D0754528

Geological Survey of Canada

Geology of Canada, no. 5

SEDIMENTARY COVER OF THE CRATON IN CANADA

edited by

D.F. Stott and J.D. Aitken

1993

This is volume D-1 of the Geological Society of America's Geology of North America series produced as part of the Decade of North American Geology project.

Available in Canada through
authorized bookstore agents and other bookstores
or by mail from

Canada Communications Group - Publishing
Ottawa, Canada K1A 0S9

and from

Geological Survey of Canada offices:

601 Booth Street
Ottawa, Canada K1A 0E8

3303-33rd Street N.W.
Calgary, Alberta T2L 2A7

100 West Pender Street
Vancouver, B.C. V6B 1R8

A deposit copy of this publication is also available for reference
in public libraries across Canada

Cat. No. M40-49/5E
ISBN 0-660-13133-1

Price subject to change without notice

Technical editors
P.J. Griffin
A.V. Okulitch

Design and layout
P.A. Melbourne
G. Léger

Cartography
GSC - Calgary and Ottawa

Cover

Writing-on- Stone Provincial Park, Alberta. Sandstone strata of the Upper Cretaceous Milk River Formation have been erosionally sculptured into hoodoos. Sweetgrass Hills in the distance, lying within Montana, U.S.A., is a laccolith of syenite yielding a K-Ar age of 48 Ma (Mid-Eocene). (Photo GSC-1992-237 by D. Leckie).

Printed in Canada

PREFACE

The Geology of North America series has been prepared to mark the Centennial of The Geological Society of America. It represents the co-operative efforts of more than 1000 individuals from academia, state and federal agencies of many countries, and industry, to prepare syntheses that are as current and authoritative as possible about the geology of the North American continent and adjacent oceanic regions.

This series is part of the Decade of North American Geology (DNAG) Project which also includes eight wall maps at a scale of 1:5 000 000 that summarize the geology, tectonics, magnetic and gravity anomaly patterns, regional stress fields, thermal aspects, seismicity, and neotectonics of North America and its surroundings. Together the synthesis volumes and maps are the first co-ordinated effort to integrate all available knowledge about the geology and geophysics of a crustal plate on a regional scale.

The products of the DNAG Project present the state of knowledge of the geology and geophysics of North America in the 1980s, and they point the way toward work to be done in the decades ahead.

From time to time since its foundation in 1842 the Geological Survey of Canada has prepared and published overviews of the geology of Canada. This volume represents a part of the seventh such synthesis and besides forming part of the DNAG Project series is one of the nine volumes that make up the latest *Geology of Canada*.

J.O. Wheeler
General Editor for the volumes
published by the
Geological Survey of Canada

A.R. Palmer
General Editor for the volumes
published by the
Geological Society of America

ACKNOWLEDGMENTS

Although the *Geology of Canada* is produced and published by the Geological Survey of Canada, additional support from the following contributors through the Canadian Geological Foundation assisted in defraying special costs related to the volume on the Appalachian Orogen in Canada and Greenland.

Alberta Energy Co. Ltd.
Bow Valley Industries Ltd.
B.P. Canada Ltd.
Canterra Energy Ltd.
Norcen Energy Resources Ltd.
Petro-Canada
Shell Canada Ltd.
Westmin Resources Ltd.

J.J. Brummer
D.R. Derry (deceased)
R.E. Folinsbee

CONTENTS

Frontispiece

A seismic-reflection survey in progress in the footwall of the McConnell thrust fault, Trans-Canada Highway at the Rocky Mountain front, west of Calgary, Alberta. In the background, Middle Cambrian limestone is superposed upon Upper Cretaceous shale and sandstone at the thrust fault. The trace of McConnell fault here marks the boundary between the Front Ranges, to the left, and the Foothills, to the right. The gentle east dip of this part of the fundamentally west-dipping thrust plane shows that the fault is folded. Photo - KGS 2388 by J.D. Aitken.

Chapter 1

INTRODUCTION

Chapter 1

INTRODUCTION

D.F. Stott and J.D. Aitken

GENERAL STATEMENT

This volume is concerned primarily with the geological development of the relatively undeformed sediments that border and cover the western and southern margins of the North American craton in Canada. It deals principally with three tectonostratigraphic provinces and their basins (Fig. 1.1): Interior Platform, Hudson Platform, and St. Lawrence Platform. These regions display parallel development at some stratigraphic levels but major differences at others.

Interior Platform, commonly known as Western Canada Basin, comprises Williston and Alberta basins, and its northern subdivision, Mackenzie Platform, which contains several smaller basins. The platform is bordered on the east by the Precambrian Canadian Shield, and on the west by the Cordilleran Orogen. It is bounded on the north and northwest by the Arctic Continental Shelf, and on the northeast by the Arctic Platform. Along its southeastern margin is Williston Basin, most of which lies south of Saskatchewan and Manitoba in the neighbouring states of Montana and North Dakota. The eastern boundary of Interior Platform is drawn along the edge of the Canadian Shield, and the platform is separated from Hudson Platform by Severn Arch and from Michigan Basin by Transcontinental Arch. In Canada, the western boundary of Interior Platform is marked by Sweetgrass Arch extending from Montana into Alberta. The term Alberta Basin is used informally for the geographic area lying between Sweetgrass Arch in the south and Tathlina Arch just north of 60° N latitude. Several sub-basins existed throughout most of Phanerozoic time, but in late Mesozoic time, some of the major elements disappeared and the prominent Rocky Mountain Trough, a foredeep, developed along the western margin. The northern part of Western Canada Basin was essentially a platform throughout much of Phanerozoic time, but at times was broken by arches and intervening sub-basins and embayments. During Late Jurassic and Cretaceous time, several sub-basins and foredeep troughs developed.

Hudson Platform lies in the central part of the Canadian Shield and comprises Moose River and Hudson Bay sedimentary basins. The smaller Moose River Basin borders and extends beneath James Bay. The large Hudson Bay Basin to the north is mainly covered by water of Hudson Bay. Rocks of the latter basin are exposed on land only in northern Hudson Bay Lowland and on Southampton, Coates, and Mansel Islands.

St. Lawrence Platform lies along the southeastern margin of the Canadian Shield and is bounded on the southeast by the Appalachian Orogen. Its history is closely linked with the histories of the Interior and Hudson platforms, for it forms part of a more extensive cratonic cover that was at times continuous with Hudson Platform to the north and Interior Platform to the northwest. St. Lawrence Platform is divided into three parts by arches extending from the Canadian Shield. The marginal areas of Michigan Basin and Allegheny Trough are separated by Frontenac Arch from Quebec Basin and Ottawa Embayment of the Ottawa-Quebec Lowland (see Chapter 10). The eastern division, separated from the western basins by Saguenay Arch, comprises Anticosti Basin, whose Paleozoic rocks outcrop on Anticosti Island and Mingan Islands and underlie the north shore of the Gulf of St. Lawrence and northwestern Newfoundland.

The Canadian Shield is basement to the mainly Phanerozoic sedimentary accumulations. For its anatomy and historical development, the reader is referred to Geology of Canada, no. 7: "Precambrian Geology of the Craton in Canada and Greenland" (Geology of Canada, no. 7, Hoffman et al., in prep.). However, because some features of the basement are known to have affected Phanerozoic sedimentation and tectonics of the platforms and basins throughout geological time, a chapter is devoted to a synthesis of our current knowledge derived from well samples of basement, gravity and aeromagnetic data, deep seismic profiles, conductivity and heat flow studies, and seismicity.

Although detailed consideration of the Cordilleran and Appalachian orogens is left to the companion volumes "Geology of the Cordilleran Orogen in Canada" (Gabrielse and Yorath, 1991) and "Geology of the Appalachian – Caledonian Orogen in Canada and Greenland" (Geology of Canada, no. 6, Williams, in prep.), parts of the depositional histories of the Western Canada Basin and St. Lawrence Platform are inextricably linked with the histories of the mountain chains that form the outer boundaries of these two elements; accordingly the salient features of the two mountain chains are outlined briefly. Detailed discussions of tectonic history and structural style are available in the other volumes.

Stott, D.F. and Aitken, J.D.
1993: Introduction; Chapter 1 in Sedimentary Cover of the Craton in Canada, D.F. Stott and J.D. Aitken (ed.); Geological Survey of Canada, Geology of Canada, no. 5, p. 1-7 (also Geological Society of America, The Geology of North America, v. D-1).

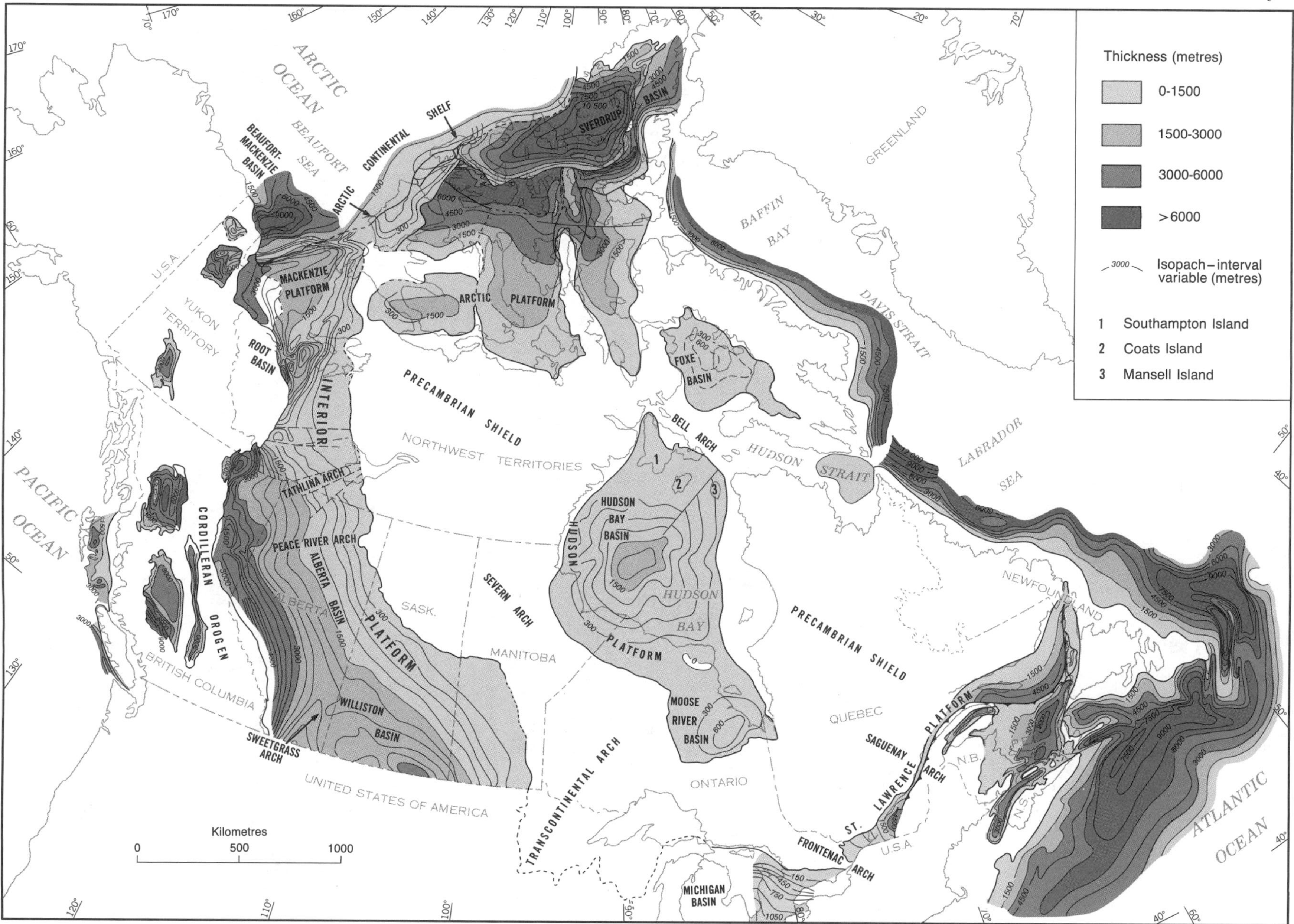

Figure 1.1. Sedimentary cover of the North American craton in Canada, showing thicknesses and distribution of unmetamorphosed sedimentary rocks. Major elements of the tectonostratigraphic provinces outlined in this volume are identified (after McCrossan and Porter, 1973).

Many similarities exist between the geological histories of the mainland platforms and Arctic Platform. The latter is dealt with in the volume "Geology of the Innuitian Orogen and Arctic Platform in Canada and Greenland" (Trettin, 1991) because its Carboniferous and later history is distinct from that of the Interior Platform.

Only a brief summary of the Quaternary history of the various regions is presented here. A more complete report is available in "Quaternary Geology of Canada and Greenland" (Fulton, 1989).

The economic geology of the Interior, Hudson, and St. Lawrence platforms is summarized with extended reports on the occurrences of energy resources, including petroleum, natural gas, and coal. The industrial minerals and groundwater resources are outlined in some detail. Deposits of metallic minerals, considered at length in the volumes on "Geology of the Cordilleran Orogen in Canada" (Gabrielse and Yorath, 1991) and "Mineral Deposits of Canada" (Geology of Canada, no. 8, Thorpe and Eckstrand, in prep.), receive only brief mention here in their stratigraphic context.

Most of the manuscripts for this volume were submitted to the editors in 1986 or earlier, and thus lack some of the new information that has come to light since that time. Because of the length of the reviewing, editing, and draughting processes, it has been impraticable to bring the volume to currency as of mid-1989, although some especially critical recent findings have been inserted.

OBJECTIVES AND SCOPE

The objectives of this volume are threefold. First, to present a balanced introduction to the regions for a geologist who has little or no previous knowledge of them. Second, as this volume is the first synthesis to be made in 20 years, to present fresh regional perspectives and new interpretations. Third, to provide a text and regional guide for earth science teachers and their students.

The text has been designed to present a concise though brief synthesis of the geology of the sedimentary cover of the North American Craton in Canada, including the mineral and fossil resources. Emphasis is given, in the light of current geological knowledge and concepts, to the tectonic and depositional history, in order to trace the geological evolution of the platforms, basins, and foredeeps. In essence, this volume modernizes and expands the syntheses dealing with the Interior, Hudson, and St. Lawrence platforms, provided in "Geology and Economic Minerals of Canada, Fifth Edition" (Douglas, 1970). Concepts of mountain building, basin development, and history have changed radically since that previous summary. The general acceptance of concepts related to plate tectonics and accretion of exotic terranes to the North American craton has resulted in major new interpretations.

The long lists of references provide entry to the vast literature dealing with the sedimentary cover of the craton. References in general are limited to recent reports in which more complete bibliographies, particularly of the older literature, may be found.

LIMITS OF THE REGION

The eastern boundary of the Interior Platform is the boundary between Phanerozoic and Precambrian rocks at the edge of the Canadian Shield (Fig. 1.1). The western boundary is arbitrary, because sedimentary rocks continuous with the strata beneath the undisturbed plains are structurally involved in the eastern Cordillera. To maintain unity and coherence, the discussion of Paleozoic sediments is carried at least to the edges of the carbonate platforms, which generally lie within the Cordilleran Orogen, but not necessarily into the deepwater basins farther west. Similarly, the discussion of sediments of the Mesozoic foreland basins includes those farthest west, which lie within the eastern Cordillera. A convenient topographic and structural boundary can be defined in the southern Cordillera along Rocky Mountain Trench, although this trench does not coincide with the margin of the Paleozoic deepwater basin. In the northern Cordillera, the edges of the carbonate platforms lie well to the northeast of Tintina Trench; there the boundary chosen is the northeastern margin of Selwyn Basin. The northern boundary is drawn along the mainland coast.

The limits of Hudson Platform are obvious (where exposed) except for the northeastern margin, which is drawn across the top of Hudson Bay along Bell Arch; the related Foxe Basin to the northeast is assigned to the Innuitian Region.

St. Lawrence Platform is readily delineated throughout most of Ontario and Quebec. It includes the Canadian portions of Michigan and Allegheny basins in the west and south, and is bordered to the north by the Canadian Shield. The southeastern boundary of the platform, the Appalachian structural front, follows St. Lawrence River and Gulf of St. Lawrence to Newfoundland.

ORGANIZATION AND EDITORIAL POLICIES

Each of the three main regions - Interior Platform, Hudson Platform, and St. Lawrence Platform, together with their basins - forms a separate part, and several chapters are devoted to each. Chapter 13, "Evolutionary Models and Tectonic Comparisons", compares and contrasts the histories of the three regions with each other and with other parts of the world.

Within the main chapters on the geology of each region, the discussion is in stratigraphic order. Most of the authors resisted the suggestion that stratigraphy be organized according to the cratonic sequences and sub-sequences of Sloss (1963), but the major sequence boundaries are recognized. Maps, sections, and other diagrams are used extensively to convey the current understanding of the geology. Summaries of the lithostratigraphy and biostratigraphy are provided, but greater emphasis is given to the interpretation of depositional environments, paleogeography, and tectonic events. The reader will find the treatment substantially more subjective than the previous editions of "Geology and Economic Minerals of Canada". Chapters or subchapters are devoted to such major topics as physiography, history of exploration, and geological investigations. Economic geology is outlined on a region-by-region basis rather than in one large, integrated chapter.

The authors were not required to follow a standard format in the treatment of geological systems, although they were requested to cover specific topics. Furthermore, the amount and quality of information vary markedly from one system to another, and from one region to another. For those systems of economic importance, a large number of publications is available; for others, limited data have been published. Some areas have been drilled extensively; other areas have few, if any, exploratory drill-holes. As a result, the discussions vary in treatment and emphasis. Because one of the main purposes of this series is to present a perspective of the tectonic development of North America, the editors have allowed the authors considerable latitude in presenting their preferred interpretations of the data, although an attempt has been made to ensure that other opinions were identified. The views expressed are those of the individual authors; other interpretations are possible and indeed published, and as indicated in the summary sections, the editors may favour interpretations differing from those of some authors.

In matters of style, the standard editorial policies of the Geological Survey of Canada have been followed throughout.

The correlation charts are fundamental contributions of this report. System and stage boundaries, as shown on the correlation charts and as used within the text, have been determined by extended consultations with many specialists and, in general, represent prevailing scientific opinions. Within rock sequences, boundaries may be ill-defined by paleontological evidence, and approximations based on lithological evidence may be difficult. The quality of paleontological control has been indicated for those units where it is available. Paleontological control in subsurface sections is sparse, hence correlations are based on lithological similarities, continuity of strata, stratigraphic position, and geophysical character. The correlation charts (in pocket) represent the present state of knowledge; correlations may change as new data become available.

Individual chapters are illustrated, to the greatest extent allowed by available space, by maps, cross-sections, and other diagrams. In addition, many photographs have been included to illustrate the character of rocks, structure, and terrain. Few up-to-date regional compilations were available and, as a result, many maps are generalized. The base map, specifically prepared for the Decade of North American Geology to provide continuity with the United States, is a modified Transverse Mercator projection. Most map diagrams have been constructed using that base, although a few have not. At the scale of publication, the discrepancies should be miminal.

This volume also includes (in pocket) a series of geological cross-sections of the Western Canada Basin. These, now redrawn and augmented, were originally prepared jointly by the Canadian Society of Petroleum Geologists and the Geological Association of Canada under the chairmanship of G.N. Wright.

ACKNOWLEDGMENTS

Responsibility for the Canadian volumes for the Decade of North American Geology was assumed from the organizational stage by the Geological Survey of Canada (GSC). The responsibility for this volume devolved upon the Institute of Sedimentary and Petroleum Geology, GSC, Calgary, Alberta, where the editors are employed. Every attempt was made to recruit knowledgeable geologists from the petroleum, coal, and mineral industry, provincial geological surveys, and the universities. The volume has been prepared and produced through the efforts of a large number of authors and contributors, reviewers, and editors. It could not have been prepared without the co-operation of consultants, companies, university departments, and governmental agencies, which provided information and also allowed for the participation of staff members.

The volume was prepared under the general direction of A.R. Palmer, Centennial Science Program Co-ordinator for the Decade of North American Geology, Geological Society of America, and J.O. Wheeler, Geological Survey of Canada, co-ordinator of the Canadian volumes. Both have provided guidance, assistance, and encouragement from the the initial stages to the finished product.

Several substantial chapters, subchapters, and sections were co-ordinated by geologists outside the Institute of Sedimentary and Petroleum Geology (ISPG). P. Fitzgerald Moore, consultant, applied, with the co-operation of Shell Canada, his extensive knowledge to the preparation of Subchapter 4D on the Devonian System. R. Burwash and E.R. Kanasewich from the University of Alberta and A.G. Green and A.M. Jessop from the GSC have provided the detailed summary of the Precambrian basement (Chapter 3). B.V. Sanford of the Continental Geoscience Division, GSC, Ottawa, prepared chapters 10 to 12 on the St. Lawrence Platform. R.D. Johnson, consultant, with N.J. McMillan, GSC, co-ordinated Subchapter 6A on Petroleum Geology, contacting many contributors within the oil industry, and P. Fitzgerald Moore, consultant, assisted with the final draft. A.M. Stalker and J.-S. Vincent of the Terrain Sciences Division, GSC, Ottawa, summarized the Quaternary history of the western plains (Subchapter 4K). W.R. Cowan, formerly of Palliser Consultants Limited, Calgary, and latterly with the Ontario Ministry of Northern Development and Mines, prepared the texts concerning the Quaternary geology of Hudson and St. Lawrence platforms (in Chapters 8 and 11). D.H. Lennox, Inland Waters/Lands Directorate of Environment Canada, provided accounts of groundwater resources in Western Canada Basin (Subchapter 6F).

Government agencies whose staff have contributed include: Alberta Geological Survey, Research Council of Alberta; Saskatchewan Geological Survey; Geological Services Branch, Manitoba Energy and Mines; Atlantic Geoscience Centre, GSC, Dartmouth, Nova Scotia; Continental Geoscience Division, GSC, Ottawa; Terrane Sciences Division, GSC, Ottawa; Cordilleran Division, GSC, Vancouver, and Pacific Geoscience Centre, GSC, Sidney, British Columbia.

Companies who have contributed information and staff time include: Gulf Canada Resources Ltd.; Petro-Canada Resources Ltd.; Shell Canada Resources Ltd.; Chevron Canada Resources Ltd.; Canadian Superior Oil Ltd.; Canterra Energy Ltd.; Canadian Hunter Exploration Ltd.

Universities whose staff have been involved include: Department of Geology, University of Alberta; Department of Geology and Geophysics, University of Calgary;

Department of Geological Sciences, University of Saskatchewan; Department of Geology, University of Toronto; Department of Geology, University of Windsor.

Several geological organizations and publications have been most generous in permitting the use of copyright material. These include: Canadian Society of Petroleum Geologists (Bulletin of Canadian Petroleum Geology); Geological Assocation of Canada and the National Research Council of Canada, (Canadian Journal of Earth Sciences); American Association of Petroleum Geologists (Bulletin); Sedimentology.

Specific acknowledgments are included in each chapter for the assistance provided by individual contributors, for photos supplied by various companies, and for those who reviewed various chapters and sections.

The volume was reviewed by T.A. Oliver, Department of Geology and Geophysics, University of Calgary, and R.G. Greggs, of Greggs and Associates. Their constructive comments and suggestions have been most helpful in improving the text.

Most of the diagrams for this volume were drafted by the Cartography Unit of the Institute of Sedimentary and Petroleum Geology, Calgary. L. MacLachlan, Head Cartographer, provided advice throughout the preparation of the original diagrams, and was instrumental in determining the final renditions. Cartographers who participated in final drafting include: J.W. Thomson, W.P. Vermette, D.J. Walter, B.H. Ortman, S. Orzeck, J.W. Waddell, B.E. Fischer, and G. Whitman. L. Wardle and G.N. Edwards of the Photomechanical Unit promptly and cheerfully processed many draft copies of maps, diagrams, and charts.

The Word Processing Unit, ISPG, was responsible for the preparation of the manuscript text and its many revisions, including the final one. The co-operation of P.L. Greener, supervisor, ensured that various drafts were accurate and available promptly. Her familiarity with the publication standards of the GSC was of great assistance to the editors. Staff who worked on the text included: H. King, M.L. Jacobs, Maureen Hill, Donna Henry, Julie Stevenson and Val Hill.

Many of the photographs were processed and printed by the Photographic Unit, ISPG. The skills of B.C. Rutley and W.B. Sharman ensured a high standard of excellence. Other institutions and companies who provided photographs include: Canada Cement Lafarge; Canadian Rock Salt Company Ltd.; Shell Canada Limited; Chevron Canada Limited, Glenbow Museum, Calgary; University of Alberta Archives.

After the retirements of D.F. Stott and J.D. Aitken, A.V. Okulitch, ISPG, served as consulting editor. He resolved admirably problems pertaining to the volume that were encountered by the cartographers and editorial staff in Ottawa.

REFERENCES

Douglas, R.J.W. (ed.)
1970: Geology and Economic Minerals of Canada; Geological Survey of Canada, Economic Geology Report No. 1, Fifth Edition, 838 p.

Fulton, R.J. (ed.)
1989: Quaternary Geology of Canada and Greenland; Geological Survey of Canada, Geology of Canada, no. 1, (also Geological Society of America, The Geology of North America, v. K-1.)

Gabrielse, H. and Yorath, C.J. (ed.)
1991: Geology of the Cordilleran Orogen in Canada; Geological Survey of Canada, Geology of Canada, no. 4 (also Geological Society of America, The Geology of North America, v. G-2.)

McCrossan, R.G. and Porter, J.W.
1973: The geology and petroleum potential of the Canadian sedimentary basins - A synthesis; in Future Petroleum Provinces of Canada, R.G. McCrossan (ed.); Canadian Society of Petroleum Geologists, Memoir 1, p. 589-720.

Sloss, L.L.
1963: Sequences in the cratonic interior of North America; Geological Society of America, Bulletin, v. 74, p. 93-114.

Trettin, H. (ed.)
1991: Innuitian Orogen and Arctic Platform: Canada and Greenland; Geological Survey of Canada, Geology of Canada, no. 3, (Also Geological Society of America, The Geology of North America, v. E.)

Authors' Addresses

J.D. Aitken
2676 Jemima Road
Denman Island, British Columbia
V0R 1T0

D.F. Stott
8929 Forest Park Drive
Sidney, British Columbia
V8L 5A7

Printed in Canada

Chapter 2

INTERIOR PLATFORM, WESTERN BASINS, AND EASTERN CORDILLERA

Subchapter 2A

INTRODUCTION TO INTERIOR PLATFORM, WESTERN BASINS, AND EASTERN CORDILLERA

D.F. Stott and J.D. Aitken

CONTENTS

GENERAL STATEMENT

The Interior Platform, underlain by Phanerozoic sedimentary rocks, is the northwestern part of the North American Craton, the stable interior region of the continent (Fig. 2A.1; see also Fig. 1.1). Essentially flat-lying rocks of the Interior Platform extend into the eastern Cordillera where they are thrust-faulted and folded (see Fig. 5.5, in pocket), but have undergone little or no metamorphism nor plutonism during the orogenic phases. This area of undeformed and deformed sedimentary rock commonly is called the Western Canada Basin. The Interior Platform is linked to Michigan Basin and St. Lawrence Platform of southeastern Canada through the central U.S.A. and merges with Arctic Platform in the north. It is separated from Hudson Platform by a broad expanse of the Canadian Shield.

The area of nearly horizontal bedrock, forming the Interior Plains physiographic province, comprises plains and plateaux covered by a thick mantle of glacial drift. It is a region of grassland, forest, and tundra, some 2 000 000 km^2 in area, that embraces parts of the prairie provinces of Manitoba, Saskatchewan, and Alberta, the northeast corner of British Columbia, and much of the western District of Mackenzie, Northwest Territories. In the southern region, clastic, carbonate, and evaporite strata of Paleozoic age rest on a basement of Precambrian crystalline rocks. The Paleozoic sediments are overlain by a relatively thick blanket of Mesozoic and Tertiary clastic rocks. The southern part of Interior Platform is the main oil-producing region of western Canada; it has been extensively drilled and its geology is well known. In the northern region, few wells penetrate the entire Phanerozoic succession, and the geology is less well known. There, lower and middle Paleozoic clastic, carbonate, and evaporite beds lie on Precambrian sedimentary and crystalline rocks and are covered by a thin veneer of Cretaceous clastic rocks. Other systems appear within the Cordillera to the west.

Stott, D.F. and Aitken, J.D.
1993: Introduction to Interior Platform, Western Basins, and Eastern Cordillera; Subchapter 2A in Sedimentary Cover of the Craton in Canada, D.F. Stott and J.D. Aitken (ed.); Geological Survey of Canada, Geology of Canada, no. 5, p. 11-13 (also Geological Society of America, The Geology of North America, v. D-1).

The eastern deformed belt of the Canadian Cordillera includes strata laid down on the western extension of the Interior Platform. It is a narrow belt, varying from 100 to 400 km in width, extending along the Rocky Mountains of western Alberta and eastern British Columbia into the District of Mackenzie and Yukon Territory. The Fold Belt (formerly known as the Rocky Mountain Fold and Thrust Belt) is characterized by Paleozoic carbonate and clastic rocks occurring in thrust plates and folds produced by pulsatory orogeny during Late Jurassic, Cretaceous, and early Tertiary times. The Rocky Mountains are bordered on the east by the Foothills Belt. This belt is underlain mainly by deformed Mesozoic and Tertiary clastic sediments, but Paleozoic carbonate strata reach the surface in the highest structures.

Within the Interior Platform, the internal structure and lateral variations reflect a long and complex history of development that involved superposition of contrasting patterns of differential subsidence and uplift. The platform is characterized by two fundamentally different phases of development. The initial phase, following a precursory period of rifting with associated sedimentation and local volcanism in very late Precambrian time, resulted in transgressive eastward onlap of the crystalline basement of the North American Craton and deposition of a continental terrace wedge. Sediments continued to be deposited on the subsiding, passive western margin of the craton from early Paleozoic time, through Devonian and Carboniferous periods, into the Mesozoic. A series of epeirogenic arches and basins developed on the platform, and subsidence of intracratonic basins was at times independent of that of the continental margin.

Deposition of the continental terrace wedge in the west ended in Late Jurassic time with the onset of convergent-margin orogeny. Eastward-migrating tectonic thickening and uplift resulted in migrating foredeeps and the accumulation of thick clastic sediments derived from the rising orogen. This phase, lasting from Middle Jurassic to Eocene time, was related in part to the accretion of allochthonous terranes to the western margin of the North American Craton and in part to intraplate convergence manifested by subduction of continental crust in the Fold and Omineca belts. During that time, the continental terrace wedge was compressed, detached from its basement, and displaced eastward over the flank of the craton.

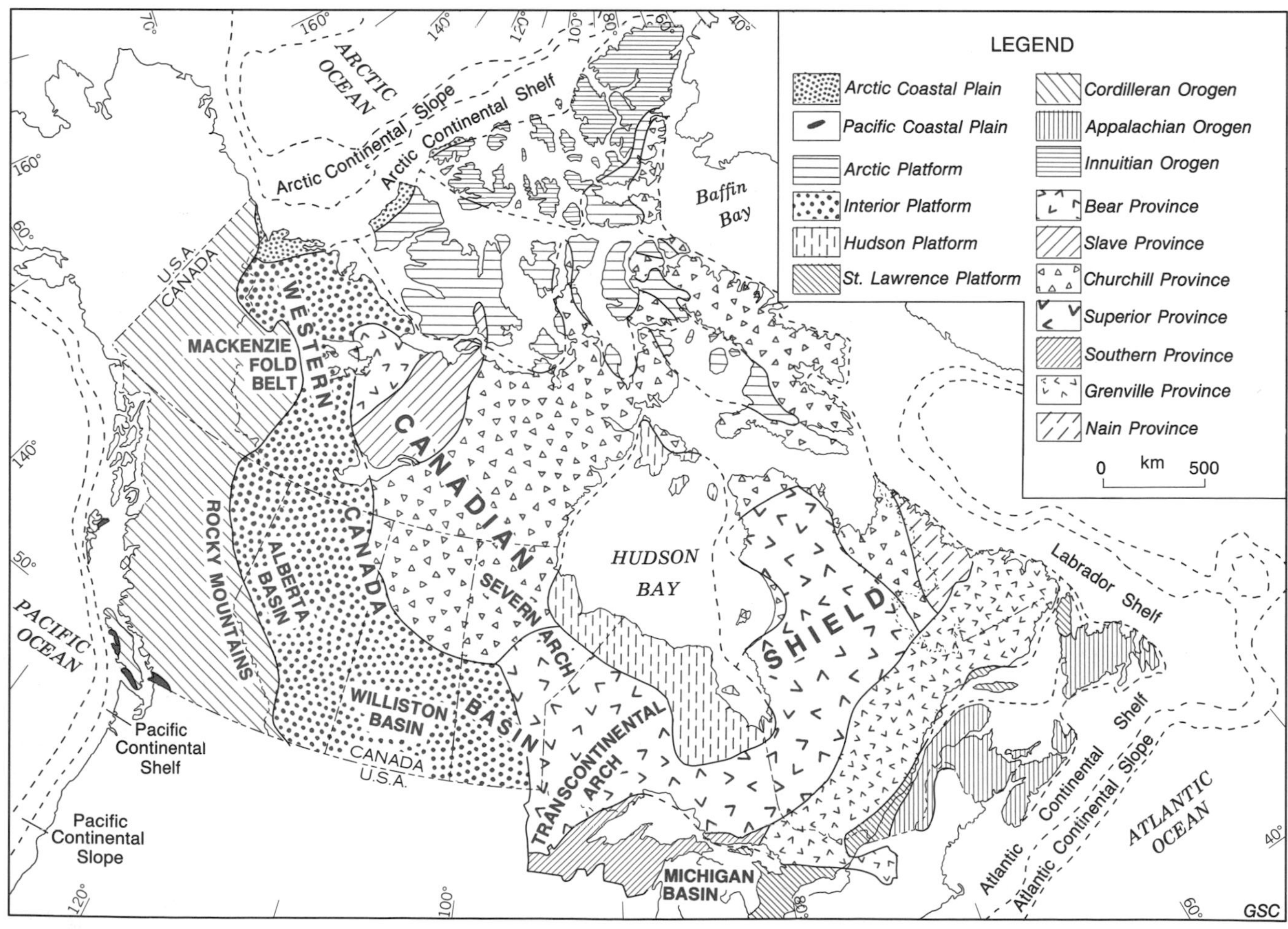

Figure 2A.1. Geological regions of Canada.

The continental interior underwent extensive erosion after early Tertiary time, and the clastic detritus was transported northward, building the Tertiary and Holocene deltas of Mackenzie River. The Mississippi Delta also received sediments from the Canadian Interior. Widespread continental glaciations of Pleistocene age affected the topography and left thick sediments throughout the region.

The Interior Platform (Western Canada Basin) is basically a simple, northeasterly tapering wedge of sedimentary rocks, more than 6 km thick in the Cordilleran Foreland Belt, extending northeasterly to the Canadian Shield. The platform is separated into several major segments by arches whose times of initial rise, duration, and subsequent history (quiescence, subsidence) are highly variable. In the southeast, the Williston Basin was a depocentre throughout much of its history, but rarely a hypsographic basin. Other areas termed basins are, in a sense, geographic areas containing the products of successive, unrelated depositional systems.

Williston Basin is a result of a combination of both structural and depositional processes. In the terms of Krumbein and Sloss (1951), it is an intracratonic basin. McCrossan and Porter (1973) termed it a cratonic centre basin. The basin is defined by the distribution of its Middle Ordovician and younger rocks. In Canada, the northeastern boundary is drawn along the edge of the Canadian Shield. The basin is separated from Hudson Platform by Severn Arch and from Michigan Basin by Transcontinental Arch. The northern limit is about 51°N latitude. In Canada, the western boundary is marked by Sweetgrass Arch, a northeasterly trending structure extending from Montana through Alberta into Saskatchewan (see Fig. 1.1). The main part of the basin lies south of Saskatchewan and Manitoba in the neighboring states of Montana and North Dakota. The basin had its inception during the Middle Ordovician and was intermittently downwarped during the Paleozoic and Mesozoic. Sediments within the basin have a maximum thickness of 5 km in U.S.A. and about 3 km on the northeast flank within southern Saskatchewan. The Canadian segment covers 572 000 km^2 and has a volume of sediments of 186 000 km^3 (Christopher et al., 1973).

The term Alberta Basin has been used informally for the area lying between Sweetgrass Arch in the south and Tathlina Arch (see Fig. 1.1) just north of 60°N latitude. The term Alberta Syncline has been used for the same element, although the synclinal character did not develop until Tertiary time. The area is essentially a monocline, dipping southwesterly from the Precambrian Canadian Shield. Within this area, Mesozoic and Paleozoic sediments, trending northwesterly, have a thickness exceeding 6 km at the edge of the deformed belt. They thin depositionally and erosionally northeastward to zero at the edge of the Shield. The volume of sediments exceeds 1 870 000 km^3 (Parsons, 1973; Torrie, 1973). In this area, a platform with several sub-basins existed through most of Paleozoic time, but in the mid- to late Mesozoic, a linear foredeep (Rocky Mountain Trough) formed along the southwestern side of the platform in response to orogeny at the western margin.

The Interior Platform lying north of 60°N latitude contains a number of sub-basins, none of which persisted throughout the Phanerozoic Eon. This northern area was essentially a platform throughout much of Paleozoic time but during the Cretaceous Period, several foreland basins and sub-basins developed. The thickest Phanerozoic deposits lie west of Mackenzie River in the mountain belt and to the north in Mackenzie Delta. Maximum thicknesses exceed 7 km, and the region contains more than 1 660 000 km^3 of sediments (de Wit et al., 1973; Gilbert et al., 1973; Kunst, 1973; Martin, 1973).

REFERENCES

Christopher, J.E., Kent, D.M., and Simpson, F.
1973: Saskatchewan and Manitoba; in Future Petroleum Provinces of Canada, R.G. McCrossan (ed.); Canadian Society of Petroleum Geologists, Memoir 1, p. 121-150.

deWit, R., Gronberg C., Richards, W.B., and Richmond, W.O.
1973: Tathlina area, southern District of Mackenzie; in Future Petroleum Provinces of Canada, R.G. McCrossan (ed.); Canadian Society of Petroleum Geologists, Memoir 1, p. 187-212.

Gilbert, D.L.F.
1973: Anderson Plain, northern District of Mackenzie; in Future Petroleum Provinces of Canada, R.G. McCrossan (ed.); Canadian Society of Petroleum Geologists, Memoir 1, p. 213-244.

Krumbein, W.C. and Sloss, L.L.
1951: Stratigraphy and Sedimentation; San Francisco, W.H. Freeman and Company, 497 p.

Kunst, H.
1973: Peel Plateau; in Future Petroleum Provinces of Canada, R.G. McCrossan (ed.); Canadian Society of Petroleum Geologists, Memoir 1, p. 245-274.

Martin, H.L.
1973: Eagle Plain Basin; in Future Petroleum Provinces of Canada, R.G. McCrossan (ed.); Canadian Society of Petroleum Geologists, Memoir 1, p. 275-306.

McCrossan, R.G. and Porter, J.W.
1973: The geology and petroleum potential of the Canadian sedimentary basins - A synthesis; in Future Petroleum Provinces of Canada, R.G. McCrossan (ed.); Canadian Society of Petroleum Geologists, Memoir 1, p. 589-720.

Parsons, W.H.
1973: Alberta; in Future Petroleum Provinces of Canada, R.G. McCrossan (ed.); Canadian Society of Petroleum Geologists, Memoir 1, p. 73-120.

Torrie, J.E.
1973: Northeastern British Columbia; in Future Petroleum Provinces of Canada, R.G. McCrossan (ed.); Canadian Society of Petroleum Geologists, Memoir 1, p. 151-186.

Author's Addresses

J.D. Aitken
2676 Jemima Road
Denman Island, British Columbia
V0R 1T0

D.F. Stott
8929 Forest Park Drive
Sidney, British Columbia
V8L 5A7

Printed in Canada

Subchapter 2B

GEOLOGICAL EXPLORATION OF THE INTERIOR PLAINS

D.F. Stott and J.D. Aitken

CONTENTS

GENERAL STATEMENT

Geological exploration has greatly influenced the development of western Canada, particularly the growth of the petroleum and natural gas, coal, and mineral industries. During the late nineteenth century when railroads were being built across the plains, settlers were encouraged by geological reports on the favorable agricultural prospects and the availability of water and fuels. The prosperity of the prairie provinces, based on their rich natural resources, would not have been achieved without the determination, enterprise, and knowledge of dedicated explorers.

This abbreviated account outlines the history of exploration and contributions of some of the notable individuals, particularly during the early years. The more recent history involves many more people, and it is not possible to acknowledge each and every one. Additional details on recent investigations may be found in the following chapters.

This subchapter has been prepared from a variety of sources in addition to original reports. It draws on a history of the Geological Survey of Canada by Zaslow (1975) and accounts of the development of the petroleum industry in western Canada by Gray (1970), de Mille (1970), Gould (1976), and a group of authors whose papers appear in Hilborn (1968).

THE ERA OF FUR TRADERS AND EXPLORERS, 1691-1870

Geological exploration in western Canada began with the earliest traders, who entered the country to obtain furs from the native Indians. These early traders were also explorers, travelling rapidly across vast areas in search of Indians with whom they might trade. The Hudson's Bay Company, established in 1670, sent traders into the western plains via Hudson Bay and its tributaries. In their travels, the traders encountered various mineral deposits and learned of others from the natives. Later, as the

Stott, D.F. and Aitken, J.D.
1993: Geological exploration of the Interior Plains; Subchapter 2B in Sedimentary Cover of the Craton in Canada, D.F. Stott and J.D. Aitken (ed.); Geological Survey of Canada, Geology of Canada, no. 5, p. 14-30 (also Geological Society of America, The Geology of North America, v. D-1).

British Government became interested in the economic potential, expeditions were mounted to investigate and report on this largely unknown land. This type of activity continued until the administration of the region was transferred from Britain to Canada in 1870.

The oldest recorded geological observations in western Canada are those made in 1691 by Henry Kelsey, the first European to visit the Plains, who was under instructions from the Hudson's Bay Company to look for mines and minerals (Whillans, 1955). He was the first to report bitumen from the Athabasca River, having been given samples by an Indian in 1719 (Morton, 1973). More than a half-century later, Peter Pond, an American fur-trader with the North West Company, observed the deposits along Athabasca River, but Alexander Mackenzie (1801), a partner in the North West Company, was the first to provide a description of this enormous bitumen deposit.

In 1789, Mackenzie, on the first of two exploring trips, travelled by canoe from Fort Chipewyan to the Arctic Ocean and reported burning coal seams near present Fort Norman. Four years later, Mackenzie (1801) travelled westward up Peace River, noting the 'bituminous substance which resembles coal' in the canyon before continuing to the Pacific Coast. Although other notable explorers, including Simon Fraser (1808), used the Peace River route, few geological observations were made until much later in the nineteenth century.

The fur traders noted the occurrences of natural resources and, in recording their explorations, produced the first maps of the region. David Thompson, while in the employ of the North West Company during the years 1792-1812, provided a remarkably accurate map of the vast region lying between 84° and 124°W longitude and 45° and 60°N latitude (Beach, 1954; Warkentin, 1967).

John Richardson, surgeon and naturalist to the First (1819-22) and Second (1825-27) Franklin expeditions and subsequent leader of his own expedition (1847-1849), forecast substantial returns from the mineral industry (Richardson, 1852). He made many geological observations and delineated with remarkable accuracy the boundary between the Precambrian Shield and the Interior Plains. Richardson's scientific observations (1823, 1851a) contrast with earlier reports which merely recorded geographic features and the occurrences of resources. He presented an extensive summary of the oil sands in the lower part of Clearwater Valley and along the Athabasca River, and described outcrops of coal and salt. His attempts at synthesis included the first geological map of all of British North America. Richardson gave an 11-lecture course on geology to the members of Franklin's party in the winter of 1825-26 at Fort Franklin, Great Bear Lake, the first geological course presented in western and northern Canada (Warkentin, 1979).

The first systematic, detailed, paleontological monograph on the Mackenzie Basin was provided by F.B. Meek (1867), on the basis of collections made by Robert Kennicott under the auspices of the Smithsonian Institution, Washington. Meek called attention to the numerous indications of petroleum along the valley of Mackenzie River. One of the Devonian localities, collected earlier by Richardson, was 'The Ramparts', a now famous, 80-m deep gorge on Mackenzie River. Plant fossils collected by Kennicott were described by the renowned Swiss paleobotanist Oswald Heer.

James Hector (Fig. 2B.1), surgeon and geologist to the British Palliser Expedition (1857-60), explored the southern plains and adjacent mountains on horseback. His stratigraphic subdivisions of the Cretaceous rocks are based on localities in southwestern Manitoba along South Saskatchewan and Souris rivers, at Cypress Hills, and in southern Alberta. Hector travelled the prairie region and the Rocky Mountains between the 49th and 54th parallels swiftly, but with remarkable results. He delineated the three great prairie levels, described the Tertiary of the Cypress Hills and Cretaceous rocks along Bow River, examined other localities in the Foothills and concluded that formations in the Rocky Mountains were of Carboniferous or Devonian age (Hector, 1863).

The Palliser Expedition (Palliser, 1863; see also Spry, 1968) had been sent out by the British Government because Rupert's Land, under the control of the Hudson's Bay Company, was still under British sovereignty. The Canadian government, unable officially to become involved in the region but concerned about developments to the

Figure 2B.1. Captain John Palliser (left) and Sir James Hector (right). Palliser was the leader of the British expedition that explored large areas of western Canada between 1857 and 1860. Hector was the geologist and also medical doctor for the expedition. Photo taken in England in late 1850s (photo - Glenbow Museum, Calgary).

Figure 2B.2. Geologists of the Geological Survey of Canada who spent many years in western Canada in late 1800s and early 1900s: **a** - A.R.C. Selwyn; **b** - G.M. Dawson; **c** - R.G. McConnell; **d** - J.B. Tyrrell; **e** - Robert Bell; **f** - D.B. Dowling; **g** - G.S. Hume (photos - Geological Survey of Canada collection).

west, responded by assisting the Assiniboine and Saskatchewan Exploring Expedition in 1858 under the leadership of Henry Youle Hind (1859). Hind, together with S.J. Dawson, had made an earlier reconnaissance of the Red River region on the Red River Exploring Expedition (S.J. Dawson, 1859). Hind's fossils were submitted to Elkanah Billings, later a geologist with the Geological Survey of Canada, who forwarded them to Meek at the Smithsonian Institution. Meek's reports (in Hind, 1859; 1876) and illustrations became the foundation for Cretaceous paleontology in the Interior Plains.

THE ERA OF DISCOVERY AND DEVELOPMENT, 1870-1914

The Government of Canada became involved in the western plains only after the transfer of Rupert's Land from the Hudson's Bay Company to Canada in 1870. Geological study of the region from 1870 to 1930 was carried out mainly by its agency, the Geological Survey. The Geological Survey of Canada, the oldest scientific organization in Canada, had been established in 1842 with William Logan as its director. At the time of the Rupert's Land transfer, the 'postage stamp' province of Manitoba was established and the remainder of the region, including the present provinces of Saskatchewan and Alberta, became the Northwest Territories. Having been given geological responsibility for this great addition to Canada, A.R.C. Selwyn (Fig. 2B.2a), then director of the Geological Survey of Canada (GSC), travelled in 1873 by Red River cart and river boat from Fort Garry (Winnipeg) to the Rocky Mountains, to assess the task with which he had been charged. Selwyn (1874) on his first visit was hindered in his reconnaissance by the lack of outcrop, saw the necessity for subsurface investigations, and recognized the need for a drilling program.

After the 1870 land transfer, one of the first requirements was to survey the Canada-United States boundary, a task given to the British North American Boundary Commission. G.M. Dawson (Fig. 2B.2b), who later became the director of the GSC, was attached for the years 1873 to 1875 as geologist-botanist to the Commission. His report (1875), which was to become a standard for the Interior Plains, contained a geological section across the region including data drawn from as much as 160 km north of the border. Dawson reviewed the occurrences of salt and lignite, both important for colonization, and stressed the petroleum potential of Devonian rocks.

With the new Canadian responsibilities for the Northwest Territories, and the addition of the provinces of Manitoba and British Columbia to confederation in 1870 and 1871 respectively, a railway link across Canada became imperative. Routes had to be surveyed and the commercial and agricultural potential of the land adjacent to the proposed line assessed. C. Horetzky and the botanist John Macoun explored the Peace River valley in 1873 as part of the Exploration and Surveys for the Canadian Pacific Railway under Sandford Fleming (1874). In 1875, the first detailed geological studies of the upper Peace River district were made by A.R.C. Selwyn (1877) accompanied by Macoun (1877). The latter was enthusiastic about the agricultural possibilities of the region, and throughout his many years of studies on the prairies continued to extol their fertility. Subsequently in 1879, G.M. Dawson (1881) and R.G. McConnell (Fig. 2B.2c) travelled from the west coast to Edmonton with another survey party, which was continuing the attempt to locate a route for the Canadian Pacific Railway. They explored the 'Grande Prairie', the western part of Peace River Plains, noting the potential for minerals and agriculture.

In the south, Dawson, accompanied again in 1881-1882 by McConnell, mapped the country in the vicinity of Bow and Belly rivers (Fig. 2B.3; Dawson, 1885). McConnell, in 1884, studied Cypress Hills and Wood Mountain farther east (1885) and described their Cretaceous and Tertiary stratigraphy and their vertebrate fossils. Dawson had been the first to discover dinosaur bones in Canada, having collected them in the vicinity of Wood Mountain. McConnell (1891, 1893) also travelled extensively through the north in the years 1887 to 1890, exploring the Peace-Athabasca country and Mackenzie Basin. He reported (1891) on the Devonian outcrops at the Ramparts (on Mackenzie River) and the petroleum potential of the area. McConnell was the first to recognize and interpret the great thrust faults in the deformed belt.

J.B. Tyrrell (Fig. 2B.2d) extended the studies by Dawson and McConnell as far as North Saskatchewan River in 1884 and 1885. In his work along Red Deer River near the present town of Drumheller, Tyrrell found dinosaur bones as well as coal. The Royal Tyrrell Museum of Paleontology at Drumheller, which opened in 1985, was named in his honour. Tyrrell, together with D.B. Dowling, explored the region between Lake Athabasca and Hudson Bay and much of Manitoba in the years between 1892 and 1902 (Tyrrell, 1893). Among Tyrrell's principal contributions were his conclusions on the Pleistocene glaciation of central Canada.

Figure 2B.3. Train of Red River carts, field party of G.M. Dawson, 1879. Prior to the coming of the railways, Red River carts were indispensable to prairie travellers. Constructed entirely of wood, they were easily repairable with buffalo leather (photo - GSC 353).

Robert Bell (Fig. 2B.2e) visited the Qu'Appelle Valley in 1873, and investigated the Manitoba Escarpment in 1874, mapping the poorly exposed Cretaceous strata in both regions. Later (1881, 1885), he summarized the occurrences of petroleum as seepages, bitumen, and staining in western Canada, including the Athabasca region.

Interest in the glacial history of the region stems from 1823 when W.H. Keating recognized that an ancient lake had covered a wide region surrounding the Red River Settlement (Winnipeg) and may have drained into the Mississippi River (Elson, 1983). Others who shared the same opinion, some of whom mapped one or more ancient strandlines, included D.D. Owen, Hind, G.K. Warren, Bell, Dawson and W.H. Winchell. Glacial Lake Agassiz was named by Warren Upham (1890) while working for the GSC. After 15 years of careful investigations, including studies in North Dakota, he published the classical United States Geological Survey Monograph on Lake Agassiz (1895).

A fledgling mineral industry made an early appearance on the prairies as the early fur traders and explorers, and later settlers, took advantage of local occurrences of salt and coal. The first mineral production in the Interior Plains was in Manitoba in 1800, when common salt was obtained from springs issuing from Paleozoic rocks near Winnipegosis. All through the nineteenth century, the Salt Plains on Slave River provided excellent salt which, according to Macoun (1877), the Hudson's Bay Company sacked each year and transported by boat or sledge to its posts across the plains.

Figure 2B.4. Drilling for coal near Roche Percée, Saskatchewan, 1880. The GSC had contracted for a number of drill holes on the southern prairies to investigate the extent of lignite deposits exposed on Souris River. After considerable trouble had been experienced in locating timber for the derrick and engine floor, this hole was drilled to depth of 295 feet (photo - GSC 433½).

Figure 2B.5. Coal mine at Lethbridge, Alberta, 1898. Coal was obtained from the Oldman Formation (photo - GSC 1636).

Coal deposits on the prairies undoubtedly were used by the traders, but the first official record of the use of coal, dated 1858, concerned coal-fired forges at Edmonton (Brown, 1985). The small mine openings near Edmonton likely were duplicated at numerous points along the river banks and coulees of the prairies (Fig. 2B.4). The first officially recorded Alberta mine was opened at Lethbridge in 1874 (Fig. 2B.5), and the next at Canmore in 1886.

With the proposed construction of a railroad to connect eastern Canada with British Columbia after the latter joined confederation in 1871, sources of coal and water became important. Selwyn (1874), during early exploration, had recognized this priority in conjunction with the need for more information on bedrock below the thick drift that covered much of the western prairies, and he instigated a drilling program to satisfy these requirements. The first well was spudded at Fort Garry (Winnipeg) in 1873, and others followed at Shoal Lake, Rat Creek (near Portage la Prairie), Fort Ellice, and Fort Pelly. From other wells drilled in the Souris Valley, Dawson (1886) predicted the existence of a buried valley, later recognized as the preglacial Missouri Valley, an important source of groundwater (Meneley et al., 1957).

Wells were drilled also by the Canadian Pacific Railway Company (CPR). In 1883, the Belle Plaine well near Regina produced the first show of natural gas in western Canada. In the same year, gas flowed from wells 60 km west of Medicine Hat (Dawson, 1886). In 1890, the CPR, drilling for coal at Medicine Hat (Fig. 2B.6), again found gas, and subsequently the town began producing gas for its own use from shallow Cretaceous sands (Landes, 1965).

The GSC extended its drilling into northern areas, siting holes at Athabasca Landing, Victoria (Fig. 2B.7), and Pelican Landing. The Pelican well obtained a large flow of natural gas in 1898 and was allowed to blow wild at

8.5 million cubic feet per day for 21 years (de Mille, 1970). The drilling program was disappointing because light gravity crude oil had not been encountered, but heavy oil had been shown to occur over a much larger area than that indicated by surface deposits, and large quantities of natural gas had been found.

In the late 1890s, the government of British Columbia gave coal-bearing land in the southeastern corner of the province to the Canadian Pacific Railway in exchange for construction of a second railroad south of the main line. The CPR in turn traded some of the land to the federal government to obtain cash to fund the construction. James McEvoy of the GSC was sent west to select a favourable area and in 1900 recommended land east of Fernie, which was to become the Dominion Coal Block.

The massive coal fields near Fernie in the east Kootenays had been known to early travellers but were not developed until the completion of the southern line of the Canadian Pacific Railway. The first mines were opened at Michel in 1898 and at nearby Morrissey in 1902, followed by others through the entire Crowsnest Pass. To the northwest, the most vigorous period of development of coal mines was from 1900 to 1910, when large operations were started in the Foothills at Nordegg, Mountain Park and Luscar (Fig. 2B.8). Mines were opened also at Drumheller in the Plains. Most of the production was used by the railways and for household heating. Farther north in British Columbia, the extensive coal deposits of the Peace River district were reported first by Dawson (1881), but were mined only for local use until recently. Lignite was produced in Saskatchewan by 1892 but was not in demand by the railway market.

The Athabasca oil sands, examined from time to time during the 1800s, continued to attract interest. After joining the federal Mines Branch in 1912, S.C. Ells made detailed topographical maps, attempted a drilling program, and tried to find commercial uses for the bituminous sands. Samples and equipment had to be hauled by scow, requiring tracking between Fort McMurray and the railhead at Athabasca Landing (Fig. 2B.9). In 1915, experimental pavement was laid by Ells in Edmonton, using Athabasca bituminous sand hauled south by sleigh and rail during the previous winter. Attempts to extract hydrocarbons from the sediments followed.

At the turn of the century, R.A. Daly was commissioned to undertake the geological examination of the boundary line (49th parallel) between Canada and the United States, from the plains westward across the mountains. His report (1912) contained an extensive discussion of the Purcell "Series" and equivalent rocks, the oldest rocks on the margin of the North American craton.

Paleontology played an important role throughout the geological studies of the western plains. In 1910 and 1911, C.D. Walcott of the Smithsonian Institution of Washington investigated remarkable Cambrian fossils of the Burgess Shale near Field, British Columbia. In 1914, F.H. McLearn started his long and distinguished career investigating Mesozoic fossils and biostratigraphy. Vertebrate fauna of the Red Deer River received much attention, particularly after 1909, when Barnum Brown made large collections for the American Museum of Natural History. C.H. Sternberg and his sons George, Charles M., and Levi were hired by the GSC to build up the Canadian collections. Later, C.M. Sternberg joined L. Lambe in Ottawa and was involved with the National Museum for many years.

Figure 2B.6. View of Medicine Hat, 1886, shortly after the railway bridge was completed over the South Saskatchewan River. A sternwheeler that operated along the river is docked against the far bank. Coal had been discovered along the river banks and drilling was undertaken to determine the extent of deposits for railway use (photo - Glenbow Museum, Calgary).

THE WAR YEARS 1914-1945

During World War I, new fuel deposits were needed for the industries of central Canada. These demands led to increased emphasis on obtaining data on stratigraphy, paleontology, and well records concerning the mineral resources of the Plains and Foothills. D.B. Dowling (Fig. 2B.2f) had been assigned at the beginning of the century to undertake a thorough investigation of the coal resources of Manitoba, Saskatchewan, Alberta, and eastern British Columbia. He produced a comprehensive report in 1914 and continued related studies until his death in 1925, when B.R. McKay took over coal investigations. Dowling, with S.E. Slipper and F.H. McLearn, published (1919) a summary of drilling records together with a series of subsurface contour maps showing the location and depths of various oil- and gas-bearing strata, as a guide to future drilling. By 1920, a good foundation had been laid for the understanding of the geological structure of the Plains. A four-year (1920-23) investigation of the Mackenzie Valley was undertaken by G.S. Hume, A.E. Cameron, D.B. Dowling, E.J. Whittaker, and M.Y. Williams. Hume (Fig. 2B.2g) continued studies of oil prospects in the southern plains for another two decades, providing much needed encouragement to the western oil and gas industry.

Although natural gas had been used at Medicine Hat since the end of the 19th century, it was used in Calgary only from 1910. After the discovery by Eugene Coste in 1912 of the Bow Island field along South Saskatchewan River, a 200-km pipeline was laid to Calgary.

The discovery in 1914 of natural gas with some associated light oil in Cretaceous sands at Turner Valley in the Foothills (Fig. 2B.10; Gould, 1976) had a profound effect on the search for petroleum. The second major development at Turner Valley occurred in 1924 when a substantial flow of gas was obtained from the Mississippian Rundle limestone. The reservoir had been drilled for naptha to be blended in automobile fuel, but natural gas had only limited use in Calgary. Between 1924 and 1931, 236 to 260 billion cubic feet of gas, one third to one half of the total reserves, was flared (Stenson, 1985). The third highlight at Turner Valley came in 1929 when substantial oil was encountered, and the fourth in 1936 when oil was discovered on the western flank of the anticlinal structure.

Edmonton had to wait until 1923 to get gas from the Viking field through a 130-km pipeline laid by Northwest Utilities (Stenson, 1985). The discovery of commercial dry

Figure 2B.7. Drilling rig at Victoria, Alberta, 1898. This rig was used in one of the early GSC attempts to locate commercial quantities of oil and gas on the prairies. This hole was abandoned at a depth of 1840 feet (photo - GSC 1615).

Figure 2B.8. View of Mountain Park, early 1920s. The town had developed adjacent to the mine, which obtained coal from the Cretaceous Luscar Group. The coal seams lie immediately west of the buildings and dip gently westward into the tree-covered hills. Mining ceased at the end of the Second World War (1945) and the town was abandoned in 1950 (photo - Glenbow Museum, Calgary).

Figure 2B.9. A line of men tracking scows on the Athabasca River, Alberta, early 1900s. Supplies were transported along the river between Fort McMurray and the railhead at Athabasca, north of Edmonton (photo - Glenbow Museum, Calgary).

Figure 2B.10. Drilling rigs at Turner Valley during early 1930s (photo - Lane Studio, Glenbow Museum, Calgary).

gas at Viking and Kinsella was followed by the discovery in 1923 of oil at Wainwright. In 1934, a large flow of natural gas was obtained near Lloydminster, followed by the discovery of heavy black oil in Cretaceous sands.

Geological exploration was by no means confined to the southern Plains and Athabasca oil sands. The Mackenzie Valley continued to be explored for conventional oil and gas. The Imperial Oil Company, already well established as Canada's largest oil company, began the first large-scale exploration in the Northwest Territories in 1919. T.A. Link, in charge of the lower Mackenzie region, recommended a drilling site near Fort Norman, and in 1920 a commercial flow of oil was yielded by the Norman Wells No. 1 well (Fig. 2B.11). The Norman Wells reservoir was recognized as a Devonian reef, the first such trap described from North America (Sproule, 1968). A small refinery was installed but had limited production until World War II and the Canol Project.

With the population increasing as settlers poured into the provinces of Alberta and Saskatchewan, which had been newly created in 1905, demands for education and research were met by the establishment of new universities. J.A. Allan became the first professor of geology at the University of Alberta in 1912. He was joined in 1923 by R.L. Rutherford, who had produced the first composite geological map of Alberta in 1920. Both were part-time employees of the federal and provincial governments and both became actively involved in the search for petroleum. In 1920 P.S. Warren had joined the staff of the Geology Department, and for half a century his name was linked to biostratigraphic studies of western Canada. Later, Warren was joined by C.R. Stelck. M.Y. Williams joined the staff of the University of British Columbia in 1921. He continued to work many summers thereafter as a stratigraphic paleontologist in northeastern British Columbia, District of Mackenzie, and southwestern

Figure 2B.11. Discovery well at Norman Wells, N.W.T., drilled by Imperial Oil Limited, 1920-1921 (photo - Glenbow Museum, Calgary).

Alberta. F.H. Edmunds of the University of Saskatchewan became involved in Quaternary geology and in exploration for petroleum around Lloydminster and in industrial minerals generally. Similarly, at the University of Manitoba R.C. Wallace and S.R. Kirk published on many stratigraphic and economic matters.

Although the provinces of Alberta and Saskatchewan had been removed from the Northwest Territories in 1905, their mineral resources were transferred to provincial jurisdiction only in 1930. Manitoba, since joining confederation in 1870 and through its enlargements in 1881 and 1912, had held the mineral rights to all crown lands. Most of the mineral rights in the Northwest Territories of 1870 resided with the federal government, although ownership of mineral rights on some land holdings was retained by the Hudson's Bay Company and also by the Canadian Pacific Railway under its 1881 charter. As a result, the management of the exploration and development of mineral resources remained largely a federal responsibility during the last third of the nineteenth century and the early part of the twentieth. In Manitoba, the Geological Services Branch, Mineral Resources Division, Department of Mines and Natural Resources issued its First Annual Report in 1931. The same year saw the establishment of the Mines Branch of the Saskatchewan Department of Natural Resources, out of which in time the Saskatchewan Geological Survey would emerge.

The Scientific and Industrial Research Council of Alberta (later Research Council of Alberta), established in 1920, undertook geological investigations, particularly in the Athabasca oil sands. In 1921, K.A. Clark accepted a position with the newly formed Council, and began a long career of research into methods of oil-sand separation. The work was carried on in co-operation with the federal Mines Branch. The unsuccessful commercial venture during the 1930s of Abasand Oil Limited was followed by a demonstration plant at Bitumount using a hot-water process developed by Clark. The plant was operated successfully in 1948 and 1949 and stimulated the major oil companies into recognizing the Athabasca oil sands as a potential source of oil for the commercial market. Recovery on a commercial basis was not achieved until much later, however, and the history becomes that of resource exploitation rather than of exploration (Carrigy, 1963).

After the transfer of mineral rights to Alberta and Saskatchewan in 1930, the Geological Survey of Canada remained active in the west as well as in the north where the federal government retained control over all lands. Cuttings and cores of wells drilled were kept by the Borings Division. The division was considerably expanded in 1930 when a new Division of Pleistocene Geology, Water Supply and Boring was established. Surficial studies during the 1930s concentrated on groundwater, and were highly important in the southern plains during the drought years when new supplies of water were located for Regina and Moose Jaw.

New techniques were applied to petroleum exploration. Aerial photography was introduced by Imperial Oil in northern Alberta and Mackenzie Valley (de Mille, 1970). In 1928, a gravity survey was tried in Turner Valley. Seismic surveys, based on techniques developed during World War I also were carried out in Turner Valley field in 1928 (Beach, 1954). In 1929, the Schlumberger brothers introduced the electric-magnetic means of investigating the electrical properties of drilling fluid and surrounding rocks in boreholes. That geophysical method was followed in 1939 and 1940 by the introduction of gamma-ray and neutron logs to measure the variation in radioactivity (Bates et al., 1982). Drilling equipment and methods also changed at the time of World War I. Whereas in the nineteenth century, derricks were made of wood, their replacement by steel structures in the twentieth century allowed for much deeper holes. The hydraulic rotary system of drilling using mud under pressure was tried at Turner Valley in 1925, but cable tool equipment still remained in use for many years after that.

After the Japanese air attack on Pearl Harbour, Hawaii, in 1941, during World War II, the possibility of Japanese attacks on Alaska and of enemy interruption of tanker traffic on the west coast caused the United States to look inland for a supply route. As a result, the precursor of the present Alaska Highway was built in Canada, linking Alaska with the south, and a pipeline was built from the Norman Wells oilfield to a refinery on the highway at Whitehorse, Yukon, a distance of some 1100 km. The construction of this pipeline, known as the Canol Project, was completed in 1944 but the pipeline was shut down in 1945 after less than 11 months of operation. The refinery was removed later to Edmonton. A summary of the geological exploration done for the Canol Project was provided by G.S. Hume and T.A. Link (1945). Other studies undertaken farther south in northeastern British Columbia along the highway route in a search for additional oil supplies include the work of C.O. Hage (1944), E.D. Kindle (1944), and M.Y. Williams (1944).

THE MODERN ERA, 1945-

The search for petroleum intensified in the 1940s, during and after World War II, and Imperial Oil undertook an extensive drilling program throughout the Plains. Finally, in 1947, after drilling 133 unsuccessful wildcat wells, Imperial found oil at Leduc in a prolific Devonian reef. Following the Leduc discovery, the search for oil concentrated on Devonian reefs. The Redwater field northeast of Edmonton was discovered in 1948, Golden Spike in 1949, and the Fenn-Big Valley trend in 1950, followed by Wizard Lake in 1951, and Acheson, Bonnie Glen, and Westerose in 1952. Mississippian oil in stratigraphic traps was discovered at Daly in the Williston Basin in Manitoba in 1951, and at Midale, Saskatchewan, in 1953. Gas was found in the Triassic of northeastern British Columbia in 1952, followed by oil discoveries in 1955. In 1953, the discovery of Pembina, the first billion-barrel field, in sandstones of the Cretaceous Cardium Formation of Alberta (Nielsen and Porter, 1984), changed the focus of oil exploration.

From 1950 on, petroleum companies explored the Western Canada Basin in great detail. Many companies systematically studied the outcrop belt, extrapolating their interpretations into the adjacent subsurface succession. Seismic lines criss-crossed the prairies and foothills, many being shot more than once. Drilling kept pace with exploration. A vast amount of geological and geophysical information was accumulated; some data such as samples, core, and well records are stored and made available to the public by the provincial agencies administering petroleum developments. Advanced computer technology has facilitated the manipulation and interpretation of these data.

Following the oil discovery at Leduc in 1947, demands increased markedly for information on the subsurface rocks of the Interior Plains and on the sediments and structures of the adjacent Rocky Mountains and Foothills. To provide the data essential for petroleum exploration, the Geological Survey of Canada opened an office in Calgary in 1950, which in 1967 became the Institute of Sedimentary and Petroleum Geology. The GSC maintained its studies in the outcrop belt, completing the reconnaissance mapping of the Rocky Mountains, Foothills, and Mackenzie-Yukon region, all of which applied directly to the Interior Platform and foreland basin. Large, helicopter-supported operations after 1955 accelerated this work, mapping being co-ordinated with stratigraphical, sedimentological, and paleontological studies. These investigations were fundamental in establishing the geological framework for a vast region of western Canada.

R.J.W. Douglas of the GSC made landmark contributions between 1950 and 1979 to the structure, stratigraphy, and tectonic evolution of the Foothills and Front Ranges of the Rocky Mountains. He provided new concepts on the origin of thrust faults, the relationships between folding and thrust faulting, and the overall kinematics of the evolution of the Foothills. A massive synthesis (1970), for which he was editor and a major contributor, documented the geology of Canada, including the regions covered in this volume.

The neglect of paleontology by the GSC during the 1930s was followed in the 1940s and 1950s by a major blossoming of activity. Continuing exploration for oil and gas demanded an improved knowledge of the rock succession, and put new emphasis on paleontological work. Among the investigations were W.A. Bell's paleobotanical studies of coal-bearing rocks. It was at this time that Hans Frebold started his long career studying Jurassic fossils; J.A. Jeletzky, Cretaceous fossils; D.J. McLaren, Devonian fossils. F.H. McLearn returned to Triassic studies, which later were continued by E.T. Tozer. L.S. Russell worked on vertebrate fossils, particularly of the Tertiary Period. New fields of research developed as microfossils became important in the study of well samples. Micropaleontology was utilized to good advantage by such workers as A.W. Nauss, R.T.D. Wickenden, C.R. Stelck, and somewhat later, J.H. Wall. Palynology is a more recent speciality, applied successfully in the plains but to a lesser degree in the more highly deformed rocks of the Foothills. Conodonts have proven extremely useful, and as in palynomorphs, their colour has been used with good results in estimating the degree of maturation of hydrocarbon-bearing rocks.

With increased drilling after 1950, the various provincial agencies became active in subsurface studies. The Manitoba Department of Mines and Natural Resources published reports on the Devonian by A.D. Baillie (1953), on the Mississippian by H.R. McCabe (1959) and on the Jurassic by D.F. Stott (1955). The Saskatchewan Department of Mineral Resources prepared several regional studies, including those on the Devonian by D.M. Kent (1963, 1968) and M.E. Holter (1969); on the Mississippian by J.G.C.M. Fuller (1956) and L.M. Fuzesy (1960); on the Jurassic by D.R. Francis (1956); and on the Cretaceous by J.E. Christopher (1974). Other studies were undertaken by the Saskatchewan Research Council, including those pertaining to Pleistocene geology, groundwater, and coal (Whitaker et al., 1978). The Research Council of Alberta (within which the Alberta Geological Survey was established recently) concentrated its work on Cretaceous rocks, including studies of oil sands by M.A. Carrigy (1966) and numerous reports on biostratigraphy by J.H. Wall and C. Singh; on coal by J.D. Campbell; on Pleistocene deposits by C.P. Gravenor and L.A. Bayrock; and on groundwater by R.N. Farvolden and J. Toth. Several professors from the University of Alberta, including P.S. Warren and C.R. Stelck, carried on collaborative programs, particularly in biostratigraphy.

The construction of modern sedimentological models for carbonate and evaporite formations overlapped and succeeded the work of establishing basic stratigraphy, and herein geologists of the petroleum industry played a conspicuous role. Regional extension in the subsurface by D.C. Pugh (1971, 1973, 1975) of large-scale Cambrian cycles recognized and interpreted by J.D. Aitken (1966, 1978) effected a new integration of Cambrian deposition. I.A. McIlreath (1977) provided new insights into the sedimentology of the Cambrian Cathedral Reef-Escarpment and its surroundings. Ordovician stratigraphy and sedimentology of Williston Basin were examined by J.W. Porter and J.G.C.M. Fuller (1964 and other papers), and of the southern part of the outcrop belt by J.D. Aitken and B.S. Norford (1967). The complex facies changes between Mackenzie Platform and Misty Creek Embayment within Cambrian to Devonian formations were examined in detail by M.P. Cecile (1982).

Because of its immense economic significance, the Devonian System of Western Canada has received the most attention in terms of depositional models. The complex, evaporite- and carbonate-dominated fill of Elk Point Basin has been the subject of many studies, among which those of J.G. McCamis and L.S. Griffith (1967), A.M. Klingspor (1969), D.G. Bebout and W.R. Maiklem (1973), N.C. Wardlaw and G.E. Reinson (1971), and G.R. Davies and S.D. Ludlam (1973) are especially worthy of note. A recent, provocative paper by G.K. Williams (1984) demonstrates that several problems of Elk Point Basin are not yet resolved and may be unresolvable. Canadian studies of Devonian reefs and their much-debated relations to enclosing sediments have provided models for reef-bearing basins world-wide. Some of the early studies of reef and off-reef facies in the Upper Devonian were made by H.R. Belyea (1964). Later subsurface studies include those by J.E. Klovan (1964), N.R. Fischbuch (1968), N.M. Sheasby (1971), and F.A. Stoakes (1980), and supporting studies of outcrop examples by J. Dooge (1966) and E.W. Mountjoy (1978). More general studies of great interpretive importance were those of T.A. Oliver and N.W. Cowper (1963) emphasizing clinoform sedimentation, P.A. Ziegler (1967) on Devonian tectonics and facies north of 60°, and the classic and still largely valid interpretation by H.G. Bassett and J.G. Stout (1968) of Devonian facies throughout the Western Canada Basin.

Important, modern contributions to Carboniferous stratigraphy and sedimentology were made by R.J.W. Douglas (1958), G. Macauley et al. (1964), R.W. Macqueen and E.W. Bamber (1968), Bamber and J.B. Waterhouse (1971), Bamber and B.L. Mamet (1978), and W.D. Stewart and R.G. Walker (1980), and (in the far north) by A.D. Graham (1973) and D.C. Pugh (1983).

Landmark papers on Permian stratigraphy and sedimentology include those of E.W. Bamber and J.B. Waterhouse (1971), I.H. Naqvi (1972), A. McGugan et al. (1964, 1968), Bamber and R.W. Macqueen (1979), and A.D. Graham (1973).

Similarly, as workers concentrated on detailed refinements of Mesozoic geology, new models were proposed for clastic sediments of the Interior Platform and foreland basin. The character of Triassic, fine-grained clastics, carbonates, and evaporites of the Interior Platform as exposed in the Rocky Mountain Foothills was documented by Barss et al. (1964) and D.W. Gibson (1974, 1975). Various aspects of Triassic sedimentology, related to the producing fields in northeastern British Columbia and west-central Alberta were discussed by A.D. Miall (1976), D.L. Barss and F.A. Montandon (1981), J.M. Bever and I.A. McIlreath (1984), and D. Cant (1984a).

The stratigraphy and sedimentology of Jurassic rocks of Williston Basin were outlined by J.E. Christopher (1966, 1974). Several recent reports deal with sedimentology in the subsurface of Alberta, including those of G.R. Davies (1983), B.J.R. Hayes (1983), and D.J. Marion (1984). The petrology and paleoenvironments of the Jurassic-Cretaceous succession in southwestern Rocky Mountains and Foothills were described by D.K. Norris (1964), J.E. Rapson (1964, 1965), D.W. Gibson (1985), L. Jansa (1972), and A.P. Hamblin and R.G. Walker (1979).

Cretaceous sequences have been studied extensively in the Western Canada Basin. In the Foothills, the detailed study by G.B. Mellon (1967) pointed out the strong influx of volcanic-derived material within the Lower Cretaceous. Related detailed studies of facies and sedimentology include those of J.R. McLean (1982) and D.F. Stott (1968 and subsequently). Many studies have been made of the Lower Cretaceous Mannville Group in local areas of the Plains, for example, that by J.C. Hopkins (1981). Recently, D. Leckie (1986) and D.J. Cant (1984b, 1986) analyzed the marine Gates sandstones and equivalent rocks of the Lower Cretaceous of northeastern British Columbia and western Alberta. The Cardium Formation, the largest oil reservoir in western Canada, has been studied intensively by R.G. Walker (1983, 1985), and A.G. Plint and their students (Plint et al., 1986). Studies of Upper Cretaceous and Tertiary strata include those of M.A. Carrigy (1971) in southern Alberta, T. Jerzykiewicz (1985) and T. Jerzykiewicz and J.R. McLean (1980) in west-central Alberta, and J. Dixon (1986) in Yukon Territory and District of Mackenzie. Similar strata in southern Saskatchewan were described by J.R. McLean (1971) and S.H. Whitaker, J.A. Irvine, and P.L. Broughton (1978).

A comprehensive summary of Tertiary rocks of the area south of 60°N latitude was published by Taylor et al. (1964). More recent studies include those of M.A. Carrigy (1971), T. Jerzykiewicz and J.R. McLean (1980), and T. Jerzykiewicz (1985). In the northern territories, J. Dixon (1986) synthesized the reports by many workers.

The development of plate tectonic concepts has significantly modified the interpretation of the development of the Western Canada Basin. Two major syntheses on the basin, published two decades ago (McCrossan and Glaister, 1964; Douglas et al., 1970) still retained models based on geosynclinal concepts in which the western Cordillera was considered to be a part of North America throughout Phanerozoic time. The application of plate tectonic concepts by J.W.H. Monger and R.A. Price (1979), J.W.H. Monger, R.A. Price, and D.J. Tempelman-Kluit (1982), R.A. Price (1981), and J.W. Porter, R.A. Price, and R.G. McCrossan (1982) showed that much of the Cordillera is a tectonic collage of terranes that were accreted to the cratonic margin after Triassic time. These concepts led directly to major revisions in the sedimentological, paleogeographic, and tectonic models for the various systems. Previously, A.W. Bally, P.L. Gordy, and G.A. Stewart (1966) had shown that the Rocky Mountain fold belt is underlain by a gently westward dipping extension of the crystalline Precambrian Shield. Recent quantitative modelling by C. Beaumont (1981) illustrated that the development of the fold-thrust belt is consistent with its being underlain by a thick lithospheric plate. The postulated opening of Canada Basin in the northwest margin of the craton (Vogt et al., 1981; Grantz and May, 1983; Sweeney, 1985; McWhae, 1986) and the formation of transcurrent faults along the western margin during Jurassic and Cretaceous time (Tempelman-Kluit, 1979; Gabrielse, 1985) have greatly altered views of the later history of the Interior Platform and foreland basins.

The origin of petroleum deposits in the Western Canada Basin has been debated extensively. Geochemistry has assumed a major role as greater emphasis has been placed on determining source, migration routes, accumulation factors, and regional variations of formation fluids. Early analyses of crude oil were carried out by the Alberta Oil and Gas Conservation Board, Alberta Research Council, Saskatchewan Department of Mineral Resources, and Fuels Research Laboratories, Ottawa. Work at the Ottawa Laboratories was carried on for many years by a group headed by D.S. Montgomery (e.g. Montgomery et al., 1974). Studies of the geochemistry of formation fluids, and data characterizing regional variations of crude oils, gases, and formation waters were provided by B. Hitchon of the Alberta Research Council (1964, 1984; Hitchon et al., 1971). Private industry now has assumed a share of this research. A recent study of geochemical and thermal maturation of petroleum in the Western Canada Basin is that of G. Deroo, B. Tissot, T.G. Powell, R.G. McCrossan, and P. Hacquebard (1977). The geochemistry of heavy oils of Alberta was summarized by Deroo, Tissot, and McCrossan in 1974. Other recent studies include those of L.R. Snowdon and T.G. Powell (1982) and G. Macauley, L.R. Snowdon, and F.D. Ball (1985) on the geochemistry of oil shale deposits.

THE LEGACY

Canada's three prairie provinces have benefitted greatly from the geological exploration that started in the early 1800s and continued to the present. Not only has a large petroleum industry developed but other mineral industries owe their beginnings to the pioneer geologists, who were interested not only in geology but in all facets of natural resources. Macoun, Dawson, McConnell, and Tyrrell, among many, reported in detail on the vegetation, fertility of the soil, availability of water and fuels, and general suitability for settlement. It was largely because of the reports of these pioneer geologists that settlers moved westward when the railroad provided access.

The production of industrial minerals, such as salt, gypsum, and building materials, has grown into major industries. The early production of salt from saline springs was replaced by that from brines encountered in deep wells. Salt was obtained from wells at Fort McMurray in Alberta, Unity in Saskatchewan, and Neepawa in Manitoba. Six operations (two in Alberta and four in Saskatchewan) now produce salt from brines. Similarly, the quarrying of building stone has developed into a major industry. During early settlement, stone quarried from the banks of the Red River was used in the construction of Lower Fort Garry, built in 1832 (Stewart, 1972). Later, Ordovician dolomitic limestone near Selkirk was used for buildings in Winnipeg. By the turn of the century, Tyndall stone had been used for the legislative buildings in Winnipeg and Regina, as well as the Parliament Buildings in Ottawa, and continues to be a favoured building stone in the western provinces. Many operations throughout the plains produce sand and gravel for concrete; gypsum is mined for building material; clays are used extensively in brick manufacture. Two recent publications summarize the present status of industrial minerals in Canada; one by Industrial Minerals (1984) and the other by the Canadian Institute of Mining and Metallurgy (Guillet and Martin, 1984). In addition, reviews of the Canadian mineral industry are published annually by Energy, Mines and Resources, Canada.

The potash deposits of Williston Basin were not discovered until 1943 when Imperial Oil drilled a well at Radville, Saskatchewan. Originally, the depth (1.1 km) was considered too great for commercial exploitation, but three years later more interest was aroused by a richer occurrence. The first commercial exploitation took place in 1953 at Patience Lake, east of Saskatoon, by the Potash Company of America. Now 10 potash mines are active in Saskatchewan, one of which uses the solution method of extraction while all others are conventional underground mines.

Lead-zinc showings at Pine Point, Northwest Territories, were examined by Robert Bell (1900), by Camsell in 1914, and by A.E. Cameron (1917), but apparently the ore was considered uneconomic because it lacked silver. However, as prices increased, Pine Point ore became valuable for its lead and zinc content alone and production began in 1964. Recent decreases in the value of base metals, however, resulted in the closing of operations at Pine Point in 1988.

At present, substantial coal operations are found both on the prairies and in the Foothills. Many of the early mines, such as Nordegg, Luscar, and Mountain Park in the Foothills and Drumheller in the Plains, were phased out in the middle of the 20th century. However, after two decades of depressed conditions, the Canadian coal industry, beginning about 1970, entered a period of rapid expansion. The development of increasing export markets for metallurgical coal and local markets for thermal coal led to the resurgence of coal exploration. In the foothills and mountains, large open-pit operations were developed in the region of Fernie Basin, in the central Alberta Foothills at Cardinal and Gregg rivers, and at Quintette and Bullmoose mountains in northeastern British Columbia. In these regions, the new developments have required the construction of new towns, new rail lines, and new infrastructure. In the Plains, expansion began after the middle of the 20th century with the introduction of large coal-fired electrical utility stations and the introduction of surface mining. Open pits are operated at Coal Valley and Wabamun west of Edmonton, in the Forestburg area east of Red Deer, Alberta, and near Estevan, Saskatchewan.

Several summaries have been prepared recently on coal resources. A unique report in atlas format (Whitaker et al., 1978), prepared under a joint federal-provincial cost-sharing agreement, presents an estimate of the magnitude and distribution of coal resources in southern Saskatchewan. The assessment is based on a drilling program, detailed geophysical and geological studies, and innovative computer techniques. A reference volume on Canadian coal deposits and methods used to exploit them was published recently by the Canadian Institute of Mining and Metallurgy (Patching, 1985).

The petroleum industry in western Canada has reached a stage of maturity, with the drilling of more than 100 000 wells defining 3000 oil pools and 9000 gas pools. Major discoveries of the last two decades include the Rainbow-Zama, Devonian reef pools and the Elmworth gas of the Deep Basin of northeastern British Columbia (Masters, 1985). Despite the drilling activity of the last 40 years, new discoveries continue to be made. Of recent interest is the West Pembina play involving pinnacle reefs in the Devonian, the Desan area of northeastern British Columbia, a Carboniferous discovery, and the Waskada area of Saskatchewan where the stratigraphic trap involves the unconformity between Paleozoic carbonates and Jurassic Watrous Formation. Full development of the giant Norman Wells pool has led to the building of a pipeline to southern markets. Given adequate oil prices, the recent oil discoveries in the Beaufort Sea also will lead to pipeline construction along the Mackenzie Corridor and improve access to markets for additional discoveries in the northern mainland.

Industrial activity resulting from the production of natural gas has been the recovery of sulphur and the development of the petrochemical industry. Sulphur has become a major commodity for export, and drilling in several areas is aimed at exploiting large accumulations of hydrogen sulphide.

The enormous potential of the Alberta Oil Sands continues to be evaluated, and developed (Carrigy, 1963, 1966; Hills, 1974). Commercial production was achieved in 1967 by Great Canadian Oil Sands, later Suncor, with the opening of a plant near Fort McMurray, which originally produced 50 000 barrels of oil per day and later was expanded to 125 000 barrels per day. A second and larger operation was begun in 1978 by Syncrude. Other plants have been proposed, but developments await more favourable economic conditions.

The heavy oils of the Cold Lake area in Alberta and the Lloydminster area at the Alberta-Saskatchewan boundary, have been subjected to numerous pilot studies. Early in 1986 an announcement was made that an upgrading facility was to be built by Husky Oil Limited to bring this resource into production.

Much of the exploration of the last four decades has been carried out by industry, and although, in general, the companies have not published large comprehensive reports of their investigations, many papers on specific topics appear in publications of the Geological Association of Canada, Canadian Society of Petroleum Geologists, Canadian Institute of Mining and Metallurgy, and others. Many of these are noted in the following chapters. Also, the companies have contributed information freely to symposia and syntheses prepared by various groups. Lengthy discussions on the hydrocarbon potential of the Western Canada Basin, Hudson Platform, and St. Lawrence Platform were provided in "The Future Petroleum Provinces of Canada - Their Geology and Potential" (McCrossan, 1973). A recent volume prepared by employees of Canadian Hunter Exploration Ltd., and dealing with "Elmworth - Case Study of a Deep Basin Gas

Field" (Masters, 1985), outlines a supergiant gas field found in a subtle trap beneath the western edge of the Peace River Plains.

As exploration increased, so did the geological community, and various organizations grew up to meet its needs. The Canadian (formerly Alberta) Society of Petroleum Geologists, starting with a modest 19 members in 1928, has grown to about 4000 (by 1987), one measure of the growth of the industry in Canada. Similarly, the Canadian Society of Exploration Geophysicists and Canadian Institute of Mining and Metallurgy developed to meet the needs of specific groups of explorationists.

Rapid accumulation of geological information, owing in large part to extensive drilling throughout the western plains after the Leduc discovery, led to several major syntheses. The first, in atlas format, "Geological History of Western Canada", edited by R.G. McCrossan and R.P. Glaister (1964) was published by the Alberta Society of Petroleum Geologists. This was followed in 1970 by the periodic update by the Geological Survey of Canada of "Geology and Economic Minerals of Canada" (Douglas, 1970), which included summaries on the geology of western Canada, Hudson Platform, and St. Lawrence Platform. Several major volumes deal with specific aspects of geology. These include volumes on the Devonian System (Oswald, 1968), Permian and Triassic systems (Logan and Hills, 1973), Cretaceous System (Caldwell, 1975), and Mesozoic rocks (Stott and Glass, 1984). Many other reports are available, including symposia volumes on Williston Basin, published periodically by the Saskatchewan Geological Society.

In summary, earth science research has had an immense social impact on the settlement of the western plains, preceding and accompanying the development of the petroleum and agricultural industries. The fossil-fuel and mineral industries continue to play a leading role throughout the region. In addition to employing large segments of the population and supporting the service industries, the petroleum industry generates substantial sums for the provincial and federal governments through royalties and other taxes.

REFERENCES

Aitken, J.D.
1966: Middle Cambrian to Middle Ordovician cyclic sedimentation, southern Rocky Mountains of Alberta; Bulletin of Canadian Petroleum Geology, v. 14, p. 405-411.
1978: Revised models for depositional Grand Cycles, Cambrian of Western Canada; Bulletin of Canadian Petroleum Geology, v. 26, p. 515-542.

Aitken, J.D. and Norford, B.S.
1967: Lower Ordovician Survey Peak and Outram formations, southern Rocky Mountains of Alberta; Bulletin of Canadian Petroleum Geology, v. 15, p. 150-207.

Baillie, A.D.
1953: Devonian System of the Williston Basin area; Manitoba Department of Mines and Natural Resources, Mines Branch Publication 52-5, 105 p.

Bally, A.W., Gordy, P.L., and Stewart, G.A.
1966: Structure, seismic data, and orogenic evolution, southern Canadian Rocky Mountains; Bulletin of Canadian Petroleum Geology, v. 14, p. 337-381.

Bamber, E.W. and Macqueen, R.W.
1979: Upper Carboniferous and Permian stratigraphy of the Monkman Pass and southern Pine Pass areas, northeastern British Columbia; Geological Survey of Canada, Bulletin 301, 27 p.

Bamber, E.W. and Mamet, B.L.
1978: Carboniferous biostratigraphy and correlation, northeastern British Columbia and southwestern District of Mackenzie; Geological Survey of Canada, Bulletin 266, 65 p.

Bamber, E.W. and Waterhouse, J.B.
1971: Carboniferous and Permian stratigraphy and paleontology, northern Yukon Territory, Canada; Bulletin of Canadian Petroleum Geology, v. 19, p. 19-250.

Barss, D.L., Best, E.W., and Meyers, N.
1964: Triassic; Chapter 9 in Geological History of Western Canada, R.G. McCrossan and R.P. Glaister (ed.); Alberta Society of Petroleum Geologists, Calgary, Alberta, p. 113-136.

Barss, D.L. and Montandon, F.A.
1981: Sukunka-Bullmoose gas fields: Models for a developing trend in the southern foothills of northeast British Columbia; Bulletin of Canadian Petroleum Geology, v. 29, p. 293-333.

Bassett, H.G. and Stout, J.G.
1968: Devonian of western Canada; in International Symposium on the Devonian System, D.H. Oswald (ed.); Alberta Society of Petroleum Geologists, v. 1, p. 717-752. (Imprint 1967).

Bates, C.C., Gaskell, T.F., and Rice, R.B.
1982: Geophysics in the Affairs of Man; Pergamon Press, Oxford, 402 p.

Beach, H.H.
1954: Our heritage in the exploration of Western Canada; Canadian Mining and Metallurgical Transactions, v. 57, p. 1-12.

Beaumont, C.
1981: Foreland basins; Geophysical Journal of the Royal Astronomical Society, v. 65, p. 291-329.

Bebout, D.G. and Maiklem, W.R.
1973: Ancient anhydrite facies and environments, Middle Devonian Elk Point Basin, Alberta; Bulletin of Canadian Petroleum Geology, v. 21, p. 287-343.

Bell, Robert
1881: On the occurrence of petroleum in the Northwest Territories, with notes on new localities; The Canadian Journal, Proceedings of the Canadian Institute, new series, v. 1, pt. 2, p. 225-230.
1885: Report on part of the basin of the Athabasca River, North-West Territory; Geological Survey of Canada, Report of Progress 1882-83-84, pt. CC, 35 p.
1900: Report on explorations in the Great Slave Lake region, Mackenzie District; Geological Survey of Canada, Summary Report for the year 1899, p. 103-110.

Belyea, H.R.
1964: Upper Devonian; Chapter 6, Part II in The Geologic Atlas of Western Canada, R.G. McCrossan and R.P. Glaister (ed.); Alberta Society of Petroleum Geologists, p. 60-88.

Bever, J.M. and McIlreath, I.A.
1984: Stratigraphy and reservoir development of shoaling-upward sequences in the Upper Triassic (Carnian) Baldonnel Formation, northeastern British Columbia (abstract); in Exploration Update '84, Program and Abstracts; Canadian Society of Petroleum Geologists - Canadian Society of Exploration Geophysicists National Convention, Calgary, Alberta, 1984, p. 147.

Brown, A.
1985: The development of the coal mining industry in Canada; in Coal in Canada, T.H. Patching (ed.); The Canadian Institute of Mining and Metallurgy, Special Volume 31, p. 5-8.

Caldwell, W.G.E. (ed.)
1975: The Cretaceous System in the Western Interior of North America; Geological Association of Canada, Special Paper 13, 666 p.

Cameron, A.E.
1917: Reconnaissance on Great Slave Lake, Northwest Territories; Geological Survey of Canada, Summary Report for 1916, p. 66-76.

Cant, D.J.
1986: Development of shoreline-shelf sand bodies in a Cretaceous epeiric sea deposit; Journal of Sedimentary Petrology, v. 54, p. 541-556.
1984a: Possible syn-sedimentary tectonic controls on Triassic reservoirs: Halfway, Doig, Charlie Lake formations, west-central Alberta (abstract); in Exploration Update '84, Program and Abstracts; Canadian Society of Petroleum Geologists - Canadian Society of Exploration Geophysicists National Convention, Calgary, Alberta, 1984, p. 45-46.
1984b: Development of shoreline-shelf sand bodies in a Cretaceous epeiric sea deposit; Journal of Sedimentary Petrology, v. 54, p. 541-556.

Carrigy, M.A. (ed.)
1963: Athabasca Oil Sands; Research Council of Alberta, K.A Clark Volume, Information Series No. 45, 241 p.
1966: Lithology of the Athabasca Oil Sands, Alberta; Research Council of Alberta, Report 18, 48 p.
1971: Lithostratigraphy of the uppermost Cretaceous (Lance) and Paleocene strata of the Alberta Plains; Research Council of Alberta, Bulletin 27, 161 p.

Cecile, M.P.
1982: The Lower Paleozoic Misty Creek Embayment, Selwyn Basin, Yukon and Northwest Territories; Geological Survey of Canada, Bulletin 335, 78 p.

Christopher, J.E.
1966: Shaunavon (Middle Jurassic) sedimentation and vertical tectonics, southwestern Saskatchewan; Billings Geological Society, Guidebook, 17th Annual Field Conference, p. 18-35.
1974: The Upper Jurassic Vanguard and Lower Cretaceous Mannville Groups of southwestern Saskatchewan; Saskatchewan Department of Mineral Resources, Report no. 151, 349 p.

Daly, R.A.
1912: Geology of the North American Cordillera at the forty-ninth parallel; Geological Survey of Canada, Memoir 38, 799 p.

Davies, G.R.
1983: Sedimentology of the Middle Jurassic Sawtooth Formation of southern Alberta; in Sedimentology of Selected Mesozoic Clastic Sequences, J.R. McLean and G.E. Reinson (ed.); The Mesozoic of Middle North America Conference, Corexpo 83, Canadian Society of Petroleum Geologists, p. 11-25.

Davies, G.R. and Ludlam, S.D.
1973: Origin of laminated and graded sediments, Middle Devonian of western Canada; Geological Society of America, Bulletin, v. 84, p. 3527-3546.

Dawson, G.M.
1875: Report on the geology and resources of the region in the vicinity of the forty-ninth parallel from the Lake of the Woods to the Rocky Mountains; British North American Boundary Commission, Dawson Brothers, Montreal, 387 p.
1881: Report on an exploration from Port Simpson on the Pacific Coast, to Edmonton on the Saskatchewan, embracing a portion of the northern part of British Columbia and Peace River country; Geology and Natural History Survey, Canada, Report of Progress 1879-80, pt. B, p. 1-177.
1885: Report on the region in the vicinity of Bow and Belly Rivers, Northwest Territories; Geological and Natural History Survey and Museum, Report of Progress 1882-83-84, pt. C, 169 p.
1886: On certain borings in Manitoba and the Northwest Territories; Royal Society of Canada, Transactions, sec. 4, p. 85-99.

Dawson, S.J.
1859: Report on the Exploration of the Country between Lake Superior and the Red River Settlement, and between the Latter Place and the Assiniboine and Saskatchewan; John Lovell, Printer, Toronto, 45 p.

de Mille, G.
1970: Oil in Canada West - the Early Years; Northwest Printing and Lithographing Ltd., Calgary, 269 p.

Deroo, G., Powell, T.G., Tissot, B., and McCrossan, R.G., with contributions by Hacquebard, P.A.
1977: The origin and migration of petroleum in the Western Canadian Sedimentary Basin, Alberta; a geochemical and thermal maturation study; Geological Survey of Canada, Bulletin 262, 136 p.

Deroo, G., Tissot, B., McCrossan, R.G., and Der, F.
1974: Geochemistry of the heavy oils of Alberta; in Oil Sands - Fuel of the Future, L.V. Hill (ed.); Canadian Society of Petroleum Geologists, Memoir 3, p. 148-167.

Dixon, J.
1986: Cretaceous to Pleistocene stratigraphy and paleogeography, northern Yukon and northwestern District of Mackenzie; Bulletin of Canadian Petroleum Geology, v. 34, p. 49-70.

Dooge, J.
1966: The stratigraphy of an Upper Devonian carbonate-shale transition between the North and South Ram rivers of the Canadian Rocky Mountains; Leidse Geologische Mededelingen, no. 39, p. 1-53.

Douglas, R.J.W.
1958: Mount Head map-area, Alberta; Geological Survey of Canada, Memoir 291, 241 p.

Douglas, R.J.W. (ed.)
1970: The Geology and Economic Minerals of Canada; Geological Survey of Canada, Economic Geology Report No. 1, Fifth Edition, 838 p.

Douglas, R.J.W., Gabrielse, H., Wheeler, J.O., Stott, D.F., and Belyea, H.R.
1970: Geology of Western Canada; Chapter VIII in Geology and Economic Minerals; of Canada, R.J.W. Douglas (ed.); Geological Survey of Canada, Economic Geology Report No. 1, Fifth Edition, p. 365-488.

Dowling, D.B.
1914: Coal fields of Manitoba, Saskatchewan, Alberta and eastern British Columbia; Geological Survey of Canada, Memoir 53, 142 p.

Dowling, D.B., Slipper, S.E., and McLearn, F.H.
1919: Investigations in the gas and oil fields of Alberta, Saskatchewan, and Manitoba; Geological Survey of Canada, Memoir 116, 89 p.

Elson, J.A.
1983: Glacial Lake Agassiz - Discovery and a century of research; in Glacial Lake Agassiz, J.T. Teller and L. Clayton (ed.); Geological Association of Canada, Special Paper 26, p. 21-41.

Fischbuch, N.R.
1968: Stratigraphy, Devonian Swan Hills reef complexes of central Alberta; Bulletin of Canadian Petroleum Geology, v. 16, p. 444-556.

Fleming, Sandford
1874: Report of progress on the exploration and surveys for the Canadian Pacific Railway, 1874; MacLean, Roger and Company, Ottawa, 249 p.

Francis, D.R.
1956: Jurassic stratigraphy of the Williston Basin area; Saskatchewan Department of Mineral Resources, Petroleum and Natural Gas Branch, Report no. 18, 69 p.

Fraser, Simon
1960: The Letters and Journals of Simon Fraser 1806-1808, W.K. Lamb (ed.), The MacMillan Company of Canada, Toronto, 292 p.

Fuller, J.G.C.M.
1956: Mississippian rocks and oilfields in southeastern Saskatchewan; Saskatchewan Department of Mineral Resources, Report 19, 72 p.

Fuzesy, L.M.
1960: Correlations and subcrops of the Mississippian strata in southeastern and south-central Saskatchewan; Saskatchewan Department of Mineral Resources, Report 51, 63 p.

Gabrielse, H.
1985: Major dextral transcurrent displacements along the Northern Rocky Mountain Trench and related lineaments in north-central British Columbia; Geological Society of America Bulletin, v. 96, p. 1-14.

Gibson, D.W.
1974: Triassic rocks of the southern Canadian Rocky Mountains; Geological Survey of Canada, Bulletin 230, 65 p.
1975: Triassic rocks of the Rocky Mountain Foothills and Front Ranges of northeastern British Columbia and west-central Alberta; Geological Survey of Canada, Bulletin 247, 61 p.
1985: Stratigraphy, sedimentology and depositional environments of the coal-bearing Jurassic-Cretaceous Kootenay Group, Alberta and British Coumbia; Geological Survey of Canada, Bulletin 357, 108 p.

Gould, E.
1976: Oil - the History of Canada's Oil and Gas Industry; Hancock House Publishers, Vancouver, 288 p.

Graham, A.D.
1973: Carboniferous and Permian stratigraphy, southern Eagle Plain, Yukon Territory, Canada; in Canadian Arctic Geology, J.D. Aitken and D.J. Glass (ed.); Geological Association of Canada - Canadian Society of Petroleum Geologists Symposium, 1973, p. 159-180.

Grantz, A. and May, S.T.
1983: Rifting history and structural development of the continental margin north of Alaska; in Studies in Continental Margin Geology, J.S. Watkins and C.L. Drake (eds.); American Association of Petroleum Geologists, Memoir 34, p. 77-99.

Gray, E.
1970: The Great Canadian Oil Patch, Toronto, Canada; Maclean-Hunter Limited, 355 p.

Guillet, G.R. and Martin, W.
1984: The Geology of Industrial Minerals in Canada; Canadian Institute of Mining and Metallurgy, Special Volume No. 29, 350 p.

Hage, C.O.
1944: Geology adjacent to the Alaska Highway between Fort St. John and Fort Nelson, British Columbia; Geological Survey of Canada, Paper 44-30, 22 p.

Hamblin, A.P. and Walker, R.G.
1979: Storm-dominated shallow marine deposits: the Fernie-Kootenay (Jurassic) transition; southern Rocky Mountains; Canadian Journal of Earth Sciences, v. 16, p. 1673-1690.

Hayes, B.J.R.
1983: Stratigraphy and petroleum potential of the Swift Formation (Upper Jurassic), southern Alberta and north-central Montana; Bulletin of Canadian Petroleum Geology, v. 31, p. 37-52.

Hector, J.
1863: Geological Report; in John Palliser, 1863, The Journals, Detailed Reports and Observations relative to the Exploration, by Captain Palliser, of that portion of British North America, which in latitude lies between the British Boundary Line and the Height of Land or Watershed of the Northern or Frozen Ocean respectively, and in longitude between the western shore of Lake Superior and the Pacific Ocean during the years 1857, 1858, 1859, and 1860; George Edward Eyre and William Spottiswoode, London, p. 216-245.

Hilborn, J.D. (ed.)
1968: Dusters and Gushers - the Canadian Oil and Gas Industry; Pitt Publishing Company Limited, Toronto, 278 p.

Hills, L.V. (ed.)
1974: Oil Sands - Fuel of the Future; Canadian Society of Petroleum Geologists, Memoir 3, 263 p.

Hind, H.Y.
1859: Northwest Territory, Reports of Progress, together with a preliminary and general report on the Assiniboine and Saskatchewan Exploring Expedition; Journal of Legislative Assembly, Province of Canada, v. 19, Appendix 36, Lovell Printer, Toronto, 210 p.

Hitchon, B.
1964: Formation Fluids; Chapter 15 in Geological History of Western Canada, R.G. McCrossan and R.P. Glaister (ed.); Alberta Society of Petroleum Geologists, p. 201-217.
1984: Geothermal gradients, hydrodynamics, and hydrocarbon occurrences, Alberta, Canada; American Association of Petroleum Geologists Bulletin, v. 68, p. 713-743.

Hitchon, B., Billings, G.K., and Klovan, J.E.
1971: Geochemistry and origin of formation waters in the Western Canada Sedimentary Basin - III. Factors controlling chemical composition; Geochimica Cosmochimica Acta, v. 35, p. 567-598.

Holter, M.E.
1969: The Middle Devonian Prairie Evaporite of Saskatchewan; Saskatchewan Department of Mineral Resources, Report 123, 133 p.

Hopkins, J.C.
1981: Sedimentology of quartzose sandstones of Lower Mannville and associated units, Medicine River Area, central Alberta; Bulletin of Canadian Petroleum Geology, v. 29, p. 12-41.

Hume, G.S. and Link, T.A.
1945: Canol geological investigations in the Mackenzie River area, Northwest Territories; Geological Survey of Canada, Paper 45-16, 87 p.

Industrial Minerals
1984: Industrial minerals in Canada - a review of recent developments; No. 200, May.

Jansa, L.
1972: Depositional history of the coal-bearing Upper Jurassic-Lower Cretaceous Kootenay Formation, southern Rocky Mountains, Canada; Geological Society of America, Bulletin, v. 83, p. 3199-3222.

Jerzykiewicz, T.
1985: Stratigraphy of the Saunders Group in the central Alberta Foothills - a progress report; in Current Research, Part B, Geological Survey of Canada, Paper 85-1B, p. 246-258.

Jerzykiewicz, T. and McLean, J.R.
1980: Lithostratigraphical and sedimentological framework of coal-bearing Upper Cretaceous and Lower Tertiary strata, Coal Valley area, central Alberta Foothills; Geological Survey of Canada, Paper 79-12, 47 p.

Kent, D.M.
1963: The stratigraphy of the Upper Devonian Saskatchewan Group of southwestern Saskatchewan; Saskatchewan Department of Mineral Resources, Report 73, 51 p.
1968: The geology of the Upper Devonian Saskatchewan Group and equivalent rocks in western Saskatchewan and adjacent areas; Saskatchewan Department of Mineral Resources, Report 99, 221 p.

Kindle, E.D.
1944: Geological reconnaissance along Fort Nelson, Liard, and Beaver Rivers, northeastern British Columbia and southeastern Yukon; Geological Survey of Canada, Paper 44-16, 19 p.

Klingspor, A.M.
1969: Middle Devonian Muskeg evaporites of western Canada; American Association of Petroleum Geologists, Bulletin, v. 53, p. 927-948.

Klovan, J.E.
1964: Facies analysis of the Redwater reef complex, Alberta, Canada; Bulletin of Canadian Petroleum Geology, v. 12, p. 1-100.

Landes, R.W.
1965: The history of exploration and oil and gas development in the Cypress Hills area of southern Alberta and Saskatchewan; in Cypress Hills Plateau, Alberta and Saskatchewan, R.L. Zell (ed.); Alberta Society of Petroleum Geologists, 15th Annual Field Conference, Guidebook, Technical Paper, p. 1-17.

Leckie, D.A.
1985: The Lower Cretaceous Notikewin Member (Fort St. John Group), northeastern British Columbia: A progradational barrier island system; Bulletin of Canadian Petroleum Geology, v. 33, p. 39-51.
1986: Petrology and tectonic significance of Gates Formation (Early Cretaceous) sediments in northeast British Columbia; Canadian Journal of Earth Sciences, v. 23, p. 129-141.

Logan, A. and Hills, L.V. (ed.)
1973: The Permian and Triassic systems and their mutual boundary; Canadian Society of Petroleum Geologists, Memoir 2, 766 p.

Macauley, G., Penner, D.G., Procter, R.M., and Tisdall, W.H.
1964: Carboniferous; Chapter 7 in Geological History of Western Canada, R.G. McCrossan and R.P. Glaister (ed.); Alberta Society of Petroleum Geologists, p. 89-102.

Macauley, G., Snowdon, L.R., and Ball, F.D.
1985: Geochemistry and geological factors governing exploitation of selected Canadian oil shale deposits; Geological Survey of Canada, Paper 85-13, p. 65.

Mackenzie, Sir Alexander
1801: Voyages from Montreal, on the River St. Lawrence through the continent of North America, to the frozen and Pacific oceans, in the years 1789 and 1793. With a preliminary account of the rise, progress and present state of the fur trade in that country, T. Cadell (ed.); London, cxxxii and 412 pages.

Macoun, John
1877: Geological and topographical notes by Professor Macoun on the Lower Peace and Athabasca rivers; Geological Survey of Canada, Report of Progress for 1875-76, p. 87-95.

Macqueen, R.W. and Bamber, E.W.
1968: Stratigraphy and facies relationships of the Upper Mississippian Mount Head Formation, Rocky Mountains and Foothills, southwestern Alberta; Bulletin of Canadian Petroleum Geology, v. 16, p. 225-287.

Marion, D.J.
1984: The Middle Jurassic Rock Creek Member in the subsurface of west-central Alberta; in The Mesozoic of Middle North America, D.F. Stott and D.J. Glass (ed.); Canadian Society of Petroleum Geologists, Memoir 9, p. 319-343.

Masters, J.A.
1985: Elmworth - Case study of a Deep Basin gas field; American Association of Petroleum Geologists, Memoir 38, 316 p.

McCabe, H.R.
1959: Mississippian stratigraphy of Manitoba; Manitoba Department of Mines and Natural Resources, Mines Branch Publication 58-1, 99 p.

McCamis, J.G. and Griffith, L.S.
1967: Middle Devonian facies relationships, Zama area, Alberta; Bulletin of Canadian Petroleum Geology, v. 15, p. 434-467.

McConnell, R.G.
1885: Report on the Cypress Hills, Wood Mountain, and adjacent country, embracing that portion of the District of Assiniboia, lying between the International Boundary and the 51st parallel and extending from longitude 106° to longitude 110°50′; Geological and Natural History Survey of Canada, Annual Report (new series), v. 1, 1885, pt. C, 85 p.
1891: Report on an exploration in the Yukon and Mackenzie Basins, Northwest Territories; Geological Survey of Canada (new series), v. 4, 1888-1889, pt. D, 163 p.
1893: Report on a portion of the District of Athabasca comprising the country between Peace River and Athabasca River north of Lesser Slave Lake; Geological Survey of Canada, Summary Report for the year 1890, pt. D, 67 p.

McCrossan, R.G. (ed.)
1973: The Future Petroleum Provinces of Canada - their Geology and Potential; Canadian Society of Petroleum Geologists, Calgary, Canada, 720 p.

McCrossan, R.G. and Glaister, R.P. (ed.)
1964: Geological History of Western Canada; Alberta Society of Petroleum Geologists, Calgary, Canada, 232 p.

McGugan, A., Roessingh, H.K., and Danner, W.R.
1964: Permian; Chapter 8 in Geological History of Western Canada, R.G. McCrossan and R.P. Glaister (ed.); Alberta Society of Petroleum Geologists, p. 103-112.

McGugan, A., Rapson-McGugan, J.F., Mamet, B.L., and Ross, C.A.
1968: Permian and Pennsylvanian biostratigraphy, and Permian depositional environments, petrography and diagenesis, southern Canadian Rocky Mountains; in Canadian Rockies, Bow River to North Saskatchewan River, Alberta, H. Hornford (ed.); Canadian Society of Petroleum Geologists, 16th Annual Field conference, Guidebook, p. 48-66.

McIlreath, I.A.
1977: Stratigraphic and sedimentary relationships at the western edge of the Middle Cambrian carbonate facies belt, Field, British Columbia; Ph.D. Thesis, Department of Geology, The University of Calgary, Calgary, Alberta.

McLean, J.R.
1971: Stratigraphy of the Upper Cretaceous Judith River Formation in the Canadian Great Plains; Saskatchewan Research Council, Geology Division, Report no. 11, 96 p.
1982: Lithostratigraphy of the Lower Cretaceous coal-bearing sequence, Foothills of Alberta; Geological Survey of Canada, Paper 80-29, 46 p.

McWhae, J.R.
1986: Tectonic history of northern Alaska, Canadian Arctic, and Spitsbergen regions since Early Cretaceous; American Association of Petroleum Geologists, Bulletin, v. 70, p. 430-450.

Meek, F.B.
1867: Remarks on the geology of the Mackenzie River, with figures and descriptions of fossils from that region, in the Museum of the Smithsonian Institution, chiefly collected by the late Robert Kennicott, Esq.; Transactions of the Chicago Academy of Science, v. 1, p. 61-114.
1976: A report on the invertebrate Cretaceous and Tertiary fossils of the upper Missouri country; United States Geological Survey, Territories Report 9, 629 p.

Mellon, G.B.
1967: Stratigraphy and petrology of the Lower Cretaceous Blairmore and Mannville groups, Alberta Foothills and Plains; Alberta Research Council, Bulletin 21, 270 p.

Meneley, W.A., Christiansen, E.A., and Kupsch, W.O.
1957: Preglacial Missouri River in Saskatchewan; Journal of Geology, v. 65, p. 441-447.

Miall, A.D.
1976: The Triassic sediments of Sturgeon Lake south and surrounding areas; in The Sedimentology of Selected Clastic Oil and Gas Reservoirs in Alberta, M. Lerand (ed.); Canadian Society of Petroleum Geologists, Calgary, Alberta, Canada, p. 25-43.

Monger, J.W.H. and Price, R.A.
1979: Geodynamic evolution of the Canadian Cordillera - progress and problems; Canadian Journal of Earth Sciences, v. 16, p. 770-791.

Monger, J.W.H., Price, R.A., and Tempelman-Kluit, D.J.
1982: Tectonic accretion and the origin of the two major metamorphic and plutonic welts in the Canadian Cordillera; Geology, v. 10, p. 70-75.

Montgomery, D.S., Clugston, D.M., George, A.E, Smiley, G.T., and Sawatzky, H.
1974: Investigation of Oils in the Western Canada Tar Belt; in Oil Sands - Fuel of the Future, L.V. Hills (ed.); Canadian Society of Petroleum Geologists, Memoir 3, p. 168-183.

Morton, A.S.
1973: A history of the Canadian West to 1870-71, being a history of Rupert's Land (the Hudson's Bay Company's Territory) and of the Northwest Territory (including the Pacific Slope); Second edition, Lewis G. Thomas (ed.); University of Toronto Press in co-operation with the University of Saskatchewan, 1039 p.

Mountjoy, E.W.
1978: Upper Devonian reef trends and configuration of the western portion of the Alberta Basin; in The Fairholme Carbonate Complex at Hummingbird and Cripple Creek, I.A. McIlreath and P.C. Jackson (ed.); Canadian Society of Petroleum Geologists, p. 1-30.

Naqvi, I.H.
1972: The Belloy Formation (Permian), Peace River area, northern Alberta and northeastern British Columbia; Bulletin of Canadian Petroleum Geology, v. 20, p. 58-88.

Nielsen, A.R. and Porter, J.W.
1984: Pembina Oil Field - in retrospect; in The Mesozoic of Middle North America, D.F. Stott and D.J. Glass (ed.); Canadian Society of Petroleum Geologists, Memoir 9, p. 1-13.

Norris, D.K.
1964: The Lower Cretaceous of the southeastern Canadian Cordillera; Bulletin of Canadian Petroleum Geology, v. 12, p. 512-535.

Oliver, T.A. and Cowper, N.W.
1963: Depositional environments of the Ireton Formation, central Alberta; Bulletin of Canadian Petroleum Geology, v. 11, p. 183-202.

Oswald, D.H. (ed.)
1968: International Symposium on the Devonian System, 1967; Alberta Society of Petroleum Geologists, 2 volumes: v. I, p. 1-1055; v. II, p. 1-1377.

Palliser, John
1863: The Journals, Detailed Reports and Observations relative to the Exploration, by Captain Palliser, of that portion of British North America, which in latitude lies between the British Boundary Line and the Height of Land or Watershed of the Northern or Frozen Ocean respectively, and in longitude between the western shore of Lake Superior and the Pacific Ocean during the years 1857, 1858, 1859, and 1860; George Edward Eyre and William Spottiswoode, London, p. 216-245.

Patching, T.H. (ed.)
1985: Coal in Canada; The Canadian Institute of Mining and Metallurgy, Special Volume 31, 327 p.

Plint, A.G., Walker, R.G., and Bergman, K.M.
1986: Cardium Formation 6: Stratigraphic framework of the Cardium in subsurface; Bulletin of Canadian Petroleum Geology, v. 34, p. 213-225.

Porter, J.W. and Fuller, J.G.C.M.
1964: Ordovician-Silurian, Part 1 - Plains; in Geological History of Western Canada, R.G. McCrossan and R.P. Glaister (ed.); Alberta Society of Petroleum Geologists, Calgary, p. 34-42.

Porter, J.W., Price, R.A., and McCrossan, R.G.
1982: The Western Canada Sedimentary Basin; Philosophical Transactions of the Royal Society of London, Series A, v. 305, p. 169-192.

Price, R.A.
1981: The Cordilleran foreland thrust and fold belt in the southern Canadian Rocky Montains; in Thrust and Nappe Tectonics, K. McClay and N.J. Price (ed.); Geological Society of London, Special Paper No. 9, p. 427-448.

Pugh, D.C.
1971: Subsurface Cambrian stratigraphy of southern and central Alberta; Geological Survey of Canada, Paper 70-10, 33 p.
1973: Subsurface Lower Paleozoic stratigraphy in northern and central Alberta; Geological Survey of Canada, Paper 72-12, 54 p.
1975: Cambrian stratigraphy from western Alberta to northeastern British Columbia; Geological Survey of Canada, Paper 74-37, 31 p.
1983: Pre-Mesozoic geology in the subsurface of Peel River map-area, Yukon Territory and District of Mackenzie; Geological Survey of Canada, Memoir 401, 61 p.

Rapson, J.E.
1964: Lithology and petrography of transitional Jurassic-Cretaceous clastic rocks, southern Rocky Mountains; Bulletin of Canadian Petroleum Geology, Special Guidebook Issue, Flathead Valley, v. 12, p. 556-586.
1965: Petrography and derivation of Jurassic-Cretaceous clastic rocks, southern Rocky Mountains, Canada; American Association of Petroleum Geologists, Bulletin, v. 49, p. 1426-1452.

Richardson, John
1823: Geognostical observations; in John Franklin, Narrative of a journey to the shores of the Polar Sea in the years 1819, 1820, 1821, and 1822; John Murray, London, Appendix I, p. 497-538.
1852: Arctic Searching Expedition: A Journal of a Boat-Voyage through Rupert's Land and the Arctic Sea, in search of the Discovery ships under command of Sir John Franklin. With an Appendix on the Physical Geography of North America; Harper and Brothers, Publishers, New York, 516 p.
1853: On some points of the physical geography of North America in connection with its geological structure; Quarterly Journal of the Geological Society, London, v. 7, p. 212-215.

Selwyn, A.R.C.
1874: Summary Report; Geological Survey of Canada, Report of Progress for 1873-74, p. 1-12.
1877: Report of exploration in British Columbia in 1875; Geological Survey of Canada, Report of Progress, 1875-76, p. 28-86.

Sheasby, N.M.
1971: Depositional patterns of the Upper Devonian Waterways Formation, Swan Hills area; Bulletin of Canadian Petroleum Geology, v. 19, p. 377-404.

Snowdon, L.R. and Powell, T.G.
1982: Immature oil and condensate-modification of hydrocarbon generation model for terrestrial organic matter; American Association of Petroleum Geologists, Bulletin, v. 66, p. 775-788.

Sproule, J.C.
1968: Exploration and discovery; in Dusters and Gushers - the Canadian Oil and Gas Industry, J.D. Hilborn (ed.); Pitt Publishing Company Limited, Toronto, p. 9-24.

Spry, M. (ed.)
1968: The papers of the Palliser expedition 1857-1860; Champlain Society, Toronto, 694 p.

Stenson, F.
1985: Waste to wealth - a history of gas-processing in Canada; Canadian Gas Processors Association/Canadian Gas Processors Suppliers Association, Calgary, 352 p.

Stewart, J.K.
1972: Minerals and Canada's Prairie Provinces; Canadian Geographical Journal, v. 84, no. 4, p. 104-115.

Stewart, W.D. and Walker, R.G.
1980: Eolian coastal dune deposits and surrounding marine sandstone, Rocky Mountain Supergroup (Lower Pennsylvanian), southeastern British Columbia; Canadian Journal of Earth Sciences, v. 17, p. 1125-1140.

Stoakes, F.A.
1980: Nature and control of shale basin fill and its effect on reef growth and termination: Upper Devonian Duvernay and Ireton formations of Alberta, Canada; Bulletin of Canadian Petroleum Geology, v. 28, p. 345-410.

Stott, D.F.
1955: Jurassic stratigraphy of Manitoba; Manitoba Department of Mines and Natural Resources, Mines Branch Publication 54-2, 78 p.

Stott, D.F. (cont.)
1963: The Cretaceous Alberta Group and equivalent rocks, Rocky Mountain Foothills, Alberta; Geological Survey of Canada, Memoir 317, 306 p.
1968: Lower Cretaceous Bullhead and Fort St. John groups between Smoky and Peace rivers, Rocky Mountain Foothills, Alberta and British Columbia; Geological Survey of Canada, Bulletin 152, 279 p.
1982: Lower Cretaceous Fort St. John Group and Upper Cretaceous Dunvegan Formation of the Foothills and Plains of Alberta, British Columbia, District of Mackenzie and Yukon Territory; Geological Survey of Canada, Bulletin 328, 124 p.

Stott, D.F. and Glass, D.J. (ed.)
1984: The Mesozoic of Middle North America; Canadian Society of Petroleum Geologists, Memoir 9, 573 p.

Sweeney, J.F.
1985: Comments about the age of the Canada Basin; in Geophysics of the Polar Regions, E.S. Husebye, G.I. Johnson, and Y. Kristoffersen (ed.); Tectonophysics, v. 114, p. 1-10.

Taylor, R.S., Mathews, W.H., and Kupsch, W.O.
1964: Tertiary; Chapter 13 in Geological History of Western Canada, R.G. McCrossan and R.P. Glaister (ed.); Alberta Society of Petroleum Geologists, Calgary; Alberta, Canada, p. 190-200.

Tempelman-Kluit, D.J.
1979: Transported cataclasite, ophiolite and granodiorite in Yukon: Evidence of arc-continent collision; Geological Survey of Canada, Paper 79-4, 27 p.

Tyrrell, J.B.
1893: Report on North-Western Manitoba with portions of the adjacent districts of Assiniboia and Saskatchewan; Geological Survey of Canada, Annual Report, 1890-91, 235 p.

Upham, Warren
1890: Report on exploration of the Glacial Lake Agassiz in Manitoba; Geological Survey of Canada, Annual Report for 1888-1889, v. 4, section E, 156 p.
1895: The Glacial Lake Agassiz; United States Geological Survey, Monograph 25, 658 p.

Vogt, P., Bernero, C., Kovacs, L., and Taylor, P.
1981: Structure and plate tectonic evolution of the marine Arctic as revealed by aeromagnetics; Oceanologica Acta, Proceedings 26th International Geological Congress, Geology of Oceans Symposium, Paris, 1980, p. 25-40.

Walker, R.G.
1983: Cardium Formation 3. Sedimentology and stratigraphy in the Garrington-Caroline area, Alberta; Bulletin of Canadian Petroleum Geology, v. 31, p. 213-230.
1985: Cardium Formation at Ricinus Field, Alberta: A channel cut and filled by turbidity currents in Cretaceous Western Interior Seaway; American Association of Petroleum Geologists, Bulletin, v. 69, p. 1963-1981.

Wardlaw, N.C. and Reinson, G.E.
1971: Carbonate and evaporite deposition and diagenesis, Middle Devonian Winnipegosis and Prairie Evaporite formations of central Saskatchewan; American Association of Petroleum Geologists, Bulletin, v. 55, p. 1759-1786.

Warkentin, J.
1967: David Thompson's geology - a document; Journal of the West, v. 6, no. 6, p. 468-490.
1979: Geological lectures by Dr. John Richardson, 1825-26; Ottawa, National Museum of Natural Science, Syllogeus, no. 22, 63 p.

Whillans, J.W.
1955: First in the West. The story of Henry Kelsey, discoverer of the Canadian Plains; Applied Arts Products Limited, Edmonton, 175 p.

Whitaker, S.H., Irvine, J.A., and Broughton, P.L.
1978: Coal resources of southern Saskatchewan: A model for evaluation methodology; Saskatchewan Research Council Report 20, Department of Mineral Resources Report 209, and Geological Survey of Canada, Economic Geology Report 30, 151 p.

Williams, G.K.
1984: Some musings on the Devonian Elk Point Basin, Alberta; Bulletin of Canadian Petroleum Geology, v. 32, p. 216-232.

Williams, M.Y.
1944: Geological investigations along the Alaska Highway from Fort Nelson, British Columbia to Watson Lake, Yukon; Geological Survey of Canada, Paper 44-28, 33 p.

Zaslow, M.
1975: Reading the Rocks - the Story of the Geological Survey 1842-1972; Macmillan, Toronto, in association with the Department of Energy, Mines and Resources and Information Canada, Ottawa, 800 p.

Ziegler, P.A.
1967: Canadian Cordillera Field Trip; Guidebook, International Symposium on the Devonian System; Alberta Society of Petroleum Geologists, Calgary, Canada, 72 p.

Authors' Addresses

J.D. Aitken
2676 Jemima Road
Denman Island, British Columbia
V0R 1T0

D.F. Stott
8929 Forest Park Drive
Sidney, British Columbia
V8L 5A7

Printed in Canada

Subchapter 2C

GEOMORPHIC DIVISIONS

D.F. Stott and R.W. Klassen

CONTENTS

INTRODUCTION

Physiographically and geologically, western Canada consists of three great parts: a core of crystalline rocks forming the Canadian (Precambrian) Shield; a surrounding band of younger, essentially flat-lying, stratified rocks; and on the western side, a rim of highly deformed sedimentary, volcanic, and igneous rocks (Fig. 2C.1, 2C.2, 2C.3). The Canadian Shield has a slightly depressed centre, presently the site of Hudson Bay and Hudson Bay Lowland. The surrounding band of undeformed Phanerozoic rocks forms the Interior Plains on the western margin and St. Lawrence Lowland on the southern margin. These areas are characterized by lowlands, plains, and plateaux, which were substantially modified by glacial action during Pleistocene time. The outer belt of deformed rock forms the magnificent mountainous terrain of the Canadian Cordillera.

The classification of physiography in western Canada has evolved since the early writings of A.R.C. Selwyn and G.M. Dawson (1884) and a classic study by R.A. Daly (1912). The physiography of British Columbia was described by S.S. Holland (1964), and that of Canada by H.S. Bostock (1970a,b). These reports have been used extensively for the following summary. As part of the current syntheses of the Geology of Canada, W.H. Mathews (1986) has modified parts of the earlier classification of Cordilleran physiography, basing his study on landsat imagery and recent reports. The physiographic divisions set forth on Matthew's geomorphic map are used in this section. A more detailed discussion of the geomorphology of the Cordillera may be found in the Cordilleran volume of this series (Gabrielse and Yorath, 1991).

CORDILLERA

The Cordillera in Canada is divided longitudinally into three great belts, the Eastern System, the Interior System, and the Western System (Bostock, 1948 and 1961; Holland, 1964). Only the Eastern System and a minor part of the

Stott, D.F. and Klassen, R.W.
1993: Geomorphic divisions; Subchapter 2C in Sedimentary Cover of the Craton in Canada, D.F. Stott and J.D. Aitken (ed.); Geological Survey of Canada, Geology of Canada, no. 5, p. 31-44 (also Geological Society of America, The Geology of North America, v. D-1).

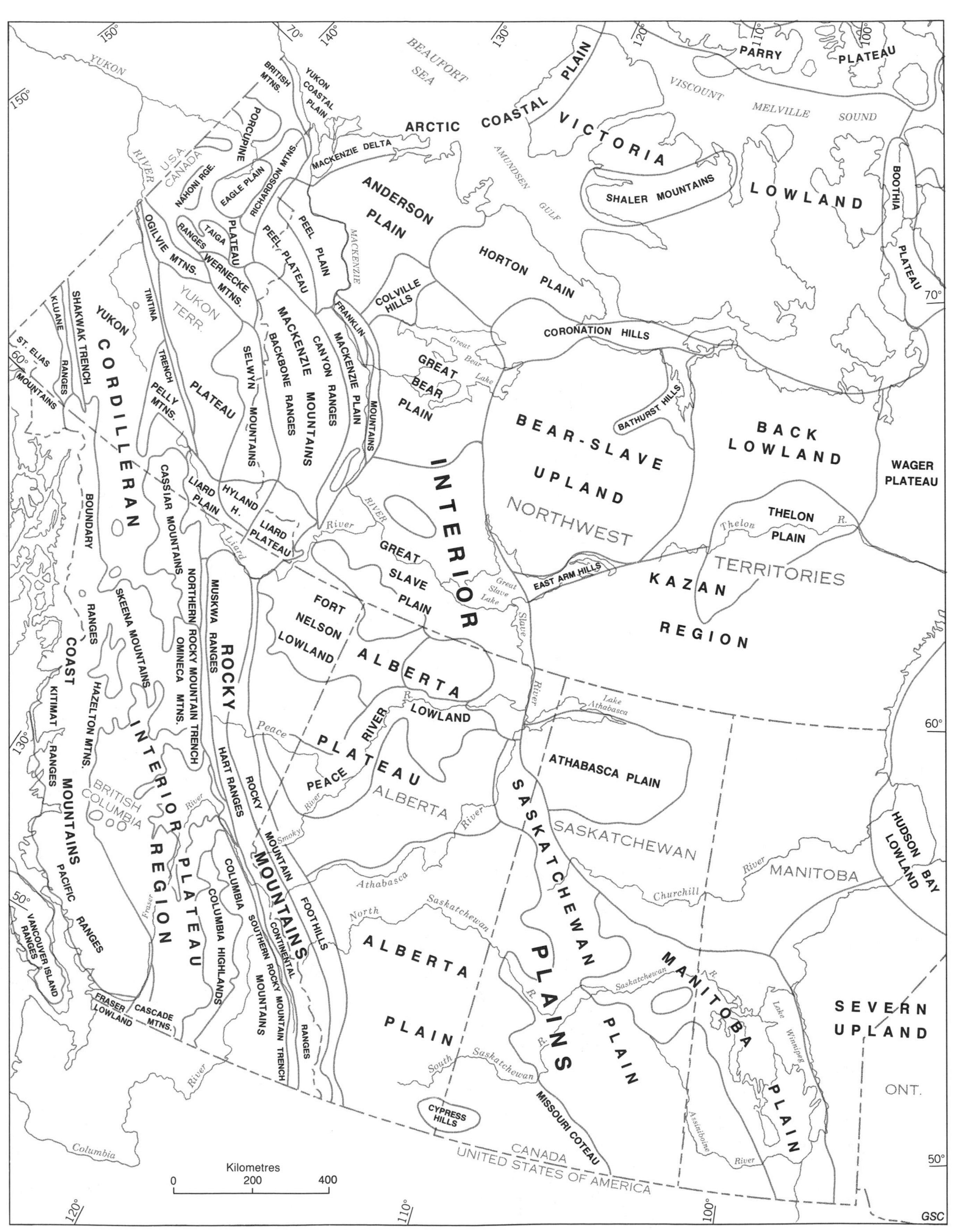
BEAUFORT SEA
ARCTIC COASTAL PLAIN
VICTORIA LOWLAND
PARRY PLATEAU
VISCOUNT MELVILLE SOUND
SHALER MOUNTAINS
BOOTHIA PLATEAU
AMUNDSEN GULF
ANDERSON PLAIN
HORTON PLAIN
CORONATION HILLS
BATHURST HILLS
BEAR-SLAVE UPLAND
BACK LOWLAND
WAGER PLATEAU
THELON PLAIN
NORTHWEST TERRITORIES
KAZAN REGION
EAST ARM HILLS
BRITISH MTNS.
YUKON COASTAL PLAIN
MACKENZIE DELTA
PORCUPINE PLATEAU
NAHONI RGE.
EAGLE PLAIN
RICHARDSON MTNS.
PEEL PLAIN
PEEL PLATEAU
TAIGA RANGES
OGILVIE MTNS.
WERNECKE MTNS.
COLVILLE HILLS
FRANKLIN MOUNTAINS
MACKENZIE PLAIN
CANYON RANGES
MACKENZIE MOUNTAINS
BACKBONE RANGES
SELWYN MOUNTAINS
GREAT BEAR PLAIN
YUKON TERR.
YUKON PLATEAU
TINTINA TRENCH
SHAKWAK TRENCH
KLUANE RANGES
ST. ELIAS MOUNTAINS
PELLY MTNS.
CORDILLERAN
INTERIOR
LIARD PLAIN
HYLAND H.
LIARD PLATEAU
CASSIAR MOUNTAINS
GREAT SLAVE PLAIN
Great Slave Lake
FORT NELSON LOWLAND
ALBERTA LOWLAND
BOUNDARY RANGES
SKEENA MOUNTAINS
NORTHERN ROCKY MOUNTAIN TRENCH
OMINECA MTNS.
MUSKWA RANGES
ROCKY MOUNTAINS
HART RANGES
PEACE RIVER PLATEAU
LOWLAND
COAST MOUNTAINS
KITIMAT RANGES
HAZELTON MTNS.
INTERIOR REGION
INTERIOR PLATEAU
BRITISH COLUMBIA
ALBERTA
ATHABASCA PLAIN
Lake Athabasca
SASKATCHEWAN PLAINS
SASKATCHEWAN
MANITOBA
HUDSON BAY LOWLAND
PACIFIC RANGES
VANCOUVER ISLAND RANGES
COLUMBIA MOUNTAINS
COLUMBIA HIGHLANDS
SOUTHERN ROCKY MOUNTAIN TRENCH
CONTINENTAL RANGES
FOOTHILLS
ROCKY MOUNTAIN TRENCH
ALBERTA PLAIN
SASKATCHEWAN PLAIN
MANITOBA PLAIN
SEVERN UPLAND
ONT.
FRASER LOWLAND
CASCADE MTNS.
CYPRESS HILLS
MISSOURI COTEAU
CANADA
UNITED STATES OF AMERICA
U.S.A. CANADA
Kilometres
0 200 400
GSC

Interior System are outlined here, because the sedimentary strata that are the focus of this volume are largely limited to those regions.

The Eastern System

The Eastern System is the easternmost of the three primary subdivisions of the Canadian Cordillera. The boundary between the Eastern System and the Interior System to the west lies along the Rocky Mountain and Tintina trenches. In Yukon Territory, part of the region east of Tintina Trench has been included at various times in the Interior System. The Eastern System is flanked on its eastern side by the Interior Plains.

The Eastern System is a mountainous belt extending northwestward from the Canada-United States boundary for 2700 km to the Arctic Ocean. It has a width of 100 km at Peace River in the Rocky Mountains, its narrowest section, and a maximum width of 400 km across the Selwyn, Mackenzie, and Franklin mountains.

Trenches

The western limit of the region discussed in this volume is defined physiographically by three great linear valleys, aligned in a nearly continuous feature trending northwesterly through the Cordilleran Region. These are the Southern and Northern Rocky Mountain trenches of British Columbia and Tintina Trench in Yukon Territory. The impressive features are narrow, straight, and steep walled. The floors range in width from about 1 km to more than 25 km, and are drained by such large rivers as the Kootenay, Columbia, Fraser, and Yukon.

Arctic Mountains

The Arctic Mountains encompass a series of mountain ranges and extensive lowlands which reflect the complexity of the underlying rocks. The **British Mountains** are the surface expression in Canada of the Romanzof Uplift, a structural feature of Tertiary age (Norris, 1974). These mountains, an eastern extension of the Brooks Range of Alaska, are highest adjacent to the Alaska-Yukon Boundary where peaks exceed 1680 m elevation. **Buckland Hills** form the northern foothills of the British Mountains. **Barn Mountains** are an eastward extension of the British Mountains (Fig. 2C.4). The **Richardson Mountains** comprise several ranges and include such structural features as the Cache Creek and Rat uplifts (Norris, 1973, 1974). These features are characterized by north-and northeast-trending folds and near-vertical faults.

The British and Barn mountains were not glaciated. Only a few peaks in the northern Richardson Mountains sustained cirque glaciers, but the mountains provided a barrier to the westward-flowing Laurentide ice-sheet. A tongue of Laurentide ice extended westward through the mountains at McDougall Pass (elev. ca. 315 m). Similarly, **Eagle Lowland** is unglaciated except in the southeast, where a lobe of the Laurentide ice-sheet extended westward across the Bonnet Plume Basin. Relief reaches a maximum of about 300 m in the highest part of the lowland, where some ridges are 800 m in elevation.

Figure 2C.1. Physiographic elements of the Interior Platform and adjacent regions. Based on divisions outlined by H.S. Bostock (GSC Map 1254A, 1970) and on modifications by W.H. Mathews (1986).

Ogilvie Mountains

The **Ogilvie Mountains**, lying east of the Tintina Trench and reaching elevations of 2200 m, are composed of sedimentary rocks intruded by granitic stocks. The southern Ogilvie Ranges supported extensive mountain glaciers during successive Pleistocene glaciations. In the northern Ogilvie Ranges, only Mount Klotz was high enough to support mountain glaciers.

Yukon Plateau

Only part of the Yukon Plateau, including Stewart Plateau, MacMillan Highland, and Ross Lowland, lies east of the Tintina Trench. In general, the central basin lies at about 1300 m and is drained by the Yukon River system. Pleistocene ice encroached from the surrounding mountains.

Selwyn Mountains

The Selwyn Mountains include the **Wernecke, Hess**, and **Logan ranges**. During the Pleistocene Epoch, the area was inundated by the Cordilleran ice-sheet, which carved the underlying bedrock into rugged, massive mountains. A few alpine glaciers remain between 2200 and 3000 m elevation.

Mackenzie Mountains

The Mackenzie Mountains consist of uplifted Proterozoic and Paleozoic clastic and carbonate rocks, which attain elevations in excess of 1800 m (Fig. 2C.5; Aitken et al., 1982). The **Backbone Ranges** in the west reach 2700 m, but the more easterly **Canyon Ranges** are more subdued (Fig. 2C.6). **Hyland Highland** is mostly less than 1500 m in elevation and is underlain mainly by shale and sandstone.

Major valleys of the Backbone Ranges were occupied by successive large Pleistocene glaciers, some of which nearly reached the mountain front. At its maximum, attained in late Wisconsinan time, the Laurentide ice-sheet impinged against the Canyon Ranges. Near the Laurentide limit, Laurentide ice-marginal features truncate moraines of older montane glaciations, and Laurentide till overlies multiple montane tills. Hyland Highland was glaciated and has a thick drift cover.

Franklin Mountains

The Franklin Mountains form a series of relatively low ranges, comprising Devonian and older carbonates and clastics. A unique feature of the **Franklin Mountains** is the reversal in topographic asymmetry that occurs along

GREAT BEAR LAKE
HORN PLATEAU
MACKENZIE RIVER
Liard R.
MACKENZIE FOLD BELT
66°
64°
62°
60°
112°
116°
120°
124°
128°

their trends. The morphological reversals are directly related to reversals in structural geometry within the ranges, which are formed by asymmetric anticlines or by steeply dipping reverse faults. To the south, the en echelon, north-trending **Liard** and **Camsell ranges** (Fig. 2C.7) have maximum elevations in the order of 2000 km. The Franklin Mountains, covered entirely by Laurentide ice during successive glaciations, profoundly affected the local directions of ice flow.

Mackenzie Plain (Fig. 2C.7) separates the Franklin Mountains from the Mackenzie Mountains and is dissected by Mackenzie River. Much of the area is underlain by bevelled, weakly resistant Cretaceous strata (Aitken et al., 1982). The plain is interrupted by local uplifts of the Franklin Mountains structural type, including Imperial Hills, West Mountain, MacKay Range, and Gambill Hills.

Rocky Mountains

The Rocky Mountains extend in a northwesterly direction for about 1300 km between the Canada-United States Boundary and Liard River (Fig. 2C.3). Their western boundary is the Rocky Mountain Trench, which is least pronounced at its northern end. The eastern front is formed by the Rocky Mountain Foothills. The Rocky Mountains are underlain by rocks, mainly sedimentary, partly metamorphic, which range from Proterozoic to Cretaceous in age. The youngest rocks are exposed in the Foothills, and progressively older rocks lie to the west.

The northern Rocky Mountains, between Liard and Peace rivers, are broadest and highest around Churchill Peak in the **Muskwa Ranges**. There, Paleozoic carbonates along east-facing scarps have been carved by glaciation, producing castellated ranges of rugged relief. Alpine glaciers occur around the higher peaks. The Canadian Rocky Mountains are narrowest at Peace River. To the south, the **Hart Ranges** are moderately rugged and few peaks have elevations that exceed 2500 m (Fig. 2C.8). The combination of lower elevation and relief, of different bedrock lithology and structure, and of reduced alpine and valley glaciation has resulted in a subdued alpine topography, which contrasts with the great relief and heights found farther south. The **Continental Ranges** include many 3000 to 3500 m peaks, including the highest peak, Mount Robson, at 3954 m. The Columbia Icefield, the largest of many alpine glaciers in the Rocky Mountains, lies within a series of high peaks on the continental divide. The Front Ranges consist of a number of longitudinal ranges whose continuity is controlled by underlying fold and fault structures (Fig. 2C.9, 2C.10). Late Pleistocene cirque glaciation has accentuated the sharpness of the east- and northeast-facing scarps. The Main Ranges are largely underlain by sedimentary rocks of Late Proterozoic and early Paleozoic age (Fig. 2C.11). The rocks occur in broad folds, forming the massive, monumental mountains, which contrast with the thrust-faulted sequences of the Front Ranges. Many large rivers head in this region: the Fraser, Columbia, and Kootenay drain westward into the Pacific; the Peace and Athabasca flow eastward and finally northward in the Mackenzie (Arctic) drainage; and the North Saskatchewan and Bow drain eastward into Hudson Bay. The southernmost ranges of the Rocky Mountains contain a great thickness of Middle Proterozoic and Paleozoic sedimentary rocks, forming massive mountains that rise to elevations of 2800 m. These are spectacularly displayed on the eastern side of Waterton Lakes National Park. Within this mountainous region, Fernie Basin is the largest of several structural depressions containing Jurassic and Cretaceous clastic sediments.

The **Foothills** are mainly linear ridges and hills of Mesozoic sandstone, with upper Paleozoic carbonates exposed in structural culminations. In northeastern British Columbia, the Foothills, with elevations of 2100 m and a relief of 1200 to 1400 m, have a mountainous character (Fig. 2C.8), which contrasts with the more subdued topography of the central and southern Foothills of Alberta. In southwestern Alberta, the gentle inclination of imbricated thrust plates accounts for the suppressed nature of the physiography of the Foothills. Except in southwestern Alberta, the Foothills were covered by eastward-flowing montane ice-sheets. Thick glacial deposits, including glacial-lacustrine sediments, were left in the valleys of major rivers and streams.

ARCTIC COASTAL PLAIN

The Arctic Coastal Plain includes the coastal terrane along the shores of the Arctic Ocean and borders the Interior Plains and Cordilleran Region. On the mainland, the Arctic Coastal Plain is represented by Mackenzie Delta and Yukon Coastal Plain. It extends beneath sea level to merge with the Arctic Continental Shelf. The plain continues unchanged seaward, reaching a depth of only 100 m at a distance of 80 km or more from the shore. There, the shelf breaks away rapidly to an irregular continental slope that drops steeply to oceanic depths beneath Beaufort Sea.

Mackenzie Delta

Mackenzie Delta is a flat, near sea-level deltaic plain developed on fine-grained unconsolidated sediments (Rampton, 1974). The delta is 220 km long and about 70 km wide, with its long axis trending north-northwest (Fig. 2C.12). It is a composite delta comprising the modern active delta and the Pleistocene delta complex of Richards Island and Tuktoyaktuk Peninsula. The delta plain is remarkable for its multitude of lakes and channels. Glacial, periglacial, and permafrost features of the region include complex outwash deposits, thermokarst depressions, patterned ground, and massive ground ice pingos (MacKay, 1962, 1971; Mackay and Stager, 1966; Stager, 1956; Rampton, 1971, 1982; Rampton and Mackay, 1971; and Pihlainen et al., 1956).

Figure 2C.2. Interior Plains between latitudes 60° and 65° North. The Interior Plains are bordered on the right by the Canadian Shield and on the left by the Mackenzie Fold Belt. Great Bear and Great Slave Lakes are remnants of much larger glacial lakes that once covered much of the region. D - areas of thick glacial drift (f - fluted, h - hummocky; Fn, Fs - Franklin Mountains (northern, southern); L - Liard Range; N - Nahanni Range. (From Satellite Image Map Sheet No. 3, Great Bear Lake, Surveys and Mapping Branch, Energy Mines and Resources.)

Figure 2C.3. Interior Plains between latitudes 49° and 57° North. The plains are bordered on the left by the Rocky Mountains and Foothills. (From Satellite Image Map Sheet No. 11, Alberta-Saskatchewan, Surveys and Mapping Branch, Energy, Mines and Resources.)

Yukon Coastal Plain

The Yukon Coastal Plain includes all flat and gently sloping plains that lie adjacent to the northeastern edge of the British, Barn, and Richardson mountains (Rampton, 1982). The plains were constructed by erosional and depositional processes but have a common base level - the Beaufort Sea. The coastal fringe is underlain by thick unconsolidated deposits. East of Firth River, lakes and ponds of thermokarst origin spot the plain. West of Herschel Island to the Alaskan boundary, the plain is formed by coastal lagoons and coalesced deltas and alluvial fans built by streams flowing from the British Mountains. The mountain fringe is primarily an erosion surface or pediment. Most of the Yukon Coastal Plain east of Firth River and also the northern edge of the adjacent mountain ranges were glaciated, presumably by early Wisconsinan ice moving west from the Mackenzie Valley (Rampton, 1982).

INTERIOR PLAINS

The Interior Plains are underlain by mainly flat-lying Proterozoic, Paleozoic, Mesozoic, and Tertiary strata. The southern part is semi-arid grassland that is extensively cultivated. The central part is tree-covered; the northern part is tundra.

The region was covered extensively by the Laurentide ice sheet and subsequently was the site of several very large glacial lakes. Glacial features were developed throughout the region and glacial deposits are widely distributed.

Northern Interior Plains

The Northern Interior Plains occupy the region between the Canadian Shield on the east and the mountains of the Cordilleran region. The southern boundary is drawn at Alberta Plateau, which rises above Liard River and Great Slave Lake (Fig. 2C.2). The region is divided into smaller physiographic units than those in the south, because of greater topographic variation. The northern limit of trees passes through the northern part of the region as does the southern limit of continuous permafrost. The effects of permafrost on topography and land forms are marked

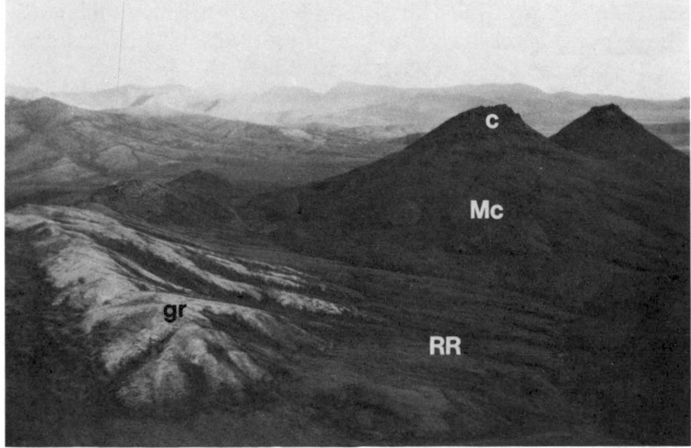

Figure 2C.4. Mount Fitton (lower left) and The Twins (middle right) on the east flank of Barn Mountains (background), northern Yukon Territory. Mount Fitton comprises Late Devonian (370 Ma) granite (gr) hosted in recessive shale of the Lower Paleozoic Road River Formation (RR). Beneath The Twins, the Road River is overlain with angular unconformity by shales of the Lower Cretaceous (Berriasian) Mcguire Formation (Mc). The Mcguire, in turn, is overlain disconformably by Albian flyschoid conglomerate (c). (Photo - D.K. Norris, ISPG 1860-40.)

Figure 2C.5. The outer fold belt of central Mackenzie Mountains, headwaters of Carcajou River, near the Canol Pipeline route, view northwestward. Upper Proterozoic to Devonian strata form the southwest limb and flat top of Stony anticline at right, and dip beneath Foran Syncline, at centre. Tawu anticline, at left, has a similar box-fold profile. Extensive pediments are developed at left. (Photo - J.D. Aitken, ISPG 3062-2.)

Figure 2C.6. Mackenzie Mountain front and Mackenzie Plain, view northward. Cambrian to Devonian, largely carbonate rocks dipping beneath Cretaceous strata of the plain are both the east-dipping limb of Gayna flexure and the northeast limb of the more local Macdougal anticline at left, which exposes Upper Proterozoic strata in its core. Carcajou River, diverted to its entrenched course along the frontal flexure by the impingement of Laurentide ice, crosses Mackenzie Plain in entrenched meanders. (Photo - M.P. Cecile, ISPG 3108-1.)

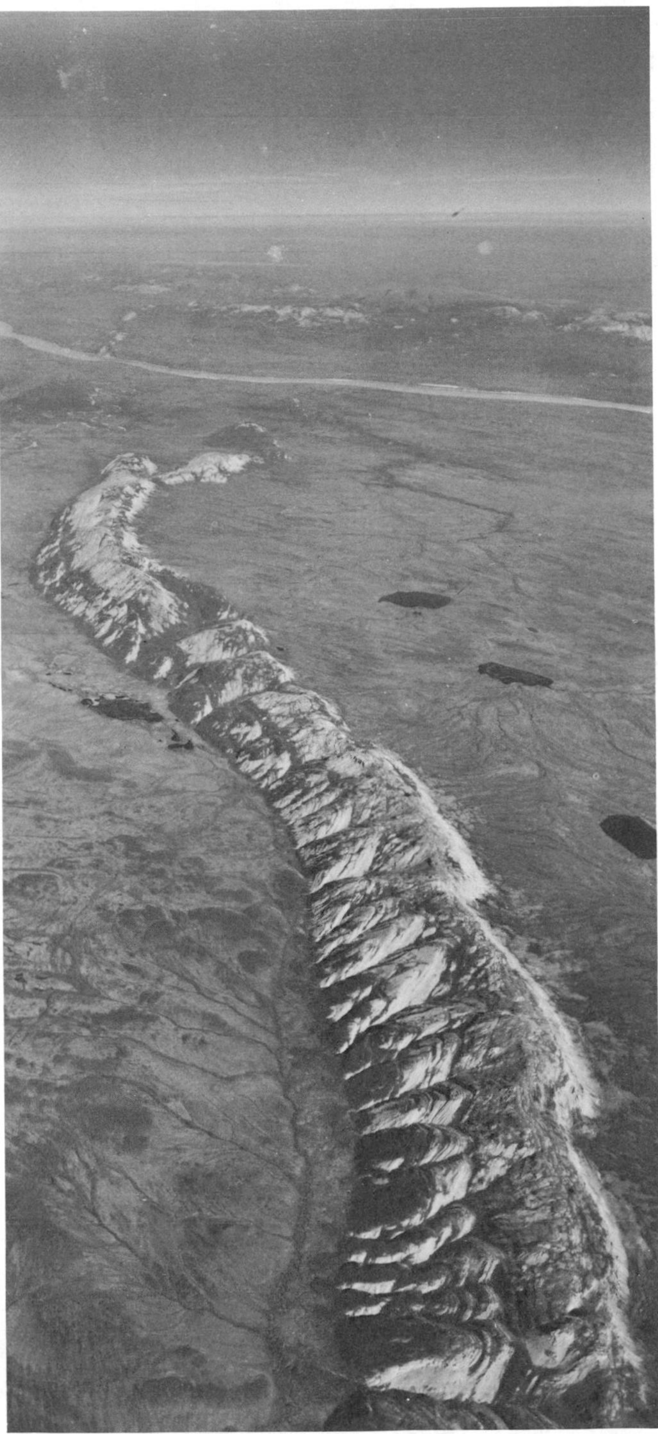

Figure 2C.7. View north along Camsell Range, Franklin Mountains and over Mackenzie Plain. Mackenzie River crosses the upper part of the photo, and the southern end of McConnell Range lies beyond the river. (R.C.A.F. Photo T12-15R.)

throughout the area by patterned ground, ground ice, pingos, oriented lakes, various solifluction structures, and many other periglacial phenomena. River valleys such as the Mackenzie, Carcajou, Mountain, Ramparts, Anderson, and Horton display active valley-wall slumping in many places because of the effects of the active layer above permafrost or enclosed ground ice.

Horton Plain is underlain by nearly flat-lying Paleozoic and Upper Proterozoic sediments, and its surface reaches elevations of 400 to 600 m, the higher parts being in the south. In the north, Upper Proterozoic strata are slightly folded and faulted and support a rolling surface of low scarps and scattered mesas. Throughout, as the streams gather size, they become entrenched 60 to 125 m below the surface. The most prominent effects of glaciation appear as extensive morainic belts, the most distinctive of which, the Melville Hills, extends westward from Horton Plain, across the southern part of Parry Peninsula. The belt consists of kames, hummocky moraine, and morainic ridges.

Anderson Plain is a broad region extending from the Arctic coast to the Mackenzie River in the southwest. It comprises a gently undulating surface, underlain by essentially flat-lying Paleozoic and Mesozoic strata and glacial drift. The region is poorly drained and covered by numerous lakes. The principal drainage is provided by the Anderson, Carnwath, and Andrew rivers. Anderson Plain is covered by a sheet of glacial drift and outwash, the main feature distinguishing it from the somewhat higher Horton Plain, where bedrock is generally exposed.

The **Colville Hills** extend northward from the Franklin Mountains and separate the Anderson and Great Bear plains. They embrace several linear, asymmetric ridges of structurally uplifted Paleozoic strata, which stand above the general level of the surrounding plains. The hills and ridges enclose hollows with several large lakes in a net-like pattern. The lower ground lies at 260 to 300 m elevation, whereas the rather sinuous ridges have summits up to 725 m.

Great Slave Plain has generally little relief and is underlain by Paleozoic strata. Low scarps of resistant carbonate strata and small shallow lakes are characteristic of much of the surface below 325 m in elevation. Its central part contains Horn Plateau, which is underlain by Cretaceous strata and is an outlier of the Alberta Plateau to the south. Great Slave Plain was completely glaciated by the Wisconsin Laurentide ice-sheet (Craig, 1965). A vast glacial lake, Glacial Lake McConnell, extended from Great Bear Lake through Great Slave Lake to Lake Athabasca. The area is almost entirely covered with a thick mantle of glacial and post-glacial deposits, including glacio-fluvial, lacustrine, alluvial, and aeolian deposits. Such glacial features as ground and end moraines and drumlinoid forms are common. Poor drainage has resulted in large areas of swamp and muskeg.

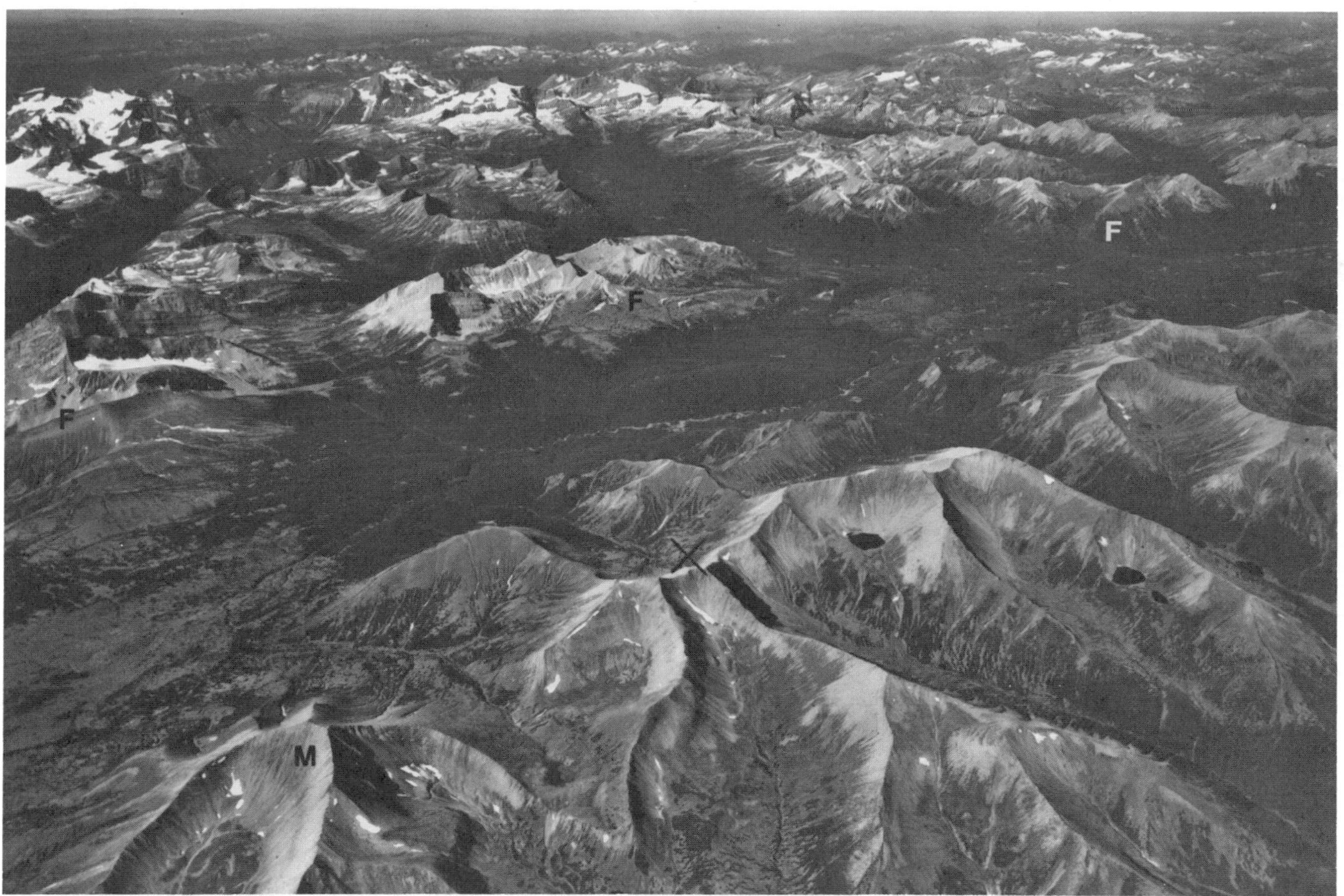

Figure 2C.8. View westward across the western Rocky Mountain Foothills to the eastern front of the Hart Ranges between the headwaters of Kakwa and Narraway rivers. Mount Minnes (M) in left foreground and ridges to right comprise Jurassic and Cretaceous strata. Fault (F) at eastern front of Paleozoic rocks dies out northwestward into folded Mesozoic rocks. Extensive alpine glaciers occur in high mountains to the west. (B.C. Photo 1204:73.)

Great Bear Plain is a very broad region of gently rolling topography that surrounds Great Bear Lake. For the most part it is underlain by poorly exposed Cretaceous strata, but lower Paleozoic carbonates occur in the western part. The surface is mainly below 300 m elevation. North of the lake, hummocky moraine is widespread. Great Bear Lake is the fourth largest lake in area on the North American continent, being in excess of 3100 km^2. It has a maximum depth of more than 450 m.

Peel Plain, lying southwest of Mackenzie River, is the northern continuation of Mackenzie Plain. It is a lowland region in the south with elevations averaging 150 m above sea level; to the north the plain rises to elevations of about 450 m. For the most part, the plain is underlain by flat-lying Cretaceous strata, but adjacent to Mackenzie River, Devonian clastic and carbonate rocks occur at the surface in structural uplifts. Glacial drift covers much of the area, and lowland areas below about 140 m are characterized by northwest-trending lakes, bogs, and swamps developed on ice-rich glaciolacustrine sediments (Hughes et al., 1973).

Peel Plateau is an upland region lying to the southwest of Peel Plain. Elevations are as much as 975 m. The plateau is an erosional remnant, capped and protected by resistant Upper Cretaceous sandstones (Aitken et al., 1982). On the south, the plateau is limited by the en echelon mountain-front flexures, in which the Cretaceous sandstones are upturned and truncated. On the north, a subdued erosional escarpment nearly 300 m high descends to Mackenzie Plain. The Arctic Red River is deeply entrenched into the plateau. The plateau was glaciated by a piedmont expansion of montane ice in pre-Wisconsin time and subsequently was covered by southeasterly derived Laurentide ice. The record of Laurentide glaciation includes extensive areas of subdued hummocky moraine, outwash deposits of sand and gravel, and northwest-trending meltwater channels.

Southern Interior Plains

Much of the rolling to nearly flat terrain and scattered uplands characteristic of the southern Interior Plains reflects preglacial fluvial erosion of the nearly flat-lying bedrock. Weakly indurated sediments of Cretaceous age

Figure 2C.9. View northwest along Front Ranges at Mountain Park, Alberta. Cambrian rocks (Ꞓ) lie above McConnell Thrust (MT) on Lower Cretaceous Luscar Group (L) of Alberta Foothills. Old coal workings are visible in lower left foreground. Note large recumbent fold in Devonian rocks at left. (Photo - J.D. Aitken, ISPG 1518-17.)

Figure 2C.11. Rocky Mountains, Main Ranges, view westward to Mount Assiniboine (elev. 3611 m) comprising Cambrian sediments and typical of the stratiform mountains of these ranges. The Simpson Pass Thrust lies in the valley in the right foreground. (Photo - J.D. Aitken, ISPG 1518-90.)

Figure 2C.10. Rocky Mountains, Front Ranges, view northward along the Flathead Range and the Alberta–British Columbia provincial boundary at Crowsnest Pass. Crowsnest Mountain, a klippe in the right distance, and the Flathead Range are underlain by the Lewis Thrust. The eastern flank of Fernie Basin lies at the extreme left. (R.C.A.F. Photo T31L-196.)

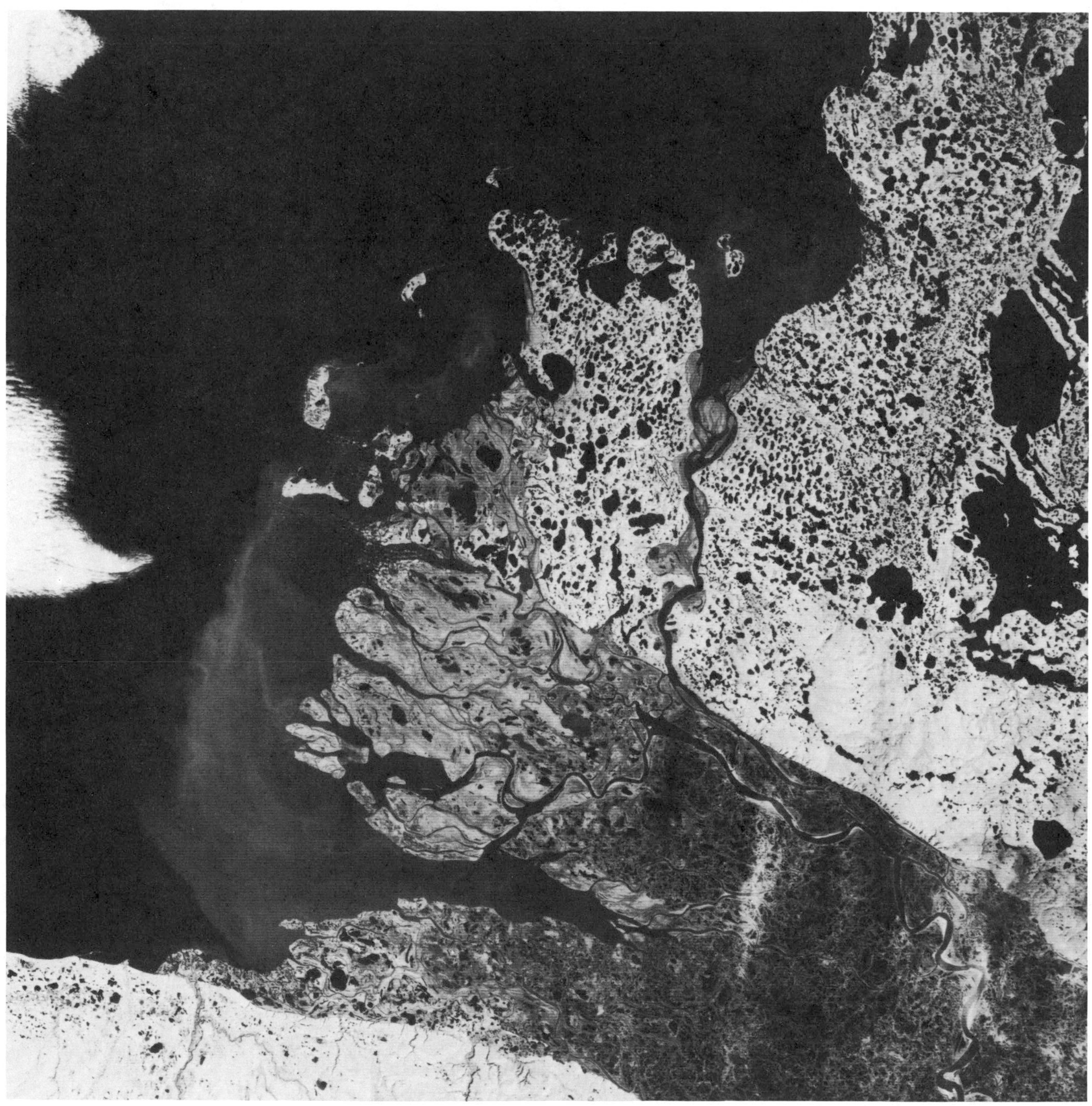

Figure 2C.12. Mackenzie Delta, Northwest Territories. A sediment plume extends into Beaufort Sea. Note the linear stream segments trending to the northwest and the northeast, and the numerous small lakes on the ancient deltaic plain of Tuktoyatuk Peninsula. (Landsat satellite image, 21 September, 1973, CCRS Photo E-1422-20190-6.)

Figure 2C.13. Alberta Plateau. View southwest across a remnant of the upland surface that has an elevation of 750 to 900 m between the Fort Nelson and Muskwa rivers. Note the scarp formed by the flat-lying Dunvegan sandstone. In the distance, anticlines of Triassic rocks lie in front of the Rocky Mountains. (R.C.A.F. Photo T27R-196.)

Figure 2C.14. The cultivated prairie of the Alberta Plain, viewed northward. Steep walls of Paleocene sediments rise above the post-glacial Bow River and its tributary gullies. Sand dunes in foreground have been covered by vegetation. (Photo -A. MacS. Stalker, 2-1-79.)

Figure 2C.15. Badland topography in uppermost Cretaceous sediments at Horseshoe Canyon, a tributary of Red Deer River, west of Drumheller, Alberta, view northward. Surface of Alberta Plain lies at the horizon. (Photo - D.F. Stott, IPSG 3086-3.)

underlie most of this region. Carbonates of Ordovician, Silurian, and Devonian age underlie the easternmost part and Tertiary sediments cap some of the uplands.

Surface elevations increase westward from 250 m at the margin of the Canadian Shield to 1200 m along the Cordilleran deformation front. Elevation increases occur abruptly along the Manitoba Escarpment and Missouri Coteau, which divide this part into three broad levels or 'steps'. The first, the Manitoba Escarpment, is formed by the erosional edge of soft Cretaceous shales overlying resistant Paleozoic carbonates. The escarpment is divided by several deep valleys into a series of morainic uplands, including Turtle Mountain, Riding Mountain, Duck Mountain, Porcupine Mountain, Pasquia Hills, and Wapawekka Hills. These have steep, east-facing slopes that mark elevation increases of about 300 m and form the boundary between the Manitoba Plain and the Saskatchewan Plain (Bostock, 1970b). The Missouri Coteau is a belt of morainic hills (hummocky moraine) that marks a rise of some 100 m from the Saskatchewan Plain to the Alberta Plain. The Coteau, well marked at the International Boundary by a line of low, rounded hills partly formed by Tertiary sediments, gradually becomes indefinite northwestward. The central and northern parts of the Interior Plains lack the distinctive natural boundaries of the southern part, and the broad elements of the bedrock topography are more dominant than in the southern part. Plateau-like interfluve areas separated by broad lowlands are designated Alberta Plateau, Peace River Lowland, and Fort Nelson Lowland.

The southern part of the Interior Plains is mainly within the Hudson Bay drainage basin whereas the northern part is within the Mackenzie drainage basin. A small part of the southwestern corner of the region drains into the Missouri basin. The drainage of the Hudson Bay basin in general occurs along east-northeast trending, preglacial drainage lines. Major river valleys in northern Alberta (Athabasca, Peace, Hay) follow northeastward courses within broad lowlands and form part of the Mackenzie River drainage. Some tributary valleys were formed as spillways and meltwater channels during deglaciation, but the main valleys follow preglacial drainage lines.

Figure 2C.16. Ancestral valley and gullied slopes on Tertiary pediments that form the southern slopes of Cypress Hills in southwestern Saskatchewan. The bevelled break in slope at the upper left is a bench that marks an earlier base level. View southerly. (Photo -R.W. Klassen, ISPG Photo 2626-22.)

Alberta Plateau (Fig. 2C.13) includes a number of extensive upland areas (Swan Hills, Pelican Mountains, Clear Hills, Birch Mountains, Caribou Mountains, Cameron Hills), 600 to 900 m in elevation, separated by the intervening Peace River and Fort Nelson lowlands some 300 to 600 m above sea level. The southernmost part of the Great Slave Plain, mostly below 300 m, extends into the northeast corner of Alberta. These broad elements of the physiography appear to reflect the preglacial landscape, although the blanket of drift may attain thicknesses of up to 150 m in the eastern lowlands and parts of uplands such as the Cameron Hills (D. Borneuf, pers. comm., 1984). Generally, however, the drift cover over the higher parts of the Alberta Plateau is less than 15 m and the topography is controlled by the underlying bedrock.

Alberta Plain (Fig. 2C.14), although locally similar to the Saskatchewan Plain, is higher, 800 to 1000 m, and includes extensive tracts of broadly rolling, drift-veneered Cretaceous bedrock plains interspersed with uplands capped by Tertiary bedrock (Wood Mountain, Cypress Hills, Neutral Hills, Hand Hills, Wintering Hills). Isolated badland areas occur along valleys in the southern part and are spectacularly developed along Red Deer and Milk rivers (Fig. 2C.15). Marked topographic contrasts are evident between the flat plateaux that form the summits of unglaciated upland (Wood Mountain, Cypress Hills) at elevations up to 1500 m in southwestern Saskatchewan and the irregular surfaces of uplands underlain by thick deposits of drift (Thickwood Hills, Mostoos Hills) in the northeastern part of the Plains. Former meltwater channels (Fig. 2C.16) and spillways generally show similar development and relationships to the regional drainage as do their downstream extensions across the Saskatchewan Plain. They are, however, cut mostly in bedrock, except where they cross or coincide with ancestral valleys.

The Porcupine Hills are a local subdivision of the Interior Plains. They are underlain mainly by gently dipping, early Tertiary strata, which form a broad, synclinal structure called the Alberta syncline. The broad, subsequent valley that separates the Porcupine Hills from the eastern Foothills is partly filled with glacial moraine, outwash, and lacustrine deposits, and was the locus of south-flowing meltwater streams during Pleistocene time.

Saskatchewan Plain consists of gently irregular till plains that merge with uplands (Moose Mountain, Touchwood Hills, Allan Hills, Wapawekka Upland, Waskesiu Upland) marked by strongly irregular, morainic terrain. Between the Manitoba Escarpment on the east and the Missouri Coteau on the west, surfaces are commonly between 400 and 800 m. The drift blanket beneath the Saskatchewan Plain is substantially thicker (30 to 120 m) and more continuous than that beneath the adjacent Manitoba and Alberta plains. As a result, local variations of the bedrock topography such as preglacial valleys are generally not reflected in the surface topography. The most striking local topographic contrasts are provided by the trench-like former meltwater channels and spillways. Bedrock is commonly exposed along the lower parts of the valley walls.

Manitoba Plain consists mostly of slightly irregular or nearly flat terrain underlain by Paleozoic carbonates dipping gently to the southwest. Till and glacio-lacustrine clays and silts of glacial Lake Agassiz blanket the bedrock in the southernmost part, but elsewhere the drift occurs mainly as a thin, discontinuous cover. Beaches of Lake Agassiz, marking successively lower water levels, are extensively developed (see Teller and Clayton, 1983). The northern part of Manitoba Plain in Manitoba and Saskatchewan includes local occurrences of low bedrock hills and cuestas of carbonate bedrock separated by gently irregular drift covered areas.

The Assiniboine River in the south and the Saskatchewan River in the north occupy shallow post-glacial channels across the Manitoba Plain. Several valleys that transect the uplands forming the Manitoba Escarpment reflect ancestral drainage lines that are largely absent over the Manitoba Plain and masked by thick drift over the Saskatchewan Plain.

REFERENCES

Aitken, J.D., Cook, D.G., and Yorath, C.J.
1982: Upper Ramparts River (106G) and Sans Sault Rapids (106H) map areas, District of Mackenzie; Geological Survey of Canada, Memoir 388, 48 p.

Bostock, H.S.
1948: Physiography of the Canadian Cordillera, with special reference to the area north of the fifty-fifth parallel; Geological Survey of Canada, Memoir 247, 106 p.
1961: Physiography and resources of the northern Yukon; Canadian Geographical Journal, v. 63, no. 4, p. 112-119.
1970a: Physiographic regions of Canada; Geological Survey of Canada, Map 1254A, Scale 1:500 000.
1970b: Physiographic subdivisions of Canada; in Geology and Economic Minerals of Canada, R.J.W. Douglas (ed.); Geological Survey of Canada, Economic Geology Report No. 1, Fifth Edition, p. 9-30.

Craig, B.G.
1965: Glacial Lake McConnell, and the surficial geology of parts of Slave River and Redstone River map-area, District of Mackenzie; Geological Survey of Canada, Bulletin 122, 33 p.

Daly, R.
1912: North American Cordillera, Forty-Ninth Parallel; Geological Survey of Canada, Memoir 38, 857 p.

Gabrielse, H. and Yorath, C.J. (ed.)
1991: Geology of the Cordilleran Orogen in Canada; Geological Survey of Canada, Geology of Canada, no. 4 (Also Geological Society of America, The Geology of North America, v. G-2).

Holland, S.S.
1964: Landforms of British Columbia - A physiographic outline; British Columbia Department of Mines and Petroleum Resources, Bulletin 48, 138 p.

Hughes, O.L., Veillette, J.J., Pilan, J., Hanley, P.T., and van Everdingen, R.O.
1973: Terrain evaluation with respect to pipeline construction; Mackenzie Transportation Corridor; Environmental-Social Committee, Northern Pipelines, Task Force on Northern Oil Development, Report No. 73-37, 74 p.

Mackay, J.R.
1962: Pingos of the Pleistocene Mackenzie Delta area; Geographical Bulletin (Canada), v. 5, no. 18, p. 21-63.
1971: Geomorphic process, Mackenzie Valley, Arctic coast, District of Mackenzie; in Report of Activities, Part A, Geological Survey of Canada, Paper 71-11, p. 189-190.

Mackay, J.R. and Stager, J.K.
1966: Thick tilted beds of segregated ice, Mackenzie Delta area, Northwest Territories; Biuletyn Peryglacjalny, no. 15, p. 39-43.

Matthews, W.H. (compiler)
1986: Physiographic map of the Canadian Cordillera; Geological Survey of Canada, Map 1701A, scale 1:5 000 000.

Norris, D.K.
1973: Tectonic styles of northern Yukon Territory and northwestern District of Mackenzie, Canada; in Arctic Geology, M.C. Ritcher (ed.); American Association of Petroleum Geologists, Memoir 19, p. 23-40.
1974: Structural geology and geological history of the northern Canadian Cordillera; Proceedings of 1973 National Convention (Calgary), Canadian Society of Exploration Geophysicists, p. 18-43.

Pihlainen, J.A., Brown, R.J.E., and Leggett, R.F.
1956: Pingo in the Mackenzie Delta, Northwest Territories, Canada; Geological Society of America, Bulletin, v. 67, p. 1119-1122.

Rampton, V.N
1971: Quaternary geology, Mackenzie Delta and Arctic Coastal Plain, District of Mackenzie; in Report of Activities, Geological Survey of Canada, Paper 71-1A, p. 173-177.
1974: Surficial geology and landforms for parts of Aklavik (107B), Blow River (117A), Demarcation Point (117C), Herschel Island (117D); Geological Survey of Canada, Open File 191, scale 1:125 000.
1982: Quaternary geology of the Yukon Coastal Plain; Geological Survey of Canada, Bulletin 317, 49 p.

Rampton, V.N. and Mackay, J.R.
1971: Massive ice and icy sediments throughout the Tuktoyaktuk Peninsula, Richards Island, and nearby areas, District of Mackenzie; Geological Survey of Canada, Paper 71-21, 16 p.

Selwyn, A.R.C. and Dawson, G.M
1884: Descriptive sketch of the physical geography and geology of the Dominion of Canada; Geological Survey of Canada, 55 p.

Stager, J.K.
1956: Progress report of the analysis of the characteristics and distribution of pingos east of the Mackenzie Delta; Canadian Geographer, no. 7, p. 13-20.

Teller, J.T. and Clayton, L. (ed.)
1983: Glacial Lake Agassiz; Geological Association of Canada, Special Paper 26, 451 p.

Yorath, C.J. and Norris, D.K.
1975: The tectonic development of the southern Beaufort Sea and its relationship to the origin of the Arctic Ocean Basin; in Canada's Continental Margins and Offshore Petroleum Exploration; Canadian Society of Petroleum Geologists, Memoir 4, p. 589-611.

Authors' Addresses

R.W. Klassen
Institute of Sedimentary and Petroleum Geology
Geological Survey of Canada
3303 - 33rd Street N.W.
Calgary, Alberta
T2L 2A7

D.F. Stott
8929 Forest Park Drive
Sidney, British Columbia
V8L 5A7

Printed in Canada

Subchapter 2D

TECTONIC FRAMEWORK

J.D. Aitken

CONTENTS

INTRODUCTION

This subchapter is concerned only with the tectonics affecting the **Phanerozoic** sedimentary cover of the Interior Platform, western basins, and eastern Cordillera and with those Upper Proterozoic deposits genetically related to them. Interpretation of the sedimentary and tectonic history of pre-Late Proterozoic time is subject to uncertainties much greater than those of Phanerozoic history, and Phanerozoic rocks are the principal focus of this volume. Precambrian tectonics as determined from stratified rocks is dealt with in Subchapter 4A, and the geology of the sparsely sampled and remotely sensed crystalline basement in Chapter 3. Furthermore, this subchapter is concerned only with those sedimentary successions that were deposited upon or in immediate connection with North America; the extremely complex history of the accreted terranes to the west is beyond the compass of this volume, and in any case is not recorded by the strata of the Western Canada Basin. The accreted terranes are fully treated by Gabrielse and Yorath (1991). In this subchapter only the most critical sources are cited; all significant literature is cited in Chapter 4.

The region of concern here has generally been treated as the Western Canada Basin, a term useful only in designating a geographic area underlain entirely by sedimentary rocks. Most of the "basin", as conventionally outlined, was a tectonic and sedimentary platform ("shelf" of some authors) during the divergent-margin phases of sedimentation, a part of it was a platform during the convergent-margin phases, and it has been a tectonic platform for most of Tertiary time. Where "Western Canada Basin" is used herein, it should be understood not to be limited to the undeformed regions, but to extend into the orogen as far as there are unmetamorphosed sedimentary rocks, and to extend from the 49th parallel to the shores of Beaufort Sea.

The organization of presentation under passive (divergent), active (convergent), and transcurrent regimes is somewhat artificial. It derives in part from an older, simpler view, based on study of the contents of the Western Canada Basin alone, in which the Paleozoic and lower

Aitken, J.D.
1993: Tectonic framework; Subchapter 2D in Sedimentary Cover of the Craton in Canada, D.F. Stott and J.D. Aitken (ed.); Geological Survey of Canada, Geology of Canada, no. 5, p. 45-54 (also Geological Society of America, The Geology of North America, v. D-1).

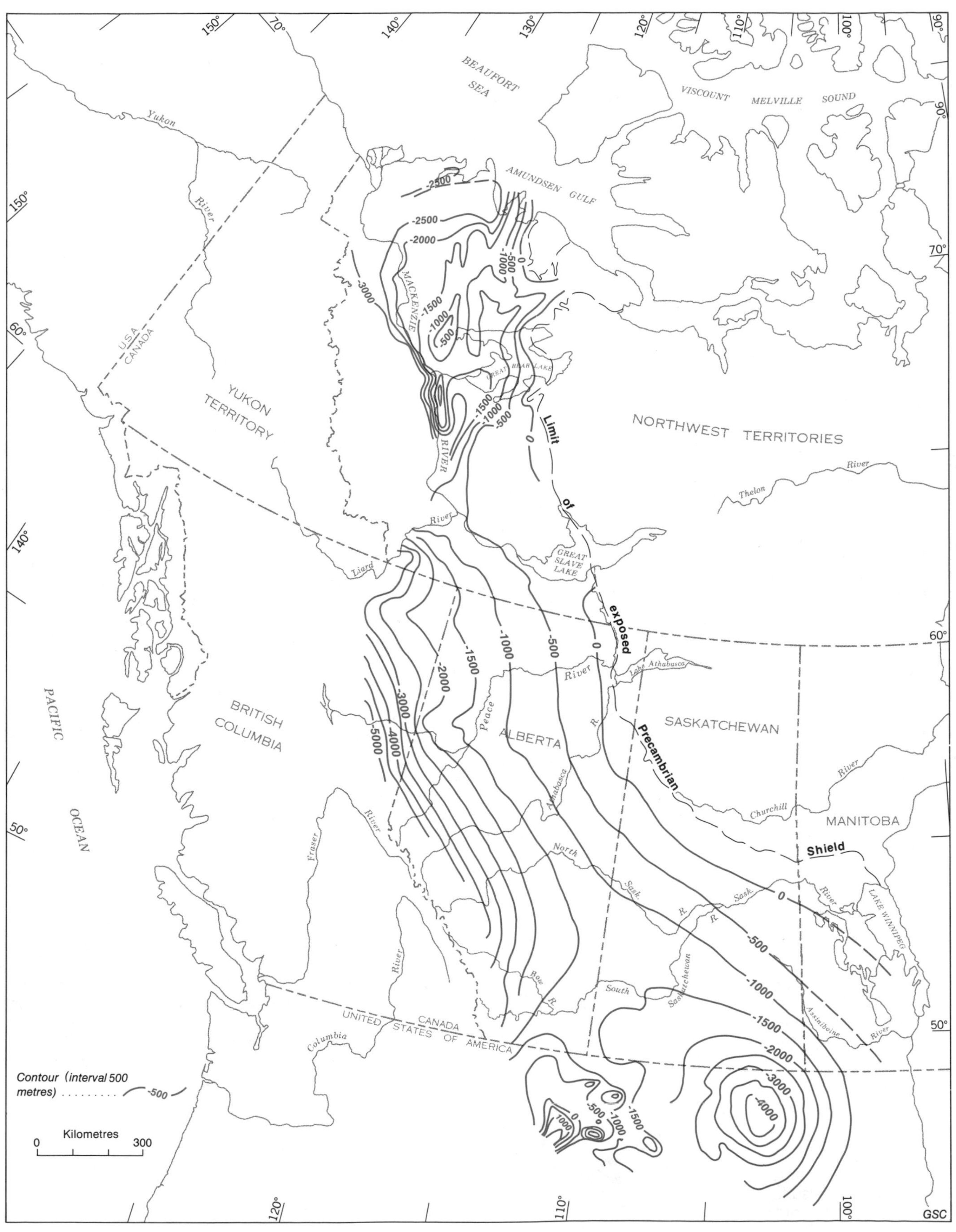
BEAUFORT SEA
VISCOUNT MELVILLE SOUND
AMUNDSEN GULF
YUKON TERRITORY
NORTHWEST TERRITORIES
BRITISH COLUMBIA
ALBERTA
SASKATCHEWAN
MANITOBA
PACIFIC OCEAN
USA CANADA
UNITED STATES OF AMERICA
CANADA
GREAT BEAR LAKE
GREAT SLAVE LAKE
Lake Athabasca
LAKE WINNIPEG
MACKENZIE RIVER
Limit of exposed Precambrian Shield
Contour (interval 500 metres) -500
Kilometres
0
300
GSC

Mesozoic successions appeared to be entirely miogeoclinal. Growing evidence for episodes of pre-Jurassic convergent tectonism that interrupted or terminated the passive regime undermines the older view. This evidence is as yet difficult to integrate with the seemingly miogeoclinal record in Western Canada Basin, however, and the organization under "passive", active, and transcurrent regimes provides a simpler, though flawed basis for discussion. The Western Canada "Basin" is seen herein as a temporally shifting mosaic of platforms, arches, basins, and sub-basins of at least two orders. In some instances different names are assigned to geographically coincident features of the divergent and convergent tectonic phases. The named elements are hybrid, in part delimited purely on tectonic criteria and in part (to divide the broadest features for ease of discussion) on geographic criteria, drawing limits where possible along some geological line. Many published names for tectonic elements have been used previously in two or more ways, and many tectonic elements have been assigned two or more names by different authors, a regrettable reality that the editors have been unable to overcome, even within the compass of this volume. In this subchapter, the interpretations are entirely those of the author, and by drawing on the manuscripts of all of the authors of the volume, the best-established and most useful names have been used (Fig. 2D.1, 2D.2, 2D.3, 2D.4).

The history of the western platforms and basins is to a large extent the history of the western continental margin (here termed the "Columbian margin" because the Columbian Orogen arose from it), and to a lesser extent the history of the ancient, Innuitian, and much younger, Beaufort Sea margins in the north. Acknowledging that a part of the distal sedimentary record of the western margin has been lost during the accretion of exotic terranes to North America and consequent deformation and erosion, that margin is drawn at the boundary with the easternmost accreted terrane, accepted here as the Slide Mountain Terrane (Klepacki and Wheeler, 1985). Debate continues, however, about which terrane is, indeed, the easternmost, in view of evidence suggesting that the Kootenay Terrane may be allochthonous (Gabrielse and Yorath, 1991). In the northern mainland, the geometry of the Innuitian margin has been disrupted during the opening of Canada Basin, and that margin is incompletely understood. The succeeding Beaufort Sea margin was created as the deposits of the older margin were disrupted (Trettin, 1991).

If the contents of the Western Canada Basin alone are considered, the history of the western margin appears simple, and to have transpired in three clearly demarcated phases, each identified by a characterizing relative motion between the North American Plate and whatever oceanic plate or plates lay to the (present) west: a divergent (passive) margin phase; a convergent (active) margin phase, and a post-convergent, transcurrent phase. In terms of sedimentation, the pivotal event was the reversal in the direction of dominant sediment transport, from westward off the Canadian Shield, during the divergent phase (latest Proterozoic to Middle Jurassic), to eastward off the rising orogen, during the convergent phase (Middle Jurassic to Eocene). Each phase had its unique suite of tectonic elements, although some elements persisted from one phase to another.

The northern margin, of which only the western and least understood part is described in this volume, was initiated by continental separation at about the same time as the Columbian margin. It underwent convergent orogeny (Ellesmerian) in Late Devonian - Early Carboniferous, but the mountain belt of that time was severed by Early Cretaceous opening of Canada Basin, and is difficult to identify, whatever its post-rotation or post-translation position. Ellesmerian folds and post-Ellesmerian orogenic deposits were again folded during late Columbian and Laramide orogeny in the northwest. Although for much of the region no folding younger than Paleocene is recorded, Miocene strata are deformed in the Mackenzie Delta region, and Pleistocene deposits along the Alaskan north shore are folded, suggesting that convergent tectonism is still in progress there.

Figure 2D.1. Structural contours on the basement (sub-Phanerozoic) surface (after Porter et al., 1982, with additions adapted from McCrossan, unpub. rep.).

Pre-Cordilleran/Innuitian Framework

The Phanerozoic and uppermost Proterozoic sediments of the Western Canada Basin were deposited upon one or other of:

(a) Lower Proterozoic and older, crystalline rocks of the Canadian Shield,

(b) Sedimentary rocks of a Middle Proterozoic, western continental margin, in part folded and metamorphosed in mid-Proterozoic, or

(c) Little-deformed, Proterozoic sedimentary and volcanic strata, in part post-dating the Proterozoic folding but in part resting on undeformed, Precambrian crystalline basement and probably pre-dating that folding.

The character and history of these rocks are outlined in Chapter 3 and Subchapter 4A.

DIVERGENT REGIMES

Columbian Margin

Inception

The Columbian margin was defined by continental separation dated as about the beginning of the Cambrian in the southern Rocky Mountains (Bond and Kominz, 1984), and the same or slightly earlier in Mackenzie and Selwyn Mountains (J.D. Aitken, this volume, Subchapters 4A, 4B). Separation had been preceded by an episode of rifting that was atypically prolonged as compared with widely accepted, current models, having its inception at 735-780 Ma. With separation, a thick continental terrace wedge began to accumulate upon the margin. From the stratigraphic record of the Western Canada Basin it would appear that sedimentation of similar style continued, with interruptions, for 400 Ma. The pre-Upper Jurassic, Phanerozoic stratigraphy of the basin has seemed, since its study began, to constitute a single wedge, although the magnitude of the

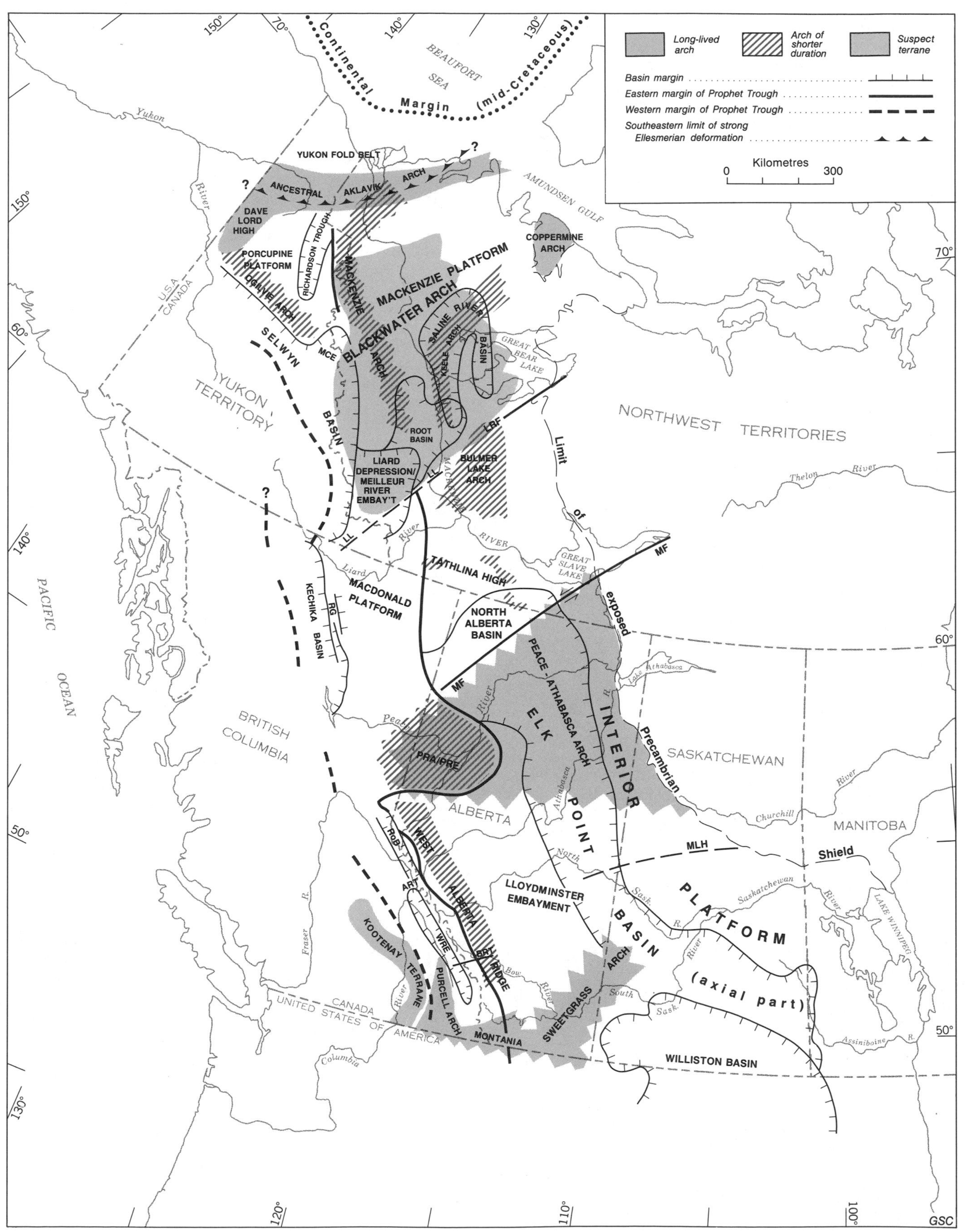

Long-lived arch
Arch of shorter duration
Suspect terrane
Basin margin
Eastern margin of Prophet Trough
Western margin of Prophet Trough
Southeastern limit of strong Ellesmerian deformation
Kilometres
0
300
Continental Margin (mid-Cretaceous)
BEAUFORT SEA
AMUNDSEN GULF
YUKON FOLD BELT
ANCESTRAL AKLAVIK ARCH
DAVE LORD HIGH
PORCUPINE PLATFORM
OGILVIE ARCH
RICHARDSON TROUGH
MACKENZIE
MACKENZIE PLATFORM
BLACKWATER ARCH
SALINE RIVER
KEELE ARCH
BASIN
COPPERMINE ARCH
GREAT BEAR LAKE
SELWYN BASIN
MCE
YUKON TERRITORY
ROOT BASIN
LIARD DEPRESSION/ MEILLEUR RIVER EMBAY'T
LBF
BULMER LAKE ARCH
NORTHWEST TERRITORIES
Limit of exposed Precambrian Shield
Thelon River
LL
TATHLINA HIGH
GREAT SLAVE LAKE
MF
KECHIKA BASIN
RG
MACDONALD PLATFORM
NORTH ALBERTA BASIN
PEACE ATHABASCA ARCH
ELK POINT BASIN
INTERIOR PLATFORM (axial part)
Lake Athabasca
PACIFIC OCEAN
BRITISH COLUMBIA
SASKATCHEWAN
PRA/PRE
ALBERTA
Churchill
MANITOBA
MLH
RoB
WEST ALBERTA RIDGE
ART
LLOYDMINSTER EMBAYMENT
KOOTENAY TERRANE
WRE
PURCELL ARCH
BRT
SWEETGRASS ARCH
MONTANIA
LAKE WINNIPEG
CANADA
UNITED STATES OF AMERICA
WILLISTON BASIN
Yukon River
Liard River
Peace River
Fraser R.
Columbia
Bow River
North Sask. R.
South Sask.
Saskatchewan River
Assiniboine R.
150°
140°
130°
120°
110°
100°
70°
60°
50°
GSC

"sub-Devonian" unconformity has caused some doubts. Most of the stratigraphy in this volume is presented from a viewpoint accepting this apparent unity as real. A major problem is that such an interpretation involves a passive-margin phase of sedimentation, Early Cambrian to Middle Jurassic, that was extraordinarily prolonged as compared with the currently prevailing, "Atlantic Margin" model. This problem may have been largely removed by the demonstration, in the Omineca Belt, of Paleozoic deformation and granitoid intrusion (Okulitch et al., 1975; Okulitch, 1985). These data could be "explained away" formerly as having come from the Kootenay Terrane, which was considered suspect (possibly non-North American), but the work of Klepacki (1985) tied the Kootenay Terrane more closely to the western margin of North America. Klepacki's work in the Omineca Belt further demonstrated at least two pre-Middle Jurassic tectonic events in Kootenay Terrane, Late Devonian to mid-Carboniferous (post-dating folding and metamorphism of lower Paleozoic strata) and Early Permian. If his interpretation of the Upper Triassic Slocan Group as a back-arc assemblage is correct, as supported by western provenance of the clastics, then a third, post-Cambrian, pre-Middle Jurassic event is recorded. The hypothesis of an allochthonous Kootenay Terrane (e.g., Silberling and Jones, 1984), except possibly its western part, is not supported by biogeographic or paleomagnetic data and requires the inference of a cryptic suture. Its only attraction now is that it accounts for the general lack of evidence for depositional responses, in Western Canada Basin, to the Paleozoic and early Mesozoic orogenic events recorded in the central Cordillera. Such evidence is not completely lacking, however. Recent work has produced evidence for a landmass and sediment source that formed the western margins of Prophet (Carboniferous) and Ishbel (Permian) troughs (Richards et al., Subchapter 4E in this volume; Henderson et al., Subchapter 4F in this volume).

The record outlined above of pre-Middle Jurassic, convergent and divergent tectonism at the western margin seems to obviate the problem of a passive-margin phase of a duration inconsistent with modern, plate-tectonic models. However, the question remains as to why no pre-Late Jurassic, orogenic clastic wedge of western derivation is known in the Western Canada Basin. This seems to require convergent deformation without the creation of a significant landmass. The missing clastic wedges remain the strongest argument for an allochthonous Kootenay Terrane, but perhaps merely demonstrate that the undeformed basin lies farther inboard of the Paleozoic continental margin than it is generally thought to be.

Figure 2D.2. Map of tectonic elements, passive regimes. MCE - Misty Creek Embayment; RG - Roosevelt Graben; RoB-Robson Basin; PRA/PRE - Peace River Arch/Peace River Embayment; WRE - White River Embayment; MLH - Meadow Lake Hingeline; LL - Liard Line; LRF - Leith Ridge Fault; MF - McDonald Fault; ART - Athabasca River Transect; BRT - Bow River Transect.

Marginal basins and promontories

The Columbian continental margin of the Early Cambrian did not follow a simple, straight or gently sinuous line, but like the margins of continents bordering the modern Atlantic Ocean, consisted of straight or arcuate segments more or less parallel to the Cordilleran trend, offset from one another by straight segments oriented at high angles to that trend. This outline is being recognized piece by piece through current work; only its most prominent features (Fig. 2D.2) will be mentioned here.

The Early Cambrian landmass and structural eminence known as Montania is recognized in the stratigraphy of the southern Rocky Mountains. At the north flank of Montania, south-trending facies belts in Cambrian strata are deflected westward. The north flank of Montania is recognized as the northeast-striking, Moyie-Dibble Creek fault system, now having the geometry of a thrust fault (P. Gordy, pers. comm.), but apparently inherited from a transform fault offsetting the ancient margin. Isopach maps show that during the Carboniferous and Permian periods, the area of the former Montania was structurally high, and had a northeast-trending depression along its northern flank.

Farther north in the Rockies, such outer-shelf features as White River Embayment, Robson Basin, and Roosevelt Graben, having margins sub-parallel to the Cordilleran trend, came into being at various times from Early Cambrian to Early Ordovician times. These appear to have originated through delayed rifting near the margin.

Near the 60th parallel, lower Paleozoic facies and isopachs and the deformation front are all offset eastward at the Liard Line, which marks an offset of some 250 km in the continental margin. This configuration is a mirror-image of the elbow in Logan's Line at the Appalachian margin. The Liard Line effectively marks the southern margin of Selwyn Basin, which is characterized by basinal deposits throughout much of the lower and middle Paleozoic. Various lines of evidence for continental separation through dextral trans-extension lead to the suggestion that Selwyn Basin originated as a large, pull-apart basin, or a sort of rhombochasm, albeit one not floored by oceanic crust.

Misty Creek Embayment, an arm of Selwyn Basin nearly aligned with Richardson Trough, shares the subsidence history of the latter (see below) and appears to be a failed attempt at a rift-connection between that feature and Selwyn Basin, or in other words, a failed rift-connection between Columbian and Innuitian margins.

Porcupine Platform is a promontory of the North American Craton. The promontory, identified as such by its geophysical characteristics and the platformal nature of the blanketing sediments, is bounded on the south by the Columbian margin and on the north by the Innuitian margin, and is strongly modified by convergent tectonism that apparently accompanied the opening of Canada Basin. Its western boundary, in eastern Alaska, is poorly understood, but may have been established in synchrony with the Columbian passive margin.

The necessary connection between the Cordilleran and Innuitian divergent margins is poorly understood. If the mostly widely accepted, current model for the opening of Canada Basin is correct, the connection has been severed by the anticlockwise rotation of northern Alaska out of the

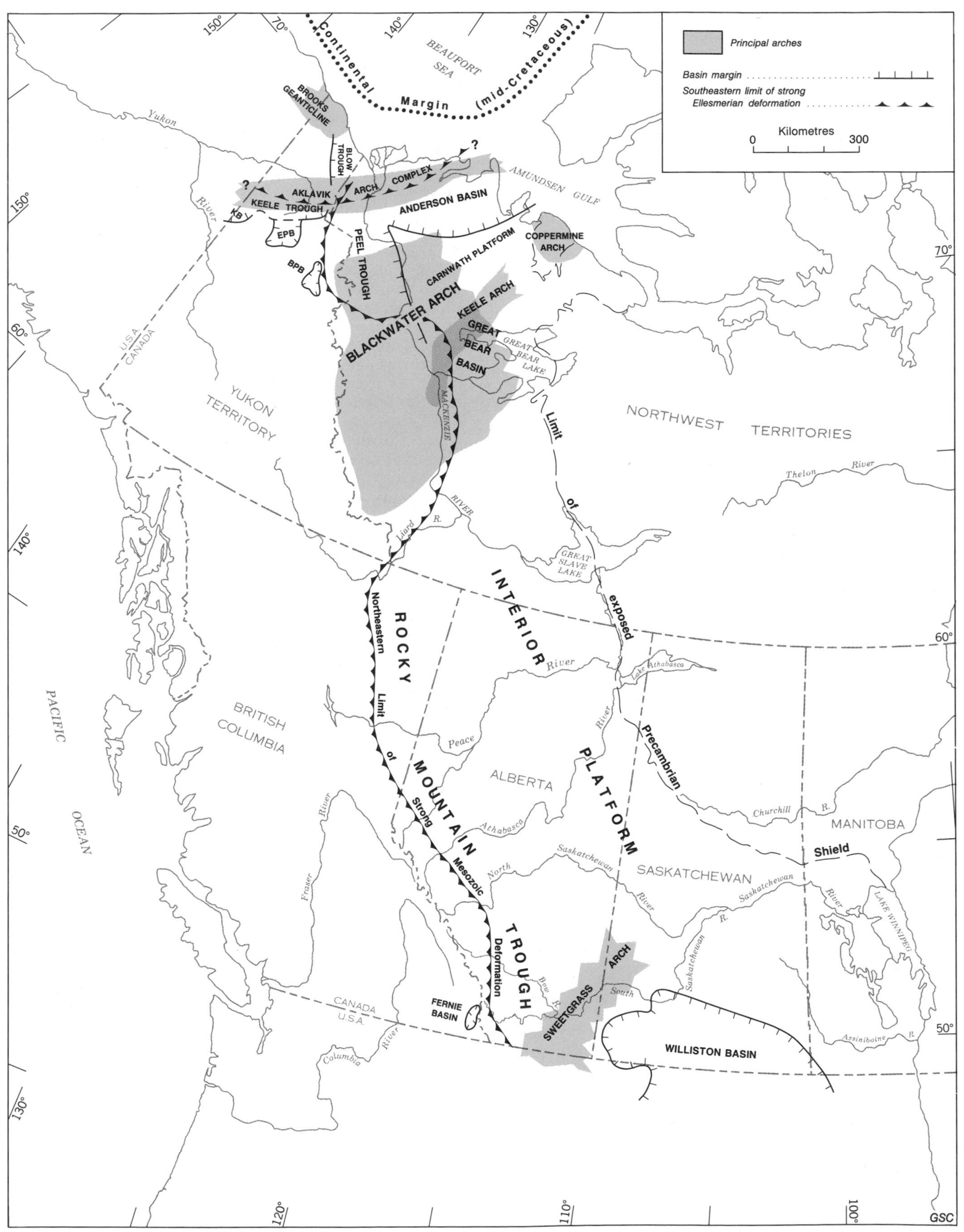
Principal arches
Basin margin
Southeastern limit of strong Ellesmerian deformation
Kilometres
0
300
Continental Margin (mid-Cretaceous)
BEAUFORT SEA
BROOKS GEANTICLINE
BLOW TROUGH
AKLAVIK ARCH COMPLEX
KEELE TROUGH
KB
EPB
BPB
ANDERSON BASIN
AMUNDSEN GULF
PEEL TROUGH
CARNWATH PLATFORM
COPPERMINE ARCH
BLACKWATER ARCH
KEELE ARCH
GREAT BEAR BASIN
GREAT BEAR LAKE
YUKON TERRITORY
NORTHWEST TERRITORIES
Limit of exposed Precambrian Shield
GREAT SLAVE LAKE
INTERIOR PLATFORM
ROCKY MOUNTAIN TROUGH
Northeastern Limit of Strong Mesozoic Deformation
BRITISH COLUMBIA
PACIFIC OCEAN
ALBERTA
SASKATCHEWAN
MANITOBA
LAKE WINNIPEG
SWEETGRASS ARCH
FERNIE BASIN
WILLISTON BASIN
CANADA U.S.A.
USA CANADA
GSC

basin in early to mid-Cretaceous time (see Grantz et al., 1990). Such rotation would have reversed the polarity of pre-Albian deposits and may account for some of the difficulties attending their interpretation.

Epicratonic arches

The history of the continental terrace wedge and the platform flanking it inboard is punctuated by the rise and subsequent stasis or subsidence of a number of epicratonic arches (Fig. 2D.2). The arches are important in that they subdivide the platforms into a number of second-order basins, act in some instances as relatively local sources of detritus, provide foundations for reef growth, and play a major role in hydrocarbon accumulation. Except for the apparent initial rise of several of these features during the Early to Middle Devonian, little of a common history is evident, and the apparent Devonian "synchrony" may not be factual. The history of epicratonic arches and basins is summarized in Figure 2D.4. It is particularly significant, in attempting to understand these features, that a number of the arches were active during both divergent- and convergent-margin phases.

The various elements of the Aklavik Arch Complex (Yorath and Norris, 1975) have had a particularly complex history, at best only partly understood. Uplift of parts of the complex took place at least by early Paleozoic time, when the complex was an intracratonic feature, and has continued intermittently into the Tertiary. Part of the complex, "Ancestral Aklavik Arch", had a clearly defined phase of activity during the Carboniferous and Permian periods. The northeastern part, Eskimo Lakes Arch, became a continental margin feature when Canada Basin opened. The elements southwest of Mackenzie River were reactivated by convergent, late-Columbian, Laramide and post-Laramide tectonism and remain structurally high today.

Epicratonic basins

In a treatment that is not highly detailed, it is necessary to include a variety of entities under the general classification of "epicratonic basin". As emphasized by Fitzgerald-Moore (Subchapter 4D in this volume), the term has been used to identify such disparate features as syndepositional tectonic basins (whether bathymetric basins or not), miogeoclines, depocentres, constructional basins defined by surrounding carbonate platforms, and post-depositional tectonic basins ("basins of preservation"). A large number of these basins and sub-basins have been identified, especially by geologists in the petroleum industry; only the more important ones are drawn in Figure 2D.2 and placed in a temporal frame of reference in Figure 2D.4.

Figure 2D.3. Map of tectonic elements, active regimes. BPB - Bonnet Plume Basin; EPB -Eagle Plains Basin; KB - Kandik Basin.

Ancient Innuitian Margin

Most knowledge of the Innuitian passive margin has been gained in the Arctic Islands. The pre-Upper Devonian stratigraphy bears many parallels to that of the Western Canada Basin, and can to a degree be projected into northern Yukon, but in the region of Porcupine Platform, Richardson Mountains, and British Mountains, it is most difficult to know whether deposits should be related to the Innuitian margin or the Cordilleran margin. The currently most popular model for the opening of Canada Basin, involving counter-clockwise rotation of northern Alaska, calls for a rotational reversal of the polarity of earlier, Innuitian-margin deposits.

Richardson Trough, an aulacogen-like feature opening to the Innuitian continental margin, began to subside in late Early Cambrian time, and continued as a tectonic and generally bathymetric trough through early and middle Paleozoic times.

Beaufort Sea Margin

The Beaufort Sea continental margin is structurally complex and difficult to interpret. The margin came into being with the opening of Canada Basin (oceanic) to the north. If this event took place in Aptian-Albian time, which is one of the likely possibilities (see Grantz et al.,1990), it coincided with strong convergence in the Cordillera and the attendant formation of foredeeps. Thus, foredeep filling and passive-margin sedimentation appear to have been in part contemporaneous, and numerous lithostratigraphic units are continuous between one and the other tectonic element.

With the opening of the oceanic Canada Basin (whatever the kinematic scenario), the northeastern part of Aklavik Arch Complex, formerly intracratonic, became a continental margin feature as Eskimo Lakes Arch, and acted as a structural "high" and a continental-margin hingeline (Yorath and Cook, 1981). Post-Laramide deformation in Romanzof Uplift provided renewed topographic relief and a nearby source of orogenic detritus.

As a consequence of progressive, northward migration of depocentres and shorelines from Cenomanian time onward, most sedimentation was taking place on the northern continental margin by Maastrichtian time. Tertiary and Quaternary depocentres have shifted progressively offshore (Dixon, 1986).

CONVERGENT REGIMES

Most of the foreland fold and thrust belt of the Columbian Orogen has undergone only a single period, albeit prolonged and pulsatory, of convergent tectonism during Phanerozoic time, namely the Columbian and Laramide orogenies extending from mid-Jurassic to early Tertiary. As described earlier in greater detail, episodes of pre-Columbian, convergent tectonism recorded in the Omineca Belt of British Columbia have left no obvious record in the Western Canada Basin at the same latitudes. The exceptional area is northern Yukon, where evidence of late Paleozoic convergent tectonism, the Ellesmerian Orogeny, is well recorded.

	Late Proterozoic	Cambrian	Ordovician	Silurian	Devonian	Carboniferous	Permian	Triassic	Jurassic	Cretaceous	Paleogene	Neogene
Great Bear Basin		0 0 0	0 0 0	0 0	0 0 0	+ +	? ?	+ + +	+ + +	+ –	–	0
Anderson Basin		0 0 0	0 0 0	0 ?	0 0 0	? ?	? ?	+ + +	+ + +	+ –	0	0
Peel Trough		0 0 0	0 0 0	0 ?	0 0 0	? ?	? ?	+ + +	+ + +	+ –	0	0
Keele/Kandik Trough						– –	+ +	? ? ?	0 0 0	– –	+	+
Blow Trough						0 0	0 ?	? ? ?	0 0 0	– –	+	+
Brooks Geanticline		? ? ?	? ? ?	? ?	0 0 +	+ 0	0 ?	? ? ?	– – –	+ +	+	+
Rocky Mtn. Trough		0 0 0	0 +? 0	? ?	0 0 0	0 0	0 0	0 0 0	0 0 –	– –	–	+
Columbian Orogen N	–	– – –	– – –	– –	– – +	+ –	– –	+ + +	+ + +	+ +	+	+
Columbian Orogen S	–	– – –	– – –	– ?	+ + +	+ –	– –	– – –	0 – –	– +	+	+
Beaufort Sea Continental Margin							? ?	? ? ?	? ? ?	+ –	–	–
Prophet/Ishbel Trough N	–	– – –	– – –	– –	– – –	– –	– –	– – –	+ + +	+ +	+	+
Prophet/Ishbel Trough S	–	– – –	– – –	– ?	+ + –	– –	– –	– – –	0 – +	+ +	+	+
Blackwater Arch		0 0 0	0 +? 0	0 0	0 0 0	? ?	? ?	+ + +	+. + +	+ 0	0	0
Yukon Fold Belt		0 0 0	0 +? 0	0 ?	0 0 +	+ 0	0 0	0 0 0	0 – –	+ +	+	+
Keele Arch		0 0 0	0 +? 0	0 +	+ 0 0	? ?	? ?	+ + +	+ + +	+ 0	0	0
Coppermine Arch		0 0 0	0 ? ?	? ?	+ 0 ?	? ?	? ?	? ? ?	+ + +	+ +	0	+
Root Basin		0 0 0	0 ? 0	0 –	– – –	+ +	? ?	+ + +	+ + +	+ +	?	0
Peace River Arch/ Embayment	+	? 0 0	? + ?	? ?	+ 0 0	– –	– ?	0 0 0	0 + –	– –	?	0
West Alberta Ridge		0 0 +	0 ? ?	? ?	+ + 0	0 0	0 0	0 0 0	0 0 –	– +	+	+
Tathlina High	+	0 0 0	? ? ?	? ?	+ + +	– +	0 0	? ? ?	? ? ?	0 0	0	+
Williston Basin		0 0 0	0 – –	– ?	– – –	– –	? ?	0 0 0	0 0 –	– –	–	+
Sweetgrass Arch		0 0 0	0 + +	?	0 0 0	– + +	? ?	+ + +	+ + +	+ +	?	+
Meadow Lk. Hingeline		0 0 0	0 – –	– ?	+ + 0	0 0	? ?	0 0 0	0 0 0	0 0	0	0
Liard Depression/ Meilleur R. Embay't.		? ? ?	– 0 –	– –	– – –	– –	? ?	+ + +	+ + +	– –	?	+
White R. Embayment	?	? – –	– – –	– ?	0 0 ?	– –	– –	? ? ?	0 – –	+ +	+	+
Purcell Arch	–	? ? 0	+ + 0	0 ?	+ ? ?	+ ?	? ?					
Robson Basin	–	– – 0	0 ? ?	? ?	? ? ?	– +	? ?					
Roosevelt Graben	–	0 – –	– + 0	0 ?	? ? ?	? ?	? ?					
Saline River Basin	0	– – 0	0 + 0	0 0	+ 0 0	? ?	? ?	+ + +	+ + +	+ –	0	0
Richardson Trough	–	0 – –	– – –	– –	– – –	0 0	0 0					
Selwyn Basin	–	– – –	– – –	– –	– 0 –	– –	– –					
Kechika Basin	–	– – –	– – –	– –	– ? ?	– –	– –					
Ogilvie Arch	+	+ + +	+ + 0	? ?	? ? ?	– –	– –					
Aklavik/Eskimo Lakes Arch	+	+ + +	0 0 0	? 0	0 0 0	0 +	+ ?	+ + +	+ + +	+ –	+	+
Dave Lord High/ Ancestral Aklavik Arch	+	0 0 0	+ + ?	? ?	+ + –	– +	+ 0	+ + +	+ 0 0	0 –	+	+
Bulmer Lake Arch	+	+ + +	+ + 0	0 ?	0 0 0							
Peace - Athabasca Arch	?	0 + +	+ + 0	0 ?	+ 0 0	0 0	0 0					
Montania	?	+ + 0	? ? ?	? ?	? ? 0	+ +	+ +	? ? ?	0 – –			
Mackenzie Arch	+	+ + 0	0 + 0	0 +	+ 0 0	? ?	? ?					
Cordilleran 'Miogeocline' N	– +	– – –	– 0 –	– –	– – –	0 0	0 0	0 0 0	0 0			
Cordilleran 'Miogeocline' S	– +	– – –	– 0 –	– ?	+ + –	– –	– –	– – –	– –			

GSC

Ellesmerian Orogeny

In northern Yukon, in Barn, Cache Creek, and Scho uplifts, northeast-trending folds affect all strata up to at least the Frasnian part of the Upper Devonian Imperial Formation. Westward-coarsening clastic strata in the folded Imperial record slightly earlier tectonic highlands to the west. The folds are bevelled by an unconformity, upon which lie strata at least as old as Viséan and possibly late Tournaisian. These relationships document an orogeny recorded in most of the Canadian Arctic Archipelago and known as the Ellesmerian Orogeny (see Trettin, 1991). The southern limit of strong Ellesmerian folding on the mainland is shown in Figures 2D.2 and 2D.3.

The region of Ellesmerian folding was re-folded during Cretaceous to early Tertiary times as part of the Columbian Orogen.

Cordilleran Orogeny

The onset of convergent tectonism at the western continental margin in Middle to Late Jurassic time recast the tectonic framework of western Canada. Thick sediments derived from the rising Columbian Orogen accumulated in a long-lived foredeep in front of the eastward-migrating deformation front, partly in continental environments, when deposition overcame subsidence, and partly in marine environments, when subsidence dominated and marine waters of the Boreal and Gulfian seas invaded the foredeep.

Earlier studies recognized numerous "orogenies" in the Central and Western Cordillera, based in part on dates of deformation and intrusion. However, it has been recognized in recent years that part of the Central, and all of the Western Cordillera consist of terranes allochthonous to North America. Interpretation of the tectonics of the Columbian Orogen has been greatly simplified by the realization that much of the deformation and igneous activity recorded to the west took place before the terranes were sutured to North America (Gabrielse and Yorath, 1991).

If the deformation and rise of the Columbian Orogen is deduced solely from the resulting record of clastic wedges, a much simpler history emerges. Analysis of this history has mainly considered the record in the Rocky Mountain Trough from the 49th parallel to the Liard River region. This has been interpreted in terms of a multi-phase Columbian Orogeny dating from Middle Jurassic to early Late Cretaceous times, followed by an apparently single-phase Laramide Orogeny dating from mid-Late Cretaceous through Paleocene times. The continued existence of a foredeep (albeit one filled mainly with fine grained deposits) between the Columbian and Laramide pulses raises the question as to whether orogeny was in fact twofold, as conventionally viewed, or essentially continuous (see Chapter 5).

Around the Mackenzie tectonic arc from the Liard River region to northern Yukon, "Columbian" and "Laramide" either are not clearly distinguished or acquire changed meanings. The first clastic wedges date only from the Aptian-Albian ages. In front of the Mackenzie Arc, the youngest preserved clastic wedge is Paleocene, and has been deformed. In the northernmost Yukon and at the Beaufort Sea margin, folding continued until at least Miocene time, and in adjacent Alaska, Pleistocene.

In Rocky Mountain Trough, the sites of major depocentres, usually marked by large delta-complexes, shifted back and forth. The boundaries of the transitory sub-basins are indistinct, however, and they have not been named or formalized.

In the structurally complex region north of the Mackenzie and Ogilvie Mountains, the Aptian and Albian strata of Blow and Keele-Kandik troughs are the earliest deposits of foredeep character, and in the Rocky Mountain framework would be classified late Columbian. The troughs do not, however, in their present configuration, occupy a typical foredeep locus, in front of and parallel to the orogen. A Laramide foredeep is difficult to recognize onshore, and certainly is not, as in Rocky Mountain Trough, superposed on the Columbian foredeep ("side-deep"?). It appears that the Laramide foredeep and the latest Cretaceous and Tertiary subsiding continental margin are one and the same.

In front (east) of the foredeeps lay the stable foreland platform, subsiding slowly. Subsidence there was largely due to sedimentary loading, facilitated by high stands of sea level during the Cretaceous Period. The platform was the site of interplay between orogen-derived and craton-derived detritus. Throughout the period of convergence, there was a progressive northeastward advance of the limit of orogen-derived sediments at the expense of craton-derived sediments. Because of slow subsidence, the Jurassic and younger sediments of the foreland platform are thin and interrupted by numerous unconformities.

Foreland arches and basins

The foreland platform is less differentiated by arches and basins than were the platforms of the passive-margin phase. Most of the features that displayed positive activity at some time between the Late Jurassic and Recent had already been established during the passive phase, e.g., Sweetgrass Arch, Keele Arch, Aklavik Arch Complex, Coppermine Arch (Fig. 2D.4). Such a history would appear to rule out any interpretation of cause-and-effect between foreland (cratonic) structures and convergence or divergence at the continental margins, but the contrary view is held by at least one of the authors of this volume.

Figure 2D.4. History of major tectonic elements: (+) - high or rising element; (O) - neutral element (platformal, or shows "regional" amount of sedimentation); (-) - differentially subsiding element, or bathymetric low; (?) - no record.

TRANSCURRENT PHASE

By Late Eocene time at latest, and perhaps as early as mid-Cretaceous, the western continental margin had become dextrally transcurrent, a condition that, except for short segments that are obliquely convergent, persists to the present. For the eastern Cordillera, this has been a time of tectonic quiet, deep erosion, and isostatic uplift due to unloading. Debate continues as to the thickness of now-eroded Paleogene strata in the Foothills and Plains; estimates of up to an average 6 to 7 km have been published (see Chapter 5). No new arches or basins formed during this tectonic phase; Sweetgrass Arch, Eskimo Lakes Arch, and possibly the area of Peace River Arch underwent slight uplift.

The Beaufort Sea continental margin has remained passive (and probably not actively divergent) through the Tertiary Period to the present. Onshore and west of Mackenzie Delta and into Alaska, on the other hand, evidence of folding as young as Miocene and even Pleistocene is recorded. This may be a response to the northward drift of Pacific ocean-floor under the transcurrent regime.

REFERENCES

Bond, G.C. and Kominz, M.A.
1984: Construction of tectonic subsidence curves for the early Paleozoic miogeocline, southern Canadian Rocky Mountains - implications for subsidence mechanisms, age of breakup, and crustal thinning; Geological Society of America Bulletin, v. 95, p. 155-173.

Dixon, J.
1968: Cretaceous to Pleistocene stratigraphy and paleogeography, northern Yukon and northwestern District of Mackenzie; Bulletin of Canadian Petroleum Geology, v. 34, p. 49-70.

Gabrielse, H. and Yorath, C.J. (ed.)
1991: Geology of the Cordilleran Orogen in Canada; Geological Survey of Canada, Geology of Canada, no. 4 (also Geological Society of America, The Geology of North America, G-2.)

Grantz, A., Johnson, L., and Sweeney, J.F. (ed.)
1990: The Arctic Ocean Region; Geological Society of America, The Geology of North America, v. L, 644 p.

Klepacki, D.W.
1985: Stratigraphy and structural geology of the Goat Range area, southeastern British Columbia; Ph.D. thesis, Massachusetts Institute of Technology, Cambridge, Massachusetts, 268 p.

Klepacki, D.W. and Wheeler, J.O.
1985: Stratigraphic and structural relations of the Milford, Kaslo and Slocan Groups, Goat Range, Lardeau and Nelson map-areas, British Columbia; in Current Research, Part A; Geological Survey of Canada, Paper 85-1A, p. 277-286.

Okulitch, A.V.
1985: Paleozoic plutonism in southeastern British Columbia; Canadian Journal of Earth Sciences, v. 22, p. 1409-1424.

Okulitch, A.V., Wanless, R.K., and Loveridge, W.D.
1975: Devonian plutonism in south-central British Columbia; Canadian Journal of Earth Sciences, v. 12, p. 1760-1769.

Silberling, N.J. and Jones, D.L.
1984: Lithotectonic terrane maps of the North American Cordillera; U.S. Geological Survey, Open File Report 84-523.

Trettin, H.P. (ed.)
1991: Innuitian Orogen and Arctic Platform of Canada and Greenland; Geological Survey of Canada, Geology of Canada, no. 3 (also Geological Society of America, The Geology of North America, v. E).

Yorath, C.J. and Cook, D.G.
1981: Cretaceous and Tertiary stratigraphy and paleogeography, northern Interior Plains, District of Mackenzie; Geological Survey of Canada, Memoir 398, 76 p.

Author's Address

J.D. Aitken
2676 Jemima Road
Denman Island, British Columbia
V0R 1T0

Printed in Canada

Chapter 3

GEOPHYSICAL AND PETROLOGICAL CHARACTERISTICS OF THE BASEMENT ROCKS OF THE WESTERN CANADA BASIN

Chapter 3

GEOPHYSICAL AND PETROLOGICAL CHARACTERISTICS OF THE BASEMENT ROCKS OF THE WESTERN CANADA BASIN

R.A. Burwash, A.G. Green, A.M. Jessop, and E.R. Kanasewich

INTRODUCTION

R.A. Burwash

Precambrian basement underlying the Interior Plains is a lateral extension of the exposed Canadian Shield. From the trend and width of most of the major tectonic units of the western part of the shield it can be inferred that they extend some distance beneath the sedimentary cover. Recognition of these units on the exposed part of the shield has been based primarily on geological mapping combined with studies of structural style, metamorphic fabric and facies, geochronology, and geophysical characteristics. In those areas of the Shield with an extensive cover of Pleistocene deposits, geophysical surveys have served to trace the boundaries of distinctive rock units. In sedimentary platforms with moderate Phanerozoic cover, geophysical maps are the only practical means of tracing the boundaries of major units. Core samples from wells drilled to basement provide information on rock assemblages, metamorphic history, geochronology, and petrophysical characteristics, but yield little of the detailed structural information that can be obtained from outcrop.

The objectives of this chapter are to present the currently available geophysical and petrological data that characterize the major tectonic units of the basement beneath the Western Canada Basin. An attempt is made to draw a map of the basement within the framework of the structural geology of North America.

Historical background

The western limit of the crystalline rocks of the Canadian Shield was first shown with reasonable accuracy on a map drawn by Sir John Richardson in 1851 (Kupsch, 1979). In his travels from Hudson Bay to the Interior Plains of western Canada, Richardson recognized the boundary between flat-lying Paleozoic rocks and underlying granites at several localities along the canoe routes developed by the fur trade.

During the mapping of the Ottawa and adjacent St. Lawrence valleys, W.E. Logan (1854) recognized an ancient series of metamorphic rocks which lay beneath the younger stratified sequences. He gave the name "Laurentian" to the granite gneisses of the complex. In the same year the word "basement" was first used in its present geological sense by Hugh Miller when referring to the Lewisian gneisses of northwestern Scotland (Murchison, 1859). The observations of Richardson and Logan, coupled with the concept of basement, formed the basis for the assumption that the Interior Plains were floored with an ancient igneous-metamorphic complex.

Within the Interior Plains of western Canada only two small areas of Precambrian granitic rocks are exposed; these are in the rims of meteorite impact structures located west of Lake Winnipeg. These limited Precambrian outcrops are in marked contrast with the number exposed in the United States, where Cenozoic tectonic activity has caused domal uplifts and basement exposures in the Black Hills and Little Belt Mountains. In the Rocky Mountains of the United States, block-faulted uplifts of crystalline basement are common. To date, only a few small fault slices of crystalline basement have been recognized in the Columbian Orogen (Evenchick et al., 1984). In the absence of Precambrian outcrops over an area of 1.8 x 10^6 km^2, reliance has been placed on data from geophysical surveys and drillholes.

Prior to 1940, basement of the Western Canada Basin had been reached only by four shallow drillholes near Fort McMurray in northeastern Alberta and one near Winnipeg (Fig. 3.1). Regional gravity surveys, started in 1945, led to the publication a decade later of the first Gravity Map of Canada (Dominion Observatory, 1957). Aeromagnetic surveys of the Western Canada Basin by the Geological Survey of Canada, started in the Leduc area in 1951, extended in several years to northeastern Alberta. The compilation of the Magnetic Anomaly Map of Canada (Geological Survey of Canada, 1967), at a scale of 1:5 000 000, made apparent for the first time the continuity of the magnetic anomaly fabric over large areas of exposed and buried Canadian Shield.

Burwash, R.A., Green, A.G., Jessop, A.M., and Kanasewich, E.R.
1993: Geophysical and petrological characteristics of the basement rocks of the Western Canada Basin; Chapter 3 in Sedimentary Cover of the Craton in Canada, D.F. Stott and J.D. Aitken (ed.); Geological Survey of Canada, Geology of Canada, no. 5, p. 55-77 (also Geological Society of America, The Geology of North America, v. D-1).

Studies of heat flow using data from deep drill holes in the Western Canada Basin were first reported by Garland and Lennox (1962). Recognition of the North American Central Plains (NACP) conductivity anomaly in southern Saskatchewan by Alabi et al. (1975) set the stage for much discussion of possible plate boundaries between cratonic blocks.

Most of the exploratory tests drilled to basement were completed in the two decades following the discovery of a large oil pool in a Devonian reef at Leduc, Alberta, in 1947. Core samples from these wells formed the basis for thesis projects by Burwash (1951, 1955) and Peterman (1962), summarized in the Geological History of Western Canada (Burwash et al., 1964). Concurrently, the Basement Rock Project of the American Association of Petroleum Geologists compiled all relevant well data and published the Basement Map of North America (Flawn, 1967).

Limitations

In the preparation of this chapter, several limitations became of concern. The southern half of the sedimentary basin (up to 60°N) has been more extensively studied geophysically and petrologically than the northern half. Even in the southern half the quality of data is far from uniform. Aeromagnetic coverage of northeastern British Columbia and Alberta west of 114°W is limited to a survey flown at an elevation of 3.5 km and line spacing of 37 km (Coles et al., 1976). The resolution of this map is much less than that of the latest Magnetic Anomaly Map of Canada (Dods et al., 1984).

Evolving petroleum exploration philosophies and thin sedimentary cover have left a number of areas with few drillholes to basement. In these areas, delineation of domain boundaries relies heavily on interpretation of geophysical surveys, without the support of direct evidence from core samples.

Acknowledgments

During the past thirty years many of the oil companies in western Canada have generously supplied core samples and information required for basement studies. Grants to R.A. Burwash and E.R. Kanasewich from the Natural Sciences and Engineering Research Council of Canada are gratefully acknowledged. The manuscript was prepared and edited with the assistance of R.W. Burwash.

LITHOSTRUCTURAL DOMAINS OF THE EXPOSED SHIELD

R.A. Burwash

The lithostructural domains (Fig. 3.2) are a synthesis of those suggested by Ayres (1978), Ermanovics and Froese (1978), Lewry et al. (1978), Sims (1980), Hoffman et al. (1982a), Fumerton et al. (1984), Bowring et al. (1984), Green et al. (1985a, b), and Hoffman (1988). Details of the geology of these domains appear elsewhere (see the companion volume, "Precambrian geology of the Craton in Canada and Greenland", Hoffman et al., in prep.). In the following synopsis, only the characteristics of the major divisions and the inter-division boundaries relevant to the Western Canada Basin are discussed.

Both Superior and Slave structural provinces (Fig. 3.2) contain numerous greenstone belts intruded by Archean mesozonal plutons and separated by linear gneissic terranes. The greenschist to lower amphibolite grade of the supracrustal rocks (Fraser et al., 1978) and the persistence of Archean K-Ar radiometric ages from biotite (Stockwell, 1982) indicate a crustal stability that satisfies a strict definition of the word "craton". The northwestern margin of Superior Province is deeply eroded to expose granulites in the Pikwitonei Subprovince (23, Fig. 3.2). The margins of Slave Province are in part onlapped by Early Proterozoic clastic wedges and in part are metamorphic fronts. Granulites are not common in Slave Province.

During the Hudsonian Orogeny (1.9-1.8 Ga) convergence of Slave and Superior cratons resulted in the formation of numerous northeast-trending lithostructural

Figure 3.1. Derrick for Alberta Government Salt Well No. 2; drilled to basement in northeastern Alberta, in 1923 (University of Alberta Archives).

domains. The Wathaman-Chipewyan batholith approximates the thermal axis of the orogen and separates regions with distinctively different lithologies (Fumerton et al., 1984). Needle Falls shear zone (14, Fig. 3.2), along the northwest margin of the batholith, is a domain boundary of regional significance (Lewry et al., 1981). In Cree Lake zone, Archean sialic basement has been remobilized with its miogeoclinal cover. Gneiss domes of relict Archean granulite occur in the Wollaston Fold Belt. Hudsonian hornblende granulite-facies metamorphism is recognized in the cover rocks infolded into the plastic

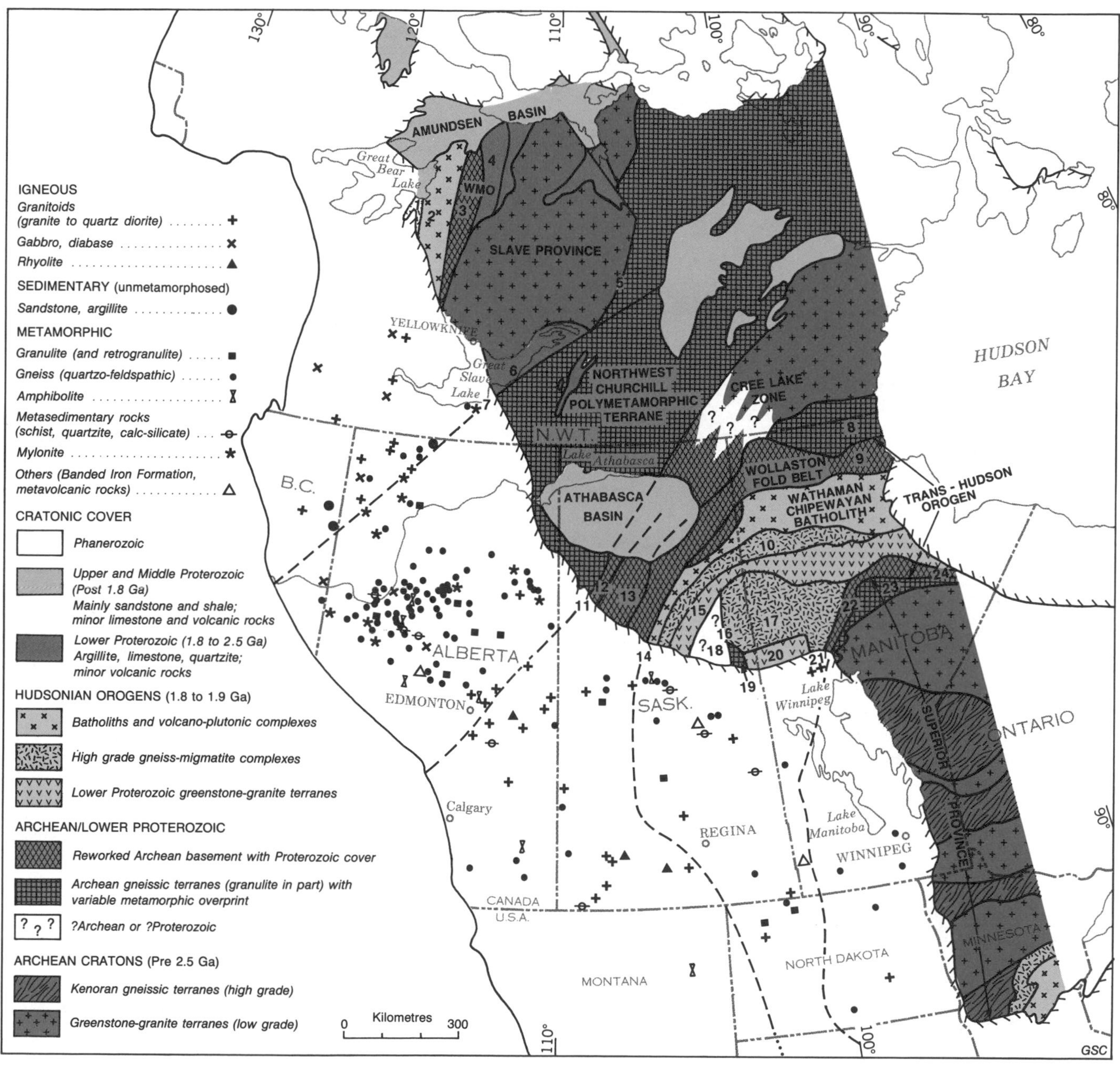

Figure 3.2. Lithostructural domains of the exposed Canadian Shield and lithology of basement cores. WMO = Wopmay Orogen; 1 = WMO, Hottah Terrane; 2 = WMO, Great Bear magmatic arc; 3 = WMO, Hepburn batholithic belt; 4 = WMO, foreland belt; 5 = Thelon Front; 6 = East Arm Fold Belt; 7 = Great Slave Lake Shear Zone; 8 = Nejanilini fold belt; 9 = Seal River fold belt; 10 = Reindeer-Southern Indian belt; 11 = Virgin River shear zone; 12 = Virgin River fold belt; 13 = Mudjatik domain; 14 = Needle Falls shear zone; 15 = LaRonge and Lynn Lake belts; 16 = Tabbernor shear zone; 17 = Kisseynew domain; 18 = Glennie Lake domain; 19 = Hansen Lake block; 20 = Flin Flon and Snow Lake belts; 21 = Setting Lake fault; 22 = Thompson Belt; 23 = Pikwitonei Subprovince; 24 = Fox River belt.

infrastructure of the Mudjatik domain (13, Fig. 3.2). An elongate batholith of Hudsonian granite parallels Virgin River shear zone (11, Fig. 3.2) (Wallis, 1970).

Between Wathaman-Chipewyan batholith and Superior Province, an ensimatic(?) eugeosyncline existed in Early Proterozoic time (Stauffer, 1984). Hudsonian compression produced the arcuate lithotectonic domains shown in Figure 3.2. Mafic to felsic volcanic sequences, greywackes, and pelites are metamorphosed to greenschist-to-granulite facies. A boundary zone of cataclasis and retrograde metamorphism separates the Trans-Hudson Orogen from Superior Province.

The polymetamorphic terrane of northwestern Saskatchewan contains a large number of blocks of granulite separated by linear zones of cataclasis and recrystallization (Beck, 1969). This tectonic style is characteristic of northwest Churchill Province to the bounding Great Slave Lake Shear Zone (7, Fig. 3.2). Granulite-facies metamorphism, circa 2.4 Ga, was documented in northwestern Saskatchewan by Koster and Baadsgaard (1970). The recrystallization is Hudsonian.

In East Arm Fold Belt (6, Fig. 3.2), carbonate and basinal clastic sequences of the Great Slave Supergroup were affected by northeast-directed thrusting but little metamorphism during Hudsonian Orogeny. Calc-alkaline laccoliths (1.86 Ga) postdate the thrusting (Bowring et al., 1984). Middle Proterozoic diabase sills are important lithological units in the belt.

Bounding Slave Craton on the west is the Early Proterozoic Wopmay Orogen (see Fig. 3.11, WMO). Its foreland belt (4, Fig. 3.2) to the east of Great Bear Lake contains sedimentary sequences correlative with those in the Great Slave Supergroup (Hoffman, 1981). West of the Asiak thrust belt, the Hepburn batholithic belt (3, Fig. 3.2) contains domes of reactivated Archean gneiss (Neilsen, 1978). The Great Bear magmatic arc (2, Fig. 3.2) is interpreted by Hoffman and Bowring (1984) as a short-lived volcano – plutonic depression (~1.9 Ga) on continental crust. Volcanic rocks, ranging from basalt to rhyolite, and nonmarine sedimentary rocks are intruded by epizonal and mesozonal calc-alkaline plutons. The poly-deformed Hottah Terrane (1, Fig. 3.2) underlies the western edge of the Great Bear magmatic arc.

Amundsen Basin of Middle and Upper Proterozoic clastic sedimentary rocks onlaps the northwestern edge of the crystalline shield. Several periods of basaltic magmatism are recorded in the basin, of which the Coppermine lavas (1.2 Ga) are the most voluminous. The contemporaneous Mackenzie diabase dykes are of widespread occurrence, both in the western part of the Canadian Shield and in Middle Proterozoic sequences of the eastern Cordillera.

MAGNETIC AND GRAVITY DATA

A.G. Green

Maps of the earth's magnetic and gravity fields are powerful tools for extrapolating our knowledge of Precambrian basement geology beneath younger sedimentary cover. High-resolution magnetic data are particularly sensitive to changes in the geology of the uppermost crust and have long been used to guide mineral exploration and geological mapping programs. On the exposed Canadian Shield, magnetic anomaly patterns commonly correlate precisely with local and regional geology (Kornik and MacLaren, 1966; Hall, 1968; Kornik, 1969, 1971; Bell, 1971a; Wilson, 1971). Many geological structures also have distinctive gravity expressions, but generally there is not the same close correspondence that is observed between surface geology and magnetic data. On the other hand, regional gravity data are useful for mapping deep structural and lithological variations.

Magnetic anomaly map

The magnetic anomaly map (Fig. 3.3) represents a composite of diverse data from a number of sources. High-quality aeromagnetic maps from the Canadian Federal-Provincial series cover the exposed Canadian Shield (Dods et al., 1984), and various maps by oil companies and universities cover the adjacent Interior Plains of Manitoba, Saskatchewan and southern Alberta (Sawatzky and Standing, 1971; Green et al., 1979, 1985a, b; PanCanadian Petroleum Ltd., pers. comm., 1984). Data from a relatively low-resolution aeromagnetic survey (37 km line spacing; Coles et al., 1976) are incorporated in two small areas of northern Saskatchewan and constitute the bulk of information available for the northwestern part of the map. South of the international boundary the map is based on the composite magnetic anomaly map of the United States (Zietz et al., 1982). Technique for matching the southern Canadian data is that of Green et al. (1979a) and details of the United States compilation are given by Zietz et al. (1982).

Most of the lithotectonic units in the Canadian Shield can be delineated on the basis of their magnetic signatures. In particular, characteristic magnetic trends or patterns are associated with Superior Province and its marginal Thompson and Fox River belts (22, 24, Fig. 3.2), the various component units of the Trans-Hudson Orogen, Northwest Churchill polymetamorphic terrane and its marginal Cree Lake zone, Slave craton and its marginal Wopmay Orogen (see Fig. 3.11). Important magnetic lineaments outline many of the bounding and internal fault systems of the lithotectonic units, including Setting Lake fault (21, Fig. 3.2) along the western margin of the Thompson Belt, Tabbernor Shear Zone (16, Fig. 3.2) within the Trans-Hudson Orogen, Needle Falls and Virgin River shear zones (14, 11, Fig. 3.2) along the eastern and western margins respectively of the Cree Lake zone and Great Slave Shear Zone along the northern margin of the Northwest Churchill polymetamorphic terrane (7, Fig. 3.2). These magnetic signatures can mostly be followed for considerable distances across the Interior Plains (Fig. 3.3).

Westerly trending magnetic anomalies reflect the Archean fabric of Superior Province and its extensions beneath the Interior Plains (Wilson and Brisbin, 1962; Kornik and MacLaren, 1966; Stockwell et al., 1968; Kornik, 1969, 1971; Bell, 1971a; Wilson, 1971; Green et al., 1979, 1985a, b). Granite/greenstone belts within Superior Province have low background magnetic fields with moderate high-amplitude elliptical anomalies across igneous rocks and iron formations. High-grade gneissic belts have broad magnetic highs. Toward the northwestern margin of Superior Province lies the Pikwitonei granulite terrane (23, Fig. 3.2) with its "bird's-eye maple pattern" of high-amplitude, ovoid-shaped magnetic anomalies (observed on the larger scale Federal-Provincial magnetic maps; Kornik and MacLaren, 1966; Kornik, 1969, 1971; Bell, 1971a; Green et al., 1979).

The magnetic fabric of Superior Province is abruptly truncated by the southwesterly to southerly trending Thompson magnetic quiet zone (see Fig. 3.11) and its extension to the south. This linear zone of low magnetization, probably the result of Proterozoic metamorphic overprinting of the Superior craton margin and its overlying supracrustal rocks, extends 1500 km from northern Manitoba to a position beneath the Interior Plains of South Dakota. Over most of its length the western edge of the magnetic low is interpreted to be a continuation of Setting Lake fault, which juxtaposes the Thompson Belt and Kisseynew domains (22, 17, Fig. 3.2).

Magnetic anomalies of the Trans-Hudson Orogen (see Fig. 3.2, 3.11) extend westerly from Hudson Bay into northern Saskatchewan and then swing to the southwest

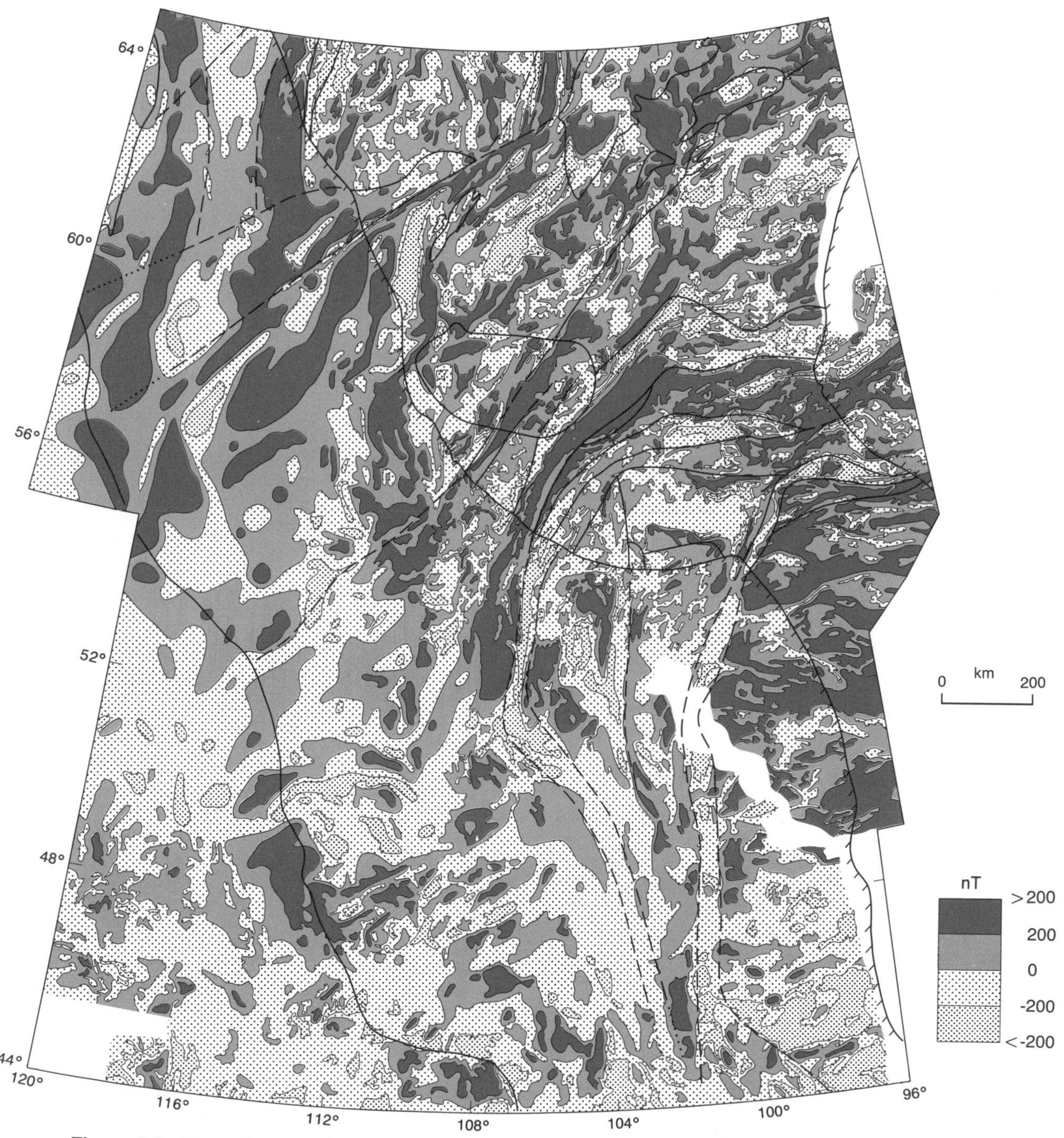

Figure 3.3. Magnetic anomaly map of Western Canada Basin and adjacent areas. Lithotectonic domains of the exposed shield from Figure 3.2 and inferred basement divisions from Figure 3.11 are superimposed.

and eventually to the south across the Interior Plains (Burwash and Culbert, 1976; Green et al., 1979, 1985a, b; Dutch, 1983). The granite/greenstone rocks of the Flin Flon and Snow Lake and LaRonge-Lynn Lake belts (20, 15, Fig. 3.2) have magnetic expressions similar to the granite/greenstone rocks of Superior Province, but the predominantly gneissic Kisseynew Domain and Reindeer-Southern Indian Belt (17, 10, Fig. 3.2) have low background magnetic fields and subdued magnetic relief. On the larger scale Federal-Provincial magnetic maps, the Tabbernor Shear Zone (16, Fig. 3.2) along the western margins of the Flin Flon and Snow Lake belts and Kisseynew Domain is delineated by a major change in style of magnetic anomalies.

At the northern and northwestern margins of the Trans-Hudson Orogen, the enormous Wathaman-Chipewyan Batholith and associated granitic bodies are represented by broad regions of magnetic high. The characteristic magnetic expressions of the Reindeer-Southern Indian Belt and the Tabbernor Shear Zone allow the Trans-Hudson Orogen to be mapped as far south as latitude 45°N (Fig. 3.3).

Within the Cree Lake zone a pattern of linear and curvilinear magnetic highs and lows extends southwesterly from the northeastern corner of the map to southern Saskatchewan and Alberta (Wallis, 1970; Burwash and Culbert, 1976; Coles et al., 1976; Green et al., 1985a, b). Along its eastern boundary a linear belt of low magnetization coincides with the Needle Falls shear zone (14, Fig. 3.2). A lineament separates terranes of contrasting magnetic fabric on opposite sides of the Virgin River shear zone (11, Fig. 3.2). Relatively intense magnetic highs occur across reworked Proterozoic metasedimentary and Archean basement rocks of the Wollaston and Virgin River (12, Fig. 3.2) fold belts and magnetic lows occur mostly across the intervening Mudjatik (13, Fig. 3.2) and the more easterly trending Seal River (9, Fig. 3.2) and Nejanilini (8, Fig. 3.2) fold belts.

South of the exposed Canadian Shield, magnetic trends of the Cree Lake zone are truncated near the eastern edge of the Cordillera and by the westerly trending magnetic low overlying the postulated Precambrian rift structure of Kanasewich et al. (1968). The strikes of magnetic anomalies on the two sides of the "rift" differ noticeably (Green et al., 1985a, b); to the north the anomalies trend south-southeasterly and to the south they trend east-southeasterly. The region south of the "rift" corresponds to the Archean Wyoming Province as outlined by Peterman (1981) on the basis of radiometric dating of basement outcrops and core samples. Beneath the Interior Plains, the eastern boundaries of the Cree Lake zone and the Wyoming Craton are not well resolved by existing data.

A general southwesterly trending magnetic anomaly pattern characterizes the reworked Proterozoic and Archean basement rocks of the Northwest Churchill polymetamorphic terrane. This pattern is broken up by some relatively intense southerly striking magnetic highs and lows about longitude 112°W, near the margin with the Interior Plains. The significance of these latter features and whether or not there is a related change in the nature of the Northwest Churchill polymetamorphic terrane beneath the Interior Plains is unclear at the present time. Along its northern boundary, Great Slave Lake Shear Zone (7, Fig. 3.2) can be traced from the Thelon Front (5, Fig. 3.2) southwestward almost to the edge of the Cordillera (Burwash and Culbert, 1976; Coles et al., 1976; Thomas et al., 1976). Magnetic anomalies on either side of the fault are sharply terminated by fault-related magnetic lineaments. Great Slave Shear Zone forms the southeastern margin of the East Arm Fold Belt (7, 6, Fig. 3.2).

In the northwestern corner of the map (Fig. 3.3), the southerly trending magnetic anomalies of the granite/greenstone and granite/gneiss terranes of Slave Province are bounded by East Arm Fold Belt and Great Slave Lake Shear Zone in the south and by southerly trending anomalies associated with Wopmay Orogen in the west. The adjacent northern Interior Plains are dominated by a pair of huge, southerly striking linear magnetic highs (Hoffman et al., 1982b). The eastern anomaly corresponds to the subsurface extension of the Great Bear Magmatic Arc (2, Fig. 3.2) of Wopmay Orogen and can be followed southward from exposures on the Canadian Shield to its intersection by the Great Slave Shear Zone (Coles et al., 1976). To the west, across the low magnetic field overlying Hottah terrane (1, Fig. 3.2), an arcuate magnetic high parallels the trend of the Mackenzie foldbelt in the north and the edge of the Cordillera in the south. Possible sources of this latter magnetic anomaly are reviewed in the section on the northern Interior Plains, later in this chapter.

Bouguer gravity anomaly map

The Bouguer gravity map (Fig. 3.4) is based on the Gravity Map of Canada (Earth Physics Branch, 1980) and the Gravity Map of the United States (Lyons and O'Hara, 1982). Many of the lithotectonic units delineated on the magnetic map also affect the gravity field. Notable Precambrian features include: westerly trending anomalies within Superior Province and their truncation near the Nelson River gravity high (4, Fig. 3.4), the westerly to southerly trending gravity gradient along the northern and western margins of the Trans-Hudson Orogen, the general southwesterly fabric of the Northwest Churchill polymetamorphic terrane and its marginal Cree Lake zone, the anomalies that almost surround Slave Province, and the southerly trending Bulmer Lake gravity high (1, Fig. 3.4). The extensive southeasterly trending anomaly pattern in the western and southwestern parts of the map is associated with the edge of the Cordillera.

One of the most studied gravity anomalies of the Canadian Shield, the Nelson River gravity high, occurs near the northwestern boundary of Superior Province with the Trans-Hudson Orogen (Innes, 1960; Wilson and Brisbin, 1962; Green et al., 1979, 1980, 1985a, b; Fountain and Salisbury, 1981). It is parallel or sub-parallel to the edge of Superior Province from Hudson Bay in the northeast to a location beneath the Interior Plains in southern Manitoba. The magnetically defined province margin corresponds to the truncation of the westerly trending gravity fabric, but the Nelson River anomaly itself seems to cut obliquely across the edge of the province. Green et al. (1985a) have suggested that the positive anomaly is a composite of three or more distinct structures generated along the same margin at different times by different boundary processes. Those anomalies overlying the Fox River Belt are caused by mafic/ultramafic dykes of Proterozoic age, those across the Pikwitonei sub-province are caused by granulites, and the anomalies within the eastern Kisseynew Domain and its probable extension to the south are caused by Proterozoic gneissic rocks. Together, the latter gravity high and the gravity low across thc adjacent Thompson Belt may constitute a paired gravity anomaly of the type described by Gibb and Thomas (1976).

A strong westerly to southerly trending gravity gradient correlates well with the western border of the Trans-Hudson Orogen. The -60mGal contour commonly outlines the border with the Cree Lake zone in the north, and the -80 and -60mGal contours run astride the most probable location of the border with the Wyoming Province in the south.

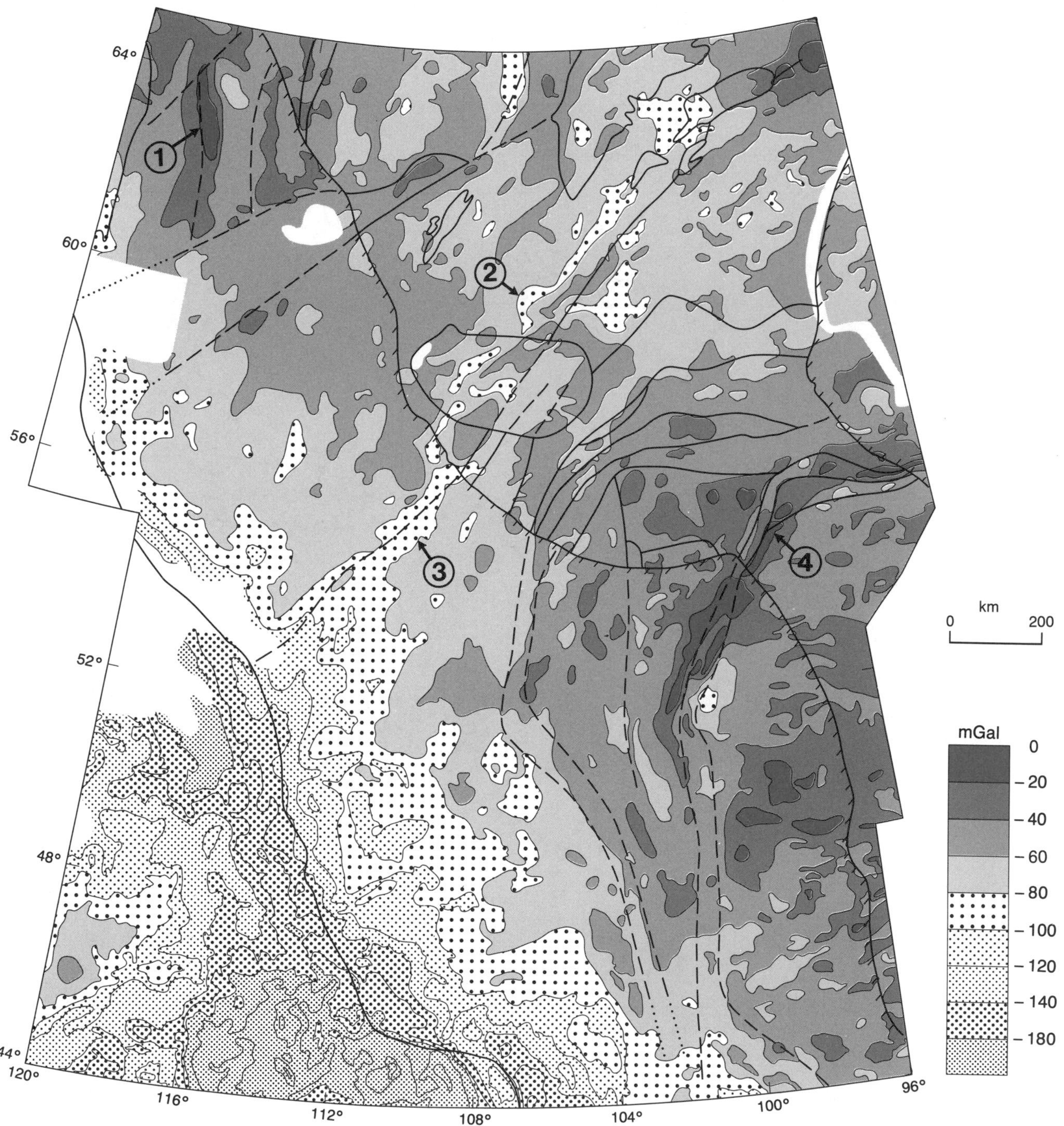

Figure 3.4. Bouguer gravity anomaly map of Western Canada Basin and adjacent areas. Lithotectonic domains of the exposed shield and inferred basement divisions from Figure 3.11 are superimposed. 1 - Bulmer Lake gravity high, 2 - Fond du Lac gravity low, 3 - Kasba Lake-Edmonton gravity low, 4 - Nelson River gravity high

Gravity anomalies within the Northwest Churchill polymetamorphic terrane and Cree Lake zone strike southwesterly. A major belt of linear gravity anomalies parallels the trend of the Virgin River shear zone and its proposed southwesterly extension towards the Cordillera (Walcott, 1968; Agarwal and Kanasewich, 1968; Wallis, 1970; Walcott and Boyd, 1971; Gibb and Halliday, 1974; Burwash and Culbert, 1976). Included in this belt are several gravity highs and the prominent Fond du Lac and Kasba Lake-Edmonton gravity lows (2, 3, Fig. 3.4). The gravity highs are underlain by high-grade metamorphic rocks and mafic/ultramafic intrusions, but the nature of the rocks associated with the gravity lows is largely undetermined because of inadequate outcrop. Walcott (1968) and Gibb and Halliday (1974) have related parts of the Fond du Lac anomaly to exposures of granitic rocks, whereas Wallis (1970) has suggested that the regional gravity lows are related to linear troughs of Proterozoic metasediments.

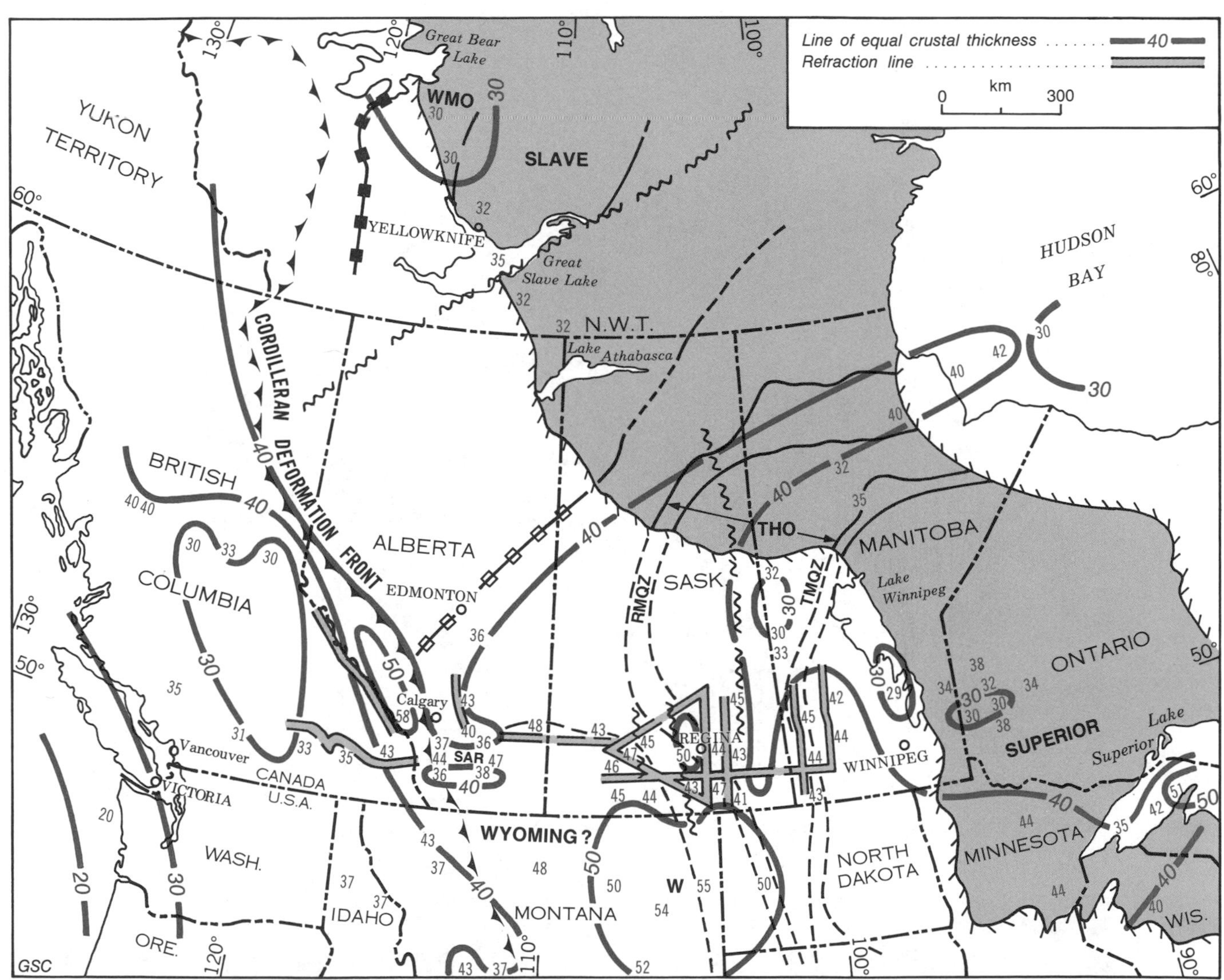

Figure 3.5. Measurements of crustal thickness in kilometres from seismic refraction and reflection studies. Major lithotectonic elements and structures (Fig. 3.11) are indicated. W = Williston basin; WMO = Wopmay Orogen; SAR = Southern Alberta Rift; THO = Trans-Hudson Orogen; RMQZ = Reindeer magnetic quiet zone; TMQZ = Thompson magnetic quiet zone. References for the data are as follows: Asada et al., 1961; Barr, 1971; Bennett et al., 1975; Berry et al., 1971; Berry and Forsyth, 1975; Chandra et al., 1972; Clowes et al., 1984; Congram, 1984; Cumming et al., 1979; DeLandro, 1981; DeLandro et al., 1982; Forsyth et al., 1974; Ganley and Cumming, 1974; Green et al., 1979; Hales and Nation, 1973; Hall and Hajnal, 1969, 1973; Hill and Pakiser, 1967; Johnson and Couch, 1970; Kanasewich, 1966, 1968; Kanasewich and Cumming, 1965; Kanasewich et al., 1985; Kazmierczak, 1980; Macrides and Kanasewich, 1984; Maureau, 1964; Mereu and Hunter, 1969; Mereu et al., 1977; Meyer et al., 1959; Richards and Walker, 1959; Smith et al., 1966; White and Savage, 1965.

Farther to the northwest, the lithotectonic units around the margins of Slave Province are delineated by a variety of gravity anomalies. A gravity low overlies the Thelon front on its eastern margin and a linear gravity high runs along the axis of East Arm Fold Belt on its southern margin. On its western margin, within Wopmay Orogen, there is a weak paired gravity anomaly adjacent to the craton (Hoffman et al., 1982b) and a stronger positive anomaly along the border between the Great Bear magmatic zone and the Hottah terrane. Finally, the high-amplitude Bulmer Lake gravity high (1, Fig. 3.4) trends southerly near the western edge of Hottah terrane (Hornal et al., 1970). This anomaly either defines the western limit of Wopmay Orogen (Meijer Drees, 1975), or represents one-half of a paired gravity anomaly within a more extensive Wopmay Orogen (Hoffman et al., 1982b).

DEEP SEISMIC PROFILES

E.R. Kanasewich

Studies made on the thickness and structure of the continental crust have been concentrated on the easily accessible southern parts of Western Canada. The quality of the data is rather uneven and it is being re-interpreted currently at several seismic laboratories. The results represented here will, no doubt, be modified with respect to local details, but the broad pattern seems to be well established.

The location of the seismic refraction lines and the crustal thicknesses are shown in Figure 3.5. The map is dominated by large regions where the continental crust is less than 40 km thick. Within these areas, significant parts thin to 30 km. These occur within the oldest part of the Canadian Shield - Superior and Slave provinces - and in part of the Intermontane Belt of southern British Columbia. In other areas, such as Williston Basin and the Rocky Mountains along the continental divide, crustal thickness approaches 55 km. This pattern of crustal thickness variation and the associated data for the thickness of the sedimentary section appears to provide evidence for vertical movements over periods of time ranging from several million years to over 1 billion years, as various basins and arches were formed. A review of the gravity anomalies, in conjunction with the seismic evidence (Kanasewich, 1966; Burwash and Krupicka, 1970; Sprenke and Kanasewich, 1983), indicates that the long-term features have their origin associated with upper mantle density variations in addition to the horizontal forces resulting from plate interactions. On a shorter time scale, long-wavelength isostatic anomalies in the northeast part of the map area indicate overcompensation, due probably to incomplete recovery of the lithosphere from the Pleistocene ice sheet. Determination of vertical movement over Williston Basin (U.S. Geodynamics Committee, 1973; Officer and Drake, 1982) shows that the area is undergoing uplift at rates of 1 to 5 mm per year.

The intracratonic Williston Basin has a well mapped crustal thickness approaching 55 km over the basin centre in eastern Montana and western North Dakota. Detailed refraction studies in southern Saskatchewan disclose smaller zones of crustal thickening, up to 50 km. The subsidence history of Williston Basin is well documented by several thousand well logs. Subsidence has occurred, with interruptions, throughout much of the Phanerozoic Era. Rocks of the Precambrian crystalline basement were eroded extensively prior to the Cambrian, leaving a smooth surface from which 10 km or more of rock had been removed. It is probably significant that, despite this peneplanation, the crust was very thick prior to the Phanerozoic deposition, which added, at most, 5 km of sediment. The cause of epeirogenesis over features such as the Williston, Denver, Michigan, and Illinois basins appears to be related to critical density anomalies, possibly at the lithosphere-asthenosphere boundary in the upper mantle (Kanasewich, 1966), but the dynamic process remains a major enigma at the present time.

Geomagnetic and magnetotelluric studies have indicated the presence of a number of electrical conductivity anomalies in Western Canada. Principal among these is the North American Central Plains (NACP) anomaly (Camfield and Gough, 1977; Alabi et al., 1975). The anomaly is not well defined (Fig. 3.6) because the station spacing is large, seldom under 100 km; towards 54°N latitude it is greater than 250 km. Measurements solely of the magnetic field preclude any depth determination except at a magnetotelluric line at 56.5°N, where Handa and Camfield (1984) placed seven stations. Here one station was over a conductive anomaly, which was

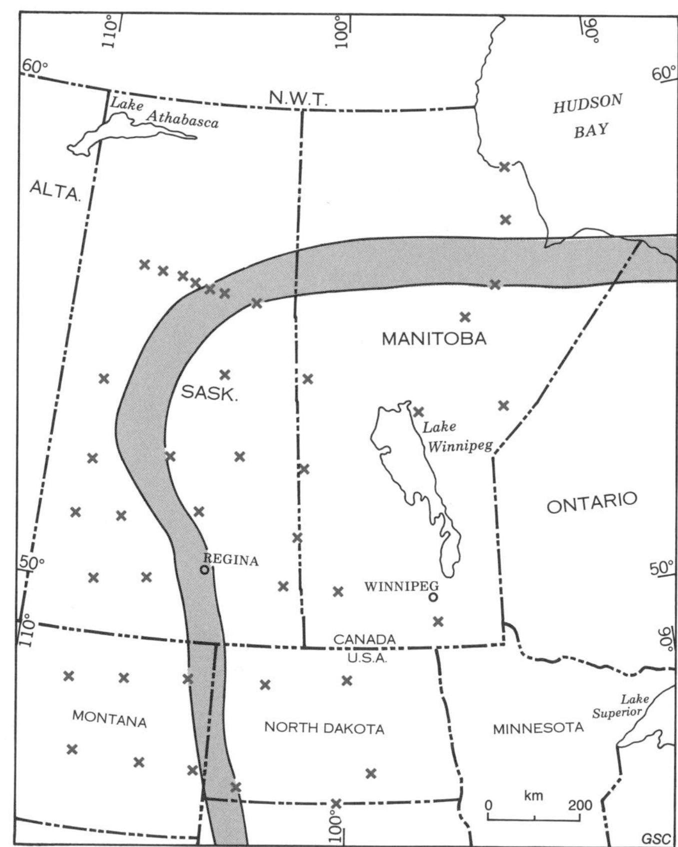

Figure 3.6. Location of the North American Central Plains conductive body (shaded area) and the stations determining its location (from Camfield and Gough, 1977; Handa and Camfield, 1984; Gupta et al., 1985).

modelled to have a resistivity of 10 ohm-metres at crustal depths of 5 to 10 km. A conductive anomaly has been detected between Gillam (56.4°N, 94.7°W) and Back, Manitoba (57.7°N, 94.2°W) near the shore of Hudson Bay (Gupta et al., 1985). However the Gillam-Back conductor is located close to surface exposures of east-trending belts of metasedimentary rocks (Fig. 3.2 and 3.6), whereas the North American Central Plains anomaly is related to more northerly structures. Another conductive zone identified through magnetotelluric studies, distinct from the North American Central Plains anomaly, has been detected by Rankin and Pascal (in press) in southern Saskatchewan at 103°W, east of Weyburn.

Detailed refraction studies near Regina, Saskatchewan support a block-faulted structure in the crystalline basement down to the Mohorovicic discontinuity. Fair evidence exists for a north-striking fault at the level of the Moho at longitude 103° in southern Saskatchewan (Kazmierczak, 1980). Another pair of north-striking faults is indicated on very good seismic broadside refraction data just west of Regina at longitudes 105° and 107°W (Macrides and Kanasewich, 1984; Kanasewich and Chiu, 1985). The fault at 105° coincides with the western edge of the Reindeer magnetic quiet zone (rmqz) and the Trans-Hudson Orogen (see Fig. 3.11) and may correlate with some parts of the North American Central Plains anomaly. Earthquake epicentres along this section (Fig. 3.7) show it to be a tectonically active zone. Another feature of the Regina region that is well established is the presence of one or more low-velocity layers. These occur at depths of 15 to 25 km in southern Saskatchewan. The velocity of the upper mantle at the Moho is between 8.0 and 8.2 km/s in southern Saskatchewan and 7.8 to 8.0 km/s in the Superior Province immediately west of the Nelson Front.

Another area where sufficient detail exists to draw some specific conclusions is in southern Alberta. The buried Cree Lake - Calgary zone is interrupted by an east-west Precambrian rift(?) ("Southern Alberta Aulacogen"), which extends from the Saskatchewan-Alberta border to

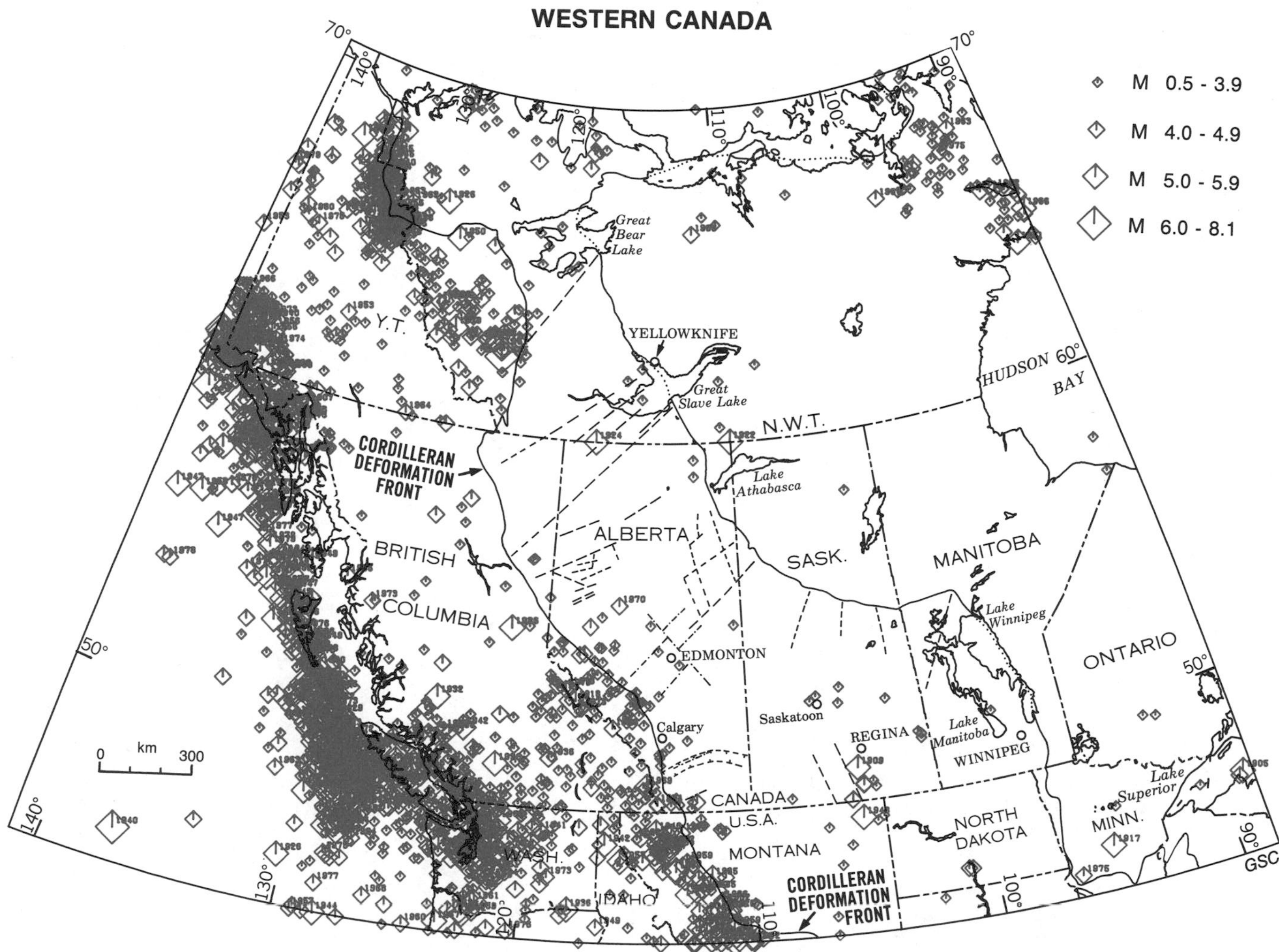

Figure 3.7. Seismic events in relation to known and inferred basement faults. Lineaments from: Aitken and Pugh, 1984; Burwash and Culbert, 1976; Garland and Bower, 1959; Green et al., 1985a; Kanasewich et al., 1968; Meijer Drees, 1975; Robinson et al., 1969; and Williams, 1973.

Kimberley, British Columbia (Fig. 3.5 and 3.7). The buried feature has been defined by two seismic reflection lines and several refraction studies. The first demonstration that deep crustal structures in this area could be mapped by the seismic reflection method used a line at 112.6°W, west of Brooks, Alberta (Kanasewich and Cumming, 1965). The most recent line of interest was obtained by Pan Canadian Petroleum (Kanasewich et al., 1985). Unlike the earlier studies, which used explosive sources, the new line of data was obtained with a Vibroseis source and common depth point profiling. The new data are from a 32 km north-south line 4 km east of Vulcan, Alberta, about 100 km west of the previous line. The results tended to confirm the hypothesis put forward by Kanasewich (1968) that the feature was a Precambrian rift valley under flat-lying Paleozoic and Mesozoic sediments. The Moho changes in depth from 47 to 38 km across a series of east-trending normal faults. Although most of the rifting occurred in Precambrian time, the new line clearly shows that the rift underwent Paleozoic and Mesozoic reactivation. Hoffman (1988) suggested that the "rift" either is, or coincides with, an Early Proterozoic suture.

The crust thickens westward to over 55 km under the Rocky Mountains and the Rocky Mountain Trench (Fig. 3.5). The northern and southern limits of this thick layer are unknown, but it is probably confined to the zone of large negative Bouguer gravity anomalies. Anomalously thin crust occurs in a northwest-trending domain between the Omineca Belt and the Coast Belt of southern British Columbia. This domain corresponds to the southern parts of the allochthonous Stikinia and Cache Creek terranes of Monger and Price (1979). Numerous seismic determinations in the Basin and Range province of the United States also show crustal thicknesses of 30 km and similar low upper mantle compressional velocities of 7.7 to 7.9 km/s. Kanasewich (1966) pointed out that the anomalous seismic thicknesses and velocities in central British Columbia are probably related to an extension of the East Pacific Rise, an oceanic spreading centre that enters the continent in the Gulf of California. Recent deep crustal seismic reflection data from project Lithoprobe suggest that Precambrian basement may extend westward as far as the zone of crustal extension.

HEAT FLOW

A.M. Jessop

As in all sedimentary covers where water may move through permeable formations, the temperature field of the Western Canada Basin is controlled by a combination of thermal conductivity and mass transport. The flow of water is driven mainly by the pronounced topographic head provided by the Rocky Mountains, with more local effects of all scales superimposed. Thus the flow pattern and the resulting temperature field are very complex.

Temperature data are available from the extensive files on oil and gas wells that are kept by the provincial governments. These data are individually of doubtful quality, but have been interpreted jointly by statistical methods to derive isotherms and temperature gradients over a wide area. Thermal conductivity is more difficult to determine, but a knowledge of typical conductivity for each rock type, combined with an analysis of the "net rock" content of each interval, provides a reasonable approximation.

A strong contrast in heat flow between Paleozoic and Mesozoic strata has been observed (Majorowicz et al., 1985; Fig. 3.8). Above the Paleozoic-Mesozoic boundary, heat flow tends to increase from the Rocky Mountains toward the exposed Canadian Shield, while below the boundary the opposite effect is observed. A great deal of local variation is superimposed on these general trends, much of which may be related to surface topography (Hitchon, 1984). In the upper part of the sediments, low temperature gradient tends to coincide with the recharge zones of relatively high land, and high gradient tends to coincide with the valleys. There is a zone of equality of heat flow between Paleozoic and Mesozoic strata; it is sinuous, but follows approximately the 2000 m isopach. This is shown as a line on Figure 3.8, but this is an approximation, and it would be preferable to think of it as a belt winding across the map. The value of heat flow in this zone is regarded as the best available estimate of the true terrestrial heat flow.

Temperature gradient or heat flow in the Precambrian basement beneath the basin is regarded as the only true test of the crustal heat flow, but measurements are lacking. The heat input to the sediments comes from the Precambrian basement, with a distribution that depends on the nature of the cratonic rocks. Some measurements of heat generation of rocks just below the interface have been made (Burwash and Cumming, 1976), but the data are sparse and unevenly distributed. The age of the Precambrian provinces is generally known, and it is possible to estimate the heat flow from averages in the exposed part of the Canadian Shield to the northwest, but the detail cannot be derived in this way. Some heat is absorbed in the chemical transformation of large hydrocarbon molecules to small ones, but this is negligible. It may thus be assumed that all heat entering the sediments from the basement eventually reaches the surface as conducted heat flow or as warm springs on the outcrops of the permeable formations.

At Regina, Saskatchewan, where a detailed and accurate temperature log is available and measurements of conductivity on the well cuttings have been made, there is evidence of upward water migration across the Mesozoic formations, a substantial change in heat flow from Mesozoic to Paleozoic strata, and downward water flow between individual units of the lowermost two formations (Jessop and Vigrass, 1989). A decline of heat flow from 75 mW/m^2 at 300 m to 51 mW/m^2 at 950 m implies an upward water flow of about 8 mm per year in this depth range. The horizontal component is not determined by this analysis, and it is probably varied and dependent on the nature of the strata, which are mainly clastic. In the Paleozoic section the heat flow is uniform at 51 mW/m^2, but there is a major aquifer system in the basal clastic unit, a zone of over 100 m of sandstone, and it cannot be assumed that heat flow in the Paleozoic strata represents heat flow in the Precambrian basement.

GEOCHRONOLOGY

R.A. Burwash

A reconnaissance program of K-Ar isotopic age determinations on whole-rock samples of basement cores from the Western Canada Basin was started in 1954. A decrease in apparent age toward the west (Shillibeer and Burwash, 1956; Burwash, 1957) can be attributed to argon loss during Mesozoic crustal loading and Cordilleran tectonism. Subsequent K-Ar dating of biotite, muscovite, and hornblende mineral separates (Fig. 3.9) confirmed that all of the basin except southern Manitoba was affected by Hudsonian metamorphism between 1.6 and 1.9 Ga (Burwash et al., 1962; Peterman, 1962). Apparent ages between 2.03 and 2.18 Ga for three hornblende samples were inferred to be "survival values" from Archean crystallization (Burwash et al., 1964). Scattered K-Ar biotite ages of less than 1.6 Ga were generally revised upward by later work (Peterman and Hedge, 1964). Biotite and hornblende separates from two diabase cores near the west side of the basin are dated as Middle Proterozoic.

Data from U-Pb isotopic analyses of zircon separates and Rb-Sr, Nd-Sm, or Pb-Pb analyses of whole-rock samples are available from only a few widely separated basement cores in the Western Canada Basin (Fig. 3.9). The peak of igneous and metamorphic activity during Early Proterozoic time occurred between 1.8 and 1.9 Ga, with a few early events pre-dating 1.9 Ga. A Late Archean (2.44 Ga) granulite facies event is documented at the edge of the basin in northeastern Alberta by a whole-rock Sm-Nd isochron (Burwash et al., 1985). Sialic crust with an age of approximately 2.9 Ga is indicated by U-Pb dating of zircons from two northwestern North Dakota localities (Peterman and Goldich, 1982); these sample localities lie within the Trans-Hudson Orogen. Eight core samples from southwestern Saskatchewan and adjacent Alberta give an average Sm-Nd crustal residence age of 2.81 Ga (Frost and Burwash, 1986). The geological relationship between the North Dakota and southwest Saskatchewan Archean terranes has not been established.

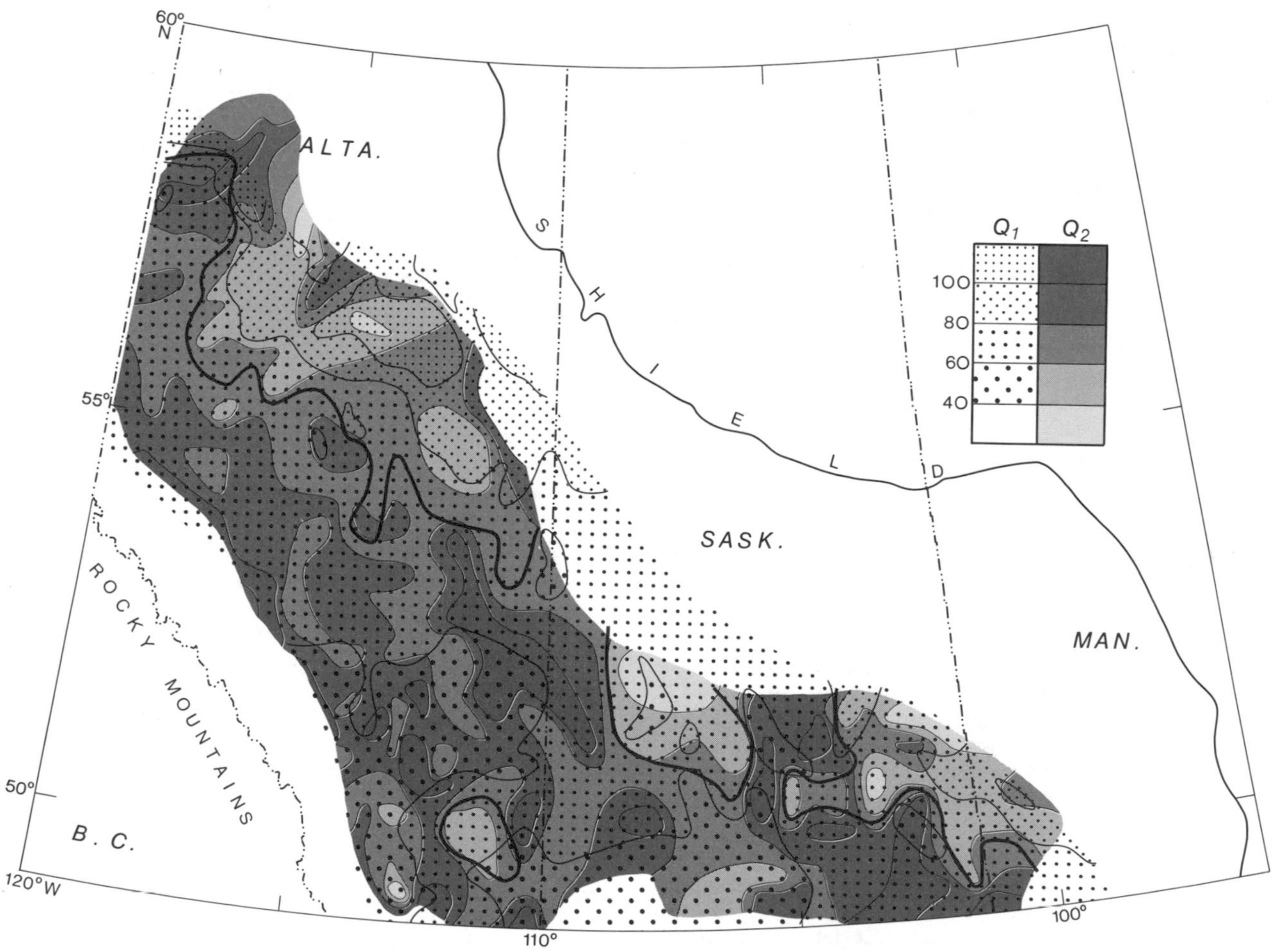

Figure 3.8. Heat flow patterns in the Western Canada Basin. Q1 - heatflow below the Paleozoic-Mesozoic boundary; Q2 - above the boundary.

LITHOSTRUCTURAL DOMAINS OF THE BURIED SHIELD

R.A. Burwash

Northern Interior Plains

The Great Bear-Great Slave Plains are bounded on the east by four major lithotectonic units of the Canadian Shield: Amundsen Basin, Wopmay Orogen, Slave Province, and East Arm Fold Belt. The adjacent units to the west are the Mackenzie Fold Belt and the Northern Rocky Mountains (Douglas and Price, 1972). A series of northeast-trending faults and north-trending geophysical anomalies define five terranes (see Fig. 3.11) underlying the Paleozoic cover.

North of the Fort Norman structure (see Fig. 3.11), formations of the Upper Proterozoic Mackenzie Mountains Supergroup flank the northwest-trending Great Bear Arch (Aitken and Pugh, 1984). Bevelling of these formations at the basal Paleozoic unconformity suggests uplift of Great Bear Arch prior to Cambrian sedimentation. The adjacent segment of the arcuate Mackenzie Arch is structurally concordant with Great Bear Arch. The Mackenzie Mountains Supergroup is equivalent to the Rae Group in the northward-dipping Amundsen Basin (Aitken and Pugh, 1984).

Between the Fort Norman structure and the Leith Ridge fault (Fig. 3.11), formations of the Hornby Bay (1.7-1.2 Ga) and Dismal Lakes groups constitute the Leith Ridge Domain, a southwestward subsurface extension of the Amundsen Basin at the Paleozoic subcrop (Aitken and Pugh, 1984). An estimated 2 km of downfaulted Middle Proterozoic strata are preserved north of the fault beneath Great Bear Lake (McGrath and Hildebrand, 1984).

The region south of the Leith Ridge fault is characterized by three major geophysical anomalies, the Great Bear magnetic high and the Bulmer Lake gravity high (see Fig. 3.3, 3.4, 3.11), and west of them, the Fort Simpson magnetic high (Fig. 3.3). The Great Bear magnetic high can be traced from the thoroughly studied Great Bear magmatic arc (see Fig. 3.11; WMO-2) (Hoffman and Bowring, 1984; Hildebrand and Bowring, 1984) south toward the Great Slave Lake Shear Zone. The intrusion of numerous mesozonal and epizonal plutons into a 100 km wide belt of thick volcanic units has produced a distinctive magnetic domain. The Hepburn belt to the east (WMO-3) and Hottah terrane to the west (WMO-1) are matched by magnetic lows. Between 120° and 122°W, a gravity high, 75 by 400 km with a relief of up to 50 mGal, can be traced from the Leith Ridge fault (LRF) to the northern margin of the Liard Block (Fig. 3.11), where it is abruptly truncated.

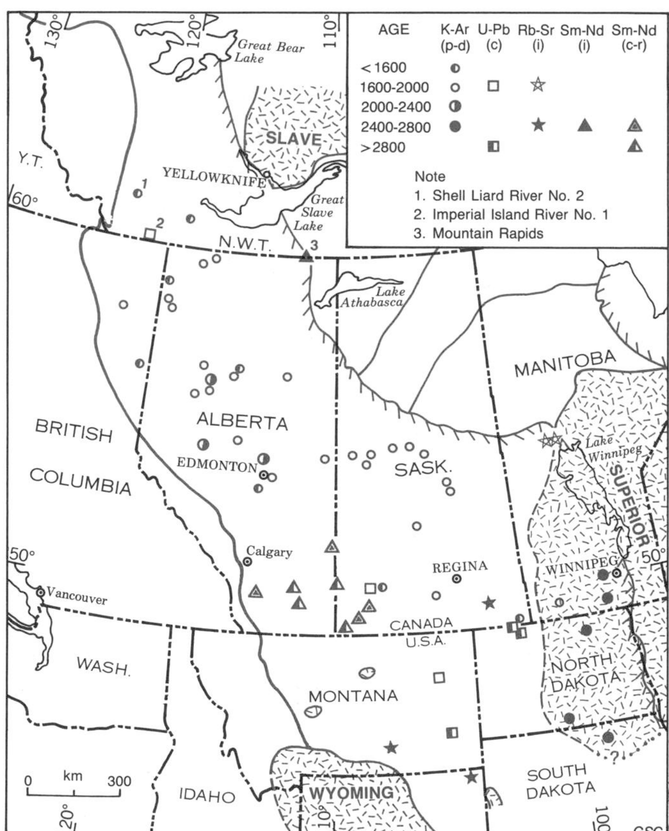

Figure 3.9. Dated basement localities in Western Canada Basin.

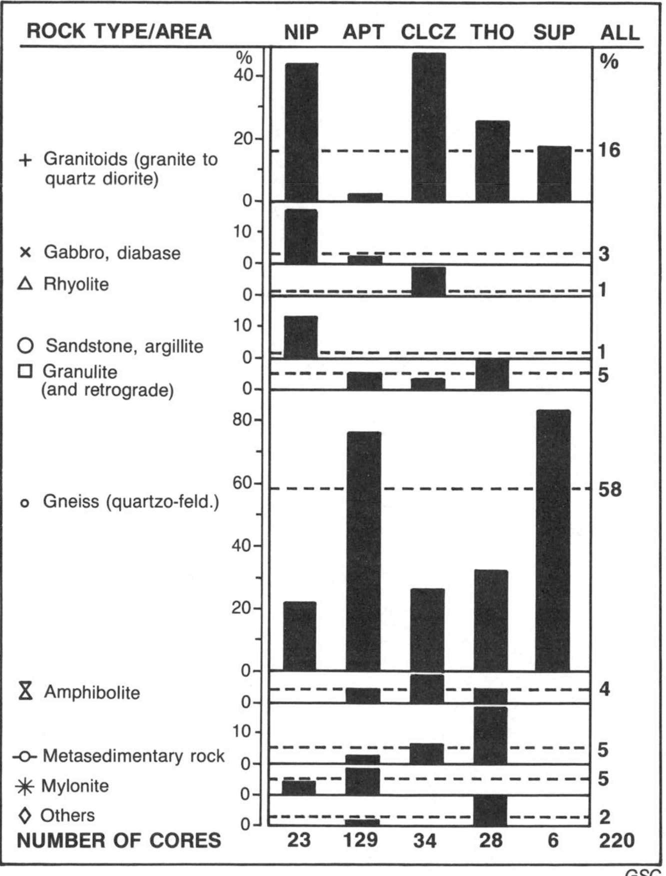

Figure 3.10. Comparative petrography of the lithostructural units of the buried shield. NIP - Northern Interior Plains; CLCZ - Cree Lake-Calgary Zone; THO - Trans-Hudson Orogen; SUP - Superior Province

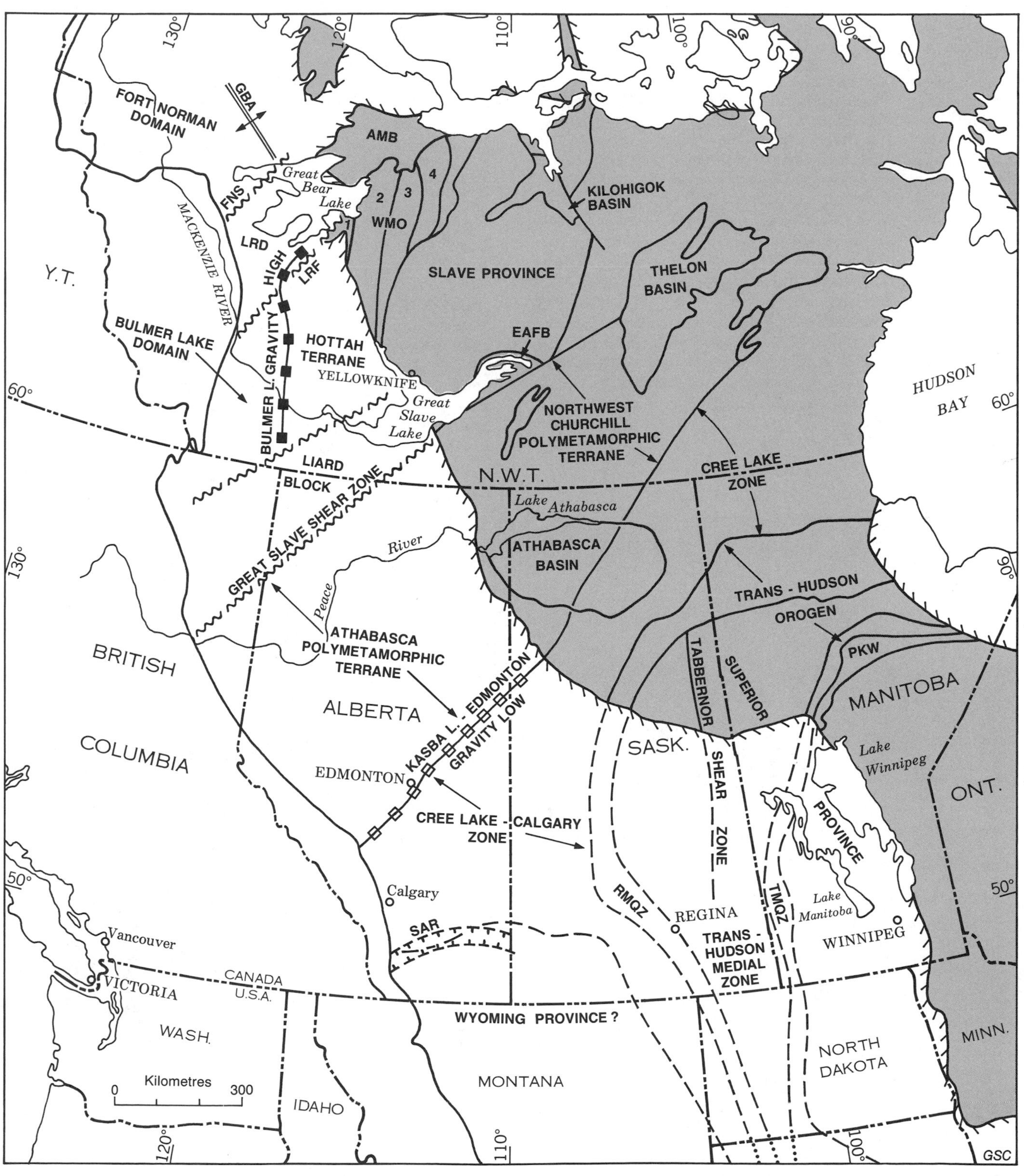
FORT NORMAN DOMAIN
GBA
AMB
Great Bear Lake
FNS
WMO
KILOHIGOK BASIN
LRD
LRF
SLAVE PROVINCE
THELON BASIN
Y.T.
MACKENZIE RIVER
BULMER L. GRAVITY HIGH
BULMER LAKE DOMAIN
HOTTAH TERRANE
EAFB
YELLOWKNIFE
Great Slave Lake
NORTHWEST CHURCHILL POLYMETAMORPHIC TERRANE
HUDSON BAY
LIARD BLOCK
N.W.T.
CREE LAKE ZONE
GREAT SLAVE SHEAR ZONE
Lake Athabasca
ATHABASCA BASIN
River
Peace
TRANS - HUDSON OROGEN
ATHABASCA POLYMETAMORPHIC TERRANE
PKW
BRITISH COLUMBIA
KASBA L. - EDMONTON GRAVITY LOW
TABBERNOR
SUPERIOR
MANITOBA
ALBERTA
SASK.
Lake Winnipeg
ONT.
EDMONTON
SHEAR ZONE
PROVINCE
CREE LAKE - CALGARY ZONE
Calgary
RMQZ
REGINA
TMQZ
Lake Manitoba
TRANS - HUDSON MEDIAL ZONE
WINNIPEG
SAR
Vancouver
VICTORIA
CANADA
U.S.A.
WASH.
WYOMING PROVINCE ?
MINN.
NORTH DAKOTA
Kilometres
0
300
IDAHO
MONTANA
GSC
130°
120°
110°
100°
90°
60°
50°

This Bulmer Lake gravity high coincides, in part, with a topographic ridge of westward-dipping Middle Proterozoic strata (Meijer Drees, 1975). Core samples from basement wells adjacent to the topographic high are diabase, suggesting that basic sills may be intercalated with the Proterozoic strata. Shell Liard River No. 2 (Fig. 3.9; Loc. 1), a projected 3500 m "basement test", encountered Proterozoic formations at a depth of 400 m. The lower 100 m of this hole were in uralitic diabase, which gave a K-Ar age (biotite) of 1100 Ma. These data support Meijer Drees' (1975) interpretation of the eastern boundary of the Bulmer Lake gravity anomaly as the "southward extension of the boundary between the Wopmay subprovince and the Coppermine homocline of the Bear Province". Hoffman (1987, 1988), on the other hand, suggested that the magnetic high may be a buried suture between Wopmay Orogen and a magmatic arc to the west, represented by the magnetic high. East of the Bulmer Lake Arch, Cambrian strata lie directly on a gneissic basement, the southward extension of Hottah terrane.

The Liard Block (Burwash and Krupicka, 1970) is a westward extension of the East Arm Fold Belt; however, the fold belt occupies only a fraction of the total length of a complex fault system, which has remained tectonically active for almost 2 Ga. The Slave-Chantry mylonite zone (Heywood and Schau, 1978) extends the Great Slave Lake Shear Zone northeastward to the Arctic Ocean at the base of the Boothia Peninsula. The aeromagnetic data of Coles et al. (1976) indicate that the zone extends across the Interior Plains to the Cordillera. Deflections of gravity, magnetic, and metamorphic trends toward the fault zone indicate dextral transcurrent movement. A seismic profile across the zone (Barr, 1971) indicates crustal thickening on the order of 4 km under the fold belt. Coles et al. (1976) suggested that the intense magnetic high east of the Thelon Front may match a similar magnetic high south of Macdonald Fault at 114°W. If this interpretation is correct, dextral displacement across the entire fault system would be of the order of 300 km. Tectonic slices of the Slave craton occur in East Arm Fold Belt (Burwash and Baadsgaard, 1962). Farther to the west, fault slices or broad arches of crystalline basement related to Wopmay Orogen are to be anticipated. Biotite granodiorite from Imperial Island River No. 1 (Loc. 2, Fig. 3.8) has been dated at 1860 ± 10 Ma (U-Pb on zircon) by S.A. Bowring (cited by Parrish in Geology of Canada, no.7, Hoffman et al., in prep.). The biotite has been chloritized and gives a K-Ar date of 850 Ma (R.A. Burwash, unpub.). The Island River well lies near the Rabbit fault, one of the multiple branches of the Great Slave fault system (Williams, 1981). Argon loss from biotite during one of the recurrent movements of the fault system could explain the discordant U-Pb and K-Ar ages. Alternatively, a post-Hudsonian regional thermal event might be involved.

Figure 3.11. The Precambrian basement of Western Canada Basin.

Lithostructural domains of the buried Shield: LRD = Leith Ridge Domain; RMQZ = Reindeer Magnetic Quiet Zone; TMQZ = Thompson Magnetic Quiet Zone.

Structural features: GBA = Great Bear Arch; FNS = Fort Norman structure; LRF = Leith Ridge fault; SAR = Southern Alberta Rift.

Lithostructural Domains of the exposed Shield: AMB = Amundsen Basin; WMO = Wopmay Orogen (subdivisions: 1 = Hottah Terrane; 2 = Great Bear Magnetic arc; 3 = Hepburn batholithic belt; 4 = Foreland Belts); EAFB = East Arm Fold Belt; PKW = Pikwitonei sub-Province

Along the western edge of the Interior Plains, a series of major magnetic highs was mapped by Coles et al. (1976). This belt of anomalies extends almost continuously from the Alaska-Yukon boundary at 66°N as a great arc following the tectonic arc, reverses curvature in northeastern British Columbia and extends into western Alberta as far south as 55°N. Coles et al. (1976) suggested that this major magnetic feature may be related to thermal enhancement of magnetic susceptibility in the crystalline basement as a result of high heat flow in the eastern Cordillera at the present time. Hoffman and Bowring (1984) and Hoffman (1987, 1988) link part of this magnetic high (Fort Simpson) to a magmatic arc during a continental collision at 1.8 to 1.9 Ga. If a dextral displacement of the order of 300 km occurred across the Great Slave fault system during a late stage of Hudsonian Orogeny, the southward extension of Wopmay Orogen would be offset 300 km west as it crossed the shear zone. The arcuate pattern of the magnetic anomalies thus may reflect either the metamorphic fabric of the buried Hudsonian crystalline basement or the imposition of present-day reheating on an old metamorphic welt.

Athabasca polymetamorphic terrane

The Athabasca polymetamorphic terrane was described by Burwash and Culbert (1976) as the Athabasca mobile zone. It is the subsurface equivalent of the northwestern zone of Churchill Province (Davidson, 1972). An important characteristic of this zone is the occurrence of numerous relict belts of metamorphic rocks of granulite facies (Fraser et al., 1978).

In establishing the boundaries of Churchill Province, Stockwell (1963) used the criterion of the last period of regional metamorphism, indicated by K-Ar age determinations mainly on biotite. He was aware that older crustal blocks were incorporated in the Churchill Province. In northwest Saskatchewan, a period of granulite-facies metamorphism, circa 2.4 Ga, was documented by Koster and Baadsgaard (1970), using K-Ar ages of hornblende. The granulites were overprinted by lower-grade Hudsonian events at 1.8 to 1.9 Ga. At Hill Island Lake, 200 km to the northeast, a granulite dome shows a histogram peak at 2450 Ma for K-Ar hornblende ages and a separate peak at 2200 Ma for biotite ages (Banks, 1980).

A comprehensive program of geological mapping, Rb-Sr and K-Ar geochronology, and structural studies in northeastern Alberta (Godfrey and Langenberg, 1978; Nielsen et al., 1981; Langenberg, 1983) established Kenoran and Hudsonian events with different P-T fields and different tectonic styles. Burwash et al. (1985) dated mafic granulites at Mountain Rapids on the Slave River (Loc. 3, Fig. 3.9) just south of 60°. A Sm-Nd isochron of 2436 ± 44 Ma is interpreted as the time of granulite metamorphism in this terrane. Aliquots of the same samples give a Rb-Sr isochron of 1898 ± 5 Ma, the time of the Hudsonian overprint.

Direct evidence for granulite-facies metamorphism in the basement is found in ten drill cores (Fig. 3.2), six of which are in the Athabasca polymetamorphic terrane. Geochronological evidence for Archean crust in this belt is given by K-Ar hornblende survival values in excess of 2.0 Ga in three other wells (Fig. 3.9).

The fabric of the domains that form the Athabasca metamorphic terrane is shown by multivariate trend surface analysis of mineralogical, chemical, and textural data (Burwash and Culbert, 1976). The three strongest R-mode factors, K-metasomatism, chloritization, and shearing, have trends between N10°W and N35°W. The chloritization factor is negative along the axis of greatest K-metasomatism. All trends terminate near the north and south boundaries of the Athabasca polymetamorphic terrane. Polynomial fitting of the data shows that along the axis of the Peace River Arch K-metasomatism extends to the western limit of drillhole sampling, confirming an earlier petrographic evaluation (Burwash and Krupicka, 1970).

Near the western limit of the northwest Churchill polymetamorphic terrane, a direct relationship is observed between the north-trending Fort Smith radiometric high (Charbonneau, 1980), a megacrystic microcline granite (Bostock, 1981), and a broad aeromagnetic low. These trends are all deflected into the Great Slave Shear Zone (Burwash and Cape, 1981). Analysis of the magnetic fabric of northeastern Alberta (Sprenke et al., 1986) gave a magnetic autocorrelogram with the major axis N10°E. This fabric can be traced southward to the boundary of the Athabasca polymetamorphic terrane, where it is abruptly truncated (Garland and Bower, 1959).

In 1970, Wallis recognized the Virgin River shear zone (11, Fig. 3.2) on the basis of a mylonite belt, elevated rhomboidal granulite blocks with dimensions of the order of 50 by 100 km, and infolded belts of Lower Proterozoic metasedimentary rocks. The granulite blocks are generally matched by positive gravity anomalies (Walcott, 1968), whereas the amphibolite-facies metasedimentary belts are gravity lows. There is marked aeromagnetic relief over the granulite fault blocks. Converging with the Virgin River shear zone near the Alberta-Saskatchewan boundary is the dominant negative gravity anomaly of Saskatchewan, the Fond du Lac low of Walcott (1968; here called the Kasba Lake-Edmonton gravity low), which can be traced southwestward to the Rocky Mountains (Walcott and Boyd, 1971). Along this gravity low occurs a series of uranium-rich granites, which Burwash (1979) interpreted as the product of anatexis along a major shear zone.

A combined geophysical-geological study of the Precambrian basement beneath central Alberta by Garland and Burwash (1959) suggested a significant change in lithology just north of Edmonton. The Kasba Lake-Edmonton gravity low (Fig. 3.11), as defined by Burwash and Culbert (1976), remains the best documented choice of a southern boundary for the northwest Churchill polymetamorphic terrane and the Athabasca polymetamorphic terrane.

Cree Lake – Calgary Zone

The reactivated Archean basement and its infolded cover as mapped on the exposed Canadian Shield can be traced into the subsurface by several persistent magnetic anomalies. Magnetite-bearing meta-arkose in the Virgin River Fold Belt (12, Fig. 3.2) has been interpreted by Wallis (1970) as the product of low- to medium-grade metamorphism of Lower Proterozoic sediments deposited in a fault trough, 15 to 50 km wide, which cuts across gneisses of granulite facies. Some of the magnetite-bearing horizons contain up to 45% iron. In the Wollaston Fold Belt, cordierite-garnet-magnetite gneisses underlie part of the broad aeromagnetic high that flanks the northwest side of the Wathaman Batholith. On the map of Coles et al. (1976), the Virgin River magnetic lineament can be extended southwestward to the Rocky Mountain Trench. The arcuate Wollaston aeromagnetic trend apparently terminates near 52°N, 108°W.

The lithologic assemblages of the Cree Lake-Calgary Zone (Fig. 3.11) in outcrop do not match those in drill cores from its subsurface extension. The Mudjatik domain (13, Fig. 3.2) appears to be eroded to the katazone, while its flanks record shallower crustal levels. Post-Hudsonian erosion of the western part of the Cree Lake-Calgary Zone to the mesozone could explain the relative abundance of unfoliated granites in the core samples (Fig. 3.2, 3.10) and the preservation of rhyolites in east-central Alberta and southwestern Saskatchewan. The accumulation of helium in basal Paleozoic sandstones above a rhyolite basement high near Swift Current, Saskatchewan, is attributed by Burwash and Cumming (1974) to a subjacent uranium-rich epizonal granitic pluton. The age of this pluton, 1.81 Ga (Rosholt et al., 1970), is late Hudsonian.

Archean basement in the Wollaston Fold Belt (Fig. 3.2), initially recognized on the basis of petrological and structural evidence (Money et al., 1970), was later confirmed by geochronology (Bell and Macdonald, 1982). The single subsurface sample of granulite facies from the Cree Lake-Calgary Zone lies along strike from an Archean charnockite complex in the Wollaston Fold Belt.

Hudsonian thermal overprinting of almost all rock units in the Cree Lake-Calgary Zone was indicated by K-Ar dating in the 1960s (Burwash et al., 1962; Wanless, 1970). K-Ar dates from biotite, hornblende, and muscovite from cores in this zone fall mainly between 1.7 and 1.8 Ga, the time of post-orogenic uplift and stabilization (Fig. 3.9). Sm-Nd isotopic analyses of eight core samples from southern Alberta and southwestern Saskatchewan give Archean crustal residence ages at all localities (Frost and Burwash, 1986). These values represent the time of separation of crust from mantle, and suggest that in the area sampled there is limited juvenile Proterozoic sial in the Cree Lake-Calgary Zone.

Trans-Hudson Orogen

In northern Saskatchewan and Manitoba the Trans-Hudson Orogen (Fig. 3.11) consists of a number of sub-parallel and generally arcuate lithotectonic domains (Lewry, 1981; Fumerton et al., 1984; Green et al., 1985a). The Wathaman-Chipewyan Batholith is a relatively homogenous body of megacrystic granite-granodiorite. It has no comagmatic early mafic phases and shows no evidence of multiple intrusions. If numerous smaller plutons are included, the composite batholithic belt is 900 km long. By analogy with Mesozoic Pacific rim belts of similar dimensions, Fumerton et al. (1984) interpreted the Wathaman-Chipewyan Batholith to be the product of plate collision.

An arcuate positive magnetic anomaly, 900 km long and up to 100 km wide, corresponds spatially with the northern region of the Wathaman Batholith and the eastern margin of the Wollaston Fold Belt. A linear belt of magnetic lows, which corresponds mainly to the Reindeer-Southern Indian belt (Green et al., 1985a), parallels the magnetic highs. The Reindeer magnetic quiet zone (RMQZ) is particularly well defined between 56°N and 52°N (Fig. 3.11). South of 52°N, the pattern of negative magnetic anomalies increases in width and becomes irregular. However, using the trends of the -60 mGal Bouguer anomaly contour and the North American Central Plains anomaly as corroborative data, the Reindeer magnetic quiet zone can be traced southward to 45°N, where it loses its identity. Beneath the Interior Plains the western margin of the Reindeer magnetic quiet zone has been chosen as the nominal boundary of the Trans-Hudson Orogen.

In contrast to the Athabasca polymetamorphic terrane and the Cree Lake-Calgary Zone, the Trans-Hudson Orogen contains only limited evidence of Archean sialic crust. The Hansen Lake block (19, Fig. 3.2), a north-trending 30 by 70 km terrane west of the Flin Flon granite-greenstone Belt, contains granulites dated at 2.4 Ga. The geological history of the adjacent Glennie Lake domain (18, Fig. 3.2) is complex. Lewry (1981) suggested that it was an Archean or Early Proterozoic microcontinent. Alternatively, it may be an older basement underlying adjacent Proterozoic terranes (Green et al., 1985a).

Basement core samples from the Trans-Hudson Orogen (Fig. 3.2 and 3.10), although limited in number, suggest several differences from the Cree Lake-Calgary Zone. If four wells in northwestern North Dakota are included, there is evidence that granulite occurs in several areas. Isotopic ages of the North Dakota granulites are Archean (Peterman and Goldich, 1982). Z.E. Peterman (pers. comm., 1983) suggested that "a major high-grade terrane may exist beneath the Williston Basin that comprises reworked Archean rocks."

Five drillholes in east-central Saskatchewan penetrated detrital sedimentary rocks (Fig. 3.2). Cores taken near Nipawin are metamorphosed banded iron-formation. Laminations on a scale of several millimetres show enrichment in either magnetite, quartz, or biotite. In the absence of documented metavolcanic rocks from this area, the Nipawin banded iron-formation and associated biotite schists are interpreted as a local sedimentary basin rather than a volcanic arc. Direct evidence of Early Proterozoic volcanism is limited in available samples to one amphibolite core on strike with the La Ronge Belt (15, Fig. 3.2) and one metavolcanic (?) core in southwestern Manitoba.

No petrological evidence was found in basement core samples to document a suture zone. If the Superior "boundary zone" is removed from the Trans-Hudson Orogen, the width of the orogen is reduced to the order of 100 km south of 50°N. Within this belt the sparsely distributed core samples are of granitic and gneissic rocks of Hudsonian age. This suggests that the east-west and north-south trending parts of the orogen must be of different fundamental character.

Superior Province

The Churchill-Superior boundary in northern Manitoba has been one of the most controversial tectonic zones in Canada (Bell, 1971b; Green et al., 1979). Because of sparse outcrop along the "nickel belt", exploration for nickel deposits near Thompson relied heavily on geophysical surveys (Zurbrigg, 1963). Linear gravity and magnetic features, in conjunction with seismic, magnetotelluric, and geochronological data, have since been used to define the Superior boundary zone (Green et al., 1985).

The most obvious anomaly associated with the boundary zone is the arcuate Nelson River gravity high (Fig. 3.4), which extends 900 km from northeastern to southwestern Manitoba. On the exposed Canadian Shield it coincides with the Archean granulite-facies gneisses of the Pikwitonei sub-province (Fig. 3.11; Weber and Scoates, 1978). The granulites have a distinctive, short-wavelength "bird's-eye maple" magnetic pattern. The northwest edge of the Pikwitonei sub-province belt has been overprinted in the Thompson structural belt by Hudsonian amphibolite-facies metamorphism, which produced a magnetic quiet zone (Green et al., 1979). The Thompson magnetic quiet zone (TMQZ), (Fig. 3.11) is slightly discordant to the gravity high, the two crossing near 54°N, 99°W.

The eastern margin of the Thompson magnetic quiet zone has been chosen as the limit of the Trans-Hudson Orogen for several reasons. On the exposed Shield it corresponds quite well to the boundary based on isotopic age determinations (Kornik and MacLaren, 1966). In the subsurface it can be traced to 45°N along a series of negative magnetic anomalies. The Thompson belt of northern Manitoba and its subsurface extension are classified as reworked Archean foreland.

The pattern of alternating belts of granite-greenstone and high-grade gneiss, which characterize the western Superior Province (Hoffman et al., 1982a), well shown on the map of metamorphic facies of the Canadian Shield (Fraser et al., 1978), can be recognized clearly on the magnetic anomaly map (Fig. 3.3). A general correlation of magnetic lows with metasedimentary and metavolcanic rocks of low metamorphic grade was observed by MacLaren and Charbonneau (1968). Variation in metamorphic grade from the centre to the edge of steeply dipping greenstone belts (Ayres, 1978), combined with faults bounding the supracrustal sequences, enhance the magnetic lineaments.

The east-trending magnetic fabric of Superior craton is sharply truncated by the Thompson magnetic quiet zone. The Hudsonian amphibolite facies overprint of the pattern of Kenoran regional metamorphism has apparently reduced the contrast in magnetic susceptibility between the various lithotectonic belts. Since the Thompson magnetic quiet zone is essentially the metamorphic front of the Trans-Hudson Orogen, relict Archean isotopic ages can be expected to occur west of the zone.

The number of drill cores available from the subsurface extension of Superior Province is inadequate to form a statistically valid petrological sample population (Fig. 3.10). The only supracrustal rock that might represent an Archean greenstone belt, an altered meta-rhyolite(?), occurs in the Thompson magnetic quiet zone near 50°N. Peterman (1962) compared the Precambrian basement rock types of Superior and Churchill (structural) provinces in Saskatchewan and Manitoba by plotting modal quartz-K-feldspar-plagioclase for all available cores of silicic plutonic rocks. K-Ar dated rocks from the adjacent exposed Canadian Shield were used to augment this sample population. He inferred that the plutons of Superior Province were predominantly granodiorite, while those of the Churchill were quartz monzonite or granite. This conclusion anticipated the use of K:Na ratios in distinguishing Archean from Proterozoic crust (Eade and Fahrig, 1971; Burwash and Krupicka, 1969).

REFERENCES

Aitken, J.D. and Pugh, D.C.
1984: The Fort Norman and Leith Ridge structures: major, buried, Precambrian features underlying Franklin Mountains and Great Bear and Mackenzie Plains; Bulletin of Canadian Petroleum Geology, v. 32, no. 2, p. 139-146.

Agarwal, R.G. and Kanasewich, E.R.
1968: A gravity investigation of the Stony Rapids area, northern Saskatchewan; Saskatchewan Department of Mineral Resources, Report 124, 42 p.

Alabi, A.O., Camfield, P.A., and Gough, D.I.
1975: The North American Central Plains conductivity anomaly; Geophysical Journal of the Royal Astronomical Society, v. 43, p. 815-833.

Asada, T., Steinhart, J.S., Rodriguez, A., Tuve, M.A., and Aldrich, L.T.
1961: The earth's crust; Carnegie Institute of Washington Yearbook, v. 60, p. 244-256.

Ayres, L.D.
1978: Metamorphism in the Superior Province of northeastern Ontario and its relationship to crustal development; in Metamorphism in the Canadian Shield; Geological Survey of Canada, Paper 78-10, p. 25-36.

Banks, C.S.
1980: Geochronology, general geology and structure of Hill Island Lake-Tazin Lake area; M.Sc. thesis, University of Alberta, Edmonton, 109 p.

Barr, K.G.
1971: Crustal refraction experiment: Yellowknife 1966; Journal of Geophysical Research, v. 76, p. 1929-1947.

Beck, L.S.
1969: Uranium deposits of the Athabasca region, Saskatchewan; Saskatchewan Department of Mineral Resources, Report No. 126, 139 p.

Bell, C.K.
1971a: Boundary geology, upper Nelson River area, Manitoba and northwestern Ontario; Geological Association of Canada, Special Paper 9, p. 11-39.
1971b: History of the Superior-Churchill boundary in Manitoba; in Geoscience Studies in Manitoba, A.C. Turnock (ed.); Geological Association of Canada, Special Paper 9, p. 5-10.

Bell, C.K. and Macdonald, R.
1982: Geochronologic calibration of the Precambrian Shield in Saskatchewan; Saskatchewan Geological Survey, Miscellaneous Report 82-4, p. 17-22.

Bennett, G.T., Clowes, R.M., and Ellis, R.M.
1975: A seismic refraction survey along the southern Rocky Mountain Trench, Canada; Bulletin of Seismological Society of America, v. 65, p. 37-54.

Berry, M.J. and Forsyth, D.A.
1975: Structure of the Canadian Cordillera from seismic refraction and other data; Canadian Journal of Earth Sciences, v. 12, p. 182-208.

Berry, M.J., Jacoby, W.R., Niblett, E.R., and Stacey, R.A.
1971: A review of geophysical studies in the Canadian Cordillera; Canadian Journal of Earth Sciences, v. 8, p. 788-801.

Bostock, H.H.
1981: A granitic diapir of batholitic dimensions at the west margin of the Churchill Province; in Current Research, Part B, Geological Survey of Canada, Paper 81-1B, p. 73-82.

Bowring, S.A., Van Schmus, W.R., and Hoffman, P.F.
1984: U-Pb zircon ages from Athapuscow aulacogen, East Arm of Great Slave Lake, Northwest Territories, Canada; Canadian Journal of Earth Sciences, v. 21, p. 1315-1324.

Burwash, R.A.
1951: The Precambrian under the central plains of Alberta; M.Sc. thesis, University of Alberta, Edmonton, 82 p.
1955: A reconnaissance of the subsurface Precambrian of the Province of Alberta, Canada; Ph.D. thesis, University of Minnesota, Minneapolis, 74 p.
1957: Reconnaissance of subsurface Precambrian of Alberta; American Association of Petroleum Geologists, Bulletin, v. 41, p. 70-103.
1979: Uranium and thorium in the Precambrian basement of western Canada. II. Petrologic and tectonic controls; Canadian Journal of Earth Sciences, v. 16, p. 472-483.

Burwash, R.A. and Baadsgaard, H.
1962: Yellowknife-Nonacho age and structural relations; in The Tectonics of the Canadian Shield, J.S. Stevenson (ed.); Royal Society of Canada, Special Publication 4, p. 22-29.

Burwash, R.A., Baadsgaard, H., and Peterman, Z.E.
1962: Precambrian K-Ar dates from the Western Canada Sedimentary Basin; Journal of Geophysical Research, v. 67, p. 1617-1625.

Burwash, R.A., Baadsgaard, H., Peterman, Z.E., and Hunt, G.H.
1964: Precambrian; in Geological History of Western Canada, R.G. McCrossan and R.P. Glaister (ed.); Alberta Society of Petroleum Geologists, Calgary, p. 14-19.

Burwash, R.A. and Cape, D.F.
1981: Petrology of the Fort Smith-Great Slave Lake radiometic high near Pilot Lake, Northwest Territories; Canadian Journal of Earth Sciences, v. 18, p. 842-851.

Burwash, R.A. and Culbert, R.R.
1976: Multivariate geochemical and mineral patterns in the Precambrian basement of western Canada; Canadian Journal of Earth Sciences, v. 13, p. 1-18.

Burwash, R.A. and Cumming, G.L.
1974: Helium source rock in southwestern Saskatchewan; Bulletin of Canadian Petroleum Geology, v. 22, p. 405-412.
1976: Uranium and thorium in the Precambrian basement of western Canada. I. Abundance and distribution; Canadian Journal of Earth Sciences, v. 13, p. 284-293.

Burwash, R.A. and Krupicka, J.
1969: Cratonic reactivation in the Precambrian basement of western Canada. I. Deformation and chemistry; Canadian Journal of Earth Sciences, v. 6, p. 1381-1396.
1970: Cratonic reactivation in the Precambrian basement of western Canada. Part II. Metasomatism and isostasy; Canadian Journal of Earth Sciences, v. 7, p. 1275-1294.

Burwash, R.A., Krupicka, J., Basu, A.R., and Wagner, P.A.
1985: Resetting of Nd whole-rock isochrons from polymetamorphic granulites, northeastern Alberta; Canadian Journal of Earth Sciences, v. 22, p. 992-1000.

Camfield, P.A. and Gough, D.I.
1977: A possible Proterozoic plate boundary in North America; Canadian Journal of Earth Sciences, v. 14, p. 1229-1238.

Chandra, N.N. and Cumming, G.L.
1972: Seismic refraction studies in western Canada; Canadian Journal of Earth Sciences, v. 9, p. 1099-1109.

Charbonneau, B.W.
1980: The Fort Smith radioactive belt, Northwest Territories; in Current Research, Part C, Geological Survey of Canada, Paper 80-1C, p. 45-57.

Clowes, R.M., Green, A.G., Yorath, C.J., Kanasewich, E.R., West, G.F., and Garland, G.D.
1984: Lithoprobe - a national program for studying the third dimension of geology; Journal of Canadian Society of Exploration Geophysicists, v. 20, p. 23-39.

Coles, R.L., Haines, G.V., and Hannaford, W.
1976: Large-scale magnetic anomalies over western Canada and the Arctic: a discussion; Canadian Journal of Earth Sciences, v. 13, p. 790-802.

Congram, A.M.
1984: A crustal refraction study in the Williston Basin of southern Saskatchewan; M.Sc. thesis, University of Saskatchewan, Saskatoon, 273 p.

Cumming, W.B., Clowes, R.M., and Ellis, R.M.
1979: Crustal structure from a seismic refraction profile across southern British Columbia; Canadian Journal of Earth Sciences, v. 16, p. 1024-1040.

Davidson, A.
1972: The Churchill Province; in Variations in Tectonic Styles in Canada, R.A. Price and R.J.W. Douglas (ed.); Geological Association of Canada, Special Paper 11, p. 381-434.

DeLandro, W.L.
1981: Deep seismic sounding in southern Manitoba and Saskatchewan; M.Sc. thesis, University of Manitoba, Winnipeg, 129 p.

DeLandro, W.L. and Moon, W.
1982: Seismic structure of Superior-Churchill Precambrian boundary zone; Journal of Geophysical Research, v. 87, p. 6884-6888.

Dods, S.D., Hood, P.J., Teskey, D.J., and McGrath, P.H.
1984: Magnetic anomaly map of Canada; Geological Survey of Canada, Map 1225A, 4th edition, 1:5 000 000.

Dominion Observatory
1957: Gravity anomaly map of Canada; 1:6 336 000.

Douglas, R.J.W. and Price, R.A.
1972: Nature and significance of variations in tectonic styles in Canada; in Variations in Tectonic Styles in Canada, R.A. Price and R.J.W. Douglas (ed.); Geological Association of Canada, Special Paper 11, p. 625-688.

Dutch, S.I.
1983: Proterozoic structural provinces in the north-central United States; Geology, v. 11, p. 478-481.

Eade, K.E. and Fahrig, W.F.
1971: Geochemical evolutionary trends of continental plates - a preliminary study of the Canadian Shield; Geological Survey of Canada, Bulletin 179, 51 p.

Earth Physics Branch
1980: Gravity Map of Canada; Gravity Map Series 80-1, Ottawa, Canada, 1:5 000 000.

Ermanovics, I.F. and Froese, E.
1978: Metamorphism of the Superior Province in Manitoba; in Metamorphism in the Canadian Shield; Geological Survey of Canada, Paper 78-10, p. 17-24.

Evanchick, C.A., Parrish, R.R., and Gabrielse, H.
1984: Precambrian gneiss and late Proterozoic sedimentation in north-central British Columbia; Geology, v. 12, p. 233-237.

Flawn, P.T.
1967: Basement map of North America, 1:5 000 000; American Association of Petroleum Geologists and United States Geological Survey.

Forsyth, D.A., Berry, M.J., and Ellis, R.M.
1974: A refraction survey across the Canadian Cordillera at 54°N; Canadian Journal of Earth Sciences, v. 11, p. 533-548.

Fountain, D.M. and Salibury, M.H.
1981: Exposed cross-sections through the continental crust: Implications for crustal structure, petrology and evolution; Earth and Planetary Science Letters, v. 56, p. 263-277.

Fraser, J.A., Heywood, W.W., and Mazurski, M.A.
1978: Metamorphic map of the Canadian Shield; Geological Survey of Canada, Map 1475A, 1:3 500 000.

Frost, C.D. and Burwash, R.A.
1986: Nd evidence for extensive Archean basement in the western Churchill Province, Canada; Canadian Journal of Earth Sciences, v. 23, p. 1433-1437.

Fumerton, S.L., Stauffer, M.R., and Lewry, J.F.
1984: The Wathaman Batholith: largest known Precambrian pluton; Canadian Journal of Earth Sciences, v. 21, p. 1082-1097.

Ganley, D.C. and Cumming, G.L.
1974: A seismic reflection model of the crust near Edmonton, Alberta; Canadian Journal of Earth Sciences, v. 11, p. 101-109.

Garland, G.D. and Bower, M.E.
1959: Interpretation of aeromagnetic anomalies in northeastern Alberta; in Fifth World Petroleum Congress, Proceedings; Geology and Geophysics, Section 1, p. 787-800.

Garland, G.D. and Burwash, R.A.
1959: Geophysical and petrological study of Precambrian of central Alberta, Canada; Bulletin of American Association of Petroleum Geologists, v. 43, p. 790-807.

Garland, G.D. and Lennox, D.M.
1962: Heat flow in western Canada; Geophysics Journal, v. 6, p. 245-262.

Geological Survey of Canada
1967: Magnetic anomaly map of Canada; Map 1255A, 1:5 000 000.

Gibb, R.A. and Halliday, D.W.
1974: Gravity measurements in southern District of Keewatin and southeastern District of Mackenzie; Gravity Map Series, Earth Physics Branch, p. 124-131.

Gibb, R.A. and Thomas, M.D.
1976: Gravity signature of fossil plate boundaries in the Canadian Shield; Nature, v. 262, p. 199-200.

Godfrey, J.D. and Langenberg, C.W.
1978: Metamorphism in the Canadian Shield of northeastern Alberta; in Metamorphism in the Canadian Shield; Geological Survey of Canada, Paper 78-10, p. 129-138.

Green, A.G., Cumming, G.L., and Cedarwell, D.
1979: Extension of the Superior-Churchill boundary zone into southern Canada; Canadian Journal of Earth Sciences, v. 16, p. 1691-1701.

Green, A.G., Hajnal, Z., and Weber, W.
1985a: An evolutionary model of the western Churchill Province and western margin of the Superior Province of Canada and the north-central United States; Tectonophysics, v. 116, p. 281-322.

Green, A.G., Stephanson, O.G., Mann, G.D., Kanasewich, E.R., Cumming, G.L., Hajnal, Z., Mair, J.A., and West, G.F.
1980: Cooperative seismic surveys across the Superior-Churchill boundary zone in southern Canada; Canadian Journal of Earth Sciences, v. 17, p. 617-632.

Green, A.G., Weber, W., and Hajnal, Z.
1985b: Evolution in Proterozoic terranes beneath the Williston Basin; Geology, v. 13, p. 624-628.

Gupta, J.C., Kurtz, R.D., Camfield, P.A., and Niblett, E.R.
1985: A geomagnetic induction anomaly from I.M.S. data near Hudson Bay, and its relation to crustal electrical conductivity in Central North America; The Geophysical Journal of the Royal Astronomical Society, v. 81, p. 33-46.

Hales, A.L. and Nation, J.B.
1973: A seismic refraction survey in the north Rocky Mountains: more evidence for an intermediate crustal layer; Geophysical Journal of the Royal Astronomical Society, v. 35, p. 381-399.

Hall, D.H.
1968: Regional magnetic anomalies, magnetic units and crustal structure in the Kenora District of Ontario; Canadian Journal of Earth Sciences, v. 5, p. 1277-1296.

Hall, D.H. and Hajnal, Z.
1969: Crustal structure of northwestern Ontario: refraction seismology; Canadian Journal of Earth Sciences, v. 6, p. 82-99.
1973: Deep seismic crustal studies in Manitoba; Bulletin Seismological Society of America, v. 63, p. 885-910.

Handa, S. and Camfield, P.A.
1984: Crustal electrical conductivity in north-central Saskatchewan: the North Central Plains anomaly and its relation to a Proterozoic plate margin; Canadian Journal of Earth Sciences, v. 21, p. 533-543.

Heywood, W.W. and Schau, M.
1978: A subdivision of the northern Churchill Structural Province; in Current Research, Part A, Geological Survey of Canada, Paper 78-1A, p. 139-143.

Hildebrand, R.S. and Bowring, S.A.
1984: Continental intra-arc depressions: A nonextensional model for their origin, with a Proterozoic example from Wopmay Orogen; Geology, v. 12, p. 73-77.

Hill, D.P. and Pakiser, L.C.
1967: Seismic refraction study of crustal structure between Nevada Test Site and Boise, Idaho; Bulletin of the Geological Society of America, 78, p. 685-704.

Hitchon, B.
1984: Geothermal gradients, hydrodynamics, and hydrocarbon occurrences, Alberta, Canada; Bulletin of American Association of Petroleum Geology, v. 68, p. 713-743.

Hoffman, P.F.
1981: Autopsy of Athapuscow Aulacogen: A failed arm affected by three collisions; in Proterozoic Basins of Canada, F.H.A. Campbell (ed.); Geological Survey of Canada, Paper 81-10, p. 97-102.
1987: Continental transform tectonics: Great Slave Lake shear zone (ca. 1.9 Ga), northwest Canada; Geology, v. 15, p. 785-788.
1988: United Plates of America, the birth of a craton: Early Proterozoic assembly and growth of Laurentia; Annual Reviews of Earth and Planetary Sciences, v. 16, p. 543-603.

Hoffman, P.F. and Bowring, S.A.
1984: Short-lived 1.9 Ga continental margin and its destruction, Wopmay Orogen, northwest Canada; Geology, v. 12, p. 68-72.

Hoffman, P.F., Card, K.D., and Davidson, A.
1982a: The Precambrian: Canada and Greenland; in Perspectives in Regional Geologic Synthesis, A.R. Palmer (ed.); Geological Society of America, The Geology of North America, Special Publication 1, p. 3-6.

Hoffman, P.F., McGrath, P.H., Bowring, S.A., Van Schmus, W.R., and Thomas, M.D.
1982b: Plate tectonic model for Wopmay Orogen consistent with zircon chronology, gravity and magnetic anomalies east of Mackenzie Mountains; Canadian Geophysical Union, 9th Annual Meeting, Abstracts, p. 13.

Hornal, R.W., Sobczak, L.W., Burke, W.E.F., and Stephens, L.E.
1970: Preliminary results of gravity surveys over the Mackenzie Basin and Beaufort Sea; Gravity Map Series, Earth Physics Branch, no. 117-119, 1:1 000 000, 12 p. text.

Innes, M.J.S.
1960: Gravity and isostasy in northern Ontario and Manitoba; Dominion Observatory, Ottawa, v. 21, no. 6.

Jessop, A.M. and Vigrass, L.W.
1989: Geothermal measurements in a deep well at Regina, Saskatchewan; Journal of Volcanology and Geothermal Research, v. 37, p. 151-166.

Johnson, S.H. and Couch, R.W.
1970: Crustal structure in the north Cascade Mountains of Washington and British Columbia from seismic refraction measurements; Bulletin Seismological Society of America, v. 60, p. 1259-1269.

Kanasewich, E.R.
1966: Deep crustal structure under the plains and Rocky Mountains; Canadian Journal of Earth Sciences, v. 3, p. 937-945.
1968: Precambrian rift: genesis of strata-bound ore deposits; Science, v. 161, p. 1002-1005.

Kanasewich, E.R. and Chiu, S.K.L.
1985: Least-squares inversion of spatial seismic refraction data; Bulletin Seismological Society of America, v. 75, p. 865-880.

Kanasewich, E.R., Clowes, R.M., and McCloughan, C.H.
1968: A buried Precambrian rift in western Canada; Tectonophysics, v. 8, p. 513-527.

Kanasewich, E.R. and Cumming, G.L.
1965: Near vertical incidence seismic reflections from the "Conrad" discontinuity; Journal of Geophysical Research, v. 70, p. 3441-3446.

Kanasewich, E.R., Phadke, S., Savage, P.J., and Kelsch, W.L.
1985: Deep crustal reflection seismic profiles near Vulcan, Alberta; Canadian Society of Exploration Geophysicists - Canadian Geophysical Union, National Convention, Abstracts, p. 704.

Kasmierczak, Z.
1980: Seismic crustal studies in Saskatchewan; M.Sc. thesis, University of Alberta, Edmonton, 134 p.

Kornik, L.J.
1969: An aeromagnetic study of the Moak Lake - Setting Lake structure in northern Manitoba; Canadian Journal of Earth Sciences, v. 6, p. 373-381.
1971: Magnetic subdivision of Precambrian rocks in Manitoba; in Geoscience Studies in Manitoba, A.C. Turnock (ed.); Geological Association of Canada, Special Paper 9, p. 51-60.

Kornik, L.J. and MacLaren, A.S.
1966: Aeromagnetic study of the Churchill-Superior boundary in northern Manitoba; Canadian Journal of Earth Sciences, v. 3, p. 547-557.

Koster, F. and Baadsgaard, H.
1970: On the geology and geochronology of northwestern Saskatchewan. I. Tazin Lake region; Canadian Journal of Earth Sciences, v. 7, p. 919-930.

Kupsch, W.O.
1979: Boundary of the Canadian Shield; in History of Concepts in Precambrian Geology, W.D. Kupsch and W.A.S. Sarjeant (ed.); Geological Association of Canada, Special Paper 19, p. 119-131.

Langenberg, C.W.
1983: Polyphase deformation in the Canadian Shield of northeastern Alberta; Alberta Geological Survey, Bulletin 45, 33 p.

Lewry, J.F.
1981: Lower Proterozoic arc-microcontinent collisional tectonics in the western Churchill Province; Nature, v. 294, p. 69-72.

Lewry, J.F., Sibbald, T.I.I., and Rees, C.J.
1978: Metamorphic patterns and their relation to tectonism and plutonism in the Churchill Province in northern Saskatchewan; in Metamorphism in the Canadian Shield; Geological Survey of Canada, Paper 78-10, p. 139-154.

Lewry, J.F., Stauffer, M.R., and Fumerton, S.
1981: A Cordilleran-type batholithic belt in the Churchill Province in northern Saskatchewan; Precambrian Research, v. 14, p. 277-313.

Logan, W.E.
1854: Report of progress for the year 1852-53; Geological Survey of Canada, p. 5-74.

Lyons, P.L. and O'Hara, N.W.
1982: Gravity anomaly map of the United States exclusive of Alaska and Hawaii; Society of Exploration Geophysicists, 1:2 500 000.

MacLaren, A.S. and Charbonneau, B.W.
1968: Characteristics of magnetic data over major subdivisions of the Canadian Shield; Geological Association of Canada, Proceedings, v. 19, p. 57-65.

Macrides, C. and Kanasewich, E.R.
1984: Seismic refraction studies over the Williston Basin; Canadian Society of Petroleum Geologists - Canadian Society of Exploration Geophysicists National Convention, Calgary, Abstracts, p. 130.

Majorowicz, J.A., Jones, F.W., Lam, H.L., and Jessop, A.M.
1985: Regional variations of heat flow differences with depth in Alberta, Canada; Geophysical Journal of the Royal Astronomical Society, v. 81, p. 479-487.

Maureau, G.T.F.R.
1964: Crustal structure in western Canada; M.Sc. thesis, University of Alberta, Edmonton, 79 p.

McGrath, P.H. and Hildebrand, R.S.
1984: An estimate, based on magnetic interpretation, of the minimum thickness of the Hornby Bay Group, Leith Peninsula, District of Mackenzie; in Current Research, Part A, Geological Survey of Canada, Paper 84-1A, p. 223-228.

Meijer Drees, N.C.
1975: Geology of the Lower Paleozoic formations in the subsurface of the Fort Simpson area, District of Mackenzie, Northwest Territories; Geological Survey of Canada, Paper 74-40, 65 p.

Mereu, R.F. and Hunter, J.A.
1969: Crustal and upper mantle structure under the Canadian Shield from Project Early Rise data; Bulletin Seismological Society of America, v. 59, p. 147-165.

Mereu, R.F., Majumdar, S.C., and White, R.E.
1977: The structure of the crust and upper mantle under the highest ranges of the Canadian Rockies from a seismic refraction survey; Canadian Journal of Earth Sciences, v. 14, p. 196-208.

Meyer, R.P., Steinhart, J.S., and Bonini, W.E.
1959: Montana Crustal Structure; in Explosion Studies of Continental Structure, J.S. Steinhart and R.P. Meyer (ed.); Carnegie Institute of Washington Publications 622, Washington, D.C., p. 305-343.

Money, P.L., Baer, A.J., Scott, B.P., and Wallis, R.H.
1970: The Wollaston Lake belt, Saskatchewan, Manitoba, Northwest Territories; in Symposium on Basins and Geosynclines of the Canadian Shield; Geological Survey of Canada, Paper 70-40, p. 171-200.

Monger, J.W.H. and Price, R.A.
1979: Geodynamic evolution of the Canadian Cordillera - progress and problems; Canadian Journal of Earth Sciences, v. 16, p. 770-791.

Murchison, R.I.
1859: On the succession of the older rocks of the northernmost counties of Scotland; Quarterly Journal of Geological Society, v. XV, p. 353-418.

Nielsen, P.A.
1978: Metamorphism of the Arseno Lake area, Northwest Territories; in Metamorphism in the Canadian Shield; Geological Survey of Canada, Paper 78-10, p. 115-122.

Nielson, P.A., Langenberg, C.W., Baadsgaard, H., and Godfrey, J.D.
1981: Precambrian metamorphic conditions and crustal evolution, northeastern Alberta, Canada; Precambrian Research, v. 16, p. 171-193.

Officer, C.B. and Drake, C.L.
1982: Epeirogenic plate movements; Journal of Geology, v. 90, p. 139-153.

Peterman, Z.E.
1962: Precambrian basement of Saskatchewan and Manitoba; Ph.D. thesis, University of Alberta, Edmonton, 317 p.
1981: Dating of Archean basement in northeastern Wyoming and southern Montana; Geological Survey of America, Bulletin, Part 1, v. 92, p. 139-146.

Peterman, Z.E. and Goldich, S.S.
1982: Archean rocks of the Churchill basement, North Dakota; in 4th International Williston Basin Symposium; Saskatchewan Geological Survey, p. 11-12.

Peterman, Z.E. and Hedge, C.E.
1964: Age of basement rocks from the Williston basin of North Dakota and adjacent area; United States Geological Survey, Professional Paper 473D, p. D100-D104.

Rankin, D. and Pascal, F.
in press: The stability of results in the magnetotelluric method; Physics of the Earth and Planetary Interiors (I.U.G.S. Proceedings).

Richards, T.C. and Walker, D.J.
1959: Measurement of the thickness of the earth's crust in the Alberta Plains of Western Canada; Geophysics, v. 24, p. 262-284.

Robinson, J.E., Charlesworth, H.A.K., and Ellis, M.J.
1969: Structural analysis using spatial filtering in Interior Plains of south-central Alberta; American Association of Petroleum Geologists, Bulletin, v. 53, p. 2341-2367.

Rosholt, J.N., Peterman, Z.E., and Bartel, A.J.
1970: U-Th-Pb and Rb-Sr ages in granite reference sample from southwestern Saskatchewan; Canadian Journal of Earth Sciences, v. 7, p. 184-187.

Sawatzky, H.B. and Standing, K.F.
1971: Regional magnetic map of southern Saskatchewan; Saskatchewan Department of Mineral Resources, 1:506 880.

Shillibeer, H.A. and Burwash, R.A.
1956: Some potassium-argon ages for western Canada; Science, v. 123, p. 938-939.

Sims, P.K.
1980: Boundary between Archean greenstone and gneiss terranes in northern Wisconsin and Michigan; in Selected Studies of Archean Gneisses and Lower Proterozoic Rocks, Southern Canadian Shield, G.B. Morey and G.N. Hanson (ed.); Geological Society of America, Special Paper 182, p. 113-124.

Smith, T.J., Steinhart, J.S., and Aldrich, L.T.
1966: Lake Superior crustal structure; Journal of Geophysical Research, v. 71, p. 1141-1172.

Sprenke, K.F. and Kanasewich, E.R.
1983: Gravity modelling and isostasy in western Canada; Journal of Canadian Society of Exploration Geophysicists, v. 18, p. 49-58.

Sprenke, K.F., Wavra, C.S., and Godfrey, J.D.
1986: The geophysical expression of the Canadian Shield of northeastern Alberta; Alberta Research Council, Bulletin, 52, 54 p.

Stauffer, M.R.
1984: Manikewan: an early Proterozoic ocean in central Canada, its igneous history and orogenic closure; Precambrian Research, v. 25, p. 257-281.

Stockwell, C.H.
1963: Second report on structural provinces, orogenies, and time-classification of rocks of the Canadian Precambrian Shield; Geological Survey of Canada, Paper 62-17, p. 123-133.
1982: Proposals for time-classification and correlation of Precambrian rocks and events in Canada and adjacent areas of the Canadian Shield. Part 1. A time-classification of Precambrian rocks and events; Geological Survey of Canada, Paper 80-19, 135 p.

Stockwell, C.H., McGlynn, J.C., Emslie, R.F., Sanford, B.V., Norris, A.W., Donaldson, J.A., Fahrig, W.F., and Currie, K.L.
1968: Superior Province; in Geology and Economic Minerals of Canada, R.J.W. Douglas (ed.); Geological Survey of Canada, Economic Geology Series Report no. 1, 5th Edition, p. 54-71.

Thomas, M.D., Gibb, R.A., and Quince, J.R.
1976: New evidence from offset aeromagnetic anomalies for transcurrent faulting associated with the Bathurst and Macdonald faults, Northwest Territories; Canadian Journal of Earth Sciences, v. 13, p. 1244-1250.

United States Geodynamics Committee
1973: U.S. Program for Geodynamics Project; National Academy of Sciences, Washington, p. 1-235.

Walcott, R.I.
1968: The gravity field of northern Saskatchewan and northeastern Alberta; Gravity Map Series of the Dominion Observatory, p. 16-20.

Walcott, R.I. and Boyd, J.B.
1971: The gravity field of northern Alberta, and part of Northwest Territories and Saskatchewan; Gravity Map Series of the Earth Physics Branch, p. 103-111.

Wallis, R.H.
1970: A geological interpretation of gravity and magnetic data, northwest Saskatchewan; Canadian Journal of Earth Sciences, v. 7, p. 858-868.

Wanless, R.K.
1970: Isotopic age map of Canada; Geological Survey of Canada, Map 1256A, 1:5 000 000.

Weber, W. and Scoates, R.F.J.
1978: Metamorphism in the northwestern Superior Province and along the Churchill-Superior boundary, Manitoba; in Metamorphism in the Canadian Shield; Geological Survey of Canada, Paper 78-10, p. 5-16.

White, W.R.H. and Savage, J.C.
1965: A seismic refraction and gravity study of the earth's crust in British Columbia; Bulletin Seismological Society of America, v. 55, p. 463-468.

Williams, G.K.
1981: Middle Devonian carbonate barrier-complex of Western Canada; Geological Survey of Canada, Open File 761, 17 p.

Wilson, H.D.B.
1971: The Superior Province of Manitoba; in Geoscience Studies in Manitoba, A.C. Turnock (ed.); Geological Association of Canada, Special Paper 9, p. 41-50.

Wilson, H.D.B. and Brisbin, W.C.
1962: Tectonics of the Canadian Shield in northern Manitoba; Royal Society of Canada, Special Publication 4, p. 60-75.

Zietz, I., Bond, K.R., Gilbert, F.P., Kirby, J.R., Riggle, J.R., and Snyder, S.L.
1982: Composite magnetic anomaly map of the United States. Part A: Conterminous United States; United States Geological Survey, Map GP-954A, 1:2 500 000, and accompanying report.

Zurbrigg, H.F.
1963: Thompson Mine geology; Transactions of Canadian Institute of Mining and Metallurgy, v. LXVI, p. 227-236.

ADDENDUM

The Glennie Domain of north central Saskatchewan, identified in Figure 3.2 by question marks, is now inferred to be underlain by Archean basement (Bickford et al., 1988), with ages around 2.5 Ga. Lower Proterozoic cover rocks were thrust over the basement during the Hudsonian Orogeny. Basement rocks of the Glennie Domain are interpreted to be the northwestern edge of the Superior craton. The width of the adjacent Trans-Hudson Orogen is thus reduced from that shown in Figure 3.11.

The various identifiable units of the Precambrian basement beneath the Western Canada Basin have recently been more sharply delineated than was possible at the time of writing, thanks to the aeromagnetic data recently made available by various petroleum exploration companies (Ross et al., 1991).

Bickford, M.E., Collerson, K.D., and Lewry, J.F.
1988: U-Pb geochronology and isotopic studies of the Trans-Hudson Orogen in Saskatchewan; in Summary of Investigations 1988, Saskatchewan Geological Survey; Miscellaneous Report 88-4, p. 35-41.

Ross, G.M., Parrish, R.R., Villeneuve, M.E., and Bowring, S.A.
1991: Geophysics and geochronology of the crystalline basement of Alberta Basin, western Canada; Canadian Journal of Earth Sciences, v. 28, p. 512-522.

Authors' Addresses

R.A. Burwash
Department of Geology
University of Alberta
Edmonton, Alberta
T6G 2E3

A.G. Green
Geological Survey of Canada
601 Booth Street
Ottawa, Ontario
K0A 1E8

A.M. Jessop
Institute of Sedimentary and Petroleum Geology
Geological Survey of Canada
3303 - 33rd Street N.W.
Calgary, Alberta
T2L 2A7

E.R. Kanasewich
Department of Physics
University of Alberta
Edmonton, Alberta
T6G 2E3

Printed in Canada

Chapter 4

STRATIGRAPHY

Subchapter 4A. PROTEROZOIC SEDIMENTARY ROCKS
J.D. Aitken

Subchapter 4B. CAMBRIAN AND LOWER ORDOVICIAN - SAUK SEQUENCE
J.D. Aitken

Subchapter 4C. ORDOVICIAN AND SILURIAN
M.P. Cecile and B.S. Norford

Subchapter 4D. DEVONIAN
P.F. Moore

Subchapter 4E. CARBONIFEROUS
B.C. Richards, E.W. Bamber, A.C. Higgins, and J. Utting

Subchapter 4F. PERMIAN
C.M. Henderson, E.W. Bamber, B.C. Richards, A.C. Higgins, and A. McGugan

Subchapter 4G. TRIASSIC
D.W. Gibson

Subchapter 4H. JURASSIC
T.P. Poulton, W.K. Braun, M.M. Brooke, and E.H. Davies

Subchapter 4I. CRETACEOUS
D.F. Stott, W.G.E. Caldwell, D.J. Cant, J.E. Christopher, J. Dixon, E.H. Koster, D.H. McNeil, and F. Simpson

Subchapter 4J. TERTIARY
D.F. Stott, J. Dixon, J.R. Dietrich, D.M. McNeil, L.S. Russell, and A.R. Sweet

Subchapter 4K. QUATERNARY
A. MacS. Stalker and J.-S. Vincent

Subchapter 4A

PROTEROZOIC SEDIMENTARY ROCKS

J.D. Aitken

CONTENTS

INTRODUCTION

Over vast areas of the undeformed platform of southwestern Canada, the sedimentary cover of the ancient North American Craton is entirely of Phanerozoic age. In northwestern Canada, on the other hand, thick Proterozoic strata intervene between Phanerozoic deposits and crystalline basement (Fig. 4A.1). Even there, the preserved sediments contain only a fraction of the record of geological events that is preserved in the Foreland Belt of the Cordillera.

Precambrian crystalline rocks beneath the sedimentary cover of the Interior Platform are dealt with in a separate chapter (Burwash et al., Chapter 3 in this volume). This brief treatment is largely a condensation of two chapters ("Middle Proterozoic", and "Upper Proterozoic") of the companion volume, "Geology of the Cordilleran Orogen in Canada" (Gabrielse and Yorath, 1991); it considers not only the Proterozoic sedimentary record of the undeformed basins and platforms, but also the more complete record preserved in the Eastern Cordillera.

H. Gabrielse and M.E. McMechan have provided welcome assistance with fact and interpretation.

STRATIGRAPHY

Gross chronostratigraphic units

Proterozoic sedimentary rocks of the Western Canada Basin may be dealt with in terms of three unconformity-bounded, major sequences (Young, 1978; Young et al., 1979).

Sequence A: ~ 1.7 to ~1.2Ga

Sequence B: ~ 1.2 to ~0.8Ga

Sequence C: ~ 0.8 to ~0.57Ga

Sequences A, B, and C provide a useful organizational framework for descriptive and most interpretive purposes. However, the bounding ages remain subject to adjustment, and because of a paucity of radiometric age determinations,

Aitken, J.D.
1993: Proterozoic sedimentary rocks; Subchapter 4A in Sedimentary Cover of the Craton in Canada, D.F. Stott and J.D. Aitken (ed.); Geological Survey of Canada, Geology of Canada, no. 5, p. 81-95 (also Geological Society of America, The Geology of North America, v. D-1).

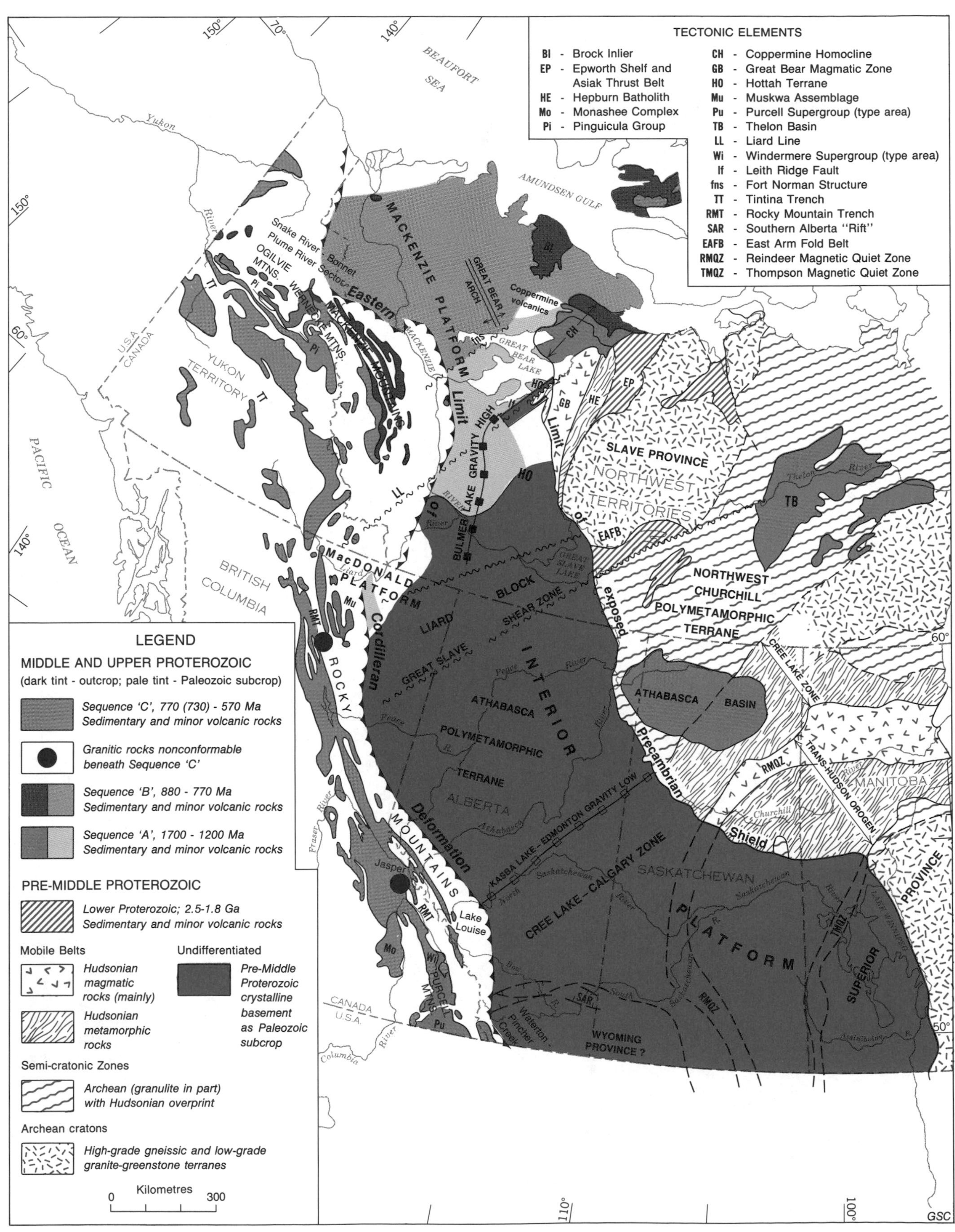
TECTONIC ELEMENTS
BI - Brock Inlier
EP - Epworth Shelf and Asiak Thrust Belt
HE - Hepburn Batholith
Mo - Monashee Complex
Pi - Pinguicula Group
CH - Coppermine Homocline
GB - Great Bear Magmatic Zone
HO - Hottah Terrane
Mu - Muskwa Assemblage
Pu - Purcell Supergroup (type area)
TB - Thelon Basin
LL - Liard Line
Wi - Windermere Supergroup (type area)
lf - Leith Ridge Fault
fns - Fort Norman Structure
TT - Tintina Trench
RMT - Rocky Mountain Trench
SAR - Southern Alberta "Rift"
EAFB - East Arm Fold Belt
RMQZ - Reindeer Magnetic Quiet Zone
TMQZ - Thompson Magnetic Quiet Zone
LEGEND
MIDDLE AND UPPER PROTEROZOIC
(dark tint - outcrop; pale tint - Paleozoic subcrop)
Sequence 'C', 770 (730) - 570 Ma
Sedimentary and minor volcanic rocks
Granitic rocks nonconformable beneath Sequence 'C'
Sequence 'B', 880 - 770 Ma
Sedimentary and minor volcanic rocks
Sequence 'A', 1700 - 1200 Ma
Sedimentary and minor volcanic rocks
PRE-MIDDLE PROTEROZOIC
Lower Proterozoic; 2.5-1.8 Ga
Sedimentary and minor volcanic rocks
Mobile Belts
Hudsonian magmatic rocks (mainly)
Hudsonian metamorphic rocks
Undifferentiated
Pre-Middle Proterozoic crystalline basement as Paleozoic subcrop
Semi-cratonic Zones
Archean (granulite in part) with Hudsonian overprint
Archean cratons
High-grade gneissic and low-grade granite-greenstone terranes
Kilometres
0
300
BEAUFORT SEA
AMUNDSEN GULF
PACIFIC OCEAN
YUKON TERRITORY
BRITISH COLUMBIA
NORTHWEST TERRITORIES
ALBERTA
SASKATCHEWAN
MANITOBA
U.S.A.
CANADA
MACKENZIE PLATFORM
GREAT BEAR ARCH
Coppermine volcanics
GREAT BEAR LAKE
GREAT SLAVE LAKE
Snake River
Plume River Sector
Bonnet
OGILVIE MTNS.
WERNECKE MTNS.
MACKENZIE MOUNTAINS
Eastern Limit of Cordilleran Deformation
BULMER LAKE GRAVITY HIGH
MacDONALD PLATFORM
LIARD BLOCK
GREAT SLAVE SHEAR ZONE
ATHABASCA POLYMETAMORPHIC TERRANE
INTERIOR PLATFORM
SLAVE PROVINCE
Limit of exposed Precambrian Shield
NORTHWEST CHURCHILL POLYMETAMORPHIC TERRANE
ATHABASCA BASIN
CREE LAKE ZONE
TRANS-HUDSON OROGEN
KASBA LAKE - EDMONTON GRAVITY LOW
CREE LAKE - CALGARY ZONE
ROCKY MOUNTAINS
Jasper
Lake Louise
PURCELL MTNS
Pincher Creek
Waterton
WYOMING PROVINCE ?
SUPERIOR PROVINCE
Thelon River
Peace River
Athabasca
Saskatchewan
Churchill
Fraser River
Columbia River
150°
140°
70°
60°
110°
100°
50°
GSC

it remains uncertain whether there is indeed overlap in depositional age between the various packages of strata assigned to a single sequence, especially within Sequence A.

Sequence A comprises the classical Purcell Supergroup of southeastern British Columbia; the Hornby Bay - Dismal Lakes succession of Coppermine Homocline; the Wernecke Supergroup of northern Yukon; and, less certainly, the Muskwa assemblage of northeastern British Columbia. Sequence B comprises the 'Mackenzie Mountains supergroup', known only from Mackenzie Mountains and northern Interior and Porcupine Platforms. Sequence C comprises the classical Windermere Supergroup of southeastern British Columbia and its demonstrated and assumed correlatives in the northern Cordillera.

Southern basin and platform

Throughout nearly all of the undisturbed southern Interior Platform (Fig. 4A.1), rocks of Middle Cambrian and younger age rest directly upon crystalline, Precambrian basement. Precambrian sedimentary rocks reached by drilling, although widespread north of the Liard line (taken here to separate northern and southern sectors) are confined to areas not far south of that line (Fig. 4A.1). North of Mackenzie River, these are sandstones and mudrocks clearly relatable to the Middle Proterozoic, Hornby Bay -Dismal Lakes succession of Coppermine Homocline (Meijer Drees, 1975; Aitken and Pugh, 1984). Southeast of Mackenzie River and northeast of Liard River, quartz arenites and coarse arkoses overlie basement over a broad swath of country (G. K. Williams, unpub. maps). These are undated, and may be of pre-Devonian, Paleozoic or Precambrian age. A little farther south, a single borehole, on passing through Silurian strata, encountered reddish dolomites assignable with some confidence to the Muskwa assemblage (Sequence A) that is exposed in the mountains to the west. Elsewhere, strata of Middle Cambrian and younger age rest on crystalline rocks belonging mainly to the Churchill Province of the Canadian Shield, but in the extreme south and southeast, to the Wyoming and Superior Provinces (Burwash et al., Chapter 3 in this volume).

The unconformity at which Cambrian strata rest upon rocks crystallized at crustal depths no later than Middle Proterozoic (1700 Ma) is profound. It is best appreciated by considering the formations that are missing there, through a combination of non-deposition and erosion, but represented in the Rocky Mountains to the west. A brief summary of these formations follows.

Figure 4A.1. Summary map showing areas of Proterozoic outcrop, principal elements of the Canadian Shield (adapted from Burwash et al., Chapter 3 in this volume), and Paleozoic subcrop in the undeformed platforms.

Classical Purcell Supergroup

The oldest, unmetamorphosed, sedimentary and volcanic strata known from the Rocky Mountains are assigned to the widely exposed and well studied Purcell Supergroup (Price, 1964; McMechan, 1981). These rocks have been penetrated by drilling only in the Waterton - Pincher Creek area of southwestern Alberta, where they are exposed in the Lewis thrust sheet and are drilled through to reach prolific Carboniferous and Devonian gas reservoirs beneath the thrust. Basement is exposed in contact with the Purcell (Belt) only at a few localities in Montana, near the eastward limit of the supergroup. Regionally, however, northwest-trending Purcell facies and isopach trends clearly cross-cut the north-to-northeast structural grain of the Churchill Province (McMechan, 1981). "Hudsonian" dates from the Churchill Province provide a maximum age of about 1700 Ma for the basal Purcell, and an intrusion dated at about 1300 Ma (Rb-Sr, whole-rock) provides an apparent younger limit (McMechan and Price, 1982). Rb-Sr, whole-rock dating of argillaceous sediments suggests that Purcell sedimentation extended from 1500 to 850 Ma, while paleomagnetic studies suggest a span of 1500 - 1250 Ma. Dating of the Purcell and sources of data were fully reviewed by McMechan and Price (1982).

The Purcell Supergroup forms a northeast-tapering wedge of sedimentary and minor volcanic rocks with the characteristics and thickness (up to 20 km) of a continental-margin succession. Evidence for a "two-sided", and thus intracratonic Belt-Purcell Basin, possibly an aulacogen, is being brought forward increasingly, however (Harrison et al., 1974; Winston et al., 1984; and papers by Winston, Grotzinger, and Slover and Winston in Roberts, 1986). McMechan and Price (1982) concluded that the Purcell was folded and cleaved in pre-Windermere time, but that view is by no means universally accepted.

The basal division consists in the southwest of thick (up to 4.2 km) clastic rocks of basinal and turbiditic character (Aldridge Formation), but in the northeast contains major intercalations of carbonate strata (Haig Brook, Tombstone Mountain, Waterton, and Altyn formations; Fermor and Price, 1983).

The lower division (not basal) of the Purcell consists mainly of clastic sedimentary rocks laid down in shallow water. Its thickness exceeds 2400 m in Canada, and reaches 4500 m in the U.S.A. It comprises the Creston and lower Kitchener formations in the west, and the Appekunny and Grinnell formations (Fig. 4A.2) in the east and along the southern front of the Rocky Mountains.

The middle (carbonate) division – middle Siyeh Formation – of the classical Purcell is of platformal aspect (Fig. 4A.2), as thin as 300 m along its northeastern limit of exposure, the Rocky Mountain front near the 49th parallel, and of more basinal aspect elsewhere. In the central and northern Purcell Mountains, and in the Hughes and Lizard Ranges of the Rockies, both facies are represented in the upper unit of the Kitchener Formation (1400 m).

Detrital, carbonate and, locally, volcanic rocks make up the upper division of the Purcell, up to 3 km thick in Canada. The central and eastern exposures share a common succession (upper Siyeh, Purcell Lava, Sheppard, Gateway, Phillips, and Roosville formations) of shallow-water to subaerial, essentially detrital rocks with one volcanic unit. In western and northern exposures, the division is thicker, contains more carbonate strata, and lacks the distinctive marker units (Purcell and Nicol Creek volcanic rocks; distinctive red quartzites of the Phillips Formation) that permit finer divisions in the south and east. The units there are the newly named Coppery Creek and La France groups (J.E. Reesor, unpub. rep.) and the overlying Mount Nelson Formation.

Northeastward thinning, coupled with changes to shoreline and subaerial facies observed in the mountains, forecasts the northeastward pinchout of the classical Purcell Supergroup. The Purcell is last seen in outcrop, unconformably overlain by Middle Cambrian Flathead Sandstone, in the Lewis thrust sheet at the mountain front. There, it is also penetrated by wells targeted on Carboniferous and Devonian gas reservoirs beneath the thrust. The rocks of the thrust sheet have undergone large-scale displacement from their site of deposition. Bally et al. (1966) speculated that a wedge-shaped feature, detected by seismic reflection profiling directly west of the southern Rocky Mountain Trench, may be the eastern pinchout of the autochthonous Purcell. The westernmost wells in the undeformed platform that penetrate the complete Cambrian section find it resting on crystalline basement, undoubtedly far east of the Purcell pinchout.

Muskwa Assemblage

The older (pre-Windermere) Proterozoic succession of the Muskwa Range of northeastern British Columbia (Fig. 4A.1) resembles lithologically the classical Purcell Supergroup and parts of the Hornby Bay - Dismal Lakes succession (Sequence A), and parts of the 'Mackenzie Mountains supergroup' (Sequence B). It is regarded here as belonging to Sequence A, because it underwent an episode of pre-Windermere folding (Preto, 1971). Like the classical Purcell, its exposures lie mainly "inboard" of the belt of Windermere strata, and its known extent along the Cordillera is limited.

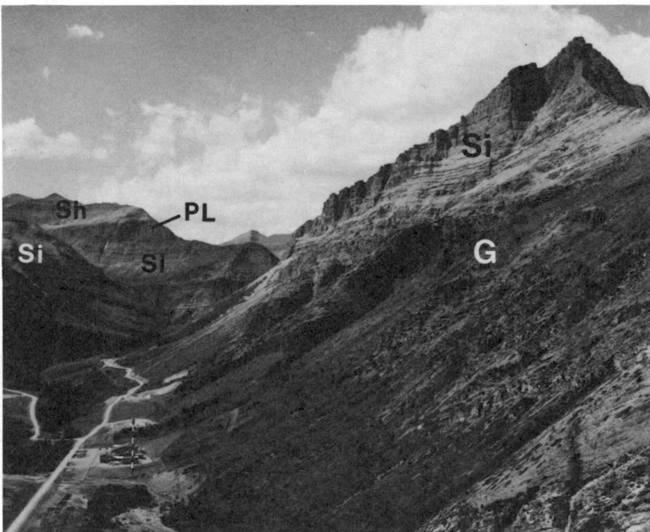

Figure 4A.2. Purcell strata of the Lewis thrust-sheet, Waterton Gas Field, Alberta. G - Grinnell Formation; Si - Siyeh Formation; PL - Purcell Lava; Sh - Sheppard Formation. The gas-production facilities in the valley bottom treat gas produced from a Carboniferous reservoir reached by drilling through the thrust-sheet (photo-Shell Canada Ltd.).

The Muskwa assemblage is a succession of seven formations, with a thickness totalling over 6 km. In upward order, these are Chischa, Tetsa, George, Henry Creek, Tuchodi, Aida, and Gataga (Bell, 1968; Taylor and Stott, 1973). They are clastic and carbonate strata mainly of deeper water origin, interrupted by two shallow-water formations. In contrast with other successions assigned to Sequence A, an overall, deepening-upward trend is apparent. The assemblage extends to the mountain front at 58 to 59 degrees north, but must thin rapidly eastward. It has been penetrated by a single well in the undeformed platform, all others in that latitude having encountered crystalline basement below Paleozoic sediments.

The Chischa, at least 940 m thick, with base concealed, is mainly shallow-water dolomite, with some orthoquartzite beds. The Tetsa (about 320 m thick) rests on the Chischa with erosional contact. It consists of deepwater mudstone, siltstone, and shale, in part carbonaceous, with turbiditic sandstones near the base. The George Formation (360-530 m) is composed of limestones that may be in part peritidal, but various kinds of resedimented beds suggest a slope or deeper water setting, albeit within the photic zone for the stromatolitic beds.

The Henry Creek Formation (about 460 m thick) comprises grey, largely carbonate-bearing mudstone with an upward-increasing content of sandstone beds. The formation lacks indicators of shallow-water deposition. The Tuchodi (about 1500 m) is characterized by crossbedded quartzites, with varying amounts of argillaceous dolomite, siltstone, and shale. The shales are black to dark grey and greenish grey, and in the east, partly red. Bell (1966) illustrated facies of the Tuchodi Formation that form a mosaic in both the geographic and stratigraphic senses. In a generally east-to-west transect, these facies are: fluvial/deltaic, mud flat/lagoonal, barrier beach/offshore bar, and offshore, shallow marine.

The Aida Formation (1200-1800 m) comprises silty, sandy, and calcareous, pale to medium grey mudstone and siltstone, with minor sandstone and limestone. A member rich in chamosite and a carbonaceous member are noteworthy. Much of the formation bears abundant evidence of turbiditic deposition and of mass movement on a depositional slope. The Gataga Formation (more than 1200 m) is dark grey to black slate, with minor silty and sandy mudstone, sandstone, and siltstone, and continues the deep-water, turbiditic aspect of the Aida.

Virtually all of the exposed Muskwa assemblage lies east of the preserved Windermere succession, which at that latitude is very thick and poorly studied and includes thick diamictites at and near its base. The Muskwa undergoes rapid, eastward bevelling at subsequent unconformities, seen as sub-Cambrian in part of the area and sub-Silurian elsewhere.

Classical Windermere Supergroup

The Windermere Supergroup extends along the Cordillera in virtually continuous outcrop, from near latitude 48°N at the Idaho-Washington boundary to near 60°N in northern

British Columbia (Fig. 4A.1). West and locally east of the southern Rocky Mountain Trench, the Windermere Supergroup intervenes between the Purcell Supergroup and basal Cambrian deposits. Its trend, north-northwest, is more northerly than that of the underlying Purcell, so that, regionally, the base of the Windermere cross-cuts the classical Purcell, to lie on crystalline basement in the central Rockies (Evenchick et al., 1984). The Windermere is missing from the Lewis thrust sheet, and there is no evidence that it was deposited anywhere as far east as the Rocky Mountain front. It is significant here as another of the first-order depositional events at and near the continental margin that are nowhere recorded in the subsurface of the platforms.

Questionable K-Ar dates on Windermere volcanic rocks from southeastern British Columbia (reviewed by Devlin et al., 1985) earlier suggested initiation of Windermere rifting, vulcanism, and sedimentation at 800 - 900 Ma. These dates are now highly suspect (Devlin et al., 1985); the 780 Ma to post-730 Ma Rb-Sr and U-Pb dates derived from northern areas (see below) appear more pertinent to the age of rifting ('Goat River orogeny'; McMechan and Price, 1982). A new, Sm-Nd date on pyroxene dates the basal Windermere, Huckleberry-Leola volcanics at 762 ± 40 Ma (W. Devlin, pers. comm., 1986). The probability of diachronous onset of rifting is acknowledged. New, U-Pb dates of about 730 Ma on zircons from granitic rocks nonconformably underlying Windermere strata at two localities adjacent to the Rocky Mountain Trench north of the latitude of Jasper (reviewed by Evenchick et al., 1984) provide a maximum age for the Windermere at those localities. The recent discovery of fossils of Ediacaran medusae in uppermost Miette Group rocks west of Jasper (Hofmann et al., 1985) demonstrates that Windermere deposition extended into latest Proterozoic time; nevertheless, a marked unconformity between Windermere and Cambrian strata has been demonstrated at many localities.

The classical Windermere (up to 9 km thick) records deposition of alternating coarse-grained and fine-grained, largely deep-water, clastic sediments in an active rift environment, interrupted in most areas by one or more episodes of carbonate sedimentation (Stewart, 1972; Lis and Price, 1976; Poulton and Simony, 1980; Teitz and Mountjoy, 1985).

Near the 49th parallel, diamictites (Toby Conglomerate) for which Aalto (1971) and Eisbacher (1981) endorsed a glacial origin, and overlying, basic volcanics (Irene) are present at the base of the succession and unconformable on strata of the classical Purcell. From the Jasper transect northward in the Rocky Mountains, amphibolite (meta-volcanics) appears here and there at or near the base of the Windermere column; diamictite (in part, but not necessarily entirely glaciogene) is common but not ubiquitous in the lower and middle parts of the succession.

In the type area of the Horsethief Creek Group, just west of the Rocky Mountain Trench (also the type area for the Windermere Supergroup), the basal tillite is again unconformable on Purcell strata, but volcanic rocks are missing (Walker, 1926). Northward plunge takes the base of the supergroup from view. One hundred kilometres north of the type area, the group, some 2000 m thick with base unknown, comprises a lower feldspathic grit unit, a middle slate, a limestone unit (not everywhere present), and an upper slate and quartzite unit (Poulton and Simony, 1980). The contact with overlying, Lower Cambrian quartzites is inferred to be unconformable, as it is in the Rocky Mountains (Devlin and Bond, 1988).

Horsethief Creek equivalents in the Rocky Mountains from about latitude 55°N southward are the Miette Group. On the Bow River transect, the base is unknown; exposed strata are a lower, Corral Creek Formation and an upper, Hector Formation (Aitken, 1969). The Corral Creek consists of grey and brown, commonly laminated slate, with a highly variable content of turbiditic, feldspathic sandstone. The Hector is largely similar to the Corral Creek, but contains a basal marker member of purple and green slate, pink, laminated, silty limestone (commonly incorporated in debris-flow breccia), and cobble conglomerate of debris-flow origin. A basic dyke and a sill (undated) intrude the Miette but not the overlying Lower Cambrian strata. Arnott and Hein (1986) have demonstrated the existence, near Lake Louise, of a submarine canyon largely filled with coarse-grained turbidites. The Hector is overlain by Lower Cambrian quartzites, at a contact that is strongly unconformable but only locally observably angular (Aitken, 1969).

The Miette Group, at and near its type area on the Athabasca River transect, is dealt with in terms of lower, middle, and upper divisions. The lower division rests nonconformably on "Hudsonian" (including reworked Archaean) Malton gneisses intruded by granite dated at about 741 Ma (U-Pb, zircon; Evenchick et al., 1984). More than 380 m thick, it consists of coarse-grained, turbiditic, gritty, feldspathic sandstone and granule conglomerate, finely crystalline, carbonaceous black limestone, silty and sandy limestone, and silty black slate. The Middle Miette, as much as 3000 m thick, is characterized by interlayered composite grit and sandstone beds in thick units, separated by silty dark green to grey pyritic slate (Fig. 4A.3). Local pebbly mudstone with limestone clasts may be equivalent to thick diamictites to the north. The Upper Miette, about 1800 m thick, consists mainly of grey to dark grey mudstone and silty argillite. Fine sandstone and siltstone

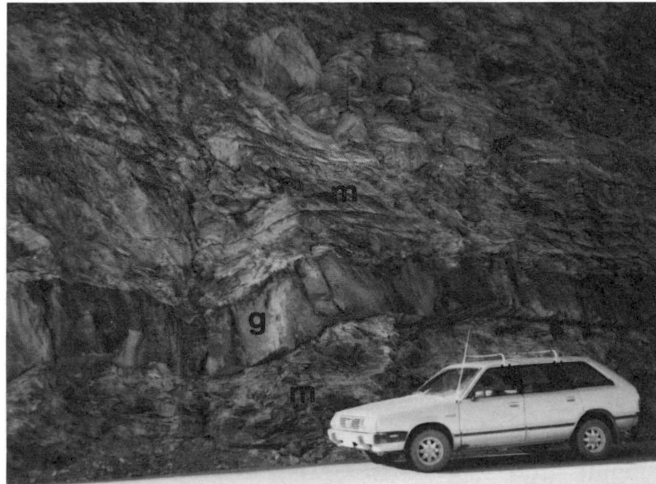

Figure 4A.3. Typical Windermere turbiditic grit (g) - deepwater mudrock (m) association, Miette Group, west of Jasper, Alberta (photo - B. Gadd).

interbeds in the lower and upper parts coarsen eastward. Lenticular interbeds (channelled?) of coarse, immature, partly conglomeratic sandstone are sparse but widespread. Here and there, the uppermost part of the Miette Group includes limestone, sandy limestone and dolomite, mudstone, and an upper sandstone, possibly in part correlative with carbonates in the upper Miette and Misinchinka groups to the north. Locally, near the Athabasca River transect, the Upper Miette is capped by platformal carbonate members up to 400 m thick (Fig. 4A.4); Ediacaran fossils occur in close association (Teitz and Mountjoy, 1985; Hofmann et al., 1985). A karstic erosion surface separates these carbonates from overlying, presumably lower Lower Cambrian quartzites.

The easternmost exposed Windermere (Miette Group) north of latitude 54°N is important in that it contains a record of volcanic activity other than basal. It comprises a unit (400+ m) at the unexposed base, of laminated, green siltite, with quartz-feldspar sandstone believed to be reworked, crystal tuff (McMechan and Thompson, 1985). An overlying unit is 300 to 900 m of diamictite with clasts ranging in size from granules to boulders (Fig. 4A.5). The clasts include extrusive rocks and felsic intrusive rocks, as well as a variety of sedimentary rocks. The next unit, about 300 m thick, consists of greenish quartz-feldspar sandstone, possibly redeposited crystal tuff, with thin beds of silty and rusty argillite. It contains olistostromal blocks of dolomite up to 400 m long. An overlying carbonate unit up to 300 m thick, and overlying rusty argillite are preserved beneath the sub-Cambrian unconformity in part of the area.

Windermere equivalents (Misinchinka Group) in the northern Rocky Mountains near the Peace River transect (latitude 56°N) are regionally metamorphosed, with metamorphic grade increasing from east to west. More than 3000 m of strata nonconformably overlie potassic granite gneiss basement dated at about 728 Ma (Evenchick et al., 1984). Basal quartzite-pebble conglomerate and quartzite are overlain by 400 m of amphibolite interbedded with quartzite and pelitic schist. These are overlain by about 400 m of pelitic schist, quartzite, marble, and amphibolite. Quartzite-cobble diamictite intervenes locally between the latter unit and overlying amphibolite, schist, marble, grit, dolomite, phyllite, and diamictite 300 m thick. The next-higher unit, 150 m of marble, limestone, and dolomite, is one of the most conspicuous units in the Upper Proterozoic succession. The uppermost strata of the Misinchinka in the region include two prominent but discontinuous limestone members 10 to 30 m thick, with an intervening diamictite unit, overlain by about 400 m of dolomitic sandstone and a thick sequence of interbedded grey and green phyllite and fine-grained sandstone (Evenchick, 1982). Opinion is divided as to whether the northern Rocky Mountain diamictites are glaciogene or not; arguments for debris-flow origin are about as convincing as those for glacial influence.

The Upper Proterozoic and Lower Cambrian 'Grit unit' of Selwyn Basin in southern and central Yukon Territory has been restudied by Gordey (in press), and named Hyland Group. These strata can be traced southward into northern Rocky Mountains. Diagnostic strata are members of feldspathic, quartz-pebble conglomerate, and purple slate. In many places, the top is marked by archaeocyathid-bearing, limestone-cobble conglomerate, quartz-pebble conglomerate, and purple slate.

The easternmost rocks of Windermere age in northern Rocky Mountains are more than 500 m of diamictite. Light-coloured dolomite clasts, in part striated, and lenses of dolomite breccia are conspicuous. The matrix is mudstone. Eisbacher (1981) inferred a relation to glaciation, while Fritz (1980b) recognized the possibility of a glacial origin but left the question open.

Figure 4A.4. Uppermost Proterozoic and Lower Cambrian strata at Mount Fitzwilliam, west of Jasper. Platformal carbonates (pale) of upper Miette Group are overlain at a karst surface by sandstones (dark) of McNaughton Formation (photo - M. Teitz).

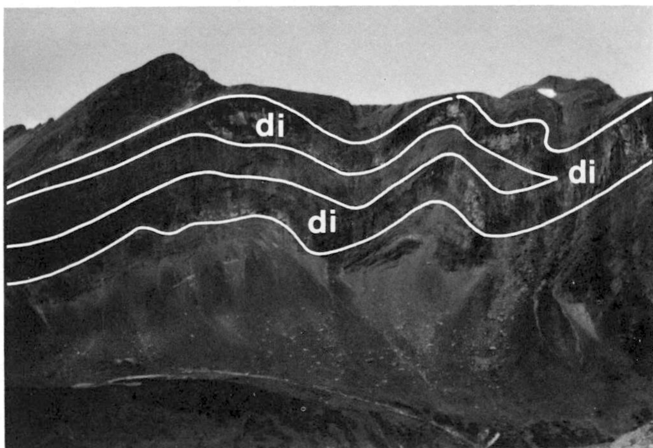

Figure 4A.5. Windermere strata 200 km northwest of Jasper. Mudrocks with thin, turbiditic sandstones are interrupted by thick, massive, cliff-forming diamictite units (di). Near-vertical fractures follow slaty cleavage (photo - M.E. McMechan, ISPG 2423-1).

Northern basins and platforms

In the deformed and undeformed Mackenzie Platform, Richardson Trough, and Porcupine Platform, basal Paleozoic strata that are generally Lower Cambrian overlie Precambrian sedimentary rocks nearly everywhere (Fig. 4A.1, 4A.6). Nevertheless, the Middle and Late Proterozoic record is punctuated with profound erosional and depositional gaps.

Hornby Bay - Dismal Lakes Succession

The oldest, unmetamorphosed, sedimentary rocks in Mackenzie Platform belong to the Hornby Bay - Dismal Lakes succession of the Coppermine Homocline (Kerans et al., 1981). The succession rests nonconformably on Lower Proterozoic (Aphebian) plutonic, volcanic, and sedimentary rocks. Sedimentation began earlier than 1660 Ma (Bowring and Ross, 1985), and was complete before the eruption of the Coppermine basalts, whose age is 1270 Ma (LeCheminant et al., 1990).

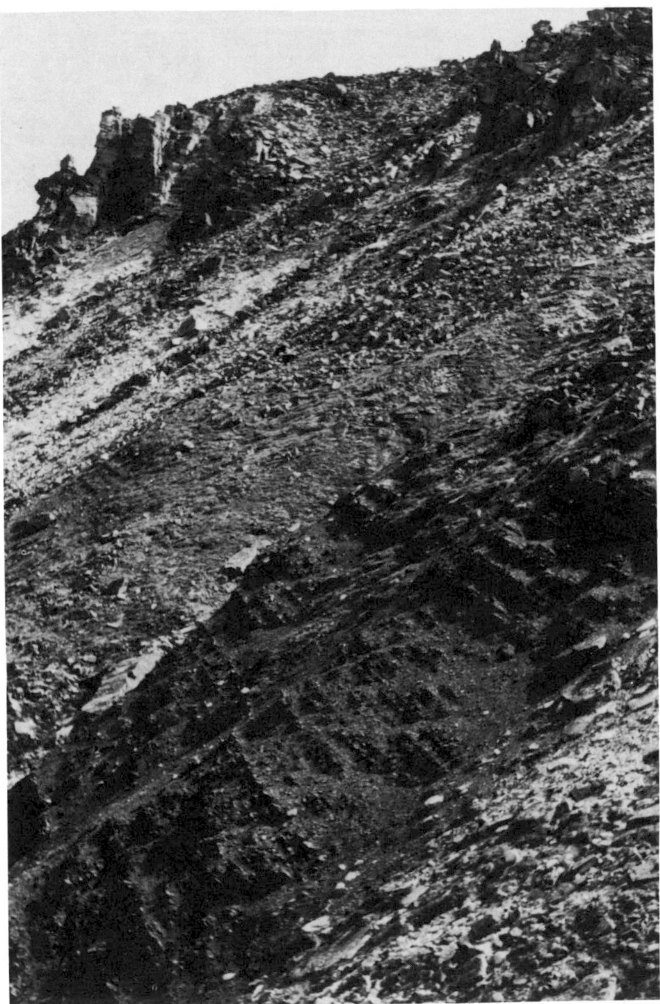

Figure 4A.6. Angular unconformity between shale and sandstone of Lone Land Formation (Dismal Lakes equivalent) and *Skolithos*-bearing sandstone of Mount Clark Formation (Lower Cambrian), Cap Mountain, District of Mackenzie (photo - J.D. Aitken, ISPG 2491-2).

The subsurface distribution of the Hornby Bay - Dismal Lakes as the subcrop to Phanerozoic strata extends only from a depositional/erosional zero-edge not far southeast of the Liard line to the Precambrian Fort Norman structure (fault). Northwest of that structure, various formations of the younger 'Mackenzie Mountains supergroup' form the subcrop. The Hornby Bay Group was deposited in a complex extensional basin. Faults were reactivated from time to time, and one local episode of volcanic activity is recorded. The Hornby Bay Group, up to 3 km thick, comprises largely red, fluvial, feldspathic sandstones, conglomerates, and mudrocks, with a significant formation of marginal- marine, carbonate strata in the upper third.

Dismal Lakes Group, more than 1 km thick, succeeds the Hornby Bay unconformably in some places and conformably in others. Deposited on a north- and west-facing shelf, it comprises mainly fluvial siliciclastic strata in its lower third, and mainly marine, carbonate strata in its upper two thirds.

The Hornby Bay - Dismal Lakes succession can be speculatively correlated with the Wernecke Supergroup of Wernecke and Ogilvie Mountains in the northern part of Yukon Territory. If the correlation is valid, the Hornby Bay - Dismal Lakes is the proximal, and the Wernecke (over 14 km thick; Delaney, 1981) the distal, continental-margin part of the depositional system. The Wernecke Supergroup underwent Precambrian folding ('Racklan') that pre-dates the 'Mackenzie Mountains supergroup' in the region.

Deep, seismic-reflection profiles of Mackenzie Platform north of Franklin Mountains (Cook, 1988a, b) reveal, unconformably beneath the undeformed, Upper Proterozoic, 'Mackenzie Mountains supergroup', a kilometres-thick succession of stratified rocks, plausibly correlated (by proximity, stratigraphic position, and seismic response), with the Hornby Bay - Dismal Lakes - Coppermine Group succession of Coppermine Homocline (Fig. 4A.7). The sub- 'Mackenzie Mountains' rocks display large-scale thrust-repetitions and long-wavelength folds (Cook, 1988a, b). Given that thick, stratified rocks (Wernecke?) are inferred on structural grounds to underlie the 'Mackenzie Mountains supergroup' in Mackenzie Mountains (Aitken et al., 1982), the stratigraphic order and the deformation ('Racklan'?) add weight to the suggested correlation between the Wernecke Supergroup and the lower part of the Coppermine Homocline succession.

Wernecke Supergroup

The Wernecke Supergroup is known only from the Wernecke and Ogilvie Mountains of northern Yukon. As described by Delaney (1981), it is over 14 km thick and has the aspect of a passive-margin succession, with two lower groups, Fairchild Lake and Quartet, consisting mainly of rather fine-grained clastic strata, and an upper, carbonate-dominated, Gillespie Lake Group. The upward facies progression thus has parallels in the classical Purcell and Hornby Bay - Dismal Lakes successions.

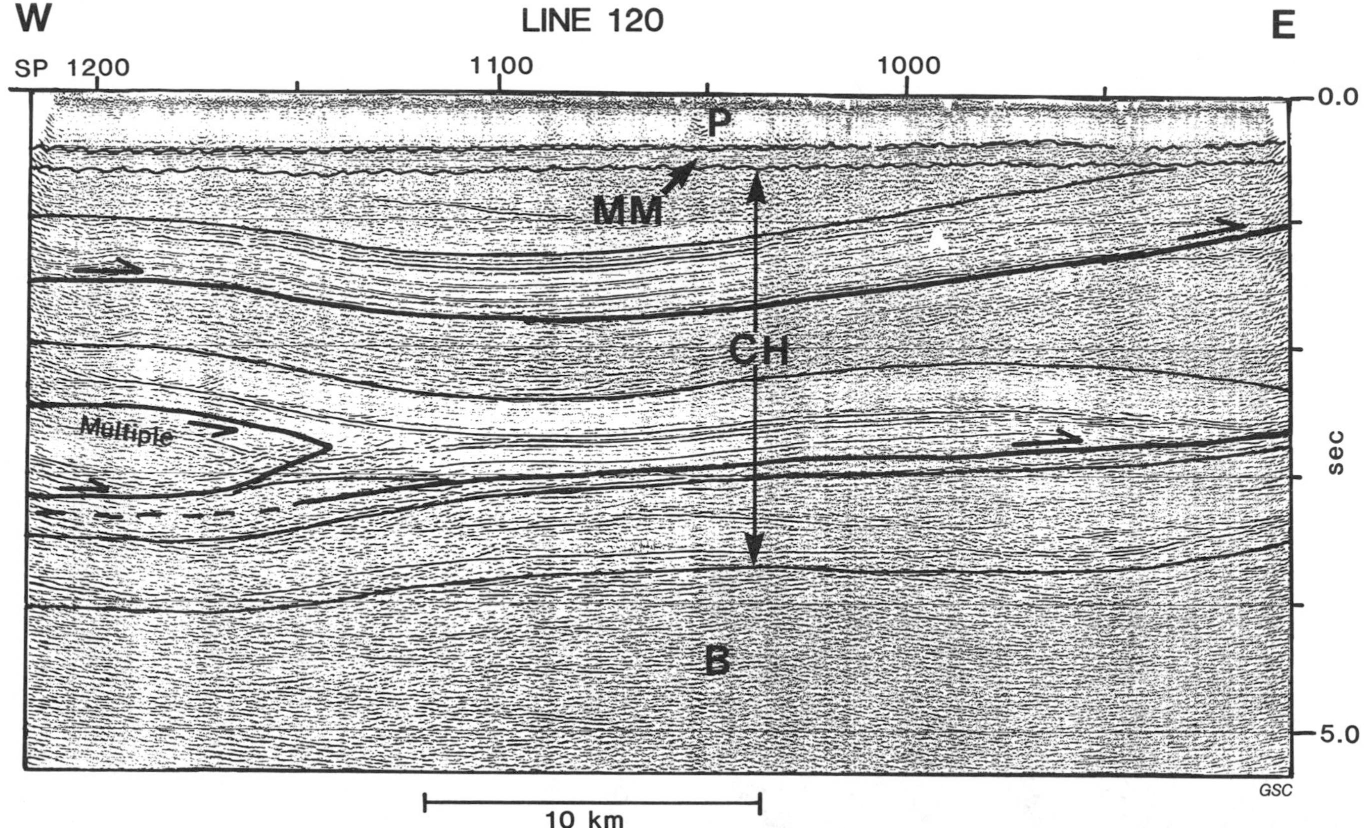

Figure 4A.7. Interpretation of an east-west, seismic reflection profile (migrated) of part of Mackenzie Platform northwest of Great Bear Lake; P - Paleozoic; MM - Mackenzie Mountains Supergroup; CH - layered, Proterozoic rocks assigned to Coppermine Homocline succession; B - crystalline basement. Light lines delimit seismically distinguishable units; heavy lines are interpreted thrust faults. Note the thrustwedge geometry at left, and at right, a footwall cutoff at 1 s, and a hanging-wall cutoff at 2.7 s. There is no vertical exaggeration for a velocity of 5.3 km/s (after Cook, 1988a).

Folds and cleavage in the Wernecke Supergroup are truncated by the base of the Pinguicula Group (see below). The Pinguicula is at youngest of early Windermere age, and the folding ('Racklan Orogeny'), therefore, is pre-Windermere. If suggestions that the Pinguicula Group correlates with the 'Mackenzie Mountains supergroup' (Young et al., 1979; Eisbacher, 1981) are correct, then the folding is pre-'Mackenzie Mountains'. This timing is suggested also (but not proved) by the fact that the 'Mackenzie Mountains supergroup', where exposed and where observed in seismic-reflection profiles (Cook, 1988a, b) has not experienced pre-Windermere folding.

Pinguicula Group

Answers to several problems of northern Cordilleran geology may lie with the little-studied and enigmatic Pinguicula Group. The typical, thickest, and most complete section of the Pinguicula is exposed in a limited area of eastern Wernecke Mountains. Thinner and less complete sequences, mainly of carbonate rocks overlying the Wernecke Supergroup in Ogilvie Mountains, are currently tentatively correlated with the Pinguicula

Figure 4A.8. Little Dal Group ('Mackenzie Mountains supergroup'), central Mackenzie Mountains. Basinal assemblage: b - basinal rhythmites; r - reef; g - Grainstone formation; gy - Gypsum formation (overlain with slight angular unconformity by Upper Cambrian strata, immediately left of photo); F - fault, occupied by diabase dyke. Quartzites of Katherine Group are covered at lower right (photo - J.D. Aitken, ISPG 1055-2).

Group. Eisbacher (1981) has provided a brief description of the 3 km-thick succession in the eastern Wernecke Mountains. He divided the Pinguicula into six informal units, A to F, of which most are carbonate formations of platformal aspect. Unit E consists of quartzite, siltstone, and dolomite, and Unit F, at the top, is particulate limestone of deeper water aspect. Volcanic rocks form the basal unit in the southern part of the area of exposure. All units thicken southward.

The base of the Pinguicula Group truncates large-scale folds in the underlying Wernecke Supergroup, and its top is overlain by conglomerate of the Rapitan Group at a contact that is locally angular. Majority opinion as published (e.g., Eisbacher, 1981; Young et al., 1979) favours correlation of the Pinguicula Group with the 'Mackenzie Mountains supergroup', but the similarities of rock-type and sequence are far from compelling. An idea at least equally attractive is correlation of the Pinguicula Group with the early Windermere 'Copper cycle' (Aitken, 1981; Coates Lake Group of Jefferson and Ruelle, 1986), as suggested earlier by Eisbacher (l978b). It may be noted that the surface distribution of the Rapitan Group at the Mackenzie-Wernecke-Richardson Mountains junction has a spur of very thick Rapitan strata pointing into the axis of Richardson Trough from the south. This suggests an early rift-depression precursory to the Paleozoic trough. The only known occurrence of thick Pinguicula Group rocks is in the same area (Fig. 4A.1), suggesting a common tectonic control. By this reasoning, the Pinguicula would be of early Windermere age and a potential 'Copper cycle' correlative.

'Mackenzie Mountains supergroup'

The 'Mackenzie Mountains supergroup' is known, from relationships in Coppermine Homocline, to overlie the Coppermine lavas, dated at 1270 Ma (LeCheminant et al., 1990). The end of an apparent polar wander path based on study of the paleomagnetism of the supergroup apparently post-dates the 'Grenville loop' and suggests a maximum age of 880 Ma (Park and Aitken, 1986). In the Mackenzie Mountains, the supergroup is cut by basic intrusions dated at about 770-780 Ma (Rb-Sr, whole rock; Armstrong et al., 1982). Four kilometres of sedimentary and minor volcanic rock was thus laid down during a period for which there is, apparently, no record in the southern Canadian Cordillera. In the subsurface Mackenzie Platform, in the frontal Mackenzie Mountains, and over parts of Porcupine Platform, the supergroup forms the subcrop to lower Paleozoic, largely Lower Cambrian strata (Fig. 4A.1). In the interior ranges of the Mackenzie Mountains, formations correlated with the Windermere Supergroup intervene. The 'Mackenzie Mountains supergroup' is notable for its consistency of lithofacies; although modest southwestward thickening is observed across Mackenzie Platform, individual formations have been traced from Mackenzie Mountains through the subsurface to outcrop in Brock Inlier, and further to outcrop as the Shaler Group of Victoria Island, in the Arctic Archipelago (Aitken et al., 1978, Young et al., l979). All formations are of platformal character throughout this great region. In Mackenzie Mountains, isopachs are arcuate and congruent with the structural arc, and display southwestward thickening (Aitken and Long, 1978), but this suggestion of a relationship to a continental margin lacks supporting evidence. Young (1981) suggested that the 'Mackenzie Mountains supergroup' may have been deposited in a "two-sided" basin.

Known stratigraphy of the supergroup begins with sandstone and mudrocks of unknown thickness (D.C. Pugh, unpub. rep.). These are overlain by grey, cherty, stromatolitic dolomite, corresponding to a member of the Glenelg Formation of Victoria Island and a unit of the type Rae Group of Coppermine Homocline. In Mackenzie Mountains, vugs in this dolomite (Map-unit H1, 400 m) contain pyrobitumen. Next are mudrocks and subordinate sandstone and carbonates of the Tsezotene Formation (750-1500 m) and correlatives, then a thick succession (700-1300 m) of orthoquartzite with minor mudrocks and carbonates, the Katherine Group (Aitken et al., 1978). The supergroup is capped by the lithologically varied, 2-km-thick, Little Dal Group (Fig. 4A.8, 4A.9, 4A.10), comprising platformal and basinal limestone and dolomite with local pyrobitumen, thick anhydrite, and gypsum, and subordinate mudrocks and sandstone (Aitken, 1981). Flows of basaltic lava are preserved locally at the top. Overlying strata are currently correlated with the Windermere Supergroup; they are in some structures the 'Copper cycle' or Coates Lake Group of Jefferson and Ruelle (1986) and in others, one or other of the formations of the Rapitan Group. The 'Mackenzie Mountains supergroup', which clearly pre-dates the rise of Mackenzie Arch, is progressively bevelled northeastward at subsequent unconformities, so that the youngest unit preserved in the undeformed part of Mackenzie Platform is the Katherine Group (Pugh, 1983, unpub. rep.).

Windermere equivalents

Significant studies of Windermere-equivalent rocks north of latitude 60°N dating from the past 10 years (Eisbacher, 1978a, 1981; Yeo, 1981; Armstrong et al., 1982; Narbonne et al., 1985; Jefferson and Ruelle, 1986; Aitken, 1988) had as a starting-point the pioneer studies of Gabrielse et al. (1973). They have not only expanded, but also changed, in part, our understanding of the age, content, and depositional setting of the northern representatives of the supergroup. Effective nomenclature has generally been established at the formational level, but some of the named groups were established prior to achievement of the present level of understanding, do not effectively identify the most significant, supra-formational packages, and do not enjoy universal acceptance. The strata north of 60° currently assigned to the Windermere Supergroup are those that underlie the base of the Cambrian (but include, in part, strata bearing Ediacaran fauna), and post-date the inception of an episode of rifting ('Hayhook orogeny' of Young et al., l979) taken to be Cordillera-long and to correspond to the poorly dated Goat River 'orogeny' of McMechan and Price (1982). Windermere formations north of 60° are generally, but not everywhere, confined to the more interior ranges of Mackenzie, Wernecke, and Ogilvie Mountains, that is, "outboard" of older Proterozoic successions. They reach the mountain front only along the Snake River - Bonnet Plume sector, and have not been encountered by exploratory drilling. A brief account of them is appropriate here because they are the only record of about 200 million years of earth history that is not recorded in the undeformed part of the platform.

The deposits of the rift phase are the 'Copper cycle' and the Sayunei and Shezal formations of the Rapitan Group. Both were deposited in a narrow, fault-bounded trough or troughs, the Rapitan units more widely, in a "steer's head" configuration. Only a few small areas of outcrop to the southwest control the width of the rift-depression(s); deposition in a single, continuous, zig-zag rift of northwest trend is therefore defensible on the evidence, whereas deposition in a broad area of extension and multiple grabens is purely speculative. Note that the Rapitan isopachs of Yeo (1981) can be re-drawn to outline a zig-zag rift.

The earliest sediments assigned to the Windermere are those of the informal 'Copper cycle' (Fig. 4A.9; Coates Lake Group of Jefferson and Ruelle, 1986), whose surface distribution is notably limited. The group commences with the Thundercloud Formation (0-300 m): volcanic-derived sandstone and conglomerate, maroon carbonate mudstone, sabkha dolomites, cyclically interbedded sandstone and shale, and local evaporites. The Thundercloud is followed gradationally by the Redstone River Formation (0-1200+ m): fanglomerates, playa redbeds, and continental anhydrite and glauberite. These are gradationally succeeded by marine-littoral, stromatolitic carbonates and overlying, slope and deeper water carbonates of the Coppercap Formation (0-300+ m). Stratabound copper deposits of large scale occur at the nonmarine to marine transition (Jefferson and Ruelle, 1986).

The Sayunei and Shezal formations (Lower and Middle Rapitan; Fig. 4A.9, 4A.10), which succeed the 'Copper cycle' display complicated relationships indicative of continued rifting. They were deposited in the same basins as the 'Copper cycle' but overstep the 'Copper cycle' to onlap the basin flanks (Yeo, 1981; J.D. Aitken, unpub. rep.). On the tops and upper flanks of the uplifted blocks, Sayunei, Shezal, or Twitya (Upper Rapitan) rest unconformably on Little Dal Formation, in many places truncating steep faults that trend northeast to north-northwest. Elsewhere, Sayunei Formation rests unconformably on 'Copper cycle', yet locally, at thick sections deposited in the most rapidly subsiding troughs, the Coppercap-Sayunei contact is gradational and conformable.

The Sayunei Formation and the Mount Berg Formation of Yeo (1981) consist of reddish and minor green and greenish grey clastic sedimentary rocks showing no signs of shallow-water sedimentation. Thickness ranges from 0 to more than 600 m. Principal lithofacies include siltstone-argillite rhythmite, lithic and feldspathic-lithic sandstone, lenticular framework conglomerate, diamictite, and massive mudstone. Jasper-hematite iron-formation occurs widely near the top of the Sayunei (Eisbacher, 1981) or at the base of the overlying Shezal (Yeo, 1981). At Snake River, this constitutes a major deposit containing six billion tons of 47.2% iron oxide (Stuart, 1963). Evidence of emplacement of sediments by mass-movement and turbidity-current processes is widespread and complicates the question of the possible glacial origin of the diamictites. Nevertheless, undoubted dropstones are widespread, and faceted and striated clasts and boulders of extra-basinal source, although not common, are present. A glacial influence on Sayunei sedimentation is generally accepted (Eisbacher, 1981; Yeo, 1981).

The Shezal Formation (0-800 m) overlies the Sayunei or older formations. It consists mainly of greenish or reddish diamictite in which indistinct, very thick bedding is visible at a distance. Intercalations of sandstone and mudstone are minor and local. The pebble- to boulder-sized megaclasts usually are almost entirely derived from the

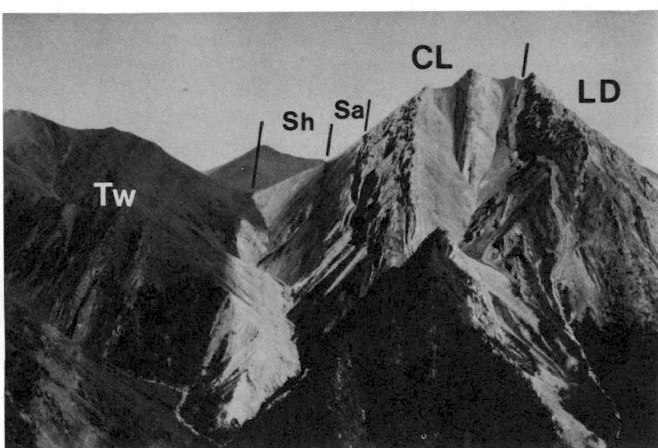

Figure 4A.9. Proterozoic strata in hanging wall of Plateau fault, central Mackenzie Mountains. LD - Little Dal Group (Gypsum, Rusty shale, and Upper carbonate formations); CL - Coates Lake Group (interpreted as basal Windermere equivalent); SA - Sayunei Formation; Sh - Shezal Formation; Tw - Twitya Formation (photo - J.D. Aitken, ISPG 1221-2).

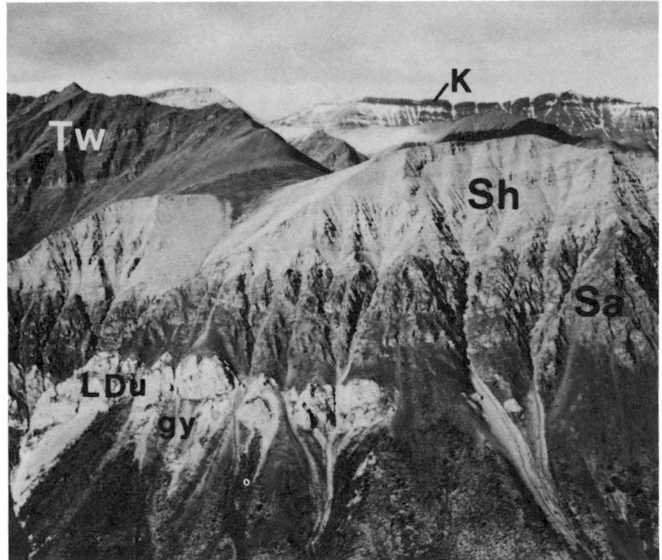

Figure 4A.10. Proterozoic strata in hanging wall of Plateau Fault, 20 km southeast of Figure 4A.9. LDu - Upper carbonate of Little Dal Group; gy - gypsum slice on minor fault; Sa - Sayunei Formation; Sh - Shezal Formation; Tw - Twitya Formation (Lower, Middle, Upper Rapitan); K - Keele Formation. Note pronounced paleotopography beneath Rapitan (photo - J.D. Aitken, ISPG 1221-3).

immediately underlying, pre-Rapitan formation, and at many localities are almost entirely carbonate rocks. Clasts of quartzite and greenstone are generally inconspicuous, those of rhyolite, granite, and gneiss rare. Carbonate and quartzite clasts rarely, but greenstone clasts generally display striations of glacial type, and at a single locality, a polished iron-stained surface is reported at the base of the Shezal (Eisbacher, 1981). Although megaclasts are almost entirely intrabasinal, variation in their type and proportions demonstrates transport from the northeast. Yeo (1981) pointed out that the nearest known source for the rhyolite, granite, and gneiss clasts is the Canadian Shield east of Great Bear Lake.

The Twitya Formation (Upper Rapitan; Fig. 4A.9, 4A.10) is more widespread and continuous than the underlying Rapitan units, and the influence of rift basins, if any, is obscure. The base of the Twitya marks the onset of sedimentation of 'Grand Cycle' style (Aitken, 1978), which characterizes the terminal Proterozoic of the region. The formation, up to 900 m thick, consists mainly of dark

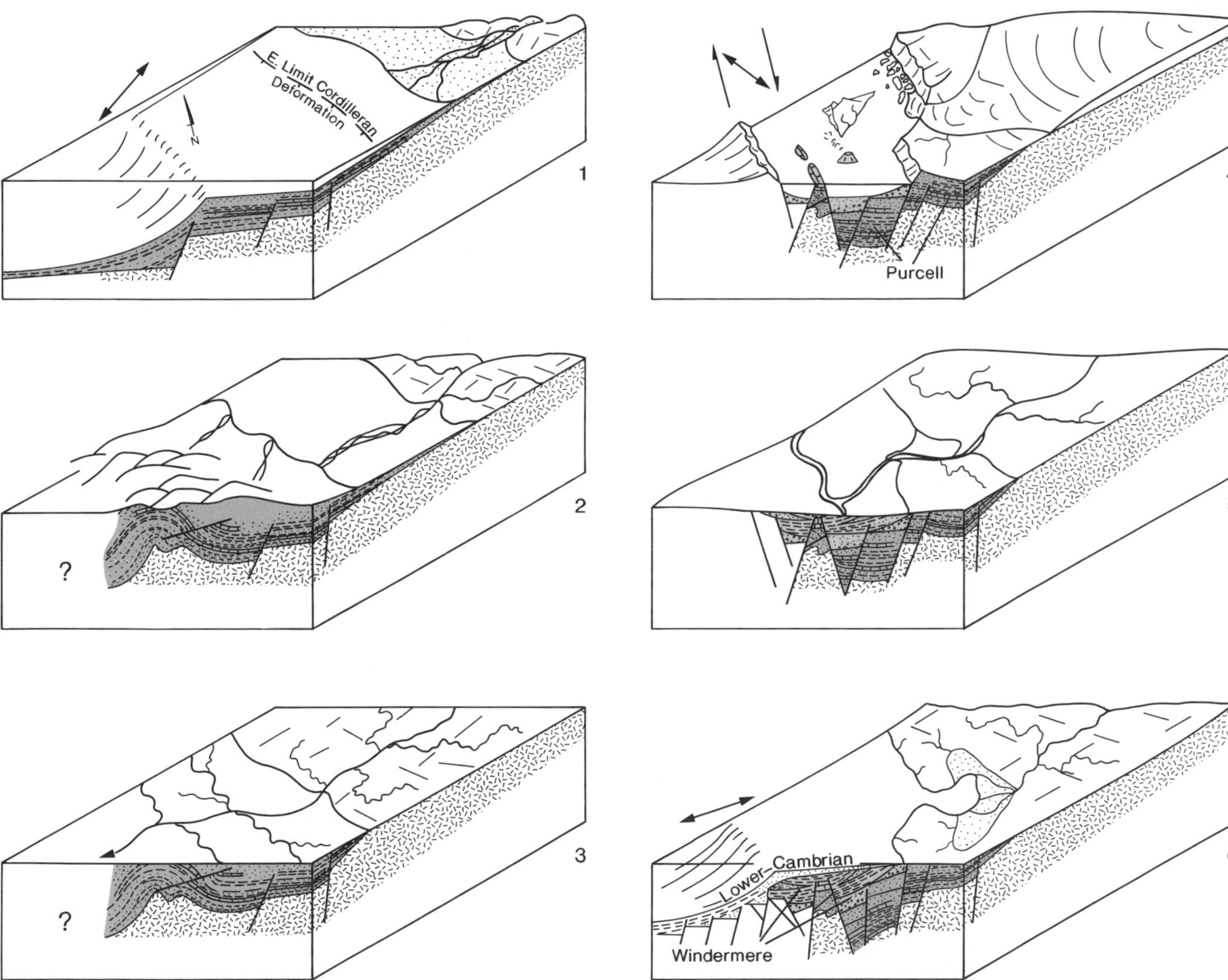

1. Middle Proterozoic (1700-1300 Ma). Deposition of Purcell Supergroup, probably as a continental terrace wedge.

2. Middle Proterozoic (ca. 1300 Ma). Folding of Purcell Supergroup (East Kootenay Orogeny).

3. Late Middle Proterozoic. Erosion and planation of Purcell Supergroup.

4. Late Proterozoic (ca. 780-730 Ma). Rifting (Goat River taphrogeny) initiating deposition of Windermere Supergroup; local volcanism; glaciation.

5. Latest Proterozoic. Uplift or eustatic fall; basinward tilting and planation.

6. Early Cambrian. Continental separation (600-550 Ma); transgression and initial deposits of the Cordilleran Miogeocline.

Figure 4A.11. Development of southern region

grey to black mudrocks, with units of finely planar-laminated and ripple cross-laminated (Bouma C?) siltstone, and graded, siltstone-shale couplets. Slump folds and slide surfaces occur here and there. At the most southwesterly exposures, packets of fine-grained, mainly quartzose sandstone appear. These display complete and incomplete Bouma sequences and apparently are turbidites introduced from the southwest.

The Twitya is capped conformably by a carbonate-dominated platform deposit, the Keele Formation (300-600 m; Fig. 4A.10), which completes the first Grand Cycle. A threefold division is more marked in the northwest than in the southeast; two members mainly of shallow-water limestone and dolomite with minor interbeds of sandstone and shale are separated by a member of mudrocks and sandstone. The lower carbonate member terminates basinward (southwestward) at a breakaway scarp that shed a sub-regional olistostrome into the basin (Eisbacher, 1978b). The upper carbonate member is locally capped by a thin diamictite, which has been interpreted as a tillite (J.D. Aitken, unpub. rep.). The tillite thickens into the basin, beyond the edge of the carbonate platform, and there becomes continuous and widespread. There, too, dropstones and rare striated clasts

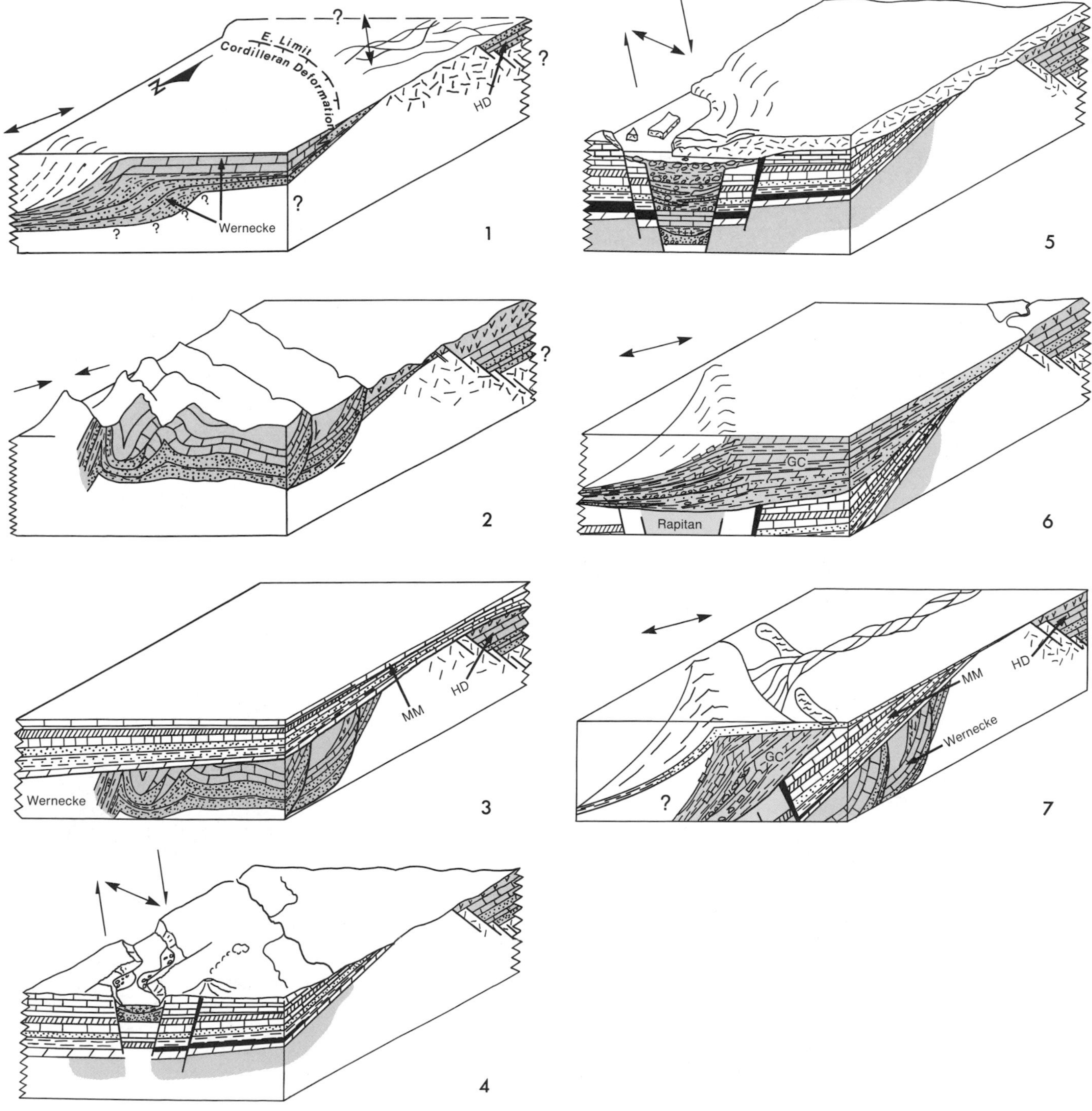

demonstrate its glacial origin. In the basin, in the absence of Keele bedded carbonates, only the tillite separates the Twitya from the overlying and lithologically similar Sheepbed Formation.

The Sheepbed Formation (up to 900 m), abruptly and disconformably succeeding the Keele, resembles the Twitya but at most exposures contains very little sandstone. At the most southwesterly exposures, however, thick, massive beds, some graded, of coarse, gritty feldspathic sandstone appear. The distribution suggests that these sandstones are turbidites of southwesterly derivation, as in the Twitya. The oldest known body fossils of the Canadian Cordillera, Ediacaran medusae, occur in the Sheepbed (G. Narbonne, pers. comm., 1988). Until recently, the Sheepbed was regarded as the youngest Proterozoic formation of the Mackenzie and Wernecke Mountains. Now, as much as 2 km of post-Sheepbed, Proterozoic strata are known.

Gradationally overlying the Sheepbed is the Gametrail Formation (up to 320 m), a carbonate unit that completes the second Grand Cycle and displays both a northeasterly, platformal aspect and a southwesterly, slope aspect. The Gametrail is abruptly succeeded by the Blueflower Formation (up to 1000 m), dark-coloured mudrocks with packets of turbiditic sandstone and limestone, debris-flow diamictites, and exotic blocks. The lowest trace fossils known in the region appear at the base of the Blueflower, and medusae and *Pteridinium* represent an Ediacaran fauna (Aitken, 1988; Hofmann, 1981).

1. Middle Proterozoic. Deposition of Wernecke Supergroup (continental terrace wedge?) and contemporaneous (?) deposition of epicratonic Hornby Bay - Dismal Lakes succession (HD; >1660-1210 Ma).
2. Middle Proterozoic. Folding and thrusting of Wernecke Supergroup (Racklan Orogeny), following eruption of Coppermine basalts (1210 Ma).
3. Late Proterozoic. Subsequent to planation, deposition of the 'platformal' Mackenzie Mountains Supergroup (MM; 880?-770 Ma).
4. Late Proterozoic. Subsequent to tilting and prolonged erosion, initiation of rifting; localized volcanism; emplacement (ca. 780 Ma) of subvolcanic intrusions (black); deposition of Coates Lake Group (red tint) in a narrow rift depression.
5. Late Proterozoic. Enlargement of rift depression, glaciation; deposition of Rapitan Group.
6. Latest Proterozoic. Widespread hinged subsidence. Three carbonate-capped Grand Cycles (GC) deposited across former rift depression. Glaciation follows first carbonate cap; Ediacaran fauna appears in first cycle.
7. Following tilting and deep erosion, transgression and deposition of basal Cambrian clastics; initiation of the Cordilleran Miogeocline.

Figure 4A.12. Development of northern region

Gradationally above the Blueflower is yet another carbonate deposit, the Risky Formation (up to 150+ m), which completes the third Grand Cycle. In eastern Wernecke Mountains, the apparent correlatives of the Risky are significantly thicker, and a medial siltstone unit contains Ediacaran fauna (Narbonne et al., 1985).

In Mackenzie Mountains, post-Sheepbed, Proterozoic formations are preserved only locally, having been erosionally bevelled eastward at a pronounced unconformity beneath the Backbone Ranges Formation. In the view of Aitken (1988), this is the expected 'sub-Cambrian unconformity'; different interpretations (notably those of W.H. Fritz), are summarized by Aitken (1988). On the basis of trace and rare body fossils, the upper member, at least, of the sandstone-dominated Backbone Ranges Formation is of Early Cambrian age (Fritz, 1980a). Resolution of the question whether the middle (carbonate) and lower members also are Cambrian awaits further work.

The three, kilometre-thick Grand Cycles (Twitya-Keele, Sheepbed-Gametrail, and Blueflower-Risky) just described are one of the most striking aspects of terminal Proterozoic sedimentation in the northern Cordillera. These display none of the lithological or distributional characteristics of rift-depression deposits. Furthermore, they each display basinward (westward to southward) thickening, and the carbonate 'caps' of the cycles display basinward facies changes and/or disappearance in the same directions. These characteristics suggest passive-margin sedimentation that commenced prior to Ediacaran time, and thus earlier than the rift/drift transition as determined from subsidence analysis in the Rocky Mountains (Bond and Kominz, 1984). Earlier commencement of passive-margin subsidence in the north would in part explain the greater craton-ward reach of Lower and Middle Cambrian deposits in Mackenzie Platform, as compared with the southern Interior Platform.

SUMMARY

The Middle and Late Proterozoic histories of the southern and northern basins and platforms (deduced entirely from studies in the mountain belt in the south, and largely from the mountain belt in the north respectively) are compared and contrasted in Figures 4A.11 and 4A.12.

Both regions have an early (post-1700 Ma, pre-1200 Ma) history of sedimentation of thick successions of continental-margin aspect (Sequence A), although evidence for a "two-sided" Belt-Purcell Basin is being brought forward increasingly. In the north, the Wernecke Supergroup underwent folding prior to deposition of sequence B (Racklan Orogeny). Evidence for pre-Windermere folding is clear in the Muskwa assemblage of northeastern British Columbia, but disputed (East Kootenay Orogeny) in the case of the classical Purcell Supergroup.

Both regions also have a Late Proterozoic (780 Ma and younger) history of sedimentation (in part glaciomarine) in rift-basin environments (Sequence C), followed by continental separation and passive-margin subsidence. The time of separation in the south was near the beginning of the Cambrian Period, and in the north, the same or possibly slightly earlier.

The regions differ principally in the presence, in the north, of a major sedimentary succession (Sequence B) that separates the Windermere-equivalent succession (C) from probable Purcell equivalents (A). This succession, 'Mackenzie Mountains supergroup', did not undergo pre-Windermere folding. It extends from the edge of Selwyn Basin completely across Mackenzie Platform to the flanks of the exposed Canadian Shield.

Thus, in the south, basal Paleozoic rocks in the undeformed basin and platform generally rest on crystalline, Proterozoic and Archaean basement, while in the north, basal Paleozoic rocks in the undeformed basin and platform generally rest on Proterozoic sedimentary rocks.

REFERENCES

Aalto, R.K.
1971: Glacial marine sedimentation and stratigraphy of the Toby Conglomerate (Upper Proterozoic), southeastern British Columbia, northwestern Idaho, and northeastern Washington; Canadian Journal of Earth Sciences, v. 8, p. 753-787.

Aitken, J.D.
1969: Documentation of the sub-Cambrian unconformity, Rocky Mountains Main Ranges, Alberta; Canadian Journal of Earth Sciences, v. 6, p. 192-200.
1978: Revised models for depositional Grand Cycles, Cambrian of the southern Rocky Mountains, Canada; Bulletin of Canadian Petroleum Geology, v. 26, p. 515-542.
1981: Stratigraphy and sedimentology of the Upper Proterozoic Little Dal Group, Mackenzie Mountains, Northwest Territories; in Proterozoic Basins of Canada, F.H.A. Campbell (ed.); Geological Survey of Canada, Paper 81-10, p. 47-72.
1988: Uppermost Proterozoic Formations, central Mackenzie Mountains, Northwest Territories; Geological Survey of Canada, Bulletin 368, 26 p.

Aitken J.D., Cook, D.G., and Yorath, C.J.
1982: Upper Ramparts River (106G) and Sans Sault Rapids (106H) map areas, District of Mackenzie; Geological Survey of Canada, Memoir 388, 48 p.

Aitken, J.D. and Long, D.G.F.
1978: Mackenzie tectonic arc - reflection of early basin configuration?; Geology, v. 6, p. 626-629.

Aitken, J.D., Long, D.G.F. and Semikhatov, M.A.
1978: Correlation of Helikian strata, Mackenzie Mountains - Brock Inlier - Victoria Island; in Current Research, Part A, Geological Survey of Canada, Paper 78-1A, p. 485-486.

Aitken, J.D. and Pugh, D.C.
1984: The Fort Norman and Leith Ridge structures: major, buried, Precambrian features underlying Franklin Mountains and Great Bear and Mackenzie Plains; Bulletin of Canadian Petroleum Geology, v. 32, p. 139-146.

Armstrong, R.L., Eisbacher, G.H., and Evans, P.D.
1982: Ages and stratigraphic-tectonic significance of Proterozoic diabase sheets, Mackenzie Mountains, northwestern Canada; Canadian Journal of Earth Sciences, v. 19, p. 316-323.

Arnott, R.W. and Hein, F.
1986: Submarine canyon fills of the Hector Formation, Lake Louise, Alberta: Late Precambrian syn-rift deposits of the proto-Pacific miogeocline; Bulletin of Canadian Petroleum Geology, v. 34, p. 395-407.

Bally, A.W., Gordy, P.L., and Stewart, G.A.
1966: Structure, seismic data and orogenic evolution of southern Canadian Rocky Mountains; Alberta Society of Petroleum Geologists, Bulletin, v. 14, p. 337-381.

Baragar, W.R.A. and Donaldson, J.A.
1973: Coppermine and Dismal Lakes map-areas; Geological Survey of Canada, Paper 71-39 (with Maps 1337A and 1338A), 20 p.

Bell, R.T.
1966: Precambrian rocks of the Tuchodi Lakes map area, northeastern British Columbia, Canada; Ph.D. thesis, Princeton University, Princeton, New Jersey, U.S.A., 138 p.
1968: Proterozoic stratigraphy of northeastern British Columbia; Geological Survey of Canada, Paper 67-68, 75 p.

Bond, G.C. and Kominz, M.A.
1984: Construction of tectonic subsidence curves for the early Paleozoic miogeocline, southern Canadian Rocky Mountains - implications for subsidence mechanisms, age of breakup, and crustal thinning; Geological Society of America Bulletin, v. 95, p. 155-173.

Bowring, S.A. and Ross, G.M.
1985: Geochronology of the Narakay Volcanic Complex: implications for the age of the Coppermine Homocline and Mackenzie igneous events; Canadian Journal of Earth Sciences, v. 22, p. 774-781.

Cook, F.A.
1988a: Proterozoic thin-skinned thrust and fold belt beneath the Interior Platform in northwest Canada; Geological Society of America, Bulletin, v. 100, p. 877-890.
1988b: Middle Proterozoic compressional orogen in northwestern Canada; Journal of Geophysical Research, v. 93, no. B8, p. 8985-9005.

Delaney, G.D.
1981: The mid-Proterozoic Wernecke Supergroup, Wernecke Mountains, Yukon Territory; in Proterozoic Basins of Canada, F.H.A. Campbell (ed.); Geological Survey of Canada, Paper 81-10, p. 1-23.

Devlin, W.J. and Bond, G.C.
1988: The initiation of the early Paleozoic Cordilleran miogeocline: evidence from the uppermost Proterozoic - Lower Cambrian, Hamill Group of Southeastern British Columbia; Canadian Journal of Earth Sciences, v. 25, p. 1-19.

Devlin, W.J., Bond, G.C., and Brueckner, H.K.
1985: An assessment of the age and tectonic setting of volcanics near the base of the Windermere Supergroup in northeastern Washington: implications for latest Proterozoic - earliest Cambrian continental separation; Canadian Journal of Earth Sciences, v. 22, p. 829-837.

Eisbacher, G.H.
1978a: Re-definition and subdivision of the Rapitan Group, Mackenzie Mountains; Geological Survey of Canada, Paper 77-35, 21 p.
1978b: Two major Proterozoic unconformities, northern Cordillera; in Current Research, Part A, Geological Survey of Canada, Paper 78-1A, p. 53-58.
1981: Sedimentary tectonics and glacial record in the Windermere Supergroup, Mackenzie Mountains, northwestern Canada; Geological Survey of Canada, Paper 80-27, 40 p.

Evenchick, C.A.
1982: Stratigraphy, structure and metamorphism in Deserter's Range, northern Rocky Mountains, British Columbia; in Current Research, Part A, Geological Survey of Canada, Paper 82-1A, p. 325-328.

Evenchick, C.A., Parrish, R.R., and Gabrielse, H.
1984: Precambrian gneiss and late Proterozoic sedimentation in north-central British Columbia; Geology, v. 12, p. 233-237.

Fermor, P. and Price, R.A.
1983: Stratigraphy of the lower part of the Belt-Purcell Supergroup (middle Proterozoic) in the Lewis thrust sheet of southern Alberta and British Columbia; Bulletin of Canadian Petroleum Geology, v. 31, p. 169-194.

Fritz, W.H.
1980a: International Precambrian - Cambrian Boundary Working Group's 1979 field study to Mackenzie Mountains, northwest Canada; in Current Research, Part A, Geological Survey of Canada, Paper 80-1A, p. 41-45.
1980b: Two Cambrian stratigraphic sections near Gataga River, northern Rocky Mountains, British Columbia; in Current Research, Part C, Geological Survey of Canada, Paper 80-1C, p. 113-119.

Gabrielse, H. and Yorath, C.J. (ed.)
1991: Geology of the Cordilleran Orogen in Canada; Geological Survey of Canada, Geology of Canada, no. 4. (Also Geological Society of America, The Geology of North America, v. G-2).

Gabrielse, H., Blusson, S.L., and Roddick, J.A.
1973: Geology of Flat River, Glacier Lake and Wrigley Lake map-areas, District of Mackenzie and Yukon Territory; Geological Survey of Canada, Memoir 366, 421 p.

Gordey, S.P.
in press: Evolution of the northern Cordilleran Miogeocline, Nahanni map area (105I), Yukon Territory and District of Mackenzie; Geological Survey of Canada, Memoir 428.

Harrison, J.E., Griggs, A.B., and Wells, J.D.
1974: Tectonic features of the Precambrian Belt Basin and their influence on post-Belt structures; United States Geological Survey, Professional Paper 866, 15 p.

Hofmann, H.J.
1981: First record of a Late Proterozoic faunal assemblage in the North American Cordillera; Lethaia, v. 14, p. 303-310.

Hofmann, H.J., Mountjoy, E., and Teitz, M.W.
1985: Ediacaran fossils from the Miette Group, Rocky Mountains, British Columbia, Canada; Geology, v. 13, p. 819-821.

Jefferson, C.W. and Ruelle, J.C.L.
1986: The Late Proterozoic Redstone Copper Belt, Mackenzie Mountains, Northwest Territories; in Mineral Deposits of the Northern Cordillera, J.A. Morin (ed.); Canadian Institute of Mining and Metallurgy, Special Volume 37, p. 154-168.

Kerans, C., Ross, G.M., Donaldson, J.A., and Geldsetzer, H.J.
1981: Tectonism and depositional history of the Helikian Hornby Bay and Dismal Lakes Groups, District of Mackenzie; in Proterozoic Basins of Canada, F.H.A. Campbell (ed.); Geological Survey of Canada, Paper 81-10, p. 152-182.

LeCheminant, A.N., Heaman, L.M., and Rainbird, R.M.
1990: Mantle plume origin for flood basalts and giant dyke swarms: 1.27 Ga Mackenzie and 0.72 Franklin igneous events, Canada; in Program and Abstracts, 2nd International Dyke Conference, Adelaide, Australia.

Lis, M.G. and Price, R.A.
1976: Large-scale block faulting during deposition of the Windermere Supergroup (Hadrynian) in southeastern British Columbia; in Current Research, Part A, Geological Survey of Canada, Paper 76-1A, p. 135-136.

McMechan, M.E.
1981: The Middle Proterozoic Purcell Supergroup in the southwestern Rocky and southeastern Purcell Mountains, British Columbia and the initiation of the Cordilleran miogeocline, southern Canada and adjacent United States; Bulletin of Canadian Petroleum Geology, v. 29, p. 583-621.

McMechan, M.E. and Price, R.A.
1982: Superimposed low-grade metamorphism in the Mount Fisher area, southeastern British Columbia - implications for the East Kootenay orogeny; Canadian Journal of Earth Sciences, v. 19, p. 476-489.

McMechan, M.E. and Thompson, R.I.
1985: Geology, southeast Monkman Pass map-area (93I, SE), B.C.; Geological Survey of Canada, Open File 1150.

Meijer Drees, N.C.
1975: Geology of the lower Paleozoic Formations in the subsurface of the Fort Simpson map-area, District of Mackenzie, Northwest Territories; Geological Survey of Canada, Paper 74-40, 65 p.

Narbonne, G.M., Hofmann, H.J., and Aitken, J.D.
1985: Precambrian-Cambrian boundary sequence, Wernecke Mountains, Yukon Territory; in Current Research, Part A, Geological Survey of Canada, Paper 85-1A, p. 603-608.

Park, J.K. and Aitken, J.D.
1986: Paleomagnetism of the Katherine Group in the Mackenzie Mountains: implications for post-Grenville (Hadrynian) apparent polar wander; Canadian Journal of Earth Sciences, v. 23, p. 308-323.

Poulton, T.P. and Simony, P.S.
1980: Stratigraphy, sedimentology and regional correlation of the Horsethief Creek Group (Hadrynian, Late Precambrian) in the northern Purcell and Selkirk Mountains, British Columbia; Canadian Journal of Earth Sciences, v. 17, p. 1708-1724.

Preto, V.A.
1971: Lode copper deposits of the Racing River - Gataga River area; in Geology, Exploration and Mining in British Columbia; British Columbia Department of Mines and Petroleum Resources, p. 75-107.

Price, R.A.
1964: The Precambrian Purcell System in the Rocky Mountains of southern Alberta and British Columbia; Bulletin of Canadian Petroleum Geology, v. 12, p. 399-426.

Pugh, D.C.
1983: Pre-Mesozoic geology in the subsurface of Peel River map-area, District of Mackenzie; Geological Survey of Canada, Memoir 401, 61 p.

Roberts, S.M. (ed.)
1986: Belt Supergroup: a guide to Proterozoic rocks of western Montana and adjacent areas; Montana Bureau of Mines and Geology, Special Publication 94, 311 p.

Stewart, J.H.
1972: Initial deposits in the Cordilleran Geosyncline: evidence of Late Proterozoic (<850 m.y.) continental separation; Geological Society of America Bulletin, v. 83, p. 1345-1360.

Stuart, R.A.
1963: Geology of the Snake River iron deposit; Department of Indian Affairs and Northern Development, Yellowknife, Open File, 18 p.

Taylor, G.C. and Stott, D.F.
1973: Tuchodi Lakes map-area, British Columbia; Geological Survey of Canada, Memoir 373, 37 p.

Teitz, M. and Mountjoy, E.W.
1985: The Yellowhead and Astoria carbonate platforms in the Late Proterozoic upper Miette Group, Jasper, Alberta; in Current Research, Part A, Geological Survey of Canada, Paper 85-1A, p. 341-348.

Walker, J.F.
1926: Geology and mineral deposits of Windermere map-area, British Columbia; Geological Survey of Canada, Memoir 148, 69 p.

Winston, Don, Woods, Marvin, and Byer, G.B.
1984: The case for an intracratonic Belt-Purcell Basin: tectonic, stratigraphic, and stable isotopic considerations; in Montana Geological Society, 1984 Field Conference and Symposium, J.D. McBane and P.B. Garrison (ed.); p. 103-118.

Yeo, G.M.
1981: The Late Proterozoic Rapitan glaciation in the northern Cordillera; in Proterozoic Basins of Canada, F.H.A. Campbell (ed.); Geological Survey of Canada, Paper 81-10, p. 25-46.

Young, G.M.
1978: Proterozoic (<1.7 b.y.) stratigraphy, paleocurrents and orogeny in North America; Egyptian Journal of Geology, v. 22, p. 45-64.
1981: The Amundsen Embayment, Northwest Territories; relevance to the Upper Proterozoic evolution of North America; in Proterozoic Basins of Canada, F.H.A. Campbell (ed.); Geological Survey of Canada, Paper 81-10, p. 203-218.

Young, G.M., Jefferson, C.W., Delaney, G.D., and Yeo, G.M.
1979: Middle and late Proterozoic evolution of the northern Canadian Cordillera and Shield; Geology, v. 7, p. 125-128.

ADDENDUM

Several important papers dealing with the terminal Proterozoic succession of the northern Cordillera have gone to press since completion of the manuscript for this chapter. These document the composition and stratigraphic range of Ediacaran fossils in the region, and the deposits of a post-Rapitan glaciation.

Aitken, J.D.
1991a: The Ice Brook Formation and post-Rapitan, Late Proterozoic glaciation, Mackenzie Mountains, N.W.T.; Geological Survey of Canada, Bulletin 404, 43 p.
1991b: Two Late Proterozoic glaciations, Mackenzie Mountains, Northwestern Canada; Geology, v. 19, p. 445-448.

Hofmann, H.J., Narbonne, G.M., and Aitken, J.D.
1990: Ediacaran remains from intertillite beds in northwestern Canada; Geology, v. 18, p. 1199-1202.

Narbonne, G.M. and Aitken, J.D.
1990: Ediacaran fossils from the Sekwi Brook area, Mackenzie Mountains, northwestern Canada; Paleontology, v. 33, Part 4, p. 945-980.

Author's Address

J.D. Aitken
2676 Jemima Road
Denman Island, British Columbia
V0R 1T0

Printed in Canada

Subchapter 4B

CAMBRIAN AND LOWER ORDOVICIAN - SAUK SEQUENCE

J.D. Aitken

CONTENTS

INTRODUCTION

General statement

The Western Canada Basin provides exceptional opportunities for the study of Cambrian and Lower Ordovician rocks. In the southwest, these rocks are part of the Cordilleran Miogeocline, the wedge of sedimentary and minor volcanic rocks deposited on a passive margin of western North America that came into being at about the beginning of the Cambrian (Fig. 4B.1). In the northeast, they form the purely sedimentary and much thinner cover of the Interior Platform. Although it is clear that a hinge line marked the transition from platform to miogeocline, the position of the hinge is difficult to locate precisely, even for small chronostratigraphic units, because of the vagaries of preservation and the uneven distribution of thickness data. In any event, the position of the hinge line was not fixed. The best approximation of its long-term average position is the eastern limit of Mesozoic deformation.

The unit chosen for analysis here is the Sauk Sequence of L.L. Sloss (1963, 1976), the record of a long period of almost continuous sedimentation on the cratons, limited below by a widespread unconformity near the base of the Cambrian and above by a widespread unconformity near the base of the Middle Ordovician. Sub-sequences I, II, and III are demarcated by relatively subtle breaks at the base of the Middle Cambrian and the base of the Franconian.

The present synthesis provides a descriptive summary, and emphasizes the striking differences seen between cross-sections from platform to miogeocline drawn through the southern Rocky Mountains, the northern Rocky Mountains, the Mackenzie Mountains, and the Richardson Mountains. It makes the point that different parts of the miogeocline, along strike from one another, differ markedly in paleotectonic regime and consequent depositional response.

The Sauk Sequence of the western basin bears a satisfactory likeness, in its patterns of depositional facies and thickness, to what the modern, 'Atlantic Margin Model'

Aitken, J.D.
1993: Cambrian and Lower Ordovician - Sauk Sequence; Subchapter 4B in Sedimentary Cover of the Craton in Canada, D.F. Stott and J.D. Aitken (ed.); Geological Survey of Canada, Geology of Canada, no. 5, p. 96-124 (also Geological Society of America, The Geology of North America, v. D-1).

would predict for a sequence beginning with the creation of a new, passive margin. Its study sheds no light, however, on the problem of why sedimentation in the Western Canada Basin persisted in a seemingly miogeoclinal mode through the Paleozoic and as late as the Middle Jurassic.

Present understanding of the Sauk Sequence in the subsurface depends heavily on studies of outcrops in the mountains and especially along the mountain front. These studies have yielded most of the important fossils and models of facies change that validate lithostratigraphic correlations in the subsurface. For this reason, information from the Foreland fold belt of the Cordillera is dealt with at some length here.

Sources

Early work (pre-1940) on the Sauk Sequence was entirely in the magnificent exposures of the southern Rocky Mountains. During the 1940s and subsequently, investigations extended eastward, as petroleum exploration opened up the deeply buried strata of the undeformed platform. At the same time, geological work in the mountains spread northward into the northern Rocky Mountains and, with the adoption of helicopters for field transport in the mid- to late 1950s, into the mountains of the Northwest Territories and Yukon Territory. Thus, most of the work on the Sauk Sequence north of Jasper is of 'first-generation' character, outlining the sequence with broad brush-strokes and identifying the major problems. Only at and south of the latitude of Jasper is 'second generation' work widespread; unfortunately, this is partly represented by unpublished theses. 'Second generation' work, although of excellent quality is barely started in the north.

The most important papers relating to the Sauk Sequence can be assigned to a 'classical era' and a 'modern era'. The 'classical era' was dominated by the work of Charles D. Walcott from 1907 to 1925; his posthumously published summary (1928) provides a key to the many paleontological and stratigraphic papers resulting therefrom. Prior to Walcott, there are only scattered references to Cambrian rocks in the Canadian Cordillera. The works of J.A. Allan (1912, 1914a, b, 1916) and those of L.D. Burling (1914, 1915, 1916a, b) gave more emphasis to the character of the rocks than those of Walcott, whose essential interest was the fossils. The 'classical era' ended with the work of Charles Deiss (1939, 1940).

The 'modern era' began with the meticulous studies of Franco Rasetti (1951, 1956) in the Rocky Mountains Main Ranges. E.W. Mountjoy (1962; Mountjoy and Aitken, 1978) was the first to map Cambrian formations in the Rocky Mountains to a modern standard. D.K. Norris and R.A. Price (1966) related the Cambrian strata near the 49th parallel to the classical sections of the Bow River transect. O.L. Slind and G.D. Perkins (1966) traced lower Paleozoic formations northward from the Jasper area. D.G. Cook (1975) established correlations at the formational level between the classical platform stratigraphy of the Rockies and the much less studied, basinal strata to the west. H.B. Whittington (1971, 1980, 1985) and co-workers (e.g., Conway Morris and Whittington, 1985) continue to work on a complete redescription and re-evaluation of the world-famous Burgess Shale fauna, based on material quarried in the 1960s and on Walcott's enormous collections.

Early works dealing in some detail with the Cambrian of the subsurface basin were those of G.O. Raasch and D.E. Campau (1957) and R.D. Hutchinson (1960). Hendrik van Hees (1959, 1964) published the first attempts at regional synthesis of the subsurface Cambrian of the Western Canada Basin; these were followed by the more detailed and complete, regional studies of D.C. Pugh (1971, 1973, 1975). Pugh (1983, unpub. rep.) went on to study the subsurface lower Paleozoic of vast areas of the Mackenzie and Porcupine platforms in the north, tying-on to the similar studies of N.C. Meijer Drees (1975).

The Sauk Sequence of the Western Canada Basin east of the Alberta-Saskatchewan border has received little attention; only the works of Lochman-Balk and Wilson (1958), W.K. Fyson (1961), R.B. Hutt (1963a, b), Porter and Fuller (1959, 1964) and H. van Hees (1959, 1964) are available.

Since 1965, W.H. Fritz has made major contributions to knowledge of the Sauk Sequence in western Canada, both south of latitude 60°N (Fritz, 1971, 1981; Fritz and Mountjoy, 1975; Aitken et al., 1972) and north of 60°, where his work (Fritz, 1972b, 1975, 1976, 1978b, 1979c and Fritz et al., 1983) provides the basic documentation for the Lower Cambrian. In addition, Fritz has reported on innumerable fossil collections submitted to him by other geologists.

The writer has devoted much of his time since 1961 to study of the Sauk Sequence in the southern Rocky Mountains (Aitken, 1966, 1968, 1969, 1971, 1978, 1981; Aitken and Greggs, 1967; Aitken and Norford, 1967; Aitken et al., 1972; Aitken, in press).

Acknowledgments

During the preparation of this subchapter, the author profited greatly from contributions of fact and interpretation freely given by the following: M.P. Cecile, W.H. Fritz, H. Gabrielse, N.C. Meijer Drees, B.S. Norford, D.K. Norris, G.C. Taylor, and G.K. Williams.

The manuscript was improved by review and constructive suggestions by H. Gabrielse and W.H. Fritz.

STRATIGRAPHIC CONTENT

This subchapter considers the record of tectonic and depositional events that took place between the earliest Cambrian – for present purposes, beginning of the Meishucun (Xing and Luo, 1984), closely approximated by the beginning of the Tommotian (see Raaben, 1969 – and about the end of Early Ordovician time (end of the Arenig). This was a time characterized, apparently worldwide, by thick and widespread sedimentation on the cratons, following and preceding briefer periods for which little or no record remains on the cratons. The depositional record, the Sauk sequence of Sloss (1963, 1976), is widely interrupted on the platforms by unconformities or disconformities at the base of the Middle Cambrian and at the base of the Franconian, which permit division of the sequence into sub-sequences, Sauk I, II, and III.

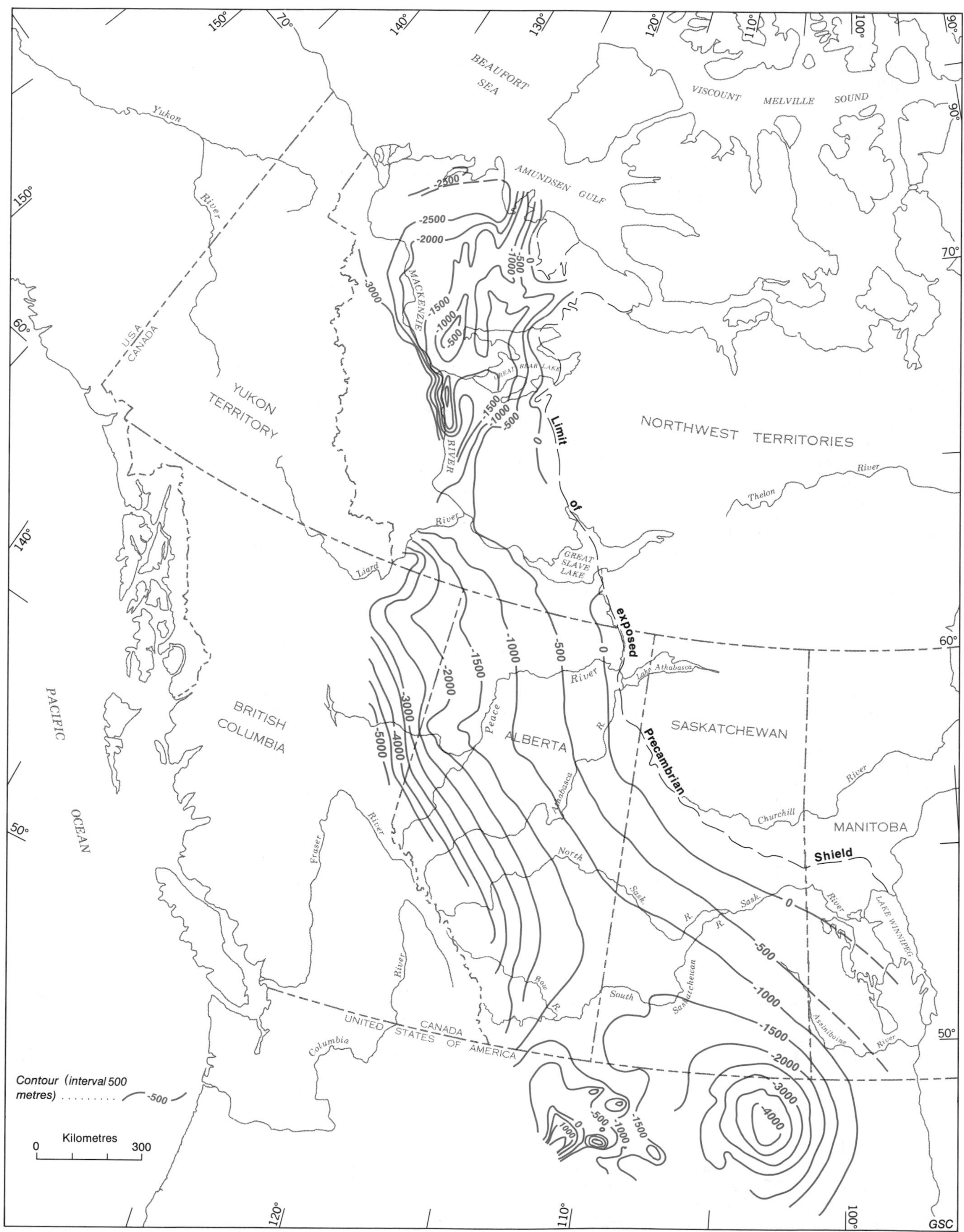

Figure 4B.1. Structural contours on the basement (pre-Phanerozoic) surface (after Porter et al., 1982; with additions adapted from McCrossan, unpub. rep.).

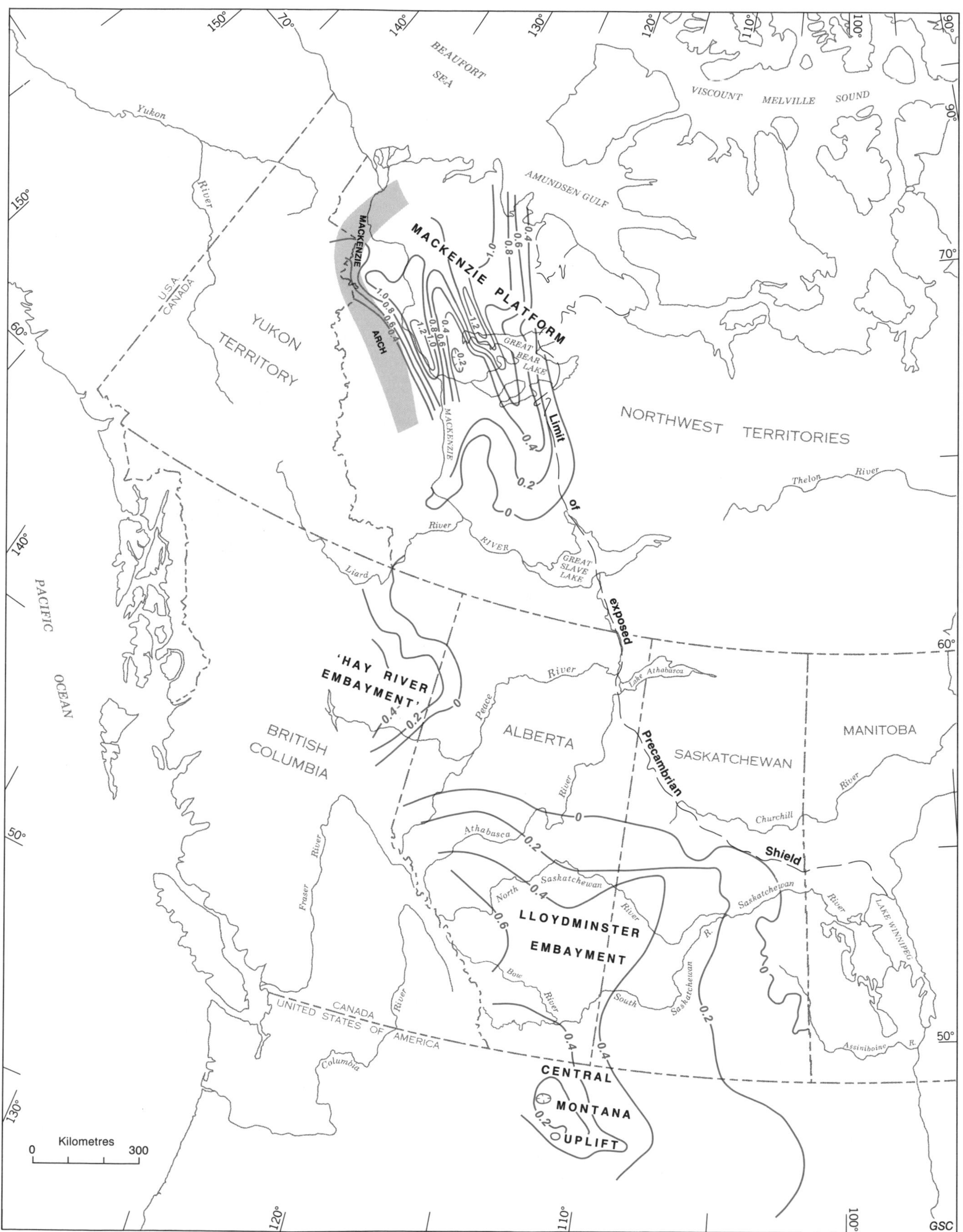

Figure 4B.2. Isopachs (interval 0.2 km), Sauk Sequence (modified from an unpublished diagram by J.W. Porter and R.G. McCrossan).

In geographic terms, this account considers the Sauk Sequence in the Interior Platform and the proximal parts of the Cordilleran Miogeocline. The platform is identified by its early Paleozoic tectonic character rather than its escape from Mesozoic deformation, and thus is considered to extend into the eastern Columbian Orogen. In order to provide an understanding of facies relationships, the treatment is extended, generally, into the Cordillera about as far as strata of platformal aspect persist. The southwestward passage of these platformal rocks, at most stratigraphic levels, into strata of basinal origin is acknowledged, but not detailed (see the companion volume, "Geology of the Cordilleran Orogen in Canada", Geology of Canada, no. 4, 1991).

Tectonic setting

Deposition of the Sauk Sequence in the southern Rocky Mountains began shortly after continental separation, an event dated at about the beginning of the Cambrian (Bond and Kominz, 1984). Similarity of the sequence of major rock-types and thickness in Lower Cambrian deposits of the Mackenzie Mountains may indicate similar timing there, but aspects of the terminal Proterozoic stratigraphy suggest the possibility of a slightly earlier commencement of drift and passive-margin subsidence (see "Proterozoic", Subchapter 4A in this volume). The onset of drift followed a precursory period of rifting, whose record, preserved only in the interior ranges of the Foreland Belt, is the Windermere Supergroup (see "Proterozoic", Subchapter 4A, and for a fuller account, "Upper Proterozoic", in Geology of Canada, no. 4, "Geology of the Cordilleran Orogen in Canada", 1991). Recognition of the ubiquitous sub-Cambrian unconformity as a 'breakup unconformity' is questionable, because an unconformity at about that level is found on all continents. The lower Paleozoic of the eastern parts of the Cordillera is a classical passive-margin succession; all chronostratigraphic units thicken westward or southwestward toward the proto-Pacific (Fig. 4B.2), and most detrital formations display proximal facies in the east and distal facies in the west (the exceptions record provenance from local, emergent "highs").

In the southern Kootenay Terrane, folds of Paleozoic age are known (Wheeler and Read, 1975), as are plutons of early Paleozoic, about Ordovician, age (Okulitch, 1985). These appear to involve rocks deposited close to North America and stand as evidence against an uninterrupted passive margin throughout the Paleozoic. On the other hand, the pre-Devonian record of the eastern Cordillera lacks evidence of vulcanism of convergent-margin type, and also lacks westward-derived clastic wedges that would be the expected result of mountain-building at the western margin. A major problem in Cordilleran geology is acknowledged, but in the absence of a documented alternative, the interpretation involving a continuously passive margin throughout Sauk Sequence deposition is adhered to in this treatment.

Overprinting this record of passive-margin subsidence due to crustal stretching, cooling, and loading by sediment and water, is a record of pulsatory, eustatic rise of sea level, culminating in Early Silurian time. As a consequence of this rise, the marine shoreline transgressed farther onto the craton than it would have in response to passive-margin processes alone, largely inundating the Canadian Shield by Late Ordovician time.

Faunas and biogeography

A reasonably satisfactory zonation of Cambrian trilobite faunas has been achieved for the North American Province (see Fig. 4B.9), although further refinements may be expected. In particular, some problems concerning the equivalence of platformal faunas and those of the deep-water, outer-detrital belt remain to be resolved in detail. Correlation with other faunal provinces is reasonably well established at some levels but less so at others (Fritz, 1981b). The Middle Cambrian has proved especially difficult, but is yielding to an attack based on the study of cosmopolitan agnostid trilobites (Fritz, 1981b).

No general agreement exists as to the paleogeographical factors responsible for faunal provinciality in the Cambrian. Effective synthesis is mainly hampered by two problems: first, many older studies of excellent paleontological quality lack accompanying lithostratigraphic detail from which depositional environments can be reconstructed according to present-day (and recently acquired) understanding, and second, reassembling the continents to their Cambrian configuration is an extremely difficult task, and one upon which there is at best only partial agreement. Nevertheless, Cambrian trilobites from epicratonic strata, worldwide, display marked provinciality.

The distribution of Cambrian faunas upon and around the North American craton (excluding paleo-European and paleo-African terranes sutured to parts of the eastern seaboard) appears fairly straightforward (Palmer, 1972). A broad core region, Palmer's "restricted shelf", is characterized by ptychoparioid trilobites almost entirely endemic to North America at the genus and species level. Agnostids are few in numbers and genera. The "restricted shelf", corresponding to the inner detrital and middle carbonate facies belts, is surrounded by a relatively narrow belt, Palmer's "shelf margin - open sea region". This region corresponds to the outer detrital facies belt, whose faunas include many agnostid and some non-agnostid genera and species, which are also found in analogous situations on other continents (see Palmer, 1973). The narrowness and under-representation of this faunal belt is partly due to tectonic shortening and metamorphic destruction of faunas in the marginal orogens.

A hint of faunal provinciality distinct from the circum-cratonic distribution sketched above can be dimly seen in North America, in the appearance of Asiatic genera and species of trilobites in the Yukon-Alaska region (but also in Nevada). The faunas from Mexico with European affinities are now known to be from an exotic terrane (A.R. Palmer, pers. comm., 1985). Palmer (1972) attempted to relate this distribution to paleolatitude, but the data in support of such an hypothesis are now seen to be less than convincing (A.R. Palmer, pers. comm., 1985).

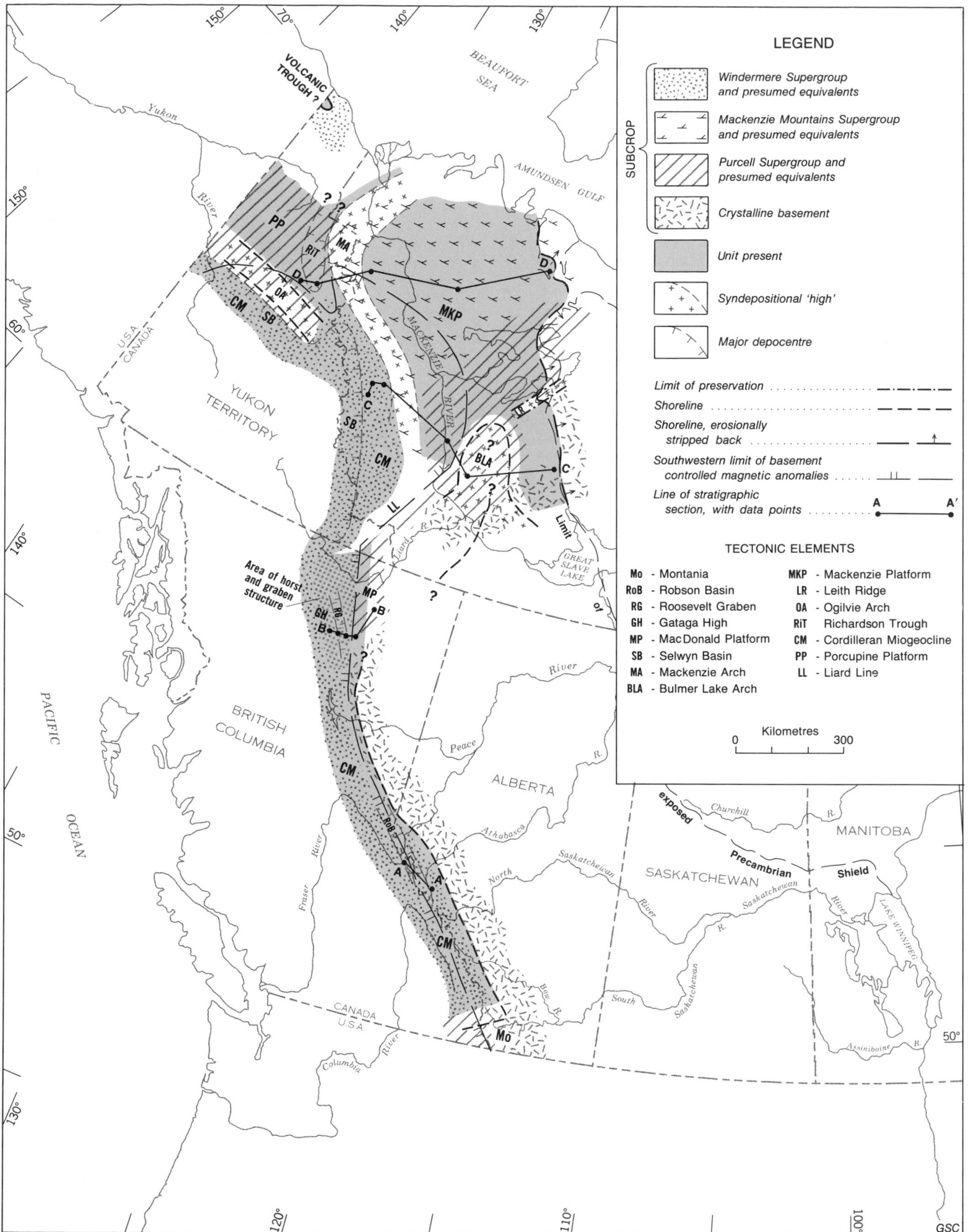

Figure 4B.3. Sauk I: subcrop, tectonic elements, shorelines.

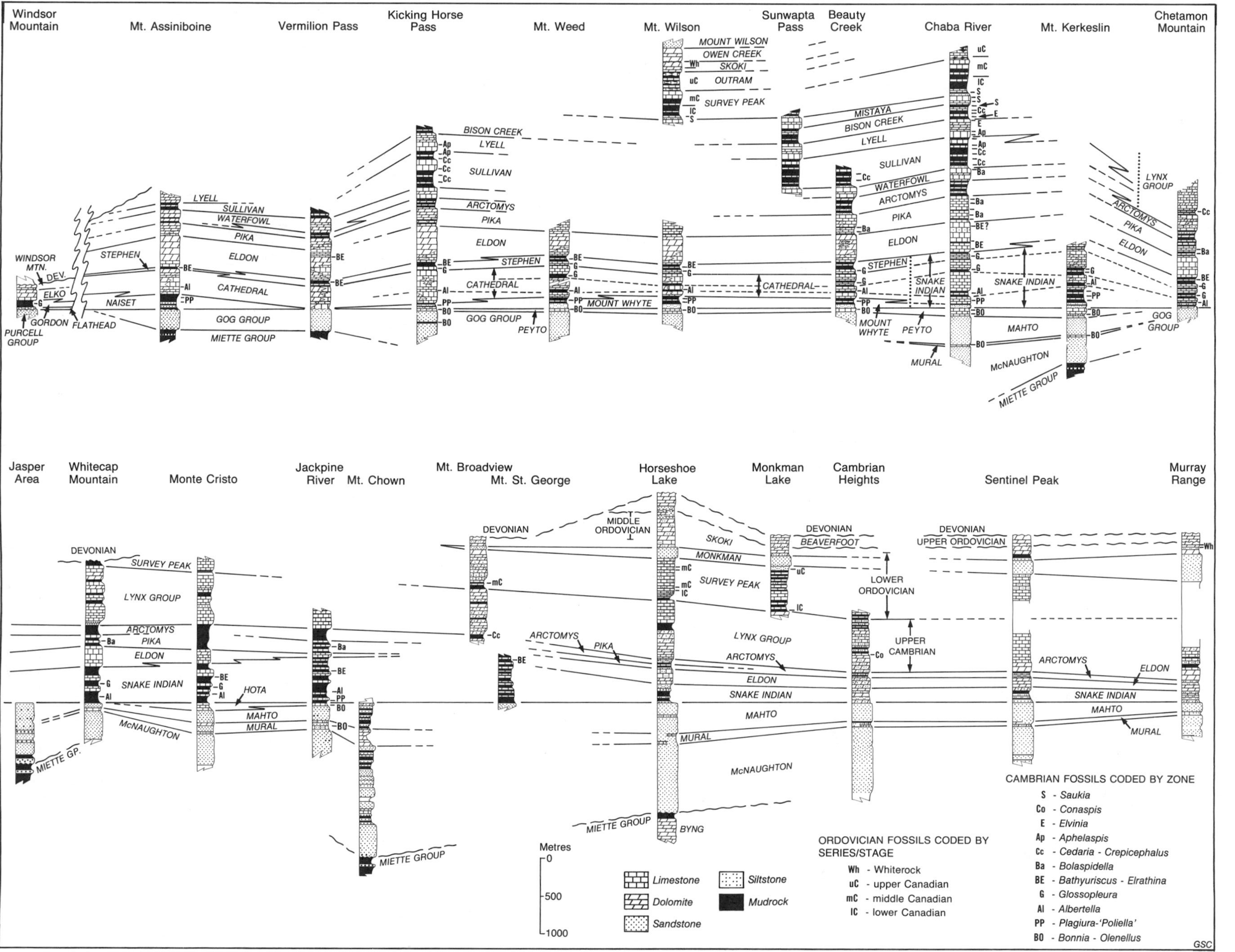
Windsor Mountain
Mt. Assiniboine
Vermilion Pass
Kicking Horse Pass
Mt. Weed
Mt. Wilson
Sunwapta Pass
Beauty Creek
Chaba River
Mt. Kerkeslin
Chetamon Mountain
WINDSOR MTN.
DEV.
ELKO
GORDON
FLATHEAD
PURCELL GROUP
NAISET
STEPHEN
LYELL
SULLIVAN
WATERFOWL
PIKA
ELDON
CATHEDRAL
GOG GROUP
MIETTE GROUP
BISON CREEK
ARCTOMYS
PEYTO
MOUNT WHYTE
MOUNT WILSON
OWEN CREEK
SKOKI
OUTRAM
SURVEY PEAK
MISTAYA
SNAKE INDIAN
MURAL
MAHTO
McNAUGHTON
LYNX GROUP
Jasper Area
Whitecap Mountain
Monte Cristo
Jackpine River
Mt. Chown
Mt. Broadview
Mt. St. George
Horseshoe Lake
Monkman Lake
Cambrian Heights
Sentinel Peak
Murray Range
DEVONIAN
MIDDLE ORDOVICIAN
MONKMAN
BEAVERFOOT
UPPER ORDOVICIAN
LOWER ORDOVICIAN
UPPER CAMBRIAN
HOTA
BYNG
MIETTE GP.
Metres
0
500
1000
Limestone
Dolomite
Sandstone
Siltstone
Mudrock
ORDOVICIAN FOSSILS CODED BY SERIES/STAGE
Wh - Whiterock
uC - upper Canadian
mC - middle Canadian
lC - lower Canadian
CAMBRIAN FOSSILS CODED BY ZONE
S - Saukia
Co - Conaspis
E - Elvinia
Ap - Aphelaspis
Cc - Cedaria - Crepicephalus
Ba - Bolaspidella
BE - Bathyuriscus - Elrathina
G - Glossopleura
Al - Albertella
PP - Plagiura-'Poliella'
BO - Bonnia - Olenellus
GSC

SAUK I SUB-SEQUENCE, LOWER CAMBRIAN

Tectonic elements

The Sauk I sub-sequence, as the earliest deposits laid down upon the young passive margin, records the most rapid subsidence rates of any of the three Sauk sub-sequences (Bond and Kominz, 1984), yet it is the Sauk I sub-sequence that advanced the least onto the craton, as predicted by current theory (Watts, 1982). The simple pattern that would be predicted from theory was modified locally by positive and negative tectonic features whose origins are as yet unexplained (Fig. 4B.3):

a. Montania, a landmass composed of Belt-Purcell sedimentary rocks, straddling the 49th parallel, was already in existence at the beginning of the Cambrian (Norris and Price, 1966).

b. Robson Basin was accommodating exceptionally thick sediments from the beginning of the Cambrian (Young, 1979).

c. Mackenzie Arch, a chord to the arc of the frontal Mackenzie Mountains, underwent the next-to-last of its several episodes of uplift prior to or during the Early Cambrian (Aitken et al., 1973).

d. Misty Creek Embayment (Cecile, 1982), an arm of Selwyn Basin flanked by platformal areas, began to subside rapidly only in the terminal Early Cambrian. Selwyn Basin itself contains the latest Proterozoic and early Paleozoic, slope and deep-water deposits of the passive margin.

e. Richardson Trough, an aulacogen-like feature opening northward to the northern continental margin, is aligned en echelon with Misty Creek Embayment and shares its history of Paleozoic subsidence (Norris, 1983), but also appears to have had an earlier history as a rift-basin in Late Proterozoic (Rapitan) time.

f. Leith Ridge, directly south of Great Bear Lake, existed during the Early Cambrian as a linear, southwest-trending, topographic ridge of granitic gneisses, controlled on its northern side by an ancient fault. Bulmer Lake Arch, a comparable feature in the subsurface to the southwest, may have stood high then, also (Meijer Drees, 1975). It underwent renewed (?) uplift prior to the Late Ordovician, however, and evidence for its earlier existence is equivocal.

South of latitude 60°N, where the breadth of the undisturbed Interior Platform is of the order of 600 km, Sauk I strata do not extend far enough toward the centre of the craton to have been penetrated by the drill, even within the disturbed belt (Fig. 4B.3). North of 60°, the breadth of the Interior Platform (from the outer edge of the disturbed belt to the edge of the exposed Canadian Shield) is about 300 km, and Sauk I deposits reach the exposed shield at least locally. This may be a consequence of the earlier onset of passive-margin subsidence north of 60°.

Figure 4B.4. Sauk Sequence: along-strike stratigraphic section, Windsor Mountain (south) to Murray Range, near Pine Pass (north). Northern half modified from Slind and Perkins (1966). Sources of data as in Figures 4B.3, 4B.8, plus Norris and Price (1966).

Stratigraphy

Southern Rocky Mountains

The best-developed section of Lower Cambrian strata in the southern Rockies is that in the vicinity of Mount Robson (Fritz and Mountjoy, 1975):

Hota Formation (80 m, top): Limestone, derived dolomite, and minor mudrocks. The limestones are skeletal at top and base, and lime mudstone is in the middle. *Bonnia - Olenellus* Zone.

Mahto Formation (380-460 m): Sandstone (quartzite), minor mudrocks; locally, archaeocyathid bioherms. Sparse fossils of *Bonnia - Olenellus* Zone.

Mural Formation (240-320 m): Limestone and derived dolomite, with a middle member of siltstone. The limestones include skeletal, intraclast, and pellet grainstone, lime mudstone, and archaeocyathid bioherms. *Bonnia - Olenellus* Zone above, *Nevadella* below.

McNaughton Formation (1500-2000 m, base): Sandstone (quartzite) and minor conglomeratic sandstone; variable amounts of mudrocks, these most abundant in the thickest sections (depositional troughs). The trace fossil *Didymaulichnus* sp. occurs immediately below the McNaughton; placement of the base of the formation is

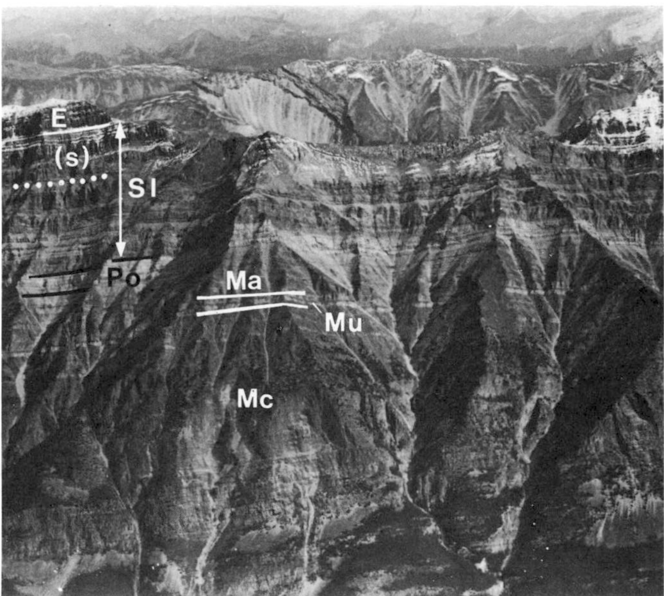

Figure 4B.5. Lower and Middle Cambrian formations exposed on Mount Kerkeslin, Jasper National Park. Mc - McNaughton Formation; Mu - Mural Formation; Ma - Mahto Formation; Po - Peyto Formation; SI - Snake Indian Formation (s - Stephen Member); E - Eldon Formation (photo - J.D. Aitken, ISPG 1518-37).

equivocal in some areas, however, and the critical fossils may lie above the 'sub-Cambrian unconformity', for which there is good evidence in the region (Young, 1979; Teitz and Mountjoy, 1985).

Southward along strike (Fig. 4B.4), the carbonate rocks of the Mural Formation thin and change facies to sandstone, and the McNaughton - Mural - Mahto triad (Fig. 4B.5) making up the northern Gog Group passes into a thick, undivided, quartzite-dominated Gog Group. The Hota equivalent, known as the Peyto Formation, persists above the quartzites to the Bow River transect near latitude 51°. South of there, the Peyto, and beneath it the upper members of the Gog clastic succession, are erosionally truncated southward beneath Middle Cambrian strata. Still farther southward, in a region in which Sauk I deposits are concealed, sub-Middle Cambrian erosion and onlap onto the Montania landmass resulted in the thinning and disappearance of the sub-sequence. Conglomerates of the Lower Cambrian Cranbrook Formation in the southern Rocky Mountain Trench attest to young relief on the northwestern flank of Montania, and suggest the possibility of Early Cambrian movement along the Moyie-Dibble Creek fault system that delineates that flank (Norris and Price, 1966).

Eastward, pinchout of the Gog Group as a whole, and within it, the Mural and Hota-Peyto carbonates, is rapid (Fig. 4B.6, A-A′). The position of the eastward pinchout of the Gog and equivalents is unknown, but is unlikely to be east of the disturbed belt. The Gog-like Flathead Sandstone in southern Alberta and Montana is Middle Cambrian in age, as are, almost certainly, Gog-like sandstones resting on crystalline basement in the subsurface of central Alberta.

Northern Rocky Mountains

In the Rocky Mountains north of the Peace - Athabasca Arch (see Fig. 4B.8), the tectonic regime and the character and distribution of Sauk I deposits are significantly different from those south of the arch. An overall westward thickening and fining in Lower Cambrian deposits is accompanied, as in the south (Fritz, 1979a, 1980b), by a concomitant increase in the amount and proportion of carbonate rocks. There is, however, an apparent record of syndepositional faulting that is lacking in the south (Fig. 4B.6, BB′).

On Macdonald Platform, along the mountain front, Lower Cambrian strata are thin, totalling as little as 110 m. They consist of orthoquartzite and subordinate siltstone. Westward, the Lower Cambrian thickens to as much as 740 m, and the proportion of quartzite is diminished by the introduction of units of shale (in part with limestone nodules), dolomite, and minor limestone (Fritz, 1979a, 1980b). *Bonnia - Olenellus* and *Nevadella* Zone fossils are present, the lowest of the latter occurring 400 m above the base.

The thickening trend apparently continues (Gabrielse, 1975; Fritz, 1980b) to the most westerly reported sections in the Rocky Mountains. A continuing westward trend of diminishing sandstone and increasing limestone and mudrock is generally observed, but some western sections consist mainly of quartzite, a suggestion of multiple sources of detritus. The base of the Lower Cambrian is strongly unconformable, resting on diamictite of Windermere age in the west, and upon the much older, Aida and Gataga formations (Purcell equivalents?) toward the mountain front.

The stratigraphy of Macdonald Platform does not continue beyond the 60th parallel on the northwesterly Rocky Mountain trend. Sedimentary facies undergo a northeastward deflection of 250 km, sub-parallel to the deflection of the structural trend into the Mackenzie arc; Lower Cambrian sections comparable in thickness and facies to those of Macdonald Platform are found in the Franklin Mountains, northeast of Mackenzie River. This deflection follows the inferred offset of the continental margin, for which a Windermere age has been argued in Subchapter 4A. Lower Cambrian strata in Selwyn Basin, which lies on a projection of the Rocky Mountains trend from Macdonald Platform, are predominantly fine-grained, basinal, clastic rocks, whose stratigraphic limits are poorly controlled. Their thickness does not exceed a few hundred metres.

Mackenzie and Franklin Mountains

In Mackenzie and Franklin Mountains, the Sauk I sub-sequence occurs in two belts separated by the broad crest of Mackenzie Arch, where Lower Cambrian strata are absent and may never have been deposited (Fig. 4B.3; 4B.6, CC′). Southwest of the arch, Lower Cambrian formations studied regionally by W.H. Fritz are limited to the Plateau Thrust Sheet and structures yet farther southwest, where they rest unconformably on three Upper Proterozoic, deepwater clastic-platform carbonate cycles, in part of Ediacaran age (Hofmann, 1981; Aitken, 1988). The succession (Fig. 4B.7) is:

Hess River Formation (basal part only; top of sub-sequence): dark mudrocks; *Bonnia - Olenellus* Zone (initial subsidence of Misty Creek Embayment).

Sekwi Formation (150-900 m): carbonate rocks; to the northeast platformal, with increasing detrital content; passing to slope deposits in the southwest. *Bonnia - Olenellus* Zone at top, *Fallotaspis* at base.

Vampire Formation (0-930 m): mainly dark-coloured, fine-grained clastic rocks; a basinal facies of the Backbone Ranges Formation (below). Acritarchs from the upper (and most persistent easterly) part correspond to the Atdabanian of Siberia (Baudet et al., 1989); the oldest known Vampire contains arthropod trace fossils and small shelly fossils whose position is near the Precambrian-Cambrian boundary (Nowlan et al., 1985).

Figure 4B.6. Sauk I: stratigraphic cross-sections (for lines of section, see Figure 4B.3). Subjacent strata: HBDL - Hornby Bay - Dismal Lakes; We - Wernecke Supergroup; Pu′ - presumed Purcell equivalent; MM - Mackenzie Mountains Supergroup; Wi - Windermere Supergroup. Tectonic elements as in Figure 4B.3. Data from Fritz (1972a, 1979a, c, 1980b); Pugh (1983 and unpub. ms.); Macqueen (1969); Aitken et al. (1973); and Aitken (in press).

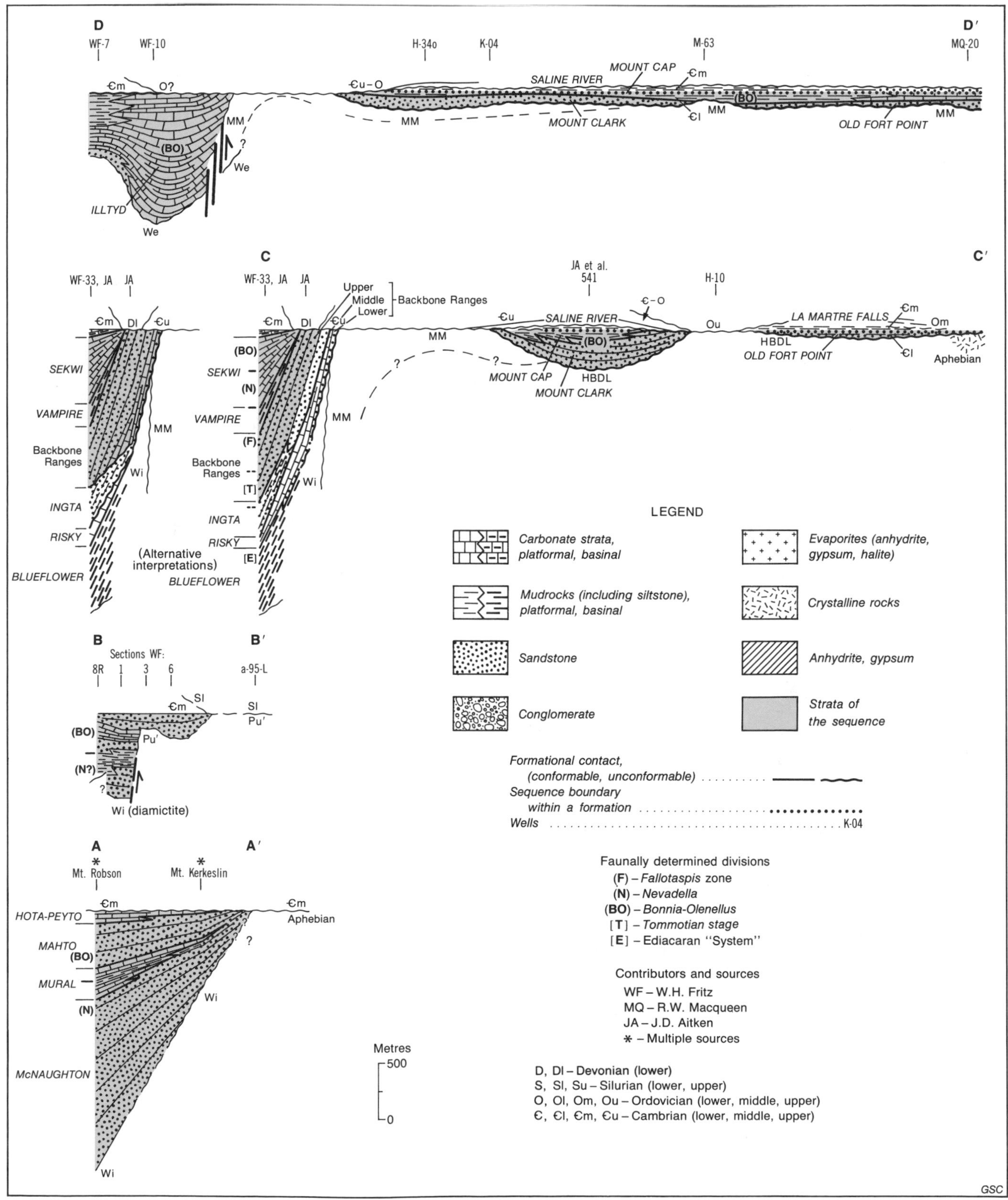

D
D′
WF-7
WF-10
H-34o
K-04
M-63
MQ-20
MOUNT CAP
SALINE RIVER
MOUNT CLARK
OLD FORT POINT
ILLTYD
C
C′
WF-33, JA JA
JA et al. 541
H-10
Upper
Middle
Lower
Backbone Ranges
LA MARTRE FALLS
HBDL
Aphebian
SEKWI
VAMPIRE
Backbone Ranges
INGTA
RISKY
BLUEFLOWER
(Alternative interpretations)
B
B′
Sections WF:
8R 1 3 6
a-95-L
Wi (diamictite)
A
A′
Mt. Robson
Mt. Kerkeslin
HOTA-PEYTO
MAHTO
MURAL
McNAUGHTON
Metres
500
0
LEGEND
Carbonate strata, platformal, basinal
Mudrocks (including siltstone), platformal, basinal
Sandstone
Conglomerate
Evaporites (anhydrite, gypsum, halite)
Crystalline rocks
Anhydrite, gypsum
Strata of the sequence
Formational contact, (conformable, unconformable)
Sequence boundary within a formation
Wells K-04
Faunally determined divisions
(F) – Fallotaspis zone
(N) – Nevadella
(BO) – Bonnia-Olenellus
[T] – Tommotian stage
[E] – Ediacaran "System"
Contributors and sources
WF – W.H. Fritz
MQ – R.W. Macqueen
JA – J.D. Aitken
* – Multiple sources
D, Dl – Devonian (lower)
S, Sl, Su – Silurian (lower, upper)
O, Ol, Om, Ou – Ordovician (lower, middle, upper)
Є, Єl, Єm, Єu – Cambrian (lower, middle, upper)
GSC

Backbone Ranges Formation (50-2200+ m, base): mainly quartz arenites with minor mudrocks and a middle carbonate member. The upper member is certainly Cambrian on the basis of its relationship to the Vampire Formation, its trace fossils, and its position above *Protohertzina* cf. *P. anabarica* (see Hofmann, 1983). Correlation of the lower two members is in dispute; these may be Ediacaran/Vendian or Cambrian.

Northeast of Mackenzie Arch, no shelly fossils older than Upper *Bonnia - Olenellus* Zone are known. There, the rocks unconformably beneath the Sauk I sub-sequence are much older than those in the Plateau Thrust Sheet. Northwest of a line through Fort Norman, various formations of the Mackenzie Mountains Supergroup subcrop, whereas southeast of that line, the subcrop is yet older sedimentary strata of the Hornby Bay and Dismal Lakes groups; the line marks a northwest-side-down, Precambrian structure (Aitken and Pugh, 1984).

The succession northeast of the Mackenzie Arch is:

Mount Cap Formation (100-200 m, top): limestone, dark grey to black shale, bioturbated, glauconitic sandstone and siltstone; upper *Bonnia - Olenellus* Zone. In some areas, the Mount Cap incorporates Middle Cambrian strata and straddles a sub-Middle Cambrian disconformity (W.H. Fritz, in Aitken et al., 1973, p. 29).

Mount Clark Formation (up to 220+ m, base): thick-bedded quartz arenites, with *Skolithos* burrows to the base.

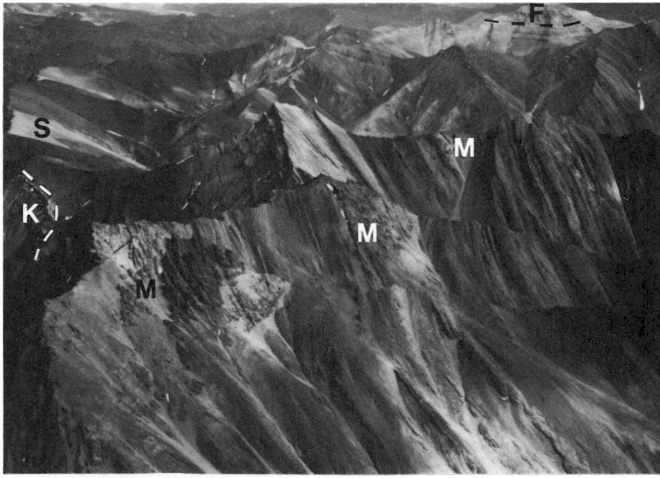

Figure 4B.7. Lower Cambrian and upper Proterozoic succession, southwest Snake River map-area, Yukon. Upper Proterozoic Keele (K) and Sheepbed (S) formations at left. Lower Cambrian quartz arenites (Backbone Ranges equivalent) are thin and inconspicuous above the Sheepbed. Higher strata are all Sekwi Formation with a higher than usual content of detrital beds (hence ledgy weathering), and several carbonate mounds (M). Cambro-Ordovician Franklin Mountain Formation (F) in right distance (photo - D.K. Norris, ISPG 2312-1).

Although the lack of diagnostic fossils in the Mount Clark leaves a degree of uncertainty, regional relationships strongly imply that, except for units that pinch out through northeastward onlap, much of the carbonate rock of the Sekwi Formation south of Mackenzie Arch has detrital correlatives northeast of the arch.

Great Bear and Anderson Plains

In contrast to their areal restriction on the Interior Platform south of the Liard Line (Fig. 4B.3), Sauk I strata are present on Mackenzie Platform northeastward to the edge of the exposed Canadian Shield. The oldest fossils recovered are from the Mount Cap Formation, and are generally of the Middle Cambrian *Glossopleura* Zone. It might be argued that the basal quartz arenites along the shield are Middle Cambrian, like the analogous Flathead Formation onlapping the Montania landmass in southern Alberta. On the other hand, a *Bonnia - Olenellus* fauna is present above thick, basal quartz arenites of the Old Fort Point Formation as far east as the Colville Hills (Macqueen and Mackenzie, 1973), and it appears probable that Lower Cambrian deposits (quartz arenites and dolomites), though thin, persist to the shield. Evidence from outcrop and the subsurface demonstrates sub-Sauk topography of appreciable relief, adjusted to the resistance of subcropping rocks.

The Mount Cap Formation is host to an undeveloped accumulation of natural gas in the Tedji Lake area. Because the gas lies beneath a seal of evaporites of the Middle (?) Cambrian Saline River Formation, its source can be no younger than Middle Cambrian.

SAUK II SUB-SEQUENCE, MIDDLE AND UPPER CAMBRIAN

Tectonic elements

The Sauk II sub-sequence is unconformable or disconformable at its base, except in basins that were subsiding rapidly at the beginning of the Middle Cambrian. Passive-margin subsidence was continuing, however, and with resumption of the sea-level rise that characterized all of early Sauk time, transgression was rapid and extensive; in the south, basal Sauk II deposits onlapped beyond the Sauk I strandline onto crystalline rocks of the Canadian Shield. Some structural elements established earlier persisted, some became extinct or were buried, and others came into being or were reactivated (Fig. 4B.8):

a. Montania was covered by marine sediments early in the Middle Cambrian, receiving clastic sediments originally and carbonates subsequently (Norris and Price, 1966).
b. Lloydminster Embayment came into being, largely through uplift of the flanking arches.
c. Robson Basin (Young, 1979) underwent its period of most rapid subsidence.
d. The Kicking Horse Rim (Aitken, 1971) came into being through localized, earliest Middle Cambrian uplift. Thereafter, it continued to be renewed through depositional processes, and to dominate lower Paleozoic depositional patterns.

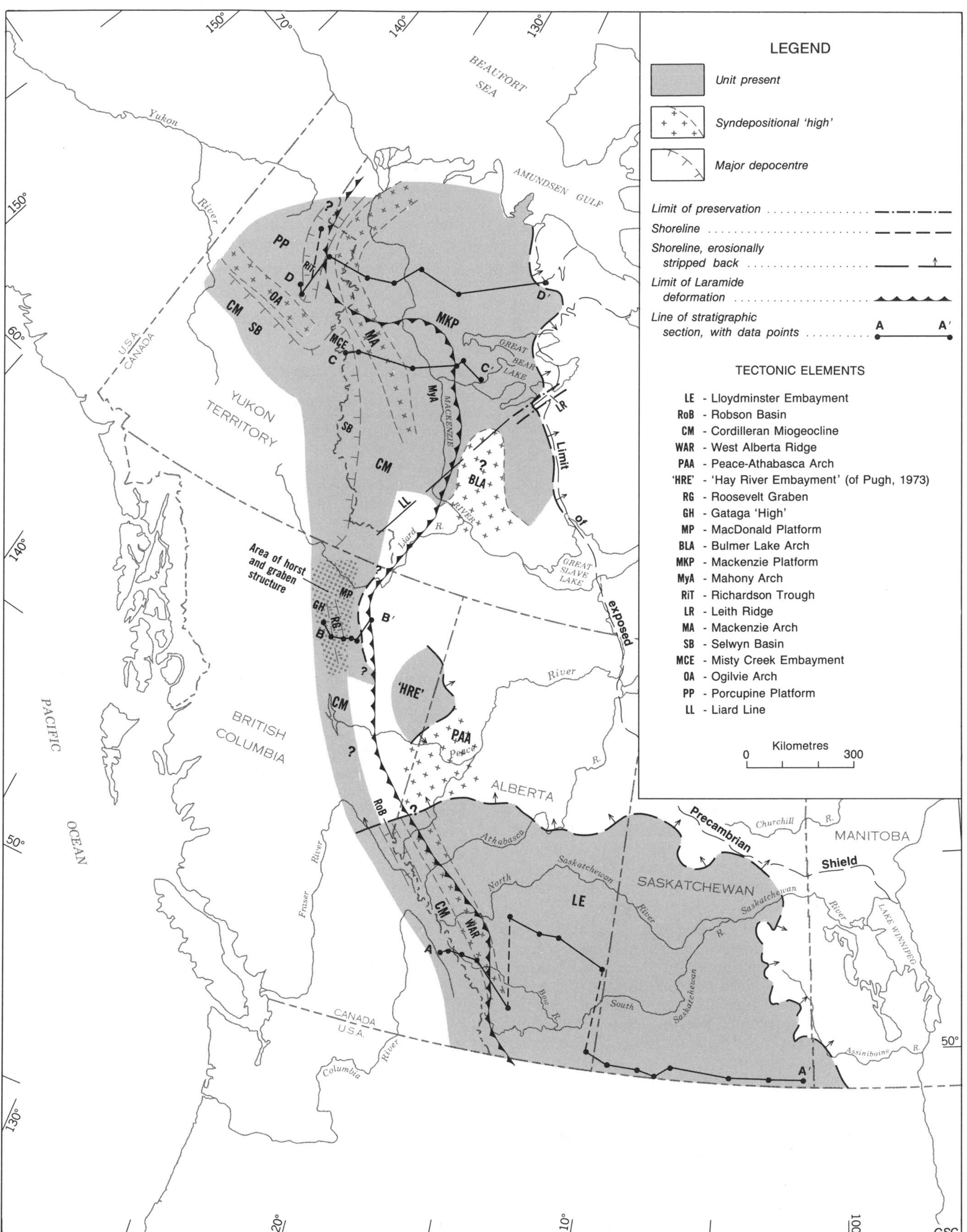

Figure 4B.8. Sauk II: tectonic elements, shorelines.

e. Initial rise of the Peace-Athabasca Arch in the Middle Cambrian is suggested by the appearance of sandstones along what, in Devonian times, became its flanks (Pugh, 1973), and by the westward swing of isopachs, both in the mountains north of Jasper and in the subsurface to the east.

f. Roosevelt Graben, in the Rocky Mountains north of the Peace-Athabasca Arch, underwent its period of most rapid subsidence; the variability of thickness and lithofacies in the Middle Cambrian of the region suggests widespread horst-and-graben tectonics.

g. Mackenzie Arch was at most a feeble source of sandy detritus during early to mid-Middle Cambrian time. If emergent at all, it was a low feature, nearly buried by Sauk I deposits. Its final phase of uplift, toward the end of the Middle Cambrian, produced a closed basin in which evaporites and redbeds of the Saline River Formation accumulated.

h. Middle Cambrian horst-and-graben tectonics apparently affected parts of Mackenzie platform (Meijer Drees, 1986, Fig. 16, 17), but lacks expression in the Sauk III sub-sequence.

j. Misty Creek Embayment and the Richardson Trough were subsiding relative to their flanking platforms. The presence of Middle Cambrian conglomerates west of Richardson Trough (W.H. Fritz, pers. comm., 1985) suggests that some fault blocks rose at that time.

k. Leith Ridge and probably the Bulmer Lake Arch continued emergent, the latter being a source of quartz sand in the Late Cambrian.

Stratigraphy

Southern Rocky Mountains

The platformal succession of the Sauk II sub-sequence is thick, virtually complete, and well-studied in the southern Canadian Rocky Mountains (Fig. 4B.4, 4B.9, 4B.10, 4B.11, 4B.14). Nomenclature established there has been successfully extended to the subsurface of the foothills belt and the undeformed basin to the east. On the other hand, the thick, penetratively deformed, sparsely fossiliferous, basinal strata to the west are lithologically monotonous and have received little study. These basinal strata present many unresolved problems, including those of their true depositional thickness and original extent.

Three concepts developed for the platform stratigraphy provide a framework within which sedimentology and paleogeography may be comprehended readily (Aitken, 1966, 1971, 1978). The first concept is that of three contiguous **facies belts**, termed inner detrital, middle carbonate, and outer detrital (Robison, 1960). Although the boundaries between these belts shifted back and forth, all three were usually present, except in those rare instances in which the middle carbonate belt was temporarily extinguished by a flood of inner-detrital sediment. The second concept is that of Grand Cycles, each spanning from one to three trilobite assemblage-zones and consisting of a sharply based formation of inner detrital character gradationally succeeded by a formation of middle-carbonate character (Fig. 4B.11, 4B.13). Grand Cycles approximate chronostratigraphic units. The third concept developed from Sauk II stratigraphy in the southern Rockies is that of the Kicking Horse Rim, a recurrent, narrow, elevated, spatially fixed carbonate shoal complex along the outer edge of the middle carbonate facies belt (carbonate platform). The rim, whenever present (as it usually was) profoundly influenced sedimentation, both in the inshore basin on the Interior Platform behind it and the open basin on the open-sea side.

Cambrian isopachs on the platform south of Jasper generally trend north-northwest, sub-parallel to the tectonic strike, but swing westward north of Jasper. Distinct, south-to-north facies changes have different expressions for different chronostratigraphic packages:

a. Middle Cambrian, except for late *Bolaspidella* time.
b. Latest Middle Cambrian, earliest Dresbachian.
c. Most of Dresbachian time.

Middle Cambrian

Platformal deposits

The Middle Cambrian deposits of the outer edge of the platform change little between Mount Assiniboine, south of Banff, and the Mount Robson area, west of Jasper. Although the Kicking Horse Rim is not as sharply definable north of Field as in its classical exposures near that place (partly because of the vagaries of exposure and preservation), the westernmost deposits of the successive carbonate platforms contain many members of peritidal facies, and lack eastward-rooted tongues of inner-detrital strata, implying proximity to the rim. A few thin units of dark green-grey argillite are probably incursions of **outer**-detrital facies.

East of the rim, a marked northward increase in the proportion of fine-grained siliciclastic units takes place. This results most notably in the passage of the Cathedral Formation from a thick (180-450 m), feature-forming unit essentially of carbonate rocks along the Bow River transect, to a succession of interleaved fine-siliciclastic and carbonate rocks along the Athabasca River transect (Fig. 4B.4). First, the mudrock-dominated Mount Whyte Formation thickens northward at the expense of the overlying Cathedral. Second, two thin (2 m and less) shaly members in the south, which are attenuated tongues of inner-detrital facies, thicken northward to destroy the monolithic character of the Cathedral. Third, the mudstone-dominated Stephen Formation thickens northward at the expense of overlying Eldon carbonates. Thus, the entire succession, Mount Whyte-Cathedral-Stephen, (up to 700 m) of the Bow River transect passes northward to Snake Indian Formation (240-550 m), a succession of alternating carbonate and shaly members of platformal character (Fig. 4B.4, 4B.5, 4B.9, 4B.11, 4B.12).

From the Bow River to the Athabasca River transect, the Eldon Formation retains its integrity as a carbonate unit, but shows some thinning (380-250 m), in part by facies

Figure 4B.9. The Cambrian succession of the Bow River transect, Rocky Mountains (graphic column). Arrows denote Grand Cycles.

CAMBRIAN AND LOWER ORDOVICIAN - SAUK SEQUENCE

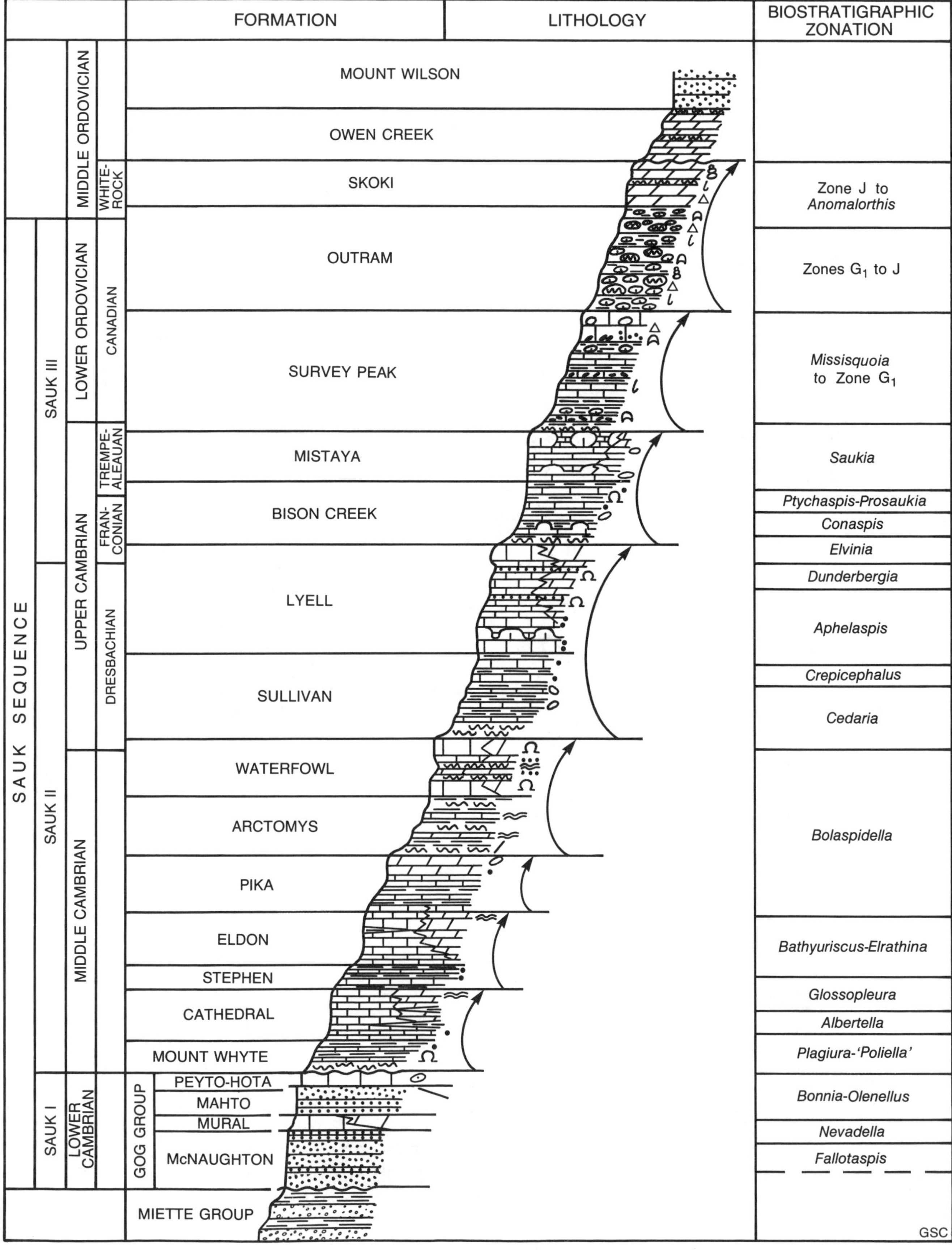

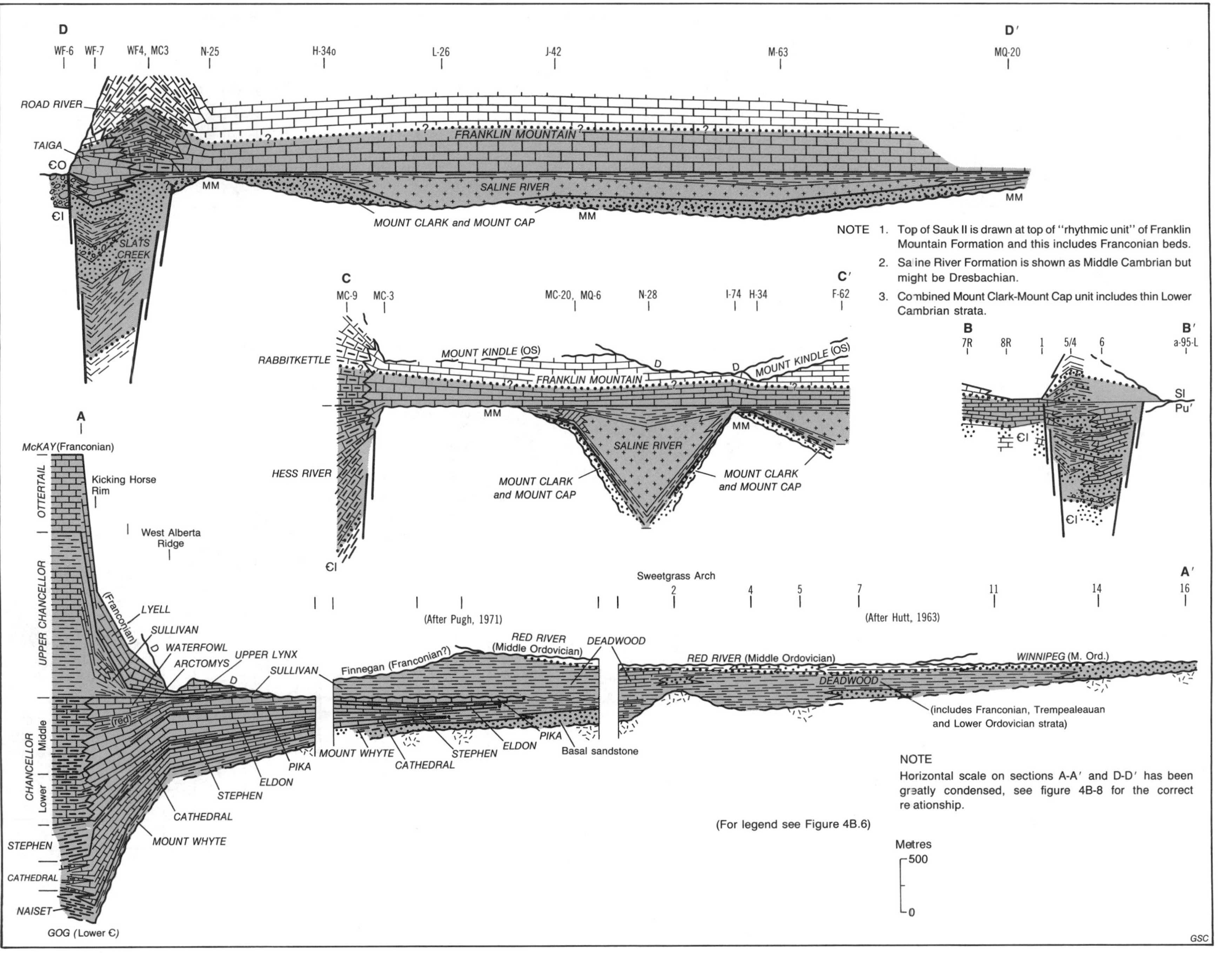
D
WF-6
WF-7
WF4, MC3
N-25
H-34a
L-26
J-42
M-63
MQ-20
D′
ROAD RIVER
TAIGA
€O
€l
SLATS CREEK
MM
FRANKLIN MOUNTAIN
SALINE RIVER
MOUNT CLARK and MOUNT CAP
NOTE 1. Top of Sauk II is drawn at top of "rhythmic unit" of Franklin Mountain Formation and this includes Franconian beds.
2. Saline River Formation is shown as Middle Cambrian but might be Dresbachian.
3. Combined Mount Clark-Mount Cap unit includes thin Lower Cambrian strata.
C
MC-9
MC-3
MC-20, MQ-6
N-28
I-74
H-34
F-62
C′
RABBITKETTLE
HESS RIVER
MOUNT KINDLE (OS)
FRANKLIN MOUNTAIN
SALINE RIVER
MOUNT CLARK and MOUNT CAP
D
B
7R
8R
1
5/4
6
a-95-L
B′
Sl
Pu′
A
Kicking Horse Rim
West Alberta Ridge
(After Pugh, 1971)
Sweetgrass Arch
(After Hutt, 1963)
A′
McKAY (Franconian)
OTTERTAIL
UPPER CHANCELLOR
CHANCELLOR
Middle
Lower
STEPHEN
CATHEDRAL
NAISET
GOG (Lower €)
LYELL
(Franconian)
SULLIVAN
WATERFOWL
ARCTOMYS
UPPER LYNX
SULLIVAN
PIKA
ELDON
STEPHEN
CATHEDRAL
MOUNT WHYTE
Finnegan (Franconian?)
RED RIVER (Middle Ordovician)
DEADWOOD
Basal sandstone
RED RIVER (Middle Ordovician)
WINNIPEG (M. Ord.)
(includes Franconian, Trempealeauan and Lower Ordovician strata)
NOTE
Horizontal scale on sections A-A′ and D-D′ has been greatly condensed, see figure 4B-8 for the correct relationship.
Metres
500
0
(For legend see Figure 4B.6)
GSC

change to shaly strata of the underlying Stephen Member of the Snake Indian. The Pika Formation maintains its character along the same route, but the shaly half-cycle of its lower part thickens markedly, and at its top near Jasper are shaly beds unknown in the south.

Changes at the level of the terminal Middle Cambrian Arctomys-Waterfowl cycle are parallel, but of different character. The Waterfowl (20-220 m) persists as a carbonate half-cycle to the latitude of Jasper, but it, and the evaporitic redbeds (20-250 m) of the Arctomys beneath, acquire a significant content of quartz sandstone beds.

Northward along the structural grain from the Athabasca River transect (Fig. 4B.4), the Middle Cambrian thins gradually, by nearly three-quarters (Slind and Perkins, 1966). As in a west-to-east traverse across the southern Rockies and plains, thinning is most marked in the lower formations, especially the Snake Indian. The Eldon (Titkana) persists at least as far as Peace River, but the Pika changes to a facies of dolomite and dolomitic sandstone with interbeds of green shale, and was included with the Arctomys by Slind and Perkins (1966). This Pika equivalent appears to pinch out south of Pine Pass (M.E. McMechan, pers. comm., 1985). The report, by Slind and Perkins (1966) of a *Cedaria* fauna from beds referred to the Arctomys, north of Jasper, indicates that an Arctomys-like redbed facies there has replaced the peritidal carbonates of the Waterfowl, and possibly a part of the overlying Lynx carbonates as well. All of these observations support the conclusions that north of Jasper, the depositional strike swings westward, as seen in the subsurface to the east (Pugh, 1975), and that clastic detritus was supplied from the northwest. An early, Middle Cambrian expression of the Peace-Athabasca Arch, albeit an arch considerably broader than its well-known expression (Peace River Arch) in the Devonian, is implied.

Basinal deposits

Knowledge of the basinal, or outer-detrital deposits west of the successive carbonate platforms is fragmentary. What is certain is that all of the Middle Cambrian, platformal carbonate lithosomes disappear abruptly, directly west of the Kicking Horse Rim. The lower Cathedral and the Eldon-Pika are last seen in a westward-thinning, ribbon-bedded lime mudstone facies of ramp origin. In contrast, the upper Cathedral terminates in a near-vertical reef-face (Fritz, 1971; McIlreath, 1977; Aitken, in press). The equivalent basinal deposits, comprised by the Chancellor Formation, consist of shale, laminated siltstone, ribbon-bedded, more or less argillaceous and silty lime mudstone, and limestone-shale couplets. Debris-flow breccias and large olistoliths are present at several levels.

Figure 4B.10. Sauk II: stratigraphic cross-sections (for lines of section, see Figure 4B.8). Subjacent strata identified as in Figure 4B.3. Data from Fritz (1978a, 1980b); Hutt (1963a, b); Pugh (1971, 1983, and unpub. ms.); Cecile (1982), Macqueen (1969, 1970); Cook (1975); Aitken et al. (1973); and Aitken (in press).

Correlation of divisions of the Chancellor Formation with their platformal equivalents is straightforward at the top, but disputed for lower levels. The shaly Sullivan Formation of the Upper Cambrian (Dresbachian) has been

Figure 4B.11. Cambrian section at Mount Loudon, southern Rocky Mountains; North Saskatchewan River at right (view upstream, northwestward). Large-scale cyclicity is strongly expressed in topography. G - Gog Group; Po - Peyto Formation; MW - Mount Whyte Formation; C - Cathedral Formation (R - Ross Lake Member, T - Trinity Lakes Member); S - Stephen Formation; E - Eldon Formation; Pa - Pika Formation; A - Arctomys Formation; W - Waterfowl Formation; Su - Sullivan Formation; L - Lyell Formation (photo - J.D. Aitken, ISPG 1518-53).

Figure 4B.12. Snake Indian Formation at Whitecap Mountain, northeast of Jasper, Alberta. G - Gog Group; E - Eldon Formation; Pa - Pika Formation. Compare the sub-Eldon strata with those in Figure 4B.11 (photo - E.W. Mountjoy, GSC 116073).

traced from the platform into continuity with the upper Chancellor. Above it, the peritidal carbonate rocks of the Lyell Formation pass westward to those of the Ottertail Formation, a record of a unique westward expansion of the carbonate (constructional) platform across former basinal areas (Cook, 1975). Cook correlated the middle Chancellor (660 m plus) with the mainly Middle Cambrian Arctomys-Waterfowl Grand Cycle, and the lower Chancellor (more than 1 km thick, base unknown) with the older Middle Cambrian formations of the platform. McIlreath (1977), on the other hand, correlated the upper middle Chancellor with the upper Eldon, plus Pika, plus Arctomys-Waterfowl of the platform, the lower middle Chancellor with the Field Member of the Eldon (a tongue of dark, outer-detrital mudrocks), and the lower Chancellor with the lower Eldon. The divisions of the Chancellor Formation referred to are fairly local entities, however, and become more difficult to delineate with distance either westward from the rim or northward or southward from the Bow-Kicking Horse transect. Fossils are extremely rare in the 'Chancellor facies', and it is not known whether equivalents of the Stephen and older Middle Cambrian formations outcrop between the Kicking Horse Rim and the Rocky Mountain Trench.

Upper Cambrian

Platformal deposits

The terminal Middle Cambrian was a time of major regression, as shown by the displacement of the pinchout of the Arctomys - Waterfowl Grand Cycle far to the west of earlier shorelines, and by a locally erosional contact between the Waterfowl and the overlying Sullivan Formation. At western localities, the base of the Dresbachian probably corresponds to a contact between unfossiliferous peritidal carbonates and overlying, richly skeletal carbonates, near the top of the Waterfowl Formation (Fig. 4B.9). The transgressive, skeletal limestones are followed by greenish shales with beds of particulate (oolitic, conglomeratic, skeletal) limestone of the Sullivan Formation (up to 425 m). The shales exposed in the Rocky Mountains are interpreted as of 'deep-water' origin, and the limestones as resedimented deposits shed eastward from a contemporaneous carbonate shoal complex, represented by the overlying and (as interpreted) laterally equivalent Lyell Formation (Aitken, 1978). Isopachs of the Sullivan Formation, showing overall westward thickening, trend north to northeast south of Jasper and swing to west-northwest through Jasper. Northward thinning there is mostly by facies change to Lyell-equivalent, Lynx carbonates. Northward disappearance of the Sullivan and the lithologically similar, Franconian Bison Creek Formation into Lynx peritidal carbonates may imply a long-lived, **carbonate** shoreline across the westward projection of Peace-Athabasca Arch, in contrast to a muddy-sandy shoreline in Saskatchewan.

Gradationally above the Sullivan, the Lyell Formation (up to 350 m), largely peritidal carbonate strata, completes the Sullivan-Lyell Grand Cycle. The lowest Franconian fossils, and the base of the Sauk III sub-sequence, appear with skeletal limestones near the top of the Lyell. At Sunwapta Pass, this intra-Lyell contact is erosional.

Basinal deposits

In contrast with Middle Cambrian chronostratigraphic units, which undergo a total change of depositional facies across the Kicking Horse Rim, Dresbachian units, while undergoing marked thickening to the west, maintain their lithological character almost to the Rocky Mountain Trench. Indeed, had not different nomenclature been coined before the correlation was recognized, the Sullivan and Lyell formations might have been so extended.

The upper Chancellor Formation corresponds to the Sullivan Formation of the platform. Not only does it belong to the same zone (*Cedaria, Crepicephalus*); its greenish grey shales (slates), subordinate, commonly thick, massive beds of limestone (skeletal and, mainly, ooid grainstones with thin conglomerates), and a siltstone member at the base are virtually identical with Sullivan rocks east of the rim. Further, a tripartite character, with a concentration of thick, massive, oolite beds in the middle, corresponds with the platform-edge section near Field, where the Sullivan contains, in its middle, a thick member of oolite. In view of the thick, massive character of the coarse-grained limestone beds in the upper Chancellor, and their lack of crossbedding, it appears that the upper Chancellor, like the Sullivan, may be a deep-water deposit with resedimented limestones.

Although fossils of the *Aphelaspis* and *Dunderbergia* Zones have not been recovered from the Ottertail Formation, it is clearly a westward and thicker (450-600+ m) continuation of the Lyell Formation of the platform. It is overlain by locally fossiliferous, Franconian beds of the lower McKay Group. Furthermore, at least in the more easterly part of its extent, it has a basal member of subtidal, thin-bedded lime mudstone and skeletal limestone, corresponding to the basal member of the Lyell. The basal member of the Ottertail is overlain by peritidal carbonate strata displaying small-scale, shallowing-upward cycles identical with those of most of the Lyell.

The westward spread of shallow-water carbonate sediments of the Lyell/Ottertail across what had earlier been a deepwater (though ensialic) basin is a unique event in the Cambrian history of the southern Canadian Rocky Mountains. It demonstrates that accelerated deposition of clastic sediments during Sullivan/upper Chancellor time brought about shallowing of the former basin to the point that autochthonous carbonate deposition could take place. A new or rejuvenated source of siliciclastic detritus is probably recorded by the basin-filling event.

Lloydminster Embayment

The Lloydminster Embayment (van Hees, 1964) is possibly the tectonic element containing the largest volume of undeformed Sauk II sediments in North America. A shallow depression on the southern Interior Platform, constrained on the north by the Peace-Athabasca Arch, on the east and southeast by the continental axis, and on the south by the Central Montana Uplift, the embayment opens westward to the thick, miogeoclinal deposits of the Rocky Mountains (Fig. 4B.8). Although to a degree a basin of preservation, the basin is outlined, not only by isopachs and erosional limits, but also by facies patterns. In particular, at several levels carbonate and fine-detrital strata in the western part of the embayment pass eastward and particularly northward to sandy facies, in a pattern nearly congruent with isopach trends. The northward increase in clasticity corresponds to that seen in Middle Cambrian formations of the Rocky Mountains. An alternative view of the embayment is that it was a region of normal subsidence in the Cambrian, framed by more slowly subsiding static, or rising elements.

Middle Cambrian

Delineation of Grand Cycles in the subsurface Middle Cambrian reveals a regular pattern that is the record of the great Cambrian transgression (Fig. 4B.10, A-A′). Each successive carbonate half-cycle (in upward order, Cathedral, Eldon, and Pika formations) persists northeastward farther than the preceeding one, and in each instance it is the terminal strata of the Grand Cycle that persist farthest in carbonate facies. Passage of carbonate formations into equivalent formations of shale and siltstone with minor carbonate beds takes place from the bottom upward, so that each carbonate formation or half-cycle is in part coeval with the shaly half-cycle it overlies (Fig. 4B.13). The record is one of prolonged expansion of successive carbonate (constructional) platforms toward the craton, behind the advancing muddy/sandy shoreline, interrupted by sudden expansions of the inner detrital belt that temporarily smothered the carbonate platforms. These detrital excursions are attributed to pulses of rapid, eustatic rise of sea level, superimposed on the overall subsidence of Lloydminster Embayment. The region east of the present mountain front was too far from the continental margin to include thermal effects; there, the subsidence was entirely downflexing due to the loads of the cooling crust outboard and the superposed sediment.

With the depositional pinchout northeastward of successive carbonate lithosomes, the entire Rocky Mountain succession, Mount Whyte-Cathedral-Stephen-Eldon-Pika passes progressively, from the bottom upward, to the equivalent Earlie Formation of glauconitic siltstone, shale, and fine-grained sandstone. This change is analogous to the northward disappearance, in the Rocky Mountains, of the great Cathedral carbonate into the Snake Indian Formation, extended to the extinction of all major carbonate lithosomes. Northward in the subsurface, however, the carbonates of the Cathedral and Eldon formations pass partly to significant members of clean, laminated, and burrowed, partly red, quartz sandstones on approaching the Peace-Athabasca Arch (Pugh, 1971, 1973).

The terminal Middle Cambrian formations of the Rocky Mountains, Arctomys and most of the Waterfowl, are not preserved east of the foothills belt. Pronounced northeastward thinning and increase in quartz sand in the mountains are premonitory of depositional pinchout of the Arctomys-Waterfowl Grand Cycle; no evidence of erosion of the top of the Pika Formation, east of that pinchout, is known. Accordingly, the Arctomys-Waterfowl records an episode of marked regression, relative to the shorelines of underlying and overlying formations. In Lloydminster Embayment, the Sullivan Formation (early Dresbachian; *Cedaria - Crepicephalus* Zone) rests upon the Pika Formation (late Middle Cambrian; *Bolaspidella* Zone) at a disconformity at which the Arctomys-Waterfowl is missing.

Upper Cambrian

As noted above, the lowest part of the Upper Cambrian, that is, the uppermost Waterfowl Formation, is missing by non-deposition east of the Foothills Belt. There, the Upper Cambrian commences with a formation of siltstone and shale with minor beds of limestone and dolomite, the Sullivan Formation (Dresbachian; *Cedaria - Crepicephalus* Zone). With the loss of all carbonate beds, the Sullivan passes eastward to the Deadwood Formation (Fig. 4B.10, A-A′). Northward also, toward the erosional zero-edge, carbonate beds disappear, and the Sullivan passes into successive facies of siltstone-shale, siltstone, and siltstone-sandstone (Pugh, 1971, 1973, 1975).

Along the Front Ranges, and eastward in the subsurface, the dolomites overlying the Sullivan Formation must be treated as the upper division of the Lynx Group, because, with the northeastward disappearance of the Bison Creek Formation (depositional pinchout ? facies change to carbonates ?), the Lyell and Mistaya formations are inseparable (Aitken and Greggs, 1967). The expected, sub-Franconian unconformity (base of the Sauk III sub-sequence), so widely recognized in the U.S.A., is hidden at a contact between carbonate units within the Lyell Formation and equivalents. In the western part of the Lloydminster Embayment, the Sullivan-Lynx contact is conformable and gradational. In central Alberta, strata apparently continous with the upper Lynx, or part thereof, were assigned to the Finnegan Formation by Pugh (1971). The Finnegan (mauve, pink, and white, micaceous, glauconitic siltstone, purple, green, and brown shale, and subordinate, partly red limestone) appears to rest

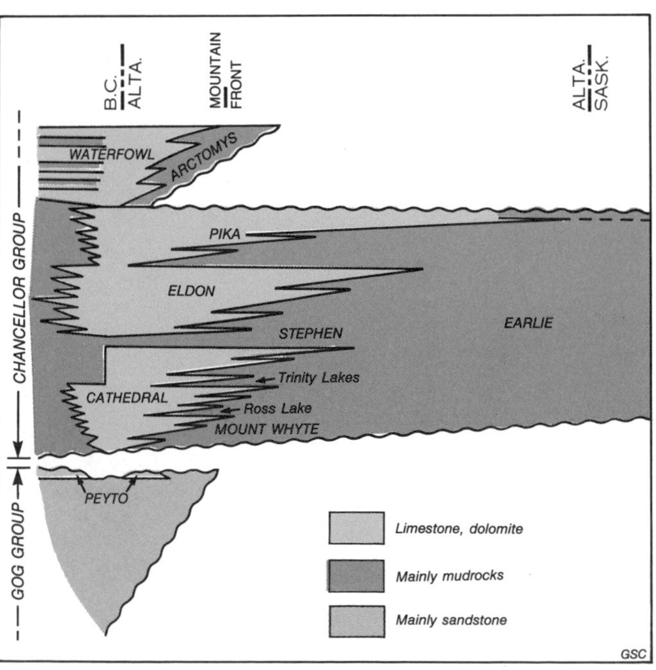

Figure 4B.13. Cartoon illustrating the advance of succesive Middle Cambrian carbonate lithosomes toward the craton. Vertical axis is time.

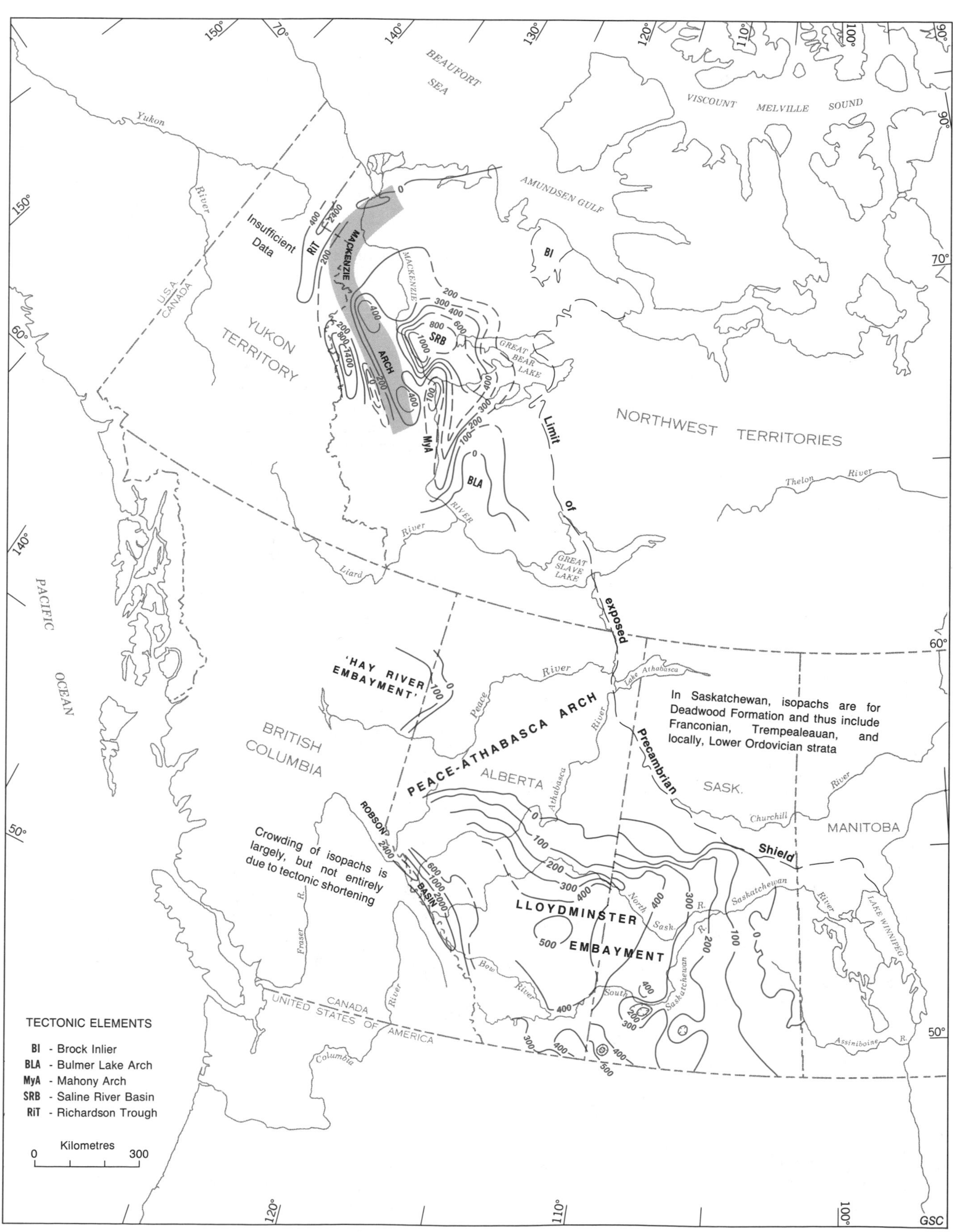

BEAUFORT SEA
VISCOUNT MELVILLE SOUND
AMUNDSEN GULF
Insufficient Data
RiT
MACKENZIE
ARCH
SRB
GREAT BEAR LAKE
BI
YUKON TERRITORY
NORTHWEST TERRITORIES
Limit of exposed Precambrian Shield
MyA
BLA
GREAT SLAVE LAKE
PACIFIC OCEAN
'HAY RIVER EMBAYMENT'
PEACE-ATHABASCA ARCH
In Saskatchewan, isopachs are for Deadwood Formation and thus include Franconian, Trempealeauan, and locally, Lower Ordovician strata
BRITISH COLUMBIA
ALBERTA
SASK.
MANITOBA
ROBSON BASIN
Crowding of isopachs is largely, but not entirely due to tectonic shortening
LLOYDMINSTER EMBAYMENT
LAKE WINNIPEG
UNITED STATES OF AMERICA
CANADA
TECTONIC ELEMENTS
BI - Brock Inlier
BLA - Bulmer Lake Arch
MyA - Mahony Arch
SRB - Saline River Basin
RiT - Richardson Trough
Kilometres
0 300
GSC

unconformably on the Deadwood Formation, and for reasons discussed below, under 'Sauk III', is tentatively considered to be post-Dresbachian.

Williston Basin

Williston Basin did not exist as such until about the beginning of Middle Ordovician (Arenig) time; Cambrian marker-defined units pass across the site of the later basin without differential thickening (van Hees, 1964). In southern Saskatchewan (later Williston Basin), Cambrian isopachs trend roughly north-south, and demonstrate thinning from about 500 m at the Alberta boundary to zero in the southwestern corner of Manitoba, on the flank of the continental axis.

Middle Cambrian

Middle Cambrian strata about 100 m thick reach western Saskatchewan as a basal sandstone (overlying Precambrian, crystalline basement) and fine-grained, detrital strata of the Earlie Formation. The top of the Earlie is defined by a geophysical log marker correlated with the top of the Pika Formation (Pugh, 1971). Pugh advocated restriction of the Deadwood Formation, which overlies the Pika in the Alberta subsurface, to Upper Cambrian strata. With the eastward disappearance of the Pika Formation, however, the top of the Middle Cambrian can be identified only by tracing a geophysical (Ra log) marker, and this marker passes into a succession of mainly fine-grained, detrital strata with minor limestone, which virtually all workers have treated as Deadwood Formation. By tracing this marker, Fuller and Porter (1962) attempted to isolate the Middle Cambrian part of the Deadwood. They showed the Middle Cambrian pinching out in central Saskatchewan, by onlap onto crystalline basement, and being overstepped eastward by the Upper Cambrian part of the Deadwood.

Upper Cambrian

Upper Cambrian strata across the later site of Williston Basin, including the Dresbachian part of the Sauk II sub-sequence, belong, with underlying Middle Cambrian and overlying Franconian, Trempealeauan, and Lower Ordovician strata, to the Deadwood Formation. Marker 'e' of van Hees (1964) may approximate the base of Franconian strata in the Richey area of Montana (Lochman-Balk and Wilson, 1967), but correlation of that marker is tenuous, and the Dresbachian part of the Deadwood has not been mapped. The Deadwood consists mainly of a green and purple, siltstone-shale lithofacies, with minor chalky limestone beds and locally important limestone members.

Figure 4B.14. Sauk II: isopachs. Isopach interval is 100 m east of the Cordillera, larger and variable in the mountains. In Saskatchewan, isopachs are for the Deadwood Formation, including Sauk III strata, hence the misfit at the Alberta-Saskatchewan border. Data from Pugh (1971, 1973, 1975, 1983 and unpub. ms.) and Aitken (in press).

Northeastward, the content of quartz sand increases and coarsens toward the depositional pinchout, which passes close to the Saskatchewan-Manitoba border at the 49th parallel (Fig. 4B.10, A-A′).

The position of the Sauk II - Sauk III contact in cross-section A-A′ of Figures 4B.10 and 4B.17 is uncontrolled and speculative, and the isopach map for the Sauk II sub-sequence in Williston Basin (Fig. 4B.14) includes all Deadwood strata, including those properly belonging to Sauk III.

Hay River Embayment

Sauk II deposits preserved in the subsurface 'Hay River Embayment' on the northern Interior Platform north of Peace-Athabasca Arch (Fig. 4B.8) barely exceed 100 m at maximum, and are limited to 'Cathedral', Eldon, and a thin remnant of 'Pika' Formation, as interpreted by Pugh (1975). While the Eldon persists there in a dolomite facies, the 'Cathedral' below and the 'Pika' above consist to such a large extent of siliciclastic rocks that the latter two terms may be inappropriate.

Northern Rocky Mountains

In the northern Rocky Mountains, a fault system that may have been initiated during the Early Cambrian underwent, during the Middle Cambrian, its period of greatest activity. Faults of the horst-and-graben system, which reached ultimate displacements of 2 km in some instances, trend northwest in part, but others, probably inherited from faults active during the Windermere rifting event, trend northeast (Taylor and Stott, 1973, map). Roosevelt Graben (Gabrielse and Taylor, 1982), the best-documented structure of the system, affected, in part, a region that was platformal during Early Cambrian and again in post-Cambrian times. Its subsidence was most marked during the Middle Cambrian (Fig. 4B.10, B-B′), and was diminishing during the Late Cambrian (see Fritz, 1979a; W.H. Fritz and H.M. Kluyver, in Gabrielse, 1975, Fig. 3).

Middle Cambrian

In Roosevelt Graben, Middle Cambrian strata exceed 800 m in thickness (Fritz, 1979a). They are mainly sandstone and siltstone, with subordinate members of dolomite and limestone and abundant units of thick, coarse, olistostromal conglomerate. In contrast, on the horst ('Gataga high') that forms the western limit of the graben, the Middle Cambrian column is only 120 to 150 m thick, and consists almost entirely of dolomite and limestone; Fritz (1979a) commented that the close spacing of successive faunizones ". . . indicates a history of slow deposition and/or intermittent removal of strata". At or near the basinward limit of exposed Middle Cambrian carbonate rocks is a chain 125 km long of elongate buildups of pale limestone, oriented north-northwest (Fritz, 1979a). The buildups form prominent topographic highs, and were apparently buried in mudrocks or shaly carbonates now eroded. The strict alignment and narrow width (less than 1 km) of the buildups suggest that they are situated along the high shoulder of a tilted fault block. In the region

Figure 4B.15. Franklin Mountain Formation (F) unconformable on Upper Proterozoic strata, Mountain River area, Mackenzie Mountains. Upper Proterozoic Little Dal Group units as follows: b - basinal assemblage; gr - grainstone formation; gy - gypsum formation (photo - J.D. Aitken, ISPG 2280-3).

generally, no clear pattern of distribution of clastic rocks is apparent, suggesting several local sources of siliciclastic detritus (horsts?).

Upper Cambrian

Differential thickening across Roosevelt Graben is less marked in Upper Cambrian deposits, indicating decelerating differential subsidence during Late Cambrian time. The Dresbachian (Sauk II) part of the Upper Cambrian graben-fill may not much exceed 250 m; it consists almost entirely of siltstone and sandstone (Fritz, 1979a). Nevertheless, faulting must have remained active; a horst-and-graben pattern is still apparent, but the 'thicks' and 'thins' in the Sauk II sub-sequence do not, in general, coincide with those of Sauk I and Sauk III (see Fritz 1979a, and W.H. Fritz and H.M. Kluyver, in Gabrielse, 1975, Fig. 3). On the 'Gataga high', condensation is even more marked than in the Middle Cambrian. At some sections, the Dresbachian carbonates are as thin as 70 m (Fritz, 1979a).

Mackenzie Mountains

In the frontal Mackenzie Mountains, except at their western termination against Richardson Mountains, the Middle Cambrian part of the Sauk II sub-sequence occurs only locally but is continuous with the generally thin Middle Cambrian of Great Bear and Anderson Plains. It reappears in the interior ranges, that is, in the Plateau thrust-sheet and similar structural positions. In the middle of the structural arc, the gap is a broad expression (100 km) of Mackenzie Arch. South of the 63rd parallel, the Middle Cambrian is of platformal facies, while to the north, only slope and deepwater facies occur southwest of Mackenzie Arch. Thus, the gap in distribution is to a large extent erosional, and records the final, pre-Dresbachian uplift of the arch (Fig. 4B.8, 4B.10, C-C′, 4B.14, 4B.15).

In contrast, the Dresbachian part of the Sauk II sub-sequence passes across the axis of Mackenzie Arch, and is continuous with coeval strata of similar, platformal, carbonate facies beneath the plains. The top of Sauk II, the base of the Franconian, is concealed within a thick succession of poorly fossiliferous dolomites of the Franklin Mountain Formation, and does not correspond to any formational contact.

Middle Cambrian

Starting with the southernmost Middle Cambrian exposures in Mackenzie Mountains, and following the tectonic arc northward and westward, facies changes with a strong, south-to-north component occur, as follows (Gabrielse et al., 1973; Cecile, 1982):

Avalanche Formation (200 - 400 m): sparsely fossiliferous, finely crystalline dolomite, silty dolomite, dolomitic siltstone and dolomitic mudstone, with crossbedding, sun-cracks, and salt hoppers.

The Avalanche Formation passes northward to:

Rockslide Formation (200 - 500 m): richly fossiliferous, platy, nodular, fine-grained argillaceous limestone and calcareous siltstone; greenish grey, locally pyritic, laminated, dolomitic sandstone, locally with salt hoppers; black shale and minor sandstone.

The Rockslide Formation passes northward to:

Hess River Formation (0 - 2500 m): thinly interbedded silty limestone and dark calcareous shale; minor sedimentary barite and phosphorite. Local, thick accumulations of sandstone-shale flysch account for the thickest sections.

This northward succession of facies is one from a platformal environment to a (largely) slope environment, to a slope-and-basin environment, and mimics the offshore succession of facies at a passive continental margin. In this instance, the strong **northward** component of change suggests that the effective margin is the segment trending northeast from latitude 60°N (Aitken and Pugh, 1984; 'Liard line' of this volume).

The redbeds and evaporites of the Saline River Formation have been considered heretofore by most authors as early Late Cambrian, because Dresbachian trilobites occur a short distance above, in an apparently conformable sequence, and because the Saline River overlies an unconformity taken to be sub-Dresbachian (Aitken et al., 1973). Here, a late Middle Cambrian age is also considered plausible, because the Saline River may be

Figure 4B.16. Sauk III: tectonic elements. Tremadocian limit, after Cecile and Norford (this volume), north of 55°, approximates Franconian limit.

CAMBRIAN AND LOWER ORDOVICIAN - SAUK SEQUENCE

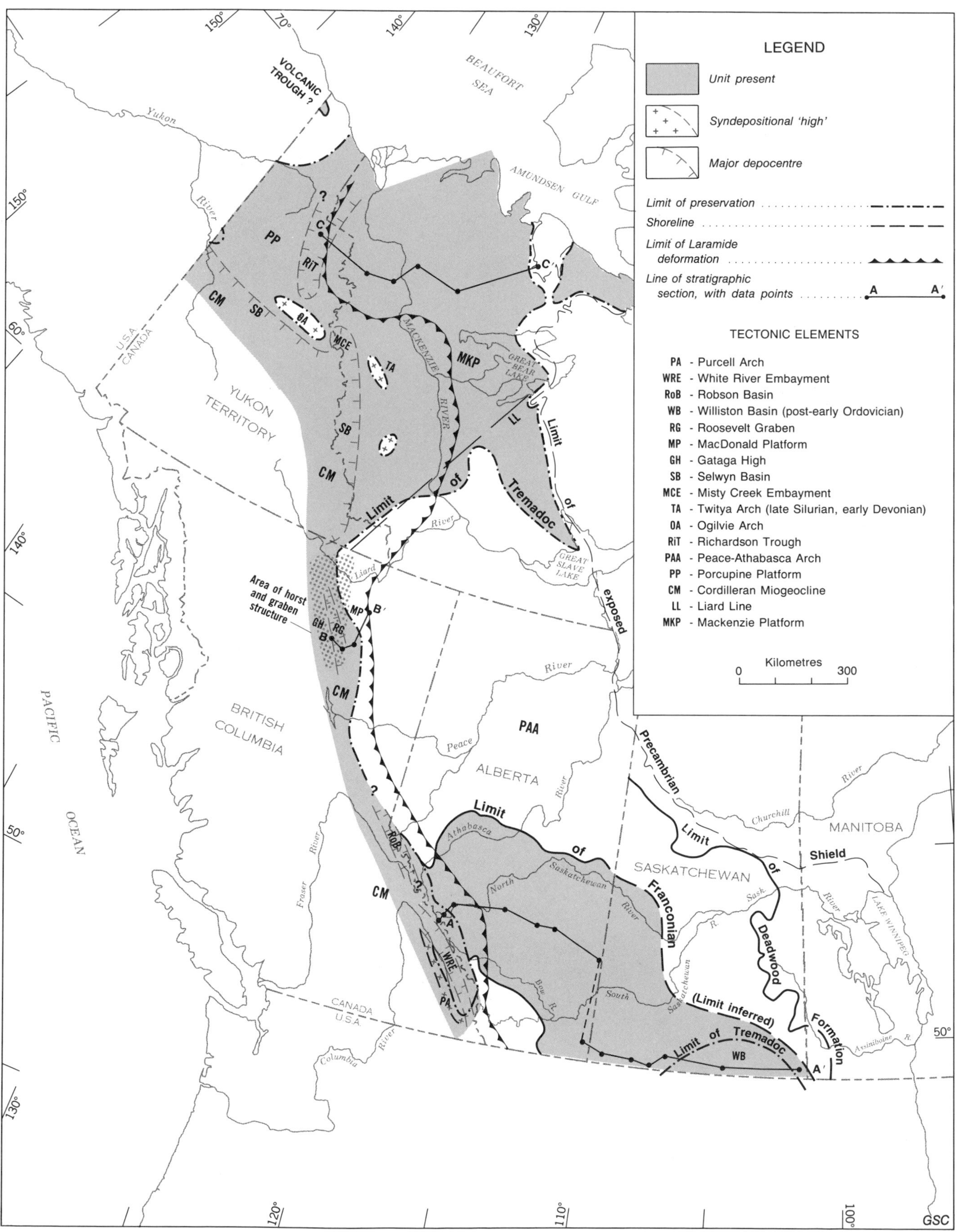

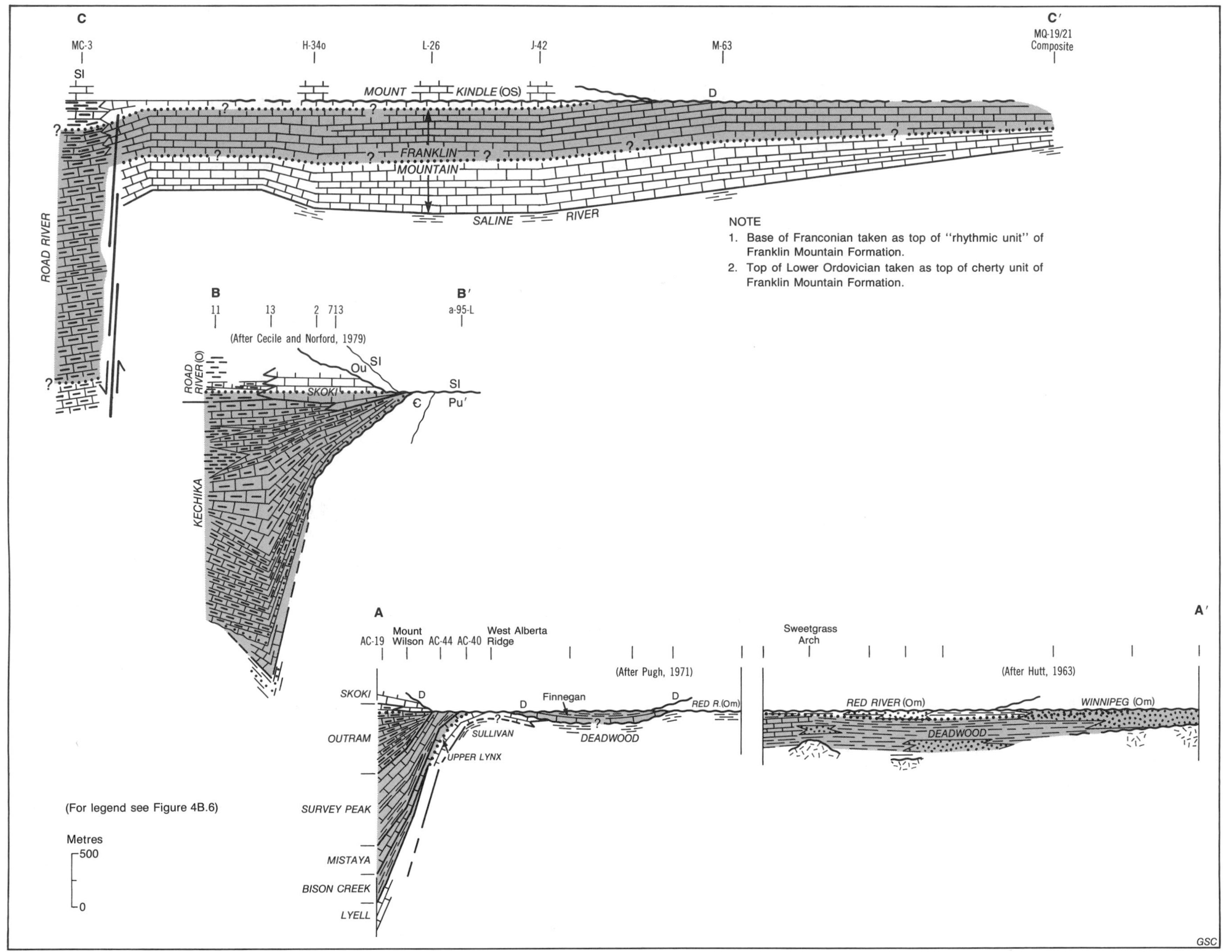
C
MC-3
SI
H-34o
L-26
J-42
M-63
C′
MQ-19/21 Composite
MOUNT KINDLE (OS)
D
FRANKLIN MOUNTAIN
SALINE RIVER
ROAD RIVER
NOTE
1. Base of Franconian taken as top of "rhythmic unit" of Franklin Mountain Formation.
2. Top of Lower Ordovician taken as top of cherty unit of Franklin Mountain Formation.
B
11
13
2
713
B′
a-95-L
(After Cecile and Norford, 1979)
ROAD RIVER (O)
Ou
SI
SKOKI
Є
Pu′
KECHIKA
A
AC-19
Mount Wilson
AC-44
AC-40
West Alberta Ridge
(After Pugh, 1971)
Sweetgrass Arch
(After Hutt, 1963)
A′
SKOKI
OUTRAM
SURVEY PEAK
MISTAYA
BISON CREEK
LYELL
Finnegan
RED R. (Om)
SULLIVAN
UPPER LYNX
DEADWOOD
RED RIVER (Om)
WINNIPEG (Om)
(For legend see Figure 4B.6)
Metres
500
0
GSC

a response to the same marked regression as that responsible for the Arctomys Formation in the southern Rockies. With an Arctomys-like facies of Dresbachian age recorded in the central Rockies (Slind and Perkins, 1966), both views may be partly correct. The outboard sill of the Saline River evaporite basin was the Mackenzie Arch in its final pulse of uplift; no comparable evaporitic facies occurs farther southwest. Saline River time witnessed more widespread, differential vertical motions as well; the principal depocentres of Saline River deposition (locally reaching thicknesses of more than 800 m) largely coincide with depocentres of the Lower and Middle Cambrian Mount Cap Formation, but are not reflected in overlying units (Meijer Drees, 1986, Fig. 16, 17).

Upper Cambrian

Dresbachian strata in Mackenzie Mountains belong to two facies: thin-bedded limestone of slope and basin origin, part of the Rabbitkettle Formation (more than 1200 m), and dolomites of platformal origin (200-900 m), part of the Broken Skull Formation in the south and of the Franklin Mountain Formation in the north. A transitional facies between platformal and slope facies has been mapped (Cecile, 1982). The formations of platformal as well as basinal facies are apparently continuous upward into the Lower Ordovician, without known break. The Franklin Mountain and Broken Skull formations pass across the crest of Mackenzie Arch, where a basal, sandstone member derived from underlying Proterozoic quartzite is normally present, and extend far eastward across Mackenzie Platform, with only subtle change of carbonate facies. Part or all of the Dresbachian portion might be missing by onlap at the crest of the arch, so far as known from existing faunal control.

Franklin Mountains and Mackenzie Platform

Sauk II deposits in Mackenzie Platform are eastward continuations of the formations exposed along the frontal Mackenzie Mountains. The Middle Cambrian part of the Mount Cap Formation thins northeastward, with progressive loss of carbonate strata and gain in the content of siliciclastic sand and silt, but persists to the edge of the exposed Canadian Shield. The major depocentre of Saline River halite, anhydrite, and redbeds underlies the broadest, northern part of Franklin Mountains, and may be responsible for the breadth and structural style of that part (Aitken et al., 1982). Nevertheless, the Saline River, greatly thinned, also reaches the edge of the Canadian Shield (Fig. 4B.10, D-D′).

Figure 4B.17. Sauk III: stratigraphic cross-sections. Tectonic elements and lines of section as in Figure 4B.16. Data from Hutt (1963a, b); Pugh (1971, 1983, and unpub. ms.); Cecile (1982); Cecile and Norford (1979); Fritz (1979a); Taylor and Stott (1973); Macqueen (1969); Aitken and Norford (1967); and Aitken (in press).

The Dresbachian part of the Sauk II sub-sequence is represented in Franklin Mountains and the plains by some lower part of the sparsely fossiliferous Franklin Mountain Formation and by the Saline River Formation, if it is Dresbachian. The basal, cyclic unit (member) of the Franklin Mountain has yielded Dresbachian trilobites; the overlying rhythmic unit has yielded Dresbachian trilobites from its base and Franconian brachiopods and echinoderms from its upper parts, and the next overlying, cherty unit has yielded Early Ordovician gastropods (Norford and Macqueen, 1975). Persistence of the three rock units eastward to the flank of Brock Inlier suggests that Dresbachian strata persist that far. The only facies change worthy of remark in these platformal deposits is a cratonward increase in the amount of secondary chert. The Franklin Mountain Formation (including Sauk III strata) thickens northeastward from about 700 m along the mountain front, to 1000 m (Pugh, 1983) to a broad depocentre beneath Peel Plateau (Fig. 4B.14). Thinning northeastward from the trough axis is evident (Fig. 4B.10, D-D′), but the true thickness along the eastern outcrop edge is unknown, because of near-horizontal bedding, low relief, and the lack of exploratory wells.

SAUK III SUB-SEQUENCE, UPPER CAMBRIAN AND LOWER ORDOVICIAN

Tectonic elements

Deposition of the Sauk III sub-sequence commenced generally at a surface of erosion, except in rapidly subsiding troughs. Locally obvious, the sub-Franconian contact is more commonly a subtle feature and is straddled by a number of formations. Through continued, passive-margin subsidence and eustatic rise of sea level, Sauk III deposits probably overstepped all earlier Cambrian formations cratonward, but have been stripped far back from their depositional limits by erosion at subsequent unconformities, most notably the sub-Devonian one (Fig. 4B.7, 4B.8, 4B.16, 4B.17).

In the southern Rocky Mountains, the tectonic pattern for Sauk III was largely a continuation of the pattern established for Sauk II. The remarkable westward expansion of carbonate deposition at the close of Sauk II time, shown by the Ottertail Formation, ended with a return to mixed sedimentation of mudrocks and limestone under basinal conditions, with the western edge of the carbonate platform, the Kicking Horse Rim, apparently near its former, Middle Cambrian, position (Aitken, 1971). A broadly synclinal belt of preserved Ordovician rocks in the southern Rockies (Norford, 1969) lies on-trend with the earlier Robson Basin and down-plunge; it may be evidence for persistence of that tectonic feature. In southeastern British Columbia, a new feature, the White River Embayment, came into being in late Early Ordovician time, and through most of the Ordovician Period received black graptolitic mud.

During Sauk III time, the Peace-Athabasca Arch appears to have affected sedimentation, though to a degree less pronounced than during Sauk II. Pronounced northward thinning is seen in Upper Cambrian deposits south of the arch. North of the arch, its effects are seen mainly in successive Ordovician units, in which the transition from eastern carbonates to western, basinal, shaly deposits is deflected to the west on approaching the long-term locus of the crest of the arch (see Cecile and Norford, Subchapter 4C in this volume). Paucity of data prohibits rigorous analysis.

Unresolved problems concern the provenance of siliciclastic detritus to both platformal and basinal facies. On the platform, especially in the Franconian and Trempealeauan, the proportion of siliciclastic as opposed to carbonate material displays no simple relationship to the cratonal shoreline, nor to the long-term locus of the outer edge of the platform. For instance, in the southern Rocky Mountains, the disappearance of the Bison Creek Formation, of inner-detrital character, into the Lynx Formation (Group) of middle carbonate character, has a strong, south-to-north component. If this records northward (eastward in the Cambrian) transport of mud in a progressively clearing current, it was in a direction opposed to the assumed northeasterly trade winds that seem to explain so much of Sauk II sedimentology (Aitken, 1978). At all levels of the Sauk III sub-sequence, clay and silt increase outboard (westward) from platformal to basinal deposits. A clear example in the southern Rocky Mountains is the eastward pinchout of tongues of the Outram Formation (dark grey, silty, deep-water shale and limestone) into Skoki platformal carbonates. Along-strike transport of mud, outboard of the carbonate belt, from unknown sources, must be assumed.

Lloydminster Embayment, which is clearly expressed in Sauk II stratigraphy, appears to have no expression in the part of the Sauk III sub-sequence remaining after sub-Ordovician and sub-Devonian erosion. No Sauk III rocks are preserved in the area of the Hay River Embayment of Sauk II time.

North of the Peace River Arch, the Roosevelt Graben persisted through the Franconian, with anomalously thick, basinal limestones along its axis, and markedly condensed sections above its shoulders, but Trempealeauan and Lower Ordovician deposits display only modest thickening and change of facies (Fig. 4B.17, B-B′).

North of latitude 60°N, the Sauk III sub-sequence consists almost entirely of carbonate rocks, not only platformal carbonates in the north and east, but equivalent, thin-bedded limestones extending far to the south and west in Selwyn Basin. Local uplifts were apparently lacking in the miogeocline, and the cratonal shoreline was remote. Mackenzie Arch was, by this time, buried. Structural differentiation, except for the persistent, platform-to-basin transition, was minimal. Misty Creek Embayment and Richardson Trough continued to subside rapidly, and Ogilvie Arch, west of the embayment, remained high, as it had since the Early Cambrian.

Stratigraphy

Southern Rocky Mountains

In the southern Rockies, the Sauk III sub-sequence commences at a subtle contact, at least locally erosional, within the uppermost beds of the Lyell Formation. There, skeletal carbonates and bioherms of cryptic origin, with a Franconian (*Elvinia* Zone) fauna, overlie barren, cyclical, peritidal carbonates. The highest occurrence of Dresbachian fossils is about 150 m lower. Above the Franconian limestones at the top of the Lyell, the succession (Aitken and Greggs, 1967; Aitken and Norford, 1967) is as follows (note the persistence of depositional Grand Cycles to the top of the Skoki Formation):

Owen Creek Formation (45-200 m, top): Peritidal dolomite, partly silty and sandy, barren. Erosional contact; karstic surface.

Skoki Formation (120-185 m): Limestone, mainly skeletal grainstone, packstone, wackestone; generally dolomitized; chert nodules. Latest Canadian and Whiterockian (*Hesperonomia, Orthidiella,* and *Anomalorthis*) Zones.

Outram Formation (170-440 m): Interbedded nodular limestone and dark brown, siliceous shale and chert; a deeper-water facies. In part, passes eastward to Skoki Formation. Canadian Zones G_1 to J; in the west, reaches Whiterockian *Orthidiella* Zone.

Survey Peak Formation: Interbedded shallow-water limestones with abundant flat-pebble conglomerates, thrombolites, and algal stromatolites, and distinctive greenish grey ('putty') shale; siltstone at base. Uppermost Cambrian *Saukia* Zone at base, overlain by Canadian Zones A to G_1.

Mistaya Formation (45-160 m): Shallow-water limestone, with abundant flat-pebble conglomerates, richly skeletal beds, and large, prominent, stromatolitic and thrombolitic bioherms. Trempealeauan *Saukia* Zone.

Bison Creek Formation (11-210 m, base): Interbedded, grey and greenish grey shale, mudstone, shaly limestone, and massive limestone, the latter richly and coarsely skeletal; algal stromatolites and thrombolites large and prominent. Franconian *Elvinia, Conaspis,* and *Ptychaspis - Prosaukia* Zones and Trempealeauan *Saukia* Zone.

The Owen Creek Formation is overlain abruptly, at an inferred unconformity, by the thick Mount Wilson Quartzite (Middle or Upper Ordovician), which records a major change in depositional regime (Fig. 4B.18).

Westward into the basin, pronounced changes take place. The Mistaya massive carbonates disappear, and they and the Bison Creek Formation pass into the lower part of the McKay Group, which also includes Survey Peak equivalents. This thick unit is lithologically somewhat similar to the Survey Peak Formation, but may be of deeper water origin. Higher, the entire Outram-Skoki-Owen Creek succession passes westward to graptolitic black shale of the Glenogle Formation (locally over 600 m). The Tipperary Quartzite (up to 175 m) occurs locally at the level of the base of the Skoki (Norford, 1969).

All or almost all of the Sauk III sub-sequence is erosionally bevelled eastward at the sub-Devonian unconformity before the mountain front is reached (Fig. 4B.16, A-A′, 4B.19). It reappears (in the subsurface) only in central Williston Basin, except for a swath west of the Alberta-Saskatchewan boundary, in which the thin Finnegan Formation (Pugh, 1971), which is probably Franconian or Trempealeauan, is preserved (see below).

Lloydminster Embayment

The Sauk III sub-sequence is represented feebly in the region outlined by Sauk II deposits as the subsurface Lloydminster Embayment. The Finnegan Formation of Pugh (1971) consists of varicoloured, micaceous, glauconitic siltstone, varicoloured shale, and subordinate, partly red limestone, nowhere exceeding 100 m and generally much less. It is apparently continuous with the upper Lynx carbonates of western Alberta, or part thereof, and on that basis is Upper Cambrian. On the other hand, the base of the Finnegan apparently truncates widely traced, subtle Ra-log markers in the underlying Deadwood Formation, including marker 'f' associated with the Cambrian-Ordovician contact in the Richey area of northeastern Montana (van Hees, 1964; Pugh, 1971). On the latter basis, the Finnegan could not be older than Early Ordovician. Because the geophysical marker units are subtly expressed, and their correlation open to question (see caveats by Porter and Fuller, 1959), and because the Finnegan, as an Ordovician unit, would be a complete anomaly in terms of regional patterns of truncation and preservation, a Franconian or Trempealeauan age is preferred here. In any event, the Finnegan is assigned to the Sauk III sub-sequence.

Williston Basin

In the Canadian part of the subsurface Williston Basin, fossil evidence for the presence of Sauk III strata is lacking. On the other hand, the Deadwood Formation in Montana appears to be a conformable succession from Dresbachian through Canadian, according to Lochman-Balk and Wilson (1967), and their maps show Deadwood beds of Early Ordovician age entering south-central Saskatchewan from Montana. Thus, in Saskatchewan, a part of the Deadwood green and purple, fine-grained, clastic beds with minor limestone belongs to Sauk III, but has not been delineated.

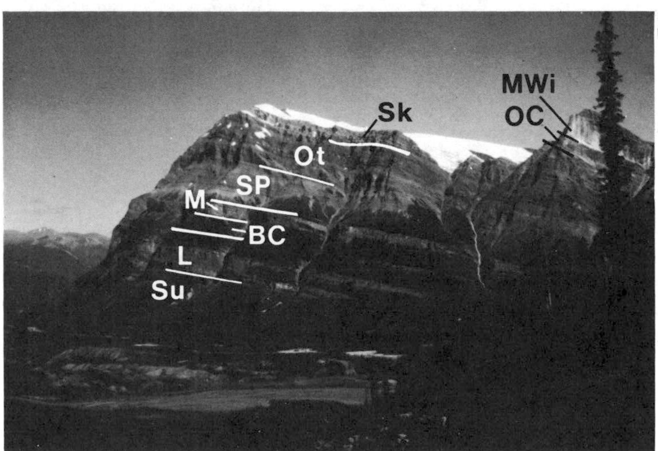

Figure 4B.18. Upper Cambrian and Ordovician section at Mount Wilson, Banff National Park. Su - Sullivan Formation; L - Lyell Formation; BC - Bison Creek Formation; M - Mistaya Formation; SP - Survey Peak Formation; Ot - Outram Formation; Sk - Skoki Formation; OC - Owen Creek Formation; MWi - Mount Wilson Quartzite (photo - J.D. Aitken, ISPG 2280-1).

Northern Rocky Mountains

In the northern Rocky Mountains, the Sauk III sub-sequence is represented mainly by the Kechika Formation (Group). The Kechika is a thick (locally over 1500 m), widely recognized and rather monotonous unit, characterized by platy, banded, and nodular lime mudstones with a variable content of interbedded shale. A basinward facies progression (Fig. 4B.17, B-B′) from shallow carbonate platform, through deeper shelf to upper slope is expressed subtly (Cecile and Norford, 1979). The basal beds of the Kechika are of Trempealeauan age at most places, but the base is diachronous, and in the inboard, platformal area and over the 'Gataga high', unconformable. Franconian strata, mainly carbonates, are locally present beneath the Kechika. The top of the Sauk III sub-sequence lies within the platformal dolomites of the Skoki Formation (Arenig to lower Caradoc). The Skoki is in intertongued contact with the Kechika, and like it, passes westward into deepwater shale and limestone of the Road River Formation (Group). Because of erosional bevelling at subsequent unconformities, notably a sub-Silurian unconformity that is strongly expressed in the region but not generally elsewhere, the Sauk III sub-sequence is not preserved eastward as far as the mountain front (Fig. 4B.17, B-B′).

Mackenzie Mountains

In the outer ranges of the Mackenzie Mountains, the Sauk III sub-sequence is represented entirely by very sparsely fossiliferous, platformal dolomites making up the upper part of the Franklin Mountain Formation in the

Figure 4B.19. The mountain front at Ghost River, west of Calgary, Alberta. C - Cathedral Formation; S - Stephen Formation; E - Eldon Formation; Pa - Pika Formation; A - Arctomys Formation ; Lx - Lynx Group; D - Devonian formations; K - Cretaceous formations (photo - J.D. Aitken, ISPG 1518-4).

north and the Broken Skull Formation in the south. The base of the sub-sequence is neither mappable nor well controlled biostratigraphically, but apparently falls within the 'rhythmic unit' of the Franklin Mountain Formation, which has yielded Dresbachian fossils from its base and Franconian fossils from its upper part (Norford and Macqueen, 1975). The 'cherty unit', uppermost of the Franklin Mountain, has yielded Early Ordovician fossils at several localities (Norford and Macqueen, 1975).

The sub-sequence has been erosionally reduced at sub-Upper Ordovician and sub-Devonian unconformities (Fig. 4B.17, C-C′), and is locally missing over paleostructural/paleotopographic highs in Mackenzie Mountains.

Southwest of Mackenzie Platform, Sauk III is represented by ribbon-bedded, silty limestones of slope origin, the Rabbitkettle Formation (up to 800 m or more). This limestone lithosome is remarkable for its off-platform extent within Selwyn Basin, but it ultimately passes to basinal mudrocks of the Road River Group (Cecile and Norford, Subchapter 4C in this volume).

Great Bear and Anderson Plains

The Sauk III sub-sequence, comprising the bulk of the Franklin Mountain Formation, persists eastward and northeastward, in a facies of increasingly stromatolitic dolomite, to its erosional limit at the edge of the Canadian Shield (Fig. 4B.17, C-C′). A sedimentological question of interest is the source and mode of transport of the immense amount of secondary silica embodied by the 'cherty unit', which is largely or entirely of Early Ordovician age. Coeval rocks in the Rocky Mountains contain comparatively miniscule amounts of chert.

Peel Plateau and Richardson Mountains

In the Richardson Trough, which had a history of subsidence from Late Proterozoic to Early Devonian times, the Sauk III sub-sequence is recorded mainly by the lower division of the Road River Group, which is lithologically similar to the Rabbitkettle Formation, and above the diachronous top of the latter, a variable thickness of dark, deepwater mudrocks of the upper Road River. In such a persistently basinal environment, neither the basal nor the upper limit of Sauk III is expressed as a detectable break in sedimentation.

The basinal, or 'trough' facies of Sauk III is not limited to the Richardson Mountains; its cratonward limit, roughly delineated by drilling, is a facies change to Franklin Mountain Formation, trending northeastward beneath Peel Plateau (Fig. 4B.16).

REFERENCES

Aitken, J.D.

1966: Middle Cambrian to Middle Ordovician cyclic sedimentation, southern Rocky Mountains of Alberta; Bulletin of Canadian Petroleum Geology, v. 14, p. 405-441.

1968: Cambrian sections in the easternmost southern Rocky Mountains and the adjacent subsurface, Alberta; Geological Survey of Canada, Paper 66-23, 96 p.

1969: Documentation of the sub-Cambrian unconformity, Rocky Mountains Main Ranges, Alberta; Canadian Journal of Earth Sciences, v. 6, p. 192-200.

Aitken, J.D. (cont.)

1971: Control of Lower Paleozoic sedimentary facies by the Kicking Horse Rim, southern Rocky Mountains, Canada; Bulletin of Canadian Petroleum Geology, v. 19, p. 557-569.

1978: Revised models for depositional Grand Cycles, Cambrian of Western Canada; Bulletin of Canadian Petroleum Geology, v. 26, p. 515-542.

1981: Generalizations about Grand Cycles; in Short Papers for the Second International Symposium on the Cambrian System, M.E. Taylor (ed.); United States Geological Survey, Open File Report 81-743, p. 8-14.

1988: Uppermost Proterozoic Formations, Central Mackenzie Mountains, Northwest Territories; Geological Survey of Canada, Bulletin 368, 26 p.

in press: Stratigraphy of the Middle Cambrian platformal succession, southern Rocky Mountains; Geological Survey of Canada, Bulletin 398.

Aitken, J.D., Cook, D.G., and Yorath, C.J.

1982: Upper Ramparts River (106 G) and Sans Sault Rapids (106 H) map areas, District of Mackenzie; Geological Survey of Canada, Memoir 388, 48 p.

Aitken, J.D., Fritz, W.H., and Norford, B.S.

1972: Cambrian and Ordovician biostratigraphy of the southern Canadian Rocky Mountains; International Geological Congress, Montreal, Guidebook XXIV A19, 57 p.

Aitken, J.D. and Greggs, R.G.

1967: Upper Cambrian formations, Southern Rocky Mountains of Alberta, an interim report; Geological Survey of Canada, Paper 66-49, 91 p.

Aitken, J.D., Macqueen, R.W., and Usher, J.L.

1973: Reconnaissance studies of Proterozoic and Cambrian stratigraphy, Lower Mackenzie River area (Operation Norman), District of Mackenzie; Geological Survey of Canada, Paper 73-9, 178 p.

Aitken, J.D. and Norford, B.S.

1967: Lower Ordovician Survey Peak and Outram formations, southern Rocky Mountains of Alberta; Bulletin of Canadian Petroleum Geology, v. 15, p. 150-207.

Aitken, J.D. and Pugh, D.C.

1984: Recurrent lithofacies and regional facies change in Middle Cambrian strata, Alberta; in Carbonates in Subsurface and Outcrop, L. Eliuk (ed.); 1984 Core Conference; Canadian Society of Petroleum Geologists, p. 33-53.

Allan, J.A.

1912: Geology of Field map area, Yoho Park, British Columbia; Geological Survey of Canada, Summary Report 1911, p. 175-187.

1914a: Rocky Mountains section between Banff, Alberta and Golden, British Columbia, along the Canadian Pacific Railway; Geological Survey of Canada, Summary Report 1912, p. 165-176.

1914b: Geology of Field map area, British Columbia and Alberta, Canada; Geological Survey of Canada, Memoir 55, 312 p.

1916: Simpson Pass to Kananaskis, Rocky Mountains Park, Alberta; Geological Survey of Canada, Summary Report 1915, p. 100-102.

Baudet, D., Aitken, J.D., and Vanguestine, M.

1989: Palynology of uppermost Proterozoic and lowermost Cambrian formations, Central Mackenzie Mountains, northwestern Canada; Canadian Journal of Earth Sciences, v. 26, p. 129-148.

Bond, G.C. and Kominz, M.A.

1984: Construction of tectonic subsidence curves for the early Paleozoic miogeocline, southern Canadian Rocky Mountains - implications for subsidence mechanisms, age of breakup, and crustal thinning; Geological Society of America Bulletin, v. 95, p. 155-173.

Burling, L.D.

1914: Early Cambrian stratigraphy in the North American Cordillera, with discussion of *Albertella* and related faunas; Geological Survey of Canada, Museum Bulletin No. 2, p. 93-129.

1915: Shallow-water deposition in the Cambrian of the Canadian Cordillera; Ottawa Naturalist, v. 29, p. 87-88.

1916a: Notes on the stratigraphy of the Rocky Mountains, Alberta and British Columbia; Geological Survey of Canada, Summary Report, 1915, p. 97-100.

1916b: The *Albertella* fauna located in the Middle Cambrian of British Columbia and Alberta; American Journal of Science, v. 42, p. 469-472.

Cecile, M.P.

1982: The Lower Paleozoic Misty Creek Embayment, Selwyn Basin, Yukon and Northwest Territories; Geological Survey of Canada, Bulletin 335, 78 p.

Cecile, M.P. and Norford, B.S.

1979: Basin to platform transition, lower Paleozoic strata of Ware and Trutch map areas, northeastern British Columbia; in Current Research, Part A, Geological Survey of Canada, Paper 79-1A, p. 219-226.

Cook, D.G.
1975: Structural style influenced by lithofacies, Rocky Mountains Main Ranges, Alberta and British Columbia; Geological Survey of Canada, Bulletin 233, 73 p.

Conway Morris, S. and Whittington, H.B.
1985: The Burgess Shale in Yoho National Park; Geological Survey of Canada, Miscellaneous Report 43, 31 p.

Deiss, C.F.
1939: Cambrian formations of southwestern Alberta and southeastern British Columbia; Geological Survey of America, Bulletin, v. 50, p. 951-1026.
1940: Lower and Middle Cambrian stratigraphy of southwestern Alberta and southeastern British Columbia; Geological Society of America, Bulletin, v. 51, p. 731-794.

Fritz, W.H.
1971: Geological setting of the Burgess Shale; in Extraordinary Fossils: North American Paleontological Convention, Proceedings Part 1 (1969), Allen Press, Lawrence, Kansas, p. 1155-1170.
1972a: Cambrian biostratigraphy, western Rocky Mountains, British Columbia; in Report of Activities, Part A, Geological Survey of Canada, Paper 72-1A, p. 209-211.
1972b: Lower Cambrian trilobites from the Sekwi Formation type section, Mackenzie Mountains, Northwestern Canada; Geological Survey of Canada, Bulletin 212, 58 p.
1975: Broad correlations of some Lower and Middle Cambrian strata in the North American Cordillera; in Report of Activities, Part A, Geological Survey of Canada, Paper 75-1A, p. 533-540.
1976: Ten stratigraphic sections from the Lower Cambrian Sekwi Formation, Mackenzie Mountains, Northwestern Canada; Geological Survey of Canada, Paper 76-22, 42 p.
1978a: Upper (carbonate) part of Atan Group, Lower Cambrian, north-central British Columbia; in Current Research, Part A, Geological Survey of Canada, Paper 78-1A, p. 7-16.
1978b: Fifteen stratigraphic sections from the Lower Cambrian of the Mackenzie Mountains, Northwestern Canada; Geological Survey of Canada, Paper 77-33, 19 p.
1979a: Cambrian stratigraphy in the northern Rocky Mountains, British Columbia; in Current Research, Part B, Geological Survey of Canada, Paper 79-1B, p. 99-109.
1979b: Cambrian stratigraphic section between South Nahanni and Broken Skull Rivers, southern Mackenzie Mountains; in Current Research, Part B, Geological Survey of Canada, Paper 79-1B, p. 121-125.
1979c: Eleven stratigraphic sections from the Lower Cambrian of the Mackenzie Mountains, Northwest Canada; Geological Survey of Canada, Paper 78-23, 19 p.
1980a: International Precambrian-Cambrian Boundary Working Group's 1979 Field Study to Mackenzie Mountains, Northwest Territories, Canada; in Current Research, Part A, Geological Survey of Canada, Paper 80-1A, p. 41-45.
1980b: Two Cambrian stratigraphic sections near Gataga River, Northern Rocky Mountains, British Columbia; in Current Research, Part C, Geological Survey of Canada, Paper 80-1C, p. 113-119.
1981a: Two Cambrian stratigraphic sections, eastern Nahanni map area, Mackenzie Mountains, District of Mackenzie; in Current Research, Part A, Geological Survey of Canada, Paper 81-1A, p. 145-156.
1981b: Cambrian biostratigraphy, southern Canadian Rocky Mountains, Alberta and British Columbia; in The Cambrian System in the Southern Canadian Rocky Mountains, M.E. Taylor (ed.); Second International Symposium on the Cambrian System, Guidebook for Field Trip 2, Denver, Colorado, 61 p.

Fritz, W.H. and Mountjoy, E.W.
1975: Lower and early Middle Cambrian formations near Mount Robson, British Columbia and Alberta; Canadian Journal of Earth Sciences, v. 12, p. 119-131.

Fritz, W.H., Narbonne, G.M., and Gordey, S.P.
1983: Strata and trace fossils near the Precambrian-Cambrian boundary, Mackenzie, Selwyn and Wernecke Mountains, Yukon and Northwest Territories; in Current Research, Part B, Geological Survey of Canada, Paper 83-1B, p. 365-375.

Fuller, J.G.C.M. and Porter, J.W.
1962: Cambrian, Ordovician and Silurian formations of the northern Great Plains, and their regional connections; Journal of Alberta Society of Petroleum Geologists, v. 10, p. 455-485.

Fyson, W.K.
1961: Deadwood and Winnipeg stratigraphy in southwestern Saskatchewan; Saskatchewan Department of Mineral Resources, Report No. 64, 37 p.

Gabrielse, H.
1975: Geology of Fort Grahame E 1/2 map area, British Columbia; Geological Survey of Canada, Paper 75-33, 28 p.

Gabrielse, H., Blusson, S.L., and Roddick, J.A.
1973: Geology of Flat River, Glacier Lake and Wrigley Lake map areas, District of Mackenzie and Yukon Territory; Geological Survey of Canada, Memoir 366, 421 p.

Gabrielse, H. and Taylor, G.C.
1982: Geological maps and cross-sections of the Cordillera from near Fort Nelson, British Columbia to Gravina Island, southeastern Alaska; Geological Survey of Canada, Open File 864.

Hofmann, H.J.
1981: First record of a Late Proterozoic faunal assemblage in the North American Cordillera; Lethaia, v. 14, p. 303-310.
1983: Early Cambrian problematic fossils near June Lake, Mackenzie Mountains, Northwest Territories; Canadian Journal of Earth Sciences, v. 20, p. 1513-1520.

Hutchinson, R.D.
1960: Middle Cambrian fossils in eastern Alberta; Journal of Alberta Society of Petroleum Geologists, v. 8, p. 137.

Hutt, R.B.
1963a: East-west cross-section of Saskatchewan; Saskatchewan Department of Mineral Resources.
1963b: North-south cross-section of Saskatchewan; Saskatchewan Department of Mineral Resources.

Lochman-Balk, C. and Wilson, J.L.
1958: Cambrian biostratigraphy in North America; Journal of Paleontology, v. 32, p. 312-350.
1967: Stratigraphy of Upper Cambrian - Lower Ordovician subsurface sequence in Williston Basin; American Association of Petroleum Geologists, Bulletin, v. 51, p. 883-917.

Macqueen, R.W.
1969: Lower Paleozoic stratigraphy, Operation Norman; in Report of Activities, Part A, Geological Survey of Canada, Paper 69-1A, p. 238-241.
1970: Lower Paleozoic stratigraphy and sedimentology, eastern Mackenzie Mountains, Northern Franklin Mountains; in Report of Activities, Part A, Geological Survey of Canada, Paper 70-1A, p. 225-230.

Macqueen, R.W. and Mackenzie, W.S.
1973: Lower Paleozoic and Proterozoic stratigraphy Mobil Colville Hills E-15 well and environs, Interior Platform, District of Mackenzie; in Report of Activities, Part B, Geological Survey of Canada, Paper 73-1B, p. 183-188.

McIlreath, I.A.
1977: Stratigraphic and sedimentary relationships at the western edge of the Middle Cambrian carbonate facies belt, Field, British Columbia; Ph.D. thesis, University of Calgary, Calgary, Alberta, 259 p.

Meijer Drees, N.C.
1975: Geology of the Lower Paleozoic formations in the subsurface of the Fort Simpson area, District of Mackenzie, Northwest Territories; Geological Survey of Canada, Paper 74-40, 65 p.
1986: Evaporitic deposits of Western Canada; Geological Survey of Canada, Paper 85-20, 118 p.

Mountjoy, E.W.
1962: Mount Robson (southeast) map area, Rocky Mountains, British Columbia; Geological Survey of Canada, Paper 61-31, 114 p.

Mountjoy, E.W. and Aitken, J.D.
1978: Middle Cambrian Snake Indian Formation (new), Jasper region, Alberta; Bulletin of Canadian Petroleum Geology, v. 26, p. 343-361.

Norford, B.S.
1969: Ordovician and Silurian stratigraphy of the southern Rocky Mountains; Geological Survey of Canada, Bulletin 176, 90 p.

Norford, B.S. and Macqueen, R.W.
1975: Lower Paleozoic Franklin Mountain and Mount Kindle formations, District of Mackenzie: Their type sections and regional development; Geological Survey of Canada, Paper 74-34, 37 p.

Norris, D.K.
1983: Geotectonic correlation chart - Porcupine Project area; Geological Survey of Canada, Chart 1532A.

Norris, D.K. and Price, R.A.
1966: Middle Cambrian lithostratigraphy of southeastern Canadian Cordillera; Bulletin of Canadian Petroleum Geology, v. 14, no. 4, p. 385-404.

Nowlan, G.S., Narbonne, G.M., and Fritz, W.H.
1985: Small shelly fossils and trace fossils near the Precambrian-Cambrian boundary in the Yukon Territory, Canada; Lethaia, v. 18, p. 233-256.

Okulitch, A.V.
1985: Paleozoic plutonism in southeastern British Columbia; Canadian Journal of Earth Sciences, v. 22, p. 1409-1424.

Palmer, A.R.
1968: Cambrian trilobites of east-central Alaska; United States Geological Survey, Professional Paper 559-B, 115 p.
1972: Problems of Cambrian biogeography; XXIV International Geological Congress, Section 7, p. 310-315.

Palmer, A.R. (cont.)
1973: Cambrian trilobites; in Atlas of Palaeobiogeography, A. Hallam (ed.); Elsevier, New York, p. 3-11.

Porter, J.W. and Fuller, J.G.C.M.
1959: Lower Paleozoic rocks of northern Williston Basin and adjacent areas; American Association of Petroleum Geologists, Bulletin, v. 43, p. 124-189.
1964: Ordovician-Silurian, Part 1 - Plains; in Geological History of Western Canada, R.G. McCrossan and R.P. Glaister (ed.); Alberta Society of Petroleum Geologists, Calgary, p. 34-42.

Porter, J.W., Price, R.A., and McCrossan, R.G.
1982: The western Canada sedimentary basin; Philosophical Transactions of the Royal Society, London, Series A, v. 305, p. 169-192.

Pugh, D.C.
1971: Subsurface Cambrian stratigraphy of southern and central Alberta; Geological Survey of Canada, Paper 70-10, 33 p.
1973: Subsurface Lower Paleozoic stratigraphy in northern and central Alberta; Geological Survey of Canada, Paper 72-12, 54 p.
1975: Cambrian stratigraphy from western Alberta to northeastern British Columbia; Geological Survey of Canada, Paper 74-37, 31 p.
1983: Pre-Mesozoic geology in the subsurface of Peel River map area, Yukon Territory and District of Mackenzie; Geological Survey of Canada, Memoir 401, 61 p.

Raaben, M.E. (ed.)
1969: Tommotskii Yarus i problema Nizhnei Granitsy Kembriya: Nauka, Moscow. (In English translation, The Tommotian Stage and the Cambrian Lower Boundary Problem; Amerind, New Delhi, 1981, 359 p.

Raasch, G.O. and Campau, D.E.
1957: Cambrian biostratigraphy of California Standard Parkland No. 4-12; Journal of Alberta Society of Petroleum Geologists, v. 5, p. 140-144.

Rasetti, F.
1951: Middle Cambrian stratigraphy and faunas of the Canadian Rocky Mountains; Smithsonian Miscellaneous Collections, v. 116, No. 5, 277 p.
1956: The Middle and Upper Cambrian of western Canada; in 20th International Geological Congress Proceedings, Part II, p. 735-750.

Robison, R.A.
1960: Lower and Middle Cambrian stratigraphy of the eastern Great Basin; Intermountain Association of Petroleum Geologists, Guidebook to the Geology of East Central Nevada, p. 43-52.

Slind, O.L. and Perkins, G.D.
1966: Lower Paleozoic and Proterozoic sediments of the Rocky Mountains between Jasper, Alberta and Pine River, British Columbia; Bulletin of Canadian Petroleum Geologists, v. 14, no. 4, p. 442-468.

Sloss, L.L.
1963: Sequences in the cratonic interior of North America; Bulletin of the Geological Society of America, v. 74, p. 93-114.
1976: Areas and volumes of cratonic sediments, western North America and eastern Europe; Geology, v. 4, no. 5, p. 272-276.

Taylor, G.C. and Stott, D.F.
1973: Tuchodi Lakes map area, British Columbia; Geological Survey of Canada, Memoir 373, 37 p.

Taylor, M.E. (ed.)
1981: The Cambrian System in the southern Rocky Mountains, Alberta and British Columbia; Second International Symposium on the Cambrian System, Guidebook for Field Trip 2, Denver, 61 p.

Teitz, M. and Mountjoy, E.W.
1985: The Yellowhead and Astoria carbonate platforms in the late Proterozoic Upper Miette Group, Jasper, Alberta; in Current Research, Part A, Geological Survey of Canada, Paper 85-1A, p. 341-348.

van Hees, H.
1959: Middle Cambrian of the southern Alberta plains; Alberta Society of Petroleum Geologists, 9th Annual Field Conference, p. 73-85.
1964: Cambrian. Part I - Plains; in Geological History of Western Canada, R.G. McCrossan and R.P. Glaister (ed.); Alberta Society of Petroleum Geologists, p. 230-312.

Walcott, C.D.
1928: Pre-Devonian Paleozoic formations of the Cordilleran provinces of Canada; Smithsonian Miscellaneous Collection, v. 75, p. 175-368.

Watts, A.B.
1982: Tectonic subsidence, flexure and global changes in sea level; Nature, v. 297, p. 469-474.

Wheeler, J.O. and Read, P.B.
1975: Geology of Lardeau (West Half), British Columbia; Geological Survey of Canada, Open File 288 (map and marginal notes).

Whittington, H.B.
1971: The Burgess Shale: history of research and preservation of fossils; North American Paleontological Convention, Chicago, 1969, Proceedings Part I, p. 1170-1201.
1980: The significance of the fauna of the Burgess Shale, Middle Cambrian, British Columbia; Proceedings of the Geological Association, v. 91, p. 127-148.
1985: The Burgess Shale; Yale University Press, New Haven, 151 p.

Xing Yusheng and Luo Huilin
1984: Precambrian-Cambrian boundary candidate, Meishucun, Jinning, Yunnan, China; Geological Magazine, v. 121, p. 143-154.

Young, F.G.
1979: The lowermost Paleozoic McNaughton Formation and equivalent Cariboo Group of eastern British Columbia: Piedmont and tidal complex; Geological Survey of Canada, Bulletin 288, 60 p.

Ziegler, A.M., Bambach, R.K., Parrish, J.T., Barrett, S.F., Gierlowski, E.H., Parker, W.C., Raymond, A., and Sepkoski, J.J., Jr.
1981: Paleozoic biogeography and climatology; in Paleobotany, Paleoecology, and Evolution, v. 2, p. 231-266.

Author's Address

J.D. Aitken
2676 Jemima Road
Denman Island, British Columbia
V0R 1T0

Printed in Canada

Subchapter 4C

ORDOVICIAN AND SILURIAN

M.P. Cecile and B.S. Norford

CONTENTS

INTRODUCTION

Geological framework

Most Ordovician and Silurian rocks in western and northwestern Canada were deposited on the North American craton and its western continental margin. Strata of these ages have also been recognized in three allochthonous terranes in the Cordillera, described in the companion volume on the Cordilleran Orogen (Gabrielse and Yorath, 1991).

The continental margin is equated with the Cordilleran Miogeocline of western Canada. The miogeocline/craton boundary is along a hinge line west of which the rate of thickening of Paleozoic strata increases significantly. On the basis of initial $^{87}Sr/^{86}Sr$ ratios for Mesozoic and younger plutonic rocks (Armstrong, 1979) that intrude them, all autochthonous Ordovician and Silurian strata, of both platformal and basinal facies, are inferred to be underlain by attenuated continental crust, and are thus in the miogeocline. Oceanic rocks are virtually unknown, and if present are either buried beneath younger strata, masked by metamorphism, and/or transported northwestward by Mesozoic and Cenozoic strike-slip faults. The only strata of this age that may be oceanic are rocks of the Kootenay Terrane in southernmost British Columbia.

The Cordilleran Miogeocline formed in Late Proterozoic to Early Cambrian time (see Bond and Kominz, 1984). It is considered to be a continental margin of Atlantic type that was affected by several periods of renewed extensional tectonism, one of which occurred during late Early Ordovician and Middle Ordovician time.

The eastern part of this ancient continental margin and adjacent craton were characterized by shallow-water, carbonate platforms and/or land areas composed mostly of exposed older Paleozoic and Precambrian stratified and crystalline rocks. In the miogeocline, dark, organic-rich, graptolitic shale, chert, and limestone accumulated during Ordovician and Early Silurian times (Fig. 4C.1. to 4C.4). Orange-brown, generally organic-free, carbonate-rich, basinal siltstone and mudstone were deposited over large areas of the miogeocline in late Early and Late Silurian time (Fig. 4C.4, 4C.5). Although most chert-facies rocks

Cecile, M.P. and Norford, B.S.
1993: Ordovician and Silurian; Subchapter 4C in Sedimentary Cover of the Craton in Canada, D.F. Stott and J.D. Aitken (ed.); Geological Survey of Canada, Geology of Canada, no. 5, p. 125-149 (also Geological Society of America, The Geology of North America, v. D-1).

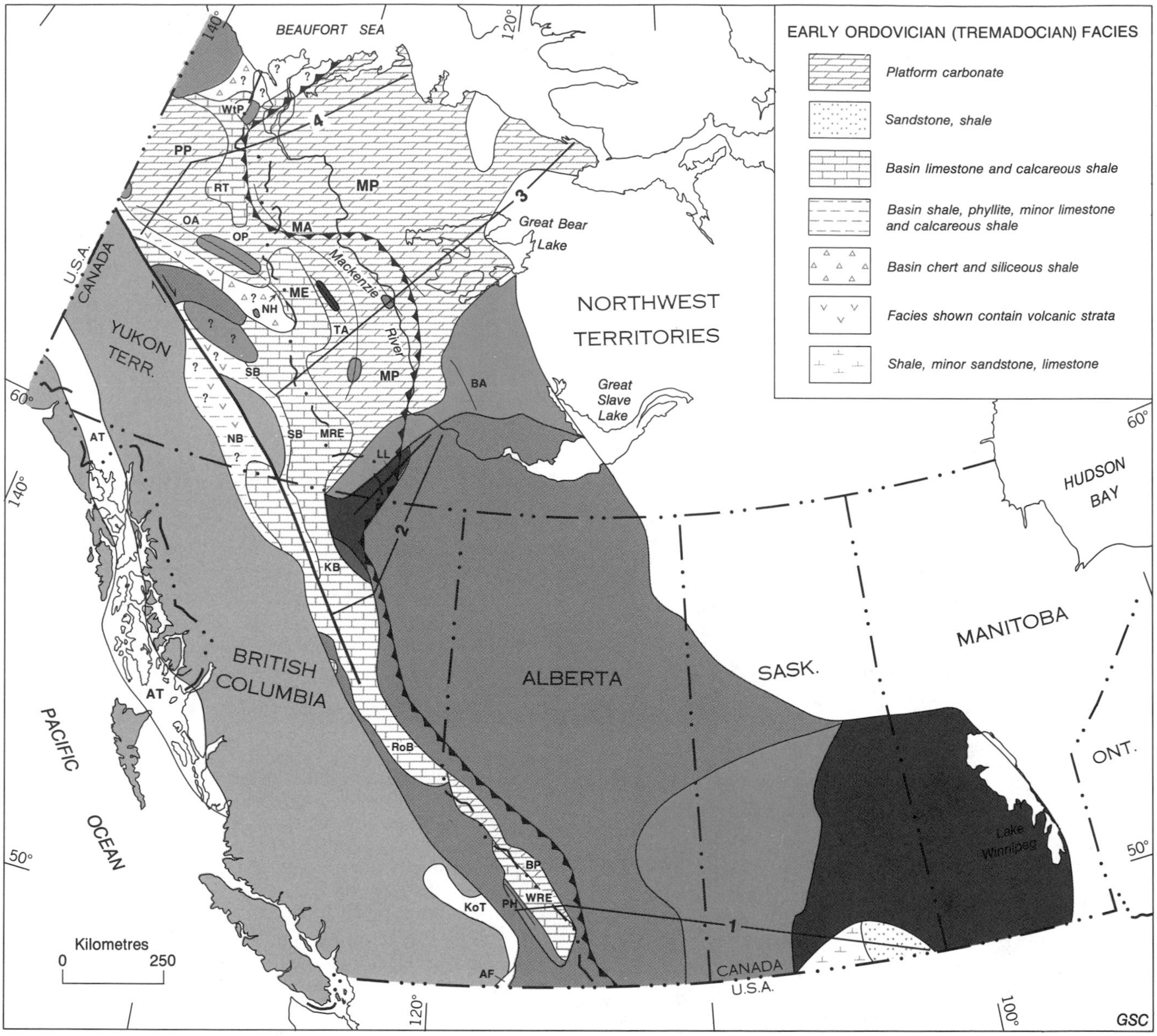

EARLY ORDOVICIAN (TREMADOCIAN) FACIES

- Platform carbonate
- Sandstone, shale
- Basin limestone and calcareous shale
- Basin shale, phyllite, minor limestone and calcareous shale
- Basin chert and siliceous shale
- Facies shown contain volcanic strata
- Shale, minor sandstone, limestone

MAJOR PALEOGEOGRAPHIC FEATURES PRESENT DURING THE TREMADOCIAN

RT	- RICHARDSON TROUGH	**KB**	- KECHIKA BASIN	**MP**	- MACKENZIE PLATFORM	**TA**	- TWITYA ARCH
SB	- SELWYN BASIN	**RoB**	- ROBSON BASIN	**BP**	- BOW PLATFORM	**BA**	- BULMER LAKE ARCH
ME	- MISTY CREEK EMBAYMENT	**WRE**	- WHITE RIVER EMBAYMENT	**WtP**	- WHITE MOUNTAINS PLATFORM	**NH**	- NIDDERY HIGH
NB	- NASINA BASIN	**PP**	- PORCUPINE PLATFORM	**OA**	- OGILVIE ARCH	**PH**	- PURCELL HIGH
MRE	- MEILLEUR RIVER EMBAYMENT	**OP**	- OGILVIE PLATFORM	**MA**	- MACKENZIE ARCH		

AF - ACTIVE FORMATION: undivided slate and units of limestone and sandstone

SUSPECT TERRANES

AT - ALEXANDER TERRANE: undivided platform and basin facies strata, minor volcanic strata

KoT - KOOTENAY TERRANE: sandstone, shale and mafic volcanics

EROSIONAL FEATURES

Tremadocian missing beneath:

- Devonian to Recent strata
- Caradocian strata
- 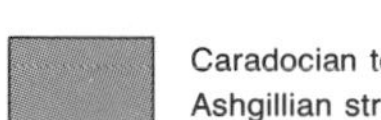Caradocian to Ashgillian strata
- Llandoverian strata
- Ludlovian strata

OTHER FEATURES

- *Arch*
- *Eastern limit of significant tectonic shortening*
- *Tintina-Northern Rocky Mountain Trench Fault System*
- *Rocks of Tremadocian age unknown*
- *Liard Line* (LL)
- *Line of section (see Fig. 7)* (2)

probably formed below the level of carbonate compensation, water depths likely were seldom more than a few hundred metres, as indicated by the fact that cherts and other basinal facies were deposited above continental crust and within a miogeocline where there is evidence of multiple periods of submarine erosion (cf. Cecile and Norford, 1979).

The northern part of the miogeocline was transitional between the Cordilleran Miogeocline and the Franklinian Miogeocline of the Arctic Islands. The transition is illustrated by the Ordovician to Silurian Richardson Trough (Fig. 4C.1), which probably was an aulacogen opening northward into the Franklinian Miogeocline (Churkin, 1975; Norris and Yorath, 1981), and was separated by only a few tens of kilometres from the Cordilleran Miogeocline across the Ogilvie Arch and Ogilvie Platform (Wheeler and Gabrielse, 1972; Churkin, 1975; Cecile, 1982). Very likely the distinctive array of arches, basin highs, troughs, and embayments that characterize the northern miogeocline is the result of crustal extension through the entire block of continental lithosphere located between the two miogeoclines (Cecile, 1986).

The boundaries of the cratonic landmass that formed the core of Ordovician to Silurian North America are enigmatic. Large areas of the present Canadian Shield were landmasses for much of that time; post-Silurian erosion around the shield has obscured their extent. Strata preserved in a few scattered outliers on the shield and in surrounding areas generally contain little or no terrigenous material, suggesting low topographic relief and slight erosion. On the other hand, Middle and Upper Ordovician, shallow-water carbonates in Williston Basin and in the southern and central Cordillera have significant interbeds and/or basal successions of terrigenous detritus. Thus, the Canadian Shield, the central core of Ordovician-Silurian North America, was an unlikely source. Instead, the most likely source was the region fringed by all these occurrences, whose central part is the Peace-Athabasca Arch (Fig. 4C.2, 4C.3). For example, in the Manitoba part of Williston Basin, the quartz sand content of the Winnipeg Formation increases northward to this general area (H.R. McCabe, pers. comm., 1985). Considering these relationships, it is reasonable to assume that the core of the Ordovician-Silurian North American craton had little local or regional topographic relief. Broad carbonate platforms likely coalesced over this core area during the Middle to Late Ordovician epochs, while a landmass of significant relief emerged in the area of Peace-Athabasca Arch. Furthermore, although broad epeiric seas may be virtually unaffected by tides, water depths and circulation patterns generally prevented deposition of evaporitic minerals.

Figure 4C.1. Descriptive paleogeography of the Early Ordovician (Tremadocian); areas and age of post-Tremadocian erosion are shown.

Tectonic and depositional terms

In this subchapter, basins, platforms, embayments, troughs, arches, and highs are defined by lithofacies distributions. Lithofacies are the direct product of ancient depositional regimes. Their original distribution is the product of the tectonic, climatic, or other controls of the time, but their present distribution is commonly greatly reduced from the former by erosion. In defining these features on the basis of lithofacies, their depositional character is emphasized first, and from that and other data, inferences are made concerning their tectonic control and preservation. In effect, when reference is made to a basin it means an area in which basinal facies were deposited, whereas a platform means an area of shallow-water carbonate or clastic facies. Depressions and linear features are tectonic features whose character are mostly inferred from isopach patterns, lithofacies patterns, and anomalies in younger tectonic structures suggesting reactivation of, or control by older structures. The miogeocline is defined by isopach patterns, the presence of thick, mainly marine, sedimentary rocks whose distribution implies a continental margin setting, and by evidence for underlying continental crust. Continental shelf areas are areas of shallow-water deposition cratonward of the miogeocline.

Because of historical precedence there is one exception to these definitions; Williston Basin does not refer to an area of basinal deposition, instead it refers to a post-Silurian tectonic depression in which Ordovician and Silurian platformal carbonates and clastic strata were preserved.

Previous work

Lower Paleozoic rocks of the southern Manitoba outcrops have been known for many years (Tyrrell, 1892; Baillie, 1952; Kupsch, 1953; Stearn, 1956; Andrichuk, 1959; McCabe, 1972a, b; Cowan, 1972), and drilling records in Williston Basin have provided details of their subsurface distribution (Porter and Fuller, 1965; Paterson, 1971; Jamieson, 1979). Sedimentological studies provide understanding of the depositional environments of many of the rock units (Kent, 1960; Roehl, 1967; Lochman-Balk and Wilson, 1967; Vigrass, 1972; Kendall, 1976, 1977).

Silurian strata in southwestern Canada were first recognized by McConnell (1887), whose terminology included the Ordovician later documented by Lapworth (1887). Walcott (1913, 1923, 1924, 1927, and 1928) made many important contributions, while Burling (1921, 1922, 1923 and 1954) provided fundamental stratigraphic information. Cairnes (1914) reported lower Paleozoic rocks on the Yukon-Alaska Boundary and Williams (1922, 1923) measured several stratigraphic sections near the Mackenzie River. The Canol Project (Hume and Link, 1945) contributed the first regional overview.

Regional syntheses (Martin, 1959; Norford, 1965; Lenz, 1972) followed rigorous stratigraphic studies of local stratotypes and systematic descriptions of stratigraphic sections (Norford, 1962, 1964, 1969; Jackson and Lenz, 1962; Slind and Perkins, 1967; Norford et al., 1967; Aitken and Norford, 1967; Aitken et al., 1972; Aitken et al., 1973; Norford and Macqueen, 1975). Detailed sedimentological and tectonstratigraphic studies began later (Jansa, 1976; Cecile and Norford, 1979; Cecile, 1982; Morrow, 1984).

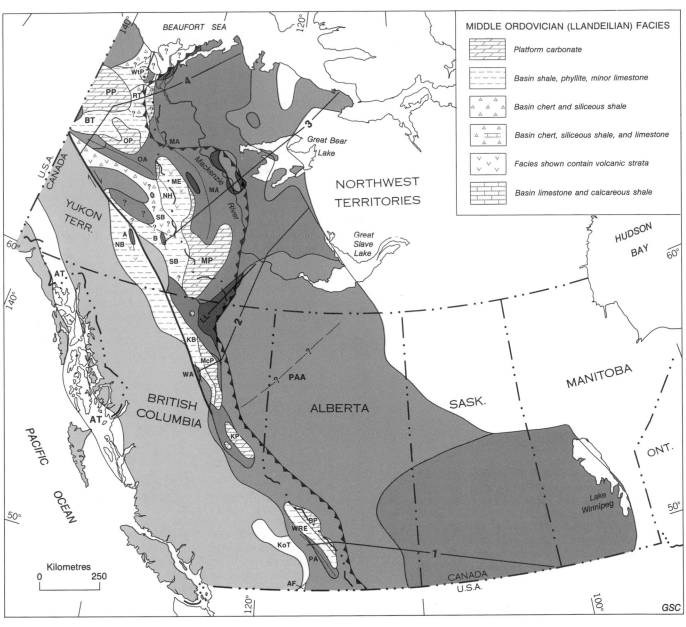

MIDDLE ORDOVICIAN (LLANDEILIAN) FACIES

- *Platform carbonate*
- *Basin shale, phyllite, minor limestone*
- *Basin chert and siliceous shale*
- *Basin chert, siliceous shale, and limestone*
- *Facies shown contain volcanic strata*
- *Basin limestone and calcareous shale*

GSC

MAJOR PALEOGEOGRAPHIC FEATURES PRESENT DURING THE LLANDEILIAN

RT	- RICHARDSON TROUGH	**WRE**	- WHITE RIVER EMBAYMENT	**KP**	- KAKWA PLATFORM	**A**	- POSITIVE AREA A
BT	- BLACKSTONE TROUGH	**WtP**	- WHITE MOUNTAINS PLATFORM	**BP**	- BOW PLATFORM	**B**	- POSITIVE AREA B
SB	- SELWYN BASIN	**PP**	- PORCUPINE PLATFORM	**OA**	- OGILVIE ARCH	**WA**	- WARE ARCH
ME	- MISTY CREEK EMBAYMENT	**OP**	- OGILVIE PLATFORM	**MA**	- MACKENZIE ARCH	**PA**	- PURCELL ARCH
NB	- NASINA BASIN	**MP**	- MACKENZIE PLATFORM	**NH**	- NIDDERY HIGH		
KB	- KECHIKA BASIN	**McP**	- MACDONALD PLATFORM	**PAA**	- PEACE - ATHABASCA ARCH		

AF - ACTIVE FORMATION: undivided slate with minor sandstone, and limestone

SUSPECT TERRANES

AT - ALEXANDER TERRANE: undivided platform and basin facies strata, minor volcanic strata

KoT - KOOTENAY TERRANE: sandstone, shale and mafic volcanics

EROSIONAL FEATURES

Llandeilian missing beneath:

Devonian to Recent strata

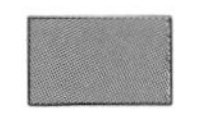
Caradocian to Ashgillian strata

Llandoverian strata

Ludlovian strata

OTHER FEATURES

Arch ....

Eastern limit of significant tectonic shortening ....

Tintina-Northern Rocky Mountain Trench Fault System ....

Rocks of Llandeilian age unknown ....

Liard Line LL

Line of section (see Fig. 7) 2

Subsurface data have been summarized by Tassonyi (1970), Meijer Drees (1974, 1975a, 1975b), and Pugh (1976, 1983, in press).

Several reports dealt with regional paleogeographic reconstructions: Ziegler (1969) and Douglas et al. (1970) dealt with all of western Canada and the Arctic; Lochman-Balk and Wilson (1967) and Porter and Fuller (1965) with the Williston Basin; Norford (1965) and Porter et al. (1982) with the southern Cordillera; and Gabrielse (1967), Norford (1964), Lenz (1972), Cecile (1982), and Pugh (1983) with the northern Cordillera.

Acknowledgments

We would like to thank D.F. Paterson, H.R. McCabe, C.J. Yorath, A.V. Okulitch, and H. Gabrielse for their critical comments and suggestions. We also benefited from discussions with S.P. Gordey, D.J. Tempelman-Kluit, L. Struik, J.O. Wheeler, P.B. Read, D.G. Cook, G.K. Williams, D.W. Morrow, A.C. Lenz, R.I. Thompson, W.F. Fritz, D.K. Norris, and D.C. Pugh.

PLATFORMS AND LAND AREAS

The Ordovician-to-Silurian continental margin of western Canada included several areas of widespread shallow-water, platform-carbonate deposition. In the northwest, White Mountains, Porcupine and Ogilvie platforms (Fig. 4C.1 to 4C.5) were present in the area between the Cordilleran and Franklinian miogeoclines. Along the Cordilleran Miogeocline and adjacent craton were, from north to south, Mackenzie, MacDonald, Kakwa, Bow, and Williston platforms. Platformal carbonates are also found on Ogilvie and Mackenzie arches and Cassiar Platform, and a positive area west of Selwyn Basin. MacDonald Platform was described by Douglas et al. (1970), and the terms Kakwa and Bow are introduced for platformal areas to the south. Although these areas are likely erosional remnants of a once continuous platform, it is convenient to name them individually.

In the north, Porcupine Platform persisted from Late Cambrian through Ordovician time, but during the Silurian it subsided with respect to adjacent platforms and was partly drowned, resulting in extensive deposition of basinal facies over older platformal facies (Fig. 4C.1 to 4C.5). In the same area, the relatively small Ogilvie Platform persisted from Late Cambrian through Ordovician and Silurian times, probably in continuity with the adjacent Ogilvie Arch. Also in the northwest, the small White Mountains Platform developed from a Cambrian and probably Early Ordovician promontory of Mackenzie Platform. In Late Ordovician and Silurian times, a trough receiving basinal facies, completely separated the White Mountains Platform from Mackenzie Platform.

Figure 4C.2. Descriptive paleogeography of the Middle Ordovician (Llandeilian); areas and age of post-Llandeilian erosion are shown.

The only known region of significant platform carbonate deposition west of the Tintina-Northern Rocky Mountain Trench is in the Cassiar Platform (Gabrielse, 1967), where Upper Ordovician carbonates of limited lateral extent are overlain by widespread Lower Silurian carbonates of the Sandpile Group. These rocks are overlain by undated carbonates (thought to be Upper Silurian to Lower Devonian), which are preserved over an even larger area. On restoration of dextral fault offsets of 400 to 750 km (Roddick, 1967; Gabrielse, 1985) the Cassiar Platform would be juxtaposed along trend with Kakwa Platform. Because carbonates of Kakwa Platform prograded westward to reach the position of Northern Rocky Mountain Trench only by Late Ordovician and Silurian time, and Cassiar Platform carbonates first began to develop in the Late Ordovician and Silurian, the stratigraphic record of the two areas is similar and supports the suggested reconstruction.

The western edges of all the platforms varied in position through time; but only Kakwa and Bow platforms prograded progressively westward from Early Ordovician to Early Silurian time (Fig. 4C.1 to 4C.4).

In Williston Basin area, Lower Ordovician strata (uppermost Deadwood Formation) are mainly marine sandstone in the east, and shale with minor limestone and sandstone in the west. Lochman-Balk and Wilson (1967) speculated that these Lower Ordovician rocks represent the beginnings of Williston Basin. Considering the history of older Cambrian strata (J.D. Aitken, Subchapter 4B in this volume), it is more likely that the shales were part of a more extensive inner detrital facies belt located between platformal carbonate in the mid-miogeocline and sandstone facies fringing the Canadian Shield.

During part of the Middle Ordovician, the site of Williston Basin was a land area exposing Precambrian, Cambrian, and Lower Ordovician rocks to erosion. Quartz sandstones of the Winnipeg Formation formed the basal deposits of the next transgression. The presence and distribution of these and slightly older and younger sandstone units indicate the existence of an Ordovician Peace-Athabasca Arch. Older (uppermost Lower Ordovician) Monkman and Tipperary quartzites developed locally within Kakwa and Bow platforms, respectively. Younger sandstones (Fig. 4C.3) preceded Late Ordovician carbonate deposition within Bow Platform (Mount Wilson Quartzite, Fig. 4C.6; Norford, 1969), southern Mackenzie Platform (Little Doctor sandstone - basal Mount Kindle; see Meijer Drees, 1975b), and Hudson Bay Platform (member 1, Portage Chute Formation; Nelson, 1963). Platformal carbonates of Late Ordovician age in MacDonald and Kakwa platforms also have either a basal sandstone or abundant sandstone interbeds (Beaverfoot Formation - unnamed Upper Ordovician carbonate; Slind and Perkins, 1967; Cecile and Norford, 1979). Some sand was carried into Kechika Basin where Upper Ordovician graptolitic shale is interbedded with quartzite (Cecile and Norford, 1979; Fig. 4C.3, 7). The widespread occurrence of these sandstones indicates that a major terrigenous source area was exposed and eroded intermittently during the Ordovician. However, as noted earlier, the scattered outliers of Middle to Upper Ordovician carbonate on the Canadian Shield lack associated sandstone, making the central parts of the North American craton an unlikely source. All the sandstone occurrences noted above fringe

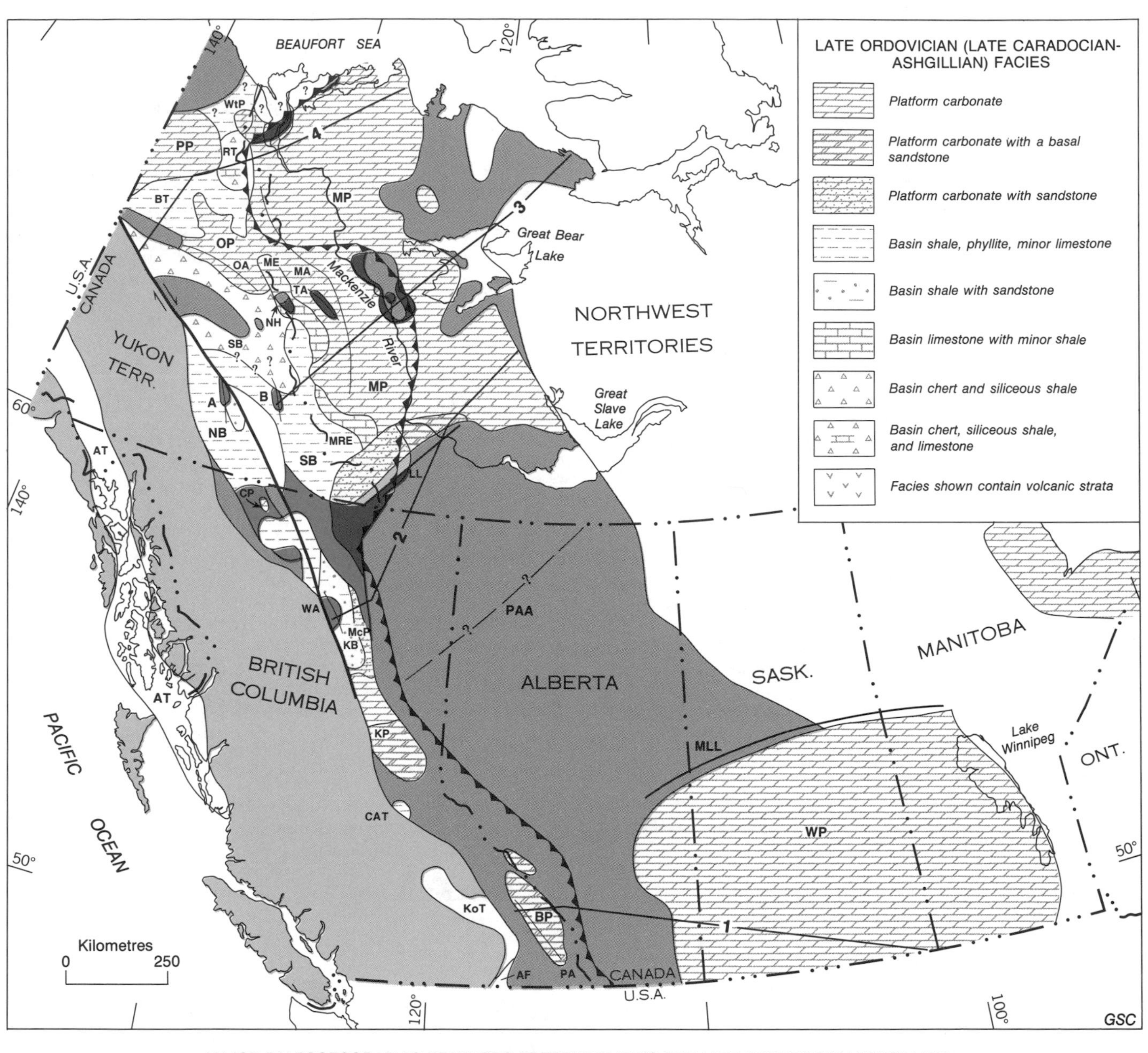

MAJOR PALEOGEOGRAPHIC FEATURES PRESENT DURING THE LATE CARADOCIAN-ASHGILLIAN

RT	RICHARDSON TROUGH	**PP**	PORCUPINE PLATFORM	**BP**	BOW PLATFORM	**TA**	TWITYA ARCH
BT	BLACKSTONE TROUGH	**MP**	MACKENZIE PLATFORM	**WP**	WILLISTON PLATFORM	**NH**	NIDDERY HIGH
SB	SELWYN BASIN	**OP**	OGILVIE PLATFORM	**WtP**	WHITE MOUNTAINS PLATFORM	**A**	POSITIVE AREA A
NB	NASINA BASIN	**CP**	CASSIAR PLATFORM	**PAA**	PEACE-ATHABASCA ARCH	**B**	POSITIVE AREA B
ME	MISTY CREEK EMBAYMENT	**McP**	MACDONALD PLATFORM	**OA**	OGILVIE ARCH	**WA**	WARE ARCH
MRE	MEILLEUR RIVER EMBAYMENT	**KP**	KAKWA PLATFORM	**MA**	MACKENZIE ARCH	**PA**	PURCELL ARCH
KB	KECHIKA BASIN						

AF - ACTIVE FORMATION: undivided slate with minor sandstone and limestone

SUSPECT TERRANES

AT - ALEXANDER TERRANE: undivided platform and basin facies strata, minor volcanic strata

KoT - KOOTENAY TERRANE: sandstone, shale and mafic volcanics

CAT - CARIBOO SUBTERRANE

EROSIONAL FEATURES

Late Caradocian to Ashgillian missing beneath:

Devonian to Recent strata

Llandoverian strata

Ludlovian strata

OTHER FEATURES

Arch

Eastern limit of significant tectonic shortening

Liard Line — LL

Meadow Lake Line — MLL

Rocks of Late Caradocian to Ashgillian unknown

Tintina-Northern Rocky Mountain Trench Fault System

Line of section (see Fig. 7) — 2

or are near an extensive area that lies between Williston and Mackenzie platforms and extends east toward Hudson Bay. This area including Peace-Athabasca Arch is essentially the one which Ketner (1968) looked to as a "long-shore drift" source of Ordovician quartzites of the western United States and southwestern Canada. At present, Devonian strata over the arch rest unconformably on Precambrian rocks. This sub-Devonian unconformity likely represents a long complex history, part of which involved erosion during the Middle and Late Ordovician epochs. The only other significant occurrences of sandstone in platformal facies are within Lower Silurian strata of MacDonald Platform, in places where the Nonda Formation rests directly on the Precambrian, and within Upper Silurian strata on Mackenzie Arch (Delorme Formation).

During Middle Ordovician to Early Silurian time, an extensive carbonate platform (Williston Platform) was present in the area conventionally designated Williston Basin. Considering the short distances and low relief between Williston Platform rocks and similar successions in the northern Hudson Platform in the east, and Bow and Kakwa platforms in the west, one can assume that all four areas were depositionally continuous.

On the platforms, Ordovician and Silurian strata typically are thickly bedded dolostone and/or limestone, representing shallow-marine to intermittently emergent conditions. Evaporites are known only from Hudson Platform and as minor interbeds within high Middle and Upper Ordovician strata (Herald Formation, Fort Garry Member of Red River Formation) of Williston Platform (Kent, 1960). In the miogeocline, most platformal carbonates display rapid transitions to basinal facies (Fig. 4C.8). Where preserved, the western edges of most of these platforms are thick successions of massive, thick-bedded, medium- to coarse-crystalline dolostone, whereas equivalent rocks farther east on the platforms are divisible into numerous stacked lithological units. Most of the platform-edge rocks lack clear evidence of framework-building organisms but have been altered extensively; much of the fossil record has been destroyed. Some silicified parts of these otherwise barren rocks contain abundant framework-building organisms, indicating a reefal origin. The transition from platform to basin was typically narrow and features spectacular foreslope deposits. One exception is the southwestern edge of Mackenzie Platform (central to southern Mackenzie Mountains), where rocks of basinal and platformal facies are interstratified in a transitional belt several tens of kilometres wide.

Ordovician and Silurian platformal carbonates hosted a variety of shallow-water flora (algae) and fauna, of which trilobites, brachiopods, corals, cephalopods, gastropods, and conodonts are biostratigraphically significant. There were, however, some significant variations stratigraphically in the relative abundance of different biota. In the Early Ordovician, platformal environments (Survey Peak and Franklin Mountain formations) were dominated by stromatolites, which occur in thin biostromes (Fig. 4C.9), or as scattered domes and biscuits. Middle Ordovician strata (Skoki Formation) contain brachiopods, sponges, and stromatolites, but they are characterized by large gastropods (*Maclurites, Palliseria*, Fig. 4C.10) and algal pisolites. Some Middle Ordovician units contain almost no fossils (e.g. Owen Creek Formation). Upper Ordovician to Lower Silurian units (Beaverfoot, Mount Kindle, Red River, Stony Mountain, Vunta formations) hosted the most diverse and abundant faunal assemblages, commonly forming large biostromes of colonial and solitary corals, large gastropods (including *Maclurites*), and large cephalopods. Upper Silurian strata (Muncho-McConnell, Delorme, Peel formations) are sparsely fossiliferous, with some fish and brachiopods (Thompson, 1989). Fish debris is found also in the Upper Ordovician of MacDonald Platform (Cecile and Norford, 1979).

Platform cycles

The platforms were the sites of several major cycles of transgression and deposition followed by regression and erosion (see Fig. 4C.7 and correlation charts, in pocket). In the north, Mackenzie Platform contains the record of three of these cycles. Cycle A (upper third of the Sauk Sequence of Sloss, 1963, 1972), resulted from widespread transgression, with deposition of thick-bedded, grey dolostone with scattered stromatolites of the Franklin Mountain Formation during Late Cambrian to Early Ordovician time. In shelf areas, deposition persisted into early Middle Ordovician time. Cycle A ended with widespread regression and erosion during the late Middle Ordovician (Fig. 4C.2, 4C.7). Cycle B (≃ Tippecanoe Sequence of Sloss, 1963, 1972) began with widespread Late Ordovician transgression, with deposition of Upper Ordovician to Lower Silurian, thick-bedded, fossiliferous, dark grey dolostone of the Mount Kindle Formation, and ended with late Early and Late Silurian (Wenlock-early Ludlow) regression and erosion (Fig. 4C.7, 4C.11). Cycle C (≃ Lower Kaskaskia Sequence of Sloss, 1963) reflects Late Silurian transgression (late Ludlow), with deposition of Upper Silurian to Lower Devonian, yellow-weathering, arenaceous dolostones of the Delorme, Peel, and Tsetso formations, followed by Early Devonian regression and minor erosion.

In the south, carbonate platforms are preserved only within parts of the Ordovician and Silurian continental shelf and within Williston Basin. In all southern areas, Cycle A is clearly recognizable. In Bow Platform, the Late Cambrian transgression was followed by deposition of Upper Cambrian to lower Middle Ordovician limestone, thrombolitic limestone, shale, pisolitic dolostone, and yellow dololutites of the Survey Peak, Skoki, and Owen Creek formations, and the cycle ended with late Middle Ordovician regression and erosion (Fig. 4C.12). Cycle A in Williston Basin area was similar except that the depositional units were sandstone and shale (Deadwood Formation), the youngest of which are late Early Ordovician.

In northeastern British Columbia, no platformal rocks equivalent to Cycle A are preserved, but the cycle can be recognized within basinal strata in Kechika Basin. Lower

Figure 4C.3. Descriptive paleogeography of the Late Ordovician (Late Caradocian-Ashgillian); areas and age of post-Late Caradocian-Ashgillian erosion are shown.

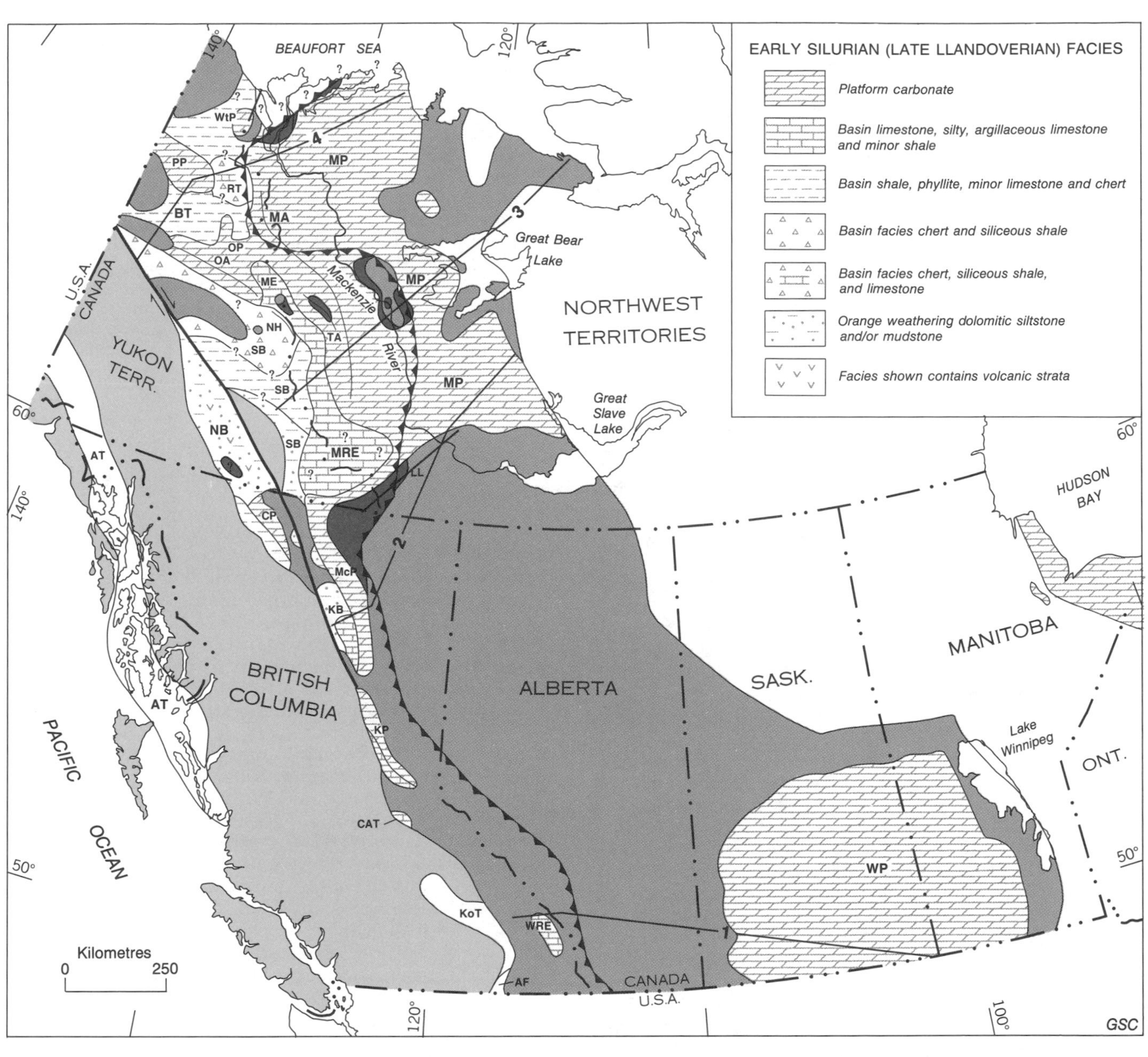

MAJOR PALEOGEOGRAPHIC FEATURES PRESENT DURING THE LATE LLANDOVERIAN

BT - BLACKSTONE TROUGH
RT - RICHARDSON TROUGH
ME - MISTY CREEK EMBAYMENT
SB - SELWYN BASIN
NB - NASINA BASIN
MRE - MEILLEUR RIVER EMBAYMENT
KB - KECHIKA BASIN
WRE - WHITE RIVER EMBAYMENT
WtP - WHITE MOUNTAINS PLATFORM
PP - PORCUPINE PLATFORM
AF - ACTIVE FORMATION: undivided slate with minor sandstone and limestone
OP - OGILVIE PLATFORM
MP - MACKENZIE PLATFORM
McP - MACDONALD PLATFORM
CP - CASSIAR PLATFORM
KP - KAKWA PLATFORM
WP - WILLISTON PLATFORM
OA - OGILVIE ARCH
MA - MACKENZIE ARCH
TA - TWITYA ARCH
NH - NIDDERY HIGH

SUSPECT TERRANES

AT - ALEXANDER TERRANE: undivided platform and basin facies strata, minor volcanic strata
CAT - CARIBOO SUBTERRANE
KoT - KOOTENAY TERRANE: sandstone, shale and mafic volcanics

EROSIONAL FEATURES

Late Llandoverian missing beneath:

Devonian to Recent strata

Ludlovian strata

OTHER FEATURES

Arch

Eastern limit of significant tectonic shortening

Tintina-Northern Rocky Mountain Trench Fault System

Rocks of Late Llandoverian age unknown

Liard Line (LL)

Line of section (see Fig. 7) (2)

Ordovician basinal facies of the Kechika Group in these areas thin dramatically and show increasing evidence of intrastratal erosion and truncation eastward (Cecile and Norford, 1979). These features suggest proximity to platformal rocks since removed by erosion just to the east.

The records of Cycles B and C in the south are much more variable. In MacDonald and Kakwa platforms, Cycle B consists of two sub-cycles. The first of these records Late Ordovician transgression and deposition (equivalents of the Mount Wilson and Beaverfoot formations: quartzite and thick-bedded grey dolostone), followed by latest Ordovician to early Early Silurian regression and erosion. The second records Early Silurian deposition (Nonda Formation) followed by erosion prior to Late Silurian (Ludlow) deposition. In Bow Platform, the latest Ordovician to early Early Silurian erosional break has not been recognized, and this area apparently records an unbroken Cycle B. The upper part of Cycle B in Bow Platform represents drowning of the platform and deposition of basinal facies (Tegart Formation; Fig. 4C.12). In Williston Basin, disconformities are subtle and related erosion is minor, but strata in the position of Cycle B can be divided into four sub-cycles, three of which are: (1) Middle Ordovician transgression and deposition (Winnipeg Formation) followed by late Middle Ordovician regression and minor erosion; (2) latest Middle Ordovician transgression and deposition, which continued into Late Ordovician time (lower Red River Formation, Yeoman Formation) before the onset of regression; and (3) Late Ordovician transgression (upper Red River, Herald, Stony Mountain, and Stonewall formations) followed by early Early Silurian regression. The fourth sub-cycle records Early Silurian transgression and deposition (Interlake Group) followed by Late Silurian regression and significant erosion. This sub-cycle is not recognized in Cycle B strata elsewhere but is considered to be the terminal part of the Tippecanoe Sequence of Sloss (1972). No strata younger than Cycle B are known in Bow Platform, and Late Silurian to Early Devonian strata in Williston Basin are missing beneath younger Devonian strata. MacDonald and Kakwa platforms, however, have a Cycle C, which is essentially identical with that found farther north: Late Silurian transgression, with late Silurian to Early Devonian deposition (Muncho-McConnell Formation) and later Early Devonian regression and erosion.

BASINS, TROUGHS, AND EMBAYMENTS

In Ordovician and Silurian time, the western Canadian miogeocline included a variety of basins, troughs, and embayments. In the northwest, the Richardson Trough virtually connected the Cordilleran and Franklinian miogeoclines. The shallow and elongate Blackstone Trough extended southwestward from the Richardson Trough. Farther south, the basinal part of the miogeocline can be divided into three parallel belts. The easternmost consists of embayments of basinal facies into carbonate platforms. From north to south these are the Misty Creek Embayment and Meilleur River Embayment and its satellite the Prairie Creek Embayment. West of the embayments there is a belt of inner basins: from north to south these are Selwyn Basin, Kechika Basin, Robson Basin, and White River Embayment. The westernmost, outer basinal belt is incompletely preserved and poorly understood. To date, only two areas of autochthonous North American strata are recognized in this belt: Nasina Basin in the north and, in the far south, the small area southeast of Trail with outcrops of the Active Formation.

Crustal shortening has affected the elements noted above, but none of the paleogeographic maps and cross-sections (Fig. 4C.1 to 4C.5, 4C.7, 4C.11) has been palinspastically restored. In the south, North American Ordovician and Silurian rocks have been transported towards the craton by as much as 200 km, the westernmost strata having been displaced the most (Price and Fermor, 1985). In the north, displacements towards the craton are as much as 100 km in the eastern shelf (Thompson, 1981; Gordey, 1981a; Cecile and Cook, 1981) and exceed 300 km in the central and western Selwyn Basin (combining the data of Cecile and Cook, 1981, and Cecile, 1984).

Although this tectonic transport towards the craton has resulted in contraction of facies belts, the relative positions of facies have not been significantly altered. On the other hand, major strike-slip displacements suggested for the Tintina-Northern Rocky Mountain Trench fault systems cannot be ignored. Reconstruction of strike-slip displacements propose right-lateral offsets ranging from 400 to 450 km (Roddick, 1967; Templeman-Kluit, 1979) to more than 750 km (Gabrielse, 1985). Restoration of strike-slip offsets of this scale significantly alters present distributions of facies and basin geometry.

Richardson Trough (Lenz, 1972; Pugh, 1983) was an elongate basin enclosed on the east, west, and south by carbonate platforms during Late Cambrian and Early Ordovician time, when it was the site of a thick succession of basinal limestone (lower Road River Formation; Fig. 4C.13). More varied and deeper basinal facies accumulated during late Early Ordovician to Late Silurian time (upper Road River Formation; Fig. 4C.14). Richardson Trough persisted as a north-south trending basin throughout its history (Fig. 4C.1 to 4C.5), but expanded laterally during the Silurian, with deposition of basinal facies over much of the earlier Porcupine Platform and Ogilvie Arch.

Blackstone Trough (Lenz, 1972) was an elongate, east-west trending Ordovician feature bounded on the south by Ogilvie Arch and on the north by Porcupine Platform. To the west, it abuts against the Tintina strike-slip fault in Alaska. Blackstone Trough was a relatively shallow depression that received a thin succession of calcareous shale and limestone (Road River Formation). It first appeared in the late Early Ordovician as a small arm of Richardson Trough, developed into a major trough during the Middle Ordovician, and expanded, along with Richardson Trough, during Late Ordovician to Silurian time when Porcupine Platform was partly submerged.

As the Richardson and Blackstone troughs expanded during the Middle Ordovician to Early Silurian, the central Richardson Trough received much more cherty sediment,

Figure 4C.4. Descriptive paleogeography of the Early Silurian (Late Llandoverian); areas and age of post-Late Llandoverian erosion are shown.

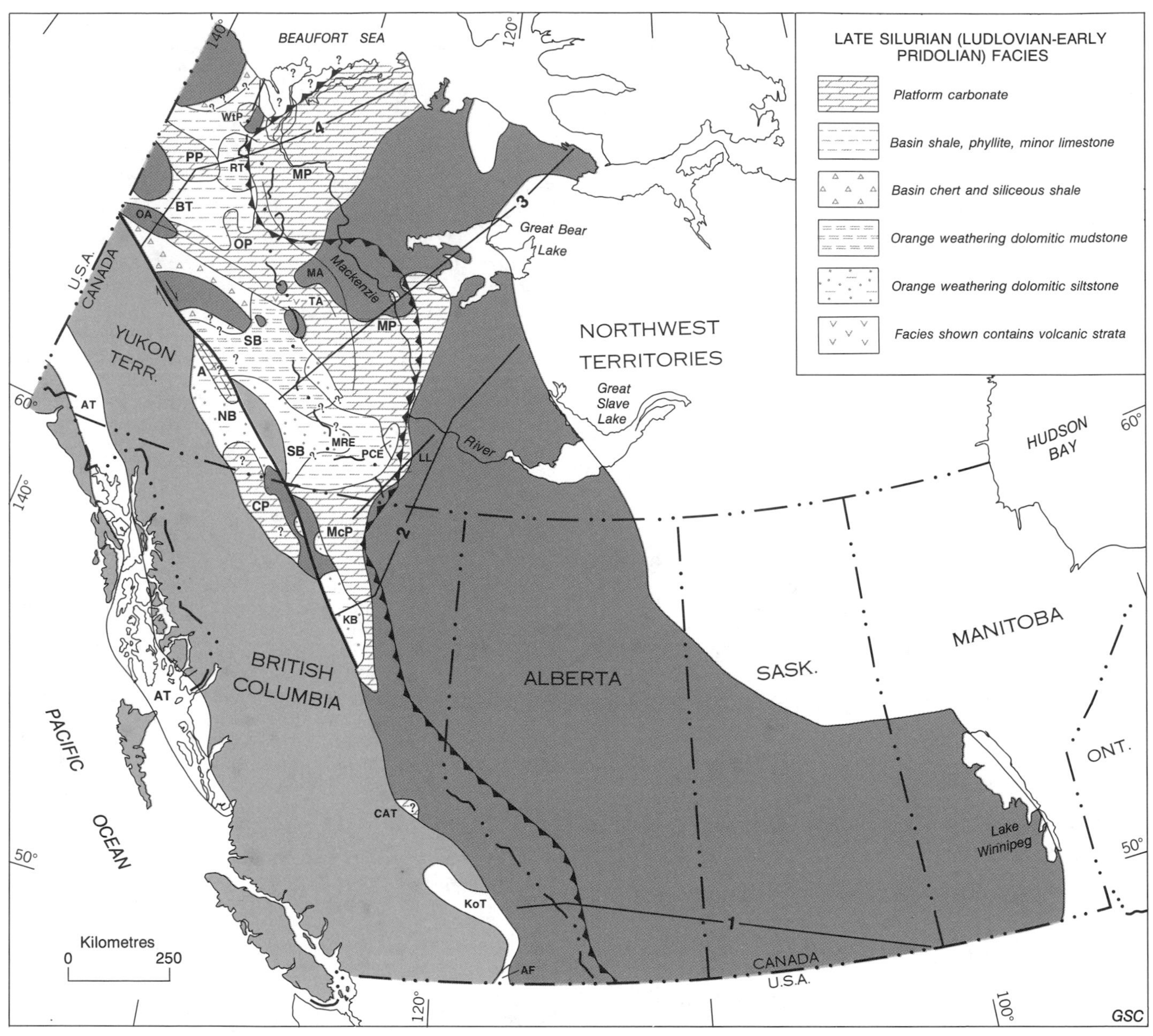

MAJOR PALEOGEOGRAPHIC FEATURES PRESENT DURING THE LUDLOVIAN-EARLY PRIDOLIAN

BT - BLACKSTONE TROUGH	**MRE** - MEILLEUR RIVER EMBAYMENT	**OP** - OGILVIE PLATFORM	**OA** - OGILVIE ARCH
RT - RICHARDSON TROUGH	**PCE** - PRAIRIE CREEK EMBAYMENT	**MP** - MACKENZIE PLATFORM	**MA** - MACKENZIE ARCH
SB - SELWYN BASIN	**WtP** - WHITE MOUNTAINS PLATFORM	**McP** - MACDONALD PLATFORM	**TA** - TWITYA ARCH
NB - NASINA BASIN	**PP** - PORCUPINE PLATFORM	**CP** - CASSIAR PLATFORM	**A** - POSITIVE AREA A
KB - KECHIKA BASIN			

AF - ACTIVE FORMATION: undivided slate with minor sandstone and limestone

SUSPECT TERRANES

AT - ALEXANDER TERRANE: undivided platform and basin facies strata, minor volcanic strata
KoT - KOOTENAY TERRANE: sandstone, shale and mafic volcanics
CAT - CARIBOO SUBTERRANE

EROSIONAL FEATURES

Ludlovian-Early Pridolian missing beneath Devonian to Recent strata

OTHER FEATURES

Arch ....

Eastern limit of significant tectonic shortening ....

Tintina-Northern Rocky Mountain Trench Fault System ....

Rocks of Ludlovian-Early Pridolian age unknown ....

Liard Line LL

Line of section (see Fig. 7) .)............................. 2

presumably indicating deeper water. The same area was also apparently the only part of Richardson and Blackstone troughs to receive carbonate-rich muds during Late Silurian time.

The Misty Creek Embayment (Cecile, 1982) was a large rectangular depression bounded on three sides by platformal carbonates (Franklin Mountain and Mount Kindle formations) and connected to the southwest with Selwyn Basin across the Niddery (submarine) High. Misty Creek Embayment developed in the late Early Cambrian, and persisted to middle Early Silurian time. In Late Cambrian and Early Ordovician time, it was the site of deposition of thick successions of thin-bedded basinal limestone (Rabbitkettle Formation) with spectacular slope-breccias and slump features in successions transitional between platform and basin (Fig. 4C.15). In the Middle Ordovician, during a period of renewed extension of the miogeocline, the embayment received a moderately thick succession of shale, chert, and basinal limestone (Duo Lake Formation) and volcaniclastic and alkalic volcanic strata (Marmot Formation). During Late Ordovician to middle Early Silurian time, a moderately thick succession of thin-bedded shaly limestone (Cloudy Formation) accumulated, locally host to patches of solitary and colonial corals and minor basic alkalic volcanic flows, tuffs, and volcaniclastics. From the Late Silurian into the Devonian, the site of Misty Creek Embayment became a shallow-water platform, at times emergent.

Richardson Trough and Misty Creek Embayment are graben-shaped. Basinal facies of Misty Creek Embayment are interstratified with thin to locally very thick successions of volcanic strata, whereas Richardson Trough contains no known Ordovician or Silurian volcanic rocks. Palinspastic restoration results in the south end of Misty Creek Embayment being rotated clockwise about an axis just northwest of the embayment to an *en échelon* alignment with Richardson Trough. Both features are located between the Franklinian and Cordilleran miogeoclines, and thus it is likely that both represent rifts that failed to open completely as the two miogeoclines developed.

Meilleur River Embayment (Bassett and Stout, 1967) was a large semi-circular embayment into the southwestern Mackenzie Platform. It was surrounded on three sides by carbonate platforms (Broken Skull, Sunblood, Whittaker, and Delorme formations) and opened westward into the southern Selwyn Basin. During the Early Ordovician the embayment was only a minor recess into Mackenzie Platform, and during the Middle Ordovician it was replaced completely by platformal carbonate strata (Sunblood Formation). In the Late Ordovician, it was re-established as a small embayment that expanded during the Silurian Period and continued to Middle Devonian time. A small satellitic embayment, the Prairie Creek Embayment, developed from the northeastern Meilleur River

Figure 4C.5. Descriptive paleogeography of the Late Silurian (Ludlovian-Early Pridolian); areas and age of post-Ludlovian-Early Pridolian erosion are shown.

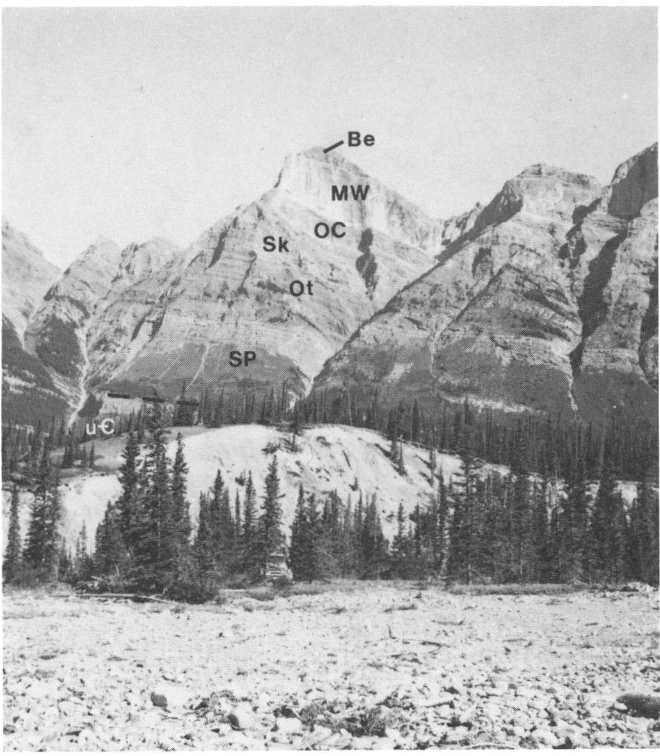

Figure 4C.6. Ordovician succession, White River Embayment - Bow Platform. UC - Upper Cambrian; SP - Survey Peak Formation; Ot - Outram Formation; Sk - Skoki Formation; OC - Owen Creek Formation; MW - Mount Wilson Quartzite; Be - Beaverfoot Formation (from Norford, 1964; photo - GSC 118948).

Embayment in Late Silurian time (Morrow, 1984). During the Early Ordovician, Late Ordovician, and Early Silurian, the Meilleur River Embayment received mainly basinal limestone and some shale (Road River units). During the Late Silurian, carbonate-rich silts were deposited in parts of Meilleur River Embayment and in Prairie Creek Embayment. Shale deposition continued elsewhere in the Meilleur River Embayment.

Selwyn Basin (Gabrielse, 1967) is the largest of the inner miogeoclinal basins. It is bounded on the north by Ogilvie Arch, on the east by Niddery High, Mackenzie Platform, and Meilleur River Embayment, and on the west by Tintina Fault. When the amount of dextral offset on the Tintina-Northern Rocky Mountain Trench fault system is clearly established, the western boundary can be revised. If either the 400 km of displacement of Roddick (1967) or the minimum of 750 km of Gabrielse (1985) is accepted, a line drawn through middle shelf positive areas A and B and Ware Arch (Figs. 4C.2, 4C.3 and below, under "Arches, Positive Areas and Submarine Highs"), after fault restoration, might make a natural boundary between the inner Selwyn and Kechika basins and the outer Nasina Basin. These positive areas were sites of Cambrian, Silurian? and Devonian carbonate buildups. To the south, Selwyn Basin merges with Kechika Basin across an arbitrary east-west boundary drawn from the southern edge of Meilleur River Embayment.

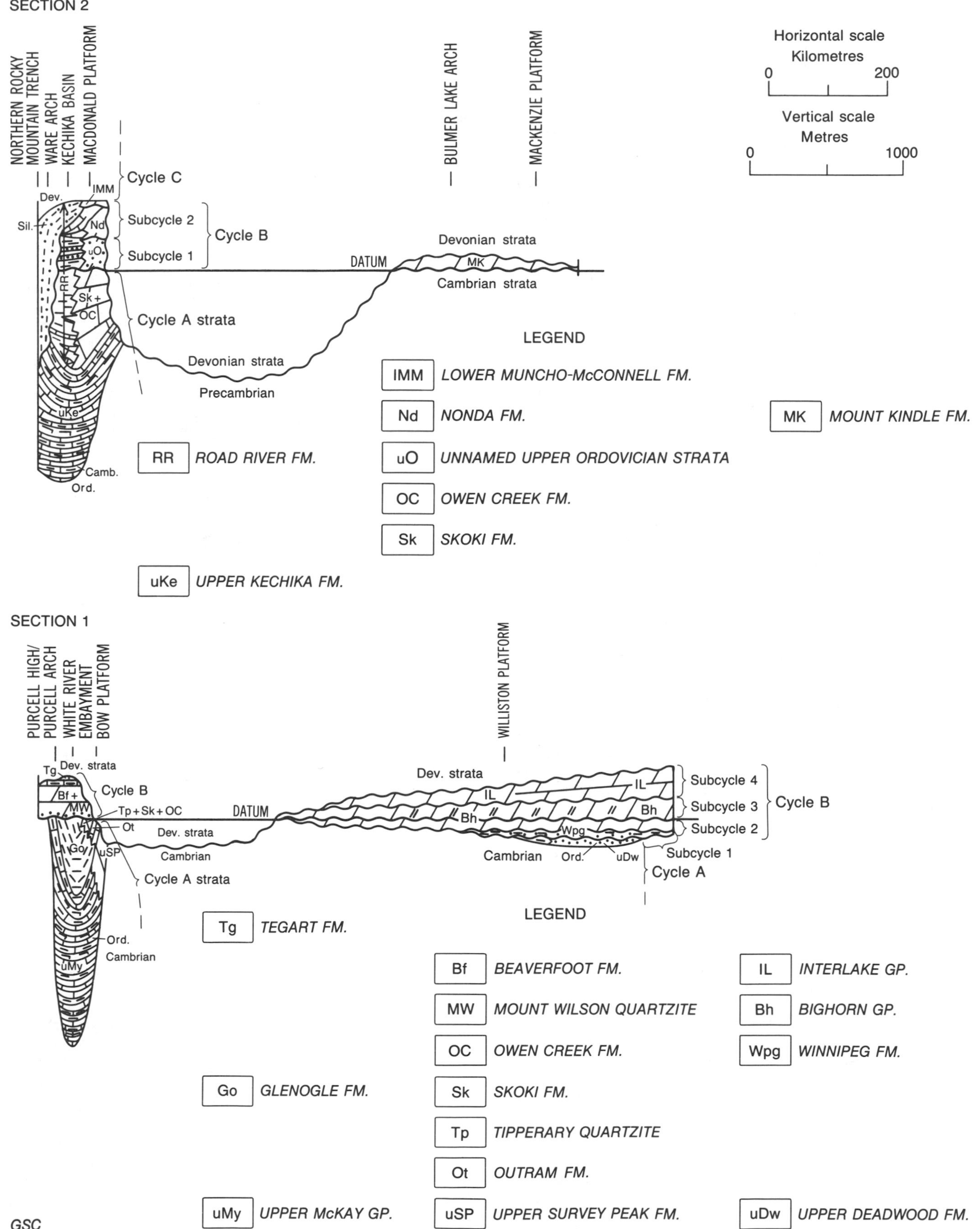

Figure 4C.7. Stratigraphic cross-sections; refer to Figures 4C.1 to 4C.5 for location. Continued on opposite page.

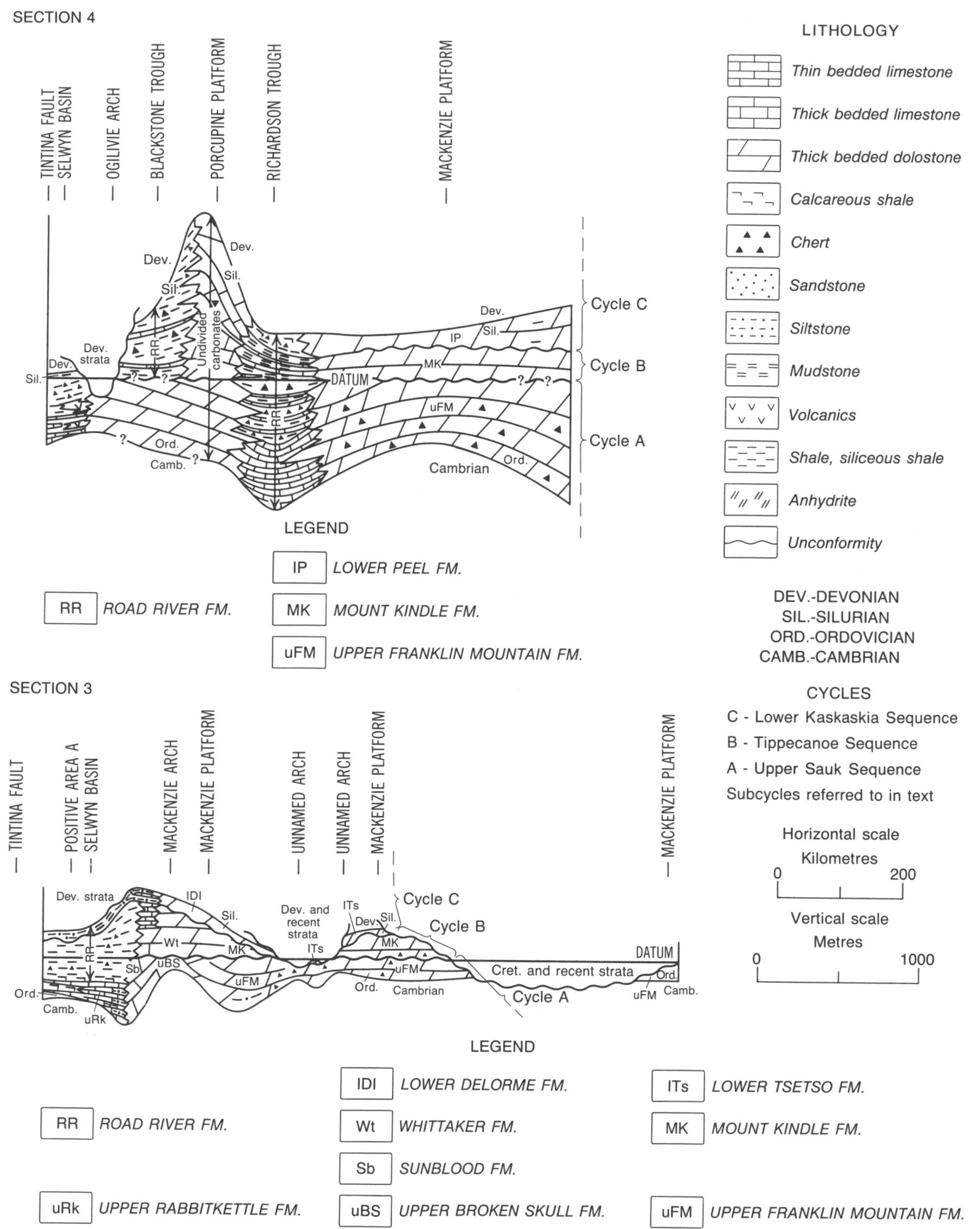
SECTION 4
TINTINA FAULT
SELWYN BASIN
OGILVIE ARCH
BLACKSTONE TROUGH
PORCUPINE PLATFORM
RICHARDSON TROUGH
MACKENZIE PLATFORM
Dev.
Sil.
Dev. strata
Undivided carbonates
RR
DATUM
IP
MK
uFM
Ord.
Camb.
Cambrian
Cycle C
Cycle B
Cycle A
LEGEND
RR ROAD RIVER FM.
IP LOWER PEEL FM.
MK MOUNT KINDLE FM.
uFM UPPER FRANKLIN MOUNTAIN FM.
LITHOLOGY
Thin bedded limestone
Thick bedded limestone
Thick bedded dolostone
Calcareous shale
Chert
Sandstone
Siltstone
Mudstone
Volcanics
Shale, siliceous shale
Anhydrite
Unconformity
DEV.-DEVONIAN
SIL.-SILURIAN
ORD.-ORDOVICIAN
CAMB.-CAMBRIAN
CYCLES
C - Lower Kaskaskia Sequence
B - Tippecanoe Sequence
A - Upper Sauk Sequence
Subcycles referred to in text
Horizontal scale
Kilometres
0
200
Vertical scale
Metres
0
1000
SECTION 3
TINTINA FAULT
POSITIVE AREA A
SELWYN BASIN
MACKENZIE ARCH
MACKENZIE PLATFORM
UNNAMED ARCH
UNNAMED ARCH
MACKENZIE PLATFORM
MACKENZIE PLATFORM
Dev. strata
lDl
Sil.
Wt
Sb
uBS
uRk
Dev. and recent strata
ITs
Cret. and recent strata
DATUM
LEGEND
lDl LOWER DELORME FM.
Wt WHITTAKER FM.
Sb SUNBLOOD FM.
uBS UPPER BROKEN SKULL FM.
RR ROAD RIVER FM.
uRk UPPER RABBITKETTLE FM.
lTs LOWER TSETSO FM.
MK MOUNT KINDLE FM.
uFM UPPER FRANKLIN MOUNTAIN FM.
GSC

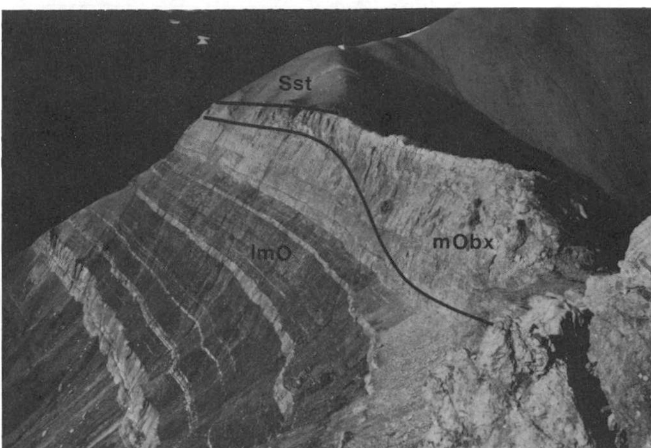

Figure 4C.8. Basinal and transitional facies, Kechika Basin and MacDonald Platform. lmO - Lower to Middle Ordovician shales and limestones, mObx - Middle Ordovician breccia, Sst - Silurian siltstone (from Cecile and Norford, 1979; photo - GSC 199405).

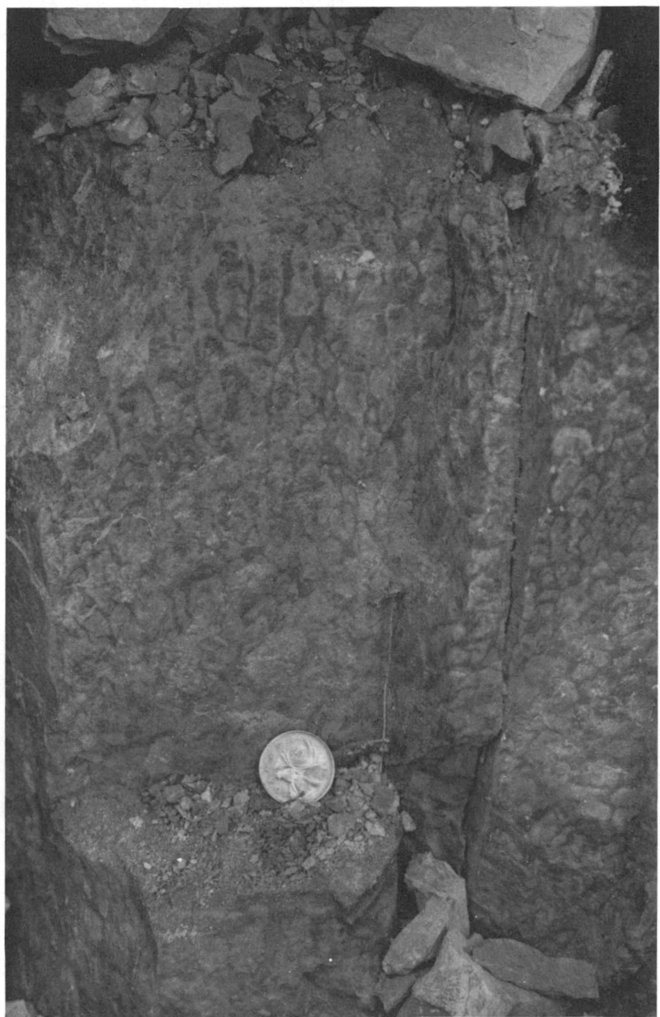

Figure 4C.9. Lower Ordovician stromatolite bioherm, west Mackenzie Arch (from Cecile, 1982; photo - GSC 199535).

Southern and northern parts of Selwyn Basin show distinctly different lithofacies patterns. From Early Ordovician to Early Silurian time, southern Selwyn Basin received a variety of basinal sediments (Rabbitkettle, Kechika, and Road River formations), including shale, chert, thin-bedded limestone, and minor basic volcanic rocks, succeeded in Late Silurian time by orange-brown, carbonate-rich mudstone and siltstone (unit of the Road River Formation). Northern Selwyn Basin was the site of shale, minor limestone, and shaly limestone deposition during the Early Ordovician. From Middle Ordovician through Late Silurian time, it was the site of a condensed, monotonous succession of chert and siliceous shale (Road River units); minor shale and basic volcanic rocks were deposited along the northern margin of the basin. Northern Selwyn Basin is close to the carbonate platform and land areas of Ogilvie Arch, and thus the lack of detritus derived from these areas and the dominance of siliceous strata in the basin are anomalous. Volcanic rocks deposited during the Early and Middle Ordovician along the northern margin perhaps acted as a barrier to basinward transport of sediment from Ogilvie Arch. These volcanics are, however, relatively thin (a few hundred metres) and occur in laterally discontinuous lenses.

Kechika Basin (Douglas et al., 1970) adjoins the south end of Selwyn Basin (at about 60°N latitude) and is bordered on the east by MacDonald Platform. Its present western limit is Cassiar Platform. The southern limit of Kechika Basin is placed at latitude 56°N along the flank of Peace River Arch, where the northern Kakwa Platform extends west of MacDonald Platform (Fig. 4C.2, 4C.3; Thompson, 1989).

Figure 4C.10. Large gastropod (*Palliseria robusta*) in Skoki carbonates of the Bow Platform (photo - ISPG 2336-1).

Like the southern Selwyn Basin, Kechika Basin received a variety of basinal sediments (Kechika and Road River formations), including shale and limestone in the Early Ordovician, shale in the Middle Ordovician, shale and locally shale and quartz sandstone in the Late Ordovician, limestone and chert in the Early Silurian, and orange-brown siltstone in the Late Silurian.

During the Early Ordovician, most of Kechika Basin was filled with a westward thickening succession of thin-bedded to nodular argillaceous limestone (Kechika Formation). The western parts of this succession are clearly basinal. The easternmost parts show an abundance of thin-bedded, framework, intraclast breccias and conglomerates, and common inter-stratal erosional surfaces. Thus, although these rocks are mainly successions of thin-bedded to nodular limestone, they are interpreted as relatively shallow-water deposits.

From Middle Ordovician to Late Silurian time, the margins of Kechika Basin showed some abrupt transitions from carbonate platform to basinal facies (Fig. 4C.8). These transitions feature abundant foreslope breccias, conglomerates, and debris flows. Some debris flows are composed entirely of platformal rocks (Fig. 4C.8). Foreslope breccias show a wide variety of textures, ranging from *in situ* breccias (Fig. 4C.15) to coarse-grained carbonate turbidites.

Sedimentation was continuous throughout Kechika Basin during most of the Ordovician Period. However, the north end sustained sub-Llandovery and sub-Ludlow erosion or non-deposition, combined with westward progradation of MacDonald Platform (Fig. 4C.3 to 4C.5). A similar pattern of erosion and carbonate platform development occurred in Cassiar Platform and adjacent basinal areas directly west of the northern Rocky Mountain Trench. Again assuming restoration of 400 to 750 km dextral offset along the Tintina-Northern Rocky Mountain Trench fault system, Cassiar Platform would be juxtaposed against Kakwa Platform, where similar patterns of erosion and platform development occurred.

Robson Basin is a basinal area on trend with Kechika Basin and White River Embayment, preserving only Lower Ordovician and older strata (Fig. 4C.1). It was a depocentre for most of the Cambrian Period.

White River Embayment was an inner basin within the southern miogeocline and essentially is defined by its area of preservation beneath the regional sub-Devonian unconformity. The trough contains a thick succession of thinly bedded and nodular limestone and shale of the Upper Cambrian and Lower Ordovician McKay Group. As in the similar Kechika Basin to the north, rocks deposited in the western parts of this basin during Early Ordovician time are of relatively deep facies, whereas equivalent strata to the east represent shallower environments and grade laterally into platformal carbonates. During late Early and Middle Ordovician time, shale and minor limestone and chert (Glenogle Formation) accumulated in White River Embayment with limestone dominant eastward adjacent to Bow Platform (Norford and Jackson, 1989); the Glenogle is laterally equivalent to carbonates of Bow Platform (Skoki and Owen Creek formations). Shallowing and an influx of quartz silt took place throughout the embayment in late Middle Ordovician time. Next, the area became emergent and was eroded prior to deposition of the upper Middle to Upper Ordovician Mount Wilson Quartzite. The trough filled during the Late Ordovician and was overstepped by the expanded Bow Platform, which extended west to the eastern flank of Purcell Arch (Norford, 1982; Fig. 4C.12; Beaverfoot Formation). Toward the close of Early Silurian time, deeper water returned with deposition of thin, wavy bedded limestone and shale (Tegart Formation, preserved only as local remnants). No Upper Silurian rocks are preserved beneath the sub-Devonian unconformity (Norford, 1969).

Nasina Basin (Templeman-Kluit, 1979) is an outer basin located west of Selwyn Basin and Cassiar Platform. The western boundary is uncertain but was beyond the limits of known Ordovician and Silurian strata. The basin received thick successions of thin-bedded limestone and shale during the Late Cambrian and Early Ordovician (mainly shale in the north and limestone in the south). In the Middle and Late Ordovician and early Early Silurian, shale deposition persisted throughout the basin, succeeded during the remainder of the Silurian by deposition of mainly orange-brown siltstone, mudstone, and some quartz sandstone (D.J. Tempelman-Kluit, pers. comm., 1985).

The eastern limits of Nasina Basin can be refined when offset on the Tintina-Northern Rocky Mountain Trench fault system is clearly established. The southern Nasina Basin would be bounded by Cassiar Platform, while in the north Nasina Basin could be defined as the area west of a straight line connecting positive areas A and B.

In the far south, in the Kootenay Arc near the United States border, the small outcrop area of the Active Formation preserves the only known occurrence of rocks of outer-basin facies considered to belong to autochthonous North America (Fig. 4C.1 to 4C.5). The Active is a poorly known, 300-to 1500-m thick succession of sooty black argillite and slate with units of limestone, arenaceous slate, argillaceous quartzite, and quartzite (Little, 1960). Early to early Middle Ordovician graptolites have been recovered. The Active Formation is laterally continuous with the Ledbetter Formation in adjacent northeastern Washington State, where Early Ordovician (upper Arenig) and Middle Ordovician (Llanvirn and Caradoc) faunas are present (Dings and Whitebread, 1965; Barnes et al., 1981). The Ledbetter Formation is considered autochthonous because underlying Cambrian rocks contain typical North American trilobite faunas.

One anomalous aspect of the basinal facies is the areal distribution of orange-brown Silurian (upper Llandovery to Pridoli) siltstone and mudstone (Road River Formation, Fig. 4C.4, 4C.5). With the exception of some mudstone, which occurs as far north as Richardson Trough, these sediments are found only over the central miogeocline area (northeastern British Columbia). The area of these detrital rocks roughly corresponds with an area of significant sub-Llandovery erosion. In comparison with other major regional unconformities, the sub-Llandovery event is anomalous. The unconformity is not known in the Mackenzie and Bow platforms, and is represented only by a minor discontinuity in the Williston Platform (sub-Interlake). Furthermore, its erosional effects are most pronounced in the MacDonald Platform-Kechika Basin parts of the miogeocline, where much of the Ordovician is missing beneath Llandovery strata that locally rest directly on Precambrian rocks. Most of the orange-brown siltstone and mudstone is younger than the

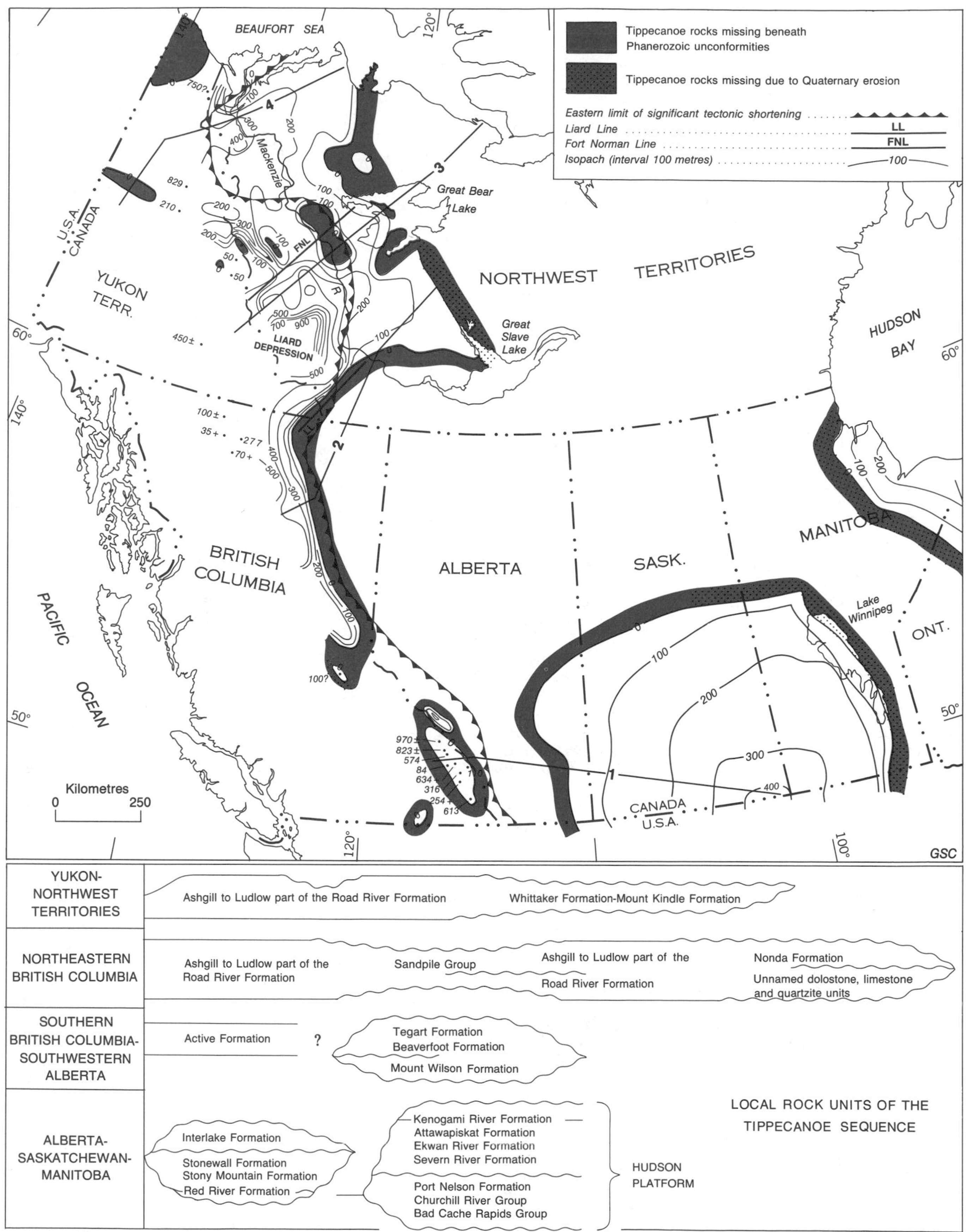

Figure 4C.11. Isopach map of the Tippecanoe sequence.

sub-Llandovery erosional event, so it is unlikely to be a product of that erosion. It is more likely that nearby younger, or continuing areas of uplift provided the detritus. Potential sources are the Peace-Athabasca Arch and areas of sub-Ludlow erosion shown in Figures 4C.1 to 4C.4.

Basinal facies and facies transitional between platform and basin contain abundant fauna and flora (algae). Transitional rocks host the most varied and abundant biota, including trilobites, brachiopods, cephalopods, gastropods, sponges, solitary and colonial corals, stromatolites, graptolites, and conodonts. Basinal facies generally host a much less abundant and varied fauna of graptolites, trilobites, the arthropod *Caryocaris*, sponge spicules, radiolarians, and small inarticulate brachiopods. Some Lower Silurian basinal strata contain solitary corals and patches of colonial corals.

Transitional-to-basin and platformal facies have different faunas. Transitional Lower Ordovician strata host conodont faunas of both Mid-Continent and North Atlantic type (Tipnis, et al., 1978; Kennedy and

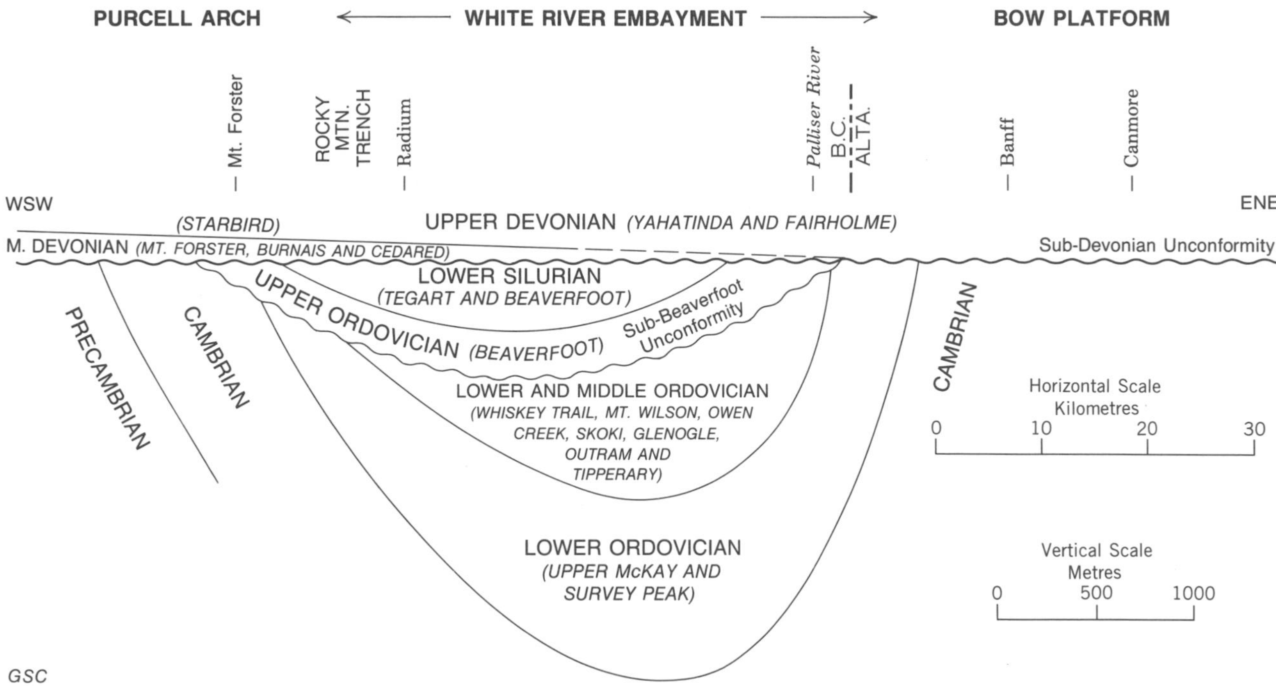

Figure 4C.12. Cross-section through White River Embayment and Bow Platform illustrating the effects of sub-Beaverfoot and sub-Devonian unconformities.

Figure 4C.13. Lower Ordovician thin-bedded basinal limestone, Richardson Trough (photo - ISPG 1862-1).

Figure 4C.14. Ordovician graptolitic shale, with resistant limestone debris flow beds, Richardson Trough (photo - D.E. Jackson, ISPG 2337-1).

Barnes, 1981). Upper Ordovician transitional facies host large irregular dolostone mounds and reefs, which contain scattered solitary corals, gastropods, and brachiopods that are typical of the *Bimuria-Dicoelosia* Fauna, known from northern California and Alaska (Potter and Cecile, 1985). Macrofossils and conodonts typical of the Upper Ordovician, western North American *Bighornia-Thaerodonta* Fauna are found in adjacent platform carbonates.

ARCHES, POSITIVE AREAS, AND SUBMARINE HIGHS

Nine major features were topographically high during all or parts of the Ordovician and Silurian periods (Fig. 4C.1 to 4C.5). Three of these, Ogilvie, Mackenzie, and Bulmer Lake arches, are found within northern platform areas. One, the Niddery (submarine) High, separates Misty Creek Embayment from Selwyn Basin, whereas four others, positive areas A and B, Ware Arch, and Purcell Arch, are within the area of basinal deposition and separate inner and outer basinal belts (see earlier text on basins). The ninth is the large Peace-Athabasca Arch complex.

The Ogilvie (Gabrielse, 1967), Mackenzie (Aitken et al., 1973; Pugh, 1983), and Bulmer Lake (Meijer Drees, 1974) arches have many features in common. Each is cored by Proterozoic strata overlain unconformably by Upper Cambrian to Lower Ordovician platform dolostone (Franklin Mountain Formation, unnamed dolostone unit). This dolostone thins over the crest of the arch. The Upper Cambrian to Lower Ordovician dolostone in turn is truncated partly, or as in the case of the Ogilvie Arch, extensively, by pre-Late Ordovician erosion. All three features are flanked by Lower and Middle Cambrian strata, which wedge out against them due to depositional onlap against the elevated arch and/or are erosionally truncated beneath the younger platform dolostones. Ogilvie and Mackenzie arches are similar in configuration and dimensions, but Mackenzie Arch is northwesterly to north-trending whereas Ogilvie Arch is westerly trending. Bulmer Lake Arch trends northwest and has little longitudinal extent. Ogilvie and Mackenzie arches are within the miogeocline and Bulmer Lake Arch is within the adjacent cratonic platform.

Niddery High (Cecile, 1982) was a positive but submarine area located between the southwestern Misty Creek Embayment and Selwyn Basin. It is characterized by a lower Paleozoic basinal succession substantially thinner than in the adjacent Misty Creek Embayment (350 m versus 1.5-3.5 km). The west boundary of Niddery High shows no significant thickness change and is defined on the change from shale, limestone, siliceous shale, and chert on the high to chert and siliceous shale westward.

Positive areas A and B, and Ware and Purcell arches are all within the outer part of the miogeocline. During the Ordovician, the histories of positive areas A and B and Ware Arch are uncertain; these are areas over which upper Lower Silurian to Upper Silurian rocks rest with apparent unconformity on Lower and some Middle Ordovician strata. Thus, although the areas must have been positive by Early Silurian time, they may have been inactive during the Ordovician. In the area of Ware Arch, this would appear to be the case. The east side of the arch shows progressive truncation of basinal Upper, Middle, and finally Lower Ordovician units in a westward direction (see Cecile and Norford, 1979; Gabrielse, 1981), implying that the arch developed only in the Silurian. Similarily, Gordey (1981b) interpreted the omission of Ordovician strata in Positive area A as erosional truncation. These observations suggest that all three areas were basinal during the Ordovician, and then were preferentially uplifted during the Silurian. During the Late Silurian, positive area A was the site of platformal carbonate deposition, while positive area B and Ware Arch continued to receive basinal orange-brown siltstone (unit of Road River Formation).

Positive areas A and B and Cassiar Platform (see earlier text) have in the past been described as parts of one continuous, positive, northerly trending linear feature, which was known as Pelly-Cassiar Platform, later shortened to Cassiar Platform (see Gabrielse, 1967; Templeman-Kluit, 1977a, b; Gordey, 1981b; Cecile, 1982). Alignment of the three areas requires restoration of 400 to 750 km of dextral offset on the Tintina-Northern Rocky Mountain Trench fault system (Roddick, 1967; Gabrielse, 1985). Facies reconstructions herein do not demonstrate continuity of these areas during any part of the Ordovician or Silurian, accordingly the use of Cassiar Platform is restricted to the area of Cassiar Mountains.

Purcell Arch (Douglas et al., 1970) is located in the southern miogeocline. Much of its stratigraphic record had been lost by erosion prior to Late Ordovician and Devonian trangressions. No Lower Ordovician, Middle Ordovician, or Silurian rocks are known.

The Peace-Athabasca Arch is a broad cratonic area bordering the Ordovician-Silurian miogeocline to the west and Williston Platform on its south margin. The extent of the landmass is poorly known, as are the location and trend of its axis. Evidence of the feature includes the peripheral

Figure 4C.15. Lower Ordovician *in situ* slope breccia, Misty Creek Embayment (from Cecile, 1982, photo - GSC 199529).

distributions of quartzites in Middle to Upper Ordovician rocks and of orange siltstones and mudstones in Silurian rocks (see earlier text).

DEPRESSIONS AND LINEAMENTS

A major trans-lithofacies depression and two major associated linear features had significant control on the thickness, preservation and, to some degree, facies distributions of Ordovician and Silurian strata.

Liard Depression (Fig. 4C.11) is located in the southern Mackenzie and adjacent mountains and is defined by abnormal thicknesses of Ordovician and Silurian strata. Isopachs of the Tippecanoe sequence indicate that this area subsided at a relatively fast rate almost continuously during the Ordovician and Silurian periods. This area of subsidence included both platformal (southwest Mackenzie, northern MacDonald) and basinal (Meilleur River Embayment) facies. The Liard Depression continued its abnormal subsidence well into Devonian time (see Bassett and Stout, 1967; Williams, 1975; P.F. Moore, Subchapter 4D in this volume).

Two major lineaments trend northeast and flank Liard Depression. Liard Line, located on the southeast side of Liard Depression, is defined by southward erosional and depositional pinchout (perhaps under the influence of Peace-Athabasca Arch) throughout the Ordovician and Silurian into the Devonian. On the north side of this linear feature, Ordovician to Devonian platformal rocks are preserved. The line is also defined by rapid northward thickening of strata (Fig. 4C.11) and transitions northward to basinal rocks. Aitken and Pugh (1984) suggested that Liard Line may be an extension of Leith Ridge Fault, which is found 200 km to the northeast. However, nothing in the facies patterns, or isopachs of Mackenzie Platform between the two features suggests that they were connected during the Ordovician and Silurian. Instead, well controlled isopachs (Meijer Drees, 1975a), which span the critical connecting area, cross this projected trend without variation. The Liard Line appears to be the result of the influence of a northeast-trending transform fault that formed, or was reactivated, during the Proterozoic-Cambrian rifting event that formed the miogeocline. The fault was likely an old northeast-trending Precambrian structure that was reactivated by crustal attenuation perpendicular to the trend of the Cordillera. It was one of several such faults used to offset crustal attenuation northeasterly into the crustal promontory between the Cordilleran and Franklinian miogeoclines. Fort Norman Line is a more subtle feature, which is essentially defined by isopach patterns along the northeast margin of the Liard Depression. It follows and projects northeastward into an

SYSTEMIC DIVISIONS	NORTH AMERICAN STAGES	CONODONT ZONES	GRAPTOLITE ZONES	TRILOBITE ZONES AND IMPORTANT BRACHIOPODS AND CORALS	EUROPEAN SERIES AND STAGES
LOWER SILURIAN (part)	ANTICOSTIAN	*kentuckyensis*	*cyphus*		LLANDOVERY SERIES (part) – RHUDDANIAN
			acinaces		
		— ? —	*atavus*	b	
	(lower part)	*nathani*	*acuminatus*	a	
UPPER ORDOVICIAN	GAMACHIAN	?	*persculptus*		ASHGILL SERIES – HIRNANTIAN
		13 – *Amorphognathus ordovicicus*	*pacificus*		
	— ? —		*complananatus ornatus*	— ? —	RAWTHEYAN
	RICHMONDIAN	12		*Bighornia-Thaerodonta*	CAUTLEYAN
			?		PUSGILLIAN
	MAYSVILLIAN		*quadri-mucronatus*	?	
		11 – *Amorphognathus superbus*		"RED RIVER" (includes much of *Cryptolithus-Anataphrus* and *Hesperorthis Oepikina* faunas)	
	EDENIAN		?		CARADOC SERIES – ONNIAN
MIDDLE ORDOVICIAN	TRENTONIAN – SHERMANIAN	10		*Whittakerites planatus*	ACTONIAN
					MARSHBROOKIAN
					LONGVILLIAN
			?	*Ceraurus mackenziensis*	SOUDLEYAN
	TRENTONIAN – KIRKFIELDIAN	9 – *Amorphognathus tvaerensis*: *Prioniodus alobatus*		*Ceraurinella necra*	
	TRENTONIAN – ROCKLANDIAN	8	— ? —	*C. longispina*	HARNAGIAN
	BLACKRIVERIAN	7 – *Prioniodus gerdae*	? *clingani*	*Ceraurus gabrielsi*	
	?	6 – *Prioniodus variabilis*	*gracilis* and *bicornis*	*Bathyurus ulu*	COSTONIAN
	CHAZYAN	*Pygodus anserinus*		*Ceraurinella nahanniensis*	LLANDEILO SERIES
		5 – *Pygodus serra*		*Bathyurus granulosus*	--- ? ---
	?		cf. *teretiusculus*	*Bathyurus nevadensis*	LLANVIRN SERIES
	WHITEROCKIAN	4 – *Eoplacognathus suecicus*	*tentaculatus*	N *Anomalorthis*	
		3		M	
		2 – *Eoplacognathus variabilis*		L *Orthidiella*	
		1 – *M. parva*	o	K	ARENIG SERIES
		P. originalis	md.		
LOWER ORDOVICIAN	IBEX SERIES	E – *Prioniodus navis*	*victoriae* l	*Hesperomena* J	
		Prioniodus triangularis			
		Prioniodus evae	*protobifidus*	H, I	
		D – *Prioniodus elegans*	*fruticosus* 3 and 4 branch	G	
			fruticosus 4 branch		
		C – *Paroistodus proteus*	*approximatus*		
		Paroistodus deltifier	*antiquus*	F	TREMADOC SERIES
				E	
				D	
				B, C	
		B – *Cordylodus angulatus*	*aureus*		
		A – *Cordylodus proavus*	*richardsoni*	A	
			tenuis		
			flab. and *Radiog.*	*Missisquoia*	
UPPER CAMBRIAN (part)	TREMPEALEAUAN	*Proconodontus*		*Saukia*	MERIONETH SERIES

GSC

Figure 4C.16. Biostratigraphic zonation of the Ordovician System. The Ordovician zonal terminology is that of Barnes et al. (1981) with modification from Lenz and McCracken (1982). For the macrofossil zonation, A-N is the Utah-Nevada terminology after Ross (1949, 1951). For the conodont zonation, A-E, 1-13 are the traditional North American assemblages.

older and perhaps ancestral structural feature, the Precambrian Fort Norman Structure (Aitken and Pugh, 1984). This feature was defined on the basis of a northwest-side-down displacement of Proterozoic rocks in the subsurface beneath the Mackenzie Platform and a major change in the level of structural detachment in the central Mackenzie Mountains (Cecile et al., 1982) that involves a major southeast-to-northwest step-down in the level of detachment within Proterozoic strata. Isopach distributions on the southwest end of the Fort Norman Line are complex. They show both subsidence on the south side of the line and subsidence in the core of Misty Creek Embayment north of the line.

VOLCANIC AND INTRUSIVE ROCKS

Numerous occurrences of volcanic rocks are known within autochthonous Ordovician and Silurian strata of western Canada. Most occurrences are in the north, firstly along the western fringe of the Mackenzie, MacDonald, and Ogilvie platforms and on the Ogilvie Arch, and secondly in the Selwyn, Kechika, and Nasina basins, near the Tintina-Rocky Mountain Trench. The documented occurrences south of latitude 56°N are an extrusive and intrusive complex interstratified with Lower Silurian strata on the western edge of Kakwa Platform (Taylor et al., 1972) and a small, pipe-like, basic intrusion cutting Middle Ordovician strata near Glacier Lake in the southern Rocky Mountains. A few mafic dykes, sills, volcanic rocks, and diatremes are associated with basinal strata of White River Embayment (Pell, 1987).

The autochthonous extrusive rocks typically occur as small, widely spaced lensoid units a few tens to hundreds of metres thick and a few kilometres long. They are highly altered and commonly contain abundant secondary minerals such as carbonate, sericite, chlorite, and leucoxene. In decreasing order of abundance, the rocks are lapilli tuffs, tuffs, volcaniclastites, flows, rare pillowed flows, and rare breccias. In areas where accumulations are relatively thick, volcanic breccia, volcaniclastic conglomerate, and a capping succession of shallow-water carbonate usually are present. In three areas where geochemical analyses have been made, the volcanics are subalkalic to alkalic basalts, and in one location, andesites. This alkalic chemistry is illustrated in some localities by the presence of euhedral to anhedral biotite phenocrysts (Goodfellow et al., 1980a, b; Cecile, 1982). Highly altered, lensoid, alkalic volcanic rocks are found interstratified with sedimentary strata. The majority of occurrences are in upper Lower to Middle Ordovician rocks, but they range as old as Proterozoic and as young as Middle Devonian. Four different types of Ordovician and Silurian intrusive rocks are known in western Canada. Dykes and sills are found mainly in association with volcanic strata. Very small mafic to ultramafic, highly altered diatremes are present in the east-central Misty Creek Embayment and in the adjacent platform (Oldershaw, 1981; McArthur et al., 1980; Godwin and Price, 1986) and in the White River Embayment (Pell, 1987). Small sodalitic syenite intrusions are found with Silurian volcanics in western Kakwa Platform (Taylor et al., 1972). Large granitoid plutons in suspect terranes just west of Kootenay Terrane in the Omineca Belt, yield poorly constrained, roughly Ordovician isotopic ages, (Okulitch, 1985).

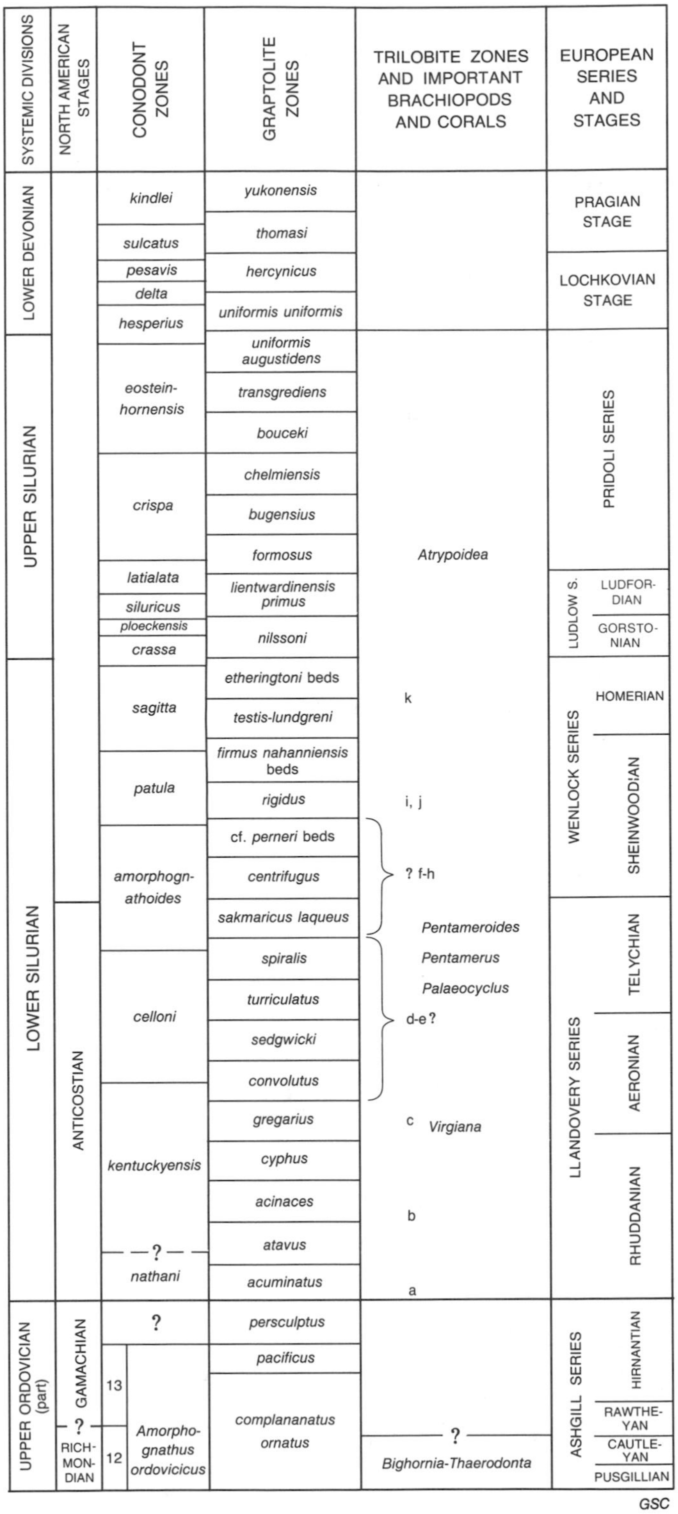

SYSTEMIC DIVISIONS	NORTH AMERICAN STAGES	CONODONT ZONES	GRAPTOLITE ZONES	TRILOBITE ZONES AND IMPORTANT BRACHIOPODS AND CORALS	EUROPEAN SERIES AND STAGES
LOWER DEVONIAN		*kindlei*	*yukonensis*		PRAGIAN STAGE
		sulcatus	*thomasi*		
		pesavis / *delta*	*hercynicus*		LOCHKOVIAN STAGE
		hesperius	*uniformis uniformis*		
UPPER SILURIAN		*eosteinhornensis*	*uniformis augustidens*		PRIDOLI SERIES
			transgrediens		
			bouceki		
		crispa	*chelmiensis*		
			bugensius		
			formosus	*Atrypoidea*	
		latialata / *siluricus*	*lientwardinensis primus*		LUDLOW S. – LUDFORDIAN
		ploeckensis / *crassa*	*nilssoni*		GORSTONIAN
LOWER SILURIAN		*sagitta*	*etheringtoni* beds	k	WENLOCK SERIES – HOMERIAN
			testis-lundgreni		
		patula	*firmus nahanniensis* beds		SHEINWOODIAN
			rigidus	i, j	
		amorphognathoides	cf. *perneri* beds	? f-h	
			centrifugus		
	ANTICOSTIAN		*sakmaricus laqueus*	*Pentameroides*	LLANDOVERY SERIES – TELYCHIAN
		celloni	*spiralis*	*Pentamerus* / d-e?	
			turriculatus	*Palaeocyclus*	
			sedgwicki		AERONIAN
			convolutus		
		kentuckyensis	*gregarius*	c *Virgiana*	
			cyphus		RHUDDANIAN
			acinaces	b	
		— ? —	*atavus*		
		nathani	*acuminatus*	a	
UPPER ORDOVICIAN (part)	GAMACHIAN	?	*persculptus*		ASHGILL SERIES – HIRNANTIAN
		13	*pacificus*		
	? / RICHMONDIAN	12 *Amorphognathus ordovicicus*	*complananatus ornatus*	? *Bighornia-Thaerodonta*	RAWTHEYAN / CAUTLEYAN / PUSGILLIAN

GSC

Figure 4C.17. Biostratigraphic zonation of the Silurian System. The Lower Silurian zonal terminology follows Lenz (1980, 1982) but uses the name *cyphus* Zone for his *gregarius* Zone and the name *gregarius* Zone for his interval *triangulatus* Zone to *argenteus* Zone. The Upper Silurian and Lower Devonian zonal terminology follows Jackson et al. (1978), modified by the placement of the *formosus* Zone in the basal Pridoli. For the macrofossil zonation, a-k are the informal divisions of the Chatterton and Perry (1984).

BIOSTRATIGRAPHIC ZONATION

Graptolites are the key fossils for intercontinental correlation of Ordovician and Silurian strata. The succession of faunas found in the Richardson Trough and Selwyn Basin is one of the most continuous sequences in the world, extending from almost the base of the Ordovician to near the top of the Lower Devonian. Shorter sequences are present in the Kechika Basin and White River Embayment. Documentation of conodont and shelly faunas has begun. Conodont zonations have been established in many localities (Tipnis et al., 1978; Kennedy and Barnes, 1981). Of the macrofauna, trilobites appear to be the most biostratigraphically sensitive (Ludvigsen, 1979, 1982; Chatterton and Perry, 1983). The interstratification of basinal and platformal facies in transitional belts provides good calibration between macrofossil, conodont, and graptolite zones (Fig. 4C.16, 4C.17).

PALEOGEOGRAPHIC AND TECTONIC SUMMARY

Autochthonous Ordovician and Silurian strata of western Canada were laid down as miogeoclinal and platformal parts of the lower Paleozoic, passive-margin succession, whose deposition began in latest Proterozoic or earliest Cambrian time. They represent mainly the Tippecanoe Sequence, and parts of the underlying, Sauk and overlying, Kaskaskia sequences. The sequence-bounding unconformities are widely, but not everywhere recognizable. Ordovician and Silurian rocks of Kootenay Terrane and Cariboo sub-terrane likely are distal facies of North American deposits, displaced substantially from their depositional sites.

The craton consists of two major areas: the eastern Mackenzie Platform in the north and Williston Basin (Platform) in the south. Their continuity is interrupted by a positive feature, Peace-Athabasca Arch, whose detailed configuration is indefinite because of post-Silurian erosion. Carbonate sedimentation was dominant.

The northern part of the miogeocline featured a complex paleogeography resulting from episodes of extensional tectonism. Troughs, embayments, and arches were developed in the northern margin, and facies belts stepped towards the craton along offsets of the continental margin caused by transform faults. The complexity of the northern part of the Cordilleran Miogeocline is thought to relate to its development on a promontory of the North American Craton between the Cordilleran and Franklinian miogeoclines.

Miogeoclinal rocks typically consist of an eastern belt of platformal carbonates with minor clastic rocks and a western complex of carbonate and siliciclastic strata laid down in slope and basinal environments. Three belts can be recognized within the western complex: a belt of embayments and troughs that extend into the adjacent carbonate platform, a belt of inner basins, and an outer belt of basins and troughs.

In the south, the miogeoclinal Bow, Kakwa, MacDonald, and Cassiar platforms (the latter in restored position) and the cratonic Williston Platform likely were continuous with one another during most of the Ordovician and Silurian. All except Cassiar Platform border the Peace-Athabasca Arch. The Kakwa and Cassiar parts formed a westward promontory that probably was related tectonically to the arch. Williston Platform was extensively exposed and eroded during the Middle Ordovician. West of the carbonate platforms, White River Embayment and Robson Basin probably formed a continuous, narrow, inner basin. To the north, the Kakwa-Cassiar promontory separated Robson and Kechika basins. White River Embayment was bounded on the west by Purcell Arch or High. West of this positive area, the Active Formation and possibly some of the the unfossiliferous units of the Kootenay Terrane, are the only representatives of Ordovician and Silurian outer-basin strata.

In the north, the tectonically complex miogeocline was transitional between the Franklinian and Cordilleran miogeoclines. Elements north and northeast of Ogilvie Arch can be considered part of the Franklinian Miogeocline. These include Porcupine, White Mountains, and northwestern Mackenzie platforms, and Richardson and Blackstone troughs. Richardson Trough was an aulacogen that opened northward into the Franklinian Miogeocline. South and east of Ogilvie Arch, all strata were deposited in the Cordilleran Miogeocline and on the adjacent craton. The eastern part of the miogeocline (Mackenzie Platform) featured shallow-water carbonate and minor clastic deposition. This large platform area was extensively exposed and eroded during Middle Ordovician and mid-Silurian times. The Middle Ordovician regression occurred during an episode of extensional tectonism. Bulmer Lake Arch and unnamed areas of sub-Upper Silurian erosional truncation have been identified in the eastern part of the miogeocline and on the adjacent craton. The western edge of Mackenzie Platform is interrupted by a large positive area, Mackenzie Arch, characterized by erosion and depositional thinning. This arch had its greatest relief in the Cambrian but persisted throughout most of the early Paleozoic and extended from the southern Mackenzie Mountains north into the subsurface east of the Richardson Mountains.

The western edge of Mackenzie Platform is indented by the Misty Creek, Prairie Creek, and Meilleur River embayments. Misty Creek Embayment is graben-like and contains alkalic volcanic strata. Meilleur River Embayment developed close to the centre of Liard Depression, which was a long-lived depocentre bounded on the southeast and northwest by lineaments inferred to be transform faults (Liard and Fort Norman lines). West of Mackenzie Platform and the embayments are the inner-miogeoclinal Selwyn Basin and farther west, Nasina Basin.

A peculiar aspect of the miogeocline is the extent and scale of Silurian erosion and truncation in the Kechika Basin and MacDonald Platform and the extensive deposition of Silurian clastic rocks in the previously eroded region. Although widespread uplift and erosion occurred at the same time in other parts of the miogeocline and adjacent craton, the scale of the truncation in the region cited implies local uplift of the central miogeocline.

ECONOMIC GEOLOGY

Oil and gas occurrences

Minor amounts of oil have been produced from Ordovician carbonate strata in the Canadian Williston Basin. In the plains of Alberta and much of British Columbia, Ordovician and Silurian rocks are mainly absent in the subsurface because of erosion at various unconformities. Ordovician and Silurian rocks are widespread in the subsurface of northern platform areas, but to date only a few scattered gas shows have been reported from these and underlying Cambrian strata (Pugh, 1983). A small number of conodont alteration index determinations (CAI) from outcrop samples indicate that Ordovician and Silurian strata in the mountains have matured beyond the "oil window". Of the very few subsurface determinations, one in the southern Eagle Plains (Porcupine Platform) indicates maturation beyond the "oil window". Two from the Franklin Mountain Formation, Norman Wells area, give CAI values of 1.5-2 and one from the same formation beneath Peel Plateau gives a value of 2-3, and another from the Peel Formation of southern Anderson Plain a value of 2, indicating maturation levels that are within the "oil window" (T.T. Uyeno, pers. comm., 1985).

Mineral occurrences

Ordovician and Silurian basinal and adjacent carbonate platformal facies have excellent base-metal potential, and host numerous showings of lead, zinc, silver, and barite. These strata are also important host rocks to metasomatic tungsten deposits found in alteration zones around Cretaceous plutons (Cecile, 1982). A number of significant phosphorite occurrences have been reported, the most important being thin phosphorite pavements on Lower Ordovician basinal limestone beds of Kechika Formation in Kechika Basin (Cecile and Norford, 1979).

Lower Silurian Road River shales of the east-central Selwyn Basin host large and relatively rich zinc-lead deposits at Howards Pass (120 million tonnes drilled and 360 million tonnes inferred at about 7% Pb-Zn; Energy, Mines and Resources, 1984; Morganti, 1979; Norford and Orchard, 1985). In addition, major lead-zinc-silver deposits in the Faro area (east side of Tintina Fault east of positive area A) are hosted by Cambrian to Ordovician strata (34 million tonnes drilled at 8% Pb-Zn and 48g/mt Ag; Energy, Mines and Resources, 1984; Morin et al., 1983).

In spite of the potential of these rocks, only the Anvil Mine at Faro and one other small mine are presently in production. This is the Brisco barite mine in the Rocky Mountain Trench south of Golden, which exploits a large vein deposit in Upper Ordovician strata (Beaverfoot Formation) of Bow Platform (Manson, 1984).

REFERENCES

Aitken, J.D. and Norford, B.S.
1967: Lower Ordovician Survey Peak and Outram formations, Southern Rocky Mountains of Alberta; Bulletin of Canadian Petroleum Geology, v. 15, p. 150-207.

Aitken, J.D. and Pugh, D.C.
1984: The Fort Norman and Leith Ridge structures: major, buried Precambrian features underlying Franklin Mountains and Great Bear and Mackenzie Plains; Bulletin of Canadian Petroleum Geology, v. 32, p. 139-146.

Aitken, J.D., Fritz, W.H., and Norford, B.S.
1972: Cambrian and Ordovician biostratigraphy of the southern Canadian Rocky Mountains; 24th International Geological Congress, Guidebook A19, 56 p.

Aitken, J.D., Macqueen, R.W., and Usher, J.L.
1973: Reconnaissance studies of Proterozoic and Cambrian stratigraphy, lower Mackenzie River area (Operation Norman), District of Mackenzie; Geological Survey of Canada, Paper 73-9, 178 p.

Andrichuk, J.M.
1959: Ordovician and Silurian stratigraphy and sedimentation in southern Manitoba; American Association of Petroleum Geologists, Bulletin, v. 43, p. 965-985.

Armstrong, R.L.
1979: Sr isotopes in igneous rocks of the Canadian Cordillera and the extent of Precambrian rocks; in Evolution of the Cratonic Margin and Related Mineral Deposits; Geological Association of Canada, Cordilleran Section, Programme and Abstracts, p. 7.

Baillie, A.D.
1952: Ordovician geology of Lake Winnipeg and adjacent areas, Manitoba; Manitoba Department of Mines, Publication 51-6, 64 p.

Barnes, C.R., Norford, B.S., and Skevington, D.
1981: The Ordovician System in Canada; International Union of Geological Sciences, Publication no. 8, 27 p.

Bassett, H.G. and Stout, J.G.
1967: Devonian of western Canada; in International Symposium on the Devonian System, v. 1, D.H. Oswald (ed.); Alberta Society of Petroleum Geologists, p. 717-752.

Bond, G.C. and Kominz, M.A.
1984: Construction of tectonic subsidence curves for the early Paleozoic miogeosyncline, southern Canadian Rocky Mountains: Implications for subsidence mechanisms, age of breakup and crustal thinning; Geological Society of America, Bulletin, v. 95, p. 155-173.

Burling, L.D.
1921: Graptolite localities of western North America, with description of two new formation names; Geological Society of America, Bulletin, v. 32, p. 127-128.
1922: A Cambro-Ordovician section in the Beaverfoot Range, near Golden, British Columbia; Geological Magazine, v. 59, p. 452-461.
1923: Cambro-Ordovician section near Mount Robson, British Columbia; Geological Society of America, Bulletin, v. 34, p. 721-747.
1954: Annotated index to the Cambro-Ordovician of the Jasper Park and Mount Robson region; Alberta Society of Petroleum Geologists, Fifth Annual Field Conference, Guide Book, p. 15-51.

Cairnes, D.D.
1914: The Yukon-Alaska International Boundary, between Porcupine and Yukon rivers; Geological Survey of Canada, Memoir 67, 161 p.

Cecile, M.P.
1982: The Lower Paleozoic Misty Creek Embayment, Selwyn Basin, Yukon and Northwest Territories; Geological Survey of Canada, Bulletin 335, 78 p.
1984: Geology of the northwest Niddery Lake map-area, Yukon; Geological Survey of Canada, Open File 1006.
1986: Lower Paleozoic embayments, troughs and arches, Northern Canadian Cordillera (abstract); Geological Survey of Canada, Abstracts for forum on Oil and Gas Activities in Canada, p. 5, 6.

Cecile, M.P. and Cook, D.G.
1981: Structural cross-section northern Selwyn and Mackenzie Mountains; Geological Survey of Canada, Open File 807.

Cecile, M.P., Cook, D.G., and Snowdon, L.R.
1982: Plateau overthrust and its hydrocarbon potential, Mackenzie Mountains, Northwest Territories; in Current Research, Part A, Geological Survey of Canada, Paper 82-1A, p. 89-94.

Cecile, M.P. and Norford, B.S.
1979: Basin to platform transition, Paleozoic strata of Ware and Trutch map-areas, northeastern British Columbia; Geological Survey of Canada, Paper 79-1A, p. 219-226.

Chatterton, B.D.E. and Perry, D.
1983: Silicified Silurian odontopleurid trilobites from the Mackenzie Mountains; Palaeontographica Canadiana, no. 1; Canadian Society of Petroleum Geologists - Geological Association of Canada, 127 p.

Churkin, M. Jr.
1975: Basement rocks of Barrow Arch, Alaska and circum-Arctic Paleozoic mobile belt; American Association of Petroleum Geologists, Bulletin, v. 59, p. 451-456.

Cowan, J.
1972: Ordovician and Silurian stratigraphy of the Interlake area, Manitoba; in Geoscience Studies in Manitoba; Geological Association of Canada, Special Publication no. 9, p. 235-241.

Dings, M.G. and Whitebread, D.H.
1965: Geology and ore deposits of the Metaline zinc-lead District, Pend Oreille County, Washington; United States Geological Survey, Professional Paper 489, 109 p.

Douglas, R.J.W., Gabrielse, H., Wheeler, J.O., Stott, D.F., and Belyea, H.R.
1970: Geology of Western Canada; in Geology and Economic Minerals of Canada, R.J.W. Douglas (ed.); Geological Survey of Canada, p. 367-546.

Energy, Mines and Resources
1984: Canadian mineral deposits not being mined in 1983; Mineral Bulletin MR198, 308 p.

Gabrielse, H.
1967: Tectonic evolution of the northern Canadian Cordillera; Canadian Journal of Earth Sciences, v. 4, p. 271-298.
1981: Stratigraphy and structure of Road River and associated strata in Ware (West Half) map-area, northern Rocky Mountains, British Columbia; in Current Research, Part A, Geological Survey of Canada, Paper 81-1A, p. 201-207.
1985: Major dextral transcurrent displacements along the Northern Rocky Mountain Trench and related lineaments in north-central British Columbia; Geological Society of America, Bulletin, v. 96, p. 1-14.

Gabrielse, H. and Yorath, C.J. (ed.)
1991: Geology of the Cordilleran Orogen in Canada; Geological Survey of Canada, Geology of Canada, no. 4. (Also Geological Society of America, The Geology of North America, v. G-2.)

Godwin, C.I. and Price, B.J.
1986: Geology of the Mountain diatreme kimberlite, north-central Mackenzie Mountains, District of Mackenzie, Northwest Territories; in Mineral Deposits of Northern Cordillera, J.A. Morin (ed.); Canadian Institute of Mining and Metallurgy, Special Volume 37, p. 298-310.

Goodfellow, W.D., Jonasson, I.R., and Cecile, M.P.
1980a: Nahanni integrated multidisciplinary pilot project geochemical studies Part 1: Geochemistry and mineralogy of shales, cherts, carbonates and volcanic rocks from the Road River Formation, Misty Creek Embayment, Northwest Territories; in Current Research, Part B, Geological Survey of Canada, Paper 80-1B, p. 149-161.
1980b: Nahanni integrated multidisciplinary pilot project geochemical studies Part 2: Some thoughts on the source, transportation and concentration of elements in shales of the Misty Creek Embayment, Northwest Territories; in Current Research; Part B, Geological Survey of Canada, Paper 80-1B, p. 163-171.

Gordey, S.P.
1981a: Stratigraphy, structure and tectonic evolution of southern Pelly Mountains in the Indigo Lake area, Yukon Territory; Geological Survey of Canada, Bulletin 318, 44 p.
1981b: Structure section across south-central Mackenzie Mountains, Yukon and Northwest Territories; Geological Survey of Canada, Open File 809.

Hume, G.S. and Link, T.A.
1945: Canol geological investigations in the Mackenzie River area, Northwest Territories and Yukon; Geological Survey of Canada, Paper 45-16, 87 p.

Jackson, D.E. and Lenz, A.C.
1962: Zonation of Ordovician and Silurian graptolites of the northern Yukon, Canada; American Association of Petroleum Geologists, Bulletin, v. 46, p. 30-45.

Jackson, D.E., Lenz, A.C., and Pedder, A.E.H.
1978: Late Silurian and Early Devonian graptolite, brachiopod and coral faunas from northwest and Arctic Canada; Geological Association of Canada, Special Publication 17, 159 p.

Jamieson, E.R.
1979: Well data and lithologic descriptions of the Interlake Group (Silurian) in southern Saskatchewan; Saskatchewan Department of Mineral Resources, Report 139, 67 p.

Jansa, L.F.
1976: Tidal deposits in the Monkman Quartzite (Lower Ordovician), northeastern British Columbia, Canada; in Tidal Deposits, R.N. Ginsburg (ed.); Springer-Verlag, New York, p. 153-161.

Kendall, A.C.
1976: The Ordovician carbonate succession (Bighorn Group) of southeastern Saskatchewan; Saskatchewan Geological Survey, Report no. 180, 186 p.
1977: Origin of dolomite mottling in Ordovician limestones from Saskatchewan and Manitoba; Bulletin of Canadian Petroleum Geology, v. 25, p. 480-504.

Kennedy, D.J. and Barnes, S.R.
1981: Conodont biostratigraphy of the Survey Peak Formation, southern Rocky Mountains, Alberta; Geological Association of Canada, Abstracts, v. 6, p. A31.

Kent, D.M.
1960: The evaporites of the upper Ordovician strata in the northern part of the Williston Basin; Saskatchewan Department of Mineral Resources, Report 46, 46 p.

Ketner, K.B.
1968: Origin of Ordovician quartzite in the Cordilleran miogeosyncline; in Geological Survey Research, United States Geological Survey, Professional Paper 600B, p. 169-177.

King, K.R.
1964: The Silurian Interlake Group in Manitoba; in Proceedings 3rd International Williston Basin Symposium, p. 51-55.

Kupsch, W.O.
1953: Ordovician and Silurian stratigraphy of Saskatchewan; Saskatchewan Geological Survey, Report 10, 62 p.

Lapworth, C.
1887: Fossils from Kicking Horse Pass; Science, v. 9, p. 320.

Lenz, A.C.
1972: Ordovician to Devonian history of the northern Yukon and adjacent District of Mackenzie; Bulletin of Canadian Petroleum Geology, v. 20, p. 321-361.
1980: Wenlockian graptolite reference section, Clearwater Creek, Nahanni National Park, Northwest Territories, Canada; Canadian Journal of Earth Sciences, v. 17, p. 1075-1086.
1982: Llandoverian graptolites of the northern Canadian Cordillera: *Petalograptus*, *Cephalograptus*, *Rhaphidograptus*, *Dimorphograptus*, Retiolitidae and Monograptidae; Life Sciences Contribution, Royal Ontario Museum, 130, 154 p.

Lenz, A.C. and McCracken, A.D.
1982: The Ordovician-Silurian boundary, northern Canadian Cordillera: graptolite and conodont correlation; Canadian Journal of Earth Sciences, v. 19, p. 1308-1322.

Little, H.W.
1960: Nelson map area, west half, British Columbia (82 F W1/2); Geological Survey of Canada, Memoir 308, 205 p.

Lochman-Balk, C. and Wilson, J.L.
1967: Stratigraphy of Upper Cambrian-Lower Ordovician subsurface sequence in Williston Basin; American Association of Petroleum Geologists, Bulletin, v. 51, p. 883-917.

Ludvigsen, R.
1979: A trilobite zonation of Middle Ordovician rocks, southwestern District of Mackenzie; Geological Survey of Canada, Bulletin 312, 99 p.
1982: Upper Cambrian and Lower Ordovician trilobite biostratigraphy of the Rabbitkettle Formation, western District of Mackenzie; Life Sciences Contributions, Royal Ontario Museum, 134, 188 p.

Manson, G.R.
1984: Brisco Barite Mine; in The Geology of Industrial Minerals in Canada, G.R. Guilet and W. Martin (ed.); Canadian Institute of Mining and Metallurgy, Special v. 29, p. 264-265.

Martin, L.J.
1959: Stratigraphy and depositional tectonics of North Yukon-Lower Mackenzie area, Canada; American Association of Petroleum Geologists, Bulletin, v. 43, p. 2399-2455.

McArthur, M.L., Tipnis, R.S., and Godwin, C.I.
1980: Early and Middle Ordovician conodont fauna from the Mountain Diatreme, northern Mackenzie Mountains; in Current Research, Part A, Geological Survey of Canada, Paper 80-1A, p. 363-368.

McCabe, H.R.
1972a: Stratigraphy of Manitoba, an introduction and review; in Geoscience Studies in Manitoba; Geological Association of Canada, Special Publication No. 9, p. 167-187.
1972b: Subsurface stratigraphic studies; in Summary of Geological Field Work, 1972; Manitoba Department of Mines and Natural Resources, Mines Branch Geological Paper 3/72, 58 p.

McConnell, R.G.
1887: Report on the geological features of a portion of the Rocky Mountains; Geological Survey of Canada, Annual Report, v. 2, pt. D, p. 1-41.

Meijer Drees, N.C.
1974: Geology of the "Bulmer Lake High", a gravity feature in the southern Great Bear Plain, District of Mackenzie; in Report of Activities, Part B, Geological Survey of Canada, Paper 74-1B, p. 274-277.
1975a: Geology of the lower Paleozoic formations in the subsurface of the Fort Simpson area, District of Mackenzie; Geological Survey of Canada, Paper 74-40, 65 p.
1975b: The Little Doctor sandstone (new sub-unit) and its relationship to the Franklin Mountain and Mount Kindle Formations in the Nahanni Range and nearby subsurface, District of Mackenzie; in Report of Activities, Part C, Geological Survey of Canada, Paper 75-1C, p. 51-57.

Morin, J.A., Grapes, K.J., and Debicki, R.L.
1983: Yukon mineral industry, 1982, an overview; in Yukon Exploration and Geology, 1982; Department of Indian and Northern Affairs, Canada, Whitehorse, p. 4-17.

Morganti, J.M.
1979: The geology and ore deposits of the Howards Pass area, Yukon and Northwest Territories: The origin of basinal sedimentary stratiform sulphide deposits; Ph.D. thesis, University of British Columbia, Vancouver, British Columbia, 327 p.

Morrow, D.W.
1984: Sedimentation in Root Basin and Prairie Creek Embayment - Siluro-Devonian, Northwest Territories; Bulletin of Canadian Petroleum Geology, v. 32, p. 162-189.

Nelson, S.J.
1963: Ordovician paleontology of the northern Hudson Bay Lowland; Geological Society of America, Memoir 90, 152 p.
Norford, B.S.
1962: The Silurian fauna of the Sandpile Group of northern British Columbia; Geological Survey of Canada, Bulletin 78, 90 p.
1964: Reconnaissance of the Ordovician and Silurian rocks of the northern Yukon Territory; Geological Survey of Canada, Paper 63-39, 139 p.
1965: Ordovician and Silurian, Part II - Cordillera; in Geological History of Western Canada, R.G. McCrossan and R.P. Glaister (ed.); Alberta Society of Petroleum Geologists, p. 42-48.
1969: Ordovician and Silurian stratigraphy of the southern Rocky Mountains; Geological Survey of Canada, Bulletin 176, 90 p.
1982: Devonian stratigraphy at the margins of the Rocky Mountain Trench, Columbia River, southeastern British Columbia; Bulletin of Canadian Petroleum Geology, v. 29, p. 540-560.
Norford, B.S., Gabrielse, H., and Taylor, G.C.
1967: Stratigraphy of Silurian carbonate rocks of the Rocky Mountains, northern British Columbia; Bulletin of Canadian Petroleum Geology, v. 14, p. 504-519.
Norford, B.S. and Jackson, D.E.
1989: Ordovician stratigraphy and graptolite faunas of the Glenogle Formation, southeastern British Columbia (abstract); Geological Association of Canada, Program with Abstracts, v. 14, p. A28.
Norford, B.S. and Macqueen, R.W.
1975: Lower Paleozoic Franklin Mountain and Mount Kindle Formations, District of Mackenzie: Their type sections and regional development; Geological Survey of Canada, Paper 74-34, 37 p.
Norford, B.S. and Orchard, M.J.
1985: Early Silurian age of rocks hosting lead-zinc mineralization at Howards Pass, Yukon Territory and District of Mackenzie; local biostratigraphy of Road River Formation and Earn Group; Geological Survey of Canada, Paper 83-18, 35 p.
Norris, D.K. and Yorath, C.J.
1981: The North American Plate from the Arctic Archipelago to the Romanzof Mountains; in The Arctic Ocean, Ocean Basin Series, v. 5, p. 37-103.
Okulitch, A.V.
1985: Paleozoic plutonism in southeastern British Columbia; Canadian Journal of Earth Sciences, v. 22, p. 1409-1424.
Oldershaw, A.E.
1981: A preliminary analysis of the Mountain and Keele Diatremes; in Department of Indian and Northern Affairs, Mineral Industry Report, 1977, Northwest Territories, E65 1981-11, p. 148-154.
Paterson, D.F.
1971: The stratigraphy of the Winnipeg Formation (Ordovician) of Saskatchewan; Saskatchewan Department of Mineral Resources, Report 140, 57 p.
Pell, J.
1987: Alkalic ultrabasic diatremes in British Columbia: Petrology, geochronology and tectonic significance; in Geological Fieldwork, 1986, British Columbia Ministry of Energy, Mines and Petroleum Resources, Paper 1987-1, p. 259-272.
Porter, J.W. and Fuller, J.G.C.M.
1965: Ordovician-Silurian, Part I, Plains; in Geological History of Western Canada, R.G. McCrossan and R.P. Glaister (ed.); Alberta Society of Petroleum Geologists, p. 34-42.
Porter, J.W., Price, R.A., and McCrossan, R.G.
1982: The Western Canada Sedimentary Basin; Philosophical Transactions of the Royal Society, London A, v. 305, p. 169-192.
Potter, A.W. and Cecile, M.P.
1985: Paleobiogeographic significance of two Late Ordovician brachiopod faunules from the Misty Creek Embayment, Selwyn Basin, Northwest Territories, Canada (abstract); Geological Society of America, Abstracts with Program, v. 17, p. 401.
Price, R.A. and Fermor, P.R.
1985: Structure section of the Cordilleran Foreland Thrust and fold belt west of Calgary, Alberta; Geological Survey of Canada, Paper 84-14.
Pugh, D.C.
1976: Cambrian stratigraphy from western Alberta to northeastern British Columbia; Geological Survey of Canada, Paper 74-37, 31 p.
1983: Pre-Mesozoic geology in the subsurface of Peel River map area, Yukon Territory and District of Mackenzie; Geological Survey of Canada, Memoir 401, 61 p.
1985: Regional stratigraphic cross-sections of Pre-Mesozoic geology, Great Bear River map-area, District of Mackenzie; Geological Survey of Canada, Open File 1176.
in press: Pre-Mesozoic geology in the subsurface of Great Bear River map area, District of Mackenzie; Geological Survey of Canada, Memoir.
Roddick, J.A.
1967: Tintina Trench; Journal of Geology, v. 73, p. 23-33.
Roehl, P.O.
1967: Stony Mountain (Ordovician) and Interlake (Silurian) facies analogues of recent low energy marine and subaerial carbonates; American Association of Petroleum Geologists, Bulletin, v. 51, p. 1979-2032.
Ross, R.J. Jr.
1949: Stratigraphy and trilobite faunal zones of the Garden City Formation, northeastern Utah; American Journal of Science, v. 247, p. 472-491.
1951: Stratigraphy of the Garden City Formation in northeastern Utah, and its trilobite faunas; Yale University, Peabody Museum, Bulletin 6, 161 p.
Slind, O.L. and Perkins, G.D.
1967: Lower Paleozoic and Proterozoic sediments of the Rocky Mountains between Jasper, Alberta and Pine River, British Columbia; Bulletin of Canadian Petroleum Geology, v. 14, p. 442-468.
Sloss, L.L.
1963: Sequences in the cratonic interior of North America; Geological Society of America, Bulletin, v. 74, p. 93-114.
1972: Synchrony of Phanerozoic sedimentary-tectonic events of the North American craton and the Russian Platform; 24th International Geological Congress, Montreal, Section 6, p. 24-32.
Stearn, C.W.
1956: Stratigraphy and paleontology of the Interlake Group and Stonewall Formation of southern Manitoba; Geological Survey of Canada, Memoir 281, 162 p.
Tassonyi, E.J.
1970: Subsurface geology, lower Mackenzie River and Anderson River area; Geological Survey of Canada, Paper 68-25, 207 p.
Taylor, G.C., Campbell, R.B., and Norford, B.S.
1972: Silurian igneous rocks in the western Rocky Mountains, northwestern British Columbia; in Report of Activities, Part A, Geological Survey of Canada, Paper 72-1A, p. 228-229.
Tempelman-Kluit, D.J.
1977a: Stratigraphic and structural relations between the Selwyn Basin, Pelly-Cassiar platform and Yukon crystalline terrane in the Pelly Mountains, Yukon; in Report of Activities, Part A, Geological Survey of Canada, Paper 77-1A, p. 223-227.
1977b: Geology of Quiet Lake (105F) and Finlayson Lake (105G) map areas, Yukon Territory (2 maps, scale 1:250 000); Geological Survey of Canada, Open File 486.
1979: Transported cataclasite, ophiolite and granodiorite in Yukon: Evidence of arc-continent collision; Geological Survey of Canada, Paper 79-14, 27 p.
Thompson, R.I.
1981: The nature and significance of large 'blind' thrusts within the northern Rocky Mountains of Canada; in Thrust and Nappe Tectonics, K.R. McClay and N.J. Price (ed.); Geological Society of London, Special Publication No. 9, p. 449-462.
1989: Stratigraphy, tectonic evolution and structural analysis of the Halfway River map area (94B), northern Rocky Mountains, British Columbia; Geological Survey of Canada, Memoir 425.
Tipnis, R.S.
1981: Early Ordovician conodont biostratigraphy and zonation of the Kechika Formation, northeastern British Columbia; Geological Association of Canada, Abstracts, v. 6, p. A-56.
Tipnis, R.S., Chatterton, B.D.E., and Ludvigsen, R.
1978: Ordovician conodont biostratigraphy of the southern District of Mackenzie, Canada; in Western and Arctic Canadian Biostratigraphy, C.R. Stelck and B.D.E. Chatterton (ed); Geological Association of Canada, Special Paper 18, p. 39-91.
Tyrrell, J.B.
1892: Report on northwestern Manitoba, with portions of the adjacent Districts of Assiniboia and Saskatchewan; Geological Survey of Canada, Annual Report for 1891, v. V, part 1E, 235 p.
Vigrass, L.W.
1972: Depositional framework of the Winnipeg Formation in Manitoba and eastern Saskatchewan; in Geoscience Studies in Manitoba; Geological Association of Canada, Special Paper 9, p. 225-234.
Walcott, C.D.
1913: Cambrian formations of the Robson Peak District, British Columbia and Alberta, Canada; Smithsonian Miscellaneous Collections, v. 57, no. 12, 17 p.
1923: Nomenclature of some post Cambrian and Cambrian Cordilleran formations (2); Smithsonian Miscellaneous Collections, v. 67, no. 8, 20 p.
1924: Geological formations of Beaverfoot-Brisco-Stanford Range, British Columbia, Canada; Smithsonian Miscellaneous Collections, v. 75, no. 1, 51 p.
1927: Pre-Devonian sedimentation in southern Canadian Rocky Mountains; Smithsonian Miscellaneous Collections, v. 75, no. 4, 27 p.

Walcott, C.D. (cont.)
1928: Pre-Devonian Paleozoic formations of the Cordilleran provinces of Canada; Smithsonian Miscellaneous Collections, v. 75, no. 5 (edited by C.E. Resser), 194 p.

Wheeler, J.O. and Gabrielse, H. (coordinators)
1972: The Cordilleran Structural Province; in Variation in Tectonic Styles in Canada, R.A. Price and R.J.W. Douglas (ed.); Geological Association of Canada, Special Paper 11, p. 1-82.

Williams, G.K.
1975: "Arnica platform dolomite", District of Mackenzie; in Report of Activities, Part C, Geological Survey of Canada, Paper 75-1C, p. 31-35.

Williams, M.Y.
1922: Exploration east of Mackenzie River between Simpson and Wrigley; Geological Survey of Canada, Summary Report 1921, Part B, p. 56-66.
1923: Reconnaissance across northeastern British Columbia and the geology of the northern extension of Franklin Mountain, Northwest Territories; Geological Survey of Canada, Summary Report 1922, Part B, p. 65-87.

Ziegler, P.A.
1969: The development of sedimentary basins in Western and Arctic Canada; Alberta Society of Petroleum Geologists, 67 p.

ADDENDUM

Since preparation of this subchapter in 1985, three important works have presented large amounts of information and have refined geological interpretations.

The first publication (Norford, 1991) includes documentation of Ordovician transport of quartz sand from the Peace-Athabasca Arch westward and southwestward into the MacDonald and Kakwa platforms and into the Kechika Basin. The sophisticated compilation of subsurface data of the second publication (Norford et al., in press) establishes that, during Ordovician and Silurian time, the Williston Platform was the site of a very subtle basin defined by facies changes and by minor thickening of several units. Anhydrites are developed in the central part of the basin within the Herald, Stony Mountain and Stonewall formations.

In the northern Yukon, recent work (Lane and Cecile, 1989) in the British Mountains has changed the interpretation of the extreme northwestern areas in Figures 4C. 1-5 and 11, designated as Ordovician and Silurian rocks having been removed by Devonian and later erosion. At the time of compilation of this Subchapter, outcrops in this area were thought to be upper Proterozoic. The new work has documented Ordovician and Silurian basin facies within the area: a very thin Ordovician unit of massive grey chert and minor greyish black siliceous shale and a thin Silurian succession of orange weathering argillites.

Lane, L.S. and Cecile, M.P.
1989: Stratigraphy and structure of the Neruokpuk Formation, northern Yukon; in Current Research, Part G; Geological Survey of Canada, Paper 89-1G, p. 57-62.

Norford, B.S.
1991: Ordovician and Silurian stratigraphy, paleogeography and depositional history in the Peace River Arch area, Alberta and British Columbia; Bulletin of Canadian Petroleum Geology, v. 38A, p. 45-54.

Norford, B.S., Haidl, F.M., Bezys, R.K., Cecile, M.P., McCabe, H.R., and Paterson, D.F.
in press: Middle Ordovician to Lower Devonian strata of the Western Canada Sedimentary Basin; in Geological Atlas of Western Canada Sedimentary Basin, G.D. Mossop and I. Shetsen (ed.); Canadian Society of Petroleum Geologists.

Authors' Addresses

M.P. Cecile
Institute of Sedimentary and Petroleum Geology
Geological Survey of Canada
3303 - 33rd Street N.W.
Calgary, Alberta
T2L 2A7

B.S. Norford
Institute of Sedimentary and Petroleum Geology
Geological Survey of Canada
3303 - 33rd Street N.W.
Calgary, Alberta
T2L 2A7

Printed in Canada

Subchapter 4D

DEVONIAN

P.F. Moore

CONTENTS

Moore, P.F.
1993: Devonian; Subchapter 4D in Sedimentary Cover of the Craton in Canada, D.F. Stott and J.D. Aitken (ed.); Geological Survey of Canada, Geology of Canada, no. 5, p. 150-201 (also Geological Society of America, The Geology of North America, v. D-1).

INTRODUCTION

Scope

The region described in this subchapter lies between the craton and the western edge of the carbonate platform built upon the Devonian continental shelf within one or two hundred kilometres of the continental margin. Sediments deposited west of the platform edge are described in the companion volume, "Geology of the Cordilleran Orogen in Canada" (Gabrielse and Yorath, 1991). The depositional record described here starts with rocks deposited in Late Silurian time following a widespread regression accompanied by epeirogenic uplift, warping, and erosion. These rocks are the first of Kaskaskia sequence (Sloss, 1963). The record ends at the base of dark radioactive shales spanning the Devonian-Mississippian boundary and marking a major transgression. In the far north, this transgression is masked by a clastic wedge originating in the Ellesmerian Orogen of the Arctic Islands and northern Alaska.

Within the region, Devonian sediments cover an area of 2 500 000 km^2. Thickness varies greatly from place to place. The thickest carbonate packets occur at the margin of the platform; close to Selwyn Basin, for example, the "Bear Rock sequence", one of the five divisions of the Devonian analyzed here, alone constitutes over 2000 m of carbonate strata. In contrast, some bathymetric basins, such as the one occupying the Mackenzie Valley region during the Givetian, were starved of sediment.

Natural resources

The Devonian System is a vast natural resource of major economic importance to Canada. The first mineral product from the Devonian in western Canada was halite from northeastern Alberta, used as table salt. The oil discoveries that sparked a quarter century of oil boom were made in a Frasnian reef and the carbonate platform draped over it at Leduc, Alberta. The Devonian contains over half of Canada's conventional oil and gas reserves, most of the sulphur, and very large potential resources of bitumen. Middle Devonian platform carbonates of the Pine Point area are hosts to large deposits of lead and zinc, and similar deposits have been discovered in the Middle Devonian carbonates of northeastern British Columbia and in the Upper Devonian of southeastern British Columbia. Potash mines in the Givetian rocks of Saskatchewan tap a giant resource of world importance. Famennian limestones at Exshaw, Alberta, supply one of the largest cement plants in the world with raw material.

The Devonian of this region has also become an important resource for scientific research. Two factors have contributed to this. One is the magnificent outcrop belt along the Front Ranges of the Canadian Rocky

Table 4D.1

Range zones of western Canadian Devonian index megafossils calibrated against standard or, where appropriate, local conodont zones and subzones or faunas. (A.E.H. Pedder, Geological Survey of Canada)

At the time the table was prepared, the Subcommission on Devonian Stratigraphy had not agreed on precise levels for the Lochkovian-Pragian, Pragian-Emsian, or Eifelian-Givetian boundaries. However, at the 1986 meeting of the subcommission, there was growing support for the lower ranges of *Eognathodus sulcatus*, *Polygnathus dehiscens*, and the first species of *Stringocephalus* as levels for these boundaries.

The letters l, m, u, lm and um in the conodont column are abbreviations for lower, middle, upper, lowermost and uppermost zones or subzones. The Middle *Polygnathus asymmetricus* Zone is shown as m_1 and m_2, because in 1971 it was first recognized in Western Canada on the basis of the incoming of *Ancyrodella gigas* (m_1). The incoming of this conodont was thought to precede that of *Palmatolepis punctata* (m_2), the definitive fossil of the zone, which had not been found in Western Canada in 1971.

The choice of zone fossils for the table is determined as much by previous usage and constraints imposed by table construction as relative merits of contending indices. Many of the older designated zone fossils are in need of taxonomic revision. Where this need is considered to be particularly acute, parentheses are used.

As in all tables of this kind, horizontal lines are intended to be isochronic, although the vertical time scale is constant throughout the table. Dashed lines are used at the top or bottom of range zones where ages of the zone boundaries are in doubt.

(See following pages for Table 4D.1).

Table 4D.1. Range zone of western Canadian Devonian index megafossils (Prepared by A.E.H. Pedder)

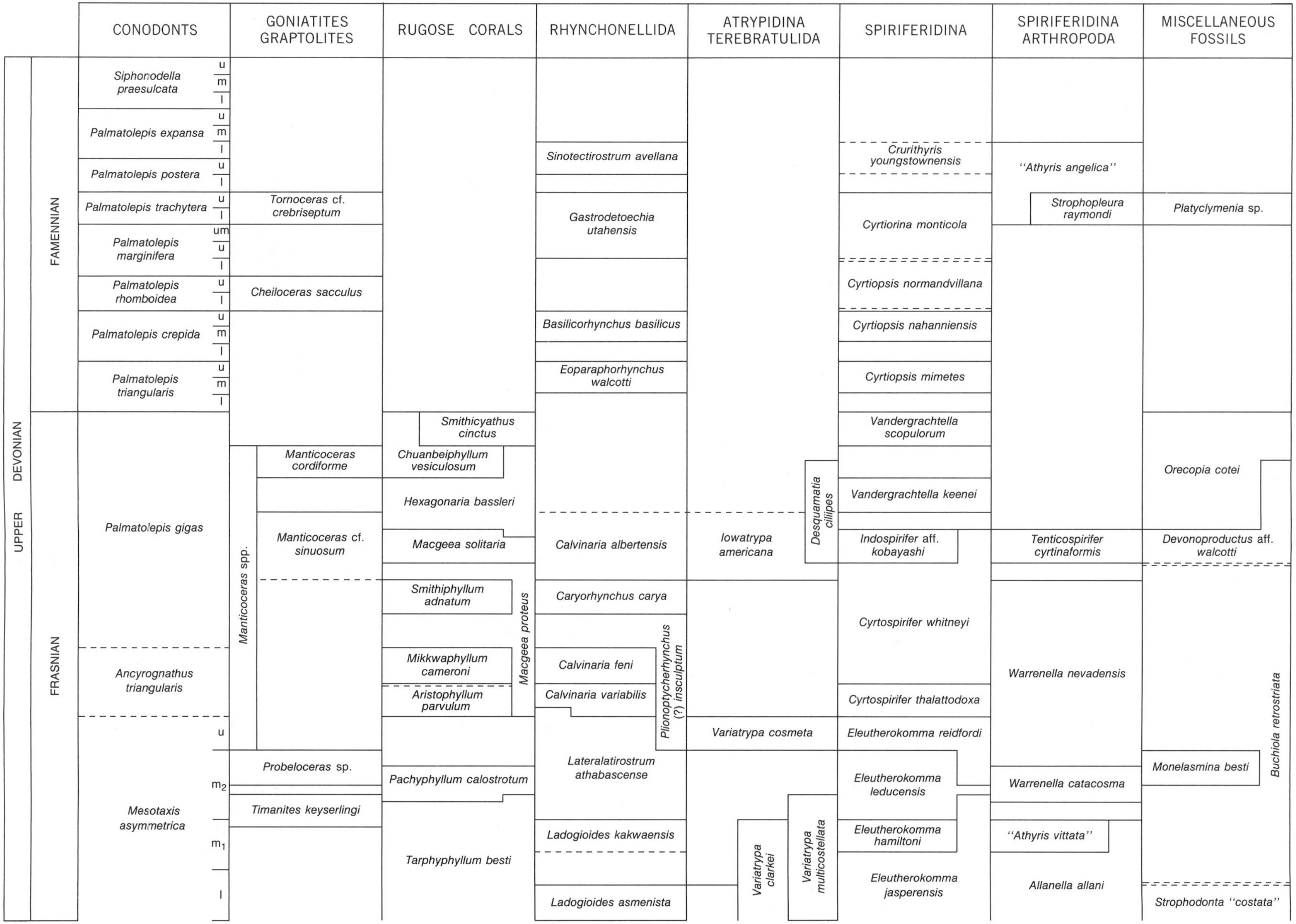

Series	Stage	CONODONTS	Subzones	GONIATITES GRAPTOLITES	RUGOSE CORALS	RHYNCHONELLIDA	ATRYPIDINA TEREBRATULIDA	SPIRIFERIDINA	SPIRIFERIDINA ARTHROPODA	MISCELLANEOUS FOSSILS
UPPER DEVONIAN	FAMENNIAN	*Siphonodella praesulcata*	u, m, l							
UPPER DEVONIAN	FAMENNIAN	*Palmatolepis expansa*	u, m, l			*Sinotectirostrum avellana*		*Crurithyris youngstownensis*	"*Athyris angelica*"	
UPPER DEVONIAN	FAMENNIAN	*Palmatolepis postera*	u, l							
UPPER DEVONIAN	FAMENNIAN	*Palmatolepis trachytera*	u, l	*Tornoceras* cf. *crebriseptum*		*Gastrodetoechia utahensis*		*Cyrtiorina monticola*	*Strophopleura raymondi*	*Platyclymenia* sp.
UPPER DEVONIAN	FAMENNIAN	*Palmatolepis marginifera*	um, u, l							
UPPER DEVONIAN	FAMENNIAN	*Palmatolepis rhomboidea*	u, l	*Cheiloceras sacculus*				*Cyrtiopsis normandvillana*		
UPPER DEVONIAN	FAMENNIAN	*Palmatolepis crepida*	u, m, l			*Basilicorhynchus basilicus*		*Cyrtiopsis nahanniensis*		
UPPER DEVONIAN	FAMENNIAN	*Palmatolepis triangularis*	u, m, l			*Eoparaphorhynchus walcotti*		*Cyrtiopsis mimetes*		
UPPER DEVONIAN	FRASNIAN	*Palmatolepis gigas*		*Manticoceras cordiforme*; *Manticoceras* cf. *sinuosum*; *Manticoceras* spp.	*Smithicyathus cinctus*; *Chuanbeiphyllum vesiculosum*; *Hexagonaria bassleri*; *Macgeea solitaria*; *Smithiphyllum adnatum*	*Calvinaria albertensis*; *Caryorhynchus carya*	*Desquamatia cilipes*; *Iowatrypa americana*	*Vandergrachtella scopulorum*; *Vandergrachtella keenei*; *Indospirifer* aff. *kobayashi*; *Cyrtospirifer whitneyi*	*Tenticospirifer cyrtinaformis*	*Orecopia cotei*; *Devonoproductus* aff. *walcotti*
UPPER DEVONIAN	FRASNIAN	*Ancyrognathus triangularis*			*Mikkwaphyllum cameroni*; *Aristophyllum parvulum*; *Macgeea proteus*	*Calvinaria feni*; *Calvinaria variabilis*; *Plionoptycherhynchus* (?) *insculptum*		*Cyrtospirifer thalattodoxa*	*Warrenella nevadensis*	*Buchiola retrostriata*
UPPER DEVONIAN	FRASNIAN	*Mesotaxis asymmetrica*	u				*Variatrypa cosmeta*	*Eleutherokomma reidfordi*		
UPPER DEVONIAN	FRASNIAN	*Mesotaxis asymmetrica*	m_2	*Probeloceras* sp.	*Pachyphyllum calostrotum*	*Lateralatirostrum athabascense*		*Eleutherokomma leducensis*	*Warrenella catacosma*	*Monelasmina besti*
UPPER DEVONIAN	FRASNIAN	*Mesotaxis asymmetrica*	m_1	*Timanites keyserlingi*	*Tarphyphyllum besti*	*Ladogioides kakwaensis*	*Variatrypa clarkei*; *Variatrypa multicostellata*	*Eleutherokomma hamiltoni*	"*Athyris vittata*"	
UPPER DEVONIAN	FRASNIAN	*Mesotaxis asymmetrica*	l			*Ladogioides asmenista*		*Eleutherokomma jasperensis*	*Allanella allani*	*Strophodonta* "*costata*"

Table 4D.1. Cont.

MIDDLE DEVONIAN	GIVETIAN	lm		*Tabulophyllum athabascense*	*Ladogioides pax*	*Atrypa scutiformis*	*Tecnocyrtina billingsi*	*Eleutherokomma impennis*	*Desquamatia independensis*
		Palmatolepis disparalis u, l		*Grypophyllum mackenziense*	*Leiorhynchus hippocastanea*	*Pseudoatrypa* (?) *percrassa*	*Emanuella vernilis*; *Warrenella timetea*		
		hermanni-cristatus							
		Polygnathus varcus u?, m, l	*Maenioceras terebratum*; *Wedekindella brilonensis*; *Cabrieroceras karpinskyi*	*Grypophyllum crickmayi*; *Temnophyllum richardsoni*; *Xystriphyllum spenceri*	*Hypothyridina cameroni*; *Stenoglossariorhynchus awokanak*; *Droharhynchia intermissa*; *Leiorhynchus castanea*	*Stringocephalus* sp. B; *Stringocephalus aleskanus*; *Ectorensselandia laevis*; *Stringocephalus chasmognathus*; *Stringocephalus glaphyrus*; *Desquamatia snakensis*	*Cyrtina "hamiltonensis"*; *"Emanuella" meristoides*	*Dechenella manitobensis*	*Rhyssochonetes aurora*; *Hypomphalocirrus manitobensis*; *Buechelia tyrrelli*; *Pseudoatrypa bremerensis*
	EIFELIAN	*Polygnathus curtigladius*	*Cabrieroceras* cf. *karpinskyi*	*Chostophyllum coniculus*; *Taimyrophyllum stirps*; *Radiastraea verrilli*	*Leiorhynchus manetoe*	*Carinatrypa dysmorphostrota*; *Variatrypa arctica*	*Warrenella kirki*; *Undispirifer compactus*	*Camsellia truncata*; *Fuscinipyge applanata*	*Dendrostella trigemme*; *Eoschuchertella adoceta*; *Spinulicosta stainbrooki*
		Steptotaxis pedderi		*"Microcyclus" multiradiatus*					
		Polygnathus costatus	*Pinacites jugleri*					*Dechenella maclareni*	
LOWER DEVONIAN	EMSIAN – DALEJAN	*Polygnathus patulus*		*Taimyrophyllum nolani*			*Warrenella quadrata*		
		Polygnathus serotinus		*Spongonaria guttata*; *Spongonaria filicata*		*Bifida ogilviensis*		*Moelleritia canadensis*	*Roemeripora* cf. *spelaeana*; *Gasterocoma* (?) *bicaula*
	EMSIAN – ZLICHOVIAN	*Polygnathus inversus*	*Teicherticeras* cf. *lissovi*	*Taimyrophyllum capax*		*Carinatina lowtherensis*	*Perryspirifer scheii*		
		Polygnathus gronbergi		*Leurelasma parca*	*Athyrhynchus susanae*		*Warrenella sekwensis*		*Megastrophia iddingsi*
		Polygnathus dehiscens	*Teicherticeras lenzi*	*Martinophyllum altiaxis*	*Nymphorhynchia pseudolivonica*	*Vagrania johnsoni*	*Elythina transversa*	*Terranovia nalivkini*	*Phragmostrophia merriami*
	PRAGIAN	*Eognathodus kindlei*	*Monograptus yukonensis*	*Werneckelasma multiseptatum*		*Vagrania* aff. *intermediafera*	*Plicocyrtina sinuplicata*		*Davidsoniatrypa johnsoni*
		Eognathodus sulcatus u, l	*Monograptus thomasi*		*Nymphorhynchia nympha*		*Plicoplasia acutiplicata*		
	LOCHKOVIAN	*Pedavis pesavis*	*Monograptus fanicus*	*Loboplasma multilobatum*	*Thliborhynchia pedderi*	*Spirigerina supramarginalis*			*Toquimaella kayi*
		Ozarkodina delta	*Monograptus hercynicus*	*Carlinastraea halysitoides*					
		Icriodus hadnagyi				*Notoparmella gilli*		*Proetus whittakeri*	
		Icriodus woschmidti	*Monograptus uniformis*	*Windelasma werneckensis*	*Ancillotoechia gutta*	*Spirigerina marginaliformis*		*Warburgella canadensis*	*Gypidula pelagica lux*; *Atrypa nieczlawiensis*

Mountains, the other the far-sighted policy of core conservation and public access to sub-surface data introduced by governmental authorities in the 1950s.

Previous work

The most recent stratigraphic syntheses of the Devonian of this region appeared in 1964 (McCrossan and Glaister), 1968 (Bassett and Stout), and 1970 (Douglas). The broad outlines of the stratigraphy laid down in these excellent publications remain unchanged. Nevertheless, during the last twenty-five years over 400 publications containing information about the Devonian rocks of this region have appeared, and it is inevitable that some changes in interpretation have been made.

Acknowledgments

The following officers of the Geological Survey of Canada made major contributions to this compilation in the form of generous provision of unpublished data, discussion, and criticism: G.K. Williams, N.C. Meijer-Drees, H. Geldsetzer, D. Morrow, G.C. Taylor, A.W. Norris, J.A. Podruski. The Table of Zones (Table 4D.1) was compiled by A.E.H. Pedder, who was consulted on all biostratigraphic matters. G.K. Williams and R.H. Workum were critical readers. Grateful acknowledgement is given to Shell Canada Resources Ltd. for the provision of computer-generated isopach maps.

Sequences and map-units

Five major depositional sequences, delimited towards the craton by discontinuities, are recognized in the Devonian of the Interior Platform (Fig. 4D.1, in pocket). For ease of reference, the sequences have been given informal names after one or more of the formal lithostratigraphic units composing them. Each sequence except the fourth is illustrated in this report by a single isopach-facies map. The fourth sequence has been divided into two map-units because an important change in basin architecture (though no evident discontinuity) occurs within it. The resulting six map layers and their approximate ages are listed below.

Palliser sequence (Famennian) 6

Beaverhill-Saskatchewan sequence
(mid-Givetian to late Frasnian):

Saskatchewan map-unit
(middle to late Frasnian) 5

Beaverhill map-unit
(mid-Givetian and early Frasnian)............. 4

Hume-Dawson sequence
(Eifelian and early Givetian) 3

Bear Rock sequence (Zlichovian to Eifelian) 2

Delorme sequence (Wenlock(?) to Pragian)............ 1

A fence diagram (Fig. 4D.1, in pocket) summarizes the Devonian system as a whole.

Within each mapped sequence there is a hierarchy of cycles of lower order. An indication of the rank of the cycle is given, where it appears helpful, by affixing a number in squared brackets. This ranking is based on an extension of the orders of sea-level cycles developed in Vail et al. (1977). Because the Sloss sequence was identified therein as a second-order cycle, the "sequences" of this report are third-order cycles or sets of third-order cycles. The smallest cycles considered are assigned to the fifth rank in conformity with the proposal of Read et al. (1984). The Pillara cycles of Australia, usually less than 10 m thick, are the type of this fifth rank (Read, 1973). Intermediate groupings are assigned to the fourth rank, although in some cases a rank intermediate between the fourth and fifth seems to be present.

Tectonic-sedimentary provinces

At any given time, the region was divisible into provinces characterized by normal, rapid, or slow sediment accumulation, by "shallow" or "deep" water, or by erosion. Areas of above-normal sediment accumulation are called depocentres. All depocentres are downwarped as a result of loading, but not all downwarps are depocentres; many are starved. Some bathymetric basins such as Richardson Trough in the Early and Middle Devonian coincide with downwarps; other bathymetric basins such as West Pembina Basin (Frasnian) are purely constructional and result from the progradation of carbonate platforms around them subsequent to sea-level rise. Treating the succession in terms of sequences and tectonic-sedimentary provinces provides a convenient framework for basin analysis. The names of most formations of scientific or economic importance have been supplied. The definitions of some are provided through specific references; the definitions of the others may be found in Alberta Society of Petroleum Geologists (1960) or Hills et al. (1981). The hierarchical level of the units (group, formation, member) has been changed from place to place as utility requires.

Sea-level change and epeirogeny

This study was undertaken at a time when it is in vogue to fit all evidence of deepening and shallowing to worldwide changes in sea level. There is a danger that this will become a dogma. The fact of eustatic sea-level change cannot be doubted, but its effect upon the cratonic sediments is always expressed in a signal mixed with other inputs from orogeny, epeirogeny, climatic change, solar energy and, upon occasion, impact by extra-terrestrial objects. This study reveals an interplay between crustal deformation and sea-level change.

Climate

Ettensohn and Barron (1981) came to the conclusion that in Late Devonian time the equator passed south of the Great Lakes, with an azimuth of 60° with respect to present true north. The 30°N paleo-parallel passed through Yukon Territory. The continent moved northward during the Devonian, but the climate remained mostly tropical to subtropical. This is in keeping with the presence of huge evaporite deposits from the southern plains (Prairie Evaporite of Givetian age) to the northern part of the mainland (Bear Rock Formation of Eifelian age). Corals and stromatoporoids flourished throughout the region, whenever salinity conditions were favourable, until the late Frasnian extinction event, the immediate cause of which may have been a worldwide lowering of temperature. The Middle and Upper Devonian reefs (Horn Plateau, Givetian; Leduc, Frasnian) seem to have had their windward sides to the present northeast, i.e., what in Devonian times was east. Trade winds with a more northerly orientation would be expected in the northern hemisphere; no explanation is offered for this deviation.

Biostratigraphy

Speculation about the correlation of cycles has been controlled by the use of all available biostratigraphic data, but this has only been cited where critical. Early biostratigraphic correlations relied mainly on brachiopods, less so on corals. In Lower Devonian formations of suitable facies, graptolites provide excellent control. Attempts to use ostracodes have been only partly successful because of the strong influence of facies. In recent years, conodont zonation has been playing a larger role and will ultimately predominate as the zonal tool, although the distribution of conodont taxa also is influenced significantly by biofacies. The Table of Zones (Table 4D.1) appears previously shows the current zonal systems based on various groups of organisms, with the conodont zonation taken as the standard. Details of correlations not given in the text can be found in Correlation Charts (in pocket) for this volume.

DELORME SEQUENCE (WENLOCK(?)-PRAGIAN)

Delorme sequence comprises a diverse assemblage of formations (Fig. 4D.3) deposited between the Mid- to Late Silurian regression and the Emsian (Early Devonian) transgression. Fish and trilobite fossils indicate that, in regions such as the southwestern District of Mackenzie, deposition of the sequence may have started as early as Wenlock time. Most of the rocks are part of a carbonate platform that once extended from the Arctic coast into northeastern British Columbia (Fig. 4D.2, in pocket). These rocks are impure, commonly silty dolomites and limestones with characteristic orange-brown hues in outcrop. In local depocentres, thick silty evaporites or solution breccias partly or wholly replace the carbonates. West of the platform, the carbonates pass within a short space to dark graptolitic shales of the Road River Formation. An entirely separate development of evaporites took place in an isolated basin in central Alberta.

Delorme sequence lies unconformably or paraconformably upon rocks ranging in age from Precambrian to Wenlock. It terminates with a major regression, revealed in the north by evidence of karstification and erosion. In the south, a thick regressive sandstone ends the sequence. This regression appears to fall at the end of T-R Cycle Ia (Johnson et al., 1985) which, in New York State, ends with a similar regressive sandstone – the Oriskany.

Tectonic-sedimentary provinces

During deposition of Delorme sequence the landmass of the Transcontinental Arch extended across Manitoba and Saskatchewan into Alberta. The only depocentre in the prairie region was Central Alberta Sub-basin, bounded on the south by Meadow Lake Hingeline and on the west by West Alberta Ridge (Fig. 4D.2, in pocket). North of it lay a landmass, which in anticipation of its future differentiation may be called the Peace-Athabasca-Tathlina High. West of the prairie region, in British Columbia and in Yukon Territory, lay the deep waters of the continental shelf occupied by Selwyn (bathymetric) Basin with its easterly gulf, Meilleur River Embayment. A broad shelf lay between the uplands and the open sea to the west. The interior portion is named Anderson-Great Slave Platform. The exterior or miogeoclinal portion has been divided for convenience of description into segments. South of Meilleur River Embayment lies

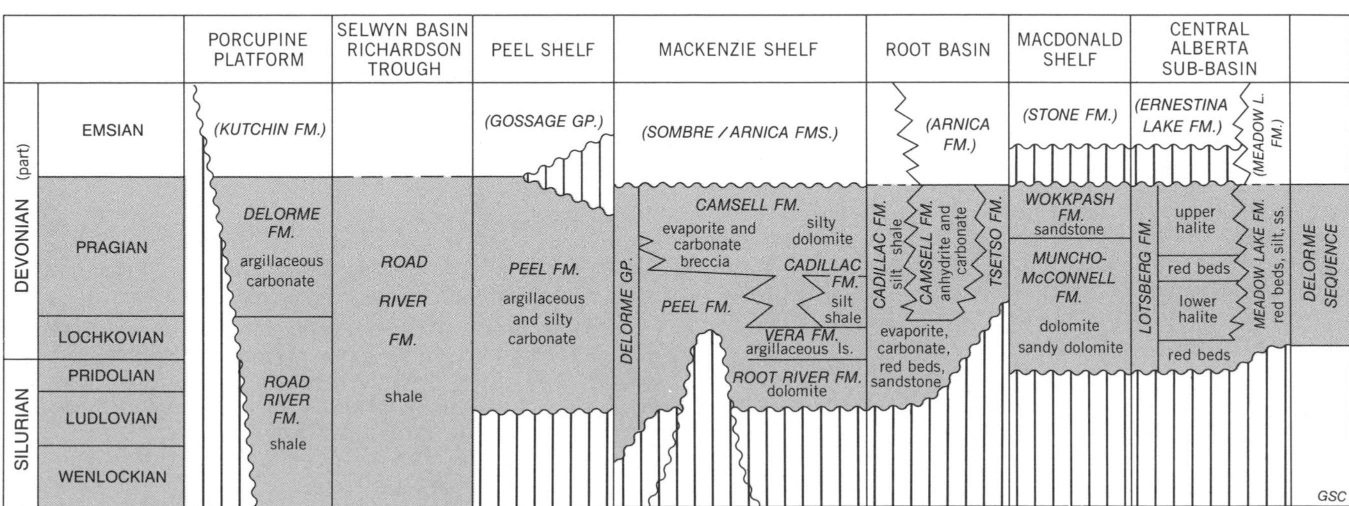

Figure 4D.3. Correlation chart of Delorme sequence.

MacDonald Platform (Gabrielse, 1967), and north of it, Mackenzie Platform. The northern part of Mackenzie Platform will be referred to here as Peel Platform.

Selwyn Basin is bounded on the north by Porcupine Platform, an element cut off from Peel Platform by Richardson Trough, a long narrow basin thought to have had a taphrogenic origin in Cambrian time. Within the northern part of Richardson Trough, a small positive feature, White Mountains Platform, was the site of persistent carbonate deposition. The core of Porcupine Platform (known as Dave Lord High) either was emergent during the whole of this sequence or was uplifted and eroded at the end of it. In any case, there seems to be no part of the sequence left upon the high, and it was not covered with sediment until the end of the next ("Bear Rock") sequence.

The most prominent positive elements influencing sedimentation during this time (Fig. 4D.2) were Keele Arch (Cook, 1975) and Redstone Arch (Gabrielse, 1967), within which Twitya Uplift (Cook and Aitken, 1978) lies at the northern end. Keele Arch, as expressed today, is the product of pre-Cretaceous uplift and erosion, but it partly coincides with an earlier positive feature referred to by Williams (1975b) as Norman Wells High. Twitya Uplift is a gentle pre-Late Silurian feature along the axis of which Mount Kindle and Whittaker formations (Ordovician-Silurian) were removed before deposition of Delorme sequence. This exhumed the sub-Upper Cambrian unconformity so that, in the core of the uplift, Delorme sequence comes to lie on rocks as old as Late Proterozoic. The influence of Redstone Arch is also felt to the south, where the sediments record shoaling (Morrow, 1984a). Aklavik Arch is a strongly faulted post-Devonian uplift, in the core of which strata down to the Proterozoic subcrop beneath the Cretaceous. The chief negative element, apart from the previously mentioned Selwyn Basin, Richardson Trough, and Central Alberta Sub-basin, was Root Basin (depocentre).

The western margin of the carbonate platform can be traced southeastward as far as Peace River, south of which it has been destroyed by Cordilleran orogeny and erosion. The margin of the platform was indented at several places. In Misty Creek Embayment (Cecile, 1982), which is mainly a lower Paleozoic feature, the volcanic Marmot Formation is interbedded with unnamed limestones to which a Devonian age has been tentatively assigned on the basis of inconclusive paleontological evidence. The entire volcanic sequence may be pre-Delorme. Meilleur River Embayment lies at about 61°N and leads, via Prairie Creek Embayment (Morrow, 1984a), into Root Basin. Some 400 km south of this, a striking facies change from carbonates to siliciclastics in the region between Halfway River and Peace River has been documented by Thompson (1976) and interpreted as an embayment or "bypass" by Taylor (1982). Taylor believed that it was one of the principal routes by which terrigenous material being eroded from the Peace River landmass reached the western basin. The siliciclastic fill of this feature, Lady Laurier Embayment, has many features in common with that of Prairie Creek Embayment. Most of the sediments it contains belong in the Bear Rock sequence, but some are Delorme equivalents (Thompson, 1976). The "embayment" interpretation depends on the palinspastic reconstruction of what is now a highly compressed anticlinorium, and is not accepted by all geologists familiar with the region.

Between Twitya Uplift and the carbonate platform margin, a local depocentre was the site of a thick evaporite deposit, now represented by breccia like that in Root Basin. Twitya Uplift may have acted as a barrier; this hypothesis is supported by the concentration of sands observed in the sediments that surround it, but on the other hand there is some suggestion of thicker sediments over it (Morrow, in press).

Porcupine Platform

On Figure 4D.2 (in pocket) some carbonate rocks are shown flanking Dave Lord High on the west side of Richardson Trough. Norris (1985) included these rocks in his Kutchin Formation, whose principal development was in Bear Rock sequence, but well logs strongly suggest a break in the succession, with Delorme Formation at the base and Kutchin Formation above. Some of these rocks are as old as Pragian (*Monograptus yukonensis* Zone).

Selwyn Basin and Richardson Trough

Selwyn Basin is an outer shelf bathymetric basin established in earlier Paleozoic times during the growth of a carbonate platform to the east. The rocks of Delorme sequence deposited in Selwyn Basin form part of Road River Formation and include Upper Silurian and Lower Devonian graptolitic shales. The formation is also present in Richardson Trough, where it forms the basinal equivalent of Peel Formation. Road River Formation consists of dark shales with associated limestone and chert, dolomite, siltstone, and sandstone, outcropping on the flanks of Richardson Mountains and to the south. The formation, over 3000 m thick, is known to range in age from Late Cambrian to Middle Devonian. The succession is of basinal character throughout and contains large debris-flow breccias near the carbonate platform margin at many localities.

Peel Platform

Subsurface strata belonging to Delorme sequence have been named Peel Formation (Pugh, 1983). Typically they are about 350 m of pale buff, pale grey, and grey-buff, aphanitic and microcrystalline dolomite, some of which is calcareous. The basal 50 m is in part argillaceous or finely silty. Peel Formation is combined in outcrop with the impure basal carbonates of the overlying Gossage Group to form the SD (Siluro-Devonian) map unit (Norris, 1985). Its buff- to orange-weathering colour makes it a distinctive unit both on the ground and on air photographs.

Both contacts of Peel Formation are unconformable, although the lower may appear paraconformable and the upper is obscured by karstification that confuses the log signature. Peel Formation is overstepped by Gossage Group on the northwest flank of Keele Arch (see cross-section 2 on Fig. 4D.1). Peel Formation has not yielded many age-diagnostic fossils, but Pugh (1983) cited three widely spaced localities that have yielded fossils of

Late Silurian age. He also mentioned a late Lochkovian to early Pragian age for beds in the middle of the suite of Road River shales and limestones thought to be equivalent to Peel Formation.

Mackenzie Platform

The deformed part of Mackenzie Platform in Mackenzie Mountains is the type region for the Delorme Formation, although the label "SD" has just as often been used. In the northern part of Mackenzie Platform, Morrow (in press) found it difficult to distinguish between Delorme Formation and the overlying Camsell Formation in many places. Indeed, he recorded several localities at which Camsell had been mapped as part of the Delorme. He has therefore erected a Delorme Group comprising an upper Camsell Formation and a lower unnamed Silurian-Devonian unit, which, in this report, is treated as a southern extension of the Peel Formation.

The Peel is most typically developed over Redstone Arch, where it consists predominantly of light yellow-weathering, argillaceous and silty, planar bedded dolomite, with a very large variety of associated dolomites, mudstones, siltstones, and sandstones. Stromatolites, small-scale ripples, and mudcracks are common in some beds, and salt casts have been recorded. The thick-bedded dolostones contain colonial corals, small hemispheroidal stromatoporoids, and amphiporoids. Fossils are sparse, but fish remains, ranging in age from Wenlock to Lochkovian (R. Thorsteinsson, pers. comm., 1985), have been found at many localities (Dineley and Loeffler, 1976). The environment of deposition over Redstone Arch ranged from intertidal to shallow subtidal.

In the southern part of Mackenzie Platform, Morrow (1984a) has erected new formations to replace the Delorme Formation in its type region. These are described in the section devoted to Root Basin, adjacent to which they have been studied in detail. The new formations probably will be traced northward in the region of the platform margin between Redstone Arch and Selwyn Basin, but revision of the stratigraphy in that area is not yet complete. Earlier work along the platform margin (Perry, 1984; Gabrielse et al., 1973) showed the presence of richly fossiliferous Ludlow to Zlichovian carbonates up to 500 m thick.

Over most of the shelf, the base of the Delorme Group appears conformable with the underlying Whittaker Formation (Silurian), but regionally it can be seen to overstep all formations down to the Upper Proterozoic. The Delorme Group interfingers westward with progressively younger shales of the Road River Formation (Perry, 1984).

Camsell Formation may be subdivided readily into a silty carbonate and an evaporite facies. On Mackenzie Platform the most characteristic facies is the former, the evaporites being confined to a small depocentre west of Twitya Uplift. The silty carbonate facies (Morrow, in press) is characterized by subtidal to intertidal hemicycles several metres thick, composed of dark carbonates grading up into lighter silty and argillaceous carbonates. The evaporitic facies will be described in the following section devoted to Root Basin.

Root Basin and Anderson-Great Slave Platform

Prairie Creek assemblage

Where Root Basin leads into Selwyn Basin a complex interfingering of facies is found in the region called Prairie Creek Embayment. These rocks have been named the Prairie Creek assemblage (Morrow, 1984a). The Prairie Creek assemblage was deposited in two phases, of which the early Cadillac phase is believed to represent the Delorme sequence. The term "Delorme Formation" was originally proposed to include what are now Root River, Vera, and Cadillac formations, forming part of the Prairie Creek assemblage. On the margin of the shelf, Camsell formation, described earlier, lies above Vera and Root River formations and forms the top of the sequence. The arrangement of these formations is shown in Figure 4D.3.

Root River Formation is a finely crystalline dolomite rimming the Prairie embayment. It is the lower of the two carbonate units in the original Delorme Formation.

Vera Formation is a limestone unit of Lochkovian age, 100-300 m thick. It consists of very thin-bedded couplets of fossiliferous limestone and argillaceous limestone. Coral- and crinoid-bearing mounds are scattered throughout its upper part. The Vera was probably deposited in an upper slope environment on the west margin of Prairie Creek Embayment.

Cadillac Formation filled the distal portion of Root Basin and Prairie Creek Embayment during the early Cadillac phase. It comprises 200 to 800 m of bright-orange-weathering siltstone and megabreccias, including conglomeratic carbonate beds that appear to be turbidites.

Tsetso Formation

The basal, mixed carbonate-evaporite-siliciclastic unit of Delorme sequence in Root Basin and the adjacent Anderson-Great Slave Platform is called Tsetso Formation (Meijer Drees, in press). Morrow and Meijer Drees (1981) reported the Tsetso as occurring at the base of the "Bear Rock" Formation, but this interpretation resulted from their extension of the name "Bear Rock" beyond its type area to include older beds interpreted here as Camsell Formation. The Tsetso is a diachronous unit that is probably continuous with the Cadillac Formation of the Delorme Group.

Camsell Formation - evaporite facies

Williams (1975b) demonstrated the equivalence of Camsell Formation to the lower of two major evaporite cycles identified in the subsurface of Root Basin. He showed two anhydrites separated by 175 m of "Arnica Platform Dolomite", forming the transgressive base of the upper cycle. The lower cycle is here assigned to Delorme sequence and the upper cycle to Bear Rock sequence. The two evaporites were mapped together as "Bear Rock" by Bassett and Stout (1968). Meijer Drees (in press) has locally done the same under the name "Fort Norman Formation". If

Arnica Formation is not well developed, it may be difficult to separate Camsell Formation from Bear Rock Formation using only geophysical logs although, according to Morrow (in press), the Camsell is consistently more silty than the overlying Bear Rock. The Camsell Formation probably exceeds 500 m in thickness, but a closer estimate would require access to seismic and dipmeter records not now available.

Johnny Hoe salt

In a small basin within the southern part of Anderson-Great Slave Platform, Camsell Formation is overlain by a thin salt named Johnny Hoe (Meijer Drees, in press).

MacDonald Platform

Muncho-McConnell and Wokkpash formations

On MacDonald Platform, Delorme sequence is represented by Muncho-McConnell and Wokkpash formations (Taylor and Mackenzie, 1970). The former is a unit of alternating medium and dark grey crystalline dolomite with a sandy base that rests disconformably on the Silurian Nonda Formation throughout much of northeast British Columbia. Thin varicoloured shale partings are present in the basal dolomites. Well rounded, frosted, fine grains of quartz are common in the upper half. Desiccation breccias have been noted. At the top there is a rapid gradation to the sandstone, dolomitic sandstone, and argillaceous dolomite of the Wokkpash Formation. The latter is probably a shoreline deposit representing the terminal regression of the sequence. It is overlain sharply, and locally disconformably, by the sandy basal beds of the Stone Formation (Bear Rock sequence).

Delorme sequence is thickest (460 m) in the most westerly outcrops. It has not been definitely recognized in any of the boreholes east of the Foothills and is assumed to have lapped out against the Peace-Tathlina-Athabasca landmass. B.S. Norford (pers. comm., 1986) measured 195 m at Clearwater Creek in Pine Pass area, but the sequence must lap out against Peace River Arch shortly to the south, for Geldsetzer (1982) found Stone Formation (Bear Rock sequence) lying directly on Ordovician strata in the Monkman Pass area. Westward, the sequence is represented by Besa River Formation (= Road River Formation). The platform margin has not been described in detail.

The age of the Muncho-McConnell and Wokkpash formations can only be determined from their position between the Nonda Formation (Llandovery) and the Stone Formation, whose earliest fossils are of the *serotinus* Zone (Dalejan or upper Emsian). Within rocks of Delorme sequence, the only datable fossils are fish identified as Early Devonian.

Lead-zinc mineralization occurs in the Muncho-McConnell at several localities (Macqueen and Thompson, 1978). It is mainly associated with breccias attributed to hydraulic fracturing caused by excess fluid pressures (Macqueen and Thompson, 1978). The main minerals are sphalerite and (or) galena, with or without pyrite or marcasite. White sparry dolomite, bitumen, and quartz are commonly present as gangue. Sulphur isotopes are consistent with a petroleum-related hydrogen sulphide source. The most important deposit is at Robb Lake (Fig. 4D.2, in pocket), where mineralization occurs in both Muncho-McConnell and Stone formations.

Central Alberta Sub-Basin of Elk Point Basin

Lotsberg Formation

Confined to the Central Alberta Sub-basin of Elk Point Basin, the Lotsberg, together with the underlying "Basal Red Beds", comprises two complete redbed-halite cycles. The present limit of salt is the result of later solution; salt probably extended far to the east (Williams, 1984). The Lotsberg has yielded no diagnostic fossils, but underlies carbonate rocks of probable Emsian age. It includes large volumes of pure rock salt, which form an important economic resource (Hamilton, 1971). The extremely low bromine content of the Lotsberg (mass fraction $<4 \times 10^{-6}$, Holser et al., 1972), and its negligible carbonate content and association with red mudstones, suggest either deposition in a continental basin with limited and only occasional access to the sea or, more probably, the dissolution of an earlier precipitate by fresh water, followed by reprecipitation. Close to Meadow Lake Hingeline, a mixture of dolomites, mudstones, and minor sandstones and limestones form the Meadow Lake Formation, the lower part of which is thought to be equivalent to the Lotsberg. Angular to rounded dolomite fragments up to 5 cm in size occur in an argillaceous matrix forming breccia conglomerates (Fuzesy, 1980). This testifies to the presence of topographic relief, an escarpment, along the hingeline coinciding with the northern erosional limit of Ordovician carbonates in Saskatchewan. Central Alberta Sub-Basin continued to sink during the Devonian (Williams, 1984); the present relief on the pre-Devonian surface is therefore greater than the paleotopographical relief at any time during the Devonian.

BEAR ROCK SEQUENCE (ZLICHOVIAN-EIFELIAN)

The Bear Rock sequence is the product of a major transgressive-regressive cycle that began in Zlichovian (Early Emsian) time and ended in mid-Eifelian (Fig. 4D.5). It includes deposits of a full spectrum of environments, from the deep waters of the open continental shelf in Yukon Territory to coastal sabkhas and supratidal deposits around the coastline in northern Alberta and adjacent areas. The first unequivocal Devonian reefs of the region occur in Bear Rock sequence.

Tectonic-sedimentary provinces

The geography of the tectonic-sedimentary elements (Fig. 4D.4, in pocket) was largely inherited from the Delorme sequence. Several elements played virtually unchanged roles viz: Richardson Trough, Selwyn Basin, Peel Platform, Mackenzie Platform, MacDonald Platform, Redstone Arch. Porcupine Platform was inundated, but a small positive feature, Dave Lord High, was emergent until late Zlichovian time. Of the elements nearer the core of the craton, Keele Arch (Norman Wells High of Williams, 1975b)

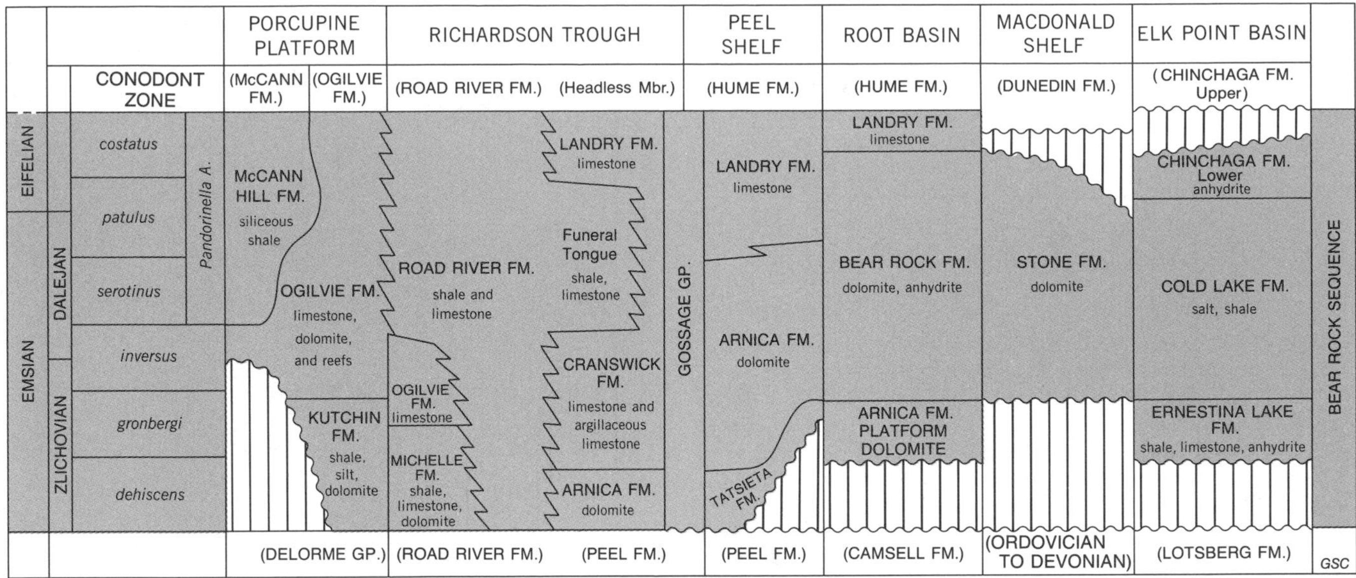

Figure 4D.5. Correlation chart of Bear Rock sequence.

was onlapped and covered by the end of sequence deposition but remained a positive feature. The strong coincidence of the arch and the reentrant of the Bear Rock Formation zero isopach is, however, mainly the effect of pre-Cretaceous uplift and erosion. Root Basin (Camsell Sub-basin of Williams, 1975b) appears to play a diminished role but was still a major depocentre in its western part.

At the beginning of Bear Rock deposition the Peace-Athabasca-Tathlina High was disrupted by the development of the North Alberta Sub-basin, a shallow (100 m) downwarp between the now distinguishable Peace-Athabasca and Tathlina arches. This basin originally extended far to the east, but has been diminished by post-Paleozoic erosion. Meadow Lake Hingeline was senescent. The Central Alberta Sub-basin owed its continued expression more to the positive influence of the Peace-Athabasca Arch than to negative activity in the basin at this time. Carbonates and evaporites were deposited in Golden Embayment, a newly formed depocentre west of West Alberta Ridge.

Porcupine Platform

Kutchin Formation

This formation comprises the sporadically argillaceous and silty dolomite and dolomitic limestone that underlie Ogilvie Formation and overlie a variety of carbonate rocks of Ordovician and Silurian ages (Norris, 1985). The type section is in the southern part of White Mountains Platform, where the Kutchin had been included formerly in the Gossage Formation. The Kutchin was not deposited on the crest of Dave Lord High; a basal conglomerate is present near the zero-edge.

The upper third or so of the Kutchin Formation contains the large, typically Emsian ostracode *Moelleritia canadensis*, tying the Kutchin closely to the overlying Ogilvie Formation, which is everywhere in concordant contact with it. It is concluded that the Kutchin Formation is a westerly representative of the lower part of Gossage Group. As mentioned under "Delorme sequence", a wedge of Delorme Formation has been removed from the Kutchin as mapped by A.W.Norris on the flanks of Dave Lord High.

Michelle Formation

This formation is a relatively thin rock unit consisting mainly of argillaceous limestone and shale, and occupying a transitional position between Road River Formation with *Monograptus yukonensis* and the Ogilvie Formation. It is areally limited but important for its rich and diverse fauna (Ludvigsen, 1970). Two-holed crinoid ossicles first appear in this formation but are more abundant in the Ogilvie. These ossicles are common in many of the formations that make up Bear Rock sequence, but are known to range upward into the Givetian. The age of the Michelle is Zlichovian; it contains a *dehiscens* Zone conodont fauna and, in its mid-part, the goniatite *Teicherticeras* (see Table 4D.1).

Ogilvie Formation

A thick carbonate (as much as 1000 m), the Ogilvie Formation, overlies the Michelle Formation in the Hart River map-area and overlies Kutchin Formation elsewhere on Porcupine Platform. It was deposited on a shallow carbonate platform, on the south side of which small bioherms about 50 m in height have been reported from at least three localities (Fig. 4D.4, in pocket; Perry et al., 1974; Clough and Blodgett, 1984). These bioherms contain Emsian fossils, including coral heads and bulbous stromatoporoids. Debris was shed from the margin of the platform into the adjacent shale basins (see below).

In Dalejan time, the outer parts of the carbonate platform were covered by deeper water crinoidal carbonates and shales. The Dalejan siliciclastic blanket is assigned to the lower part of the McCann Hill Chert on the west side of the carbonate platform and to the Road River Formation (the strata formerly known as Prongs Creek) on the east side. This drowning phase perhaps represents the start of T-R Cycle Ic. Carbonate sedimentation continued in the interior of the platform until late in the Hume-Dawson sequence. This part of Ogilvie Formation will therefore be described in a later section.

Selwyn Basin and Richardson Trough

Road River Formation and Funeral Tongue

Norris (1985) estimated that the Devonian portion of the Road River Formation (see Delorme sequence) may amount to 1000 m. The thickness to be assigned to Bear Rock sequence is uncertain. During Bear Rock deposition there was a marked development of slope deposits. Large exotic blocks of limestone containing two-holed crinoids occur in the Road River Formation (Lenz, 1972; Perry et al., 1974; Macqueen, 1974) and are also known east of Snake River at the carbonate platform margin. Dark argillaceous limestones in the Snake River area show widespread evidence of gravity deformation. They overlie carbonate rocks of Delorme sequence on the margins of the basin and represent an incursion of deeper water sediments over the platform. The argillaceous limestones and overlying shales are thus analogous to the Funeral Tongue several hundred kilometres to the south, which advanced over the margin of the platform into Root Basin at this time.

Cranswick Formation

The facies transitional between Road River shales and the platform carbonates of Gossage Group is assigned to the Cranswick Formation in the Snake River region, where it is 135 m thick. It comprises an upper Zlichovian limestone facies and an overlying, more argillaceous facies partly of Dalejan age (Pedder and Klapper, 1977).

Peel Platform

Gossage Group

The Gossage Group (Gossage Formation of Tassonyi, 1969) is an open platform carbonate succession consisting, from base to top, of Tatsieta, Arnica, and Landry formations. It lies unconformably upon the Peel Formation, which was uplifted and karstified before being re-submerged. The deposition of a silty and argillaceous basal unit on the karstified surface confuses the interpretation of geophysical logs. In constructing the maps and cross-sections for this study, the boundary was placed at the base of a widespread, strongly radioactive zone. The Gossage Group is overlain conformably by the Hume Formation of Hume-Dawson sequence.

The Cranswick Formation (see above) can be traced into Gossage Group, and therefore may be as old as Zlichovian. The uppermost beds contain the rare conodont *Pandorinellina* sp. A, which may be as young as *costatus* Zone (Eifelian).

Figure 4D.6. Bear Rock breccia, frontal Mackenzie Mountains, Powell Creek, Northwest Territories (65°16′N; 128°46′W) (photo - A.E.H. Pedder, ISPG 2582-1).

Tatsieta Formation

The unit formerly known as the Lower Limestone Member of the Gossage Formation is now known as Tatsieta Formation (Pugh, 1983); however, the boundaries chosen by Pugh at the type locality do not coincide exactly with those chosen by Tassonyi (1969). The formation is about 60 m thick in the central part of Peel Platform, where it consists of pale, micrograined limestone with waxy shale partings and minor argillaceous and silty dolomite with green hues. The base of the formation is strongly radioactive. In some sections the formation contains micropyritic shale and this no doubt accounts for the orange-weathering colour observed in outcrop sections just south of Peel Shelf, where it is included in the "SD Unit". The formation, locally thickening to 100 m, wedges out eastward against Keele Arch. It has been extended westward by Pugh across Richardson Trough to embrace rocks which, in the present work, are included in the Kutchin and, possibly, Michelle formations.

Arnica Formation

Over much of Peel Platform the former "dolomite member of the Gossage Formation" is a readily mappable unit with fairly consistent thickness. Raising it to the status of formation within Gossage Group is probably justified, although there are regions where the relationship between dolomite and limestone units is clearly one of lateral facies change rather than vertical succession. For example (Fig. 4D.1, cross-section no. 1, in pocket), just west of the facies change to Bear Rock Formation, the entire carbonate succession is dolomite. On the other hand, just east of Richardson Trough the entire section is limestone. The Arnica attains a thickness of 250 m in the north-central part of Peel Platform. In the subsurface (Tassonyi, 1969) it consists almost entirely of brown and buff, locally calcareous dolomites. It exhibits some porosity, and bleeding oil was observed in a drill core.

Landry Formation

Formerly known as the "Pellet Limestone Member" of the Gossage Formation, the Landry Formation is 230 m thick in the type well, where it consists of light-coloured, occasionally argillaceous limestone with aphanitic or pellet texture. The formation was deposited in a more restricted platform environment than the underlying Arnica. A few intercalated dolomite beds locally display intergranular porosity, which has yielded a little gas on test.

Bear Rock Formation

At the type locality of Bear Rock, on Mackenzie River close to where it crosses Keele Arch, the Bear Rock formation is a very distinctive, coarse dolomite and limestone breccia (Morrow and Meijer Drees, 1981). It is now generally agreed that this breccia (Fig. 4D.6) is the surface and shallow subsurface expression of equivalent interbedded carbonate and anhydrite in the subsurface. Bassett and Stout (1968) used the name "Bear Rock Formation" in a very loose sense to include all formations deposited at about this time, i.e., in the sense in which Bear Rock sequence is used here. In this study, following Tassonyi (1969), it is restricted to the evaporitic facies.

Bear Rock Formation replaces Arnica Formation laterally by facies change, both southward and eastward from Peel Platform. In parts of Peel Platform it is overlain by a tongue of Landry Formation included by Bassett (1961) in Bear Rock Formation. The thickness of the formation in the subsurface of Norman Wells area is 285 m. A description is deferred to the discussion of Root Basin where it has been cored. The formation was laid down in a restricted marine shelf lagoon.

Mackenzie Platform

Arnica, Sombre, Landry and Bear Rock formations

During Bear Rock deposition the influence of Redstone Arch was diminished. A blanket of carbonate rock, typically 500 m thick, comprising Arnica Formation (or its western facies, Sombre Formation) and Landry Formation, and giving way to evaporitic Bear Rock Formation eastward, was laid down. The term "Gossage Group" is not used in this tectonic sedimentary province, where surface geological mapping predominates. Terrigenous material is generally absent from these rocks in contrast to those of the underlying Delorme sequence (Morrow, in press), although the Sombre Formation is silty in parts.

Arnica Formation is a dark brownish grey, bituminous, fine- to medium-crystalline dolomite characteristically developed in cycles [5] from less than a metre to several metres thick, in which dark, *Amphipora*-bearing beds grade up into lighter, stromatolitic, fenestral dolomites. These cycles represent fluctuations from subtidal to tidal environments. A higher order cyclicity [4] is observed in some areas. In the northern part of the shelf, the Arnica contains numerous limestone interbeds; the top of the last massive dolomite is taken as the top of the formation. Arnica Formation extends eastward under the evaporites of Bear Rock Formation as a tongue called Arnica Platform Dolomite (Williams, 1975b).

Sombre Formation is a tripartite dolomite consisting of a thick, banded, lower member, a thin, very dark, middle member containing stromatoporoids and corals, and an upper, lighter coloured member. It is about 1000 m thick north of Meilleur River Embayment; farther north it becomes indistinguishable from Arnica Formation.

Landry Formation is typically represented in this sedimentary province by couplets of resistant, pelletal lime wackestone and recessive, lime mudstone with very thin planar to wavy bedding. The couplets range from 0.5 to 3 m in thickness and may represent aggradational cycles [5] (Morrow, in press). In the western part of the shelf, the Landry facies appeared early; a Zlichovian to Dalejan age, the same as the type Cranswick, was provided by one faunal collection.

Westward from Mackenzie Platform the carbonates pass into the Road River Formation. A shaly tongue, mapped as Funeral Formation, enters Root Basin to the extent indicated on Figure 4D.4, (in pocket) and may represent the start of T-R Cycle Ic. The main transition from Arnica dolomite to Funeral shale is rather abrupt (Noble and Ferguson, 1971), but the precise relationships are obscured by a diagenetic facies described below.

The evaporites and breccias of Bear Rock Formation, with which a capping Landry tongue is customarily included for mapping purposes, extend onto Mackenzie Platform over the Carcajou Salient (Aitken et al., 1982) of Norman Wells High (Keele Arch, Fig. 4D.4, in pocket), but the formation has its main development to the east and will be described more fully under "Anderson-Great Slave Platform and Root Basin".

Manetoe-Presqu'ile Dolomite

Cutting across Arnica Formation in the southwestern part of Mackenzie Platform and extending to the basin margin, is a mass of late diagenetic, white, very coarsely crystalline, cavity-filling and replacement dolomite. It has been mapped as Manetoe "Formation" but is more correctly referred to as a diagenetic facies. In the Beaver River and Pointed Mountain gas fields of southeastern Yukon Territory and northeastern British Columbia (Fig. 4D.7, in pocket), Manetoe dolomite (replacing Dunedin Formation of the Hume-Dawson sequence) is host to gas accumulations in anticlinal traps (Snowdon, 1977). It is found farther south on MacDonald Platform invading both Stone and Dunedin formations (Macqueen and Taylor, 1974). Morrow (1984b) found Manetoe dolomites to be similar to Mississippi Valley-type dolomites in the curvature of the crystals, their stoichiometry, carbon isotope values, and bulk strontium contents. Manetoe dolomite differs, however, in having a lower iron and ^{18}O content. In certain wells (e.g., Fig. 4D.1, section 3, in pocket) a body of unusually radioactive dolomite is found associated with Manetoe dolomite; the origin of this is not known. East of the main occurrences of Manetoe dolomite, a similar irregular body, known as Presqu'ile "Formation", has replaced the upper part of Hume-Dawson sequence.

MacDonald Platform

Stone Formation

South of Meilleur River Embayment, Arnica Formation is replaced by Stone Formation, a light-coloured, fine- to medium-crystalline dolomite containing breccias. Sandstone commonly occurs at its base but is in some places hard to distinguish from the underlying Wokkpash Formation of Delorme sequence. Basal lag deposits include pebbles of barite. The Stone Formation is nearly 600 m thick in the northern part of the shelf, but to the south, where it approaches Peace-Athabasca Arch, it thins and becomes sandy. It has been traced along the west side of the arch in Monkman Pass area, where it thins from 570 m to zero in just over 100 km (Geldsetzer, 1982). Eastward, the Stone thins and passes into the lower Chinchaga Formation, composed of anhydrite and dolomite and analogous to the Bear Rock Formation to the north. Over Fort Nelson High the section is reduced to 12 m of sandstone. Westward, the Stone passes into Road River equivalent, here called Besa River Formation (Taylor et al., 1975). The nature of the transition has not been reported.

In southern MacDonald Platform the upper and lower boundaries of the Stone Formation are unconformable, but to the north the upper contact becomes gradational. Corals collected 410 m above the base near Monkman Pass are Zlichovian or Dalejan in age. Conodonts from various levels in the Stone have yielded ages from *serotinus* Zone to *costatus* Zone, confirming its equivalence to part of the Gossage Group (Geldsetzer, 1982).

Several types of breccia characterize the Stone Formation. In a northern area extending from north of Muncho Lake to south of Tuchodi Lakes, tabular, largely conformable sheets of dolomite breccia up to 4 m thick and several hundreds of metres long occur throughout the formation (Morrow, 1975). Another type of breccia occurrence, cutting the overlying Dunedin Formation as well as the Stone, is called a buttress structure. Such structures are composed of rubble breccia similar to those observed in present day sink-holes; this led Morrow to attribute them to processes now active in the Florida Aquifer. He ascribed the solution and brecciation to the sub-Watt Mountain disconformity at the end of Hume-Dawson sequence.

Breccias associated with the Robb Lake mineral deposit (Fig. 4D.4, in pocket) occur in both Stone and Muncho-McConnell formations and have already been described. In the same region, some galena and smithsonite mineralization in the Stone Formation is associated with limestone pseudobreccias (Macqueen and Taylor, 1974). Irregular masses of the Manetoe-Presqu'ile diagenetic facies also occur in this region but chiefly invade the overlying sequence.

Anderson-Great Slave Platform and Root Basin

To the east of Mackenzie Platform, Anderson-Great Slave Platform represents the interior of the pericratonic marine platform during Bear Rock deposition and is characterized by the dominance of evaporitic over carbonate deposits. It is transected by Keele Arch (in its earlier manifestation as Fort Norman High) over which rocks of the sequence thin. Arnica Formation occupies the crest of the arch shown in Cross-section 2 (Fig. 4D.1, in pocket), while Bear Rock Formation was deposited on either side; on the southwestward plunge-out, Bear Rock covers the arch. In the outcrop belt, which is characterized by sinkholes, the Bear Rock is represented by a chaotic calcareous and gypsiferous breccia of angular to slightly rounded dolomite fragments from sand size to 15 m in diameter; anhydrite is never seen.

South of Keele Arch, Root Basin contains thick Bear Rock anhydrite. A well there penetrated the following succession (Meijer Drees, 1980):

Funeral Formation (a shaly facies of Landry Formation)	607 m
Bear Rock Formation:	
Dolomitic Member	35 m
Evaporitic Member	1443 m
Arnica Formation Platform Dolomite	>159 m

(The measurement is drilled interval, not thickness).

In this well, the Bear Rock comprises numerous upward-shoaling evaporitic cycles [5], 1 to 5 m thick and consisting of a basal pebble bed overlain by anhydrite and dolomite. The character of the basal bed is consistent with a storm origin.

South of Root Basin lies Willow Lake Basin, a depocentre which, although much less pronounced, may have been bathymetrically deeper at the start of sequence deposition. Arnica Platform Dolomite in Root Basin merges into Ernestina Lake Formation in Willow Lake Basin, followed, in the centre of the basin, by a halite assigned to Cold Lake Formation. This is the northernmost of a string of small halite deposits on either side of Tathlina and Peace-Athabasca arches. Above the Cold Lake Formation, and separated from it by a thin layer of redbeds, lies an extension southwards of Bear Rock Formation. Eastward

Figure 4D.8. La Loche Formation, Stony Island, Slave River, northern Alberta (59°35′N; 111°26′W). The lower hammer head marks the contact with Precambrian crystalline basement; the upper marks the contact with the Fitzgerald Formation. (photo - A.E.H. Pedder, ISPG 940-257).

from Willow Lake Basin, Bear Rock sequence is represented at outcrop by basal redbeds (Mirage Point Formation) overlain by dolomites and evaporites of Chinchaga Formation within which the Ebbutt unconformity that ends the sequence may be concealed.

Elk Point Basin

Elk Point Basin expanded during the course of Bear Rock deposition to form the North Alberta and Central Alberta sub-basins, which are connected via the east side of Tathlina High to Willow Lake Basin. Meadow Lake Hingeline was still active, confining deposition to its northern side. Peace-Athabasca Arch remained positive.

The sequence consists of Ernestina Lake and Cold Lake formations overlain, in North Alberta Basin, by a thin layer of lower Chinchaga (= Bear Rock) Formation. Ernestina Lake Formation consists of a basal red dolomitic shale 3 to 4 m thick followed by 10 m of limestone containing ostracodes and a cap of about 6 m of anhydrite. The overlying Cold Lake Formation consists of up to 50 m of halite with 6 m of basal red shale. The salt is confined to the sub-basins and is not found over the arch. Cold Lake salt, like Lotsberg salt, has an abnormally low bromine content and is likewise thought to indicate dissolution by fresh water and reprecipitation. Near Meadow Lake Hingeline, Bear Rock sequence is represented by the upper part of Meadow Lake Formation, a mixed clastic and carbonate coastal facies.

The ostracode *Welleria meadowlakensis*, typical of Ernestina Lake Formation, is also found in the Fitzgerald Formation, a dolomite and dolomitic limestone overlying La Loche Formation – the basal Paleozoic sandstone and regolith – on Slave River (Fig. 4D.8). Ernestina Lake and Fitzgerald, as defined by Norris (1963), are synonyms. The Fitzgerald has also yielded the giant ostracode *Moelleritia* and the coral *Planetophyllum planetum*, the association strongly suggesting a Zlichovian-Dalejan age (A.E.H. Pedder, pers. comm., 1985). The *Welleria* fauna is considered by Braun (1977) to be a facies equivalent of his DM3 Assemblage found in the Funeral Formation on Snake River.

Golden Embayment

Between West Alberta Ridge and a Purcell landmass, whose dimensions are unknown, to the west, there was a deep embayment whose connection with the open sea lay near the present town of Golden, British Columbia. The basal Devonian unit, Cedared Formation, consists of well-bedded dolomites (in part detrital) and dolomitic quartz sandstones passing vertically and laterally into Burnais Formation, an evaporite-carbonate-shale facies including mineable deposits of gypsum. The Cedared contains charophyte oogonia, fish remains, brachiopods, gastropods, and ostracodes thought to indicate an Eifelian age (Norford, 1981). It is probable that a part at least of the Cedared and Burnais formations belongs to Bear Rock sequence.

Ebbutt break

The upper limit of Bear Rock sequence is marked locally by an unconformity called the Ebbutt break ("break" signifying a lacuna), which is most evident in the region of Peace-Athabasca and Tathlina arches. G.K. Williams (pers. comm., 1985) thought he detected the Ebbutt break over Keele Arch, within the evaporitic succession normally mapped as Bear Rock Formation. North of the arch, W.B. Brady and G.O. Raasch (unpub. rep.) postulated a lacuna on faunal grounds. In neither case is the evidence conclusive. Westward, toward the carbonate platform margin, it is generally agreed that the boundary between Bear Rock and Hume-Dawson sequences is transitional.

HUME-DAWSON SEQUENCE (EIFELIAN-LOWER GIVETIAN)

Deposition of the Hume-Dawson sequence started with a major transgression, which for the first time in the Devonian crossed Meadow Lake Hingeline and extended beyond the present outcrop of the Devonian on the shores of Lakes Manitoba and Winnipegosis. It ended with a regression and, in the central part of the region, emergence (Watt Mountain unconformity). The sequence is punctuated by lacunae (Fig. 4D.9, see also Correlation Charts in pocket), the evidence for which is strongest in the region of Tathlina Arch, suggesting that the effects of eustasy have been modified by uplift. Over most of the region, the sequence begins with *pedderi* Zone (Eifelian) and ends with middle *varcus* Zone (Givetian), spanning T-R Cycles Id, Ie, If, and part of IIa (Johnson et al., 1985), but it starts earlier (Cycle Ic) on MacDonald Platform.

Tectonic-sedimentary provinces

The tectonic-sedimentary framework of Hume-Dawson sequence (Fig. 4D.7, in pocket) is less differentiated than that of the preceding Bear Rock sequence. Porcupine Platform was gradually drowned during the transgression; only a few small reefal areas (Ogilvie Formation) survived to the end. Richardson Trough and Selwyn Basin continued as negative regions receiving argillaceous sediment (Road River Formation). That part of Yukon Territory and District of Mackenzie lying east of the shale basins can be treated as a single tectonic-sedimentary province, Northern Open Shelf, occupied by a carbonate platform that is remarkably uniform, both in thickness and lithology (Hume Formation). MacDonald Platform is continuous with the Northern Open Shelf, but the deposition of a transgressive carbonate sequence (Dunedin Formation) began earlier there than on Mackenzie Platform.

Across the region, in a belt stretching from Great Slave Lake to the Cordillera in the latitude of Peace River, a great carbonate barrier complex was built (Keg River Barrier). It comprises an assemblage of rock units occupying a discrete tectonic-sedimentary province that includes Tathlina Arch. At its southwest end, the barrier approaches Peace River Arch, to which it is anchored as to the abutment of a dam – the simile is apt, for the barrier kept open-marine waters out of Elk Point Basin during its periods of evaporative drawdown (Maiklem, 1971) or even dessication.

Behind Keg River Barrier, and flanked on the southwest by West Alberta Ridge, an elongated downwarp (Elk Point Basin) extended across northern Alberta and Saskatchewan into western Manitoba and northwestern North Dakota. This basin was at first the site of mixed carbonate-siliciclastic-redbed deposition and later of carbonates and evaporites. Immediately behind the barrier lay Black Creek Basin, separated from the rest of Elk Point Basin by Hay River Bank. West of West Alberta Ridge, which persisted as a landmass throughout the sequence, Golden Embayment continued to receive sediment.

Porcupine Platform

Ogilvie Formation

The widespread Emsian portion of Ogilvie Formation was much reduced in areal extent by the rise in sea level at the beginning of sequence deposition, although sedimentation continued in the interior regions of the platform without recorded interruption. Perry et al. (1974) noted the presence in the Ogilvie Formation of the *Warrenella kirki* and *Leiorhynchus castanea* zones overlain by *Stringocephalus* spp., indicating late Eifelian and early Givetian ages. The *Stringocephalus* spp., which allow correlation of this part of the Ogilvie Formation to the reef foundation facies of the Ramparts Formation, are known from three localities (Fig. 4D.7, in pocket) where early Givetian reefs were constructed after the rest of the platform had been drowned.

McCann Hill Chert

Black, fissile, siliceous shale of the McCann Hill Chert (commonly oxidized and brick red in outcrop) continued to onlap the west side of the platform throughout Hume-Dawson deposition. The upper part appears to be homotaxial with the Horn River Group, which was deposited on the open marine platform to the east. The formation is over 200 m thick in southwestern Ogilvie Mountains, over 300 m on the eastern flank of Nahoni Range, and nearly 400 m in a subsurface section between the Givetian reefs (Norris, 1985). One of the few diagnostic fossils recovered from the McCann Hill Chert of this region is *Agoniatites* cf. *A. fulguralis* (Whidborne), which provided a Givetian dating.

The McCann Hill Chert and Horn River Group in this region and in the adjacent Richardson Trough were deposited in a starved basin. Stratal discordance, observed at some outcrops, is thought to be the result of submarine erosion; lacunae in the biostratigraphic record are a natural concomitant of such conditions. Bassett and Stout (1968) interpreted the unconformities within the shales as the product of subaerial erosion, a view maintained by A.W. Norris (1985).

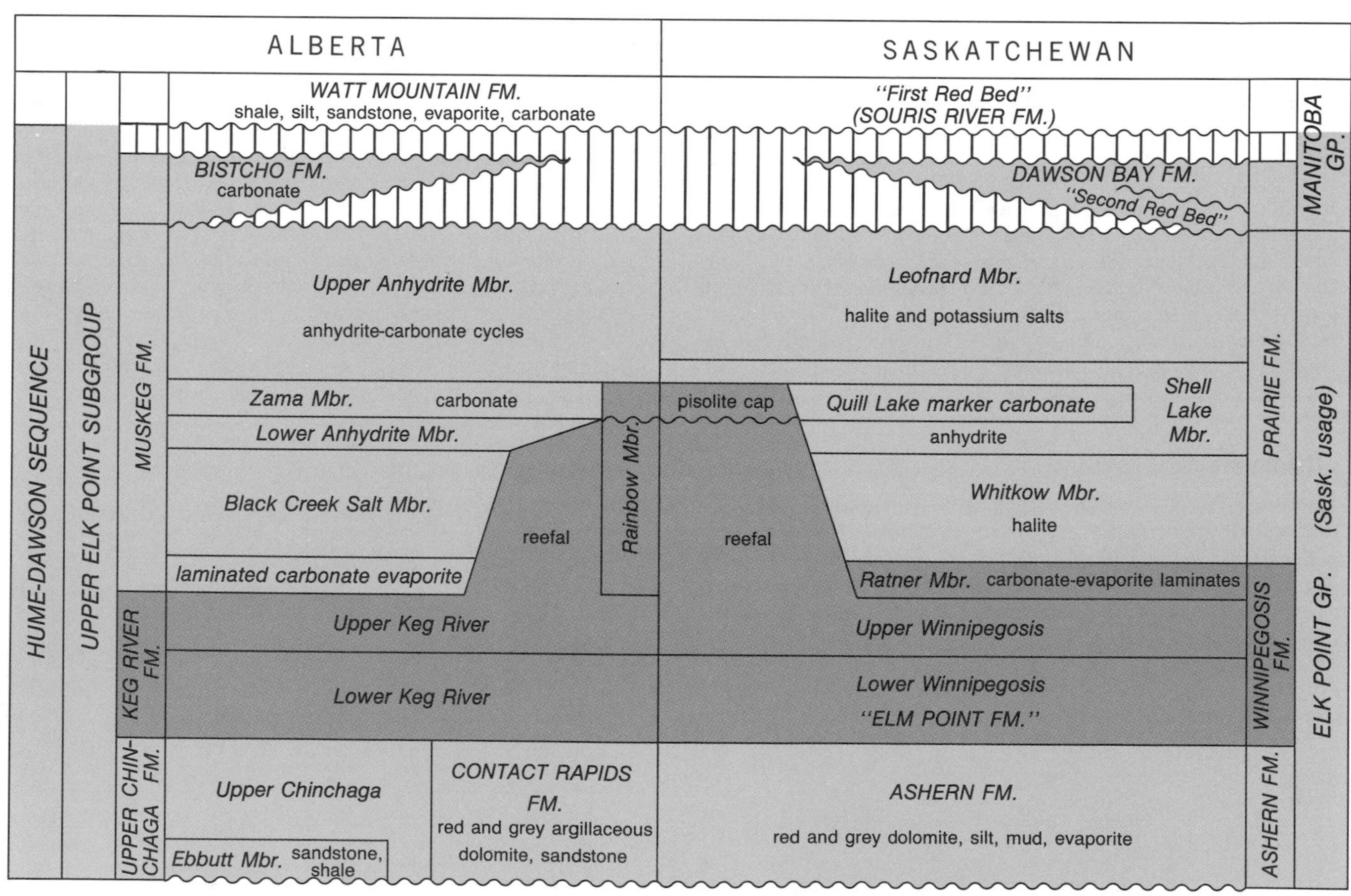

Figure 4D.9. Comparative nomenclature of Hume-Dawson sequence in Alberta and Saskatchewan.

Northern Open Shelf

Hume Formation

The northern part of the open shelf tectonic- sedimentary province includes the type locality of the Hume Formation (Bassett, 1961). At Hume River, the formation is 122 m thick and consists, in ascending order, of 81 m of limestone and calcareous shale, 27 m of very fossiliferous limestone, 5 m of shaly limestone and calcareous shale and 9 m of fossiliferous limestone. Correlation by physical tracing and biostratigraphy has shown the Hume to be equivalent to the Nahanni Formation (Hage, 1945), or jointly to the redefined Nahanni Formation (Douglas and Norris, 1961) and an underlying, more argillaceous, Headless Formation. Nahanni and Headless are therefore treated as members of the Hume Formation. Their mutual relationship is diachronous. The Hume is composed of alternating, wedge-shaped bodies of more and less argillaceous limestone. G.K. Williams (pers. comm., 1985) has shown that these clinothems dip west-northwest.

The contact of the Hume Formation on the underlying Gossage Group is in places transitional but usually sharp, though concordant. Toward the craton interior, a lacuna develops and increases, but over most of the Northern Open Shelf it appears to be negligible. The occurrence of *Moelleritia canadensis* just a few metres below Hume Formation east of Aklavik Arch suggested to W.B. Brady and G.O. Raasch (unpub. rep.) that there is a lacuna in that region. However, though typically Emsian, *M. canadensis* also occurs in the base of Dunedin Formation on MacDonald Shelf, above Stone Formation with *serotinus* to *costatus* zone conodonts, suggesting that it ranges into the Eifelian. Hume Formation can be traced at least as far north as Campbell Uplift, where limestone mapped as "Cranswick" by Norris (1985) contains a typical Hume fauna (A.E.H. Pedder pers. comm., 1984). Farther north, the reservoir carbonate in two oil wells on Aklavik Arch is probably Hume Formation, but this is a faulted region and the oil has migrated into the Hume from a Cretaceous source (Paul Brooks, pers. comm., 1986). The Hume maintains a rather uniform thickness of between 100 and 200 m over most of the region, but thickens to nearly 400 m near the platform margin northeast of Meilleur River Embayment.

Horn River Group

Hume Formation is succeeded by the Horn River Group (Pugh, 1983), a wedge of shale, containing reefs locally, that thins from over 400 m at its eastern outcrop to 20 m in the west (Fig. 4D.1, Section 2, in pocket). The lowest unit of the group is the Bluefish Member of Hare Indian Formation, a dark, bituminous, highly radioactive shale containing, in the lower 5 to 10 cm, concentrations of tentaculitids, styliolinids, plant fragments, bivalves, and inarticulate brachiopods. Conodonts belonging to the *"Spathognathodus" ormistoni* faunal unit (earliest Givetian) occur in the Bluefish Member (Chatterton, 1978).

Overlying the Bluefish Member is the upper member of Hare Indian Formation, a basin-margin clinothem sequence dominated by a succession of greenish grey shale, marl, and medium-grey calcisiltite, ranging in thickness from zero to over 200 m (at the mouth of Mountain River).

Ramparts Formation, which follows at certain localities, consists of a carbonate platform (reef foundation) upon which are reefal build-ups. From faunal evidence it appears that, whereas much of the foundation belongs in Hume-Dawson sequence, all the reef growth occurred during deposition of the ensuing Beaverhill-Saskatchewan sequence. Near its northern limit, where it outcrops at The Ramparts of Mackenzie River (see Fig. 4D.13, in pocket), Ramparts foundation consists of 45 m of bedded limestone with "floating" stromatoporoid fragments, overlain by the reefal build-up. The bedded limestone contains three faunas: at the base is *Stringocephalus asteius* and *Rensselandia (Ectorensselandia) laevis*; *Stringocephalus aleskanus* occurs about 25 m higher; and *Leiorhynchus hippocastanea* at the top, just beneath the reef (Crickmay, 1970). This gives an age range from middle or lower *varcus* to lower *disparilis* Zone. At the southern end of the reef foundation, where it outcrops on tributaries to Mountain River (Braun, 1977), the *Stringocephalus*-bearing beds are in a more argillaceous facies known as the "Shale Ramp" (Muir et al., 1984), which is followed immediately by a biostromal foundation containing a *disparilis* conodont fauna. *Leiorhynchus hippocastanea* has not been found in this area. At the top of the Shale Ramp, a layer of large pyrite nodules is associated with hardgrounds, which might possibly represent the Watt Mountain unconformity. At Norman Wells, the argillaceous limestones of the "Shale Ramp" give way to shales and are included in the Hare Indian Formation. *Grypophyllum mackenziense*, an upper *disparilis* Zone coral that occurs in the Ramparts reef, has been found in the reef foundation at Powell Creek (Lenz and Pedder, 1972), suggesting that reef initiation was later there than at The Ramparts.

East of Norman Wells, the Hume is unchanged but the overlying Hare Indian thickens towards what must have been the source of the siliciclastic basin fill. This is an example of a basin constructed within the shelf interior by advancing carbonate platforms. All such basins seem to have filled with wedge-shaped bodies or clinothems in which siliciclastic and carbonate sediments alternate. Other examples will be described from the Waterways (Beaverhill Group) and Ireton formations (Woodbend Group).

West of Little Chicago on lower Mackenzie River, part of the Hare Indian Formation is a fossiliferous, clean quartz sandstone that falls in the *aleskanus* and *hippocastanea* zones (MacKenzie et al., 1975). Sandstone influx in Alberta did not take place until later.

South from Norman Wells, Hume Formation continues unchanged until just north of Keg River Barrier. In that region (Law, 1971) the lower part of the Hume changes through a dolomite facies (Willow Lake Formation) to an evaporitic (sabkha) facies (upper Chinchaga Formation). The line of facies change is shown on Figure 4D.7, in pocket. The upper part of the Hume continues south to become Keg River Formation (locally known as "Lonely Bay" from its outcrop on Great Slave Lake).

Horn Plateau Reefs

In the upper Mackenzie River area, small, high-profile reefs of Horn Plateau Formation ("pinnacles") grew on the drowned Hume platform in front of the Keg River Barrier. Fuller and Pollock (1972) showed that there were two periods of reef growth at the Fawn Lake reef followed by lengthy emergence. The first stage consisted of about 50 m of build-up of crinoidal and stromatoporoidal calcarenites, capped by "calcified coral-rock". Vopni and Lerbekmo (1971) interpreted most of the reef as a biolithite with reef-flat calcarenites at the top. Contemporaneous calcirudites accumulated on the flanks. "Erosion surfaces" (?hardgrounds) formed both at the top of the low reef stage and about 15 m below it. Many of the reefs encountered by drilling in this region grew no higher than the "low reef" stage. Others, like the Fawn Lake reef, continued to build to a second stage. This is construed as evidence of a drop in sea level after the first stage, followed by a rapid rise that drowned some of the first-stage reefs. After second-stage growth, fissures were opened at Fawn Lake down to 90 m below the present-day top of the reef. This has been interpreted as evidence of another sea-level fall, although it is possible for fissures to open in reefs without emergence.

The reef of both the high-level and low-level stages contains *Spathognathodus brevis* and *Polygnathus linguiformis mucronatus*. Reef flank deposits thought to be equivalent to the first stage yielded *Ancyrognathus walliseri*. These data suggest a *varcus* age for both low- and high-level stages and in general a correlation with the Rainbow Member of Keg River Formation (see "Black Creek Basin"). The fissures contain "conodont forms referable to the *Schmidtognathus hermanni – Polygnathus cristatus* zone" (Fuller and Pollock, 1972), and thus belonging to the Beaverhill- Saskatchewan sequence. Flank deposits lying on top of those already described also contain this fauna. Johnson et al. (1985) concluded that the mid-Horn Plateau break was equivalent to the Watt Mountain break: this interpretation is rejected here. As will be demonstrated in the discussions of Black Creek Basin and Interior Elk Point Basin that follow, reef growth was interrupted everywhere, recording a fall in sea level at about the middle of the Hume-Dawson sequence. This fall was no doubt accentuated by evaporative draw-down behind the Keg River Barrier, but it was most probably triggered by a drop in sea level outside the barrier to a level below the sills of the channels that passed through it. The fall is here named the Black Creek Event, and the simplest and most plausible explanation of the mid-Horn Plateau break is to equate it with that event. The later lowering of sea level, supposed to have opened the fissures, is equated with the sub-Watt Mountain unconformity. The sea returned in *hermanni-cristatus* time, the reefs were drowned again, and the fissures filled with sediment.

Off-reef deposits

Dark shales of the Horn River Formation surround the reefs. The basin in which Horn Plateau reefs grew up was starved of clastic influx until the clinothems of the Frasnian Fort Simpson Formation downlapped westward into the basin. The fondothems of Fort Simpson Formation are included for mapping purposes in the Horn River Formation, which is thus a highly condensed deposit. These relationships have analogues in the Clarke Lake and Norman Wells areas (Williams, 1983). The Horn River attains a thickness of nearly 200 m in outcrops just west of Great Slave Lake.

Selwyn Basin and Richardson Trough

Shales assigned to the Road River Formation continued to accumulate in the western basins during Hume-Dawson deposition (Fig. 4D.7, in pocket). At the westerly shale-out of Hume Formation in the southeast part of Meilleur River Embayment (Noble and Ferguson, 1971), the Nahanni Member, 225 m thick, exhibits two prograding tongues of richly fossiliferous limestone with an intercalated lime mudstone and a basal shale. The top of the Nahanni falls in the zone of *Leiorhynchus castanea*, characteristic of the basal part of the Hare Indian Formation on Peel Platform. The two prograding tongues fall in the upper Hume zone of "*Billingsastrea verrilli*" or *Carinatrypa dysmorphostrota* (see Table 4D.1). The underlying beds, assigned to the Headless Member, comprise a shale-limestone couplet 80 m thick lying within the lower Hume *Eoschuchertella adoceta* Zone. The limestones display a wide variety of fossil assemblages comprising brachiopods, corals, trilobites, and stromatoporoids; the matrix of these boundstones, rudstones, and floatstones is almost entirely micritic, indicating low-energy conditions. Elsewhere the formation is simply a lime mudstone, usually with some admixture of clay, and rarely porous. The pure, resistant carbonates of the Nahanni Member pass laterally into argillaceous

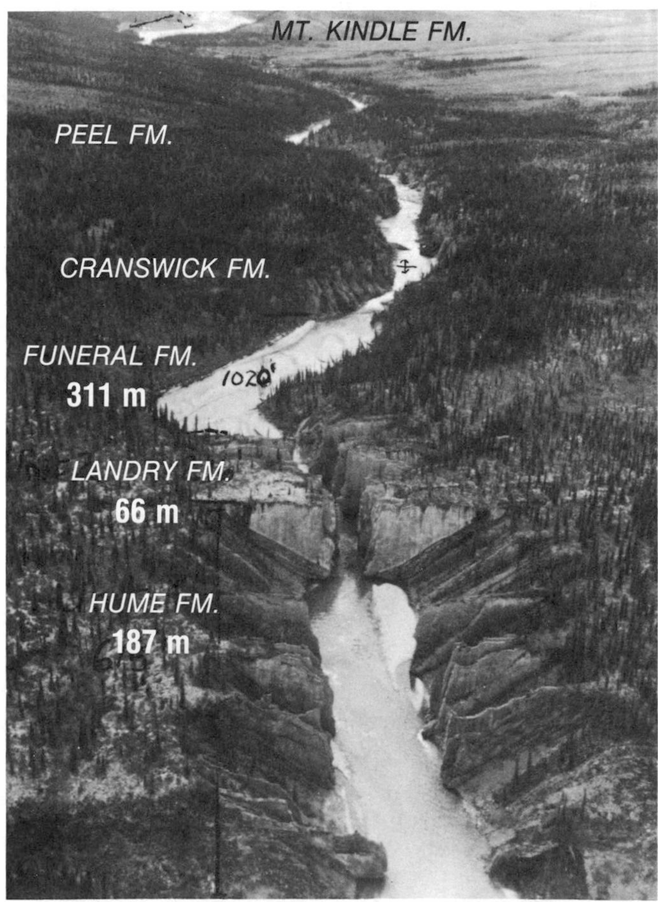

Figure 4D.10. Devonian formations of the frontal Mackenze Mountains, river 14 km east of Snake River, N.W.T.; view southward (photo - P.F. Moore and D.W.R. Wilson).

limestones and shale over a distance of less than 1500 m. No barrier developed at the margin, but there seems to have been a definite break in slope on the ramp ("epeiric slope" of Noble and Ferguson, 1971). At another Hume shale-out, 650 km to the northwest (Fig. 4D.10) the full set of Hume fossil zones is represented by Headless facies, 187 m of richly fossiliferous limestone and argillaceous limestone comprising about 10 couplets. This unit becomes increasingly shaly into the basin until, west of Snake River, it is mainly calcareous shale, there called Mount Baird Formation by Norris (1985).

MacDonald Platform

Dunedin Formation

Hume Formation can be traced into the Dunedin Formation of MacDonald Platform, the region between Keg River Barrier and the western shale-out (Fig. 4D.7, in pocket). But whereas the Hume seems to be the product of a transgressive-regressive cycle, the Dunedin was claimed by Morrow (1978) to be entirely transgressive, being composed of a dolomitic wackestone shoreface facies overlain by a grainstone-wackestone offshore facies. At the northern end of the platform, however, the "Middle Devonian", mapped as Dunedin, bears witness to sabkha conditions (Snowdon, 1977).

The base of the Dunedin is transitional to Stone Formation in the northwestern part of MacDonald Platform, near the British Columbia-Yukon Territory border, where it is nearest the carbonate platform margin. Collections of two-holed crinoids (Morrow, 1978), the giant ostracode *Moelleritia canadensis* (W.B. Brady and G.O. Raasch, unpub. rep.; Taylor and Mackenzie, 1970) and conodonts (Chatterton, 1978) from Dunedin Formation in the northern two-thirds of MacDonald Platform indicate an early Eifelian age, older than the Hume. Raasch (in W.B. Brady and G.O. Raasch, unpub. rep.) concluded that there is a major, though as yet physically undetected, lacuna within the Dunedin. A different explanation favoured here is that transgression began earlier in this region than elsewhere, possibly as a result of differential subsidence. Toward Peace River Arch, the basal contact of the Dunedin becomes sharp and unconformable (Taylor and Mackenzie, 1970). Thin sandstones (equated with the Ebbutt sandstones of the Tathlina Arch area) appear at the base in the southern third of the shelf.

The top of the Dunedin Formation becomes younger toward Peace River Arch. In the northwest it may be as old as *pedderi* Zone (*Eoschuchertella* cf. *adoceta*), while to the south it gets as young as *curtigladius* Zone (*Leiorhynchus castanea*). It then merges with the Keg River Barrier.

Besa River Formation is the name applied on MacDonald Platform to a Middle Devonian to Carboniferous extension of Road River Formation that overlies Middle Devonian and younger carbonates. The Besa River fills the stratigraphic position occupied to the northeast by Horn River Group and the Upper Devonian Fort Simpson Formation jointly. Pelzer (1966), using mineralogical zones, showed the Horn River portion of the Besa River Formation thinning from 210 m at the Keg River Barrier to less than 30 m in the mountain outcrops.

Keg River Barrier

The entire barrier complex is here assigned to the Keg River Formation, except where a distinct Bistcho Formation ("Sulphur Point" of some authors) can be recognized. For reasons given earlier, the Presqu'ile dolomite is no longer treated as a formation. Pine Point and Sulphur Point formations, which outcrop on the shores of Great Slave Lake, have not been extended into the subsurface because in neither case is common usage in accord with the stratotype. The barrier is mapped in Fig. 4D.7 (in pocket; for details, see Williams, 1981a, 1981b; McAdam, 1983).

During the Ebbutt regression that terminated the underlying Bear Rock sequence, the entire barrier belt was above sea level. During the pre-Watt Mountain regression that terminated the Hume-Dawson sequence, Elk Point Basin to the south dried up, but the open shelf tectonic-sedimentary province did not.

The barrier crosses several cratonic sags and swells, the largest of which is Tathlina Arch. Above the latter, the barrier front has been obliterated by pre-Watt Mountain erosion (Williams, 1981a), but farther west it is intact. The barrier is indented from the open-shelf side by the deep Cordova and Utahn embayments, the latter containing Horn Plateau-type reefs. Both the reefs and the sinuous margin of the barrier are reservoirs for gas. In most cases a dual reservoir has been created by a Slave Point reef superimposed from the succeeding sequence. In some cases both Keg River and Slave Point formations have been metasomatically replaced by Presqu'ile dolomite.

The barrier runs south from Evie Lake (Fig. 4D.7, in pocket) in the subsurface and outcrops in the thrust-faulted foothills, where it has been included in the Dunedin Formation ("Pine Point" Formation of Taylor and Mackenzie, 1970; Morrow, 1978). In this region, debris-flow breccia from the barrier have been traced thousands of metres into the Besa (Road) River shales (Macqueen and Thompson, 1978). South of Peace River the back-reef evaporite basin closes and the barrier loses its identity in a rather poorly exposed belt of outcrops. At the southern end of MacDonald Platform, the Dunedin thins southward from 250 m to 60 m and oversteps Stone Formation onto Ordovician quartzites forming the west flank of Peace River Arch (Geldsetzer, 1982). In that area, the lower part of the Dunedin is siltstone to coarse-grained sandstone and the upper part limestones with *Stringocephalus*. Eastward-directed paleocurrent indicators characterize the basal sandstones, suggesting the presence of a westerly landmass, possibly a salient of Peace River Arch.

Upper Chinchaga Formation

The basal unit of Hume-Dawson sequence in the barrier complex is the upper Chinchaga Formation, equivalent in age to the lower part of the Hume Formation. It begins with the mid-Chinchaga sandstones (Belyea, 1971), equated with the Ebbutt Member of the Northern Open Shelf. The sandstones are followed by evaporites and dolomites of sabkha origin, forming a diachronous blanket about 50 m thick under the barrier belt except where paleotopographic eminences remained as islands.

Keg River Formation

As the sea deepened, the Chinchaga evaporites gave way to the open marine carbonates of Keg River Formation. The Keg River is normally divided into a lower, blanket-like, crinoid-bearing limestone member and an upper member of reefs and their off-reef equivalents. The core of the barrier complex (Fig. 4D.11) is the organic barrier facies consisting of very fossiliferous dolomite, including fragments of frame-builders such as massive and tabular stromatoporoids, dendroid stromatoporoids, and corals. Skall (1975) regarded this as a cementation reef. Northward, toward the open shelf, a detrital lime-sand facies passes, via an argillaceous off-reef facies, into the *Tentaculites* facies – a dark brown-to-black, argillaceous and very bituminous limestone. This unit, containing *Leiorhynchus castanea*, can be correlated with the Evie Lake Member of the Cordova Embayment and with the Bluefish Member of Hare Indian Formation. Evie Lake Member is overlain by Buffalo River shale, a dark green to bluish grey, fissile, limy shale that correlates with the upper member of the Hare Indian Formation of the open shelf and with the Otter Park Formation found in front of the barrier to the west. Hare Indian Formation equivalents thin away from the barrier and eventually merge with the overlying Canol Formation (= Muskwa) to form the undifferentiated Horn River Formation (see Fig. 4D.17). Therefore, despite the introduction of a complete set of new formation names in the barrier region, the geology of the fore-reef turns out to be very similar to that in the Norman Wells area. Behind (south of) the organic barrier facies, back-reef dolomite, some parts of which show relict stromatolites, other parts of which contain *Amphipora*, is transitional to the Muskeg anhydrites.

Bistcho Formation

Above the Muskeg Formation, up to 95 m of interbedded dark limestones containing gastropods and *Amphipora* and fine-grained, light-coloured, micritic dolomite is assigned to Bistcho Formation. It overlies Muskeg Formation discordantly and is locally replaced by Presqu'ile dolomite. Over the central part of the barrier, the Bistcho Formation consists, where not completely replaced by Presqu'ile dolomite, of light cream to white limestone. It seems concordant with the underlying Keg River Formation, but the distribution of lithofacies at the contact between Keg River and Bistcho formations "offers permissive evidence for an intervening period of non-deposition and partial subaerial exposure" (Rhodes et al., 1984). The Bistcho does not outcrop at or even near Sulphur Point, but many authors refer to it as "Sulphur Point Formation", a term abandoned by Skall (1975). When McCamis and Griffith (1967) erected Bistcho "Member" they considered it a facies variant of Muskeg Formation, but Tranchant (1975) showed that at least part of the solution of Black Creek Salt took place between the deposition of the Upper Anhydrite Member of the Muskeg and that of the Bistcho, so that there must be a break between the two. This evidence, together with the "permissive evidence" quoted above, has led to the Bistcho being given formation status. The Bistcho includes a thin wedge of carbonates "for which the name Windy

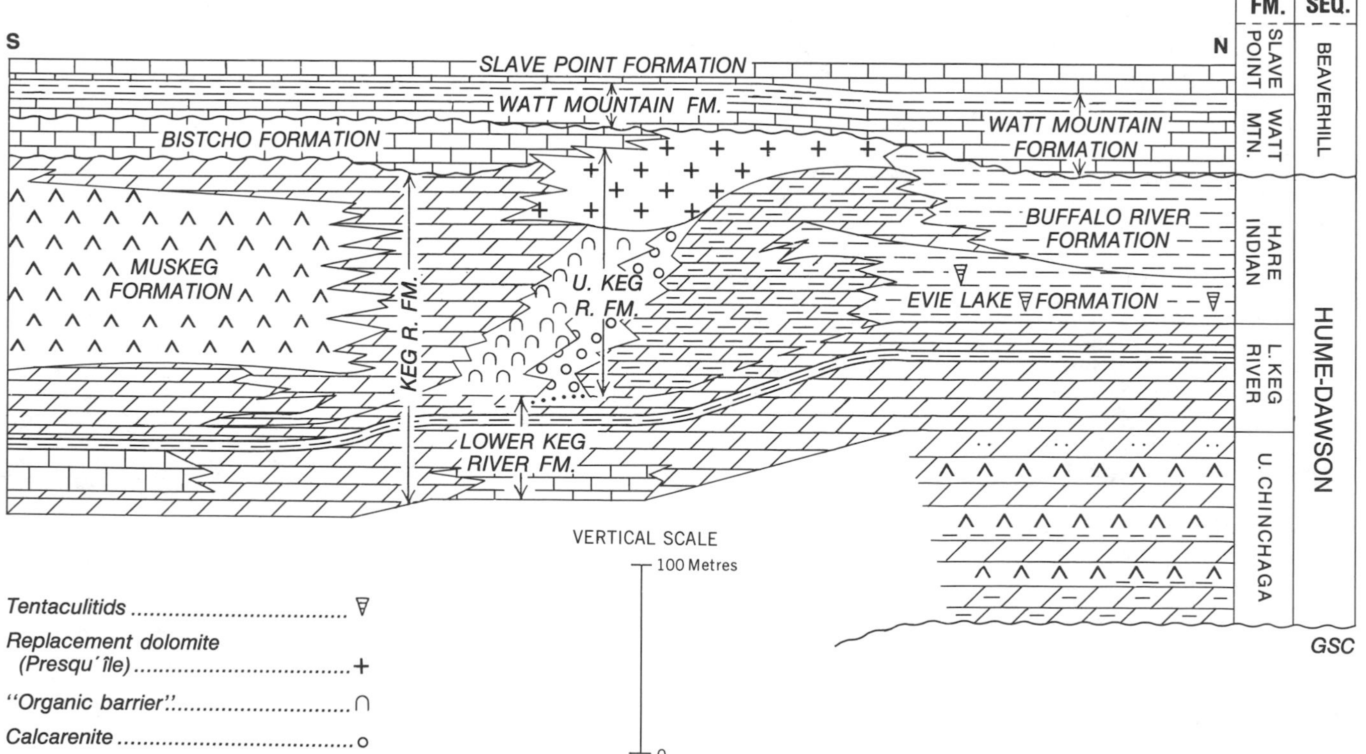

Figure 4D.11. Section across Keg River Barrier near Pine Point.

Point will be proposed" (Rhodes et al., 1984). The outcrop of Windy Point Formation on Great Slave Lake appears to be the place where Kindle and others found *Stringocephalus burtini* (Norris, 1965). Emergence, probably dating from the end of middle *varcus* time, took place following Bitscho deposition. This event was marked by a combination of eustatic lowering and epeirogenic differentiation so that, when siliciclastics of the Watt Mountain Formation began to be deposited, an entirely new paleogeographic configuration prevailed. The Watt Mountain is the basal unit of the Beaverhill-Saskatchewan sequence, to be described below.

The Pine Point lead-zinc orebodies on the south shore of Great Slave Lake are mainly hosted by paleokarst features in the upper part of the barrier complex. Karst networks coincide with two areas of preferential Presqu'ile dolomite development that mainly replaces Bistcho ("Sulphur Point") Formation. Other orebodies occur within the Keg River Formation (Rhodes et al., 1984). Fluid-inclusion data from sphalerite and dolomite and the organic geochemistry of the bitumens contained in the host rocks indicate that some mineralization and dolomitization events represent thermal anomalies with respect to the host rocks (Macqueen and Powell, 1983).

Black Creek Basin

Upper Chinchaga Formation

The upper Chinchaga Formation is part of the formally defined Lower Elk Point Sub-group, but the group boundaries do not coincide with the sequence boundaries. At the base is a detrital zone in which quartz sandstones are especially abundant near Tathlina and Peace River Highs. The detrital zone or "Ebbutt Member" is traceable by gamma-ray log over most of the basin. The overlying beds consist of about 50 m of anhydrite and red beds. The formation thus comprises a complete cycle [4].

Keg River Formation

When deeper water returned to the area as part of the Hume-Dawson transgression, the evaporitic sabkha deposits of Chinchaga Formation gave way abruptly to subtidal, argillaceous and bituminous limestones and dolomites with a fauna of crinoids, corals, and tentaculitids, forming the lower Keg River member. A degree of hypersalinity is recorded in its upper part by the high boron content of the clay residues (Dumestre, 1969). The lower Keg River is very probably a source rock for oil trapped in the overlying reefs. A strong gamma-ray marker separates lower Keg River member from the basal, crinoidal part of upper Keg River member, during the deposition of which normal marine waters returned, and swarms of mounds dotted the sea floor. Reefs developed on these mounds in a number of sub-basins. Within each sub-basin the reefs had a slightly different growth history. Between the reefs a much thinner layer of non-fossiliferous, dolomitic laminites (containing later secondary anhydrite) was deposited. This layer, rich in organic matter, is also a probable source rock for the oil in reef reservoirs.

The most important sub-basins in which reefs developed are Rainbow and Zama, where the reefs, and the carbonate beds draped over them, are highly productive of oil. The reefs are assigned to "Rainbow Member" and "upper Keg River member" respectively.

Keg River Barrier was only intermittently effective in restricting circulation and, as the transgression continued, the reefs grew as they did during the same period in the region to the north. However, after the drop in sea level recorded in the Horn Plateau reefs and here called the "Black Creek Event", Black Creek Basin was cut off from the open sea and evaporative draw-down took place. Between the reefs, carbonate-anhydrite laminites were deposited, followed by a halite bed, the Black Creek Salt Member of Muskeg Formation. The salt has been largely removed by later solution from most of the sub-basins except Rainbow. The interpretations offered here agree with those of Bebout and Maiklem (1973), but a different interpretation, with two periods of salt deposition, that at Zama being earlier than the salt at Rainbow, has been proposed by Tranchant (1975). During the period of evaporative draw-down, during which laminites and halite were deposited, the inter-reef areas remained subaqueous but the tops of the reefs were exposed to vadose alteration. Contrary views postulating sabkha or playa origins for the laminites and non-vadose diagenesis of the reefs have been expressed (Fuller and Porter, 1969; Husain and Warren, 1985; Schmidt, 1971).

Following deposition of the Black Creek Salt, freshening of the waters in the inter-reef and bank margin areas led to deposition, first of sulphates (lower Anhydrite member of Muskeg Formation, Fig. 4D.9) and then of laminar unfossiliferous dolomites (Zama Member of Muskeg). In Zama sub-basin, upper Keg River reef is overlain either directly by Zama Member, an important oil reservoir, or by lower Anhydrite member followed by Zama Member. In Rainbow sub-basin, the upper part of the Rainbow Member is the product of shoaling over the reefs during the deposition of the dolomite-anhydrite couplets which, in other sub-basins, are assigned to Zama Member of the Muskeg Formation.

The phase of reef growth that took place after the Black Creek event was referred to by Schmidt et al. (1977) as the "second cycle", but Bebout and Maiklem (1973) treated it, in effect though not explicitly, as a "third" cycle since they had already postulated a lowering of sea level in the early stages of mound growth, prior to the Black Creek event, on the timing - if not the intensity - of which all are agreed. This last cycle [4] of reef growth consists of about fifteen carbonate cycles [5], represented by carbonate-evaporite couplets in off-reef areas.

The Keg River reefs are commonly called "pinnacles", from their appearance on cross-sections with greatly exaggerated vertical scale. Their shape is, in fact, like that of an inverted soup plate with sides sloping up to 30 degrees and heights of about 200 m. They are mostly small, with areas in the range of 2 to 350 ha, but some are as large as 2000 ha. The distribution of biofacies in the reefs is interpreted as evidence of an atoll-like build-up (Elloy, 1972) in spite of the low proportion of frame builders preserved in the reef rock. Marine cementation played an important part in stabilizing the reef (Schmidt et al., 1977) but this does not necessarily exclude the activity of frame-builders.

Muskeg Formation

In Black Creek Basin, the Muskeg Formation comprises Black Creek Salt Member, lower Anhydrite member, Zama Member, upper Anhydrite member, and Bistcho Member (Fig. 4D.9). All except the latter were partly contemporaneous or in reciprocal relation with reef growth, as explained above. The formation will receive fuller treatment in the discussion of the interior part of Elk Point Basin. About twenty cycles [5] succeed the fifteen that locally were accompanied by reef-building.

Interior Elk Point Basin

The interior part of Elk Point Basin was separated from Black Creek Basin by the Hay River carbonate bank. Previously published regional maps have mostly been of total Elk Point Group and thus emphasize the early differentiation of this basin into sub-basins controlled by Peace-Athabasca Arch and Meadow Lake Hingeline (Fig. 4D.4, in pocket). The influence of the arch continued to be noticeable at the start of Hume-Dawson sequence; subsidence, and hence reef growth, was much greater north of it than over it. By Muskeg-Prairie time, however, the influence of the arch and hingeline had so diminished that the basin may be treated as one. The sequence of deposits is remarkably uniform from one end of the interior basin to the other, but this uniformity is obscured by the coexistence of two sets of formation names (Fig. 4D.9). Note that the correlation between Bistcho Formation and Dawson Bay Formation is not firmly established. In order to simplify discussion, the Saskatchewan nomenclature will be used throughout interior Elk Point Basin.

Figure 4D.12. Argillaceous, red- and pale-orange-weathering dolomite of the Ashern Formation in quarry 7.2 km west of Mulvihill, Manitoba, overlying dolomite breccia of the Middle Interlake Group. Staff graduated in feet and tenths (photo - A.W. Norris, GSC 170397).

Ashern Formation

Lying above Ordovician and Silurian beds south of Meadow Lake Hingeline is the basal unit of Hume-Dawson sequence, known as the Ashern Formation (Fig. 4D.12). It consists of red and grey microcrystalline dolomite with minor quartz silt, and contains some nodular anhydrite. The redbeds occupy the lower paleotopography and are overlain everywhere by a continuous layer of slightly argillaceous grey beds. The average thickness in Saskatchewan is 15 m, and the maximum 30 m in a minor depocentre southwest of Saskatoon (Paterson, 1973). Drillholes near the outcrop in Manitoba penetrate a full section of about 13 m. Thicknesses of up to 50 m occur in west-central North Dakota, where halite was deposited (Lobdell, 1984). North of the halite occurrence, breccia and other evidence of exposure occur between Ashern and Winnipegosis formations (Perrin, 1982). Thus the Ashern comprises a complete cycle [4] and is not merely the transgressive phase of the overlying Winnipegosis Formation.

The Ashern Formation contains rare crinoid ossicles and calcispheres but has yielded no age-diagnostic macrofossils, although some have mistakenly been attributed to it. Rare conodonts suggesting a late Eifelian age have been recovered from beds transitional to the overlying Winnipegosis (Norris et al., 1982) and corroborate correlation of the Ashern with upper Chinchaga Formation.

Contact Rapids Formation

North of Meadow Lake Hingeline, the Ashern thickens and passes into the Contact Rapids Formation, which also has a lower red unit (10 m) and an upper grey unit (35 m) of argillaceous dolomite. On the northeast and southeast flanks of Peace River Arch, thick feldspathic sandstones (a tongue of La Loche Formation) occupy this interval.

Winnipegosis Formation and the lower part of the Prairie Formation

The Winnipegosis is the extension of Keg River Formation into the interior Elk Point Basin, where it undergoes no significant change except in name. In general architecture, the Winnipegosis consists of (1) a bank complex passing landward to siliciclastics adjacent to the landmasses of Peace River, West Alberta Ridge, and in the United States, Sweetgrass Arch; (2) several starved, constructional basins within which numerous small-diameter reefs grew to elevations of up to 100 m above the sea floor. Earlier models depicted a central basin with large banks and small patch reefs intermixed, but, where seismic reflection profiles have been released (Gendzwill, 1978), it appears that most of these isolated banks were artifacts of contouring without seismic control. Most wells that penetrate the Winnipegosis have been drilled on seismic anomalies and thus bias the sampling. Without seismic information, the reef basins cannot be accurately mapped and their outlines on Figure 4D.7 (in pocket) are schematic. Although a field on the Peace-Athabasca Arch produces from the bank complex, oil and gas are rare in the Winnipegosis Formation of Canada. A few basinal reefs near the International Boundary have had significant oil production. Over large areas of the basin, the bituminous laminites, thought to be the principal source, may not have reached maturity.

Winnipegosis Formation, including Elm Point Formation, which is a facies variant of its lower member, ranges in age from *curtigladius* Zone to lower *varcus* Zone, i.e., from late Eifelian to early Givetian (Norris et al., 1982).

Growth of the Winnipegosis reefs took place in stages analogous to those described in Black Creek Basin, i.e., foundation, build-up, exposure (Black Creek event). In contrast to Black Creek Basin, no further carbonate deposition seems to have taken place over the reefs after the Black Creek event, presumably because of higher salinities in the interior. The history of reef growth may be quite complicated in detail. For example, the carbonates of the foundation stage are punctuated by a varying number of tentaculitid-rich shale layers and capped by a shallow-water oncolite rock, recording marked oscillation in water depth (Jones, 1965; Perrin, 1982).

The build-up stage has been interpreted as inorganically cemented (Wilson, 1984), but Perrin (1982) produced extensive evidence of sediment-binding activity at the reef margins, including encrusting activity by blue-green algae, *Thamnopora*, stromatoporoids, and chaetitids. She suggested that some of the basinal reefs were atolls, with a framework margin and a lagoonal interior characterized by packstone and wackestones with algae, *Amphipora*, calcispheres, and ostracodes.

Carbonate laminae with bituminous partings forming Unit I of the Ratner Member (Wardlaw and Reinson, 1971) were deposited in the inter-reef areas contemporaneously with reef growth. Laminae have been traced for thousands of metres by matching profiles in well cores (Davies and Ludlam, 1973). The hypersalinity that occurred during the Black Creek event may have given rise to the overlying anhydrite laminae (Unit II of Ratner Member), either through the direct precipitation of anhydrite or, more probably, by diagenetic alteration of existing carbonate laminae. Dolomitization and replacement by anhydrite occurred on the flanks of reefs and banks at this time. The laminites were followed by halite of the Whitkow Member. The Whitkow and the Black Creek Member of Black Creek Basin have been equated because they occupy the same position relative to the other members of the reef assemblage. Corrigan (1975), however, concluded from his correlations that the lowest preserved salt of the Prairie Formation north of Peace-Athabasca Arch was deposited later than the lowest preserved salt south of it.

At the maximum lowering of sea level the upper parts of the reefs underwent vadose diagenesis. The result was what Wardlaw and Reinson (1971) called the Winnipegosis Pisolite Cap and Perrin (1982) called the Pisolite Dolomite Facies. Wilson (1984), considered the objects described by others as vadose pisoliths to be oncoliths and made no comment on the evidence of downward growth and polygonal fitting adduced by Wardlaw and Reinson.

Adjacent to the tops of the reefs and overlying the Whitkow Member is an outwash of interbedded pellet packstone, stromatolitic mudstone, pisolitic grainstone-packstone, and *Amphipora* wackestone (Quill Lake Marker Beds). These beds are interbedded with the anhydrite of the Shell Lake Member. They are interpreted as the product of a post-Black Creek sea-level rise, and are thus analogous to the Zama Member of Black Creek Basin.

Prairie Formation – Leofnard Member

The Prairie Formation (or "Prairie Evaporite") of the interior Elk Point Basin is equivalent to the Muskeg Formation of Black Creek Basin. In contrast to the Muskeg, which consists of anhydrite and dolomite, the Prairie Formation consists mainly of halite. The lower part of the formation is contemporaneous with reef growth and has been described in that context. The upper part, known as the Leofnard Member (Jordan, 1968), contains valuable deposits of potassium salts. The Leofnard, with a maximum thickness of 210 m in east-central Alberta, is divisible into a number of regionally correlatable cycles (Klingspor, 1969; Busson, 1974). The contrast between the starved-basin-fill character of the lower Prairie Formation and the widespread cycles of the Leofnard is striking. The present distribution of the Leofnard salts is mainly the result of solution that started in the Devonian and continues to the present day, as evidenced by salt springs. The Saskatoon earthquake of 1909 was attributed to collapse of strata as a result of salt removal. Removal of salt by solution has played an important role in creating structural closures in the overlying strata.

The potash deposits of the Leofnard Member in southern Saskatchewan are thought to be the largest deposits of high-grade potash in the world (Fuzesy, 1982). They occur within the uppermost 60 m of the member, at depths of over 900 m. They extend a few kilometres into Manitoba and southward at increasing depth into North Dakota and northeastern Montana. Individual potash-rich beds are widespread and consistent in grade and thickness. Local thickening occurs where carnallite, rather than sylvite, is the dominant potash mineral. Four zones of halite-bearing potash minerals have been distinguished. From the base up, these are the Esterhazy, White Bear, Belle Plaine, and Patience Lake sub-members. These sub-members are commonly between 6 m and 15 m thick and separated by 1 to 45 m of halite with minor potassium-rich layers. The potash members and interbeds thin southwesterly. The Patience Lake sub-member was truncated by pre-Dawson Bay erosion. Beneath the unconformity there is a secondary cap of halite up to 30 m thick, known as the "salt-back", the presence of which is essential to keep underground water out of the mines.

Dawson Bay Formation

An impure red dolomite ("Second Red Bed") followed by a shallowing-upward cycle [4] of carbonates capped by an evaporite succeeds the Prairie Formation and is called Dawson Bay Formation (Dunn, 1982). Its base is an unconformity, beneath which marker beds of the Prairie Formation are truncated (Klingspor, 1969; Fuzesy, 1982). The upper boundary of the Dawson Bay is placed beneath the "First Red Bed" of the Souris River Formation and may be disconformable. There was emergence at the end of Dawson Bay deposition and a sea-level rise at the start of

Souris River deposition; whether or not there was accompanying epeirogenesis is uncertain, because of correlation problems to be discussed later.

The Second Red Bed Member, average thickness 4 m, consists of a reddish brown to greenish grey dolomite mudstone with minor quartz and clay; it is veined with halite and contains anhydrite nodules. It represents a transgressive deposit in which the effects of an advancing sea on a highly soluble substrate produced a mixture of carbonate, evaporite, and siliciclastic sediment. At the outcrop in Manitoba the Second Red Bed is embraced by the Mafeking Member which, although part of Dawson Bay Formation, in fact includes the residuum of Prairie Formation from which the soluble salts have been leached. Dunn (1982) showed that there was a lacuna, perhaps accompanied by earth movement, between the Second Red and the overlying Burr Member.

The Burr Member, average thickness 20 m, is characterized by bioturbated dolomitic limestone with corals, bryozoans and crinoids, and numerous hardgrounds. A low-energy shelf environment is indicated.

The Neely Member maintains a thickness of 15 m to 18 m from central Saskatchewan to Manitoba, where it is rather more argillaceous. Unfossiliferous, argillaceous limestone at the base is followed by a bituminous limestone with gastropods, stromatoporoids, corals, and *Stringocephalus*. A higher energy shoal is indicated. Still further shallowing led to the restriction of the fauna to ostracodes and small gastropods and eventually to algal stromatolites with anhydrite and evidence of emergence at the top of the member.

The uppermost member is the Hubbard Evaporite, a halite bed attaining a maximum thickness of 19 m. Its occurrence is limited to a depocentre east of the third meridian (106°). Its position in the sequence and the admixture of red dolomitic mudstone, anhydrite, and cryptocrystalline dolomite suggests deposition in coastal lagoons.

Dawson Bay Formation at outcrop is divided into three members (B, C, D) above the redbeds (Norris et al., 1982). The exact manner in which they relate to the subsurface members has not been worked out. Fossil collections from the outcrop include corals (Pedder, 1977), brachiopods and conodonts (Norris et al., 1982), and ostracodes (Braun and Mathison, 1982). The conodonts from all four outcrop members were assigned by Uyeno (in Norris et al., 1982) to the Middle *varcus* Subzone. The ostracode assemblages are characteristic of the Hare Indian/Ramparts transition in the Norman Wells area. The corals indicate a position in the *Rensselandia (Ectorensselanida) laevis* Zone (see Table 4D.1). Brachiopods include a form of *Rhyssochonetes aurora* similar to one in the middle part of the Keg River Formation (Evie Member). *Stringocephalus* does not occur in the outcrop belt, possibly because of unsuitable facies. In the subsurface, Braun found ostracode assemblages that occur in the Ramparts Formation of the Norman Wells area. The youngest of these assemblages is also found in the base of the Swan Hills Formation (see Beaverhill-Saskatchewan sequence, below). Despite these facts, Braun and Mathison (1982, p. 45) concluded that "In determining the upper age limit of Dawson Bay Formation, it follows that its ostracode fauna must be older, and at the most, barely equivalent to the lowermost Slave Point of northern Alberta . . . "

The Dawson Bay Formation thins westward and becomes unrecognizable somewhere in eastern Alberta. The interpretation adopted here is that it is equivalent to Bistcho Formation in the Keg River Barrier belt. Both contain *Stringocephalus* and show local discordance with the underlying evaporites. However, other correlations are possible, e.g., with the siliciclastic interval at the base of the Watt Mountain Formation or with the lacuna beneath it. An attempt has been made by some geologists (Williams, 1984; Johnson et al., 1985) to correlate Dawson Bay Formation with Slave Point Formation. This is hard to reconcile (1) with the fact that middle *varcus* Zone conodonts have been recovered from all four members of the Dawson Bay at outcrop (Norris et al., 1982), (2) with the presence of *Stringocephalus* in the subsurface, or (3) with the age of the beds overlying the Dawson Bay at outcrop, which contain the brachiopod *Atrypa independensis*, characteristic of the Slave Point (Norris et al., 1982).

Golden Embayment

Harrogate Formation

A unit of fossiliferous, Eifelian limestone, dolomite, shaly and nodular limestone, mudstone, and shale is named Harrogate Formation (Norford, 1981). In Lussier syncline it is 90 to 140 m thick. At one locality, the Harrogate passes up into rocks assigned to the Beaverhill-Saskatchewan sequence with no physical break but with a considerable biostratigraphical hiatus. Toward the postulated Purcell landmass to the west, the Harrogate passes to thick (circa 530 m) littoral or non-marine mudstone and dolomite without age-diagnostic fossils (Mount Forster Formation).

BEAVERHILL-SASKATCHEWAN SEQUENCE (UPPER GIVETIAN-UPPER FRASNIAN)

Beaverhill-Saskatchewan sequence comprises the rocks, spanning late Givetian to late Frasnian time, deposited between the sub-Watt Mountain unconformity and the sub-Sassenach unconformity. Transgressive conditions prevailed in the early part of the sequence, during which cyclic changes in water depth promoted two episodes of reef growth (Ramparts-Swan Hills, Leduc). In mid-Frasnian time, carbonate ramps with downslope reefs (Zeta Lake) prograded into the basins. Finally, in late Frasnian time, a regression brought nearshore and coastal sediments westward into Saskatchewan. The reefs of the Beaverhill-Saskatchewan sequence together comprise the largest reservoir of Devonian oil in Canada.

No important regional break occurs within this sequence, although local unconformities appear in both basinal and nearshore successions. It is divided, for purposes of illustration and discussion, into two map-units named after the lithostratigraphic groups that most closely match their boundaries: Beaverhill map-unit and Saskatchewan map-unit.

BEAVERHILL MAP-UNIT

Beaverhill map-unit was deposited while the sea advanced over the former sites of West Alberta Ridge and Tathlina Arch and up the flanks of Peace River Arch (Fig. 4D.13, in pocket); the evidence of its advance eastward onto the craton has been removed by later erosion. This transgression forms part of a North America-wide inundation named the Taghanic Onlap by Johnson (1970). The earliest conodont fauna recovered from the sequence came from the lower part of the Souris River Formation in Manitoba. It cannot be assigned precisely to any established zone, but conceivably may belong to the upper *P. varcus* Subzone or the *S. hermanni – P. cristatus* Zone or both (Norris et al., 1982). On these tenuous grounds the base of the sequence has been assumed to coincide roughly with the base of the *hermanni – cristatus* Zone. The map-unit thus comprises the upper part of T-R Cycle IIa and T-R Cycles IIb and IIc. The reference area for Beaverhill map-unit is in central Alberta, where it consists of two cycles [3] (Fig. 4D.14). The first cycle begins with a siliciclastic unit (Watt Mountain Formation) and ends at the top of the Slave Point Formation (which acts locally as the foundation for Swan Hills reefs). The beginning of the second cycle records a rapid sea-level rise, which promoted reef-building. As the rate of rise slowed, infill processes became dominant, and the argillaceous Waterways Formation prograded over basin and reef alike, except in the most positive areas flanking West Alberta Ridge. Finally, another blanket carbonate (Cooking Lake Formation) spread over all but the most basinal areas. Cooking Lake Formation is formally part of the overlying Woodbend Group and has been included with Saskatchewan map-unit, although at least that part of it below the Elstow Member belongs in the Beaverhill cycle.

It has not proved possible to match the cycles in Alberta with those in Saskatchewan, although broad faunal correlation has been established. There is a rather drastic faunal change, especially in the ostracodes, between the two cycles that make up Beaverhill map-unit. Although this has been attributed to the presence of an erosional unconformity (Braun, 1977; Norris and Uyeno, 1981), a more likely cause was the drowning of previously existing barriers to migration by rising sea level during the second cycle. This is in accord with Johnson's (1970) concept of "the end of North American provinciality". There is no coincidence between these events and the Givetian/Frasnian boundary, which falls within the Calmut Member of Waterways Formation (Fig. 4D.14; Uyeno, 1974).

The base of Beaverhill map-unit is well-marked in central Alberta. Eastward into Saskatchewan and Manitoba there is disagreement as to whether it should be chosen at the First Red Bed (this study) or the Second Red Bed (Williams, 1984; Johnson et al., 1985). Northward,

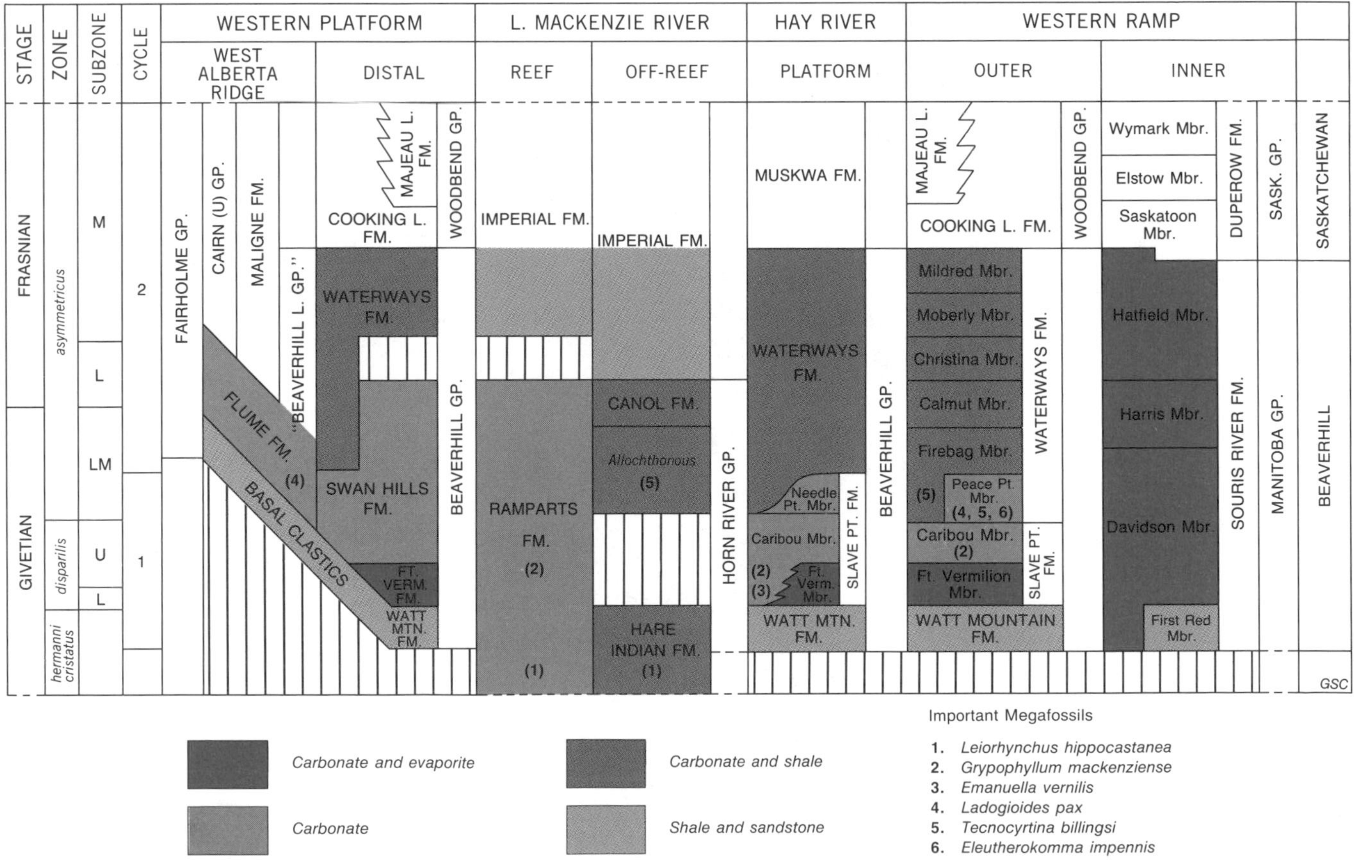

Figure 4D.14. Correlation chart of Beaverhill map-unit.

beyond Keg River Barrier, the boundary is lost in the condensed sediments of Horn River Formation, which were deposited in the starved basin that covered most of the Northwest Territories at this time. Only in the lower Mackenzie Valley area, where a siliciclastic wedge (Hare Indian Formation) had entered the basin earlier from the northeast and become the site of reef building, is the boundary discernable (though still disputed in detail). The "natural" top of the second Beaverhill cycle [3] lies beneath extremely widespread dark shales (Duvernay, Muskwa, and Canol formations), which represent the initial deposits of a deepened sea. Exceptions occur where reefs grew in response to the drowning (Leduc Formation) or where the waters were already deep and distant enough for the new influx of fondoform sediment to merge its identity with the old (the domain of Horn River Formation).

Tectonic-sedimentary provinces

Beaverhill sedimentation took place in the following provinces:

1) Eastern Ramp - originally fringing the west side of the Transcontinental Arch; having now preserved two facies belts
   (a) inner ramp (evaporitic)
   (b) outer ramp (limestone/shale);
2) Western platform and reefs - fringing Peace River Arch and West Alberta Ridge and eventually covering the latter;
3) Lower Mackenzie River reefs;

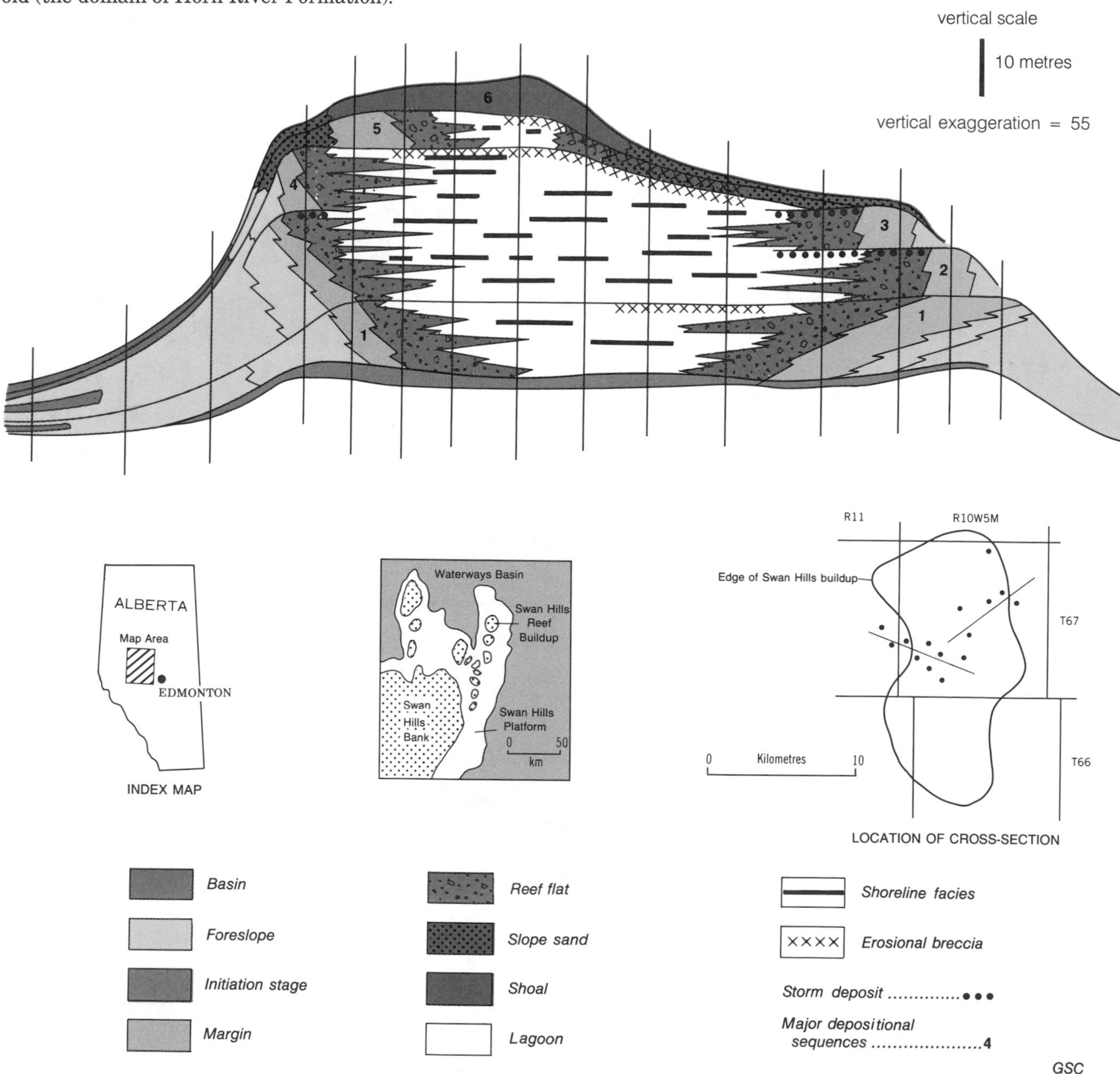

Figure 4D.15. Stratigraphic cross-section through the Swan Hills reef buildup (from C. Viau, 1983).

4) Hay River Platform - with a reefal rim overlying the shelf-margin barrier of the previous cycle; and
5) Northern Starved Basin.

Western Platform and Reefs

Watt Mountain and Yahatinda formations

The Western Platform was initiated on the south flank of Peace River Arch and on both flanks of West Alberta Ridge. The upper beds of the platform completely cover West Alberta Ridge and reduce Peace River Arch to an island. The end of the previous cycle had witnessed an emergence of these landmasses, accompanied by tectonic uplift and faulting (Kramers and Lerbekmo, 1967), and the consequent spread of a siliciclastic apron around them comprising Watt Mountain and Yahatinda formations, which include fluviatile, shoreline, and coastal lagoon sediments. Watt Mountain Formation flanks Peace River Arch. Its southern development records sedimentation on a coastal plain (Jansa and Fischbuch, 1974) and includes fluviatile sandstones of the Gilwood Member, which attain 18 m in thickness but are commonly about 5 m thick. The rivers depositing Gilwood sandstone flowed east and south to the sea, where barrier-face sands developed along the shore and formed reservoirs for large volumes of oil at Nipisi (Shawa, 1969) and Mitsue (Christie, 1971).

Fort Vermilion, Flume and Cairn formations

Gilwood sandstone bears witness to a pluvial period. Arid conditions soon returned together with the beginning of a transgression, causing coastal evaporite deposition to take place in lagoons and sabkhas on top of the former coastal plain. Anhydrites, with interbedded peloidal limestone, dolomite, and shale, up to about 11 m in thickness, make up Fort Vermilion Formation. The sea continued to deepen and a carbonate platform developed on the flanks of Peace River Arch and West Alberta Ridge, where the reefal east margin is gas-bearing. In the mountain outcrops this platform is called Cairn Formation; paleontological evidence (Maurin and Raasch, 1972) shows that it is a diachronous unit that becomes younger towards the crest of the arch, covering West Alberta Ridge by middle *asymmetricus* time. A lower cherty member, equivalent in age to the Calmut Member of Waterways Formation, is separately mapped as Flume Formation. The Flume and the overlying basal beds of the upper Cairn (often loosely referred to as "Cairn" by a tacit redefinition) together form the foundation for reefs and banks of the succeeding (Saskatchewan) map-unit. Reefal dolomites continuous with the Cairn also flank the Peace River Arch and contain a few small oil accumulations.

Figure 4D.16. Ramparts buildup at Powell Creek, frontal Mackenzie Mountains, N.W.T. Ca - Canol; BR - Bear Rock; Hu - Hume; HI - Hare Indian; Im - Imperial; MK - Mount Kindle; Ra - Ramparts; a - allochthonous beds (photo - R.A. McLean, ISPG 2660-1).

Swan Hills Formation

In the bight between Peace River Island and West Alberta Ridge, an extension of the fringing platform was built out over the drowned delta. Soon drowned itself, this platform became the foundation for reefs. Some completed their growth during deposition of Beaverhill map-unit (Swan Hills reefs), others, nearer the positive area of the ridge, continued to grow through deposition of the next map-unit (Leduc Formation reefs built on the Western Platform). Swan Hills reefs contain major reserves of oil. Jansa and Fischbuch (1974) have traced the development of the Judy Creek region from coastal plain to drowned reef. At the start, a carbonate blanket about 6 m thick spread evenly across the region. Upon this blanket there was differentiation into a marginal reef bank complex, whose location was controlled by the morphology of the underlying Gilwood coastal plain, and an interior shelf facies. These together formed the rimmed platform. Later, the platform was drowned and became the foundation for reef growth. The foundation comprises two cycles, the first containing *Grypophyllum mackenziense*, most characteristic of the upper *disparilis* Subzone, elsewhere found in the Slave Point Formation. An important faunal break occurs just above the base of the build-up stage. A cycle-break at that horizon marks the exit of the last "Middle Devonian" ostracode and seven species of stromatoporoid. Norris and Uyeno (1983) correlated this break with that observed between the Slave Point and the Peace Point Member of Waterways Formation.

The foundation and superposed platform reefs grew in pulses referred to here as cycles [4]. The number of cycles recognized in a reef depends on the criteria used to identify them; Figure 4D.15 is a typical interpretation. The principal processes contributing to reef cyclicity are: (1) lateral facies shift arising from rare reef-destroying storms; (2) lateral facies shift arising from the re-establishment of reef growth on an exposed and possibly eroded surface at the start of a new punctuated aggradational cycle; (3) shoaling cycles [5] unrelated to changes in the rate of sea-level change (autocycles). Facies-shift stages have chiefly been documented on the reef margins; simple shoaling autocycles with limited lateral extent (Wong and Oldershaw, 1980) are characteristic of the reef interior. There is no necessary correlation between the two. Besides facies shift, the following criteria have been used to delimit cycles by various investigators: hardgrounds, extinction of stromatoporoid species (Fischbuch, 1968); shoaling-upward sequences; redox cycles (which do not always coincide with the shoaling); cement paragenesis (Viau and

Oldershaw, 1984); vadose silt and cement (Havard and Oldershaw, 1976); and the presence of a layer of green clay at the base of many cycles.

Green "clays", with various, even dominant, admixtures of dolomite-, lime- and quartz-silt, in some instances containing finely disseminated pyrite, have been noted, particularly at the turning point of cycles, in nearly every Middle and Upper Devonian sequence in western Canada. Havard and Oldershaw (1976) attributed them to karst processes: Wendte and Stoakes (1982) argued that they had an external source and postulated storms as the transporting agent. Here they are considered eolianites. This explains their occurrence at the end of a shoaling sequence when carbonate sedimentation is brought to a halt and extensive areas of dolomitic sabkhas are exposed in the upwind landmasses. If the surface on which they are deposited is karstified, the dust will enter the cavities; if it is not, the dust will form a layer, which may be reworked during the subsequent transgression.

Waterways Formation

Interbedded argillaceous limestone and shale clinothems of Waterways Formation, prograding from the Eastern Ramp, downlapped onto the Swan Hills build-ups and eventually covered them (Sheasby, 1971). A glauconite-crinoid-brachiopod-rich calcarenite occurs at the base of some of the downlapping units (Jansa and Fischbuch, 1974). Sheasby's (1971) diagrams show interbedding of reef and off-reef material throughout the vertical growth of the reef; it is more likely, however, that the build-ups stood in clear water until at least the end of the first platform cycle (which is covered by a thin layer of dark grey open marine shale), or later. The sudden shift of the reef margin southwestward in the early stages of reef-building may be connected with the main arrival of clay. Waterways progradation did not reach the reef area until foundation construction had begun and probably not until the first cycle of foundation construction had ended.

Lower Mackenzie River Reefs

Ramparts and Canol formations

Discussion flows easily from the Western Platform to the lower Mackenzie River area because of the affinity between the two in their early development. Although initiated as early as *varcus* Zone in the northern province, active foundation-building was proceeding during *hermanni – cristatus* Zone and part of *disparilis* Zone in both provinces (Uyeno, 1978; Lenz and Pedder, 1972). The build-up of reefs (Fig. 4D.17), here included in Ramparts Formation ("Kee Scarp"), followed a pattern of cycles similar to those

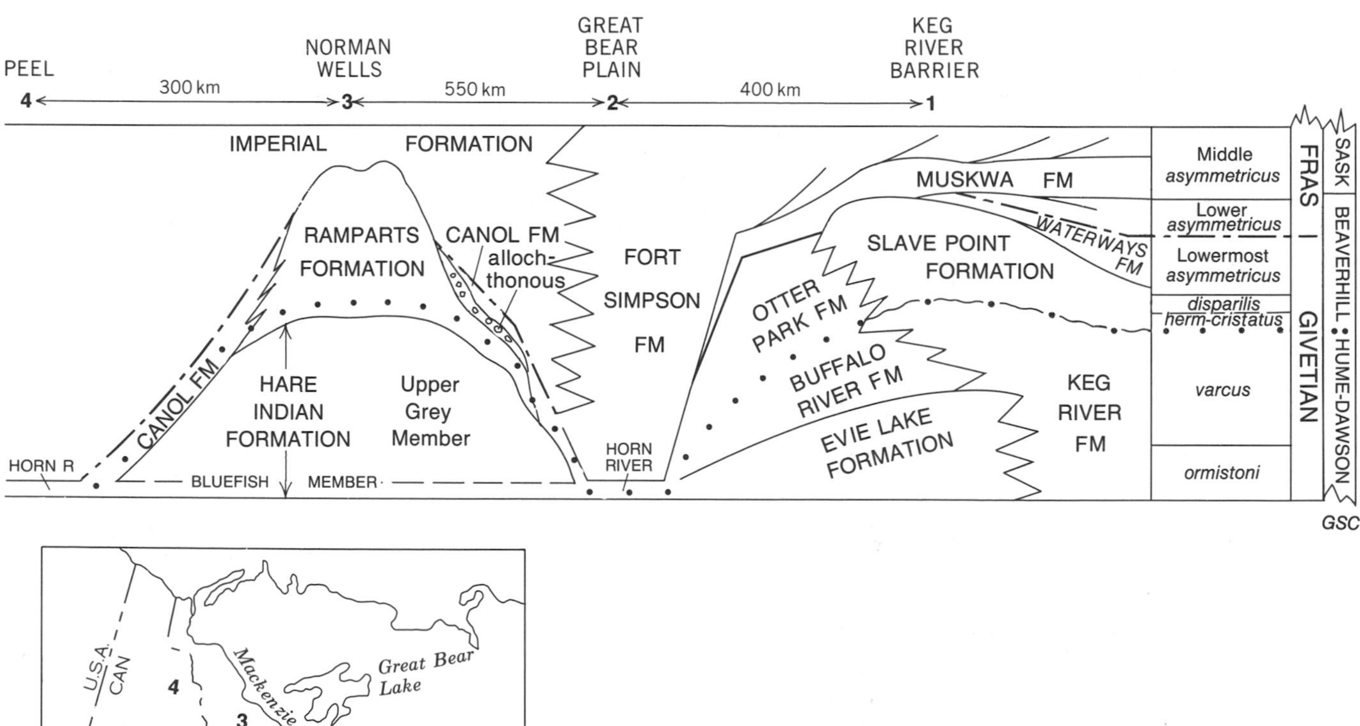

Figure 4D.17. Stratigraphic relationships in Northern Starved Basin during the Givetian Age.

described from the Swan Hills reefs, the number of interpreted cycles being dependent on the investigator (compare Muir et al., 1984; Fischbuch, 1984). The foundation in each of the Ramparts reefs consists of a biostromal unit 55 to 65 m in thickness; it is followed by 90 to 160 m of build-up. The reefs have boundstone margins in which tabular encrusting stromatoporoids alternate with stromatoporoid rubble. The reef interiors contain cycles [5] of sub-tidal and tidal-flat deposits in which *Amphipora* is common. One of the reefs, at Norman Wells, is a major oilfield containing estimated initial recoverable reserves of $40x10^6$ m^3.

At Powell Creek (Fig. 4D.13, 4D.16), the flanking beds contain a good ostracode (Braun, 1966) and conodont (Uyeno, 1978) fauna. The tongue of Ramparts Formation at the base, as already mentioned, spans upper *varcus* and *disparilis* zones. Overlying it are allochthonous beds (reef debris) with a lowermost *asymmetricus* conodont fauna and the brachiopod *Tecnocyrtina billingsi*, a fossil characteristic of the Firebag Member of Waterways Formation (Fig. 4D.14, 4D.17). These are followed by the Canol Formation, dark platy and siliceous shales, which are generally poorly fossiliferous but with concretions bearing a lowermost to lower *asymmetricus* fauna characteristic of the lower part of the Waterways. Canol shale surrounds the reefs and interfingers with the flanking beds (Muir et al., 1984). It does not everywhere cover the reefs, which at their crests are surmounted by Imperial Formation (Fig. 4D.17). The off-reef beds thin westward, so that the Canol comes to lie upon a remnant of Hare Indian Formation. These relationships will be described later in this subchapter in connection with the Northern Starved Basin.

Figure 4D.18. Micritic limestone, about 6 m thick, of the Point Wilkins Member of Souris River Formation, highway roadcut 1.6 km south of Red Deer River bridge, near Dawson Bay, Lake Winnipegosis (photo - A.W. Norris, GSC 170480).

Eastern Ramp - Inner Facies Belt

Souris River Formation

The Eastern Ramp (Fig. 4D.13, in pocket) presumably passed eastward into a siliciclastic fringe around the Transcontinental Arch. No evidence for that remains in Canada, where pre-Mesozoic erosion has truncated the Beaverhill map-unit well to the west of its shoreline. The most easterly sediments, the Point Wilkins (Fig. 4D.18) and Sagemace members of Souris River Formation, outcropping on Lakes Manitoba and Winnipegosis, are open-marine limestones about 80 m thick (Norris et al., 1982). These rocks represent only the lower part of Beaverhill map-unit. In the subsurface, the Souris River (Lane, 1964) has been divided into three members. The lower, Davidson Member, a complete shallowing-upward cycle [4] (Price and Ball, 1971), is very similar to the Dawson Bay Formation. Following the basal redbeds, the carbonate passes from a brachiopod-crinoid facies to a stromatoporoid facies and then evaporites. At the depocentre of the halite (Fig. 4D.13) the member is 100 m thick. The middle, Harris Member, rather uniform in thickness at 30 to 40 m, introduces a succession of lenticular, shoal-water evaporitic cycles [4] a few metres thick, which characterize the succession in the area throughout deposition of Beaverhill and most of Saskatchewan map-units. The upper, Hatfield Member, about 40 to 50 m thick, is similar to the Harris Member in lithology, but can be mapped separately with the aid of well logs. The top of the Souris River Formation coincides with the top of the Waterways Formation at Waterways. In the southern townships there can be a slight discrepancy, depending on whether a rock-defined top or a marker-defined top is selected.

The inner ramp facies can be traced across southern Alberta into southeastern British Columbia, where it is represented by about 100 m of Hollebeke Formation (Price, 1965). The lower member, consisting of dolomite that is locally silty or argillaceous, contains sedimentary breccias that Price interpreted as relicts of former evaporites. The upper member consists of lime wackestone with sporadic solution breccias and stromatolites. Stromatoporoids, tabulate corals, and *Amphipora* are common in some beds; fossils indicate correlation with the upper Waterways.

Eastern Ramp - Outer Facies Belt

Slave Point and Waterways formations

Above the Slave Point Formation, the outer facies belt of the Eastern Ramp is entirely composed of open marine sediments. This distinguishes it from the inner facies belt whose cycles typically end with evaporites. At the base of the succession, Watt Mountain Formation is represented by a thin siliciclastic zone, not recognizable in outcrop. It is succeeded by a few metres of Slave Point Formation comprising a lower anhydrite (Fort Vermilion Member) and an upper shaly calcarenitic limestone (Caribou Member; Richmond, 1965). The upper surface of the Slave Point has been interpreted as an unconformity in the outcrop area (Norris and Uyeno, 1983), but the outcrop is disturbed by later solution collapse and the evidence is not conclusive.

Waterways Formation is divided into six members (Crickmay, 1957; Norris, 1973; Uyeno, 1974; Norris and Uyeno, 1981). At the base, Peace Point Member is a greenish grey, calcareous shale and argillaceous limestone, 0 to 6 m thick, preserved in low spots on the Slave Point surface. It contains a rich brachiopod fauna including *Ladogioides pax*, *Eleutherokomma impennis*, and *Tecnocyrtina billingsi*, which indicate an important change in the fauna from the Slave Point, but not necessarily a lacuna. The Peace Point yielded an *in situ* conodont assemblage probably of lowermost *asymmetricus* age. Peace Point Member has not been recognized in the subsurface, where it would be difficult to distinguish from Slave Point Formation. The succeeding and partly equivalent Firebag Member is a dark shale with a fauna similar to that of the Peace Point. It passes upward to pure limestone of the Calmut Member ("Calumet" of Norris), correlatable by well logs over thousands of square kilometres.

Christina Member represents a return to shale deposition and, together with the overlying Moberly Member, forms a second shale-limestone couplet that is widely traceable. Moberly Member locally swells into a mound-like feature several kilometres across (Fig. 4D.1, Section 5). The uppermost unit, Mildred Member, is more shaly and can be considered the base of a couplet that is completed with the Cooking Lake Formation of the Saskatchewan map-unit.

Hay River Platform

Watt Mountain, Slave Point and Waterways formations

North of Peace River Arch, Slave Point Formation developed into a reef-rimmed platform. Its interior lies under the present drainage basin of Hay River, which gives the platform its name. The history of the region starts with emergence at the end of Hume-Dawson deposition. Appreciable erosion took place along Tathlina Arch on the north side of the platform, where a karst surface developed (Williams, 1981a). Peace River Arch on the south side of the platform was uplifted, as discussed earlier, and Watt Mountain Formation was deposited on a fluvial plain flanking it (Rottenfusser and Oliver, 1977). The Watt Mountain thins from 60 m close to the arch to a metre or so of radioactive clay at the top of the karstified surface at distant localities. Fluvial and coastal sandstones of Gilwood Member of the Watt Mountain Formation are derived from the gneissic terrane of Peace River Arch, as demonstrated by their progressive loss of feldspar away from it. Limestone beds were deposited in fresh-water lakes on the fluvial plain, and fish remains are common in the coastal and estuarine deposits. At the east end of Peace River Arch, the rising sea of Beaverhill-Saskatchewan time advanced over a rugged gneissic landscape. Feldspathic sandstones and fringing reefs were deposited around the buried hills and became important reservoirs for oil.

As with the Western Platform, the onset of aridity caused deposition of the anhydritic Fort Vermilion Member of Slave Point Formation (Fort Vermilion is given formation status only where it underlies Swan Hills Formation). These anhydrites extend to the outcrop belt on the east, where they merge into a brecciated unit that includes the underlying Muskeg Formation. As sea level rose, evaporites onlapped the coastal plain followed by open-marine carbonates constructing a reef-rimmed platform (Caribou Member of Slave Point Formation). A short period of exposure is recorded in the platform interior (Crawford, 1972) before renewed deepening, as a result of westward tilting, put an end to carbonate deposition. Over much of its length the reefal rim follows the underlying Keg River Barrier. In many localities both formations have been replaced by coarse white dolomite of the Presqu'ile diagenetic facies. The rim hosts major reserves of gas in northeast British Columbia.

Where the platform margin outcrops at two localities in the mountains, organic growth managed to keep up with relative rise in sea level to create reefs of the Swan Hills Formation (Fig. 4D.13). Back from the rim, very low reefs or biostromes (Needle Lakes Member of Slave Point Formation; Richmond, 1965) developed during the final deepening stage, providing traps for gas at Botha River and the oil show at Melville River. At the east end of Peace River Arch a number of oil pools in the Slave Point may also be reefal, but local dolomitization plays a large part in trap formation.

During the infill stage, Waterways Formation was deposited as a prograding wedge derived from a (now eroded) easterly source. As interpreted here, the wedge, which thins to zero behind the reefal margin, downlapped at an extremely low angle onto the drowned surface of the Slave Point on Hay River Platform (Fig. 4D.17). On the other hand, Griffin (1965) saw the relationship as one of facies change between Slave Point and Waterways, while Bassett and Stout (1968) thought that the lacuna between Slave Point Formation and Waterways Formation was erosional.

In summary, Hay River Platform has a well developed foundation stage followed by a poorly developed build-up stage, except at the platform margin where thick reefs developed; the infill stage is represented by a thin wedge of limestone and shale.

Northern Starved Basin

Horn River Formation

Northern Starved Basin is the tectonic-sedimentary province lying outboard of Hay River Platform during Beaverhill deposition. It extends northward for over 1000 km and is interrupted only by the Lower Mackenzie reefs. At the southern end of the basin delineated by the margin of Hay River Platform (Fig. 4D.11), clinoform calcareous shales of the Otter Park Formation, overlying the Buffalo River and Evie Lake shales of the Hume-Dawson sequence, appear to be the slope equivalent of Slave Point Formation (Williams, 1983). The Otter Park is overlain by the dark radioactive fondoform shales of Muskwa Formation. Within a few kilometres of the platform margin, the entire packet, comprising the shales of both the Hume-Dawson sequence and the Beaverhill map-unit, has condensed to a radioactive shale about 30 m thick, named Horn River Formation, which continues northward around the west side of the Lower Mackenzie reefs as far as Richardson Trough. This packet of shale is exposed in the region between Norman Wells and Richardson Trough where, at several localities, it can be seen to comprise a microfossil-rich layer of Bluefish Member (Hare Indian Formation) overlain by siliceous platy shale typical of the Canol Formation. At localities as distant from each other as English Chief River (Braun, 1977) and Snake River (Fig. 4D.13), the Horn River with the above characteristics lies upon the thin *castanea* Zone, indicating that there has been no erosion beneath it, although Braun (1977) reported a significant faunal hiatus in boreholes east of English Chief River. In Richardson Trough, where the underlying strata are in shale facies, condensation is extreme, and what appears to be Horn River Formation has been found within a few decametres of dated Silurian shales. Occasional angular discordances within this packet (Norris, 1985) are interpreted as the products of submarine erosion (Pray, 1969; Pugh, 1983) or downlap, although some geologists (Bassett and Stout, 1968; Norris, 1985) have considered them evidence of subaerial erosion consequent upon uplift.

Pugh (1983) used the term "Horn River Group" (Fig. 4D.17) for the gross packet, including all of Hare Indian Formation, Ramparts Formation, and Canol Formation. West of Richardson Trough, a similar packet of siliceous shale, dated Givetian, has been assigned to the Chert and Shale member of McCann Hill Chert (Clough and Blodgett, 1984). It is probably continuous with the Horn River.

SASKATCHEWAN MAP-UNIT

This map-unit is named after the Saskatchewan Group with whose limits it coincides in Saskatchewan, although it includes some younger beds elsewhere. The unit corresponds roughly with TR Cycles IIc and IId of Johnson et al. (1985), but the base is a datum in the middle of a minor cycle, convenient for subsurface studies. Over most of the region, the top of the unit coincides with the extinction event at the end of the Frasnian (McLaren, 1982), which resulted in the virtual disappearance of the principal frame builders and radically changed the character of carbonate sedimentation for the remainder of the Devonian. Although locally there is a minor unconformity between the top of the unit and younger Devonian strata, there is no physical evidence of a terminal "catastrophe". In Saskatchewan and Manitoba, the Frasnian-Famennian boundary is believed to occur within the siliciclastic beds of the overlying Palliser sequence. Early deposition in this unit in central Alberta was characterized by deepening water and reefal build-up, later deposition by the progradation of marginal carbonate platforms with accompanying clinoformal basin fill. Throughout all but the very latest Frasnian time it seems that the sea continued its transgression of the Transcontinental Arch (interrupted, of course, by minor regressive pulses).

Tectonic-sedimentary provinces

Deposition of Saskatchewan map-unit was characterized by interplay between two regions (Fig. 4D.19, in pocket): a northern siliciclastic province dominated by detrital influx, and a southern region where carbonate deposition continued to follow the pattern established in the Middle Devonian. The northern siliciclastic wedge penetrated deep into the southern region during the deepening phase and then retreated. The southern region may be divided into an Eastern Platform and a Western Reef Domain (Fig. 4D.20). During the deepening phase the latter was

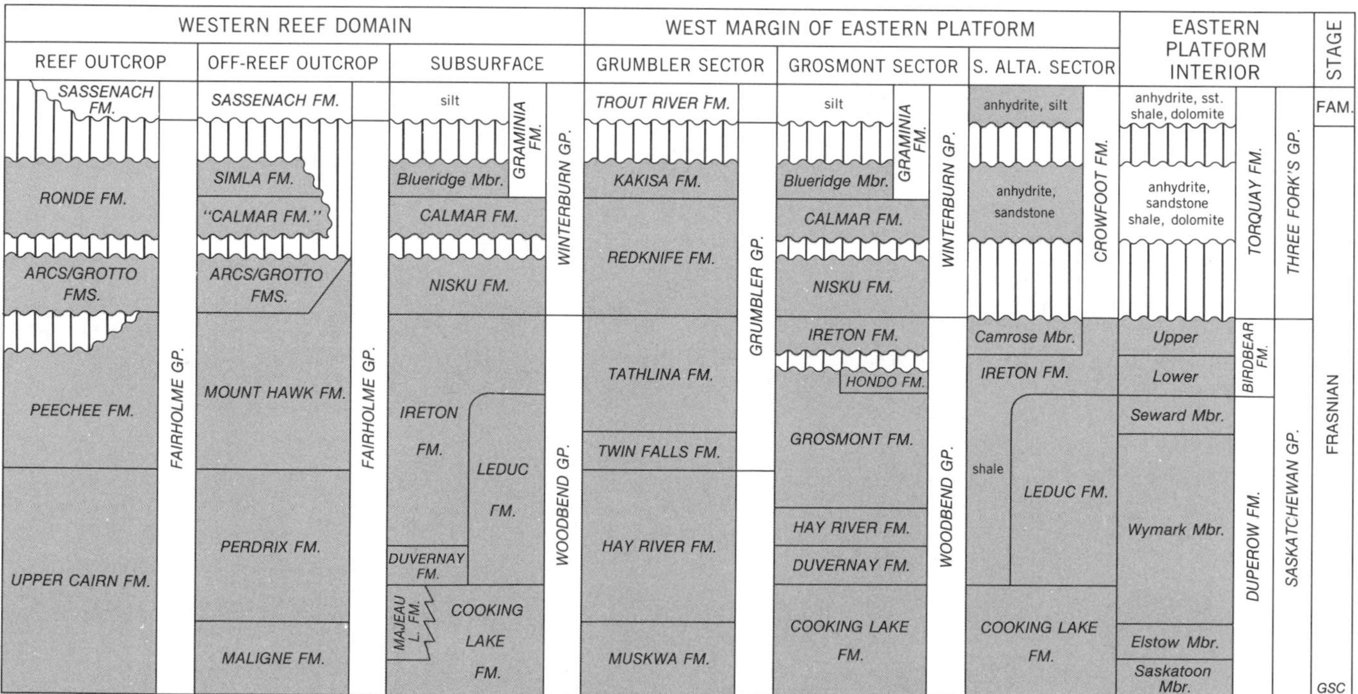

Figure 4D.20. Correlation chart of Saskatchewan map unit.

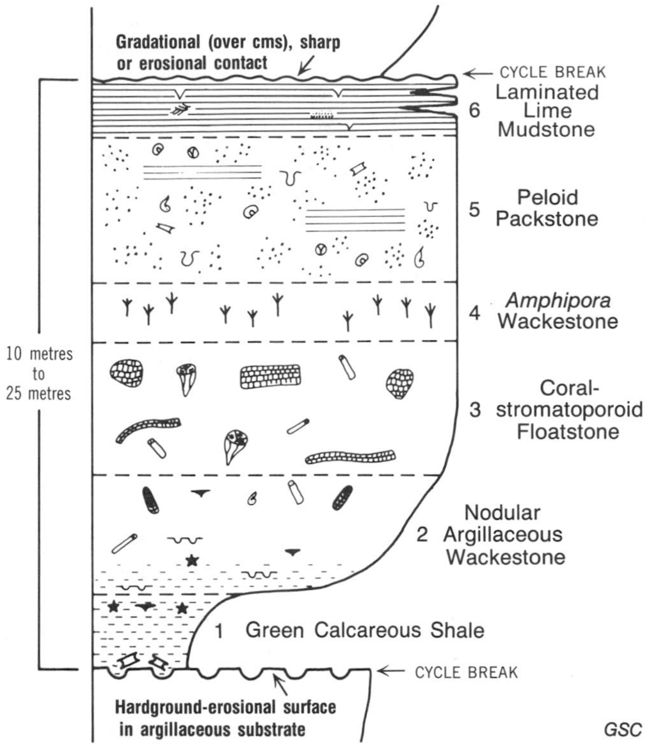

Figure 4D.21. Complete Grosmont cycle (from W.G. Cutler, 1983).

differentiated into platform reefs and shale-filled basins, while shallow-water carbonate-evaporite cycles continued to be deposited over the former. During the progradational phase, the Western Reef Domain was covered by a carbonate ramp with downslope reefs developed in relict basins, while the Eastern Platform continued to be the site of carbonate-evaporite cycles until finally covered with nearshore sediments having a high siliciclastic content. The map (Fig. 4D.19) shows the paleogeography of Alberta and British Columbia at the time of maximum differentiation, before deposition of the progradational and regressive sediments.

Eastern Platform Margin

The western limit of the eastern platform is preserved in three sectors, the South Alberta, Grumbler and Grosmont sectors. The platform itself extends into Saskatchewan, where it is truncated beneath the sub-Mesozoic unconformity.

South Alberta Sector

The South Alberta platform (or "shelf") margin has generally been assumed to be of the accretionary rimmed shelf type (Belyea, 1958), but its later development as a ramp-like progradation has been proved in at least six boreholes. Section 7 (Fig. 4D.1) crosses the margin at one of these sites; the situation is also illustrated in Figure 4D.24. The marginal rocks are classified

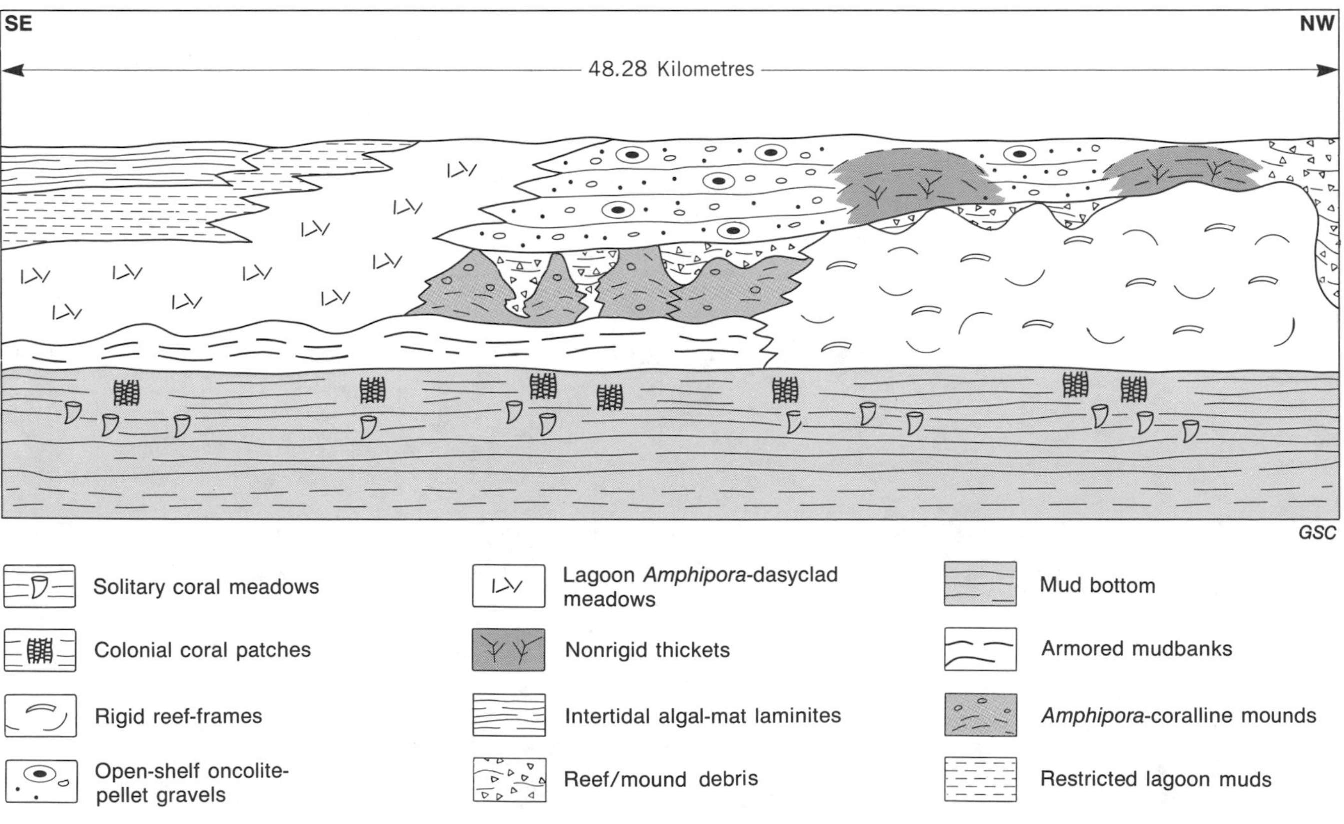

Figure 4D.22. Schematic distrubtion of facies in the Alexandra reef-complex (Grumbler Group) (from E.R. Jamieson, 1971).

stratigraphically with those of the adjacent Western Reef Domain, i.e., Leduc Formation overlain by a thin tongue of basinal shales of the Ireton Formation, covered by Camrose Member (of Ireton Formation) and Crowfoot Formation.

Leduc Formation is a structureless grey crystalline dolomite; it is thought to be "reefal", but the original fabric is largely obliterated. Camrose Member is a carbonate tongue that can be traced eastward into basal Birdbear Formation. Crowfoot Formation is a thin sandy anhydrite that represents the condensed shoreward facies of Winterburn Group (Belyea, 1958) deposited during the terminal regressive phase.

When the South Alberta platform margin first developed, it was as a linear rim with a reef-rimmed basin behind it. A hypersalinity event put an end to reef growth and filled the basin with salt (Belyea, 1958, Fig. 2). Blanket carbonate and anhydrite sedimentation then covered the salt and buried the evidence for this brief episode. Section 7 (Fig. 4D.1, in pocket) crosses a remnant of the salt and one of the reefs. Most of the salt was subsequently removed by solution.

Grumbler and Grosmont Sectors

The Grumbler-Grosmont margin is a distally steepened ramp in the interior of which evaporitic lagoons developed and over which siliciclastic sediments from time to time were spread; shale tongues thicken westward and thin sandstones apparently had an eastern source. In the subsurface, this ramp comprises, in ascending order, Grosmont Formation, Ireton Formation, and Nisku Formation. It was undoubtedly anchored to a platform of Cooking Lake Formation to the east, but this connection has been eroded and, when first encountered west of the subcrop, the Grosmont has prograded over a tongue of Hay River shale (usually called "Lower Ireton" by drillers).

Figure 4D.23. Louise Falls Member of Hay River Formation and Alexandra Member of Twin Falls Formation, Louise Falls, Hay River, N.W.T. (60°30′N; 110°14′W). The river is falling over the Louise Falls Member. The cliff in the background, about 35 m high, is formed by the Alexandra Member (photo - A.E.H. Pedder, ISPG 2582-3).

Grosmont Formation (Cutler, 1983) is composed of rather thick (10-25 m) shoaling-upward cycles [5] (Fig. 4D.21) grouped westward into clinothems. In the interior, Hondo Formation consists of anhydrite and dolomite up to 100 m thick, deposited in a shelf lagoon. The Grosmont contains important reserves of bitumen (estimated at 50 x 10^9 m^3) trapped up-dip along a subcrop margin beneath the sub-Cretaceous unconformity (Harrison, 1986).

The Nisku Formation is a thin platformal carbonate overlying the Grosmont, from which it is separated by a tongue of marine Ireton shale. It represents the progradational phase and therefore extends west of the mapped limit of the Eastern Platform.

Traced northward, the Grosmont-Nisku package emerges from beneath Mesozoic cover to outcrop as part of Grumbler Group (Belyea and McLaren, 1962). The group comprises, in ascending order, Twin Falls, Tathlina, Redknife, and Kakisa formations, of which the first two are equivalent to the Grosmont. Whereas Grosmont Formation is largely dolomitized, Grumbler Group is mainly limestone with considerable amounts of quartz silt and sandstone in some of the beds. Outcrops along Hay River (Fig. 4D.23; Jamieson, 1971) add a horizontal dimension to the shoaling-upward sequence described by Cutler (1983). Figure 4D.22 shows the complexity of the facies pattern; it is typical of this group that the abundant stromatoporoid and coral patch reefs seldom build up to heights of more than a few metres. Redknife and Kakisa formations, the deposits of the progradational phase, extend west of the mapped Eastern Platform margin. North of Great Slave Lake, the platform swings eastward and has been removed at the sub-Cretaceous unconformity, but the prograding carbonates, almost overwhelmed by sands and silts, are preserved in a narrow belt along Mackenzie River. The carbonates include coralliferous biostromes with Grumbler fossils.

Eastern Platform Interior

The platform interior is covered by a succession of carbonates, evaporites, and minor shales belonging to the Saskatchewan Group, comprising Duperow and Birdbear formations (Kent, 1968). The Duperow is further subdivided into named members (Fig. 4D.20) in southwestern Saskatchewan, and into numbered units in southeastern Saskatchewan (Dunn, 1975). Part of the overlying Torquay and Lyleton formations is Frasnian; these formations represent a final regressive pulse of Saskatchewan map-unit, but for convenience of mapping are included in the overlying Palliser sequence.

Duperow Formation

This unit, notable for its cycles of shoaling carbonates capped by evaporites, overlies Beaverhill map-unit (Wilson, 1967; Dunn, 1975; Kissling and Ehrets, 1984). Agreement on where the "natural" cycle-breaks occur is limited and the number of lower order cycles [4-5] is in dispute. Dunn's four units of the Duperow are cycles [4] (average thicknesses 4.4, 7.5, 2, and 8.1 m in upward order) containing 8, 7, 6, and 6 lower order [5] cycles. Wilson had four cycles in Unit 2 in place of Dunn's seven (in the same well). Kissling and Ehrets had five "Units" in place of

Dunn's four; they claimed that the observed pattern arose from the interaction of two independent controls, one on bathymetry and one on salinity (climate?). The final member of Dunn's ideal cycle and of Kissling and Ehrets' salinity cycle is a green, pyritic, argillaceous dolomite - one of the green "clays" interpreted here as eolianites.

At the base of the Duperow is the Saskatoon Member, 27 m thick at the type locality, consisting mainly of microcrystalline limestone, dolomitic limestone, and dolomite, with interbedded anhydrite in the top 10 m. Elstow Member, which follows, is a thin, widespread shale with a diverse ostracode fauna (Braun and Mathison, 1982) recording marine transgression; it can be traced into the shale in the middle of the Cooking Lake Formation (Western Reef Domain). The numerous anhydritic layers of the succeeding Wymark Member show that it was the most restricted unit of the sequence. Dunn's Unit 3 (upper Wymark) terminates with a widespread halite (Flat Lake Evaporite), directly beneath which is the richest source-rock of the Saskatchewan Group so far identified (Osadetz and Snowdon, 1986). The halite deposit described in connection with the South Alberta Platform Margin was also laid down in early Wymark time. Zones rich in sporomorphs attributed to marine phytoplankton have been used for correlation (Kent, 1968). The uppermost of the Duperow members is the Seward, a shale and carbonate unit whose lower part was deposited during a marine transgression (Braun and Mathison, 1982).

Birdbear Formation

Birdbear Formation comprises a complete shoaling cycle [3-4] that begins with a tongue of open-marine argillaceous deposits which pinches out to the east in central Saskatchewan. This marine pulse correlates with the base of T-R Cycle IId, the most transgressive unit on Johnson et al.'s (1985) qualitative eustatic curve. The open-marine sediments pass upward to shoal-water carbonates. In southeastern Saskatchewan, where the average thickness of the formation is 20 m, the terminal deposits are cycles [5] of alternating sulphates and pelletal lime muds. The Birdbear thickens to 50 m in western Saskatchewan and then passes westward into Ireton Shale and the Camrose Tongue in southeastern Alberta (Fig. 4D.20). At the end of Birdbear deposition the region was emergent; subaerial weathering and solution probably removed some of the upper sulphate and carbonate layers and developed the brecciated and conglomeratic regoliths included in the base of the next sequence (Nichols, 1970).

The Saskatchewan Group does not contain important oil reservoirs, but the Hummingbird structure of southeastern Saskatchewan produces oil from 8 m of fine crystalline, vuggy dolomite of Birdbear Formation (and also from a Lower Carboniferous reservoir). It is interesting as a trap developed by the multistage solution and collapse of salt in the underlying Prairie Formation (Smith and Pullen, 1967).

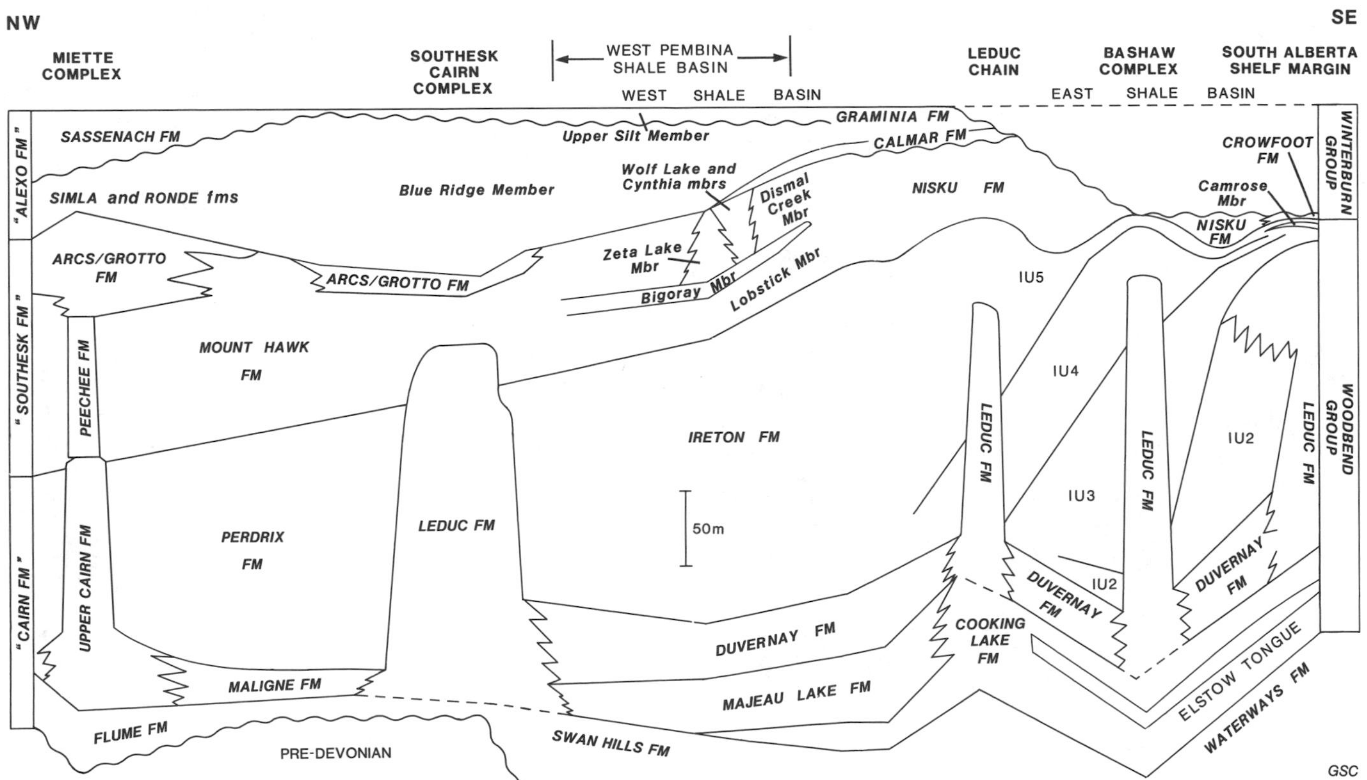

Figure 4D.24. Section through middle and upper Frasnian reefs in central Alberta.

Western Reef Domain

This is the most important sedimentary province in western Canada, from an economic standpoint. It was the site of the historic Leduc oil discovery in 1947, the mainstay of the petroleum industry for the next 30 years, and a source of renewed surprises during the subsequent decade.

Reefs and reef-rimmed banks, ranging in area from one or two square kilometres to thousands of square kilometres, grew in a shelf sea lying between Peace River Island to the north and the Grumbler-Grosmont-South Alberta platform margin to the east and south (Fig. 4D.19, in pocket). Description of the succession is complicated by subsurface nomenclature different from that used in studies of outcrop (Fig. 4D.24); neither is ideally adapted to present interpretations of the stratigraphic relationships. In the following overview, separate treatment is given to (1) the reefs of the Jasper, West, and East shale basins, (2) the bank complexes, and (3) the fringing reef of Peace River Island. To simplify the nomenclature, the formations have been grouped into "assemblages" rather than described unit by unit as in preceding sections.

Reefs of the Shale Basins

The **Cooking Lake** assemblage comprises the reef foundations and their basinal equivalents. At the base of the map-unit, Cooking Lake Formation was deposited as a carbonate platform complex of rather uniform thickness (20-50 m) covering the eastern part of the domain. It consists of two superposed carbonate banks separated by an open-marine argillaceous bed (Elstow Tongue). Published lists of Elstow fossils from its outcrop in a Saskatchewan potash shaft are being revised, and support this correlation (A.E.H. Pedder, pers. comm., 1984). The Cooking Lake platform provided the foundation for reef growth in the next stage. The platform was apparently reef-rimmed on its west side where it passes abruptly into Majeau Lake shale in the West Shale Basin (equivalent to Maligne Formation in Jasper Basin).

In the western part of the Western Reef Domain, in the region overlying the recently covered West Alberta Ridge (Fig. 4D.13), foundation building began earlier than in the eastern part; a foundation had already been established by the end of Beaverhill map-unit deposition. This early foundation is represented by Swan Hills Formation in subsurface and Flume Formation in outcrop. The foundation of the western reefs is complex. At Miette Reef (Fig. 4D.25, see inset, Fig. 4D.19, in pocket), two stages of Flume Formation followed by a biostrome at the base of the upper Cairn were emplaced before rising sea level started the upward growth of the reefs (Noble, 1970). The foundation stage of the upper Cairn is thought to be coeval with Cooking Lake Formation.

The **Leduc assemblage** developed as rapid deepening stimulated growth of banks and platform reefs throughout the Western Reef Domain. The reefs are assigned to Leduc Formation in the subsurface and to the upper Cairn and Peechee formations in outcrop. The build-up stage produced linear arrangements of reefs such as the Bashaw Complex and the Leduc chain (Fig. 4D.25, see inset, Fig. 4D.19, in pocket) and isolated reefs such as Redwater on the east side and Miette on the west side. A number of small satellite reefs have been discovered adjacent to the large reefs in East Shale Basin. The type Leduc Formation is about 240 m thick, but in down-dip reefs such as Strachan and Ricinus (both classified as "Giant Gas Fields") reef height exceeds 275 m (Hriskevich et al., 1980).

Leduc Formation of the producing reefs is usually completely dolomitized, and the only subdivision recognized is into a lower "dark reef" and an upper "light reef" facies corresponding to Cairn and Peechee formations in outcrop. Where the original textures are preserved, as at Redwater, Alberta, Leduc reefs display multiple growth stages like those of the Swan Hills reefs (F.A. Stoakes and

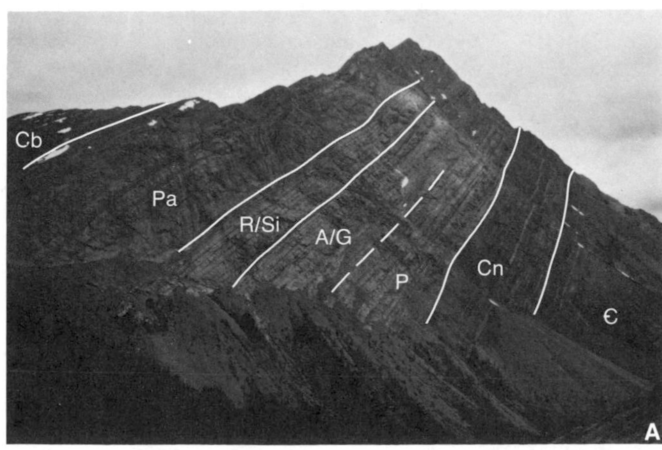

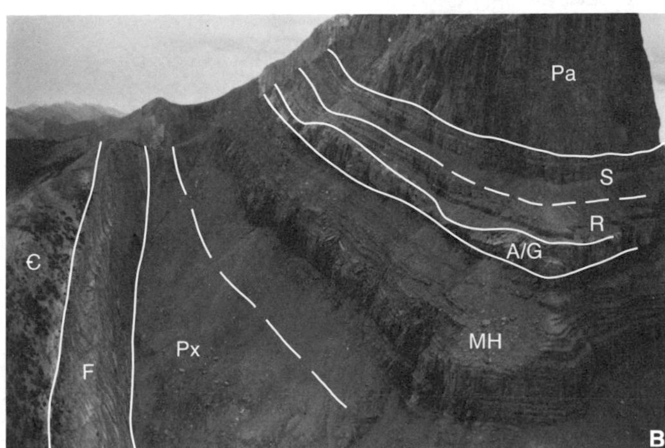

Figure 4D.26. Reef (**A**) and off-reef (**B**) facies at Roche Miette and vicinity, Alberta. Photo A is within the reef (west side of Slide Creek at 53°03′N; 117°14′W). Photo B is off the reef on the northeast shoulder of Roche Miette (53°10′N; 117°55′W). Cb - Carboniferous; Pa - Palliser; S - Sassenach (missing over reef); R/Si - Ronde-Simla; A/G - Arcs/Grotto; P - Peechee; MH - Mount Hawk; Px - Perdrix; F - Flume; Cn - Cairn; C - Cambrian. Thickness from top of Cambrian to base of Palliser is 440 m at A and 460 m at B (photo - R.H. Workum, ISPG 2462-8; interpretation slightly modified from R.H. Workum, pers. comm., 1986).

S. Creaney, pers. comm., 1984). The episodic filling of the basin, to be described, is attributed to sea-level fluctuation and was coeval with reef growth; this lends support to the interpretation of staged reef build-up. Such an interpretation has not emerged from studies of the reefs in outcrop; but this was also the case with Ramparts reefs until recently. At Redwater, Klovan (1964) demonstrated a mid-Leduc step-back on the windward side, followed by deposition of Duvernay Formation over the previous reef flat and subsequent re-advance of the reef. This step-back may be evidence of a mid-Leduc deepening. A distinct step forward of Peechee Formation over Cairn Formation occurs at Burnt Timber (Jackson and Harrison, 1982). In several subsurface reefs, step-backs on the windward (northeast to east) side have been found by seismic surveys and successfully drilled for oil and gas. At Golden Spike, west of Leduc, a crude subdivision into four stages was proposed by McGillivray and Mountjoy (1975). Reef-interior cycles [5] of Pillara type have been described in Miette Reef Complex, where twenty-eight cycles 2 to 5 m in thickness were identified, some terminating with well documented exposure surfaces (Mountjoy and Burrowes, 1982; Fig. 4D.26). Workum (1985) reported an intra-reef unconformity at a horizon between Cairn and Peechee formations, but this has not been adequately documented. Reef growth terminated through burial by detrital sediment (north of Redwater, for example) or possibly because lowered sea level exposed the reefs. On the west side of the Ancient Wall Complex there is evidence of emergence and karstification of Peechee Formation at the top of the main reef (H.H.J. Geldsetzer, pers. comm., 1986).

Dysaerobic conditions prevailed between the growing reefs. In these deeper waters, the bituminous, argillaceous limestones of Duvernay and basal Perdrix formations were deposited. In East Shale Basin, the Duvernay appears to be a diachronous fondofacies of Grosmont Formation, but in West Shale Basin, markers within the Duvernay are traceable for over 100 km and seem isochronous (Mountjoy, 1980). Deposition was synchronous with reef growth; the shales contain reef detritus. These rocks are rich in organic matter and are believed to be the principal source rocks of the Western Reef Domain.

The **Ireton assemblage** comprises those fine-grained, mixed calci- and siliciclastic rocks that prograded into the spaces between the growing reefs of Leduc assemblage. The early deposited clinothems of East Shale Basin can be traced up-slope into the platform carbonates of Grosmont Formation in the east. Later deposited clinothems can be traced up into Nisku Formation. Ireton clinothems are composed of clay minerals, fine quartz grains, and carbonate in varying proportions, and terminate, where the substrate is carbonate, in submarine hardgrounds (Stoakes, 1980). They prograde from the east and appear to downlap westward over and around the reefs. In the region of the Grumbler-Grosmont ramp, infill started early in the history of reef growth and the northern extension of the Leduc reef trend was overlapped when the reefs had built up to a mere 70 m (compared to over 200 m at Leduc). This demonstrates the contemporaneity of at least the eastern part of Ireton Formation with reef growth. On the west side of West Shale Basin, log correlation shows that clinothems of Ireton or Mount Hawk facies, traceable into Nisku equivalents, downlap against the lower flanks of Leduc reefs on the windward side of the Windfall-Marlborough complex (N.C. Meijer Drees, pers. comm., 1985). Whether the reefs were still growing or moribund by the time the Ireton sediments arrived is not known, but in this region most of the reef growth clearly predates the infill. An important observation is that part of what is mapped as Ireton Formation in West Shale Basin is of the same age as Nisku Formation in East Shale Basin.

The episodic progradation of the basin-fill clinothems can be explained by fluctuations in relative sea level. A shelf-to-basin relief of about 100 m has been deduced (Stoakes, 1980). At least one of the shallowing stages may have been at the end of Peechee deposition, as mentioned above. An earlier low stage might have given rise to the debris flows mentioned below.

During periods of lower sea level, carbonate sediment reworked from the surface of the platforms was added to the continuing supply of terrigenous material brought into the basin by contour currents from a more distant source, to create packets of sediment, which Workum (1985) called "basinal sediment with no shelf correlatives". Subsurface studies in the western part of Western Shale Basin have revealed similar packets of "extra" sediment, which defy log correlation to the basin margins.

An important phenomenon observed in outcrop and occasionally reported from subsurface cores (Hriskevich et al., 1980) is extensive sheets of debris extending from the reef into the basinal sediments. These include megabreccias, the larger blocks measuring 50 m across, which seem to be derived from the collapse of the reef margin (Cook et al., 1972). These mostly occur at the Perdrix-Mount Hawk boundary and have been found up to 5 km from the reefs; finer rudites derived from the mobilization of early cemented calcarenites in the foreslope facies also occur (Hopkins, 1977).

As the bathymetric basins created by Leduc reef growth became filled, a new sedimentary regime, the **Nisku assemblage** appeared, characterized by the progradation of carbonate ramps from east and west. The easterly ramp is Nisku Formation and the westerly ramp Arcs/Grotto Formation. In each case there is a proximal, largely biostromal carbonate passing downslope into a mixed carbonate and shale facies assigned to Mount Hawk Formation. Both the eastern and western carbonate facies are characterized by a dark, coralliferous lower part and a lighter, less fossiliferous upper part. Workum (1982) extended the name Nisku into the outcrop region.

Where Nisku Formation was deposited above Leduc reefs, differential compaction in the intervening Ireton shales has produced drape structures that provide a secondary, and locally principal, trap for oil and gas.

Nisku Formation passes landward (southeast) into siltstones and anhydrites of Crowfoot Formation (Belyea, 1958). Westward into West Shale Basin it becomes split by an argillaceous tongue (Bigoray Member) into an upper Dismal Creek Member (Machel, 1983) and a lower Lobstick Member (Chevron, 1979). This tripartite package forms the "Nisku Outer Shelf" and delimits a reduced West Shale Basin called West Pembina Basin (Fig. 4D.19). At the margin of this shelf a porous reefal facies assigned to Zeta Lake Member is developed locally. Meekwap field is an example of production from such a reefal margin in a situation where the underlying Lobstick Member is argillaceous and usually included in Ireton Formation.

Off the edge of the platform, Dismal Creek Member is replaced by isolated downslope reefs of Zeta Lake Member and by an off-reef, shale-carbonate couplet comprising Cynthia and Wolf Lake members. The carbonate stage of either Bigoray or Lobstick Member may act as foundation for reef growth. Although formally referred to Nisku Formation, these slope and basin members would be assigned more properly to Mount Hawk Formation. In the mountain outcrops, such a succession, complete with small reefs, occurs at several localities. Zeta Lake Member may build up to 100 m in thickness; the reefs average 1500 m in diameter and are prolific producers of oil. The lower part of the reefs consists of impure lime mudstone, floatstone, and grainstone, and in a few places of bindstone and bafflestone (Machel, 1983). The upper part is distinguished by a much greater variety of macrofossils including stromatoporoids; encrusting renalcids and stromatolites are common in some beds. Intertonguing with off-reef sediments suggests that there was generally very little relief between reef and off-reef. This is also the relationship observed in mountain outcrops of downslope reefs, for example at Winnifred Pass, where a relief of 15 m was estimated (Desbordes and Maurin, 1974). Both the Zeta Lake and the outcropping reefs contain stromatactis and are considered to be mud mounds, although corals are abundant in the former and rare in the latter.

The **Calmar-Blueridge assemblage** comprises the deposits of a single cycle [4] lying locally above an unconformity. A relative fall in sea level occurred at the end of the Nisku depositional stage; Workum (1985) described the contact with the overlying Calmar Formation as unconformable and attributed dolomitization and the development of porosity to exposure at that time. There is no evidence, however, that the basinal areas were emergent. As the region began once more to submerge, an influx of fine quartz sand and silt was deposited together with carbonates and evaporites as Calmar Formation. In West Pembina Basin, the Calmar changes facies to argillaceous limestone; in the Jasper area, through which Figure 4D.24 is drawn, the Calmar has tentatively been included in the silty Ronde Formation. Clean, variegated sandstones associated with carbonate breccias occur farther south in the outcrop belt of the Bow River valley, and have been included in the portmanteau "Alexo Formation" by field geologists.

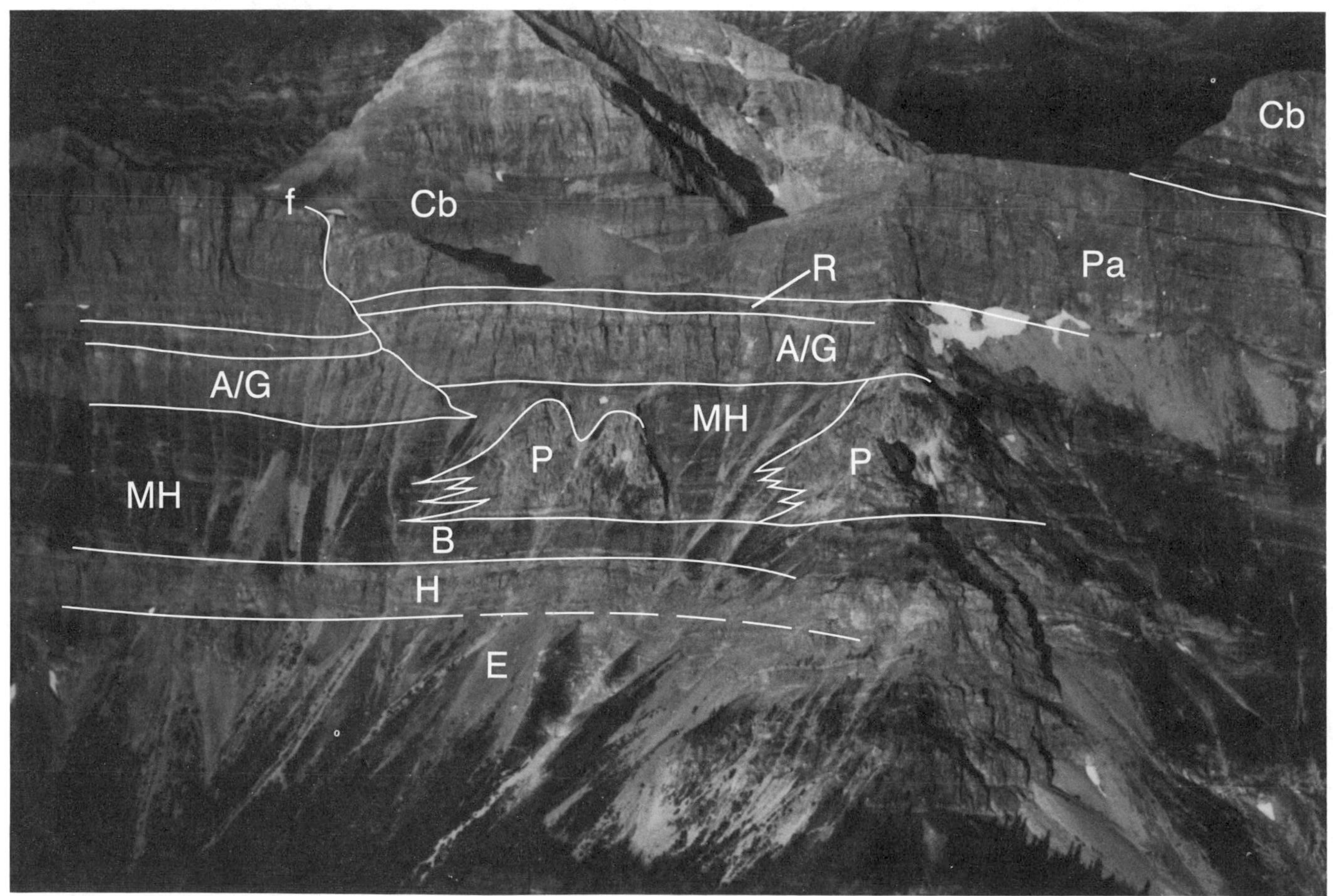

Figure 4D.27. Bioherms just off the southwest margin of the Fairholme Bank at South Lost Creek (49°25′N; 114°36′W). Cb - Carboniferous; Pa - Palliser; R - Ronde; A/G - Arcs/Grotto; MH - Mount Hawk; P - Peechee; B - Borsato; H - Hollebeke; E - Elko (Cambrian); f - fault. Thickness from the top of the Cambrian to the base of the Palliser is 430 m (photo and interpretation - R.H. Workum, ISPG 2462-4).

Carbonate sedimentation resumed with the deposition of Blue Ridge "Formation" (formally a member of Graminia Formation, the upper member of which belongs in the next sequence) and its westerly, surface equivalent, Simla Formation. These formations form a prism of sediment whose depocentre, containing a small deposit of halite, is displaced northwesterly from that of the Nisku. Simla Formation can be followed west and north of the limits of the diagram (Fig. 4D.24) into the Grande Cache region (Geldsetzer, 1982) where it overlies Mount Hawk Formation. It is there a limestone about 70 m thick, so crowded with stromatoporoids and corals that it may be called a biostrome. On faunal grounds, the Simla can be correlated with Kakisa Formation of the Grumbler Group (Fig. 4D.20). Log correlations carried south from Kakisa Formation tie it to the Blue Ridge Formation to complete the "loop".

Bank complexes

Carbonate banks with reefal margins (e.g., Cairn-Southesk and Fairholme complexes) cover thousands of square kilometres in the western part of the domain. These banks are transected by deep embayments which have a northeast orientation when palinspastically restored (Fig. 4D.25, see inset, Fig. 4D.19, in pocket). Biohermal structures have been found not only at the margins of the banks (Price, 1965 and Fig. 4D.27) but also in their interior (Desbordes and Maurin, 1974), showing that at times they were completely open to marine circulation.

Southesk-Cairn Complex consists of at least two, and possibly three discrete banks. A section measured at Coronation Mountain (Mountjoy, 1978), near the assumed western margin of bank development, has about 730 m of Fairholme Group carbonates, the thickest section known.

Fairholme Complex is a vast carbonate bank about 400 m thick and over 300 km in length. It is completely exposed in several mountain ranges. The overall appearance is of a lower, dark-coloured, bedded dolomite about 200 m thick (Cairn Formation), overlain by an equal thickness of less well bedded, light-coloured dolomite (Southesk Formation) and topped by a yellowish, recessive-weathering, silty and sandy layer (Alexo Formation). In detail it shows great complexity. At Canmore, for instance (Desbordes and Maurin, 1974), the Cairn is

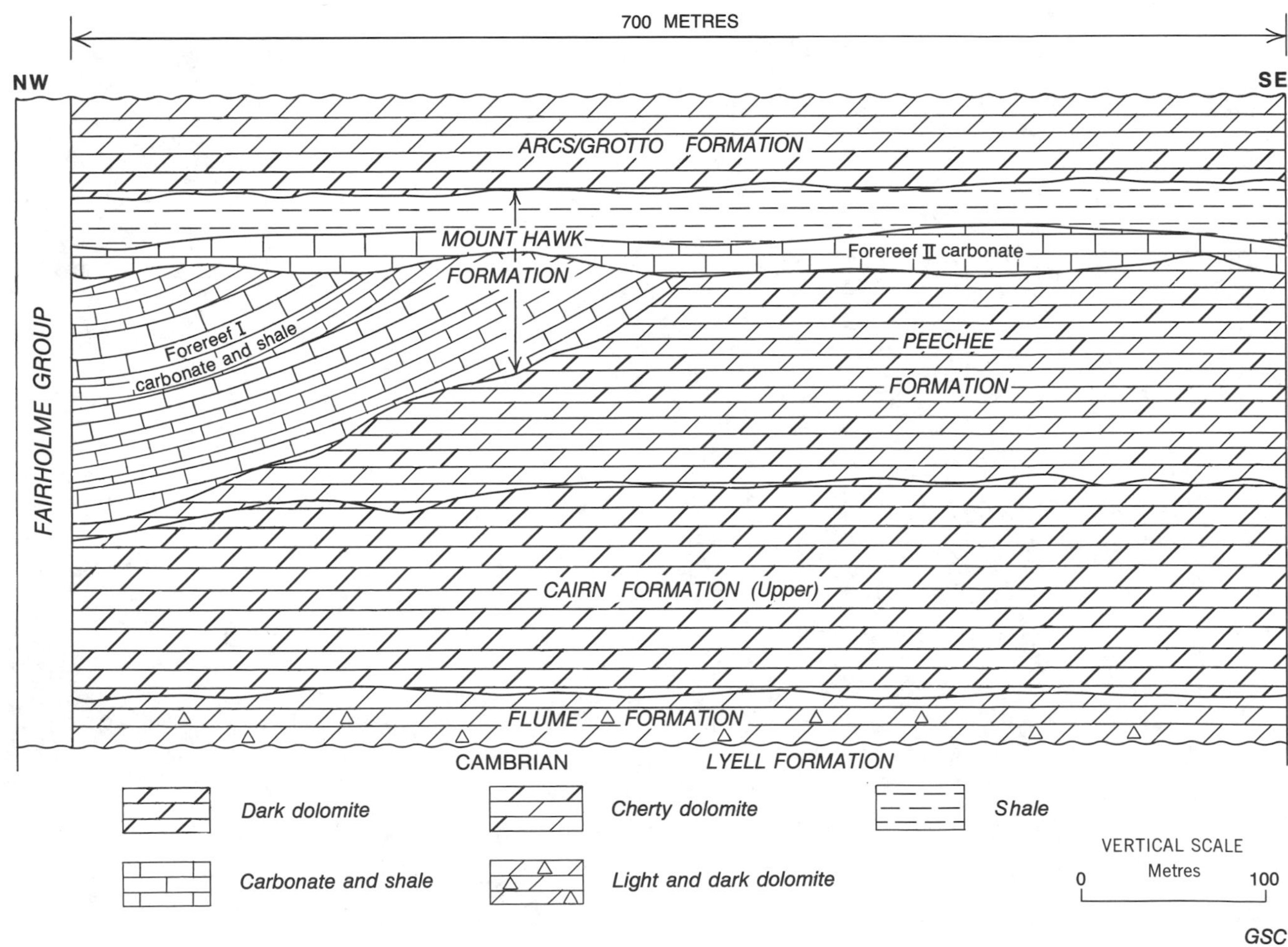

Figure 4D.28. Diagrammatic, true-scale section through the bank margin at Cripple Creek, Alberta. The Peechee Formation is a stratigraphic reef. Forereef I may either interfinger with, or onlap, the Peechee. Bedding traced from a photograph by R.H. Workum.

divided into a lower dark cherty member (Flume Formation) and an upper dark organic member. The upper member, containing crinoids and bulbous stromatoporoids, is interpreted as a product of the deep- ening event that started the Beaverhill-Saskatchewan sequence. Bioherms grew upon it but only attained a thickness of 30 m, presumably because of its position in the bank interior. The bioherms are overlain by 50 m of biostromes with varying faunal content (coral, gastropod, crinoid, lamellar stromatoporoid). Ferruginous hardgrounds occur and, near the top, there is an open-marine intercalation with goniatites in addition to corals and crinoids. These open-marine beds might be equated with the mid-Leduc deepening and possibly the early Seward (member of Duperow Formation) marine incursion, but there is insufficient faunal evidence to decide the matter.

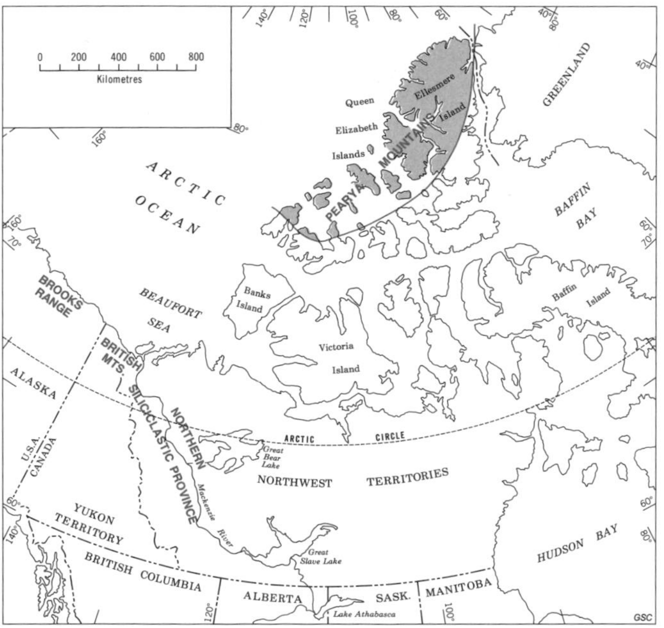

Figure 4D.29. Extent of the Pearya Mountains, Late Devonian-Early Mississippian.

The overlying Southesk Formation comprises 200 m of coarsely crystalline, light grey dolomite with occasional evidence of *Amphipora* and other fossils. Alexo Formation, also fully exposed near Canmore, Alberta, consists of 50 m of silty dolomite, siltstones, and (solution?) breccias with a distinct diastem in the middle. The lower green- and purple-stained siltstones are presumably Calmar Formation and the upper part may include Graminia Formation.

Most of the interest in the Fairholme Complex has focused on its margins, where the carbonate platform passes into argillaceous deposits of the Perdrix and Mount Hawk formations. This transition can be seen at the southwest side of the bank near the Flathead well (Fig. 4D.25, see inset, Fig. 4D.19, in pocket), where large bioherms are developed in the Peechee Formation (Price, 1965 and Fig. 4D.27). Because of differences in stratigraphic terminology it is more instructive to describe the situation on the northwest flank at Hummingbird and Cripple creeks. The following account (and Fig. 4D.28) draws on the work of Dooge (1966) and Workum (1978) at Cripple Creek and on comments by Mountjoy (1980). Because of imperfect exposure, there is lack of agreement in interpretation. The foundation of reef growth is Cairn Formation, comprising 144 m of dark, fossiliferous, crystalline dolomite lying with slight angular discordance upon Cambrian strata. The lower part is distinguished as Flume Formation (34 m), characterized by the development of chert; it is early Frasnian in age and thus belongs in the Beaverhill map-unit. The upper part consists of a lower, *Amphipora*-stromatoporoid assemblage followed in some places by a coral-brachiopod assemblage. Upon this foundation three major facies were emplaced: a reef, forereef carbonates, and off-reef shales and carbonates. Peechee Formation forms a stratigraphic reef 122 m thick, consisting of light and dark biostromal dolomites, possibly with some bioherms. It either interfingers with, or more probably is onlapped by, forereef, bedded carbonates and shales with original dip (unnamed unit). A 34-metre tongue of forereef carbonates overlies the reef margin, probably as the result of a step-back. The forereef strata grade into basinal dark shales (Perdrix). The dark shales are overlain by greenish grey shales and argillaceous limestones (Mount Hawk), a richly fossiliferous tongue (20 m) of which passes over the reef

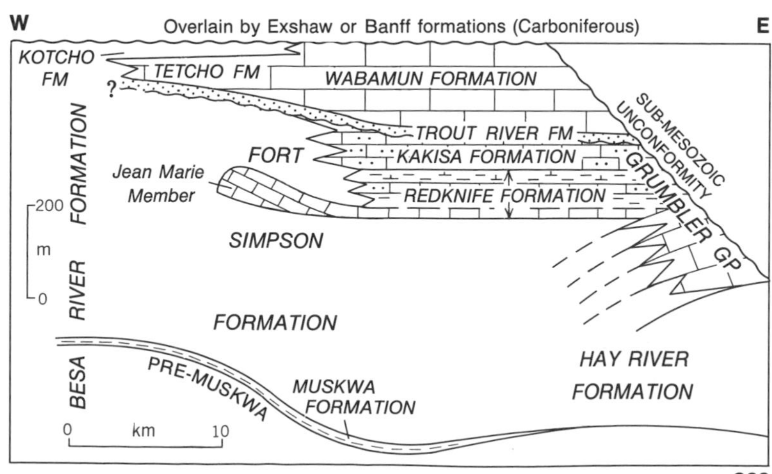

Figure 4D.30. Shale-out of Wabamun Formation and Grumbler Group in northern Alberta.

margin, penetrates into the interior of the bank, and records a marine incursion. The Arcs/Grotto and Alexo formations follow. Neither this transition nor perhaps any of those reported to date lie on the fully windward side of a bank or reef, where the zone between reef and basinal sediments might be much narrower.

Peace River Fringing Reefs

The Peace River Fringing Reefs are not illustrated in Figure 4D.20 but are crossed by Section 7 (Fig. 4D.1, in pocket). A discontinuous barrier reef forms a horseshoe about the windward side and flanks of Peace River Island. The reef is separated from the island by a coastal plain and lagoon complex of sandstones and shales. On the north flank of the island in the Worsley Field, an interesting reservoir has been discovered where reefs and interbedded sandstones both produce gas. The updip part of the barrier reef has not yielded the large hydrocarbon reserves one might expect. The problem may lie with the seal, as the Ireton incursion did not penetrate far towards the land; also, there may have been no route by which the reservoirs could have been charged by the supposed source rocks (Duvernay Formation).

Northern Siliciclastic Province

The region lying between Peace River Island and the northern limits of the study area was a major depocentre into which more than 2000 m of sediment was poured during Late Devonian time. At the start of the period, a landmass in the northern Arctic islands (Fig. 4D.29) was flanked on its southern side by a coastal plain and marine shelf (Embry and Klovan, 1976). As a result of steady regression, this coastline moved southward during Late Devonian time. At the end of the Devonian, the clastic wedge on the south flank of the landmass was folded by the Ellesmerian Orogeny, which also affected the regions of Brooks Range and British Mountains. The description of this province embraces both Frasnian and Famennian sediments, because they can only be distinguished paleontologically.

In the southern part of the province, near Great Slave Lake, Grumbler Group (Frasnian) and Wabamun Formation (Famennian), consisting of slightly silty and sandy carbonates deposited on a shallow marine shelf (Jamieson, 1967), pass westward into open marine shales of the Fort Simpson Formation (Fig. 4D.30), whose clinothems (Williams, 1975a) advanced over fondoform dark shales of Muskwa Formation. East of the Muskwa pinchout (Fig. 4D.1, Section 4), some of these clinothems appear to downlap onto Waterways Formation, the plane of discordance being interpreted as a submarine hiatal surface.

North from Great Slave Lake, although coral patch reefs survived until the end of Frasnian time, siliciclastic strata dominate, and Imperial Formation replaces Grumbler Group and Wabamun Formation. The Imperial is in a shallow marine facies with interbedded limestone in its easterly outcrops but changes facies westward to deeper

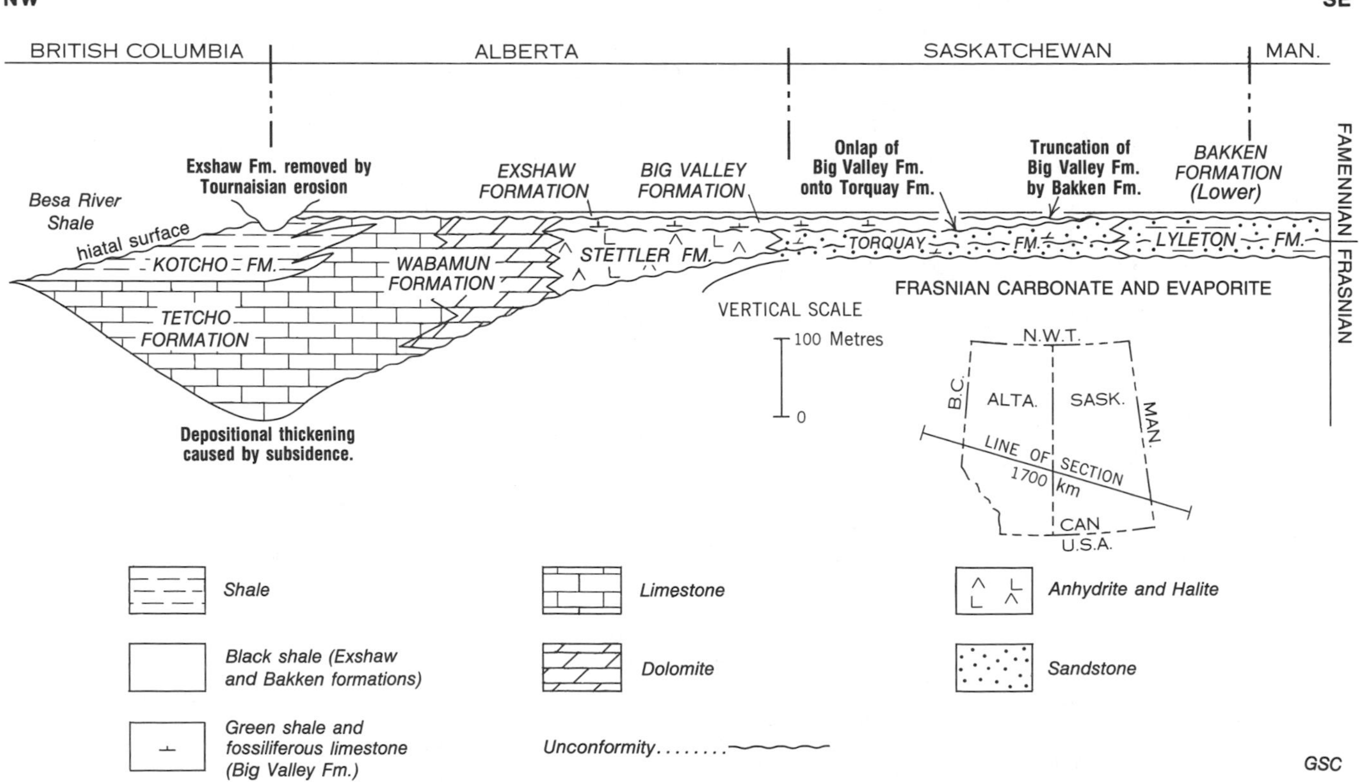

Figure 4D.32. Changes in facies and thickness across the Famennian shelf (Palliser sequence).

water sediments. The patch reefs do not reach Arctic Red River, although a wedge of shelf sandstone and shale has prograded over the underlying turbidites there. At Trail River most of the formation is a flysch facies dominated by turbidites (Hills and Braman, 1978). Still farther west, in Ogilvie Mountains, the corresponding interval is either missing or represented by part of the dark shale, limestone, and chert of the McCann Hill Chert. Current indicators and westward- and southward-diminishing grain size suggest that the source of the turbidites lay to the north. In southern Mackenzie Mountains, 520 m of slate and argillite represent some of the Middle and all of the Upper Devonian (Gabrielse et al., 1973). Frasnian brachiopods like those found in Jean-Marie Member of Redknife Formation are found about 275 m above the base. The beds 390 to 450 m above the base are Famennian. Still farther south, in northeastern British Columbia, equivalent beds can be identified in Besa River Formation (Road River Formation) by tracing clay-mineral zones (Pelzer, 1966). About 300 m of dark shale represents the Upper Devonian, the upper 100 m being Famennian.

The upper 500 m of siliciclastics exposed on Trail River is much coarser sandstone, rich in chert pebbles. These are included in Tuttle Formation (Pugh, 1983), a map-unit that may contain conglomerates of several different ages. At least some of the Tuttle represents coarse siliciclastics derived from the uplift of the ancestral Brooks Range and British Mountains, and is therefore related to the Kekiktuk Formation of Alaska; the latter formation, however, is Lower Carboniferous whereas the coarse siliciclastics on Trail River begin within the Famennian (Braman, 1981). Most of the Tuttle Formation appears to be Lower Carboniferous. It is noteworthy that the younger Famennian beds in the Mackenzie Valley region (e.g., at Imperial River, Wrigley, Carlson Lake) are shaly and calcareous in contrast to the sandy, shallow-water sediments that preceded them. A deepening event seems indicated; it is attributed to isostatic tilting of the foreland as a consequence of Ellesmerian orogeny and the infilling of a foreland basin, in other words the same phenomenon that caused coarse sediments to be deposited in the west. Several granitic plutons in the northern part of Yukon Territory are dated as Late Devonian, i.e., Ellesmerian (Gordey, 1987).

Frasnian-Famennian boundary

The one generally accepted feature of the Frasnian-Famennian boundary is that it is characterized by a drastic faunal and floral break. It marked the end of most species of rugose coral (Pedder, 1982) and of all the rich stromatoporoid fauna except *Labechia*. As mentioned earlier, the consequences for carbonate sedimentation were profound, as most of the reef-builders were wiped out at a stroke. The cause of this is likely to have been climatic (a mild precursor of the "nuclear winter"), possibly as a result of a bolide (i.e., meteorite) impact as suggested by McLaren (1982). If McLaren is right, there is no need to infer an unconformity to explain the paleontological data. Bassett and Stout (1968) postulated a major unconformity between the stages and referred to "peripheral erosion of the upper flanks of the basin" during the Late Frasnian regression. The conglomerate they used to support their hypothesis, however, has been subsequently identified as the Sharp Mountain conglomerate of Early Cretaceous age (D.K. Norris, 1981). There does not seem to be any physical evidence of an unconformity at the top of the Frasnian in the northern region.

Paleontological data are insufficient to support the construction of separate Frasnian and Famennian isopach maps, but where separate thickness values are available they have been posted (Fig. 4D.19, 4D.31). The control north of Mackenzie Mountains is mainly from palynomorphs; in the mountains it is from macrofossils. Where both sets of data are available, as at Powell Creek, they are in agreement (Braman, 1981; Lenz and Pedder, 1972).

PALLISER SEQUENCE (FAMENNIAN)

Over most of Alberta the base of Palliser sequence is marked by a silty or sandy zone more or less coincident with the base of the Famennian. The sequence ends at the widespread black shale at the base of the Exshaw Formation. The Devonian-Carboniferous boundary occurs within the Exshaw or its exact equivalent, the lower and middle members of the Bakken Formation (Richards et al., Subchapter 4E in this volume). In Saskatchewan and Manitoba the Frasnian-Famennian boundary occurs within a poorly fossiliferous siliciclastic succession and is not readily identified; the unit mapped as Palliser sequence therefore includes some Frasnian sediments there. In the Northern Siliciclastic Province, Famennian sediments were treated together with Frasnian in the previous section and will receive no further discussion; individual thicknesses of the Famennian, as determined paleontologically, are shown in Figure 4D.31 (in pocket).

During deposition of Palliser sequence there was, as with the Saskatchewan map-unit, a northern siliciclastic province and a southern carbonate-dominated platform province. However, the regression that started in late Frasnian time continued during the Famennian, shifting the facies belts to the northwest thus allowing Famennian nearshore and terrestrial sediments to be spared the pre-Mesozoic erosion that removed most of their Frasnian counterparts (Compare Fig. 4D.19 and 4D.31, in pocket).

Southern carbonate-dominated Platform Province

In spite of mountain building in the far north and the influx of molassic deposits into that region, the Famennian remained a time of relative quiet in the south. A distally steepened ramp extended from Manitoba to the southern District of Mackenzie. It comprised six facies belts spanning the change from terrestrial to open-marine deposition (Fig. 4D.32):

1) redbeds: Lyleton Formation
2) mixed clastic-evaporite-carbonate: Torquay Formation
3) evaporite: Stettler Formation
4) carbonate: Palliser-Wabamun formations
5) carbonate-to-shale transition: Tetcho and Kotcho formations
6) distal shale facies: Besa River (Road River) Formation

In addition there are thin units at the top and bottom of the succession, which will be described where appropriate.

Redbeds

The Lyleton Formation includes the Frasnian-Famennian redbeds in Manitoba; the boundary between the two stages has not been determined paleontologically. The Lyleton consists of soft, orange-red, calcareous shale and red, dolomitic sandstone, and ranges up to 60 m in thickness; anomalous thicknesses occur where deposition was concurrent with the removal of underlying Prairie Formation salt by solution. On geophysical logs it cannot be separated from the overlying Devonian-to-Carboniferous Bakken Formation and is generally mapped with it (McCabe, 1971).

Mixed clastic-evaporite-carbonate beds

The Torquay Formation (Christopher, 1961) is the lowest unit of the Three Forks Group in Saskatchewan. It consists of about 50 m of dolomitic and anhydritic, red-brown and green siltstone and shale. It overlies Birdbear Formation sharply and is disconformably overlain by either Big Valley Formation or Bakken Formation, the other two formations of the Three Forks Group. It grades by facies change into Lyleton Formation on the east and Stettler Formation on the west. The Torquay contains as many as nine regolithic beds 1 to 5 m thick, some of which can be traced across southern Saskatchewan. One major intraformational discontinuity, which may represent the Frasnian/Famennian boundary, occurs at about one-third the height of the formation and is succeeded by a shale bed marking a possible transgression. Two additional important discontinuities have been recognized higher in the formation (Christopher, 1961). The formation represents a shoreward condensation of several carbonate cycles deposited during the general regression that took place towards the end of the Devonian Period.

Big Valley Formation has its type locality in central Alberta, where it consists of 10 to 25 m of normal marine limestone. Toward the southeast it changes to a splintery, green, calcareous shale containing ostracodes and crinoid ossicles. The arrangement of the microfossil zones (Lethiers, 1978) suggests progressive eastward onlap over Torquay Formation. The ostracode zonation also suggests an unconformity beneath the overlying Exshaw shale (lower shale of Bakken Formation); the shingled arrangement of the zones is attributed to bevelling rather than offlap. Big Valley Formation contains the *Strophopleura raymondi* fauna (Crickmay, 1956), which farther west in Alberta and also in Montana occurs in the *trachytera* Zone.

Evaporitic belt

Stettler Formation ("Potlatch" on old logs) consists of up to 210 m of anhydrite and dolomite with local halite. In southeastern Alberta it overlies the thin, silty and sandy zone (called Crowfoot Formation) that probably contains the Frasnian-Famennian boundary. Farther north, the Stettler may be separated from Nisku Formation by a breccia zone considered by Storey (1970) to be sedimentary and indicative of a basal Famennian transgression. The Stettler is overlain disconformably by Big Valley Formation, the two units together making up Wabamun Group. Log correlations show discordance of dip between markers in the Stettler and the base of the overlying Big Valley. This is attributed by Storey (1970) to pre-Big Valley erosion; the collapse of beds as a result of anhydrite solution gives the unconformity an exaggerated appearance.

At the western margin of the evaporitic belt, a mid-Stettler transgression from the west introduced Crossfield Member, which includes massive dolomite, shoal lime sands, stromatoporoid mud banks (gas reservoir), and tidal/intertidal beds. The stromatoporoid is probably *Labechia palliseri* Stearn (1961), single specimens of which have been found at five scattered outcrops of Costigan Member near the top of the Palliser Formation. Stearn observed that such survival under environmental stress is typical of primitive forms - *Labechia* originated in the Ordovician.

The Crossfield Member is about 35 m thick south of Calgary and is overlain by about 100 m of Stettler evaporites representing the return of sabkha deposits in a regressive phase. These create a regional stratigraphic trap for sour gas (Workman and Metherell, 1969; Eliuk, 1984). There is no evidence of transgression in the evaporite beds east of Calgary; hence the transgression may have been caused by tilting rather than eustasy. The evaporite cover extends some distance westward (Fig. 4D.31, in pocket); gypsum is found at one locality in the Front Ranges, and in the Limestone Mountain area there are breccias that probably represent dissolved evaporites.

Carbonate belt

Within the carbonate belt, there are two sub-belts: the inner one, next to the evaporite belt, is dominated by dolomite, the outer one is mainly limestone with local zones of dolomitization providing reservoirs for gas (these may be fault- or fracture-controlled). The sheet of carbonate is called Wabamun Formation in the subsurface and Palliser Formation (which has priority) in outcrop. In addition to the fracture-controlled dolomites, at least one biostrome produces hydrocarbons in the Normandville region on the south flank of Peace River Island (Fig. 4D.19, in pocket). The reservoir is a stromatoporoid (*Labechia*)-algal boundstone (Nishida et al., 1985). Other important gas reserves occur at shallow depth at the updip eroded edge of the Wabamun beneath Cretaceous cover.

At the base of Palliser/Wabamun Formation there is a thin silt unit, not independently named and usually incorporated in Alexo Formation (outcrop) or Graminia Formation (subsurface). No evidence of unconformity beneath the silt has been disclosed by subsurface mapping, but absence in some outcrop areas of the uppermost Frasnian brachiopod zone (*Vandergrachtella scopulorum*) has been attributed to pre-Famennian erosion (Maurin and Raasch, 1972). Westward this silt grades to sand and becomes thicker, as Sassenach Formation (dated Famennian by brachiopods). A thick wedge of Sassenach fills the void that remained in Jasper Basin at the end of the last sequence (Mountjoy, 1980). A possible source for these western sandstones is in Ordovician quartzite of the

Cordillera. Another sandstone with a similar stratigraphic position has been called Broadwood Quartzite at Elko in southeastern British Columbia, where it is 70 m thick. Famennian *Cyrtiopsis mimetes* has been reported from these quartzites (Price, 1965); a horn coral, rare in the Famennian, has also been reported near the top (Gibson, 1978). Northward from Jasper, Sassenach thins to zero and Palliser lies directly on Simla Formation with disconformity (Geldsetzer, 1982).

Palliser Formation is widely divisible into three units. The basal unit is usually a dolomite about 30 m thick, which can be followed from the mountains into the subsurface where it is locally gas-bearing. The middle unit is a dolomitic limestone mottled with thalassinoid burrows. The two lower units (dolomite, mottled limestone) together comprise Morro Member of the Palliser and average 200 to 220 m in thickness. The upper unit, Costigan Member, 25 to 60 m thick, is not burrowed but is well-bedded to platy. It carries a rich fauna of brachiopods and cephalopods; large orthocones are quite common. *Strophopleura raymondi* links it with Big Valley Formation. In spite of its laminar appearance, which at first suggests a tidal deposit, the diversity of the fauna leads to the conclusion that it represents a deeper environment than the burrowed zone. Thus Costigan Member and Big Valley Formation represent a rise in relative sea level, possibly coeval with the deepening noted in the mid-Famennian units of the Northern Siliciclastic Province.

North of Elko, in southeastern British Columbia, Costigan Member is 161 m thick, has a basal conglomerate, and lies unconformably on Morro Member (Gibson, 1978). The latter is 452 m thick and consists of burrowed limestones with a medial argillaceous limestone band. The youngest dated conodont fauna, collected near the base of the Costigan, is lower *marginifera* Zone (H.H.J. Geldsetzer, pers. comm., 1985). An equally great thickness of Palliser has been observed at Three Isle Lake, about 100 km to the north (A.E.H.Pedder, pers. comm., 1985).

The top of Costigan Member is generally knife-sharp and highly silicified, with irregular black chert nodules; the silica probably came from the overlying Exshaw Formation, which is siliceous and volcanic in places. The base of the Exshaw locally consists of a lag deposit containing phosphatic grains including fish teeth, and even logs of wood. Locally, where volcanic ash was deposited at the base of the Exshaw, its initial bed is a sanidine sandstone with abundant zircon.

Carbonate-shale and shale belts

In the carbonate-shale and shale belts of northern Alberta and southern District of Mackenzie, Trout River Formation is the local manifestation of the basal Famennian sandstones and siltstones. It lies on Kakisa Formation, which is Frasnian. Many of the Frasnian formations are sandy and silty; in this region the influx of siliciclastics began long before the extinction event. Trout River Formation is limited to the *Eoparaphorhynchus* Zone (Sartenaer, 1969), unlike Sassenach Formation, which extends higher, into the *Basilicorhynchus* Zone. Trout River conodonts fall in the lower to middle *crepida* Zone (Klapper and Lane, 1985).

Above the basal sandstone, a tongue of Palliser limestone (Tetcho Formation) is succeeded by Kotcho Formation, mainly shale. The boundary between the two formations is diachronous, Kotcho gradually increasing in thickness at the expense of Tetcho westward until the sequence is entirely shale (Besa River Formation). Pelzer (1966) used mineralogical analyses to trace the Exshaw Formation and some lower markers into Besa River Formation and demonstrate the great reduction in thickness that takes place basinward.

The carbonate-shale belt strikes south and outcrops in the area of Kakwa Lake (Fig. 4D.31, in pocket). There, the Palliser is divided by Geldsetzer (1982) into an upper shaly unit and a lower mottled limestone unit (interpreted here as Kotcho and Tetcho formations respectively). Locally, Exshaw Formation is absent, and shales of the Banff Formation with a Tournaisian palynoflora lie directly upon the Kotcho with a *marginifera* Zone fauna (H.H.J. Geldsetzer, pers. comm., 1985). A rather similar situation is illustrated in the west panel of section 4 (Fig. 4D.1). Although at 427 m the Kakwa Lake section is four times the thickness in the borehole on section 4, this is not inconsistent with its situation on the westerly prolongation of a thick trend centred south of the buried Peace River Island. North from Kakwa Lake the Palliser section thins rapidly so that, within 150 km, Besa River shales, possibly Early Carboniferous in age, appear to lie directly upon Simla Formation.

The positions of boundaries between carbonate, carbonate-shale, and shale belts are quite subjective and depend on the criteria used. In Figure 4D.31 the boundary between carbonate and carbonate-shale is simply an arbitrary cut-off between Wabamun Formation (undivided) and Kotcho and Tetcho formations.

Pre-Exshaw unconformity?

The end of Palliser sequence lies at the base of Exshaw Formation although recent conodont studies (Richards et al., Subchapter 4E in this volume) have confirmed that the Devonian-Carboniferous boundary falls within the Exshaw. Over almost the entire region covered by the Exshaw in Alberta, perfect parallelism prevails between it and the underlying Palliser or Wabamun Formation, in outcrop as in the subsurface. Wherever brachiopods have been recovered from the uppermost Palliser or Wabamun, the fauna is characterized by *Strophopleura raymondi* (Crickmay, 1956) and falls into Sartenaer's (1969) *Gastrodetoechia* Zone and the *trachytera* conodont zone. In mountain outcrops north of Jasper, the topmost Palliser has yielded conodont faunas ranging from *marginifera* to *expansa* zones (A. Higgins, pers. comm., 1986). Where diagnostic conodonts have been recovered from the lower part of Exshaw Formation in the same region, they usually fall in *expansa* Zone. The conclusions drawn from these data are that there are no strata missing by erosion beneath Exshaw Formation and that any hiatus was of short duration. The situation may be different in Saskatchewan, where the platform interior has already been shown to be a region of multiple disconformities. Although some might doubt the power of ostracodes to provide so fine a zonation, Lethiers (1978) concluded that there is a lacuna beneath Bakken Formation representing a large part of "Strunian" time. In the upper Mackenzie

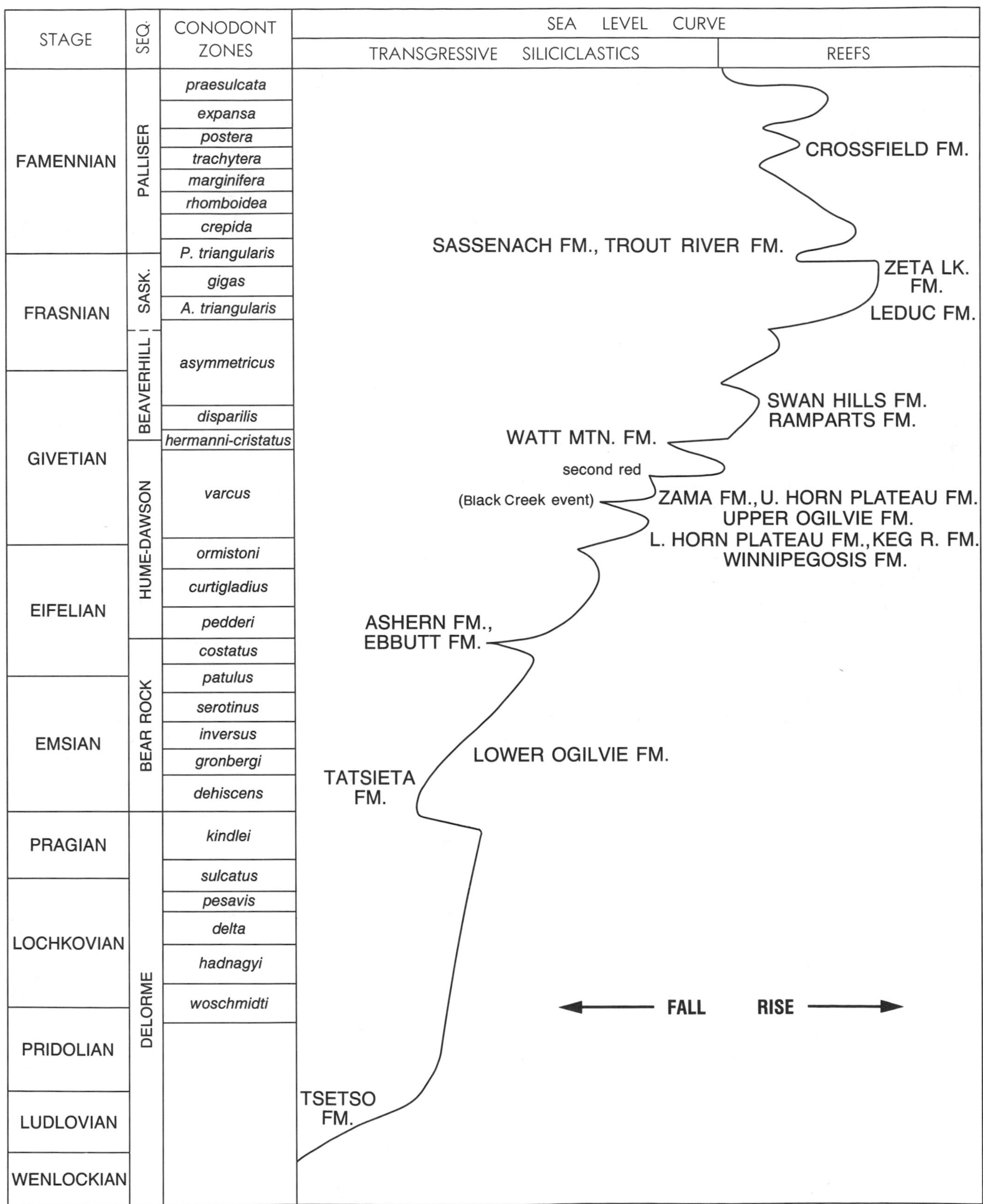

Figure 4D.33. Devonian transgression with shallowing and deepening cycles (adapted from Johnson et al., 1985).

River Valley region, the uppermost Famennian *Sinotectirostrum avellana* Zone (former *Athyris angelica* Zone), which is absent in Alberta, occurs in the Imperial Formation. This is probably the result of facies change, not of unconformity, although Pedder (Table 4D.1) shows the *Sinotectirostrum* Zone lying over the *Gastrodetoechia* Zone.

In the region of northeastern British Columbia in which the Exshaw is absent, the Banff Formation (Besa River Formation), which may be as young as Viséan in places, lies on Palliser of *marginifera* Zone or older, i.e., on strata older than those generally underlying the Exshaw. It is believed that the (still rather confused) stratigraphic relationships in this region are the product of two distinct causes: (1) basinal (submarine) unconformities within both Upper Devonian and Lower Carboniferous strata, producing lacunae in the record, (2) post-Exshaw (early Tournaisian) uplift and erosion that cut down into the underlying Palliser in places (Richards et al., Subchapter 4E in this volume). There is no reason to infer a pre-Exshaw unconformity of any greater magnitude here than elsewhere.

CONCLUSIONS

Transgression, regression, and tectonism

Devonian sedimentation in the Western Canada Basin took place mainly during a long period of advance of the sea over the craton (transgression), which had begun in mid-Silurian time. A shorter period of withdrawal (regression) began in late Frasnian time. The final episode, taking place during the last few thousand years of the Devonian, was renewed transgression. This broad pattern is illustrated by the curve on Figure 4D.33 and follows closely the envelope of the sea-level curve published by Johnson et al. (1985).

The order in which the main areas were inundated was:

Mackenzie Platform (Late Silurian);

Northern Alberta to Meadow Lake Hinge (Emsian);

Saskatchewan and Manitoba (Eifelian);

West Alberta Ridge (Frasnian);

Peace River Arch (Famennian).

This does not reflect the steady centripetal movement that would follow from a simple tectonic model of marginal thermal subsidence and consequent crustal flexure (Watts, 1982). The timing of the flexure is in any case too late to be explained by the thermal evolution of a passive margin. Movement of localized tectonic elements was superimposed upon the tilting of the craton. For example, downwarping took place across Meadow Lake Hingeline. Peace River Arch during the Middle Devonian was an active positive element with evidence of faulting. Its post-Devonian history, on the other hand, was one of subsidence and continued faulting. The shoreline continued to transgress over the Peace River landmass during the Famennian while regressing in Manitoba and Saskatchewan. Synsedimentary movement of West Alberta Ridge is less well documented but may have taken place. Tathlina Arch shows evidence of movement during the Devonian Period. Keele and Redstone arches also had a long history of movement. Porcupine and White Mountains platforms appear to have been passive positive elements, but this may be the result of imperfect knowledge. Aklavik Arch had a history of activity in both pre- and post-Devonian time, but no conclusive evidence has been found for its control over Devonian sedimentation. A comparison of the maps of each sequence shows a greater tectonic-sedimentary differentiation in the early sequences than in the later ones; this is mainly because elevated areas were degraded and covered.

The overall transgression was punctuated by regressive events, which resulted in emergence and the deposition of sandstones around the paleotopographic "highs" with exposed crystalline rocks or quartzite. These regressive events, whose timing is shown in Figure 4D.32, delimit the sequences used as mapping units. No major diastrophism has been inferred until the end of the period - and then only in the Arctic. The great unconformity postulated by some workers between the "Middle" and "Upper" Devonian seems to have been the product of "millenialist" thinking. The same applies to the terminal Frasnian unconformity, where it is indeed rather surprising to find any break at all if the ultimate cause of the faunal extinction was a bolide impact (unless it shook the earth to its very core). Changes in the balance of erosion and deposition would be a natural product of climatic change.

The question as to whether global sea-level rise was concurrent with regional flexure and subsidence cannot be answered from the study of one margin alone and is beyond the scope of this paper. Eustasy alone certainly cannot account for the sediments being thousands of metres thicker on the outer than on the inner part of the craton.

Tectonic control over reef development in Alberta has been postulated by many investigators (Goodman, 1956; Sikabonyi and Rodgers, 1959, Fig. 12; Andrichuk, 1961; Martin, 1967, Fig. 7; Keith, 1970; Greggs, 1987; Mountjoy, 1987). Goodman's anomaly was almost certainly the result of salt solution rather than faulting. On the other hand there is little doubt that the topographical features upon and around which certain Slave Point reefs developed were influenced by Devonian faulting. Viau (1987) has mapped a detailed correspondence between sedimentological/ morphological patterns **within** the reef at Swan Hills and a set of southwest-trending offsets believed due to faulting; but this is a phenomenon of different order. No convincing evidence has been advanced for fault control of the Leduc reefs other than their supposed arrangement along certain lineaments. The Rimbey-Meadowbrook chain may look rectilinear, but when the Clyde extension (under the Grosmont platform) is added to the picture it becomes clearly arcuate. Stoakes and Wendte (1987) have argued convincingly for a purely sedimentological explanation of the alignments.

Reefs and source rocks

The rise in relative sea level that was the immediate cause of the transgression proceeded in pulses. When these had the right rate, reefs were engendered. A slow-quick-slow tempo of sea-level rise (the Cretic foot of classic verse) favours reef growth and the deposition of source rocks for oil. During the first slow phase a biostromal foundation is laid down. During the phase of rapid rise the reefs build up, both organic framework builders and inorganic, early cementation seemingly playing a part in consolidating the structure. During this phase, the inter-reef areas are starved of mineral matter and oxygen and become the sites of sediments rich in organic matter; with burial, these will become source rocks for oil to charge the reefs. Finally, as sea-level rise slows again, up-building ceases, calcarenite shoals cap the reefs, and the basin fills by lateral accretion. The main periods of reef-building in response to rapid increase in water depth (Fig. 4D.33) occurred during the Bear Rock, Hume-Dawson, and Beaverhill-Saskatchewan sequences i.e., during Emsian, Eifelian, Givetian-early Frasnian, and middle Frasnian times, respectively. The late Frasnian, Zeta Lake reefs had a different setting and history. There are no true reefs in the Famennian because only one primitive genus of stromatoporoid, *Labechia*, seems to have been available for their construction. *Labechia* did, however, take part in mound-building during the mid-Famennian deepening.

Basin-fill

Most of the basin-filling clinothems seem to have advanced from east to west, probably as a result of trade wind-induced marine circulation. Thus reefs on the eastern side of the basin could build out, even on their windward flanks, over allochthonous material piled against them by these currents; but reefs located farther west could only build out on their leeward sides over their own debris. An important and perhaps principal source of siliciclastic sediment lay to the north, but an eastern cratonic source must also have existed to provide the coarser sandstones that are found interbedded in the Frasnian deposits of both the Grumbler-Grosmont province and the shelf facies of the Northern Siliciclastic province (Fig. 4D.19, in pocket). The heavy-mineral studies that would differentiate these sources remain to be done.

Aggradational cycles

It has been widely assumed that lithological cycles of various orders correspond to cycles of global sea-level change of similar orders. There is no agreement on the extent to which this is the case in the Devonian of western Canada. The conclusion reached here is that the observed cycles are polygenetic. The presence of autocycles in the interior of Swan Hills reefs is accepted. The conglomerate-based punctuations in Bear Rock Formation reported from Root Basin may be storm deposits. Salinity variations, possibly caused by climatic cycles, have been invoked to explain aspects of the Duperow cycles. Muskeg and Prairie formations display fourth-order cycles that seem to be controlled by rhythmic changes in water depth; whether this was a global effect or confined to Elk Point Basin is undetermined. In short, it has not been demonstrated that eustatic cycles of lower than the third order can consistently overprint the other contributory and perhaps local causes of cyclicity.

Remaining problems

Numerous problems remain to be solved in the Devonian stratigraphy of western Canada. There is as yet no acceptable model that will explain the tectonic-sedimentary evolution of the region in terms of plate tectonics. Problems in applying existing models were mentioned above.

As the layer-cake model of basin fill gives way to one (based on oceanic analogues; Lukin et al., 1982) of clinothems advancing over hiatal surfaces, radical revision of established correlations is under way. Decisive evidence may come from conodont biostratigraphy, but it is also very desirable that high-quality seismic surveys from the basinal areas should be published, in order to disclose the distribution of wedge-shaped bodies, downlap relationships, and local intrabasinal unconformities. Problems such as reef to off-reef relationships need additional field and subsurface work in the light of new theory.

The problems of diagenesis and of migration of hydrocarbons and mineral-bearing fluids have fallen outside the scope of this review, but offer unlimited opportunities for valuable research.

RESOURCES

Devonian rocks of Western Canada contain important resources of crude oil, natural gas, sulphur, crude bitumen, carbon dioxide, potash, common salt, gypsum, lead, zinc, and limestone; some phosphate also occurs.

Crude oil and natural gas

Discoveries of conventional crude oil to 1985 totalled 2.9×10^9 m^3 oil in place, which represents about 37% of all the oil in place discovered in the Western Canada Basin. Natural gas discoveries amount to 1.4×10^{12} m^3 of raw gas reserves.

No significant reserves have been found in beds older than Eifelian. The principal reservoir formations, grouped in order of decreasing age, are listed below, together with the initial oil in place of the largest oil pool (in 10^6 m^3) and the initial raw gas in place of the largest gas pool (in 10^9 m^3) in parenthesis [ ]:

Eifelian:

Nahanni Member of Hume Formation in structural trap (Beaver River Field) [0:7]

Lower Givetian:

Upper Keg River Formation (including Rainbow Member and carbonates of Muskeg Formation draped over Keg River reefs) [41:50]

Upper Givetian:

Gilwood Sandstone Member of Watt Mountain Formation and "Granite Wash" of various ages adjacent to Peace River Arch [129:13]

Upper Givetian and lower Frasnian:

Slave Point Formation (platform margin and atolls) [<5:62]

Swan Hills Formation [303:106]

Ramparts (Kee Scarp) Formation [40:0]

Middle and upper Frasnian:

Leduc Formation [207:52]

Nisku Formation (sensu stricto) [60:7]

Zeta Lake Member (of "Nisku" Formation) [<10:0]

Famennian:

Wabamun Formation (including Crossfield Member) [<1:47]

Sulphur

Sulphur is recovered from sour gas. An important source at moderate depth is the Crossfield Member of the Wabamun Formation; other major reserves of sulphur occur in the Wabamun within the folded belt and in deep reservoirs of the Leduc, Swan Hills, and Slave Point formations. Estimated initial reserves from gas already discovered amount to 122 Mt, of which about half has already been extracted.

Crude bitumen

The principal reserve of crude bitumen is in the Grosmont Formation (Frasnian), in which it is estimated that 50×10^9 m^3 may be stored. Techniques of recovery are being investigated.

Carbon dioxide gas

At one time considered a useless gas detracting from the value of any associated hydrocarbon gases, carbon dioxide has now found application in enhanced oil recovery technology. The largest reserves of naturally occurring carbon dioxide discovered in Canada to date are in the folded belt of southeast British Columbia, where the Palliser Formation (Famennian) is one of the two main reservoirs.

Potash

Potash is recovered from the Leofnard Member of the Prairie Formation in Saskatchewan. Production in 1982 from nine shaft mines and one solution mine totalled 7 Mt.

Gypsum

Gypsum is mined from the Burnais Formation of probable Eifelian age in Golden Embayment (Fig. 4D.7). Reserve data are not available but an average of 0.5 Mt/a has been mined in recent years.

Common salt

Both the Leofnard Member of the Prairie Formation and the Lotsberg Formation contain huge reserves of halite from which common salt is extracted by brining or as a by-product of the conventional mining of potash. Production in 1983 was 8.6 Mt.

Lead and zinc

The largest ore field occurs on the south side of Great Slave Lake in karst networks within upper Keg River Formation (Givetian) that has been dolomitized to the Presqu'île diagenetic facies. By 1984 52.8 Mt of 3% Pb and 6.7% Zn had been produced, with remaining reserves of 23.3 Mt of 2.7% Pb and 6.3% Zn (Rhodes et al., 1984). Mining operations ceased in 1988. Other occurrences of lead and zinc are found in the Cordillera. In northeastern British Columbia these occur in the Stone and Dunedin formations (Middle Devonian). In southeastern British Columbia and southwestern Alberta the showings have been in the Palliser Formation (Famennian).

Limestone

The Palliser Formation (Famennian) in the front ranges of the Alberta Rocky Mountains provides abundant limestone suitable for the production of cement. About 1.7 Mt of limestone were quarried for cement manufacture in 1985.

Phosphate

Phosphate occurs in the black shale member of the Exshaw Formation (uppermost Devonian) but mostly in low grades at inaccessible locations.

REFERENCES

Aitken, J.D., Cook, D.G., and Yorath, C.J.
1982: Upper Ramparts River and Sans Sault Rapids (106H) map areas, District of Mackenzie; Geological Survey of Canada, Memoir 388, 48 p.

Alberta Society of Petroleum Geologists
1960: Lexicon of Geologic Names in the Western Canada Sedimentary Basin and Arctic Archipelago; Alberta Society of Petroleum Geologists, Calgary, Alberta, 380 p.

Andrichuk, J.M.
1961: Stratigraphic evidence for tectonic and current control of Upper Devonian reef sedimentation, Duhamel area, Alberta, Canada; American Association of Petroleum Geologists, Bulletin, v. 45, no. 5, p. 612-632.

Bassett, H.G.
1961: Devonian stratigraphy, central Mackenzie River region, Northwest Territories, Canada; in Geology of the Arctic, G.O. Raasch (ed.); v. 1, University of Toronto Press, p. 481-498.

Bassett, H.G. and Stout, J.G.
1968: Devonian of Western Canada; in International Symposium on the Devonian System, Calgary, D.H. Oswald (ed.); Canadian Society of Petroleum Geologists, v. 1, p. 717-752. (Imprint 1967).

Bebout, D.G. and Maiklem, W.R.
1973: Ancient anhydrite facies and environments, Middle Devonian Elk Point Basin, Alberta; Bulletin of Canadian Petroleum Geology, v. 21, p. 287-343.

Belyea, H.R.
1958: Devonian sediments in southern Alberta and correlations with northwestern Montana; Billings Geological Society Guidebook, 9th Annual Field Conference, p. 49-56.
1971: Middle Devonian tectonic history of the Tathlina Uplift, southern District of Mackenzie and northern Alberta, Canada; Geological Survey of Canada, Paper 70-14, 38 p.

Belyea, H.R. and McLaren, D.J.
1962: Upper Devonian formations, southern part of Northwest Territories, northeastern British Columbia and northwestern Alberta; Geological Survey of Canada, Paper 61-29, 74 p.

Braman, D.R.
1981: Upper Devonian-Lower Carboniferous miospore biostratigraphy of the Imperial Formation, District of Mackenzie and Yukon; Ph.D. thesis, University of Calgary, Calgary, Alberta, 377 p.

Braun, W.K.
1966: Stratigraphy and microfauna of Middle and Upper Devonian formations, Norman Wells area, Northwest Territories, Canada; Neues Jahrbuch für Geologie und Paläontologie Abhandlungen, v. 125, p. 247-264.
1977: Usefulness of ostracodes in correlating Middle and Upper Devonian rock sequences in Western Canada; in Western North America: Devonian, M.A. Murphy, W.B.N. Berry, and C.A. Sandberg (ed.); University of California Riverside Campus Museum, Contribution 4, p. 65-79.

Braun, W.K. and Mathison, J.E.
1982: Ostracodes as a correlation tool in Devonian studies of Saskatchewan and adjacent areas; in 4th International Williston Basin Symposium, October 5-7, 1982, Regina, Saskatchewan, J.E. Christopher and J. Kaldi (ed.); Saskatchewan Geological Society, Special Publication no. 6, p. 43-50.

Busson, V.G.
1974: Interpretation et synthèse des données de sondages de l'Upper Elk Point (Dévonien Moyen du Canada occidental); Genèse des evaporites et rapports avec les carbonates; in Report of Activities, Part A, Geological Survey of Canada, Paper 74-1A. p. 291-295.

Cecile, M.P.
1982: The Lower Paleozoic Misty Creek Embayment, Selwyn Basin, Yukon and Northwest Territories; Geological Survey of Canada, Bulletin 335, 78 p.

Chatterton, B.D.E.
1978: Aspects of late Early and Middle Devonian conodont biostratigraphy of western and northwestern Canada; in Western and Arctic Biostratigraphy, C.R. Stelck and B.D.E. Chatterton (ed.); Geological Association of Canada, Special Paper 18, p. 161-231.

Chevron Standard Exploration Staff
1979: The geology, geophysics and significance of the Nisku Reef discoveries, West Pembina area, Alberta, Canada; Bulletin of Canadian Petroleum Geology, v. 27, p. 326-359.

Christie, H.H.
1971: Mitsue oil field: a rich stratigraphic trap; Proceedings 8th World Petroleum Congress, Moscow, v. 2, Applied Science Publishers Ltd., London, p. 269-274.

Christopher, J.E.
1961: Transitional Devonian-Mississippian formations of Southern Saskatchewan; Saskatchewan Department of Mineral Resources, Report 66, 103 p.

Clough, J.G. and Blodgett, R.G.
1984: Lower Devonian basin to shelf carbonates in outcrop from the western Ogilvie Mountains, Alaska and Yukon Territory; in Carbonates in Subsurface and Outcrop, L. Eliuk (Chairperson); Canadian Society of Petroleum Geologists 1984 Core Conference, October 18-19, 1984, Calgary, Alberta, Canadian Society of Petroleum Geologists, p. 57-81.

Cook, D.G.
1975: The Keele Arch - a pre-Devonian and Pre-Late Cretaceous paleo-upland in the northern Franklin Mountains and Colville Hills; in Report of Activities, Part C, Geological Survey of Canada, Paper 75-1C, p. 243-246.

Cook, D.G. and Aitken, J.D.
1978: Twitya Uplift - a pre-Delorme phase of the Mackenzie Arch; in Current Research, Part A, Geological Survey of Canada, Paper 78-1A, p. 383-388.

Cook, H.E., Daniel, P.N., Mountjoy, E.W., and Pray, L.C.
1972: Allochthonous carbonate debris flows at Devonian bank ("reef") margins Alberta, Canada; Bulletin of Canadian Petroleum Geology, v. 20, p. 439-497.

Corrigan, V.A.F.
1975: The evolution of a cratonic basin from carbonate to evaporitic deposition and the resulting stratigraphic and diagenetic changes, Upper Elk Point Subgroup, northeastern Alberta; Ph.D. thesis, University of Calgary, Calgary, Alberta, 328 p.

Crawford, F.D.
1972: Facies analysis and depositional environments in the Middle Devonian Fort Vermilion and Slave Point formations of northern Alberta; Masters thesis, University of Calgary, Calgary, Alberta, 99 p.

Crickmay, C.H.
1956: The Palliser-Exshaw contact; Guide Book for the 6th Annual Field Conference of the Alberta Society of Petroleum Geologists, p. 56-58.
1957: Elucidation of some Western Canada Devonian formation; C.H. Crickmay, Calgary, 17 p.
1970: Ramparts, Beavertail and other Devonian formations; Bulletin of Canadian Petroleum Geology, v. 18, p. 67-79.

Cutler, W.G.
1983: Stratigraphy and sedimentology of the Upper Devonian Grosmont Formation; Bulletin of Canadian Petroleum Geology, v. 31, p. 282-325.

Davies, G.R. and Ludlam, S.D.
1973: Origin of laminated and graded sediments, Middle Devonian of Western Canada; Geological Society of America, Bulletin, v. 84, p. 3527-3546.

Desbordes, B. and Maurin, A.F.
1974: Trois examples d'étude du Frasnien d'Alberta Canada; Compagnie Française des Pétroles, Notes et Memoires 11, p. 293-336.

Dineley, D.L. and Loeffler, E.J.
1976: Ostracoderm faunas of the Delorme and associated Siluro-Devonian formations, Northwest Territories, Canada; Paleontological Association Special Papers on Paleontology No. 18, 214 p.

Dooge, J.
1966: The stratigraphy of an Upper Devonian carbonate-shale transition between the North and South Ram Rivers of the Canadian Rocky Mountains; Leidse Geologische Mededelingen, no. 39, p. 1-53.

Douglas, R.J.W. (ed.)
1970: Geology and Economic Minerals of Canada; Geological Survey of Canada, Economic Geology Report No. 1 (Fifth Edition), 838 p.

Douglas, R.J.W. and Norris, D.K.
1961: Camsell Bend and Root River map-areas, District of Mackenzie, Northwest Territories; Geological Survey of Canada, Paper 61-13, 36 p.

Dumestre, A.
1969: Relations entre hydrocarbures et environment évaporitique à Rainbow Alberta (Canada); Association Française des Techniciens du Petrole, Revue no. 194, p. 29-46.

Dunn, C.E.
1975: The Upper Devonian Duperow Formation in southeastern Saskatchewan; Saskatchewan Department of Mineral Resources, Report 179, 151 p.
1982: Geology of the Middle Devonian Dawson Bay Formation in the Saskatoon Potash Mining District; Saskatchewan Energy and Mines, Report 194, 117 p.

Eliuk, L.
1984: A hypothesis for the origin of hydrogen sulphide in Devonian Crossfield Dolomite, Wabamun Formation, Alberta, Canada; in Carbonates in Subsurface and Outcrop, L. Eliuk (Chairperson); Canadian Society of Petroleum Geologists 1984 Core Conference, October 18-19, 1984, Calgary, Alberta, Canadian Society of Petroleum Geologists, p. 245-289.

Elloy, R.
1972: Réflexions sur quelques environments récifaux du Paléozoïque; Bulletin du Centre de Recherches, Pau- SNPA, v. 6, p. 1-105.

Embry, A. and Klovan, J.E.
1976: The Middle-Upper Devonian clastic wedge of the Franklinian Geosyncline; Bulletin of Canadian Petroleum Geology, v. 24, p. 485-639.

Ettensohn, F.R. and Barron, L.S.
1981: Depositional model for the Devonian-Mississippian black shale sequence of North America: a tectono-climatic approach; United States, Department of Energy DOE/METC/12040-2, 83 p.

Fischbuch, N.R.
1968: Stratigraphy, Devonian Swan Hills reef complexes of central Alberta; Bulletin of Canadian Petroleum Geology, v. 16, p. 444-556.
1984: Facies and reservoir analysis, Kee Scarp Formation, Norman Wells area, Northwest Territories; Geological Survey of Canada, Open File 1116.

Fuller, J.G.C.M. and Pollock, C.A.
1972: Early exposure of Middle Devonian reefs, southern Northwest Territories, Canada; Proceedings of the 24th International Geological Congress, Montreal, Section 6, p. 144-155.

Fuller, J.G.C.M. and Porter, J.W.
1969: Evaporites and carbonates; two Devonian basins of Western Canada; Bulletin of Canadian Petroleum Geology, v. 17, p. 182-193.

Fuzesy, A.
1982: Potash in Saskatchewan; Saskatchewan Department of Energy and Mines Report 181, 44 p.

Fuzesy, L.M.
1980: Geology of the Deadwood (Cambrian) Meadow Lake and Winnipegosis (Devonian) formations in west-central Saskatchewan; Saskatchewan Mineral Resources Report 210, 64 p.

Gabrielse, H.
1967: Tectonic evolution of the northern Canadian Cordillera; Canadian Journal of Earth Sciences, v. 4, p. 271-298.

Gabrielse, H., Busson, S.L., and Roddick, J.A.
1973: Geology of Flat River, Glacier Lake and Wrigley Lake map areas, District of Mackenzie and Yukon Territory; Geological Survey of Canada, Memoir 366, 121 p.

Gabrielse, H. and Yorath, C.J. (ed.)
1991: Geology of the Cordilleran Orogen in Canada; Geological Survey of Canada, Geology of Canada, no. 4. (Also Geological Society of America, The Geology of North America, v. G-2.)

Geldsetzer, H.H.J.
1982: Depositional history of the Devonian succession in the Rocky Mountains southwest of the Peace River Arch; in Current Research, Part C, Geological Survey of Canada, Paper 82-1C, p. 55-64.

Gendzwill, D.J.
1978: Winnipegosis mounds and Prairie Evaporite Formation of Saskatchewan; American Association of Petroleum Geologists, Bulletin, v. 62, p. 73-86.

Gibson, G.
1978: Geology of the Munroe-Alpine-Boivin carbonate-hosted zinc occurrences, Rocky Mountain Front Ranges, southeastern British Columbia; British Columbia Ministry of Energy, Mines and Petroleum Resources, Preliminary Map No. 46, 1:25 000.

Goodman, A.J.
1956: Comment on "Time of Migration" by W.C. Gussow; Alberta Society of Petroleum Geologists, Journal, v. 4, no. 8, p. 184-185.

Gordey, S.P.
1987: Devono-Mississippian sedimentation and tectonism in the Canadian Cordillerian Miogeocline; Second International Symposium on the Devonian System, Program and Abstracts, Canadian Society of Petroleum Geologists, Calgary, Alberta, p. 98.

Greggs, R.G.
1987: A structural model for the formation of carbonate 'buildups' (reefs); Second International Symposium on the Devonian System, Program and Abstracts, p. 103.

Griffin, D.L.
1965: The Devonian Slave Point, Beaverhill Lake and Muskwa formations; British Columbia Department of Mines and Petroleum Resources, Bulletin 50, 90 p.

Hage, C.O.
1945: Geological reconnaissance along the lower Liard River, Northwest Territories, Yukon and British Columbia; Geological Survey of Canada, Paper 45-22, 33 p.

Hamilton, W.N.
1971: Salt in east-central Alberta; Alberta Research Council, Bulletin 29, 53 p.

Harrison, R.
1986: Stratigraphy, sedimentology, and bitumen potential of the Upper Devonian Grosmont Formation, Northern Alberta (abstract); Canadian Society of Petroleum Geologists, Reservoir, v. 13, no. 1, p. 1-3.

Havard, C. and Oldershaw, A.
1976: Early diagenesis in back-reef sedimentary cycles, Snipe Lake reef complex, Alberta; Bulletin of Canadian Petroleum Geology, v. 24, p. 27-69.

Hills, L.V. and Braman, D.R.
1978: Sedimentary structures of the Imperial Formation, northwest Canada; in Canadian Society of Petroleum Geologists Core Conference, 1978, A.F. Embry (compiler), Display Summaries, p. 35-37.

Hills, L.V., Sangster, E.V., and Suneby, L.B. (ed.)
1981: Lexicon of Canadian Stratigraphy, Volume 2, Yukon Territories and District of Mackenzie; Canadian Society of Petroleum Geologists, Calgary, Canada, 240 p.

Holser, W.T., Wardlaw, N.C., and Watson, D.W.
1972: Bromide in salt rocks: extraordinarily low content in the Lower Elk Point salt, Canada; in Geology of Saline Deposits, G. Richter-Bernburg (ed.); Proceedings of the Hanover Symposium 1968 (Earth Sciences 7), Unesco, Paris, p. 69-73.

Hopkins, J.C.
1977: Production of foreslope breccia by differential submarine cementation and downslope displacement of carbonate sands, Miette and Ancient Wall buildups, Devonian, Canada; Society of Economic Paleontologists and Mineralogists, Special Publication 25, p. 155-170.

Hriskevich, M.E., Faber, J.M., and Langton, J.R.
1980: Strachan and Ricinus West gas fields, Alberta, Canada; in Giant Oil and Gas Fields of the Decade 1968-1978, M.T. Halbouty (ed.); American Association of Petroleum Geologists, Memoir 30, p. 315-327.

Husain, M. and Warren, J.K.
1985: Origin of laminae in Holocene-Pleistocene evaporitic sequence of salt-flat playas, West Texas and New Mexico (Abstract); American Association of Petroleum Geologists, Bulletin, v. 69, no. 2, p. 268.

Jackson, P.C. and Harrison, R.S. (ed.)
1982: Hummingbird Reef Complex, Field Trip Guidebook; American Association of Petroleum Geologists Annual Convention, June 27-30, 1982, Calgary, Alberta, Canadian Society of Petroleum Geologists, Trip no. 7, 91 p.

Jamieson, E.R. (Compiler)
1967: Field Trip A-11 Guidebook - Upper Devonian outcrops, Hay River area; International Symposium on the Devonian System, Calgary, 36 p.

Jamieson, E.R.
1971: Paleoecology of Devonian reefs in western Canada; Proceedings of the North American Paleontological Convention, Chicago, 1969, Proceedings Part J, Allen Press, Lawrence, Kansas, p. 1300-1340.

Jansa, L.F. and Fischbuch, N.R.
1974: Evolution of a Middle and Upper Devonian sequence from a clastic coastal plain complex into overlying carbonate reef complex and banks, Sturgeon-Mitsue area, Alberta; Geological Survey of Canada, Bulletin 234, 105 p.

Johnson, J.G.
1970: Taghanic onlap and the end of North American Devonian provinciality; Geological Society of America, Bulletin, v. 81, p. 2077-2106.

Johnson, J.G., Klapper, G., and Sandberg, C.A.
1985: Devonian eustatic fluctuations in Euramerica; Geological Society of America, Bulletin, v. 96, p. 567-587.

Jones, L.
1965: The Middle Devonian Winnipegosis Formation of Saskatchewan; Saskatchewan Department of Mineral Resources, Report 98, 101 p.

Jordan, S.P.
1968: Will Zama be duplicated at Quill Lake, Saskatchewan?; Oilweek, v. 19, no. 2, p. 10-12.

Keith, J.W.
1970: Tectonic control of Devonian reef sedimentation, Alberta (abstract); American Association of Petroleum Geologists, Bulletin, v. 54, no. 5, p. 854.

Kent, D.M.
1968: The geology of the Upper Devonian Saskatchewan Group and equivalent rocks in western Saskatchewan and adjacent areas; Saskatchewan Department of Mineral Resources, Report 99, 221 p.

Kissling, D.L. and Ehrets, J.R.
1984: Depositional Models for the Duperow and Birdbear Formations: implications for correlation and exploration; Oil and Gas in Saskatchewan, J.A. Lorsong et al. (ed.); Saskatchewan Geological Society, Special Publication No. 7, (Addendum), unpaginated.

Klapper, G. and Lane, H.R.
1985: Upper Devonian (Frasnian) conodonts of the *Polygnathus* biofacies, Northwest Territories, Canada; Journal of Paleontology, v. 59, no. 4, p. 904-951.

Klingspor, A.M.
1969: Middle Devonian Muskeg evaporites of Western Canada; American Association of Petroleum Geologists, Bulletin, v. 53, p. 927-948.

Klovan, J.E.
1964: Facies analysis of the Redwater Reef Complex, Alberta, Canada; Bulletin of Canadian Petroleum Geology, v. 12, p. 1-100.

Kramers, J.W. and Lerbekmo, J.E.
1967: Petrology and mineralogy of Watt Mountain Formation, Mitsue-Nipisi area, Alberta; Bulletin of Canadian Petroleum Geology, v. 15, p. 346-378.

Lane, D.M.
1964: Souris River Formation in southern Saskatchewan; Saskatchewan Department of Mineral Resources, Report 92, 72 p.

Law, J.
1971: Regional Devonian geology and oil and gas possibilities, Upper Mackenzie River area; Bulletin of Canadian Petroleum Geology, v. 19, p. 437-484.

Lenz, A.C.
1972: Ordovician to Devonian history of northern Yukon and adjacent District of Mackenzie; Bulletin of Canadian Petroleum Geology, v. 20, p. 321-361.

Lenz, A.C. and Pedder, A.E.H.
1972: Lower and Middle Paleozoic sediments and paleontology of Royal Creek and Peel River, Yukon, and Powell Creek Northwest Territories; International Geological Congress 24th Session, Montreal, Guidebook Field Excursion A14, 43 p.

Lethiers, F.
1978: Ostracodes du Dévonien terminal de la Formation Big Valley, Saskatchewan et Alberta; Palaeontographica, Abteilung A, 162, Leiferung 3-6, p. 81-143, Stuttgart, Oktober 1978.

Lobdell, F.K.
1984: Age and depositional history of the Middle Devonian Ashern Formation in the Williston Basin, Saskatchewan and North Dakota, J.A. Lorsong and M.A. Wilson (ed.); Saskatchewan Geological Society, Special Publication No. 7, p. 5-12.

Ludvigsen, R.
1970: Age and fauna of the Michelle Formation, northern Yukon Territory; Bulletin of Canadian Petroleum Geology, v. 18, p. 407-429.
Lukin, A. Ye., Chizhova, V.A., Alekseyeva, L.P., Larchenov, A. Ya.
1982: Relationship between wedge-shaped geologic bodies and biostratigraphic units; Doklady Akademii Nauk SSSR, v. 266, p. 1212-1215. (English translation)
Machel, Hans-G.
1983: Facies and diagenesis of some Nisku buildups and associated strata, Upper Devonian, Alberta, Canada; in Carbonate Buildups - a Core Workshop (SEPM Core Workshop No. 4), P.M. Harris (ed.); Society of Economic Paleontologists and Mineralogists, Tulsa, Oklahoma, p. 144-181.
Mackenzie, W.S., Pedder, A.E.H., and Uyeno, T.T.
1975: A Middle Devonian Sandstone Unit, Grandview Hills area, District of Mackenzie; in Report of Activities, Part A, Geological Survey of Canada, Paper 75-1A, p. 547-552.
Macqueen, R.W.
1974: Lower and middle Paleozoic studies, northern Yukon Territory (106E, F, L); in Report of Activities, Part A, Geological Survey of Canada, Paper 74-1A, p. 323-326.
Macqueen, R.W. and Powell, T.G.
1983: Organic geochemistry of the Pine Point lead zinc ore field and region, Northwest Territories, Canada; Economic Geology, v. 78, p. 1-25.
Macqueen, R.W. and Taylor, G.E.
1974: Devonian stratigraphy, facies changes and zinc-lead mineralization, southwestern Halfway River area (94B) northeast British Columbia; in Report of Activities, Part A, Geological Survey of Canada, Paper 74-1A, p. 327-331.
Macqueen, R.W. and Thompson, R.I.
1978: Carbonate-hosted lead-zinc occurrences in northeastern British Columbia with emphasis on the Robb Lake Deposit; Canadian Journal of Earth Sciences, v. 15, p. 1737-1762.
Maiklem, W.R.
1971: Evaporitive drawdown - a mechanism for water-level lowering and diagenesis in the Elk Point Basin; Bulletin of Canadian Petroleum Geology, v. 19, p. 485-501.
Martin, R.
1967: Morphology of some Devonian reefs in Alberta: a paleogeomorphological study; in International Symposium on the Devonian System, D.H. Oswald, (ed.), Alberta Society of Petroleum Geologists, Calgary, Alberta, v. 2, p. 365-385.
Maurin, A.F. and Raasch, G.O.
1972: Early Frasnian stratigraphy, Kakwa-Cecilia Lakes British Columbia, Canada; Compagnie Française des Pétroles, Notes et Mémoires No. 10, Paris, 80 p.
McAdam, K.A.
1983: Devonian subsurface correlations, northeast British Columbia; (set of 7 sections) British Columbia, Ministry of Energy, Mines and Petroleum Resources.
McCabe, H.R.
1971: Stratigraphy of Manitoba, an introduction and review; Geological Association of Canada, Special Paper No. 9, p. 167-187 (1972).
McCamis, J.G. and Griffith, L.S.
1967: Middle Devonian facies relationships, Zama area, Alberta; Bulletin of Canadian Petroleum Geology, v. 15, p. 434-467.
McCrossan, R.G. and Glaister, R.P. (ed.)
1964: Geological history of Western Canada; Alberta Society of Petroleum Geologists, Calgary, Alberta, 232 p.
McGillivray, J.G. and Mountjoy, E.W.
1975: Facies and related reservoir characteristics, Golden Spike Reef Complex, Alberta; Bulletin of Canadian Petroleum Geology, v. 23, p. 753-809.
McLaren, D.J.
1982: Frasnian-Famennian extinctions; in Geological Implications of Impacts of Large Asteroids and Comets on the Earth, L.T. Silver and P.H. Schultz (ed.); Geological Society of America, Special Paper 190, p. 477-484.
Meijer Drees, N.C.
1980: Description of the Hume, Funeral, and Bear Rock formations in the Candex et al. Dahadinni M43A well, District of Mackenzie; Geological Survey of Canada, Paper 78-17, 26 p.
in press: The Devonian succession in the subsurface of the Great Slave and Great Bear Plains, Northwest Territories; Geological Survey of Canada, Bulletin 393.
Morrow, D.W.
1975: The Florida Aquifer: a possible model for a Devonian paleoaquifer in northeastern British Columbia; in Report of Activities, Part B, Geological Survey of Canada, Paper 75-1B, p. 261-266.
1978: The Dunedin Formation - a transgressive shelf carbonate sequence; Geological Survey of Canada, Paper 76-12, 35 p.
1984a: Sedimentation in Root Basin and Prairie Creek embayment - Siluro-Devonian, Northwest Territories; Bulletin of Canadian Petroleum Geology, v. 32, p. 162-189.
1984b: The Manetoe Facies - a gas-bearing late diagenetic dolomite of the Northwest Territories, Canada; in Carbonates in Subsurface and Outcrop, L. Eliuk (chairperson); Canadian Society of Petroleum Geologists 1984 Core Conference, October 18-19, 1984, Calgary, Alberta, Canadian Society of Petroleum Geologists, p. 55-56.
in press: The Silurian-Devonian sequence of the northern part of the Mackenzie Shelf, Northwest Territories; Geological Survey of Canada, Bulletin.
Morrow, D.W. and Meijer Drees, N.C.
1981: The Early to Middle Devonian Bear Rock Formation in the type section and other surface sections, District of Mackenzie; in Current Research, Part A, Geological Survey of Canada, Paper 81-1A, p. 107-114.
Mountjoy, E.W.
1978: Upper Devonian reef trends and configuration of the western portion of the Alberta Basin; in The Fairholme Carbonate Complex at Hummingbird and Cripple Creek, I.A. McIlreath and P.C. Jackson (ed.); Canadian Society of Petroleum Geologists, p. 1-30.
1980: Some questions about the development of Upper Devonian carbonate buildups (reefs), western Canada; Bulletin of Canadian Petroleum Geology, v. 28, p. 315-344.
1987: Controls on reef development Devonian of Western Canada Sedimentary Basin (abstract); in Second International Symposium on the Devonian System, Program and Abstracts, Canadian Society of Petroleum Geologists, Calgary, Alberta, p. 174.
Mountjoy, E.W. and Burrowes, G.
1982: Upper Devonian Miette Reef Complex, Jasper National Park, Alberta; International Association of Sedimentologists, 11th Annual Congress on Sedimentology, Hamilton, Excursion 27A, 57 p.
Muir, I., Wong, P., and Wendte, J.
1984: Devonian Hare Indian-Ramparts (Kee Scarp) evolution, Mackenzie Mountains and subsurface Norman Wells, Northwest Territories: Basin-fill and platform reef development; in Carbonates in Subsurface and Outcrop, L. Eliuk (ed.); Canadian Society of Petroleum Geologists 1984 Core Conference, October 18-19, 1984, Calgary, Alberta, Canadian Society of Petroleum Geologists, p. 82-102.
Nichols, R.A.H.
1970: The petrology and economic geology of the Upper Devonian Birdbear Formation, Saskatchewan; Saskatchewan Department of Mineral Resources, Report 125, 93 p.
Nishida, D.K., Murray, J.W., and Stearn, C.W.
1985: Stromatoporoid-algal-facies hydrocarbon traps in Upper Devonian (Famennian) Wabamun Group, north-central Alberta, Canada (abstract); American Association of Petroleum Geologists, Bulletin, v. 69, p. 293.
Noble, J.P.A.
1970: Biofacies analysis, Cairn Formation of Miette Reef Complex (Upper Devonian) Jasper National Park, Alberta; Bulletin of Canadian Petroleum Geology, v. 18, p. 493-543.
Noble, J.P.A. and Ferguson, R.D.
1971: Facies and faunal relations at edge of early Mid-Devonian Carbonate Shelf, South Nahanni River area, Northwest Territories; Bulletin of Canadian Petroleum Geology, v. 19, p. 570-588.
Norford, B.S.
1981: Devonian stratigraphy at the margins of the Rocky Mountain Trench, Columbia River, southeastern British Columbia; Bulletin of Canadian Petroleum Geology, v. 29, p. 540-560.
Norris, A.W.
1963: Devonian stratigraphy of northeastern Alberta and northwestern Saskatchewan; Geological Survey of Canada, Memoir 313, 168 p.
1965: Stratigraphy of Middle Devonian and older Palaeozoic rocks of the Great Slave Lake region, Northwest Territories; Geological Survey of Canada, Memoir 323, 180 p.

Norris, A.W. (cont.)
1973: Paleozoic (Devonian) geology of northeastern Alberta and northwestern Saskatchewan; in Guide to the Athabasca Oil Sands Area, M.A. Carrigy and J.W. Kramers (ed.); Alberta Research Council, Information Series 65, p. 16-61.
1985: Stratigraphy of Devonian outcrop belts in northern Yukon Territory and northwestern District of Mackenzie (Operation Porcupine area); Geological Survey of Canada, Memoir 410, 81 p.

Norris, A.W. and Uyeno, T.T.
1981: Stratigraphy and paleontology of the lowermost Upper Devonian Slave Point Formation on Lake Claire and the lower Upper Devonian Waterways Formation of northeastern Alberta; Geological Survey of Canada, Bulletin 334, 53 p.
1983: Biostratigraphy and paleontology of Middle-Upper Devonian boundary beds, Gypsum Cliffs area, northeastern Alberta; Geological Survey of Canada, Bulletin 313, 65 p.

Norris, A.W., Uyeno, T.T., and McCabe, H.R.
1982: Devonian rocks of the Lake Winnipegosis Lake Manitoba outcrop localities; Geological Survey of Canada, Memoir 392, 280 p.

Norris, D.K.
1981: Geology Old Crow, Yukon Territory; Geological Survey of Canada, Map 1518A, 1:250 000.

Osadetz, K.G. and Snowdon, L.R.
1986: Petroleum source rock reconnaissance of southern Saskatchewan; in Current Research, Part A, Geological Survey of Canada, Paper 86-1A, p. 609-617.

Paterson, D.F.
1973: Computer plotted isopach and structure maps of the Devonian Formations in Saskatchewan; Saskatchewan Department of Mineral Resources, Report 164, 15 p.

Pedder, A.E.H.
1977: Corals of the Lower/Middle and Middle/Upper Devonian boundary beds of northern and western Canada; in Western North America: Devonian, M.A. Murphy, W.B.N. Berry, and C.A. Sandberg (ed.); University of California Riverside Campus Museum, Contribution 4, p. 99-106
1982: The rugose coral record across the Frasnian/Famennian boundary; Geological Society of America, Special Paper 190, p. 485-489.

Pedder, A.E.H. and Klapper, G.
1977: Fauna and correlation of the type section of the Cranswick Formation (Devonian), Mackenzie Mountains, Yukon; in Report of Activities, Part B, Geological Survey of Canada, Paper 77-1B, p. 227-234.

Pelzer, E.E.
1966: Mineralogy, geochemistry and stratigraphy of Besa River Shale, British Columbia; Bulletin of Canadian Petroleum Geology, v. 14, p. 273-321.

Perrin, N.A.
1982: Environments of deposition and diagenesis of the Winnipegosis Formation (Middle Devonian), Williston Basin, North Dakota; in 4th International Williston Basin Symposium, J.E. Christopher and J. Kaldi (ed.); Saskatchewan Geological Society, Special Publication No. 6, p. 51-66.

Perry, D.G.
1984: Brachiopoda and biostratigraphy of the Silurian-Devonian Delorme Formation in the District of Mackenzie; Royal Ontario Museum Life Sciences, Contributions No. 138, 150 p.

Perry, D.G., Klapper, G., and Lenz, A.C.
1974: Age of the Ogilvie Formation (Devonian), northern Yukon, based primarily on the occurrence of brachiopods and conodonts; Canadian Journal of Earth Sciences, v. 11, p. 1055-1097.

Pray, L.C.
1969: Basin sloping submarine (?) unconformities at margins of Paleozoic banks, West Texas and Alberta (Abstracts); Geological Society of America, Special Paper 121, p. 243.

Price, L.L. and Ball, N.L.
1971: Stratigraphy of Duval Corporation potash shaft No.1, Saskatoon, Saskatchewan; Geological Survey of Canada, Paper 70-71, 107 p.

Price, R.A.
1965: Flathead map area, British Columbia and Alberta; Geological Survey of Canada, Memoir 336, 221 p.

Pugh, D.C.
1983: Pre-Mesozoic geology in the subsurface of Peel River map area, Yukon Territory and District of Mackenzie; Geological Survey of Canada, Memoir 401, 61 p.

Read, J.F.
1973: Paleo-environments and paleogeography, Pillara Formation (Devonian), Western Australia; Bulletin of Canadian Petroleum Geology, v. 21, p. 344-394.

Read, J.F., Grotzinger, J.P., Bova, J.A., and Koerschner, W.F.
1984: Models for generation of 5th order carbonate cycles; Geological Society of America, Abstracts with Programs, p. 631.

Rhodes, D., Lantos, E.A., Lantos, J.A, Webb, R.J., and Owens, D.C.
1984: Pine Point orebodies and their relationship to the stratigraphy, structure, dolomitization and karstification of the Middle Devonian Barrier Complex; Economic Geology, v. 79, no. 5, p. 991-1055.

Richmond, W.O.
1965: Paleozoic stratigraphy and sedimentation of the Slave Point Formation, southern Northwest Territories and northern Alberta; Ph.D. thesis, Stanford University, Stanford, California, 565 p.

Rottenfusser, D.D. and Oliver, T.T.
1977: Depositional environment and petrology of the Gilwood Member north of the Peace River Arch; Bulletin of Canadian Petroleum Geology, v. 25, p. 907-928.

Sartenaer, P.
1969: Late Upper Devonian (Famennian) rhynchonellid brachiopods from Western Canada; Geological Survey of Canada, Bulletin 169, 219 p.

Schmidt, V.
1971: Diagenesis of Keg River bioherms, Rainbow Lake, Alberta (abstract); American Association of Petroleum Geologists, Bulletin, v. 54, p. 868.

Schmidt, V., McDonald, D.A., and McIlreath, I.A.
1977: Growth and diagenesis of Middle Devonian Keg River cementation reefs, Rainbow Field, Alberta; in The Geology of Selected Carbonate Oil, Gas and Lead-Zinc Reservoirs in Western Canada (supplement), I.A. McIlreath and R.D. Harrison (ed.); Canadian Society of Petroleum Geologists, p. 1-21.

Shawa, M.S.
1969: Sedimentary history of the Gilwood Sandstone (Devonian) Utikama Lake area, Alberta, Canada; Bulletin of Canadian Petroleum Geology, v. 17, p. 392-409.

Sheasby, N.M.
1971: Depositional patterns of the Upper Devonian Waterways Formation, Swan Hills area; Bulletin of Canadian Petroleum Geology, v. 19, p. 377-404.

Sikabonyi, L.A. and Rodgers, W.J.
1959: Paleozoic tectonics and sedimentation in the northern half of the West Canadian Basin; Journal of the Alberta Society of Petroleum Geologists, v. 7, no. 9, p. 193-216.

Skall, H.
1975: The paleoenvironment of the Pine Point lead-zinc district; Economic Geology, v. 70, p. 22-47.

Sloss, L.L.
1963: Sequences in the cratonic interior of North America; Geological Society of America, Bulletin, v. 74, p. 93-113.

Smith, D.G. and Pullen, J.R.
1967: Hummingbird structure of southeast Saskatchewan; Bulletin of Canadian Petroleum Geology, v. 15, p. 468-482.

Snowdon, D.M.
1977: Beaver River Gas Field: a fractured carbonate reservoir; in The Geology of Selected Carbonate Oil, Gas and Lead-Zinc Reservoirs in Western Canada, I.A. McIlreath and R.D. Harrison (ed.); Canadian Society of Petroleum Geologists, p. 1-18.

Stearn, C.W.
1961: Devonian stromatoporoids from the Canadian Rocky Mountains; Journal of Paleontology, v. 35, no. 5, p. 932-948.

Stoakes, F.A.
1980: Nature and control of shale basin fill and its effect on reef growth and termination: Upper Devonian Duvernay and Ireton Formations of Alberta, Canada; Bulletin of Canadian Petroleum Geology, v. 28, p. 345-410.

Stoakes, F.A. and Wendte, J.C.
1987: The Woodbend Group; in Devonian Lithofacies and Reservoir Styles in Alberta, F.F. Krause, and O.G. Burrowes, (ed.); 13th Canadian Society of Petroleum Geologists Core Conference and Second International Symposium on the Devonian System (Calgary, 1987), Canadian Society of Petroleum Geologists, Calgary, Alberta, p. 153-170.

Storey, T.P.
1970: Evaporite basin configuration - structural versus sedimentary interpretation: in Third Symposium on Salt, J.L. Rau and L.F. Dellwig (ed.); Northern Ohio Geological Society Inc., Cleveland, Ohio, v. 1, p. 8-19.

Tassonyi, E.J.
1969: Subsurface geology, lower Mackenzie River and Anderson River areas, District of Mackenzie; Geological Survey of Canada, Paper 68-25, 207 p.

Taylor, G.C.
1982: Geological guide to the central and southern Rocky Mountains of Alberta and British Columbia; Canadian Society of Petroleum Geologists Field Trip No. 4, Canadian Society of Petroleum Geologists/American Society of Petroleum Geologists, Calgary, Alberta, 40 p.

Taylor, G.C. and Mackenzie, W.S.
1970: Devonian stratigraphy of northeastern British Columbia; Geological Survey of Canada, Bulletin 186, 62 p.

Taylor, G.C., Macqueen, R.W., and Thompson, R.I.
1975: Facies changes, breccias and mineralisation in Devonian rocks of Rocky Mountains, northeast British Columbia; in Report of Activities, Part A, Geological Survey of Canada, Paper 75-1A, p. 577-585.

Thompson, R.I.
1976: Some aspects of stratigraphy and structure in the Halfway River map area (94B), British Columbia; in Report of Activities, Part A, Geological Survey of Canada, Paper 76-1A, p. 471-477.

Tranchant, J-C.
1975: Essai de synthèse des données de sondage de l' "Upper Elk Point" (dévonien moyen) des champs de Rainbow et de Zama (nord-ouest de l'Alberta): chronologie des formations "Muskeg" et "Upper Keg River"; in Report of Activities, Part B, Geological Survey of Canada, Paper 75-1B, p. 303-307.

Uyeno, T.T.
1974: Conodonts of the Waterways Formation (Upper Devonian) of northeastern and central Alberta; Geological Survey of Canada, Bulletin 232, 93 p.
1978: Devonian conodont biostratigraphy of Powell Creek and adjacent areas, western District of Mackenzie; in Western and Arctic Canadian Biostratigraphy, C.R. Stelck and B.D.E. Chatterton (ed.); Geological Association of Canada, Special Paper 18, p. 233-257.

Vail, P.R. Mitchum, R.M., Jr., and Thompson, S., III.
1977: Seismic stratigraphy and global changes in sea level; in Seismic Stratigraphy - Applications to Hydrocarbon Exploration, C.E. Payton (ed.); American Association of Petroleum Geologists, Memoir 26, p. 99-116.

Viau, C.A.
1983: Depositional sequences, facies and evolution of the Upper Devonian Swan Hills Reef buildup, central Alberta, Canada; in Carbonate Buildups - a Core Workshop, Paul M. Harris (ed.), Society of Exploration Paleontologists and Mineralogists, Core Workshop No. 4, p. 112-143.
1987: Structural control on the sedimentological development of the Swan Hills Formation, Swan Hills field, Alberta, Canada; in 13th Canadian Society of Petroleum Geologists Core Conference and Second International Symposium on the Devonian System, Canadian Society of Petroleum Geologists, Calgary, Alberta, Program and Abstracts, p. 235.

Viau, C.A. and Oldershaw, A.E.
1984: Structural controls on sedimentation and dolomite cementation in the Swan Hills Formation, Swan Hills Field, Central Alberta; in Carbonates in Subsurface and Outcrop, L. Eliuk (ed.); Canadian Society of Petroleum Geologists, p. 103-131.

Vopni, L.K. and Lerbekmo, J.R.
1971: Sedimentology and ecology of the Horn Plateau Formation; a Middle Devonian coral reef, Northwest Territories, Canada; Geologische Rundschau, v. 61, no. 2, p. 626-646.

Wardlaw, N.C. and Reinson, G.E.
1971: Carbonate and evaporite deposition and diagenesis Middle Devonian Winnipegosis and Prairie Evaporite formations of central Saskatchewan; American Association of Petroleum Geologists Bulletin, v. 55, p. 1759-1786.

Watts, A.B.
1982: Tectonic subsidence, flexure and global changes in sea level; Nature, v. 297, p. 469-474.

Wendte, J.C. and Stoakes, F.A.
1982: Evolution and corresponding porosity distribution of the Judy Creek reef complex, Upper Devonian, central Alberta; in Canada's Giant Hydrocarbon Reserves, W.G. Cutler (ed.); Canadian Society of Petroleum Geologists/American Association of Petroleum Geologists, Core Conference, Calgary, Alberta, p. 63-81.

Williams, G.K.
1975a: The Hay River Formation and its relationship to adjacent formations, Slave River map area, Northwest Territories; Geological Survey of Canada, Paper 75-12, 17 p.
1975b: Arnica Platform Dolomite (NTS 85, 95, 96, 105, 106), District of MacKenzie; in Report of Activities, Part C, Geological Survey of Canada, Paper 75-1C, p. 31-35.
1981a: Middle Devonian carbonate barrier complex of Western Canada; Geological Survey of Canada, Open File 761 (2 cross-sections, 6 maps 1:1 000 000).
1981b: Dolomitization pattern of the Keg River barrier complex, Yukon; Geological Survey of Canada, Open File 818 (4 maps 1:500 000).
1983: What does the term "Horn River Formation" mean?; Bulletin of Canadian Petroleum Geology, v. 31, p. 117-122.
1984: Some musings on the Devonian Elk Point Basin, Alberta; Bulletin of Canadian Petroleum Geology, v. 32, p. 216-232.

Wilson, J.L.
1967: Carbonate-evaporite cycles in Lower Duperow Formation of Williston Basin; Bulletin of Canadian Petroleum Geology, v. 15, p. 230-312.

Wilson, N.
1984: The Winnipegosis Formation of south-central Saskatchewan; Saskatchewan Geological Society, Special Publication No. 7, p. 13-16.

Wong, P.K. and Oldershaw, A.E.
1980: Causes of cyclicity in reef interior sediments, Kaybob Reef, Alberta; Bulletin of Canadian Petroleum Geology, v. 28, p. 411-424.

Workman, L.E. and Metherell, R.G.
1969: Geology Crossfield East and Lone Pine Creek gas fields, Alberta; Bulletin of Canadian Petroleum Geology, v. 17, p. 92-108.

Workum, R.H.
1978: Cripple Creek, a leeward Leduc reef margin; in The Fairholme Carbonate Complex at Hummingbird and Cripple Creek, I.A. McIlreath and P.C. Jackson (ed.); Canadian Society of Petroleum Geologists, p. 74-87.
1982: The pattern of Devonian stratigraphy Alberta Rocky Mountains: in Major Controls on Devonian Stratigraphy and Sedimentation; Canadian Society of Petroleum Geologists. Conference in honour of Dr. Helen R. Belyea, sponsored by The Canadian Society of Petroleum Geologists, March 18-19, 1982, Calgary, Alberta, unpaginated.
1985: Relationship between Devonian shelf carbonates and basin fill, Alberta Rocky Mountains (abstract); Reservoir, Canadian Society of Petroleum Geologists, v. 12, no. 3, p. 1-2.

ADDENDUM

No attempt has been made to bring the report up to date with material published after 1985, but the following two reports by the author may be referred to for supplementary information on the distribution of reefs and their regional setting (Moore, 1988) and on the petrography of the reservoir rocks (Moore, 1989).

Moore, P.F.

1988: Devonian reefs in Canada and some adjacent areas; in Reefs, Canada and Adjacent Areas, H.H.J. Geldsetzer, N.P. James and G.E. Tebbutt (ed.); Canadian Society of Petroleum Geologists, Memoir 13, p. 367-390.

1989: The Lower Kaskaskia sequence - Devonian; in Western Canada Sedimentary Basin: A Case History, B.D. Ricketts (ed.); Canadian Society of Petroleum Geologists, p. 139-164.

Author's Address

P.Fitzgerald Moore
2003 - 8 Street S.W.
Calgary, Alberta
T2T 2Z5

Printed in Canada

Subchapter 4E

CARBONIFEROUS

B.C. Richards, E.W. Bamber, A.C. Higgins, and J. Utting

CONTENTS

INTRODUCTION

The Carboniferous System in Western Canada Basin (Fig. 4E.1-4E.3) is a thick succession of strata deposited on the downwarped and downfaulted western margin of the ancestral North American plate, the central to western cratonic platform, and southern Yukon Fold Belt. This succession, representing the upper Kaskaskia sequence and lower Absaroka sequence of Sloss (1963), comprises two main lithofacies assemblages. The lower assemblage is basinal shale and generally thickens southwestward or basinward (see Fig. 4E.9-4E.13). Upward and northeastward, it passes into an upper assemblage of platform and ramp carbonates (Fig. 4E.4, 4E.5, 4E.9-4E.13) and sandstone-dominated siliciclastic facies, deposited in deep-water slope to continental settings. Both assemblages consist of numerous formations, some separated by regional disconformities. Subaerial erosion during the Late Carboniferous, Permian, and subsequent periods removed large parts of the succession, particularly in the Interior Plains, the region west of the Rocky Mountain Front Ranges, and the Cordillera between southwestern District of Mackenzie and northern Yukon Territory. Where the Carboniferous remains, it is generally unconformably overlain by either Permian or Mesozoic strata.

Carboniferous formations are preserved in two main regions. The southern one, which includes much of the eastern Cordillera and southern to western Interior Plains, extends from southwestern Manitoba to southwestern District of Mackenzie. The northern area includes the eastern Cordillera of northern Yukon Territory and northwestern District of Mackenzie (Fig. 4E.1). Between these regions, erosional remnants are present in the Mackenzie and Selwyn Mountains of east-central Yukon and west-central District of Mackenzie. The Lower Carboniferous is most complete and best exposed in the southwestern part of the southern region, and the Upper Carboniferous is best represented in northern Yukon (Fig. 4E.1, 4E.2, 4E.3).

Richards, B.C., Bamber, E.W., Higgins, A.C., and Utting, J.
1993: Carboniferous; Subchapter 4E in Sedimentary Cover of the Craton in Canada, D.F. Stott and J.D. Aitken (ed.); Geological Survey of Canada, Geology of Canada, no. 5, p. 202-271 (also Geological Society of America, The Geology of North America, v. D-1).

Previous work

Geological studies on the Carboniferous in Western Canada Basin date back to 1887 (McConnell) and 1924 (Kindle). The current lithostratigraphic nomenclature and the depositional origins of major lithofacies have been established by numerous studies, particularly those of Douglas (1958), Macauley (1958), Halbertsma (1959), Edie (1958), Christopher (1961), Scott (1964), Macqueen and Bamber (1967, 1968), Bamber and Waterhouse (1971), Macqueen et al. (1972), Graham (1973), Bamber and Mamet (1978), and Pugh (1983). Lithostratigraphic syntheses were written by Macauley et al. (1964), Douglas et al. (1970), and Bamber et al. (1984). Important biostratigraphic reports include those of Petryk et al. (1970), Bamber and Waterhouse (1971), Mamet (1976), Bamber and Mamet (1978), Sando and Mamet (1981), Crasquin (1984), Sando and Bamber (1985), Mamet et al. (1986), and Carter (1987). Excellent accounts of previous work were given by Norris (1965), Macqueen and Bamber (1968), Macqueen and Sandberg (1970), Bamber and Waterhouse (1971), Fuzesy (1973), and Bamber and Mamet (1978).

Acknowledgments

Several geologists of the Geological Survey of Canada, including S.P. Gordey, M.E. McMechan, J.W.H. Monger, D.K. Norris, K.G. Osadetz, and R.I. Thompson contributed to this compilation by providing unpublished data, criticism, and discussion. We thank W.J. Sando of the United States Geological Survey for discussions and information about uppermost Devonian and Carboniferous stratigraphy, sedimentation, and tectonism in Montana. J.A. Dolph of Gulf Canada Resources Limited and R. McWhae of Petro-Canada Incorporated provided maps of the Carboniferous System in Western Canada Basin, and B.L. Mamet of the University of Montreal provided biostratigraphic data. J.C.L. Ruelle, W. Styan, and J. Kaldi of Shell Canada Limited read an early draft of the manuscript.

A.C. Higgins, J. Utting, and E.W. Bamber wrote the biostratigraphic section of this report, E.W. Bamber prepared the section on northern Yukon Territory, and B.C. Richards compiled the remainder of the manuscript.

TECTONIC ELEMENTS

During the Carboniferous Period, the principal tectonic elements in Western Canada Basin were Prophet Trough, Peace River Embayment, the cratonic platform, Williston Basin, and Yukon Fold Belt (Fig. 4E.1).

The name Prophet Trough is introduced here for the downwarped and down-faulted western margin of the North American plate of uppermost Devonian and Carboniferous times. Prophet Trough was apparently continuous with the Antler Foreland Basin (Fig. 4E.6; Poole, 1974; Nilsen, 1977) of the western United States, and extends from southeastern British Columbia to Yukon Fold Belt in northern Yukon and Alaska. The downwarped belt has been called the Cordilleran Miogeocline (Monger and Price, 1979; Monger and Ross, 1984; Gordy et al., 1987). This may be inappropriate, however, as the trough had an elevated western rim, and a Late Devonian to Early Carboniferous foreland basin may have developed in Prophet Trough. Douglas (1970, Fig. 8-23) called the southern part of this element Alberta Trough, the central section Liard Trough, and the northern part Peel Basin. This three-fold division is abandoned, as a continuous depression occupied the region. The origin of Prophet Trough is uncertain. Some workers (Tempelman-Kluit, 1979: Struik, 1987; Gordey et al., 1987) believed it resulted from Late Devonian and Carboniferous rifting and separation of terranes from western North America. Extension, recorded by block faulting, did occur in part of the trough, but data summarized in this report suggest that the Antler Orogeny or a related contractional event influenced Tournaisian and Viséan subsidence in its southern part at least. Compression during the latest Devonian to earliest Carboniferous Ellesmerian Orogeny caused downwarping in northernmost Prophet Trough.

A broad hinge zone, marking a point at which water depths and sedimentation rates increased quite rapidly basinward, formed the boundary between Prophet Trough and the cratonic platform to the east. The location of this hinge, which migrated basinward, is best established for the late Tournaisian (Tn3) and early Viséan (V1) (Fig. 4E.1). Between southwestern District of Mackenzie and northern Yukon, the location of the hinge is uncertain because the Carboniferous rocks occur only as isolated erosional remnants.

The nature and location of the western margin of Prophet Trough is largely unknown because most of the Carboniferous succession deposited in the western part has been eroded. Most of the Cordillera west of the trough comprises allochthonous terranes that coalesced with North America during the Mesozoic Era. South of northern Yukon, the western boundary of Prophet Trough was an elevated rim, at least locally exposed during Early Carboniferous time. Remnants of the rim are preserved mainly in the pericratonic Kootenay Terrane of Monger and Berg (1987). Strata in lower and eastern parts of the Kootenay Terrane (see Fig. 4E.6B, 4E.25) generally have affinities with the off-shelf lower Paleozoic succession of the Western Canada Basin. The western and upper parts of the terrane include Upper Devonian and Carboniferous siliciclastics derived partly from outboard sources (Gordey et al., 1987), Devonian to Lower Carboniferous calc-alkaline plutons and volcanics (Mortensen et al., 1987; Mortensen and Jilson, 1985; Evenchick et al., 1984), and felsic to intermediate volcanics (Mortensen, 1982). The Kootenay Terrane appears to have been marginal to North America - hence pericratonic - and was apparently bounded on the west by oceanic crust of the Upper Devonian to Triassic Slide Mountain Terrane (J.O. Wheeler, pers. comm., 1988). Components of the western rim of Prophet Trough may also be locally preserved in the Cassiar Terrane, which consists of displaced continental margin rocks formed inboard of Kootenay Terrane.

Part of the uplift that occurred on the rim may have resulted from block faulting related to continental rifting and separation along the western margin of North America (Tempelman-Kluit, 1979; Gordey et al., 1987, Struik, 1987). In the western United States, the western boundary of the Antler Foreland Basin was the Antler Orogenic Belt, which apparently developed east of a back-arc basin (Poole, 1974; Poole and Sandberg, 1977; Nilsen, 1977). The Antler or a related orogenic belt possibly extended into

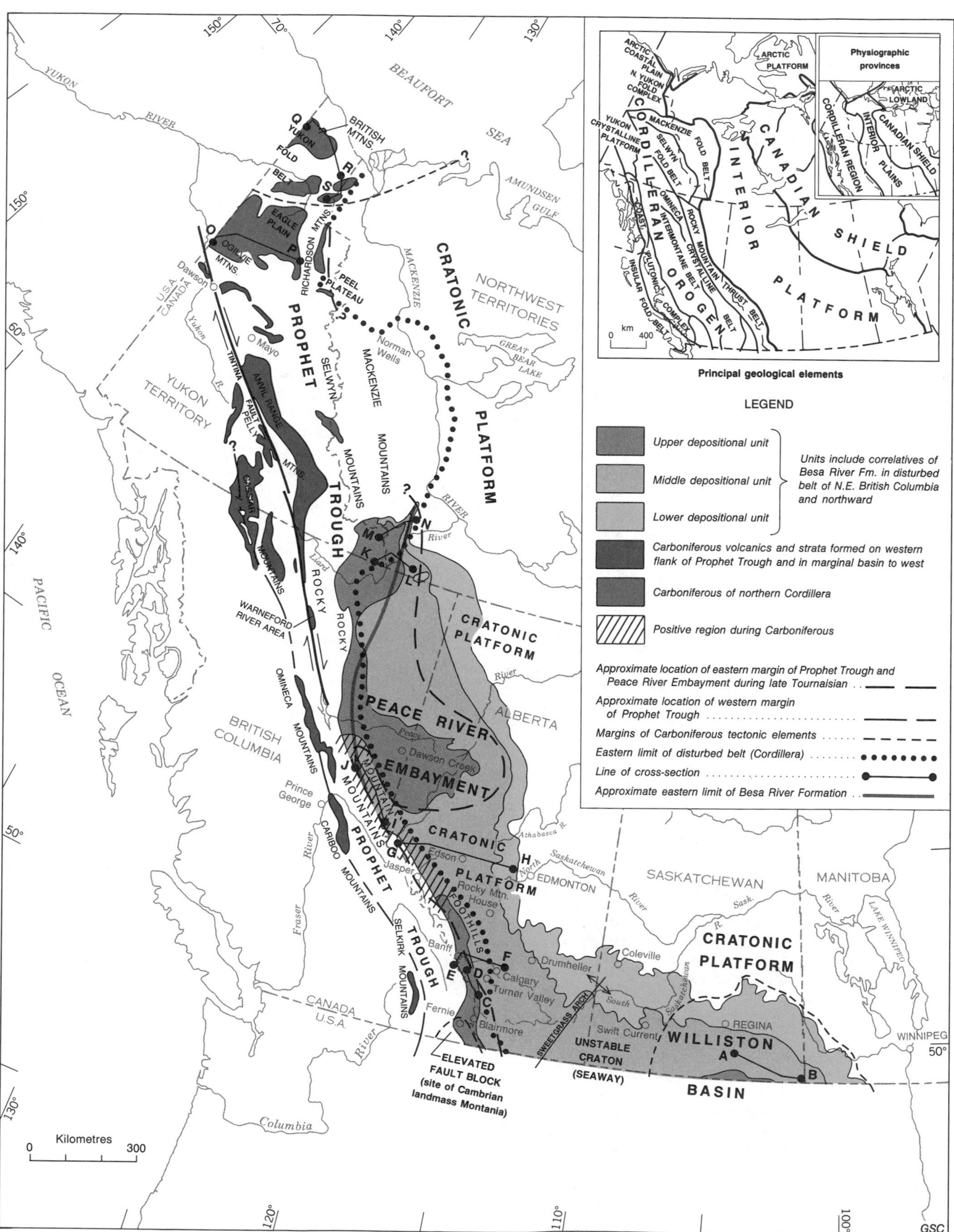
Principal geological elements
Physiographic provinces
LEGEND
Upper depositional unit
Middle depositional unit
Lower depositional unit
Units include correlatives of Besa River Fm. in disturbed belt of N.E. British Columbia and northward
Carboniferous volcanics and strata formed on western flank of Prophet Trough and in marginal basin to west
Carboniferous of northern Cordillera
Positive region during Carboniferous
Approximate location of eastern margin of Prophet Trough and Peace River Embayment during late Tournaisian
Approximate location of western margin of Prophet Trough
Margins of Carboniferous tectonic elements
Eastern limit of disturbed belt (Cordillera)
Line of cross-section
Approximate eastern limit of Besa River Formation
PROPHET TROUGH
CRATONIC PLATFORM
PEACE RIVER EMBAYMENT
WILLISTON BASIN
UNSTABLE CRATON (SEAWAY)
SWEETGRASS ARCH
ELEVATED FAULT BLOCK (site of Cambrian landmass Montania)
YUKON FOLD BELT
EAGLE PLAIN
BRITISH MTNS.
OGILVIE MTNS.
RICHARDSON MTNS.
PEEL PLATEAU
SELWYN MOUNTAINS
MACKENZIE MOUNTAINS
TINTINA FAULT
ANVIL RANGE
PELLY MTNS.
CASSIAR MOUNTAINS
ROCKY MOUNTAINS
WARNEFORD RIVER AREA
OMINECA MOUNTAINS
CARIBOO MOUNTAINS
SELKIRK MOUNTAINS
FOOTHILLS
BEAUFORT SEA
AMUNDSEN GULF
GREAT BEAR LAKE
PACIFIC OCEAN
NORTHWEST TERRITORIES
YUKON TERRITORY
BRITISH COLUMBIA
ALBERTA
SASKATCHEWAN
MANITOBA
USA CANADA
Dawson
Mayo
Norman Wells
Dawson Creek
Prince George
Edson
Jasper
EDMONTON
Rocky Mtn. House
Banff
Calgary
Turner Valley
Drumheller
Coleville
Fernie
Blairmore
Swift Current
REGINA
WINNIPEG
Kilometres
0
300
GSC

southeastern British Columbia (Høy, 1977; Monger and Price, 1979; Monger and Ross, 1984; Lethiers et al., 1986), forming part of the western rim of Prophet Trough. The presence of Upper Devonian and Lower Carboniferous calc-alkaline plutons extending in a narrow belt from southeastern British Columbia into Alaska suggests uplift from plate convergence and eastward-directed subduction along the western rim of Prophet Trough.

The western rim of Prophet Trough was bounded on the southwest by a marginal basin, represented by oceanic mafic volcanics, ultramafic rocks, and bedded chert of the Slide Mountain Terrane (Fig. 4E.25). The basin lay northeast (cratonward) of probable arc terranes, and may have been partly a back-arc basin resembling that west of the Antler belt (Monger and Price, 1979; Monger and Ross, 1984). In southeastern British Columbia, Quesnellia - an allochthonous terrane preserving an Upper Paleozoic volcanic-arc assemblage (Harper Ranch sub-terrane) - lies west of the rim of Prophet Trough and the Slide Mountain Terrane (Monger and Ross, 1984). To the north, middle Paleozoic arcs may have developed west of the marginal basin but have not been unequivocally identified. During the Late Carboniferous and Permian, island arcs, preserved in the allochthonous Stikine Terrane of western Canada (Fig. 4E.25) and related terranes to the south and northwest, probably lay to the southwest (Miller, 1987).

In northwestern Alberta and northeastern British Columbia, Peace River Embayment (Douglas et al., 1970) was part of Prophet Trough during the early Tournaisian (Tn1) and early middle Tournaisian (early Tn2) (Fig. 4E.1). The arcuate embayment, a broad re-entrant into the cratonic platform, had a poorly to moderately well-defined southern margin at approximately 54°45'N and a poorly defined opposing side with a northwesterly trend. During the middle and late Tournaisian (late Tn2 and Tn3), the northern margin of the embayment became well defined as the eastern margin of Prophet Trough stepped basinward. Peace River Embayment still opened into the trough, but was a distinct tectonic element. The depositional and structural axis of the embayment has an easterly trend and coincides approximately with that of the Late Devonian Peace River Arch. Extensive block faulting, accompanied by regional flexural subsidence, produced the embayment, which is characterized by an anomalously thick Lower Carboniferous succession and by numerous northeasterly and northwesterly striking normal faults (Sikabonyi and Rodgers, 1959). The paleogeography of Peace River Embayment is not well established.

Prophet Trough and adjacent areas on the cratonic platform included several subordinate tectonic elements (Fig. 4E.1). In southernmost Canada and northern Montana, the trough periodically included a positive area on the site of the Cambrian landmass Montania (see Douglas et al., 1970). This area, which was episodically positive during the late Paleozoic, is characterized by a relatively thin Carboniferous succession. It was bounded on the north side by a northeast-striking trough or half-graben(?), which occurred in approximately the same area as the Cambrian Eager Trough (see Douglas et al., 1970) and persisted into the Permian (Henderson et al., Subchapter 4F in this volume). The younger trough contains an anomalously thick upper Paleozoic succession and was possibly bordered on the south by syndepositional basement faults. In the Rockies of west-central Alberta and east-central British Columbia, disconformities and trends of Lower Carboniferous lithofacies indicate the presence of a low, unnamed, northwest-trending positive element with at least three phases of relative uplift, extending from south of Jasper into the Peace River Embayment. The first positive phase was earliest middle Tournaisian (Tn2), the second, late Tournaisian (Tn3) and possibly early Viséan (V1), and the third and most extensive, late Viséan to latest Carboniferous.

Mesozoic and early Tertiary overthrusting has moved most of the Prophet Trough succession eastward, with the western deposits displaced to the greatest extent. In the south, the western succession has been thrust from 100 to 200 km eastward (Norris, 1965; Monger and Ross, 1984).

The western cratonic platform extended northward from the United States to the Ellesmerian and Yukon fold belts. It included Williston Basin and a broad northeasterly trending uplift in the vicinity of the Laramide Sweetgrass Arch (Fig. 4E.1). Williston Basin, centred in North Dakota, is a syndepositional and post-depositional basin extending (in Canada) from southwestern Manitoba to southwestern Saskatchewan. This flexural basin (Nisbet and Fowler, 1984) also occupies a vast region in eastern Montana and the Dakotas (Sheldon and Carter, 1979). During Tournaisian time, Williston Basin was connected to Prophet Trough and the Antler Foreland Basin by a broad seaway, extending from southeastern Alberta to Wyoming. Subsequently, Williston Basin became a topographic basin as a low, broad, uplift (Sweetgrass Arch of Douglas et al., 1970, Fig. VII-33) developed across the seaway in southeastern Alberta and north-central Montana.

The Yukon Fold Belt of northern Yukon and Alaska (Bell, 1973) resulted from the Famennian to early(?) Tournaisian (Tn1) Ellesmerian Orogeny, named and best studied in the Canadian Arctic Archipelago (Thorsteinsson and Tozer, 1970; Trettin, 1973, 1991). According to the most widely accepted of several hypotheses, the Yukon Fold Belt was apparently continuous with the Ellesmerian Fold Belt of the Canadian Arctic Archipelago, and lay northwest of

Figure 4E.1. Map of Carboniferous units subcropping beneath Permian and Mesozoic formations in Western Canada Basin, Carboniferous tectonic elements, and lines of cross-section. See Figure 4E.2 for formational composition of lower, middle and upper depositional units and Figure 4E.3 for Carboniferous formations of northern Cordillera.

Figure 4E.2. Correlation of Carboniferous lithostratigraphic units, southwestern Manitoba to southwestern District of Mackenzie, with standard chronostratigraphic units and Carboniferous zonal schemes. Dashed lines indicate nature of contact uncertain; question marks indicate position of lines uncertain. **(See following pages)**

SYSTEM
SERIES
FORAMINIFERAL ZONES
DEPOSITIONAL UNIT
SW DISTRICT OF MACKENZIE
S. MACKENZIE FOLD BELT, 61°00′N TO 61°15′N
NW SE
SUBSURFACE E. DISTURBED BELT AND W. INTERIOR PLAINS, 60°05′N
W E
SUBSURFACE, PEACE RIVER EMBAYMENT, NW ALBERTA AND NE B.C., INTERIOR PLATFORM AND EASTERN DISTURBED BELT
NW SE
ROCKY MTN. FRONT RANGES, EAST-CENTRAL B.C., 54°20′N TO 55°10′N
NW SE
ROCKY MTN. FRONT RANGES, WEST-CENTRAL ALBERTA AND EAST-CENTRAL B.C., 53°30′ TO 54°20′N
W E
ROCKY MTNS., EASTERN FRONT RANGES, BOW VALLEY (EXSHAW) TO 53°30′N
NW SE
LOWER CRETACEOUS FORT ST. JOHN GP.
PERMIAN KINDLE and FANTASQUE fms.
PERMIAN BELLOY FORMATION
PERMIAN BELCOURT FORMATION
PERMIAN BELCOURT FM.
TRIASSIC
PERMIAN (mainly JOHNSTON CANYON FM.)
CARBONIFEROUS
UPPER CARBONIFEROUS
LOWER CARBONIFEROUS
DEVONIAN
MOSCOVIAN (Part)
BASHKIRIAN
SERPUKHOVIAN
VISEAN
TOURNAISIAN
FAMEN-NIAN (Part)
UPPER (V3)
MIDDLE (V2)
LOWER (V1)
UPPER (TN3)
MIDDLE (TN2)
LOWER (TN1)
TN1A
TN1B
>22
22
21
20
19
18
17
16S
16-I
15
14
13
12
11
10
9
8
7
pre 7
unzoned
UPPER
MIDDLE
LOWER
BESA RIVER
MATTSON FORMATION
GOLATA FM.
RUNDLE GROUP
FLETT FORMATION
Meilleur Member
Jackfish Gap Member
Tlogotsho Member
PROPHET FORMATION
CLAUSEN FORMATION
YOHIN FORMATION
BANFF? FM.
BESA RIVER FM.
no data
IMPERIAL? FM.
GOLATA FM.
Prophet FM.
FORMATION F
CLAUSEN FM.
PEKISKO FM.
BANFF FORMATION
Member D
Member A
EXSHAW FM. (Black Shale Unit)
BESA R. FM.
KOTCHO FM.
STODDART GROUP
TAYLOR FLAT FM.
KISKATINAW FORMATION
DEBOLT FORMATION
Elkton Member
SHUNDA FM.
BANFF FM.
Member B
Mbr. A
Siltstone Unit of Exshaw
EXSHAW FM.
Black Shale Unit
WABAMUN FM.
MT. HEAD FM.
TURNER VALLEY FM.
PROPHET FM.
FM. F
PEKISKO FM.
BESA RIVER FM.
BANFF FM.
EXSHAW FM.
PALLISER FM.
MOUNT HEAD FM.
Salter Mbr.
Baril Member
Wileman Mbr.
Upper Member
Middle Mbr.
Elkton Mbr.
LIVINGSTONE FM.
SHUNDA FM.
FM. F
PEKISKO FM.
Mbr. C
Member B
Mbr. A
TOBERMORY FM.
STORELK FM.
TYRWHITT FM.
undivided
SPRAY LAKES GP.
Todhunter Member
ETHERINGTON FORMATION
Carnarvon Member
Marston Member
Loomis Member
Salter Member
Baril Mbr.
Wileman Mbr.
Upper Mbr.
Middle Mbr.
Elkton Member
MOUNT HEAD FM.
RUNDLE GROUP
Member C
Siltstone Unit
Black Shale Unit
FOSSIL OCCURRENCES: Foraminifers
Conodonts
Corals
Brachiopods
Ostracodes

SW ALBERTA	WILLISTON BASIN AREA		DEPOSITIONAL UNIT	FORAMINIFERAL ZONES (Mamet and Skipp, 1970)	CONODONT ZONES (after Sandberg et al., 1978; Lane et al., 1980)	CORAL ZONES (Sando and Bamber, 1985)	BRACHIOPOD ZONES (Carter, 1987)	OSTRACODE ZONES (Crasquin, 1984)	SERIES	SYSTEM
EASTERN TO WESTERN FRONT RANGES, 50°30′N TO BOW VALLEY (W–E)	SOUTH-CENTRAL TO SE SASKATCHEWAN (W–E)	SW MANITOBA (W–E)								
PERMIAN JOHNSTON CANYON FORMATION	LOWER JURASSIC WATROUS and AMARANTH fms.	LOWER JURASSIC AMARANTH FM.								

SW Alberta: SPRAY LAKES GP.: KANANASKIS FM.; TOBERMORY FM.; STORELK FM.; TYRWHITT FM. RUNDLE GROUP: ETHERINGTON FORMATION (Todhunter Member; undivided); MOUNT HEAD FM. (Carnarvon Member; Opal Mbr.; Marston Member; Loomis Member; Salter Member; Baril Mbr.; Wileman Mbr.); LIVINGSTONE FORMATION; PEKISKO FM. BANFF FM. (Mbr. F; FM. F; undivided; Mbr. E; Member B; Mbr. A). EXSHAW FM. (Siltstone Unit; Black Shale Unit). PALLISER FM.

South-central to SE Saskatchewan: L. BIG SNOWY GP.: KIBBEY FM. (Age upper boundary not established). MADISON GROUP: CHARLES FM.; MISSION CANYON FM.; LODGEPOLE FORMATION. THREE FORKS GROUP (part): BAKKEN FM. (Upper Shale Mbr.; Middle Sandstone Member; Lower Shale Mbr.). BIG VALLEY FM.

SW Manitoba: MADISON GROUP: CHARLES FM.; MISSION CANYON FORMATION; LODGEPOLE FM. undivided (Flossie L. Mbr.; Whitewater L. Mbr.; Virden Mbr.; Scallion Mbr.; Routledge Sh.). THREE FORKS GROUP (part): BAKKEN FM. (Upper Shale Mbr.; Middle Sandstone Member).

DEPOSITIONAL UNIT	FORAMINIFERAL ZONES	CONODONT ZONES	CORAL ZONES	BRACHIOPOD ZONES	OSTRACODE ZONES	SERIES		SYSTEM	
UPPER	>22	unzoned	unzoned	unzoned	unzoned	MOSCOVIAN (Part)		UPPER CARBONIFEROUS	CARBONIFEROUS
UPPER	? 22 ? 21 ?	unzoned	unzoned	unzoned	unzoned	BASHKIRIAN		UPPER CARBONIFEROUS	CARBONIFEROUS
UPPER	20	unzoned	unzoned	unzoned	unzoned	BASHKIRIAN		UPPER CARBONIFEROUS	CARBONIFEROUS
UPPER	19	unzoned	VI	unzoned	unzoned	SERPUKHOVIAN		LOWER CARBONIFEROUS	CARBONIFEROUS
UPPER	18	unzoned	VI	unzoned	unzoned	SERPUKHOVIAN		LOWER CARBONIFEROUS	CARBONIFEROUS
UPPER	17	?	V B	unzoned	unzoned	SERPUKHOVIAN		LOWER CARBONIFEROUS	CARBONIFEROUS
UPPER	16S	*Cavusgnathus*	V A	unzoned	? M105	UPPER (V3)	VISEAN	LOWER CARBONIFEROUS	CARBONIFEROUS
UPPER	16-I	*Cavusgnathus*	V A	unzoned	M105	UPPER (V3)	VISEAN	LOWER CARBONIFEROUS	CARBONIFEROUS
UPPER	15	*Cavusgnathus*	IV	unzoned	M105	UPPER (V3)	VISEAN	LOWER CARBONIFEROUS	CARBONIFEROUS
UPPER	14	*Cavusgnathus*	IV	unzoned	?	UPPER (V3)	VISEAN	LOWER CARBONIFEROUS	CARBONIFEROUS
MIDDLE	13	*Cavusgnathus*	III D / C	unzoned	M104	MIDDLE (V2)	VISEAN	LOWER CARBONIFEROUS	CARBONIFEROUS
MIDDLE	12	*Cavusgnathus*	III B	unzoned	M104	MIDDLE (V2)	VISEAN	LOWER CARBONIFEROUS	CARBONIFEROUS
MIDDLE	11	? *texanus*	III A	unzoned	?	LOWER (V1)	VISEAN	LOWER CARBONIFEROUS	CARBONIFEROUS
MIDDLE	10	*texanus*	III A	unzoned	M103	LOWER (V1)	VISEAN	LOWER CARBONIFEROUS	CARBONIFEROUS
MIDDLE	9	*anchoralis-latus*	II B	? *Avonia minnewankensis-Marginata burlingtonensis* ?	M103	UPPER (TN3)	TOURNAISIAN	LOWER CARBONIFEROUS	CARBONIFEROUS
MIDDLE	8	*typicus*	II B	*Stegacanthia* cf. *bowsheri-Marginata fernglenensis* ?	? M102	UPPER (TN3)	TOURNAISIAN	LOWER CARBONIFEROUS	CARBONIFEROUS
MIDDLE	7	*isosticha*-UPPER *crenulata*	II B / A	? *Calvustrigis rutherfordi* ?	? M101	MIDDLE (TN2)	TOURNAISIAN	LOWER CARBONIFEROUS	CARBONIFEROUS
LOWER	? pre 7	LOWER *crenulata*	I	unzoned	?	MIDDLE (TN2)	TOURNAISIAN	LOWER CARBONIFEROUS	CARBONIFEROUS
LOWER	unzoned	? *sandbergi*	I	unzoned	unzoned	LOWER (TN1) TN1B	TOURNAISIAN	LOWER CARBONIFEROUS	CARBONIFEROUS
LOWER	unzoned	*duplicata*	I	unzoned	unzoned	LOWER (TN1) TN1B	TOURNAISIAN	LOWER CARBONIFEROUS	CARBONIFEROUS
LOWER	unzoned	*sulcata*	? unzoned	unzoned	unzoned	LOWER (TN1) TN1A	TOURNAISIAN	LOWER CARBONIFEROUS	CARBONIFEROUS
LOWER	unzoned	*praesulcata*	unzoned	unzoned	unzoned	FAMENNIAN (Part)		DEVONIAN	
LOWER	unzoned	? *expansa*	unzoned	unzoned	unzoned	FAMENNIAN (Part)		DEVONIAN	

GSC

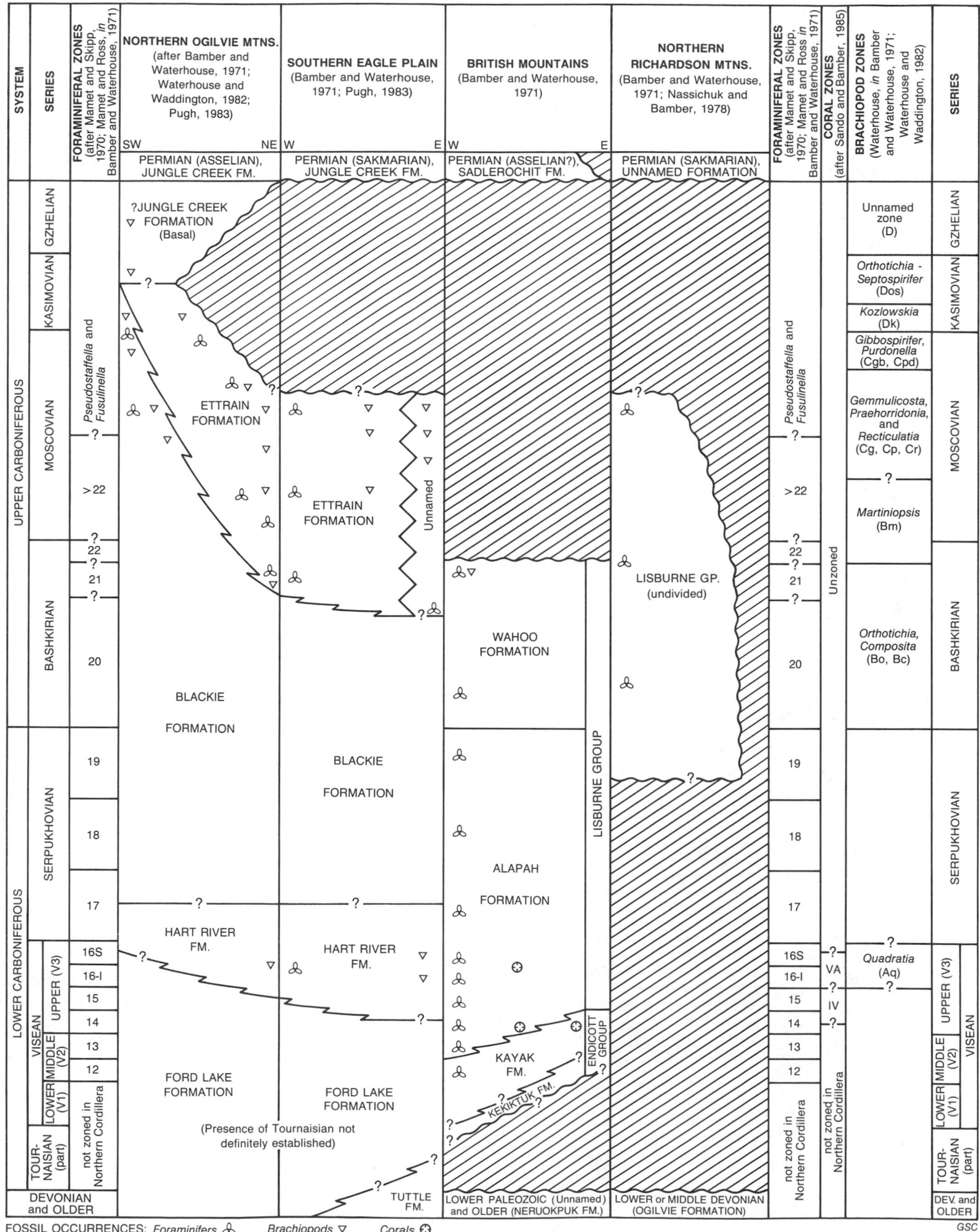

SYSTEM
SERIES
FORAMINIFERAL ZONES (after Mamet and Skipp, 1970; Mamet and Ross, in Bamber and Waterhouse, 1971)
NORTHERN OGILVIE MTNS. (after Bamber and Waterhouse, 1971; Waterhouse and Waddington, 1982; Pugh, 1983)
SOUTHERN EAGLE PLAIN (Bamber and Waterhouse, 1971; Pugh, 1983)
BRITISH MOUNTAINS (Bamber and Waterhouse, 1971)
NORTHERN RICHARDSON MTNS. (Bamber and Waterhouse, 1971; Nassichuk and Bamber, 1978)
CORAL ZONES (after Sando and Bamber, 1985)
BRACHIOPOD ZONES (Waterhouse, in Bamber and Waterhouse, 1971; Waterhouse and Waddington, 1982)
SW
NE
W
E
PERMIAN (ASSELIAN), JUNGLE CREEK FM.
PERMIAN (SAKMARIAN), JUNGLE CREEK FM.
PERMIAN (ASSELIAN?), SADLEROCHIT FM.
PERMIAN (SAKMARIAN), UNNAMED FORMATION
UPPER CARBONIFEROUS
LOWER CARBONIFEROUS
GZHELIAN
KASIMOVIAN
MOSCOVIAN
BASHKIRIAN
SERPUKHOVIAN
VISEAN
UPPER (V3)
MIDDLE (V2)
LOWER (V1)
TOUR-NAISIAN (part)
DEVONIAN and OLDER
Pseudostaffella and Fusulinella
>22
22
21
20
19
18
17
16S
16-I
15
14
13
12
not zoned in Northern Cordillera
?JUNGLE CREEK FORMATION (Basal)
ETTRAIN FORMATION
Unnamed
BLACKIE FORMATION
HART RIVER FM.
FORD LAKE FORMATION
(Presence of Tournaisian not definitely established)
TUTTLE FM.
WAHOO FORMATION
ALAPAH FORMATION
LISBURNE GROUP
KAYAK FM.
KEKIKTUK FM.
ENDICOTT GROUP
LOWER PALEOZOIC (Unnamed) and OLDER (NERUOKPUK FM.)
LISBURNE GP. (undivided)
LOWER or MIDDLE DEVONIAN (OGILVIE FORMATION)
Unzoned
VA
IV
Unnamed zone (D)
Orthotichia - Septospirifer (Dos)
Kozlowskia (Dk)
Gibbospirifer, Purdonella (Cgb, Cpd)
Gemmulicosta, Praehorridonia, and Recticulatia (Cg, Cp, Cr)
Martiniopsis (Bm)
Orthotichia, Composita (Bo, Bc)
Quadratia (Aq)
DEV. and OLDER
FOSSIL OCCURRENCES: Foraminifers
Brachiopods
Corals
GSC

the southern Arctic Islands. Early Cretaceous counterclockwise rotation brought the Yukon Fold Belt to its present position (Sweeney et al., 1978; Nilsen, 1981; Ziegler, 1988).

CARBONATE DEPOSITIONAL MODELS

The Carboniferous carbonate buildups in the Western Canada Basin are mainly poorly differentiated platforms and ramps (see Fig. 4E.4, 4E.5). Platforms are large buildups with a subhorizontal top and relatively abrupt shelf margins, where sediment is deposited in high-energy settings. The shelf margin is separated from the main shoreline by a broad, relatively low-energy, protected shelf, where deposition occurs in the shallow subtidal and intertidal zones. Ramps are large buildups that prograde away from positive areas and down gentle regional slopes. They lack an obvious break in slope, and the environments of highest energy lie close to the main shoreline (Wilson, 1975).

SOUTHEASTERN MANITOBA TO SOUTHWESTERN DISTRICT OF MACKENZIE

The Carboniferous succession from southwestern Manitoba to southwestern District of Mackenzie is divided, on the basis of lithology and depositional history, into lower, middle, and upper depositional units (Fig. 4E.1). Figure 4E.2 shows the lithostratigraphic units that constitute these principal divisions, which resemble those of Macauley et al. (1964). From east-central British Columbia to southwestern District of Mackenzie, the three units overlie and pass basinward into a fourth depositional unit, the Besa River Formation (Fig. 4E.2, 4E.11-4E.13).

Lower depositional unit

The lower depositional unit comprises carbonates and siliciclastics of the Bakken, Exshaw, Lodgepole, Banff, and Yohin formations. The Bakken constitutes the upper Three Forks Group of Christopher (1961), and the Lodgepole, the lower Madison Group (see Sando and Dutro, 1974). In most areas, the lower unit, which is part of the Kaskaskia sequence of Sloss (1963), disconformably overlies upper Famennian strata and is overlain by the less widely distributed middle depositional unit. In most of Williston Basin, much of the southern Cordillera, and on the Interior Plains of southernmost Alberta, the top of the lower unit becomes younger basinward as the middle unit grades into it. Elsewhere, the boundary between these two units is generally abrupt and commonly a minor disconformity. East and north of the subcrop edge of the middle unit, the lower unit is unconformably overlain by Mesozoic strata.

In the lower depositional unit, lowest Carboniferous deposits lie in the Exshaw and Bakken formations, but where these formations are absent or incompletely developed, the oldest Carboniferous strata are in the Banff Formation (Fig. 4E.2). Upper Devonian strata occurring in the Exshaw, Bakken, and Banff formations are discussed with the Carboniferous because they are relatively thin, separated from underlying Famennian strata by either widespread disconformities or hiatal surfaces, and lithologically indistinguishable from overlying Carboniferous strata.

The Bakken and Lodgepole formations were deposited in Williston Basin and on the unstable craton of southeastern Alberta and southwestern Saskatchewan, while the Exshaw, Banff, and Yohin formed on the western cratonic platform and in Prophet Trough, which included the poorly defined Peace River Embayment. Prior to deposition of the lower Lodgepole, subaerially exposed Devonian strata on the Central Montana High (a prominent Devonian element) extended from Wyoming into central and northern Montana (Craig, 1972; Smith, 1977). As a consequence, the connection between Williston Basin and Prophet Trough/Antler Foreland Basin was a relatively narrow seaway extending from southern Canada into northern Montana. During sedimentation of the lower Lodgepole, the Central Montana High was rapidly transgressed (Craig, 1972; Smith, 1977), and a broad, commonly deep seaway developed between southern Alberta and Wyoming.

Bakken Formation

The widely distributed Bakken Formation (Nordquist, 1953) comprises two black shale members separated by an arenaceous member (Fig. 4E.2). It is of late Famennian to middle Tournaisian (Tn2) age, and paleontological evidence (conodonts and palynomorphs) suggests that the Devonian-Carboniferous boundary is in the middle member. The Bakken occurs from southwestern Manitoba to southeastern Alberta and in western North Dakota and northeastern Montana. The lower and upper units of the Exshaw Formation in Alberta and British Columbia are stratigraphically and lithologically equivalent to the lower and middle Bakken respectively. The Bakken-Exshaw boundary has been arbitrarily placed along Sweetgrass Arch in southeastern Alberta by Macqueen and Sandberg (1970). In Canada, the Bakken is generally less than 30 m thick and is thickest in south-central Saskatchewan.

The lower boundary of the Bakken Formation is probably either a minor disconformity or a hiatal surface in most areas. According to McCabe (1959) and Christopher (1961), the Bakken of Manitoba and southeastern Saskatchewan east of 102°40'W disconformably overlies the Famennian Torquay Formation. Elsewhere in southeastern Saskatchewan and in south-central Saskatchewan, the Bakken abruptly but conformably(?) overlies the upper Famennian Big Valley Formation (Christopher, 1961). In most of southwestern Saskatchewan and in southeastern Alberta, the contact is disconformable. (Macqueen and Sandberg, 1970; Christopher, 1961).

Figure 4E.3. Correlation of Carboniferous lithostratigraphic units in northern Yukon Territory and northwestern District of Mackenzie, with standard chronostratigraphic units and Carboniferous zonal schemes. Dashed lines indicate nature of contact uncertain; question marks indicate position of lines uncertain.

In most of the area, the Bakken is overlain conformably by the Lodgepole Formation (Fig. 4E.2, 4E.8; Christopher, 1961; Macqueen and Sandberg, 1970). In western Saskatchewan, however, the upper contact of the Bakken is disconformable (Kent, 1974, 1984).

The lower black shale member of the Bakken is present throughout most of the region but is absent in easternmost Saskatchewan and most of southwestern Manitoba (Christopher, 1961; McCabe, 1959). Its absence is generally attributed to erosion; however, a facies or colour change may make it indistinguishable from underlying formations. The lower Bakken generally closely resembles the partly coeval black shale unit of the Exshaw. Near the northern edge of the Bakken in Saskatchewan, however, it consists of red, green, and variegated shales.

In most areas, the middle Bakken conformably overlies the lower Bakken, but in easternmost Saskatchewan and in most of southwestern Manitoba, it disconformably overlies the Torquay Formation (Christopher, 1961; McCabe, 1959). Lithofacies of the middle Bakken, generally 6 to 18 m thick, occur in most areas where the Bakken is preserved but are absent in westernmost Saskatchewan (Kent, 1974, 1984) and near the erosional edge of the formation. The middle Bakken comprises massive to bioturbated and crossbedded sandstone and siltstone with subordinate shale and silty dolostone; ooid lime grainstone is present locally in Saskatchewan (McCabe, 1959; Christopher, 1961; Penner, 1958).

The upper shale member is generally present south and west of the subcrop belt of the middle Bakken, but is locally absent in southwestern Saskatchewan (Kent, 1974) and poorly developed to absent in southeastern Alberta. This member is lithologically and stratigraphically equivalent to the lower black shale of the Banff Formation in Alberta (Macqueen and Sandberg, 1970) and the Cottonwood Canyon Member of the Lodgepole Formation in southwestern Montana. The nature of the abrupt contact between the upper and middle Bakken is uncertain. Macqueen and Sandberg (1970) interpreted this contact to be a minor disconformity in southern Canada; in addition conodont data suggest it is a disconformity in North Dakota (Hays, 1985) and in eastern Saskatchewan. The upper member, generally less than 3.7 m thick, resembles the lower Bakken and comprises brownish black, anomalously

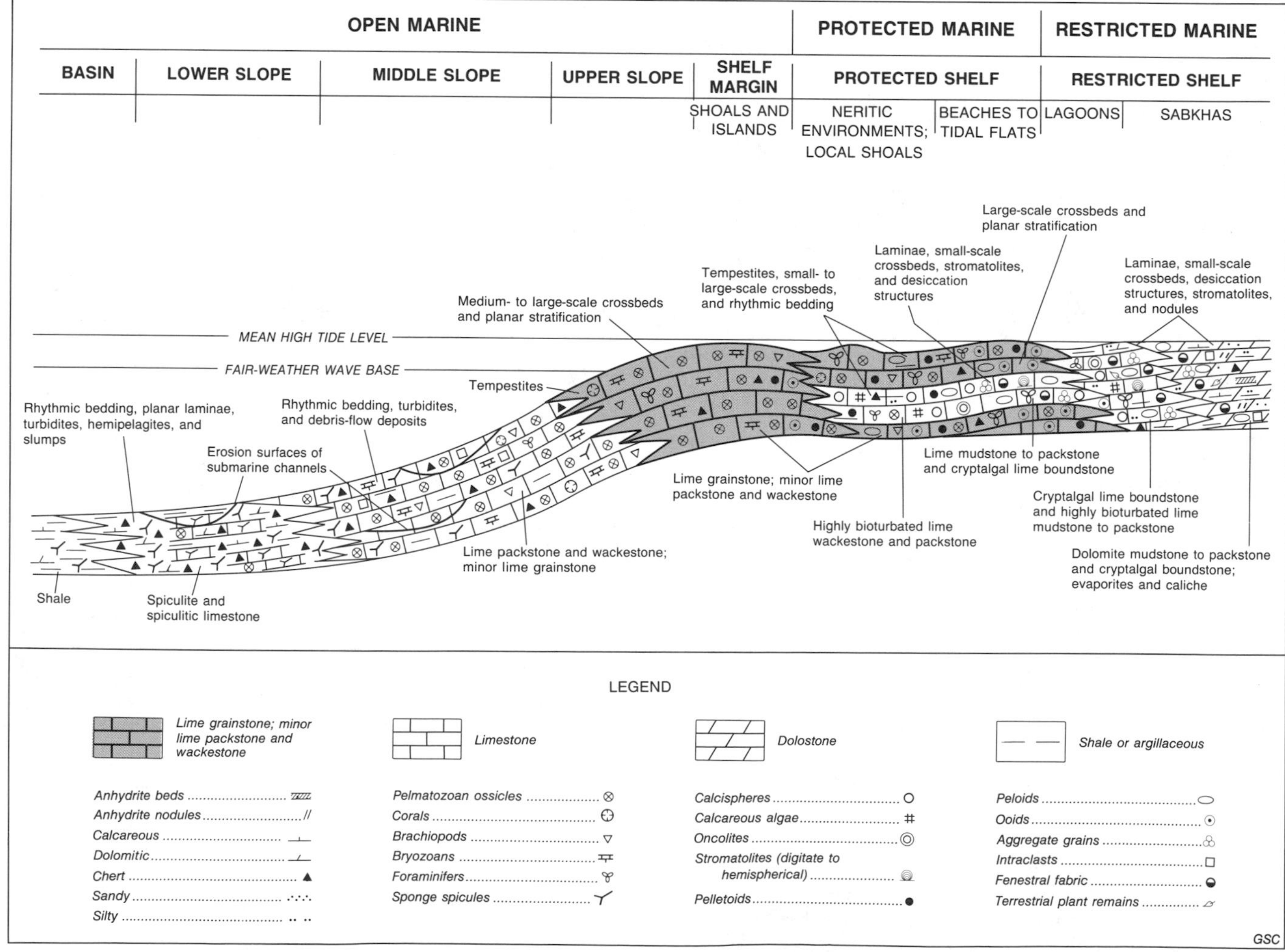

Figure 4E.4. Generalized depositional model of an Early Carboniferous carbonate platform.

radioactive shale that has a high organic content. Red shale is widespread near its northern erosional edge (Christopher, 1961).

Black shale of the upper and lower Bakken was deposited in offshore-marine settings and at moderate water depths (below storm-wave base) in the anaerobic zone of an extensive, chemically- and density-stratified basin. This is indicated by the paucity of benthonic fossils and by the pyritic black shale containing diverse, offshore-marine conodont faunas. Similar conclusions were reached for these deposits in North Dakota (Hays, 1985). The red to variegated shale of the upper and lower members probably resulted either from deposition in shallow, well-oxygenated water or from the diagenetic oxygenation of black shale subsequent to the early Tournaisian. Transgressions are recorded by the presence of the lower and upper Bakken shales above shallow marine to supratidal deposits.

The middle Bakken, deposited in relatively well-oxygenated water, and partly at very shallow depths, records shallowing after deposition of the lower Bakken. This is indicated by the widespread occurrence of wave- and current-formed crossbedding and by the local presence of ooid lime grainstone. The light colour, in addition to the presence of abundant trace fossils and some body fossils, records deposition in the aerobic zone.

Exshaw Formation

Sedimentation of the Exshaw Formation (Warren, 1937; Macqueen and Sandberg, 1970) occurred during that of the lower and middle Bakken Formation in late Famennian to middle(?) Tournaisian (Tn2) time. The Exshaw, comprising black shale overlain by siltstone, sandstone, and limestone, occurs on most of the western Interior Plains from southern Alberta into southwestern District of Mackenzie. In the eastern Cordillera it is generally present from 49°00′N to 52°30′N, but from 52°30′N to 54°25′N, it occurs only locally because of erosion prior to deposition of the Banff Formation (Fig. 4E.2, 4E.9-4E.12; Harker and McLaren, 1958). The Exshaw, which is 49 m thick at its type section near Canmore, Alberta (Macqueen and Sandberg, 1970), is generally between 7 and 50 m thick and thickens southwestward.

In most areas, the Exshaw overlies upper Famennian strata with probable disconformity (Macqueen and Sandberg, 1970; Harker and McLaren, 1958), but its basal contact may be locally conformable in the Rockies and on the northern Interior Plains. The lower contact is subplanar and commonly overlain by thin phosphatic deposits indicative of slow sedimentation.

In the southern Peace River Embayment and southward, the Exshaw is generally disconformably overlain by the Banff Formation (Harker and McLaren, 1958; Macqueen and Sandberg, 1970). To the north, however, it is mainly gradationally overlain by the Banff and thins northward as it grades laterally into that formation (Macauley, 1958; Richards, 1989). From southern Peace River Embayment northward, the Exshaw passes westward into black shale of the Besa River Formation (Fig. 4E.2, 4E.12).

At its type locality, the Exshaw comprises a lower, 10 m thick unit of black shale, gradationally overlain by an upper, 39 m thick unit of siltstone and silty limestone (Macqueen and Sandberg, 1970). Elsewhere in the south, both divisions are generally present, but in most of the

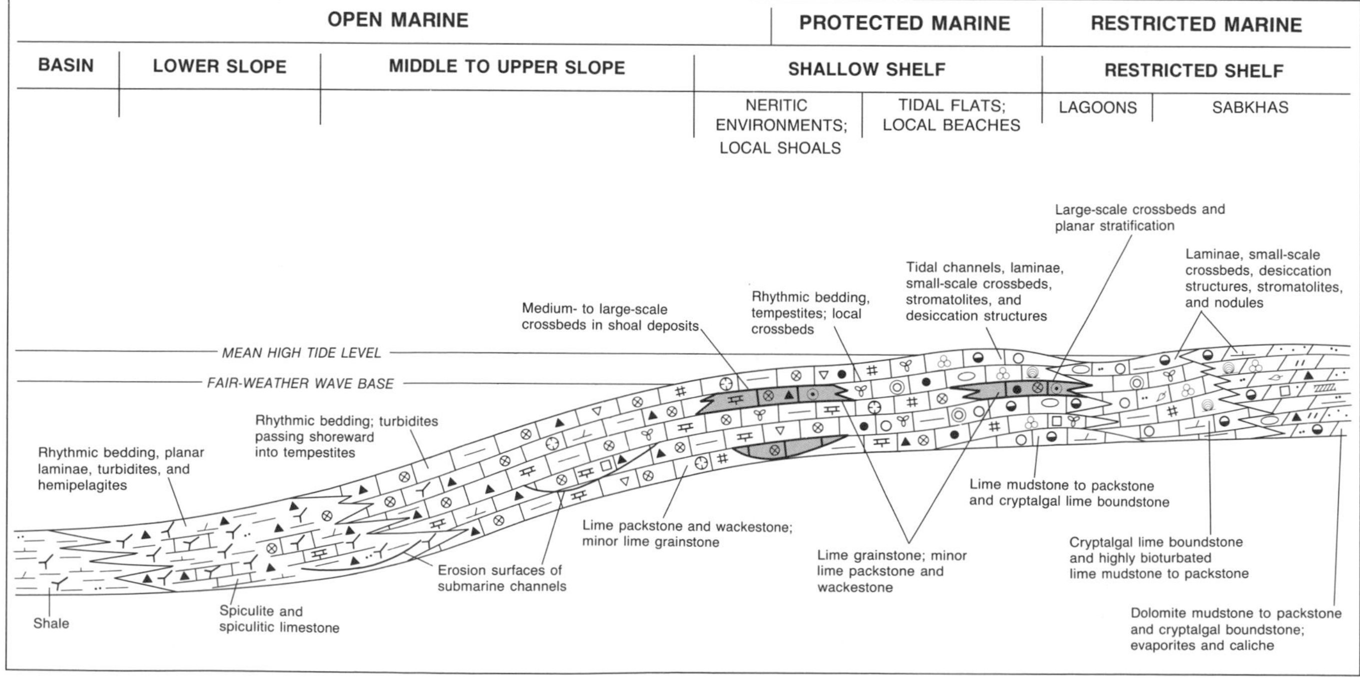

Figure 4E.5. Generalized depositional model of an Early Carboniferous carbonate ramp. See Figure 4E.4 for legend.

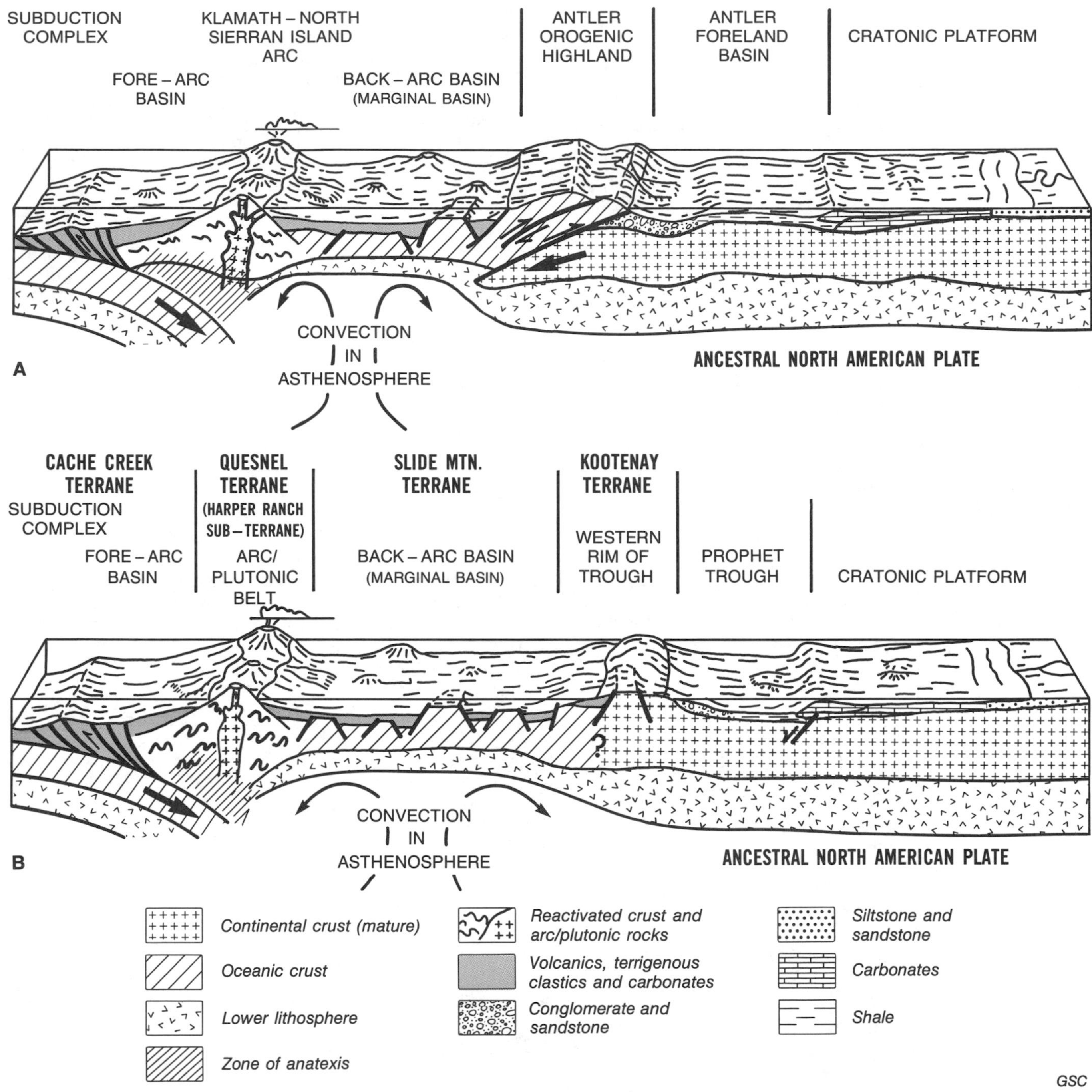

Figure 4E.6. Hypothetical generalized models showing relationships between Late Devonian to Carboniferous island arcs and the western margin of ancestral North America. Model A, representing western United States during the Antler Orogeny, shows a continental margin influenced by compression and back-arc thrusting (after Poole and Sandberg, 1977); it may also represent parts of western Canada during the Tournaisian and Viséan. Model B may represent part of western Canada during Late Devonian and Carboniferous times. It shows widespread extension on the contineтal margin and in an adjoining back-arc basin; the arc may include crust rifted from the continent. Arrows show relative sense of motion.

central Rockies, only the shale unit occurs (Harker and McLaren, 1958). In southern Peace River Embayment, two divisions are usually present, but farther north the upper Exshaw passes laterally into the Banff (Macauley, 1958).

The lower Exshaw consists of anomalously radioactive, brownish black shale that has a high organic carbon content and generally lacks benthonic fossils. Bentonite beds and marine tuff are present locally (Chatellier, 1984; Richards and Higgins, 1988).

Sparsely fossiliferous, calcareous, and dolomitic siltstone with subordinate silty limestone constitute the upper Exshaw in much of the southern Rockies. In the mountains, primary sedimentary structures other than subplanar bedding are rare, but the trace fossil *Taenidium* sp. (= *Scalarituba* of Macqueen and Sandberg, 1970) is abundant. To the east, the upper Exshaw comprises grey shale grading up into siltstone, sandstone, silty limestone, and skeletal to ooid lime grainstone.

The black shale unit of the Exshaw was deposited in the moderately deep water (below storm-wave base) of an anaerobic basin. Such deposition is indicated by planar-laminated, black, pyritic shale; diverse, open-marine conodont assemblages; and the absence of wave- and current-formed structures. A similar interpretation was made by Pelzer (1966). Transgression is recorded by the occurrence of the shale above shallow-neritic to supratidal deposits of the Famennian Palliser/Wabamun Formation.

The upper Exshaw records regression and deposition in moderately deep- to shallow-marine environments. The common occurrence of *Taenidium* sp., characteristic of outer neritic to bathyal environments (Chamberlain, 1978) and the absence of wave-formed structures suggests that most of the unit in the Rockies was deposited at moderate depth (below storm wave base). Much of this unit east of the mountains was, however, probably deposited in shallow-marine environments as it locally contains ooid grainstone and crossbedding. The upper Exshaw is lighter coloured, coarser grained, and contains more benthonic fossils than the lower Exshaw, thereby recording improved circulation in the basin.

Lodgepole Formation

The widely distributed Lodgepole Formation (Collier and Cathcart, 1922), of middle and late Tournaisian (Tn2 to Tn3) age, was deposited after the Bakken and Exshaw formations and contemporaneously with the Banff (Fig. 4E.2, 4E.7, 4E.8). The Lodgepole, which is mainly limestone, occurs from southwestern Manitoba to southeastern Alberta (Fig. 4E.1), and is widely distributed in North and South Dakota and Montana (Roberts, 1979; Sheldon and Carter, 1979). It includes the Souris Valley beds (Macauley et al., 1964; Kent, 1984), most of the Strathallen beds (Kent, 1974), and locally parts of several other informal marker-defined units. In Canada, the southwestward-thickening Lodgepole is mainly between 150 and 180 m thick.

In Canada, the Lodgepole overlies the Bakken and is overlain either conformably by the Mission Canyon or unconformably by Jurassic strata. In Manitoba and easternmost Saskatchewan, the Lodgepole-Mission Canyon contact is commonly abrupt, but to the west it is gradational. Also, the upper Lodgepole becomes younger southwestward (basinward) because the overlying Mission Canyon grades laterally into it. The Banff-Lodgepole boundary is arbitrarily placed at the axis of the Sweetgrass Arch and at the 49th parallel.

In southwestern Manitoba east of 101°W, the Lodgepole is divided into the Scallion, Virden, Whitewater Lake, and Flossie Lake members, in ascending order (Stanton, 1958; McCabe, 1959, 1963). Elsewhere in Canada, it has not been formally divided.

The Scallion Member, 61 m thick at its type section in Manitoba (McCabe, 1959), comprises shale and carbonates deposited during a period of rising sea level. In Canada, it conformably overlies the Bakken, but in North Dakota, the Scallion conformably overlies the Bakken in the west and disconformably overlies pre-Bakken strata in the east (Bjorlie, 1979). Chert-rich skeletal lime mudstone grading upward to pelmatozoan lime wackestone constitutes most of the Scallion, but in the east, pelmatozoan lime packstone and grainstone occur in its upper part. The lower Scallion in the east includes the Routledge shale (0 to 30 m thick) in Manitoba and the Carrington shale facies (0 to 27 m thick) in North Dakota. The western Scallion thins basinward and is overlain by a westward-thickening succession of basinal shale and lime mudstone that constitutes part of the undivided Lodgepole. According to Bjorlie (1979), the Scallion carbonates are mainly of slope and shallow-marine shelf origin, and the Routledge and Carrington were probably deposited in marine environments that lay landward of a barrier. The barrier consisted of lime mudstone to wackestone and may have included Waulsortian mounds. Transgression is recorded by deposition above a subaerial disconformity and onlapping stratigraphic relationships within the Scallion. Shallowing upward is indicated by a transition from basinal shale and slope lime mudstone to shallow-marine grainstone.

The Virden and conformably overlying Whitewater Lake are cyclic members that overlie the Scallion and grade westward into argillaceous slope limestone of the undivided lower Lodgepole. Ooid and skeletal lime grainstone and packstone intercalated with shale and argillaceous limestone constitute the regressive, lower parts of these members (McCabe, 1959; Stanton, 1958). Their upper parts, which indicate deepening, comprise pelmatozoan lime grainstone, packstone, and wackestone with subordinate ooid lime grainstone. Lithofacies in these members are characteristic of shallow-marine, platform and ramp environments (Figs. 4E.4, 4E.5); moreover their presence above the Scallion records shallowing.

The Flossie Lake Member, 65.8 m thick at its type section in Manitoba (McCabe, 1963), conformably overlies the Whitewater Lake and passes basinward into argillaceous slope(?) limestone of the upper Souris Valley beds. In the west, it conformably underlies the Tilston beds of the Mission Canyon Formation, whereas in the east it unconformably underlies Mesozoic beds. Most of the cyclic Flossie Lake comprises micritic to argillaceous and cherty, skeletal limestone (probably mudstone and wackestone), grading upward and eastward to intercalated argillaceous and clean pelmatozoan limestone, probably packstone and grainstone. The uppermost Flossie Lake is a marker horizon comprising red to grey shale and argillaceous limestone abruptly overlain by Tilston ooid and skeletal grainstone of shelf-margin to protected-shelf origin.

Deposition in upper slope to protected- and restricted-marine shelf environments (Fig. 4E.4, 4E.5) is suggested by the lithology and facies trends of the Flossie Lake Member.

In westernmost Manitoba, Saskatchewan, and southeastern Alberta, the Lodgepole Formation is mainly cherty limestone and resembles the undivided Banff Formation of southern Alberta and the lower Scallion of Manitoba (Figs. 4E.2, 4E.8). Deposits resembling the Virden and Whitewater Lake members are generally absent, but ooid lime grainstone occurs locally at a similar stratigraphic position near the northern limit of the Lodgepole (Kent, 1974). Chert is common (MacDonald, 1956) and becomes more abundant westward. The undivided Lodgepole becomes less argillaceous, thicker bedded, lighter coloured, coarser grained, and more fossiliferous upward and toward the north and east. Thickness trends of units within the Lodgepole Formation and Madison Group suggest that much of the Lodgepole was deposited as clinothems inclined at low angles toward the west or southwest.

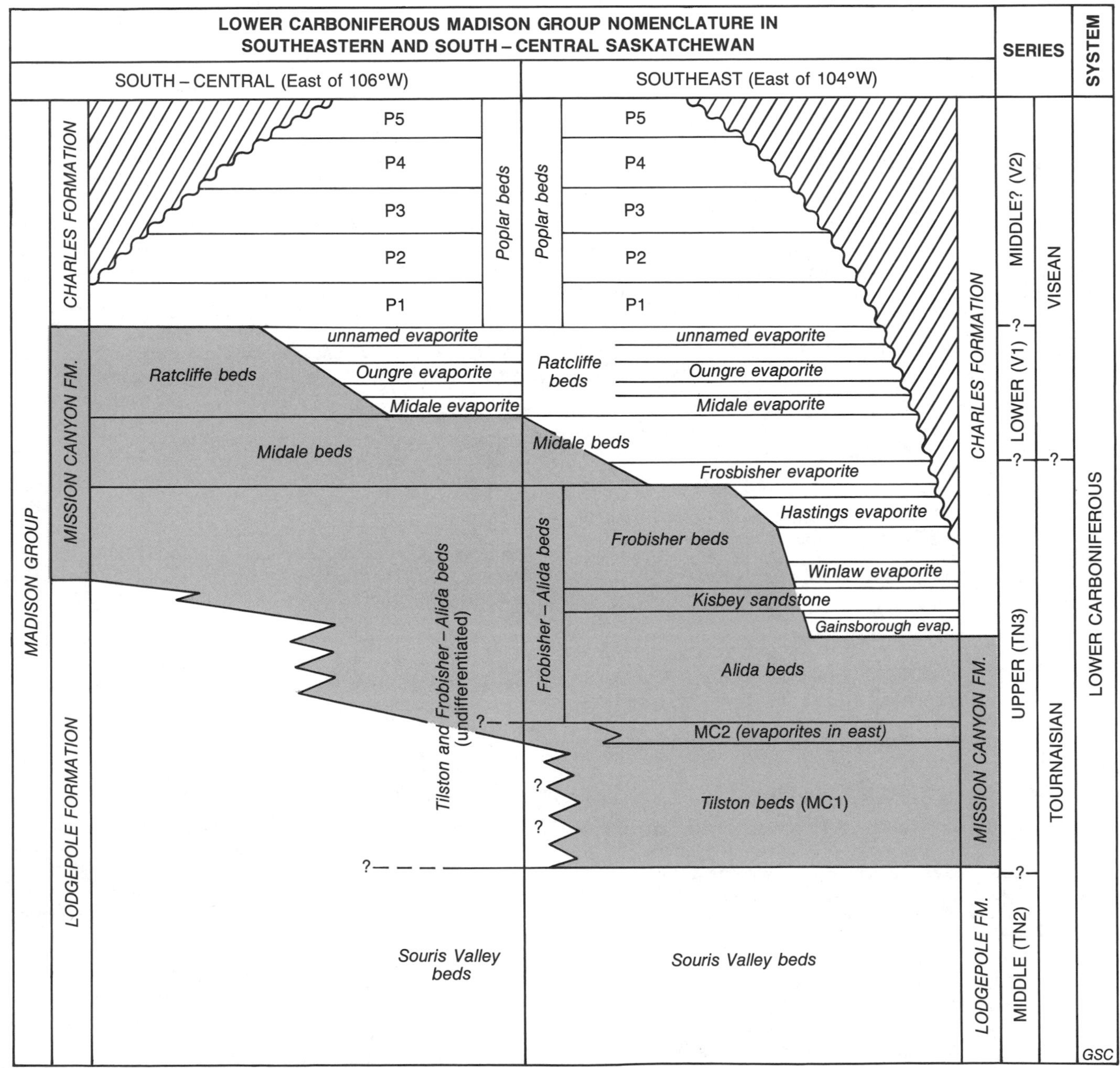

Figure 4E.7. Formal and informal lithostratigraphic units in Carboniferous Madison Group of southeastern and south-central Saskatchewan (after Fuzesy, 1973).

In Saskatchewan, Alberta, and Montana, the basal Lodgepole is commonly a thin unit of glauconitic, pelmatozoan lime wackestone and mudstone (Kent, 1974; Smith 1977; Sando and Dutro, 1974). Smith suggested that the Montana deposits accumulated in relatively shallow water during the transgression of the Central Montana High by the Madison Sea.

The glauconitic beds are overlain by a thick sequence of dark, cherty, argillaceous to bituminous lime mudstone with subordinate shale (Kent, 1984; MacDonald, 1956; Smith, 1977). The lime mudstone, commonly rich in spicules, is planar laminated to thin bedded. In central Montana, where Waulsortian mounds, consisting largely of lime mudstone and submarine cement and lacking an organic framework, are locally developed, the mudstone is rhythmically bedded (Cotter, 1965; Smith, 1977, 1982). This lithofacies formed mainly in anaerobic, moderately deep-water, starved-basin environments (Wilson, 1969; Smith, 1977; Kent, 1984).

The starved-basin deposits grade upward, northward, and eastward to lighter coloured, cherty, bioturbated, and mainly nonargillaceous lime mudstone (Kent, 1974, 1984). These deposits, which are commonly spiculitic and locally contain other skeletal material, represent deposition in the dysaerobic zone (zone within the pycnocline; calcareous epifauna lacking but resistant infauna present) on a low-gradient slope (Kent, 1984).

Upward and toward the paleoshoreline in the north and east, the lime mudstone acquires beds of skeletal lime wackestone, packstone, and grainstone (MacDonald, 1956; Kent, 1974, 1984). Pelmatozoan ossicles predominate, but bryozoans and brachiopods are locally important. Skeletal lime packstone and grainstone become more abundant as these deposits, in turn, pass upward and toward the paleoshoreline to shelf-margin lime grainstone of the Mission Canyon Formation. The carbonates of the upper Lodgepole were, therefore, probably deposited chiefly in middle- to upper-slope environments (Fig. 4E.4).

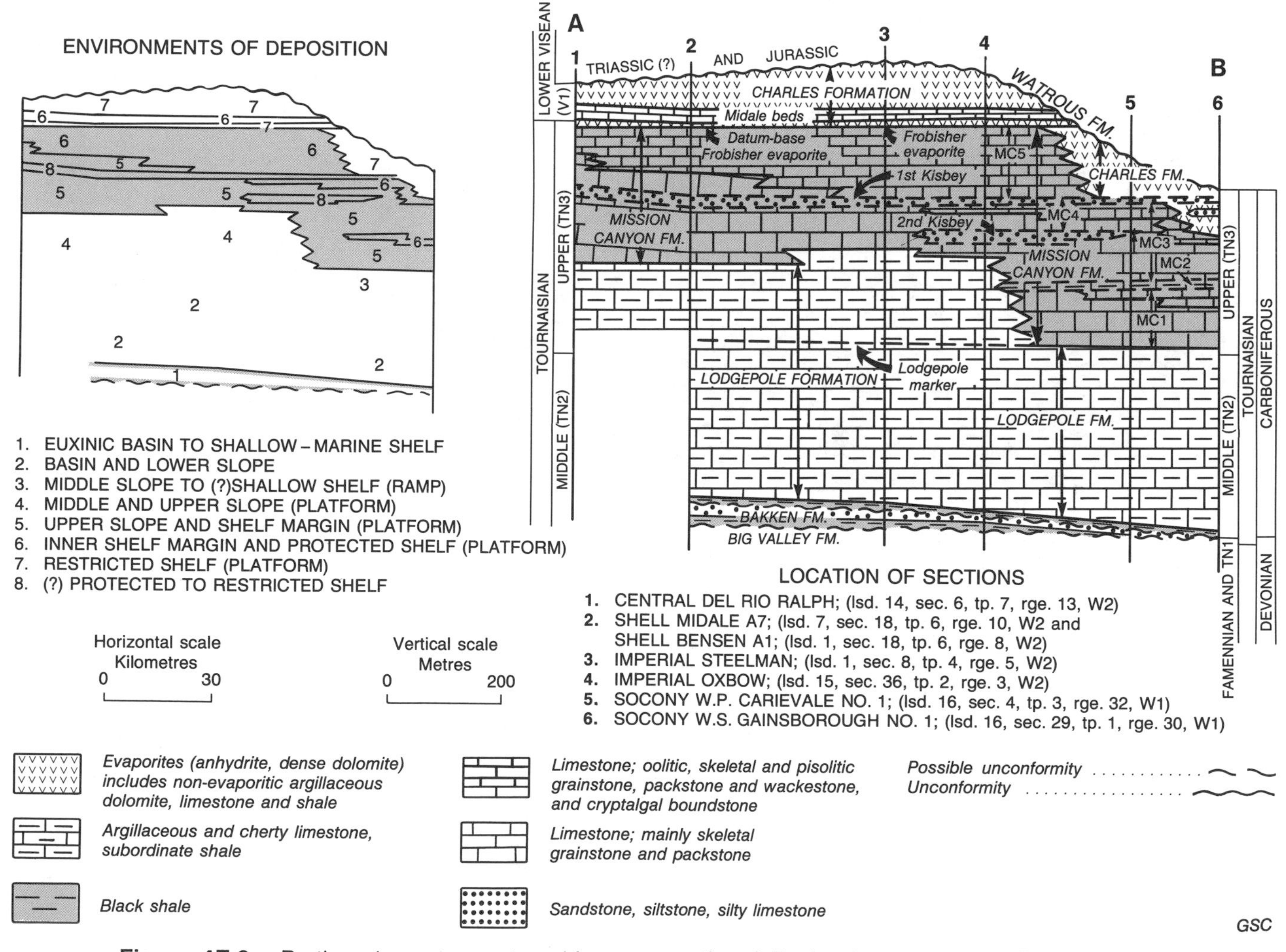

Figure 4E.8. Partly schematic stratigraphic cross-section A-B showing uppermost Devonian and Carboniferous units in eastern and central Williston Basin, southeastern Saskatchewan. See Figure 4E.1 for line of section (after Edie, 1958).

West and southwest of Manitoba, the Lodgepole records a major transgression followed by a long period of progradation and shallowing. Following deposition of the basal, transgressive glauconite beds in relatively shallow water, continued transgression and deepening resulted in the establishment of deep-water, starved-basin conditions over wide areas and deposition of the bituminous, basinal lime mudstone and shale of the lower Lodgepole (Smith, 1977; Sheldon and Carter, 1979). Overlying, middle- to upper-slope limestone in the upper Lodgepole marks an end to the deep-water, starved basin. It also records the onset of southward and westward progradation of shallow-shelf carbonates across Williston Basin.

Banff Formation

Sedimentation of the Banff Formation (Kindle, 1924; Warren, 1927), of latest Famennian(?) to late Tournaisian (Tn3) age, partly coincided with that of the Lodgepole Formation. The Banff, consisting of shale, carbonates, and sandstone, extends from southeastern British Columbia to southwestern District of Mackenzie (Fig. 4E.1, 4E.2, 4E.9-4E.12). This eastward-thinning formation ranges from less than 150 m on the Plains of west-central Alberta to more than 500 m in southwestern District of Mackenzie.

The Banff generally overlies the Exshaw Formation, but it disconformably overlies the Palliser in the eastern Cordillera from 52°30′N to 54°25′N, where most of the

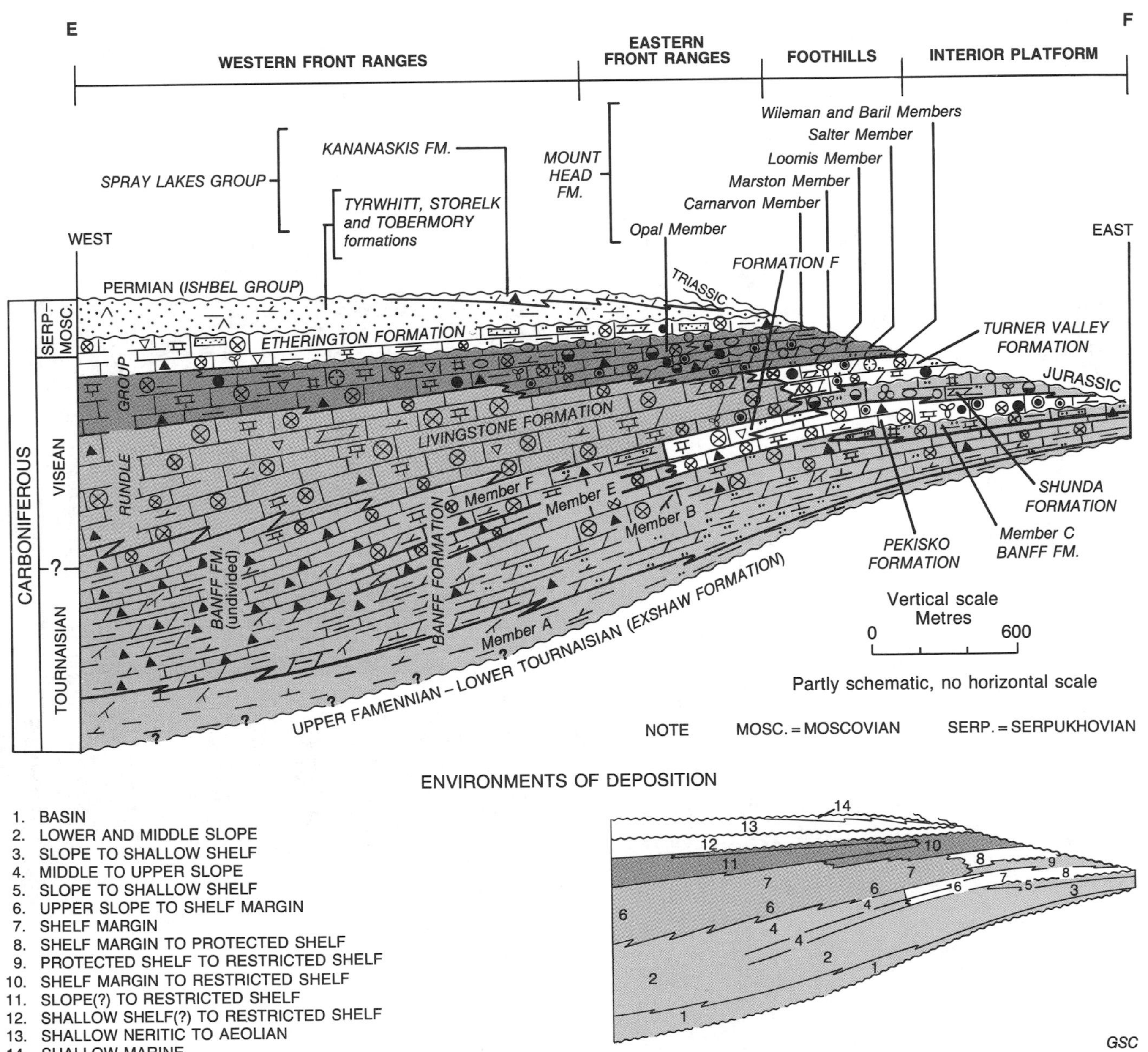

Figure 4E.9. Partly schematic, non-palinspastic stratigraphic cross-section E-F showing Carboniferous units of southwestern Alberta. See Figure 4E.1 for line of section and Figure 4E.10 for legend.

Exshaw was removed prior to Banff deposition. It locally unconformably overlies the Fammenian Wabamun Formation, coarse-grained Devonian siliciclastics, and Precambrian plutonic rocks on the axis of the Devonian Peace River Arch in west-central Alberta. A thin unit of arkose commonly constitutes the basal Banff where it unconformably overlies the Precambrian and Palliser. North of 55°N, the Banff passes basinward (generally southwestward) into the Besa River Formation (Fig. 4E.2, 4E.11-4E.13).

The Pekisko and Livingstone formations generally overlie the Banff, except in British Columbia north of 55°N, where the Banff is commonly overlain by formation F (see Fig. 4E.16). In most areas, the boundary between the Banff and overlying Pekisko is sharp and erosional. However, on the Plains south of Calgary and in the eastern Cordillera south of 51°15'N, it is commonly gradational. The Banff is conformably overlain by the Livingstone in the southeastern Cordillera and on the Plains of southernmost Alberta, and in these areas grades eastward and northward into the lower part of the Livingstone Formation (Fig. 4E.2, 4E.9). Northeast of the subcrop edges of the overlying Carboniferous formations, Mesozoic strata unconformably overlie the Banff.

The middle and upper Banff (members E and F; Fig. 4E.2, 4E.9) pass laterally into the Pekisko Formation and formation F toward the north on the Plains of southernmost Alberta, and toward the northeast in the Cordillera south of 52°15'N. In these areas, the boundary with the Pekisko and formation F is the basinward limit of the thick lime grainstone units characteristic of the Pekisko.

In a regional study in progress, Richards and Higgins have determined that the Pekisko of southern and southwestern Alberta passes basinward into formation F and the Banff rather than into the Livingstone as suggested by Macqueen and Bamber (1967), Macqueen et al. (1972), and Chatellier (1983, 1984). Strata in the eastern Front Ranges west of Calgary contain the same conodont zones as the Pekisko and part of the basal Shunda to the northwest and east (Fig. 4E.2), and are assigned here to the Pekisko and formation F rather than to the middle and upper Banff as previously done by other workers. Similar correlations were proposed earlier by Moore (1958). Richards and Higgins have also concluded that the western part of the basal Shunda passes basinward into the Banff through formation F.

For this report, part of the Banff is divided into six informal members (A to F) (Fig. 4E.2, 4E.9-4E.12, 4E.14). Members A, B, and D are equivalent to the lower shale unit, middle carbonate unit, and upper clastic unit of Macauley (1958), respectively. Member C is partly correlative with member D. The E and F members are equivalent to parts of the middle and upper Banff respectively of Macqueen and Bamber (1967).

Member A, chiefly shale, overlies the Palliser and Exshaw formations. It occurs in most areas where the Banff is preserved but is locally absent on the Interior Plains south of 51°00'N. The thickness of this basinward-thickening member ranges from less than 8 m in southwestern Alberta to 490 m in southwestern District of Mackenzie. Basal member A is commonly a thin black shale unit that resembles the upper Bakken Formation and occurs at a similar stratigraphic level. Turbiditic siltstone and silty carbonates interbedded with shale are commonly present higher in the member. Sandstone and siltstone units occur locally in the north. Member A usually grades upward and northeastward into slope deposits of member B, but in the northwest, member A is gradationally overlain by member D.

Member A was deposited in deep to moderately deep water (Richards, 1989) as indicated by deep-water conodont assemblages, lithology, and facies relationships. Southwestward-dipping clinoforms, evident in some seismic profiles of the Banff in northeastern British Columbia (Chatellier, 1983, 1984), show that part of member A formed at depths of 350 m or more. A regional transgression and marked deepening are recorded by the local presence of these deposits above a subaerial erosion surface and by their widespread occurrence above relatively shallow-water facies in the upper Exshaw.

Member B, which generally overlies member A and consists of carbonates, occurs from southern Alberta to northeastern British Columbia. It is gradationally overlain by, and grades laterally to, either member C or D; however, where these members are absent, member B is overlain by either member E or the Rundle Group. Member B, mainly 50 to 250 m thick, generally constitutes either the middle Banff or most of the lower and middle Banff.

Member B is cherty to argillaceous spiculite, dolostone, lime mudstone, and wackestone, which commonly grade upward and northeastward into cherty lime packstone and grainstone. Shale is also locally common. Toward the north and northeast on the Interior Plains, siltstone and sandstone become locally abundant. In the southern to central Rockies and foothills, lower member B consists of laminae of possible hemipelagic origin and rhythmically bedded turbidite-like beds. Southwestern occurrences of upper member B comprise rhythmically bedded pelmatozoan lime wackestone and packstone. To the northeast, upper member B is commonly dominated by pelmatozoan lime grainstone showing medium- to large-scale crossbedding.

Member B is a shallowing-upward hemicycle formed in environments ranging from basin and slope to shelf margin, protected shelf, and shallow shelf (Fig. 4E.4, 4E.5, 4E.9, 4E.10). Shallowing upward is recorded by a transition from hemipelagites and turbidites, representing basin and slope environments, to tempestites (storm deposits) and crossbedded lime grainstone of shelf-margin to protected-shelf and shallow-shelf origin. In upper member B of west-central Alberta and east-central British Columbia, shelf-margin environments are represented by thick, widespread units of crossbedded pelmatozoan grainstone, which pass northeastward into facies of protected-shelf aspect in members B and C.

Member C, which is normally less than 100 m thick and consists of limestone, dolostone, and shale, gradationally overlies and passes southwestward into upper member B. The member is widely distributed on the Interior Plains and in the eastern Cordillera from southern Alberta to east-central British Columbia. Member C is, however, commonly absent or poorly developed on the Plains south of Calgary and in the Cordillera to the southwest. Toward the north and northeast, this member grades into

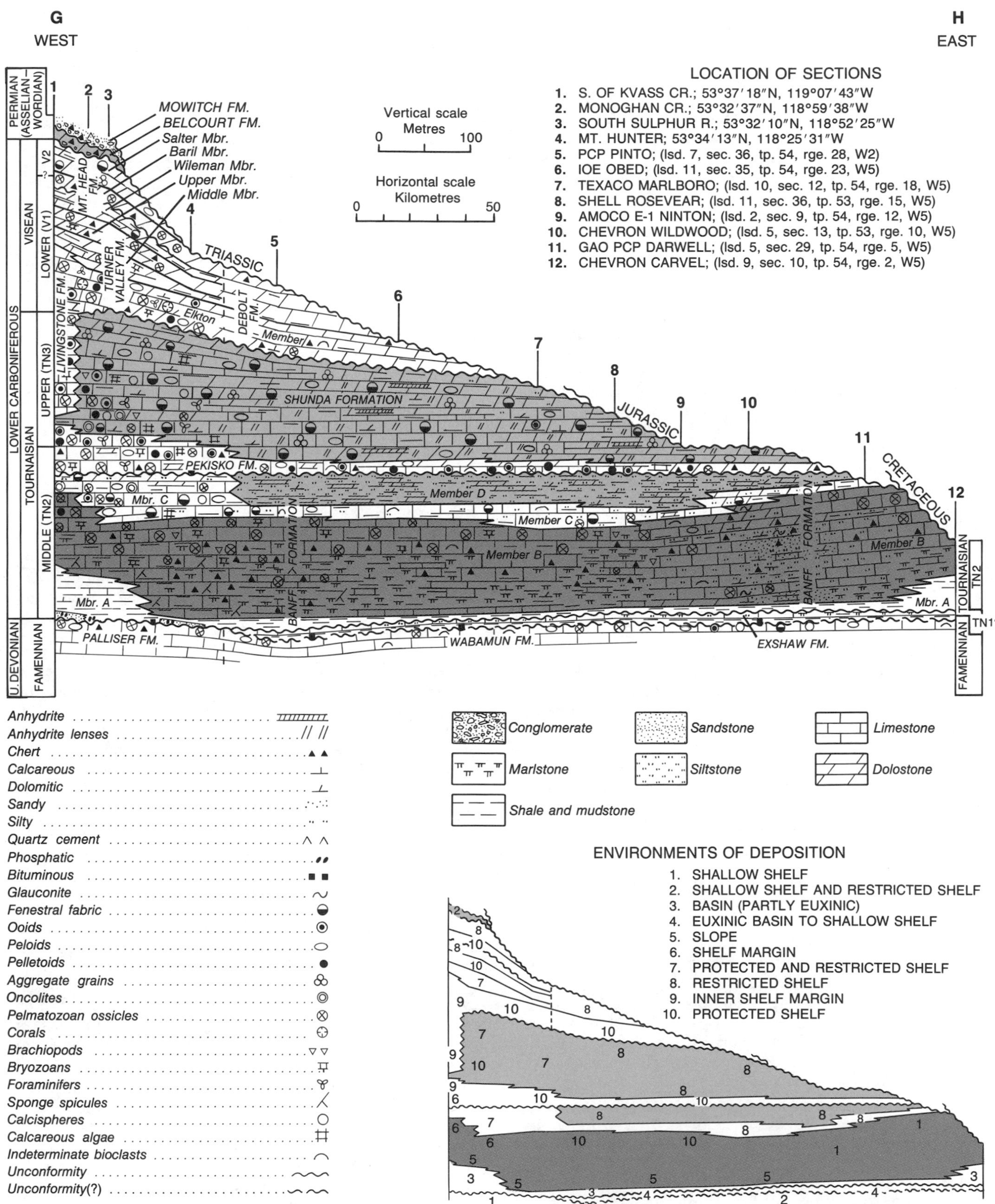

Figure 4E.10. Partly schematic, non-palinspastic stratigraphic cross-section G-H showing uppermost Devonian and Lower Carboniferous units of west-central Alberta. See Figure 4E.1 for line of section.

member D, which commonly overlies it as well. Where the latter is absent, member C is generally overlain by the Rundle Group.

In the southwest, member C comprises cherty, bryozoan-pelmatozoan lime wackestone, packstone, and grainstone with subordinate shale, marlstone, and dolostone. Rhythmically bedded deposits resembling turbidites and tempestites generally predominate. Bedding is locally truncated by large-scale submarine paleochannels, and some grainstone shows crossbedding. These lithofacies grade northeastward into fenestral, cryptalgal boundstone and algal-peloid wackestone, which are commonly associated with mixed-skeletal limestone, shale, and silty microcrystalline dolostone. East of the Rockies, siltstone and sandstone are moderately common.

Member C was deposited mainly on the restricted shelf, shallow shelf, and protected shelf, and locally on the slope. On the Interior Plains and in part of the easternmost Cordillera, widespread deposition on the restricted shelf and on tidal flats of either the protected shelf or shallow shelf is recorded by the occurrence of cryptalgal boundstone and associated rock types. Southwestward, shallow-neritic environments (above fair-weather wave base) are represented by crossbedded grainstone and algal-peloid to mixed-skeletal lime wackestone and packstone. Some of these neritic deposits pass basinward to shelf-margin grainstone of upper member B, thereby indicating deposition on the protected shelf of a platform. However, most probably formed on the shallow-shelf of a ramp because they pass directly basinward into rhythmically bedded slope deposits of members C and B.

Member D occurs on the Plains from southern Alberta to southwestern District of Mackenzie and locally in the Rocky Mountain Foothills of Alberta. Only locally developed south of Calgary, it generally overlies either member B or C, but in the District of Mackenzie and immediately adjacent areas to the south, it gradationally overlies member A. Member D ranges in thickness from less than 40 m in the south to more than 135 m in the north. Interbedded shale, siltstone, silty to sandy dolostone, and sandstone predominate, but skeletal limestone and other carbonates are also developed.

In most of member D, shallow-marine to supratidal deposition is recorded by wave-and current-formed crossbedding, the local occurrence of fenestral carbonates and desiccation structures (Martin, 1967, 1969, Richards, 1989), and the relationship to coeval restricted-shelf carbonates in member C. However, turbidite-like beds, suggesting deposition in slope environments, are included in lower member D in the District of Mackenzie and in adjacent areas to the south. The latter deposits grade southwestward into basinal shale of member A and the Besa River Formation.

Member E, which generally overlies member B and underlies member F, is present in the eastern Cordillera of southwestern Alberta (Fig. 4E.2, 4E.9) and the western part of the southern Plains north of 49°30′N. On the Plains, member E grades northward, and in the Cordillera northeastward into the Pekisko Formation. Member E passes basinward into the undivided Banff. Cherty, spiculitic, pelmatozoan lime packstone and wackestone with subordinate grainstone, dolostone, and spiculite constitute the member. Toward the south and southwest this resistant unit becomes finer grained and more cherty. Member E contains rhythmically bedded, turbidite-like beds, and was deposited in moderately deep-water slope environments basinward of shelf-margin to upper-slope carbonates of the Pekisko.

Member F overlies member E and occurs in the same areas. It is mainly a southwestern correlative of formation F, but upper member F grades cratonward into the lower Livingstone Formation (Fig. 4E.2, 4E.9). This member is more argillaceous than member E, closely resembles formation F, and coarsens upward. It comprises rhythmically bedded, spiculitic dolostone, marlstone, and cherty to dolomitic skeletal lime wackestone and packstone. Pelmatozoan lime grainstone, silty carbonates, and siltstone commonly prevail in its upper part.

Member F was deposited in moderately deep-water slope settings, as indicated by the common occurrence of numerous turbidite-like beds. It also passes toward the paleoshoreline into slope deposits of formation F and upper-slope to shelf-margin facies of the Livingstone. In upper member F, shallowing is recorded by an upward increase in grainstone.

Most of the Banff Formation is undivided on the Interior Plains south of 49°45′N, in the western Front Ranges west of Calgary, and in the Rockies south of 50°30′N. The undivided Banff, which overlies either the Exshaw Formation or member A, comprises laminated to thin-bedded, spicular chert and cherty spicule lime mudstone to wackestone, which grade upward into medium-bedded chert and cherty skeletal limestone. Shale units are locally present, and skeletal lime grainstone occurs in its upper part. Laminae in this unit resemble hemipelagites and distal turbidites, whereas the beds commonly resemble distal to proximal turbidites.

The undivided Banff was deposited in basin and slope environments, as indicated by the predominance of dark, fine-grained strata lying well basinward of coeval shelf margins of the Pekisko and Livingstone. In the Cordillera of southeastern British Columbia and southern Alberta, marked basinward thinning of some units in the undivided Banff suggests clinothems facing southward or southwestward (W. Styan, Shell Canada Ltd., pers. comm., 1986).

Carbonate platform and ramp deposits prograded southward and southwestward over basinal shale during deposition of members B to F. These buildups developed mainly south of Peace River Embayment and are absent from District of Mackenzie. Strata in upper member B locally record poorly developed platforms; but lithofacies deposited contemporaneously with those of the Pekisko and lower Livingstone formed on widespread platforms. At these three stratigraphic levels, platforms are indicated by slope lithofacies passing northeastward into mainland supratidal facies through shelf-margin and protected-shelf facies. However, the apparent occurrence of slope deposits passing directly eastward into shallow-shelf and mainland facies (Fig. 4E.5, 4E.9, 4E.10) indicates that ramps predominated during sedimentation of lower to middle member B, middle to upper member C, and part of member F.

The eastern Banff, comprising members A to D, records a major transgression and subsequent regression. Transgression is recorded by the basinal lithofacies of

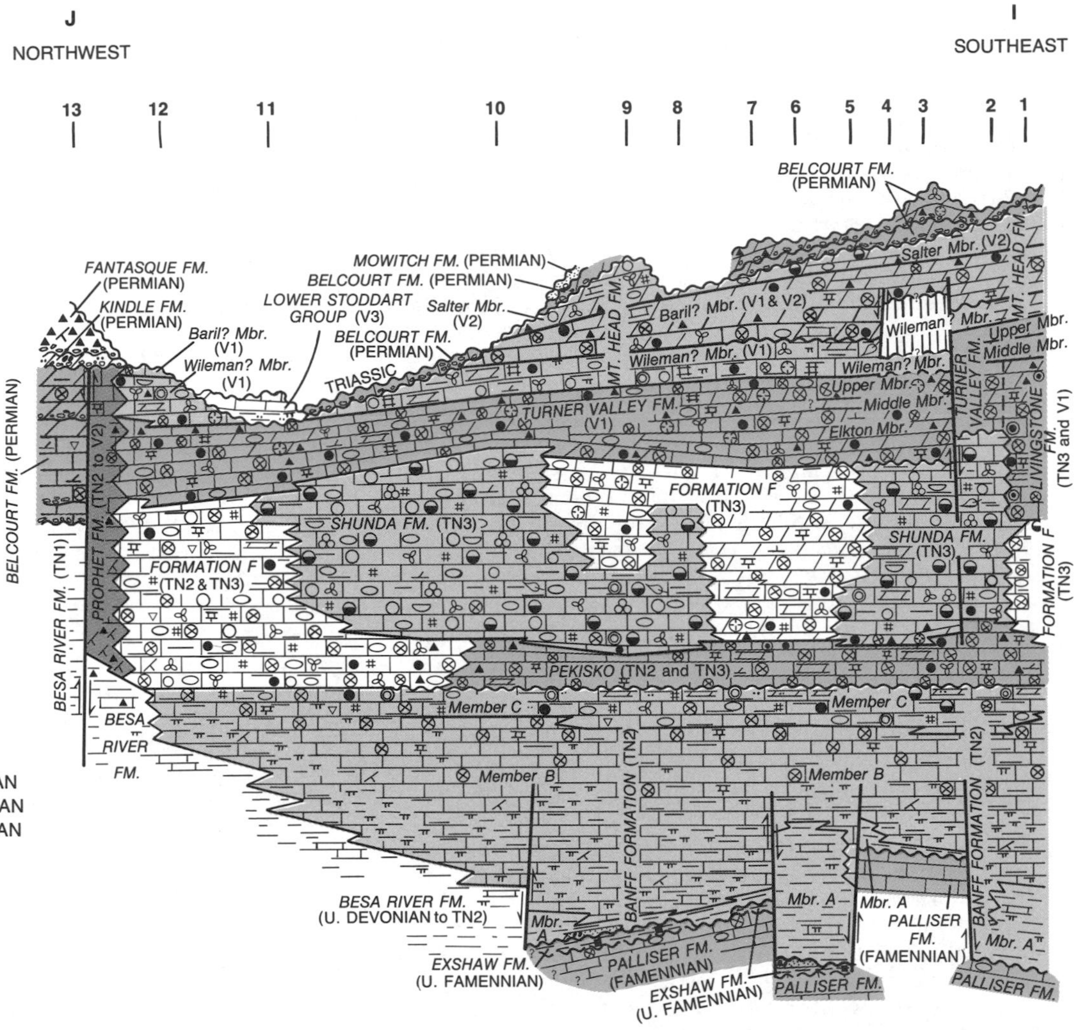

J
NORTHWEST

I
SOUTHEAST

13 12 11 10 9 8 7 6 5 4 3 2 1

BELCOURT FM. (PERMIAN)
FANTASQUE FM. (PERMIAN)
KINDLE FM. (PERMIAN)
MOWITCH FM. (PERMIAN)
BELCOURT FM. (PERMIAN)
LOWER STODDART GROUP (V3)
Salter Mbr. (V2)
BELCOURT FM. (PERMIAN)
Baril? Mbr. (V1)
Wileman? Mbr. (V1)
TRIASSIC
MT. HEAD FM.
Baril? Mbr. (V1 & V2)
Salter Mbr. (V2)
Wileman? Mbr.
MT. HEAD FM.
Upper Mbr.
Middle Mbr.
Wileman? Mbr. (V1)
Wileman? Mbr.
Upper Mbr.
Middle Mbr.
Elkton Mbr.
TURNER VALLEY FM. (V1)
TURNER VALLEY FM.
LIVINGSTONE FM. (TN3 and V1)
BELCOURT FM. (PERMIAN)
PROPHET FM. (TN2 to V2)
BESA RIVER FM. (TN1)
SHUNDA FM. (TN3)
FORMATION F (TN2 & TN3)
FORMATION F (TN3)
SHUNDA FM. (TN3)
FORMATION F (TN3)
PEKISKO (TN2 and TN3)
Member C
Member C
BESA RIVER FM.
Member B
Member B
BANFF FORMATION (TN2)
BANFF FORMATION (TN2)
Mbr. A
Mbr. A
Mbr. A
Mbr. A
BESA RIVER FM. (U. DEVONIAN to TN2)
EXSHAW FM. (U. FAMENNIAN)
PALLISER FM. (FAMENNIAN)
EXSHAW FM. (U. FAMENNIAN)
PALLISER FM.
PALLISER FM. (FAMENNIAN)
PALLISER FM.

Hiatus
Spiculite
Limestone
Dolostone
Marlstone
Sandstone
Conglomerate
Shale and mudstone

V3 = UPPER VISEAN
V2 = MIDDLE VISEAN
V1 = LOWER VISEAN
TN3 = UPPER TOURNAISIAN
TN2 = MIDDLE TOURNAISIAN
TN1 = LOWER TOURNAISIAN
SERP. = SERPUKHOVIAN

Horizontal scale
Kilometres
0 20

Vertical scale
Metres
0 100

Chert
Calcareous
Dolomitic
Silty
Phosphatic
Bituminous
Fenestral fabric
Ooids
Peloids
Pelletoids
Aggregate grains
Oncolites
Pelmatozoan ossicles
Corals
Brachiopods
Bryozoans
Ostracodes
Foraminifers
Sponge spicules
Calcispheres
Calcareous algae
Terrestrial plant remains
Unconformity(?)
Unconformity
Possible normal faults

LOCATION OF SECTIONS

1. JARVIS LAKES; 54°06′21″N, 120°12′27″W
2. NW OF MT. HANINGTON; 54°10′46″N, 120°14′45″W
3. MEOSIN MTN.; 54°17′08″N, 120°18′55″W
4. MUINOK MTN.; 54°20′03″N, 120°23′19″W
5. SW OF BELCOURT LAKE; 54°21′54″N, 120°29′58″W
6. RED DEER CR.; 54°27′00″N, 120°35′14″W
7. MT. BECKER; 54°31′02″N, 120°38′46″W
8. ONION LAKE; 54°37′57″N, 120°45′34″W
9. FELLERS CR.; 54°42′19″N, 120°53′06″W
10. ALBRIGHT RIDGE; 54°48′15″N, 121°17′37″W
11. MT. PAULSON; 55°06′32″N, 121°48′23″W
12. WATSON PEAK; 55°13′58″N, 122°05′07″W
13. LEAN – TO CR.; 55°04′33″N, 121°59′40″W

ENVIRONMENTS OF DEPOSITION

1. EUXINIC BASIN
2. BASIN (PARTLY EUXINIC)
3. SLOPE
4. UPPER SLOPE AND SHELF MARGIN (PLATFORM)
5. UPPER SLOPE AND SHALLOW SHELF (RAMP)
6. SHALLOW SHELF (RAMP)
7. PROTECTED SHELF AND RESTRICTED SHELF (PLATFORM)
8. SHELF MARGIN AND PROTECTED SHELF (PLATFORM)
9. SHALLOW SHELF AND RESTRICTED SHELF (RAMP)
10. INNER SHELF MARGIN (PLATFORM)
11. PROTECTED SHELF (PLATFORM)
12. RESTRICTED SHELF
13. SLOPE TO SHALLOW SHELF (RAMP IN PART)
14. BASIN? TO SHALLOW SHELF

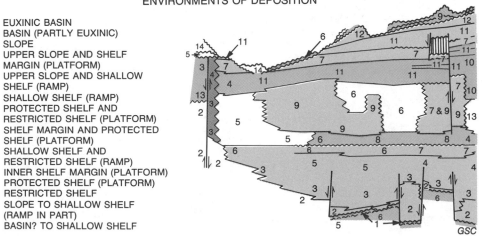

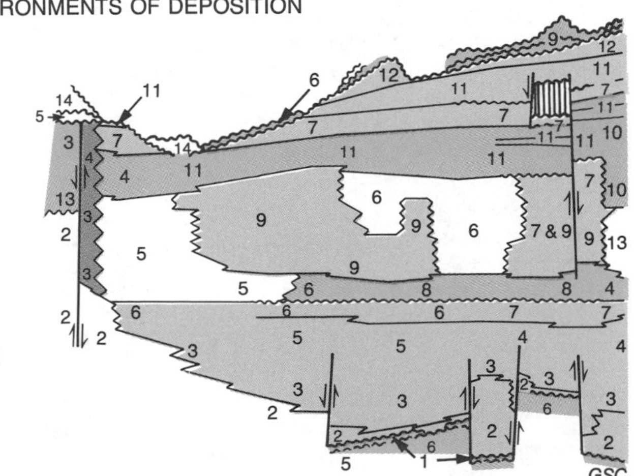

member A and lower B. Subsequent shallowing is recorded in member B and locally in member D by an upward transition from basin deposits to shallow-marine facies. The regression culminated with sedimentation of supratidal deposits in members C and D. Members E and F were deposited during a subsequent transgressive-regressive hemicycle, which is well recorded by coeval lithofacies of the Pekisko and overlying formation F and basal Shunda to the northeast.

Yohin Formation

The lower and middle Tournaisian (upper Tn1 to Tn2) Yohin Formation (Harker 1961) was deposited contemporaneously with member D of the Banff Formation, and occurs in the southern Mackenzie Fold Belt between 60°55′N and 61°45′N (Fig. 4E.2, 4E.12, 4E.13). This formation, consisting mainly of siltstone and sandstone, ranges in thickness from 81 m in the east to 157 m in the west (Richards, 1989).

In most areas, the Yohin gradationally overlies and passes basinward into shale of the Besa River Formation and is conformably overlain by shale of the Clausen Formation. However, in the northeast it gradationally overlies a shale-dominated unit that is equivalent to the lower Banff Formation and has been provisionally called Banff(?) by Richards (1989).

A graded-bed lithofacies forms the lower one-third or less of the eastern Yohin; at Jackfish Gap (Fig. 4E.13) and farther westward it constitutes most of the formation. This facies, which grades into basinal shale of the Besa River and Banff(?), comprises siltstone and sandstone with subordinate skeletal limestone and shale. The predominant deposits are sharp-based, rhythmically stratified beds and laminae that vary from ungraded to normally graded. At Jackfish Gap and westward, these deposits occur with large-scale paleochannels and are probably turbidites. To the east, where they are associated with the crossbedded-sandstone lithofacies, they are interpreted as tempestites. In the east, the graded-bed lithofacies was probably deposited in relatively shallow water (near wave base), either on a gentle slope or on the basinward margin of a marine shelf. To the west, it was deposited in deeper water slope and toe-of-slope settings (Richards, 1989).

A crossbedded-sandstone lithofacies, which overlies and passes westward into the graded-bed lithofacies, thickens eastward and forms most of the upper part of the Yohin northeast of Jackfish Gap. It occurs in coarsening-upward sequences of siltstone and sandstone. Small-scale crossbedding predominates in most sequences, but medium- to large-scale crossbedding is common in the upper part of some. This lithofacies was deposited in shallow-marine environments (above storm wave base) on a shelf that prograded westward (Richards, 1989).

The Yohin and underlying Besa River and Banff(?) formations form a shallowing-upward megasequence recording a transition from deposition of moderately deep-water shale to that of shallow-marine, crossbedded sandstone. It correlates with a similar, coeval, shoaling sequence in the Banff Formation on the Interior Plains.

Figure 4E.11. Partly schematic, non-palinspastic stratigraphic cross-section I-J showing Carboniferous units of east-central British Columbia. Locations of Carboniferous normal faults are based on changes in thickness and lithology between stratigraphic sections; actual faults not observed. See Figure 4E.1 for line of section.

Depositional summary

The Bakken and Exshaw formations comprise fine-grained siliciclastics deposited in euxinic-basin to shallow-neritic environments during late Famennian and early to earliest middle Tournaisian (Tn1 to earliest Tn2) time. Facies of the Exshaw and coeval deposits in the lower and middle Bakken and lower part of the northern Banff Formation (Fig. 4E.2) record regional transgression followed by shallowing. The latter culminated with the subaerial erosion of most of the Exshaw and part of the underlying Palliser Formation in the central Rockies and Foothills. To the east, part of the Exshaw and the middle Bakken may also have been eroded. Middle Tournaisian (Tn2) black shale in the upper Bakken, and correlatives in the basal Banff south of 56°N record the initial phase of a second regional transgression.

The overlying Banff and Lodgepole formations are heterogeneous assemblages of carbonate platform and ramp deposits and siliciclastics. Shale, spiculite, and fine-grained carbonates in the lower parts of these sequences record the continuation of the second transgression and the establishment of widespread, moderately deep-water, basin environments. The transgression was followed, in the middle Tournaisian (Tn2), by shallowing and basinward progradation of slope to supratidal carbonates and siliciclastics. On most of the stable cratonic platform and in eastern Peace River Embayment, this trend culminated with deposition of the regressive uppermost Banff and its correlatives during late middle Tournaisian (late Tn2) time. However, on the unstable craton of southernmost Alberta and in most of Williston Basin and Prophet Trough, the trend culminated during the late Tournaisian (Tn3) and Viséan with sedimentation of the middle depositional unit. In southern Prophet Trough, progradation and shallowing were interrupted by a latest middle Tournaisian (Tn2, upper zone 7) regional transgression. The latter resulted in deposition of slope deposits of lower member F of the Banff and correlatives in lower formation F to the east.

Middle depositional unit

The middle depositional unit (Figs. 4E.1, 4E.2), which constitutes part of the upper Kaskaskia sequence of Sloss (1963), is principally carbonates. It comprises the Mission Canyon and Charles formations of the Madison Group and all of the Rundle Group (Warren, 1927; Douglas, 1958) except the Etherington Formation. This depositional unit generally overlies the lower depositional unit, but from east-central British Columbia to southwestern District of Mackenzie, western deposits of the unit overlie and pass basinward into the Besa River Formation (Fig. 4E.2, 4E.11-4E.13). In most areas, the middle depositional unit is unconformably overlain by either Permian or Mesozoic strata. It is, however, overlain by the sandstone-dominated, upper depositional unit in south-central Saskatchewan, on part of the westernmost Interior Plains, and over wide areas in the eastern Cordillera. The contact between the middle and upper depositional units may be conformable in Saskatchewan, but it is generally disconformable from southwestern Alberta to the southwestern part of Peace River Embayment. Farther north, the middle depositional unit is abruptly, but usually conformably, overlain by the upper unit at a contact that becomes older northwestward.

Deposition of the Rundle Group occurred in Prophet Trough and Peace River Embayment and on the western cratonic platform, whereas that of the Mission Canyon and Charles formations took place in Williston Basin and on the unstable craton to the west (Fig. 4E.1). During late middle and early late Tournaisian (late Tn2 to early Tn3) time, the connection between Williston Basin and Prophet Trough/Antler Foreland Basin was a seaway extending from southern Alberta into northern Colorado. This connection progressively narrowed, and during early to middle Viséan (V1 to V2) time, Williston Basin was occupied by a restricted to semi-restricted sea.

Mission Canyon Formation

After sedimentation of the middle Tournaisian (Tn2) Lodgepole strata, the Mission Canyon Formation (Collier and Cathcart, 1922) was deposited in Williston Basin and to the west (Fig. 4E.1, 4E.2, 4E.7, 4E.8). In Canada, upper Tournaisian (Tn3) to lower Viséan (V1) carbonates of the Mission Canyon occur from southwestern Manitoba to southeastern Alberta. To the south, they occur in North Dakota, South Dakota, Montana, and Wyoming (Sando and Dutro, 1974; Sheldon and Carter, 1979; Roberts, 1979). The Mission Canyon is separated from stratigraphic equivalents in the lower Rundle Group of southern Alberta by arbitrary boundaries at 49°N and along the Sweetgrass Arch. The Mission Canyon is 90 m thick at its type section in Montana (Sando and Dutro, 1974). In Canada, it is thickest in south-central Saskatchewan and generally less than 180 m. In Williston Basin, it has been divided into informal, marker-defined units called beds (Fig. 4E.7; Fuzesy, 1973; Sheldon and Carter, 1979).

In Canada, the Mission Canyon Formation conformably overlies the Lodgepole Formation. In southwestern Manitoba and in southern Saskatchewan east of 106°W, the Mission Canyon is gradationally overlain by the Charles Formation (Fig. 4E.2, 4E.7, 4E.8). The Mission Canyon-Charles contact, which in most places underlies the lowest anhydrite unit above the MC2 of the Madison Group, becomes younger toward the west and southwest. Beyond the erosional edge of the Charles Formation, the sub-Mesozoic disconformity truncates the Mission Canyon toward the east, north, and west.

The Mission Canyon Formation comprises limestone with subordinate dolostone, shale, and sandstone. Anhydrite is locally present, mainly in the lower part (MC2 of Tilston beds) in Manitoba and easternmost Saskatchewan. Dolostone becomes more abundant upward and eastward, both in the formation as a whole and within the numerous constituent fining-upward hemicycles. Sandstone occurs mainly in the Kisbey sandstone of the Frobisher-Alida beds, and shale and argillaceous carbonates are common as thin, widespread markers.

Regional and local studies, including those of Edie (1958), McCabe (1959), Fuzesy (1966, 1973), Kent (1984), and Lindsay and Roth (1982), have demonstrated that the Mission Canyon was deposited in upper-slope to restricted-shelf environments similar to those represented by Figure 4E.4. Lime grainstone and packstone containing abundant pelmatozoan fragments generally constitute the upper-slope and shelf-margin deposits. The upper-slope deposits occur mainly in the lower part of the Mission Canyon west of 102°00′W and contain turbidite-like beds, load casts, and slump structures. Toward the west and southwest, this facies grades into deeper water slope and basin carbonates of the Lodgepole Formation. Upward and toward the northern and eastern paleoshorelines, the upper-slope deposits pass into shelf-margin, pelmatozoan grainstone. The latter, which locally shows medium- to large-scale crossbedding, grades upward and toward the paleoshorelines to a mixed-skeletal facies formed on the shelf margin (Kent, 1984) and a peloid-skeletal lithofacies deposited on the protected shelf (Lindsay and Roth, 1982). Both the mixed-skeletal and the peloid-skeletal facies generally grade upward and toward the paleoshoreline into an ooid lithofacies. Deposition of the ooid facies occurred on shoals along the inner part of the shelf margin (Kent, 1984) and on tidal bars and barrier beaches along the main coast (Lindsay and Roth, 1982). Ooid facies of the shelf margin underlie and grade eastward or northward into carbonates deposited in relatively open-marine settings on the protected shelf. Those deposited on the main coast (inner protected shelf) grade upward and toward the paleoshoreline into restricted-shelf facies. Protected-shelf deposits, which range from skeletal and ooid grainstone to peloid limestone and cryptalgal boundstone, varied with the degree of restriction on the protected shelf and the distance from the shelf margin. The restricted-shelf deposits, although present in the upper part of the Mission Canyon Formation, are best developed in the coeval lower Charles Formation. They include evaporites, bioturbated lagoonal limestone containing abundant peloids and calcareous algae, and intertidal to supratidal carbonate deposits with fenestral fabric and desiccation structures.

The Mission Canyon Formation overall is a regressive succession that prograded westward and southwestward over deeper water Lodgepole facies. At several levels in the Mission Canyon and lower Charles, low-order transgressions are recorded by the occurrence of ooid and skeletal limestone above evaporites and other restricted-shelf deposits.

Charles Formation

The upper Tournaisian (Tn3) to middle(?) Viséan (V2?) Charles Formation (Seager, 1942), which conformably overlies and grades basinward into the Mission Canyon Formation (Fig. 4E.2, 4E.7, 4E.8), is widely distributed in Williston Basin. In Canada, this formation of carbonates, halite, and anhydrite occurs in southwestern Manitoba and in southern Saskatchewan east of 106°W. To the south, it extends into central and eastern Montana and into western North Dakota (Sheldon and Carter, 1979; Roberts, 1979). The Charles is thickest in southernmost Saskatchewan and mainly less than 185 m. Toward the east, north, and west, it thins to zero beneath a sub-Mesozoic regional unconformity.

In Manitoba and easternmost Saskatchewan, the Charles Formation includes most of the Frobisher-Alida beds (McCabe, 1959; Fig. 4E.7). To the west, it comprises, in ascending order, parts of the Midale and Ratcliffe beds, and the Poplar beds (see Wegelin, 1984; Fuller, 1956; Fuzesy, 1960). These marker-defined units, beginning with the Frobisher-Alida beds in the east and progressing upward to the top of the Ratcliffe beds in the west, pass southwestward into the Mission Canyon Formation.

The Charles Formation is overlain conformably(?) by the Carboniferous Kibbey Formation and unconformably by Mesozoic strata. In Saskatchewan, the common occurrence of halite and anhydrite beds immediately below the Kibbey suggests that this boundary is conformable. Maughan and Roberts (1967) and Kent (1974, Fig. 11)

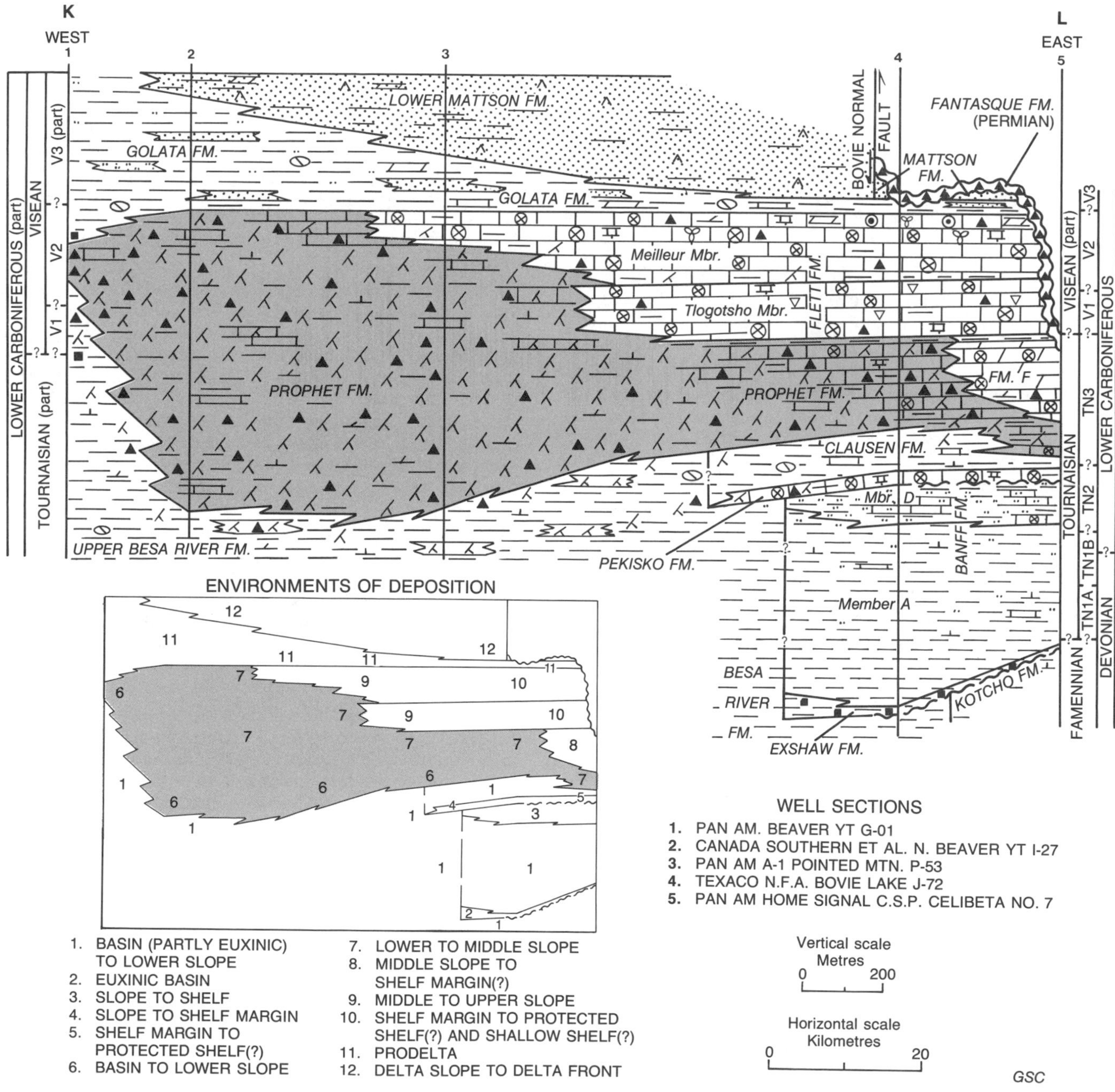

Figure 4E.12. Partly schematic, palinspastic stratigraphic cross-section K-L showing Lower Carboniferous units of southwestern District of Mackenzie and southeastern Yukon Territory. See Figure 4E.1 for line of section and Figure 4E.13 for legend.

reported local intertonguing of the Charles and Kibbey formations, and Roberts (1979) interpreted the contact to be conformable in most of Montana. However, Sheldon and Carter (1979), Sando et al. (1975), and Dutro (1979) have interpreted the Charles-Kibbey boundary in Williston Basin to be a regional subaerial disconformity. North of the erosional edge of the Kibbey, the Charles is overlain by strata that are mainly of either Triassic or Jurassic age (Poulton, 1984).

The cyclic Charles Formation comprises two principal lithofacies assemblages. The lower assemblage contains marginal-marine evaporites, whereas the upper one is dominated by basinal evaporites (Sheldon and Carter, 1979).

In the lower assemblage (Frobisher-Alida beds to the top of the Ratcliffe beds), restricted-shelf evaporites and associated deposits grade basinward (southwestward and southward), through transitional protected-shelf deposits, into shelf-margin carbonates of the Mission Canyon Formation. Evaporites of this assemblage are mainly microcrystalline dolostone and anhydrite that are associated with cryptalgal boundstone, peloid and skeletal limestone, and shale (McCabe, 1959; Fuzesy, 1973; Fuzesy, 1960; Hartling et al., 1982). In the east, siltstone and sandstone (Kisby sandstone) occur as well. The protected-shelf deposits are ooid and mixed-skeletal grainstone and peloid limestone. They are developed mainly in the Mission Canyon Formation but are also commonly present between evaporite units of the Charles.

The upper lithofacies assemblage, which comprises the Poplar beds, consists of evaporites, shale, and associated carbonates that extend across the centre of the basin (Fuzesy, 1973; Sheldon and Carter, 1979). Evaporites in the lower part of this assemblage are mainly anhydrite and microcrystalline dolomite, whereas halite generally predominates in the upper part (unit P5, Fig. 4E.7). Halite also becomes more abundant toward the centre of the basin, where its cumulative thickness exceeds 90 m. Associated carbonates, formed in hypersaline to near normal marine water, include ooid and skeletal limestone. In the upper assemblage, the youngest definite marine beds are of middle Viséan (V2, foraminiferal zone 12) age (B.L. Mamet, pers. comm., 1986). Overlying strata in the assemblage may have been deposited principally in hypersaline lacustrine settings (W.J. Sando, pers. comm., 1986).

Several transgressive-regressive sequences constitute the Charles Formation (Fuller, 1956; McCabe, 1959). They have a lower, transgressive, ooid or skeletal carbonate facies overlain by one or more regressive evaporite units. Below the Poplar beds, the evaporitic component of each successive sequence extends farther basinward than that of its predecessor, but this trend is not evident in the Poplar beds.

During sedimentation of the lower Charles Formation, Williston Basin contained a semi-restricted sea that was connected to the Antler Foreland Basin until the early middle Viséan (V2) by the Central Montana Trough (Roberts, 1979; Sheldon and Carter, 1979). The lower lithofacies assemblage of the Charles and coeval deposits in the Mission Canyon Formation record a major episode of progradation and general expulsion of this sea. The upper lithofacies assemblage records periodic restriction of Williston Basin and precipitation of basinal evaporites.

Pekisko Formation

The Pekisko Formation (Douglas, 1958; Penner, 1958), deposited in Prophet Trough and Peace River Embayment and on the western cratonic platform, is widely distributed from southern Alberta to southwestern District of Mackenzie (Fig. 4E.2, 4E.9-4E.12, 4E.15, 4E.17). This formation, consisting of middle to upper Tournaisian (upper Tn2 to lower Tn3) limestone and dolostone (Macqueen et al., 1972), is present on most of the western Interior Plains north of 49°45′N but is commonly absent and poorly developed on the Plains north of 55°00′N. It is also generally present in the foothills from 49°45′N to 55°N and in the eastern Rockies from 50°45′N to 55°N. The Pekisko is thickest (40 to 120 m) in the Foothills and eastern Front Ranges of southwestern Alberta.

The Pekisko Formation generally overlies and grades basinward into the Banff Formation. In the Cordillera of southwestern Alberta, it generally passes southwestward into the Banff, but on the Plains south of 50°45′N, this transition takes place southward. In the District of Mackenzie and adjacent areas to the south, the Pekisko locally overlies and grades southwestward into the Besa River Formation.

Pekisko lithofacies underlie the Shunda Formation over wide areas of the Interior Plains of western Alberta and in the eastern Cordillera between 50°30′N and 54°50′N. In these areas, the Pekisko generally thins northeastward as it grades into the Shunda Formation (Fig. 4E.2, 4E.9, 4E.10). Formation F overlies the Pekisko in the following areas: part of the Interior Plains of southernmost Alberta, parts of the eastern Cordillera from southern Alberta to east-central British Columbia, Peace River Embayment, and in most of the region north of the embayment. The Pekisko generally thins southwestward as its upper part passes into formation F. Along the southern margin of Peace River Embayment, however, the Pekisko thins as it passes northward into formation F (Fig. 4E.11). The Clausen Formation gradationally overlies the Pekisko in the District of Mackenzie and in adjacent areas of northeastern British Columbia (Fig. 4E.12). East of the erosional edges of the Clausen, formation F, and Shunda, the Pekisko is unconformably overlain by Mesozoic strata.

Three principal lithofacies constitute the Pekisko Formation: bryozoan-pelmatozoan, peloid-skeletal, and ooid-skeletal (Speranza, 1984). In most areas these lithofacies, commonly arranged in multibed hemicycles, were deposited on carbonate platforms, but along the southern margin of Peace River Embayment they appear to have been deposited on a carbonate ramp (Speranza, 1984; Beauchamp et al., 1986; Richards, 1989). The Pekisko facies assemblage generally occurs as a sequence that becomes finer and less resistant upward (Fig. 4E.15), but near its basinward depositional limit in southern Alberta and locally elsewhere, the Pekisko coarsens upward.

The bryozoan-pelmatozoan lithofacies constitutes most of the Pekisko Formation near its southwestern and southern depositional limits and makes up the lower one-third or more of the formation over wide areas to the east and north. In these areas, it generally forms broad, northwest-striking belts (Fig. 4E.9-4E.11). Along the southern margin of Peace River Embayment, however, it occurs as large, shoal-like deposits separated by the peloid-skeletal lithofacies (Beauchamp et al., 1986). Chert becomes more abundant basinward. In western outcrops, large-scale, submarine paleochannels (Bamber et al., 1981; Speranza, 1984) and turbidite-like beds are common. Elsewhere, the bryozoan-pelmatozoan facies comprises massive beds, deposits showing medium- to large-scale crossbedding, and beds with diffuse planar stratification. Toward the northeastern paleoshoreline, units of this facies thin and grade into the peloid-skeletal lithofacies. In southern and southwestern Alberta, the bryozoan-pelmatozoan lithofacies grades basinward into deeper water slope deposits of either the Banff Formation or

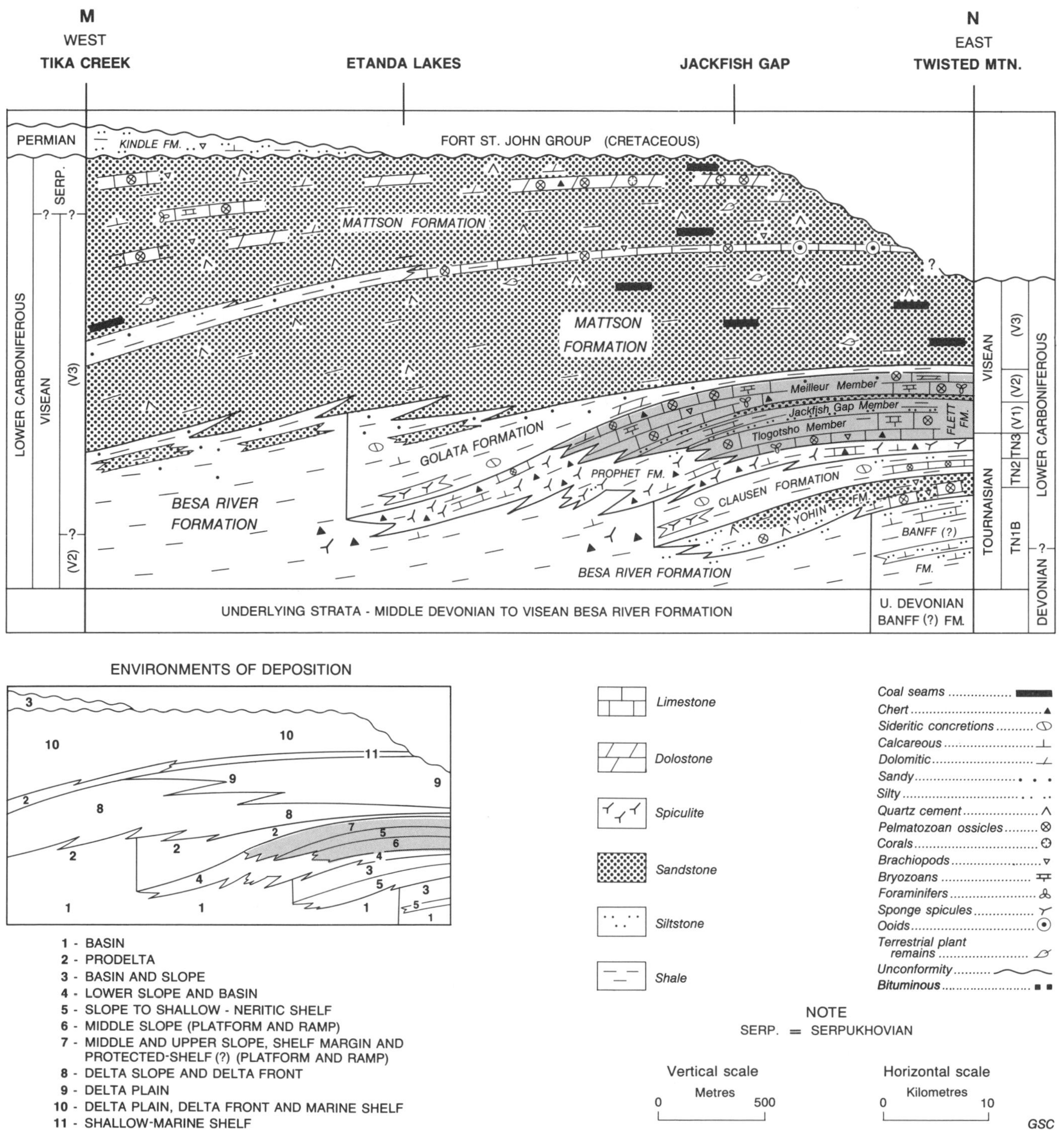

Figure 4E.13. Partly schematic, palinspastic stratigraphic cross-section M-N showing Lower Carboniferous units in eastern Cordillera of southwestern District of Mackenzie. See Figure 4E.1 for line of section.

formation F. From east-central British Columbia to the District of Mackenzie, it passes basinward into either slope deposits of formation F and the Prophet Formation or into the slope and basin deposits of the Besa River Formation.

Deposition of the bryozoan-pelmatozoan lithofacies occurred mainly on the shelf margin and outer protected shelf of carbonate platforms, but partly on the upper slope and shallow shelf (Fig. 4E.4, 4E.5, 4E.9-4E.12). Widespread deposition in the first two environments is indicated by high-energy deposits, crossbedding, and predominance of open-marine fossils. At western localities, upper-slope deposition is indicated by turbidite-like grainstone beds passing basinward into the finer grained slope carbonates of other formations. Shoal-like occurrences of the bryozoan-pelmatozoan facies in the southern part of the Peace River Embayment have been interpreted as the shallow-shelf deposits of a ramp (Beauchamp et al., 1986).

The peloid-skeletal lithofacies, which constitutes a major part of the Pekisko Formation in and northeast of the eastern Cordillera, overlies and grades basinward to the bryozoan-pelmatozoan lithofacies. It also underlies and passes northeastward into either the ooid-skeletal facies of the Pekisko Formation or the protected- to restricted-shelf deposits of the Shunda. Peloid-pelletoid-skeletal lime grainstone with subordinate lime packstone to wackestone, and microcrystalline dolostone constitute the peloid-skeletal lithofacies. Chert and bioturbation are common in the packstone, wackestone, and dolostone. Small- to large-scale crossbedding is locally common.

Deposition of the peloid-skeletal lithofacies occurred principally in relatively open-marine, protected-shelf environments on platforms, as indicated by its relationship to coeval shelf-margin and restricted-shelf deposits. The diverse environments of the protected shelf are also indicated by the joint occurrence of open-marine and restricted-marine fossils, numerous peloids (small, micritic, subrounded allochems, irrespective of origin) and pelletoids (micritized allochems), and deposits ranging from crossbedded grainstone to bioturbated wackestone. However, along the southern margin of Peace River Embayment this facies was probably deposited on the shallow shelf of a ramp because it lies between large, shoal-like deposits formed in that setting.

The ooid-skeletal lithofacies commonly constitutes most of the Pekisko Formation near its erosional zero edge on the Interior Plains, and part of the upper Pekisko to the southwest (Fig. 4E.9, 4E.10). Upward and toward the paleoshoreline, this lithofacies, which is mainly ooid-skeletal lime grainstone, grades into restricted-shelf carbonates of the Shunda Formation.

Most of the ooid-skeletal lithofacies was probably deposited in high-energy environments along the landward part of the protected shelf because it generally lies between the peloid-skeletal facies of the Pekisko and restricted-shelf deposits of the Shunda. The widespread existence of grainstone with abundant ooids is also indicative of deposition in beach and other extensive high-energy settings. Some occurrences, however, originated on the inner part of the shelf margin because the facies locally lies directly on shelf-margin deposits of the bryozoan-pelmatozoan lithofacies.

In most areas, the Pekisko and overlying Shunda formations resulted from a transgressive-regressive hemicycle. Transgression is recorded by the presence of skeletal and ooid grainstone of the Pekisko above shallow-marine siliciclastics and restricted-shelf carbonates of the Banff Formation (Fig. 4E.9, 4E.12, 4E.15, 4E.17). This event did not establish deep-water environments on most of the cratonic platform, but did so in Prophet Trough, particularly after the initial phase of the transgression. Above the thin western part of the Pekisko Formation, this deepening is recorded by slope-to-basin deposits of the Clausen Formation and formation F, which are coeval with the middle to upper Pekisko to the northeast. On the Interior Plains, the regression is recorded by an upward transition from shelf-margin grainstone in the lower part of the Pekisko to restricted-shelf facies in the Shunda.

Clausen Formation

The middle to upper Tournaisian (Tn2 to Tn3) Clausen Formation (Harker, 1961) is a shale that conformably overlies the Yohin and Pekisko formations (Fig. 4E.2, 4E.12, 4E.13) and was deposited in Prophet Trough. It outcrops in the southern Mackenzie Fold Belt of the southwestern District of Mackenzie and occurs on the western Interior Plains to the southeast. The eastward-thinning Clausen, ranging in thickness from less than 30 m on the Plains to 193 m in the Mackenzie Fold Belt (Richards, 1989), generally underlies and grades eastward into slope carbonates and spiculite of the Prophet Formation. From 61°00′N to 61°25′N, where it overlies the Yohin Formation, the lower Clausen Formation locally contains a thin limestone unit that correlates with the Pekisko.

The Clausen Formation, a tongue of the shale lithosome constituting most of the Besa River Formation, is mainly dark shale deposited in moderately deep-water basin environments. Units of a spiculite lithofacies deposited in basin to lower-slope environments are also moderately common, and become more abundant upward and eastward. Lime grainstone and packstone of possible upper-slope to shelf origin occur locally near the base of the formation (Richards, 1989).

Transgression and marked deepening in Prophet Trough are recorded by the occurrence of Clausen basinal shale above shallow-marine deposits of the Yohin and Pekisko formations. Where the Clausen overlies the Pekisko, the transgression began during sedimentation of the latter and culminated with that of the Clausen. Elsewhere, it commenced with sedimentation of the basal Clausen.

Formation F

Formation F ("Shunda" Formation of Bamber and Mamet, 1978; Beauchamp et al., 1986; Richards, 1989), which was deposited contemporaneously with the Pekisko and Shunda formations, occurs from 49°45′N in southwestern Alberta to the southwestern District of Mackenzie (Fig. 4E.2, 4E.9, 4E.11, 4E.12, 4E.16). From southern Alberta to the southern part of the Peace River Embayment, middle to upper Tournaisian (Tn2 to Tn3) formation F is developed mainly in the eastern Cordillera.

To the north, it is present in the western Interior Plains. This formation, which generally thins eastward and consists chiefly of carbonates, is thickest in Peace River Embayment, where it is commonly more than 250 m thick.

Formation F gradationally overlies the Pekisko Formation from southern Alberta to the southern and eastern parts of Peace River Embayment. In most of the embayment, where correlatives of the Pekisko form a resistant limestone unit in the lower part of formation F (Fig. 4E.2, 4E.11, 4E.16), the latter abruptly overlies the Banff Formation. Farther north, formation F generally overlies the Pekisko in the east, and at its western depositional limit from east-central British Columbia to southwestern District of Mackenzie, formation F overlies and passes basinward into the Prophet Formation.

From southwestern Alberta to northeastern British Columbia, eastern occurrences of formation F generally underlie and pass northeastward into the Shunda Formation. In the southern part of Peace River Embayment, however, the transition into the Shunda generally takes place toward the south and southwest (Beauchamp et al., 1986) and locally in this area formation F overlies the Shunda. West of the depositional limit of the Shunda in the Cordillera of southwestern Alberta and east-central British Columbia, the Turner Valley and Livingstone formations abruptly to gradationally overlie formation F. Farther north, in northeastern British Columbia and southwestern District of Mackenzie, the Debolt and Flett formations, respectively, generally overlie western occurrences of formation F.

South of the Peace River Embayment and commonly in its southern part, formation F overlies the Pekisko Formation and is a moderately recessive unit that generally coarsens and becomes more resistant upward (Fig. 4E.9, 4E.11). Rhythmically bedded, dolomitic and cherty, skeletal lime wackestone and packstone predominate, but skeletal grainstone is commonly present in the upper part of formation F and in the east. Most limestone beds in this unit, which locally shows clinoforms and large-scale, submarine paleochannels, are massive, partly bioturbated, and grade upward to marlstone. Many limestone beds resemble turbidites and tempestites. In southwestern Alberta, these deposits pass basinward into slope carbonates occurring in members E and F of the Banff Formation. Toward the paleoshoreline, they grade into upper-slope and shelf-margin deposits of the Pekisko, and shallow-marine Shunda facies.

In the region discussed above, formation F generally comprises slope facies with subordinate shallow-marine facies and records a regional transgressive-regressive hemicycle, also recorded to the north. Deposition on slopes is indicated by regional facies relationships (Fig. 4E.9, 4E.11) and widespread occurrence of turbidite-like beds containing open-marine fossils. Transgression is recorded by the presence of slope packstone and wackestone above upper-slope, shallow-shelf, and shelf-margin grainstone of the Pekisko. In the uppermost part of formation F, shallow-marine microfossils, fenestral cryptalgal boundstone, grainstone, and crossbedding record shallow-water sedimentation and regression.

In most of Peace River Embayment, lower formation F, which commonly becomes finer and less resistant upward, was deposited contemporaneously with the Pekisko Formation and comprises cherty skeletal lime wackestone and packstone with subordinate grainstone. The bedding and sedimentary structures resemble those of formation F to the south, but probable Waulsortian mounds are locally present (Morgan and Jackson, 1970; Davies et al., 1988). Seismic profiles reveal basinward-dipping clinoforms on the northeastern flank of the embayment. Toward the axis of the embayment and generally toward the west, lower formation F grades into deeper water slope and basin lithofacies of the Prophet Formation. In the southern part of Peace River Embayment, the lower part of formation F grades southward into shallow-shelf grainstone of the Pekisko Formation, but elsewhere in the embayment, it grades northeastward into upper-slope and shelf-margin grainstone of that unit.

The lower part of formation F (Pekisko correlative) within the embayment was deposited in shallow-marine to slope environments. In the south, algal packstone and grainstone in the basal part of this interval record shallow-marine deposition. Widespread slope sedimentation is indicated by the occurrence of clinoforms and turbidite-like beds deposited basinward of shelf-margin and shallow-shelf grainstone of the Pekisko. The upward transition from shallow-marine facies in the Banff and basal formation F to slope deposits records the onset of a regional transgression.

In the region northwest of Peace River Embayment and above the Pekisko correlative within the embayment, formation F is mainly a progradational, coarsening-upward sequence with several sub-sequences (Fig. 4E.12, 4E.16). The main sequence is thickest and best developed in the west. There it either overlies spiculite of the Prophet Formation or has a basal shale- and marlstone-dominated interval recording basin to lower-slope sedimentation and the culmination of the transgression discussed above. The sequence comprises argillaceous to cherty lime wackestone grading upward to peloid-skeletal and ooid-skeletal lime grainstone through pelmatozoan grainstone and packstone. In the lower part of this sequence in the west, rhythmically bedded, turbidite- and tempestite-like beds of packstone and wackestone predominate. In the southern part of the Peace River Embayment, these commonly grade southward and southwestward into shallow-marine to supratidal facies of the Shunda Formation without passing through a grainstone-dominated facies (Fig. 4E.11). To the north, moderately thick and widespread grainstone units commonly occur in the upper part of the sequence in the west and at progressively lower stratigraphic positions toward the northeast. Where the basal strata are grainstone, they locally merge with the shelf-margin to protected-shelf deposits of the Pekisko Formation. Most grainstone units in this part of formation F pass northeastward into protected- and restricted-shelf facies of the Shunda. Toward the axis of the embayment and southwestward, facies in the main sequence (Shunda correlative) generally grade into lower slope and basin deposits of the Prophet Formation (Fig. 4E.11, 4E.12). Seismic profiles of the lower part of the sequence in the eastern part of the Peace River Embayment reveal southwestward-dipping clinoforms.

Above its transgressive lower facies, the sequence described above (Shunda correlative) was deposited under shallowing-upward conditions, as indicated by the upward transition from basin deposits to those characteristic of shelf-margin, shallow-shelf, and protected-shelf environments. Widespread slope deposition is indicated by the occurrence of clinoforms and turbidite-like beds containing open-marine fossils.

The uppermost part of formation F, at western occurrences in the axis of the Peace River Embayment and northward, is generally a westwardly thickening shale-dominated unit. It is of basinal aspect, and probably records the early phase of a transgression that occurred during deposition of the basal Turner Valley Formation to the south (Richards, 1989).

Formation F is interpreted to be mainly carbonate platform deposits, because it generally passes eastward into Pekisko shelf-margin deposits and Shunda facies deposited in protected- to restricted-marine settings on platforms. In the southern part of the Peace River Embayment, where slope and shallow-shelf facies of formation F grade directly into shallow- to restricted-shelf deposits of the Shunda, it forms part of a ramp.

Shunda Formation

The upper Tournaisian (Tn3) Shunda Formation (Stearn, 1956), deposited principally on the cratonic platform, extends from southwestern Alberta to northeastern British Columbia (Fig. 4E.2, 4E.9, 4E.10, 4E.11, 4E.17; Macqueen et al., 1972). It consists mainly of carbonates, and is widely distributed on the Interior Plains north of 50°N. The Shunda, which also occurs in the Foothills from 50°N to 55°10′N and in the Front Ranges between 51°15′N and 55°10′N, is up to 170 m thick in the eastern Front Ranges and thins relatively rapidly southwestward.

In most areas, the Shunda conformably overlies the Pekisko Formation; however, in part of the southern Peace River Embayment and over wide areas to the north, it gradationally overlies formation F. Near its subcrop edge in northwestern Alberta, the Shunda also locally overlies the Banff Formation.

Between the Peace River Embayment and Calgary, the Shunda grades southwestward into the Pekisko, Turner Valley, and Livingstone formations and formation F (Fig. 4E.2, 4E.9, 4E.10, 4E.11); farther south it grades southward and southwestward into these formations. Along the margins of the embayment and northward, the Shunda grades basinward into formation F.

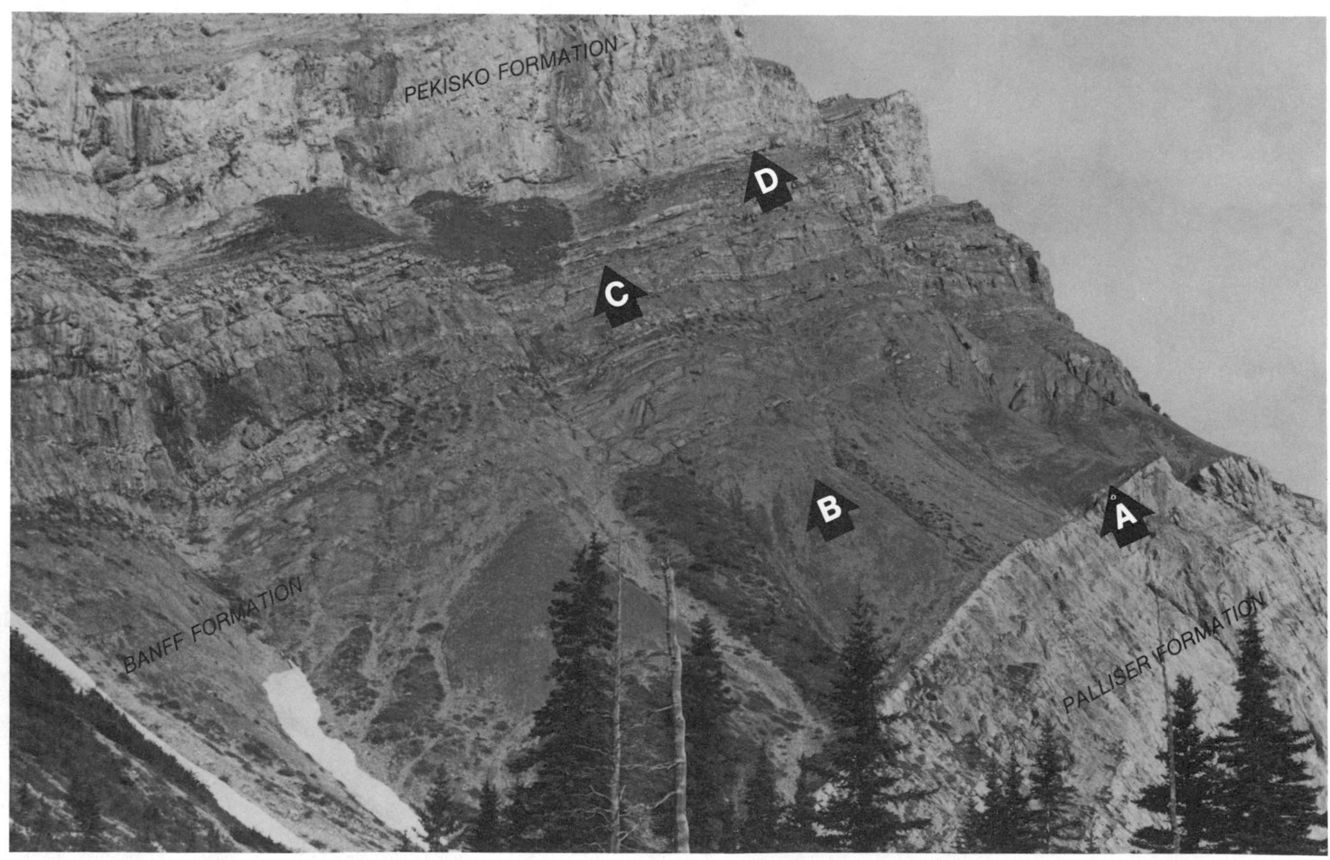

Figure 4E.14. Palliser, Banff (162 m thick) and Pekisko formations, head of South Sulphur River, Rocky Mountain Front Ranges, west-central Alberta. A - disconformable contact between Palliser (limestone) and member A (basinal shale) of Banff; B - base of member B (slope to shelf-margin carbonates) of Banff; C - base of member C (protected- and restricted-shelf deposits) of Banff; D - erosional base of Pekisko (shelf-margin grainstone). View is to northwest (photo - B.C. Richards, ISPG 1854-20).

The Shunda Formation is generally abruptly and commonly disconformably overlain by the Turner Valley Formation (Fig. 4E.2, 4E.17) in the Foothills and eastern Front Ranges south of 55°10′N and on the southwestern Plains. Near its basinward depositional limit, however, its contact with the Turner Valley is gradational. To the north, the Shunda is conformably(?) overlain by the Debolt Formation, but east of the subcrop edge of the Debolt and Turner Valley, it is overlain by Jurassic and Cretaceous strata.

Several lithofacies constitute the cyclic Shunda Formation (Macqueen et al., 1972; Bamber et al., 1981). The principal ones are the mixed-skeletal facies, algal-wackestone facies, fenestral-carbonate facies, and silty-dolostone facies. Of these, the first three form most of the formation in the Cordillera and on the western Plains. Although commonly present in the Cordillera, the silty-dolostone lithofacies is more abundant and locally predominates in the east.

The mixed-skeletal lithofacies commonly constitutes the lower 5 to 30 m of the Shunda Formation near its southwestern depositional limit. This facies also commonly forms the upper 5 m or more of the Shunda and the lower parts of shallowing-upward sequences within the formation. Peloid- and pelletoid-bearing, mixed-skeletal lime wackestone and packstone predominate. Eastward-thinning tongues of peloid-skeletal and ooid lime grainstone are common where the western part of the Shunda grades laterally into the Livingstone and Turner Valley formations. Beds of the mixed-skeletal lithofacies, which is commonly rhythmically bedded, are mainly massive and bioturbated, but stromatolites, graded storm beds, mud cracks, and crossbedding occur locally. The mixed-skeletal lithofacies chiefly overlies or grades basinward into shelf-margin to protected-shelf grainstone of the Pekisko, Turner Valley, Livingstone, and formation F. Upward and toward the paleoshoreline, it grades into either the algal-wackestone or the fenestral-carbonate facies of the Shunda.

The algal-wackestone lithofacies comprises foraminifer-algal lime wackestone with subordinate lime packstone and dolostone. This facies, which is dominantly subplanar bedded, locally shows large-scale paleochannels of tidal origin. It is commonly either interbedded with or grades

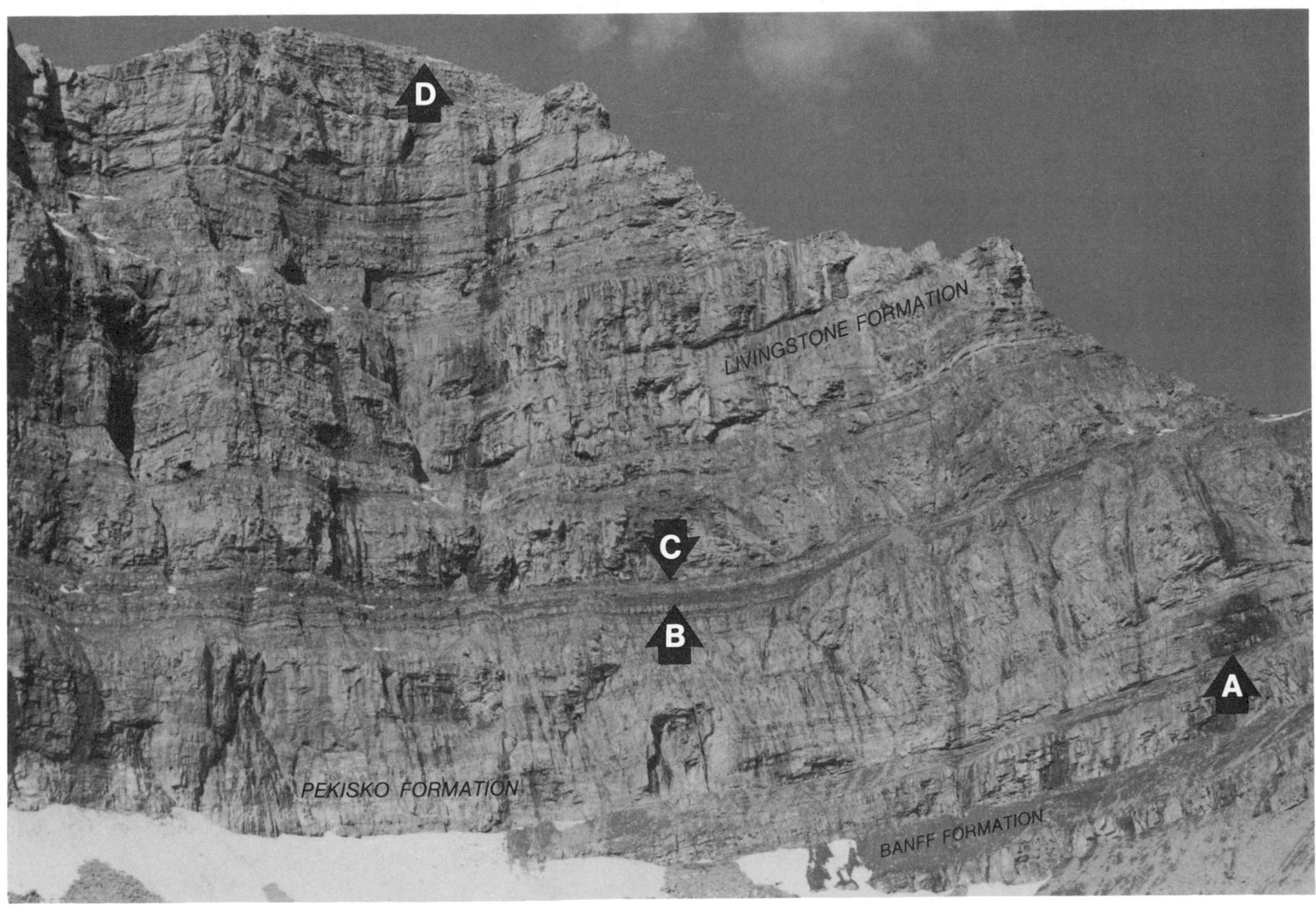

Figure 4E.15. Upper Banff Formation and lower Rundle Group, south of Kvass Creek, Rocky Mountain Front Ranges, west-central Alberta. A - top of Banff (basin deposits grading upward into restricted-shelf deposits) and base of Pekisko Formation (shelf-margin grainstone); B - base of Shunda Formation (14 m thick, protected-shelf deposits) near western depositional limit; C - base of Livingstone Formation (shelf-margin grainstone); D - top of Livingstone. View is to west (photo - B.C. Richards, ISPG 1988-27).

up into the fenestral-carbonate lithofacies. Both the algal-wackestone lithofacies and the other Shunda facies are commonly abruptly overlain by shale.

The fenestral-carbonate lithofacies comprises fenestral, cryptalgal lime and dolomite boundstone, which are mainly subplanar bedded. The main allochems are: peloids, aggregate grains, calcareous algae, and pisolites. It commonly either forms the top of shallowing-upward sequences or grades up into the silty-dolostone lithofacies interbedded with anhydrite, solution-collapse breccia, and siltstone. Where the fenestral-carbonate lithofacies occurs in the basal Shunda Formation, it overlies and passes basinward into protected-shelf deposits of the Pekisko Formation.

The silty-dolostone lithofacies comprises silty, microcrystalline dolostone that locally contains desiccation structures and small-scale crossbedding. This lithofacies, which is commonly planar laminated, grades into the fenestral-carbonate facies and a subordinate Shunda lithofacies of dolomitic siltstone. In the eastern subsurface, the silty-dolostone lithofacies commonly forms the lower part of hemicycles consisting of dolostone and shale.

Except for the mixed-skeletal lithofacies, most Shunda lithofacies were deposited on the restricted shelf and on inner parts of the shallow shelf and protected shelf. This is shown by the common occurrence of evaporites, structures and fabrics diagnostic of subaerial exposure, and abundant restricted-marine fossils. Moreover, these facies commonly either overlie or occur cratonward of Pekisko ooid grainstone that was deposited on the inner part of the shelf margin and on mainland beaches of the protected shelf. The presence of these Shunda lithofacies above the more open marine facies of the Pekisko and formation F records a major regression.

The mixed-skeletal lithofacies, particularly in the lower and upper parts of the Shunda Formation, was probably deposited in relatively open-marine environments on the shallow shelf and protected shelf. This is suggested by the predominance of bioclasts and sedimentary structures characteristic of these settings, the rarity of features indicative of subaerial exposure, and the position of the facies basinward of restricted-shelf deposits.

The Shunda Formation and coeval parts of the Pekisko, Livingstone, and Turner Valley formations, and formation F are interpreted to be mainly carbonate-platform deposits, but carbonate-ramp deposits occur locally. Deposition on platforms is indicated by slope carbonates grading northward to northeastward into shelf-margin facies that lie basinward of protected-shelf deposits. In the southern part of Peace River Embayment,

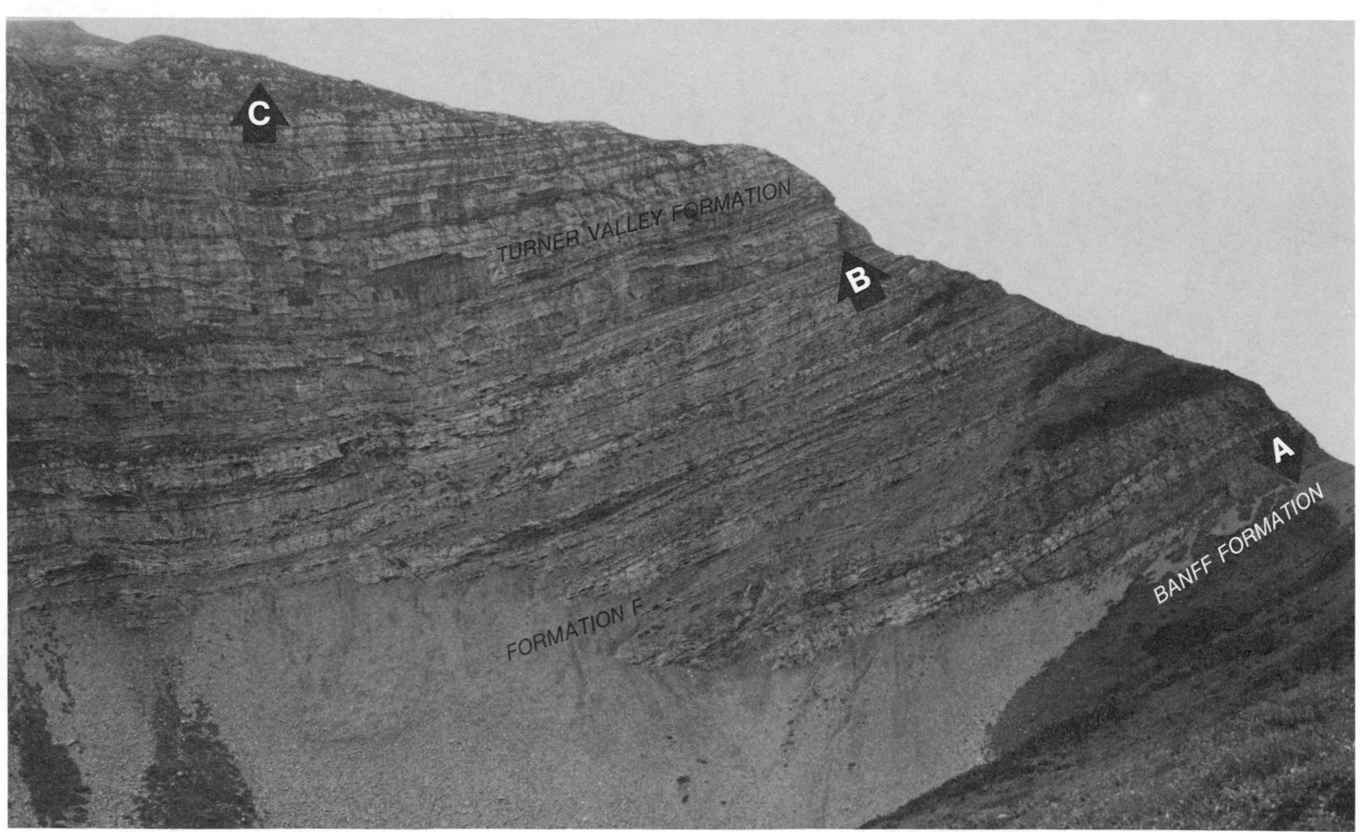

Figure 4E.16. Lower Carboniferous units at Watson Peak, southwestern rim of Peace River Embayment, Rocky Mountains, east-central British Columbia. A - top of Banff Formation and base of formation F (160 m thick, shallow-shelf carbonates passing upward into slope carbonates); B - base of Turner Valley Formation (slope deposits passing upward into shelf-margin carbonates); C - base of Mount Head Formation (protected- and restricted-shelf deposits). View is to northwest (photo - B.C. Richards, ISPG 2457-1).

however, the Shunda and its correlatives constitute part of a ramp. In that area, slope and shallow-shelf deposits in formation F grade southward to southwestward into Shunda facies deposited on the inner part of the shallow shelf. Intermediate shelf-margin facies are not present.

Livingstone Formation

The middle Tournaisian (Tn2) to middle Viséan (V2) Livingstone Formation (Douglas, 1958), which passes northeastward into the Shunda and Turner Valley formations and lower to middle Mount Head Formation, extends from southern Alberta to east-central British Columbia (Fig. 4E.2, 4E.9, 4E.10, 4E.15). Comprising mainly grainstone, it occurs from 49°N to 51°30'N in the Rocky Mountains and Foothills of southwestern Alberta and southeastern British Columbia. It also underlies the western Plains south of 50°00'N (Macauley et al., 1964), and is locally present in the eastern Rockies from 51°30'N to 54°10'N. Most of the Livingstone was deposited in Prophet Trough, but in the southern Plains it was deposited on the cratonic platform. The thickness of the northeastward-thinning Livingstone is chiefly from 275 to 425 m in the Cordillera and less than 130 m on the Plains.

Boundaries of the Livingstone Formation are mainly diachronous (Fig. 4E.2, 4E.9, 4E.10). In most areas, it conformably overlies the Banff Formation, but in the east, it conformably overlies the western Shunda and formation F. Southwestward in the eastern Cordillera and southward on the southernmost Plains, the basal Livingstone becomes younger as it grades basinward into the upper part of either the Banff or formation F. Where it overlies the Shunda, the basal Livingstone Formation becomes younger northward or northeastward as it grades into the latter. The upper Livingstone, which is overlain by the Mount Head Formation and locally by the middle Turner Valley Formation, becomes younger southwestward, where the middle to upper Turner Valley, followed by the lower and middle Mount Head, grades basinward into the Livingstone. In the east, the Livingstone-Mount Head contact is sharp and at least locally erosional.

A bryozoan-pelmatozoan lithofacies constitutes most of the Livingstone Formation, but an ooid-skeletal lithofacies and several other facies occur.

In the first lithofacies, which comprises bryozoan-pelmatozoan lime grainstone with subordinate dolomitized grainstone, chert is locally common and becomes more abundant basinward. Most beds are massive or have diffuse internal stratification, but medium- to very large-scale crossbedding is common in the eastern Front Ranges and in the upper Livingstone Formation to the southwest. In the eastern Front Ranges, both this facies and the ooid-skeletal lithofacies commonly constitute the lower part of thick fining-upward hemicycles capped by packstone to microcrystalline dolostone. Southwestward in the Cordillera and southward on the southernmost Plains, the bryozoan- pelmatozoan lithofacies of the lower Livingstone grades into slope packstone and wackestone of the upper Banff (Fig. 4E.9). In addition, the bryozoan-pelmatozoan facies becomes finer grained, rhythmically bedded, and intercalated with basinward-thickening units of cherty wackestone and packstone. The wackestone and packstone, which commonly form the lower part of coarsening-upward sequences capped by grainstone, closely resemble slope deposits of the Banff Formation and are, therefore, probably of slope origin. Northeastward in the Cordillera and northward on the southern Plains, the bryozoan-pelmatozoan lithofacies passes into shelf-margin to protected- and restricted-shelf carbonates in the Shunda, Turner Valley, and lower to middle Mount Head formations (Fig. 4E.9, 4E.10). This transition commonly occurs through the ooid-skeletal lithofacies of the Livingstone.

A shelf-margin origin for most of the bryozoan-pelmatozoan lithofacies in the southern Plains, eastern Front Ranges, and the upper Livingstone Formation to the west is indicated by its position between coeval slope and protected- to restricted-shelf facies, its high-energy deposits and crossbedding, and the predominance of open-marine fossils. An upper-slope origin for the bryozoan-pelmatozoan lithofacies in part of the lower Livingstone of the southern Plains and in most of the lower to middle Livingstone of the central to western Front Ranges is recorded by the nature of coeval adjacent lithofacies, rhythmic bedding, and absence of shallow-water indicators.

In the ooid-skeletal lithofacies, which consists of ooid-bryozoan-pelmatozoan lime grainstone, medium- to very large-scale crossbedding is common. Rhythmically bedded tempestites occur locally. The ooid-skeletal lithofacies commonly gradationally overlies and passes basinward into the shelf-margin, bryozoan-pelmatozoan deposits. It also passes upward and toward the paleoshoreline into protected- to restricted-shelf lithofacies of the Shunda, Turner Valley, and Mount Head formations. The ooid-skeletal lithofacies was deposited on the inner shelf margin, as shown by the sedimentary structures, allochems, and the nature of adjacent, coeval lithofacies.

In the eastern Front Ranges, the bryozoan-pelmatozoan and ooid-skeletal lithofacies of the Livingstone Formation are commonly associated with a lithofacies comprising cherty, skeletal wackestone and packstone. The latter occurs as basinward-thinning units in the upper parts of grainstone-dominated, fining-and shallowing-upward sequences. The packstone and wackestone intervals in the eastern part of the Livingstone resemble protected-shelf deposits in the Shunda and Turner Valley formations, and are possibly of similar origin.

Where it overlies the Banff and formation F, the Livingstone is a progradational, shallowing-upward succession recording a transition from slope to shelf-margin environments. In southernmost Alberta, it records marked shallowing of a deeper water environment that occupied the seaway connecting Prophet Trough and Williston Basin. Similarily, the Livingstone records general shallowing of southeastern Prophet Trough.

The great thickness of the Livingstone shelf-margin deposits in the eastern Cordillera indicates that the shelf margin was relatively stationary from late Tournaisian (foraminiferal zone 8) to early Viséan (lower zone 11) time. Subsequently, the shelf-margin lithofacies, followed by protected- and restricted-shelf deposits of the Mount Head, prograded rapidly basinward.

Turner Valley Formation

The lower Viséan (V1) Turner Valley Formation (Douglas, 1958; Penner, 1958), which was deposited principally on the cratonic platform, occurs from southwestern Alberta to east-central British Columbia (Fig. 4E.2, 4E.9-4E.11, 4E.16, 4E.17). Comprising mainly carbonates, it is present from 50°N to about 55°15'N in the Foothills and part of the Plains of western Alberta. It also outcrops extensively in the Front Ranges from 51°15'N to 55°15'N. The southwestward-thickening Turner Valley is mainly between 50 and 120 m thick (Rupp, 1969; Macqueen et al., 1972).

In most areas, the Turner Valley abruptly overlies the Shunda Formation; however, in part of east-central British Columbia and in southernmost Alberta, it gradationally overlies and passes basinward into formation F. North of 50°45'N, the Turner Valley Formation passes southwestward, and in southernmost Alberta southward, into the Livingstone. In west-central Alberta north of 53°15'N, the Turner Valley passes cratonward into the lower Debolt Formation; an arbitrary nomenclatural boundary separates the two. The Turner Valley Formation is overlain mainly by the Mount Head Formation, and the sharp contact between them is at least locally a minor disconformity. East of the erosional zero edge of the Mount Head in southwestern Alberta, the Turner Valley is unconformably overlain by Mesozoic strata.

The Turner Valley Formation was divided into the Elkton Member, middle dense member, and upper porous member by Penner (1958) and into informal units Mt1 to Mt3 by Rupp (1969). Three members are present in most areas (Fig. 4E.2). The lowest of these, the Elkton, extends farthest eastward, whereas the upper two extend farthest basinward. The resistant Elkton, 42.7 m thick at its type section, comprises grainstone-and packstone-dominated units that have sharp bases and grade upward into thin recessive units of restricted-marine aspect. Microcrystalline dolostone, fenestral cryptalgal boundstone, and peloid-skeletal packstone and wackestone constitute the latter. The moderately recessive middle member, generally 20 to 35 m thick, comprises numerous hemicycles that resemble those of the Elkton Member but are thinner and contain a higher proportion of restricted-marine carbonates. Shale and marlstone also occur in the middle member, which commonly contains an upper interval dominated by silty microcrystalline dolostone. The upper member, generally 15 to 40 m thick, consists of one or more resistant units of grainstone and packstone, which commonly pass upward into carbonates

Figure 4E.17. Lower Carbonifeorus units, Monoghan Creek, Rocky Mountain Front Ranges, west-central Alberta. A - top of Banff Formation (basin deposits grading upward into restricted-shelf carbonates) and base of Pekisko Formation (44 m thick, shelf-margin and protected-shelf carbonates); B - base of Shunda Formation (protected-and restricted-shelf carbonates); C - base of Turner Valley Formation (shelf-margin to restricted-shelf carbonates); D - base of Mount Head Formation (shelf-margin to restricted-shelf carbonates). View is to northwest (photo - B.C. Richards, ISPG 1988-87).

of restricted-marine aspect. Bryozoan-pelmatozoan, peloid-skeletal, and ooid-skeletal lithofacies constitute the resistant, grainstone- and packstone-dominated units. At many localities, the chert-rich Turner Valley Formation has been entirely dolomitized.

The bryozoan-pelmatozoan lithofacies, which closely resembles that of the Pekisko and Livingstone formations, is commonly present in southwestern occurrences of the Turner Valley Formation and in the basal part of fining-upward hemicycles to the northeast. Units of this lithofacies are principally crossbedded, cherty, bryozoan-pelmatozoan grainstone. They generally thicken southwestward and appear to pass into similar units of the Livingstone Formation. In east-central British Columbia, where this facies grades basinward into finer grained slope deposits of the Prophet Formation, rhythmically bedded, turbidite-like beds predominate (Fig. 4E.11, 4E.16). Upward and northeastward, the bryozoan-pelmatozoan lithofacies grades into the peloid-skeletal and ooid-skeletal lithofacies of the Turner Valley Formation.

Most of the bryozoan-pelmatozoan lithofacies was deposited on the shelf margin, as indicated by its similarity to deposits of that origin in the Pekisko and Livingstone. In east-central British Columbia, deposition on the upper slope is recorded by the presence of turbidite-like beds passing basinward into Prophet slope facies.

In the peloid-skeletal facies, which constitutes most of the Turner Valley Formation, chert-rich, peloid-pelletoid-skeletal grainstone to wackestone predominate. Beds showing planar stratification and large-scale crossbedding are common in western occurrences. Elsewhere, beds are mostly massive, locally resemble storm deposits, and are subplanar bedded. This lithofacies generally grades upward and northeastward into Turner Valley facies of restricted-marine aspect; however, it also commonly passes both basinward, upward, and northeastward into the ooid-skeletal lithofacies.

Sedimentation of the peloid-skeletal lithofacies occurred in low- to moderately high-energy, protected-shelf environments, as it closely resembles Pekisko peloid- skeletal lithofacies deposited in these settings. The occurrence of

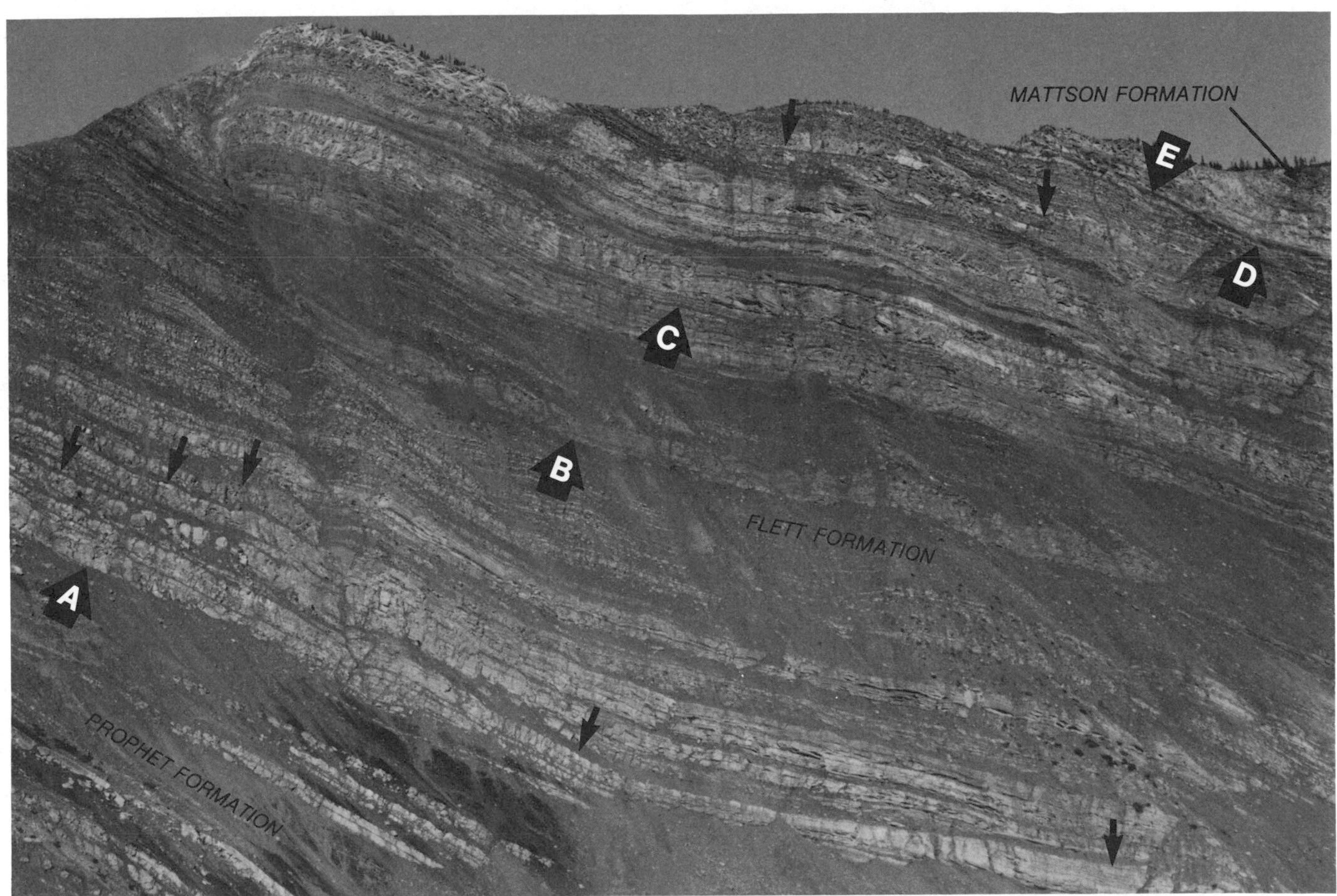

Figure 4E.18. Lower Carboniferous units, Jackfish Gap, Liard Ranges, southwestern District of Mackenzie. A - top of Prophet Formation (basin and lower-slope deposits) and base of Tlogotsho Member (middle-slope deposits) of Flett; B - base of Jackfish Gap Member (80 m thick, slope to shelf siliciclastics and carbonates) of Flett; C - base of Meilleur Member (middle-slope to shelf-margin carbonates) of Flett; D - Golata Formation (8 m thick, prodelta shale); E - base of Mattson Formation (deltaic sandstone). Small arrows indicate erosion surfaces of submarine paleochannels. View is to northeast (photo - B.C. Richards, ISPG 798-171).

the lithofacies cratonward of the main belt of coeval shelf-margin deposits also records deposition on the protected shelf.

The ooid-skeletal lithofacies, which closely resembles that of the Livingstone Formation, consists of massive to crossbedded ooid-skeletal grainstone. It is best developed in the thick southwestern part of the Turner Valley Formation, where it is commonly contiguous with the ooid-skeletal facies of the Livingstone. The lithofacies is also locally well developed where grainstone of the Turner Valley Formation grades into restricted-marine carbonates.

Southwestern occurrences of the ooid-skeletal lithofacies probably originated on the inner shelf-margin, as they are high-energy deposits formed directly landward of coeval, shelf-margin, bryozoan-pelmatozoan lithofacies. Most other occurrences were deposited in high-energy settings of the inner protected shelf, as they overlie the peloid-skeletal lithofacies and pass upward and toward the paleoshoreline into restricted-marine facies.

Deposits of restricted-marine aspect intercalated with the three facies discussed above were deposited on the restricted shelf and on relatively restricted parts of the protected shelf. This is exemplified by the presence of fenestral cryptalgal boundstone, stromatolites, anhydrite nodules, and numerous restricted-marine fossils.

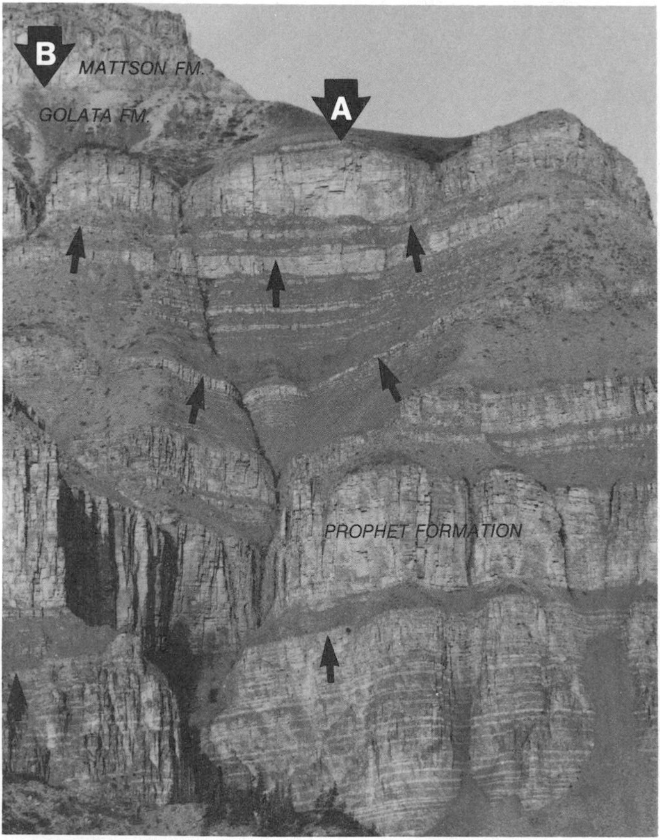

Figure 4E.19. Lower Carboniferous units at Ram Creek, north side of Tlogotsho Range, southwestern District of Mackenzie. A - top of Prophet Formation (lower and middle slope deposits) and base of Golata Formation (78 m thick, prodelta shale); B - base of Mattson Formation (deltaic sandstone). Small arrows indicate erosion surfaces of submarine paleochannels. View is to southwest (photo - B.C. Richards, ISPG 1756-83).

The Turner Valley Formation records two main transgressions, an intervening regional regression, and numerous minor transgressive-regressive hemicycles. Open-marine grainstone of the Elkton Member above restricted-marine facies of the Shunda Formation records the first regional transgression. Subsequent regression is recorded by the restricted-marine, middle Turner Valley. Overlying lithofacies of protected-shelf origin in the upper part of the Turner Valley Formation record a second widespread transgression. Minor transgressive-regressive events are recorded by thin shallowing-upward sequences.

Mount Head Formation

Carbonates of the lower to upper Viséan (V1 to V3) Mount Head Formation (Douglas, 1958) are widely distributed in the eastern Cordillera and occur locally on the Interior Plains (Fig. 4E.2, 4E.9-4E.11). They overlie the Turner Valley Formation in the east and the Livingstone Formation to the west. In the Cordillera, the Mount Head Formation extends from southeastern British Columbia and southwestern Alberta to 55°15′N in east-central British Columbia (Oswald, 1964; Macqueen and Bamber, 1968; Beauchamp et al., 1986). On the Interior Platform north of 53°15′N, lithological and stratigraphic equivalents of the Mount Head have been included in the Debolt Formation by Macauley et al. (1964); an arbitrary nomenclatural boundary separates the two. The Mount Head was deposited on the western cratonic platform and in southern Prophet Trough and the southwestern part of Peace River Embayment. This northeastward- thinning unit is generally between 75 and 250 m thick in the Rocky Mountains, and normally less than 75 m thick to the east.

The Mount Head Formation is overlain disconformably by the Etherington Formation in most of the southern Rocky Mountains and in part of the adjacent Foothills, but in the southwest the contact with the Etherington may be conformable (Fig. 4E.2, 4E.9, 4E.20). The sub-Etherington disconformity is commonly a solution surface that has rundkarren (large corrosion features developed beneath soil) and is overlain by nodular calcrete. North and east of the erosional edge of the Etherington Formation, Permian and Mesozoic strata unconformably overlie the Mount Head in most areas.

Part of the Mount Head Formation was divided by Douglas (1958) into the Wileman, Baril, Salter, Loomis, Marston, and Carnarvon members. A seventh member, the Opal, was erected by Macqueen and Bamber (1968) for western correlatives of the Marston and lower to middle Carnarvon members. These members are widely distributed, but cannot be easily identified in most of the eastern Rocky Mountains from 51°45′N to 53°30′N, the central to eastern Foothills, and the Interior Plains (Macqueen et al., 1972; McGugan, 1984). Only the lower three members are present from 53°30′N to 55°15′N.

The recessive Wileman Member, which overlies the Livingstone and Turner Valley formations, ranges in thickness from 7.6 m in the east to 25 m in the west, and is chiefly silty, microcrystalline dolostone (Macqueen and Bamber, 1968). Siltstone, shale, marlstone, anhydrite, and peloid-skeletal grainstone and wackestone occur as well.

The basal contact of the Wileman Member is commonly a minor disconformity, but near its southwestern depositional limit, it is gradational. This member is abruptly and commonly disconformably overlain by the Baril Member.

A restricted-shelf origin for most of the Wileman Member is recorded by its evaporites and related solution-collapse breccias, an impoverished biota dominated by calcareous algae and foraminifers, and the common occurrence of structures and fabrics indicative of subaerial exposure and very shallow water. Moreover, the Wileman grades basinward into Livingstone lithofacies deposited on the protected shelf and shelf margin.

The resistant, southwestward-thickening Baril Member, which overlies the Wileman and ranges in thickness from 11 to 39 m, is chiefly pelletoid-skeletal and skeletal-ooid lime grainstone with subordinate dolostone. In western sections, cherty, pelletoid-skeletal wackestone and packstone also occur and become more abundant upward. In general, the abundance of ooids increases northeastward, whereas that of bryozoans and pelmatozoan fragments increases southwestward. In most areas, the Baril Member is abruptly and disconformably(?) overlain by the Salter Member, but western facies of the upper Baril grade northeastward into the latter.

In most of the Baril Member, sedimentation in moderately high-energy, protected-shelf environments on a carbonate platform is indicated by small-to large-scale crossbedding, and by the widespread ooids, pelletoids, and foraminifers. Moreover, the Baril passes southwestward into coeval shelf-margin lithofacies of the Livingstone Formation and northeastward into restricted-shelf dolostone of the undivided Mount Head Formation.

The recessive Salter Member, ranging in thickness from 29 m in the east to 67 m in the west, generally overlies the Baril Member, but in the southwest it gradationally overlies the Livingstone Formation. Microcrystalline dolostone, fenestral cryptalgal boundstone, and foraminifer-algal lime wackestone and packstone predominate. Skeletal-ooid grainstone, peloid-bryozoan-pelmatozoan grainstone to wackestone, evaporites, and solution-collapse breccia occur locally. At western localities, chert is very abundant. Near its southwestern depositional limit, the Salter is conformably overlain by the Loomis Member, but elsewhere, the contact is abrupt and commonly disconformable.

Most of the Salter Member, like the Wileman, records restricted-shelf sedimentation. Ooid grainstone and peloid-skeletal grainstone, packstone, and wackestone in the western part of the Salter Member indicate sedimentation in protected-shelf environments.

The resistant, northeastward-thinning Loomis Member is generally 20 to 100 m thick and predominantly lime grainstone (Macqueen and Bamber, 1968). In most areas, this member overlies the Salter, but near its southwestern depositional limit, the Loomis conformably overlies shelf-margin, bryozoan-pelmatozoan grainstone of the Livingstone Formation. Southwestern occurrences of the Loomis generally comprise an ooid-skeletal lithofacies resembling that of the eastern part of the Livingstone Formation, but an ooid facies predominates locally. To the northeast, the Loomis Member comprises a pelletoid-ooid-skeletal lithofacies and an ooid facies. Most beds in the Loomis are massive, but some show large-scale crossbedding and subplanar stratification. The Loomis, which is abruptly and disconformably(?) overlain by the Marston Member in the northeast, is conformably overlain by the Opal Member in the southwest. By analogy with units of similar lithology and relationships (Pekisko, Turner Valley, and Baril), the Loomis was deposited in shelf-margin environments and in relatively open-marine, high-energy, protected-shelf environments.

Deposits of the recessive Marston Member (18 to 67 m thick) overlie the Loomis Member and pass southwestward into the Opal Member (Fig. 4E.2, 4E.9; Macqueen and Bamber, 1968). Microcrystalline dolostone with subordinate fenestral cryptalgal boundstone, ooid to skeletal limestone, shale, and marlstone constitute the rhythmically bedded Marston. Several shallowing-upward hemicycles, comprising skeletal to ooid limestone grading upward to dolostone and shale, are generally present. The member is abruptly, and possibly disconformably, overlain by the Carnarvon Member. Macqueen et al. (1972) interpreted the Marston to be chiefly lagoon and sabkha deposits.

The Opal Member thickens southwestward, and is 161 m thick at its type locality southwest of Calgary. It overlies the Loomis, passes northeastward into the Marston and lower to middle Carnarvon, and is conformably(?) overlain by the Carnarvon (Macqueen et al., 1972).

The moderately resistant lower and middle parts of the Opal Member consist of pelletoid-skeletal and ooid-skeletal lime grainstone with subordinate marlstone and cherty, pelletoid-skeletal lime packstone and wackestone. They were deposited mainly on either the protected shelf to shelf margin of a platform or the shallow shelf of a ramp. This origin is suggested by the joint occurrence of abundant open-marine and restricted-marine fossils, abundant pelletoids, and the common presence of medium- to large-scale crossbedding. It is also suggested by the occurrence of these deposits southwest of coeval restricted-shelf facies in the Marston and Carnarvon members.

The recessive upper part of the Opal Member comprises shale and marlstone that are rhythmically interbedded with fenestral, cryptalgal boundstone and peloid-skeletal lime wackestone and packstone. In southwestern occurrences of this subdivision and in its lower part to the east, the joint occurrence of open- and restricted-marine fossils suggests deposition in relatively open- marine environments on either the shallow shelf of a ramp or the protected shelf of a platform. Elsewhere, deposition on the restricted shelf and the inner part of either the protected shelf or shallow shelf are recorded by the predominance of fenestral, cryptalgal boundstone and foraminifer-algal wackestone.

Resistant lithofacies of the Carnarvon Member overlie the Marston Member in the east and the Opal Member to the southwest, and are overlain by the Etherington Formation. The lower and middle parts of the Carnarvon of eastern sections grade southwestward into the Opal Member (Macqueen et al., 1972). Peloid-foraminifer-algal and mixed-skeletal lime wackestone to packstone constitute most of the Carnarvon, but fenestral cryptalgal boundstone is common in the eastern Front Ranges. Most

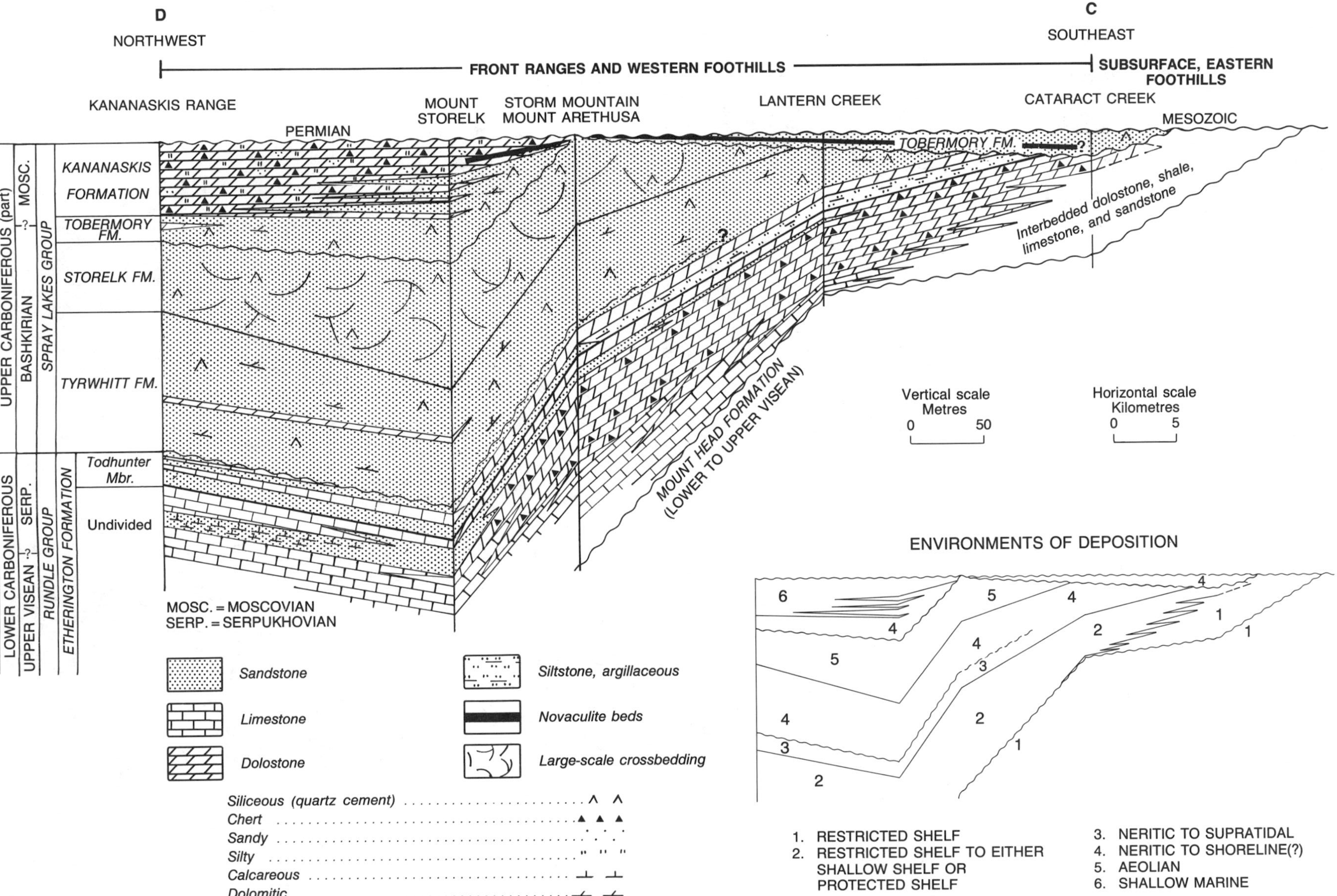

Figure 4E.20. Partly schematic stratigraphic cross-section C-D showing uppermost Rundle Group and Spray Lakes Group of southwestern Alberta. See Figure 4E.1 for line of section (after Scott, 1964).

limestone beds are rhythmically interbedded with thin shale beds. At its type locality southwest of Calgary, the Carnarvon Member is 47 m thick.

In most of the Carnarvon Member, fenestral fabric or abundant calcareous algae and foraminifers record deposition on the restricted shelf and relatively restricted parts of either the protected shelf of a platform or shallow shelf of a ramp. Southwestern occurrences comprise bioturbated peloid-skeletal limestone characteristic of relatively open-marine, shallow-shelf, and protected-shelf settings.

The undivided Mount Head Formation comprises sparsely fossiliferous microcrystalline dolostone with subordinate shale, anhydrite, siltstone, carbonate breccias, and limestone (Bamber et al., 1981; McGugan, 1984). Deposition on the restricted shelf is indicated by the presence of cryptalgal laminae, evaporites, and evaporite solution-collapse breccias.

Three regional transgressions and four important regressions are recorded in the Mount Head Formation. The transgressions are indicated by the occurrence of protected-shelf grainstone of the Baril, Loomis and middle Opal members above restricted-shelf deposits of the Wileman, Salter and lower Opal members, respectively. Restricted- shelf lithofacies of the Wileman and Salter, which overlie shelf-margin to protected-shelf grainstone, resulted from the first two regressions. During the Salter regression, erosion and karst-plain development, which produced the regional disconformity above the Mission Canyon Formation of Montana, occurred on the cratonic platform in western Montana and possibly southeastern Alberta (Sando, 1988; Sando et al., 1969). The third and fourth regressions are recorded by restricted-marine deposits of the Marston, and upper Opal to Carnarvon, respectively.

During sedimentation of the Mount Head Formation, marked shallowing and regression occurred in the western part of the seaway connecting Prophet Trough and Williston Basin. On the cratonic platform of western Montana, and possibly southeastern Alberta, the seaway site was exposed to middle and late(?) Viséan (V2 and V3?) erosion. Regression and pronounced shallowing also occurred throughout eastern to central Prophet Trough of southeastern British Columbia and southwestern Alberta.

Debolt Formation

During deposition of the Mount Head Formation, carbonates of the Debolt Formation (Macauley, 1958) were deposited in Prophet Trough and on the western cratonic platform (Fig. 4E.2). The lower to upper Viséan (lower V1 to lower V3) Debolt occurs beneath the western Interior Plains and in part of the Foothills from 53°15′N in west-central Alberta to 60°N in northeastern British Columbia (Macauley et al., 1964). This formation, 244 m thick at its type section in west-central Alberta (Macauley, 1958), thickens slowly southwestward before thinning rapidly near its southwestern depositional limit.

The Debolt Formation presents several stratigraphic problems. At the type section and over wide areas to the west and south, it includes stratigraphic and lithological equivalents of the Mount Head and Turner Valley formations and is currently separated from them by arbitrary nomenclatural boundaries. Most of the Debolt of British Columbia does not closely resemble the type Debolt, which consists mainly of restricted-marine carbonates. Instead, it closely resembles the stratigraphically equivalent Flett Formation of the southwestern District of Mackenzie, and is separated from it by a strictly nomenclatural boundary at 60°N. The upper part of the Debolt Formation of British Columbia is also stratigraphically and lithologically equivalent to part of the upper Prophet Formation and is separated from it by an arbitrary nomenclatural boundary near the eastern margin of the Cordillera.

Beneath the Interior Plains, the Debolt Formation generally conformably overlies the Shunda Formation in the east and formation F in the west. However, on the extreme western part of the Plains and in the Foothills, it overlies and passes southwestward into the Prophet Formation.

The Debolt Formation is abruptly, and at least locally, disconformably overlain by the Golata Formation in most of the Peace River Embayment and near its depositional limit to the northwest (Fig. 4E.2). The hiatus between these two formations is probably greatest in the southwestern part of the embayment, where upper Viséan (V3, foraminiferal zone 15) strata of the Golata (Bamber and Mamet, 1978) appear to overlie lower middle Viséan (lower V2) strata of the Debolt. Locally on the southern flank of the embayment, the Debolt is disconformably overlain by the Kiskatinaw Formation. Elsewhere, the Debolt is unconformably overlain by Permian and Mesozoic strata.

Macauley (1958) divided the Debolt Formation into lower (102.7 m), and upper members (141.7 m), and Law (1981) subdivided it into five informal members. Beneath the Plains of west-central Alberta, the basal carbonate unit of the Debolt closely resembles the correlative Elkton Member in the Turner Valley Formation, and has been named accordingly (Fig. 4E.2, 4E.10; Young, 1967).

The Debolt Formation comprises limestone and dolostone with subordinate shale, marlstone, anhydrite, siltstone, and sandstone. Limestone becomes more abundant southwestward, whereas the proportion of dolostone increases upward and toward the east, where it commonly predominates. Shale and marlstone, present as partings and beds in most of the formation, are usually most abundant in the middle part of the lower Debolt and in the lower part of the upper Debolt. Anhydrite occurs mainly in the upper Debolt of Alberta, where it is commonly a major component.

Little has been published about the lithofacies and sedimentology of the Debolt Formation. Published data (Bamber and Mamet, 1978; Law, 1981) and data from well logs suggest that most of this unit comprises carbonate-platform lithofacies deposited in slope to restricted-shelf environments, but carbonate ramp lithofacies may be present.

Near its southwestern depositional limit in northeastern British Columbia, most of the lower Debolt and part of the upper Debolt consist of probable slope lithofacies. These are well-bedded, chert-rich, spicule-bryozoan-pelmatozoan lime wackestone, packstone, and grainstone that are rhythmically intercalated with shale and marlstone. Southwestward, these deposits grade into coeval cherty spiculite and spicule lime packstone and wackestone of the Prophet Formation, deposited in lower- to middle-slope environments (Fig. 4E.19). Upward and northeastward within the Debolt Formation, slope facies grade into deposits characteristic of shelf-margin environments.

Debolt lithofacies of shelf-margin aspect occur mainly in northeastern British Columbia, where they are best developed in the upper Debolt in the west and the lower Debolt in the east. To the south, the Debolt was deposited cratonward of the main belt of shelf-margin facies, which is preserved in the Turner Valley, and locally the Livingstone Formation (Fig. 4E.10, 4E.11). The outer shelf-margin deposits of the Debolt Formation are mainly bryozoan-pelmatozoan lime grainstone. Upward and northeastward, this grainstone commonly grades to ooid-skeletal and ooid lime grainstone characteristic of inner shelf-margin environments. In much of the middle part of the Debolt, the grainstone units are thin, and some may be isolated shoals formed on the shallow shelves of ramps rather than on the shelf margins of platforms. The shelf-margin deposits grade northeastward into carbonate lithofacies of protected- to restricted-shelf aspect.

Deposits of protected- to restricted-shelf aspect constitute part of the Debolt Formation in northeastern British Columbia and most of this formation in Alberta. The lithofacies of probable protected-shelf origin comprise peloid-skeletal lime wackestone, packstone, and grainstone containing bioclasts diagnostic of open- and restricted-marine settings. Lithofacies of probable restricted-marine origin are dominated by microcrystalline dolostone that is commonly silty to sandy and interbedded with anhydrite and limestone containing restricted-marine bioclasts. Siltstone, sandstone, and shale are generally interbedded with the carbonates, particularly in the lower part of the upper Debolt.

The Debolt Formation records at least one main transgressive-regressive hemicycle. In Alberta, an early Viséan (early V1) transgression is recorded by the presence of protected-shelf carbonates of the basal part of the Debolt above the restricted-shelf carbonates of the Shunda Formation. The subsequent regression, recorded by anhydrite and restricted-shelf carbonates preserved in the upper Debolt of west-central Alberta and east-central British Columbia, appears to have coincided with sedimentation of the regressive Wileman and Salter members of the Mount Head Formation to the southwest (Fig. 4E.2). To the north, middle and upper Viséan (upper V2 and lower V3) strata of the upper Flett, Golata, and lower Mattson formations record a second transgressive-regressive event, and the upper Debolt of northeastern British Columbia probably records this event as well (Richards, 1989).

Flett Formation

Sedimentation of the Flett Formation (Harker, 1961; Richards, 1989) occurred in Prophet Trough during latest Tournaisian (Tn3) to latest middle Viséan (late V2) time (Fig. 4E.2, 4E.12, 4E.13, 4E.18). The Flett, which is separated from the partly coeval Debolt Formation by a strictly nomenclatural boundary at 60°N, comprises mainly limestone and occurs in southwestern District of Mackenzie. There, it outcrops in the southern part of the Mackenzie Fold Belt and underlies the western part of the Interior Platform. The southwestward-thinning Flett is generally between 120 and 350 m thick.

In most areas, the Flett Formation conformably overlies the Prophet Formation, but on the western Interior Platform, it locally overlies formation F. Southwestern occurrences of the Flett grade basinward into the Prophet, with the upper Flett extending farthest. The Flett is abruptly but conformably overlain by the Golata Formation in the Mackenzie Fold Belt and on the extreme eastern Interior Platform. East of the subcrop edge of the Golata, it is unconformably overlain by Cretaceous and locally by Permian strata.

Richards (1989) divided the Flett Formation into the Tlogotsho, Jackfish Gap, and Meilleur members, in ascending order. The resistant lower and upper members, mainly limestone, have almost the same distribution as the Flett. In contrast, the recessive Jackfish Gap Member, which comprises siltstone, sandstone, shale, mudstone, and limestone, is confined to the Mackenzie Fold Belt north of 60°55′N.

A mixed-skeletal lime-packstone lithofacies and a lime-grainstone lithofacies constitute most of the Flett Formation. Siltstone, sandstone, shale and mudstone are commonly present.

Cyclic deposits of the mixed-skeletal lime-packstone lithofacies constitute most of the Tlogotsho Member, the lower and western Meilleur Member, and the limestone beds in the Jackfish Gap Member. Skeletal lime packstone and wackestone, rhythmically intercalated with marlstone, generally predominate. Most beds are probably turbidites, but deposits of storm and debris-flow origin are present. Large-scale submarine paleochannels are common. The mixed-skeletal lime-packstone lithofacies generally grades southwestward into lower and middle slope deposits of the Prophet Formation (Fig. 4E.19). Upward and eastward, it grades into the lime-grainstone lithofacies of the Flett. Most of the mixed-skeletal lime-packstone lithofacies has been interpreted as middle-slope deposits.

Most of the lime-grainstone lithofacies occurs in the upper part of the Meilleur Member, but it is also present in the eastern part of the Tlogotsho Member south of 60°45′N. Bryozoan-pelmatozoan lime grainstone and packstone predominate. Medium-to large-scale crossbedding and beds interpreted as storm deposits are moderately common, and large-scale submarine paleochannels truncate the bedding at some localities. The lime-grainstone lithofacies has been interpreted to be of upper-slope and shelf-margin origin.

The siltstone and sandstone lithofacies occurs in the Jackfish Gap Member, where it generally forms coarsening-upward sequences underlain and overlain by shale and mudstone lithofacies (Fig. 4E.13, 4E.18). Abundant trace fossils and wave- and current-formed crossbedding occur in the sandstone and siltstone. This lithofacies has been interpreted as slope to shallow-neritic deposits.

The shale and mudstone lithofacies is present mainly in the middle Jackfish Gap Member, but it also constitutes part of the upper Tlogotsho, the lower Meilleur, and the uppermost Meilleur in the northeast. It was probably deposited at depths ranging from near to well below fair-weather wave base.

The Flett Formation is a progradational succession recording two transgressions and an intervening regression. Because of the fluctuations in water depth, two carbonate buildups comprising platform deposits with subordinate ramp deposits developed north of 60°55′N, while one formed to the south. The first transgression resulted in deposition of the carbonates of the Tlogotsho Member north of 60°55′N and the coeval lower Tlogotsho to the south. A subsequent break in buildup growth occurred north of 60°55′N. This break, resulting from regression, is recorded by the shallow-marine sandstone and siltstone of the Jackfish Gap Member (Fig. 4E.13). During siliciclastic deposition in the north, deposition of the upper Tlogotsho carbonates continued to the south. North of 60°55′N, the initial phase of a second transgression is recorded by the rapid transition from shallow-neritic sandstone in the Jackfish Gap Member to moderately deep-water limestone and shale in the overlying basal Meilleur Member. To the south it is recorded by the transition from shallow-water carbonates in the upper Tlogotsho to shale and spicule-rich deposits in the basal Meilleur. After the initial phase of the second transgression, carbonates of the Meilleur were deposited under shallowing-upward conditions.

Prophet Formation

The Prophet Formation (Sutherland, 1958), deposited in Prophet Trough and Peace River Embayment, extends from east-central British Columbia to the southwestern part of the District of Mackenzie (Fig. 4E.2, 4E.11-4E.13, 4E.19). In this region, the middle Tournaisian (Tn2) to upper Viséan (lower V3) Prophet occurs in the Foothills and eastern Rocky Mountains and on part of the westernmost Interior Plains. It comprises spiculite and carbonates with subordinate shale that were deposited contemporaneously with most of the Rundle Group to the east. The Prophet, which is thickest in southwestern District of Mackenzie, ranges in thickness from more than 770 m in the east to less than 75 m in the west.

Boundaries of the Prophet Formation are diachronous, with the upper part extending farthest basinward and the lower beds farthest toward the paleoshoreline (Fig. 4E.2, 4E.11-4E.13). Southwestern occurrences either gradationally overlie the Clausen Formation or overlie and grade basinward into the Besa River Formation. To the northeast, the Prophet Formation generally grades upward and northeastward into the Pekisko, Debolt, and Flett formations and formation F. South of the axis of Peace River Embayment, however, it grades southeastward into the Turner Valley and Mount Head formations and formation F (Fig. 4E.11). In the southwest, the Prophet is generally overlain conformably by the Golata Formation, but to the northeast it is overlain conformably by the formations of the Rundle Group noted above and unconformably by Permian strata.

A spiculite and spicule-lime-packstone lithofacies, a dark-shale lithofacies, and a mixed-skeletal lime-packstone lithofacies constitute the Prophet Formation (Richards, 1989). These lithofacies form a progradational, coarsening- and shallowing-upward succession. In parts of northeastern British Columbia, the Prophet lithofacies assemblage has been divided, in ascending order, into members A, B, and C (Sutherland, 1958; Bamber and Mamet, 1978).

The rhythmically bedded spiculite and spicule-lime-packstone lithofacies (Fig. 4E.19) generally constitutes most of the lower to middle Prophet in the northeast and most of the formation in the southwest. It consists of chert-rich spiculite, spicule lime packstone and wackestone, which occur as hemipelagic deposits, laminated to medium-bedded turbidites, and platy bioturbated deposits. Bedding is locally truncated by slump scars and large-scale submarine paleochannels. Toward the southwest, this lithofacies grades into basinal shale of the Besa River Formation. Upward and northeastward it grades into carbonate slope deposits occurring in the Prophet mixed-skeletal lime-packstone lithofacies and in other formations of the Rundle Group. Richards (1989) interpreted the spiculite and spicule-lime-packstone lithofacies as having been deposited on the lower slopes of carbonate platforms and ramps.

Units of the mixed-skeletal lime-packstone lithofacies generally constitute most of the upper part of the Prophet Formation in the east. They also occur locally as southwestward-thinning tongues in other parts of the formation. This lithofacies consists of beds of chert-rich lime wackestone and packstone rhythmically intercalated with marlstone. Turbidites predominate, but some thin debris-flow deposits are possibly present. In some areas, the mixed-skeletal lime-packstone lithofacies, which is a southwestward extension of similar lithofacies in the Flett and Debolt formations, contains numerous large-scale submarine paleochannels (Fig. 4E.19). Toward the southwest, the lithofacies generally grades into the spiculite and spicule-lime-packstone lithofacies of the Prophet. The mixed-skeletal lithofacies is considered to be mainly platform deposits of middle-slope origin.

The dark-shale lithofacies, most abundant in the lower and northern parts of the Prophet Formation, occurs principally as southwestward-thickening tongues that pass basinward into the Besa River Formation. This lithofacies consists of black shale deposited in moderately deep-water basin environments.

Depositional summary

The middle depositional unit, which was deposited during late middle Tournaisian (late Tn2) to early late Viséan (early V3) time, comprises carbonate-platform lithofacies with subordinate carbonate-ramp lithofacies and basinal to supratidal siliciclastics. In most of this depositional unit, the slope lithofacies were deposited on low-gradient slopes; evidence for submarine escarpments is absent. In most of Prophet Trough and Peace River Embayment and on parts of the western cratonic platform, the dominantly shallow-marine platform and ramp carbonates of the Rundle Group prograded basinward (mainly southwestward) over deeper-water shale and carbonates of the Banff and Besa River formations. In the seaway developed on the unstable craton of southernmost Alberta, the Rundle prograded southward over the deeper-water Banff deposits. At the same time in Williston Basin, carbonate-platform facies of the Mission Canyon Formation and lower Charles Formation prograded basinward (southwestward to westward) over deeper water carbonates of the Lodgepole Formation. Elsewhere, particularly on most of the stable cratonic platform, carbonates of the Rundle Group were deposited above shallow-marine facies of the upper Banff Formation.

The middle depositional unit, particularly in Williston Basin, Peace River Embayment, and Prophet Trough, is a shallowing-upward succession overall, although it formed during several transgressive-regressive hemicycles.

Most of the Prophet Trough deposits and those formed in Williston Basin and the western part of the Peace River Embayment record continuation of the general progradational, shallowing-upward trend established during sedimentation of the upper part of the lower depositional unit. At the base of the adjacent cratonic-platform succession, however, and in the eastern parts of the Prophet Trough and Peace River Embayment succession, the Pekisko Formation and coeval deposits in the Clausen Formation and formation F record a major regional transgression. That event, which occurred during the latest middle to earliest late Tournaisian (latest Tn2 to earliest Tn3) time, was followed by several regional regressions and transgressions.

The first of the major regressions occurred during sedimentation of the Shunda Formation and its correlatives during late Tournaisian (Tn3) time. Subsequent regional regressions resulted in deposition of the Wileman, Salter, Marston, and Carnarvon members of the Mount Head Formation and their correlatives in the Debolt and Flett formations and upper Madison Group. During the Salter regression, most of the cratonic platform of western Montana was exposed, resulting in an extensive karst plain that probably extended into southeastern Alberta.

After the Pekisko transgression, the main regional transgressions occurred during latest Tournaisian (latest Tn3) to earliest Viséan (earliest V1) time and in late middle Viséan (late V2) time. The latest Tournaisian event resulted in deposition of the lower Turner Valley Formation and its correlatives found in the lower Debolt and Flett formations and the Midale beds of the Mission Canyon Formation. The middle Viséan event is recorded by the Loomis Member of the Mount Head Formation and by correlatives in the upper parts of the Debolt and Flett formations.

Subsequent to deposition of the Mount Head and Debolt formations, at least part of the region south of the axis of Peace River Embayment was subaerially exposed. This produced the widespread but minor disconformity between the Mount Head and Etherington formations in southwestern Alberta and the disconformity between the Debolt Formation and the Stoddart Group in the southwestern part of the embayment.

Upper depositional unit

The upper depositional unit consists of the Etherington Formation, Stoddart Group (Halbertsma, 1959), Golata and Mattson formations, Spray Lakes Group (McGugan and Rapson, 1963), and minor occurrences of the Kibbey Formation of the Big Snowy Group (Scott, 1935a, b) (Fig. 4E.1, 4E.2). This depositional unit, comprising sandstone with subordinate carbonates and shale, overlies and passes basinward into the Besa River Formation from east-central British Columbia into southwestern District of Mackenzie. In most areas, the upper depositional unit is unconformably overlain by Permian strata, but east and north of the subcrop edge of the latter, it is unconformably overlain by Triassic to Lower Cretaceous strata.

Deposition of the Kibbey Formation took place in Williston Basin, whereas the remainder of the upper depositional unit was deposited in Prophet Trough, Peace River Embayment and, at least locally, on the western part of the stable cratonic platform. Peace River Embayment was well developed during late Viséan (V3) to Serpukhovian time, but its existence during Late Carboniferous time is undocumented because most of the region lacks known post-Serpukhovian strata. During deposition of the Kibbey Formation, the connection between Williston Basin and Prophet Trough/Antler Foreland Basin was a relatively narrow, east-trending trough in Montana (Rawson, 1969; Sando et al., 1975).

The upper depositional unit constitutes parts of the upper Kaskaskia and lower Absaroka sequences of Sloss (1963). In southwestern Alberta and southeastern British Columbia, the inter-sequence boundary is the regional disconformity separating the Etherington Formation from the overlying Spray Lakes Group.

Kibbey Formation

In Canada, the sandstone- and siltstone-dominated Kibbey Formation (Weed, 1899), which conformably(?) overlies the Charles Formation, is confined to the extreme southern part of central Saskatchewan (Fig. 4E.1, 4E.2). To the south, the upper middle(?) Viséan (V3) to Serpukhovian(?) Kibbey is widely distributed in Montana and in western North Dakota (Easton, 1962; Maughan and Roberts, 1967; Sheldon and Carter, 1979). In Saskatchewan, the Kibbey is generally unconformably overlain by Mesozoic strata, but locally (Tp. 1, Rge. 17, W2) it appears to be conformably overlain by shale and marlstone of the Serpukhovian(?) Otter Formation, which is the middle formation of the Big Snowy Group of Montana and North Dakota. The Kibbey of Saskatchewan is generally less than 45 m thick and thins rapidly northward to zero; however it is locally up to 83 m thick. The age of the Kibbey in Williston Basin has not been determined, but in southwestern Montana it probably is middle to late Viséan (zone 13 to 15) (Sando et al., 1985; Wardlaw and Pecora, 1985).

The Kibbey Formation of Saskatchewan comprises three members, but in most areas only the lower member has been preserved. The lower member (up to 43 m thick) consists of red siltstone with subordinate interbedded reddish sandstone and shale and minor dolostone and anhydrite. The middle member, or Ray Member of Rawson (1968), is about 6 m thick. It abruptly overlies the lower member and comprises limestone grading upward into anhydrite. Lime mudstone predominates in the Ray, but ooid grainstone is moderately common in Montana and North Dakota (Rawson, 1969). Red sandstone with subordinate red siltstone and shale constitute the upper member, which is up to 33 m thick and gradationally overlies the Ray.

A continental to marginal-marine origin for most of the Kibbey Formation in Canada is suggested by the predominance of redbeds, lack of marine fossils, and presence of interbeds of anhydrite and carbonates. Lithofacies trends in the formation, particularly in the Ray Member, suggest that Williston Basin was periodically occupied by a restricted to semi-restricted sea (Rawson, 1969; Sando et al., 1975).

Lithofacies of the Kibbey Formation together with the shallow-marine deposits of the Otter Formation and slightly deeper water marine deposits of the overlying Heath Formation (upper formation in Big Snowy Group of Montana) indicate a regional regression and subsequent transgression (Roberts, 1979, p. 241). The transgression was succeeded by late Serpukhovian (foraminiferal zone 19) regression and subaerial erosion (Sando et al., 1975, Fig. 13; Easton, 1962; Maughan and Roberts, 1967).

Etherington Formation

After sedimentation of the Mount Head Formation, the upper Viséan (V3) to Serpukhovian Etherington Formation (Douglas, 1958) was deposited, partly contemporaneously with the Big Snowy Group of Montana, in Prophet Trough and on the western cratonic platform (Fig. 4E.2, 4E.9, 4E.20). Comprising carbonates and siliciclastics, the Etherington occurs in the southern part of the Rocky Mountains and part of the western Foothills from 49°00'N to approximately 52°30'N (Macauley et al., 1964; Scott, 1964). It is 58 m thick at its type section in the western Foothills southwest of Calgary, and thickens to over 250 m toward the west and southwest. The Etherington thins gradually northwestward to zero below a sub-Bashkirian disconformity, and in the east it thins rapidly eastward beneath sub-Bashkirian and younger unconformities.

In most areas, the Etherington Formation overlies the Mount Head Formation and is overlain disconformably by the Bashkirian Tyrwhitt Formation, but east of the erosional edge of the latter, it is locally overlain disconformably by the Bashkirian Tobermory Formation (Scott, 1964). East of the erosional edge of the Tobermory, the Etherington is unconformably overlain by Mesozoic strata.

Deposits of the Todhunter Member (Norris, 1957, 1965), 27 m thick at its type locality in southeastern British Columbia, generally constitute the upper part of the Etherington Formation. The remainder of the formation has not been formally divided.

Limestone and dolostone with subordinate shale, sandstone, and siltstone constitute the heterogeneous Etherington Formation (Fig. 4E.9, 4E.20). The proportion of siltstone and sandstone increases upward and eastward, whereas limestone and dolostone become more abundant downward and southwestward. Shale is generally most abundant in the basal Etherington and in the east.

Most of the Etherington Formation below the Todhunter Member, particularly in the eastern Front Ranges, comprises multiple fining-upward hemicycles. The latter, which have sharp and commonly erosional bases, thicken basinward and range in thickness from less than 2 m to more than 20 m. Where completely developed, they comprise a resistant lower lithofacies assemblage that grades upward and northeastward into a recessive upper assemblage. The lower assemblage comprises mixed-skeletal to ooid lime grainstone and packstone with subordinate wackestone. These lower deposits, which generally become more dolomitic and finer grained upward, vary from massive and bioturbated to planar stratified and crossbedded. The upper assemblage comprises microcrystalline dolostone and green to maroon shale with subordinate lime wackestone, packstone, and grainstone, cryptalgal boundstone, siltstone, and sandstone. Wave-formed crossbedding and planar laminae are common in the sandstone and some carbonates. Terra-rossa breccia (breccia in paleosol), nodular calcrete, chert nodules, desiccation structures, and solution surfaces with rundkarren are commonly present.

The hemicycles were deposited in open- to restricted-marine environments during transgressive-regressive events. In the lower part of each hemicycle, the allochems and sedimentary structures generally record deposition in relatively open-marine, shallow-neritic (above storm wave base) to intertidal beach environments on either the protected shelf of a platform or the shallow shelf of a ramp. In contrast, deposits in the upper parts of the hemicycles are diagnostic of lagoon, sabkha, and other restricted-shelf environments.

In some eastern sections, the hemicycles described above occur with sequences consisting mainly of sandstone and shale. The latter, which coarsen and become more resistant upward, are generally dominated by bioturbated or crossbedded sandstone. Sequences of this type generally result from sedimentation in shallow-neritic to shoreline environments.

The Todhunter Member consists of sparsely fossiliferous sandstone with subordinate siltstone, dolostone, and limestone. Where completely developed, the Todhunter comprises a recessive lower siltstone and sandstone unit, a middle division of sandstone with sandy limestone and dolostone, and an upper sandstone-siltstone unit (Scott, 1964). In the sandstone and siltstone units, scattered marine fossils, redbeds, and abundant crossbedding suggest deposition in shallow-neritic to supratidal environments. The limestone and dolostone, which are commonly fossiliferous, are characteristic of shallow, open- to restricted-marine environments. Like deposits of the lower part of the Etherington Formation, those of the Todhunter Member commonly constitute hemicycles of transgressive-regressive origin.

The Etherington Formation, which records the first major Carboniferous influx of sand-sized siliciclastics into southern Prophet Trough, indicates a major regression and subsequent transgressive-regressive hemicycle. Widespread regression is recorded by the sub-Etherington disconformity and by overlying calcrete and associated restricted-shelf deposits of the basal Etherington. Overlying upper Viséan (V3, foraminiferal zone 16s) to lower Serpukhovian lithofacies that are dominantly of open-marine aspect record the subsequent transgression. The final major regression is recorded by sandstone of the Todhunter Member and the overlying sub-Bashkirian erosional disconformity.

Golata Formation

The middle(?) to upper Viséan (uppermost V2 to lower V3) Golata Formation (Halbertsma, 1959) is widely distributed from northwestern Alberta to southwestern District of Mackenzie and southeastern Yukon Territory (Fig. 4E.2, 4E.12, 4E.13, 4E.19). In this region, the Golata, which comprises shale with subordinate sandstone and carbonates, occurs in the eastern Cordillera and beneath the western Interior Plains. This formation, which forms the basal part of the Stoddart Group south of 58°30′N, was deposited mainly in Peace River Embayment and Prophet Trough. In the east, the Golata is commonly less than 15 m thick, but toward the west it thickens to more than 520 m.

The Golata Formation, a tongue of the shale lithosome constituting most of the Besa River Formation to the west, overlies the Debolt and Flett formations in the east and the Prophet Formation to the west. In most areas, the Golata is overlain by, and passes northeastward into, the Kiskatinaw Formation south of 58°30′N and Mattson Formation to the northwest. The boundary is commonly conformable and diachronous, becoming younger southwestward, but it is erosional in most of Peace River Embayment. East of the erosional edge of the Kiskatinaw and Mattson, Permian and Mesozoic strata unconformably overlie the Golata.

Shale and mudstone with subordinate sandstone, limestone, and dolostone constitute most of the Golata Formation (Richards, 1989); anhydrite and coal occur locally in Alberta (Halbertsma, 1959). Most of the shale and mudstone were deposited in prodelta environments and in other delta-related marine to marginal-marine environments. Some deposits in the lower part of the western Golata formed basinward of carbonate-platform lithofacies preserved in the upper Prophet, Flett, and Debolt formations (Fig. 4E.12). Shale and mudstone in the upper Golata grade upward and northeastward into delta and delta-related sandstone of the Kiskatinaw and Mattson formations.

The presence of the Golata above local unconformities and shelf-margin to restricted-shelf carbonates of the Rundle Group records a transgression and the onset of a subsequent regression. Golata rocks also record the initial phase of a major period of siliciclastic sedimentation culminating with the westward progradation of delta-plain and delta-related deposits in the overlying lower Kiskatinaw Formation and Mattson Formation.

Kiskatinaw and Taylor Flat formations

During and after sedimentation of the upper part of the Golata Formation, the Kiskatinaw Formation and overlying Taylor Flat Formation (Halbertsma, 1959) were deposited in northern Alberta and British Columbia (Fig. 4E.1, 4E.2). The upper Viséan (V3) Kiskatinaw is mainly sandstone, whereas the uppermost Viséan(?) to Serpukhovian Taylor Flat comprises carbonates with subordinate fine-grained siliciclastics (Bamber and Mamet, 1978; Mamet, 1976, Fig. 2). These two formations, which constitute most of the Stoddart Group, occur principally in the axial region of Peace River Embayment but are also present from 56°30′N to 58°30′N (Fig. 4E.1; Macauley et al., 1964). There, they occur in the eastern Cordillera and locally beneath the western Interior Plains. An arbitrary boundary, placed in the disturbed belt at 58°30′N, separates the Kiskatinaw and Taylor Flat formations from the partly coeval and lithologically similar Mattson Formation to the north.

The Kiskatinaw and Taylor Flat formations vary greatly and abruptly in thickness because of block faulting (partly syndepositional) and post-depositional, differential erosion, but general thinning trends also occur. Near Fort St. John, British Columbia, the Kiskatinaw is 296 m thick and the Taylor Flat 282 m thick (Halbertsma, 1959). Toward the southwest in the Foothills south of 55°40′N, these formations and the entire Stoddart Group are truncated by a regional unconformity (Fig. 4E.1). The erosion appears to have occurred chiefly during the Late Carboniferous event discussed below. Toward the south, east, and north, the Kiskatinaw/Taylor Flat succession is truncated by a sub-Asselian (Permian) unconformity.

The Kiskatinaw Formation, which grades basinward into the Besa River Formation, generally overlies the Golata Formation in the east and the Besa River southwest of the basinward limit of the Golata (Fig. 4E.2; Halbertsma, 1959; Bamber and Mamet, 1978). In the southwestern part of the Peace River Embayment area, however, the Kiskatinaw locally disconformably overlies the Debolt Formation. In most of Peace River Embayment, the boundary between the Kiskatinaw and overlying Taylor Flat is relatively abrupt and commonly unconformable. North of the embayment, and locally within it, these two formations are difficult to separate. The Taylor Flat is unconformably overlain by Permian beds and locally by Triassic beds (Macauley et al., 1964; Bamber and Mamet, 1978). Eastward, Permian strata unconformably overlie the Kiskatinaw Formation.

Sandstone (mainly quartzarenite) with subordinate shale, skeletal limestone, dolostone, and siltstone constitute the Kiskatinaw (Halbertsma, 1959; Bamber and Mamet, 1978). This formation, which closely resembles the Mattson Formation (Fig. 4E.2, 4E.21), shows marked lateral and vertical lithofacies variations, with southwestward increase in the proportion of shale, siltstone, and carbonates.

Silty and sandy, skeletal to locally oolitic limestone and dolostone constitute most of the Taylor Flat Formation (Halbertsma, 1959). Sandstone, shale, and siltstone beds are commonly present, however, and are locally abundant enough to cause difficulty in distinguishing this formation from the Kiskatinaw.

Little has been published about the sedimentology of the Kiskatinaw and Taylor Flat formations. A brief examination of subsurface data, however, suggests that the Kiskatinaw consists mainly of deltaic and delta-related deposits, whereas the Taylor Flat comprises nondeltaic marine lithofacies. A deltaic origin for part of the Kiskatinaw, at least in the axial region of Peace River Embayment, is suggested by the association of marine deposits and abundant sharp-based sequences comprising sandstone grading upward into shale. These latter sequences, which have gamma-ray signatures and deposits characteristic of estuary and fluvial channel-fills, occur in Alberta and easternmost British Columbia. They are best developed and most abundant in the lower part of the Kiskatinaw Formation and do not appear to be present in the Taylor Flat. Toward the west and the top of the Kiskatinaw, coarsening- and shallowing-upward sequences of delta-front or shallow-neritic to intertidal aspect become predominant. The coarsening-upward sequences generally comprise shale and siltstone grading upward into sandstone, but limestone and dolostone are present in some. Sequences similar to the latter also are found in the Taylor Flat Formation, where they are associated with numerous thin carbonate units that are partly of restricted-shelf aspect. The carbonate units in the Taylor Flat do not form well-developed carbonate ramps and platforms.

The Kiskatinaw/Taylor Flat succession records a major regression and subsequent transgression. Regression is indicated by the estuarine, fluvial and related deltaic facies of the lower Kiskatinaw, which prograded southwestward over prodelta deposits of the Golata and Besa River formations. The onset of transgression is recorded in the upper part of the Kiskatinaw by an upward increase in the ratio of marine to fluvial and related nonmarine facies. Its continuation is recorded by the overlying, carbonate-dominated Taylor Flat.

Mattson Formation

The thick, upper Viséan (V3) to Serpukhovian Mattson Formation (Patton, 1958) of northwestern Canada was deposited contemporaneously with the Stoddart Group to the south (Fig. 4E.2, 4E.12, 4E.13, 4E.21). This sandstone-dominated formation, deposited in Prophet Trough, extends from 58°30'N in northeastern British Columbia to 61°25'N in southwestern District of Mackenzie (Harker, 1963; Braman and Hills, 1977). In most of this region, the Mattson is preserved as a down-faulted succession lying west of the extensive, north-striking Bovie normal fault system. The Mattson is generally less than 15 m thick east of Bovie normal fault and thickens westward, attaining over 1410 m in the southern Mackenzie Fold Belt (Richards, 1989). Douglas and Norris (1959, 1960) and Harker (1961, 1963) divided the formation into three informal members.

The Mattson Formation overlies and grades basinward to the Golata Formation in the northeast and the Besa River Formation to the southwest. In most areas, the Mattson thins northeastward below a sub-Permian unconformity. North of the erosional limit of the Permian strata at about 60°40'N, Cretaceous strata unconformably overlie the Mattson.

Figure 4E.21. Type section of Mattson Formation, Jackfish Gap, Liard Range, southwestern District of Mackenzie. A - base of Mattson (1009 m thick, deltaic sandstone) and top of Golata Formation (8 m thick, prodelta shale). Carbonates of Flett Formation underlie Golata. View is to northeast (photo - B.C. Richards, ISPG 1281-3).

Sandstone (principally quartzarenite and subchertarenite according to the classification of Folk, 1974) with subordinate shale, dolostone, and limestone constitute the Mattson Formation. Shale becomes more abundant southwestward, and limestone and dolostone occur in the middle to upper Mattson. Thin coal seams also occur, principally north of 59°30′N, in the lower to middle part of the eastern Mattson.

The Mattson Formation, which prograded southwestward or basinward over prodelta shale of the Golata and upper Besa River formations, consists mainly of lithofacies deposited in deltaic and related environments during several delta cycles (Richards, 1989). Most of the Mattson deltas can be classified as fluvially dominated, wave- and tide-influenced deltas of lobate form. Thick Platte-type braided-stream deposits consisting of sandstone with medium-scale tabular crossbedding are common in the delta-plain facies assemblage. Coal, shale, and sandstone in coarsening-upward sequences of overbank and delta-front origin are associated with the braided-stream deposits. Some deposits of meandering streams are also present. Toward the southwest, the delta-plain deposits, which are best developed in the eastern Mattson (Fig. 4E.21) of the District of Mackenzie, grade into delta-front deposits. The latter include tidal-channel deposits but are mainly bioturbated to crossbedded and planar-stratified, coarsening- and shallowing-upward sequences. The sequences comprise sandstone and subordinate shale deposited at depths ranging from below storm wave base to shoreline. Farther southwestward, the delta-front lithofacies grade into delta-slope deposits of the Mattson and prodelta shale of the Golata. Sandstone turbidites constitute most of the delta-slope succession, which overlies and grades basinward into prodelta shale of the Golata and Besa River formations.

Limestone and dolostone in the Mattson Formation are mainly skeletal wackestone, packstone, and grainstone, but skeletal-ooid lime grainstone occurs locally. The carbonates, occurring mainly as units less than 10 m thick, commonly overlie deltaic and delta-related siliciclastics, thereby recording numerous minor transgressions. These carbonates, which do not constitute well-defined platforms or ramps, were probably deposited mainly in shallow-neritic to shoreline environments.

Lithofacies of the Mattson Formation record a major regression and subsequent transgression. Regression is recorded by the widespread occurrence of fluvial lithofacies that prograded well basinward of shallow-marine deposits in the underlying Viséan carbonate succession (Fig. 4E.13). The distribution of coal and fluvial deposits in the Mattson indicates that maximum regression occurred during the late Viséan (late V3; lower foraminiferal zone 16s). Subsequently, a gradual transgression took place. The latter is recorded by uppermost Viséan (V3) to Serpukhovian shallow-marine sandstone, shale, and carbonates that onlapped the western delta-plain deposits. In the northeast, however, fluvial sedimentation continued intermittently into the early part of Serpukhovian time. Most sand in the Mattson had a northern sedimentary provenance, probably the Ellesmerian Fold Belt and adjacent uplifted areas to the south of it.

Tyrwhitt, Storelk, and Tobermory formations

Deposition of the Bashkirian to lowest Moscovian Tyrwhitt, Storelk, and Tobermory formations (Scott, 1964) occurred in southern Canada after the erosional event that followed sedimentation of the Etherington Formation (Fig. 4E.2, 4E.9, 4E.20). Representing the base of the Absaroka sequence of Sloss (1963), this succession was deposited in Prophet Trough and on the westernmost part of the cratonic platform. It constitutes most of the Spray Lakes Group and outcrops in the Rocky Mountains of southeastern British Columbia and southwestern Alberta (south of 52°45′N). This southwestward-thickening succession, which is the Misty Formation of Norris (1965), is locally present also in the western Foothills of southern Alberta (Scott, 1964; Norris, 1965; Stewart and Walker, 1980). At the type locality in southeastern British Columbia, the Tyrwhitt, Storelk, and Tobermory formations are 85 m, 93 m, and 66 m thick, respectively (Scott, 1964).

The succession comprising the Tyrwhitt, Storelk, and Tobermory formations disconformably overlies the Etherington Formation and in most areas is overlain conformably by the Kananaskis Formation. It is, however, locally overlain unconformably by Jurassic and Triassic strata in the east (Figs. 4E.2, 4E.20). In most areas, the Tyrwhitt overlies the Etherington. The sub-Tyrwhitt disconformity, which correlates with the regional unconformity between the Big Snowy and Amsden groups of central Montana (Maughan and Roberts, 1967; Sando et al., 1975, Fig. 13), marks the top of the Kaskaskia sequence of Sloss (1963). Although abrupt, the Tyrwhitt-Storelk contact is probably conformable (Stewart and Walker, 1980; Scott, 1964). The boundary between the Tobermory and the underlying Storelk and Tyrwhitt is an unconformity that truncates underlying strata eastward to bring the Tobermory into contact with the Etherington. Toward the southwest, the upper Tobermory grades into the overlying Kananaskis.

The Tyrwhitt, Storelk, and Tobermory formations were deposited in environments ranging from neritic to eolian (Stewart and Walker, 1980) and comprise sandstone (chiefly quartzarenite) with subordinate dolostone and minor shale and limestone. Most of the Tyrwhitt is bioturbated sandstone deposited below fair-weather wave base. Associated deposits are sandstone units that display medium-scale crossbedding and were probably deposited by storm-generated currents. The conformably overlying Storelk Formation is dominated by sandstone that shows large- to very large-scale crossbedding and was deposited principally in a coastal dune environment. Deposits of the disconformably overlying Tobermory resemble those of the Tyrwhitt and were deposited in a similar setting. Carbonate beds are, however, more abundant in the Tobermory, which also has a higher ratio of crossbedded to bioturbated sandstone. Both the Tyrwhitt and Tobermory contain coarsening-upward sequences but lack overall coarsening- and shallowing-upward trends.

Two transgressions and an intervening regression are recorded by the succession. The first transgression is recorded by the presence of marine Tyrwhitt lithofacies above a regional erosional disconformity, and subsequent regression is recorded by the overlying aeolian Storelk

Formation. The second regional transgression is recorded by the occurrence of neritic Tobermory lithofacies above the post-Storelk erosional disconformity. This transgression culminated with deposition of carbonates that constitute the overlying Kananaskis Formation (Fig. 4E.2).

Serpukhovian to upper Viséan (V3) and possibly older Carboniferous strata on the southern cratonic platform were exposed to subaerial erosion during late Bashkirian time. The unconformity between the Bashkirian to lowest Moscovian Tobermory Formation and the Etherington Formation records the event (Fig. 4E.20).

Scott (1964) suggested that most sand in this succession came from Paleozoic strata on the cratonic platform to the east. Much of the sand, however, was probably derived from subaerially exposed Lower Carboniferous strata in east-central British Columbia and west-central Alberta, where the Rundle and Stoddart groups were subjected to deep erosion. Some sand possibly came from the west as well, because similar partly coeval sandstone in the Quadrant Formation of southwestern Montana apparently had a western provenance (see Maughan, 1975).

Kananaskis Formation

Subsequent to deposition of the Tobermory Formation, the Moscovian Kananaskis Formation (McGugan and Rapson, 1961) was deposited (Fig. 4E.2, 4E.9, 4E.20). This youngest known Carboniferous deposit in Western Canada Basin south of northern Yukon Territory, was deposited in Prophet Trough and possibly on the western cratonic platform.

The Kananaskis, the youngest formation of the Spray Lakes Group, conformably overlies the Tobermory Formation and unconformably underlies the Permian Ishbel Group (McGugan and Rapson, 1961; McGugan et al., 1968; Scott, 1964; Norris, 1965). This formation is preserved south of 52°10′N in the Rocky Mountains of southeastern British Columbia and southwestern Alberta. There, the Kananaskis is up to 51 m thick and thins eastward because of both stratigraphic condensation and erosion prior to deposition of the overlying Lower Permian (Asselian to Artinskian) Ishbel Group.

Most of the Kananaskis Formation is silty and sandy, microcrystalline dolostone that commonly grades into dolomitic siltstone and sandstone. Chert nodules and thin beds of chert are very common, and some novaculite beds occur. Intraformational chert breccias are characteristic of the upper part of the Kananaskis in some eastern sections. Toward the west, these breccias become less common and pass into facies containing unbrecciated chert (McGugan and Rapson, 1961; Scott, 1964). A marine origin for most of the Kananaskis is recorded by the common occurrence of fusulinids, brachiopods and other marine fossils in the chert. Marine carbonates of the Kananaskis Formation record the culmination of a transgression that commenced with deposition of marine sandstone in the underlying Tobermory Formation.

Substantial pre-Permian erosion occurred from southwestern Alberta to east-central British Columbia, with maximum erosion in the northwest. In the Cordillera north of 51°00′N, a disconformity below Bashkirian deposits of the Spray Lakes Group truncates the upper Viséan (V3) to Serpukhovian Etherington Formation northwestward (Fig. 4E.2). Southwestward truncation of pre-Moscovian Carboniferous deposits also took place. From 54°30′N to 55°35′N, the upper Viséan (V3) to lower Serpukhovian Stoddart Group occurs on the western Interior Platform. Toward the southwest, first the Stoddart Group and then formations in the underlying Rundle Group are truncated by unconformities (Fig. 4E.1). These stratigraphic relationships indicate that southwestward truncation occurred subsequent to early Serpukhovian and prior to Permian time.

Depositional summary

The upper depositional unit, which was deposited from late Viséan (V3) to early Moscovian time, comprises delta and delta-related deposits, non-deltaic(?) siliciclastics, and carbonates. Sandstone-dominated delta and delta-related deposits, which constitute most of the Kiskatinaw and Mattson formations, prograded southwestward (basinward) over basinal to prodelta shale of the Golata and upper Besa River formations. Non-deltaic(?) sandstone of neritic to continental origin constitutes most of the Tyrwhitt, Storelk, Tobermory, and Kibbey formations, and the upper part of the Etherington Formation. Also, non-deltaic(?), shallow-marine sandstone grading southwestward into probable slope facies constitutes part of the Taylor Flat Formation. Carbonate lithofacies predominate in the lower Etherington, Kananaskis and most of the Taylor Flat formations, and form parts of the Kibbey and upper Mattson. Carbonates of the Etherington and Kananaskis formations are probably of carbonate platform or ramp origin, but those of the other formations do not constitute well-defined carbonate buildups. In general, the carbonate lithofacies of Prophet Trough and the western cratonic platform are of restricted-shelf origin in the east and of open-marine origin to the southwest.

The upper depositional unit records the culmination of the general regressive trend that predominated during sedimentation of the middle depositional unit. It is cyclical, however, and also records numerous minor and some major transgressions and regressions.

In the Big Snowy Group of central Williston Basin, the lower Kibbey Formation appears to represent the culmination of a regional regression, whereas the upper Kibbey and overlying Otter and Heath formations indicate a subsequent transgression. This transgression was followed by late Serpukhovian regression and subaerial erosion.

In Prophet Trough and Peace River Embayment and on the adjacent stable craton, the first major regression occurred during late Viséan (foraminiferal zones 14 through 16i) time. Deltaic and related lithofacies of the upper Golata and overlying lower and middle Mattson Formation record this event in the north. In Peace River Embayment deltaic siliciclastics of the upper Golata and lower Kiskatinaw formations resulted from the regression, which followed a late Viséan transgression indicated by the lower Golata. Farther south, the regression is recorded by the disconformity between the Mount Head and Etherington formations and by restricted-shelf carbonates in the upper part of the middle depositional unit and lower Etherington. The regression was followed by a gradual transgression, interrupted by numerous minor regressions, during deposition of upper Viséan (foraminiferal zone 16s) to middle Serpukhovian (foraminiferal zone 18) carbonates and sandstone in the Etherington and Taylor Flat formations and the upper Kiskatinaw and Mattson formations.

Three subsequent regional regressions and two intervening transgressions occurred in southern Prophet Trough and on the adjacent cratonic platform. Regression is indicated by Serpukhovian (foraminiferal zones 18 and 19) sandstone in the Todhunter Member of the Etherington Formation and by the overlying sub-Tyrwhitt disconformity. Subsequent transgression is recorded by Bashkirian, shallow-marine sandstone of the Tyrwhitt Formation. The onset of the second regression is recorded by eolian sandstone of the Bashkirian Storelk Formation and its culmination by the regional disconformity below the overlying upper Bashkirian to lower Moscovian Tobermory Formation. Bashkirian shallow-marine siliciclastics in the Tobermory resulted from the initial phase of the second transgression, which culminated with deposition of lower Moscovian carbonates of the Kananaskis Formation. A final regression, which exposed all of the Western Canada Basin from southwestern District of Mackenzie southward, took place after early Moscovian time and prior to Asselian (Early Permian) time.

Besa River Formation

The Middle Devonian to upper Viséan (V3) Besa River Formation (Kidd, 1963) is widely distributed from 54°45′N in east-central British Columbia to southeastern Yukon Territory and southwestern District of Mackenzie (Pelzer, 1966; Bamber and Mamet, 1978; Richards, 1989). Consisting mainly of shale, it occurs in the eastern Cordillera and on the western Interior Plains (Fig. 4E.2, 4E.11-4E.13). The Besa River, up to 1655 m thick, is generally thickest in the Foothills and on the extreme western part of the Interior Plains.

In most areas, the Besa River Formation conformably overlies Middle Devonian strata (Pelzer, 1966; Bamber and Mamet, 1978) and is conformably overlain by Tournaisian to upper Viséan (V3) strata of the lower, middle, and upper depositional units (Fig. 4E.2, 4E.11-4E.13). In east-central and northeastern British Columbia it is locally disconformably overlain by Permian strata (Bamber et al., 1968; Bamber and Macqueen, 1979). Eastward, Carboniferous strata of the Besa River Formation pass into coarser siliciclastics and carbonates (Rundle Group and Yohin, Kiskatinaw, Taylor Flat, and Mattson formations) and into shale of the Exshaw, Banff, Clausen, and Golata formations.

A dark-shale lithofacies constitutes most of the Besa River Formation, but a spiculite lithofacies occurs in the upper part from east-central British Columbia northward. In addition, a sandstone-and-shale facies is present in the upper Besa River of northeastern British Columbia and southeastern Yukon Territory and in the lower Besa River of the Rocky Mountains from 55°00′N to 55°50′N.

The dark-shale lithofacies, which generally underlies and grades eastward into the other two facies, is mainly black shale, which commonly contains abundant sponge spicules and radiolarians. It was deposited in deep water (Pelzer, 1966), basinward of carbonate slope deposits and deltaic to nondeltaic sandstone-dominated successions (Richards, 1989).

Spiculite, bedded chert, and spicule-rich carbonates constitute the spiculite lithofacies. These deposits, which are commonly intercalated with the dark-shale facies, generally pass upward and eastward into slope deposits of the upper Banff and Prophet formations. Most of the spiculite lithofacies is preserved as basinward-thinning tongues deposited in poorly oxygenated, moderately deep water at the toes of carbonate platforms and ramps.

The shale-and-sandstone facies consists of silty to sandy shale and mudstone interbedded with subordinate siltstone and sandstone turbidites and slump deposits. In the upper part of the Besa River Formation, where the strata grade upward and eastward into deltaic and nondeltaic slope siliciclastics of the Yohin and Mattson formations and the Stoddart Group, deposition took place in deep, poorly oxygenated water. Occurrences of shale and sandstone in the middle(?) Besa River of east-central British Columbia contain locally abundant terrestrial plant fragments and are probably of latest Devonian age. These deposits, locally associated with plant-bearing polymictic conglomerate units (Muller, 1967, p. 80), are probably of deep-water origin (below storm wave base) as suggested by the presence of coarse-grained turbidites and debris-flow beds interbedded with black shale. The middle(?) Besa River occurrences probably had a western source, possibly the rim of Prophet Trough, because the region to the east was the site of carbonate and black, marine shale deposition and would not have been a suitable source.

In northeastern British Columbia and northward, much of the Besa River Formation was deposited in a starved basin, as indicated by marked basinward depositional thinning in the formation (Pelzer, 1966; Richards, 1989). Similar evidence for starved-basin conditions is shown by some correlatives of the Besa River, particularly the Prophet (Fig. 4E.12, 4E.13). Starved-basin conditions were established during the Late Devonian and persisted into late Viséan (V3) time.

MACKENZIE AND SELWYN MOUNTAINS, EAST-CENTRAL YUKON AND WEST-CENTRAL DISTRICT OF MACKENZIE

In the west-central Mackenzie Mountains and Selwyn Mountains, Carboniferous formations are much less continuously preserved than to the south (Fig. 4E.1). The Carboniferous succession (Fig. 4E.26) begins with the Tournaisian upper Earn Group (Campbell, 1967; Gordey et al., 1987), which rests unconformably on Famennian strata of the lower Earn Group (Gordey et al., 1982). The upper Earn (Prevost Formation) is principally shale and siltstone that is of moderately deep-water aspect and resembles prodelta deposits of the Besa River Formation. Thick submarine-channel fills of westerly derived sandstone and conglomerate are also locally present in the lower part of this unit and resemble coeval conglomerate and sandstone deposits in the Tuttle Formation on the flanks of the Richardson Mountains to the north (Fig. 4E.3, 4E.22).

The upper Earn Group is abruptly and possibly disconformably overlain by the upper Tournasian to upper Viséan (TN3 to V3) sandstone-dominated Tsichu Formation, partly correlative with the Mattson Formation to the south (Gordey et al., 1982; 1987). The Tsichu, up to 200 m thick, is at least partly of shallow-marine origin and appears to grade westward into a deeper water shale facies. Above the Tsichu, the Carboniferous succession comprises shale and interbedded limestone grading upward into sandy skeletal limestone with subordinate sandstone. This upper, unnamed interval, up to 600 m thick, is probably mainly of late Viséan (V3) to Bashkirian age. It is overlain unconformably by Permian strata.

The age and depositional history of the succession in the Mackenzie and Selwyn mountains are not well established. This succession is significant, however, because it indicates that Carboniferous strata of shallow to moderately deep-water origin were deposited over extensive areas between southwestern District of Mackenzie and northern Yukon Territory, although none are preserved.

NORTHERN CORDILLERA – NORTHERN YUKON TERRITORY AND NORTHWESTERN DISTRICT OF MACKENZIE

An essentially continuous succession of Viséan to Moscovian carbonates and siliciclastics is preserved beneath the regional sub-Permian disconformity in the northern Cordillera north of 64°30′N (Fig. 4E.3, 4E.22, 4E.23). Younger Carboniferous deposits occur only in the Ogilvie Mountains, where Kasimovian to Gzhelian strata are present and sedimentation may have been locally continuous across the Carboniferous/Permian boundary (Fig. 4E.22; Waterhouse and Waddington, 1982). Tournaisian strata have been definitely identified only in southwestern Peel Plateau (Bamber and Waterhouse, 1971). They may, however, be present on the western flank of the southern Richardson Mountains (Pugh, 1983).

The Carboniferous strata of the northern Cordillera comprise two separate, but closely related successions. The more completely preserved succession consists of Tournaisian(?) to Gzhelian strata, which occur mainly south of 67°N, in the Ogilvie Mountains and Eagle Plain and on the flanks of the Richardson Mountains (Fig. 4E.3, 4E.22). It was deposited in northern Prophet Trough and possibly on the western cratonic platform. The second succession consists principally of Viséan to Moscovian strata that onlap, from Prophet Trough, northward onto the Yukon Fold Belt and possibly eastward onto the cratonic platform. It lies north of 68°00′N in a southeast-trending, discontinuous belt extending from the Alaska/Yukon border into northwestern District of Mackenzie (Fig. 4E.3, 4E.23). Both successions comprise a lower interval dominated by terrigenous clastics and an upper one consisting mainly of carbonates. The two successions originally formed part of a continuous depositional complex, but their distribution apparently has been altered by counter-clockwise rotation of northern Yukon and Alaska during the Mesozoic Era (Sweeney et al., 1978; Nilsen, 1981). They are now separated by the northeast-trending, late Paleozoic Ancestral Aklavik Arch (see Henderson et al., Subchapter 4F in this volume; Bamber and Waterhouse, 1971). Along the axis of the arch, Carboniferous strata are absent, and Permian strata lie unconformably on Devonian and older rocks.

Southern assemblage

The lower part of the southern succession comprises Tournaisian(?) to Bashkirian siliciclastics and carbonates of the Tuttle, Ford Lake, Hart River, and lower Blackie formations (Fig. 4E.3, 4E.22). These are generally overlain by Bashkirian to Kasimovian(?) carbonates of the Ettrain Formation, which grade southward and southwestward into shale and carbonates of the upper Blackie Formation. In the Ogilvie Mountains, Kasimovian(?) to Gzhelian terrigenous clastics of the basal Jungle Creek Formation overlie the Ettrain and constitute the top of the Carboniferous System. In most areas, the southern succession is overlain by Permian strata, but north of the erosional edge of the Permian, it is truncated northward beneath a regional sub-Cretaceous unconformity (Pugh, 1983).

Tuttle Formation

At the base of the southern succession, thick, coarse-grained siliciclastics of the Tuttle Formation (Pugh, 1983) rest conformably on the Upper Devonian Imperial Formation in eastern Eagle Plain and on the flanks of the Richardson Mountains. The Tuttle is mainly Late Devonian, but may contain strata as young as Tournaisian in its upper part. It consists of chert-rich conglomerate and sandstone with subordinate siltstone and shale, and attains a thickness of 1400 m or more. The upper Tuttle becomes finer grained as it passes upward and southwestward into basinal shale and siltstone of the Ford Lake Formation. The characteristics and regional lithofacies relationships of the Tuttle suggest deposition in deltaic and related environments. Lithofacies trends and grain-size variations indicate that terrigenous clastics of the Tuttle had a northern to northeastern provenance, probably the Yukon and Ellesmerian fold belts.

Ford Lake Formation

The Upper Devonian to upper Viséan (V3) Ford Lake Formation (Fig. 4E.3, 4E.22) comprises up to 1000 m of siliciclastics, which are distributed over a wide area, including the west flank of the Richardson Mountains, southern Eagle Plain, and the Ogilvie Mountains (Pugh, 1983). Tournaisian strata have not been identified, but they may be present, because there is no definite evidence for a disconformity at the base of the Carboniferous succession. The Ford Lake consists of dark grey, silty, pyritic shale and siltstone with subordinate sandstone, conglomerate, and silty limestone. In the northeast, the Ford Lake conformably overlies the Tuttle Formation. The upper Ford Lake grades northeastward into spiculitic carbonates of the lower Hart River Formation.

The lithology and lithofacies relationships of the Ford Lake Formation record deposition in a moderately deep-water, basinal environment similar to that represented by the Besa River Formation. Basinal shale and siltstone of the lower Ford Lake onlap eastward and northward over deltaic facies of the Tuttle Formation, thereby indicating a transgression during latest Devonian and Early Carboniferous time. Viséan sandstone and conglomerate occurring in the upper part of the eastern Ford Lake Formation form wedges and lenses that thin southward or basinward (Graham, 1973) and may have been deposited partly in delta-slope environments. During deposition of the Ford Lake Formation, Viséan strata of the Keno Hill Quartzite (R. Thompson, pers. comm., 1987), in central Yukon near Dawson and Mayo, were deposited on the southwestern flank of Prophet Trough (Fig. 4E.1). The presence of the Keno Hill, formerly considered to be of Cretaceous age (Green, 1972), indicates that an extensive exposed region lay southwest of the trough during Viséan time and supplied a large volume of quartz sand.

Unnamed correlatives of the eastern Ford Lake Formation occur in southwestern Peel Plateau (Bamber and Waterhouse, 1971, p. 50). They consist of up to 1300 m of middle(?) Tournaisian (Tn2) and Viséan, cyclic, marine and non-marine shale, siltstone, sandstone, and coal, which are unconformably overlain by Cretaceous shale and sandstone.

Hart River Formation

The oldest carbonates in the southern succession are in the upper Viséan (V3) to Serpukhovian Hart River Formation (Bamber and Waterhouse, 1971). The formation lies chiefly in southern Eagle Plain, where it is up to 691 m

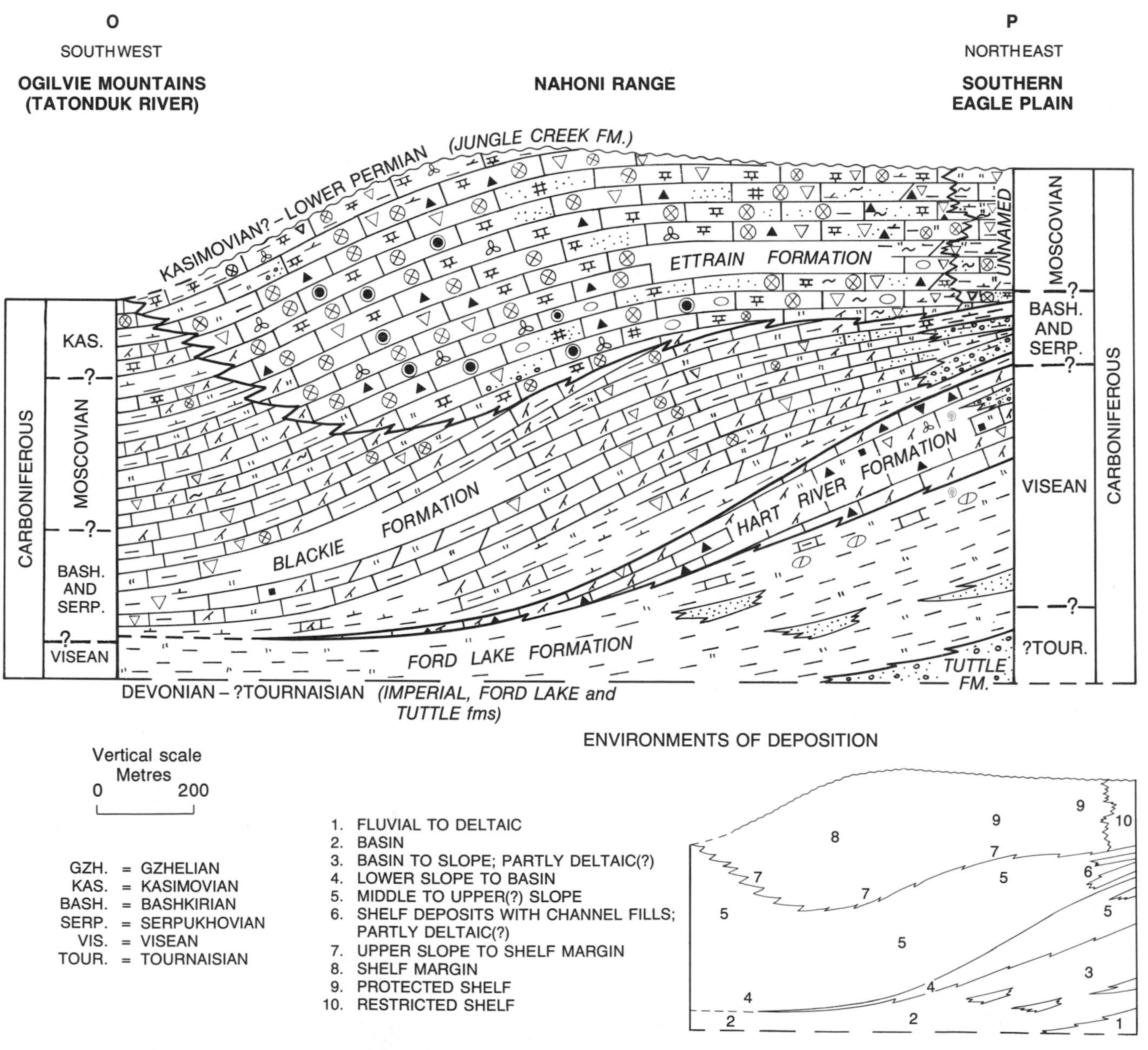

Figure 4E.22. Partly schematic stratigraphic cross-section O-P showing Carboniferous units of northwestern Yukon Territory. See Figure 4E.1 for line of section and Figure 4E.23 for legend.

thick (Pugh, 1983). It thins southwestward into the Ogilvie Mountains as it progrades over and passes laterally into basinal shale and siltstone of the upper part of the Ford Lake Formation (Fig. 4E.3, 4E.22). Most of the Hart River comprises thinly laminated, cherty spiculite and spicule lime packstone with subordinate sandstone, siltstone, and calcareous shale, but lime grainstone occurs locally in eastern outcrops. Thick, discontinuous lenses and wedges of sandstone grading into chert-rich conglomerate are also present in the east-central Eagle Plain, and thin rapidly toward the south and west (e.g., Chance Sandstone Member; Martin, 1972).

Graham (1973) has demonstrated with seismic reflection profiles that the carbonates and siliciclastics of the Hart River Formation prograded southwestward and were deposited in shelf, slope, and basin environments. The discontinuous sandstone and conglomerate units of east-central Eagle Plain appear to have had a northeastern provenance, and some are possibly submarine channel-fills.

Blackie Formation

Siliclastics and carbonates of the lower Serpukhovian to Kasimovian(?) Blackie Formation (Pugh, 1983) occur in the southwestern Eagle Plain, Keele Range, and Ogilvie Mountains, where they attain a thickness of more than 700 m. The Blackie (Fig. 4E.3, 4E.22), which underlies and grades northeastward into shelf carbonates of the Ettrain Formation, extends northeastward as a thin tongue (Unit 2 of Bamber and Waterhouse, 1971) separating the Ettrain from the underlying Hart River Formation. In the southwest, the Blackie comprises multiple hemicycles that consist of dark grey, calcareous shale and siltstone, and silty, spiculitic limestone. Within the northeastern tongue, there are discontinuous bodies of sandy limestone and calcareous sandstone. Thick, sharp-based, fining-upward units of chert-pebble conglomerate and sandstone occur in the easternmost outcrops (Eagle Plain).

Most of the Blackie Formation was deposited under shallowing-upward conditions in basin to slope environments, as indicated by the low-energy deposits and regional lithofacies relationships. Shale and spiculitic carbonates of the lower Blackie onlap eastward and northward over the slope and shelf deposits of the Hart River Formation, recording an abrupt influx of terrigenous clastics and possibly local deepening during early Serpukhovian time. Some conglomerate and sandstone units in the Blackie of the Eagle Plain are probably channel-fill deposits that had a northeastern provenance (Pugh, 1983). The environment in which the fill sequences developed has not been determined. Southwest of the correlative Ettrain shelf carbonates, upper Blackie lithofacies indicate that slope to basinal sedimentation continued until Kasimovian(?) time.

Ettrain Formation

Bashkirian to Kasimovian(?) carbonates of the Ettrain Formation, more than 600 m thick, cap the Carboniferous succession throughout much of the northern Ogilvie Mountains and Keele Range (Fig. 4E.3, 4E.22; Bamber and Waterhouse, 1971). They are also present in a narrow belt extending eastward across southern Eagle Plain (Pugh, 1983). The Ettrain thins southward and westward as it grades into the upper part of the Blackie Formation. It also thins beneath the sub-Permian disconformity toward eastern Eagle Plain, where the youngest Ettrain strata are of Moscovian age. The disconformity, and general absence of post-Moscovian Carboniferous deposits resulted from regional, latest Carboniferous to earliest Permian regression and subaerial erosion. In the northwestern Ogilvie Mountains, however, Kasimovian(?) strata of the Ettrain are overlain by sandstone, conglomerate, and shale of probable latest Carboniferous age (Kasimovian and Gzhelian) in the basal Jungle Creek Formation. The Ettrain/Jungle Creek contact may be locally conformable in this area. Siliciclastics of the Jungle Creek were derived from the northeast-trending Ancestral Aklavik Arch (see Henderson et al., Subchapter 4F in this volume).

Ettrain carbonates are mainly cherty, echinoderm-bryozoan and ooid lime grainstone, and mixed-skeletal lime packstone. Foraminifers, calcareous algae, and brachiopods are locally abundant. In the east, quartz and chert sand and silt are commonly present in the carbonates.

The lithological character and facies relationships (Bamber and Waterhouse, 1971) of the Ettrain Formation suggest that it is a shallowing-upward succession deposited principally in protected-shelf, shelf-margin, and upper-slope environments (Fig. 4E.22). In eastern outcrops of Eagle Plain, near the western flank of the Richardson Mountains, unnamed correlatives of the upper Ettrain contain abundant silty dolostone and siltstone characteristic of restricted-shelf environments. The Ettrain prograded westward and southward over the slope and basin deposits of the Blackie Formation, recording a major regional regression.

Northern assemblage

The northern Carboniferous succession outcrops in the British, Barn, and northern Richardson Mountains and comprises Tournaisian(?) to Moscovian deposits (Fig. 4E.3, 4E.23, 4E.24). This succession includes the Endicott and Lisburne groups, first described in northern Alaska (Bowsher and Dutro, 1957; Tailleur et al., 1967). It lies unconformably on Precambrian to Devonian rocks, and its base ranges in age from Tournaisian(?) or early Viséan (V1) in the British Mountains to Bashkirian in the northern Richardson Mountains. The succession extends into the Brooks Range of northern Alaska, where it is similar in lithology and origin (Wood and Armstrong, 1975; Armstrong and Mamet, 1977). In the British and Barn Mountains, Tournaisian(?) and Viséan terrigenous clastics and subordinate carbonates of the Endicott Group (Kekiktuk and Kayak formations) constitute the lower part of the succession. These lower strata are conformably overlain by carbonates of the Viséan to Bashkirian Lisburne Group (Alapah and Wahoo formations), which are disconformably overlain by Lower Permian strata. Carboniferous and Permian strata are truncated northeastward beneath regional sub-Triassic and sub-Jurassic unconformities (Norris, 1983).

Kekiktuk Formation

Chert-pebble conglomerate and sandstone of the Kekiktuk Formation (Brosgé et al., 1962) form a laterally discontinuous unit at the base of the northern succession

(Fig. 4E.3, 4E.23). With a thickness of a few to more than 50 m, they rest with angular unconformity on the Proterozoic and Lower Paleozoic Neruokpuk Formation in the British Mountains and on unnamed Ordovician and Silurian siliciclastics and chert in the Barn Mountains (Norris, 1981, 1983). The Kekiktuk has not been dated in Canada, but is conformably overlain by Viséan strata of the Kayak Formation and is, therefore, assumed to be of Early Carboniferous age. A fluvial origin has been demonstrated by Nilsen (1981) for the Kekiktuk in Alaska.

Kayak Formation

Middle(?) and upper Viséan (V2? and V3) siliciclastics and subordinate carbonates of the Kayak Formation (Bowsher and Dutro, 1957) gradationally overlie the Kekiktuk Formation (Fig. 4E.3, 4E.23). Where the Kekiktuk is not developed, the Kayak rests with angular unconformity on the Proterozoic and Lower Paleozoic Neruokpuk Formation. The upper Kayak grades both upward and westward into shelf carbonates of the Alapah Formation.

In the British Mountains, where the Kayak attains a thickness of more than 350 m, it is a deepening-upward succession similar to that in the Brooks Range of Alaska (see Nilsen, 1981). Thin beds of sandstone, shale, and coal at the base pass upward and westward into dark grey, calcareous shale with subordinate argillaceous, silty, mixed-skeletal lime packstone and wackestone. The proportion of limestone increases upward as the Kayak grades into the Alapah Formation. Marine shale and limestone of the upper Kayak pass eastward into shale, carbonaceous sandstone, and coal outcropping in the Barn Mountains (Bamber and Waterhouse, 1971).

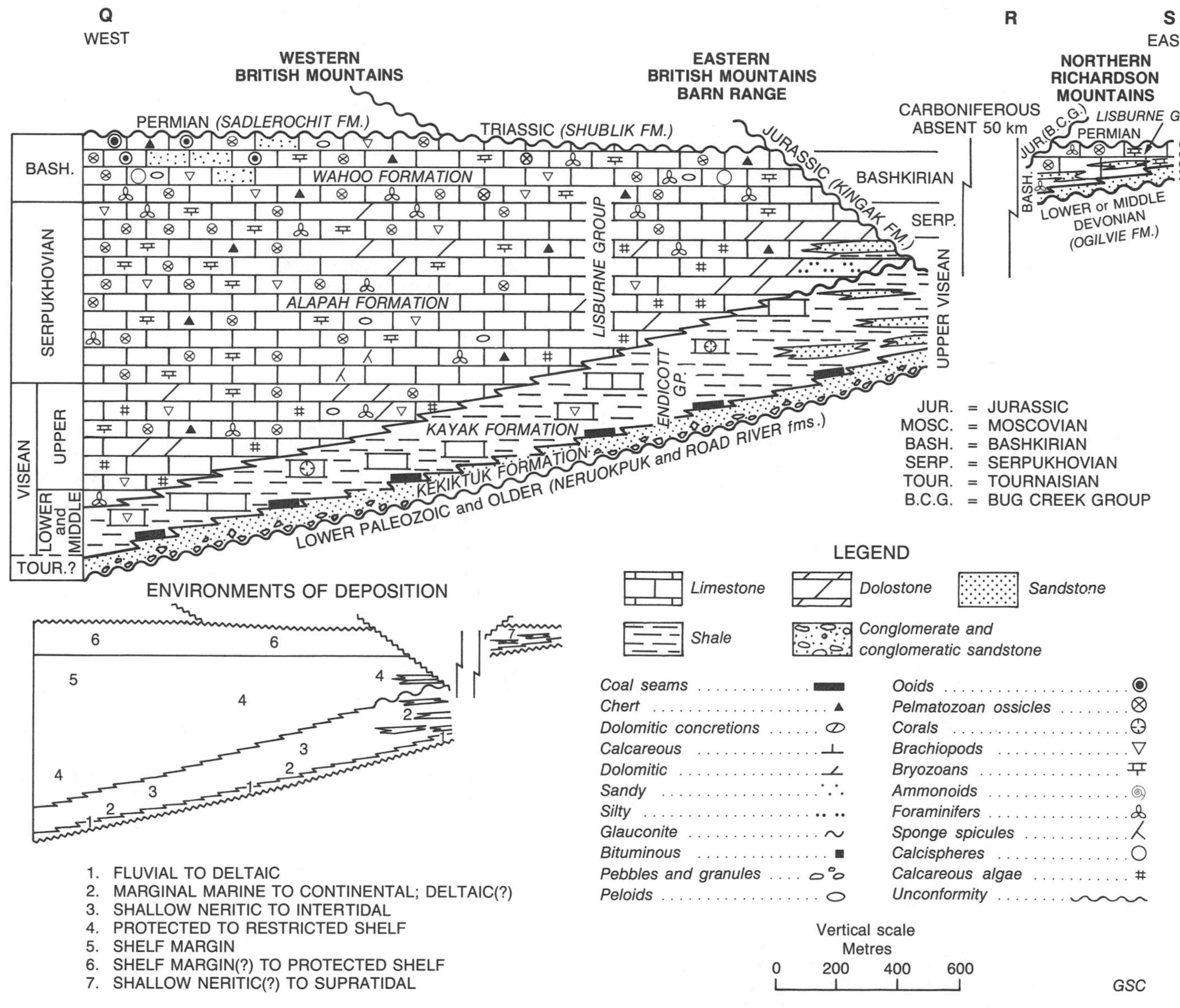

Figure 4E.23. Partly schematic stratigraphic cross-section Q-R-S showing Carboniferous units of northern Yukon Territory and northwestern District of Mackenzie. See Figure 4E.1 for line of section.

Regional lithofacies trends indicate that the Kayak Formation of the Barn Mountains, and the diachronous, coal-bearing siliciclastics of the basal Kayak to the west were deposited in marginal-marine to continental environments that succeeded the fluvial(?) to deltaic(?) environments represented by the underlying Kekiktuk Formation. Kayak shale and carbonates occurring above and westward of the basal and eastern facies were probably deposited in shallow-neritic (above storm wave base) to intertidal environments. The latter lay on the landward side of a protected shelf on which coeval carbonates of the lower Alapah Formation were accumulating.

Alapah Formation

The middle Viséan (V2) to Serpukhovian Alapah Formation (Fig. 4E.3, 4E.23; Bowsher and Dutro, 1957) is the thickest, most extensively preserved Carboniferous carbonate unit of northern Yukon. It is more than 1300 m thick in western British Mountains and thins northeastward. Its thickness and lithological character in the Barn Mountains have not been determined (Bamber and Waterhouse, 1971). In most areas, the Alapah, which overlies the Kayak Formation, is conformably(?) overlain by the Wahoo Formation. North and east of the erosional edge of the latter, Mesozoic strata unconformably overlie the Alapah.

In the British Mountains, the lower Alapah Formation is mainly lime mudstone and wackestone, with abundant calcareous algae. Mixed-skeletal lime packstone and grainstone, containing numerous echinoderm fragments, bryozoans, and foraminifers, dominate the upper part. Grainstone is particularly abundant in the southwest. In the northeastern British Mountains, the Alapah contains abundant finely crystalline dolostone, calcareous algae, and fine-grained, terrigenous clastics. These lithofacies relationships suggest that lower and northeastern strata of the Alapah were deposited in restricted- to protected-shelf environments, with shelf-margin environments developing to the southwest during deposition of the upper Alapah.

Wahoo Formation

The northern Carboniferous succession is capped by Bashkirian limestone of the Wahoo Formation (Fig. 4E.3, 4E.23; Brosgé et al., 1962). In most areas, these carbonates, which conformably(?) overlie the Alapah Formation, are disconformably overlain by siliciclastics and carbonates of the Lower Permian Sadlerochit Formation. North and northeast of the erosional edge of the latter, Mesozoic strata unconformably overlie the Wahoo. The Wahoo ranges in thickness from approximately 130 to 225 m in the British Mountains.

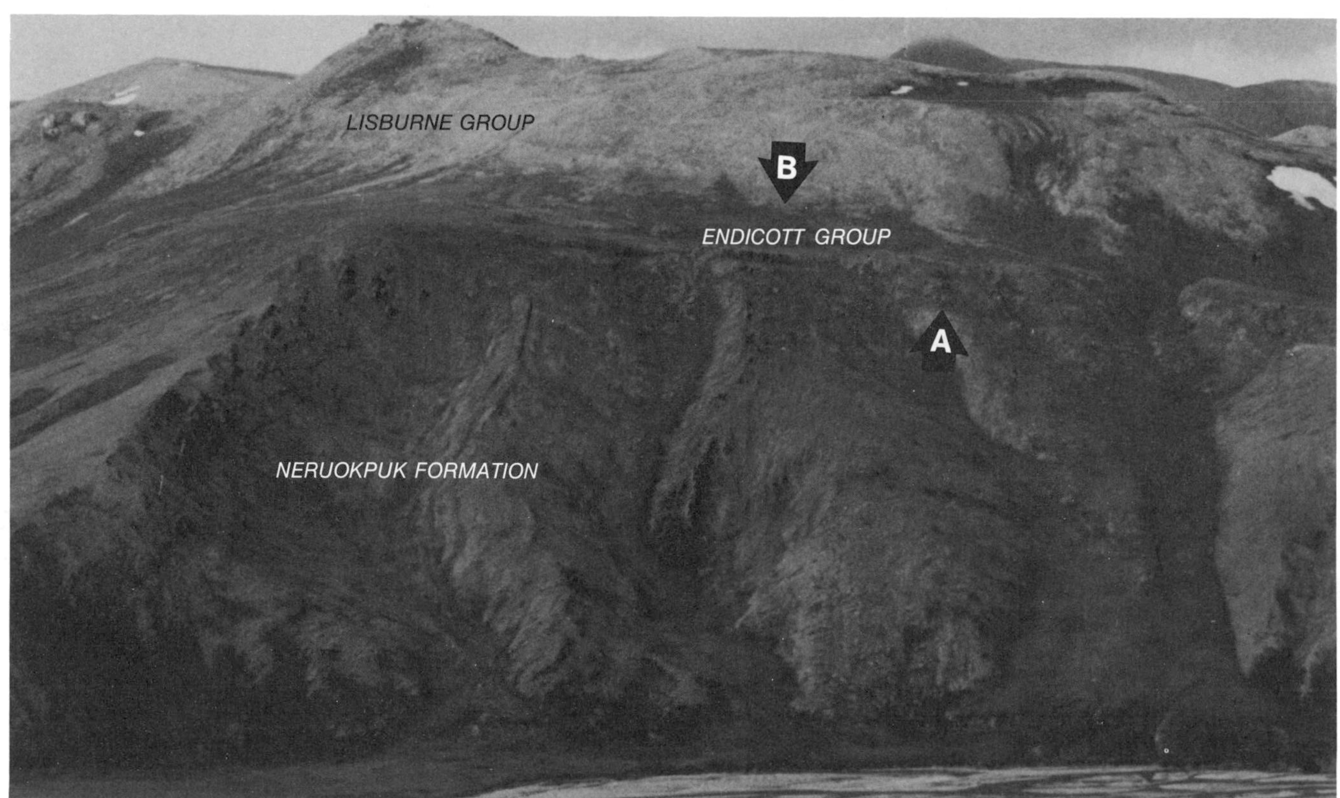

Figure 4E.24. Carboniferous units, British Mountains, northern Yukon Territory. A - base of Endicott Group (Kekiktuk and Kayak formations) and angular unconformity on steeply dipping stata of Lower Paleozoic and Precambrian Neruokpuk Formation; B - base of Lisburne Group (carbonates of Alapah and Wahoo formations). View is to northwest (photo - D.K. Norris, GSC 662-51).

In the British Mountains, the Wahoo Formation is chiefly lime grainstone and packstone rich in echinoderm, bryozoan, and brachiopod fragments. Foraminifers occur throughout, and peloids and calcispheres are locally abundant. Ooid and skeletal-ooid grainstone dominate the upper part in the northwest. In the Barn Mountains, the Wahoo contains skeletal grainstone and packstone of unknown thickness (Bamber and Waterhouse, 1971). The Wahoo carbonate lithofacies and allochems suggest deposition in protected-shelf to shelf-margin environments.

Eastern Lisburne equivalents

East of the Barn Mountains, in the southern part of Rapid Depression (Norris, 1983), the Lisburne Group was removed by regional erosion prior to Jurassic deposition. Farther east, in the northern Richardson Mountains (northwestern District of Mackenzie), the Carboniferous comprises Bashkirian to Moscovian shelf carbonates and subordinate sandstone of the Lisburne Group (undivided) (Fig. 4E.3, 4E.23). These deposits, which resemble those of the Lisburne to the west, disconformably overlie Middle Devonian rocks (Nassichuk and Bamber, 1978) and constitute the most easterly part of the onlapping, northern Carboniferous succession. They occur as erosional remnants beneath the sub-Permian disconformity on the north flank of the Ancestral Aklavik Arch. Prior to late Paleozoic (post-Moscovian) and Mesozoic erosion, however, they were probably continuous with the Lisburne Group to the west. Carboniferous strata of the northern Richardson Mountains are truncated eastward toward the Interior Platform and southward toward the axis of the Ancestral Aklavik Arch.

Depositional summary

Carboniferous depositional patterns and lithofacies relationships in the northern Cordillera record northward regional transgression and onlap of the Yukon Fold Belt, and possible eastward onlap of the cratonic platform. Coarse-grained siliciclastics derived from the Yukon Fold Belt, and possibly from the Ellesmerian highlands to the northeast, constitute the Kekiktuk and Tuttle formations, the basal part of the Kayak Formation, and units of conglomerate and sandstone within the Ford Lake, Hart River, and Blackie formations. The major, transgressive-regressive hemicycles that affected the more southerly parts of the Western Canada Basin have not been recognized in the northern Cordillera.

In the southern area, the onset of regional transgression is marked by Tournaisian(?) basinal shale of the Ford Lake Formation above deltaic deposits of the Tuttle Formation. Carbonates of the overlying Hart River, Blackie, and Ettrain formations were deposited in slope to restricted-shelf environments, which were established in late Viséan (V3) time and persisted throughout the Eagle Plains/Ogilvie Mountains area until Kasimovian(?) time. An increase in the supply of siliciclastics from the northeast is recorded by conglomerate and sandstone in the northeastern tongue of the Blackie Formation. These terrigenous clastics probably also indicate a Serpukhovian to Bashkirian regression. Shelf-carbonate deposition in the southern part of the area was terminated in latest Carboniferous (Kasimovian?) time by a major regression, represented by the regional disconformity below the Kasimovian(?) to Lower Permian Jungle Creek Formation. Local uplift to the north along the Ancestral Aklavik Arch is recorded by conglomerate and sandstone of the Kasimovian(?) to Gzhelian, basal Jungle Creek in the northern Ogilvie Mountains.

In the northernmost part of Yukon Territory, the onset of regional transgression is recorded in the western British Mountains by the uppermost Tournaisian(?) to lower Viséan (uppermost Tn3 to V1), marginal-marine deposits of the basal Kayak Formation. Stratigraphic relationships in the British, Barn, and northern Richardson mountains indicate that transgression and onlap continued toward the east and north until at least Moscovian time. The Carboniferous rocks in the Brooks Range of northeastern Alaska also record this transgression (Brosgé et al., 1962; Bamber and Waterhouse, 1971; Wood and Armstrong, 1975). As in the south (Ogilvie Mountains/Eagle Plains area), sedimentation in the northern area was terminated by a major Late Carboniferous or earliest Permian regression, represented by the sub-Permian disconformity.

BIOSTRATIGRAPHY

The biostratigraphy of the Lower Carboniferous of the Western Canada Basin is now established at a reconnaissance level for several fossil groups. Zonations are available for conodonts, foraminifers, corals, ostracodes, and for a small part of the succession, brachiopods. The earlier Lower Carboniferous zonations were based on brachiopods and corals, which were mainly from the eastern Cordillera of southwestern Alberta and southeastern British Columbia (Brown, 1952; Harker and Raasch, 1958; Nelson, 1961). These two groups are of limited value for correlations with standard successions outside of North America. Also, they occur mainly in shallow-marine lithofacies and, therefore, cannot be used for detailed local correlations between shallow- and deep-marine successions. The first detailed correlation between western Canada and standard Lower Carboniferous successions in Europe was achieved by application of the international foraminiferal zonation of Mamet and Skipp (1970). Biostratigraphic data based on foraminifers, corals, and conodonts were subsequently combined to form an integrated zonal scheme applicable to the eastern Cordillera and Interior Platform of Canada and the western United States (Sando and Bamber, 1985). Recent, unpublished conodont studies for the Lower Carboniferous rocks of western Canada support and amplify established international correlations. In addition, the broad paleoecological distribution of the conodonts allows precise dating of some slope and basin lithofacies and facilitates correlation between shallow-marine and deeper-water lithofacies. Carboniferous zonal schemes used in the study area and their relationships to standard chronostratigraphic units are summarized in Figures 4E.2 and 4E.3.

The Upper Carboniferous succession from central Yukon southward (Fig. 4E.2) is generally thin and incomplete and contains few fossils. Consequently, its biostratigraphy is poorly known in most areas. In northern Yukon Territory, however, where the Upper

Carboniferous succession is richly fossiliferous and relatively complete, detailed zonations based on brachiopods and foraminifers have been established (Bamber and Waterhouse, 1971).

Conodont zonation

The conodont sequence in the Lower Carboniferous succession of the Western Canada Basin has not been described in detail; moreover, few accounts of conodont distribution have been published. Baxter (1972) and Baxter and von Bitter (1979) have applied a modified zonal scheme, which was proposed by Collinson et al. (1971) and is based on Mississippi Valley faunas and sections. The scheme used in the present account (Fig. 4E.2) is more widely applied, being applicable to Canada, the United States, and areas outside of North America. The uppermost Devonian conodont zonation used here was proposed by Ziegler (1962) and modified by Ziegler and Sandberg (1984). Lower Carboniferous siphonodellid-bearing horizons are zoned using the scheme of Sandberg et al. (1978), and Lower Carboniferous post-siphonodellid strata are zoned using the scheme of Lane et al. (1980).

The Devonian/Carboniferous boundary has not been precisely located within western Canada. Late Devonian and earliest Carboniferous conodonts have been collected from the Exshaw, Bakken, Yohin, and Banff(?) formations. However, several internationally recognized conodont zones that normally span the Devonian/Carboniferous boundary are either absent or have not been recognized. For example, the *praesulcata* and *sulcata* zones, which are respectively of latest Devonian and earliest Carboniferous age, have not been found, and the Lower Carboniferous *sandbergi* Zone has not been located. The basal Exshaw at its type section near Canmore, Alberta, has yielded a specimen of *Palmatolepis* referable to either *P. perlobata* or *P. rugosa* (Macqueen and Sandberg, 1970) and assignable to the upper Famennian *expansa* Zone. Richer and more varied faunas belonging to the *expansa* Zone have been collected from the Exshaw in the Alberta subsurface and in the Rocky Mountains of east-central British Columbia (Richards and Higgins, 1988).

Carboniferous conodonts are not common in the Exshaw Formation, but their distribution indicates that the Devonian/Carboniferous boundary occurs within the formation at some localities in southwestern Alberta. For example, at the type section, siphonodellids including *S. cooperi* were reported by Macqueen and Sandberg (1970) from the top of the black shale unit. *S. cooperi* first appears in the Upper *duplicata* Zone and continues upward into the Upper *crenulata-isosticha* Zone (Sandberg et al., 1978), thereby indicating an Early Carboniferous (late Tn1b to Tn2) age. *S. cooperi* has also been reported from the overlying siltstone unit of the Exshaw near the type area (Macqueen and Sandberg, 1970). There are insufficient biostratigraphic data from the Exshaw at its type section and other southern localities to determine the presence or absence of a disconformity at the Devonian/Carboniferous boundary. Moreover, there is no unequivocal evidence for the presence of the lower Tournaisian *praesulcata, sulcata, duplicata*, or *sandbergi* zones. In east-central British Columbia, Upper Devonian faunas of the *expansa* Zone lie either at or near the top of the Exshaw. These faunas occur directly below strata of the Banff Formation that contain the middle Tournaisian (Tn2) Lower *crenulata* Zone, thereby indicating the presence of a disconformity at the top of the Exshaw.

The conodont situation in the Exshaw Formation and basal part of the Banff Formation parallels that of the Bakken Formation (Fig. 4E.2) in Saskatchewan and North Dakota (Hayes, 1985). The lower Bakken belongs to the *expansa* Zone and the upper Bakken to the Lower *crenulata* Zone, but the presence or absence of intervening zones has not been proven.

Some of the conodont faunas can be used as indicators of depositional environment. In Western Canada, the shallow-water and deep-water faunas recognized by Gutschick and Sandberg (1983, Fig. 7D) occur in strata judged on other criteria to be of shallow- or deep-water origin. In general, the stratigraphic ranges of both faunas are sufficiently well known, and contain enough species in common, to allow correlation with standard chronostratigraphic units and between facies.

duplicata *Zone*

The base of the *duplicata* Zone is defined by the first occurrence of *Siphonodella duplicata* morphotype 1 of Sandberg et al. (1978). Characteristic conodont faunas include *Pseudopolygnathus primus, Patrognathus variabilis, Pseudopolygnathus dentilineatus, Polygnathus communis communis, P. inornatus, P. longiposticus*, and *Bispathodus aculeatus aculeatus*. The zone has been recognized by Higgins (in Richards, 1989) only in the Banff(?) Formation and overlying lower Yohin Formation of southwestern District of Mackenzie.

Lower crenulata *Zone*

The Lower *crenulata* Zone is widely present in the lower Banff Formation of western Alberta and eastern British Columbia. It has been locally recognized in the lower Yohin Formation of southwestern District of Mackenzie and the upper Bakken and lower Lodgepole formations of southeastern Saskatchewan. The zone is characterized by the presence of siphonodellids, including *Siphonodella crenulata, S. cooperi, S. quadruplicata*, and *S. obsoleta* together with *Pseudopolygnathus marginatus* and *Elictognathus bialatus*.

Upper crenulata-isosticha *Zone*

The Upper *crenulata-isosticha* Zone has been recognized in the upper Banff Formation and lower Pekisko Formation at numerous localities in the Rocky Mountains of west-central Alberta and east-central British Columbia. Farther south, in the Cordillera east of Banff, Alberta, it has been found locally in the Pekisko Formation and in the upper Banff below the western Pekisko. In southeastern Saskatchewan, the Upper *crenulata-isosticha* Zone has been locally recognized in the Lodgepole Formation. The zone is characterized by *Siphonodella isosticha, S.* cf. *isosticha, Patrognathus variabilis, Anchignathodus penescitulus, Pseudopolygnathus multistriatus,*

Polygnathus communis carina, Gnathodus delicatus, Gnathodus punctatus, and locally a species of *Clydagnathus*.

typicus *Zone*

In the Rocky Mountains of east-central British Columbia and west-central Alberta, the *typicus* Zone is found in the upper Pekisko Formation, formation F, and the lower Shunda Formation. Farther south, in the Rocky Mountains near Canmore, this zone has been locally recognized in the lower to upper parts of the Pekisko Formation, formation F, and basal Livingstone Formation. It is characterized by the disappearance of the siphonodellids and by the presence of *Gnathodus typicus, Eotaphrus bultyncki, Gnathodus semiglaber, Gnathodus cuneiformis*, and *Pseudopolygnathus oxypageus*.

anchoralis-latus *Zone*

The *anchoralis-latus* Zone occurs in upper formation F and the lower Turner Valley Formation in the Rocky Mountains of east-central British Columbia (55°13′N). Farther south (54°45′N to Banff area), it has been recognized in formation F and in the Shunda, lower Turner Valley, and Livingstone formations. The zone is characterized by *Eotaphrus burlingtonensis, Eotaphrus bultyncki, Polygnathus communis carina*, and *Gnathodus typicus* morphon 2 of Lane et al. (1980).

texanus *Zone*

In the Rocky Mountains of east-central British Columbia and west-central Alberta, the *texanus* Zone is in the Turner Valley Formation. Farther south, in the Banff area, it has been recognized in the middle and upper parts of the eastern Livingstone Formation. Species characterizing this zone include *Taphrognathus varians, Anchignathodus scitulus, Polygnathus mehli, Cloghergnathus* spp., and a variety of species of the genus *Apatognathus*. The genus *Apatognathus* is distributed throughout the Lower Carboniferous succession, but it is particularly common in this zone, where some of its species may be of stratigraphic importance. The genus *Cloghergnathus* is restricted to the upper part of the *texanus* Zone, and its presence may be useful in dividing this zone into lower and upper parts.

Cavusgnathus *Zone*

The *Cavusgnathus* Zone has been found in the Mount Head Formation at many localities and locally in the Etherington Formation near Banff. Characteristic species include *Cavusgnathus unicornis, C. characta*, and *Vogelgnathus campbelli*.

Foraminiferal zonation

Fifteen foraminiferal zones have been recognized in the Carboniferous System of the Western Canada Basin. They include zones pre- 7 to 22 of Mamet and Skipp (1970), which range in age from Tournaisian to Bashkirian (Fig. 4E.2, 4E.3). Details of the zones and their distribution have been summarized by Bamber and Mamet (1978), Mamet and Bamber (1979), and Mamet et al. (1986). Foraminifers are generally common and diversified in limestone of upper-slope to restricted-shelf origin (Fig. 4E.4, 4E.5) but rare to absent in deeper water deposits. The association of foraminifers with conodont faunas that extend into deeper water lithofacies allows correlation between shelf and slope to basin deposits and closer correlation between foraminiferal and conodont zonal schemes than was previously possible.

Upper Carboniferous foraminifers occur mainly in the northern part of Yukon Territory, where Mamet and Ross (in Bamber and Waterhouse, 1971) identified zones 20 to 22, and Ross (1967) obtained Moscovian fusulinaceans. Early Moscovian fusulinaceans also occur in the Kananaskis Formation of the Banff region (McGugan et al., 1968; Ross and Bamber, 1978).

Brachiopod zonation

The first brachiopod zonation of the Lower Carboniferous of western Canada was that of Brown (1952), who described three zones from the Banff Formation and one from the lower part of the Mount Head Formation in the Rocky Mountains east of Jasper, Alberta. Parts of the upper Rundle Group and the overlying Spray Lakes Group in southwestern Alberta were zoned by Raasch (1958). His zonation was extended downward to the base of the Banff Formation by Harker and Raasch (1958), who included corals as zonal indices. Nelson (1958, 1960, 1961) described 15 zones, based on both brachiopods and corals, which extended from the base of the Banff to the top of the Etherington Formation in the southern Canadian Rocky Mountains.

The most recent study of the Early Carboniferous brachiopods is that of Carter (1987), which includes a zonation of the Banff Formation and lower Rundle Group in southwestern Alberta. Carter recognized a lower assemblage zone that occurs in the upper Banff of east-central British Columbia and west-central Alberta, and in the Pekisko Formation near Canmore, Alberta. Two overlying range zones lie within the upper Banff and formation F in the Front Ranges of southwestern Alberta.

Early and Late Carboniferous brachiopods were described and illustrated from northern Yukon Territory, where Waterhouse (in Bamber and Waterhouse, 1971) recognized five zones ranging in age from Viséan to Gzhelian. This zonal scheme was modified by Waterhouse and Waddington (1982). Bashkirian brachiopods from southwestern Alberta were illustrated by Bamber and Copeland in Bamber et al. (1970).

Coral zonation

Corals are common in some middle-slope to shallow-marine carbonates of the uppermost Banff Formation and overlying Rundle Group, and were used in joint brachiopod-coral zonal schemes erected by Harker and Raasch (1958) and Nelson (1960, 1961). Coral zonation data by Bamber (in Macqueen and Bamber, 1968) and Macqueen et al. (1972) contributed to a major publication on the coral zonation of the Lower Carboniferous of western North America (Sando and Bamber, 1985). In that report, six Oppel zones and

11 subzones were erected. Five of the zones and nine subzones, ranging in age from middle Tournaisian (Tn2) to Serpukhovian, are recognized in Canada. The zonation applies to the eastern Cordillera and Interior Plains and is integrated with the foraminiferal zonation of Mamet and Skipp (1970).

Ostracode zonation

Latest Devonian ostracodes have been described from subsurface samples from Alberta and Saskatchewan by Lethiers (1978) and Coquel et al. (1976). From the Exshaw Formation and lower part of the Bakken Formation, they described ostracodes belonging to the *Shemonaella* Zone of Strunian (latest Devonian) age. No ostracode faunas have been recovered from earliest Carboniferous strata of the Exshaw and Bakken formations.

The first comprehensive study of Early Carboniferous ostracodes was that of Green (1963), who described faunas from the Banff Formation in the Rocky Mountains of southern Alberta. He recognized two major faunal associations within the Banff. Earlier, Loranger (1958) recognized three broad ostracode zones in the Lower Carboniferous strata of western Canada, and Copeland (1960) noted the occurrence of ostracodes in the Banff Formation near Blairmore, southern Alberta.

The most recent comprehensive study of Early Carboniferous ostracodes is that of Crasquin (1984). Using faunas from five sections in the Rocky Mountains of southwestern Alberta and east-central British Columbia, she established five ostracode zones extending from the lower Banff Formation to the lower Etherington Formation. Ostracodes are most common in shallow-water lithofacies and may be absent from many deeper water deposits.

Ammonoids

In western Canada, Carboniferous ammonoids are generally rare and known mainly from Lower Carboniferous strata. Their distribution has been summarized by Mamet and Bamber (1979), who showed that few genera have been identified to specific level. Detailed comparison with Europe and the United States, at the species level, has been made only in the upper Viséan strata of northern Yukon Territory (Bamber and Waterhouse, 1971) and in northeastern British Columbia (Bamber et al., 1968), where *Goniatites crenistria* and *Goniatites granosus* have been identified.

Late Carboniferous goniatites are known only from northern Yukon Territory, where a Late Carboniferous (Kasimovian?) ammonoid, *Metapronorites stelcki* was described from the Ogilvie Mountains (Nassichuk, 1971).

Pollen and spores

No detailed palynological zonation is available for the Carboniferous System of western Canada, but Carboniferous microfloras have been studied from several parts of the succession.

In southwestern District of Mackenzie, Van der Zwan and Walton (1981) investigated the variation and vertical distribution of the *Cyrtospora cristifera* morphon in the Upper Devonian to Lower Carboniferous deposits of the Banff, Yohin, Clausen, and Flett formations. McGregor (in Richards, 1989) recorded spores, including *Retispora lepidophyta*, of latest Devonian age (Tn1a or early Tn1b) in the lower to middle Banff. Similar assemblages occur to the south in the Exshaw Formation and lower to middle Bakken Formation (Coquel et al., 1976). In southwestern District of Mackenzie, spores of earliest Carboniferous age (upper Tn1b and Tn2) are found in the upper Banff Formation and in the Yohin Formation. J. Utting (in Richards, 1989) records *Auroraspora macra, Cyrtospora cristifera, Densosporites anulatus, Densosporites spitsbergensis, Dibolisporites distinctus, Grandispora echinata, Spelaeotriletes* sp. cf. *S. pretiosus,* and *Vallatisporites banffensis* in the assemblages. Many of these species continue into the overlying Clausen Formation (Tournaisian Tn2-Tn3) in which *Tripartites incisotrilobus* occurs. In the middle to upper parts of the overlying Flett Formation (Viséan, V1 to V2), *Lycospora pusilla* and *Murospora aurita* appear. Assemblages from the overlying Golata Formation resemble those that Staplin (1960) reported from the type section in west-central Alberta.

Hacquebard and Barss (1957), Sullivan (1965), and Braman and Hills (1977) studied microfloras from the Mattson Formation of southwestern District of Mackenzie. In the lower to middle Mattson (Viséan, V3), species of *Convolutispora, Densosporites, Diatomozonotriletes, Knoxisporites, Lycospora, Murospora*, and *Punctatisporites* commonly occur with rare specimens of *Rotaspora fracta*. *Grandispora spinosa* and *Schulzospora* spp. appear in the upper Mattson. According to J. Utting (in Richards, 1983) the presence in the uppermost Mattson of *Costatascyclus crenatus* and *Florinites* spp. suggests a Serpukhovian age.

Pollen and spores from the upper Blackie Formation (Bashkirian to Kasimovian?) and the overlying basal Jungle Creek Formation (Lower Permian) in the Tatonduk River section, northern Yukon Territory, were summarized by Bamber and Barss (1969), who assigned an Early Permian age to the assemblages. Subsequently, a Late Carboniferous age was assigned to the Blackie and lowest Jungle Creek in the same section, on the basis of marine faunas (Bamber and Waterhouse, 1971; Waterhouse and Waddington, 1982). Recent work on Upper Carboniferous assemblages from the northern part of Yukon Territory (Ettrain Formation and correlatives) indicates the presence of a varied microflora. From the lower Ettrain upward to the lower Jungle Creek, there are gradual quantitative changes in the assemblages (J. Jerzykiewicz, pers. comm., 1986). For example, the relative proportion of spores (especially triletes) gradually decreases upwards, whereas the proportion of pollen (especially monosaccates and disaccates) increases. The stratigraphically important genus *Vittatina* appears in the upper Ettrain and upper Blackie formations. No dramatic qualitative or quantitative changes occur at the Carboniferous/Permian boundary in the lower part of the Jungle Creek Formation.

Paleobiogeography

Differences in the degree of endemicity of different fossil groups, and differences of opinion among authors render impossible a unified summary of the provincial position of western Canada throughout the Carboniferous Period. Nevertheless, indications of the position are beginning to emerge.

Mamet and Skipp (1979) recognized three foraminiferal faunal realms for the Lower Carboniferous. A Tethyan Realm extends from North Africa through northern Europe, the Russian Platform, Malaya, and Vietnam to northwestern and possibly southeastern Australia. The Tethyan faunas are rich, highly diversified, and show a high degree of endemism. A Taimyr/Alaska Realm includes the Taimyr, Lena, and Kolyma areas, and the Omolon Massif in the Soviet Union, and the arctic of North America. Faunas of this second realm show lower diversity than those in the Tethyan realm and little endemism. The third realm is the Kuznetz/North American Realm, which is intermediate between the first two.

Distributions of other Early Carboniferous faunas in the Western Canada Basin suggest that they occurred in realms similar to those outlined above. For example, Carter (1987) records great similarity between brachiopods of the Banff Formation in the southern Rocky Mountains in Canada, and those of Kuznetz. In contrast, Late Carboniferous brachiopods from northern Yukon Territory were shown by Waterhouse (in Bamber and Waterhouse, 1971) to be more closely similar to those of the Urals than to other North American faunas.

Sando et al. (1977), recognized a Western Interior Coral Province that included the western United States and western Canada, but excluded northern and southeastern Alaska, northern Yukon Territory, northwestern Washington, central Oregon, and northern California. Corals of the western Interior Province are more closely related to those of northern Yukon Territory and northern Alaska than to those elsewhere in North America. No comparison was made with areas outside North America.

Conodont faunas were essentially cosmopolitan across the equatorial belt in Carboniferous time. There are, however, differences in diversity between North American faunas and those of Europe, North Africa, China, and the Russian Platform and the Urals. In areas other than northern Yukon Territory, Early Carboniferous conodont faunas of North America are less diverse than on other continents. Late Carboniferous conodont faunas of Yukon Territory have not been studied, but those of the Canadian Arctic Islands are closer in both generic and specific composition to those of the Russian Platform and the Urals than to those of the rest of North America.

Crasquin (1984) recognized three Lower Carboniferous ostracode provinces. The widespread *Bairdia brevis* Province occurs in the eastern Cordillera and Interior Plains of North America, northern Siberia, Kazakhstan, and northwestern Europe. It is divided into: a subprovince extending over the Interior Plains and eastern Cordillera of North America; a second subprovince that includes northern Siberia and Kazakhstan; and a third subprovince that covers northwestern Europe. The second province, the *Graphiadactyllis* Province, is found only in the Appalachian region of North America. The third province, the *Hollinella* Province, is in North Africa, western Turkey, and the Moscow Basin.

What emerges from this survey is the unity of Early Carboniferous faunas in the eastern Cordillera and Western Interior Plains both in the United States and Canada. This unity is shown by the foraminifers, corals, brachiopods, ostracodes, and conodonts. These faunas, which differ, in varying degrees, from those of equivalent age in eastern Canada and northwestern Europe, show greater affinities with faunas from the Kuznets Basin. Early Carboniferous faunas of northern Yukon Territory are not well known; however, the known faunas show affinity with those of both the Alaska-Taimyr region and more southerly faunas of western North America.

Early Carboniferous microfloras from northern Yukon to southwestern District of Mackenzie are similar to those described by Playford (1962, 1963) from Spitsbergen, and may be assigned to the *Lophozonotriletes* Region of Van der Zwan (1981). This was the tropical humid belt of north Euramerica and extended from the Soviet Union to Spitsbergen. Towards the south, evaporites become abundant, indicating hot arid conditions. Microfloras from Atlantic Canada have more in common with western European microfloras (Utting, 1980, 1987). The latter belong to the *Vallatisporites* Region of Van der Zwan (1981), which included the subtropical dry belt of south Euramerica from eastern Canada, Ireland, Great Britain, and Denmark.

TECTONIC HISTORY

Latest Famennian (*expansa* conodont zone) to earliest middle Tournaisian (earliest Tn2, *sandbergi* conodont zone)

During latest Famennian to earliest middle Tournaisian time, regional subsidence and local rifting, followed by local uplift, occurred in most of the Western Canada Basin. In northernmost Yukon Territory and northwestern District of Mackenzie, however the Yukon Fold Belt was developing during late stages of the Ellesmerian Orogeny. Prophet Trough was forming but not clearly differentiated from the cratonic platform and included the poorly defined Peace River Embayment. In the south, the intracratonic Williston Basin was connected to Prophet Trough by a seaway developed on the unstable craton of southern Alberta and north-central Montana. Precambrian basement rocks and coarse-grained Devonian siliciclastics, remnants of the formerly extensive Peace River Arch (Douglas et al., 1970; P.F. Moore, Subchapter 4D in this volume) remained as islands in northeastern British Columbia and northwestern Alberta, where the Peace River Embayment was beginning to develop from block faulting and flexural subsidence in eastern Prophet Trough. The islands persisted until deposition of the lower Banff Formation. In the western United States, back-arc(?) thrusting during the Antler Orogeny emplaced the Roberts Mountain Allochthon on the western side of the craton (Fig. 4E.6). The orogenic belt and related Antler Foreland Basin to the east extended from southern California to central Idaho and possibly into southern British Columbia (Poole, 1974; Dickinson, 1977; Nilsen, 1977; Johnson and Pendergast, 1981).

South of northern Yukon Territory, widespread subsidence occurred in Prophet Trough and on the cratonic platform. The subsidence, of both flexural and fault-block character, is recorded by the upper Famennian to lowest middle Tournaisian (lowest Tn2) Exshaw and Bakken formations and partly coeval deposits in the Besa River Formation and northern part of the Banff Formation (Fig. 4E.2). The basal deposits of the upper Earn Group in east-central Yukon Territory and west-central District of Mackenzie (Fig. 4E.26) may record this event as well. In eastern Prophet Trough, subsidence was greatest from the axis of the western part of Peace River Embayment into southwestern District of Mackenzie and southeastern Yukon Territory. Slow subsidence, possibly followed by minor uplift, apparently occurred on the cratonic platform and in Williston Basin. This is suggested by the veneer of deposits present and by the occurrence of regional disconformities below the Banff and upper Bakken formations (Fig. 4E.2, 4E.8, 4E.10). Part of the deepening and subsequent shallowing recorded by the Exshaw Formation and its correlatives may have resulted from eustatic events, as Late Devonian glaciations have been documented in South America (Caputo, 1985). In the western United States, emplacement of the Roberts Mountain Allochthon greatly restricted the epicontinental sea, creating widespread, euxinic, marine conditions (Poole, 1974). A related event may have produced the euxinic environment in which black shale of the Exshaw Formation was deposited.

Paleontological and lithostratigraphical evidence (Orchard, 1985; Klepacki and Wheeler, 1985) indicate that uplifted and folded, subaerially exposed, lower Paleozoic strata of Kootenay Terrane (Fig. 4E.25) lay on the southwestern side of Prophet Trough in southeastern British Columbia. Farther north, uppermost Devonian to Lower Carboniferous conglomerates occur locally in the Cariboo and Omineca Mountains of east-central British Columbia, Cassiar Mountains, and the Warneford River area of northeastern British Columbia (Struik, 1981; Sutherland Brown, 1963; Monger and Ross, 1984; Gordey et al., 1987), and the Selwyn Mountains of central Yukon Territory (Gordey et al., 1982) (Fig. 4E.26). Their presence also indicates that uplifted, subaerially exposed terrain was at least locally present either along or adjacent to the western rim of Prophet Trough. Sandstone and conglomerate in the lower Besa River Formation of east-central British Columbia provide additional evidence.

Marine tuff (rich in quartz and K feldspar) and bentonite are found locally in the black shale unit of the Exshaw Formation in southwestern Alberta and east-central British Columbia. Their presence records latest Devonian volcanism that was probably related to the calc-alkaline plutonism occurring in Kootenay Terrane on the western rim of Prophet Trough.

After deposition of the regressive upper Exshaw and middle Bakken formations, widespread differential uplift took place in southern Prophet Trough and on the extreme western side of the southern cratonic platform, while subsidence continued to the northwest. In the eastern Rocky Mountains from Canmore, Alberta, to 52°30′N, the uplift is recorded by northwestward erosional thinning of the Exshaw. Farther north in the Rocky Mountains, but south of latitude 55°N, the Exshaw was completely eroded. The underlying Famennian Palliser Formation was also partly to completely removed from this northern positive area (Figs. 4E.2, 4E.10; Geldsetzer, 1982).

The Ellesmerian Orogeny is recorded in northern Yukon Territory and northern Alaska by the Yukon Fold Belt (Bell, 1973; Norris, 1973; Nilsen, 1981) and the deposits of northern Prophet Trough. Convergence between North America and another plate was the probable cause (Trettin, 1991); however, the nature of the tectonic element that converged with North America is uncertain. The main sedimentological record for the orogeny in northern Yukon Territory is the thick flysch of the Upper Devonian Imperial Formation, overlying conglomeratic deposits of the Tuttle Formation, and shale and sandstone of the Ford Lake Formation (Fig. 4E.3, 4E.22). Northern Prophet Trough resembles a foreland basin, because it contains this thick succession of terrigenous clastics and resulted partly from synorogenic subsidence.

Early middle (early Tn2, lower *crenulata* conodont zone) to earliest late Tournaisian (earliest Tn3, basal foraminiferal zone 8)

Lower middle to lowest upper Tournaisian strata record local block faulting and two periods of major flexural subsidence. Both episodes of subsidence were followed by periods characterized by low to moderate subsidence rates. The principal tectonic elements were those of the previous period. During deposition of the Pekisko Formation in late middle and early late Tournaisian time (late Tn2 and early Tn3), however, Peace River Embayment became differentiated from Prophet Trough as the eastern margin of the trough stepped basinward in the region north of the embayment axis.

The first period of subsidence occurred during early middle Tournaisian (early Tn2) time and is recorded by transgressive deeper water deposits of the lower Banff Formation and coeval deposits in the Lodgepole and Besa River formations (Fig. 4E.2). Marked subsidence took place in most of Prophet Trough, central Williston Basin, and on the western cratonic platform. In eastern to central Prophet Trough, which was not clearly demarcated from the cratonic platform, subsidence was greatest in the northeast-trending half-graben(?) north of the site of the Cambrian land mass Montania and in Peace River Embayment. In southwestern District of Mackenzie, however, subsidence rates in the trough were lower than during the previous period, as indicated by shallowing upward trends in the Yohin and Banff formations. The unstable craton of north-central Montana and southern Alberta subsided substantially, forming a moderately deep seaway between Williston Basin and Prophet Trough/Antler Foreland Basin. Block faulting accompanied subsidence of the seaway in Montana (Smith, 1982, p. 53). Trends on structural contour maps and pronounced local changes in the thickness and lithology of the Banff Formation (Fig. 4E.11; Sikabonyi and Rodgers, 1959; Lavoie, 1958) suggest that subsidence in Peace River Embayment, which was not a well defined topographic feature at this time, also resulted partly from extensive normal faulting. Following the downwarping, regional

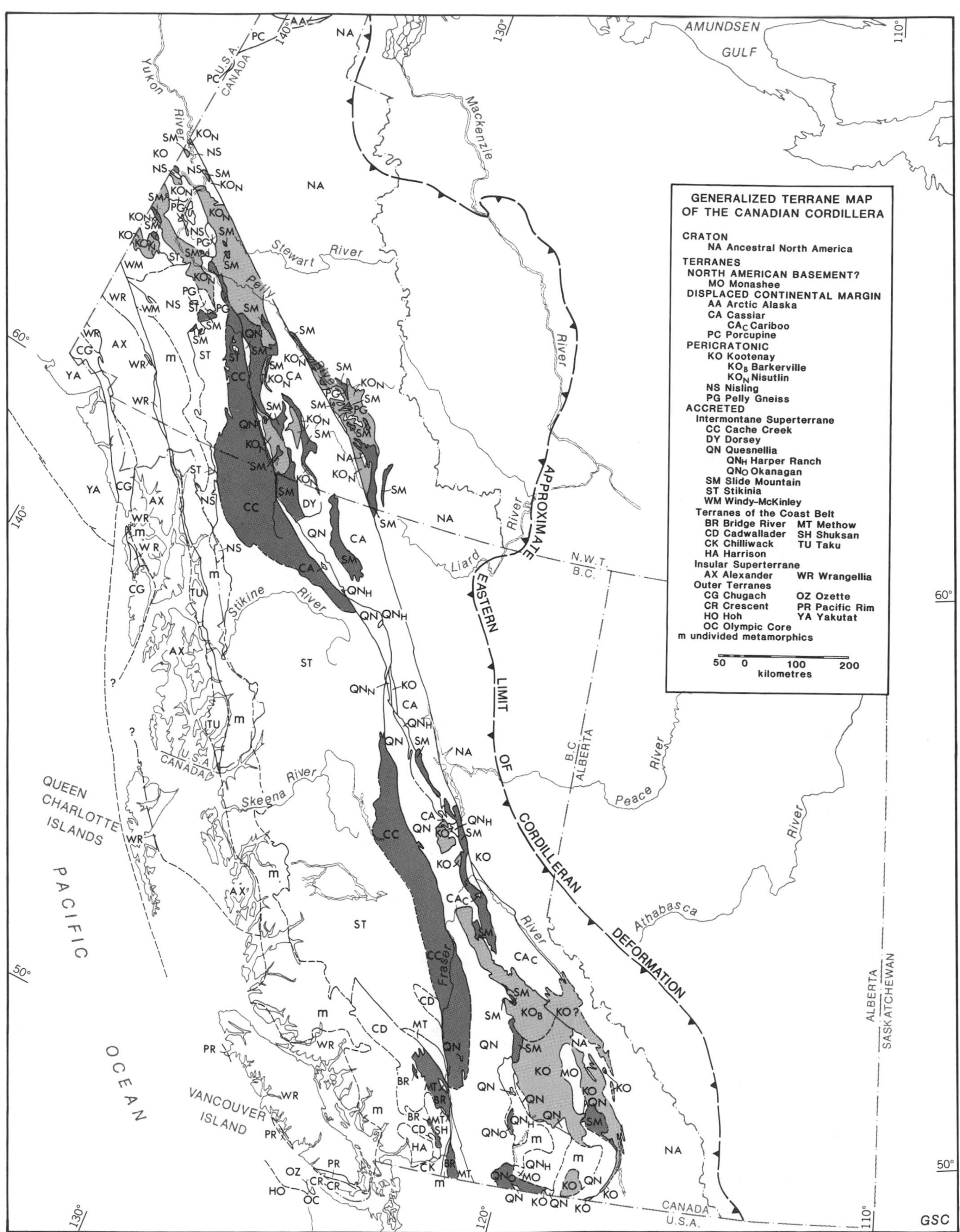

Figure 4E.25. Simplified lithotectonic terrane map of Western Canada (modified from Wheeler et al., 1988).

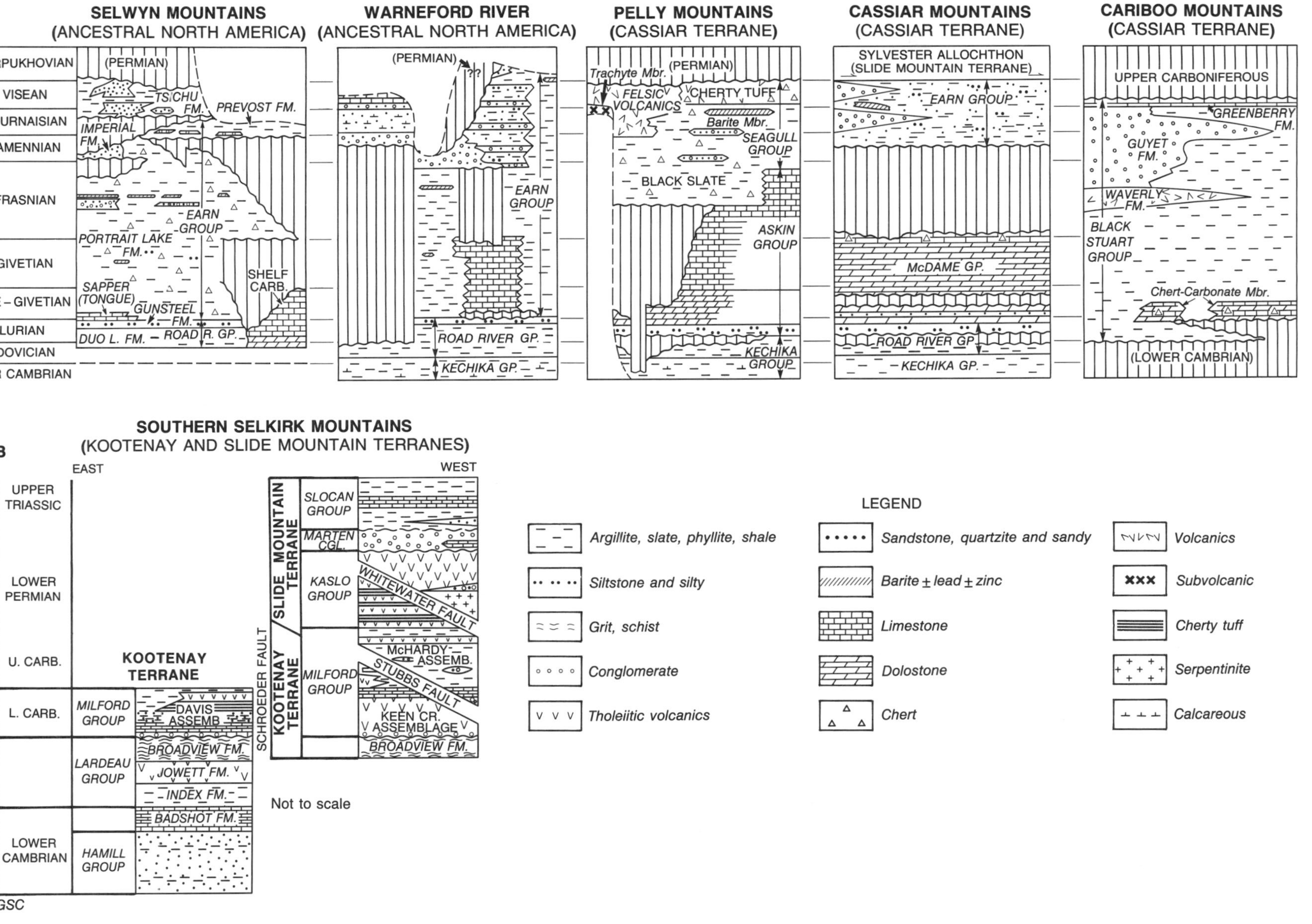

Figure 4E.26. Stratigraphic relationships and correlation of Devonian and Carboniferous units in lithotectonic terranes of the eastern Cordillera. A (after Gordey et al., 1987) illustrates deposits of western Prophet Trough; B (after Klepacki and Wheeler, 1985) shows the Milford Group of southeastern British Columbia. The Davis and Keen Creek assemblages of the Milford were deposited on the western rim of Prophet Trough. The McHardy assemblage of the Milford and the overlying Kaslo Group formed in a marginal basin directly to the west.

subsidence rates decreased, as indicated by a shallowing upward trend, which continued until late middle Tournaisian (late Tn2) time.

The second major episode of subsidence occurred during late middle to earliest late Tournaisian time (late Tn2 to earliest Tn3, upper foraminiferal zone 7 to lower zone 8) and is recorded by the Pekisko Formation and partly coeval strata in the Clausen, Banff, and Lodgepole formations and formation F (Fig. 4E.2). Marked subsidence occurred in central Williston Basin and in most of Prophet Trough, with maximum subsidence in the trough taking place in the same areas as during the previous episode and also in southwestern District of Mackenzie. The unstable craton (seaway) of southernmost Alberta and northern Montana continued to subside, but the stable cratonic platform did not subside appreciably. As a consequence of the differential subsidence, Peace River Embayment and the boundary between Prophet Trough and the cratonic platform became relatively well defined.

Subsidence also took place in Prophet Trough between southwestern District of Mackenzie and the Yukon Fold Belt during early middle to earliest late Tournaisian time (early Tn2 to earliest Tn3) or possibly slightly later. In the Selwyn Mountains, this is indicated by the deep-water character of the lower upper Earn Group. Farther north, in the Ogilvie Mountains/southern Eagle Plain region, marked subsidence is recorded by superposition of the Ford Lake shale on the Tuttle Formation (Fig. 4E.2).

Most of the Yukon Fold Belt was exposed during early middle to earliest late Tournaisian time. Gradual northeastward onlap was probably beginning in the south (Fig. 4E.23), suggesting slow subsidence.

Lower Carboniferous marine strata preserved in the Kootenay, Cassiar, and Slide Mountain terranes of the Omineca Belt (Fig. 4E.25, 4E.26) (Gordey et al., 1987; Monger, 1977; Monger and Ross, 1984), record marked subsidence in western Prophet Trough and the marginal basin to the west. These deposits in the Kootenay and Cassiar terranes include conglomerate (Struik, 1981; Orchard and Struik, 1985; Monger and Ross, 1984), indicative of subaerially exposed terrane on or adjacent to the western rim of the trough. The conglomerate, best represented by the Guyet Formation in the Cariboo Mountains (Sutherland Brown, 1963; Struik, 1981) and upper Earn Group of the Warneford River area and Cassiar Mountains (Fig. 4E.26), was generally deposited as eastward thinning clastic wedges in shallow to moderately deep marine settings (Gordey et al., 1987) of western Prophet Trough. In the Kootenay Terrane of the southern Selkirk Mountains, conglomerate and associated deposits of the Milford Group (Fig. 4E.25, 4E.26) record eastward onlap of a broad uplifted region. This positive feature, partly developed on the lower Paleozoic Purcell Arch (see Douglas et al., 1970), formed part of the western rim of Prophet Trough. Lithofacies in the lower part of the onlap succession are of middle Tournaisian age or older in the west and of early Serpukhovian age in the east (Orchard, 1985; Klepacki and Wheeler, 1985). Bedded chert and tholeiitic pillow basalt of the Slide Mountain Terrane record deep-water oceanic conditions west of the rim of Prophet Trough (Struik and Orchard, 1985; Struik, 1987).

At this time in the western United States, a flysch trough developed in the Antler Foreland Basin directly east of the Antler Orogenic Belt. Marked foreland subsidence and deepening, coinciding with rising global sealevel, occurred in the foreland basin east of the flysch (Poole, 1974; Johnson and Pendergast, 1981; Gutschick et al., 1980). Tectonic loading may have caused subsidence in southern Canada as well.

Early late Tournaisian (early Tn3) through middle Viséan (V2)

From early late Tournaisian to latest middle Viséan time, moderate to low rates of flexural subsidence apparently predominated, but local block faulting and uplift occurred. Williston Basin and Peace River Embayment were well developed, and the boundary between Prophet Trough and the cratonic platform was moderately well defined. Paleontological and lithostratigraphic evidence from the Milford Group in the Kootenay Terrane indicates that part of the western rim of southern Prophet Trough was still exposed. In the Cariboo Mountains to the northwest, however, the presence of the upper Tournaisian to lower Viséan (Tn3 to V1) Greenberry limestone above Tournaisian conglomerate of the Guyet Formation (Orchard and Struik, 1985) indicates that part of the western rim in east-central British Columbia was transgressed. Part of the rim possibly was still exposed farther to the northwest, where it apparently supplied gravel for the Earn Group of the Cassiar Mountains (Fig. 4E.26) and sand for the Viséan Keno Hill quartzite near Dawson, west-central Yukon Territory.

In most areas, subsidence rates were substantially lower than during earliest late Tournaisian time (earliest Tn3). In Williston Basin and in the succession of southern to central Prophet Trough, this is indicated by the presence of a relatively continuous sequence (upper Madison Group, Livingstone, lower Mount Head, Debolt, and Flett formations) showing marked basinward progradation of shallow-marine carbonates over basin and slope deposits (Lodgepole, Banff, Clausen, Prophet, and Besa River formations) (Fig. 4E.8, 4E.9, 4E.12, 4E.13). Although lower than during the previous period, subsidence rates in most of the trough remained moderately high into early Viséan time (V1), as recorded by thick successions. On the stable cratonic platform, slow subsidence is recorded by a succession (Shunda, Turner Valley, lower Mount Head, and Debolt formations) dominated by shallow-marine to supratidal carbonates but lacking major unconformities. In northern Yukon Territory, regional subsidence resulted in the slow northeastward onlap of the Yukon Fold Belt by the Endicott Group (Fig. 4E.23).

Subsidence was greatest in central Williston Basin, in the axial region of Peace River Embayment, and in Prophet Trough. In the latter, subsidence rates increased southwestward and were greatest in the northeast-trending trough or half-graben(?) north of the site of the Cambrian landmass Montania (Fig. 4E.1), and in southwestern District of Mackenzie and environs. Marked subsidence also occurred in northern Prophet Trough as recorded by thick basin deposits of the Ford Lake Formation (Fig. 4E.22). In Peace River Embayment, trends on structural contour maps and abrupt changes in thickness (Lavoie, 1985; Sikabonyi and Rodgers, 1959) suggest subsidence accompanied by widespread normal faulting.

Several areas were uplifted slightly. Deposits of this age in the Livingstone and lower Mount Head formations are anomalously thin southwest of Fernie, southeastern British Columbia, suggesting that the Montania block was slightly positive. On the unstable craton of southeastern Alberta and environs, a broad uplift in the vicinity of Sweetgrass Arch rose and restricted Williston Basin. To the northwest, local uplift and erosion are recorded by the disconformity between the Debolt Formation and overlying Stoddart Group in the southwestern part of Peace River Embayment. Farther west, in the eastern Rocky Mountains of British Columbia (54°25′N to 55°05′N), the Rundle Group records westward shallowing. Thus a northwest-trending positive feature was present in that part of the western area in which substantial uplift and erosion had occurred during earliest middle Tournaisian time (earliest Tn2).

In the western United States, uplift occurred in the Antler Orogenic Belt, which was the source of coarse clastics in a flysch succession in the western Antler Foreland Basin. East of the flysch, the foreland basin subsided substantially, with great subsidence occurring during latest Tournaisian time (latest Tn3). Middle Viséan (V2, foraminiferal zone 12/13 boundary) epeirogenic uplift and a drop in sealevel exposed most of the cratonic platform east of the foreland basin to karstification (Gutschick et al., 1980). In southern Canada, the epeirogenic event may be recorded by the regressive upper Salter Member of the Mount Head Formation and the disconformity below the overlying Loomis Member (Fig. 4E.2).

Late Viséan (V3) to early late Serpukhovian (lower foraminiferal zone 19)

The block faulting and slow regional subsidence that characterized the previous period continued until early late Serpukhovian time. Block faulting was more widespread, however, and there was a massive influx of terrigenous clastics. Volcanism, continuing into the Late Carboniferous, was common in the west. Most of the principal tectonic elements of the previous period persisted, but Peace River Embayment became more clearly differentiated from Prophet Trough as the eastern hinge of the trough stepped basinward in the region north of the embayment. Paleontological and stratigraphic evidence (Orchard, 1985) from the Milford Group (Fig. 4E.26), deposited on the western rim of southern Prophet Trough and to the west, indicate that part of the area that constituted a subaerially exposed rim in southernmost British Columbia during most of Early Carboniferous time subsided and was transgressed from the west. Rifting accompanied the transgression (D. Klepacki, pers. comm., 1986). In addition, the apparent absence of post-Viséan, westerly derived siliciclastics in Prophet Trough indicates that most of its western rim was transgressed in pre-Serpukhovian time. In the western United States, the Antler Orogenic Belt continued to supply terrigenous clastics to the Antler Foreland Basin (Poole, 1974; Johnson and Pendergast, 1981).

Lithofacies and isopachs of the Kibbey Formation (Fig. 4E.2) and overlying formations in the Big Snowy Group indicate that Williston Basin subsided slowly, but more rapidly than the surrounding cratonic platform, until early late Serpukhovian time (Sheldon and Carter, 1979; Sando et al., 1975). The basin was generally well developed in Montana and North Dakota, but probably extended only into southernmost Canada.

Elsewhere in the Western Canada Basin, but south of northern Yukon Territory, the upper depositional unit and its correlatives (Fig. 4E.2) record slow subsidence rates on the western cratonic platform and moderate to high rates in Prophet Trough and Peace River Embayment. Flexural downwarping was at least locally accompanied by block faulting. Thickness and lithofacies trends indicate that subsidence was greatest in the northeast-trending half-graben(?) north of the Cambrian landmass Montania, in the axis of the Peace River Embayment, and in Prophet Trough in southwestern District of Mackenzie and areas directly adjacent to the south and west. In Prophet Trough, subsidence generally increased westward.

Extensive block faulting accompanied sedimentation of the upper depositional unit in and possibly north of Peace River Embayment (Lavoie, 1958; Sikabonyi and Rodgers, 1959). In the axis of the embayment, late Viséan (V3) block faulting is indicated by numerous, marked, local thickness changes in the Kiskatinaw Formation where it underlies the Taylor Flat Formation. These changes occur across normal faults. Carboniferous block faulting also took place in the embayment either during or after deposition of the Taylor Flat. The thickness of the latter, where it is unconformably overlain by Permian (lower Asselian) strata, varies greatly across normal faults. In northeastern British Columbia and southwestern District of Mackenzie, block faulting, which occurred prior to deposition of the Lower Permian sediments and may have been partly syndepositional, is indicated by the pronounced westward increase in the thickness of the Mattson Formation across the Bovie normal fault (Fig. 4E.12).

The upper Viséan (V3) to lower upper Serpukhovian succession from southwestern District of Mackenzie southward contains numerous shallowing-upward hemicycles and diastems. Most of these probably resulted from eustatic events rather than an oscillatory style of tectonism, because there were numerous, relatively minor fluctuations in global sea level at that time (Ramsbottom, 1973, 1977; Ross and Ross, 1985).

In northern Yukon Territory and northwestern District of Mackenzie, differential flexural subsidence apparently predominated. Subsidence rates must have been moderately high in the southwest (northern Ogilvie Mountains and Eagle Plain), where moderately deep-water deposits of the Blackie Formation were deposited (Fig. 4E.22). To the north, slower subsidence is suggested by the occurrence of shallow-marine lithofacies (Kayak and Alapah formations) that slowly onlapped the Yukon Fold Belt northeastward (Fig. 4E.23).

Submarine volcanic rocks, preserved in the Milford Group, were deposited on the western rim of southern Prophet Trough, and record substantial extension-related subsidence. Viséan volcanics in the Cassiar Terrane of the Pelly Mountains (Fig. 4E.26) apparently record local (?) submarine volcanism in western Prophet Trough.

Submarine volcanism, which continued into the Permian Period, was widespread in the marginal basin to the west, as recorded by the Slide Mountain Terrane (Monger, 1977; Monger and Ross, 1984; Struik and Orchard, 1985).

A major influx of northerly to northeasterly derived terrigenous clastics is represented by upper Viséan and Serpukhovian sandstone in the south (Fig. 4E.2, 4E.9, 4E.12, 4E.13) and by coeval terrigenous deposits in the northern Cordillera (Fig. 4E.3, 4E.22, 4E.23). Numerous minor fluctuations in global sea level took place during the interval. Consequently, either a change in sediment dispersal patterns or uplift in northern source areas, rather than a major eustatic drop in sea level, is conceived as the main cause of the influx. The Ellesmerian Fold Belt in the Arctic Islands may have been the main source of the clastics (Richards, 1989); that region was undergoing extensive block faulting and had high relief (Kerr, 1981).

Latest Serpukhovian (upper foraminiferal zone 19) through Bashkirian

Latest Serpukhovian and Bashkirian time was characterized by periods of broad differential uplift and erosion in the south and gradual subsidence in northern Yukon Territory and northwestern District of Mackenzie. Most of the main tectonic elements of the previous period persisted, but Peace River Embayment may not have been present, as most of that region lacks documented occurrences of Upper Carboniferous strata and was subjected to post-early Serpukhovian uplift and erosion. The boundary between Prophet Trough and the cratonic platform was poorly defined. An unnamed, northwest-trending positive area developed from the southwestern part of Peace River Embayment to southern Prophet Trough (Fig. 4E.1). A massive influx of westerly(?) derived quartz sand occurred in southern Prophet Trough (Fig. 4E.20), while deposition of northeasterly derived terrigenous clastics continued in the north (Fig. 4E.22).

During latest Serpukhovian time, a major eustatic drop in sea level took place (Saunders and Ramsbottom, 1986), and the entire Williston Basin area was subjected to subaerial erosion and possibly some uplift (Sando et al., 1975). Erosion was greatest on the basin margins and on the surrounding cratonic platform (Sheldon and Carter, 1979). The Williston Basin area of Canada lacks stratigraphic evidence for a Late Carboniferous depositional basin, but the extensive Wyoming Basin developed to the south (Sando et al., 1975).

During latest Serpukhovian (upper foraminiferal zone 19) and Bashkirian time, two main episodes of uplift and erosion occurred on the southwestern cratonic platform and in southern Prophet Trough. The first of these took place during latest Serpukhovian and possibly continued into early Bashkirian time (lower foraminiferal zone 20). It is recorded by the regional disconformity that underlies the Spray Lakes Group and truncates the Etherington Formation northwestward (Fig. 4E.2, 4E.20). Part of the erosion probably resulted from the mid-Carboniferous eustatic event referred to above, but the northwestward truncation of the Etherington indicates differential uplift. The second period of uplift, during late Bashkirian time, is represented by the sub-Tobermory disconformity, which truncates the Storelk, Tyrwhitt, and Etherington formations eastward (Fig. 4E.20; Scott, 1964). Both events probably influenced the region between the northern zero edge of the Spray Lakes Group and 54°N, but post-Serpukhovian erosion has removed most of the record.

Late Carboniferous uplift and erosion also took place farther northwestward, but south of the axis of Peace River Embayment. This is recorded by the hiatus between Permian strata and underlying middle Viséan (V2) and lower Tournaisian (Tn1b) strata (Fig. 4E.1, 4E.2, 4E.11). The southwestward truncation of the Stoddart and Rundle Groups indicates that uplift increased southwestward. Normal faulting probably accompanied the uplift, because the amount of pre-Permian erosion increases abruptly southwestward in the Front Ranges at 55°10′N.

Bashkirian sandstone of the Spray Lakes Group and similar coeval sandstone of the Quadrant Formation in southwestern Montana may record uplift on the western side of Prophet Trough/Antler Foreland Basin. The great thickness of the Quadrant sandstone in western Montana, its rapid eastward depositional thinning, and its overlapping and intertonguing with eastern carbonates suggest an extensive source area to the west or northwest (Wardlaw and Pecora, 1985, Figs. 1, 3; Maughan, 1975). Part of that source area may have been the Copper Basin Highland of east-central Idaho, which resulted from uplift of the Antler flysch belt (Skipp et al., 1979). Alternatively, the main provenance of both units may have been the Antler Belt or the uplifted region northwest of the present northern erosional edge of the Spray Lakes Group, where the Rundle and Stoddart groups were subjected to deep erosion. The Antler Orogenic Belt, which was undergoing initial transgression, may have been broken up and partly rejuvenated by an event (Dickinson, 1977) related to uplift of the Ancestral Rocky Mountains. The main uplifts and extensional basins, southeast and east of the Antler Belt, were caused by sinistral intraplate shearing during the Ouachita-Marathon Orogeny (Kluth and Coney, 1981; Budnick, 1986).

In northern Yukon Territory and northwestern District of Mackenzie, gradual, differential, flexural subsidence predominated. Subsidence rates must have been moderately rapid in the southwest (northern Ogilvie Mountains and Eagle Plain), because moderately deep-water environments, represented by the Blackie Formation (Fig. 4E.22), prevailed. Conglomerate and sandstone units in the easternmost Blackie may have resulted from both the mid-Carboniferous eustatic drop and uplift in the Ellesmerian Fold Belt. Farther north (in the British and Barn mountains), low subsidence rates are recorded by the predominance of shallow-marine lithofacies (upper Alapah and Wahoo Formation) that onlapped the Yukon Fold Belt northeastward (Fig. 4E.23).

Moscovian through Gzelian

Little is known of Moscovian to Gzhelian tectonism south of the northern part of Yukon Territory. The Moscovian Kananaskis Formation of the southern Rocky Mountains (Figs. 4E.2, 4E.20) records regional transgression in southern Prophet Trough. Westward deepening suggests that the transgression resulted at least partly from regional subsidence. The absence of post-Moscovian deposits and the presence of a major sub-Lower Permian unconformity suggests that appreciable post-Moscovian subsidence did not take place; instead, there was periodic differential uplift.

In northern Yukon Territory and northwestern District of Mackenzie, regional subsidence was followed by differential uplift and erosion. The subsidence that characterized the previous period continued into Moscovian time and is recorded by Moscovian marine lithofacies onlapping toward the northeast and east (Fig. 4E.23). Subsidence rates were probably lower than during Bashkirian time because marked regional shallowing occurred. In the northern Ogilvie Mountains, and possibly in part of the region to the north and east, subsidence continued into Kasimovian(?) time, as recorded by the upper Blackie and Ettrain formations (Fig. 4E.22). During Kasimovian(?) and Gzhelian time, the Ancestral Aklavik Arch, which was well developed by Early Permian time, started to rise, as indicated by Kasimovian(?) to Gzhelian conglomerate in the basal Jungle Creek Formation. Broader deformation may also have taken place, with uplift increasing northward and eastward as suggested by the pre-Permian differential erosion of the Upper Carboniferous succession (Fig. 4E.22, 4E.23).

RESOURCES

The Carboniferous succession of the Western Canada Basin contains substantial accumulations of oil and gas, ranking third in total recoverable reserves, after the Devonian and Cretaceous. Natural gas constitutes approximately 90% of the Carboniferous hydrocarbon reserves. Sulphur is next in economic importance, and limestone is the other main resource. Potentially economic coal seams are present.

Oil and gas traps

All of the producible oil and gas reserves are in Lower Carboniferous strata; moreover most occur in formations of the middle depositional unit east of the Rocky Mountains (Fig. 4E.1, 4E.2). Dolomitized lime grainstone and packstone are the main reservoir rocks, but a wide range of other carbonate rock types, and sandstone, contain oil and gas pools. Most of the reserves are in unconformity-related traps and in traps that are primarily structural (Podruski et al., 1988). The unconformity traps lie along the northern and eastern erosional subcrop edges of Lower Carboniferous formations (Smith, 1980; Boreski, 1978; Young, 1967). They resulted from regional, late Paleozoic and Mesozoic tilting and concomitant erosion, which caused formations to dip southward in Manitoba and Saskatchewan and southwestward beneath the Interior Platform of Alberta and British Columbia. Seals for the unconformity traps are mainly formed by overlying Mesozoic strata. In much of Williston Basin, oil is not found directly at the unconformity, but is trapped down dip by anhydrite-occluded porosity. Underlying and overlying beds of restricted-marine carbonates and evaporites commonly form part of the seals. The structural traps are formed by Laramide folds and thrust faults in the foothills (Hennessey et al., 1982; Gordy et al., 1975), fault blocks and related folds in the Peace River Embayment region, normal faults beneath the western Interior Platform, structures resulting from dissolution of Devonian salt in Williston Basin (Kent, 1984, Fuzesy, 1973), and a suspected meteorite impact structure (Sawatzky, 1972) in Williston Basin. Reservoir seals are similar to those of the unconformity traps, but also include overlying Permian strata in the Peace River region, and fault zones. Stratigraphic traps related to depositional facies or diagenetic changes occur, but tectonism commonly played some role in their development.

Oil

About 18% of the proven conventional oil reserves of the Western Canada Basin occur in the Lower Carboniferous succession. It is the principal oil exploration target in the Canadian part of Williston Basin, but is of lesser importance elsewhere. Approximately 94% of the oil is in the middle depositional unit (Podruski et al., 1988).

East of Sweetgrass Arch, most oil pools are found in unconformity-related traps in the Williston Basin of southwestern Manitoba and southeastern Saskatchewan (Kent, 1984). Little of this oil occurs west of 105°00'W and south of the Coleville area, because of late Paleozoic and Mesozoic erosion. In Manitoba, most oil pools are in the Lodgepole Formation, from which production started during 1951 (McCabe, 1963). In Saskatchewan, the first Paleozoic oil was discovered during 1951 in the middle Bakken Formation at Coleville (Reasoner and Hunt, 1958). Since the early 1950s, the main focus of exploration has been on unconformity- and structure-related traps of southeastern Saskatchewan, where most reserves are in shelf carbonates of the Mission Canyon Formation and lower Charles Formation (Fig. 4E.2, 4E.7; Smith, 1980; Fuzesy, 1966, 1973; Podruski et al., 1988). Additional reserves are found in sandstone of the middle Bakken Formation and shelf carbonates of the Lodgepole Formation (Christopher, 1961; Kent, 1984). The source is believed to be mainly bituminous shale and carbonates of the Lodgepole, but the oil in some pools probably migrated from the Bakken (Osadetz and Snowdon, 1986a, b).

Most of the recoverable oil reserves in Carboniferous rocks west of Sweetgrass Arch are in southern Alberta, where the Turner Valley Formation (Fig. 4E.2), mainly fossiliferous shelf grainstone and packstone, is the most important producing unit (Podruski et al., 1988). Major fields in the formation are Harmattan East, Harmattan-Elkton, Sundre (Hemphill, 1957), and Turner Valley. Oil pools in the Turner Valley Formation, with the exception of the two in the Turner Valley Field, lie in unconformity-related traps. The gas cap of the latter Carboniferous field was discovered in 1924, and the oil leg in 1936 (Gordy et al., 1975). This field, the first substantial discovery of conventional oil in the Western Canada Basin, occurs in a faulted overthrust anticline (Gallup, 1954) and is the only major oil field in the gas-dominated foothills belt. The Pekisko Formation, mainly shelf grainstone, is the second most important producing unit in southern Alberta. Most of the reserves occur in unconformity traps between Drumheller and Rocky Mountain House, but the Pekisko-edge trend extends from Sweetgrass Arch to southern Peace River Embayment. Major fields are Twining (Boreski, 1978), Three Hills Creek, and Sylvan Lake. Laramide faults and anticlines form traps in grainstone of the Livingstone Formation on the Sweetgrass Arch (Podruski et al., 1988), where the main fields are Del

Bonita (Humphreys, 1955), Red Coulee, and Regan. Small oil pools are in unconformity-related traps in Banff carbonates from Sweetgrass Arch to the southern flank of the Carboniferous Peace River Embayment.

In the gas-dominated Peace River region of northwestern Alberta and northeastern British Columbia, oil is entrapped in distributary channel sandstone of the Kiskatinaw Formation and in dolostone of the Debolt Formation. The Debolt pools are in fault-related traps of the Blueberry Field (Law, 1981). On the northeast flank of the embayment, pools were recently discovered in mud mounds of Waulsortian type in the Pekisko Formation and formation F (Davies et al., 1988).

Farther northwest in northeastern British Columbia, several small oil pools lie in shelf carbonates of the Pekisko and Debolt formations, and formation F. They are found in stratigraphic, structural, and unconformity traps from the Pekisko subcrop edge to the foothills belt. This trend, discovered during 1983, includes the Desan and Tooga fields.

Beneath Eagle Plain in northern Yukon Territory, oil accumulations occur in structural traps developed in sandstone of the Hart River Formation (Fig. 4E.3, 4E.22; Graham, 1973) but have not been produced.

Gas

Most of the proven gas reserves are found in carbonates of the middle depositional unit. The major fields are in southwestern Alberta, but some commercial reserves occur in Williston Basin and other areas. In Williston Basin, most gas occurs in solution in oil of the Mission Canyon Formation and lower Charles Formation; no gas caps or fields of unassociated gas are known (Beach, 1968). In the foothills of Alberta, gas is produced from structural traps in the Turner Valley and Livingstone formations. Fields in the trend, discovered during 1924 at Turner Valley, include the giant Waterton (Gordy et al., 1977) and Turner Valley fields, and Savanna Creek, Jumping Pound West, and Jumping Pound (Rupp, 1969; Davidson, 1975). With the exception of the Turner Valley Field, most gas in this trend is not associated with oil. East of the foothills, reserves are mainly in unconformity-related traps of the Turner Valley and Pekisko formations. Unconformity traps of the Turner Valley and partly equivalent Elkton Member of the Debolt (Fig. 4E.2) extend from the Calgary region to Edson, the largest accumulations being the giant Edson and Brazeau River fields. The first major discovery in this trend was the Harmattan-Elkton Field in 1954 (Prather and McCourt, 1968). In most fields of this trend, gas is not associated with oil; important exceptions are the Harmattan-Elkton and neighboring fields. East of this trend, the main gas reserves occur in the Pekisko Formation. The proven reserves in this unit are predominantly nonassociated gas, and the main fields are Three Hills Creek and Minnehik-Buck Lake. Important exceptions include the Sylvan Lake and Twining North fields, where the gas overlies oil. In Peace River Embayment and northwestward, numerous small pools lie in structural traps in dolostone of the Debolt Formation (Law, 1981) and sandstone of the Kiskatinaw Formation. In northern Yukon Territory, structural traps in sandstone of the Hart River Formation contain significant gas pools (Graham, 1973) that have not been produced.

Sulphur

Sulphur, derived from hydrogen sulphide, is an important resource in the Carboniferous succession, which contains about 25% of the Western Canada Basin's proven reserves. Most of the hydrogen sulphide is produced from sour gas pools in structural traps in the Turner Valley and Livingstone formations of southwestern Alberta. Important producing fields include Waterton, Jumping Pound, and Jumping Pound West (Gordy et al., 1977; Davidson, 1975).

Coal

Coal seams, ranging in thickness from less than 0.1 m to 1.4 m, are common in the Mattson Formation of southwestern District of Mackenzie and are locally present in the Mattson of southeastern Yukon Territory and northeastern British Columbia. Most seams are not areally extensive, and those with economic potential occur only in eastern outcrops of the lower and middle Mattson. The reflectances (0.71 to 1.14 Romax) and rank (mainly high volatile A or B) are relatively low, indicating only moderate depth of burial and/or a moderate geothermal gradient (A. Cameron, pers. comm., 1982). Hacquebard and Barss (1957) presented chemical and petrographic data from a coal seam in the lower Mattson at Jackfish Gap.

In northern Yukon Territory, the Kayak Formation contains coal seams in and near the British and Barn mountains and in unnamed Tournaisian and Viséan correlatives of the Ford Lake Formation in Peel Plateau (Bamber and Waterhouse, 1971). The thickest known seam, about 5.5 m thick, occurs in the Kayak southeast of the Barn Mountains. It consists of good quality, anthracite thermal coal with less than 7 per cent ash and average sulphur content of 0.5 per cent (Cameron et al., 1986).

Limestone

Limestone, used in the manufacture of Portland cement and lime, is a major resource in the eastern Rocky Mountains of Alberta. Relatively pure, high-calcium limestone of the Rundle Group has been quarried in the Bow Valley east of Canmore and in Crowsnest Pass near Blairmore, southwestern Alberta. Production in the Canmore area began in 1906 and continues today at the Continental Lime Limited quarry on Grotto Mountain. In Crowsnest Pass, where production commenced in 1909, several quarries have been operated and abandoned, but production continues at the quarries of Summit Lime Works Limited (Goudge, 1946; Holter, 1976).

REFERENCES

Armstrong, A.K. and Mamet, B.L.
1977: Carboniferous microfacies, microfossils, and corals, Lisburne Group, Arctic Alaska; United States Geological Survey, Professional Paper 849, 144 p.

Bamber, E.W. and Barss, M.S.
1969: Stratigraphy and palynology of a Permian section, Tatonduk River, Yukon Territory; Geological Survey of Canada, Paper 68-18, p. 1-39.

Bamber, E.W., Bolton, T.E., Copeland, M.J., Cummings, L.M., Frebold, H., Fritz, W.H., Jeletzky, J.A., McGregor, D.C., McLaren, D.J., Norford, B.S., Norris, A.W., Sinclair, G.W., Tozer, E.T., and Wagner, F.J.E.
1970: Biochronology: standard of Phanerozoic time; in Geology and economic minerals of Canada, R.J.W. Douglas (ed.); Geological Survey of Canada, Economic Geology Report No. 1, p. 593-674.

Bamber, E.W. and Macqueen, R.W.
1979: Upper Carboniferous and Permian stratigraphy of the Monkman Pass and southern Pine Pass areas, northeastern British Columbia; Geological Survey of Canada, Bulletin 301, 27 p.

Bamber, E.W., Macqueen, R.W., and Ollerenshaw, N.C.
1981: Mississippian stratigraphy and sedimentology, Canyon Creek (Moose Mountain), Alberta; in Field Guide to Geology and Mineral Deposits, Calgary '81 Annual Meeting, R.I. Thompson and D.G. Cook (ed.); Geological Association of Canada, Mineralogical Association of Canada, Canadian Geophysical Union, p. 177-194.

Bamber, E.W., Macqueen, R.W., and Richards, B.C.
1984: Facies relationships at the Mississippian carbonate platform margin, western Canada; in Part 3: Sedimentology and Geochemistry, E.S. Belt and R.W. Macqueen (ed.); Neuvième Congrès International de Stratigraphie et de Géologie du Carbonifère, 1979; Compte Rendu, v. 3, p. 461-478.

Bamber, E.W. and Mamet, B.L.
1978: Carboniferous biostratigraphy and correlation, northeastern British Columbia and southwestern District of Mackenzie; Geological Survey of Canada, Bulletin 266, 65 p.

Bamber, E.W. and Waterhouse, J.B.
1971: Carboniferous and Permian stratigraphy and paleontology, northern Yukon Territory, Canada; Bulletin of Canadian Petroleum Geology, v. 19, p. 29-250.

Bamber, E.W., Taylor, G.C., and Procter, R.M.
1968: Carboniferous and Permian stratigraphy of northeastern British Columbia; Geological Survey of Canada, Paper 68-15, 25 p.

Baxter, S.
1972: Conodont biostratigraphy of the Mississippian of western Alberta and adjacent British Columbia, Canada; Ph.D. thesis, Ohio State University, Columbus, Ohio, 185 p.

Baxter, S. and von Bitter, P.H.
1979: Conodont succession in the Mississippian of southern Canada; in Biostratigraphy, P.K. Sutherland and W.L. Manger (ed.); Neuvième Congrès International de Stratigraphie et de Géologie du Carbonifère, 1979; Compte Rendu, v. 2, p.253-264.

Beach, F.K.
1968: Geology of natural gas - Saskatchewan Paleozoic; in Natural Gases of North America, B.W. Beebe and B.F. Curtis (ed.); American Association of Petroleum Geologists, Memoir 9, p. 1285-1287.

Beauchamp, B., Richards, B.C., Bamber, E.W., and Mamet, B.L.
1986: Lower Carboniferous lithostratigraphy and carbonate facies, upper Banff Formation and Rundle Group, east-central British Columbia; in Current Research, Part A; Geological Survey of Canada, Paper 86-1A, p. 627-644.

Bell, J.S.
1973: Late-Paleozoic orogeny in the northern Yukon; in Canadian Arctic Geology, J.D. Aitken and D.J. Glass (ed.); Proceedings of the Symposium on the geology of the Canadian Arctic, Geological Association of Canada-Canadian Society of Petroleum Geologists Symposium, 1973, p. 23-28.

Bjorlie, P.F.
1979: The Carrington shale facies (Mississippian) and its relationships to the Scallion subinterval in central North Dakota; North Dakota Geological Survey, Report of Investigation no. 67, 46 p.

Boreski, C.V.
1978: Sedimentation and diagenesis of the Twining Oil Field, Alberta; M.Sc. thesis, University of Calgary, Calgary, Alberta, 163 p.

Bowsher, A.L. and Dutro, J.T., Jr.
1957: The Paleozioc section in the Shainin Lake area, central Brooks Range, Alaska; United States Geological Survey, Professional Paper 303-A, 39 p.

Braman, D.R. and Hills, L.V.
1977: Palynology and paleoecology of the Mattson Formation, northwest Canada; Bulletin of Canada Petroleum Geology, v. 25, p. 582-630.

Brosgé, W.P., Dutro, J.T., Jr., Mangus, M.D., and Reiser, H.N.
1962: Paleozoic sequence in eastern Brooks Range, Alaska; American Association of Petroleum Geologists Bulletin, v. 46, p. 2174-2198.

Brown, R.A.C
1952: Carboniferous stratigraphy and paleontology in the Mount Greenock area, Alberta, Canada; Geological Survey of Canada, Memoir 264, 119 p.

Budnick, R.T.
1986: Left-lateral intraplate deformation along the Ancestral Rocky Mountains; implications for late Paleozoic plate motions; Tectonophysics, v. 132, p. 195-214.

Cameron, A.R., Norris, D.K., and Pratt, K.C.
1986: Rank and other compositional data on coals and carbonaceous shale of the Kayak Formation, northern Yukon Territory; in Current Research, Part B; Geological Survey of Canada, Paper 86-1B, p. 665-670.

Campbell, R.B.
1967: Reconnaissance geology of Glenlyon map area, Yukon Territory; Geological Survey of Canada, Memoir 352, 92 p.

Caputo, M.V.
1985: Late Devonian glaciation in South America; Palaeogeography, Palaeoclimatology, Palaeoecology, v. 51, p. 291-317.

Carter, J.L.
1987: Lower Carboniferous brachiopods from the Banff Formation of western Alberta; Geological Survey of Canada, Bulletin 378, 183 p.

Chamberlain, C.K.
1978: Recognition of trace fossils in cores; in Trace Fossil Concepts, P.B. Basan (ed.); Society of Economic Paleontologists and Mineralogists, Short course No. 5, Oklahoma City, p. 133-183.

Chatellier, J.Y.
1983: Sedimentology of the Mississippian Banff Formation in southwestern Alberta mountains and plains; M.Sc. thesis, University of Calgary, Calgary, Alberta, 159 p.
1984: Sedimentologie de la Formation Banff et des formations à son toit et son mur (Dévonien supérieur et Carbonifère inferieur, Alberta, Canada); Thèse de troisieme cycle, Mémoires des Sciences de la Terre, Université Pierre et Marie Curie, Académie de Paris, Paris, France, 171 p.

Christopher, J.E.
1961: Transitional Devonian-Mississippian formations of southern Saskatchewan; Saskatchewan Mineral Resources Report 66, 103 p.

Collier, A.J. and Cathcart, S.H.
1922: Possibility of finding oil in laccolithic domes south of the Little Rocky Mountains, Montana; United States Geological Survey, Bulletin 736-F, p. 171-178.

Collinson, C., Rexroad, C.B., and Thompson, T.L.
1971: Conodont zonation of the North American Mississippian; in Symposium on Conodont Biostratigraphy, W.C. Sweet and S.M. Bergstrom (ed.); Geological Society of America, Memoir 127, p. 353-394.

Copeland, M.J.
1960: A Kinderhook microfauna from Crowsnest Pass, Alberta; Royal Society of Canada Transactions, ser. 3, v. 55, p. 37-43.

Coquel, R., Loboziak, S., and Lethiers, F.
1976: Répartition de quelques ostracodes et palynologie à la limite Dévono-Carbonifère dans l'ouest Canadien; Actes du 101e Congrès National des Societés Savantes, Lille, 1976, Sciences, fasc. I, p. 69-84.

Cotter, E.
1965: Waulsortian-type carbonate banks in the Mississippian Lodgepole Formation of central Montana; Journal of Geology, v. 73, p. 881-888.

Craig, L.C.
1972: Mississippian System; in Geologic Atlas of the Rocky Mountain Region, W.W. Mallory (ed.); Rocky Mountain Association of Geologists, p. 100-110.

Crasquin, S.
1984: Ostracodes du Dinantien: Systématique, Biostratigraphie, Paléoécologie (France, Belgique, Canada); Thèse de troisième cycle, l'Université des Sciences et Techniques de Lille, Lille, France, 2 volumes, 238 p.

Davidson, J.
1975: Jumping Pound and Sarcee gas fields; in Structural Geology of the Foothills between Savanna Creek and Panther River, S.W., Alberta, Canada, H.J. Evers and J.E. Thorpe (ed.); Canadian Society of Petroleum Geologists-Canadian Society of Exploration Geophysicists, Exploration Update '75, Guidebook, p. 30-34.

Davies, G.R., Edwards, D.E., and Flach, P.
1988: Lower Carboniferous (Mississippian) Waulsortian reefs in the Seal area of north-central Alberta; in Reefs, Canada and Adjacent Areas, H.H.J. Geldsetzer, N.P. James, and G.E. Tebbutt (ed.); Canadian Society of Petroleum Geologists, Memoir 13, p. 643-648.

Dickinson, W.R.
1977: Paleozoic plate tectonics and the evolution of the cordilleran continental margin; in Paleozoic Paleogeography of the Western United States, J.H. Stewart, C.H. Stevens, and A. Eugene Fritsche (ed.); Pacific Section, Society of Economic Paleontologists and Mineralogists, Pacific Coast Paleogeography Symposium 1, p. 137-155.

Douglas, R.J.W.
1958: Mount Head map-area Alberta; Geological Survey of Canada, Memoir 291, 241 p.

Douglas, R.J.W. (ed.)
1970: Geology and Economic minerals of Canada; Geological Survey of Canada, Economic Geology Report no. 1, 838 p.

Douglas, R.J.W., Gabrielse, H., Wheeler, J.O., Stott, D.F., and Belyea, H.R.
1970: Geology of Western Canada; in Geology and Economic Minerals of Canada, R.J.W. Douglas (ed.); Geological Survey of Canada, Economic Geology Report no. 1, p. 366-488.

Douglas, R.J.W. and Norris, D.K.
1959: Fort Liard and La Biche map-areas, Northwest Territories and Yukon; Geological Survey of Canada, Paper 59-6, 23 p.
1960: Virginia Falls and Sibbeston Lake map-areas, Northwest Territories 95F and 95G; Geological Survey of Canada, Paper 60-19, 26 p.

Dutro, J.T., Jr.
1979: Introduction; in Carboniferous of the Northern Rocky Mountains, J.T. Dutro, Jr. (ed.), Field Trip no. 15, Ninth International Congress of Carboniferous Stratigraphy and Geology; American Geological Institute, p. 1-4.

Easton, W.H.
1962: Carboniferous formations and faunas of central Montana; United States Geological Survey, Professional Paper 348, 126 p.

Edie, R.W.
1958: Mississippian sedimentation and oil fields in southeastern Saskatchewan; in Jurassic and Carboniferous of Western Canada, A.J. Goodman (ed.); American Association of Petroleum Geologists, Allan Memorial Volume, p. 331-363.

Evenchik, C.A., Parrish, R.R., and Gabrielse, H.
1984: Precambrian gneiss and late Proterozoic sedimentation in north-central British Columbia; Geology, v. 12, p. 233-237.

Folk, R.L.
1974: Petrology of Sedimentary Rocks; Hemphill Co., Austin, Texas, 182 p.

Fuller, J.G.C.M.
1956: Mississippian rocks and oilfields in southeastern Saskatchewan; Saskatchewan Department of Mineral Resources, Report 19, 72 p.

Fuzesy, L.M.
1960: Correlation and subcrops of the Mississippian strata in southeastern and south-central Saskatchewan; Saskatchewan Department of Mineral Resources, Report 51, 63 p.
1966: Geology of the Frobisher-Alida beds; Saskatchewan Department of Mineral Resources, Report 104, 59 p.
1973: The geology of the Mississippian Ratcliffe beds in south-central Saskatchewan; Department of Mineral Resources, Saskatchewan Geological Survey, Report 163, 63 p.

Gallup, W.B.
1954: Geology of Turner Valley oil and gas field, Alberta, Canada; in Western Canada Sedimentary Basin, L.M. Clark (ed.); American Association of Petroleum Geologists, Ralph Leslie Rutherford Memorial Volume, p. 397-414.

Geldsetzer, H.H.J.
1982: Depositional history of the Devonian succession in the Rocky Mountains southwest of the Peace River Arch; in Current Research, Part C; Geological Survey of Canada, Paper 82-1C, p. 55-64.

Gordey, S.P., Abbott, J.G., and Orchard, M.J.
1982: Devono-Mississippian (Earn Group) and younger strata in east-central Yukon; in Current Research, Part B; Geological Survey of Canada, Paper 82-1B, p. 93-100.

Gordey, S.P., Abbott, J.G., Tempelman-Kluit, D.J., and Gabrielse, H.
1987: "Antler" clastics in the Canadian Cordillera; Geology, v. 15, p. 103-107.

Gordy, P.L., Frey, F.R., and Norris, D.K.
1977: Geological guide for the CSPG 1977 Waterton-Glacier Park field conference; Canadian Society of Petroleum Geologists, 93 p.

Gordy, P.L., Frey, F.R., and Ollerenshaw, N.C.
1975: Road log Calgary - Turner Valley - Jumping Pound - Seebe; in Structural Geology of the Foothills between Savanna Creek and Panther River, S.W. Alberta, Canada, H.J. Evers and J.E. Thorpe (ed.); Canadian Society of Petroleum Geologists-Canadian Society of Exploration Geophysicists, Exploration Update '75, Guidebook, p. 37-61.

Goudge, M.F.
1946: Limestones of Canada, their occurrence and characteristics; Canada Department of Mines and Resources, Mines and Geology Branch, Report 811, 233 p.

Graham, A.D.
1973: Carboniferous and Permian stratigraphy, southern Eagle Plain, Yukon Territory, Canada; in Canadian Arctic Geology, J.D. Aitken and D.J. Glass (ed.); Geological Association of Canada-Canadian Society of Petroleum Geologists Symposium, 1973, p. 159-180.

Green, L.H.
1972: Geology of Nash Creek, Larsen Creek, and Dawson map-areas, Yukon Territory; Geological Survey of Canada, Memoir 364, 157 p.

Green, R.
1963: Lower Mississippian ostracodes from the Banff Formation, Alberta; Research Council of Alberta, Bulletin 11, p. 1-237.

Gutschick, R.C. and Sandberg, C.A.
1983: Mississippian continental margins of the conterminous United States; in The Shelf Break: Critical Interface on Continental Margins, D.J. Stanley and G.T. Moore (ed.); Society of Economic Paleontologists and Mineralogists, Special Publication no. 33, p. 79-96.

Gutschick, R.C., Sandberg, C.A., and Sando, W.J.
1980: Mississippian shelf margin and carbonate platform from Montana to Nevada; in Paleozoic Paleogeography of the West-central United States, Rocky Mountain Paleogeography Symposium 1, T.D. Fouch and E.R. Magathan (ed.); Rocky Mountain Section, Society of Economic Paleontologists and Mineralogists, p. 111-128.

Hacquebard, P.A. and Barss, M.S.
1957: A Carboniferous spore assemblage in coal from the South Nahanni River area, Northwest Territories; Geological Survey of Canada, Bulletin 40, 63 p.

Halbertsma, H.L.
1959: Nomenclature of Upper Carboniferous and Permian strata in the subsurface of the Peace River area; Journal of Alberta Society of Petroleum Geologists, v. 7, p. 109-118.

Harker, P.
1961: Summary of Carboniferous and Permian formations, south-western District of Mackenzie; Geological Survey of Canada, Paper 61-1, 9 p.
1963: Carboniferous and Permian rocks, southwestern District of Mackenzie; Geological Survey of Canada, Bulletin 95, 91 p.

Harker, P. and McLaren, D.J.
1958: The Devonian-Mississippian boundary in the Alberta Rocky Mountains; in Jurassic and Carboniferous of Western Canada, A.J. Goodman (ed.); American Association of Petroleum Geologists, John Andrew Allan Memorial Volume, p. 244-259.

Harker, P. and Raasch, G.O.
1958: Megafaunal zones in the Alberta Mississippian and Permian; in Jurassic and Carboniferous of Western Canada, A.J. Goodman (ed); American Association of Petroleum Geologists, John Andrew Allan Memorial Volume, p. 216-231.

Hartling, A., Brewster, A., and Posehn, G.
1982: The geology and hydrocarbon trapping mechanisms of the Mississippian Oungre zone (Ratcliffe beds) of the Williston Basin; in Fourth International Williston Basin Symposium, J.E. Christopher and J. Kaldi (ed.); Saskatchewan Geological Society, Special Publication no. 6, p. 217-223.

Hayes, M.D.
1985: Conodonts of the Bakken Formation (Devonian and Mississippian), Williston Basin, North Dakota; The Mountain Geologist, v. 22, p. 64-77.

Hemphill, C.R.
1957: History and development of the Sundre, Westward Ho, and Harmattan oil fields; Bulletin of Canadian Petroleum Geology, v. 5, p. 232-247.

Hennessey, W.J., Bamber, E.W., and Norris, D.K.
1982: Geology of the Plateau Mountain area; Canadian Society of Petroleum Geologists, American Association of Petroleum Geologists, Annual Convention June 27-30, 1982, Field Trip Guidebook, Trip no. 9, 42 p.

Holter, M.E.
1976: Limestone resources of Alberta; Alberta Research Council, Economic Geology Report 4, 91 p.

Høy, T.
1977: Stratigraphy and structure of the Kootenay Arc in the Riondel area, southeastern British Columbia; Canadian Journal of Earth Sciences, v. 14, p. 2301-2315.

Humphreys, J.T.
1955: Del Bonita area - southern Alberta; in Sweetgrass Arch - Disturbed Belt, Montana, P.J. Lewis (ed.); Billings Geological Society, Guidebook, 6th Annual Field Conference, p. 189-194.

Johnson, J.G. and Pendergast, A.
1981: Timing and mode of emplacement of the Roberts Mountain allochthon, Antler Orogeny; Geological Society of America Bulletin, v. 92, p. 648-658.

Kent, D.M.
1974: A stratigraphic and sedimentologic analysis of the Mississippian Madison Formation in southwestern Saskatchewan; Department of Mineral Resources, Saskatchewan Geological Survey, Report no. 141, 85 p.

Kent, D.M.
1984: Depositional setting of Mississippian strata in southeastern Saskatchewan: a conceptual model for hydrocarbon accumulation; in Oil and Gas in Saskatchewan, J.A. Lorsong and M.A. Wilson (ed.); Saskatchewan Geological Society, Special Publication no. 7, p. 19-30.

Kerr, J.W.
1981: Evolution of the Canadian Arctic Islands; a transition between the Atlantic and Arctic oceans; in The Ocean Basins and Margins, A.E.M. Nairn, M. Churkin Jr., and F.G. Stehli (ed.); v. 5, The Arctic Ocean, p. 105-199.

Kidd, F.A.
1963: The Besa River Formation; Bulletin of Canadian Petroleum Geology, v. 11, p. 369-372.

Kindle, E.M.
1924: Standard Paleozoic section of Rocky Mountains near Banff, Alberta; Pan-American Geologist, v. 42, p. 113-124.

Klepacki, D.W. and Wheeler, J.O.
1985: Stratigraphic and structural relations of the Milford, Kaslo and Slocan groups, Goat Range, Lardeau and Nelson map areas, British Columbia; in Current Research, Part A; Geological Survey of Canada, Paper 85-1A, p. 277-286.

Kluth, C.F. and Coney, P.J.
1981: Plate tectonics of the ancestral Rocky Mountains; Geology, v. 9, p. 10-15.

Lane, H.R., Sandberg, C.A., and Ziegler, W.
1980: Taxonomy and phylogeny of some Lower Carboniferous conodonts and preliminary standard post-*Siphonodella* zonation; Geologica et Palaeontologica, v. 14, p. 117-164.

Lavoie, D.H.
1958: The Peace River Arch during Mississippian and Permo-Pennsylvanian time; Journal of the Alberta Society of Petroleum Geologists, v. 6, p. 69-73.

Law, J.
1981: Mississippian correlations, northeastern British Columbia, and implications for oil and gas exploration; Bulletin of Canadian Petroleum Geology, v. 29, p. 378-398.

Lethiers, F.
1978: Ostracodes du Devonien Terminal de la Formation Big Valley, Saskatchewan et Alberta; Palaeontographica, Abteilung A, Band 162, p. 81-143.

Lethiers, F., Braun, W.K., Crasquin, S., and Mansy, J.L.
1986: The Strunian event in western Canada with reference to ostracode assemblages; Annales de la Société Géologique de Belgique, v. 109, p. 149-157.

Lindsay, R.F. and Roth, M.S.
1982: Carbonate and evaporite facies, dolomitization and reservoir distribution of the Mission Canyon Formation, Little Knife field, North Dakota; in Fourth International Williston Basin Symposium, J.E. Christopher and J. Kaldi (ed.); Saskatchewan Geological Society, Special Publication no. 6, p. 153-179.

Loranger, D.M.
1958: Mississippian micropaleontology applied to the western Canada basin; in Jurassic and Carboniferous of Western Canada, A.J. Goodman (ed.); American Association of Petroleum Geologists, John Andrew Allan Memorial Volume, p. 232-243.

Macauley, G.
1958: Late Paleozoic of Peace River area, Alberta; in Jurassic and Carboniferous of Western Canada, A.J. Goodman (ed.); American Association of Petroleum Geologists, John Andrew Allan Memorial Volume, p. 289-308.

Macauley, G., Penner, D.G., Procter, R.M., and Tisdall, W.H.
1964: Carboniferous; in Geological History of Western Canada, R.G. McCrossan and R.P. Glaister (ed.); Alberta Society of Petroleum Geologists, p. 89-102.

MacDonald, G.H.
1956: Subsurface stratigraphy of the Mississippian rocks of Saskatchewan; Geological Survey of Canada, Memoir 282, 46 p.

Macqueen, R.W. and Bamber, E.W.
1967: Stratigraphy of Banff Formation and lower Rundle Group (Mississippian), southwestern Alberta; Geological Survey of Canada, Paper 67-47, 37 p.
1968: Stratigraphy and facies relationships of the Upper Mississippian Mount Head Formation, Rocky Mountains and Foothills, southwestern Alberta; Bulletin of Canadian Petroleum Geology, v. 16, p. 225-287.

Macqueen, R.W., Bamber, E.W., and Mamet, B.L.
1972: Lower Carboniferous stratigraphy and sedimentology of the southern Rocky Mountains; 24th International Geological Congress, Montreal, Quebec, Guidebook, Field Excursion 17, 62 p.

Macqueen, R.W. and Sandberg, C.A.
1970: Stratigraphy, age, and inter-regional correlation of the Exshaw Formation, Alberta Rocky Mountains; Bulletin of Canadian Petroleum Geology, v. 18, p. 32-66.

Mamet, B.L.
1976: An atlas of microfacies in Carboniferous carbonates of the Canadian Cordillera; Geological Survey of Canada, Bulletin 255, 131 p.

Mamet, B.L. and Bamber, E.W.
1979: Stratigraphic Correlation Chart of the lower part of the Carboniferous, Canadian Cordillera and Arctic Archipelago; in Palaeontological Characteristics of the Main Subdivisions of the Carboniferous, S.V. Meyen, V.V. Menner, E.A. Reitlinger, A.P. Rotai, and M.N. Solovieva (ed.); Huitième Congrès International de Stratigraphie et de Géologie Carbonifère, 1975; Compte Rendu, v. 3, p. 37-49.

Mamet, B.L., Bamber, E.W., and Macqueen, R.W.
1986: Microfacies of the Lower Carboniferous Banff Formation and Rundle Group, Monkman Pass map-area, northeastern British Columbia; Geological Survey of Canada, Bulletin 353, 93 p.

Mamet, B.L. and Skipp, B.A.
1970: Preliminary foraminiferal correlations of Early Carboniferous strata in the North American Cordillera; in Colloque sur la Stratigraphie du Carbonifère; Les Congrès et Colloques de l'Université de Liege, v. 55, p. 327-348.
1979: Lower Carboniferous Foraminifera-paleogeographical implications; in Regional Carboniferous Biostratigraphy of Modern Continents, S.V. Meyen, V.V. Menner, E.A. Reitlinger, A.P. Rotai, and M.N. Solovieva (ed.); Huitième Congrès International de Stratigraphie et de Géologie Carbonifère, 1975; Compte Rendu, v. 2, p. 48-66.

Martin, H.L.
1967: Mississippian subsurface geology, Rocky Mountain House area, Alberta; Geological Survey of Canada, Paper 65-27, 14 p.
1969: The Banff and Exshaw formations in the Wabamun Lake area, Alberta (83G); Geological Survey of Canada, Paper 69-7, 24 p.
1972: Upper Paleozoic stratigraphy of the Eagle Plain basin, Yukon Territory; Geological Survey of Canada, Paper 71-14, 54 p.

Maughan, E.K.
1975: Montana, North Dakota, northeastern Wyoming, and northern South Dakota; in Paleotectonic Investigations of the Pennsylvanian System in the United States, E.D. Mckee and E.J. Crosby (coordinators); United States Geological Survey Professional Paper 853, Part 1, p. 270-293.

Maughan, E.K. and Roberts, A.E.
1967: Big Snowy and Amsden groups and the Mississippian-Pennsylvanian boundary in Montana; United States Geological Survey, Professional Paper 554-B, 27 p.

McCabe, H.R.
1959: Mississippian stratigraphy of Manitoba; Province of Manitoba Department of Mines and Natural Resources, Publication 58-1, 99 p.
1963: Mississippian oil fields of southwestern Manitoba; Province of Manitoba Department of Mines and Natural Resources, Publication 60-5, 31 p.

McConnell, R.G.
1887: Report on the geological structure of a portion of the Rocky Mountains; Geological Survey of Canada, Annual Report 1886, New Series, v. 1, Part D, p. 1D-41D.

McGugan, A.
1984: Carboniferous and Permian Ishbel Group stratigraphy, North Saskatchewan Valley, Canadian Rocky Mountains, western Alberta; Bulletin of Canadian Petroleum Geology, v. 32, no. 4, p. 372-381.

McGugan, A. and Rapson, J.E.
1961: Stratigraphy of the Rocky Mountain Group (Permo-Carboniferous), Banff area, Alberta; Journal of Alberta Society of Petroleum Geologists, v. 9, p. 73-106.
1963: Permian stratigraphy and nomenclature, western Alberta and adjacent regions; in Sunwapta Pass Area, D.E. Jackson (ed.); Edmonton Geological Society Guidebook, 5th Annual Field Trip, p. 52-64.

McGugan, A., Rapson-McGugan, J.F., Mamet, B.L., and Ross, C.A.
1968: Permian and Pennsylvanian biostratigraphy, and Permian depositional environments, petrography and diagenesis, southern Canadian Rocky Mountains; in Canadian Rockies, Bow River to North Saskatchewan River, Alberta, H. Hornford (ed.); Canadian Society of Petroleum Geologists 16th Annual Field Conference, guidebook, p. 48-66.

Miller, M.M.
1987: Dispersed remnants of a northeast Pacific fringing arc: upper Paleozoic terranes of Permian McCloud faunal affinity, western U.S.; Tectonics, v. 6, p. 807-830.

Monger, J.W.H.
1977: Upper Paleozoic rocks of the western Canadian Cordillera and their bearing on Cordilleran evolution; Canadian Journal of Earth Sciences, v. 14, p. 1832-1859.

Monger, J.W.H. and Berg, H.C.
1987: Lithotectonic terrane map of western Canada and southeastern Alaska; United States Geological Survey, Miscellaneous Field Studies Map, Report no. MF-1874-B, 12 p., 1:2 500 000.

Monger, J.W.H. and Price, R.A.
1979: Geodynamic evolution of the Canadian Cordillera-progress and problems; Canadian Journal of Earth Sciences, v. 16, p. 770-791.

Monger, J.W.H. and Ross, C.A.
1984: Upper Paleozoic volcanosedimentary assemblages of the western North American craton; in Part 3: Sedimentology and Geochemistry, E.S. Belt and R.W. Macqueen (ed.); Neuvième Congrès International de Stratigraphie et de Géologie du Carbonifère, 1979, Compte rendu, v. 3, p. 219-228.

Moore, P.F.
1958: Late Paleozoic stratigraphy in the Rocky Mountains and Foothills of Alberta - a critical historical review; in Jurassic and Carboniferous of Western Canada, A.J. Goodman (ed.); American Association of Petroleum Geologists, John Andrew Allan Memorial Volume, p. 145-176.

Morgan, G.R. and Jackson, D.E.
1970: A probable 'Waulsortian' carbonate mound in the Mississippian of northern Alberta; Bulletin of Canadian Petroleum Geology, v. 18, p. 104-112.

Morin, G.
1981: Mississippian porosity trends; confidential report, Applied Geoscience and Technology Consultants Ltd., phase III, v. 1, p. 63.

Mortensen, J.K.
1982: Geological setting and tectonic significance of Mississippian felsic metavolcanic rocks in the Pelly Mountains, southeastern Yukon Territory; Canadian Journal of Earth Sciences, v. 19, p. 8-22.

Mortensen, J.K. and Jilson, G.A.
1985: Evolution of the Yukon-Tanana Terrane: evidence from southeastern Yukon Territory; Geology, v. 13, p. 806-810.

Mortensen, J.K., Montgomery, J.R., and Fillipone, J.
1987: U-Pb zircon, monazite, and sphene ages for granitic orthogneiss of the Barkerville terrane, east-central British Columbia; Canadian Journal of Earth Sciences, v. 24, p. 1261-1266.

Muller, J.E.
1967: Pine Pass area; in Report of Activities, Part A; Geological Survey of Canada, Paper 67-1A, p. 77-80.

Nassichuk, W.W.
1971: An Upper Pennsylvanian ammonoid from the Ogilvie Mountains, Yukon Territory; in Contributions to Canadian Paleontology, Geological Survey of Canada, Bulletin 197, p. 79-84.

Nassichuk, W.W. and Bamber, E.W.
1978: Middle Pennsylvanian biostratigraphy, eastern Cordillera and Arctic Islands, Canada - a summary; in Western and Arctic Canadian Biostratigraphy, C.R. Stelck and B.D.E. Chatterton (ed.); Geological Association of Canada, Special Paper 18, p. 395-413.

Nelson, S.J.
1958: Brachiopod zones of the Mount Head and Etherington formations, southern Canadian Rockies; Royal Society of Canada, Transactions, v. 52, sec. 4, p. 45-53.
1960: Mississippian lithostrotionid zones of the southern Canadian Rocky Mountains; Journal of Paleontology, v. 34, p. 107-126.
1961: Mississippian faunas of western Canada; Geological Association of Canada, Special Paper 2, p. 1-39.

Nilsen, T.H.
1977: Paleogeography of Mississippian turbidites in south-central Idaho; in Paleozoic Paleogeography of the Western United States, J.H. Stewart, C.H. Stevens, and A.E. Frische (ed.); Pacific Section, Society of Economic Paleontologists and Mineralogists, Pacific Coast Paleogeography Symposium 1, p. 275-299.
1981: Upper Devonian and lower Mississippian redbeds, Brooks Range, Alaska; in Sedimentation and Tectonics in Alluvial Basins, A.D. Miall (ed.); Geological Association of Canada, Special Paper 23, p. 187-219.

Nisbet, E.G. and Fowler, C.M.R.
1984: A review of some models for the formation of the Williston Basin; in Oil and Gas in Saskatchewan, Proceedings of a Symposium, Regina, 1984, J.A. Lorsong and M.A. Wilson (ed.); Saskatchewan Geological Society, Special Publication no. 7, p. 1-3.

Nordquist, J.W.
1953: Mississippian stratigraphy of northern Montana; in Little Rocky Mountains - Montana, southwestern Saskatchewan, J.M. Parker (ed.); Billings Geological Society, Guidebook, 4th Annual Field Conference, p. 68-82.

Norris, D.K.
1957: The Rocky Mountain Succession at Beehive Pass, Alberta; Journal of Alberta Society of Petroleum Geologists, v. 5, p. 248-254.
1965: Stratigraphy of the Rocky Mountain Group in the southeastern Cordillera of Canada; Geological Survey of Canada, Bulletin 125, 82 p.

Norris, D.K. (cont.)
1973: Tectonic styles of northern Yukon Territory and northwestern District of Mackenzie, Canada; in Arctic Geology, M.G. Pitcher (ed.); American Association of Petroleum Geologists, Memoir 19, p. 23-40.
1981: Blow River and Davidson Mountain, Yukon Territory-District of Mackenzie; Geological Survey of Canada, Map 1516A, 1:250 000.
1983: Geotectonic correlation chart 1532A - Operation Porcupine project area; Geological Survey of Canada.

Orchard, M.J.
1985: Carboniferous, Permian and Triassic conodonts from the central Kootenay Arc, British Columbia: constraints on the age of the Milford, Kaslo and Slocan groups; in Current Research, Part A; Geological Survey of Canada, Paper 85-1A, p. 287-300.

Orchard, M.J. and Struik, L.C.
1985: Conodonts and stratigraphy of upper Paleozoic limestones in Cariboo gold belt, east-central British Columbia; Canadian Journal of Earth Sciences, v. 22, p. 538-552.

Osadetz, K.G. and Snowdon, L.R.
1986a: Petroleum source rock reconnaissance of southern Saskatchewan; in Current Research, Part A, Geological Survey of Canada, Paper 86-1A, p. 609-617.
1986b: Speculation on the petroleum source rock potential of the Lodgepole Formation (Mississippian) of southern Saskatchewan; in Current Research, Part B; Geological Survey of Canada, Paper 86-1B, p. 647-651.

Oswald, D.H.
1964: Mississippian stratigraphy of southeastern British Columbia; Bulletin of Canadian Petroleum Geology, Flathead Valley Guidebook Issue, v. 12, p. 452-459.

Patton, W.J.H.
1958: Mississippian succession in South Nahanni River area, Northwest Territories; in Jurassic and Carboniferous of Western Canada, A.J. Goodman (ed.); American Association of Petroleum Geologists, John Andrew Allan Memorial Volume, p. 309-326.

Pelzer, E.E.
1966: Mineralogy, geochemistry and stratigraphy of the Besa River Shale, British Columbia; Bulletin of Canadian Petroleum Geology, v. 14, p. 273-321.

Penner, D.G.
1958: Mississippian stratigraphy of southern Alberta plains; in Jurassic and Carboniferous of Western Canada, A.J. Goodman (ed.); American Association of Petroleum Geologists, John Andrew Allan Memorial Volume, p. 160-288.

Petryk, A.A., Mamet, B.L., and Macqueen R.W.
1970: Preliminary foraminiferal zonation, Rundle Group and uppermost Banff Formation (Lower Carboniferous), southwestern Alberta; Bulletin of Canadian Petroleum Geology, v. 18, p. 84-103.

Playford, G.
1962: Lower Carboniferous microfloras of Spitsbergen Part I: Paleontology, v. 5, pt. 3, p. 550-618.
1963: Lower Carboniferous microfloras of Spitsbergen Part II: Paleontology, v. 5, pt. 4, p. 619-678.

Podruski, J.A., Barclay, J.E., Hamblin, A.P., Lee, P.J., Osadetz, K.G., Procter, R.M., and Taylor, G.C.
1988: Conventional oil reserves of western Canada (light and medium density), part 1, resource endowment; Geological Survey of Canada, Paper 87-26, 149 p.

Poole, F.G.
1974: Flysch deposits of Antler Foreland Basin, Western United States; in Tectonics and Sedimentation, W.R. Dickinson (ed.); Society of Economic Paleontologists and Mineralogists, Special Publication no. 22, p. 58-82.

Poole, F.G. and Sandberg, C.A.
1977: Mississippian paleogeography and tectonics of the western United States; in Paleozoic Paleogeography of the Western United States, J.H. Stewart, C.H. Stevens, and A.E. Fritsche (ed.); Pacific Section, Society of Economic Paleontologists and Mineralogists, Pacific Coast Paleogeography Symposium 1, p. 67-85.

Poulton, T.P.
1984: The Jurassic of the Canadian western interior, from 49°N to Beaufort Sea; in The Mesozoic of Middle North America, D.F. Stott and D.J. Glass (ed.); Canadian Society of Petroleum Geologists, Memoir 9, p. 15-41.

Prather, R.W. and McCourt, G.B.
1968: Geology of gas accumulations in Paleozoic rocks of Alberta plains; in Natural Gases of North America, B.W. Beebe and B.F. Curtis (ed.); American Association of Petroleum Geologists, Memoir 9, p. 1238-1284.

Pugh, D.C.
1983: Pre-Mesozoic geology in the subsurface of Peel River map area, Yukon Territory and District of Mackenzie; Geological Survey of Canada, Memoir 401, 61 p.

Raasch, G.O.
1958: Upper Paleozoic section at Highwood Pass, Alberta; in Jurassic and Carboniferous of western Canada, A.J. Goodman (ed.); American Association of Petroleum Geologists, John Andrew Allan Memorial Volume, p. 190-215.

Ramsbottom, W.H.C.
1973: Transgressions and regressions in the Dinantian; a new synthesis of British Dinantian stratigraphy; Proceedings of the Yorkshire Geological Society, v. 39, p. 567-607.
1977: Major cycles of transgression and regression (mesothems) in the Namurian; Proceedings of the Yorkshire Geological Society, v. 41, p. 261-291.

Rawson, R.R.
1968: The Kibbey Limestone of the Williston basin and central Montana; Wyoming Geological Association, Earth Science Bulletin, v. 1, p. 35-47.
1969: Petrographic analysis of the "Kibbey Limestone"; in Eastern Montana Symposium - The Economic Geology of Eastern Montana and Adjacent Areas, Billings; Montana Geological Society 20th Annual Conference, p. 165-177.

Reasoner, M.A. and Hunt, A.D.
1958: Structure of Coleville-Buffalo Coulee area, Saskatchewan; in Jurassic and Carboniferous of Western Canada, A.J. Goodman (ed.); American Association of Petroleum Geologists, John Andrew Allan Memorial Volume, p. 391-405.

Richards, B.C.
1983: Uppermost Devonian and Lower Carboniferous stratigraphy, sedimentation, and diagenesis, southwestern District of Mackenzie and southeastern Yukon Territory; Ph.D. thesis, University of Kansas, Lawrence, Kansas, 373 p.
1989: Uppermost Devonian and Lower Carboniferous stratigraphy, sedimentation, and diagenesis, southwestern District of Mackenzie and southeastern Yukon Territory; Geological Survey of Canada, Bulletin 390, 135 p.

Richards, B.C. and Higgins, A.C.
1988: Devonian-Carboniferous boundary beds of the Palliser and Exshaw formations at Jura Creek, Rocky Mountains, southwestern Alberta; in Devonian of the World, N.J. McMillan, A.F. Embry, and D.J. Glass (ed.); Canadian Society of Petroleum Geologists, Memoir 14, v. 3, p. 397-410.

Roberts, A.E.
1979: Northern Rocky Mountains and adjacent plains region; in Paleotectonic Investigations of the Mississippian System in the United States, Part I. Introduction and Regional Analysis of the Mississippian System; United States Geological Survey, Professional Paper 1010, p. 221-247.

Ross, C.A.
1967: Late Paleozoic Fusulinacea from northern Yukon Territory; Journal of Paleontology, v. 41, p. 709-725.

Ross, C.A. and Bamber, E.W.
1978: Middle Carboniferous and Early Permian Fusulinaceans from the Monkman Pass area, Northeastern British Columbia; in Contributions to Canadian Paleontology; Geological Survey of Canada, Bulletin 267, p. 25-41.

Ross, C.A. and Ross, J.R.P.
1985: Late Paleozoic depositional sequences are synchronous and worldwide; Geology, v. 13, p. 194-197.

Rupp, A.W.
1969: Turner Valley Formation of the Jumping Pound area, foothills, southern Alberta; Bulletin of Canadian Petroleum Geology, v. 17, p. 460-485.

Sandberg, C.A., Ziegler, W., Leuteritz, K., and Brill, S.M.
1978: Phylogeny, speciation, and zonation of *Siphonodella* (Conodonta, Upper Devonian and Lower Carboniferous); Newsletters in Stratigraphy, v. 7, p. 102-120.

Sando, W.J.
1988: Madison Limestone (Mississippian) paleokarst: A geologic synthesis; in Paleokarst, N.P. James and P.W. Choquette (ed.); Springer-Verlag, New York, p. 256-277.

Sando, W.J. and Bamber, E.W.
1985: Coral-zonation of the Mississippian System in the Western Interior Province of North America; United States Geological Survey, Professional Paper 1334, 61 p.

Sando, W.J. and Dutro, J.T., Jr.
1974: Type sections of the Madison Group (Mississippian) and its subdivisions in Montana; United States Geological Survey, Professional Paper 842, 22 p.

Sando, W.J. and Mamet, B.L.
1981: Distribution and stratigraphic significance of foraminifera and algae in well cores from Madison Group (Mississippian), Williston Basin, Montana; United States Geological Survey, Bulletin 1529-F, 12 p.

Sando, W.J. and Bamber, E.W., and Armstrong, A.K.
1977: The zoogeography of North American Mississippian corals; in Second International Symposium on Corals and Fossil Coral Reefs; Bureau de Recherches Géologiques et Minières Mémoire 89, p. 175-184.

Sando, W.J., Gordon, M., Jr., and Dutro, J.T., Jr.
1975: Stratigraphy and geologic history of the Amsden Formation (Mississippian and Pennsylvanian) of Wyoming; United States Geological Survey, Professional Paper 848-A, 83 p.

Sando, W.J., Mamet, B.L., and Dutro, J.T., Jr.
1969: Carboniferous megafaunal and microfaunal zonation in the northern Cordillera of the United States; United States Geological Survey, Professional Paper 613-E, 29 p.

Sando, W.J., Sandberg, C.A., and Perry, W.S., Jr.
1985: Revision of Mississippian stratigraphy, northern Tendoy mountains, southwest Montana; United States Geological Survey Bulletin 1656, Part A. 10 p.

Saunders, W.B. and Ramsbottom, W.H.C.
1986: The mid-Carboniferous eustatic west; Geology, v. 14, p. 208-212.

Sawatzky, H.B.
1972: Viewfield - a producing fossil crater?; Journal of the Canadian Society of Exploration Geophysicists, v. 8, p. 22-40.

Scott, D.L.
1964: Pennsylvanian stratigraphy; Bulletin of Canadian Petroleum Geology, v. 12, Flathead Valley Guidebook Issue, p. 460-493.

Scott, H.W.
1935a: Upper Mississippian and Lower Pennsylvanian stratigraphy in Montana (abstract); Geological Society of America Proceedings, 1934, p. 367.
1935b: Some Carboniferous stratigraphy in Montana and northwestern Wyoming; Journal of Geology, v. 43, p. 1011-1032.

Seager, O.A.
1942: Test on Cedar Creek Anticline, southeastern Montana; Bulletin of American Association of Petroleum Geologists, v. 26, p. 861-864.

Sheldon, R.P. and Carter, M.D.
1979: Williston Basin Region; in Paleotectonic Investigations of the Mississippian System in the United States; Part 1. Introduction and Regional Analysis of the Mississippian System; United States Geological Survey, Professional Paper 1010, p. 249-271.

Sikabonyi, L.A. and Rodgers, W.J.
1959: Paleozoic tectonics and sedimentation in the northern half of the West Canadian Basin; Journal of the Alberta Society of Petroleum Geologists, v. 7, p. 193-216.

Skipp, B., Hoggan, R.D., Schleicher, D.L., and Douglas, R.C.
1979: Upper Paleozoic carbonate bank in east-central Idaho - Snaky Canyon, Bluebird Mountain, and Arco Hills formations, and their paleotectonic significance; United States Geological Survey, Bulletin 1486, 78 p.

Sloss, L.L.
1963: Sequences in the cratonic interior of North America; Geological Society of America Bulletin, v. 74, p. 93-114.

Smith, D.L.
1977: Transition from deep- to shallow-water carbonates, Paine Member, Lodgepole Formation, central Montana; in Deep-water Carbonate Environments, H.E. Cook and P. Enos (ed.); Society of Economic Paleontologists and Mineralogists, Special Publication No. 25, p. 187-201.
1982: Waulsortian bioherms in the Paine Member of the Lodgepole Limestone (Kinderhookian) of Montana, U.S.A.; in Symposium on the Paleoenvironmental Setting and Distribution of the Waulsortian Facies, K. Bolten, H.R. Lane, and D.V. LaMone (ed.); El Paso Geological Society and University of Texas at El Paso, El Paso, Texas, p. 51-64.

Smith, S.R.
1980: Petroleum geology of the Mississippian Midale beds, Benson Oil Field, southeastern Saskatchewan; Saskatchewan Mineral Resources, Report 215, 98 p.

Speranza, A.
1984: Sedimentation and diagenesis of the Pekisko Formation (Mississippian), Canyon Creek, Alberta; Master of Science thesis, University of Waterloo, Waterloo, Ontario, 278 p.

Stanton, M.S.
1958: Stratigraphy of the Lodgepole Formation, Virden-Whitewater area, Manitoba; in Jurassic and Carboniferous of Western Canada, A.J. Goodman (ed.); American Association of Petroleum Geologists, John Andrew Allan Memorial Volume, p. 372-390.

Staplin, F.L.
1960: Upper Mississippian plant spores from the Golata Formation, Alberta, Canada; Palaeontographica, Abteilung B, Band 107, p. 1-40.

Stearn, C.W.
1956: Type section of the Shunda Formation; Journal of the Alberta Society of Petroleum Geologists, v. 4, p. 237-239.

Stewart, W.D. and Walker, R.G.
1980: Eolian coastal dune deposits and surrounding marine sandstones, Rocky Mountain Supergroup (Lower Pennsylvanian), southeastern British Columbia; Canadian Journal of Earth Sciences, v. 17, p. 1125-1140.
Struik, L.C.
1981: A re-examination of the type area of the Devono-Mississippian Cariboo Orogeny, central British Columbia; Canadian Journal of Earth Sciences, v. 18, p. 1767-1775.
1987: The ancient western North American margin: an alpine rift model for the east-central Canadian Cordillera; Geological Survey of Canada, Paper 87-15, 19 p.
Struik, L.C. and Orchard, M.J.
1985: Late Paleozoic conodonts from ribbon chert delineate imbricate thrusts within the Antler Formation of the Slide Mountain Terrane, central British Columbia; Geology, v. 13, p. 794-798.
Sullivan, H.J.
1965: Palynological evidence concerning the regional differentiation of Upper Mississippian floras; Pollen et spores, v. 7, no. 3, p. 539-563.
Sutherland, P.K.
1958: Carboniferous stratigraphy and rugose coral faunas of northeastern British Columbia; Geological Survey of Canada, Memoir 295, 177 p.
Sutherland Brown, A.,
1963: Geology of the Cariboo River area, British Columbia; British Columbia Department of Mines and Petroleum Resources, Bulletin 47, 60 p.
Sweeney, J.F., Irving, E., and Geuer, J.W.
1978: Evolution of the Arctic Basin; in Arctic Geophysical Review, J.F. Sweeney (ed.); Energy, Mines and Resources Canada, Publications of the Earth Physics Branch, v. 45, no. 4, p. 91-100.
Tailleur, I.L., Brosge, W.P., and Reiser, H.N.
1967: Palinspastic analysis of Devonian rocks in northwestern Alaska; in International Symposium on the Devonian System, Calgary, 1967, D.H. Oswald (ed.); Alberta Society of Petroleum Geologists, v. 1, p. 1345-1361.
Tempelman-Kluit, D.J.
1979: Transported cataclasite, ophiolite and granodiorite in Yukon: evidence of arc-continent collision; Geological Survey of Canada, Paper 79-14, 27 p.
Thorsteinsson, R. and Tozer, E.T.
1970: Geology of the Arctic Archipelago; in Geology and Economic Minerals of Canada, R.J.W. Douglas (ed.); Geological Survey of Canada, Economic Geology Report no. 1, p. 548-590.
Trettin, H.P.
1973: Early Paleozoic evolution of northern parts of Canadian Arctic Archipelago; in Arctic Geology, M.G. Pitcher (ed.); American Association of Petroleum Geologists, Memoir 19, p. 57-75.
1991: Middle Devonian to Early Carboniferous deformation, metamorphism, and granitic plutonism, northern Ellesmere and Axel Heiberg islands; Chapter IIX, Part F; in Geology of the Innuitian Orogen and Arctic Platform of Canada and Greenland, H.P. Trettin (ed.); Geological Survey of Canada, Geology of Canada, no. 3. (Also Geological Society of America, The Geology of North America, v. E.) (1992)
Utting, J.
1980: Palynology of the Windsor Group (Mississippian) in a borehole at Stewiacke, Shubenacadie Basin, Nova Scotia; Canadian Journal of Earth Science, v. 17, p. 1031-1045.
1987: Palynostratigraphic investigations of the Albert Formation (Lower Carboniferous) of New Brunswick, Canada; Palynology, v. 11, p. 73-96.
Van der Zwan, C.J.
1981: Palynology, phytogeography and climate of the Lower Carboniferous; Paleogeography, Palaeoclimatology, Palaeoecology, v. 33, p. 279-310.
Van der Zwan, C.J. and Walton, H.S.
1981: The *Cyrtospora cristifer* morphon: inclusion of *Cornispora varicornata* and *C. monocornata*; Review of Palaeobotany and Palynology, v. 33, p. 139-152.
Wardlaw, B.R. and Pecora, W.C.
1985: New Mississippian-Pennsylvanian stratigraphic unit in southwest Montana and adjacent Idaho; United States Geological Survey, Bulletin 1656, Part B, 9 p.
Warren, P.S.
1927: Banff area, Alberta; Geological Survey of Canada, Memoir 153, 94 p.
1937: Age of the Exshaw shale in the Canadian Rockies; American Journal of Science, Ser. 5, v. 33, p. 454-457.
Waterhouse, J.B. and Waddington, J.
1982: Systematic descriptions, paleoecology and correlations of the late Paleozoic subfamily Spiriferellinae (Brachiopoda) from the Yukon Territory and the Canadian Arctic Archipelago; Geological Survey of Canada, Bulletin 289, 72 p.
Weed, W.H.
1899: Description of the Fort Benton quadrangle, Montana; United States Geological Survey, Geological Atlas Folio 55, 7 p.
Wegelin, A.
1984: Geology and reservoir properties of the Weyburn Field, southeastern Saskatchewan; in Oil and Gas in Saskatchewan, Proceedings of a Symposium, Regina 1984, J.A. Lorsong and M.A. Wilson (ed.); Saskatchewan Geological Society, Special Publication no. 7, p. 71-82.
Wheeler, J.O., Brookfield, A.J., Gabrielse, H., Monger, J.W.H., Tipper, H.W., and Woodsworth, G.J.
1988: Terrane map of the Canadian Cordillera; Geological Survey of Canada, Open File 1894, Scale 1:2 000 000.
Wilson, J.L.
1969: Microfacies and sedimentary structures in "deeper water" lime mudstones; in Depositional Environments in Carbonate Rocks, G.M. Friedman (ed.); Society of Economic Paleontologists and Mineralogists, Special Publication No. 14, p. 4-19.
1975: Carbonate facies in geologic history: Springer-Verlag, New York, 471 p.
Wood, G.V. and Armstrong, A.K.
1975: Diagenesis and stratigraphy of the Lisburne Group limestones of the Sadlerochit mountains and adjacent areas, northeastern Alaska; United States Geological Survey, Professional Paper 857, 47 p.
Young, F.G.
1967: Elkton reservoir of Edson gas field, Alberta; Bulletin of Canadian Petroleum Geology, v. 15, p. 50-64.
Ziegler, P.A.
1988: Laurussia - the Old Red Continent; in Devonian of the World, N.J. McMillan, A.F. Embry, and D.J. Glass (ed.); Canadian Society of Petroleum Geologists, Memoir 14, p. 15-48.
Ziegler, W.
1962: Taxonomie und Phylogenie Oberdevonischer Conodonten und ihre stratigraphische Bedeutung; Hessische Landesamt Bodenforschung Abteilung 38, p. 1-166.
Ziegler, W. and Sandberg, C.A.
1984: *Palmatolepis*-based revision of upper part of standard Late Devonian conodont zonation; in Pander Society Symposium on Conodont Biofacies and Provincialism, D.L. Clark (ed.); Geological Society of America, Special Paper 196, p. 143-178.

ADDENDUM

Since the manuscript was revised in 1988, numerous papers about the Carboniferous succession of the Western Canada Sedimentary Basin have either been published or are in press. The most significant of these works are listed below. Most of the papers support and amplify the data and interpretations presented in this chapter, but some formational assignments and major interpretations in Richards et al. (in press) differ substantially from those given herein.

Bamber, E.W., Henderson, C.M., Jerzykiewicz, J., Mamet, B.L., and Utting, J.
1989: A summary of Carboniferous and Permian biostratigraphy, northern Yukon Territory and northwestern District of Mackenzie; in Current Research, Part G; Geological Survey of Canada, Paper 89-1G, p. 13-21.
Barclay, J.E.
1988: The Lower Carboniferous Golata Formation of the Western Canada Basin, in the context of sequence stratigraphy; in Sequences, Stratigraphy, Sedimentology: Surface and Subsurface, D.P. James and D.A. Leckie (ed.); Canadian Society of Petroleum Geologists, Memoir 15, p. 1-14.

Barclay, J.E., Krause, F.F., Campbell, R.I., and Utting, J.
1990: Dynamic casting and growth faulting: Dawson Creek graben complex, Carboniferous-Permian Peace River Embayment, Western Canada; Bulletin of Canadian Petroleum Geology, v. 38A, p. 115-148.

Henderson, C.M.
1989: The lower Absaroka Sequence: Upper Carboniferous and Permian, Chapter 10; in Western Canada Sedimentary Basin, A Case History, B.D. Ricketts (ed.); Canadian Society of Petroleum Geologists, p. 203-217.

Higgins, A.C., Richards, B.C., and Henderson, C.M.
1991: Conodont biostratigraphy and paleoecology of the uppermost Devonian and Carboniferous of the Western Canada Sedimentary Basin; in Ordovician to Triassic Conodont Paleoecology of the Canadian Cordillera, M.J. Orchard and A.D. McCracken (ed.); Geological Survey of Canada, Bulletin 417, p. 215-251.

O'Connell, S.C.
1990: The development of the Lower Carboniferous Peace River Embayment as determined from Banff and Pekisko Formation depositional patterns; Bulletin of Canadian Petroleum Geology, v. 38A, p. 93-114.

Playford, G. and McGregor, D.C.
in press: Miospores and organic-walled microphytoplankton of Devonian-Carboniferous boundary beds (Bakken Formation), southern Saskatchewan; a systematic and stratigraphic appraisal; Geological Survey of Canada, Bulletin.

Potter, J., Richards, B.C., and Cameron, A.R.
in press: The petrology and origin of coals from the Lower Carboniferous Mattson Formation, southwestern District of Mackenzie, Canada; International Journal of Coal Geology.

Richards, B.C.
1989: Upper Kaskaskia Sequence: uppermost Devonian and Lower Carboniferous, Chapter 9; in Western Canada Sedimentary Basin, A Case History, B.D. Ricketts (ed.); Canadian Society of Petroleum Geologists, p. 165-201.

Richards, B.C., Henderson, C.M., Higgins, A.C., Johnston, D.I., Mamet, B.L., and Meijer Drees, N.C.
1991: The Upper Devonian (Famennian) and Lower Carboniferous (Tournaisian) at Jura Creek, southwestern Alberta; in A Field Guide to the Paleontology of the Southwestern Canada, Canadian Paleontology Conference 1, Vancouver 1991, P.L. Smith (ed.); Paleontology Division, Geological Association of Canada, p. 34-81.

Richards, B.C., Barclay, J.E., Bryan, D., Hartling, A., Henderson, C.M., Hinds, R.C., and Trollope, F.H.
in press: Carboniferous; in Geological Atlas of the Western Canada Sedimentary Basin, G.D. Mossop and I. Shetsen (ed.); Canadian Society of Petroleum Geologists and Alberta Research Council, Calgary.

Sando, W.J., Bamber, E.W., and Richards, B.C.
1990: The rugose coral *Ankhelsasma* – index to Viséan (Lower Carboniferous) shelf margin in the western interior of North America; in Shorter Contributions to Paleontology and Stratigraphy; United States Geological Survey, Bulletin 1895, p. B1-B29.

Utting, J.
1989: Thermal maturity of Lower Carboniferous rocks in northern Yukon Territory; in Current Research, Part G; Geological Survey of Canada, Paper 89-1G, p. 101-104

1991: Lower Carboniferous miospore assemblages from the Hart River Formation, northern Yukon; in Contributions to Canadian Paleontology; Geological Survey of Canada, Bulletin 412, p. 81-99.

Authors' Addresses

B.C. Richards
Institute of Sedimentary and Petroleum Geology
Geological Survey of Canada
3303-33rd Street, N.W.
Calgary, Alberta
T2L 2A7

E.W. Bamber
Institute of Sedimentary and Petroleum Geology
Geological Survey of Canada
3303-33rd Street, N.W.
Calgary, Alberta
T2L 2A7

A.C. Higgins
9-Treglof Court
Oatlands Drive
Weybridge, Surrey
England KT13 9JG

J. Utting
Institute of Sedimentary and Petroleum Geology
Geological Survey of Canada
3303-33rd Street, N.W.
Calgary, Alberta
T2L 2A7

Printed in Canada

Subchapter 4F

PERMIAN

C.M. Henderson, E.W. Bamber, B.C. Richards, A.C. Higgins, and A. McGugan

CONTENTS

INTRODUCTION

Permian rocks are preserved throughout most of the eastern Cordillera and locally, in the Peace River-Liard River area, on the Interior Platform (Fig. 4F.1, 4F.2, 4F.3). They are absent from the remainder of the Interior Platform and from most of the Mackenzie Mountains through truncation at several disconformities beneath Mesozoic strata.

Permian sediments were deposited mainly along the margin of the North American plate in a Permian depositional basin, here named Ishbel Trough, extending from the 49th parallel to the Ancestral Aklavik Arch in northern Yukon Territory. The trough occupies approximately the same position as the Carboniferous Prophet Trough (Richards et al., Subchapter 4E in this volume). In the Peace River-Liard River area, Permian sediments were also deposited on the western cratonic platform. The western margin of Ishbel Trough has not been identified. It (Fig. 4F.2, 4F.3) may have been in the area presently occupied by Cariboo Terrane (Struik and Orchard, 1985; Struik, 1986), which lacks Permian strata and may have been partly subaerially exposed during the Permian Period. A marginal basin to the west of Cariboo Terrane (Barkerville Subterrane of Kootenay Terrane; Monger and Price, 1979) contains a remnant of Permian strata (Sugar Limestone). The eastern margin of Ishbel Trough was a broad hinge-line along the western margin of the cratonic platform. In northeastern British Columbia and southwestern District of Mackenzie the hinge coincided with a zone of normal faulting, but generally the position of the eastern trough margin is unclear, because of eastward truncation of Permian strata (Fig. 4F.2, 4F.3). The Permian succession, which corresponds to part of the Absaroka sequence of Sloss (1963), consists mainly of marine, supratidal to basinal siliciclastics and silty to sandy carbonates. It lacks the extensive carbonate buildups that characterize the underlying Carboniferous succession and has no major basinal shale accumulations or well-defined continental deposits.

Henderson, C.M., Bamber, E.W., Richards, B.C., Higgins, A.C., and McGugan, A.

1993: Permian; Subchapter 4F in Sedimentary Cover of the Craton in Canada, D.F. Stott and J.D. Aitken (ed.); Geological Survey of Canada, Geology of Canada, no. 5, p. 272-293 (also Geological Society of America, The Geology of North America, v. D-1).

A regional unconformity above strata ranging in age from Early Silurian to latest Carboniferous marks the base of the Permian System (see Fig. 4F.1; McGugan et al., 1968; Bamber and Macqueen, 1979). In all areas a disconformity separates Permian from Triassic and younger strata, with the result that no Permian rocks younger than Wordian are preserved. From the 49th parallel to southeastern Yukon Territory, a regional, intra-Permian disconformity separates a thin, upper Artinskian to Wordian assemblage of chert and siliciclastics from underlying Asselian to lower Artinskian siliciclastics and carbonates. This disconformity also may be present in the Permian strata of northern Yukon and northwestern District of Mackenzie, within a thick Asselian to Wordian succession on the flanks of the northeast-trending Ancestral Aklavik Arch (see Fig. 4F.16; Bamber and Waterhouse, 1971).

Previous work

The stratigraphic succession and current formational nomenclature for the Permian System in Western Canada Basin have been established by numerous studies, particularly those of McGugan and Rapson (1963a),

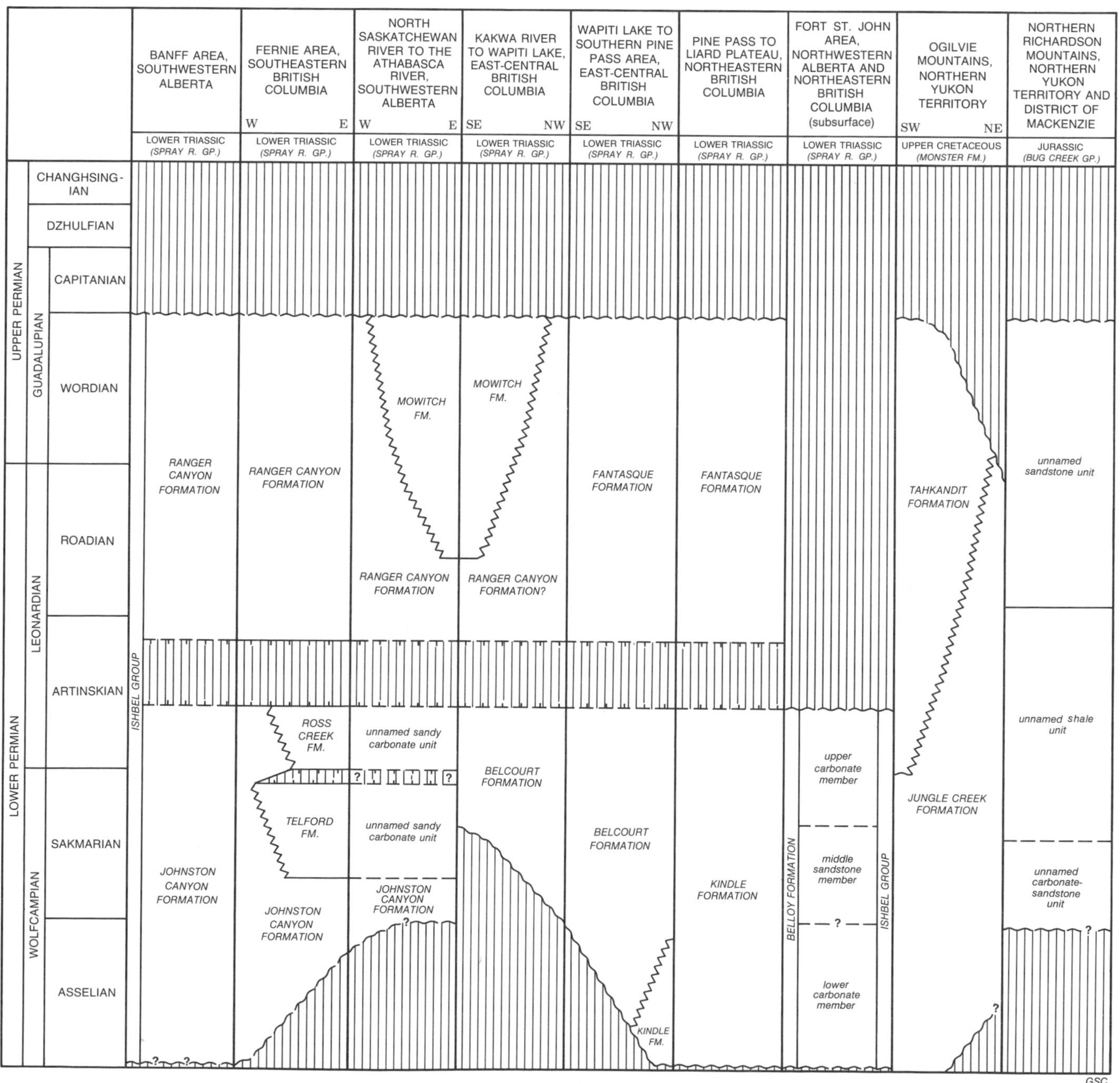

Figure 4F.1. Correlation of Permian formations, Eastern Cordillera and Interior Platform.

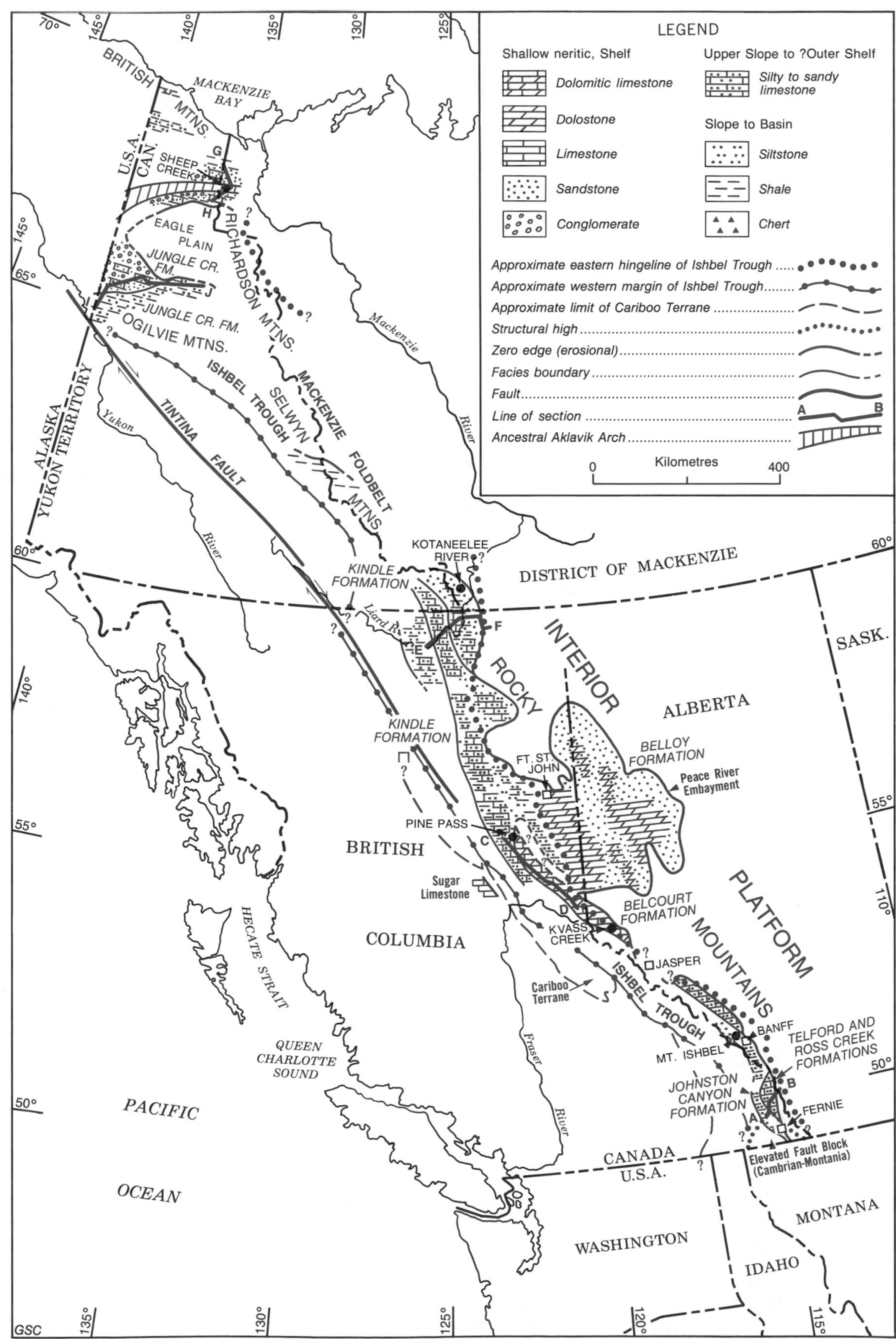

Figure 4F.2. Map of Asselian to Lower Artinskian facies (not palinspastically restored).

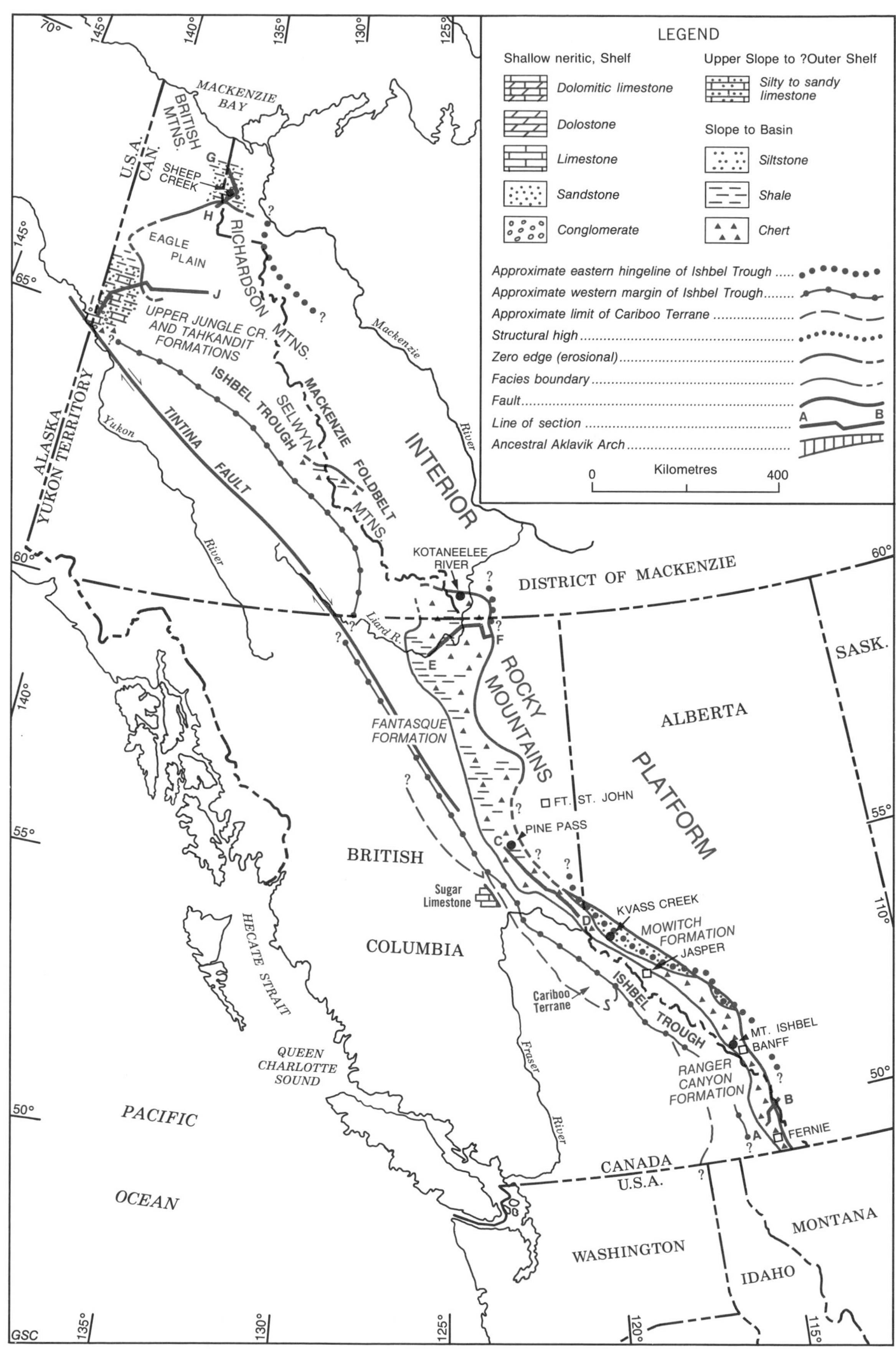

Figure 4F.3. Map of Upper Artinskian to Wordian facies (not palinspastically restored).

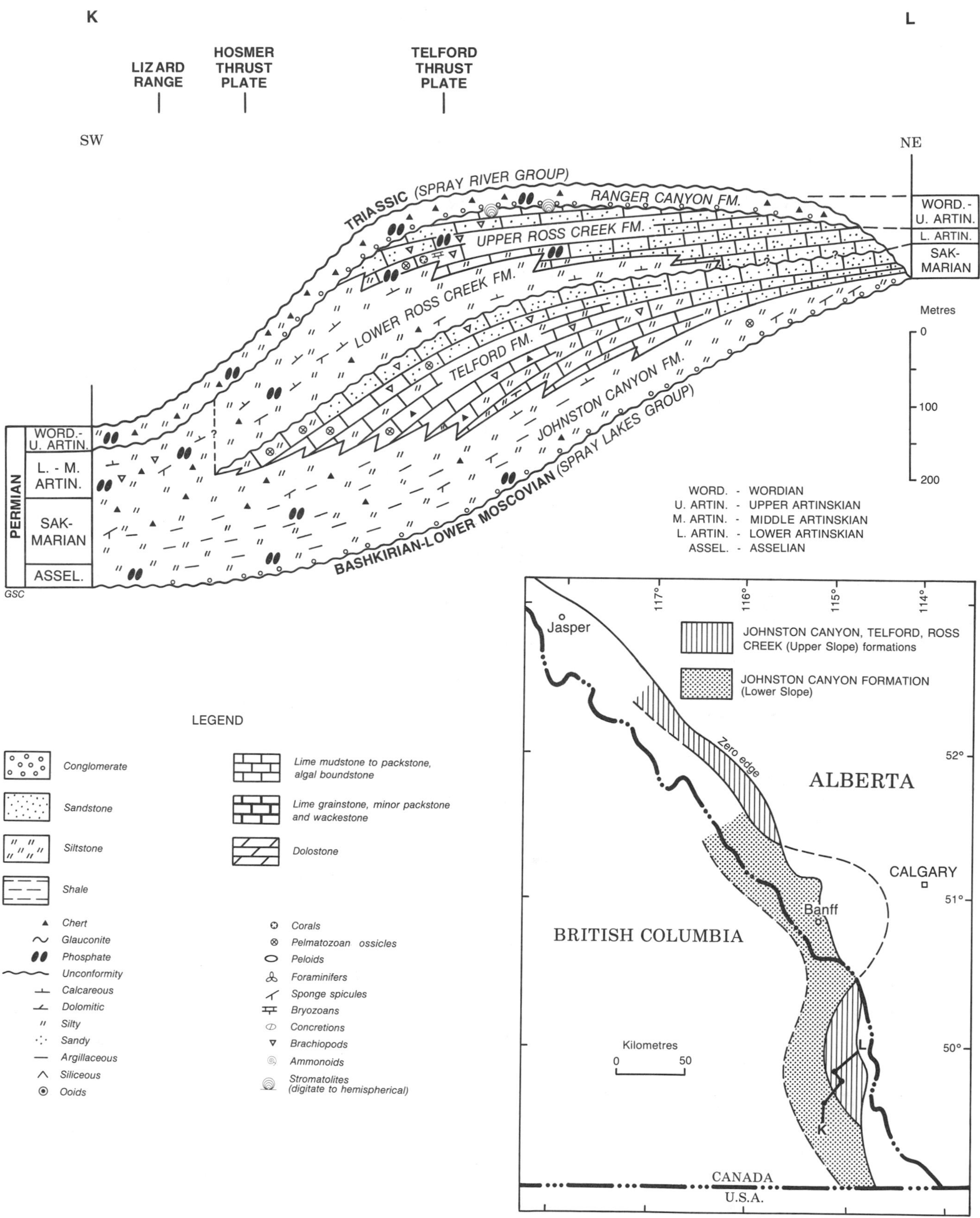

Figure 4F.4. Stratigraphic cross-section of Permian Ishbel Group, southern Rocky Mountains (see Fig. 4F.2, A-B).

McGugan and Rapson-McGugan (1976), MacRae and McGugan (1977), and McGugan (1984) for the southern Canadian Rocky Mountains; Halbertsma (1959), Naqvi (1972), and Bamber and Macqueen (1979) for northeastern British Columbia; and Bamber and Waterhouse (1971) for northern Yukon Territory and northwestern District of Mackenzie. Regional lithostratigraphic syntheses were prepared by McGugan et al. (1964) and Douglas et al. (1970). Important biostratigraphic reports include Logan and McGugan (1968), McGugan et al. (1968), MacRae and McGugan (1977), Bamber and Copeland (1970), Henderson and McGugan (1986), Jansonius (1962), Nelson and Johnson (1968), Nassichuk (1969, 1971), Ross and Bamber (1978), and Bamber and Waterhouse (1971).

Acknowledgments

Several geologists of the Geological Survey of Canada, including S.P. Gordey, J.W.H. Monger, and D.K. Norris, contributed to this compilation by providing unpublished data and helpful criticism and discussion. The manuscript was critically read by R.W. Macqueen.

C.M. Henderson and A.C. Higgins wrote the biostratigraphic section. C.M. Henderson and A. McGugan prepared the section on the southern Rocky Mountains and the Peace River Embayment, E.W. Bamber compiled the northern Cordillera section, and E.W. Bamber and B.C. Richards prepared the section on the Cordillera of east-central and northeastern British Columbia.

Figure 4F.5. Basal Permian transgressive sequence, Telford thrust plate, 40 km north of Fernie, British Columbia (49°45'N; 115°02'W); recessive phosphatic siltstone of Johnston Canyon Formation (JC) (60 m thick) conformably overlain by resistant, sandy carbonates of Telford Formation (Te); Johnston Canyon disconformably overlies Carboniferous Spray Lakes Group (SL); view to southwest (photo - A. McGugan).

SOUTHERN ROCKY MOUNTAINS AND FOOTHILLS (SOUTHWESTERN ALBERTA AND SOUTHEASTERN BRITISH COLUMBIA)

The Permian succession, referred in this area to the Ishbel Group (McGugan and Rapson, 1963a), is generally thin and is dominated by slope to basinal facies. It is confined to the Front Ranges from southeastern British Columbia to the Jasper area (Athabasca River) (Fig. 4F.1, 4F.2, 4F.3, 4F.4) and is absent from the adjacent subsurface.

Permian biostratigraphic and facies relationships of the region are best understood in southeastern British Columbia (Telford Thrust Plate), where the succession is much thicker (460 m) than elsewhere (MacRae and McGugan, 1977; Henderson and McGugan, 1983). There, the succession below the intra-Permian disconformity begins with the Johnston Canyon Formation (Fig. 4F.5), which is about 60 m thick and consists of a basal conglomerate of chert and phosphate pebbles overlain by phosphatic, argillaceous siltstone and nodular to bedded, spicular chert of slope to basinal origin. Eastward, the formation onlaps Carboniferous strata of Bashkirian to early Moscovian age. The Telford Formation (Sakmarian) progrades westward over the Johnston Canyon Formation (Fig. 4F.4) and coarsens upward from slope to possible outer shelf deposits. It consists of approximately 240 m of silty to sandy carbonates with subordinate, interbedded, spicular chert and pelmatozoan-brachiopod wackestone in the lower part. In the middle part, a thin, mixed-skeletal carbonate unit containing large, solitary corals, large brachiopods, and phosphatic material is developed locally. Brachiopod wackestone and medium- to coarse-grained

Figure 4F.6. Contact between Telford Formation (resistant, sandy carbonates-Te) and overlying Ross Creek Formation (lower, recessive, phosphatic siltstone-lRCr, grading upward into resistant, sandy carbonates-uRCr); Telford Thrust plate, 40 km north of Fernie, British Columbia (49°45'N; 115°02'W); view to northwest; approximately 110 m of Ross Creek Formation exposed (photo - A. McGugan).

calcareous sandstone with *Zoophycos* occur in the upper part. The Telford is recognized only in southeastern British Columbia (McGugan and Rapson, 1963a), although possibly correlative carbonates are exposed between Banff and Jasper, in Alberta (Fig. 4F.2). A disconformity at the top of the Telford Formation is indicated by the absence of one conodont zone that is present in a more basinal section. This disconformity may account for the lack of any preserved shallow shelf facies at the top of the Telford. The Telford is disconformably overlain by lowest Artinskian (lower Leonardian) basinal to slope facies of the lower Ross Creek Formation (Fig. 4F.6), including phosphatic and calcareous siltstone, and minor black, spicular chert with a thin conglomeratic unit locally present at the base. The northward extent of this disconformity is uncertain, since it has not been recognized definitely elsewhere. The upper Ross Creek sediments, which prograded basinward (southwestward) over the lower, include lower to middle Artinskian silty and sandy carbonates, phosphatic siltstone, and sandstone, with abundant thin-shelled brachiopods. A mixed-skeletal carbonate unit, 4 m thick, containing solitary corals, productid brachiopods, echinoderms, bryozoans, and oolites, occurs locally near the base of the upper Ross Creek Formation, suggesting deposition in the upper slope and possibly the outer shallow shelf of a ramp (Richards et al., Subchapter 4E in this volume; Fig. 4E.5). Stromatolites at the top indicate deposition in the intertidal to supratidal zones. The Ross Creek Formation (lower and upper), like the Telford, is known only from southeastern British Columbia, where it has an average thickness of 150 m (McGugan and Rapson, 1963a). Unnamed, correlative carbonates between Banff and Jasper are discussed below.

A westerly Permian facies (Fig. 4F.4, 4F.7) is thinner and characterized by black, spicular chert, platy, phosphatic siltstone, and minor pelmatozoan wackestone. Rhythmic bedding and abundant phosphorite and the absence of shallow-marine sedimentary structures suggests slope or basinal depositional environments. It has been referred to the Johnston Canyon Formation in the Banff area. There, in the type section (McGugan and Rapson, 1963a), the Johnston Canyon consists of 45 m of rhythmically bedded, resistant, phosphatic siltstone and recessive, shaly siltstone. It unconformably overlies the Upper Carboniferous Spray Lakes Group (Bashkirian to lower Moscovian) and is disconformably overlain by the upper Artinskian to Wordian Ranger Canyon Formation. The Johnston Canyon thins rapidly eastward to a few centimetres of conglomerate. Conodont studies suggest that at least part of the Johnston Canyon in the Banff area is correlative with the Ross Creek Formation of southeastern British Columbia (Fig. 4F.1). Strata in the Hosmer Thrust Plate and Lizard Range sections (Fig. 4F.4; MacRae and McGugan, 1977) also contain conodonts that indicate correlation with the nearby Telford and Ross Creek formations. The Hosmer and Lizard Range sections are presently in juxtaposition with the Telford thrust plate sequence but were deposited farther west.

The most easterly succession in the southern Rocky Mountains begins with the Johnston Canyon Formation (Sakmarian), which consists of phosphatic siltstone and

Figure 4F.7. Type section of Ishbel Group; west flank of Mt. Ishbel, Sawback Range, near Banff, Alberta (51°15′N; 115°47′W); Johnston Canyon Formation (rhythmically bedded, resistant, phosphatic siltstone and recessive, argillaceous siltstone-JC) and Ranger Canyon Formation (blue-grey chert-RCa) separated by regional intra-Permian disconformity; view to southeast; approximately 10 m of Ranger Canyon Formation shown (photo - C.M. Henderson).

Figure 4F.8. Belcourt (Be) and Mowitch (Mo) formations, near Kvass Creek, west-central Alberta (53°37′18″N; 119°07′43″W); Belcourt Formation separated by regional disconformity from underlying, Lower Carboniferous Mount Head Formation (MH) (thickness shown about 70 m); Lower Triassic (Ѣ) shale and siltstone on skyline; view to southwest (photo - B.C. Richards, ISPG 1988-8).

minor fusulinid-bearing wackestone lenses, suggesting deposition on the slope to outer shallow shelf of a ramp. Between Banff and Jasper (Fig. 4F.1, 4F.2), the Johnston Canyon is overlain by a 15 to 30 m succession of sandy carbonates (McGugan and Rapson, 1963b) that resembles the Telford and Ross Creek formations to the south and a carbonate-rich facies of the Kindle Formation to the north. Conodont studies indicate that these thin eastern facies are in part correlative with the Ross Creek (lower Artinskian). Eastward thinning of the Permian succession is related to a combination of onlap and truncation beneath the intra-Permian disconformity.

Strata above the intra-Permian disconformity are remarkably consistent in lithology and thickness from southeastern British Columbia to the Jasper area in western Alberta (Fig. 4F.3; McGugan et al., 1964). Over most of this area the interval comprises a relatively thin but widespread sequence of blue-grey chert, silicified sandstone, and phosphatic siltstone of the Ranger Canyon Formation (Fig. 4F.7). It is a slope to basinal facies, as judged by the presence of rhythmic bedding and abundant phosphatic material, the predominance of spicules, and the absence of shallow-water sedimentary structures and fossils. A basal unit of very thin, but laterally persistent,

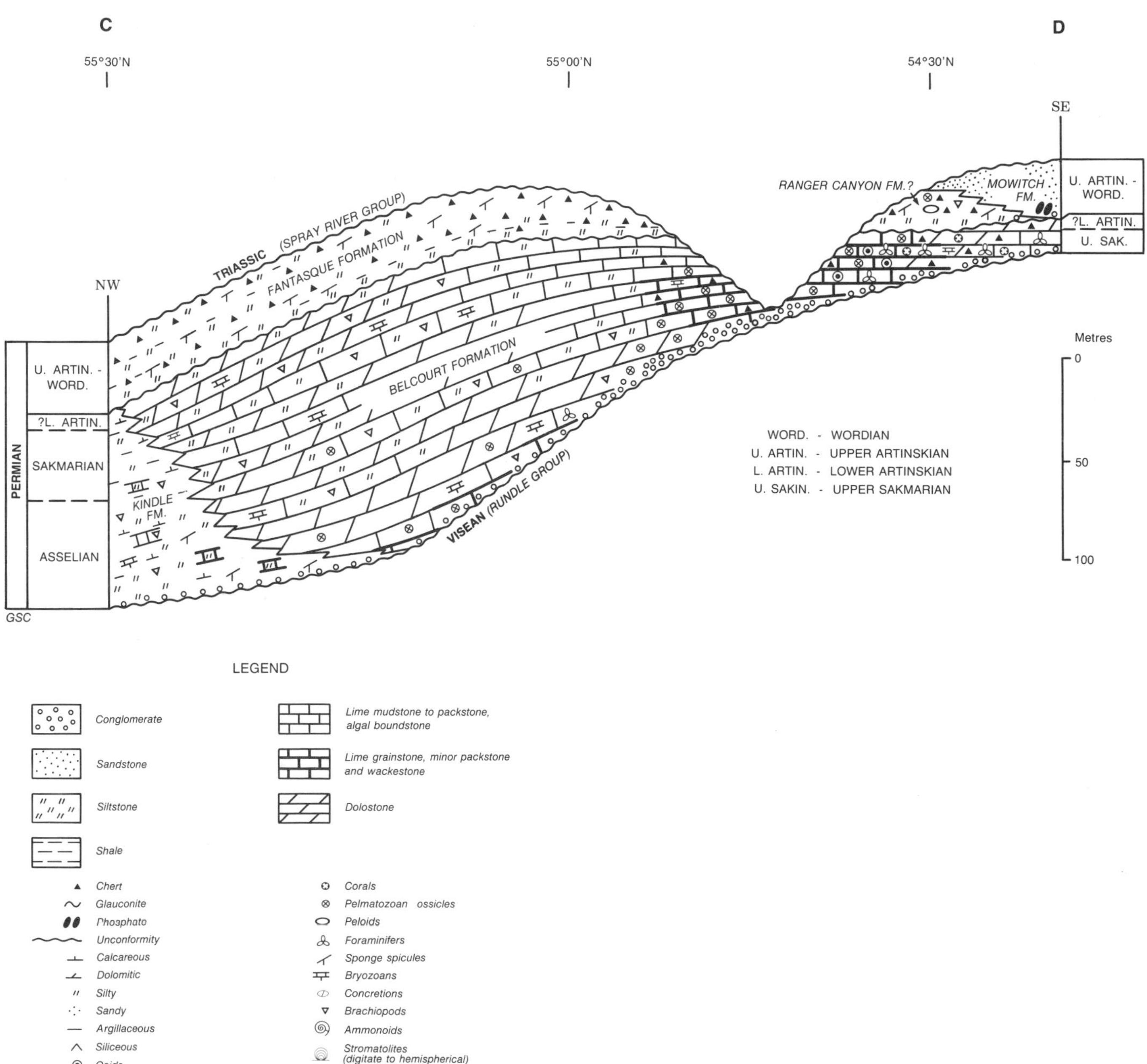

Figure 4F.9. Stratigraphic cross-section (schematic) of Permian formations, central Rocky Mountains and Foothills, east-central British Columbia (see Fig. 4F.2, C-D).

phosphatic chert-pebble conglomerate, containing inarticulate brachiopods, sponge spicules, phosphatized bone fragments and *Helicoprion* sp., onlaps the intra-Permian disconformity. In the type area at Banff, the thickness of the Ranger Canyon ranges from less than 1 m to 45 m, but is generally about 10 m (McGugan and Rapson, 1963a). In the Jasper area, the formation grades into an eastern facies of medium-bedded, brown, marine, quartzose sandstone of the Mowitch Formation (Fig. 4F.8). There, the Mowitch conformably overlies and possibly interfingers with the Ranger Canyon (Fig. 4F.9; McGugan, 1984). The ages of these two formations are poorly known, but data from conodont and rare brachiopod occurrences indicate late Artinskian to Wordian. Significant stratigraphic condensation has been postulated for this thin interval (MacRae and McGugan, 1977), because of the considerable time span that it represents.

Three major transgressions (see Fig. 4F.16) are represented by the deeper water sediments of the Johnston Canyon Formation (Asselian-early Sakmarian), the lower Ross Creek Formation (early Artinskian), and the Ranger Canyon Formation (?late Artinskian-Wordian). Major regressions are suggested by the shallower water sediments of the Telford and upper Ross Creek formations. This succession is similar to that in the Cassia Mountains of Idaho, where a corresponding sequence of transgressions and regressions has been recognized (Mytton et al., 1983).

CENTRAL AND NORTHERN ROCKY MOUNTAINS, MACKENZIE FOLD BELT, AND INTERIOR PLATFORM (WEST-CENTRAL ALBERTA TO SOUTHWESTERN DISTRICT OF MACKENZIE)

From west-central Alberta to southwestern District of Mackenzie, the Permian succession rests disconformably on Viséan to Serpukhovian strata and contains abundant, thin, slope to basinal facies similar to those in the southern Rocky Mountains. In addition, extensive shallow-neritic (above normal wave-base) to supratidal facies are developed in the eastern outcrop belt and adjacent subsurface (Fig. 4F.1, 4F.2).

The strata below the intra-Permian disconformity have been most thoroughly studied in the Rocky Mountains south of Pine Pass in east-central British Columbia (McGugan and Rapson-McGugan, 1976; Bamber and Macqueen, 1979). There (Fig. 4F.9), Lower Permian carbonates of the Belcourt Formation rest disconformably on Viséan carbonates (Rundle Formation) and shale (Besa River Formation). Temporal and facies relationships within the diachronous deposits of the Belcourt Formation indicate regional eastward transgression during Early Permian time (Fig. 4F.9, 4F.16; Bamber and Macqueen, 1979). A widespread limestone conglomerate (up to 6 m thick), containing pebbles of upper Tournaisian and Viséan chert and carbonate, forms the base of the Belcourt. In the northwestern part of the area the conglomerate is overlain by relatively thick (44-138 m), Asselian to ?lower Artinskian carbonates consisting mainly of silty, mixed-skeletal wackestone and packstone and finely crystalline dolostone. The base of the Belcourt becomes progressively younger southeastward along the outcrop belt, where the basal conglomerate is directly overlain by upper Sakmarian (Sterlitamakian), fusulinid-and coral-bearing, shallow-neritic to supratidal carbonates. Within this eastern facies, which contains the type section of the Belcourt Formation (Forbes and McGugan, 1959), shallow-neritic, oolitic and skeletal grainstone grades eastward into thin (3.8-10 m), planar-bedded, microcrystalline dolomite, characteristic of intertidal to supratidal environments. This shallow-marine carbonate belt extends farther southeastward into west-central Alberta (Fig. 4F.2; McGugan et al., 1964), where it exhibits similar facies relationships. Lithofacies trends within the Belcourt record deposition on a carbonate ramp (Richards et al., Subchapter 4E in this volume; Fig. 4F.4).

The Permian succession of the Interior Platform to the east, in the subsurface of west-central Alberta and northeastern British Columbia, is assigned to the Belloy Formation, which consists of three informal members – lower carbonate, middle sandstone, and upper carbonate (Halbertsma, 1959). The thickness of the formation varies from a zero edge in the east to more than 180 m near the Foothills south of Fort St. John. Isopachs for the Permian strata demonstrate northwest trends (Sikabonyi and Rodgers, 1959), which are related to facies distribution (McGugan et al., 1964) and appear to parallel pre-Belloy or syndepositional block-faulting trends, with carbonates developed on the highs (Fig. 4F.2). The Belloy Formation

Figure 4F.10. Kindle Formation, Tika Creek, southeastern Yukon Territory (60°43'N; 124°53'W); rhythmically bedded, calcareous siltstone and shale of upper part of formation. Bar represents 1.5 m (photo - B.C. Richards, ISPG 798(67)).

is separated from the correlative Belcourt Formation by an area within which Permian strata are unobserved and possibly absent. This area may have been high during Permian time, judging by the marked contrast between facies on either side of it and the presence, directly to the west, of an onlapping, shallow-marine succession (Belcourt Formation) with a thick basal conglomerate (Fig. 4F.9). It cannot be discounted, however, that this area may be merely an artifact related to lack of well control.

The base of the Belloy Formation is marked by phosphatic chert conglomerate. In the area east of Fort St. John, this conglomerate is overlain by the lower carbonate member, which consists of very fine grained, glauconitic sandstone, fossiliferous white chert, and dolomitic limestone. The middle sandstone member consists of coarsening-upward, glauconitic, very fine to medium grained sandstone with abundant phosphate, dominantly in the form of fish debris. The upper carbonate

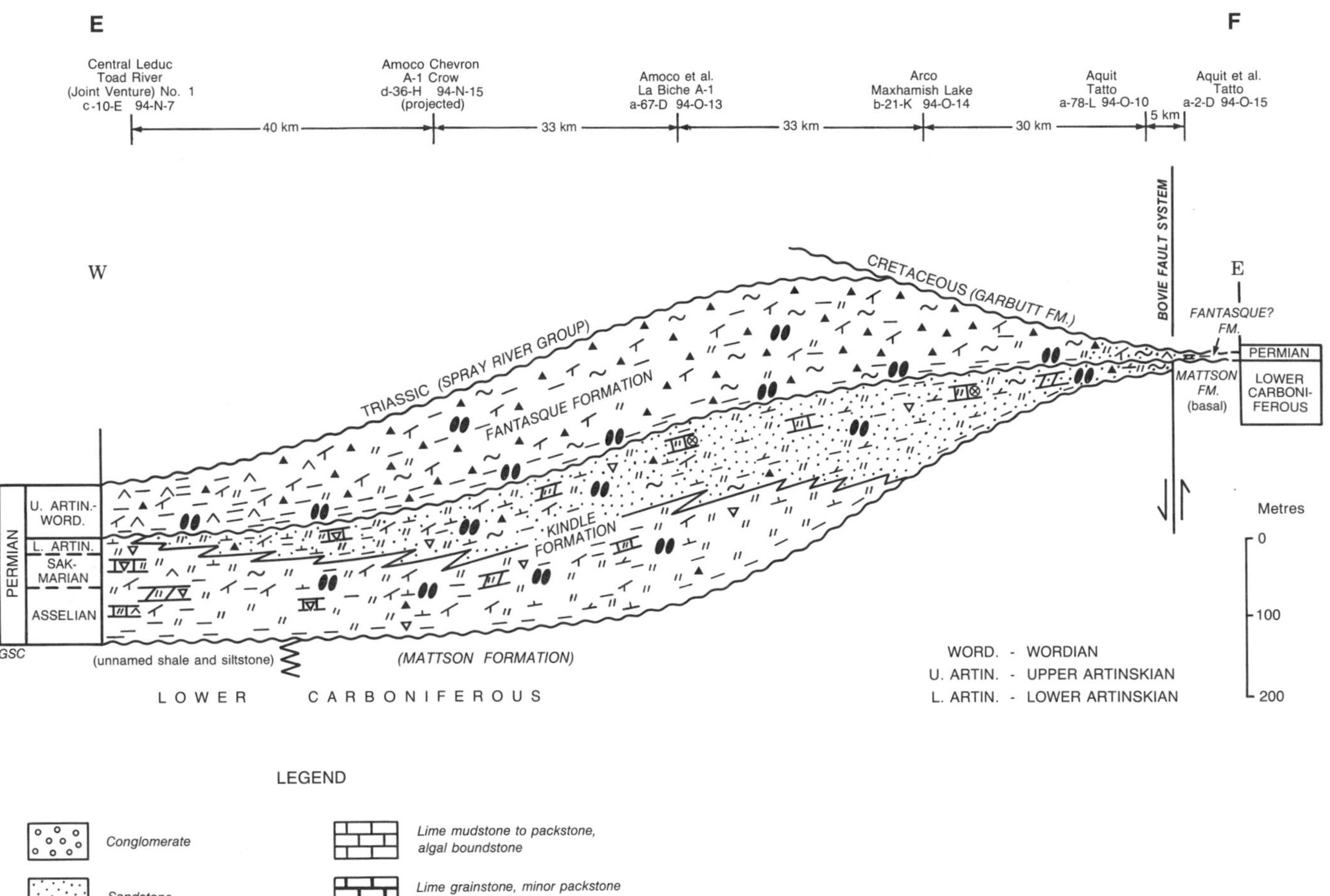

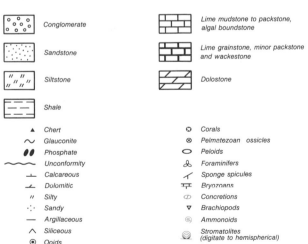

Figure 4F.11. Stratigraphic cross-section (schematic) of Permian formations, Interior Platform northeastern British Columbia (see Fig. 4F.2, E-F).

member consists of minor, fine-grained, glauconitic, quartzose sandstone, dolomitic limestone and, in the type section, brachiopod-rich, coarsely crystalline, calcareous dolostone. This eastern, shallow-neritic assemblage is similar to facies within the Belcourt Formation to the west and the Kindle Formation to the northwest. South of Fort St. John, the Belloy comprises mainly slope to basinal siltstone, a facies similar to part of the Kindle Formation and markedly different from the Belcourt Formation.

The Belloy Formation appears to correlate with only those formations below the intra-Permian disconformity (Fig. 4F.1), suggesting that Fantasque or Ranger Canyon equivalents are absent from the Interior Platform (Naqvi, 1972). Conodont studies indicate that the upper part of the middle sandstone member is correlative with the lower to middle Telford Formation (?Asselian to lower Sakmarian), suggesting that the upper carbonate member may be correlative with the upper Telford and Ross Creek formations (upper Sakmarian to lower Artinskian). Age determinations for the uppermost Belloy are based on palynomorphs that indicate a Leonardian and possible Guadalupian age (Jansonius, 1962), and on ammonoids that indicate an early Artinskian age (Nassichuk, 1969; pers. comm., 1985).

North of Pine Pass, in the northern Rocky Mountains, southern Mackenzie Fold Belt, and Interior Platform, strata correlative with the Belcourt Formation are assigned to the Kindle Formation (Fig. 4F.10; Bamber et al., 1968; Laudon and Chronic, 1949). The Kindle consists of Asselian to ?lower Artinskian, basinal to shallow-neritic siliciclastics and silty carbonates resting disconformably on Carboniferous, Viséan to Serpukhovian strata (Prophet and Mattson formations, Stoddart Group; Richards et al., Subchapter 4E in this volume). As in the area south of Pine Pass, facies relationships indicate regional, eastward transgression during earliest Permian time (Fig. 4F.11), but detailed age relationships within the Kindle have not been determined. At its type section in the western part of the outcrop belt, (Laudon and Chronic, 1949), the Kindle comprises basinal to slope siltstone, shale, and subordinate silty carbonates, with a maximum thickness of 205 m. The formation becomes phosphatic, glauconitic, and highly calcareous eastward in the outcrop belt. There, it comprises siltstone units with subordinate calcareous sandstone and shale, which were deposited in slope environments. These deposits have features typical of turbidites, such as rhythmic bedding and sharp-based beds that are commonly graded. Their occurrence directly above the subaerial erosion surface at the base of the Kindle records the establishment of moderately deep marine environments during the initial Permian transgression (see Fig. 4F.16). An increase in the proportion of calcareous sandstone toward the top of the formation suggests shallowing during regional regression. Sandstone increases to the east also, as the slope deposits of the outcrop belt pass laterally into an eastern, progradational, slope to shallow-neritic facies consisting of coarsening-upward hemicycles of shale, siltstone, and crossbedded sandstone. This facies, similar to one in the contemporaneous Belloy Formation (Fig. 4F.1), is present in the western Interior Platform of northeastern British Columbia and easternmost Mackenzie Fold Belt, where the Kindle reaches a maximum thickness of 133 m and thins rapidly eastward to its subcrop edge (Fig. 4F.2, 4F.11).

An unnamed, Lower Permian shale and sandstone unit, correlative with the Kindle Formation, is present to the north, in west-central Mackenzie Mountains and Selwyn Mountains (Gordey et al., 1982). In the intervening area (Fig. 4F.2), Mesozoic strata rest disconformably on Carboniferous and older rocks. Facies relationships within the northern succession have not been described, and the nature of the basal contact with the Carboniferous succession has not been determined.

The interval above the intra-Permian disconformity consists mainly of slope to basinal deposits, comprising rhythmically bedded, spicular chert, shale, and siltstone of the Fantasque Formation (Fig. 4F.12). This thin, widespread unit, which closely resembles and is correlative with the Ranger Canyon Formation of the southern Rocky Mountains, extends from the Pine Pass area (Fig. 4F.3) into the Interior Platform and southern Mackenzie Fold Belt. A thin lag deposit of phosphate and chert nodules and

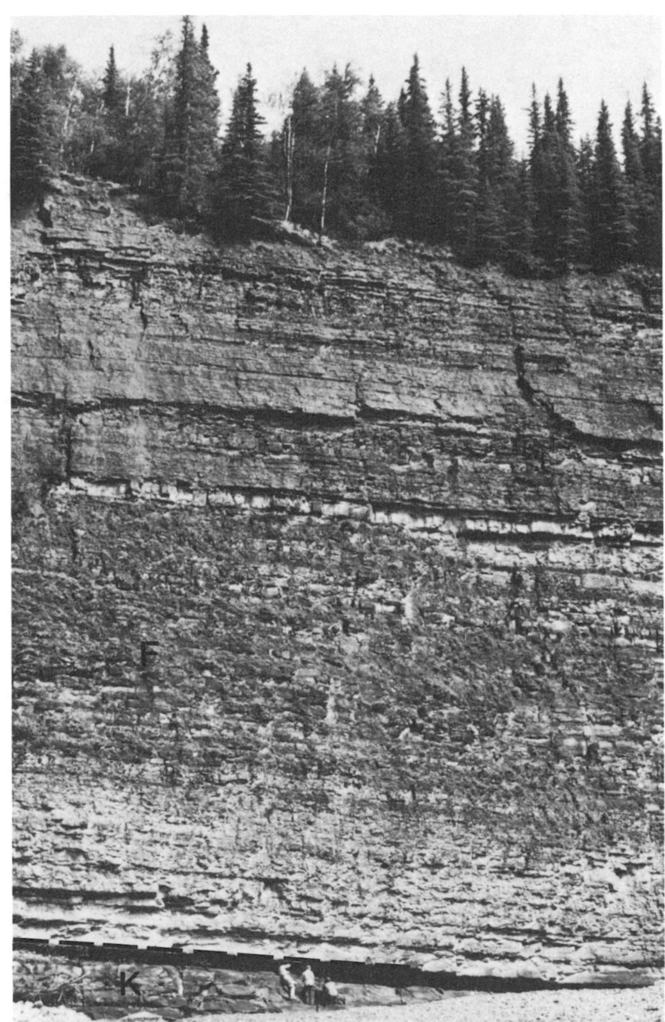

Figure 4F.12. Fantasque Formation, Kotaneelee River, southwest District of Mackenzie (60°30'N; 124°05'W); Fantasque (F) rests disconformably on Kindle Formation (K); view to north. Persons at base of exposure provide a scale (photo - B.C. Richards, ISPG 2378(3)).

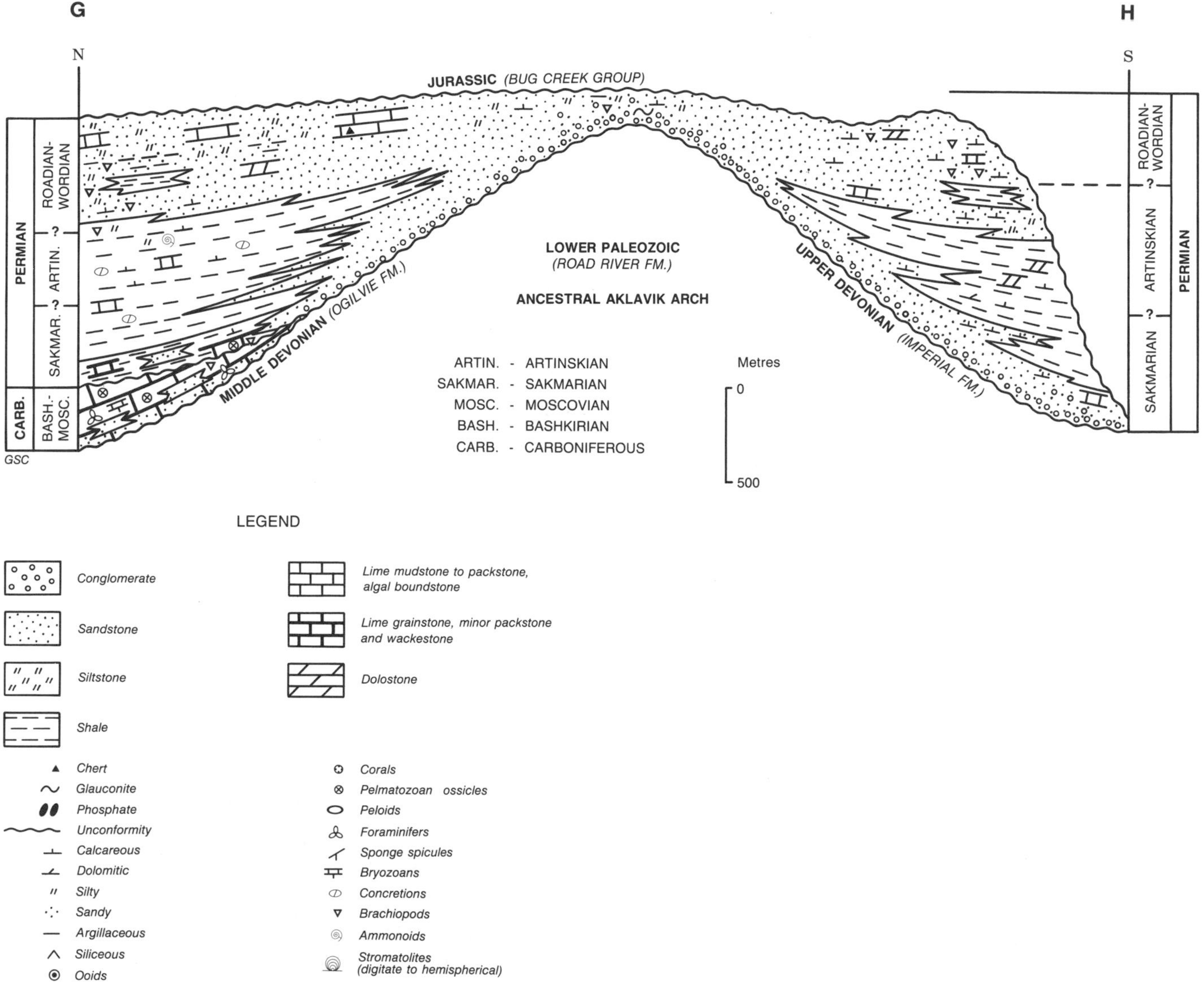

Figure 4F.13. Stratigraphic cross-section (schematic) of unnamed Permian formations, northern Richardson Mountains (see Fig. 4F.2, G-H).

pebbles, which accumulated during a second, regional, eastward transgression (see Fig. 4F.16), marks the disconformity (hiatal surface?) at the base of the Fantasque throughout most of the area. In easternmost sections of the western Interior Platform, a regolith consisting of partly chertified, brecciated sandstone occurs at this level. Thin beds and laminae of shale and siltstone form a minor part of the formation in eastern sections and are abundant near the base (Fig. 4F.9, 4F.11). Phosphate pellets and nodules occur locally, and glauconite is very abundant in the chert of eastern sections. The proportion of siliciclastics increases westward as the Fantasque grades laterally into basinal shale and mudstone. A unit of chert and shale resembling the Fantasque Formation caps the Permian succession to the north, in west-central Mackenzie Mountains and Selwyn Mountains (Gordey et al., 1982). Lateral facies relationships within this unit have not been determined. In east-central British Columbia, eastern equivalents of the Fantasque are light grey, silicified, skeletal carbonates interbedded with fine-grained quartz sandstone. This facies, assigned to the Ranger Canyon Formation (McGugan and Rapson-McGugan, 1976; Bamber and Macqueen, 1979), was deposited in shallow-neritic to intertidal environments. It appears to grade southeastward into intertidal to shallow-neritic, cherty, phosphatic chert arenite and conglomerate of the Mowitch Formation, that occupies the entire interval above the Belcourt Formation in easternmost outcrops of east-central British Columbia and west-central Alberta (McGugan et al., 1964).

NORTHERN CORDILLERA - OGILVIE MOUNTAINS TO BEAUFORT SEA (NORTHERN YUKON TERRITORY AND NORTHWESTERN DISTRICT OF MACKENZIE)

Permian rocks of the northern Cordillera are preserved mainly in the Ogilvie Mountains and southern Eagle Plain, the northern Richardson Mountains, and the British Mountains (Fig. 4F.1, 4F.2, 4F.3). Asselian to Wordian marine strata occur along the flanks of the Ancestral Aklavik Arch in facies belts trending northeast, approximately at right angles to those of the underlying Carboniferous strata, as noted by Bamber and Waterhouse (1971).

Marine transgression toward the axis of the arch began in Asselian time (see Fig. 4F.16) and continued, with minor breaks, until Wordian time, when the arch was covered by marine sediments. This depositional pattern is best illustrated by facies relationships in the northern Richardson Mountains (Fig. 4F.13), where the Sakmarian to Wordian succession adjacent to the arch consists of three unnamed units (Fig. 4F.1; Bamber and Waterhouse, 1971) - a basal unit of sandstone and conglomerate (100-200 m), a middle unit of calcareous and silty shale with thin carbonate beds and nodules (200-?700 m), and an upper unit of calcareous sandstone with subordinate siltstone and shale (250-800 m). The basal unit, unconformable on Devonian strata (Fig. 4F.14), contains Sakmarian faunas on the flanks of the arch. It becomes progressively younger toward the central part of the arch, where it is directly overlain by ?upper Artinskian to Wordian siliciclastics and rests with angular unconformity on rocks as old as Early Silurian. At all levels, coarse-grained, shallow-marine siliciclastics on both flanks of the arch pass outward, away from the axis, into laterally equivalent, finer grained siliciclastics and carbonates. On the north side of the arch, Permian strata extend into the northernmost Richardson Mountains, where shallow water carbonates and associated ?hydrozoan mounds are developed near the base. South of the arch and eastward toward the Interior Platform, the Permian succession is truncated within a short distance beneath a regional sub-Jurassic disconformity.

To the southwest, in northern Eagle Plain, Permian strata are absent as a result of truncation beneath a regional sub-Cretaceous unconformity (Fig. 4F.2, 4F.3; Pugh, 1983). In northern Ogilvie Mountains near the Yukon-Alaska border, a thick, varied assemblage of Asselian to Wordian siliciclastics, carbonates, and chert is present on the south flank of the Ancestral Aklavik Arch. The nature and magnitude of the sub-Permian disconformity there have not been clearly demonstrated (Waterhouse and Waddington, 1982), but deposition may have been continuous locally across the Carboniferous-Permian boundary. Gzhelian and lower Asselian conglomerate, sandstone, and sandy carbonates of the lower Jungle Creek Formation (137-220 m) rest on Upper Carboniferous (Moscovian or younger) carbonates and siliciclastics of the Ettrain and Blackie formations (Fig. 4F.15). The upper Jungle Creek Formation (290 to more than 440 m) comprises Asselian to Roadian sandstone, shale, siltstone, and subordinate carbonates. It is overlain by conglomeratic, glauconitic lime grainstone and spicular chert of the Artinskian to Wordian Tahkandit Formation (29-410 m). Abundant chert pebbles and beds of chert-pebble conglomerate occur in the limestone of the lower Tahkandit. The Jungle Creek-Tahkandit contact, which marks a change from shelf slope to shallow-neritic depositional environments, appears to be disconformable, but its relationship to well developed disconformities in the southern Cordillera (Fig. 4F.16) is unknown. Near the axis of the Ancestral Aklavik Arch to the northwest, in adjacent, east-central Alaska, the Permian consists of Roadian to Wordian conglomerate, sandstone, and lime grainstone of the Tahkandit and Step formations. These rest unconformably on Precambrian to Lower Carboniferous strata (Brabb and Churkin, 1969). Thus, the base of the Permian in the northern Ogilvie Mountains becomes younger toward the axis of the arch, as it does in the northern Richardson Mountains.

Southward and southeastward, away from the arch, the shallow-neritic assemblage of the northern Ogilvie Mountains grades into spicular carbonates, chert, and shale of slope origin (?Blackie Formation; Pugh, 1983). In southern Eagle Plain, to the east, only the Jungle Creek Formation is preserved beneath the sub-Cretaceous unconformity (Fig. 4F.2, 4F.15). The Jungle Creek consists of fine-grained sandstone, siltstone, shale, and subordinate carbonates, and appears to pass southward, through the

Figure 4F.14. Permian siliciclastics on Sheep Creek, northern Richardson Mountains (67°40'N; 136°20'W); unnamed basal sandstone (bss) and conglomerate resting with angular unconformity on Upper Devonian Imperial Formation (Im) (foreground); middle, unnamed shale unit (ms) on skyline saddle (middle background); unnamed, upper sandstone unit (uss) on skyline (upper right background); view to southwest along south flank of Ancestral Aklavik Arch (photo - D.K. Norris, GSC 662-126).

eastern Ogilvie Mountains, into an undifferentiated succession of slope to basinal shale, siltstone, and silty carbonates (Graham, 1973).

The Permian succession is poorly known near the axis of the Ancestral Aklavik Arch and on its north flank in the central and western Yukon Territory. In the British Mountains, well to the north of the axial area, Lower Permian shale, sandstone, and minor carbonates of the Sadlerochit Formation, up to 200 m thick, are preserved locally as erosional remnants above Upper Carboniferous, Bashkirian carbonates of the Lisburne Group (Bamber and Waterhouse, 1971). Permian facies relationships in this northern area are unstudied.

BIOSTRATIGRAPHY

Permian strata of the eastern Cordillera have been dated previously by ammonoids, brachiopods, fusulinaceans, and palynomorphs (Bamber and Copeland, 1970). Zonal schemes have been proposed for brachiopods (northern Yukon Territory) and fusulinaceans (northeastern British Columbia); however, the geographic separation of these studies make interregional correlations difficult. Furthermore, these faunas are largely restricted to shallow shelf facies and are therefore difficult to relate to faunas in slope and basinal facies of southern British Columbia and Alberta. Recent conodont studies have dated these slope and basinal

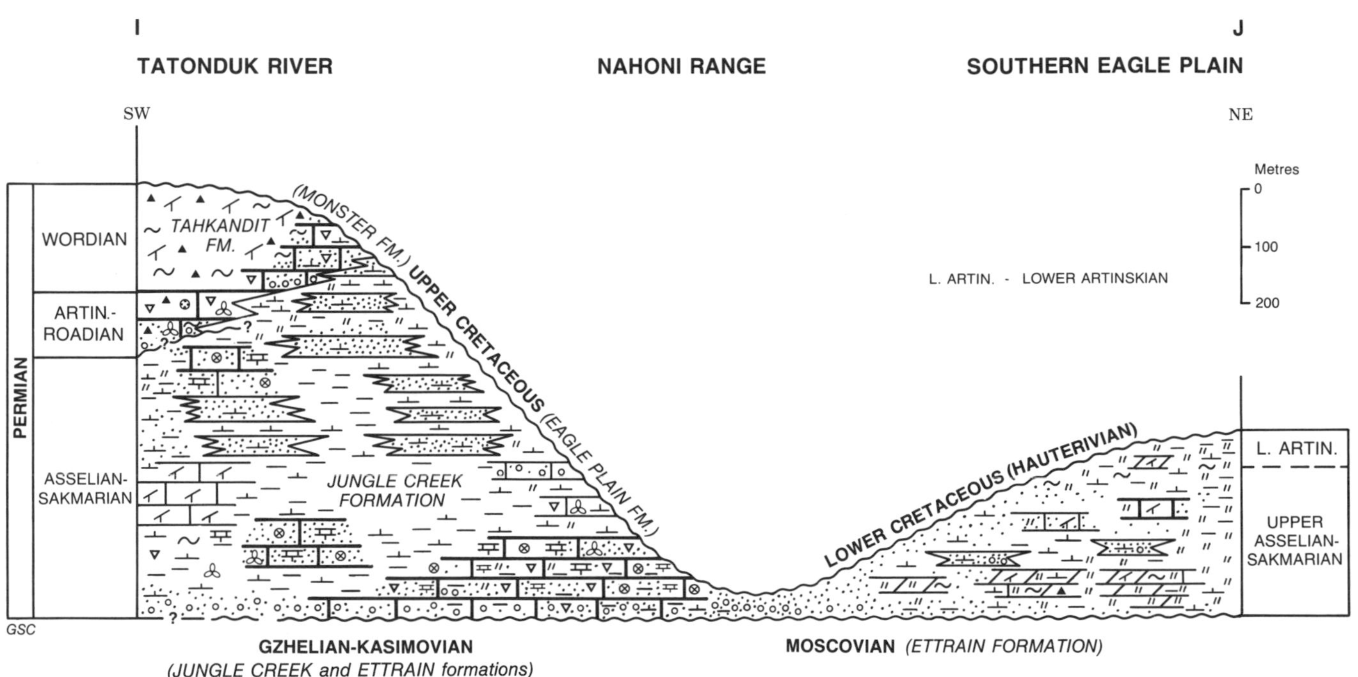

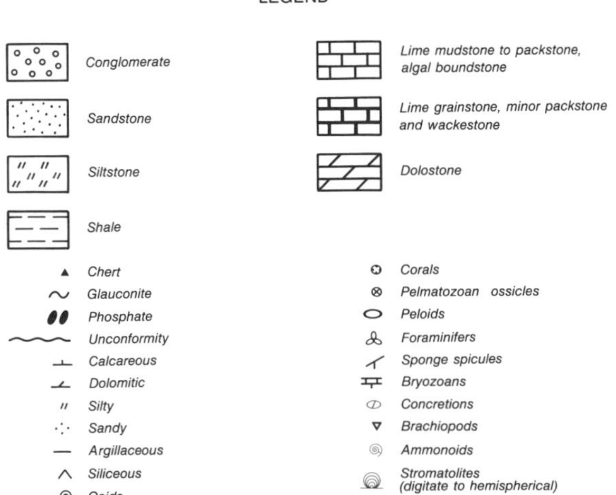

Figure 4F.15. Stratigraphic cross-section (schematic) of Permian formations, Ogilvie Mountains and Eagle Plain (see Fig. 4F.2, I-J).

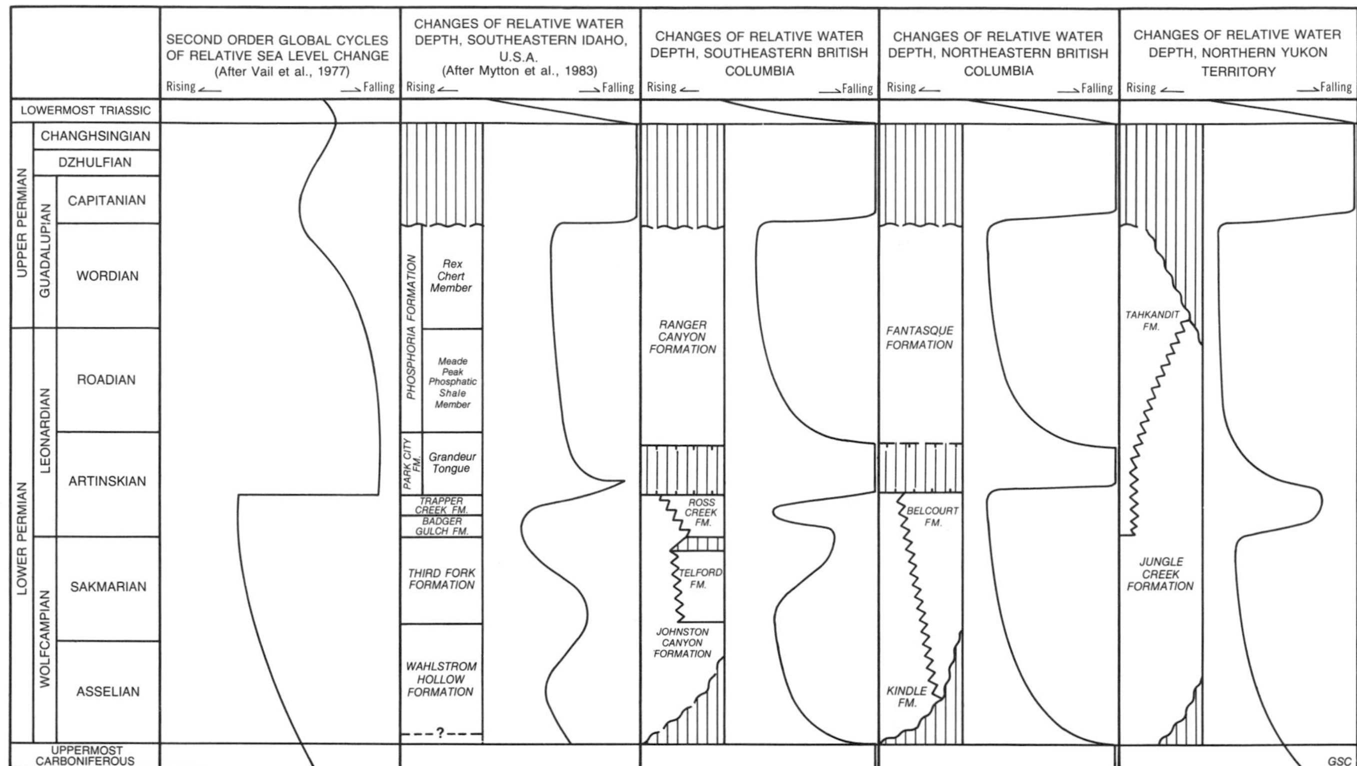

Figure 4F.16. Comparison of water depth changes as reflected by vertical facies variations during the Permian, Eastern Cordillera, with second-order relative sea-level cycles proposed by Vail et al. (1977).

sediments (Henderson and McGugan, 1986). Isolated samples from other regions indicate that some species found in southern British Columbia and Alberta occur in several correlative facies, which will provide the link for integration with paleontological data from other fossil groups.

Previously published zonations are discussed only briefly in the following account. Conodont zones that were not discussed by Henderson and McGugan (1986) are dealt with in more detail. The zonal schemes applied to western Canadian stratigraphy are summarized in Figures 4F.17, 4F.18, and 4F.19.

Conodont zonation

Reconnaissance conodont studies from the southern Canadian Rocky Mountains (Henderson and McGugan, 1983, 1985, 1986) indicate the presence of at least eight zones (Fig. 4F.17), three of which are based on single localities and were not discussed in Henderson and McGugan (1986).

Strata immediately above the sub-Permian disconformity contain, in addition to definite Permian conodonts, *Neognathodus* spp., which are either homeomorphs of Bashkirian (Carboniferous) species or reworked from Bashkirian strata.

Idiognathodus ellisoni *Zone*

This zone is present in the Johnston Canyon and lower Telford formations at the Telford thrust plate (Henderson and McGugan, 1986) and in the lower to middle Belloy Formation in the subsurface of west-central Alberta. It has been assigned a Virgilian to Tastubian (early Sakmarian) age range by Clark et al. (1979) in the Great Basin, southwest U.S.A.

Streptognathodus elongatus *Zone*

This zone occurs in the upper Telford Formation at the Telford thrust plate (Henderson and McGugan, 1986). Clark et al. (1979) have shown that *S. elongatus* occurs above the range of *Idiognathodus ellisoni* in southwest U.S.A., indicating a latest Tastubian? to Sterlitamakian (late Sakmarian) age.

Adetognathus paralautus – Sweetognathus *cf.* whitei *Zone*

This zone, which has not been described formally, was recognized from the basinal facies of the Johnston Canyon Formation and includes *Adetognathus* sp., *Neogondolella bisselli*, and *Sweetognathus* cf. *whitei*. It appears to correlate with an assemblage that includes *Neogondolella bisselli, Streptognathodus artinskiensis*, and *S. elongatus*

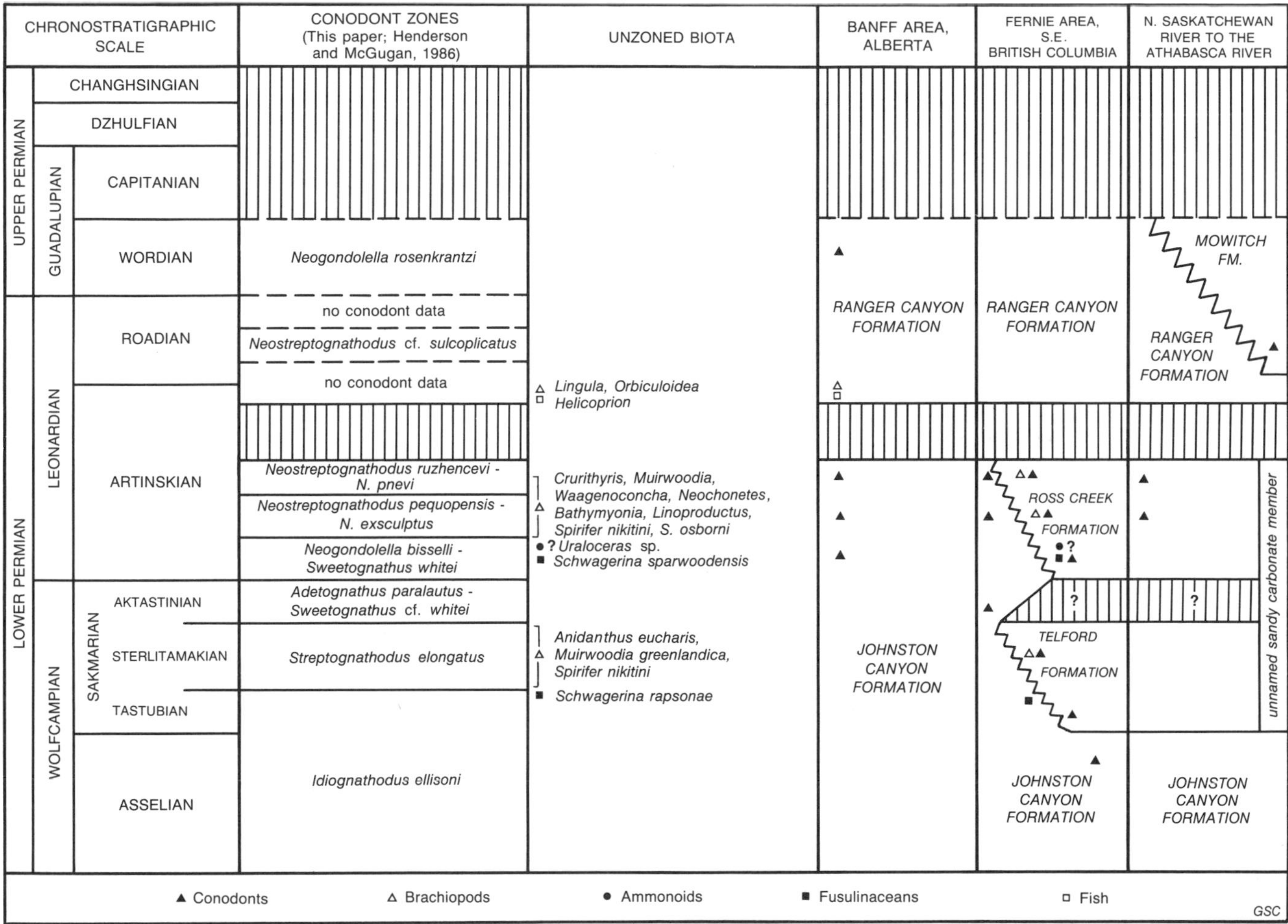

Figure 4F.17. Biostratigraphy of Eastern Cordillera, southwestern Alberta and southeastern British Columbia.

and is dated as Aktastinian (early Artinskian) by Movshovich et al. (1979), who correlate this zone with the uppermost Tensleep Formation of Wyoming (Rhodes, 1963), which yielded *Sweetognathus whitei* and *Streptognathodus* spp. Another probably correlative assemblage was recovered from the Harper Ranch beds, near Kamloops, British Columbia (Orchard, 1984) and includes *Adetognathus paralautus, Diplognathodus* sp., *Hindeodus minutus, Neogondolella bisselli,* and *Sweetognathus* cf. *whitei*. Brachiopods, conodonts, and fusulinids from this interval in the Ural Mountains, U.S.S.R. (Movshovich et al., 1979) have greater affinity with late Sakmarian (Sterlitamakian) than with Artinskian (Baigendzhinian) faunas. With this in view, Waterhouse (1976) included the Aktastinian horizon within the Sakmarian, and further stated that the "classical" Artinskian of the Arti region is entirely Baigendzhinian.

*Addendum: *Sweetognathus* cf. *whitei* is now recognized as *S. inornatus*.

Neogondolella bisselli – Sweetognathus whitei *Zone*

This zone is recognized in the lower Ross Creek Formation (Henderson and McGugan, 1986). Considering the argument for the age of the previous zone, the *bisselli-whitei* zone is regarded as early Artinskian (Baigendzhinian). This zone was established in the southwest U.S.A. by Clark and Behnken (1971), who assigned a late Wolfcampian age. However, it is recognized in the basal 15 m of the Skinner Ranch Formation of west Texas (Behnken, 1975), which is Leonardian by definition. Thus the Sakmarian/Artinskian boundary, as recognized herein, is equivalent to the Wolfcampian/Leonardian boundary of southwestern U.S.A.

Neostreptognathodus pequopensis – N. exsculptus *Zone*

This zone, present in the lower part of the upper Ross Creek Formation (Henderson, 1986), was established in southwest U.S.A. as the *N. pequopensis* zone by Clark (1974) and Behnken (1975), who assigned it to the Leonardian Stage. Movshovich et al. (1979) recognized the zone in Artinskian strata of the Ural Mountains. The distribution of *N. exsculptus* almost totally coincides with that of *N. pequopensis* at sections in the Telford thrust plate.

Neostreptognathodus ruzhencevi – N. pnevi *Zone*

This zone, which occurs in the upper part of the upper Ross Creek Formation (Henderson and McGugan, 1986), was established as the *N. ruzhencevi* zone by Movshovich et al. (1979) in upper Artinskian strata of the Ural Mountains.

Disconformably overlying the *ruzhencevi-pnevi* zone in southern British Columbia and Alberta is a condensed sequence referable to the Ranger Canyon and Mowitch formations. These formations have been dated as uppermost Artinskian to Wordian on the basis of *Helicoprion* and rare brachiopod occurrences. Conodonts have been recovered from only one locality in the Mowitch Formation between Banff and Jasper. All specimens belong to *Neostreptognathodus* cf. *sulcoplicatus*, which has been assigned an early Roadian age (Clark et al., 1979). A single location from the upper part of the Ranger Canyon Formation near Banff, Alberta, yielded species of *Neogondolella*, including *N. rosenkrantzi*, which is assigned to the Wordian *N. rosenkrantzi* zone (Wardlaw and Collinson, 1979).

Brachiopod zones

Zonation on the basis of brachiopods has been achieved only in northern Yukon Territory and northwestern District of Mackenzie (Fig. 4F.19; Bamber and Waterhouse, 1971; Sarytcheva and Waterhouse, 1972; Waterhouse and Waddington, 1982; Nelson, 1961, 1962; Nelson and

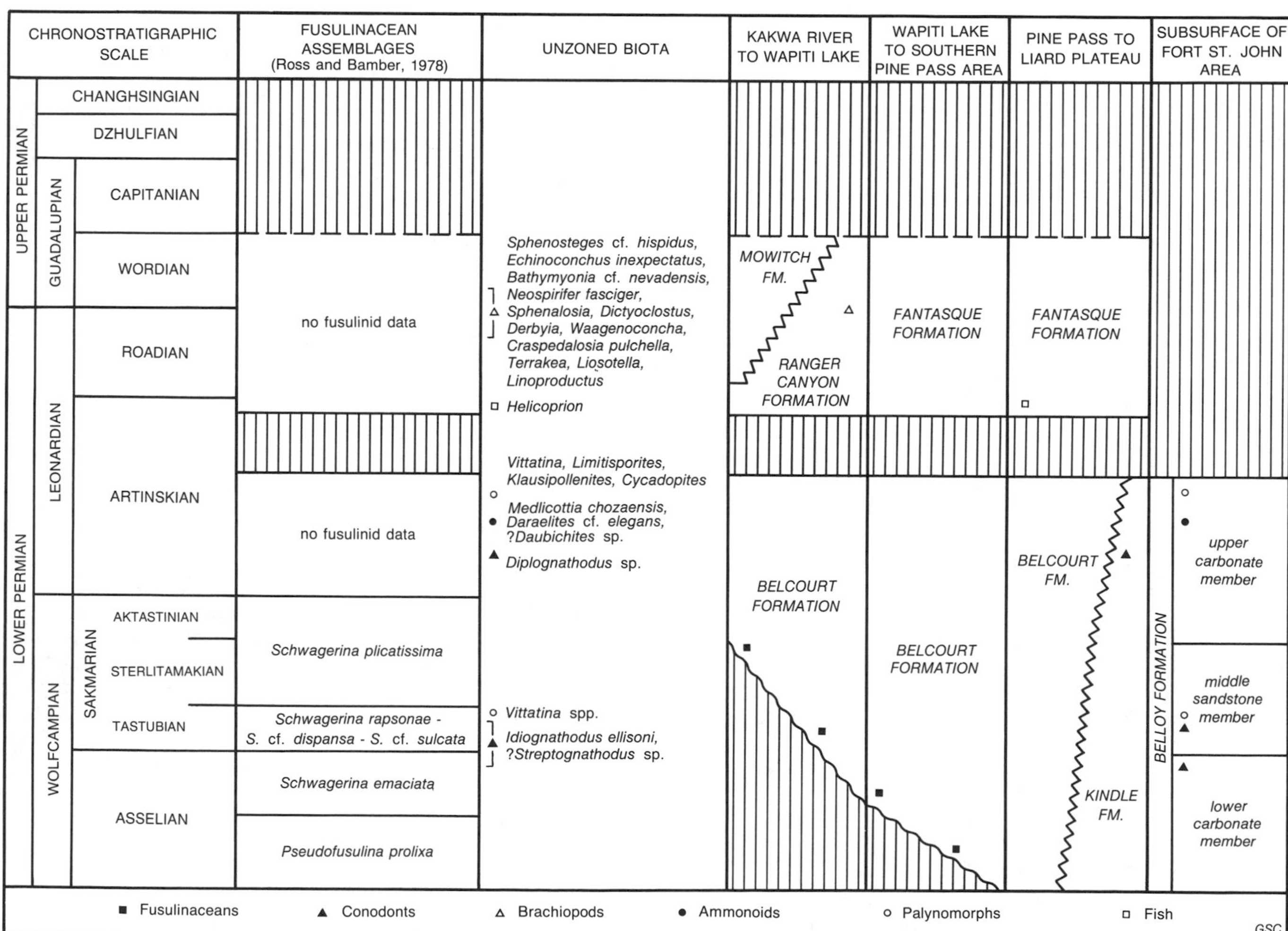

Figure 4F.18. Biostratigraphy of Eastern Cordillera and Interior Platform, west-central Alberta to southwestern District of Mackenzie.

Johnson, 1968). In other parts of the eastern Cordillera, brachiopods occur only locally and have not been studied systematically (Logan and McGugan, 1968; MacRae and McGugan, 1977).

In the northern Ogilvie Mountains of the Yukon Territory, the Jungle Creek Formation contains seven zones. These include three Asselian zones (*Kochiproductus-Attenuatella; Orthotichia;* and *Attenuatella-Tomiopsis*) and four Sakmarian (Tastubian to Aktastinian) zones (*Yakovlevia; Attenuatella; Tornquistia;* and *Jakutoproductus*). Deposition in this area was probably continuous across the Carboniferous/Permian boundary, but away from the northern Ogilvie Mountains several zones are missing at the boundary (Bamber and Waterhouse, 1971; Waterhouse and Waddington, 1982). The lower Tahkandit Formation of the Tatonduk River area includes two zones (*Antiquatonia* and *Sowerbina*) that are correlated with the Baigendzhinian (Artinskian). The middle and upper Tahkandit Formation contains three zones, *Pseudosyrinx* and *Thuleproductus* (Roadian) and *Cancrinelloides* (Wordian). In the northern Richardson Mountains four zones have been recorded, including the *Yakovlevia, Neochonetes, Lissochonetes,* and *Cancrinelloides* zones. The *Neochonetes* and *Lissochonetes* zones are correlative with the *Pseudosyrinx* and *Thuleproductus* zones of the Ogilvie Mountains, respectively.

Fusulinacean assemblages

Fusulinid data are available from several areas but are most complete in northeastern British Columbia, in the shallow shelf facies of the Belcourt Formation (Fig. 4F.18), where Ross and Bamber (1978) recognized several assemblages. These include the *Pseudofusulina prolixa* assemblage (lower to middle Asselian), the *Schwagerina emaciata* assemblage (middle to upper Asselian), the *S. rapsonae-S.* cf. *dispansa-S.* cf. *sulcata* assemblage (lower Sakmarian), and the *S. plicatissima* assemblage (upper Sakmarian). In southern British Columbia, *S. rapsonae* was reported (McGugan, 1963) from a carbonate lens near the top of the Johnston Canyon

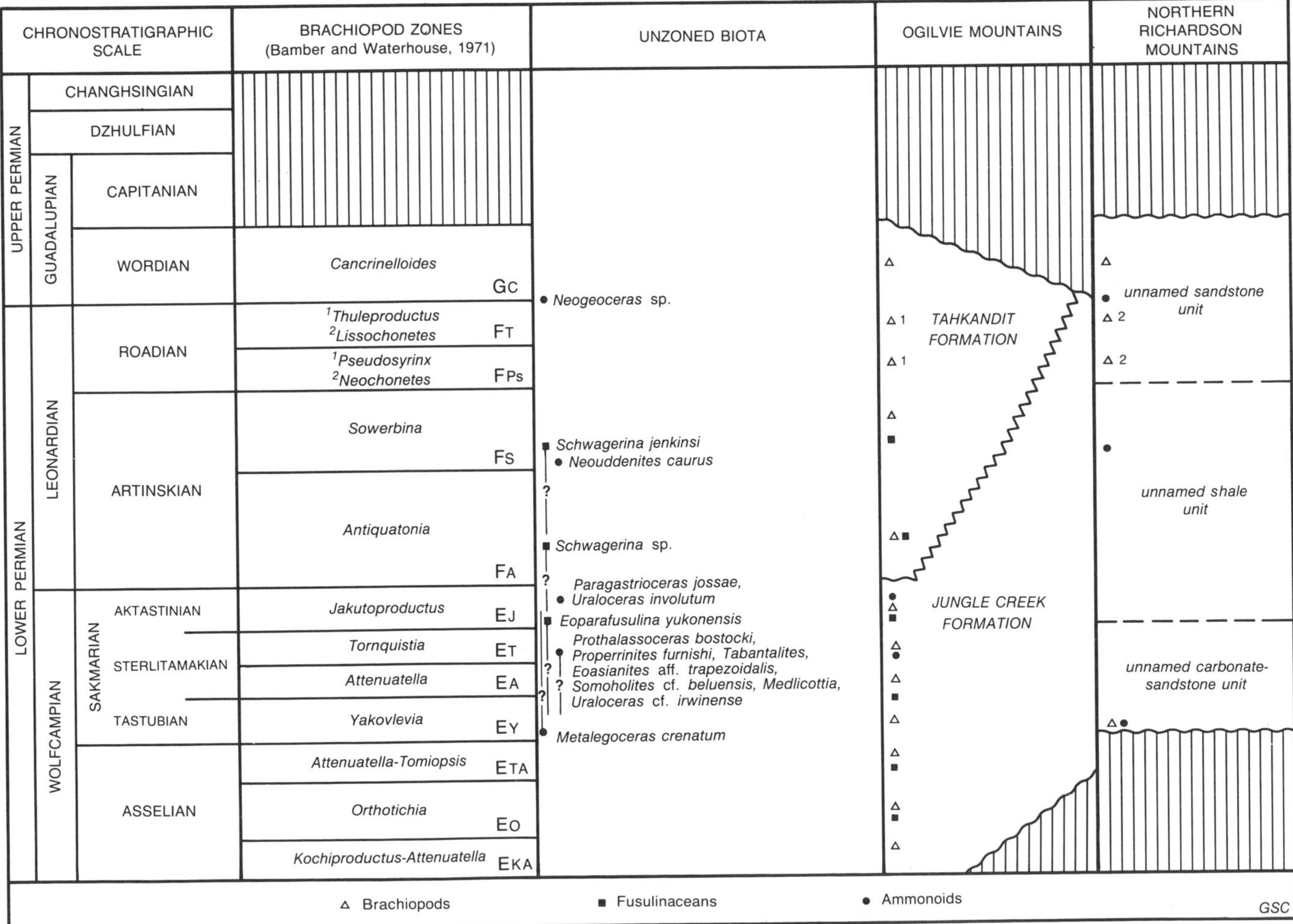

Figure 4F.19. Biostratigraphy of Northern Cordillera, Ogilvie Mountains to Beaufort Sea, northern Yukon Territory and northwestern District of Mackenzie. Symbols for unzoned biota indicate actual stratigraphic position with respect to brachipod zones. Attached vertical lines indicate possible age ranges.

Formation. McGugan (1983) described *S. sparwoodensis* from the lower Ross Creek Formation, which has been dated as early Artinskian on the basis of conodonts. In the northern Yukon Territory, fusulinacean assemblages ranging from Asselian to Artinskian were reported by Ross (1967, 1969). Local correlations based on fusulinaceans agree with the brachiopods, but some discrepancies occur in the correlations to standard chronostratigraphic units.

Ammonoids

Ammonoid occurrences are few but valuable for worldwide correlation. Occurrences in the northern Yukon Territory were reported by Nassichuk et al. (1965) and Nassichuk (1971). An assemblage from the Jungle Creek Formation near Peel River, including *Prothalassoceras bostocki, Properrinites furnishi, Tabantalites bifurcatus, Eoasianites* aff. *trapezoidalis, Somoholites* cf. *beluensis, Medlicottia* sp., and *Uraloceras* cf. *irwinense*, was assigned an early Sakmarian age by Nassichuk (1971); however, brachiopods from the same horizon indicate a late Sakmarian age. Nassichuk (1971) described a second assemblage from the upper Jungle Creek or lower Tahkandit formation, which included *Paragastrioceras jossae* and *Uraloceras involutum*. This fauna was correlated with the Baigendzhinian (Artinskian), but brachiopods from this same interval were assigned an Aktastinian age by Bamber and Waterhouse (1971). Nassichuk et al. (1965) indicated a Wordian age for *Neogeoceras* sp., which was recovered from a section in the northern Richardson Mountains. Bamber and Waterhouse (1971) indicated that this locality was within or above the *Lissochonetes* zone. Other ammonoid faunas from the northern Richardson Mountains include *Metalegoceras crenatum* (Aktastinian; but associated brachiopods indicate Tastubian) and *Neouddenites caurus* (Artinskian). Ammonoids from the Belloy Formation of northeastern British Columbia include *Medlicottia chozaensis, Daraelites* cf. *elegans*, and *Daubichites* sp. Nassichuk (1969) assigned this fauna to the early Artinskian. A single specimen of *Uraloceras* sp. was recovered from an isolated block near Fernie, southeastern British Columbia, and was considered by Nassichuk (1971) to be either Sakmarian or Artinskian. The matrix suggests that it may have originated from the lower Ross Creek Formation (lower Artinskian on the basis of conodonts).

Palynomorphs

Jansonius (1962) suggested a Leonardian and possibly Gualalupian age for the upper part of the Belloy Formation in the subsurface of west-central Alberta on the basis of a microfloral assemblage dominated by species of *Vittatina*. *Limitisporites, Klausipollenites*, and *Cycadopites* are also present in the upper Belloy (J. Jansonius in Naqvi, 1972). Comparison with Lower Permian assemblages in Sverdrup Basin of the Canadian Arctic Archipelago suggests a correlation of the upper Belloy with strata dated as late Asselian to Sakmarian (Utting, 1985; pers. comm., 1986). Barss (in Bamber and Barss, 1969) illustrated a great variety of monosaccate, and striate and non-striate disaccate and polyplicate pollen from the Lower Permian Jungle Creek Formation (middle recessive unit) of northern Yukon Territory. The assemblages are difficult to differentiate from those occurring in the upper part of the underlying Upper Carboniferous units. There are no dramatic qualitative changes at the Carboniferous/Permian boundary, but rather gradual quantitative changes with monosaccate and disaccate pollen becoming increasingly common (J. Jerzykiewicz, pers. comm., 1986).

Paleogeographic implications of the faunas and floras

As for the Carboniferous Period (Richards et al., Subchapter 4E in this volume), differences in the degree of endemism of different fossil groups and differences of opinion among authors render impossible a unified summary of the correct provincial position of western Canada throughout the Permian Period.

Most of the conodont faunas described herein are cosmopolitan, but there are some differences in diversity and degree of endemism between the southern Canadian faunas and those described from southwestern U.S.A., China, Japan, the Urals of the U.S.S.R., and the Arctic Islands of Canada. Ammonoids are generally cosmopolitan, but those described herein show greatest affinity with assemblages of the Russian platform and the Urals. Changes in the character of northern Yukon brachiopod faunas indicate widespread, rhythmic, climatic fluctuations in the Permian Period (Bamber and Waterhouse, 1971). Both brachiopods and fusulinaceans of this area are of boreal aspect. The brachiopods are more closely linked with those of the Urals than with those of the southwestern United States, and indicate possible paleolatitudes of 40 to 45 degrees for the northern Yukon Territory. Similarly, pollen and spore assemblages of this area have more in common with those of Sverdrup Basin of the Canadian Arctic Archipelago and Pechora Basin of the northeast European part of Russia than with those known from the central U.S.A. (Utting, 1985; pers. comm., 1986).

TECTONIC AND DEPOSITIONAL HISTORY

The Permian of the Eastern Cordillera and Interior Platform belongs to the Absaroka stratigraphic sequence, established for North America by Sloss (1963). Typical Absaroka features exhibited by the Canadian succession include: presence of oscillatory, transgressive-regressive sedimentary packages (Fig. 4F.16); punctuation of the succession by interregional disconformities (Fig. 4F.1); and predominance of siliciclastics over carbonates (Fig. 4F.2), with facies distribution partly controlled by high-angle block faulting, resulting in parallel or en-echelon facies trends. In contrast to other North American Absaroka successions, however, evaporites, mineralogically immature siliciclastics, and coal are absent.

Permian sediments of continental-margin character accumulated along the margin of the North American Plate from southern Alberta to northern Yukon Territory. Permian oceanic, volcanic, and sedimentary rocks occurring farther west (Eastern Assemblage of Monger, 1977) are in tectonic contact with the autochthonous North American strata. The oceanic assemblage was imbricated both before and after emplacement of granites in the Early Permian, events unrecorded in the autochthonous assemblage (see the companion volume, "Geology of the

Cordilleran Orogen in Canada", Gabrielse and Yorath, 1991). Juxtaposition of the oceanic and continental-margin assemblages during the Mesozoic records the closure of an oceanic basin of unknown breadth. Structural, paleomagnetic, and paleontological evidence from other upper Paleozoic, volcanic-sedimentary assemblages in the Western Cordillera indicates that they were widely dispersed at their time of deposition, and were brought together and accreted to the western margin of North America during Mesozoic time (Monger, 1977).

During latest Carboniferous and earliest Permian time, regional subaerial erosion, recorded by the sub-Permian disconformity, occurred throughout the Eastern Cordillera and Interior Platform. Widespread subsidence and transgression occurred during Asselian and Sakmarian time, resulting in the establishment of relatively deep-water marine depositional conditions (below storm wave base) in Ishbel Trough (Fig. 4F.2).

The Peace River Embayment in Permian time (Douglas et al., 1970), a deeply downwarped and downfaulted part of the cratonic platform in northeastern British Columbia and northwestern Alberta (Fig. 4F.2), appears to represent a rejuvenation of the Early Carboniferous embayment which occupied approximately the same area (Richards et al., Subchapter 4E in this volume). The detailed paleogeography of the Permian embayment is poorly known. Its northern depositional boundary had a northwest trend and its southern boundary trended east-west. Facies distribution, thickness variations (Belloy Formation), and local occurrences of thick basal Permian limestone boulder conglomerate (Belcourt Formation) suggest block faulting on a northwest trend within the embayment and in the area of the present Rocky Mountains of east-central British Columbia. The anomalously thick Permian succession in the Telford thrust plate of southeastern British Columbia records deposition in a northeast-trending trough of probable fault origin. This trough was on the northern boundary of an elevated fault block, the site of Cambrian Montania (Douglas et al., 1970) in southernmost Ishbel Trough, south of Fernie, British Columbia. On the high block, a thin sequence of lowest Artinskian strata rests unconformably on clastics and carbonates of the Upper Carboniferous Kananaskis Formation, suggesting that the area was high during at least Asselian and Sakmarian times. In northern Yukon and northwest District of Mackenzie, Lower Permian depositional patterns and structural relationships at the sub-Permian disconformity indicate local uplift and probable block faulting in latest Carboniferous and earliest Permian times along the Ancestral Aklavik Arch.

Early Permian subsidence was followed, in Artinskian time, by extensive regression (Fig. 4F.16). A recurrence of block faulting at this time is indicated by relationships at the intra-Permian disconformity. On the Interior Platform of northeastern British Columbia and southwestern District of Mackenzie, the Fantasque Formation rests disconformably on units of different ages on either side of the Bovie Lake fault system (Fig. 4F.11). In the disturbed belt, from west-central Alberta to northeastern British Columbia, the Belcourt Formation exhibits abrupt thickness changes beneath its disconformable contact with the Fantasque and Mowitch formations. In addition, thick boulder conglomerate in the basal Mowitch Formation records late Artinskian local uplift in several areas. The Artinskian regressive phase, which corresponds with an Artinskian (mid-Leonardian) global drop in relative sea level shown by Vail et al. (1977), was followed by a second episode of widespread transgression. Marine sedimentation continued in Ishbel Trough until at least Wordian time (Fig. 4F.1, 4F.3) and was followed by a period of regression and non-deposition, marked by the North America-wide, sub-Triassic disconformity.

REFERENCES

Bamber, E.W. and Barss, M.S.
1969: Stratigraphy and palynology of a Permian section, Tatonduk River, Yukon Territory; Geological Survey of Canada, Paper 68-18, 39 p.

Bamber, E.W. and Copeland, M.J.
1970: Carboniferous and Permian faunas; in Biochronology: Standard of Phanerozoic Time; Chapter XI in Geology and Economic Minerals of Canada, R.J.W. Douglas (ed.); Geological Survey of Canada, Economic Geology Report No. 1, p. 623-632.

Bamber, E.W. and Macqueen R.W.
1979: Upper Carboniferous and Permian stratigraphy of the Monkman Pass and southern Pine Pass areas, northeastern British Columbia; Geological Survey of Canada; Bulletin 301, 27 p.

Bamber, E.W. and Waterhouse J.B.
1971: Carboniferous and Permian stratigraphy and paleontology, northern Yukon Territory, Canada; Bulletin of Canadian Petroleum Geology, v. 19, no. 1, p. 29-250.

Bamber, E.W., Taylor, G.C., and Procter, R.M.
1968: Carboniferous and Permian stratigraphy of northeastern British Columbia; Geological Survey of Canada, Paper 68-15, 25 p.

Behnken, F.H.
1975: Leonardian and Guadalupian (Permian) conodont biostratigraphy in western and southwestern United States; Journal of Paleontology, v. 49, p. 284-315.

Brabb, E.E. and Churkin, M., Jr.
1969: Geologic map of the Charley River Quadrangle, east-central Alaska; United States Geological Survey, Miscellaneous Geologic Investigations, Map 1-573, 1:250 000.

Clark, D.L.
1974: Factors of Early Permian conodont paleoecology in Nevada; Journal of Paleontology, v. 48, p. 710-720.

Clark, D.L. and Behnken, F.H.
1971: Conodonts and biostratigraphy of the Permian; in Symposium on Conodont Biostratigraphy, W.C. Sweet and S.M. Bergstrom (ed.); Geological Society of America, Memoir 127, p. 415-439.

Clark, D.L., Carr, T.R., Behnken, F.H., Wardlaw, B.R., and Collinson, J.W.
1979: Permian conodont biostratigraphy in the Great Basin; Brigham Young University Geology Studies, p. 143-147.

Douglas, R.J.W., Gabrielse, H., Wheeler, J.O., Stott, D.F., and Belyea, H.R.
1970: Geology of western Canada; Chapter VIII in Geology and Economic Minerals of Canada, R.J.W. Douglas (ed.); Geological Survey of Canada, Economic Geology Report No. 1, p. 365-488.

Forbes, C.L. and McGugan, A.
1959: A Lower Permian fusulinid fauna from Wapiti Lake, B. C.; Journal of the Alberta Society of Petroleum Geologists, v. 7, no. 2, p. 33-42.

Gabrielse, H. and Yorath, C.J. (ed.)
1991: Geology of the Cordilleran Orogen in Canada; Geological Survey of Canada, Geology of Canada no. 4. (Also Geological Society of America, The Geology of North America, v. G-2.)

Gordey, S.P., Abbott, J.G., and Orchard, M.J.
1982: Devono-Mississippian (Earn Group) and younger strata in east-central Yukon; in Current Research, Part B, Geological Survey of Canada, Paper 82-1B, p. 93-100.

Graham, A.D.
1973: Carboniferous and Permian stratigraphy, southern Eagle Plain, Yukon Territory, Canada; in Proceedings of the Symposium on the Geology of the Canadian Arctic, J.D. Aitken and D.J. Glass (ed.); Geological Association of Canada - Canadian Society of Petroleum Geology, p. 89-102.

Halbertsma, H.L.
1959: Nomenclature of Upper Carboniferous and Permian strata in the subsurface of the Peace River area; Journal of the Alberta Society of Petroleum Geologists, v. 7, no. 5, p. 109-118.

Henderson, C.M. and McGugan, A.
1983: Permian conodonts from the Canadian Rocky Mountains; Geological Society of America, Abstracts with programs, v. 15, no. 5, p. 410.

Henderson, C.M. and McGugan, A. (cont.)
1985: Permian conodont biostratigraphy, Canadian Rocky Mountains; Geological Society of America, Abstracts with programs, v. 17, p. 224.
1986: Permian conodont biostratigraphy of the Ishbel Group, southwestern Alberta and southeastern British Columbia; Contributions to Geology, University of Wyoming, v. 24, p. 219-235.

Jansonius, J.
1962: Palynology of Permian and Triassic sediments, Peace River area, western Canada; Palaeontographica, Abteilung B, Band 110, p. 35-98.

Laudon, L.R. and Chronic, B.J.
1949: Paleozoic stratigraphy along Alaska Highway in northeastern British Columbia; Bulletin of American Association of Petroleum Geologists, v. 33, p. 189-222.

Logan, A. and McGugan, A.
1968: Biostratigraphy and faunas of the Permian Ishbel Group, Canadian Rocky Mountains; Journal of Paleontology, v. 37, no. 3, p. 1123-1139.

MacRae, J. and McGugan, A.
1977: Permian stratigraphy and sedimentology - southwestern Alberta and southeastern British Columbia; Bulletin of Canadian Petroleum Geology, v. 25, no. 4, p. 752-766.

McGugan, A.
1963: A Permian brachiopod and fusulinid fauna from the Elk Valley, British Columbia, Canada; Journal of Paleontology, v. 37, p. 621-627.
1983: A new species of *Schwagerina* from the Lower Permian of southeastern British Columbia; Journal of Foraminiferal Research, v. 13, p. 242-246.
1984: Carboniferous and Permian Ishbel Group Stratigraphy, North Saskatchewan Valley, Canadian Rocky Mountains, western Alberta; Bulletin of Canadian Petroleum Geology, v. 32, no. 4, p. 372-381.

McGugan, A. and Rapson-McGugan, J.E.
1976: Permian and Carboniferous stratigraphy, Wapiti Lake area, northeastern British Columbia; Bulletin of Canadian Petroleum Geology, v. 24, no. 2, p. 193-210.

McGugan, A. and Rapson, J.E.
1963a: Permian stratigraphy and nomenclature, western Alberta and adjacent regions; Edmonton Geological Society Guidebook, 5th Annual Field Conference, p. 52-64.
1963b: Permo-Carboniferous stratigraphy between Banff and Jasper, Alberta; Bulletin of Canadian Petroleum Geology, v. 11, no. 2, p. 150-160.

McGugan, A., Rapson-McGugan, J.E., Mamet, B.L., and Ross, C.A.
1968: Permian and Pennsylvanian biostratigraphy, and Permian depositional environments, petrography and diagenesis, southern Canadian Rocky Mountains; in Canadian Rockies, Bow River to North Saskatchewan River, Alberta, H. Hornford (ed.); Alberta Society of Petroleum Geologists, 16th Annual Field Conference Guidebook, September, 1968, p. 48-66.

McGugan, A., Roessingh, H.K., and Danner, W.R.
1964: Permian; in Geological History of Western Canada, R.G. McCrossan and R.P. Glaister (ed.); Alberta Society of Petroleum Geologists, p. 103-112.

Monger, J.W.H.
1977: Upper Paleozoic rocks of the western Cordillera and their bearing on Cordilleran evolution; Canadian Journal of Earth Sciences, v. 14, no. 8, p. 1832-1859.

Monger, J.W.H. and Price, R.A.
1979: Geodynamic evolution of the Canadian Cordillera - progress and problems; Canadian Journal of Earth Sciences, v. 16, no. 3, p. 770-791.

Movshovich, E.V., Kozur, H., Pavlov, A.M., Pnev, V.P., Polozova, A.N., Chuvashov, B.N., and Bogoslovskaya, M.R.
1979: Conodont complexes of the Lower Permian in Priuralye and problems of the correlation of Lower Permian deposits; Conodonts of the Urals and their stratigraphic significance; Ural'skii Nauchnyi Tsentr., U.S.S.R. Academy of Sciences, No. 145, p. 94-131.

Mytton, J.W., Morgan, W.A., and Wardlaw, B.R.
1983: Stratigraphic relationships of Permian units, Cassia Mountains, Idaho; Geological Society of America, Memoir 157, p. 281-303.

Naqvi, I.H.
1972: The Belloy Formation (Permian), Peace River area, northern Alberta and northeastern British Columbia; Bulletin of Canadian Petroleum Geology, v. 20, no. 1, p. 58-88.

Nassichuk, W.W.
1969: Permian ammonoids from the Belloy Formation, northeastern British Columbia; Geological Survey of Canada, Bulletin 182, p. 113-122.
1971: Permian ammonoids and nautiloids, southeastern Eagle Plain, Yukon Territory; Journal of Paleontology, v. 45, p. 1001-1021.

Nassichuk, W.W., Furnish, W.M., and Glenister, B.F.
1965: The Permian ammonoids of Arctic Canada; Geological Survey of Canada, Bulletin 131, p. 1-56.

Nelson, S.J.
1961: Permo-Carboniferous of the northern Yukon Territory; Journal of the Alberta Society of Petroleum Geologists, v. 9, p. 1-9.
1962: Horridonid brachiopods as horizon indicators, Permo-Pennsylvanian of the Yukon Territory; Journal of the Alberta Society of Petroleum Geologists, v. 10, p. 192-197.

Nelson, S.J. and Johnson, C.E.
1968: Permo-Pennsylvanian brachythyrid and horridonid brachiopods from the Yukon Territory, Canada; Journal of Paleontology, v. 42, p. 715-746.

Orchard, M.J.
1984: Early Permian conodonts from the Harper Ranch beds, Kamloops area, southern British Columbia; in Current Research, Part B, Geological Survey of Canada, Paper 84-1B, p. 207-215.

Pugh, D.C.
1983: Pre-Mesozoic geology in the subsurface of Peel River map-area, Yukon Territory and District of Mackenzie; Geological Survey of Canada, Memoir 401, 61 p.

Rascoe, B., Jr. and Baars, D.L.
1972: Permian System; in Geological Atlas of the Rocky Mountain Region, United States of America, W.W. Mallory, R.R. Mudge, V.E. Swanson, D.S. Stone, and W.E. Lumb (ed.); Rocky Mountain Association of Geologists, A.P. Hirschfeld Press, Denver, p. 143-165.

Rhodes, F.H.T.
1963: Conodonts from the topmost Tensleep Sandstone of the eastern Big Horn Mountains, Wyoming; Journal of Paleontology, v. 37, p. 401-408.

Ross, C.A.
1967: Late Paleozoic Fusulinacea from northern Yukon Territory; Journal of Paleontology, v. 41, p. 709-725.
1969: Upper Paleozoic Fusulinacea: *Eowaeringella* and *Wedekindellina* from Yukon Territory and giant *Parafusulina* from British Columbia; Geological Survey of Canada, Bulletin 182, p. 129-134.

Ross, C.A. and Bamber, E.W.
1978: Middle Carboniferous and Early Permian fusulinaceans from the Monkman Pass area, northeastern British Columbia; in Contributions to Canadian Paleontology; Geological Survey of Canada, Bulletin 267, p. 25-41.

Sarytcheva, T.G. and Waterhouse, J.B.
1972: Description of brachiopods of family Retariidae from the Permian of northern Canada; Paleontological Journal, USSR Academy of Sciences, v. 6, p. 501-513.

Sikabonyi, L.A. and Rodgers, W.J.
1959: Paleozoic tectonics and sedimentation in the northern half of the West Canadian Basin; Journal of the Alberta Society of Petroleum Geologists; v. 7, no. 9, p. 193-216.

Sloss, L.L.
1963: Sequences in the cratonic interior of North America; Bulletin of the Geological Society of America, v. 74, p. 93-114.

Struik, L.C.
1986: Imbricated terranes of the Cariboo gold belt with correlations and implications for tectonics in southeastern British Columbia; Canadian Journal of Earth Sciences, v. 23, p. 1047-1061.

Struik, L.C. and Orchard, M.J.
1985: Late Paleozoic conodonts from ribbon chert delineate imbricate thrusts within the Antler Formation of the Slide Mountain terrane, central British Columbia; Geology, v. 13, p. 794-798.

Utting, J.
1985: Preliminary results of palynological studies of the Permian and lowermost Triassic sediments, Sabine Peninsula, Melville Island, Canadian Arctic Archipelago; in Current Research, Part B, Geological Survey of Canada, Paper 85-1B, p. 231-238.

Vail, P.R., Mitchum, R.M., Jr., and Thompson, S. III
1977: Seismic stratigraphy and global changes of sea level, Part 4: Global cycles of relative changes of sea level; in Seismic Stratigraphy - Applications to Hydrocarbon Exploration, C.E. Payton (ed.); American Association of Petroleum Geologists, Memoir 26, p. 83-97.

Wardlaw, B.R. and Collinson, J.W.
1979: Youngest Permian conodont faunas from the Great Basin and Rocky Mountains; Brigham Young University Geology Studies, v. 26, p. 151-164.

Waterhouse, J.B.
1976: World correlations for Permian marine faunas; Department of Geology, University of Queensland papers, v. 7, 232 p.

Waterhouse, J.B. and Waddington, J.
1982: Systematic descriptions, paleoecology and correlations of the Late Paleozoic subfamily Spiriferellinae (Brachiopoda) from the Yukon Territory and the Canadian Arctic Archipelago; Geological Survey of Canada, Bulletin 289, 73 p.

ADDENDUM

Research activity subsequent to production of this manuscript has required modification to some of the data and interpretations presented herein. The reader is referred to the following for these changes:

Henderson, C.M.
1989: Absaroka Sequence - The Lower Absaroka Sequence: Upper Carboniferous and Permian; in Western Canada Sedimentary Basin - A Case History, B.D. Ricketts (ed.); Canadian Society of Petroleum Geologists, Special Publication no. 30, Calgary, Alberta, p. 203-217.

Authors' Addresses

C.M. Henderson
Department of Geology and Geophysics
University of Calgary
Calgary, Alberta
T2N 1N4

E.W. Bamber
Institute of Sedimentary and Petroleum Geology
Geological Survey of Canada
3303 - 33rd Street N.W.
Calgary, Alberta
T2L 2A7

B.C. Richards
Institute of Sedimentary and Petroleum Geology
Geological Survey of Canada
3303 - 33rd Street N.W.
Calgary, Alberta
T2L 2A7

A.C. Higgins
9 - Treglof Court
Oatlands Drive
Weybridge, Surrey
England, KT 13 9JG

A. McGugan (retired)
1157 Rolman Cr.
RR#2 Cobble Hill
British Columbia
V0R 1L0

Printed in Canada

Subchapter 4G

TRIASSIC

D.W. Gibson

CONTENTS

INTRODUCTION

General statement

Triassic rocks of the Interior Platform and eastern Cordillera of Canada have long been recognized as an interesting and variable sequence of marine strata occupying an elongate belt that extends from the United States border on the south to 69°N latitude and Beaufort Sea on the north (Fig. 4G.1, 4G.2). Triassic deposits are best developed in Western Canada Basin of British Columbia and Alberta, where they occupy three main physiographic provinces, Rocky Mountains on the west, Rocky Mountain Foothills, and Interior Plains to the east. Triassic rocks also occur in Liard River area of southern Yukon Territory and within northern Yukon Territory and District of Mackenzie in the British, Barn, Richardson, Selwyn, Wernecke, and northwestern Ogilvie Mountains. However, because much more published information on Triassic rocks in Western Canada Basin is available, and because of the greater development and economic importance of the system there, most of this report will be directed toward the region south of latitude 60°N.

Triassic rocks of the Rocky Mountains, Foothills, and Interior Plains comprise over 1200 m (Fig. 4G.1) of westward-thickening, siliciclastic and carbonate rocks and lesser amounts of evaporites. Contained marine faunas range in age from Griesbachian to Norian. The Triassic rocks of northern Yukon Territory display similar lithofacies but lack evaporites. The strata form a marine wedge deposited along a topographically low, tectonically stable continental shelf and shoreline, a continuation of Permian conditions at the passive western margin of the North American Craton. They form part of what is referred to as the Absaroka sequence within the cratonic interior of North America (Sloss, 1963). Deposition along the shelf preceded the accretion of the allochthonous terranes, which today characterize much of the central and western Cordillera. Far offshore to the west and southwest in Panthalassa were volcanic islands, archipelagos, shoals, and banks with intervening deep-water basins and troughs (Tozer, 1982b). Some of these islands and archipelagos were fringed with corals and other fauna characteristic of shallow warm water. It is the rocks of these offshore areas

Gibson, D.W.
1993: Triassic; Subchapter 4G in Sedimentary Cover of the Craton in Canada, D.F. Stott and J.D. Aitken (ed.); Geological Survey of Canada, Geology of Canada, no. 5, p. 294-320 (also Geological Society of America, The Geology of North America, v. D-1).

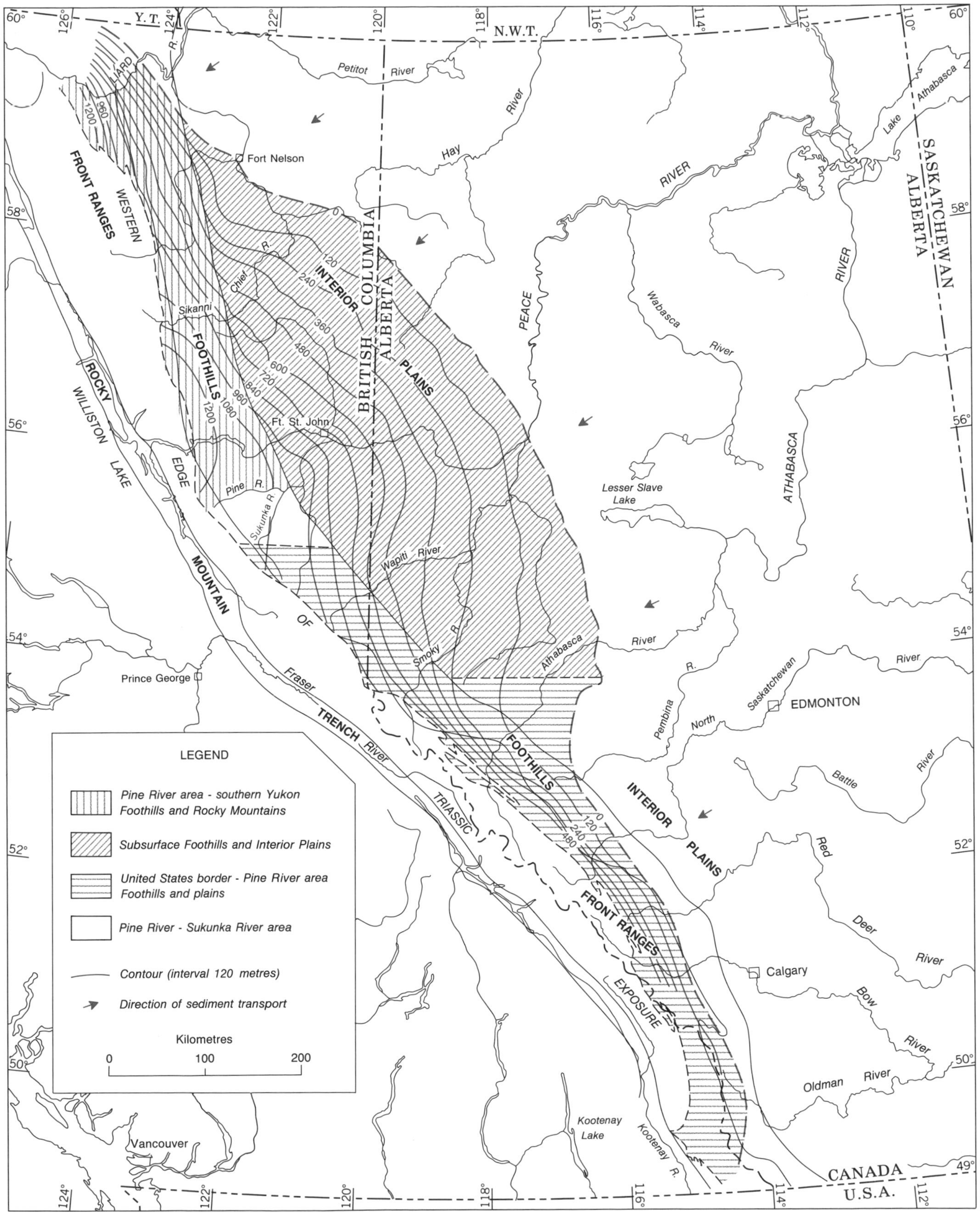

Figure 4G.1. Index map of Eastern Cordillera and Western Interior Plains, Alberta and British Columbia, illustrating geographic names, important physiographic and nomenclature boundaries, distribution, isopachs and direction of sediment transport of Triassic rocks within Western Canada Basin (modified from Barss et al., 1964).

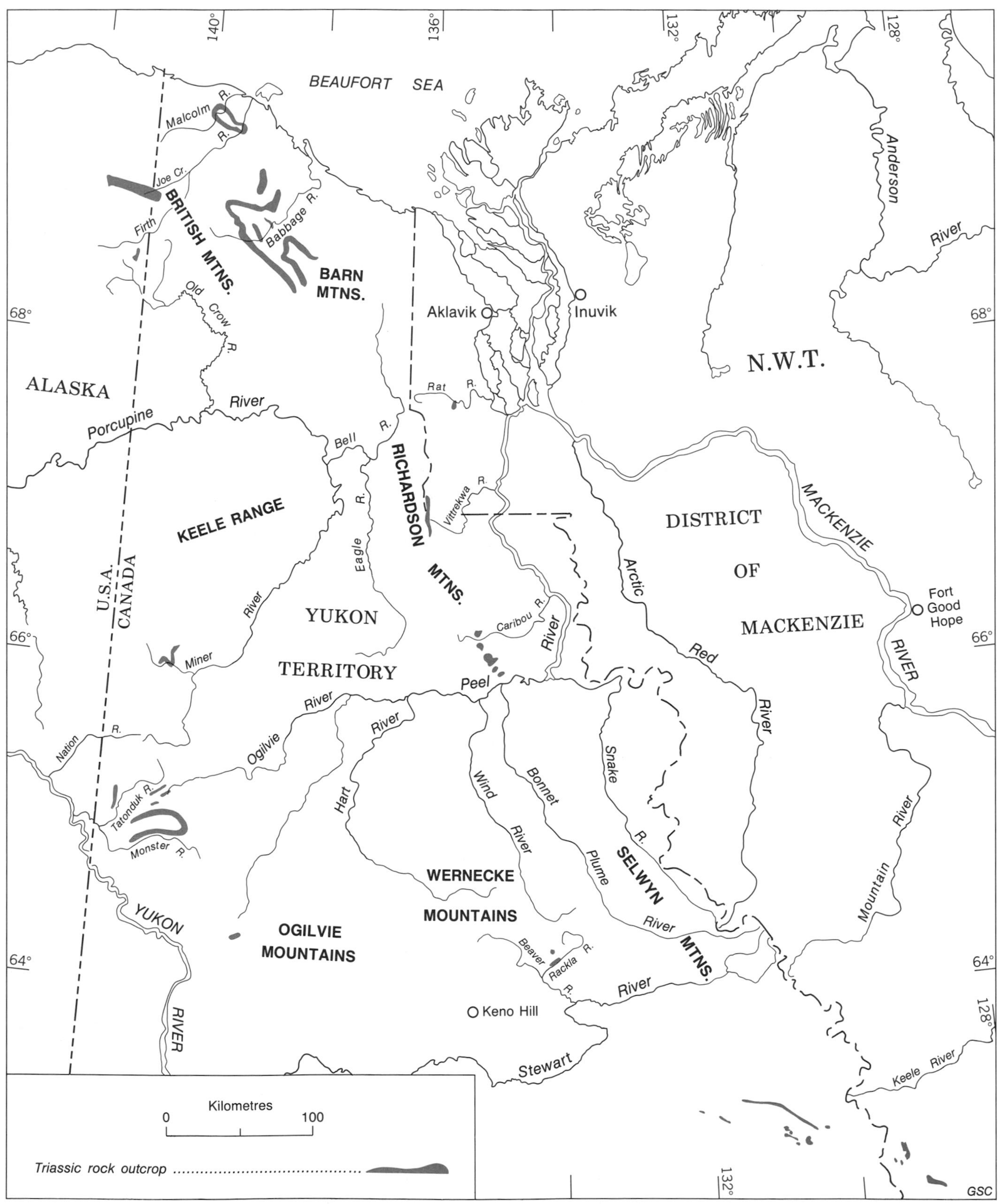

Figure 4G.2. Index map of Triassic rock occurrences, northern Yukon Territory and District of Mackenzie (modified from Mountjoy, 1967).

that now constitute the exotic terranes of the central and western Cordillera. To date, evidence for nearby tectonism or volcanism in Triassic rocks of the Western Canada Basin is lacking.

Three main nomenclatural systems have been adopted for Triassic rocks in Western Canada Basin (Fig. 4G.3). Strata within the Rocky Mountains and Foothills between the Pine-Sukunka rivers area and the United States border (49°N latitude), and strata in the subsurface of the adjacent Alberta Foothills and Plains south of Township 58 are called the Spray River Group (Sulphur Mountain and Whitehorse formations). Exposures in the Rocky Mountains and Foothills between the Pine-Sukunka rivers area of British Columbia and the Liard River area of the southern Yukon Territory are characterized by a different lithofacies and nomenclature and are referred to, in ascending order, as the Grayling and Toad formations and the Schooler Creek Group (Liard, Charlie Lake, Ludington, Baldonnel, Pardonet, and Bocock formations). The economically important oil- and gas-bearing Triassic rocks in the subsurface of the Foothills and Plains of Alberta and British Columbia north of Township 58, in ascending order, are the Daiber Group (Montney and Doig formations) and Schooler Creek Group (Halfway, Charlie Lake, Baldonnel, and Pardonet formations). Each of the three regions and systems of nomenclature will be discussed separately.

Data base

Much of the following information concerning Triassic lithostratigraphy and biostratigraphy in Western Canada Basin and northern Yukon Territory and District of Mackenzie has been obtained from published reports by Gibson (1965, 1968a, 1968b, 1969, 1970, 1971a, 1971b, 1972, 1974, 1975) and Tozer (1967, 1984), and by Mountjoy (1967). Information on Triassic rocks of the subsurface of the Foothills and Plains has largely been obtained from unpublished reports by J.A. Dolph of Gulf Canada Resources and I.A. McIlreath of Petro-Canada Resources.

Because of text limitations, it is not possible to discuss the many previous Triassic contributions by other authors. These are cited by the authors named above. Of importance, however, is the report by D.L. Barss, E.W. Best, and N. Meyers published in 1964, a synthesis of all published Triassic surface and subsurface information in western Canada available to the end of 1962. This paper still serves as an important reference to complement the information presented in the following text. Additional important investigations concerning Triassic rocks, not mentioned above, have been published by Miall (1976), Barss and Montedan (1981), Bever and McIlreath (1984), and Cant (1984, 1986).

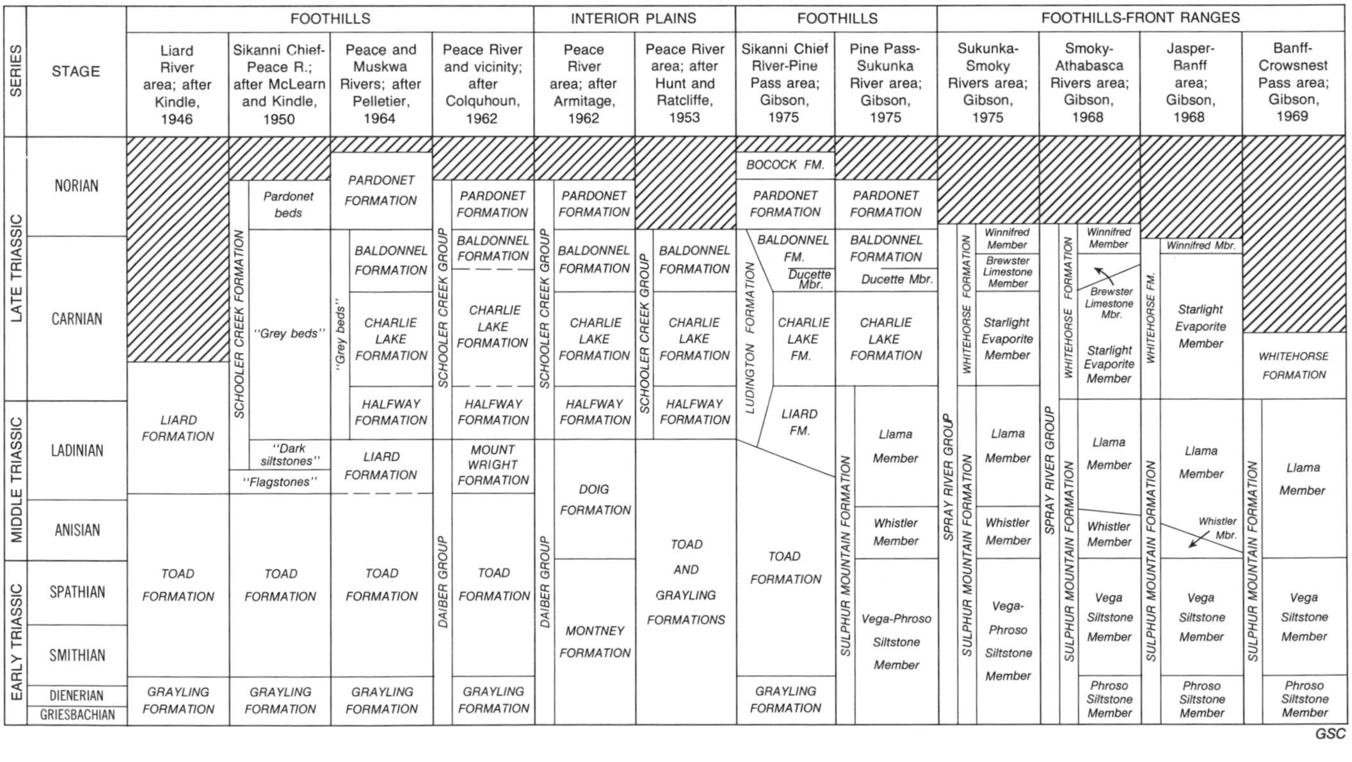

SERIES	STAGE	FOOTHILLS				INTERIOR PLAINS		FOOTHILLS		FOOTHILLS-FRONT RANGES			
		Liard River area; after Kindle, 1946	Sikanni Chief-Peace R.; after McLearn and Kindle, 1950	Peace and Muskwa Rivers; after Pelletier, 1964	Peace River and vicinity; after Colquhoun, 1962	Peace River area; after Armitage, 1962	Peace River area; after Hunt and Ratcliffe, 1953	Sikanni Chief River-Pine Pass area; Gibson, 1975	Pine Pass-Sukunka River area; Gibson, 1975	Sukunka-Smoky Rivers area; Gibson, 1975	Smoky-Athabasca Rivers area; Gibson, 1968	Jasper-Banff area; Gibson, 1968	Banff-Crowsnest Pass area; Gibson, 1969
LATE TRIASSIC	NORIAN		SCHOOLER CREEK FORMATION: Pardonet beds	PARDONET FORMATION	SCHOOLER CREEK GROUP: PARDONET FORMATION	SCHOOLER CREEK GROUP: PARDONET FORMATION		BOCOCK FM.; PARDONET FORMATION	PARDONET FORMATION				
	CARNIAN		"Grey beds"	"Grey beds": BALDONNEL FORMATION; CHARLIE LAKE FORMATION	BALDONNEL FORMATION; CHARLIE LAKE FORMATION	BALDONNEL FORMATION; CHARLIE LAKE FORMATION	SCHOOLER CREEK GROUP: BALDONNEL FORMATION; CHARLIE LAKE FORMATION	LUDINGTON FORMATION; BALDONNEL FM.; Ducette Mbr.; CHARLIE LAKE FM.	BALDONNEL FORMATION; Ducette Mbr.; CHARLIE LAKE FORMATION	WHITEHORSE FORMATION: Winnifred Member; Brewster Limestone Member; Starlight Evaporite Member	WHITEHORSE FORMATION: Winnifred Member; Brewster Limestone Mbr.; Starlight Evaporite Member	WHITEHORSE FM.: Winnifred Mbr.; Starlight Evaporite Member	WHITEHORSE FORMATION
MIDDLE TRIASSIC	LADINIAN	LIARD FORMATION	"Dark siltstones"; "Flagstones"	HALFWAY FORMATION; LIARD FORMATION	HALFWAY FORMATION; MOUNT WRIGHT FORMATION	HALFWAY FORMATION	HALFWAY FORMATION	LIARD FM.	SULPHUR MOUNTAIN FORMATION: Llama Member	SPRAY RIVER GROUP; SULPHUR MOUNTAIN FORMATION: Llama Member	SPRAY RIVER GROUP; SULPHUR MOUNTAIN FORMATION: Llama Member	SULPHUR MOUNTAIN FORMATION: Llama Member	SULPHUR MOUNTAIN FORMATION: Llama Member
	ANISIAN					DAIBER GROUP: DOIG FORMATION			Whistler Member	Whistler Member	Whistler Member	Whistler Mbr.	
EARLY TRIASSIC	SPATHIAN	TOAD FORMATION	TOAD FORMATION	TOAD FORMATION	DAIBER GROUP: TOAD FORMATION		TOAD AND GRAYLING FORMATIONS	TOAD FORMATION	Vega-Phroso Siltstone Member	Vega-Phroso Siltstone Member	Vega Siltstone Member	Vega Siltstone Member	Vega Siltstone Member
	SMITHIAN					MONTNEY FORMATION							
	DIENERIAN	GRAYLING FORMATION	GRAYLING FORMATION	GRAYLING FORMATION	GRAYLING FORMATION			GRAYLING FORMATION			Phroso Siltstone Member	Phroso Siltstone Member	Phroso Siltstone Member
	GRIESBACHIAN												

GSC

Figure 4G.3. Correlation chart of Triassic lithostratigraphic units, Western Canada Basin (modified from Gibson, 1975).

Acknowledgments and text responsiblities

This report represents the co-operative effort of five contributors. J.A. Dolph, Gulf Canada Resources, contributed information and data concerning the Triassic of the subsurface of the Foothills and Plains of Alberta and British Columbia north of Township 58. I.A. McIlreath, Petro-Canada Resources, provided data on the subsurface Baldonnel Formation. E.T. Tozer, Geological Survey of Canada provided information on the biostratigraphy and biochronology for all areas. S.P. Gordey, Geological Survey of Canada, provided information on the stratigraphy of the Yukon Territory. The author co-ordinated and integrated all contributions for the report, and assumed responsibility for the Triassic outcrop belt in the remainder of the Western Canada Basin and northern Yukon Territory and District of Mackenzie.

Many colleagues from the petroleum industry and the Geological Survey of Canada have contributed to the satisfactory completion of this report, through personal discussions and information provided through published and unpublished geological reports. To these individuals and companies the writer is grateful.

Special acknowledgment and appreciation is extended to Gulf Canada Resources and Petro-Canada Resources for so generously providing the time and technical support in the preparation of their respective contributions, and most importantly, for providing access to and release of much heretofore unpublished geological information.

The helpful comments and suggested improvements by critical reader D.L. Barss are gratefully appreciated.

FAUNA, FLORA, AND BIOCHRONOLOGY

The biochronological scale and terminology used in this report (Fig. 4G.4) is taken from Tozer (1984). The stratigraphic position of diagnostic faunas in the various formations and members in Western Canada Basin is summarized in Figures 4G.5, 4G.6 and 4G.7. Details on the occurrence and collection localities are given in Tozer (1967, 1979). Most of the age determinations are based on cosmopolitan ammonoids and thin-shelled pelagic bivalves (*Posidonia, Daonella, Halobia, Eomonotis, Monotis*). Brachiopods and benthonic bivalves are locally abundant in Ladinian and Carnian strata. Conodonts have also been found at many levels (Mosher, 1973; Orchard, 1983). Fish remains are fairly common in the Lower Triassic (Schaeffer and Mangus, 1976), ichthyosaur bones in the Middle and Upper Triassic. Palynomorphs are known from the subsurface Triassic (Jansonius, 1962).

The cosmopolitan character of the faunas facilitates correlation with other parts of Canada. The faunal record extends from the Upper Griesbachian to the lower Upper Norian (*Cordilleranus* Zone). Earliest and latest Triassic faunas (Lower Griesbachian, uppermost Norian) are unknown.

Late Triassic benthonic faunas of Western Canada Basin differ from those of the accreted terranes of the Western Cordillera, although there are sufficient cosmopolitan pelagic forms to permit correlation. The faunas of the accreted terranes include corals and thick-shelled bivalves unknown in Western Canada Basin. These fossils evidently indicate that the rocks were deposited in warmer water, presumably at a lower latitude, than those of Western Canada Basin. Today the Triassic rocks of the accreted terranes and those of Western Canada Basin are at comparable latitudes. This juxtaposition is attributed to substantial northward tectonic movement of the accreted terranes since Triassic time.

LITHOSTRATIGRAPHY

The distribution of Triassic rocks in Western Canada Basin and northern Yukon Territory and District of Mackenzie is illustrated in Figures 4G.1 and 4G.2. The differences in nomenclature adopted for the northern and southern areas of the basin reflect differences in the exposed rocks. A transitional region between Pine and Sukunka rivers has been chosen as the boundary between the northern and southern nomenclature (Fig. 4G.1, 4G.3; Gibson, 1972, 1975). Similarly, different nomenclature is necessary for Triassic units in the subsurface. Township 58 was selected as a convenient boundary between the Triassic nomenclature of the north and that of the south. The

SERIES	STAGE	SUBSTAGE	FAUNA NUMBER	ZONES
UPPER TRIASSIC	NORIAN	Upper Norian		*Choristoceras crickmayi*
			27	*Cochloceras amoenum*
			26	*Gnomohalorites cordilleranus*
		Middle Norian (Alaunian)	25	*Himavatites columbianus*
			24	*Drepanites rutherfordi*
		Lower Norian	23	*Juvavites magnus*
			22	*Malayites dawsoni*
			21	*Stikinoceras kerri*
	CARNIAN	Upper Carnian (Tuvalian)	20	*Klamathites macrolobatus*
			19	*Tropites welleri*
			18	*Tropites dilleri*
		Lower Carnian (Julian)	17	*Austrotrachyceras obesum*
			16	*Trachyceras desatoyense*
MIDDLE TRIASSIC	LADINIAN		15	*Frankites sutherlandi*
			14	*Maclearnoceras maclearni*
			13	*Meginoceras meginae*
			12	*Progonoceratites poseidon*
			11	*Eoprotrachyceras subasperum*
	ANISIAN	Upper Anisian (Illyrian)	10	*Frechites chischa*
			9	*Frechites deleeni*
		Middle Anisian (Pelsonian)	8	*Anagymnotoceras varium*
		Lower Anisian (Aegean)	7	*Lenotropites caurus*
LOWER TRIASSIC	SPATHIAN		6	*Keyserlingites subrobustus*
				"Olenikites" pilaticus
	NAMMALIAN	Smithian	5	*Wasatchites tardus*
			4	*Euflemingites romunderi*
		Dienerian	3	*Vavilovites sverdrupi*
			2	*Proptychites candidus*
	GRIESBACHIAN	Upper Griesbachian (Ellesmerian)	1	*Proptychites strigatus*
				Ophiceras commune
		Lower Griesbachian (Gangetian)		*Otoceras boreale*
				Otoceras concavum

GSC

Figure 4G.4. Triassic time scale (modified from Tozer, 1984). The faunal zones are located by number on Figures 4G.5, 4G.6 and 4G.7.

subsurface stratigraphic units of the southern region are discussed with those of the southern outcrop belt. Triassic rocks of northern Yukon Territory and District of Mackenzie are discussed under a separate heading.

Western Canada Basin

Rocky Mountain Front Ranges, Foothills and Plains, United States border to Pine River

Spray River Group

The Spray River Group of the Rocky Mountain Foothills and Front Ranges of Alberta and British Columbia is divisible into two distinctive and contrasting formations, a lower, Sulphur Mountain and an upper, Whitehorse (Fig. 4G.8). Each formation displays internal lithological variation that generally facilitates subdivision into members. The Spray River Group ranges in thickness from an erosional pinch-out in the east to a maximum of 850 m near Smoky River in the west (Fig. 4G.1). It disconformably overlies chert, cherty dolostone, and sandstone of the Permian Ishbel Group throughout most of the region, but in some Eastern Foothills localities it disconformably overlies cherty dolostone of the Mississippian Rundle Group. The Spray River Group is disconformably overlain by the Jurassic Fernie Formation (Fig. 4G.9, 4G.10, 4G.11).

The Sulphur Mountain Formation comprises dark grey to predominantly rusty brown weathering calcareous and dolomitic siltstone, sandstone, and shale, and silty to sandy limestone and dolostone, ranging in thickness from an erosional eastern pinch-out to 556 m near the Pine-Sukunka rivers area in the west (Gibson, 1975). In most areas the Sulphur Mountain can be readily divided into four members, in ascending order: Phroso Siltstone, Vega Siltstone, Whistler, and Llama members (Fig. 4G.3). In part of the Front Ranges and Foothills of the Athabasca-Bow River region, the Vega Siltstone Member is characterized by an additional unit called the Mackenzie Dolomite Lentil (Fig. 4G.3, 4G.9). This distinctive unit can

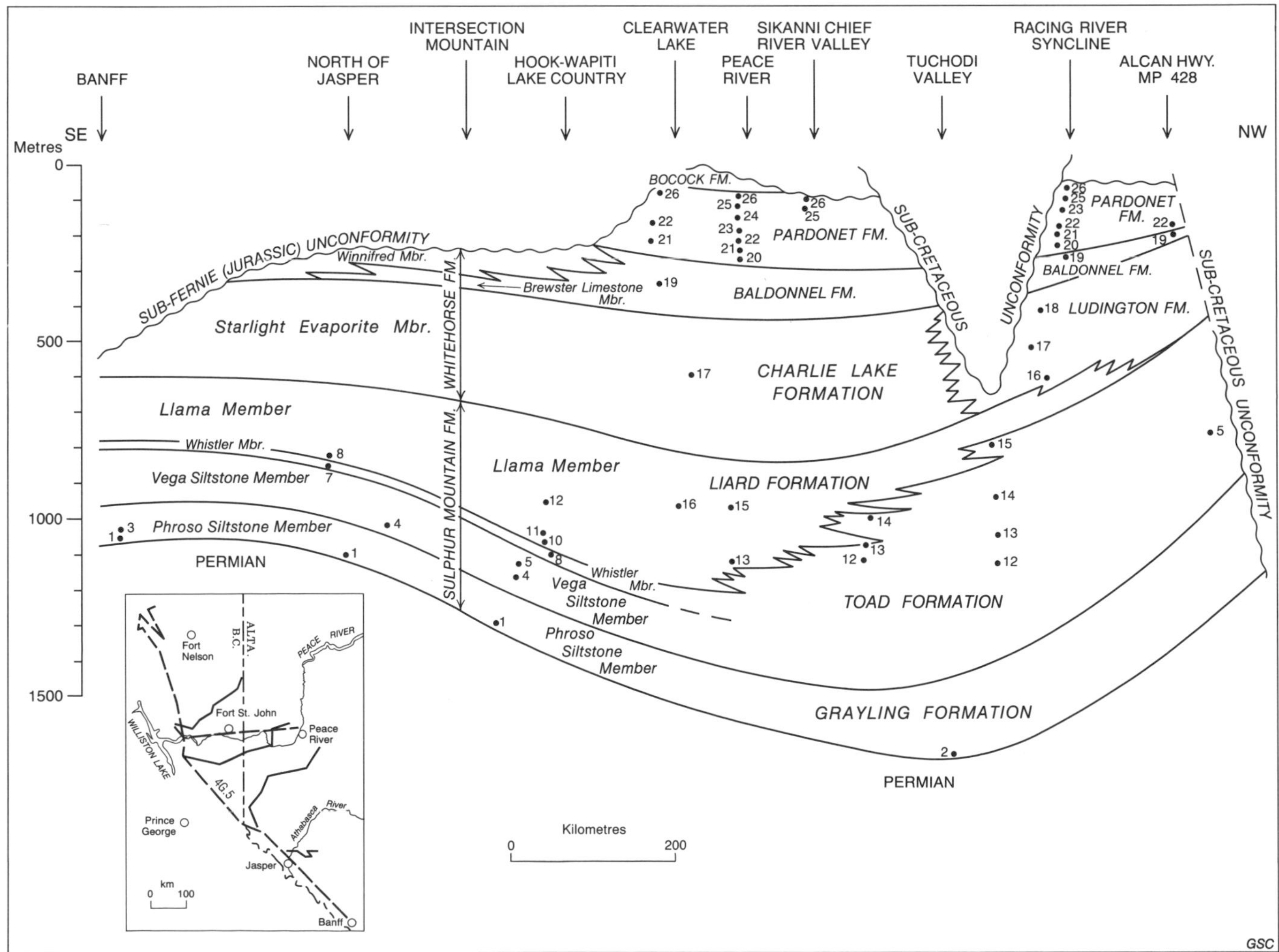

Figure 4G.5. Schematic longitudinal section, Banff, Alberta to Milepost 428, Alaska Highway, northeastern British Columbia. Numbers refer to ammonoid faunal zones listed in Figure 4G.4 (prepared by E.T. Tozer).

also be recognized in the subsurface of the Plains to the east (Gibson, 1974). Between Smoky and Pine rivers, the Phroso and Vega Siltstone members lose their distinctive lithological character and accordingly, must be grouped as a single undifferentiated member of the Sulphur Mountain Formation.

The Phroso Siltstone Member comprises recessive, shaly to flaggy weathering siltstone and silty shale and, less commonly, very fine grained sandstone (Fig. 4G.8, 4G.9, 4G.10). In some areas south of Bow River, Phroso strata are characterized by thin interbeds of very calcareous siltstone. Bedding is difficult to recognize and accordingly, the strata appear as massive or shaly weathering, finely laminated siltstone. The Phroso Member ranges in thickness from a minimum of 30 m in the east to more than 240 m in the southwest near Sparwood, British Columbia (Gibson, 1974).

The Phroso Siltstone Member rests unconformably on the Permian Ishbel Group (Fig. 4G.9), or, in the more easterly sections of the Foothills and in the subsurface, on the Carboniferous Rundle Group. The contact with the overlying Vega Siltstone Member is conformable and generally abrupt, but at some localities gradational.

The Phroso Siltstone Member is sparsely fossiliferous. To date, a few bivalves, poorly preserved ammonoids, and indeterminate fish fragments have been recovered from the lower and upper facies of the member. *Claraia stachei* Bittner indicates an Early Triassic, Griesbachian age.

The overlying Vega Siltstone Member comprises a distinctive rusty brown weathering sequence of cyclically bedded siltstone, silty limestone, and shale (Fig. 4G.8, 4G.9, 4G.10). Locally, beds of very fine grained sandstone and sandy to silty dolostone are intercalated. In some areas the Vega Siltstone is characterized by distinctive, light grey- to buff-weathering dolostone and dolomitized coquina lentils. In the Front Ranges and Foothills from Cadomin to near Banff, one such dolostone forms a distinctive marker, the Mackenzie Dolomite Lentil. Other distinctive thin, light grey-weathering dolostones occur stratigraphically higher in the Vega Member (Gibson, 1969, 1974). Barss et al. (1964) have also recorded dolostone lithofacies in the Sulphur Mountain Formation of the subsurface in Alberta.

In outcrop, the Vega Siltstone Member ranges in thickness from a minimum of 50 m near the Athabasca River to a maximum of 360 m in southeastern British Columbia (Gibson, 1974).

The Vega Siltstone Member is abruptly overlain by either the Whistler Member or the Llama Member (Fig. 4G.9). Ammonoids identified from the Vega Siltstone include *Vavilovites sverdrupi* (Tozer), indicating the Dienerian *sverdrupi* Zone, and *Euflemingites cirratus* (White) of Smithian age (*romunderi* Zone, see Fig. 4G.4, 4G.5).

The Mackenzie Dolomite Lentil occurs in the Cadomin and Cascade River areas of Alberta, and in the adjacent subsurface of the Plains (Gibson, 1974). The dolostone lentil is conspicuous in the lower part of the Vega Siltstone Member (Fig. 4G.3, 4G.9). It comprises a relatively thin sequence of light grey to yellowish grey, silty to sandy dolostone and dolomitized coquina, locally with interbeds of siltstone and very fine grained sandstone. The dolostone is commonly vuggy and porous and, in the subsurface of the Plains, is a potential hydrocarbon reservoir. The Mackenzie Dolomite Lentil ranges in thickness from a depositional pinch-out to a maximum of 50 m (Gibson, 1974).

The Vega-Phroso Siltstone Member comprises dark brownish grey to rusty brown weathering siltstone, finely crystalline to skeletal limestone, shale, and minor quartz sandstone (Fig. 4G.12). Between Smoky and Pine rivers, division into two members is impractical because of interfingering between the two characteristic rock types over several hundred metres. Unlike the separate members south of Smoky River, the Vega-Phroso to the north contains limestone beds, some of which contain whole and fragmented bivalve molluscan and brachiopod shells. Fossil-bearing limestone concretions are found in many areas. Near Wapiti Lake, very fine to fine-grained quartz sandstone forms a distinctive orange-brown weathering marker facies. In outcrop, the member ranges in thickness from 80 m to 270 m.

The Vega-Phroso Member unconformably overlies one or other of several formations in different parts of the area: the Permian Fantasque, Belcourt, and Mowitch formations and the Lower Carboniferous Prophet Formation. The contact with the overlying Whistler Mountain is generally sharp and abrupt.

Ammonoids and bivalves from the Vega-Phroso Siltstone Member indicate an age ranging from Early Triassic Griesbachian to Spathian.

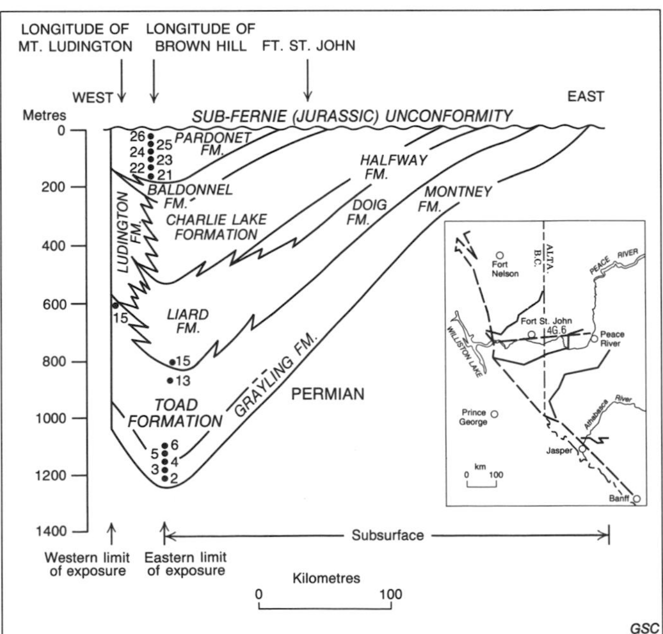

Figure 4G.6. Schematic cross-section illustrating relationships between surface and subsurface Triassic stratigraphy and nomenclature, Rocky Mountain Foothills and Interior Plains between Peace River-Williston Lake and area east of Fort St. John, northeastern British Columbia. Numbers refer to ammonoid faunal zones listed in Figure 4G.4 (prepared by E.T. Tozer).

Recessive, dark grey to black-weathering silty dolostone, siltstone, silty limestone, shale, and locally phosphatic sandstone and pebble conglomerate, recognized as the Whistler Member (Fig. 4G.12) and in the subsurface as part of the Doig Formation (Fig. 4G.3, 4G.10), form a most distinctive and widespread unit of the Sulphur Mountain Formation. Its distribution south of Bow River and in the Foothills east of Jasper is limited (Fig. 4G.9, 4G.10; Gibson, 1974). The eastward and southward thinning and disappearance are attributed in part to syndepositional thinning and in part to an inferred facies change to the Llama Member. Maximum thickness is 85 m near Sunkunka River.

The Whistler Member appears conformable above resistant-weathering siltstones of the Vega Siltstone Member at most localities. However, a widely occurring, basal phosphatic pebble conglomerate and the abrupt change in lithology between the two members may indicate a minor unconformity or diastem. The contact with the overlying Llama Member is conformable and generally distinct.

The Whistler Member contains ammonoids indicating the *caurus* and *varium* Zones (Lower and Middle Anisian; see Fig. 4G.4, 4G.5).

The uppermost, Llama Member of the Sulphur Mountain Formation comprises a generally cliff-forming sequence of siltstone, dolostone, silty and skeletal limestone, and locally, sandstone and shale (Fig. 4G.8, 4G.9, 4G.12), ranging in thickness from 3 m in the east near Athabasca River to a maximum of 356 m near Sukunka River.

The contact between the Llama Member and the overlying Whitehorse and Charlie Lake formations is abrupt in some areas and gradational in others. Between Pine and Sukunka rivers, the Llama Member is overlain

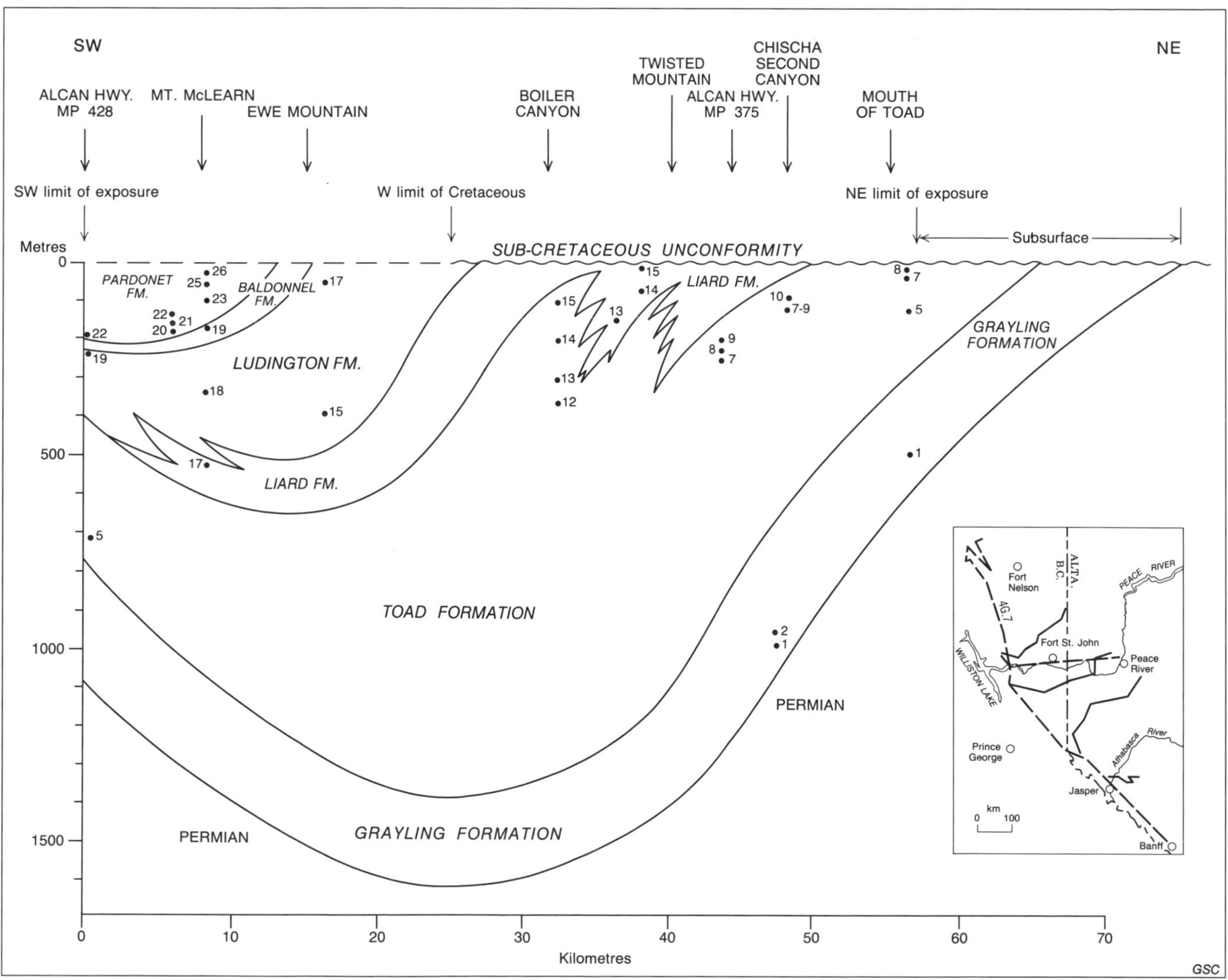

Figure 4G.7. Schematic cross-section of Triassic strata, Milepost 428, Alaska Highway to Liard River-Nelson Forks area, northeastern British Columbia. Numbers refer to ammonoid faunal zones listed in Figure 4G.4 (prepared by E.T. Tozer).

gradationally by the Charlie Lake Formation. At most western localities of the Front Ranges between Athabasca River and southeastern British Columbia, the contact is abrupt. At most localities in the eastern Foothills and in the subsurface of the Plains, the contact is gradational. Barss et al. (1964) suggested the possibility of a disconformity between the Sulphur Mountain and Whitehorse formations in this eastern area; however, no convincing evidence has yet been found to confirm the suggestion.

The Llama Member, one of the most fossiliferous units of the Triassic succession, contains ammonoid faunas of the *chischa* Zone (Anisian) and the overlying *subasperum, meginae, maclearni* and *sutherlandi* zones (Ladinian; see Fig. 4G.4, 4G.5).

The Whitehorse Formation of the Spray River Group, ranging in thickness from an erosional pinch-out in the east to 500 m near Smoky River in the west, is divisible into three members, in ascending order: Starlight Evaporite Member, Brewster Limestone Member and Winnifred Member (Fig. 4G.3, 4G.8). The Brewster Member does not extend into the region south of Athabasca River. In the Cadomin-Banff area, the Starlight Evaporite Member is characterized by a distinctive quartzitic sandstone called the Olympus Sandstone Lentil. The members of the Whitehorse Formation are not recognizable in the subsurface of the Foothills and Plains. The Whitehorse Formation gradationally overlies the Sulphur Mountain Formation and is disconformably overlain by the Jurassic Fernie Formation.

The Starlight Evaporite Member comprises variegated, buff, yellow, light to medium grey and reddish-brown weathering dolostone and limestone, sandstone, siltstone, and intraformational and/or solution breccia. Locally, the breccia contains beds and lenses of gypsum, which in some areas attain a thickness up to 44 m (Fig. 4G.9, 4G.10). The breccias form a diagnostic facies of the member, particularly in the thicker western localities. Abrupt facies changes are characteristic of the Starlight Evaporite, and bedding or lithological correlation between closely spaced sections in the same locality is generally not possible, except for the Olympus Sandstone Lentil.

The Starlight Evaporite Member ranges in thickness from an erosional pinch-out in the eastern Foothills and Plains subsurface to more than 400 m in the Front Ranges of Banff National Park (Gibson, 1974).

The Starlight Evaporite Member is conformably but abruptly overlain in areas north of Athabasca River by strata of the Brewster Limestone Member (Fig. 4G.9, 4G.10). In the Foothills and subsurface of the Plains to the east, and the area between Athabasca and Bow rivers where the Brewster Member is absent because of a probable facies change, the Starlight Evaporite is conformably overlain by strata of the Winnifred Member. In parts of the Foothills north of Bow River and throughout most of the area to the south, the member is disconformably overlain by the Jurassic Fernie Formation.

Rarity of fossils reflects the shallow-water, evaporitic depositional environment of the member. On the basis of apparent equivalence to the Charlie Lake Formation to the north, the Starlight Evaporite Member is assumed to be of Late Triassic, Carnian age.

Between Athabasca and Bow rivers, the Starlight Evaporite Member is characterized by a distinctive, lenticular sandstone unit called the Olympus Sandstone Lentil (Fig. 4G.3; Gibson, 1974). It consists of cliff-forming, well indurated sandstone, with local intercalations of sandy dolostone, attaining a maximum thickness of 140 m.

The Brewster Limestone Member comprises cliff-forming limestone with minor amounts of dolostone and intraformational carbonate breccia (Fig. 4G.8, 4G.9, 4G.10). The member thickens from east to west, from a depositional feather edge to 62 m. As noted above, the distribution of the Brewster Member in the outcrop belt is limited; south of Athabasca River and in the eastern Foothills, it is absent, probably through facies change. The member contains bivalves of the *Mysidioptera poyana* Zone of the Peace River district and is therefore of Carnian age (Gibson, 1968a).

The contact of the Brewster with the overlying Winnifred Member is sharp and conformable and placed at a prominent lithological break.

The Winnifred Member, the uppermost unit of the Whitehorse Formation, consists of dolostone and limestone with lesser interbeds of dolomitic siltstone, very fine grained sandstone, and intraformational conglomerate in some western localities (Fig. 4G.8, 4G.9, 4G.10). In the Smoky River region, it contains at some localities a distinctive redbed facies similar in appearance to that in the underlying Starlight Member (Gibson, 1968a, 1974). The Winnifred ranges in thickness from an erosional pinch-out (Front Ranges and Foothills south of Bow River) to a maximum of 225 m in the Front Ranges. It is disconformably overlain by the Jurassic Fernie Formation.

Identifiable fossils are rare. Because of stratigraphic position above the Brewster Member, the Winnifred is interpreted to be of Late Triassic age.

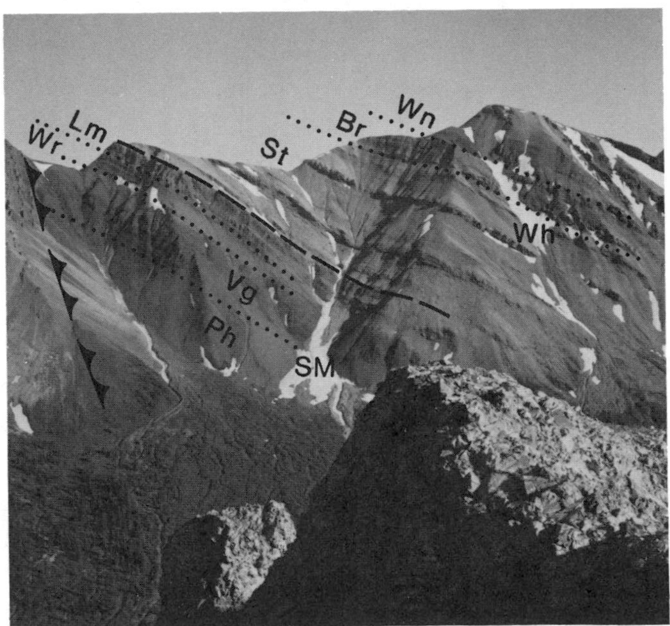

Figure 4G.8. Sulphur Mountain (SM) and Whitehorse (Wh) formations, Mowitch Creek, northern Jasper National Park, view southeastward. Members: Ph - Phroso; Vg - Vega; Wr - Whistler; Lm - Llama; St - Starlight; Br - Brewster; Wn - Winnifred (photo - D.W. Gibson, GSC 7-3-62).

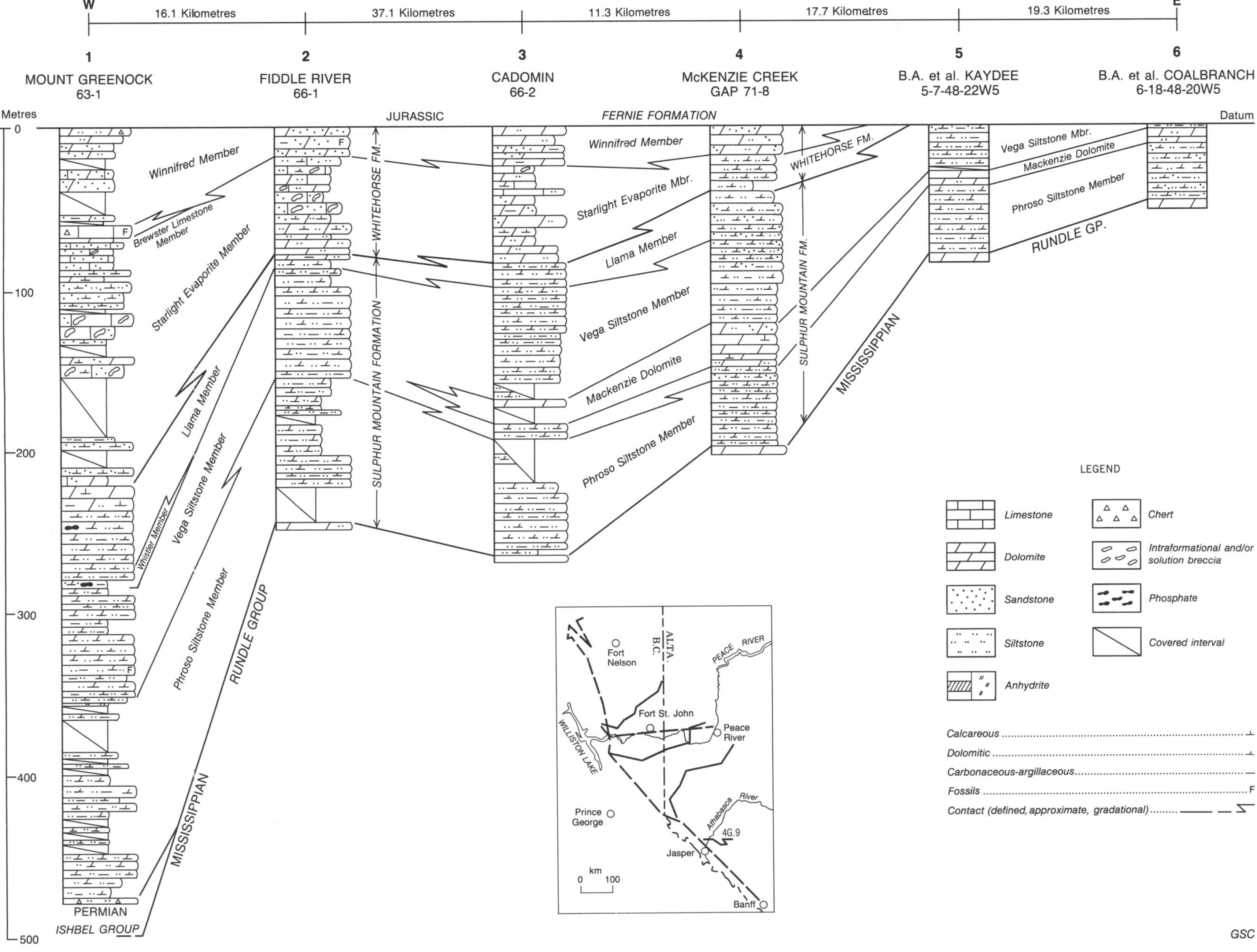

Figure 4G.9. Stratigraphic cross-section of Triassic units, Rocky Mountain Front Ranges and Foothills between Athabasca and Brazeau rivers, Alberta (after Gibson, 1974).

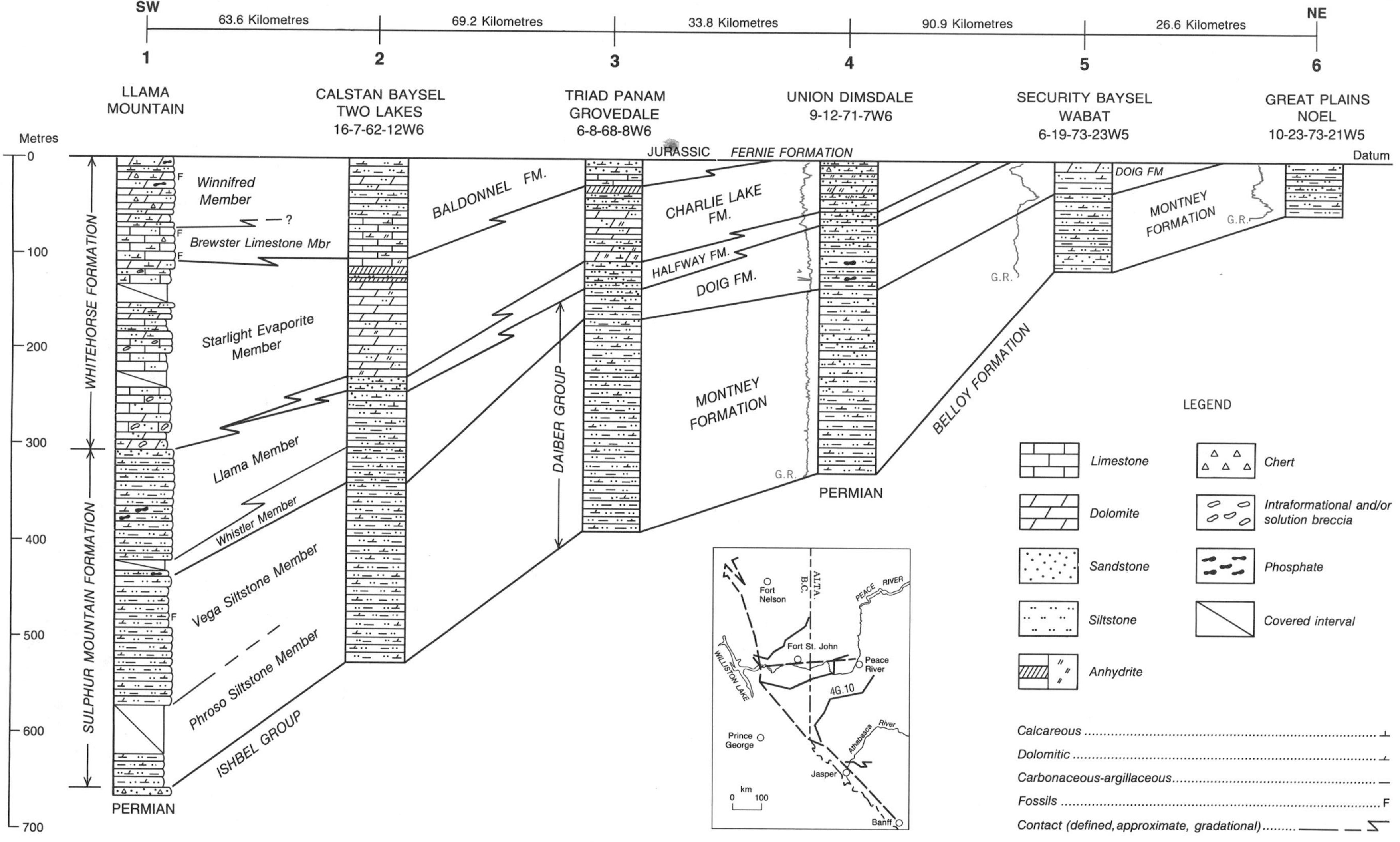

Figure 4G.10. Stratigraphic cross-section of Triassic units, Rocky Mountain Front Ranges, Foothills and Interior Plains between Smoky River and Valleyview area, Alberta.

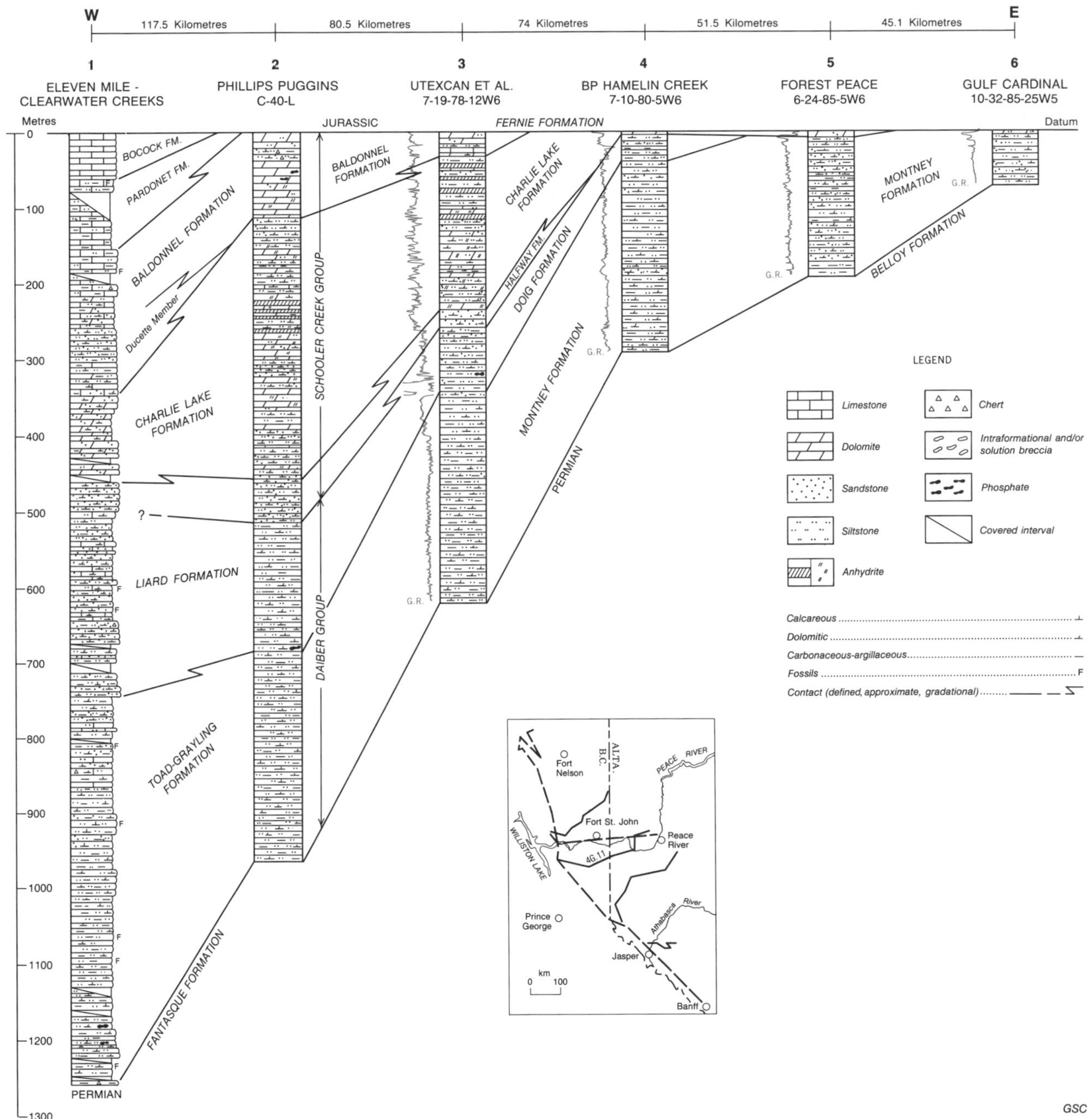

Figure 4G.11. Stratigraphic cross-section of Triassic units, Rocky Mountain Foothills and Interior Plains, northeastern British Columbia and western Alberta.

Rocky Mountain Foothills and Front Ranges, Pine River area to southern Yukon Territory

Triassic rocks in the outcrop belt of the Rocky Mountain Foothills and Front Ranges between Pine River and the Liard River area of southern Yukon Territory can be divided into seven formations, in ascending order: Grayling, Toad, Liard, Charlie Lake, Baldonnel, Pardonet, and Bocock (Fig. 4G.3). In the western Foothills north of Peace River, the Ludington Formation is a lateral equivalent of the Liard, Charlie Lake, and Baldonnel formations. South of, and near Williston Lake the Baldonnel is divided into two units, the lower of which is called Ducette Member.

The Grayling Formation at the base comprises recessive weathering siltstone, shale, and lesser limestone, dolostone, and very fine grained sandstone (Fig. 4G.13). In outcrop it ranges in thickness from a minimum of 35 m near Halfway River (Gibson, 1975) to more than 395 m near Liard River (Pelletier, 1963). The Grayling disconformably overlies the Permian Fantasque Formation (Fig. 4G.11,

4G.13). The contact with the overlying Toad Formation is gradational. Fossils are rare in the Grayling. *Claraia stachei* Bittner, indicating a Griesbachian age, is known from Liard and Dunedin rivers. Dienerian (*candidus* Zone) ammonoid faunas have been collected from the formation (see Fig. 4G.4, 4G.7).

The Toad Formation (Fig. 4G.13) comprises dark grey weathering, calcareous siltstone (locally phosphatic), silty limestone, shale, and minor amounts of dolostone and sandstone, ranging in thickness from a minimum of 158 m near Williston Lake to more than 825 m near Halfway River (Gibson, 1975). The formation displays pronounced thickening from east to west, partly attributable to depositional thickening and partly attributable to an eastward facies change to Liard Formation (see Fig. 4G.5, 4G.7; Gibson, 1971a, 1975).

The contact of the Toad Formation with the overlying Liard and Ludington formations is gradational in some places and abrupt in others. Between Pine River and Williston Lake, the contact is gradational. An abrupt but conformable contact between Toad and Liard formations occurs in the eastern Foothills north of Williston Lake. In the extreme western Foothills north of Williston Lake, and possibly in the Liard River and southern Yukon Territory area, the Toad has a similarly abrupt but conformable contact with medium grey weathering sandstone, siltstone, and sandy to silty limestone of the Ludington Formation (Fig. 4G.3, 4G.14). In an earlier study, Gibson (1975) recorded a pebble to cobble conglomerate unit up to 1.5 m thick at the base of the Ludington Formation at two localities near Graham and Halfway rivers, suggesting the possibility of a disconformable contact between the two formations. However, recent work in these areas indicates that the conglomerate facies is confined to possible subaqueous debris-flow channels, which locally occur near and at the base of the formation. Accordingly, convincing evidence in support of the disconformity is lacking. Additional work is necessary to resolve the contact relationship. In the extreme eastern Foothills belt of the Liard River-Alaska Highway area, the Toad Formation is erosionally and disconformably overlain by sandstone, siltstone, and shale of the Lower Cretaceous Fort St. John Group (Stott, 1982).

The Toad Formation contains ammonoid and bivalve faunas ranging in age from Dienerian to Ladinian (see Fig. 4G.4, 4G.5, 4G.6, 4G.7). No single section yields all faunal zones of the formation. Details on the faunal zonation are provided by Tozer (1967). The faunas clearly indicate that the Toad-Liard boundary is diachronous, older in the east than in the west. In eastern sections the Toad-Liard boundary approximately coincides with that between the Anisian and Ladinian stages. Farther west, beds as young as Late Ladinian (*sutherlandi* Zone) occur in rocks regarded as Toad.

The Liard Formation consists of sandstone, siltstone, and minor amounts of dolostone (Fig. 4G.11). Interbeds of skeletal limestone occur at many localities. South of Williston Lake, finely crystalline limestone forms a conspicuous component of the Liard. The variation in reported thickness may be due in part to difficulty in placing the contact with the overlying Charlie Lake Formation, or the difficulty in some areas of placing the

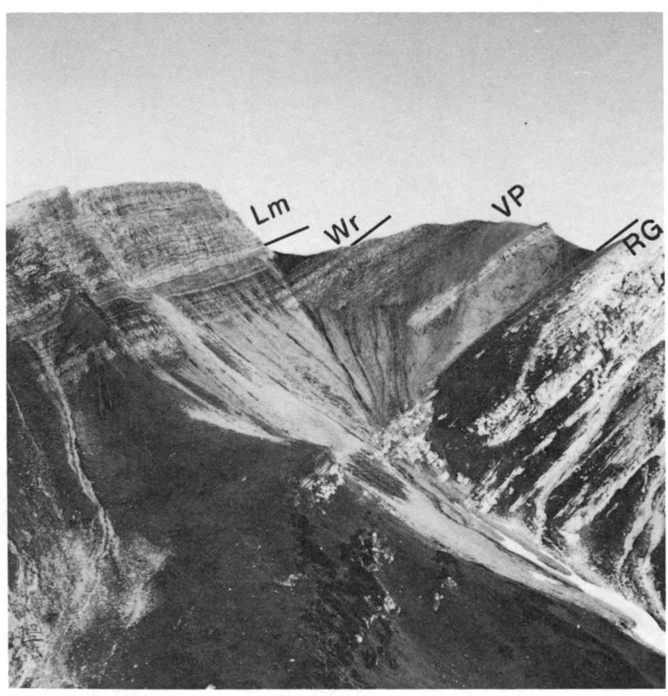

Figure 4G.12. Vega-Phroso Siltstone (VP), Whistler (Wr) and Llama (Lm) members of the Sulphur Mountain Formation, underlain by Lower Carboniferous Rundle Group (RG), Meosin Mountain, west-central Alberta. Note resistant sandstone band near base of Vega-Phroso Siltstone Member. View is northward. (photo - D.W. Gibson, GSC 6-7-70).

Figure 4G.13. Permian Fantasque Formation (Fa) and Triassic Grayling (Gr) and Toad (Td) formations, Pine River-Williston Lake area, northeastern British Columbia. View is southeastward. (photo - D.W. Gibson, GSC 1457 17).

contact between the Toad and Liard formations. The Liard Formation ranges in thickness from an erosional pinch-out in the northeast to a maximum of 417 m near Pine River (Gibson, 1975).

Throughout most of the outcrop belt, the Liard Formation is gradationally overlain by strata of the Charlie Lake Formation. In some areas the contact is difficult to pinpoint because of a similarity in lithology between the two formations. In the extreme western Foothills, the Liard is interpreted to grade laterally into strata of the Ludington Formation (see Fig. 4G.5, 4G.6, 4G.7). To the northeast, in the eastern Foothills, the Liard Formation is erosionally and disconformably overlain by Lower Cretaceous rocks of the Fort St. John Group (Stott, 1982).

The macrofauna of the Liard Formation includes *Nathorstites* and other ammonoids, bivalves, and brachiopods. The ammonoids prove the presence of the *meginae, maclearni*, and *sutherlandi* zones (see Fig. 4G.4). Their distribution shows that Liard rocks of eastern sections were deposited contemporaneously with Toad and Ludington rocks farther west (see Fig. 4G.7).

The Charlie Lake Formation, like the Starlight Evaporite Member, comprises a variable sequence of buff to yellow and grey to orange-brown weathering sandstone, siltstone, sandy limestone, dolostone, and minor amounts of intraformational and solution breccia (Fig. 4G.11). In the subsurface of the Plains to the east, it contains thick intervals and interbeds of anhydrite and in addition, contains several unconformities. However, the unconformities have not been recorded in strata of the outcrop belt. The Charlie Lake ranges in thickness from a northeastern erosional pinch-out in the eastern Foothills and subsurface of the Plains to a maximum of 405 m near Williston Lake (Gibson, 1975).

The contact of the Charlie Lake Formation with the overlying Baldonnel Formation north of Williston Lake is gradational. Between Williston Lake and Sukunka River, the contact is abrupt. Eastward and northeastward across the Foothills and the western Plains, the Charlie Lake is overlain by the Baldonnel and farther east is overlain erosionally and disconformably by the Jurassic Fernie Formation.

The Charlie Lake contains few fossils, but the deeper water correlative facies of the western Foothills and Front Ranges, the Ludington Formation, contains a fauna of probable Ladinian to Carnian age (Gibson, 1971a, 1975).

The Ludington Formation consists of siltstone, very fine to fine grained sandstone, and skeletal limestone (Fig. 4G.14). It occurs in the extreme western Foothills north of Williston Lake, extending north at least as far as Liard River. Strata similar in appearance and age to the Ludington have been identified by Gabrielse (1977) in a narrow northward-elongate belt of about 60 km^2 in the western Muskwa Ranges of northern British Columbia (Fig. 4G.1, lat. 58°). In an area between the headwaters of Halfway and Graham rivers, the Ludington is characterized by an unusual occurrence of skeletal limestone, which consists of a dense coquina of whole and fragmented pelecypod and brachiopod shells and crinoid columnals, forming thin lenticular beds and pods, and as distinctive thin to thick "shell banks", the latter ranging in thickness from 4 m to 150 m west of Mount Laurier (Gibson, 1975). The formation ranges in measured thickness from a minimum of 500 m at the type locality at Mount Ludington, to approximately 960 m near Mount Laurier to the northwest (Pelletier, 1964; Gibson, 1975). Thickness trends of the Ludington Formation toward the Liard River area to the north are not known.

Contact relationships with overlying formations are uncertain throughout most of the areas studied. Overlying strata in many areas have been eroded. At Mount Ludington, strata typical of the Baldonnel Formation were not observed, and Ludington strata are overlain by limestones and siltstones of the Pardonet Formation (Fig. 4G.14). The absence of Baldonnel strata at this locality is due to a lateral facies change into strata typical of the Ludington (Gibson, 1971a). The Ludington Formation in outcrop has not been observed in contact with the Liard Formation in the Mount Ludington-Mount Laurier area. However, in the Liard River-Alaska Highway area to the north and possibly elsewhere, the Ludington Formation is in an interfingering contact relationship with the underlying Liard Formation, as illustrated in Figures 4G.5, 4G.6 and 4G.7. As noted previously, the Ludington forms an abrupt, but conformable contact with the underlying Toad Formation in some western exposures.

The fauna of the Ludington Formation includes ammonoids and species of *Halobia*. Late Ladinian (*sutherlandi* Zone), Early Carnian (*obesum* Zone), and Late Carnian (*dilleri* and *welleri* zones) faunas have been recognized (see Fig. 4G.5, 4G.6).

The Baldonnel Formation throughout most areas of the Foothills comprises pale-weathering, cliff-forming limestone and dolostone, with less common siltstone and sandstone (Fig. 4G.11). However, between Williston Lake and Sukunka River, the Baldonnel is characterized by an additional lithofacies at the base, consisting of dark-weathering siltstone, sandstone, limestone, and dolostone called the Ducette Member. Its upper contact with the main carbonate facies of the Baldonnel is gradational. The Ducette Member ranges in thickness in the Foothills from a depositional feather-edge to a maximum of 118 m near Pine River. The Baldonnel Formation, including the Ducette Member, ranges in thickness from an erosional pinch-out in the subsurface of the Plains to the east, to a maximum of 146 m south of Williston Lake (Gibson, 1975).

The contact of the Baldonnel Formation with the overlying Pardonet Formation is generally abrupt and distinct. Toward the east, the Baldonnel is disconformably overlain by the Jurassic Fernie Formation.

An ammonoid fauna of Late Carnian age (*welleri* Zone) has been recovered from strata probably equivalent to the Baldonnel Formation in the Toad River area of northeastern British Columbia (E.T. Tozer, pers. comm., 1984). The Ducette Member, because of its stratigraphic position and relationship to the main facies of the Baldonnel Formation and underlying Charlie Lake Formation, also is of Late Triassic, Carnian age.

The Pardonet Formation consists of dark-weathering limestone, calcareous and dolomitic siltstone, and minor shale (Fig. 4G.11, 4G.15). Much of the limestone is skeletal and consists of whole and fragmented shells of the bivalves *Halobia* and *Monotis*, which generally form dense coquinas and falsely resemble cryptalgal laminites. The Pardonet

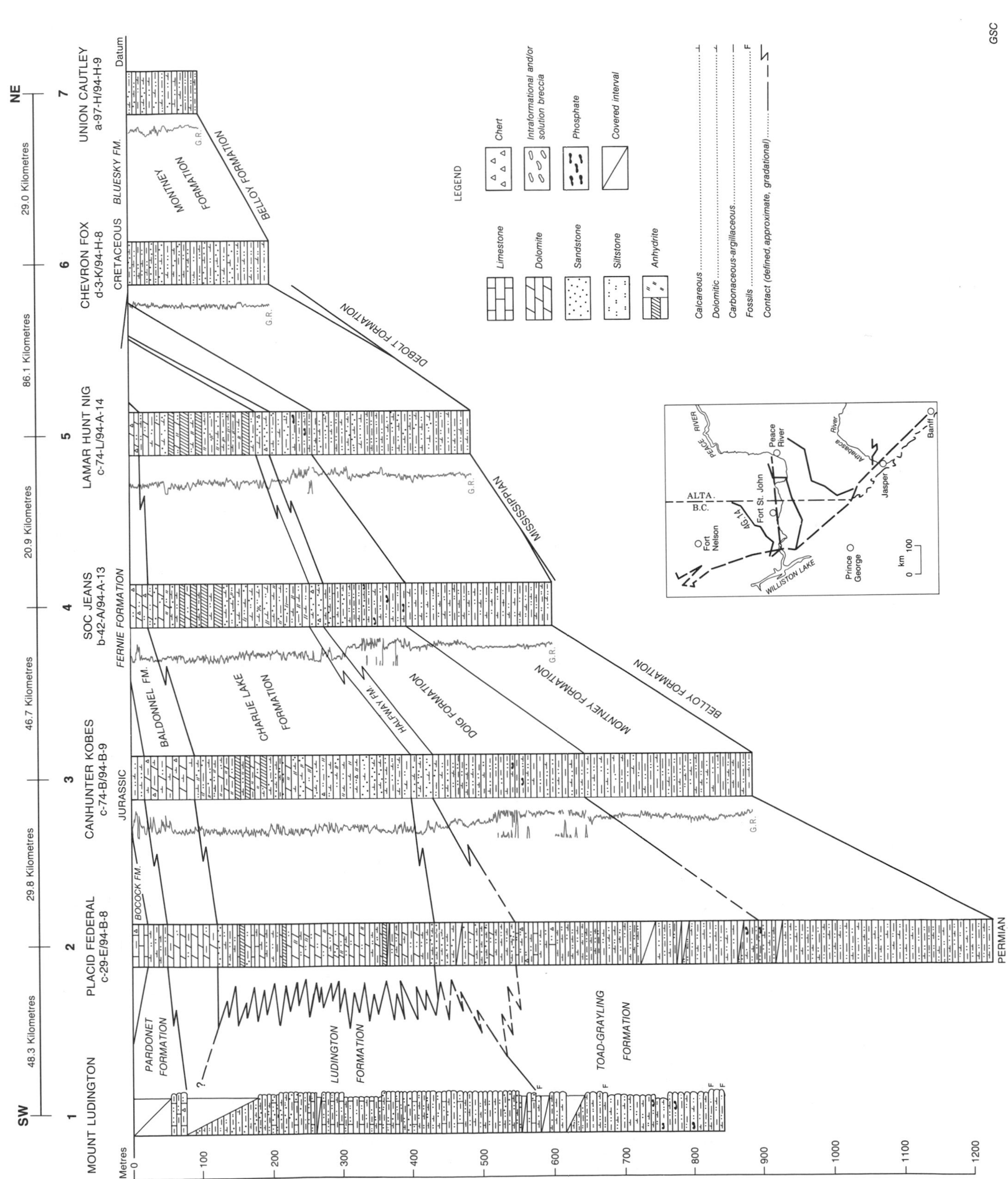
SW
NE
48.3 Kilometres
29.8 Kilometres
46.7 Kilometres
20.9 Kilometres
86.1 Kilometres
29.0 Kilometres
1
MOUNT LUDINGTON
2
PLACID FEDERAL
c-29-E/94-B-8
3
CANHUNTER KOBES
c-74-B/94-B-9
4
SOC JEANS
b-42-A/94-A-13
5
LAMAR HUNT NIG
c-74-L/94-A-14
6
CHEVRON FOX
d-3-K/94-H-8
7
UNION CAUTLEY
a-97-H/94-H-9
Metres
0
100
200
300
400
500
600
700
800
900
1000
1100
1200
Datum
JURASSIC
CRETACEOUS
PERMIAN
MISSISSIPPIAN
FERNIE FORMATION
BLUESKY FM.
BOCOCK FM.
PARDONET FORMATION
BALDONNEL FM.
CHARLIE LAKE FORMATION
LUDINGTON FORMATION
TOAD-GRAYLING FORMATION
HALFWAY FM.
DOIG FORMATION
MONTNEY FORMATION
BELLOY FORMATION
DEBOLT FORMATION
G.R.
LEGEND
Limestone
Dolomite
Sandstone
Siltstone
Anhydrite
Chert
Intraformational and/or solution breccia
Phosphate
Covered interval
Calcareous
Dolomitic
Carbonaceous-argillaceous
Fossils
Contact (defined, approximate, gradational)
PEACE RIVER
Peace River
Athabasca River
ALTA.
B.C.
Fort Nelson
Fort St. John
4G.14
WILLISTON LAKE
Prince George
Jasper
Banff
km
0
100
GSC

Formation in outcrop ranges in thickness from an erosional pinch-out south of Sukunka River to a maximum of 137 m south of Peace River. The Pardonet does not extend very far eastward into the Plains nor south of Sukunka River because of pre-Jurassic erosion.

Throughout most of northeastern British Columbia, the Pardonet is unconformably overlain by dark grey, recessive shale, shaly siltstone, and limestone of the Jurassic Fernie Formation. However, between Williston Lake and Pine River, the Pardonet is overlain abruptly, and possibly disconformably, by the Bocock Formation (Fig. 4G.3, 4G.11, 4G.15).

Ammonoids and *Halobia* are abundant in the Pardonet Formation. The upper beds of the Pardonet are characterized by *Monotis subcircularis* Gabb and *Monotis ochotica* (Keyserling) of Late Norian age. The oldest faunas in the formation are Late Carnian (*macrolobatus* Zone), the youngest, Late Norian (*cordilleranus* Zone) (see Fig. 4G.4, 4G.5, 4G.6, 4G.7). At one locality, the former site of Ne-Parle-Pas Rapids on Peace River (now part of Williston Lake Reservoir), the Upper Norian *Monotis* beds are overlain by strata that possibly represent the *amoenum* Zone.

In the westernmost Foothills between Williston Lake and Pine River, the uppermost Triassic strata are a cliff-forming light grey weathering limestone called the Bocock Formation (Fig. 4G.11, 4G.15). The formation ranges in thickness from an erosional feather-edge to a maximum of 63 m (Gibson, 1975). Strata of the Bocock have not been observed in outcrop north of Peace River; however, they have been identified in boreholes immediately north of Williston Lake (Gibson, 1971a). Conodonts from the Bocock Formation (M.J. Orchard, in Tozer, 1982a), indicate a Late Norian age.

The contact with the overlying Jurassic Fernie Formation is distinct and disconformable (Fig. 4G.15).

Subsurface, Rocky Mountain Foothills and Interior Plains, northeastern British Columbia and west-central Alberta

Daiber Group

The subsurface Daiber Group of the Foothills and Plains north of Township 58 is divided into two formations, a lower, Montney and an upper, Doig (Armitage, 1962; Fig. 4G.3). The group ranges in thickness from an erosional pinch-out in the Alberta Plains to more than 600 m in the Foothills. The Daiber Group rests unconformably upon rocks of Permian or Carboniferous age. It is overlain by strata of the Halfway Formation of the Schooler Creek Group. In the eastern subsurface of the Plains, the contact with the Halfway Formation is unconformable (Armitage, 1962).

Figure 4G.14. Stratigraphic cross-section of Triassic units, Rocky Mountain Foothills and Interior Plains, northeastern British Columbia.

The Montney Formation consists of siltstone and shale (Fig. 4G.10, 4G.11, 4G.14). Along the eastern depositional margins, however, the Montney is characterized by hydrocarbon-bearing sandstone and dolomitized coquina-sandstone interbeds, the latter occurring at several levels. For example, a basal sandstone is the earliest depositional unit of the Triassic succession in the Peace River area of western Alberta (Barss et al., 1964), occupying erosional lows on the underlying carbonates. Grain size in the basal sandstone increases eastward toward the depositional and/or erosional edge. The dolomitized coquina-sandstone facies is also an eastern basin margin facies to the south (Miall, 1976). The coquina-sandstone lithofacies is progradational, with the lowest units forming in the south and east and succeeding ones developing toward the north and west. The Mackenzie Dolomite Lentil, discussed previously, represents a similar coquina facies within the Sulphur Mountain Formation (Fig. 4G.9). The coquina-sandstone units in the Montney Formation of the Peace River area occur within the lower 105 m. The coquinas are leached; their moldic porosity forms excellent hydrocarbon reservoirs. The associated sandstones have generally poor porosity. The Montney ranges in thickness from an eastern erosional edge to more than 450 m in the Foothills.

The Montney Formation unconformably overlies the Permian Belloy Formation in some areas and the Carboniferous Debolt in others. The Montney is abruptly overlain by phosphatic shale of the basal Doig Formation (Fig. 4G.3, 4G.10, 4G.11, 4G.14).

The Montney Formation, because of its equivalence to the Vega Siltstone and Phroso Siltstone members of the Sulphur Mountain Formation, is interpreted to be of Early Triassic age (Fig. 4G.3).

The Doig Formation consists of siltstone, with developments of very fine grained sandstone and dolomitized coquina along the eastern margin of deposition (Fig. 4G.10, 4G.11, 4G.14). Throughout much of the region, the base of the Doig is characterized by nodular phosphate in thicknesses up to 9 m. The phosphate concentration provides a strong gamma-ray signal marking the base of the formation. The phosphate content decreases westward, so that no equivalent phosphatic unit can be recognized with any degree of certainty in surface exposures of the Toad Formation. However, the phosphate unit, although thin, still characterizes the base of the Whistler Member of the Sulphur Mountain Formation to the south. Along the eastern depositional margin, it is difficult to distinguish between Doig coquina-sandstone units and overlying coquina-sandstone units of the Halfway Formation, because of stratigraphic convergence and the presence of local unconformities. In the Grande Prairie and adjacent areas of Alberta and northeastern British Columbia, the Doig Formation is characterized by large channel-fill sandstone units 30 to 55 m thick, which display easterly to northerly trends (Cant, 1986). The Doig thins from more than 120 m in the Foothills, to an eastern erosional edge (Armitage, 1962). An ammonoid indicating the *caurus* Zone (Lower Anisian, see Fig. 4G.4) was recovered from the formation of the Guardian well (7-7-80-12-W6), near Pouce Coupe, British Columbia.

The Doig Formation is overlain by strata of the Halfway Formation. The contact is unconformable in the east but gradational in the west.

Schooler Creek Group

The Schooler Creek Group in the subsurface of the Foothills and Plains consists of four formations, in ascending order, Halfway, Charlie Lake, Baldonnel, and Pardonet (Fig. 4G.3). The Schooler Creek ranges in thickness from an eastern erosional feather-edge to more than 600 m in the Foothills to the west. It is unconformably overlain by the Jurassic Fernie Formation in most areas but by the Cretaceous Fort St. John Group in some northern areas.

The subsurface Halfway Formation in the Foothills and Plains of northeastern British Columbia and west-central Alberta comprises very fine to medium grained sandstone, with lesser amounts of finely crystalline dolostone (Fig. 4G.10, 4G.11, 4G.14). In addition, along the eastern depositional margin dolomitized coquinas are developed locally (Cant, 1986). The Halfway ranges in thickness from an eastern erosional pinch-out to more than 120 m in the west.

Halfway strata throughout most of the Plains and Foothills area form a "blanket" sandstone lithofacies. However, near the eastern depositional pinch-out (Fig. 4G.16), the Halfway Formation is composed of a series of linear, discontinuous, sandstone and coquinoid sandstone outliers. In some areas these outliers consist of reworked Doig sandstone and coquina and, accordingly, are difficult to differentiate from those of the Doig Formation. The discontinuous sandstone-coquina outliers rest unconformably upon the Doig strata and are commonly prime oil producers.

The Halfway Formation is conformably and abruptly overlain by the Charlie Lake Formation in the east; in the west the contact is gradational.

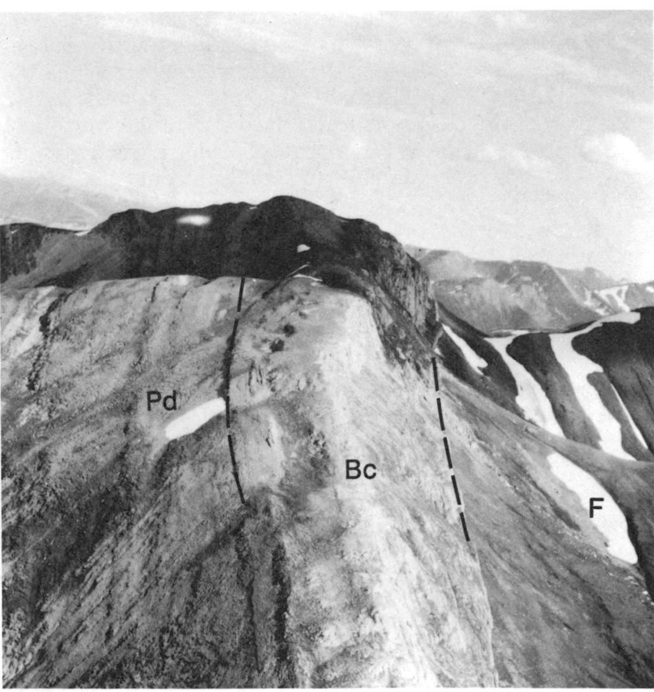

Figure 4G.15. Pardonet (Pd) and Bocock (Bc) formations and Jurassic Fernie (F) Formation, Pine River-Williston Lake area, northeastern British Columbia. View is southeastward. (photo - D.W. Gibson, GSC 145716A).

In the subsurface, the Charlie Lake Formation consists of intercalated siliciclastic, carbonate and evaporite rocks (Fig. 4G.10, 4G.11, 4G.14). Intraformational and collapse or solution breccias are common in some areas. The siliciclastics consist of shale, siltstone, and very fine to medium grained sandstone. Redbeds are best developed toward the eastern margin of deposition. Laterally extensive, microcrystalline to cryptocrystalline dolostone and skeletal limestone and dolostone are common. Anhydrite, gypsum, and minor salt are typical of the subsurface Charlie Lake Formation, and form persistent markers, particularly in the central and western areas of the Foothills and Plains.

The Charlie Lake Formation contains three formally defined members, in ascending order, Inga, Coplin, and Boundary Lake. In addition, as many as 10 local and regionally significant stratigraphic marker units have been recognized and delineated by the British Columbia Department of Mines and Petroleum Resources and several petroleum exploration companies (Fig. 4G.17). The formation contains several unconformities, which are most evident along the eastern margin of deposition but can also be recognized elsewhere in northeastern British Columbia. The "Coplin unconformity" has regional significance, and serves to informally divide the formation into two divisions. In discussing similar unconformities in the Wembley Field, Cant (1986, p. 337) suggested that such discontinuities do not indicate major periods of uplift and erosion. Rather, he proposed that they developed as the result of sea-level fluctuations and local tectonism related to the Peace River Arch.

The Charlie Lake Formation ranges from an erosional eastern edge to more than 425 m in the Foothills. It is abruptly overlain by the Baldonnel Formation in most areas. In the eastern regions it is unconformably overlain by strata of the Jurassic Fernie Formation and in the northern regions by strata of the Cretaceous Fort St. John Group (Fig. 4G.14).

The Baldonnel Formation of the subsurface of the Foothills and Plains comprises dolostone and minor limestone (Fig. 4G.10, 4G.11, 4G.14). Medium grey chert lenses and nodules and chert-filled vugs are conspicuous in some areas. The Baldonnel can be subdivided into several shallowing-upward sequences, characterized by subtidal mudstone and wackestone followed by shallow subtidal to lower intertidal packstone and grainstone, and capped by upper intertidal to supratidal cryptalgal laminite, thin-bedded anhydrite, and rare red siltstone (Bever and McIlreath, 1984). Similar sequences have been recognized by Barss and Montandon (1981, Fig. 4G.18). The formation ranges in thickness from an erosional pinch-out (Fig. 4G.16), to more than 80 m in the Foothills.

The Baldonnel Formation in the Foothills generally is overlain gradationally, and in some places abruptly, by the Pardonet Formation. However, throughout most areas of the Plains and eastern Foothills, it is unconformably overlain by strata of the Jurassic Fernie Formation.

The uppermost formation of the Schooler Creek Group in the subsurface is restricted mainly to the area of the Foothills. The Pardonet Formation has recently been divided into four informal stratigraphic units, which are, in ascending order, highly radioactive, dark

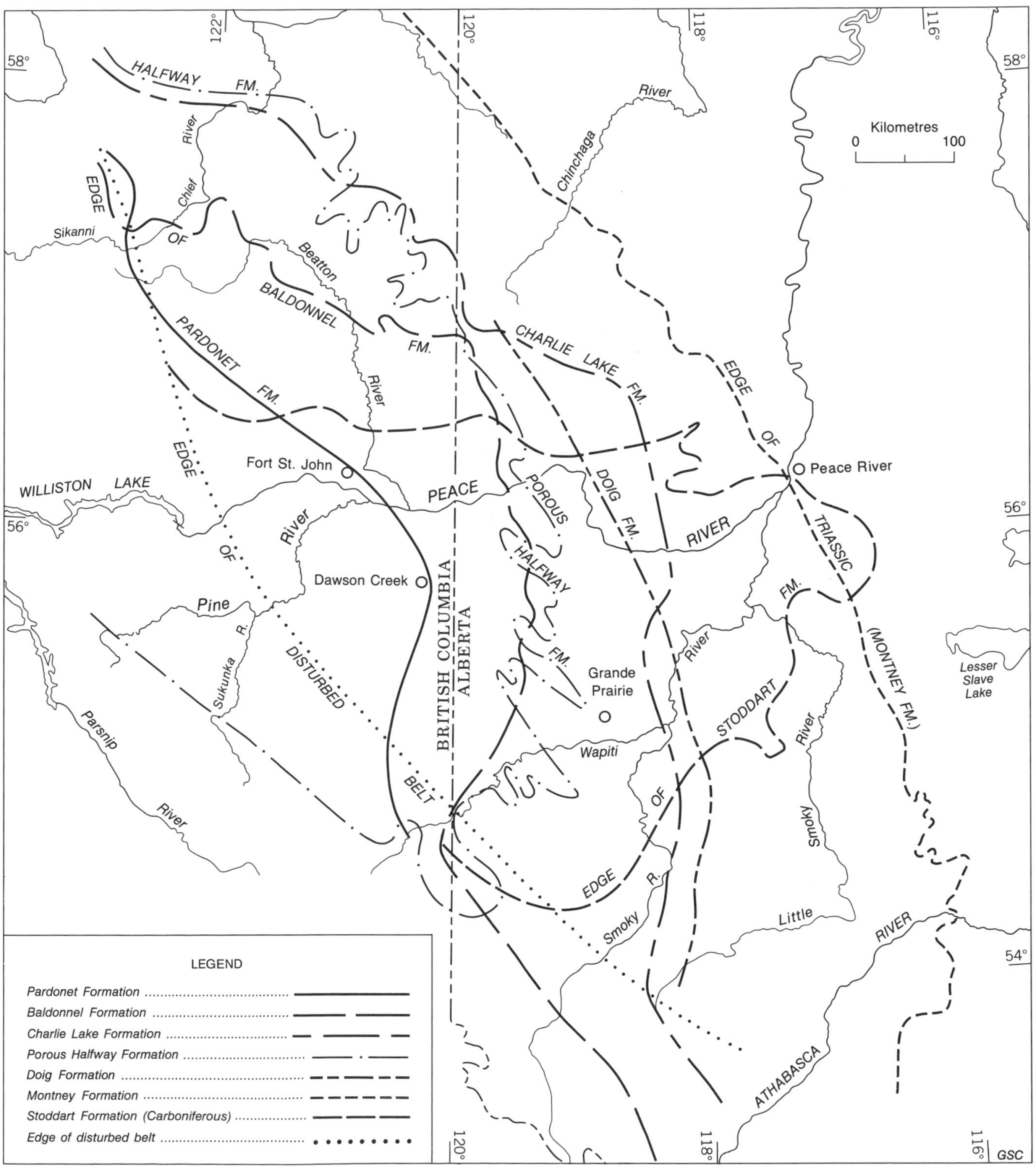

Figure 4G.16. Distribution of important subsurface Triassic units, northeastern British Columbia and west-central Alberta (Gulf Canada Resources).

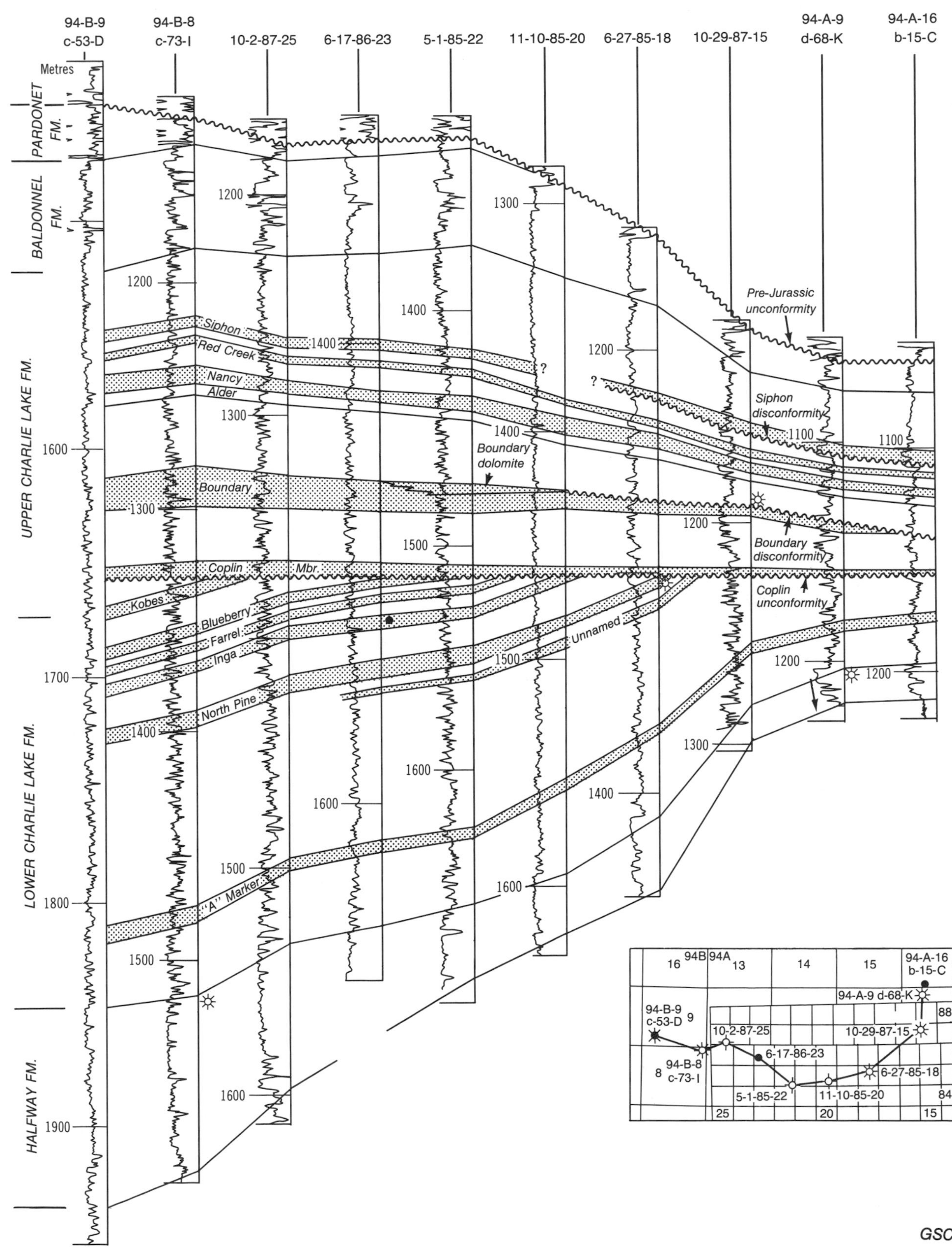

Figure 4G.17. Nomenclature and stratigraphic relationships, subsurface Charlie Lake Formation, northeastern British Columbia and west-central Alberta (after Torrie, 1973).

grey to black, bituminous, dolomitic, pyritic, argillaceous siltstone; argillaceous silty limestone and silty shale; relatively clean limestone or dolostone interbedded with silty argillaceous limestone; and clean arenaceous limestone or dolostone (Fig. 4G.18; Barss and Montandon, 1981). The Pardonet Formation in the subsurface ranges in thickness from an erosional pinch-out (Fig. 4G.16), to more than 50 m.

The Pardonet is unconformably overlain by the Jurassic Fernie Formation.

Northern Yukon Territory and District of Mackenzie

Triassic rocks in northern Yukon and District of Mackenzie occur in several widely separate areas in the British, Barn, Richardson, Ogilvie, Wernecke, and Selwyn mountains (Fig. 4G.2). To date, however, only two formations have been named and described: the Permian-Triassic Sadlerochit Formation and the Lower to Upper Triassic Shublik Formation (Mountjoy, 1967). Recently Gordey et al. (1981) described an unnamed Triassic sequence from Selwyn Mountains.

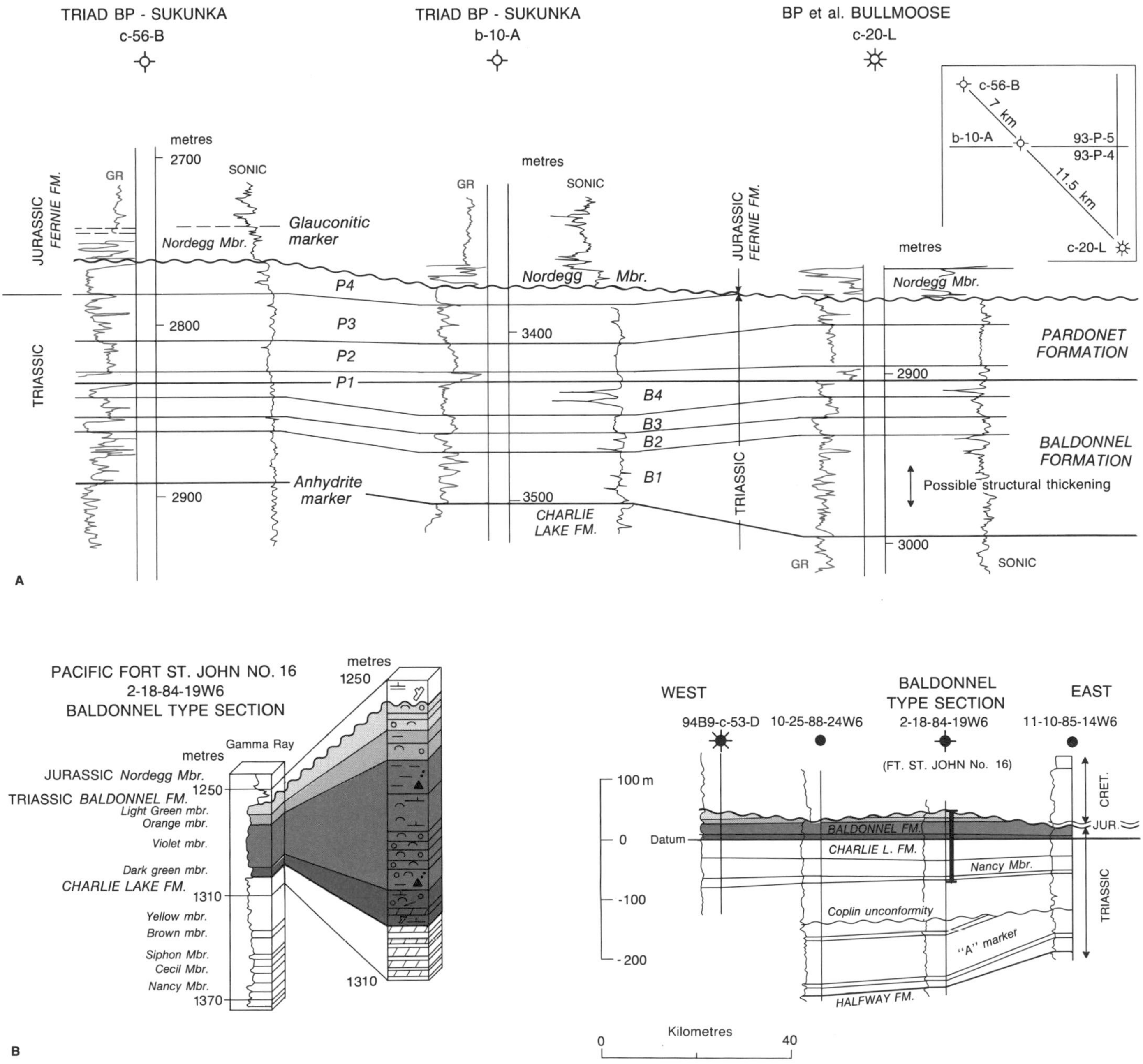

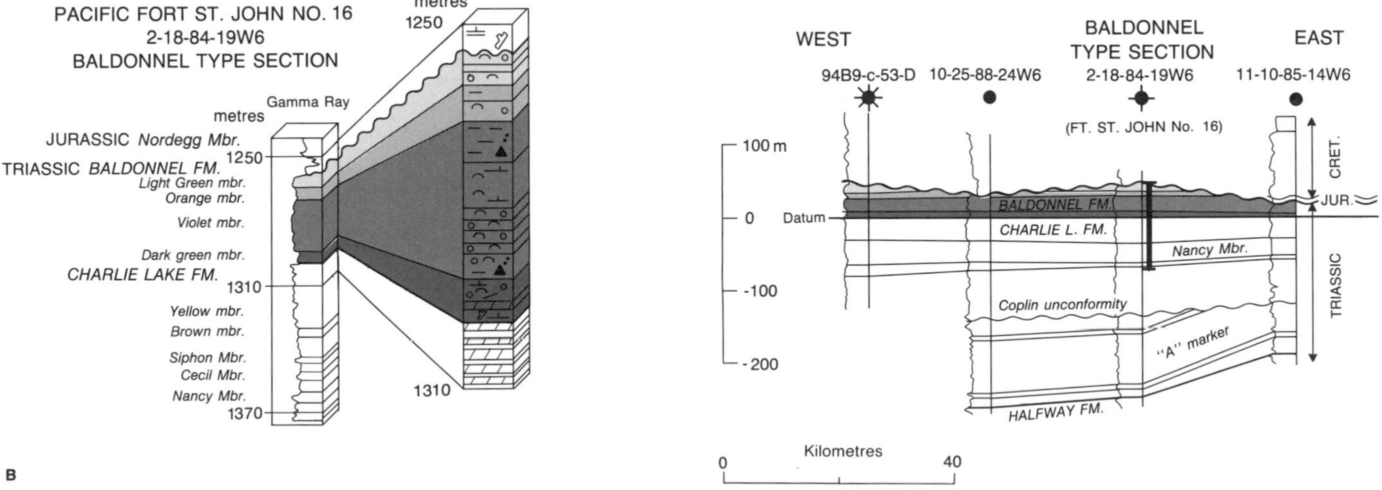

Figure 4G.18. Local lithostratigraphic subdivisions, subsurface Pardonet, Baldonnel and Charlie Lake formations, northeastern British Columbia (A - After Barss and Montandon, 1981, B - Petro-Canada Ltd.).

The Sadlerochit Formation of northern Yukon Territory is limited to an area just east of the Alaska-Yukon border in the British Mountains (Fig. 4G.2). It consists of two members, the upper of which, the Ivishak Member, is Triassic in age (Fig. 4G.19). The lower, called the Echooka Member, is Permian. The Ivishak Member consists of brown sandstone and grey shale (Mountjoy, 1967), and ranges in thickness from an erosional edge to 130 m.

The Ivishak Member is underlain abruptly, and presumably unconformably, by interbedded silty shale and sandstone and minor carbonates of the Echooka Member and overlain unconformably by the Shublik Formation (Fig. 4G.19; Mountjoy, 1967).

The Shublik Formation occurs mainly in three widely separated regions of northern Yukon Territory; the British and Barn mountains, Richardson Mountains, and northwestern Ogilvie Mountains (Fig. 4G.2). The Shublik comprises a regionally variable sequence of siltstone, sandstone, shale, and silty and skeletal limestone, with minor quartz conglomerate near the base in the British Mountains. The limestone is a conspicuous component of the formation (Fig. 4G.20), and consists of coquinas in some areas, silty limestone and coarse-grained skeletal calcarenites in others. The Shublik is variable in thickness, ranging from an erosional pinch-out to values in excess of 120 m (Mountjoy, 1967).

Near the Alaska-Yukon border, the Shublik Formation conformably overlies the Lower Triassic Ivishak Member of the Sadlerochit Formation. In the British Mountains, it overlies the Carboniferous Lisburne Group and Kayak Formation and middle and lower Paleozoic Road River Formation. Strata equivalent to the Shublik Formation unconformably overlie Permian and Carboniferous clastic rocks and the Devonian Imperial Formation in Richardson Mountains, and the Permian Tahkandit Formation in

			North central Brooks Range, Alaska (composite)	Eastern Brooks Range, Joe Creek, Yukon and Alaska	British and western Barn Mountains	Northern Richardson Mountains	Central and southern Richardson Mountains	Western Ogilvie Mountains
Overlying formations			*KINGAK FM.* (JURASSIC)	*KINGAK FM.* (L. JURASSIC)	*Shale* (L. JURASSIC)	*Sandstones* (JURASSIC, Jeletsky, 1967)	UPPER JURASSIC	UPPER CRET.
TRIASSIC	UPPER	RHAETIAN						
		NORIAN	*SHUBLIK FM.*	*SHUBLIK FM.*	*SHUBLIK FM.*		*Unnamed limestone*	*Unnamed shale and argillaceous limestone*
		CARNIAN				?	?	?
	MIDDLE	LADINIAN	*Shale Mbr.* ?			*Nonmarine clastics*		
		ANISIAN				?		
	LOWER	SPATHIAN						
		SMITHIAN						
		DIENERIAN	*SADLEROCHIT FM.*					
		GRIESBACHIAN	*Ivishak Mbr.*	*Ivishak Mbr.*	*IMPERIAL FM.*			
Underlying formations			*LOWER SADLEROCHIT FM.* / *LISBURNE GP.*	*Echooka Mbr.* SADLEROCHIT FORMATION (PERMIAN)	*LISBURNE GP.* / *Kayak shale* / *NERUOKPUK FORMATION*	*Permian clastics* / *IMPERIAL FM.* (DEVONIAN)	*Mississippian clastics*	*TAHKANDIT FM.* / *Permian shales*

GSC

Figure 4G.19. Correlation chart of Triassic strata, northern Yukon Territory (after Mountjoy, 1967).

Figure 4G.20. Limestones of the Upper Triassic Shublik Formation (TS) unconformably overlying Lower Devonian shales and chert of the Road River Formation (DR), and overlain by Jurassic (J) shales, lower Firth River area, northern Yukon Territory (photo - D.K. Norris, ISPG 2421-206).

Ogilvie Mountains. The Shublik Formation is unconformably overlain by the Jurassic Kingak Formation in the British Mountains, an unnamed Jurassic sandstone sequence in the Richardson Mountains, and Upper Cretaceous and/or Tertiary strata in the Ogilvie and Wernecke mountains.

The Shublik Formation in the British and Barn mountains has yielded the pelecypod *Monotis ochotica* (Keyserling) of Late Triassic, Late Norian age. In the southern Ogilvie Mountains, isolated occurrences of Triassic siltstones and limestones of probable Smithian to Late Norian age have been reported (Tempelman-Kluit, 1970). The Triassic in this area is probably unconformably underlain by the Permian Tahkandit Formation and unconformably overlain by mid-Jurassic strata.

Triassic rocks have recently been described from isolated exposures in Selwyn Mountains of east-central Yukon Territory and District of Mackenzie (Gordey et al., 1981). They comprise a thick succession of siltstone, sandstone, and shale up to 750 m thick. Meagre conodont collections indicate a Smithian to Norian age.

Correlation

Correlation of Triassic formations and members between various areas of the outcrop belt and the subsurface is illustrated by the stratigraphic cross-sections of Figures 4G.9, 4G.10, 4G.11, 4G.14, and by Figures 4G.3 and 4G.19. The correlation between some surface and subsurface units of the Foothills and Plains of northeastern British Columbia and west-central Alberta is still speculative, because of the absence of adequate fossil and lithological control. All available well logs and core and chip samples from the eastern Foothills and adjacent Plains have not to date been systematically examined and related to field exposures. Furthermore, fossil preservation in most members of the Triassic succession in the Jasper-Banff-Crowsnest Pass area is generally poor, and accordingly, correlation of southern members with those to the north and in the subsurface is based mainly upon lithology and homotaxis or relative stratigraphic position.

Recognition and correlation of subsurface units in some areas of the Plains remain speculative. For example, in the Hamelin Creek and Forest Peace wells included in Figure 4G.11, the dolostone capping the Triassic succession is interpreted by some to be equivalent to the Charlie Lake Formation as shown. Others, however, interpret the dolostone to be equivalent to the newly discovered "Worsley/Tangent Dolomite", a hydrocarbon-bearing unit, which is younger than the Charlie Lake. Additional work is necessary to resolve this and other subsurface correlation problems.

For details concerning the regional correlation of formations and members discussed in this report, the interested reader is referred to reports by Barss et al. (1964), Mountjoy (1967), Tozer (1967), and Gibson (1975).

HISTORY AND ENVIRONMENTS OF DEPOSITION

Western Canada Basin

The following synthesis concerning the history and paleoenvironments of deposition in Western Canada Basin is summarized from reports by Hunt and Ratcliffe (1959), Colquhoun (1960), Clark (1961), Armitage (1962), Barss et al. (1964), Pelletier (1963, 1965), Sharma (1969), Mothersill (1968), Gibson (1968a, 1971a, b, 1974, 1975), Nelson (1970), Torrie (1973), Miall (1976), Tozer (1982b), Cant (1984), and Bever and McIlreath (1984), and from personal communication with colleagues of the Geological Survey of Canada, Gulf Canada Resources Limited, and Petro Canada Limited.

Triassic rocks form a westward-thickening wedge of predominantly marine sediments deposited along a tectonically stable continental shelf and shoreline of the passive western margin of the North American Craton, inherited from the Permian Period. Barss et al. (1964) outlined the shelf and suggested that the eastern, erosional edge of the Triassic succession closely approximates the original, eastern shoreline. Some early plate-tectonic models for western Canada (Monger et al., 1972; Tozer, 1982b) cast doubt, however, on the idea of a deep trough or volcanic archipelago adjoining the shelf to the west (Barss et al., 1964; Gibson, 1968a, 1971a, b, 1974, 1975; Nelson, 1970). Tozer (1982b) suggested that the Triassic sedimentary and volcanic rocks of the central and western Cordillera accumulated in a volcanic archipelago as much as 5000 km distant from the Triassic continental shelf of western Canada, thereby suggesting that the shelf deposits faced an open sea. The island-arc rocks were considered to be unrelated to the shelf deposits, but rather to belong to a belt of exotic terranes, which were accreted to the North American Craton during early Middle Jurassic time (Monger et al., 1972). However, recent work in southeastern British Columbia by Klepacki (1985) showed that Triassic volcanics lie unconformably on Paleozoic rocks, which, in turn, lie on the pericratonic Kootenay Terrane. These relationships suggest that part of the allochthonous rocks may have been deposited not far from ancestral North America. Nevertheless, there is little or no evidence of a western source for the Triassic sediments of the Rocky Mountains and Foothills.

Early Triassic Epoch

Phroso Siltstone, Vega Siltstone, Vega-Phroso Siltstone members, Grayling, Lower Toad, and Montney formations

Most of the Lower Triassic clastic sediments were probably derived from sources to the east and northeast (Fig. 4G.1). These sources were likely of low relief and mature topography developed in sedimentary rocks. This is suggested by the concentration of quartz in relation to other detrital minerals, the general absence of rock fragments and ferromagnesian minerals, the occurrence of westerly dipping crossbeds, the westerly thickening stratigraphic units, and the westerly decrease in insoluble residue content of some members of the Sulphur Mountain Formation.

Prior to Triassic deposition the Permian shelf and coastal plain areas were subjected to a period of marine regression and uplift. Succeeding deposition took place in the relatively deep, eastward-shallowing open-shelf marine environment, of an easterly transgressing sea in which the Phroso Siltstone, Vega Siltstone, and Vega-Phroso Siltstone members of the Sulphur Mountain Formation, and the Grayling, Lower Toad, and Montney formations were laid down.

Coarse-grained siltstones, sandstones, and silty and skeletal dolostones of the Plains were deposited in a shallow-water, river-dominated, deltaic environment (Miall, 1976). The coarser grained sediments are interpreted by Miall to represent delta-front sands, perhaps laid down in subaqueous distributary channel or distributary mouth bar depositional environments. The dolostones and other thin carbonate facies of the Sulphur Mountain and Montney formations represent interdeltaic and interdistributary shell-bank environments (Miall, 1976). Alternatively, these plains strata may be interpreted as shallow-marine shelf or lower shoreface sediments deposited away from direct fluvial processes. The absence of associated coal and carbonaceous coaly strata and specific deltaic configurations are not in keeping with the hypothesis of a river-dominated delta. However, such facies associations and deltaic configurations may have been generated landward but subsequently removed by post-Early Triassic erosion. Additional work is necessary to resolve the problem.

Farther west, in the outcrop belt of the Front Ranges and Foothills, Lower Triassic rocks consist of finer grained, thinner bedded siltstone and shale at the base, grading upward into slightly coarser grained units of alternating siltstone and silty shale. The siltstones generally increase in frequency and thickness toward the top. Sedimentary structures consist of regular to wavy laminations in the thinner bedded and finer grained strata, possible hummocky crossbedding in some of the thicker bedded strata, small-scale crossbedding, and ripple marks. Sole marks such as groove casts, flute casts, bounce casts, and load casts occur in some siltstones of the Vega Siltstone Member and some facies of the Grayling and Toad formations. Graded bedding is rare. The distinctive alternation of resistant and recessive weathering beds in much of the succession, and the associated sedimentary structures, suggest deposition of the coarser beds by turbidity or storm-generated offshore currents. The occurrence of fine regular to wavy lamination in the thinner less resistant more carbonaceous beds, and the occurrence of thin shelled pelagic bivalves and probable radiolarians in some strata of the Grayling and lower Toad formations, indicate deposition in a deep water distal shelf environment. However, recent work by Nelson (1982) in Yukon Delta of Alaska has demonstrated the occurrence, in recent sediments, of alternating silts and muds containing sedimentary structures typical of so-called deep-water turbidity currents. The recent sediments of Yukon Delta, however, are found in shallow water at depths of less than 20 m and distances in excess of 100 km offshore.

Available evidence suggests that Lower Triassic sediments of Western Canada Basin were deposited in a deltaic to distal shelf environment, in an initially transgressive (Phroso Siltstone, Grayling, Lower Montney) but subsequently regressive sea (Vega Siltstone, Lower Toad, Upper Montney), influenced by turbidity and/or storm generated currents.

Middle Triassic Epoch

Whistler Member, Middle Toad and lower Doig formations

Following the Early Triassic regression, some areas were subjected to a brief interval of nondeposition and/or erosion, as indicated by the abrupt change between Lower and Middle Triassic facies, and the occurrence in many areas of granular phosphate and phosphatic pebble conglomerate.

The Middle Triassic epoch began with transgression. Eastward shallowing is indicated by pinch-out of the basal Middle Triassic Whistler Member, and eastward coarsening in the Llama Member and Doig Formation. In the Foothills and Front Ranges south of Pine River, lower Middle Triassic sediments of the Whistler Member are characterized by thin beds with fine, regular to lenticular lamination, a high concentration of organic carbonaceous-ferruginous matter, and in some areas, thin, skeletal and silty limestone interbeds. Shells are mostly unfragmented, suggesting deposition below normal wave base. The dark grey to black colour and high content of organic matter may suggest that deposition took place in an oxygen-deficient environment. Furthermore, the thin development of the member and the thin-bedded character of the strata suggest that the basin may have been receiving a low input of detrital sediment. The easternmost Llama Member and lower Doig Formation display a notable decrease in organic and ferruginous matter and lighter colouration.

Equivalent strata north of the Sukunka-Pine rivers area exhibit characteristics similar to those of the Lower Triassic Series, and therefore, probably were deposited in a similar depositional environment.

Llama Member, Liard, Halfway, and upper Doig formations

Following deposition of the Whistler Member and laterally equivalent facies, and during Llama-Liard-upper Doig-Halfway deposition, the western shelf was again subjected to regression. In the eastern Foothills and Plains, the sands, silts, and associated sediments of the progradational succession are interpreted as probable shallow marine shelf, barrier beach, shoreface, and tidal channel deposits. The conspicuous increase in average detrital grain size compared to that in the underlying members and formations, the presence of well-rounded silt and sand-sized detrital siliciclastic and carbonate grains, the occurrence of crossbedding in some sandstones, the lack of micaceous minerals and the high concentration of fragmented shells in some areas, all support the interpretation of a shallow-water, high-energy depositional environment. Furthermore, lingulid and orbiculid shells and shell fragments are common in Llama and Liard strata, and all strata are in part highly bioturbated.

In the western Foothills, the Liard-Llama deposits are characterized by strata suggestive of deposition below normal wave base. In this environment the sandstones and siltstones are fine grained, somewhat carbonaceous and argillaceous, and in some areas contain interbeds of silty and skeletal limestone with ammonoids and bivalves. In the Foothills of the Sikanni Chief River area, the shallow-water Liard sandstones and siltstones grade westward into deeper water strata of the Toad Formation. Cant (1986, p. 334), in describing the Halfway sandstone of the Wembley Field, described it as a shelf to shoreline sequence, dominated by tide- and wave-influenced, beach-barrier deposits. He suggested the Halfway coquinas were wave built, shell bank, channel lags, on locally developed shell beaches.

During the later stages of the Middle Triassic Epoch, the climate appears to have changed, from temperate with normal precipitation, to arid or semi-arid with reduced or little precipitation. This change is reflected in the strata of the upper Llama Member and the Liard Formation, and also in the upper Doig and Halfway formations. Strata from this interval become less carbonaceous, less silty and sandy, and more dolomitic and calcareous. Carbonates become a conspicuous component of the strata, as individual beds and as cement. These Middle Triassic strata are succeeded by shallow-water evaporitic strata of the Charlie Lake and Whitehorse formations, pointing to continuing and increasing aridity. This climatic change appears to have occurred throughout much of western North America, as shown by the occurrence of extensive Triassic "redbeds" in the western United States.

Late Triassic Epoch

Starlight Evaporite Member, Charlie Lake and Ludington formations

Following deposition of the Llama Member, Liard, upper Doig, and Halfway formations, the seas continued to shallow, accompanied by the development of shallow shelves, coastal bars and inlets, restricted hypersaline lagoons, dunes, and extensive tidal flat areas. The postulated depositional environment resembles in some ways the environments existing today along the Texas Gulf Coast, Gulf of California, and the Trucial coast of the Persian Gulf.

The carbonates, redbeds, evaporites, and some breccias of the Charlie Lake Formation and Starlight Evaporite Member of the Whitehorse Formation may have formed in coastal inlets, restricted lagoons, tidal flat, and sabkha environments. The barrier island and coastal shoreface sands of the Halfway Formation restricted water circulation, and coupled with the prevailing arid climate, resulted in the precipitation of gypsum and anhydrite. During deposition of the shallow-water sediments, the eastern part of the basin appears to have been subjected to episodes of erosion during continued westward downwarping, a feature responsible for the development of the Coplin and other small unconformities, and stratigraphic truncation of lower Charlie Lake units (Fig. 4G.16).

The Ludington Formation is composed of strata indicative of deposition under much less restricted conditions in a deep water outer shelf environment. However, small- to medium-scale crossbedding and fragmentation and rounding of bioclasts in some of the calcareous sandstones and limestones in some areas suggest wave and/or current activity. Deposition generally took place below normal wave base. The thick "shell bank" coquina facies of the Ludington (Gibson, 1975) represent deeper water, shelf-slope detrital, channel-fill accumulations of predominantly bivalve and lesser brachiopod shells.

An arid to semi-arid climate at the beginning of Late Triassic time has been postulated above. In such a climate, the transporting capacity of river systems and the input of detritals to the sea would be diminished, promoting at times the deposition of non-clastic sediments.

Brewster Limestone and Winnifred members, and Baldonnel Formation

The depositional environment changed from predominantly shallow-water, restricted marine, to slightly deeper water, more open marine in most areas during Brewster Limestone-Winnifred-Baldonnel deposition.

The Baldonnel Formation in the subsurface of the Plains and eastern Foothills is characterized by several shallowing-upward sequences, each bounded by hiatal surfaces, recording a series of minor transgressions and regressions during Baldonnel deposition. Shallowing-upward cycles, although present in some areas, are not common in the outcrop belt. The fragmented and well-rounded nature of the bioclasts and grains is suggestive of deposition in a relatively high-energy environment. In the western Foothills of northeastern British Columbia, Baldonnel strata contain more organic carbonaceous matter and less fragmented bioclastic material than equivalent beds to the east, suggesting deposition in a relatively deeper water, probably less oxygenated environment. The change from the evaporitic, Starlight-Charlie Lake depositional regime to the open-sea regime (normal marine) of the Baldonnel Formation and Brewster Limestone Member is due to transgression. South of Sukunka River, the Brewster Limestone is overlain by the Winnifred Member of the Whitehorse Formation. The Winnifred is characterized by silty and sandy carbonates, rare sandstones, and locally, by intraformational and/or solution breccias and redbeds, thus the transgression did not eliminate evaporitic conditions everywhere. The environments of Winnifred deposition are considered to be relatively deeper and less restricted than those postulated for the underlying Starlight Evaporite Member and Charlie Lake Formation.

Pardonet and Bocock formations

Throughout the Foothills and parts of the Plains north of Sukunka River (Township 72), the Baldonnel Formation is overlain by silty limestones and calcareous siltstones of the Pardonet Formation. These strata are interpreted to have been deposited in a relatively deep-water depositional environment below storm wave-base in a restricted, probably euxinic environment. Evidence is provided by the intact pelecypod shells, which in many areas form dense coquinas, and by the fetid sulphurous odour and large concentration of carbonaceous-argillaceous matter.

Following deposition of the Pardonet Formation, the seas again regressed, with most of the region subjected to pre-Jurassic erosion. However, between Williston Lake and Pine River, a shallow-water, marine embayment remained within which limestones of the Bocock Formation were deposited. Alternatively, the Bocock may have been deposited over a much broader region of northeastern British Columbia and subsequently preserved only in the Williston Lake-Pine River area as a pre-Jurassic erosional remnant. The well-rounded fragmented shells and locally carbonate breccia clasts, are suggestive of a high-energy environment.

Northern Yukon Territory and District of Mackenzie

Information on Triassic environments and history of deposition in northern Yukon Territory and District of Mackenzie is sparse. Mountjoy (1967) provided an interpretation for most Triassic environments of the region, which included marine and nonmarine; however, all Triassic rocks of the northern Yukon Territory and District of Mackenzie are now interpreted to be of marine origin (D.K. Norris, pers. comm., 1985). The Lower Triassic Sadlerochit Formation was interpreted by Mountjoy (1967) as a shallow-water sandstone, siltstone, and shale facies, the sediment having been derived from the underlying Neruokpuk Formation. The Upper Triassic Shublik Formation is characterized by rapid lateral and vertical facies variations. These strata were deposited along a northwest-trending stable shelf in a nearshore, relatively shallow-water environment. Descriptions by Mountjoy (1967) suggest that the coarser grained and conglomeratic strata record beach and shoreface depositional environments, and the finer grained siltstones and shales deeper water deposition in a restricted environment. Clastic sediments of the Shublik Formation came from local sources. The limestones, some of which consist of bivalve coquinas, accumulated away from the influx of terrigenous sediment, in a shallow-water shelf environment.

In southern Ogilvie Mountains, lithology and faunas suggest an environment of quiet marine sedimentation at moderate depths (Tempelman-Kluit, 1970). The Triassic strata of the Selwyn Mountains are interpreted to represent subtidal deposition on a shallow marine shelf (S.P. Gordey, pers. comm., 1985).

ECONOMIC GEOLOGY

Oil and Gas

Triassic rocks of northeastern British Columbia and northwestern Alberta contain substantial proven and potential reserves of oil and gas. The system in the Peace River region alone accounts for approximately 5% of the discovered initial reserves of the conventional oil reserves of Western Canada Basin (Podruski, et al., 1988, p. 50). Currently, production is from the Halfway, Charlie Lake, and Baldonnel formations of the Schooler Creek Group, with lesser but economically significant production from the Doig and Montney formations of the Daiber Group. In the Foothills of west-central Alberta, minor hydrocarbon production has been obtained from the Sulphur Mountain Formation. To date, most production is from stratigraphic traps; however, recently renewed emphasis has been directed toward the discovery of structural traps in the Foothills, particularly traps involving the Halfway Formation (Torrie, 1973; Podruski et al., 1988).

Although potential hydrocarbon sources, the Sulphur Mountain, Grayling, Toad, and subsurface Montney formations are commonly too fine grained, non-porous, and impermeable to serve as potential hydrocarbon reservoirs. However, the eastern shoreline facies of these formations are commonly characterized by porous sandstone and coquinoid dolostone, which are hydrocarbon producers (Miall, 1976; Podruski et al., 1988). Limited production has been obtained from the Mackenzie Dolomite Member of the Sulphur Mountain Formation in the Cadomin-Mountain Park area of Alberta. In recent years, the upper part of the Doig Formation of the Daiber Group has yielded new hydrocarbon discoveries in channel sandstones and coquina "banks" near the contact with the Halfway Formation (Podruski et al., 1988).

In the eastern Foothills and Plains, the subsurface Halfway Formation serves as an excellent reservoir for oil. Toward the west, however, the stratigraphically equivalent Liard Formation in the Foothills belt is in most areas impermeable and nonporous. The overlying Charlie Lake Formation contains several excellent sandstone and carbonate reservoirs in many areas of the Alberta and British Columbia Plains. The laterally equivalent Ludington Formation of the extreme western Foothills between Halfway and Graham rivers, contains lenticular coquina beds, pods, and "shell banks", some of which are up to 150 m thick. These carbonate rocks are stained by hydrocarbons, are porous and permeable, and under favorable geological conditions could serve as excellent hydrocarbon reservoirs. The Starlight Evaporite Member of the Whitehorse Formation in the Foothills and Front Ranges south of Sukunka River (Township 72) contains many porous and permeable horizons, but to date has not yielded hydrocarbons. Unfortunately, the member thins rapidly eastward, and in those western regions in which Starlight strata are thick, the Triassic succession is near surface and severely folded and faulted, and thus of reduced hydrocarbon reservoir potential.

The Baldonnel Formation of northeastern British Columbia is a major gas producer (Bever and McIlreath, 1984). Production is obtained from porous skeletal dolostones in structural and stratigraphic traps. Equivalent facies in outcrop of the westernmost Foothills are mainly limestone displaying no visible porosity or permeability. The Brewster Limestone and Winnifred members in the Foothills and Front Ranges to the south are nonporous and impermeable and rate as poor exploration targets.

The Pardonet Formation is very carbonaceous and exhibits a strong, petroliferous, fetid odour. The formation produces hydrocarbons in the Sukunka River area (Barss and Montandon, 1981), but the strata to the west are very fine grained siltstones and shales and effectively nonporous. The rocks may, however, be potential hydrocarbon source rocks for overlying or underlying reservoirs. The youngest Triassic formation in the Foothills, the Bocock, appears in outcrop to display the character of an excellent petroleum reservoir. The Bocock Formation is capped by impermeable Fernie Formation shales and limestones.

Triassic rocks of northern Yukon Territory and District of Mackenzie outcrop or occur only slightly below the present surface, and therefore, do not appear to be potential hydrocarbon reservoirs (Mountjoy, 1967).

Phosphate

Collophane, a cryptocrystalline mineral of the apatite group, is found in most Triassic strata, although generally in quantitatively insignificant concentrations. The most economically promising phosphate horizons are those at the base of the Whistler Member of the Sulphur Mountain Formation, and within the lower Toad Formation of northeastern British Columbia. Within certain stratigraphic intervals of the Vega-Phroso and Whistler members of the Sulphur Mountain Formation, and the Toad Formation, phosphate particles occur in concentrations up to 71% by weight, with P_2O_5 content up to 30% by weight (Gibson, 1975). The phosphate forms as thin conglomerate and sandstone beds (up to 15 cm) at or near the base of the Sulphur Mountain members, and thin lenses and nodules (up to 10 cm) dispersed in the lower Toad and uppermost Grayling formations. In the latter formations, the phosphate is distributed vertically over several metres of strata. Near the Alaska Highway and Tetsa River, thin beds of granular phosphate lie in the middle part of the Toad Formation. Rare occurrences of phosphate are found in the upper Winnifred Member of the Whitehorse Formation, adjacent to the contact with the overlying Jurassic Fernie Formation. Phosphate is also in the subsurface Doig Formation, although probably in uneconomic concentrations.

Gypsum

Gypsum occurs sparingly in the Starlight Evaporite Member of the Whitehorse Formation of the Foothills and Front Ranges of Alberta, as thin to thick beds, lenses, or pods, interbedded with dolostone and solution or collapse breccia, in units ranging from a few centimetres to several metres thick. Most of the gypsum is found in the thicker Triassic exposures of the Rocky Mountain Front Ranges. At one locality near Jasper, gypsum interbedded with thin dolostones attains a thickness of 44 m (Gibson, 1968b, 1974). Gypsum has not been observed in the outcrop belt of British Columbia, but anhydrite is a common component of the equivalent Charlie Lake Formation in the subsurface of the easternmost Foothills and Plains of northeastern British Columbia and west-central Alberta.

REFERENCES

Armitage, J.H.
1962: Triassic oil and gas occurrences in northeastern British Columbia, Canada; Journal of the Alberta Society of Petroleum Geologists, v. 10, no. 2, p. 35-56.

Barss, D.L., Best, E.W., and Meyers, N.
1964: Triassic; Chapter 9 in Geological History of Western Canada, R.G. McCrossan and R.P. Glaister (ed.); Alberta Society of Petroleum Geologists, Calgary, Alberta, p. 113-136.

Barss, D.L. and Montandon, F.A.
1981: Sukunka-Bullmoose gas fields: models for a developing trend in the southern foothills of northeast British Columbia; Bulletin of Canadian Petroleum Geology, v. 29, no. 3, p. 293-333.

Bever, J.M. and McIlreath, I.A.
1984: Stratigraphy and reservoir development of shoaling-upward sequences in the Upper Triassic (Carnian) Baldonnel Formation, northeastern British Columbia; in Exploration Update '84, Program and Abstracts; Canadian Society of Petroleum Geologists - Canadian Society of Exploration Geophysicists National Convention, Calgary, Alberta, 1984, p. 147.

Cant, D.
1984: Possible syn-sedimentary tectonic controls on Triassic reservoirs: Halfway, Doig, Charlie Lake formations, west-central Alberta; in Exploration Update '84, Program and Abstracts; Canadian Society of Petroleum Geologists - Canadian Society of Exploration Geophysicists National Convention, Calgary, Alberta, 1984, p. 45-46.
1986: Hydrocarbon trapping in the Halfway Formation (Triassic), Wembley Field, Alberta; Bulletin of Canadian Petroleum Geology, v. 34, no. 3, p. 329-338.

Clark, D.R.
1961: Primary structures of the Halfway sand in the Milligan Creek oilfields, British Columbia; Journal of the Alberta Society of Petroleum Geologists, v. 9, no. 4, p. 109.

Colquhoun, D.J.
1960: Triassic stratigraphy of western central Canada; Ph.D. dissertation, University of Illinois, Urbana, Illinois, 160 p.

Gabrielse, H.
1977: Geology of Toodoggone (NTS 94E) and Ware (NTS 94F, W1/2) map areas, British Columbia; Geological Survey of Canada, Open File 483.

Gibson, D.W.
1965: Triassic stratigraphy near the northern boundary of Jasper National Park, Alberta; Geological Survey of Canada, Paper 64-9, 144 p.
1968a: Triassic stratigraphy between the Athabasca and Smoky Rivers of Alberta; Geological Survey of Canada, Paper 67-65, 105 p.
1968b: Triassic stratigraphy between the Athabasca and Brazeau Rivers of Alberta; Geological Survey of Canada, Paper 68-11, 84 p.
1969: Triassic stratigraphy of the Bow River-Crowsnest Pass region, Rocky Mountains of Alberta and British Columbia; Geological Survey of Canada, Paper 68-29, 48 p.
1970: Triassic stratigraphy, Pine Pass area, northeastern British Columbia; Edmonton Geological Society, Field Conference Guidebook, p. 23-38.
1971a: Triassic stratigraphy of the Sikanni Chief River-Pine Pass region, Rocky Mountain Foothills, northeastern British Columbia; Geological Survey of Canada, Paper 70-31, 105 p.
1971b: Triassic petrology of Athabasca - Smoky River region, Alberta; Geological Survey of Canada, Bulletin 194, 59 p.
1972: Triassic stratigraphy of the Pine Pass-Smoky River area, Rocky Mountain Foothills and Front Ranges of British Columbia and Alberta; Geological Survey of Canada, Paper 71-30, 108 p.
1974: Triassic rocks of the southern Canadian Rocky Mountains; Geological Survey of Canada, Bulletin 230, 65 p.
1975: Triassic rocks of the Rocky Mountain Foothills and Front Ranges of northeastern British Columbia and west-central Alberta; Geological Survey of Canada, Bulletin 247, 61 p.

Gordey, S.P., Wood, D., and Anderson, R.G.
1981: Stratigraphic framework of southeastern Selwyn Basin, Nahanni map area, Yukon Territory and District of Mackenzie; in Current Research, Part A, Geological Survey of Canada, Paper 81-1A, p. 395-398.

Hunt, A.D. and Ratcliffe, J.D.
1959: Triassic stratigraphy, Peace River area, Alberta and British Columbia, Canada; Bulletin of the American Association of Petroleum Geologists, v. 43, p. 563-582.

Jansonius, J.
1962: Palynology of Permian and Triassic sediments, Peace River area, Western Canada; Paleontographica, Abteilung B, Band 110, p. 35-98.

Klepacki, D.W.
1985: Stratigraphy and structural geology of the Goat Range area, southeastern British Columbia; Ph.D. thesis, Massachusetts Institute of Technology, Cambridge, Massachussetts, 268 p.

Miall, A.D.
1976: The Triassic sediments of Sturgeon Lake south and surrounding areas; in The Sedimentology of Selected Clastic Oil and Gas Reservoirs in Alberta, M. Lerand (ed); Canadian Society of Petroleum Geologists, Calgary, Alberta, Canada, p. 25-43.

Monger, J.W.H., Souther, J.G., and Gabrielse, H.
1972: Evolution of the Canadian Cordillera: A plate-tectonic model; American Journal of Science, v. 272, p. 577-602.

Mosher, L.C.
1973: Triassic conodonts from British Columbia and the northern Arctic Islands; in Contributions to Canadian Paleontology; Geological Survey of Canada, Bulletin 222, p. 141-148.

Mothersill, J.S.
1968: Environments of deposition of the Halfway Formation, Milligan Creek area, British Columbia; Bulletin of Canadian Petroleum Geology, v. 16, no. 2, p. 180-199.

Mountjoy, E.W.
1967: Triassic stratigraphy of northern Yukon Territory; Geological Survey of Canada, Paper 66-19, 44 p.

Nelson, C.H.
1982: Modern shallow-water graded sand layers from storm surges, Bering Shelf: a mimic of Bouma sequences and turbidite systems; Journal of Sedimentary Petrology, v. 52, no. 2, p. 537-545.

Nelson, S.J.
1970: The face of time, the geological history of Western Canada; Alberta Society of Petroleum Geologists, Calgary, Alberta, p. 111-113.

Orchard, M.
1983: *Epigondolella* populations and their phylogeny and zonation in the Upper Triassic; in Taxonomy, Ecology and Identity of Conodonts, Proceeding of ECOS III, Lund, 1982; Fossils and Strata, no. 15, p. 177-192.

Pelletier, B.R.
1960: Triassic stratigraphy, Rocky Mountain Foothills, northeastern British Columbia 94J and 94K; Geological Survey of Canada, Paper 60-2, 32 p.
1963: Triassic stratigraphy of the Rocky Mountains and Foothills, Peace River district, British Columbia; Geological Survey of Canada, Paper 62-26, 43 p.
1964: Triassic stratigraphy of the Rocky Mountain Foothills between Peace and Muskwa Rivers, northeastern British Columbia; Geological Survey of Canada, Paper 63-33, 89 p.
1965: Paleocurrents in the Triassic of northeastern British Columbia; in Primary Sedimentary Structures and their Hydrodynamic Interpretation, G.V. Middleton (ed.); Society of Economic Paleontologists and Mineralogists, Special Publication 12, p. 233-245.

Podruski, J.A., Barclay, J.E., Hamblin, A.P., Lee, P.J., Osadetz, K.G., Procter, R.M., and Taylor, G.C.
1988: Conventional oil resources of western Canada (light and medium), Part I: Resource Endowment; Geological Survey of Canada, Paper 87-26, p. 1-125.

Schaeffer, R. and Mangus, M.
1976: An Early Triassic fish assemblage from British Columbia; Bulletin American Museum of Natural History, v. 156, Article 5, New York, p. 515-564.

Sharma, G.D.
1969: Paragenetic evolution in Peejay Field, British Columbia, Canada; Mineralium Deposita, v. 4, p. 346-354.

Sloss, L.L.
1963: Sequences in the cratonic interior of North America; Bulletin of Geological Society of America, v. 74, p. 93-114.

Stott, D.F.
1982: Lower Cretaceous Fort St. John Group and Upper Cretaceous Dunvegan Formation of the Foothills and Plains of Alberta, British Columbia, District of Mackenzie and Yukon Territory; Geological Survey of Canada, Bulletin 328, 124 p.

Tempelman-Kluit, D.J.
1970: Stratigraphic and structure of the "Keno Hill Quartzite" in Tombstone River-Upper Klondike River map areas, Yukon Territory (116 B/7, B/8); Geological Survey of Canada, Bulletin 180, 102 p.

Torrie, J.E.
1973: Northeastern British Columbia; in Future Petroleum Provinces of Canada, R.G. McCrossan (ed.); Canadian Society of Petroleum Geologists, Memoir 1, p. 151-186.

Tozer, E.T.
1967: A standard for Triassic time; Geological Survey of Canada, Bulletin 156, 103 p.

Tozer, E.T. (cont.)
1979: Latest Triassic ammonoid faunas and biochronology, Western Canada; in Current Research, Part B; Geological Survey of Canada, Paper 79-1B, p. 127-135.
1982a: Late Triassic (Upper Norian) and earliest Jurassic (Hettangian) rocks and ammonoid faunas, Halfway River and Pine Pass map areas, British Columbia; in Current Research, Part A, Geological Survey of Canada, Paper 82-1A, p. 385-391.
1982b: Marine Triassic faunas of North America: their significance for assessing plate and terrane movements; Geologische Rundschau, v. 71, no. 3, p. 1077-1104.
1984: The Trias and its ammonoids: the evolution of a time scale; Geological Survey of Canada, Miscellaneous Report 35, 171

ADDENDUM

In addition to the important Triassic references listed under 'Data base' and 'History and Environments of Deposition', recent reports by Gibson and Barclay (1989) and Gibson and Edwards (1990) should now be included.

Gibson, D.W. and Barclay, J.E.
1989: Middle Absaroka sequence: the Triassic table craton; in Western Canada Sedimentary Basin, A Case History, B.D. Ricketts (ed.); Canadian Society of Petroleum Geologists, Calgary, 1989, p. 219-231.

Gibson, D.W. and Edwards, D.E.
1990: An overview of Triassic stratigraphy and depositional environments in the Rocky Mountain Foothills and western Interior Plains, Peace River Arch area, northeastern British Columbia; in Geology of the Peace River Arch, S.C. O'Connell and J.S. Bell (ed.); Bulletin of Canadian Petroleum Geology, v. 38A, p. 146-158.

Author's Address

D.W. Gibson
Institute of Sedimentary and Petroleum Geology
Geological Survey of Canada
3303 - 33rd Street N.W.
Calgary, Alberta
T2L 2A7

Printed in Canada

Subchapter 4H

JURASSIC

T.P. Poulton, W.K. Braun, M.M. Brooke, and E.H. Davies

CONTENTS

INTRODUCTION

The Jurassic System in western Canada is of interest particularly for the evidence it brings to bear on the early history of the mid-Jurassic to mid-Cretaceous Columbian orogeny, and for its economic significance. Its sediments contain the record of eustatic and epeirogenic events related partly to the early stages of the opening of the North Atlantic Ocean in the east, and the history of collision of allochthonous terranes in the west with the westerly drifting continent. Superimposition of the effects of the early phases of the Columbian Orogeny and the related foredeep on the older pre-orogenic sediments of Western Canada Basin, and the demise of Williston Basin as a depocentre are seen in the Jurassic succession of the western Plains and Rocky Mountains. Late events of the Ellesmerian tectonic phase in Brooks-Mackenzie Basin are recorded in northern Yukon Territory and adjacent Northwest Territories.

Jurassic sedimentary rocks form significant hydrocarbon reservoirs and source rocks in western Canada. The immense coal reserves of the southern Canadian Rocky Mountains occur in an Upper Jurassic-Lower Cretaceous succession; the Jurassic of southern Manitoba hosts two of Canada's major gypsum mines; and some of the phosphate deposits, currently subeconomic, in southeastern British Columbia are Jurassic in age.

Acknowledgments and responsibilities

W.K. Braun and M.M. Brooke provided the discussions of ostracodes and Foraminifera in this report, and the microfossil zonations presented in Table 4H.1. E.H. Davies contributed the palynological discussions and the zonation presented in Table 4H.1. The remainder of the report is the responsibility of T.P. Poulton, and is modified from a previous compilation (Poulton, 1984). The manuscript has been read critically by J.E. Christopher, W.K. Braun, and R. Hall.

D.F. Stott, D.W. Gibson, N.C. Ollerenshaw, D.K. Norris, M. McMechan, R. Hall, R.L. Christie, and A. Sweet were particularly helpful in providing data for British Columbia

Poulton, T.P., Braun, W.K., Brooke, M.M., and Davies, E.H.
1993: Jurassic; Subchapter 4H in Sedimentary Cover of the Craton in Canada, D.F. Stott and J.D. Aitken (ed.); Geological Survey of Canada, Geology of Canada, no. 5, p. 321-357 (also Geological Society of America, The Geology of North America, v. D-1).

Table 4H.1. Correlation chart of key macrofossils, microfossils and palynomorphs. Vertical lined pattern indicates hiatuses

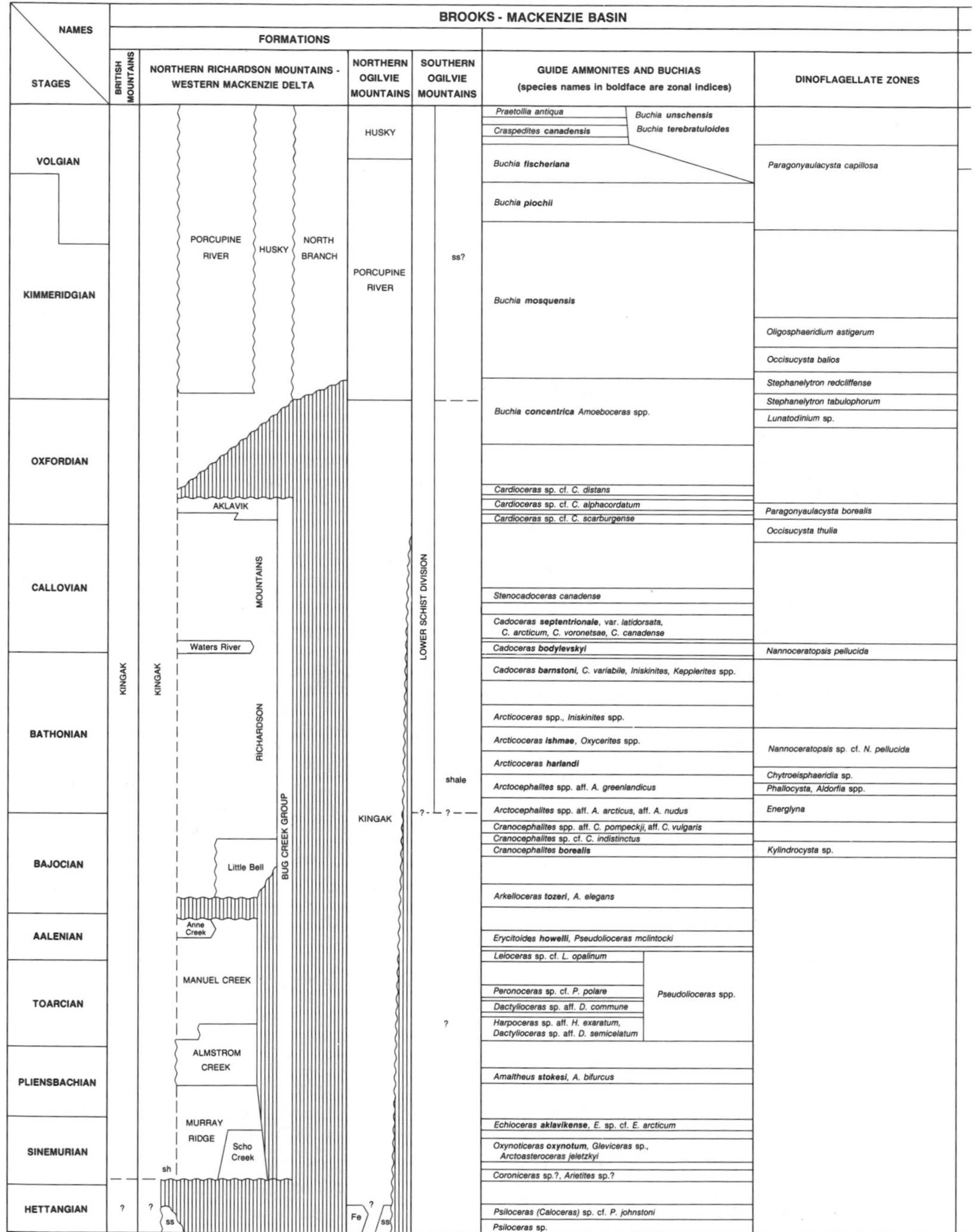

STAGES	GUIDE AMMONITES AND BUCHIAS (species names in boldface are zonal indices)	DINOFLAGELLATE ZONES
VOLGIAN	*Praetollia antiqua*; *Craspedites* ***canadensis***; *Buchia* ***unschensis***; *Buchia* ***terebratuloides***; *Buchia* ***fischeriana***; *Buchia* ***piochii***	*Paragonyaulacysta capillosa*
KIMMERIDGIAN	*Buchia* ***mosquensis***	*Oligosphaeridium astigerum*; *Occisucysta balios*; *Stephanelytron redcliffense*
OXFORDIAN	*Buchia* ***concentrica*** *Amoeboceras* spp.; *Cardioceras* sp. cf. *C. distans*; *Cardioceras* sp. cf. *C. alphacordatum*; *Cardioceras* sp. cf. *C. scarburgense*	*Stephanelytron tabulophorum*; *Lunatodinium* sp.; *Paragonyaulacysta borealis*
CALLOVIAN	*Stenocadoceras canadense*; *Cadoceras* ***septentrionale***, var. *latidorsata*, *C. arcticum*, *C. voronetsae*, *C. canadense*; *Cadoceras* ***bodylevskyi***	*Occisucysta thulia*; *Nannoceratopsis pellucida*
BATHONIAN	*Cadoceras* ***barnstoni***, *C. variabile*, *Iniskinites*, *Kepplerites* spp.; *Arcticoceras* spp., *Iniskinites* spp.; *Arcticoceras* ***ishmae***, *Oxycerites* spp.; *Arcticoceras* ***harlandi***; *Arctocephalites* spp. aff. *A. greenlandicus*; *Arctocephalites* spp. aff. *A. arcticus*, aff. *A. nudus*	*Nannoceratopsis* sp. cf. *N. pellucida*; *Chytroeisphaeridia* sp.; *Phallocysta*, *Aldorfia* spp.; *Energlyna*
BAJOCIAN	*Cranocephalites* spp. aff. *C. pompeckji*, aff. *C. vulgaris*; *Cranocephalites* sp. cf. *C. indistinctus*; *Cranocephalites* ***borealis***; *Arkelloceras* ***tozeri***, *A. elegans*	*Kylindrocysta* sp.
AALENIAN	*Erycitoides* ***howelli***, *Pseudolioceras mclintocki*; *Leioceras* sp. cf. *L. opalinum*	
TOARCIAN	*Peronoceras* sp. cf. *P. polare*; *Dactylioceras* sp. aff. *D. commune*; *Harpoceras* sp. aff. *H. exaratum*, *Dactylioceras* sp. aff. *D. semicelatum*; *Pseudolioceras* spp.	
PLIENSBACHIAN	*Amaltheus* ***stokesi***, *A. bifurcus*	
SINEMURIAN	*Echioceras* ***aklavikense***, *E.* sp. cf. *E. arcticum*; *Oxynoticeras* ***oxynotum***, *Gleviceras* sp., *Arctoasteroceras jeletzkyi*; *Coroniceras* sp.?, *Arietites* sp.?	
HETTANGIAN	*Psiloceras* (*Caloceras*) sp. cf. *P. johnstoni*; *Psiloceras* sp.	

Table 4H.1. Cont.

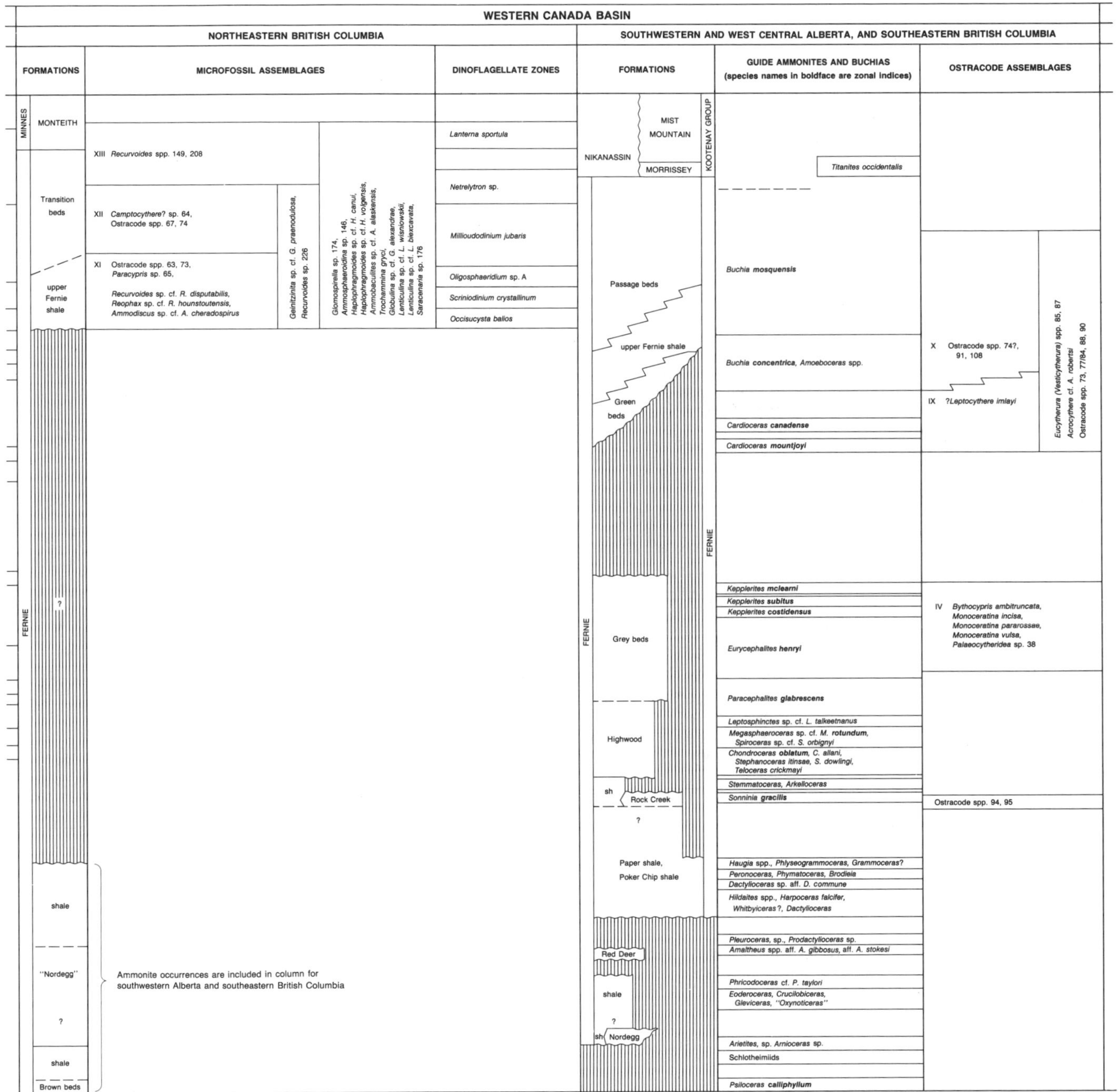

Table 4H.1. Cont.

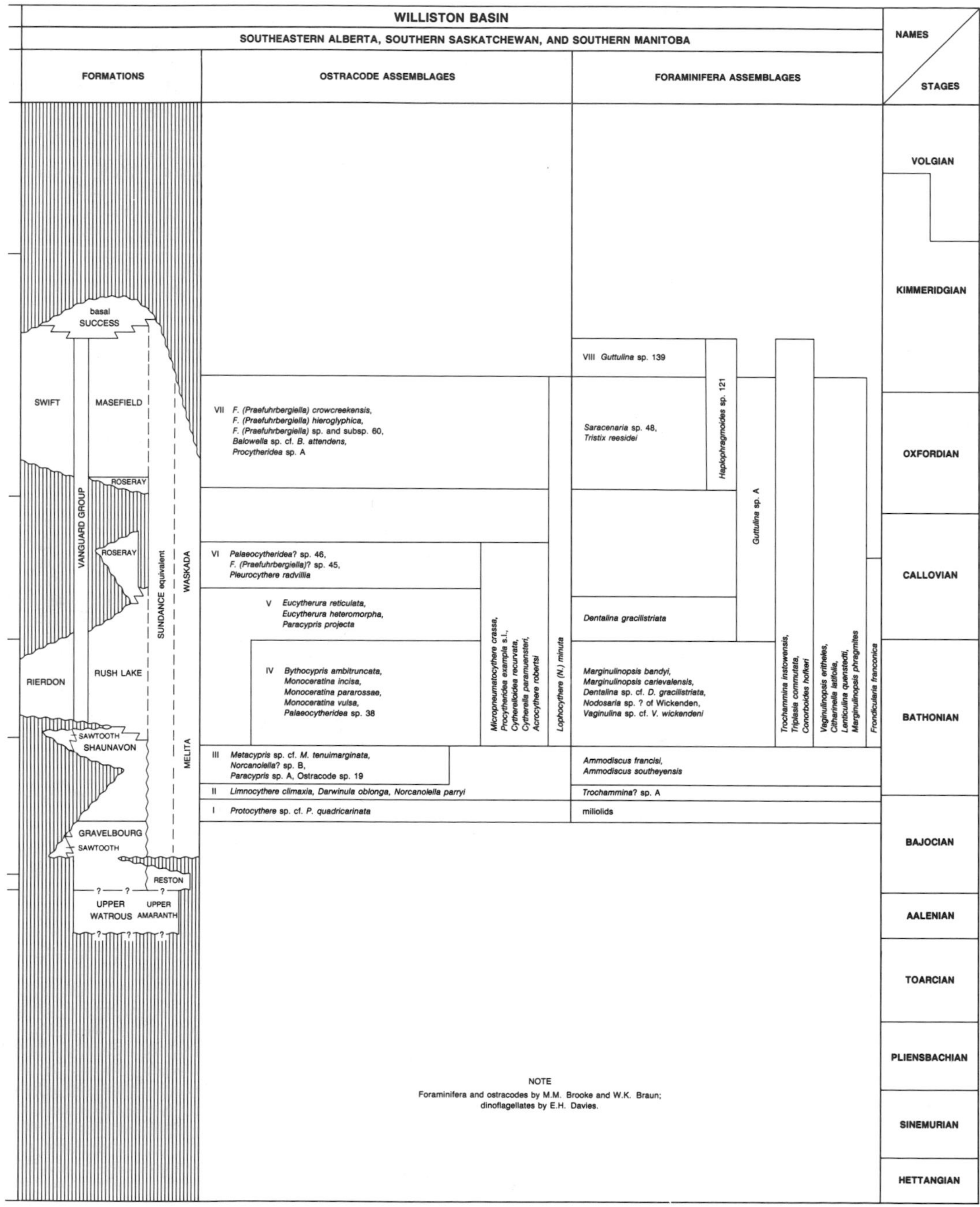

and western Alberta; D.K. Norris, J.A. Jeletzky, F.G. Young, and J. Dixon have provided unpublished information on the geology of northern Yukon Territory. J.E. Christopher, S. Anderson, H. McCabe, L.L. Price, and K.G. Osadetz provided some recent, unpublished information on Williston Basin.

DEPOSITIONAL BASINS AND TECTONIC ELEMENTS

Different sequences of Jurassic rocks occur in three major epicratonic depositional settings: in Brooks-Mackenzie Basin (see Balkwill et al., 1983) in northern and central Yukon Territory and adjoining parts of the Northwest Territories (Fig. 4H.1, 4H.2); on the western margin of the craton (Alberta Basin) and in the succeeding foredeep, the Rocky Mountain Trough (Fig. H.3), in western Alberta and eastern British Columbia; and in Williston Basin of southeastern Alberta, southern Saskatchewan, and southern Manitoba.

The Jurassic and older systems in the western Cordillera, which contain volcanic and plutonic packages of the several allochthonous terranes accreted to the continent, are treated in a companion volume of this series, "Geology of the Cordilleran Orogen in Canada" (Gabrielse and Yorath, 1991).

Brooks-Mackenzie Basin

Jurassic rocks are well exposed in the northern Ogilvie Mountains (Poulton, 1982) and in the northern Richardson Mountains (Poulton et al., 1982), where they are especially fossiliferous. They also have been penetrated in wells drilled in Mackenzie Delta (Young et al., 1976; Poulton, 1978a; Dixon, 1982a, b). West and northwest of Richardson Mountains and north of latitude 68°N, they are, in general, poorly exposed, tectonically disturbed, and poorly fossiliferous.

Although Jurassic fossils from the northern Yukon Territory and adjacent Mackenzie Delta area are among the earliest collected in Canada (Meek, 1859), their Jurassic age was not correctly established until a century later (Frebold, 1960). Mapping of surface exposures and stratigraphic studies by the Geological Survey of Canada and oil companies since about 1955, in large part related to oil and gas exploration, have resulted in our present understanding of the Jurassic. Recent compilations of data, interpretations, and bibliographies have been given by Poulton et al. (1982), Poulton (1982) and Dixon (1982a, b). The presence of Jurassic rocks, a southerly extension of the northerly facies, in a narrow band across central Yukon Territory, was not known until ammonites were reported in 1962 by Green and Roddick, and this belt remains poorly known. Upper Jurassic rocks in northern Anderson Plain were first identified by Brideaux and Fisher (1976).

Brooks-Mackenzie Basin (Balkwill et al., 1983) comprises that part of the Ellesmerian tectonic regime (Lerand, 1973) that occupied what is now northern Yukon Territory, part of the adjacent Northwest Territories, northern Alaska and adjacent, offshore areas of Beaufort Sea.

Many tectonic models have been proposed for the origin of the Arctic Ocean basin and the evolution of its margins, but none explain all of the known geology. Currently popular is a hypothesis involving counterclockwise rotation of the terrane underlying northern Alaska and northwestern Yukon Territory, about a pivot point in northeastern Yukon (e.g. Grantz et al., 1979). This terrane, prior to rotation, would have lain off the coast of northern Canada. Rifting that initiated rotation may have begun in the Early Cretaceous, so that Jurassic rocks would have been rotated and displaced, according to the hypothesis. The degree to which post- or syn-rift, transcurrent faulting (Norris, 1983) has additionally displaced Jurassic rocks is unknown. If sufficient transcurrent movement is assumed, the pre-rotation, depositional site of the northern Alaska-northern Yukon Jurassic rocks would have lain off the present shelf west and northwest of Banks Island. Thus, the northern basin margin or series of highlands in the Jurassic in northern Alaska, northern Yukon, and northwestern District of Mackenzie may have been parts of originally southeastern basin margins that rifted away from what is now the southeastern margin of the Beaufort Sea. This rifting may account for the discontinuous nature of the highlands and for the fault-bounded character of some of them. Additionally, the Jurassic sediments of northern Alaska and northwestern Yukon Territory themselves would originally have formed part of a continuous belt of sediments filling what is now a gap between Prince Patrick Island to the northeast and the Tuktoyuktuk Peninsula-Mackenzie Delta area to the southwest. The absence of significant coarse clastic wedges derived from highlands north of Brooks-Mackenzie Basin in its present location would indicate the absence during the Jurassic of a major river system draining northward between Sverdrup Basin and the Richardson Mountains portion of Brooks-Mackenzie Basin. The Jurassic shales presently preserved in a graben on Banks Island (Miall, 1979) represent a Late Jurassic, cratonward-transgressive extension of a more complete Jurassic sedimentary wedge that would have been deposited to their west and northwest.

Neither the character of the Jurassic sediments nor that of their bordering highlands contradicts the rift-rotation hypothesis, although Poulton (1982) pointed out the absence of Cretaceous volcanics in the proposed pivot and rift zones as an argument against it. In any case, in the absence of a more than merely hypothetical reconstruction, present-day locations and cardinal compass directions are used in connection with the descriptions of this report.

Further to the question of palinspastic reconstruction, the Jurassic rocks of northern Yukon Territory do not contain firm evidence to support or reject the transcurrent movement on the order of 200 to 500 km required by the rift-rotation hypothesis. Movement may have been concentrated along a structural feature that may be a continuation into Canada of the Kaltag fault of Alaska (Norris, 1974, 1983). Continuity of the basin across the fault throughout much of the sedimentary column, including the Paleozoic, has been deduced from the similarity of both lithological and paleontological characteristics and successions on either side of the fault (see Poulton, 1982).

The present northern and northeastern margins of Brooks-Mackenzie Basin in Canada are poorly known. In adjacent Alaska, the northern margin in Early and Mid-Jurassic time was a highland called Barrow Arch (Detterman, 1970; Rickwood, 1970). In Canada, a northern highland is indicated also, but it was probably not a source of any significant amount of sediment. Evidence includes the onlap of probable Upper Jurassic shales onto Precambrian rocks northward beyond the limits of Lower and Middle Jurassic, Triassic and Paleozoic sequences at Spring River in northernmost Yukon Territory (D.K. Norris, 1972, 1974), and the abrupt disappearance northward in Mackenzie Delta of first the Lower and Middle Jurassic, and then the Upper Jurassic rocks against a series of fault-bounded(?) uplifts (Dixon, 1982a, b). It is not at all clear whether these enigmatic highlands were

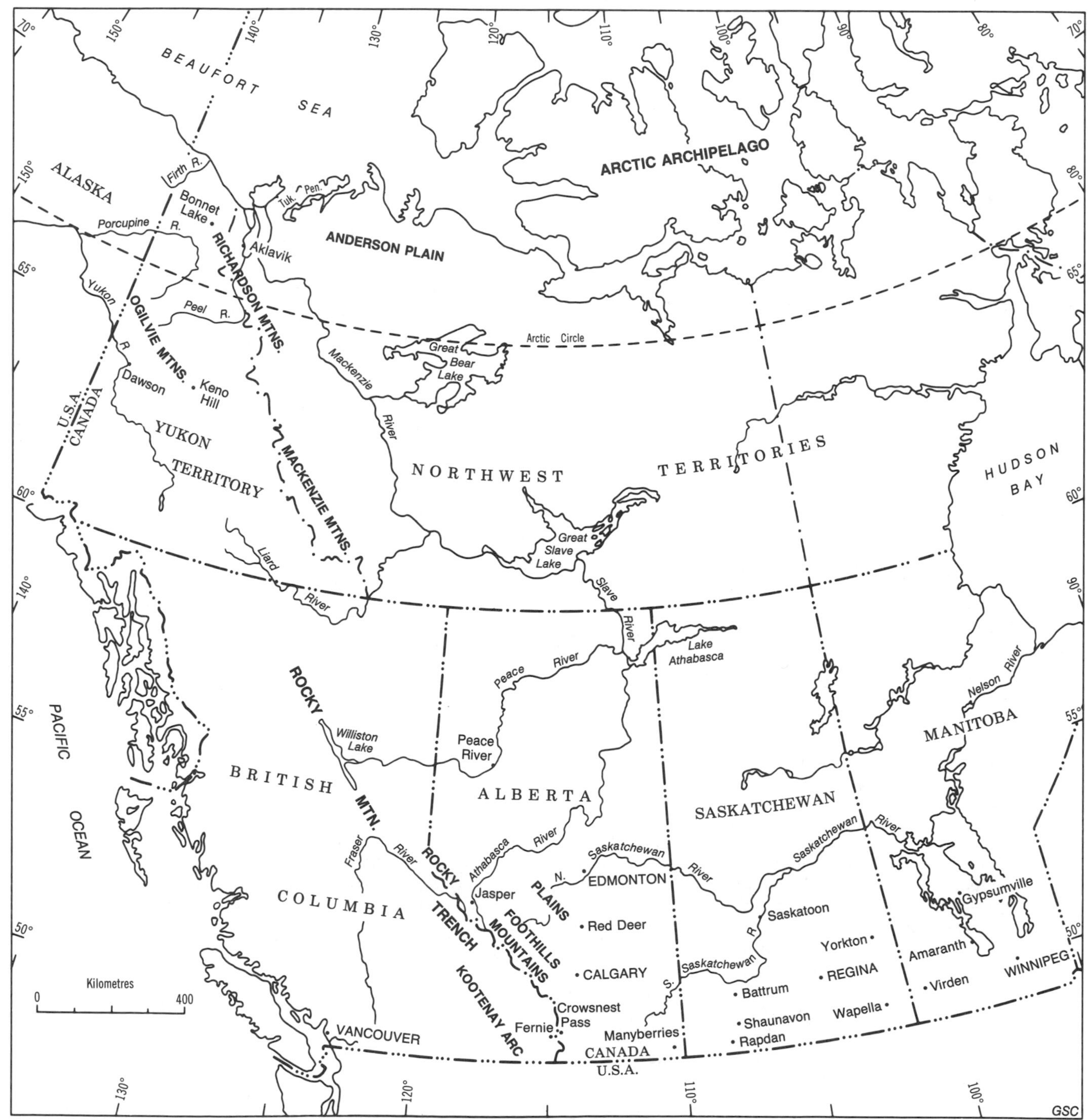

Figure 4H.1. Index map of western Canada.

part of a continuous highland area, or if they were simply individual elements of a more extensive series of en echelon uplifts that characterized the unstable cratonic margins in much of Arctic Canada. Unlike the Jurassic rocks in northern Alaska (Rickwood, 1970; Detterman, 1973; Detterman et al., 1975), no coarse clastic rocks occur in the Kingak Formation at Spring River to indicate a northern sediment source there, and a thick sequence of fossiliferous Kingak shales farther west at Firth River lies between that locality and northern Alaska. The source of some turbidites in the Upper Pliensbachian strata near Firth River is unknown. In northern Yukon Territory and northern Mackenzie Delta, Upper Jurassic shales are spread across the older basin margins, lying upon whatever uplifted terrane was present previously (Poulton, 1982).

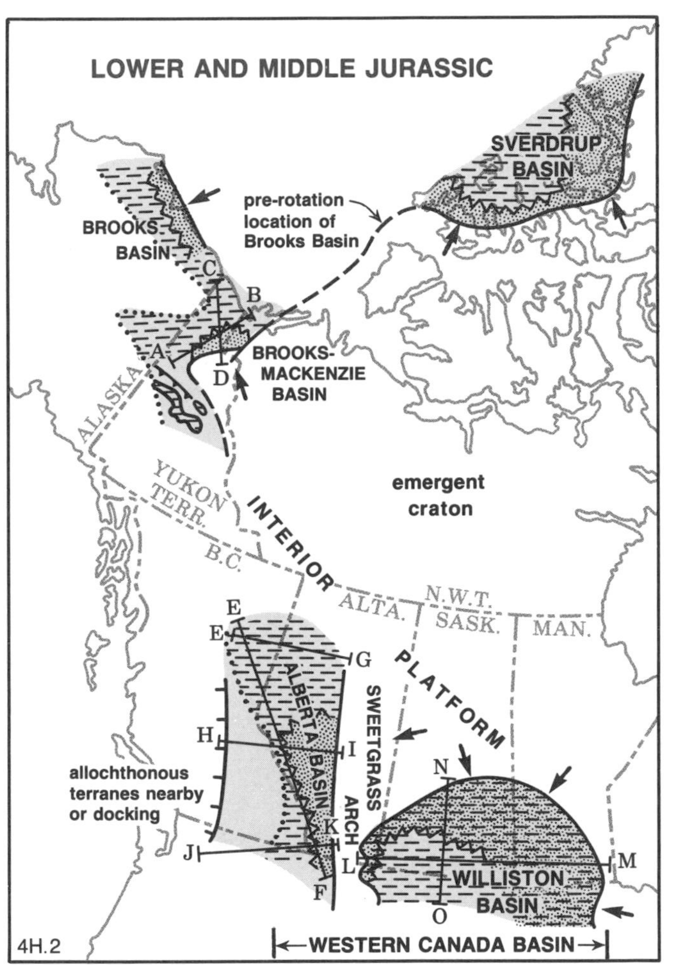

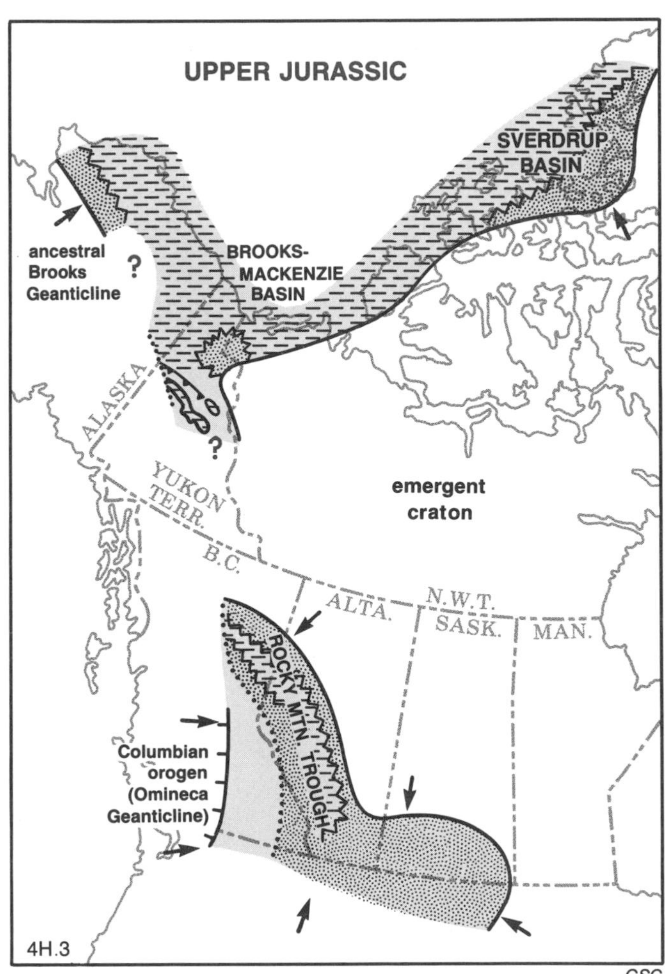

LEGEND

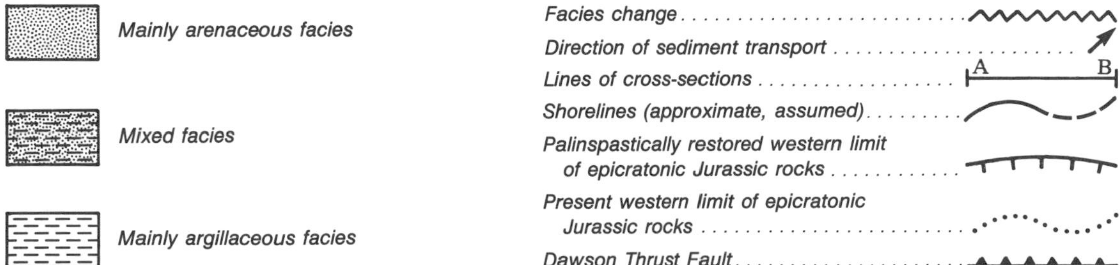

Figures 4H.2, 3. Tectonic setting of western Canada in Early and Middle (Fig. 4H.2), and Late (Fig. 4H.3) Jurassic times. Lines of cross-sections (Fig. 4H.5, 4H.6, 4H.12-15, 4H.19, 4H.20) are shown in Figure 4H.2. The major reorganization accompanying early phases of Columbian Orogeny in southern British Columbia and proposed rotation of northern Alaska is shown. Facies are not restored palinspastically.

The southeastern margin of Brooks-Mackenzie Basin coincided with the Eskimo Lakes Fault Zone in the Mackenzie Delta area, where Lower Jurassic to Oxfordian rocks were deposited against high-standing blocks of upper Paleozoic rocks. Upper Jurassic rocks were deposited across this fault zone as a consequence of a regional transgressive episode, and are preserved in various places on the adjacent parts of the craton as far east as the northern Anderson Plain (Brideaux and Fisher, 1976). The source of the clastic sediments was from the southeast, and both local and distant sources were apparently present (Poulton, 1982; Poulton et al., 1982). Southeastward, the Jurassic sedimentary wedge becomes coarser grained and more proximal in aspect, as it thins and becomes less complete.

The southeastern basin margin trended southwestward across northern Yukon Territory to the northern Ogilvie Mountains (Poulton, 1982; Poulton et al., 1982). The basin margin must then have curved southward and eastward, in the vicinity of Ogilvie Mountains, to account for the location of the basinal, argillaceous rocks of the "Lower Schist Division" (Green and Roddick, 1962) in central Yukon Territory. A west-jutting promontory of the craton was thus present, around which deposition took place (Poulton and Tempelman-Kluit, 1982).

The western and southwestern basin margins during Early and Middle Jurassic time are completely unknown, apparently having been displaced by subsequent large-scale movement along Tintina and other transcurrent faults. No westerly source of detritus, and hence no evidence for orogenic highlands in the west, is known in the Canadian part of the basin until the Cretaceous Period (e.g. Young, 1973; Poulton, 1982). In northwestern Alaska, however, Upper Jurassic turbidites derived from the south demonstrate uplift of volcanic terranes in that direction (Campbell, 1967), initiation there of the Brookian tectonic regime, and the initial subsidence of Colville Trough.

The positions and orientations of a series of mid-Jurassic through Early Cretaceous uplifts on the northwestern margin of the continent (e.g. Miall, 1979; Poulton, 1982; Poulton et al., 1982; Dixon, 1982a, b) are consistent with them being high blocks along the unstable cratonic margin, if the hypothesis of rotation of northern Alaska away from Arctic Canada is correct. Alternatively they may represent jostled blocks in a strike-slip regime. The Jurassic age of two of the uplifts, White Mountains Uplift (Poulton et al., 1982) and Tununuk High (Dixon, 1982a, b), as judged by stratigraphic relations, suggests initiation of the tectonic activity during Jurassic time.

Alberta Basin and Rocky Mountain Trough

Jurassic rocks are extensively exposed in the Rocky Mountains and Foothills of western Canada, and are penetrated by wells in the Foothills and the adjoining Plains. Discovery of phosphate deposits and vast coal reserves in southwestern Alberta and southeastern British Columbia near the end of the last century led to local studies and the first stratigraphic nomenclature. Regional mapping followed, in part inspired by exploration for oil and gas as well as coal, and resulted in recognition of Jurassic rocks as far north as northeastern British Columbia. The Jurassic column in the subsurface in the eastern part of this area still is poorly understood, although it hosts significant oil and gas reserves. The most useful regional summaries are by Springer et al. (1964) and Stott (1970).

The Jurassic records the changes in tectonic and depositional regimes of Western Canada Basin caused by the earliest pulses of the Columbian Orogeny. A tectonically quiet region with relatively stable environments of deposition resulting in a thin sequence of shallow-water epicratonic sedimentary rocks was abruptly transformed by Late Jurassic time into a highly mobile region, with orogenic uplift in the west and an associated narrow, arcuate foredeep, Rocky Mountain Trough, to its east.

The Late Jurassic increase in subsidence rates, recording initiation of the foredeep, is shown strikingly in the burial history curve of pre-Jurassic units (Fig. 4H.4). Latest Jurassic, western source areas are clearly indicated by the presence of conglomerates in the westernmost Upper Jurassic sections in northeastern and southwestern British Columbia (Gibson, 1985; Stott, 1984).

The tectonic character of the western edge of the North American continent during the Early and Middle Jurassic is enigmatic. The prevailing, "Atlantic margin" model (see, for example, Bond and Kominz, 1984, and references

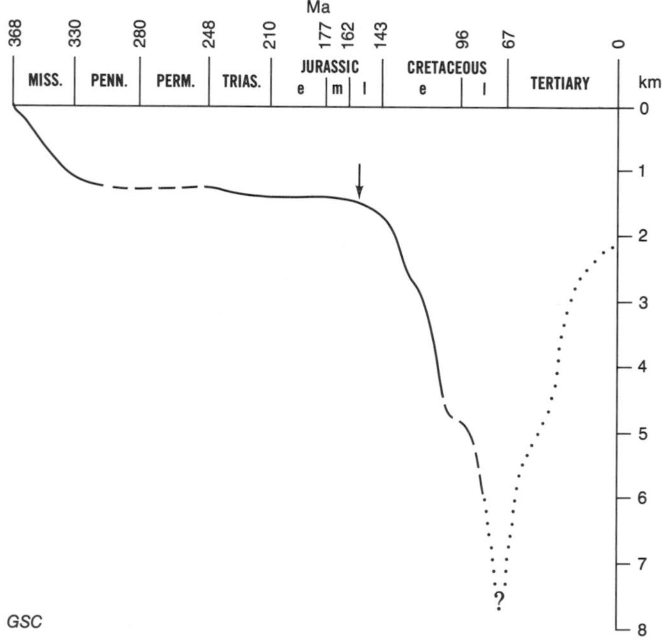

Figure 4H.4. Burial history curve of the Devonian-Carboniferous Exshaw Formation at Fernie, southeastern British Columbia. Note the abrupt increase in subsidence rate (indicated by arrow) initiating the foreland basin (Rocky Mountain Trough) associated with loading by thrust faulting in the adjacent Cordillera in the Late Jurassic. The formation thicknesses are not corrected for compaction effects so that the depths are minimum estimates. Curves drawn for localities farther east show younger (Cretaceous) initiation of subsidence of the foredeep (modified from R.A. Price and P. Fermor, unpub. data).

therein), predicts cessation of passive-margin subsidence 200 million years at most after separation. Jurassic subsidence thus cannot be (according to that model) simply a continuation of the subsidence of "one sided basin" style set in motion by continental separation at about the beginning of the Cambrian Period (Bond and Kominz, 1984). From other evidence, the Alberta Basin may have been influenced by convergent plate tectonic activity occurring to the west.

Radiometric evidence for tectonism and uplift related to suturing of parts of the allochthonous terranes in southeastern British Columbia during the Middle Jurassic Epoch (Archibald et al., 1983) bears on this problem. The exotic terranes and superterranes were in part amalgamations of earlier accreted smaller terranes and included lands of volcanic-arc origin (Beddoe-Stephens and Lambert, 1981) that developed off the west coast of North America. They may have been transported to their present locations from more southerly sites by northward movement oblique to the west coast of North America or by transcurrent faulting along the coast, arriving by mid-Jurassic time (Tipper, 1984). The widespread, pelletal phosphorite beds at the base of the Fernie Formation in southeastern British Columbia may indicate proximity to an open ocean with upwelling currents to the west (Sheldon, 1964). These phosphorite beds may thus define an older limit (Sinemurian) to the time at which exotic terranes came to lie off the west coast, changing a "one-sided" to a "two-sided" basin. In order that accretion and associated uplift be accomplished by some time in the Middle Jurassic, it seems likely that initial docking may have occurred at the time that lower (but not lowest) parts of the Fernie Formation were being laid down. That the sites of major Jurassic intrusion and uplift of southeastern British Columbia lie opposite the depocentre of epicratonic Middle Jurassic sediments (Highwood Member, Grey Beds and equivalent units, Fernie Formation) may not be a coincidence. Summaries of the current interpretations regarding accretion of western allochthonous terranes and the related plutonic belt at the suture zone are given by Monger et al. (1982), Tipper (1984), and (Gabrielse and Yorath, 1991).

A few bentonite layers in the Lower and Middle Jurassic rocks, particularly in the 'Nordegg Member' of northeastern British Columbia and the Highwood Member of southeastern British Columbia indicate some volcanic activity, presumably to the west. Late Jurassic vulcanism may also be indicated by the peculiar chemistry of tonsteins in the latest Jurassic coal measures of southeastern British Columbia.

Whatever lay to the west of the epicratonic sedimentary prism in Early and Middle Jurassic times can no longer be seen north of about latitude 52°N, because of its subsequent removal by large-scale transcurrent northward movements and/or erosion related to uplift of the Columbian Orogen. Westernmost exposures of Upper Jurassic units in southeastern and northeastern British Columbia include conglomerates that indicate proximity to the early Columbian tectonic highlands, which shed detritus eastward. For the most part the Upper Jurassic detritus is thought to have been derived from the older sedimentary rocks uplifted in the western thrust sheets of the incipient Rocky Mountains (e.g. Rapson, 1964, 1965; Jansa, 1972; Gibson, 1977, 1985). Earlier and possibly more significant uplifts in the western source areas are indicated in the south compared to the north, according to the model of Hamblin and Walker (1979), in which sedimentation is supposed to have prograded in a northerly direction, and according to the regional tectonic model of Eisbacher (1981). Radiometric studies of batholiths in the Kootenay Arc (Archibald et al., 1983) give evidence of Late Jurassic and Early Cretaceous uplift of source terranes there, and plate collisions in that area may have created highlands as early as Middle Jurassic time.

Late Cretaceous and Tertiary (Laramide) thrusting resulted in easterly transport of the western facies of Jurassic rocks in southwestern Alberta and southeastern British Columbia, and their near juxtaposition against more easterly facies. Some unknown amount of northerly strike-slip movement associated with the thrusting may have occurred also.

Only farther east did sediments continue to accumulate in the stable environments of the eastern basin margin. The eastern shorelines are almost entirely unknown, owing to erosion of the marginal facies prior to Cretaceous time, but they appear to have migrated widely in position during the Jurassic Period because of sea-level fluctuations and the influences of local, low-amplitude tectonic elements. Of principal interest are the pre-Late Jurassic uplift of a broad area straddling what earlier had been Peace River Embayment, and the Middle Jurassic inundation of Sweetgrass Arch in southeastern Alberta. In the Late Jurassic, some detritus may have been transported northward into the foredeep and its eastern margin from adjacent parts of the U.S.A. to the south and southeast (e.g. Stelck et al., 1972).

Williston Basin

Jurassic rocks preserved in Williston Basin (Fig. 4H.2) are penetrated by many wells in southern Alberta, Saskatchewan, and Manitoba. As oil and gas exploration progressed, authors of regional studies either adopted the American nomenclature, which was commonly used at that time (e.g. Weir, 1949; Schmitt, 1953; Francis, 1956, 1957; Peterson, 1957), or established their own local nomenclature (e.g. Hadley and Milner, 1953; Milner and Thomas, 1954; Stott, 1955). In more recent detailed studies (e.g. Christopher, 1964, 1974), some of the relationships have been further refined, and additional names established, reflecting the complexity of the Jurassic sequences. The strata are essentially undisturbed. Only a few ammonites and other useful megafossils have been recovered, and microfossil and palynological zonations and correlations provide a biostratigraphic framework tied to ammonite-based control sections that outcrop in the northern United States and southwestern Alberta. Excellent regional summaries of the subsurface Jurassic sequences have been published by Springer et al. (1964), Stott (1970), and Christopher (1984a).

Williston Basin is that part of the Interior Platform that subsided differentially from Ordovician through Jurassic times about a depocentre in northeastern Montana and northwestern North Dakota. Approximately the northern half of the basin lies in Canada. The northern and northeastern basin margins remained relatively stable through Jurassic time, except for fluctuations caused by sea-level changes. Minor variations in clastic sediment input indicate some local influences.

The northwestern margin, formed by Sweetgrass Arch, was progressively inundated through Middle Jurassic time. Eventually, in the Late Jurassic, sediments from southern parts of the Columbian Orogen reached as far east as Williston Basin. This last event coincided more or less with a slight increase in the subsidence rate of Williston Basin and modest expansion of its limits. In latest Jurassic or earliest Cretaceous time, the basinal configuration was lost, and the American part of the basin was transformed into an uplifted area, from which detritus may have been shed northward into Canada, and upon which nonmarine clastic sediments of the Morrison Formation were deposited.

BIOSTRATIGRAPHY AND PALEOBIOGEOGRAPHY

The dating and correlation of the marine Jurassic formations are based mainly on ammonites, which occur sporadically and rarely in most sequences but prolifically at some horizons, on species of the bivalve genus *Buchia*, and on some palynomorphs in the Upper Jurassic strata. For biostratigraphic correlations, foraminifers and ostracodes are also important, and are invaluable in such areas as Williston Basin and the Mackenzie Delta, where only subsurface samples are available. Bivalves such as *Gryphaea, Inoceramus*, and various pectinids as well as belemnites, provide useful data for correlation of some units. Nonmarine beds, such as the Mist Mountain Formation of the Kootenay Group, are dated on the basis of spores, pollen, and to a lesser extent, plant macrofossils. Other fossil groups, among them brachiopods, gastropods, crinoids, scaphopods, serpulids, and vertebrates occur sporadically; most of them are not evaluated as yet and are of auxiliary biostratigraphic value only. Some of the southern Canadian subsurface units are dated by lithological correlation with their American counterparts, whose better established ages and faunas have been summarized comprehensively by Imlay (1980).

According to current interpretations, Jurassic rocks of western Canada can be placed between about paleolatitudes 45° and 70°N, on a globe whose climatic belts were presumably less strongly differentiated than they are today. Jurassic climates of the Western Interior were cooler than those of southern United States or southern Europe, as indicated by the scarcity of limestone deposits and the absence of reef-forming organisms.

Many of the significant macrofossils have been illustrated in atlases by Frebold (1964a, 1970a, b), and the ammonite data have been summarized in reviews by Frebold and Tipper (1970), Poulton (1978b), Westermann (1980), Callomon (1984), Taylor et al. (1984), and Hall (1984).

A proposed zonation of much of the Jurassic column of the entire region based on terrestrial spores and pollen and marine dinoflagellates and acritarchs was published by Pocock (1970, 1972). Major reports on the microfaunas of selected regions and their biostratigraphic value have been published by Wall (1960) and Brooke and Braun (1972, 1981).

Lower Jurassic series and fossils

The base of the Jurassic System is disconformable throughout the Alberta and Brooks-Mackenzie basins.

The oldest dated Jurassic rocks are of Early Hettangian age, characterized by *Psiloceras calliphyllum*. They overlie Upper Triassic sedimentary rocks in northeastern British Columbia (Tozer, 1982). Probable Early Hettangian *Psiloceras* species and bivalves were described from northern Yukon Territory by Frebold and Poulton (1977). Parts of all the other Lower Jurassic stages are indicated in Brooks-Mackenzie and Alberta basins by sporadic ammonite occurrences (Warren, 1931; Frebold, 1951, 1957, 1960, 1966, 1969, 1970b, 1975 1976; Frebold et al., 1967; Table 4H.1). The Early Jurassic ammonites mainly represent groups that can be readily correlated with northwestern European index species.

Autochthonous Lower Jurassic sequences in western Canada have not yielded any diagnostic microfauna so far, except in Brooks-Mackenzie Basin, where a few species occur (mentioned by K. Leskiw in Poulton et al., 1982).

Early Jurassic palynomorphs from Brooks-Mackenzie Basin were discussed by A.P. Audretsch (in Poulton et al., 1982), and some from other areas of western Canada by Pocock (1970, 1972). Those of the Brooks-Mackenzie Basin have closer affinities to those of offshore eastern Canada than to those of the Canadian Arctic Islands (Davies, 1983; Davies and Norris, 1980).

Middle Jurassic series and fossils

Aalenian sediments, including Lower Aalenian strata with *Leioceras opalinum*, are known with certainty only in Brooks-Mackenzie Basin, where they overlie lithologically similar Toarcian strata gradationally. No Aalenian fossils are known with certainty from the Alberta Basin. The other Middle Jurassic stages, except for Middle and Upper Callovian, are more or less well represented by ammonites throughout western Canada (Whiteaves, 1889; Buckman, 1929; McLearn, 1924, 1927, 1928, 1929, 1932; Warren, 1932, 1947; Frebold, 1957, 1960, 1961, 1963, 1964b, 1976; Frebold et al., 1967; Westermann, 1964; Paterson, 1968; Hall and Westermann, 1980; Hall and Stronach, 1981; and Hall, 1984).

The Middle Jurassic ammonites at several levels show extreme provincialism in western Canada, and detailed correlations between Brooks-Mackenzie, Alberta, and Williston basins are hampered by the scarcity of faunas in common (e.g., Imlay, 1965; Callomon, 1984; Taylor et al., 1984). The Brooks-Mackenzie Basin represents a region of mixed boreal and northeast Pacific faunas. In the Aalenian, they are *Leioceras, Pseudolioceras*, and *Erycitoides*, but without *Tmetoceras*. In the Bajocian, *Arkelloceras* and *Cranocephalites* are characteristic boreal genera. In contrast, certain minor elements of the Bathonian ammonite fauna are eastern Pacific in affinity, such as the genus *Iniskinites*, or cosmopolitan such as *Phylloceras* and *Cadomites*, but the fauna is dominated by the boreal Cardioceratidae. The mixed fauna is explained by the location of the basin at the northwestern corner of the continent in Jurassic time (Poulton, 1982). The boreal Middle and Late Bathonian ammonite succession at

Salmon Cache Canyon on Porcupine River (Poulton, 1987) is of particular importance because it provides the basis for a northwestern Canadian zonation. The Early Callovian *Cadoceras* and *Kepplerites* faunas are essentially boreal or sub-boreal in affinity.

Most of the Bajocian ammonites of Alberta and Williston basins are closely related to European species (see Hall and Westermann, 1980; Hall, 1984). Bathonian species, in contrast, seem to be a mixture of endemic forms, which are entirely unknown elsewhere (e.g. *Paracephalites*), of eastern Pacific forms (e.g. *Iniskinites, Eurycephalites*), and of European or cosmopolitan forms (e.g. *Leptosphinctes*) (Callomon, 1984). Additionally, some widespread boreal elements, such as *Kepplerites*, appear in these areas, and are consequently of singular importance for inter-basinal correlations.

Middle Jurassic microfossils have been described from Williston Basin by Wickenden (1933), Loranger (1955), Wall (1960), Guliov (1967), and Brooke and Braun (1972), the last authors producing a comprehensive biostratigraphic zonal scheme. Microfossils from southeastern British Columbia were described by Weihmann (1964), and from Brooks-Mackenzie Basin by K. Leskiw (in Poulton, et al., 1982).

Characteristic microfossils of the Bajocian to Lower Bathonian Upper Gravelbourg and Shaunavon formations of the Williston Basin include species of brackish-water Ostracoda, Charophyta, and a few marginal-marine Foraminifera. They are endemic, except for a few ostracodes of northwestern European affinity. Three successive assemblages (I-III) were recognized (Brooke and Braun, 1972).

Another microfauna (assemblage IV) established by Brooke and Braun (1972) characterizes the Rierdon and Rush Lake formations and the Grey Beds of the Fernie Formation, both of Middle Bathonian through Early Callovian age approximately. This prominent marine fauna is an endemic Western Interior fauna, with some boreal elements and rarer species with northwestern European affinities.

Palynomorphs from the Middle Jurassic were discussed by Pocock (1962, 1970, 1972), and some from Brooks-Mackenzie Basin were described by A.P. Audretsch (in Poulton et al., 1982). The latter assemblages in the Bajocian and Bathonian strata are unlike those of Canadian Arctic Islands, resembling more closely those of Williston and Alberta basins, as well as those of offshore Eastern Canada, northwestern Europe, and eastern Greenland. The Callovian dinoflagellates of western Canada more closely resemble those of the Canadian Arctic Islands, characterized by *Nannoceratopsis pellucida* and *Glomodinium* species, but also with the circum-North Atlantic *Ctenidodinium* association.

Upper Jurassic series and fossils

Lower Oxfordian strata, with *Cardioceras* species, are present in apparently continuous, poorly fossiliferous sequences in Brooks-Mackenzie Basin, where the next older ammonites are Early Callovian, and in Alberta and Williston basins, where older fossils are Late Bathonian or Early Callovian in age. The remaining Upper Jurassic is marine and nearly complete, as indicated by a succession of *Buchia* species in seemingly continuous sequences. Various species of *Buchia* and a few of the rare ammonites of Late Jurassic age have been described by Jeletzky (1965, 1966, 1973, 1984), who erected a zonation for Western and Arctic Canada. The other Late Jurassic marine faunas have been described by Whiteaves (1903), Frebold (1957), Frebold et al. (1959, 1967), and Poulton and Tempelman-Kluit (1982). These boreal faunas are a product of the southward expansion of the marine Boreal Realm in the late Middle Jurassic.

Microfossils of Oxfordian to Volgian age from Rocky Mountain Trough of northeastern British Columbia were described and zoned by Brooke and Braun (1981; assemblages I-III, corresponding to XI-XIII of this report); some of Callovian to Oxfordian age from Williston Basin by Brooke and Braun (1972; assemblages V-VIII, corresponding to V-VIII of this report); and those of Kimmeridgian and younger Late Jurassic age of Brooks-Mackenzie Basin by Hedinger (in press). In the more southerly parts of Rocky Mountain Trough, some calcareous Foraminifera and Ostracoda at the base of the Upper Jurassic pile (Green beds of Fernie Formation) are comparable to those of northwestern Europe and offshore Eastern Canada. Those of the younger Upper Jurassic rocks of Rocky Mountain Trough and Brooks-Mackenzie Basin are closely similar and are essentially boreal in affinity. This change in paleobiogeographic affinity of the microfauna in approximately Early Kimmeridgian time conforms with filling or elevation of the sedimentary basins of the Western Interior area, so that southerly and/or possible easterly connections were lost. The change also indicates subsidence of the northerly connected Rocky Mountain Trough adjacent to the incipient Columbian Orogen.

The Late Jurassic (Oxfordian) microfauna of Williston Basin is essentially endemic.

The Late Jurassic palynomorphs from Rocky Mountain Trough and Williston Basin have been described by Rouse (1959), Gussow (1960), Pocock (1962, 1964, 1970, 1972, 1980), and Davies and Poulton (1986). Those from northeastern British Columbia have strong affinities to floras from the Brooks-Mackenzie Basin and from the Canadian Arctic Islands, whereas those from more southerly areas have closer affinities to the North Atlantic areas. The Late Jurassic palynomorphs from Brooks-Mackenzie Basin have been described or recorded by Brideaux (1971, 1976, 1977a, b), Brideaux and Myhr (1976), Brideaux and Fisher (1976), Pocock (1967, 1970, 1972, 1976), and Fensome (1987).

Spores and pollen, still undescribed, are the principal tools for dating and correlating the nonmarine beds of the lower Kootenay Group in southwestern Alberta and southeastern British Columbia. Plant macrofossils have also been used for biostratigraphy in the Upper Jurassic of Rocky Mountain Trough (Berry, 1929; Bell, 1944, 1946, 1956; Brown, 1946).

The presence of Upper Jurassic strata in Canadian parts of Williston Basin is indicated by palynomorphs and microfossils. The completeness of the record there is still uncertain.

The Jurassic-Cretaceous boundary is in dispute worldwide because of failure to recognize, as yet, reliable, widespread marker species (see Jeletzky, 1984; Zeiss, 1983, 1984). For present purposes, the base of the *Buchia okensis* bivalve zone is taken to mark the base of the Cretaceous in the boreal marine successions of western Canada, following Jeletzky (1965).

The Jurassic-Cretaceous boundary seems to be transitional in the Rocky Mountain Trough, falling within the upper part of the nonmarine Mist Mountain Formation of southern regions and within the marine Nikanassin and Monteith formations of more northerly areas. In Brooks-Mackenzie Basin, it lies within the Kingak and Husky formations. The contact is defined by *Buchia* species, dinoflagellates, and microfossils in marine facies, and by spores and pollen in nonmarine facies. Ammonites are rare. The nature and position of the Jurassic-Cretaceous boundary in Williston Basin remain uncertain. In some areas, a disconformity within the lower part of the Mannville Group appears to separate Upper Jurassic from Lower Cretaceous rocks.

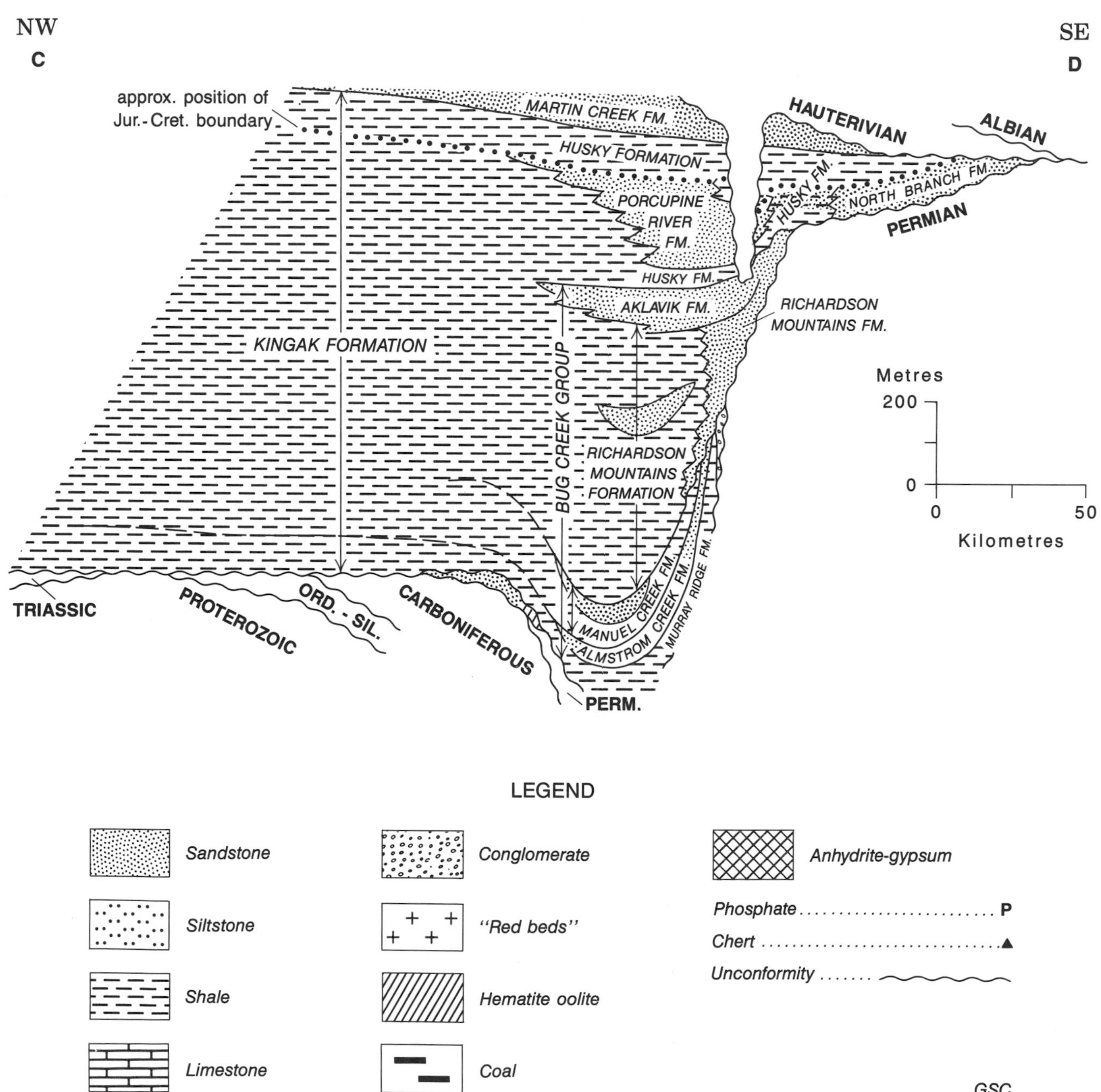

Figure 4H.5. Generalized northwest-southeast stratigraphic cross-section (C-D) across southeastern edge of Brooks-Mackenzie Basin, northern Yukon Territory. Line of section shown in Figure 4H.2.

BROOKS-MACKENZIE BASIN

Lithostratigraphy

The Jurassic rocks of northeastern and west-central Yukon Territory and adjacent Northwest Territories are autochthonous. They comprise in the east a series of marine, probably easterly derived, mainly shelf sandstone complexes (Bug Creek Group, Husky, Porcupine River, North Branch formations). These sandstones grade to outer shelf shale and siltstone facies (Kingak Formation) to the north, west, and southwest (Fig. 4H.5, 4H.6), becoming generally thicker, proportionately finer grained, and more complete in terms of the preserved record in those directions. No unequivocal nonmarine facies are preserved. The sedimentary facies and indications of point sources suggest that the sediments were deposited entirely in non-deltaic, marine environments along-shore from the mouth of a major river. No significant sediment introduction from basin margins in other directions is indicated.

The Jurassic rocks of northwestern Yukon Territory (Kingak Formation), together with those of adjacent northern Alaska, may well have been rotated counter-clockwise, and translated southwestward, away from an original depositional site off Banks Island, (Grantz et al., 1979; Norris, 1983). If this hypothesis is correct they were deposited to the northeast of the northeasternmost Jurassic sediments preserved on Tuktoyaktuk Peninsula. In any event, they are discussed here in the context of their presently preserved location.

The sub-Jurassic surface, formed on Precambrian, Paleozoic, and Triassic rocks, was apparently nearly peneplaned by latest Permian and/or Triassic erosion. In Aklavik Range, the early Jurassic basin margin was controlled by relief associated with faults of the Eskimo Lakes Fault Zone (Poulton et al., 1982).

The basal Jurassic rocks onlapping the craton become progressively younger eastward and southeastward. They range in age from Hettangian near Bonnet Lake, to late Middle or Late Jurassic southeast of Aklavik, where the Upper Jurassic rocks overstepped the Eskimo Lakes Fault Zone, which formed the Early and Middle Jurassic basin margin. The transition from Jurassic to Cretaceous strata lies within shales of the Husky and Kingak formations.

The Upper Jurassic rocks have a general sediment point-source and sand-shale distribution similar to that of the older Jurassic rocks, but are spread over a broader area, and their distribution apparently is not controlled by faulting.

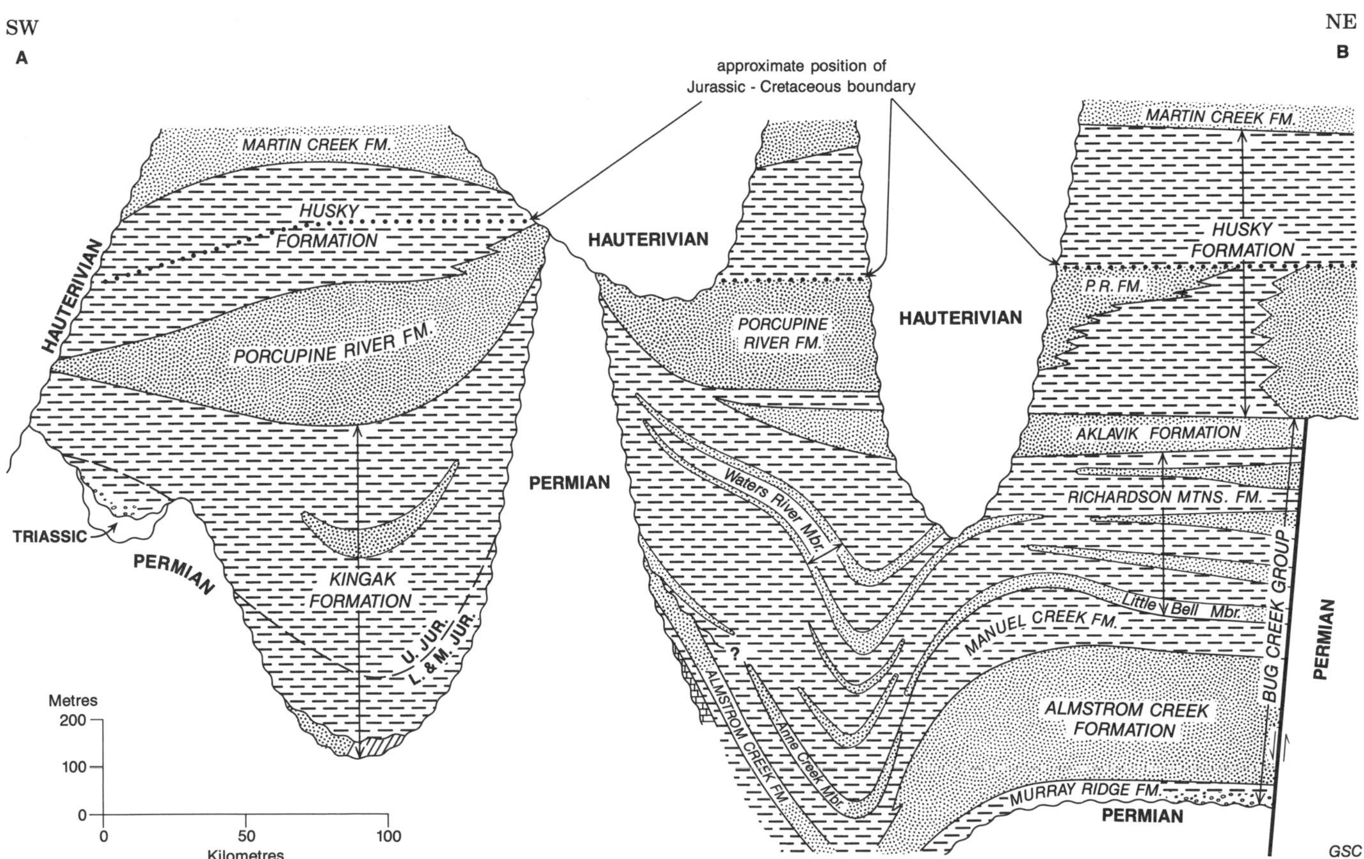

Figure 4H.6. Southwest-northeast stratigraphic cross-section (A-B) along southeastern edge of Brooks-Mackenzie Basin, northern Yukon Territory. Line of section shown in Figure 4H.2. The thicknesses and relationships in the southwestern half are poorly known and largely speculative. Those shown are based partly on data of Jeletzky (1975, 1977, 1980). Other parts of the diagram draw on data from Dixon (1982a, b and unpub. data) and Young (1975). For legend see Figure 4H.5.

Western basinal facies

The argillaceous sequence to the northwest and west, from northern Yukon Territory south as far as the northern Ogilvie Mountains, comprises most of the Kingak Formation, defined in northeastern Alaska (Leffingwell, 1919; Detterman et al., 1975; Fig. 4H.7). Because of structural complications, its thickness is uncertain but appears to exceed 1200 m in some places. Parts of every Jurassic stage except Sinemurian are identified by characteristic genera of ammonites or species of *Buchia* in the monotonous Kingak shale, and the Sinemurian is well represented at the base of the adjacent basin-margin sandstone succession. Basal sandstones of Hettangian age occur near Bonnet Lake (Frebold and Poulton, 1977) and Lower Jurassic chamosite oolites and *Gryphaea*-rich siltstones are found in the same area. The last may correlate with similar beds at the base of the Bug Creek Group in its western localities (below shales with Early Sinemurian *Coroniceras*?).

The upper parts of the Kingak Formation are equivalent in age and similar lithologically to the shales of the Husky Formation. They are characterized in many places by large, buff-weathering, siliceous and dolomitic septarian concretions, similar to those of the equivalent Ringnes Formation of the Arctic Archipelago (Balkwill et al., 1977) and the upper Fernie shales of southwestern Alberta and northeastern British Columbia (Poulton, 1984).

The name Kingak is applied also to a predominantly shale-siltstone unit in northern Ogilvie Mountains, although it is geographically close to and lithologically similar to the Glenn Shale, which was established by Brabb (1969) in adjacent Alaska. The rocks in northern Ogilvie Mountains were presumably deposited in physical continuity with the Kingak farther north and are similar to it, especially to its upper parts of Late Jurassic age. Fossils are rare but indicate Early through Late Jurassic ages (Poulton, 1982). The Kingak Formation in Ogilvie Mountains contains sandstone units in its upper part that seem to be distal equivalents of the Porcupine River Formation of west-central Richardson Mountains.

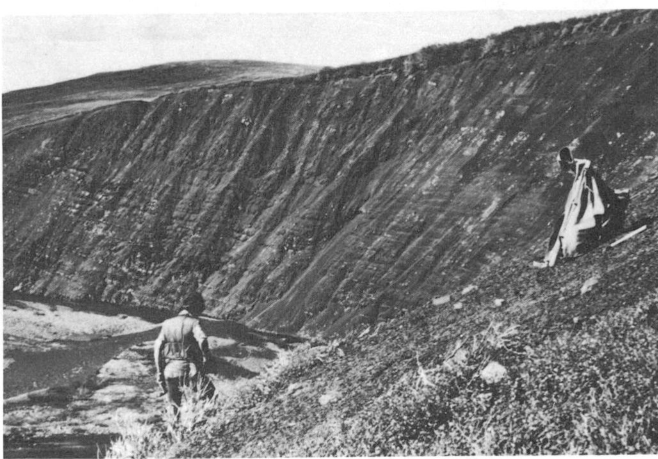

Figure 4H.7. Kingak concretionary shales and siltstones, along Caribou Creek, southeast of Trout Lake, northern Yukon Territory. The beds shown are not well dated owing to scarcity of fossils, but Bathonian ammonites occur in approximately the same stratigraphic position in the next bluffs to the south (photo - T.P. Poulton, ISPG 2445-12).

Figure 4H.8. Metamorphosed shales and siltstones of the 'Lower Schist Division' (L. Sch), forming most of the slope, overthrust by quartzites of the Paleozoic Keno Hill Quartzite (KHQ); east side of Dempster Highway south of North Fork Pass, central Yukon Territory. Sandstone beds in the upper part of the 'Lower Schist' resemble distal tongues of the Porcupine River Formation on the extreme western edge of the northern Richardson Mountains, northern Yukon Territory (photo - T.P. Poulton, ISPG 2445-7).

Along the outcrop belt in northern Ogilvie Mountains, the basal Kingak unit varies from one locality to another, ranging from hematitic oolite to sandstone, to thin conglomerate, and the thickness of these seemingly pre-Upper Oxfordian rocks is highly variable. In places along the eastern side of northern Ogilvie Mountains, tentatively dated Upper Jurassic shales lie directly on Paleozoic rocks. Elsewhere shales are present that are thinly developed, poorly dated, and lacking the characteristic concretions of the Upper Jurassic Husky sequence; these therefore are considered to be pre-Late Jurassic in age. The differences in thickness of the probable pre-Upper Jurassic shales may attest to deposition on an irregular surface associated with deeply indented Early and Middle Jurassic shorelines, across which the present outcrop belt provides a more or less straight transect.

Shales with Middle and Late Jurassic ammonites in a narrow belt across central Yukon Territory, from about Dawson City in the west to beyond Keno Hill in the east, have been called the 'Lower Schist Division' (Green, 1972; Frebold et al., 1967; Tempelman-Kluit, 1970; Poulton and Tempelman-Kluit, 1982; Fig. 4H.8). The shales occur in a series of thrust sheets, which have been transported northward so that the original depositional site is not precisely known. However, the lithological succession of these Jurassic rocks resembles that in the southerly parts of Brooks-Mackenzie Basin (Frebold et al., 1967; Poulton and Tempelman-Kluit, 1982), some 180 km to the north-northwest. For this reason, they are treated as a southerly extension of Brooks-Mackenzie sequence. The similarity of the sequences in northern and central Yukon Territory extends to some marine sandstones in the upper part of the 'Lower Schist Division', which closely resemble distal tongues of the Porcupine River Formation of northern Yukon Territory, where they extend out into the Kingak Shale.

Eastern basin-marginal facies

Sinemurian to Lower Oxfordian rocks in most of the northern Richardson Mountains form the Bug Creek Group, and are composed mainly of sandstone and siltstone, interbedded with argillaceous tongues of the Kingak Formation (Poulton et al., 1982).

Figure 4H.9. Upper part of the Manuel Creek Formation (MC) (Toarcian and Aalenian shale with minor sandstone beds), abruptly overlain by bluff-forming sandstone of the Little Bell Member (LB) (Bajocian; 24 m thick) of the Richardson Mountains Formation (RM). Higher beds are interbedded sandstones, shales and siltstones of the Richardson Mountains Formation (Late Bajocian through Early Oxfordian age), and the sequence is capped by resistant sandstone bluffs of the Oxfordian Aklavik Formation (A). Murray Ridge, Northwest Territories (photo - T.P. Poulton, ISPG 2445-11).

The Bug Creek clastic wedge was limited along its depositional strike to an area between the present-day northernmost Ogilvie Mountains and the northeastern Mackenzie Delta. It disappears southwestward by thinning and lateral gradation to shale. To the northeast, it terminates abruptly against Tununuk High along a northwest-trending locus near Reindeer Channel, presumably along a Late Jurassic fault. The overlying Upper Jurassic and Lower Cretaceous Husky Formation was deposited across that structure (Dixon, 1982a, b).

The basin margin was complicated by local uplifts and depressions, some of which must have been fault-bounded. One small uplift, 'White Mountains Uplift' (Poulton and Callomon, 1976; Poulton et al., 1982), from which Lower Jurassic sediments were stripped prior to Late Bajocian deposition, was possibly a precursor of the more extensive Cretaceous Cache Creek Uplift, which forms part of the Rapid Fault Array (Young et al., 1976; Jeletzky, 1980).

Except for local basin-margin sandstones and conglomerates of the Scho Creek Member in the northern Richardson Mountains and thin basal lag deposits elsewhere, the lowest Bug Creek rocks, the Murray Ridge Formation, are open-shelf shales and siltstones. Articulated crinoid remains and *Chondrites* attest to poorly oxygenated bottom conditions and lack of currents. The formation thickens gradually westward from a minimum of about 30 m in the eastern northern Richardson Mountains. The Murray Ridge Formation grades upward to shelf sandstone of the Almstrom Creek Formation, which was deposited by storm and/or tidal agencies, and is mainly Pliensbachian and Early Toarcian in age. Maximum thickness of the Almstrom Creek of over 300 m is found near the headwaters of Bell and Big Fish rivers, beyond which the sandstone grades rapidly westward into Kingak shale.

The overlying Manuel Creek Formation is a blanket of shale of Toarcian and Aalenian age, thickening gradually westward from about 30 m in the central northern Richardson Mountains (Fig. 4H.9). Laterally equivalent rocks to the southwest, near the headwaters of Waters River, are somewhat thicker and include significant sandstone units, the lower of which are tongues of the Almstrom Creek Formation. An upper sandstone unit, the Anne Creek Member, is a shoal development associated with a regression that culminated in a period of stillstand and negligible sedimentation in Early Bajocian time. This last event and the succeeding transgression, during which the rate of sedimentation was slow also, is represented by a thin blanket sandstone, the Little Bell Member of the Richardson Mountain Formation (Fig. 4H.9).

The Early and early Middle Jurassic southeastern basin margin was determined in part, at least, by faults of the Eskimo Lakes Fault Zone. The margin probably was close to the present-day southeastern limits of Lower Jurassic and lowest Middle Jurassic rocks, although the latter were partly controlled by pre-Late Bajocian erosion. Uplift of the southeastern basin margin more or less coincided with the elevation of White Mountains Uplift, and a corresponding depression of adjacent Waters River depocentre.

Bajocian through Lower Oxfordian shelf shales, siltstones, and sandstones of the Richardson Mountains Formation were deposited as a major complex that is thicker and more argillaceous toward the north and west (Fig. 4H.9). In the depocentre near the head of Waters River, some thick Lower Callovian sandstone units of the Waters River Member represent shoals or possibly barrier islands. The upper part of this succession, represented by the Aklavik Formation (Fig. 4H.10), is a prograded shelf sandstone of Early Oxfordian age. It may include some nonmarine beds in the southernmost localities. The Richardson Mountains and Aklavik formations together thicken westward to more than 700 m, beyond which the sandstones disappear into the Kingak shale.

Upper Oxfordian and younger Jurassic rocks exhibit a depositional geometry basically similar to that of the underlying Bug Creek Group but are transgressive over a broader area onto the craton. Their facies indicate easterly or east-southeasterly sediment sources near the head of Vittrekwa River. The sandstones extend basinward in shallow marine facies, grading into Husky shale to the north and Kingak shale to the west and south.

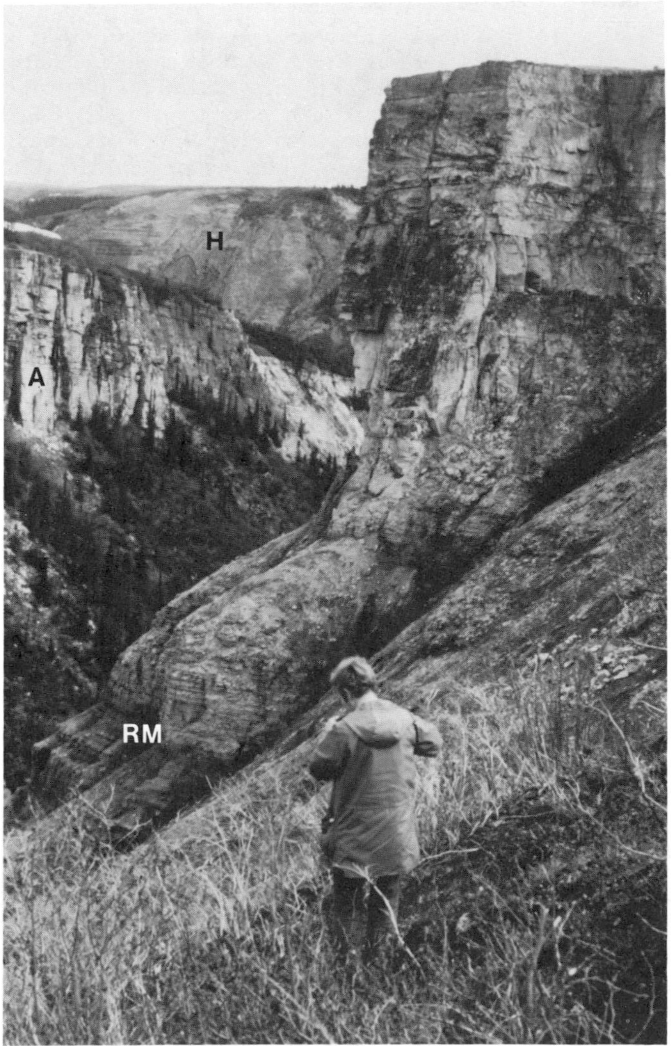

Figure 4H.10. Bluff-forming sandstones of the Aklavik Formation (A) (Oxfordian) gradationally overlying interbedded siltstones and sandstones of the Richardson Mountains Formation (RM) (Bajocian through Oxfordian). The upper surface of the Aklavik is overlain abruptly by soft fissile black shales of the lower Husky Formation (H). Martin Creek, Northwest Territories (photo - T.P. Poulton, ISPG 2445-10).

The Upper Jurassic units reach a maximum thickness of about 400 m in a depocentre near Salmon Cache Canyon on Porcupine River. Most of the thickening is in the Porcupine River marine sandstone (Jeletzky, 1977). Its main development lay in a northeast-trending broad band extending from northern Ogilvie Mountains in the southwest to the headwaters of Big Fish River in the northeast. Shelf sandstones of very shallow marine origin to the southeast, included in the North Branch Formation (Jeletzky, 1967; Fig. 4H.11), are the most proximal facies preserved. The sandstones pass northward into the arenaceous member of the marine Husky Formation in the Mackenzie Delta area and northernmost Richardson Mountains (Jeletzky, 1967, 1975; Dixon, 1982a, b; Braman, 1985).

The Husky sequence is transgressive eastward in Mackenzie Delta area beyond the fault-controlled limits of older Jurassic rocks, resting on Paleozoic and Proterozoic rocks in Kugpik O-13 well in northern Mackenzie Delta, and in Aklavik F-17 well near Aklavik (Dixon, 1982a, b). It was deposited as far east as northern Anderson Plain, where the sequence is included in the mainly Lower Cretaceous Langton Bay Formation (Brideaux and Fisher, 1976), and equivalent shales overlie Paleozoic rocks on Banks Island (Miall, 1979; Balkwill et al., 1983).

Basin history

The succession along the southeastern margin of Brooks-Mackenzie Basin records five major regressive events. The Sinemurian and Pliensbachian Murray Ridge and Almstrom Creek formations, the Oxfordian upper Richardson Mountains and Aklavik formations, and the overlying Upper Jurassic Husky, Porcupine River, and North Branch formations represent coarsening and

Figure 4H.11. Lower and middle parts of the North Branch Formation (Upper Jurassic) at its type section, North Branch of Vittrekwa River, Northwest Territories. The prominent, light-coloured sandstone bed is about 2.5 m thick (photo - T.P. Poulton, ISPG 2445-13).

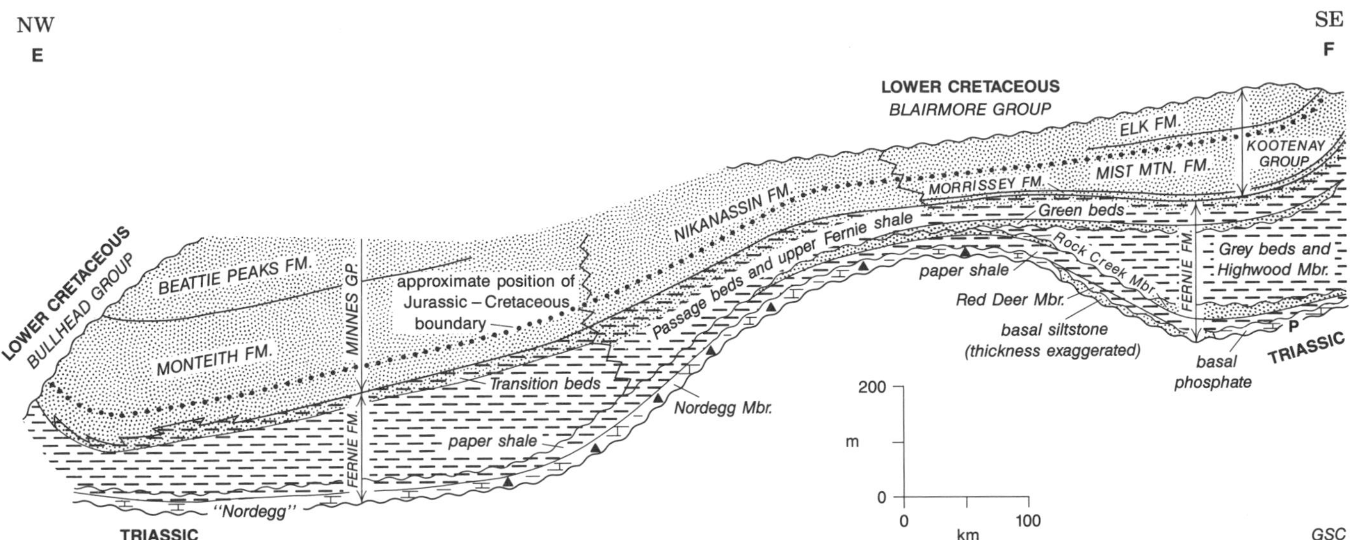

Figure 4H.12. Northwest-southeast stratigraphic cross-section (E-F) along western parts of the Alberta Basin and superjacent Rocky Mountain Trough in western Alberta and eastern British Columbia. Data from many sources, particularly Stott (1967a) in northeastern British Columbia. Line of section shown in Figure 4H.2. For legend see Figure 4H.5.

shallowing-upward, basin-fill successions, each culminating in significant sandstone deposits. Regressive episodes resulting in only minor sandstone development are the shoaling-upward Toarcian and Aalenian Manuel Creek Formation, and the Bathonian and Lower Callovian, lower and middle parts of the Richardson Mountains Formation. The Aalenian or Early Bajocian maximum regression more or less coincided with erosion on the basin margin and development of the White Mountains Uplift and Waters River depocentre on the shelf. The upper beds of the Almstrom Creek Formation, intertongued with Toarcian shales, and the basal sandstone of the Richardson Mountains Formation, which is Early Bajocian in the west and Late Bajocian in the east, are transgressive facies.

Deposition of each of the regressive packages was preceded by cratonward transgressions, which are remarkable for the consistency with which progressive onlap can be seen in the preserved record. The depocentres of the Lower, Middle, and Upper Jurassic show a progressive southerly advance associated with transgression in that direction. The locus of maximum sandstone deposition for each series correspondingly shifted southward within the general Waters River area, from near the head of the Waters River (western northern Richardsons) in the Middle and possibly also Early Jurassic (Poulton et al., 1982) to about Salmon Cache Canyon on Porcupine River in Late Jurassic time. The absence of evidence for a Jurassic, western source of sediment stands against any interpretation of tectonic lands to the west (e.g. Jeletzky, 1975).

ALBERTA BASIN AND ROCKY MOUNTAIN TROUGH

Lithostratigraphy

The pre-Jurassic surface of much of the Interior Platform appears to have been a relatively smooth peneplane, sloping gently to the west, and not significantly channelled. However, Lackie (1958) described southwest-trending sub-Jurassic channels reconstructed from Jurassic isopachs in the Peace River area of northern Alberta and northeastern British Columbia. The Jurassic succession overlies a regional unconformity, which truncates successively older beds eastward onto the craton, Triassic and Permian in the west, and Carboniferous in the east. The sub-Cretaceous unconformity, and unconformities within and below the Jurassic strata merge eastward with various major unconformities of Triassic and Paleozoic age. Where Mesozoic rocks overlie Carboniferous carbonates, karst occurs commonly, and a 'residual zone' of breccia and various sandstones is characteristic at the contact.

Lower and Middle Jurassic rocks are easterly derived shales and minor quartzose sandstones of the lower and middle Fernie Formation (Frebold, 1957; Hall, 1984; Stronach, 1984; Poulton, 1984). Upper Jurassic rocks seem to be mainly products of the early phases of the Columbian Orogeny, presumably derived from uplifts to the west, southwest, or south. Some local channel-fill sandstone units near the eastern erosional edge of the Jurassic succession in south-central Alberta are from cratonal sources (Hopkins, 1981) and may be Late Jurassic in age. Certain quartzose eastern facies of the lower Monteith Formation in northeastern British Columbia may also be craton-derived (Stott, 1984; Poulton, 1984; Fig. 4H.12-15). The continuity of individual sandstones, and their respective ages, along the complex eastern basin margin are still incompletely resolved.

Although the oldest Jurassic unit of the Fernie Formation in an extreme western locality in northeastern British Columbia is Hettangian, the basal beds over most of the area are Sinemurian or Pliensbachian in age. Toarcian or younger shales were more widespread than the Sinemurian units, indicating progressive but episodic transgression of the craton. These and the younger Jurassic rocks are limited in their present eastward extent by pre-Cretaceous or pre-Late Jurassic erosion.

To the east, northeast, and north, the Jurassic sequence thins from maximum thicknesses greater than 1000 m in the west. It disappears onto the craton partly by depositional thinning of some units, but mainly by erosional truncation below unconformities within the Jurassic itself and below the Lower, but not lowest, Cretaceous Blairmore, Bullhead, and Mannville groups. The facies and thickness trends in the Lower Jurassic beds indicate that they were originally deposited more extensively northward over the craton, and have been

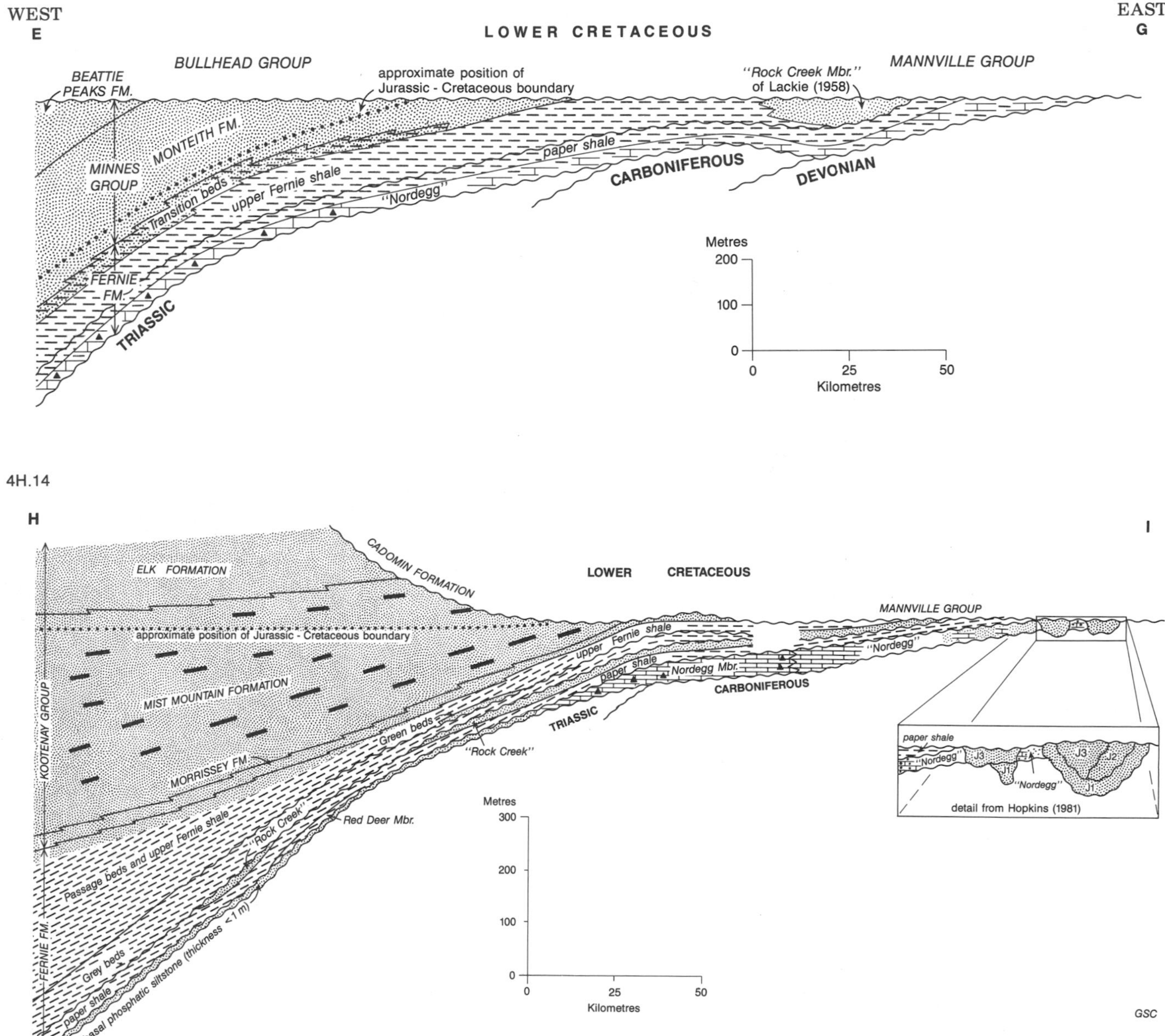

Figures 4H.13, 14, 15. Generalized west-east (E-G, H-I), and southwest-northeast (J-K) stratigraphic cross-sections across Alberta Basin and the superjacent Rocky Mountain Trough; (E-G) adapted from Stott (1967a, b) and Lackie (1958); eastern part of (H-I) from Hopkins (1981); (J-K) partly from Gibson (1977, 1979); remainder from many sources. Relationships of upper beds in central part of H-I, upper beds in eastern parts of E-G, and lower and middle beds in eastern parts of J-K remain unproven. Lines of sections shown in Figure 4H.2. The western, tectonically telescoped, facies have been restored palinspastically to show the pre-deformation relationships, following the restoration model of Norris (1971). For legend see Figure 4H.5.

stripped back from their depositional limits by subsequent erosion. All of the Lower and Middle Jurassic units are thin, reaching a maximum thickness of about 200 m, of which most is accounted for by the Bathonian Grey Beds.

In southernmost Alberta, Middle and Upper Jurassic units are continuous with those of the Ellis Group of the northern part of Williston Basin in southeastern Alberta and western Montana, south of and across Sweetgrass Arch.

Cratonic facies

Hettangian strata, shales and limestones with *Psiloceras calliphyllum* (Tozer, 1982), are known at the base of the Fernie Formation at a single locality, near Williston Lake, northeastern British Columbia. These, together with the exceptional thickness and completeness of the underlying Triassic section, indicate the continued presence of the Peace River Embayment, a depocentre initiated in the Late Devonian persisting through earliest Jurassic time.

Over most of the remainder of the area, a widespread, banded, phosphatic limestone-shale unit, with a thin shale or a 'residual zone' of grit or breccia at its base, forms the basal Jurassic unit. In west-central Alberta, about 125 km northwest of Calgary, the basal unit, the Nordegg Member, is characterized by banded and nodular chert (Spivak, 1949; Stott et al., 1968; Hall, 1984; Fig. 4H.16). Eastward, its equivalents include phosphorite, limestone and dolomite, and shallow marine quartz-chert sandstone facies in central Alberta (Bovell, 1979; Rall, 1980; Hopkins, 1981). To the west in the Foothills of central Alberta, the chert disappears (Ollerenshaw, 1968) and eventually the entire unit is absent at Bighorn Creek. A thin, richly fossiliferous or coquinoid shallow marine limestone bed ('*Oxytoma* bed' of Frebold, 1953 in part) lies at the top of the Nordegg Member in many places. Its arietitid ammonites indicate an Early Sinemurian age (Frebold, 1969).

Calcareous or mainly shaly beds with Early Sinemurian *Arnioceras* (Warren, 1931; Frebold, 1957) in the Crowsnest Pass area of southwestern Alberta are not significantly cherty but contain sub-economic phosphate deposits (Telfer, 1933; Christie, 1979a). To the southeast, for instance at Adanac Mines and Hastings Ridge and to the northeast in the Livingstone Range of the southern Foothills, a thin, phosphatic, calcareous sandstone forms the basal unit of the Fernie Formation. It contains many bivalves and gastropods locally, as well as some plesiosaur and ichthyosaur remains (Nicholls, 1976).

Sinemurian fossils are not known with certainty from northeastern British Columbia, those ammonites identified previously as '*Arnioceras*' having been re-identified as the Late Pliensbachian genus *Amaltheus* by Frebold (1970b). Nevertheless, the lower beds of what is called the 'Nordegg Member' in this region (Stott, 1967a, b, 1969) may be Sinemurian judging by their lithological

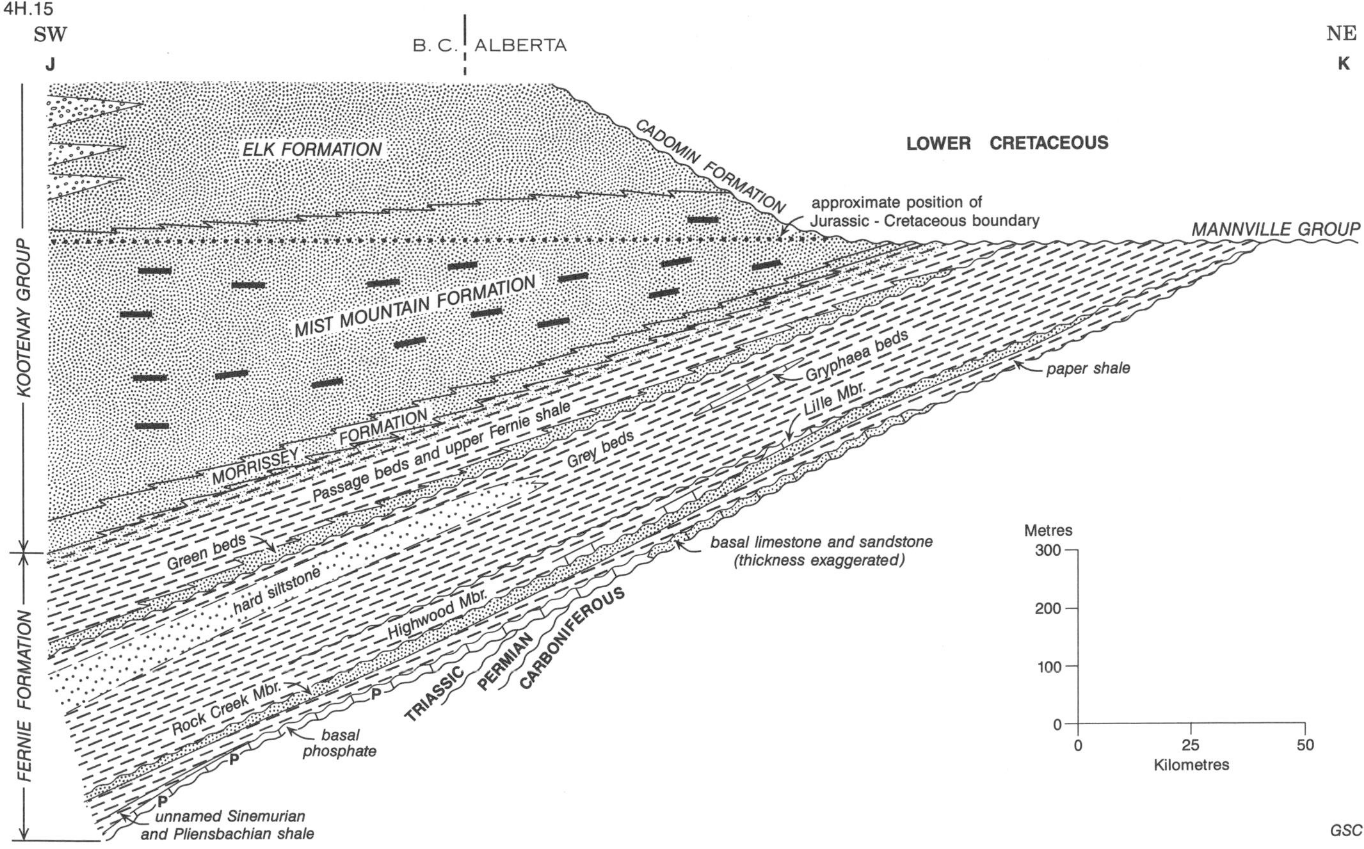

Figures 4H.13, 14, 15. Cont.

similarity with the Nordegg Member farther south. To the east in the subsurface of northeastern British Columbia, a basal Jurassic, speckled, calcareous shale unit (Stott, 1967a, b; Lackie, 1958; Hamilton, 1962) may be their equivalent.

Upper Sinemurian and lowermost Pliensbachian shales occur in southeastern British Columbia (Frebold, 1969). Siltstones with Late Pliensbachian fossils in west-central Alberta were named the Red Deer Member by Frebold (1969). Quartz-chert sandstones, probably of Early Jurassic age, which partially fill channels cut through the Nordegg Member into Lower Carboniferous rocks just northwest of Red Deer (Rall, 1980; Hopkins, 1981), may be an eastern equivalent. A pre-Late Pliensbachian erosional surface below the Red Deer Member in west-central Alberta may well reflect the same channelling event.

Figure 4H.16. Recessive lower part (lN) and resistant upper part (uN) of the Nordegg Member limestones and cherts (Sinemurian), overlying Triassic Whitehorse Formation (TW), between Cadomin and Mountain Park, west-central Alberta. The bluff-forming upper part of the Nordegg is about 10 m thick (photo - T.P. Poulton, ISPG 2445-2).

A regional euxinic black shale of Toarcian age, the Poker Chip Shale of Spivak (1949) or Paper Shale of Frebold (1957), overlies Upper Pliensbachian or Sinemurian beds paraconformably. This unit forms the base of the thin Jurassic sequence at Canyon Creek just southwest of Calgary, indicating overstep of older beds by the Toarcian shale.

Middle Jurassic beds are known with certainty only from about latitude 54°N southward, the area to the north presumably having undergone Middle or post-Middle Jurassic uplift.

Lower Bajocian quartzose sandstones and siltstones, the Rock Creek Member (Warren, 1934; Frebold, 1957, 1976), overlie the Toarcian beds with apparent conformity in southwestern Alberta and southeastern British Columbia. The ammonite genus *Sonninia* occurs in it at some localities (Frebold, 1976; Hall and Stronach, 1982; Hall, 1984). Locally developed, thin calcareous grits and coquinas in southernmost southeastern British Columbia, in places with the bivalve *Myophorella argo* (Crickmay) and belemnites, are probably equivalents of the Rock Creek. The Lille Member (McLearn, 1927; Frebold, 1957), a thin bivalve coquina, known only from one locality south of Grassy Mountain in the Crowsnest Pass area, overlies the Rock Creek and is also presumably Bajocian in age.

Younger Lower Bajocian shales in the southern areas (Highwood Member of Hall and Stronach, 1982; Stronach, 1984) contain *Chondroceras* (*Defonticeras*), *Stemmatoceras*, and other ammonites near their base. Calcareous beds with abundant belemnites (including 'Belemnite Zone' of Hume, 1930 and 'Bajocian Limestone' of Stronach, 1984) are characteristic in the lower part, as are thin phosphatic and bentonitic horizons. The Highwood Member overlaps eastward the older Bajocian sandstones and directly overlies Toarcian shales at Canyon Creek just southwest of Calgary.

The basal transgressive beds of the Highwood shale are correlative with the base of the Sawtooth Formation in northwest Montana and presumably also with the Sawtooth in the subsurface of southern Alberta, as judged from their ammonite content. Higher beds of the Highwood Member at Rock Creek contain Late Bajocian and Bathonian ammonites (Hall and Stronach, 1982; Hall, 1984). The Bajocian strata near Jasper comprise interbedded sandstones and shales with *Stemmatoceras* and *Arkelloceras* at their top (Frebold, 1957; Westermann, 1964), which may indicate continuous sedimentation in this region across the hiatus that separates the Rock Creek and Highwood members farther south. Other sandstones in west-central Alberta have been assigned to the Rock Creek Member and may also be Bajocian in age (Marion, 1984). They have coquinas at their base in some places. It is not clear whether or not these sandstones are physically continuous with the Rock Creek Member of southern Alberta.

Bathonian shales and siltstones, the Grey Beds in part (Frebold, 1957, 1958, 1963), are regionally extensive in southwestern Alberta. A disconformity below the Grey Beds at many localities is suggested by the appearance of the typical 'Rierdon' microfauna (Assemblage IV of Brooke and Braun, 1972) directly above the 'Bajocian Limestone' of the Highwood Member (of Stronach, 1984).

The Grey Beds unit is a hard, grey siltstone in southeastern British Columbia, possibly indicating a sediment source toward the south or southwest. Buff-weathering siltstones and limestone concretionary bands with many bivalves (e.g., '*Corbula munda* beds' and '*Gryphaea* beds'; McLearn, 1929; Frebold, 1957, 1963) occur in southern Alberta. The latter units contain the ammonites *Paracephalites* and *Eurycephalites* (i.e. *Warrenoceras*) faunas respectively. Callovian strata are not known to be represented in Alberta Basin, unless by the undated ammonite *Imlayoceras* that lies in a zone of large concretions in the Grey Beds of the Jasper area. The Grey Beds are similar lithologically to the equivalent Rierdon Formation in southeastern Alberta. However, the microfauna of the Grey Beds (Assemblage IV of Brooke and Braun, 1972) indicates that only equivalents of the lower to middle part of the Rierdon Formation of Saskatchewan, which is considered to be of Bathonian age, are present in it.

The thick Middle Jurassic succession in southeastern British Columbia and southwestern Alberta contrasts with the absence or a near absence of rocks of this age in northeastern British Columbia. This difference is partly due to erosion below the Oxfordian-Kimmeridgian beds north of about latitude 54°N, but may also indicate subsidence of the southern areas in response to Middle Jurassic convergent tectonism in southeastern British Columbia. In particular, a Middle Jurassic tectonic event may be indicated by the still incomplete evidence for a southern or southwestern source area for the Grey Beds, and by reported western sources for equivalents in southwestern Alberta (Hall and Stronach, 1982).

A thin, shallow-marine, concretionary, chamositic (berthierine) sandstone-siltstone unit named the Green Beds (McLearn, 1927) represents the Oxfordian at many places in the southeastern Rocky Mountains, Foothills, and adjacent Plains (Frebold et al., 1959). Characteristic fossils include the bivalve *Buchia concentrica*, the ammonites *Cardioceras* and *Goliathiceras*, and a diverse microfauna of calcareous and arenaceous Foraminifera and some ostracodes still to be described. The 'Upper Fernie Shales' of the foregoing area, strongly glauconitic at the base in some places (Marion, 1984), presumably represent parts of the same interval in a less condensed facies. Undated 'glauconitic' sandstones in the subsurface of Peace River area (Stott, 1967a, b) may represent the Green Beds (Poulton, 1984), whereas Lackie (1958) had earlier assigned them to the Rock Creek Member. The upper Fernie shales of the western Peace River area in northeastern British Columbia contain minor 'glauconite' intervals even in their lower parts (Stott, 1967a, b).

Study of the microfauna from the Fernie shales in northeastern British Columbia by Brooke and Braun (1981) and of the microflora (Davies and Poulton, 1986) indicates a hiatus below the Oxfordian or Kimmeridgian beds, encompassing nearly all, if not all, of the Middle Jurassic interval. The absence of undisputed Middle Jurassic fossils or characteristic rock types in the entire area north of about latitude 54°N, and discoveries of the ammonite *Amoeboceras*(?) and the bivalve *Buchia concentrica* a few metres above the 'Nordegg Member' west of Grande Cache, north of Jasper, suggest the regional extent of this hiatus. Uplift of the northeastern British Columbia region prior to deposition of the Upper Jurassic rocks may represent rejuvenation of Peace River Arch, following a long period of general subsidence, from Late Devonian through earliest Jurassic time, which localized the Peace River Embayment depocentre.

The Green Beds are probably a complex of condensed offshore basal transgressive sandstones deposited after immediately pre-orogenic or earliest orogenic regression and erosion. The degree to which older Jurassic beds were removed is erratic along depositional strike, as is the presence of the Green Beds facies. Thickness variations of the upper Fernie beds from about 10 to as much as 400 m are probably due to the presence of significant relief on the sub-Late Jurassic surface as much as to differential foundering of the developing foredeep of the incipient Columbian Orogen.

Either the Green Beds and their shaly equivalents to the west and north in some places, or the probably partly equivalent and partly younger Passage Beds in other places, are the basal unit overlying a regional unconformity that truncates underlying Jurassic beds progressively eastward.

In southernmost Alberta, the 'glauconitic' chert-pebble bearing beds of the Swift Formation, as well as certain 'glauconitic' chert-pebble beds with belemnites in the Middle Vanguard of southern Saskatchewan (Milner and Thomas, 1954; see also Frebold et al., 1959), are lithologically similar to the Green Beds and seem to be approximately equivalent.

Near the easternmost erosional edge of Jurassic rocks in south-central Alberta, some local channel-fill sandstones may be Late Jurassic in age (Hopkins, 1981). Other eastern arenaceous facies of the Upper Jurassic succession are included in the basal beds of the Mannville Group in some places in southwestern Saskatchewan (Christopher, 1974; 1984a, b) and easterly quartzose facies of the Monteith Formation in northeastern British Columbia (Stott, 1984). Together with the so-called 'Rock Creek' sandstones (Lackie, 1958) in northeastern British Columbia, treated here as possibly Oxfordian in age, these sandy facies may indicate easterly or northeasterly sediment sources not directly related to Columbian uplifts to the west. These various units may be only locally preserved remnants of what were much more widespread eastern shoreline and shelf facies, mostly removed by sub-Cretaceous erosion.

Orogenic facies

Most of the thick Upper Jurassic package of western Alberta and eastern British Columbia is generally thought to be derived from western uplifted terranes and to reflect the change in tectonic regimes arising from the early Columbian Orogeny. The generally coarsening-upward, basin-fill succession, initiated about Kimmeridgian time, consists of shale in the lower part ('Upper Fernie Shale') overlying a basal craton-derived sandstone (Green Beds), and a succeeding unit of shales and siltstones with upwardly increasing amounts of sandstone. These last are assigned to the Passage Beds of the Fernie Formation (Fig. 4H.17) and the Kootenay Group in the south (McLearn, 1927; Gibson et al., 1983), the 'Transition beds' and lower Monteith Formation in the north (Stott, 1967a, b, 1969), and the lower Nikanassin Formation in central western Alberta (Mackay, 1929). Some beds contain as much as 15 per cent detrital chert, and abundant detrital mica on bedding surfaces further distinguishes this upper package from the lower, mainly quartzose assemblages. In southwestern Alberta, the upper Fernie shales have been included in the Ribbon Creek Member by Stronach (1984). At Ribbon Creek southwest of Calgary, they contain a microfauna that also characterizes the Green Beds and Passage Beds, indicating probable lateral equivalence of these various units. Stronach (1984) also included in the Ribbon Creek Member shales with large concretions that underlie the Green Beds. These are treated together with the Grey Beds in earlier parts of this subchapter. An undated turbiditic siltstone unit, the Pigeon Creek Member (Crockford, 1949; Norris, 1957; Hall, 1984), may have been deposited in the foredeep in early Late Jurassic time.

Sedimentological studies of the Passage Beds of southern Alberta indicate a shallowing-upward foredeep sequence of basin-fill with transport directions from the south, southwest, and southeast (Hamblin and Walker, 1979). If the original source of the detritus lay to the west, then presumably secondary transport mechanisms along the basin axis were involved. The overlying unit in the southwest is the massive quartzose basal Kootenay sandstone (Morrissey Formation of Gibson, 1979; Gibson et al., 1983; Gibson, 1985). A Portlandian (Volgian approximately) age was originally suggested on the basis of a single, large, poorly understood ammonite *Titanites occidentalis* (Jeletzky in Newmarch, 1953; Frebold, 1954, 1957; Westermann, 1966; Callomon, 1984) and another fragment is now known (Westermann in Hamblin and Walker, 1979). An approximately Portlandian age is supported by a recent study of Kootenay palynomorphs by A.R. Sweet (in Poulton, 1984), who placed the Jurassic/Cretaceous boundary just below the top of the Mist Mountain Formation, which overlies the Morrissey.

Figure 4H.17. Shallowing- and coarsening-upward sequence in the upper Fernie Passage Beds (Late Oxfordian through Kimmeridgian), section overlooking Alexander Creek, Crowsnest Pass, southwestern Alberta (photo - T.P. Poulton, ISPG 2445-9).

The Morrissey is a shallow-water, coastal beach-dune-barrier and/or delta-front sheet sand complex (see Jansa, 1972; Gibson, 1977; Hamblin and Walker, 1979; Gibson et al., 1983), indicating that the southern part of the Rocky Mountain Trough had been filled by this time. Generally northward or northwest-trending paleocurrents and paleoslopes for the basal sandstone in southern Alberta are related to either northerly progradation, or longshore drift of sediment whose primary westerly sources are inferred but not proven. Stelck et al. (1972) have suggested that some of the sediment was derived from the east or southeast, from the area formerly occupied by Williston Basin that had been recently uplifted.

The Morrissey is overlain by the Mist Mountain Formation (Gibson, 1979), which straddles the Jurassic-Cretaceous boundary. The Mist Mountain is a diverse, mainly nonmarine, fluvio-deltaic, variably quartzose and chert-rich, sandstone-siltstone-shale succession with economically important coal deposits (Jansa, 1972; Gibson, 1977, 1985; Gibson et al., 1983; Hamblin and Walker, 1979; Fig. 4H.18). Chert-pebble conglomeratic beds occur in the western localities (Gibson, 1977, 1985), indicating proximity to a source terrane.

The coal measures of the Mist Mountain Formation contain thin clay layers known as tonsteins. The origin of tonsteins worldwide is a subject of much study and uncertainty, and seems to be different from one example to another. Those of the Mist Mountain Formation are composed of kaolinite (Meriaux, 1972; Grieve, 1984); there is no montmorillonite. High barium and yttrium contents may possibly indicate a volcanic origin, however. This is the only possible indication of volcanic activity in the mobile belt that must have lain west of the foredeep.

The absence of any significant amounts of detritus other than that which could have been derived from older sedimentary rocks led Gibson (1985) to conclude that the Kootenay sediments were eroded from uplifted sedimentary terranes east of the Rocky Mountain Trench. However, the abundant detrital mica on some bedding planes, and kaolinite in some intervals of the Upper Jurassic regionally, may represent products eroded from metamorphic and igneous terranes as well.

North of the North Saskatchewan River, the Kootenay Group grades into marine cherty sandstones of the Nikanassin Formation (Gibson, 1978, 1979). In northeastern British Columbia, the marine, mainly quartzose Monteith sandstone (Stott, 1967a, 1975) occurs at the base of the Minnes Group, which is mainly Early Cretaceous in age, but whose lowest part is latest Jurassic. Western facies appear to have been derived from uplifted terranes to the west (Stott, 1975), and M. McMechan (in Poulton, 1984) reported westerly changes to nonmarine facies. Conglomerates west of Mt. Minnes in western Kakwa River area between Jasper and Williston Lake contain volcanic and foliated greenschist clasts (M. McMechan and D.F. Stott in Poulton, 1984). Stott (1984) suggested that the more easterly facies of the Monteith may be craton derived.

Basin history

Episodes of transgression set the stage for deposition of each of the following packages: the Lower Hettangian shales; the Sinemurian Nordegg argillaceous package; the Upper Pliensbachian Red Deer Member; the Toarcian shales; the Lower Bajocian Highwood shales, limestones, and sandstones; the Bathonian basal Grey Beds; and the post-'*Gryphaea* beds' shales of the Grey Beds.

Major regressive packages include: the Toarcian shales through the Bajocian sandstones; the Bajocian Highwood shales; the Bathonian Grey Beds; and the post-'*Gryphaea* beds' shales of the Grey Beds. Major erosional episodes with little preserved sediment are those succeeding deposition of the Sinemurian(?) *Oxytoma* beds, the Upper Pliensbachian Red Deer siltstone member, and the Bathonian and Callovian(?) shales.

Following a period of uplift and erosion, most pronounced in northeastern British Columbia and northwestern Alberta, the Oxfordian Green Beds were deposited, apparently as a diachronous series of transgressive sandstones basal to the succeeding foredeep-fill facies. The foredeep facies, of Oxfordian through latest Jurassic age, consist of the upper Fernie shale and Passage Beds. The overlying Kootenay sandstones indicate deposition of the basal marginal marine sandstone at about sea level, and overlying coal-bearing sandstones in nonmarine environments adjacent to rising highlands.

WILLISTON BASIN

Lithostratigraphy

Jurassic rocks in Williston Basin have a maximum thickness of about 450 m (Fig. 4H.19, 4H.20). The Mesozoic section rests unconformably on Carboniferous, Devonian, or, in the extreme east, lower Paleozoic strata. Thickness variations of the lower part of the Jurassic sequence indicate deposition on a surface of considerable local relief, the most extreme (about 250 m) in Manitoba (Stott, 1955). Thickness variation in the lower beds is related to structure in the sub-Jurassic rocks (Francis, 1956; Kent, 1974; Christopher, 1974, 1984a), to channels, to thinning over cuestas formed by the Lower Carboniferous carbonates on the sub-Mesozoic erosional surface, to relief on a surface created by meteorite impacts (Sawatzky, 1974), and to collapse resulting from solution of Paleozoic salts. The lower units were deposited in two distinct sub-basins, the Watrous sub-basin in southern Saskatchewan and the Amaranth sub-basin in southern Manitoba, separated by an arch aligned along the provincial boundary (Christopher, 1984a, b). A ridge of Paleozoic carbonate rocks extending westward across south-central Saskatchewan from the northern end of this arch separated a small, narrow depositional trough to its north from the main part of Williston Basin as well.

The northwestern margin of the basin, formed by Sweetgrass Arch, was transgressed in Middle Jurassic time. The succession on the arch is similar to that of adjacent Montana (see Peterson, 1981), from where its formational nomenclature is taken; in ascending order, the Sawtooth, Rierdon, and Swift formations of the Ellis Group. These units grade into the Fernie Formation farther west in southernmost Alberta. Endemic microfauna in the Upper Callovian(?) to Lower Oxfordian strata of Williston Basin indicate the strong influence exerted by Sweetgrass Arch even when the arch was the site of marine sedimentation.

The middle part of the Jurassic succession, a sequence of shales and sandstones, indicates a change of dominant sediment sources from the east and north in Middle Jurassic time to the west and southwest in Late Jurassic time. Initiation of significant sediment supply from the western orogen more or less coincided with expansion and accelerated subsidence of the basin, as well as a southerly shift of the depocentre into northern United States. The basin configuration was lost about Kimmeridgian time when the American area became the site of deposition of the nonmarine Morrison Formation. The Jurassic rocks are unconformably overlain by the Lower Cretaceous Mannville sandstones.

Lower formations

The oldest Mesozoic rocks in Williston Basin are part of a widespread undated unit of dolomitic and evaporitic, at least partly eolian, mainly red, silty shales and siltstones. These beds, comprising the lower Watrous Formation of

Figure 4H.18. Mist Mountain Formation (latest Jurassic or earliest Cretaceous) exposed just west of Mount Gibraltar, above old Burns coal mine near head of Sheep River, west of Turner Valley, southwestern Alberta. Several hundred metres of interbedded sandstones, siltstones, shales and minor coal are exposed. View is towards west (photo - T.P. Poulton, ISPG 2445-1).

Saskatchewan (Milner and Thomas, 1954) and the lower Amaranth Formation of Manitoba (Stott, 1955; Barchyn, 1982, 1984), overlie Paleozoic carbonates with regional unconformity (Fig. 4H.19, 4H.20). They are probably the result of mainly nonmarine deposition of locally derived sediments, with occasional marine flooding in two sub-basins, one in southern Manitoba and the other in southern Saskatchewan (Barchyn, 1982; Christopher, 1984a).

Francis (1956) and Carlson (1968) considered that the Watrous Formation in Canada contained both Triassic and Jurassic rocks. They correlated, on the basis of lithology, the lower part of the Watrous with the Lower Triassic, upper part of the Spearfish Formation in the United States. American authors have recognized Permian, Triassic, and Jurassic redbeds that may be equivalent in part to the Watrous Formation (Dow, 1967; McLachlan, 1972; Peterson, 1972). In recent regional studies in North Dakota (S. Anderson in Poulton, 1984), one or more major unconformities interpreted from geophysical logs are thought to exist within the succession.

Overlying beds in Canada, also undated, are anhydrite interbedded with minor carbonates, variegated or brown to grey shale, and sandstone, of the upper Watrous or upper Amaranth formations. Economic gypsum deposits are mined from this latter unit in southwestern Manitoba, where it lies near the surface.

The upper parts of the Watrous and Amaranth formations are apparently gradational, through a unit of evaporites, into the overlying Bajocian carbonate and shale units, which also contain minor evaporites. This suggests a simple transgressive basin-flooding succession conformable with the Bajocian beds, and thus an early Middle Jurassic age (e.g. Stott, 1955; Barchyn, 1982). This interpretation gains support from the presence of evaporites and red siltstones in beds in the same stratigraphic position in the United States, the lower Piper, Gypsum Spring, and Nesson formations, which may be of Bajocian age (Imlay, 1980). An alternative interpretation, that the Watrous is at least partly Sinemurian in age, was proposed by Poulton (1984) on the basis of a facies correlation with Sinemurian phosphates and cherty

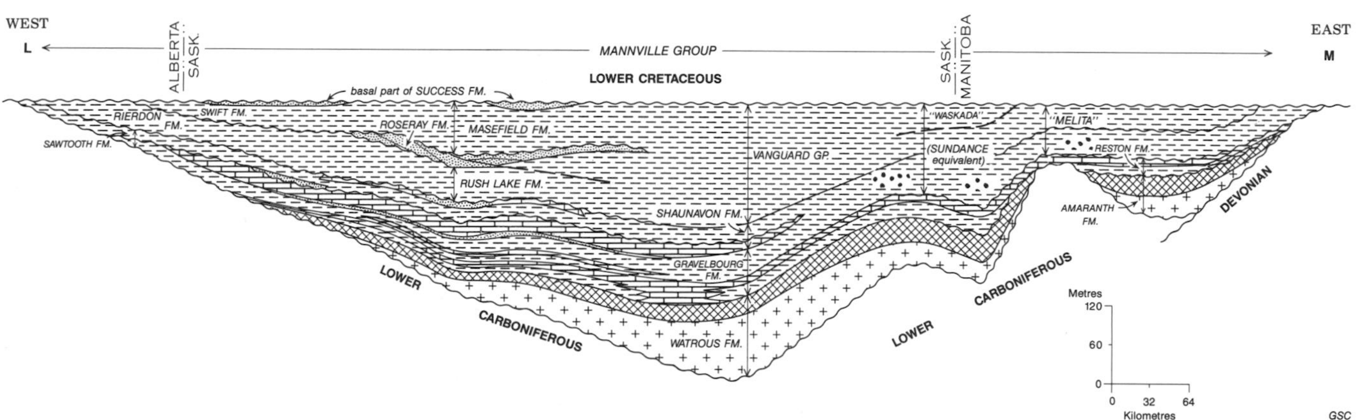

Figure 4H.19. West-east stratigraphic cross-section (L-M) across northern edge of Williston Basin in southern plains of Canada; modified from Francis (1956), with additional data mainly from Stott (1955), Weir (1949) and Christopher (1964, 1974). Line of section shown in Figure 4H.2. For legend see Figure 4H.5.

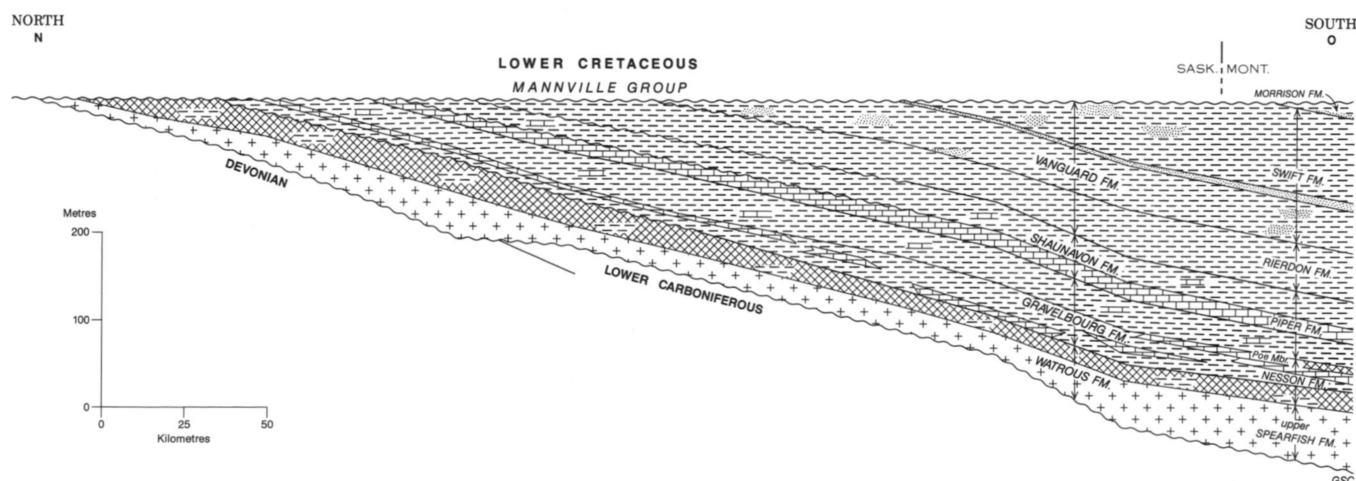

Figure 4H.20. North-south stratigraphic cross-section (N-O) through north-central edge of Williston Basin in southern Sakskatchewan, modified from Francis (1956) and showing correlation with formations in the U.S.A. Line of section shown in Figure 4H.2. For legend see Figure 4H.5.

limestones in the Rocky Mountains. This follows a model of laterally associated phosphorites, cherts, evaporites, carbonates, and redbeds proposed by Sheldon et al. (1967) and McKelvey et al. (1967).

Disturbed Lower Carboniferous and Devonian carbonate rocks, sandwiched between lower Watrous beds on the rim of an interpreted meteorite crater in southeastern Saskatchewan, were considered to date the impact as contemporaneous with the Triassic or Lower Jurassic redbed deposition (Sawatzky, 1974). Another crater in southern Manitoba, partially filled by Jurassic evaporites, may be of the same age (Ballantyne and McCabe, 1984).

The overlying carbonates and commonly green or variegated shales with minor sandstone were assigned to the Reston Formation in Manitoba (Stott, 1955), and to the lower Gravelbourg Formation in Saskatchewan (Milner and Thomas, 1954; Christopher, 1984a). They have been correlated with the Bajocian upper Gypsum Spring Formation of Wyoming (Francis, 1956) and the upper Nesson of North Dakota (Nordquist, 1955). Stott (1955) considered that a chert and breccia zone at the lower contact in Manitoba marked an unconformity.

Westerly thinning of the evaporite and carbonate units, and overlap of the former by the latter in southwestern Saskatchewan, indicate approach to the depositional limits of those units and progressive transgression of the basin margin. No such variation is seen to the north, indicating that the basin was originally more extensive in that direction. In Manitoba, the carbonate overlaps the underlying evaporite unit in an easterly direction. The original northeastern basin margin is not known, the near coincidence of zero edges of the Lower or Middle Jurassic(?) redbed units and the Middle Jurassic evaporite and carbonate units resulting from pre-Cretaceous erosion.

The arch straddling the Manitoba-Saskatchewan border was transgressed by an upper oolitic member of the carbonate unit at its southern extremity (e.g., Virden area; Stott, 1955), and by 45 m of sub-Rierdon sandstone to its north near Yorkton-Langenberg in southeastern Saskatchewan (L.L. Price in Poulton, 1984). The top of the carbonate unit is thought to be an unconformity and is represented locally by a zone of bluish chalcedony (Vigrass, 1952), which may be a hardground.

Overlying units, at least partly Bajocian in age, are mainly green-grey, calcareous shale, characterized by interbeds of arenaceous, partly oolitic carbonate and fossiliferous calcareous sandstone in the western part of the basin. They were called the Piper Formation by Francis (1956) and are now included in the upper Gravelbourg and Shaunavon formations (Milner and Thomas, 1954). Channel sandstones and internal unconformities characterize the upper part of the upper Gravelbourg in the east. Its upper contact is thought to be transitional (Christopher, 1984a), except in Manitoba (Stott, 1955), and over the arch at the Saskatchewan-Manitoba border.

A massive unit of lime mudstone and oolite forms the lower part of the Shaunavon Formation (middle Piper equivalent). Its characteristic buff colour, restricted fossil assemblage, and consistent character, thickness, and seismic response make it a prominent regional marker (Christopher, 1984a).

The complex facies changes and depositional environments in the shallow marine limestones and sandstones of the upper Shaunavon Formation and their relation to local paleotopography in southwestern Saskatchewan have been described by Christopher (1964, 1966a, b). The carbonate units become less common and more discontinuous to the east, grading to varicoloured, dominantly shaly facies of the Melita Formation (lower Sundance equivalent) in southeastern Saskatchewan and Manitoba. Regionally extensive, as well as local, channel-fill sandstones are increasingly common northward and eastward, indicating approach to shorelines and sediment sources in those directions (Stott, 1955; Francis, 1956; Christopher, 1984a). An unconformity occurs at the base of the upper Shaunavon, which is also characterized by internal discontinuities (Christopher, 1964, 1984a).

On the northwestern side of Williston Basin in southeastern Alberta, the upper Gravelbourg and Shaunavon interval thins and limestone becomes insignificant toward shorelines along Sweetgrass Arch. The most westerly extensive, though discontinuous, sandstone and shale unit, called the Sawtooth Formation (Weir, 1949; Davies, 1983), is equivalent to the Shaunavon and upper Gravelbourg. It has basal quartzitic sandstone beds and contains chert pebbles and belemnite-bearing conglomerates with oysters of the genus *Gryphaea* (Thompson and Crockford, 1958).

Upper formations

Upper units are mainly shales of the Vanguard Group. A consistent marker on electric logs ('Rierdon Shoulder' of American geologists) at a limestone or calcareous siltstone bed (e.g., Hadley and Milner, 1953; Francis, 1956; Thompson and Crockford, 1958; Hayes, 1983) differentiates upper and lower parts of the Vanguard in the central parts of Williston Basin, the lower parts being mainly calcareous ('Rierdon'), the upper being partly sandy and micaceous ('Swift'). The mica content of the upper units may reflect the same events as it does elsewhere in Western Canada Basin - the uplift and erosion of early Columbian metamorphic terranes to the west - although clastic rocks derived from the craton in Williston Basin are also consistently micaceous.

A quartzose sandstone unit, the Roseray Formation, within the Vanguard Group in southwestern Saskatchewan, led Christopher (1974) to subdivide the group into the Rush Lake, Roseray, and Masefield formations. The Roseray Formation comprises a complex of local southward and southeastward accreted marine sandstone wedges, which Christopher (1974, 1984a) considered to be clinothem equivalents of the Rierdon shales in the basin centre to the east and southeast. Thus the Rierdon shale in the basin centre may be equivalent only to the upper part of the Rierdon in northern Montana and may rest disconformably on a surface of nondeposition in a starved basin.

The upper Rierdon-Rush Lake sequence contains a microfossil assemblage (V of Brooke and Braun, 1972) younger than that recovered from the homotaxial Grey Beds of the Fernie Group to the west, possibly being as young as Early Callovian in age in Saskatchewan and north-central Montana.

The Masefield shale lies unconformably on the Rierdon and most of the Roseray, but a thin upper part of the Roseray, which contains a microfauna with some elements in common with the Masefield (assemblage VI of Brooke and Braun, 1972; VII of this report), may be basal to the Masefield sequence and overlie the unconformity.

The Rierdon transgressive event on the northeast side of the basin is recorded by shales that overlie the 45 m-thick basal Jurassic sandstone on the arch straddling the Saskatchewan-Manitoba border (L.L. Price in Poulton, 1984). The Melita and Waskada formations in southern Manitoba are not well dated. The latter may be equivalent to some part of the Roseray-Masefield interval.

Some 'glauconitic' basal Swift sandstones and shales with chert pebbles in southern Alberta (Thompson and Crockford, 1958; Hayes, 1983), like the similar but perhaps slightly older Oxfordian Green Beds of the Fernie Formation to the west, are basal transgressive deposits following the pre-Oxfordian regressive and erosional event. The sandstones of the Swift Formation in southern Alberta ('Ribbon sand' in part; Weir, 1949; Hayes, 1983) seem to be distal, shallow-marine equivalents of the first clearly southwesterly derived sediments related to early Columbian orogenic events, but they do not extend far into the Canadian parts of Williston Basin. These beds are presumably equivalent to the upper third of the Roseray Formation and the Masefield Formation in southwestern Saskatchewan (Hayes, 1983).

The Masefield shale is overlain in southwestern Saskatchewan by the thin, kaolinitic, and quartzose sandstone that was called the Success Formation (basal Mannville Group) by Christopher (1974). The contact appears to be gradational in many places (Hayes, 1983), but Christopher (1974, 1984b) considered it to be in general disconformable and to be characterized by channel relief. Palynomorphs confirm the Late Jurassic age suggested by Christopher (1974) for the lower ('S1') unit of the Success, and 'glauconite' indicates its marine origin. The precise age and depositional environments of immediately overlying sandstones (upper Success) remain undetermined.

The lower Success sandstone appears to occur as locally preserved erosional remnants of what was a more widespread thin sandstone, the last marine deposits of what is called Williston Basin. In adjacent U.S.A., the former basin became the site of deposition of the nonmarine Morrison Formation. The Morrison was presumably partly equivalent to, and partly a source of sediment for, the basal Success of southwestern Saskatchewan and eastern shelf/shoreline facies of the Rocky Mountain Trough. The entire region seems to have experienced extensive post-Kimmeridgian, pre-Cretaceous erosion, followed by Early Cretaceous deposition of sandstones of the remainder of the Mannville Group. The significance of the unconformity within the lower part of sandstones that are traditionally assigned to the Mannville led Hayes (1983) to distinguish the lower Success Unit (S1) and to include it in the underlying Vanguard Group.

In southeastern Saskatchewan, an oil-producing sandstone at the base of the Mannville Group, called the Wapella Sand (Kreis, 1985), may also be of Jurassic age, and correspond to the Success Formation of southwestern Saskatchewan. However, it is entirely undated, and the nature of the Jurassic-Cretaceous boundary in that area is not understood.

Basin history

Regressive basin-fill packages characterized by shale and topped by shallow-marine limestone or sandstone are indicated in the lower Gravelbourg and Reston formations (perhaps Early Bajocian in age), the upper Gravelbourg, lower Shaunavon, and Sawtooth formations (mid-Bajocian through Early Bathonian approximately), the upper Shaunavon (Middle Bathonian approximately), the Rierdon, or Rush Lake to Roseray formations (Late Bathonian through Callovian or Early Oxfordian), and the Masefield or Swift to basal Success formations (Oxfordian and younger). Erosion associated with regressive events preceded deposition of the upper Gravelbourg, Reston, and Rierdon (Sundance equivalent) formations in the east; and of the upper Shaunavon, Sawtooth, Rush Lake, and Rierdon, and the Masefield and Swift formations in the west.

Progressively more extensive transgression on the northeast side of Williston Basin is indicated at perhaps Early Bajocian and about Middle Bajocian times by overlap of underlying units by the Reston and Melita formations respectively, and at about Late Bathonian time by overlap of Rierdon - equivalent shales over underlying sandstones. On the northwest side of the basin, transgressive events mark the base of the Gravelbourg, Sawtooth, upper Shaunavon, Rierdon, and Swift formations. Some vertical movement on faults may have occurred about Oxfordian time, modifying the basin configuration in southwestern Saskatchewan. During or following Kimmeridgian time, virtually all of the area formerly occupied by Williston Basin became emergent.

HISTORICAL SUMMARY

The Jurassic of the western interior of Canada reflects the change in depositional-tectonic regime from a relatively quiet, perhaps "passive" continental margin ("one-sided basin") to one, involved in early phases of Columbian orogeny, that was "two-sided" and highly active. This history is now seen to be more complex and difficult to interpret than has been generally realized. First, according to current models, the styles of subsidence and sedimentation, resembling those of a passive (divergent) margin, are too young by approximately 200 million years to be directly inherited from the early Paleozoic, rifted margin. Second, the sedimentation style of the Lower and Middle Jurassic beds does not appear to reflect the convergent tectonism that was occurring to the west, as a complex of allochthonous terranes approached and docked against North America. The latter events are recorded by the radiometric ages of plutonic and metamorphic rocks of the central Cordillera rather than by sedimentary rocks. A puzzling lag of some 30 million years appears to be recorded between suturing of the new terranes to the old continent on the one hand, and the creation of a foredeep and its filling with detritus eroded from tectonic lands to the west, on the other. These problems of the timing of Jurassic events are among the major problems of Cordilleran geology.

The Jurassic record of autochthonous northwestern North America, as of the other continents, is one of progressive, if intermittent, flooding of the cratonic margins (e.g., Hallam, 1981; Vail and Todd, 1981; Fig. 4H.21). The same is true for southern areas of Alberta Basin until about Kimmeridgian time, when sediments derived from the uplifted Columbian Orogen to the west and southwest filled the depositional basins, leading to widespread Late Jurassic regression. The depositional record for Jurassic time in autochthonous western Canada is summarized graphically in Figures 4H.22 to 4H.32.

The general transgressive trend, characteristic of the earlier half of the Zuni Sequence, was accomplished by at least eight major transgressive events. They are recognized either by extension of marine deposits beyond the limits of older ones, or by upward change from coarse to finer clastic rocks that are inferred to indicate such transgressions. These events occurred about earliest Hettangian, earliest Sinemurian, latest Sinemurian, late Pliensbachian, early Toarcian, late Early Bajocian, Late Bathonian, and earliest Oxfordian times (Fig. 4H.21). More are recognized in Williston Basin, presumably because of the more complete succession and the shallower environments of deposition that dominate the preserved record there.

Most of the Jurassic sedimentary packages throughout the area consist of clastic rocks, some in coarsening upward cycles, which are essentially regressive in character. At least four major regressive episodes culminated in erosion even in basinward areas. They are indicated by sediments representing early Pliensbachian, latest Aalenian to earliest Bajocian, and late Callovian times (Poulton, 1988). The latest Triassic – earliest Jurassic sequence indicates a major regression at that time also.

The record of Jurassic transgressions and regressions is generally similar throughout autochthonous western Canada, given the uncertainties of interpretation of the sedimentary record and of correlation. A few of these events correspond with some of the 'worldwide eustatic' events postulated by Hallam (1978) and Vail and Todd (1981) (shown by asterisks in Figure 4H.21), but pronounced discrepancies reflect the tectonic events of the western Canada basins. Poulton (1988) recognized four major transgressive-regressive events that can be correlated between Western Canada, Williston, Arctic, and east coast, offshore basins. They are approximately earliest Jurassic, latest Pliensbachian-early Toarcian, late Aalenian-early Bajocian, and early Oxfordian in age.

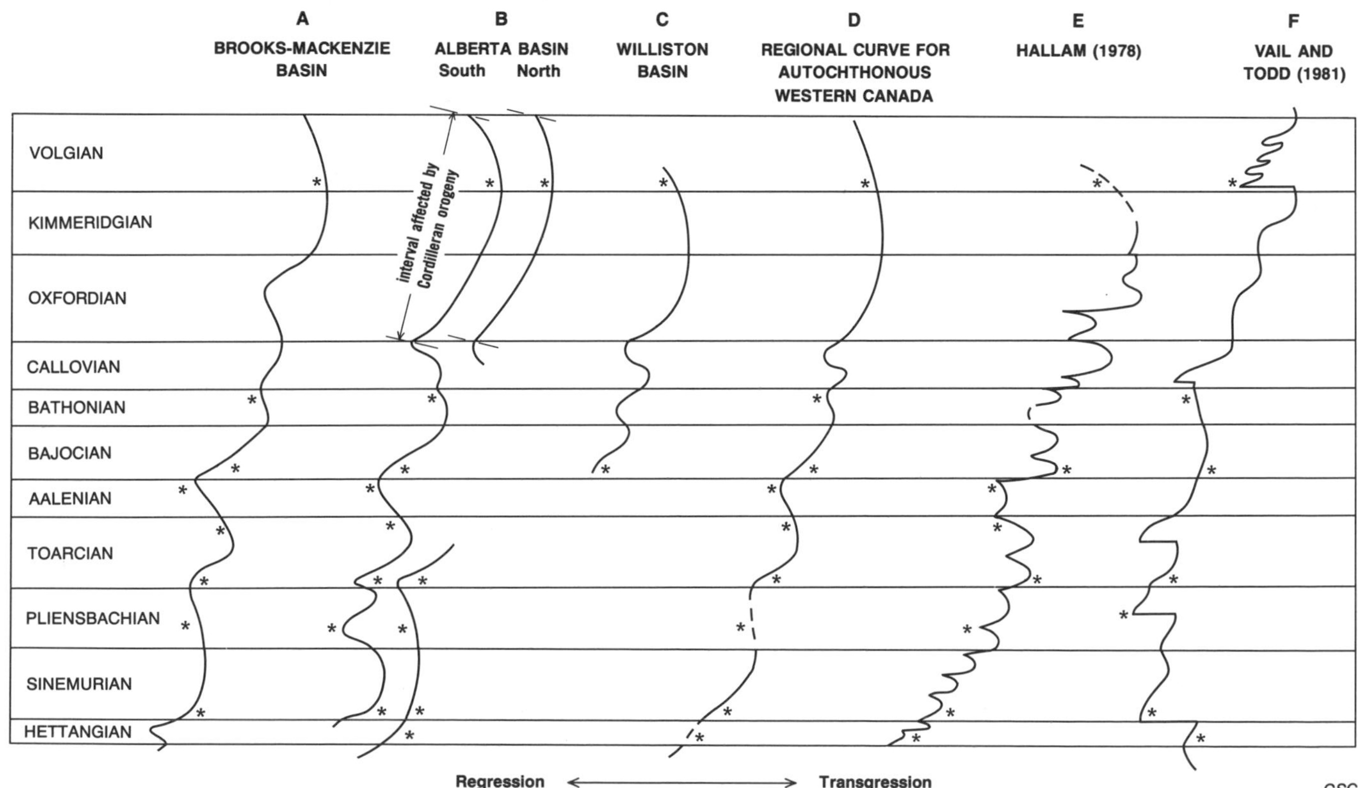

Figure 4H.21. Relative sea-level fluctuation curves for (A) Brooks-Mackenzie Basin, (B) Alberta Basin, and (C) Williston Basin. They are summarized (D) and compared with the 'worldwide eustatic' curves of (E) Hallam (1978) and (F) Vail and Todd (1981). Points of correspondence with the 'eustatic' curves of Hallam and Vail and Todd are indicated by asterisks on the left of the lines (regression) or on the right (transgression). Similar curves were published by Poulton (1988).

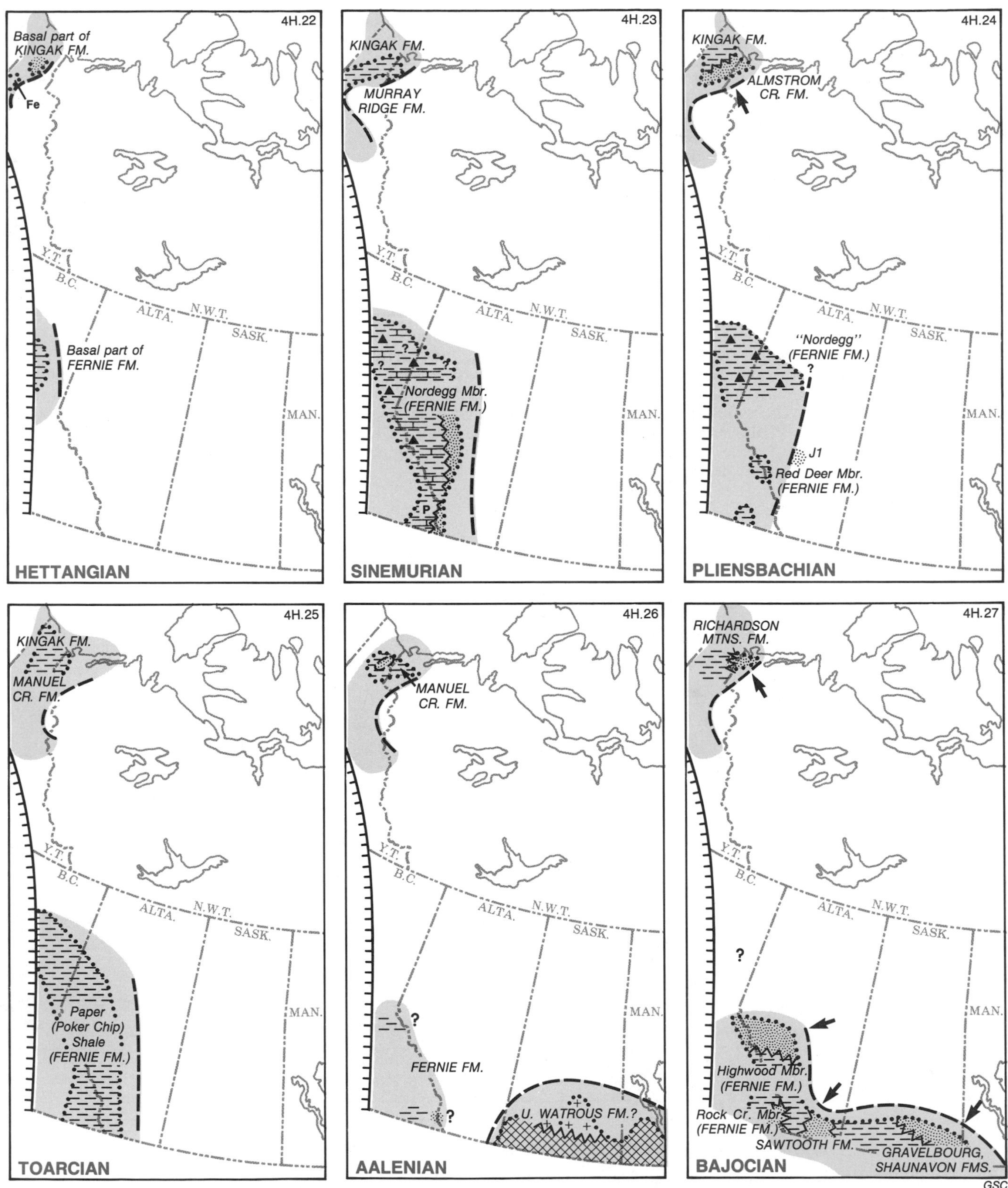

Figures 4H.22-32. Generalized facies distribution maps for selected intervals of Jurassic time. Colour indicates area of marine deposition. Legend for all figures follows Figure 4H.32.

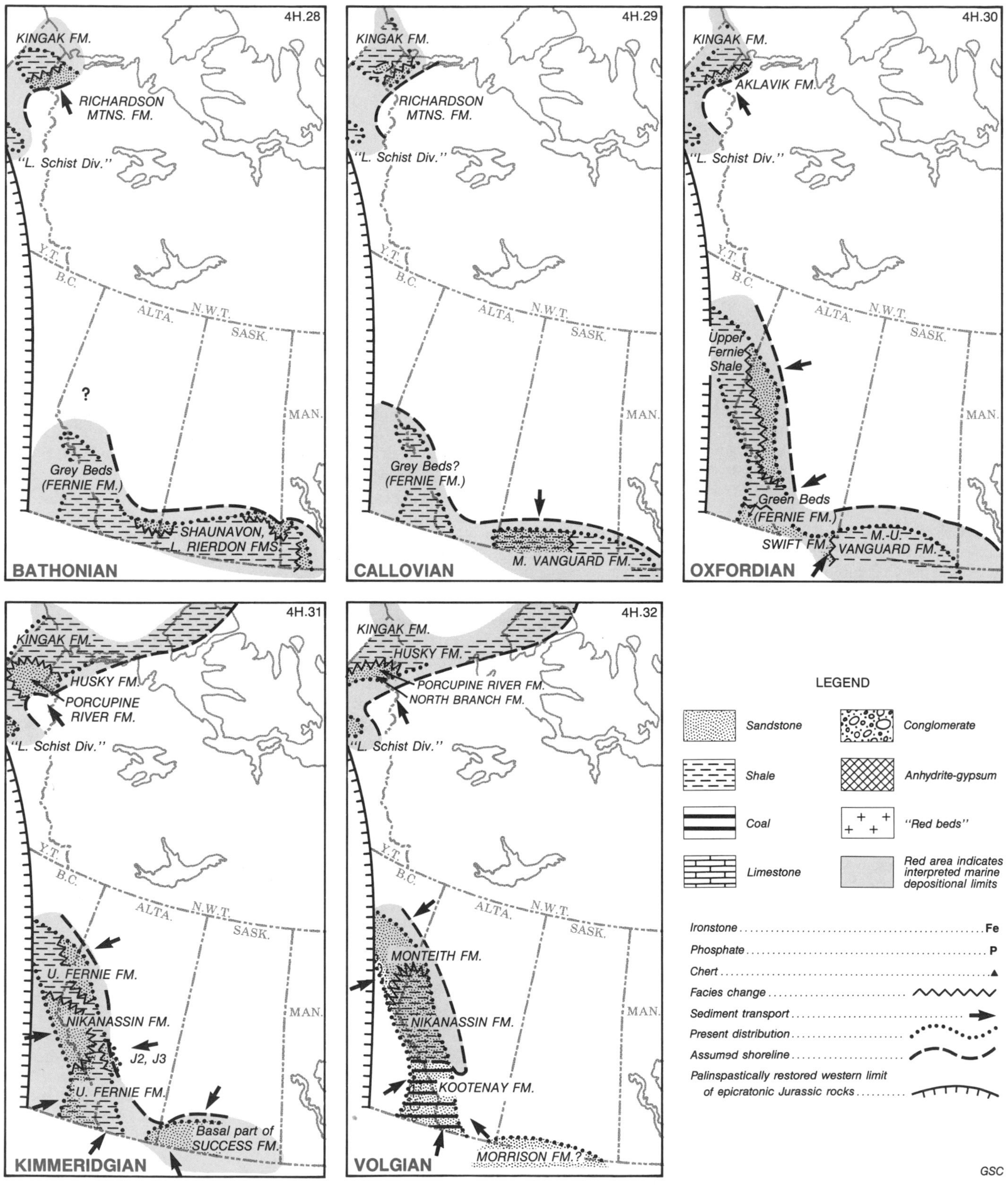

Figures 4H.22-32. Cont.

Regional considerations

Differences in the preserved record from one area to another can be explained to a large extent by differences in the quantity of detritus available for deposition, or by differences in the rate of subsidence of the basins. Major events that are more localized include the renewed Middle Jurassic subsidence of the earlier established Williston Basin, the deposition of thick Pliensbachian and Middle Jurassic sandstones in Brooks-Mackenzie Basin, and the deposition of major craton-derived Upper Jurassic sandstone units in Brooks-Mackenzie Basin and that part of Rocky Mountain Trough that underlies northeastern British Columbia.

A generally greater proportion of terrigenous clastic material as compared to chemical sediments characterizes Brooks-Mackenzie Basin, presumably reflecting both the greater influx of clastic detritus from the northern interior area of Canada, and the higher paleolatitudes of deposition.

The sandstone wedges of Brooks-Mackenzie Basin were presumably derived ultimately from uplifted regions of the craton to the southeast. That part of the craton in the general vicinity of northwestern Alberta and southeastern Yukon Territory may have yielded significant volumes of Upper Jurassic sediment to the south and southwest as well.

Other regional differences are clearly related to local tectonic influences. The most important is the Late Jurassic uplift of the Columbian Orogen, associated with subsidence of the adjacent foredeep, and deposition of orogen-derived sediments in it. The transgressive aspect of the Upper Jurassic record of Brooks-Mackenzie Basin was essentially eustatically controlled; however, the renewed influx of sandy detritus may be a response to Cordilleran tectonic events. Increased subsidence and sedimentation rates in Williston Basin also seem to coincide with the early Columbian tectonic events. Deposition in Williston Basin of micaceous clastics derived from southwestern sources in northwestern United States (Swift Formation) preceded the otherwise similar events represented by the Passage Beds and Kootenay Group elsewhere in Alberta. A progressively northerly migration of orogenic source areas of the early phases of the Columbian Orogen may be indicated by the general northward younging of the micaceous clastic wedges derived from them (Poulton, 1984). An interesting tectonic model relating this north-south variation in timing of sedimentation in the foredeep to large-scale dextral offsets of uplifted source areas along strike-slip faults on the western side of the Omineca Belt of the orogen was proposed by Eisbacher (1981).

Other local tectonic activity that affected the depositional basins includes the minor Middle and Late Jurassic uplifts and depressions on the unstable margins of Brooks-Mackenzie Basin, the greater cratonward extension of Lower Jurassic marine rocks in the northern part of Alberta Basin than in its southern part, and the Middle or early Late Jurassic uplift of the Peace River area of Alberta Basin in northeastern British Columbia. The local preservation of Upper Triassic and lowest Jurassic rocks in northeastern British Columbia indicates either the presence of a local depocentre there, or the presence of a structural depression in which they were preserved from sub-Sinemurian erosion.

ECONOMIC GEOLOGY

Oil and gas

Jurassic rocks of Alberta, southern Saskatchewan, and Manitoba are significant host rocks for oil and gas deposits. Jurassic reservoirs account for about 4% of western Canada's total initial reserves of oil and 1.4% of its marketable gas (Wallace-Dudley, 1983a, b).

Alberta Basin

Oil and gas occur in interfluvial erosional remnants of a thin dolomitic sandstone-limestone unit, correlated with the Lower Jurassic Nordegg Member of the Fernie Formation, in the Gilby-Medicine River-Sylvan Lake fields of central Alberta. The hydrocarbons are trapped below the impermeable 'Poker Chip Shale'. Oil occurs also below and between probable Middle and Upper Jurassic channel-fill kaolinitic sandstones, as well as below the Lower Cretaceous, basal Mannville, Ellerslie unit (see Hopkins, 1981). Some of the Jurassic channel sandstones are oil-producing also. Natural gas is produced also in the Lower Jurassic unit in the Paddle River-Whitecourt-Goodwin areas northwest of Edmonton.

Hydrocarbons have been discovered in sandstones, probably equivalent to the Bajocian Rock Creek Member of the Fernie Formation, in the area of Rocky Mountain House just west of Red Deer (Marion, 1984) and also west of Edmonton (J. Barclay in Podruski et al., 1986).

Some of the gas in the 'Nikanassin Formation' of northeastern British Columbia may be hosted in Jurassic parts of the formation. Most, if not all, is in Cretaceous rocks, however.

An upper sandstone unit of the thin Middle Jurassic Sawtooth Formation of southern Alberta produces oil in the Conrad and Grand Forks fields (see G. Davies, 1983). Some oil is also produced from lower sandstones. Uppermost Sawtooth impermeable shales, or overlying Rierdon shales, formed barriers to hydrocarbon migration updip toward the crest of Sweetgrass Arch.

Some of the oil in southeastern Alberta, mainly hosted in the Lower Cretaceous Mannville Group, is present in the Upper Jurassic 'Ribbon Sand unit' of the Swift Formation in the Flat Coulee and Manyberries fields (see Hayes, 1983). It is trapped in discontinuous small sandstone bodies by impermeable shales and other fine-grained nonmarine sediments overlying it and adjacent to it.

Gas also occurs in several fields in southeastern Alberta, where the Sawtooth Formation in the Black Butte field is the biggest producer.

The Toarcian 'Poker Chip Shale' and the approximately Kimmeridgian 'Upper Fernie Shale' or lower Passage beds of the Fernie Formation are possibly significant hydrocarbon source beds in western Canada, but no confirmative geochemical information is available.

Williston Basin

The main productive district comprises a belt of 21 oil pools in southwestern Saskatchewan from Battrum in the north to Rapdan in the south. The reservoirs are in the Middle and Upper Jurassic or Lower Cretaceous units on the northwest side of Williston Basin. Early Cretaceous hydrocarbon migration from Paleozoic source limestones upward through local fracture systems on the west side of Shaunavon syncline has been postulated (Christopher, 1984a, b). Thin intervals of the upper Watrous and Gravelbourg formations are oil sources (Type II marine). They are probably too thin to be of great significance, and anomalies regarding their thermal maturity are not understood (Osadetz and Snowdon, 1986).

Shoreline and nearshore sandstones and coquinas of the Middle Jurassic upper Shaunavon Formation contain oil in small fields in the southern half of the oil belt in southwestern Saskatchewan. In the Leitchville-Shaunavon area, the lower Shaunavon limestone also produces oil. The pools are perhaps not so much localized by their position on the irregular western margin of Williston Basin as by a zone of flexure and fracturing on the west side of Shaunavon Syncline, where small fault blocks were active prior to, during, and following Shaunavon deposition. Post-Shaunavon, approximately Late Jurassic uplift of Williston Basin, forming Swift Current Platform in the west, may have played a role in the fracture and migration patterns. Hydrocarbons were trapped at stratigraphic pinch-out edges of discrete sandstone units encompassed by interbedded impermeable Shaunavon shales and carbonates and below the overlying Rierdon shales. The stratigraphy, structural geology, and petroleum geology of these fields are complex, each having its own unique combination of characteristic features (see Christopher, 1964, 1966b, 1984a; Kent, 1974).

The Middle or Upper Jurassic Roseray and Upper Jurassic or Lower Cretaceous Success sandstones contain oil in small fields in the northern half of the oil belt of southwestern Saskatchewan. Some of the reservoirs are formed by the western up-dip pinch-out of porous facies of the Roseray Formation against the eastern flank of Sweetgrass Arch. Others are erosional mesas, buttes, and interfluvial highs preserved below and between valley-fill sediments of the overlying Lower Cretaceous Cantuar Formation, which formed permeability barriers to hydrocarbon migration (Christopher, 1974, 1975, 1984a, b).

Oil is present at several levels in the Jurassic beds of southeastern Saskatchewan, in Wapella, Moosomin, and Red Jacket fields. Deposition of the Wapella Sand, possibly of Jurassic age, at the base of the Mannville Group, may be related to a local structural arch (Kreis, 1985).

The sandstone-siltstone redbed facies of the Triassic or Lower Jurassic, lower Amaranth Formation contains oil in the Waskada-Pierson area of southwestern Manitoba. The oil occurs in systems of interconnected permeable zones in laterally and vertically varied basal and higher sandstone units. Late Cretaceous hydrocarbon migration upward and updip from Lower Carboniferous strata was blocked by impermeable beds within the lower Amaranth itself, related to updip facies changes (Barchyn, 1982, 1984).

Coal

The Upper Jurassic and Lower Cretaceous Mist Mountain Formation of the Kootenay Group is a major source of low- to medium-volatile bituminous coal in western Canada. Currently six mines are operating, all in the Fernie and Crowsnest Pass areas of southeastern British Columbia, and a large number of other unexploited prospects and abandoned mines exist in that area and in southwestern Alberta. Total inferred resources for this formation comprise 5635 megatonnes, 13% of Canada's inferred coal resources (Smith, 1989).

Phosphate

Phosphatic strata, currently subeconomic, have long been known in the Lower Jurassic Nordegg Member and its equivalents, and in the Middle Jurassic lower Highwood Member of the Fernie Formation of western Alberta and eastern British Columbia (Telfer, 1933; Christie, 1979a). Minor amounts of phosphate are also present in the equivalent Middle Jurassic, basal Richardson Mountains Formation ('*Arkelloceras* beds') in Brooks-Mackenzie Basin (Poulton et al., 1982).

Gypsum

Gypsum is mined at Amaranth, Gypsumville, and until recently at Silver Plains, in southern Manitoba, where the widespread Lower or lower Middle Jurassic upper Amaranth anhydrite lies near the surface and has been converted to gypsum by groundwater.

REFERENCES

Archibald, D.A., Glover, J.K., Price, R.A., Farrar, E., and Carmichael, D.M.

1983: Geochronology and tectonic implications of magmatism and metamorphism, southern Kootenay Arc and neighbouring regions, southeastern British Columbia, Part I: Jurassic to mid-Cretaceous; Canadian Journal of Earth Sciences, v. 20, p. 1891-1913.

Balkwill, H.R., Cook, D.G., Detterman, R.L., Embry, A.F., Håkansson, E., Miall, A.D., Poulton, T.P., and Young, F.G.

1983: Arctic North America and northern Greenland; in Phanerozoic of the World II, Mesozoic, A, A.E.M. Nairn and A. Moullade (ed.); Elsevier Scientific Publication Co., p. 1-31.

Balkwill, H.R., Wilson, D.G., and Wall, J.H.

1977: Ringnes Formation (Upper Jurassic) Sverdrup Basin, Canadian Arctic Archipelago; Bulletin of Canadian Petroleum Geology, v. 25, p. 1115-1144.

Ballantyne, B. and McCabe, H.

1984: Manitoba crater revealed; GEOS, v. 13, p. 10-13.

Barchyn, D.

1982: Geology and hydrocarbon potential of the lower Amaranth Formation, Waskada-Pierson area, southwestern Manitoba; Mineral Resources Division, Manitoba Department of Energy and Mines, Geological Report GR 82-6, 30 p.

1984: The Waskada lower Amaranth (Spearfish) oil pool, southwestern Manitoba: a model for Spearfish exploration in Saskatchewan; in Oil and Gas in Saskatchewan, J.A. Lorsang and M.A. Wilson (ed.); Saskatchewan Geological Society, Special Publication No. 7, p. 103-112.

Beddoe-Stephens, B. and Lambert, R. St. J.

1981: Geochemical, mineralogical, and isotopic data relating to the origin and tectonic setting of the Rossland volcanic rocks, southern British Columbia; Canadian Journal of Earth Sciences, v. 18, p. 858-868.

Bell, W.A.

1944: Use of some fossil floras in Canadian stratigraphy; Royal Society of Canada, Transactions, 3rd Series, v. 38, Sec. 4, p. 1-13.

1946: Age of the Canadian Kootenay formation; American Journal of Science, v. 244, p. 513-526.

1956: Lower Cretaceous floras of western Canada; Geological Survey of Canada, Memoir 285, 331 p.

Berry, E.W.

1929: The Kootenay and lower Blairmore floras; National Museum of Canada, Bulletin 58, Geological Series No. 50, p. 28-54, pl. IV-VIII.

Bond, G.C. and Kominz, M.A.
1984: Construction of tectonic subsidence curves for the early Paleozoic miogeocline, southern Canadian Rocky Mountains: Implications for subsidence mechanisms, age of breakup, and crystal thinning; Geological Society of America, Bulletin, v. 95, p. 155-173.

Bovell, G.R.L.
1979: Sedimentation and diagenesis of the Nordegg Member in central Alberta; M. Sc. thesis, Queen's University, Kingston Ontario, 176 p.

Brabb, E.E.
1969: Six new Paleozoic and Mesozoic formations in east-central Alaska; United States Geological Survey, Bulletin 1247-I, p. I1-I26.

Braman, D.R.
1985: The sedimentology and stratigraphy of the Husky Formation in the subsurface, District of Mackenzie, Northwest Territories; Geological Survey of Canada, Paper 83-14, 24 p.

Brideaux, W.W.
1971: Jurassic to Cretaceous assemblages (palynomorphs); in Biostratigraphic Determinations of Fossils from the Subsurface of the Yukon Territory and the District of Mackenzie, B.S. Norford, M.S. Barss, W.W. Brideaux, T.P. Chamney, F.H. Fritz, W.S. Hopkins, J.A. Jeletzky, A.E.H. Pedder, and T.T. Uyeno (ed.); Geological Survey of Canada, Paper 71-15, p. 5-7.
1976: Jurassic, Cretaceous and Tertiary assemblages (palynomorphs); in Biostratigraphic Determinations from the Subsurface of the Districts of Franklin and Mackenzie and the Yukon Territory, W.W. Brideaux, D.R. Clowser, M.J. Copeland, J.A. Jeletzky, B.S. Norford, A.W. Norris, A.E.H. Pedder, A.R. Sweet, R. Thorsteinsson, T.T. Uyeno and J. Wall (ed.); Geological Survey of Canada, Paper 75-10, p. 2-4.
1977a: Taxonomy of Upper Jurassic-Lower Cretaceous microplankton from the Richardson Mountains, District of Mackenzie, Canada; Geological Survey of Canada, Bulletin 281, 89 p.
1977b: Occurrence charts for microplankton species and range charts of selected taxa in the Banff-Aquitaine-Arco Rat Pass K-35 well and the I.O.E. Stoney Core Hole F-42; Geological Survey of Canada, Open File 403.

Brideaux, W.W. and Fisher, M.J.
1976: Upper Jurassic-Lower Cretaceous dinoflagellate assemblages from Arctic Canada; Geological Survey of Canada, Bulletin 259, 53 p.

Brideaux, W.W. and Myhr, D.W.
1976: Lithostratigraphy and dinoflagellate cyst succession in the Gulf Mobil Parsons N-10 well; in Report of Activities, Part B, Geological Survey of Canada, Paper 76-1B, p. 235-250.

Brooke, M.M. and Braun, W.K.
1972: Biostratigraphy and microfauna of the Jurassic System of Saskatchewan; Saskatchewan Department of Mineral Resources, Report No. 161, 83 p.
1981: Jurassic microfaunas and biostratigraphy of northeastern British Columbia and adjacent Alberta; Geological Survey of Canada, Bulletin 283, 69 p.

Brown, R.W.
1946: Fossil plants and Jurassic-Cretaceous boundary in Montana and Alberta; American Association of Petroleum Geologists, Bulletin, v. 30, p. 238-248.

Buckman, S.S.
1929: Mesozoic paleontology of Blairmore region, Alberta, Jurassic Ammonoidea; National Museum of Canada, Bulletin 58, Geological Series 50, p. 1-27, pl. I-III.

Callomon, J.H.
1984: A review of the biostratigraphy of the post-Lower Bajocian Jurassic ammonites of western and northern North America; in Jurassic-Cretaceous Biochronology and Paleogeography of North America, G.E.G. Westermann (ed.); Geological Association of Canada, Special Paper 27, p. 143-174.

Campbell, R.H.
1967: Areal geology in the vicinity of the Chariot Site, Lisburne Peninsula, northwestern Alaska; United States Geological Survey, Professional Paper 395, 71 p.

Carlson, C.E.
1968: Triassic-Jurassic of Alberta, Saskatchewan, Manitoba, Montana, and North Dakota; American Association of Petroleum Geologists, Bulletin, v. 52, p. 1969-1983.

Christie, R.L.
1979a: Phosphorite in sedimentary basins of western Canada; in Current Research, Part B, Geological Survey of Canada, Paper 79-1B, p. 253-258.

Christopher, J.E.
1964: The Middle Jurassic Shaunavon Formation of southwestern Saskatchewan; Saskatchewan Department of Mineral Resources, Report No. 95, 95 p.
1966a: The Middle Jurassic Shaunavon Formation of southwestern Saskatchewan. Lithologic descriptions of selected cored sections; Saskatchewan Department of Mineral Resources, Report No. 110, 308 p.

Christopher, J.E. (cont.)
1966b: Shaunavon (Middle Jurassic) sedimentation and vertical tectonics, southwestern Saskatchewan; Billings Geological Society, Guidebook, 17th Annual Field Conference, p. 18-35.
1974: The Upper Jurassic Vanguard and Lower Cretaceous Mannville groups of southwestern Saskatchewan; Saskatchewan Department of Mineral Resources, Report No. 151, 349 p.
1984a: Depositional patterns and oil field trends in the lower Mesozoic of the northern Williston Basin, Canada; in Oil and Gas in Saskatchewan, J.A. Lorsang and M.A. Wilson (ed.); Saskatchewan Geological Society, Special Publication No. 7, p. 83-102.
1984b: Stratigraphy and tectonic setting of the Lower Cretaceous Mannville Group: northern Williston Basin region, Canada; in The Mesozoic of Middle North America, D.F. Stott and D. Glass (ed.); Canadian Society of Canadian Petroleum Geologists, Memoir 9, p. 109-126.

Crockford, M.B.B.
1949: Geology of the Ribbon Creek area, Alberta; Research Council of Alberta, Report 52, p. 1-67.

Davies, E.H. and Norris, G.
1980: Latitudinal variations in encystment modes and species diversity in Jurassic dinoflagellates; in The Continental Crust and its Mineral Deposits, D.W. Strangway (ed.); Geological Association of Canada, Special Paper 20, p. 361-373.

Davies, E.H. and Poulton, T.P.
1986: Late Jurassic dinoflagellate cysts and strata, northeastern British Columbia; in Current Research, Part B, Geological Survey of Canada, Paper 86-1B, p. 519-537.

Davies, G.R.
1983: Sedimentology of the Middle Jurassic Sawtooth Formation of southern Alberta; in Sedimentology of Selected Mesozoic Clastic Sequences, J.R. McLean and G.E. Reinson (ed.); The Mesozoic of Middle North America Conference, Corexpo 83, Canadian Society of Petroleum Geologists, p. 11-25.

Detterman, R.L.
1970: Sedimentary history of Sadlerochit and Shublik Formations in northeastern Alaska; in Proceedings of the Geological Seminar on the North Slope of Alaska, Pacific Section; American Association of Petroleum Geologists, p. 0-1 to 0-13.
1973: Mesozoic sequence in Arctic Alaska; in Proceedings of the 2nd International Arctic Symposium, M.G. Pitcher (ed.); American Association of Petroleum Geologists, Memoir 19, p. 376-387.

Detterman, R.L., Reiser, H.N., Brosgé, W.P., and Dutro, J.T., Jr.
1975: Post-Carboniferous stratigraphy, northeastern Alaska; United States Geological Survey, Professional Paper 886, 46 p.

Dixon, J.
1982a: Jurassic and Lower Cretaceous subsurface stratigraphy of the Mackenzie Delta - Tuktoyaktuk Peninsula, Northwest Territories; Geological Survey of Canada, Bulletin 349, 52 p.
1982b: Upper Oxfordian to Albian geology, Mackenzie Delta, Arctic Canada; in Arctic Geology and Geophysics, A.F. Embry and H.R. Balkwill (ed.); Canadian Society of Petroleum Geologists, Memoir 8, p. 29-42.

Dow, W.G.
1967: The Spearfish Formation in the Williston Basin of western North Dakota; North Dakota Geological Survey, Bulletin 52, 28 p.

Eisbacher, G.H.
1981: Late Mesozoic-Paleogene Bowser Basin molasse and Cordilleran tectonics, western Canada; in Sedimentation and Tectonics in Alluvial Basins, A.D. Miall (ed.); Geological Association of Canada, Special Paper 23, p. 125-151.

Fensome, R.A.
1987: Taxonomy and biostratigraphy of schizaealean spores from the Jurassic-Cretaceous boundary beds of the Aklavik Range, District of Mackenzie; Palaeontolographica Canadiana, no. 4, 49 p.

Francis, D.R.
1956: Jurassic stratigraphy of the Williston Basin area; Saskatchewan Department of Mineral Resources, Report No. 18, 69 p.
1957: Jurassic stratigraphy of Williston Basin area; Bulletin of the American Association of Petroleum Geologists, v. 41, p. 367-398.

Frebold, H.
1951: Contributions to the paleontology and stratigraphy of the Jurassic System in Canada; Geological Survey of Canada, Bulletin 18, p. 1-54.
1953: Correlation of the Jurassic formations of Canada; Geological Society of America Bulletin, v. 64, p. 1229-1246.
1954: Stratigraphic and paleogeographic studies in the Jurassic Fernie Group; Alberta Society of Petroleum Geologists, News Bulletin, v. 2, p. 1-2.
1957: The Jurassic Fernie Group in the Canadian Rocky Mountains and Foothills; Geological Survey of Canada, Memoir 287, 197 p.

Frebold, H. (cont.)
1958: Stratigraphy and correlation of the Jurassic in the Canadian Rocky Mountains and Alberta Foothills; in Jurassic and Carboniferous of Western Canada, A.J. Goodman (ed.); American Association of Petroleum Geologists, John Andrew Allan Memorial Volume, p. 10-26.
1960: The Jurassic faunas of the Canadian Arctic: Lower Jurassic and lowermost Middle Jurassic ammonites; Geological Survey of Canada, Bulletin 59, 33 p.
1961: The Jurassic faunas of the Canadian Arctic: Middle and Upper Jurassic ammonites; Geological Survey of Canada, Bulletin 74, 43 p.
1963: Ammonite faunas of the upper Middle Jurassic beds of the Fernie Group in western Canada; Geological Survey of Canada, Bulletin 93, 33 p.
1964a: Illustrations of Canadian fossils, Jurassic of western and Arctic Canada; Geological Survey of Canada, Paper 63-4, 107 p.
1964b: The Jurassic faunas of the Canadian Arctic; Cadoceratinae; Geological Survey of Canada, Bulletin 119, 27 p.
1966: Upper Pliensbachian beds in the Fernie Group of Alberta; Geological Survey of Canada, Paper 66-27, 9 p.
1969: Subdivision and facies of Lower Jurassic rocks in the southern Canadian Rocky Mountains and Foothills; Proceedings of the Geological Association of Canada, v. 20, p. 76-89.
1970a: Marine Jurassic faunas; in Geological and Economic Minerals of Canada, R.J.W. Douglas (ed.); Geological Survey of Canada, Economic Geology Report No. 1, p. 641-648.
1970b: Pliensbachian ammonoids from British Columbia and southern Yukon; Canadian Journal of Earth Sciences, v. 7, p. 435-456.
1975: The Jurassic faunas of the Canadian Arctic, Lower Jurassic ammonites, biostratigraphy and correlations; Geological Survey of Canada, Bulletin 243, 24 p.
1976: The Toarcian and lower Middle Bajocian beds and ammonites in the Fernie Group of southeastern British Columbia; Geological Survey of Canada, Paper 75-39, 33 p.

Frebold, H. and Poulton, T.P.
1977: Hettangian (Lower Jurassic) rocks and faunas, northern Yukon Territory; Canadian Journal of Earth Sciences, v. 14, p. 89-101.

Frebold, H., Mountjoy, E.W., and Reed, R.
1959: The Oxfordian beds of the Jurassic Fernie Group, Alberta and British Columbia; Geological Survey of Canada, Bulletin 53, p. 1-47.

Frebold, H., Mountjoy, E.W., and Tempelman-Kluit, D.
1967: New occurrences of Jurassic rocks and fossils in central and northern Yukon; Geological Survey of Canada, Paper 67-12, 35 p.

Frebold, H. and Tipper, H.W.
1970: Status of the Jurassic in the Canadian Cordillera of British Columbia, Alberta and Southern Yukon; Canadian Journal of Earth Sciences, v. 7. p. 1-21.

Gabrielse, H. and Yorath, C.J. (ed.)
1991: Geology of the Cordillera Orogen in Canada; Geological Survey of Canada, Geology of Canada no. 4. (Also Geological Society of America, The Geology of North America, vol. G-2).

Gibson, D.W.
1977: Sedimentary facies in the Jura-Cretaceous Kootenay Formation, Crowsnest Pass area, southwestern Alberta and southeastern British Columbia; Bulletin of Canadian Petroleum Geologists, v. 25, p. 767-791.
1978: The Kootenay-Nikanassin lithostratigraphic transition, Rocky Mountain Foothills of west-central Alberta; in Current Research, Part A, Geological Survey of Canada, Paper 78-1A, p. 379-381.
1979: The Morrissey and Mist Mountain formations - newly defined lithostratigraphic units of the Jura-Cretaceous Kootenay Group, Alberta and British Columbia; Bulletin of Canadian Petroleum Geology, v. 27, p. 183-208.
1985: Stratigraphy, sedimentology and depositional environments of the coal-bearing Jurassic-Cretaceous Kootenay Group, Alberta and British Columbia; Geological Survey of Canada, Bulletin 357, 108 p.

Gibson, D.W., Hughes, J.D., and Norris, D.K.
1983: Stratigraphy, sedimentary environments and structure of the coal-bearing Jura-Cretaceous Kootenay Group, Crowsnest Pass area, Alberta and British Columbia; Canadian Society of Petroleum Geologists, Conference - The Mesozoic of Middle North America, Field Trip Guidebook No. 4, 58 p.

Grantz, A., Eittrem, S., and Dinter, D.A.
1979: Geology and tectonic development of the continental margin north of Alaska; Tectonophysics, v. 59, p. 263-291.

Green, L.H.
1972: Geology of Nash Creek, Larsen Creek and Dawson map areas, Yukon Territory; Geological Survey of Canada, Memoir 364, 157 p.

Green, L.H. and Roddick, J.A.
1962: Dawson, Larsen Creek and Nash Creek map areas, Yukon Territory; Geological Survey of Canada, Paper 62-7, 20 p.

Grieve, D.A.
1984: Tonsteins: possible stratigraphic correlation aids in East Kootenay coalfields (82G/15, 82J/2); British Columbia Ministry of Energy, Mines and Petroleum Resources, Geological Fieldwork, 1983, Paper 1984-1, p. 36-41.

Guliov, P.
1967: Jurassic microfossils from a locality in southeastern Saskatchewan; Saskatchewan Department of Mineral Resources, Report No. 105, Part II, p. 41-45.

Gussow, W.C.
1960: Jurassic-Cretaceous boundary in western Canada and late Jurassic age of the Kootenay; Transactions of the Royal Society of Canada, v. LIV, Ser. III, p. 45-64.

Hadley, H.D. and Milner, R.L.
1953: Stratigraphy of Lower Cretaceous and Jurassic, northern Montana - southwestern Saskatchewan; in Little Rocky Mountains - Montana, Southwestern Saskatchewan, J.M. Parker (ed.); Billings Geological Society, Guidebook, Fourth Annual Field Conference, p. 85-86.

Hall, R.L.
1984: Lithostratigraphy and biostratigraphy of the Fernie Formation (Jurassic) in the southern Canadian Rocky Mountains; in The Mesozoic of Middle North America, D.F. Stott and D.J. Glass (ed.); Canadian Society of Petroleum Geologists, Memoir 9, p. 233-247.

Hall, R.L. and Stronach, N.J.
1981: First record of Late Bajocian (Jurassic) ammonites in the Fernie Formation, Alberta; Canadian Journal of Earth Sciences, v. 18, p. 919-925.
1982: A guidebook to the Fernie Formation of southern Alberta and British Columbia, 1st Field Conference, Calgary, Circum-Pacific Jurassic Research Group; International Geological Correlation Program No. 171, 48 p.

Hall, R.L. and Westermann, G.E.G.
1980: Lower Bajocian (Jurassic) cephalopod faunas from western Canada and proposed assemblage zones for the Lower Bajocian of North America; Palaeontographica Americana, v. 9, no. 52, p. 1-87.

Hallam, A.
1978: Eustatic cycles in the Jurassic; Palaeogeography, Palaeoclimatology, Palaeoecology, v. 23, p. 1-32.
1981: A revised sea-level curve for the early Jurassic; Journal of Geological Society, London, v. 138, p. 735-743.

Hamblin, A.P. and Walker, R.G.
1979: Storm-dominated shallow marine deposits: the Fernie Kootenay (Jurassic) transition, southern Rocky Mountains; Canadian Journal of Earth Sciences, v. 16, p. 1673-1690.

Hamilton W.
1962: The Jurassic Fernie Group in northeastern British Columbia; Edmonton Geological Society, Fourth Annual Field Trip Guidebook, p. 46-56.

Hayes, B.J.R.
1983: Stratigraphy and petroleum potential of the Swift Formation (Upper Jurassic), southern Alberta and north-central Montana; Bulletin of Canadian Petroleum Geology, v. 31, p. 37-52.

Hedinger, A.S.
in press: Late Jurassic (Oxfordian-Portlandian) Foraminifera from the Aklavik Range, District of Mackenzie, Northwest Territories; Geological Survey of Canada, Bulletin.

Hopkins, J.C.
1981: Sedimentology of quartzose sandstones of Lower Mannville and associated units, Medicine River area, central Alberta; Bulletin of Canadian Petroleum Geology, v. 29, p. 12-41.

Hume, G.S.
1930: The Highwood-Jumpingpound Anticline, with notes on Turner Valley, New Black Diamond, and Priddis Valley structures, Alberta; Geological Survey of Canada, Report for 1929, pt. B, p. 1-24.

Imlay, R.W.
1965: Jurassic marine faunal differentiation in North America; Journal of Paleontology, v. 39, p. 1023-1038.
1980: Jurassic paleobiogeography of the coterminous United States in its continental setting; United States Geological Survey, Professional Paper 1062, 134 p.

Jansa, L.
1972: Depositional history of the coal-bearing Upper Jurassic-Lower Cretaceous Kootenay Formation, southern Rocky Mountains, Canada; Geological Society of America, Bulletin, v. 83, p. 3199-3222.

Jeletzky, J.A.
1965: Late Upper Jurassic and early Lower Cretaceous fossil zones of the Canadian Western Cordillera, British Columbia; Geological Survey of Canada, Bulletin 103, 70 p.
1966: Upper Volgian (Late Jurassic) Ammonites and Buchias of Arctic Canada; Geological Survey of Canada, Bulletin 128, 51 p.

Jeletzky, J.A. (cont.)
1967: Jurassic and (?) Triassic rocks of the eastern slope of Richardson Mountains, northwest District of Mackenzie, 106M and 107B (parts of); Geological Survey of Canada, Paper 66-50, 171 p.
1973: Biochronology of the marine boreal latest Jurassic, Berriasian and Valanginian in Canada; in The Boreal Lower Cretaceous, R. Casey and P.F. Rawson (eds.); Geological Journal, Special Issue No. 5, Seal House Press, Liverpool, p. 41-80.
1975: Jurassic and Lower Cretaceous paleogeography and depositional tectonics of Porcupine Plateau, adjacent areas of northern Yukon and those of Mackenzie District, Northwest Territories; Geological Survey of Canada, Paper 74-16, 52 p.
1977: Porcupine River Formation; a new Upper Jurassic sandstone unit, northwest Yukon Territory; Geological Survey of Canada, Paper 76-27, 43 p.
1980: Lower Cretaceous and Jurassic rocks of McDougall Pass area and some adjacent areas of north-central Richardson Mountains, northern Yukon Territory and northwestern District of Mackenzie, Northwest Territories (NTS 116P/9 and 116P/10): a reappraisal; Geological Survey of Canada, Paper 78-22, 35 p.
1984: Jurassic-Cretaceous boundary beds of western and Arctic Canada and the problem of the Tithonian-Berriasian stages in the Boreal Realm; in Jurassic-Cretaceous Biochronology and Paleogeography of North America, G.E.G. Westermann (ed.); Geological Association of Canada, Special Paper 27, p. 175-255.

Kent, D.M.
1974: The relationship between hydrocarbon accumulations and basement structural elements in the northern Williston Basin; in Fuels: a Geological Appraisal, G.R. Parslow (ed.); Saskatchewan Geological Society, Special Publication No. 2, p. 63-80.

Kreis, L.K.
1985: Notes on the Jurassic-Cretaceous petroleum-producing zones in the Wapella-Moosomin area, southeastern Saskatchewan; in Summary of Investigations 1985; Saskatchewan Geological Survey, Saskatchewan Energy and Mines, Miscellaneous Report 85-4, p. 172-179.

Lackie, J.H.
1958: Subsurface Jurassic of the Peace River area; in Jurassic and Carboniferous of Western Canada, A.J. Goodman (ed.); American Association of Petroleum Geologists, John Andrew Allan Memorial Volume, p. 85-97.

Leffingwell, E. de K.
1919: The Canning River region, northern Alaska; United States Geological Survey, Professional Paper 109, 251 p.

Lerand, M.
1973: Beaufort Sea; in The Future Petroleum Provinces of Canada: their Geology and Potential, R.G. McCrossan (ed.); Canadian Society of Petroleum Geologists, Memoir 1, p. 315-386.

Loranger, D.M.
1955: Palaeogeography of some Jurassic microfossil zones in the south half of the Western Canada Basin; Proceedings of the Geological Association of Canada, v. 7, pt. 1, p. 31-60.

MacKay, B.R.
1929: Brûlé Mines coal area, Alberta; Geological Survey of Canada, Summary Report, 1928, pt. B, p. 1-29.

Marion, D.J.
1984: Sedimentology of the Middle Jurassic Rock Creek Member in the subsurface of west-central Alberta; in The Mesozoic of Middle North America, D.F. Stott and D.J. Glass (ed.); Canadian Society of Petroleum Geologists, Memoir 9, p. 319-343.

McKelvey, V.E., Williams, J.S., Sheldon, R.P., Cressman, E.R., Cheney, T.M., and Swanson, R.W.
1967: The Phosphoria, Park City and Shedhorn formations in Western Phosphate field; in Anatomy of the Western Phosphate Field, L.A. Hale (ed.); Salt Lake City, Intermountain Association of Geologists, 15th Annual Field Conference, p. 15-33.

MacLachlan, M.E.
1972: Triassic System; in Geologic Atlas of the Rocky Mountain Region; Rocky Mountain Association of Geologists, Denver, p. 166-176.

McLearn, F.H.
1924: New pelecypods from the Fernie Formation of the Alberta Jurassic; Transactions of the Royal Society of Canada, 3d Ser., Sec. 4, v. 18, p. 39-61.
1927: Some Canadian Jurassic faunas; Royal Society of Canada, Transactions, Ser. 3, v. 21, Sec. 4, p. 61-73.
1928: New Jurassic Ammonoidea from the Fernie Formation, Alberta; Geological Survey of Canada, Bulletin No. 49, Geological Series No. 48, Contributions to Canadian Palaeontology, p. 19-22, Pl. IV-VIII.
1929: Mesozoic paleontology of Blairmore region, Alberta; National Museum Canada, Bulletin 58, Geological Series 50, Preface, p. v-vi; Stratigraphic Paleontology, p. 80-107.
1932: Three Fernie Jurassic Ammonoids; Transactions of the Royal Society of Canada, 3d Ser., v. 26, Sec. iv, p. 111-116, Pl. i-v.

Meek, F.B.
1859: Remarks on the Cretaceous fossils collected by Professor Henry Y. Hind, on the Assiniboine and Saskatchewan exploring expedition, with descriptions of some new species, Northwest Territory; Reports of Progress, together with a preliminary and general report on the Assiniboine and Saskatchewan exploring expedition, Henry Youle Hind, Toronto, p. 182-185.

Meriaux, E.
1972: Les tonsteins de la veine de charbon no 10 (Balmer) à Sparwood Ridge dans le bassin de Fernie (Colombie Britannique); in Report of Activities, Part B, Geological Survey of Canada, Paper 72-1B, p. 11-22.

Miall, A.D.
1979: Mesozoic and Tertiary geology of Banks Island, Arctic Canada: the history of an unstable craton margin; Geological Survey of Canada, Memoir 387, 235 p.

Milner, R.L. and Thomas, G.E.
1954: Jurassic System in Saskatchewan; in Western Canada Sedimentary Basin, L.M. Clark (ed.); American Association of Petroleum Geologists, Ralph Leslie Rutherford Memorial Volume, p. 250-267.

Monger, J.W.H., Price, R.A., and Tempelman-Kluit, D.J.
1982: Tectonic accretion and the origin of the two major metamorphic and plutonic welts in the Canadian Cordillera; Geology, v. 10, p. 70-75.

Newmarch, C.B.
1953: Geology of the Crowsnest Coal Basin with special reference to the Fernie area; British Columbia Department of Mines, Bulletin No. 33, 107 p.

Nicholls, E.L.
1976: The oldest known North American occurrence of the Plesiosauria (Reptilia: Sauropterygia) from the Liassic (Lower Jurassic) Fernie Group, Alberta, Canada; Canadian Journal of Earth Sciences, v. 13, p. 185-188.

Nordquist, J.W.
1955: Pre-Rierdon Jurassic stratigraphy in northern Montana and Williston Basin; in Sweetgrass Arch-Disturbed Belt, Montana, P.J. Lewis (ed.); Billings Geological Society Guidebook, Sixth Annual Field Conference, p. 96-106.

Norris, D.K.
1957: Canmore, Alberta; Geological Survey of Canada, Paper 57-4, 8 p.
1971: The geology and coal potential of the Cascade Coal Basin; in A Guide to the Geology of the Eastern Cordillera along the Trans Canada Highway between Calgary, Alberta and Revelstoke, British Columbia, I.A.R. Halladay and D.K. Mathewson (ed.); Alberta Society of Petroleum Geologists, Calgary, p. 25-39.
1972: Structural and stratigraphic studies in the tectonic complex of northern Yukon Territory, north of Porcupine River; in Report of Activities, Part B, Geological Survey of Canada, Paper 72-1B, p. 91-99.
1974: Structural geometry and geological history of the northern Canadian Cordillera; in Proceedings of the 1973 National Convention, A.E. Wren and R.B. Cruz (eds.); Canadian Society of Exploration Geophysicists, p. 18-45.
1983: Porcupine virgation - the structural link among the Columbian, Innuitian and Alaskan orogens; Geological Association of Canada/Mineralogical Association of Canada, Joint Annual Meeting, University of Victoria, May 11-13, 1983; Geological Association of Canada, Program with Abstracts, v. 8, p. A51.

Ollerenshaw, N.C.
1968: Preliminary account of the geology of Limestone Mountain map area, Alberta; Geological Survey of Canada, Paper 68-24, p. 1-37.

Osadetz, K.G. and Snowdon, L.R.
1986: Petroleum source rock reconnaissance of southern Saskatchewan; in Current Research, Part A, Geological Survey of Canada, Paper 86-1A, p. 609-617.

Paterson, D.F.
1968: Jurassic megafossils of Saskatchewan with a note on Charophytes; Saskatchewan Department of Mineral Resources, Geological Sciences Branch, Sedimentary Geology Division, Report No. 120, 135 p.

Peterson, J.A.
1957: Marine Jurassic of northern Rocky Mountains and Williston Basin; Bulletin of the American Association of Petroleum Geologists, v. 41, no. 3, p. 399-440. Reprinted 1958; in Jurassic and Carboniferous of Western Canada, A.J. Goodman (ed.); American Association of Petroleum Geologists, John Andrew Allan Memorial Volume, p. 100-141.
1972: Jurassic System; in Geologic Atlas of the Rocky Mountain Region; Rocky Mountain Association of Geologists, Denver, p. 177-189.
1981: General stratigraphy and regional paleostructure of the western Montana overthrust belt; Montana Geological Society, 1981 Field Conference and Symposium Guidebook to southwest Montana, p. 5-35.

Pocock, S.A.J.
1962: Microfloral analysis and age determination of strata at the Jurassic-Cretaceous boundary in the western Canada plains; Palaeontographica, Abteilung B, Band 111, p. 1-95.
1964: Palynology of the Kootenay Formation at its type section; Bulletin of Canadian Petroleum Geology, v. 12, Special Guide Book Issue - Flathead Valley, p. 500-511.
1967: The Jurassic-Cretaceous boundary in northern Canada; Review of Paleobotany and Palynology, v. 5, p. 124-136.
1970: Palynology of the Jurassic sediments of western Canada: Part I, terrestrial species; Palaeontographica, Abteilung B, Band 130, p. 12-72, 73-136.
1972: Palynology of the Jurassic sediments of western Canada: Part 2, Marine species; Palaeontographica, Abteilung B, Band 137, p. 85-153, Pl. 22-29.
1976: A preliminary dinoflagellate zonation of the uppermost Jurassic and lower part of the Cretaceous, Canadian Arctic, and possible correlation in the western Canada Basin; Geoscience and Man, v. XV, p. 101-114.
1980: Palynology at the Jurassic-Cretaceous boundary in North America; Palynology Congress, Lucknow (1976-77), v. 2, p. 377-385.

Podruski, J.A., Barclay, J.E., Hamblin, A.P., Lee, P.J., Osadetz, K.G., Procter, R.M., Taylor, G.C., Conn, R.F., and Christie, J.A.
1988: Conventional oil resources of western Canada (light and medium); Geological Survey of Canada, Paper 87-26, 149 p.

Poulton, T.P.
1978a: Correlation of the Jurassic Bug Creek Formation in the subsurface of the Mackenzie Delta, District of Mackenzie; in Current Research, Part C, Geological Survey of Canada, Paper 78-1C, p. 39-42.
1978b: Pre-Late Oxfordian Jurassic biostratigraphy of northern Yukon and adjacent Northwest Territories; Geological Association of Canada, Special Paper 18, p. 445-471.
1982: Paleogeographic and tectonic implications of Lower and Middle Jurassic facies patterns in northern Yukon Territory and adjacent Northwest Territories; in Arctic Geology and Geophysics, A.F. Embry and H.R. Balkwill (ed.); Canadian Society of Petroleum Geologists, Memoir 8, p. 13-27.
1984: Jurassic of the Canadian Western Interior, from 49°N Latitude to Beaufort Sea; in The Mesozoic of Middle North America, D.F. Stott and D.J. Glass (ed.); Canadian Society of Petroleum Geologists, Memoir 9, p. 15-41.
1987: Boreal Middle Bathonian to Lower Callovian (Jurassic) ammonites, zonation and correlation, Salmon Cache Canyon, Porcupine River, Northern Yukon; Geological Survey of Canada, Bulletin 358, 155 p.
1988: Major interregionally correlatable events in the Jurassic of western interior, Arctic and eastern offshore, Canada; in Sequences, Stratigraphy, Sedimentology: Surface and Subsurface, D.P. James and D.A. Leckie (ed.); Canadian Society of Petroleum Geologists, Memoir 15, p. 195-206.

Poulton, T.P. and Callomon, J.H.
1976: Major features of the Lower and Middle Jurassic stratigraphy of northern Richardson Mountains, northeastern Yukon Territory and northwestern District of Mackenzie; in Report of Activities, Part B, Geological Survey of Canada, Paper 76-1B, p. 345-352.

Poulton, T.P. and Tempelman-Kluit, D.J.
1982: Recent discoveries of Jurassic fossils in the Lower Schist Division of central Yukon; in Current Research, Part C, Geological Survey of Canada, Paper 82-1C, p. 91-94.

Poulton, T.P., Leskiw, K., and Audretsch, A.P.
1982: Stratigraphy and microfossils of the Jurassic Bug Creek Group of northern Richardson Mountains, northern Yukon and adjacent Northwest Territories; Geological Survey of Canada, Bulletin 325, 137 p.

Rall, R.D.
1980: Stratigraphy, sedimentology and paleotopography of the Lower Jurassic in the Gilby-Medicine River Fields, Alberta; M.Sc. thesis, University of Calgary, Calgary, Alberta, 142 p.

Rapson, J.E.
1964: Lithology and petrography of transitional Jurassic-Cretaceous clastic rocks, southern Rocky Mountains; Bulletin of Canadian Petroleum Geology, v. 12, Special Guidebook Issue - Flathead Valley, p. 556-576.
1965: Petrography and derivation of Jurassic-Cretaceous clastic rocks, southern Rocky Mountains, Canada; Bulletin of the American Association of Petroleum Geologists, v. 49, pt. II, no. 9, p. 1410-1425.

Rickwood, F.K.
1970: The Prudhoe Field; in Proceedings of the Geological Seminar on the North Slope of Alaska, Pacific Section; American Association of Petroleum Geologists, p. L1-L11.

Rouse, G.E.
1959: Plant microfossils from Kootenay Coal Measures strata of British Columbia; Micropaleontology, v. 5, p. 303-324.

Sawatzky, H.G.
1974: Astroblemes in the Williston Basin; in Fuels: A Geological Appraisal, G.R. Parslow (ed.); Saskatchewan Geological Society, Special Publication No. 2, p. 95-118.

Schmitt, G.T.
1953: Regional stratigraphic analysis of Middle and Upper marine Jurassic in northern Rocky Mountains - Great Plains; Bulletin of the American Association of Petroleum Geologists, v. 37, no. 2, p. 355-393.

Sheldon, R.P.
1964: Paleolatitudinal and paleogeographic distribution of phosphorite; United States Geological Survey, Professional Paper 501-C, p. C106-C113.

Sheldon, R.P., Maughan, E.K., and Cressman, E.R.
1967: Sedimentation of rocks of Leonard (Permian) age in Wyoming and adjacent states; in Anatomy of the Western Phosphate Field, L.A. Hale (ed.); Salt Lake City, Intermountain Association of Geologists, 15th Annual Field Conference, p. 1-13.

Smith, G.G.
1989: Coal resources of Canada; Geological Survey of Canada, Paper 89-4, 146 p.

Spivak, J.
1949: Jurassic sections in foothills of Alberta and northeastern British Columbia; Bulletin of the American Association of Petroleum Geologists, v. 33, p. 533-546: Reprinted 1954; in Western Canada Sedimentary Basin, L.M. Clark (ed.); American Association of Petroleum Geologists, Ralph Leslie Rutherford Memorial Volume, p. 219-232.

Springer, G.D., MacDonald, W.D., and Crockford, M.B.B.
1964: Jurassic; Chapter 10 in Geological History of Western Canada, R.G. McCrossan and R.P. Glaister (ed.); Alberta Society of Petroleum Geologists, Calgary, p. 137-155.

Stelck, C.R., Wall, J.H., Williams, G.D., and Mellon, G.B.
1972: The Cretaceous and Jurassic of the foothills of the Rocky Mountains of Alberta; XXIV International Geological Congress, Montreal, Guidebook, Field Excursion A20, 51 p.

Stott, D.F.
1955: Jurassic stratigraphy of Manitoba; Manitoba Department of Mines and Mineral Resources, Publication 54-2, 78 p.
1967a: Fernie and Minnes strata north of Peace River, foothills of northeastern British Columbia; Geological Survey of Canada, Paper 67-19 (Part A), 58 p.
1967b: Jurassic and Cretaceous stratigraphy between Peace and Tetsa Rivers, northeastern British Columbia; Geological Survey of Canada, Paper 66-7, 73 p.
1969: Fernie and Minnes strata north of Peace River, foothills of northeastern British Columbia; Geological Survey of Canada, Paper 67-19 (Part B), 132 p.
1970: Mesozoic stratigraphy of the Interior Platform and Eastern Cordilleran Orogen; in Geology and Economic Minerals of Canada, R.J.W. Douglas (ed.); Geological Survey of Canada, Economic Geology Report 1, p. 438-445.
1975: The Cretaceous System in northeastern British Columbia; in The Cretaceous System in the Western Interior of North America, W.G.E. Caldwell (ed.); Geological Association of Canada, Special Paper 13, p. 441-467.
1984: Cretaceous sequences of the foothills of the Canadian Rocky Mountains; in The Mesozoic of Middle North America, D.F. Stott and D.J. Glass (ed.); Canadian Society of Petroleum Geologists, Memoir 9, p. 85-108.

Stott, D.F., Gibson, D.W., and Ollerenshaw, N.C.
1968: Mesozoic and Cenozoic rocks between Bow and Brazeau Rivers, Alberta; Guidebook - Central Canadian Rockies and foothills of Alberta, 16th Annual Field Conference 1968, Alberta Society of Petroleum Geologists, p. 67-105.

Stronach, N.J.
1984: Depositional environments and cycles in the Jurassic Fernie Formation, southern Canadian Rocky Mountains; in The Mesozoic of Middle North America, D.F. Stott and D.J. Glass (ed.); Canadian Society of Petroleum Geologists, Memoir, p. 43-68.

Taylor, D.G., Callomon, J.H., Hall, R., Smith, P.L., Tipper, H.W., and Westermann, G.E.G.
1984: Jurassic ammonite biogeography of western North America - the tectonic implications; in Jurassic-Cretaceous Biochronology and Paleogeography of North America, G.E.G. Westermann (ed.); Geological Association of Canada, Special Paper 27, p. 121-142.

Telfer, L.
1933: Phosphate in the Canadian Rockies; Canadian Institute of Mining and Metallurgy, Bulletin, v. 260, p. 566-605.

Tempelman-Kluit, J.
1970: Stratigraphy and structure of the "Keno Hill Quartzite" in Tombstone River - Upper Klondike River map areas, Yukon Territory; Geological Survey of Canada, Bulletin 180, 102 p.

Thompson, R.L. and Crockford, M.B.B.
1958: The Jurassic subsurface in Southern Alberta; in Jurassic and Carboniferous of Western Canada, A.J. Goodman (ed.); American Association of Petroleum Geologists, Tulsa, John Andrew Allan Memorial Volume, p. 52-64.
Tipper, H.W.
1984: The allochthonous Jurassic-Cretaceous terranes of the Canadian Cordillera and their relation to correlative strata of North American craton; in Jurassic-Cretaceous Biochronology and Paleogeography of North America, G.E.G. Westermann (ed.); Geological Association of Canada, Special Paper 27, p. 113-120.
Tozer, E.T.
1982: Late Triassic (Upper Norian) and earliest Jurassic (Hettangian) rocks and ammonoid faunas, Halfway River and Pine Pass map-areas, British Columbia; in Current Research, Part A, Geological Survey of Canada, Paper 82-1A, p. 385-391.
Vail, P.R. and Todd, R.G.
1981: Northern North Sea Jurassic unconformities, chronostratigraphy and sea-level changes from seismic stratigraphy; in Petroleum Geology of the Continental Shelf of Northwest Europe, L.V. Illing and G.D. Hobson (ed.); Heyden, London, p. 216-235.
Vigrass, L.W.
1952: Jurassic stratigraphy of southern Saskatchewan; M.Sc. thesis, University of of Saskatchewan, Saskatoon, Saskatchewan, 71 p.
Wall, J.H.
1960: Jurassic microfaunas from Saskatchewan; Saskatchewan Department of Mineral Resources, Report No. 53, 229 p.
Wallace-Dudley, K.
1983a: Gas pools of western Canada; Geological Survey of Canada, Map 1558A, 1:1 013 760.
1983b: Oil pools of western Canada; Geological Survey of Canada, Map 1559A, 1:1 013 760.
Warren, P.S.
1931: A Lower Jurassic fauna from Fernie, British Columbia; Transactions of the Royal Society of Canada, Ser. 3, Sec. IV, v. 25, p. 105-111, Pl. I.
1932: A new pelecypod fauna from the Fernie Formation, Alberta; Transactions of the Royal Society of Canada, Ser. 3, Sec, IV, v. 26, p. 1-36, Pl. 1-5.
1934: Present status of the Fernie shale; American Journal of Science, Ser. 5, v. 27, p. 56-70.
1947: Description of Jurassic ammonites from the Fernie Formation, Alberta; in Geology of Highwood-Elbow Area, Alberta, J.A. Allan and J.L. Carr (ed.); Research Council of Alberta, Report No. 49, p. 67-74, Pl. I-VII.
Weihmann, I.
1964: Stratigraphy and microfauna of the Jurassic Fernie Group, Fernie Basin, Southeast British Columbia; Bulletin of Canadian Petroleum Geology, v. 12, Special Guidebook Issue – Flathead Valley, p. 587-599.
Weir, J.D.
1949: Marine Jurassic Formations of southern Alberta Plains; Bulletin of the American Association of Petroleum Geologists, v. 33, p. 547-563. Reprinted 1954; in Western Canada Sedimentary Basin, L.M. Clark (ed.); American Association of Petroleum Geologists, Ralph Leslie Rutherford Memorial Volume, p. 233-249.
Westermann, G.E.G.
1964: Occurrence and significance of the Arctic *Arkelloceras* in the Middle Bajocian of the Alberta foothills (Ammonitina, Jurassic); Journal of Paleontology, v. 38, p. 405-409.
1966: The holotype (plastotype) of ?*Titanites occidentalis* Frebold from the Kootenay sandstone (Upper Jurassic) of southern British Columbia; Canadian Journal of Earth Sciences, v. 3, p. 623-625.
1980: Ammonite biochronology and biogeography of the Circum-Pacific Middle Jurassic; in The Ammonoidea, M.R. House and J.R. Senior (ed.); Academic Press, London and New York, Systematics Association Special Volume No. 18, p. 459-498.
Whiteaves, J.F.
1889: Fossils from the Rocky Mountains three miles north of the east end of Devils Lake; Geological Survey of Canada, Contributions of Canadian Paleontology, v. 1, pt. 2, p. 163-172.
1903: Description of a species of *Cardioceras* from the Crowsnest Coal Fields; Ottawa Naturalist, v. 17, p. 65-67.
Wickenden, R.T.D.
1933: Jurassic Foraminifera from wells in Alberta and Saskatchewan; Transactions of the Royal Society of Canada, Sec. IV, p. 157-170, Pl. I-II.
Young, F.G.
1973: Mesozoic epicontinental, flyschoid, and molassoid depositional phases of Yukon's north slope; in Canadian Arctic Geology, J.D. Aitken and D.J. Glass (ed.); University of Saskatchewan, Saskatoon, May 1973, Canadian Society of Petroleum Geologists, p. 181-202.
1975: Stratigraphic and sedimentologic studies in northeastern Eagle Plain, Yukon Territory; in Report of Activities, Part B, Geological Survey of Canada, Paper 75-1B, p. 309-319.
Young, F.G., Myhr, D.W., and Yorath, C.J.
1976: Geology of the Beaufort-Mackenzie Basin; Geological Survey of Canada, Paper 76-11, 65 p.
Zeiss, A.
1983: Zur frage der Aquivalenz der Stufen Tithon/Berrias/Wolga/Portland in Eurasien und Amerika: Ein Beitrag zur Klörung der weltweiten Korrelation der Jura-/Kreide-Grenzschichten im marinen Bereich; Zitteliana, v. 10, p. 427-438.
1984: Comments (on paper by Jeletzky, 1984; see this reference list); in Jurassic-Cretaceous Biochronology and Paleogeography of North America, G.E.G. Westernmann (ed.); Geological Association of Canada, Special Paper 27, p. 250-253.

ADDENDUM

The basal Jurassic (Sinemurian) strata of southwesternmost Alberta and southeastern British Columbia include a phosphorite unit as thick as 10 m. The requirement of "West coast type" phosphorite depositional models for access to upwelling ocean currents places limits on how close allochthonous terranes could have lain off the west coast of the continent in Sinemurian times (Poulton and Aitken, 1989).

The basal Jurassic Nordegg black chert and limestone member, and an equivalent 40 m thick carbonate platform, in west central Alberta is replaced in northwestern Alberta and northeastern British Columbia by a platy argillaceous limestone facies. It is an important hydrocarbon source rock (Riediger et al., 1990), similar and perhaps party equivalent to the Pliensbachian Red Deer Member. A major sea-level fall prior to its deposition would be consistent with its general lowstand distribution pattern relative to the adjacent and older(?) Nordegg carbonate platform. The limits and general character of the carbonate platform and of the argillaceous facies to its north have been outlined by Poulton et al. (1991).

Palynological studies lend further support to the hypothesis that the Rock Creek and other Middle Jurassic sandstones and shales of west central Alberta are absent in northwestern Alberta below a regional sub-Upper Jurassic unconformity. A characteristic glauconitic sandstone (Niton B) overlies the unconformity, which has been well documented in central Alberta by Losert (1990). Aalenian strata, comprising interbedded sandstones and shales of the upper Poker Chip Shale of northwestern Alberta, have been identified for the first time in the Western Canada Basin, on the basis of palynology (Poulton et al., 1991). An interval of micaceous sandstones and interbedded shales that are either in the lower part of the Rock Creek Member or the upper part of the Poker Creek Shale along its eastern preserved limit in central Alberta also contain Aalenian palynomorphs (G. Dolby, unpub. data).

Sandstone-rich strata in southeastern Saskatchewan equivalent to the Middle Jurassic Upper Gravelbourg and Shaunavon formations have been distinguished as the Red Jacket Formation (Kreis, 1988). The 'Rierdon Shoulder' is equivalent to the top of one of the clinothem sandstone units of the roseray Formation, rather than a marker at the top of the Rierdon Formation (Christopher, 1988).

Most recent summaries of the Jurassic, including access to references, are by Poulton (1989a) for the northern Yukon area and by Poulton (1989b) for the Western Canada Basin. The major Early Jurassic and Aalenian invertebrate macrofossils for northern Yukon and adjacent Northwest Territories have been illustrated and described (Poulton, 1991).

REFERENCES

Christopher, J.E.
1988: Note on an Upper Jurassic section in southeastern Saskatchewan and its relationship to adjacent regions; in Summary of Investigations 1988; Saskatchewan Geological Survey, Miscellaneous Report 88-4, p. 196-201.

Kreis, L.K.
1988: The Red Jacket Formation of southeastern Saskatchewan; in Summary of Investigations 1988, Saskatchewan Geological Survey; Miscellaneous Report 88-4, p. 211-223.

Losert, J.
1990: The Jurassic-Cretaceous boundary units and associated hydrocarbon pools in the Niton Field, west-central Alberta; Alberta Research Council, Open File Report 1990-1, 41 p.

Poulton, T.P.
1989a: Current status of Jurassic biostratigraphy and stratigraphy, northern Yukon and adjacent Mackenzie Delta; in Current Research, Part G; Geological Survey of Canada, Paper 89-1G, p. 25-30.

Poulton, T.P. (cont.)
1989b: Upper Absaroka to Lower Zuni: the transition to the foreland basin; in Western Canada Sedimentary Basin. A case history, B.D. Ricketts (ed.), Canadian Society of Petroleum Geologists, p. 233-247.
1991: Hettangian through Aalenian (Jurassic) guide fossils and biostratigraphy, northern Yukon and adjacent Northwest Territories; Geological Survey of Canada, Bulletin 410.

Poulton, T.P. and Aitken, J.D.
1989: The Lower Jurassic phosphorites of southeastern British Columbia and terrane accretion to western North America; Canadian Journal of Earth Sciences, v. 26, p. 1612-1616.

Poulton, T.P., Tittemore, J., and Dolby, G.
1991: Jurassic strata, northwestern (and west central) Alberta and northeastern British Columbia; Bulletin of Canadian Petroleum Geology, v. 38A, p. 159-175.

Riediger, C.L., Fowler, M.G., Snowdon, L.R., Goodarzi, F., and Brooks, P.W.
1990: Source rock analysis of the Lower Jurassic "Nordegg Member" and oil-source rock correlations, northwestern Alberta and northeastern British Columbia; Bulletin of Canadian Petroleum Geology, v. 38A, p. 236-249.

Authors' Addresses

T.P. Poulton
Institute of Sedimentary and Petroleum Geology
Geological Survey of Canada
3303 - 33rd Street N.W.
Calgary, Alberta
T2L 2A7

W.K. Braun
Department of Geological Sciences
University of Saskatchewan
Saskatoon, Saskatchewan
S7N 0W0

M.M. Brooke
Department of Geological Sciences
University of Saskatchewan
Saskatoon, Saskatchewan
S7N 0W0

E.H. Davies
Branta Biostratigraphy Ltd.
410 Elbow Park Lane SW
Calgary, Alberta
T2S 2S6

Printed in Canada

Subchapter 4I

CRETACEOUS

D.F. Stott, W.G.E. Caldwell, D.J. Cant, J.E. Christopher, J. Dixon, E.H. Koster, D.H. McNeil, and F. Simpson

CONTENTS

INTRODUCTION

Cretaceous rocks extend through the mid-continent from northern Yukon Territory and District of Mackenzie southward along Mackenzie and Rocky Mountains and eastward across the Plains of Alberta, Saskatchewan, and Manitoba to the Canadian Shield. They underlie Mackenzie Delta and Plains in the District of Mackenzie. Exposures of Cretaceous rocks are found in the southern Plains along many of the major river systems. One of the celebrated localities is at Dinosaur Park, southeastern Alberta, designated a World Heritage Site. Elsewhere in the Plains, Cretaceous rocks are mainly covered by Pleistocene drift, except in the Cypress Hills and along the Manitoba Escarpment on the eastern side of the basin.

Many investigations of Cretaceous rocks have been carried out over a period of more than 125 years. Studies of oil sands, heavy oil areas, and coal-bearing areas, regional and economic studies, and detailed surface and subsurface studies have contributed to understanding the history of these rocks. More than 100 000 boreholes have been drilled in the southern part of the Western Canada Basin. Well control is relatively sparse in the District of Mackenzie where, too, outcrops are lacking in large areas. Drilling in the Eagle Plains, Mackenzie Delta, and along the Mackenzie Valley has provided much needed data. Recent coal exploration in the foothills and mountains of British Columbia and in the Plains of Alberta and Saskatchewan has contributed greatly to the overall knowledge of the Cretaceous rocks.

Tectonic elements and sub-elements

The Cretaceous was a time of subduction and accretion of exotic terranes to the western margin of the North American continent, accompanied by mountain building and the associated development of foredeeps and in the north, the opening of the oceanic Canada Basin. The deformed, mountainous terrane of western Canada, defined as the Cordilleran Orogen (Fig. 4I.1) by Wheeler (1970), is a product of several episodes of deformation and

Stott, D.F., Caldwell, W.G.E., Cant, D.J., Christopher, J.E., Dixon, J., Koster, E.H., McNeil, D.H., and Simpson, F.
1993: Cretaceous; Subchapter 4I in Sedimentary Cover of the Craton in Canada, D.F. Stott and J.D. Aitken (ed.); Geological Survey of Canada, Geology of Canada, no. 5, p.358-438 (also Geological Society of America, The Geology of North America, v. D-1).

consists of several structural belts. The orogen is composed of a mobile, uplifted core of granitic and medium- to high-grade metamorphic rocks, flanked by belts in which the tectonic transport has been directed principally away from the core zone. The eastern belt, the Columbian Orogen (Wheeler, 1970), originated mainly during mid-Jurassic to early Tertiary times. The supracrustal rocks involved in the orogen are a Proterozoic to upper Jurassic sequence of miogeoclinal and platformal aspect succeeded by an upper Mesozoic-lower Tertiary clastic

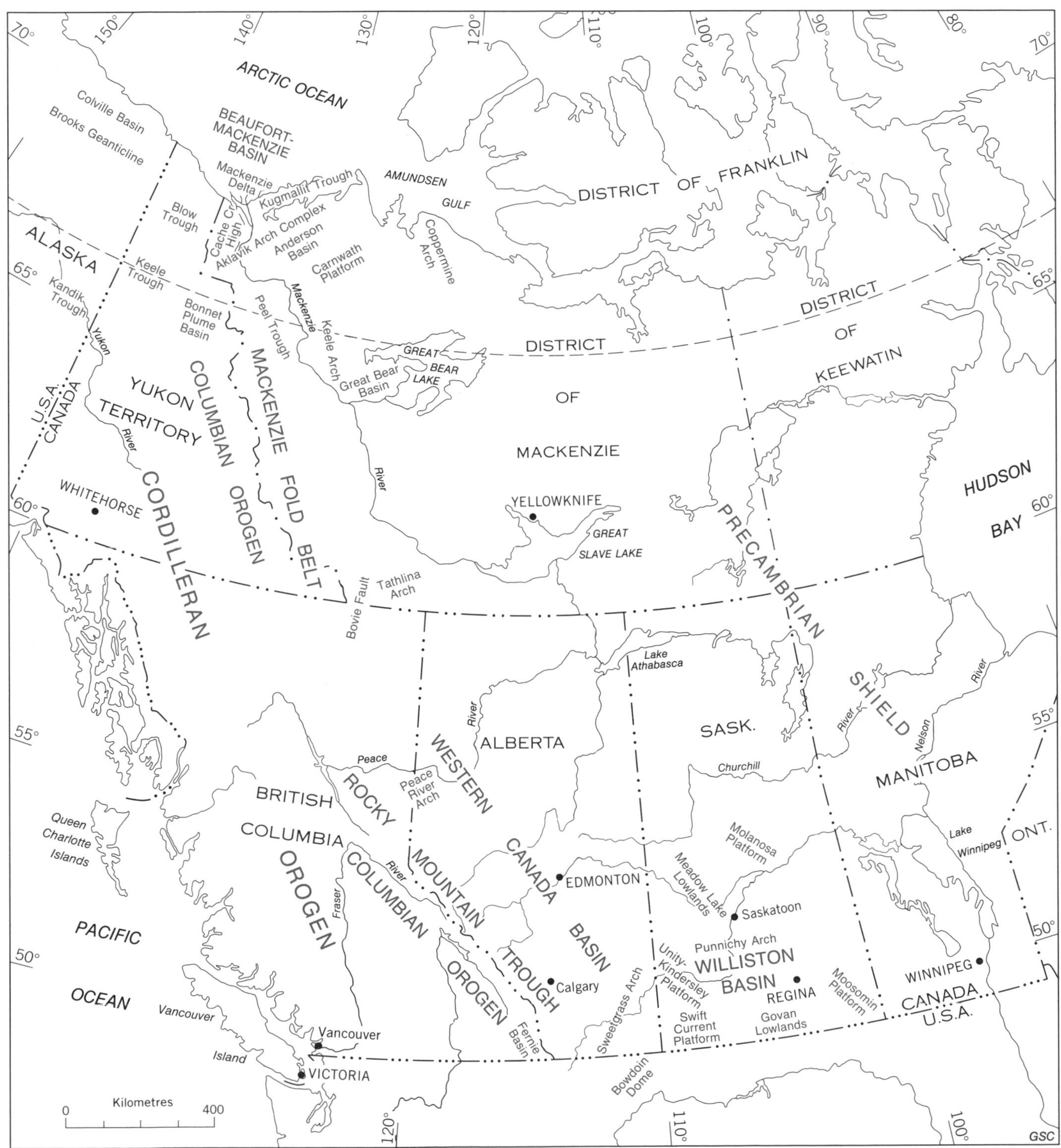

Figure 4I.1. Index map of tectonic elements that were active during the Cretaceous Period.

wedge sequence. The orogen in Canada is over 2000 km long and 400 km wide. It extends from beyond Canada's southern border northwestward into Alaska as a sinuous belt marked by eastward-convex salients. It is bounded on the east by the Interior Platform and on the north by the Arctic Continental Shelf. The eastern boundary is marked by the easternmost folds or faults of the Rocky Mountain Foothills, the Mackenzie Mountains and the Franklin Mountains.

The Cretaceous also witnessed one of the most widespread inundations of the North American Craton. Sediments were deposited in the broad, low-lying Western Canada Basin between the Canadian Shield on the east, the emerging Columbian Orogen on the west (Fig. 4I.1), and the developing Arctic Ocean to the north. During the Cretaceous, Western Canada Basin was the northwestern part of the Western Interior Basin of North America; it was flooded at various times during the period by seas that extended southward from Boreal regions and also northward from the Gulf of Mexico.

Sedimentation in subsiding foredeeps flanking the Columbian Orogen and Brooks Range Geanticline continued from Jurassic through Cretaceous into Tertiary times. In northern Yukon Territory, Blow Trough extended southward, merging into the southwestward- trending Keele-Kandik Trough. Peel Trough, a southeasterly trending foredeep, bordered the rising Mackenzie Mountains. A number of uplifted terranes in the northern regions had varying affects on rates of deposition, facies, and distribution of Cretaceous sediments. These include Coppermine and Keele arches, and Aklavik Arch Complex, of which Eskimo Lakes Arch is the largest component. Aklavik Arch was bordered by Kugmallit Trough to the northwest and by Anderson Basin to the southeast. To the south, Great Bear Basin lies between Keele Arch and the Canadian Shield.

Along the western side of the southern part of Western Canada Basin, the linear Rocky Mountain Trough trapped sediments shed by the Columbian Orogen, whereas on the eastern margin, sediments were derived from the Canadian Shield and its Paleozoic cover. Several prominent arches separated Western Canada Basin into local sub-basins. In the southeast, Punnichy Arch, possibly formed in response to solution of Paleozoic evaporites, emphasized the northern limits of the earlier Williston Basin. To the west, Sweetgrass Arch separated the Williston Basin from Rocky Mountain Trough. Farther north, Peace River Arch, a positive element during Devonian and older times but a negative element, Peace River Embayment, during Carboniferous and Permian times, continued to influence Cretaceous deposition. An "axial high", extending from northeastern British Columbia to Swift Current Platform (possibly a tectonic forebulge), was flanked to the west by the foreland basin and to the east by a depression caused by subsurface leaching of Devonian salt.

Paleotopography of the pre-Mannville surface is the result of pre-Albian erosion, which produced a strongly developed drainage pattern over large parts of the region. Ridges of Paleozoic, mainly Devonian and Carboniferous, rocks extend northwestward from Saskatchewan through Alberta into British Columbia.

In the southwestern Rocky Mountains, Fernie Basin is a large, Laramide (Late Cretaceous-Tertiary) syncline in which is preserved a thick succession of Jurassic and Cretaceous sediments.

Tectono-stratigraphic assemblages

The Cretaceous successions of Western Canada Basin record geological events of continental as well as local scale. In order to develop a comprehensive and integrated history of the basin, its paleogeography, and tectonic develoment, the Cretaceous successions are described in the context of tectono-stratigraphic assemblages (Fig. 4I.2, 4I.3).

Upper Jurassic and Berriasian to Hauterivian rocks record the early phases of the Columbian Orogeny. The assemblage includes Berriasian progradational deposits and a major transgressive-regressive couplet of Valanginian age and, only in the northern District of Mackenzie region, a well developed, Hauterivian cycle. Erosion during Hauterivian time in the south resulted in the development of a regional unconformity. The Berriasian-Hauterivian succession equates with the upper part of the Zuni II and the lower part of Zuni III sequences of Sloss (1963) and in part to the Transgressive-Regressive sequence, T_1-R_1 of Kauffman (1977c; see Fig. 4I.49).

The Barremian to Middle Cenomanian assemblage, deposited during the second phase of the Columbian Orogeny, comprises a thick sequence of marine to alluvial sediments deposited in foredeeps along the eastern flanks of the rising Columbian Orogen. Early events are best recorded in the northern basins, but a series of transgressive-regressive cycles are best documented in the region of Peace River. The assemblage is equivalent to the upper part of the Zuni III sequence of Sloss (1963) and to the Transgressive-Regressive sequences T_4-R_4 to T_5-R_5 of Kauffman (1977c).

The Upper Cenomanian to Maastrichtian assemblage records marine conditions within the basin until at least Late Campanian time and includes Laramide molasse in front of the rising Columbian Orogen. This assemblage is equivalent to the Zuni IV sequence of Sloss (1963) and the Transgressive-Regressive Cycles T_6-R_6 to T_{10}-R_{10} of Kauffman (1977c).

Discussion of the Cretaceous succession in the following pages is organized by geographic regions, five in number, coinciding to some extent with major tectonic elements: northern basins (including all of Western Canada Basin north of latitude 60°N); Rocky Mountain Foothills; Alberta Plains; Saskatchewan Plains; and Manitoba Escarpment. The northern basins of Yukon Territory and District of Mackenzie contain the sediments of the western foredeeps and of the eastern sub-basins, which are separated from the former by arches or elevated platforms.

The Rocky Mountain Foothills coincide with the western foredeep, or Rocky Mountain Trough, where subsidence and sedimentation rates were at a maximum. Sediments were derived from tectonically active western uplands; they are dominantly of shallow-water and littoral, coastal plain, and alluvial character. Alberta includes the west-central zone of Kauffman (1977b), axial basin of Kauffman (1985), or west-median trough of McNeil and Caldwell (1981). Deposits are predominantly dark mudstones with numerous sand tongues from western sources. Saskatchewan lies between the west-median or axial trough and the eastern platform, including the "hinge zone" of Kauffman (1977b) or the east-median hinge of McNeil and Caldwell (1981). Subsidence and sedimentation rates were moderate to low. The sedimentary sequence comprises mainly clays, shales, and rare carbonates and is relatively thin and broken by disconformities. The Manitoba Escarpment lies within the eastern stable platform zone of Kauffman (1977b) or the eastern platform of McNeil and Caldwell (1981). This is a broad zone of slow subsidence and low sedimentation. Sediments are predominantly fine-grained terrigenous clastics, calcareous shales, and rare calcarenites, implying lack of a major eastern source. These sequences also are thin and much interrupted by disconformities.

Previous work and principal references

Initial geological studies of Cretaceous rocks in western Canada include those of James Hector (1863) of the Palliser Expedition, Henry Hind (1859) of the Assiniboine and Saskatchewan Exploring Expedition, and S.J. Dawson (1859) of the Red River Expedition. Early surface exploration was linked to scientific surveys along the American boundary, plans for a coast-to-coast railroad, and investigations on the possibility of agricultural development and the availability of coal and water. The pioneer geologist, G.M. Dawson (1881), in the late nineteenth century, made remarkably accurate forecasts of the great economic potential of the area of the Athabasca oil sands and of the large reserves of coal in the southern plains.

Between the turn of the century and the end of the first World War (1918), the emphasis on the exploration of new country changed to more detailed investigations of local areas as resources were located. Thus, drilling for water and coal led to the discovery of the first major gas field at Medicine Hat in 1904, and was followed by other gas discoveries. The first Cretaceous oil was discovered in Alberta at Turner Valley in 1914, and surface geological work led to the discovery of other small oil fields in east-central Alberta.

Between the two world wars (1918 to 1939), geological investigations proceeded at a slow but steady pace. Studies of the Cretaceous System accelerated during the Second World War, when the need for new petroleum resources was high.

The discovery of oil in Devonian reefs at Leduc in 1947 marked the beginning of an era of rapid expansion of geological studies and widespread use of geophysical techniques. Discovery in 1953 of a world-class pool of

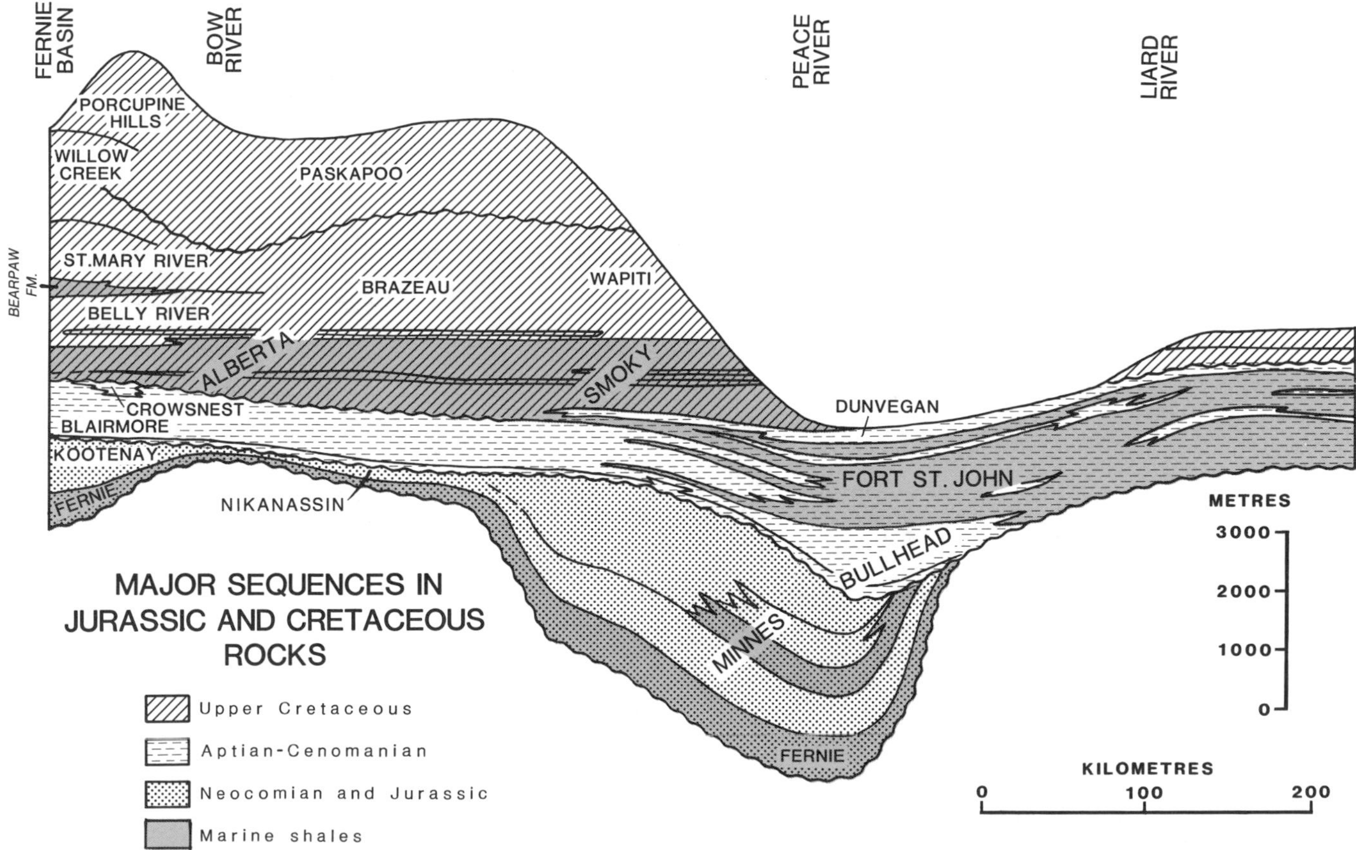

Figure 4I.2. Diagrammatic longitudinal section of the three clastic wedges deposited during Late Jurassic to early Tertiary time in the Rocky Mountains and Foothills between Fernie Basin and Liard River. The Paskapoo and Porcupine Hills formations at the top of the uppermost sequence are of Tertiary age (from Stott, 1982).

Upper Cretaceous light crude oil at Pembina (Nielson and Porter, 1984) gave impetus to detailed investigations and a lively interest in Cretaceous rocks since then.

Subsequent to the Canol Project (Hume, 1954) in the northern Yukon and District of Mackenzie in 1942-1945, during the Second World War, most of the regional geological work on Cretaceous rocks was accomplished during three reconnaissance operations of the Geological Survey of Canada: Operation Mackenzie (Stott, 1960), Operation Norman (Yorath and Cook, 1981), and Operation Porcupine (Norris, 1963). The studies of Mountjoy and Chamney (1969), Tassonyi (1969), Yorath and Cook (1981), and Aitken et al. (1982) provide the most comprehensive summaries of Cretaceous-Tertiary stratigraphy of Peel Plateau and the region to the east. The maps of Norris and stratigraphic studies by Jeletzky (1958, 1960, 1961, 1965, 1971c, 1972, 1974, 1975a, 1975b, 1980b), Young (1971, 1972, 1973a, 1973b, 1975a, 1975b, 1977), Young et al. (1976) and Mountjoy (1967) provide basic data for northern Yukon Territory and adjacent areas. The works of Young et al. (1976) and Dixon (1982a, 1986a) are the most comprehensive for Mackenzie Delta. Additionally, regional syntheses and summaries have been published by Lerand (1973), Miall (1973), and Balkwill et al. (1983).

The Kootenay succession in the southern Rocky Mountains and Foothills has been summarized by Norris (1964), Jansa (1972), Hamblin and Walker (1979), and Gibson (1977b, 1979, 1985). The equivalent Minnes Group farther north has been described by Stott (1967a, 1975). The overlying Blairmore Group has been outlined by Norris (1964), Mellon (1967), and McLean (1982), and the Alberta and Smoky groups, by Stott (1963, 1967b). The Upper Cretaceous nonmarine sequences have been described by Tozer (1956), Carrigy (1970), Kramers and Mellon (1972), Jerzykiewicz and McLean (1980), and Jerzykiewicz (1985). The Cretaceous System in the Foothills was summarized by Stott (1984).

The Mannville succession of the Plains has been outlined by Nauss (1945), Carrigy (1959), Glaister (1959), Mellon and Wall (1963), Williams (1963), Singh (1964), Mellon (1967), Cant (1983, 1984), Jackson (1985), Maycock (1967), Vigrass (1968), and Christopher (1974, 1984). The Upper Cretaceous Colorado and Montana groups have been described by Simpson (1975, 1984), Shaw and Harding (1949), McLean (1971), and Caldwell (1968). The Cretaceous succession along Manitoba Escarpment was outlined by McNeil and Caldwell (1981) and McNeil (1984).

Early investigations of Upper Cretaceous sediments were followed by more detailed subsurface correlations (Elliot, 1960), in part because of renewed interest in coal resources (Campbell, 1974; Hughes, 1984) and hydrocarbons (Myhr and Meijer Drees, 1976; Shouldice, 1979). Recent work has been concerned with clastic facies models (Walker, 1983a, b, 1985; Reinson, 1984) and problems of the Cretaceous-Tertiary boundary (Lerbekmo et al., 1979a, b). In the remote northern plains, the mostly nonmarine Upper Cretaceous succession remains relatively undifferentiated (Merrill and Buchwald, 1965; Kramers and Mellon, 1972).

Interpretations of Cretaceous stratigraphy, sedimentology, and paleogeography are still evolving. Syntheses dealing with Western Canada Basin, or large

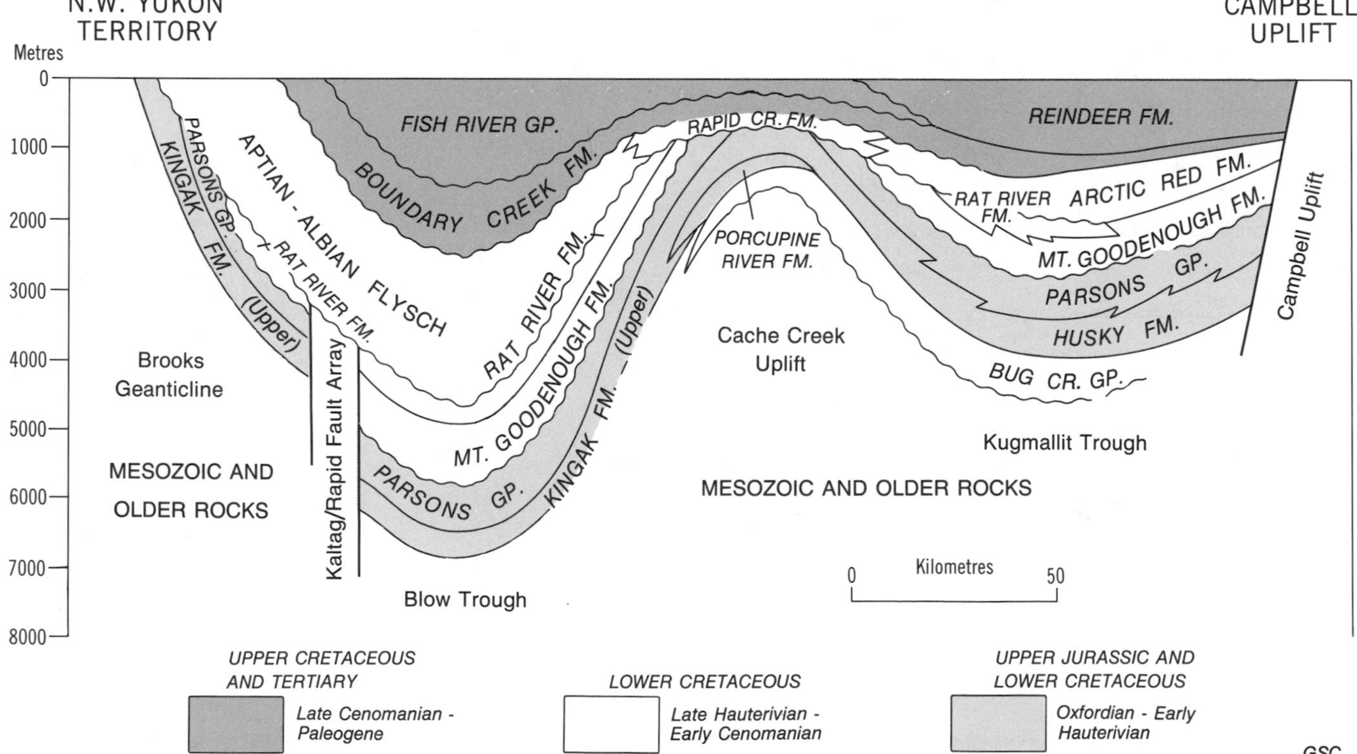

Figure 4I.3. Diagrammatic cross-section of the three clastic wedges deposited during Late Jurassic to early Tertiary time in Blow and Kugmallit troughs (adapted from Balkwill et al., 1983).

parts of it, include reports by McLearn and Kindle (1950), Rudkin (1964), Williams and Burk (1964), Springer et al. (1964), Zeigler (1969), Douglas et al. (1970), Stott (1984), Williams and Stelck (1975), Caldwell (1984a), and Christopher (1984). Concepts relating to tectonism, sedimentation, and basin history during the Cretaceous Period have been developed by Price and Mountjoy (1970), Monger and Price (1979), Eisbacher et al. (1974), Douglas et al. (1970), Porter et al. (1982), Monger et al. (1982), Williams and Stelck (1975), Caldwell (1984a), Stott (1984), and Jeletzky (1977, 1978).

The principal biostratigraphic reports dealing with Cretaceous macrofossils are those of McLearn (1933, 1937, 1944), Landes (1940), Warren and Stelck (1955, 1958a, 1969), Jeletzky (1964, 1968, 1971a, 1971b, 1980a), Tozer (1956b), Gill and Cobban (1973), Riccardi (1983), and Jeletzky and Stelck (1981).

Cretaceous megafloras were described by Bell (1949, 1956, 1963). Palynological studies of the Kootenay Formation include those of Rouse (1959) and Pocock (1964) and of the Mannville and equivalent strata by Pocock (1962, 1964, 1976), Singh (1964, 1971, 1975), Vagvolgyi and Hills (1969), Playford (1971), and Burden (1984). Palynological studies of Albian to Cenomanian rocks of Alberta were made by Norris (1967) and of Upper Cretaceous rocks by Snead (1969), Srivastava (1970), Gunther and Hills (1972), Brideaux (1971), Sweet (1978b), and Singh (1983).

The micropaleontology of Lower Cretaceous successions has been studied by Stelck et al. (1956), Sutherland and Stelck (1972), Stelck (1975a), Anan-York and Stelck (1978), Koke and Stelck (1984, 1985), and that of Upper Cretaceous successions by Wall (1960, 1967), Stelck and Wall (1954, 1955), Stelck and Wetter (1958), Given and Wall (1971), North and Caldwell (1964, 1970, 1975), Caldwell (1968), and McNeil and Caldwell (1981). A major overview of the biostratigraphy of Cretaceous foraminiferal faunas is that of Caldwell et al. (1978).

In spite of unusually rich vertebrate assemblages, a zonal scheme beyond the resolution of the host formations has not been developed (Russell, 1975). Recent work on land vertebrate fossil assemblages includes that of Dodson (1983), Fox (1975, 1984), and Currie (1982).

Radiometric dates for parts of the Cretaceous succession have been determined by Folinsbee et al. (1963, 1965), Obradovich and Cobban (1975), and Lerbekmo and Coulter (1984, 1985a, b).

References to all investigations of Cretaceous rocks of the western basin are not possible in a summary such as this, but lists of pertinent publications may be found in the references noted.

Responsibilities and acknowledgments

This synthesis of Cretaceous history has been undertaken by contributors who are actively investigating the Cretaceous System of western Canada. It relies heavily on recent studies and compilations, all of which have incorporated data obtained from earlier investigations and interpretations. It represents the co-operation and generosity of earth-science departments of several universities, many petroleum companies, and agencies of provincial and federal governments.

The Introduction and Summary of Economic Geology were prepared by D.F. Stott, Geological Survey of Canada, who also was responsible for the overall co-ordination of this subchapter. Biostratigraphy and Basin History were the responsibility of W.G.E. Caldwell, formerly of the Department of Geological Sciences, the University of Saskatchewan, and more recently, at the University of Western Ontario. He had the co-operation and assistance of G. Norris and R.A. Fensome, University of Toronto, A.R. Sweet, Geological Survey of Canada, and D.A. Russell, National Museum of Natural Sciences. The section on the Beaufort-Mackenzie area was prepared by J. Dixon, Geological Survey of Canada. The description of the Rocky Mountain region was written by D.F. Stott. D. Cant of the Alberta Geological Survey (more recently, Geological Survey of Canada) summarized the Mannville succession of Alberta, and J.E. Christopher of Saskatchewan Energy and Mines the equivalent succession in Saskatchewan. The Upper Cretaceous nonmarine and marine rocks of the basin were summarized by E. Koster, formerly of the Alberta Geological Survey, and the Tyrrell Museum of Palaeontology and more recently, Ontario Science Centre. F. Simpson, Department of Geology, University of Windsor, contributed the discussions on the Colorado Group of Alberta and Saskatchewan and the Montana Group of Saskatchewan. D.H. McNeil, Geological Survey of Canada, summarized the Cretaceous of the Manitoba Escarpment on the eastern side of the Western Canada Basin.

J.M. Porter, Canadian Superior Oil Limited, has made available regional maps prepared by himself, R.G. McCrossan, formerly of Esso Resources Canada Limited, and D.G. Cook, Geological Survey of Canada. Canterra Energy Ltd. (formerly Aquitaine Company of Canada Ltd.) provided a map of the pre-Cretaceous bedrock surface. M. Ranger, of Gulf Canada Resources, has made available his maps showing the distribution and thickness of Cretaceous Mannville rocks. Use has been made of a report prepared by R. de Wit for Applied Geoscience and Technology (AGAT) Consultants Limited on the Nikanassin of northeastern British Columbia. Canadian Hunter Exploration Limited provided a manuscript copy of its AAPG Memoir on the Deep Basin of Alberta.

C.R. Stelck of the University of Alberta and J.H. Wall of the Geological Survey of Canada critically reviewed the manuscript. Their intimate knowledge of the Cretaceous System of western Canada has contributed significantly to improvement of the text.

BIOSTRATIGRAPHY

Biogeography and paleogeography

Western Canada Basin was part of the Cretaceous Western Interior Basin of North America, an epeiric seaway of continental proportions; it stretched some 4800 km from the Beaufort Sea to the Gulf of Mexico, and some 1600 km from Utah to Iowa. In the north, the seaway may have been narrower. During much of the Late Cretaceous it may have been linked to the circum-polar ocean by way of Hudson and Baffin bays (Williams and Stelck, 1975).

Geographic variations in the faunas and floras inhabiting the seaway and its margins may be attributed to two main factors. Firstly, biotic elements were of both

STAGES		SELECTED MOLLUSCAN ZONES	FORAMINIFERAL ZONES	FORAMINIFERAL SUBZONES
PALEOCENE			Nonmarine deposits	
MAASTRICHTIAN		[Dinosaurs (*Triceratops*)] *Baculites grandis* *Baculites baculus* *Baculites eliasi* *Baculites jenseni* *Baculites reesidei*	*Haplophragmoides excavata* *Anomalinoides* sp.	
CAMPANIAN		*Baculites cuneatus* *Baculites compressus* *Didymoceras cheyennense* *Exiteloceras jenneyi* *Didymoceras stevensoni* *Baculites gregoryensis* *Baculites perplexus* *Baculites asperiformis* *Baculites maclearni* *Baculites obtusus* *Baculites* sp. *Scaphites hippocrepis*	*Haplophragmoides fraseri* *Eoeponidella linki* *Glomospira corona* *Trochammina ribstonensis*	*Praebulimina kickapooensis* *Gaudryina bearpawensis* *Dorothia* cf. *D. smokyensis* *Quinqueloculina sphaera* *Spiroplectammina sigmoidina*
SANTONIAN		*Desmoscaphites bassleri* *Desmoscaphites erdmanni* *Clioscaphites choteauensis* *Clioscaphites vermiformis* *Scaphites depressus*	*Globigerinelloides* sp. *Bullopora laevis*	*Heterohelix* cf. *H. reussi* *Gavelinella henbesti*
CONIACIAN		*Scaphites ventricosus* *Scaphites preventricosus*	*Trochammina* sp.	
TURONIAN		*Collignoniceras woollgari* *Inoceramus labiatus* *Watinoceras reesidei*	*Pseudoclavulina* sp. *Hedbergella loetterlei*	*Whiteinella aprica* *Clavihedbergella simplex*
CENOMANIAN		*Dunveganoceras hagei* *Dunveganoceras* cf. *D. parvum* *Dunveganoceras albertense* *Dunveganoceras* cf. *D. conditum* *Acanthoceras athabascense* *Beattonoceras beattonense*	*Flabellammina gleddiei* *Verneuilinoides perplexus* *Textularia alcesensis*	*Haplophragmoides spiritense* *Ammobaculites pacalis* *Gaudryina irenensis* *Ammobaculites gravenori*
ALBIAN		*Neogastroplites maclearni* *Neogastroplites muelleri* *Neogastroplites cornutus* *Neogastroplites haasi* *Inoceramus comancheanus* *Stelckiceras liardense* *Gastroplites allani* *Gastroplites kingi* *Pseudopulchellia pattoni* *Arcthoplites macconnelli* *Arcthoplites irenense* *Lemuroceras* cf. *L. indicum* *Cleoniceras* cf. *C. subbaylei* *Pachygrycia* spp.	*Miliammina manitobensis* *Haplophragmoides gigas* *Gaudryina nanushukensis*	*Haplophragmium swareni* *Haplophragmoides postis goodrichi* *Verneuilina canadensis* *Ammobaculites wenonahae* *Ammobaculites* sp. *Haplophragmoides multiplum* *Marginulinopsis collinsi*– *Verneuilinoides cummingensis* *Trochammina mcmurrayensis* *Rectobolivina* sp.
APTIAN		Absent *Aucellina* of *aptiensis-causasica* gp.	Foraminifera poorly known. Zones not yet established.	
NEOCOMIAN	BARREMIAN	*Crioceratites* cf. *C. lardii* *Crioceratites* cf. *C. nolani* *Oxyteuthis* cf. *O. jasikowi*		
NEOCOMIAN	HAUTERIVIAN	*Craspedodiscus* cf. *C. discofalcatus* *Simbirskites* cf. *S. kleini* Unnamed zone Marine rocks unknown		
NEOCOMIAN	VALANGINIAN	*Buchia inflata* *Buchia* cf. *B. keyserlingi*		
NEOCOMIAN	BERRIASIAN	*Buchia* n. sp. aff. *B. volgensis* *Buchia okensis*		
TITHONIAN		*Buchia fischeriana* *Buchia piochii*		

GSC

northern and southern derivation, and although many elements migrated deeply into the basin to form a broad and fluctuating zone of overlap that became an important endemic centre (Kauffman, 1973, 1977b, 1984b), biogeographic provincialism was maintained. Secondly, a northward-declining palaeotemperature gradient, substantiated by oxygen-isotope measurements, influenced the distribution of biota. Northward decline in the number and variety of many biotic elements may be attributable to palaeolatitude. Such an interpretation finds additional support in an observed northward decline in the proportion and purity of carbonate sediments deposited when the Western Interior Sea was most widespread. The existence of two biogeographic provinces, one of northern or "Boreal" ancestry and the other of southern or Gulf Coast ancestry, long has been recognized (for example, Jeletzky, 1971a, b). These now are formalized as the Northern Interior Subprovince, extending south into Wyoming, and the Southern Interior Subprovince, extending north into Colorado. Both belong to the North American Province and lie within the global North Temperate Realm (Kauffman, 1973). Recently, an intervening Central Interior Subprovince, spanning the approximate latitudinal extent of Wyoming (Kauffman, 1984b), was introduced and corresponds to the southern part of the Western Interior Endemic Centre. Western Canada Basin lay within the Northern Interior Subprovince, and only in Late Albian time, when northern and southern gulfs became confluent, and at such later times as the seaway expanded to unusually wide limits, did southern biotic elements and their descendents form components of the subprovincial biota (Kauffman, 1984b).

Zones and stages

Many fossil groups are becoming better known within Canada and hold potential for stratigraphic zonal schemes, but the molluscs continue to offer the most complete, detailed, and reliably tested scheme for regional biostratigraphic and chronostratigraphic analysis. Ammonites and inoceramid bivalves have proved to be key fossils in effecting long-range correlation. Other components of the biota are measured against the standard molluscan zones (Fig. 4I.4).

Although the North American Commission on Stratigraphic Nomenclature (1983) suggests a clear distinction between zones (biostratigraphic divisions) and stages (chronostratigraphic divisions), the two are intimately related, as is evident from the molluscan zones. These zones form the basis of a relative time-scale used in the basin, and boundaries between stages are drawn in the original d'Orbignyan sense to coincide with boundaries between zones. Some of these boundaries are marked by important extinction events (Kauffman, 1984a). For various reasons, the international stage boundaries have not yet been reliably established in many parts of the Western Canada Basin. Remoteness and geographical restriction of some sequences, limited exposure in a heavily glaciated terrain, paucity of molluscan fossils, and disconformities or paraconformities all have acted as constraints (Caldwell and North, 1984).

Berriasian through Barremian rocks are preserved only at the northern limits of the Western Interior Basin (Jeletzky, 1964, 1968, 1971c, 1973; Young et al., 1976; Yorath and Cook, 1981), and Berriasian and Valanginian rocks (possibly with some Upper Barremian rocks) occur in the local Peace River basin (Stott, 1967a, 1969, 1975). At these sites, the paleontological criteria for stage boundaries are weak, and although Jeletzky (1984) now has identified the Tithonian-Berriasian (Jurassic-Cretaceous) boundary, many of the higher boundaries are only tentative (Jeletzky, 1968, 1973). Aptian rocks also are present in most northerly parts of the Western Interior Basin, and some younger Aptian rocks may be preserved in the Peace River basin. However, the base of the stage can be placed only within a broad zone named for *Aucellina* of the *aptiensis-caucasica* group. These aucellinids may be a reliable guide to the upper limit of the Aptian in this region, but in other continents they have been reported also in Lower Albian rocks (Kauffman, 1979).

The Albian is well represented in the Western Canada Basin (Caldwell, 1984a; Yorath and Cook, 1981; Dixon, 1986a). The precise position of the base of the stage is uncertain; it may coincide with *Pachygrycia* spp. (Jeletzky and Stelck, 1981), an ammonite recorded only in the District of Mackenzie. The thick Albian section in the Peace River district has yielded a sequence of ammonite and foraminiferal faunas that is without peer in the entire Western Interior Basin. The upper boundary of the Albian has been placed by Warren and Stelck (1969) between the subzone of Neogastroplites maclearni and that of *N. septimus*. The *Neogastroplites* Zone is succeeded by the Early Cenomanian ammonite zones of *Irenicoceras bahani* and *Beattonoceras beattonense*, which are confined to the Peace River sequence. Farther east, a hiatus may mark the Albian-Cenomanian boundary.

Upper Cretaceous sections in the western foredeep of the southern part of Western Canada Basin are relatively complete; sections on the eastern platform are markedly less complete, less so even than those outlined for the eastern platform in the United States by Kauffman et al. (1977). Jeletzky (1968) presented the biostratigraphic basis for delineating the principal stage boundaries between the top of the Cenomanian and the top of the Santonian, and his interpretation, with subsequent revisions and with the addition of data on foraminiferal distribution relative to these boundaries, has been summarized by Caldwell and North (1984).

The Campanian-Maastrichtian boundary, at the base of the *B. reesidei* Zone, (Cobban in Obradovich and Cobban, 1975; compare Jeletzky, 1968), can be drawn precisely in southwestern Saskatchewan (Caldwell, 1968). To the east, the boundary beds have been eroded from large tracts of the basin, but to the west, the boundary may be traced using both baculitid ammonites and the associated foraminiferal faunas almost to the Rocky Mountain Foothills (Fig. 4I.5).

Marine waters withdrew from the northern portion of the Western Interior Basin in Maastrichtian times, and the Maastrichtian-Paleocene (Cretaceous-Tertiary) boundary is based on terrestrial fossils, particularly pollen and spores.

Figure 4I.4. Molluscan and foraminiferal zones and subzones of the Cretaceous System in Western Canada Basin (W.G.E. Caldwell).

Molluscs

Molluscs, including ammonites, belemnites, bivalves, and gastropods all feature in zonal systems of the Western Interior Basin, but of these, the ammonites (craspeditids, ancyloceratids, gastroplitinids, acanthoceratids, collignoniceratids, scaphitids, and baculitids particularly) determine and identify most zones. The ammonites have been studied to the point that their lines of phyletic development and their paths of migration are reasonably well understood. Most component zones in the biostratigraphic framework are lineage zones or phylozones, and the species that discriminate them are mostly short-ranging chronospecies; for example, the twenty baculitid zones between the upper Lower Campanian and the lower Lower Maastrichtian span only 11 million years as determined from K-Ar isotopic ages (Gill and Cobban, 1973). Among other molluscan groups, the benthonic bivalves warrant special mention; their dispersal and biostratigraphic potential may have been seriously underestimated (Kauffman, 1975). Inoceramids probably rank second to ammonites as diagnostic elements of zonal assemblages, and ostreids and such genera as the pectinaceid *Buchia* and the lucinaceid *Thyasira* have demonstrable biostratigraphic value at different levels within the Cretaceous (Kauffman, 1965, 1977a; Jeletzky, 1964, 1968, 1970, 1971a, b, 1973, 1984). The molluscs diversified and thrived within the basin to the extent that their fossilized remains permit construction of a framework of biozones and chronozones as refined as that established in the Gulf Coast or in the classical outcrop belts of Europe.

Differences in the biogeographical subprovinces find expression in somewhat different zonal schemes for the northern and the southern portions of the Western Interior Basin. The sequence of molluscan zones that applies in the southern portion has been compiled and refined largely by W.A. Cobban of the United States Geological Survey. The sequence of molluscan zones applied in the northern portion of the basin has been compiled and refined largely by J.A. Jeletzky of the Geological Survey of Canada. The basis for the biostratigraphic framework and a comparison of the framework with that prevailing in the United States and in Europe is contained in Jeletzky (1968), variations in the framework with geographical region in Jeletzky (1970), and the palaeobiogeographic implications of many of the zonal ammonites in Jeletzky (1971 a, b). Because the northern arm of the Western Interior Sea transgressed continuously, if in pulsatory fashion, into the northern portion of the basin from the onset of Early Cretaceous time and the southern arm of the sea did not transgress extensively into the southern portion of the basin until latest Middle or earliest Late Albian time, the marine Lower Cretaceous sequence is much more complete in Canada than it is in the United States. Jeletzky's zonal scheme has the advantage, therefore, of being applicable to almost the entire Cretaceous System. Cobban's zonal scheme, however, is more refined and simpler. Caldwell (1968, 1982) demonstrated that Cobban's detailed sequence of Upper Campanian-Lower Maastrichtian zones based on baculitid ammonites is as applicable in southern Canada as it is in the northern Interior Plains of the United States. Kauffman (1979) outlined a zonal framework for the Arctic and Western Interior that draws on Cobban's and Jeletzky's schemes and supplements them, particularly with data on inoceramids and other bivalves. Illustrations of the more important molluscan species found in the Canadian Western Interior Basin may be found in Jeletzky's (1964, 1970) reviews of molluscan biostratigraphy and in other references quoted herein.

Jeletzky (1984) reviewed the biostratigraphy of the Jurassic-Cretaceous boundary beds and, although the work concentrates on western British Columbia, much of his analysis is applicable to the Western Interior. The Tithonian-Berriasian boundary is drawn between the *Buchia terebratuloides* (*sensu lato*) Zone and the *B. okensis* (*sensu stricto*) Zone, and these zones contain other species of *Buchia* and species of craspeditid ammonites. Succeeding Berriasian and Valanginian beds are included in a succession of zones characterized by species of *Buchia* and associated craspeditids (Jeletzky, 1973). Marine rocks of the Hauterivian lack diagnostic fossils in northern Canada, although some Upper Hauterivian beds in the Richardson Mountains and Porcupine Plateau contain the Boreal ammonite *Simbirskites* (Jeletzky, 1968). Consequently, Hauterivian strata are identified principally by stratigraphic position. Marine rocks of the Barremian from the same areas are zoned on various ancyloceratid ammonites associated with belemnites and bivalves.

Rocks in the northeastern Richardson Mountains, Porcupine Plateau, and probably also in Peace River Basin contain bivalves of the *Aucellina aptiensis-caucasica* group, which may be restricted to the Upper Barremian and Aptian, and also, species of the ancyloceratid *Tropaeum* have been found in the Richardson Mountains.

Marine Albian deposits are much more widespread than those of Berriasian to Aptian age. Revisions by Jeletzky (1980a) and Jeletzky and Stelck (1981) of many of the biostratigraphically important Early and Middle Albian ammonites have profound implications for the concepts of the Albian and Early Cenomanian ammonite faunas. The new desmoceratid genus *Pachygrycia* Jeletzky and Stelck, defines the lowest Albian zone and may mark the base of the Lower Albian. *Beudanticeras (Grantziceras) affine* species group ranges from the top of the *Pachygrycia* spp. Zone through the Lower Albian accompanied by species of *Arcthoplites* Spath (*Lemuroceras* of Jeletzky, 1968) and *Freboldiceras* Imlay (Stelck and Kramers, 1980). Within the northern basin, the lower part of the Middle Albian is characterized by *Beudanticeras (Grantziceras) affine* and associated species, but without accompanying species of Arcthoplites. The upper zones of the Middle Albian are distinguished by such gastroplitinid ammonites (Jeletzky, 1980a) as *Pseudopulchellia pattoni* Imlay, *Gastroplites* spp., and *Stelckiceras liardense* (Whiteaves). Successive species of *Neogastroplites* McLearn (Warren and Stelck, 1969), a genus endemic to the Western Interior Basin, serve as the basis for a series of subzones through the Upper Albian.

Lower Cenomanian rocks in the Peace River district are unique insofar as they contain ammonites that may be used to establish a zonal sequence. Basal Cenomanian beds are marked by the last of the sequential species of *Neogastroplites*, (Warren and Stelck, 1969), and succeeding beds by the distinct genera *Irenicoceras* Warren and Stelck and *Beattonoceras* Warren and Stelck, whose phyletic roots are not fully understood (Jeletzky, 1980a). Middle and Upper Cenomanian rocks in the northern Western Interior

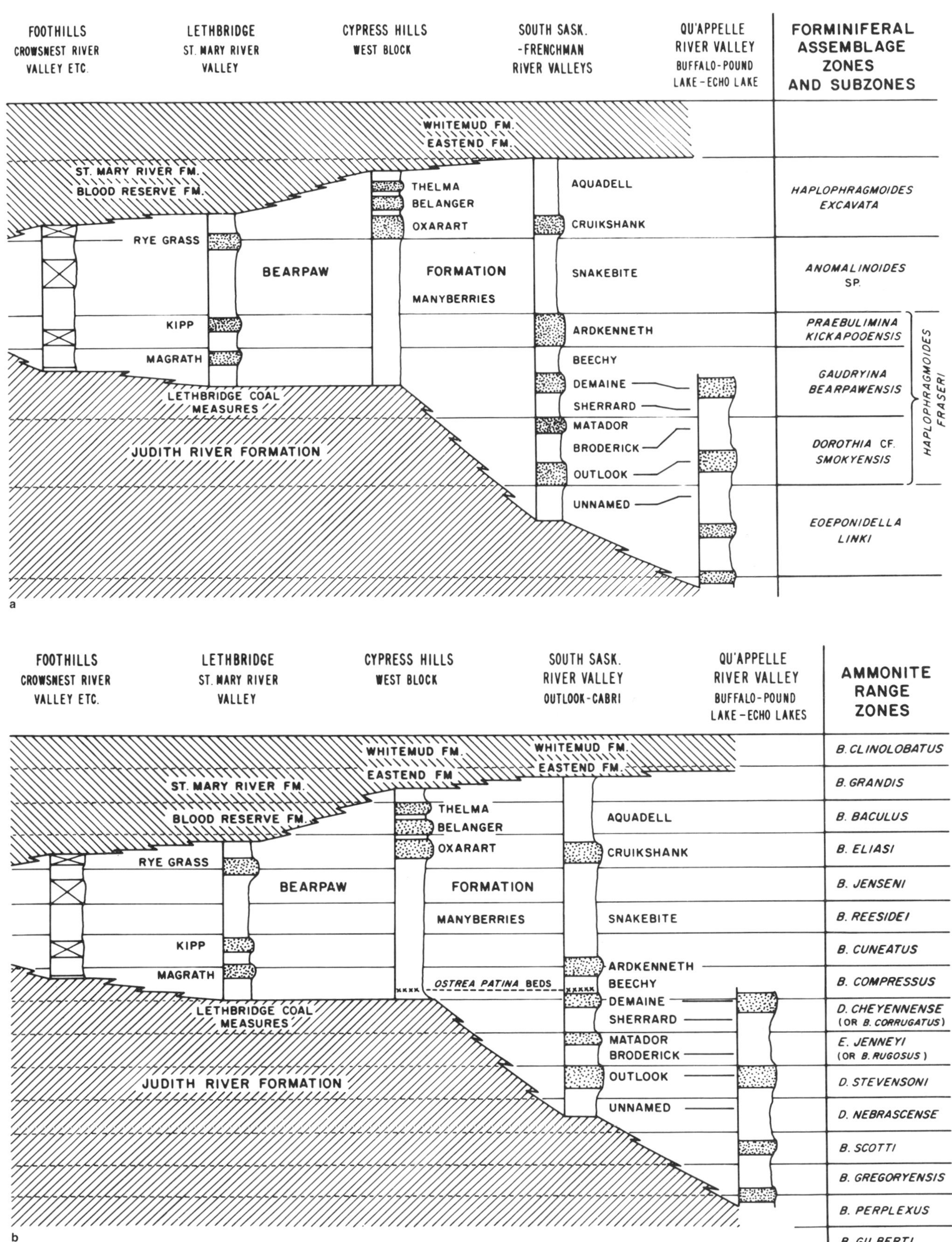

Figure 4I.5. Five composite reference sections of the Bearpaw Formation set in a relative time frame of Late Campanian-Early Maastrichtian ammonites (a) and foraminifera (b) (from Caldwell, 1982).

Basin are distinguished by strongly tuberculate acanthoceratid ammonites, in association with the bivalves *Inoceramus dunveganensis* McLearn and I. *rutherfordi*, followed by successive species of *Dunveganoceras* Warren and Stelck.

The acanthoceratids range into the Lower Turonian, where they are represented by *Watinoceras* Warren in association with the widespread bivalve *Mytiloides labiatus* (Schlotheim). The upper part of the Lower Turonian and the remainder of the Turonian are distinguished by collignoniceratid ammonites. The detailed succession of collignoniceratid and scaphitid zones, identified by Cobban in the Turonian rocks of the United States, has not yet been recognized in Canada (Cobban in Obradovich and Cobban, 1975).

Species of *Scaphites* are the dominant Coniacian and Santonian zone fossils and are accompanied by distinctive species of *Inoceramus*. Most of Cobban's clioscaphitid zones in the United States occur also in Canada. The uppermost Santonian throughout the Western Interior Basin is distinguished by *Desmoscaphites* Reeside, with a single zone in the northern part of the basin and two in the south.

The scaphitid ammonites continue to be important indices in the Campanian and Maastrichtian. Species of two lineages of baculitid ammonites designate nineteen of twenty-five Campanian and Lower Maastrichtian zones established in the United States but applicable also in Canada. The first line belongs to an endemic stock, which developed between Early Campanian and Early Maastrichtian time. The second originated in the Gulf Coast (Gill and Cobban, 1973) and distinguishes four Upper Campanian zones.

Foraminifera

Benthonic and planktonic foraminifers have been used to establish a Cretaceous zonal framework for Western Canada Basin (Fig. 4I.4, 4I.6; Caldwell et al., 1978). Currently the zones span Albian to Early Maastrichtian. The foraminiferal zonal scheme for the southern Plains includes brief comment on the regional stratigraphy, an historical outline of studies foundational to the zonal scheme, and a systematic treatment of seventeen zones with twenty-two subzones.

The diversity of Cretaceous planktonic foraminifers becomes reduced from the Gulf Coast through the Western Interior Basin, almost certainly in response to cooler paleoclimate with increasing palaeolatitude. Diversity within the Canadian Interior is too low (keeled forms having been almost totally eliminated) and stratigraphic distribution too irregular for even a modified version of the

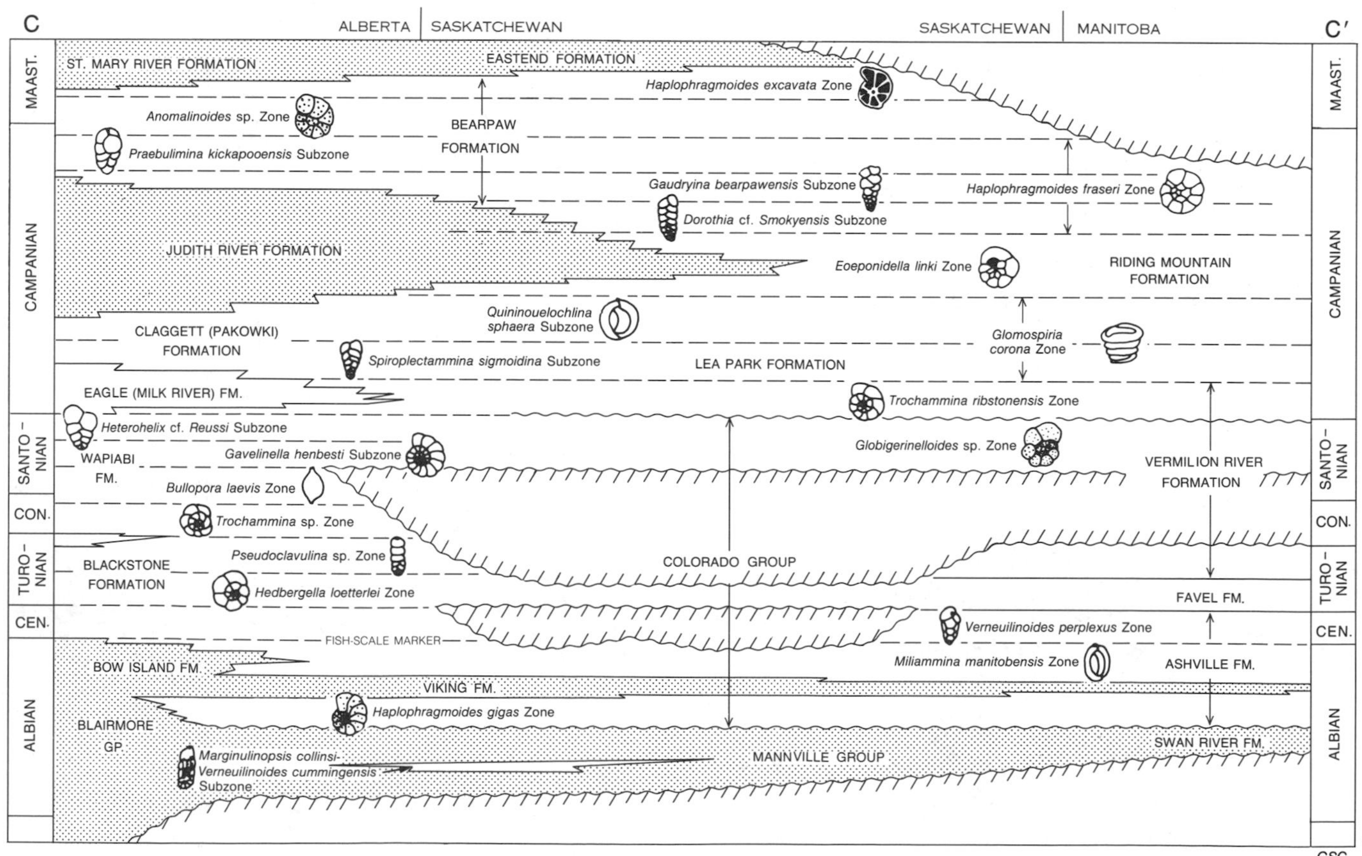

Figure 4I.6. Schematic cross-section through the southern plains of Alberta, Saskatchewan and Manitoba, showing the distribution of some foraminiferal zones and subzones and exposing the magnitude of the hiatuses represented by two pre-Campanian disconformities (from Caldwell, 1982).

planktonic zones of Pessagno (1967) to be applied. Planktonic foraminifers dominate only two faunas, and these are believed to have been carried northward in warm-water transgressions during latest Cenomanian-Early Turonian and Coniacian-Santonian times (Kauffman, 1984b).

Benthonic foraminifers are much more common and, although these foraminifers fall far short of the ammonites and inoceramid bivalves in terms of biostratigraphic resolution, the zonal framework based upon them seems to be applicable over a vast region. Detailed phyletic lines have not been well established. Many stocks evolved slowly relative to the ammonites and for that reason give rise to zones with markedly greater stratigraphic spans. Whereas Cobban (in Obradovich and Cobban, 1975) distinguished sixty ammonite zones and Kauffman (1977a) eighty-eight ammonite-bivalve zones in the Upper Albian through Lower Maastrichtian, Caldwell et al. (1978) distinguished only twenty-four foraminiferal zones and subzones for the same interval. Furthermore, whereas many of the ammonite zones are interval (range) zones, founded upon chronospecies, the benthonic (and also the planktonic) foraminiferal zones are mostly assemblage zones, more specifically Oppel zones, based upon associations of morphospecies. As such, the foraminiferal zones are inherently less reliable guides to chronostratigraphy. For two main reasons, however, the foraminiferal zonal scheme has been important to the Cretaceous stratigraphy of the southern Interior Plains of Canada. Firstly, within the Western Interior Basin, the number and variety of molluscs, like the planktonic foraminifers, become reduced from south to north and in parts of the northern basin few zone molluscs may be present. The foraminiferal zonal scheme then provides a viable alternative to the molluscan scheme in interpreting stratigraphic relationships, for example in the outcrop belt of the Manitoba Escarpment (McNeil and Caldwell, 1981; McNeil, 1984). Secondly, much of the Canadian segment of the basin is blanketed by Pleistocene surficial deposits. Interpretation of the bedrock stratigraphy, therefore, is much more dependent on cored sections from the subsurface. Generally these reveal fewer molluscs but yield more numerous and better-preserved foraminifers than weathered outcrop sections.

Little has been published on Early Cretaceous foraminifera in the District of Mackenzie. The Berriasian-Albian faunas of the Richardson Mountains and Valanginian faunas in the Peace River basin are entirely benthonic and are quite diverse, some assemblages being agglutinated-walled species, and others a mixture of agglutinated-and calcareous-walled species.

Between fifteen and twenty families of foraminifers are known in the Albian of the southern Interior Plains of Canada, with ataxophragmiids, lituolids, textulariids, trochamminids, and nodosariids being most common. Onc of the most distinctive faunas is that of the Lower Albian *Marginulinopsis collinsi-Verneuilinoides cummingensis* Subzone of the *Gaudryina nanushukensis* Zone, a fauna with a relatively high content of calcareous-walled species (anomalinids, nodosariids, polymorphinids, and spirillinids). Information on this diverse suite, as it occurs in the Moosebar Formation of Alberta, has been provided by Wall (in McLean and Wall, 1981) and in the Mannville Group in the Lloydminster heavy oil fields (Alberta-Saskatchewan borderland) by Caldwell (1984b). Koke and Stelck (1984) have substantiated the validity of the *A. wenonahae* Subzone and the *H. gigas* assemblage. On the eastern side of the basin, McNeil and Caldwell (1981) have traced the *H. gigas* Zone and the *Verneuilina canadensis* Subzone along the length of the Manitoba Escarpment. At a still higher stratigraphic level, Stelck and Hedinger (1983) have given added definition to the *Haplophragmium swareni* Subzone, thereby contributing to locating and tracing the Albian-Cenomanian (Lower-Upper Cretaceous) boundary. Modification of age assignments of the foraminiferal zones are summarized by Caldwell and North (1984).

Three zonal assemblages occur in the uppermost part of the Albian and Cenomanian. Lower Turonian substages contain representatives of ten families, and ataxophragmiids, lituolids, textulariids, and trochamminids continue to be the most common components, with planktonic heterohelicids making an appearance in the *V. perplexus* Zone and turrilinids in the *F. gleddiei* Zone. The *T. alcesensis* and *F. gleddiei* zones remain known only from the Peace River district. Their absence eastwards signals the presence of disconformities (Fig. 4I.6). The distribution of the foraminiferal zones in combination with the apparent absence of neogastroplitid ammonites suggests the presence of a major hiatus, embracing the Albian-Cenomanian boundary in the Manitoba Escarpment (Caldwell, 1984a; Caldwell and North, 1984).

Two zones, *Hedbergella loetterlei* and *Globigerinelloides* sp., respectively of Early Turonian and Santonian or possibly Early Campanian age, are dominated by planktonic heterohelicids, planomalinids, and rotaliporids, although they also carry a benthonic component of anomalinids and turrilinids. Both have exceedingly widespread geographical distributions and, in contrast to so many of the older foraminiferal assemblages belonging to the Northern Interior Subprovince, may be linked to the Southern Interior faunas.

Three zonal assemblages separating the two planktonic assemblages are composed entirely of benthonic elements, and become increasingly diverse in upward sequence, but are not easily extended far to the east (McNeil and Caldwell, 1981).

The Campanian and Lower Maastrichtian foraminiferal zones are composed of agglutinated-walled or mixed agglutinated- and calcareous-walled benthonic species, with only rare planktonic associates. Twenty-six families are represented in these six zonal assemblages, of which eighteen are composed of calcareous-walled elements. The *G. corona* and *Anomalinoides* sp. assemblages are by far the most diversified and contain the highest proportions of calcareous-walled elements (McNeil and Caldwell, 1981). Uppermost Campanian and lowest Maastrichtian rocks are eroded from most of the escarpment and the eastern plains or are unexposed. Farther west, between Saskatchewan and the Rocky Mountain Foothills, the upper four zones were found in the Bearpaw transgressive-regressive cycle (Caldwell, 1982; Fig. 4I.5).

Dinoflagellates

Dinoflagellates form an important part of the marine, post-Triassic fossil record. In terms of diversity, they reached their acme between the Late Jurassic and Paleogene. In Cretaceous sequences, therefore, they are abundant and diverse, and they constitute valuable biostratigraphic indices. Most Cretaceous dinoflagellates belong primarily to the gonyaulacalean and peridinialean stocks. At the base of the system, gonyaulacalean dinoflagellates are dominant; species of the *Achomosphaera* Evitt-*Spiniferities* Mantell complex appear for the first time in uppermost Jurassic or lowermost Cretaceous strata. Gonyaulacalean cysts continue to be diverse and numerous throughout the Cretaceous, amongst which, chorate genera such as *Oligosphaeridium* Davey and Williams, *Surculosphaeridium* Davey et al., and *Cordosphaeridium* Eisenack are prominent and biostratigraphically valuable. Ceratiacean dinoflagellates and their allies (e.g. *Muderongia* Cookson and Eisenack, *Phoberocysta* Millioud and *Cyclonephelium* Deflandre and Cookson appear initially in the Upper Jurassic, increase in number and diversity in the Lower Cretaceous, and decline in diversity through the Upper Cretaceous. *Pareodinia* Deflandre, *Batioladinium* Brideaux, and their relatives are important in the lowest Cretaceous strata but decrease in importance upwards in the section and have not been widely reported above the Barremian. The peridinialean stock, which is of major importance in Upper Cretaceous and Tertiary strata, is first recognized in Lower Cretaceous (or possibly uppermost Jurassic) strata. Earliest representatives, mostly from the Hauterivian or Barremian, have combination archeopyles. Some of these genera continue throughout the Cretaceous, but Late Cretaceous peridinialean dinoflagellate assemblages are dominated by genera with simple intercalary archeopyles (e.g. *Chatanigiella* Vozzhennikova and *Isabelidinium* Lentin and Williams).

In Western Canada Basin, Cretaceous dinoflagellate biostratigraphy is less well known (Singh, 1964, 1971; Brideaux, 1971, 1977; McIntyre, 1974; Brideaux and McIntyre, 1975) than that of offshore eastern Canada (Williams, 1975; Bujak and Williams, 1978; Barss et al., 1979), and of Sverdrup Basin and environs of the Arctic Archipelago (Brideaux and Fisher, 1976; Pocock, 1976; Davies, 1983). Although the differences between zonal schemes in these three regions are due in part to the differences in the biostratigraphic techniques and taxonomic emphases of various workers, notable discrepancies in composition and diversity between the regional assemblages may accord with palaeolatitudinal setting and, therefore, may be paleoclimatologically controlled (Davies and Norris, 1980). Brideaux (1977) reported that Lower Cretaceous dinoflagellate assemblages from Richardson Mountains contain elements common to both Scotian Shelf and the Arctic Archipelago, together with distinctive new species and elements common to Alaskan assemblages.

The Hauterivian-Barremian sequence is not well studied outside the Scotian Shelf, but Aptian-Albian dinoflagellate assemblages of Western Interior Basin have been studied by Brideaux and McIntyre (1975) in the Horton River area, Singh (1964, 1971) in Alberta, and Brideaux (1971) in Alberta. In these upper Lower Cretaceous strata, some gonyaulacalean elements are characteristic, ceratacean elements are prominent, and peridinialean dinoflagellates become common for the first time.

McIntyre (1974) distinguished three divisions in a Santonian-Maastrichtian sequence of the Horton River area, and Nichols and Tschudy (1983) distinguished three interval zones, identified by species of *Alterbia* and *Chatangiella*, in the discontinuous Cenomanian-Campanian sequence of the central and northern Rocky Mountains in the United States. Provincialism is again in evidence amongst dinoflagellates from the Upper Cretaceous Series. For example, the Campanian dinoflagellates of North America can be separated paleogeographically into Atlantic seaboard and Western Interior assemblages (Lentin and Williams, 1980).

Biostratigraphic correlation of Cretaceous strata across North America using dinoflagellates is still at an early stage of development and correlations should become better established with a more complete knowledge of taxonomy, paleogeography, and paleoecology.

Terrestrial palynomorphs, exclusive of megaspores

Earliest Cretaceous spore-pollen floras are characterized by rapid diversification of pteridophyte genera, which originated during the Late Jurassic (Fensome, 1983). Hepaticalean genera are common for the first time. Certain pteridophyte and hepaticalean species appear at, or close to, the Jurassic-Cretaceous boundary, and these can be used for precise correlation in the Northwest Territories (Fensome, 1987). The rate of diversification slowed after Valanginian time, although the microfloras continued to be highly diverse. Angiosperms did not appear until Barremian time.

Early Cretaceous assemblages tend to be dominated by gymnosperms and rarely pteridophytes. Climatic and localized paleoecological controls seem to have been important in determining the dominant elements, which, therefore, are of little biostratigraphic value.

Pre-Albian terrestrial microfloras are not yet widely known. Burden (1984) examined diverse microfloral assemblages from the lower Mannville Group of Alberta and Montana, established three zones, and suggested a Valanginian to earliest Albian age-span for that succession. His dates indicate it to be in part older than the age assigned by Singh (1964, 1971) and Pocock (1976). Fensome and Norris (1982) compared Aptian-Cenomanian terrestrial assemblages from the Scotian Shelf with those of Alberta (Singh, 1964, 1971) and found the pteridophyte microfloras especially to be closely similar.

Detailed work in the eastern United States (Doyle, 1969; Doyle and Robbins, 1977) has demonstrated the biostratigraphic utility of early angiosperm pollen, and this work now is being applied to Aptian-Albian sections in western Canada (Singh, 1971, 1975, 1983). Small tricolpate angiosperm pollen first appear in the Middle Albian rocks of western Canada.

Cenomanian terrestrial palynofloras are markedly changed from those preceding (Singh, 1983). Schizaeaceous striate spores continued to diversify, and the Late Albian radiation of gymnospermous *Rugubivesiculites*

Pierce continued. Angiosperm pollen, however, show the most notable changes: the Albian-Cenomanian boundary is marked by the earliest appearances of small psilate triangular tricolporate pollen, large thick-walled angiosperm pollen in permanent tetrads (Singh, 1975) and large *Liliacidites* Couper; and in the Middle Cenomanian Substage, a number of new species make their appearance.

No definitive changes took place in the terrestrial flora of the midcontinental United States during the Cenomanian and Turonian stages (Nichols et al., 1982). Nichols et al. recorded the earliest triporate pollen (both *Proteacidites* Anderson and that belonging to the Normapolles) in the Coniacian and the oldest *Aquilapollenites* Rouse in the earliest Campanian. In Western Canada a single species of triporate pollen, of oculate pollen (*Azonia*) and a possible precursor to triprojectate pollen are found in the latest Turonian (McIntyre and Sweet, 1986). It is not until the Santonian that well developed triprojectate pollen, referable to the genus *Aquilapollenites*, occurs (Jarzen and Norris, 1975, with age interpretations based on Caldwell et al., 1978), following which triprojectate pollen underwent rapid radiation throughout western Canada, northwestern United States, and Siberia. Terrestrial floral provincialism in North America appears to be linked to climate and to geographic isolation of eastern from western North America by the Late Cretaceous epeiric seaway (Srivastava, 1981; Herngreen and Khlonova, 1981) superimposed upon an overall circumpolar provincialism (Samolovich, 1977).

In western North America, several zones based on pollen and spore assemblages are recognized in the Campanian-Maastrichtian sequence. In these assemblages, triprojectate, oculate, and other angiosperm pollen comparable to that of extant families are common in the mid-latitudes (Srivastava, 1970; Jarzen and Norris, 1975; Norris et al., 1975; Nichols et al., 1982). In northern Canada, as in Alaska and northeastern Siberia, a similar suite of angiosperm pollen is present but with a greater diversity of oculate pollen in the southern arctic and *Expressipollis* pollen in the northern arctic regions (Felix and Burbridge, 1973; McIntyre, 1974; Wiggins, 1976; Herngreen and Khlonova, 1981). Interpretations by Hickey et al. (1983) of these Late Cretaceous floras as being markedly diachronous from higher to lower latitudes now are considered to be erroneous and to have been based on incorrect and incomplete magnetostratigraphic data (Norris and Miall, 1984).

The highly diverse Maastrichtian palynofloras of the Western Interior Basin appear to be subtropical (Srivastava, 1970), and the floras may have analogues in parts of southeast Asia and Indomalaysia today (Jarzen, 1978). The Maastrichtian palynofloras are succeeded by Early Paleocene palynofloras dominated by fern and moss spores and later by taxodiaceous and cupressaceous pollen with significant quantities of juglandaceous and betulaceous pollen (Leffingwell, 1970; Nichols, 1973; Nichols and Ott, 1978; Fredericksen and Christopher, 1978). These floral changes at the Cretaceous-Tertiary boundary appear to be transcontinental in extent and associated with the extinction of the dinosaurs (Srivastava, 1972; Russell, 1983b; Russell and Singh, 1978; Lerbekmo et al., 1979b).

Megaspores

Although megaspores are strictly the larger, functionally female spores of heterosporous plants, the term is used also for spores of undetermined affinity having a size greater than 200 m. The seeds *Costatheca* and *Spermatites* also commonly occur in megaspore preparations from rocks of Albian and Late Cretaceous age.

Cretaceous megaspores have an affinity with either lycopods or the heterosporous ferns. Species of azonate and zonate genera have been reported by Sweet (1978a) from the Kootenay Group of latest Jurassic to earliest Cretaceous age, by Singh (1964, 1971, 1983) from Lower Cretaceous and Cenomanian formations in north-central Alberta, by Speelman and Hills (1980) from the Campanian Pakowki and Judith River formations of southern Alberta, and by Gunther and Hills (1972) from the Campanian and Maastrichtian Brazeau Formation in the Rocky Mountain Foothills of Alberta. Worldwide, species of azonate genera are usually long-ranging and therefore of limited biostratigraphic value. Restrictions in the stratigraphic range of some zonate species are documented by Tschudy (1976).

Like those of the angiosperms, the development and radiation of the heterosporous ferns took place during the Cretaceous, wherein changes in the morphology of the megaspore and associated structures suggest that two discrete lineages evolved rapidly. One lineage, referred to the Marsileaceae, reached its peak of diversity during Early to mid-Cretaceous times (Fig. 4I.7). Species of *Arcellites* have been recorded by Singh (1964, 1971, 1983) from Lower Cretaceous and Cenomanian rocks in Alberta; species of *Molaspora* by Singh (1983) from the Middle Cenomanian Dunvegan Formation of northern Alberta, and by Campbell and Untergasser (1972) and Gunther and Hills (1972) from Campanian and Maastrichtian rocks in more southerly parts of Alberta.

The second lineage of heterosporous ferns, referred to the Salvinaceae, also evolved during the Early Cretaceous Epoch, provided the genera *Ricinospora* and *Balmeisporites* be included in the family. Maximum radiation was not achieved until the Late Cretaceous Epoch (Fig. 4I.7). Species of this lineage are relatively short-ranging, making them the most significant components biostratigraphically of the Cretaceous megaspore assemblage. Records of salvinaceous megaspores from the Canadian Interior Plains include those of *Balmeisporites* in the Middle to Upper Albian rocks by Singh (1971); of *Ariadnaesporites, Balmeisporites*, and *Ricinospora* in Cenomanian rocks by Singh (1983); of *Ariadnaesporites* in Campanian rocks by Hills and Jensen (1966), Gunther and Hills (1970), and Campbell and Untergasser (1972), and of that genus, together with *Azolla, Azollopsis*, and *Ghoshispora (Balmeisporites)*, in Campanian rocks by Speelman and Hills (1980); of the above genera in Campanian rocks and, with the exception of *Ariadnaesporites*, in Maastrichtian rocks by Gunther and Hills (1972); of *Azollopsis* in Campanian and Maastrichtian rocks by Sweet and Hills (1974); of *Ghoshisopora* in Maastrichtian rocks by Srivastava and Binda (1969) and Srivastava (1971, 1978); and of *Azolla* in Maastrichtian rocks by Srivastava (1968), Snead (1969), Jain (1971), Sweet and Hills (1976), and Sweet (1978b).

Vertebrates

Stratigraphically, the most useful vertebrate-bearing sediments in Western Canada Basin are those of the Fish-Scale Marker. Rich in the disassociated bones and scales of fishes, the marker provides a convenient regional lithostratigraphic guide to the approximate position of the Lower-Upper Cretaceous boundary in western Canada (Warren and Stelck, 1969). The Fish-Scale strata are a thin northern extension of the Mowry Formation of the Western Interior Basin of the United States (Vuke, 1984).

A series of "land mammal ages" (*sensu* Savage and Russell, 1983) has been established in terrestrial facies, largely on the basis of tooth evolution (Russell, 1975; Lillegraven et al., 1979). Associated microvertebrate remains of fishes, amphibians, turtles, champsosaurs, lizards, snakes, and crocodiles have not provided data of comparable biostratigraphic importance. The mammalian assemblage from the Milk River Formation (Early Campanian) resembles those from younger Cretaceous strata in the abundance of multituberculates and marsupials. However, the Milk River contains triconodont and symmetrodont survivors from much older faunas. The Judith River (Oldman) assemblage (middle Campanian) lacks these ancient forms, and the multituberculate and marsupialian components are intermediate in dental morphology between those from the Milk River and Scollard-Frenchman assemblages. The Late Campanian-Early Maastrichtian mammals of the St. Mary River Formation appear most closely related to those of the Scollard-Frenchman formations. The latter assemblages are from the youngest Cretaceous sediments in the western interior region, and appear to represent faunas of greater

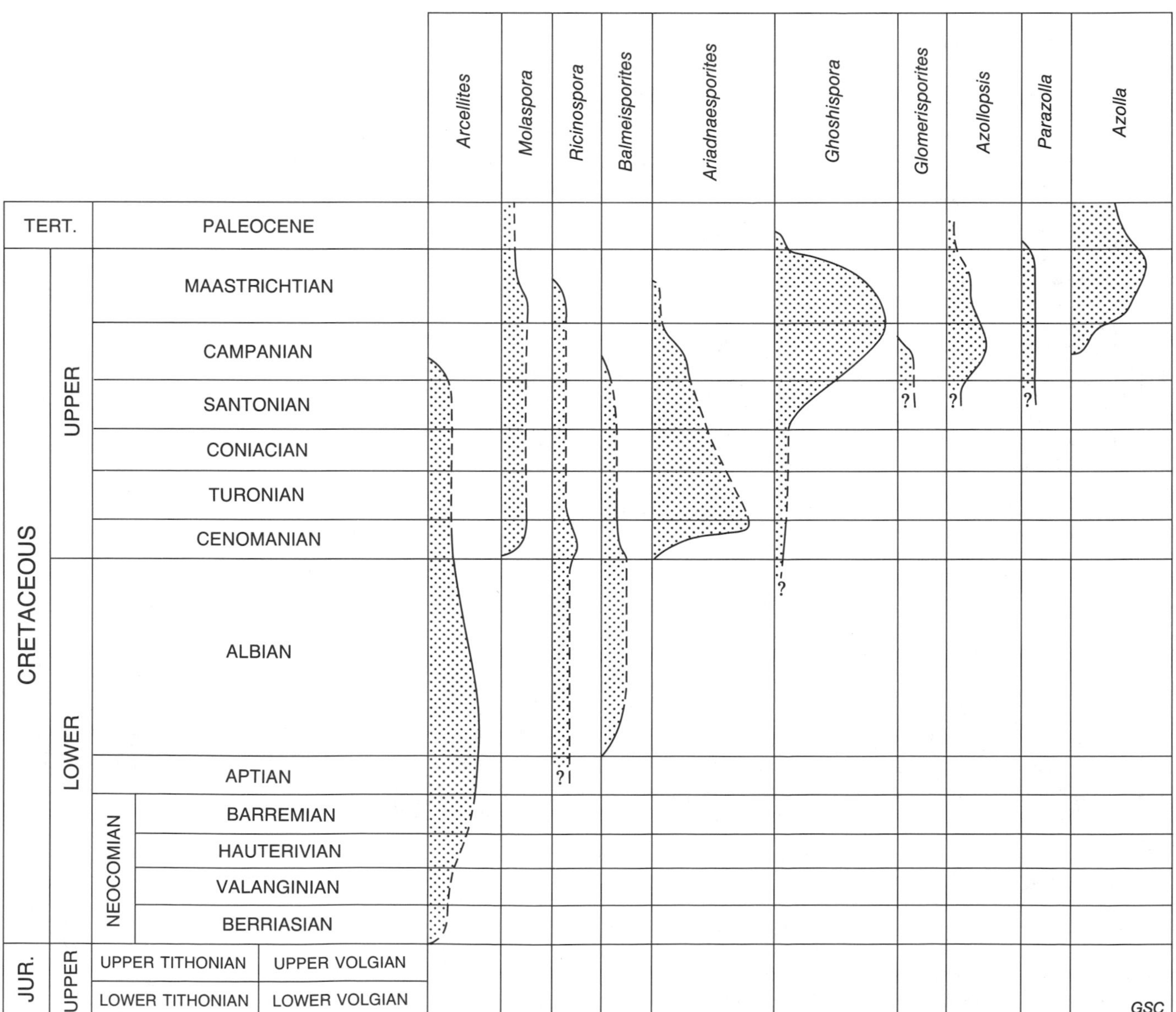

Figure 4I.7. Range chart of Cretaceous megaspores, Western Canada Basin (A.R. Sweet).

diversity. Multituberculates were numerous, and marsupials were abundant and diversified. Insectivores, lemuroid primates, and condylarths also are represented.

Dinosaurs probably evolved as rapidly as mammals (Russell, 1984a, 1984b), but the taxonomy of the dinosaurs is based on gross skeletal rather than dental morphology, and well-preserved skeletal material is usually necessary to make generic identifications (the limits of intra-specific variation are poorly established in dinosaurs). Dinosaurian assemblages seem to have been strongly affected by local environmental conditions (Russell and Chamney, 1967; Beland and Russell, 1978). No dinosaurs equivalent in age to Wealden dinosaurs in England are known from North America. The footprint assemblage from the Aptian-Albian Gething Formation (Currie and Sarjeant, 1979; Currie, 1983), demonstrates that hadrosaur-like ornithopods dominated the faunas of some Early Cretaceous environments in which flowering plants were as yet rare or absent. The formation also contains tracks of shore birds. Skeletal remains of Aptian-Albian dinosaurs are unknown in Canada, but in the United States dromaeosaurids (small raptorial carnivores), sauropods (long-necked elephantoid herbivores), hypsilophodonts (small bipedal herbivores) and nodosaurs (armoured forms lacking tail clubs) have been recovered, along with other dinosaurian taxa (Ostrom, 1970; Langston, 1974).

The dinosaurian record in Cenomanian through Santonian strata is poor, but significant change evidently occurred in the composition of dinosaurian faunas. Younger Cretaceous skeletal assemblages are dominated by tyrannosaurs (large carnivores), hadrosaurs (duck-billed dinosaurs), ceratopsians (horned dinosaurs), and ankylosaurids (armoured forms with tail clubs). Sauropods have not been identified in Canada, and the animals may have preferred relatively warmer environments to the south (Sloan, 1970). In contrast, Upper Cretaceous sediments in the Canadian prairies have produced a greater number and variety of dinosaur skeletons than any other region of the continent, or of the globe. Remains of more than 500 dinosaurs have been collected from three formations of Campanian-Maastrichtian age (Russell, 1984b).

The middle Campanian Judith River (Oldman) Formation records a broad alluvial plain that extended nearly 500 km between the western mountains and the interior seaway. Thirty-seven species of dinosaurs are recognized among the more than 300 specimens collected from this formation near Dinosaur Provincial Park, east-central Alberta (Koster and Currie, 1987). The specimens document the largest number of dinosaurian genera known from any single area of outcrop of terrestrial Mesozoic strata (Russell, 1983a, 1984b). At least ten varieties were relatively small, weighing between 50 and 100 kg. The remaining genera ranged between three and four tonnes in adult weight, and 4 to 9 m in length. The fauna was unusual in the diversity of spikes and crests that ornamented the skulls of the ceratopsians and of the numerically dominant hadrosaurs. Some of the small carnivores (troodontids) and omnivores (ornithomimids) possessed relatively the largest brains known in dinosaurs (Russell, 1983b).

The Upper Campanian-Lower Maastrichtian Horseshoe Canyon Formation of central Alberta does not contain the abundance of fish and small vertebrate remains found in Dinosaur Provincial Park. In the coaly lower part of the unit, the fauna is dominated by hadrosaurine (flat-headed) hadrosaurs, whereas the upper channel facies contains an assemblage resembling that of Dinosaur Provincial Park, with abundant lambeosaurine (hollow-crested) hadrosaurs (Russell and Chamney, 1967). The ceratopsian component features more long-frilled forms than those of older assemblages.

The terminal Cretaceous dinosaurian assemblage of Late Maastrichtian age has been identified in the Scollard and Frenchman formations over a vast area of the southern Canadian plains, and in equivalent strata as far south as central Colorado in the United States. In spite of their great geographic extent, these strata are poorly exposed and have not been well sampled in Canada. Many dinosaurs (including such familiar genera as *Tyrannosaurus, Triceratops*, and *Ankylosaurus*) were much larger than related forms from the Judith River and Horseshoe Canyon formations, and specimens of *Triceratops* were larger than those from coeval strata farther south. The apparent abundance of protoceratopsids and ankylosaurids and the scarcity of hadrosaur remains in the Scollard Formation in central Alberta, may have been an ecological effect of the proximity of the western mountains.

Late Cretaceous dinosaur assemblages from the Western Interior of North America resemble contemporaneous assemblages from eastern Asia more closely than those from Europe, Africa, and South America. Oviraptorosaurs (egg-eating dinosaurs), dromaeosaurs, tyrannosaurs, and ankylosaurs are known only from the land masses bordering the north Pacific Ocean, whereas sauropods commonly dominate dinosaur assemblages from the other three regions (Molnar, 1980). Some of these differences between the assemblages from western North America and eastern Asia may be due to a humid, maritime climate in the former area, and a semi-arid continental climate in the latter (Fox, 1978).

Marine vertebrate assemblages seem to be similar in northern and southern regions of Canada (Russell, 1967a; Bardack, 1968). Hesperornithiform diving birds occur in both, as do mosasaurs which evolved rapidly during Late Cretaceous time (Russell, 1967b). The mosasaurs can be readily compared to specimens recovered from stratotypic marine sections in Europe. Likewise, the plesiosaurs, which seem to have favoured more northerly waters, may be sufficiently common in both regions to be of some biostratigraphic usefulness.

The disappearance of the dinosaurs coincided with the terminal Cretaceous extinctions, the causes of which are currently the subject of vigorous debate (compare Alvarez et al., 1982; McLaren, 1983; and Van Valen, 1984). At present the evidence of any of the groups of vertebrates involved in the extinction having experienced previous decline is ambiguous. A layer of clay a few centimetres thick has been identified in southern Canada (Sweet and Hills, 1984; Nichols et al., 1986) as containing the Cretaceous-Tertiary boundary. The clay carries evidence suggestive of an asteroid impact and profound ecological disturbance (Wolbach et al., 1985). This stratum is apparently a chronostratigraphic datum of inter-continental extent in both marine and continental facies (Lerbekmo and Coulter, 1984, 1985a, b; Lerbekmo, 1985).

LITHOSTRATIGRAPHY

Introduction

In the south, the northeasterly tapering wedge of Upper Jurassic to Paleocene rocks is synorogenic with the Columbian and Laramide orogenies. The wedge consists of clastic deposits of western provenance in contrast with the non-orogenic, shallow-water clastic and carbonate sediments of the eastern platform. The synorogenic sediments accumulated in northeasterly migrating foredeeps (Bally et al., 1966) as the continental lithosphere subsided in response to the increased load produced by northeasterly displacement and tectonic thickening of overlying supracrustal rocks (Price, 1981). In the north, Upper Jurassic to Aptian sediments were craton-derived, whereas from Albian time onward, the sediments came from the Cordilleran orogens and accumulated in the flanking foredeeps. One large allochthonous terrane, the Intermontane Superterrane, formed by the amalgamation of several foreign terranes, had collided with North America by mid-Jurassic time, causing compression and uplift of the miogeoclinal sediments of the Omineca Belt (Monger and Price, 1979; Price, 1981; Monger et al., 1982). Continued convergence and the resulting deformation continued well into the Cretaceous and is referred to as the Columbian Orogeny.

During later Cretaceous time, the western margin of North America was again affected by renewed convergence of large composite terranes (Price, 1981; Monger et al., 1982). Miogeoclinal sediments along the Omineca Belt and the Upper Jurassic-Lower Cretaceous foreland strata were detached from the craton and thrust eastward as imbricate slices and décollement folds (Price, 1981). These tectonic movements, extending from Late Cretaceous into Tertiary time and referred to as the Laramide Orogeny, resulted in the deposition of another clastic wedge in the foreland basin.

Berriasian-Barremian sequence

The lowermost Cretaceous sediments are closely related to those of the preceding Jurassic sequence and to the development of a foredeep along the eastern margin of the Columbian Orogen. The orogenic belt lies along the boundary between the ancient continental terrace wedge (miogeocline) of North America and a large composite terrane that collided with, and was accreted to, North America in about mid-Jurassic time (Monger et al., 1982). The orogen became the dominant source of the sediments that were deposited in the trough, at least in the south, thereby resulting in reversal of the direction of sediment transport from that of earlier Phanerozoic deposits. A major transgression in Oxfordian time advanced over older Jurassic strata.

Berriasian to Barremian paleogeography of the Yukon-Mackenzie region was characterized by a broad shelf with cratonic source areas to the southeast and east (Young, 1973a; Dixon, 1982a), and no indication of a western source. In contrast, lowermost Cretaceous rocks in the Peace River region of Western Canada Basin are restricted to a narrow, northwesterly trending trough, which apparently received substantial deposits from both craton and orogen (Stott, 1984).

Northern basins

Berriasian

Berriasian strata, spanning the *Buchia okensis*, *B. uncitoides*, and *B. volgensis* zones, are fully preserved only in northernmost Yukon Territory and northern Richardson Mountains, and beneath the Mackenzie Delta. They include the red-weathering and upper members of the Husky Formation (*sensu* Jeletzky, 1967) and their equivalent in the subsurface (the upper member of the Husky Formation, *sensu* Dixon, 1982a), the uppermost Kingak Formation, and the Martin Creek Formation (Fig. 4I.8). Pre-McGuire (Valanginian) erosion removed the Martin Creek Formation and parts of the uppermost Husky Formation in the southern Richardson Mountains. West of Nahoni Range, in the Alaskan part of Kandik Basin, uppermost beds of the Glenn Shale (Brabb, 1969) appear to be equivalent to the upper Husky and Martin Creek formations. East of the Richardson Mountains and Mackenzie Delta, Berriasian strata are unknown.

Isopach and facies trends of the Husky and Martin Creek formations show no major change in basin configuration from that of latest Jurassic time. Canoe Depression and Kugmallit Trough continued to be depocentres. Tununuk High and Eskimo Lakes Arch were positive regions, the latter forming the southeastern basin margin (Dixon, 1982a, b, 1986a).

The Husky to Martin Creek succession, in excess of 400 m thick in Kugmallit Trough, coarsens upward from shale to fine-grained sandstone. The Husky shale probably was deposited in a mid- to outer-shelf environment. In northwesternmost Yukon, bioturbated, fine-grained sandstones interbedded with mudstones in the Martin Creek Formation indicate a mid- to outer-shelf setting, whereas on the eastern slopes of Richardson Mountains, an abundance of hummocky cross-stratified sandstone reflects nearshore to inner-shelf environments. In the Mackenzie Delta, the Husky-Martin Creek succession is interpreted as a prograding-shelf to barrier-island sequence (Dixon, 1982a, b).

By Late Berriasian time, maximum progradation had taken place, and beach and near-shore sediments in the east grade westward into inner, mid- and outer-shelf sediments. A southward-directed embayment may have extended into Anderson Basin, with the shoreline skirting the northern end of Eskimo Lakes Arch. Facies trends indicate an easterly to southeasterly source.

By latest Berriasian and, possibly, earliest Valanginian time, uplift occurred along Rat Uplift-Eagle Arch, and probably along Eskimo Lakes Arch. Erosion of Berriasian strata occurred south of that hinge zone.

Valanginian-?Middle Hauterivian

Valanginian to middle Hauterivian strata include McGuire and Kamik formations of the Parsons Group, which have a wide distribution throughout Porcupine Plateau and Mackenzie Delta (Fig. 4I.8). The depositional basin was similar in extent to that of previous depositional episodes, with two major depocentres, Canoe Depression and Kugmallit Trough, evident. Regional facies and isopach trends indicate a main source area to the south and southeast, and a depositional edge not far from the present erosional edge (Dixon, 1986a). Marked thickening immediately basinward of Eskimo Lakes Fault Zone provides evidence of growth faults along the Eskimo Lakes Arch. The arch appears to have been a major source of clastic sediments. No equivalent strata are found east of Tuktoyaktuk Peninsula.

McGuire strata, whose lower part is dated as Early to Middle Valanginian by *Buchia keyserlingi* and *B. crassa* (Jeletzky, 1961), are predominantly shales with interbedded siltstone and intensely bioturbated, very fine grained sandstone. These beds represent outer- to mid-shelf deposits, which grade upward to mid- and inner-shelf sediments. The initial marine sedimentation was short lived and rapidly succeeded by a phase of alluvial deposition centred over the Mackenzie Delta. Fault movement and uplift of Eskimo Lakes Arch are believed to have initiated this change in depositional pattern.

In Mackenzie Delta, the lower one-third of the Kamik Formation comprises alluvial deposits (Myhr and Young, 1975; Dixon, 1982a, b). On the western slopes of the northern Richardson Mountains, equivalent strata containing sedimentary structures of strandline to nearshore origin are marine, and farther west, inner-shelf to mid-shelf sandstones are cross-stratified and bioturbated. The overall setting during deposition of the lower Kamik Formation was that of a broad shelf over much of northern Yukon Territory, with a delta-complex in the present location of the Mackenzie Delta. An east-to-west change from littoral to shelf sedimentation can be identified in the series of thick, coarsening-upward and thickening-upward cycles of the upper two-thirds of the Kamik Formation of northern Yukon Territory (Dixon, 1986a).

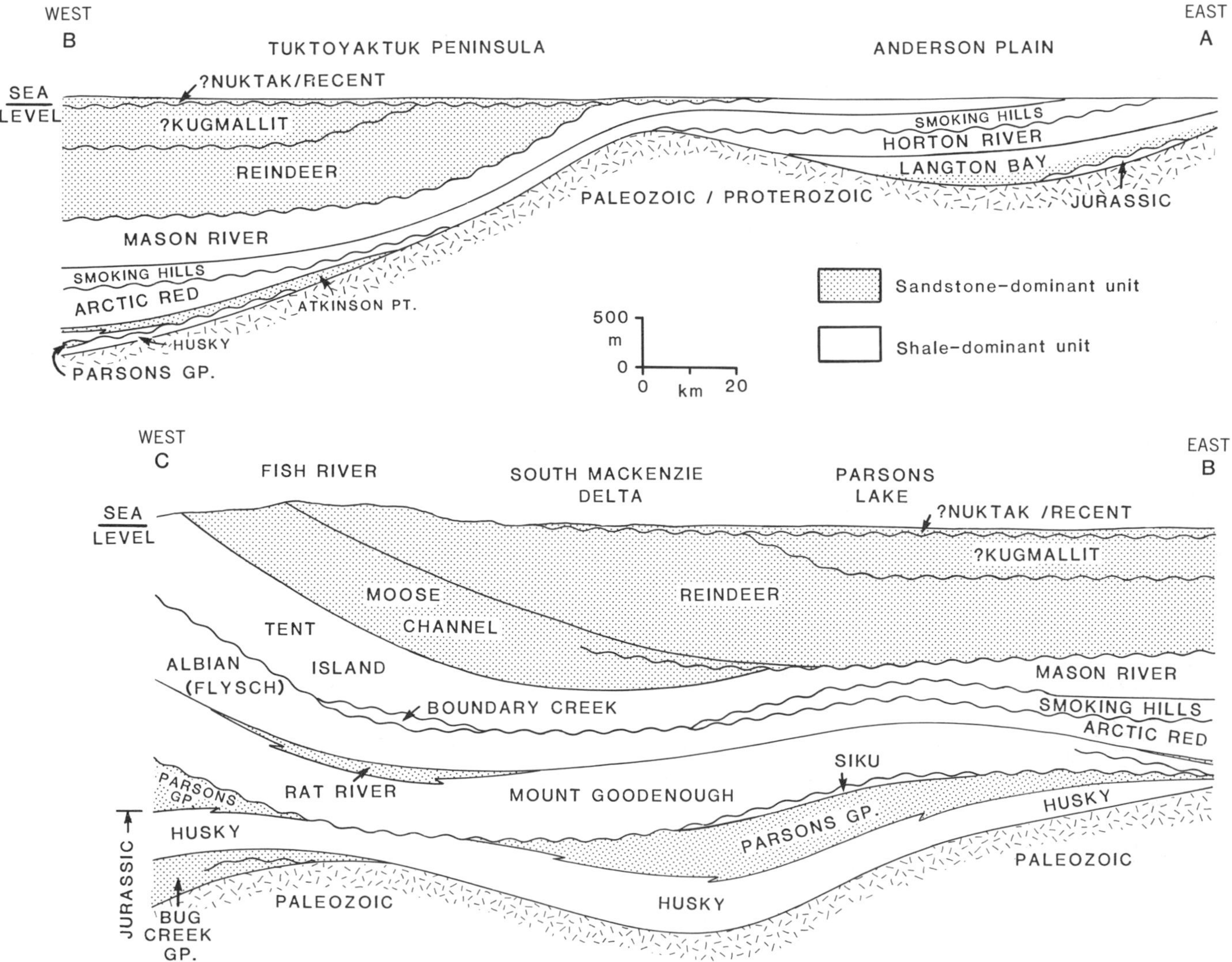

Figure 4I.8. Diagrammatic cross-section of Cretaceous deposits, from Fish River to Anderson Plain across Mackenzie Delta (J. Dixon). For line of section, see Figure 4I.15.

Latest Middle or earliest Late Hauterivian uplift resulted in a regional unconformity at the base of the Mount Goodenough Formation. Eskimo Lakes Arch, Cache Creek Uplift, and Tununuk High were elevated and in part subaerially eroded.

Upper Hauterivian-Barremian

Upper Hauterivian to Barremian strata include the Mount Goodenough Formation and the lower part of the Rat River Formation (Fig. 4I.8). These strata are up to 700 m thick in Kugmallit Trough, and upwards of 1500 m thick in the Bonnet Lake area (Jeletzky, 1971c). The Mount Goodenough Formation is dated by Jeletzky (1958, 1960, 1961) as Late Hauterivian to Barremian and possibly as young as Aptian, and the Rat River, Aptian and possibly Late Barremian.

Transgressive beds at the base of the Mount Goodenough Formation, a few metres to 100 m thick and varying from thin conglomerate to thick sandstone, were deposited across the previously uplifted terrane. Younger, shaly beds are interpreted to be offshore, shelf deposits. Sandstones in the basal Rat River Formation (Fig. 4I.9) display bioturbated beds, wave-ripple lamination, and hummocky cross-stratification, characteristic of nearshore inner-shelf sedimentation.

Coarse detritus from southwest of the Mackenzie Delta built the Late Hauterivian Barrier River-Stony Creek deltaic lobe (Jeletzky, 1975b). The shoreline is presumed to have had a trend similar to that of the underlying McGuire and Kamik formations, that is, extending along Eskimo Lakes Arch and across the western edge of Peel Plateau. These sands grade northwestward into fine deposits in Canoe Depression. The trends suggest that Kugmallit Trough and Canoe Depression filled from the south. Some evidence indicates that uplift of the basin margin had occurred by the close of Barremian time.

Figure 4I.9. Light grey, cliff-forming sandstone and dark grey shale and siltstone of the Rat River Formation, overlain disconformably by recessive shale and sandstone of the Martin House Formation. Lower Stoney Creek, on the east flank of the Richardson Mountains, northwestern District of Mackenzie (photo - D.K. Norris, ISPG 662-69).

Within the Foreland Belt between northern Yukon Territory and northeastern British Columbia, Upper Jurassic to Barremian rocks are unknown. However, reworked Jurassic palynomorphs have been recovered from Aptian and younger formations in the outer Mackenzie Fold Belt (Yorath and Cook, 1981).

Rocky Mountain Trough (Rocky Mountain Foothills)

Berriasian to Valanginian

Berriasian to Valanginian sediments, recording an early phase of the Columbian Orogeny, were deposited in the elongate Rocky Mountain Trough, which, during its early history, was bordered on the southwest by deltas (Stott, 1984) and was deepest and subsided most rapidly in the region of Peace River. Large quantities of detritus were deposited along the western margin. As the southern part of the trough filled with carbonaceous sediments, the sea retreated northward, but persisted through Berriasian and most of Valanginian time in the Peace River region. A connection with Tyaughton Trough on the western side of the geanticlinal welt has been suggested by Warren and Stelck (1958b), Jeletzky and Tipper (1968), and Jeletzky (1968, 1971a, 1971b), but the embayment may have extended from the present Yukon region. Hauterivian-Barremian rocks are largely missing owing to nondeposition or pre-Aptian erosion.

In northeastern British Columbia, marine Jurassic Fernie shale is transitionally overlain by the uppermost Jurassic to lowermost Cretaceous Minnes Group (Fig. 4I.10). This succession grades southeastward in the Alberta Foothills into the Nikanassin Formation, which in turn grades into the Kootenay Group of the southern foothills and mountains.

The Minnes Group (Ziegler and Pocock, 1960; Stott, 1967a, 1975) is most complete between Berland River in the northern Alberta Foothills and Halfway River in northeastern British Columbia. The group, over 2 km thick in Carbon Creek Basin, does not extend far into Peace River Plains nor beyond Muskwa River in the Foothills. Four formations are included in the Minnes Group near Peace River - in ascending order, Monteith, Beattie Peaks, Monach, and Bickford (Mathews, 1947; Stott, 1981). The Minnes is dominantly marine north of Pine River, but southward, the upper three formations grade laterally into nonmarine, coal-bearing beds of the Gorman Creek Formation (Fig. 4I.10, 4I.11).

Basal beds of the Monteith Formation contain *Buchia* spp. from the *B. piochi*, *B. fischeriana*, and *B. terebratuloides* Zones of Tithonian/Volgian age. The Lower Berriasian Zone of *B. okensis* is found in the lower middle part of the formation, and ammonites and *Buchia* from the middle and upper part are assigned to the Upper Berriasian *B. uncitoides* Zone. The lower part of the formation is similar to the alternating units of fine-grained sandstone and dark grey shale of the Nikanassin Formation near Athabasca River. The Monteith attains a maximum thickness of 600 m in western sections but decreases eastward to less than 300 m. Coarse quartzose sandstone at Peace River is part of an eastern fluvial-deltaic system, which introduced detritus from the craton. In the centre of the trough, sediments exhibit features of turbidity flows. Those beds grade southwestward into fluvial-deltaic, channel sandstone and carbonaceous beds, suggesting proximity to the western margin of the trough. Evidence for a western source is found in conglomerates of the Monteith, which contain pebbles and cobbles of foliated quartzites, volcanics and radiolarian and other cherts. This evidence implies the emplacement of oceanic terranes in the Omineca Belt by latest Jurassic or earliest Cretaceous time.

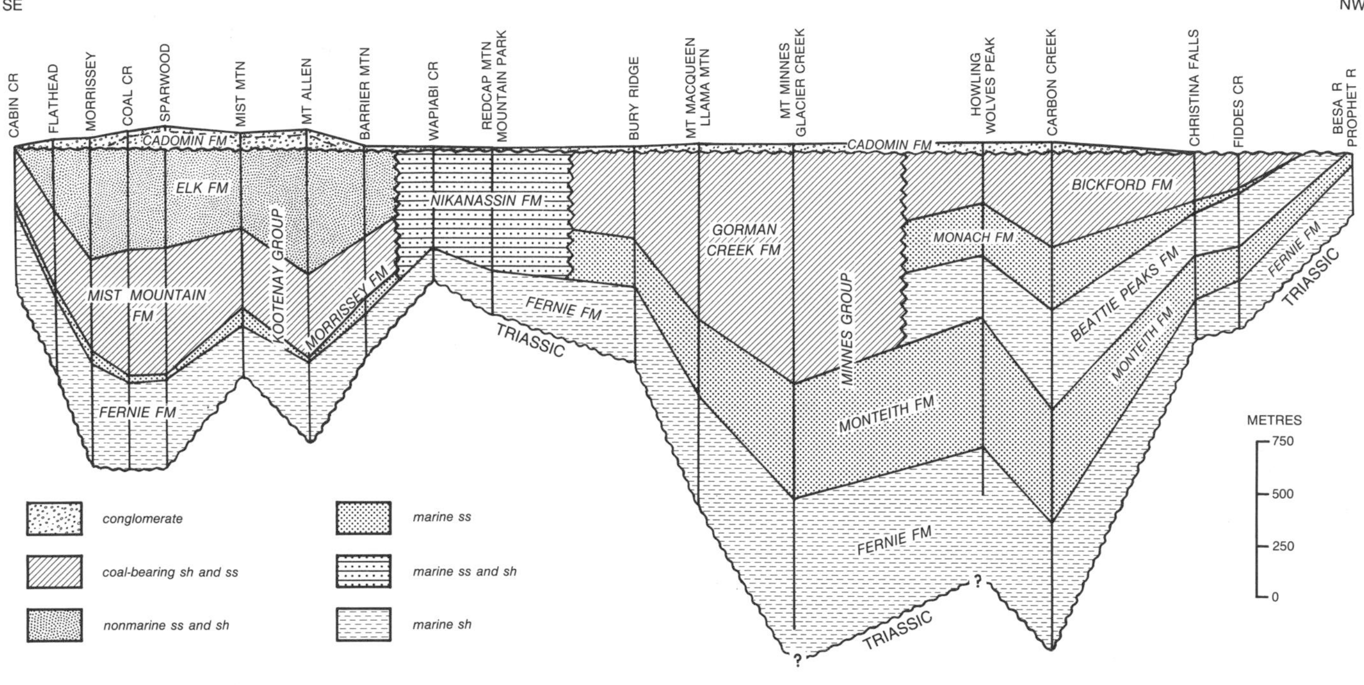

Figure 4I.10. Schematic section parallel to basin axis, illustrating the first, or oldest, clastic wedge, comprising the Fernie Formation and Kootenay Group of southern Rocky Mountains, Nikanassin Formation of central Alberta Foothills, and Minnes Group of northeastern British Columbia (adapted from Stott, 1984, and Gibson, 1985).

Figure 4I.11. Mesozoic succession at Stinking Springs, between Torrens and Kakwa rivers, foothills of northern Alberta. View south. Ŧ - Triassic; F - Fernie Formation; Mon - Monteith Formation; GC - Gorman Creek Formation (photo - D.F. Stott, ISPG 2438-23).

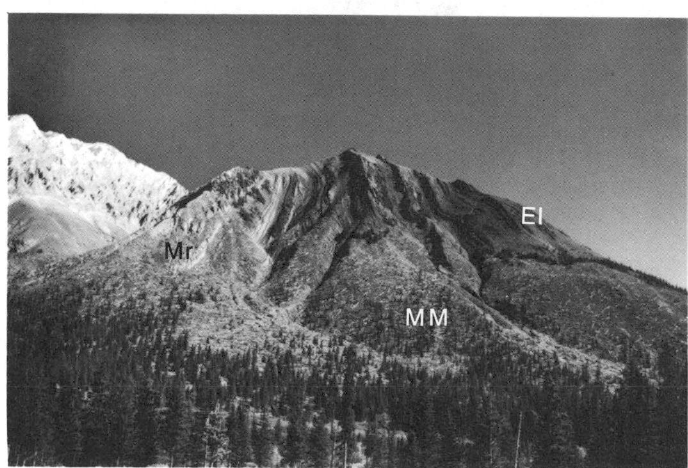

Figure 4I.12. Kootenay Group at the type section of the Mist Mountain Formation, near Mist Mountain in Highwood Pass area, Alberta. View north. Mr - Morrissey Formation; MM - Mist Mountain Formation; El - Elk Formation. Coal seams occur within many of the recessive talus-covered intervals (photo - D.W. Gibson, ISPG 2454-6).

The overlying marine mudstone and siltstone of the Beattie Peaks Formation contain *Amundiptychites* and *Ringnesiceras* of earliest Late Valanginian age. These prodeltaic to mid-basin sediments grade westward and southward into fine-grained, silty sandstone, marking the western margin of the Valanginian embayment. The shoreline trends southeastward along the western Foothills between Peace and Sukunka rivers, and more easterly south of there. In the subsurface, a significant eastward thinning takes place.

Argillaceous to quartzose sandstones of the Monach Formation, dated as Late Valanginian by the *Buchia* n. sp. aff. *inflata-Buchia bulloides* fauna, have a distribution similar to that of the quartzose sandstone of the Monteith and appear to have developed along the eastern margin of the embayment. The overlying Bickford Formation, 240 to 300 m thick, consists mainly of interbedded sandstone and mudstone, but also includes carbonaceous strata and thin coal seams, particularly in the region around Pine River.

To the south, the Gorman Creek Formation, with a maximum thickness of 1.35 km, is a succession of coal-bearing beds (Stott, 1981). The lower part probably represents fluvial plain deposits, whereas the coarse sandstone and conglomerate in the upper part are more typical of channel sediments.

In the Athabasca region, the Nikanassin Formation is poorly dated; fossils of Kimmeridgian-Tithonian/Volgian age are reported only from the basal beds (Mountjoy, 1962). The formation, 300 to 600 m thick in the west and truncated eastward, consists of alternating fine-grained sandstone and dark mudstone. The lower shaly part probably represents an offshore facies, although thin carbonaceous beds occur in the upper part. Unlike the Kootenay to the south, the Nikanassin lacks commercial coal seams.

Although partly of Jurassic age, the Kootenay Group of the southern region (Fig. 4I.10, 4I.12) includes Lower Cretaceous strata in the more westerly and thickest sections of the Fernie Basin, where it attains a maximum thickness of 1.1 km. The basal Kootenay is probably Late Jurassic (Tithonian/Volgian), an age based mainly on a single large ammonite, identified and described by Jeletzky (in Newmarch, 1953) and Frebold (1957). Bell (1956) considered the Kootenay flora to be of Neocomian-Barremian age. Microfloral studies by Rouse (1959) and Pocock (1964) indicated a Jurassic age, although Pocock suggested some strata could be Early Cretaceous. Recent palynological studies by Sweet (in Gibson, 1985) also provide evidence for an Early Cretaceous age of the upper part of the group.

The Kootenay sediments are considered to have been derived from western thrust sheets of the rising Rocky Mountains or from farther west in central British Columbia (Norris, 1964, 1971; Price, 1965; Rapson, 1964, 1965; Mellon, 1967; Jansa, 1972; Gibson, 1977b, 1985).

Basal Kootenay beds, the Morrissey Formation, are a coarsening-upward sequence of massive, cliff-forming sandstone (Fig. 4I.10; Gibson, 1979, 1985). These sandstones are part of a wave-dominated delta shoreline. Their maximum thickness of 80 m contrasts markedly with the much greater thickness of Monteith marine sandstone of northeastern British Columbia. The overlying Mist Mountain Formation (Fig. 4I.12) consists of interbedded sandstone and mudstone, rare conglomerate, and economically important bituminous to semianthracitic coal, deposited on a fluvial plain. The overlying Elk Formation is somewhat similar to the underlying Mist Mountain but lacks commercial coal seams. In Fernie Basin, it includes thick beds of chert-pebble conglomerate, deposited within an alluvial-fan to braidplain environment.

Throughout the Rocky Mountains and Foothills, the Kootenay-Nikanassin-Minnes sequence is bevelled easterly to an erosional edge beneath conglomerate and conglomeratic sandstone of the Cadomin Formation (Fig. 4I.13, 4I.14). In westernmost sections, little or no hiatus may exist between the Kootenay and Blairmore groups of Fernie Basin (Rapson, 1965; Gibson, 1977b, 1985) and between the Minnes and Bullhead groups of northeastern British Columbia (Stott, 1981). The unconformity in northeastern British Columbia occurs between beds of Late Valanginian and others presumably of Barremian-Aptian age. No record of Hauterivian marine rocks is known. The unconformity may record significant decrease in tectonism and an isostatic readjustment of the craton.

East-median trough (Saskatchewan)

Uppermost Jurassic-lowermost Cretaceous

Four main, non-contiguous, regional bodies of Jurassic-Cretaceous sandstone and mudstone occur in Saskatchewan (see Fig. 4I.23; Christopher, 1974, 1975, 1980, 1984). The two southern bodies, assigned to the Success Formation, are located in the southwest and southeast, where they form northward extensions of similar bodies in Montana and North Dakota. In general, they overlie the Jurassic Vanguard Group. The two northerly bodies, referred to as Jura-Cretaceous Sands, are

Figure 4I.13. Unconformity at base of massive cliffs of Cadomin conglomerate (Cd) underlain by sandstone and mudstone of the Gorman Creek Formation (GC) of the Minnes Group, and overlain by the Gething Formation (Ge). View eastward to the southern flank of Mount Gorman along the British Columbia-Alberta border (photo - D.F. Stott, ISPG 2438-28).

located more centrally; they overlie the Lower Carboniferous Madison Formation in the west and the Upper Devonian Torquay, Birdbear, and Duperow formations in the east.

The Success Formation of southwestern Saskatchewan, ranging in age from ?Kimmeridgian to 'Neocomian', is characterized by white kaolinitic, sideritic, and locally cherty, quartzose sandstone and mudstone. It is up to 105 m thick. In the Kindersley region of southwestern Saskatchewan, white sandstone overlies an irregular erosion surface on the Madison Limestone and has been called informally the "Detrital", the "Residual Zone" (White, 1969), or more formally the Deville Formation (Badgley, 1952; Maycock, 1967). The Success of central eastern Saskatchewan consists of quartzose, coarse- to very fine-grained sandstone with siltstone and kaolinitic mudstone.

These Success sand bodies are remnants of a continuous unit that blanketed the region of southern and central Saskatchewan and eastern Manitoba (Christopher, 1974, 1984). The depositional slope was southward and southwestward from the Canadian Shield. The general depositional environment was fluvial- lacustrine, with some shallow marine indications in the extreme south of the province. The upper Success may be coeval with the Lower Cretaceous Lakota Formation of Williston Basin, and the lower part with the Jurassic-Cretaceous Morrison Formation of Fergus-Great Falls Basin and perhaps the Jurassic Swift sandstone of Sweetgrass Arch.

Barremian-Middle Cenomanian sequence

Early events of this sequence are most completely documented in the northern basins, where a major tectonic reorganization commenced in late Early Cretaceous time with a rising western highland and a north-south trending foredeep.

Strong subsidence and widespread transgression in Western Canada Basin occurred during Albian time, although a break in deposition is apparent in the southern part of the basin after Early Albian time. Subsequently, renewed flooding brought about the joining of Boreal and Gulfian seas.

Northern Basins

Aptian

In the northern region, Aptian transgressions spread southward. The resulting embayment, apparently of limited extent although its southern boundaries are unknown, was bordered by a large fluvial apron whose

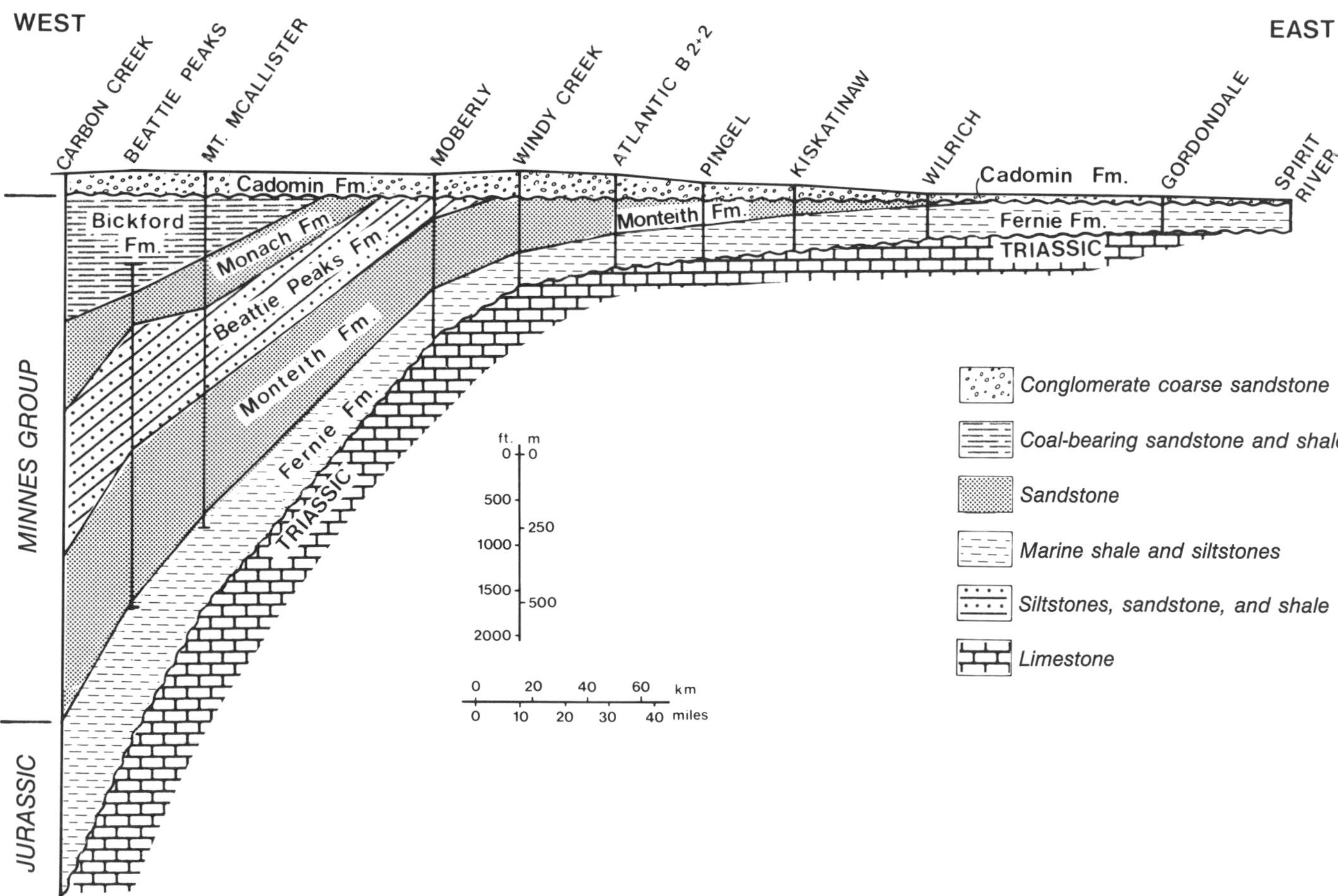

Figure 4I.14. Schematic section illustrating unconformable relationships of Minnes Group and Cadomin Formation at the latitude of Peace River (from D.F. Stott, 1982).

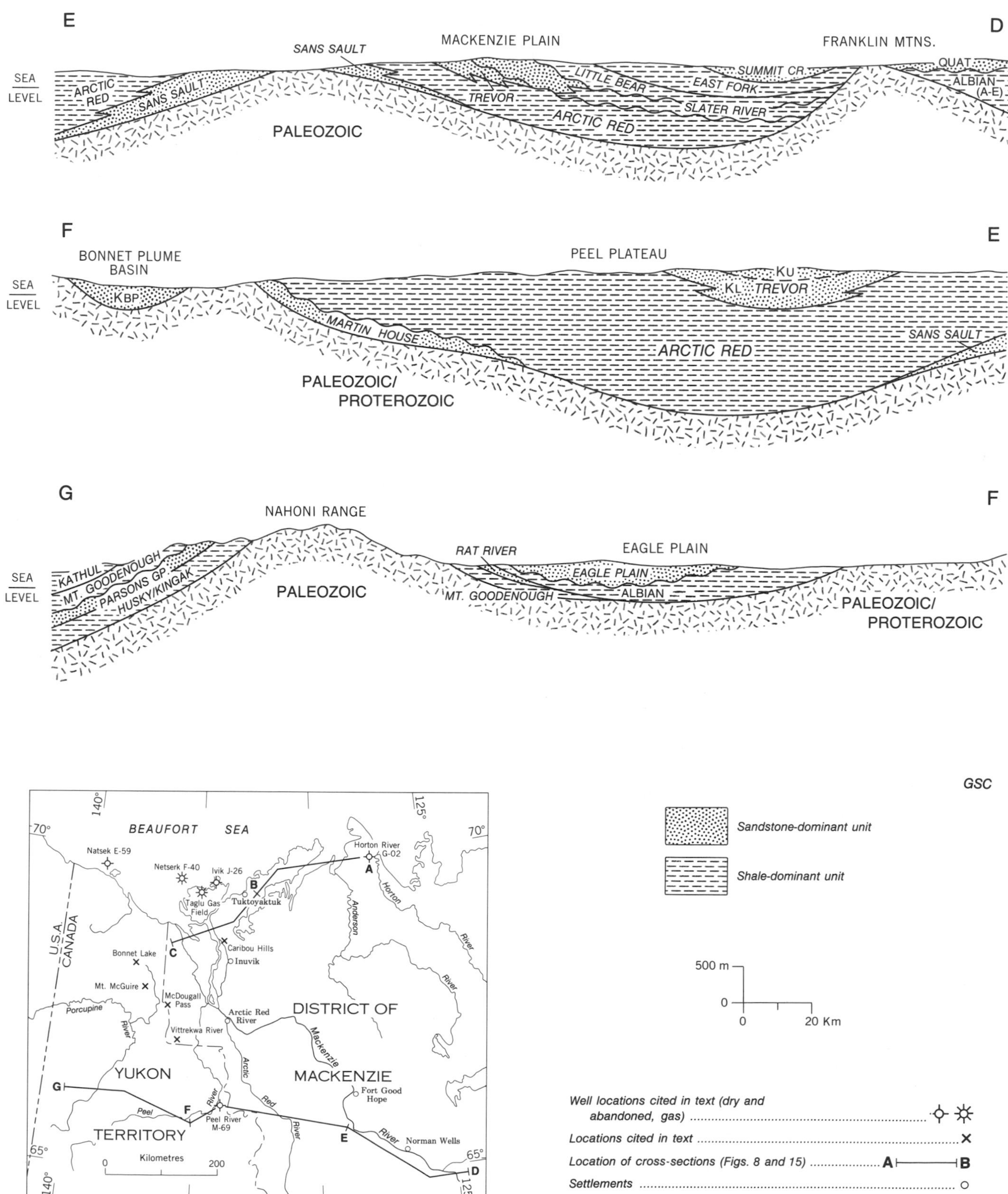

Figure 4I.15. Diagrammatic cross-section of Cretaceous deposits from Nahoni Range, Yukon Territory, to Franklin Mountains, District of Mackenzie (J. Dixon).

boundaries extended eastward beyond Anderson Basin. A foredeep along the rising western Cordillera was initiated in the region of Peel Trough.

On the eastern flank of the embayment, nonmarine sediments were deposited over a broad area extending from Coppermine Arch, across the southern Anderson Basin and Carnwath Platform, at least to the eastern edge of Peel Trough (Yorath and Cook, 1981, 1984). The Gilmore Lake Member of the Langton Bay Formation (Fig. 4I.8) grades northwestward from coarse conglomerate on the flanks of Coppermine Arch through point-bar sandstone to estuarine siltstone and mudstone in Anderson Plain. Upper Gilmore Lake deltaic sandstone and coal were deposited along the margin of the embayment. Across Carnwath Platform, thin, discontinuous, fine-to coarse-grained, crossbedded Gilmore Lake sandstone fills sub-Cretaceous paleodepressions, with relief as much as 100 m. Marine siltstone and mudstone on the eastern side of the seaway are included in the Crossley Lakes Member, but were not deposited over Eskimo Lakes Arch on the northwestern flank of the fluvial basin. A small fan-delta, part of the Atkinson Point Formation, prograded off the northwestern flank of the arch, and a true depositional edge of the Atkinson Point is preserved in the subsurface of Tuktoyaktuk Peninsula (Dixon, 1979).

Sands of the upper Rat River Formation were deposited over much of the central shelf of the Aptian embayment (Fig. 4I.9). Between Vittrekwa River and Aklavik Range, the Rat River consists of numerous sandstones, up to 60 m thick, alternating with similarly thick, dark, silty, bioturbated mudstones. In the northern Richardson Mountains, the formation is 200 to 300 m thick and contains several coarsening- and thickening-upward cycles (Young, 1978), in which the sandstones are characterized by hummocky cross-stratification, wave-ripple laminations, and bioturbated beds. In the region of Peel Trough to the southeast, the lower Martin House Formation consists of similar marine mudstone and siltstone, probably of mid- to outer-shelf origin.

Uppermost Aptian-Albian

Southerly and westerly source areas became dominant as the Columbian Orogen and Brooks Range Geanticline shed sediment into foreland troughs and basins during Aptian and Albian time (Fig. 4I.1). These extended from northern Yukon Territory into Alaska, as Blow and Keele-Kandik troughs (Young et al., 1976) or Rapid Depression (Norris, 1983). On the western flanks of Blow Trough, Lower Albian conglomerates rest unconformably on Jurassic to Hauterivian shales. To the southwest, the Lower Albian Sharp Mountain Formation rests on Carboniferous, Permian, and Jurassic rocks. Although there was significant uplift of Brooks Range Geanticline, there is little evidence to suggest uplift in the region of the Richardson or Mackenzie mountains prior to the Albian transgression.

Latest Aptian or earliest Albian transgression is represented by the Glauconite Member of the Martin House Formation, the upper Crossley Lakes Member of the Langton Bay Formation, the basal glauconitic sandstone of the Arctic Red Formation, and the lower sandstone of the Sans Sault Formation (Fig. 4I.15).

Thicknesses of Albian strata are highly variable. As much as 4 km of Lower Albian strata are recorded in Blow Trough (Young, 1973b), yet directly to the east, on the flanks of Cache Creek Uplift, they thin to about 400 m (Young, 1972). Likewise, Arctic Red strata in Kugmallit Trough are 600 m thick but on Eskimo Lakes Arch thin to less than 100 m. Southward in Peel Trough, Albian strata are up to 1600 m thick. In Eagle Plain, Albian shale is about 400-600 m thick, and in Kandik Basin, the Kathul Greywacke is at least 450 m (Brabb, 1969). In Anderson Basin on the eastern side of the embayment, the Horton River Formation is generally less than 200 m.

In Blow Trough, the Albian flyschoid sequence (Fig. 4I.16) includes a lower shale unit, a conglomerate and sandstone unit, an upper shale, and a turbiditic sandstone and shale unit. The lower shale, lying below beds containing Early Albian *Pachygrycia* sp. is as much as 730 m thick in the centre of the trough, thins eastward, and is not present at Fish River. On the western side of Blow and Keele-Kandik troughs, conglomeratic and sandy deposits, 2100 m thick (Young, 1972, 1973b; Young et al., 1976), were deposited as submarine fans or in submarine channels. Their lateral equivalents to the north are finer grained sandstone and shale arranged in Bouma sequences typical of outer fan and basin-plain environments. The Lower Albian coarse-grained basinal deposits fine upward, indicating either a curtailment of the supply of coarse sediments, or a switch in the locus of fan deposition to another part of the basin. The location of the Lower Albian Sharp Mountain sandstones between the deep-water sediments of Blow Trough to the north and Kandik Basin

Figure 4I.16. Alternating rusty shale and siltstone of the unnamed Albian flyschoid succession on the Arctic Coastal Plain north of Barn Mountains, northern Yukon Territory (photo - D.K. Norris, ISPG 662-26).

to the southwest suggests that they may be a continuation of those deep-water sediments (Dixon, 1986a). If they are shallow marine, as suggested by Jeletzky (1975b), then they represent shelf deposits of the western land mass.

To the east, the upper part of the flyschoid sequence grades eastward into an unusual phosphatic iron-formation on Cache Creek High, named the Rapid Creek (Young and Robertson, 1984). Those beds thin from 4 km in Blow Trough to only 60 m over Cache Creek High. Phosphate grains are composed primarily of rare minerals such as satterlyite, arrojadite, and ormanite, which suggest an original calcium-deficient composition. The deposition of iron and magnesium phosphates as well as apatite is strongly indicated, and this condition is unique for marine phosphorites. This most northerly known phosphorite (75°N paleolatitude) probably resulted from cold, northeast-flowing currents upwelling on the western flank of Cache Creek High. Farther east at Mackenzie Delta, the iron formation changes to dark grey shale of the Arctic Red Formation.

During Early Albian time, much of the eastern area was a shelf, covered by mud and silt of the Arctic Red and Horton River formations and unnamed Albian beds of Eagle Plains. Local sand facies, such as the basal sandstone of the Arctic Red Formation on Eskimo Lakes Arch and the Sans Sault Formation on Keele Arch, are associated with shoals. The deepening Peel Trough was filled with a basal glauconitic sand and overlying thick, concretionary, fossiliferous and silty mud of the Arctic Red Formation, which disconformably overlie strata of the Upper Devonian Imperial Formation between the Mackenzie Mountain front and Mackenzie River. The glauconitic sandstone and lower shales have yielded Early Albian *Pachygrycia* sp., and the uppermost shale, Middle Albian *Gastroplites* sp. Farther east, Sans Sault sandstone and mudstone also bear *Pachygrycia*, providing additional evidence of the transgression of the Albian Clearwater Sea from Boreal regions into midwestern Canada (Jeletzky and Stelck, 1981). Fine-grained sand and mud were deposited in Great Bear Basin, and marine shales of the upper Crossley Lake and Horton River formations were deposited on the broad shelf area extending eastward across Anderson Basin.

By late Middle Albian time, Peel Trough was receiving coarse clastic sediments from the south and southwest, represented by the marine Trevor Formation (Yorath and Cook, 1981). Shelf muds and silts are presumed to have been deposited over much of the northern area and Early Albian troughs are believed to have begun to fill by the end of the Albian. Presumably by Late Albian time, coarse clastics were deposited over a much wider area.

Regression during Middle to Late Albian time affected most of the northern basins, ending the earlier Albian flooding. A regional unconformity separates Albian from Upper Cretaceous rocks throughout much of the northern foreland belt. Although Yorath and Cook (1981) suggested

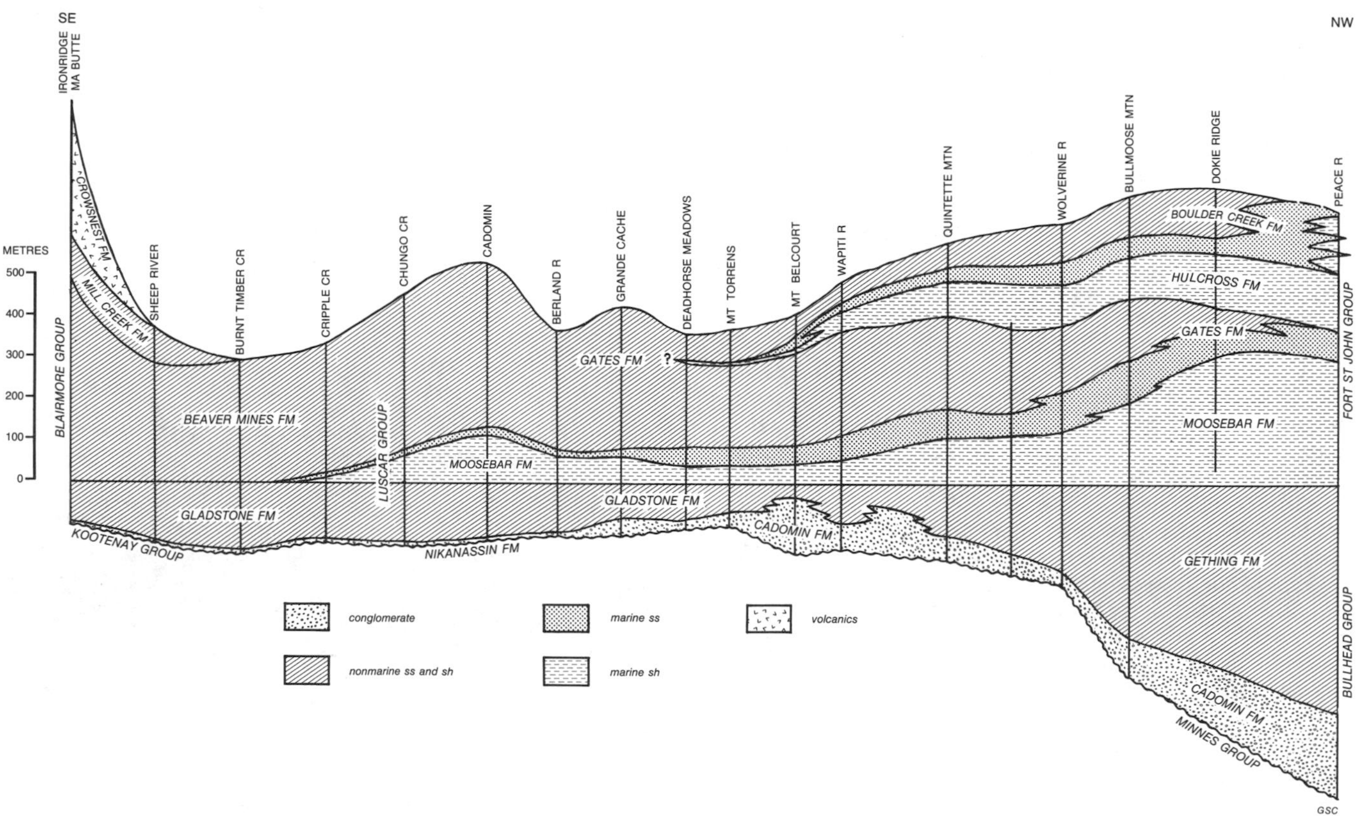

Figure 4I.17. Schematic section along Rocky Mountain Foothills, showing the lower part of the second clastic wedge, comprising the Blairmore, Luscar, Bullhead and lower Fort St. John groups, Alberta and northeastern British Columbia (adapted from Stott, 1982, and McLean, 1982).

that deposition in the region of Peel Trough was continuous, G.K. Williams (pers. comm., 1988) reported that deposition may have been disrupted in Late Albian time by movement of local blocks related to Keele Arch.

Rocky Mountain Trough (Rocky Mountain Foothills)

Barremian-Aptian

Columbian orogenic uplift in the Cordilleran region was accompanied by subsidence of the Rocky Mountain Trough, which trapped coarse clastic sediments and allowed the gradual incursion of marine water from boreal regions into Western Canada Basin. The Barremian-Aptian to Lower Cenomanian clastic wedge comprises the Blairmore, Luscar, Bullhead, and Fort St. John groups and the Dunvegan Formation (Fig. 4I.2, 4I.17) and is 450 m to 2.1 km thick in the west, thinning to 325 m or less in the plains. In northeastern British Columbia, the wedge is bounded by unconformities. Paleogeographic reconstruction of the Blairmore-Fort St. John succession reveals extensive flooding of Western Canada Basin and shifts in shoreline positions.

The Aptian-earliest Albian embayment recorded by the Gething Formation did not extend much farther south than the Peace River (see Fig. 4I.50c; Stott, 1984) where a large delta of an ancestral Peace River formed the southwestern shore. Other deltas were present on the eastern side, which was flanked by ridges of Paleozoic rocks (Rudkin, 1964). The Boreal sea was bordered to the south by estuarine to lacustrine environments (McLean and Wall, 1981), and by an extensive, poorly drained fluvial plain.

In northeastern British Columbia, the Bullhead Group (Stott, 1968, 1973, 1982) lies above strata of Middle to Late Valanginian age and below Fort St. John shales of Early Albian age. It seems likely that the Bullhead Group is no older than Barremian and is possibly as young as earliest Albian. The Gething macroflora is comparable to the "Lower Blairmore" flora (Bell, 1956; Stott, 1968, 1973). The sequence is about 3 km thick in the western Foothills at Peace River (Fig. 4I.18). The base is marked by the distinctive Cadomin Formation, a conglomerate consisting predominantly of chert pebbles, which in northeastern British Columbia was deposited in a piedmont-alluvial environment. A coal-bearing facies, included in the Gething Formation and forming an extensive delta at Peace River, grades northward into fine-grained sandstone, which in turn grades into marine mudstone and siltstone. Those beds record an Aptian to possibly earliest Albian transgression, the initial incursion of a Boreal sea after a major period of erosion.

In the central Foothills, the Gething coal-bearing facies gives way to alluvial sands and shales with point-bar and floodplain deposits of the lower Gladstone Formation (Fig. 4I.17; Mellon, 1967; McLean, 1982). Its "Lower Blairmore" flora is of Aptian to Middle Albian age. The upper Gladstone in the southern Foothills includes calcareous shale and limestone, variously designated as the Ostracode Zone or Calcareous Member (Loranger, 1951; Glaister, 1959), which contain fresh-water bivalves, gastropods, and ostracodes, which are replaced northward by a brackish-water fauna (Mellon and Wall, 1963; Mellon, 1967; McLean and Wall, 1981; Finger, 1983).

Albian

During Early Albian time the Columbian Orogeny was renewed and the foredeep migrated eastward. Molasse and coal deposits are characteristic. The boreal sea advanced southward and eventually joined, in latest Middle Albian time, with the northward expanding Gulfian Sea (Stelck et al., 1956; Stelck, 1958; Jeletzky, 1971a, b).

The Albian Fort St. John Group of northeastern British Columbia comprises four transgressive-regressive cycles (Fig. 4I.17, 4I.19; Stott, 1968, 1982). The pattern of deposition was influenced, in part, by movement of Peace

Figure 4I.18. Gething Formation, comprising numerous cycles of carbonaceous mudstone, coal and argillaceous sandstone; type locality downstream from W.A.C. Bennett Dam at head of Peace River Canyon. View southwestward (photo - D.F. Stott, ISPG 2438-17).

Figure 4I.19. Bullhead and Fort St. John strata on north face of Mount Belcourt, Monkman Pass map-area, British Columbia. GC - Gorman Creek; Cd - Cadomin; Ge - Gething; Mo - Moosebar; Ga - Gates; Hu - Hulcross; BC - Boulder Creek. View southward from Mount Hamelin (photo - D.F.Stott, ISPG 2438-22).

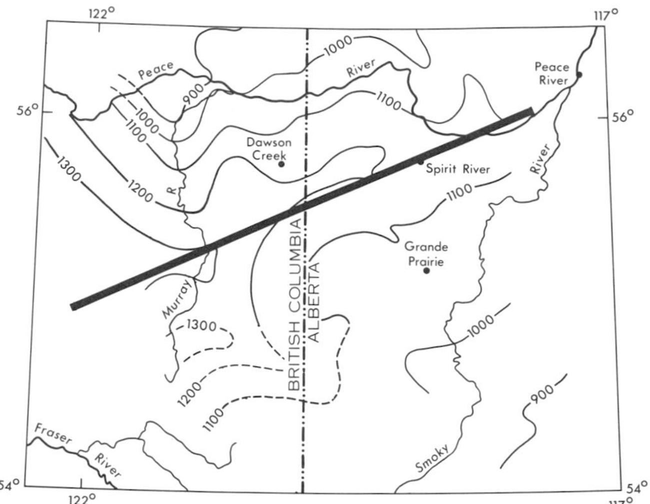

Figure 4I.20. Isopachs of the combined Moosebar and Gates formations, illustrating the northeasterly trending deltaic lobe, which parallels the northeast trend of the ancient Peace River Arch (thick line) (Stott, 1968). Similar patterns are found in other cycles of the second wedge.

River Arch. In the central and southern Foothills, equivalent strata comprise alluvial shale and sandstone of the upper Luscar and Blairmore groups (Mellon, 1967; McLean, 1982; Langenberg and McMechan, 1984).

The Early Albian transgression extended southward well into present Alberta (McLean and Wall, 1981), and extensive coal deposits formed landward of the embayment (see Fig. 4I.50d). The basal marine shales, about 400 m thick, are included in the Garbutt, basal Buckinghorse, and Moosebar formations. In the north basal beds lie within the *Pachygrycia* Zone (Jeletzky and Stelck, 1981) and farther south, possibly within the younger *Arcthoplites* Zone. Foraminiferal assemblages are assigned to the lower part of the *Gaudryina nanushukensis* Zone (Caldwell et al., 1978), with the *Marginulinopsis collinsi-Verneuilinoides cummingensis* Subzone marking peak transgression and the ensuing regression. Delta-front sandstones occur in the Bulwell Member of the Scatter Formation at Liard River. A series of shorelines within the Gates Formation at Peace River have an easterly trend (Stott, 1968, 1982; Leckie and Walker, 1982; Cant, 1984). Shorelines and maximum depositional accumulation parallel the trend of the ancient Peace River Arch (Fig. 4I.20). Extensive coal deposits landward of the shoreline sandstones are included in the Grande Cache Member of the Gates Formation in the

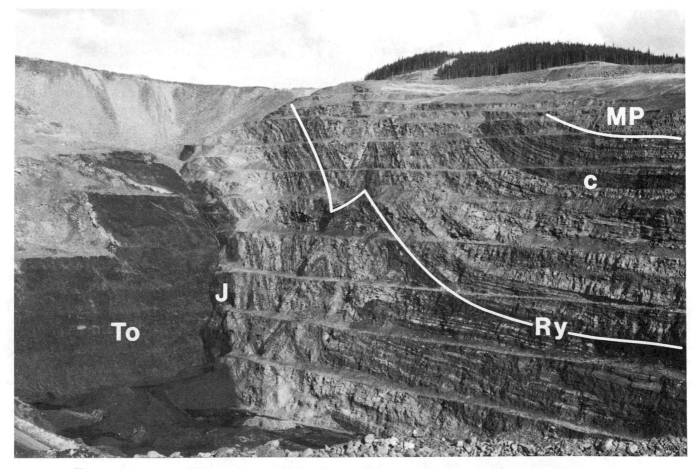

Figure 4I.21. Luscar Group in east wall of the Cardinal River coal pit, south of Athabasca River in central Alberta Foothills. To - Torrens Member; J - Jewel seam; Ry - Ryder Seam and c - coal and coaly shale of Grande Cache Member; MP - Mountain Park Member (photo - C.W. Langenberg).

northern Alberta Foothills. The Gates contains the Lower Blairmore-Luscar-Gething flora, and fauna of the Lower Albian *Arcthoplites* Zone (Stott, 1968, 1982).

Deposits of the second cycle, approximately 400 m thick and of Middle Albian age, are similar to those of the previous cycle, although the sea transgressed only as far south as Smoky River in the Foothills. This transgression is recorded in the Wildhorn shales of the Scatter Formation at Liard River, the middle Buckinghorse shales of the Halfway-Muskwa area, and the Hulcross shales south of Peace River (see Fig. 4I.50e). The regression is recorded in the delta-front and prodeltaic sandstone and shale of the Tussock Member of the Scatter Formation at Liard River and in the shoreline sandstone and fluvial sediments of the Boulder Creek Formation south of Peace River. The shoreline facies, like those of the previous cycle, lie south of the ancient Peace River Arch (Leckie, 1985; Cant, 1984). The upper part of the Hulcross Formation is assigned to the zone of *Pseudopulchellia pattoni*, and the Boulder Creek Formation and Cadotte Member of the Peace River Plains lie mainly within the *Gastroplites kingi* Subzone. The *Haplophragmoides multiplum* foraminiferal fauna (Stelck et al., 1956) characterizes the Hulcross Formation and its equivalents. The Boulder Creek and Cadotte sandstones lie within the *Ammobaculites* sp. Subzone of the *G. nanushukensis* Zone (Caldwell et al., 1978). Middle Albian strata are not known elsewhere and presumably much of the southern part of Western Canada Basin was the site either of nonmarine deposition or of erosion.

In the central Alberta Foothills, beds equivalent to the Bullhead and lower Fort St. John Group are included in the Luscar Group (Fig. 4I.17; Langenberg and McMechan, 1984). The Gladstone Formation (Mellon, 1967; McLean, 1982), overlying a thin Cadomin conglomerate, comprises fining-upward sequences of sandstone, shale, and minor coal seams. The Gates Formation, overlying thin marine shales of the Moosebar, is divided into three members: the basal, Torrens sandstone, which was deposited in a high energy, shoreline environment; the Grande Cache Member, characterized by thick coal seams (Fig. 4I.21); and the upper, Mountain Park Member comprising greenish, fine- to medium-grained sandstones, which are interpreted as fluvial channel deposits.

The Beaver Mines Formation of the Blairmore Group (Fig 4I.17; Mellon, 1967; McLean, 1982) of the southern Alberta Foothills contains the Lower Blairmore flora and is equivalent to the upper Luscar and lower Fort St. John groups. It consists of nonmarine varicoloured shales, and green feldspathic sandstone with abundant volcanic detritus and chloritic cement. Several beds of conglomerate containing igneous pebbles occur throughout the Blairmore Group. Those from the McDougall-Segur conglomerate (Douglas, 1950) have yielded K-Ar ages between about 113 and 174 Ma (Norris et al., 1965). The source for these clasts was almost certainly in the Omineca Belt. Coal beds are absent, in contrast with the abundant coal in equivalent beds to the north and also with the underlying Kootenay Group.

Upper Albian shales of northeastern British Columbia, included in the Hasler, Buckinghorse, and Lepine formations, mark the linkage of Boreal and Gulfian seas. In northeastern and central Alberta the Joli Fou shale and Viking sandstone developed during an early phase of the Late Albian cycle. The distribution of Viking sandstone reveals the positive influence of Peace River Arch (see Fig. 4I.50f; Koke and Stelck, 1984, 1985). The distinctive *Haplophragmoides gigas* foraminiferal assemblage is found within the Joli Fou (Stelck, 1958; Koke and Stelck, 1984, 1985). In southern Alberta, related beds are included in the nearshore Bow Island sandstones and nonmarine Mill Creek Formation (Fig. 4I.17). The Mill Creek comprises sandstone and green and red mottled shale, which grade upward into trachytic tuff and agglomerate of the Crowsnest Formation (Mellon, 1967). The upper Blairmore or Mill Creek Albian flora, a predominantly dicotyledonous association, occurs in both the sedimentary and pyroclastic beds (Bell, 1956).

Fine-grained, regressive marine, nearshore sandstones of the Goodrich and Sikanni formations of northeastern British Columbia, lying within the *Neogastroplites* Zone, post-date the Viking sandstone. The microfauna is assigned to the *Miliammina manitobensis* Zone (Stelck, 1975a; Caldwell et al., 1978). A prominent subsurface unit, the Fish-Scale marker, generally assumed to mark the Albian-Cenomanian boundary (Warren and Stelck, 1969), lies within, or just above, the Goodrich and Sikanni sandstones in their westernmost exposures. These sandstones represent a sinuous shoreline that extended from near Kakwa River to the Mackenzie River. This northwesterly trend more or less parallels later Laramide structures, contrasting with the northeasterly trending shorelines of the earlier cycles. Beds of latest Albian age apparently are not present in the southern Foothills of Alberta. The volcanic activity recorded in the Crowsnest Formation of the Crowsnest Pass region is equated with the Vaughn Member of the Blackleaf Formation (Cobban et al., 1959, 1976), which underlies the Fish-Scale sandstone. Thus, the western margin of the basin was marked by shorelines in the north, nondeposition or erosion in the middle, and volcanic activity in the south.The last Albian transgression is recorded by Cruiser and Sully marine shales, which grade upward into the Cenomanian (earliest Late Cretaceous) Dunvegan Formation (Fig. 4I.22). The base of the succession falls within the *Textularia alcesensis* Zone (Caldwell et al., 1978). The Dunvegan Formation contains fauna tentatively correlated with the Middle? to Upper Cenomanian *Acanthoceras* Zone (Jeletzky, 1968), a megaflora dated by Bell (1963), and a microflora dated by Singh (1983) as Cenomanian. The Dunvegan consists mainly of conglomeratic piedmont-alluvial plain facies in the Liard region, alluvial-deltaic sediments between there and Peace River, and delta-front and prodeltaic sediments to the south (see Fig. 4I.50g; Stott, 1982). The apparent arc required to connect the trends of the Monster Formation in Eagle Plains with those of the Dunvegan of northeastern British Columbia and Alberta (Fig. 4I.50g) suggests that they may parallel the Late Cretaceous structures of the Mackenzie Mountains. Plutons of mid-Cretaceous age (90-110 Ma) in the Selwyn Mountains (Gabrielse et al., 1973) may be at the centre of the uplift that shed the Monster-Dunvegan sediments. Depositional trends that parallel those of the later Laramide tectonic trends are not without precedent; the earlier Goodrich and Sikanni sandstones (Late Albian) of northeastern British Columbia also provide examples.

In the Liard region of British Columbia, the hiatus at the top of the Dunvegan Formation (see Fig. 4I.2, 4I.37) spans part of Cenomanian and all of Turonian and Coniacian time (Stott, 1963). Although much of the Cenomanian and Turonian are represented by marine shale in the northern Foothills of Alberta, the post-Blairmore hiatus in the central and southern Foothills, increasing in magnitude southward, appears to represent much of Late Albian and some of Cenomanian time. At Cadomin and Cripple Creek, Late Albian faunas are present in the lower Blackstone Formation. Farther south at Ghost River, a Middle Cenomanian fauna directly overlies the Blairmore Group. In the Crowsnest Pass, the disconformity is no less pronounced, as a Late Cenomanian fauna is found in the basal Blackstone (Wall, 1967; Warren and Stelck, 1955).

West-median or axial trough (Alberta)

?Aptian-Lower Albian

The Mannville Group and equivalents of Alberta lie unconformably on units ranging in age from earliest Cretaceous in the west to Devonian in the east (Fig. 4I.23). The nature of these underlying rocks, particularly differences in their resistance to erosion and resultant paleotopographic patterns, have influenced Mannville deposition.

Primitive angiosperm pollen indicate that the base of the Mannville Group is no older than latest Barremian, and is probably earliest Aptian in most localities (Singh, 1964; Pocock, 1976), although remnants of Valanginian sandstone may be present in some areas (Burden, 1984). The middle Mannville transgressive unit is well dated as Early Albian (McLean and Wall, 1981) by foraminifera and ostracodes. The upper Mannville is dated as latest Early Albian on the basis of ammonites (Stelck and Kramers, 1980), pollen (Singh, 1964), and foraminifera (Caldwell, 1984b).

Figure 4I.22. Massive conglomerate beds of Dunvegan Formation, Akue Creek east of Muskwa River (photo - D.F. Stott, GSC 128458).

The Mannville Group, ranging from 60 to 900 m in thickness (Fig. 4I.24), has been subdivided in different areas into various subgroups, formations, and members (Fig. 4I.25). For the purpose of regional interpretation, the group can be divided into three units.

The lowest unit is patchy and nonmarine over much of the basin (Fig. 4I.26). In northeastern Alberta, the basal 30 to 60 m, a well-sorted, quartzose sandstone, is designated McMurray Formation. Where impregnated with heavy oil, these sandstones constitute the Athabasca Oil Sands (Fig. 4I.27). The correlative Cadomin and Gething formations in the Peace River Plains of northwestern Alberta are eastward extensions of conglomerate, conglomeratic sandstone, and coal-bearing beds of the Foothills. In central Alberta, the equivalent Ellerslie Formation (Basal Quartz Sandstone) is about 60 m of quartzose sandstone with shale and minor coal. A basal residual detritus formed of chert and other rock fragments, known as the Deville Member, occurs locally. In east-central Alberta, the basal quartzose sandstone is termed the Dina Member. In southern Alberta, the lower Mannville comprises the Cutbank, a medial member that is commonly referred to as the "Sunburst", and a capping "Calcareous" member (Hayes, 1986).

The thickness of the lowest unit is largely a function of the relief on the erosional surface (Fig. 4I.26). The most prominent features, trending northwest, are produced by ridges of resistant Paleozoic carbonates and intervening valleys of recessive shales. The Mannville valleys of Alberta have been named, from east to west, the St. Paul, Edmonton, and Spirit River channels. These are separated by Wainwright Ridge (Williams, 1963) and Fox Creek Escarpment (McLean, 1977b). The erosional relief was up to 100 m.

Following uplift and erosion, sedimentation began filling the valleys (Fig. 4I.28; Hayes, 1986). Locally, remnants of Jurassic sediments may remain (Hopkins, 1981), but in most areas nonmarine lower Mannville clastics lie directly on Paleozoic rocks. The Cadomin conglomerate is confined to the area west of the Fox Creek Escarpment (McLean, 1977b). Farther east, fluvial systems of the lower Mannville flowed northwest, depositing sand, even as erosion continued in the ridges of Paleozoic rocks. Each of the Gething, Gladstone, and McMurray formations occupies a major valley system and features large-scale meandering stream deposits (Taylor and Walker, 1984; Mossop and Flach, 1983). Northward, estuarine facies appear, particularly in the upper parts of these units. In southern Alberta, valleys were filled with nonmarine sediments (Dalhousie, Moulton, Cutbank, "Sunburst", Detrital, and Ellerslie), which eventually overlapped the intervening ridges. Dominantly nonmarine sedimentation was brought to an end by a decreased input of clastic sediments combined with a southerly transgression of the Boreal Sea.

The middle Mannville marine to marginal-marine sediments were deposited during the Early Albian transgression from the north. The unit is in general a thin, probably diachronous, sheet of sandstone, shale, and minor limestone. In the Athabasca region of northeastern Alberta, the basal Wabiskaw Member of the Clearwater Formation is a well sorted, glauconitic sandstone, underlain by calcareous shale, correlative with the Ostracode Zone of central Alberta. The equivalent Bluesky Formation of the Peace River region also is characterized by glauconitic sandstone and shale. The Ostracode Member (=Calcareous Member) of central Alberta (Wanklyn, 1985) includes calcareous shale and siltstone with abundant ostracodes. Farther east, the Cummings Member is a marine shale that includes equivalents of the Ostracode Zone (Loranger, 1951) of southern Alberta. In some places, beds of the middle unit are transitional from those of the lower unit, and the two are difficult to separate (e.g., Bluesky-Gething in the subsurface). In other places, upper beds of the middle unit form a minor regressive tongue (e.g., Cummings sandstone), and more closely resemble the overlying unit. Locally, boundaries between the units are somewhat arbitrary.

The transgressive beds are widespread and thin. The major northwesterly trends still reflect the underlying surface of Paleozoic rocks, but the smaller northeasterly ones represent sands projecting from the edge of subtle Paleozoic ridges. These littoral bodies probably formed at breaks in slope caused by differential compaction of the lower Mannville.

As the sea moved southward, it advanced most rapidly up the axes of the buried valleys. Over large areas, thin sheets of sand were laid down in estuarine, shoreface, and shallow marine environments (Bluesky, Wabaskaw, Glauconitic, Cummings; Fig. 4I.29). Locally, small, regressive, coarsening-upward sequences are present where the shore prograded. Where transgression encountered a slightly elevated area, a thicker, more extensive, littoral sand body was developed. The glauconitic sandstone of the Hoadley barrier trend (Chiang, 1985) is an example. The transgression also resulted in a lower gradient of river systems in the southern region, the formation of large lakes, which were partly flooded by seawater as the transgression proceeded, and development of brackish water conditions in which the Calcareous Member and Ostracode Zone limestones were deposited. It reached into southernmost Alberta and possibly into Montana.

The upper Mannville unit represents another pulse of clastic input, with marine sediments in the north, transgressive-regressive, shoreline to shallow marine deposits in central Alberta, and nonmarine sediments in the south. Thus, in the Athabasca region, Loon River marine shales grade into the Clearwater shales and Grand Rapids sandstones. To the west in the Peace River region, equivalent marine and nonmarine strata are included in the Wilrich, Falher, and Notikewin members of the Spirit River Formation. Beds equivalent to the younger Peace River Formation (Harmon, Cadotte) are not recognized within the Mannville to the east, where a hiatus apparently occurs at the top of the group. The upper Mannville Group is characterized by sand-shale couplets, some of which are capped by coal beds. The upper Mannville (Glaister, 1959) of the central and southern parts of the basin includes all beds between the Ostracode (Calcareous) Member and the base of overlying Colorado marine shales. It is nonmarine in the south, with the exception of some of the sands included in the Glauconitic sandstone at the base.

Upper Mannville sediments thicken markedly northwestward (Fig. 4I.30). In east-central Alberta, the northwesterly thinning trend subtly reflects the underlying erosional surface. Very locally in the Lloydminster area, a few small remnants of Paleozoic carbonate are onlapped by upper Mannville sediments (O'Connell, 1984). The huge anomaly in the northwest reflects subsidence in the area of the former Peace River Arch.

Initially, shales (Moosebar, Wilrich, Clearwater) were deposited during late Early Albian time in relatively deep water in the west (Leckie and Walker, 1982; McLean and Wall, 1981) and shallower water in the east. As the sands prograded northward, marginal marine, then nonmarine conditions were established. Because of variations in sediment supply, many minor transgressions and regressions occur within the framework of the overall regression. At maximum transgression, the shoreline extended east-west across central Alberta (Fig. 4I.31) from the Deep Basin at Elmworth (Cant, 1983, 1984; Jackson, 1985) to the Fort McMurray area (Kramers, 1982), then curved southward toward Lloydminster (Putnam, 1982). In each of those areas, small-scale (10-50 m thick) transgressive-regressive shoreline cyclothems are displayed. In western Alberta, the Spirit River Formation (Cant, 1984) includes five shoreline sequences. In east-central Alberta similar, though fewer, cycles are recognized in the Grand Rapids Formation. Likewise, in the Lloydminster area, the Lloydminster, Rex, General Petroleum, Sparky, and Colony sands were formed by those shoreline fluctuations.

In the northwest, the Middle Albian Harmon shale and the Cadotte-Paddy shoreline complex reflect another transgressive-regressive pulse. In the east, the interval is represented by a hiatus, probably because subsidence was far less than in the west.

Middle Albian-Cenomanian

During Middle to Late Albian time, Lower Cretaceous nonmarine beds of the western and southern margins of Western Canada Basin were progressively overlapped by marine sediments of Boreal and Gulfian seas, which invaded the region and eventually joined. In the Plains, marine beds above the Mannville Group are assigned to the Colorado Group (Rudkin, 1964; Williams and Burk, 1964). The group occurs mainly in subsurface, but equivalent beds are found in outcrops of the Alberta Group in the adjacent Foothills of Alberta and the upper Fort St. John and Smoky groups in northeastern British Columbia and northwestern Alberta.

Distribution of Colorado facies in the Alberta Plains was controlled by continuing downwarp on the western basin margin. Reactivation of such northeast-trending elements as Sweetgrass and Peace River arches as well as smaller scale, linear basement features with northeast and northwest orientation influenced facies distribution. The paleotopography of the sub-Mannville unconformity is mimicked by compaction-drape folds in dominantly argillaceous parts of the succession, and may have influenced configuration of some of the shelf sands.

The marine Colorado Group rests unconformably on the Mannville Group. A distinct microfloral break at the Grand Rapids (Mannville)-Joli Fou contact (Norris, 1967) probably represents the interval from early Middle Albian to early Late Albian time. The Colorado succession was laid down as a series of transgressive-regressive cycles, which equate with the Skull Creek, Greenhorn, and Niobrara cycles of Kauffman (1967, 1969). The group is divided into upper and lower subgroups, commonly at the base of the Fish-Scale Marker (Albian-Cenomanian), although Simpson (1975) made the division in Saskatchewan at the base of the Second Speckled Shale (Turonian).

In the southern Plains of Alberta, the first transgression, represented by Joli Fou shales, is related to the northward advance of the Gulfian sea. The Joli Fou

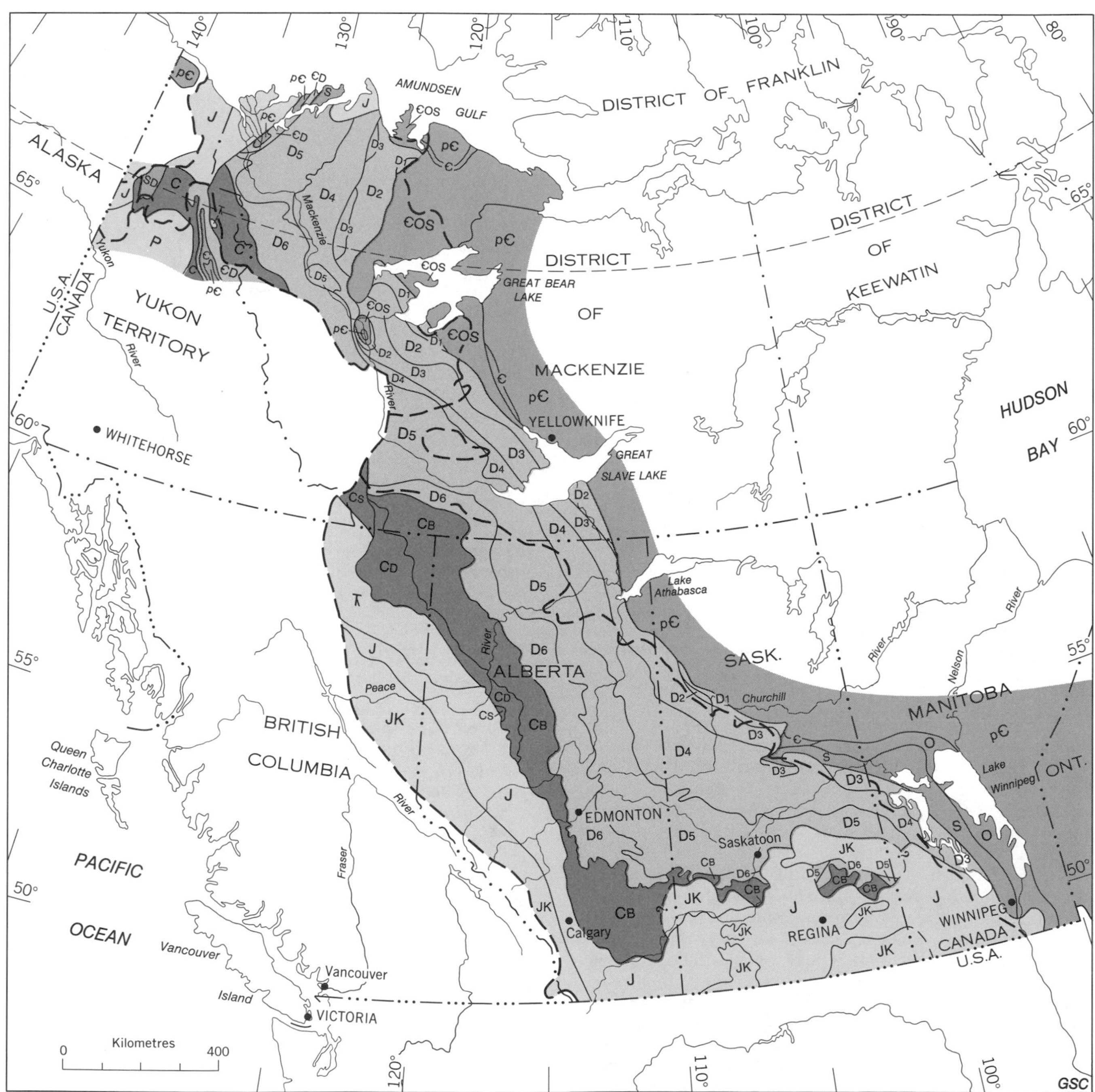

Figure 4I.23. Subcrop map of rocks beneath the unconformity at the base of the Cadomin Formation, Mannville Group and equivalent rocks (Barremian-Aptian) in Western Canada Basin.

comprises dark grey, marine shales, 20 to 30 m thick, equivalent to the Skull Creek Shale of the United States. No equivalent beds appear to be present in the central or northern Foothills of Alberta. In southeastern Alberta, the Cessford Sand (Basal Colorado Sand), 6 m thick and occurring at the base of the Joli Fou, may reflect winnowing of siliciclastic debris on Sweetgrass Arch.

In the southwestern part of the basin the Joli Fou is replaced by the Bow Island Formation, 175 m thick, consisting of shaly sandstone and subordinate conglomerate in three composite sandstone bodies, dominated by transgressive-regressive, coarsening-m upward sequences. The Bow Island is similar in composition to the upper Blairmore of the mountains, and is tuffaceous in its upper part (Glaister, 1959). Bentonitic clays and argillaceous sandstone containing orange-red clinoptilolite (Red Speck Zone), between the First and Second Bow Island sands, correlate with the Vaughn Member of the Blackleaf Formation of Montana (Cobban et al., 1976) and with the Crowsnest Formation.

The Bow Island is replaced northeastward by a shallow-marine sandstone unit, the Viking Formation. The Viking, 20 to 50 m thick and marking a shoreline lying in central Alberta well east of the Foothills, becomes finer grained and more argillaceous eastward across the basin.

The Viking Formation of south-central Alberta is characterized by a coarsening-upward sequence of bioturbated sandstone and shale, fine-grained sandstone, and chert-pebble conglomerate and pebbly sandstone (Hein et al., 1986; Leckie, 1986b). The sequence represents a progradational, shoreface sequence, followed by a lowstand in sea level accompanied by channel cut-and-fill. The upper part represents a transgressive, shelf complex in which sandstone and conglomerate were deposited and reworked. The authigenic mineral and porosity trends exhibited in the sandstone reflect depth control of diagenetic processes (Reinson and Foscolos, 1986). These processes have produced (1) internal diagenetic alterations resulting from compaction-related effects, and (2) external

LEGEND

JK	*Jurassic-Cretaceous*
J	*Jurassic*
Ѫ	*Triassic*
P	*Permian*
CS	*Carboniferous: Stoddart*
CD	*Carboniferous: Debolt*
CB	*Carboniferous: Banff*
D6	*Devonian: Palliser, Famennian*
D5	*Devonian: Saskatchewan, middle and upper Frasnian*
D4	*Devonian: Beaverhill, upper Givetian-lower Frasnian*
D3	*Devonian: Hume-Dawson*
D2	*Devonian: Bear Rock*
D1	*Devonian: Delorme*
S	*Silurian*
O	*Ordovician*
Є	*Cambrian*
pЄ	*Precambrian*
C	*Carboniferous*
SD	*Silurian-Devonian*
ЄOS	*Cambrian-Ordovician-Silurian*
ЄO	*Cambrian-Ordovician*
ЄD	*Cambrian-Devonian*

Limit of Cretaceous ....

mineralogical reactions in adjacent shales, which are imprinted in the authigenic minerals of the sandstones via mass transfer of chemical components.

The Viking is overlain by an unnamed but relatively thick sequence termed the 'Upper Shale Unit', whose top is placed at the base of the Fish-Scale Marker. This shale unit, included within the Big River Formation in Saskatchewan (Simpson, 1975), is about 100 m thick near the Fourth Meridian and thins westward as the Viking and Bow Island thicken. The Fish-Scale Marker, a zone of abundant fish remains, is less than 30 m thick over most of the Plains; the base of the zone forms a distinctive resistivity marker, which is recognized throughout most of Western Canada Basin. In southwestern Alberta, a fine-grained sandstone, several metres thick, has been called the Fish Scale Sandstone.

Lower Colorado beds in the Athabasca region of northeastern Alberta are included in the Joli Fou and Pelican formations and in the lower part of the Labiche Formation. The Joli Fou Formation yields the distinctive *Haplophragmoides gigas* foraminiferal assemblage of earliest Late Albian age, a reliable guide to the transgressive deposits of the Kiowa-Skull Creek Cyclothem. These beds are younger than the Harmon shale and Cadotte sandstone of northeastern British Columbia and northwestern Alberta (Stelck, 1958; Brideaux, 1971; Singh, 1971). Beds equivalent to the Joli Fou are recognized in the basal

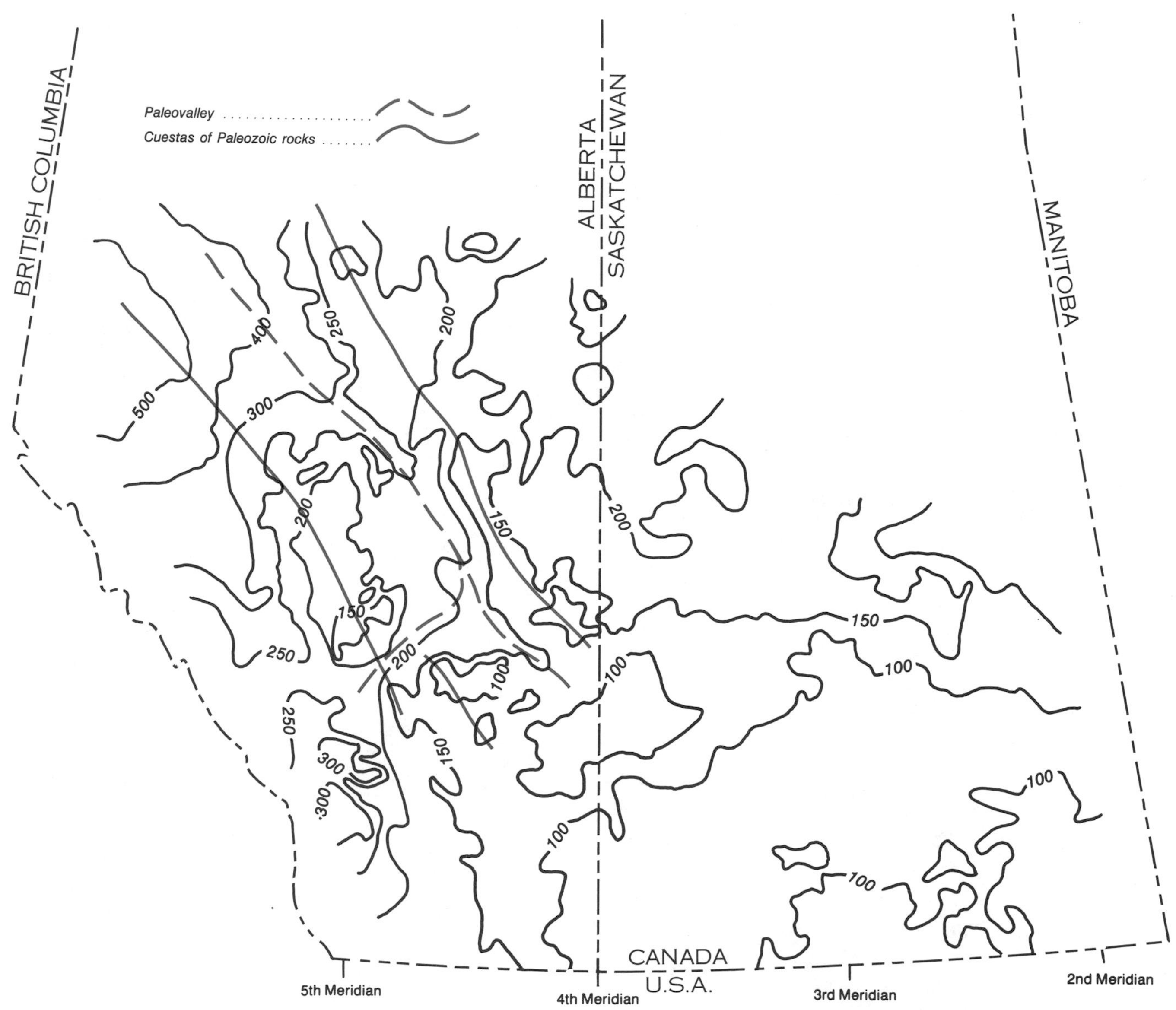

Figure 4I.24. Isopachs (metres) of the Mannville Group and equivalent rocks in the southern part of the Western Canada Basin (D. Cant). Note the northwestward thickening.

Shaftesbury Formation on Peace River in northeastern British Columbia (Koke and Stelck, 1985). Glauconite, marine fossils, and coaly material in the Pelican sediments in northeastern Alberta suggest a mixed marine and nonmarine depositional environment.

The last Early Cretaceous transgression is recorded in northeastern British Columbia and northwestern Alberta by the shales between the Fish-Scale Marker within the Shaftesbury Formation and the Dunvegan Formation. These contain the Late Albian *Miliammina manitobensis* microfauna. To the west in the Foothills, equivalent beds lie at or near the top of the Goodrich and Sikanni sandstones. During the regressive phase, the Dunvegan developed as an extensive delta on the northwestern margin of the basin between Peace and Liard rivers. Its numerous sandstones grade laterally eastward and southward into marine shale (Burk, 1963; Stott, 1982) and westward and northwestward to deltaic and alluvial sediments with abundant conglomerate and thin coal.

East-median hinge (Saskatchewan)

Aptian-Lower Albian

Mannville strata unconformably overlie rock units ranging in age from Jurassic-Cretaceous in southern Saskatchewan to Devonian in central Saskatchewan, to Cambro-Ordovician in east-central Saskatchewan and western Manitoba (Fig. 4I.23). Because of marked erosional relief on the subjacent surface, the thickness of Mannville Group is irregular, ranging from less than 25 m to more than 550 m. Regional thickness conforms to that of the classical wedge, decreasing from Rocky Mountain Trough in the west to a post-Tertiary erosional feather edge that trends southeastward across northwestern Alberta, central Saskatchewan, and southwestern Manitoba (Rudkin, 1964).

More specifically, Mannville distribution shows a broad platform region, occupying southern Saskatchewan and southeastern Alberta. Central to this region, an irregular belt of thin deposits forms a gentle arc whose east-west portion, called Punnichy Arch (Christopher, 1980), terminates in the west against the north-south Swift Current-Sweetgrass Arch complex. North and northwest of the platform, thickness variation is related to cuesta-controlled drainage on the Devonian topography. Whereas it may be inferred that the Mannville represents infill of an irregular topographic surface, tectonic overprinting was operative even in the south. For example, Swift Current-Sweetgrass and Punnichy arches are structural elements that were active during the Mannville depositional episode.

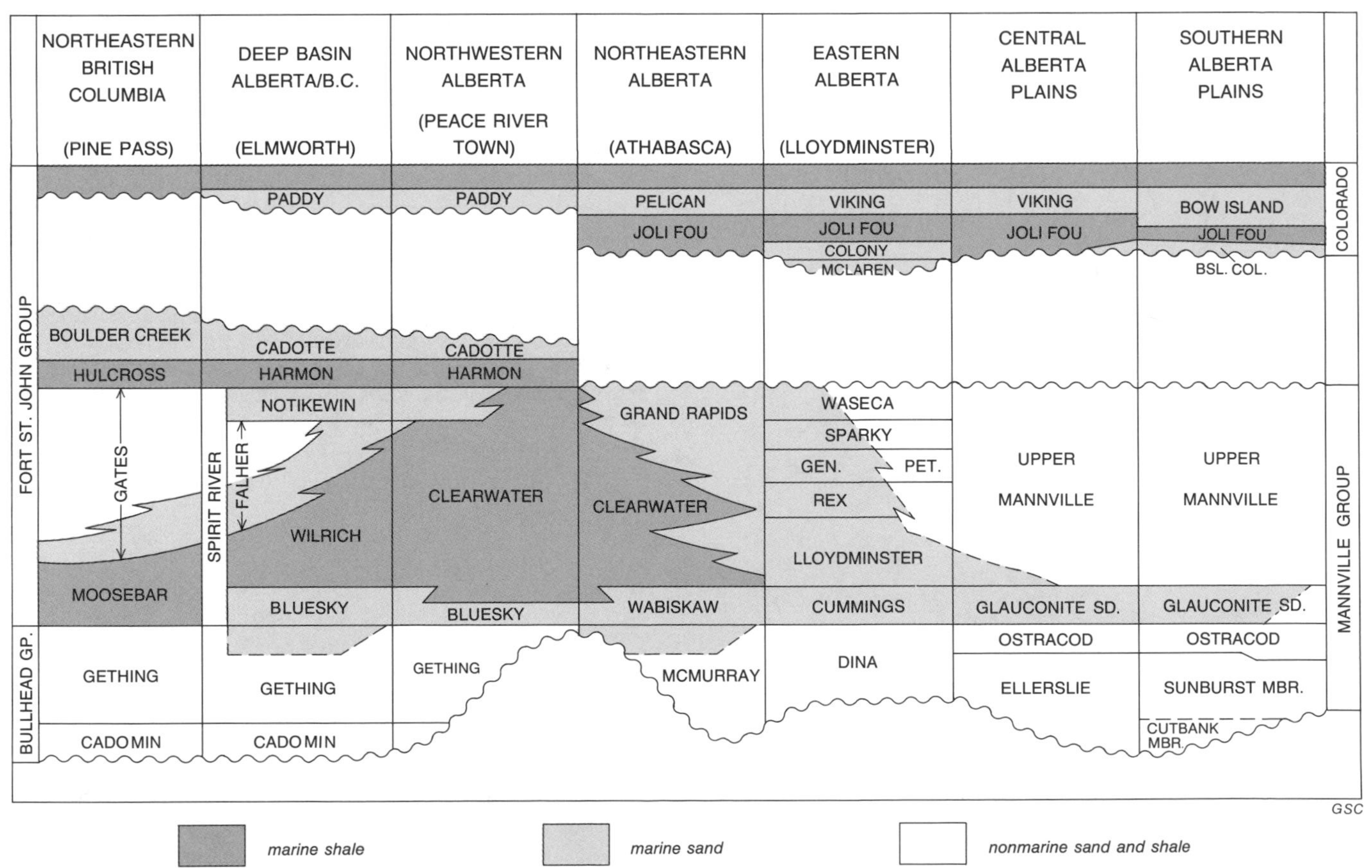

Figure 4I.25. Lower Cretaceous correlations and general facies from northeastern British Columbia to southern Alberta, showing the intertonguing of marine and nonmarine sediments (modified from Jackson, 1985).

The age of the Mannville Group in Saskatchewan is largely determined by its correlation with the Mannville Group of Alberta, which is considered to be Barremian to Early Albian in age (Singh, 1964; Pocock, 1976; McLean and Wall, 1981; Stelck and Kramers, 1980; Caldwell, 1984b).

Mannville nomenclature varies from region to region. However, an important reference area of outcrops is the drainage basin of the Athabasca, Christina, and Clearwater Rivers in east-central Alberta and west-central Saskatchewan, the Athabasca oil sand area. There, the Mannville is divided into the McMurray, Clearwater, and Grand Rapids formations.

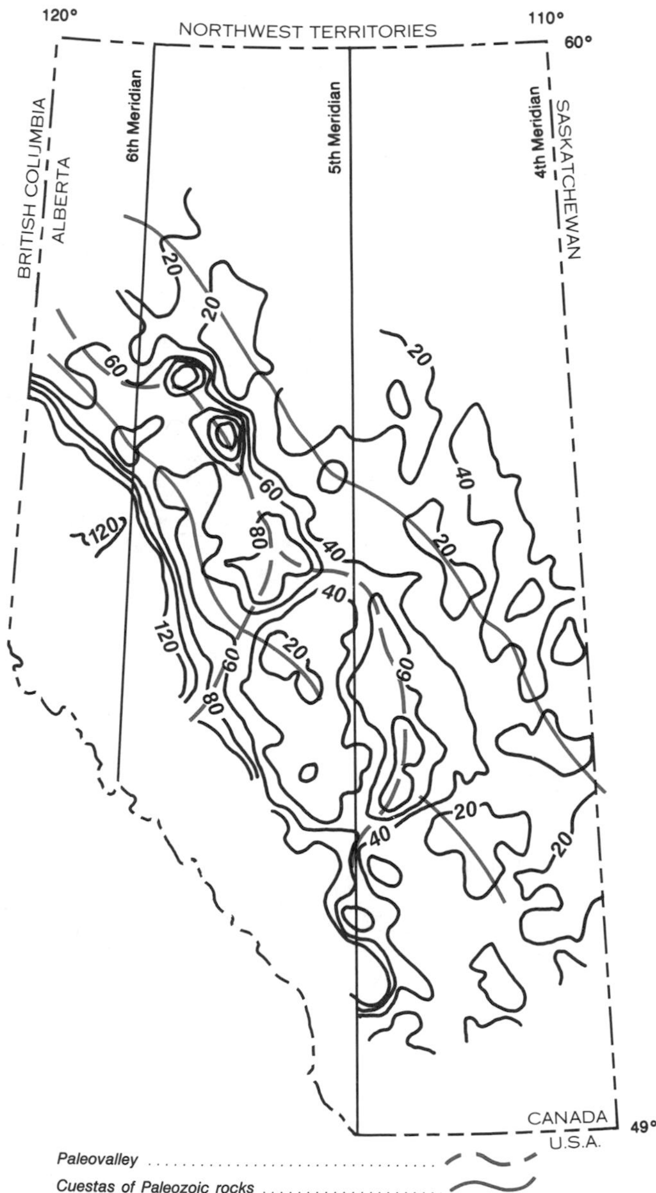

Figure 4I.26. Isopachs (metres) of the lower Mannville Group, Alberta (D. Cant). Note the thickening in paleovalleys and thinning over the adjacent ridges.

Figure 4I.27. Bituminous sands of the McMurray Formation overlying Devonian limestone (D). Epsilon cross-stratified sands overlie thick-bedded sands and underlie argillaceous sands. Clearwater shale (Cl) exposed at the top of cliff. Athabasca River, Alberta (photo - Alberta Geological Survey).

Because of their overall fluvial character and basal position below the widespread Wabiskaw-Cummings marine shale marker, the sandstones equivalent to the McMurray Formation are traceable into southern Saskatchewan. Similar sandstones are included in the Dina Formation of the Lloydminster region of west-central Alberta and adjacent Saskatchewan.

The Clearwater Formation generally comprises marine, interbedded, grey sandstones and shales that interfinger with onshore, lagoonal, and distributary facies on rises that separate broad embayments and narrower estuaries. The Grand Rapids Formation is a glauconitic-chloritic facies that feathers southward and southeastward into the grey facies typical of the Clearwater (Christopher, 1974, 1984). In the Lloydminster region, the combined facies form cyclic sandstone-shale couplets representative of virtually the full range of deposits associated with a gentle but irregularly shelving shore. Environments ranged from backshore-lagoonal to beach and offshore; from fluvial-deltaic to tidal-estuarine and tidal flats.

The Mannville Group of the Lloydminster region comprises nine units: Dina, Cummings, Lloydminster, Rex, General Petroleum, Sparky, Waseca, McLaren, and Colony (Fig. 4I.32). To the south, most of these units, apart from the Cummings shale, become poorly defined, because the overall character of the formations changes from dominantly marine in the north and northwest, to dominantly fluvial in the south. In southwestern Saskatchewan, the condensed Mannville section is represented in the chloritic and glauconitic Cantuar Formation. In eastern Saskatchewan and western Manitoba, the dark grey Clearwater mudstones and sandstones interfinger with white kaolinitic quartzose sandstones of the Swan River Formation, derived from the Canadian Shield.

The Mannville sequence is terminated by one or more hiatuses and erosional intervals at or about the contact with the Joli Fou Formation. The marine Harmon shale and Cadotte sandstone of northwestern Alberta, forming the upper part of the Peace River Formation, are absent in eastern Alberta. Likewise the 45-m thick, marine Pense Formation of southern Saskatchewan is truncated to a feather edge in central Saskatchewan. More significantly, the Pense Formation itself lies disconformably on the Waseca and older members where the underlying Mannville is eroded to its very base across Punnichy Arch of southern Saskatchewan (Fig. 4I.33; Christopher, 1984).

The development of the Mannville depositional basin in Saskatchewan began in post-Jurassic time with uplift of the Sweetgrass-Swift Current platform complex by as much as 600 m. The deep dissection of Jurassic beds on this upland developed a topographic grain that was radial to the north and northeast in southeastern Alberta and southwestern Saskatchewan. However, the cuesta-dominated landforms on the Devonian carbonates trended predominantly northwestward. Valley widths ranged from less than 1 km in the southern uplands to more than 10 km down valley. In Saskatchewan, the main valley system is called the Assiniboia (Christopher, 1975).

Most of the basal Dina-Ellerslie sediments represent products of sheet wash, surface creep, and near-source streams, and are derived from Jurassic and Jurassic-Cretaceous outcrops of the interfluves and uplands. The early sediments were augmented and succeeded by other detritus introduced by regional stream systems flowing from the weathered Canadian Shield in the north and east and from the Paleozoic uplands in the south and west. The sediments were spread out along and across the valleys as channel cut-and-fill, quartzose sands and as coalescent alluvial-fan sediments, creating lacustrine basins, lagoons, and marshes. An initial invasion of the Boreal Sea advanced up the valleys and into shallow inlets and estuaries in which the later sediments were reworked.

The Moosebar-Clearwater invasion was a major pulse that advanced into the heads of the embayments in southern Saskatchewan and Alberta, and in the southeast penetrated Williston Basin in North Dakota. Successive

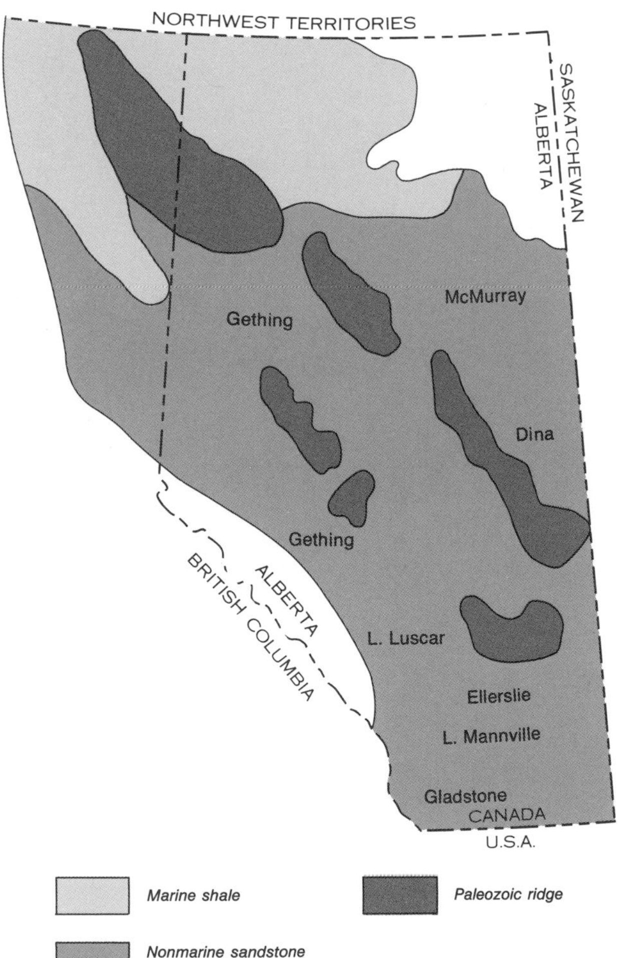

Figure 4I.28. Paleogeographic map of the lower Mannville Group and equivalent rocks in Alberta, showing the extensive distribution of nonmarine sediments surrounding ridges of Paleozoic carbonate rocks (D. Cant). Compare with Figure 4I.26.

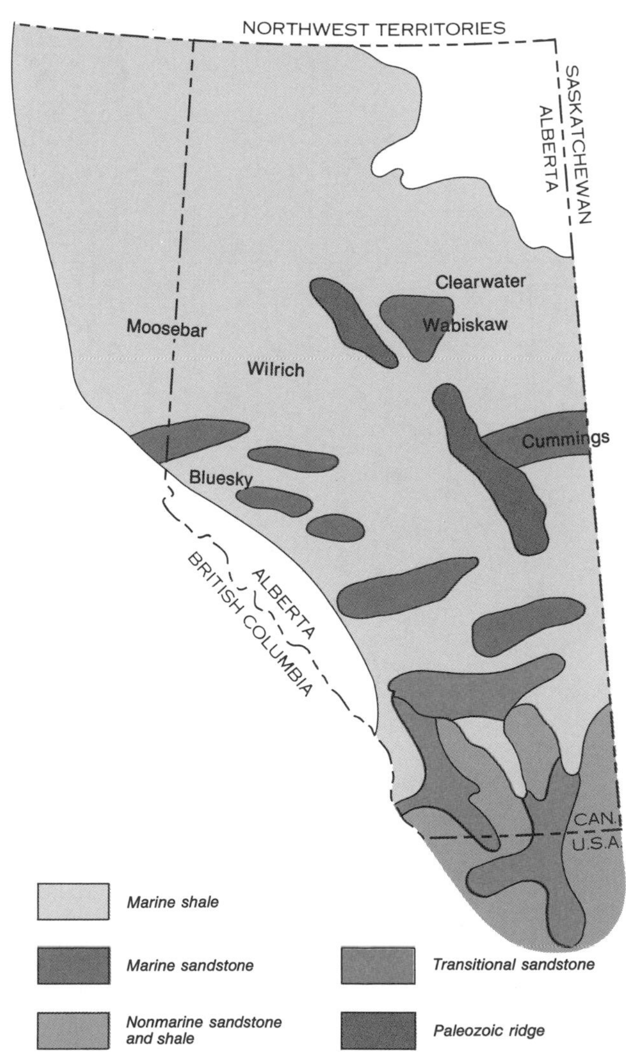

Figure 4I.29. Paleogeographic map of the middle Mannville Group and equivalent rocks in Alberta, showing the distribution of marine, transitional, and nonmarine sediments, offshore bars, and ridges of Paleozoic rocks associated with the southward marine advance (D. Cant).

Clearwater pulses occurred within a regressional mode even as the southern region subsided under the negative influence of Williston Basin and its western offshoot, Fergus-Great Falls Basin of Montana. The deeper parts of the seaway lay above the sub-Mannville depressions and valleys, and the shallow parts corresponded to the buried crests and highs.

Towards the end of Mannville time, accelerated downwarp of Williston Basin uptilted its northern rim at Punnichy Arch, while permitting entry of a precursor Gulfian Sea. Contemporaneously, the basin northwest of the arch and extending into central Alberta also deepened. Sediments eroded from the Mannville on Punnichy Arch were swept south onto an episodically subsiding Swift Current Platform, and redeposited there as the Pense Formation.

Middle Albian to Cenomanian

In Saskatchewan, Middle Albian to Cenomanian marine strata, commonly included in the lower Colorado Group (Fig. 4I.34; Rudkin, 1964; Williams and Burk, 1964; Simpson, 1975, 1979, 1982, 1984), are dominantly argillaceous with subordinate shaly sandstone and siltstone and minor occurrences of conglomerate. These beds are known mainly from subsurface data because exposures are largely restricted to a few scattered locations, notably in Pasquia Hills along the northern erosional edge.

Facies of the lower Colorado Group were controlled by basement structure and solution-generated collapse features of strong relief, and by drape over sub-Mannville topography.

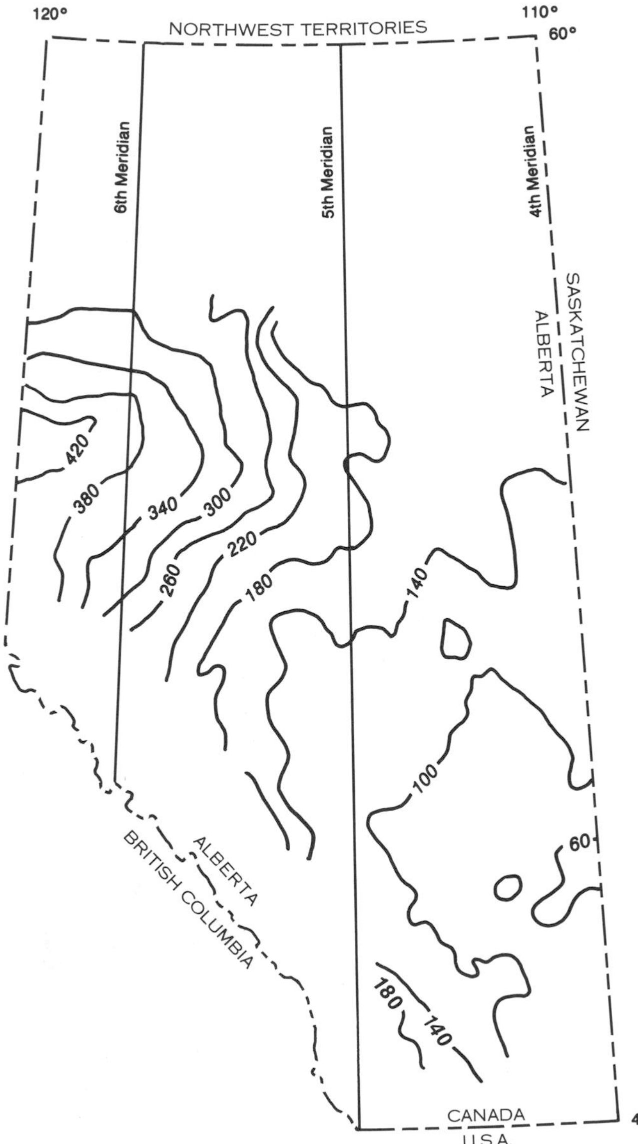

Figure 4I.30. Isopachs (metres) of the upper Mannville Group in Alberta (D. Cant). Note the northwestward thickening and the absence of paleovalleys, which were evident in the lower Mannville.

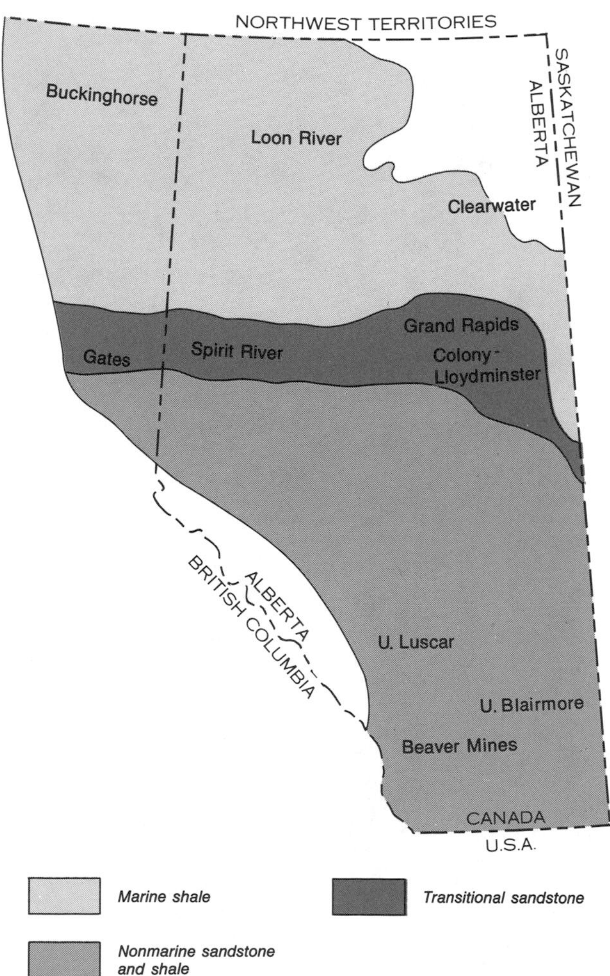

Figure 4I.31. Paleogeographic map of the upper Mannville Group and equivalent rocks in Alberta, showing the east-west trends of shorelines during the general progradation northward of nonmarine sediments (D. Cant). Compare with Figure 4I.30.

The boundary between Lower and Upper Cretaceous strata is generally assumed to be at the base of the Fish-Scale Marker, a widespread phosphatic unit some 18 to 30 m below the top of the Second White-Speckled Shale. However Caldwell (1984b) has shown that, in Manitoba Escarpment and across most of the southern Interior Plains, the base of the Fish-Scale Marker lies much lower, within the Upper Albian. He considered that the Albian-Cenomanian boundary in Saskatchewan lies within a hiatus that spans a good portion of Late Albian and all of Early Cenomanian time.

Colorado Group rocks below the Second White-Speckled Shale, about 215 m thick in southwestern Saskatchewan, are divided from the top downward into the Big River, Viking, and Joli Fou formations (Simpson, 1975, 1984). In central Saskatchewan, the dominantly argillaceous Joli Fou and Big River formations display a northerly increase in sandstone and siltstone, whereas the intervening sandy Viking Formation feathers out.

The Joli Fou Formation, up to 61 m, consists of non-calcareous shale with minor sandstone of the Kiowa-Skull Creek Marine Cycle (Kauffman, 1967; Caldwell, 1984a) and lies within the *Haplophragmoides gigas* Zone of early Late Albian age (Caldwell et al., 1978). In central Saskatchewan, basal beds consist of up to 37 m of interbedded glauconitic sandstone and non-calcareous mudstone of the Spinney Hill Sandstone (Edwards, 1960; Simpson, 1982), laid down on the eastern shelf as a broad tongue, oriented at right angles to the depositional strike. This sandstone is interpreted as estuarine. A discontinuous sandstone unit, 8 m thick and about 12 m above the base of the Joli Fou in southwestern Saskatchewan, is referred to the Cessford Sand. In parts of southeastern Alberta and southwestern Saskatchewan, a thin discontinuous, sandstone unit, resting disconformably on the Mannville Group, is termed the Basal Colorado Sandstone.

Regression led to the formation of the Bow Island-Viking wedge on the western shelf, and the Newcastle and Flotten Lake sands on the eastern shelf. The Viking Formation, 9 to 21 m thick in west-central Saskatchewan (Simpson, 1982), is made up of shaly sandstone and subordinate conglomerate. It is dominated by coarsening-upward sequences considered to have

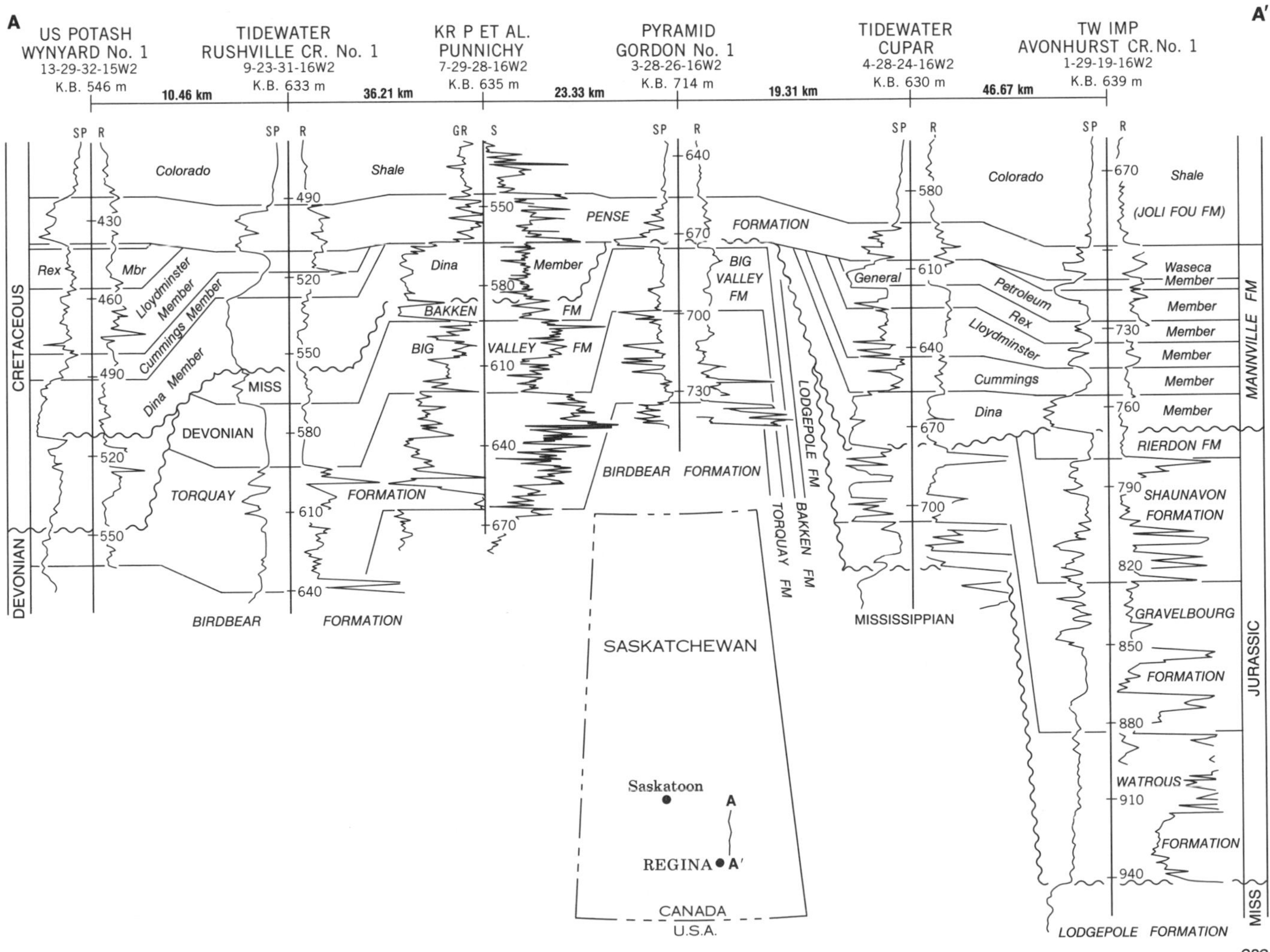

Figure 4I.32. North-south stratigraphic cross-section of the Punnichy Arch, southeastern Saskatchewan, Range 16W2, Townships 32-19 (after Christopher, 1980).

developed as a series of sand ridges similar to those formed in modern shelf settings. In southwestern Saskatchewan, the single sandstone unit of the Viking Formation gives way to the Second and Third Bow Island sands (Jones, 1961; Evans, 1970), which are composite, multistorey sandstone beds separated by mudstone and shale. The Sweetgrass Arch appears to have limited their eastward spread. The Punnichy Arch performed a similar function in the eastern shelf region during deposition of the Newcastle succession. The latter is characterized by

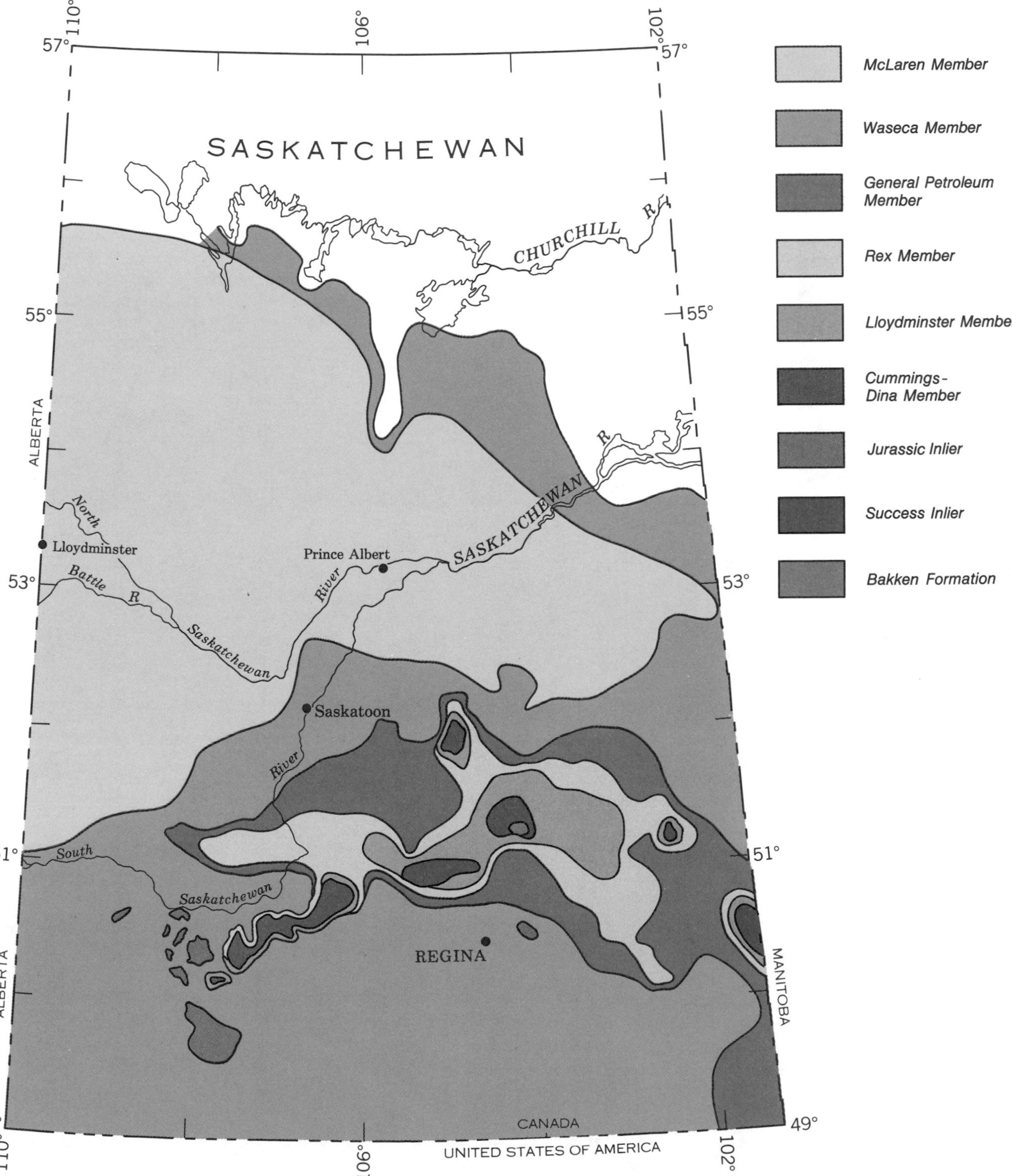

Figure 4I.33. Paleogeological map of the pre-Pense erosion surface, Saskatchewan (after Christopher, 1980).

pronounced thickness changes over short distances, with northward passage from stacked, multi-story sandstone bodies into a single sandstone unit. Similarly, a southwestward-thinning wedge of sandstones about 21 m thick, known as the Flotten Lake sand, occurs in west-central Saskatchewan.

The Big River Formation, 150 m in the southwest, is made up of non-calcareous mudstone with minor interbedded sandstone and coarse-grained siltstone. The unit incorporates the Fish-Scale Marker, which thins northeastward from 21 m in southern Saskatchewan to 1.5 m or less at the northern erosional limit. The sandy part of the formation, occurring below the Fish-Scale Marker in southwestern Saskatchewan and correlative with the First Bow Island sand of southeastern Alberta, is termed the Spikes sandstone. In west-central Saskatchewan, a southwestward-thinning wedge of shaly sandstone and scarce conglomerate, the St. Walburg sandstone, occurs below the Fish-Scale Marker. The Okla sandstone, an approximately equivalent unit, is 8 m thick in east-central Saskatchewan and pinches out to the south. The shale above the Viking sand and below the Fish-Scale Marker contains the *Miliammina manitobensis* Zone (Caldwell et al., 1978).

Eastern Platform (Manitoba Escarpment)

Albian to Cenomanian

At Manitoba Escarpment, which marks the eastern erosional edge of Cretaceous strata, approximately 600 m of Albian to Maastrichtian sediments, interrupted by numerous unconformities, were deposited in terrestrial, transitional, and marine environments (Fig. 4I.35; McNeil and Caldwell, 1981; McNeil, 1984). The section consists of sand, shale, and calcareous sediments, which are collectively diagnostic of the "eastern facies belt" (sensu Tourtelot, 1962), extending from the area of Manitoba Escarpment southward into Nebraska and Kansas. In terms of the regional structural zones for the Western Interior Cretaceous Basin (Kauffman, 1977b), the eastern facies belt was deposited mainly in the "eastern platform zone", which at times of peak marine transgression was the site of chalky shale, marl, or limestone deposition.

The Cretaceous record begins with terrestrial to marginal marine terrigenous clastic sediments of the upper Middle to lower Upper Albian Swan River Formation, which rests unconformably on Jurassic and older strata. The formation varies markedly in thickness from near 0 to 120 m, reflecting the underlying valley-interfluve topography (Christopher, 1975; McNeil and Caldwell, 1981). The lower Swan River, consisting mainly of grey silts and clays, in part kaolinitic, was deposited in fluvial and paludal environments marginal to the boreal seaway. The upper Swan River, comprising fine, well-sorted sand, richly glauconitic in places and often with black clay interbeds, represents littoral and shoreface deposition. Upper Swan River strata were dated palynologically by Playford (1971) as late Middle Albian, possibly to earliest Late Albian in age. He correlated the Swan River beds with the Harmon and Cadotte members of western Alberta, although no Mannville beds of equivalent age are known between the Swan River and Harmon-Cadotte sections (Norris, 1967; Singh, 1964, 1975). Alternatively, Stelck (1958) proposed that the Swan River sediments may represent an early onlap of the Gulfian Joli Fou sea.

The Ashville Formation, 50 to 100 m thick, comprising Skull Creek, Newcastle, Westgate, and Belle Fourche members, rests with sharp, probably disconformable contact on the Swan River along most of Manitoba Escarpment. Eastward transgression of the Late Albian sea is recorded in the diachronous rise of the Ashville-Swan River contact and by the disappearance eastward of the lower part of the foraminiferal Zone of *Haplophragmoides gigas* (McNeil, 1984). Marine shale deposition (Skull Creek Member) was terminated in Late Albian time by basin-wide shoaling and subsequent deposition of several closely contemporaneous sands derived from both the western and eastern flanks of the basin. The easterly derived Newcastle sandstone of the Ashville Formation is recognized discontinuously along Manitoba Escarpment and into the subsurface of eastern Saskatchewan (McNeil and Caldwell, 1981). Transgressive marine sedimentation was re-established as the boreal Mowry Sea spread far to the south. This event is recorded in the dark shale, 8 to 30 m thick, of the Westgate Member, which carries foraminifers of the *Miliammina manitobensis* Zone.

The latest Albian to Cenomanian Belle Fourche Member of the Ashville Formation is a 45-m-thick shale that diminishes northwestward to 10 or 15 m in east-central Saskatchewan. The lower few metres of the Belle Fourche contain quartzose silts of the Fish-Scale Marker beds, which represent regional shoaling. These silts were probably derived from the east and have been correlated with the D Sandstone, which extends from the arenaceous Dakota Group into the Graneros Shale of the Denver Basin (McNeil, 1984).

The upper Belle Fourche carries the calcarenitic *Ostrea beloiti* beds and an associated marker bentonite believed by McNeil and Caldwell (1981) to be equivalent to the "X" bentonite of the Graneros Shale in the Colorado-Kansas area. Oyster-bearing calcarenite is a consistent feature of the black shales of the upper Belle Fourche and equivalents along the eastern facies belt and signals the transition to the peak period of marine transgression (McNeil, 1984).

Upper Cenomanian-Maastrichtian sequence

The Cenomanian to Campanian history of Western Canada Basin is marked by widespread expansion of the seaway. A regional unconformity, found throughout much of the basin, separates Lower from Upper Cretaceous strata, forming a major tectonostratigraphic boundary. Late Campanian to Maastrichtian phases of the Laramide Orogeny resulted in widespread molassic deposits.

In the northern basins, Cenomanian and younger strata record the gradual northward migration of depocentres and shorelines, such that by Paleocene time most deposition was on the continental margin of the Beaufort Sea. The rising Cordillera and Porcupine Platform supplied vast quantities of sediment to form major deltaic depositional complexes.

Turonian to Campanian epineritic sands in the south are associated with several transgressive-regressive couplets within a dominantly shale sequence. Campanian to Maastrichtian alluvial-deltaic sediments subsequently prograded eastward across the basin, terminating the marine cycle.

Northern basins

Cenomanian-Turonian

Deposition at the beginning of Early Cretaceous time in the northern basins was restricted mainly to the western troughs. In Kandik Trough, the Monster Formation, its lowest beds dated as Cenomanian on the basis of dinoflagellates (McIntyre in Ricketts, 1988), records the northward to northeastward progradation of a coastal fan-delta complex. A thin marine mudstone at the base of the formation is rapidly succeeded by littoral and nonmarine strata. Along the southern edge of Eagle Plain, the Eagle Plain Formation, over 700 m thick, consists of sandstone and shales (Fig. 4I.15). The association of Cenomanian megaflora and *Inoceramus* ex gr. *dunveganensis* suggests deposition in lacustrine and fluvial deltaic to marine environments and proximity to an oscillating shoreline (Mountjoy, 1967). To the north, Eagle Plain strata are interpreted to be nearshore, inner-shelf deposits. In Blow Trough to the northwest, and in Beaufort-Mackenzie Basin, organic-rich, bentonitic and ferruginous Boundary Creek shales of Cenomanian to latest Turonian age lie unconformably on older Cretaceous rocks (Jeletzky, 1960; Young et al., 1976).

The entire area of Anderson Basin, Carnwath Platform, and the northern part of Keele Arch may have been emergent during Late Albian to Turonian time (Yorath and Cook, 1981, 1984), although Cenomanian strata at an isolated locality on the southern edge of Anderson Basin suggests such strata were once present over a much wider area. In Peel Trough, the possibility exists that the Trevor Formation, as described by Yorath and Cook (1981), lies below or includes an unconformity representing some part of Late Albian to Cenomanian time. A regional unconformity appears to separate Lower and Upper

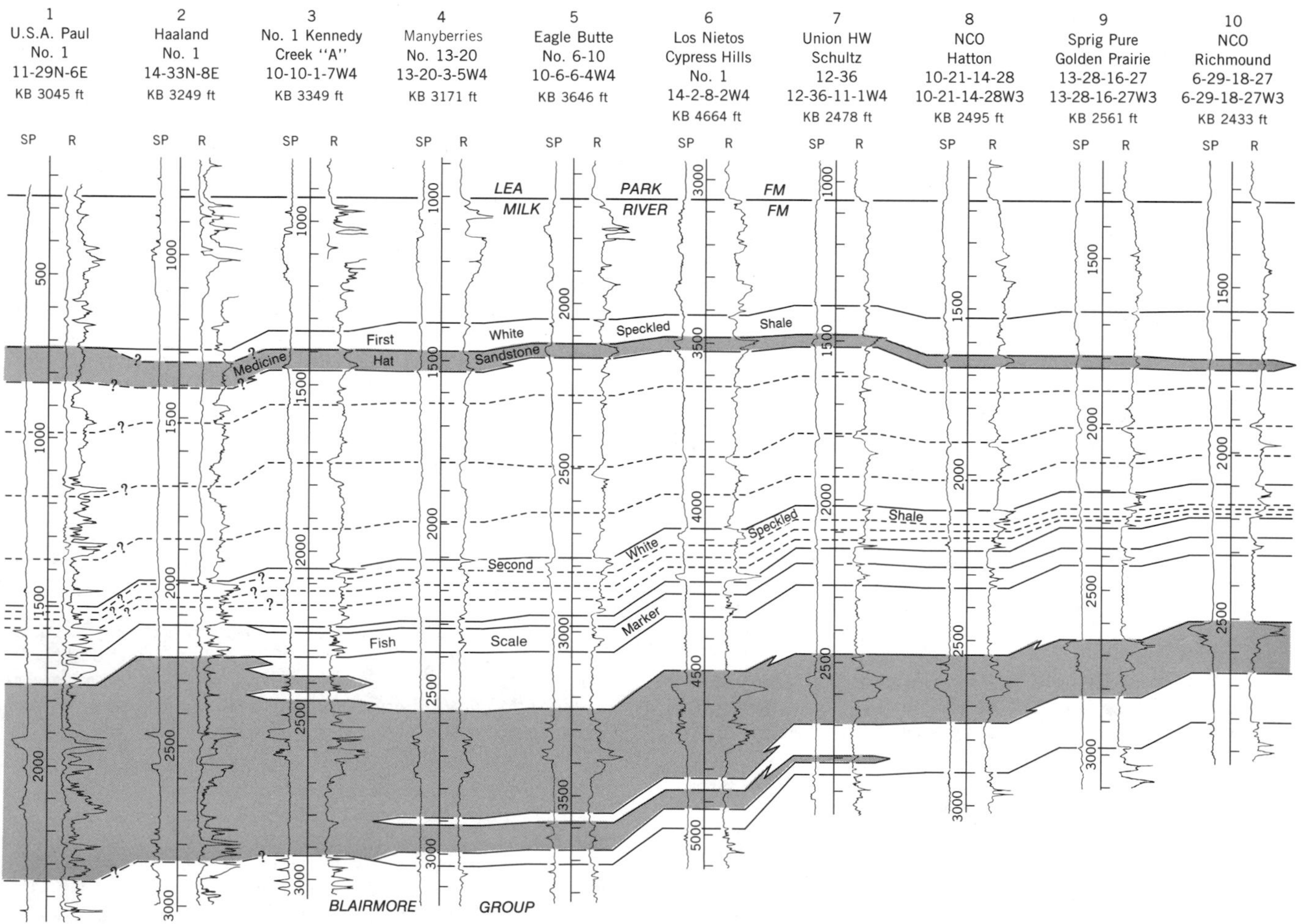

Figure 4I.34. South-north stratigraphic cross-section showing main Colorado sandstone bodies of the western shelf, north-central Montana, southeastern Alberta, and western Saskatchewan (after Simpson, 1975).

Cretaceous strata throughout most of the District of Mackenzie. In southern Peel Trough, shale of the Slater River Formation forms the basal Cretaceous unit and is apparently of Late Cretaceous age.

A sequence of Albian shales (800 m) in Great Bear Basin is poorly known but records subsidence, at least in Middle Albian time, and may have been continuous with the Arctic Red and Trevor strata of Peel Trough (Yorath and Cook, 1981). Upper Albian and Turonian rocks are also present in the area. The sequence is truncated by Campanian to Maastrichtian strata. Outcrops of Cenomanian-Turonian strata are insufficient to provide an adequate outline of basin history.

Coniacian-Campanian

During Coniacian to Campanian time, an extensive shallow-water shelf received coarse detritus from the rising Cordillera to the south, and the nonmarine to near-shore marine upper Eagle Plain Formation was deposited along the northern flank of the orogen. To the southeast, Peel Trough continued as a depocentre, and Anderson Basin as a low-energy, dysaerobic, shelf environment.

Upper Coniacian to Lower Campanian strata of the upper Eagle Plain, Smoking Hills, and Little Bear formations are poorly exposed and of limited areal extent (Figs. 4I.8, 4I.15). Thicknesses are not great, the Smoking Hills being only about 100 m and the Little Bear over most of its distribution less than 300 m (Yorath and Cook, 1981). No Coniacian beds are recognized in Great Bear Basin, although Upper Campanian shale is present.

The Smoking Hills Formation of Mackenzie Delta and Anderson Plain (Fig. 4I.8), dated as Santonian to Campanian on the basis of palynomorphs (McIntyre, 1974), consists of dark grey to black, bituminous, marine shale, rich in organic carbon at its base. The Little Bear Formation, overlying the Slater River shale in Peel Trough, also has yielded Santonian to Campanian palynomorphs (Brideaux, 1971; see also Aitken and Cook, 1974), and is thickest to the southwest and thinnest toward Keele Arch.

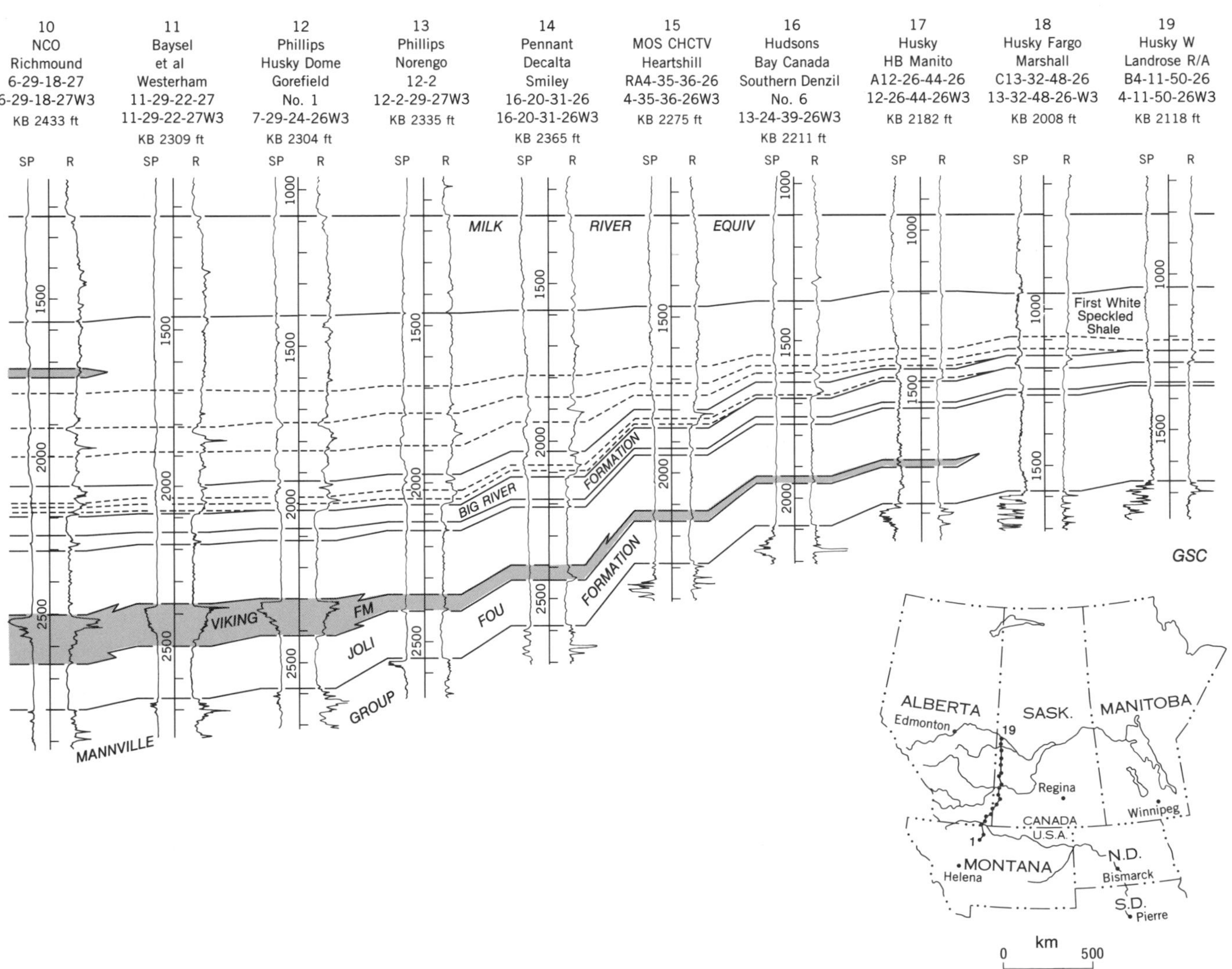

Figure 4I.34. Cont.

It is dominated by nonmarine coarse clastics at the eastern end of Peel Trough but becomes partly marine to the northeast and east.

A major unconformity separates the Cenomanian to Turonian Boundary Creek strata of Mackenzie Delta from Maastrichtian Tent Island strata whereas, in Anderson Basin, the contact between the Santonian to Campanian Smoking Hills and Maastrichtian Mason River Formation is sharp but shows no evidence of erosion. In Peel Trough, the contact between the Little Bear and overlying East Fork Formation is disconformable and erosional in places (Yorath and Cook, 1981). Great Bear Basin may have been emergent during part of Coniacian to Campanian time. Anderson Basin on the other hand appears to have remained below sea level.

Campanian to Paleocene

During Campanian-Maastrichtian time, marine muds were deposited in the Boreal Sea across much of northern Yukon Territory and adjacent areas to the east. Shelf muds of the Tent Island, East Fork, and Mason River formations (Fig. 4I.8) represent the transgressive phase. Basins were filled rapidly as alluvial sediments of the Moose Channel, Summit Creek, and Bonnet Plume formations spread northward and eastward from the rising Cordillera.

At the base of the Tent Island Formation in northern Yukon Territory, a distinct conglomerate-sandstone unit, the Cuesta Creek Member (Fig. 4I.36) was originally described as alluvial deposits (Holmes and Oliver, 1973; Young, 1975a). These beds have features more characteristic of sediment gravity flow deposits and may have been deposited on a submarine fan. Prograding over the Cuesta Creek beds are prodelta muds that form the bulk of the Tent Island Formation. In Anderson Basin, equivalent strata are represented by shelf muds of the Mason River Formation, which has no basal sandy unit.

To the southwest in Peel Trough, East Fork marine shales (Yorath and Cook, 1981) are interbedded with fine-grained argillaceous sandstone and siltstone. As well, at the east end of Peel Trough, locally derived sands of the East Fork Formation were deposited on the flank of Keele Arch. To the west in Bonnet Plume Basin, nonmarine coal-bearing sediments were deposited in an intermontane

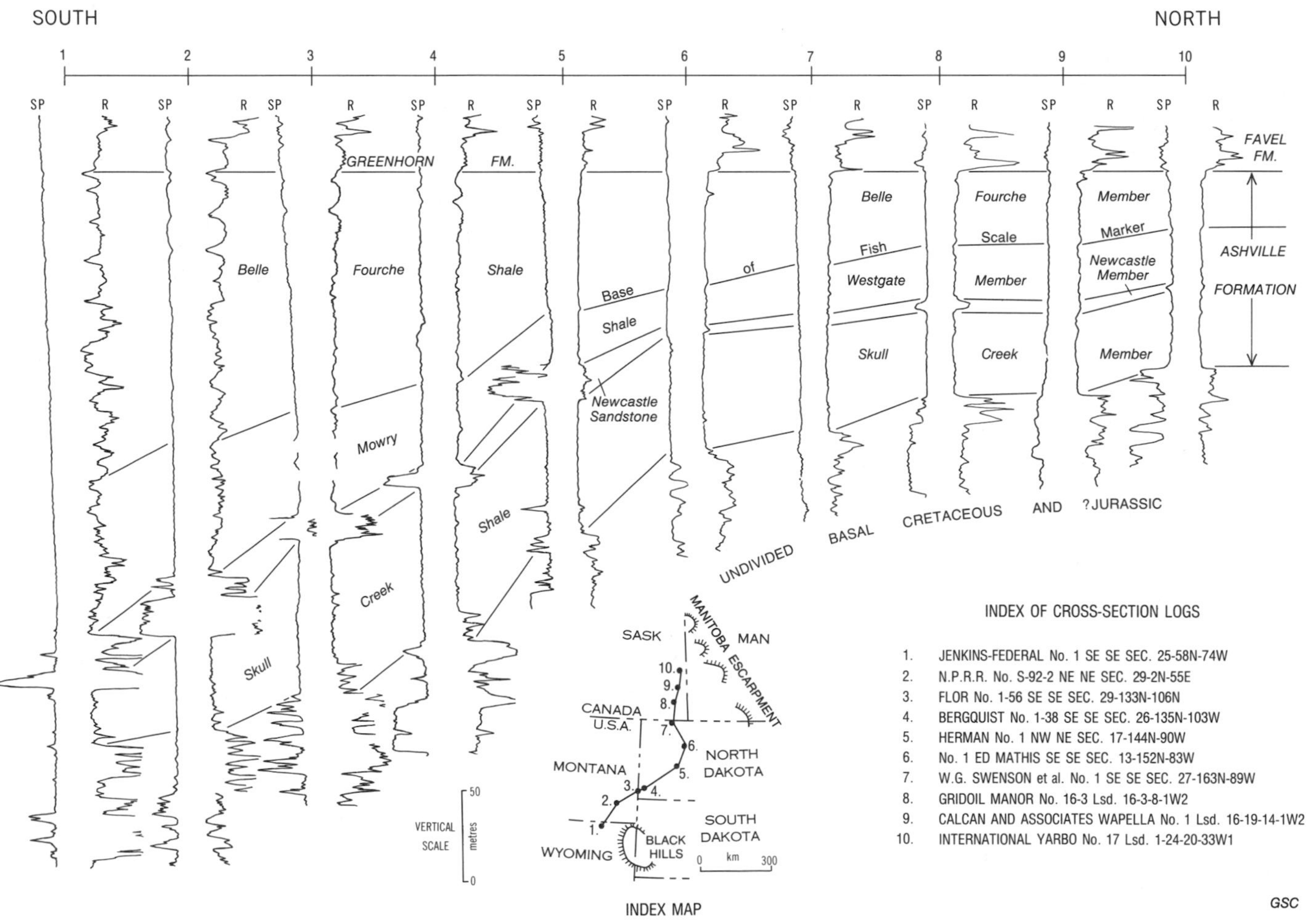

Figure 4I.35. South-north stratigraphic cross-section of Lower Cretaceous units, Wyoming, Montana, North Dakota and Saskatchewan (after McNeil and Caldwell, 1981).

basin (Norris and Hopkins, 1977). The river system(s) in the Bonnet Plume Basin probably carried sediment into the marine western end of Peel Trough.

In Peel Trough, Upper Maastrichtian to Paleocene conglomerates, sandstones, ash beds, and low-grade coal, are included in the Summit Creek Formation (Yorath and Cook, 1981; Cameron et al., 1986), which overlies the East Fork shales (Fig. 4I.15). The sediments appear to be part of a large alluvial fan derived from the Mackenzie Mountains. Unnamed rocks of equivalent age in Great Bear Basin are poorly exposed, and thicknesses are not known.

By the close of Maastrichtian and beginning of Paleocene time, coarse clastic sediments of the Moose Channel, Summit Creek, and upper Bonnet Plume formations were prograding into the basins. A gradual transition from shelf and prodeltaic muds to coarse-grained, nearshore, and deltaic sands is reported from most areas (Holmes and Oliver, 1973; Young, 1975a; Yorath and Cook, 1981; B.D. Ricketts, pers. comm., 1985), although local erosional contacts have been reported. A large delta complex, or series of complexes, at least 1000 m thick along the north slope of Yukon Territory and represented by the Moose Channel Formation, continued through the Paleocene Epoch. To the southeast nonmarine strata of the Bonnet Plume Formation accumulated in an intermontane basin. Farther east, the nonmarine Summit Creek Formation filled the final vestige of the foredeep. A shallow-water, marine embayment probably extended from northern Yukon Territory, southeast across Peel Plateau and western Anderson Basin to the Norman Wells area. This embayment gradually filled and the shoreline migrated northward. Paleocene strata are believed to have been eroded from the area of the embayment.

To the northwest, evidence for uplift, erosion, and the development of an unconformity at the close of Moose Channel deposition is limited to the Natsek E-56 well and Caribou Hills. Reflection seismic data in the Natsek area indicate a hiatus between the Moose Channel (Fish River sequence) and younger units (Dixon et al., 1985). At Caribou Hills, the succeeding Reindeer Formation rests unconformably on Smoking Hills (Santonian-Campanian) strata.

Figure 4I.36. Cuesta Creek Member of the Tent Island Formation, Fish River, northern Richardson Mountains. Conglomerate and sandstone occupy a channel eroded into turbiditic sandstone and shale. Pale outcrops in distance are shales of the Tent Island Formation (photo - J. Dixon, ISPG 2236-6).

Rocky Mountain Trough

Cenomanian to Maastrichtian

Upper Cretaceous marine strata attain a maximum thickness of 4 km in the western Foothills (Fig. 4I.37) but thin rapidly to less than 1 km in the plains. The succession lies unconformably on the Blairmore and Luscar groups in the southern and central Foothills of Alberta (Fig. 4I.38), and on the Cenomanian Dunvegan Formation in northeastern British Columbia. The post-Dunvegan hiatus north of Peace River represents all of Turonian time and possibly Coniacian as well. In the central and northern Foothills of Alberta and extending into the Peace River region, the succession is more continuous, and the upper contact is conformable and transitional. The Upper Cretaceous beds are overlain throughout the Foothills by Paleocene sediments assigned to the upper Willow Creek, Porcupine Hills, and Paskapoo formations.

The boreal invasion that spread southward into the Western Interior Basin of the United States in Late Albian time extended westward into the present Foothills region of southern Alberta in late Cenomanian to Turonian time. The marine Alberta and Smoky groups (Fig. 4I.37), of Late Cenomanian to Campanian age and equivalent to the Colorado and Montana groups of the Plains, include sideritic and calcareous shale, siltstone, and fine-grained sandstone (Stott, 1963, 1967b), and range in thickness from 600 to 1400 m. The sequence occurs along the length of the Foothills to a northern erosional limit near Peace River. The upper part of this succession and overlying alluvial beds reappear in the broad Liard Syncline north of the Alaska Highway. Two megacycles are recognized.

The shales of the first megacycle are included in Upper Albian to Turonian Blackstone and Kaskapau formations of the Alberta and Smoky groups respectively (Stott, 1963, 1967b). Upper Cenomanian (*Dunveganoceras* Zone) basal shales of the Blackstone and Kaskapau are present in most of the Foothills region south of Pine River, but near Peace River, the Doe Creek and Pouce Coupe sandstones (Warren and Stelck, 1940) and the Howard Creek sandstone (Stelck and Wall, 1954) are nearshore equivalents. The succeeding, Turonian, Vimy calcareous shales mark the peak transgression. They lie within the zone of *Prionocyclus woollgari* and are equivalent in part to the Second Speckled Shale of the Plains (Williams and Burk, 1964) and part of the Greenhorn Formation of the Western Interior of the United States. Those beds are overlain by fetid mudstone of the Haven Member and sideritic to glauconitic mudstone and siltstone of the Opabin Member. The Turonian shales grade northward near Pine River into such littoral and near-shore sediments as the Wartenbe and Tuskoola sandstones (Stelck, 1955; Stott, 1967b). To the north of Peace River, no equivalent sediments are known, and Turonian time is represented by the hiatus at the top of the Dunvegan Formation.

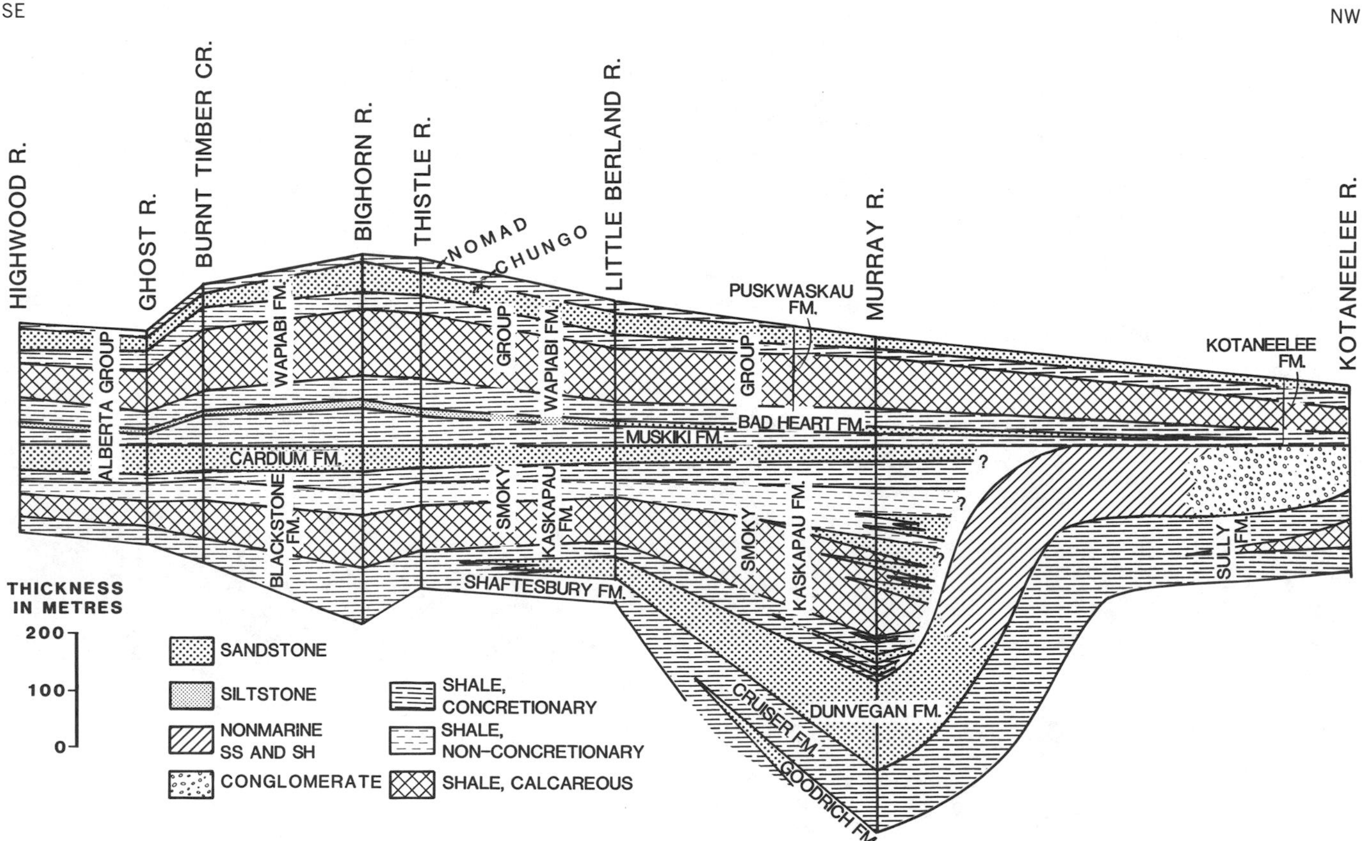

Figure 4I.37. Schematic section along Rocky Mountain Foothills, illustrating the marine phase of the third clastic wedge, recorded in the Alberta and Smoky groups (modified from Stott, 1967b).

The pulsatory regressive phase of this first megacycle is found in the thin, Turonian, Cardium Formation (Fig. 4I.39), which includes sediments ranging from deltaic and lagoonal to neritic (Stott, 1963, 1967b; Nielsen and Porter, 1984; Krause and Nelson, 1984; Wright and Walker, 1981; Walker, 1983a, b). Littoral to neritic, well-sorted sandstone is found in the basal (Ram) and uppermost (Sturrock) members. Lagoonal and marsh deposits, assigned to the Moosehound Member, developed on the western margin. Coarse-grained, conglomeratic sandstone of the Baytree Member is present along the northern outcrop belt near Dawson Creek.

The second megacycle is represented by the Wapiabi and Puskwaskau shales (Fig. 4I.37) of Turonian to Campanian age. The basal, Muskiki, shales lie within the uppermost Turonian zone of *Scaphites preventricosus* and *Inoceramus deformis* Meek. In the Liard region, equivalent beds are found in the Kotaneelee Formation (Stott, 1960). The oldest fauna known from the Kotaneelee being of Early Santonian age, the marine advance may not have extended into the Liard region until somewhat later than in the Foothills to the south. An Early Santonian (*Scaphites depressus* Zone) regression resulted in deposition of the Bad Heart and similar sandstone across northeastern British Columbia from the Foothills to the junction of Smoky and Peace rivers. A later deepening of the basin is indicated by shales of the Santonian Dowling Member, and return to open-basin conditions is recorded in the calcareous shales of the Thistle Member.

Eastward Late Santonian progradation is recorded by the Hanson mudstone and Chungo sandstone, although several minor reversals occurred (Stott, 1963, 1967b; Lerand, 1983). To the south and east, the Chungo merges with the lower part of the Belly River Formation of the southern Foothills and the Milk River sandstone of the Plains. The development of this wedge of coarse clastics indicates an early phase of Laramide mountain building to the southwest (see Fig. 4I.50j).

Dark grey marine shales of the Nomad Member, generally less than 35 m thick, can be traced along almost the entire length of the Foothills. They represent a short-lived transgression related to the Early to Mid-Campanian Claggett Cycle (Lea Park and Pakowki formations) of the eastern basin.

The widespread and prolonged, Late Cretaceous marine deposition in Western Canada Basin was brought to an end as alluvial, coarse-grained sandstone, and interbedded mudstone, variously assigned to the Belly River, Brazeau, and Wapiti formations, were deposited. The Belly River Formation and its equivalents (Tozer, 1956b; Carrigy, 1971) form a thick clastic wedge that extends from the Foothills into Saskatchewan. In the southern Foothills and central Alberta, the Belly River, over 600 m thick and consisting of green sandstone, shale, and some coal, is overlain by the Campanian to Maastrichtian Bearpaw marine shale (Wall and Rosene, 1977; Caldwell et al., 1978). The Bearpaw Formation represents a significant readvance of the seaway during

Figure 4I.38. Unconformable contact between Blackstone Formation (Bl) and Luscar Group (Lu), tributary of Brazeau River, Alberta (photo - D.F. Stott).

Late Campanian time (see Fig. 4I.50k). Although this seaway extended to the west in the Crowsnest region, it did not extend into the present central and northern Foothills of Alberta, nor did it extend northwestward much beyond Edmonton.

In the southern Alberta Foothills, the Bearpaw is overlain by 700 to 1000 m of sandstone and shale, dominantly alluvial with the exception of the basal part, assigned to the St. Mary River Formation (Tozer, 1956b; Carrigy, 1971). The younger Willow Creek Formation, lying transitionally above the St. Mary River, is about 1.2 km thick, and consists of sandstone with varicoloured shales that grade upward into buff-weathering sandstone and brown shale. No hiatus is recognized within the Willow Creek although the upper part contains a typical Paleocene terrestrial invertebrate fauna.

In the central Foothills, the lower beds of the Brazeau Formation, equivalent to the Belly River Formation, include pebble conglomerate, and the upper 1.6 km consists of greenish sandstone, shale, some tuff, and thin coal seams. The Brazeau is overlain by the Entrance conglomerate, the coal-bearing Coalspur Formation, and thick deposits of the Paleocene Paskapoo Formation (Jerzykiewicz and McLean, 1980; Jerzykiewicz, 1985). The Wapiti Formation farther north is similar to the Brazeau Formation (Kramers and Mellon, 1972). These thick uppermost Cretaceous and Tertiary clastic sediments record continued pulses of Laramide Orogeny. Mountain building followed by uplift of the craton to the north resulted in retreat of the seaway from the continental interior and the dominance of alluvial environments.

West-median or axial trough (Alberta)

Cenomanian to Santonian

The record of Cenomanian to Santonian deposition on the Plains is found in the upper Colorado Group. The succession includes two coccolithic marker units, the Second (lower) and First (upper) White-Speckled Shales, which mark the peak transgression respectively of the Turonian Greenhorn and Santonian Niobrara cyclothems. The Colorado is overlain by marine shale of the Lea Park Formation throughout much of the region, but in southeastern Alberta, it is overlain gradationally by Milk River sandstone. Maximum thickness is in excess of 400 m. The predominant lithology is dark grey marine shale, with some prominent sandstones and thinner beds of limestone and basal conglomerate.

In central and southern Alberta, the First and Second White-Speckled shales are about 110 m and 60 m thick respectively. These sediments contain more silt and mud than the chalky facies of the eastern plains. They are separated by as much as 60 m of unnamed, noncalcareous shale equivalent to the Cardium sandstone of the Foothills and the Carlile Shale of the United States. The calcareous shales are bituminous, with intercalated shaly chalk and chalky, skeletal calcarenite deposited under anoxic conditions. The Second Specks Sandstone, 25 m thick, may have formed by intensified winnowing of bottom sediments. In the Lloydminster district of east-central Alberta, lower beds of dark, non-calcareous, siliceous shale are succeeded by a single unit, up to 30 m thick, of white-speckled shale correlative with the First White-Speckled shale farther south.

Along the western edge of the Plains, the two calcareous units are separated by the Cardium sandstone, the principal producing reservoir of the Pembina Oil Field, Canada's largest (Nielsen and Porter, 1984). The Cardium marks the termination of Late Turonian regression and the Greenhorn Cycle. In the southern and central plains, the Cardium consists of shaly sandstone, well-washed sandstone and conglomerate with interbedded mudstone. In the Pembina area (Krause and Nelson, 1984), the lower part is characterized by a coarsening-upward sequence

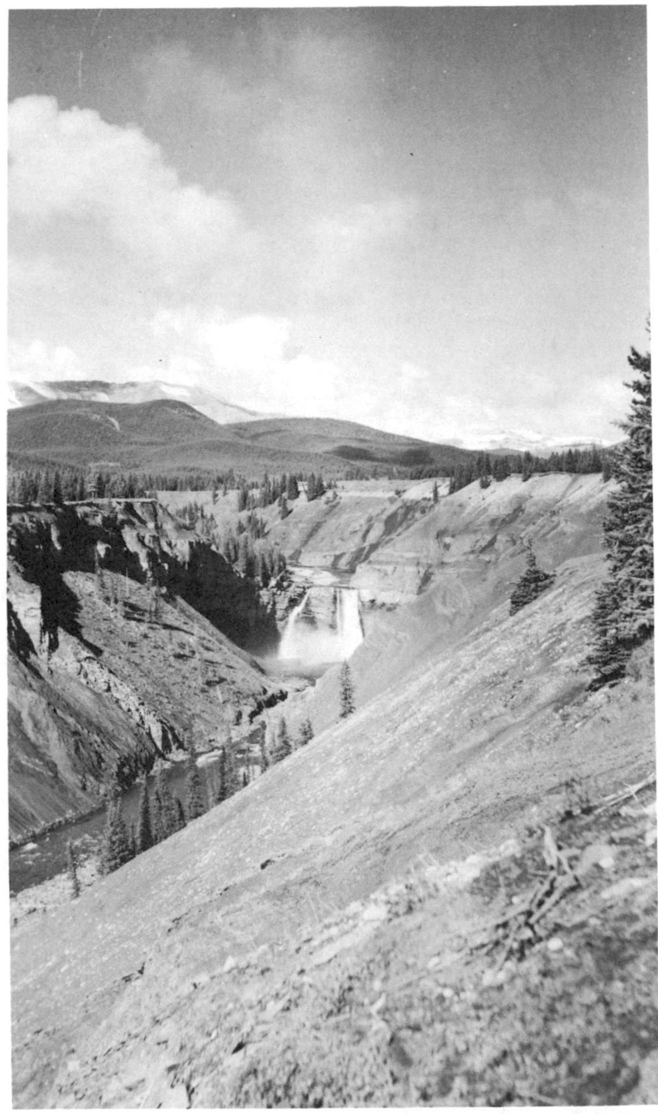

Figure 4I.39. Alberta Group at Ram Falls near North Saskatchewan River, Alberta, view westward. Falls formed by Sturrock Member, Cardium Formation, on west flank of anticlinal fold. Shales above falls are part of the Muskiki Member, Wapiabi Formation and are overlain by pale coloured glacial deposits. Ram Range in background is an anticlinal inlier of Paleozoic rocks (photo - D.F. Stott, GSC 1326-14).

containing bioturbated to distinctly bedded sandstone and shale, capped by cherty conglomerate. This lithofacies is typical of a marine shelf affected by frequent storm events. The upper part contains bioturbated and finely laminated, black shale and siltstone, with pebble stringers and a conglomerate bed at top. Two distinctly different types of rock bodies are recognized by Plint et al. (1986) within the Cardium Formation in the subsurface of Alberta, from Kakwa River in the north to the Caroline-Garrington and Ricinus areas in the south. The first consists of upward-coarsening sequences terminated by erosional surfaces. These sequences are composed essentially of bioturbated mudstone at the base and sandstone with or without bioturbated mudstone toward the top. The second type consists of deposits that immediately overlie the erosional surfaces and occur in scours up to 20 m deep. They consist of conglomeratic bodies with minor sandstone and mudstone. The upward-coarsening sequences were interpreted as being deposited on an offshore shelf. The conglomeratic sediments were considered to be deposited during episodes of lowered sea level that resulted in erosion of shoreline and offshore sediments. Sand bodies in the Ricinus, Caroline, and Garrington fields occur as patchy sheet sands (Walker, 1983a, b; 1985). A well preserved ichnofauna (Pemberton and Frey, 1984) is indicative of offshore, quiet, to nearer shore, higher energy conditions.

The Medicine Hat Sandstone of southeastern Alberta lies about 30 m below the top of the First White-Speckled Shale. It consists of alternating layers of calcareous and non-calcareous sandstone and is noted for the giant natural gas field at Medicine Hat.

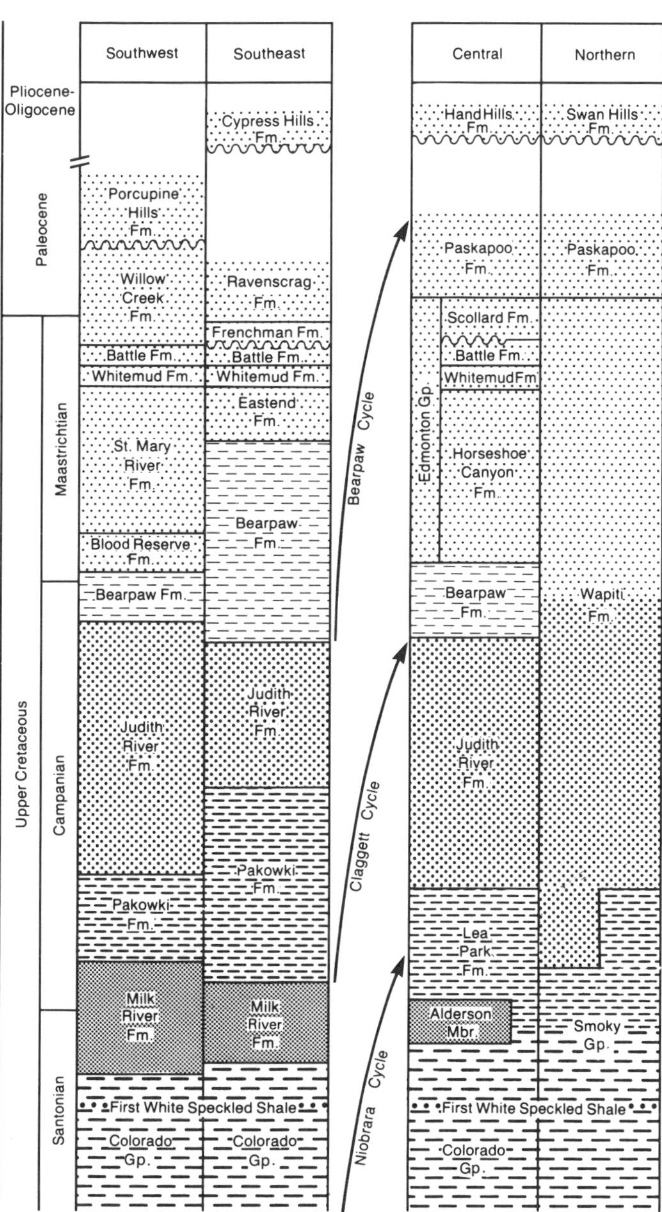

Figure 4I.40. Stratigraphic nomenclature for post-Colorado formations of the Alberta plains, in relation to the transgressive-regressive cycles recognized in the Western Canada Basin (E.H. Koster).

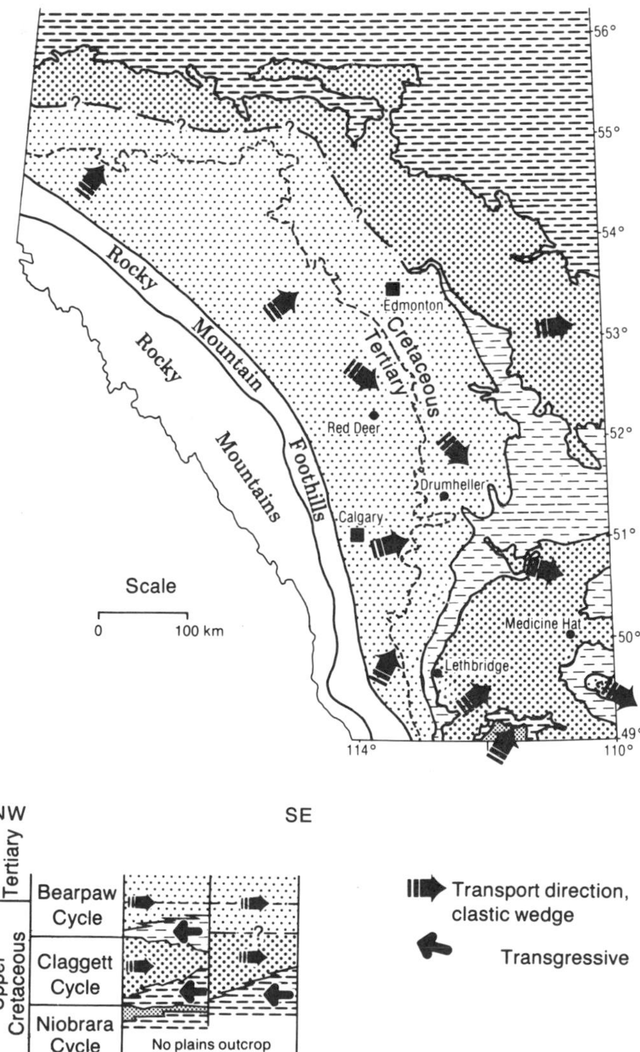

Figure 4I.41. Surface distribution of the rock units overlying the Colorado Group in the Alberta Plains, showing mean transport directions for alluvium in clastic wedges (E.H. Koster). See also Figure 4I.40.

Within the basal Kaskapau shales of the Peace River region, three sandstones, Doe Creek, Pouce Coupé, and Howard Creek (Stelck and Wall, 1955) are related to Late Cenomanian shorelines, which appear to be more closely aligned with the preceding Dunvegan delta than the later Laramide structural trends. Although the White-Speckled Shale markers are recognized within the Kaskapau and Puskwaskau formations (Burk, 1963), abundant silt in calcareous shale farther west obscures the white specks. The Cardium sandstone is poorly developed in the Peace River Plains, but the younger Bad Heart sandstone, about 6 m thick, is extensive. The latter contains abundant ooids similar to those found in iron-rich Cretaceous beds in the Clear Hills of northern Alberta (Mellon, 1962). Those beds mark a minor regression in Santonian (*Scaphites depressus*) time that is recognized along the length of the Foothills. The overlying Puskwaskau Formation, 125 to 200 m thick, contains concretionary and calcareous shales, the latter being equivalent to the First White-Speckled Shale of the Niobrara Cyclothem.

Upper Santonian-Maastrichtian

Uppermost Cretaceous strata of the Alberta Plains (Fig. 4I.40) consist of three eastward-thinning clastic wedges, each related to a Cordilleran orogenic pulse, intercalated with finer grained marine sequences. North of Edmonton, paleogeographic changes were not so pronounced and the record is one of a single marine to continental sequence. Over most of the Plains, strata usually appear horizontal in outcrop.

The post-Colorado or Montana Group (Weimer, 1960; Williams and Burk, 1964) refers to Upper Santonian, Campanian, and Maastrichtian strata above the first

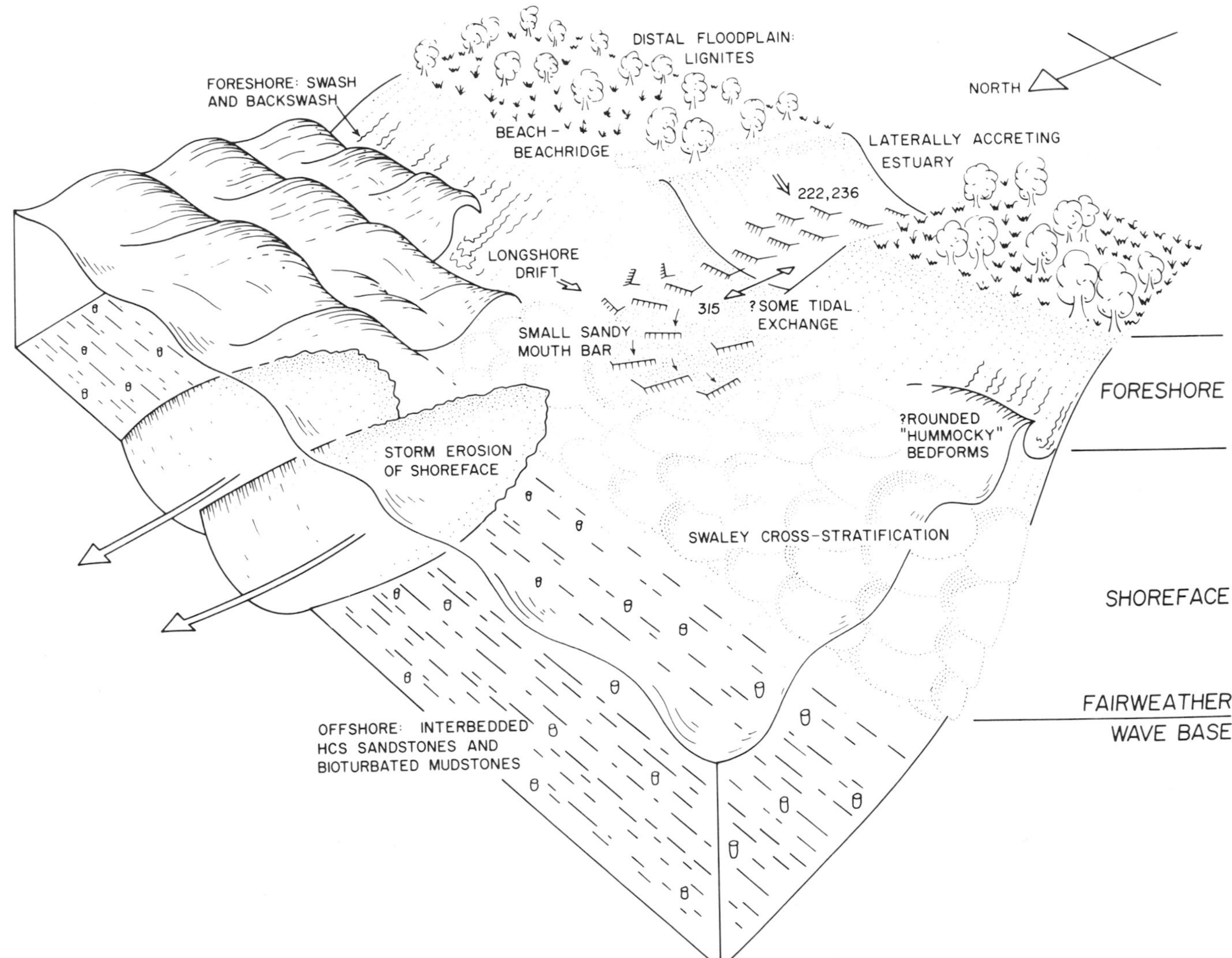

Figure 4I.42. Model of progradational shoreline processes at the margin of the Milk River clastic wedge at Writing-on-Stone Provincial Park, Alberta (after McCrory and Walker, 1986). The Telegraph Creek Member is represented by sediments below fair-weather wave base, the Virgelle Member by foreshore and shoreface deposits, and the Deadhorse Coulee Member by subaerial and channel deposits of the coastal plain.

White-Speckled Shale (Fig. 4I.40), one of several datums commonly used in regional subsurface studies. Transgressive-regressive cycles of this group (Weimer, 1960; Kauffman, 1967, 1969; McGookey et al., 1972), each spanning between 5 and 18 million years (Kauffman, 1977c; Lerand, 1982a), include the regressive phase of the Niobrara (Late Santonian to earliest Campanian) and all of the Claggett (mid-Campanian) and Bearpaw (Late Campanian, Maastrichtian, and Paleogene) cycles.

Gradual retreat of the Niobrara Sea across southern Alberta is recorded by the 90 m thick, upward-coarsening Milk River Formation. Marine conditions persisted in central Alberta and Saskatchewan, such that the Milk River correlates with the nearshore Chungo sandstone (Lerand, 1982b) in the Foothills, and upper parts of the marine Smoky Group and Lea Park Formation in the northern and eastern Plains, respectively.

The Milk River Formation is divided into Telegraph Creek, Virgelle, and Deadhorse Coulee members. Northeastward these members lose their identity and merge with the offshore Alderson Member of the Lea Park Formation. Finely interbedded shale and hummocky cross-stratified sandstone of the Telegraph Creek suggest deposition below fairweather wave base (Fig. 4I.42). Coarser sediments of the Virgelle Member form the main cliff at Writing-on-Stone Provincial Park, south of Medicine Hat. An erosional surface separates the basal clean shoreface sand with swaley cross-stratification and *Ophiomorpha* from an upper, whitish sandstone interval. In the latter, abundant crossbedding occurs within channels attributed to laterally migrating estuaries (McCrory and Walker, 1986). Whereas Meijer Drees and Myhr (1981) interpreted the Deadhorse Coulee sequence of coaly shales and thin lignite seams, interrupted by thin sandstones, as representing a lagoon affected by storm washovers, McCrory and Walker envisaged a distal floodplain with crevasse splays. Fox (1975, 1984) has described Early Campanian snakes and various mammals, including marsupials, from this unit. The northwest-trending alignment of the clean, thick sandstone bodies in the Virgelle Member (Meyboom, 1960), including those of the gas-producing fields at Alderson and Bantry (Herbaly, 1974), may reflect either coastal shoals produced by longshore currents or a coastal plain channel system.

Progradation of the Milk River sequence (Fig. 4I.42) and of its offshore correlative, the Alderson Member, was halted by the Claggett transgression. The Alderson consists of sub-littoral shale and sandstone. The abrupt change to marine shales of the Pakowki Formation is marked by a conspicuous deflection on resistivity logs termed the 'Eagle shoulder' by McLean (1971).

The Pakowki, up to 280 m thick northeast of Medicine Hat (Meyboom, 1960), consists mainly of shales similar to those in the correlative Wapiabi of the Foothills (Walker and Hunter, 1982). The uppermost part of the Pakowki exposed in Milk River canyon yielded an inner-neritic assemblage of bivalves, gastropods, foraminifera, and

Figure 4I.43. North side of Milk River Canyon, Alberta, near the U.S. border, showing the cyclic nature of the marine-continental transition at the base of the Judith River (Foremost) Formation. Section is approximately 60 m thick (photo - E.H. Koster).

ostracodes (Ogunyomi and Hills, 1977). An upward increase in megaspores, consistent with shallowing, also takes place (Speelman and Hills, 1980).

Northeastward regression of the Pakowki shoreline was punctuated by subsidiary transgressive reversals, causing the marine shale facies to be intercalated with progradational sand. The formational names Foremost and Oldman have been applied in the southern Alberta plains to this basal transitional unit and the overlying nonmarine succession, respectively, but McLean (1971, 1977a) has recommended that all strata between the marine phases of the Claggett and Bearpaw cycles be designated as the Judith River Formation. Five cycles are defined in Milk River canyon (Fig. 4I.43); each represents a sequence from offshore, to littoral/barrier island, to lagoonal salt- and fresh-water marsh (Ogunyomi and Hills, 1977). From gamma-ray geophysical logs, the basinward strata are interpreted as stacked upward-coarsening sequences (Rosenthal, 1982). The northeastward pinchout of each terminal sand oversteps the underlying ones, thus showing the incremental, diachronous nature of the regression (McLean, 1971). Thin seams of sub-bituminous coal cap the upper cycles and collectively constitute the McKay and Taber coal zones (Holter and Chu, 1978).

The younger (Oldman) part of the Judith River Formation accumulated as an extensive coastal plain (McLean, 1971; Rahmani and Lerbekmo, 1975) under a warm, moist paleoclimate with pronounced seasonality (Dodson, 1971; Beland and Russell, 1978). The uppermost 60 to 100 m of the Judith River is extensively exposed in badlands at Dinosaur Provincial Park, east-central Alberta (Koster, 1984; Koster and Currie, 1987). There, the strata contain an unrivalled fossil assemblage of dinosaurs and other Late Cretaceous vertebrates (Currie, 1985; Dodson, 1983, 1985), and primarily for this reason the park was designated a UNESCO World Heritage Site in 1979. The variable form and facies of paleochannel sequences are best explained in terms of estuarine conditions over lower reaches of the coastal-plain drainage (Koster and Currie, 1987). Articulated skeletons in the park, of which over 300 have so far been recovered, indicate stranding and quick burial after flotation from a nearby death site. Bone beds occur at amalgamation surfaces in multi-storey channel sequences.

The coastal plain succession is capped by the Lethbridge coal zone which, in turn, is overlain by *Ophiomorpha*-bearing tidal deposits. This sequence of only 10 to 15 m heralds the Bearpaw inundation. A line linking the edge of the Foothills west of Calgary, north through Pembina to Edmonton, marks the maximum limit of the Bearpaw transgression northwestward across Alberta, but evidence of marine strata is found as far north as Buffalo Head Hills (Wall and Singh, 1975). Caldwell (1968) concluded that this terrestrial-marine transition exhibits greater diachronism than hitherto thought, with beds of the Bearpaw Formation grading laterally westward into coarser facies of the clastic wedges above and below. The nonmarine sequence in the northern Plains beyond this marine tongue is termed the Wapiti Formation (Kramers and Mellon, 1972). Including the overlying Paleocene, this sequence totals approximately 1600 m (Williams and Burk, 1964).

East of Red Deer, the lower Bearpaw Formation, about 145 m thick (Given and Wall, 1971), consists of two sandstone and three shale members. According to Habib (1981), the alternating sandstone and shale facies originated as an eastward series of prograding river-dominated delta lobes. Each lobe is recorded by an upward-coarsening cycle from prodelta shales to sandier deposits of distributary mouth bars, and finally to coal seams accumulated in delta-top swamps. Riccardi's (1983) map of Bearpaw localities at which scaphitid ammonoids have been collected suggests that these were restricted to the more offshore zone across central and southeast Alberta. Preservation of more than twenty bentonites in the Bearpaw shale across extreme southern Alberta (Link and Childerhose, 1931) implies volcanic activity upwind and deposition below storm wave-base. East of Calgary, the lower 31 m of the Bearpaw unit contains a foraminiferal fauna of shallow marine origin (Wall et al., 1971). The succeeding 87 m of beds, transitional into the Horseshoe Canyon Formation (Fig. 4I.44), resembles several of

Figure 4I.44. Bearpaw marine shale (Be) gradationally overlain by Horseshoe Canyon Formation (HC), south of East Coulee on Red Deer River, Alberta (photo - D.W. Gibson, ISPG 532-33).

Figure 4I.45. Horseshoe Canyon (HC), Whitemud (Wm) and Battle (Bt) formations at Elnora on Red Deer River, Alberta. Note light grey weathering Kneehills Tuff (Knt) in the Battle Formation (photo - D.W. Gibson, GSC 532-48).

Habib's cycles; their foraminifera indicate a salt-marsh environment (Wall, 1976). In the southwestern plains, the upper transition zone is known as the Blood Reserve Formation; it has become a reference unit in the development of facies models for barrier beach and tidal inlet sequences (Reinson, 1984).

The onset of fully continental conditions, accompanying final regression, has been intensively studied in the Drumheller badlands. The Edmonton Group includes, in upward order, the Horseshoe Canyon (230-270 m), Whitemud (3-8 m), and Battle (3-11 m) formations (Fig. 4I.45), and at the top, the coal-bearing Scollard Formation. The status of the Scollard unit, which contains the Cretaceous-Tertiary boundary (Lerbekmo et al., 1979b; Lerbekmo and Coulter, 1985a, b), has been particularly problematic (see also Irish, 1970; Gibson, 1977a; L.S. Russell, 1983). The Edmonton Group is succeeded by the Paskapoo Formation. In the southwestern plains, post-Bearpaw stratigraphy mainly comprises the St. Mary River (Horseshoe Canyon equivalent) followed by the Willow Creek and Porcupine Hills formations (Carrigy, 1970).

The Upper Cretaceous to lower Tertiary nonmarine succession along the Red Deer River (Fig. 4I.45), upstream from Drumheller, was considered by Gibson (1977a) to contain channel sands of braided-river origin. Others (Waheed, 1983; Hughes, 1984; Nurkowski and Rahmani, 1985) have concluded that deposition occurred in meandering systems whose channels trended east-southeast and were as much as 2 km wide. The principal coal seams are located in the interfluve areas. A long history of coal mining (Campbell, 1964) led to a numerical scheme, supplemented in some cases by local place names, of up to twelve coal seams within the dinosaur-bearing Horseshoe Canyon Formation and a further two in the Scollard Formation (Gibson, 1977a).

The Whitemud Formation, which thickens southeastward to the Cypress Hills, is a barren alluvial interval mostly consisting of crossbedded sandstone and interbedded clay units. The overlying Battle Formation, a distinctive regional marker (Irish and Havard, 1968; Carrigy, 1971), consists of bentonitic mudstone with silicified megaspores (Binda and Srivastava, 1968) and contains three horizons of volcanic tuff (Gibson, 1977a). The age of this interval, Kneehills Tuff Zone, has been determined to be 65 Ma (Shafiqullah et al., 1964).

The succeeding Scollard Formation is a product of extensive freshwater peat swamps. The Nevis coal seam marks the point of dinosaurian extinction, a significant microfloral change, and the end of the Cretaceous Period (Lerbekmo et al., 1979b). In southeastern Alberta, the Frenchman Formation, comprising bentonitic clays and crossbedded sandstone, lies unconformably on the Battle and below the lowest Ravenscrag coal seam (Irish, 1965). The Frenchman has yielded remains of *Triceratops* and is equivalent to the upper part of the Edmonton Group and St. Mary River Formation.

Above the Edmonton Group and on the higher western plains, the Paskapoo Formation, up to 950 m thick, represents the development of extensive alluvial systems (Carrigy, 1971). In the Cypress Hills, Paskapoo-equivalent strata appear to comprise the relatively thin, fluviatile Ravenscrag Formation whose sediments were derived from western Montana (Misko and Hendry, 1979). Outliers of the uppermost Cretaceous and Paleocene units across the plains are typically capped by braidplain gravels of Oligocene to Pliocene age (Vonhof, 1965).

East-median hinge (Saskatchewan)

Cenomanian to Maastrichtian

In Saskatchewan, Upper Cretaceous shale and subordinate sandstone and siltstone are included in the upper Colorado and Montana groups. The Colorado Group is overlain transitionally by sandstone and siltstone of the Milk River Formation and equivalent shales of the Lea Park Formation, which form the basal units of the Montana Group. The Colorado Group is commonly divided into two subgroups at the Fish-Scale Marker, a widespread phosphatic unit, but Simpson (1975) drew the boundary at the higher, Second White-Speckled Shale.

Shales of the Big River Formation, at the base of the upper Colorado Group, mark the early phase of the Greenhorn Marine Cycle. The St. Walburg and Okla sandstones were deposited on the eastern shelf, and fish remains were dispersed across most of the area. The widespread influence of storm-surge-generated suspension currents as agents of sediment dispersal is evidenced by ubiquitous sandstone-mudstone couplets, current markings, and simple grading of siliciclastic grains and fish debris.

The Second White-Speckled Shale, about 60 m thick and lying above the Big River Formation, is bituminous, calcareous shale and mudstone with intercalated shaly chalk and chalky, skeletal calcarenite. These beds are characterized by the Lower Turonian Zone of *Hedbergella loetterlei* (Caldwell et al., 1978). The Phillips Sandstone (Second Speckled Sandstone of southeastern Alberta) occurs about 6.5 m below the top of the Second Speckled Shale and is as much as 38 m thick near the International Boundary. The overlying calcareous shales and mudstones are known locally as the Greenhorn Lime.

The Second White-Speckled shales are overlain by up to 60 m of non-calcareous shale, correlative with the Morden Formation of Manitoba Escarpment, Cardium sandstone of the Alberta Foothills, and upper Carlile Shale of the northern and central United States. The Bowdoin Sandstone, 61 m thick, consisting of sandstone and siltstone, is found in the southwest above the Second Speckled Shale. The non-calcareous shales are absent in central Saskatchewan (Simpson, 1975; McNeil and Caldwell, 1981) but extend through southern Saskatchewan into eastern Alberta. North and Caldwell (1975) concluded that the disappearance of the shales was due to a disconformity beneath the First Speckled Shale, and McNeil and Caldwell (1981) concluded that the disconformity terminated the Greenhorn Cyclothem.

The First White-Speckled Shales, similar to the older, Second Speckled Shale, consist of calcareous shale and chalky calcarenite, and are part of the Niobrara Marine Cycle (McNeil and Caldwell, 1981; McNeil, 1984). The *Globigerinelloides* sp. Zone is coincident with these shales (Caldwell et al., 1978), which are included in the Niobrara Formation along the Manitoba Escarpment. The Martin Sandy Zone, about 60 m thick near the border with the United States, occurs within the shale sequence, which also includes the Medicine Hat Sandstone, as much as 14 m thick, about 30 m below the top. Gravity anomalies delineate the northern and eastern limits of the Medicine Hat Sandstone at the Hatton pool of southwestern Saskatchewan; coincidence with an elevated basement block seems likely.

Strata above the First White-Speckled Shale are almost entirely argillaceous in eastern Saskatchewan, but in the west incorporate interbedded sandstones as eastward-thinning tongues. These sediments, forming the bedrock over much of southern Saskatchewan, are known from abundant subsurface data and numerous exposures. The succession is dominantly marine, and the interbedded sandstones commonly reflect eastward passage from continental to transitional and marine depositional settings. These beds, assigned to the post-Colorado group (Williams and Burk, 1964) and to the Montana Group (Simpson, 1975, 1984), rest unconformably on the First White-Speckled Shale in eastern Saskatchewan, the hiatus being that recognized between the Niobrara and Pierre formations over most of North Dakota and South Dakota (Cobban and Reeside, 1952). The contact appears to be conformable and gradational in the southwest. The succession is overlain disconformably by sandstone, siltstone, and clay of the uppermost Cretaceous Frenchman Formation, about 760 m thick in the Missouri Coteau region.

In south-central Saskatchewan, shales of the post-Colorado or Montana Group are divided into Lea Park and Bearpaw formations, separated by sandstones of the Judith River Formation (McLean, 1971; or Belly River Formation of some authors).

In southwestern Saskatchewan and southeastern Alberta, Lea Park shales are replaced by shaly sandstones of the Milk River Formation and overlying shales of the Pakowki Formation. Lea Park shale is marine, 300 m thick, glauconitic with ferruginous concretions in the lower part and bentonitic in its upper part. The laterally equivalent Milk River sandstones mark a regressive influx of relatively coarse detritus across the western shelf. The Milk River comprises about 100 m of monotonous,

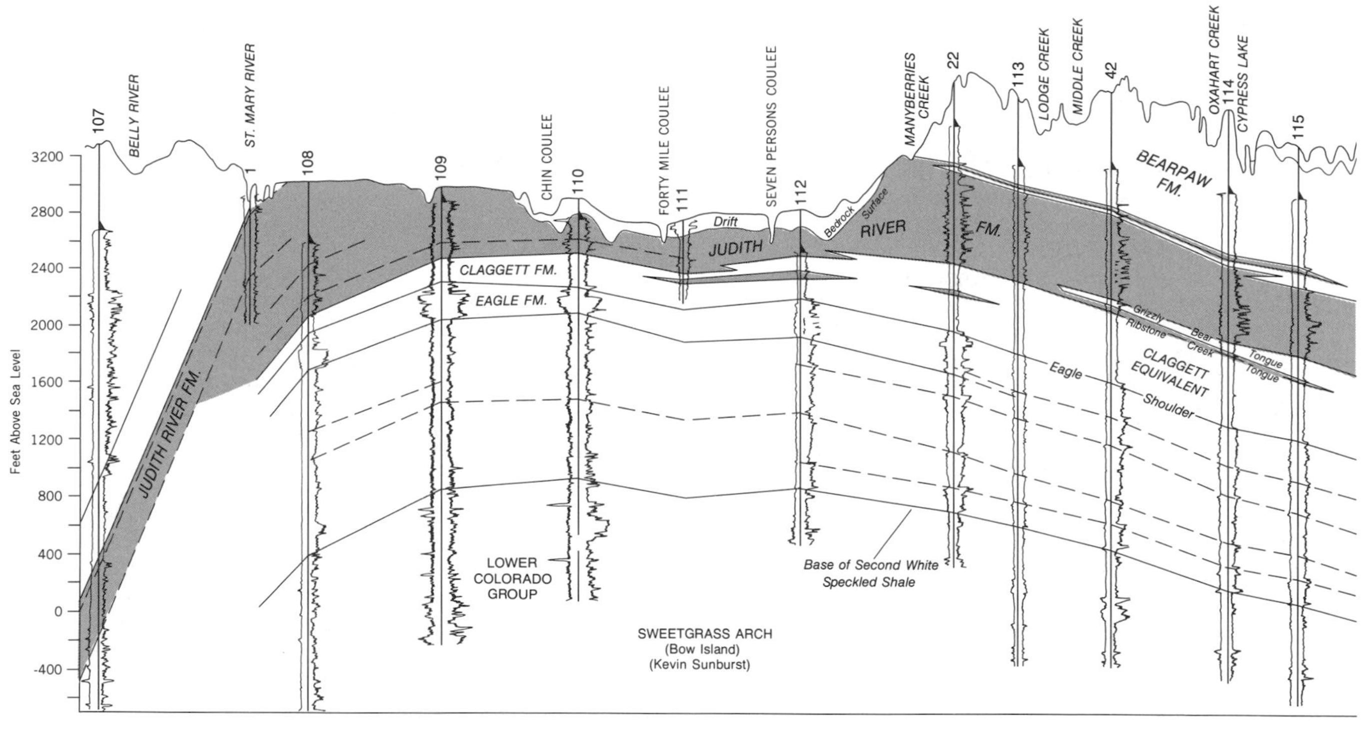

INDEX OF CROSS-SECTION LOGS

107 *Sinclair Baysel McLeod 4-24-8-26-W4 GL 3286*
1 *U of S Lethbridge NE-9-2-7-22-W4 GL 2930*
108 *Mobil Oil C.P.R. Wilson No. 5-4 4-5-8-20-W4 KB 3039*
109 *Mid Continent No. 4 11-1-8-17-W4 GL 2995*
110 *Dominion Legend Province No. 1 2-15-7-13-W4 KB 2914*
111 *U of S Foremost NW-13-1-9-11-W4 GL 2750*
112 *British Dominion Etzikom No. 4 6-30-6-8-W4 KB 2845*
22 *CEG East Manyberries No. 6-32 6-32-5-4-W4 KB 3724*
113 *Anglo Socony et al. Bain No. 12-3 3-12-5-3-W4 KB 3616*
42 *Canadian Delhi et al. Cypress Hills No. 1 9-9-6-30-W3 KB 3690*
114 *B.A. Co-op Calvan Cypress Lake No. 2-26 2-26-6-27-W3 KB 3576*
115 *B.A. Co-op Calvan Cypress Lake No. 12-34 12-34-5-25-W3 KB 3305*
116 *B.A. Eastend Arendt 3-15-5-23-W3 KB 3346*
117 *Tidewater White Mud Crown No. 8-9 8-9-5-20-W3 KB 3154*
118 *Imperial Tidewater Chambery No. 5-20 5-20-5-18-W3 KB 3365*
119 *B.A. Climax Foss 8-31-4-16-W3 KB 3026*
120 *Sohio McCarty and Coleman Hillandale No. 1 4-11-5-14-W3 KB 2806*
121 *Socony Sohio Burnbrae No. 32-7 7-32-5-12-W3 KB 3103*
122 *Amerada Shell Crown S-H No. 3-10 3-10-6-10-W3 KB 2655*
84 *International Helium Wood Mountain No. 10-3 10-3-5-8-W3 KB 2799*
123 *Shell Wood Mountain No. 12 16-35-3-6-W3 KB 3269*
124 *White Rose Wood Mountain No. 12-19 12-19-4-3-W3 KB 2997*
125 *Socony Sohio Canopus No. 25-12 12-25-3-2-W3 KB 3221*
126 *Socony Sohio Lisieux No. 1 5-7-4-29-W2 KB 2761*
127 *Pacific Mobil Harptree No. 1-28 1-28-3-26-W2 KB 2749*
128 *Mobil Oil Sohio Coal Mine Lake No. 18-9 9-18-4-22-W2 KB 2585*
129 *Shell Salt Lake No. A-10-21 10-21-4-20-W2 KB 2461*
130 *Jordan et al. Cecilia Lake 3-3-4-18-W2 KB 2466*

Figure 4I.46. Stratigraphic cross-section of Judith River (Belly River) Formation across southern Alberta and Saskatchewan, illustrating eastward gradation of sandstone to shale (after McLean, 1971).

bioturbated shaly sandstone and siltstone; its lower part includes shale and siltstone, assigned in Montana to the Telegraph Creek Formation.

In southern and central Alberta and southwestern Saskatchewan, ten members of alternating nonmarine and marine strata were recognized within the Belly River Formation by Shaw and Harding (1949; see also Nicols and Wyman, 1969). The five main sandstone tongues (Brosseau, Victoria, Ribstone Creek, Birch Lake, and Oldman, in order of decreasing age), are separated in vertical sequence by shale units (Shandro, Vanesti, Grizzly Bear, Mulga, from oldest to youngest). McLean (1971) assigned the sandstones to the Judith River Formation and the intertonguing shales to the laterally equivalent Lea Park. This sequence becomes entirely continental westward and marine eastward, decreasing from about 300 m in the southwest to 15 m in central Saskatchewan (Fig. 4I.46). It represents a large number of coalescing deltas (Shaw and Harding, 1949; McLean, 1971).

The Bearpaw Formation, 275 m thick in the South Saskatchewan River Valley, is divisible into eleven members (Caldwell, 1968), six of silty clay (unnamed, Broderick, Sherrard, Beechy, Snakebite, and Aquadell) alternating with five of silty sand (Outlook, Matador, Demain, Ardkenneth, and Cruickshank) (Fig. 4I.5). Ammonite and foraminiferal faunas of Late Campanian to earliest Maastrichtian age are recognized (Caldwell, 1968; North and Caldwell, 1970). Bentonites provide isotopic ages of 70 to 74.5 Ma (Folinsbee et al., 1960, 1961). Sedimentary features and fossils suggest that epineritic conditions prevailed over the Bearpaw seabed and water depth probably never exceeded 50 m. The sandy units interbedded with Bearpaw shale record the pulsatory advance of the Bearpaw sea.

The succeeding regression is recorded in the Eastend and Whitemud formations, which were laid down in transitional and continental conditions respectively. In the Missouri Coteau and in the Cypress Hills, Bearpaw shales are succeeded by sandstones of the Eastend Formation

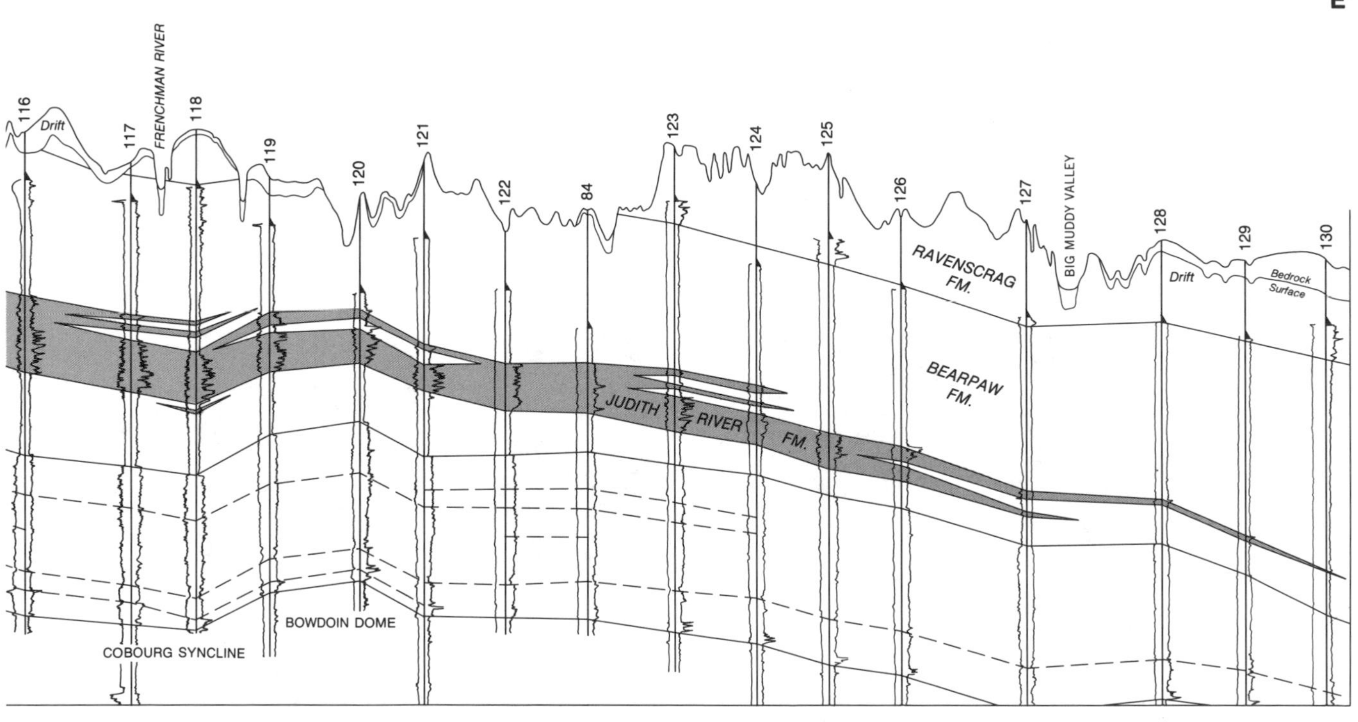

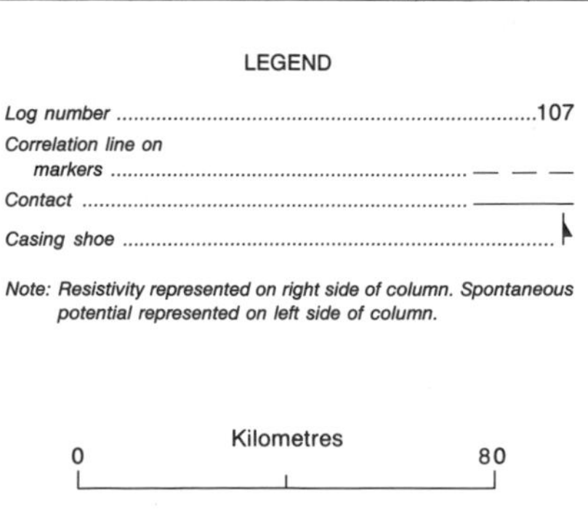

Figure 4I.46. Cont.

(less than 30 m), kaolinitic mudstones of the Whitemud Formation (up to 23 m), and clays of the Battle Formation (about 8 m).

Eastern Platform (Manitoba Escarpment)

Cenomanian to Maastrichtian

Chalk-speckled shale, shaly limestone, and calcarenite of the uppermost Cenomanian to Middle Turonian Favel Formation overlie the Ashville Formation in Manitoba Escarpment. The Favel, made up of the Keld and Assiniboine Members, is about 45 m thick in southern Manitoba, but decreases markedly to about 11 m in east-central Saskatchewan, where its uppermost beds are affected by an unconformity (McNeil and Caldwell, 1981).

The Favel records a peak marine transgression during Turonian time, and its calcareous planktonic organisms (foraminiferal Zone of *Hedbergella loetterlei*) reveal an incursion of warm Gulfian water. Because shorelines were distant and terrigenous input was minimal, the Favel has a high component (30-60%) of biogenic carbonate sediment.

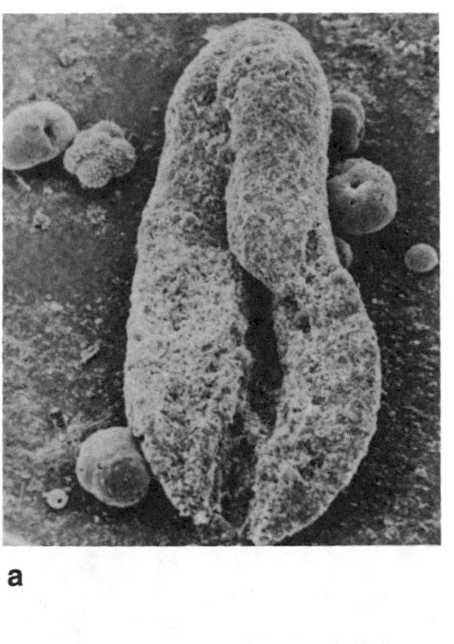

a

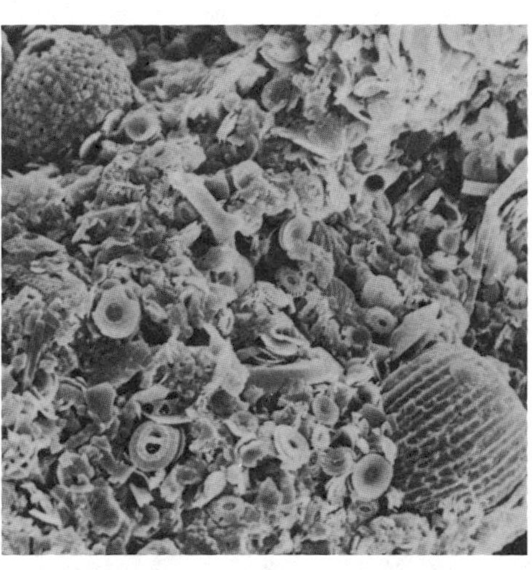

b

c

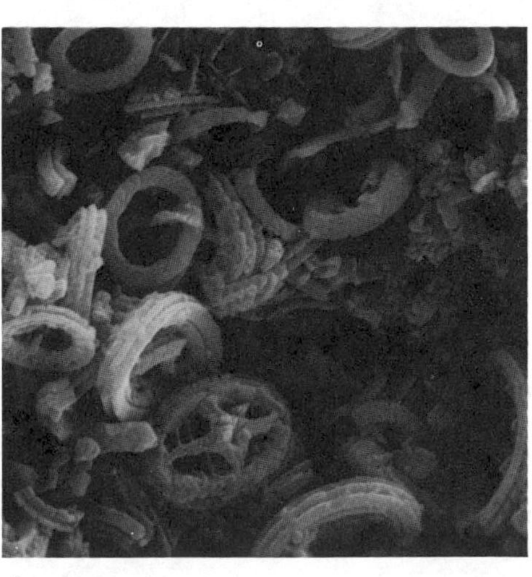

d

a. Modern fecal pellet (length 1.8 mm) of pelagic copepod.

b. Enlargement of the coccolith-rich interior of the pellet illustrated in a. (a and b published with permission of S. Honjo, Woods Hole Oceanographic Institute, Massachusetts.)

c. Longitudinal cross-section of fecal pellet (length 1.9 mm) in argillaceous limestone of the Favel Formation.

d. Enlarged view of coccolith-stuffed interior of the fecal pellet in c.

Figure 4I.47. Comparison of Recent and Late Cretaceous fecal pellets (after D.H. McNeil, 1984).

A rich molluscan assemblage also is present. The *Mytiloides labiatus* beds occur in the Keld Member as well as in the Laurier Limestone Beds, which also yield *Collignoniceras woollgari* and other ammonites and bivalves. The Marco Calcarenite, a unit of the Assiniboine Member, yields bivalves and belemnites. Coccoliths are disseminated throughout the Favel and the equivalent Second (Lower) White-Speckled Shale of the Interior Plains in 1 to 2 mm sized "white specks" of chalk. The white specks are coccolith-stuffed fecal pellets of pelagic organisms (Fig. 4I.47).

The overlying late Middle Turonian Morden Shale is remarkably uniform and carbonaceous. It represents a regional regressive phase that is terminated at a disconformity. The formation yields the ammonite *Prionocyclus hyatti* (Stanton) and foraminifera thought to be indicative of the zones of *Pseudoclavulina* sp. and *Trochammina* sp. (McNeil and Caldwell, 1981).

Conditions of peak marine transgression were quickly re-established following the Late Turonian regression, as indicated by the lithologies and faunas of the 20- to 70-m thick Niobrara Formation, which McNeil and Caldwell (1981) divided into calcareous shale and chalky members. Uppermost Turonian to lowest Campanian chalky calcareous sediments were deposited along the eastern platform from eastern Colorado and western Kansas to Manitoba Escarpment. Planktonic foraminifers of the Zone of *Globigerinelloides* sp. record the northward spread of warm temperate or possibly sub-tropical marine waters at least as far as 54°N latitude in latest Santonian to earliest Campanian time. Like the earlier transgressive Favel Formation, the Niobrara is characterized by abundant coccolith-stuffed fecal pellets, commonly referred to as "white specks" in the correlative First (Upper) White-Speckled Shale of the Interior Plains. The Niobrara Formation is terminated in Manitoba Escarpment by an unconformity.

The lower Lower Campanian to Maastrichtian Pierre Shale unconformably overlies the Niobrara Formation and consists of five members: Gammon Ferruginous, Pembina, Millwood, Odanah, and an unnamed shale. McNeil and Caldwell (1981) have interpreted the Pierre as a product of a long oscillatory regression. The Gammon Ferruginous Member is a dark, blocky-fracturing shale containing ferruginous concretions, which yield *Baculites minerensis* (Landes). The unit is about 50 m thick in the subsurface of southwestern Manitoba, but is absent by unconformity at the Manitoba Escarpment, except in the area of Riding Mountain, where it infilled an erosional depression on the Niobrara Formation (McNeil, 1984).

Dark, marine Pembina shales rest conformably on the Gammon Ferruginous Member in the subsurface of southwestern Manitoba, but overstep the Gammon towards Manitoba Escarpment, where they rest unconformably on the Niobrara Formation. The Pembina Member contains a striking sequence of 20 to 30 bentonite beds contemporaneous with the late Early Campanian Zone of *Baculites obtusus*.

The Pembina Member is overlain by the lower Upper Campanian Millwood Member, which consists of greenish clays, silts, and thin marls yielding rich foraminiferal assemblages diagnostic of the *Glomospira corona* and *Eoeponidella linki* zones and the ammonite zones of *Baculites gregoryensis* and *B. scotti*. An abundance of planktonic foraminifers in the marls at the base of the Millwood indicates conditions of open-marine sedimentation and near-peak transgression. The Millwood changes southeasterly from 150 m of silty clay at its type locality near the Saskatchewan-Manitoba border to 15 m of shales and thin marls at Pembina Mountain at the international border (McNeil, 1984). The facies change marks the transition from the median facies belt of fine clastics to the calcareous eastern facies, sensu Tourtelot (1962), which is best developed in equivalent shales, calcareous shales, and chalks of the Gregory and DeGrey members of the Pierre Shale of South Dakota.

The Millwood passes gradationally into the overlying, 150 m thick, olive-grey siliceous shale of the Odanah Member. The siliceous lithotype is unique in the Manitoba section and extends no farther than eastern Saskatchewan and northern North Dakota. The origin of the silica is not definitely known, but McNeil and Caldwell (1981) considered it biogenic. Manganese concretions are also characteristic (Beck, 1974). The Odanah yields the largely agglutinated late Late Campanian foraminiferal assemblage of the *Haplophragmoides fraseri* Zone and contains in its lower beds ammonites suggestive of the *Didymoceras nebrascense* Zone.

The uppermost, unnamed member of the Pierre Shale (McNeil and Caldwell, 1981), or the Coulter Member (Bamburak, 1978), consists of 20 to 55 m of fine-grained clayey silt with fine sands in the upper beds. The member is presumed to be Maastrichtian in age, but little is known of its fossil content and regional correlation.

The Pierre Shale is conformably overlain by approximately 45 m of Maastrichtian sands of the Boissevain Formation in southern Manitoba (McNeil and Caldwell, 1981). The sands are crossbedded, medium grained, and quartzose, and contain interbeds of silt and clay, and ironstone concretionary layers. In general terms, the Boissevain represents marginal marine to possibly terrestrial environments, associated with the final retreat of the Cretaceous epicontinental seas from this area in late Maastrichtian time. The formation is sharply and probably unconformably overlain by Tertiary clays, silts, sands, and lignites of the Turtle Mountain Formation.

SOME ASPECTS OF BASIN EVOLUTION: A SYNOPSIS

Tectonic framework

The Western Canada Basin is bounded to the west by the Columbian-Laramide Orogen and to the east by the Canadian Shield; the oceanic Canada Basin forms its northern boundary. During the Cretaceous, the Western Canada Basin was an integral part of the Western Interior Basin of North America, an elongate trough of truly continental proportions (Fig. 4I.48).

Basin and orogen are inextricably linked in their history. They share a common lithospheric foundation in the Precambrian basement, which not only underlies the basin but continues westward beneath the 80 km width of the Rocky Mountains. Mechanically coupled by this foundation, basin and orogen grew in geodynamic partnership. Late Jurassic through Eocene subsidence of the basin may be attributed to isostatic adjustment in response to crustal loading by eastward-displaced thrust sheets stacked along the western margin, and related elastic downwarping that extended eastward well beyond the limit of the fold-thrust belt. As the margin of the Cordillera migrated eastwards with growth of the Rocky Mountains, so also in response did the foredeep of the flanking basin. The asymmetric form and to some extent the width of the basin then may be related to the magnitude of load, both tectonic (the total thickness of stacked thrust sheets) and depositional (the thickness of synorogenic clastic sediments). The response to these loads was affected by the mechanical properties of the underlying continental crust (Bally et al., 1966; Price, 1973; Beaumont, 1981). Quantitative modelling of the behavior of the lithospheric plate beneath the basin in response to the superimposed loads has shown that evolution of basin and orogen is linked by the growth of a lithospheric flexure. Subsidence of the basin may be readily interpreted as an expression of crustal thickening within the orogen exceeding the rate of mass wasting from it. Further, regional uplift and denudation of the basin may be interpreted as consequences of deep and widespread, post-orogenic erosion of the fold-thrust belt and the foreland-basin fill that had depressed the crust. These movements of the basin are consistent with it being underlain by "an old, cool, thick, lithospheric plate" (Beaumont, 1981).

The Western Canada Basin in Late Jurassic and subsequent times was, therefore, a tectonically generated foreland basin, which provided a huge collecting ground for vast quantities of detritus shed by the developing eastern Cordillera. The foreland basin developed on a broad belt of the western craton - a belt in which various arches (for example, Peace River and Sweetgrass) and flanking basins (for example, Williston Basin and Peace River Embayment) had undergone repeated movements. During the existence of the foreland basin, some of these more localized cratonic structures were essentially dormant, while others continued to undergo relative movements of variable magnitude (Stelck, 1975b; Porter et al., 1982).

The tectonic evolution of the northernmost part of Western Canada Basin involved two principal elements, an extensional regime (in part during the formation of oceanic crust in Canada Basin) and the compressional regime of the Columbian Orogen. Not until Aptian times is there evidence for the development of foredeeps associated with orogeny in the northern Cordillera. During the Aptian and Albian, a complex interplay of extensional and compressional forces acted upon these northern areas. From Aptian time onward, orogeny in the northern Cordillera and post-rift subsidence of the Arctic Ocean continental margin were the principal tectonic activities.

Evolution in concert of the basins and the orogen is underscored dramatically by the relationship between deformation and sedimentation. Episodes of orogenesis, variable in intensity and in degree of local and regional expression, were recurrent throughout the Cretaceous, so much so that it is difficult, both spatially and temporally, to distinguish between waning phases of the Columbian Orogeny and premonitory phases of the succeeding Laramide Orogeny. In general, however, there was a period of relative quiescence between Cenomanian and Santonian times, so that episodes of deformation to the end of the Albian Age conveniently may be dubbed as "Columbian" and those of the Campanian and Maastrichtian ages as "Laramide". Following each episode of deformation, the newly displaced and elevated folds and thrust sheets underwent erosion and the resulting detritus was swept rapidly into the flanking basins to be deposited as deltaic outgrowths, which rapidly became compounded into marginal alluvial plains. These commonly extended across the foredeeps to give way ultimately to thinning sheets of sand, silt, and mud deposited beyond the prograding shorelines. Examples abound. Some of the most continuous and strong late Columbian movements affected what is now northeastern British Columbia in Albian time. They are recorded in the basin by at least four progradational sequences of continental to marine conglomerates and sandstones wedged between the same number of marine shales (Stott, 1984). Intensifying Laramide movements affected what is now southeastern British Columbia and southwestern Alberta in Campanian time. They are recorded in the basin by the huge deltaic sandstone wedges of the Eagle or Milk River and Judith River formations (McLean, 1971), which extended as far east as Saskatchewan.

Unlike regions in the United States where depositional patterns within the foreland basin were influenced by foreland uplifts and plutons, evidence of the effect of similar structures on Cretaceous deposition in the Canadian foreland basin is not so well documented nor so evident. In Canada, laterally continuous facies, both nonmarine and marine, are characteristic of the broad, Cretaceous foreland basin. However, Cretaceous depositional patterns in the

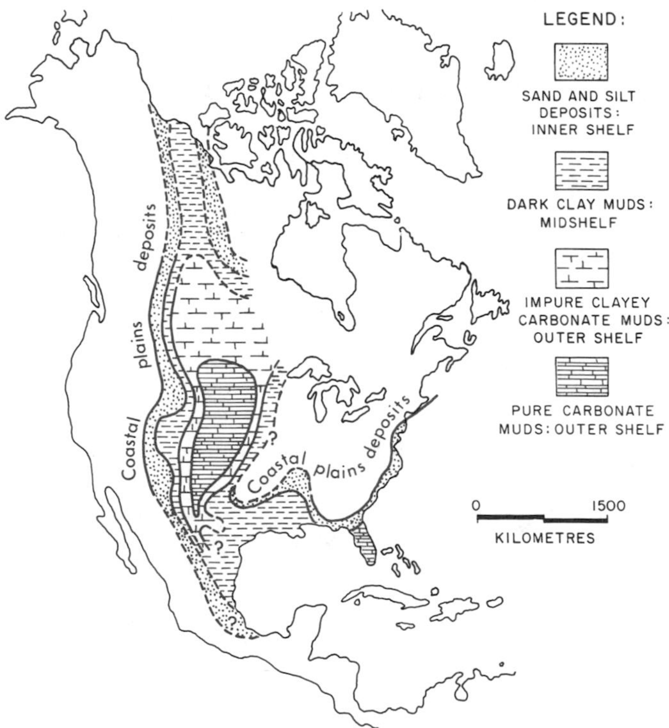

Figure 4I.48. Generalized distribution of sediments in the Western Interior Basin during the Early Turonian maximum transgression of the Greenhorn Sea (after McNeil and Caldwell, 1982).

United States have been linked to a series of lineaments attributed to pre-existing structural trends that date from the Precambrian and to continued fragmentation of the North American craton (Weimer et al., 1983; Weimer, 1984; Dickinson et al., 1988). In contrast, there appears to be little tectonic partitioning of the foreland basin in Canada. Some recurrent movements on basement blocks, particularly in the Peace River region (Stott, 1982) and in Williston Basin (Christopher, 1984), may have influenced depositional patterns. Possibly, more detailed studies may define unrecognized effects on Cretaceous deposition of recurrent movements of basement fault blocks within the Canadian foreland basin.

Igneous activity

Subduction of the Farallon and Kula plates in the Pacific Ocean and continuing compression of the exotic terranes against the cratonic margin of North America manifested themselves during the Cretaceous, not only in the progressive eastward deformation of the sedimentary pile, but also in plutonism and vulcanism; and just as the episodes of deformation put their stamp on the sedimentary sequences of the flanking foreland basin, so did those of igneous activity. The record of contemporary vulcanism may be reconstructed more accurately and more completely by interpreting the stratigraphy of the basin than by identifying volcanic centres and their products within the Cordillera.

Cretaceous plutonism seems to have been associated mainly with the waning phases of Columbian Orogeny and to have found expression, in part, in batholiths of granodiorite and associated rocks emplaced at high levels within the Omineca Belt (Gabrielse and Reesor, 1974; Woodsworth et al., 1991). Some of the intrusions dated as Early Cretaceous represent only the terminal pulses of multi-phase batholiths, the history of which extends to much earlier Mesozoic time (see, for example, Tipper et al., 1974, 1981; Woodsworth et al., 1991). By Albian time, an uplifted Omineca Belt probably had become unroofed to expose its metamorphic core, if not the batholiths themselves (Stott, 1968). Certainly many post-Aptian, eastward-tapering sandstone formations contain significant proportions of detritus derived from crystalline rocks of the western hinterland. In some sandstones, detrital components eroded from both plutonic and metamorphic source rocks may be identified with assurance; in others, the plutonic component, although probably present, may be discriminated with less confidence. The Lower Albian Gates Formation of the Foothills and Plains of northeastern British Columbia (Leckie, 1986a) and the Upper Campanian Judith River Formation of the plains of southern Alberta and Saskatchewan (McLean, 1971) - two sandstones contrasted in age and location within the basin - illustrate these points. Attributing components of the sandy detritus to the erosion of plutonic and metamorphic source rocks is not difficult, but identifying precisely which rocks supplied these components is quite another matter. Localizing the provenance of the detritus is complicated (1) by uncertainty as to when particular batholiths of the Omineca Belt became unroofed; (2) by dextral strike-slip displacement since mid-Cretaceous times of many hundreds of kilometres along the Rocky Mountain Trench-Tintina fault system (Tempelman-Kluit, 1979; Gabrielse, 1985); and (3) by recognition that sediment dispersal, controlled by the direction of drainage and palaeoslope inclination, may not bear a predictable relationship to the trend of the orogenic front, either locally or regionally. Some of these factors are considered by Leckie (1986a) and Rahmani and Lerbekmo (1975).

Whereas plutonism seems mainly to have accompanied the waning phases of the Columbian Orogeny in the Early Cretaceous, vulcanism persisted throughout the greater part of the period and gives signs of having intensified with early Laramide pulses toward the end of the Late Cretaceous. Thick piles of Cretaceous volcanic rocks are known to exist deep within the Intermontane Belt in central British Columbia, but with the exception of the Crowsnest volcanic remnants, little direct evidence persists of more easterly volcanic centres bordering the basin. Sedimentary sequences within the basin, however, carry a record of vulcanism, mainly of Albian and Campanian-Maastrichtian age, expressed in two contrasted types of volcanic product - detritus eroded directly from accumulations of lava and tephra, and ash blown from erupting volcanoes. Both were carried into the basin to accumulate as stratified marine or terrestrial deposits.

Mellon (1964) attributed the high proportion of volcanic lithic fragments and other detritus in the middle Blairmore and upper Mannville groups (see also Williams, 1963) to volcanic centres along the western margin of the basin in southeastern British Columbia. Altered ash beds (bentonites) in the Gething, Moosebar, and Hulcross sequence (Kilby, 1985) in northeastern British Columbia testify to contemporaneous volcanic eruption. These occurrences demonstrate the establishment of active centres of vulcanism by Early Albian time. Bentonite and tuff beds, volcanic rock fragments, and some crystalline minerals, including fresh sanidine, in the Upper Campanian Judith River Formation of the Alberta and Saskatchewan Plains is attributed by McLean (1971) to contemporaneous volcanic centres in the Elkhorn Mountains of Montana and possibly in Alberta and British Columbia. Similarly Byers (1969) has attributed much of the sandy detritus in the Lower Maastrichtian upper Eastend Formation of southwestern Saskatchewan to fast mechanical weathering of mountains of freshly extruded volcanic rocks in western Montana, and much of the sediment of the overlying Whitemud Formation to slow chemical weathering and leaching of the same volcanic pile once it had become worn down. The persistence of volcanic influence is supported by Misko and Hendry's (1979) identification of the Elkhorn Mountains and even younger Adel Mountains as sources of the lithic fragments and other grains of volcanic origin present in the Upper Maastrichtian Frenchman Formation and Paleocene Ravenscrag Formation of southern Saskatchewan.

Bentonites, the most easily recognized and recurrent indices of contemporaneous vulcanism, commonly contain minerals (such as sanidine) suitable for obtaining isotopic ages by the K-Ar technique and may be treated as "instantaneous" deposits and used as markers in lithostratigraphic and chronostratigraphic correlation. Folinsbee et al. (1961, 1963, 1965) dated a number of (mainly) Campanian-Maastrichtian bentonites, and McNeil and Caldwell (1981) used the Cenomanian 'X' bentonite to effect international correlation of the Belle Fourche Member of the Ashville Formation. On a more limited scale, McNeil and Caldwell used a remarkable concentration of about thirty bentonites to achieve an exceptionally refined correlation of the Lower Campanian lower Pembina Member of the Pierre Shale between Manitoba and North Dakota.

The geographic spread and thickness of some volcanic ash layers, such as 'X' bentonite, attest to the highly explosive (Vulcanian or Plinian) character of the eruptions. The frequency and violence of the eruptions may have influenced climate (Axelrod, 1981). The concentration of bentonite beds in the Campanian-Maastrichtian rocks of the southern Interior Plains suggests that vulcanicity intensified with Laramide deformation, and given that volcanic ash beds generally decrease in number and thickness away from their generating centres, and that these centres were located in the Cordillera, the distribution of these same bentonite beds points to prevailing westerly winds across the basin (compare Gordon, 1973; Lloyd, 1982).

Transgressions and regressions

The flooding of the Western Canada Basin was achieved through progressively more extensive, pulsatory transgression in the later part of the Early Cretaceous. The seas may not have occupied much of the basin until Aptian time, and the extent of southward transgression even then is uncertain. By Albian time, however, the sea had swept through all of the Canadian portion of the basin and far beyond. Confluence of the strongly transgressive northern sea and weakly transgressive southern sea to form the continuous Western Interior Sea took place in latest Middle or earliest Late Albian time in the region of southern Colorado. Once established, the epicratonic seaway persisted, with only the briefest of interruptions, from Late Albian until well into Maastrichtian time. During the Late Cretaceous, the shorelines fluctuated markedly in response to continuing transgressive-regressive pulses. Inundation reached a maximum in Early Turonian and again during latest Santonian to earliest Campanian time.

According to Kauffman (1977b, 1979, 1984b), ten discrete, transgressive-regressive pulses were controlled by fluctuations in global sea level. They are believed to correspond to third-order cycles of Vail et al. (1977a, b). Whereas eustasy has been accepted by some as the controlling factor (e.g. Hancock, 1975; Hancock and Kauffman, 1979), others have argued that, in general, any eustatic influence would have been dwarfed and overprinted by local and regional tectonic influence (e.g. Jeletzky, 1977, 1978). Caldwell (1984a) concluded that transgressions and regressions were influenced more or less by the effect of global tectonic activity on world-wide sea level and by local and regional tectonic activity on the form of the Western Interior Basin, and that it is impossible as yet to segregate a single dominant factor.

The Cretaceous cyclothemic sedimentary record offers a convenient framework within which to discuss the systemic stratigraphy. In the American portion of the Western Interior Basin, Kauffman (1977b, 1979, 1984b) has recognized the latest six (T_5-R_5 to T_{10}-R_{10}) of his ten Cretaceous marine cycles held to be eustatically controlled. All but the tenth have been named and Caldwell (1984a) described their manifestation within the Canadian portion of the basin (Fig. 4I.49; see also Stott, 1984; Dixon, 1986a). The extended section of marine Lower Cretaceous rocks preserved in the Western Canada Basin does not yet permit a comparable evaluation of the earliest (T_1-R_1 to T_4-R_4) of Kauffman's cycles. It does indicate that earlier marine cycles are recorded in the basin, and it provides impressive detail on the pulsatory character of the Kiowa-Skull Creek (T_5-R_5) cycle.

Berriasian through Valanginian marine rocks are confined to northern Porcupine Plateau, Richardson Mountains, and Mackenzie River delta in the District of Mackenzie and to the Peace River region in northeastern British Columbia (Fig. 4I.50a,b; Young et al., 1976; Dixon, 1986b; Stott, 1967a, 1969, 1981). Facies distribution in the Peace River region suggested to Caldwell (1984a) the possibility of a major marine Valanginian cycle (Beattie Peaks), involving southward transgression and northward regression, which may be related to Kauffman's T_1-R_1 cycle (Fig. 4I-50b); Stott (1984) suggested two cycles, one of Late Jurassic-Berriasian (upper Fernie and Monteith formations) and a second, the Valanginian Beattie Peaks. In Richardson Mountains and beneath Mackenzie Delta, two T-R cycles occur within Kauffman's T_1-R_1; a Berriasian cycle (Husky to Martin Creek formations) and a Valanginian to ?Middle Hauterivian cycle (McGuire to Kamik formations; Dixon, 1986a). The dominantly Barremian Mount Goodenough Formation may have been deposited during an episode corresponding to Kauffman's T_3-R_3 cycle. Compared to the "Neocomian" marine rocks, Aptian Rat River and equivalent rocks of Porcupine Plateau and Anderson Plain have a wider distribution (Dixon, 1986a; Yorath and Cook, 1981, 1984). Combined with the evidence of the thick Cadomin and Gething progradational deposits in Peace River Basin, which contains molluscs dated as probably Aptian or at youngest Early Albian, the possibility of an extensive Aptian marine cycle related to Kauffman's T_4-R_4 cycle must be entertained (Fig. 4I.50c). In Barremian through Aptian times, however, more southerly parts of the Western Interior Basin received mainly continental sediments and were drained to the northwest by vast river systems, which have been sketched by Williams (1963), McLean (1977b) and Christopher (1980), and thus do not record the foregoing T-R cycles.

In Canada, the Kiowa-Skull Creek marine cycle (T_5-R_5) is revealed to have been strongly pulsatory from latest Aptian or earliest Albian to Late Albian time. Three distinct phases are distinguished (Caldwell, 1984a). The first is found in the Early Albian Clearwater couplet, with transgressive peak recorded in the Clearwater and Moosebar formations of Alberta and northeastern British Columbia (Fig. 4I.50d). The second is recorded in the Middle Albian Hulcross couplet (more restricted than the Clearwater), with transgressive peak recorded in the Wildhorn Member of the Scatter Formation, Hulcross Formation, and Harmon Member of the Peace River Formation in northeastern British Columbia and northwestern Alberta (Fig.4I.50e). The third pulse produced the dominantly Late Albian Joli Fou or Skull Creek Couplet, with transgressive peak recorded within the Shaftesbury Formation in northeastern British Columbia, the Joli Fou Formation of Alberta and Saskatchewan, and the Skull Creek Member of the Ashville Formation in Manitoba (Fig. 4I.50f). Only the earliest of these transgressions is reasonably well documented in the north; it resulted in deposition of sediments now included in the Arctic Red, Sans Sault, and Horton River formations (Yorath and Cook, 1984), even though the later Skull Creek transgression was much more extensive within the basin as a whole and led to initial establishment of the Western Interior Seaway.

The Early Turonian Greenhorn marine cycle (T_6-R_6) saw the Western Interior Sea expanded to its widest limits (Fig. 4I.50h). The transgressive phase brought warmer Gulfian waters into the Canadian basin, and with them, molluscan and planktonic foraminiferal immigrants. Calcareous deposits, in the form of coccolith-bearing white-speckled shales, became widespread. The foraminiferal and molluscan fossils permit precise correlation of the white-speckled shales with the Bridge Creek Limestone Member of the upper Greenhorn Formation, the Fairport Chalk Member of the Carlile Formation in the mid-basin of the United States, and the Vimy Member of the Blackstone and Kaskapau formations of the Foothills. Organic-rich shales were deposited over much of northern Canada during this cycle.

During the Coniacian-earliest Campanian Niobrara marine cycle (T_7-R_7), the marine basin re-attained its widest limits (Fig. 4I.50i), although if Kauffman's (1967) model of a marine cyclothem be applied to identify the time of the transgressive peak, that may have been later in the northern basin than in the southern basin. In the District of Mackenzie, the expanding sea re-invaded the Anderson Basin to deposit organic-rich muds now included in the Smoking Hills Formation (Yorath and Cook, 1984). Equivalent calcareous shales in the southern Foothills are included in the Thistle Member of the Wapiabi and Puskwaskau formations. In the eastern part of the basin, white-speckled muds and chalky marls, rich in planktonic foraminifers and coccolith-bearing fecal pellets, were deposited widely under environmental conditions similar to those prevailing during the Greenhorn marine cycle.

The last two marine cycles recorded in the southern Canadian portion of the Western Interior Basin, the Claggett and the Bearpaw, are clearly expressed only on the western flank. The Claggett marine cycle seems to have reached its transgressive peak in Early Campanian time (Fig. 4I.50j), the Bearpaw in latest Campanian to earliest Maastrichtian time (Fig. 4I.50k). The transgressive phase of the Claggett cycle is most faithfully recorded in the non-calcareous shales of the lower Lea Park and Claggett (Pakowki) formations of Alberta and Saskatchewan, and the Nomad Member of the Foothills;

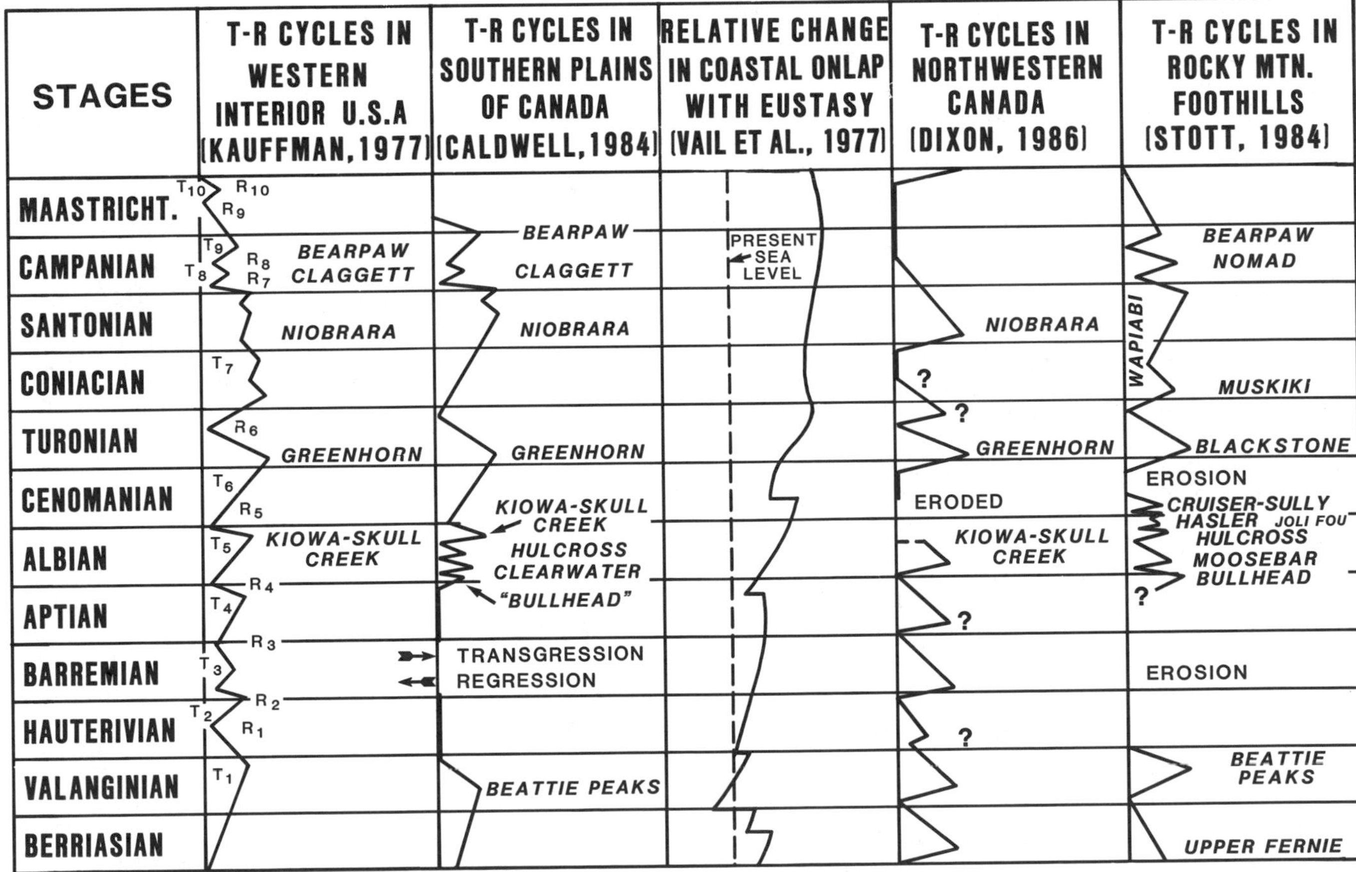

Figure 4I.49. Transgressive-regressive (T-R) couplets and cycles in the southern Interior Plains of Canada (after Caldwell, 1984a).

the regressive phase appears in the nonmarine and marine sandstones of the Belly River Formation of the southern Rocky Mountain Foothills and Judith River Formation of the Plains. For the Bearpaw cycle, the transgressive phase culminates in the Snakebite and equivalent members of the Bearpaw Formation, and the regressive phase in the largely nonmarine sandstones and mudstones that cap the Cretaceous sequence of the plains (Fig. 4I.50l; the Blood Reserve and St. Mary River formations of Alberta; Eastend, Whitemud, Battle, and Frenchman formations of Saskatchewan; and Boissevain and Turtle Mountain formations of Manitoba). The Claggett and Bearpaw cycles have not been recognized in northeastern British Columbia and adjacent parts of the District of Mackenzie.

In the northern part of the Western Canada Basin, Kauffman's T_{10}-R_{10} cycle is probably represented by the Tent Island and Moose Channel formations and equivalent strata. Maximum transgression during this cycle occurred during Late Maastrichtian time and regression continued in Early Paleocene time.

Facies distribution

Southern Western Canada Basin

Facies patterns within Western Canada Basin south of latitude 60°N were controlled by the form of the basin and by the transgressive-regressive cycles, which were determined by interaction of tectonic and eustatic movements. Each of the four longitudinal sectors of the basin, described previously had its own characteristic facies. The western foredeep contains the coarsest grained terrigenous clastic sediments shed from the developing Rocky Mountains. The west-median trough trapped finer sands, silts, and muds. The east-median hinge is covered by muds, and the eastern platform is characterized by chalky marlstones and muds. The basin was never static; its limits expanded and shrank in response to local and regional tectonic activity and transgression and regression, and the facies belts migrated accordingly.

Kauffman (1967, 1969), working in the Western Interior of the United States, recognized a dozen distinct lithofacies that form repetitive cyclothemic sequences. These lithofacies range from marginal-marine sandstones, through increasingly finer grained terrigenous clastic sediments to off-shore massive limestones, chalky limestones, and chalks. Carbonate rocks were poorly developed in the cooler waters of the Canadian part of the basin, and there is no counterpart in any marine cycle in Canada of the three most calcareous facies, including chalks and massive limestones, which formed in the warmer southern basin. McNeil (1984), in applying this scheme in the Manitoba Escarpment, showed that transgressive phases of the Kiowa-Skull Creek cycle ranged from marginal marine sandstone to dark non-calcareous shale, and the regressive phase from dark non-calcareous shale to platy, shaly sandstone and siltstone. Other transgressive sequences are more complete, culminating, for example, in shaly limestone at the top of the Favel Formation and chalky shale in the upper Niobrara Formation. Some sequences are abbreviated by unconformities. The unconformity-bounded Greenhorn sequence begins and ends in sandy and silty shales, and the unconformity- bounded Niobrara sequence, in non-calcareous shales.

The western foredeep, and to some extent the west-median trough, are interpreted as responses to tectonic loading in the adjacent eastern Cordillera. When rates of sedimentation exceeded rates of subsidence, the western foredeep rapidly filled with marine sediments, followed by vast alluvial deposits, their tops marked in places by disconformities. In the trough, where the rate of subsidence generally outstripped sedimentation, deposition was more continuous and disconformities are rare. The hinge was a broad, fluctuating zone in which sedimentation was moderately continuous except where local cratonic structures underwent short-lived uplift, as for example in Saskatchewan. The eastern platform lay adjacent to the craton and during transgressions was submerged, with sediments being deposited over a broad area; during regressions it was exposed to subaerial or submarine erosion. Sedimentation was discontinuous and increasingly so eastwards, so that along Manitoba Escarpment and in the adjoining eastern plains most formations and many members are unconformity-bounded.

A number of distinctive lithotypes have been recognized, and the lithotypes of the transgressive hemicyclothem are commonly repeated in reverse order through the regressive hemicyclothem. The lithotypes grade laterally as well as vertically into each other and are more or less diachronous. The key to unravelling the facies

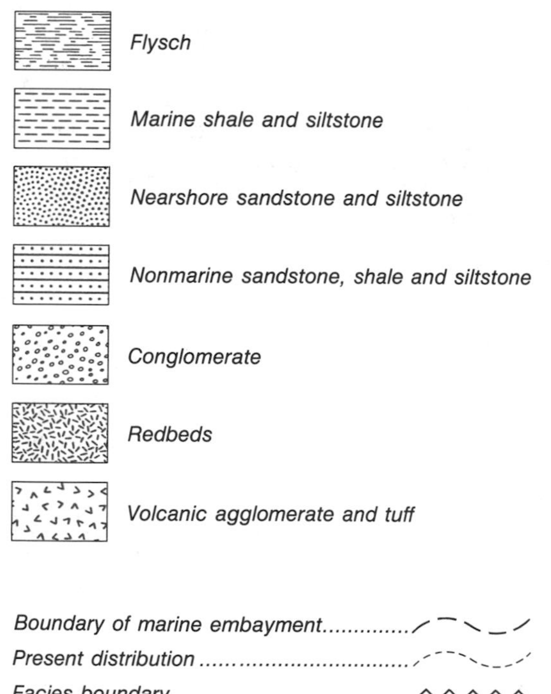

Figure 4I.50. Paleogeographic maps showing distribution of marine, transitional and nonmarine environments in the Western Canada Basin during the Cretaceous Period. Dashed lines show eastern and western boundaries of the seaway; patterns cover areas of known Cretaceous sediments (D.F. Stott).

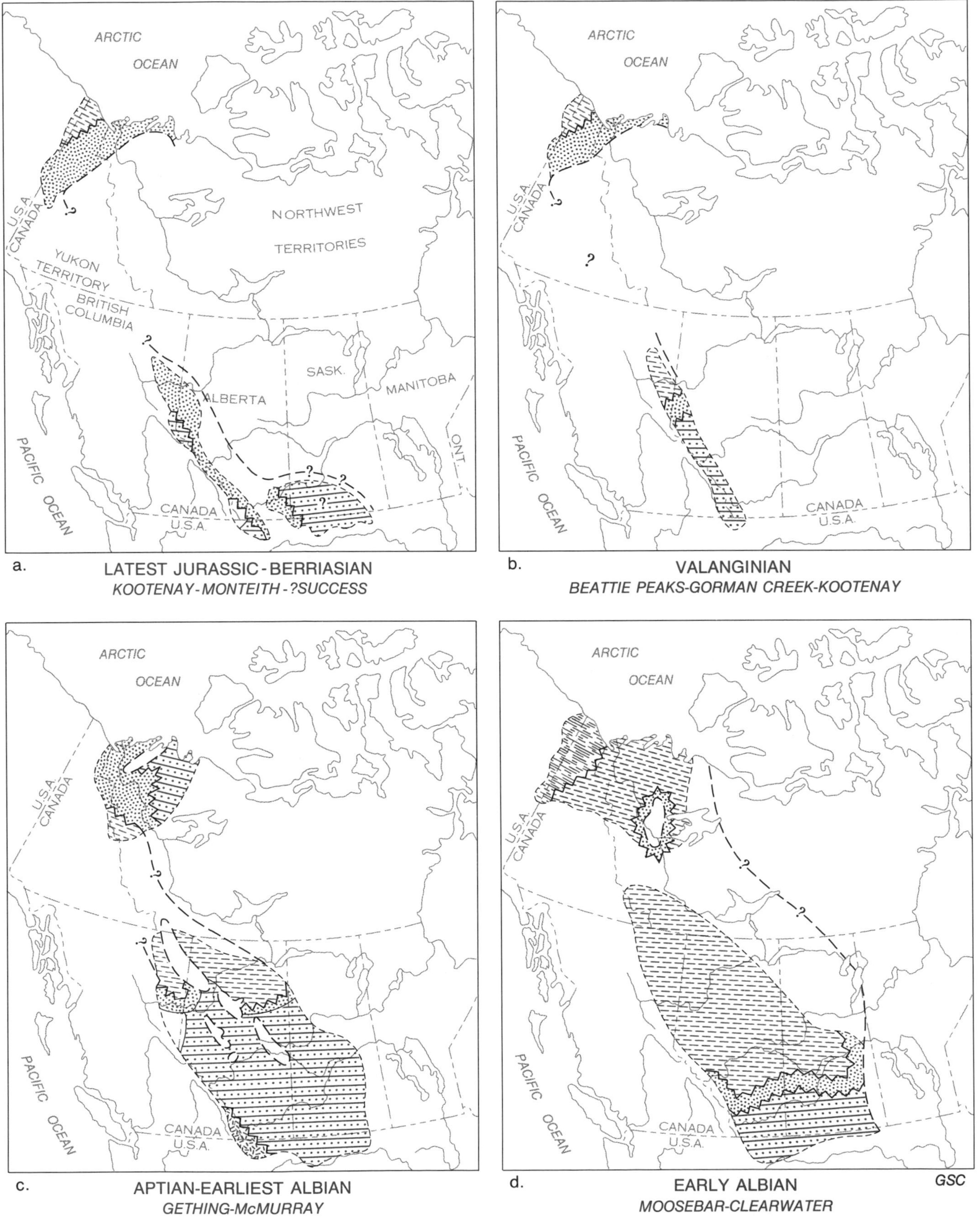
ARCTIC
OCEAN
NORTHWEST
TERRITORIES
U.S.A.
CANADA
YUKON
TERRITORY
BRITISH
COLUMBIA
ALBERTA
SASK.
MANITOBA
ONT.
PACIFIC
OCEAN
CANADA
U.S.A.
a.
LATEST JURASSIC-BERRIASIAN
KOOTENAY-MONTEITH-?SUCCESS
b.
VALANGINIAN
BEATTIE PEAKS-GORMAN CREEK-KOOTENAY
c.
APTIAN-EARLIEST ALBIAN
GETHING-McMURRAY
d.
EARLY ALBIAN
MOOSEBAR-CLEARWATER
GSC

Figure 4I.50. Cont.

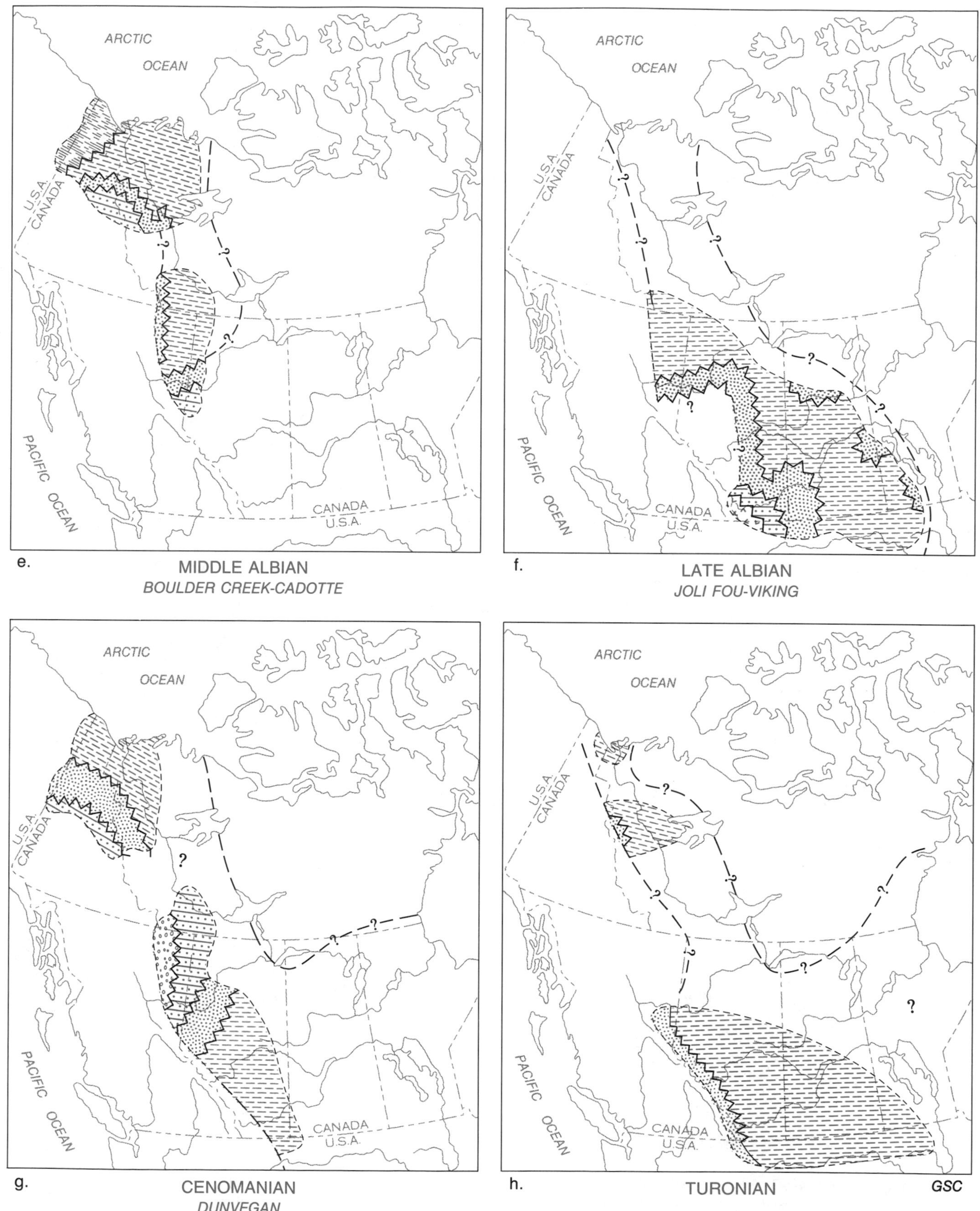
ARCTIC
OCEAN
U.S.A.
CANADA
PACIFIC
OCEAN
CANADA
U.S.A.
e.
MIDDLE ALBIAN
BOULDER CREEK-CADOTTE
f.
LATE ALBIAN
JOLI FOU-VIKING
g.
CENOMANIAN
DUNVEGAN
h.
TURONIAN
GSC

Figure 4I.50. Cont.

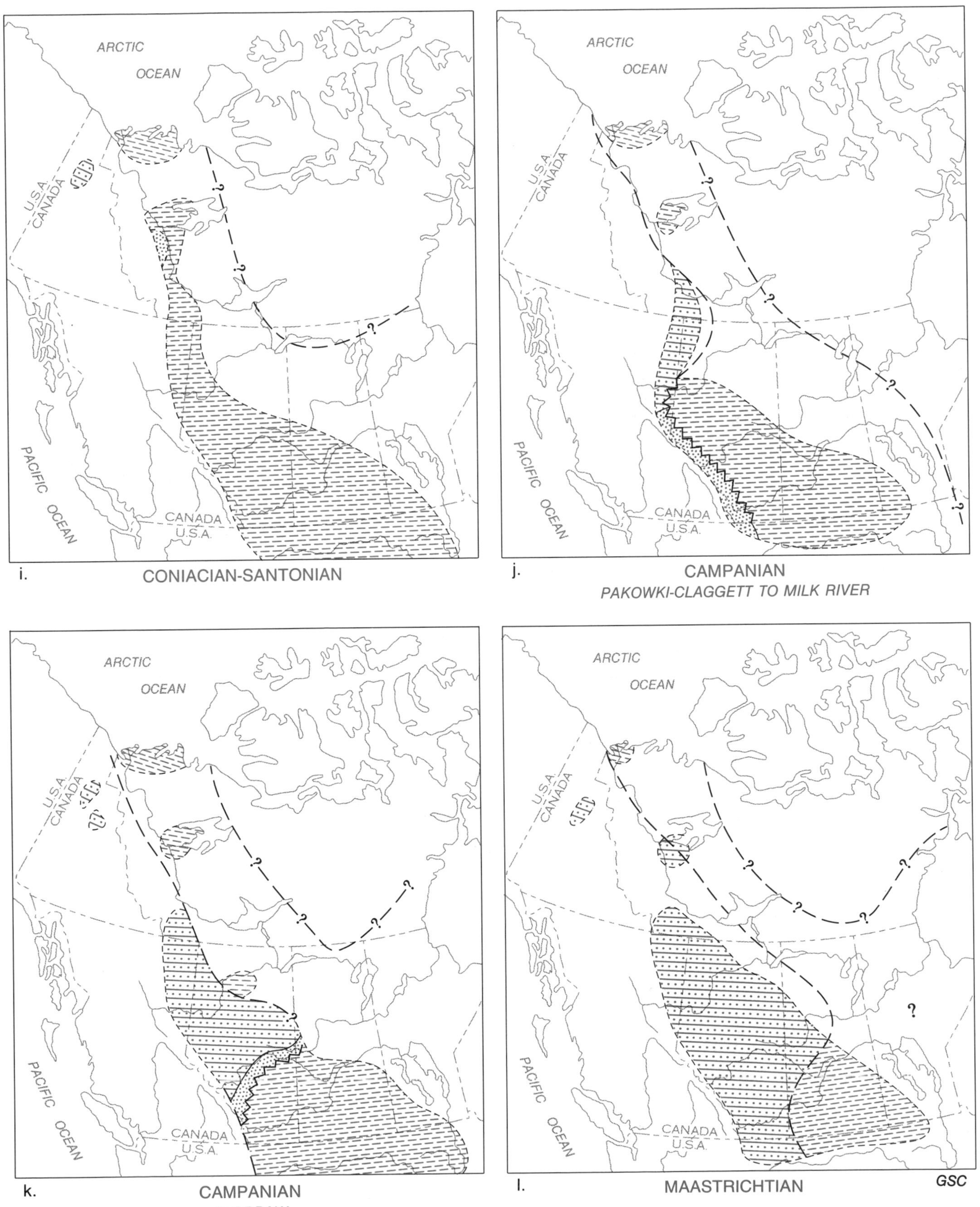
ARCTIC
OCEAN
U.S.A.
CANADA
PACIFIC
OCEAN
CANADA
U.S.A.
i.
CONIACIAN-SANTONIAN
j.
CAMPANIAN
PAKOWKI-CLAGGETT TO MILK RIVER
k.
CAMPANIAN
BEARPAW
l.
MAASTRICHTIAN
GSC

Figure 4I.50. Cont.

patterns has lain in remarkably refined chronocorrelation, using a relative time-scale based on about sixty-five molluscan and thirty foraminiferal biozones. Integration of that relative time-scale with an absolute time-scale based on about fifty volcanic ash beds, has produced a powerful biostratigraphic-geochronometric yardstick for the measurement of Cretaceous time (see Caldwell, 1982). Newer chemostratigraphic and magnetostratigraphic methods of correlation are now being developed and applied to resolving facies relationships, commonly as a supplement to conventional biostratigraphic methods. Thus, some Upper Maastrichtian to Paleocene continental sections in Alberta, where they contain the global iridium spike (Lerbekmo and St. Louis, 1986), and their counterparts in Saskatchewan and North Dakota, have proven amenable to magnetostratigraphic correlation (Lerbekmo et al., 1979a; Lerbekmo, 1985; Lerbekmo and Coulter, 1984, 1985a, b). The method has offered a check on the reliability of using the distribution of certain key palynomorphs (for example, the disappearance of *Aquilapollenites*) to identify the Cretaceous-Tertiary boundary, and has exposed disconformities, diachronism, facies relationships, and contrasted rates of sedimentation in the formations enclosing the boundary.

Northern basins

Facies distribution in the northern basins was controlled by forces similar to those in the southern part of Western Canada Basin; however, an additional tectonic regime must be considered, that of extensional tectonics until Albian time. From Late Hauterivian onward, an oceanic basin (the present day Canada Basin) was forming and, according to one model (although others have been proposed), northern Alaska-northern Yukon Territory was rotating in an anticlockwise direction away from the Arctic Islands.

Transgressive-regressive cycles contain fewer facies types than those described by Kauffman (1969) and are characterized by a basal transgressive interval, generally coarse clastics but silty to sandy shale in places, which varies in thickness from a few centimetres to 100 m or more. Shales of the regressive (progradational) phase usually overlie transgressive beds abruptly. These become progressively siltier up-section and interbeds of sandstone become more common. The upper part of the regressive interval is generally marine to nonmarine sandstone, with thin shale interbeds, and local coal in nonmarine intervals. This ideal cycle with optimum preservation of facies is characteristic of basin-centre deposition. Toward the basin margin, sandstone may occupy the complete cycle, and erosion of part or the whole of a cycle is not uncommon. Toward the basin centre, the sandstone facies may be completely replaced by shale, but there is usually a silty to sandy facies equivalent.

Berriasian to Middle Hauterivian facies belts are characterized by an eastern platform area in which nonmarine beds at the basin margin grade westward and northwestward to shelf sands and muds (Fig. 4I.50a). Subsidence and sedimentation rates tended to be low in these areas. At the approximate northwest flank of the Aklavik Arch Complex, sediments thicken dramatically. This was a zone of rifting, with higher rates of subsidence and sedimentation. On the northwest side of the rift margin, structural, and possibly bathymetric troughs developed in which middle to outer shelf deposits accumulated. During the Berriasian to Middle Hauterivian, an east to west/northwest change from continental to nearshore to shelf deposits can be documented. By Late Hauterivian, the progressive development of Canada Basin resulted in significant changes in paleogeography and facies. While the eastern platform and central troughs persisted, the eastern end of the Brooks Range Geanticline began to influence sedimentation. During the Late Hauterivian transgression, sandy beds appear to have been deposited along the eastern flank of the geanticline as well as on the eastern platform. The implications are that a "two-sided" basin may have existed at this time. During the Barremian and Aptian, evidence for a western source terrane is lacking.

A significant expansion of depositional limits occurred after Aptian transgression, when the eastern platform was inundated and became the site of continental and nearshore marine deposition (Fig. 4I.50c). A lack of reliably dated Aptian strata in Blow and Keele-Kandik troughs negates any attempt to extrapolate facies to the west. If the lower part of the Martin House Formation has been correctly dated as Aptian, the shales of this formation could indicate the development of a minor foredeep in front of the western end of Mackenzie Arc.

Albian time saw the culmination of seafloor spreading in Canada Basin, and the resulting convergence led to uplift of the Brooks Range Geanticline, which shed coarse sediments into the adjacent Blow and Keele-Kandik troughs. Also during Albian time, orogeny in the Cordillera was producing uplifted terranes that shed sediments northward into Peel Trough. The eastern platform area became an areally extensive, epicratonic shelf on which muds and siltstones were deposited (Fig. 4I.50d, e). Little or none of this detritus was of eastern provenance.

From Cenomanian time until the present, the Cordillera has been the principle source of clastic sediments. The stratigraphy records progressive northward migration of foredeeps until Late Maastrichtian-Early Paleocene time, when deposition on the continental margin of Canada Basin became pre-eminent (Fig. 4I.50g-l). Late Cretaceous facies tend to be arranged in belts subparallel to the mountain front.

Paleoclimatological considerations

From palaeomagnetic measurements, Couillard and Irving (1975), Irving (1979) and Van der Voo (1981) refined the position of the Cretaceous pole, and from that constructed a grid of paleolatitude for the North American continent. At maximum extent, the Western Interior Basin reached from paleolatitude 30° to paleolatitude 75°, the northern (Canadian) portion of the basin occupying the higher 25° of that range. Assuming normal polarity, the Cretaceous basin trended north-northeast.

Although the Western Interior Basin occupied much the same latitudinal range in Cretaceous times as it does today, the climate was warmer and more uniform, as judged by the distribution of biotas within the basin (for example, rudistid reefs near the Gulf Coast, bread-fruit trees within the modern Arctic Circle). This paleobiological evidence is now further substantiated by paleotemperature measurements using oxygen isotopes (see Kauffman, 1977b, 1984b). Thus, for example, Forester et al. (1977) obtained $\delta^{18}O$ values of +0.06 to -2.14 from well-preserved ammonite shells in the Campanian-Maastrichtian Bearpaw Formation of southwestern Saskatchewan. These values translate into temperatures of about 17° to 27°C, with an average of 23.2°C ± 3.2°C (corrected for an ice-free Earth to about 20°C) for the Bearpaw sea at about palaeolatitude 55°. More recent studies of stable isotopes suggest that the $\delta^{18}O$ (and $\delta^{13}C$) values obtained throughout the basin are too negative, possibly because of variations in salinity of the epicratonic seas, and thus the precise temperatures of the seas are in doubt. It may be concluded, however, that temperatures throughout the basin were considerably higher than those prevailing today. Warmer and more equable though Cretaceous climates were (Barron, 1983), there were still distinct differences in the biotas that lived along the length of the seaway. Northward decrease in diversity almost certainly reflects northward decrease in temperature. These factors then help explain biogeographical provincialism, not only the Boreal and Tethyan provinces (Jeletzky, 1971a, b) but the Northern, Central, and Southern Interior Subprovinces of the Western Interior Basin (Kauffman, 1973, 1977b, 1979, 1984b).

When the Western Interior Seaway was established in Late Albian time, elements of the northern and southern subprovincial faunas migrated for great distances along the strait. Thus, at peak transgression of the Kiowa-Skull Creek cycle, both foraminifers and molluscs from the Gulf Coast and the southern portion of the basin spread as far as Alberta and Saskatchewan (Caldwell, 1982). Likewise, the Greenhorn and Niobrara cycles brought pulses of warmer waters (possibly by as much as 5°C; Kauffman, 1977b), to which the fauna adapted, into the northern part of the basin. The planktonic *Hedbergella loetterlei* foraminiferal fauna, found in the Lower Turonian Favel, lower White-Speckled Shale, and lower Blackstone Formation, and the dominantly planktonic *Globigerinelloides* fauna, found in the Santonian Niobrara, upper White-Speckled Shale, and upper Wapiabi Formation across the Canadian part of the Western Interior Seaway, forcefully illustrate this point. Keeled planktonic species, paleoenvironmental indicators of warm waters, encroached as far north as an oblique line between the Pasquia Hills of northeastern Saskatchewan and the Rocky Mountain Foothills of southwestern Alberta - a line essentially parallel to paleolatitude - during the Niobrara transgression (McNeil and Caldwell, 1981; McNeil, 1984). The Greenhorn and Niobrara transgressions were marked also by a spread of calcareous lithofacies northward into the Canadian portion of the Western Interior Seaway. A good proportion of southern, calcareous-walled foraminiferal taxa in the middle Lea Park and middle Bearpaw formations of Saskatchewan and adjoining regions, and even a stray rudistid in the lower Bearpaw, suggest similar warm-water invasions during the Claggett and Bearpaw cycles (Caldwell and Evans, 1961; Caldwell et al., 1978). The northward migration of some southern groups of foraminifers and molluscs into the Canadian portion of the Western Interior Basin has been portrayed graphically by Kauffman (1984b).

The Western Interior epicratonic seaway is without modern analogue, which contributes to the difficulty of reconstructing atmospheric and hydrospheric circulation patterns for the entire basin, far less its northern part, and for assessing the influence they may have exercised on palaeoclimates. Parrish et al. (1984) provided a summary of the factors controlling air and water circulation, the various attempts to model these qualitatively and quantitatively, and the discrepancies and similarities among the results obtained. Large-scale models, dealing with aspects of air and water circulation and applicable to the Canadian portion of the basin, have been proposed by Gordon (1973), Kauffman (1975), Barron and Washington (1982), Lloyd (1982), and Parrish and Curtis (1982).

ECONOMIC GEOLOGY

Potential mineral resources contained in Cretaceous rocks of western Canada were first recognized over two hundred years ago by the earliest explorers. Peter Pond, in 1778, noted the tarry deposits of the oil sands along Athabasca River. Alexander Mackenzie (1801) also commented on the bitumen deposits and in 1798, discovered the coal deposits of the Peace River canyon. As stated by Masters (1985), the Cretaceous rocks of the Western Canada Basin contain the greatest accumulation of hydrocarbons of any group of rocks in the world. Furthermore, many parts of the basin include extremely large deposits of coal.

Hydrocarbons

The immense oil sands of the McMurray Formation, the heavy oils of the Lloydminster area, the Upper Cretaceous oil of the Pembina Field, natural gas of the Elmworth area, the Upper Cretaceous oil shales throughout the basin, and the thousands of oil and gas pools that occur through the Western Canada Basin hold vast hydrocarbon resources. In Northwest Territories and Yukon Territory, only limited resources in Cretaceous strata have been identified. The hydrocarbon resources of western Canada, outlined briefly in the following discussion, are described in greater detail in Subchapter 6A of this volume.

Petroleum

The southern Cretaceous basin is a simple homocline rising eastward toward the Canadian Shield. However, its eastern rim has a dip reversal along a 1000 km anticlinal axis (Masters, 1985), which was produced in part by structural drape over the eastern edge of the Devonian beds, which have undergone salt collapse. Part of the structure may be related to the foreland bulge. The structure was a barrier and trap to hydrocarbons migrating up-dip. The sub-Albian unconformable surface is mainly Paleozoic carbonates and has a relief of 200 m. These carbonates, beneath the main oil-sands deposits in east-central Alberta, are host to large petroleum accumulations, much of which may have been derived from Cretaceous sources (Deroo et al., 1974; Deroo et al., 1977).

The Mannville Group of Alberta, one of the world's richest hydrocarbon-bearing units, is of great interest because of the enormous amounts of bitumen in the oil sands and of natural gas in its deeper, western portion. Geochemical analyses (Deroo et al., 1977; Welte et al.,

1982; Moshier and Waples, 1985) indicate that on average, Mannville shales have a total organic content of 1.3%, and vitrinite reflectance values of less than 0.7% (Moshier and Waples, 1985). These rocks are therefore somewhat immature over most of the province. Only in the west, where Tertiary burial was deeper, is maturity higher. Welte et al. (1982) have reported vitrinite reflectances values up to 1.0% there. The kerogen in that area is dominantly woody, making the area gas prone. They have concluded also that some Devonian sources have contributed to the bitumen of the Cretaceous oils sands of Alberta.

In the Northwest Territories, the most prospective area for petroleum exploration of Cretaceous reservoirs underlies the Tuktoyaktuk Peninsula and the southern Mackenzie Delta. The faulted northwest margin of the Eskimo Lakes Arch provides the structural setting for accumulations in small horsts and faulted anticlines.

Oil sands

The Mannville sandstones of the Athabasca-Peace River region in northeastern Alberta are among the largest known oil reservoirs in the world, and are estimated by Alberta Energy Resources Conservation Board (1983) to contain more than 198 x 10^9 m^3 of bitumen in place. The oil, having a gravity of 6-10° API, behaves as a solid unless heated. The bitumen is considered to have originated and migrated as liquid oil, but once emplaced was water washed, degraded by bacterial action, and possibly oxidized by inorganic processes (Deroo et al., 1977). The five main deposits are Athabasca, Peace River, Wabasca, Cold Lake, and Buffalo Head Hills. About 10 per cent of the deposits is amenable to open-pit mining and the remainder to other types of recovery processes using steam injection and other techniques. The Athabasca deposits at Fort McMurray lie within fluviatile, lacustrine, and lagoonal sands of the McMurray Formation, lying on the highly irregular surface of eroded Devonian rocks. The Peace River deposit, lying along the west side of a Mississippian ridge, consists of the Cadomin, Gething, and Bluesky formations, but extends upward into the Grand Rapids Formation (Kramers, 1982). At Cold Lake, trapping was controlled by the northward pinchout of marine Mannville sands along east-west trending shorelines.

Heavy oils

Heavy oils, found near Lloydminster along the Alberta-Saskatchewan border, occur within fluvial-intertidal sediments forming cyclic, sand-shale sequences within the Mannville Group. Parts of the Lloydminster deposits were influenced by shale-filled channels cut into producing sands. Gravity of the oil ranges from 13 to 18 degrees API. Total heavy oil in place is estimated to be 5200 x 10^6 m^3 at high confidence level, 7460 x 10^6 m^3 at average expectation level, and 10 630 x 10^6 m^3 at the speculative level (Procter et al., 1984).

Conventional oil

Numerous small oil fields occur in the Mannville Group, the reservoirs being related to shoreline sediments and channel fills. Oil has been discovered in more than 500 pools containing a total of 240 x 10^6 m^3 of recoverable oil.

The Pembina Field, producing from the Upper Cretaceous Cardium Formation, is areally one of the largest known in North America. The history of development and the reservoirs and reserves of the Pembina Field were recently reviewed by Nielsen and Porter (1984) and the detailed sedimentology by Krause and Nelson (1984). Cardium sands in the Garrington, Caroline, and Ricinus areas of Alberta have been described by Walker (1983a, b, 1985) and in a larger region of Alberta by Plint et al. (1986). The original proven recoverable reserves of the Pembina Oil Field were 290 x 10^6 m^3, of which the Cardium sandstone contributes 250 x 10^6 m^3 (Nielsen and Porter, 1984).

Virtually all of the main sandstone units of the Colorado Group of the Alberta Plains yield commercial production of hydrocarbons. Particularly noteworthy are those in the Viking Formation at Joarcam. The Viking-Bow Island reservoirs occur as elongate sand deposits, trending northwesterly in central Alberta.

Widespread production of hydrocarbons is obtained from Judith River (Belly River) reservoirs in southern Alberta.

Oil and condensate have been recovered from four Cretaceous tests in the Mackenzie Delta area. These were in strata of the Atkinson Point Formation and Parsons Group, at Atkinson M-25, Tuk M-09, Kamik D-48, and Kugpik O-13. Reserves in these four discoveries are small by frontier standards; the largest probably does not exceed 5 x 10^6 m^3 of recoverable oil.

Natural gas

Some 8500 x 10^9 m^3 of natural gas are estimated to be present within Cretaceous sediments of the "Deep Basin", along the northeastern front of the Rocky Mountains (Masters, 1985). These sediments include a thick sequence of coal and organic-rich shales, which are still actively generating hydrocarbons (Welte et al., 1982). Gas-saturated sands of very low porosity and ultra-low permeability ("tight gas sands") grade updip into porous, water-saturated sandstones. The trap is not tightly sealed, so that gas is constantly migrating upward as it is generated (Welte et al., 1982; Geis, 1985; Varley, 1984). The reservoirs are found in the Falher and Notikewin shoreline sands, as well as in the Bluesky, Gething, and Cadomin formations. The coals of the Falher and equivalent Gates Formation are regarded as source beds.

About 5% of Western Canada's recoverable gas reserves are present in the low-permeability reservoirs of the Upper Cretaceous Medicine Hat Sandstone and Milk River sandstones of southeastern Alberta and southwestern Saskatchewan.

Some 3000 gas fields produce from thin Cretaceous sandstones at shallow depth, and have a combined reserve of 481 x 10^9 m^3. In Alberta, gas is reported from Dunvegan, Edmonton, Brazeau, and Wapiti strata, and is produced from many of the Cardium and Viking fields. In Saskatchewan, natural gas reservoirs are found in the Ribstone Creek Tongue of the Judith River Formation and in the Spinney Hill, St. Walburg, and Phillips sandstones of the Colorado Group.

In the Mackenzie Delta area, only one accumulation of gas, the Parsons Gas Field, has been discovered in Cretaceous strata. The Parsons Group is the reservoir, and gas is trapped in a complexly faulted anticline. Two structural culminations divide the field into North and South Parsons fields. Recoverable reserves are estimated at 62.3 x $10^9 m^3$.

Oil shales

Oil shales occur within the Upper Cretaceous Favel and Boyne formations along the Manitoba Escarpment (Macauley et al., 1985); both shales contain 3 to 4% organic carbon. These shales, being widely distributed across central Saskatchewan and into northern Alberta, contain large volumetric reserves of shale oil but give only modest yields. Equivalent shales, including the Smoking Hills Formation, occur in northernmost District of Mackenzie and Yukon Territory. Thus, they form the largest areal deposit in Canada, and probably the largest potential resources of in situ shale oil. Each of the two oil shale zones in Manitoba ranges in thickness from 30 to 45 m. They are thinner in Saskatchewan but thicker (35 to 60 m) in Alberta. The oil shales are classified as marine, Type II, mixed and immature. The potential oil resources are vast. In one small area at Pasquia Hills in Manitoba, in situ oil has been estimated to be 200 x 10^6 m^3.

Coal

Coal deposits of the Cretaceous are widely distributed in western Canada, from the Foothills and southern Rocky Mountains to the central Interior Plains. Coal rank, petrography, and resources are outlined by A.R. Cameron in Subchapter 6B.

Rocky Mountains and Foothills

In the Crowsnest Pass-Fernie area of southwestern Alberta and southeastern British Columbia, coal deposits occur in the Kootenay Group of Early Cretaceous age (Gibson, 1977b, 1979, 1985). The coal is mostly bituminous and locally semi-anthracite. Within the Mist Mountain Formation, seams are as thick as 18 m and are laterally extensive. To the north of Fernie Basin, coal is less abundant, although locally some thick seams have good economic potential and some have been mined until recently at Canmore on Bow River.

In the central Foothills, coal was mined for many years at Mountain Park, Cadomin, and Luscar, but those mines were closed prior to 1956. Coal is once again being mined from the Lower Cretaceous Luscar Group near Cadomin. To the east, coal in the Upper Cretaceous and Tertiary Saunders Group is being mined at Sterco and Robb.

In the northern Foothills, the major coal deposits occur within the Gates and Gething formations and are presently being exploited at Grande Cache in Alberta and near Tumbler Ridge in British Columbia (Hoffman and Jordan, 1984). Total coal resources of the northern Foothills region may be greater than 8 x 10^9 tonnes.

Interior Plains

Coal deposits of the Alberta Plains occur mostly in Upper Cretaceous formations, the most important being the Foremost, Oldman, and Edmonton. Most are ranked as high-volatile bituminous. Coal seams in the Foremost are mined at Taber and Milk River and in smaller mines as far east as Medicine Hat. Coal seams of the Oldman are mined at Lethbridge and Magrath. Sub-bituminous coal from the Edmonton Group is found in central Alberta, being produced at Wabamun, Castor, and Drumheller. Coal seams in the Horseshoe Canyon Formation have been explored in the Dodds-Round Hill area near Camrose (Hughes, 1984).

In Saskatchewan, lignite beds at the base of the Ravenscrag Formation nearly coincide with the Cretaceous-Tertiary boundary. Major deposits are those at Estevan, Willow Bunch, Wood Mountain, and Cypress Hills (Whitaker et al., 1978). Total resources are estimated to be 1.6 x 10^9 tonnes of measured coal, 3.0 x 10^9 tonnes indicated, 3.57 x 10^9 tonnes demonstrated, and 27.0 x 10^9 tonnes inferred.

Coals are present in the Upper Cretaceous strata of the Great Bear and Bonnet Plume basins.

Industrial minerals

In Manitoba, Cretaceous strata exposed along or near the Manitoba Escarpment include siliceous, carbonaceous, kaolinitic, and illitic shales, which are used in producing bricks. Non-swelling, calcium bentonite near the base of the Pembina Member is of value as an absorbent clay and is the only commercial source in production in Canada. Calcareous shale of the Boyne Member is suitable for cement.

In Saskatchewan, clays of commercial and industrial significance are obtained from the Upper Cretaceous Whitemud Formation and include plastic stone clays, ball clays, and refractory clays. Swelling, sodium bentonite from the Cretaceous Bearpaw Formation is marketed to the iron-ore pelletizing industry, foundries, and oil-well drilling mud companies. It also finds a market as a sealant for reservoirs and dugouts.

In Alberta, Cretaceous shales are used in structural clay products plants at Medicine Hat and Redcliff, and at a brick plant in Edmonton. Shales of the Bearpaw, Judith River, and Wapiabi formations are used in cement and lightweight aggregate manufacture and for mineral wool insulation. Swelling bentonite from the Bearpaw and Horseshoe Canyon Formation is quarried near Rosalind and used for drilling muds and foundry clay, and as a sealer.

High-silica sands occur at the base of the Swan River, Mannville, and McMurray formations.

Additional information on industrial minerals of western Canada is provided in Subchapter 6D of this volume.

Gem stone

Ammonites from the Campanian Bearpaw Formation of southern Alberta yield gem-quality aragonite. The mineral, known commercially as ammolite, is being used in the production of fine jewellery.

REFERENCES

Aitken, J.D. and Cook, D.G.
1974: Carcajou Canyon map area (96D), District of Mackenzie, Northwest Territories; Geological Survey of Canada, Paper 74-13, 28 p.

Aitken, J.D., Cook, D.G., and Yorath, C.J.
1982: Upper Ramparts River (106G) and Sans Sault Rapids (106N) map area, District of Mackenzie; Geological Survey of Canada, Memoir 388, 48 p.

Alberta Energy Resources Conservation Board
1983: Alberta's Reserves of crude oil, oil sands, gas, natural gas liquid, and sulfur; Energy Resources Conservation Board ST 84-18, unpaged.

Alvarez, L.W., Asaro, F., and Michel, H.V.
1982: Current status of the impact theory for the terminal Cretaceous extinction; in Geological Implications of Impacts of large Asteriods and Comets on the Earth, L.T. Silver (ed.); Geological Society of America, Special Paper 190, p. 305-316.

Anan-Yorke, R. and Stelck, C.R.
1978: Microfloras from Upper Albian *Neogastroplites* Zone, Sikanni Chief River, northeastern British Columbia; in Western and Arctic Canadian Biostratigraphy, C.R. Stelck and B.D.E. Chatterton (ed.); Geological Association of Canada, Special Paper 18, p. 473-493 (1979).

Axelrod, D.K.
1981: Role of volcanism in climate and extinction; Geological Society of America, Special Paper 185, 59 p.

Badgley, P.C.
1952: Notes on the subsurface stratigraphy and oil and gas geology of the Lower Cretaceous series in central Alberta; Geological Survey of Canada, Paper 52-11, 12 p.

Balkwill, H.R., Cook, D.G., Detterman, R.L., Embry, A.F., Håkansson, E., Miall, A.D., Poulton, T.P., and Young, F.G.
1983: Arctic North America and northern Greenland; in Phanerozoic of the World II, Mesozoic A, A.E.M. Nairn and A. Moullade (ed.); Elsevier Scientific Publishing Co., p. 1-31.

Bally, A.W., Gordy, P.L., and Stewart, G.A.
1966: Structure seismic data and orogenic evolution, southern Canadian Rocky Mountains; Bulletin of Canadian Petroleum Geology, v. 14, p. 337-381.

Bamburak, J.D.
1978: Stratigraphy of the Riding Mountain, Boissevain and Turtle Mountain formations in the Turtle Mountain area, Manitoba; Manitoba Department of Mines, Resources and Environmental Management, Geological Survey, Geological Report 78-2, 47 p.

Bardack, D.
1968: Fossil vertebrates from the marine Cretaceous of Manitoba; Canadian Journal of Earth Sciences, v. 5, p. 145-153.

Barron, E.J.
1983: A warm, equable Cretaceous: The nature of the problem; Earth Science Review, v. 19, p. 305-338.

Barron, E.J. and Washington, W.M.
1982: Cretaceous climate: a comparison of atmospheric simulations with the geologic record; Palaeogeography, Palaeoclimatology, Palaeoecology, v. 40, p. 103-133.

Barss, M.S., Bujak, J.P., and Williams, G.L.
1979: Palynological zonation and correlation of sixty-seven wells, eastern Canada; Geological Survey of Canada, Paper 78-24, 118 p.

Beaumont, C.
1981: Foreland basins; Geophysical Journal of the Royal Astronomical Society, v. 65, p. 291-329.

Beck, L.S.
1974: Geological investigations in the Pasquia Hills area; Saskatchewan Department of Mineral Resources, Saskatchewan Geological Survey, Report No. 158, 16 p.

Beland, P. and Russell, D.A.
1978: Paleoecology of Dinosaur Provincial Park (Cretaceous), Alberta, interpreted from the distribution of articulated vertebrate remains; Canadian Journal of Earth Sciences, v. 15, p. 1012-1024.

Bell, W.A.
1949: Uppermost Cretaceous and Paleocene floras of western Alberta; Geological Survey of Canada, Bulletin 13, 231 p.
1956: Lower Cretaceous floras of western Canada; Geological Survey of Canada, Memoir 285, 331 p.

Bell, W.A. (cont.)
1963: Upper Cretaceous floras of the Dunvegan, Bad Heart, and Milk River formations of western Canada; Geological Survey of Canada, Bulletin 94, 76 p.

Binda, P.L. and Srivastava, S.K.
1968: Silicified megaspores from upper Cretaceous beds of southern Alberta, Canada; Micropaleontology, v. 14, p. 105-113.

Brabb, E.E.
1969: Six new Paleozoic and Mesozoic formations in east-central Alaska; United States Geological Survey, Bulletin 1274-1, 25 p.

Brideaux, W.W.
1971: Palynology of the Lower Colorado Group, central Alberta, Canada: I, Introductory remarks, geology and microplankton studies; Palaeon-tographica, Section B, v. 135, p. 53-114.
1977: Taxonomy of Upper Jurassic-Lower Cretaceous microplankton from the Richardson Mountains, District of Mackenzie, Canada; Geological Survey of Canada, Bulletin 281, 89 p.

Brideaux, W.W. and Fisher, M.J.
1976: Upper Jurassic-Lower Cretaceous dinoflagellate assemblages from Arctic Canada; Geological Survey of Canada, Bulletin 259, 53 p.

Brideaux, W.W. and McIntyre, D.J.
1975: Miospores and microplankton from Aptian-Albian rocks along Horton River, District of Mackenzie, Canada; Geological Survey of Canada, Bulletin 252, 85 p.

Bujak, J. and Williams, G.L.
1978: Cretaceous palynostratigraphy of offshore southeastern Canada; Geological Survey of Canada, Bulletin 297, 19 p.

Burden, E.T.
1984: Terrestrial palynomorph biostratigraphy of the lower part of the Mannville Group (Lower Cretaceous), Alberta and Montana; in The Mesozoic of Middle North America, D.F. Stott and D.J. Glass (ed.); Canadian Society of Petroleum Geologists, Memoir 9, p. 249-269.

Burk, C.F. Jr.
1962: Upper Cretaceous structural development of the southern flank of the Peace River Arch; Journal of the Alberta Society of Petroleum Geologists, v. 10, no. 5, p. 223-227.
1963: Structure, isopach and facies maps of Upper Cretaceous marine successions, west-central Alberta and adjacent British Columbia; Geological Survey of Canada, Paper 62-31, 10 p.

Byers, P.N.
1969: Mineralogy and origin of the Eastend and Whitemud formations of south-central and southwestern Saskatchewan and southeastern Alberta; Canadian Journal of Earth Sciences, v. 6, p. 317-334.

Caldwell, W.G.E.
1968: The Late Cretaceous Bearpaw Formation in the South Saskatchewan River valley; Saskatchewan Research Council, Geology Division, Report 8, 89 p.
1982: The Cretaceous System in the Williston basin - a modern appraisal; in Proceedings of the Fourth International Symposium on the Williston Basin, J.E. Christopher and D.M. Kent (ed.); Saskatchewan Geological Society, Regina, p. 295-312.
1984a: Early Cretaceous transgressions and regressions in the southern Interior Plains; in The Mesozoic of Middle North America, D.F. Stott and D.J. Glass (ed.); Canadian Society of Petroleum Geologists, Memoir 9, p. 173-203.
1984b: Some regional palaeogeographic considerations and their implications for the Mannville Group in the Lloydminster heavy-oil fields; in Proceedings of the Saskatchewan Heavy-Oil Conference, Saskatoon, University of Saskatchewan, p. 29-55.

Caldwell, W.G.E. and Evans, J.K.
1961: A Cretaceous rudist from Canada, and a redescription of the holotype of *Ichthyosarcolites coraloidea* (Hall and Meek); Journal of Paleontology, v. 37, p. 615-620.

Caldwell, W.G.E. and North, B.R.
1984: Cretaceous stage boundaries in the southern Interior Plains of Canada; Bulletin of the Geological Society of Denmark, v. 33, p. 57-69.

Caldwell, W.G.E., Stelck, C.R., and Wall, J.H.
1978: A foraminiferal zonal scheme for the Cretaceous System in the Interior Plains of Canada; in Canadian Biostratigraphy, C.R. Stelck and B.D.E. Chatterton (ed.); Geological Association of Canada, Special Paper 18, p. 495-575 (1979).

Cameron, A.R., Norris, D.K., Ricketts, B.D., and Sweet, A.R.
1986: Geology and coal resource potential, Summit Creek Formation, Fort Norman area, Northwest Territories; in Program and Abstracts, Canada's Hydrocarbon Reserves for the 21st Century; Canadian Society of Petroleum Geologists, p. 31.

Campbell, J.D.
1964: Catalogue of coal mines on the Alberta plains; Research Council of Alberta, Preliminary Report 64-3, 140 p.
1974: Coal resources, Taber-Manyberries area, Alberta; Alberta Research Council, Report 74-3, 140 p.

Campbell, J.D. and Untergasser, B.
1972: Two new megaspore species from the continental Upper Cretaceous (Campanian-Maastrichtian) of the Alberta Plains; Canadian Journal of Botany, v. 50, p. 2553-2557.

Cant, D.J.
1983: Spirit River Formation - A stratigraphic-diagenetic gas trap in the Deep Basin of Alberta; American Association of Petroleum Geologists, Bulletin, v. 67, p. 577-587.
1984: Development of shoreline-shelf sand bodies in a Cretaceous epeiric sea deposit; Journal of Sedimentary Petrology, v. 54, no. 2, p. 541-556.

Carrigy, M.A.
1959: Geology of the McMurray Formation; Research Council of Alberta, Memoir 1, 130 p.
1970: Proposed revision of the boundaries of the Paskapoo Formation in the Alberta plains; Bulletin of Canadian Petroleum Geology, v. 18, p. 156-165.
1971: Lithostratigraphy of the uppermost Cretaceous (Lance) and Paleocene strata of the Alberta Plains; Research Council of Alberta, Bulletin 27, 161 p.

Chiang, K.K.
1985: The giant Hoadley Gas Field, south-central Alberta; in Elmworth - Case Study of a Deep Basin Gas Field, J.A. Masters (ed.); American Association of Petroleum Geologists, Memoir, p. 297-313.

Christopher, J.E.
1974: The Upper Jurassic Vanguard and Lower Cretaceous Mannville Group of southwestern Saskatchewan; Saskatchewan Department of Mineral Resources, Report No. 151, 349 p.
1975: The depositional setting of the Mannville Group (Lower Cretaceous) in southwestern Saskatchewan; in The Cretaceous System in the Western Interior of North America, W.G.E. Caldwell (ed.); The Geological Association of Canada, Special Paper 13, p. 523-552.
1980: The Lower Cretaceous Mannville Group of Saskatchewan - A Tectonic Overview; in Lloydminster and Beyond, Geology of Mannville Hydrocarbon Reservoirs, L.S. Beck, J.E. Christopher, and D.M. Kent (ed.); Saskatchewan Geological Society, Special Publication 5, p. 3-32.
1984: The Lower Cretaceous Mannville Group, northern Williston Basin region, Canada; in The Mesozoic of Middle North America, D.F. Stott and D.J. Glass (eds.); Canadian Society of Petroleum Geologists, Memoir 9, p. 109-126.

Cobban, W.A., Erdman, C.E., Lemke, R.W., and Maughan, E.K.
1959: Colorado Group on Sweetgrass Arch, Montana; Bulletin of American Association of Petroleum Geologists, v. 43, no. 12, p. 2786-2796.
1976: Type sections and stratigraphy of the members of the Blackleaf and Marias River formations (Cretaceous) of the Sweetgrass Arch, Montana; United States Geological Survey, Professional Paper 974, 66 p.

Cobban, W.A. and Reeside, J.B., Jr.
1952: Correlation of the Cretaceous formations of the Western Interior of the United States; Bulletin of Geological Society of America, v. 63, p. 1011-1044.

Couillard, R. and Irving, E.
1975: Palaeolatitude and reversals: evidence from the Cretaceous Period; in The Cretaceous System in the Western Interior of North America, W.G.E. Caldwell (ed.); Geological Association of Canada, Special Paper 13, p. 21-29.

Currie, P.J.
1982: Hunting dinosaurs in Alberta's huge bonebed; Canadian Geographic Journal, v. 10, no. 4, p. 34-39.
1983: Hadrosaur trackways from the Lower Cretaceous of Canada; Acta Palaeontologica Polonica, v. 28, p. 63-73.
1985: Small theropods of Dinosaur Provincial Park, Alberta (abstract); Geological Society of America, Abstracts with Program, v. 17, no. 4, p. 215.

Currie, P.J. and Sarjeant, W.A.S.
1979: Lower Cretaceous dinosaur footprints from the Peace River Canyon, British Columbia, Canada; Paleogeography, Paleoclimatology, Paleoecology, v. 28, p. 103-115.

Dawson, G.M.
1881: Report on an exploration from Port Simpson on the Pacific Coast, to Edmonton, on the Saskatchewan, embracing a portion of the northern part of British Columbia and the Peace River country; Geological and National History Survey of Canada, Report of Progress 1879-80, Pt. B, p. 1-177.

Dawson, S.J.
1859: Report on the exploration of the country between Lake Superior and the Red River Settlement, and between the latter place and the Assiniboine and Saskatchewan; John Lovell, Toronto, 45 p.

Davies, E.H.
1983: The dinoflagellate Oppel-zonation of the Jurassic-Lower Cretaceous sequence in the Sverdrup Basin, Arctic Canada; Geological Survey of Canada, Bulletin 359, 59 p.

Davies, E.H. and Norris, G.
1980: Latitudinal variations in encystment modes and species diversity in Jurassic dinoflagellates; in The Continental Crust and Its Mineral Deposits, D.W. Strangway (ed.); Geological Association of Canada, Special Paper 20, p. 362-373.

Deroo, G., Powell, T.G., Tissot, B., and McCrossan, R.G.
1977: The origin and migration of petroleum in the Western Canada sedimentary basin, Alberta; Geological Survey of Canada, Bulletin 262, 136 p.

Deroo, G., Tissot, B., McCrossan, R.G., and Der, F.
1974: Geochemistry of heavy oils of Alberta; in Oil Sands - Fuel of the Future, L.V. Hills (ed.); Canadian Society of Petroleum Geologists, Memoir 3, p. 148-167.

Dickinson, W.R., Klute, M.A., Hayes, M.J., Janecke, S.U., Lundin, E.R., McKittrick, M.A., and Olivares, M.D.
1988: Paleogeographic and paleotectonic setting of Laramide sedimentary basins in the central Rocky Mountains region; Geological Society of America, Bulletin, v. 100, no. 7, p. 1023-1039.

Dixon, J.
1979: The Lower Cretaceous Atkinson Point Formation (new name) on the Tuktoyaktuk Peninsula, Northwest Territories: a coastal fan-delta to marine sequence; Bulletin of Canadian Petroleum Geology, v. 27, p. 163-182.
1982a: Jurassic and Lower Cretaceous subsurface stratigraphy of the Mackenzie Delta - Tuktoyaktuk Peninsula, Northwest Territories; Geological Survey of Canada, Bulletin 349, 52 p.
1982b: Sedimentology of the Neocomian Parsons Group in the subsurface of the Mackenzie Delta area, Arctic Canada; Bulletin of Canadian Petroleum Geology, v. 30, p. 9-28.
1986a: Cretaceous to Pleistocene stratigraphy and paleogeography, northern Yukon and northwestern District of Mackenzie; Bulletin of Canadian Petroleum Geology, v. 34, no. 1, p. 49-70.
1986b: Comments on the stratigraphy, sedimentology and distribution of the Albian Sharp Mountain Formation, northern Yukon; in Current Research, Part B, Geological Survey of Canada, Paper 86-1B, p. 375-428.

Dixon, J., Dietrich, J.R., McNeil, D.H., McIntyre, D.J., Snowdon, L.R., and Brooks, P.
1985: Geology, biostratigraphy and organic geochemistry of Jurassic to Pleistocene strata, Beaufort-Mackenzie area, northwest Canada; Canadian Society of Petroleum Geology, Course Notes, 65 p.

Dodson, P.
1971: Sedimentology and taphonomy of the Oldman Formation (Campanian), Dinosaur Provincial Park, Alberta; Paleogeography, Paleoclimatology, Paleoecology, v. 10, p. 21-74.
1983: A faunal review of the Judith River (Oldman) Formation, Dinosaur Provincial Park, Alberta (Canada); Delaware Valley Paleontological Society, The Mosasaur, v. 1, p. 89-118.
1985: Studies of dinosaur paleoecology by repeated microfaunal sampling in Campanian sediments, southern Alberta (abstract); Geological Society of America, Abstracts with Programs, v. 17, no. 4, p. 216.

Douglas, R.J.W.
1950: Callum Creek, Langford Creek, and Gap map areas, Alberta; Geological Survey of Canada, Memoir 255, 124 p.

Douglas, R.J.W., Gabrielse, H., Wheeler, J.O., Stott, D.F., and Belyea, H.R.
1970: Geology of Western Canada; Chapter VIII in Geology and Economic Mineral of Canada, R.J.W. Douglas (ed.); Geological Survey of Canada, Economic Geology Report No. 1, Fifth Edition, p. 365-488.

Doyle, J.A.
1969: Cretaceous angiosperm pollen of the Atlantic Coastal Plain and its evolutionary significance; Journal of the Arnold Arboretum, v. 50, p. 1-35.

Doyle, J.A. and Robbins, E.T.
1977: Angiosperm pollen zonation of the continental Cretaceous of the Atlantic coastal plain and its application to deep wells in the Salisbury Embayment; Palynology, v. 1, p. 43-78.

Edwards, R.G.
1960: Cretaceous Spinney Hill sand in west-central Saskatchewan; Journal of Alberta Society of Petroleum Geologists, v. 8, p. 141-153.

Eisbacher, G.H., Carrigy, M.A., and Campbell, R.B.
1974: Paleodrainage pattern and late-orogenic basins of the Canadian Cordillera; in Tectonics and Sedimentation, W.R. Dickinson (ed.); Society of Economic Paleontologists and Mineralogists, Special Publication 22, p. 143-166.

Elliot, R.H.J.
1960: Subsurface correlation of the Edmonton Formation; Journal of the Alberta Society of Petroleum Geologists, v. 8, p. 324-337.

Evans, W.E.
1970: Imbricate linear sandstone bodies of Viking Formation, Dodsland-Hoosier area of southwestern Saskatchewan, Canada; American Association of Petroleum Geologists, Bulletin, v. 54, p. 469-486.

Felix, C.J. and Burbridge, P.P.
1973: A Maastrichtian-age microflora from Arctic Canada; Geoscience and Man, v. 7, p. 1-29.

Fensome, R.A.
1987: Taxonomy and biostratigraphy of schizaealean spores from the Jurassic-Cretaceous boundary beds of the Aklavik Range, District of Mackenzie; Palaeontographica Canadiana No. 4, Canadian Society of Petroleum Geologists and Geological Association of Canada, 48 p.

Fensome, R.A. and Norris, G.
1982: Palynostratigraphic comparison of Cretaceous of the Moose River basin, Ontario, with marginal marine assemblages from the Scotian Shelf and Alberta; in Geoscience Research Grant Program; Summary of Research, 1981-1982, E.G. Pye (ed.), Ontario Geological Survey, Miscellaneous Paper 103, p. 37-42.

Finger, K.L.
1983: Observations on the Lower Cretaceous ostracode zone of Alberta; Bulletin of Canadian Petroleum Geology, v. 31, p. 326-337.

Folinsbee, R.E., Baadsgaard, H., and Cumming, G.L.
1963: Dating volcanic ash beds (bentonites) by the K-Ar method; National Academy of Sciences/National Research Council Publication 1075, p. 70-82.

Folinsbee, R.E., Baadsgaard, H., and Lipson, J.
1960: Potassium-argon time scale: Pre-Quaternary absolute age determination; International Geological Congress XXI, Session Report, Pt. 3, , p. 717.
1961: Potassium-argon dates of Upper Cretaceous ash falls, Alberta, Canada; Annals of the New York Academy of Science, v. 91, p. 352-359.

Folinsbee, R.E., Baadsgaard, H., Cumming, G.L., Nascimbene, G., and Shafiqullah, M.
1965: Late Cretaceous radiometric dates from the Cypress Hills of western Canada; Alberta Society of Petroleum Geologists, 15th Annual Field Conference, Guidebook, Pt. 1, p. 162-174.

Forester, R.W., Caldwell, W.G.E., and Oro, F.H.
1977: Oxygen and carbon isotopic study of ammonites from the Late Cretaceous Bearpaw Formation in southwestern Saskatchewan; Canadian Journal of Earth Sciences, v. 14, no. 9, p. 2086-2100.

Fox, R.C.
1975: Fossil snakes from the upper Milk River Formation (Upper Cretaceous), Alberta; Canadian Journal of Earth Sciences, v. 12, p. 1557-1563.
1978: Upper Cretaceous terrestrial vertebrate stratigraphy of the Gobi Desert (Mongolian People's Republic) and western North America; Geological Association of Canada, Special Paper 18, p. 577-594.
1984: A primitive, "obtuse-angled" symmetrodont (Mammalia) from the Upper Cretaceous of Alberta, Canada; Canadian Journal of Earth Sciences, v. 21, p. 1204-1207.

Frebold, H.
1957: The Jurassic Fernie Group in the Canadian Rocky Mountains and Foothills; Geological Survey of Canada, Memoir 287, 197 p.

Frederiksen, N.O. and Christopher, R.A.
1978: Taxonomy and biostratigraphy of Late Cretaceous and Paleogene triatrite pollen from South Carolina; Palynology, v. 2, p. 113-145.

Gabrielse, H.
1985: Major dextral transcurrent displacements along the Northern Rocky Mountain Trench and related lineaments in north-central British Columbia; Geological Society of America Bulletin, v. 96, p. 1-14.

Gabrielse, H., Blusson, S.L., and Roddick, J.A.
1973: Geology of Flat River, Glacier Lake, and Wrigley Lake map areas, District of Mackenzie and Yukon Territory; Geological Survey of Canada, Memoir 366, 153 p.

Gabrielse, H. and Reesor, J.E.
1974: The nature and setting of granitic plutons in the central and eastern parts of the Canadian Cordillera; Pacific Geology, v. 8, p. 109-138.

Gibson, D.W.
1977a: Upper Cretaceous and Tertiary coal-bearing strata in the Drumheller-Ardley region, Red Deer River valley, Alberta; Geological Survey of Canada, Paper 76-35, 41 p.
1977b: Sedimentary facies in the Jura-Cretaceous Kootenay Formation, Crowsnest Pass area, southwestern Alberta and southeastern British Columbia; Bulletin of Canadian Petroleum Geology, v. 25, no. 4, p. 767-791.
1979: The Morrissey and Mist Mountain formations - newly defined lithostratigraphic units of the Jura-Cretaceous Kootenay Group, Alberta and British Columbia; Bulletin of Canadian Petroleum Geology, v. 27, no. 2, p. 183-208.
1985: Stratigraphy, sedimentology and depositional environments of the coal-bearing Jurassic-Cretaceous Kootenay Group, Alberta and British Columbia; Geological Survey of Canada, Bulletin 357, 108 p.

Gies, R.M.
1985: Case history for a major Alberta Deep Basin Gas Trap; The Cadomin Formation; in Elmworth - Case study of a Deep Basin gas field, J.A. Masters (ed.); American Association of Petroleum Geologists, Memoir 38, p. 115-140.

Gill, J.R. and Cobban, W.A.
1973: Stratigraphy and geologic history of the Montana Group and equivalent rocks, Montana, Wyoming, and North and South Dakota; United States Geological Survey, Professional Paper 776, 37 p.

Given, M.M. and Wall, J.H.
1971: Microfauna from the Upper Cretaceous Bearpaw Formation of south-central Alberta; Bulletin of Canadian Petroleum Geology, v. 19, p. 504-546.

Glaister, P.
1959: Lower Cretaceous of southern Alberta and adjoining areas; American Association of Petroleum Geologists, Bulletin, v. 43, no. 3, p. 590-640.

Gordon, W.A.
1973: Marine life and ocean surface currents in the Cretaceous; Journal of Geology, v. 81, p. 269-284.

Gunther, P.R. and Hills, L.V.
1970: Heterospory in *Ariadnaesporites*; Pollen et Spores, v. 12, p. 123-130.

Gunther, P.R. and Hills, L.V.
1972: Megaspores and other palynomorphs of the Brazeau Formation (Upper Cretaceous), Nordegg area, Alberta; Geoscience and Man, v. 4, p. 29-48.

Habib, A.G.E.
1981: Geology of the Bearpaw Formation in south-central Alberta; M.Sc. thesis, University of Alberta, 103 p.

Hamblin, E.P. and Walker, R.G.
1979: Storm-dominated shallow marine deposits: the Fernie-Kootenay (Jurassic) transition, southern Rocky Mountains; Canadian Journal of Earth Sciences, v. 16, p. 1673-1690.

Hancock, J.M.
1975: The sequence of facies in the Upper Cretaceous of northern Europe compared with that in the Western Interior; in The Cretaceous System in the Western Interior of North America, W.G.E. Caldwell (ed.); Geological Association of Canada, Special Paper 13, p. 83-118.

Hancock, J.M. and Kauffman, E.G.
1979: The great transgressions of the Late Cretaceous; Journal of Geological Society of London, v. 136, p. 175-186.

Hayes, B.J.R.
1986: Stratigraphy of the basal Cretaceous Lower Mannville Formation, southern Alberta and north-central Montana; Bulletin of Canadian Petroleum Geology, v. 34, no. 1, p. 30-48.

Hector, James
1863: Geological report; in John Palliser, 1863, The Journals, Detailed Reports and Observations relative to the Exploration, by Captain Palliser, of that portion of British North America, which in latitude lies between the British Boundary Line and the Height of Land or Watershed of the Northern or Frozen Ocean respectively, and in longitude between the western shore of Lake Superior and the Pacific Ocean during the years 1857, 1858, and 1860; George Edward Eyre and William Spottiswoode, London, p. 216-245.

Hein, F.J., Dean, M.E., Delure, A.M., Grant, S.K., Robb, G.A., and Longstaffe, F.J.
1986: The Viking Formation in the Caroline, Garrington and Harmattan East Fields, western south-central Alberta: Sedimentology; Bulletin of Canadian Petroleum Geology, v. 34, no. 1, p. 91-110.

Herbaly, E.L.
1974: Petroleum geology of Sweetgrass Arch, Alberta; American Association of Petroleum Geologists, Bulletin, v. 58, p. 2227-2244.

Herngreen, G.F.W. and Khlonova, A.F.
1981: Cretaceous microfloral provinces; Pollen et Spores, v. 23, p. 441-555.

Hickey, L.J., West, R.M., Dawson, M.R., and Choi, D.K.
1983: Arctic terrestrial biota: paleomagnetic evidence of age disparity with mid-northern latitudes during the Late Cretaceous and early Tertiary; Science, v. 221, p. 1153-1156.

Hills, L.V. and Jensen, E.
1966: *Capulisportes longiprocessum* n. sp., a possible marker plant spore from the Belly River Formation (Campanian) of Alberta, Canada; Canadian Journal of Earth Sciences, v. 3, p. 413-417.

Hind, H.Y.
1859: Northwest Territory, Reports of Progress, together with a preliminary and general report on the Assiniboine and Saskatchewan Exploring Expedition; Journal of Legislative Assembly, Province of Canada, v. 19, Appendix 36, Lovel Printer, Toronto, 210 p.

Hoffman, G.L. and Jordan, G.R.
1984: Coalfields of the Northern Foothills of the Canadian Rocky Mountains; in The Mesozoic of Middle North America, D.F. Stott and D.J. Glass (ed.); Canadian Society of Petroleum Geologists, Memoir 9, p. 541-549.

Holmes, D.W. and Oliver, T.A.
1973: Source and depositional environments of the Moose Channel Formation, Northwest Territories; Bulletin of Canadian Petroleum Geology, v. 21, p. 435-478.

Holter, M.E. and Chu, M.
1978: Geology and coal resources of southeastern Alberta; Alberta Research Council, Open File Report 1978-13, 23 p.

Hopkins, J.C.
1981: Sedimentology of quartzose sandstones of Lower Mannville and associated units, Medicine River area, central Alberta; Bulletin of Canadian Petroleum Geology, v. 29, p. 12-41.

Hughes, J.D.
1984: Geology and depositional setting of the Late Cretaceous, Upper Bearpaw and lower Horseshoe Canyon formations in the Dodds-Round Hill coalfields of central Alberta - a computer-based study of closely spaced exploration data; Geological Survey of Canada, Bulletin 361, 81 p.

Hume, G.S.
1954: The lower Mackenzie river area, Northwest Territories and Yukon; Geological Survey of Canada, Memoir 273, 118 p.

Irish, E.J.W.
1965: Notes to accompany the geological map, Cypress Hills region, Alberta and Saskatchewan; in 15th Annual Field Conference Guidebook, Alberta Society of Petroleum Geologists, p. 15-22.
1970: The Edmonton Group of south-central Alberta; Bulletin of Canadian Petroleum Geology, v. 18, p. 125-155.

Irish, E.J.W. and Havard, C.J.
1968: The Whitemud and Battle formations ("Kneehills Tuff Zone") - a stratigraphy marker; Geological Survey of Canada, Paper 67-63, 51 p.

Irving, E.
1979: Paleopoles and paleolatitudes of North America and speculations about displaced terranes; Canadian Journal of Earth Sciences, v. 16, p. 669-694.

Jackson, P.C.
1985: Paleogeography of the Lower Cretaceous Mannville Group of western Canada; in Elmworth - Case Study of a Deep Basin Gas Field, J.A. Masters (ed.); American Association of Petroleum Geologists, Memoir 38, p. 49-78.

Jain, R.K.
1971: Pre-Tertiary records of Salviniaceae; American Journal of Botany, v. 58, p. 487-496.

Jansa, L.
1972: Depositional history of the coal-bearing Upper Jurassic-Lower Cretaceous Kootenay Formation, southern Rocky Mountains, Canada; Geological Society of America Bulletin, v. 83, p. 3199-3222.

Jarzen, D.M.
1978: Some Maastrichtian palynomorphs and their phytogeographical and paleoecological implications; Palynology, v. 2, p. 29-38.

Jarzen, D.M. and Norris, G.
1975: Evolutionary significance and botanical relationships of Cretaceous angiosperm pollen in the western Canadian interior; Geoscience and Man, v. 11, p. 47-60.

Jeletzky, J.A.
1958: Uppermost Jurassic and Cretaceous rocks of Aklavik Range, northeastern Richardson Mountains, Northwest Territories; Geological Survey of Canada, Paper 58-2, 84 p.
1960: Uppermost Jurassic and Cretaceous rocks, east flank of Richardson Mountains between Stony Creek and lower Donna River, Northwest Territories; Geological Survey of Canada, Paper 59-14, 31 p.
1961: Uppermost Jurassic and Lower Cretaceous rocks, west flank of Richardson Mountains between the headwaters of Blow River and Bell River, Yukon Territory; Geological Survey of Canada, Paper 61-9, 42 p.
1964: Illustrations of Canadian Fossils, Lower Cretaceous marine index fossils of western and Arctic Canada; Geological Survey of Canada, Paper 64-11, 101 p.
1965: Late Upper Jurassic and early Lower Cretaceous fossil zones of the Canadian western Cordillera, British Columbia; Geological Survey of Canada, Bulletin 103, 70 p.
1967: Jurassic and (?)Triassic rocks of the eastern slopes of Richardson Mountains, northwestern District of Mackenzie; Geological Survey of Canada, Paper 66-50, 171 p.
1968: Macrofossil zones of the marine Cretaceous of the Western Interior of Canada and their correlation with the zones and stages of Europe and Western Interior of the United States; Geological Survey of Canada, Paper 67-72, 66 p.

Jeletzky, J.A. (cont.)
1970: Cretaceous macrofaunas; in Geology and Economic Minerals of Canada, R.J.W. Douglas (ed.); Geological Survey of Canada, Economic Geology Report 1, p. 649-662.
1971a: Marine Cretaceous biotic provinces and paleogeography of western and Arctic Canada: illustrated by a detailed study of ammonites; Geological Survey of Canada, Paper 70-22, 92 p.
1971b: Marine Cretaceous biotic provinces of western and Arctic Canada; Proceedings of the North American Paleontological Convention, September, 1969, Pt. 1, p. 1638-1659, Allen Press, Inc., Kansas.
1971c: Stratigraphy, facies and paleogeography of Mesozoic rocks of northern and west-central Yukon; in Report of Activities Part A, Geological Survey of Canada, Paper 71-1A, p. 203-221.
1972: Stratigraphy, facies and paleogeography of Mesozoic and Tertiary rocks of northern Yukon and northwest District of Mackenzie, Northwest Territories (NTS-197B, 106M, 117A, 116O N1/2); in Report of Activities Part A, Geological Survey of Canada, Paper 72-1A, p. 212-215.
1973: Biochronology of the marine Boreal latest Jurassic, Berriasian and Valanginian in Canada; in The Boreal Lower Cretaceous, R. Casey and P.F. Rawson (eds.); Geological Journal, Special Issue No. 5, p. 41-80.
1974: Contribution to the Jurassic and Cretaceous geology of northern Yukon Territory and District of Mackenzie, Northwest Territories; Geological Survey of Canada, Paper 74-10, 23 p.
1975a: Jurassic and Lower Cretaceous paleogeography and depositional tectonics of Porcupine Plateau, adjacent areas of northern Yukon and those of Mackenzie District; Geological Survey of Canada, Paper 74-16, 52 p.
1975b: Sharp Mountain Formation (new): A shoreline facies of the Upper Aptian-Lower Albian Flysch Division, eastern Keele Range, Yukon Territory (NTS 117O); in Report of Activities Part B, Geological Survey of Canada, Paper 75-1B, p. 237-244.
1977: Causes of Cretaceous oscillations of sea level in western and Arctic Canada and some general geotectonic implications; in Mid-Cretaceous Events, K. Kanmera (ed.); Paleontological Society of Japan, Special Paper 21, p. 233-246.
1978: Causes of Cretaceous oscillations of sea level in western and Arctic Canada and some general geotectonic implications; Geological Survey of Canada, Paper 77-18, 44 p.
1980a: New or formerly poorly known, biochronologically and paleo-biogeographically important gastroplitinid and cleoniceratinid (Ammonitida) taxa from Middle Albian rocks of mid-western and Arctic Canada; Geological Survey of Canada, Paper 79-22, 63 p.
1980b: Lower Cretaceous and Jurassic rocks of McDougall Pass area and some adjacent areas of north-central Richardson Mountains, northern Yukon Territory and northwestern District of Mackenzie, Northwest Territories (NTS-116P/9 and 116P/10): A reappraisal; Geological Survey of Canada, Paper 78-22, 35 p.
1984: Jurassic-Cretaceous boundary beds of western and Arctic Canada and the problem of the Tithonian-Berriasian stages in the Boreal Realm; in Jurassic-Cretaceous Biochronology and Paleogeography of North America, G.E.G. Westermann (ed.); Geological Association of Canada, Special Paper 27, p. 175-255.

Jeletzky, J.A. and Stelck, C.R.
1981: *Pachygrycia*, a new *Sonneratia*-like ammonite from the Lower Cretaceous (earliest Albian?) of northern Canada; Geological Survey of Canada, Paper 80-20, 25 p.

Jeletzky, J.A. and Tipper, H.W.
1968: Upper Jurassic and Cretaceous rocks of Taseko Lakes map area and their bearing on the geological history of southwestern British Columbia; Geological Survey of Canada, Paper 67-54, 218 p.

Jerzykiewicz, T.
1985: Stratigraphy of the Saunders Group in the central Alberta Foothills - a progress report; in Current Research, Part B, Geological Survey of Canada, Paper 85-1B, p. 246-258.

Jerzykiewicz, T. and McLean, J.R.
1980: Lithostratigraphical and sedimentological framework of coal-bearing Upper Cretaceous and Lower Tertiary strata, Coal Valley area, central Alberta Foothills; Geological Survey of Canada, Paper 79-12, 47 p.

Jones, H.L.
1961: The Viking Formation in southwestern Saskatchewan; Saskatchewan Department of Mineral Resources, Report 65, 79 p.

Kauffman, E.G.
1965: Middle and Late Turonian oysters of the *Lopha lugubris* group; Smithsonian Miscellaneous Collections, v. 148, no. 6, 93 p.
1967: Coloradoan macroinvertebrate assemblages, central Western Interior United States; in Paleoenvironments of the Cretaceous Seaway in the Western Interior - A Symposium, E.G. Kauffman and H.E. Kent (ed.); Colorado School of Mines Publication, p. 67-143.

Kauffman, E.G. (cont.)
1969: Cretaceous marine cycles of the Western Interior; Mountain Geologist, v. 6, p. 227-245.
1973: Cretaceous Bivalvia; in Atlas of Paleobiogeography, A. Hallam (ed.); Elsevier Scientific Publishing Company, New York, p. 353-383.
1975: Dispersal and biostratigraphic potential of Cretaceous benthonic Bivalvia in the Western Interior; in The Cretaceous System in the Western Interior of North America, W.G.E. Caldwell (ed.); Geological Association of Canada, Special Paper 13, p. 163-194.
1977a: Evolutionary rates and biostratigraphy; in Concepts and Methods of Biostratigraphy, E.G. Kauffman and J.E. Hazel (ed.); Stroudsburg, PA., Dowden, Hutchinson and Ross, p. 109-140.
1977b: Geological and biological overview: Western Interior Cretaceous basin; Mountain Geologist, v. 14, p. 75-99.
1977c: Upper Cretaceous cyclothems, biotas and environments, Rock Canyon anticline, Pueblo, Colorado; Mountain Geologist, v. 14, p. 129-152.
1979: Cretaceous; in Treatise on Invertebrate Paleontology, Part A, Introduction; Geological Society of America and University of Kansas Press, p. A418-487.
1984a: The fabric of Cretaceous marine extinctions; in Catastrophes and Earth History, W.A. Berggren and J.A. van Couvering (ed.); Princeton University Press, Princeton, New Jersey, p. 151-246.
1984b: Paleobiogeography and evolutionary response dynamic in the Cretaceous Western Interior seaway of North America; in Jurassic-Cretaceous Biochronology and Biogeography of North America, G.E.G. Westermann (ed.); Geological Association of Canada, Special Paper 27, p. 273-306.
1985: Cretaceous evolution of the Western Interior Basin of the United States; in SEPM Field Trip Guidebook No. 4, 1985 Midyear Meeting, Golden, Colorado; Society of Economic Paleontologists and Mineralogists, p. iv-xiii.

Kauffman, E.G., Cobban, W.A., and Eicher, D.L.
1977: Albian through lower Coniacian strata and biostratigraphy, Western Interior United States: Proceedings, 2nd International Conference, Middle Cretaceous Working Group, International Geological Correlation Programme, Special Volume; Annales Musée Histoire Naturelle, Nice, France, v. 4, p. XXIII.1-XXIII.52.

Kilby, W.E.
1985: Tonstein and bentonite correlations in northeast British Columbia (93O, P, I; 94A); British Columbia Ministry of Energy, Mines and Petroleum Resources, Geological Fieldwork, 1984, Paper 1985-1, p. 257-277.

Koke, K.R. and Stelck, C.R.
1984: Foraminifera of the *Stelckiceras* Zone, basal Hasler Formation (Albian), northeastern British Columbia; in The Mesozoic of Middle North America, D.F. Stott and D.J. Glass (ed.); Canadian Society of Petroleum Geologists, Memoir 9, p. 271-279.
1985: Foraminifera of a Joli Fou Shale equivalent in the Lower Cretaceous (Albian) Hasler Formation, northeastern British Columbia; Canadian Journal of Earth Sciences, v. 22, p. 1299-1313.

Koster, E.H.
1984: Sedimentology of a foreland coastal plain: Upper Cretaceous Judith River Formation at Dinosaur Provincial Park; Canadian Society of Petroleum Geologists, Field Trip Guidebook, 115 p.

Koster, E.H. and Currie, P.J.
1987: Upper Cretaceous coastal plain sediments at Dinosaur Provincial Park, southeast Alberta; in Geological Society of America Centennial Field Guide Volume 2, Rocky Mountain Section, S.S. Beus (ed.); Geological Society of America, p. 9-14.

Kramers, J.W.
1982: Grand Rapids Formation, north-central Alberta: An example of nearshore sedimentation in a high energy, shallow, inland sea; American Association of Petroleum Geologists, Bulletin, v. 66, p. 589-590.

Kramers, J.W. and Mellon, G.B.
1972: Upper Cretaceous-Paleocene coal-bearing strata, northwest-central Alberta plains: Proceedings, First Geological Conference on Western Canadian Coal; Research Council of Alberta, Information Series, no. 60, p. 109-124.

Krause, F.F. and Nelson, D.A.
1984: Storm event sedimentation: Lithofacies association in the Cardium Formation, Pembina Oilfield area, west-central Alberta, Canada; in The Mesozoic of Middle North America, D.F. Stott and D.J. Glass (ed.); Canadian Society of Petroleum Geologists, Memoir 9, p. 485-511.

Landes, R.W.
1940: Paleontology of the marine formations of the Montana Group; in Geology of the Southern Alberta Plains, L.S. Russell and R.W. Landes (ed.); Geological Survey of Canada, Memoir 221, Pt. 2, p. 129-217.

Langenberg, C.W. and McMechan, M.E.
1984: Lower Cretaceous Luscar Group of the northern and north-central Foothills of Alberta; Bulletin of Canadian Petroleum Geology, v. 33, p. 1-11.

Langston, W., Jr.
1974: Nonmarine Comanchean tetrapods; Geoscience and Man, v. 8, p. 77-102.

Leckie, D.A.
1985: The Lower Cretaceous Notikewin Member (Fort St. John Group), northeastern British Columbia: A progradational barrier island system; Bulletin of Canadian Petroleum Geology, v. 33, no. 1, p. 39-51.
1986a: Petrology and tectonic significance of Gates Formation (Early Cretaceous) sediments in northeast British Columbia; Canadian Journal of Earth Sciences, v. 23, p. 129-141.
1986b: Tidally influenced, transgressive shelf sediments in the Viking Formation, Caroline, Alberta; Bulletin of Candian Petroleum Geology, v. 34, no. 1, p. 111-125.

Leckie, D.A. and Walker, R.G.
1982: Storm- and tide-dominated shorelines in Cretaceous Moosebar-Lower Gates interval - Outcrop equivalents of Deep Basin gas trap in western Canada; American Association of Petroleum Geologists, Bulletin, v. 66, p. 138-157.

Leffingwell, H.A.
1970: Palynology of the Lance (Late Cretaceous) and Fort Union (Paleocene) formations of the type Lance area (Wyoming); Geological Society of America, Special Paper 127, p. 1-64.

Lentin, J.K. and Williams, G.L.
1980: Dinoflagellate provincialism with emphasis on Campanian peridiniaceans; American Association of Stratigraphic Palynologists, Contribution Series, no. 7, 47 p.

Lerand, M.
1973: Beaufort Sea; in The Future Petroleum Provinces of Canada, R.G. McCrossan (ed.); Canadian Society of Petroleum Geologists, Memoir 1, p. 315-386.
1982a: Sedimentology of Upper Cretaceous fluviatile, deltaic and shoreline deposits: Trap Creek - Lundbreck area; American Association of Petroleum Geologists, Annual Convention, Calgary, 1982, Canadian Society of Petroleum Geologists, Field Trip Guidebook No. 8, 123 p.
1982b: Chungo (sandstone) Member, Wapiabi Formation at Mount Yamnuska, Alberta; in Clastic Units of the Front Ranges, Foothills and Plains between Field, British Columbia and Drumheller, Alberta, R.G. Walker (ed.); 11th International Congress on Sedimentology, Hamilton, 1982, Guidebook, Excursion 21A, p. 96-116.
1983: Sedimentology of the Chungo (sandstone) Member, Wapiabi Formation at Mount Yamnuska; in Sedimentology of Jurassic and Upper Cretaceous Marine and Nonmarine Sandstones, Bow Valley, The Mesozoic of Middle North America, M.M. Lerand, M.E. Wright, and A.P. Hamblin (ed.); Canadian Society of Petroleum Geology, Field Trip Guidebook No. 7, p. 39-76.

Lerbekmo, J.F.
1985: Magnetostratigraphic and biostratigraphic correlation of Maastrichtian to early Paleocene strata between south-central Alberta and southwestern Saskatchewan; Bulletin of Canadian Petroleum Geology, v. 33, no. 2, p. 213-226.

Lerbekmo, J.F. and Coulter, K.C.
1984: Magnetostratigraphic and biostratigraphic correlations of Late Cretaceous to Early Paleocene strata between Alberta and North Dakota; in The Mesozoic of Middle North America, D.F. Stott and D.J. Glass (ed.); Canadian Society of Petroleum Geologists, Memoir 9, p. 313-317.
1985a: Late Cretaceous to early Tertiary magnetostratigraphy of a continental sequence, Red Deer River valley, Alberta, Canada; Canadian Journal of Earth Sciences, v. 22, p. 567-583.
1985b: Magnetostratigraphic and lithostratigraphic correlation of coal seams and contiguous strata, upper Horseshoe Canyon and Scollard formations (Maastrichtian to Paleocene), Red Deer Valley, Alberta; Bulletin of Canadian Petroleum Geology, v. 33, no. 3, p. 295-305.

Lerbekmo, J.F. and St. Louis, R.M.
1986: The terminal Cretaceous iridium anomaly in the Red Deer Valley, Alberta, Canada; Canadian Journal of Earth Sciences, v. 23, p. 120-124.

Lerbekmo, J.F., Evans, M.E., and Baadsgaard, H.
1979a: Magnetostratigraphy, biostratigraphy and geochronology of Cretaceous-Tertiary boundary sediments, Red Deer valley; Nature, v. 279, no. 5708, p. 26-30.

Lerbekmo, J.F., Singh, C., Jarzen, D.M., and Russell, D.A.
1979b: The Cretaceous-Tertiary boundary in south-central Alberta - a revision based on additional dinosaurian and microfloral evidence; Canadian Journal of Earth Sciences, v. 16, p. 1866-1869.

Lillegraven, J.A., Kielan-Jaworowska, Z., and Clemens, W.A.
1979: Mesozoic Mammals; The First Two-Thirds of Mammalian History; University of California Press, Berkeley, California, 311 p.
Link, T.A. and Childerhose, A.J.
1931: Bearpaw Shale and contiguous formations in Lethbridge area, Alberta; in Stratigraphy of Plains of Southern Alberta, Donaldson Bogart Dowling Memorial Symposium; American Association of Petroleum Geologists, p. 99-114.
Lloyd, C.R.
1982: The mid-Cretaceous Earth: paleogeography; ocean circulation and temperature; atmospheric circulation; Journal of Geology, v. 90, p. 393-413.
Loranger, D.M.
1951: Useful Blairmore microfossil zone in central and southern Alberta, Canada; in Western Canada Sedimentary Basin, Rutherford Memorial Volume, L.M. Clark (ed.); American Association of Petroleum Geologists, Bulletin, v. 35, p. 2348-2367. Reprinted 1954, American Association of Petroleum Geologists, p. 182-203.
Macauley, G., Snowdon, L.R., and Ball, F.D.
1985: Geochemistry and geological factors governing exploitation of selected Canadian oil shale deposits; Geological Survey of Canada, Paper 85-13, 65 p.
Mackenzie, A.
1801: Voyages from Montreal on the River St. Lawrence, through the continent of North America, to the Frozen and Pacific Oceans in the years 1789 and 1793; London, 412 p. (Facsimile published 1981, M.G. Hurtig Ltd., Edmonton, Alberta.)
Masters, J.A.
1985: Lower Cretaceous oil and gas in western Canada; in Elmworth - Case Study of a Deep Basin Gas Field, J.A. Masters (ed.); American Association of Petroleum Geologists, Memoir 38, p. 1-34.
Mathews, W.H.
1947: Geology and coal resources of the Carbon Creek, Mount Bickford map area; British Columbia Department of Mines, Bulletin 24, 27 p.
Maycock, I.D.
1967: Mannville Group and associated Lower Cretaceous clastic rocks in southwestern Saskatchewan; Saskatchewan Department of Mineral Resources, Report No. 96, 108 p.
McCrory, V.L.C. and Walker, R.G.
1986: A storm and tidally influenced prograding shoreline - Upper Cretaceous Milk River Formation of southern Alberta, Canada; Sedimentology, v. 33, p. 47-60.
McGookey, D.P., Haun, J.D., Hale, L.A., Goodell, H.G., McCubbin, D.G., Weimer, R.J., and Wulf, G.R.
1972: Cretaceous system; in Geologic Atlas of the Rocky Mountain Region, W.W. Mallory (ed.); Rocky Mountain Association of Geologists, p. 190-228.
McIntyre, D.J.
1974: Palynology of an Upper Cretaceous section, Horton River, District of Mackenzie, Northwest Territories; Geological Survey of Canada, Paper 74-14, 56 p.
McIntyre, D.J. and Sweet, A.R.
1986: Late Turonian marine and nonmarine palynomorphs from the Cardium Formation, north-central Alberta Foothills, Canada (abstract); Proceedings of the American Association of Stratigraphic Palynologists, El Paso, Texas, 254 p.
McLaren, D.J.
1983: Bolides and biostratigraphy; Geological Society of America Bulletin, v. 94, p. 313-324.
McLean, J.R.
1971: Stratigraphy of the Upper Cretaceous Judith River Formation in the Canadian Great Plains; Saskatchewan Research Council, Geology Divison Report No. 11, 96 p.
1977a: Lithostratigraphic nomenclature of the upper Cretaceous Judith River Formation in southern Alberta: philosophy and practice; Bulletin of Canadian Petroleum Geology, v. 25, p. 1105-1114.
1977b: The Cadomin Formation; stratigraphy, sedimentology, and tectonic implications; Bulletin of Canadian Petroleum Geology, v. 25, p. 792-827.
1982: Lithostratigraphy of the Lower Cretaceous coal-bearing sequence, Foothills of Alberta; Geological Survey of Canada, Paper 80-29, 46 p.
McLean, J.R. and Wall, J.H.
1981: Early Cretaceous Moosebar Sea in Alberta; Bulletin of Canadian Petroleum Geology, v. 29, p. 334-377.
McLearn, F.H.
1933: The ammonoid genera *Gastroplites* and *Neogastroplites*; Royal Society of Canada, Transactions, 3rd series, v. 27, sec. 4, p. 13-26.
1937: The fossil zones of the Upper Cretaceous Alberta shale; Royal Society of Canada, Transactions, 3rd series, v. 31, sec. 4, p. 111-120.
1944: Revision of the Lower Cretaceous of the Western Interior of Canada; Geological Survey of Canada, Paper 44-17, 2nd edition, 1945, 14 p.
McLearn, F.H. and Kindle, E.D.
1950: Geology of northeastern British Columbia; Geological Survey of Canada, Memoir 259, 239 p.
McNeil, D.H.
1984: The eastern facies of the Cretaceous System in the Canadian Western Interior; in The Mesozoic of Middle North America, D.F. Stott and J.D. Glass (ed.); Canadian Society of Petroleum Geologists, Memoir 9, p. 145-171.
McNeil, D.H. and Caldwell, W.G.E.
1981: Cretaceous rocks and their foraminifera in the Manitoba Escarpment; Geological Association of Canada, Special Paper 21, 439 p.
Meijer Drees, N.C. and Myhr, D.W.
1981: The Upper Cretaceous Milk River and Lea Park formations in southeastern Alberta; Bulletin of Canadian Petroleum Geology, v. 29, p. 42-74.
Mellon, G.B.
1962: Petrology of Upper Cretaceous oolitic iron-rich rocks from northern Alberta; Economic Geology, v. 57, p. 921-940.
1964: Discriminatory analysis of calcite- and silicate-cemented phases of the Mountain Park Sandstone; Journal of Geology, v. 72, no. 6, p. 786-809.
1967: Stratigraphy and petrology of the Lower Cretaceous Blairmore and Mannville groups, Alberta Foothills and Plains; Alberta Research Council, Bulletin 21, 270 p.
Mellon, G.B. and Wall, J.H.
1963: Correlation of the Blairmore Group and equivalent strata; Bulletin of Canadian Petroleum Geology, v. 11, no. 4, p. 396-409.
Merrill, W.M. and Buchwald, C.E.
1965: Gross sedimentary facies in uppermost Cretaceous and lower Tertiary sediments, west-central Alberta; American Association of Petroleum Geologists, Bulletin, v. 49, no. 3, p. 350.
Meyboom, P.
1960: Geology and groundwater resources of the Milk River Sandstone in southern Alberta; Research Council of Alberta, Memoir 2, 89 p.
Miall, A.D.
1973: Regional geology of northern Yukon; Bulletin of Canadian Petroleum Geology, v. 21, p. 81-116.
Misko, R.M. and Hendry, H.E.
1979: The petrology of sands in the uppermost Cretaceous and Palaeocene of southern Saskatchewan: a study of composition influenced by grain size, source area and tectonics; Canadian Journal of Earth Sciences, v. 16, p. 38-49.
Molnar, R.
1980: Australian late Mesozoic terrestrial tetrapods: some implications; Memoires de la Societe Geologique de France, no. 139, p. 131-143.
Monger, J.W.H. and Price, R.A.
1979: Geodynamic evolution of the Canadian Cordillera - progress and problems; Canadian Journal of Earth Sciences, v. 16, p. 770-791.
Monger, J.W.H., Price, R.A., and Tempelman-Kluit, D.J.
1982: Tectonic accretion and the origin of the two major metamorphic and plutonic welts in the Canadian Cordillera; Geology, v. 10, no. 2, p. 70-75.
Moshier, S.O. and Waples, D.W.
1985: Quantitative evaluation of the Lower Cretaceous Mannville Group as the source rock for Alberta's Oil Sands; American Association of Petroleum Geologists, Bulletin, v. 69, p. 161-172.
Mossop, G.D. and Flach, P.D.
1983: Deep channel sedimentation in the Lower Cretaceous McMurray Formation, Athabasca Oil Sands, Alberta; Sedimentology, v. 30, p. 493-510.
Mountjoy, E.W.
1962: Mount Robson (southeast) map area, Rocky Mountains of Alberta and British Columbia; Geological Survey of Canada, Paper 61-31, 114 p.
1967: Upper Cretaceous and Tertiary stratigraphy, northern Yukon and northwestern District of Mackenzie; Geological Survey of Canada, Paper 66-16, 70 p.
Mountjoy, E.W. and Chamney, T.P.
1969: Lower Cretaceous (Albian) of the Yukon: Stratigraphy and foraminiferal subdivisions, Snake and Peel Rivers; Geological Survey of Canada, Paper 68-26, 71 p.
Myhr, D.W. and Meijer Drees, N.C.
1976: Geology of the southeastern Alberta Milk River gas pool; in The Sedimentology of Selected Clastic Oil and Gas Reservoirs in Alberta, M.M. Lerand (ed.); Canadian Society of Petroleum Geologists, p. 96-125.
Myhr, D.W. and Young, F.G.
1975: Lower Cretaceous (Neocomian) sandstone sequence of Mackenzie Delta and Richardson Mountains area; in Report of Activities Part C, Geological Survey of Canada, Paper 75-1C, p. 247-266.

Nauss, A.W.
1945: Cretaceous stratigraphy of Vermilion area, Alberta, Canada; American Association of Petroleum Geologists, Bulletin, v. 29, p. 1605-1629.

Newmarch, C.B.
1953: Geology of the Crowsnest Coal Basin with special reference to the Fernie area; British Columbia Department of Mines, Bulletin No. 33, 107 p.

Nichols, D.J.
1973: North American and European species of Momipites ("*Engelhardtia*") and related genera; Geoscience and Man, v. 7, p. 103-117.

Nichols, D.J. and Ott, H.L.
1978: Biostratigraphy and evolution of the *Momipites-Caryapollenites* lineage in the early Tertiary in the Wind River Basin, Wyoming; Palynology, v. 2, p. 94-112.

Nichols, D.J, Jacobson, S.R., and Tschudy, R.H.
1982: Cretaceous palynomorph biozones for the central and northern Rocky Mountains region of the United States; in Geological Studies of the Cordilleran Thrust Belt, R.B. Powers (ed.); Rocky Mountain Association of Geologists (2 v.), p. 721-733.

Nichols, D.J., Jarzen, D.M., Orth, C.J., and Oliver, P.Q.
1986: Palynological and iridium anomalies at the Cretacous-Tertiary boundary, south-central Saskatchewan; Science, v. 231, p. 714-717.

Nicols, R.A.H. and Wyman, J.M.
1969: Interdigitation versus arbitrary cutoff: Resolution of an Upper Cretaceous stratigraphic problem, Western Saskatchewan; American Association of Petroleum Geologists, Bulletin, v. 53, no. 9, p. 1880-1893.

Nielsen, A.R. and Porter, J.
1984: Pembina - in retrospect; in The Mesozoic of Middle North America, D.F. Stott and D.J. Glass (ed.); Canadian Society of Petroleum Geologists, Memoir 9, p. 1-14.

Norris, D.K.
1963: Operation Porcupine; in Summary of Research: Field, 1962, Geological Survey of Canada, Paper 63-1, p. 17-19.
1964: The Lower Cretaceous of the southeastern Canadian Cordillera; Bulletin of Canadian Petroleum Geology, v. 12, p. 512-535.
1971: The geology and coal potential of the Cascade coal basin; in A guide to the Geology of the Eastern Cordillera along the Trans Canada Highway between Calgary, Alberta and Revelstoke, British Columbia, I.A.R. Halladay and D.H. Mathewson (ed.); Alberta Society of Petroleum Geologists, Calgary, p. 25-39.
1983: Operation Porcupine Project area; Geological Survey of Canada, Geotectonic Correlation Chart 1532A.

Norris, D.K. and Hopkins, W.S., Jr.
1977: The geology of the Bonnet Plume Basin, Yukon Territory; Geological Survey of Canada, Paper 76-8, 20 p.

Norris, D.K., Stevens, R.D., and Wanless, R.K.
1965: K-Ar age of igneous pebbles in the McDougall-Segur conglomerate, southeastern Canadian Cordillera; Geological Survey of Canada, Paper 65-26; 11 p.

Norris, G.
1967: Spores and pollen from the lower Colorado Group (Albian-?Cenomanian) of central Alberta; Palaeontographica, Band 120, Abteilung B, Lieferung 1-4, p. 72-115.

Norris, G., Jarzen, D.M., and Awai-thorne, B.V.
1975: Evolution of the Cretaceous terrestrial palynoflora in western Canada; Geological Association of Canada, Special Paper 13, p. 333-364.

Norris, G. and Miall, A.
1984: Arctic terrestrial biota: paleomagnetic evidence of age disparity with mid-northern latitudes during the Late Cretaceous and Early Tertiary (Discussion); Science, v. 224, p. 174-175.

North, B.R. and Caldwell, W.G.E.
1964: Foraminifera from the Cretaceous Lea Park Formation in south-central Saskatchewan; Saskatchewan Research Council, Geology Division, Report 5, 43 p.
1970: Foraminifera from the Late Cretaceous Bearpaw Formation in the South Saskatchewan River valley; Saskatchewan Research Council, Geology Division, Report 9, 117 p.
1975: Foraminiferal faunas in the Cretaceous System of Saskatchewan; in The Cretaceous System in the Western Interior of North America, W.G.E. Caldwell (ed.); Geological Association of Canada, Special Paper 13, p. 303-331.

North American Commission on Stratigraphic Nomenclature
1983: North American Stratigraphic Code; American Association of Petroleum Geologists, Bulletin, v. 67, p. 841-875.

Nurkowski, J.R. and Rahmani, R.A.
1985: An Upper Cretaceous fluvio-lacustrine coal-bearing sequence, Red Deer area, Alberta, Canada; in Sedimentology of Coal and Coal-bearing Sequences, R.A. Rahmani and R.M. Flores (ed.); International Association of Sedimentologists, Special Publication 7, p. 163-176.

Obradovich, J.D. and Cobban, W.A.
1975: A time-scale for the Late Cretaceous of the Western Interior of North America; in The Cretaceous System of the Western Interior of North America, W.G.E. Caldwell (ed.); Geological Association of Canada, Special Paper 13, p. 31-54.

O'Connell, S.C.
1984: Mid-Mannville depositional systems and reservoir geology in the Chauvin and Chauvin-South Fields, Lloydminster heavy oil trend, Alberta (abstract); Canadian Society of Petroleum Geologists, National Convention, Program and Abstracts, p. 43.

Ogunyomi, O. and Hills, L.V.
1977: Depositional environments, Foremost Formation (Late Cretaceous), Milk River area, southern Alberta; Bulletin of Canadian Petroleum Geology, v. 25, p. 929-968.

Ostrom, J.H.
1970: Stratigraphy and paleontology of the Cloverly Formation (Lower Cretacous) of the Bighorn Basin area, Wyoming and Montana; Peabody Museum of Natural History, Yale University, Bulletin, v. 35, 234 p.

Parrish, J.T. and Curtis, R.L.
1982: Atmospheric circulation, upwelling, and organic-rich rocks in the Mesozoic and Cenozoic Eras; Palaeogeography, Palaeoclimatology, Palaeoecology, v. 40, p. 31-66.

Parrish, J.T., Gaynor, G.C., and Swift, D.J.P.
1984: Circulation in the Cretaceous Western Interior Seaway of North America, a review; in The Mesozoic of Middle North America, D.F. Stott and D.J. Glass (ed.); Canadian Society of Petroleum Geologists, Memoir 9, p. 221-231.

Pemberton, S.G. and Frey, R.W.
1984: Ichnology of storm-influenced shallow marine sequence: Cardium Formation (Upper Cretaceous) at Seebe, Alberta; in The Mesozoic of Middle North America, D.F. Stott and D.J. Glass (ed.); Canadian Society of Petroleum Geology, Memoir 9, p. 281-304.

Pessagno, E.A.
1967: Upper Cretaceous planktonic foraminifera from the western Gulf Coastal Plain; Paleontographica Americana, v. 5, no. 37, p. 245-445.

Playford, G.
1971: Palynology of basal Cretaceous (Swan River) strata of Saskatchewan and Manitoba; Palaeontology, v. 14, Pt. 4, p. 533-565.

Plint, A.G., Walker, R.G., and Bergman, K.M.
1986: Cardium Formation 6; stratigraphic framework of the Cardium in subsurface; Bulletin of Canadian Petroleum Geology, v. 34, p. 213-225.

Pocock, S.A.J.
1962: Microfloral analysis and age determination of strata at the Jurassic-Cretaceous boundary in the western Canada plains; Palaeontographica, section B, v. 111, p. 1-95.
1964: Palynology of the Kootenay Formation at its type section; Bulletin of Canadian Petroleum Geology, Special Guidebook Issue, Flathead valley, v. 12, p. 501-512.
1976: A preliminary dinoflagellate zonation of the uppermost Jurassic and lower part of the Cretaceous, Canadian Arctic, and possible correlation in the western Canada basin; Geoscience and Man, v. XV, p. 101-114.

Porter, J.W., Price, R.A., and McCrossan, R.G.
1982: The Western Canada Sedimentary Basin; Philosophical Transactions of the Royal Society of London, Series A, v. 305, p. 169-192.

Price, R.A.
1965: Flathead map area, British Columbia and Alberta; Geological Survey of Canada, Memoir 336, 221 p.
1973: Large-scale gravitational flow of supracrustal rocks, southern Canadian Rockies; in Gravity and Tectonics, K.A. de Jong and R. Scholten (ed.); John Wiley and Sons, New York, p. 491-592.
1981: The Cordilleran foreland thrust and fold belt in the southern Canadian Rocky Mountains; in Thrust and nappe tectonics, K. McClay and N.J. Price (eds.); Geological Society of London, Special Paper No. 9, p. 427-448.

Price, R.A. and Mountjoy, E.
1970: Geologic structure of the Canadian Rocky Mountains between Bow and Athabasca Rivers - a progress report; Geological Association of Canada, Special Paper No. 6, p. 7-25.

Procter, R.M., Taylor, G.C., and Wade, J.A.
1984: Oil and natural gas resources of Canada 1983; Geological Survey of Canada, Paper 83-31, 59 p.

Putnam, P.E.
1982: Aspects of the petroleum geology of the Lloydminster heavy oil fields, Alberta and Saskatchewan; Bulletin of Canadian Petroleum Geology, v. 30, p. 81-111.

Rahmani, R.A. and Lerbekmo, J.F.
1975: Heavy-mineral analysis of Upper Cretaceous and Paleocene sandstones in Alberta and adjacent areas of Saskatchewan; in The Cretaceous System in the Western Interior of North America, W.G.E. Caldwell (ed.); Geological Association of Canada, Special Paper 13, p. 607-632.

Rapson, J.E.
1964: Lithology and petrography of transitional Jurassic-Cretaceous clastic rocks, southern Rocky Mountains; Bulletin of Canadian Petroleum Geology, Special Guidebook Issue, Flathead valley, v. 12, p. 556-586.
1965: Petrography and derivation of Jurassic-Cretaceous clastic rocks, southern Rocky Mountains, Canada; American Association of Petroleum Geologists, Bulletin, v. 49, p. 1426-1452.

Reinson, G.E.
1984: Barrier Island and associated strand-plain system; in Facies Models, 2nd edition, R.G. Walker (ed.); Geoscience Canada, Reprint Series 1, p. 119-140.

Reinson, G.E. and Foscolos, A.E.
1986: Trends in sandstone diagenesis with depth of burial, Viking Formation, southern Alberta; Bulletin of Canadian Petroleum Geology, v. 34, no. 1, p. 126-152.

Riccardi, A.C.
1983: Scaphitids from the upper Campanian-lower Maastrichtian Bearpaw Formation of the Western Interior of Canada; Geological Survey of Canada, Bulletin 354, 103 p.

Ricketts, B.D.
1988: The Monster Formation: A coastal fan system of Late Cretaceous age, Yukon; Geological Survey of Canada, Paper 86-14, 27 p.

Rosenthal, L.
1982: Depositional environments and paleogeography of the Upper Cretaceous Wapiabi and Belly River formations in southwestern Alberta (abstract); 11th International Congress on Sedimentology, Hamilton 1982, Abstracts of Papers, p. 159.

Rouse, G.E.
1959: Plant microfossils from Kootenay coal measures strata of British Columbia; Micropaleontology, v. 5, p. 303-324.

Rudkin, R.A.
1964: Lower Cretaceous; Chapter 11, in Geological History of Western Canada, R.G. McCrossan and R.P. Glaister (ed.); Alberta Society of Petroleum Geologists, Calgary, Alberta, p. 156-168.

Russell, D.A.
1967a: Cretaceous vertebrates from the Anderson River, Northwest Territories; Canadian Journal of Earth Sciences, v. 4, p. 21-38.
1967b: Systematics and morphology of American mosasaurs; Peabody Museum of Natural History, Yale University, Bulletin 23, 241 p.
1983a: A Canadian dinosaur park; Terra 21, p. 3-9.
1983b: Exponential evolution: implications for intelligent extraterrestrial life; Advances in Space Research, v. 3, p. 95-103.
1984a: The gradual decline of the dinosaurs: fact or fallacy?; Nature, v. 307, p. 360-361.
1984b: A check list of the families and genera of North American dinosaurs; Syllogeus 53, 35 p.

Russell, D.A. and Chamney, T.P.
1967: Notes of the biostratigraphy of dinosaurian and microfossil faunas in the Edmonton Formation (Cretaceous), Alberta; National Museum of Canada, Natural History Paper 35, 22 p.

Russell, D.A. and Singh, C.
1978: The Cretaceous-Tertiary boundary in south-central Alberta - a reappraisal based on dinosaurian and microfloral extinctions; Canadian Journal of Earth Sciences, v. 15, p. 284-292.

Russell, L.S.
1975: Mammalian faunal succession in the Cretaceous System of western North America; in The Cretaceous System in the Western Interior of North America, W.G.E. Caldwell (ed.); Geological Association of Canada, Special Paper 13, p. 137-161.
1983: Evidence for an unconformity at the Scollard-Battle contact, Upper Cretaceous strata, Alberta; Canadian Journal of Earth Sciences, v. 20, p. 1219-1231.

Samolovich, S.R.
1977: A new outline of the floristic zoning of the Northern Hemisphere in the Late Senonian; Paleontological Journal, v. 11, no. 3, p. 366-371. (English translation from Russian).

Savage, D.E. and Russell, D.A.
1983: Mammalian Paleofaunas of the World; London, Addison-Wesley, 432 p.

Shafiqullah, M., Folinsbee, R.E., Baadsgaard, H., Cumming, G.L., and Lerbekmo, J.F.
1964: Geochronology of Cretaceous-Tertiary boundary, Alberta, Canada; Proceedings of the 22nd International Geological Congress, India, Part 3, Section 3, p. 1-20.

Shaw, E.W. and Harding, S.R.L.
1949: Lea Park and Belly River formations of east-central Alberta; Bulletin of American Association of Petroleum Geologists, v. 33, p. 487-499.

Shouldice, J.R.
1979: Nature and potential of Belly River gas sand traps and reservoirs in western Canada; Bulletin of Canadian Petroleum Geology, v. 27, p. 229-241.

Simpson, F.
1975: Marine lithofacies and biofacies of the Colorado Group (middle Albian to Santonian) in Saskatchewan; in The Cretaceous System in the Western Interior of North America, W.G.E. Caldwell (ed.); Geological Association of Canada, Special Paper 13, p. 553-587.
1979: Low-permeability gas reservoirs in marine, Cretaceous sandstones of Saskatchewan: 1. Project outline and rationale; in Summary of Investigations, 1979, J.E. Christopher and R. MacDonald (ed.); Saskatchewan Geological Survey, Saskatchewan Mineral Resources, Miscellaneous Report 79-10, p. 174-180.
1982: Sedimentology, palaeoecology and economic geology of Lower Colorado (Cretaceous) strata, west-central Saskatchewan; Saskatchewan Energy and Mines, Report No. 150, 183 p.
1984: Hydrocarbons potential of low-permeability marine Cretaceous sandstones of Saskatchewan; in The Mesozoic of Middle North America, D.F. Stott and D.J. Glass (eds.); Canadian Society of Petroleum Geologists, Memoir 9, p. 513-532.

Singh, C.
1964: Microflora of the Lower Cretaceous Mannville Group, east-central Alberta; Alberta Research Council, Bulletin 15, 239 p.
1971: Lower Cretaceous microfloras of the Peace River area, northwestern Alberta; Alberta Research Council, Bulletin 28, 542 p.
1975: Stratigraphic significance of early angiosperm pollen in the mid-Cretaceous strata of Alberta; in The Cretaceous System in the Western Interior of North America, W.G.E. Caldwell (ed.); Geological Association of Canada, Special Paper, no. 13, p. 364-389.
1983: Cenomanian microfloras of the Peace River area, northwestern Alberta; Alberta Research Council, Bulletin 44, 322 p.

Sloan, R.E.
1970: Cretaceous and Paleocene terrestrial communities of western North America; Proceedings of the First North American Paleontological Convention, Part E, p. 427-453.

Sloss, L.L.
1963: Sequences in the cratonic interior of North America; Bulletin of the Geological Society of America, v. 74, p. 93-113.

Snead, R.G.
1969: Microfloral diagnosis of the Cretaceous-Tertiary boundary, central Alberta; Research Council of Alberta, Bulletin 25, 148 p.

Speelman, J.D. and Hills, L.V.
1980: Megaspore paleoecology: Pakowki, Foremost and Oldman formations (Upper Cretaceous), southeastern Alberta; Bulletin of Canadian Petroleum Geology, v. 28, p. 522-541.

Springer, G.C., MacDonald, W.D., and Crockford, M.B.B.
1964: Jurassic; Chapter 10, in Geological History of Western Canada, R.G. McCrossan and R.G. Glaister (ed.); Alberta Society of Petroleum Geologists, p. 137-155.

Srivastava, S.K.
1968: *Azolla* from the Upper Cretaceous Edmonton Formation, Alberta, Canada; Canadian Journal of Earth Sciences, v. 5, p. 915-919.
1970: Pollen biostratigraphy and paleoecology of the Edmonton Formation (Maastrichtian), Alberta, Canada; Palaeogeography, Palaeoclimatology, Palaeoecology, v. 7, p. 221-276.
1971: Systematic revision of the genus *Styx* Norton et Hall, 1967; Review of Palaeobotany and Palynology, v. 11, p. 297-309.
1972: Paleoecology of pollen genera *Aquilapollenites* and *Mancicorpus* in Maastrichtian deposits of North America; Proceedings, 24th Session, International Geological Congress, Montreal, 1972, v. 7 (Paleontology), p. 111-120.
1978: The Cretaceous megaspore genus *Ghoshispora*; Palaeontographica, Abteilung B, v. 167, p. 175-184.
1981: Evolution of Upper Cretaceous phytogeoprovinces and their pollen flora; Review of Palaeobotany and Palynology, v. 35, p. 155-173.

Srivastava, S.K. and Binda, P.L.
1969: Megaspores of the genus *Balmeisporites* from the Upper Cretaceous of Alberta and Saskatchewan, Canada; Revue de Micropaléontogie, v. 11, no. 4, p. 205-209.

Stelck, C.R.
1955: Cardium Formation of the Foothills of northeastern British Columbia; Canadian Mining and Metallurgical Bulletin, v. XLVIII, p. 266-273. Also Transactions, v. LVIII, 1955, p. 132-139.
1958: Stratigraphic position of the Viking Sand; Journal of the Alberta Society of Petroleum Geologists, v. 6, no. 1, p. 2-7.

Stelck, C.R. (cont.)
1975a: The Upper Albian *Miliammina manitobensis* Zone in northeastern British Columbia; in The Cretaceous System in the Western Interior of North America, W.G.E. Caldwell (ed.); The Geological Association of Canada, Special Paper 13, p. 251-275.
1975b: Basement control of Cretaceous sand sequences in western Canada; in The Cretaceous System in the Western Interior of North America, W.G.E. Caldwell (ed.); Geological Association of Canada, Special Paper 13, p. 427-440.
Stelck, C.R. and Hedinger, A.S.
1983: Foraminifera of the lower part of the Sully Formation (Upper Albian), northeastern British Columbia; Canadian Journal of Earth Sciences, v. 20, p. 1248-1259.
Stelck, C.R. and Kramers, J.W.
1980: *Freboldiceras* from the Grand Rapids Formation of north-central Alberta; Bulletin of Canadian Petroleum Geology, v. 28, p. 509-521.
Stelck, C.R. and Wall, J.H.
1954: Kaskapau Foraminifera from Peace River area of western Canada; Research Council of Alberta, Report 68, 38 p.
1955: Foraminifera of the Cenomanian *Dunveganoceras* Zone from Peace River area of western Canada; Alberta Research Council, Report 70, p. 1-62.
Stelck, C.R. and Wetter, R.E.
1958: Lower Cenomanian foraminifera from Peace River area, western Canada; Alberta Research Council, Bulletin 2, Pt. I, p. 1-35.
Stelck, C.R., Bahan, W.G., and Martin, L.J.
1956: Middle Albian foraminifera from Athabasca and Peace River drainage areas of western Canada; Alberta Research Council, Report 75, 60 p.
Stott, D.F.
1960: Cretaceous rocks in the region of Liard and Mackenzie Rivers, Northwest Territories; Geological Survey of Canada, Bulletin 63, 36 p.
1963: The Cretaceous Alberta Group and equivalent rocks, Rocky Mountain Foothills, Alberta; Geological Survey of Canada, Memoir 317, 306 p.
1967a: The Fernie and Minnes strata north of Peace River, Foothills of northeastern British Columbia; Geological Survey of Canada, Paper 67-19, Part A, 58 p.
1967b: The Cretaceous Smoky Group, Rocky Mountain Foothills, Alberta and British Columbia; Geological Survey of Canada, Bulletin 132, 133 p.
1968: Lower Cretaceous Bullhead and Fort St. John Groups between Smoky and Peace Rivers, Rocky Mountain Foothills, Alberta and British Columbia; Geological Survey of Canada, Bulletin 152, 279 p.
1969: Fernie and Minnes strata north of Peace River Foothills of northeastern British Columbia; Geological Survey of Canada, Paper 67-19, Part B, 132 p.
1973: Lower Cretaceous Bullhead Group between Bullmoose Mountain and Tetsa River, Rocky Mountain Foothills, northeastern British Columbia; Geological Survey of Canada, Bulletin 219, 228 p.
1975: The Cretaceous System in northeastern British Columbia; in The Cretaceous System in the Western Interior of North America, W.G.E. Caldwell (ed.); Geological Association of Canada, Special Paper 13, p. 441-467.
1981: Bickford and Gorman Creek, two new formations of the Jurassic-Cretaceous Minnes Group, Alberta and British Columbia; in Current Research, Part B, Geological Survey of Canada, Paper 81-1B, p. 1-9.
1982: Lower Cretaceous Fort St. John Group and Upper Cretaceous Dunvegan Formation of the Foothills and Plains of Alberta, British Columbia, District of Mackenzie and Yukon Territory; Geological Survey of Canada, Bulletin 328, 124 p.
1984: Cretaceous sequences of the foothills of the Canadian Rocky Mountains; in The Mesozoic of Middle North America, D.F. Stott and D.J. Glass (ed.); Canadian Society of Petroleum Geologists, Memoir 9, p. 85-107.
Sutherland, C.D. and Stelck, C.R.
1972: Foraminifera from the Cretaceous *Neogastroplites* Zone, Moberly Lake, British Columbia; Bulletin of Canadian Petroleum Geology, v. 20, no. 3, p. 549-582.
Sweet, A.R.
1978a: Jurassic and Cretaceous megaspores; American Association of Stratigraphic Palynologists, Contribution Series No. 5B, p. 1-30.
1978b: Palynology of the Ravenscrag and Frenchman formations; in Coal Resources of Southern Saskatchewan: A Model for Evaluation Methodology, S.H. Whitaker, N.A. Irvine, and P.L. Braughton (ed.); Geological Survey of Canada, Economic Geology Report 30, p. 29-39.
Sweet, A.R. and Hills, L.V.
1974: A detailed study of the genus *Azollopsis*; Canadian Journal of Botany, v. 52, p. 1625-1642.
Sweet, A.R. and Hills, L.V. (cont.)
1976: Early Tertiary species of *Azolla* subg. *Azolla* sect. *Kremastospora* from western and Arctic Canada; Canadian Journal of Botany, v. 54, p. 334-351.
1984: A palynological and sedimentological analysis of the Cretaceous-Tertiary boundary, Red Deer River Valley, Alberta, Canada; Sixth International Palynological Conference, Calgary, Abstracts, p. 160-161.
Tassonyi, E.J.
1969: Subsurface geology, lower Mackenzie River and Anderson River area, District of Mackenzie; Geological Survey of Canada, Paper 68-25, 207 p.
Taylor, D.R. and Walker, R.G.
1984: Depositional environments and paleogeography in the Albian Moosebar Formation and adjacent fluvial Gladstone and Beaver Mines formations, Alberta; Canadian Journal of Earth Sciences, v. 21, p. 698-714.
Tempelman-Kluit, D.J.
1979: Transported cataclasite, ophiolite and granodiorite in Yukon: evidence of arc-continent collision; Geological Survey of Canada, Paper 79-14, 27 p.
Tipper, H.W., Campbell, R.B., Taylor, G.C., and Stott, D.F.
1974: Parsnip River, British Columbia; Geological Survey of Canada, Map 1424A, 1:250 000.
Tipper, H.W., Woodsworth, G.J., and Gabrielse, H.
1981: Tectonic assemblage map of the Canadian Cordillera; Geological Survey of Canada, Map 1505A, Scale 1:2 000 000.
Tourtelot, H.A.
1962: Preliminary investigation of the geologic setting and chemical composition of the Pierre Shale, Great Plains region; United States Geological Survey, Professional Paper 390, 74 p.
Tozer, E.T.
1956: Uppermost Cretaceous and Paleocene nonmarine molluscan faunas of western Alberta; Geological Survey of Canada, Memoir 280, 125 p.
Tschudy, R.H.
1976: Stratigraphic distribution of species of the megaspore genus *Minerisporites* in North America; United States Geological Survey, Professional Paper 743-E, p. E1-E11.
Vagvolgyi, A. and Hills, L.V.
1969: Microflora of the Lower Cretaceous McMurray Formation of northeast Alberta; Bulletin of Canadian Petroleum Geology, v. 17, no. 2, p. 155-181.
Vail, P.R., Mitchum, R.M., and Thompson, S.
1977b: Seismic stratigraphy and global changes of sea level, Part 4, Global cycles of relative changes of sea level; in Seismic Stratigraphy; Applications to Hydrocarbon Exploration, C.E. Payton (ed.); American Association of Petroleum Geologists, Memoir 26, p. 83-97.
Vail, P.R. and Thompson, S.
1977a: Seismic stratigraphy and global changes of sea level, Part 3: Relative changes of sea level from coastal onlap; in Seismic Stratigraphy - Applications to Hydrocarbon Exploration, C.E. Payton (ed.); American Association of Petroleum Geologists, Memoir 26, p. 63-81.
Van der Voo, R.
1981: Paleomagnetism of North America: a brief review; in Paleoreconstruction of the Continents, Geodynamics Series, v. 2, Washington, D.C. and Boulder, Colorado, M.W. McElhinny and D.A. Valencio (ed.); American Geophysical Union and Geological Society of America, p. 159-176.
Van Valen, L.
1984: Catastrophes, expectations, and the evidence; Paleobiology, v. 10, p. 121-137.
Varley, C.J.
1984: The Cadomin Formation: A model for the Deep Basin type gas trapping mechanism; in The Mesozoic of Middle North America, D.F. Stott and D.J. Glass (ed.); Canadian Society of Petroleum Geologists, Memoir 9, p. 471-484.
Vigrass, L.W.
1968: Geology of Canadian heavy oil sands; Bulletin of American Association of Petroleum Geologists, v. 52, p. 1984-2000.
Vonhof, J.A.
1965: The Cypress Hills Formation and its reworked deposits in southwestern Saskatchewan; in 15th Annual Field Conference Guidebook, Alberta Society of Petroleum Geologists, Part 1, p. 142-161.
Vuke, S.M.
1984: Depositional environments of the Early Cretaceous Western Interior Seaway in southwestern Montana and the northern United States; in The Mesozoic of Middle North America, D.F. Stott and D.J. Glass (ed.); Canadian Society of Petroleum Geologists, Memoir 9, p. 127-144.

Waheed, A.
1983: Sedimentology of the coal-bearing Bearpaw-Horseshoe Canyon formations (Upper Cretaceous), Drumheller area, Alberta, Canada; M.Sc. thesis, University of Toronto, Toronto, Ontario, 161 p.

Walker, R.G.
1983a: Cardium Formation 2. Sand-body geometry and stratigraphy in the Garrington-Caroline-Ricinus area, Alberta - "The Ragged Blanket" model; Bulletin of Canadian Petroleum Geology, v. 31, no. 1, p. 14-26.
1983b: Cardium Formation 3. Sedimentology and stratigraphy in the Garrington-Caroline area, Alberta; Bulletin of Canadian Petroleum Geology, v. 31, p. 213-320.
1985: Cardium Formation at Ricinus Field, Alberta: A channel cut and filled by turbidity currents in Cretaceous Western Interior Seaway; American Association of Petroleum Geologists, Bulletin, v. 69, no. 11, p. 1963-1981.

Walker, R.G. and Hunter, D.
1982: Transition, Wapiabi to Belly River Formation at Trap Creek and Highwood River, Alberta; in Clastic Units of the Front Ranges, Foothills and Plains in the Area between Field, British Columbia and Drumheller, Alberta, R.G. Walker (ed.); 11th International Congress on Sedimentology, Hamilton, 1982, International Association of Sedimentologists, Guidebook to Excursion 21A, p. 61-71.

Wall, J.H.
1960: Upper Cretaceous foraminifera from the Smoky River area, Alberta; Research Council of Alberta, Bulletin 6, 43 p.
1967: Cretaceous foraminifera of the Rocky Mountain Foothills, Alberta; Alberta Research Council, Bulletin 20, 185 p.
1976: Marginal marine foraminifera from the Late Cretaceous Bearpaw-Horseshoe Canyon transition, southern Alberta, Canada; Journal of Foraminiferal Research, v. 6, no. 3, p. 193-201.

Wall, J.H. and Rosene, R.K.
1977: Upper Cretaceous stratigraphy and micropaleontology of the Crowsnest Pass-Waterton area, southern Alberta Foothills; Bulletin of Canadian Petroleum Geology, v. 25, no. 4, p. 842-867.

Wall, J.H. and Singh, C.
1975: A Late Cretaceous microfossil assemblage from the Buffalo Head Hills, north-central Alberta; Canadian Journal of Earth Sciences, v. 12, p. 1157-1174.

Wall, J.H., Sweet, A.R., and Hills, L.V.
1971: Paleoecology of the Bearpaw and contiguous Upper Cretaceous formations in the C.P.O.G. Strathmore well, southern Alberta; Bulletin of Canadian Petroleum Geology, v. 19, no. 3, p. 691-702.

Wanklyn, R.P.
1985: Stratigraphy and depositional environments of the Ostracode Member of the McMurray Formation (Lower Cretaceous: Late Aptian-Early Albian) in west central Alberta; Ph.D. thesis, University of Colorado, Boulder, Colorado, 289 p.

Warren, P.S. and Stelck, C.R.
1940: Cenomanian and Turonian faunas in the Pouce Coupé District, Alberta and British Columbia; Transactions of the Royal Society of Canada, 3rd series, sec. 4, v. 34, p. 143-153.
1955: New Cenomanian ammonites from Alberta; Alberta Research Council, Report 70, p. 63-80.
1958a: Lower Cenomanian Ammonoidea and Pelecypoda from Peace River area, western Canada; Alberta Research Council, Geology Division, Bulletin 2, Pt. II, p. 36-51.
1958b: The Nikanassin-Luscar hiatus in the Canadian Rockies; Transactions of the Royal Society of Canada, 3rd series, v. 52, sec. 4, p. 55-62.
1969: Early *Neogastroplites*, Fort St. John Group, western Canada; Bulletin of Canadian Petroleum Geology, v. 17, no. 4, p. 529-547.

Weimer, R.J.
1960: Upper Cretaceous stratigraphy, Rocky Mountain area; American Association of Petroleum Geologists, Bulletin, v. 44, p. 1-20.
1984: Relation of unconformities, tectonics, and sea-level changes, Cretaceous of Western Interior, U.S.A.; in Interregional Unconformities and Hydrocarbon Accumulation, J.S. Schlee (ed.); American Association of Petroleum Geologists, Memoir 36, p. 7-35.

Weimer, R., Emm, J.J., Farmer, C.L., Anna, L.O., Davis, T.L., and Kidney, R.L.
1983: Tectonic influences on sedimentation, Early Cretaceous, east flank Powder River Basin, Wyoming and South Dakota; Colorado School of Mines Quarterly, v. 77, no. 4, p. 1-61.

Welte, D.H., Schaefer, R.G., Radke, M., and Weiss, H.M.
1982: Origin, migration and entrapment of natural gas in Alberta Deep Basin, Part 1 (abstract); American Association of Petroleum Geologists, Bulletin, v. 66, p. 642.

Wheeler, J.O.
1970: Summary and Discussion; in Structure of the Southern Canadian Cordillera, J.O. Wheeler (ed.); Geological Association of Canada, Special Paper No. 6, p. 155-166.

Whitaker, S.H., Irvine, J.A., and Broughton, P.L.
1978: Coal Resources of southern Saskatchewan: a model for evaluation methodology; Geological Survey of Canada, Economic Geology Report 30, 151 p.

White, W.I.
1969: Geology and petroleum accumulations of the North Hoosier area, west central Saskatchewan; Department of Mineral Resources, Geological Sciences Branch, Sedimentary Geology Division, Report No. 133, 37 p.

Wiggins, V.D.
1976: Fossil oculate pollen from Alaska; Geoscience and Man, v. 15, p. 51-76.

Williams, G.D.
1963: The Mannville Group (Lower Cretaceous) of central Alberta; Bulletin of Canadian Petroleum Geology, v. 11, p. 350-368.

Williams, G.D. and Burk, C.F.
1964: Upper Cretaceous; Chapter 12 in The Geological History of Western Canada, R.G. McCrossan and R.P. Glaister (ed.); Alberta Society of Petroleum Geologists, p. 169-189.

Williams, G.D. and Stelck, C.R.
1975: Speculations on the Cretaceous paleogeography of North America; in The Cretaceous System in the Western Interior of North America, W.G.E. Caldwell (ed.); Geological Association of Canada, Special Paper 13, p. 1-20.

Williams, G.L.
1975: Dinoflagellate and spore stratigraphy of the Mesozoic-Cenozoic, offshore eastern Canada; Geological Survey of Canada, Paper 74-30, v. 2, p. 107-161.

Wolbach, W.S., Lewis, R.S., and Anders, E.
1985: Cretaceous extinctions: evidence for wildfires and search for meteoritic material; Science, v. 230, p. 167-170.

Woodsworth, G.J, Anderson, R.G., Armstrong, R.L., with contributions by Struik, L.C. and Van der Heyden, P.
1991: Plutonic regimes; Chapter 15 in Geology of the Cordilleran Orogen in Canada, H. Gabrielse and C.J. Yorath (ed.); Geological Survey of Canada, Geology of Canada no. 4. (Also Geological Society of America, The Geology of North America, no. G-2).

Wright, M.E. and Walker, R.G.
1981: Cardium Formation (Upper Cretaceous) at Seebe, Alberta - storm-transported sandstones and conglomerates in shallow marine depositional environments below fair-weather wave base; Canadian Journal of Earth Sciences, v. 18, p. 795-809.

Yorath, C.J. and Cook, D.G.
1981: Cretaceous and Tertiary stratigraphy and paleogeography, Northern Interior Plains, District of Mackenzie; Geological Survey of Canada, Memoir 398, 76 p.
1984: Mesozoic and Cenozoic depositional history of the northern Interior Plains of Canada; in The Mesozoic of Middle North America, D.F. Stott and D.J. Glass (ed.); Canadian Society of Petroleum Geologists, Memoir 9, p. 69-83.

Young, F.G.
1971: Mesozoic stratigraphic studies, northern Yukon Territory and northeastern District of Mackenzie; in Report of Activities Part A, Geological Survey of Canada, Paper 71-1A, p. 245-247.
1972: Cretaceous stratigraphy between Blow and Fish Rivers, Yukon Territory; in Report of Activities Part A, Geological Survey of Canada, Paper 72-1A, p. 229-235.
1973a: Mesozoic epicontinental flyschoid and molassoid depositional phases of Yukon's North Slope; in Canadian Arctic Geology, J.D. Aitken and D.J. Glass (ed.); Geological Association of Canada - Canadian Society of Petroleum Geologists, p. 181-201.
1973b: Jurassic and Cretaceous stratigraphy between Babbage and Blow Rivers, Yukon Territory; in Report of Activities Part A, Geological Survey of Canada, Paper 73-1A, p. 277-281.
1975a: Upper Cretaceous stratigraphy, Yukon Coastal Plain and northwestern Mackenzie Delta; Geological Survey of Canada, Bulletin 249, 83 p.
1975b: Stratigraphic and sedimentologic studies in northeastern Eagle Plain, Yukon Territory; in Report of Activities Part B, Geological Survey of Canada, Paper 75-1B, p. 309-323.
1977: The mid-Cretaceous flysch and phosphatic ironstone sequence, northern Richardson Mountains, Yukon Territory; in Report of Activities Part C, Geological Survey of Canada, Paper 77-1C, p. 67-74.
1978: Geological and geographical guide to the Mackenzie Delta area; Canadian Society of Petroleum Geologists, 158 p.

Young, F.G., Myhr, D.W., and Yorath, C.J.
1976: Geology of the Beaufort-Mackenzie Basin; Geological Survey of Canada, Paper 76-11, 63 p.

Young, F.G. and Robertson, B.T.
1984: The Rapid Creek Formation: an Albian flysch-related phosphatic iron formation in northern Yukon Territory; in The Mesozoic of Middle North America, D.F. Stott and D.J. Glass (ed.); Canadian Society of Petroleum Geologists, Memoir 9, p. 361-372.

Zeigler, W.H.
1969: The development of sedimentary basins in western and Arctic Canada; Alberta Society of Petroleum Geologists, 89 p.

Zeigler, W.H. and Pocock, S.A.J.
1960: The Minnes Formation; Second Annual Field Conference Guidebook, Edmonton Geological Society, p. 43-71.

ADDENDUM

A study of the crystalline basement of the Alberta Basin by Ross et al. (1991), based on aeromagnetic and gravity data combined with U-Pb zircon and monazite geochronology, showed that the geophysical signature of tectonic domains in the Canadian Shield can be traced nearly to the Foothills. The basement of the Alberta Basin was divided into 22 distinct domains.

New information of the sequence stratigraphy of the Beaufort-Mackenzie Basin was provided by Dixon and Dietrich (1988) and in a subsequent publication by Dixon et al. (1992).

Many of the papers in James and Leckie (1988) pertain to Cretaceous sediments of the Western Canada Basin, interpreting them in the light of current geological concepts, with a strong emphasis on sequence stratigraphy. Sedimentology of the Bluesky sandstones was interpreted by Moslow and Pemberton (1988), O'Connell (1988), and Oppelt (1988). Linear estuarine conglomerates of the Gates Formation of the Foothills of northeastern British Columbia were described by Carmichael (1988). Cyclic marine sedimentation of the Luscar Group and Spirit River Formation were outlined by Macdonald et al. (1988). The Cadotte Member was discussed by Hayes (1988) and Rahmani and Smith (1988). The McMurray Formation and Wabiskaw Member were the subject of three papers by Keith et al. (1988), Ranger et al. (1988), and Ranger and Pemberton (1988). Two papers by Rosenthal (1988) and Strobl (1988) discussed the Lower Cretaceous Glauconite Formation of Alberta; another by Banerjee and Davies (1988) provided a lithostratigraphic and palynostratigraphic analysis of the underlying Ostracode Zone of central Alberta. The environments of the Grand Rapids Formation were interpreted from its ichnofossils by Beynon et al. (1988). The Doe Creek Member of Kaskapau Formation was described by Wallace-Dudley and Leckie (1988). The economically important Cardium and Viking sandstones were the subject of reports by Bergman and Walker (1988), Eyles and Walker (1988), Pattison (1988), Plint et al. (1988), Power (1988), Raddysh (1988), and Vossler and Pemberton (1988). Late Turonian marine and nonmarine palynomorphs were documented from Cardium Formation by Sweet and McIntyre (1988). Geochemical studies included an oil-oil and oil-source rock correlation of Lower Cretaceous oils by Feinstein et al. (1988), and a description of the use of biomarker geochemistry to identify variable biodegradation levels in Lower Cretaceous oil sands by Brook et al. (1988).

The Lower Cretaceous Minnes Group of northeastern British Columbia was documented by Stott (in press) in a report which establishes a stratigraphic framework, outlines the correlations and facies, and provides an interpretation of depositional environments.

Several studies have provided new information on various units within the Mannville Group. The diagenetic history and sedimentology of the Clearwater Formation, which contains heavy oil, was described by Hutcheon et al. (1989), based on petrographic, X-ray, SEM and isotopic examination of core samples. Hydrocarbon-producing quartzose sandstone within the Sunburst Member in southeastern Alberta were shown by Farshori and Hopkins (1989) to be deposited in fluvial and brackish water environments in narrow elongate ribbons and broad sheets, respectively. The nature of the pre-Cretaceous unconformity and the dominant controls on hydrocarbon accumulation of the Cantuar Formation of southeastern Saskatchewan were illustrated by Putnam (1989).

Stelck (1991) added to the documentation of the Albian foraminiferal succession of northeastern British Columbia.

The allostratigraphy of the Viking Formation in Willesden Green area was defined by Boreen and Walker (1991) who also provided an interpretation of the depositional environments.

The river- and wave-dominated deltaic sediments of the Upper Cretaceous Dunvegan Formation were expanded upon by Bhattacharya (1988) and by Battacharya and Walker (1991a,b).

The Coniacian-Santonian Muskiki and Marshybank formations of the Foothills and subsurface were described in detail by Plint (1990) and Plint and Norris (1991) and the Bad Heart Formation by Plint et al. (1990).

The Upper Cretaceous (Santonian) Medicine Hat Formation was described as a bacteriogenic gas reservoir by Hankel et al. (1989).

New stratigraphic and sedimentological data were obtained by Eberth et al. (1990) from the Judith River Formation of west-central Saskatchewan and the succession yielded the first nonmarine Campanian vertebrate fossils from that area. The Judith River Formation of Dinosaur Provincial Park, Alberta, was described by Wood (1989).

REFERENCES

Bergman, K.M. and Walker, R.G.
1988: Formation of Cardium erosion surface E5, and associated deposition of conglomerate; Carrot Creek Field, Cretaceous Western Interior Seaway, Alberta; in Sequences, Stratigraphy, Sedimentology; Surface and Subsurface, D.P. James and D.A. Leckie (ed.); Canadian Society of Petroleum Geologists, Memoir 15, p. 15-24.

Banerjee, I. and Davies, E.H.J.
1988: An integrated lithostratigraphic and palynostratigraphic study of the Ostracode Zone and adjacent strata in the Edmonton Embayment, central Alberta; in Sequences, Stratigraphy, Sedimentology; Surface and Subsurface, D.P. James and D.A. Leckie (ed.); Canadian Society of Petroleum Geologists, Memoir 15, p. 261-274.

Beynon, B.M., Pemberton, S.G., Bell, D.D., and Logan, C.A.
1988: Environmental implications of ichnofossils from the Lower Cretaceous Grand Rapids Formation, Cold Lake Oil Sands Deposit; in Sequences, Stratigraphy, Sedimentology; Surface and Subsurface, D.P. James and D.A. Leckie (ed.); Canadian Society of Petroleum Geologists, Memoir 15, p. 275-290.

Bhattacharya, J.
1988: Autocyclic and allocyclic sequences in river- and wave-dominated deltaic sediments of the Upper Cretaceous, Dunvegan Formation, Alberta: core examples; in Sequences, Stratigraphy, Sedimentology; Surface and Subsurface, D.P. James and D.A. Leckie (ed.); Canadian Society of Petroleum Geologists, Memoir 15, p. 25-32.

Bhattacharya, J. and Walker, R.G.
1991a: Allostratigraphic subdivision of the Upper Cretaceous Dunvegan, Shaftesbury, and Kaskapau formations in the northwestern Alberta subsurface; Bulletin of Canadian Petroleum Geology, v. 39, p. 145-164.

Bhattacharya, J. and Walker, R.G. (cont.)
1991b: River- and wave-dominated depositional systems of the Upper Cretaceous Dunvegan Formation, northwestern Alberta; Bulletin of Canadian Petroleum Geology, v. 39, p. 165-191.

Boreen, T. and Walker, R.G.
1991: Definition of allomembers and their facies assemblages in the Viking Formation, Willesden Green area, Alberta; Bulletin of Canadian Petroleum Geology, v. 39, p. 123-144.

Brooks, P.W., Fowler, M.G., Macqueen, R.W., and Mathison, J.E.
1988: Use of biomarker geochemistry to identify variable biodegradation levels, Cold Lake Oils Sands (Fort Kent Area), Alberta; in Sequences, Stratigraphy, Sedimentology; Surface and Subsurface, D.P. James and D.A. Leckie (ed.); Canadian Society of Petroleum Geologists, Memoir 15, p. 529-536.

Carmichael, S.M.M.
1988: Linear estuarine conglomerate bodies formed during a mid-Albian marine transgression; "Upper Gates" Formation, Rocky Mountains Foothills of north-eastern British Columbia; in Sequences, Stratigraphy, Sedimentology; Surface and Subsurface, D.P. James and D.A. Leckie (ed.); Canadian Society of Petroleum Geologists, Memoir 15, p. 49-62.

Dixon, J. and Dietrich, J.R.
1988: The nature of depositional and seismic sequence boundaries in Cretaceous-Tertiary strata of the Beaufort-Mackenzie Basin; in Sequences, Stratigraphy, Sedimentology; Surface and Subsurface, D.P. James and D.A. Leckie (ed.); Canadian Society of Petroleum Geologists, Memoir 15, p. 63-72.

Dixon, J., Dietrich, J.R., and McNeil, D.H.
1992: Upper Cretaceous to Holocene sequence stratigraphy of the Beaufort-Mackenzie and Banks Island areas, northwest Canada; Geological Survey of Canada, Bulletin 407, 90 p.

Eberth, D.A., Braman, D.R., and Tokaryk, T.T.
1990: Stratigraphy, sedimentology and vertebrate paleontology of the Judith River Formation (Campanian) near Muddy Lake, west-central Saskatchewan; Bulletin of Canadian Petroleum Geology, v. 38, p. 387-406.

Eyles, C.H. and Walker, R.G.
1988: "Geometry" and facies characteristics of stacked shallow marine sandier-upwards sequences in the Cardium Formation at Willesden Green, Alberta; in Sequences, Stratigraphy, Sedimentology; Surface and Subsurface, D.P. James and D.A. Leckie (ed.); Canadian Society of Petroleum Geologists, Memoir 15, p. 85-96.

Farshori, M.Z. and Hopkins, J.C.
1989: Sedimentology and petroleum geology of fluvial and shoreline deposits of the Lower Cretaceous Sunburst Sandstone Member, Mannville Group, southern Alberta; Bulletin of Canadian Petroleum Geology, v. 37, p. 371-388.

Feinstein, S., Brooks, P.W., Fowler, M.G., Snowdon, L.R., and Williams, G.K.
1988: Families of oils and source rocks in the central Mackenzie Corridor: a geochemical oil-oil and oil-source rocks correlation; in Sequences, Stratigraphy, Sedimentology; Surface and Subsurface, D.P. James and D.A. Leckie (ed.); Canadian Society of Petroleum Geologists, Memoir 15, p. 543-552.

Hankel, R.C., Davies, G.R., and Krouse, H.R.
1989: Eastern Medicine Hat gas field; a shallow, Upper Cretaceous, bacteriogenic gas reservoir of southeastern Alberta; Bulletin of Canadian Petroleum Geology, v. 37, p. 98-112.

Hayes, B.J.R.
1988: Incision of a Cadotte Member paleovalley-system at Noel, British Columbia – evidence of a Late Albian sea level fall; in Sequences, Stratigraphy, Sedimentology; Surface and Subsurface, D.P. James and D.A. Leckie (ed.); Canadian Society of Petroleum Geologists, Memoir 15, p. 97-106.

Hutcheon, I., Abercrombie, H.J., Putnam, P., Gardner, R., and Krouse, H.R.
1989: Diagenesis and sedimentology of the Clearwater Formation at Tucker Lake; Bulletin of Canadian Petroleum Geology, v. 37, p. 83-97.

James, D.P. and Leckie, D.A. (ed.)
1988: Sequences, Stratigraphy, Sedimentology; Surface and Subsurface; Canadian Society of Petroleum Geologists, Memoir 15, 586 p.

Keith, D.A.W., Wightman, D.M., Pemberton, S.G., MacGillivray, J.R., Berezniuk, T., and Berhane, H.
1988: Sedimentology of the McMurray Formation and Wabiskaw Member (Clearwater Formation), Lower Cretaceous, in the central region of the Athabasca Oil Sands Area, northeastern Alberta; in Sequences, Stratigraphy, Sedimentology; Surface and Subsurface, D.P. James and D.A. Leckie (ed.); Canadian Society of Petroleum Geologists, Memoir 15, p. 309-324.

Macdonald, D.E., Langenberg, C.W., and Stobl, R.S.
1988: Cyclic marine sedimentation in the Lower Cretaceous Luscar Group and Spirit River Formation of the Alberta Foothills and Deep Basin; in Sequences, Stratigraphy, Sedimentology; Surface and Subsurface, D.P. James and D.A. Leckie (ed.); Canadian Society of Petroleum Geologists, Memoir 15, p. 143-154.

Moslow, T.F. and Pemberton, S.G.
1988: An integrated approach to the sedimentological analysis of some Lower Cretaceous shoreface and delta front sandstone sequences; in Sequences, Stratigraphy, Sedimentology; Surface and Subsurface, D.P. James and D.A. Leckie (ed.); Canadian Society of Petroleum Geologists, Memoir 15, p. 373-386.

O'Connell, S.C.
1988: The distribution of Bluesky Facies in the region overlying the Peace River Arch, northwestern Alberta; in Sequences, Stratigraphy, Sedimentology; Surface and Subsurface, D.P. James and D.A. Leckie (ed.); Canadian Society of Petroleum Geologists, Memoir 15, p. 387-400.

Oppelt, H.
1988: Sedimentology and ichnology of the Bluesky Formation in northeastern British Columbia; in Sequences, Stratigraphy, Sedimentology; Surface and Subsurface, D.P. James and D.A. Leckie (ed.); Canadian Society of Petroleum Geologists, Memoir 15, p. 401-416.

Pattison, S.A.J.
1988: Transgressive, incised shoreface deposits of the Burnstick Member (Cardium "B" Sandstone) at Caroline, Crossfield, Garrington and Lochend; Cretaceous Western Interior Seaway, Alberta, Canada; in Sequences, Stratigraphy, Sedimentology; Surface and Subsurface, D.P. James and D.A. Leckie (ed.); Canadian Society of Petroleum Geologists, Memoir 15, p. 155-167.

Plint, A.G.
1990: An allostratigraphic correlation of the Muskiki and Marshybank formations (Coniacian-Santonian) in the Foothills and subsurface of the Alberta Basin; Bulletin of Canadian Petroleum Geology, v. 38, p. 288-306.

Plint, A.G. and Norris, B.
1991: Anatomy of a ramp margin sequence; facies successions, paleogeography, and sediment dispersal patterns in the Muskiki and Marshybank formations, Alberta Foreland Basin; Bulletin of Canadian Petroleum Geology, v. 39, p. 18-42.

Plint, A.G., Norris, B., and Donaldson, W.S.
1990: Revised definitions for the Upper Cretaceous Bad Heart Formation and associated units in the Foothills and Plains of Alberta and British Columbia; Bulletin of Canadian Petroleum Geology, v. 38, p. 78-88.

Plint, A.G., Walker, R.G., and Duke, W.L.
1988: An outcrop to subsurface correlation of the Cardium Formation in Alberta; in Sequences, Stratigraphy, Sedimentology; Surface and Subsurface, D.P. James and D.A. Leckie (ed.); Canadian Society of Petroleum Geologists, Memoir 15, p. 167-184.

Power, B.A.
1988: Coarsening-upwards shoreface and shelf sequences: examples from the Lower Cretaceous Viking Formation at Joarcam, Alberta, Canada; in Sequences, Stratigraphy, Sedimentology; Surface and Subsurface, D.P. James and D.A. Leckie (ed.); Canadian Society of Petroleum Geologists, Memoir 15, p. 185-194.

Putnam, P.E.
1989: Geological controls on hydrocarbon entrapment within the Lower Cretaceous Cantuar Formation, Wapella Field, southeastern Saskatchewan; Bulletin of Canadian Petroleum Geology, v. 37, p. 389-400.

Raddysh, H.K.
1988: Sedimentology and "geometry" of the Lower Cretaceous Viking Formation, Gilby A and B Fields, Alberta; in Sequences, Stratigraphy, Sedimentology; Surface and Subsurface, D.P. James and D.A. Leckie (ed.); Canadian Society of Petroleum Geologists, Memoir 15, p. 417-430.

Rahmani, R.A. and Smith, D.G.
1988: The Cadotte Member of northwestern Alberta: a high energy barred shoreline; in Sequences, Stratigraphy, Sedimentology; Surface and Subsurface, D.P. James and D.A. Leckie (ed.); Canadian Society of Petroleum Geologists, Memoir 15, p. 431-438.

Ranger, M.J. and Pemberton, S.G.
1988: Marine influence on the McMurray Formation in the Primrose area, Alberta; in Sequences, Stratigraphy, Sedimentology; Surface and Subsurface, D.P. James and D.A. Leckie (ed.); Canadian Society of Petroleum Geologists, Memoir 15, p. 439-450.

Ranger, M.J., Pemberton, S.G., and Sharpe, R.J.
1988: Lower Cretaceous examples of a shoreface-attached marine bar complex: the Wabiskaw "C" Sand of northeastern Alberta; in Sequences, Stratigraphy, Sedimentology; Surface and Subsurface, D.P. James and D.A. Leckie (ed.); Canadian Society of Petroleum Geologists, Memoir 15, p. 451-462.

Rosenthal, L.
1988: Wave-dominated shorelines and incised channel trends; Lower Cretaceous Glauconite Formation, west-central Alberta; in Sequences, Stratigraphy, Sedimentology; Surface and Subsurface, D.P. James and D.A. Leckie (ed.); Canadian Society of Petroleum Geologists, Memoir 15, p. 207-221.

Ross, G.M., Parrish, R.R., Villeneuve, M.E., and Bowring, S.A.
1991: Geophysics and geochronology of the crystalline basement of the Alberta Basin, western Canada; Canadian Journal of Earth Sciences, v. 28, p. 512-522.

Stelck, C.R.
1991: Foraminifera of the middle to upper Albian transition (Lower Cretaceous), northeastern British Columbia; Canadian Journal of Earth Sciences, v. 28, p. 561-580.

Stott, D.F.
in press: Fernie Formation and Minnes Group (Jurassic and Lowermost Cretaceous), Northern Foothills of Alberta and British Columbia; Geological Survey of Canada, Bulletin.

Strobl, R.S.
1988: The effects of sea-level fluctuations on prograding shorelines and estuarine valley-fill sequences in the Glauconitic Member, Medicine River Field and adjacent area; in Sequences, Stratigraphy, Sedimentology; Surface and Subsurface, D.P. James and D.A. Leckie (ed.); Canadian Society of Petroleum Geologists, Memoir 15, p. 221-236.

Sweet, A.R. and McIntyre, D.J.
1988: Late Turonian marine and nonmarine palynomorphs from the Cardium Formation, north-central Alberta Foothills, Canada; in Sequences, Stratigraphy, Sedimentology; Surface and Subsurface, D.P. James and D.A. Leckie (ed.); Canadian Society of Petroleum Geologists, Memoir 15, p. 499-516.

Vossler, S. and Pemberton, S.G.
1988: Ichnology of the Cardium Formation (Pembina Oilfield); implications for depositional sequence stratigraphic interpretations; in Sequences, Stratigraphy, Sedimentology; Surface and Subsurface, D.P. James and D.A. Leckie (ed.); Canadian Society of Petroleum Geologists, Memoir 15, p. 237-253.

Wallace-Dudley, K.E. and Leckie, D.A.
1988: Preliminary observations on the sedimentology of the Doe Creek Member, Kaskapau Formation, in the Valhalla Field, northwestern Alberta; in Sequences, Stratigraphy, Sedimentology; Surface and Subsurface, D.P. James and D.A. Leckie (ed.); Canadian Society of Petroleum Geologists, Memoir 15, p. 485-496.

Wood, J.M.
1989: Alluvial architecture of the Upper Cretaceous Judith River Formation, Dinosaur Provincial Park, Alberta, Canada; Bulletin of Canadian Petroleum Geology, v. 37, p. 169-181.

Authors' Addresses

D.F. Stott
8929 Forest Park Drive
Sidney, British Columbia
V8L 5A7

W.G.E. Caldwell
Room 319, Stevenson Lawson Building
University of Western Ontario
London, Ontario
N6A 5B8

D.J. Cant
Institute of Sedimentary and Petroleum Geology
3303 - 33rd Street N.W.
Calgary, Alberta
T2L 2A7

J.E. Christopher
252 Coldwell Road
Regina, Saskatchewan
S4R 4L2

J. Dixon
Institute of Sedimentary and Petroleum Geology
Geological Survey of Canada
3303 - 33rd Street N.W.
Calgary, Alberta
T2L 2A7

E.H. Koster
Ontario Science Centre
770 Don Mills Road
Don Mills, Ontario,
M3C 1T3

D.H. McNeil
Institute of Sedimentary and Petroleum Geology
Geological Survey of Canada
3303 - 33rd Street N.W.
Calgary, Alberta
T2L 2A7

F. Simpson
Department of Geology
University of Windsor
Windsor, Ontario
N9B 3P4

Printed in Canada

Subchapter 4J

TERTIARY

D.F. Stott, J. Dixon, J.R. Dietrich, D.H. McNeil, L.S. Russell, and A.R. Sweet

CONTENTS

INTRODUCTION

The depositional environments existing at the end of the Cretaceous Period continued into the Paleocene Epoch as the Laramide Orogeny continued to affect the western margin of the basin. Thick sequences of sediments derived from the Cordilleran Orogen were deposited in the Beaufort Sea in the north, while continental Paleocene beds were widely distributed in the south, spreading eastward from the rising Rocky Mountains across the Interior Platform.

The thickest Tertiary sediments in the southern part of Western Canada Basin are preserved along the eastern side of the Rocky Mountain Foothills and within the adjacent Alberta Syncline (Fig. 4J.1). These strata originally extended much more widely across the Interior Plains, but occur now only as erosional remnants, the largest being the Cypress Hills of southern Alberta and Saskatchewan (Fig. 4J.2) and their eastern extension along the Saskatchewan-United States border from Wood Mountain to Turtle Mountain in Manitoba. Tertiary sediments of the eastern Canadian Cordillera and Interior Plains are almost wholly continental, contrasting with the thick marine and nonmarine section encountered in the region of Mackenzie Delta and Beaufort Sea.

Tertiary sediments are extensively developed in Mackenzie Delta, Arctic Coastal Plain, and on the Arctic Continental Shelf. Limited exposures are found in the Caribou Hills on the eastern side of Mackenzie Delta (Fig. 4J.3), along the western margin of the delta, in the Richardson Mountains, and in a small area near Babbage River on Yukon Coastal Plain. Other occurrences are found in Bonnet Plume Basin, Brackett Basin near Norman Wells, and adjacent Great Bear Basin.

Tectonic elements

The early phases of the Laramide Orogeny rejuvenated the mountain belts of the Columbian Orogen formed during the earlier, Late Jurassic–Early Cretaceous orogeny, and resulted in the development of the Rocky Mountain Thrust

Stott, D.F., Dixon, J., Dietrich, J.R., McNeil, D.H., Russell, L.S., and Sweet, A.R.
1993: Tertiary; Subchapter 4J in Sedimentary Cover of the Craton in Canada, D.F. Stott and J.D. Aitken (ed.); Geological Survey of Canada, Geology of Canada, no. 5, p. 439-465 (also Geological Society of America, The Geology of North America, v. D-1).

Belt, the Mackenzie Fold Belt, and the Northern Yukon Fold Complex. The Laramide phase began during the Late Cretaceous Epoch, continued during the early Tertiary, and terminated during the Eocene when Cretaceous and early Tertiary sequences were involved in the deformation of the eastern Rocky Mountains and Foothills. The final orogenic phase in the north was somewhat later, ending during the Miocene. The Upper Eocene, Oligocene, and Miocene beds of the south are post-orogenic, whereas only Miocene and younger beds in Mackenzie Delta are post-orogenic.

Sediments derived from the Columbian Orogen were transported eastward across Western Canada Basin. The short-lived Cannonball Sea, recorded only in southernmost Manitoba, may have had connections with the Arctic Ocean to the north via Hudson Bay and to the south with the Gulf of Mexico (Fox and Olsson, 1969). During mid-Tertiary time, erosion was dominant on the Interior Plains, resulting in the development of widespread surfaces of low relief. In contrast, continued deposition at the northern

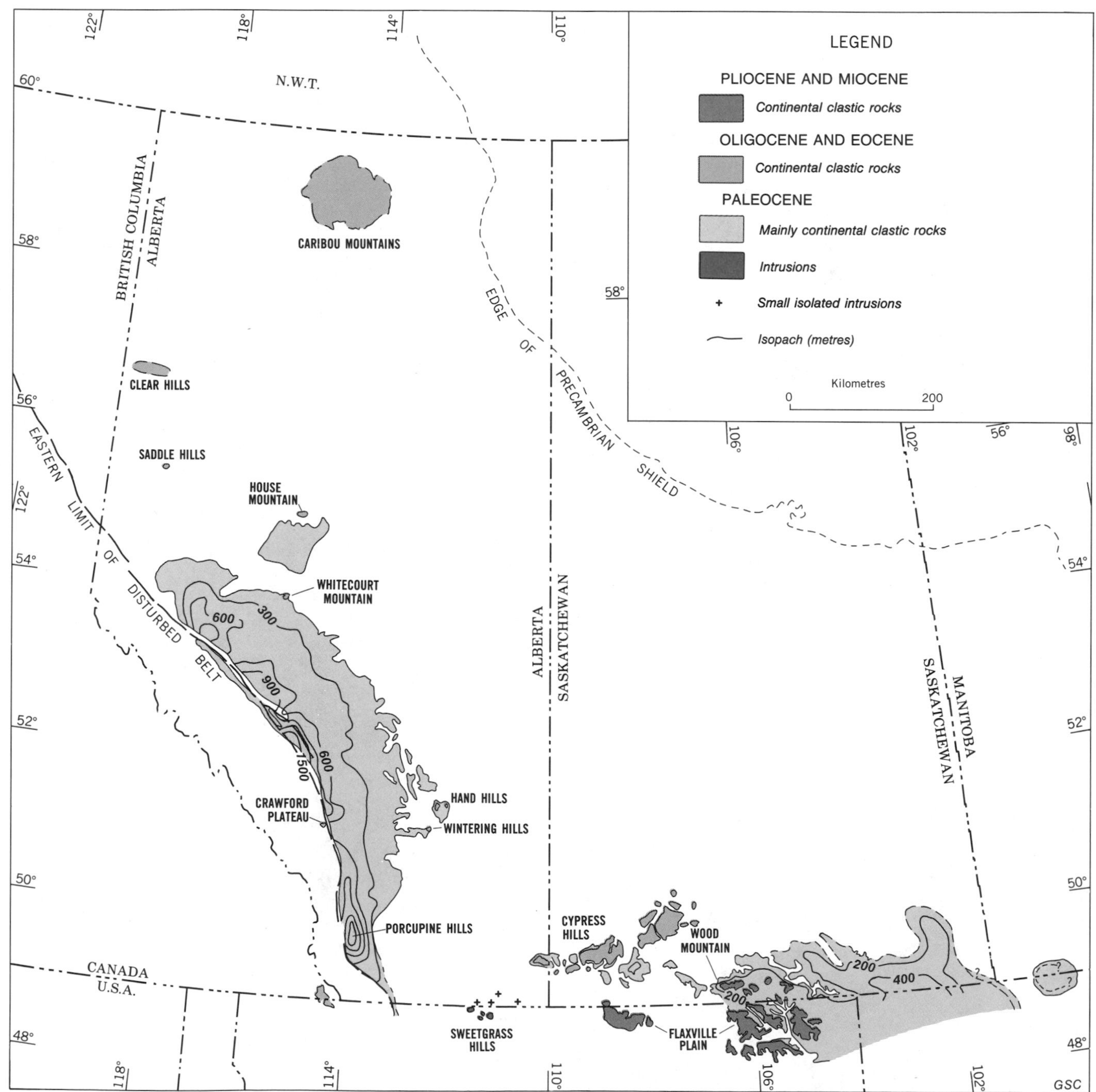

Figure 4J.1. Isopachs and lithology of Tertiary strata in Western Canada Basin (after Taylor et al., 1964).

Figure 4J.2. View of Ravenscrag Butte, Cypress Hills, Saskatchewan, exposing sections of Whitemud (Wm), Battle (Bt), Frenchman (Fr) and Ravenscrag (R) beds (photo G.M. Furnival, GSC 87157).

Figure 4J.3. West-facing scarp of Caribou Hills at eastern margin of Mackenzie Delta, exposing the upper part of the Reindeer Formation (Re) (lower bare ridge), the white clay unit' (wc) of Price et al. (1980) (upper bare spurs) and basal gravel beds of the Beaufort(?)/Nuktak Formation (Be) (photo L.L. Price, ISPG 1015-20).

continental margin resulted in development of the thick clastic succession mainly preserved in the Beaufort-Mackenzie Basin.

Regional and local structures related to collapse following dissolution and removal of salts from the Devonian Prairie Evaporite, occur in Saskatchewan. Kent (1968), Holter (1969), and Christiansen (1967) related distribution of sediments and development of structures to salt dissolution, and Whitaker (1978) discussed some of the structures involving Tertiary sediments. Broughton (1977) related variation of coal thicknesses in Tertiary lignites of Saskatchewan to salt dissolution.

Data base

Although many thousands of wells have been drilled through nonmarine Tertiary and Upper Cretaceous strata of western Canada, few samples or cores of the Tertiary sequence are available for study, and little attention has been given to these rocks by oil-company geologists. However, geophysical logs of the beds below well-casing depths of approximately 200 m are available. In addition, companies exploring for near-surface coal have drilled, sampled, and cored the Tertiary strata extensively in many areas of Saskatchewan and Alberta and to a limited extent in the District of Mackenzie.

In contrast, the Tertiary succession of the Beaufort-Mackenzie Basin is a prime petroleum-exploration target, and data from approximately 230 wells and many thousands of kilometres of reflection seismic records are available.

Previous work

Tertiary rocks of southwestern Alberta have been described in some detail by Douglas (1950), Tozer (1956), and Carrigy (1971), and of southeastern British Columbia by Price (1965), McMechan (1981), and Jones (1969). Similar Tertiary beds in the central foothills have been outlined recently by Jerzykiewicz and McLean (1980), and Jerzykiewicz (1985). The more current reports on the Tertiary of the plains are by Irish (1965), Vonhof (1965), and Russell (1974). Comprehensive summaries were published by Taylor et al. (1964) and Douglas et al. (1970). Subsurface correlations have been published by Elliot (1960), Ower (1960), and Irish and Havard (1968). Whitaker et al. (1978) presented a comprehensive report on Paleocene strata of southern Saskatchewan.

Reconnaissance work in northern Yukon Territory and District of Mackenzie was undertaken during Operation Porcupine (Mountjoy, 1967; Norris, 1976) and later related studies (Norris and Hopkins, 1977). Studies in the region of the Richardson Mountains, Mackenzie Delta, and Beaufort Sea include those of Young (1975a, b), Young et al. (1976), Johnson et al. (1976), Hawkings and Hatlelid (1975), Dixon (1981), Willumsen and Coté (1982), Young and McNeil (1984), and Deitrich et al. (1985). The Tertiary stratigraphy and paleogeography of the northern Interior Plains of the District of Mackenzie were described by Yorath and Cook (1981). These reports were synthesized by Dixon (1986).

Plant fossils have been useful in Tertiary biostratigraphy. The most comprehensive review of the macroflora of the Tertiary System of western Canada was provided by Bell (1949). Palynological investigations have multiplied rapidly in the last decade and provide the best biostratigraphical control. The Tertiary microfloras of the central Plains were studied by Snead (1969), Srivastava (1970), and Russell and Singh (1978). Sweet (1978) provided a new zonation of Tertiary coal areas in Saskatchewan. Eliuk (1969), Gunther and Hills (1972), and Sweet (in Jerzykiewicz and McLean, 1980) have reported on Tertiary strata of the Athabasca region in the central Foothills. Rouse and Srivastava (1972) illustrated a Paleocene microflora from Bonnet Plume Basin in Yukon Territory.

The most definitive work on the macro-invertebrate fauna of the Tertiary System of the southern Plains and Foothills was provided by Tozer (1956). Other studies include those of Russell (1967b). Microfaunal studies include that of Bamburak (1973, 1978). Freshwater algae and ostracodes in the Willow Creek and Porcupine Hills strata were described by Germundson (1965). Insects of the Paskapoo Formation of the Red Deer Valley were illustrated by Kevan and Wighton (1981).

A large and varied vertebrate fauna has been recovered from parts of the Tertiary succession in Alberta and Saskatchewan; much of the early work was undertaken by L.M. Lambe, C.M. Sternberg, and L.S. Russell. A bibliography of Cretaceous and Tertiary vertebrates from western Canada was published by Fox (1970). Russell and Churcher (1972) provided lists of the vertebrate faunas found in the Paleocene Paskapoo Formation, the Upper Eocene Swift Current, the Lower Oligocene Cypress Hills, and the Miocene Wood Mountain formations. More recent reports include those of Storer (1975a, 1975b, 1976, 1978a, 1978b, 1978c, 1984), Krause (1977, 1978), and Fox (1984).

In the western Northwest Territories, the main reports on micropaleontology include those of Young and McNeil (1984) and on palynology, those of Brideaux (1975, 1976), Staplin (1976), Wilson (1978), Ioannides and McIntyre (1980), McNeil et al. (1982), and Norris (1986).

Responsibilities and acknowledgments

This compilation of Tertiary history is based to a considerable degree on early investigations of the southern Plains and to on-going studies in the region of Beaufort-Mackenzie Basin. Many of the more recent geological studies on the plains are related to the appraisal of coal resources by the Alberta Geological Survey, the Saskatchewan Research Council, the Saskatchewan Geological Survey, and the Geological Survey of Canada. Much of the information concerning Mackenzie Delta and Beaufort Sea is derived from current exploration for petroleum and natural gas and reflects studies by the exploration companies and by the Geological Survey of Canada.

The discussion of the lithostratigraphy of the Mackenzie Delta region and northern basins was prepared by J. Dixon and J.R. Dietrich. The biostratigraphy of the northern basins was summarized by D.H. McNeil. The palynology of the southern plains was the responsibility of A.R. Sweet, who also contributed to the discussion of geological boundaries. The summary of the vertebrate record was written by Loris S. Russell of the Royal Ontario Museum, an active investigator for more than 50 years.

The remainder of the text was prepared by D.F. Stott, who was responsible for the overall co-ordination of this subchapter.

This report has been reviewed by L.V. Hills, University of Calgary, and D.J. McIntyre, Geological Survey of Canada. Their comments have been most useful in improving the text.

BIOSTRATIGRAPHY

Northern basins

Micropaleontology of Beaufort-Mackenzie Basin

Tertiary floras and faunas of the Beaufort-Mackenzie Basin are preserved in complex sequences of terrestrial and marine terrigenous clastic sediments (Fig. 4J.4). The paleontological record reflects a style of sedimentation that was dominated by major episodes of delta formation characterized by relatively rapid rates of deposition, massive reworking at times, and complex facies changes. The Tertiary record was also influenced by a climate that progressively deteriorated from warm temperate in the Paleocene and Eocene epochs to boreal in the Pliocene Epoch.

The Cretaceous-Tertiary boundary occurs approximately at the Tent Island-Moose Channel formational contact (Fig. 4J.4). Low-diversity assemblages of pollen, spores, and foraminifers characterize the Lower to Middle Paleocene Moose Channel Formation. In contrast, rich assemblages of palynomorphs, including abundant fungal spores, occur in the Upper Paleocene to Lower Eocene Reindeer Formation (Ioannides and McIntyre, 1980). The uppermost Reindeer yields the marginal-marine dinoflagellates *Wetzeliella (Apectodinium) hormomorpha* Deflandre and Cookson and *W. articulata* Eisenack, which date the Reindeer-Richards contact at approximately the boundary between the Early and Middle Eocene. Although the Reindeer is largely terrestrial in origin, marine units do occur, as indicated by the nannofossils and by the low-diversity assemblages of agglutinated foraminifers (Young and McNeil, 1984). In

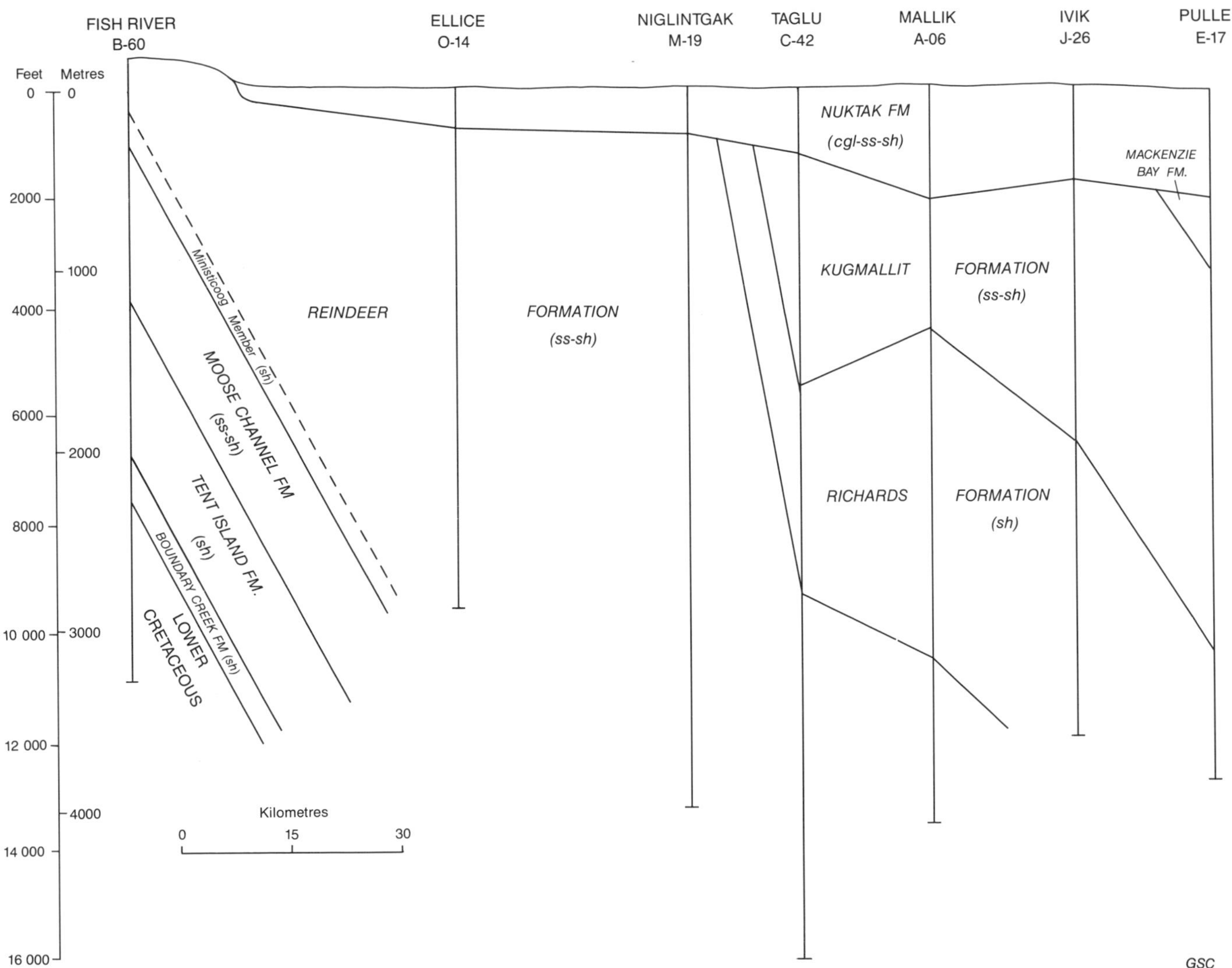

Figure 4J.4. Diagrammatic cross-section, Tertiary subsurface, Mackenzie Delta (J. Dixon).

the outer Mackenzie Delta and Beaufort Sea areas, the Upper Eocene to Lower Oligocene mudstone of the neritic to bathyal Richards Formation yields agglutinated foraminifers dominated by species of *Bathysiphon, Haplophragmoides, Cyclammina*, and *Recurvoides*.

Climatic cooling is first recognized, in the Kugmallit Formation, by the disappearance of a variety of fungal spores and hyphae, bryophytes, and pteridophytes, and thermophylic angiosperms including *Ulmus, Tilia, Quercus, Castanea*, which are common in the underlying Richards Formation (Norris, 1982). The Kugmallit Formation is terrestrial throughout much of its extent under Mackenzie Delta but intertongues progressively with marine sediments to the north beneath Beaufort Sea, where it carries a mixed assemblage of calcareous and agglutinated foraminiferal species. The calcareous component, which is better developed in the overlying Mackenzie Bay Formation, includes *Turrilina alsatica* Andreae (Oligocene) and can be correlated with microfaunas of the North Atlantic and northwest Europe (McNeil et al., 1982), indicating a well-established circulation between the Arctic and Atlantic oceans.

An amelioration of the climate occurred in Late Oligocene to Middle Miocene time, as indicated by palynomorphs of the marine Mackenzie Bay Formation (Norris, 1982). This event may also be recorded in the terrestrial Miocene (?) Beaufort Formation on nearby Banks Island, where Hills et al. (1974) recorded the walnut *Juglans eocineria* (indicative of a temperate climate). A fairly rich calcareous foraminiferal assemblage characterizes the Mackenzie Bay Formation; *Turrilina alsatica* Andreae and *Asterigerina guerichi* (Franke) occur in succession in the formation (Young and McNeil, 1984). In the offshore area, the upper Mackenzie Bay yields an assemblage of dinoflagellates (e.g., *Cannosphaeropsis* and *Nematosphaeropsis*), which correlate with upper Middle to Upper Miocene oceanic microplankton zones (Bujak and Davies, 1981).

The Mackenzie Bay Formation is overlain by the seismically defined Upper Miocene Akpak sequence, which yields a calcareous benthic foraminiferal assemblage characterized by numerous species of *Cibicidoides* and unilocular genera. A regional unconformity terminates the Akpak sequence and in many areas of Beaufort Sea also truncates the upper Mackenzie Bay Formation. The unconformity corresponds approximately to a eustatic drop in sea level and cooler climatic conditions, as determined by the appearance of comparatively modern arctic floras and faunas above the unconformity (McNeil et al., 1982). For example, the cold-water *Elphidium* spp. foraminiferal assemblage of Young and McNeil (1984) is widely recognized in the upper Nuktak Formation of Mackenzie Delta and in the upper Iperk group of Beaufort Sea. The important index species, *Cibicides grossus* ten Dam and Reinhold, occurs in the lower half of the Iperk group. Its last appearance datum approximates the onset of continental glaciation, dated at about 2.5 Ma in the eastern Arctic of North America (Feyling-Hanssen et al., 1983).

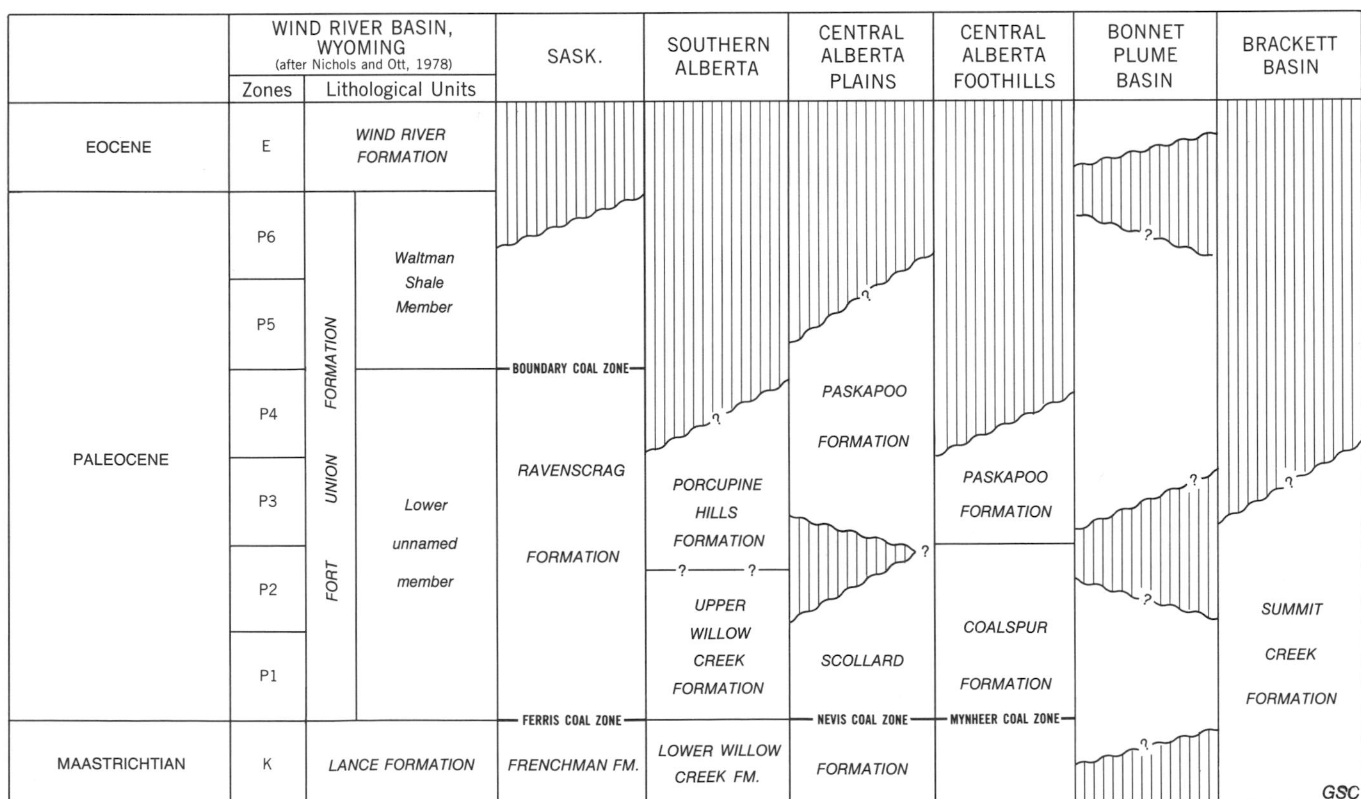

Figure 4J.5. Palynological zonation of Paleocene strata of Western Canada Basin (A.R. Sweet, based on zonation established by Nichols and Ott, 1978).

Western Canada Basin

Mollusca

According to Tozer (1956), who provided a detailed study of the nonmarine molluscan shells collected by various workers since 1880, Paleocene strata contain a rich molluscan fauna, which he designated as the Paskapoo fauna. It includes unionids, sphaeriids, and gastropods referable to both aquatic and terrestrial genera. The Paskapoo unionids are similar to those of preceding faunas, but the aquatic gastropods, almost without exception, are not known in the older faunas of Alberta. *Viviparus retusus* (Meek and Hayden), *V. planolatere* Russell, *Lioplacodes nebrascensis* (Meek and Hayden), and *Valvata bicincta* Whiteaves are characteristic and widespread components of this fauna. The most striking feature of the Paskapoo fauna is the terrestrial element, of which *Grangerella mcleodensis* (Russell), *Oreohelix thurstoni* (Russell), and *Dimorphoptychia douglasi* (Tozer) are most widespread and distinctive. The fauna provides an excellent criterion for the recognition of deposits of Paleocene age.

Macroflora

One of the early comprehensive reports on Paleocene flora was by Berry (1930), who described the flora from Ravenscrag beds of southern Saskatchewan and correlated it with the Fort Union flora of Montana. He commented that it indicated a distinctly temperate flora of probably northern origin. Later, latest Cretaceous and Paleocene floras of western Alberta were outlined and illustrated by Bell (1949) in a study which remains definitive. He indicated that the upper part of the Willow Creek Formation of the Foothills is Paleocene but considered lower beds to be of Late Cretaceous age. Post-Brazeau beds farther north also were considered to be definitely of Paleocene age. The substantial collections of Paskapoo flora were considered to be approximately the same age as the Ravenscrag Formation of southeastern Alberta and Saskatchewan. Equisetums, ferns, and ginkgos are rather rare, as are monocotyledons, and the flora is dominantly coniferous and dicotyledonous. Of the conifers, *Sequoiites* and *Cryptomerites* are dominant, and of 20 dicotyledons, the most abundant and widespread genera are *Platanus, Trochodendroides, Celestrinites*, and *Viburnum*. The assemblage was interpreted as a temperate forest one, with a moderate admixture of plants from bushland, forest border, and undergrowth, and with a few forms of aquatic and riparian habit.

A Paleocene floral assemblage from Genesee, Alberta, containing 14 species, was illustrated and described by Chandrasekharam (1974) from beds occurring above a coal seam correlated with the Ardley Coal Seam of east-central Alberta. A warm temperate, humid, winter-dry and summer-wet climate was inferred from floristic and vegetational analysis.

Palynology

Floristic changes within the Paleocene can be recognized and used in its zonation (Fig. 4J.5; Snead, 1969; Srivastava, 1970; Rouse and Srivastava, 1972; Nichols and Ott, 1978; Sweet, 1978; Jarzen, 1982; Sweet and Hills, 1984).

The diverse Late Maastrichtian flora, dominated by spores and angiosperm pollen, contrasts sharply with the low diversity of the Early Paleocene flora, which is dominated initially by spores but subsequently by gymnosperms. The floral change occurs between the Frenchman and Ravenscrag formations, within the Scollard Formation at the base of the Nevis coal zone, within the Coalspur beds at the base of the Mynheer coal zone, and at the base of a tuffaceous coal-bearing sequence in the Summit Creek Formation of the Tertiary Hills in the Mackenzie region.

In eastern Saskatchewan, the youngest portion of the Ravenscrag Formation, that part above the Boundary coal zone, can be correlated with Zones P5 and P6 of the Paleocene Fort Union Formation of the Wind River Basin in Wyoming (Fig. 4J.5; Nichols and Ott, 1978). This correlation is based on the diversity within *Momipites-Caryapollenites* lineage and the entrance of *Pistillipollenites mcgregorii* Rouse into the flora. Much of the Paskapoo Formation appears to encompass the mid-Paleocene zones P3 and P4 of Nichols and Ott (1978). However, to the west, Upper Paleocene strata correlative to Zone P5 and possibly P6 can be recognized by the occurrence of *Pistillipollenites mcgregorii*, for example in the Obed Marsh coal deposit. The associated pollen of the *Momipites-Caryapollentites* lineage is less diverse than in the Upper Paleocene of Wyoming.

An assemblage referable to Zone P3 was recovered from one sample located near the stratigraphically highest, preserved part of the Porcupine Hills Formation of southwestern Alberta. This occurrence suggests that the Porcupine Hills is correlative with the Paskapoo Formation, rather than being younger as indicated by Carrigy (1971).

A palynological zonation for the Eocene and Oligocene of western and northern Canada has been outlined by Rouse (1977). Unpublished data indicate an Eocene or possibly younger age for strata at the extreme top of the Bonnet Plume Formation in Bonnet Plume Basin, as determined by the presence of pollen of *Carya, Juglans, Momipites, Pterocarya*, and *Tilia*. No palynological information is available on the Tertiary gravels of the southern Plains area.

Tertiary vertebrates

The Tertiary vertebrate assemblages of western Canada represent the northern extension of an outstanding record of Tertiary vertebrates in the western United States, which permits reconstruction of the physical and biological history of Tertiary time throughout western North America and facilitates comparisons with western Europe.

Fossil vertebrates have been found in Ravenscrag strata in the Souris Valley of southern Saskatchewan (Krause, 1977) and at Cypress Hills in southeastern Alberta (Krishtalka, 1973). The faunas are representative of Late Paleocene time, as typified by the Tiffanian and Clarkforkian stages of Colorado, Wyoming, and Montana (Fig. 4J.6). Among the mammals, the multituberculates make up a large portion of the faunas. Important but less numerous are the insectivores and the condylarths (primitive ungulates). Marsupials, so very abundant in Late Cretaceous time, are rare. The orders Rodentia, Perissodactyla, and Artiodactyla have yet to make their appearance, although their presence at the beginning of the Eocene sequence in Wyoming suggests that advance guards will eventually be found. A single mammalian jaw fragment found in the Ravenscrag on the eastern flank of the Cypress Hills has teeth almost identical with those of a condylarth from the Lower Paleocene (Puercan) of New Mexico.

Fossil vertebrates have been found in Paskapoo strata from Bow Valley in the south to Swan Hills in the north, but the richest occurrences are in the Red Deer Valley east of Red Deer city (Simpson, 1927; Krause, 1978; Fox, 1984). The mammalian fauna is similar to that of the Ravenscrag, with multituberculates and a variety of insectivores, but other groups are conspicuous, such as lemuroid primates and the relatively large pantotheres. Comparison of the Paskapoo mammals with those of the Paleocene faunas of the western United States shows that the age of the Red Deer Valley and Swan Hills occurrences is Late Paleocene (Tiffanian), whereas those of the Bow River area are somewhat older, probably Middle Paleocene (Torrejonian).

Beginning in Late Eocene time, a new mode of deposition changed the nature of the Tertiary record in western Canada. The oldest fossil fauna from these later Tertiary deposits is the Swift Current local fauna (Russell, 1965; Storer, 1978b). The presence of such modern groups as rodents, rabbits, perissodactyls, and selenodont artiodactyls provides a striking contrast with the Paleocene mammalian faunas. But primitive insectivores (leptictids) and condylarths (*Hyopsodus*) are hold-overs from the Early Eocene. Comparison with mammalian faunas in the Tertiary strata of the western United States shows a very close correlation with that in the upper part of the Uinta Formation (Myton local fauna) of Utah, which is regarded as the type of the Late Eocene (Uintan) mammalian assemblages. South of Swift Current, gravels and sands yield fossil mammals of Oligocene age in the upper part of the section.

PERIOD	SERIES	NORTH AMERICA LOCAL CLASSIFICATION	SEQUENCE IN SOUTHERN SASKATCHEWAN
TERTIARY	PLIOCENE	BLANCAN	
		HEMPHILLIAN	
	MIOCENE	CLARENDONIAN	
		BARSTOVIAN	WOOD MOUNTAIN
		HEMINGFORDIAN	
	OLIGOCENE	ARIKAREEAN	
		WHITNEYAN	
		ORELLAN	
		CHADRONIAN	CYPRESS HILLS
	EOCENE	DUCHESNEAN	
		UINTAN	SWIFT CURRENT
		BRIDGERIAN	
		WASATCHIAN	
	PALEOCENE	CLARKFORKIAN	RAVENSCRAG
		TIFFANIAN	
		TORREJONIAN	
		DRAGONIAN	
		PUERCAN	

GSC

Figure 4J.6. Zonation of Tertiary vertebrates, southern part of Western Canada Basin (L.S. Russell).

At the eastern end of Cyress Hills, a small mammalian fauna has been obtained northwest of Southfork station (Storer, 1984). The main sources of Cypress Hills fossils are farther west, in the valleys of Conglomerate Creek and its tributaries, notably Calf Creek (Russell, 1934, 1972; Storer, 1978c). The predominant mammalian fossils are the skeletal remains of the large perissodactyls called brontotheres (*'Titanotheres'*). Also present are teeth and bones of rhinoceroses, horses (*Mesohippus*), antelopes, camels and the selenodont artiodactyls such as oreodonts. Less common but interesting are small dog-like carnivores, a large creodont (*Hemipsalodon*), and the bunodont artiodactyls called entelodonts ('giant pigs'). Most remarkable is the presence of small multituberculates, the last known survivors of a characteristic Mesozoic group that was long supposed to have died out in Paleocene time.

The Cypress Hills mammalian fauna is a northern extension of part of the rich White River assemblage of the Black Hills region of South Dakota and Wyoming. The presence of brontotheres permits correlation with the Chadron Formation, the lowest division of the White River Group. More precisely, the Early Oligocene Cypress Hills fauna indicates that the Southfork locality represents very early Chadronian time, in contrast to the Calf Creek occurrences, which seem to be of mid-Chadronian age.

At one area west of Conglomerate Creek, the uppermost beds of the Cypress Hills Formation have yielded teeth of horses and camels of genera that are different from those of the Calf Creek fauna and indicate a Middle Miocene (Hemingfordian) age (Storer, 1975a). These and other fossil mammals of the Cypress Hills Formation show that deposition occurred at intervals from Late Eocene to Middle Miocene time.

The Wood Mountain Formation of southern Saskatchewan has yielded a rich and varied vertebrate fauna (Sternberg, 1930; Storer, 1975b). Among the mammals, the most conspicuous are the three-toed horse *Merychippus* and the antelope *Cosoryx*. Rodent remains are abundant in places, including representatives of the squirrel, prairie-dog, and beaver families. Rabbits are also present. The carnivores represent the dog, cat, and weasel families. The most striking element of the fauna is the four-tusked mastodon (*Gomphotherium*) one of the earliest proboscideans in the Tertiary record of North America. Comparison of the Wood Mountain fauna with those of the western States, notably Nebraska, permits correlation with the Middle Miocene (Barstovian). This fauna is of special interest as a transition between the earlier Tertiary assemblages and the Pleistocene and modern faunas of western North America. The Wood Mountain Formation contains mammalian fossils older than those of the Flaxville Gravel of Montana (Late Miocene; Storer, 1969, 1978a).

Mammalian teeth, notably those of horses and rodents, indicate a Middle Miocene age for the lower deposits of the Hand Hills gravels of eastern Alberta, and an Early Pleistocene age for the upper beds (Storer, 1976, 1978a).

Fossils occur within the Kishenehn Formation at a number of localities in the Flathead Valley on the Canadian and American sides of the International Boundary. The fossils consist of terrestrial gastropods and the teeth and jaws of small mammals, including marsupials, rodents, insectivores, rabbits, and a selenodont artiodactyl. Comparison of the Kishenehn mammals with faunas from Saskatchewan, Montana, and Utah reveals that the fossils are about on the boundary between uppermost Eocene and lowest Oligocene (Duchesnean or early Chadronian) ages.

Cretaceous–Tertiary boundary

The boundary between the Cretaceous and Paleocene has been controversial for many years, having been placed indefinitely within a formation or sequence of rocks, or placed arbitrarily at a distinct lithologic change (Russell, 1950, 1965; Bell, 1949; Kupsch, 1956; Jeletzky, 1960). Recent palynological, paleontological, and paleomagnetic studies have resulted in more precise placement of the boundary in some areas.

No hiatus is recognized within the Willow Creek Formation in the southern Foothills, although the lower part contains Late Cretaceous nonmarine invertebrates and the upper part contains a typical Paleocene fauna (Tozer, 1956). A problem remains as to whether a widespread unconformity exists in the central Alberta Plains as maintained by Allan and Sanderson (1945), Elliot (1960), and Russell (1950, 1965), or whether local channelling is responsible for removal of the upper part of the Scollard Formation (Ower, 1960; Campbell, 1962; Gibson, 1977). However, the controversial break in sedimentation occurs at the contact between the Scollard and Paskapoo formations some distance above the Cretaceous-Tertiary boundary as presently recognized (Fig. 4J.7).

In the Athabasca region, the Cretaceous-Tertiary boundary was placed, on the basis of palynology (Eliuk, 1969; Gunther and Hills, 1972; Sweet, in Jerzykiewicz and McLean, 1980) within the Coalspur beds between the Brazeau and Paskapoo formations. A more recent palynological study (Jerzykiewicz et al., 1984) reveals that a diverse angiosperm flora is replaced by a dominant assemblage of gymnosperm pollen and miospores near the base of the Mynheer coal of the Coalspur Formation. Analysis of this level for iridium shows an anomaly across 6 cm of the coal, reaching a maximum of about 6 ppb in the lowest 2 cm (Fig. 4J.8).

In the Plains, the Cretaceous-Tertiary boundary was placed at a floral break at the top of the Nevis Seam within the Scollard Formation of the Red Deer River valley

ERA	PERIOD	EPOCH	SOUTHERN ALBERTA	NORTHEASTERN MONTANA	SOUTHERN SASKATCHEWAN	NORTHWESTERN NORTH DAKOTA	SOUTHWESTERN MANITOBA	CENTRAL ALBERTA	CENTRAL FOOTHILLS
CENOZOIC	QUAT.	PLEISTOCENE	GLACIAL DRIFT	GLACIAL DRIFT	GLACIAL DRIFT	GLACIAL DRIFT	GLACIAL DRIFT	GLACIAL DRIFT	GLACIAL DRIFT
	TERTIARY	PLIOCENE			EMPRESS				
		MIOCENE		FLAXVILLE	WOOD MOUNTAIN			HAND HILLS	
		OLIGOCENE			CYPRESS HILLS	WHITE RIVER			
		EOCENE			SWIFT CURRENT CREEK	GOLDEN VALLEY			
		PALEOCENE	PORCUPINE HILLS WILLOW CREEK	SENTINEL BUTTE FT. UNION: TONGUE RIVER, LEBO, TULLOCK	RAVENSCRAG	SENTINEL BUTTE FT. UNION: TONGUE RIVER, LUDLOW, CANNONBALL	TURTLE MOUNTAIN CANNONBALL	PASKAPOO SCOLLARD	PASKAPOO COALSPUR
MESOZOIC	CRETACEOUS	UPPER CRETACEOUS	WILLOW CREEK BATTLE (KNEEHILLS) WHITEMUD ST. MARY RIVER BLOOD RESERVE BEARPAW JUDITH RIVER (BELLY RIVER) CLAGGETT (PAKOWKI)	HELL CREEK COLGATE FOX HILLS BEARPAW JUDITH RIVER CLAGGETT	FRENCHMAN BATTLE WHITEMUD EASTEND BEARPAW JUDITH RIVER LEA PARK RIDING MOUNTAIN	HELL CREEK COLGATE FOX HILLS PIERRE	BOISSEVAIN RIDING MOUNTAIN	SCOLLARD EDMONTON: BATTLE, WHITEMUD, HORSESHOE CANYON BEARPAW JUDITH RIVER / BELLY RIVER LEA PARK	COALSPUR BRAZEAU WAPIABI

GSC

Figure 4J.7. Correlation chart of uppermost Cretaceous and Tertiary formations in southern Saskatchewan and adjacent areas.

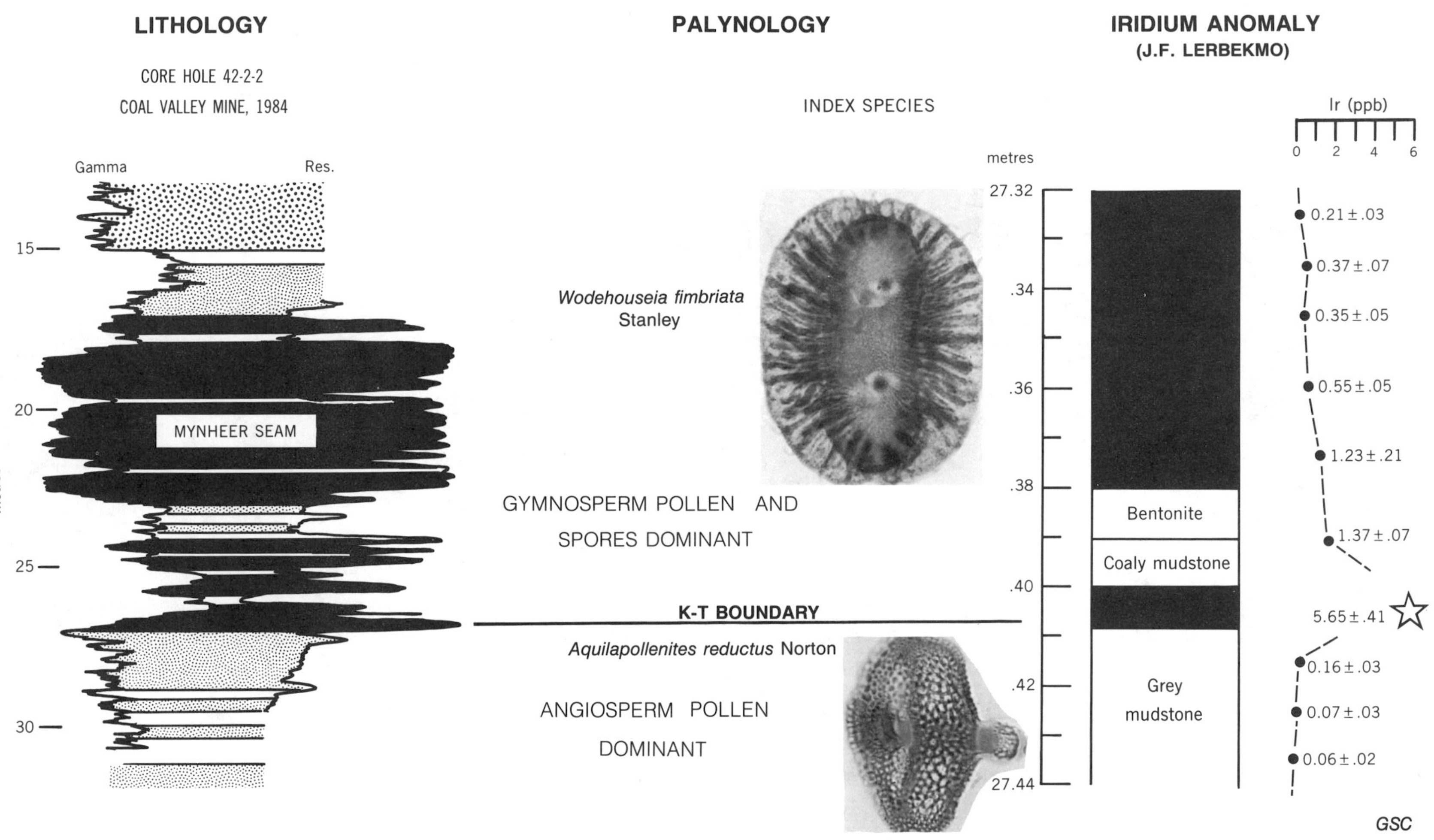

Figure 4J.8. Cretaceous-Tertiary boundary at base of Mynheer Seam, Coalspur Formation, as defined by palynology and iridium anomaly (J.F. Lerbekmo).

(Fig. 4J.9; Snead, 1969; Russell and Singh, 1978; Srivastava, 1970; Lerbekmo et al., 1979b). Latest Cretaceous palynomorphs become extinct at the seam, and no dinosaur remains are known to occur above the Nevis Seam (Lerbekmo et al., 1979b). A floral break marked by an upward decrease in diversity and abundance of the *Aquilapollenites* complex and associated angiosperms generally characterizes the boundary. An abrupt change to *Laevigatosporites* dominance, followed by an increase in gymnosperm pollen in basal Paleocene strata was reported by Sweet and Hills (1984), who placed the boundary at the base of the Nevis coal, which coincides with the position of an iridium anomaly (Lerbekmo and Coulter, 1985). Integrated magnetostratigraphic and biostratigraphic data for continental Cretaceous-Tertiary boundary sediments in Alberta allow a correlation with recognized seafloor magnetic anomalies 29 and 30, and the boundary occurs near the top of a reversed polarity zone shown to be 29r (Lerbekmo and Coulter, 1984, 1985). Eleven K-Ar dates on bentonite sanidines obtained near this boundary yield a mean age of 63.1 Ma (Lerbekmo et al., 1979a).

In Saskatchewan, the boundary was known to lie somewhere in the upper part of the Frenchman Formation or in the lower part of the Ravenscrag Formation (Fig. 4J.1; Russell, 1950, 1965; Kupsch, 1956; Jeletzky, 1960). The floral assemblages of the Frenchman-Ravenscrag succession were studied by Sweet (1978), who recognized two assemblages. The lower, obtained from the upper Frenchman Formation, contains angiosperm pollen considered to be of Late Maastrichtian age. The second, obtained from the basal Ravenscrag and dominated by gymnosperm pollen, was assigned an Early Paleocene age. These palynological studies suggest that the occurrence of the first coal bed at the base of the Ravenscrag Formation is nearly coincident with the Cretaceous-Tertiary boundary. The extinction of many of the distinctive taxa characteristic of the Maastrichtian Age is assumed to mark the Cretaceous-Tertiary boundary. Following the extinctions, an increase in the relative abundance of simple tricolpate, and triporate pollen more characteristic of temperate climates occurred within a dominantly gymnosperm assemblage.

Farther east, the Maastrichtian Boissevain Formation of southern Manitoba is overlain by the Turtle Mountain Formation. Bamburak (1973, 1978) recognized foraminifera of the Paleocene Cannonball sea from shales within the succession.

Figure 4J.9. Cretaceous-Tertiary boundary (arrows) in Red Deer Valley near Drumheller, Alberta. The boundary occurs within the Scollard Formation of the Edmonton Group, at the base of the Nevis Coal Seam (photo A.R. Sweet, ISPG 2454-5).

The approximate position of the Cretaceous-Tertiary boundary in Yukon Territory and Northwest Territories is known within Brackett Basin, Bonnet Plume Basin and the Yukon Coastal Plain. Near Fort Norman, in Brackett Basin, the boundary is placed within the Summit Creek Formation near the base of a coal, conglomerate, and tuff-bearing unit (Bihl, 1973; Wilson, 1978; Brideaux, in Yorath and Cook, 1981; Ricketts, 1985). In Bonnet Plume Basin, the change from Maastrichtian to Lower Paleocene occurs near the top of the Bonnet Plume Formation at the base of conglomeratic and coal-bearing beds (Rouse and Srivastava, 1972; Long, 1978).

The Cretaceous-Tertiary boundary in Yukon Coastal Plain and Mackenzie Delta has been variously placed. Young (1975a, b) and Wilson (1978) considered the boundary to be at the base of an interval of coal and conglomerate within the Aklak Member of the Reindeer Formation. More recent palynological evidence favors a boundary position at the top of the Tent Island Formation or possibly within the basal beds of the Moose Channel Formation (D.J. McIntyre and A.R. Sweet, pers. comm., 1985).

LITHOSTRATIGRAPHY

Northern basins and Mackenzie Delta

Upper Cretaceous strata record the gradual northward migration of depocentres and shorelines, such that by Paleocene time most deposition was in Beaufort Sea. The rising Cordillera and Yukon Fold Complex supplied vast quantities of sediment to form major deltaic depositional complexes. Over 11 km of Upper Cretaceous and Cenozoic sediments accumulated beneath Beaufort Sea (Fig. 4J.10), which is flanked to the southeast by the northern Interior Platform and to the southwest by the northern Cordillera.

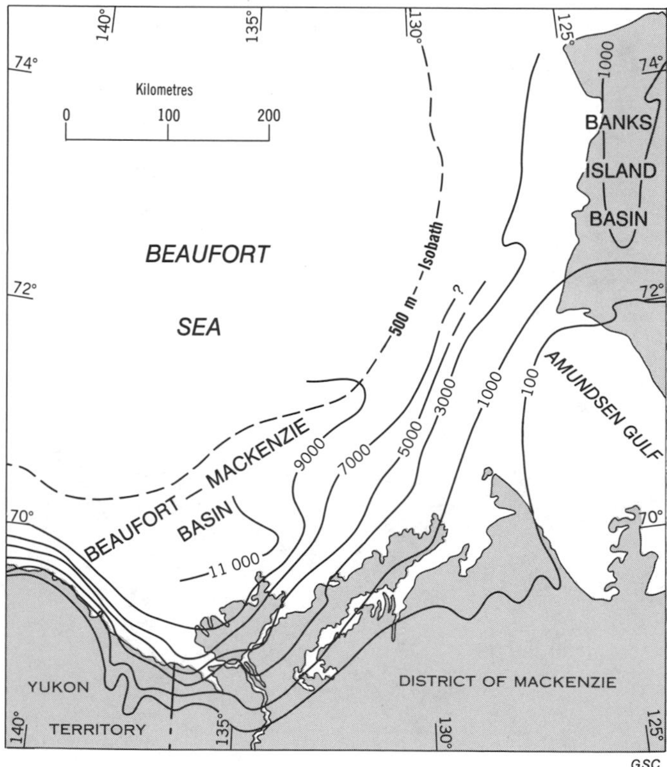

Figure 4J.10. Isopachs of Upper Cretaceous and Tertiary sediments, Beaufort-Mackenzie Basin (J. Dixon and J.R. Dietrich).

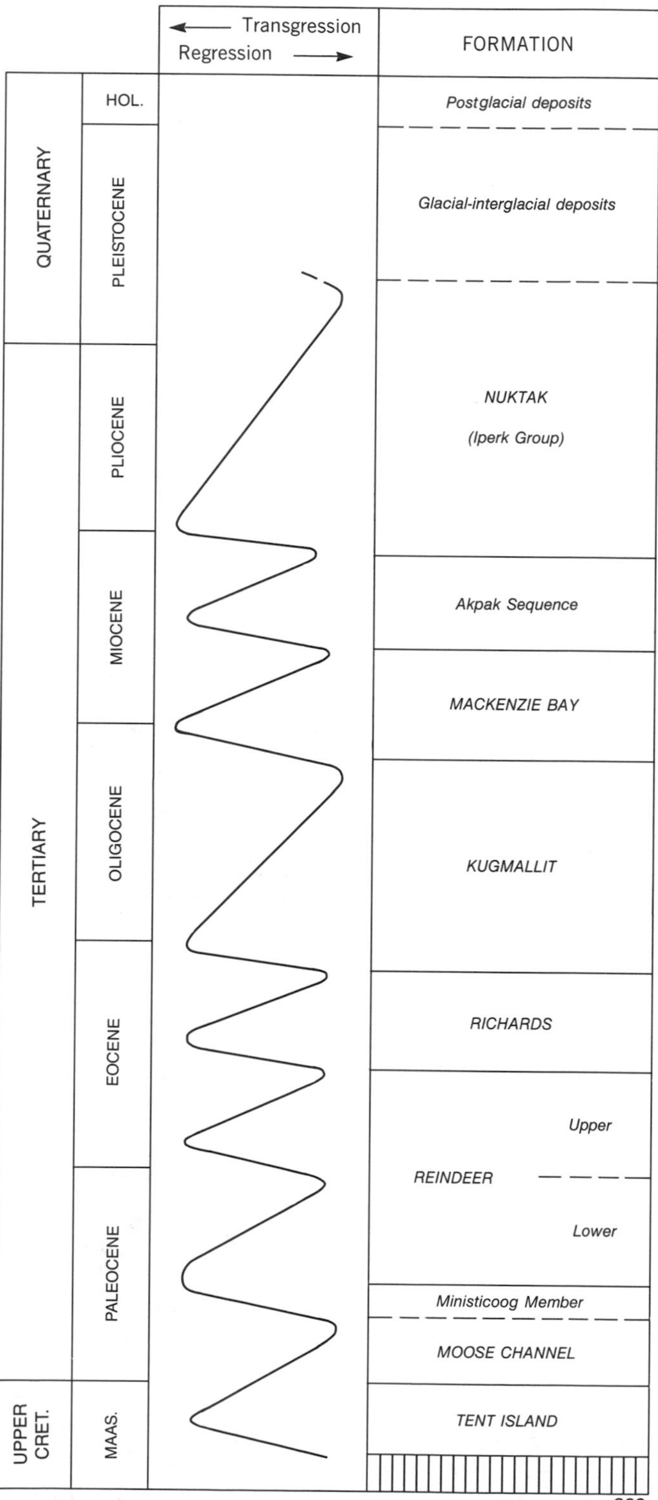

Figure 4J.11. Transgressive-regressive cycles in Tertiary strata of the Beaufort-Mackenzie Basin (J. Dixon).

By latest Miocene time, the final orogenic pulse was coming to an end, resulting in a major regional unconformity. The succeeding Pliocene-Pleistocene succession forms a structurally simple, prograding wedge of sediment under the Beaufort Shelf.

In the Mackenzie Delta area, the interaction of eustasy and distinct tectono-sedimentary pulses is reflected by large-scale, regressive, deltaic or alluvial cycles, which gradually built seaward and outward to form the southern continental shelf of the Beaufort Sea. In landward areas, tectonic uplifts brought previously deposited molassic wedges and older rocks to the surface where they were eroded and recycled into the next younger clastic complex. The complexes (Fig. 4J.11) represent regressive cycles and record substantial shifts of environments, ranging from alluvial plain through delta plain to prodelta and mid-basin.

Paleocene-Middle Eocene

Paleocene to Middle Eocene rocks include the Moose Channel and Reindeer formations (Fig. 4J.4). These strata have limited outcrop in the northern Richardson Mountains and Caribou Hills but are extensively developed in the subsurface of Mackenzie Delta and Tuktoyaktuk Peninsula. The Moose Channel Formation of the Fish River area west of Mackenzie Delta comprises over 2700 m of chert conglomerate, sandstone, mudstone, and coal-bearing sediments, interpreted as fluvial, delta-plain, nearshore, and prodelta deposits (Holmes and Oliver, 1973; Young, 1975a). Within Mackenzie Delta, lower Moose Channel strata are predominantly deltaic, although the upper beds of the Ministicoog Member are predominantly marine mudstone and siltstone and mark the beginning of another major phase of deltaic sedimentation represented by the overlying Reindeer Formation. In outcrop, only the lower part of the Reindeer Formation is preserved, whereas complete sections are present in the subsurface. In the Yukon Coastal Plain, the lower beds, included in the Aklak Member by Young (1975a), are characterized by coal seams and a high proportion of alluvial deposits (Fig. 4J.12). Under Mackenzie Delta the Reindeer Formation is dominated by prodelta to delta plain, coarsening-upward cycles throughout its entire thickness (Dixon, 1981). The Reindeer succession is rich in temperate terrestrial palynomorphs and yields sparse, brackish water foraminifers of the *Saccammina-Trochammina* spp. assemblage.

To the south in Bonnet Plume Basin, the upper Bonnet Plume Formation includes a succession of conglomerate, sandstone, shale, and lignite (Mountjoy, 1967), considered to be Maastrichtian to Paleocene and possibly as young as Eocene in age (Rouse and Srivastava, 1972). Beds of Paleocene age are restricted to the upper conglomerate and lignitic unit. These sediments were deposited in a depression between the Knorr and Deslauriers faults, an assemblage of nearly vertical, north-trending right-lateral strike-slip faults (Norris and Hopkins, 1977).

Maastrichtian to Paleocene conglomerate, sandstone, ash beds, and lignite occur in Brackett Basin, a small basin adjacent to the Mackenzie Mountains (Yorath and Cook, 1981; Ricketts, 1985). These sediments, included in the Summit Creek Formation, represent the development of synorogenic molasse in the form of a broad alluvial fan that prograded eastward during Laramide uplift of the Mackenzie Mountains. The ash beds may have been derived from eruption of the Carmacks volcanics, which yield K-Ar dates ranging between 67.9 and 73.1 Ma (Grond et al., 1984), consistent with a Maastrichtian age. In the eastern Great Bear Basin, local nonmarine and transitional Maastrichtian-Paleocene sediments mark the end of marine conditions.

Middle Eocene-Lower Oligocene

The Richards Formation of Middle Eocene to Early Oligocene age consists of approximately 2000 m of mudstone beneath Mackenzie Delta (Fig. 4J.3). The Richards Formation is a prodelta-shelf deposit in which gravity-flow beds have been identified (Glaister and Hopkins, 1974). The formation is marked by the *Haplophragmoides* spp. agglutinated foraminiferal assemblage and, in the lower part, by a rich zone of dinoflagellates. The upper part of the Richards Formation

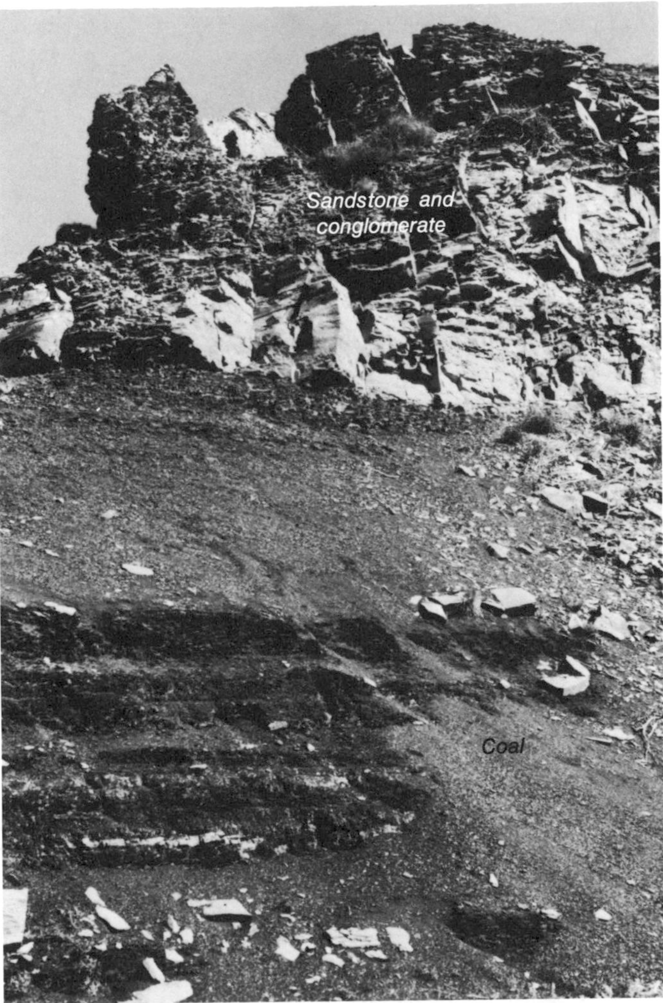

Figure 4J.12. Coal, mudstone, conglomerate and sandstone, in upward succession; Aklak Member of the Reindeer Formation, Aklak Creek, northern Richardson Mountains (photo F.G. Young, ISPG 2454-3).

probably represents the northeastward-migrating toe of a major upper Paleogene deltaic wedge that is included in the Kugmallit Formation. Landward equivalent facies have not been preserved over most of the delta because of extensive pre-Nuktak erosion.

The Kugmallit Formation occurs under Mackenzie Delta; the thickest occurrence is in the Ivik area, where 2500-3000 m of strata are present. In that area, and in some wells to the north, a major unconformity has been identified within the lower part of the Kugmallit Formation. Kugmallit strata in Mackenzie Delta and nearshore areas consist of numerous prodelta- to delta-front and delta-plain, coarsening-upward cycles (Young and McNeil, 1984). In the uppermost few hundred metres of the formation, thick sandstone units are more typical of delta-plain environments. A rich terrestrial palynoflora, dated as Oligocene, occurs in the Kugmallit Formation (Norris, 1986). Late Oligocene transgression curtailed deltaic sedimentation and the succeeding prodelta mudstone of the Mackenzie Bay Formation rests abruptly on Kugmallit beds.

Unconsolidated conglomerate, mud, and brown peat of Late Eocene age were described from the Old Crow Basin, a structural depression in northern Yukon Territory (Hopkins and Norris, 1974). The palynological assemblage suggests a temperate to warm temperate climate.

Upper Oligocene-Middle Miocene

Mudstones of the Mackenzie Bay Formation (Young and McNeil, 1984) of Late Oligocene to Middle Miocene age, are generally thin, 600 m or less, in the nearshore area. The Mackenzie Bay beds are marine, prodeltaic to shelf and slope sediments. No nonmarine strata have been recognized. The Mackenzie Bay Formation carries cool-temperate to boreal terrestrial palynomorphs and the neritic *Cibicides* spp. foraminiferal assemblage, which is rich in calcareous benthonic species. Pre-Nuktak erosion truncated the Mackenzie Bay Formation to the south, the erosional edge occurring just seaward of Mackenzie Delta.

In Mackenzie Delta and nearshore areas, the Mackenzie Bay Formation is the youngest succession preserved under the pre-Nuktak unconformity (Fig. 4J.4). However, farther offshore an additional stratigraphic unit, the Upper Miocene Akpak sequence, lies above the Mackenzie Bay beds (Dietrich et al., 1985).

Upper Miocene to Pleistocene

In Mackenzie Delta, the term Beaufort Formation has been used for at least two successions. Young and McNeil (1984) regarded the Beaufort Formation as a Miocene unit that underlies and is, in part, laterally equivalent to the Mackenzie Bay Formation. Hea et al. (1980) and Willumsen and Coté (1982) used the term Beaufort for the younger Plio-Pleistocene succession that Young and McNeil called the Nuktak Formation and Jones et al. (1980) termed the Iperk Group. It remains debatable whether or not the Beaufort Formation, as used and correlated by Young and McNeil, is laterally equivalent to the Mackenzie Bay and is separable from the upper beds of the Kugmallit Formation. The Mackenzie Bay Formation is considered herein to rest directly on Kugmallit strata, and the Beaufort Formation is not identified as a separate unit under Mackenzie Delta.

Beds of Pliocene to Pleistocene age are represented by the Nuktak Formation (Young and McNeil, 1984; the Beaufort Formation, Iperk Formation or Iperk sequence of other authors), gravels tentatively identified as the Beaufort in Anderson Plain (Yorath and Cook, 1981), and the Herschel Island Formation (Johnson et al., 1976) on the Yukon Coastal Plain. These beds rest with marked erosional unconformity on older strata, and the unconformity represents the termination of tectonic activity.

The Nuktak consists of a lower, nonmarine gravel member and an upper, marine mud member and contains the cool water, inner shelf, *Elphidium* spp. foraminiferal assemblage. On Tuktoyaktuk Peninsula, equivalent strata are dominated by gravels, as are presumed equivalent beds in Anderson Plain. These gravels and the similar deposits in Old Crow Plain are interpreted as alluvial deposits. The Nuktak Formation under Mackenzie Delta is between 100 and 300 m thick, thickening seaward.

The stratigraphic relationship between Nuktak and Herschel Island Formations is not clear, but the two are probably equivalent. The Herschel Island Formation, a mixture of nonmarine and marine beds, is about 60 m thick.

Pleistocene glacio-fluvial deposits that occur throughout much of the northern mainland are described in more detail in the companion volume on Quaternary geology (Fulton, 1989).

Off-shore Beaufort-Mackenzie Basin

In the off-shore Beaufort-Mackenzie Basin, Late Cretaceous to Recent sedimentation was, and still is, dominated by deltaic processes, resulting in a series of thick, generally northward prograding, deltaic complexes. On-shore terminology (Young and McNeil, 1984; Dixon, 1986) is less appropriate basinward where lithological differentiation is less apparent in successions dominated by prodeltaic to basinal mudstone and siltstone, and additional strata are present. Eleven transgressive-regressive sequences have been identified using reflection seismic profiles, well data, and, to a limited extent, outcrop data (Fig. 4J.13, 4J.14; Dietrich et al., 1985). The sequence boundaries are usually subaerial unconformities at the basin margins and basinward, may represent submarine unconformities. In many parts of the basin, the sequences are disrupted by large-scale, syndepositional listric faults.

The Fish River sequence, of Maastrichtian to Paleocene age, is identified with certainty only on the basin margins. It is probably 4000 to 6000 m thick in the Natsek area, and appears from seismic data to have a depocentre in the western Beaufort Sea. The known succession of the lower part of the Fish River is characterized by prodelta mud and silt of the Tent Island Formation and is overlain by delta-plain and delta-front sediments of the Moose Channel Formation.

The Reindeer sequence, of Paleocene to Early Eocene age, includes progradational prodelta and shelf muds (Ministicoog Member of Moose Channel Formation) and delta-plain and delta-front facies (Reindeer Formation). Seismic stratigraphic analysis has shown that strata encompassed by the Reindeer Formation contain a major unconformity (Dixon et al., 1985) that is angular in the western Beaufort Sea but disconformable in the central Beaufort-Mackenzie Delta area. This division into lower and upper successions is accompanied by a shift in the locus of deltaic sedimentation. The main depocentre of the Reindeer sequence appears to have been located in the Ellice area where 4000 m of strata may occur. Along the southern part of Tuktoyaktuk Peninsula, the lower sequence comprises nonmarine and marginal marine sands and gravels, which grade northeastward into prodeltaic muds. The upper sequence is centred over Mackenzie Bay-Ellice Island and is about 2000 m thick. Deltaic sediments are dominant in the depocentre, with prodeltaic muds in the western Beaufort Sea. The basinward extent of Reindeer strata is unknown.

The Richards sequence, of Middle to Late Eocene age, is 1200 to 1500 m thick in the Ivik area, and up to 2000 m thick in fault-bounded synclines in the Ellice Island area. Mud and silt are dominant, although some local sand and gravel are present. A Late Eocene period of uplift and/or sea-level fall led to erosion of the Richards succession at the basin margin and paleo-shelf edge; consequently Richards strata may be very thin or absent in many parts of the outer Beaufort Shelf.

The Kopanoar sequence, of Late Eocene and/or Early Oligocene age, is identified only in the farthest offshore wells. It is separated from older strata by an erosional unconformity and overlain conformably but abruptly by the Kugmallit sequence. Where penetrated, the sequence rarely exceeds 1000 m, although thicknesses of up to 2500 m are indicated by seismic data. The Kopanoar sequence consists mainly of silt and mud of deep-water origin, with local sandy intervals.

The Kugmallit sequence, ranging in age possibly from Late Eocene to Middle or earliest Late Oligocene, is present under much of Beaufort Shelf. It is as much as 3000 m thick in the vicinity of North Issungnak L-86. Interbedded sand and mud of delta-front origin occur in the depocentre, but the succession grades landward into delta-plain deposits and basinward into slope and basinal muds.

The Mackenzie Bay sequence, of Late Oligocene to Middle Miocene age, rests abruptly on Kugmallit strata and is present under most of the Beaufort Shelf. It ranges from 362 m at Netserk B-44 to as much as 2500 m thick in the western Beaufort Sea. The known lithotypes are dominantly mud and silt.

The Akpak sequence, probably of Middle to early Late Miocene age, is present throughout most of the basin. It is thickest in the western Beaufort Sea where it is about 2500 m thick. Where penetrated, the Akpak is predominantly mud and silt with a few thin sand beds.

The Iperk sequence, of Pliocene to Pleistocene age, is a thick, lens-shaped deposit whose thickness varies from a few tens of metres at the basin margin to about 5000 m at its depocentre northeast of Kenalooak J-94 well. It is a relatively unstructured, simple progradational sequence which lies with marked unconformity on older beds.

The youngest identifiable seismic sequence is the Shallow Bay, assumed to have been deposited during or after the last major glaciation. It may include sediments as old as latest Pleistocene but is mostly Holocene. The sequence rests erosionally on older strata. It consists of mud, silt, sand, and gravel, and includes sediments of the modern delta.

The Beaufort-Mackenzie Basin is discussed in its broader context within the Arctic region in the companion volume by Grantz et al. (1990).

Western Canada Basin

In Western Canada Basin, the final marine incursion, the short-lived Paleocene Cannonball Sea of Gulfian origin, extended along an epeirogenic depression that extended northward through North Dakota into Manitoba. Marine deposition was soon terminated as continental sediments spread eastward. Farther west, widespread coal deposition was followed by extensive development of alluvial fans across the foreland basin. Late-phase tectonic movements deformed Willow Creek and Paskapoo strata and produced the latest thrust plates of the Foothills. The Sweetgrass Arch underwent renewed uplift, which was accompanied by small, mid-Eocene intrusions. Post-orogenic normal faults developed along the western flank of the Canadian Rocky Mountains, with resultant accumulation of sediments within the down-faulted blocks. During mid-Tertiary time, epeirogenic uplift was dominant, erosion was extensive, peneplains were produced, and elements of the present drainage system were established. Subsequent erosion removed Tertiary sediments from much of the eastern area, and the main preserved sequences are concentrated along the Foothills and in a series of erosional remnants extending along the Canada-United States border from Cypress Hills in Alberta to Turtle Mountain in Manitoba (Fig. 4J.2). Only scattered remnants of the late alluvial deposits are preserved in the southern Interior Plains.

Paleocene

Continental sedimentation continued without apparent interruption from Cretaceous to Paleocene time along the eastern side of the Columbian Orogen. Continental deposits of Paleocene age are widespread in southern Saskatchewan and much of southern and central Alberta, extending into the Foothills and Rocky Mountains. Thick accumulations of plant debris later became extensive commercial coal seams. The Paleocene in the Interior Plains ranges in thickness from more than 1500 m in the Porcupine Hills of southwestern Alberta to a few hundred metres at Turtle Mountain.

The Willow Creek Formation of the southern Alberta Foothills consists of grey sandstone and variegated shales overlain by buff-weathering beds (Douglas, 1950; Carrigy, 1971). It contains two molluscan faunas (Tozer, 1956); the older is correlated with that in the Upper Cretaceous Edmonton Group and the younger with that in the Paleocene Paskapoo and Ravenscrag formations. The Willow Creek, more than 1200 m thick near Castle River, thins northeastward, the lower part merging with the upper Edmonton. The upper part is bevelled beneath the overlying Porcupine Hills Formation (Douglas, 1950).

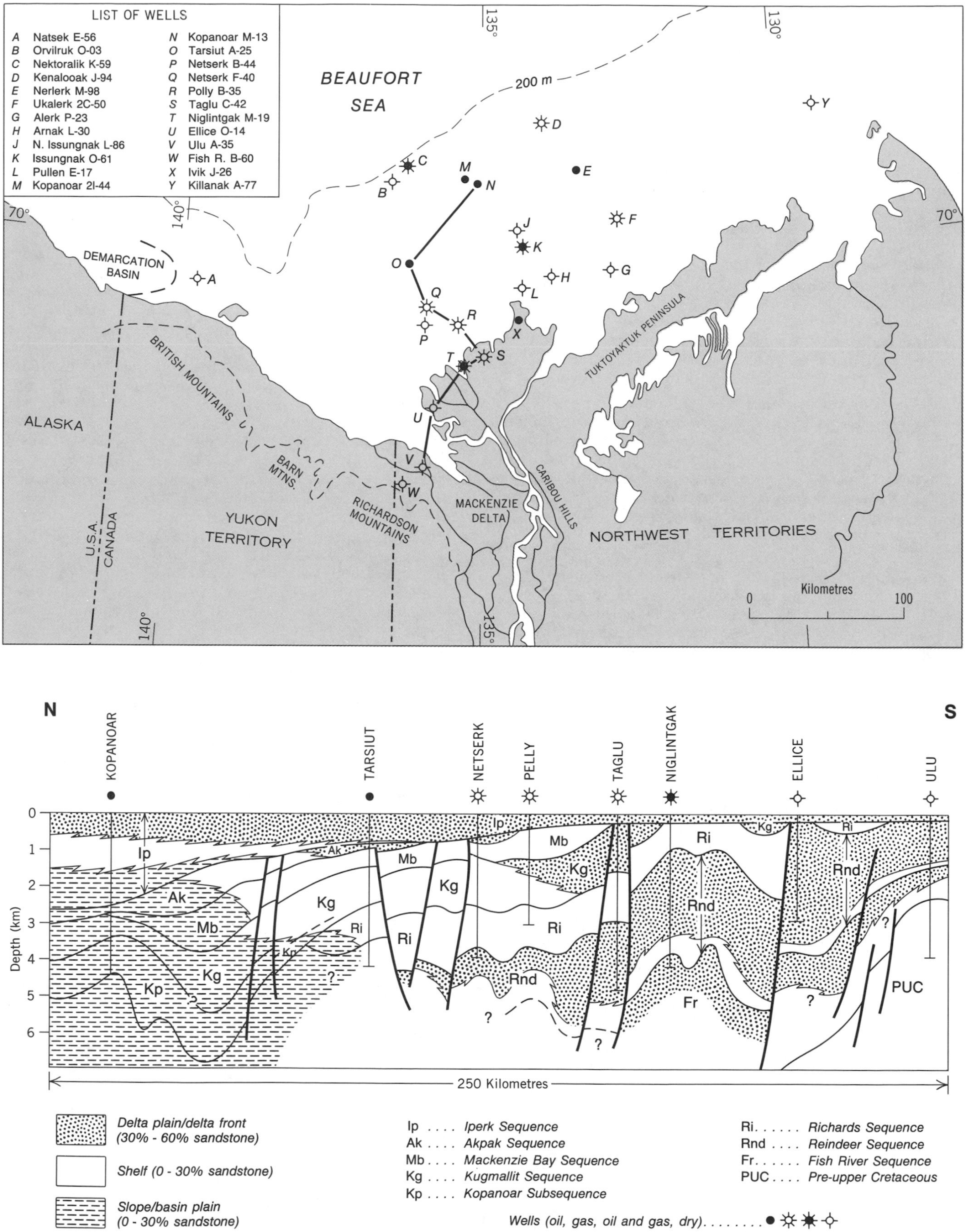

LIST OF WELLS

A	Natsek E-56	N	Kopanoar M-13
B	Orvilruk O-03	O	Tarsiut A-25
C	Nektoralik K-59	P	Netserk B-44
D	Kenalooak J-94	Q	Netserk F-40
E	Nerlerk M-98	R	Polly B-35
F	Ukalerk 2C-50	S	Taglu C-42
G	Alerk P-23	T	Niglintgak M-19
H	Arnak L-30	U	Ellice O-14
J	N. Issungnak L-86	V	Ulu A-35
K	Issungnak O-61	W	Fish R. B-60
L	Pullen E-17	X	Ivik J-26
M	Kopanoar 2I-44	Y	Killanak A-77

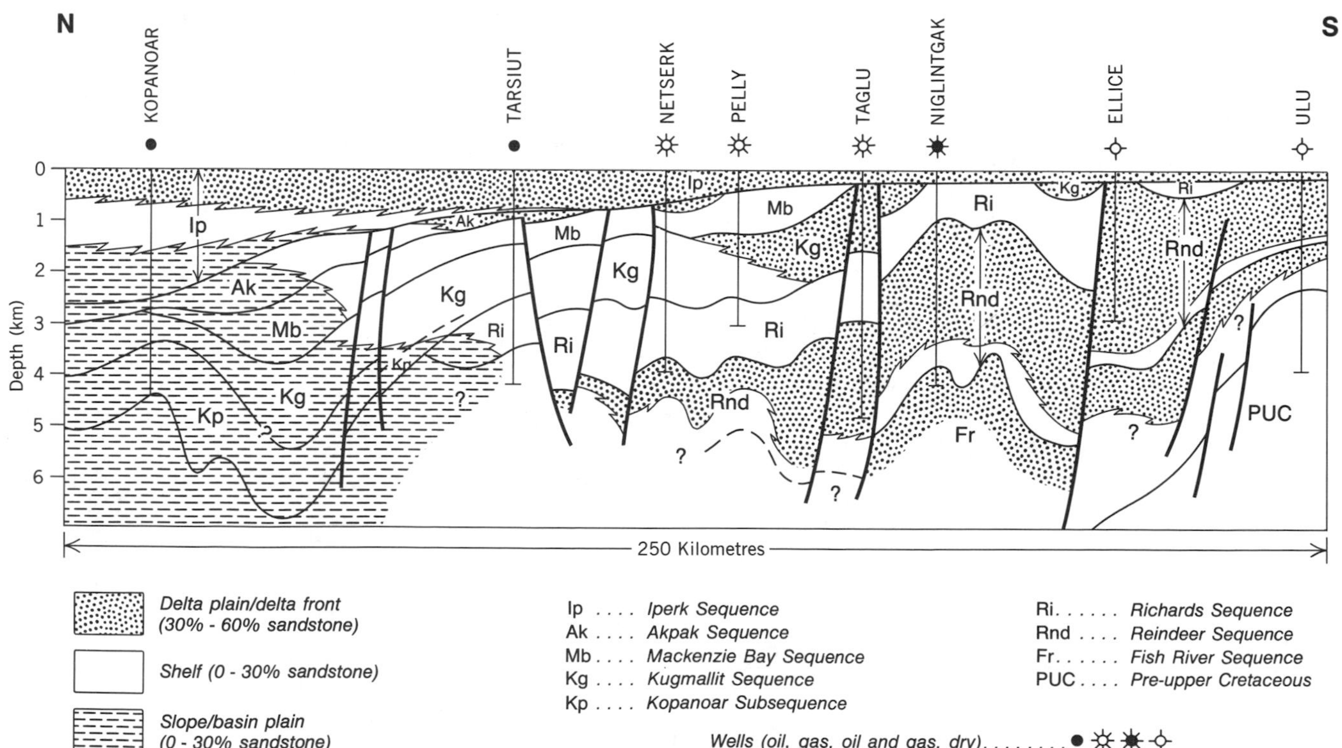

Figure 4J.13. Schematic cross-section through the Beaufort-Mackenzie Basin (from Dietrich et al., 1985).

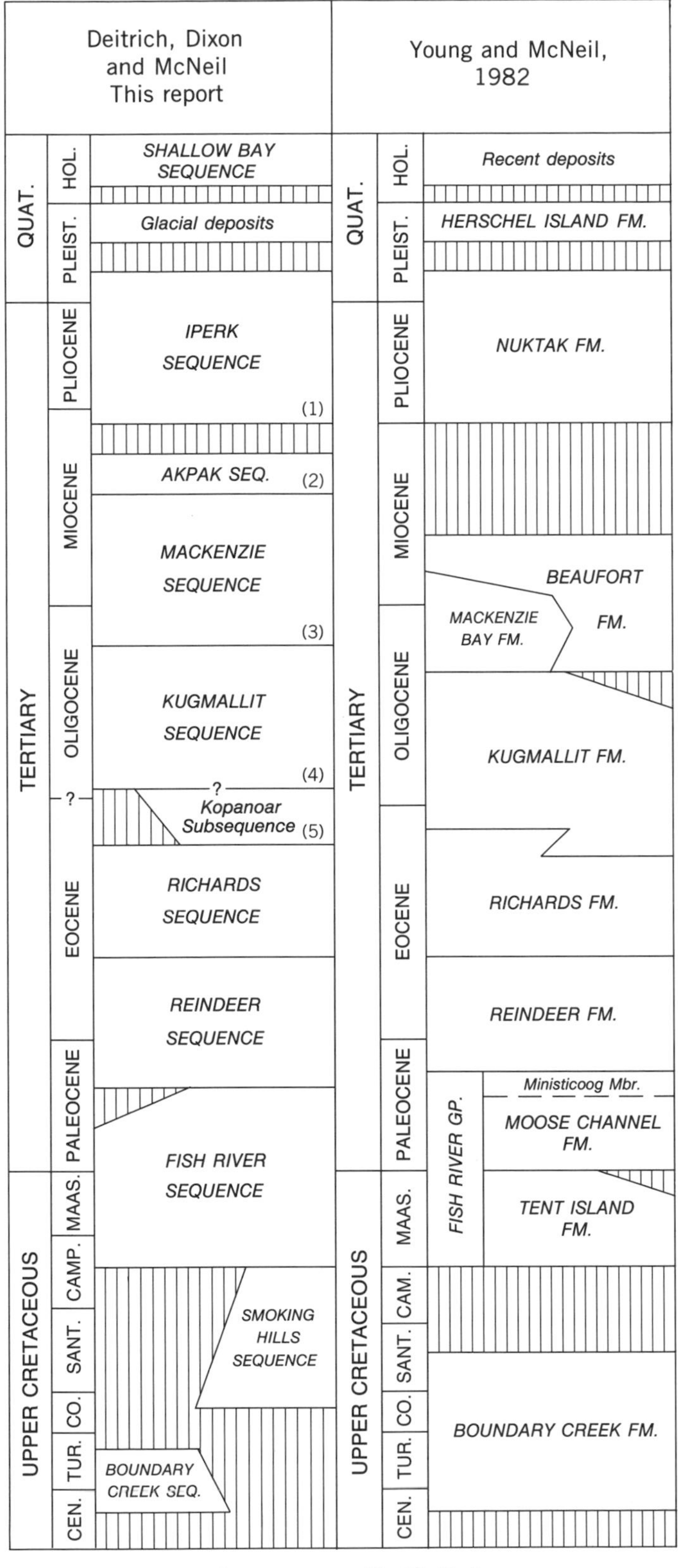

Deitrich, Dixon and McNeil This report
Young and McNeil, 1982
QUAT.
HOL.
PLEIST.
TERTIARY
PLIOCENE
MIOCENE
OLIGOCENE
EOCENE
PALEOCENE
UPPER CRETACEOUS
MAAS.
CAMP.
SANT.
CO.
TUR.
CEN.
SHALLOW BAY SEQUENCE
Glacial deposits
IPERK SEQUENCE
(1)
AKPAK SEQ. (2)
MACKENZIE SEQUENCE
(3)
KUGMALLIT SEQUENCE
(4)
?
Kopanoar Subsequence (5)
RICHARDS SEQUENCE
REINDEER SEQUENCE
FISH RIVER SEQUENCE
SMOKING HILLS SEQUENCE
BOUNDARY CREEK SEQ.
CAM.
Recent deposits
HERSCHEL ISLAND FM.
NUKTAK FM.
BEAUFORT FM.
MACKENZIE BAY FM.
KUGMALLIT FM.
RICHARDS FM.
REINDEER FM.
FISH RIVER GP.
Ministicoog Mbr.
MOOSE CHANNEL FM.
TENT ISLAND FM.
BOUNDARY CREEK FM.

Numbers refer to sequences identified in the Kopanoar M-13 well (Dixon et al., 1984) .... (4)

Figure 4J.14. Stratigraphic terminology used for Upper Cretaceous, Teritiary and Quaternary strata in the Beaufort-Mackenzie Basin (from Dietrich et al., 1985).

The Porcupine Hills Formation includes about 1000 m of coarse-grained, massive crossbedded sandstone and calcareous bentonitic shale. In the type area, the Porcupine Hills contains abundant freshwater molluscs of Paleocene age (Tozer, 1956).

The Paskapoo Formation is more than 3000 m thick in the central Foothills and thins eastward across the central Interior Plains to an erosional edge. It contains massive, crossbedded, coarse-grained sandstone, bentonitic shales, lignite, and conglomerate. Carrigy (1971) argued that the Paskapoo was overlain by the Porcupine Hills Formation, but that relationship is not supported by the mid- or Late Paleocene age determinations established from well-preserved molluscan fauna (Tozer, 1956), mammalian remains (Russell, 1958; Fox, 1984; Krause, 1978), and palynomorphs (A.R. Sweet, pers. comm., 1986). The thin, lenticular Entrance conglomerate in the Athabasca River region has been considered to mark the base of the Paskapoo in the central Foothills, but Jerzykiewicz and McLean (1980) placed the boundary above the younger Coalspur coal beds, now included in the Coalspur Formation (Jerzykiewicz, 1985). On the basis of palynological studies, the Cretaceous-Paleocene boundary occurs at the base of the Mynheer coal within the lower Coalspur beds (Jerzykiewicz et al., 1984; Jerzykiewicz and Sweet, 1986). Primate fossils (Krause, 1978) permit correlation of Paskapoo strata of southwestern Saskatchewan, southeastern and central Alberta, and the Swan Hills of north-central Alberta. The source of volcanic detritus in Paleocene formations of the Foothills and Plains was identified by Carrigy (1971) as volcanic rocks in the interior of British Columbia and the source of chert, dolomite, and limestone, as Paleozoic carbonates and cherts of the thrust plates of the Rocky Mountains.

In Red Deer Valley near Drumheller, the uppermost Cretaceous-lowermost Paleocene beds are included in the Scollard Formation (Fig. 4J.8), originally designated as a member of the Paskapoo Formation (Irish, 1970; Holter et al., 1975) but since included as a formation in the Edmonton Group (Gibson, 1977). The formation, having a maximum thickness of 85 m, consists of sandstone, shale, and coal. Two main coal zones are recognized, the Nevis and Ardley (lower Ardley of some authors; Campbell, 1967; Holter et al., 1975; Gibson, 1977). Gibson suggested that the coarser sediments were deposited by an extensive system of braided rivers and that the finer mud and carbonaceous material were deposited in swamps.

In the Cypress Hills along the Alberta-Saskatchewan border, a cyclic succession of Paleocene lignite, sand, silt, and clay are included in the Ravenscrag Formation (Fig. 4J.15; Furnival, 1946; Irish, 1965; Whitaker et al., 1978). The base is drawn below the lowest commercial coal seam and appears to be conformable with the underlying dinosaur-bearing Frenchman Formation. Owing to variable erosion the Ravenscrag is thickest in the Estevan area of southern Saskatchewan. The volcanic material of the Frenchman and Ravenscrag formations was likely derived from the Elkhorn Mountain Volcanics and the Adel Mountains Volcanics of western Montana (Misko and Hendry, 1979). Uplift of the Front Ranges of the Rocky Mountains may account for the increased carbonate and quartz in the sands of the Ravenscrag Formation. A large and diverse assemblage of vertebrate fossils in the Paleocene sediments is considered to be equivalent of the Tiffanian and Clarkforkian stages (Upper Paleocene) of Colorado, Montana, and Wyoming (Krause, 1977).

The Turtle Mountain Formation of Manitoba is an eastward equivalent of the Ravenscrag Formation. It consists of shale, sandstone, and lignite beds, having a maximum thickness of about 120 m. Foraminifera provide evidence of the Paleocene Cannonball sea (Bamburak, 1973, 1978; Fox and Olsson, 1969). Conditions fluctuated from shallow marine to continental, with swamps in which the plant material accumulated.

Eocene-Oligocene

In upper Flathead Valley of the southern Rocky Mountains, the Kishenehn Formation of latest Eocene to earliest Oligocene (Duchesnean to early Chadronian) age (Russell, 1964) was deposited in a structural basin along the down-thrown side of the Flathead normal fault. Two main facies (Price, 1965; Jones, 1969; McMechan, 1981) are conglomerates and breccias adjacent to the fault derived locally from the Paleozoic and Precambrian rocks of the upthrown side, and a basinal facies of fine clastic material, marl, and lignite. Sedimentation took place simultaneously with displacement on the fault, shown by the sequence of clasts, which is the inverse of the normal stratigraphic order. About 300 m of sediments are present. The Kishenehn lies with angular unconformity on strata deformed by the Laramide Orogeny, and establishes an upper limit for dating the underlying structures of the southern Rocky Mountains in Canada.

A sequence of sedimentary rocks along Parsnip River in the Rocky Mountain Trench of northeastern British Columbia, has yielded a palynomorph assemblage correlative to assemblages of late Early Oligocene (Chadronian) age (Hopkins et al., 1972). The Oligocene sediments are mildly deformed relative to the Rocky Mountains, conforming with evidence from farther south that the time of most intense orogenic activity occurred earlier. The deformation may relate to post-Laramide, strike-slip faulting.

The Laramide Orogeny, which ended during the Eocene Epoch, had resulted in major uplift and profound deformation of the Rocky Mountains, accompanied by thrusting and folding within the Foothills, and associated downwarping of the foredeep basin. Farther east, the Sweetgrass Arch acquired its present form. A minette offshoot related to the Sweetgrass laccoliths of northern Montana and dyke equivalents in southern Alberta has yielded a K-Ar age of 48 Ma (mid-Eocene; Baadsgaard et al., 1961), and was considered as evidence of the late age of the Laramide Orogeny, but the intrusions may be unrelated to convergent tectonism.

Following the uplift of the Rocky Mountains during the Laramide Orogeny, they and the adjacent plains underwent intense erosion. Estimates based on the coal rank of near-surface coal (Hacquebard, 1977; Nurkowski, 1984) and from shale compaction (Magara, 1976) indicate that 1 to 2 km of sediments were removed by erosion during the Tertiary Period. Hitchon (1984), in a study of geothermal gradients, indicated that in the most westerly areas, 2.7 to 3 km of sediments were removed. As pointed out by Beaumont et al. (1985), when overthrusting ceased, denudation of the mountains resulted in isostatic adjustments, uplift, and erosion of the foreland basin. Only insignificant quantities of the enormous amount of transported debris are preserved in the Plains. Upper Eocene to Oligocene coarse gravels are preserved on the Interior Plains as far east as Saskatchewan, where they are only few metres to a few tens of metres thick. These younger fluvial deposits were eroded during subsequent uplift and stream incision, which left only remnants.

The Oligocene Epoch marked the beginning of a change to cooler and dryer climate, which culminated in the ice ages of Pleistocene time. By Oligocene time, horses, rhinoceroses, camels, canids, and felids were making their appearance.

Aggradation across the former foredeep and the platform to the east during Late Eocene and Oligocene time left relatively thin deposits, represented now by the local Eocene Swift Current beds and the Oligocene Cypress Hills conglomerate. The Upper Eocene to Middle Miocene Cypress Hills Formation is composed of fluviatile conglomerate, gravel, sand, silt, and clay, capping the Cypress Hills in southeastern Alberta and southwestern Saskatchewan. The pebbles and boulders are primarily quartzite with some chert. Near the Alberta-Saskatchewan border, gravel is the major constituent, whereas farther east the formation becomes sandier and gravel is only a minor component (Vonhof, 1965; Storer, 1978a). The formation is over 100 m thick on the western end of the Cypress Hills, but thins to a few metres at the eastern end (Irish, 1965). Vonhof demonstrated that the sands and gravels were fluvial sediments deposited in northeasterly trending valleys, and were derived from a source in the Belt (Purcell) Supergroup of northwestern Montana. Gravel deposits south of Swift Current in Saskatchewan resemble the Cypress Hills gravels and are of Late Eocene age (Russell, 1965; Storer, 1978b). The extensive alluvial plains of Eocene and Oligocene times subsequently underwent uplift and intense erosion.

The Middle Miocene Wood Mountain Formation, capping the Wood Mountain Upland in Saskatchewan, is composed of sand and gravel, with some interbedded clay. Vonhof (1965) established that there are compositional differences between the Cypress Hills Formation and Wood Mountain Formation, the latter containing a greater percentage of chert and cherty sandstone. Vertebrate fauna include amphibians and reptiles (Holman, 1976), and mammals (Storer, 1975b). The Flaxville Gravel of northern Montana is younger, being Late Miocene in age (Storer, 1969, 1978a).

The Swan Hills gravels of northwestern Alberta are as much as 5 m thick and consist almost entirely of water-worn pebbles of pure white quartzite (Allan, 1919; Russell, 1967b). They may be correlated with gravels capping the Cypress Hills and Hand Hills. Other remnants of fluvial deposits representing erosion of the Rocky Mountains are the gravels of Whitecourt Mountain, Saddle Hills, Clear Hills, and Caribou Mountains.

Other fluvial deposits are known from various places across Alberta. The Hand Hills Formation, consisting of shale, marl, and thin conglomerate beds, occupies a small area on the summit of the Hand Hills east of Drumheller. Storer (1976, 1978a) collected vertebrate remains of Middle Miocene and Early Pleistocene age.

Dating of the Hand Hills gravels, according to Russell, has some important implications on the late Cenozoic history of the Canadian plains. The older gravels are of about the same age as the Wood Mountain deposits, but unlike those occurrences, there is no older plateau from which the Hand Hills sediments could have been derived. Instead, they must have resulted from erosion of quartzite and chert in the re-elevated Rocky Mountains. Apparently general uplift occurred in Middle Miocene time, involving both the eastern Cordillera and the adjacent high plains. An Early Pleistocene age for the upper Hand Hills gravel implies an episode of minor erosion and redeposition of material from the Miocene deposits. This was followed by profound erosion during later Pleistocene and Recent time. The fossiliferous gravels of the Hand Hills are about 150 m above the present prairie land surface a few miles to the north. Other upland remnants in Alberta are capped by Tertiary or Pleistocene gravels. Vast amounts of terrane have been carved away from around these remnants during Quaternary time.

In southern Saskatchewan, interbedded gravel, sand, silt, and clay, considered to be of Pliocene to Quaternary age, occur between till and bedrock in several ancient river valleys (Whitaker and Christiansen, 1972). These sediments, included in the Empress Group, were laid down as fluvial, lacustrine, and colluvial deposits prior to and during glaciation.

LARAMIDE OROGENY AND BASIN DEVELOPMENT: A SUMMARY

Tectonics and structure

The major uplift and deformation in the Canadian Rocky Mountains referred to as the Laramide Orogeny commonly has been considered to be of early Tertiary age (Taylor et al., 1964). Douglas et al. (1970) dated it as Late Cretaceous to Early Oligocene, and Price and Mountjoy (1970) emphasized that the hiatus separating Paleocene clastic wedge deposits of the western plains from Upper Eocene-Lower Oligocene sediments marks a major reduction in the supply of detritus and the end of thrusting in the Rocky Mountains, rather than its climax. The Upper Eocene-Lower Oligocene sediments are associated with a new and different phase of deformation involving extensional faulting, during which block fault structures were superimposed on older thrust and fold structures. Price and Mountjoy (1970), on the basis of the clastic sequence, concluded that structures in the Main Ranges probably began to develop as early as Late Jurassic time, the western Front Ranges structures may have emerged in the Late Jurassic and Early Cretaceous, and the Front Ranges in Late Cretaceous time. Three main pulses of clastic sedimentation can be recognized in the foreland basin of the Rocky Mountains (Stott, 1984); Late Jurassic to Valanginian, Aptian-Albian-Cenomanian, and Late Cenomanian to Paleocene. The first two are considered as phases of the Columbian Orogeny and the last as the Laramide Orogeny.

The earlier concept of an early Tertiary Laramide Orogeny that produced the Rocky Mountains as an event separate from an earlier (Columbian) orogeny that was restricted to the western Cordillera has been largely destroyed by the development of plate tectonic theories. Those theories postulate subduction and the accretion of allochthonous terranes along the western margin of the North American craton, with accompanying compression, thrusting, and shortening of the continental terrace wedge and foreland basin. The first plate-tectonic models (Dercourt, 1972; Monger et al., 1972) of the Canadian Cordillera have undergone considerable modifications as continuing studies provide new geophysical and geological data that outline the extreme complexity of the Cordillera. The basic configuration of the Cordillera was established between Middle Jurassic and Late Cretaceous times as a result of accretion of exotic terranes, arc magmatism related to subduction in the western Cordillera, and compression in the eastern Cordillera (Monger and Price, 1979). Major dextral transcurrent displacements, dated as mid-Cretaceous and Late Cretaceous to Eocene or Oligocene, occur well east of but sub-parallel with the continental margin along the Northern Rocky Mountain Belt and in the Northern Yukon Fold Complex (Gabrielse, 1985). These relationships suggest that the movement of the exotic terranes may have been oblique to the continental margin.

The eastern Cordillera has features in common with typical alpine belts - a core zone of metamorphic rocks (the Omineca Belt), and a marginal zone (the Rocky Mountains) characterized by essentially surficial or thin-skinned deformation above a décollement (Wheeler et al., 1974). The Rocky Mountains, involving 'pre-orogenic' sedimentary rocks as old as Middle Precambrian, contrast with the marginal zone of the Alps where only Mesozoic rocks are involved in the Cenozoic deformation.

The geological structure, tectonic evolution, and geotectonic significance of the Canadian Rocky Mountains were discussed by Price (1981), and of the Northern Yukon Fold Complex by Norris and Yorath (1981). Beaumont's (1981) model of the Alberta foreland basin involving tectonic loads continuously advancing on the craton in a series of pulses between Late Jurassic and Eocene times is consistent with observations.

The clastic wedges of western Canada developed in two stages, in Middle Jurassic to Early Cretaceous, and Late Cretaceous to Paleocene times. The Upper Proterozoic to Middle Jurassic continental terrace wedge was compressed and displaced over the western margin of the craton. Part of the supracrustal cover was scraped off the craton and incorporated in the eastward-advancing stack of imbricate fault slices. Isostatic depression of the continental lithosphere in response to the tectonic loading produced the foreland basin, which was filled with detritus from the rising mountains.

The maximum age of deformation in the southern Canadian Rocky Mountains during the Late Cretaceous Epoch is established by the emplacement of the McConnell and Lewis thrust sheets onto Campanian sediments. In the central Rocky Mountains north of Jasper, movement along the Back Range-Snake Indian fault system occurred earlier than the deposition of Upper Cretaceous-Paleocene Sifton conglomerates along the Northern Rocky Mountain Trench, but postdates Late Jurassic or earlier cleavage formation and metamorphism (McMechan, 1987). In the Foothills of Alberta, the thick upper Upper Cretaceous and Paleocene formations are involved in the easternmost structures. The minimum age of deformation in the south is given by unconformable overlap by the Upper Eocene to Lower Oligocene Kishenehn Formation (Price, 1965).

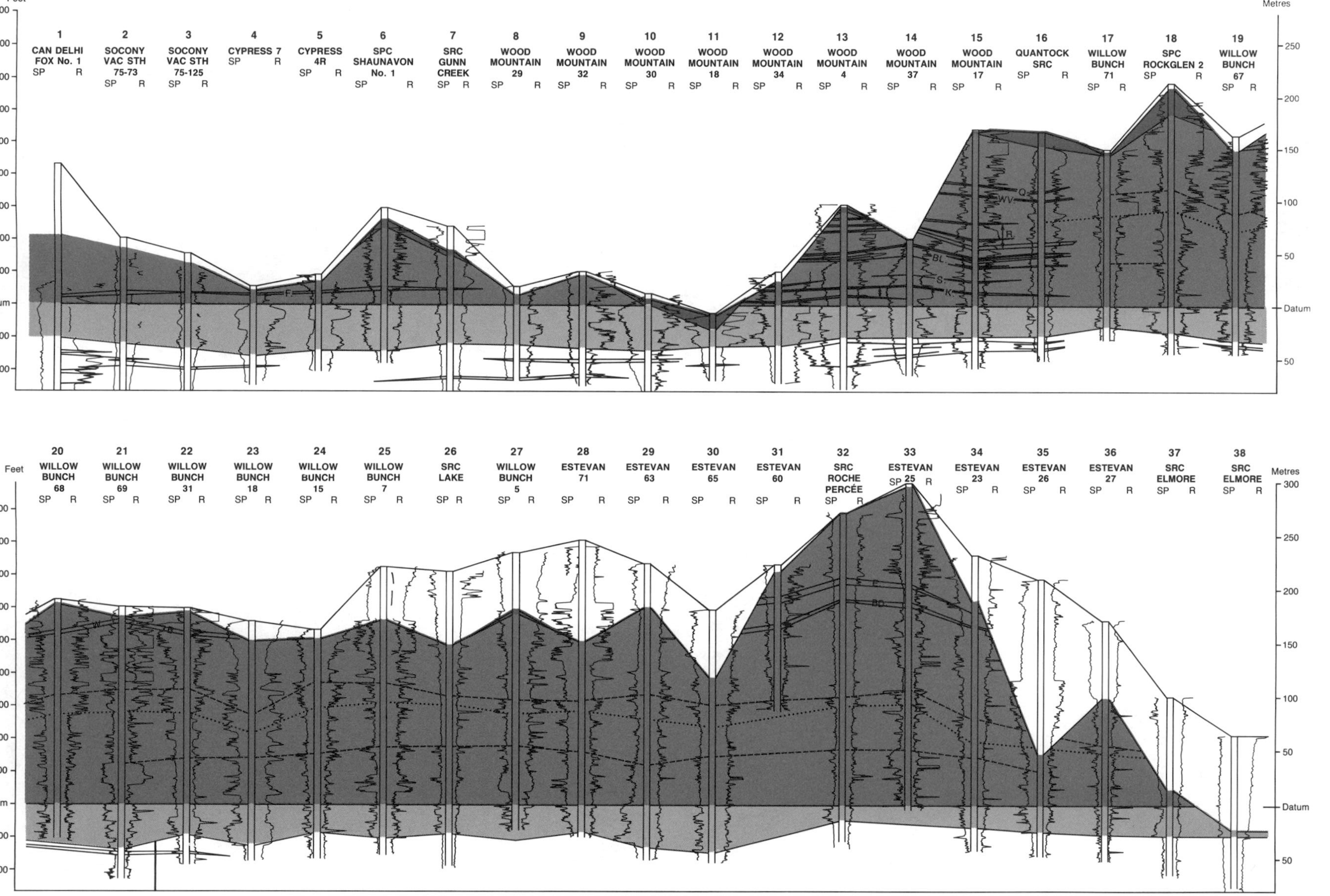

Figure 4J.15. Regional cross-section of Tertiary strata from Cypress Hills, Alberta and Saskatchewan, across Missouri Coteau to Turtle Mountain, Manitoba (from Whitaker et al., 1978).

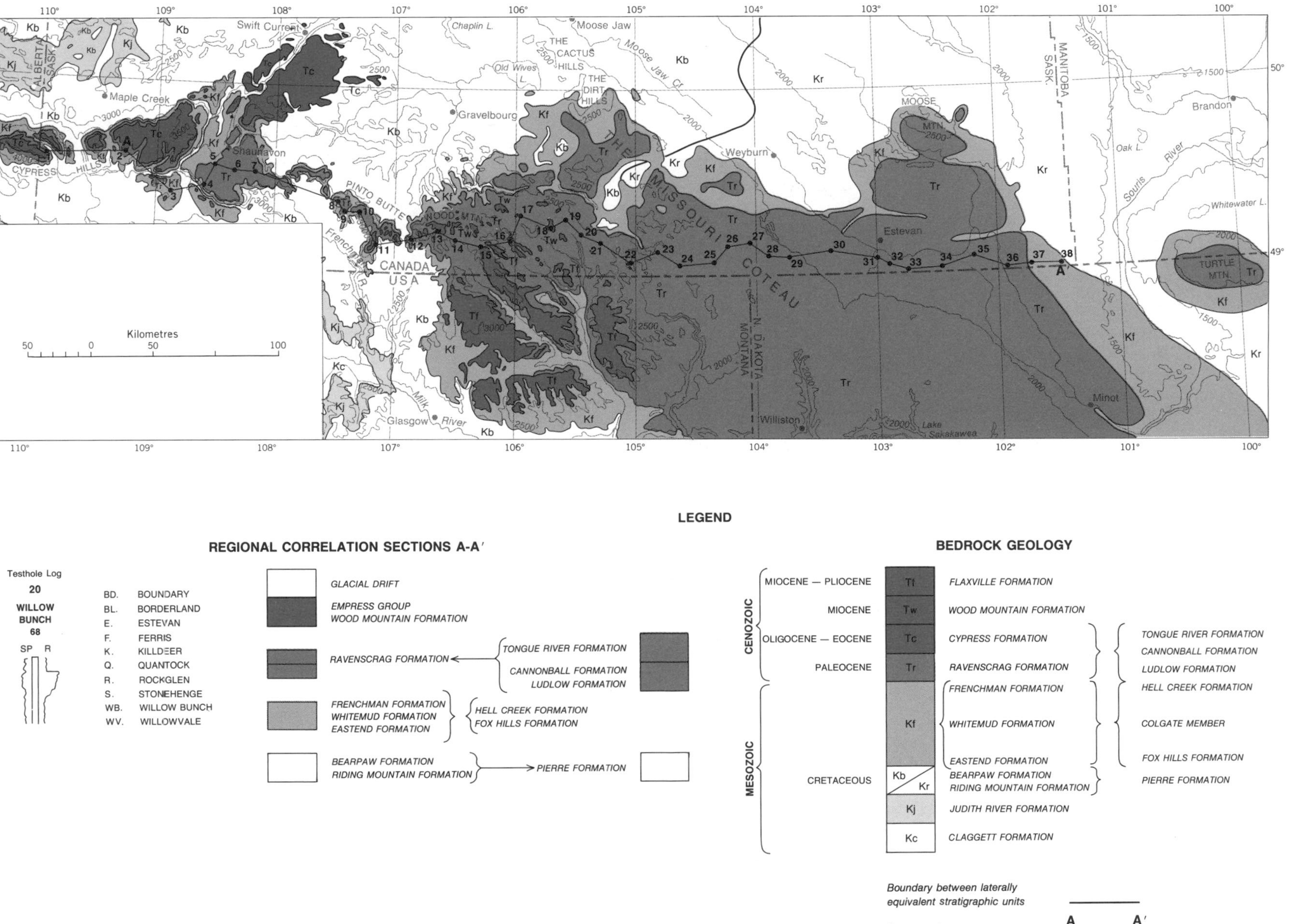

Figure 4J.15. Cont.

The southern Rocky Mountains and Foothills are formed of several large, west-dipping thrust sheets, each of which has been displaced eastwards many kilometres. The strata within the thrust sheets are broken by many closely spaced, subparallel, steeply dipping faults of smaller displacement, and the folds are complex and disharmonious. In the extreme south, total displacement is about 160 km, representing shortening of 50% (Bally et al., 1966; Price, 1981), and the thrust belt, as indicated by seismic data, is allochthonous above an undeformed, gently west-dipping basement of Precambrian crystalline rock. Many of the more westerly structures are considered to have formed in the Columbian Orogeny and to have been modified slightly in the Laramide Orogeny.

The structure in the central Rocky Mountains and Foothills is dominated by thrust faults, although folds are widespread and developed in conjunction with thrusting (Price and Mountjoy, 1970). In the region between the Bow and Athabasca rivers, shortening is estimated to be in the order of 200 km. Shortening farther north at latitude 53°15′N is about 140 km (Campbell et al., 1982).

Northward beyond Peace River, the structure changes from thrust-dominated to fold-dominated (Thompson, 1979, 1981). Although the deformation in the north is thin-skinned like that in the southern Rocky Mountains and Foothills, much of the west-to-east displacement has occurred on nonsurfacing or 'blind' faults, and the thick stack of imbricate thrusts of the southern region is not developed at the surface. Shortening across the northern Rocky Mountains and Foothills decreases from approximately 70 km at Peace River (McMechan, 1987) to 50 km at 58°30′N (Gabrielse and Taylor, 1982).

The Laramide Orogeny, dated as Late Cretaceous to Paleocene, is considered to be the final and most important tectonic event in the Mackenzie and Franklin mountains (Aitken et al., 1982). The structures of the region are north- to west-trending arcuate, large-scale folds and subordinate contractional faults. Four subregions of contrasting tectonic style are recognized. Each of these subregions has a unique stratigraphic succession, which, by controlling the level of structural detachments, has dictated the tectonic style.

The Northern Yukon Fold Complex, forming the structural and stratigraphic link between the structures of the Mackenzie Arc and those of northern Alaska, also was affected by the Laramide Orogeny. Both the history and the structures are complex, as rocks were deformed by orogenies during Precambrian, Mid-Paleozoic, Late Cretaceous, and Tertiary times (Norris and Yorath, 1981). A belt of Cretaceous thrust faults on the south flank of the Brooks Range in Alaska, comparable in style to those of the Rocky Mountains, continues southeastward in Yukon Territory. North- and northeast-trending, vertical to near-vertical faults, some with dextral strike-slip displacement, were active in the region of the Richardson Mountains during the Late Cretaceous Epoch and very probably in Tertiary time. Some of the movements are attributed to the opening of the Canada Basin (for additional detail, see Gabrielse and Yorath, 1991 and Trettin, 1991.

Recently, the Northern Rocky Mountain Trench Fault and a number of other prominent faults have been shown to belong to a system of dextral transcurrent faults that dates from the middle Cretaceous to Late Eocene or Oligocene (Gabrielse, 1985). Offsets of shelf to off-shelf facies boundaries in lower Paleozoic rocks indicate a cumulative displacement of at least 750 km, and probably greater than 900 km.

In the southern Rocky Mountains, several post-orogenic normal faults are considered to flatten downward and merge with pre-existing thrusts (Bally et al., 1966). The Upper Eocene-Lower Oligocene Kishenehn Formation was deposited in a graben bounded by the Flathead Fault that postdates the Laramide, Lewis Thrust-Sheet. Lack of offset of older transverse structures precludes any major strike-slip displacement along the Southern Rocky Mountain Trench, although normal faults are known to border it (Leech, 1966).

A more detailed review of the structure and tectonic evolution of the Canadian Cordillera as a whole is provided by Gabrielse and Yorath (1991).

ECONOMIC GEOLOGY

Sandstone of the Porcupine Hills (Paskapoo) Formation was used extensively in Calgary and Edmonton for many of the early buildings, some of which are now designated as Provincial and National Heritage sites.

The Ravenscrag Formation is a source of plastic kaolinitic clays. Other Tertiary clays are used locally in structural clay products. The semiconsolidated Tertiary quartzitic gravels of the Swift Current and Cypress Hills areas and other smaller erosional remnants are used directly or crushed as aggregate. Clay resources of the plains were surveyed by Crockford (1951), and more recently clays and gravels of individual provinces were outlined by Hamilton (1984), Guliov (1984), and Bannatyne (1984).

The Paleocene Ravenscrag Formation of Saskatchewan is a major source of thermal coal, which is mined at Estevan (Fig. 4J.15; Whitaker et al., 1978). Nineteen coal zones are recognized in the Estevan area, of which only five zones are shallow enough to be of immediate interest. To the west, six coal zones are recognized in the Willowbunch and Wood Mountain areas, but only two coal zones are recognized in the Cypress area. Correlation of coal zones from area to area is tenuous because erosion has removed the coal-bearing strata between the four areas.

In central Alberta, coal seams up to 5 m thick occur in the Paleocene Scollard Formation (Campbell and Almadi, 1964; Campbell, 1967). All of the major coal-bearing seams above the Kneehills Marker Horizon (Battle Formation) were combined in the Ardley Coal Zone (Campbell, 1967). The Ardley coal zone is subdivided into three main units, designated from base to top as the Lower Ardley 'A' (Nevis or No. 13 seam), Lower Ardley 'B' (Ardley or No. 14 seam), and Upper Ardley (Holter et al., 1975). The Lower Ardley 'B' seam is the most prominently developed. The Upper Ardley is composed of a number of thin seams and is of economic interest only in the region of Red Deer. Ardley coal has been mined in Three-Hills-Trochu, Ardley, and Wizard Lake areas and is presently being stripped at Wabamun Lake west of Edmonton for two thermal plants. The Coalspur Formation (Jerzykiewicz, 1985) contains major Tertiary coal seams in the vicinity of Athabasca River east of Jasper. The Val d'Or and Mynheer seams are mined at Coal Valley near Hinton, and are probably equivalent to the Ardley Zone. Other coal deposits are mined from the Obed Coal zone of the Paskapoo Formation.

In Bonnet Plume Basin in Yukon Territory, more than a billion tonnes of lignite may be preserved in the upper Bonnet Plume Formation (Long, 1978).

In the Tertiary of the Beaufort-Mackenzie Basin, naphthenic oils and condensates have been generated from terrestrially derived organic matter in source rocks juxtaposed with the reservoir at reflectance levels of 0.4 to 0.6%R° (Snowdon and Powell, 1982). The source for these early oils and condensates is considered to be resinite occurring dispersed in coal fragments.

The Reindeer Formation of Mackenzie Delta has yielded natural gas and minor quantities of oil in Taglu Gas Field (Bowerman and Coffman, 1975; Hawkings and Hatlelid, 1975). Discoveries of natural gas and oil have been made in other Tertiary deltaic sands within the Beaufort-Mackenzie Basin.

REFERENCES

Aitken, J.D., Cook, D.G., and Yorath, C.J.
1982: Upper Ramparts River (106G) and Sans Sault Rapids (106H) map areas, District of Mackenzie; Geological Survey of Canada, Memoir 388, 48 p.

Allan, J.A.
1919: Geology of the Swan Hills in Lesser Slave Lake district, Alberta; Geological Survey of Canada, Summary Report 1918, pt. C, p. 7-13.

Allan, J.A. and Sanderson, J.O.G.
1945: Geology of Red Deer and Rosebud sheets, Alberta; Research Council of Alberta, Report 13, 115 p.

Baadsgaard, H., Folinsbee, R.W., and Lipson, J.
1961: Potassium-argon dates of biotites from Cordilleran granites; Bulletin of the Geological Society of America, v. 72, p. 689-702.

Bally, A.W., Gordy, P.L., and Stewart, G.A.
1966: Structure, seismic data and orogenic evolution of southern Canadian Rockies; Bulletin of Canadian Petroleum Geology, v. 14, p. 337-381.

Bamburak, J.D.
1973: The Upper Cretaceous and Paleocene stratigraphy of Turtle Mountain, Manitoba; M.Sc. thesis, University of Manitoba, Winnipeg, Manitoba, 110 p.
1978: Stratigraphy of the Riding Mountain, Boissevain and Turtle Mountain formations in the Turtle Mountain area, Manitoba; Manitoba Department of Mines, Mineral Resources Division, Geological Report 78-2, 47 p.

Bannatyne, B.B.
1984: Manitoba: Industrial minerals in Canada - a review of recent developments; Industrial Minerals, p. 89-93.

Beaumont, C.
1981: Foreland basins; Geophysical Journal, Royal Astronomical Society, v. 65, p. 291-329.

Beaumont, C., Boutilier, R., Mackenzie, A.S., and Rullkötter, J.
1985: Isomerization and aromatization of hydrocarbons and the paleothermometry and burial history of Alberta Foreland Basin; American Association of Petroleum Geologists, Bulletin, v. 69, no. 4, p. 546-566.

Bell, W.A.
1949: Uppermost Cretaceous and Paleocene floras of western Alberta; Geological Survey of Canada, Bulletin 13, 231 p.

Berry, E.W.
1930: Fossil plants from the Cypress Hills of Alberta and Saskatchewan; National Museum of Canada, Bulletin 63, p. 15-28.

Bihl, G.
1973: Palynostratigraphic investigation of Upper Maastrichtian and Paleocene strata near Tate Lake, Northwest Territories; Ph.D. thesis, University of British Columbia, Vancouver, British Columbia, 53 p.

Bowerman, J.N. and Coffman, R.C.
1975: The geology of the Taglu gas field in the Beaufort Basin, Northwest Territories; in Canada's Continental Margins, C.J. Yorath, E.R. Parker, and D.J. Glass (ed); Canadian Society of Petroleum Geologists, Memoir 4, p. 649-669.

Brideaux, W.W.
1975: Cretaceous and Tertiary assemblages (palynomorphs); in Biostratigraphic determinations of fossils from the subsurface of the districts of Franklin and Mackenzie; Geological Survey of Canada, Paper 74-39, p. 2-5.
1976: Jurassic, Cretaceous and Tertiary assemblages (palynomorphs); in Biostratigraphic determinations from the subsurface of the districts of Franklin and the Yukon Territory; Geological Survey of Canada, Paper 75-10, p. 2-4.

Broughton, P.L.
1977: Origin of coal basins by salt solution; Nature, v. 270, p. 420-423.

Bujak, J.P. and Davies, E.H.
1981: Neogene dinoflagellate cysts from the Hunt Dome Kopanoar M-13 well, Beaufort Sea, Canada; Bulletin of Canadian Petroleum Geology, v. 29, p. 420-425.

Campbell, J.D.
1962: Boundaries of the Edmonton Formation in the central Alberta Plains; Journal of the Alberta Society of Petroleum Geologists, v. 10, p. 308-319.
1967: Ardley coal zone in the Alberta Plains: central Red Deer River Area; Research Council of Alberta, Report 67-1, 28 p.

Campbell, J.D. and Almadi, I.S.
1964: Coal occurrences of the Vulcan-Gleichen area, Alberta; Research Council of Alberta, Report 64-2, 58 p.

Campbell, J.D., Mountjoy, E.W., and Struik, L.C.
1982: Structural cross-section through south-central Rocky and Cariboo Mountains to the Coast Range; Geological Society of Canada, Open File 844.

Carrigy, M.A.
1971: Lithostratigraphy of the uppermost Cretaceous (Lance) and Paleocene strata of the Alberta Plains; Research Council of Alberta, Bulletin 27, p. 161.

Chandrasekharam, A.
1974: Megafossil flora from the Genesee Locality, Alberta, Canada; Palaeontographica Abteilung B, v. 147, Lieferung 1-3, p. 1-41.

Christiansen, E.A.
1967: Collapse structures near Saskatoon, Saskatchewan, Canada; Canadian Journal of Earth Sciences, v. 4, p. 757-767.

Crockford, M.B.
1951: Clay deposits of Elkwater Lake area of Alberta; Alberta Research Council of Alberta, Report 61, 102 p.

Dercourt, J.
1972: The Canadian Cordillera, the Hellenides, and the sea-floor spreading theory; Canadian Journal of Earth Sciences, v. 9, p. 709-743.

Dietrich, J.R., Dixon, J., and McNeil, D.H.
1985: Sequence analysis and nomenclature of Upper Cretaceous to Holocene strata in the Beaufort-Mackenzie Basin; in Current Research, Part A, Geological Survey of Canada, Paper 85-1A, p. 613-628.

Dixon, J.
1981: Sedimentology of the Eocene Taglu delta, Beaufort-Mackenzie basin: example of a river-dominant delta; Geological Survey of Canada, Paper 80-11, 11 p.
1986: Cretaceous to Pleistocene stratigraphy and paleography, northern Yukon and northwestern District of Mackenzie; Bulletin of Canadian Petroleum Geology, v. 34, p. 47-70.

Dixon, J., Dietrich, J.R., McNeil, D.H., McIntyre, D.J., Snowdon, L.R., and Brooks, P.
1985: Geology, biostratigraphy, and organic chemistry of Jurassic to Pleistocene strata, Beaufort-Mackenzie area, northwest Canada; Canadian Society of Petroleum Geology, Course notes, 65 p.

Douglas, R.J.W.
1950: Callum Creek, Langford Creek and Gap map areas, Alberta; Geological Survey of Canada, Memoir 225, 124 p.

Douglas, R.J.W., Gabrielse, H., Wheeler, J.O., Stott, D.F., and Belyea, H.R.
1970: Geology of western Canada; in Geology and Economic Minerals of Canada, Fifth Edition, R.J.W. Douglas (ed.); Geological Survey of Canada, Economic Report No. 1, Chapter VIII, p. 366-488.

Elliot, R.H.
1960: Subsurface correlation of the Edmonton Formation; Journal of the Alberta Society of Petroleum Geologists, v. 8, p. 324-337.

Eliuk, L.S.
1969: Correlation of the Entrance Conglomerate, Alberta, by palynology; M.Sc. thesis, University of Alberta, Edmonton, Alberta.

Feyling-Hanssen, R.W., Funder, S., and Petersen, K.S.
1983: The Lodin Elv Formation: a Plio-Pleistocene occurrence in Greenland; Bulletin of the Geological Society of Denmark, v. 31, p. 81-106.

Fox, R.C.
1970: A bibliography of Cretaceous and Tertiary vertebrates from western Canada; Alberta Society of Petroleum Geologists, v. 18, no. 2, p. 263-281.

Fox, R.C. (cont.)
1984: First North American record of the Paleocene primate Saxonella; Journal of Paleontology, v. 58, p. 892-894.

Fox, S.K., Sr. and Olsson, R.K.
1969: Danian planktonic foraminifera from the Cannonball Formation in North Dakota; Journal of Paleontology, v. 16, p. 660-673.

Fulton, R.J.
1989: Quaternary Geology of Canada and Greenland; Geological Survey of Canada, Geology of Canada, no. 1, 839 p., (Also Geological Society of America, The Geology of North America, v. K-1.)

Furnival, G.M.
1946: Cypress Lake map area, Saskatchewan; Geological Survey of Canada, Memoir 242, 161 p.

Gabrielse, H.
1985: Major dextral transcurrent displacements along the Northern Rocky Mountain Trench and related lineaments in north-central British Columbia; Geological Society of America, Bulletin, v. 96, no. 1, p. 1-14.

Gabrielse, H. and Taylor, G.C.
1982: Geological maps and cross sections of the northern Canadian Cordillera from southwest of Fort Nelson, British Columbia to Gravina Island, southeastern Alaska; Geological Survey of Canada, Open File Report 864.

Gabrielse, H. and Yorath, C.J. (ed.)
1991: Geology of the Cordilleran Orogen in Canada; Geological Survey of Canada, Geology of Canada no. 4, (Also Geological Society of America, The Geology of North America, v. G-2.)

Germundson, R.K.
1965: Stratigraphy and micropaleontology of some late Cretaceous-Paleocene continental formations, Western Interior, North America; Ph.D. thesis, University of Missouri, Rolla, Missouri, 224 p.

Gibson, D.W.
1977: Upper Cretaceous and Tertiary coal-bearing strata in the Drumheller-Ardley region, Red Deer River Valley, Alberta; Geological Survey of Canada, Paper 76-35, 41 p.

Glaister, R.P. and Hopkins, J.C.
1974: Turbidity-current and debris-flow deposits; in Use of Sedimentary Structures for Recognition of Clastic Environments, M.S. Shawa (ed.); Canadian Society of Petroleum Geologists, p. 23-40.

Grantz, A., Johnson, L., and Sweeney, J.F. (ed.)
1990: The Arctic Ocean Region; in Geological Society of America, The Geology of North America, v. L, 644 p.

Grond, H.C., Churchill, S.J., Armstrong, R.L., Harakal, J.E., and Nixon, G.T.
1984: Late Cretaceous age of the Hutshi, Mount Nansen, and Carmacks groups, southwestern Yukon Territory and northwestern British Columbia; Canadian Journal of Earth Sciences, v. 21, p. 554-548.

Guliov, P.
1984: Saskatchewan; in Industrial Minerals in Canada - A Review of Recent Developments; Industrial Minerals, p. 93-101.

Gunther, P.R. and Hills, L.V.
1972: Megaspores and other palynomorphs of the Brazeau Formation (Upper Cretaceous), Nordegg area, Alberta; Geoscience and Man, v. 4, p. 29-48.

Hacquebard, P.A.
1977: Rank of coal as an index of organic metamorphism for oil and gas in Alberta; in The Origin and Migration of Petroleum in the Western Canada Sedimentary Basin, G. Deroo, T.G. Powell, B. Tissot, and R.G. McCrossan (ed.); Geological Survey of Canada, Bulletin 262, p. 11-22.

Hamilton, W.N.
1984: Alberta; in Industrial Minerals in Canada - A Review of Recent Developments; Industrial Minerals, p. 105-111.

Hawkings, T.J. and Hatlelid, W.G.
1975: The regional setting of the Taglu field; in Canada's Continental Margins and Offshore Petroleum Exploration, C.J. Yorath, E.R. Parker, and D.J. Glass (ed.); Canadian Society of Petroleum Geologists, Memoir 4, p. 633-647.

Hea, J.P., Arcuri, J., Campbell, G.R., Fraser, I., Fuglem, M.O., O'Bertos, J.J., Smith, D.R., and Zayat, M.
1980: Post-Ellesmerian basins of Arctic Canada: Their depocentres, rates of sedimentation and petroleum potential; in Facts and Principles of World Oil Occurrences, A.D. Miall (ed.); Canadian Society of Petroleum Geologists, Memoir 6, p. 447-488.

Hills, L.V., Klovan, J.E., and Sweet, A.R.
1974: *Juglans eocinerea* n. sp., Beaufort Formation (Tertiary), southwestern Banks Island, Arctic Canada; Canadian Journal of Botany, v. 52, p. 65-90.

Hitchon, B.
1984: Geothermal gradients, hydrodynamics, and hydrocarbon occurrences, Alberta, Canada; American Association of Petroleum Geologists, Bulletin, v. 68, p. 713-743.

Holman, J.A.
1976: Cenozoic Herpetofaunas of Saskatchewan; in ATHLON Essays in Palaeontology in Honour of Loris Shano Russell, C.S. Churcher (ed.); Royal Ontario Museum, Life Sciences Miscellaneous Publications, p. 80-92.

Holmes, D.W. and Oliver, T.A.
1973: Source and depositional environments of the Moose Channel Formation, Northwest Territories; Bulletin of Canadian Petroleum Geology, v. 21, p. 435-478.

Holter, M.E.
1969: The Middle Devonian Prairie Evaporite of Saskatchewan; Saskatchewan Department of Mineral Resources, Report 123, 113 p.

Holter, M.E., Yurko, J.R., and Chu, M.
1975: Geology and coal reserves of the Ardley Coal Zone of central Alberta; Research Council of Alberta, Report 75-7, 41 p.

Hopkins, W.S., Jr. and Norris, D.K.
1974: An occurrence of Paleogene sediments in the Old Crow structural depression, northern Yukon Territory; in Report of Activities, Part A, Geological Survey of Canada, Paper 74-1, p. 315-316.

Hopkins, W.S., Jr., Rutter, N.W., and Rouse, G.E.
1972: Geology, paleoecology, and palynology of some Oligocene rocks in the Rocky Mountain Trench of British Columbia; Canadian Journal of Earth Sciences, v. 9, p. 460-470.

Ioannides, N.W. and McIntyre, D.J.
1980: A preliminary palynological study of the Caribou Hills outcrop section along the Mackenzie River, District of Mackenzie, Canada; in Current Research, Part A, Geological Survey of Canada, Paper 80-1A, p. 197-208.

Irish, E.J.W.
1965: Notes to accompany the geological map, Cypress Hills region, Alberta and Saskatchewan; in 15th Annual Field Conference Guidebook, Alberta Society of Petroleum Geologists, p. 15-22.
1970: The Edmonton Group of south-central Alberta; Bulletin of Canadian Petroleum Geology, v. 18, p. 125-155.

Irish, E.J.W. and Havard, C.J.
1968: The Whitemud and Battle formations ("Kneehills Tuff Zone") - a stratigraphic marker; Geological Survey of Canada, Paper 67-63, 51 p.

Jarzen, D.M.
1982: Angiosperm pollen from the Ravenscrag Formation (Paleocene), southern Saskatchewan, Canada; Pollen et Spores, v. 24, p. 119-155.

Jeletzky, J.A.
1960: Youngest marine rocks in western interior of North America and the age of the *Triceratops*-beds, with remarks on comparable dinosaur-bearing beds outside North America; International Geological Congress, Proceedings, sec. 5, p. 25-40.

Jerzykiewicz, T.
1985: Stratigraphy of the Saunders Group in the central Alberta Foothills - A progress report; in Current Research, Part B, Geological Survey of Canada, Paper 85-1B, p. 247-258.

Jerzykiewicz, T., Lerbekmo, J.F., and Sweet, A.R.
1984: The Cretaceous-Tertiary Boundary, central Alberta Foothills (abstract); in 1984 Canadian Paleontology and Biostratigraphy Seminar, Geological Association of Canada, Programme with Abstracts, p. 4.

Jerzykiewicz, T. and McLean, J.R.
1980: Lithostratigraphical and sedimentological framework of coal-bearing Upper Cretaceous and Lower Tertiary strata, Coal Valley area, central Alberta Foothills; Geological Survey of Canada, Paper 79-12, 26 p.

Jerzykiewicz, T. and Sweet, A.R.
1986: The Cretaceous-Tertiary boundary interval: I. Stratigraphy, Coalspur Formation, central Alberta Foothills, Canada; Canadian Journal of Earth Sciences, v. 23, p. 1356-1374.

Johnson, B., Calverley, A.E., and Pendlebury, D.C.
1976: Pleistocene foraminifera from northern Yukon, Canada; in Paleocecology and Biostratigraphy, C.T. Schafer and B.R. Pelletier (ed.); Part B, First International Symposium on Benthonic Foraminifera of Continental Margins; Maritime Sediments, Special Publication No. 1, p. 393-400.

Jones, P.B.
1969: The Tertiary Kishenehn Formation, British Columbia; Bulletin of Canadian Petroleum Geology, v. 17, p. 234-246.

Jones, P.B., Brache, J., and Lentin, J.K.
1980: The geology of the 1977 offshore hydrocarbon discoveries in the Beaufort-Mackenzie Basin, Northwest Territories; Bulletin of Canadian Petroleum Geology, v. 28, p. 81-102.

Kent, D.M.
1968: The geology of the Upper Devonian Saskatchewan Group and equivalent rocks in western Saskatchewan and adjacent areas; Saskatchewan Department of Mineral Resources, Report 99, 221 p.

Kevan, D.K.M. and Wighton, D.C.
1981: Paleocene orthopteroids from south-central Alberta, Canada; Canadian Journal of Earth Sciences, v. 18, p. 1824-1837.

Krause, D.W.
1977: Paleocene multituberculates (Mammalia) of the Roche Percée local fauna, Ravenscrag Formation, Saskatchewan, Canada; Palaeontographica, A, v. 159, p. 1-36.
1978: Paleocene primates from western Canada; Canadian Journal of Earth Sciences, v. 15, p. 1250-1271.

Krishtalka, L.
1973: Late Paleocene mammals from the Cypress Hills, Alberta; The Museum, Texas Tech University, Special Publications, No. 2, p. 1-77.

Kupsch, W.O.
1956: Cretaceous-Tertiary transition in southwestern Saskatchewan (abstract); XX International Geological Congress, Resumenes de los Trabajos Presenta dos, p. 29.

Leech, G.B.
1966: The Rocky Mountain Trench; in The World Rift System, T.N. Irvine (ed.); Geological Survey of Canada, Paper 66-14, p. 307-329.

Lerbekmo, J.F. and Coulter, K.C.
1984: Magnetostratigraphic and biostratigraphic correlations of Late Cretaceous to Early Paleocene strata between Alberta and North Dakota; in The Mesozoic of Middle North America, D.F. Stott and D.J. Glass (ed.); Canadian Society of Petroleum Geologists, Memoir 9, p. 313-317.

Lerbekmo, J.F. and Coulter, K.C.
1985: Late Cretaceous to early Tertiary magnetostratigraphy of a continental sequence: Red Deer Valley, Alberta, Canada; Canadian Journal of Earth Sciences, v. 22, p. 567-583.

Lerbekmo, J.F., Evans, M.E., and Baadsgaard, H.
1979a: Magnetostratigraphy, biostratigraphy and geochronology of Cretaceous-Tertiary boundary sediments, Red Deer Valley; Nature (London), v. 279, p. 26-30.

Lerbekmo, J.F., Singh, C., Jarzen, D.M., and Russell, D.A.
1979b: The Cretaceous-Tertiary boundary in south-central Alberta; Canadian Journal of Earth Sciences, v. 16, p. 1866-1869.

Long, D.G.F.
1978: Lignite deposits in the Bonnet Plume Formation, Yukon Territory; in Current Research, Part A, Geological Survey of Canada, Paper 78-1A, p. 399-401.

McMechan, M.E.
1987: Stratigraphic and structure, Mount Selwyn area, Rocky Mountains, northeastern British Columbia; Geological Survey of Canada, Paper 85-28, 34 p.

McMechan, R.
1981: Stratigraphy, sedimentology, structure and tectonic implications of the Oligocene Kishenehn Formation, Flathead Valley Graben, southeastern British Columbia; Ph.D. thesis, Queen's University, Kingston, Ontario, 327 p.

McNeil, D.H., Ioannides, N.S., and Dixon, J.
1982: Geology and biostratigraphy of the Dome Gulf et al. Ukalerk C-50 well, Beaufort Sea; Geological Survey of Canada, Paper 80-31, 17 p.

Magara, K.
1976: Thickness of removed sedimentary rocks, paleopore pressure, and paleotemperature, southwestern part of Western Canada Basin; American Association of Petroleum Geologists, Bulletin, v. 60, p. 554-565

Misko, R.M. and Hendry, H.E.
1979: The petrology of sands in the uppermost Cretaceous and Palaeocene of southern Saskatchewan, a study of composition influenced by grain size, source area, and tectonics; Canadian Journal of Earth Sciences, v. 16, p. 38-49.

Monger, J.W.H. and Price, R.A.
1979: Geodynamic evolution of the Canadian Cordillera - progress and problems; Canadian Journal of Earth Sciences, v. 16, p. 770-791.

Monger, J.W.H., Souther, J.G., and Gabrielse, H.
1972: Evolution of the Canadian Cordillera: a plate-tectonic model; American Journal of Science, v. 272, p. 577-602.

Mountjoy, E.W.
1967: Upper Cretaceous and Tertiary stratigraphy, northern Yukon and northwestern District of Mackenzie; Geological Survey of Canada, Paper 66-16, 70 p.

Nichols, D.J. and Ott, H.L.
1978: Biostratigraphy and evolution of the *Momipites-Caryapollenites* lineage in the Early Tertiary in the Wind River Basin, Wyoming; Palynology, v. 2, p. 93-112.

Norris, D.K.
1976: Structural and stratigraphic studies in the northern Canadian Cordillera; in Report of Activities, Part A, Geological Survey of Canada, Paper 76-1A, p. 457-466.

Norris, D.K. and Hopkins, W.S., Jr.
1977: The geology of the Bonnet Plume Basin, Yukon Territory; Geological Survey of Canada, Paper 76-8, 20 p.

Norris, D.K. and Yorath, C.J.
1981: The North American Plate from the Arctic Archipelago to the Romanzof Mountains; in The Ocean Basins and Margins, A.E.M. Nairn, M. Churkin Jr. and F.G. Stehli (ed.); Chapter 3, Plenum Press, New York and London, p. 37-103.

Norris, G.
1982: Spore-pollen evidence for early Oligocene high-latitude cool climatic episode in northern Canada; Nature, v. 297, p. 387-389.
1986: Systematic and stratigraphic palynology of Eocene to Pliocene strata in Imperial Nuktak C-22 well, Mackenzie Delta region, District of Mackenzie, Northwest Territories; Geological Survey of Canada, Bulletin 340, 89 p.

Nurkowski, J.R.
1984: Coal quality, coal rank variation and its relation to reconstructed overburden, Upper Cretaceous and Tertiary Plains coals, Alberta, Canada; American Association of Petroleum Geologists, Bulletin, v. 68, no. 3, p. 285-295.

Ower, J.R.
1960: The Edmonton Formation; Journal of Alberta Society of Petroleum Geologists, v. 8, p. 309-323.

Price, L.L., McNeil, D.H., and Ioannides, N.W.
1980: Revision of the Tertiary Reindeer Formation in the Caribou Hills, District of Mackenzie; in Current Research, Part B, Geological Survey of Canada, Paper 80-1B, p. 179-184.

Price, R.A.
1965: Flathead map area, British Columbia and Alberta; Geological Survey of Canada, Memoir 336, p. 221.
1981: The Cordilleran foreland thrust and fold belt in the southern Canadian Rocky Mountains; in Thrust and Nappe Tectonics, K.R. McClay and N.J. Price (ed.); The Geological Society of London, Special Publication 9, p. 427-448.

Price, R.A. and Mountjoy, E.W.
1970: Geologic structure of the Canadian Rockies between Bow and Athabasca Rivers - A preliminary report; in Structure of the Southern Canadian Cordillera, J.O. Wheeler (ed.); The Geological Association of Canada, Special Paper No. 6, p. 7-26.

Ricketts, B.D.
1985: Possible plinian eruptions of Paleocene age in central Yukon: evidence from volcanic ash, Norman Wells area, Northwest Territories; Canadian Journal of Earth Sciences, v. 22, p. 473-479.

Rouse, G.E.
1977: Paleogene palynomorph ranges in Western and Northern Canada; in Contributions of Stratigraphic Palynology, v. 1, Cenozoic Palynology; American Association of Stratigraphic Palynologists, Contribution Series, no. 5A, p. 48-65.

Rouse, G.E. and Srivastava, S.K.
1972: Palynological zonation of Cretaceous and Early Tertiary rocks of the Bonnet Plume Formation, northeastern Yukon, Canada; Canadian Journal of Earth Sciences, v. 9, p. 1163-1179.

Russell, D.A. and Singh, C.
1978: The Cretaceous-Tertiary boundary in south-central Alberta - a reappraisal based on dinosaurian and microfloral extinctions; Canadian Journal of Earth Sciences, v. 15, p. 284-292.

Russell, L.S.
1934: Revision of the Lower Oligocene vertebrate fauna of the Cypress Hills, Saskatchewan; Royal Canadian Institute, Transactions, v. 20, pt. 1, p. 49-67.
1950: Correlation of the Cretaceous-Tertiary transition in Saskatchewan and Alberta; Bulletin of Geological Society of America, v. 61, p. 27-42.
1958: Paleocene mammal teeth from Alberta; Bulletin of the National Museum of Canada, Bulletin 1947, p. 96-101.
1964: Kishenehn Formation: Field Conference Guide Book Issue; Bulletin of Canadian Petroleum Geology, v. 12, p. 536-543.
1965: Tertiary mammals of Saskatchewan Part I: The Eocene fauna; Royal Ontario Museum, Life Science Contribution 67, p. 1-33.
1967a: Unionidae from the Cretaceous and Tertiary of Alberta and Montana; Journal of Paleontology, v. 41, p. 1116-1120.
1967b: Palaeontology of the Swan Hills area, north-central Alberta; Royal Ontario Museum, Life Sciences Contribution 71, p. 1-31.
1972: Tertiary mammals of Saskatchewan Part II: The Oligocene fauna, non-ungulate orders; Royal Ontario Museum, Life Sciences Contribution 84, p. 1-97.
1974: Fauna and correlation of the Ravenscrag Formation (Paleocene) of southwestern Saskatchewan; Royal Ontario Museum, Life Sciences Contribution 102, p. 1-53.

Russell, L.S. and Churcher, C.S.
1972: Vertebrate palaeontology, Cretaceous to Recent, Interior Plains, Canada; XXIV International Geological Congress, Montreal, Guidebook, Field Excursion A21, 46 p.

Simpson, G.G.
1927: Mammalian fauna and correlation of the Paskapoo Formation of Alberta; American Museum Novitates, no. 268, 10 p.

Snead, R.G.
1969: Microfloral diagnosis of the Cretaceous-Tertiary boundary, central Alberta; Research Council of Alberta, Bulletin 25, 148 p.
Snowdon, L.R. and Powell, T.G.
1982: Immature oil and condensate - modification of hydrocarbon generation model for terrestrial organic matter; American Association of Petroleum Geologists, Bulletin, v. 66, p. 775-588.
Srivastava, S.K.
1970: Pollen biostratigraphy and paleoecology of the Edmonton Formation (Maestrichtian), Alberta, Canada; Palaeogeography, Palaeoclimatology, Palaeoecology, v. 7, p. 221-276.
Staplin, F.L.
1976: Tertiary biostratigraphy, Mackenzie Delta region, Canada; Bulletin of Canadian Petroleum Geology, v. 24, p. 117-136.
Sternberg, C.M.
1930: Miocene gravels in southern Saskatchewan; Royal Society of Canada, Transactions, Ser. 3, v. 24, Sec. 4, p. 29-30.
Storer, J.E.
1969: An Upper Pliocene neohipparion from the Flaxville Gravels, northern Montana; Canadian Journal of Earth Sciences, v. 6, no. 4, p. 791-794.
1975a: Middle Miocene mammals from the Cypress Hills, Canada; Canadian Journal of Earth Sciences, v. 12, no. 3, p. 520-522.
1975b: Tertiary mammals of Saskatchewan Part III: The Miocene fauna; Royal Ontario Museum, Life Sciences Contribution 103, p. 1-134.
1976: Mammals of the Hand Hills Formation, southern Alberta; in ATHLON Essays on Palaeontology in Honour of Loris Shano Russell, C.S. Churcher (ed.); Royal Ontario Museum, Life Sciences Miscellaneous Publications, p. 186-209.
1978a: Tertiary sands and gravels in Saskatchewan and Alberta: correlation of mammalian faunas; in Western and Arctic Canadian Biostratigraphy; Geological Association of Canada, Special Paper 18, p. 595-602.
1978b: Rodents of the Swift Current Creek local fauna (Eocene, Uintan) of Saskatchewan; Canadian Journal of Earth Sciences, v. 15, no. 10, p. 1673-1674.
1978c: Rodents of the Calf Creek local fauna (Cypress Hills Formation, Oligocene, Chadronian) Saskatchewan; Saskatchewan Museum of Natural History, Natural History Contributions, no. 1, p. 1-54.
1984: Fossil mammals of the Southfork local fauna (early Chadronian) of Saskatchewan; Canadian Journal of Earth Sciences, v. 21, p. 1400-1405.
Stott, D.F.
1984: Cretaceous sequences of the Foothills of the Canadian Rocky Mountains; in The Mesozoic of Middle North America, D.F. Stott and D.J. Glass (ed.); Canadian Society of Petroleum Geologists, Memoir 9, p. 85-107.
Sweet, A.R.
1978: Palynology of the Ravenscrag and Frenchman formations; in Coal Resources of Southern Saskatchewan: A Model for Evaluation Methodology; Geological Survey of Canada, Economic Geology Report 30, p. 29-38.
Sweet, A.R. and Hills, L.V.
1984: A palynological and sedimentological analysis of the Cretaceous-Tertiary boundary, Red Deer River Valley, Alberta, Canada (abstract); Sixth International Palynological Conference, Calgary 1984, p. 160.
Taylor, R.S., Mathews, W.H., and Kupsch, W.O.
1964: Tertiary; in Geological History of Western Canada, R.G. McCrossan and J. Porter (ed.); Alberta Society of Petroleum Geologists, p. 190-194.
Thompson, R.I.
1979: A structural interpretation across part of the northern Rocky Mountains, British Columbia, Canada; Canadian Journal of Earth Sciences, v. 16, p. 1228-1241.
1981: The nature and significance of large 'blind' thrusts within the northern Rocky Mountains of Canada; in Thrust and Nappe Tectonics, K.R. McClay and N.J. Price (ed.); Geological Society of London, Special Publication 9, p. 449-462.
Tozer, E.T.
1956: Uppermost Cretaceous and Paleocene nonmarine molluscan faunas of western Alberta; Geological Survey of Canada, Memoir 280, 125 p.
Trettin, H.P. (ed.)
1991: Geology of the Innuitian Orogen and Arctic Platform of Canada and Greenland; Geological Survey of Canada, Geology of Canada, no. 3. (Also Geological Society of America, The Geology of North America, v. E.)
Vonhof, J.A.
1965: The Cypress Hills Formation and its reworked deposits in southwestern Saskatchewan; in 15th Annual Field Conference Guidebook; Alberta Society of Petroleum Geologists, p. 142-161.
Wheeler, J.O., Charlesworth, H.A.K., Monger, J.W.H., Muller, J.A., Price, R.A., Reesor, J.E., Roddick, J.A., and Simony, P.S.
1974: Western Canada; in Mesozoic-Cenozoic orogenic belts, A.M. Spencer (ed.); Geological Society of London, Special Publication 4, p. 592-623.
Whitaker, S.H.
1978: Regional Geology; in Coal Resources of Southern Saskatchewan: A Model for Evaluation Methodology, S.H. Whitaker, J.A. Irvine, and P.L. Broughton (ed.); Geological Survey of Canada, Economic Geology Report 30, p. 21-27.
Whitaker, S.H. and Christiansen, E.A.
1972: The Empress Group in southern Saskatchewan; Canadian Journal of Earth Sciences, v. 9, p. 353-360.
Whitaker, S.H., Irvine, J.A., and Broughton, P.L.
1978: Coal Resources of Southern Saskatchewan: A Model for Evaluation Methodology; Geological Survey of Canada, Economic Geology Report 30, 151 p.
Willumsen, P.S. and Coté, R.P.
1982: Tertiary sedimentation in the southern Beaufort Sea, Canada; in Arctic Geology and Geophysics, A.F. Embry and H.R. Balkwill (ed.); Canadian Society of Petroleum Geologists, Memoir 8, p. 43-53.
Wilson, M.A.
1978: Palynology of three sections across the uppermost Cretaceous/Paleocene boundary in the Yukon Territory and District of Mackenzie, Canada; Palaeontographica, B, v. 166, p. 99-183.
Young, F.G.
1975a: Upper Cretaceous stratigraphy, Yukon Coastal Plain and northwestern Mackenzie Delta; Geological Survey of Canada, Bulletin 249, p. 83.
1975b: Stratigraphic and sedimentologic studies in northeastern Eagle Plain, Yukon Territory; in Report of Activities, Part B, Geological Survey of Canada, Paper 75-1B, p. 309-323.
Young, F.G. and McNeil, D.H.
1984: Cenozoic stratigraphy of the Mackenzie Delta, Northwest Territories; Geological Survey of Canada, Bulletin 336, 63 p.
Young, F.G., Myhr, D.W., and Yorath, C.J.
1976: Geology of the Beaufort-Mackenzie Basin; Geological Survey of Canada, Paper 76-11, 63 p.
Yorath, C.J. and Cook, D.G.
1981: Cretaceous and Tertiary stratigraphy and paleogeography, northern Interior Plains, District of Mackenzie; Geological Survey of Canada, Memoir 398, 76 p.

ADDENDUM

The sequence stratigraphy of the Beaufort-Mackenzie Basin was expanded upon by Dixon et al. (1992). The Reindeer sequence was raised in status to a supersequence and the informal lower and upper units recognized within it were named the Aklak and Taglu sequences, respectively. The Kopanoar sequence/subsequence was recognized to be the lowermost beds of the Kugmallit sequence. Several parasequences were recognized in the Kugmallit Formation of the Nipterk structure by James and Baxter (1988).

Three new members were designated in the Paskapoo Formation of the central Alberta Plains (Demchuk and Hills, 1991).

Magnetostratigraphic analysis was integrated with palynological studies and mammal data in a study of the continental Paleocene Scollard and Paskapoo formations in the Red Deer Valley, Alberta (Lerbekmo et al., 1992). Five palynological zones, corresponding closely to zones P1 to P5 in Wyoming, were recognized from the K-T boundary upward through the upper Scollard and overlying Paskapoo formations. The Paskapoo Formation contains 13 mammal sites belonging to the Tiffanian Ti3 and Ti3 North American Land Mammal Ages. The Paleocene upper portion of the Scollard Formation contains part of magnetozones 29r, 29, 28r, and 28. The Paskapoo Formation embodies a normal magnetozone between two partial reversed magnetozones, which are interpreted to be 26r, 26 and 25r, on the basis of the mammal and palynofloral assemblages. Therefore, there is a hiatus between the Scollard and Paskapoo formations which in this area represents chrons 27r, 27 and part of 26r. This hiatus of 2-3 million years increases to 3-4 million years in the Scollard Canyon area. The sub-Paskapoo disconformity agrees in age with a major Paleocene unconformity attributed to eustatic sea level fall, and may be related to withdrawal of the Cannonball Sea.

REFERENCES

Denchuk, T.D. and Hills, L.V.
1991: A re-examination of the Paskapoo Formation in the central Alberta Plains; the designation of three new members; Bulletin of Canadian Petroleum Geology, v. 39, p. 270-282.

Dixon, J., Dietrich, J.R., and McNeil, D.H.
1992: Upper Cretaceous to Holocene sequence stratigraphy of the Beaufort-Mackenzie and Banks Island areas, northwest Canada; Geological Survey of Canada, Bulletin 407, 90 p.

James, D.P. and Baxter, A.J.
1988: Stratigraphy and sedimentology of the Kugmallit Formation, Nipterk Structure; Beaufort-Mackenzie Basin, Canada; in Sequences, Stratigraphy, Sedimentology; Surface and Subsurface, D.P. James and D.A. Leckie (ed.); Canadian Society of Petroleum Geologists, Memoir 15, p. 117-136.

Lerbekmo, J.F., Demchuk, T.D., Evans, M.E., and Hoye, G.S.
1992: Magnetostratigraphy and biostratigraphy of the continental Paleocene of the Red Deer Valley, Alberta, Canada, Bulletin of Canadian Petroleum Geology, v. 40, p. 24-35.

Authors' Addresses

D.F. Stott
8929 Forest Park Drive
Sidney, British Columbia
V8L 5A7

J. Dixon
Institute of Sedimentary and Petroleum Geology
Geological Survey of Canada
3303 - 33rd Street N.W.
Calgary, Alberta
T2L 2A7

J.R. Dietrich
Institute of Sedimentary and Petroleum Geology
Geological Survey of Canada
3303 - 33rd Street N.W.
Calgary, Alberta
T2L 2A7

D.H. McNeil
Institute of Sedimentary and Petroleum Geology
Geological Survey of Canada
3303 - 33rd Street N.W.
Calgary, Alberta
T2L 2A7

L.S. Russell
Royal Ontario Museum
100 Queens Park
Toronto, Ontario
M5S 2C6

A.R. Sweet
Institute of Sedimentary and Petroleum Geology
Geological Survey of Canada
3303 - 33rd Street N.W.
Calgary, Alberta, Canada
T2L 2A7

Printed in Canada

Subchapter 4K

QUATERNARY

A. MacS. Stalker and J.-S. Vincent

CONTENTS

INTRODUCTION

The Quaternary spans approximately the last two million years (Fig. 4K.1). In Canada, it is associated with glaciation, and the Interior Plains display a glacial record that may be unrivalled for detail and completeness. This review gives a synopsis of the period; a fuller account is found in Chapter 2 (Klassen, 1989; Vincent, 1989) of the Quaternary volume of this series (Fulton, 1989). That account also includes additional references, figures, and tables.

The Quaternary record of the Interior Plains is relatively rich, but still very incomplete. First, although much is known about the glaciations and some of the interglaciations, most of nonglacial time is probably unrecorded. Second, much of the region, and especially parts of northern Alberta, northeast British Columbia, and southern parts of the Northwest Territories, have not been mapped. This lack of information renders correlation between the better known southern Plains and the region near the Arctic coast extremely difficult. Third, the available information indicates little about climate and environment during most of the Quaternary Period.

Preglacial setting

During the first half of the Tertiary, erosion of the newly risen Cordillera provided a flood of silt and sand that was spread as a thick blanket over much of the western Interior Plains. During the second half of that period, the Plains rose relative to sea level, erosion began to predominate over deposition, and much of the Tertiary cover and some of the underlying Cretaceous beds were stripped away. This process, later interrupted by Quaternary glaciation, produced a well-developed drainage system characterized by deep and extremely broad valleys. It also produced the prominent, commonly gravel-capped, erosional remnants found scattered across the plains, such as the Cypress Hills of southeast Alberta (Fig. 4K.2) and southwest Saskatchewan and the Caribou Hills east of Mackenzie Delta. Except for the valleys, erosion remnants, and prominent cuestas such as the Missouri Coteau, the Plains

Stalker, A. MacS. and Vincent, J.-S.
1993: Quaternary; Subchapter 4K in Sedimentary Cover of the Craton in Canada, D.F. Stott and J.D. Aitken (ed.); Geological Survey of Canada, Geology of Canada, no. 5, p. 466-482 (also Geological Society of America, The Geology of North America, v. D-1).

surface was fairly smooth, but otherwise little is known about it. The prolonged exposure of the surface may have produced thick soils, but their remnants are rare.

GLACIATION

Major ice sheets spread onto the Interior Plains several times during the Quaternary (Fig. 4K.1., 4K.2). Those glaciers that advanced from the western mountains are referred to as Cordilleran, and those that came from the Canadian Shield as Laurentide.

On the Interior Plains, drift from old glaciers, especially Cordilleran ones, occurs on many uplands, where it lay beyond the reach of younger, but thinner, glaciers. In general, however, most of the old drift and the thickest Quaternary deposits are found in the preglacial valleys and, to a lesser extent, in the interglacial and interstadial ones, where they were somewhat protected from erosion by subsequent glaciers (Fig. 4K.3). Most of that drift is

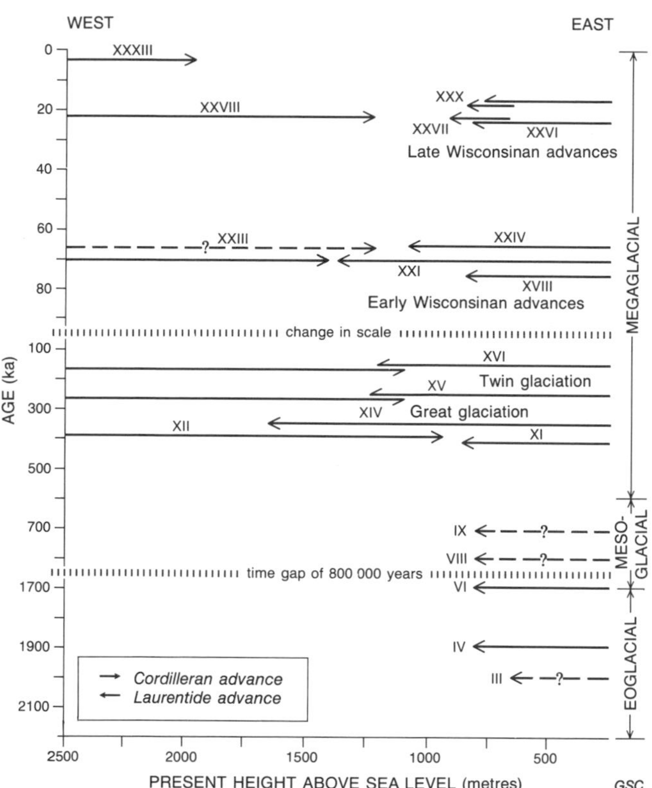

Figure 4K.1. Glacial record of the southwestern Interior Plains and Foothills of Canada, showing approximate times of advances, approximate altitudes reached, and interrelationships of the various Cordilleran and Laurentide glaciers, along with their assumed positions on an absolute time scale. Identification numbers III to XXXIII for the glacial advances correspond to those used in Chapter 2 of the Quaternary Volume (Fulton, 1989). Ages given for many of the glaciations, and especially for advances III and VIII to XVI, are extremely uncertain. Cordilleran equivalents of advances III, IV, VI, VIII and IX have not been recognized. Note gap in time scale between 1.7 Ma and 0.8 Ma, and change in scale after 100 ka.

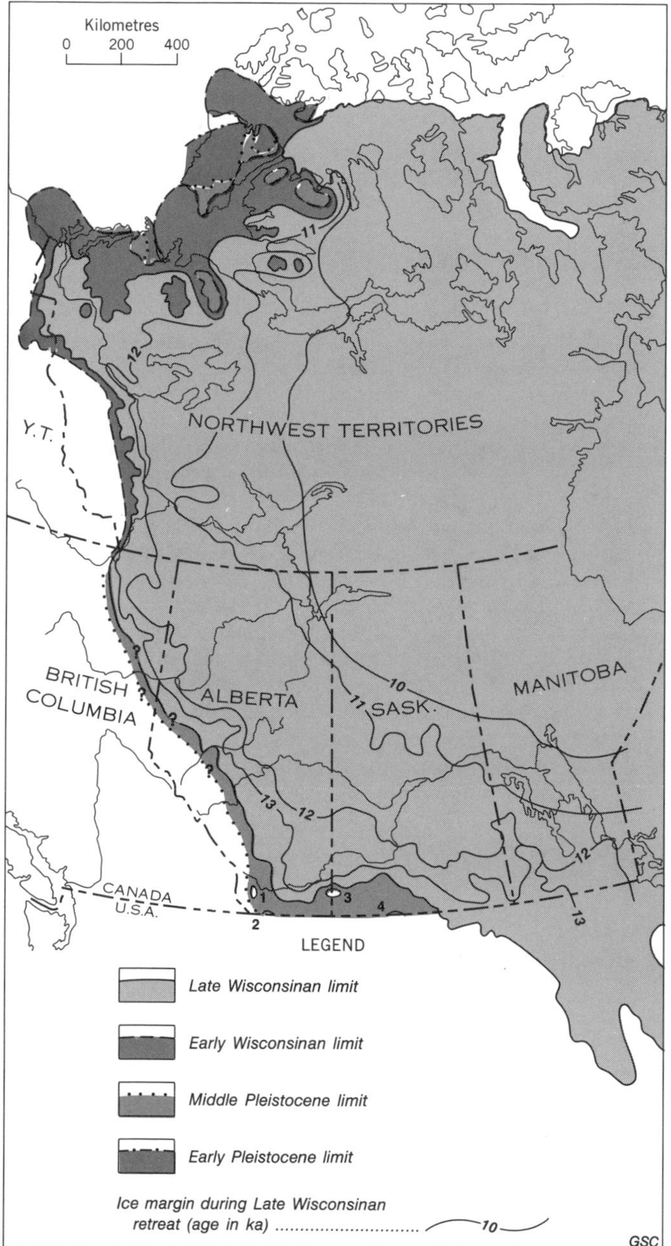

Figure 4K.2. Limits of Laurentide ice for successive Pleistocene glaciations, and successive positions of the Laurentide ice-margin during final retreat. Boundaries shown in British Columbia are highly speculative, and the Middle Pleistocene boundary shown for Alberta and British Columbia is placed at the limit of crystalline stones introduced by ice from the Canadian Shield. Because of fusion of the Cordilleran and Laurentide glaciers into a single ice-mass on the southwestern Plains during Early Wisconsinan time, the Early Wisconsinan limits cannot be shown there. Nonglaciated uplands shown in the southwestern Plains are: 1 Porcupine Hills; 2 Del Bonita area; 3 Cypress Hills; 4 Wood Mountain area.

Laurentide, but some in the western part of the Plains is solely Cordilleran or both Cordilleran and Laurentide. The latter chronicles the interaction between the two sets of glaciers. In general, the Laurentide record is by far the better known, partly because Laurentide drift has received more study, contains fossils between its tills, and is better preserved in preglacial valleys. Most of the Cordilleran drift was laid down in mountain valleys, where it was readily destroyed as subsequent valley glaciers scoured those valleys down to bedrock. As a result, typically only drift of the last local glacier is preserved or, in far downvalley segments, drift of glaciers that reached well beyond the limits of the younger ones.

Local controls

Major factors affecting glaciation and producing distinctive glacial features on the Interior Plains were type of bedrock, regional topography, especially surface slope, position of the Plains on the continent, and climate. The bedrock is predominantly sedimentary, poorly consolidated, and typically rather impermeable, and generally dips gently west-southwest, opposed to the northeasterly regional slope. These factors affected the manner of glacial erosion and deposition, determined subglacial drainage, and induced widespread deformation and contortion of the bedrock by glaciers advancing from east or north. Further, except for pieces of coal, concretions, petrified wood, and fossilized bone, most of the bedrock could only supply fine material to the glaciers, and as a result the till of the area typically is clayey and silty, with few stones. Most of the stones that are present came from the harder rocks of the Canadian Shield and its adjoining Paleozoic belt, and the mountains and foothills.

Topography in the southern Plains, in turn, strongly influenced the advance and retreat of the glaciers and drainage from them. Because of the general westward rise of the surface, eastern glaciers could advance westward only after considerable thickening of the ice on the low-lying eastern Plains. Their westward spread was slow, and those glaciers that did not reach the necessary thickness were restricted to the eastern Plains. This blockage of the glaciers by higher ground, combined with the unconsolidated nature of the bedrock, helped produce the thrusting, folding, and drag effects on bedrock and till now so prominently displayed in many parts of the Plains. The rise to the west also blocked surface drainage coming from the ice, inducing the formation of large lakes and producing the large meltwater spillways whose trenches are such outstanding features of the Plains. In the northern Interior Plains the situation was different, because the glaciers generally moved downhill from the Canadian Shield onto lower areas centred on Great Slave and Great Bear lakes and Mackenzie River basin. Ice thrusting and development of glacial lakes were common, nevertheless, particularly on the Arctic Coastal Plain.

Figure 4K.3. Twin Cliffs Bluff, on the northeast bank of the South Saskatchewan River about 5 km north of Medicine Hat, Alberta. The river has cut through the fill of its preglacial valley to expose four till sheets overlying silt and sand laid down during the approach of the first glacier in the region (photo A.MacS. Stalker, GSC 203193-Z).

Figure 4K.4. Hummocky moraine about 45 km south of Drumheller, Alberta (photo J.-S. Vincent, GSC 200737-F).

The Interior Plains lie towards the centre of the continent and far from the ameliorating effects of the Pacific and Atlantic oceans. They thus have a continental climate, with cold winters and warm summers. In addition, precipitation is low because they lie in the rain shadow of the Cordillera. These factors obviate against local accumulation of ice, hence the glaciers that affected the Plains rose either in the western mountains or on the Canadian Shield. As a result, the Plains was a region where glaciers terminated, rather than one with centres of dispersion, and a region where glaciers stagnated, forming immense areas of hummocky moraine, a distinctive feature of the Plains (Fig. 4K.4). Further, because these glaciers were nearing their limits of advance, glacial deposition was more important than erosion, thereby preserving earlier deposits.

Extent of glaciation

Non-glaciated areas

Laurentide glaciers covered most of the southern Interior Plains and all of the northern Interior Plains at one time or another during the Quaternary. A few small, high areas (e.g. Cypress Hills, Fig. 4K.2 and interfluve areas near the United States border and the Foothills), did escape glaciation by both Laurentide and Cordilleran ice. Their total area is small, perhaps 1000 km^2, and they did suffer some indirect effects, for they were subjected to permafrost and elevated groundwater tables during some of the glaciations and to wind action and loess deposition during glacier retreat. As a result, their original surfaces and soils have been strongly modified and contorted. Certainly, too, the fauna and flora were severely affected, and perhaps destroyed in some of them.

Glaciated areas

In the northern Interior Plains, Quaternary Laurentide glaciers, likely originating west of Hudson Bay, repeatedly advanced west and northwest to the Cordilleran mountain front and at one time or another covered all the plains. Because of limited glaciation in the Mackenzie and Richardson mountains, Laurentide ice rarely coalesced there with Cordilleran ice, as it did in the southern Plains. Exceptionally, Laurentide ice coalesced locally with valley glaciers. Large unglaciated areas remain in the mountainous southwestern District of Mackenzie and northern Yukon Territory.

In addition to the small unglaciated areas in the southern Interior Plains, Laurentide glaciers also missed some areas near the mountain front, mainly in the Foothills and upper reaches of the mountain valleys. Many of these areas, in turn, were overrun by Cordilleran glaciers (Fig. 4K.1, nos. XXIII, XXVIII, XXXIII). In addition, during several glaciations, the glaciers from the two sources overlapped east of the region of solely Cordilleran drift, giving a belt containing both drifts (Fig. 4K.1, nos. XII, XIV, XV, XVI). In most exposures showing overlap the Laurentide drift lies above the corresponding Cordilleran drift, in some instances with intervening lake or stream deposits. This relationship indicates that Cordilleran ice advanced first, having to flow only a short distance down slope, whereas the Laurentide glaciers had to advance upslope and much farther. In most instances the Cordilleran glaciers had even begun active retreat, and some receded considerably, before Laurentide ice arrived. This retreat probably was not caused by increased melting; more likely it resulted from lack of nourishment as expanding ice-sheets farther west in the Cordillera intercepted more and more of the moisture brought eastward by winds from the Pacific before it could reach the eastern slopes of the mountains, while growing Laurentide ice masses effectively cut off supplies from other directions.

During one glaciation, apparently the Early Wisconsinan (Fig. 4K.1, no. XXI), the two ice masses fused into a single entity that covered nearly all the Interior Plains. On the southwestern Plains, the piedmont glaciers formed by the Cordilleran valley glaciers combined with the western part of the Laurentide Ice Sheet, and the joint ice mass flowed southeastward. Little is known about the flow pattern farther north, but on the southwestern plains the flow of many of the ice streams from the valley glaciers can be traced far onto thc Plains by their long erratic trains (Fig. 4K.5). Many of the valley glaciers evidently headed west of the Continental Divide and were fed largely by the massive mountain glacier that had developed over the western Cordillera. Probably only the availability of this major source of nourishment prevented the starvation and retreat that had affected the Cordilleran glaciers during earlier glaciations.

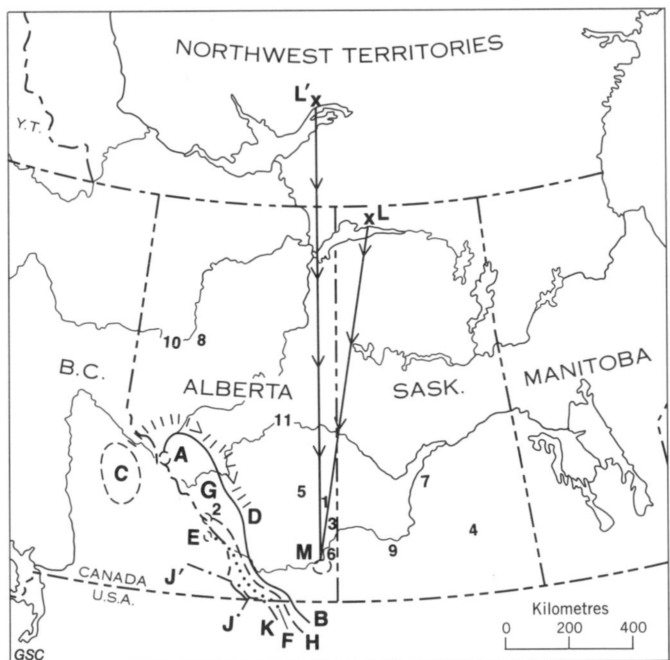

Figure 4K.5. Erratics trains and sources of some erratics found on the southwestern Interior Plains: A-B Foothills Erratics Train (Stalker, 1956); C-D Athabasca Valley Erratics Train (Roed et al., 1967); E-F McNeill (Twin Rivers) Erratics Train (dolomites); G-H vuggy dolomite; J-K shale from Purcell Supergroup(?), with possible source areas J and J′; L-M shallow water oolitic iron formation, with possible sources L and L′. Movement L-M may be Late Wisconsinan, the other movements most likely took place in Early Wisconsinan time. Numbers refer to the sites, listed in Figure 4K.6, where mammalian taxa have been recovered. Photograph shows the Okotoks erratic southwest of Calgary, Alberta (photo N.C. Ollerenshaw).

The Quaternary Record of the Southern Interior Plains

Preglacial Saskatchewan Gravels and Sands

The Saskatchewan Gravels and Sands consist of alluvium laid down in the large valleys of the Plains before Quaternary glacial ice disrupted local drainage. Their deposition started long before the Quaternary Period, and continued well into it; in the higher parts of the southwestern Plains nearly two-thirds of Quaternary time had passed before glacier advances XII and XIV (Fig. 4K.1) terminated fluvial deposition. Because glaciers did not cover all regions at the same time, the upper boundary of the unit is strongly diachronous.

Silt undoubtedly forms the largest component of the Saskatchewan Gravels and Sands, but the gravel, being more conspicuous and of greater economic value, has attracted most attention. Because the rivers that laid the unit down headed in the western mountains, most of the coarser material originated there. Its lack of crystalline Shield material distinguishes this alluvium from river deposits laid down after the first glaciation of any area.

The unit is an important source of gravel, sand, and groundwater. It has also yielded vertebrate and other fossils that give an inkling of the environment and fauna at time of its deposition (Fig. 4K.6, Kansan and Yarmouthian columns).

Glaciation

Glaciers undoubtedly were present on the southern Plains prior to two million years ago, though no trace of them remains. The ice age probably began with glaciers forming in mountain cirques and valley heads, and at times coalescing to flow downvalley through the Foothills to the western edge of the Plains. Also, near the Canadian Shield, small, early Laurentide glaciers may have approached the plains from the northeast. However, traces of such glaciers have not been found.

The glacial record presented here is based on sections examined from the Wellsch Valley, north of Swift Current, Saskatchewan, to the Crowsnest Valley on the British Columbia-Alberta boundary. Cordilleran and Laurentide deposits are assigned numbers used throughout this report (Fig. 4K.1). The record is very incomplete. Records of some advances may be completely obliterated; other Laurentide advances may not have reached the sites studied in the central plains. Further, most of the nonglacial record is missing, particularly for the middle part of the Quaternary Period.

The record is discussed under five headings or divisions, based on extent and timing of glaciations. These divisions are: the Eoglacial interval, or the time of the first recorded glaciers, which lasted from about 2.0 Ma to 1.7 Ma; the Mesoglacial or middle interval, which lasted from about 1.7 Ma to 0.6 Ma and so spanned about half of Quaternary time, but which on the Plains shows glacial activity only towards its end; the Megaglacial interval, or time of large glaciers, which lasted from about 0.6 Ma to 8.5 ka; the warm Altithermal interval, also known either as Climatic Optimum or Hypsithermal, an interlude from about 8.5 to 4.5 ka that typically lacked glaciation; and the Neoglacial interval from about 4.5 ka to the present, which saw minor renewal of glacial activity in the mountain valleys.

Eoglacial interval

The glacial record of the southern Plains starts with three moderately strong glaciations represented by advances III, IV, and VI (Fig. 4K.1). These are all Laurentide; their Cordilleran equivalents are not known. These advances were strong enough to reach central Saskatchewan but not the Medicine Hat area of Alberta. Evidently advance III was less extensive than the other two, and less is known about it, for its till is recorded only from boreholes. The stones in the tills from advances IV and VI tend to disintegrate readily, attesting to prolonged weathering and exposure to groundwater. These two tills appear to have borne thick soils, further indicating long spells of interglacial weathering.

At Wellsch Valley Site, an important sequence of silt, clay, and minor sand beds intervenes between the tills of advances IV and VI. That sequence contains, near its top, a volcanic ash bed of Cascade origin. These intertill beds, which were laid down in lakes and streams and by wind, have yielded (Fig. 4K.6, Nebraskan column) an extensive mammal fauna (Stalker and Churcher, 1982) described later. These fossils, particularly the rodents, indicate an age of about 1.8 Ma. In addition, the sequence encompasses a period of strong paleomagnetic instability including several reversals, probably part of the Olduvai Polarity Event of roughly 1.9 to 1.7 Ma (Mankinen and Dalyrmple, 1979), corroborating the age indicated by the fossils. This age determination is of extreme importance, for it is the basis for assigning ages of between 2.0 and 1.7 Ma to the Eoglacial interval.

Mammoths (Fig. 4K.6, No. 39) and certain small rodents in the loess and lake deposits (Fig. 4K.6, no. 16, 18, 19, 20) indicate that the sequence was deposited following what probably was the most conspicuous of all mammoth and microtine dispersal events from Asia into North America (C.A. Repenning, pers. comm., 1983). This dispersal required establishment of the Bering Land Bridge from Asia, and that in turn probably required a

Figure 4K.6. North American time spans of some of the mammalian taxa recovered from the Quaternary deposits of the Interior Plains. Modified after Stalker and Churcher (1982). Site names (located on Figure 4K.5): 1 Bindloss; 2 Cochrane; 3 Empress Elevators; 4 Fort Qu'Appelle (Bliss); 5 Hand Hills; 6 Medicine Hat area; 7 Saskatoon area, 8 Watino; 9 Wellsch Valley; 10 West Peace River; 11 General Edmonton area, but definite locations unknown; 12 Arctic Coastal Plain. Note: ages of deposits from which bones recovered at Hand Hills (site 5) and Edmonton area (site 11) are unknown, but are assumed to be Sangamon; these ages are indicated on the chart by ? for taxa numbers 5 and 69. The fauna represented in those two areas has an extremely long range of possible ages.

QUATERNARY

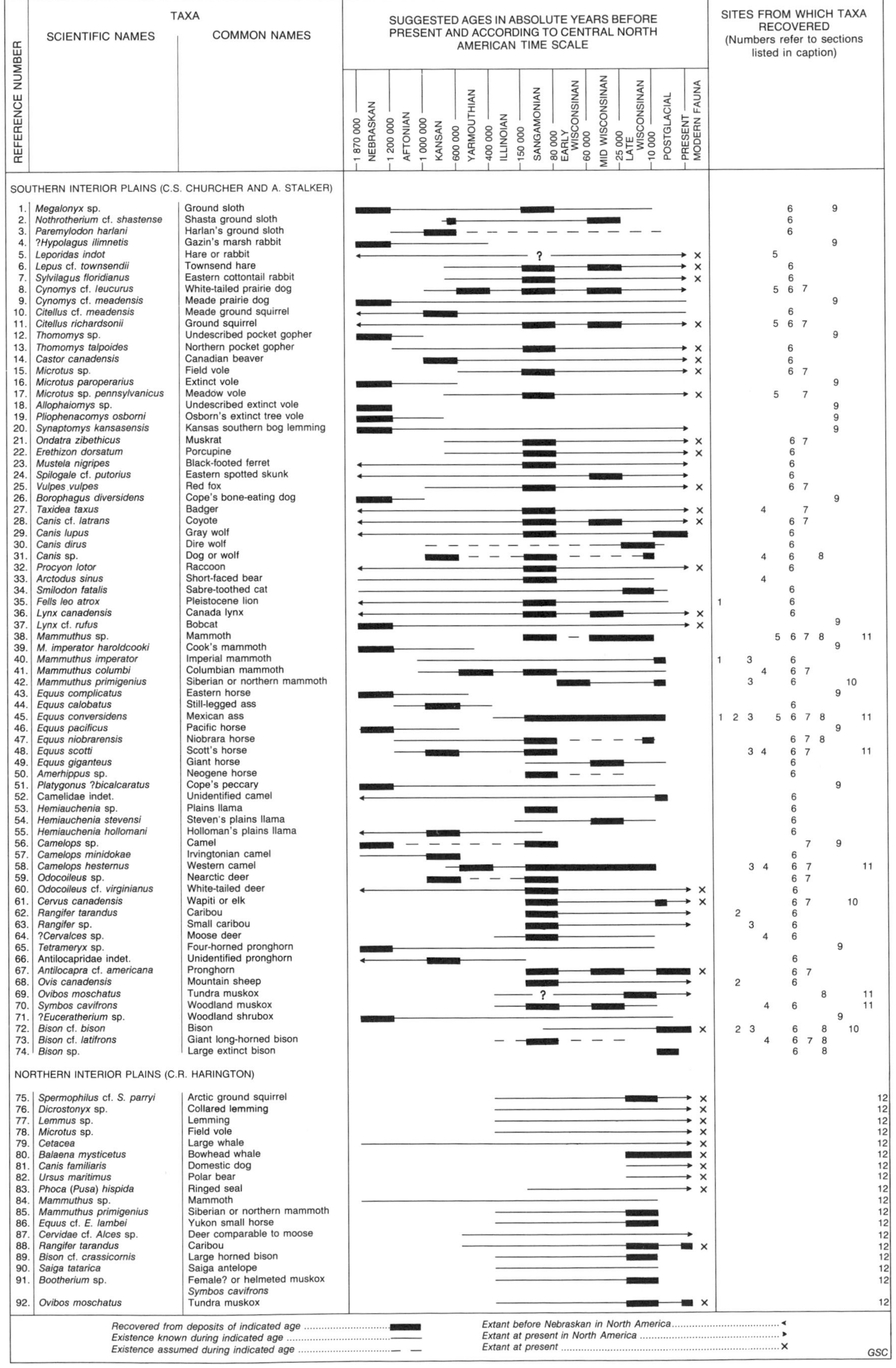

REFERENCE NUMBER	TAXA: SCIENTIFIC NAMES	TAXA: COMMON NAMES	Extant at present	SITES FROM WHICH TAXA RECOVERED (Numbers refer to sections listed in caption)
SOUTHERN INTERIOR PLAINS (C.S. CHURCHER AND A. STALKER)				
1.	*Megalonyx* sp.	Ground sloth		6 9
2.	*Nothrotherium* cf. *shastense*	Shasta ground sloth		6
3.	*Paremylodon harlani*	Harlan's ground sloth		6
4.	*?Hypolagus ilimnetis*	Gazin's marsh rabbit		9
5.	*Leporidas indot*	Hare or rabbit	×	5
6.	*Lepus* cf. *townsendii*	Townsend hare	×	6
7.	*Sylvilagus floridianus*	Eastern cottontail rabbit	×	6
8.	*Cynomys* cf. *leucurus*	White-tailed prairie dog		5 6 7
9.	*Cynomys* cf. *meadensis*	Meade prairie dog		9
10.	*Citellus* cf. *meadensis*	Meade ground squirrel		6
11.	*Citellus richardsonii*	Ground squirrel	×	5 6 7
12.	*Thomomys* sp.	Undescribed pocket gopher		9
13.	*Thomomys talpoides*	Northern pocket gopher	×	6
14.	*Castor canadensis*	Canadian beaver	×	6
15.	*Microtus* sp.	Field vole	×	6 7
16.	*Microtus paroperarius*	Extinct vole		9
17.	*Microtus* sp. *pennsylvanicus*	Meadow vole	×	5 7
18.	*Allophaiomys* sp.	Undescribed extinct vole		9
19.	*Pliophenacomys osborni*	Osborn's extinct tree vole		9
20.	*Synaptomys kansasensis*	Kansas southern bog lemming		9
21.	*Ondatra zibethicus*	Muskrat	×	6 7
22.	*Erethizon dorsatum*	Porcupine	×	6
23.	*Mustela nigripes*	Black-footed ferret		6
24.	*Spilogale* cf. *putorius*	Eastern spotted skunk		6
25.	*Vulpes vulpes*	Red fox	×	6 7
26.	*Borophagus diversidens*	Cope's bone-eating dog		9
27.	*Taxidea taxus*	Badger	×	4 7
28.	*Canis* cf. *latrans*	Coyote	×	6 7
29.	*Canis lupus*	Gray wolf		6
30.	*Canis dirus*	Dire wolf		6
31.	*Canis* sp.	Dog or wolf		4 6 8
32.	*Procyon lotor*	Raccoon	×	6
33.	*Arctodus sinus*	Short-faced bear		4
34.	*Smilodon fatalis*	Sabre-toothed cat		6
35.	*Fells leo atrox*	Pleistocene lion		1 6
36.	*Lynx canadensis*	Canada lynx	×	6
37.	*Lynx* cf. *rufus*	Bobcat	×	9
38.	*Mammuthus* sp.	Mammoth		5 6 7 8 11
39.	*M. imperator haroldcooki*	Cook's mammoth		9
40.	*Mammuthus imperator*	Imperial mammoth		1 3 6
41.	*Mammuthus columbi*	Columbian mammoth		4 6 7
42.	*Mammuthus primigenius*	Siberian or northern mammoth		3 6 10
43.	*Equus complicatus*	Eastern horse		9
44.	*Equus calobatus*	Still-legged ass		6
45.	*Equus conversidens*	Mexican ass		1 2 3 5 6 7 8 11
46.	*Equus pacificus*	Pacific horse		9
47.	*Equus niobrarensis*	Niobrara horse		6 7 8
48.	*Equus scotti*	Scott's horse		3 4 6 7 11
49.	*Equus giganteus*	Giant horse		6
50.	*Amerhippus* sp.	Neogene horse		6
51.	*Platygonus ?bicalcaratus*	Cope's peccary		9
52.	Camelidae indet.	Unidentified camel		6
53.	*Hemiauchenia* sp.	Plains llama		6
54.	*Hemiauchenia stevensi*	Steven's plains llama		6
55.	*Hemiauchenia hollomani*	Holloman's plains llama		6
56.	*Camelops* sp.	Camel		7 9
57.	*Camelops minidokae*	Irvingtonian camel		6
58.	*Camelops hesternus*	Western camel		3 4 6 7 11
59.	*Odocoileus* sp.	Nearctic deer		6 7
60.	*Odocoileus* cf. *virginianus*	White-tailed deer	×	6
61.	*Cervus canadensis*	Wapiti or elk	×	6 7 10
62.	*Rangifer tarandus*	Caribou		2 6
63.	*Rangifer* sp.	Small caribou		3 6
64.	*?Cervalces* sp.	Moose deer		4 6
65.	*Tetrameryx* sp.	Four-horned pronghorn		9
66.	Antilocapridae indet.	Unidentified pronghorn		6
67.	*Antilocapra* cf. *americana*	Pronghorn	×	6 7
68.	*Ovis canadensis*	Mountain sheep		2 6
69.	*Ovibos moschatus*	Tundra muskox		8 11
70.	*Symbos cavifrons*	Woodland muskox		4 6 11
71.	*?Euceratherium* sp.	Woodland shrubox		9
72.	*Bison* cf. *bison*	Bison	×	2 3 6 8 10
73.	*Bison* cf. *latifrons*	Giant long-horned bison		4 6 7 8
74.	*Bison* sp.	Large extinct bison		6 8
NORTHERN INTERIOR PLAINS (C.R. HARINGTON)				
75.	*Spermophilus* cf. *S. parryi*	Arctic ground squirrel	×	12
76.	*Dicrostonyx* sp.	Collared lemming	×	12
77.	*Lemmus* sp.	Lemming	×	12
78.	*Microtus* sp.	Field vole	×	12
79.	*Cetacea*	Large whale	×	12
80.	*Balaena mysticetus*	Bowhead whale	×	12
81.	*Canis familiaris*	Domestic dog	×	12
82.	*Ursus maritimus*	Polar bear	×	12
83.	*Phoca (Pusa) hispida*	Ringed seal	×	12
84.	*Mammuthus* sp.	Mammoth		12
85.	*Mammuthus primigenius*	Siberian or northern mammoth		12
86.	*Equus* cf. *E. lambei*	Yukon small horse		12
87.	*Cervidae* cf. *Alces* sp.	Deer comparable to moose		12
88.	*Rangifer tarandus*	Caribou	×	12
89.	*Bison* cf. *crassicornis*	Large horned bison		12
90.	*Saiga tatarica*	Saiga antelope		12
91.	*Bootherium* sp.	Female? or helmeted muskox *Symbos cavifrons*		12
92.	*Ovibos moschatus*	Tundra muskox	×	12

Recovered from deposits of indicated age ▬
Existence known during indicated age —
Existence assumed during indicated age – –
Extant before Nebraskan in North America ◂
Extant at present in North America ▸
Extant at present ×

glaciation of at least moderate extent. On the basis of assumptions made about its size and its timing, that glaciation most likely was the one of advance III.

Despite their moderate size, ice advances of the Eoglacial interval must have had devastating effects on the Plains. They ended the prolonged, fairly stable conditions of the latter part of the Tertiary, during which drainage, soils, fauna, and flora had developed in concert. The glaciers disrupted drainage and diverted major rivers, filled some of the preglacial valleys, and cut new channels and spillway valleys. They buried or swept away the pre-existing soils, left a chaotic landscape and, through climatic changes, decimated the flora and fauna, even far beyond the regions they overran. The first glacier (advance III, Fig. 4K.1) caused the original southward diversion of South Saskatchewan River from its preglacial course in Saskatchewan, and also changed the courses of the North Saskatchewan and many other rivers.

Mesoglacial interval

The Mesoglacial interval is long and important. However, information about it is meagre and restricted mainly to deposits of the few glaciers that occurred towards its end, and to the younger part of the Saskatchewan Gravels and Sands, whose deposition continued in western regions until its close. The only other nonglacial deposit known from that time is at the Wellsch Valley Site; it consists of about 1 m of lacustrine and eolian silt, sand, and clay. This deposit has not been dated and lacks fossils. Cordilleran events during the Mesoglacial interval are unknown.

Most of Mesoglacial time was nonglacial, and only towards its end do signs appear of glacial activity with two moderate-sized Laurentide ice advances (nos. VIII and IX of Fig. 4K.1). There may have been other, but weaker, Laurentide glaciers which, like their Cordilleran counterparts, left no traces. Advances VIII and IX, like the Eoglacial advances, reached the Wellsch Valley Site but failed to reach the Medicine Hat region where deposition of Saskatchewan Gravels and Sands continued unabated in the valley of the South Saskatchewan river. Little is known about those advances, and their deposits have been identified solely at Wellsch Valley. The age assignments of 0.8 and 0.7 Ma (Fig. 4K.1) are highly speculative, but if incorrect are more likely to be younger rather than older. The freshness of the stones in their tills is in marked contrast to the strong weathering seen on those in the tills from the earlier advances IV and VI, implying a long interlude between advances VI and VIII.

Most of Mesoglacial time was, in all probability, a time of relative stability on the Plains, with conditions much like those of the present. Its length should have allowed development, in addition to soils, of a well-integrated drainage system with deep, broad valleys. Likely little will ever be known of the environment during this interval. However, fauna and flora once again returned from their glacially enforced sojourns, though probably with diminished diversity, and had time to again become well established while evolution continued.

Megaglacial interval

The Megaglacial interval, best known in southern Alberta, was the most striking part of the Quaternary ice age. Its glaciers were so much thicker and more extensive than in preceding episodes that, in regions affected solely by those strong glaciers, this episode was for a long time considered to constitute the complete ice age. During this interval Laurentide ice first reached southern Alberta, and for the first time Cordilleran glaciation left a significant record. On the southern Plains and in the eastern Cordillera, the severest four or five glaciations of the Quaternary are recorded with the first known overlap of Cordilleran and Laurentide glaciers. It is the first interval in which the Cordilleran and Laurentide records can be combined.

Incorportation of mountain outwash into the upper part of the Saskatchewan Gravels and Sands, together with Laurentide advance XI, signalled the start of the main part of the Megaglacial interval. Advance XI was the most extensive glaciation to that time on the southern Plains, but its drift has been recognized only in a restricted area along the South Saskatchewan valley of southern Alberta, and evidently was destroyed elsewhere by subsequent glaciers. The ice seems to have covered the Medicine Hat region but failed to reach Lethbridge. Otherwise, this advance remains largely an enigma, and whether it represents a separate glaciation or was merely the first pulse of the succeeding Labuma glacier (Stalker, 1983) remains unknown; it is shown as the latter (Fig. 4K.1) and therefore is assigned an age of about 0.4 Ma. Advance XII, which laid down the Albertan drift (Dawson and McConnell, 1896), was probably its Cordilleran equivalent. This Cordilleran glacier may also have lasted throughout the early part of Laurentide advance XIV.

The Great Glaciation

The outstanding event of the Megaglacial interval, and probably of the whole ice age on the southern Plains, was the ensuing 'Great Glaciation' of advance XIV and its corresponding Cordilleran advance XII (Stalker and Harrison, 1977). Either then or during Early Wisconsinan advance XXI, both the Cordilleran and Laurentide ice masses reached their greatest dimensions ever. However, farther north on the western Arctic Islands, the most extensive glacial advance apparently was much earlier (Vincent, 1984). During the Great Glaciation, ice flowed down the mountain valleys to coalesce on the plains and form a large piedmont glacier. It also sent lobes far down each of the major preglacial valleys on the Plains, an indication that its advance was unobstructed by Laurentide ice. It laid down the Albertan Till, the most extensive of all the Cordilleran till sheets.

Laurentide advance XIV reached the mountain front unobstructed by Cordilleran ice. This, along with the fact that stream and lake sediments commonly separate these Cordilleran and Laurentide tills, shows substantial retreat of the Cordilleran ice before Laurentide ice reached the western Plains, though the two ice masses may have met locally in front of the major mountain valleys. This Laurentide glacier apparently was more extensive, and thicker, and reached higher in the mountains and foothills than any other. It blocked drainage as it advanced, forming large proglacial lakes in the river valleys, and it finally ended all deposition of the Saskatchewan Gravels and

Sands. During its retreat proglacial lakes again formed, and many of the large spillways found in the plains probably date from then, even though they were reoccupied and modified during subsequent glaciations.

The age of advance XIV is uncertain. At Medicine Hat, the river and lake sediments laid down directly before this advance are normally magnetized, indicating a Brunhes Polarity Epoch age of <730 ka, whereas vertebrate fossils from slightly lower in the section are of Yarmouthian Age (Stalker and Churcher, 1982). Also at Medicine Hat, sediments gave a somewhat older fission track date of 435 ka from a bed of volcanic ash (Westgate et al., 1978); that data is the basis for the age estimates (Fig. 4K.1) of about 350 ka for this advance and 400 ka for advance XI; on the other hand vertebrate evidence would tend to place them earlier, but still later than 600 ka.

The Twin Laurentide Glaciers

Laurentide advances XV and XVI, two glaciations during the middle part of the Megaglacial interval, intervened between the two most extensive glaciations of the Pleistocene Epoch, advances XIV and XXI, both of which greatly overshadowed them. Nevertheless, they surpassed in extent all the others, and so rank as third and fourth most extensive of all the Laurentide glaciations. They are here referred to as the 'Twin Glaciers', for they were of about equal size - it is not known which was larger - and seem to have shared many characteristics. Little is known about them, because the area they covered was completely overrun by the subsequent Early Wisconsinan ice, which left only scattered remnants of their drifts in buried valleys. They are so alike that some authorities consider their tills to represent the same glaciation (Alley, 1973); here, however, they are considered to represent two distinct glaciations, largely because the till of advance XV overlies bedrock in the oldest set of interglacial valleys, whereas that of advance XVI does not appear before the second set, in which it overlies either bedrock or the till of advance XV. Undoubtedly, there were corresponding Cordilleran advances, but little record exists.

The 'Twin Glaciations' are considered to be of Illinoian Age, and are given estimated ages of 250 ka and 150 ka. The latter age is assigned because it was the last major glaciation before the Sangamon Interglaciation of about 100 ka, and the former because it lies between that glaciation and advance XIV with its estimated age of 350 ka.

Sangamonian Interglaciation

The Sangamonian is considered to be the last Quaternary interglaciation, and it separated the final, or Wisconsinan glaciations from those of the Megaglacial interval. Conditions during the Sangamonian Interglaciation probably resembled those prevailing in the other interglaciations of the last 500 ka. On the Plains this ice-free interval is estimated to have lasted from before 125 ka to about 85 ka, or far longer than postglacial time. Rivers were re-established to form a drainage network better integrated than the present one, according to limited evidence. The interval also gave ample time for soils to form, and locally its deposits have yielded abundant vertebrate fauna (Harington, 1978; Stalker and Churcher, 1982). Nevertheless, those deposits do not give an accurate view of the Sangamonian Interglaciation, for they were laid down during the last part of the Sangamonian Interglaciation, only after valleys had been deeply incised and broadened. The interglaciation may have had some long, warm spells, but most of the deposits appear to have developed during cooler episodes, and some possibly were laid down when glaciers were again forming to the east, north, and west.

Wisconsinan Glaciation

The Wisconsinan Glaciation, because of its younger age, the great expanse and good exposures of its deposits, and its economic importance, is much better known than earlier glaciations. However, information about it, and especially about its first two-thirds, remains meagre. The stage consists of two major substages, the Early and Late Wisconsinan, (Fig. 4K.1; advances XVIII, XXI, XXIII, XXIV, XXVI, XXVII, XXVIII, XXX), and one long, nonglacial interval. The Middle Wisconsinan interval probably lasted from earlier than 60 ka to about 25 ka, or for more than half of Wisconsinan time. The various advances exhibited in southern Alberta for each glacial episode represent pulses or subdivisions of the main Laurentide advances. Major ice withdrawals separated those pulses; for example, lake and stream deposits intervene locally between glacial beds, and bones and wood are found between advances XVIII and XXI (Stalker and Churcher, 1982). Eastward, the three Laurentide pulses of each episode evidently belong to one major Early Wisconsinan and one major Late Wisconsinan advance, each represented by one till. However, as those Early and Late Wisconsinan glaciers withdrew northeastward, further pulsations of the ice-margins took place.

The Early Wisconsinan encompassed either the most extensive or second most extensive glaciation of the plains, depending upon the correct assignment of high-altitude deposits in southwest Alberta. Probably during the middle pulse (advance XXI; Fig. 4K.1) the Cordilleran and Laurentide glaciers fused into one major ice-mass that swept south far into the United States. Most of the erratic trains (Fig. 4K.5) developed at that time, and the combined glacier received vast quantitites of Cordilleran ice. This is adjudged to have been the only time the two ice masses combined, and also the only time the valley glaciers received such large incursions of ice from farther west.

Evidence is lacking for whatever Cordilleran and Laurentide glacial activity took place during the long Middle Wisconsinan substage. Altogether, this substage is poorly represented by deposits, and most of those recognized are probably from its cool and humid final stages, perhaps laid down while Wisconsinan glaciers developed in distant regions. They are, therefore, unrepresentative of the interval. The best record comes from the Medicine Hat region, where deposits ranging from at least 41 ka to 24 ka contain a well-developed vertebrate fauna, along with minor quantities of wood and pollen, none of which indicates much about paleoenvironments. As in the Sangamonian Interglaciation, a well-developed drainage system evolved towards the end of the Middle Wisconsinan interval, and conditions probably resembled those of today.

The cooling trend towards the end of the Middle Wisconsinan interval signalled the coming of Late Wisconsinan glaciation. Apparently the middle Laurentide pulse (advance XXVII) spread farthest. Here it is considered to have advanced to the boundary shown for Late Wisconsinan ice in Figure 4K.2, but many United States and some Canadian workers (Christiansen, 1979; Clayton and Moran, 1982; Mickelson et al., 1983; Shetsen, 1984) considered it to have been more extensive than Early Wisconsinan ice, and to have spread well into Montana. Certainly it did reach the limit shown in Figure 4K.2, and so Late Wisconsinan ice is responsible for most of the drift mantle of the Interior Plains.

Cordilleran events during the Late Wisconsinan are less well known, although undoubtedly there were many minor advances and retreats of valley glaciers (Rutter, 1972; Alley, 1973; Stalker and Harrison, 1977; Jackson, 1980a, b). It is doubtful that Cordilleran ice advanced much beyond the mountain front, and apparently no contact was made between Cordilleran and Laurentide glaciers.

The Late Wisconsinan, Laurentide Ice Sheet on the Interior Plains is considered to have reached its maximum at about 18 to 20 ka, and to have begun active retreat from the higher regions at 16 to 14 ka.

Late Wisconsinan Deglaciation

Late Wisconsinan ice retreated completely from the Interior Plains between 16 and 10 ka (Fig. 4K.2). That retreat from regions near the mountains to the Hudson Bay area was depicted on a map by Prest (1969), and the broad interpretations of the deglaciation have not changed much since that time, though chronology and ice-margin positions have been refined. The main controversy still centres on the position of the ice margin at maximum advance. If the more northerly limit (Fig. 4K.2) is correct, then the features to the south and west are Early Wisconsinan or older in age. Another controversy centres around the date at which Late Wisconsinan ice began to retreat from southern Manitoba and adjacent parts of the United States. Teller and Fenton (1980), Clayton and Moran (1982), and Mickelson et al. (1983) inferred that deglaciation took place several thousand years later than was advocated by Elson (1967), Prest (1969), Klassen (1975), and Christiansen (1979). The earlier date advocated by the latter authors is adhered to here.

During Late Wisconsinan time, the waning valley glaciers retreated to approximately their present positions, probably with some fluctuations en route. The retreat of the massive Laurentide glacier was much more complex. That glacier retreated downslope to the east or northeast, but with many halts and readvances. Its margin was lobate because of topography and uneven retreat, with some sectors melting back faster than others, or with local advances taking place while other parts of the margin were retreating. Retreat resulted largely by downmelting at the ice surface, but also, especially where the glacier was fronted by glacial lakes, considerable retreat was caused by melting at the glacier margin. Much ice-stagnation took place along broad marginal zones, the resulting slow disintegration of the ice producing vast areas of hummocky moraine, kames, outwash plains, and other associated features.

Marginal positions of the glacier during its retreat (Fig. 4K.2) are based largely on altitudes, drainage patterns, locations of proglacial lakes, spillways, and moraines. Ages are determined largely by radiocarbon dating, which introduces uncertainties because few samples are directly associated with ice-margin positions, but rather come from below or above the drift. In some instances the discrepancy may be several thousand years. The distance between the 12 and 10 ka isochrones on the eastern plains indicates an average ice-margin retreat of about 200 m per year towards the northeast (Fig. 4K.2). However, irregular spacing and northward flexures of the isochrones suggest uneven rates of retreat.

Much of the retreat on the southeastern Plains was directly associated with large glacial lakes, in front of which retreat was particularly rapid. Lake Agassiz reached its greatest expanse at about 10 ka and subsequently fell rapidly and drained from its southern and western reaches while spreading northward behind the retreating ice. Upon final disintegration of the ice in northeastern Manitoba at about 8 ka (Dredge and Cowan, 1989), water from Hudson Bay (the Tyrrell Sea) inundated the Hudson Bay Lowland, and mixed with the remnants of Lake Agassiz.

Altithermal interval

Final disintegration of the ice on the eastern Plains was hastened by a strong warming trend, culminating in the warm Altithermal interval. It was the most important post-Wisconsinan interval, witnessing drastic effects on fauna, flora, and early man, (Stalker, 1969, 1970). It is still not known whether it consisted of one long, warm span or of two or more short ones separated by cooler spells (Stalker, 1970, 1973), and temperature and precipitation changes remain conjectural. It was a time of strong wind action, and most of the sand dunes found on the Plains originated then. Widespread loess settled as a thin blanket over much of the Prairies, adding greatly to the fertility of the soils.

The beginning of the Altithermal interval is estimated at about 9 ka (Fig. 4K.1). Certainly it was considerably before 6.7 ka, for Mazama Ash of that age commonly is found towards the base of loess that, in turn, overlies a well developed soil that probably formed partly during the Altithermal interval. Its end is better established at about 4.5 ka.

Neoglacial interval

The Quaternary Period ends with the short Neoglacial interval. Conditions became cooler and perhaps damper, and dune formation and loess deposition largely ended, probably because the more humid climate encouraged vegetation. The surface stabilized with formation of the modern soil. The main features of the topography of the Plains were established previously, however, with even the trunk rivers incised to almost their present levels. In the mountains, valley glaciers reformed and advanced, at times beyond their present positions, but no evidence exists that any of them reached the Plains or even the Foothills.

The Quaternary record of the northern Interior Plains

Sediments in the northern Interior Plains, like those of the southern Interior Plains, provide a long Quaternary record. Most knowledge has been gained since the early 1970s, with most work a response to needs related to oil and gas exploration. Major regional syntheses include those of Mackay (1958, 1963), Craig (1965), Hughes (1972, 1987), Hughes et al. (1981), Rampton (1982, 1988), and Catto (1986). The longest records are on the flanks and foothills of the Mackenzie and Richardson mountains and on the Arctic Coastal Plain. In the latter, the record is difficult to unravel for three reasons. Firstly, mainly nonglacial sediment suites are stacked without intervening glacial or marine sediments, which elsewhere serve as marker units. Secondly, widespread ice thrusting along the Beaufort Sea coast has made it difficult to establish the correct relative position of sediments. Finally, widespread thermokarst activity and the growth of massive bodies of ground ice has strongly disturbed the sediments. Even though distinct events can be identified and their relative timing established, their absolute age is rarely known. In many areas, various authors have mapped what they considered to be the ice-sheet limits during distinct glacial stades or stages (Fig. 4K.2). Because of the lack of reliable chronological data and detailed mapping in certain critical areas, together with conflicting interpretations, there is no agreement on the age of recognized glacial limits in the northern Plains, as is also the case in the south. Indeed, one worker's glacial limit can be a still-stand or readvance position to a colleague. Obviously the limits shown in Figure 4K.2 are tentative.

Preglacial record

On the northern Interior Plains, sediments that predate those of the earliest glaciation are present in the foothills of the Richardson Mountains and on the Yukon Coastal Plain. In river valleys of the foothills, preglacial gravels found directly above bedrock and beneath glacial and nonglacial drift (Catto, 1986) contain no Shield-derived clasts and, therefore, are probably reworked from Tertiary alluvium. On the Yukon Coastal Plain, major pediments (Rampton, 1982) record periglacial conditions but have been little studied.

Early Pleistocene

Early Pleistocene time spans the interval between 2 Ma and 730 ka. In the South Nahanni River basin an ice advance before 350 ka, termed the First Canyon Glaciation, is documented by Ford (1976). In the Snake, Peel, and Rat River basins, fluvial gravel with Shield-derived clasts probably also records an old glacial event in the foothills of the Richardson Mountains. In the lower Horton River area, Fulton and Klassen (1969) have reported at least three tills lying above Tertiary gravels in an area that probably lies outside at least the Late Wisconsinan glacial limit. Paleomagnetic investigation of this sequence by Vincent and Barendregt (unpublished data) has revealed that the lowermost tills are magnetically reversed, thus indicating an Early Pleistocene age for the glaciation responsible for their deposition. On the basis of these data, Laurentide ice probably extended onto the Plains and up the flanks of the Mackenzie and Richardson mountains several times during Early Pleistocene time.

Middle Pleistocene glacial and nonglacial record

Middle Pleistocene time spans the interval between 730 ka (Brunhes/Matuyama boundary) and 128 ka (beginning of Sangamonian Stage). Sediments in several localities of northern Interior Plains record glacial and nonglacial events that likely occurred during this period. During one strong glaciation, Laurentide ice advanced to a height of about 1500 m on the south flank and 1400 m on the north flank of the Mackenzie Mountains. In the South Nahanni River basin, one of these glacial advances (termed Clausen Glaciation) possibly was Illinoian in age (Ford, 1976; Harmon et al., 1977).

Along Snake River, in the foothills of the Richardson Mountains, Catto (1986) identified a Laurentide till under fluvial sediments and younger glacial deposits. In the fluvial deposits, wood was dated at >39ka (GSC 3697 – the identification number refers to radiocarbon-dated sample analyzed by the Geological Survey of Canada). In the Rat River basin, sediments deposited in a lake dammed by Laurentide ice standing to the east also record a Middle Pleistocene ice advance, for they are overlain by organic-bearing fluvial sediments considered to be of interglacial origin and by younger glacial deposits.

On the north flank of Caribou Hills along the East Channel of the Mackenzie River, and also at the south end of Eskimo Lakes, fluvial sands, silts, and clays, interbedded with peat containing wood, are exposed. Radiocarbon and uranium/thorium ages on the organic deposits have given non-finite ages [more than 200 ka (UQT-149 and 188; J.S. Vincent, unpublished data) in the case of the U/Th age determination]. These deposits were probably laid down by the ancestral Mackenzie River and record a Middle Pleistocene interglaciation for which there is little other evidence (Vincent, 1989).

Thick sands with minor gravel and clay underlie till of probable Early Wisconsinan age in the Tuktoyaktuk Coastlands, including islands fronting Mackenzie Delta, Tuktoyaktuk Peninsula, and in areas surrounding Eskimo Lakes and Liverpool Bay. Rampton (1988) proposed a lithostratigraphic framework for these sediments, some of which may date from the Sangamonian Stage. The succession starts at the base of the Kendall sediments and Hooper clay, found west of the East Channel of Mackenzie River and probably on Nicholson Peninsula. These consist of fossiliferous marine clays, silts, and sands. Overlying Kidluit beds, consisting of fluvial, crossbedded, grey sands with much organic detritus, are exposed between Pelly Island and Liverpool Bay. Finally, the Kittigazuit Formation, consisting of brown, most likely glaciofluvial deltaic sediments, is well exposed on the south side of Liverpool Bay and on Richards Island. All of these sediments were laid down on the coastal plain by the ancestral Mackenzie River and other rivers farther east, either as outwash when pre-Wisconsinan glaciers melted or between glaciations when sea level was higher.

A record of a Middle Pleistocene drift exists on Bathurst Peninsula, where ice of the Mason River Glaciation (Rampton, 1988) is recorded on the southwestern portion of the peninsula (Fig. 4K.2). Deposits from Mason River ice advance underlie interglacial deposits and clearly predate the younger Wisconsinan Toker Point Stade ice advance and thus provide probably the best record of a Middle Pleistocene glaciation in the northern Interior Plains. During ice retreat, marine waters covered the area to at least 50 m above present sea level.

As indicated by the several sites described above, a long record of Middle Pleistocene glacial and nonglacial events is documented in the northern Interior Plains. Unfortunately, it has been impossible to correlate the different deposits reliably and thus reconstruct the history of the region.

Sangamonian interglaciation

In the Rat River basin, organic-rich fluvial sediments record a boreal forest to open tundra environment with a climate warmer than today (Catto, 1986). These may date from the last interglaciation. On Yukon Coastal Plain, deposits underlying drift of probable Early Wisconsinan age are widespread (Rampton, 1982). These consist of thick and complex sequences of sands, bedded silts, clays, and gravels, commonly ice-thrusted and deformed, laid down in rivers, estuaries, and shallow to moderately deep seas. Also along the lower reaches of some of the rivers flowing across the coastal plain, unoxidized sands and gravels are widespread over oxidized sands and gravels with ice wedge casts. Both the oxidized and unoxidized materials, with paleosols, probably are the fluvial equivalents of the thick marine and estuarine deposits mentioned above. Although most probably date from the Sangamonian Stage, some of these sediments may be older.

Complex sequences of bedded sands, silts, and clays with soil horizons and ice wedge casts, found on Bathurst Peninsula, were possibly deposited in alluvial plains, in perimarine environments, and in tundra ponds during the last interglaciation. Also of likely interglacial age are thick driftwood mats that are found on low benches, up to about 7 m above sea level (a.s.l.). These record a time of higher sea level than today, when the Mackenzie River basin had to be ice free in order to produce the driftwood (Vincent, 1989).

Wisconsinan Glaciation

Most Quaternary geologists agree as to which glacial deposits are of Wisconsinan age in the northern Interior Plains, but disagree on their positions within the Wisconsinan. Some contend that the strongest Wisconsinan advance was during the Early Wisconsinan Subage and that Late Wisconsinan ice was less extensive; this is the point of view portrayed in Figure 4K.2. On the other hand, others consider that the maximum Wisconsin advance occurred in the Late Wisconsinan Subage, and was followed locally by a readvance. In this report the Early and Late Wisconsinan histories are discussed separately, with alternate interpretations of age of the glacial limits given where controversy exists. In the absence of unequivocal age control, age assignments are generally tentative and very much depend on the biases of individual authors.

Early Wisconsinan

The limit proposed for Early Wisconsinan ice advance is shown in Figure 4K.2. During this time Bathurst Peninsula, Tuktoyaktuk Peninsula, and Brock Upland are believed to have escaped glaciation. In the northeastern Interior Plains, an ice lobe advanced in Amundsen Gulf from the mainland, impinging on the coast and merging with northwestward flowing ice from Great Bear Lake area. These glaciers probably were too thin to cover Brock Upland and Bathurst Peninsula (Vincent, 1989). Cliffs on the east coast of Bathurst Peninsula effectively blocked further westerly progress of the ice lobe (likely an ice shelf) in Amundsen Gulf (Franklin Bay Stade of Rampton, 1988), whereas the ice flowing northwesterly from Great Bear Lake did not progress much north of West River and was restricted to the west of the Anderson River Basin (Toker Point Stade of Rampton, 1988; Fig. 4K.7). Farther west, ice from north of Great Bear Lake reached the Beaufort Sea at the mouth of Mackenzie River and impinged on the southern part of Tuktoyaktuk Peninsula. West of Mackenzie River, the ice advanced from the Mackenzie Valley into Beaufort Sea and impinged on the Yukon Coastal Plain. The limit and deposits of this event, termed the Buckland Glaciation, have been documented by Rampton (1982). The glacial limit falls from about 1000 m southwest of Aklavik, in the Richardson Mountains, to near sea level west of Herschel Island. A major stillstand or readvance, termed the Sabine Phase of Buckland Glaciation, occurred during deglaciation. The western limits of this presumably Early Wisconsinan glacier on the lower slopes of the Mackenzie and Richardson mountains are controversial (Hughes et al., 1981; Prest, 1984). They are not readily traced on the steep slopes where significant changes in ice thicknesses caused only limited changes in extent. As there are indications that Laurentide ice reached farthest along the mountain front, at least locally, in the latter part of the Wisconsinan Glaciation (Hughes et al., 1981; Catto, 1986), this area is discussed below.

The Early Wisconsinan age assignment for this ice advance is based on radiocarbon dates obtained from the Arctic Coastal Plain. Fragile willow twigs in crossbedded terrace sands on Bathurst Peninsula, which postdate the glacial advance, have been dated at 33.8 ka (GSC 1974; Rampton, 1988). Also marine shells coming from littoral deposits overlying till laid down during this advance are too old for radiocarbon dating (Vincent, 1989).

Middle Wisconsinan

Deposits that can clearly be assigned to the interstadial period preceding the Late Wisconsinan Subage are rare. Deposits dated at 33.8 ka (GSC 1974; Rampton, 1988) occur on Bathurst Peninsula, sediments with organic matter yielding finite radiocarbon ages have been found in a terrace on Malcolm River on the Yukon Coastal Plain, and a date of 36.9 ka (GSC 2422; Hughes et al., 1981) has been obtained from wood in fluvial and deltaic sediments underlying till in Bonnet Plume Basin.

Late Wisconsinan

The Late Wisconsinan glacial limit shown in Figure 4K.2 is the minimum position (see Prest, 1984). When the ice was at this position significant areas of the mainland south of Beaufort Sea and Amundsen Gulf were free of ice. Ice flowed westward from the Keewatin Sector of the Laurentide Ice Sheet in the upper Mackenzie River basin to the Mackenzie Mountains, where it was deflected north along Mackenzie River and south up Liard River.

Between the Richardson Mountains and Anderson Plain, ice flowed north to northwest down the Mackenzie Valley. West of Mackenzie River and north of the

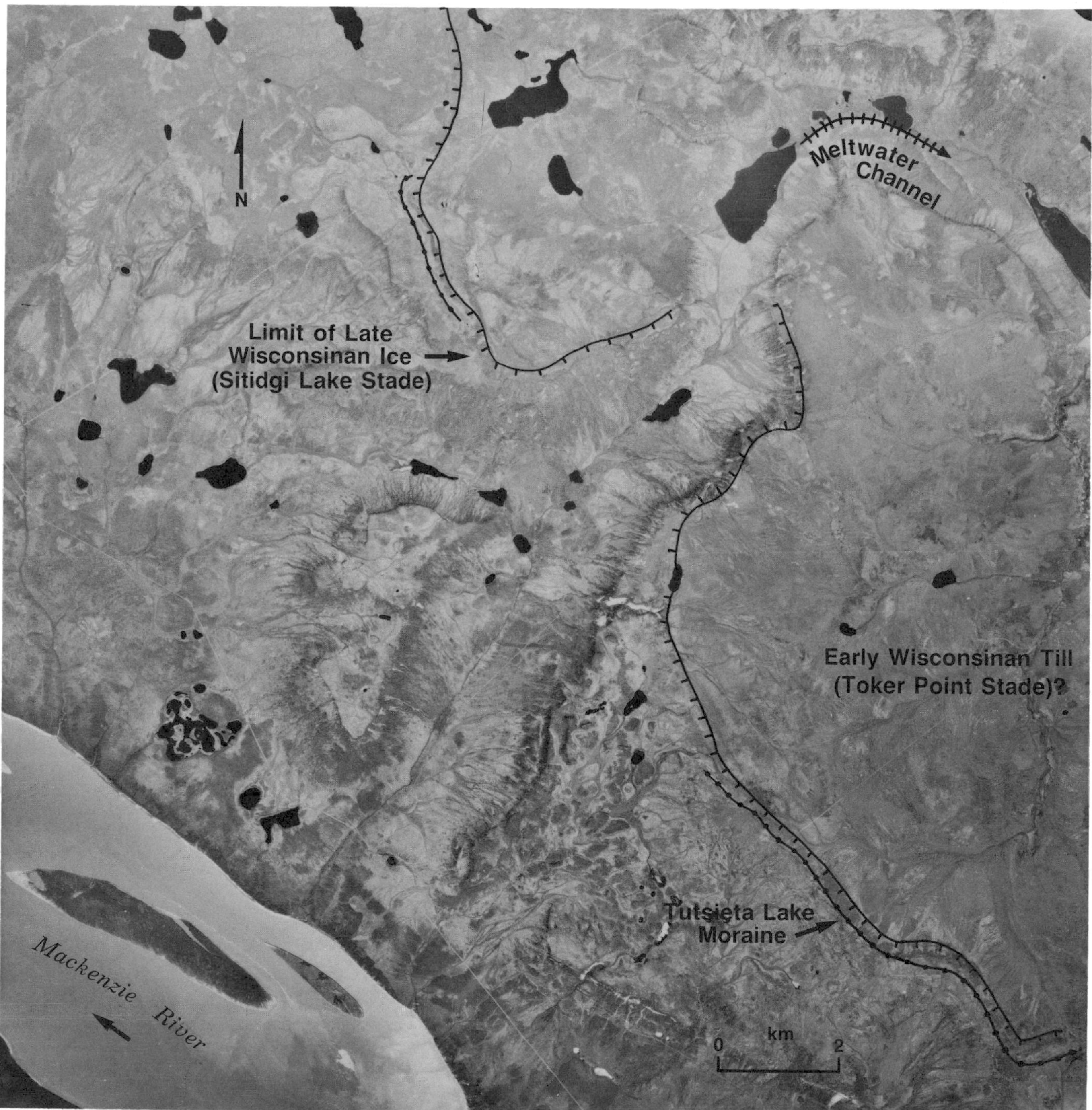

Figure 4K.7. Limit of Laurentide ice advance during the Late Wisconsinan Sitidgi Lake Stade (note the terminal moraine) and surface covered during the Early? Wisconsinan Toker Point Stade. Area southeast of Inuvik, N.W.T. (photo A-21585-191, National Air Photo Library, Department of Energy, Mines and Resources).

Mackenzie Mountains, its limit is placed where Hughes (1972) initially mapped the limit of Wisconsinan Laurentide Ice. North of latitude 68°, it follows the margin of Late Wisconsinan glaciation (Sitidgi Lake Stade) shown by Rampton (1982, 1988). East of Mackenzie River the limit follows a well developed terminal moraine, named the Tutsieta Lake Moraine (Hughes, 1987), which coincides with the Sitidgi Lake Stade limit of Rampton (Fig. 4K.7). Hughes et al. (1981) and Hughes (1987) did not agree that this is the Late Wisconsinan limit, but considered it a stillstand during retreat of the Laurentide glacier in Late Wisconsinan time. They placed the Late Wisconsinan limit farther west and north at the farthest glacial margin, where Rampton (1988) placed the Early Wisconsinan Toker Point Stade limit on the Arctic Coastal Plain. Hughes et al. (1981) and Catto (1986) placed the Late Wisconsinan margin in this position to explain Late Wisconsinan lakes in the unglaciated part of Yukon Territory, presumably created by blocked drainage from the Porcupine River basin through McDougall Pass in the Richardson Mountains. In addition, in the Bonnet Plume drainage basin, well west of the Late Wisconsinan limit (Fig. 4K.2), allochthonous wood from below till gave a radiocarbon date of 36.9 ka (GSC 2422; Hughes et al., 1981), indicating that the overlying till (Hungry Creek Glaciation of Hughes et al., 1981) was of Late Wisconsinan age. This extreme position of Late Wisconsinan ice is, seemingly, not compatible with data from the Arctic Coastal Plain, which corroborates the limit of Figure 4K.2. Having abundant ice on the east flank of the Cordillera creates problems with Arctic Coastal Plain reconstructions, whereas less ice on the Arctic Coastal Plain creates problems in the reconstruction of events on the east side of the Cordillera. This dilemma remains unsolved, and Wisconsinan ice may have advanced into the area either during two distinct stades (Early Wisconsinan and Late Wisconsinan) or reached its maximum during early Late Wisconsinan and developed a second well marked position during a Late Wisconsinan stillstand or readvance.

East of Mackenzie River and south of Amundsen Gulf, the ice advanced northwesterly from Great Bear Lake. Topographic depressions, such as the upper part of the Anderson River and Hornaday River valleys and Dolphin and Union Strait, along with higher plateaus in upper Horton River basin and south of Brock Upland, directed the ice flow.

In the upper Mackenzie River basin the ice generally retreated northeastward and eastward (Craig, 1965), allowing glacial lakes dammed in mountain valleys at the glacial maximum to spread eastward. Glacial Lake McConnell enveloped much of the Liard and southern Mackenzie River basin and extended south of latitude 60°N in the valleys of Peace and Slave rivers, and into the basins of Great Slave, Great Bear, and Athabasca lakes. Radiocarbon dates from the lower Mackenzie River basin indicate that ice retreat was underway at 13 ka (Rampton, 1988). Organic detritus overlying glacial lacustrine sediments near the mouth of the Great Bear River, dated at 10.6 ka (GSC 2328, Hughes, 1987), gives the oldest minimum age for deglaciation in the upper Mackenzie River area.

East of Mackenzie River and south of Amundsen Gulf, the ice front generally retreated southeasterly towards the Canadian Shield (Fig. 4K.2), while building spectacular systems of end moraines and leaving widespread hummocky moraine and ice-contact deposits. In the Hornaday River basin, a radiocarbon date of 10.8 ka (GSC 1139; Lowdon et al., 1971) on peaty moss from a pingo provides a minimum age for deglaciation. As the ice withdrew, the sea flooded the coastal areas. Marine limit, which is at about 145 m above present sea level at the mouth of Coppermine River, decreases gradually towards the west. West of Parry Peninsula no evidence of emergence exists.

In the Beaufort Sea, north of the Mackenzie Delta and Tuktoyaktuk Peninsula, a continuous record of Late Wisconsinan and Holocene submergence exists (Forbes, 1980). Hill et al. (1985) have prepared a sea-level curve showing a relative rise of sea level in the Beaufort Sea of 140 m since possibly 27 ka.

Holocene

Since deglaciation, geomorphic processes have been active. As in most regions, rivers are the chief landscape modifier. Mass movements on slopes are also important. Gelifluction and creep are common on all slopes with fine-grained materials, as shown by the widespread presence of stripes and solifluction lobes and sheets. Rapid mass movements are less common, but active-layer detachment failures in the poorly lithified Cretaceous shales are important, as are rock falls and debris slides in the mountains. On the Arctic Coastal Plain, frost processes have caused sorting, mixing, heaving, tension cracking, and shattering of both consolidated and unconsolidated surface materials. Patterned ground with hummocks, high- and low-centered polygons, mudboils, and other features reflect this frost activity. The growth or decay of ground-ice bodies, such as massive segregated ice, possibly remnant glacier ice, ice wedges, and pingo ice, has substantially altered large tracts of land. Coastal processes that modify sea shores and shores of the larger lakes, as well as limited eolian and human activity, have contributed to a lesser extent in modifying these northern landscapes.

PALEOECOLOGY

Knowledge about the Quaternary paleoecology of the Canadian Interior Plains is sketchy. Most of the flora and fauna was destroyed during the successive glaciations. Sediments are preserved from only limited parts of nonglacial intervals, and those probably misrepresent general conditions, for they are largely from the cool, final stages, in some cases deposited even while glaciers were forming and advancing elsewhere.

Both flora and fauna were vastly richer and more varied at the beginning of Quaternary time than at the end, despite the intermittent introduction of different species of animals, and perhaps plants, from Asia via Bering Land Bridge. Each large glacier extinguished most life on the Great Plains, and each withdrawal saw the re-introduction of genera and species from the south, fewer in number than previously. Depletion of taxa was general throughout the Quaternary Period, but the greatest impoverishment came during the 'Great Extinction' towards its end. Loss of plant taxa probably was greater than the loss of animal taxa, partly because plants could not repopulate the region as readily.

Flora

Most knowledge about the floral history of the Interior Plains comes from pollen studies (Ritchie, 1976, 1984; Mott, 1973; Ritchie and Yarranton, 1978; see also Klassen, 1989). Most is known about the latter part of Late Wisconsinan time and the Holocene, partly because deposits from those times are more common and better studied, but also because that time span is within the range of radiocarbon dating.

In general, the studies show a predominantly spruce forest along the ice margin in southern Manitoba during Late Wisconsinan time. This forest spread north behind the retreating glacier, and also covered the southeastern part of the plains between 14 and 12 ka. As climate warmed, grassland replaced much of the forest, which by the peak of the Altithermal episode about 6.5 ka was largely restricted to its present ranges in the Canadian Shield and northern Saskatchewan and Alberta.

In the northern Interior Plains much is known about the vegetation changes in the lower Mackenzie River basin during Late Wisconsinan and Holocene times, but little about other areas and earlier Quaternary times (Ritchie, 1984). From about 30 to 14 ka, the unglaciated northern Yukon Territory was cold and the vegetation resembled the modern herb fell-field communities. As Late Wisconsinan ice retreated, the northern Yukon Territory at about 14 ka and the lower Mackenzie valley at about 12 ka became warmer and wetter, with an increase in dwarf birch indicating a shift from herb to shrub tundra. Between about 11 and 9 ka, the area was much warmer and the tree line migrated northward. About 5.5 ka the forest started to retreat, and by 4 ka BP the climate had cooled subtantially and in many areas shrub tundra, similar to that of today, became established.

Fauna

During the last 20 years, vertebrate fossils - chiefly mammal - have become the basis of correlation between sections and determination of relative ages of Quaternary deposits on the southern Interior Plains. They have also established that the surficial deposits on the Plains represent a long record.

Most of the Quaternary fossil taxa identified for both the southern and northern Plains are listed in Figure 4K.6, which also shows their time ranges. Harington (1978) listed by area the vertebrate faunas described in Canada and gave brief descriptions of the vertebrate sites and past environments. A chart by Stalker and Churcher (1982) depicts the fossils recovered from the Wellsch Valley and Medicine Hat regions, indicates their ages, and shows their North American time ranges.

At the onset of Quaternary glaciation, the fauna had had a prolonged, uninterrupted term of development, and was well adapted to the environment. It was rich and diverse and, although somewhat impoverished in relation to earlier times, it remained so until the Great Extinction in Late Wisconsinan time. There were constant changes, but extinctions were largely offset by evolution and by introduction of taxa from Asia via the Bering Land Bridge and, in the early part of the Quaternary Period, from South America via the Isthmus of Panama. In general, the record is better for those animals that lived along rivers or browsed on brush in the large valleys than for those that lived on the high plains or foothills. Bones from the larger animals and those that lived during the cooler parts of the interglacials were also more readily preserved, and the samples are biased towards these.

The oldest, comprehensive Quaternary fauna yet found on the plains comes from Wellsch Valley (Fig. 4K.6, no. 1, 4, 9, 12, 16, 18, 19, 20, 26, 37, 39, 43, 46, 51, 56, 65, 71). It dates from about 1.75 Ma and is a fairly typical preglacial fauna for the southern Plains, especially for species that lived in the large valleys. Large animals included ground sloths, horses, and camels; mammoths, which had just previously reached North America across the Bering Land Bridge, were starting to speciate and become firmly established. Among mid-size animals were early rabbits, peccaries, pronghorns, and shrub-oxen. Carnivores were represented only by the bobcat and *Borophagus* - an early, heavily built dog. Rodents, especially voles, were common, but many were primitive types.

The next relatively complete fauna, found in the lowest beds in the Medicine Hat region, is about 1 million years younger, being laid down in Kansan and Yarmouthian time, some 600 thousand years to 500 thousand years ago. That was still before glaciers had reached that region, thus the animals had continued their evolution without being directly affected by glaciers. However, the deposits show that drastic changes had taken place since Wellsch Valley time, with most genera now represented by different species. The fauna (Fig. 4K.6, no. 2, 3, 8, 10, 14, 31, 41, 44, 48, 55, 57, 58, 59, 66) was dominated by camels, horses, elephants, and ground sloths and included beaver, deer, pronghorn, and prairie-dog; the last may indicate a warmer climate than now. *Borophagus* was extinct, and other carnivores and rodents are poorly represented.

The next fauna, considered to be Sangamonian and dated at 120 to 80 ka, or about 400 thousand years younger than that just described, is found near Medicine Hat, Saskatoon, and Fort Qu'Appelle. It is the most complete fauna yet recognized (Fig. 4K.6, Sangamonian column), and again shows drastic changes from the preceding one, having some modern aspects. Bison, recent migrants from Asia, were becoming abundant, and elephants and horses, the latter new species, were still prominent. Camels, again new species, were still abundant though apparently declining relative to the other animals. Deer, wapiti, elk, caribou, mountain sheep, porcupine, muskrat, racoon, and hare appear for the first time, while ground sloth, rabbit, prairie-dog, and ground squirrel were still competing successfully. Carnivores are well represented for the first time, with lion, lynx, fox, and ferret resident in the region. Short-faced bear and muskox are found at Fort Qu'Appelle, and badger both there and at Saskatoon.

The fauna continued to resemble the Sangamonian assemblage until the end of Wisconsinan time, when most of the large species disappeared during the 'Great Extinction' and the fauna assumed its modern, but greatly impoverished, aspect. This change occurred between 15 and 9 ka. Rodents, rabbits, hares, beavers, smaller carnivores, pronghorns, mountain sheep, bison, and some of the caribou and elk survived, as did the smaller carnivores such as fox, coyote, ferret, skunk, bobcat, and lynx. However, of the formerly dominant taxa only the bison remained abundant; the elephants, horses, camels, and ground sloth were gone, as were the lions, sabre-toothed cats, and dire wolves that preyed on them. The only major changes since then have been the introduction of horses, cattle, and other domestic animals by man, and the near extinction of bison.

Climate

During the Quaternary, the climate fluctuated drastically on the Interior Plains. Long cold spells, accompanied by glacier development, alternated with the interglacials, when the climate at times was warmer than now. Few details are known; neither the amount of precipitation nor the relative lengths of the glaciations and interglaciations. Even the vertebrate record gives little information, for most of the animals, particularly the large ones, could tolerate a broad range of conditions.

Undoubtedly, the climate was continental during the interglaciations, with cold winters and warm, or even hot, summers - at least in the southern Plains. Though there were spells of climate warmer than now, intervals of cooler and perhaps more humid climate probably were more common. Further cooling, particularly during the summers, probably accompanied by enhanced precipitation, led to glacier advance. However, temperatures need not have been lower nor precipitation higher during the major glaciations than during the lesser ones, for the extent of glaciers may have depended more on duration of glacial conditions than on their intensity.

ECONOMIC GEOLOGY

Events of the Quaternary have affected practically all facets of Plains life, from impoverishment of fauna and flora to enrichment of soils. Their most obvious economic affects have been on water supply, agriculture and forestry, and supply of construction materials.

Water supply

Over broad regions of the Interior Plains, bedrock is a poor source of water, with quantity limited and quality poor due to a high content of dissolved matter including salts. Further, over broad areas surface water is not available or is an undependable source. As a result, glacial drift has been a major, though not entirely satisfactory, source of water, and undoubtedly settlement of large parts of the southern Plains would not have taken place, or would have been long deferred, without the groundwater available in drift. In general, this water is of better quality than either the surface or bedrock water. The tills typically yield small quantities of water sufficient for household and minor farm use, whereas the interglacial and surface silt, sand, and gravel deposits give larger supplies. The best potential resources, however, are the gravels and sands found near the bottoms of the buried preglacial and interglacial valleys. These have been little used to date, and their potential remains unknown. In the more northerly northern plains groundwater is permanently frozen, but ample surface water is available.

Agriculture and forestry

The important agricultural and forestry industries of the Interior Plains are based on the fertile soils, which are mostly developed on Quaternary deposits. The earlier soils developed on bedrock typically would have been impoverished in elements such as calcium and sulphur, and overrich in sodium and certain salts. The advancing glaciers remedied these faults by spreading till, which, together with glacial-lake sediments and Holocene loess, forms most of the parent material. Without the Pleistocene drift and Holocene loess, the southern Plains would not be as fertile and would remain largely unsettled.

Permeability and good subsurface drainage are also necessary for successful forestry and agriculture, for without them water would lie near the surface until it evaporated. Resulting problems would include, on the one hand rapid run off and erosion, and on the other hand, waterlogging and salinity. Such problems are widespread today where bedrock is shallow, as in southeastern Alberta. Though some bedrock formations are coarse enough to allow good drainage, for instance the Porcupine Hills Formation and parts of the Edmonton and Paskapoo formations, in general only the parent material laid down during the Quaternary Period, with its better permeability and drainage, has allowed development of economic forests and a viable farming industry.

The climate of the Northern Plains is unfavourable for large-scale forestry and agriculture. There, soil problems are largely associated with permafrost and large masses of ground ice, making construction of buildings, roads, and pipelines difficult and costly.

Construction materials

Alluvium (the preglacial gravels and sands) laid down during the last half of the Tertiary is a valuable source of sand and gravel in parts of the Interior Plains. In most areas, however, such supplies are lacking and resort must be had to Quaternary deposits. In the southern Plains the best and largest of these deposits are found in the preglacial, and to a lesser extent the interglacial, valleys, but those deposits commonly are too deep to be economic. As a result, ice-contact deposits such as kames, eskers, and outwash, and the alluvium of modern rivers supply much of the construction materials. Though those deposits are commonly poorly sorted and contain weak or rotted stones, ease of access and development largely compensate for those deficiencies.

In the northern Interior Plains granular materials generally can be recovered from widespread outwash and ice-contact deposits, nearshore glaciolacustrine sediments, and preglacial and Quaternary alluvium. On the Arctic Coastal Plain, construction resources are scarce because of the large amount of ground ice and the general lack of material coarser than sand.

REFERENCES

Alley, N.F.
1973: Glacial stratigraphy and the limits of the Rocky Mountain and Laurentide ice sheets in southwestern Alberta, Canada; Bulletin of Canadian Petroleum Geology, v. 21, p. 153-177.

Catto, N.R.
1986: Quaternary sedimentology and stratigraphy, Peel Plateau and Richardson Mountains, Yukon and Northwest Territories; Ph.D. thesis, University of Alberta, Edmonton, Alberta, 728 p.

Christiansen, E.A.
1979: The Wisconsinan glaciation of southern Saskatchewan and adjacent areas; Canadian Journal of Earth Sciences, v. 16, p. 913-938.

Clayton, L. and Moran, S.R.
1982: Chronology of Late Wisconsinan glaciation in middle North America; Quaternary Science Reviews, v. 1, p. 55-82.

Craig, B.G.
1965: Glacial Lake McConnell, and the surficial geology of parts of Slave River and Redstone River map areas, District of Mackenzie; Geological Survey of Canada, Bulletin 122, 33 p.

Dawson, G.M. and McConnell, R.G.
1896: Glacial deposits of southwestern Alberta in the vicinity of the Rocky Mountains; Geological Survey of America, Bulletin, v. 7, p. 31-66.

Dredge, L.A. and Cowan, W.R.
1989: Quaternary geology of the southwestern Canadian Shield; in Quaternary Geology of Canada and Greenland, R.J. Fulton (ed.); Geological Survey of Canada, Geology of Canada, no. 1. (Also Geological Society of America, The Geology of North America, v. K-1.)

Elson, J.A.
1967: Geology of glacial Lake Agassiz; in Life, Land and Water, W.J. Mayer-Oakes (ed); Occasional Papers, Department of Anthropology, University of Manitoba, No. 1, University of Manitoba Press, Winnipeg, p. 37-95.

Forbes, D.L.
1980: Late Quaternary sea levels in the southern Beaufort Sea; in Current Research, Part B, Geological Survey of Canada, Paper 80-1B, p. 75-87.

Ford, D.C.
1976: Evidences of multiple glaciation in South Nahanni National Park, Mackenzie Mountains, Northwest Territories; Canadian Journal of Earth Sciences, v. 13, p. 1433-1445.

Fulton, R.J. and Klassen, R.W.
1969: Quaternary geology, Northwest District of Mackenzie; in Report of Activities, Part A, Geological Survey of Canada, Paper 69-1A, p. 193-194.

Fulton, R.J. (ed.)
1989: Quaternary Geology of Canada and Greenland; Geological Survey of Canada, Geology of Canada, no. 1., 839 p. (Also Geological Society of America, The Geology of North America, v. K-1.)

Harington, C.R.
1978: Quaternary vertebrate faunas of Canada and Alaska and their suggested chronological sequence; National Museum of Natural Sciences, Syllogeus, no. 15, 105 p.

Harmon, R.S., Ford, D.C., and Schwarcz, H.P.
1977: Interglacial chronology of the Rocky and Mackenzie Mountains based upon 230Th-234U dating of calcite speleothems; Canadian Journal of Earth Sciences, v. 14, p. 2543-2552.

Hill, P.R., Mudie, P.J., Moran, K., and Blasco, S.M.
1985: A sea-level curve for the Canadian Beaufort Shelf; Canadian Journal of Earth Sciences, v. 22, p. 1383-1393.

Hughes, O.L.
1972: Surficial geology of northern Yukon Territory and northwestern District of Mackenzie, Northwest Territories; Geological Survey of Canada, Paper 69-36, 11 p.
1987: The Late Wisconsinan Laurentide glacial limits of northwestern Canada: the Tutsieta Lake and Kelly Lake phases; Geological Survey of Canada, Paper 85-25, 19 p.

Hughes, O.L., Harington, C.R., Janssens, J.A., Matthews, J.V., Jr., Morland, R.E., Rutter, N.W., and Schweger, C.E.
1981: Upper Pleistocene stratigraphy, paleoecology, and archaeology of the northern Yukon interior, Eastern Bergina; 1. Bonnet Plume Basin; Arctic, v. 34, p. 329-365.

Jackson, L.E., Jr.
1980a: Quaternary stratigraphy and history of the Alberta portion of the Kananaskis map area (82J) and its implications for the existence of an ice free corridor during Wisconsinan time; Canadian Journal of Anthropology, v. 1, p. 9-10.
1980b: Glacial history and stratigraphy of the Alberta portion of the Kananaskis Lakes map area; Canadian Journal of Earth Sciences, v. 17, p. 459-477.

Klassen, R.W.
1975: Quaternary geology and geomorphology of Assiniboine and Qu'Appelle valleys of Manitoba and Saskatchewan; Geological Survey of Canada, Bulletin 228, 61 p.
1989: Quaternary geology of the southern Interior Plains; in Quaternary Geology of Canada and Greenland, R.J. Fulton (ed.); Geological Survey of Canada, Geology of Canada, no. 1. (Also Geological Society of America, The Geology of North America, v. K-1.)

Lowdon, J.A., Robertson, I.M., and Blake, W., Jr.
1971: Geological Survey of Canada radiocarbon dates XI; Radiocarbon, v. 13, p. 255-324.

Mackay, J.R.
1958: The Anderson River map area, Northwest Territories; Canada Department of Mines and Technical Surveys, Geographical Branch, Memoir 5, 137 p.
1963: The Mackenzie delta area, Northwest Territories; Canada Department of Mines and Technical Surveys, Geographical Branch, Memoir 8, 202 p.

Mankinen, E.A. and Dalrymple, G.B.
1979: Revised paleomagnetic polarity time scale for the interval 0-5 m.y. BP; Journal of Geosphysical Research, v. 84, p. 615-626.

Mickelson, D.M., Clayton, L., Fullerton, D.S., and Barns, H.W.
1983: The Late Wisconsin glacial record of the Laurentide ice sheet in the United States; in Late-Quaternary Environments of the United States, H.E. Wright, Jr. (ed); v. 1; The Late Pleistocene, J.C. Porter (ed); University of Minnesota Press, Minneapolis, p. 3-37.

Mott, R.J.
1973: Palynological studies in central Saskatchewan - Pollen stratigraphy from lake sediment sequences; Geological Survey of Canada, Paper 72-49, 18 p.

Prest, V.K.
1969: Retreat of Wisconsin and recent ice in North America; Geological Survey of Canada, Map 1257A, 1:5 000 000.
1984: The Late Wisconsinan glacier complex; in Quaternary Stratigraphy of Canada-A Canadian Contribution to IGCP Project 24, R.J. Fulton (ed.); Geological Survey of Canada, Paper 84-10, p. 21-36.

Rampton, V.N.
1982: Quaternary geology of the Yukon Coastal Plain; Geological Survey of Canada, Bulletin 317, 49 p.
1988: Quaternary geology of the Tuktoyaktuk coastlands, Northwest Territories; Geological Survey of Canada, Memoir 423, 98 p.

Ritchie, J.C.
1976: The Late Quaternary vegetational history of the western interior of Canada; Canadian Journal of Botany, v. 54, no. 15, p. 1793-1818.
1984: Past and present vegetation of the far northwest of Canada; University of Toronto Press, Toronto, 251 p.

Ritchie, J.C. and Yarranton, G.A.
1978: Patterns of change in the late-Quaternary vegetation of the western interior Canada; Canadian Journal of Botany, v. 56, no. 17, p. 2177-2183.

Roed, M.A., Mountjoy, E.W., and Rutter, N.W.
1967: The Athabasca Valley Erratics Train; Canadian Journal of Earth Sciences, v. 4, p. 625-632.

Rutter, N.W.
1972: Geomorphology and multiple glaciation in the area of Banff, Alberta; Geological Survey of Canada, Bulletin 206, 54 p.

Shetsen, I.
1984: Application of till pebble lithology to the differentiation of glacial lobes in southern Alberta; Canadian Journal of Earth Sciences, v. 21, p. 920-933.

Stalker, A.MacS.
1956: The Erratics Train; Foothills of Alberta; Geological Survey of Canada, Bulletin 37, 28 p.
1969: Quaternary stratigraphy in southern Alberta; Report II: sections near Medicine Hat, Alberta; Geological Survey of Canada, Paper 69-26, 28 p.
1970: A probable Late Pinedale terminal moraine in Castle River Valley, Alberta; Reply; Geological Society of America Bulletin, v. 81, p. 3775-3778.
1973: Surficial geology of the Drumheller area, Alberta; Geological Survey of Canada, Memoir 370, 122 p.
1983: Quaternary stratigraphy in southern Alberta; Report III: The Cameron Ranch Section; Geological Survey of Canada, Paper 83-10, 20 p.

Stalker, A.MacS. and Churcher, C.S.
1982: Ice Age deposits and animals from the southwestern part of the Great Plains of Canada; Geological Survey of Canada, Miscellaneous Report No. 31.

Stalker, A.MacS. and Harrison, J.E.
1977: Quaternary glaciation of the Waterton-Castle River region of Alberta; Bulletin of Canadian Petroleum Geology, v. 25, no. 4, p. 882-906.

Teller, J.T. and Fenton, M.M.
1980: Late Wisconsinan glacial stratigraphy and history of southeastern Manitoba; Canadian Journal of Earth Sciences, v. 17, p. 19-35.

Vincent, J-S.
1984: Quaternary stratigraphy of the western Canadian Arctic Archipelago; in Quaternary Stratigraphy of Canada - a Canadian Contribution to IGCP Project 24, R.J. Fulton (ed.); Geological Survey of Canada, Paper 84-10, p. 87-100.
1989: Quaternary geology of the northern Interior Plains; in Quaternary Geology of Canada and Greenland, R.J. Fulton (ed.); Geological Survey of Canada, Geology of Canada no. 1. (Also Geological Society of America, The Geology of North America, v. K-1.)

Westgate, J.A., Briggs, N.D., Stalker, A.MacS., and Churcher, C.S.
1978: Fission-track age of glass from tephra beds associated with Quaternary vertebrate assemblages in the southern Canadian Plains; Geological Society of America, Abstracts with programs, v. 10, p. 514-515.

Authors' Addresses

A. MacS. Stalker
Geological Survey of Canada
601 Booth Street
Ottawa, Ontario
K1A 0E8

J.-S. Vincent
Geological Survey of Canada
601 Booth Street
Ottawa, Ontario
K1A 0E8

Printed in Canada

Chapter 5

TECTONIC EVOLUTION AND BASIN HISTORY

CHAPTER 5

TECTONIC EVOLUTION AND BASIN HISTORY

J.D. Aitken

INTRODUCTION

The sedimentary cover of the North American craton in western Canada (see Fig. 5.5, in pocket) is in large part related to processes at the continental margins. It is in part the proximal deposits of continental terrace wedges, laid down on subsiding, divergent margins (miogeoclines), and in part the deposits of foreland basins depressed by the load of crustal and sedimentary rocks tectonically thickened at convergent margins. Evidence from the Omineca Belt of the Cordillera suggests that it may, in late Paleozoic to mid-Jurassic times, have been a back-arc basin, but that record cannot be read from the sediments of the Western Canada Basin. Certain disconformities and instances of stratigraphic condensation are probably related to the flexural bulge inboard of the foreland basin. An analogous, though less pronounced flexural bulge due to the **sediment** load at and inboard of divergent margins is expected, but has nowhere been identified beyond question.

On the other hand, the view of the authors, based on current information presented in this volume, is that the development of circumscribed Phanerozoic basins and arches well within the craton (Fig. 2D.2, 2D.3, 2D.4, 5.1) was independent of activity, whether convergent or divergent, at the continental margins. Contemporaneity of some events deep inside the craton with events at the margins may be apparent, rather than real, and plausible mechanisms for linkage of epicratonic movements with activity at the margins have not been proposed. The several arches and basins that were active during both convergent and divergent phases provide the strongest support for our view, which is not shared by all contributing authors.

In this chapter, only hypotheses and interpretations are cited as to source; observations are developed more completely in the various sections of Chapter 4, with appropriate citations.

Aitken, J.D.
1993: Tectonic evolution and basin history; Chapter 5 in Sedimentary Cover of the Craton in Canada, D.F. Stott and J.D. Aitken (ed.); Geological Survey of Canada, Geology of Canada, no. 5, p. 483-502 (also Geological Society of America, The Geology of North America, v. D-1).

ANCIENT PASSIVE CONTINENTAL MARGINS

Cordilleran margins

The deposits of the ancient western margins of North America are now hundreds of kilometres inboard of the Pacific littoral, and separated from that coast by a collage of exotic terranes (see the companion volume by Gabrielse and Yorath, 1991.)

Pre-Cordilleran margin

The oldest unmetamorphosed deposits are those of the Middle Proterozoic Purcell Supergroup of the Purcell and southern Rocky mountains (1500-1300 Ma; McMechan, 1981), and the plausibly correlative Wernecke Supergroup of northern Yukon Territory (Delaney, 1981). Price and Hatcher (1983) and Gabrielse and Yorath (1991) regarded the Purcell-Wernecke rocks as deposits of a Cordilleran continental margin, in the same sense that the lower Paleozoic rocks are so regarded. They are treated herein, preferably, as belonging to a pre-Cordilleran tectonic cycle. These deposits, whether strictly contemporaneous or not, are mere remnants of formerly much more extensive successions, disrupted and partly removed at the subsequent, Cordilleran margin (the margin from whose deposits the Columbian Orogen arose). The long-standing interpretation (especially by Canadian geologists) of the classical Purcell as a passive-margin assemblage is currently under attack (especially by students of the U.S. northwest), who increasingly see evidence for a western, bounding landmass.

The northwest-trending Purcell Supergroup is truncated by the base of the succeeding, more northerly-trending, Upper Proterozoic Windermere Supergroup, considered here to be the oldest deposits of the Cordilleran margin sensu stricta.

The Wernecke Supergroup is exposed in the Wernecke and Ogilvie mountains. From structural arguments it is assumed to be present at depth in the Mackenzie Mountains; furthermore, deep seismic-reflection profiles of the Mackenzie Platform reveal a kilometres-thick, layered succession underlying the Mackenzie Mountains Supergroup at an angular unconformity (Cook, 1988a, b). These remotely sensed, deformed rocks can only correspond to the Hornby Bay - Dismal Lakes - Coppermine

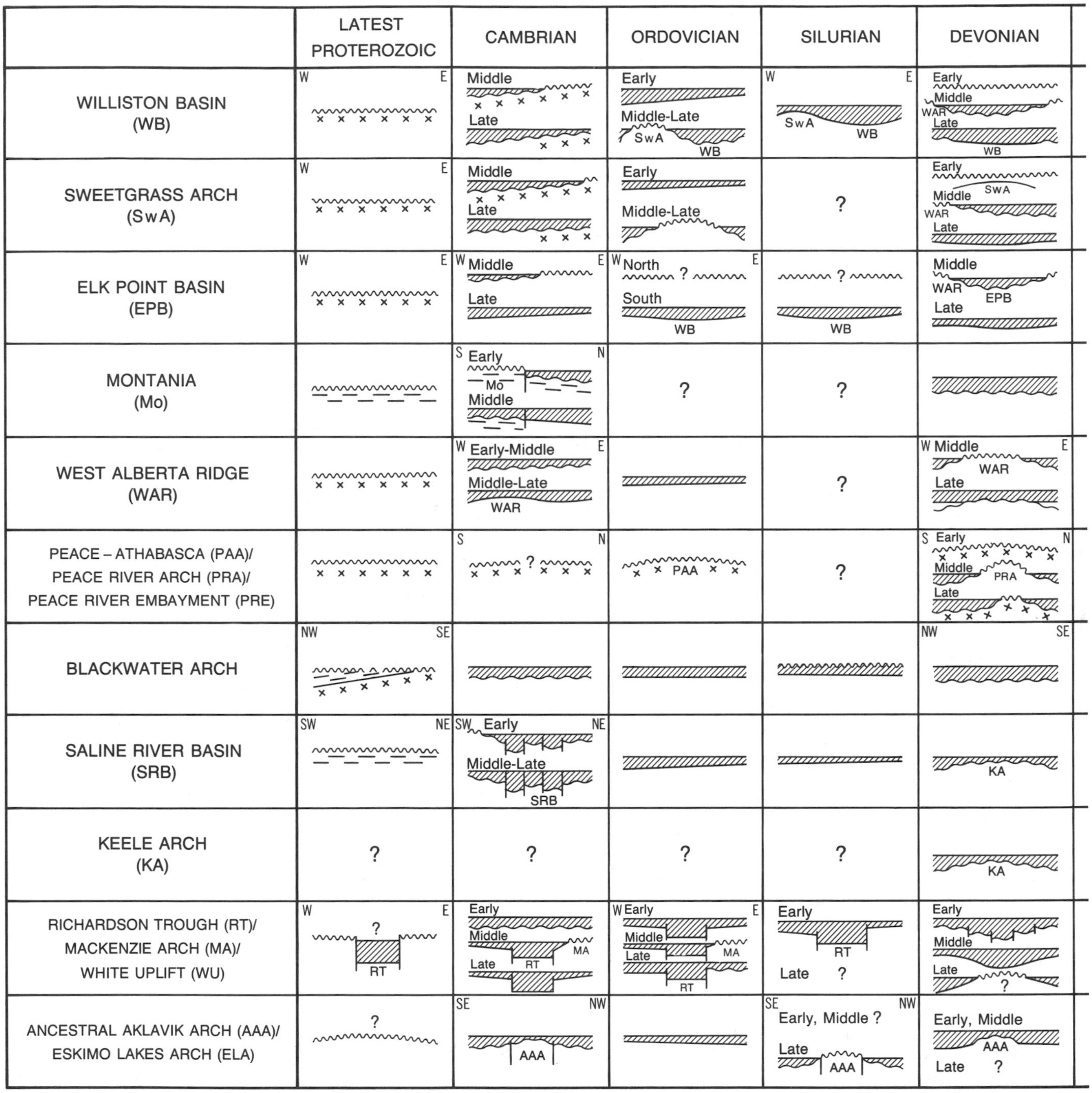

	LATEST PROTEROZOIC	CAMBRIAN	ORDOVICIAN	SILURIAN	DEVONIAN
WILLISTON BASIN (WB)	W E	Middle Late	Early Middle-Late SwA WB	W E SwA WB	Early Middle WAR Late WB
SWEETGRASS ARCH (SwA)	W E	Middle Late	Early Middle-Late	?	Early Middle SWA WAR Late
ELK POINT BASIN (EPB)	W E	W Middle E Late	W North ? E South WB	? WB	Middle WAR EPB Late
MONTANIA (Mo)		S Early N Mo Middle	?	?	
WEST ALBERTA RIDGE (WAR)		W Early-Middle E Middle-Late WAR		?	W Middle E WAR Late
PEACE – ATHABASCA (PAA)/ PEACE RIVER ARCH (PRA)/ PEACE RIVER EMBAYMENT (PRE)		S N ?	PAA	?	S Early N Middle PRA Late
BLACKWATER ARCH	NW SE				NW SE
SALINE RIVER BASIN (SRB)	SW NE	SW Early NE Middle-Late SRB			KA
KEELE ARCH (KA)	?	?	?	?	KA
RICHARDSON TROUGH (RT)/ MACKENZIE ARCH (MA)/ WHITE UPLIFT (WU)	W E ? RT	Early Middle MA Late RT	W Early E Middle Late MA RT	Early RT Late ?	Early Middle Late ?
ANCESTRAL AKLAVIK ARCH (AAA)/ ESKIMO LAKES ARCH (ELA)	?	SE NW AAA		SE NW Early, Middle ? Late AAA	Early, Middle AAA Late ?

Figure 5.1. Sheet of cartoons showing development of Phanerzoic basins and arches.

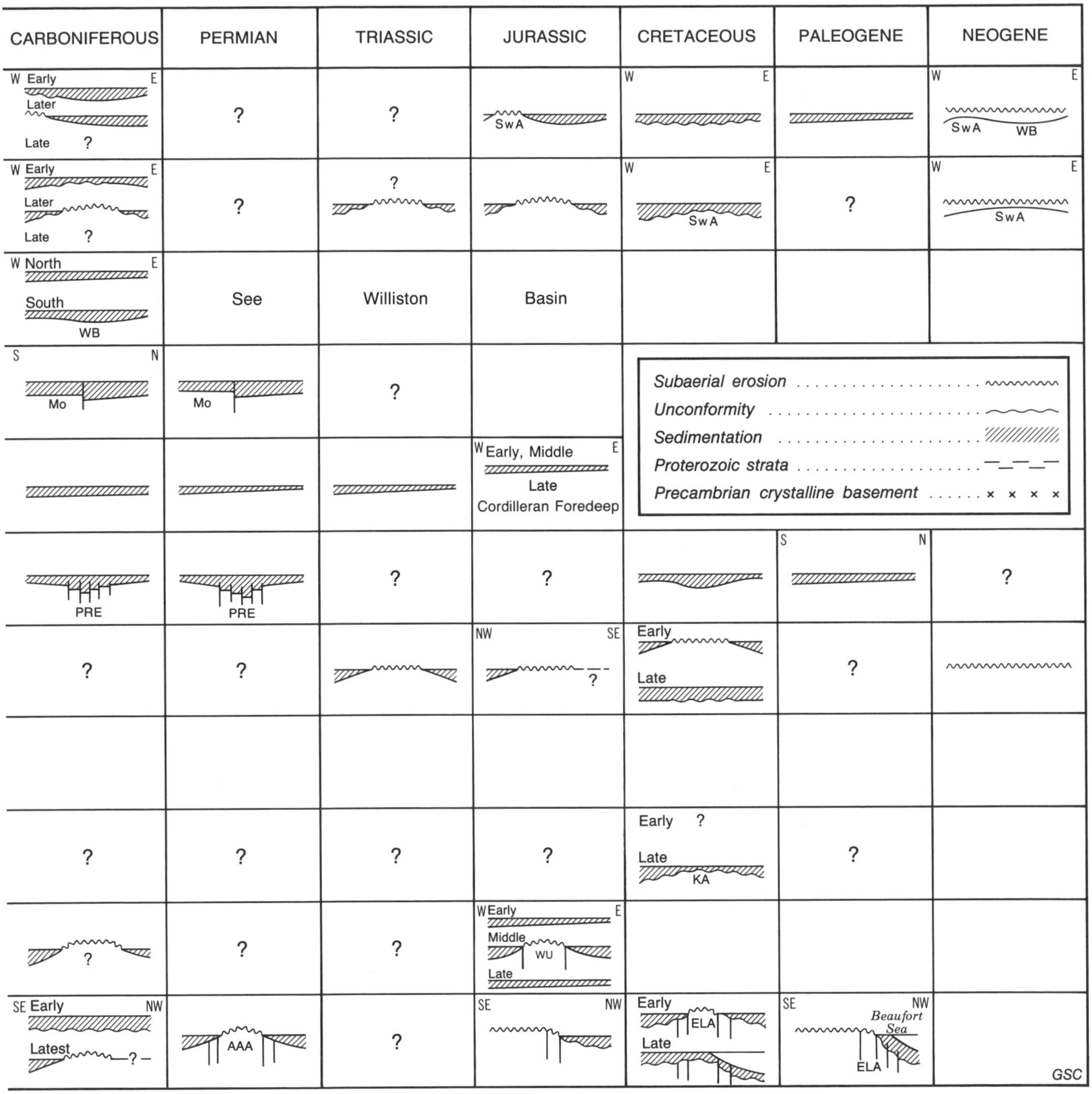

CARBONIFEROUS	PERMIAN	TRIASSIC	JURASSIC	CRETACEOUS	PALEOGENE	NEOGENE
W Early E Later Late ?	?	?	SwA	W E		W E SwA WB
W Early E Later Late ?	?	?		W E SwA	?	W E SwA
W North E South WB	See	Williston	Basin			
S N Mo	Mo	?				
			W Early, Middle E Late Cordilleran Foredeep			
PRE	PRE	?	?		S N	?
?	?		NW SE ?	Early Late	?	
?	?	?	?	Early ? Late KA	?	
?	?	?	W Early E Middle WU Late			
SE Early NW Latest ? –	AAA	?	SE NW	Early ELA Late	SE NW Beaufort Sea ELA	GSC

Subaerial erosion ....

Unconformity ....

Sedimentation ....

Proterozoic strata ....

Precambrian crystalline basement ...... × × × ×

Figure 5.1. Cont.

lavas succession of the Coppermine Homocline, or to the probably equivalent Wernecke Supergroup; available isotopic dates are permissive of such equivalence. If the two successions are in fact equivalent, the Hornby Bay and Dismal Lakes Groups appear to have been deposited in an extensional basin that joined the miogeocline of Wernecke deposition at a high angle. None of these rocks are known south of the Liard line (Fig. 4D.2). Although the distribution of exposures prevents a firm conclusion, it appears that the Wernecke Supergroup, at deposition, wrapped around the Porcupine promontory of the craton, and was truncated subsequently, probably by rifting of Windermere age.

Cordilleran margin

An episode of rifting that post-dated Purcell-Wernecke deposition by nearly 500 million years established the fundamental outline of the continental margin (Fig. 2D.1) that persisted until the arrival of the first allochthonous terrane in Middle Jurassic time. The Windermere Supergroup, consisting largely of clastic rocks of deep-water origin, is the sedimentary response to the Late Proterozoic rifting. In the Mackenzie and Ogilvie mountains, rifting is dated by volcanic and sub-volcanic, basic rocks with isotopic ages of 760 to 780 Ma, but at two localities near the Rocky Mountain Trench (near latitudes 53° and 57°N), Windermere strata rest nonconformably on granitic rocks dated at 720 to 735 Ma. These data suggest southward-younging diachroneity of rifting.

Although the overall path of rifting established the Cordilleran tectonic trend, the rifted margin was stepped and indented in detail. Several lines of evidence suggest that the rifting was in part dextrally transcurrent or transtensional (e.g., Eisbacher, 1977). Individual rift-basins tend to be limited by faults aligned north-northwest to west-northwest, parallel to the strike-slip trend (according to position in the non-linear fold and thrust belt), or northeast to north, perpendicular to the direction of maximum tension under dextral shear. In the Rocky Mountains, this orientation of potential tensional openings was close to the northeast trend of major strike-slip faults in the basement (R.A. Burwash et al., Chapter 3 in this volume; Geological Survey of Canada, 1987), and accordingly is well developed in such structures as the Moyie-Dibble Creek and St. Mary faults and possibly also the Southern Alberta Aulacogen (Kanasewich et al., 1968), if the last is of appropriate age. This point is valid even though the geometry of the faults has been strongly modified by Mesozoic, contractional reactivation. At and north of latitude 60°, a long-lived fault or family of faults, initiated near the end of Early Proterozoic time, is similarly oriented and appears to have controlled a major offset in the margin, referred to here as the Liard line (Aitken and Pugh, 1984). The distribution of deposits of Windermere and early Paleozoic age is difficult to reconcile with a continental margin created by simple extension. On the other hand, the hypothesis of dextral transtensional rifting, with a transcurrent component of several hundred kilometres, provides an attractive explanation for the marked widening of the belt of Windermere outcrop (assumed to bear a relationship to original, depositional extent) north of the Liard line, and the Late Proterozoic precursor to the Selwyn Basin as a pull-apart basin, albeit one floored by continental, not oceanic crust.

Continental separation and thermally driven, initial passive-margin subsidence of the Cordilleran Miogeocline in the southern Rocky Mountains date from about the beginning of the Cambrian, as judged by subsidence curves (Bond and Kominz, 1984). In the Mackenzie and Wernecke mountains, separation may have been earlier, as suggested by the onset there of sedimentation characterized by depositional Grand Cycles (Subchapters 4A, 4B in this volume) slightly before the beginning of Ediacaran time. This would be consistent with the earlier onset of rifting there, as suggested by isotopic age determinations.

If the ubiquitous "sub-Cambrian unconformity" is a "breakup unconformity" (and this is equivocal, the earliest Cambrian being a time of extreme regression, worldwide), it has puzzling features. Wherever information is adequate, the unconformity bevels progressively older units cratonward. Nowhere is described an unconformity that down-cuts oceanward, as would be predicted by a model of a thermally elevated continental margin at breakup. Here the dominant 'Atlantic margin' model does not appear to fit the facts. New evidence shows, however, that the parautochthonous, North American, continental-margin succession extends farther west (e.g., Klepacki, 1985) than many have thought previously. Possibly the "thermal shoulder" predicted by the "Atlantic margin model" lies near or within the core zone of either the Paleozoic or the Mesozoic orogen, or both, and cannot be recognized stratigraphically.

It is not yet resolved whether the early Paleozoic, passive-margin subsidence was a direct sequel to Windermere rifting, or whether a second rifting event was involved. This question is raised by the long delay (up to 200 million years) between the onset of Windermere rifting and continental separation. Bond et al. (1985) argued for a distinct, younger rifting event as the immediate precursor to latest Proterozoic or earliest Cambrian continental separation. The evidence for this event is, however, scarcely more impressive than that for minor episodes of faulting and vulcanism in the miogeocline, recorded in strata as young as Ordovician. Reflecting on the analogy with the Triassic graben of the Atlantic margin, which pre-dated separation there by about 100 million years, the favoured interpretation is that of a single, prolonged rifting event, rather than two.

Marginal basins and promontories

Some rifting activity clearly persisted, although apparently with diminished energy, into early Paleozoic time, as documented by several late-initiated basins, demonstrably or plausibly fault-bounded, that interrupt the orderly, southwestward thickening of the miogeoclinal deposits (Fig. 2D.2).

In the north, Richardson Trough, opening to the Innuitian margin, and Misty Creek Embayment, opening to the Cordilleran margin, are aligned *en echelon* to one another, in the manner of a failed attempt to rift the Porcupine Platform away from North America. Both troughs contain a deep-water sedimentary fill, and Misty

Creek Embayment contains submarine volcanic rocks. Marked differential subsidence commenced in both during latest Early Cambrian time, but unusually thick Windermere strata exposed at the south end of the Richardson Trough and the north end of the Misty Creek Embayment suggest the possibility of a Late Proterozoic, precursor rift-basin.

Porcupine Platform northeast of Tintina Fault and west at least to the Alaska boundary has Proterozoic and lower and middle Paleozoic successions that are relatable to parautochthonous successions in the Wernecke and Mackenzie mountains. It was a promontory or northwestern shoulder of the craton prior to the emplacement of allochthonous terranes. The Proterozoic and lower and middle Paleozoic, partly platformal strata of the Brooks Range, in Alaska, appear to belong to an extremely long, attenuated extension of the Porcupine Platform. This appearance is misleading if the range, or its core of pre-Carboniferous strata, is allochthonous, as several different tectonic scenarios suggest (see Grantz et al., 1990).

Robson Basin, aligned parallel to the Cordilleran margin, contains Lower and Middle Cambrian deposits that are anomalously thick but of platformal facies.

Roosevelt Graben, apparently also aligned parallel to the Cordilleran margin, underwent its most marked differential subsidence, witnessed by both thickness and facies, in the Middle Cambrian. Activity waned thereafter, and the effect seen in Lower Ordovician deposits is slight.

White River Embayment, another margin-parallel element, began to subside differentially and markedly in Early Ordovician time, as documented by thickness and facies, but by Late Ordovician time was again apparently subsiding slowly, in harmony with the surrounding miogeocline.

Liard Depression is recorded as a marked "thick" in isopachs for the Tippecanoe Sequence, but differential subsidence there had little or no effect upon sedimentary facies. The same feature as seen in the stratigraphy of the Kaskaskia Sequence, where it affects facies as well as thickness, is called Meilleur River Embayment. Cambrian stratigraphy of the region of the embayment is not well known, and a pre-Tippecanoe effect cannot be ruled out. Open to the west, Liard Depression/Meilleur River Embayment may well reflect an embayment in the Cordilleran margin, although the stratigraphic effects seen must lie inboard of the actual continent-ocean transition.

The structurally high block known as Montania, at the 49th Parallel, first appears in the stratigraphic record as an Early Cambrian landmass. Montania is delimited on the north by the Moyie and Dibble Creek faults, which have the present geometry of thrusts (P. Gordy, pers. comm., 1985). Cambrian facies in the Cordillera undergo a westward excursion at the latitude of Montania, whose Cambrian cover resembles the succession beneath the plains to the northeast. These relationships, and the fact that some northeast-striking faults of the region, for example the St. Mary fault, underwent large-scale vertical displacements during Late Proterozoic Windermere deposition, suggest that the north flank of Montania was related to a transform fault on which the continental margin was offset, so that Montania was further from the margin than the country immediately to the north. Isopachs show that the area of Montania was again high during Carboniferous and Permian times, and was flanked on the north side by a marked structural trough in which anomalously thick sediments accumulated.

Data on the basin within Mackenzie Platform that contains the upper Middle or lower Upper Cambrian evaporites of the Saline River Formation are insufficient for a complete understanding. The timing of subsidence, the shape of the basin, and thickness anomalies in the underlying Lower and Middle Cambrian strata suggest a faulted basin and a genetic association with the Cordilleran margin. On the other hand, its extremely inboard position might suggest that it is a cratonic basin. Its duration was fairly brief, and in early Late Cambrian time it had stabilized as part of the Mackenzie Platform.

Depositional platforms

Although far removed from the site of the western continental margin, the Cambrian through Middle Jurassic platform deposits clearly bear a relationship to that margin:

a. Early in the Paleozoic, given allowance for second-order regressive events, onlap of the Canadian Shield was progressive, resulting in complete or near-complete submergence of the shield by Late Ordovician - Early Silurian time (Beaverfoot - Mount Kindle - Stony Mountain formations).
b. Nearly all chronostratigraphic units thicken westward or southwestward, in the absence of interference from cratonic arches and basins.
c. A progression from proximal facies in the northeast to distal in the southwest is ubiquitous, except in the vicinity of cratonic arches.
d. Except for transects through transient "highs", e.g., West Alberta Ridge, Mackenzie Arch, all of which have platformal deposits outboard, the platforms appear one-sided, with no internal, sedimentary evidence for a western/southwestern shoreline. All evidence suggestive of land to the west is found in the Omineca Belt (see "Cordilleran Active Margins; Kootenay Terrane", below).

The deposits of the platforms are too far removed from the continental margin to have been deposited under a regime of thermal subsidence. They record, instead, a self-perpetuating, though decelerating process of outer-platform downwarping caused by the load of sediment and water superposed upon the thermally subsiding crust (Sleep, 1971; McKenzie, 1978), leading to a sediment load farther inboard on the downwarp, further downwarping, and so on (Watts and Ryan, 1976). Thus, the northeastward limit of loading, subsidence, and further loading migrated progressively inward from the newly formed continental margin. A major interruption of sedimentation in mid-Ordovician time is recorded in western Canada as on all continents (Sloss, 1976). This interruption, necessarily eustatic, did not put an end to the advance of the shoreline. It took place only 60 to 70 million years after separation, when effective, thermally driven subsidence at the margin would continue for tens of millions of years, according to current models (Sleep, 1971; McKenzie, 1978). The processes just described were augmented by the overall, undoubtedly eustatic, early Paleozoic transgression. A long-standing, major problem in the geology of western Canada, namely the extraordinarily long duration of sedimentation of passive-margin aspect, requires examination.

Duration of Cordilleran "divergent" regime

Viewed solely from Western Canada Basin, the Cordilleran Miogeocline is again anomalous, as compared with the prevailing, "Atlantic margin" model, in the apparently remarkable duration of its passive-margin-like behaviour. Most of the stratigraphy in this volume is presented as if that appearance reflected reality. The geophysical models that dominate present thought (e.g., Sleep, 1971; McKenzie, 1978) predict that thermal subsidence and subsidence-dependent sedimentation on downwarps should become negligible 120 million years after separation. Yet the Cordilleran Miogeocline apparently subsided (again, as viewed from the Western Canada Basin) in passive-margin fashion for nearly three times as long, from Early Cambrian to Middle Jurassic times. The answer to the problem may lie in the recent demonstration of a complicated history for the Kootenay Terrane (Fig. 2D.2), especially its western part. The eastern part is now well established as parautochthonous, or at least pericratonic (Klepacki, 1985; Klepacki and Wheeler, 1985). The western part, overlying a zone of talc schist that may be a dismembered ophiolite, has undergone post-Early Cambrian, pre-Late Devonian deformation, metamorphism, and intrusion (Okulitch et al., 1975; Okulitch, 1985). The western part has also undergone Late Carboniferous and Early Permian rifting, and Late Triassic tectonic-arc activity (Klepacki, 1985; Klepacki and Wheeler, 1985). If the Kootenay Terrane is entirely pericratonic, these events imply several rejuvenations of the continental margin, but if not, some or even all of the pre-Jurassic deformation and igneous activity may have occurred offshore North America. The question of the paleogeography of the Kootenay Terrane, especially its western part, relative to North America, is dealt with at length by Gabrielse and Yorath (1991). Under whatever paleogeographic scenario, evidence remains to suggest that the entity commonly referred to as the Cordilleran Miogeocline, was, in a broad sense, a back-arc basin (or at least, a "two-sided" basin) at least twice, prior to the onset of Columbian orogeny. The new sedimentary evidence for a landmass west of the Carboniferous Prophet Trough (B.C. Richards et al., Subchapter 4E in this volume) supports this interpretation, and the prolonged subsidence of the "Cordilleran Miogeocline" is seen to have been only partly of passive-margin character.

The problem remains as to why or how the strata of the Western Canada Basin fail almost totally to record any hint of the recurring tectonic activity to the west. Sparse volcanic rocks, generally of alkaline affinity, are found locally interbedded with strata of Early Cambrian to Middle Devonian age, the older occurrences being the more widespread. They occur mostly outboard of the depositional platforms, or at the platform margins, and probably record continued or renewed extension. The alkaline, Middle Devonian, Ice River intrusive complex and Early Silurian, alkaline, intrusive and extrusive rocks at the outer edge of the Kakwa Platform are of similar position and interpretation. Sanidine crystals in the Lower Carboniferous Exshaw Formation probably record volcanic activity to the west. In the Rocky Mountains, the absence of any pre-Late Jurassic, westerly derived clastic wedge is anomalous, and remains, perhaps, the best argument for an allochthonous Kootenay Terrane. On the other hand, these anomalies perhaps merely demonstrate that the Western Canada Basin lies farther inboard of the continental margin than generally perceived.

Innuitian margin

The ancient Innuitian passive margin of northern Yukon Territory is so fragmented by Mesozoic and Tertiary structures that it is almost unrecognizable. One is forced to assume that the developmental scenario for that margin, worked out in the Arctic Archipelago (Trettin, 1991), can be projected southwestward into northern Yukon Territory. Evidence from the Arctic Islands is that passive-margin subsidence of the Innuitian margin, like that of the Cordilleran margin, commenced in latest Proterozoic or earliest Cambrian time. The aulacogen-like Richardson Trough of northern Yukon, with similar timing, opened northward to the early Paleozoic Innuitian continental margin. Recent work by M.P. Cecile (pers. comm., 1988) has revealed that most or all of the Neruokpok Formation of northern Yukon, previously considered Proterozoic, is a fossiliferous, lower Paleozoic succession with marked similarities to that of Selwyn Basin. Thus, the facies relationships of the platformal strata of the northern Mackenzie and Porcupine Platforms are best viewed as in continuity with deep-water deposits of the northern margin, rather than the western.

Evidence of late Paleozoic tectonism is well preserved in the north, in contrast to the southern Cordillera, where little evidence of such tectonism suffered by the outboard part of the miogeocline is preserved. Late Devonian and Early Carboniferous, Ellesmerian folds of northeast trend, and a related clastic wedge, are preserved in the Barn, Cache Creek, and Scho uplifts. The Ellesmerian episode was brief, however, and the region had returned to hinged subsidence of passive-margin style as early as Tournaisian (early Early Carboniferous). The similarities with parts of the tectonic histories of the Kootenay Terrane and Sverdrup Basin are obvious.

Cordilleran – Innuitian junction

A final problem is the history and configuration of the junction between the Cordilleran and Innuitian continental margins. The opening of Canada Basin (Early Cretaceous?) has certainly broken the connection that necessarily existed, by either rotation or translation of large crustal blocks. The several, mutually incompatible models that have been proposed for the opening of the basin are reviewed by Grantz et al. (1990). The model involving rotation of northern Alaska (Brooks Range to the continental margin) counter-clockwise out of Canada Basin currently receives the most widespread support, but is not without problems. If the rotational scenario is correct, and the Paleozoic core of the Brooks Range lay offshore the

Arctic Islands, then the Paleozoic continental margin had a shoulder or corner, rather than a promontory, near the northern Yukon-Alaska boundary.

Epicratonic arches and basins

Semantics - The term "(cratonic) arch" (synonymous with "uplift" and "high") denotes structural elements that subsided less rapidly than their surroundings, as well as elements that in fact rose, relative to some deep-earth datum. The former kind of cratonic arch is recognized as areas in which chronostratigraphic units are anomalously thin, the latter as areas in which chronostratigraphic units present in surrounding regions are missing, implying the former presence of land, either during or subsequent to deposition of the missing units. As pointed out by P.F. Moore (Subchapter 4D in this volume), the term "basin" has been used to refer to such diverse and overlapping entities as syndepositional tectonic basins (whether bathymetric basins or not), miogeoclines, depocentres, constructional basins defined by surrounding carbonate platforms, and post-depositional tectonic basins ("basins of preservation"). The platforms of this volume, justifiably so named because of the blanket character of their mainly shallow-water deposits, are basins from a viewpoint on the Canadian Shield; except where elevated in the mountain belts, most of their sedimentary contents now lies at greater depths than when deposited.

"Epicratonic basin" is used here in the sense of a cratonal area undergoing **differential** tectonic subsidence during the period of sedimentation considered. In this sense, a part of the Hudson Platform was Hudson Bay Basin during parts of its history, although it is largely a post-depositional, tectonic basin, containing mainly platformal sediments and mainly delimited by the rise of surrounding arches. Of course, depocentres correlate strongly with tectonic basins whether syndepositional or post-depositional.

Common causation? - The near-synchroneity of initial differential subsidence in three of the epicratonic basins (early Middle Ordovician in the Williston and Michigan basins, early Late Ordovician in the Hudson Bay Basin) suggests common causation, but the suggestion may be misleading; marked subsidence ended at grossly different times, early Tertiary in the Williston, Early Carboniferous in the Michigan, and Late Devonian in the Hudson Bay basins. Furthermore, as pointed out by Bally and Snelson (1980) among others, it is largely the rise (or failure to subside) of cratonic arches that demarcates the several basins from one another. Sanford et al. (1985) emphasized the (apparent) synchrony of activity on numerous cratonal arches. However, the apparent synchroneity is considered here as largely a record of eustatic rises and falls of sea level. Furthermore, the subsequent history of the various "synchronous" arches, following initial appearance, is so disparate as to cast doubt on common causation (Fig. 2D.3, 5.1).

Epicratonic arches

The Mackenzie Arch had a complex history, as yet incompletely known. It was not present during deposition of the 'Mackenzie Mountains supergroup' (<800, >780 million years). It is debatable for the period of Windermere deposition (780-570 Ma), but probably lay inside (northeast of) the belt of Windermere rift-depressions. It underwent uplift prior to latest Early Cambrian time and again toward the end of Middle Cambrian time; these uplifts might be considered peripheral-bulge effects due to loading by thick Lower and Middle Cambrian deposits outboard, but locally thick strata of the same age inboard do not fit the flexural model, and the complexity of thickness patterns in the Saline River Formation (latest Middle and/or earliest Late Cambrian) calls for tectonic (horst-and-graben?) activity inboard. The main arch was finally and permanently buried in early Late Cambrian time, but the Redstone Arch, active early in the Paleozoic, and Twitya Uplift, a Late Silurian feature, may be regarded as locally reactivated parts of the Mackenzie Arch.

Sweetgrass Arch is first evident in the Middle Ordovician, when its rise coincided with the initial differential subsidence of the Williston Basin. Never, apparently, an important source of detritus until the Tertiary, it persisted to influence distribution and thickness of sediments as late as Late Jurassic time, and the preservational distribution of strata as young as Upper Cretaceous. It is prominently expressed in the basement surface today. The arch diverges too much from the Cordilleran trend for it to be a peripheral bulge, and the simple "bulge" hypothesis (Beaumont, 1981) fails to explain its Paleozoic history.

The Peace River Arch arose as a landmass in Early Devonian time (its earlier history is uncertain) from part of the older (since Cambrian at least) and much more extensive Peace-Athabasca Arch. By Late Devonian time, it had become inactive, and by earliest Carboniferous, the area of the former arch had become one of accelerated subsidence, the Peace River Embayment, which had filled with sediment by Late Carboniferous time. Late Paleozoic block faulting took place, and Permian strata are anomalously thick in the area. Isopachs of Lower Cretaceous deposits suggest the possibility that the area of the arch during their deposition may have been a horst and adjacent half-graben. The arch remains a "high" in the basement surface today. Because of its areal restriction and its trend, perpendicular to the continental margin, the arch is not plausibly related to processes at the margin. Similarly, Walcott's (1970) isostatic hypothesis for basement uplifts failed to explain the history of the Peace River Arch. The Peace-Athabasca Arch appears to have been a broad, topographic ridge on crystalline basement rocks at the inception of passive-margin subsidence - an initial requirement for arch uplift resulting from differential sedimentary loading, according to Walcott's hypothesis. The distribution of quartz sand in Ordovician deposits (M.P. Cecile and B.S. Norford, Subchapter 4C in this volume) may record such uplift by differential loading by lower Paleozoic strata, but the delay until the Peace River Arch *sensu stricto* was strongly uplifted during or prior to the Early Devonian is unexplained. Similarly unexplained by the isostatic hypothesis, as Walcott (1970) acknowledged, is the Carboniferous and Permian subsidence of the arch, which, in view of the greater sediment load on its flanks, would appear to be anti-isostatic.

The West Alberta Ridge, which plausibly connects at its northern end with Peace River Arch, had a distinctly different history. An early expression of the ridge is seen in depositional thinning of lower Upper Cambrian units, but this behaviour (flexural bulge caused by sedimentary loading outboard?) was merely "less negative", and did not give rise to a landmass. The record is uncertain until the sub-Devonian (sub-Givetian?) unconformity, which deeply bevels the axis of the ridge. Fluvial channels filled with red beds testify to a topographic feature of appreciable relief that defined the southwest flank of Elk Point Basin. This episode is clearly too retarded, relative to thick, early Paleozoic deposition outboard, and the scale of uplift excessive, for it to be viewed as a flexural-bulge response to that loading. Upper Devonian and Carboniferous formations pass across the site in facies and thicknesses normal for the region, and the ridge did not undergo subsequent block faulting.

Blackwater Arch is the name newly applied here to a very broad feature of the northern platforms active in Triassic to early Cretaceous times and possibly manifested in the Early Carboniferous (Fig. 2D.2). Its existence is manifested in the first instance by a marked westward excursion of the zero edges of Carboniferous, Permian, Triassic, Jurassic, and pre-Aptian Cretaceous deposits, from near the 60th parallel on the south to about the latitude of Eagle Plains Basin (Fig. 2D.3) on the north. This preservational configuration dates activity of the arch only as post-Devonian and pre-Aptian, but depositional facies at several stratigraphic levels provide evidence of a positive feature active intermittently or continously over a long period:

a. Following a reconnaissance account of Triassic deposits of northern Yukon Territory, Mountjoy (1967) observed, ". . . most of the region south of the British Mountains appears to have been moderately positive during Triassic time and probably formed a peninsula of land extending west and northwest from the craton to the east." This statement corresponds to the concept of Blackwater Arch, except that the "peninsula" is visualized here as having an axis oriented west-southwest. Further, Pelletier (1964) published maps of Triassic deposits of northeastern British Columbia showing paleocurrents directed south-westward between latitudes 56° and 59°N and south-ward between latitudes 59° and 60°N. In the same belt, shorelines change from a northeast to an east-west orientation.
b. In the general vicinity of Blow Trough, Jurassic deposits reveal shorelines and a source of detritus to the southeast (T.P. Poulton, Subchapter 4H in this volume), evidence of the north flank of Blackwater Arch. In the south, Jurassic deposits are missing in the Plains and eastern Cordillera north of latitude 58°, and the rocks that might have contained evidence of the presence of the arch have been eroded. On the other hand, this distribution might be viewed as evidence of the maximum southward expansion of the arch's southern flank.
c. North and south of Blackwater Arch, as drawn in Figure 2D.2, Cretaceous formations as old as Berriasian are present. Across the arch, the earliest Cretaceous deposits, resting everywhere on pre-Carboniferous deposits, are Late Aptian or younger.

A suggestion of the presence of an arch in pre-Triassic times is seen in the facies and configuration of Lower Carboniferous deposits near the Yukon-Northwest Territories boundary just north of latitude 60°. These, including much sandstone, display northward shoaling facies, and facies belts undergo a marked swing to the west (B.C. Richards et al., Subchapter 4E in this volume). Overlying Permian strata are of deepwater facies. The facies of Carboniferous and Permian deposits at their northeastern erosional limits off the southwesterly plunge of the arch provide evidence against a pre-Triassic Blackwater Arch, however. These are of deepwater facies, and are viewed by B.C. Richards et al. and C.M. Henderson et al. (Subchapters 4E and 4F in this volume) as deposits of the Prophet and Ishbel troughs, subsequently uplifted and eroded over the Blackwater Arch.

Blackwater Arch is unique among the major epicratonic arches in its limited amount of cumulative uplift; Upper Devonian strata are preserved over much of its postulated extent. This seems to imply that, instead of rising markedly, the arch simply failed to subside along with the country to the north and south, from Carboniferous through Early Cretaceous times.

Unlike the previously described arches, Keele Arch (whose outline is drawn very differently by different authors) formed over a region that had been host, in late Middle to early Late Cambrian time, to both a marked depocentre (the Saline River evaporite basin) and a local "high". Excluding basins "regurgitated" by Mesozoic compression, it is the only feature described here to have undergone inversion from basin to arch. Its post-Cambrian, pre-Devonian history is uncertain; sedimentological evidence for an emergent landmass is lacking. The arch is well defined by the "sub-Devonian unconformity"; Lower Devonian deposits, thin over the arch, rest on formations as old as Middle Cambrian, but the former topographic "high" so recorded is not more precisely dated. The arch is next seen at an unconformity beneath Upper Cretaceous (Turonian to Campanian-Maastrichtian) strata that rest on lower Paleozoic units along the crest of the arch. It remains structurally high today. Because convergent tectonism was underway in the Cordillera when the arch re-manifested itself in the Cretaceous, some might view this as a response to orogeny, were it not for the clear Devonian and earlier expression of the same feature.

The Aklavik Arch Complex is a northeasterly trending, linear *en echelon* array of paleostructural and, from time to time, paleotopographic features crossing the apex of the modern Mackenzie Delta. Some elements at least appeared as positive features in early Paleozoic time, and the complex had a profound effect on the depositional and erosional history of the region during both the passive-margin and active-margin phases. In the northeast, manifested as the Eskimo Lakes Arch, it is a continental-margin feature, yet its southwestward extension in Yukon Territory, which possibly extends as far as Alaska as the early Paleozoic feature called the Dave Lord High, is intracratonic. It appears that in the northeast, rifting attendant upon the opening of Canada Basin (according to whatever scenario) reactivated the faults on which the arch had been originally uplifted; thus an intracratonic feature acquired a marginal position.

Elements of Aklavik Arch Complex have undergone upward movement during different phases of Columbian, Laramide, and possibly post-Laramide convergent tectonism, and are structurally high today.

Epicratonic basins

The origins of the epicratonic basins of western Canada are complex, and typically involve, in addition to differential subsidence, isolation or definition by cratonic arches and paleotopography.

The Saline River Basin (Fig. 2D.2, 5.1) is poorly understood, and as mentioned earlier, may be fundamentally a marginal basin. It would not be considered a possible cratonic basin except for its inboard position. In addition to its rapid differential subsidence in late Middle and/or earliest Late Cambrian time while receiving thick redbeds and evaporites, it may also have subsided differentially in late Early and earlier Middle Cambrian time. The basin, whose outline strongly suggests fault-controlled subsidence, was limited on the southwest by the Mackenzie Arch, on the southeast by a broad expression of the Peace-Athabasca-Tathlina Arch, on the east by the Canadian Shield and on the northwest (the apparent connection with the ocean) by a broad, unnamed swell that might better be regarded as normally subsiding platform. Post-Cambrian differential subsidence is undocumented, and in the Devonian, a large area of the basin became the Keele Arch.

Lloydminster Embayment in the broadest sense, that is, the element containing thick Cambrian strata in southern Alberta and southern Saskatchewan, is best regarded, not as an embayment in any genetic sense, but as a normally subsiding region of the early Paleozoic platform. Its nearly east-trending northern flank, which creates the "embayment" outline, is the southern flank of the Cambrian through the Silurian Peace-Athabasca Arch. Its eastern/southeastern limit is the continental axis in the U.S.A. The arches, not the embayment, form the anomaly.

Williston Basin may be the simplest cratonic basin in western Canada, because its history is mainly one of simple (though unexplained), differential subsidence. During early Paleozoic until Middle Ordovician times, the region of the subsequent basin was part of the normally subsiding platform; informal, marker-defined units in the Deadwood Formation reveal no differential thickening. Modest differential subsidence began in early Middle Ordovician time, contemporaneously with the first uplift of Sweetgrass Arch, which forms the western flank of the basin. The tectonic feature identified in Devonian geology as the Meadow Lake Hingeline marks the northern flank of the basin as conventionally drawn, and probably underwent south-side-down movement during the Ordovician and Silurian periods, premonitory to its marked movement of opposite sense in latest Silurian - Early Devonian times. Ordovician and Silurian strata in the basin are entirely of platformal character; it is not clear whether their thinning toward the margins is depositional or the result of successive unconformities. The basin behaved essentially as a platform through the Devonian, but differential thickening of individual Carboniferous stratigraphic units, and Carboniferous deeper-water facies, record renewed subsidence. Williston Basin was a bathymetric basin and depocentre during parts of the Jurassic. Selective preservation of Cretaceous and Paleocene deposits over the basin centre records modest Mesozoic and Cenozoic differential subsidence, but the timing of different episodes is uncertain. At least some episodes of uplift of the Sweetgrass Arch correspond with episodes of subsidence in the Williston Basin.

Elk Point Basin (Fig. 2D.2, 5.1) exemplifies a polygenetic, cratonic basin. The Early Devonian sea, transgressing southward from the Mackenzie and MacDonald platforms, entered an elongate topographic depression confined on the northeast by the Canadian Shield (partly thinly veneered by lower Paleozoic strata), and on the southwest by the West Alberta Ridge and the Peace River Landmass (Arch). Southeastward marine advance, inundating the Peace-Athabasca Arch for the first time, was stopped by a low escarpment developed at the Meadow Lake Hingeline by a reversal of the earlier hinge movement. This reverse movement (Early Devonian?) necessarily involved subsidence of the broad, Peace-Athabasca Arch. Although **en bloc** subsidence of the arch is well displayed by its progressive inundation during the Devonian, differential subsidence also played a role in the definition of the Central Alberta and North Alberta sub-basins of Elk Point Basin. In this instance again, subsidence is apparently linked with uplift in an adjoining region (Peace River Arch). The recurrently evaporitic character of the early fill of the basin was dictated by two factors, bathymetry and the aridity evidenced throughout the Devonian succession in Western Canada. The sill of the evaporite basin, namely the Keg River Barrier (reef complex) was constructional, but Peace River Arch on the southwest and Tathlina Arch on the northeast, newly risen in the Early Devonian, provided topographic abutments for the barrier. By middle and late Frasnian time, West Alberta Ridge had been inundated and differential subsidence of the Elk Point Basin had ended; the region (including the area of the former Peace-Athabasca Arch) had acquired the aspect of a normally subsiding platform, open westward to the ocean. The record in Western Canada Basin suggests that this character persisted until the onset of Columbian convergent tectonism in Middle Jurassic time. The record of deformation, vulcanism, and erosion in the Omineca Belt, on the other hand, suggests the existence of western landmasses at least twice during Devonian to Middle Jurassic times, but it is not yet proven that these were part of North America.

Basement composition: influence on subsidence and uplift

Many geologists appear to believe that the age and composition of ancient, thoroughly cratonized, crystalline, continental basement influences or even dictates the tendency of that basement to rise or subside in the absence of later-imposed loads, not only for 100-200 million years after cratonization, but at any later time. The Phanerozoic history of the marginal and epicratonic basins of Canada provides little support for that view. **Local** correspondences between up/down behaviour and basement provinces or boundaries between basement provinces are rejected as supporting "basement control", because the hypothesis would predict a near-1:1 correspondence throughout the length of basement belts. The only convincing example of "basement control" is coincidence of a belt of K-metasomatized (at 1.8 Ga) basement of the Churchill Structural Province of the Canadian Shield with the broad, early (Cambrian to earliest Devonian?) Peace-Athabasca Arch (Burwash and Krupicka, 1970; Burwash and Culbert, 1976). The areally restricted Peace River Arch of Devonian times, however, occupied only a small part of the K-metasomatized belt.

On the other hand, the first-order basins appear independent of basement age and character:

a. The Middle Proterozoic Hornby and Athabasca basins of the Canadian Shield lie athwart basement province boundaries (but only small parts of those boundaries).

b. Williston Basin sprawls across part of the boundary between Churchill and Superior structural provinces without apparent effect (although this boundary coincides neatly with Cape Henrietta Maria Arch separating Hudson Bay and Moose River basins; see A.W. Norris (Chapter 8 in this volume and Fig. 7.1).

The positions and histories of basement arches similarly lack explanation by basement age and composition, in nearly all cases:

a. West Alberta Ridge lies across projected and aeromagnetically revealed basement trends.

b. At the southwestern edge of the Canadian Shield, the axis of the Trans-Hudson Orogen is nicely aligned with the Sweetgrass Arch. Southwestward in the subsurface, however, the axis curves far to the east of the arch; "basement control" must be denied.

c. Tathlina Arch, in a region of northerly basement trends, was an equidimensional element for most of its existence. When it became briefly elongate, during Middle Devonian time, its trend was east-west.

To put the case against "basement control" another way, there are vast areas of the Canadian Shield whose basement age and composition are similar to those of the Hudson Bay, Williston, and Athabasca basins. If "basement control" reflects reality, why did these areas not subside also ?

Sanford et al. (1985) published maps of the exposed and buried craton showing a grid-like arrangement of intersecting positive and negative elements. The degree of speculative projection of data embodied in these maps does not appear justified; however, if the maps do, indeed, reflect reality, the positive and negative elements as outlined completely ignore the gross structure of the Canadian Shield.

On the other hand, examples abound of faults of various ages that are aligned with ancient faults or the strike of basement gneissosity. Particularly conspicuous are ancient northeast-trending faults in the crystalline basement of Western Canada (Geological Survey of Canada, 1987; Burwash et al., Chapter 3 in this volume). These appear to have controlled, in part, the path of continental separation as the western margin was formed, and appear to be responsible for the numerous examples of northeasterly oriented Devonian trends and carbonate platform edges.

None of the foregoing should be taken as a denial of the hypothesis that the response of the basement (lithosphere) to loading is strongly influenced by basement character, especially basement age. Rigidity increases as the lithosphere ages, cools, and thickens, and this increase can influence profoundly the flexural response to sedimentary loading, as at continental margins, and tectonic loading, as in the orogen-foredeep couple (see Beaumont, 1981).

The origin of epicratonic basins and arches remains a mystery. Following a brief overview of published hypotheses, Bally and Snelson (1980) drew much the same conclusion: "The models that have been proposed do not satisfy the observations, and knowledge we have of the initiation of cratonic basins is not sufficient to provide useful answers".

Beaufort Sea margin

The Beaufort Sea margin adjacent to the northern Cordillera came into being with the formation of Canada Basin. Because Aptian to Middle Albian formations are relatable to Cordilleran foredeeps (Peel Trough, Blow-Keele-Kandik troughs) and foreland platforms in front of them, the stratigraphic record suggests that deposition related to the new (Beaufort Sea) margin commenced in about Cenomanian time, following a major hiatus. Since that time, deposition at the margin has been of progradational character overall (Young et al., 1976; Dixon, 1986).

The major positive features affecting deposition at the Beaufort Sea margin belong to a loose, northeast-trending chain of syndepositional "highs" referred to as the Aklavik Arch Complex and including Eskimo Lakes Arch and Cache Creek Uplift, among others. The complex is marginal to the Beaufort Sea southwestward as far as Mackenzie River or possibly Cache Creek Uplift, but farther southwest lies within the continent. Earlier uplift of various parts of the complex took place, not only in Early Cretaceous time, but as far back as the early Paleozoic (e.g., Dave Lord High, Ancestral Aklavik Arch). Thus, the complex is not inherently marginal, but its northeastern part appears to have provided a favorable path for rifting associated with the opening of the Canada Basin. Laramide and post-Laramide convergent tectonism may have reactivated some elements of the Aklavik Arch Complex.

ANCIENT ACTIVE CONTINENTAL MARGINS

Ellesmerian active margin

Northern and northwestern Canada underwent, in Late Devonian and Early Carboniferous times, a rather brief episode of convergent tectonism known, because it is well recorded in the Arctic Archipelago, as the Ellesmerian Orogeny. In northern Yukon Territory, northeast-trending folds affecting all formations up to at least the Frasnian part of the Upper Devonian Imperial Formation are only locally preserved and visible. A Devono-Carboniferous clastic wedge, in part conglomeratic, is of northwestern derivation. Graben along the Yukon-Northwest Territories boundary farther south, near latitude 64°, are filled with cherty, Devono-Carboniferous "black clastics" derived from the northwest.

By Late Carboniferous time, the region affected by Devono-Carboniferous tectonism had resumed flexural subsidence, and continued thus until late Early Cretaceous (Aptian or Albian) time.

Cordilleran active margins

Kootenay Terrane

The earliest record of Phanerozoic convergent tectonism in the southern Canadian Cordillera is found in the Kootenay Terrane of southeastern British Columbia. The fact that that tectonism has left at most a feeble record in the sediments of the Western Canada Basin calls into question Kootenay Terrane as a North American or pericratonic entity. The question is not yet fully resolved.

The succession of Proterozoic and lower Paleozoic formations of the eastern Kootenay Terrane is clearly relatable to that of Purcell Anticlinorium, and on that basis appears to be North American or pericratonic. In western Kootenay Terrane, folded, metamorphosed, and intruded, lower Paleozoic strata are unconformably overlain by Upper Devonian to Upper Carboniferous strata (Okulitch, 1985). The problem is that the rocks of concern may be separated from North American rocks by a cryptic suture (Gabrielse and Yorath, 1991). Thus, the late Paleozoic orogeny (Antler, Ellesmerian) may not have involved the British Columbia part of the Cordilleran margin of North America. These questions await final resolution.

In western Kootenay Terrane and also in the Sylvester Allochthon of northern British Columbia and southern Yukon Territory, there is a record of Permo-Triassic (Sonoman?) thrust imbrication. The convergent tectonism apparently occurred prior to obduction of the deformed terrane onto the North American continent (Gabrielse and Yorath, 1991).

Clastic rocks of western derivation (though no thick, clastic wedge) occur in the Carboniferous succession of Prophet Trough (B.C. Richards et al., Subchapter 4E in this volume) and the Permian succession of Ishbel Trough (C.M. Henderson et al., Subchapter 4F in this volume). These record episodes during which the "miogeocline" was instead a two-sided basin.

In western Kootenay Terrane, upper Triassic volcanic arenites pass westward to proximal volcanic rocks. In the Cariboo area, a parallel relationship is seen in Upper Triassic and Lower Jurassic strata. Thus, a back-arc basin lay at, or offshore from the western continental margin at those times (Gabrielse and Yorath, 1991).

Western Canada Basin

The pivotal event in the Phanerozoic history of Western Canada Basin is associated with the onset of opening of the North Atlantic Ocean and relative westward drift of the continent. The diachroneity of the commencement of Atlantic seafloor spreading (Late Jurassic off Nova Scotia, Late Cretaceous north of Newfoundland) is not clearly reflected in parallel diachroneity of convergent tectonism in the Columbian Orogen; sea-floor spreading was also vigorously in progress in the Pacific Basin, and oriented independently of the Atlantic spreading centre. New lithosphere had to be, and was accomodated by subduction of oceanic lithosphere and destruction of the Cordilleran continental terrace wedge. Deformation of the terrace wedge took place in a generally west-to-east progression. The rising orogen provided both a tectonic load, depressing the continental crust in a foredeep, and a highland source for copious detritus, part of which was trapped in the foredeep (see Fig. 5.5, in pocket).

Deformation in the Foreland Belt was essentially thin-skinned, involving detachment of the sedimentary cover from undeformed crystalline basement (Fig. 5.2, 5.3). In the Rocky Mountains, in transects characterized by a "skeleton" of thick carbonate formations, for example, the Bow River transect (e.g., Price and Fermor, 1982), the dominant structures are thrust faults controlled to a large degree by the plane of bedding, and folds are relatively unimportant. Where the elements of the carbonate "skeleton" are thinner, giving way to fine-grained clastics, as along and north of the Athabasca River transect, thrust faults still play an important role, but tight folds become prominent also (e.g., Mountjoy, 1962). Slices of basement were detached and transported northeastward with the sediments only here and there in the Rocky Mountains, but basement involvement was more widespread in the Omineca Belt to the west (e.g., Price and Fermor, 1982; Campbell et al., 1982). In the Mackenzie Mountains, structure at the surface is characterized by long, broad folds, commonly of box-fold character, and a few large thrust faults (Gabrielse et al., 1973; Aitken et al., 1982). Important detachment surfaces must be present at depth. Crystalline basement is nowhere known to be involved. In the northern part of the Franklin Mountains, the most obvious surface structures lie above a regional detachment in Cambrian evaporites, but in the narrower belt of the southern Franklin Mountains, a much deeper detachment (subtly active also to the north) dominates, as judged by the involvement of the evaporites in the surface structures. Structural style is fully described and interpreted in the companion volume, "Geology of the Cordilleran Orogen in Canada" (Gabrielse and Yorath, 1991).

The lithosphere of the foreland was old, and consequently thick and cool. South of the MacDonald Fault (Fig. 2D.1), it was the subsurface continuation of the Churchill Structural Province of the Canadian Shield, and possibly, in part, the Wopmay Orogen (see R.A. Burwash et al., Chapter 3 in this volume). It is not clear how completely the Archean terranes within the Churchill Province were reworked on a full-crustal scale, but even if reworking was intense it was complete by 1.8 Ga, more than one and one-half billion years before the onset of the Cordilleran orogeny. North of the MacDonald Fault, crystalline basement is less easily related to established provinces of the exposed shield, but there again, relict Archean isotopic ages and Hudsonian-aged overprint have been identified. Nothing is known of the age and composition of the crystalline basement north of the Liard Line, but the uniform, platformal character of the sediments of the Proterozoic Mackenzie Mountains Supergroup (>770, <1200 Ma) and the thin-skinned character of deformation in underlying strata (Cook, 1988a, b) imply a stable, and hence old crust. These characteristics of the mid-Mesozoic lithosphere predict rigidity, and hence a foredeep that would be wide and shallow, as compared with the response to be expected from the same load applied to a younger, thinner lithosphere (Beaumont, 1981).

Timing

Two ideas that dominate current thinking about the deformational history of the eastern Cordillera were published in 1970. The first of these is that the westward-coarsening, Mesozoic and early Tertiary clastic wedges of the Rocky Mountain Front Ranges and Foothills and the western Plains date Rocky Mountain deformation, and that their eastward progression records eastward advance of the deformation front (Price and Mountjoy, 1970). The second is that the clastic-wedge record demonstrates a pause in deformation between the deposition of a stack of coarse clastic wedges in Late Jurassic to early Late Cretaceous times, the product of the Columbian Orogeny, and another, simpler coarse clastic wedge laid down in mid-Late Cretaceous to Paleocene times, the product of the Laramide Orogeny *sensu stricto* (Douglas et al., 1970). The latter idea was not fully expounded in its first presentation, but soon came to be widely understood in the terms expressed here.

While it undoubtedly is based on real phenomena, and continues to have value for communication, the Columbian-Laramide two-orogeny scenario becomes less clear-cut as data accumulate. The Laramide clastic wedge indeed has the integrity to be expected of the product of a single orogeny or orogenic pulse, and is present to a greater or lesser degree along the front of the eastern Cordillera. The Columbian "clastic wedge", on the other hand, is a

Figure 5.2. Rocky Mountains Front Ranges at Lake Minnewanka, near Banff; view northwestward. Near the photo centre, pale, cliff-forming Upper Devonian Palliser limestones (uD) lie upon dark, non-resistant Jurassic Fernie shales (F) at the southwest-dipping Inglismaldie Thrust. Above the Palliser lie Lower Carboniferous formations, the Banff (Ba) and Rundle Group (Ru), then dark-weathering Upper Carboniferous and Permian Rocky Mountain Group and Triassic clastics (uC-T̄R), capped by pale, Triassic Whitehorse Formation (Wh). At near-extreme left, the north end of the Canmore Coal Basin and Cascade River are underlain by non-resistant, Jurassic Fernie and Jura-Cretaceous Kootenay formations (JK), overthrust from the west by Devonian and Carboniferous carbonates (DC) at the Rundle Thrust. Beneath the Inglismaldie Thrust (right, northeast), a dark, relatively non-resistant section (P-J), extending from Rocky Mountain Group to Fernie Formation, again rests upon Lower Carboniferous and older carbonates. Note the dominance of southwestward dip, characteristic of the Front Ranges. (photo - Chevron Canada Resources Limited).

complex entity, embodying the deposits of deltas distinct from one another in both space and time (Fig. 5.4), and evidencing at least three phases or pulses of orogeny, as follows.

As interpreted from the conventional viewpoint that relates each **coarse** clastic wedge to a pulse of plate convergence, a first phase, in Late Jurassic and earliest Cretaceous times, is represented by a depocentre with maximum thickness in the Peace River region and extending, although thinning somewhat, southward to the 49th parallel. Northward thinning from the Peace River is apparent, but the succession is truncated by erosion at the Muskwa River. Blow and Peel troughs (foredeeps) had not yet developed, and indications there of a western source of sediment during this period are lacking.

A second phase, of Aptian and Albian age, is inferred from coarse clastic-wedge deposits from the 49th parallel to northern Yukon Territory. The molasse in southwestern Alberta is relatively thin. In central Alberta, the clastic wedge is characterized by east-trending marine shorelines that record infilling, by the outbuilding of transverse, subaerial deltas, of a north-northwest-trending tectonic trough. The depositional tectonics of the region north of the Mackenzie Arc is extremely complicated and puzzling. In the east, Peel Trough was a simple foredeep in front of the Mackenzie Mountains. Farther west, Keele-Kandik Trough has the characteristics of a foredeep in front of the east-west fold belt of the Ogilvie Mountains, but the relative timing and relationships between the north-south Nahoni fold belt and the trough it disrupts are unclear. Like Keele-Kandik Trough, Blow Trough received thick

Figure 5.3. Rocky Mountains Foothills north of Burmis, Alberta; view northwestward. In the foreground, resistant ribs of Jurassic and Cretaceous sandstones outline large-scale folds. On the distant right skyline, Moose Mountain Dome, a major structural culmination of the Foothills, brings resistant, pale-weathering Carboniferous carbonates to the surface. At upper left, at the mountain front, pale, resistant, Paleozoic carbonates are superposed on Mesozoic clastics above the west-dipping Livingstone and McConnell thrust faults. The entrenched Oldman River cuts through tightly folded Mesozoic strata at right centre. (photo - Chevron Canada Resources Limited).

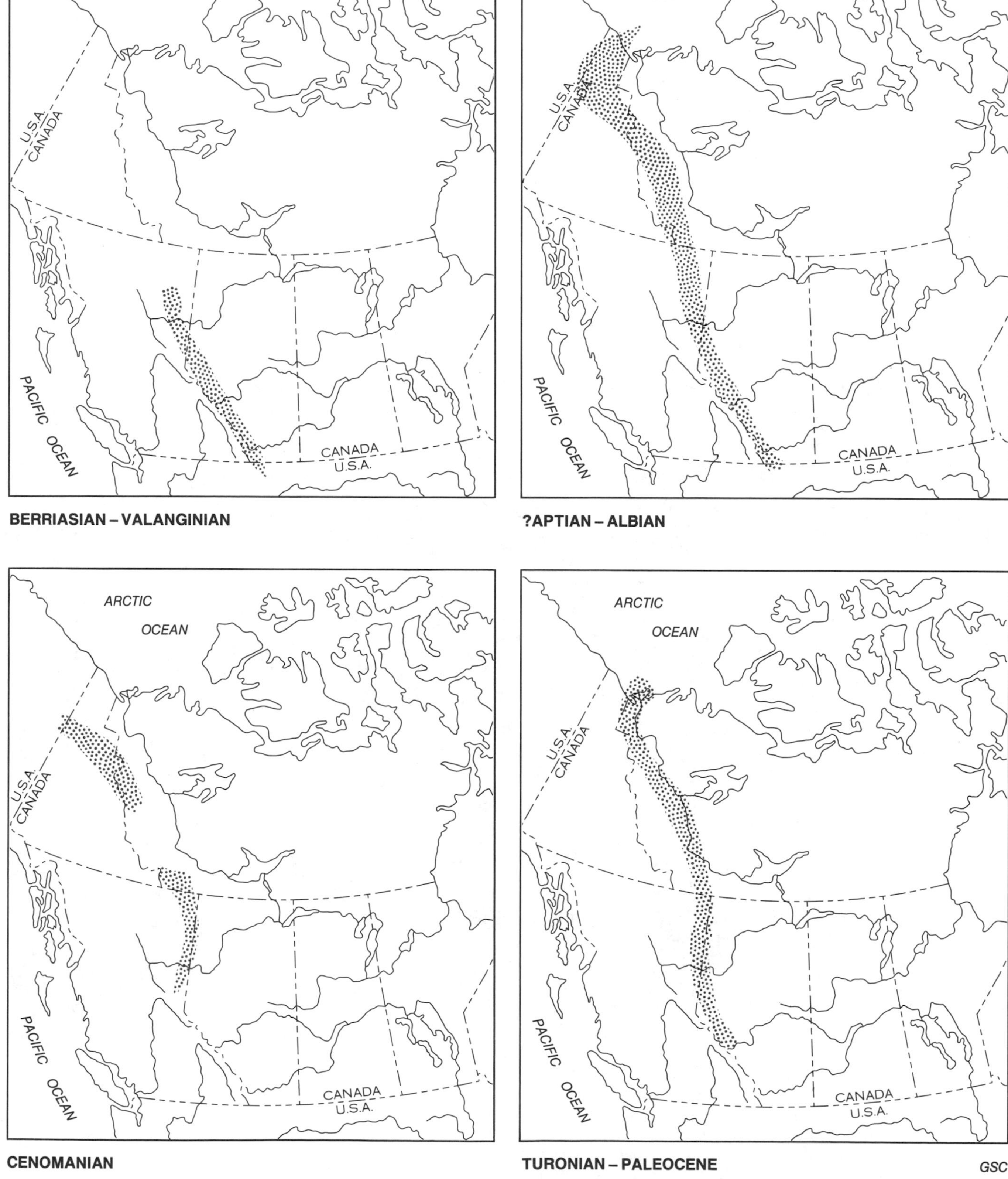

Figure 5.4. Maps of successive Columbian and Laramide foredeep depocentres.

flysch deposits during late Aptian and Albian times, but as a "foredeep" it is anomalous. Much of it is now caught between two relatively stable platforms, Mackenzie Platform with its flanking Aklavik Arch Complex, and Porcupine Platform. Various parts of the southwestern Aklavik Arch Complex, already with a long history of positive behaviour, may have behaved in a sense as forebulges in front of the troughs, but none of this is clear at present, partly because of severe, post-Albian, convergent and transcurrent dismemberment of the mid-Cretaceous tectonic elements. The need and opportunity for further study are obvious.

The third phase, in Late Albian to Early and mid-Cenomanian times, resulted in the connection of the Boreal and Gulfian seas. Its termination is represented by the thick, coarse molasse of the Dunvegan deltas in the regions of Peace and Liard rivers, and by similar deposits of the Monster Formation in northern Yukon Territory. Granitic plutons in the Selwyn Mountains, dated at 90 to 110 Ma, cut pre-existing folds (Gabrielse et al., 1973) and appear to be related to this phase of convergence. In the Foothills south of Peace River and beneath the Plains, on the other hand, a depositional and erosional hiatus is found.

The depocentres of the second and third phases display a marked correspondence to reentrants in the deformation front, as pointed out by Eisbacher et al. (1974), and suggest structural control of the debouchements of major rivers from the orogen. In several instances, large modern rivers emerge from the orogen at structural reentrants which are the sites of ancient deltas.

The Campanian to Paleocene, Laramide clastic wedge, like the second-phase Columbian wedge, is rather evenly distributed along the front of the orogen, but is feebly developed in Peel Trough, where its onset was slightly earlier than elsewhere. Large subaerial deltas extended across Alberta and Saskatchewan prior to the last recorded withdrawal of seas from the foredeep.

The decades-old division of Cordilleran foreland convergent deformation into two phases, Columbian and Laramide, is based mainly on the time-spans of Kootenay-Blairmore and Belly River - Paskapoo coarse clastic wedges, and the period separating them (late Cenomanian through Santonian), characterized by deposition of rather thick mudrocks in the west, but lacking a coarse clastic wedge. The period of fine-grained sedimentation has been interpreted conventionally as the record of the pause in convergence between Columbian and Laramide orogenies. But the conventional interpretation does not account for the persistence of a foredeep (**thick** mudrocks). A recently published hypothesis, like most good ideas obvious once pointed out, turned the conventional interpretation of coarse clastic wedges upside-down. Blair and Bilodeau (1988) inferred that during episodes of plate convergence, the rapidly loaded foredeep should be at its most pronounced, and that coarse detritus from the orogen should be trapped close to the deformation front. Conversely, it is during the post-convergence episodes of relaxation and uplift, with a decaying foredeep, that coarse detritus can be transported far into the foreland basin. Application of these ideas to the clastic-wedge history of the Alberta Trough would yield four episodes of convergence: Oxfordian to latest Kimmeridgian or Tithonian/Volgian; Aptian-Albian (most pronounced in the Northwest Territories and Yukon Territory); Late Cenomanian through Santonian; and late Paleocene. It must be noted that the record of the first convergent episode is the sub-Oxfordian unconformity, interpreted by Poulton (Subchapter 4H in this volume) as a forebulge (the inferred foredeep deposits, lying to the west, are not preserved). Conversely, the foredeep and relaxation-phase deposits of the final convergence (deformation of Paleocene strata) are not preserved because there has been no subsequent tectonic loading and consequent burial. This reinterpretation of the clastic-wedge record is more in harmony with the inferred timing of periods of accelerated convergence of allochthonous terranes deduced from the record in the Omineca Belt of the Cordillera (Gabrielse and Yorath, 1991), and with the stratigraphic positions of the regional unconformities in the foreland basin successions (D.F. Stott et al., Subchapters 4I, 4J, in this volume).

Cessation of convergent tectonism in the southern Rocky Mountains is reasonably well dated. Paleocene strata are folded in the Foothills, and the Upper Eocene and Oligocene Kishenehn Formation, deposited against the scarp of the Flathead normal fault, records the onset of extensional conditions. In the central part of the Mackenzie Arc, Paleocene beds are folded also, but the absence of preserved, post-Paleocene strata leaves uncertain the question of a younger limit to convergent deformation. The question as to persistence, if any, of the foredeep following the cessation of convergence is an interesting one that yields conflicting answers. A study of burial history in southern Alberta, based on vitrinite reflectance (England and Bustin, 1986) suggested that on average, 6 to 7 km of rock has been eroded. The inferred load would have been thrust sheets and/or Paleocene and Eocene sediments. Well established structural models deny the likelihood of extension of high-level thrust sheets east of the structural "triangle zone", whose eastern limit is the axis of the Alberta Syncline (a structure present in Mesozoic but not underlying Paleozoic formations). The inferred load suggested by reflectance data must, therefore, have been Paleocene and Eocene sediments, and such a thickness implies an active foredeep. The methodology of the study by England and Bustin (1986) is currently under attack, however; their approach may have yielded excessive values for vanished cover. A lingering, post-convergence foredeep would imply that, at the cessation of convergence, the leading edge of the tectonic load was not isostatically compensated, and continued to subside and to depress the foredeep with it.

On the other hand, the sedimentary record of distant areas suggests the prompt elimination of an active foredeep with the cessation of convergence, if the foredeep is viewed as an effective trap for coarse, orogenic sediment. Detritus from the southern frontal Rocky Mountains and from the Foothills and Plains that is not preserved in the foredeep or the foreland platform, must have exited to the continental margin via an ancestral Mississippi River drainage. The sedimentary record of the Gulf Coast shows that the building of large, sand-rich deltas into the Gulf Coast Basin did not begin until Eocene time (various sources cited by Landes, 1970). This can be read as dating the time at which coarse detritus from the orogen was first able to escape the trap of the Cordilleran foredeep. A supporting argument is found in the total absence of

Eocene strata directly in front of the orogen from the United States border to Yukon Territory; in an active Eocene foredeep, sediments would be expected to be depressed to a preservable level somewhere. If this approach, yielding an earlier date for demise of the foredeep than that suggested by the vitrinite reflectance approach, is valid, it implies that the leading edge of the Rocky Mountain thrust sheets was close to isostatic equilibrium during its final advance.

Tectonics outside the orogen

In comparison with the passive-margin phase, differentiation of the craton, now the foreland, was subdued during the convergent phase.

The behaviour of the several foreland arches and basins displays no obvious temporal relationship to phases of convergence in the orogen. The Sweetgrass Arch, first expressed in Middle Ordovician time, markedly affected the thickness and facies of Jurassic strata, but its effects in the Cretaceous are seen only in the thinning of Upper Cretaceous deposits (Williams and Burk, 1964). It underwent its final uplift in the Tertiary. Similarly, the Bowdoin Dome, a feature centred in Montana, is markedly revealed by Upper Cretaceous isopachs.

The area of the Devonian Peace River Arch, which had displayed mostly negative behaviour since the beginning of the Carboniferous, and was a depocentre for the Columbian clastic wedges, may have undergone brief uplifts in Middle Jurassic and Middle to Late Albian times, judging by the strong east-west trends developed in Boulder Creek and Viking sandstones.

Great Bear Basin appears to have a two-phase history. During Early and Middle Albian times, it may have been an undifferentiated part of the foreland platform, receiving detritus from the craton rather than the orogen. It then appears to have been isolated as a (tectonic) basin by reactivation of Keele Arch. A younger succession, nonmarine in large part and Campanian to Paleocene in age, rests unconformably on the Albian strata and is preserved only in the centre of the basin. For this phase, Great Bear Basin is a post-depositional, "basin of preservation". The relatively localized tectonic load of the northern Franklin Mountains may be responsible for the younger depression.

Aklavik Arch Complex, with a history of positive behaviour established in early Paleozoic time or earlier, was active during Columbian and Laramide orogenies, but not in a simple way. Its northeastern part was reactivated during rifting attendant upon the opening of Canada Basin in about mid-Cretaceous time, and again, possibly in response to transcurrent faulting, in Late Cretaceous-Early Tertiary times. Southwest of the Mackenzie River, various parts of the complex were again uplifted at various times from the Jurassic onward, and remain structurally high today. It is by no means clear which of these later movements were essentially forebulge effects, which were of the nature of "pop-up" movements of blocks caught between strike-slip faults, and which were simply renewed uplifts driven by the same forces that had established the complex during the passive-margin phase.

Brooks Range – Colville Trough

The British Mountains of northern Yukon Territory are the eastward continuation and termination of the Brooks Range of Alaska, a tectonic element that presents maddening difficulties of interpretation, summarized in part by, for example, Nilsen et al. (1980) and Mull (1982). The range consists of a core or anticlinorium deformed in Late Devonian to Early Carboniferous times (Ellesmerian Orogeny), overlain (mainly on the flanks) by an Upper Carboniferous through Middle Jurassic, platformal succession and a thick, Upper Jurassic and Lower Cretaceous clastic succession. According to several different tectonic scenarios, the core was not in its present orientation or position relative to North America when formed. North of the Brooks Range lies the Colville Trough, with a thick fill of Upper Jurassic and Neocomian clastics and Albian flysch. Colville Trough may be either a successor basin relative to the Ellesmerian deformation (e.g., Mull, 1982), or a foredeep (e.g., Grantz et al., 1979). The range and trough were both deformed in Late Jurassic to Tertiary times, largely by northward-directed thrust faults. The presence of obducted ophiolites at the southern margin of the range, the polarity of the deformation, and the position of Colville Trough (as a foredeep) appear to link the Brooks Range and the British Mountains with the Columbian Orogen. According to this reasoning, the range was in existence and in more or less its present position from Late Jurassic time onward, as a source of detritus. If, however, the core of the range was rotated away from a position northwest of the Arctic Islands during the opening of Canada Basin, and the Mesozoic deformation and the obducted ophiolites are related to that rotation, then the Brooks Range is not Cordilleran, and was not in its present position as a source of detritus until Albian time or later.

TRANSCURRENT MARGIN

The western continental margin had evolved to the dextral transcurrent mode that characterized it for most of the Tertiary by no later than Late Eocene time, as judged by the dating of the extensional Flathead fault in southeastern British Columbia. Under this tectonic regime, the eastern Cordillera was essentially quiet, and undergoing erosion and consequent isostatic uplift due to unloading. A study of burial history based on vitrinite reflectance (England and Bustin, 1986) concluded that, on average, 6 to 7 km of rock has been eroded from the western plains. This conclusion calls for inferred deposition of at least 4.5 km (and perhaps as much as 7.5 km!) of orogenic, Paleogene sediments younger than the youngest preserved Paleocene. Projection of old erosional surfaces allows a calculation of 450 to 1500 m of Neogene erosion, at a rate an order of magnitude lower than that calculated for the Paleogene strata (England and Bustin, 1986). Few western Canadian stratigraphers can accept comfortably the high rates of Paleogene deposition and erosion implied by the vitrinite reflectance data.

Folding of strata as young as Miocene offshore from the north coast of Yukon Territory, and even Pleistocene along the northern coast of Alaska, took place during the regime that is transcurrent relative to most of the Cordillera. Canada Basin is aseismic and therefore no longer opening.

It appears, therefore, that the very young folding along the western Beaufort Sea coast must be fundamentally related to the Pacific continental margin, and be a response to the convergent component (relative to Alaska) of the northward drift of the northeastern Pacific seafloor.

Outside the Cordillera, no new arches or basins came into being during the transcurrent-margin phase. Of the cratonic and foreland structures, only Sweetgrass Arch and, possibly, the area of Peace River Arch underwent slight uplift. Bonnet Plume Basin, whose fill includes beds of Maastrichtian to as young as Eocene age, appears to be a pull-apart basin developed as a consequence of strike-slip faulting.

REFERENCES

Aitken, J.D., Cook, D.G., and Yorath, C.J.
1982: Upper Ramparts River (106G) and Sans Sault Rapids (106H) map areas, District of Mackenzie; Geological Survey of Canada, Memoir 388, 48 p.

Aitken, J.D. and Pugh, D.C.
1984: The Fort Norman and Leith Ridge structures, major, buried, Precambrian features underlying Franklin Mountains and Great Bear and Mackenzie Plains; Bulletin of Canadian Petroleum Geology, v. 32, p. 139-146.

Bally, A.W. and Snelson, S.
1980: Realms of subsidence; in Facts and principles of world petroleum occurrence; A.D. Miall (ed.); Canadian Society of Petroleum Geologists, Memoir 6, p. 9-94.

Beaumont, C.
1981: Foreland basins; Geophysical Journal of the Royal Astronomical Society, v. 65, p. 291-329.

Blair, T.C. and Bilodeau, W.L.
1988: Development of tectonic cyclothems in rift, pull-apart, and foreland basins: Sedimentary response to episodic tectonism; Geology, v. 16, p. 517-520.

Bond, G.C., Christie-Blick, M., Kominz, M.A., and Devlin, W.J.
1985: An Early Cambrian rift to post-rift transition in the Cordillera of western North America; Nature, v. 315 (6032), p. 742-746.

Bond, G.C. and Kominz, M.A.
1984: Construction of tectonic subsidence curves for the early Paleozoic miogeocline, southern Canadian Rocky Mountains - implications for subsidence mechanisms, age of breakup, and crustal thinning; Geological Society of America Bulletin, v. 95, p. 155-173.

Burwash, R.A. and Culbert, R.R.
1976: Multivariate geochemical and mineral patterns in the Precambrian basement of western Canada; Canadian Journal of Earth Sciences, v. 13, p. 1-18.

Burwash, R.A. and Krupicka, J.
1970: Cratonic reactivation in the Precambrian basement of western Canada. Part II. Metasomatism and isostasy; Canadian Journal of Earth Sciences, v. 7, p. 1275-1294.

Campbell, R.B., Mountjoy, E.W., and Struik, L.C.
1982: Structural cross-section through south-central Rocky and Cariboo mountains to the Coast Range; Geological Survey of Canada, Open File 844.

Coles, R.L., Haines, G.V., and Hannaford, W.
1976: Large-scale magnetic anomalies over western Canada and the Arctic: a discussion; Canadian Journal of Earth Sciences, v. 13, p. 790-802.

Cook, F.A.
1988a: Proterozoic thin-skinned thrust and fold belt beneath the Interior Platform in northwest Canada; Geological Society of America Bulletin, v. 100, p. 877-890.
1988b: Middle Proterozoic compressional orogen in northwestern Canada; Journal of Geophysical Research, v. 93, no. B8, p. 8985-9005.

Delaney, G.D.
1981: The mid-Proterozoic Wernecke Supergroup, Wernecke Mountains, Yukon Territory; in Proterozoic Basins of Canada, F.H.A.Campbell (ed.); Geological Survey of Canada, Paper 81-10, p. 1-23.

Dixon, J.
1986: Cretaceous to Pleistocene stratigraphy and paleogeography, northern Yukon and northwestern District of Mackenzie; Bulletin of Canadian Petroleum Geology, v. 34, p. 49-70.

Douglas, R.J.W., Gabrielse, H., Wheeler, J.O., Stott, D.F., and Belyea, H.R.
1970: Geology of Western Canada; Chapter VIII, in Geology and Economic Minerals of Canada, R.J.W. Douglas (ed.); Geological Survey of Canada, Economic Geology Report No. 1, p. 365-488.

Eisbacher, G.H.
1977: Tectono-stratigraphic framework of the Redstone Copper Belt, District of Mackenzie; in Report of Activities, Part A, Geological Survey of Canada, Paper 77-1A, p. 229-234.

Eisbacher, G.H., Carrigy, M.A., and Campbell, R.B.
1974: Paleodrainage pattern and late-orogenic basins of the Canadian Cordillera; in Tectonics and Sedimentation, W.R. Dickinson (ed.); Society of Economic Paleontologists and Mineralogists, Special Publication 22, p. 143-166.

England, T.D.J. and Bustin, R.M.
1986: Thermal maturation of the Western Canadian Sedimentary Basin south of the Red Deer River: (I) Plains; Bulletin of Canadian Petroleum Geology, v. 34, p. 71-90.

Gabrielse, H., Blusson, S.L., and Roddick, J.A.
1973: Flat River, Glacier Lake and Wrigley Lake map areas (95E, L, M), District of Mackenzie and Yukon Territory; Geological Survey of Canada, Memoir 366, 421 p.

Gabrielse, H. and Yorath, C.J. (ed.)
1991: Geology of the Cordilleran Orogen in Canada; Geological Survey of Canada, Geology of Canada no. 4. (Also Geological Society of America, The Geology of North America, v. G.2.)

Geological Survey of Canada
1987: Magnetic anomaly maps of Canada, fifth edition; Geological Survey of Canada, Map 1255A, Scale 1:5 000 000.

Grantz, A, Ittreim, S., and Dinter, D.A.
1979: Geology and tectonic development of the continental margin north of Alaska; Tectonophysics, v. 59, p. 263-291.

Grantz, A., Johnson, L., and Sweeney, J.F. (ed.)
1990: The Arctic Ocean Region; Geological Society of America, The Geology of North America, v. L, 644 p.

Kanasewich, E.R., Clowes, R.M., and McCloughan, C.H.
1968: A buried Precambrian rift in western Canada; Tectonophysics, v. 8, p. 513-537.

Klepacki, D.W.
1985: Stratigraphy and structural geology of the Goat Range area, southeastern British Columbia; Ph.D. thesis, Massachusetts Institute of Technology, Cambridge, Massachusetts, 268 p.

Klepacki, D.W. and Wheeler, J.O.
1985: Stratigraphic and structural relations of the Milford, Kaslo and Slocan Groups, Goat Range, Lardeau and Nelson map-areas, British Columbia; in Current Research, Part A; Geological Survey of Canada, Paper 85-1A, p. 277-286.

Landes, K.K.
1970: Petroleum Geology of the United States; Wiley-Interscience, Toronto, 571 p.

McKenzie, D.P.
1978: Some remarks on the development of sedimentary basins; Earth and Planetary Science Letters, v. 40, p. 25-32.

McMechan, M.E.
1981: The Middle Proterozoic Purcell Supergroup in the southwestern Rocky and southeastern Purcell mountains, British Columbia, and the initiation of the Cordilleran Miogeocline, southern Canada and adjacent United States; Bulletin of Canadian Petroleum Geology, v. 29, p. 583-621.

Mountjoy, E.W.
1962: Mount Robson (southeast) map-area, Rocky Mountains of Alberta and British Columbia; Geological Survey of Canada, Paper 61-31, 114 p.
1967: Triassic stratigraphy of northern Yukon Territory; Geological Survey of Canada, Paper 66-19, 44 p.

Mull, C.G.
1982: Tectonic evolution and structural style of the Brooks Range, Alaska: an illustrated summary; in Geologic Studies of the Cordilleran Thrust Belt; Rocky Mountain Association of Geologists, Denver, p. 1-46.

Nilsen, T.H., Brosge, W.P., Moore, T.E., Dutro, J.T., Jr., and Balin, D.F.
1980: Significance of the Endicott Group for tectonic models of the Brooks Range; in The United States Geological Survey in Alaska: accomplishments during 1980, W.L. Coonrad (ed.); United States Geological Survey Circular 844, p. 28-32.

Okulitch, A.V.
1985: Paleozoic plutonism in southeastern British Columbia; Canadian Journal of Earth Sciences, v. 22, p. 1409-1424.

Okulitch, A.V., Wanless, R.K., and Loveridge, W.D.
1975: Devonian plutonism in south-central British Columbia; Canadian Journal of Earth Sciences, v. 12, p. 1760-1769.

Pelletier, B.R.
1964: Triassic stratigraphy of the Rocky Mountain Foothills between Peace and Muaskwa rivers, northeastern British Columbia; Geological Survey of Canada, Paper 63-33, 89 p.

Price, R.A. and Fermor, P.R.
1982: Structural section of the Cordilleran thrust and fold belt west of Calgary, Alberta; Geological Survey of Canada, Open File 882.

Price, R.A. and Hatcher, R.D., Jr.
1983: Tectonic similarities in the evolution of the Alabama-Pennsylvania Appalachians and the Alberta-British Columbia Canadian Cordillera; in Contributions to the Tectonics and Physics of Mountain Chains, R.D. Hatcher, Jr., H. Williams and I. Zeitz (ed.); Geological Society of America, Memoir 158, p. 149-160.

Price, R.A. and Mountjoy, E.W.
1970: Geologic structure of the Canadian Rocky Mountains between Bow and Athabasca Rivers - a progress report; in Structure of the Southern Canadian Cordillera, J.O. Wheeler (ed.); Geological Association of Canada, Special Paper No. 6, p. 7-25.

Sanford, B.V., Thompson, F.J., and McFall, G.H.
1985: Plate tectonics - a possible controlling mechanism in the development of hydrocarbon traps in southwestern Ontario; Bulletin of Canadian Petroleum Geology, v. 33, p. 52-71.

Sleep, N.H.
1971: Thermal effects of the formation of Atlantic continental margins by continental breakup; Geophysical Journal of the Royal Astronomical Society, v. 24, p. 325-350.

Sloss, L.L.
1976: Areas and volumes of cratonic sediments, western North America and eastern Europe; Geology, v. 4, p. 272-276.

Trettin, H.P. (ed.)
1991: Geology of the Innuitian Orogen and Arctic Platform of Canada and Greenland; Geological Survey of Canada, Geology of Canada, no. 3. (Also Geological Society of America, The Geology of North America, v. E.)

Walcott, R.I.
1970: Isostatic response to loading of the crust in Canada; Canadian Journal of Earth Sciences, v. 7, p. 2-13.

Watts, A.B. and Ryan, W.B.F.
1976: Flexure of the lithosphere and continental margin basins; Tectonophysics, v. 36, p. 25-44.

Williams, G.D. and Burk, C.F.
1964: Upper Cretaceous; Chapter 12 in Geological History of Western Canada, R.G. McCrossan and R.P. Glaister (ed.); Alberta Society of Petroleum Geologists, p. 169-189.

Young, F.G., Myhr, D.W., and Yorath, C.J.
1976: Geology of the Beaufort-Mackenzie Basin; Geological Survey of Canada, Paper 76-11, 65 p.

Author's Address

J.D. Aitken
2676 Jemima Road
Denman Island
British Columbia
V0R 1T0

Printed in Canada

Chapter 6

ECONOMIC GEOLOGY

Subchapter 6A

PETROLEUM

R.D. Johnson and N.J. McMillan

CONTENTS

INTRODUCTION

Petroleum has been discovered in commercial quantities in two major sedimentary basins of the Western Interior: Western Canada Basin and Beaufort-Mackenzie Basin. Western Canada Basin comprises the Phanerozoic sedimentary wedge lying between the Canadian Shield and the Cordillera; it includes the thrust-faulted rocks of the Foothills Belt but not the Rocky, Mackenzie, and Ogilvie mountains. This vast basin extends from the 49th parallel northward beyond the Arctic Circle (Fig. 6A.1). Western Canada Basin is divided into the following regions: Alberta Basin, Williston Basin, Foothills Belt, and Northern Basins. Beaufort-Mackenzie Basin straddles the Arctic coastline north of Western Canada Basin; a summary description is given in this subchapter but its geology and economic potential are described more fully in the companion volume on the Arctic Region (Grantz et al., 1990).

The distribution of oil and natural gas in western Canada south of latitude 60°N is shown on two pocket figures (Maps 1558A, 1559A), which may be used to locate fields referred to in the text. The maps are coloured to show the geological age of the reservoirs. Marginal notes contain statistical data on the largest pools, including Initial Volume In Place (IVIP), initial recoverable oil (Initial Established Reserves (IER)), pool area, net pay, porosity, and discovery year. Pie diagrams show the initial volume and the initial reserves, and the initial reserves by age of reservoir (see also Fig. 6A.2, 6A.3). For current statistical data on reserves and production for individual fields, the annual summaries published by the provincial regulatory agencies and the Canada Oil and Gas Lands Administration should be consulted.

The history of petroleum exploration in the Western Canada Basin from pioneer days to 1987 can be divided conveniently into six phases, the first three of which partially overlap.

1. The "pioneer" phase (Beach and Irwin, 1940): Prior to 1914 discoveries were made either by accident while drilling for water or by drilling for petroleum near surface seepages. Numerous accidental discoveries of

Johnson, R.D. and McMillan, N.J.
1993: Petroleum; Subchapter 6A in Sedimentary Cover of the Craton in Canada, D.F. Stott and J.D. Aitken (ed.); Geological Survey of Canada, Geology of Canada, no. 5, p. 505-562 (also Geological Society of America, The Geology of North America, v. D-1).

natural gas in water wells are recorded from the 1880s, particularly in southern Alberta. However, the only field of significance resulting from such discoveries is the Medicine Hat Gas Field, which was connected by pipeline to Calgary in 1912.

Drilling near seepages with, or without, geological advice resulted in substantial gas shows at Pelican Rapids in northern Alberta (1896), with some oil recovery in 1904, and in 1901 minor oil shows in Cameron Creek, now within Waterton National Park. During the "pioneer" years, the only discovery of importance was the large oil field at Norman Wells along the Mackenzie River in 1920.

2. The "anticlinal" phase: This period of field geology focused on anticlinal structures in the Foothills, resulting in the discovery in 1914 of oil in Cretaceous rocks of the Turner Valley Anticline. Further drilling on this anticline resulted in the discovery in 1923 of a second productive gas and condensate zone in Lower Carboniferous limestone. A third major event was the discovery, by a step-out well, of oil on the west flank of the structure in 1936.

   The addition of seismic surveys to Foothills exploration techniques eventually revolutionized the geological interpretation of the structures and opened the door to discoveries in deeper anticlines lacking overlying surface expression. Many large sour gas fields have resulted from this technique, the first of which was Jumping Pound (1944).
3. The early phase of subsurface geology and seismic reflection surveys on the Plains: During this phase many holes were drilled over a wide area of the Plains, at first without success except for the discovery of some heavy oil and natural gas for which there was no market. This state of affairs changed dramatically in 1947 with the discovery of Devonian oil at Leduc, Alberta in the Nisku Formation, followed, within a few months, by the discovery of much greater quantities in the underlying Leduc reef.
4. The first post-Leduc decade (1947-1957): The discovery of Leduc brought geologists and geophysicists to Alberta from all over the world and ushered in the post-Leduc era, dominated by the ever-increasing sophistication of the seismic reflection tool and the refinement of subsurface geology, borehole geophysical surveys, and testing methods. By the end of the first post-Leduc decade the potential of nearly all the important oil and gas zones had been demonstrated by significant discoveries. The key discoveries are listed stratigraphically, together with the date when the first significant discovery in the zone was made (Table 6A.1).
5. The second post-Leduc decade (1958 to 1967): During this period, important discoveries of prolific new oil zones in Middle Devonian strata were added to the already productive Upper Devonian zones (Table 6A.2).
6. The final phase, which brings the history of petroleum exploration in the Western Canada Basin to the time of writing, covers the third and fourth post-Leduc decades (1968-1977; 1978-1987): This 20-year period has yielded few new discoveries of major importance, with the exception of the non-conventional Cretaceous gas play in the Elmworth area discovered in 1968 (Masters, 1984). Other discoveries included oil in the Permian Belloy Formation in 1970 and in the Devonian Zeta Lake reefs of the West Shale Basin in 1977, which opened up a brief flurry of new activity. New plays in previously established productive zones attracted most exploratory effort and yielded the most success during this period. At the same time exploration started in the Beaufort-Mackenzie Basin, where the first important finds of oil occurred in 1970, followed by a major gas discovery in 1971.

Table 6A.1. Key discoveries of the first post-Leduc decade (to 1957) in Western Canada Basin

Series/System Formation	Field	Date
Upper Cretaceous		
Belly River	Willesden Green	1956
Cardium Formation	Pembina	1953
Lower Cretaceous		
Viking Formation	Joarcam	1949
Jurassic		
Shaunavon Formation	Dollard Trend (Sask)	1953
J Sands	Medicine River	1956
Triassic		
Boundary Lake Formation	Boundary Lake (NE B.C.)	1955
Halfway sandstone	Milligan Creek	1957
Carboniferous		
Turner Valley subcrop	Harmattan	1955
Pekisko subcrop	Twining	1952
Charles and Mission Canyon subcrop	Midale	1953
Lodgepole subcrop	Daly	1951
Upper Devonian		
Wabamun (Crossfield	Okotoks (sour gas)	1951
Member)	Leduc	1947
Nisku platform	Leduc	1947
Leduc Reef		
Middle Devonian		
Swan Hills Formation	Swan Hills	1957
Slave Point/Presq'ile	Clarke Lake (gas)	1957

Table 6A.2. Key oil discoveries of the second post-Leduc decade (to 1967) in Western Canada Basin

Series/Formation	Field	Date
Middle Devonian		
Slave Point and Granite wash	Red Earth	1958
Gilwood Sandstone Member of Watt Mountain Formation	Mitsue	1964
Keg River Formation reefs	Rainbow	1965
Keg River sandstones	Utikama Lake	1963

Crude bitumen deposits in Cretaceous sands and Devonian carbonates are a major resource and were the earliest to be discovered. In 1788, explorer Peter Pond reported the presence of oil seeping from the Athabasca Oil Sands (Layer, 1958). Since that time more deposits have been discovered at Cold Lake, Wabasca, Peace River, and in the "carbonate triangle" (see Fig. 6A.54). Research into extraction methods, a major concern of the Alberta Research Council since its inception, came to fruition in 1967 with open-pit mining by Great Canadian Oil Sands (later Suncor), followed by Syncrude in 1978. Subsurface extraction has been successful recently at the Imperial Oil Enterprises plant at Cold Lake, Alberta.

This report describes the petroleum **resources** that have been discovered in the region. From a geological point of view the most significant measure of the resource is the **Initial Volume In Place (IVIP).** This measure is used where available (some agencies do not publish data on IVIP). That portion of the IVIP that is recoverable under given economic and technical conditions is referred to as the **Initial Established Reserves (IER).** All statistics refer to 1986 unless otherwise stated. A concluding section of this report is devoted to Initial and Remaining Established reserves and to estimates of the undiscovered resources.

Within the Western Canada Basin an IVIP of over 8400 x 10^6 m^3 of light and medium crude oil has been discovered. The largest volumes are found in the Devonian beds (36%), followed closely by those in the Cretaceous rocks (34%). The Initial Established Reserves of the Devonian are, however, much greater than those of the Cretaceous because of higher recovery factors.

Heavy oil (density >900 kg/m^3) is concentrated in the Cretaceous System, with a minor quantity (8%) in the Carboniferous System. However, heavy oil volumes reported by regulatory agencies, totalling about 2680 x 10^6 m^3, represent only part of the geological resource; Procter et al. (1984) estimated that the IVIP of heavy oil was in the range 5200-7400-10 630 x 10^6 m^3 (high confidence - average expectation - speculative estimate). The distribution by region of the combined light, medium, and heavy oils is shown in Figure 6A.2a.

Crude bitumen is found only in Alberta, in Cretaceous sandstones ("oil sands") and in Paleozoic carbonates. The Alberta Energy Resources Conservation Board (1987) estimated 26 700 x 10^6 m^3 in place in the mineable areas and 1679 x 10^6 m^3 in the sites being developed underground. Initial Established Reserves for the combined categories are 5370 x 10^6 m^3. As in the case of heavy oil, these volumes represent only a small fraction of the geological resource.

The regional distribution of the natural gas resources, totalling over 5400 x 10^9 m^3, is shown in Figure 6A.2b. Cretaceous reservoirs contain the largest amounts (47%), followed by Devonian (28%). Natural gas liquids are a valuable by-product of gas processing; the heavier fractions referred to as "pentanes plus" add at least 17% to the reserves of crude oil and equivalents.

Sulphur is a valuable by-product of sour gas processing in both Alberta and British Columbia and also of crude bitumen upgrading in Alberta. The Initial Established Reserves (at end 1986) for sulphur were 206 Mt for Alberta and 15 Mt for British Columbia.

The Beaufort-Mackenzie Basin contains large volumes of oil and gas in Cretaceous and Tertiary rocks, but exploration is in the very early stages.

As a result of enlightened government conservation regulations, logs, cores, and cuttings generated by the drilling of exploration and development wells have been placed in the public domain. This has encouraged a liberal policy of publication, so that documentation on most of the plays is available. Important basin compilations have also been published, notably volumes edited by McCrossan and Glaister (1964), Douglas (1970), and McCrossan (1973).

Many different types of traps have been identified (Table 6A.3). The most important traps for conventional oil and gas are reefs and reef complexes built on drowned platforms during the Devonian Period. This trap type (Rd) is followed closely in importance by the pinchout of marine shelf sandstones (Ls3) of Cretaceous age. Subcrop traps (Ub) are particularly important for Lower Carboniferous and Triassic oil, but the large Lower Carboniferous gas reserves are trapped in anticlines carried on low-angle thrusts (Sa2). The largest accumulation of crude bitumen appears to be trapped by a bitumen plug (Ld2); the bitumen trapped in carbonates is in a subcrop trap (Ub1), but most of the heavy oils are in lithological traps within fluviatile sediments (Ls1). Traps in the Beaufort-Mackenzie Basin are mainly of types Sf1 (traps against synsedimentary listric faults), Sa1 (anticlinal traps above listric normal faults), and Sd5 (drape folds over mud diapirs). Other trap types are of considerable geological interest and of some commercial importance but have not made a major impact on resource volumes.

ACKNOWLEDGMENTS

This subchapter has benefited from input of a large number of geologists with individual experience and expertise on specific accumulations. They include C.V. Campbell, A.R. Clark, J. Christopher, J.R. Dietrich, J. Dixon, P.L. Gordy, A.D. Graham, R.S. Harrison, D.M. Kent, J.E. Klovan, J.W. Kramers, G. Macauley, L. Matwe, K.A. McAdam, T.A. Oliver, J. Plante, R.M. Procter, P.E. Putnam, M. Raicar, D.K. Smith, J.C. Wendte, and P.K. Wong. The principal authors have substantially modified the input of others and are alone responsible for errors and omissions.

Organizations supporting the involvement of professional staff in this paper include: Alberta Geological Survey, British Columbia Ministry of Energy, Mines and Petroleum Resources, Canadian Hunter Explorations Ltd., Chevron Canada Resources Ltd., Esso Resources Canada Limited, ICG Resources Ltd., Husky Oil Operations Ltd., PanCanadian Petroleum Limited, R.P.I. Ltd., R.D. Johnson and Associates Ltd., Saskatchewan Geological Survey, Shell Canada Resources Limited, University of Calgary, University of Regina, and Western Oil Maps Ltd.

This text has drawn extensively on a report by R.M. Procter, G.C. Taylor and J.A. Wade (1984) on "Oil and natural gas resources of Canada, 1983", and on a report by J.A. Podruski, J.E. Barclay, A.P. Hamblin, P.J. Lee, K.G. Osadetz, R.M. Procter, and G.C. Taylor (1988) entitled "Conventional oil resources of Western Canada (light and medium)".

Special acknowledgment is made to P.F. Moore, who carried out the final revision of this subchapter, provided much helpful advice and knowledge, made available his classification of traps, and served as a consulting editor.

CONVENTIONAL OIL AND GAS

Alberta Basin

The Alberta Basin is bounded by Tathlina Arch on the north and by Sweetgrass Arch on the south (Fig. 6A.1). Tathlina Arch is a Devonian feature that was finally covered during the Givetian Age; Sweetgrass Arch was moderately mobile during Paleozoic time and became a pronounced post-Paleozoic anteclise. Within the basin, the Peace River Arch was active during the Devonian Period and not completely covered until the beginning of the Early Carboniferous Epoch, at which time it began to change to a syneclise (Peace River Embayment of Richards et al., Subchapter 4E in this volume). Its continuation southward as the West Alberta Arch was covered in Frasnian time; the axis of this arch was subsequently involved in the Rocky Mountain thrusts and is thus excluded from the basin, but the upper Middle Devonian reef-margined carbonate platforms that mantled its east flank are important reservoirs in the Foothills and adjacent Plains.

The sedimentary fill of the Alberta Basin forms a wedge thickening from its outcrop edge along the Precambrian (Canadian) Shield on the east to 6 km in and adjacent to the Foothills (Fig. 5.5). This wedge consists of a preorogenic Cambrian to Middle Jurassic platformal sequence overlain by a synorogenic Upper Jurassic to Paleogene foredeep sequence. The Paleozoic rocks are marine shelf carbonates, with lesser amounts of evaporites, sandstones, and shales: the Mesozoic rocks are dominantly siliciclastic.

The platformal sequence contains numerous important unconformities that merge as the Canadian Shield is approached; most of them exert some control over trapping conditions. The principal Paleozoic unconformities occur beneath Devonian and Permian strata. The principal Mesozoic unconformities are those beneath the Triassic and Jurassic systems and the Cretaceous Aptian Stage.

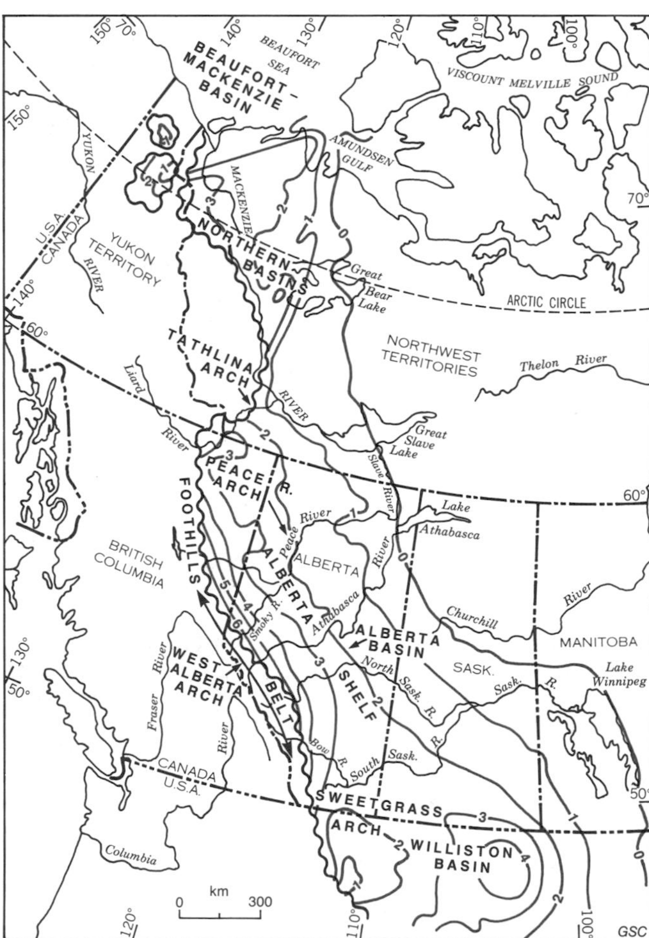

Figure 6A.1. Total basin-fill of Western Canada Basin, showing thickness (km) of unmetamorphosed sedimentary rocks (adapted from McCrossan and Porter, 1973).

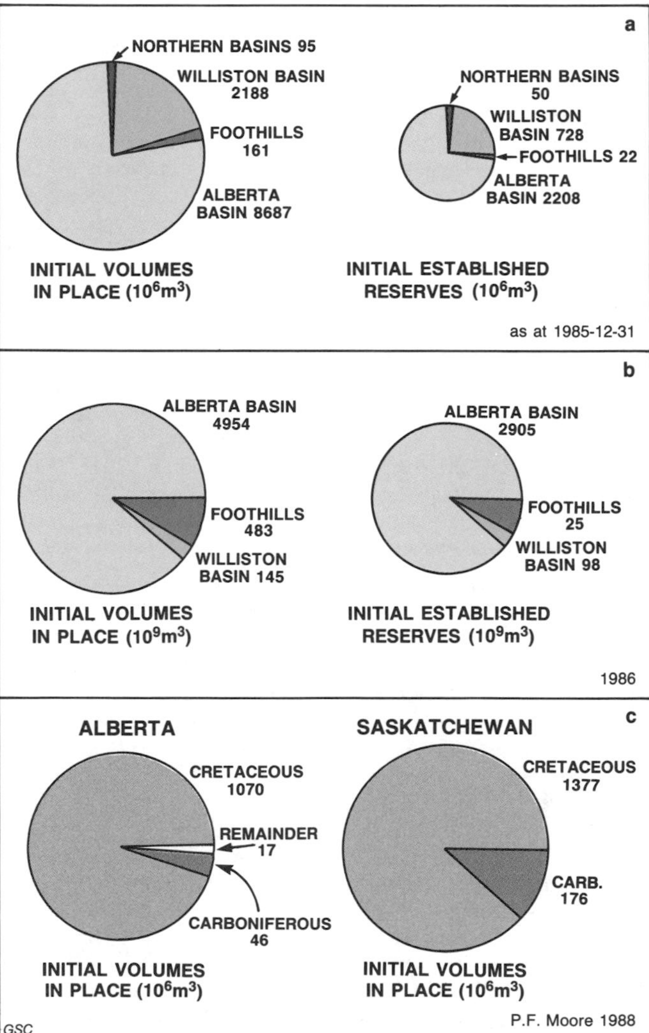

Figure 6A.2. Distribution of hydrocarbon volumes among the regions of Western Canada Basin: a) light and medium oil; b) gas; c) heavy oil.

The richest petroleum source rocks so far identified are: the Middle Devonian Muskeg laminites and basal Waterways shales; the Upper Devonian Duvernay Formation; the Upper Devonian to Lower Carboniferous Exshaw Formation; and the Upper Cretaceous Second White Specks zone. However, good source rocks occur at other horizons throughout the succession. Thermal maturity ranges from immature in the shallow gas areas of the northern and eastern parts of the basin, to overmature in the Foothills and adjacent areas.

The most important reservoirs are: Devonian biogenic carbonates; erosional edges of the Devonian and Carboniferous carbonate units at the sub-Mesozoic unconformity; Lower Cretaceous fluviatile and estuarine sands; and Upper Cretaceous marine shelf sands. Other prolific reservoirs include Devonian sandstones forming an apron around the Peace River Arch, and Triassic algal carbonates in northeastern British Columbia.

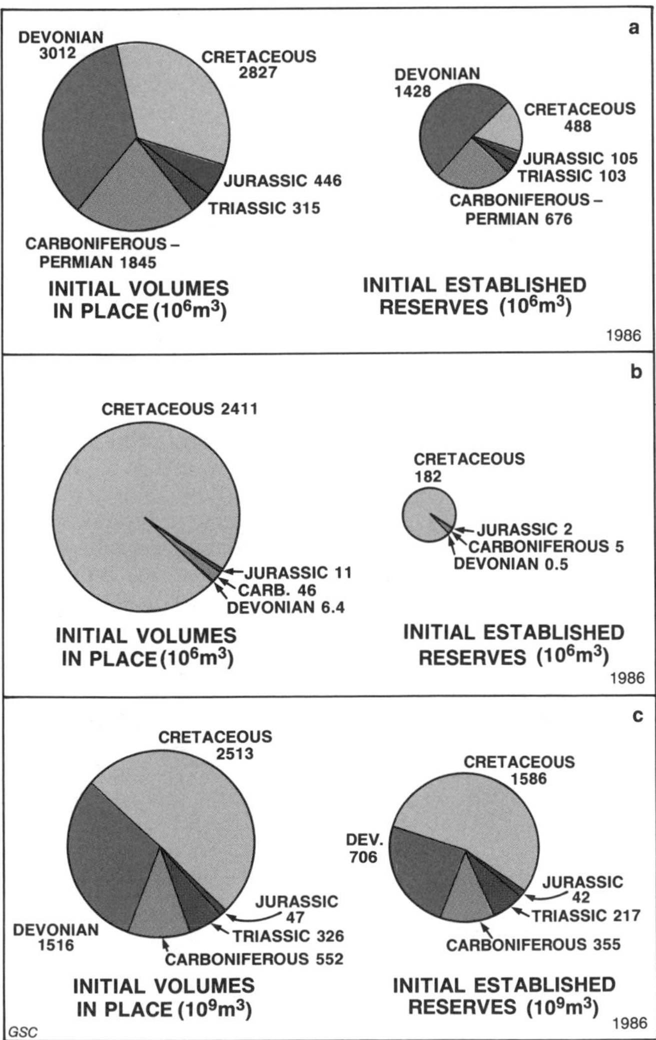

Figure 6A.3. Distribution of hydrocarbon volumes by geological system in Alberta Basin: a) light and medium oil; b) heavy oil; c) gas.

Over 100 000 wells have been drilled in Alberta Basin and nearly 20 000 gas and 3000 oil pools have been discovered. Most of these pools are very small (for instance, fewer than 10% of the gas pools contain more than $300 \times 10^6 m^3$ of gas) and the resource volumes depend heavily on the largest pools. The distribution of oil and gas by geological formation is shown in Figure 6A.3a. Production to 1986 was approximately $1630 \times 10^6 m^3$ of conventional oil and $1520 \times 10^9 m^3$ of gas.

Devonian

The Devonian System is the principal oil producer in the Alberta Basin. A stratigraphic diagram (Fig. 6A.4) of the Devonian succession indicates by means of symbols the chief oil- and gas-producing units. A map (Fig. 6A.5) outlines the principal reefs. The most important trap types and fields are described for each of the five groups into which the succession is divided (Fig. 6A.4). The Devonian System contains a disproportionally large fraction of the recoverable reserves owing to its excellent reservoir properties and the lighter than average density of the Devonian oils (Fig. 6A.3a).

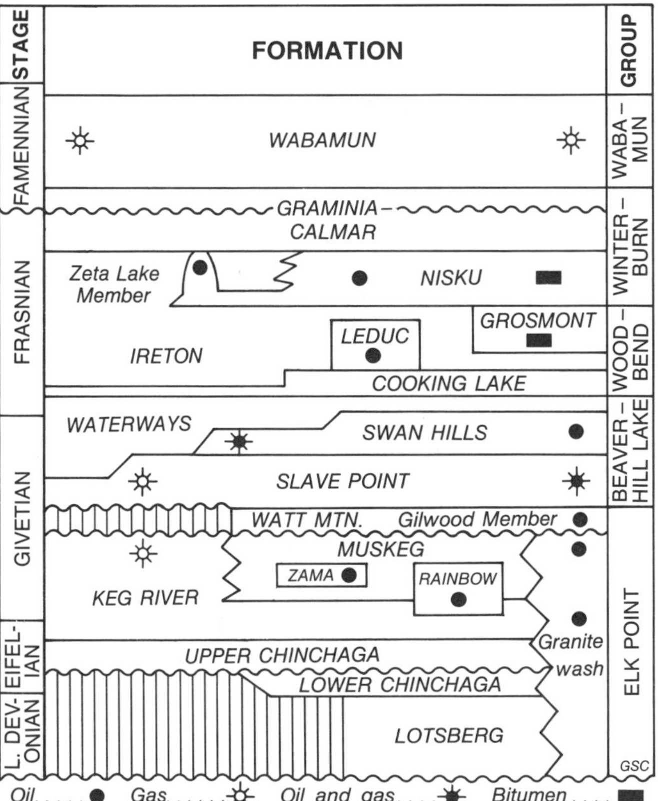

Figure 6A.4. Principal producing zones in the Devonian System of the Alberta Basin.

Elk Point Group

Middle and possibly Lower Devonian sandstone reservoirs are found in an apron flanking the Peace River Arch and its northeasterly extension (Fig. 6A.6). Minor oil accumulations occur in sandstones collectively known as "Granite Wash" but locally taking the name of the carbonate-evaporite formation with which they are correlated (e.g. Chinchaga, Keg River).

The Red Earth Oil Field (Century, 1967) provides examples of traps in buttress sands (Trap Type Ua2: Table 6A.3). The most important sandstone fields of this age are at Mitsue and Nipisi on the eastern flank of the Peace River Arch. The reservoirs are in channel and barrier face sandstones (Trap Type Ls2) of the Givetian Gilwood Member, which shale-out up-dip into marine shales. At Nipisi (Alcock and Benteau, 1976) the reservoir, lying at a depth of 1700 m, is a poorly to moderately well-sorted feldspathic sandstone, with an average porosity of 13.5% and a net pay of 5.2 m, containing oil with an IVIP of 114 x 10^6 m^3. The best reservoirs are in channel sandstones of the delta-plain where cementation is least complete. Sandstones of the delta-front facies (barrier face sands) are well cemented by anhydritic and siliceous cements, creating permeability barriers.

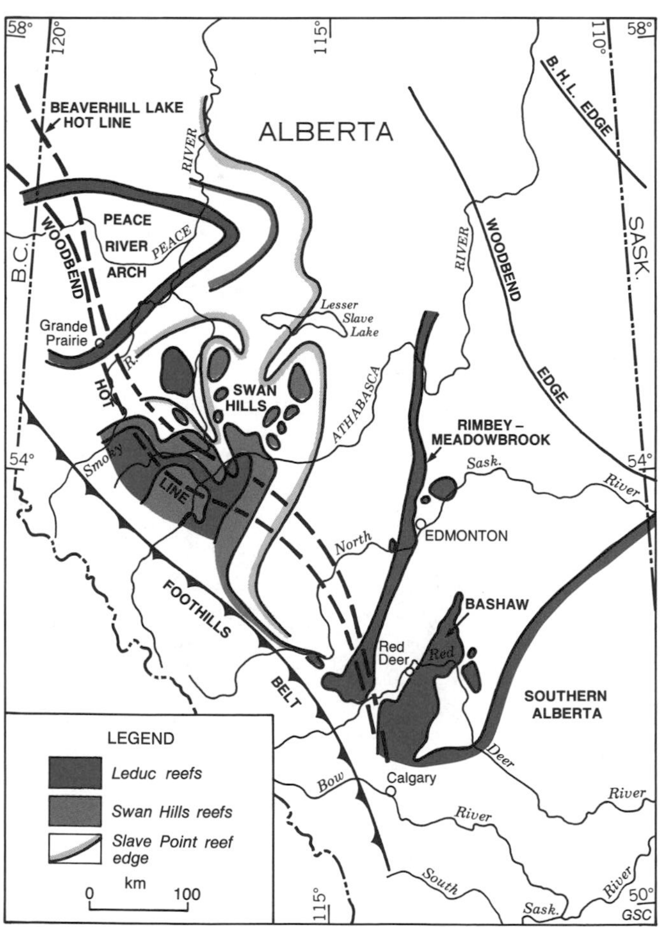

Figure 6A.5. Distribution of principal Devonian carbonate complexes of Alberta Basin (J.A. Podruski).

The Keg River Formation formed a barrier reef behind which developed the Elk Point Basin, a periodically isolated, evaporitic inland sea. Lagoonal sub-basins within Elk Point Basin (e.g., Rainbow, Fig. 6A.7; Shekilie and Zama basins) contain swarms of "pinnacle" reefs. In unexaggerated sections some of these reefs have the profile of an inverted soup-plate (Elloy, 1972), with an areal extent ranging typically from 16 ha to 260 ha: others are more conical. Oil was discovered in these reefs (Trap Type Rd4) in 1965. The main producing horizon is the Rainbow Member of Keg River Formation, which occurs as atolls and mounds resting on the fine-grained carbonate foundation of the lower Keg River (Langton and Chin, 1968). The Rainbow reefs appear to be ecological (Elloy, 1972) but were strongly affected by early cementation (Schmidt et al., 1977). The reservoir contains numerous facies, including reef-flank debris with corals, brachiopods, and a variety of stromatoporoids; a reef-margin facies with bulbous stromatoporoids and corals; interior laminites; and a cap of shoaling cycles, which is equivalent to the Zama Member in the Shekilie-Zama Basin. The larger reefs reach a height of 250 m. Light oil is produced from depths of about 1800 m. Porosity ranges from 4% to 13% and permeability from 50 mD to 780 mD. Overlying the reefs are a few small pools in drape traps (Trap Type Sd3) within Muskeg Formation and Bistcho Formation (also called "Sulphur Point").

Keg River Barrier Reef is largely replaced by a white hydrothermal dolomite known as Presqu'ile facies. This facies is host to gas acculumations in dolomitized reefs lying outboard of the barrier (e.g., at Yoyo Field, where the reservoir is called "Pine Point" and has net pay of 45.7 m with a porosity of 9.3% containing gas with an IVIP of 50 x 10^9 m^3).

Beaverhill Lake Group

In northern Alberta, the Slave Point Formation formed a platform with a reefal rim that, over much of its length, coincides in position with the Keg River Barrier Reef. Where this occurs, the Slave Point is commonly dolomitized to Presqu'ile facies. This has created gas reservoirs in the Clarke Lake Gas Field along the barrier reef margin in British Columbia and in isolated reefs such as Kotcho Gas Field.

At Clarke Lake (Fig. 6A.8) the porosity resulting from early and intermediate dolomitization is predominantly intercrystalline, averaging 10 to 12% and reaching as high as 17%; the permeability is less than 100 mD. Later dolomitization increased porosity somewhat, but the voids are in part filled by late precipitated dolomite (Gray and Kassube, 1963). The gas is trapped by a facies change to impermeable limestone (Trap Type Ld1). Average net pay is 35.4 m and depth to production is about 1950 m. The estimated IVIP of gas is 62 x 10^9 m^3.

On the northeast side of the Peace River Arch, the Slave Point Formation formed fringing biostromes buttressing Precambrian granite islands (Trap Type Rf). Similar reefs of more "shoreward" position are probably equivalent to the Swan Hills Formation, though still classified as "Slave Point" by some geologists. None of the contained oil pools has an IVIP greater than 10 x 10^6 m^3. A well documented example from the older set of pools is Slave Field, which contains a light oil. "The principal porosity types observed

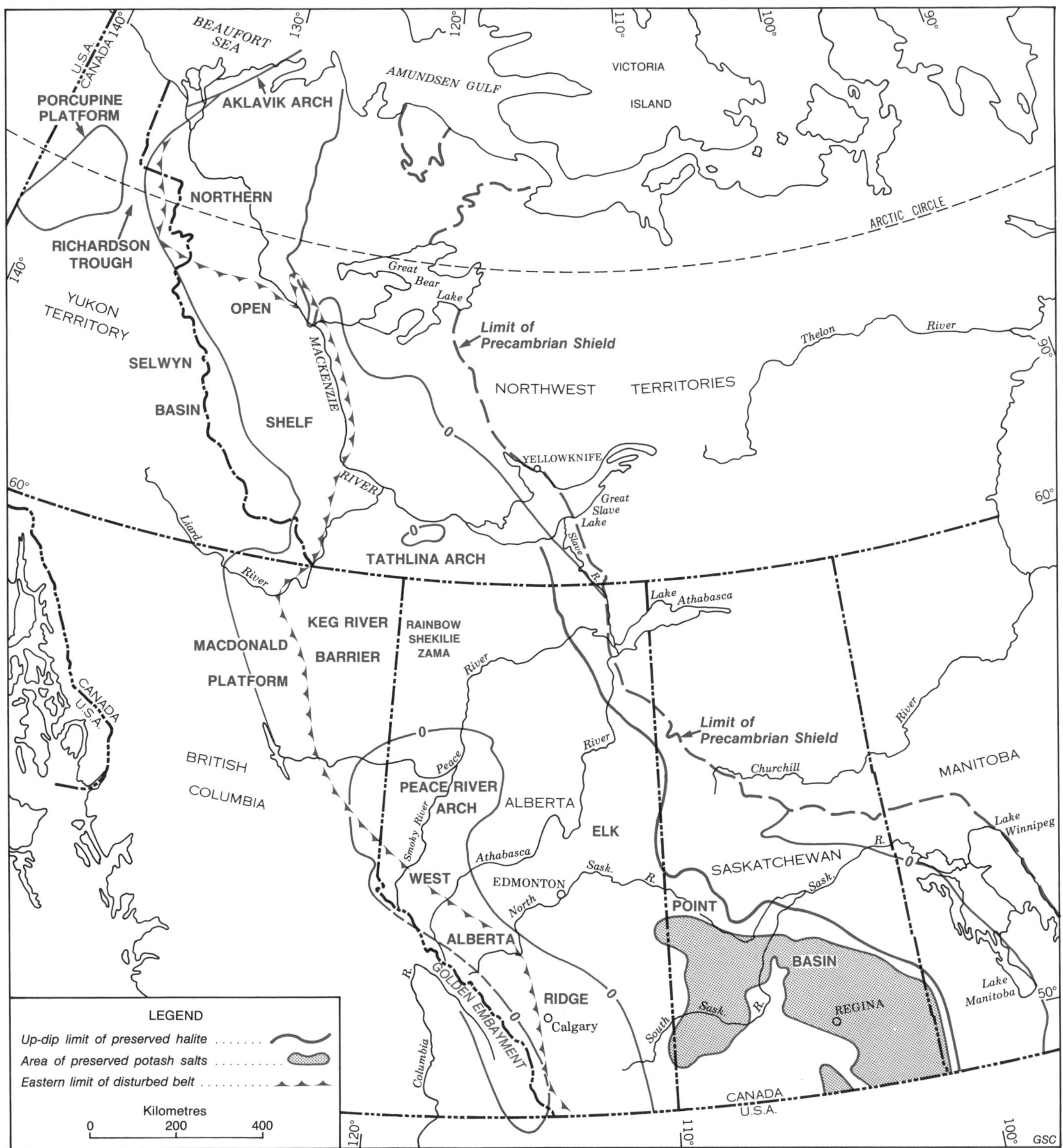

Figure 6A.6. Regional configuration of Elk Point Basin during Middle Devonian time, showing emergent land areas of Peace River Arch, Tathlina Arch and West Alberta Ridge. (adapted from Moore, Subchapter 4D, in this volume).

in the Slave Field reservoir are moldic porosity developed as a result of the leaching of fossil elements . . . and intercrystalline porosity developed between dolomite crystals in the dolomite matrix" (Dunham et al., 1983).

South of Peace River Arch, the Slave Point Formation forms the foundation for a step-wise buildup of platform reefs (Fig. 6A.9). The combined foundation and superstructure are included in the Swan Hills Formation, one of the principal oil and gas producers of the Alberta Basin. Swan Hills Field is the largest oil accumulation in this play, with an IVIP greater than 400×10^6 m^3.

The Swan Hills reef complexes (Hemphill et al., 1970), mainly limestone, are commonly more than 10 km across, range in thickness from 60 to 120 m, and contain oil columns up to 60 m. The reefs lie at depths between 2500 and 3000 m. As illustrated by a stratigraphic cross-section through the Judy Creek "A" pool (Fig. 6A.10; Wendte and Stoakes, 1982), Swan Hills reefs comprise a number of depositional stages with the number dependent on the criteria adopted. These complexes of organic and bioclastic limestone have original intraskeletal, interskeletal, and intergranular porosity. Dolomite cements are present and have occluded some of the porosity, but despite diagenetic effects a striking correlation has been observed between porosity distribution and facies (Viau, 1983). Downdip from the oil-bearing limestone reefs, gas has been discovered in the margin of an extensive platform developed on the east flank of the West Alberta Arch. The gas field at Kaybob South has an IVIP of over 100×10^9 m^3.

Table 6A.3. Principal trap types of western Canada

(From P.F. Moore, unpublished data)	
R	Reef traps.
Rd	Reefs and reef complexes on drowned carbonate platforms. Subtypes: 1) Margins and crests of platform reefs; 2) Barrier reefs; 3) Satellite reefs; 4) Lagoonal bioherms; 5) Traps against interior channels.
Rp	Reefs on prograding carbonate platforms. Subtypes: 1) Interior; 2) Marginal; 3) Basin slope.
Rf	Fringing reefs and biostromes.
U	Unconformity-related traps.
Ua	Traps situated above the unconformity. Subtypes: 1) Transgressive marine sandstone; 2) Buttressing islands, "arches" and monadnocks; 3) Valley fill; 4) Valley side.
Ub	Traps situated beneath the unconformity (subcrop traps). Subtypes: 1) Traps controlled by paleogeomorphology of unconformity; 2) Traps requiring a seat seal.
L	Lithologic traps (facies change).
Ls	Sedimentation controlled traps. Subtypes: 1) Fluviatile environment; 2) Littoral system (deltaic and shoreline); 3) Marine shelf (siliciclastic); 4) Carbonate platform; 5) Deep water sands.
Ld	Diagenetically controlled traps. Subtypes: 1) Pore enhancement or reduction by alteration of rock; 2) Pore reduction by bitumen plugging.
S	Structural traps.
Sd	Drape folds (not over folded basement). Subtypes: 1) Over a fault at depth; 2) Over buried topography; 3) Over buried reef; 4) Over salt remnant; 5) Over a mud diapir.
Sf	Fault traps. Subtypes: 1) Against synsedimentary listric faults; 2) Against overthrust listric faults; 3) Against normal faults; 4) Against reverse faults.
Sa	Anticlinal traps. Subtypes: 1) Above listric faults in a synsedimentary (tensional) setting; 2) In an overthrust or other compressional setting.
M	Miscellaneous trap types. Subtypes: 1) In structures caused by astroblemes; 2) In sands trapped against mud diapirs; 3) In thickened pods related to salt solution.

Forty kilometres farther south, Rosevear Gas Field is trapped by a channel across the platform margin (Trap Type Rd1-Rd5). This gas play was extended still farther south to Caroline where very large additional reserves were discovered in 1986. At Rosevear the reservoir is dolomite, the dolomitization being apparently controlled by the same intra-shelf channel that controls the trap.

Woodbend Group

The Woodbend Group of Alberta Basin provides three major trap types (Fig. 6A.5). From west to east these are: oil and gas pools trapped in reefs of the Western Reef Domain (P.F. Moore, Subchapter 4D in this volume); oil pools trapped in reefs of the Eastern Platform; and bitumen trapped in the subcrop of the Nisku Formation and Grosmont complex (described under "Non-conventional Oil and Gas").

The oil fields of the Western Reef Domain include Sturgeon Lake with an IVIP greater than 50 x 10^6 m^3 and Simonette with an IVIP of about 20 x 10^6 m^3. Both trap oil in the updip margins of reef complexes in the Leduc Formation (Trap Type Rd1). An outboard satellite patch reef yields oil at Sturgeon Lake (Trap Type Rd3). Gas fields of the Leduc Formation include Westerose South, Strachan, and Ricinus West (Hriskevich et al., 1980), each with an IVIP over 40 x 10^9 m^3. These deep (4200 m) fields have a high (7 to 9%) hydrogen sulphide content; the sulphur constitutes a major economic resource.

The Strachan and Ricinus West reefs (Fig. 6A.11) are ellipsoidal, oriented northwest, and measure about 11 x 4 km, with relief of up to 275 m. Although heavily dolomitized, the reefs still exhibit an organic framework of massive stromatoporoids and colonial corals, an external talus facies and an interior facies of skeletal rudites, arenites, and muds that was apparently open to marine

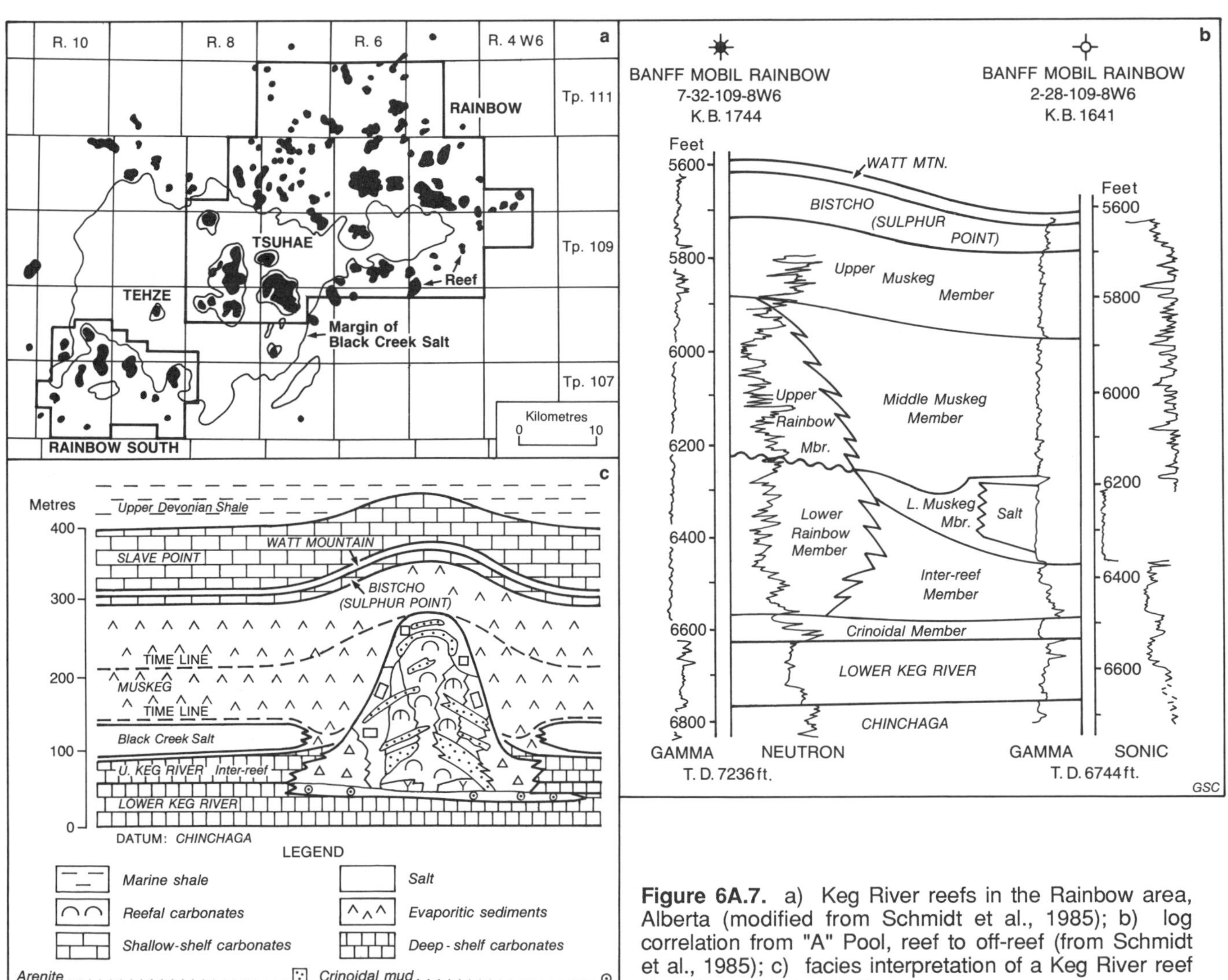

Figure 6A.7. a) Keg River reefs in the Rainbow area, Alberta (modified from Schmidt et al., 1985); b) log correlation from "A" Pool, reef to off-reef (from Schmidt et al., 1985); c) facies interpretation of a Keg River reef (from Barss et al., 1970).

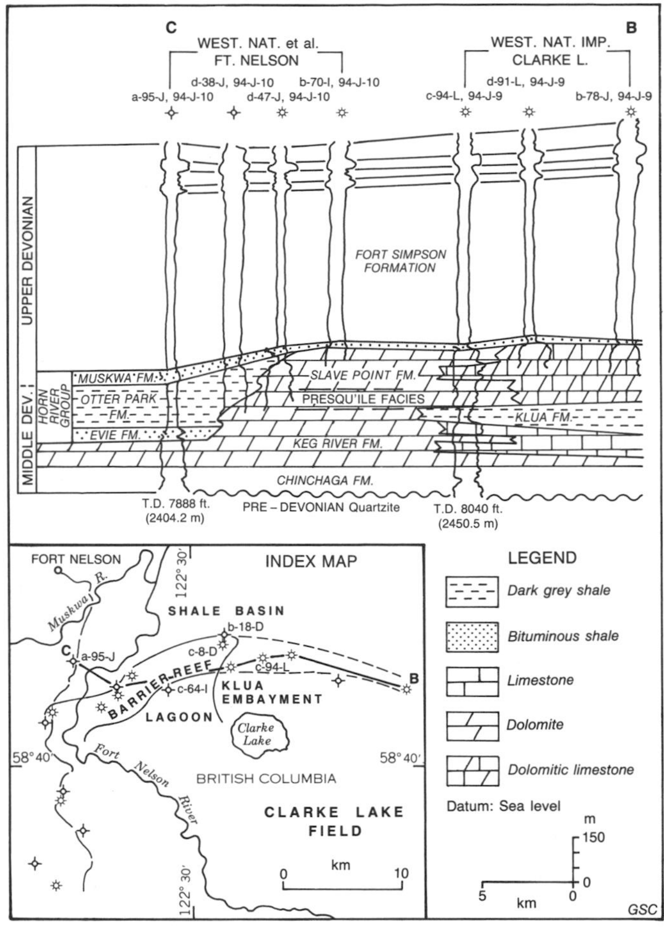

Figure 6A.8. Section across the Clarke Lake (Middle Devonian) Gas Field (from Gray and Kassube, 1963).

circulation at all times. Dolomitization has significantly increased the primary porosity, especially where the primary matrix was altered while fossil material remained calcite. The most favourable reservoir characteristics are found in the organic reef and proximal areas of the forereef and backreef facies. The poorest is found in the lower forereef areas. Porosities in the Strachan Field range from 2 to 10% and permeabilities from 1 to 40 mD.

The overwhelming majority of the oil fields of the Eastern Platform (Fig. 6A.5) are found in reefs of the Leduc Formation. Most of these fields are in reefs along two linear trends, the Rimbey-Meadowbrook trend and the Bashaw-Malmo-Duhamel trend, but the biggest single field, Redwater, is an isolated reef complex. Characteristically, these fields produce from the crest or updip margins of Leduc reefs (Trap Type Rd1). In the Bashaw Complex, channels across the reef flat also create traps (Type Rd5). Satellite pinnacle and tabular reefs (e.g. Golden Spike) occur outboard from the main reef complexes (Trap Type Rd3).

Net pay in the major pools ranges from 10 m to over 135 m; porosity from 5 to 15%. The oil is light and recovery factors high (about 0.50), hence the leading role played by Leduc reefs in the oil economy. The IVIP at Redwater is estimated at 207 x 10^6 m^3. Eleven of the Leduc pools on the Eastern Platform have IVIPs of over 10 x 10^6 m^3. All of the gas in this region is associated with oil, but the reserves are of the same order of magnitude as those of the gas fields in the Western Reef Domain.

Growth of the Leduc reef complexes took place during the rising sea-level stage of Woodbend deposition. Buildup was episodic, and sedimentary stages similar to those documented in the Swan Hills reefs have been observed in the relatively undolomitized reefs at Redwater and Golden Spike (Fig. 6A.12). All the other producing reefs are dolomitized. The Duvernay Formation, a bituminous and limy shale deposited under anoxic conditions between the

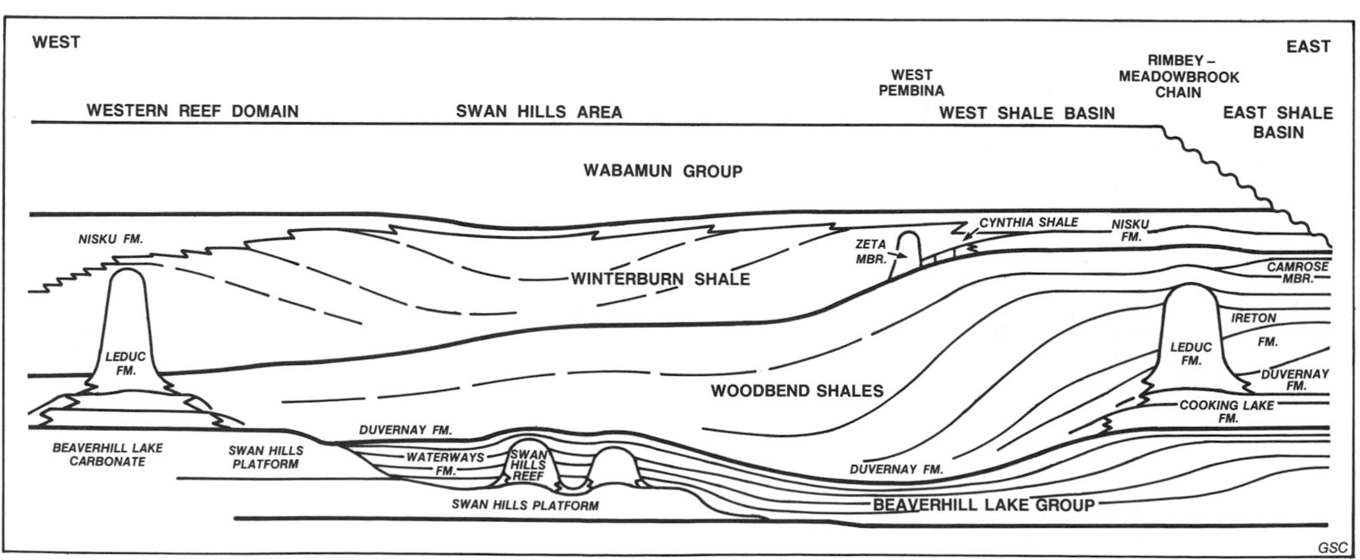

Figure 6A.9. Stratigraphic section across Alberta Basin showing reefs and off-reef cyclic sediments (adapted from Stoakes and Wendte, 1987).

reefs, is a rich source rock and is thought to have generated most of the oil and gas. The reefs were covered and sealed by shales of the Ireton Formation during the progradational phase of Woodbend deposition.

At Redwater (Klovan, 1964), the best porosity occurs in the organically bound reefs and in the coarse skeletal calcarenites and calcirudites of the talus slopes. Of Golden Spike, Walls (1983, p. 449) reported that:

(1) Porosity in the margin, particularly in reef flank facies is, with few exceptions, consistently low. Tabular stromatoporoids and detrital coral facies illustrate some of the lowest porosity in the reef due to pore occlusion by cementation and deposition of marine internal sediments. Argillaceous skeletal rubble facies generally contain low primary porosity due to a matrix (mud) supported fabric. (2) Reef facies illustrate both solution enlarged and cement reduced primary pores, as well as unaltered primary porosity. (3) Porosity in the reef interior is variable. Subtidal skeletal sand flat deposits contain some of the highest porosity in the reef complex, due to preservation of original interparticle and fenestral porosity, solution of early interparticle and fenestral porosity, solution of early cements, and solution enlargement of primary pores.

Winterburn Group

Within the Winterburn Group, only the Nisku Formation is of economic importance. It forms a broad, shallow, marine carbonate platform complex which extends eastward across Alberta and Saskatchewan. In central Alberta, four types of trap contain major oil reserves and limited gas reserves: (1) structural closure due to drape over Leduc reefs (Type Sd3); (2) up-dip regional facies changes from porous, biostromal carbonate to non-porous evaporite (Type Ls4); (3) local, isolated, porous patch reefs within the non-porous carbonate and evaporitic facies (Type Rp1); (4) reefs on the margins and basinward slopes of a prograding carbonate platform (Types Rp2 and Rp3).

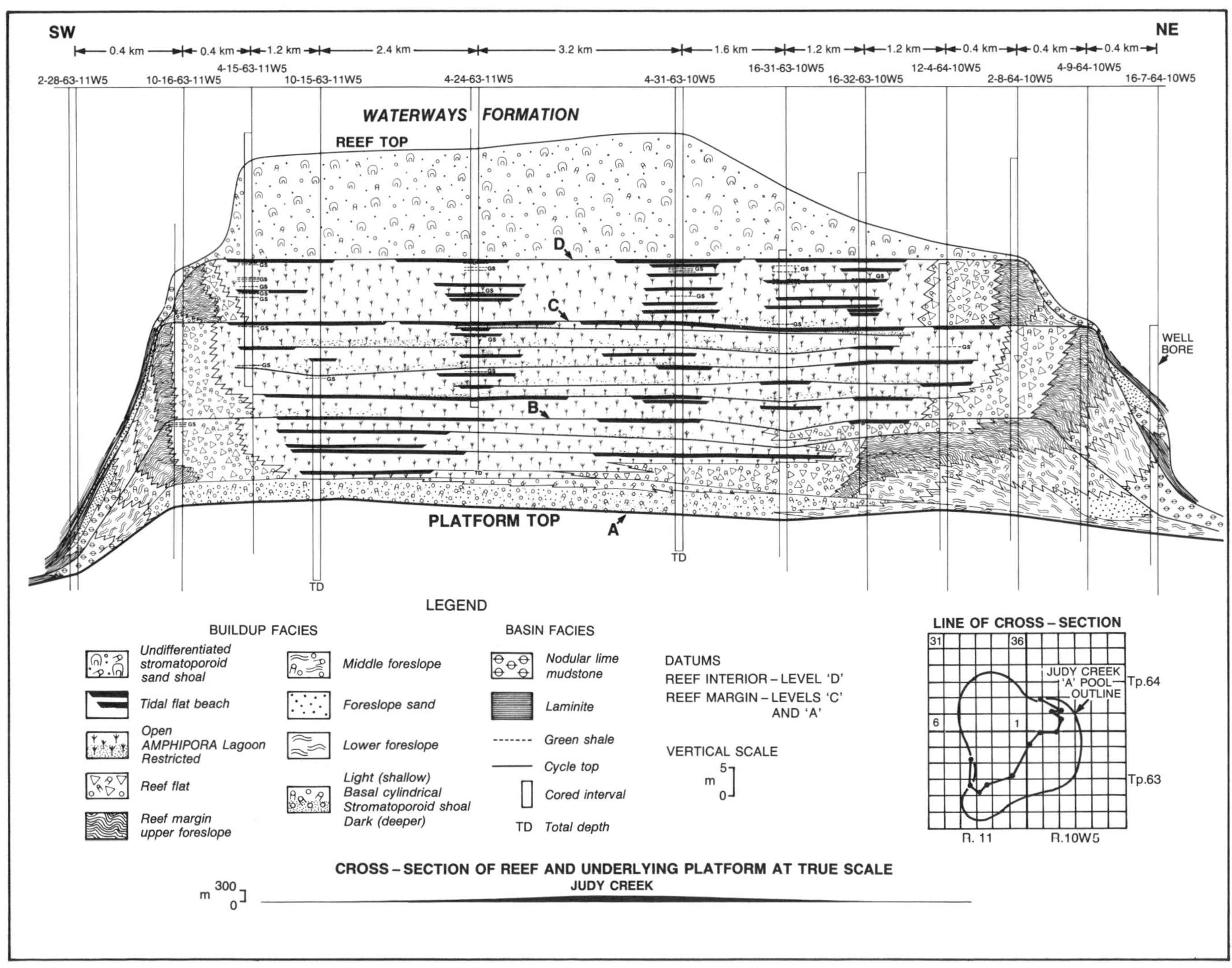

Figure 6A.10. Section of Swan Hills reef at Judy Creek Field. Tops of upward-shallowing megacycles are labelled A, B, C, D. Each megacycle ended abruptly when the depth of water suddenly increased (modified from Wendte and Stoakes, 1982).

Nisku reservoirs overlying Leduc fields of the Rimbey-Meadowbrook reef chain north of Bonnie Glen consist of dolomite with good vuggy and fracture porosity. Major oil pools include Leduc, Fenn-Big Valley, and Clive. East of the Rimbey-Meadowbrook trend, the upper part of the Nisku Formation grades to anhydrite but the lower Nisku and the underlying coral-stromatoporoid-rich dolomite of the Camrose Formation of the Woodbend Group form reservoirs above the Bashaw (Leduc Formation) Complex. Porosity is partly filled with irregularly distributed anhydrite. Pay zones vary to a maximum of 40 m; porosity is generally from 3 to 5% and the density of the oil ranges from 825 to 875 kg/m^3.

The western edge of the carbonate platform (Fig. 6A.13) is marked by a transition from shallow carbonates to deeper water, basinal sediments like those of the underlying Ireton Formation. At the margin west of Swan Hills, the upper part of the Nisku Formation is reefal, and oil is trapped in it at the Meekwap Field (Fig. 6A.14) by facies change up-dip into impermeable limestone and shale (Trap Type Rp2). Dolomitization has enhanced primary porosity and has created vuggy and channel porosity (Cheshire and Keith, 1977). The oil zone, at a depth of 2300 m, averages about 10 m in thickness; porosity is 8%, permeability 50 mD, and the oil has a density of 865 kg/m^3. The IVIP of oil is estimated at about 9 x 10^6 m^3. Small isolated Nisku reefs occurring basinward of the outer shelf edge in west-central Alberta reach a thickness of approximately 120 m (Trap Type Rp3). Unlike the shelf rocks, which are extensively dolomitized, some of the isolated reefs are limestone. They contain major oil reserves and significant gas reserves. Since 1977, some 50 separate pools have been discovered in the West Pembina area (Chevron Standard Ltd., 1979).

Significant gas accumulations are found in the Nisku Formation of the Western Reef Domain where the basin is deeper and gas prone (e.g., the Obed D-2A gas pool with an IVIP of 7 x 10^9 m^3).

Wabamun Group

Oil and gas are produced from lithological, subcrop, structural, and diagenetic traps within the Wabamun Group, a carbonate shelf complex. Sour gas is produced from the Crossfield dolomite in a belt of fields lying between Okotoks and Olds in Alberta (Fig. 6A.15). The gas occurs in the intercrystalline and pin-point porosity of fine- to medium-crystalline dolomites, which grade eastward up-dip into impermeable dolomite and anhydrite. The porosity has been developed in a stromatoporoid biostrome, probably through the agency of acid gases (Eliuk, 1984). This is considered a trap of Type Ld1-Rp2. Crossfield alone has an IVIP of 47 x 10^9 m^3 at a depth of 2590 m. These important gas fields contain up to 40% H_2S by volume.

A number of small fields and discoveries produce gas from the up-dip erosional edge of the Wabamun Group at the sub-Cretaceous unconformity (Trap Type Ub1). The largest gas pool is Flat with an IVIP of 3.5 x 10^9 m^3. Small oil fields of this trap type occur at the north end of the Rimbey-Meadowbrook reef chain, where drape over the Leduc reefs may enhance the traps (e.g., St. Albert-Big Lake with an IVIP of less than 3 x 10^6 m^3). The hydrocarbons occur in sugary, crystalline dolomites with good intercrystalline porosity, and in vuggy brecciated dolomite and limestone. Porosity has been increased by leaching and weathering of the dolomite at the erosion surface.

Gas with no associated oil or water occurs in porous vuggy dolomite in the lower part of the Wabamun Group at Simonette, Pine Creek, and South Sturgeon Lake. The gas is trapped in the dolomite against overlying fine-grained, tightly cemented Wabamun limestone (Trap Type Ld1).

Combination lithological-structural traps occur in the southern Peace River Arch area, where faults have displaced the Wabamun. The Blood Field in southern Alberta is in a faulted and folded structure in which numerous fractures result in high productivity.

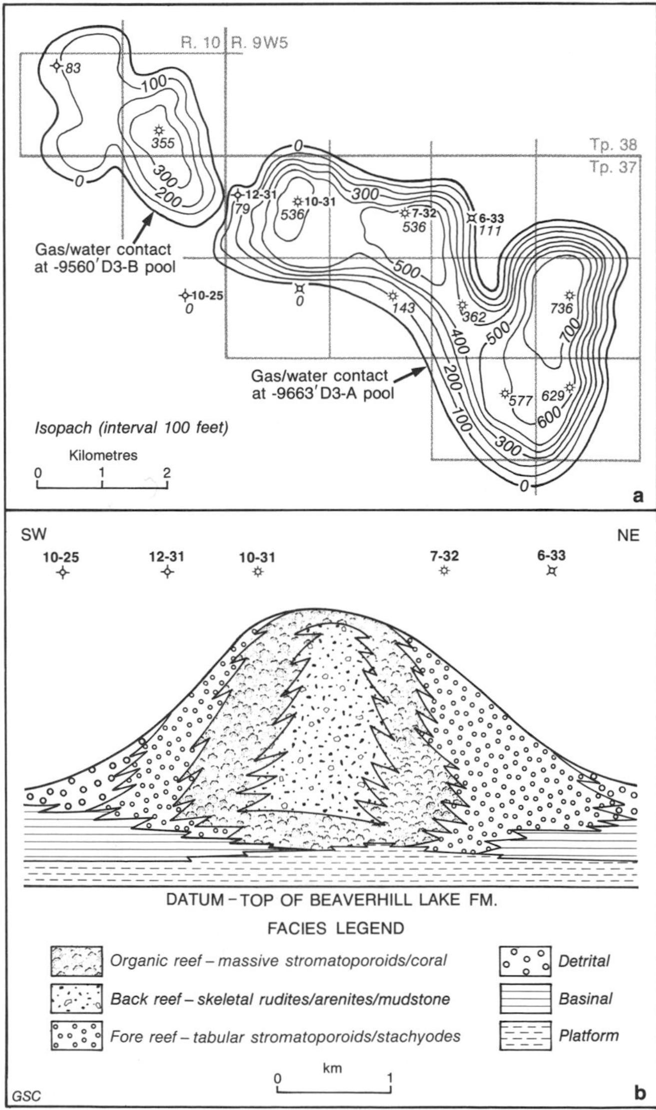

Figure 6A.11. a) isopachs of net pay in Strachan D3-A and D3-B pools in the Leduc Formation of the Western Reef Domain; b) facies interpretation of the reef (from Hriskevich et al., 1980).

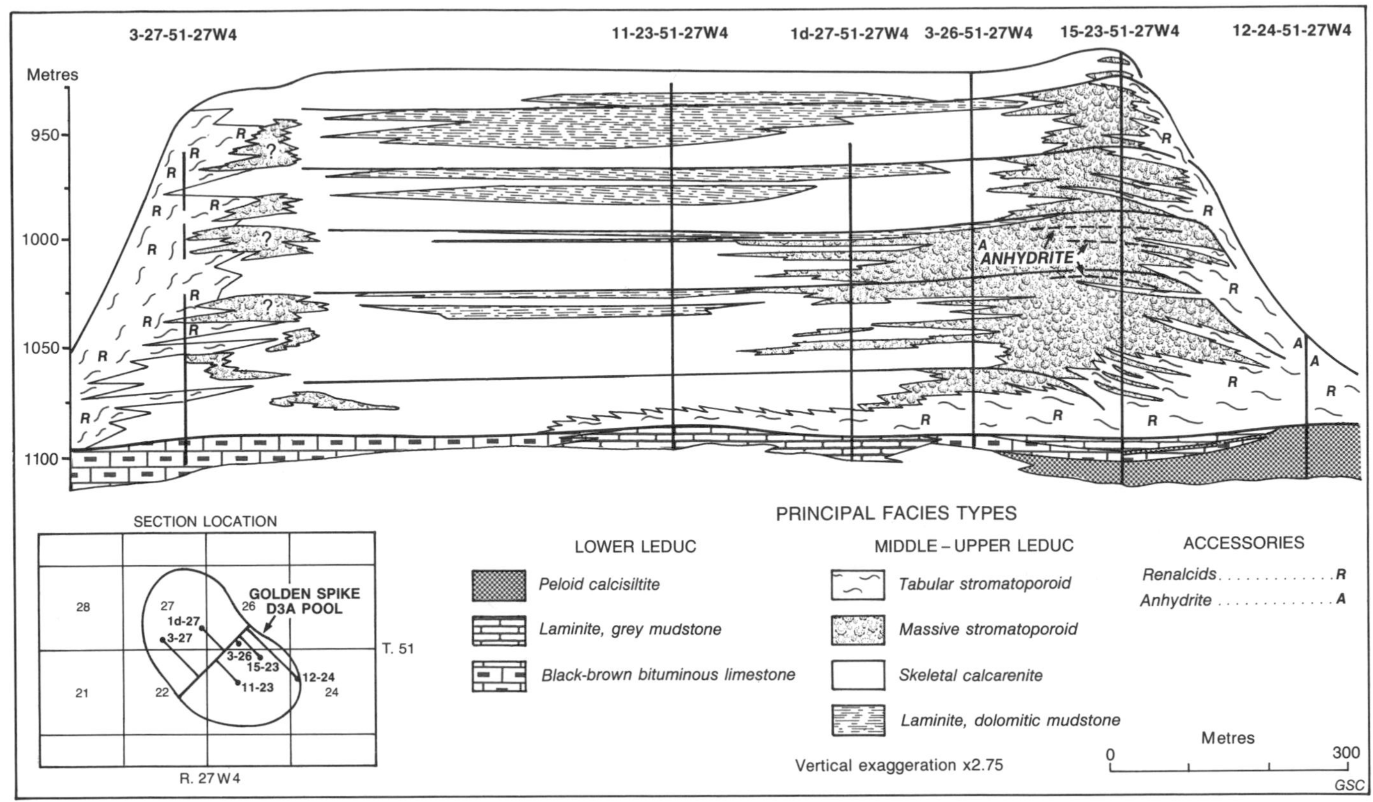

Figure 6A.12. Section of Leduc reef at Golden Spike (after McGillivray, 1977).

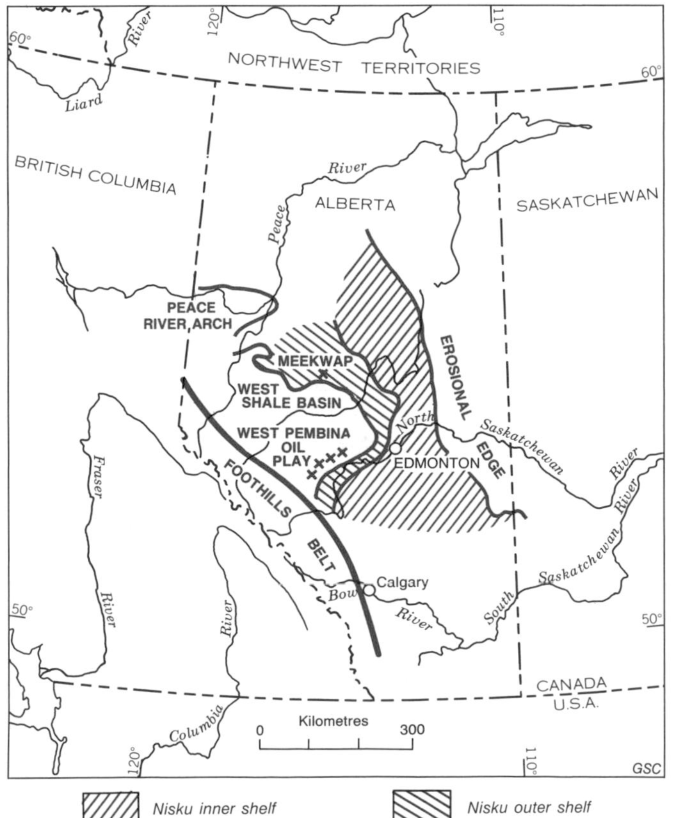

Figure 6A.13. Paleogeography during Nisku and equivalent deposition in Alberta Basin (adapted from Chevron Standard Ltd., 1979).

Carboniferous

The petroleum and natural gas resources of the Carboniferous System of the Alberta Basin are third in importance after those of the Devonian and Cretaceous systems. A stratigraphic diagram (Fig. 6A.16) of the Carboniferous (and Permian) succession identifies the chief oil- and gas-producing zones. The Carboniferous rocks account for 9% of the IVIP of light and medium oil in the Alberta Basin and 17% of the gas.

During the Carboniferous Period, the Peace River Embayment that succeeded Peace River Arch had a different geological history to that of the Alberta Shelf and so is treated separately. The Desan area of northeastern British Columbia and the Sweetgrass Arch area of southern Alberta are also given separate mention.

On the Alberta Shelf (Fig. 6A.1) nearly all of the oil and gas is in subcrop traps of types Ub1 and Ub2, but some is in lithological traps of type Ld1, where dolomite changes updip to impermeable limestone. Production is obtained from porous zones in the following formations: Banff (middle carbonate member), Pekisko, Shunda, and Turner Valley (Elkton Member).

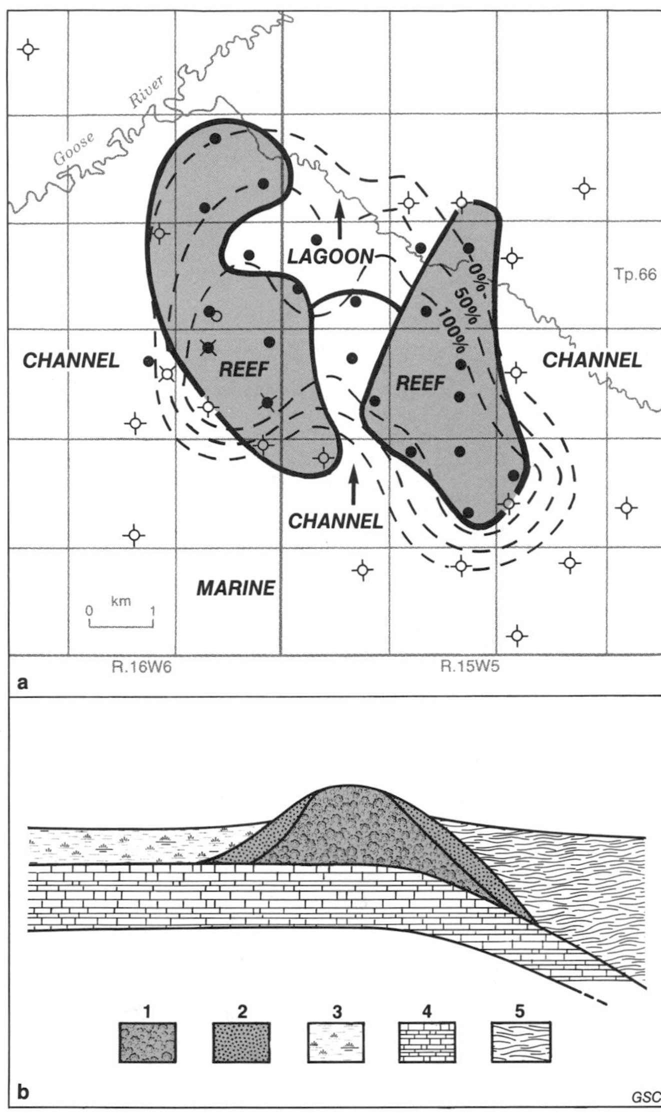

Figure 6A.14. a facies distribution and dolomitization (%) in the Nisku Formation of Meekwap Oil Field; b facies interpretation of Meekwap Field. Legend: 1 Reef: contains thin laminar stromatoporoids and *Renalcis*. 2 Reef slope: contains laminar stromatoporoids in lime-wacke to grainstone groundmass. 3 Lagoon: contains *Amphipora* in grainstone and packstone groundmass. 4 Platform: contains *Amphipora* and corals in wackestone. 5 Open marine: lime- and argillaceous lime-wackestone and mudstone with rare brachiopod and gastropod debris (adapted from Cheshire and Keith, 1977).

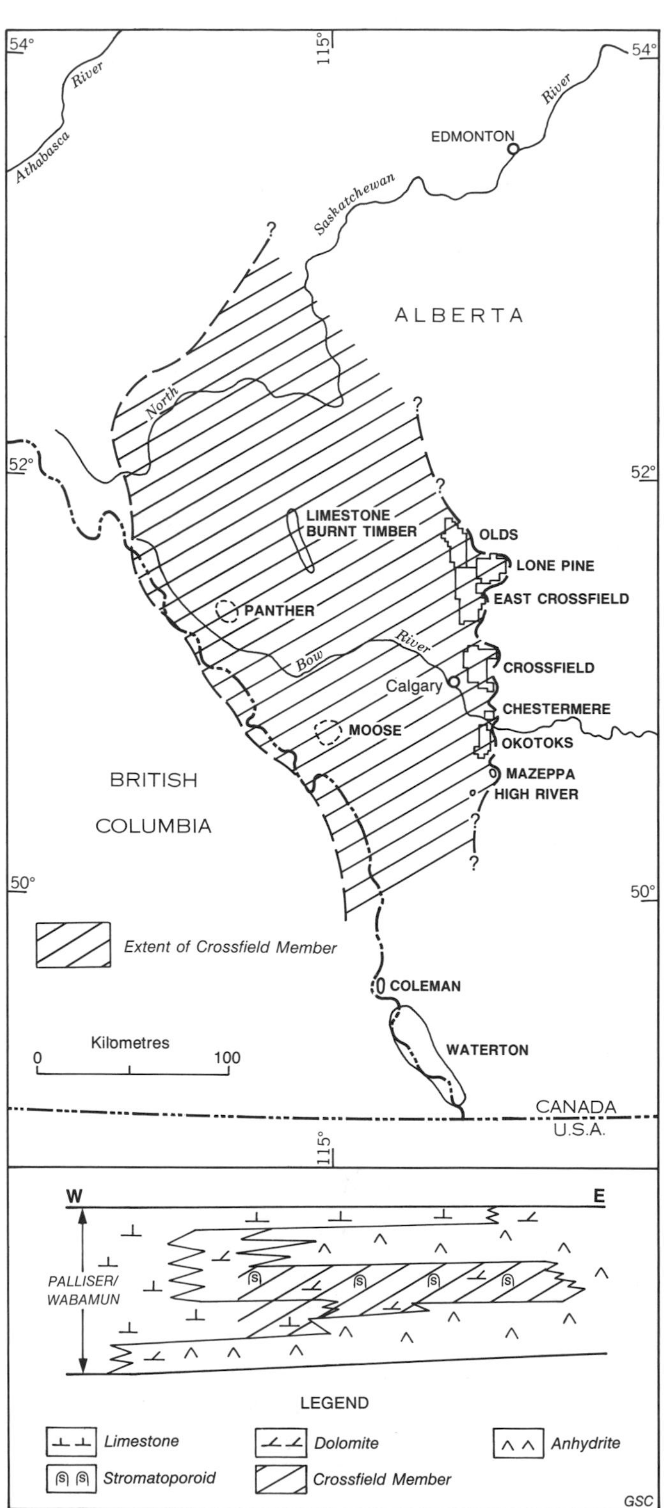

Figure 6A.15. Wabamun gas fields of the Olds-Okotoks trend (after L.S. Eliuk, 1984).

The Banff Formation contains numerous scattered oil and gas pools along the erosional edge of the middle carbonate member in the Westerose South and Cherhill-Highvale areas of west-central Alberta. None of the Banff pools is of major importance.

The Pekisko Formation consists mainly of skeletal limestone, locally dolomitized along the erosional edge. Most of the oil occurs in the Twining-Sylvan Lake-Medicine River area and most of the gas in the Minnehik-Buck Lake (Fig. 6A.17), Three Hills, and Whitecourt areas. The Twining pool with an IVIP of nearly 90 x 10^6 m^3 is the largest Carboniferous oil accumulation in the Alberta Basin; however, the Initial Established Reserves are only 7 x 10^6 m^3. This field (Williams and McNeil, 1976) has an average pay thickness of 14.5 m at a depth of 1645 m and a porosity of 6%; the oil density is about 875 kg/m^3.

EPOCH/AGE		PEACE RIVER EMBAYMENT	ALBERTA SHELF
LATE PERMIAN	CHANGHSINGIAN		
	DZHULFIAN		
	CAPITANIAN		
	WORDIAN		
EARLY PERMIAN	ROADIAN		
	ARTINSKIAN	BELLOY ✷	BELLOY ●
	SAKMARIAN		
	ASSELIAN		
LATE CARB.	GZHELIAN		
	KASIMOVIAN		
	MOSCOVIAN		
	BASHKIRIAN		
EARLY CARBONIFEROUS	SERPUKHOVIAN	STODDART GROUP: TAYLOR FLAT; KISKATINAW ✷; GOLATA	
	VISEAN	RUNDLE GROUP: DEBOLT ✷	RUNDLE GROUP: TURNER VALLEY (Upper, Middle, Elkton) ✷
	TOURNAISIAN	SHUNDA; "UNIT F"; PEKISKO ✷; BANFF; Siltstone; EXSHAW	SHUNDA; LIVINGSTONE ✷; PEKISKO ✷; BANFF ●; Siltstone; EXSHAW
LATE DEV.	FAMENNIAN	Black shale; WABAMUN	Black shale; WABAMUN

GSC

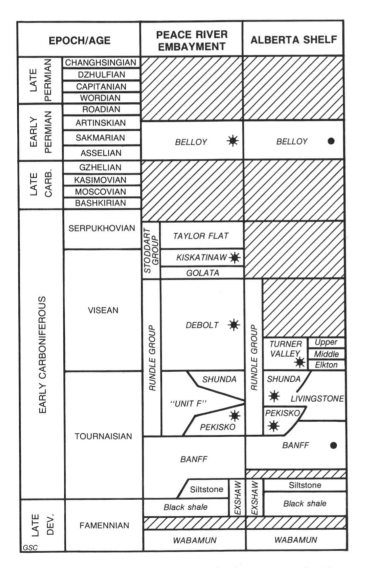

Figure 6A.16. Principal producing zones in the Carboniferous and Permian systems of Alberta Basin. (symbols described in Figure 6A.4) (adapted from Podruski et al., 1988).

Williams and McNeil (1976) also reported that the Pekisko Formation is almost entirely limestone in the area and varies from lime-mudstone through wackestone to packstone and, less commonly, grainstone. Porosity is generally intergranular but also includes some vugs and fractures. Best porosity is generally found in the crinoidal and oolitic zones.

The Elkton Member of the Turner Valley Formation is the most important Carboniferous producing zone for both oil and gas (Fig. 6A.18). Traps are formed partly where porous dolomite changes up-dip to impermeable limestone and partly where the Elkton is truncated at the sub-Mesozoic unconformity and sealed by impermeable overlying beds (Trap Type Ub-Ld1). Reserves are in leached and dolomitized encrinite (crinoidal limestone). The largest gas accumulation is at Edson with an IVIP of 60 x 10^9 m^3, but currently the largest Initial Established Reserves are at the Brazeau River Field. Edson (Young, 1967) produces sour gas from subcropping crystalline dolomite at a drilling depth of about 2850 m; average pay is 7.1 m and porosity 11.5%. The largest oil pools are Harmattan East, Harmattan-Elkton, and Sundre, production being obtained at depths of 3000 to 3500 m. Pay thickness in these fields ranges from 10 to 20 m and porosity from 5 to 10%. The oil density is in the range 850 to 875 kg/m^3.

In the Peace River region, oil and gas are trapped, mostly in small pools, in the Pekisko, Debolt, and Kiskatinaw formations. In the Seal area, heavy oil is trapped in Waulsortian reefs of the Pekisko Formation (Trap type Rp3). Porosity in these reefs is mainly of primary intergranular, shelter, and intraskeletal types (Davies et al., 1987). The Blueberry Oil Field and a number of small gas fields produce from the Debolt Formation. The reservoirs are generally cherty bioclastic limestone partly replaced by dolomite, with intercrystalline and solution cavity porosity. The fields occur along the up-dip eroded margin of the Debolt Formation; accumulation is partly the result of erosion and sealing of the beds at the pre-Mesozoic erosion surface and partly due to porosity pinchout from porous dolomite to limestone (Trap Type Ub - Ld1). The Blueberry Oil Field lies along a northwest-trending anticline, offset by younger faults, and is thus a combination stratigraphic-structural trap (Type Ub-Sa2); it has an IVIP of over 10 x 10^6 m^3. Oil at Eaglesham, Normandville, and George is structurally trapped by faults that developed during the Early Carboniferous in the region of the Peace River Embayment (Trap type Sf3).

Kiskatinaw traps are of the lithological pinchout type. In the Josephine Oil Pool the reservoir is a deltaic splay sandstone (Trap Type Ls2). At the Boundary Lake South Gas Pool the sandstone is marine (Kendall, 1969) (Trap Type Ls3-Sa2).

Early in 1983 an oil discovery was made at a depth of only 600 m in Lower Carboniferous limestones at Desan, in the Helmet area of northeastern British Columbia. The oil at Desan appears to be trapped in porous zones isolated within the strata, unrelated to an unconformity. Multiple reservoirs occur in zones 1 to 36 m thick in crinoidal limestones of the Debolt, Shunda, and Pekisko formations. Although the discovery well was encouraging and the IVIP is large, the deliverability of individual wells is low.

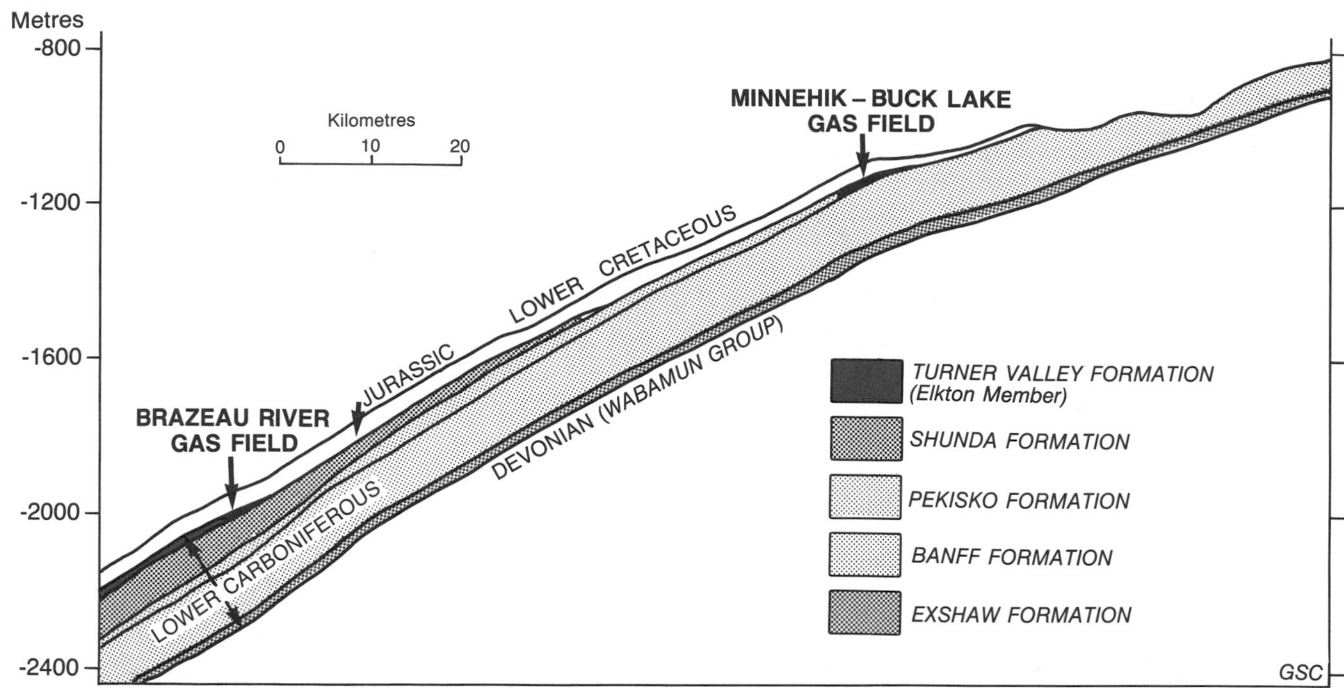

Figure 6A.17. Diagrammatic cross-section of Lower Carboniferous formations from Brazeau River Field to Minnehik-Buck Lake Field, Alberta (from H. Martin, in Little et al., 1970).

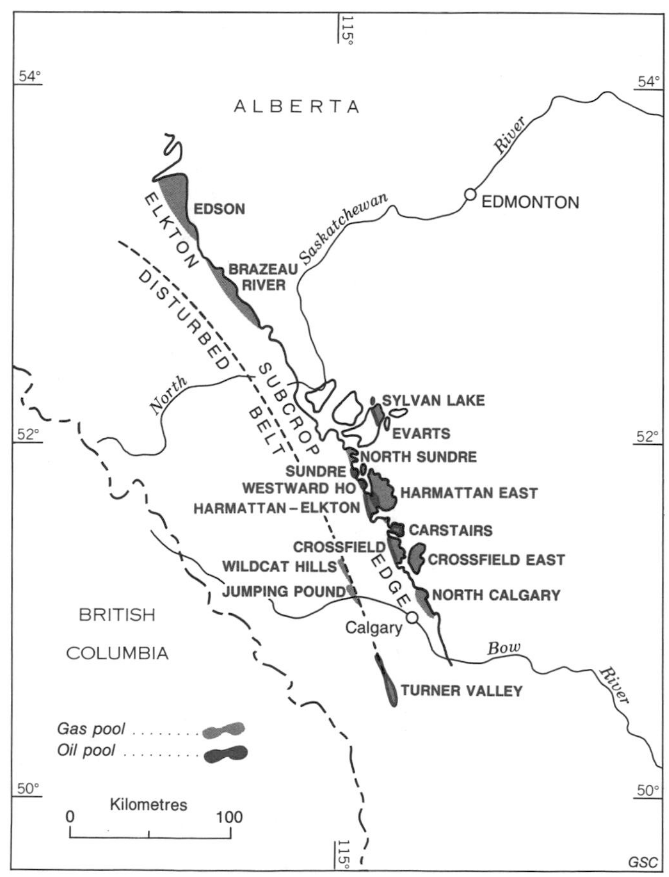

Figure 6A.18. Oil and gas production from Lower Carboniferous Elkton Member of the Turner Valley Formation, western Alberta (from Young, 1967).

AGE			PEACE RIVER SUBSURFACE		OIL AND GAS
JURASSIC			*NIKANASSIN*		
			FERNIE		
TRIASSIC	LATE	NORIAN		*Worsley dolomite*	
			SCHOOLER CREEK GROUP	*PARDONET*	
		CARNIAN		*BALDONNEL*	☼
				CHARLIE LAKE	
				Boundary	●
				Inga	●
	MIDDLE	LADINIAN		*HALFWAY*	✸
		ANISIAN	DAIBER GROUP	*DOIG*	●
				?	
	EARLY	SPATHIAN		*MONTNEY*	●
		SMITHIAN			
		DIENERIAN			
		GRIESBACHIAN			

GSC

Figure 6A.19. Principal producing zones of the Triassic System in Alberta Basin (symbols described in Fig 6A.4) (modified from Podruski et al., 1988).

Medium to heavy oil is trapped along Sweetgrass Arch in anticlinal and fault structures in the Banff and Livingstone formations at fields such as Claresholm, Del Bonita, and Red Coulee.

Permian

Permian rocks, because of their limited thickness and distribution, contain only minor hydrocarbon accumulations. Oil is trapped in the Belloy marine sandstone in fault and fold structures in the area of the Peace River Embayment in northeastern British Columbia and northwestern Alberta. The most important pools are in the Eagle and Stoddart areas, where the reservoirs are porous sandstone interbedded with dense limestone and cherty limestone. The largest discovered pool is Stoddart West, with an IVIP of over 20×10^6 m^3.

In west-central Alberta, oil is produced from outliers of very porous Belloy sandstone in the Virginia Hills area. The trap is at the sub-Mesozoic unconformity and is seat-sealed by Shunda Formation (Trap Type Ub2).

Triassic

Hydrocarbon production from Triassic rocks in Alberta Basin is obtained from northeastern British Columbia and from the adjacent Peace River region of Alberta. A stratigraphic diagram (Fig. 6A.19) of the Triassic formations identifies the chief oil- and gas-producing zones. The Triassic rocks contain approximately 5% of the IVIP of light and medium oil in the Alberta Basin and 7% of the gas. Three quarters of the IVIP of oil and a third of the gas in British Columbia is Triassic. More than 90% of Triassic oil production is from the Halfway Formation and the Inga and Boundary Lake members of the Charlie Lake Formation. The largest reserves are in the Inga and Boundary Lake fields. Most of the remaining production is from small pools with relatively thin pay zones at various Charlie Lake horizons. All Triassic oils are relatively light; Halfway and Inga oil lies in the range 800 to 830 kg/m^3 and Boundary Lake averages 855 kg/m^3. The Baldonnel Formation contains the major portion of established reserves of Triassic gas,

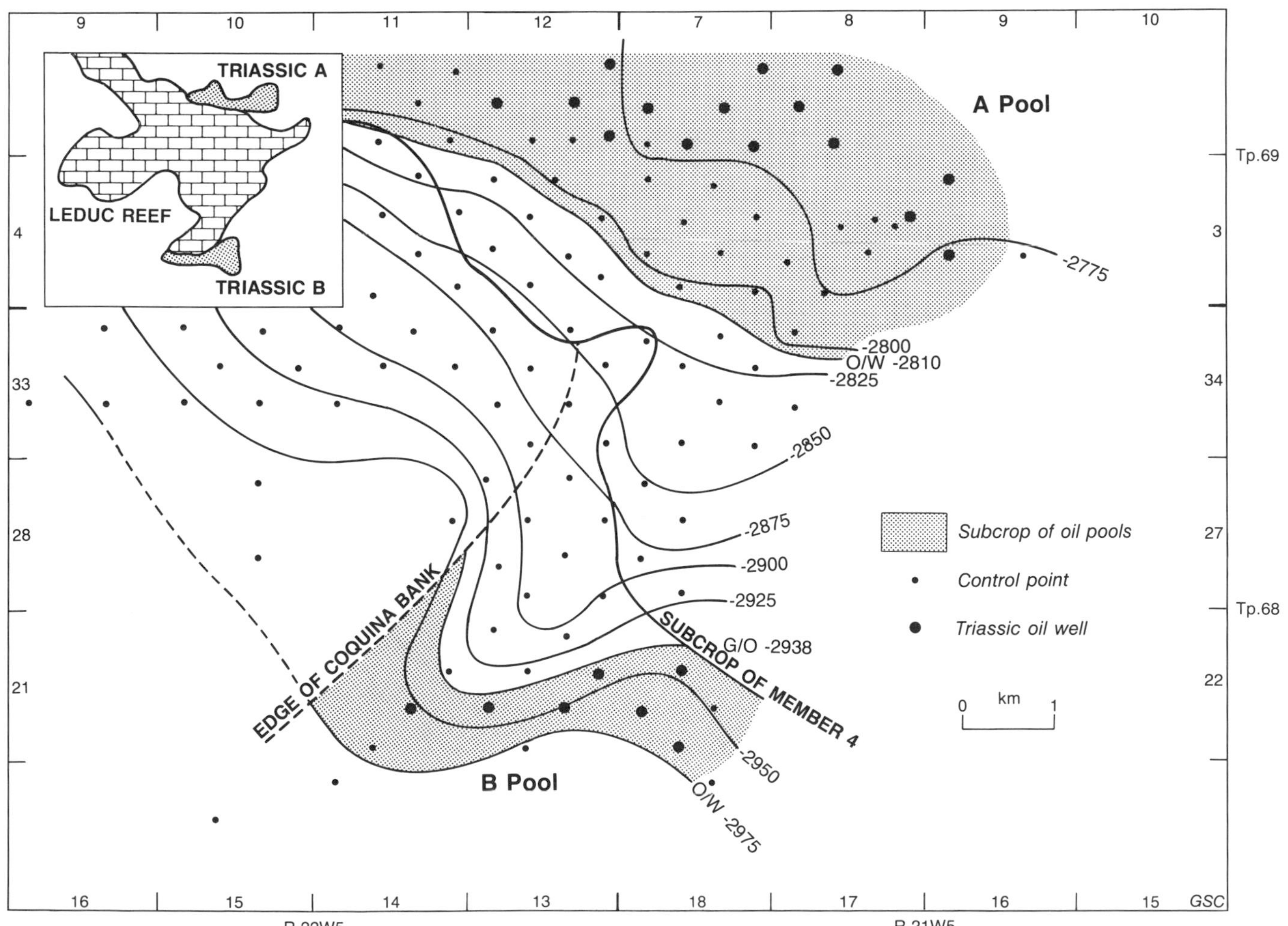

Figure 6A.20. Trapping conditions in Sturgeon Lake South Triassic A and B pools. Inset shows the relationship of the Triassic oil pools to the underlying Upper Devonian (Leduc Formation) reef. Structure contours (in feet, subsea) are drawn on the sub-Jurassic unconformity (adapted from Miall, 1976).

which is relatively sweet and dry; the Halfway Formation contains most of the rest. The principal gas fields are centred around Fort St. John.

The Triassic accumulations of Alberta Basin are found in a wide variety of traps, which are commonly complex. Northeastern British Columbia is a region where a belt of Laramide folds lies in front of the overthrust belt; thus many traps in the western part of the region are controlled, or at least modified, by anticlinal structures.

The main producing zones are described in ascending order. The principal reservoir rock type in the Montney Formation is a sandy and silty coquinoid dolomite that occurs in banks, up to 36 m thick, whose location was controlled by paleostructural highs (Wallace-Dudley et al., 1982). The Montney Formation contains one major oil accumulation at the Kaybob South Triassic "A" Pool with an estimated IVIP of 35×10^6 m^3. Trapping at the Sturgeon Lake South "B" Pool (Fig. 6A.20; Miall, 1976) occurs by a combination of subcrop beneath Jurassic Nordegg Formation and northwesterly facies change from coquina to siltstone and shale (Combination Trap Type Ls3-Ub3). A buried Woodbend reef apparently exerted some paleostructural control on facies. The dolomite coquina has an average porosity of 13%, mainly from vugs and fossil moulds, but the IVIP is only just over 1×10^6 m^3 at a depth of 1500 m.

The Doig Formation produces from a number of very small pools, mostly in anticlinal structures. The exception is the Doig "A" Pool at Spirit River, which is in a paleogeomorphologic trap (Type Ub1) beneath an intra-Triassic unconformity (Aukes and Webb, 1986) (Fig. 6A.21). The reservoir consists of interbedded sandstones and coquinas deposited in tidal inlets. Both primary and secondary solution porosity are present and are described as good to excellent. This oil pool has an IVIP of over 7×10^6 m^3.

The Halfway Formation has the most varied trapping conditions. For example, in the Jedney-Bubbles area, the Halfway Formation yields gas from a series of north-trending anticlines parallel to the eastern front of the Foothills Belt, whereas within Boundary Lake field, the Halfway traps are partly controlled by high-angle basement faults (Trap Type Sf3 or Sf4). The Peejay-Milligan area contains the largest oil pools (e.g., Peejay Unit 2 with an IVIP of nearly 9×10^6 m^3). Most of the Halfway oil in this area is stratigraphically trapped in discontinuous, northwest-trending offshore bars (Mothersill, 1968) of sand and coquina deposited on the undulating erosional surface of the Doig Formation (Trap Type Ua1). Dolomitized pelecypod coquina provides the best porosity and permeability, which deteriorate westward as the clay content and degree of cementation increase. The oil density ranges from 900 kg/m^3 near the subcrop edge to 780 kg/m^3 adjacent to the Foothills Belt.

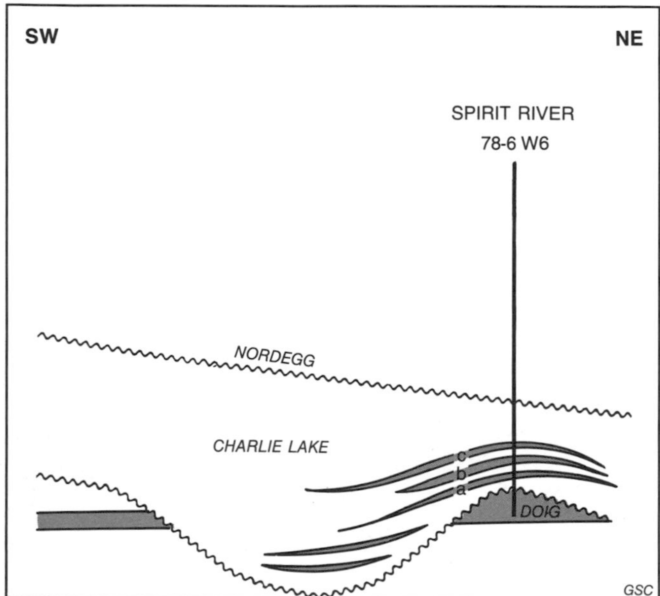

Figure 6A.21. Trapping conditions in the Doig Formation of Spirit River pool, northwestern Alberta. Lenses a, b and c are productive, stromatolitic dolomites in the Charlie Lake Formation (after Aukes and Webb, 1986).

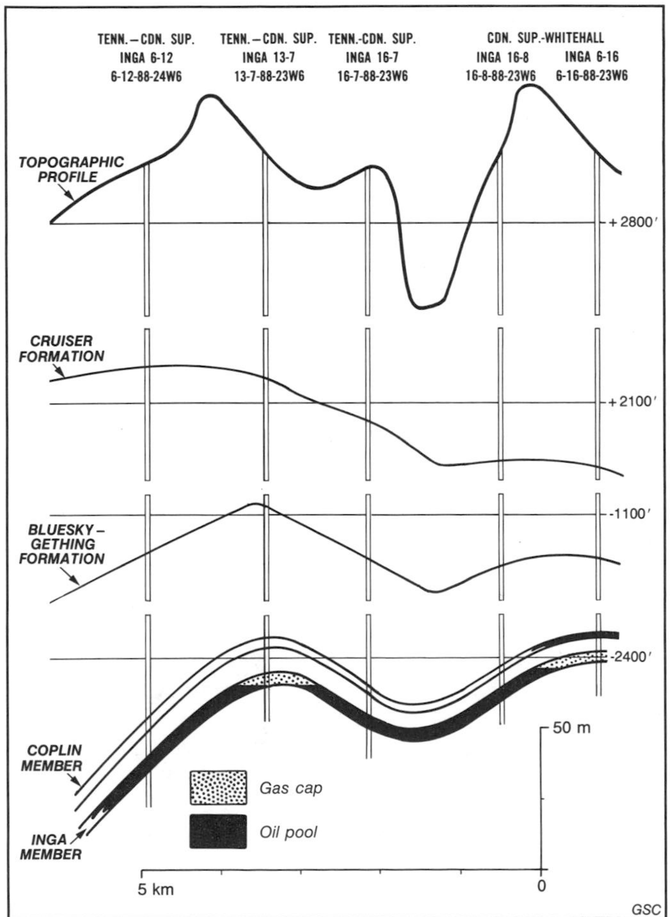

Figure 6A.22. Structural cross-section across Inga Field showing anticlinal trap (after Fitzgerald and Peterson, 1967).

Other thin Halfway sandstones occur southwest of the discontinuous trend in continuous shallow marine deposits with shorelines oriented northwesterly. The largest pool in this play is the Halfway "B" Pool in the Wembley Field of north-central Alberta (Halton, 1981). At Wembley, oil and gas is trapped where Halfway sandstone and coquina thin and pinch out northeastward. Most of the porosity is secondary from the leaching of cement, fossils, and some sand grains. The pinchout is interpreted by Barclay and Leckie (1986) as resulting from lithological change in a tidal inlet and barrier island shoreface environment (Trap Type Ls2-3). On the other hand, Cant (1986) attributed the pinchout to synsedimentary downwarping accompanied by onlapping deposition of the Halfway. These events may be related to tensional deformation associated with the ancient Peace River Arch/Embayment and specifically to a major horst structure directly beneath Wembley Field. The pool has an IVIP of almost 6×10^6 m^3.

Charlie Lake Formation contains numerous producing zones of which the Boundary Lake carbonate is the most important, although the Inga Member and other sandstone zones are also significant. In the Inga Field, a trap associated with Laramide fold structures in Inga Member quartzose sandstone (Fig. 6A.22) has an IVIP of approximately 15×10^6 m^3. "The western edge and part of the eastern boundary of the pool [are] determined by the loss of porosity resulting from increasing dolomite cementation and secondary anhydrite" (Fitzgerald and Peterson, 1967, p. 65); this is therefore a combination trap of type Sa2-Ld1.

EPOCH	ALBERTA BASIN: MEDICINE RIVER	ALBERTA BASIN: SWEETGRASS ARCH	WILLISTON BASIN
EARLY CRET.			?
LATE JURASSIC	J3 ✷		Upper Success ●
	?		?
	?		Lower Success ●
	J2 ✷	SWIFT ●	VANGUARD GROUP: MASEFIELD
			VANGUARD GROUP: ROSERAY ●
MIDDLE JURASSIC		RIERDON	VANGUARD GROUP: RUSH LAKE
		SAWTOOTH ✷	SHAUNAVON ●
			GRAVELBOURG
	ROCK CREEK ✷		WATROUS/ AMARANTH ●
	Poker Chip Shale		
EARLY JURASSIC	Red Deer		?
	Shale		
	NORDEGG AND J1 ✷		

GSC

Figure 6A.23. Principal producing zones of the Jurassic System in the Alberta and Williston basins (symbols described in Figure 6A.4) (based on information in Hopkins, 1981; Podruski et al., 1988).

The largest Charlie Lake oil accumulation is at the Boundary Lake Field in the Boundary Lake Member, a cyclic unit of stromatolitic, pelmicritic-micritic limestone, and dolomite (Roy, 1972). The oil reservoir is a buried erosional remnant beneath an intraformational unconformity; it is overlain by redbeds (Trap Type Ub1). Average net pay in this pool is only 2.8 m, but porosity averages over 20% and the estimated IVIP is about 70×10^6 m^3. The member is also involved in an anticline developed along a high-angle fault that strikes northeast. Gas is trapped at the crest of the anticline and downdip in closure provided by minor faults. Gas is also trapped at the erosional edge of the member.

The Baldonnel Formation is a major gas producer. Accumulations occur in porous skeletal dolostones within structural and stratigraphic traps (Bever and McIlreath, 1984). Gas production is obtained from the Fort St. John, Nig Creek, Blueberry, Beg, Jedney, Bubbles, and Laprise Creek fields. Laprise Creek is the largest pool, with an IVIP of 26×10^9 m^3. Trapping in the Laprise area is a combination of: anticlinal structure, which determines the gas-water contact; sub-Cretaceous erosion, which cut down through the Jurassic cover and removed part of the Baldonnel reservoir; and facies change to impermeable carbonate (Armitage, 1962). The trap is thus a combination type Sa2-Ub1-Ld1.

Jurassic

Jurassic rocks are host only minor oil and gas reserves in Alberta Basin, accounting for less than 1% of the IVIP of oil and 1% of the gas. A stratigraphic diagram of the Jurassic formations locates the chief oil- and gas-producing zones (Fig. 6A.23).

Three main producing zones are the Nordegg sandstone facies, the Rock Creek Member of the Fernie Formation, and the "J2-J3" sandstones, which may be equivalent to the Roseray and lower Success sandstones of Williston Basin (Hopkins, 1981). On Sweetgrass Arch, minor oil and gas production is obtained from the Sawtooth and Swift formations.

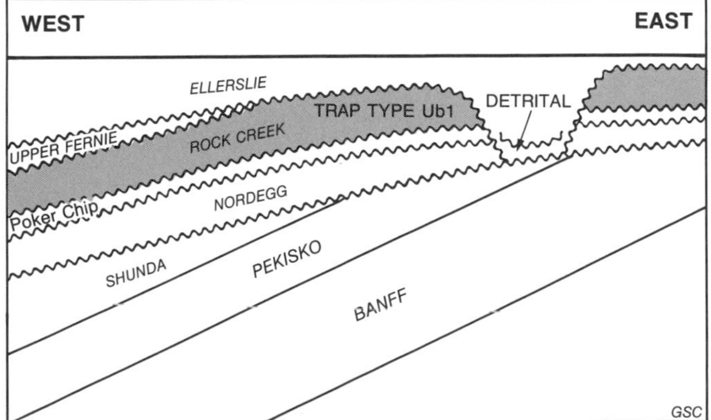

Figure 6A.24. A trap in sandstone of the Rock Creek Member, controlled by paleotopography at the sub-Cretaceous unconformity, west-central Alberta (from Marion, 1984).

The Nordegg dolomitic sandstones occur as eroded headlands at the eastern margin of the Jurassic prism. Oil and gas are found in unconformity-related traps of Type Ub1. Oil pools of this type are found in the Gilby-Medicine River-Sylvan Lake fields of central Alberta. The reservoirs are covered by the impermeable "Poker Chip Shale" or by other Jurassic or Lower Cretaceous units, and laterally sealed by Jurassic or Cretaceous valley-fill. Gas is produced from similar beds in the Paddle River-Whitecourt-Goodwin areas northwest of Edmonton.

The Rock Creek Member produces oil and natural gas in the area of Rocky Mountain House (Marion, 1984) (Fig. 6A.24). Sub-Cretaceous erosion left paleotopographic highs capped by Rock Creek sandstone; these highs were subsequently covered by impermeable Cretaceous sediments (Trap Type Ub1). All of the pools so far discovered in this play are small, one of the largest being the Niton F Pool with an IVIP of 6.6×10^6 m^3 of oil with 11.4×10^9 m^3 associated gas.

The J2-J3 sandstones fill paleovalleys in the Gilbey-Medicine River area and produce from traps of Type Ua4 or Ua3 (Fig. 6A.25). The stratigraphy of this area is complicated, and it has proved difficult to assign reserves to specific zones; even the overlying Cretaceous Ellerslie sandstones are sometimes confused with the Jurassic sandstones. With that proviso, an IVIP of about 50×10^6 m^3 has been allocated to oil pools in this play, the largest being the Gilby B Pool with over 12×10^6 m^3. The Gilby Jurassic Gas Pool has an IVIP of 6×10^9 m^3 produced from a fine-grained calcareous sandstone averaging 5 m in pay thickness with a porosity of 14% at a depth of 2120 m (Cheesman, 1969).

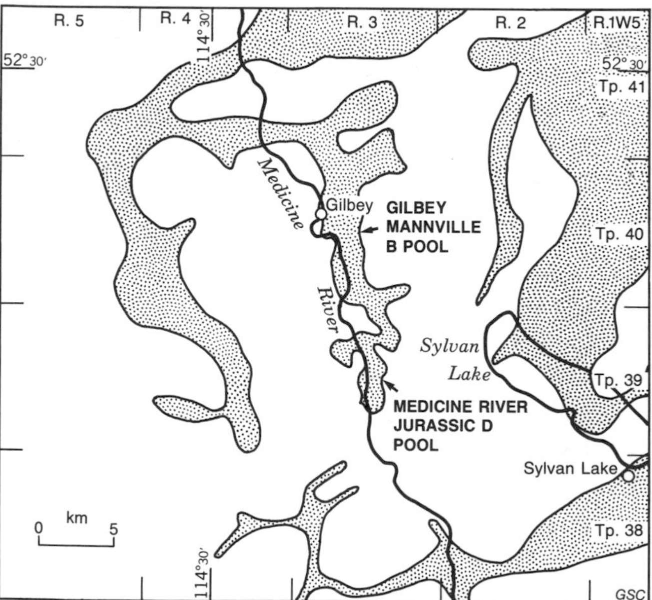

Figure 6A.25. Pre-Mannville subcrop map of the Medicine River and Gilby areas, central Alberta. Stippled pattern indicates subcrop of the Banff Formation in the bottom of valleys excavated through the Carboniferous cover by Mesozoic river systems. The two oil pools shown are reservoirs in valley-filling quartzose sandstones (from Hopkins, 1981).

In the Conrad and Grand Forks fields of southern Alberta, the discontinuous distribution of the Sawtooth sandstone is controlled by topography on the underlying unconformity surface (Davies, 1983). In that area, impermeable shales of the Sawtooth or overlying Rierdon Formation formed a seal along the western flank of the Sweetgrass Arch, trapping some small pools of heavy oil. Sawtooth sandstones also produce gas in southeastern Alberta, the largest pool being Black Butte, with an IVIP of less than 1×10^9 m^3.

Some oil is found in discontinuous sandstone of the Upper Jurassic "Ribbon Sand" of the Swift Formation in the Flat Coulee and Manyberries fields (Hayes, 1983).

Cretaceous

Although the in-place volume of Cretaceous hydrocarbons exceeds that of any other system in Alberta Basin, most of it is in the form of crude bitumen; much of the conventional oil is heavy and recovery factors are generally lower than in Devonian rocks. Hence the Cretaceous succession contains only 22% of the recoverable light and medium crude oil, although it contains 41% of the IVIP. These facts should be borne in mind when evaluating the IVIPs of oil that are mentioned in the following account. Gas IVIP totals 2500×10^9 m^3, comprising some 47% of the gas in the basin; this does not include any volume allocated to the gas contained in the tight sandstones of the Deep Basin, which is discussed, together with the crude bitumen reserves, under "Non-conventional Oil and Gas".

The Cretaceous rocks of Alberta Basin are divided into groups, which vary according to geographic locality. It is convenient to use the central plains of Alberta as a standard and to discuss the reservoirs under the headings of Mannville, Colorado, and Belly River groups. The formation names and principal producing zones are shown in Figure 6A.26. The Mannville Group is a mixed continental and marine succession. It contains the major non-conventional heavy oil and bitumen reserves as well as large amounts of conventional oil and natural gas. The Colorado Group represents the marine incursion of Late Albian to Santonian time and contains the largest single oil field in Canada. Continental conditions returned with the Belly River Group, which contains only minor reserves.

Mannville Group (Aptian to Albian)

The paleogeography of the Mannville Group is shown in Figure 6A.27. The group is divided into a Lower and an Upper Subgroup, which are described separately. The Lower Mannville in the south and west starts with a lower unit comprising continental, fluvial, and coastal plain deposits, which gradually filled the depressions in the landscape left by erosion at the end of Neocomian time. The central and northern parts of the basin were invaded by the boreal sea, with attendant deposition of estuarine and shallow marine deposits. Floodplain deposits of the medial unit of the Lower Mannville Subgroup are succeeded by an upper unit comprising lacustrine sediments of the "Calcareous" or "Ostracod" member. Most of the oil and gas occurs in the lower unit, of which the Cutbank Formation is typical on the Sweetgrass Arch, the Ellerslie Formation in central Alberta, and the Bullhead Group in the northwestern part of the basin.

Over 300 conventional oil pools have been discovered in the Lower Mannville Subgroup, 40 of which have IVIP of over 1 x 10^6 m^3. The traps are mainly in fluvial deposits above the unconformity (Types Ua3 and Ua4; Jackson, 1984). These constitute one of the most widely occurring plays in western Canada, with potential for many more discoveries. The largest in-place volumes of light and medium oil are in the Wayne-Rosedale Basal Quartz B Pool, which has an IVIP of nearly 10 x 10^6 m^3 of oil with a density of 880 kg/m^3 (Podruski et al., 1988). The Sunburst sandstone in this pool is typically medium grained, quartzose, angular to sub-angular, very well sorted, and friable, with an average gross thickness of over 20 m (Erickson and Crewson, 1959). The setting is that of a valley cut into the sub-Cretaceous rocks, but the local cause of entrapment is clay plugging of the sandstone (Trap Type Ls1). A similar trap in the Grand Forks Field produces 900 kg/m^3 oil from a Lower Mannville channel sand, 6 km in length and 1200 m in average width. The sand is almost completely bounded by nonporous sandy shales. The IVIP

EPOCH / AGE	NORTHEASTERN BRITISH COLUMBIA AND WEST CENTRAL ALBERTA	NORTHERN ALBERTA	CENTRAL ALBERTA	SOUTHERN ALBERTA SWEETGRASS ARCH	WESTERN SASKATCHEWAN AND WILLISTON BASIN
LATE CRETACEOUS	WAPITI	(hatched)	BELLY RIVER ●	BELLY RIVER ☼	BELLY RIVER (JUDITH RIVER)
	(SMOKY GROUP) PUSKWASKAU		Shale	MILK RIVER ☼	MILK RIVER (EAGLE)
				(COLORADO GROUP) First White Specks	(COLORADO GROUP) First White Specks
	Badheart			MEDICINE HAT ☼	MORDEN
	Muskiki			Shale	
	CARDIUM		(COLORADO GROUP) CARDIUM ●		
	KASKAPAU		Second White Specks	Second White Specks ☼	Second White Specks
	DUNVEGAN		Shale	Shale	BELLE FOURCHE
ALBIAN	(FORT ST. JOHN GROUP) SHAFTESBURY		Fish Scale Sandstone	Fish Scale Sandstone ●	
			Shale	Shale	
	PADDY ☼	PELICAN	VIKING ✷	BOW ISLAND ☼	WESTGATE
	(hatched)	JOLI FOU	JOLI FOU		VIKING ●
	CADOTTE ☼	(hatched)	(hatched)	Basal Colorado ✷	JOLI FOU
	HARMON			(hatched)	
	NOTIKEWIN	CLEARWATER/ GRAND RAPIDS	(MANNVILLE GROUP) UPPER BLAIRMORE/ CLEARWATER	(MANNVILLE GROUP) UPPER BLAIRMORE	(MANNVILLE GROUP) COLONY ●
	FAHLER ☼				McLAREN ●
					WASECA ●
					SPARKY ●
					GEN. PETROLEUM ●
					REX
	WILRICH				LLOYDMINSTER ●
	BLUESKY ☼	BLUESKY/WABISKAW ●	Glauconitic Sst. ✷	Glauconitic Sst.	CUMMINGS ●
APTIAN	(BULLHEAD GROUP) GETHING ☼	GETHING/McMURRAY	Ostracod Zone ●	Ostracod Zone	DINA ●
			ELLERSLIE	SUNBURST ●	
	CADOMIN ☼	CADOMIN		CUTBANK/TABER ●	
NEO-COMIAN	(hatched)	(hatched)	J3?	(hatched)	(hatched)

GSC

Figure 6A.26. Principal producing zones of the Cretaceous System in Alberta Basin (symbols described in Figure 6A.4).

was estimated to be 16×10^6 m^3 (Berry, 1974). Other large oil fields in southeastern Alberta, including Bantry and Cessford, occur in the Lower Mannville on the east flank of the Alberta Syncline and the northwest flank of the Sweetgrass Arch. The porosity is in basal conglomerates and overlying, quartzose sands. Traps are dominantly lithological, although some are related to structure and drape over eroded Paleozoic subcrop.

In the Bellshill Lake area of east-central Alberta, northwesterly trending sand-filled channels in the Dina interval reach thicknesses of 95 m (Fig. 6A.28). When filled with clean sands, these channels commonly form excellent reservoirs. The Bellshill Lake Oil Pool produces from sandstone with average thickness of 8.6 m and porosity of nearly 26%; it has an IVIP of 41×10^6 m^3. The Niton Basal Quartz B Pool, north of Pembina, contains

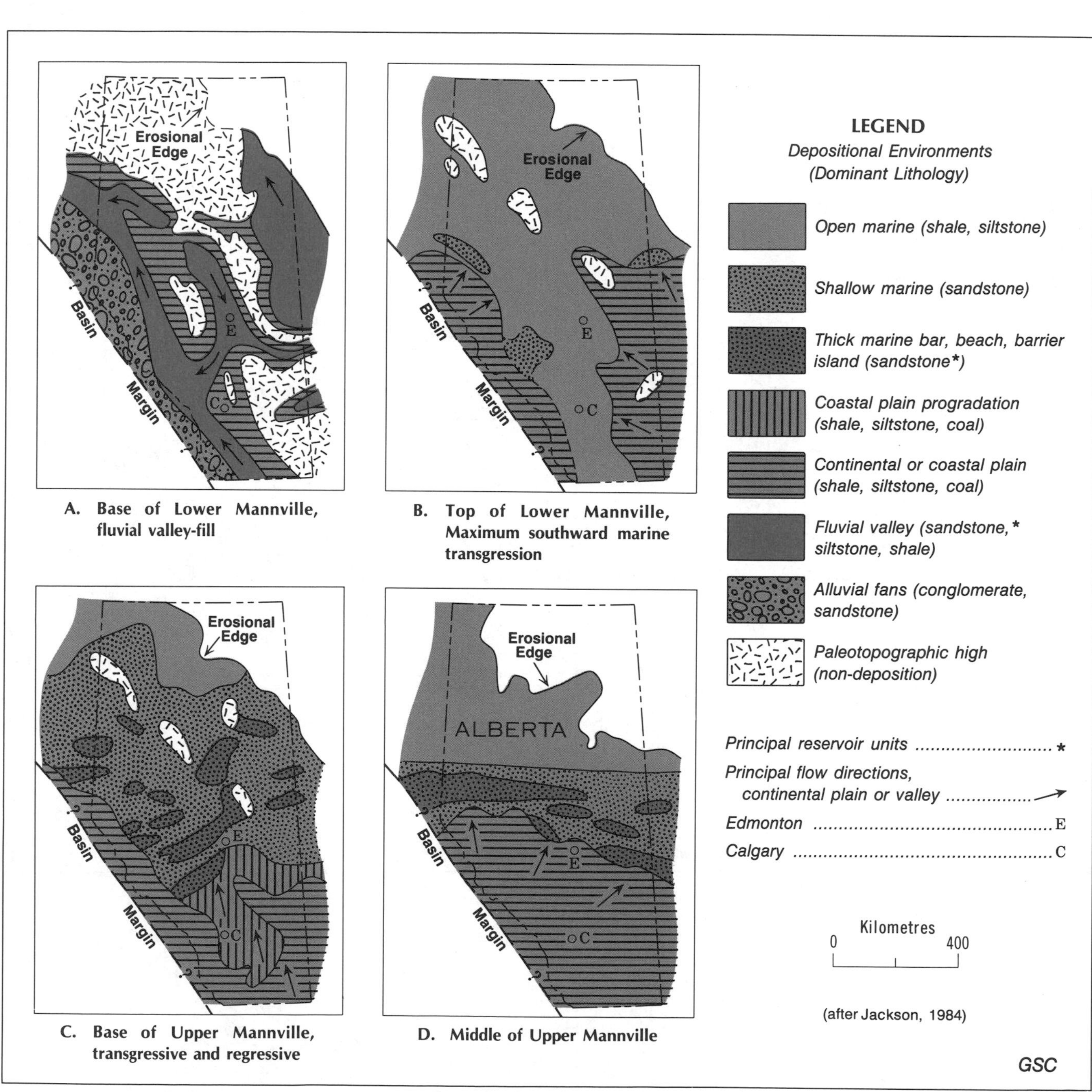

Figure 6A.27. Mannville Group paleogeography. Diagrams illustrate northwest-trending valley systems (A); inundation by southward directed marine transgression (B); and regression with basin filled from south to north (C, D) (from Podruski et al., 1988).

substantial reserves of oil and nearly 8 x 10^9 m^3 of associated gas. Natural gas occurs in hundreds of small pools in the Lower Mannville; one of the larger pools is the Alexander Field in central Alberta, where a thick channel-fill of Basal Quartz sandstone trends northwesterly. Gas, with an IVIP of 4 x 10^9 m^3, is trapped by shale-out updip to the northeast (Trap Type Ls2). The most substantial accumulations of gas in Lower Mannville equivalents are found in the Bullhead Group of the northwestern plains. In Alberta, the Edson Field produces from Gething strata deposited in a deep, east-west trending re-entrant eroded in Lower Carboniferous rock. In northeastern British Columbia, the Rigel Field produces from chert conglomerates of the Cadomin Formation above the unconformity, whereas the Buick Creek Field produces from Jurassic to lowermost Cretaceous quartzose sands in erosional remnants beneath the sub-Albian unconformity.

Many small pools occur in the brackish to marine Ostracod Member of the Mannville Group, where estuarine, lacustrine, brackish bay, and coastal to deltaic sands are common at the southern extremity of the Lower Cretaceous marine basin. The largest oil pool in this category is the Pembina E Pool, with an IVIP of 3.7 x 10^6 m^3.

The Upper Mannville succession was deposited by a deltaic system, advancing from the south, which cut deeply down into the underlying lacustrine sediments of the Ostracod Member. Fluvial and near-shore to shoreline sandstones in the Upper Mannville are locally productive over most of central and southern Alberta. Conventional oil is produced from a Mannville beach deposit in the Kaybob Field, near Edmonton. In southern Alberta, channels were mapped by Herbaly (1974), barrier islands were described from the Jenner Field by Holmes and Rivard (1976), and beach deposits in the Suffield area by Tilley and Longstaffe (1982). Offshore bars were identified in the Countess Field by McCoy and Moritz (1982) and distributary channels were described from Little Bow (Hopkins et al., 1982).

The Hoadley Gas Field (Chiang, 1985) is an extensive gas condensate field covering approximately 3885 km^2 in south-central Alberta (Fig. 6A.29). The producing zone in the Lower Cretaceous Glauconitic member (Upper Mannville) comprises 7.6 to 24.4 m of sandstone. The sand was deposited as an extensive marine barrier bar complex with a width of about 24 km and length of more than 200 km. The middle and southwestern portion of the barrier bar is entirely saturated with wet gas trapped laterally by impermeable shale and updip by shale-filled tidal channels (Trap Type Ls2). The field is estimated to contain an ultimate potential reserve of 170 to 200 x 10^9 m^3 of marketable gas and 55 to 64 x 10^6 m^3 of natural gas liquids. Other large gas accumulations include the Glauconitic pools of the Pembina and Bigoray fields.

High gas productivity is found in some lenses of coarse-grained sandstone and conglomerate in the Deep Basin (see Fig. 6A.58) of western Alberta and northeastern British Columbia. Reservoirs are found in Falher and Notikewin shoreline sandstones and conglomerates as well

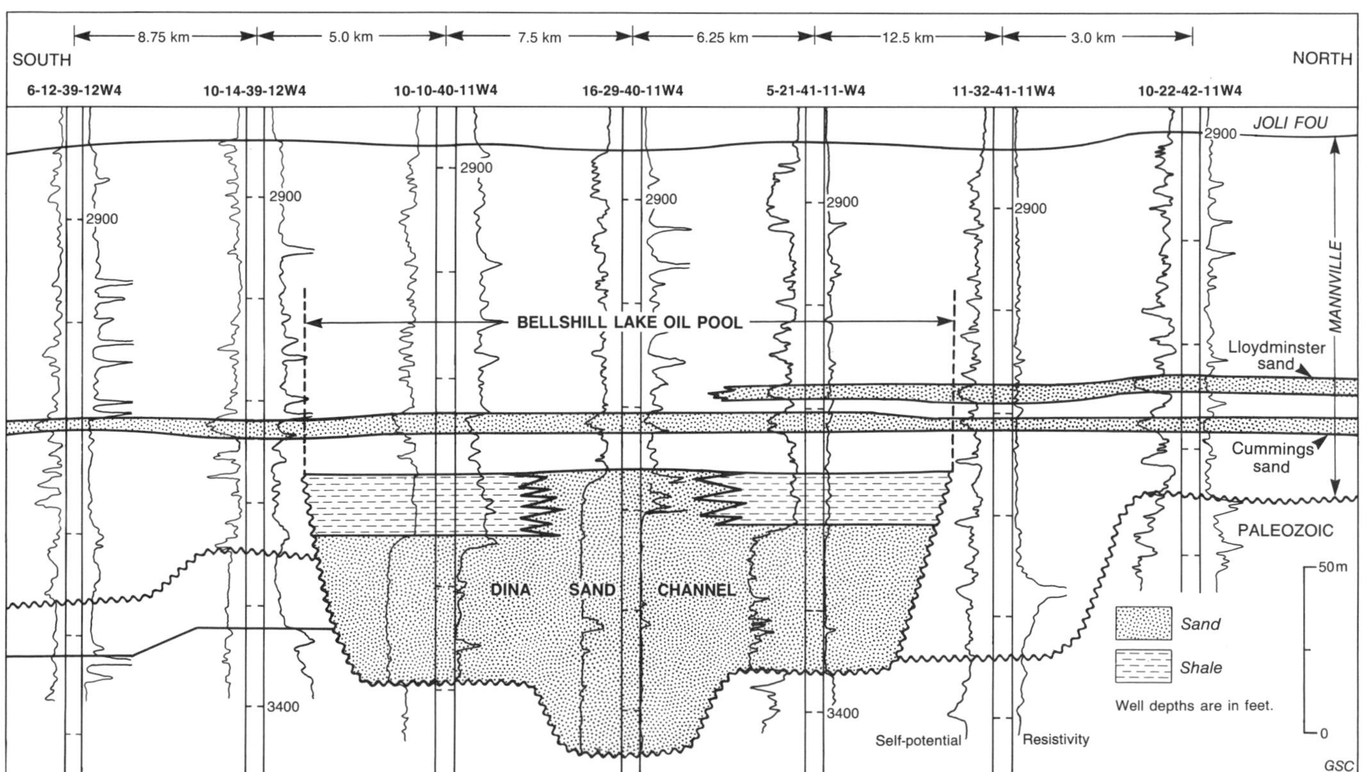

Figure 6A.28. Stratigraphic cross-section of a Mannville channel-fill of the Bellshill Lake Field in east-central Alberta (from Gross, 1980).

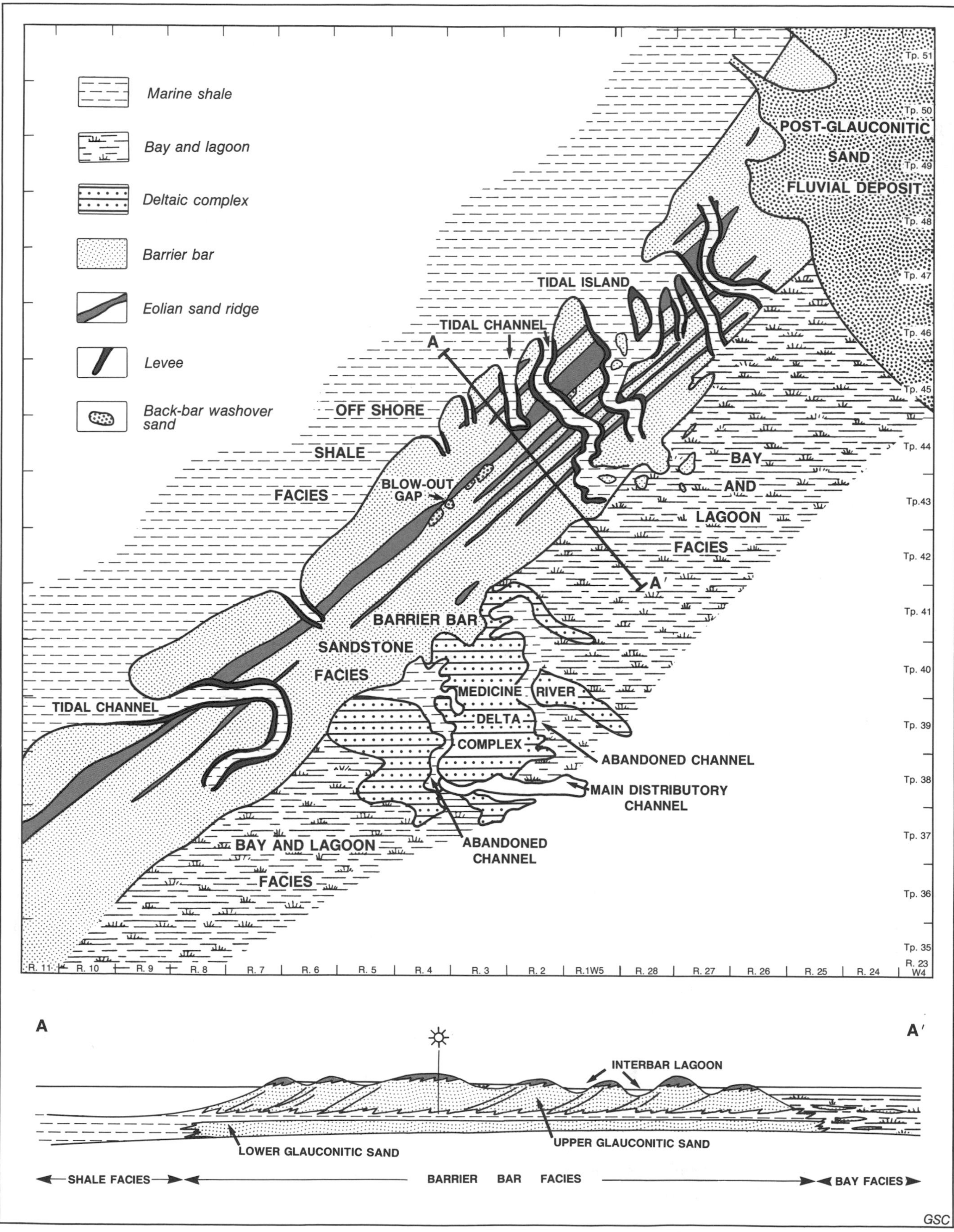

Marine shale
Bay and lagoon
Deltaic complex
Barrier bar
Eolian sand ridge
Levee
Back-bar washover sand
POST-GLAUCONITIC SAND FLUVIAL DEPOSIT
TIDAL ISLAND
TIDAL CHANNEL
A
A′
OFF SHORE SHALE FACIES
BLOW-OUT GAP
BAY AND LAGOON FACIES
BARRIER BAR SANDSTONE FACIES
TIDAL CHANNEL
MEDICINE RIVER DELTA COMPLEX
ABANDONED CHANNEL
MAIN DISTRIBUTORY CHANNEL
ABANDONED CHANNEL
BAY AND LAGOON FACIES
Tp. 51
Tp. 50
Tp. 49
Tp. 48
Tp. 47
Tp. 46
Tp. 45
Tp. 44
Tp.43
Tp. 42
Tp. 41
Tp. 40
Tp. 39
Tp. 38
Tp. 37
Tp. 36
Tp. 35
R. 11
R. 10
R. 9
R. 8
R. 7
R. 6
R. 5
R. 4
R. 3
R. 2
R.1W5
R. 28
R. 27
R. 26
R. 25
R. 24
R. 23 W4
A
A′
INTERBAR LAGOON
LOWER GLAUCONITIC SAND
UPPER GLAUCONITIC SAND
SHALE FACIES
BARRIER BAR FACIES
BAY FACIES
GSC

as in the Bluesky marine sandstones. The Elmworth Falher A-1 Pool has been assigned an IVIP of 8.6 x 10^9 m^3, but it is possible that the producing zone could draw additional reserves from the enclosing tight sands.

To sum up the trapping conditions of the Mannville Group, lithological trap types dominate. There are, however, some structural influences. Mannville sands directly overlying an unconformity are commonly influenced structurally and depositionally by the paleomorphology of the erosional surface of underlying Paleozoic rocks. Drape traps of Type Sd2 are known, although not volumetrically important. Also, structures caused by differential compaction over more deeply buried Devonian reefs provide traps in the Mannville in a few places. Some minor Mannville hydrocarbons are associated with faults on Peace River Arch and the Sweetgrass Arch. Recently, Hopkins (1987) described the accumulation of the reservoir sandstone in the Upper Mannville C Pool, Berry Field, as taking place during synsedimentary faulting or downwarping, possibly caused by the solution of Devonian salt at depth.

Colorado Group and post-Colorado formations

The Colorado Group contains the largest conventional oil reserves in the Cretaceous System, and also very large accumulations of natural gas, although these lie at shallow depths and have low reservoir pressures. The Colorado was deposited during the period of maximum marine incursion, when continental deposits were restricted to the edges of the basin. The basin was mainly the site of shale deposition with thin interbeds of marine sandstone and minor conglomerate. In central Alberta, two major sandstone units, the Viking and Cardium formations, provide the important reservoirs. In southeastern Alberta, the Medicine Hat sandstone and the Second White Specks sandstone are important additional gas reservoirs. In northwestern Alberta and northeastern British Columbia, the Dunvegan sandstones, interposed between the Viking equivalents and the Cardium, provide reservoirs for oil and gas.

The reservoirs are all terrigenous clastic sediments, mainly of fine-grained sand, with minor conglomerate. Any sandstone or conglomerate unit within or in contact with thick marine shale is likely to contain oil or gas. Pay zones are generally thin, averaging 6 to 7 m. In order to accommodate large reserves, the reservoirs must therefore be continuous over exceptionally wide areas, as are the Cardium Formation at Pembina and the Medicine Hat sandstone near Medicine Hat. The traps are, with few exceptions, simple updip lithological pinchouts (mainly Type Ls3), because the sequence is structurally undisturbed and dips uniformly southwest at one-half degree. The oil of the Colorado Group, above the heavier oil of the basal sands, is mostly in the range of 825 to 875 kg/m^3. About 80% of the oil in this part of the section is contained in the Cardium Formation, and about two-thirds of that occurs in the Pembina Field. Associated gas is present in every oil reservoir, but by far the largest gas accumulations occur at low pressure in extensive and continuous sandstones in southern Alberta. Non-associated natural gas is mostly methane with very little sulphur or liquid.

Basal Colorado Sandstone

The basal Colorado sandstone is restricted to the southern part of Alberta. The sandstone is generally fine to medium grained but contains chert-pebble stringers in places. At the only important oil field, Cessford, the reservoir averages 3 m in thickness and contains oil with a density of 905 kg/m^3. The trap is a combination of lithological pinchout and structural drape caused by compaction over paleotopography eroded in Lower Carboniferous rocks (Trap Type Ls2-Sd2). Gas accumulations are found at Cessford, Hussar, and Countess.

Viking-Bow Island-Peace River formations

A typical Viking section consists of a coarsening-upward sequence of fine sandstone capped disconformably by a conglomerate. The sandstones were deposited on a marine shelf and then channelled during regression; the conglomerates and associated sheet sands were deposited during the transgression accompanying a subsequent rise in sea level (Fig. 6A.30) (Leckie, 1986). Pay zones generally average no more than 1 to 3 m in thickness, with porosity about 20%. A progressive reduction in primary intergranular porosity with increasing depth of burial is evident from the shallow Provost reservoirs through to the deeply buried Caroline reservoir (Reinson and Foscolos, 1986).

More than 125 lithologically trapped pools produce from the Viking sandstone in central and eastern Alberta. Most of the Viking sandstone oil and gas pools are arranged in belts with a pronounced northwesterly trend (Fig. 6A.31). The northern belt from Provost to Westlock contains large amounts of oil and associated gas. Provost Field, for example, has an IVIP of oil of 82 x 10^6 m^3 with 32 x 10^9 m^3 associated raw gas. The Viking A Gas Pool in the Viking-Kinsella Field, also on this trend, has an IVIP of 27 x 10^9 m^3. South of Red Deer, the Viking, or approximately equivalent Bow Island Formation, is almost exclusively gas-bearing.

Three of the most important Viking reservoirs of central Alberta, Gilby, Joarcam, and Joffre, have similar shapes and dimensions. They are elongate, northwest-striking bodies, from 30 to 50 km in length, and 2 to 3 km in width, thought to be related to ancient shorelines. Reservoir rocks are fine- to medium-grained sandstones, commonly capped by chert-pebble conglomerate. All of these fields are simple lithological traps related to an updip facies change within a uniformly dipping succession (Type Ls3). Pay thicknesses in these fields range up to 10 m, and average about 3 m. Field areas range from 50 to 80 km^2. Oils at Gilby, Joarcam, and Joffre range from 820 to 830 kg/m^3. The Joarcam Field has an IVIP of 40 x 10^6 m^3.

Non-linear, irregularly shaped sand bodies also form oil reservoirs in the Viking Formation. In the Hamilton Lake Field (Lerand and Thompson, 1976), the fine-grained sandstone, interbedded in part with siltstone, is less favourably developed than in the linear fields. Maximum

←

Figure 6A.29. Facies and paleogeography of the Hoadley barrier bar complex and the Medicine River delta-complex in central Alberta (from Chiang, 1985).

pay is only 3 m, and average pay less than 1.5 m; Hamilton Lake oil is 875 kg/m^3. A few traps are related to structural drape caused by compaction over paleotopography on the sub-Cretaceous surface (Type Sd2; e.g. Bindloss).

In northern Alberta and northeastern British Columbia, the Peace River Formation, in a similar stratigraphic position but older than the Viking, contains gas in such fields as Pouce Coupé, Gordondale, and Dawson Creek.

Dunvegan Formation

The Dunvegan Formation is a deltaic and shoreline complex consisting of a number of thick, stacked sandstones interbedded with shale and siltstone, thinning eastward to zero in west-central Alberta. The largest oil pool is Valhalla I with an IVIP of over 13 x 10^6 m^3. At Valhalla, the overlying Doe Creek sandstone is productive also.

Second White Specks sandstone

The Second White Specks is a coccolith-rich marl that is extremely widespread. Locally, in southern Alberta, it includes thin sheet sandstones that contain large reserves of low-pressure gas. The Alderson and Suffield fields have, between them, an IVIP of some 30 x 10^9 m^3 within the Second White Specks sandstone.

Cardium Formation

Over 25 separate oil pools occur in the Cardium Formation of west-central Alberta (Fig. 6A.32). Most of these, however, contain only 1% or less of the oil in place of the Pembina Field. Pembina is the only "giant" oilfield, excluding the oil sands and heavy oils, amongst the Cretaceous pools in the Western Canada Basin.

The Cardium oil-producing sands and conglomerates are completely enveloped by marine shales. Within the Cardium, coarsening-upward sequences are composed essentially of bioturbated mudstones and extensive sheet sandstones. In many fields, conglomeratic bodies overlie erosion surfaces, filling scours and forming important reservoirs (Fig. 6A.33; Plint et al., 1986). The succession of rock types is similar to that in the Viking fields (compare Fig. 6A.30). Many of the sands are characterized by relatively poor porosities and permeabilities. However, wherever the Cardium sandstones have adequate porosity and permeability, oil has been discovered. The reservoirs are usually filled with oil and there are no gas fields of any consequence; virtually all gas produced is solution gas from the oil reservoirs. The overpressured gas-solution drive mechanism suggests isostatic rebound as a result of post-Eocene denudation.

The Pembina Oil Field (Nielsen and Porter, 1984), discovered in 1953, is Canada's largest, with an estimated IVIP of 1180 x 10^6 m^3; it encompasses an area of more than 2331 km^2. Its Initial Established Reserves amounted to

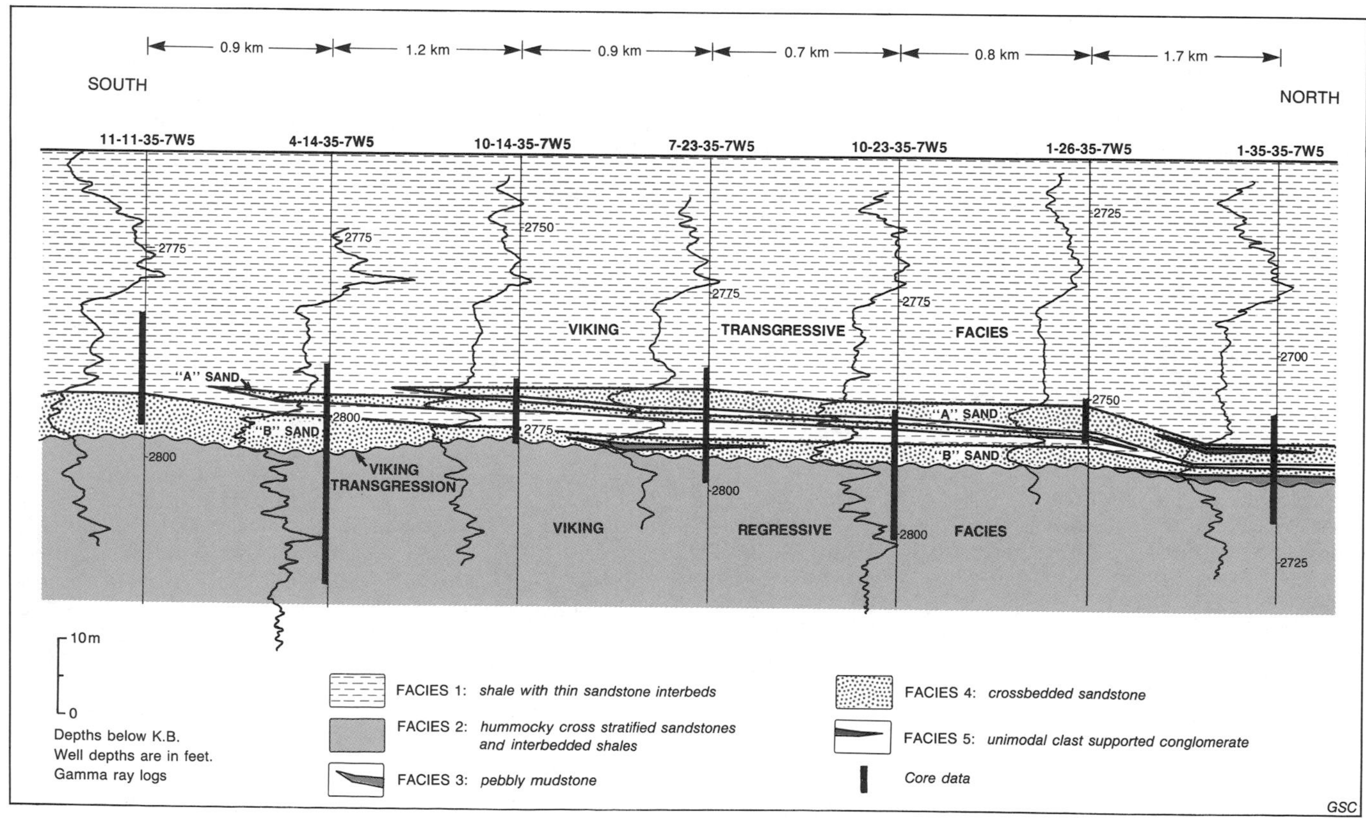

Figure 6A.30. Cross-section of the Viking Formation in the Caroline Field. All the wells on the section are oil-producing (adapted from D. Leckie, 1986).

290×10^6 m^3 from 66 separate pools. The Cardium sandstone is the principal producing reservoir and contributes 250×10^6 m^3 to this total, with the remainder from deeper pools. The Cardium at Pembina is a succession of regressive units of fine-grained sandstone, each generally capped by transgressive conglomerate ranging from a few centimetres to 5 to 6 m thick (Krause and Nelson, 1984). Pay zones average 6 m, with a maximum of 12.5 m. Porosity is 12 to 15% and oil density 830 kg/m^3. No gas cap or active water drive is present, and there has been no biodegradation of the crude by invasion of meteoric water. The updip limit of the field is defined by the 'shale-out' or termination of the reservoir sands, and the zero edge extends from the vicinity of Calgary to beyond Pembina. The westward extent of production is limited by loss of porosity resulting from burial diagenesis; this is a consequence of post-migration synorogenic loading of the foreland basin. The diagenetic seal is parallel to a subtle structural hinge.

Smaller Cardium fields have the same general characteristics as Pembina, but the pay zones are thinner (Walker, 1983a, 1983b, 1985). A few fields, like Crossfield and Garrington, display a linearity and narrowness similar to that of the Viking Formation of the Gilby, Joarcam, and Joffre fields. Cardium fields, such as Ricinus (Walker, 1985), Cym-Pem, and Carrot Creek, produce from conglomerates up to 15 m thick, deposited within large erosional scours (Fig. 6A.33).

Medicine Hat sandstone

The Medicine Hat sandstone (Meijer Drees and Myhr, 1981; McCrory and Walker, 1986) is one of three thin (1-2 m), shallow (350-500 m), low pressure, low permeability, fine-grained Upper Cretaceous gas-bearing sandstones in southern Alberta and southwestern Saskatchewan (Male and Pacholko, 1982). The Medicine Hat sandstone, the Second White Specks sandstone, and the Milk River sandstone have similar characteristics and collectively account for about 70% of the total gas reserves in the Upper Cretaceous beds of Alberta. Twenty-seven gas fields are recognized in a belt between Drumheller, Alberta, and Maple Creek, Saskatchewan: the pools east of the Saskatchewan boundary are mentioned in the section devoted to Williston Basin. Reserves per well are small and the flow rates of individual wells decline rapidly, but because of the large areal extent of each reservoir, the total reserves are large. In Medicine Hat Field, for example, the Medicine Hat sandstone had an IVIP of about 80×10^9 m^3, of which 50×10^9 m^3 is producible, over 470 000 ha. The

Figure 6A.31. Viking fields in Alberta (from Hein et al., 1986; modified from Amajor, 1980).

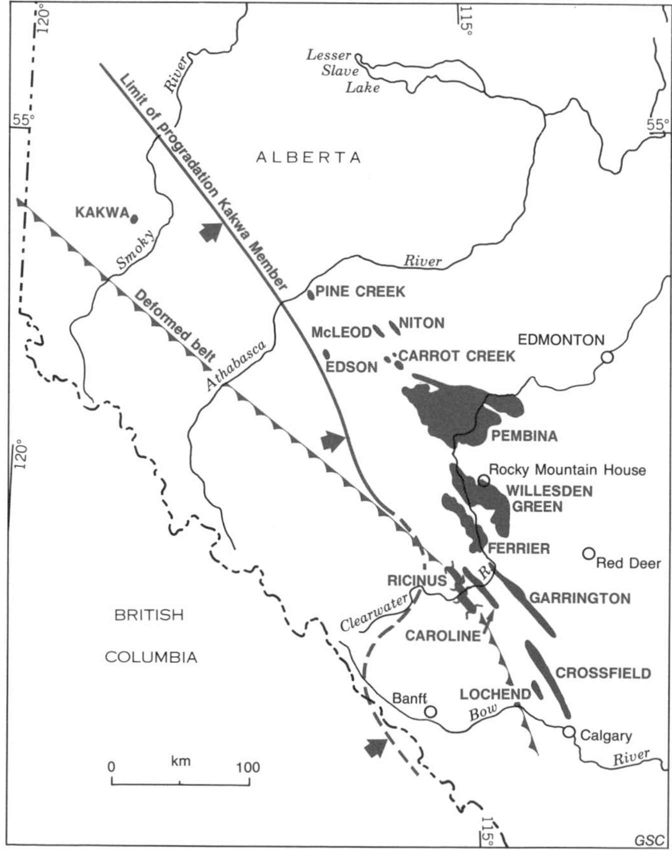

Figure 6A.32. Main Cardium fields in subsurface of Alberta Basin (adapted from Plint et al., 1986).

gas in all of the reservoirs has little or no sulphur and is dry, being 95% methane and 4% nitrogen, with traces of other constituents. Rice and Shurr (1980) showed that gas in the equivalent Upper Cretaceous succession of north-central Montana formed through the breakdown of organic matter in the associated shales by anaerobic bacteria.

Milk River Formation

The Milk River reservoir lies above the top of the Colorado Group, and is partly equivalent to the Lea Park Formation of eastern Alberta. The Milk River gas of southeastern Alberta occurs on the flanks and crest of Bow Island Arch, downdip from porous, fresh water-saturated Virgelle sandstone to the south. The gas is present in sandy shale of the Alderson Member, transitional between a sandy nearshore facies, Milk River, and a shaly offshore facies, Lea Park. Good porosity and permeability are restricted to the thin, laminated sandstone beds within the shales.

An estimated 142 x 10^9 m^3 of recoverable gas is present in the Milk River sandstone of eastern Alberta and southwestern Saskatchewan (Meijer Drees and Myhr, 1981). In the Medicine Hat Field alone, the Milk River gas reservoir has an IVIP of 60 x 10^9 m^3, of which 40 x 10^9 m^3 is producible, over an area of 370 000 ha. The reservoir characteristics and deliverability of these shallow sands are similar to those of the Medicine Hat gas sand.

Belly River Group

Apart from a single Tertiary pool, the basal Belly River sands form the youngest hydrocarbon-producing reservoirs in Alberta and Saskatchewan (Shouldice, 1979). Oil production is restricted to a relatively small area southwest of Edmonton, but gas is encountered throughout the broad area of distribution of the Belly River sands. Hydrocarbons are trapped in up-dip pinchouts of shoreface (Trap Type Ls2) and fluvial (Trap Type Ls1) sandstones and in gentle structures involving widespread sands. The shoreface and structural types are the most important. The largest oil accumulation in shoreface sands is the Pembina-Keystone B Pool with an IVIP of 29 x 10^6 m^3; the largest of the fluvial sand accumulations is the Ferrier A Pool with an IVIP of 4 x 10^6 m^3. The Herronton Pool southeast of Calgary appears to be an offshore bar (Trap Type Ls3), flanked on the east by marine shale. In the Atlee, Bindloss, and Medicine Hat regions, Belly River strata are gently draped in closed structures over buried hills on the sub-Mesozoic erosional surface (Trap Type Sd2). Belly River sands of the Alberta Basin have estimated IVIPs of 143 x 10^6 m^3 and 70 x 10^9 m^3 of oil and gas respectively, but the reservoirs have low recovery factors and poor deliverability and are not a major resource.

Tertiary

A single gas pool, Bigoray, east of the Pembina Field, represents the discovered hydrocarbons in Tertiary strata. The gas is in Paskapoo sand and may have migrated from

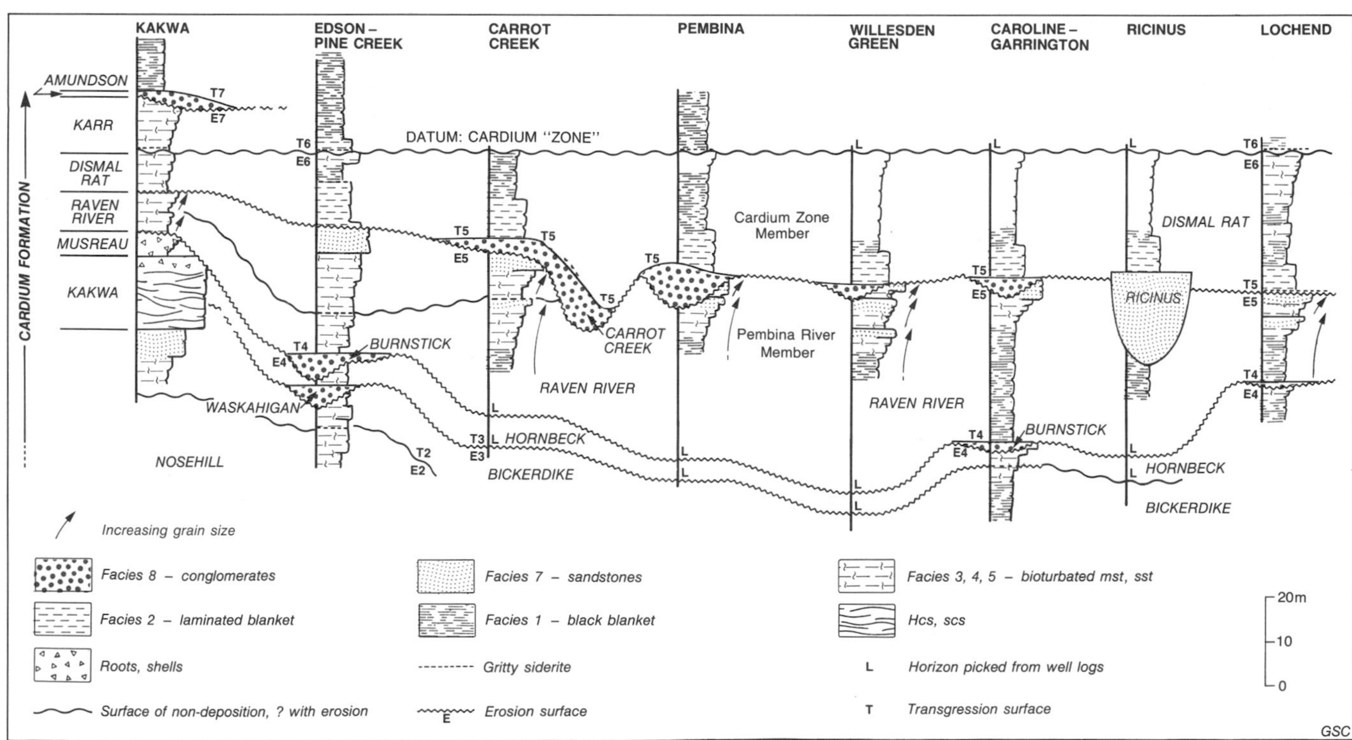

Figure 6A.33. Correlation of measured core sections from Cardium fields. Terminology of Krause and Nelson (1984) in lower case (from Plint et al., 1986).

AGE	ZONE	WILLISTON BASIN	
CARBONIFEROUS			
	BIG SNOWY	KIBBEY	
	MADISON	CHARLES	●
		MISSION CANYON	●
		LODGEPOLE	●
DEVONIAN	THREE FORKS	BAKKEN	●
		BIG VALLEY	
		TORQUAY	
	SASK	BIRDBEAR	●
		DUPEROW	
	MANITOBA	SOURIS RIVER	
		DAWSON BAY	
	ELK POINT	PRAIRIE EVAPORITE	
		WINNIPEGOSIS	●
		ASHERN	
SILURIAN			
		INTERLAKE	
ORDOVICIAN		STONY MOUNTAIN	
		RED RIVER	●
		WINNIPEG	
CAMBRIAN		DEADWOOD	
		GSC	

Figure 6A.34. Principal producing zones of Paleozoic rocks of Williston Basin (symbols described in Figure 6A.4).

a downdip source. The Tertiary rocks in Western Canada Basin are entirely continental, shallow, and thermally immature, and are not considered very prospective.

Williston Basin

The Canadian portion of Williston Basin occupies the southern part of Saskatchewan and the southwestern corner of Manitoba (Fig. 6A.1). The geological limits of the basin are debatable because it had different geological expressions at different times. The limits of the basin in this report were chosen for statistical convenience: oil and gas reserve data for the selected area are those for Manitoba and Areas II, III, and IV of Saskatchewan (Saskatchewan Energy and Mines, 1987). The Cretaceous heavy oils of the Lloydminster region are treated separately under "Non-conventional oil and gas". The region lies between the International Boundary and the northeastern outcrop limits of Phanerozoic sediments. Its western boundary is the Alberta-Saskatchewan boundary north to Township 39 and thence along an imaginary line running northeast to Township 61.

The sedimentary fill of Williston Basin forms a lens with its depocentre in the United States. The northern part of the basin, lying in Canada, covers approximately 565 000 km^2, and contains 23% of the IVIP of light and medium oil in Western Canada Basin, 8% of the heavy oil, and 3% of the gas. The cumulative production from the basin to December 1986 was 320 x 10^6 m^3 of oil and 32 x 10^9 m^3 of gas. The stratigraphic columns of the Canadian part of Williston Basin show the principal oil- and gas-producing zones (Fig. 6A.23, 6A.34). The greater part of the oil production (Fig. 6A.35) comes from Lower Carboniferous rocks on the east flank and Jurassic rocks toward the west flank of the basin. A third producing area, of less importance, lies near the north flank in the Coleville area. The Lower Carboniferous and Jurassic productive strata are of shallow marine, carbonate-evaporite facies and littoral, mixed carbonate-siliciclastic facies with minor evaporites. The principal trapping control on the Lower Carboniferous accumulations is the sub-Mesozoic unconformity: the Jurassic traps are mainly lithological pinchouts.

Although more than 20 000 wells have been drilled in Williston Basin in Canada, they have been heavily concentrated in the areas of established plays; large parts of the basin, and deeper zones, remain very lightly explored. Potential petroleum source rocks have been identified in Ordovician, Devonian, Carboniferous, Jurassic, and Cretaceous carbonates and shales (Osadetz and Snowdon, 1986a, 1986b; Brooks et al., 1987). Potential reservoir rocks are common throughout the section, and include Cambro-Ordovician sandstones, Lower Carboniferous carbonates, and Jurassic and Cretaceous sandstones.

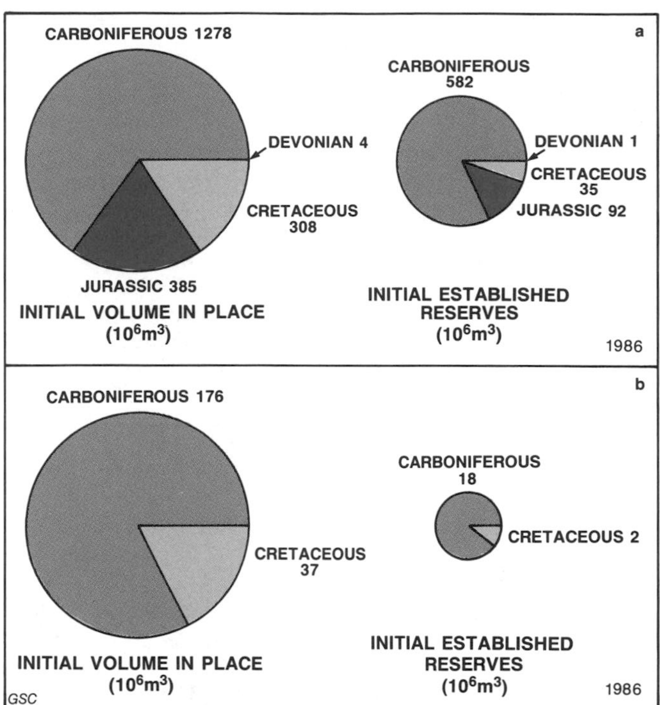

Figure 6A.35. Distribution of hydrocarbon volumes by geological system in Williston Basin: a) light and medium oil; b) heavy oil.

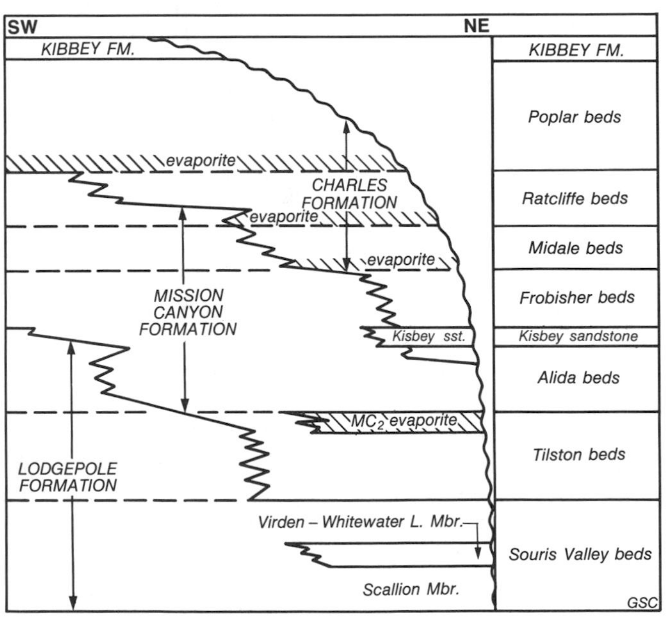

Figure 6A.36. Nomenclature of the informal "beds" of the Carboniferous Madison Group (from Podruski et al., 1988).

Pre-Devonian

The first commercial oil production from the Ordovician rocks of Saskatchewan was obtained in 1957 at Hummingbird (A.C. Kendall, 1976). This proved to be a single-well pool, as was the case with two further discoveries at Lake Alma (1963) and nearby Beaubier.

Devonian

In spite of the presence of numerous Middle Devonian reefs similar to those that have yielded large discoveries in the United States portion of the Williston Basin, there had, until 1986, been very little production from the Devonian rocks of the Canadian portion. In 1986 Home Oil drilled the first of a number of hydrocarbon-bearing reefs of the Winnipegosis Formation in the Tableland area of southeastern Saskatchewan. The well flowed 223 m^3/d of 840 kg/m^3 oil (Martindale and Orr, 1988).

Upper Devonian pools in the Birdbear Formation, including Hummingbird and Kisbey, are in Type M3 traps due to salt-solution collapse (Halabura, 1982). The main pay zones are found within dolomitic intertidal bank packstone and wackestone and in thin microsucrosic dolomite. Porosity in the zones is primarily solution-enlarged mouldic, interparticle, and channel.

Carboniferous

The principal Carboniferous pools lie at depths ranging from 800 m near the Manitoba border to 2000 m in southwestern Saskatchewan. The principal producers are Lower Carboniferous carbonates. Bakken pools are exceptional in producing from sandstone and in being mainly located on the northwestern subcrop.

Bakken sandstone (basal Carboniferous)

Production in the Coleville area of western Saskatchewan is obtained from the Bakken Formation of Devonian and earliest Carboniferous age (Hornford, 1985). Although the formation is mostly black shale, the productive middle member is a fine-grained quartzose sandstone. Traps (Type Ub2) occur along the east-west subcrop belt beneath Cretaceous Mannville strata. Producing fields include Fusilier, North Hoosier, and Coleville. Coleville has an estimated IVIP of 49 x 10^6 m^3 at about 825 m depth, but the recovery factor is only 15%. The oil density is 845 kg/m^3.

The Bakken pools in the Roncott and Rocanville fields of southeastern Saskatchewan are trapped in domes (Type M3) created by the differential removal of salt from the underlying Middle Devonian Prairie Evaporite (Christopher et al., 1971).

Lower Carboniferous carbonates

The Lower Carboniferous carbonate fields are distributed over 27 200 km^2 of southeastern Saskatchewan and southwestern Manitoba, but most are concentrated in the Weyburn-Midale and Steelman areas (Christopher et al., 1973), which hold more than half of the recoverable reserves. The oil density ranges from 835 to 935 kg/m^3. The rocks are formally included in the Lodgepole, Mission

Canyon, and Charles formations, but an informal subdivision into "beds" prevails in the oilfields (Fig. 6A.36). These beds, dipping southward, subcrop beneath the Mesozoic in a series of concentric arcs (Fig. 6A.37).

Certain general principles govern the trapping conditions in all of the subcrop belts (Kent, 1985). The most common type of trap (Type Ub-Ld1) is one in which the pool occurs beneath the sub-Mesozoic unconformity and is sealed by diagenesis (anhydritization) beneath the unconformity surface (Fig. 6A.38, case B). The seat-seal may or may not be essential to trap morphology (cases A and C), depending on the amount of paleogeomorphic relief and diagenesis.

Some traps are entirely lithological, formed either by textural variations such as grainstone or packstone interfingering with lime mudstone; diagenetic differences such as limestone grading laterally to dolomite; or changes in cementation. The proximity of these lithological traps to the unconformity may be coincidental or may be the result of hydrodynamic conditions at the unconformity enhancing sealing capacities. The porosity in most Lower Carboniferous reservoir rocks of southern Saskatchewan has been modified by diagenetic processes into three types of secondary porosity: intercrystal, cavity, and mixed (Kent et al., 1982). The original porosity was greatly reduced by cementation, neomorphism, and compaction, and enhanced by dolomitization and solution.

Before describing the main features of the individual subcrop belts, mention should be made of two less important trapping mechanisms. A few small pools occur in drape structures (Trap Type M3), probably originating in multi-stage salt solution. Oil is obtained from single or multiple zones, and local traps are commonly formed by porosity-permeability pinch-outs and fractures. In an unusual occurrence, oil is trapped at Viewfield in a structure interpreted as an astrobleme (Trap Type M1; Sawatsky, 1975; Fig. 6A.39).

Oil is trapped along the subcrop edge of sublittoral and shelf-margin carbonates of the Lodgepole Formation (Souris Valley beds). Paleogeomorphic highs on the sub-Mesozoic erosional surface are important in localizing accumulations at the subcrop, but drape folds also exert strong control at Daly and elsewhere in this belt (Trap Type Ub1-Sd4). The Daly Pool has been assigned an IVIP of 24×10^6 m^3, but recovery factors in the Lodgepole fields are very low (about 13%) The rocks are sealed by conformably overlying impermeable carbonates of the Lodgepole Formation and by strata above the sub-Mesozoic unconformity. At the unconformity, the seal is either a zone of plugging and metasomatic replacement of carbonates by anhydrite at the top of the Carboniferous, or a cap of impermeable anhydrite and shale of the overlying Triassic (?) to Jurassic Amaranth Formation.

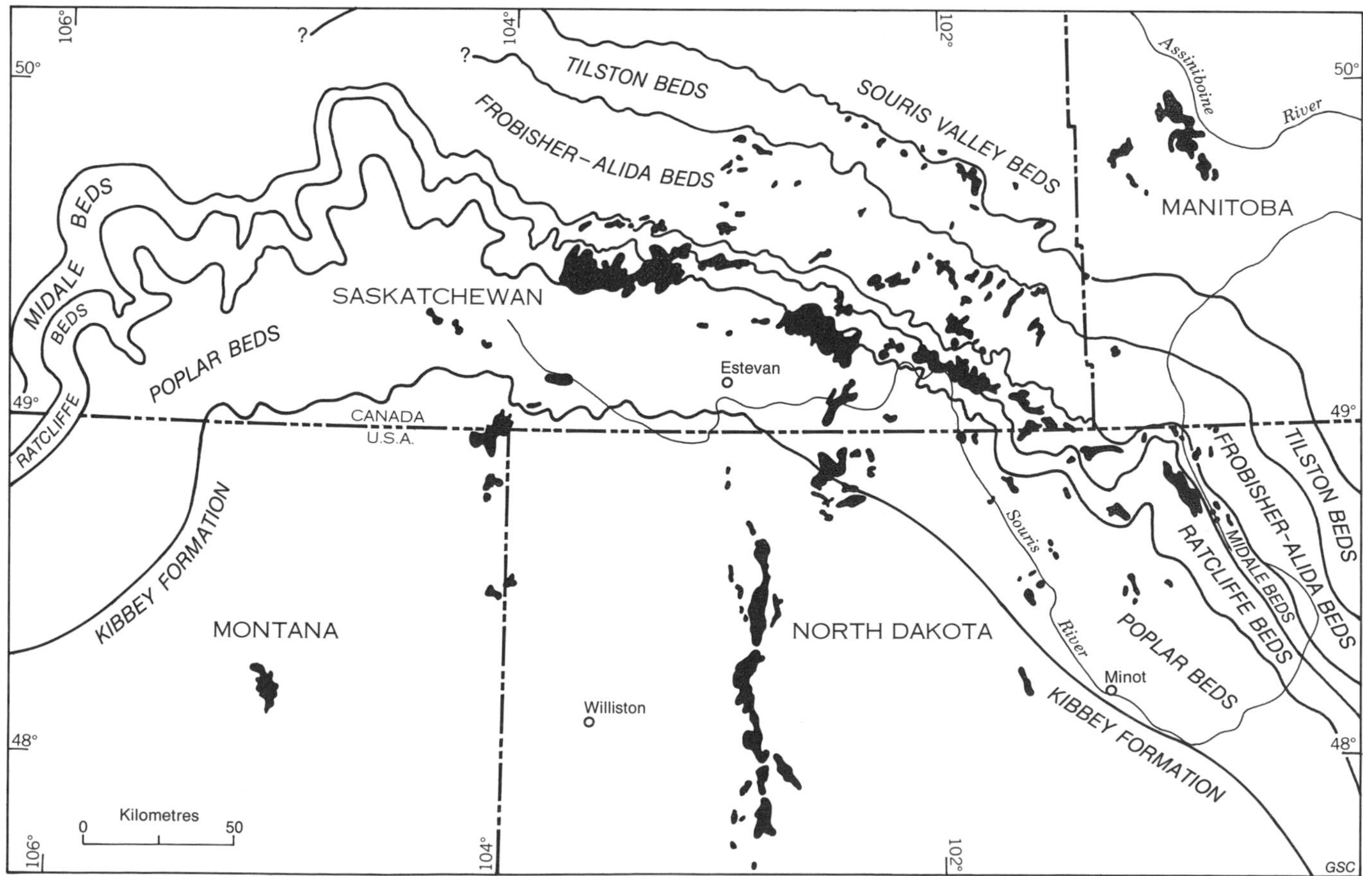

Figure 6A.37. Pre-Mesozoic subcrop pattern and distribution of Carboniferous oilfields in Williston Basin (from Kent, 1985).

The Parkman Field, with an estimated IVIP of 24×10^6 m^3 of 855 kg/m^3 oil, is unique in having a continuous oil column through the Souris Valley beds (Lodgepole) and the Tilston beds (Mission Canyon). There are 20 pools in the combined Tilston and Souris Valley subcrop play.

Similarly, shelf-margin carbonates of the Alida and lower Frobisher beds and sublittoral carbonates associated with supratidal evaporites of the upper Frobisher are oil-bearing. Oil is trapped in the Frobisher beds of the Innes Field within oolitic-pelletoidal packstone (Crabtree, 1982); this is one of the largest pools, having an IVIP of 10×10^6 m^3. There are 78 pools in the trend, including Hastings, Willman, and Workman.

The Midale beds form the most important producing belt, containing the Weyburn U Pool, assigned an IVIP of 117×10^6 m^3, and seven other pools with IVIPs greater than 10×10^6 m^3. The oil reservoirs of the Midale beds are coarsely crystalline vuggy dolomite and fractured, bioturbated, calcareous dolomite (Kaldi, 1982). Lithological traps downdip from the unconformity (Type Ls4-Ub2) are typical of the Midale, Weyburn, and Steelman fields (Illing et al., 1967). The reservoirs include micro- to fine-crystalline dolomite (porosity 20 to 35%, permeability 5 to 50 mD) and fragmental and pelletoid limestone with a variably dolomitized matrix (porosity 10 to 20%, permeability 0.5 to 10 mD). Permeability decreases updip to 0.3 to 3 mD in non-dolomitized chalky lime-mudstone or skeletal lime-mudstone. Similar decreases in permeability along strike in carbonate rocks control the lateral extent of the fields.

Pools in the Ratcliffe interval include those along the Hummingbird trend trapped in structures commonly associated with multistage salt solution (Type M^3); the largest pool is Flat Lake with an IVIP of less than 10×10^6 m^3. In the Oungre trend, early diagenetic changes form traps (Ld1-Ls4) in shallow shelf dolomites and limestones (Hartling et al., 1982).

Jurassic

Sandstone and siltstone of the redbed facies of the Triassic or Lower Jurassic, lower Amaranth Formation produce oil in the Waskada-Pierson area of southwestern Manitoba. The reservoir sands overlie the erosional edge of Lower Carboniferous strata. Hydrocarbon migration upward from Carboniferous sources and across the unconformity was apparently blocked by impermeable beds within the lower Amaranth Formation (Barchyn, 1984). The Waskada Lower Amaranth Oil Pool had IER of 3×10^6 m^3 (IVIP are not published for Manitoba).

Twenty-one Jurassic oil fields are found in southwestern Saskatchewan along a north-trending belt 70 km east of Maple Creek (Christopher, 1984). Reservoirs are in Middle and Upper Jurassic sandstones of the Shaunavon, Roseray, and Success formations (Fig. 6A.23). There are two contiguous productive belts, the Shaunavon trend and the Roseray-Success trend. Oil migration from Paleozoic (Christopher, 1984) or Jurassic (van Delinder, 1984) sources through a system of faults and fractures has been postulated. Alternatively, the Jurassic oil may have migrated from adjacent Jurassic source beds (Wilson, 1986).

Hydrocarbons are trapped in sandstones and carbonates of the Shaunavon Formation by impermeable shale of the Shaunavon and also by overlying Rierdon shales. Shaunavon oil reservoirs are found in northeasterly oriented beach and shoal sandstone bodies and northwest-trending channel-fills of sand and shell debris. The beach deposits are 5 to 20 m thick, 3 to 7 km wide, and 15 to 20 km long, and form the main reservoirs at Rapdan, Whitemud, Bone Creek, Instow, and Dollard.

The Roseray and Success producing trend overlaps and extends north and northeast of the Shaunavon trend. The Upper Jurassic Roseray Formation is dominated by fine-grained, well sorted, quartzose sandstone intercalated with marine shale, deposited as lensoid clinobeds with

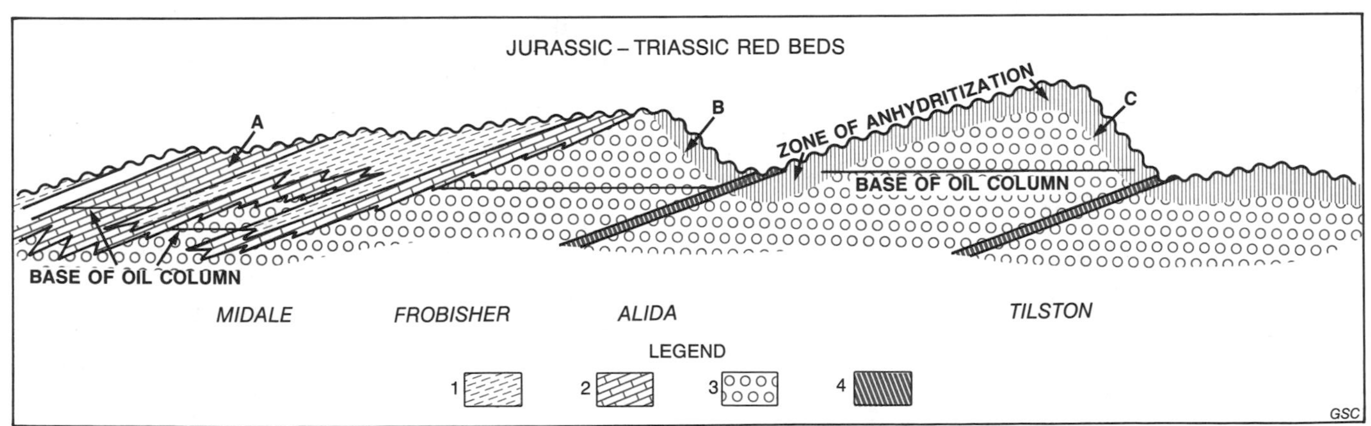

Figure 6A.38. Schematic cross-section of oil traps at the sub-Mesozoic unconformity in Williston Basin. At A the trap is dependent on the seat-seal (Trap Type Ub2-Ld1); at B it is modified by diagenesis (Trap Type Ub2-Ld1); at C it is independent of seat-seal (Trap Type Ub1-Ld1). Rock types: 1 - low permeability siliciclastics; 2 - dolomitized lime mudstone with skeletal and peloidal wackestone; 3 - oolitic, peloidal and crinoidal wackestones, grainstones, and packstones; 4 - low permeability carbonates (adapted from Kent, 1985).

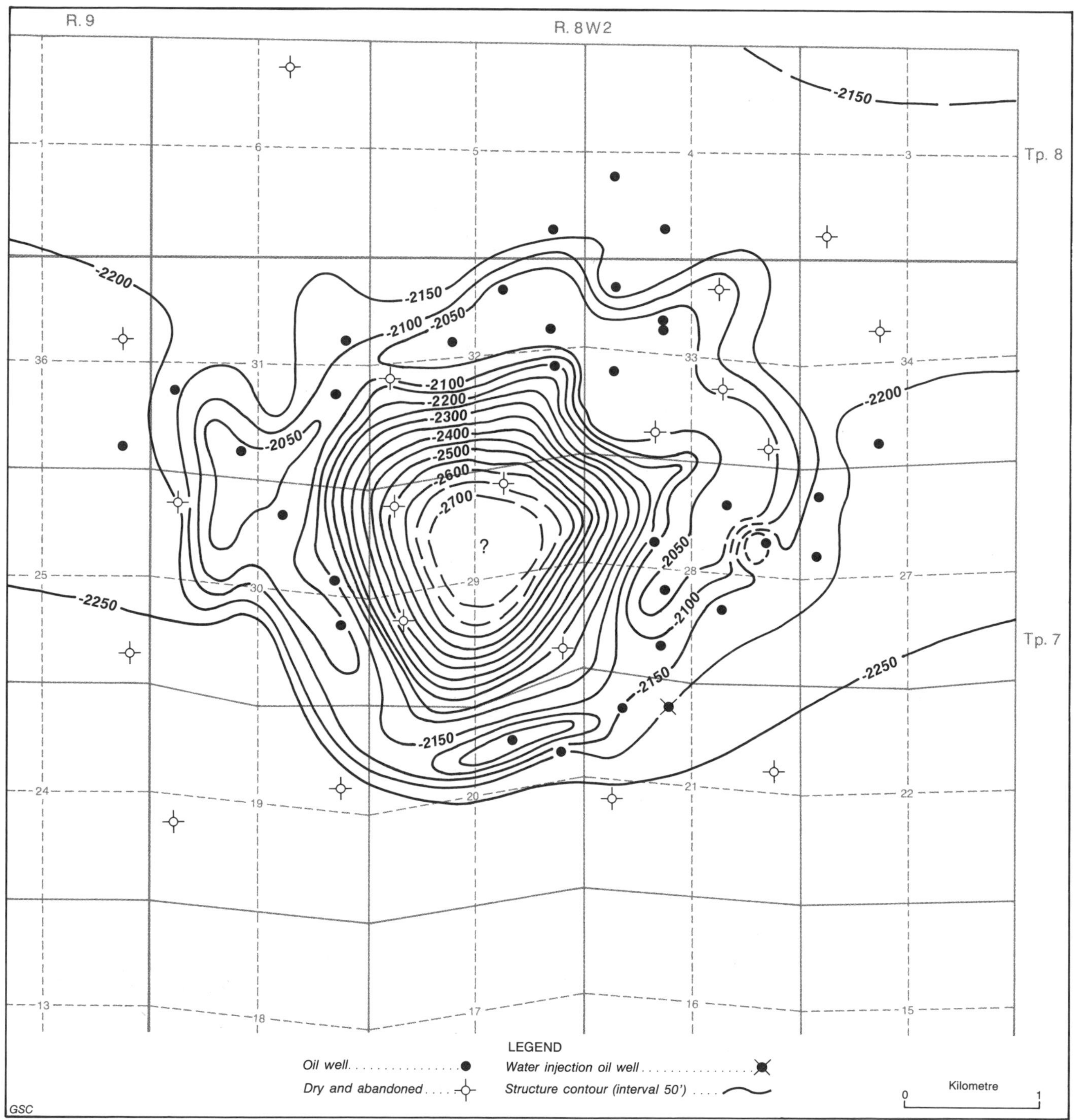

Figure 6A.39. Map of the Viewfield Oil Field where oil is trapped in a structure interpreted as an astrobleme (Trap Type M1). Structure contours on Lower Carboniferous horizon (after Sawatsky, 1975).

southeasterly slope (Fig. 6A.40). The lenses of sandstone change facies to shale both up and down slope. The overlying Success Formation of Late Jurassic and Early Cretaceous age is a kaolinitic quartzose sandstone ranging in grain size from very fine to coarse. Both sandstones are deeply dissected by pre-Cantuar erosion at an intra-Cretaceous unconformity. Where the Cantuar valley fill is impermeable it acts as a seal. At Gull Lake and Suffield, oil is trapped in the updip western facies edge of the Roseray Formation and by Cretaceous valley-fill in a combination lithological-unconformity trap (Type Ls3-Ub1). Other fields, including Premier, Verlo, Hazlet, Fosterton, Cantuar, Success, and Battrum, occur in paleogeomorphic traps (Type Ub1) having the form of buttes and mesas buried by less permeable Cretaceous sediments (Christopher, 1984). Production in all fields is from both Roseray and Success sandstones, though the proportions vary greatly. The Battrum pool is the largest in the trend and has been assigned an IVIP of 29 x 10^6 m^3 with 23% recoverable.

Oil is also found in Jurassic sands of southeastern Saskatchewan in Wapella, Moosomin, and Red Jacket fields. Accumulations of the Wapella sand may be related to local structure (Kreis, 1985).

Cretaceous

The most important hydrocarbon accumulations in the Cretaceous rocks (Fig. 6A.26) of Williston Basin, excluding the heavy oils of the Lloydminster area, are (1) the light oils and non-associated gas of the Viking Formation found in the Coleville region, and (2) an extension of a trend from the Alberta border near Hatton of low-pressure, Upper Cretaceous natural gas fields, which include a number of Viking gas pools.

Throughout much of southern Saskatchewan, basal Cantuar sands of the Mannville Group are in contact with older reservoirs. Migration of oil from the older strata into Cantuar beds has occurred. Known accumulations of this type are predominantly in sediments filling valleys carved in the pre-Cantuar erosion surface (Trap Type Ua3). They are not economically important.

Virtually all significant Cretaceous reserves of crude oil are in the Viking sands of west-central Saskatchewan (Fig. 6A.26). This complex extends east-west, and consists of reservoirs (Trap Type Sd2) structurally controlled by drape over erosional cuestas developed on the Lower Carboniferous Madison Limestone, and blanketed by Cretaceous Mannville and Joli Fou strata. Two possible origins are suggested for Viking petroleum: indigenous petroleum driven from the enclosing Colorado shales by the northeasterly flow of formation waters; and cross-formation flow from Paleozoic source beds (Simpson, 1984). There are seven fields, but over 90% of the recoverable reserves is concentrated in the Dodsland, Smiley-Dewar, and Eureka fields. The Dodsland Viking Eagle Lake Unit has an IVIP of 13 x 10^6 m^3. The IVIP of non-associated gas along this trend is about 17 x 10^9 m^3. Gas with an IVIP of 14 x 10^9 m^3 is found in the Viking Formation of the Hatton area, where the accumulation conditions are similar to those of the Upper Cretaceous gas sands described below.

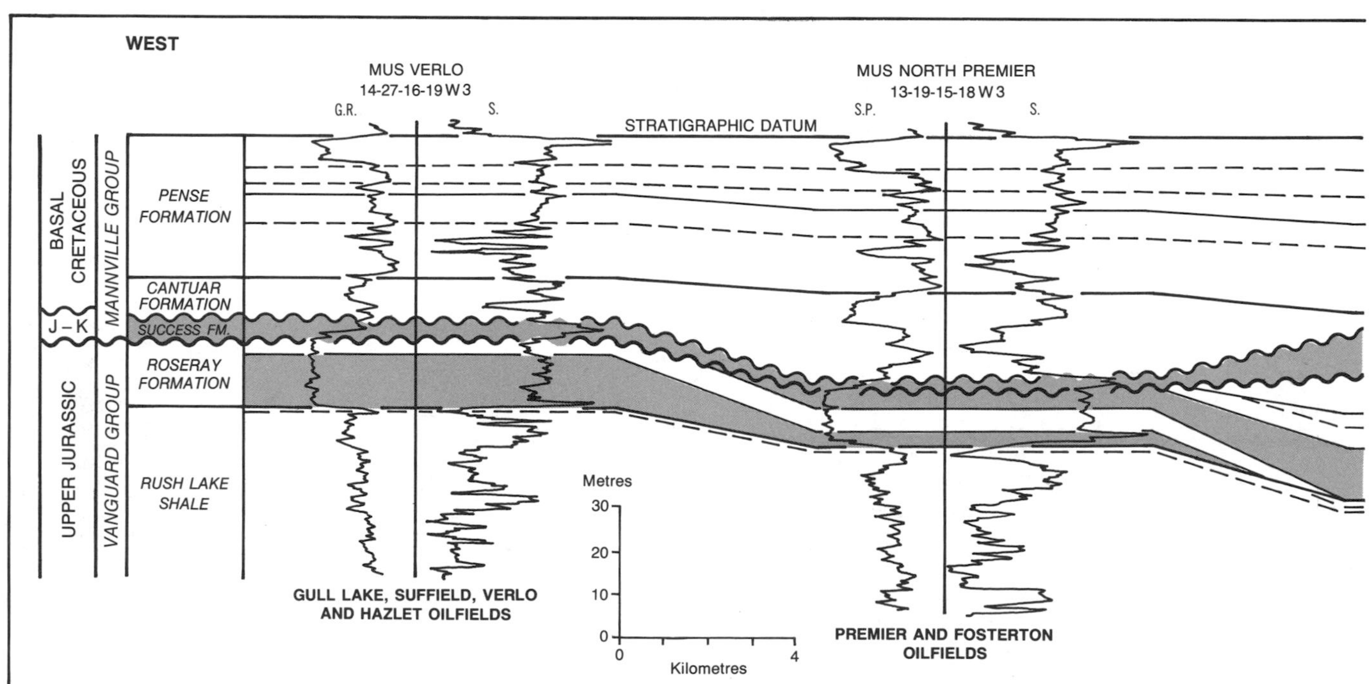

Figure 6A.40. Stratigraphic section of Jurassic and basal Cretaceous formations across the producing trend in western Saskatchewan (modified from Christopher, 1974).

The principal gas-producing formations of the Upper Cretaceous strata in Saskatchewan are the Second White Specks and Medicine Hat sands of the Colorado Group and the Milk River Formation. These reservoirs are mainly argillaceous sandstone and siltstone, with porosities less than 30%; the Medicine Hat sandstone averages 16% and the Milk River generally less than 20% (Simpson, 1984). Permeabilities range between 1 and 250 mD in the Medicine Hat sandstone and are commonly less than 1 mD in the Milk River sandstone. The major structural influences on hydrocarbon accumulations are (1) northeast-trending regional uplifts, (2) northeast- and northwest-trending linear features outlining basement blocks, (3) solution-generated collapse features associated with the Middle Devonian Prairie Evaporite, and (4) drape folds over paleotopography at the sub-Cretaceous unconformity (Simpson, 1984). The estimated IVIP of the gas in the Upper Cretaceous sandstones is 46×10^9 m^3.

Minor gas production is obtained from other thin sandstones within the Cretaceous succession, including the Spinney Hill Sandstone, St. Walburg Sandstone, and Ribstone Creek Tongue of the Judith River Formation of west-central Saskatchewan, and the Phillips Sandstone of the southwest.

Foothills Belt

The Foothills Belt lies between Alberta Basin and the Rocky Mountains. It extends approximately 1000 km north from the International Boundary to the southern District of Mackenzie, covering an area of 190 000 km^2 (Fig. 6A.41). Little exploration has taken place in the mountains, and the only production of any significance comes from part of the Waterton Field that has been overridden by the Lewis Thrust Sheet.

All the important accumulations in the Foothills are trapped in overthrust anticlines and monoclines (Trap Types Sa2 and Sf2). Except for the Turner Valley Oil Field, foothills fields produce natural gas, most of which is sour. Volumes of gas by geological system are shown in Figure 6A.42. Hydrogen sulphide from the Foothills sour gas fields account for 20 Mt (about 20%) of the established sulphur reserves of Western Canada remaining in 1987.

Three producing sectors are recognized. The Alberta Foothills sector, extending from the International Boundary to Athabasca River and producing mainly from Lower Carboniferous carbonates, contains most of the larger pools and reserves. The Peace River sector, embracing the Inner Foothills belt of British Columbia from Athabasca River to Liard River, produces mainly from Upper Triassic carbonates. The Liard sector, spanning the boundary between British Columbia and the District of Mackenzie, contains two large Middle Devonian gas fields. The "Foothills" status of the Liard sector could be challenged, but it is convenient to include it in this section.

Alberta Foothills

Nearly all the gas reserves in the Alberta Foothills occur south of the North Saskatchewan River (Fig. 6A.41). Although many structures are found north of this river, the character of the Carboniferous reservoir rock deteriorates northward, and no large accumulations have been found.

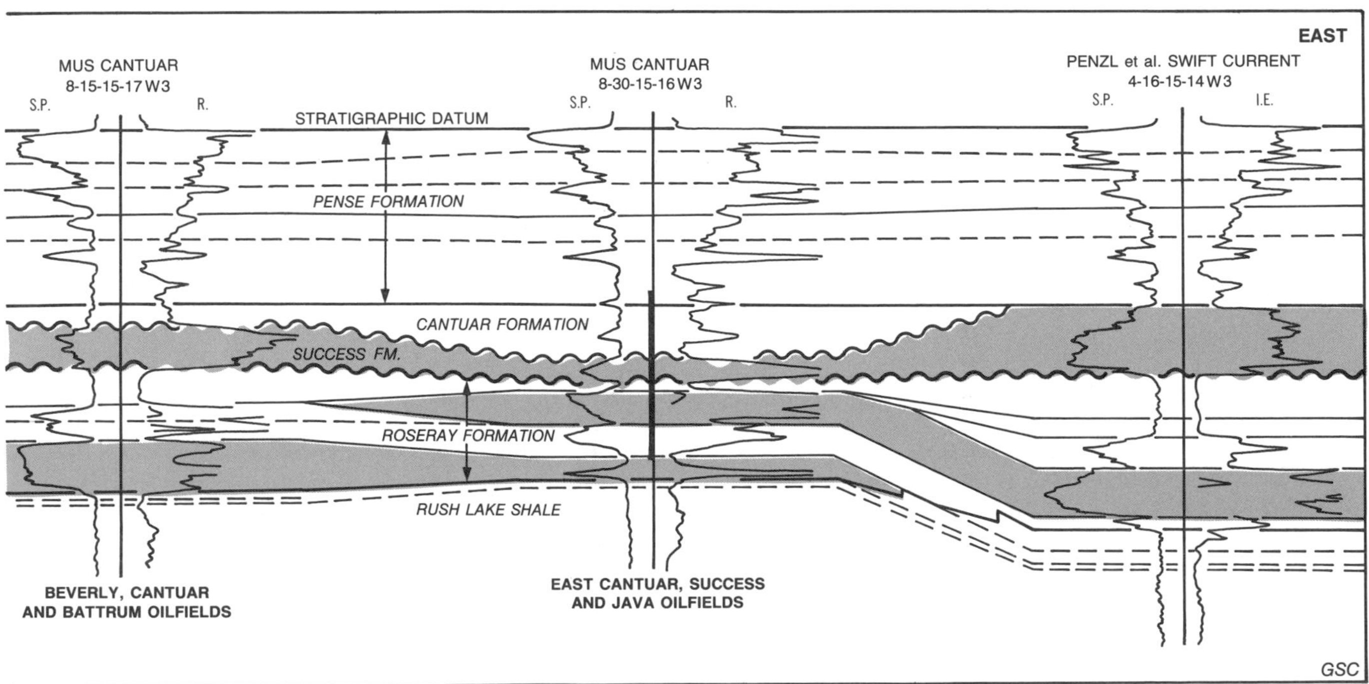

Figure 6A.40. Cont.

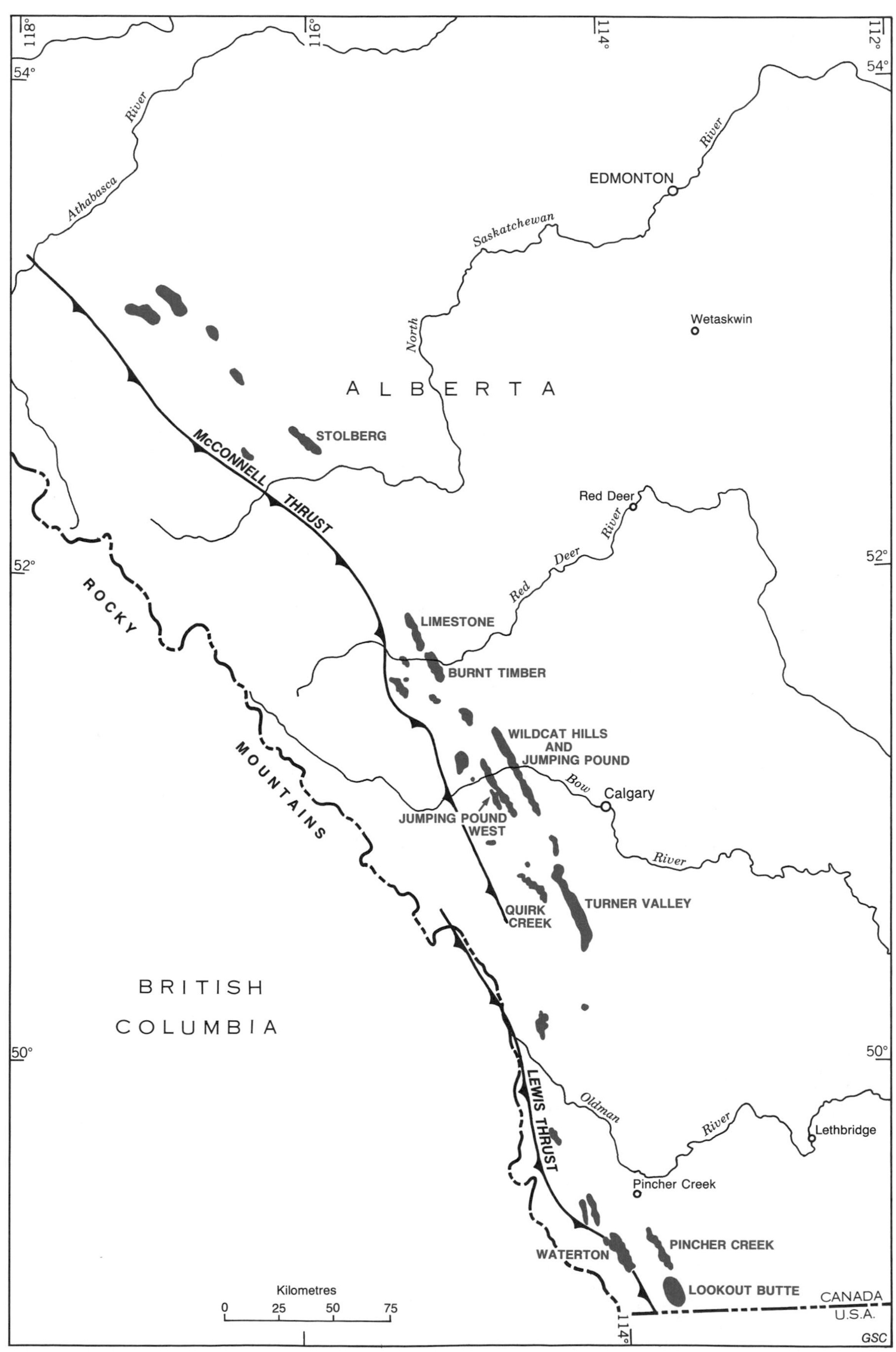
118°
116°
114°
112°
54°
54°
52°
52°
50°
50°
Athabasca
River
EDMONTON
River
Saskatchewan
North
Wetaskwin
A L B E R T A
STOLBERG
McCONNELL THRUST
Red Deer
Red Deer River
ROCKY
MOUNTAINS
LIMESTONE
BURNT TIMBER
WILDCAT HILLS AND JUMPING POUND
Bow
Calgary
JUMPING POUND WEST
River
QUIRK CREEK
TURNER VALLEY
BRITISH COLUMBIA
LEWIS THRUST
Oldman
River
Lethbridge
Pincher Creek
PINCHER CREEK
WATERTON
LOOKOUT BUTTE
Kilometres
0 25 50 75
CANADA
U.S.A.
114°
GSC

The Alberta Foothills may be subdivided into an outer and an inner belt, which differ in the number of superposed Paleozoic imbricates and the stratigraphic level of the décollement.

In the Outer (eastern) Foothills nearly all structural traps are single or imbricate, commonly folded, thrust sheets of Lower Carboniferous carbonates. Most sheets do not contain Devonian carbonates. The thrust planes pass into bedding planes within Banff shale near the base of the Lower Carboniferous section. Trap size is determined by reservoir development in carbonate rocks of the Lower Carboniferous Rundle Group and length of the truncating thrust fault. Marine shales of the Jurassic Fernie Formation form an effective seal. Nearly all the traps are closed eastward against thrust faults, but eastward-facing anticlinal limbs above the thrust also play a role. Horizontal displacements of individual thrusts at the level of the Carboniferous strata are generally small, up to a few kilometres. Vertical displacements can range up to hundreds of metres. All reservoirs in the structures are filled to capacity with gas except the Turner Valley structure, which has gas, oil, and water, and the Highwood structure, which is mainly water-filled. Residual solid hydrocarbons in many of the gas pools suggest that the structures at one stage of thermal or structural history may have contained oil, which later was either decomposed thermally or displaced by migrating gas. The fields in the eastern Foothills share many characteristics and differ only by the amount of thrust displacement, vertical uplift above the regional gradient, and reservoir development. At the time of its discovery in 1914 the Turner Valley Field (Fig. 6A.43; Gallup, 1975) produced light oil and large volumes of gas from the Lower Cretaceous Blairmore Group. In 1924, deeper drilling encountered large volumes of wet gas from porous carbonates of the Lower Carboniferous Turner Valley Formation. Subsequently, prolific crude oil production with a low gas/oil ratio was achieved in 1936. The Pincher Creek and Lookout Butte fields, typical of the Outer Foothills, have thrust imbricates of porous Rundle carbonates raised above the regional structural gradient (Gordy and Frey, 1977). Similarly, the Jumping Pound (Rupp, 1969) and Sarcee fields are contained in relatively simple thrust plates involving only Lower Carboniferous and younger strata (Davidson, 1975). Other fields in the Outer Foothills include Whiskey Creek (Stein, 1977), Burnt Timber, and Stolberg.

The Inner (western) Foothills are characterized by stacked thrust sheets that provide multiple targets. The deeper thrust sheets are generally emplaced in a step-like manner, with displacements of up to tens of kilometres. Their emplacement folded the overlying but previously emplaced thrust sheets. Variation in the number and thickness of thrust sheets stacked in the Inner Foothills gives rise to prominent axial culminations and depressions. Both simple fold and fault closures occur, but not all closed structures are hydrocarbon-bearing or filled to capacity, probably because of the complex maturation and migration history of the hydrocarbons or a lack of seals. The principal reservoirs are in Lower Carboniferous strata, but gas is found also in the Upper Devonian Wabamun Group and Nisku Formation at Waterton and Panther River respectively. The Carboniferous reservoirs exhibit fine intergranular and vuggy porosity in dolomite or vuggy

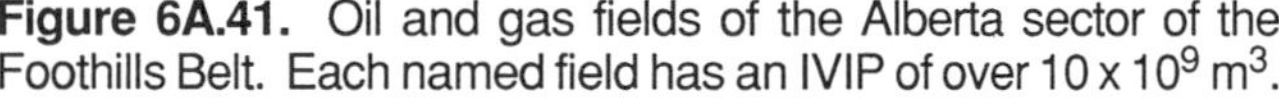

Figure 6A.41. Oil and gas fields of the Alberta sector of the Foothills Belt. Each named field has an IVIP of over 10 x 10^9 m^3.

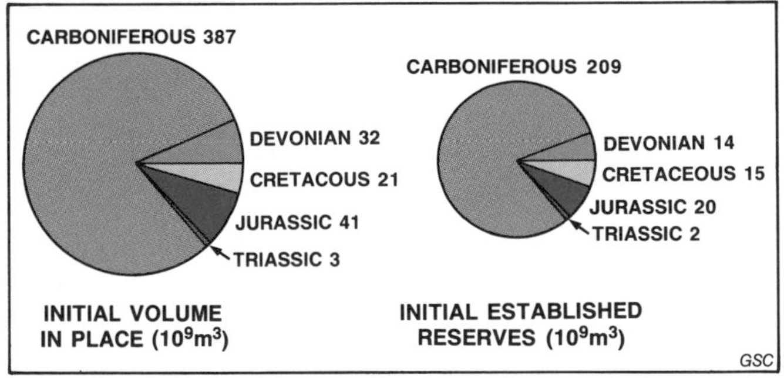

Figure 6A.42. Distribution of natural gas volumes by geological system in the Foothills Belt.

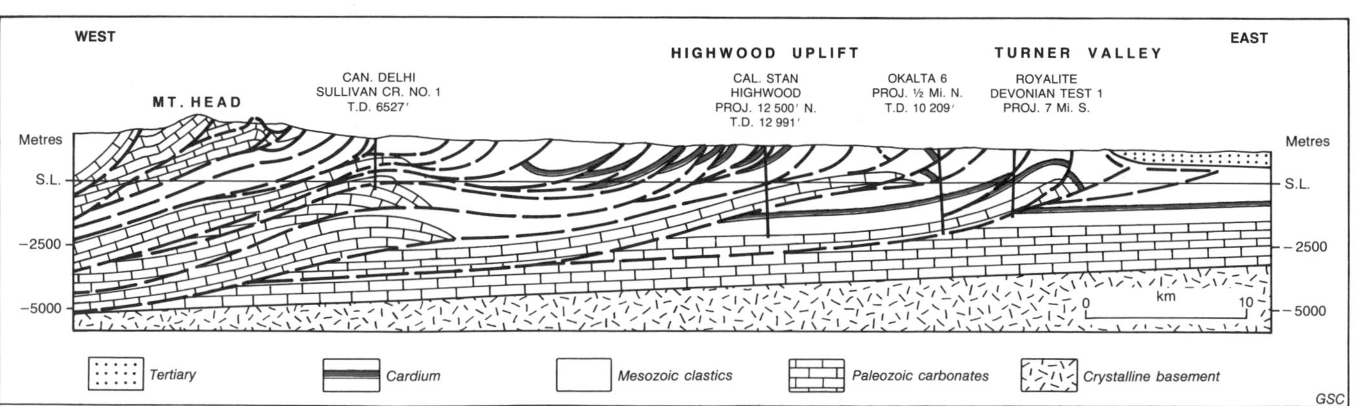

Figure 6A.43. Schematic structural cross-section from Rocky Mountains to Plains across Highwood uplift and Turner Valley in southern Alberta (from Bally et al., 1966).

Figure 6A.44. Natural gas processing plant, Waterton, Alberta (photo Shell Canada Resources Ltd.).

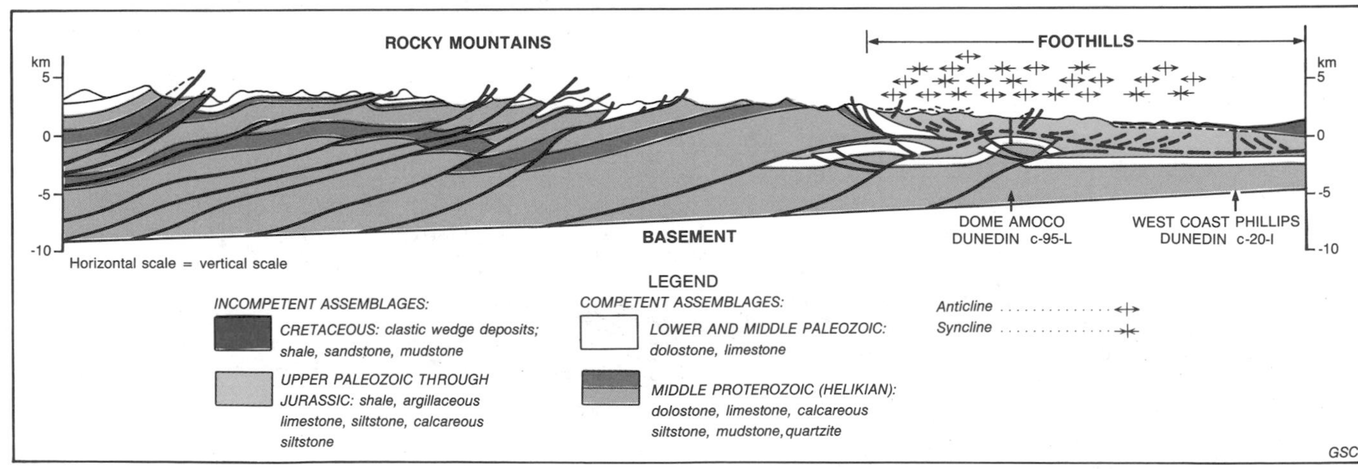

Figure 6A.45. Structural section across the Rocky Mountains and Foothills Belt north of Peace River, B.C. In contrast to the overthrust pattern of the Alberta foothills, this sector is characterized by blind thrusts and broad folds (after R.I. Thompson, 1989).

porosity in crinoidal limestone. Much of the permeability is the result of fracturing of the brittle carbonates. Reservoirs are at depths of 3000 to 4000 m.

The Waterton Field (Gordy and Frey, 1977; Fig. 6A.44) is one of Canada's largest, with an IVIP of 117 x 10^9 m^3 and an estimated ultimate gas reserve of 62 x $10^9 m^3$. The structure is a complex stack of at least three major thrust sheets involving Paleozoic carbonates. Folds with locally overturned beds and many subsidiary faults carrying intermediate thrust slices complicate the structural picture. The Savanna Creek structure consists of four separate slices of Lower Carboniferous carbonate rocks stacked and folded in a doubly-plunging anticline (Hennessey, 1975). Other fields in this belt include Coleman, Moose Mountain (Ower, 1975), Panther River, Limestone, and Nordegg.

A thin organic-rich shale, the Exshaw Formation (Bakken equivalent) at the base of the Lower Carboniferous succession, is considered to be the major source rock for the petroleum of the Foothills. Maturation and expulsion of hydrocarbons probably began in Jurassic time in the deeper part of the basin to the west and then migrated eastward in conjunction with the early thrusting and sedimentation in foredeeps that developed ahead of the advancing thrust front. Thus, the present day accumulations are probably the result of gas migrating from earlier formed traps that were destroyed or modified by subsequent thrust events and post-orogenic uplift.

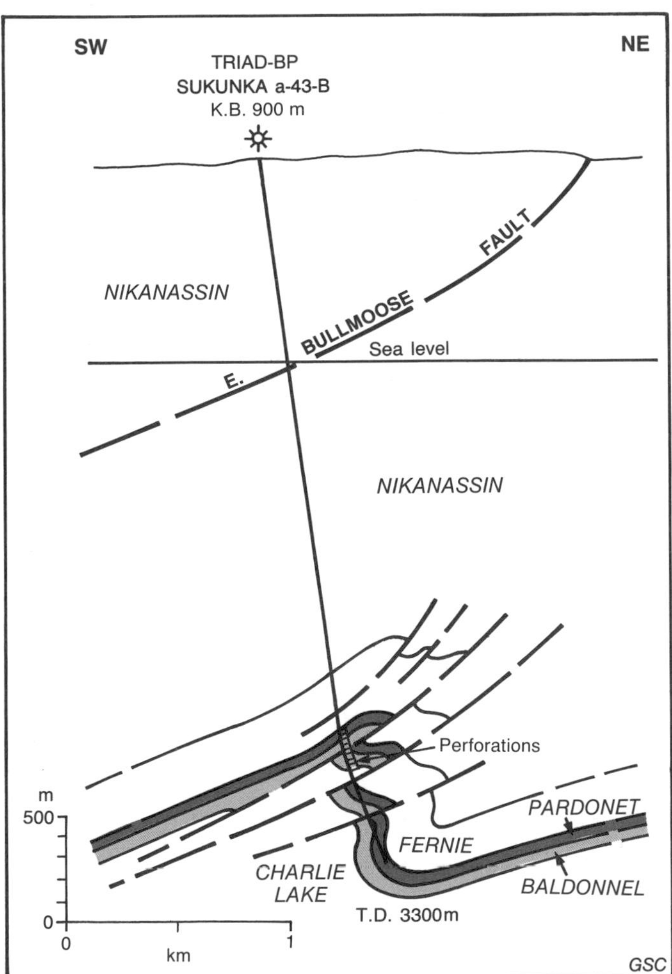

Figure 6A.46. Structure of the Sukunka Field of northeastern British Columbia (from Barss and Montandon, 1981).

In the southern part of the Alberta Foothills, coal-rank studies indicate that thermal maturation increases from east to west (Norris, 1971; Hacquebard and Donaldson, 1974). Coal-rank studies in the northern part of the Alberta Foothills (Kalkreuth and McMechan, 1984) suggest, on the contrary, that in that region the level of thermal maturation of Triassic and Carboniferous carbonates decreases from dry gas-overmature under the Outer Foothills to oil preservation-wet gas under the western part of the Inner Foothills.

Gas composition varies from dry and sweet to very sour. The H_2S component generally ranges from 1 to 6% in Lower Carboniferous reservoirs, but concentrations as high as 33% are found in producing fields such as Waterton. The Devonian reservoirs generally have higher concentrations of H_2S, with the maximum concentration encountered being over 95% at Panther River (Eliuk 1984). Sulphur has become a significant commercial by-product of gas production and is the main source of income in several fields.

Peace River Foothills

Hydrocarbon accumulations have been found in a series of structures in the Inner Foothills of northeastern British Columbia as far north as Prophet River. The region sometimes referred to as the Outer Foothills of this sector has been included in the discussion of the Alberta Basin because the structural style is different from that of the Outer Foothills of Alberta: it comprises a region of the Plains where Laramide folds extend beyond the limit of the surface overthrust belt. The numerous thrust faults with large displacements, which are the rule in the Alberta Foothills, give way in the Peace River Foothills to folds of large amplitude, blind thrusts, and thrusts with small displacements (Fig. 6A.45; Thompson, 1981). Reservoirs are found in rocks ranging from Early Carboniferous to Cretaceous in age, but most of the production is obtained from Triassic strata (Barss and Montandon, 1981).

Eight substantial gas accumulations, with reserves of about 27 x $10^9 m^3$, occur in thrust-faulted structural traps south of Pine River. Compared to those in Alberta, the structures have greater fold complexity at the leading edge of the thrust sheets (Fig. 6A.46). In the Sukunka-Bullmoose gas fields, the main reservoir comprises carbonates of the Triassic Pardonet and Baldonnel formations deposited in supratidal to shallow shelf environments. Net pay may be over 90 m. Porosities are generally less than 5%, but fractures created by folding permit good to excellent drainage. Initial well deliverability is usually very good, ranging from 350 x 10^3 m^3/d to 1500 x 10^3 m^3/d. Hydrogen sulphide and carbon dioxide concentrations are high, ranging from about 25% to more than 45%.

Liard Foothills

The northernmost gas fields of the Foothills Belt, Beaver River and Pointed Mountain, lie on the southeast flank of the Mackenzie Mountains near the junction of the boundaries between British Columbia, Yukon Territory, and the District of Mackenzie. Both are anticlines that are broad and relatively simple at surface but have thrust-faulted cores of Middle Devonian carbonates. Closure is controlled by the faulted anticlines. The porosity is low, but the reservoir drainage is enhanced by fractures. The Beaver River Field (D.M. Snowdon, 1977) produces from an extensively fractured reservoir (Fig. 6A.47), whose porosity and permeability are largely the result of extensive diagenetic and tectonic alteration. The reservoir is 670 m thick but has a weighted average porosity of only 2.09%. The IVIP of the Beaver River Gas Field has been estimated at 7.4 x 10^9 m^3 (British Columbia Energy Mines and Petroleum Resources, 1985). No IVIP is available for Pointed Mountain, but the IER is estimated at 2 x 10^9 m^3 (Energy Mines and Resources Canada, 1986).

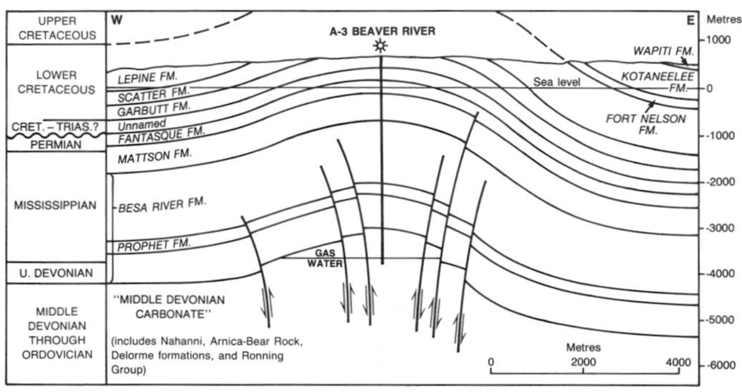

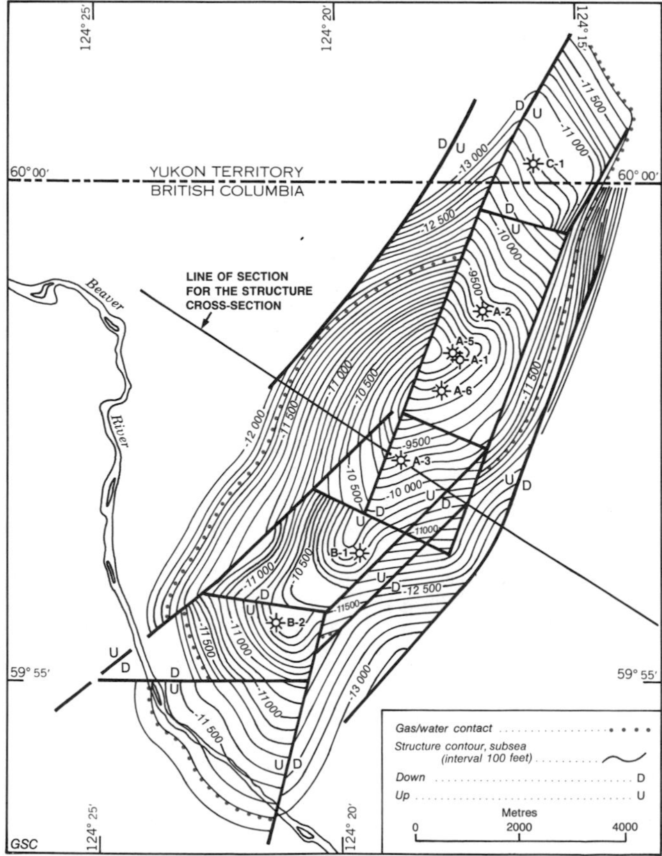

Figure 6A.47. Structural interpretation of the Beaver River anticline and its gas pool (from D.M. Snowdon, 1977).

Northern Basins

The Northern Basins region is that portion of Western Canada Basin extending north from Tathlina Arch near the 60th parallel to the southern boundary of the Beaufort-Mackenzie Basin. The region has an area of approximately 500 000 km^2. Numerous depocentres developed in this region during the course of geological history, but those formed in one major sequence seldom persisted into the next sequence. Consequently the areal subdivisions (Fig. 6A.48) used to describe the petroleum resources of this region have a certain artificiality and can hardly be described as "basins". The first of these subdivisions is the Upper Mackenzie River area (Law, 1971), which includes Tathlina Arch (a positive tectonic feature during Middle Devonian time) and several important Devonian depocentres underlying Great Slave Plain. The second region, the Lower Mackenzie area, lying north of Fort Norman and east of the Richardson Mountains and drained by Anderson River, Arctic Red River and part of Peel River, is underlain by a thick wedge of Paleozoic sediments unconformably covered by a thinner wedge of Cretaceous clastics. The third, the Eagle Fold Belt lying west of the Richardson Mountains, is a modern physiographic intermontane basin preserving an economically important wedge of Upper Paleozoic sediments. The fourth, Old Crow Basin, receives only brief comment as it is entirely untested by the drill. The northeasterly part of Aklavik Arch, including the Eskimo Lakes block, is considered part of the Beaufort-Mackenzie Basin and is described later. Generalized stratigraphic columns are shown in Figure 6A.49.

The earliest drilling in the region took place near oil seepages along the Mackenzie River, resulting in the discovery of the Norman Wells oil field in 1920. This is the only major oil accumulation so far discovered in the region. By the end of 1986 about 730 wells had been drilled in the Northern Basins, most of them since 1960. No other fields have been brought into production.

Upper Mackenzie River area

The Upper Mackenzie River area contains numerous structural elements. Some of the anticlines and thrust-faulted structures on the west side of the area have been drilled without encouraging results, but most exploratory drilling has been concentrated in the Great Slave Plain and over Tathlina Arch. Early exploration was

Figure 6A.48. Northern Basins region of Western Canada Basin and Beaufort-Mackenzie Basin.

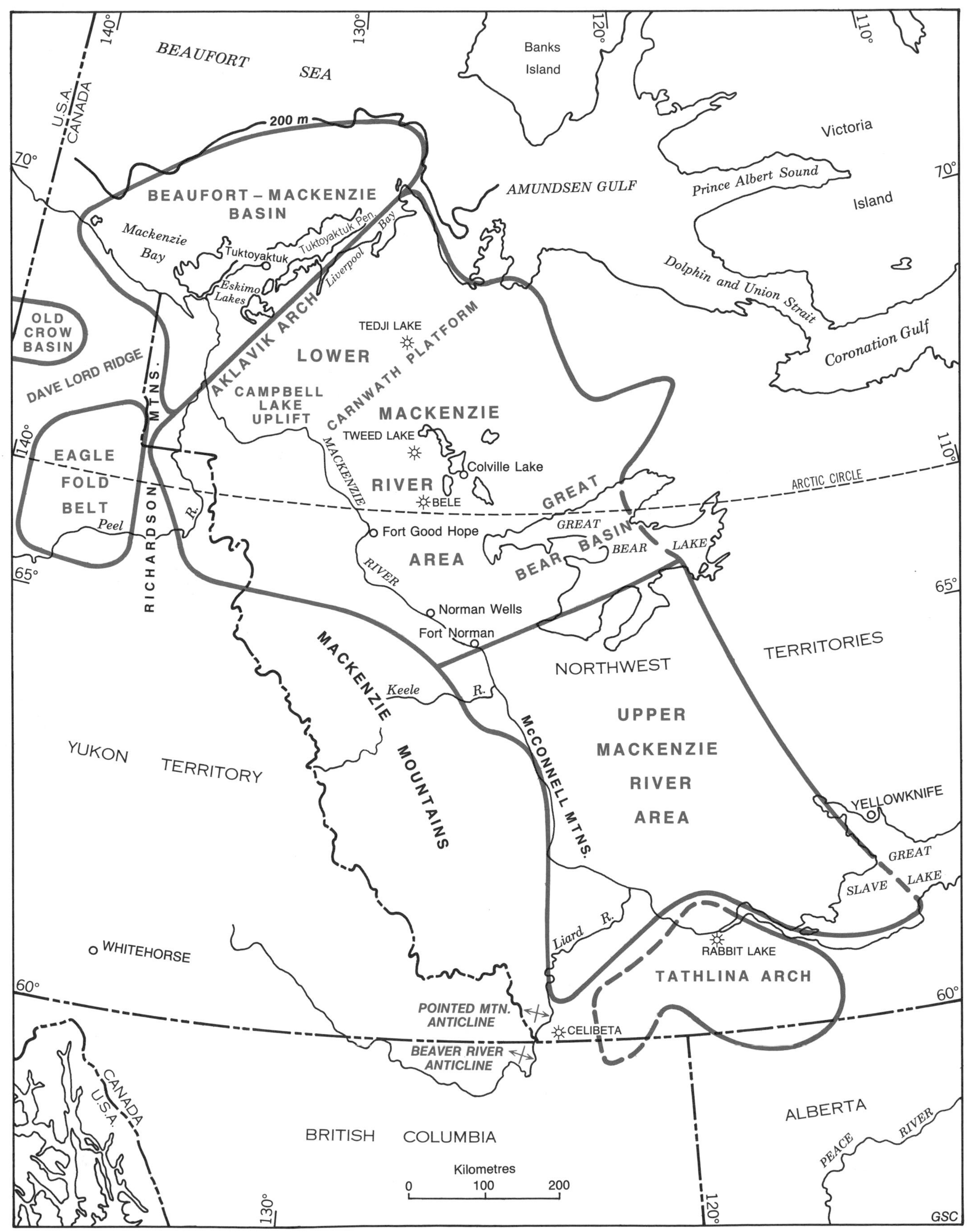
BEAUFORT SEA
U.S.A.
CANADA
140°
130°
120°
110°
70°
65°
60°
Banks Island
Victoria Island
Prince Albert Sound
AMUNDSEN GULF
Dolphin and Union Strait
Coronation Gulf
200 m
BEAUFORT – MACKENZIE BASIN
Mackenzie Bay
Tuktoyaktuk
Tuktoyaktuk Pen.
Liverpool Bay
Eskimo Lakes
OLD CROW BASIN
DAVE LORD RIDGE
AKLAVIK ARCH
CARNWATH PLATFORM
TEDJI LAKE
LOWER MACKENZIE RIVER AREA
CAMPBELL LAKE UPLIFT
TWEED LAKE
Colville Lake
BELE
EAGLE FOLD BELT
Peel
R.
RICHARDSON MTNS.
MACKENZIE RIVER
ARCTIC CIRCLE
Fort Good Hope
GREAT BEAR BASIN
GREAT BEAR LAKE
Norman Wells
Fort Norman
NORTHWEST TERRITORIES
MACKENZIE MOUNTAINS
Keele R.
McCONNELL MTNS.
UPPER MACKENZIE RIVER AREA
YUKON TERRITORY
YELLOWKNIFE
GREAT SLAVE LAKE
Liard R.
RABBIT LAKE
TATHLINA ARCH
WHITEHORSE
POINTED MTN. ANTICLINE
CELIBETA
BEAVER RIVER ANTICLINE
CANADA
U.S.A.
BRITISH COLUMBIA
ALBERTA
PEACE RIVER
Kilometres
0
100
200
GSC

drawn to the tar seeps on the north shore of Great Slave Lake and to seeps associated with lead-zinc showings (later developed as Pine Point mine) on the south shore. By 1986, about 180 wells had been drilled in an area of 65 000 km^2. The lower Paleozoic rocks under the Great Slave Plain form a west-dipping homoclinal wedge within which broad uplifts, normal faults, and Givetian (upper Middle Devonian) reefs provide drillable anomalies (Law, 1971; de Wit et al., 1973). The Tathlina Arch was differentiated from the broad landmass lying south of it early in the Devonian Period. Like the Peace River Arch, it localized carbonate shoals and reefs during later Devonian time. Again like the Peace River Arch, its subsidence during Late Carboniferous time was accompanied by clastic deposition. Mid-Devonian uplift of Tathlina Arch resulted in the partial truncation of the Slave Point Formation on its crest. The northern limit of the Slave Point, which has been one of the principal drilling targets, is thus partly controlled by erosion and partly by a barrier reef front. Post-Devonian, probably Laramide, normal faulting with a northeasterly trend is locally mappable at the surface. Some of the faulting reactivated Precambrian faults. Numerous shallow exploratory holes have been drilled over Tathlina Arch.

In the Rabbit Lake gas field, consisting of two shut-in wells, the Sulphur Point reservoir is a stromatoporoid-coral limestone with good intergranular porosity. The accumulation occurs on the crest of an anticline on the northwest, upthrown side of a small northeast-trending normal fault. G.K. Williams (1977) indicated that this fault does not extend up to the level of the reservoir (Trap Type is Sd1 - drape fold over deeper faulted structure). The Celibeta structure is probably of Laramide age (G.K. Williams, 1977) and is the site of a one-well gas field probably of the same trap type. Several other small gas discoveries have been made.

In the Great Slave Plain north of Tathlina Arch, the principal targets have been the small high-profile platform reefs of the Horn Plateau Formation (Keg River equivalents). No significant discoveries have been made.

Lower Mackenzie River area

This is a vast region with numerous oil and gas seepages, especially near Norman Wells and Fort Good Hope. It contains the large Norman Wells oil field (Devonian) and several Cambrian gas strikes.

The region as a whole is underlain by a thick wedge of carbonates ranging in age from Cambrian to Middle Devonian and dipping gently westward. This is followed by a wedge of Upper Devonian and Carboniferous clastics. Both wedges are truncated and covered by a thinner layer of Cretaceous clastics. The region is divided structurally into a belt of thrust-faulted anticlines adjacent to the Mackenzie Mountains and a broad homocline to the north and east of it. The homocline is transected by the Keele paleo-arch and by a north-trending belt of Laramide folds and faults in the Colville Lake area. Keele Arch is a large, ancient structure, about 120 km wide, that extends north for 300 km from the junction of Keele and Mackenzie rivers (Cook, 1975). Initially developed prior to the Devonian Period, the arch was reactivated prior to Late Cretaceous time. Basal Cambrian sandstone and younger Ordovician to Devonian rocks occur on its flanks and are overlapped by Cretaceous sediments. In spite of its favourable trapping configuration, no significant shows have been reported.

Potential source rocks exist in shales of Paleozoic and Cretaceous age, including Ordovician to Devonian graptolitic shales and Middle and Upper Devonian bituminous shales; Proterozoic shales may also have potential. The discoveries and important shows in this area are described in ascending stratigraphic order.

In 1974, gas was discovered in basal Cambrian sandstone near Tedji Lake in the Colville Hills (Davis and Willott, 1978). A thin pay zone of about 5 m occurs at depth of 1150 m. The trap is relatively small, but reserves are estimated to be about 2.5 x 10^9 m^3 (Procter et al., 1984). Further discoveries were made 90 km and 135 km south of Tedji Lake in 1985 and 1986 respectively. The reservoir is Lower Cambrian, Old Fort Island sandstone and is sealed by shales of the Mount Cap Formation which are assumed to be the source beds. However, underlying Proterozoic shales may still have potential to generate dry gas (Snowdon and Williams, 1986). Traps of both structural and stratigraphic type may be present.

At some localities west of Norman Wells, dolomites of the lower Paleozoic Franklin Mountain and Mount Kindle formations contain intervals of vuggy porosity

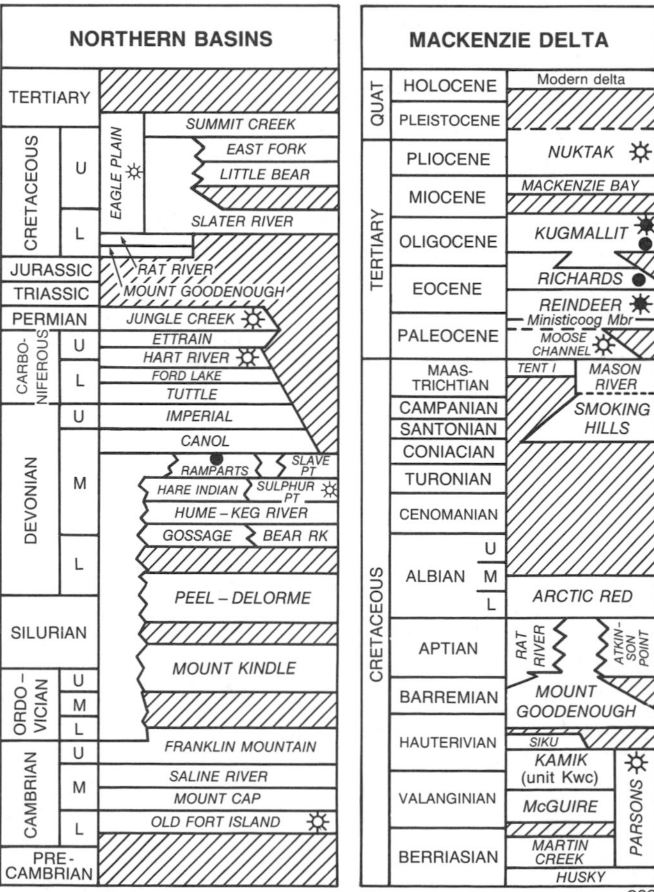

Figure 6A.49. Principal producing zones of the Northern Basins region and of Beaufort-Mackenzie Basin (symbols described in Figure 6A.4) (after Procter et al., 1984).

(Aitken et al., 1982). Large flows of water on drillstem test confirm the existence of significant permeability. A strong showing of heavy oil (935 kg/m^3) from Upper Franklin Mountain cherty dolomite near Fort Norman is, however, believed to have its source in the Cretaceous Slater River Formation (Yorath and Cook, 1981). In general, the lower Paleozoic carbonates lack the lithological differentiation and shale interbeds that would makes them attractive targets. The exception is the western margin of the carbonate platform where it passes fairly abruptly to shales of the Road River Formation (Kunst, 1973). Unfortunately this facies change occurs along the flank of the Richardson Mountains and deep within the Mackenzie Mountains.

The Lower Devonian succession is not very different from underlying Cambrian to Silurian strata, but Middle Devonian strata show greater differentiation into large reef complexes and shale basins. Minor reefs occur on top of the Hume Formation and more important ones in the Ramparts Formation. Both types are associated with oil seepages. The Norman Wells field (Fig. 6A.50, 6A.51),

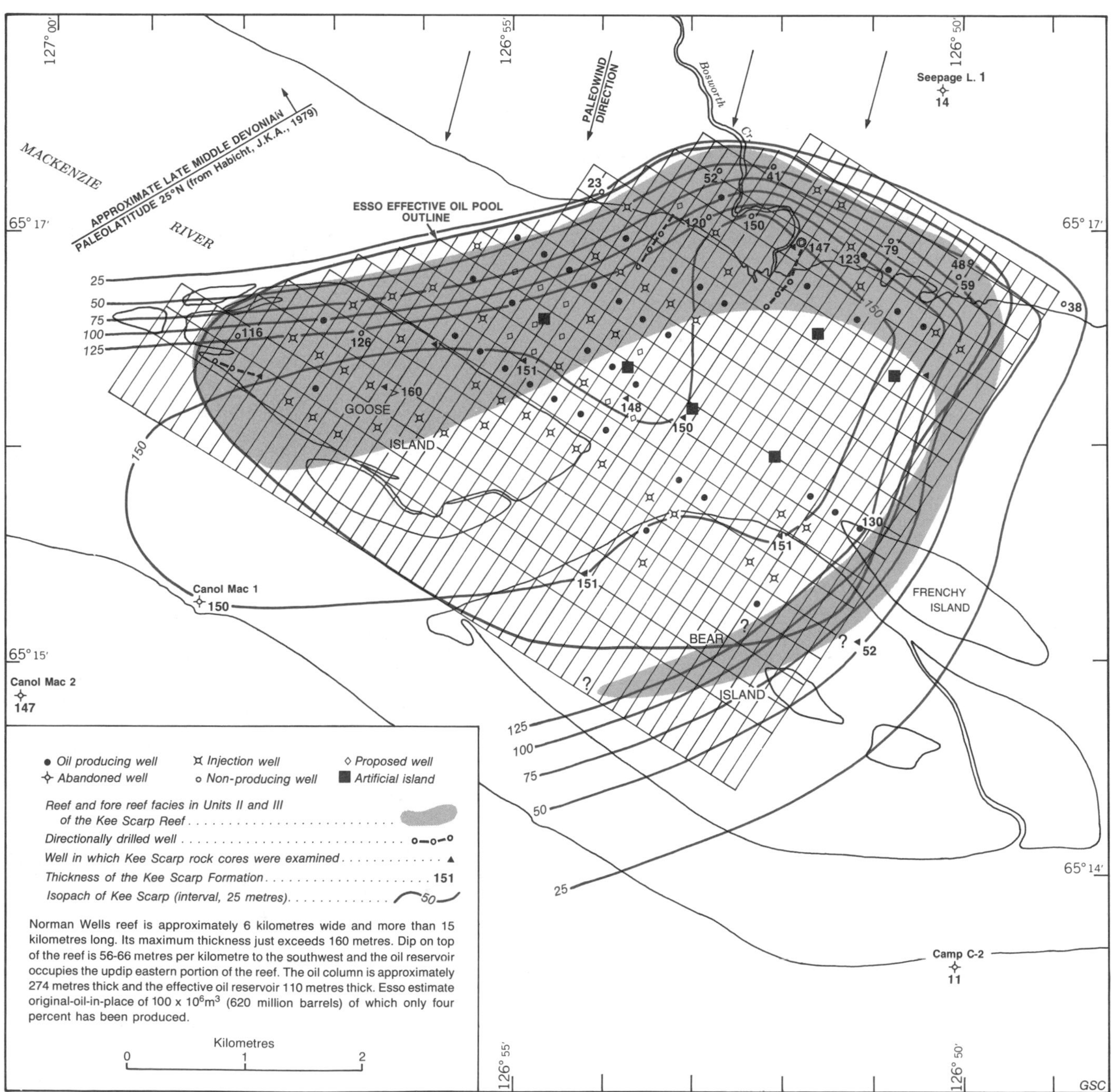

Figure 6A.50. Isopach map of the Middle Devonian Kee Scarp Formation, Norman Wells Oil Field, District of Mackenzie (adapted from Fischbuch, 1984).

which produces from one of these reefs, lies beneath Mackenzie River at the southern end of the Lower Mackenzie River area, about 50 km south of the Arctic Circle (Fischbuch, 1984). Oil is trapped in the updip periphery of a Middle Devonian (Givetian) subsurface reef complex of the Ramparts (Kee Scarp) Formation (Trap Type Rd1). This field had an estimated IVIP of $100 \times 10^6\ m^3$. Stratigraphic relationships similar to those within this reservoir are evident in a large Ramparts reef complex that is well exposed in a 30-km long sector of the Mackenzie Mountain front, 100 km west of Norman Wells (Muir et al., 1984). Where penetrated in other areas, the Ramparts Formation generally has had poor porosity.

Cretaceous rocks of the region northeast of the Mackenzie River have low potential for the production of hydrocarbons (Yorath and Cook, 1981) because they are not deeply buried and are widely exposed. Little Bear and East Fork sandstones pinch out eastward into shales but have poor porosity and low potential as reservoirs. Although oil and gas seeps occur on Carnwath Platform, Cretaceous rocks are immature in that area and the hydrocarbon source must be in Paleozoic rocks. In parts of Great Bear Basin, the basal Cretaceous sand is known to be porous and to occur at shallow depths. Cretaceous sandstone may have more potential southwest of Mackenzie River, where a small show of medium oil was recovered (Aitken et al., 1982).

Eagle Fold Belt

Twenty four wells have been drilled in this belt, resulting in three gas discoveries and one gas and oil discovery in Carboniferous sandstones, but the region is far from markets and no development has taken place. A significant show was reported from the Cretaceous beds. Long, north-trending anticlines, some with associated thrust faults, are mapped in Cretaceous beds at the surface, and their culminations have been tested by several wells. Evidence of hydrocarbons has not been found in the Cambrian to Middle Devonian carbonate sequence, although porosity, characteristically filled with fresh water, has been encountered. Probable source beds for the hydrocarbons are the Devonian-to-Carboniferous bituminous shales of the basinal facies and organic-rich limestones of the slope facies (Martin, 1973).

Cretaceous rocks lie with profound unconformity upon Permian and Carboniferous rocks, which are truncated northward. The deepest producing zone is the Carboniferous Hart River Formation, mainly composed of spicular limestone, but locally also containing conglomeratic sandstone bodies that provide good reservoirs. Chance Sandstone Member, up to 200 m thick, is the most important of these, having yielded oil and gas in three tests. The best well was Chance J-19, with 44 m of gas pay and 13 m of oil pay. During the deposition of the Hart River Formation a mixed carbonate-siliciclastic platform was prograding southward across the region. The clinoforms are clearly visible on seismic sections; deep-water sandstones at their distal terminations have yielded gas in two tests, the best being East Chance C-18 where wet gas flowed from 44 m of sandstone at $160 \times 10^3\ m^3/d$. The well bottomed in the sand, whose full thickness is thus not known (Graham, 1973). These deep-water sandstones have poor porosity and permeability.

Hart River Formation is overlain by the Permian Jungle Creek Formation, consisting of very high-energy, nearshore to non-marine, fine- to coarse-grained and conglomeratic sandstones in the north, which grade to low-energy marine shales to the south. Porosity, locally over 10%, has been reported to deteriorate westward as well as to the south; one well contained 16 m of gas pay.

The Cretaceous succession is also prospective in the Chance area, significant recoveries of hydrocarbon having been obtained from a prominent lower Upper Cretaceous sandstone.

Old Crow Basin

No wells have been drilled in Old Crow Basin, although up to 2500 m of Permian and Mesozoic rocks may be present in addition to an undetermined thickness of Carboniferous limestone overlying Proterozoic or lower Paleozoic rocks. Lawrence (1973) speculated that petroleum source beds, as well as reservoirs, occur in the sedimentary column.

Beaufort-Mackenzie Basin

The Beaufort-Mackenzie Basin, 55 000 km^2 in area, comprises the onshore Mackenzie Delta, Tuktoyaktuk Peninsula, and the adjacent Beaufort Sea to the 200-m isobath. The region has been the site of numerous oil and gas discoveries (Fig. 6A.52). The petroleum geology of the Beaufort-Mackenzie region was outlined by Lerand (1973)

Figure 6A.51. Recently completed production platforms of the Norman Wells Oil Field, in Mackenzie River. The town of Norman Wells, the initial wells, and the refinery are on the north bank, at left. Beyond, the reservoir formation outcrops in Kee Scarp (limestone quarry) (photo - J.D. Aitken, ISPG 3062-14).

and the petroleum potential of offshore areas is discussed in more detail in the companion volume on "The Arctic Ocean Region" (Grantz et al., 1990).

The Aklavik Arch Complex, forming the southeastern margin of the Beaufort-Mackenzie Basin, is characterized by down-to-basin block faulting (Yorath and Norris, 1975; Young et al., 1976). The arch consists of a number of components arranged in a right-hand en-echelon sequence. The stratigraphic succession along the arch is disrupted by five angular unconformities and five disconformities, which attest to prolonged tectonic activity extending from late Proterozoic to Tertiary times. Various components of the complex were intermittently and independently active. One of these, Eskimo Lakes Arch, extends northeast from Campbell Lake Uplift into the offshore area adjacent to the northeastern Tuktoyaktuk Peninsula.

The Beaufort-Mackenzie basin is a pericratonic basin containing as much as 12 km of Mesozoic and Cenozoic clastics (Fig. 6A.53). Reservoirs occur in several upper Mesozoic and Cenozoic sandstones that were deposited in a series of migrating deltaic wedges and associated turbiditic deposits. Initial exploration of the Beaufort-Mackenzie Basin has shown it to be a significant hydrocarbon area. The hydrocarbons are mainly indigenous within Tertiary rocks, but late Cretaceous and Jurassic sources have also been identified (Snowdon and Powell, 1982; and D.F. Stott et al., Subchapter 4J in this volume).

Exploratory drilling was initiated in the area in 1962 with two exploratory tests near Liverpool Bay, but it was not until 1965, with the spudding of the Reindeer D-27 well, that the presence of a thick Tertiary and Mesozoic sequence was established. This well and the next four drilled were abandoned. The first hydrocarbon discovery, which flowed oil from Lower Cretaceous sandstones, was made in 1970 at Atkinson Point (H-25). The first gas discovery was made at Taglu G-33 in 1971, recovering gas and condensate from Eocene sandstones (Bowerman and Coffman, 1975). Gas condensate was discovered in the Parsons Lake Field that same year (Langhus, 1980). Several more small oil and gas fields were discovered onshore before exploratory drilling moved offshore in 1973. The Adgo (F-28) well, drilled from an artificial island in 1974, was the first offshore discovery of oil and gas. By 1976, drilling activity had moved into deeper waters of the Beaufort Shelf using ice-strengthened drillships. The Nektoralik (K-59) well discovered the first oil and gas in the outer shelf area. Continued exploration

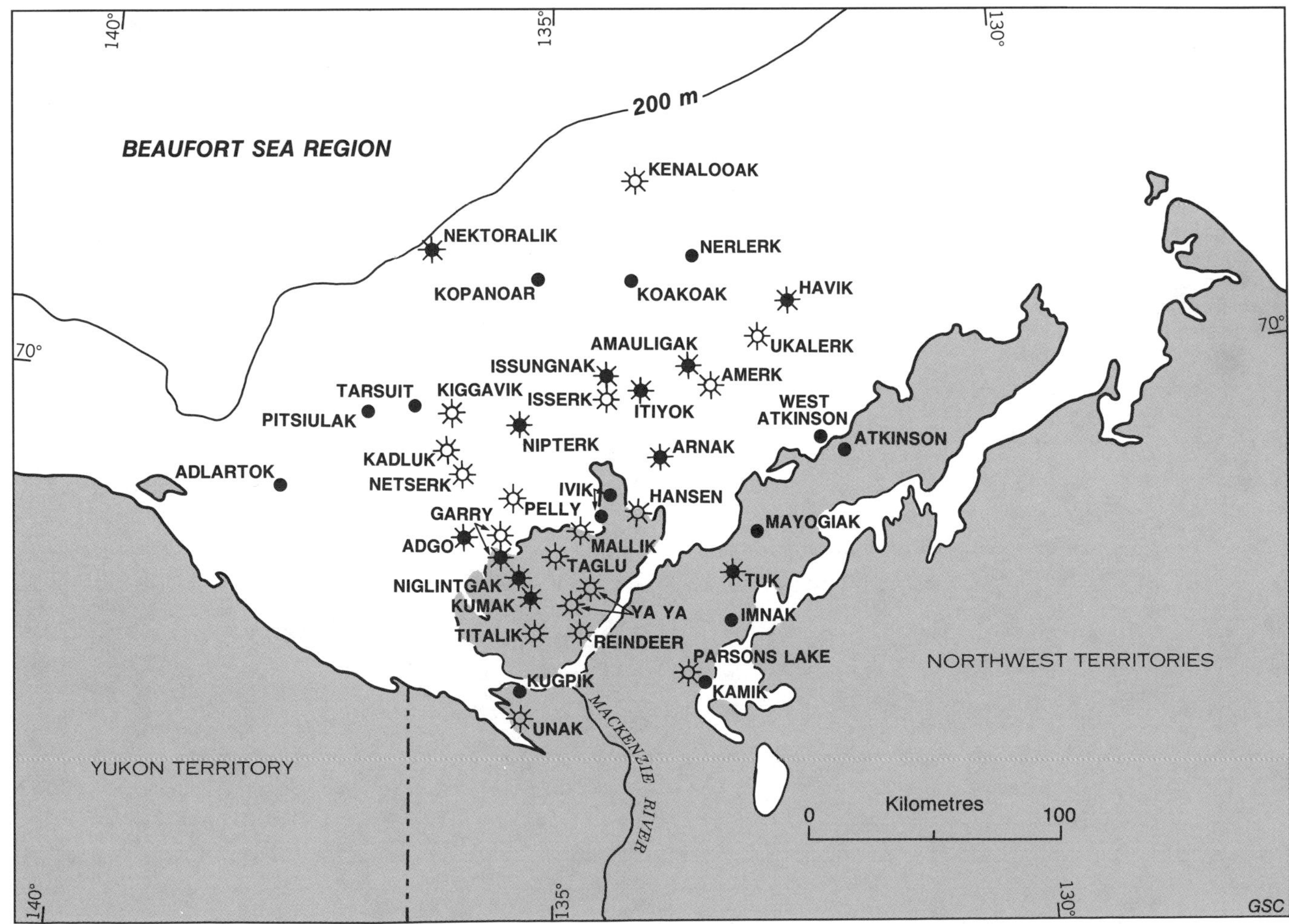

Figure 6A.52. Significant hydrocarbon discoveries of Beaufort-Mackenzie Basin.

into the 1980s led to major discoveries of oil at Kopanoar and Tarsiut and oil and gas at Issungnak and Amauligak. The Amauligak (J-44) discovery well was drilled from a mobile drilling platform.

To 1988, 237 wells have been drilled, 78 of which have been offshore. This has resulted in the discovery of 47 significant hydrocarbon fields, 21 gas, 17 oil, and 9 oil and gas (Fig. 6A.52). The discoveries include several near-giant gas fields, including Parsons Lake (Langhus, 1980), Taglu (Bowerman and Coffman, 1975; Hawkins and Hatlelid, 1975), and Issungnak, with estimated recoverable gas reserves of 62 x 10^9 m^3, 68 x 10^9 m^3, and 71 x 10^9 m^3 respectively. Amauligak, the largest oil field delineated to date, has recoverable oil estimated by Dixon et al. (1988) at greater than 80 x 10^6 m^3. Further delineation drilling is required to evaluate the potentially large oil accumulations at Kopanoar, Koakoak, and Adlartok.

Hydrocarbons are trapped in a variety of stratigraphic units within and peripheral to the central Beaufort-Mackenzie Basin. With the exception of the Devonian carbonate reservoirs at Mayogiak and West Atkinson, which contain migrated oil of Upper Cretaceous type, all of the discovered hydrocarbons occur in Tertiary and Lower Cretaceous sandstones. The most prolific units include the Lower Cretaceous Parsons Group, the Lower Eocene Reindeer sequence and the Oligocene Kugmallit sequence. The undiscovered hydrocarbon resource potential for the region is believed to remain largely in Paleogene strata beneath the central and western Beaufort shelf area.

A regional cross-section (Fig. 6A.53) from Parsons Lake to the shelf edge illustrates the general stratigraphic and structural features typical of the central Beaufort-Mackenzie Basin. The oil- and gas-bearing sand-stones were deposited in a variety of prograding depositional environments, including: lower delta-plain, delta-front, nearshore, shelf and submarine-fan settings. Each sequence has complex facies reflecting the shifting positions of the shelf break and deltaic distributary channels. Rapid subsidence and deposition and consequent undercompaction of sediments initiated large-scale syndepositional structures including listric fault-bounded, rollover anticlines, rotated fault blocks, and mud diapirs, as well as structurally complex combinations of faults and diapirs. Compressional folds and reverse faults occur in the western part of the basin. The shelf and turbidite sand reservoirs occur in mudstone-dominated successions, which provide effective seals for the generally isolated sand bodies. Turbidite sand reservoirs in the far offshore fields are significantly overpressured. Stratigraphic traps may be important components in several of the hydrocarbon accumulations in the basinal sediments. Hydrocarbon accumulations in the sand-dominated, deltaic strata commonly have seals related to unconformities, and are overlain by fine-grained sediments of a transgressive depositional episode (e.g., lower Richards seals upper Reindeer at Taglu).

The hydrocarbons discovered in the Beaufort-Mackenzie Basin are believed to be derived from Tertiary, Cretaceous, and Jurassic sources. Tertiary hydrocarbons occur mainly in offshore areas but were discovered at Taglu (Hawkins and Hatlelid, 1975) and Niglintgak fields. The Tertiary petroleum deposits are characterized by (1) a gas/oil ratio lower than expected for a typical Tertiary delta complex, (2) properties that indicate generation of oil and gas at lower levels of thermal alteration than predicted by the conventional hydrocarbon generation model, and (3) properties that indicate derivation from terrestrially dominated organic matter rather than marine organic matter (Snowdon, 1980a; Snowdon and Powell, 1982; Powell and Snowdon, 1983). Recent work by Brooks (1986) indicates that the dominant and possibly sole source of the Tertiary oils has been the Eocene Richards Formation. Cretaceous hydrocarbons, identified from the Atkinson H-25 and IOE Magogiak J-17 wells, were considered to be derived from the Upper Cretaceous "Bituminous Zone" (Bruce and Parker, 1975; Snowdon, 1980b; Creaney, 1980). Gas in the South and North Parson Lakes fields is believed to have been generated in the Jurassic Husky Formation (Langhus, 1980).

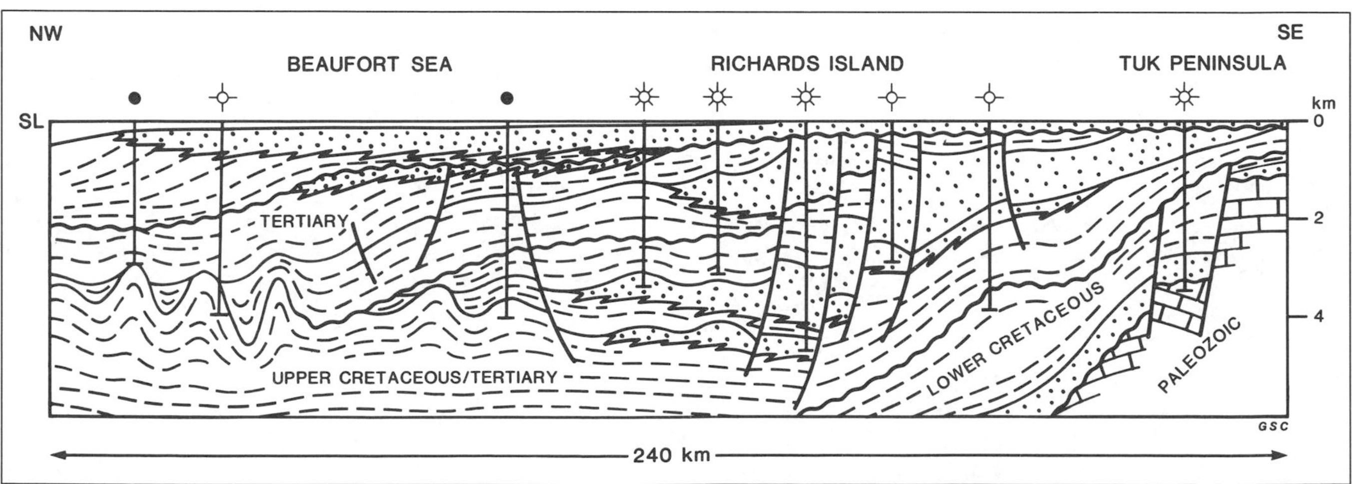

Figure 6A.53. Schematic stratigraphic and structural relationships of the central Beaufort-Mackenzie Basin (from Procter et al., 1984).

NON-CONVENTIONAL OIL AND GAS

Non-conventional hydrocarbons include crude bitumen in sands or carbonate rocks; shale oil; and gas in reservoirs defined as "tight". Crude bitumen is a viscous mixture, composed mainly of hydrocarbons heavier than pentanes, that is not recoverable at a commercial rate through a well by means of conventional oil-field technology. There is a transition from crude bitumen to heavy crude oils (oils with a density greater than 900 kg/m^3). Some heavy crude oils are produced conventionally while others, which may be from the same deposits, are being exploited with non-conventional production technology, in which case they are included in the "non-conventional heavy oil" category. "Tight" gas occurs downdip from water-saturated reservoirs in rocks of very low porosity with ultra-low permeability. The in-place volumes of non-conventional hydrocarbons dwarf those of the conventional category, but how much of this volume will become commercial will depend on the technology developed for its extraction and future energy prices. Volumetric statistics for heavy oils are complicated by differing criteria between provinces and by the fact that reserves are only allocated to areas that are currently under development.

Bituminous sands

Lower Cretaceous bituminous sands, commonly known as "oil sands", containing bitumen that ranges in density from 965 to over 1010 kg/m^3, occur in a discontinuous trend extending from the Peace River deposit in northern Alberta through Wabasca and Athabasca south to Cold Lake (Fig. 6A.54). This belt of oil-sand deposits, with an estimated IVIP of 268 x 10^9 m^3 of bitumen (Alberta Energy Resources Conservation Board, 1987), probably contains the largest accumulation of oil in the world. The bitumen must be extracted either by open-pit mining, or by in-situ thermal stimulation techniques at depth.

Of the four major deposits of oil sand (Fig. 6A.54) Athabasca is by the far the largest and contains 74% of the bitumen in place; Cold Lake is next with 17%; Peace River and Wabasca are estimated to hold 6% and 3% respectively. The Alberta Energy Resources Conservation Board has estimated that 5.2 x 10^9 m^3 of the remaining bitumen in place is in surface-mineable oil sands, all of which occur in the Athabasca deposit.

The major differences between oil-sands bitumen and conventional crude oil are the result of biodegradation, water washing, oxidation, and differential movement of lighter fractions of the original oil as it migrated into near-surface reservoirs near the edge of the Western Canada Basin (Deroo and Powell, 1978; Deroo et al., 1977). The bitumen has a high density and a high viscosity (105 to 107 mPa.s) at reservoir temperatures. Variations in bitumen properties are found throughout the deposits, both laterally and vertically. With increased depth, bitumen of the oil sands grades to heavy oil. Although the origin of the bitumen by degradation of lighter oil is now accepted by most petroleum geologists, the source of the oil is still controversial. Some geochemical studies have suggested that all of the oil originated from Lower Cretaceous source rocks deeper in the basin, but this is not consistent with the volumetric calculations of Moshier and Waples (1985), which call for additional sources.

The oil-sand deposits occur either as continuous reservoirs, such as the McMurray sands of the Athabasca deposits, Bluesky and Gething sands of the Peace River deposits, and the large Clearwater sand reservoir of the Cold Lake deposit, or in a series of stacked reservoirs separated by impermeable shales, such as the smaller

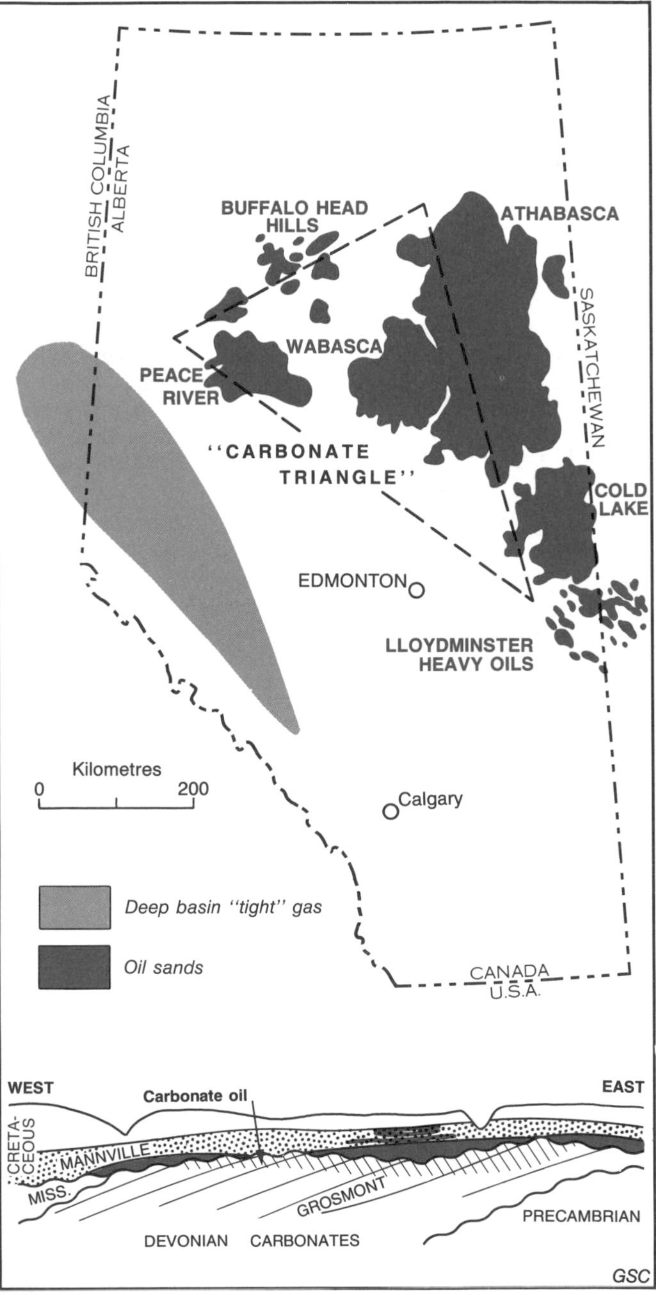

Figure 6A.54. Distribution of Cretaceous and Devonian non-conventional hydrocarbon deposits, including bituminous sands ("oil sands"), heavy oil deposits, Deep Basin "tight" gas, and the "Carbonate Triangle" within which occur deposits of crude bitumen trapped in Devonian carbonate rocks. The schematic cross-section depicts relationship of the Mannville Group to underlying Paleozoic rocks (adapted from Procter et al., 1984).

Grand Rapids reservoirs in the Wabasca and Cold Lake deposits. Trapping mechanisms may be either stratigraphic or structural. The steeply dipping oil/water contact along the east side of the large Athabasca deposit is believed to be caused by a bitumen plug formed as the oil was degraded near the surface (Trap Type Ld2).

Most of the reserves in the Athabasca area are contained within the Lower Cretaceous McMurray Formation, a nonmarine sequence of uncemented sands and shales overlying the eroded surface of Devonian limestones (Flach, 1984; Hills, 1974; Mossop, 1980). The thickness of the formation, a function of relief on the underlying unconformity, ranges from over 150 m in the centre of the deposit to zero in the west, where it pinches out against a paleotopographic ridge of Devonian limestone. In the Athabasca deposit, thickness of oil-saturated sand ranges up to 35 m, the average being from 5 to 15 m. Bitumen content of up to 0.18 (mass fraction) is found in the richest portions of the mining area of the Athabasca deposit, but more average "rich" mass fractions are in the 0.10 range. The average pore-volume bitumen saturation ranges from 0.50 to 0.70. Approximately half of the oil-sands reservoirs have an underlying water zone and some have an overlying water zone that may be the result of a gas cap being replaced by water.

In the Wabasca oil-sand deposit, the Grand Rapids Formation represents a progradational clastic sequence (Kramers, 1974, 1982; Keeler, 1980; Lennox and Lerand, 1980). The formation, up to 90 m thick, contains three sandstone units separated by marine shale and siltstone. The deposit is found at depths ranging from 76 to 610 m. Oil density ranges from 1014 to 1029 kg/m^3 and the sulphur content from 4.5 to 5.5%. The regional dip is to the southwest, with local reversals leading to closure. Tilting of the reservoirs may have taken place after degradation of the oil. The oil-impregnated sands have a relatively low montmorillonite content compared to the barren sands; clay content must be considered in in-situ recovery schemes.

In the Peace River deposit, oil is pooled in the sands of the Gething and Bluesky formations (Jardine, 1974; Rottenfusser, 1982), which were deposited on an eroded surface of Jurassic and Paleozoic strata. Maximum relief on the unconformity is about 45 m. The oil-bearing section thins from about 120 m and pinches out against a ridge of Lower Carboniferous rock to the northeast (Trap Type Ua2). The thickest and richest oil sands occur in the updip part of the trap. Porosity in the sands ranges from 16 to 30%, with an average of 24%. Permeability in the cleaner sands may reach 1 darcy. Oil density ranges from 915 to 1000 kg/m^3.

The Cold Lake oil sands in east-central Alberta cover more than 9000 km^2. Distribution of the sands was controlled by the topography of the pre-Mannville surface, and oil accumulations are controlled by depositional facies (Minken, 1974; Kemp, 1984). The Clearwater sands, having an aggregrate thickness of about 50 m, form the richest reservoir and are the target of most pilot operations in the area, but the Grand Rapids Formation has an equal volume of in-place oil.

Production from the oil sands is constantly increasing as pilot plants are expanded to commercial scale. The crude bitumen from the mining projects is upgraded and marketed as "synthetic crude oil". The crude bitumen from the other projects is temporarily marketed unprocessed at present. At the end of 1986, synthetic oil production from the Athabasca oil sands was from two operating surface mines, Suncor and Syncrude, both near Fort McMurray. The surface-mineable area is situated along the valley of Athabasca River, where overlying Cretaceous formations have been eroded, leaving the oil sands within 50 m of the surface. Separation of bitumen from the sand is achieved by the Clark Hot-Water Process. Production of synthetic crude in 1986 was 14 x 10^6 m^3 (Alberta, Energy Resources Conservation Board, 1987). Ninety percent of the oil-sands reserves lie below 50 m, too deep to be surface mined. Either an in-situ process or an underground mining method will be required to recover the oil. An experimental underground facility has been constructed from which holes will be drilled upward into the oil sands. After injection of steam into the sands the bitumen is expected to flow down into a gathering system in the horizontal tunnels. The major in-situ thermal recovery processes either in production or being field-tested in other areas are: (1) cyclic steam stimulation (huff and puff), where injection and production are carried out in the same well; (2) steam flood or steam drive, where a pattern of injection and production wells is used; and (3) combustion, where the formation is ignited to raise the temperature and increase fluidity. By 1986 there were 31 experimental schemes in operation, with an annual production of 2.1 x 10^6 m^3 of crude bitumen (Alberta, Energy Resources Conservation Board, 1987).

At Cold Lake, Esso Resources obtains commercial production from the oil sands of the Clearwater Formation with a successful application of the cyclic steam stimulation process. This facility has gone through six phases of expansion and another four have been approved. Several other facilities are producing from the Cold Lake deposit (e.g., at Wolf Lake, British Petroleum operates an in-situ oil-sands production facility, which is to be upgraded to 3800 m^3/d in 1989). A commercial facility is under construction to exploit the Peace River oil-sand deposits.

Bituminous carbonates

Crude bitumen accumulations requiring non-conventional recovery methods are found in Devonian and Carboniferous carbonate rocks that subcrop beneath the Alberta oil sands. The accumulations occur in an area that has been termed the "carbonate triangle" (Fig. 6A.54). Much of the crude bitumen has been found in the Devonian Grosmont Formation, which has the best reservoir properties for potential development, but the Nisku Formation also contains substantial volumes of bitumen. The Alberta Energy Resources Conservation Board (1987) has estimated that in-place volumes amount to 60 x 10^9 m^3 (Grosmont and Nisku), and Outtrim and Evans (1978) have estimated much greater amounts within the total carbonate section.

The Grosmont Formation is a broad, multi-stage, shallow marine, carbonate complex, approximately 160 km in width and 500 km in length, averaging 150 m in thickness (Cutler, 1983; Harrison, 1982). The southern and western limits of the formation are at the transition from shallow platform facies to deeper basinal equivalents. The formation is largely dolomitized and passes northward into limestones of the Grumbler Group. The updip eastern margin has been erosionally truncated and is covered by Cretaceous sandstones.

Economic interest in the Grosmont Formation is primarily focused on the eastern subcrop margin. Large quantities of bitumen, as well as a number of gas pools, are trapped along this updip edge at depths of 200 to 500 m. The bitumen is contained within a belt approximately 50 km wide and 300 km long. The chemistry, viscosity, and origin of the bitumen are similar to those of the younger overlying Cretaceous oil sands (Hoffmann and Strausz, 1986). Porosity within the Grosmont varies in character and distribution and is almost entirely diagenetic rather than primary in nature. Intercrystalline porosity is most important, and is developed pervasively through much of the Grosmont section. In the richest horizons, pore- volume bitumen saturation is commonly 0.70 to 0.80. The bitumen density is approximately 1020 kg/m^3. Geochemical studies confirm the suggestion that this deposit is genetically related to the Athabasca oil sands (Hoffmann and Strausz, 1986), but the Grosmont bitumen has been subjected to a greater degree of biodegradation and water washing. Volumetric studies (Moshier and Waples, 1985) indicate that the Cretaceous Mannville shales of Western Canada Basin cannot be the major source rocks of the oil sand and Grosmont bitumen accumulations of northern Alberta.

Cyclic (huff and puff) steam injection has been the dominant recovery method utilized in the pilot projects (Harrison, 1986).

Non-conventional heavy oil fields

Accumulations of heavy oil occur over an area of approximately 28 000 km^2 in the subsurface of east-central Alberta and west-central Saskatchewan, around Lloydminster (Fig. 6A.54, 6A.55; Orr et al., 1977; Beck et al., 1980). These accumulations are the down-dip, southern extension of the bituminous sand deposits of northeastern Alberta. The oils are lighter than the bitumen and some of the deposits can be produced by conventional methods, although with very low primary recovery. In recent years, thermal recovery pilot projects have demonstrated considerably higher recoveries, and oil produced

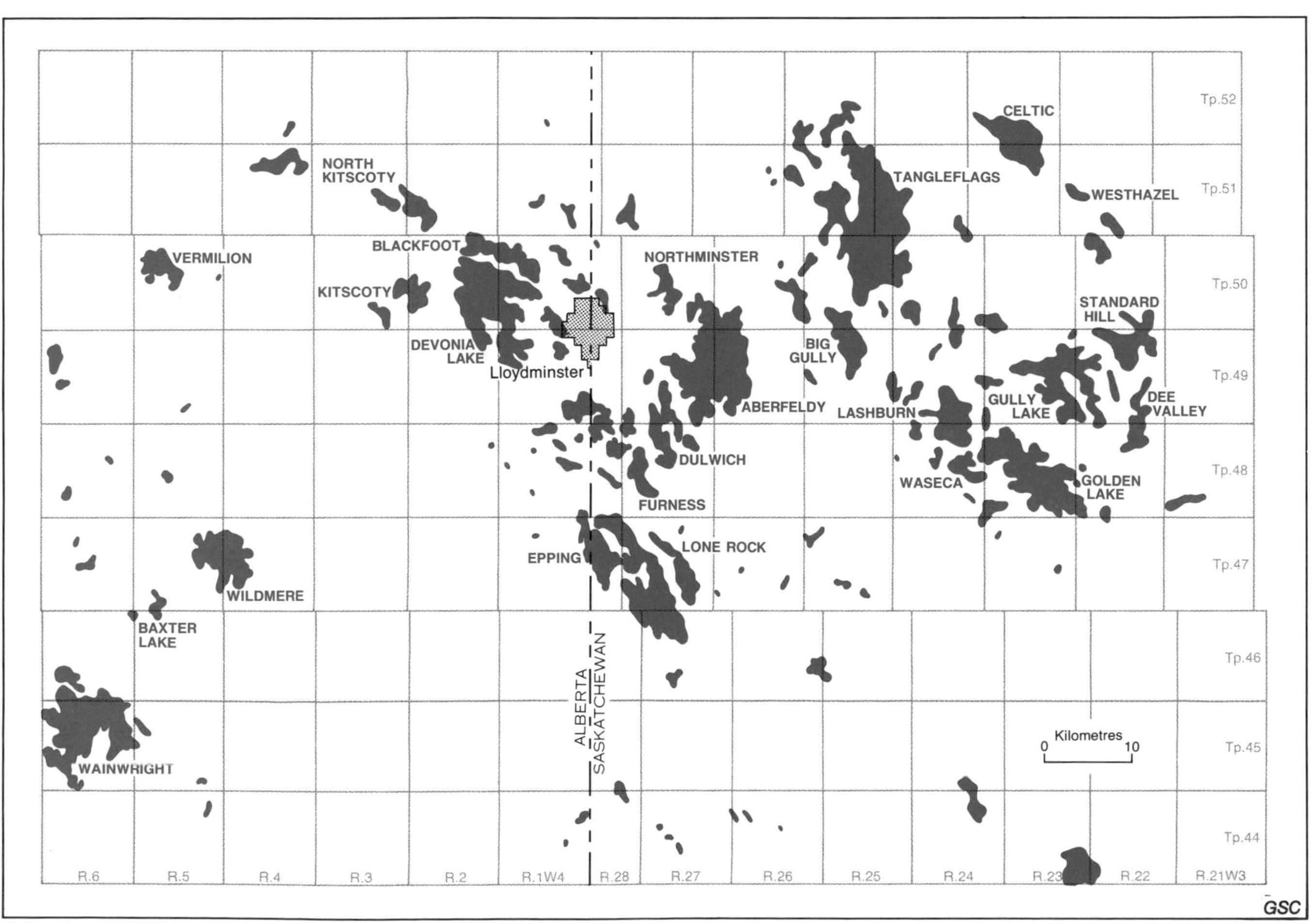

Figure 6A.55. Distribution of heavy oil fields in Alberta and Saskatchewan (from Maccagno and Watson, 1980).

with this technology is classified as non-conventional. Within the Lloydminster heavy oil area, Husky Oil Operations Ltd. is the largest producer, with production in 1985 averaging about 4800 m^3/d, which amounts to about 50% of the Saskatchewan heavy oil production and about 25% of the Alberta production (Berry, 1985). In Saskatchewan, seven experimental projects in the Aberfeldy, Golden Lake, Tangleflags, and Pikes Peak fields are designed to evaluate different enhanced oil-recovery processes in various reservoirs. These projects include five fireflood projects and two steam-injection projects, one being steam drive and the other being cyclic steam or "huff and puff".

Regionally, oil is trapped by a major antiform trending northwest through Lloydminster, Cold Lake, and Fort Athabasca. The antiform results from reversal of the regional southwesterly dip by collapse caused by leaching of the underlying Middle Devonian Prairie evaporites (Orr et al., 1977). Local hydrocarbon traps are created by: minor structural rollovers (Trap Type Sd), updip seals formed by shale-filled channels (Trap Type Ls1; Fig. 6A.56), localized sand-filled channels (Trap Type Ls1), and sand pinchouts against paleotopographic highs (Trap Type Ua2).

Reservoirs in the multiple sands of the Mannville Group are influenced by paleotopography and structure on the pre-Mannville unconformity surface (Putnam, 1980, 1982). Widely developed channel-fill deposits and a succession of cyclical marine deposits in the Upper Mannville over much of the area have produced a highly complex and discontinuous series of reservoirs, which occur at an average depth of 550 m (Beck et al., 1980; Smith et al., 1984). Porosity of the mostly unconsolidated, very fine- to fine-grained sands is 30 to 35%, with permeability up to 6 D. The oil column is in the order of 2 to 8 m, rarely up to 30 m. The oil density ranges from 905 to 1000 kg/m^3 and averages 965 kg/m^3. Viscosity ranges from 100 to over 10 000 mPa.s, sulphur content is 3.5 to 4%, solution gas-oil ratio approximates 10 (m^3/m^3), and the shrinkage factor is in the order of 0.96 to 0.99. Initial pore-volume oil saturation is from 0.75 to 0.80 and primary production is 3 to 8% of oil in place.

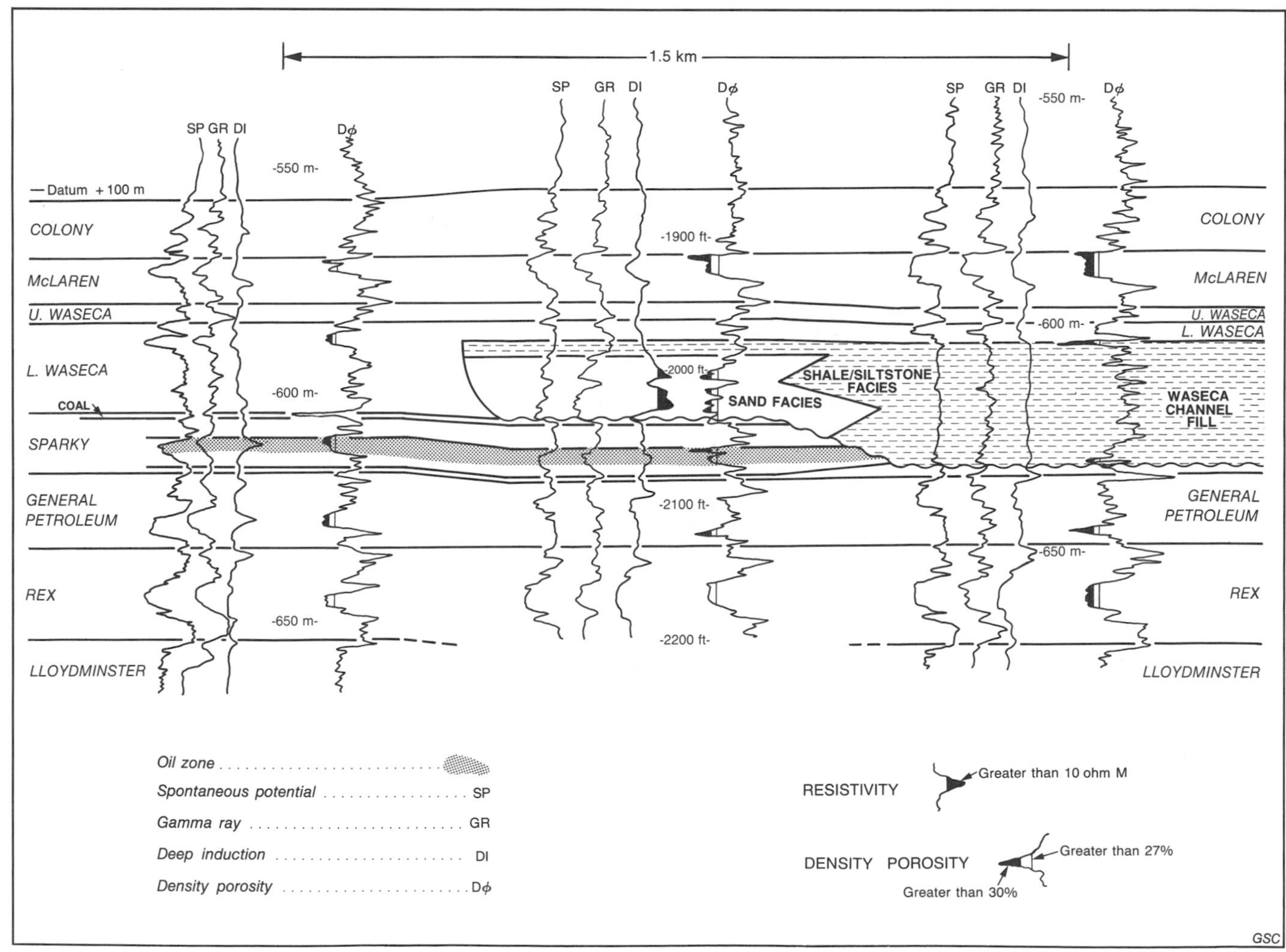

Figure 6A.56. Typical trap in the heavy oil area of the Mannville Group (Trap Type Ls1), illustrated by Freemont Field (from S.R. Smith et al., 1984).

Geochemically, Lloydminster crude oils may be characterized as being asphaltic, highly naphthenic, moderate- to high-sulphur oils (Deroo et al., 1974; Deroo and Powell, 1978). The lighter aromatics have been removed as a result of water-washing and biodegradation, although they have been transformed to a lesser degree than the bitumens of the Cold Lake and other oil-sand deposits. Deroo et al. (1977) concluded, on the basis of organic chemical composition, that the source of the Lloydminster heavy oil is the same as that for more conventional Lower Cretaceous crudes, that is, Lower Cretaceous shales found in the deeper parts of Alberta Basin. However, since their conclusions about the origin of the bitumen have been challenged on both volumetric and geochemical grounds (see above), this interpretation is also open to question. Coal vitrinite reflectance levels are only 0.4%, well below the minimum maturation level for oil.

The Mannville sands of the Lloydminster area are estimated to contain from 5.2 to 10.6 x 10^9 m^3 of oil in place (Procter et al., 1984). Large fields in Saskatchewan include Aberfeldy (Maccagno and Watson, 1980; Smith, 1984) and Tangleflags (M. Wilson, 1984). About 40% of the production on the Alberta side of the Lloydminster region has come from the lighter oils of the Wainwright, Chauvin, and Kinsella fields. Reservoirs are found in the Colony, McLaren, Waseca, Sparky, General Petroleum (GP), Lloydminster, and Cummings sands. The Sparky is the most prolific producer; the Cummings and Colony have been minor contributors.

Oil shales

Oil shales that may at some future date have economic potential occur in Devonian and Cretaceous strata of western Canada (Fig. 6A.57; Macauley, 1984; Macauley et al., 1985).

In the Norman Wells and Fort Good Hope areas of the northern Interior Platform, potential oil shales are found in the Devonian Canol Formation, a marine unit up to 100 m thick, which overlies the Ramparts (Kee Scarp) reef and which possibly provided a source for the oil in the reef reservoir.

Oil shales occur in the Upper Cretaceous Favel and Niobrara (Boyne) formations and are widely distributed across the prairies from Manitoba into Alberta, ranging in thickness from 20 to 60 m. In outcrop along the Manitoba Escarpment, each formation is 30 to 45 m thick. Throughout much of the area, they are separated by 40 m of barren shale of the Morden Formation. Both oil shale zones contain "white specks" (coccoliths and foraminiferal debris).

The total organic carbon (TOC) is about 7% along the outcrop belt, although it averages only 3 to 4% over the entire area. The organic matter is immature and marine (Type II mixed). Zones of lesser organic carbon may contain the most humic material. Oil from the shales at Pasquia Hills at the north end of the Manitoba Escarpment is of low quality (990 kg/m^3), highly aromatic, and hydrogen deficient.

Although the volume of oil shale is large, yields are modest. Average yields for each 1% organic carbon range from 3.5 to 4.5 kg/t over the total interval of each oil shale unit, but increase up to 6 kg/t in beds where Type II kerogen predominates. An area mapped on the north flank of Pasquia Hills, close to the Manitoba-Saskatchewan border, has an optimum yield of 40 to 50 kg/t from the Boyne Formation. The IVIP of the oil in this area is estimated to be at least 200 x 10^6 m^3 at mineable localities (Macauley et al., 1985).

In the Northwest Territories, oil shales of the Cretaceous Smoking Hills and Boundary Creek formations are lithologically similar and stratigraphically equivalent to those of the Favel and Niobrara (Boyne) formations and are classed as marine, Type II mixed deposits. The Smoking Hills deposit is considered immature, whereas the Boundary Creek beds have been shown (L.R. Snowdon, 1980b) to be a likely source for some oil in Mackenzie Delta.

Deep Basin "tight" gas

Natural gas, apparently in very large volume, is contained in Lower Cretaceous strata along the axis of the Alberta Syncline, east of the Foothills in west-central Alberta and northeastern British Columbia. Within the 4500 m thick clastic wedge of this region, termed the "Deep Basin" (Fig. 6A.54), every sandstone in the Mesozoic succession is gas saturated (Fig. 6A.58). The IVIP was estimated by Masters (1984) to be as large as 8500 x 10^9 m^3. The gas occurs in sandstones of extremely low permeability, downdip from permeable water-bearing sands in shallower areas of the basin.

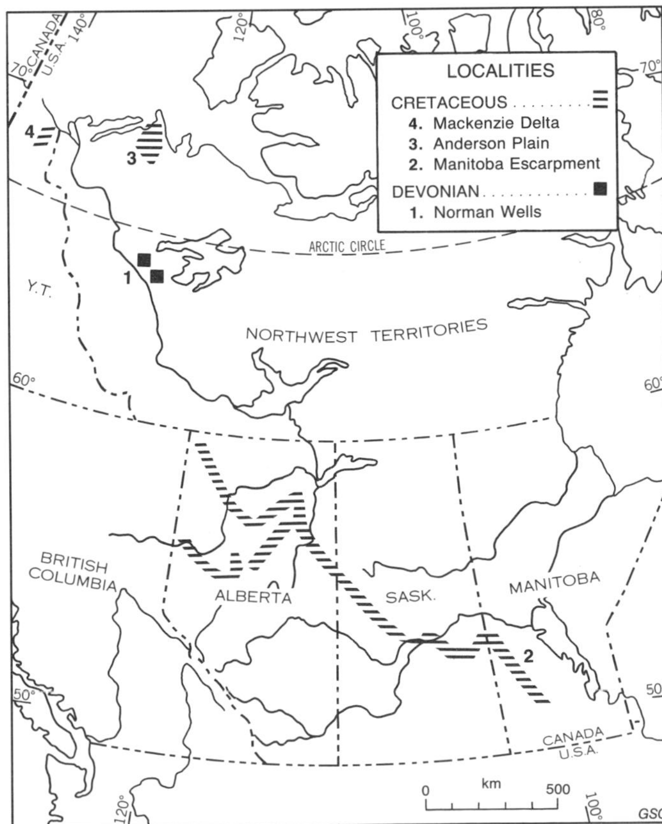

Figure 6A.57. Principal oil shale deposits of Western Canada Basin (modified from Macauley et al., 1985).

Analysis of the Elmworth Gas Field in the Deep Basin of west-central Alberta indicates that Mannville equivalents are the principal source rock (Welte et al., 1984). The strata are extremely rich in Type III kerogen (terrestrial plants), ideal for gas generation. Maturity ranges from 0.8% to 1.8% Ro, well within the gas generation "window". Much of the gas appears to originate in the numerous Cretaceous coal seams. Hydrocarbons are still being actively generated in the Deep Basin, but because of low permeability, migration is probably dominated by diffusion. The pervasive Deep Basin gas traps are considered to be dynamic systems (Welte et al., 1982, 1984; Gies, 1985). Gas generated downdip migrates slowly updip, ultimately escaping into water-saturated, shallower strata of higher permeability, where it is in part caught by conventional traps. Gas escaping from Deep Basin traps is replaced by gas that is continually generated within the basin. If hydrocarbon generation ceased, these pervasive gas traps probably would slowly lose gas by diffusion and be filled with water.

Production in the area has so far been entirely by conventional technology from more porous and permeable lenses. The feasibility of producing the "tight" gas is speculative and no reserves are assigned to it directly by the Alberta Energy Resources Conservation Board, although the interbedded conventional reservoirs have been allocated a slightly augmented volume of gas. The Board expects technological advance and improved prices to make exploitation of "tight" gas viable during the 21st century.

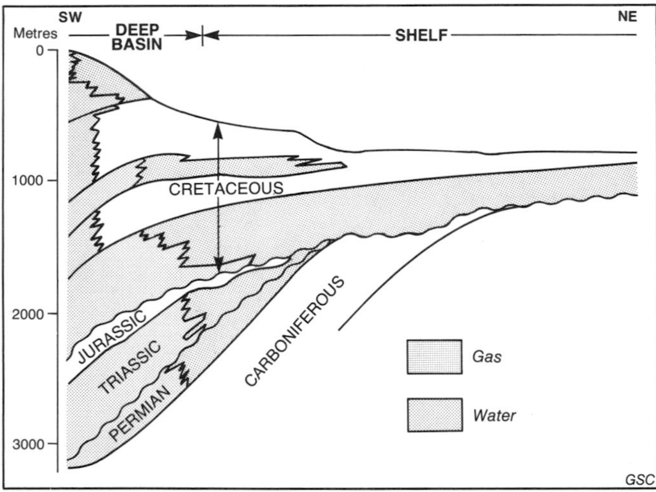

Figure 6A.58. Schematic section of gas trapped below water by permeability barriers, Deep Basin of western Alberta. The shaded areas, containing water, generally represent the porous, siliciclastic parts of the sedimentary column (after Procter et al., 1984).

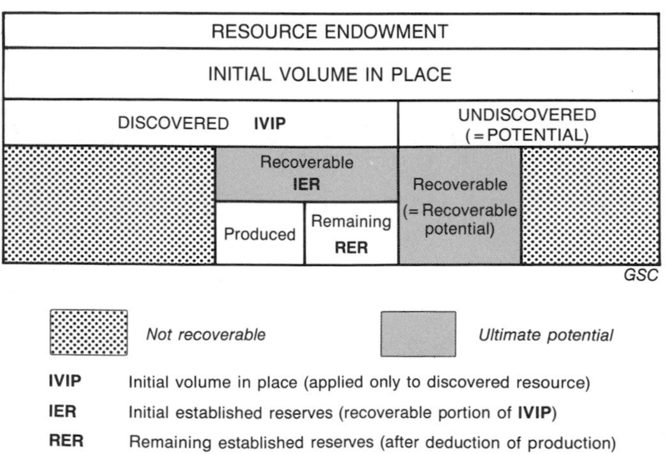

Figure 6A.59. Explanation of terms used in the discussion of resource endowment and reserves.

RESERVES AND POTENTIAL

Previous sections refer to discovered resources in terms of Initial Volumes In Place (IVIP). In this summary, estimates are given of the volumes of discovered hydrocarbons that remain to be produced and on the probable size of undiscovered resources.

The **Remaining Established Reserves (RER)** are those left after subtracting cumulative production from the Initial Established Reserves (IER). The **Potential** is defined as an estimate of the undiscovered resources, whether or not they will eventually be discovered or, if discovered, be economically exploitable. This usage follows that of the Geological Survey of Canada and should not be confused with the **Ultimate Potential**, which is used by the Interprovincial Advisory Committee on Energy to refer to an estimate of the IER + additions to existing pools + future discoveries. Therefore, to compare Potential to Ultimate Potential it is necessary to estimate the recoverable portion of the former and then add it to the IER (Fig. 6A.59).

Table 6A.4. Conventional crude oil and equivalent.

	IER/10^6 m^3	RER/10^6 m^3	Life Index
British Columbia	84	19	9.3
Alberta	2162	635	11.9
Saskatchewan	446	112	9.6
Manitoba	n/a	9	11.3
Mainland Territories	n/a	51	36.0
Beaufort/Mackenzie	n/a	193	n/a

IER - Initial Established Reserves

RER - Remaining Established Reserves

Note 1. Sources of data:

British Columbia Energy, Mines and Petroleum Resources 1986.

Alberta Energy Resources Conservation Board 1987.

Saskatchewan Energy and Mines 1987.

Manitoba's designated oil pools, Dec. 31, 1987. (Informal circular).

Canada Oil and Gas Lands Administration Annual Report 1986.

Note 2. Life indices (in years) calculated by dividing RER by production for the year 1986.

Table 6A.5. Natural Gas (marketable)

	IER/10^6 m^3	RER/10^6 m^3	Life Index
British Columbia	411	227	33
Alberta	3026	1720	25
Saskatchewan	106	68	29
Mainland Territories*	23	23	n/a
Beaufort/Mackenzie	n/a	293	n/a

*includes Pointed Mountain gas field which lies within Foothills Belt.
IER - Initial Established Reserves
RER - Remaining Established Reserves
Note 1. Sources of data:
British Columbia Energy, Mines and Petroleum Resources 1986.
Alberta Energy Resources Conservation Board 1987.
Saskatchewan Energy and Mines 1987.
Manitoba's designated oil pools, Dec. 31, 1987. (Informal circular).
Canada Oil and Gas Lands Administration Annual Report 1986.
Note 2. Life indices (in years) calculated by dividing RER by production for the year 1986.

Tables 6A.4 and 6A.5 show the IER and RER for conventional crude oil and natural gas grouped by reporting agency. Table 6A.6 shows published estimates of Ultimate Potential for natural gas, crude oil, and pentanes. These data indicate that in the Alberta and Williston basins the greater part of the discovered resources of conventional crude oil and natural gas, recoverable under current economic and technical conditions, has already been produced. It is to be expected that there will be further discoveries and that technological innovations will result in a greater proportion of the IVIP being recoverable. However, the amount that will be recovered at any given time will be highly sensitive to changes in price. Hence econometric studies (Uhler, 1986; Conn and Christie, 1988) are necessary for an evaluation of the Ultimate Potential.

A recent forecast for Alberta (Energy Resources Conservation Board, 1987) contains the following statement:

"By 2010 conventional oil production is forecast to be about only one-third of the 1985 rate. During the 25-year period, cumulative production is forecast to be some 825 million cubic metres . . . which is some 70 per cent of the current recoverable reserves plus expected reserve additions. By the end of the forecast period, the Alberta conventional oil industry will be in its later stage of development.

The situation is somewhat different with respect to gas. The production rate in 2010 is predicted to be about two-thirds the current rate but today's production is substantially less than what could be achieved with better market opportunities. Cumulative production during the twenty-five year forecast period is forecast to be some 75 per cent of current reserves plus reserve additions. By 2010 gas productive capability will be in the decline phase

Table 6A.6. Estimates of remaining ultimate potential

	Gas	Oil	Pentanes+	Synthetic
Alberta (1985-12)[1]	2910	1.2	0.200	41.5
British Columbia[2]	1280			
Mainland Territories[3]	312	0.095		
Beaufort-Mackenzie[3]	2151	1.5		

Estimates for the missing elements are not currently available but are not likely to be much greater than the RERs reported in Tables 6A.4 and 6A.5.

[1] Alberta's Energy Resources – a summary (ERCB ST 86-32).

[2] Gas Resources of N.E. British Columbia: K.E. Wallace-Dudley, P.J. Lee, and R.M. Procter. Geological Survey of Canada, Open File 817 (1982). Based on an estimated recoverable potential of 900 $10^9 m^3$ and "ultimate" remaining reserves of use of "ultimate" differs from that of the provincial agencies.

[3] Canada Oil and Gas Lands Administration Annual Report 1986.

but production will likely continue at substantial rates for several decades into the 21st century. Indeed, during this period the Board expects that gas reserves known to exist in tight formations will become viable for development as a result of technological advances and improved prices."

Podruski et al. (1988) estimated that the recoverable potential (see definition above) for the light and medium crude oil of the Western Canada Basin is 570 x 10^6 m^3. Procter et al. (1984) concluded that the reserves of "rich" heavy oil deposits, having thicker than average beds, high oil saturation, and no underlying water, have an average expectation of 3010 x 10^6 m^3.

The data base for a satisfactory forecast is not yet adequate. Nevertheless, it seems clear that by the end of the next half century most of Canada's oil and gas supply will either have to be found in the frontier areas or produced from "non-conventional" resources.

REFERENCES

Aitken, J.D., Cook, D.G., and Yorath, C.J.
1982: Upper Ramparts River (106G) and Sans Sault Rapids (106H) map areas, District of Mackenzie; Geological Survey of Canada, Memoir 388, 48 p.

Alberta Energy Resources Conservation Board
1987: Alberta's reserves of crude oil, oil sands, gas, natural gas liquids and sulphur; Reserves to December 31, 1986; Report ST87-18.

Alcock, F.G. and Benteau, R.I.
1976: Nipisi field - a Middle Devonian clastic reservoir; in The Sedimentology of Selected Clastic Oil and Gas Reservoirs in Alberta, M.M. Lerand (ed.); Canadian Society of Petroleum Geologists, p. 1-24.

Amajor, L.C.
1980: Chronostratigraphy, depositional patterns and environmental analysis of subsurface Lower Cretaceous (Albian) Viking reservoir sandstones in central Alberta and part of southwestern Saskatchewan; Ph.D. thesis, University of Alberta, Edmonton, Alberta, 330 p.

Armitage, J.H.
1962: Triassic oil and gas occurrences in northeastern British Columbia, Canada; Journal of the Alberta Society of Petroleum Geologists, v. 10, no. 2, p. 35-56.

Aukes, P.G. and Webb, T.K.
1986: Triassic Spirit River Pool northwestern Alberta; in Core Conference 1986, N.C. Meijer Drees (ed.); Canadian Society of Petroleum Geologists, p. 3-1 to 3-34.

Bally, A.W., Gordy, P.L., and Stewart, G.A.
1966: Structure, seismic data, and orogenic evolution of southern Canadian Rocky Mountains; Bulletin of Canadian Petroleum Geology, v. 14, no. 3, p. 337-381.
Barchyn, D.
1984: The Waskada Lower Amaranth (Spearfish) oil pool, southwestern Manitoba: A model for Spearfish exploration in Manitoba; in Oil and Gas in Saskatchewan, J.A.Lorsong and M.A.Wilson (ed.); Saskatchewan Geological Society, Special Publication, no. 7, p. 103-111.
Barclay, J.E. and Leckie, D.A.
1986: Tidal inlet reservoirs of the Triassic Halfway Formation, Wembley Region, Alberta; in Core Conference 1986, N.C. Meijer Drees (ed.); Canadian Society of Petroleum Geologists, p. 4-1 to 4-6.
Barss, D.L., Copland, A.B., and Ritchie, W.D.
1970: Geology of Middle Devonian reefs, Rainbow area, Alberta, Canada; in Geology of Giant Petroleum Fields, M.T. Halbouty (ed.); American Association of Petroleum Geologists, Memoir 14, p. 19-49.
Barss, D.L. and Montandon, F.A.
1981: Sukunka-Bullmoose gas fields: Models for a developing trend in the southern foothills of northeast British Columbia; Bulletin of Canadian Petroleum Geology, v. 29, no. 3, p. 293-333.
Beach, F.K. and Irwin, J.L.
ca.1940: The history of Alberta oil; Issued by the Department of Lands and Mines. Published by the Publicity and Travel Bureau, Edmonton, Alberta.
Beck, L.S., Christopher, J.E., and Kent, D.M. (ed.)
1980: Lloydminster and beyond: Geology of Mannville hydrocarbon reservoirs; Saskatchewan Geological Society, Special Publication no. 5, 268 p.
Berry, A.D.
1974: A note on the discovery and development of the Grand Forks Cretaceous oil field, southern Alberta; Bulletin of Canadian Petroleum Geology, v. 22, no. 3, p. 325-339.
Berry, H.J.
1985: Husky's heavy oil experience in the Lloydminster area; in Proceedings of the Saskatchewan Heavy Oil Conference, November 28-30, 1974, University of Saskatchewan; The University of Saskatchewan Printing Services, Saskatoon, p. 300-308.
Bever, J.M. and McIlreath, I.A.
1984: Stratigraphy and reservoir development of shoaling-upward sequences in the Upper Triassic (Carnian) Baldonnel Formation, northeastern British Columbia (abstract); in Exploration Update '84; Canadian Society of Petroleum Geologists - Canadian Society of Exploration Geophysicists National Convention, Calgary, Alberta, 1984, Program and Abstracts, p. 147.
Bowerman, J.N. and Coffman, R.C.
1975: The geology of the Taglu gas field in the Beaufort Basin, N.W.T.; in Canada's Continental Margins and Offshore Petroleum Exploration, C.J. Yorath, E.R. Parker and D.J. Glass (ed.); Canadian Society of Petroleum Geologists, Memoir 4, p. 649-662.
British Columbia Energy Mines and Petroleum Resources
1985: Engineering and Geological Reference Book; Petroleum Resources Division, 2 v., 1025 p.
Brooks, P.W.
1986: Unusual biological marker geochemistry of oils and possible source rocks, offshore Beaufort-Mackenzie Delta, Canada; in Advances in Organic Geochemistry, 1985, Part I, Petroleum Geochemistry, D. Leythaeuser and J. Rullkotter (ed.); Proceedings of 12th International Meeting on Organic Geochemistry, Julich, West Germany, September 1985, Organic Geochemistry, v. 10, nos. 1-3, p. 401-406.
Brooks, P.W., Osadetz, K.G., and Snowdon, L.R.
1987: Origin of compositional differences amongst oils from the Hummingbird Field (Paleozoic), southeast Saskatchewan; in Current Research, Part A, Geological Survey of Canada, Paper 87-1A, p. 331-348.
Bruce, C.J. and Parker, E.R.
1975: Structural features and hydrocarbon deposits in the Mackenzie Delta; Proceedings of the Ninth World Petroleum Congress, Tokyo, v. 2, p. 251-261.
Cant, D.J.
1986: Hydrocarbon trapping in the Halfway Formation (Triassic), Wembley field, Alberta; Bulletin of Canadian Petroleum Geology, v. 34, no. 3, p. 329-338.
Century, J.R.
1967: Oil fields of Alberta Supplement 1966; Alberta Society of Petroleum Geologists, 136 p.
Cheesman, B.
1969: Gilby Field, Jurassic; in Gas Fields of Alberta, L.H. Larson (ed.); Alberta Society of Petroleum Geologists, p. 179.
Cheshire, S.G. and Keith, J.W.
1977: Meekwap field - a Nisku (Upper Devonian) shelf edge reservoir; in The Geology of Selected Carbonate Oil, Gas and Lead-zinc Reservoirs in Western Canada, I.A. McIlreath and R.D. Harrison (ed.); Canadian Society of Petroleum Geologists, p. 107-108.
Chevron Standard Limited
1979: The geology, geophysics and significance of the Nisku reef discoveries, West Pembina area, Alberta, Canada; Bulletin of Canadian Petroleum Geology, v. 27, no. 3, p. 326-359.
Chiang, K.K.
1985: The giant Hoadley gas field, south-central Alberta; in Elmworth - Case Study of a Deep Basin Gas Field, J.A. Masters (ed.); American Association of Petroleum Geologists, Memoir 38, p. 297-314.
Christopher, J.E.
1974: The Upper Jurassic Vanguard and Lower Cretaceous Mannville groups of southwestern Saskatchewan; Saskatchewan Energy and Mines, Report no. 151, 349 p.
1984: Depositional patterns and oil field trends in the lower Mesozoic of the northern Williston Basin, Canada; in Oil and Gas in Saskatchewan, J.A. Lorsong and M.A. Wilson (ed.); Saskatchewan Geological Society, Special Publication no. 7, p. 83-102.
Christopher, J.E., Kent, D.M., and Simpson, F.
1971: Hydrocarbon potential of Saskatchewan; Saskatchewan Energy and Mines, Report no. 157, 47 p.
1973: Saskatchewan and Manitoba; in The Future Petroleum Provinces of Canada -Their Geology and Potential, R.G. McCrossan (ed.); Canadian Society of Petroleum Geologists, Memoir 1, p. 121-149.
Cook, D.G.
1975: The Keele Arch - a pre-Devonian and pre-Late Cretaceous paleo-upland in the northern Franklin Mountains and Colville Hills; in Report of Activities, Part C, Geological Survey of Canada, Paper 75-1C, p. 243-246.
Conn, R.F. and Christie, J.A.
1988: Conventional oil resources of Western Canada, Part II: Economic analysis; Geological Survey of Canada, Paper 87-26, p. 127-141.
Crabtree, H.T.
1982: Lithologic types, depositional environment, and reservoir properties of the Mississippian Frobisher Beds, Innes field, southeastern Saskatchewan; in Fourth International Williston Basin Symposium, J.E. Christopher and J. Kaldi (ed.); Saskatchewan Geological Society, Special Publication no. 6, p. 203-210.
Creaney, S.
1980: The organic petrology of the Upper Cretaceous Boundary Creek Formation, Beaufort-Mackenzie Basin; Bulletin of Canadian Petroleum Geology, v. 28, p. 112-129.
Cutler, W.G.
1983: Stratigraphy and sedimentology of the Upper Devonian Grosmont Formation, northern Alberta; Bulletin of Canadian Petroleum Geology, v. 31, no. 4, p. 282-325.
Davidson, J.
1975: Jumping Pound and Sarcee gas fields; in Structural Geology of the Foothills between Savanna Creek and Panther River, S.W. Alberta, Canada, H.J. Evers and J.E. Thorpe (ed.); Canadian Society of Petroleum Geologists - Canadian Society of Exploration Geophysicists, Exploration Update '75 - Guidebook, p. 30-34.
Davies, G.R.
1983: Sedimentology of the Middle Jurassic Sawtooth Formation of southern Alberta; in Sedimentology of Selected Mesozoic Clastic Sequences, J.R. McLean and G.E. Reinson (ed.); Corexpo 83, Canadian Society of Petroleum Geologists, p. 11-25.
Davies, G.R., Edwards, D.E., and Flach, P.
1987: Geology and reservoir characteristics of Mississippian Waulsortian reefs in the Seal area of Alberta (abstract); Canadian Society of Petroleum Geologists, Reservoir, v. 14, no. 8, p. 1-2 .
Davis, J.W. and Willott, R.
1978: Structural geology of the Colville Hills; Bulletin of Canadian Petroleum Geology, v. 26, no. 1, p. 105-122.
Deroo, G. and Powell, T.G.
1978: The oil sands deposits of Alberta; Their origin and geochemical history; in Oil and Oil Shale Chemistry, O.P. Strausz and E.M. Lown (ed.); Springer Verlag, p. 11-32.
Deroo, G., Tissot, B., McCrossan, R.G., and Der, F.
1974: Geochemistry of the heavy oil of Alberta; in Oil Sands Fuel of the Future, L.V. Hills (ed.); Canadian Society of Petroleum Geology, Memoir 3, p. 148-189.
Deroo, G., Powell, T.G., Tissot, B., and McCrossan, R.G.
1977: The origin and migration of petroleum in the Western Canadian Sedimentary Basin, Alberta - a geochemical and thermal maturation study; Geological Survey of Canada, Bulletin 262, 136 p.

de Wit, R., Gronberg, E.C., Richards, W.B., and Richmond, W.O.
1973: Tathlina area, District of Mackenzie; in The Future Petroleum Provinces of Canada - Their Geology and Potential, R.G. McCrossan (ed.); Canadian Society of Petroleum Geologists, Memoir 1, p. 187-212.

Dixon, J., Morrell, G.R., Dietrich, J.R., Procter, R.M., and Taylor, G.C.
1988: Petroleum resources of the Mackenzie Delta - Beaufort Sea; Geological Survey of Canada, Open File 1926, 74 p.

Douglas, R.J.W. (ed.)
1970: Geology and economic minerals of Canada; Geological Survey of Canada, Economic Geology Report No. 1, fifth edition, 838 p.

Dunham, J.B., Crawford, G.A., and Panasiuk, W.
1983: Sedimentology of the Slave Point Formation (Devonian) at Slave Field, Lubicon Lake, Alberta; in Carbonate Buildups, a Core Workshop, P.M. Harris, (ed.); Society of Economic Paleontologists and Mineralogists, p. 73-111.

Eliuk, L.S.
1984: A hypothesis for the origin of hydrogen sulphide in Devonian Crossfield Member dolomite, Wabamun Formation, Alberta, Canada; in Carbonates in Subsurface and Outcrop, 1984 C.S.P.G. Core Conference, L.S. Eliuk (ed.); Canadian Society of Petroleum Geologists, p. 245-289.

Elloy, R.
1972: Réflexions sur quelques environments récifaux du Paléozoique: Bulletin du Centre de Recherches, Pau-SNPA, v. 6, p. 1-105.

Energy, Mines and Resources Canada
1986: Canadian minerals yearbook 1985 Review and outlook; Mineral Report 34.

Energy, Mines and Resources Canada-Indian and Northern Affairs Canada
1987: Canada Oil and Gas Land Administration, Annual Report 1986, 41 p.

Erickson, R.H. and Crewson, J.S.
1959: Wayne oil field, Alberta; in Moose Mountain-Drumheller, 9th Annual Field Conference Guidebook, G.H. Austen (ed.); Alberta Society of Petroleum Geologists, p. 158-163.

Fischbuch, N.R.
1984: Facies and reservoir analysis, Kee Scarp Formation, Norman Wells area, Northwest Territories; Geological Survey of Canada, Open File 1116, 46 p., 3 sections, map.

Fitzgerald, E.L. and Peterson, D.J.
1967: Inga oil field, British Columbia; Bulletin of Canadian Petroleum Geology, v. 15, no. 1, p. 65-81.

Flach, P.D.
1984: Oil sands geology - Athabasca deposit north; Geological Survey Department, Alberta Research Council, Bulletin 46, iv + 31 p.

Gallup, W.B.
1975: A brief history of the Turner Valley oil and gas field; in Structural Geology of the Foothills between Savanna Creek and Panther River, S.W. Alberta, Canada, H.J. Evers and J.E. Thorpe (ed.); Exploration Update '75, Canadian Society of Petroleum Geologists - Canadian Society of Exploration Geophysicists, Guidebook, p. 12-17.

Gies, R.M.
1985: Case history for a major Alberta Deep Basin gas trap: The Cadomin Formation; in Elmworth - Case Study of a Deep Basin Gas Field, J.A. Masters (ed.); American Association of Petroleum Geologists, Memoir 38, p. 115-140.

Gordy, P.L. and Frey, F.R.
1977: Some notes on the gas fields of southwestern Alberta; in Geological Guide for the CSPG 1977 Waterton-Glacier Park Field Conference, P.L. Gordy, F.R. Frey, and D.K. Norris (ed.); Canadian Society of Petroleum Geologists, Calgary, p. 83-89.

Graham, A.D.
1973: Carboniferous and Permian stratigraphy, southern Eagle Plain, Yukon Territory, Canada; in Proceedings of the Symposium on the Geology of the Canadian Arctic, J.D. Aitken and D.J. Glass (ed.); Geological Association of Canada-Canadian Society of Petroleum Geologists, Special Paper, p. 159-180.

Grantz, S., Johnson, L., and Sweeney, J.F. (ed.)
1990: The Arctic Ocean Region; Geological Society of America, The Geology of North America, v. L, 644 p.

Gray, F.F. and Kassube, J.R.
1963: Geology and stratigraphy of Clarke Lake gas field, British Columbia; American Association of Petroleum Geologists Bulletin, v. 47, no. 3, p. 467-483.

Gross, A.A.
1980: Mannville channels in east-central Alberta; in Lloydminster and Beyond: Geology of Mannville Hydrocarbon Reservoirs, L.S. Beck, J.E. Christopher, and D.M. Kent (ed.); Saskatchewan Geological Society, Special Publication no. 5, p. 33-63.

Hacquebard, P.A. and Donaldson, J.R.
1974: Rank studies of coals in the Rocky Mountains and Inner Foothills belt, Canada; in Carbonaceous Materials as Indicators of Metamorphism, R.S. Dutcher, P.A. Hacquebard, J.M. Schopf, and J.A. Simon (ed.); Geological Society of America, Special Paper 153, p. 75-94.

Halabura, S.
1982: Depositional environment of the Upper Devonian Birdbear Formation, Saskatchewan; in Fourth International Williston Basin Symposium, J.E. Christopher and J. Kaldi (ed.); Saskatchewan Geological Society, Special Publication no. 6, p. 113-124.

Halton, E.
1981: Wembley field: facies variations in the Halfway Formation; in Environmental Analysis from Core and Field Sample Studies, F.A. Stoakes (ed.); Core Conference 1981, Canadian Society of Petroleum Geologists, p. 9.

Harrison, R.S.
1982: The Grosmont Project: evaluation of a bitumen-bearing Paleozoic carbonate in northern Alberta; in Canada's Giant Hydrocarbon Reservoirs, W.G. Cutler (ed.); Core Conference 1982, Canadian Society of Petroleum Geologists, p. 55-61.
1986: Stratigraphy, sedimentology, and bitumen potential of the Upper Devonian Grosmont Formation, northern Alberta (abstract); Canadian Society of Petroleum Geologists, Reservoir, v. 13, no. 1, p. 1-3.

Hartling, A., Brewster, A., and Posehn, G.
1982: The geology and hydrocarbon trapping mechanisms of the Mississippian Oungre Zone (Ratcliffe Beds) of the Williston Basin; in Fourth International Williston Basin Symposium, J.E. Christopher and J. Kaldi (ed.); Saskatchewan Geological Society, Special Publication no. 6, p. 217-223.

Hawkings, T.J. and Hatlelid, W.G.
1975: The regional setting of the Taglu field; in Canada's Continental Margins and Offshore Petroleum Exploration, C.J. Yorath, E.R. Parker, and D.J. Glass (ed.); Canadian Society of Petroleum Geologists, Memoir 4, p. 633-647.

Hayes, B.J.R.
1983: Stratigraphy and petroleum potential of the Swift Formation (upper Jurassic), southern Alberta and north-central Montana; Bulletin of Canadian Petroleum Geology, v. 31, no. 1, p. 37-52.

Hemphill, C.R., Smith, R.I., and Szabo, F.
1970: Geology of Beaverhill Lake reefs, Swan Hills area, Alberta; in Geology of Giant Petroleum Fields, M.T. Halbouty (ed.); American Association of Petroleum Geologists, Memoir 14, p. 50-90.

Hennessey, W.J.
1975: A brief history of the Savanna Creek gas field; in Structural Geology of the Foothills between Savanna Creek and Panther River, S.W. Alberta, Canada, H.J. Evers and J.E. Thorpe (ed.); Exploration Update '75, Canadian Society of Petroleum Geologists-Canadian Society of Exploration Geophysicists, Guidebook, p. 18-21.

Herbaly, E.L.
1974: Petroleum geology of Sweetgrass Arch, Alberta; American Association of Petroleum Geologists Bulletin, v. 58, no. 1, p. 2227-2244.

Hills, L.V. (ed.)
1974: Oil Sands Fuel of the Future; Canadian Society of Petroleum Geologists, Memoir 3, 263 p.

Hoffmann, C.F. and Strausz, O.P.
1986: Bitumen accumulation in Grosmont platform complex, Upper Devonian, Alberta, Canada; American Association of Petroleum Geologists Bulletin, v. 70, no. 9, p. 1113-1128.

Holmes, I.G. and Rivard, Y.A.
1976: A marine barrier island bar, Jenner field, southeastern Alberta; in The Sedimentology of Selected Clastic Oil and Gas Reservoirs in Alberta, M.M. Lerand (ed.); Canadian Society of Petroleum Geologists, p. 44-61.

Hopkins, J.C.
1981: Sedimentology of quartzose sandstones of lower Mannville and associated units, Medicine River area, central Alberta; Bulletin of Canadian Petroleum Geology, v. 29, no. 1, p. 12-41.
1987: Contemporaneous subsidence and fluvial channel sedimentation: Upper Mannville "C" Pool, Berry Field (Lower Cretaceous of Alberta); American Association of Petroleum Geologists, Bulletin, v. 71, no. 3, p. 334-345.

Hopkins, J.C., Hermanson, S.W., and Lawton, D.C.
1982: Morphology of channels and channel-sand bodies in the Glauconitic Sandstone Member (Upper Mannville), Little Bow Area, Alberta; Bulletin of Canadian Petroleum Geology, v. 30, no. 4, p. 274-285.

Hornford, H.T.
1985: An analysis of the petroleum geology of the Bakken Formation of the Williston Basin of western Canada; Geological Survey of Canada, Open File 1123, 9 p., 17 maps and cross-sections.

Hriskevich, M.E., Faber, J.M., and Langton, J.R.
1980: Strachan and Ricinus West gas fields, Alberta, Canada; in Giant Oil and Gas Fields of the Decade 1968-1978, M.T. Halbouty (ed.); American Association of Petroleum Geologists, Memoir 30, p. 315-327.

Illing, L.V., Wood, G.V., and Fuller, J.G.C.M.
1967: Reservoir rocks and stratigraphic traps in non-reef carbonates; in Origin of Oil, Geology and Geophysics; Seventh World Petroleum Congress, Proceedings, v. 2, London, Elsevier Publishing Co., p. 487-499.

Jackson, P.C.
1984: Paleogeography of the Lower Cretaceous Mannville Group of western Canada; in Elmworth - Case Study of a Deep Basin Gas Field, J.A. Masters (ed.); American Association of Petroleum Geologists, Memoir 38, p. 49-77.

Jardine, D.
1974: Cretaceous oil sands of western Canada; in Oil Sands Fuel of the Future, L.V. Hills (ed.); Canadian Society of Petroleum Geologists, Memoir 3, p. 50-67.

Kaldi, J.
1982: Reservoir properties, depositional environments and diagenesis of the Mississippian Midale Beds, Midale field, southeastern Saskatchewan; in Fourth International Williston Basin Symposium, J.E. Christopher and J. Kaldi (ed.); Saskatchewan Geological Society, Special Publication No. 6, p. 211-216.

Kalkreuth, W. and McMechan, M.E.
1984: Regional pattern of thermal maturation as determined from coal-rank studies, Rocky Mountain Foothills and Front Ranges north of Grande Cache, Alberta - Implications for petroleum exploration; Bulletin of Canadian Petroleum Geology, v. 32, no. 3, p. 249-271.

Keeler, R.G.
1980: Lower Cretaceous (Mannville Group) Grand Rapids Formation, Wabasca A oil sand deposit area, northeast Alberta; in Lloydminster and Beyond: Geology of Mannville Hydrocarbon Reservoirs, L.S. Beck, J.E. Christopher, and D.M. Kent (ed.); Saskatchewan Geological Society, Special Publication no. 5, p. 96-131.

Kemp, E.M.
1984: Cold Lake Project: the resource, the particular production techniques, and related challenges; in Proceedings of the Saskatchewan Heavy Oil Conference, November 28-30, 1984, University of Saskatchewan; The University of Saskatchewan Printing Services, Saskatoon, p. 181-196.

Kendall, A.C.
1976: The Ordovician carbonate succession (Bighorn Group) of south-eastern Saskatchewan; Saskatchewan Energy and Mines, Report no. 180, 185 p.

Kendall, G.H.
1969: Boundary Lake South field - Kiskatinaw Formation; in Gas Fields of Alberta, L.H. Larson (ed.); Alberta Society of Petroleum Geologists, p. 72-73.

Kent, D.M.
1985: Mississippian reservoirs in Williston Basin; International Congress on Carboniferous Stratigraphy and Geology (9th: 1979), Compte Rendu, v. 4, p. 99-110.

Kent, D.M., Leibel, R.J., and Eriyagama, S.C.
1982: Pore systems and reservoir quality in Mississippian carbonates and their relationship to hydrocarbon reservoirs in southern Saskatchewan; in Fourth International Williston Basin Symposium, J.E. Christopher and J. Kaldi (ed.); Saskatchewan Geological Society, Special Publication no. 6, p. 191-201.

Klovan, J.E.
1964: Facies analyses of the Redwater reef complex, Alberta, Canada; Bulletin of Canadian Petroleum Geology, v. 12, no. 1, p. 1-100.

Kramers, J.W.
1974: Geology of the Wabasca A oil sand deposit (Grand Rapids Formation); in Oil Sands Fuel of the Future, L.V. Hills (ed.); Canadian Society of Petroleum Geologists, Memoir 3, p. 68-83.
1982: Grand Rapids Formation, north-central Alberta: an example of nearshore sedimentation in a high energy, shallow, inland sea (abstract); American Association of Petroleum Geologists Bulletin, v. 66, no. 5, p. 589-590.

Krause, F.F. and Nelson, D.A.
1984: Storm event sedimentation; Lithofacies association in the Cardium Formation, Pembina, west-central Alberta, Canada; in The Mesozoic of Middle North America, D.F. Stott and D.J. Glass (ed.); Canadian Society of Petroleum Geologists, Memoir 9, p. 485-511.

Kreis, L.K.
1985: Notes on the Jurassic-Cretaceous petroleum-producing zones in the Wapella-Moosomin area, southeastern Saskatchewan; in Summary of Investigations 1985, Saskatchewan Geological Survey, Saskatchewan Energy and Mines, Miscellaneous Report 85-4, p. 172-179.

Kunst, H.
1973: Peel Plateau; in The Future Petroleum Provinces of Canada - Their Geology and Potential, R.G. McCrossan (ed.); Canadian Society of Petroleum Geologists, Memoir 1, p. 245-273.

Langhus, B.G.
1980: Generation and migration of hydrocarbons in the Parsons Lake area, N.W.T., Canada; in Facts and Principles of World Petroleum Occurrences, A.D. Miall (ed.); Canadian Society of Petroleum Geologists, Memoir 6, p. 523-534.

Langton, J.R. and Chin, G.E.
1968: Rainbow Member facies and related reservoir properties, Rainbow Lake, Alberta; Bulletin of Canadian Petroleum Geology, v. 16, no. 1, p. 104-143.

Law, J.
1971: Regional Devonian geology and oil and gas possibilities, Upper Mackenzie River area; Bulletin of Canadian Petroleum Geology, v. 19, no. 2, p. 437-486.

Lawrence, J.R.
1973: Old Crow Basin; in The Future Petroleum Provinces of Canada - Their Geology and Potential, R.G. McCrossan (ed.); Canadian Society of Petroleum Geologists, Memoir 1, p. 307-314.

Layer, D.B.
1958: Characteristics of major oil and gas accumulations in the Alberta Basin; in Habitat of Oil, L.G. Weeks (ed.); American Association of Petroleum Geologists, p. 113-128.

Leckie, D.
1986: Tidally influenced, transgressive shelf sediments in the Viking Formation, Caroline, Alberta; Bulletin of Canadian Petroleum Geology, v. 34, no. 1, p. 111-125.

Lennox, T.R. and Lerand, M.M.
1980: Geology of an in situ pilot project, Wabasca oil sands deposit, Alberta (abstract); in Lloydminster and Beyond: Geology of Mannville Hydrocarbon Reservoirs, J.S. Beck, J.E. Christopher, and D.M. Kent (ed.); Saskatchewan Geological Society, Special Publication no. 5, p. 267-268.

Lerand, M.M.
1973: Beaufort Sea; in The Future Petroleum Provinces of Canada - Their Geology and Potential, R.G. McCrossan (ed.); Canadian Society of Petroleum Geologists, Memoir 1, p. 315-386.

Lerand, M.M. and Thompson, D.K.
1976: Provost field-Hamilton Lake pool; in Joint Convention on Enhanced Recovery, W.J.F. Clark and G. Huff (ed.); Core Conference 1976, Calgary, Canadian Society of Petroleum Geologists-Petroleum Society of Canadian Institute of Mining, p. B1-B34.

Little, H.W., Belyea, H.R., Stott, D.F., Latour, B.A., and Douglas, R.J.W.
1970: Economic minerals of western Canada; in Geology and Economic Minerals of Canada, fifth edition, R.J.W. Douglas (ed.); Geological Survey of Canada, Economic Geology Report no. 1, Chapter IX, p. 490-546.

Macauley, G.
1984: Geology of the oil shale deposits of Canada; Geological Survey of Canada, Paper 81-25, 65 p.

Macauley, G., Snowdon, L.R., and Ball, F.D.
1985: Geochemistry and geological factors governing exploitation of selected Canadian oil shale deposits; Geological Survey of Canada, Paper 85-13, 65 p.

Maccagno, H.J. and Watson, M.D.
1980: Geology of the forty acre Aberfeldy steam pilot; in Lloydminster and Beyond: Geology of Mannville Hydrocarbon Reservoirs, J.S. Beck, J.E. Christopher, and D.M. Kent (ed.); Saskatchewan Geological Society, Special Publication No. 5, p. 149-176.

Male, W.H. and Pacholko, R.R.
1982: Upper Cretaceous gas reservoirs of the Suffield military range, southeastern Alberta; in Canada's Giant Hydrocarbon Reservoirs, W.G. Cutler (ed.); Core Conference, Canadian Society of Petroleum Geologists, p. 95-108.

Marion, D.J.
1984: The Middle Jurassic Rock Creek Member and associated units in the subsurface of west-central Alberta; in The Mesozoic of Middle North America, D.F. Stott and D.J. Glass (ed.); Canadian Society of Petroleum Geologists, Memoir 9, p. 319-343.

Martin, H.L.
1973: Eagle Plain Basin, Yukon Territory; in The Future Petroleum Provinces of Canada - Their Geology and Potential, R.G. McCrossan (ed); Canadian Society of Petroleum Geologists, Memoir 1, p. 275-306.

Martindale, W. and Orr, N.E.
1988: Middle Devonian Winnipegosis reefs of the Tableland area, S.E. Saskatchewan (abstract); Canadian Society of Petroleum Geologists, Reservoir, v. 15, no. 3, p. 1-2.

Masters, J.A.
1984: Lower Cretaceous oil and gas in western Canada; in Elmworth - Case Study of a Deep Basin Gas Field, J.A. Masters (ed.); American Association of Petroleum Geologists, Memoir 38, p. 1-33.

McCoy, A.W. III and Moritz, C.A.
1982: Countess oil field, south central Alberta; Case history in finding a stratigraphic trap; Oil and Gas Journal, v. 80, no. 44, p. 95-98.

McCrory, V.L.C. and Walker, R.G.
1986: A storm - and tidally-influenced prograding shoreline - Upper Cretaceous Milk River Formation of southern Alberta, Canada; Sedimentology, v. 33, no. 1, p. 47-60.

McCrossan, R.G. (ed.)
1973: The future petroleum provinces of Canada - their geology and potential; Canadian Society of Petroleum Geologists, Memoir 1, 720 p.

McCrossan, R.G. and Glaister, R.P. (ed.)
1964: Geological history of western Canada, Alberta Society of Petroleum Geologists, Calgary, 232 p.

McCrossan, R.G. and Porter, J.W.
1973: The geology and petroleum potential of the Canadian sedimentary basins - a synthesis; in The Future Petroleum Provinces of Canada - Their Geology and Potential, R.G. McCrossan (ed.); Canadian Society of Petroleum Geologists, Memoir 1, p. 589-720.

McGillivray, J.G.
1977: Golden Spike D3A Pool; in The Geology of Selected Carbonate Oil, Gas and Lead-zinc Reservoirs in Western Canada, I.A. McIlreath and R.D. Harrison (ed.); Canadian Society of Petroleum Geologists, p. 67-88.

Meijer Drees, N.C. and Myhr, D.W.
1981: The Upper Cretaceous Milk River and Lea Park formations in southeastern Alberta; Bulletin of Canadian Petroleum Geology, v. 29, no. 1, p. 42-74.

Miall, A.D.
1976: The Triassic sediments of Sturgeon Lake South and surrounding areas; in The Sedimentology of Selected Clastic Oil and Gas Reservoirs in Alberta, M.M. Lerand (ed.); Canadian Society of Petroleum Geologists, p. 25-43.

Minken, D.F.
1974: The Cold Lake oil sands; geology and reserve estimate; in Oil Sands Fuel of the Future, L.V. Hills, (ed.); Canadian Society of Petroleum Geologists, Memoir 3, p. 84-99.

Moshier, S.O. and Waples, D.W.
1985: Quantitative evaluation of the Lower Cretaceous Mannville Group as the source rock for Alberta's Oil Sands; American Association of Petroleum Geologists Bulletin, v. 69, no. 2, p. 161-172.

Mossop, G.D.
1980: Facies control on bitumen saturation in the Athabasca oil sands; in Facts and Principles of World Petroleum Occurrence, A.D. Miall (ed.); Canadian Society of Petroleum Geologists, Memoir 6, p. 609-632.

Mothersill, J.S.
1968: Environments of deposition of the Halfway Formation, Milligan Creek area, British Columbia; Bulletin of Canadian Petroleum Geology, v. 16, no. 2, p. 180-199.

Muir, I., Wong, P., and Wendte, J.
1984: Devonian Hare Indian - Ramparts (Kee Scarp) evolution, Mackenzie Mountains and subsurface Norman Wells, N.W.T.: Basin fill and platform-reef development; in Carbonates in Subsurface and Outcrop, L. Eliuk (ed.); 1984 Core Conference, Canadian Society of Petroleum Geologists, p. 82-101.

Nielsen, A.R. and Porter, J.W.
1984: Pembina - in retrospect; in The Mesozoic of Middle North America, D.F. Stott and D.J. Glass (ed.); Canadian Society of Petroleum Geologists, Memoir 9, p. 1-13.

Norris, D.K.
1971: The geology and coal potential of the Cascade Coal Basin; in A Guide to the Geology of the Eastern Cordillera along the Trans-Canada Highway between Calgary, Alberta and Revelstoke, British Columbia, I.A.R. Halladay, and D.H. Matthewson (ed.); Alberta Society of Petroleum Geologists, p. 25-39.

Orr, R.D., Johnston, J.R., and Manko, E.M.
1977: Lower Cretaceous geology and heavy-oil potential of the Lloydminster area; Bulletin of Canadian Petroleum Geology, v. 25, no. 6, p. 1187-1221.

Osadetz, K.G. and Snowdon, L.R.
1986a: Petroleum source rock reconnaissance of southern Saskatchewan; in Current Research, Part A, Geologial Survey of Canada, Paper 86-1A, p. 609-617.
1986b: Speculation on the petroleum source rock potential of portions of the Lodgepole Formation (Mississippian) of southern Saskatchewan; in Current Research, Part B, Geological Survey of Canada, Paper 86-1B, p. 647-651.

Outtrim, C.P. and Evans, R.G.
1978: Alberta's oil sands reserves and their evaluation; in The Oil Sands of Canada-Venezuela, D.A. Redford and A.G. Winestock (ed.); Canadian Institute of Mining and Metallurgy, Special Volume no. 17, p. 36-66.

Ower, J.
1975: The Moose Mountain structure, birth and death of a folded fault play; in Structural Geology of the Foothills between Savanna Creek and Panther River, S.W. Alberta, Canada, H.J. Evers and J.E. Thorpe (ed.); Exploration Update '75, Canadian Society of Petroleum Geologists - Canadian Society of Exploration Geophysicists, Guidebook, p. 22-29.

Plint, A.G., Walker, R.G., and Bergman, K.M.
1986: Cardium Formation 6. Stratigraphic framework of the Cardium in subsurface; Bulletin of Canadian Petroleum Geology, v. 34, no. 2, p. 213-225.

Podruski, J.A., Barclay, J.E., Hamblin, A.P., Lee, P.J., Osadetz, K.G., Procter, R.M., and Taylor, G.C.
1988: Conventional oil resources of Western Canada (light and medium). Part I: Resource Endowment; Geological Survey of Canada, Paper 87-26, p. 1-125.

Powell, T.G. and Snowdon, L.R.
1983: A composite hydrocarbon generation model: Implications for evaluation of basins for oil and gas; Erdol und Kohle Erdgas Petrochemie, v. 36, p. 163-170.

Procter, R.M., Taylor, G.C., and Wade, J.A.
1984: Oil and natural gas resources of Canada - 1983; Geological Survey of Canada, Paper 83-31, 59 p.

Putnam, P.E.
1980: Fluvial deposition within the upper Mannville of west-central Saskatchewan: stratigraphic implications; in Lloydminster and Beyond: Geology of Mannville Hydrocarbon Reservoirs, J.S. Beck, J.E. Christopher, and D.M. Kent (ed.); Saskatchewan Geological Society, Special Publication no. 5, p. 197-216.
1982: Aspects of the petroleum geology of the Lloydminster heavy oilfield, Alberta and Saskatchewan; Bulletin of Canadian Petroleum Geology, v. 30, no. 2, p. 81-111.

Reinson, G.E. and Foscolos, A.E.
1986: Trends in sandstone diagenesis with depth of burial, Viking Formation, southern Alberta; Bulletin of Canadian Petroleum Geology, v. 34, no. 1, p. 126-152.

Rice, D.R. and Shurr, G.W.
1980: Shallow, low-permeability reservoirs of northern Great Plains - assessment of their natural gas resources; American Association of Petroleum Geologists Bulletin, v. 64, no. 7, p. 969-987.

Rottenfusser, B.
1982: Facies of the Gething Formation, Peace River oil sands deposit; in Canada's Giant Hydrocarbon Reservoirs, W.G. Cutler (ed.); Core Conference, Canadian Society of Petroleum Geologists, p. 47-53.

Roy, K.J.
1972: The Boundary Member: a buried erosional remnant of Triassic age in northeastern British Columbia; Bulletin of Canadian Petroleum Geology, v. 20, no. 1, p. 27-57.

Rupp, A.W.
1969: Turner Valley Formation of the Jumping Pound area, foothills southern Alberta; Bulletin of Canadian Petroleum Geology, v. 17, no. 4, p. 460-485.

Saskatchewan Energy and Mines
1987: Reservoir annual 1985; Miscellaneous Report 86-1.

Sawatsky, H.B.
1975: Astroblemes in Williston Basin; American Association of Petroleum Geologists Bulletin, v. 59, no. 4, p. 694-710.

Schmidt, V., McDonald, D.A., and McIlreath, I.A.
1977: Growth and diagenesis of Middle Devonian Keg River cementation reefs, Rainbow Field, Alberta; in The Geology of Selected Carbonate Oil, Gas and Lead-zinc Reservoirs in Western Canada, I.A. McIlreath and R.D. Harrison (ed.); Canadian Society of Petroleum Geologists, p. 1-21.
1980: Growth and diagenesis of Middle Devonian Keg River cementation reefs, Rainbow Field, Alberta; in Carbonate Reservoir Rocks, R.B. Halley and R.G. Loucks (ed.); Society of Economic Paleontologists and Mineralogists Workshop No. 1, Denver, Colorado, 1980, p. 43-63.

Schmidt, V., McIlreath, I.A., and Budwill, W.E.
1985: Origin and diagenesis of Middle Devonian pinnacle reefs encased in evaporites, "A" and "E" pools, Rainbow Field, Alberta; in Carbonate Petroleum Reservoirs, P.O. Roehl and P.W. Choquette (ed.); Casebooks in Earth Science, Springer Verlag, New York, N.Y., p. 141-160.

Shouldice, J.R.
1979: Nature and potential of Belly River gas sand traps and reservoirs in western Canada; Bulletin of Canadian Petroleum Geology, v. 27, no. 2, p. 229-241.

Simpson, F.
1984: Potential for additional hydrocarbon recovery from the Colorado and Montana groups (Cretaceous) of Saskatchewan; in Oil and Gas in Saskatchewan, J.A. Lorsong and M.A. Wilson (ed.); Saskatchewan Geological Society, Special Publication no. 7, p. 211-244.

Smith, S.R.
1984: The Lower Cretaceous Sparky Formation, Aberfeldy steamflood pilot, Saskatchewan: A wave-dominated delta?; in The Mesozoic of Middle North America, D.F. Stott and D.J. Glass (ed.); Canadian Society of Petroleum Geologists, Memoir 9, p. 413-429.

Smith, S.R., van Hulten, F.F.N., and Young, S.D.
1984: Distribution of the Sparky Formation heavy oil fields within the Lloydminster sub-basin; in Oil and Gas in Saskatchewan, J.A. Lorsong and M.A. Wilson (ed.); Saskatchewan Geological Society, Special Publication no. 7, p. 149-168.

Snowdon, D.M.
1977: Beaver River gas field: a fractured carbonate reservoir; in The Geology of Selected Carbonate Oil, Gas and Lead-zinc Reservoirs in Western Canada, I.A. McIlreath and R.D. Harrison (ed.); Canadian Society of Petroleum Geologists, p. 1-18.

Snowdon, L.R.
1980a: Resinite - a potential petroleum source in the Upper Cretaceous/Tertiary of the Beaufort-Mackenzie Basin; in Facts and Principles of World Petroleum Occurrence, A.D. Miall (ed.); Canadian Society of Petroleum Geologists, Memoir 6, p. 509-521.
1980b: Petroleum source potential of the Boundary Creek Formation, Beaufort-Mackenzie Basin; Bulletin of Canadian Petroleum Geology, v. 28, p. 46-58.

Snowdon, L.R. and Powell, T.G.
1982: Immature oil and condensate; modification of hydrocarbon generation model for terrestrial organic matter; American Association of Petroleum Geologists Bulletin, v. 66, no. 6, p. 775-788.

Snowdon, L.R. and Williams, G.K.
1986: Thermal maturation and petroleum source potential of Cambrian and Proterozoic rocks in the Mackenzie Corridor; Geological Survey of Canada, Open File 1367, 15 p.

Stein, E.M.
1977: The Turner Valley Formation at Whiskey Creek, a Mississippian carbonate reservoir rock; in The Geology of Selected Carbonate Oil, Gas and Lead-zinc Reservoirs in Western Canada, I.A. McIlreath and R.D. Harrison (ed.); Canadian Society of Petroleum Geologists, p. 109-124.

Stoakes, F.A. and Wendte, J.C.
1987: The Woodbend Group; in Devonian Lithofacies and Reservoir Styles in Alberta, F.F. Krause and O.G. Burrowes (ed.); 13th CSPG Core Conference and Display, Second International Symposium on the Devonian System, Canadian Society of Petroleum Geologists, Calgary, p. 153-170.

Thompson, R.I.
1981: The nature and significance of large "blind" thrusts within the northern Rocky Mountains of Canada; in Thrust and Nappe Tectonics, K.R. McClay and N.J. Price (ed.); Geological Society of London, Special Publication 9, p. 449-462.
1989: Stratigraphy, tectonic evolution and structural analysis of the Halfway River map area (94B), northern Rocky Mountains, British Columbia; Geological Survey of Canada, Memoir 425, 119 p.

Tilley, B.J. and Longstaffe, F.J.
1982: Sedimentology and mineralogy of Suffield heavy oil sands (Lower Cretaceous), southeastern Alberta (abstract); American Association of Petroleum Geologists Bulletin, v. 66, no. 5, p. 636.

Uhler, R.S.
1986: The potential supply of crude oil and natural gas reserves in the Alberta Basin; Economic Council of Canada, 90 p.

van Delinder, D.G.
1984: Source of oils in Cretaceous fields of southern Saskatchewan; in Oil and Gas in Saskatchewan, J.A. Lorsong and M.A. Wilson (ed.); Saskatchewan Geological Society, Special Publication no. 7, p. 113-118.

Viau, C.
1983: Depositional sequences, facies and evolution of the Upper Devonian Swan Hills buildup, Central Alberta, Canada; in Carbonate Buildups - a Core Workshop, P.M. Harris (ed.); Society of Economic Paleontologists and Mineralogists Core Workshop No. 4, p. 112-143.

Walker, R.G.
1983a: Cardium Formation 2. Sand-body geometry and stratigraphy in the Garrington-Caroline-Ricinus area, Alberta - "The Ragged Blanket" model; Bulletin of Canadian Petroleum Geology, v. 31, no. 1, p. 14-26.

Walker, R.G., (cont.)
1983b: Cardium Formation 3. Sedimentology and stratigraphy in the Garrington-Caroline area, Alberta; Bulletin of Canadian Petroleum Geology, v. 31, no. 4, p. 213-230.
1985: Cardium Formation at Ricinus Field, Alberta; a channel cut and filled by turbidity currents in Cretaceous Western Interior Seaway; American Association of Petroleum Geologists, Bulletin, v. 69, no. 11, p. 1963-1981.

Wallace-Dudley, K.E., Lee, P.J., and Procter, R.M.
1982: Gas resources of northeast British Columbia; Geological Survey of Canada, Open File 817, 19 p.

Walls, R.A.
1983: Golden Spike Reef Complex, Alberta; in Carbonate Depositional Environments, P.A. Scholle, D.G. Bebout, and C.H. Moore (ed.); American Society of Petroleum Geologists, Memoir 33, p. 445-453.

Welte, D.H., Schaefer, R.G., Radke, M., and Weiss, H.M.
1982: Origin, migration and entrapment of natural gas in Alberta Deep Basin: Part 1 (abstract); American Association of Petroleum Geologists Bulletin, v. 66, no. 5, p. 642.

Welte, D.H., Schaefer, R.G., Stoessinger, W., and Radke, M.
1984: Gas generation and migration in the Deep Basin of western Canada; in Elmworth - case history of a Deep Basin Gas Field, J.A. Masters (ed.); American Association of Petroleum Geologists, Memoir 38, p. 35-47.

Wendte, J.C. and Stoakes, F.A.
1982: Evolution and corresponding porosity distribution of the Judy Creek Reef Complex, Upper Devonian, central Alberta; in Canada's Giant Hydrocarbon Reservoirs, W.G. Cutler (ed.); Canadian Society of Petroleum Geologists, p. 63-81.

Williams, E.P. and McNeil, D.J.
1976: The Twining field, Alberta - Rundle A Pool (Pekisko Formation); in Core Conference 1976, Joint Convention on Enhanced Recovery, W. Clark and G. Huff (ed.); Petroleum Society of Canadian Institute of Mining - Canadian Society of Petroleum Geologists, p. D1-D16.

Williams, G.K.
1977: The Celibeta structure compared with other basement structures on the flanks of the Tathlina High, District of Mackenzie; in Report of Activities, Part B, Geological Survey of Canada, Paper 77-1B, p. 301-310.

Wilson, R.C.L.
1986: A reconnaissance sedimentological study of the Middle Jurassic Shaunavon Formation, southwestern Saskatchewan; Geological Survey of Canada, Open File 1299, 40 p. + 14 pl.

Wilson, M.
1984: Depositional environments of the Mannville Group (Lower Cretaceous) in the Tangleflags area, Saskatchewan; in Oil and Gas in Saskatchewan, J.A. Lorsong and M.A. Wilson (ed.); Saskatchewan Geological Society, Special Publication no. 7, p. 119-134.

Young, F.G.
1967: Elkton reservoir of Edson gas field, Alberta; Bulletin of Canadian Petroleum Geology, v. 15, no. 1, p. 50-64.

Young, F.G., Myhr, D.W., and Yorath, C.J.
1976: Geology of the Beaufort-Mackenzie Basin; Geological Survey of Canada, Paper 76-11, 65 p.

Yorath, C.J. and Cook, D.G.
1981: Cretaceous and Tertiary stratigraphy and paleogeography, northern Interior Plains, District of Mackenzie; Geological Survey of Canada, Memoir 398, 76 p.

Yorath, C.J. and Norris, D.K.
1975: The tectonic development of the southern Beaufort Sea and its relationship to the origin of the Arctic Ocean Basin; in Canada's Continental Margins and Offshore Petroleum Potential, C.J. Yorath, E.R. Parker, and D.J. Glass (ed.); Canadian Society of Petroleum Geologists, Memoir 4, p. 589-611.

Authors' Addresses

R.D. Johnson
R.D. Johnson and Associates Ltd.
200, 409 - 8 Avenue S.W.
Calgary, Alberta
T2P 1E3

N.J. McMillan
Institute of Sedimentary and Petroleum Geology
Geological Survey of Canada
3303 - 33 Street N.W.
Calgary, Alberta
T2L 2A7

Printed in Canada

Subchapter 6B

COAL

A.R. Cameron

CONTENTS

INTRODUCTION

Over 95% of Canada's known coal resources are found in the sedimentary basins of the Interior Platform and eastern Cordillera (Fig. 6B.1); a minor amount is found in the Moose River Basin of the Hudson Platform. This distribution is shown in Table 6B.1, which summarizes coal resources for all of Canada. Coal also occurs in the platforms and basins of western Northwest Territories, eastern and northern Yukon Territory and the Arctic Islands, but low exploration activity has limited calculation of regional resources. The coals range in rank from lignite to anthracite and in age from Early Carboniferous to Tertiary.

An interesting pattern of resource distribution by rank is shown by Table 6B.1. A large proportion is subbituminous and lignite - valuable as fuels for power generation and possible future feedstocks for gasification and liquefaction. High-volatile bituminous coals constitute a relatively small percentage of total resources. On the other hand, Canada is well endowed with medium- and low-volatile bituminous coals, and these are of special interest to the metallurgical market. Of Canada's total resources of immediate interest probably 25% are in the medium/low-volatile bituminous category, virtually all of which are in Alberta and British Columbia.

Coal in western Canada has gone through periods of fluctuating fortunes paralleling the experience of the industry elsewhere in North America. During the late 1950s and early 1960s, production decreased markedly and many mines closed. In the late 1960s and the decade of the 70s, a resurgence in mining took place due to increased exports of metallurgical coal and greater use for electricity generation. Active exploration marked this period and a number of new properties were developed. In the 1980s, exploration declined, though Canadian production continued to grow from 1984 through 1987 (Table 6B.2). All production, except that from Nova Scotia and New Brunswick, has come from the geographic areas covered in this volume.

Cameron, A.R.
1993: Coal; Subchapter 6B in Sedimentary Cover of the Craton in Canada, D.F. Stott and J.D. Aitken (ed.); Geological Survey of Canada, Geology of Canada, no. 5, p. 563-598 (also Geological Society of America, The Geology of North America, v. D-1).

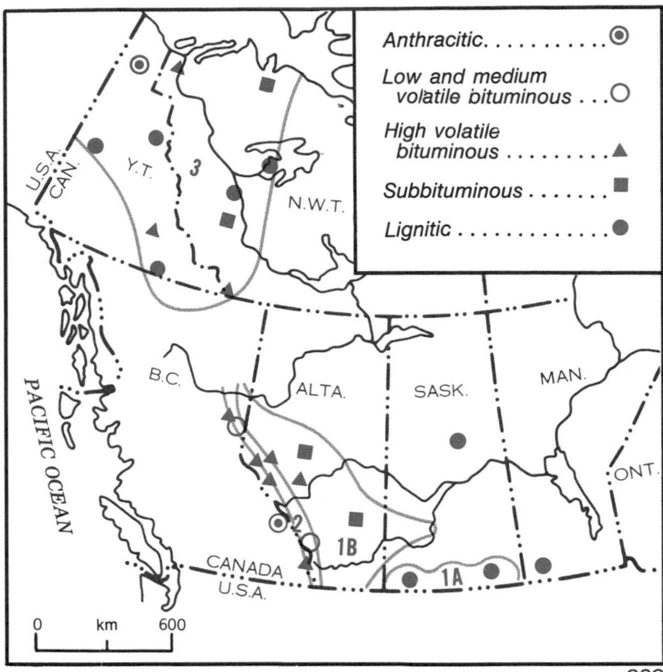

Figure 6B.1. Index map for coal-bearing areas of the Interior Platform and eastern Cordillera in Canada (modified from Bielenstien et al., 1979).: 1 Plains and Outer Foothills A. Williston Basin B. Alberta Syncline; 2 Cordillera: Rocky Mountains and Inner Foothills; 3 Northern Interior Platform and Cordillera

Canada's coal resources have been reviewed and re-calculated periodically. An important older report is that by MacKay (1947). More recent general references on resources and quality are periodic reports published by the Federal Department of Energy, Mines and Resources, as for example Bielenstein et al. (1979) and Romaniuk and Naidu (1987). Also important are periodic reports by the Provincial Government of Alberta, such as Alberta Energy Resources Conservation Board (ERCB) Report ST 87-31. Another general reference is the Canadian Institute of Mining and Metallurgy volume "Coal in Canada" edited by T.H. Patching (1985). The most recent reference on Canada's coal resources has been compiled by Smith (1989).

In the last 15 years two important developments in coal geology and resource evaluation have occurred. The first is increased emphasis on the sedimentology of coal-bearing strata and the second, the use of computer modelling, not only to assess quantity and quality of coal but also to assess the geometry and facies characteristics of all units in a coal-bearing sequence. Examples of papers from Canada dealing with the first development are those of Gibson (1977, 1985a, b). An example of a study dealing with computer modelling is that of Hughes (1984).

The discussion of coal deposits in this subchapter follows a division based on geography and geological differences (see Figure 6B.1). These divisions are:

1. Plains and Outer Foothills - Alberta, British Columbia, and Saskatchewan (Late Cretaceous-Tertiary)
2. Inner Foothills and Rocky Mountains - Alberta and British Columbia (Late Jurassic-Early Cretaceous)
3. Eastern and northern Yukon Territory and western Northwest Territories (Carboniferous to Tertiary).

Coal occurrences in the Moose River Basin of Ontario are discussed in Chapter 9.

Compositional characteristics of the coals discussed in this chapter are defined in several ways. Proximate and ultimate analyses and determination of heating value are routinely made to assess quality and to classify by rank according to specifications of the American Society for Testing and Materials (ASTM, 1979). Data from such analyses are cited and rank expressed in ASTM terms. In addition, quality data are reported from petrographic analyses. These include reflectance measurements for rank definition and determination of maceral content. Macerals in coal are the analogs of minerals in inorganic rocks. Coal seams differ in the maceral compositions, and because they do, data on distribution can be used to correlate seams. Maceral composition can be used also to analyze depositional environments in the peat swamp, for example, conditions of good versus poor preservation of plant material. Maceral content also may be related to parent-plant communities. Finally, because they differ chemically, macerals can be used to define technological behaviour, that is, coking, gasification, or liquefaction performance.

PLAINS AND OUTER FOOTHILLS - ALBERTA, BRITISH COLUMBIA, AND SASKATCHEWAN

Williston Basin - Southern Saskatchewan

A belt of lignite-bearing lower Tertiary strata of the Ravenscrag Formation underlies about 25 900 km^2 (Guliov, 1972) in southern Saskatchewan (Fig. 6B.2), and small erosional remnants occur also in southwestern Manitoba and southeastern Alberta. The Ravenscrag, on the northern flank of Williston Basin, is the northern extension of the important coal-bearing Fort Union Group of North Dakota and Montana. The main reference on the geology and coal resources of the Ravenscrag is the report by Whitaker et al. (1978). Earlier references include McLearn (1929, 1930), Fraser et al. (1935), Furnival (1950), and Holter (1972).

Stratigraphy

The Paleocene Ravenscrag Formation is overlain by younger Tertiary and glacial deposits and underlain by the Eastend, Whitemud, and Frenchman formations which, in ascending order, separate it from the marine Bearpaw Formation. In Saskatchewan the Ravenscrag is entirely nonmarine, consisting of poorly consolidated sands, silts, clays, and coal of fluvial-deltaic origin. Coal occurrence is described in terms of "zones", which according to Irvine (1978a) denote related seams appearing in proximity to one another if not joined as one unit. Thus a zone contains one or more seams with interseam clastic strata.

Whitaker et al. (1978) discussed the Ravenscrag and its contained coal resources in terms of four map-areas, Cypress, Wood Mountain, Willow Bunch, and Estevan (Fig. 6B.2). Representative sections show the Ravenscrag to be an eastward-thickening wedge with an eastward increase in the number of coal zones (Fig. 6B.3). Net coal thickness and zone thickness data are summarized in Table 6B.3. Continuous zones from Cypress to Estevan are unlikely, though coals occur at comparable stratigraphic positions in the various areas (Whitaker et al., 1978). Thus the Ferris and Anxiety Butte zones at Cypress are at the same stratigraphic level as the Killdeer and Stonehenge zones at Wood Mountain, the Landscape zone at Willow Bunch, and the Auburnton , Woodley, and Elcott zones at Estevan.

Whitaker (1978) found at least two geophysical log markers that are useful for stratigraphical delineation. One of these, between the Estevan and Boundary zones at Estevan, corresponds approximately to a palynological break established by Sweet (1978).

Table 6B.1. Summary of Canada's coal resources (from Smith, 1989)

Coal Region	Coal Rank	IMMEDIATE INTEREST (megatonnes)			FUTURE INTEREST (megatonnes)			
		measured	indicated	inferred	measured	indicated	inferred	speculative
COASTAL BRITISH COLUMBIA								
- Vancouver Island	h-mvb	35	80	200	-	-	300	-
- Queen Charlotte Islands	lvb-an	-	-	10	-	-	-	-
	h-mvb	-	15	10	-	-	-	-
	lig-sub	-	-	50	-	-	-	500
INTERMONTANE BRITISH COLUMBIA								
- Northern District	lvb-an	100	500	1000	-	-	-	4000
	h-mvb	30	50	100	-	-	-	100
- Southern District	sub-hvb	40	120	340	-	-	-	-
	lig-sub	450	320	270	-	-	-	-
ROCKY MOUNTAINS AND FOOTHILLS								
- Front Ranges								
- East Kootenay	h-mvb	1390	1320	4040	-	2700	-	-
- Crowsnest	m-lvb	265	140	510	-	200	-	-
	h-mvb	330	170	630	-	-	-	-
- Cascade	lvb-an	240	120	455	-	210	-	-
- Panther River-Clearwater	lvb-an	-	-	-	15	15	700	-
- Inner Foothills								
- Southern District	m-lvb	635	320	1145	-	245	-	-
	h-mvb	150	75	275	-	-	-	-
- Northern District	m-lvb	1115	2385	6270	-	100	-	-
- Outer Foothills	sub-hvb	830	740	1955	-	200	-	-
PLAINS								
- Mannville Group	lig-sub	-	35	100	-	-	30	-
- Belly River/Edmonton/Wapiti	sub-hvb	1240	585	1860	-	820	-	-
	lig-sub	11 860	4935	16 575	-	14 115	-	-
- Paskapoo	sub-hvb	120	60	175	-	25	-	-
- Ravenscrag	lig-sub	1445	2680	3440	165	3910	23 510	-
- Deep coal	sub-hvb	-	-	-	1200	4000	50 000	85 000
HUDSON BAY LOWLAND								
- Onakawana	lig-sub	170	10	-	-	(no available estimates)		-
ATLANTIC PROVINCES	h-mvb	345	365	770	-	1500	215	-
NORTHERN CANADA								
- Yukon Territory and District of Mackenzie	lvb-an	-	-	90	-	(no available estimates		-
	h-mvb	-	-	150	-	of resources of future		-
	sub-hvb	-	-	350	-	interest for this region)		-
	lig-sub	-	-	2290	-	-	-	-
- Arctic Archipelago	sub-hvb	-	-	-	-	500	550	4500
	lig-sub	-	-	-	-	7000	7500	31 000
TOTALS	lvb-an	340	620	1555	15	225	700	4000
	m-lvb	2015	2845	7925	-	545	-	-
	h-mvb	2280	2075	6175	-	4200	515	100
	sub-hvb	2230	1505	4680	1200	5545	50 550	89 500
	lig-sub	13 925	7980	22 725	165	25 025	31 040	31 500

Structure

The structure of the southern Saskatchewan coal fields is relatively uncomplicated. Whitaker (1978) mentioned as major tectonic features the Roncott/Ceylon platform, Bowdoin Dome, and the associated downwarps of the Radville, Hummingbird, and Killdeer troughs and the Coburg Syncline (Fig. 6B.2). Depocentres were controlled by regional subsidence associated with the overall tectonics of the epicratonic Williston Basin. Superimposed on this, however, is more local subsidence, related to salt solution in the underlying Prairie Evaporite, which affected coal formation in the Radville, Coronach, Killdeer, and Hummingbird troughs (Broughton, 1979). Minor discontinuities and faults have been noted in all fields. At least some are related to ice-thrusting, as exemplified by an apparent vertical displacement of 18 m in two coal zones of the Estevan area (Irvine, 1978b).

Rank and other quality parameters

The major Ravenscrag coals are of lignite A rank according to ASTM specifications. A summary (Table 6B.4) of data relating to coal quality is condensed from more extensive tables published by Dyck et al. (1980). Many more samples than sample sites are indicated in each area, because zones were sampled in increments at each site. All samples were from reverse circulation drilling. Mean heat values on the equilibrium moisture, mineral-matter free basis range from 8299 Btu/lb (19.3 MJ/kg) for the Estevan zone to 7492 Btu/lb (17.4 MJ/kg) for the Ferris zone. Heat values on the equilibrium moisture basis suggest higher rank coal in the Estevan area, but correction for mineral matter reduces the difference between areas. The Estevan and Souris zones in the Estevan area and the Willow Bunch zone in the Willow Bunch field show the lowest mean ash contents. Mean sulphur values are all below 1%, with the three Estevan zones showing the lowest contents.

Other compositional data on Saskatchewan lignites are of a petrographic nature. Determination of reflectivities on samples along an east-west transect from Estevan to Wood Mountain showed gradually decreasing values in a westerly direction (Cameron and Marconi, 1979; Cameron, 1991). These values, which correspond to decreasing rank, conform with indications of slightly lower rank westward from Estevan, as suggested by Table 6B.4 and mentioned by Dyck et al. (1980). The stratigraphically lower seams in the Wood Mountain-Willow Bunch areas have lower reflectances than stratigraphically higher seams in the

Table 6B.2. Coal production[1] by province (Romaniuk and Naidu, 1987; Statistics Canada, 1988)

Province	1984 10^3 Tonnes	1985 10^3 Tonnes	1986 10^3 Tonnes	1987 10^3 Tonnes
Nova Scotia	3 093	2 800	2 695	2 925
New Brunswick	564	560	490	533
Saskatchewan	9 918	9 672	8 281	10 020
Alberta	23 052	24 712	24 519	25 739
British Columbia	20 775	22 994	20 359	21 990
Totals	57 402	60 738	56 344	61 207

[1] Clean coal from mines with preparation plants plus raw coal from mines without such plants.

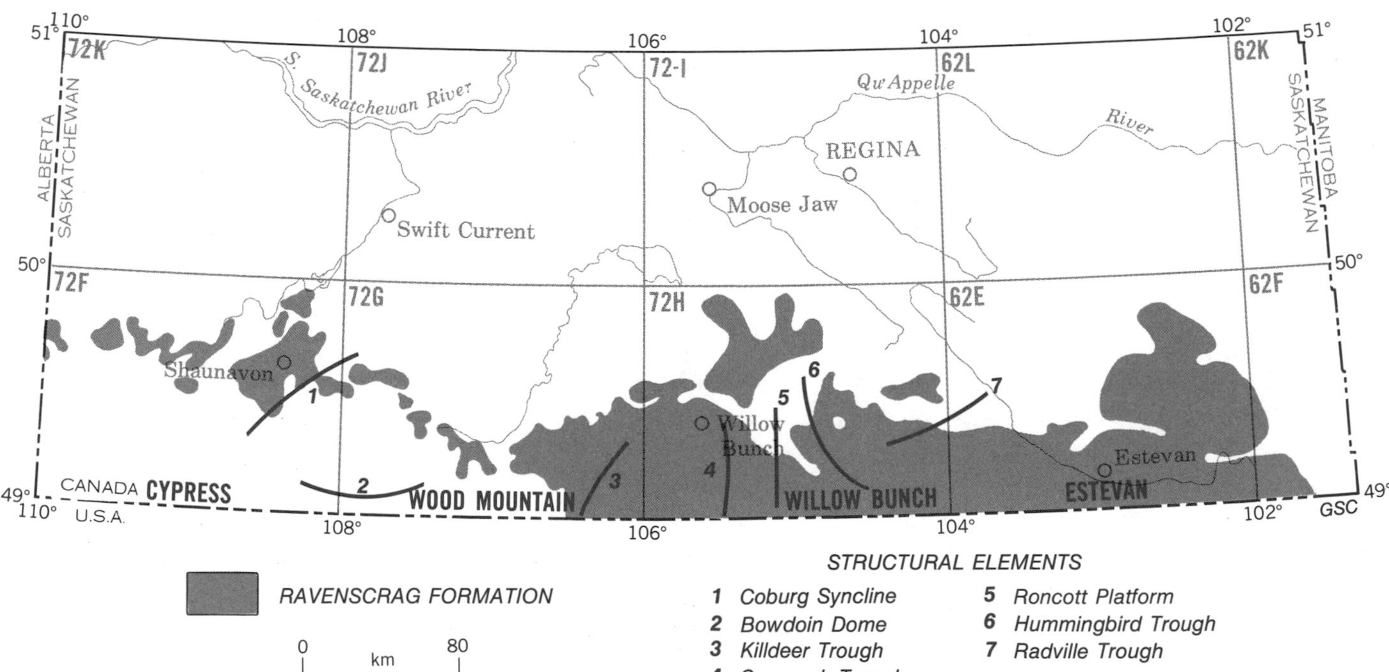

Figure 6B.2. Distribution and structure of Ravenscrag Formation in southern Saskatchewan (modified from Whitaker et al., 1978).

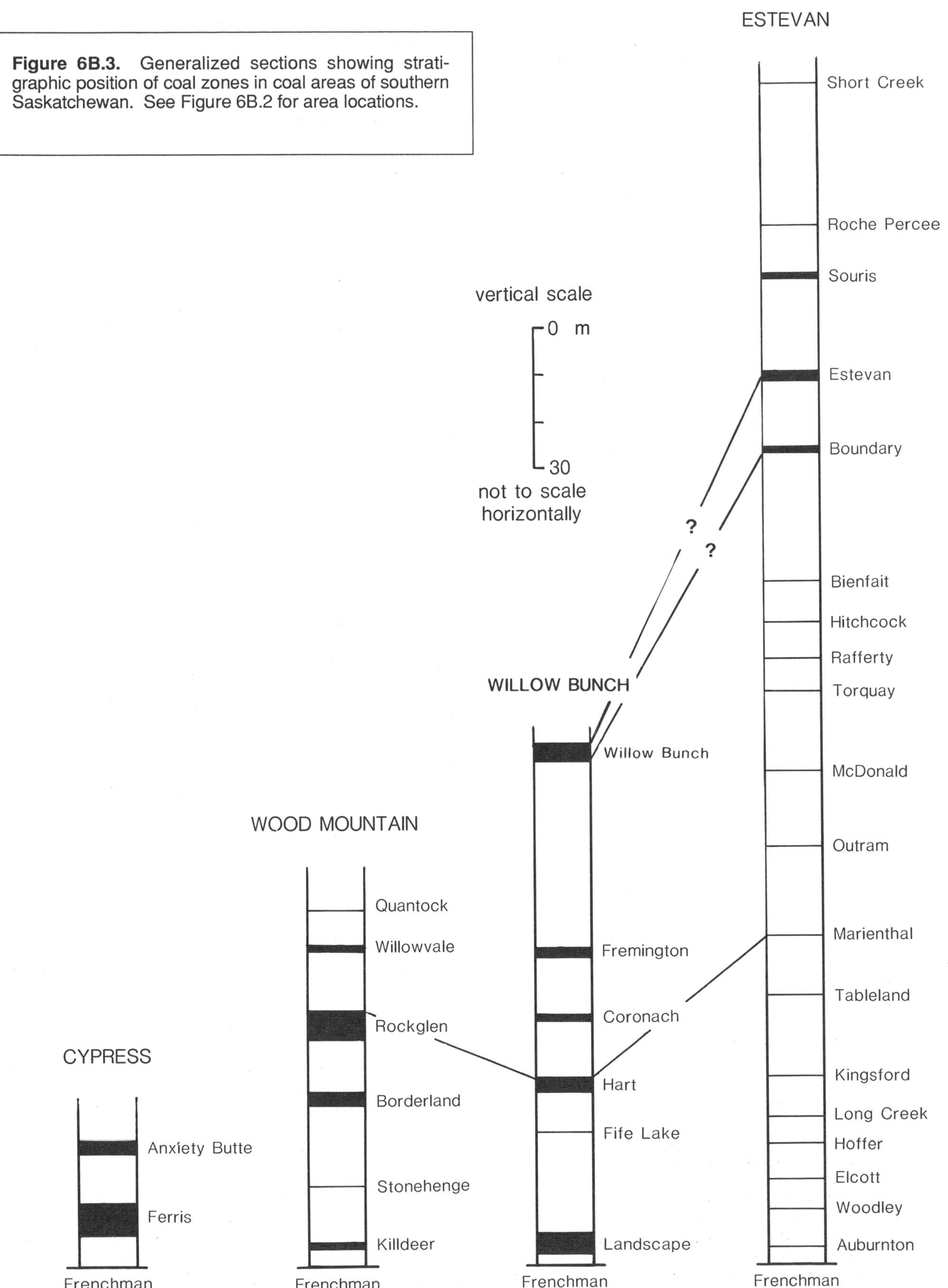

Figure 6B.3. Generalized sections showing stratigraphic position of coal zones in coal areas of southern Saskatchewan. See Figure 6B.2 for area locations.

Table 6B.3. Summary of data on zone and net coal thicknesses in Ravenscrag Formation of southern Saskatchewan; Data from Whitaker et al. (1978)

Area	Coal Zone	Number of Seams	Average Thickness of Zone (m)	Average Net Thickness of Coal in Zone (m)
Cypress	Anxiety Butte	2	2.5	1.6
	Ferris	3	6.3	2.9
	Quantock	1	1.1	1.0
	Willowvale	1	1.0	0.9
Wood Mountain	Rock Glen	2	6.1	2.2
	Borderland	2	2.9	2.3
	Stonehenge	2	2.5	1.7
	Killdeer	2	1.5	1.2
	Willow Bunch	2	3.4	2.6
	Fremington	1	1.6	1.2
	Coronach	1	1.4	1.1
Willow Bunch	Hart	2	3.6	3.1
	Fife Lake	1	1.4	1.1
	Landscape	3	4.4	2.1
	Short Creek	2	2.0	1.6
	Roche Percee	2	1.7	1.5
	Souris	2	1.5	1.3
	Estevan	3	4.2	2.9
	Boundary	3	2.6	1.7
	--*			
	Bienfait	1	0.7	0.7
	Hitchcock	1	2.5	0.7
Estevan	Rafferty	2	1.8	1.0
	Torquay	1	2.2	1.0
	McDonald	3	2.7	1.0
	Outram	3	2.2	0.8
	Marienthal	3	3.2	1.2
	Tableland	3	3.2	1.4
	Kingsford	3	4.1	1.5
	Long Creek	1	1.7	0.9
	Hoffer	2	1.6	1.0
	Elcott	2	1.1	0.6
	Woodley	1	0.8	0.7
	Auburnton	2	1.7	0.8

*----Boundary between shallow and deep coal zone in Estevan area (after Irvine 1978b).

Estevan area, suggesting possibly greater depth of burial at one time in the latter area or a higher heat flow regime as postulated by Majorowicz and Jessop (1981).

Significant differences in maceral content occur with depth, at least in the Estevan area, involving a change in the relative proportions of two huminite macerals, ulminite and densinite. Ulminite is a constituent that still shows variably preserved evidence of plant cell structure, whereas densinite is more attrital. In the Estevan area nearly all the seams with high ulminite and low densinite lie in the lower part of the section, whereas the majority of samples from the Estevan, Souris, and Roche Percée zones show higher densinite/ulminite ratios (Cameron, 1978). Different environments and/or different plant communities are thus indicated for the uppermost seams in the Estevan section. It may be significant that Sweet (1978) found a floral break at approximately the change in petrological character, that is, between the Boundary and Estevan zones.

Resources, production

Whitaker et al. (1978) provided data on coal resources that are summarized in Table 6B.5. Quantities are reported in measured, indicated, and inferred categories. They are also classified as "Resources of immediate interest" and two categories defined as "Resources of future interest", one at depths of less than 46 m and one at depths greater than 46 m (deep coal). The table presents these data for each of the four main coal areas. The following criteria for resource classification are taken from Irvine (1978a)[1] and identify resource categories in Table 6B.5:

A. Measured resources: coal located 0.25 miles (0.4 km) or less from a borehole.

B. Indicated resources: coal located 0.5 mile (0.8 km) or less, but more than 0.25 mile (0.4 km) from a borehole.

[1] In 1989 a revised set of standards for reporting coal resources was published: Hughes, J.D., Klatzel-Mudry, L. and Nikols, D.J. 1989: A standardized coal resource/reserve reporting system for Canada; Geological Survey of Canada, Paper 88-21, p. 17.

Table 6B.4. Compositional data on important Ravenscrag coal zones. Data represent mean values on weight per cent basis (modified from Dyck et al., 1980)

Chemical Parameters	Boundary Zone 3 sites; 8 samples	Estevan Zone 17 sites; 153 samples	Souris Zone 9 sites; 29 samples	Hart Zone 15 sites; 129 samples	Willow Bunch Zone 19 sites; 114 samples	Borderland Zone 16 sites; 94 samples	Rock Glen Zone 10 sites; 62 samples	Ferris Zone 9 sites; 55 samples
Equilibrium moisture (EQM)	25.9	28.9	29.3	27.7	27.8	24.9	23.4	28.4
Ash (EQM)	23.4	15.1	14.3	19.4	16.3	23.1	28.0	22.6
Volatile matter (EQM)	24.5	26.5	29.0	27.5	30.3	26.5	25.5	26.1
Fixed carbon (EQM)	26.2	29.5	27.5	25.5	25.6	25.5	23.2	24.9
Sulphur (EQM)	0.5	0.5	0.6	0.8	0.8	0.9	0.9	0.6
Calorific value (EQM)	14.2	15.4	14.7	13.7	14.7	13.5	12.6	13.5
Calorific value (EQM; MMF)	19.3	18.6	17.6	17.6	17.9	18.2	18.4	17.4

EQM - Equilibrium moisture basis
MMF - Mineral matter free
Calorific value - MJ/kg (Megajoules/kilogram)
Coversion: Btu/pound to MJ/Kg = Btu x 0.002326
MJ/Kg to Btu/pound = MJ/Kg x 429.923

C. Inferred resources: coal located more than 0.5 mile (0.8 km) but less than 1.5 mile (2.4 km) from a borehole.

Two categories of deep coal are identified, demonstrated and inferred, according to the following criteria:

Demonstrated resources: includes measured and indicated coal located 0.5 mile (0.8 km) or less from a borehole.

Inferred Resources: coal more than 0.5 mile (0.8 km) but less than or equal to 3 miles (4.8 km) from a borehole.

The detailed resource tables in Whitaker et al. (1978) indicate that, in terms of total resources of immediate interest, the Hart zone is the most important, followed by the Willow Bunch, Borderland, and Ferris zones. In terms of "measured resources" in the immediate interest category, the Hart zone is again the most important, followed by Willow Bunch, Estevan, and Ferris.

Present production is exclusively from surface mines, of which four are in the Estevan area and one near Poplar River in the Willow Bunch field. Most coal is used for power generation at mine-site power stations. A small amount is shipped as steam coal to Manitoba and northwestern Ontario. Lignite in Saskatchewan apparently was produced commercially first in 1890, when 200 tonnes were mined. Since then production has increased steadily even during the depression years of the 1930s. Production for 1987 was 10.02 million tonnes (Statistics Canada, 1988).

Alberta Basin – Alberta, British Columbia, and Saskatchewan

Upper Cretaceous to Lower Tertiary coal-bearing rocks underlie much of the Plains and outer Foothills of central and southern Alberta and extend into Saskatchewan and British Columbia (Fig. 6B.4). The most important resources are in Alberta. Large parts of this area are underlain also by the Lower Cretaceous Mannville Group containing large coal resources that are uneconomic at present (Yurko, 1976).

In southern Alberta the Cretaceous-Tertiary beds are the Foremost and Oldman formations of the Belly River Group (Judith River of McLean, 1971), in central Alberta the Horseshoe Canyon and Scollard formations of the Edmonton Group, in the Foothills the Brazeau and Coalspur formations of the Saunders Group, and in the north the Wapiti Formation.

The coals range in rank from lignite to high-volatile bituminous and have been mined since at least 1873, although never more than on a small scale in some districts. At present eight mines are active, all of them surface operations.

Stratigraphy

The Upper Cretaceous-Lower Tertiary coalmeasures belong to the westward thickening sedimentary wedge of the foreland basin.

Table 6B.5. Coal resources of the Ravenscrag Formation (in 10^6 tonnes)* (from Whitaker et al., 1978)

	Measured	Indicated	Demonstrated	Inferred 0.81-2.41 km** (0.5-1.5 miles)	Inferred 2.41-4.83 km** (1.5-3.0 miles)
Resources of <u>immediate interest:</u>					
Estevan	310	497		438	
Willow Bunch	749	1 044		1 420	
Wood Mountain	277	733		1 114	
Cypress	162	406		466	
Subtotals	1 500	2 700		3 400	
Resources of future interest; at less than 45.7 m (150 ft) depth:					
Estevan	41	99		126	
Willow Bunch	68	107		168	
Wood Mountain	45	119		323	
Cypress	8	22		45	
Subtotals	200	300		700	
Resources of future interest; at greater than 45.7 m (150 ft) depth:					
Estevan			420	2 190	4 683
Willow Bunch			1 597	5 239	4 981
Wood Mountain			1 329	4 211	1 132
Cypress			221	320	96
Subtotals			3 600	12 000	10 900
TOTALS	1 700	3 000	3 600	16 100	10 900

* Subtotals and totals rounded off.
** See text under heading "Resources, Production; Williston Basin-Saskatchewan" for significance of these distances.

In southern Alberta, the Belly River Group (Judith River) is divided into Foremost and overlying Oldman formations, both of which contain coal seams. Toward the Foothills, separate units cannot be recognized and the Belly River is treated as a formation. Although the Belly River extends into Saskatchewan (Fig. 6B.4), the most important coal deposits are in the Lethbridge-Taber area of southern Alberta, with smaller occurrences in the Foothills. Thickness of the Belly River ranges from 1370 m in the Beaver Mines area of southwestern Alberta to about 350 m at Lethbridge and 30 m west of Regina (McLean, 1971; Wall and Rosene, 1977). To the northwest (Foothills) and north (Plains), the Belly River passes into the Brazeau Formation and Wapiti Formation respectively.

Coal seams in the Foremost and Oldman formations are relatively thin; Campbell (1974) reported thicknesses of 0.9 to 2.1 m for individual seams including partings. The Alberta Energy Resources Conservation Board Report (1987) indicated maximum aggregate thicknesses of 2.4 m for the Foremost and 1.9 for the Oldman, with a maximum of 5 seams in each formation. In the Foothills, coal occurs near the top of the undivided Belly River (Douglas, 1950).

The Edmonton Group includes those rocks containing coal that lie above the marine Bearpaw Formation and beneath continental Paskapoo beds. The Edmonton Group underlies much of southern Alberta (Fig. 6B.4) with good exposures in Red Deer Valley. It is correlative with part of the Wapiti Formation northwest of Edmonton and part of the Saunders Group of the Foothills. Thickness of the Edmonton Group ranges from 335 m to 760 m exclusive of the Scollard beds (Irish, 1970). Gibson (1977) included the Scollard and found average thickness in Red Deer Valley of 336 m for the Edmonton Group.

In Red Deer Valley, 15 seams or coal zones occur in the Edmonton Group, 13 of them in the Horseshoe Canyon Formation, as shown in the composite section of Figure 6B.5. In terms of past mining history, the most important seams are the Nos. 1, 2, 5, 7, 11, and 12. Figure 6B.5 indicates substantial variation in seam thickness, although many seams are continuous and can be traced over considerable distances. Maximum thickness reported by Gibson was 3.3 m for both the Nos. 1 and 7 seams. Steiner et al. (1972) correlated seams in Red Deer Valley with those of the Edmonton area, and reported average thickness of coals in the Horseshoe Canyon Formation of 1.3 m, with maximum thickness of 7.9 m. The uppermost seam in the section, the No. 14 seam, is also known as the Ardley or "Big" seam (Holter et al., 1975). It is probably the most important coal bed in the Edmonton Group and is being mined near Wabamun Lake west of Edmonton (where its thickness is much greater than shown in Fig. 6D.5). The group in its upper part straddles the Cretaceous-Tertiary boundary. According to Sweet and Hills (1984) the boundary, based on palynological evidence, is at the base of or within the Nevis (No. 13) coal zone.

The Saunders Group (Fig. 6B.4) comprises the whole sequence of nonmarine, in part coal-bearing, rocks lying above the Alberta Group in the central Alberta Foothills (Allan and Rutherford, 1924; Jerzykiewicz and McLean, 1980). The group has a maximum thickness of 3600 m and consists of three coal-bearing units, the Brazeau, Coalspur, and Paskapoo formations (Jerzykiewicz and McLean, 1980; Jerzykiewicz, 1985). The Brazeau Formation contains only a few thin seams, which, at least in the Chungo Creek area, have a maximum thickness of only 0.3 m (Jerzykiewicz, pers. comm., 1984). The seams in the Coalspur Formation are more important economically, and have their maximum development in the Coal Valley and Brazeau River areas. A series of sections from this region (Fig. 6B.6) indicates seam nomenclature and thickness. The most important seams are the Val D'Or and Mynheer, best developed in the Mercoal and Coal Valley areas Fig. 6B.8. Thickness of the former ranges from 9.6 to 20.0 m and of the latter 5.5 to 20.0 m, including partings. Seams in the Coalspur Formation increase in number and thickness northward from Blackstone River and appear to be correlatable with the Ardley coal zone of the Plains. Recent studies by Jerzykiewicz et al. (1984) show that the Cretaceous- Tertiary boundary is at or near the base of the Mynheer seam. The youngest Saunders coal zone is in the Paskapoo Formation and is identified as the Obed coal zone containing up to three seams, with a maximum aggregate thickness of 5.4 m (Alberta Energy Resources Conservation Board, 1987).

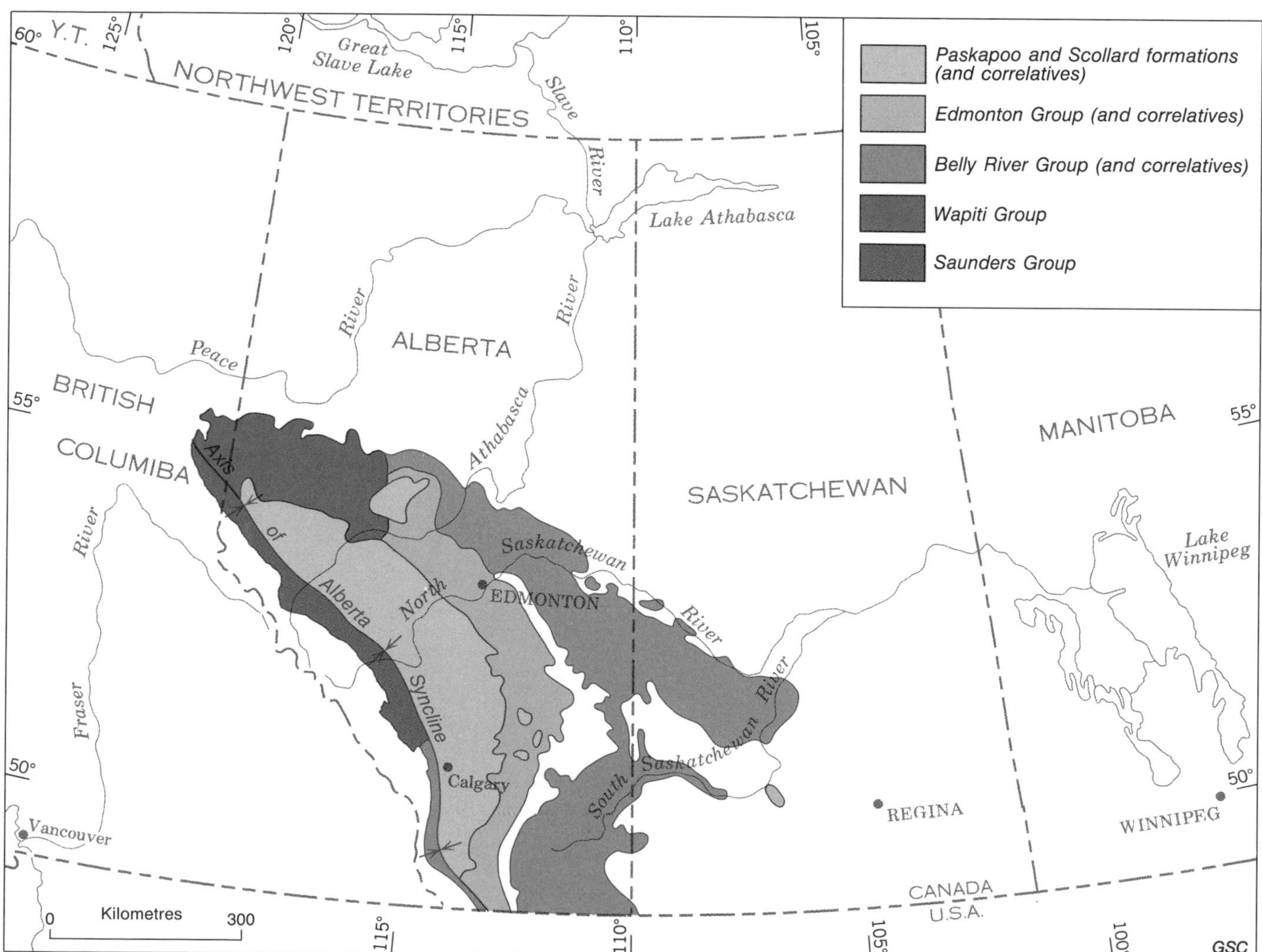

Figure 6B.4. Distribution of Upper Cretaceous and Tertiary coal-bearing rocks in the Plains and outer Foothills (modified from Canada - Department Energy, Mines and Resources, 1982; Alberta; ERCB, 1987).

The Wapiti Formation occupies the northern part of the coal area of Alberta (Fig. 6B.4). Near the city of Edmonton, the Wapiti merges with Belly River, Bearpaw, and Edmonton beds. The Wapiti is underlain by marine rocks of the Smoky Group and overlain by Paskapoo strata and glacial deposits. Geology and coal potential of the Wapiti Formation were described by Kramers and Mellon (1972) and also by Allan and Carr (1946). The Wapiti thins eastward due to depositional wedging and erosion. Kramers and Mellon (1972) estimated thicknesses of about 760 m near the eastern outcrop margin and 1370 m near the Foothills. Coal seams occur throughout the Wapiti section (Fig. 6B.7). The most important zone, possibly equivalent to the Ardley zone and containing several seams of mineable thickness, is about 30-60 m above the regionally persistent Kneehills Tuff horizon in the upper part of the Wapiti. Recent observations (Dawson and Kalkreuth, 1989) indicate that a considerable thickness of Wapiti strata, including the equivalent of the Ardley zone, exists above the top of the section shown in Figure 6B.7.

Structure and tectonics

Three major tectonic elements have controlled the deposition, rank, and present-day distribution of coal-bearing sequences in Alberta. These elements in order of importance are (1) the Rocky Mountain Trough, (2) the Cordilleran Fold Belt, and (3) Sweetgrass Arch. The Fold Belt was the source of thousands of metres of sediments that accumulated in the Rocky Mountain Trough. Repeatedly, deltaic and alluvial environments in the foreland were optimal for the development of extensive peat swamps preserved as the numerous coal seams of the succession. Sweetgrass Arch is a positive element in southeastern Alberta and may have been active in Late Cretaceous time (Williams and Burk, 1964), though not necessarily emergent (McLean, 1971). Pre-Belly River beds are exposed on the crest of the arch, with Belly River and younger sediments arranged concentrically around its margin. The regional structure for these Plains and Foothills rocks is that of an asymmetric syncline with a

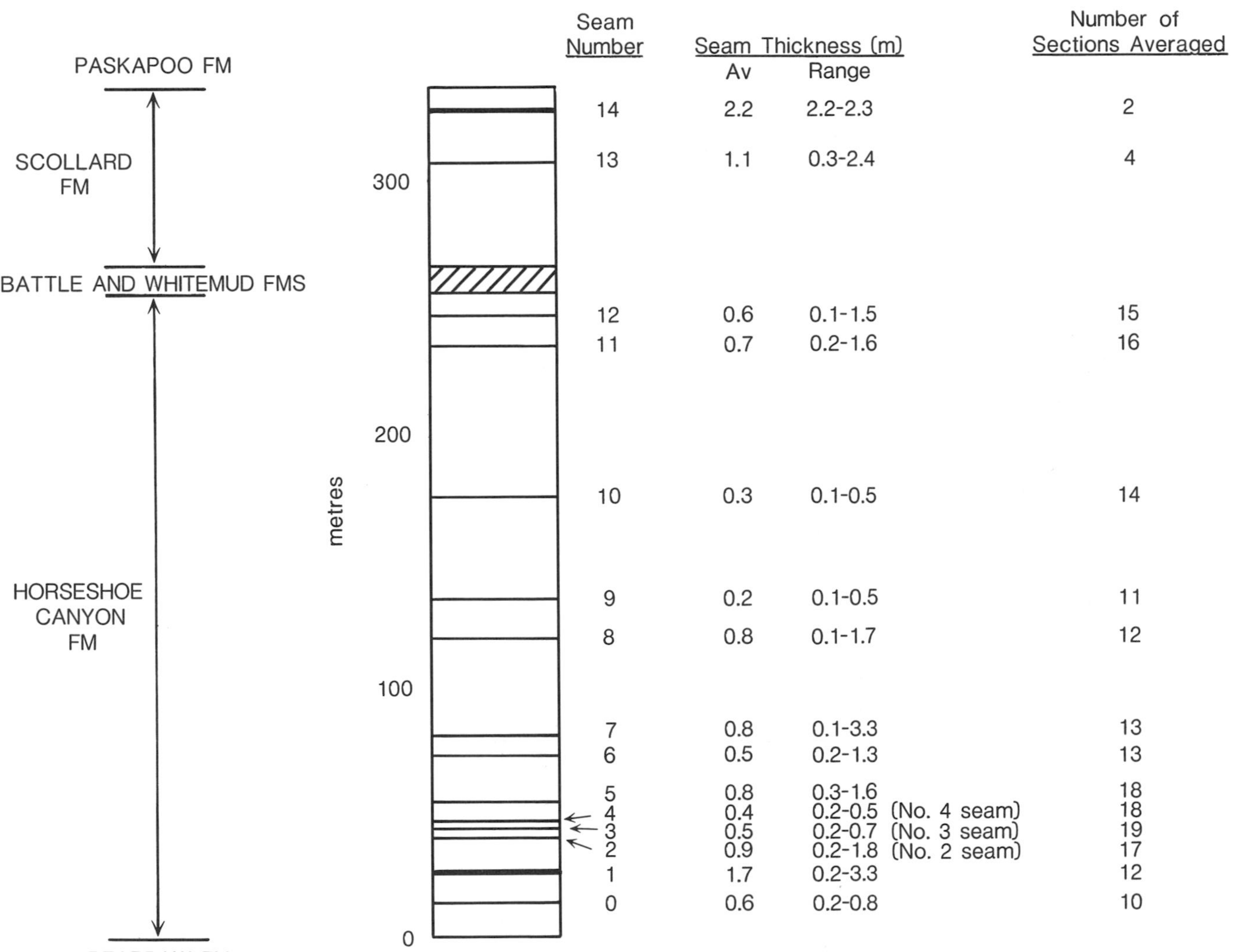

Figure 6B.5. Composite section of Edmonton Group in Red Deer Valley, Alberta, showing stratigraphic position of coal seams (data from Gibson, 1977).

broad, gently dipping and essentially undeformed eastern flank and a narrow, steep to overturned western flank associated with numerous faults.

Bedrock deformation caused by Pleistocene glaciation is evident at a number of localities. Campbell (1974) reported examples from southeastern Alberta, and others have been observed in the Coal Valley area of the central Foothills and in the Plains of central Alberta (Fig. 6B.9). Such deformation affects areas as large as a mine property and has produced vertical displacements of as much as 60 m (F.M. Dawson, pers. comm., 1985).

Rank and other quality parameters

The coals of the Alberta Plains and outer Foothills range in rank from lignite to high-volatile bituminous. Some representative chemical and heating value analyses are given in Table 6B.6 (localities shown in Fig. 6B.10). In Alberta, rank increases in a westward direction. For example, the rank in Belly River coals is subbituminous in the southeast, high-volatile C bituminous near Lethbridge, and high-volatile A/B bituminous in the southern Foothills. Rank relationships in the Plains have been explored by Hacquebard (1977) and Nurkowski (1984).

Sulphur content of these coals is generally low, usually less than 1% (Alberta Energy Resources Conservation Board, 1987). Nurkowski (1984) showed that average content for Belly River coals is about 0.66%, for the Horseshoe Canyon 0.47%, and the Scollard 0.39%.

Steiner et al. (1972) summarized much of the chemical compositional data and published a series of maps showing regional patterns in the distribution of heating value, fixed carbon, moisture, ash, and sulphur. The plot for heating value (Fig. 6B.11) shows increase toward the mountains, with lignitic coals along the eastern margin of the basin and bituminous coals in the west. The data for Figure 6B.11 were determined on near-surface coals and represent all four of the major coal-bearing successions (Belly River, Edmonton, Saunders, and Wapiti).

Maceral and reflectance data on 10 subbituminous coals from the Plains (Fig. 6B.10) are summarized in Table 6B.7 (Parkash et al., 1984). The data show a difference

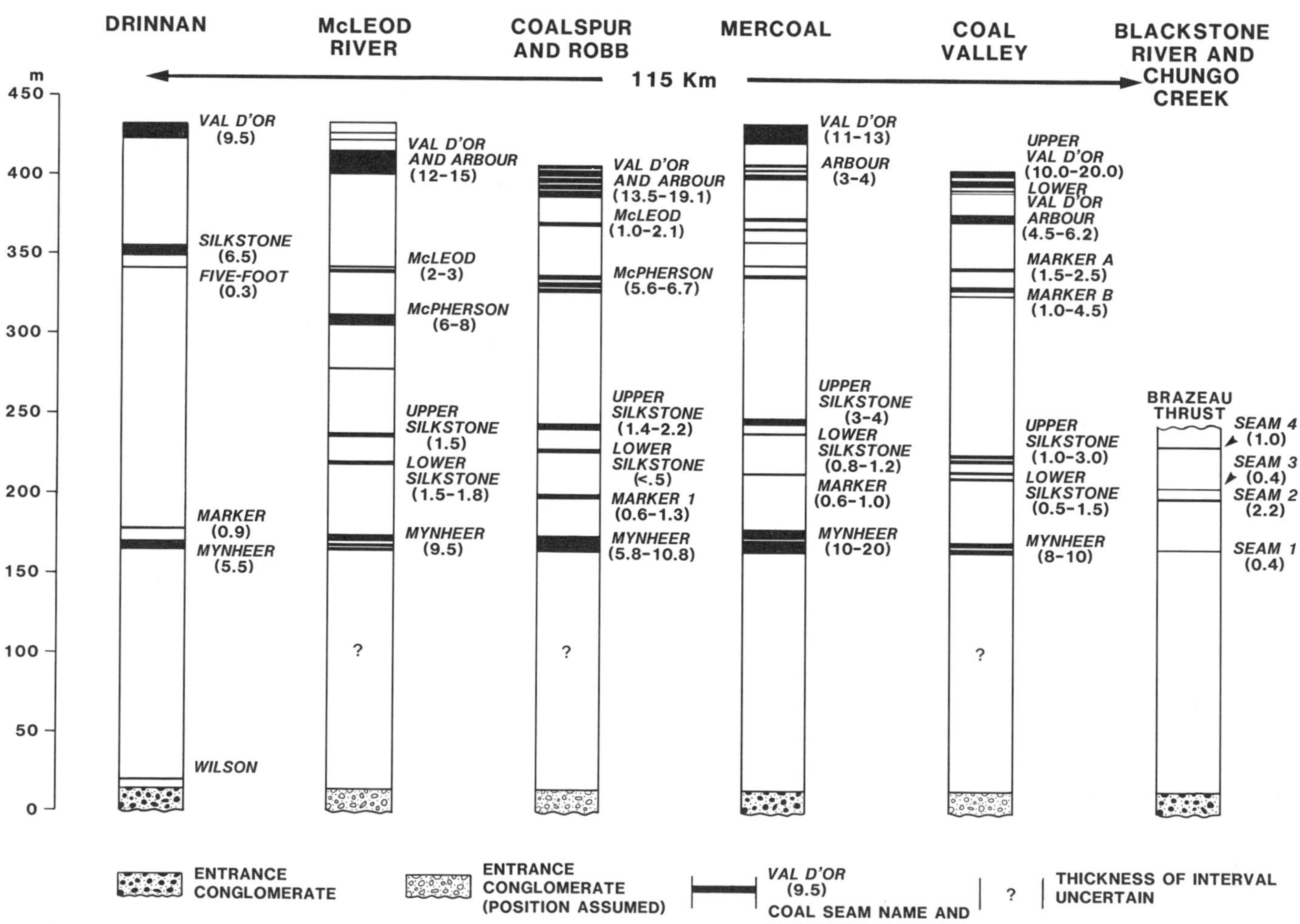

Figure 6B.6. Sections of Coalspur Formation in central Alberta Foothills showing stratigraphic position of coal seams (after Jerzykiewicz, 1985).

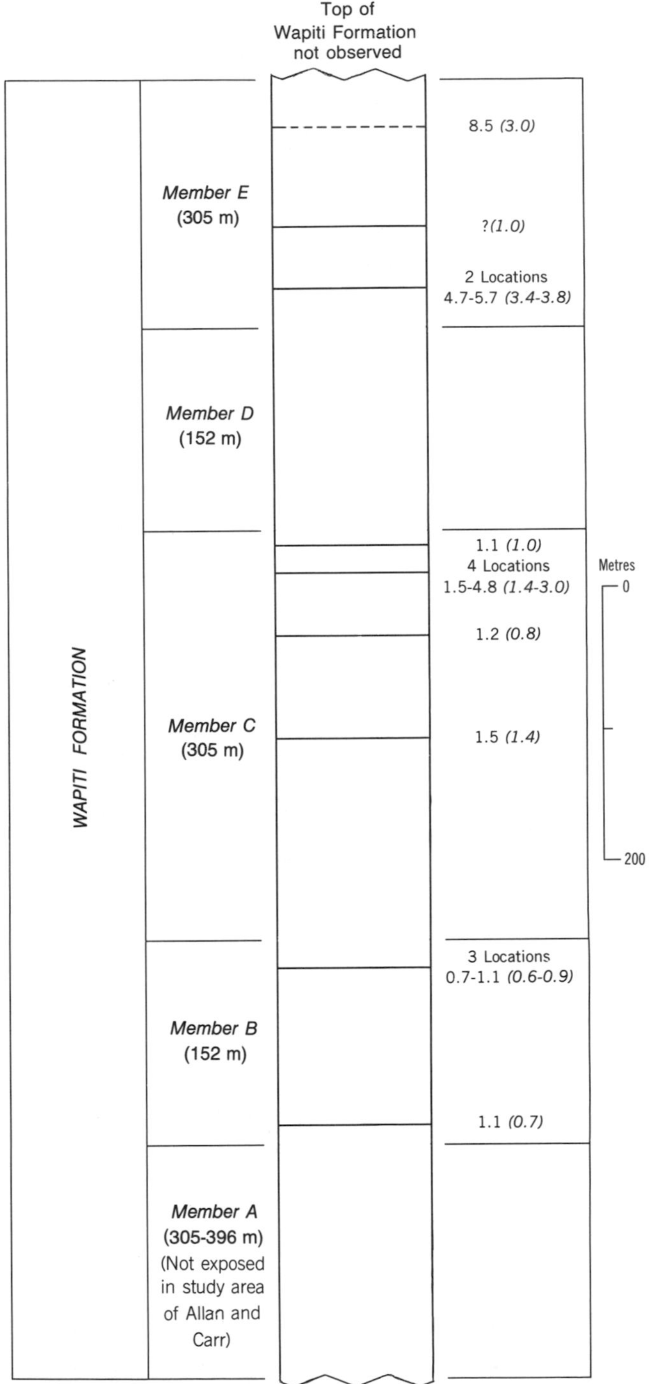

Figure 6B.7. Section of Wapiti Formation, Wapiti-Cutbank area, Alberta, showing stratigraphic position and thickness of coal seams (modified from Allan and Carr, 1946). Numbers on right side of column indicate coal zone thickness and aggregate coal thickness, the latter in brackets. All measurements from one location unless otherwise stated.

in maceral distribution between Scollard and Horseshoe Canyon coals. Except for the Vesta sample, the Horseshoe Canyon coals have significantly higher contents of huminite (vitrinite). Maceral data obtained by Hacquebard and Birmingham (1973) on the Val D'Or and Mynheer seams of the Coalspur Formation show average vitrinite plus liptinite contents of 65 and 63% respectively. These seams are believed to correlate with seams of the Scollard Formation, and their compositional characteristics agree with the above-stated observation regarding maceral distribution in the Scollard coals vis-a-vis those of the Horseshoe Canyon Formation.

Resources and production

Established resource as used by the Alberta Energy Resources Conservation Board (1987) means coal in the ground delineated by exploration, with a minimum seam thickness of 0.6 m and in some instances to a maximum depth of 600 m.

Total established resources for all four of the coal-bearing successions are estimated to be about 56 gigatonnes (56 billion tonnes) (Alberta Energy Resources Conservation Board, 1987). Table 6B.8 summarizes by formation much of the data presented by the ERCB. For each formation, data are given individually for the most important fields, and the remainder of the fields and deposits are grouped together as a single entry. The number of seams used in the calculations is specified in one column; "average aggregate thickness" represents cumulative coal thicknesses used in the compilation. The Horseshoe Canyon and Scollard formations contain by far the largest established resources. The most important fields in terms of resources are Ardley, Westaskiwin, and Tofield-Dodds, in that order.

In terms of rank distribution most of the established resources are in the subbituminous category. A rank breakdown of the data in Table 6B.8 is as follows:

High-volatile bituminous 4631 megatonnes

High-volatile/subbituminous 706 megatonnes

Subbituminous 49 720 megatonnes

Lignite 814 megatonnes

In addition, an isolated lignite occurrence in the Lower Cretaceous McMurray Formation named the Firebag Field has been identified by the ERCB north of Fort McMurray (township 98, ranges 5-6). Estimated in-place resources are 160 megatonnes.

Most of the coal resources discussed are at fairly shallow depths (surface to 610 m, see Table 6B.8). However, all of the coal-bearing sequences have a westward dip and contain coal at greater depths. Williams and Murphy (1981) estimated quantities on these deep resources utilizing data from deep borings in the Alberta syncline. Table 6B.9 summarizes their findings. About half of the resources estimated are believed to be in the Mannville Group. A comparison of Tables 6B.8 and 6B.9 shows some overlap in depth categories less than 610 m.

Coal production from the Plains and outer Foothills is confined to the Edmonton Group and Coalspur Formation, and has more than doubled since 1978 (Table 6B.10). Nearly all production is from eight mines (1986), all of them

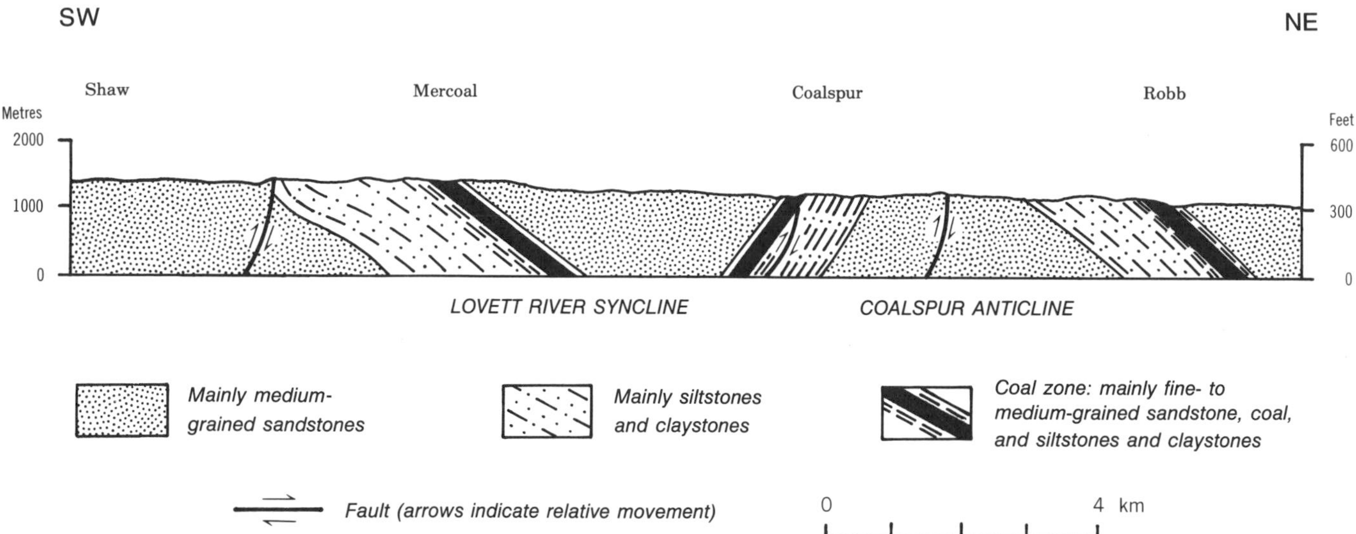

Figure 6B.8. Folding of Coalspur Formation in the Coal Valley area, Alberta (after Jerzykiewicz and McLean, 1980).

Figure 6B.9. Superficial deformation, probably glacial, in a coal seam of Edmonton Group, central Alberta Plains (photo G. Smith, GSC 2338-1).

open-pit. Much of the coal produced is used for power generation in Alberta, although a substantial amount is shipped out of that province.

INNER FOOTHILLS AND ROCKY MOUNTAINS – ALBERTA AND BRITISH COLUMBIA

General statement

Some of Canada's most valuable coal assets occur in a narrow belt in the Rocky Mountains and Foothills of Alberta and British Columbia that extends some 1000 km from the United States border to north of the Peace River in British Columbia (Fig. 6B.12). The belt comprises a number of separate areas of coal-bearing strata ranging in age from Late Jurassic (Portlandian) to Early Cretaceous (Albian). It is appropriate to discuss this belt in two parts: a southern area, where coal-bearing strata belong to the Kootenay Group, and a northern part where coal is found in the Minnes, Blairmore-Luscar-Bullhead, and Fort St. John groups (Fig. 6B.12). All of the northern units except the Minnes are younger than the Kootenay. The younger Dunvegan Formation in the north also contains coal, but the resource has not been evaluated. A voluminous literature on the coal-related geology of the belt has accumulated; much of it has been reviewed and synthesized by Gibson (1985a), Stott (1984), Stott et al. (Subchapter 4I in this volume), and Horachek (1985). Tectonic and structural aspects have been discussed by Price (1961), Norris (1964), Price and Mountjoy (1970), and Thompson (1979).

Because they occur in a foreland fold and thrust belt, the coals commonly are highly deformed (Fig. 6B.13). Such deformation creates geological problems in seam correlation, and production problems relating to coal quality and mineability. Underground operations pose special difficulties. Locally, the deformation has been advantageous, in that tectonic thickening of seams presents opportunities to mine large tonnages in small areas.

Coal ranks range from high-volatile bituminous to semianthracite, with a high proportion in the premium coking medium- and low-volatile bituminous category. Much of the production is used for coking.

Southern area – Kootenay Group

Stratigraphy

The Kootenay Group has been described in detail by Norris (1959, 1971) and more recently by Gibson (1985a), who divided it into Morrissey, Mist Mountain, and Elk formations, in ascending order. Kootenay beds outcrop in a series of parallel narrow bands (Fig. 6B.12), a reflection of thrust faults that have repeatedly brought slices of Kootenay to the surface along west-dipping fault planes, making them accessible for exploitation of the contained coal. The Kootenay conformably overlies marine beds of the Fernie Formation and is overlain unconformably by continental beds of the Blairmore Group. The environment of deposition was a prograding delta. To the east, the Kootenay wedges out due to non-deposition or post-Kootenay erosion and does not extend much beyond the Foothills. To the north, it passes into the partly marine beds of the Nikanassin Formation near North Saskatchewan River. Nearly all of the coals occur in the Mist Mountain Formation, with a few thin seams in the Elk Formation (Fig. 6B.14).

The Kootenay Group has maximum thickness of 1100 m on Sparwood Ridge in the Fernie Basin. In the Fernie Basin 14 major seams occur and in the Elk Valley as many as 40 seams have been noted (Gibson, 1985a). Representative sections indicating stratigraphic position and thickness of coal seams show that the most laterally persistent and thickest (up to 18 m) seams are found in the lower part of the Mist Mountain Formation (Fig. 6B.14). The number of seams and, to some extent, thickness of individual seams decrease from west to east and also from south to north.

Table 6B.6. Chemical compositional data on Plains and Foothills coals

Position on Index Map (Fig. 6B.10)	Area	Formation	Rank (ASTM)	Heating[1] Value MJ/kg	Moisture	Ash	Volatile Matter	Sulphur
					Basis: As Rec'd* or Equil. Moisture data: wt.percent			
1	Halcourt	Wapiti	HVC	28.6	12.2*	5.4	34.0	0.4
2	Morinville	Wapiti	Sub. C	21.2	27.1	6.8	28.1	0.5
3	Morinville	Wapiti	Sub. C	21.5	23.8	9.8	28.7	0.3
4	Thorhild-Abee	Wapiti	Lig. A	18.9	33.0*	4.4	30.0	0.3
5	Obed	Paskapoo	HVC	29.2	8.0*	13.0	36.0	0.5
6	Coal Valley	Coalspur	HVC	27.9	8.1*	10.3	27.5	0.3
7	Mayerthorpe	Scollard	Sub. B	23.5	19.8	10.7	26.6	0.5
8	Alix	Scollard	Sub. B	23.9	18.8*	8.8	28.3	0.3
9	Wabumun (Whitewood)	Scollard	Sub. B	22.1	20.2	11.1	25.8	0.2
10	Wabamun (Highvale)	Scollard	Sub. B	22.2	23.8	12.6	27.3	0.2
11	Wetaskiwin	Scollard	Sub. A	25.1	17.7*	7.6	29.0	0.4
12	Battle River (Diplomat)	Horseshoe Canyon	Sub. C	21.3	25.6	6.3	33.4	0.4
13	Battle River (Vesta)	Horseshoe Canyon	Sub. C	21.3	23.8	8.7	34.8	0.3
14	Sheerness	Horseshoe Canyon	Sub. C	21.2	24.2	10.8	32.7	0.5
15	Drumheller	Horseshoe Canyon	Sub. B	24.4	16.5	9.1	31.5	0.5
16	Lethbridge	Oldman	HVC	27.5	11.5	17.7	29.4)	0.5[2]-0.7
17	Lethbridge	Oldman	HVC	27.4	12.1	11.3	31.2)	
18	Redcliff	Foremost	Sub. C	21.4	25.7	6.4	27.2	0.5
19	Taber	Foremost	Sub. A	26.3	15.3	7.6	32.0	0.6[3]-1.2
20	Pincher Creek	Belly River	HVB	31.2	6.8*	14.7	34.5	0.8
21	Pekisko	Belly River	HVB	32.2	5.8*	7.1	36.8	0.6

[1] Heating values - moist - mineral-matter free basis
[2] Range of sulphur values for coals in Lethbridge area
[3] Range of sulphur values for coals in Taber area
Data from: Stansfield and Lang (1944), Campbell (1964, 1974), ERCB (1987)

The top of the Morrissey is Portlandian in age (Frebold, 1957), and palynological evidence indicates that the Jurassic-Cretaceous boundary is close to the Mist Mountain-Elk contact (A.R. Sweet, pers. comm., 1984).

Tectonics and structure

The structural deformation of the Kootenay Group is part of the telescoping of much of the Paleozoic and Mesozoic successions of the southern Canadian Rocky Mountains by a series of thrust faults of the Late Cretaceous-early Tertiary Laramide Orogeny. Palinspastic reconstructions indicate eastward displacement of 150 km for some sections (Gibson, 1985a). A number of important normal faults are apparently younger than the thrusts. A typical west-east cross-section (Fig. 6B.15) through the north end of the Fernie Basin (Fig. 6B.12) and extending into the Plains shows five major thrusts as well as one major normal fault. Folds are another component of the deformation and are the important structural element on the mine scale, in at least some basins (Norris, 1971). The incompetent coal seams commonly show a high degree of shearing, especially in folds and near faults. Shearing is an integral part of intraseam transport of the coal, producing thickened and thinned areas. The highly sheared coal produces large amounts of fines during mining.

Norris (1964, 1966) made detailed studies of the structural fabric in Kootenay coals and associated strata and listed bedding, joints and cleats, extension, contraction, and wrench faults, and locally drag folds as the principal mesoscopic elements. Extension faults are numerous, commonly occurring in swarms, though usually stratigraphic displacement is small. On the other hand, contraction faults, though less numerous, generally have more stratigraphic displacement and as a result are more troublesome in mining.

The individual areas of Kootenay coals (Fig. 6B.12) may be grouped as follows: (a) occurrences in which the dominant structural element is a syncline, examples of which are the Cascade (Canmore-Mt. Allan area), Elk Valley (Fig. 6B.16), and Fernie basins, and (b) occurrences dominated structurally by westward-dipping homoclines closely adjacent to major thrusts. Examples of the latter type are the Blairmore and Coleman fields, lying in Alberta just east of the Fernie Basin. In each type, smaller-scale folding and faulting may be quite significant on the local level. Structural geometry determines exploration and mining activity. Coals preserved on the flanks of synclines were favoured targets for early mining, whereas the thrust faults have brought multiple slices of Kootenay strata to surface, creating attractive sites for exploration.

Rank and other quality parameters

Typical proximate analysis data, heating values, and sulphur contents on Kootenay coals from Alberta and British Columbia are presented in Table 6B.11. Rank ranges from high-volatile bituminous to semianthracite. Sulphur content is normally quite low, often under 0.5%, while ash content tends to be somewhat higher than North American Carboniferous coals of the same rank. The relationship of these characteristics and others to coking potential is discussed below.

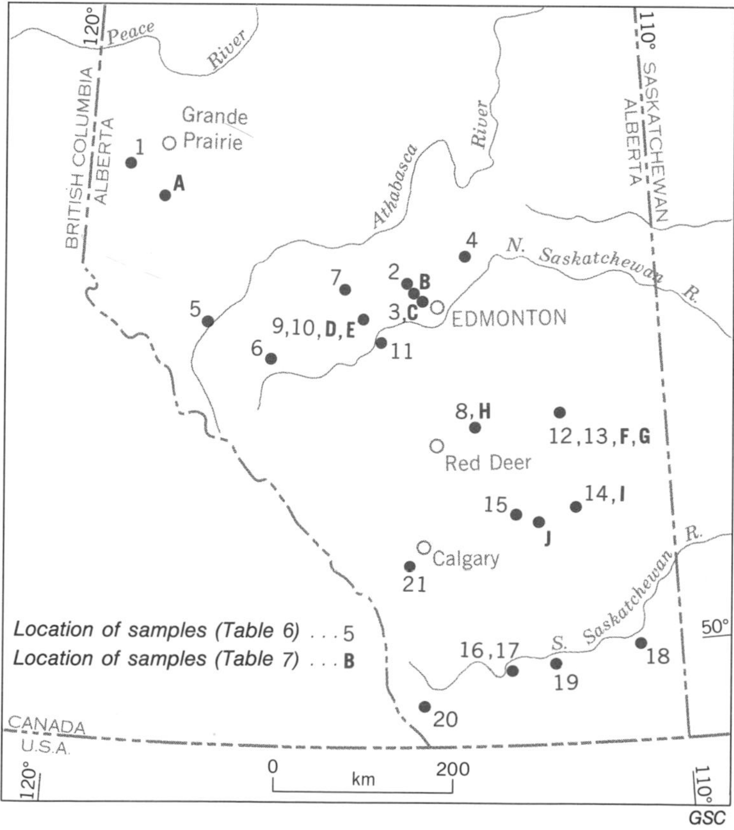

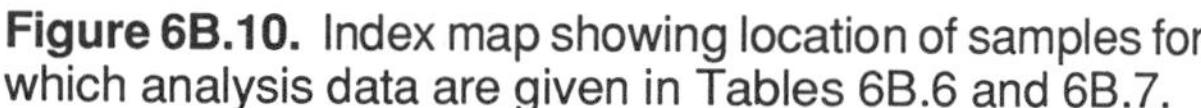
Figure 6B.10. Index map showing location of samples for which analysis data are given in Tables 6B.6 and 6B.7.

Rank has been determined by reflectance measurements on many Kootenay coals (Hacquebard and Donaldson, 1974; Graham et al., 1977; Bustin, 1982; Gibson, 1985a; Hughes and Cameron, 1985; Pearson and Grieve, 1985). A number of representative Kootenay sections with reflectance data show values ranging from 0.61 to 2.50 (Romax) (Fig. 6B.17). The pattern of rank distribution is not random. The highest values occur toward the north, in sections N, near Canmore, and O, near Barrier Mountain, while the lowest occur in sections J and L in Elk Valley. A decrease in reflectance also occurs from west to east; for example, the basal seam at Sparwood (section D) measures 1.53 Romax, while the basal seam at Grassy Mountain (section E) is only 1.19. Rapid lateral changes in reflectance are not uncommon, as exemplified by sections I and J, only a few kilometres distant from one another on opposite sides of Elk Valley.

Lateral variations in rank and variations in rank gradient (% change in Ro/100 m) in specific sections have been attributed to regional changes in geothermal gradient and to variation in depth of burial by younger strata. The influence of additional tectonic burial by the overriding thrusts has to be considered and is a problem currently being investigated. Initial results from some areas suggest that such an effect may be less than expected. The effect of groundwater movement on heat flow in these disturbed belts must also be considered in evaluating rank distribution (Majorowicz et al., 1985). In the Canmore-Mount Allan area, Norris (1971) suggested that the enhanced rank might be the result of igneous intrusion at depth adjacent to the Rocky Mountain Trench (the restored, pre-orogenic position of the strata). Studies by Pearson and Grieve (1985) have indicated that a substantial amount of the coalification in the Fernie Basin is post-deformational.

Maceral distribution in Kootenay coals appears to follow a consistent pattern. In a number of sections in Elk Valley and Fernie Basin, Cameron (1972) found that coals in the bottom part of the Mist Mountain Formation tended to have low vitrinite and high inertinite and semi-inertinite contents. In the upper part of the formation the proportions are reversed. This pattern of maceral distribution is illustrated in Figure 6B.18.

Resources and production

In 1986, five surface mines and one subsurface mine were producing coal from the Kootenay Group, all of them in British Columbia (Romaniuk and Naidu, 1987). The last

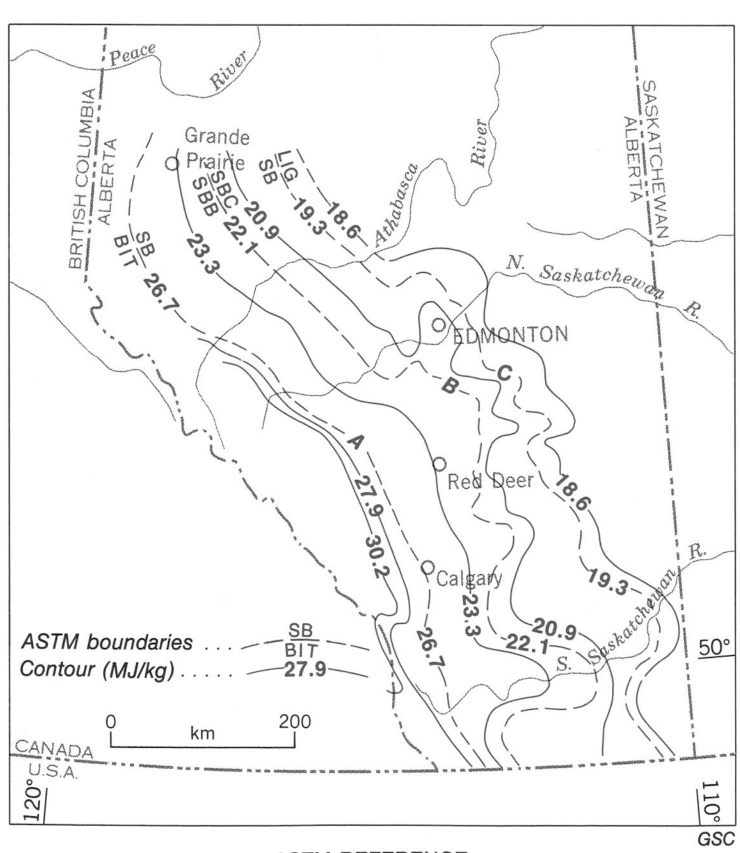

Figure 6B.11. Isopleths of heating value for near surface coals in Alberta, showing westward increase. Data on moist mineral-matter-free basis (modified from Steiner et al., 1972).

Table 6B.7. Petrographic composition of Alberta subbituminous coals (from Parkash et al., 1984)

Sample No.	Location and Formation	ASTM rank	Volume percent maceral composition (mineral matter free)					Huminite reflectance (random) (%)
			Huminite	Liptinite	Semi-fusinite	Fusinite	Other inertinites	
A	Smoky Tower[1] Wapiti?	Sb A[2]	70.0	6.4	8.6	4.3	10.7	0.51 ± 0.05
B	Egg Lake Horseshoe Canyon	Sb C	92.4	4.2	0.8	0.4	2.2	0.33 ± 0.04
C	Starkey Horseshoe Canyon	Sb C	91.5	5.8	1.5	0.6	0.6	0.33 ± 0.03
D	Wabamun Scollard	Sb B	65.0	6.0	15.0	3.0	11.0	0.36 ± 0.02
E	Highvale Scollard	Sb B	74.6	2.1	22.2	1.1	nil	0.48 ± 0.04
F	Forestburg Horseshoe Canyon	Sb C	92.5	3.8	0.6	1.9	1.2	0.39 ± 0.04
G	Vesta Horseshoe Canyon	Sb C	73.3	0.3	8.9	3.3	14.1	0.43 ± 0.04
H	Heatburg Scollard	Sb B	74.8	1.0	9.6	2.1	12.5	0.44 ± 0.04
I	Sheerness Horseshoe Canyon	Sb C	92.7	5.3	1.2	0.4	nil	0.42 ± 0.03
J	East Coulee Horseshoe Canyon	Sb B	91.9	4.9	0.6	1.4	1.2	0.46 ± 0.03

[1]likely occurs in lateral equivalent of Scollard Formation
[2]subbituminous A, B, C

Kootenay operation in Alberta ceased in 1983. The coals produced range from high-volatile to low-volatile bituminous, and both thermal and metallurgical products are shipped. Three operations are multiple seam developments with upwards of 11 seams being mined at once (Fording River Mine; Romaniuk and Naidu, 1987). Production statistics for the Kootenay are given in Table 6B.12.

Coal resources in the Kootenay Group for British Columbia total 52 000 megatonnes in measured, indicated, and inferred categories (Table 6B.13; Bielenstein et al., 1979). Of this total, Romaniuk and Naidu (1984) estimate that 1762 megatonnes are mineable, that is, coal for which feasibility studies have been done and which can be mined using current technology and given current economic conditions.

In Alberta, established resources (ERCB, 1984) in the Kootenay Group amount to 3550 megatonnes (Table 6B.13). Of these, 780 megatonnes are in the reserve category, that is, coal in the ground and potentially mineable under present conditions. The most important fields in Alberta in terms of established resources are Canmore (810 megatonnes), Blairmore-Coleman (700 megatonnes), and Oldman River (610 megatonnes).

Northern area – Luscar, Minnes, Bullhead, and Fort St. John groups

Stratigraphy

The northern segment of the Rocky Mountains and Foothills coal belt contains important coal resources in the Luscar, Bullhead, and Fort St. John groups. The underlying Minnes Group also contains coal seams, though of less significant volume. These beds outcrop extensively from Mountain Park in west-central Alberta to Prophet River in northeastern British Columbia, a distance of some 645 km (Fig. 6B.12).

Published information on these Lower Cretaceous rocks has been summarized by Stott (1968, 1972a, b, 1975, 1981, 1984) and Stott et al. (Subchapter 4I in this volume). Gibson (1985b) has reported on the Gething Formation at Carbon Creek, British Columbia.

The coal measures in the northern segment belong to three sequences, each with intercalated marine sediments. The lowest sequence consists of the Minnes Group, underlain by marine beds of the Fernie Formation and overlain unconformably by the Cadomin conglomerate. South of Pine River the upper three formations of the Minnes Group pass laterally into the Gorman Creek Formation, which includes thin coal seams (Stott, 1984).

Table 6B.8. Summary of established coal resources in Alberta Plains and outer Foothills (Data from ERCB Report, 1987)

Formation	Field	No. of[1] Seams	Avg.[2] Aggregate Thickness Metres	Area km²	Max. Depth Metres	Coal in[3] place (megatonnes)	Rank[4] (ASTM)
	Medicine Hat	5	2.1	304	204	770	Sub. C
Foremost	6 other fields or deposits	1-4	0.9-2.4	553	32-121	997	Sub. C-A
	Totals	1-5	0.9-2.4	857	32-204	1770	Sub. C-A
	Lethbridge	5	1.9	220	280	536	HVC
Oldman	4 other fields or deposits	1-3	1.1-1.6	312	66-190	576	sub C - HVC
	Totals	1-5	1.1-1.9	532	66-280	1100	sub. C-HVC
	Tofield-Dodds	5	3.6	849	185	4119	Sub. C
	Battle River	4	2.3	1221	165	3742	Sub. C
Horseshoe Canyon	11 other fields or deposits	1-9	1.1-5.2	2375	32-446	11 496	Sub. C - HVC
	Totals	1-9	1.1-5.2	4445	32-446	19 400	Sub. C - HVC
	Ardley	4	3.3	1622	292	7562	Sub. B/A
	Wetaskiwin	10	3.5	1226	266	5956	Sub. B/A
Scollard	8 other fields or deposits	3-10	1.7-5.1	2717	74-265	12 863	Sub. C - A
	Totals	3-10	1.7-5.1	5565	74-292	26 400	Sub. C - A
Brazeau	2 deposits	1	1.7-3.1	11	152-610	40	HVC
	McLeod River	24	13.3	121	589	2473	HVC
Coalspur	4 other fields or deposits	1-10	1.8-7.4	138	200-610	1202	HVC/B
	Totals	1-24	1.8-13.3	259	200-610	3700	HVC/B
Obed Coal Zone	2 fields	3	2.9-5.4	81	91-138	380	HVC
	Morinville	4	2.2	724	149	2150	Sub. C
Wapiti	4 other fields or deposits	2-6	1.1-1.6	437	22-96	814	Lig. A - Sub. C
	Totals	2-6	1.1-2.2	1161	22-149	3000	Lig. A - Sub. C
Undiff. Paskapoo	Musreau Lake	4	2.2	50	147	130	Sub. A - HVC

[1]Maximum number of seams in field.
[2]Average coal thickness.
[3]Total resources (underlined numbers) rounded off.
[4]Lig. A = lignite A; sub. A, B, C = subbituminous A, B, C, HV B, C = high volatile B and C bituminous.

The coal-bearing part of the Minnes has its best development in the area between Sukunka and Berland rivers (Stott, 1972a). The Minnes Group reaches a maximum thickness of 2100 m between Smoky and Peace rivers.

The next younger sequence with major coal resources is the Gething Formation overlying widespread Cadomin conglomerates of the Bullhead Group, jointly equivalent to the lower part of the Blairmore Group in the southern Alberta Foothills. Deposition of the Gething Formation took place within a deltaic complex extending from Peace River southeastward into Alberta (Stott, 1972b). The most favourable environments for coal accumulation occurred on the southern margin of the complex (Fig. 6B.19). Maximum thickness for the Gething is about 1050 m at Carbon Creek; representative sections are shown in Figure 6B.20.

Coal seams of the Gething are few in number and mostly thin south of Smoky River, but to the north, the number and thickness of seams increase. The best occurrences of coal in the Gething seem to be in the area from Bullmoose Mountain north to Peace River Canyon. Near Bullmoose Mountain (70 km north of Monkman Pass, Fig. 6B.12) at least 13 seams with thicknesses ranging up to 6 m have been exposed. In the Carbon Creek area, Gibson (1985b) noted over 50 seams, though few were thicker than 2 m. In the Peace River Canyon, also, about 50 seams occur in the Gething. In the Sukunka River area (100 km northwest of Monkman Pass, Fig 6B.12), the Chamberlain and Skeeter seams in the upper part of the formation may be laterally extensive. Thicknesses are on the order of 3.0 to 5.5 m for these seams (Wallis and Jordan, 1974). North of Peace River, the number of seams in the Gething decreases, and few occur north of Halfway River.

The Moosebar Formation is the basal unit of the Fort St. John Group. It is overlain by the Gates Formation, which contains important seams in its Grande Cache Member from Smoky River in Alberta north to Bullmoose Mountain in British Columbia. The Gates has a maximum thickness of 266 m at Mount Belcourt but thins northward toward Peace River and becomes increasingly marine. Toward the south, the Gates passes into the upper part of the Luscar Group (Langenberg and McMechan, 1985) with important coal seams at Luscar, Cadomin, and Mountain Park. In Figure 6B.21, the areal distribution of the coal-bearing part is superimposed on an isopach map of the combined Moosebar and Gates. Like the Gething, the Gates is a deltaic deposit and the distribution of coal is similar. Typical stratigraphic sections of the Gates are shown in Figure 6B.22.

Gates coals have been mined at Mountain Park and Nordegg in Alberta and presently are being mined at Luscar, Gregg River, and Grande Cache in Alberta and at Quintette and Bullmoose Mountain (near Monkman Pass) in British Columbia. A few thin coal seams have been reported in the Boulder Creek Formation. Other coals may occur in the Dunvegan Formation, especially between Murray and Pine rivers (Stott, 1972a).

Tectonics and structure

The tectonic setting in which these Lower Cretaceous strata accumulated was similar to that which prevailed for the Kootenay Group farther south, that is, a foredeep lying east of the emergent parts of the Columbian Orogen.

Table 6B.9. Estimated deep coal resources (in megatonnes) for Alberta Plains (modified from Williams and Murphy, 1981)

Seam Thickness	Level of[1] Confidence	Depth from Surface (metres) 301-500	501-1000	Total
>2m	Measured	1 220	2 370	3 590
	Indicated	3 930	7 330	11 260
	Inferred	50 200	81 600	131 800
<2m	Measured	1 290	1 520	2 810
	Indicated	4 250	4 940	9 190
	Inferred	45 400	55 100	100 500

[1] Measured <400m from borehole; Indicated 400-800m from borehole; Inferred 800-2400m from borehole

Table 6B.10. Coal production (in megatonnes) from Plains and Outer Foothills, Alberta (Data from ERCB Rpt. ST 88-29, 1988)

Year	Production	Year	Production
1978	9.59	1983	18.14
1979	12.81	1984	19.85
1980	14.80	1985	21.13
1981	16.52	1986	22.03
1982	18.27	1987	23.00

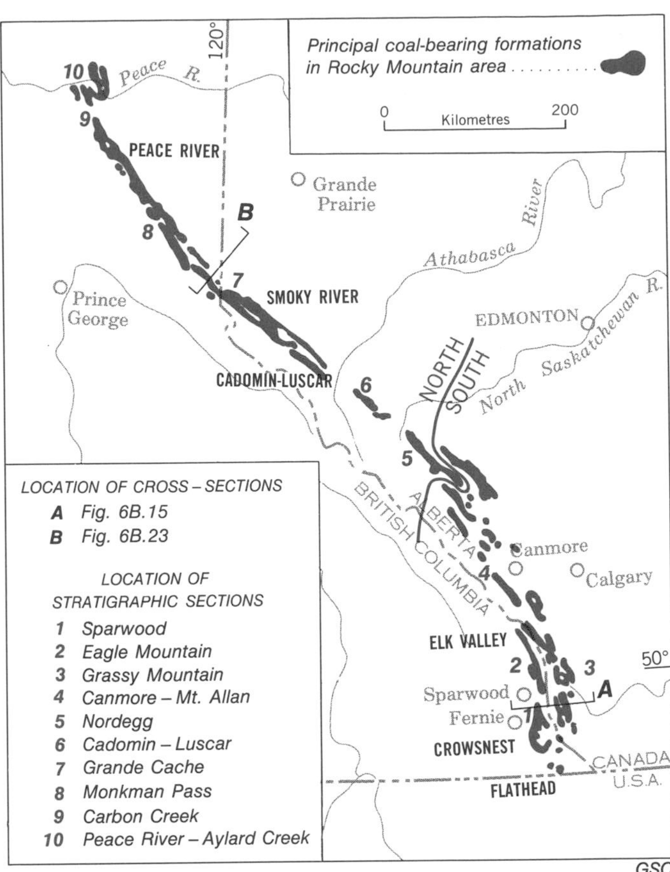

Figure 6B.12. Distribution of Upper Jurassic-Lower Cretaceous coal-bearing rocks in foothills and mountains; division into north and south regions (modified from Romaniuk and Naidu, 1987).

Figure 6B.13. Disharmonic folding in coal seam of Kootenay Group at Grassy Mountain, Alberta (photo D.K. Norris, GSC 201686).

However, the internal stratigraphic relationships of the coal-bearing stratal package in the north are more complex in that several marine transgressions interrupted dominant continental accumulation, the latter including large peat-swamp deposits. In the south, the only period of peat-swamp development is represented by the Mist Mountain Formation of the Kootenay Group.

The overall impact of the Laramide Orogeny differs from south to north. Thrust faults are present in the north but folds are regionally more important. This change in structural style from south to north is gradual, with first a decrease in magnitude of displacement on the faults and then a decrease in the actual number visible in surface exposures. For example, a typical cross-section northwest

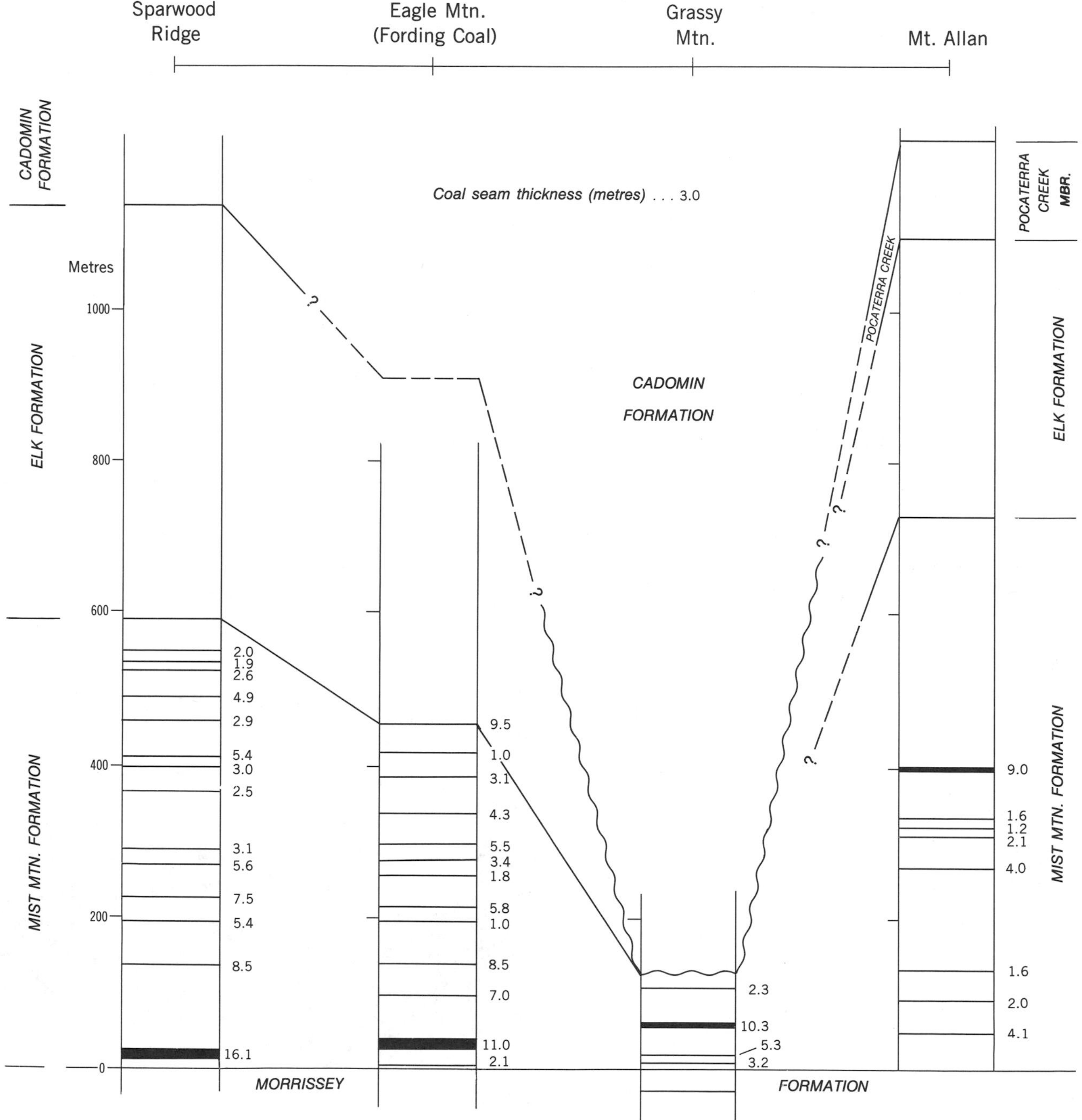

Figure 6B.14. Representative sections of Kootenay Group emphasizing stratigraphic position and thickness (m) of coal seams (see Fig. 6B.12 for section locations). Note the decrease in number of seams thicker than 1 m at Mount Allan (data from Gibson, 1985a; Gibson et al., 1983; D.K. Norris, pers. comm., 1985).

of Grande Cache indicates the existence at depth of a number of thrust faults. However, displacement on any of these is relatively small and many do not extend to surface (Fig. 6B.23). Box and chevron folds are much more common in the north than in the south. An examples of these is shown in Figure 6B.24. According to Hoffman and Jordan (1984) the northern Foothills coal area, at least from Grande Cache north, is essentially one large anticlinorium, within which many large to small folds occur, usually arranged en échelon. Some of the larger folds can be traced for 50 km or more. The dip of fold limbs ranges from very shallow to overturned, though most are between 20° and 50°. Many folds are cut by faults, including many small-scale thrusts.

Mineability of coal seams is largely controlled by the structural style. In the north, many of the potentially exploitable coal deposits are in box folds where the cores of the folds involve relatively large areas of gently dipping to nearly horizontal beds. Examples of such folds are the anticlines at Mount Babcock, Mount Duke, and (much more complex) Grande Cache. Examples of synclinal deposits are Quintette and Luscar. Locally, shearing has destroyed primary sedimentary features, such as bedding, particularly in the cores of folds. Such sites are also often the loci of tectonically thickened seams and thus are favoured targets for exploration.

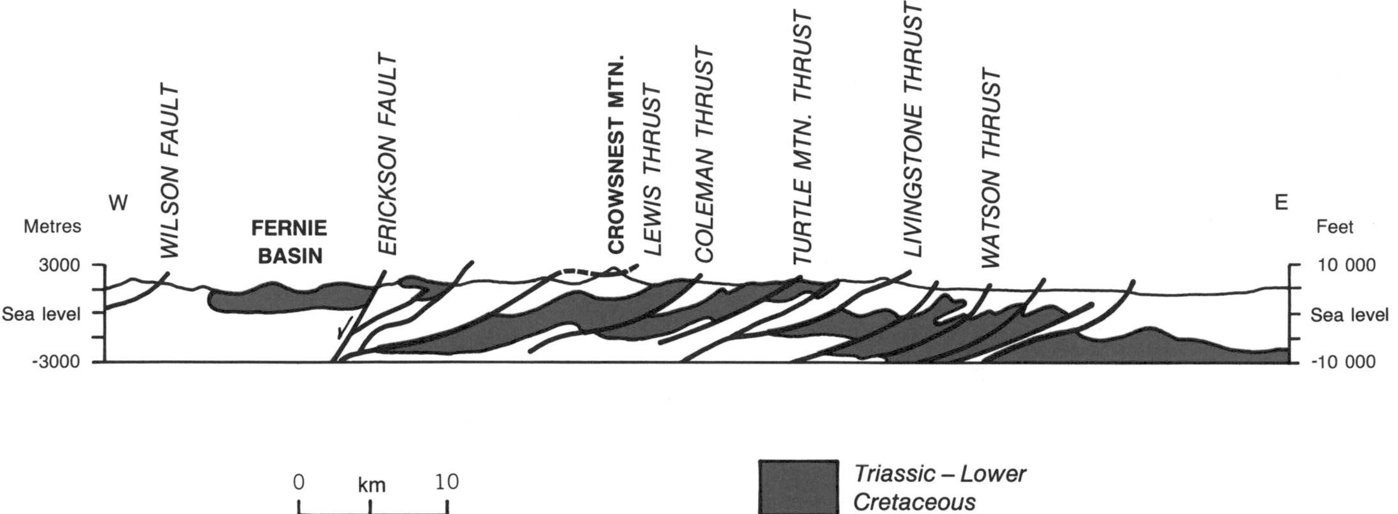

Figure 6B.15. Cross-section through the Rocky Mountains and foothills near Sparwood, British Columbia and Coleman, Alberta; see Section A, Figure 6B.12 for location (modified from Monger and Preto, 1972).

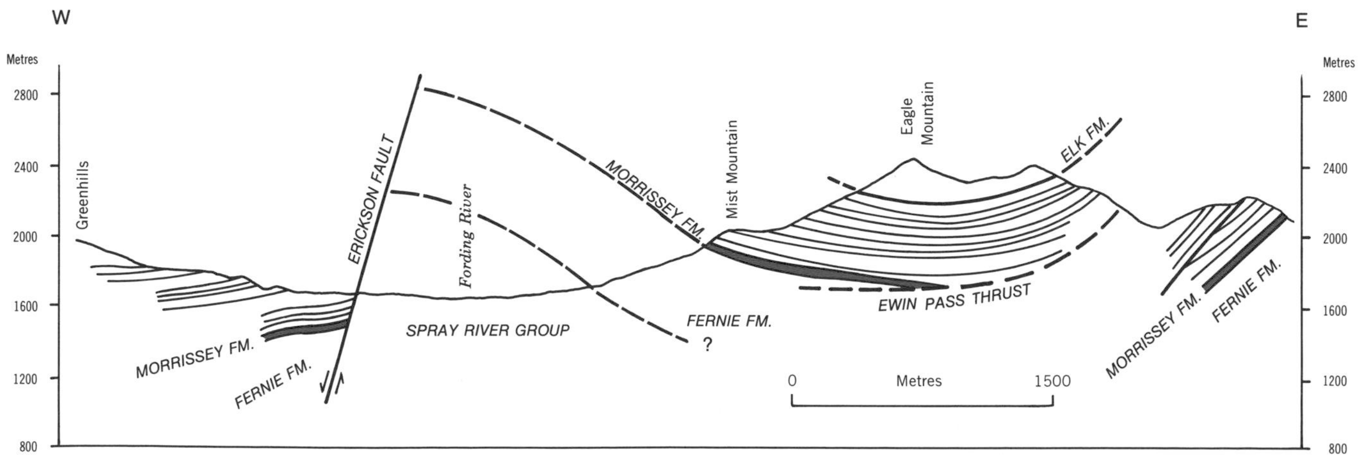

Figure 6B.16. Cross-section through Kootenay Group on Fording Coal property (Eagle Mountain), Elk Valley, British Columbia. Coal seams shown by lines roughly parallel to highlighted Morrissey Formation (see Fig. 6B.12 for location) (modified from Gibson et al., 1983).

Table 6B.11. Representative compositional data on coals of the Kootenay Group (Data on raw coal, as received basis)

Province	Area	No. of Seams	Weight Per Cent Moisture	Ash	Volatile Matter	Fixed Carbon	Sulphur	MJ/kg	Rank
Alberta	Adanac	1	1.4	23.5	23.9	51.2	0.4	25.1	MV
Alberta	Beaver Mines	2	3.0-3.9	11.7-16.4	29.6-32.8	47.4-54.3	0.9-1.6	27.1-29.7	HV
Alberta	Bellevue	2	1.5-2.5	16.5-16.9	25.3-25.6	54.8-56.4	0.5-0.6	28.0-28.5	MV
Alberta	Blairmore	2	2.5	15.0	23.4	59.1	0.5	29.0	MV
Alberta	Cascade	14	1.1-5.8	3.3-35.4	11.2-24.0	43.5-83.3	0.4-1.0	31.8-33.0	MV-SA
Alberta	Highwood	at least 4	1.5-9.4	10.7-27.0	11.9-20.8	53.4- 72.6		24.6-30.3	LV-SA
Alberta	Oldman	1	3.9	10.9	22.8	62.4	0.7	30.5	MV
British Columbia	Coal Mtn.	1	1.6-5.1	14.0-18.3	21.0-24.5	56.3-60.9	0.2-0.3	27.0-28.7	MV
British Columbia	Elk Valley Eagle Mtn.	?	11.7	12.1	20.5	55.7	26.3	0.5	MV
British Columbia	Elk Valley Greenhills	4	1.6-8.1	5.9-25.9	17.4-23.5	51.7-63.4	0.3-0.6	24.9-30.7	MV
British Columbia	Elk Valley E.V.-holes[1]	22	2.4-12.6	4.3-34.7	17.2-31.5	37.5-65.6	0.3-1.2	19.6-32.9	HV-MV
British Columbia	Line Creek	4	1.6-4.2	13.8-37.8	16.0-21.3	44.6-60.8	0.2-0.7	21.4-29.6	LV
British Columbia	Sparwood	8	0.6-2.1	6.6-16.0	16.6-32.7	55.5-68.4	0.2-0.6	29.3-34.0	HV-MV

[1]see Graham et al. (1977) for location of holes

a. MJ/kg = heating value (megajoules/kilogram), HV, MV, LV = high volatile, medium volatile, and low volatile bituminous; SA - semianthracite

b. Sources for data:
Stansfield and Lang (1944)
Swartzman and Tibbetts (1955)
Nicolls (1952)
Bonnell et al. (1983)
ERCB (1987)

Rank and other quality parameters

Representative chemical analyses and heating values for Gates and Gething coals of northeast British Columbia and adjacent parts of Alberta are given in Table 6B.14. The predominant ranks are medium- to low-volatile bituminous. Sulphur contents are low, most being below 1%.

Rank variations have been investigated through vitrinite reflectance by Hacquebard and Donaldson (1974), Karst and White (1980), Kalkreuth (1982), and Kalkreuth and McMechan (1984). Karst and White noted that reflectance values for a restricted horizon near the top of the Gething Formation increase from east to west, reaching a peak in the Foothills, then decrease westward. This pattern contrasts with that of the Kootenay Group in which the highest ranks usually occur in the most westerly exposures. Kalkreuth and McMechan suggested that the coals of highest rank do not occur in the most western exposures (Fig. 6B.25) because they were not as deeply buried nor for so long a time as those farther east near the axis of the Alberta syncline. Their study also has important implications for the maturity of hydrocarbons in this area.

Coals from the northern segment of the Foothills (Fig. 6B.26) show a wide range in maceral distribution, with vitrinite content ranging from less than 40 to almost 100%. Data for some of the major seams being mined at Grande Cache and Luscar show that at least some of these northern coals tend to be inertinite rich, as are some Kootenay coals.

Two other characteristics of the maceral composition of the northern coals are worthy of note. Kalkreuth (1982) found that a number of coals showed anomalously low reflectances when plotted in rank-depth profiles. He related the low reflectances to the liptinite content of these coals and demonstrated that the reflectance of the vitrinite was depressed, apparently by infiltration into the vitrinite of bituminoid materials derived from the liptinite. Furthermore, many of the Minnes coals contain a rather distinctive liptinite tentatively believed to be of algal origin. Both in hand specimen and microscopically these coals have the same appearance as the "needle" coals of the Kootenay Group. Because the occurrences in the Kootenay and the Minnes are approximately at the same stratigraphic position, it is possible that the vertically restricted occurrence of this special variety of liptinite may be a useful stratigraphic index horizon on a regional scale.

Resources and production

Resource estimates for northern Foothills coals are given in Table 6B.15. Total measured resources for northeastern British Columbia are 996 megatonnes and the approximately synonymous term "established resources" for Alberta is given as 3445 megatonnes. According to Alberta

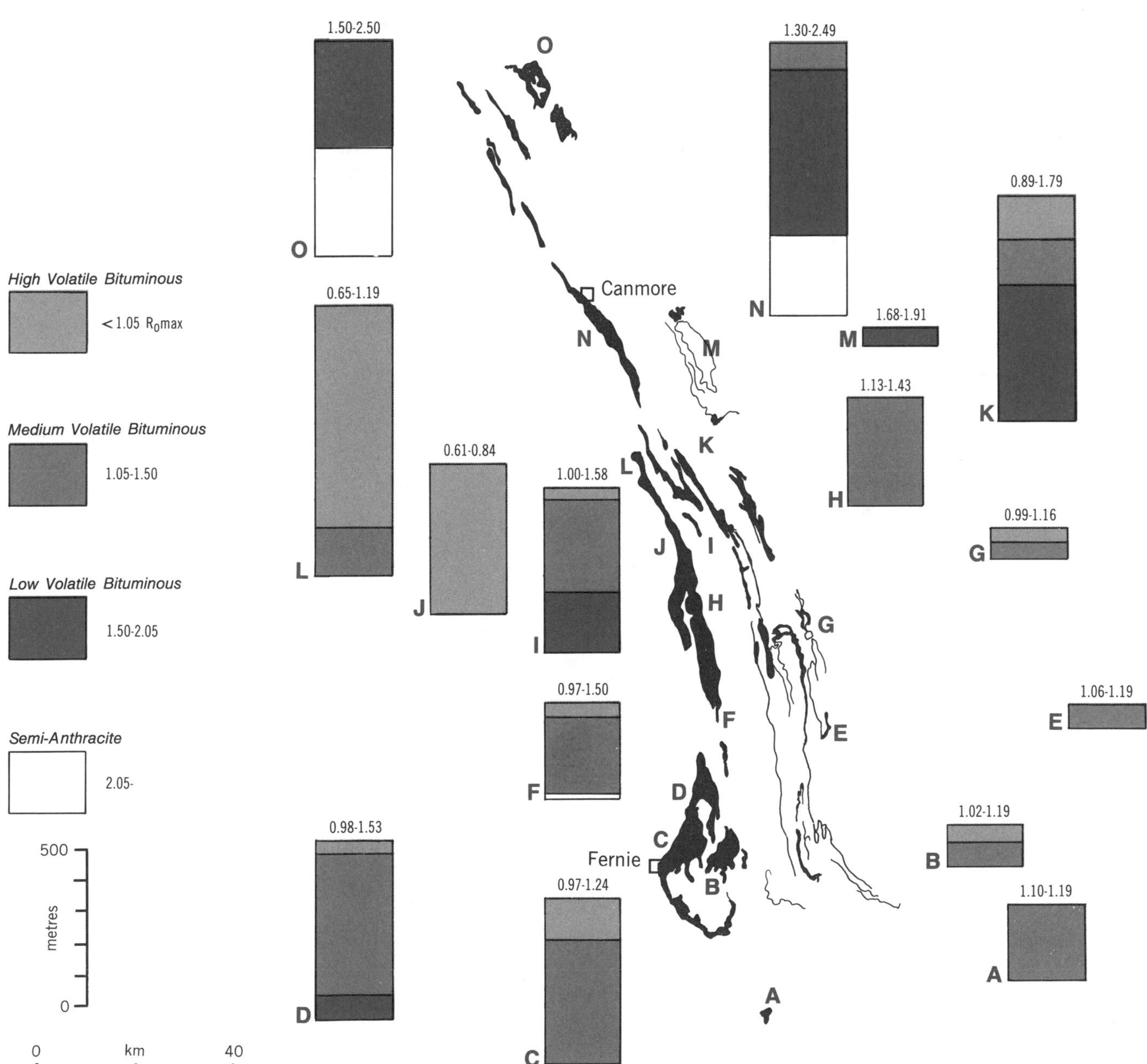

Figure 6B.17. Rank (reflectance) distribution in representative Kootenay sections. Reflectance range for individual sections given at top of each section.

ERCB (1987) the three most important fields in Alberta are Smoky River, Kakwa River, and Ram River, with 940, 580, and 410 megatonnes in place respectively.

Coal from Lower Cretaceous sequences in northeastern British Columbia and west-central Alberta has been mined since the early 1900s, with nearly all production from Alberta and from the Gates Formation or equivalent beds. After the Second World War, mining gradually decreased and finally ceased in 1959, temporarily as it turned out. In 1970 mining resumed, with two fundamental differences; the main market demand was for metallurgical coal and nearly all production was from large surface mines.

Coals from the Gething Formation have been mined to date only on a small scale. For a number of years small mines near the Peace River produced from this formation, but these are now closed.

Coal from the northern segment is currently being mined at five different places, Luscar, Gregg River (near Cadomin), and Grande Cache in Alberta and Quintette and

Table 6B.12. Preliminary 1986 production statistics for Kootenay Group coal (from Romaniuk and Naidu, 1987)

Company	Area	Mine Output 10^3 tonnes
Byron Creek	Coal Mountain	1052
Westar	Sparwood	5474
Westar	Greenhills	2839*
Fording Coal	Fording River	7524
Crowsnest Res.	Line Creek	2314
		TOTAL: 19 203

* Includes 25 kilotonnes from underground operation; all other production is surface

Table 6B.13. Coal resources in Kootenay Group (Data modified from Bielenstein et al., 1979; ERCB, 1984)

Province	Resources of Immediate Interest (Megatonnes)			Rank
	Measured	Indicated	Inferred	
Alberta	3550[1]	-	-	HV - SA[2]
British Columbia	6300	9400	36 300	HV - LV

[1] Data from ERCB (1987). Figure quoted is "Established Resources" according to ERCB criteria.
[2] HV, LV = high volatile and low volatile bituminous respectively; SA = semianthracite

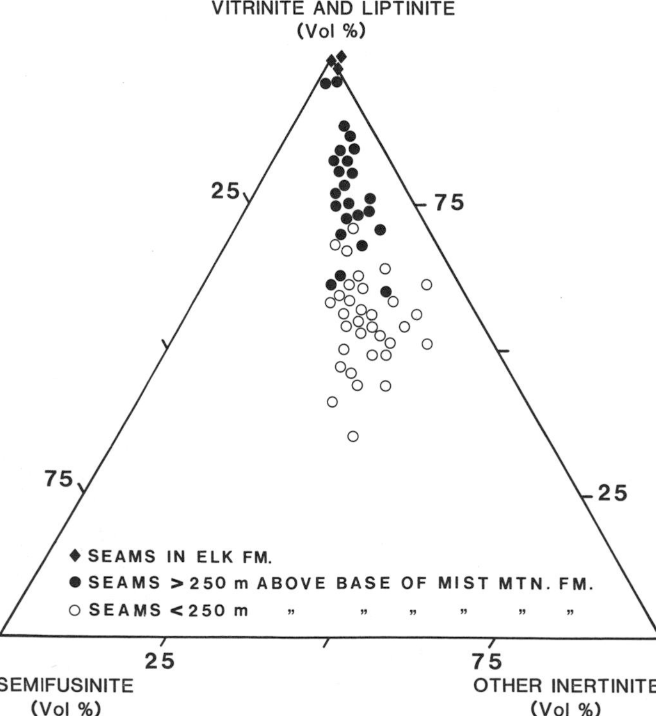

Figure 6B.18. Relationship between maceral distribution (mineral-free) and stratigraphic position of coal seams, Mist Mountain Formation (after Cameron and Kalkreuth, 1982).

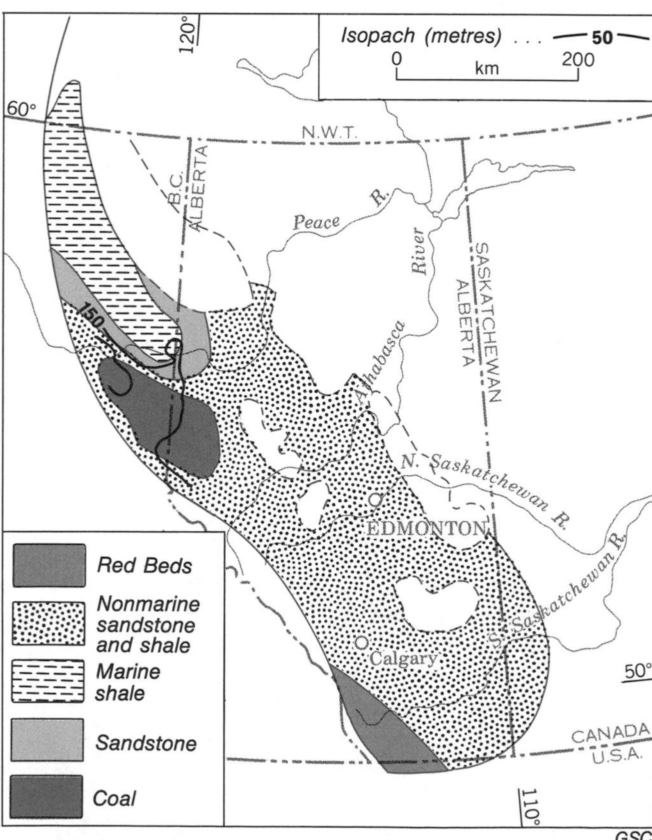

Figure 6B.19. Isopachs and depositional facies, Gething Formation and equivalents, showing coal-bearing area (modified from Stott, 1984).

Bullmoose Mountain (near Monkman Pass) in British Columbia (Fig. 6B.12). Most of the output is from surface operations; some underground production is obtained at Grande Cache. Preliminary production figures for 1986 are given for these areas (Table 6B.16).

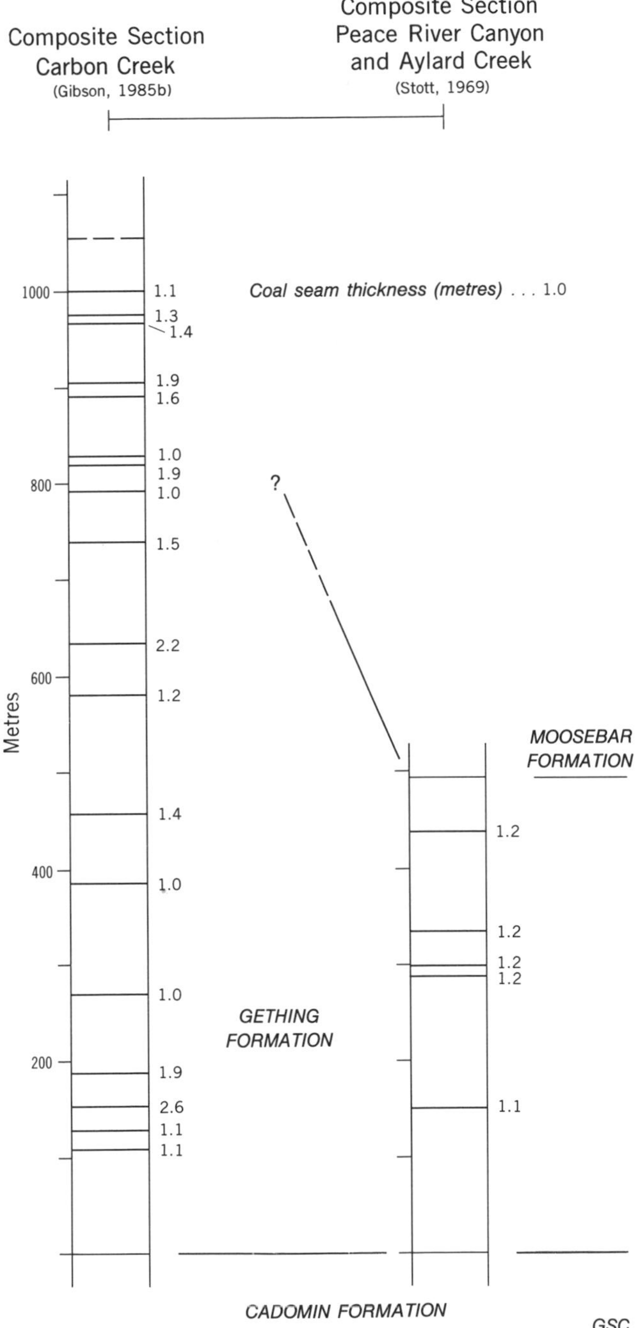

Figure 6B.20. Representative Gething sections showing stratigraphic position and thickness of seams greater than 1 m. Gething-Moosebar contact not observed at Carbon Creek but believed to be close to top of measured section (see Fig. 6B.12 for location of sections) (modified from Stott, 1969; Gibson, 1985b).

Coking properties of Foothills and Mountain coals

The rank of many of these coals is optimal for the coking industry. Because they differ somewhat in their properties from other coals used in the coking trade, including the Carboniferous coals of eastern North America, it is useful to examine some of the differences.

To evaluate a coal for coking potential, the characteristics of rank, ash and sulphur content and maceral distribution are important. Fluidity as measured by the Gieseler test is routinely quoted, and strength of resulting coke is a key property. Maceral distribution is important in coking because some macerals (vitrinite and liptinite) are reactive and others are not (inertinite). In terms of maceral composition, many western Canadian coals are characterized by high inertinite content, although the nature of some of these inertinite constituents may be different from that reported for coals in other parts of the world (Koensler, 1980; Cameron, 1984; Pearson and Pearson, 1983). Western Canadian coals have significantly lower sulphur and somewhat higher ash content than competitive fuels from other parts of the world.

A comparison of fluidity of western Canadian coals with that of eastern Carboniferous coals of similar rank (Fig. 6B.27) shows that the western coals have lower

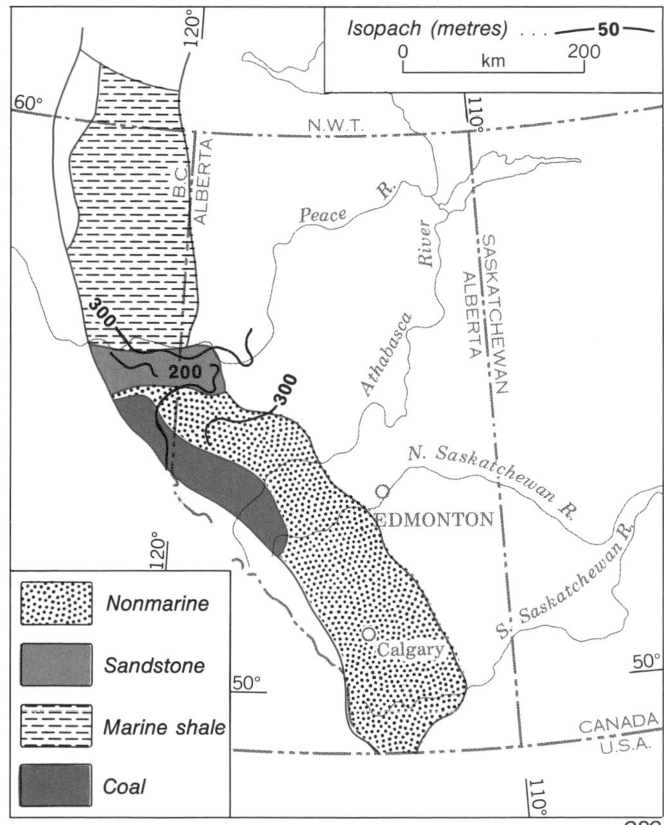

Figure 6B.21. Isopachs and depositional facies, combined Moosebar and Gates formations and equivalents, showing coal-bearing area (modified from Stott, 1984).

values than their Appalachian counterparts. Although a minimum amount of fluidity is necessary in a coking coal, the quality of coke does not increase in linear fashion with ever higher fluidities in the parent coal (Gray et al., 1978). Therefore, a number of Western Canadian coals can be rated as having good coking potential, though they have significantly lower fluidities than Appalachian coals of comparable rank.

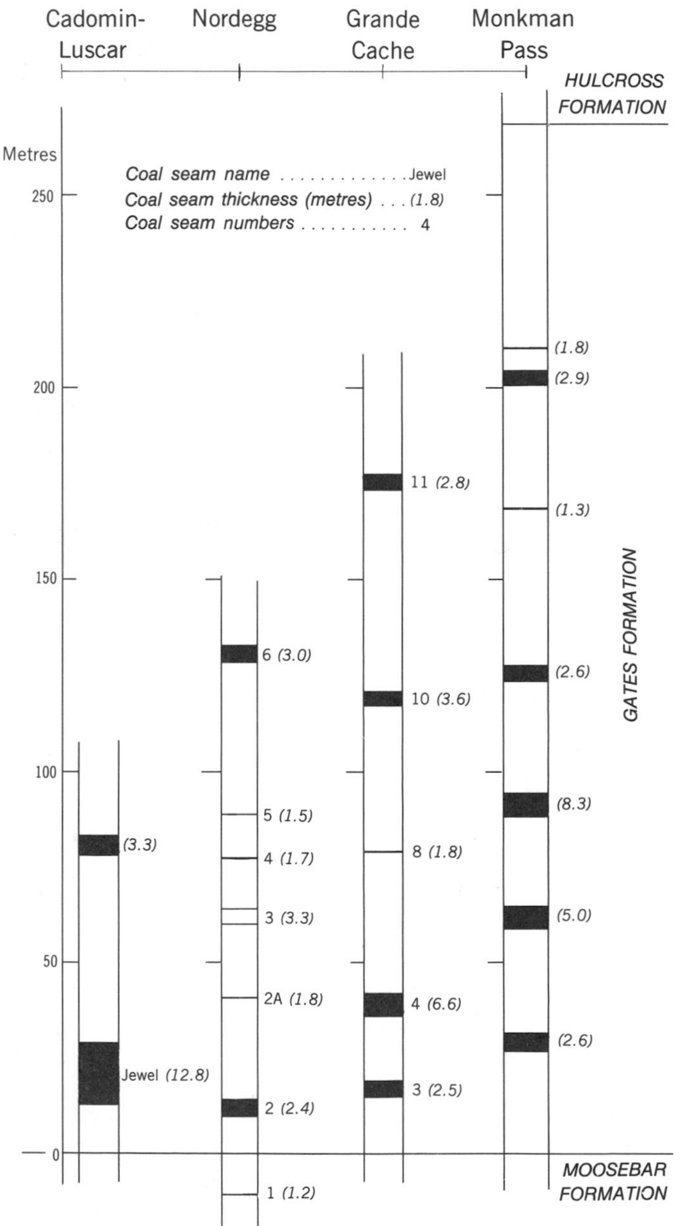

Figure 6B.22. Representative sections of the Gates Formation emphasizing coal seams designations (where established) and thickness (m) in brackets. Sections modified from Horachek (1985) and Schiller et al. (1983), with additional data from Kalkreuth (pers. comm., 1985). See Figure 6B.12 for locations.

Figure 6B.23. Structure section in northern segment, Mountains and Foothills coal belt. Compare structural style with that shown in Figure 6B.15. See Section B Figure 6B.12 for location (modified from Kalkreuth and McMechan, 1984).

An important physical property of coke is strength, measured by a variety of empirical tests. In North America, the ASTM Tumbler Test is commonly used, and stability factor values of 50 or more, derived from this test, are indicative of good quality coke. Whether a coal has or has not the potential to make this threshold is strongly influenced by its rank and maceral composition. From the position of the isostability lines in Figure 6B.28 it is clear that western Canadian coals have the potential to produce good cokes, despite relatively high inert and semi-inert contents, because many of them fall in the optimum rank range between 1.1 and 1.5 reflectance.

YUKON TERRITORY AND DISTRICT OF MACKENZIE

In Yukon Territory and western Northwest Territories, coal occurs in at least ten areas, and in strata ranging in age from Early Carboniferous to Tertiary (Fig. 6B.29). Lateral extent of the coal-bearing measures at nearly all locations is poorly known, and rarely have individual seams been traced beyond the limit of specific outcrops. Quoted resource data should therefore be viewed with these reservations in mind.

Some of the deposits have been mined on a small scale for local consumption, and during the 1970s a moderate amount of exploration was carried out in some areas to evaluate economic potential. On the basis of present information, the most important deposits are those of the Fort Norman, Mackenzie Delta, Bonnet Plume, Dawson, Ross River, and Watson Lake areas.

Much relevant data on these deposits have been summarized by Ricketts (1984). The following discussion of specific occurrences is organized by geographic area (Fig. 6B.29). Locations of important exposures are given in Table 6B.17 and compositional data appear in Table 6B.18.

A. Liard River - Nahanni River area.

1. Thin seams (maximum thickness 0.4 m) of subbituminous coal occur in the Cretaceous Wapiti Formation; estimated resources are 240 million tonnes (Lord, 1983).
2. At least two seams in the Mattson Formation (Early Carboniferous) are exposed in the Liard and Kotaneelee ranges (Harker, 1963; Richards, 1983). Maximum seam thickness is 1.8 m and the rank is mainly high-volatile bituminous. Estimated resources are 154 million tonnes (Lord, 1983).

B. Watson Lake Area.

Lignite to subbituminous coal of Tertiary age occurs along the Rancheria and Frances rivers. In one locality at least 5 seams have been exposed with thicknesses

Figure 6B.23

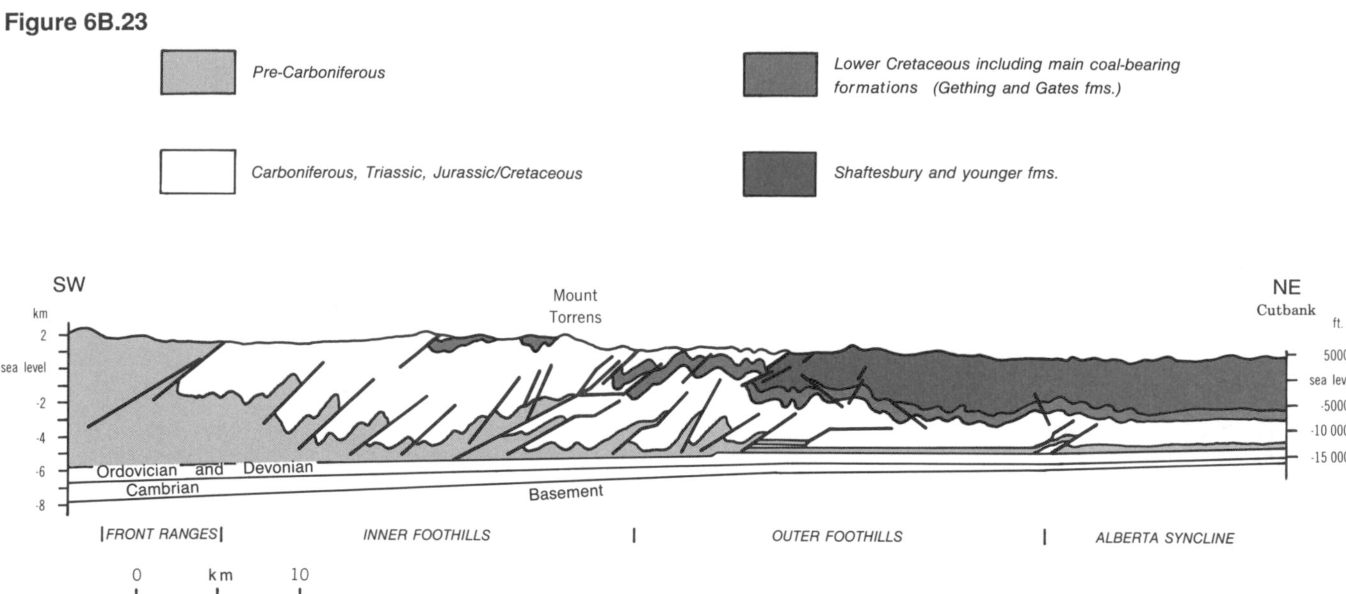

Figure 6B.24. Aerial view of high wall at Cardinal River Mine of Luscar Ltd. showing box fold (syncline) (left centre) with relatively flat beds in core of fold and steeply inclined flanks (photo Luscar Ltd.).

Table 6B.14. Representative compositional data on coals of Gates and Gething formations

Province	Area	Formation	No. of Seams	Weight Per Cent Moisture	Ash	Volatile Matter	Fixed Carbon	Sulphur	MJ/kg	Rank
Alberta	Nordegg	Gates	2	0.5-2.1	6.0-12.6	15.6-22.9	69.0-74.0	0.5	31.7-33.3	MV-LV
Alberta	Mtn. Park	Gates	3	0.8-4.9	9.2-16.2	24.2-30.4	55.3-60.3	0.2-0.6	26.7-32.2	HV-MV
Alberta	Brûlé	Gates	1?	1.0	13.9	18.8	66.3	0.4	30.6	LV
Alberta	Luscar	Gates	2?	0.8-3.3	8.8-30.1	18.0-22.4	50.2-65.5	0.2-0.3	23.9-31.4	MV-LV
Alberta	Smoky River	Gates	3	0.5-2.7	7.2-19.5	15.5-20.6	60.0-72.6	0.3-0.8	28.0-32.8	LV
British Columbia	Bullmoose-Quintette	Gates		6.0	20.0-30.0	21.0-27.0	45.0-55.0	0.3-0.5	21.0-26.0	MV
British Columbia	Monkman (clean)	Gates	?	8.0	7.9	21.4	62.7	0.5	?	MV
British Columbia	Peace River	Gething	7	1.1-11.0	4.1-14.9	18.7-23.9	52.4-73.6	0.6-1.0	23.2-33.6	MV-LV
British Columbia	Sukunka	Gething	1	0.6	5.0	23.7	70.7	0.5	?	MV

[1] MJ/kg (Megajoules/kilogram) = heating value, HV, MV, LV = high volatile, medium volatile and low volatile bituminous respectively
[2] all data on as-received basis and on raw coal unless specified as clean
[3] Sources for data: Stansfield and Lang (1944); Nicolls (1952); Romaniuk and Naidu (1987); ERCB (1987); Bonnell et al. (1983); Schiller et al. (1983); Birmingham and Cameron (1970, 1971)

ranging from 0.4 to 2.1 m (Hughes and Long, 1980). On Coal River 90 km east of Watson Lake, three seams of similar rank and thickness have been described (Thomas, 1943).

C. Ross River Area.

In this area coal occurs in outcrop near the town and along Lapie River. The best exposure shows 5 seams with a maximum thickness of 2.7 m. Although the Ross River and Dawson area coals are of the same age (Tertiary), the former are considerably higher in rank, being in the high-volatile to low-volatile bituminous range. Resource estimates based on the 2.7 m thick seam are 1 million tonnes (Hughes and Long, 1980).

D. Tintina Trench (Dawson area).

Lignite to subbituminous coal occurs over a distance of 96 km along the Yukon River and tributaries. Coal was mined sporadically at some sites such as Coal Creek, where four coal zones with a maximum thickness of 12.9 m have been found (Hughes and Long,1980).

E. Bonnet Plume Basin.

Thick seams ranging in rank from lignite to high-volatile C bituminous are found in the Bonnet Plume Formation (Cretaceous-Paleocene). Seams up to 11 m thick occur in both upper and lower parts of the formation, which may underlie much of the 1800 km^2 area of the basin. Estimated resources are 200 million tonnes (Long, 1978; Norris and Hopkins, 1977).

Table 6B.15. Coal resources in Lower Cretaceous strata of northeastern British Columbia and west-central Alberta (Data modified from Bielenstein et al., 1979; ERCB, 1987)

Province	Resources of Immediate Interest (Megatonnes) Measured	Indicated	Inferred	Rank[2]
Alberta	3400[1]	-	-	HV - LV
British Columbia	1000	500	7700	MV - LV

[1] Data from ERCB (1987). Figure quoted is "Established Resources" according to ERCB criteria.
[2] HV, MV, LV = high volatile, medium volatile, low volatile bituminous respectively

Table 6B.16. Preliminary 1986 production statistics from Gates Formation for northeastern British Columbia and west-central Alberta (data from Romaniuk and Naidu, 1987)

Area	Surface (10^3 tonnes)
Luscar	1900
Grande Cache	1204[1]
Gregg River	2533
Quintette	9919[2]
Bullmoose Mountain	2514[2]

[1] Some underground production.
[2] Some coal for thermal market.

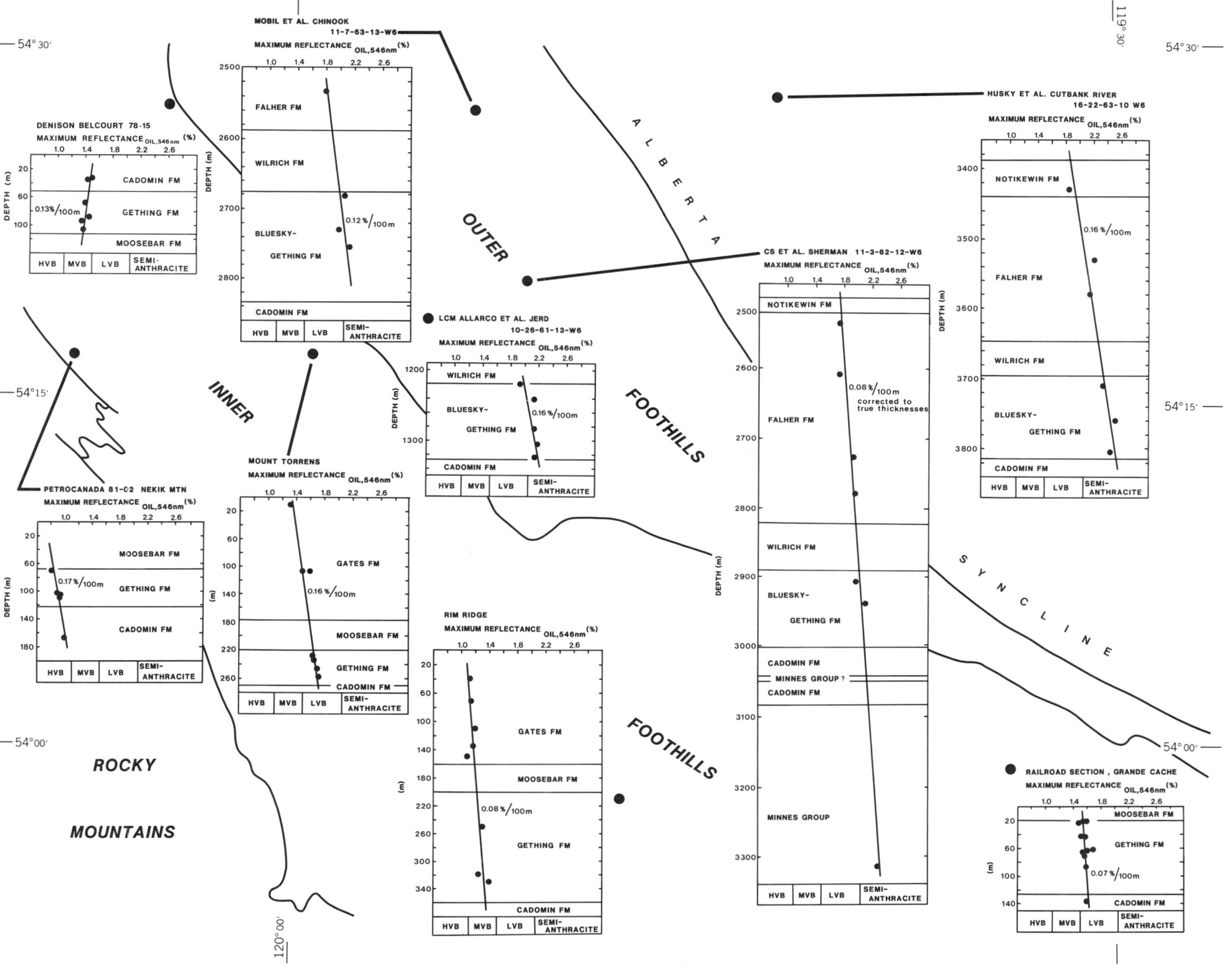

Figure 6B.25. Reflectance data for various coal-bearing sections in northeast British Columbia and adjacent parts of Alberta (from Kalkreuth and McMechan, 1984).

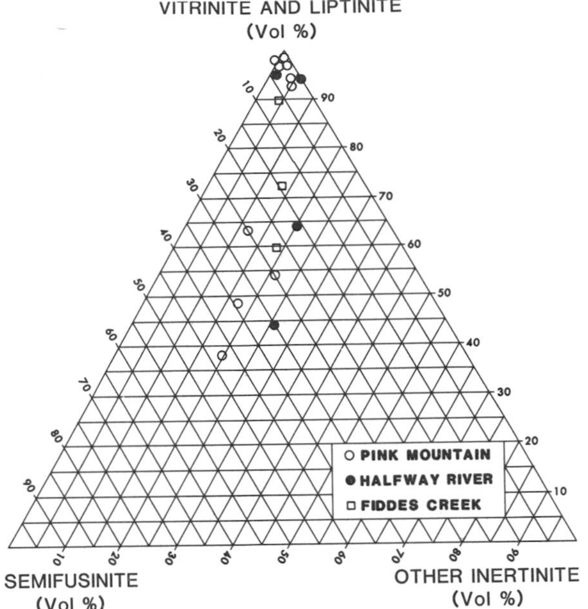

Figure 6B.26. Maceral composition (mineral-free) for Gething and Gates coals at three localities in northeast British Columbia and west central Alberta (from Kalkreuth, 1982).

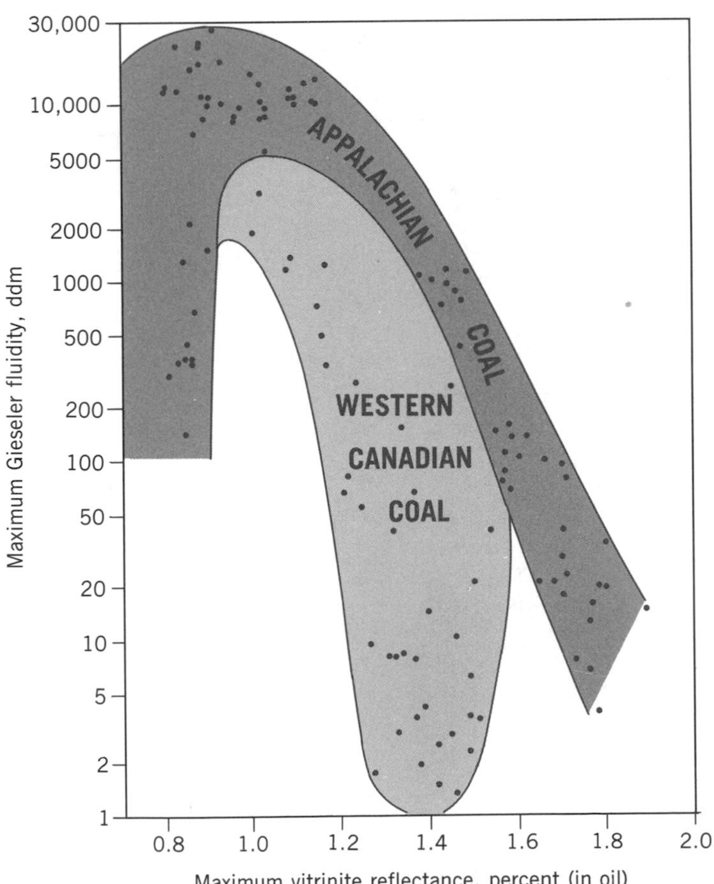

Figure 6B.27. Plot of Gieseler fluidity versus vitrinite reflectance, Appalachian and western Canadian coals (Paulencu and Readyhough, 1974). Fluidity expressed in dial divisions/ minute (ddm).

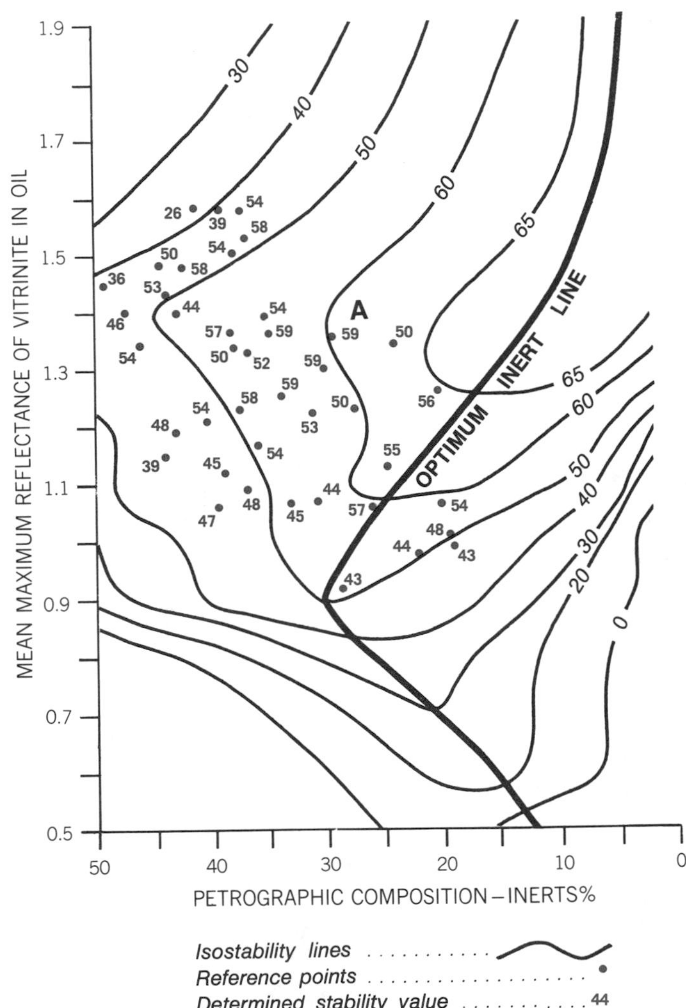

Figure 6B.28. Predicted and measured coke stabilities for some western Canadian coals. Predicted values based on maceral compostion and reflectance. The data are superimposed on isostability curves by Pearson (1980) based on work of Schapiro et al. (1961). Points are plotted according to predicted values; appended numbers are measured values; for example for Point A predicted value is 60 and measured stability is 59.

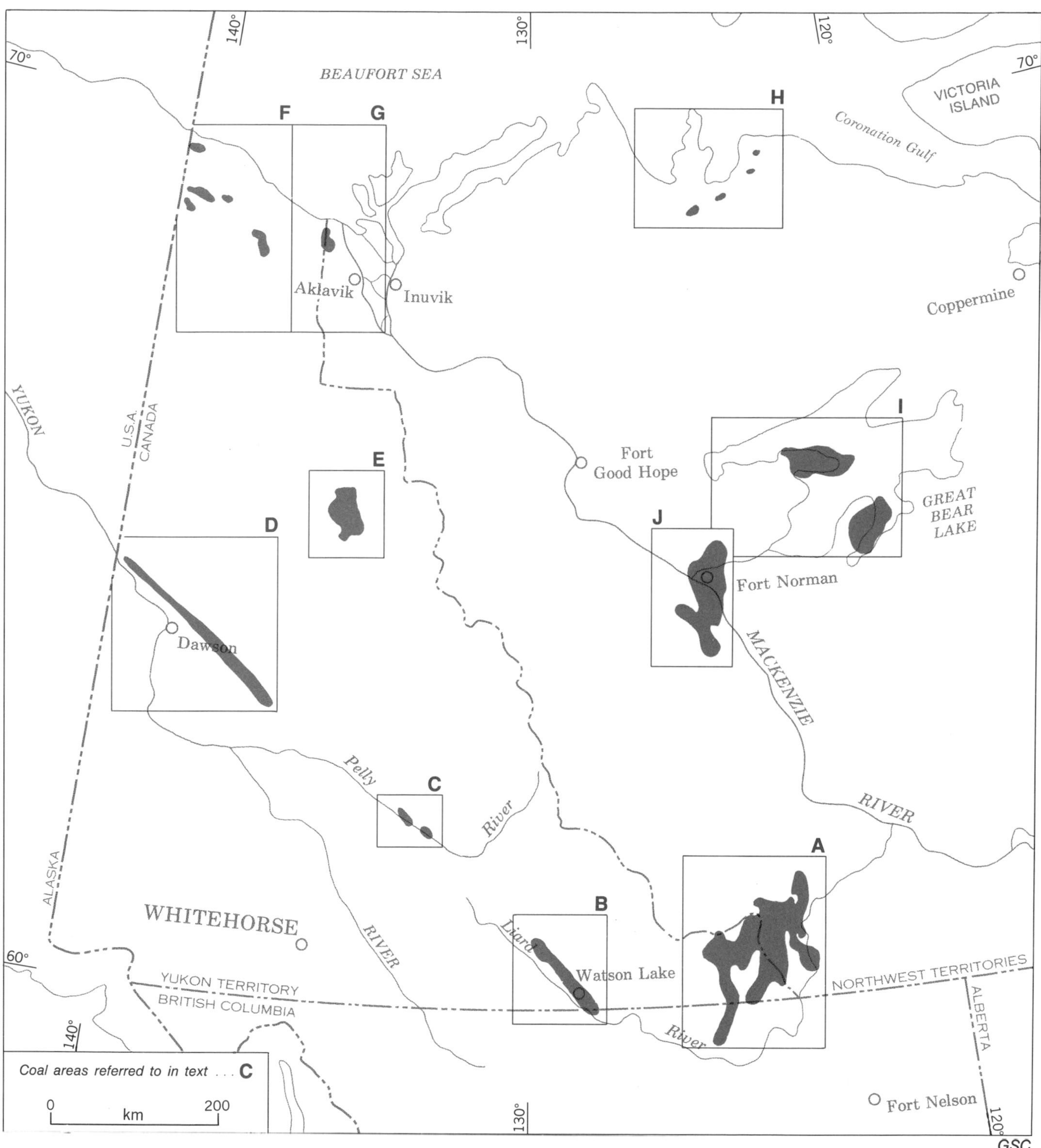

Figure 6B.29. Index map for coal areas in the western District of Mackenzie and Yukon Territory north of and including Tintina Trench. Coal areas, referred to in text are: A Liard River - Nahanni River; B Watson Lake; C Ross River; D Tintina Trench (Dawson); E Bonnet Plume Basin; F British Mountains - Coastal Plains; G Mackenzie Delta; H Franklin Bay - Darnley Bay; I Great Bear Lake; J Fort Norman. Tint indicates areas underlain by known or potential coal-bearing rocks (modified from Energy, Mines and Resource Map **Canada - Coal**, 1982.)

Table 6B.17. Data for selected coal exposures in Yukon Territory and Mackenzie District (NWT)

Area (see Figure 6B.29)	Age	Formation	NTS Map	Location of Exposures (Lat. and Long.)
A.	L. Cret.	Wapiti	95B/3, 4	60°15'N, 123°30'W
	E. Carboniferous	Mattson	95C/16	60°50'N, 124°25'W
	E. Carboniferous	Mattson	95F/1	61°10'N, 124°20'W
	E. Carboniferous	Mattson	95G/4	61°06'N, 123°55'W*
B.	E. Tertiary	unnamed	105A/2, 3	60°08'N, 129°15'W*
	Eocene	unnamed	95D/3	60°05'N, 127°05'W*
C.	Eocene?	unnamed	105F/15	61°58'N, 132°35'W
D.	Eocene	unnamed	116B/2	64°08'N, 138°56'W*
	Eocene	unnamed	116C/8, 9, 10	64°27-38'N, 140°07-40'W*
E.	Cret.-E. Tert.	Bonnet Plume	106E/6, 7	65°15'N, 135°00'W*
F.	E. Carboniferous	Kayak	117C/8	69°20'N, 140°42'W*
	E. Carboniferous	Kayak	117A/10	68°36'N, 137°55'W*
	E. Cret.	Kamik	116P/15	67°55'N, 136°57'W
	L. Cret.-E. Tert.	Reindeer	117A/14	68°53'N, 138°02'W
	Oligocene?	unnamed	117A/3	68°06'N, 138°20'W
G.	L. Cret.-Tert.	Reindeer	117A/9	68°42'N, 136°18'W*
	L. Cret.-Tert.	Reindeer	107B/3, 6	68°15'N, 134°42'W
	E. Cret.	Kamik	107B/4	68°12'N, 135°25'W
H.	E. Cret.	Langton Bay	97B/15	68°51'N, 125°24'W
	E. Cret.	Langton Bay	97D/6, 11	68°30'N, 122°45'W*
I.	Cretaceous	unnamed	96H/13	65°55'N, 121°45'W
	Cretaceous	unnamed	96I/4	66°10'N, 120°50'W*
J.	Cret.-Paleocene	Summit Creek	96C/5, 12	64°30'N, 125°45'W
	Cret.-Paleocene	Summit Creek	96C/11, 14	64°45'N, 125°20'W*

*Locations for which compositional data are listed in Table 6B-18

Abbreviations: E = Early, L = Late, Cret. = Cretaceous, Tert. = Tertiary

F. British Mountains - Coastal Plains Area.

Coal in this area occurs in strata of four different ages: in the Kayak Formation (Early Carboniferous), in the Kamik Formation (Early Cretaceous), in the Reindeer Formation (Late Cretaceous-Tertiary), and in unnamed Oligocene beds. Maximum seam thicknesses and ranks are: Kayak Formation, 5.0 m - semianthracite-anthracite (Cameron et. al., 1988); Kamik Formation, 0.2 m - high-volatile bituminous to semianthracite; Reindeer Formation, 1.5 m - high-volatile bituminous; Oligocene, 12.0+ m - lignite (Norris, 1974).

G. Mackenzie Delta Area.

Coal occurs in the Reindeer Formation (Late Cretaceous-Tertiary) at three localities. At the Coal Mine Lake site high-volatile bituminous coal was mined from 1942 to 1960. Here three seams with thicknesses of 3.1, 7.0, and 3.7 m are present. Lignitic to sub-bituminous coal of the Reindeer Formation has also been reported in the Caribou Hills east of the delta (Young, 1975).

H. Franklin Bay-Darnley Bay Area.

Coal has been reported at four localities over a distance of 128 km in the Langton Bay Formation (Early Cretaceous). Maximum reported thickness is 3 m and the rank is lignite to subbituminous (Yorath et al., 1975).

I. Great Bear Lake Area.

Lignite is exposed in Cretaceous strata at several places along the shore. Maximum reported thickness is 5.6 m (Yorath and Cook, 1981).

J. Fort Norman Area.

Numerous coal exposures are found in the Summit Creek Formation (Late Cretaceous-Paleocene). Two major coal zones, with cumulative coal of 4 and 11 m, occur west of Tate Lake. These zones can be traced for 45 km to the vicinity of Fort Norman on Mackenzie River. The rank is lignite to subbituminous. In a few places rank has increased, apparently because of burning of coal beds in outcrop. (Yorath and Cook, 1981; Sweet et al., 1989).

Table 6B.18. Summary of analytical data on coals of Yukon Territory and western District of Mackenzie (from Ricketts, 1984)

			Proximate Analysis (wt. per cent as rec'd except H : dry)							
Area (see Figure 6B.29)	NTS Map	Lat. Long.	Moisture	Ash	Volatile Maher	Fixed Carbon	Sulphur	Cal.[1] value MJ/kg	Ro[2]	Rank[3] by ASTM (as rec'd)
A.	95G/4	61°06'N 123°55'W	3.3	3.6	34.9	58.2		29.2		HV C/B
B.	95D/3	60°05'N 127°05'W	2.2-3.8	22.0-47.5	6.1-10.1	38.4-69.9				Lignite
B.	105A/2,3	60°08'N 129°15'W	14.0-49.3	4.7-41.5	18.1-39.6	8.5-25.8		5.8-17.4	0.11-0.32	Lignite
C.	105F/15	61°58'N 132°35'W	3.4-21.8	10.4-31.9	18.4-28.4	28.6-58.1	0.23-0.80	13.28-29.66	1.06-2.03	HVA-SA
D.	116B/2	64°08'N 138°56'W	24.9-33.6	9.6-16.8	25.4-31.5	30.4-36.5	tr.-0.6	16.5-19.0		Lig./Sub. C.
D.	116C/8	64°27'N 140°07'W	16.8-20.2	8.3-26.9	26.1-35.8	26.2-36.4	0.5-1.5	14.6-18.2	0.46-0.54	Sub. C/B
D.	116C/9	64°33'N 140°28'W	27.5-31.4	6.6-11.4	31.4-32.6	29.7-34.4	0.7-1.1	15.5-18.3	0.36-0.46	Lignite
D.	116C/10	64°35'N 140°35'W	19.0	14.9	30.7	35.4	0.5	17.9	0.74	Sub. C
E.	106E/6, 7	65°15'N 135°00'W	5.6	13.5	35.1	45.8	0.5	23.6		Lig./HVC
F.	117C/8	69°20'N 140°42'W	1.8	6.4	10.0	81.9	0.7		2.89-3.27 (Romax)	SA/A
F.	117A/10	68°36'N 137°55'W	2.0-2.1	2.0	9.5-14.0	82.0-86.4		33.2-33.5		LV/SA
G.	117A/9	68°42'N 136°18'W	6.2-7.2	11.6-14.4	36.1-38.9	42.3-43.3	0.2-0.7	25.3-25.7	0.53-0.75	HV C/B
H.	97D/11	69°30'N 122°45'W		17.0-46.1	20.8-36.0	33.1-47.0	0.7-1.6	15.4-24.4		Sub. B
I.	96I/4	66°10'N 120°50'W	32.2-48.9	4.2-22.8	20.4-49.5	19.4-28.8	0.1-0.9	10.4-14.5		Lignite
J.	96C/11, 12, 14	64°45'N 125°20'W						17.7-24.7	0.31-1.23	lig. to bit.

[1]calorific value in megajoules/kilogram
[2]Ro random = percent random reflectance in oil
[3]Lig. = lignite; sub. B,C = subbituminous B and C; HV B, C = high volatile bituminous B and C; LV = low volatile bituminous; SA and A = semianthracite and anthracite

REFERENCES

Alberta Energy Resources Conservation Board (ERCB)
1987: Reserves of Coal - Province of Alberta; ERCB Report ST 87-31.
1988: Alberta Coal Industry - Annual Statistics; ERCB Report ST 88-29, 45 p.

Allan, J.A. and Carr, J.L.
1946: Geology and coal occurrences of Wapiti-Cutbank area, Alberta; Research Council of Alberta, Report no. 48, 43 p.

Allan, J.A. and Rutherford, R.L.
1924: Geology along the Blackstone, Brazeau and Pembina Rivers in the Foothills belt, Alberta; Alberta Scientific and Industrial Research Council, Report no. 9, 53 p.

American Society for Testing and Materials (ASTM)
1979: Annual Book of ASTM Standards, Part 26, Gaseous Fuels; Coal and Coke; Atmospheric Analysis. ASTM, 1916 Race Street, Philadelphia, Pa., p. 187-462.

Bielenstein, H.U., Chrismas, L.P., Latour, B.A., and Tibbetts, T.E.
1979: Coal resources and reserves of Canada; Department of Energy, Mines and Resources, Canada, Report EMR 79-9, 37 p.

Birmingham, T.F. and Cameron, A.R.
1970: Petrography of the Nos. 4, 10 and 11 seams in the No. 2 Mine area of the Smoky River coalfield, Alberta; Geological Survey of Canada, Internal Report TR 83E/14-1, 34 p.
1971: Petrography of the Chamberlain seam in the Sukunka River area, British Columbia; Geological Survey of Canada, Internal Report TR 93P/4-3, 19 p.

Bonnell, G.W., Janke, L.C., and Romaniuk, A.S.
1983: Analysis directory of Canadian commercial coals - supplement No. 5; CANMET Report 84-1E, 223 p.

Broughton, P.L.
1979: Tectonic controls on sand distribution in the Late Cretaceous-Tertiary coal basins of southern Saskatchewan; in Summary of Investigations 1979; Saskatchewan Geological Survey, Miscellaneous Report 79-10, p. 214-221.

Bustin, R.M.
1982: The effect of shearing on the quality of some coals in the southeastern Canadian Cordillera; Canadian Institute of Mining and Metallurgy, Bulletin, v. 75, p. 76-83.

Cameron, A.R.
1972: Petrography of Kootenay coals in the upper Elk River valley and Crowsnest areas; Research Council of Alberta, Information Series No. 60, p. 31-45.
1978: Petrographic characteristics of lignites from the Estevan area; in Coal Resources of Southern Saskatchewan: A Model for Evaluation Methodology, S.H. Whitaker, J.A. Irvine, and P.L. Broughton (ed.); Geological Survey of Canada Economic Report 30, Saskatchewan Department of Mineral Resources Report 209, Saskatchewan Research Council Report 20, p. 41-48.
1984: Comparison of reflectance data for various macerals from coals of the Kootenay Group, southeastern British Columbia; in Symposium on the Geology of Rocky Mountain Coal; North Dakota Geological Society, Proceedings 1984, Publication 84-1, p. 61-75.
1991: Regional patterns of reflection in lignites of the Ravenscrag Formation, Saskatchewan, Canada; Organic Geochemistry, v. 17, p. 223-242.

Cameron, A.R. and Kalkreuth, W.D.
1982: Petrological characteristics of Jurassic-Cretaceous coals in the Foothills and Rocky Mountains in western Canada; Proceedings 5th Symposium on the Geology of Rocky Mountain Coal; Utah Geological and Mineral Survey, Bulletin 118, p. 163-167.

Cameron, A.R. and Marconi, L.G.
1979: Reflectance measurements on ulminite from Saskatchewan lignites; Geological Society of America, Abstracts with Programs, v. 11, p. 397.

Cameron, A.R., Boonstra, C., and Pratt, K.C.
1988: Compositional characteristics of anthracitic coals in the Hoidahl Dome area, northern Yukon Territory; in Current Research, Part D, Geological Survey of Canada, Paper 88-1D, p. 67-74.

Campbell, J.D.
1964: Catalogue of coal mines of the Alberta Plains; Research Council of Alberta Report 64-3, 140 p.
1974: Coal resources, Taber-Manyberries area, Alberta; Alberta Research Council, Report 74-3, 140 p.

Canada-Department of Energy, Mines and Resources
1982: Canada-Coal (Map); The National Atlas of Canada, 5th edition.

Dawson, F.M. and Kalkreuth, W.
1989: Preliminary results of a continuing study of the stratigraphic context, distribution and characteristics of coals in the Upper Cretaceous to Paleocene Wapiti Group, northwestern Alberta; in Contributions to Canadian Coal Geoscience, Geological Survey of Canada, Paper 89-8, p. 43-48.

Douglas, R.J.W.
1950: Callum Creek, Langford Creek and Gap map areas, Alberta; Geological Survey of Canada, Memoir 255, 124 p.

Dyck, J.H., McKenzie, C.T., Tibbetts, T.E., and Whitaker, S.H.
1980: Quality and occurrence of coal in southern Saskatchewan, Canada, v. 1; Saskatchewan Research Council and CANMET, Report ERP/ERL 80-84, 117 p.

Fraser, F.J., McLearn, F.H., Russell, L.S., Warren, P.S., and Wickenden, R.T.D.
1935: Geology of southern Saskatchewan; Geological Survey of Canada, Memoir 176, 137 p.

Frebold, H.
1957: The Jurassic Fernie Group in the Canadian Rocky Mountains and Foothills; Geological Survey of Canada, Memoir 287, 197 p.

Furnival, G.M.
1950: Cypress Lake map-area, Saskatchewan; Geological Survey of Canada, Memoir 242, 161 p.

Gibson, D.W.
1977: Upper Cretaceous and Tertiary coal-bearing strata in the Drumheller-Ardley region, Red Deer Valley, Alberta; Geological Survey of Canada, Paper 76-35, 41 p.
1985a: Stratigraphy, sedimentology and depositional environments of the coal-bearing Jurassic-Cretaceous Kootenay Group, Alberta and British Columbia; Geological Survey of Canada, Bulletin 357, 108 p.
1985b: Stratigraphy and sedimentology of the Lower Cretaceous Gething Formation, Carbon Creek coal basin, northeastern British Columbia; Geological Survey of Canada, Paper 80-12, 29 p.

Gibson, D.W., Hughes, J.D., and Norris, D.K.
1983: Stratigraphy, sedimentary environments and structure of the coal-bearing Jura-Cretaceous Kootenay Group, Crowsnest Pass area, Alberta and British Columbia; Field trip no. 4, The Mesozoic of Middle North America, May 1983, Calgary, Alberta; Canadian Society of Petroleum Geologists, 80 p.

Graham, P.S., Gunther, P.R., and Gibson, D.W.
1977: Geological investigations of the coal-bearing Kootenay Formation in the subsurface of the upper Elk River valley, British Columbia; in Report of Activities, Part B, Geological Survey of Canada, Paper 77-1B, p. 203-210.

Gray, R.J., Goscinski, J.S., and Schoenberger, R.W.
1978: Selection of coals for coke-making; Preprint of paper at joint conference of Iron and Steel Society (AIME) and Society of Mining Engineers (AIME), Pittsburgh, Pa., October 3, 1978, 49 p.

Guliov, P.
1972: Lignite coal resources of Saskatchewan; Research Council of Alberta, Information Series No. 60, p. 165-172.

Hacquebard, P.A.
1977: Rank of coal as an index of organic metamorphism for oil and gas in Alberta; Chapter 3 in The Origin and Migration of Petroleum in the Western Canadian Sedimentary Basin, Alberta, G. Deroo, T.G. Powell, B. Tissot, and R.G. McCrossan (ed.); Geological Survey of Canada, Bulletin 262, p. 11-22.

Hacquebard, P.A. and Birmingham, T.F.
1973: Petrography of the Val D'Or and Mynheer Seams, Coalspur Coal area, Alberta; Geological Survey of Canada, Technical Report No. 83-F/2-1, 21 p.

Hacquebard, P.A. and Donaldson, J.R.
1974: Rank studies of coals in the Rocky Mountains and Inner Foothills Belt, Canada; in Carbonaceous materials as indicators of metamorphism, R.R. Dutches, P.A. Hacquebard, J.M. Schopt, and J.A. Simon (ed.); Geological Society of America, Special Paper 153, p. 75-94.

Harker, P.
1963: Carboniferous and Permian rocks, southwestern District of Mackenzie; Geological Survey of Canada, Bulletin 95, 91 p.

Hoffman, G.L. and Jordan, G.R.
1984: The coalfields of the northern foothills of the Canadian Rocky Mountains; in The Mesozoic of Middle North America, D.F. Stott and D.J. Glass (ed.); Canadian Society of Petroleum Geologists, Memoir 9, p. 541-549.

Holter, M.E.
1972: Coal seams of the Estevan area, southern Saskatchewan; in Proceedings First Geological Conference on Western Canadian Coal, G.B. Mellon, J.W. Kramers, and E.J. Seagel (ed.); Alberta Research Council, Information Series, No. 60, p. 173-184.

Holter, M.E., Yurko, J.O., and Chu, M.
1975: Geology and coal reserves of the Ardley Coal Zone of central Alberta; Alberta Research Council, Report 75-7, 41 p.

Horachek, Y.
1985: Geology of Alberta coal; in Coal in Canada, T. Patching (ed.); Canadian Institute of Mining and Metallurgy, Special Volume 31, p. 115-133.

Hughes, J.D.
1984: Geology and depositional setting of the Late Cretaceous, Upper Bearpaw and Lower Horseshoe Canyon Formations in the Dodds-Round Hill coalfield of central Alberta - A computer-based study of closely-spaced exploration data; Geological Survey of Canada, Bulletin 361, 81 p.

Hughes, J.D. and Cameron, A.R.
1985: Lithology, depositional setting and coal rank-depth relationships in the Jurassic-Cretaceous Kootenay Group at Mount Allan, Cascade coal basin, Alberta; Geological Survey of Canada, Paper 81-11, 41 p.

Hughes, J.D. and Long, D.G.F.
1980: Geology and coal resource potential of Early Tertiary strata along Tintina Trench, Yukon Territory; Geological Survey of Canada, Paper 79-32, 21 p.

Irish, E.J.W.
1970: The Edmonton Group of south-central Alberta; Bulletin of Canadian Petroleum Geology, v. 18, p. 125-155.

Irvine, J.A.
1978a: Data base construction and resource evaluation methodology; in Coal Resources of Southern Saskatchewan: A model for evaluation methodology, S.H. Whitaker, J.A. Irvine, and P.L. Broughton (ed.); Geological Survey of Canada Economic Report 30, Saskatchewan Department of Mineral Resources Report 209, Saskatchewan Research Council Report 20, p. 89-106.
1978b: Estevan map area; in Coal Resources of Southern Saskatchewan; A model for Evaluation Methodology, S.H. Whitaker, J.A. Irvine, and P.L. Broughton (ed.); Geological Survey of Canada Economic Report 30, Saskatchewan Department of Mineral Resources Report 209, Saskatchewan Research Council Report 20, p. 49-67.

Jerzykiewicz, T.
1985: Stratigraphy of the Saunders Group in the central Alberta Foothills - a progress report; in Current Research, Part B, Geological Survey of Canada, Paper 85-1B, p. 247-258.

Jerzykiewicz, T. and McLean, J.R.
1980: Lithostratigraphical and sedimentological framework of coal-bearing Upper Cretaceous and Lower Tertiary strata, Coal Valley area, central Alberta foothills; Geological Survey of Canada, Paper 79-12, 47 p.

Jerzykiewicz, T., Lerbekmo, J.F., and Sweet, A.R.
1984: The Cretaceous-Tertiary boundary, central Alberta foothills (abstract); in Programme with Abstracts; Canadian Paleontology and Biostratigraphy Seminar, Ottawa, September 28-30, 1984, p. 4.

Kalkreuth, W.D.
1982: Rank and petrographic composition of selected Jurassic-Lower Cretaceous coals of British Columbia, Canada; Bulletin of Canadian Petroleum Geology, v. 30, p. 112-139.

Kalkreuth, W.D. and McMechan, M.E.
1984: Regional pattern of thermal maturation as determined from coal-rank studies, Rocky Mountain Foothills and Front Ranges north of Grande Cache, Alberta - implications for petroleum exploration; Bulletin of Canadian Petroleum Geology, v. 32, p. 249-271.

Karst, R. and White, G.
1980: Coal rank distribution within the Bluesky-Gething stratigraphic horizon of northeastern British Columbia; Geological Field Work 1979, British Columbia Ministry of Energy, Mines and Petroleum Resources, Paper 1980-1, p. 103-107.

Koensler, W.
1980: Das Verhalten des Inertinits westkanadischer Kreidekohlen bei der Verkokung; Dissertation, Rheinisch - Westfalische Technische Hochschule, Aachen, Federal Republic of Germany, 118 p.

Kramers, J.W. and Mellon, G.B.
1972: Upper Cretaceous-Paleocene coal-bearing strata, northwest-central Alberta plains; Research Council of Alberta, Information Series No. 60, p. 109-124.

Langenberg, C.W. and McMechan, M.E.
1985: Lower Cretaceous Luscar Group (revised) of the northern and north-central Foothills of Alberta; Bulletin of Canadian Petroleum Geology, v. 33, p. 1-11.

Long, D.G.F.
1978: Lignite deposits in the Bonnet Plume Formation, Yukon Territory; in Current Research, Part A, Geological Survey of Canada, Paper 78-1A, p. 399-401.

Lord, C.C.
1983: Nahanni Region; Department of Indian and Northern Affairs, Mineral Industry Report 1979, Northwest Territories, EGS-1983-9, p. 245-265.

MacKay, B.R.
1947: Coal reserves of Canada; Reprint of Chapter 1 and Appendix A of Report of the Royal Commission on Coal, 1946, and Four Supplementary Maps Relating to Estimates of Coal Reserves; Ottawa, Kings Printer and Controller of Stationery, 113 p.

Majorowicz, J.A. and Jessop, A.M.
1981: Regional heat flow patterns in the western Canadian sedimentary basin; Tectonophysics, v. 74, p. 209-238.

Majorowicz, J.A., Rahman, M., Jones, F.W., and McMillan, N.J.
1985: The paleogeothermal and present thermal regimes of the Alberta Basin and their significance for petroleum occurrences; Bulletin of Canadian Petroleum Geology, v. 33, p. 12-21.

McLean, J.R.
1971: Stratigraphy of the Upper Cretaceous Judith River Formation in the Canadian Great Plains; Saskatchewan Research Council, Report No. 11, 96 p.

McLearn, F.H.
1929: Southern Saskatchewan; Geological Survey of Canada, Summary Report 1928, Part B, p. 30-44.
1930: Stratigraphy, clay and coal deposits of southern Saskatchewan; Geological Survey of Canada, Summary Report 1929, Part B, p. 48-63.

Monger, J.W.H. and Preto, V.A.
1972: Geology of the southern Canadian Cordillera; Guidebook for Excursions AO3-CO3; XXIV International Geological Congress, Montreal, 87 p.

Nicolls, J.H.H.
1952: Analyses of Canadian coals and peat fuels; Canada Department of Mines and Technical Surveys, Mines Branch Report 831, 409 p.

Norris, D.K.
1959: Type section of the Kootenay Formation, Grassy Mountain, Alberta; Journal of the Alberta Society of Petroleum Geologists, v. 7, p. 223-237.
1964: Microtectonics of the Kootenay Formation near Fernie, British Columbia; Bulletin of Canadian Petroleum Geology, v. 12, p. 383-398.
1966: The mesoscopic fabric of rock masses about some Canadian coal mines; International Society of Rock Mechanics, v. 1, p. 191-198.
1971: The geology and coal potential of the Cascade coal basin; in A Guide to the Geology of the Eastern Cordillera along the Trans-Canada Highway between Calgary, Alberta and Revelstoke, British Columbia, I.A.R. Halladay and D.H. Mathewson (ed.); Alberta Society of Petroleum Geologists, p. 25-39.

Norris, D.K. (cont.)
1974: Structural and stratigraphic studies in the northern Canadian Cordillera; in Report of Activities, Part A, Geological Survey of Canada, Paper 74-1, Part A, p. 343-349.

Norris, D.K. and Hopkins, W.S., Jr.
1977: The geology of the Bonnet Plume Basin, Yukon Territory; Geological Survey of Canada, Paper 76-8, 20 p.

Nurkowski, J.R.
1984: Coal quality, coal rank variation and its relation to reconstructed overburden, Upper Cretaceous and Tertiary plains coals, Alberta, Canada; American Association of Petroleum Geologists, Bulletin, v. 68, p. 285-295.

Parkash, S., du Plessis, M.P., Cameron, A.R., and Kalkreuth, W.D.
1984: Petrography of low rank coals with reference to liquefaction potential; International Journal of Coal Geology, v. 4, p. 209-234.

Patching, T.H. (ed.)
1985: Coal in Canada; Canadian Institute of Mining and Metallurgy, Special Volume 31, 327 p.

Paulencu, H.N. and Readyhough, P.J.
1974: Interpreting coal properties for utilization in commercial coke-making; Proceedings of Symposium on Coal Evaluation, Alberta Research Council, Information Series 76, p. 108-133.

Pearson, D.E.
1980: The quality of western Canadian coking coal; Canadian Institute of Mining and Metallurgy, Bulletin, v. 73, p. 1-15.

Pearson, D.E. and Grieve, D.A.
1985: Rank distribution and coalification pattern, Crowsnest coalfield, British Columbia, Canada; Proceedings, 9th International Congress of Carboniferous Stratigraphy and Geology, v. 4, p. 600-608.

Pearson, D.E. and Pearson, J.S.
1983: The reactivity of inertinites in coking coals from British Columbia (abstract); Canadian Institute of Mining and Metallurgy, Bulletin, v. 76, no. 857, Program of Smithers, B.C. Meeting, p. 52.

Price, R.A.
1961: Fernie map-area, east half, Alberta and British Columbia; Geological Survey of Canada, Paper 61-24, 65 p.

Price, R.A. and Mountjoy, E.W.
1970: Geologic structure of the Canadian Rocky Mountains between Bow and Athabasca Rivers - a progress report; The Geological Association of Canada, Special Paper 6, p. 7-39.

Richards, B.C.
1983: Uppermost Devonian and Lower Carboniferous stratigraphy, sedimentation and diagenesis, southwestern District of Mackenzie and southeastern Yukon Territory; Ph.D. thesis, University of Kansas, Lawrence, Kansas, 373 p.

Ricketts, B.D.
1984: A coal index for Yukon and District of Mackenzie, Northwest Territories; Geological Survey of Canada, Open File Report 1115, 82 p.

Romaniuk, A.S. and Naidu, H.G.
1984: Coal mining in Canada; Energy, Mines and Resources Canada, CANMET Report 83-20E, 46 p.
1987: Coal mining in Canada: 1986; Energy, Mines and Resources Canada, CANMET Report 87-3E, 44 p.

Schapiro, N., Gray, R.J., and Eusner, G.R.
1961: Recent developments in coal petrography: Proceedings of blast furnace, coke oven and raw materials committee; American Institute of Mining Engineering, v. 20, p. 89-112.

Schiller, E.A., Santiago, S.P., and Plachner, W.A.
1983: Monkman coal property, British Columbia; Canadian Institute of Mining and Metallurgy Bulletin, v. 76, p. 65-71.

Smith, G.G.
1989: Coal resources of Canada; Geological Survey of Canada, Paper 89-4, 146 p.

Stansfield, E. and Lang, W.A.
1944: Coals of Alberta: their occurrence, analysis and utilization; Research Council of Alberta, Report No. 35, 174 p.

Statistics Canada
1988: December 1987 coal and coke statistics; Catalogue no. 45-002, v. 66, no. 12, 16 p.

Steiner, J., Williams, G.D., and Dickie, G.J.
1972: Coal deposits of the Alberta plains; Research Council of Alberta, Information Series No. 60, p. 85-108.

Stott, D.F.
1968: Lower Cretaceous Bullhead and Fort St. John groups between Smoky and Peace rivers, Rocky Mountain Foothills, Alberta and British Columbia; Geological Survey of Canada, Bulletin 152, 279 p.
1969: The Gething Formation at Peace River Canyon, British Columbia; Geological Survey of Canada, Paper 68-28, 30 p.
1972a: Cretaceous stratigraphy, northeastern British Columbia; Research Council of Alberta, Information Series No. 60, p. 137-150.

Stott, D.F. (cont.)
1972b: The Cretaceous Gething Delta, northeastern British Columbia; Research Council of Alberta, Information Series No. 60, p. 151-163.
1975: The Cretaceous system in northeast British Columbia; in The Cretaceous System in the Western Interior of North America, W.G.E. Caldwell (ed.); Geological Association of Canada, Special Paper No. 13, p. 441-467.
1981: Bickford and Gorman Creek, two new formations of the Jurassic-Cretaceous Minnes Group, Alberta and British Columbia; in Current Research, Part B; Geological Survey of Canada, Paper 81-1B, p. 1-9.
1984: Cretaceous sequences of the Foothills of the Canadian Rocky Mountains; in The Mesozoic of Middle North America, D.F. Stott and D.J. Glass (ed.); Canadian Society of Petroleum Geologists, Memoir 9, p. 85-107.

Swartzman, E. and Tibbetts T.E.
1955: Analysis directory of Canadian coals, Supplement No. 1; Mines Branch, Canada Department of Mines and Technical Surveys Report No. 850, 81 p.

Sweet, A.R.
1978: Palynology of the Ravenscrag and Frenchman formations; in Coal Resources of Southern Saskatchewan: A Model for Evaluation Methodology, S.H. Whitaker, J.A. Irvine, and P.L. Broughton (ed.); Geological Survey of Canada Economic Report 30, Saskatchewan Department of Mineral Resources Report 209, Saskatchewan Research Council Report 20, p. 29-39.

Sweet, A.R. and Hills, L.V.
1984: A palynological and sedimentological analysis of the Cretaceous-Tertiary boundary, Red Deer Valley, Alberta, Canada; Abstract, Sixth International Palynological Conference, Calgary, 1984, p. 60-61.

Sweet, A.R., Ricketts, B.D., Cameron, A.R., and Norris, D.K.
1989: An integrated analysis of the Brackett Coal Basin, Northwest Territories; in Current Research, Part G, Geological Survey of Canada, Paper 89-1G, p. 85-99.

Thomas, L.O.
1943: Mineral possibilities of areas adjacent to the Alaska Highway, Part 1, Yukon section; Canadian Institute of Mining and Metallurgy Transactions, v. XLVI, p. 375-401.

Thompson, R.I.
1979: A structural interpretation across part of the northern Rocky Mountains, British Columbia, Canada; Canadian Journal of Earth Sciences, v. 16, p. 1228-1241.

Wall, J.H. and Rosene, R.K.
1977: Upper Cretaceous stratigraphy and micropaleontology of the Crowsnest Pass - Waterton area, southern Alberta foothills; Bulletin of Canadian Petroleum Geology, v. 25, p. 842-867.

Wallis, G.R. and Jordan, G.R.
1974: The stratigraphy and structure of the Lower Cretaceous Gething Formation of the Sukunka River coal deposit in British Columbia; Canadian Institute of Mining and Metallurgy Bulletin, v. 67, p. 142-147.

Whitaker, S.H.
1978: Regional geology; in Coal Resources of Southern Saskatchewan: A Model for Evaluation Methodology, S.H. Whitaker, J.A. Irvine, and P.L. Broughton (ed.); Geological Survey of Canada Economic Report 30, Saskatchewan Department of Mineral Resources Report 209, Saskatchewan Research Council Report 20, p. 21-27.

Whitaker, S.H., Irvine, J.A., and Broughton, P.L.
1978: Coal resources of southern Saskatchewan: A model for evaluation methodology; Geological Survey of Canada Economic Report 30, Saskatchewan Department of Mineral Resources Report 209, Saskatchewan Research Council Report 20, 156 p.

Williams, G.D. and Burk, C.F.
1964: Upper Cretaceous; Chapter 12 in Geological History of Western Canada, R.G. McCrossan and R.P. Glaister (ed.); Alberta Society of Petroleum Geologists, p. 169-189.

Williams, G.D. and Murphy, M.C.
1981: Deep coal resources of the Interior Plains, estimated from petroleum borehole data; Geological Survey of Canada, Paper 81-13, 15 p.

Yorath, C.J. and Cook, D.G.
1981: Cretaceous and Tertiary stratigraphy and paleogeography, northern Interior Plains, District of Mackenzie; Geological Survey of Canada, Memoir 398, 76 p.

Yorath, C.J., Balkwill, H.R., and Klassen, R.W.
1975: Franklin Bay and Malloch Hill map areas, District of Mackenzie; Geological Survey of Canada, Paper 74-36, 35 p.

Young, F.G.
1975: Upper Cretaceous stratigraphy, Yukon coastal plain and Northwestern Mackenzie Delta; Geological Survey of Canada, Bulletin 249, 83 p.

Yurko, J.R.
1976: Deep Cretaceous coal resources of the Alberta Plains; Alberta Research Council Report 75-4, 47 p.

Author's Address

A.R. Cameron (retired)
Institute of Sedimentary and Petroleum Geology
Geological Survey of Canada
3303 - 33 Street N.W.
Calgary, Alberta
T2L 2A7

Printed in Canada

Subchapter 6C

GEOTHERMAL ENERGY

A.M. Jessop

Geothermal sources in sedimentary basins are of moderate to low temperature, are associated with flat-lying or gently dipping sediments, extend over long distances, and take the shape of the layered host rocks. They contrast with geothermal fields on continental margins associated with subduction zones or of rift zones associated with crustal extension; these are the result of vigorous hydrothermal systems feeding on volcanic heat sources, are relatively localized, generally small, irregular in shape, and of diverse character.

Geothermal heat may be recoverable from almost any part of the sedimentary section because mobile water is present to some extent in all sediments with the exception of halite and other evaporite beds. However, the amount of water, its temperature, chemical character, pressure, and potential production rate may vary considerably from place to place, even in the same aquifer. Porosity and permeability of the rock are factors that affect the pressure drawdown and the lifetime of any production well. Thus, the economic geothermal potential must be individually evaluated at any proposed location.

Temperatures in sedimentary aquifers range from near zero to 150°C and possibly higher in very deep basins. The lower limit of utility is about 40°C, which corresponds to a depth of about 500 to 1500 m, depending on surface temperature and geothermal gradient. Geothermal gradient in sediments is often distorted by hydrological effects, and conductive heat is disturbed by convective heat transfer. Areas of relatively high surface elevation act as recharge zones, depressing subsurface temperature, and low areas act as discharge zones with an associated high temperature (Hitchon, 1984). The depth and temperature at which useful resources are to be found at any location depends on the intended use.

Exploration for geothermal energy in the Western Canada Basin has benefitted from the large amount of drilling by the petroleum industry. In fact, it is unlikely that geothermal exploration could be economically worthwhile alone, despite the enormous size of the resource. Well-logs provide abundant temperature data, but bottom-hole temperatures are taken at the time of maximum disturbance, and commonly the data are of doubtful quality. Temperature logs, intended to show anomalies such as zones of setting cement, also are run at times of maximum disturbance, and calibration of the thermometer system is often neglected. However, numerous bottom-hole temperature measurements may, by statistical treatment, give a general picture of temperature distribution and the major regional anomalies. A large file of such data has been used to show variations in temperature gradient and heat flow, both laterally and vertically, and very large anomalies have been found in western Canada (e.g., Majorowicz et al., 1985).

A well on the campus of the University of Regina, drilled for geothermal production but used only for research, has yielded good data on the detailed thermal state of the sediments and on the problems of temperature measurement (Jessop and Vigrass, 1989). The studies show that temperature gradient at any location depends on distribution of conductivity and on the flow regime of water through the sediments, particularly flow parallel to the temperature gradient, which is normally vertical and perpendicular to the strata. Vertical variation of heat flow provides a powerful tool for inferring small components of vertical water flow.

In the Western Canada Basin between latitudes of 49°N and 60°N, the total heat in water, above a threshold temperature of 60°C and for reinjection at 30°C, is estimated to be 16 x 10^{21}J. That resource is about 500 times greater than the energy in the Canadian oil reserves as of December, 1983. The figure may be significantly increased by considering the heat in the solid rock, which may be partially recovered by the circulating water, or by using heat pumps and thus employing lower threshold and reinjection temperatures. However, the presence of the energy does not imply that it is economically feasible to exploit more than a small portion of it.

The most promising areas for geothermal resources are the deep, western side of the basin, stretching from southern Alberta to northeastern British Columbia, and the northern flank of Williston Basin in Saskatchewan. Jones et al. (1985) have shown that above the Paleozoic surface, at shallower depths, water temperatures greater than 60°C will occur in areas west of the Calgary-Swan Hills-Grande Prairie line (Fig. 6C.1). Above the Precambrian surface, at greater depths, water temperatures greater than 60°C occur in almost the entire west half of Alberta (Fig. 6C.2). Significant high-temperature anomalies are found in British Columbia and in the Weyburn area of Saskatchewan, with small anomalies in the Hinton-Edson area of Alberta (Majorowicz et al., 1986; Jones et al., 1984).

Jessop, A.M.
1993: Geothermal energy; Subchapter 6C in Sedimentary Cover of the Craton in Canada, D.F. Stott and J.D. Aitken (ed.); Geological Survey of Canada, Geology of Canada, no. 5, p. 599-600 (also Geological Society of America, The Geology of North America, v. D-1).

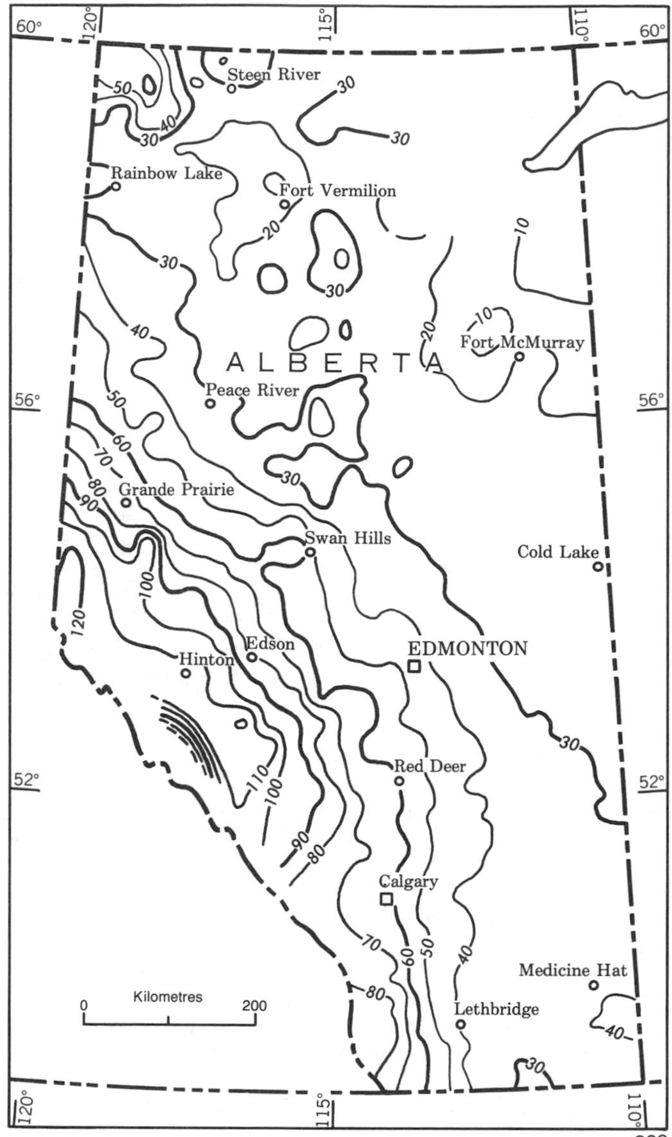

Figure 6C.1. Temperature (C°) distribution at the sub-Mesozoic unconformity in Alberta (from Jones et al., 1985). Contours are based on bottom-hole data from logs, corrected for disturbance caused by drilling and mud circulation.

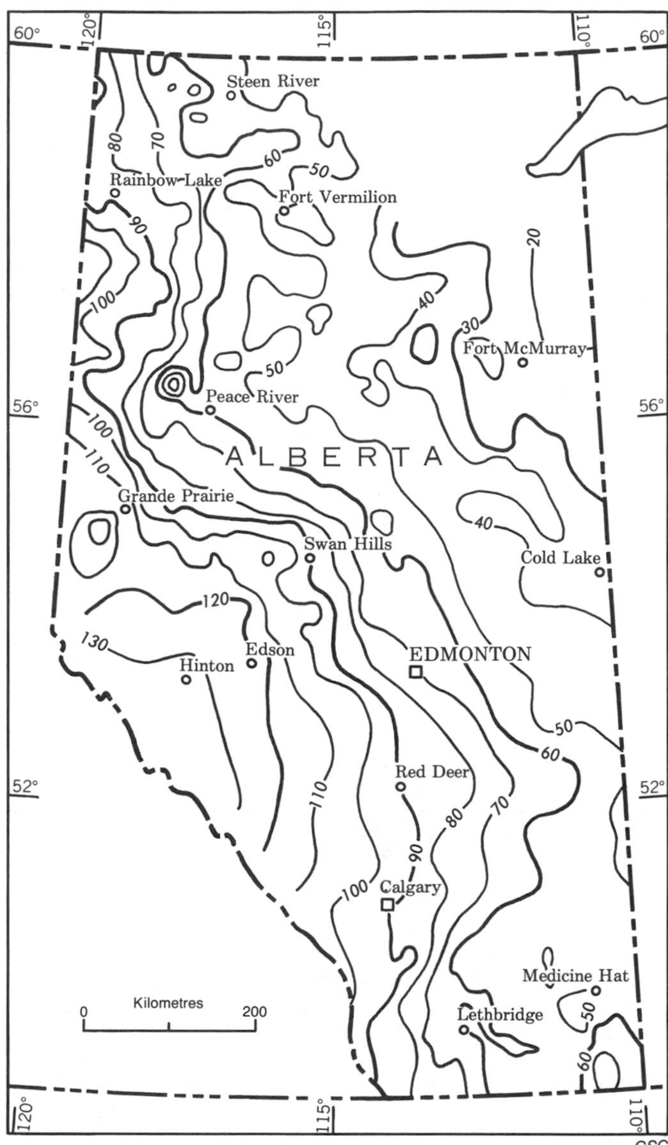

Figure 6C.2. Temperature (C°) distribution at the Precambrian basement surface in Alberta (from Jones et al., 1985). Contours are based on bottom-hole data from logs, corrected for disturbance caused by drilling and mud circulation.

REFERENCES

Hitchon, B.
1984: Geothermal gradients, hydrodynamics, and hydrocarbon occurrences, Alberta, Canada; American Association of Petroleum Geologists, Bulletin, v. 68, p. 713-743.

Jessop, A.M. and Vigrass, L.W.
1989: Geothermal measurements in a deep well at Regina, Saskatchewan; Journal of Volcanology and Geothermal Research, v. 37, p. 151-166.

Jones, F.W., Kushigbar, C., Lam H.-L., Majorowicz, J.A., and Rahman, M.
1984: Estimates of terrestrial thermal gradients and heat flow variations with depth in the Hinton - Edson area of the Alberta basin derived from petroleum bottom-hole temperature data; Geophysical Prospecting, v. 32, p. 1111-1130.

Jones, F.W., Lam, H.-L., and Majorowicz, J.A.
1985: Temperature distributions at the Paleozoic and Precambrian surfaces and their implications for geothermal energy recovery in Alberta; Canadian Journal of Earth Sciences, v. 22, p. 1774-1780.

Majorowicz, J.A., Jones, F.W., and Jessop, A.M.
1986: Geothermics of the Williston Basin in Canada in relation to hydrodynamics and hydrocarbon occurrences; Geophysics, v. 51, no. 3, p. 767-779.

Majorowicz, J.A., Jones, F.W., Lam H.-L., Linville, A., and Nguyen, C.D.
1985: Topography and subsurface temperature regime in the Western Canada Sedimentary Basin: implications for low-enthalphy geothermal energy recovery; Geothermics, v. 14, p. 175-187.

Author's Address

A.M. Jessop
Institute of Sedimentary and Petroleum Geology
Geological Survey of Canada
3303 - 33 Street N.W.
Calgary, Alberta
T2L 2A7

Printed in Canada

Subchapter 6D

INDUSTRIAL MINERALS OF THE WESTERN CANADA BASIN

P. Guliov

CONTENTS

INTRODUCTION

The Western Canada Basin is a storehouse of vast industrial mineral wealth. All of the Phanerozoic systems are represented in the basin and most of them have contributed or have the potential of contributing useful industrial mineral commodities. Producing deposits and significant occurrences are scattered throughout the more populous southern region of the basin (Fig. 6D.1, 6D.2). Commodities of low value such as sand and gravel, brick clay, and expandable clays tend to be relatively common and widespread. Development of these therefore tends to be concentrated around urban centres close to markets. Others, such as refractory clays, gypsum, cement raw materials, and lime are processed or used in manufacturing end products, whose added value permits transportation over relatively long distances. Some commodities, potash and sulphur for example, have gained global significance.

Production of industrial minerals in the Western Canada Basin accounts for about 65% of the total value of Canadian production, of which about 75% is contributed by sulphur and potash (Table 6D.1). Reviews of industrial minerals have been published recently for Manitoba (Bannatyne, 1984a; Davies et al., 1962; Fogwill, 1983), for Saskatchewan (Guliov, 1984; Saskatchewan Energy and Mines, 1983) and for Alberta (Hamilton, 1976, 1984). An earlier summary of the industrial minerals of the western provinces was published by Douglas (1970).

Acknowledgments

Much of the information incorporated in this section has been derived from Industrial Minerals in Canada, which contains a series of reviews prepared by provincial agencies. The author has drawn substantially on reports prepared by W.N. Hamilton, Alberta Geological Survey, on Alberta, by B. Bannatyne, Department of Mines, Resources and Environment Management, on Manitoba, and from his own report for Saskatchewan.

Guliov, P.
1993: Industrial minerals of Western Canada Basin; Subchapter 6D in Sedimentary Cover of the Craton in Canada, D.F. Stott and J.D. Aitken (ed.); Geological Survey of Canada, Geology of Canada, no. 5, p. 601-611 (also Geological Society of America, The Geology of North America, v. D-1).

Others who have contributed to the development of this subchapter include D. Edwards and D. Scafe of the Alberta Geological Survey, D. Watson and R. Gunter of the Manitoba Department of Mines, Resources and Environment Management, and J.E. Christopher of the Saskatchewan Department of Energy and Mines.

SAND, GRAVEL, AND OTHER AGGREGATE MATERIALS

Exploitable deposits of sand and gravel are widespread in the Western Canada Basin. Those of glacial origin are of prime importance. The most widespread and common are glacial outwash deposits occurring in numerous localities throughout the basin. Glaciolacustrine sands and gravels deposited as extensive beaches are important sources in southern Manitoba and parts of eastern Saskatchewan. Other significant sources include deltaic deposits, kame-eskerine complexes, and morainal deposits. Some contain sufficient proportions of coarse material to be suitable for crushed aggregate production. Some deposits, particularly those in parts of western Manitoba and eastern Saskatchewan, have a deleterious content of shale pebbles derived from Cretaceous shales in the uplands regions of Riding Mountain, Duck Mountain, Porcupine Hills, and Pasquia Hills.

Preglacial sand and gravel deposits are of local importance in Alberta and Saskatchewan. Deposits formed in preglacial bedrock channels are usually below economic depth and are far more important as sources of groundwater. A few, such as a large deposit near Edmonton, are significant sources of excellent quality aggregate. Several bedrock highs in southwestern Saskatchewan and southeastern Alberta are capped by semiconsolidated Tertiary quartzitic gravels, which locally are used directly or as crushed aggregate. These include Eocene-Oligocene gravels of the Swift Current and Cypress Hills areas, Miocene gravels of the Wood Mountain Formation in Saskatchewan, and Pliocene gravels of the Hand Hills and Wintering Hills of Alberta.

Many of the glacial-lake deposits throughout the plains area are important sources of clays suitable for lightweight aggregate manufacture. These are discussed briefly under Clay and Clay Products.

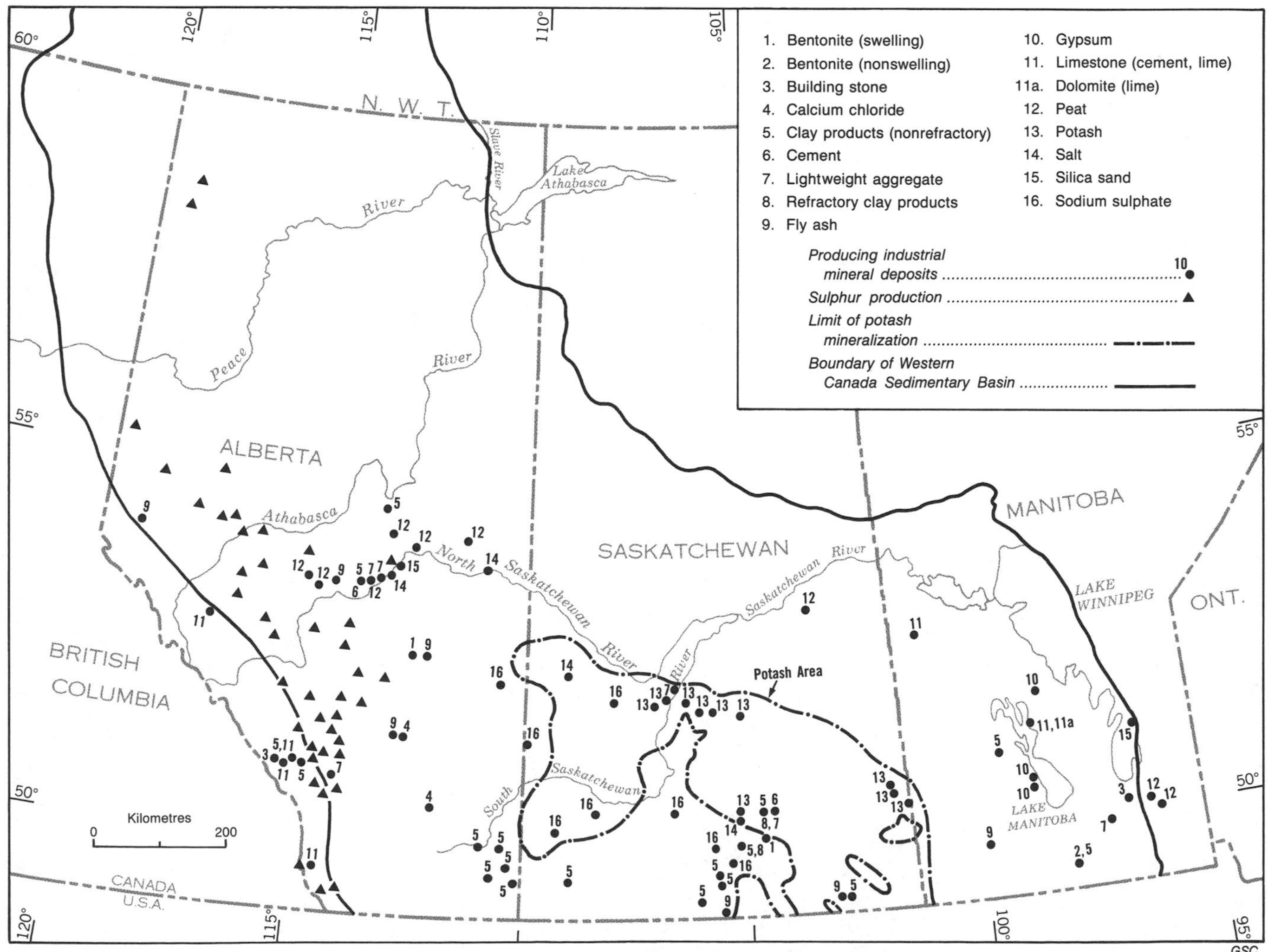

Figure 6D.1. Producing industrial mineral deposits.

LIMESTONE, DOLOMITE, LIME, AND CEMENT

The Middle Devonian Elm Point Formation and the Middle to Upper Devonian Souris River Formation of southeastern and west-central Manitoba respectively are sources of high-quality limestone for the lime and cement industries (Stonehouse, 1982). Production takes place near Faulkner and at Steep Rock (on Lake Manitoba) for a local lime plant and a portland cement plant near Winnipeg. The lime plant also utilizes high-purity Silurian dolomite of the Interlake Formation for a high-magnesia lime. Limestone of the Souris River Formation is quarried near Mafeking for cement production at Regina. The Devonian Dawson Bay Formation contains large reserves of limestone at or near the surface in Manitoba (Bannatyne, 1975).

Although the Souris River Formation extends westward into Saskatchewan, it subcrops beneath the Lower Cretaceous sandstone of the Mannville Formation and is not known to be at economic depth.

In the area of the south end of Pinehouse Lake, Saskatchewan, the Meadow Lake Formation of Middle Devonian age occurs about 70 m below surface. The formation occupies the same stratigraphic position as the Elm Point in Manitoba and contains about 10 m of high-calcium limestone apparently derived through de-dolomitization processes. Although not presently economic, the deposit remains of interest for future development.

Paleozoic formations along the Rocky Mountains of western Alberta contain large reserves of commercial and potentially commercial limestones (Holter, 1976). Economic deposits are restricted to Cambrian, Devonian, and Lower Carboniferous strata. The principal sources are the Devonian Palliser Formation and the Lower Carboniferous Livingstone and Mount Head formations. Quarries are operated at four localities in the Rocky Mountains and supply the raw material for a large cement and lime industry. Two quarries near Cadomin produce from the Palliser Formation, with grades of up to 96% $CaCO_3$. Quarries at Exshaw, in the Bow Valley, produce

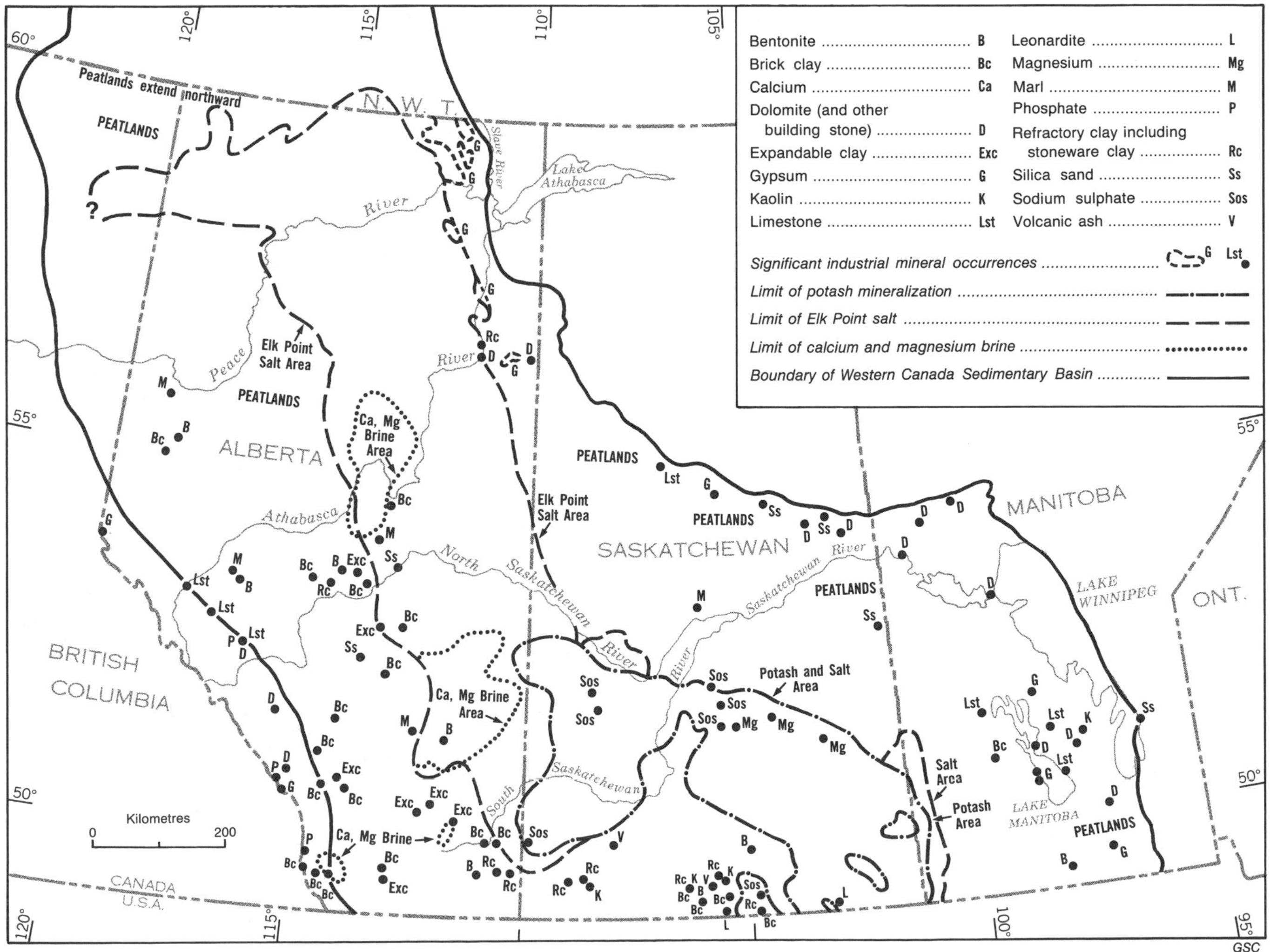

Figure 6D.2. Significant industrial mineral occurrences.

from the Palliser and Livingstone formations (Fig. 6D.3). The Cambrian Eldon Formation was a former source of commercial limestone in the same area. In the Crowsnest Pass, the Livingstone and Mount Head formations are quarried for local lime production. A small amount of dolomite is also produced from the Devonian Fairholme Group in the area.

Limestones in the Upper Devonian Waterways Formation are exposed near the Precambrian Canadian Shield margin and along the Athabasca and Clearwater river valleys in northeastern Alberta. Despite the large reserves and good grade of these deposits, development appears to be precluded by the vast resources of limestones in the Cordilleran region marginal to the Western Canada Basin and near large markets.

Surface and near-surface Paleozoic dolomites are widespread in south-central and west-central Manitoba, in the lowlands of east-central Saskatchewan, and in northeastern Alberta. Except for a limited production of high-magnesia lime from Silurian dolomites in Manitoba the use and potential use of these materials appears to lie in the building-stone category. They are, therefore, dealt with in the next section.

Calcareous shale of the Boyne Member of the Cretaceous Vermilion River Formation is similar in composition to typical cement rock. Sample analyses from the Babcock area of Manitoba and Pasquia Hills in Saskatchewan indicate substantial thicknesses with a CaO content of 30 to 40% and MgO less than 1%. This shale was used in Manitoba for natural cement production between 1898 and 1924.

Table 6D.1. Industrial Mineral Production and Value of Sales 1987

	Production tonnes (x 1000)	Values of Sales ($ x 1000)
Cement	1 524+	196 719+
Clay products	N/A	22 654
Gypsum	380	6 060
Lime*	N/A	25 856
Peat Moss	164	27 366
Potash (K_2o)	6 449	670 739
Quartz (Silica sand)	N/A	7 039+
Salt	1 914	35 880
Sand and gravel	70 959	213 908
Sodium sulphate	258+	24 756
Stone (all types)	4 020	17 434
Sulphur*	6 584	604 999

\+ Indicates additional production or sales for which data are not available.

* Includes Foothills and Rocky Mountains production in Alberta.

N/A Data not available.

BUILDING STONE

The building stone industry is concentrated in the eastern part of the basin, in Manitoba (Prud'homme, 1982), with minor production in Alberta. The widely known Tyndall stone, a tapestry-mottled dolomitic limestone from the Ordovician Red River Formation, is quarried at Garson, Manitoba. This stone is prized for the decorative value of its distinct mottled appearance and the contained fossils. The major products are cut slabs and random ashlar blocks.

Numerous exposures of Ordovician Red River and Stony Mountain dolomites and less commonly, Silurian dolomites of the Interlake Formation, occur in the Cumberland-Deschambault-Amisk lowlands of east-central Saskatchewan (Kupsch, 1952) and throughout the Interlake region of Manitoba. The most accessible are the Red River dolomites along Saskatchewan's Highway 106 and Highways 6 and 391 in Manitoba. The stone varies from grey to brown to mottled and various combinations of grey, brown, red, and yellow. Preliminary evaluation of Ordovician dolomite in Saskatchewan indicates the stone has good potential as a building material. Unlike Tyndall stone many of these dolomites take a good polish and would be classified as marbles by the stone industry.

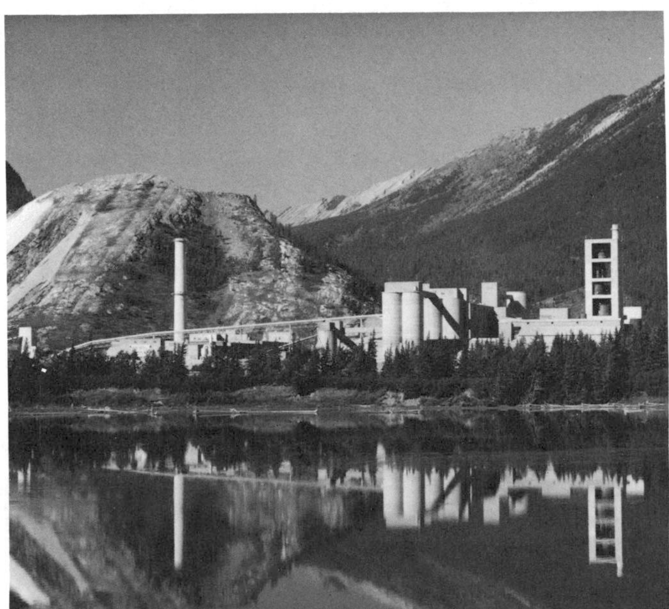

Figure 6D.3. Cement plant at Exshaw, Alberta, on the Bow River. The quarry face to the left of the rocky knoll behind the plant is formed by bedding surfaces of the Devonian Palliser Formation. The dip slope on the ridge to the right is formed by the Etherington and Mount Head formations of the Carboniferous Rundle Group. The Exshaw Thrust lies in the valley of Exshaw Creek behind the plant (photo - Canada Cement Lafarge Ltd.)

In Alberta, one of the main building stones is a fine-grained, medium-grey, flaggy siltstone from the Triassic Spray River Formation, quarried in the Bow River Valley near Canmore. The material is marketed for use as a rough building and decorative stone. Another building stone, the Tertiary Paskapoo sandstone, was used in many of the original buildings in Calgary and Edmonton.

Dolomites of the Middle Devonian Methy Formation occur marginal to the Canadian Shield in northeastern Alberta. These are too remote for consideration as building materials in the forseeable future.

GYPSUM

Major gypsum deposits are exploited only in the eastern part of the Western Canada Basin, in southern Manitoba. The region supplies much of the market in the prairie provinces. Gypsum is mined from open pits in the Jurassic Amaranth Formation 16 km north of Amaranth. Deposits are about 6 m thick and lie beneath 6 m of glacial overburden. At Gypsumville, 120 km to the northeast, Jurassic Amaranth deposits have been preserved within the 25 km diameter Lake St. Martin impact crater. There the gypsum is mined by open pit method from isolated ridges rising above swampy country. Anhydrite and glauberite are found with the gypsum at a depth of 33 to 38 m near the base of the deposit. Production at Silver Plains, about 50 km south of Winnipeg, also from the Jurassic beds, took place between 1964 and 1975. The best grades in the Jurassic rocks of Manitoba range from 90 to 95% gypsum (Bannatyne, 1984b).

Other significant deposits within the Western Canada Basin are in northeastern Alberta. The Middle Devonian Prairie Evaporite subcrops in the region and is exposed along some of the major river valleys. Salt beds of the formation were leached out, leaving thick anhydrite strata, which were subsequently altered to gypsum. A deposit underlying the Clearwater River valley in Alberta ranges from 9 to 15 m thick at depths ranging from near-surface to 90 m for a distance of nearly 30 km (Hamilton, 1976). The average grade is 84% gypsum. A similar deposit is postulated to underlie the Athabasca River valley 96 km north of Fort McMurray. Extensive Devonian gypsum deposits are present in Wood Buffalo National Park, northern Alberta, where the Cretaceous overburden is absent. About 25 m of solid gypsum is exposed at Gypsum Cliffs along a 22 km section of the Peace River. This deposit is probably among the largest and highest grade in the world.

Two significant gypsum deposits occur in the Cordilleran region of western Alberta. The most accessible is a 24-m thick bed within the Upper Devonian Fairholme Group exposed on Mount Invincible in the Kananaskis Lakes area, 128 km southwest of Calgary. Grab samples from the steeply dipping beds assayed 90 to 92% gypsum. Detailed investigation and development of the deposit are precluded by the high recreational value of the region. The other deposit, of which two exposures have been reported (Govett, 1961), lies in the Triassic Whitehorse Formation on the Continental Divide, directly west of Edmonton and within 65 km of rail. The deposit appears to be large, but the grade and extent are difficult to assess owing to structural complexities.

Large deposits of gypsum in the Middle Devonian Burnais Formation are being mined in quarries near Windermere, British Columbia, on the western side of the Rocky Mountains.

PHOSPHATE

Phosphate-bearing zones are known to occur in the basal Lower Carboniferous Exshaw Formation, in the upper part of the Permian and Upper Carboniferous Rocky Mountain Group, and in the basal and middle parts of the Jurassic Fernie Formation in the Rocky Mountain region of Alberta. The zones are low grade, discontinuous, thin, and generally steeply dipping. Although not presently economic, the phosphate beds represent future potential if sufficient changes in the economics of traditional sources occur (Hamilton, 1976).

CLAY AND CLAY PRODUCTS

Clay and shale are widespread throughout much of the Western Canada Basin. Most accessible deposits are Mesozoic or younger. The most useful materials are found in nonmarine bedrock formations.

The most important source of the better grades of clay in Canada is the Upper Cretaceous Whitemud Formation (Worcester, 1950). The formation extends from the Cypress Hills area of southeastern Alberta eastward at least as far as Big Muddy Valley, Saskatchewan, and is confined to within about 100 km of the United States border. The formation is exposed in many localities around the margins of uplands or in valley walls capped by overlying Tertiary sediments. More than one period of post-depositional erosion has complicated the distribution pattern, and the formation is now present as a dissected, discontinuous bed and a few scattered erosional outliers.

The Whitemud Formation varies in thickness but locally may reach about 30 m. It is subdivided broadly into a lower kaolinitic sand member and an upper plastic clay member. The lower member contains up to 50 or 60% kaolin in south-central Saskatchewan. The kaolin is being investigated for use as a filler for paper production. Successful pilot-scale testing in Regina may lead to the construction of a full-scale (160 000 tonne/year) operation near Wood Mountain, Saskatchewan, by the early 1990s.

The upper unit contains the important clays, including white and light-burning ball clay, stoneware clay, and fireclay. Products derived from these clays include refractory, common, and face brick, including white decorative brick, functional and decorative pottery and other ceramics, sewer pipe, and vitrified flue linings. The clays are believed to have been derived from *in situ* weathering of a feldspathic sand. It is likely that reworking, redeposition, and concentration of kaolinite took place prior to the end of Whitemud (Late Cretaceous) time. The clays of the Whitemud Formation tend to increase in refractoriness eastward from the Cypress Hills area. Refractory brick is produced from Whitemud clays in the Claybank area of southern Saskatchewan.

The Paleocene Ravenscrag Formation of southern Saskatchewan is also a source of plastic kaolinitic clays (Guliov, 1984). Some production is from the Willow Bunch Member occurring high on the slopes of the dissected uplands south of the Big Muddy valley. These clays are, on the average, much less refractory than Whitemud clays, although some are of low duty refractoriness. Most can be used for earthenware and for stoneware products in the lower temperature range.

Clays of stoneware and fireclay quality underlie the Athabasca oil sands in certain areas north of Fort McMurray.

Bloating clays suitable for the production of lightweight aggregate (expanded clay) are widespread throughout the Western Canada Basin. In Manitoba and Saskatchewan, montmorillonite-rich, glacio-lacustrine clays are presently used for lightweight aggregate production near Winnipeg, Regina, and Saskatoon. In Alberta, marine shales of the Upper Cretaceous Bearpaw Formation and nonmarine shales from the Cretaceous Judith River and Tertiary Paskapoo formations are suitable for lightweight aggregate (Hamilton, 1976). In Saskatchewan, glacial lake clays near Regina, Saskatoon, and Unity are suitable for lightweight aggregate. Production normally takes place near the larger urban centres, as transportation is a large factor in the selling price.

Structural clay products, particularly common and face brick, have been made in the region from a variety of clays, including those of the Jurassic and Cretaceous beds in Manitoba (Bannatyne, 1984b), Tertiary beds in Saskatchewan and Alberta, and Cretaceous and Quaternary beds throughout the region.

BENTONITE

Bentonite is a common constituent in Upper Cretaceous sedimentary rocks of the Western Canada Basin.

Sodium or swelling bentonite is associated commonly but not exclusively with the marine shales of the Bearpaw Formation in the south-central portion of the basin (Alberta and Saskatchewan) and the mainly nonmarine Horseshoe Canyon Formation in central Alberta (Scafe, 1975). Swelling bentonites also occur in the Paleocene Ravenscrag Formation in southern Saskatchewan. Despite the numerous occurrences, economic deposits of sodium bentonite are not common. Only two mines currently are operating, one in the Battle River Valley, near Rosalind, Alberta (Horseshoe Canyon Formation), and another near Truax, Saskatchewan (Bearpaw Formation). The bentonite is marketed for use in foundries, for pelletizing, for sealing reservoirs, and as drilling mud. The quality and yield of the product is comparable to the well known Wyoming bentonite.

Calcium or nonswelling bentonite is a relatively rare commodity in the region. The only production in Canada is from the marine Upper Cretaceous Pembina Member of the Vermilion River Formation near Morden, Manitoba. The bentonite occurs in six main seams with a combined recoverable thickness of 75 cm, separated by organic black shale layers totalling 25 cm (Bannatyne, 1984b). The material is used as a decolourizing and absorbing agent for mineral and vegetable oils, as cat litter, and as a binder for feed pellets and foundry sand. Other calcium bentonite occurrences are known in the Battle and Ravenscrag formations of southern Saskatchewan and in the Bearpaw and Horseshoe Canyon formations of south-central Alberta.

SILICA

The principal sources of silica sand within the Western Canada Basin are Ordovician, Lower Cretaceous, and post-glacial Quaternary deposits (Boyd, 1982b).

One of North America's purest silica sands is located on Black Island in Lake Winnipeg. The sand is produced from a bed of loosely consolidated quartz sandstone 10 to 15 m thick at the base of the Ordovician Winnipeg Formation. It consists of well rounded and frosted quartz grains with 95 to 98% SiO_2, less than 1% heavy minerals, and minor limonite and kaolin (Bannatyne, 1984a). Annual production of washed sand of -20 to +100 mesh is about 100 000 tonnes and grades up to 99.7% SiO2. About half of the production is shipped to Redcliff, Alberta, for use in glass products. The remainder is used as foundry and filter bed sand, for sandblasting, and many other uses. Similar but thinner deposits in the same formation occur along the northeastern edge of the basin in east-central Saskatchewan. The accessible deposits there are capped by up to 3 m of dolomite of the Ordovician Red River Formation.

Lower Cretaceous silica sands are widespread and lie at or near surface in many localities fringing the northern edge of the basin in the Swan River or equivalent Mannville formations in Manitoba and Saskatchewan and in the Shaftesbury and McMurray formations of northwestern and northeastern Alberta, respectively. Potential sources of Lower Cretaceous high-silica material include the Swan River area in Manitoba, Red Deer River and Wapawekka Lake-Nipekamew River, Saskatchewan (Babey, 1955; Langford, 1973), and Fort McMurray and Peace River, Alberta. The Red Deer River deposits have been at the development stage for several years.

The Fort McMurray sands are a by-product of oil-sand mining and processing and are produced in quantities far exceeding western Canadian requirements; fines would have to be eliminated for glass production. Silica grades in these and other Lower Cretaceous sands generally range from about 90 to 98%.

Figure 6D.4. Sulphur plant at Elkton, Alberta. The plant recovers sulphur from sour gas produced from Carboniferous beds (photo - B.C. Rutley, GSC 3015-15.)

Post-glacial silica sands, deposited as surficial dunes, are scattered over many areas in the western region of the basin. They are best known in Alberta, where many have been investigated and one is exploited. As silica sands these deposits are of a low grade and would probably be more aptly termed "industrial sands". A large deposit near Bruderheim, 35 km northeast of Edmonton, Alberta is easily upgraded for use in lower grade glass products. The sand is now used mainly in the production of glass fibre insulation. One deposit of post-glacial sand near Beausejour, Manitoba, was mined and used as feed for a local glass plant until 1913. Tailings produced from a heavy-mineral recovery operation in glacial outwash north of Edmonton, Alberta, are being developed as a source of silica.

SULPHUR

The Alberta portion of the Western Canada Basin ranks as a major world producer of sulphur (Boyd, 1982a). Practically all of the sulphur is a by-product of sour natural-gas production (Fig. 6D.4). The hydrogen sulphide concentration in Western Canadian sour gases is generally between 3 and 20%, but in some gas fields it exceeds 50%. Sulphur production in 1987 totalled about 6.6 million tonnes, of which a small proportion results from the processing of the Athabasca tar sands in northeastern Alberta. Proved recoverable reserves of sulphur are about 283 million tonnes. A small amount of by-product sulphur is produced at the heavy-oil upgrader in Regina, Saskatchewan.

POTASH

Middle Devonian evaporite beds underlie a large part of the Western Canada Basin. They were deposited within the Elk Point Basin, which extended from northwestern Alberta through southern Saskatchewan, into southwestern Manitoba, North Dakota, and Montana. An early stage of evaporite deposition is represented by the thick and extensive halite beds of the lower Elk Point Group. These are confined almost entirely to Alberta. Later, during late Elk Point time, evaporitic conditions became far more widespread, and the thick salt sequence known as the Prairie Formation was deposited throughout much of the basin. The upper part of the Elk Point Group is more than 200 m thick in places and occurs at depths ranging from 400 m to 2750 m. In Saskatchewan and an adjacent part of Manitoba, the upper 60 m of the Prairie Formation includes potash-bearing beds in four members, in ascending order, the Esterhazy, White Bear, Belle Plaine, and Patience Lake members (A. Fuzesy, 1982; L.M. Fuzesy, 1983; Barry, 1982a). The members are commonly 6 to 15 m thick, are separated by 1 to 45 m of halite and contain one or more individual potash beds as much as 7 m thick. The members and interbeds generally become thinner toward the southwest.

Figure 6D.5. Continuous mining machine at work in a potash mine, Saskatchewan (photo - Potash Corporation of Saskatchewan.)

A small area of potash mineralization occurs in eastern Alberta, between Provost and the Battle River.

Potash in the Esterhazy, Belle Plaine, and Patience Lake members is mined in Saskatchewan. Nine mines are of the conventional underground type (Fig. 6D.5) and one produces by a solution technique. Ore grade averages between 21 and 27% K_2O and in some areas exceeds 30%. Ore usually consists of halite, sylvite, and insoluble minerals, but in some areas includes carnallite. Each member is carnallitic along its margins and sylvinitic in the central areas.

Manitoba potash deposits are at the development stage and mining is scheduled to commence near Russell in the 1990s.

Reserve estimates vary from 4.5 to 14.0 billion tonnes K_2O equivalent recoverable by conventional mining and from 42 to 62.6 billion tonnes recoverable by solution mining. Regardless of the variance in published reserve figures, it is well demonstrated that the potash resources in the Western Canada Basin are unquestionably among the richest and largest known in the world.

Potash production peaked in 1980 and 1984 at 7.3 and 7.6 million tonnes K_2O respectively, and bottomed in 1982 and 1986 at 5.2 and 6.0 million tonnes K_2O respectively. The industry appears to be recovering from its most recent slump, and production is expected to equal or surpass previous peak levels by 1989.

SODIUM SULPHATE

Natural sodium sulphate deposits have formed in numerous alkaline lakes with restricted drainage in the south-central region of the Western Canada Basin (Barry, 1982c; Last and Schweyen, 1983; Tomkins, 1954). The region has the dry climate and seasonally low temperatures required for concentration of brine and precipitation of sodium sulphate (Fig. 6D.6). The lake brines are primarily of the Na-Mg-SO_4 type, which give rise to mirabilite ($Na_2SO_4 \cdot 10H_2O$) and, to a lesser extent, thenardite (Na_2SO_4), bloedite ($Na_2 \cdot Mg(SO_4)_2 \cdot 4H_2O$), and epsomite ($Mg SO_4 \cdot 7H_2O$). Solutes derived from leaching of the surrounding terrain and from inflowing brackish or brine springs undergo evaporative concentration during warm, dry summers. Where solute concentration is sufficient the salts precipitate with declining ambient temperature in the autumn. Some lakes remain as brine while others precipitate a crystal deposit, which may be temporary or may add to a permanent crystal bed. Typical thicknesses of permanent crystal beds among the alkaline lake deposits of western Canada are commonly 1 to 5 m. Maximum thicknesses in some lakes may be in excess of 30 m. The crystal beds are never pure but are usually admixed or interlayered with material such as clay, silt, sand, and organic material.

In Alberta, the Metiskow Lake deposit contains about 2.7 million tonnes of anhydrous sodium sulphate and is the only one in the province presently being exploited. In Saskatchewan there are 21 deposits, in each of which original reserves exceed 500 000 tonnes. Most are in excess of 1 million tonnes and one-third of them are in excess of 3 million tonnes. A total original reserve of 51 million tonnes is attributed to the 21 deposits. There are 11 other deposits in Saskatchewan with resources ranging from 100 000 to 500 000 tonnes each.

Five companies, four in Saskatchewan and one in Alberta, produce about 500 000 tonnes of anhydrous sodium sulphate annually from seven plants. Most of the saltcake product is used in kraft-paper production. A small amount is purified to 99.7% Na_2SO_4 for use in the detergent industry. Experimentation is now in progress in Saskatchewan to combine sodium sulphate with potash to produce an economically viable potassium sulphate fertilizer product for chloride-sensitive crops.

SALT

Much of the Middle Devonian of the Western Canada Basin is characterized by evaporites, including several halite deposits. The most widespread and economically most significant of these belong to the Elk Point Group (Barry, 1982b).

Figure 6D.6. Sulphate plant, Chaplin, Saskatchewan. Evaporation reservoirs lie adjacent to the plant, with the alkali lake lying beyond (photo - P. Guliov.)

Salt of the lower Elk Point is very thick and extensive, and is confined largely to Alberta. The remarkable purity of the halite is probably the result of a complex history of solution and recrystallization.

The upper Elk Point salt is the well known Prairie Formation renowned for its rich potash ore in the southeastern part of the basin. The salt beds are extremely extensive, continuing in a broad belt from the Peace River area in northwestern Alberta to the southwestern corner of Manitoba. The salt lies at depths ranging from about 200 m in northern Alberta to 2750 m in southeastern Saskatchewan. In the northern part of the basin, about 200 km northeast of Edmonton, the cumulative thickness of lower and upper Elk Point salt reaches 425 m. In Saskatchewan and a small part of southwestern Manitoba the Prairie Formation is commonly about 150 m thick but in some areas exceeds 200 m. As described earlier the upper 60 m of the Prairie Evaporite in Saskatchewan is potash-bearing over most of its areal extent.

Although an enormous quantity of salt is brought to surface during potash mining operations in Saskatchewan most of it is destined for the spoil piles, together with other impurities. Six operations across the plains area can be considered as producers of salt. Five of these, two in Alberta and three in Saskatchewan, are brining operations, producing salt for the chemical industry, domestic and agricultural use, water softening, and ice-control. One salt plant, near Belle Plaine, Saskatchewan, uses waste brine from a nearby solution potash mining operation as its feedstock. The sixth producer is a potash mine supplying some by-product salt for control of road ice. Total salt production in the plains area of western Canada in 1987 was about 2 million tonnes.

Thick Middle Devonian salt deposits are also used indirectly at many locations. Large caverns are formed by controlled solution of halite and used as underground storage for petroleum products, especially natural gas.

CALCIUM CHLORIDE

Formation waters of deep Devonian reservoirs in several areas of Alberta contain high concentrations of calcium and magnesium. In some waters calcium exceeds 60 000 mg/l and magnesium exceeds 9 000 mg/l (Hitchon and Holter, 1971). Extraction of calcium chloride from the brines takes place at Drumheller, Brooks, and two localities near Slave Lake. The brines are also enriched in bromine, iodine, and other metallic chlorides. Calcium chloride-rich brine is also produced in Saskatchewan by the PCS Cory potash mine near Saskatoon. The brine seeps into the underground mine workings from overlying Devonian carbonates and is collected and pumped to the surface. The calcium chloride content is about 28%.

FLY ASH

Coal-fired power plants are the principal source of fly ash in the region. There are eight such plants in total, three in Saskatchewan, four in Alberta, and two in Manitoba. All except the Manitoba plants, which use Saskatchewan lignite, are mine-mouth coal consumers. The plants produce a large amount of fly ash, and only a very small

Figure 6D.7. Aerial photograph of the Julius peat bog in southeastern Manitoba. Scrubby trees on better-drained ground outline the bog. The Canadian Pacific Railway crosses the centre of the bog, providing convenient transportation of the peat produced (National Air Photo Library A21992-156.)

proportion of this is utilized by industry. Its most significant application, to date, is as a pozzolanic additive to concrete. It has also been used successfully on an experimental basis as filler in asphalt road base. Other potential uses include mineral wool and structural products. Very fine, 20 to 200 micron-size, vitrified and gas-filled particles of fly ash, known as cenospheres, have a potential use as a coolant in steel mills and as a replacement for some feldspars.

A small tonnage of bottom and fly ash accumulates as a by-product of the oil sands extraction process in Alberta. The ash results from the combustion of by-product bitumen coke in the thermal coking of tar. The ash contains up to 3.5% vanadium and 1.2% nickel.

PUMICITE

Pumicite deposits are common in the southern and central regions of the Western Canada Basin (Crawford, 1955). They range in age from Late Cretaceous to Quaternary. The pumicite is commonly partially weathered to bentonite, although a few deposits exhibit little evidence of alteration. Thicknesses range from a few centimetres to over 6 m. The most significant deposits are in southeastern Alberta and southern Saskatchewan.

In Alberta, a bentonite bed in the Upper Cretaceous Bearpaw Formation of the Cypress Hills area is 1.5 to 3 m thick and passes laterally into volcanic ash and ashy bentonite.

Deposits in Saskatchewan are better known and perhaps economically more promising. The three largest are the Rockglen and Duncairn deposits, both of Eocene-Oligocene age, and the St. Victor deposit of the Paleocene Ravenscrag Formation. The Duncairn and Rockglen pumicites are the finest grained (93-99% in the -200 mesh size), whereas the St. Victor material is slightly coarser. Chemical analyses indicate they are derived from acid or intermediate magmas.

Potential uses of the major pumicite deposits include lightweight structural products, ceramic glazes, enamels, mineral wool, pozzolans, and abrasive cleaners.

MAGNESIUM

The Middle Devonian Prairie Formation within the potash belt contains numerous carnallite-rich bodies. These occur largely between the South Saskatchewan River and the Assiniboine River. The richest are associated with the Belle Plaine and Patience Lake members (Fuzesy, 1983). In the Quill Lakes region of Saskatchewan, for example, the Belle Plaine Member contains extensive carnallite bodies averaging 14 m in thickness and up to 90% carnallite. The magnesium content of carnallite ($KCl \cdot MgCl_2 \cdot 6H_2O$) is 8.75 per cent.

Of possible future interest as a magnesium source are the Devonian formation brines of central and southern Alberta. The magnesium content is more than 9000 mg/l, several times that of sea water.

Within the Rocky Mountains south of Mount Assiniboine, magnesite was quarried until 1991 from the Middle Cambrian Cathedral Formation. Ore was supplied to a magnesium plant at High River, Alberta. Magnesium is used in aluminum alloys, castings and wrought products.

PEAT

Extensive peatlands in the Western Canada Basin occur in a broad region bordering the southern margin of the Canadian Shield. The peatland belt extends from southeastern Manitoba through central Saskatchewan, throughout much of northern Alberta, and into northeastern British Columbia and the western Northwest Territories. The region contains more than 20 million hectares of peatlands.

Resources in the peatlands of Manitoba and Saskatchewan range from dominantly horticultural peat to dominantly fuel peat. There appears to be an increase in the frequency of fuel peat occurrences westward. This is apparent in Saskatchewan, where humified sedge peats of fuel grade dominate in western regions (Troyer, 1985). Peatlands in Alberta, northeastern British Columbia, and the western Northwest Territories are poorly known.

Peatland development, excluding agricultural land use, is restricted to horticultural peat production in Manitoba, Saskatchewan, and Alberta (Fig. 6D.7). Production in 1987 was 60 000 tonnes in Manitoba, 74 000 in Alberta, and 21 000 in Saskatchewan. Most of the peat is of the sphagnum type and is marketed mainly in the United States.

Recent work in west-central Saskatchewan has demonstrated the technical feasibility of extruded fuel-peat sod production. Large-scale development of fuel peatlands (by a power generating facility, for example) could provide cheap raw material for the production of a variety of solid domestic and industrial fuels and other carbon products (Saskmont Engineering, 1984).

REFERENCES

Babey, W.J.
1955: The Red Deer River Silica Sand Deposit of east-central Saskatchewan; Saskatchewan Department of Mineral Resources, Report of Investigation, no. 7, 29 p.

Bannatyne, B.B.
1975: High-calcium limestone deposits of Manitoba; Manitoba Department of Mines, Resources and Environmental Management, Mineral Resources Division Publication no. 75-1, 103 p.
1984a: Manitoba; in Industrial Minerals in Canada – a review of recent developments; Industrial Minerals, May, 1984, p. 89-93.
1984b: Gypsum in Canada; in The Geology of Industrial Minerals in Canada; Canadian Institute of Mining and Metallurgy, Special v. 29, p. 163-166.

Barry, G.S.
1982a: Potash; in Review, Canadian Mineral Industry, 1982; Canada Department of Energy, Mines and Resources, p. 31.1-31.12
1982b: Salt; in Review, Canadian Mineral Industry, 1982; Canada Department of Energy, Mines and Resources, p. 35.1-35.7.
1982c: Sodium sulphate; in Review, Canadian Mineral Industry, 1982; Canada Department of Energy, Mines and Resources, p. 39.1-39.5.

Beck, L.S.
1974: Geological investigations in the Pasquia Hills area; Saskatchewan Department of Mineral Resources, Report 158, 16 p.

Boyd, B.W.
1982a: Sulphur; in Review, Canadian Mineral Industry, 1982; Canada Department of Energy, Mines and Resources, p. 42.1-42.9.
1982b: Silica; in Review, Canadian Mineral Industry, 1982; Canada Department of Energy, Mines and Resources, p. 37.1-37.4.

Crawford, G.S.
1955: Pumicite in Saskatchewan; Saskatchewan Department of Mineral Resources, Report 16, 35 p.

Davies, J.F., Bannatyne, B.B., Barry, G.S., and McCabe, H.R.
1962: Geology and Mineral Resources of Manitoba; Manitoba Department of Mines and Natural Resources, 190 p.

Douglas, R.J.W. (ed.)
1970: Geology and Economic Minerals of Canada; Geological Survey of Canada, Economic Geology Report no. 1, 838 p.

Fogwill, W.D.
1983: Mining in Manitoba - Past and present; in Canadian Institute of Mining and Metallurgy Directory, v. 17, p. 75-91.

Fuzesy, A.
1982: Potash in Saskatchewan; Saskatchewan Energy and Mines Report 181, 44 p.

Fuzesy, L.M.
1980: Geology of the Deadwood (Cambrian), Meadow Lake and Winnipegosis (Devonian) formations in west-central Saskatchewan; Saskatchewan Energy and Mines Report 210, 64 p.

Fuzesy, L.M. (cont.)
1983: Petrology of the Middle Devonian Prairie evaporite potash ore in Saskatchewan; in Summary of Investigations 1983; Saskatchewan Geological Survey, Saskatchewan Department of Energy and Mines, Miscellaneous report 83-4, p. 138-143.

Govett, G.J.S.
1954: Occurrence and stratigraphy of some gypsum and anhydrite deposits in Alberta; Alberta Research Council Bulletin 7, 62 p.

Guliov, P.
1984: Saskatchewan; in Industrial Minerals in Canada - A Review of Recent Developments; Industrial Minerals, May, 1984, p. 93-105.

Hamilton, W.N.
1976: Industrial Minerals: Alberta's uncelebrated endowment; in Eleventh Industrial Minerals Forum; Montana Bureau of Mines and Geology, Special Publication 74, p. 17-35.
1984: Alberta; in Industrial Minerals in Canada - A Review of Recent Developments; Industrial Minerals, May, 1984, p. 105-111.

Hitchon, B. and Holter, M.E.
1971: Calcium and magnesium in Alberta brines; Alberta Research Council, Economic Geology Report 1, 39 p.

Holter, M.E.
1976: Limestone resources in Alberta; in Eleventh Industrial Minerals Forum; Montana Bureau of Mines and Geology, Special Publication 74, p. 37-50.

Kupsch, W.O.
1952: Ordovician and Silurian stratigraphy of east-central Saskatchewan; Saskatchewan Department of Mineral Resources Report 10, 62 p.

Langford, F.F.
1973: The geology of the Wapawekka area, Saskatchewan; Saskatchewan Department of Mineral Resources Report 147, 36 p.

Last, W.M. and Schweyen, T.N.
1983: Sedimentology and geochemistry of saline lakes of the Great Plains; in Hydrobiologia no. 105, p. 245-262.

Prud'homme, M.
1982: Stone; in Review, Canadian Mineral Industry, 1982; Canada Department of Energy, Mines and Resources, p. 41.1-41.13.

Saskatchewan Energy and Mines
1983: Mineral Statistics Yearbook, 1982, Miscellaneous Report 83-3, 111 p.

Saskmont Engineering Company Ltd.
1984: Buffalo Narrows fuel peat demonstration project; Saskatchewan Energy and Mines, Open File Report 84-22.

Scafe, D.W.
1975: Alberta Bentonites; Alberta Research Council, Economic Geology Report 2, 19 p.

Stonehouse, D.H.
1982: Lime; in Review, Canadian Mineral Industry, 1981; Canada Department of Energy, Mines and Resources, p. 25.1-25.5.

Tomkins, R.V.
1954: Natural sodium sulphate in Saskatchewan; Saskatchewan Department of Mineral Resources, Report 6, 71 p.

Troyer, R.
1985: Peat resources in Saskatchewan; Saskatchewan Energy and Mines Report 218, 74 p.

Worcester, W.G.
1950: Clay resources of Saskatchewan; Saskatchewan Department of Mineral Resources Report 7, 198 p.

ADDENDUM

SILICA SAND:

The Black Island silica sand in Manitoba is undergoing pilot testing at Selkirk for the production of silicon metal. Favourable results may be beneficial to the local silica market. One of the traditional markets for the sand was for glass container production at Redcliffe, Alberta. The glass plant closed down in 1989. Total production in the basin in 1989 was an estimated 450 000 tonnes.

SULPHUR:

Ultra sour gas containing 90 per cent H_2S from the Bearberry sour gas field in Alberta is undergoing pilot testing for sulphur recovery. Daily sulphur capacity of all sour gas and tar sands operations in Alberta is about 32 000 tonnes.

POTASH:

Since initial writing one conventional potash mine has been lost to flooding but was later successfully coverted to a solution mining operation.

SODIUM SULPHATE:

Declining demand and the closure of three sodium sulphate operations have reduced the annual production level to about 274 000 tonnes by 1991. However, research into the use of sodium sulphate for the production of soda ash and potassium sulphate fertilizer could assist in raising and stabilizing production levels. Total production in 1989 was 369 500 tonnes of anhydrous product.

SALT:

Rising chloralkali chemical demand for an expanding pulp industry may give rise to additional salt brining operations particularly in Alberta. The current salt production capacity in the Western Canada Basin is 2.08 million tonnes per annum.

Author's Address

P. Guliov
Saskatchewan Geological Survey
Saskatchewan Department of Energy and Mines
1914 Hamilton Street
Regina, Saskatchewan
S4P 4V4

CALCIUM CHLORIDE:

Recent investigations have identified calcium concentrations ranging up to nearly 120 000 mg/L in Devonian Souris River and Dawson Bay Formation brines in Saskatchewan. Analyses of potash brine inflows indicate bromine and iodine levels up to 28 500 and 806 mg/L respectively.

MAGNESIUM:

Brine resource studies in Saskatchewan have identified magnesium concentrations up to 13 700 mg/L in the Souris River-Dawson Bay formations and in excess of 51 000 mg/L in the Winnipegosis Formation.

PEAT:

There are three major horticultural peat producers in the Western Canada Basin producing from about 12 bogs. Combined production represents about 28 per cent of Canadian output.

GEMSTONES:

Fossil ammonites from the Bearpaw Formation in Alberta are a source of gem material commercially known as "ammolite". The material is the inner or nacreous pearly layer of the shells of two species of *Placenticeras*. The material is being marketed worldwide. The two species, *P. meeki* and *P. intercalare*, are also present in the Bearpaw of Saskatchewan.

Recent descoveries of diamondiferous kimberlites in the Fort à la Corne area of Saskatchewan have triggered a staking rush which has spread southward and westward into Alberta and the Northwest Territories. No economic diamond deposits have been found to date but exploration is still in the initial stages and optimism is high.

Printed in Canada

Subchapter 6E

METALLIC DEPOSITS

J.D. Aitken and D.F. Stott

CONTENTS

INTRODUCTION

The main metallic mineral deposits, with the important exception of Pine Point, lie within the eastern Cordillera and most are not being produced currently. Ferrous deposits of sedimentary origin and placer gold, both of low commercial interest at present, occur within the Plains. The metallic mineral deposits are outlined only briefly; additional detail may be found in the companion volumes on "Geology of the Cordilleran Orogen in Canada" (Gabrielse and Yorath, 1991) and "Mineral Deposits of Canada" (Geology of Canada, no. 8, Thorpe and Eckstrand, in prep.).

LEAD – ZINC – SILVER

Stratabound, lead-zinc deposits of Mississippi Valley type occur at several places in the Western Canada Basin. The only one of these to achieve large production to date is at Pine Point, in the undeformed, platform succession; the others are in the Fold Belt of the Columbian Orogen. The shared characteristics of these deposits are carbonate host rocks, generally dolomitized and commonly brecciated, and proximity to equivalent, shaly facies. Pine Point and Monarch-Kicking Horse are associated with reef complexes.

Pine Point

R.W. Macqueen

Located on the south shore of Great Slave Lake, Northwest Territories, the Pine Point property contains 87 known sphalerite- and galena-bearing deposits hosted by carbonate rocks of Middle Devonian age, within an area approximately 65 km long by 24 km wide (Rhodes et al., 1984). Since production began in 1964, 36 of these deposits, all but two mined as open pits, have yielded more than 58.2 million tonnes of ore averaging 3.0% lead and 6.7% zinc (Rhodes et al., 1984). The Middle Devonian host rocks at Pine Point form the exposed part of a carbonate barrier complex, which is marginal to the Elk Point evaporite basin

Aitken, J.D. and Stott, D.F.
1993: Metallic deposits; Subchapter 6E in Sedimentary Cover of the Craton in Canada, D.F. Stott and J.D. Aitken (ed.); Geological Survey of Canada, Geology of Canada, no. 5, p. 612-615 (also Geological Society of America, The Geology of North America, v. D-1).

to the south and the Mackenzie carbonate-shale basin to the north. Precambrian rocks, approximately 300 m below the base of the Middle Devonian carbonate succession, are the locus of northeast-trending faults of, or associated with, the Great Slave Shear Zone. These faults may be responsible for relief of the barrier during sedimentation and for post-depositional flexure of the carbonate host rocks, recognized by study of persistent stratigraphic markers at two levels (E shale, Amoco marker; Rhodes et al., 1984). The stratigraphy of the property is exceptionally well known through systematic study of some 600 000 m of core from more than 10 000 exploration and development diamond-drill holes.

Lead-zinc orebodies are of two types: prismatic, with vertical dimensions greater than horizontal, and of the highest ore grade; and tabular, with horizontal dimensions greater than vertical, and of lower grade (Rhodes et al., 1984). Most orebodies mined to date are prismatic. Lead-zinc orebodies occur as open space fillings within cavities of probable karst origin (Rhodes et al., 1984). Karst development is believed to have been initiated in mid- to late Givetian (late Middle Devonian) time, and to have continued during Carboniferous and Permian time. Eight major stages of diagenesis were recognized by Krebs and Macqueen (1984), with dissolution and dolomitization being important prerequisites to the development of sulphide mineralization. Fluid-inclusion studies indicate that ore fluids were highly saline brines and that precipitation temperatures of coarse crystalline sphalerite, far less abundant than the ubiquitous colloform sphalerite, ranged from $\simeq$ 50°C to $\simeq$ 100°C. Through studies of various types of organic matter and bitumen from the property and region, Macqueen and Powell (1983) and Powell and Macqueen (1984) were able to demonstrate that the orebodies represent thermal anomalies with respect to the host rocks, and that organic matter-sulphate reactions can account for all the sulphide required to precipitate the orebodies.

The timing of sulphide precipitation is poorly constrained. Lead-isotope data suggest an age of about 310 Ma, whereas meagre paleomagnetic data indicate ages ranging between 232 and 280 Ma. Many aspects of the origin of these and similar Mississippi Valley type lead-zinc deposits would be less disputed if mineralization ages were known with greater confidence. Meanwhile, the Pine Point deposits appear to represent the happy coincidence of 1) metal-bearing brines, probably travelling as chloride complexes, 2) suitable large-scale pre-existing host-rock pore space originating through subaerial karst development; and 3) a locally available sulphide source responsible for ore precipitation, developed through organic matter-sulphide reactions. The role of faults associated with the Great Slave Shear Zone may also be important in karst development, flexure of the property, and control of fluid migration from depth. Broadly similar deposits are known at Robb Lake, northeastern British Columbia (see below), but the search for Pine Point-like deposits elsewhere in the relatively undisturbed Paleozoic rocks of the Interior Platform has not yielded success.

Mining operations at Pine Point ceased in 1988.

Monarch and Kicking Horse Mines

The Monarch and Kicking Horse Mines, near Field, British Columbia, exploited a number of lenticular orebodies in a reefal facies of the Middle Cambrian Cathedral Formation that changes facies nearby to argillaceous limestone and shale. The coarse dolomitization of the host rocks is a sub-regional characteristic, and hence not necessarily related to mineralizing fluids. The Zn/Pb ratio of the ore is high, and Ag content low; mining was economically feasible only during World War II, under a wartime subvention on metal prices. (References: Ney, 1954; Aitken and McIlreath, 1984).

Gayna River

Several mineralized bodies of Mississippi Valley type ("orebodies"), lying in the upper reaches of Gayna River, Mackenzie Mountains, were drilled extensively in the late 1970s. The host formation is the informal Grainstone formation of the Upper Proterozoic Little Dal Group. The host grainstones and cryptalgal carbonates are dolomitized; mineralization is associated with brecciation and subsequent influx of white, sparry dolomite. The area of economic interest coincides with part of a belt of large cryptalgal reefs in the underlying basinal assemblage; sphalerite, but no significant orebodies, occurs in the reefs. Possible reserves are estimated as over 50 million tonnes of 5% combined Pb-Zn. The Zn/Pb ratio is high, and the Ag content low. Because of its remote situation, and the existence of larger, higher-grade and more accessible deposits in Selwyn Basin, further development of the property is unlikely to take place for some time. (References: Hewton, 1982; Aitken, 1981).

Robb Lake area

Several occurrences of Pb-Zn mineralization, hosted by Paleozoic, mainly Devonian carbonate rocks, are known from the Rocky Mountains of northeastern British Columbia. Of these, the best explored are the deposits near Robb Lake. At Robb Lake, mineralization is found in broadly conformable breccias in the Stone and Muncho-McConnell formations. Sparry dolomite forms the breccia matrix. Reported resources are 5.5 x 10^6 tonnes of 7.3% combined lead-zinc. (References: Macqueen and Thompson, 1978).

COPPER

Redstone Copper Belt

Stratabound copper mineralization occurs widely in the central Mackenzie Mountains at the contact between evaporitic, clastic redbeds of the Redstone River Formation and peritidal carbonates of the overlying Coppercap Formation (basal Windermere Supergroup). The most promising of the known deposits, at Coates Lake, is a bed 1 m thick, with drill-indicated "reserves" of 37 million tonnes averaging 3.92% Cu and 11.3 g/t Ag ("about two average porphyry copper deposits" [J.C.L. Ruelle, pers. comm., 1979]). Elsewhere in the belt, drill-indicated thicknesses of up to 52 m, with comparable or higher grades, have been encountered. (References: Ruelle, 1982; Eisbacher, 1977; Jefferson, 1978; Jefferson and Ruelle, 1986).

have been encountered. (References: Ruelle, 1982; Eisbacher, 1977; Jefferson, 1978; Jefferson and Ruelle, 1986).

Lode copper, Racing - Gataga rivers area

Copper-bearing quartz-carbonate veins are widespread in the Racing-Gataga rivers area of the Rocky Mountains in northeastern British Columbia. The veins cut phyllite, slate, and impure carbonate rocks of the Middle (?) Proterozoic Aida and Gataga formations in an area at least 80 km long and 50 km wide. Only one of the many claim groups has given rise to commercial production; the Magnum Mine produced 352 146 tons of ore grading about 3.3% copper. The ore shoots (chalcopyrite) occupy only short portions of much more extensive, essentially barren veins. On the evidence of Preto (1971), and the assumption that the widespread, post-ore diabase dykes are related to the onset of Windermere rifting, the mineralization is not only Precambrian but pre-Windermere. (References: Preto, 1971).

GOLD

Placer gold has been recovered from rivers in northeastern British Columbia and Alberta for more than a century. In British Columbia, Peace River between Hudson Hope and Fort St. John was worked in the latter part of the 19th century, but the total production for the Peace River Division from 1900 to 1947 is reported to be only slightly more than 100 kg. Several operations have been undertaken above Peace River Canyon, and although small amounts of gold were recovered, the larger operations were not profitable.

In Alberta, placer gold has been produced from South Saskatchewan, Peace, MacLeod, Athabasca, Bow, Oldman, and other rivers, but the North Saskatchewan River has been the chief source. The total amount of gold recovered from the North Saskatchewan River from 1887 to 1981 was about 1000 kg. Production was 133 kg in 1980, but has declined steadily since then. (References: Giusti, 1986; Halferdahl, 1965).

IRON

Two large sedimentary iron deposits, neither yet economically exploitable, occur in Western Canada Basin. They are the Upper Proterozoic Snake River deposit, in the northwestern Mackenzie Mountains, and the Upper Cretaceous Clear Hills deposit, in the undeformed Interior Platform of west-central Alberta.

Snake River deposit

Near the debouchement of Snake River from the Mackenzie Mountains into Peel Plateau, in northeastern Yukon Territory, occurs the world's largest single iron "orebody", six billion tons of 47.2% iron. The "orebody" is a thick, rich part of a regional, banded, jaspilitic iron formation near the top of the Upper Proterozoic Sayunei Formation (Windermere Supergroup). Despite the immense size of the deposit, its remoteness and some beneficiation problems related to fine grain-size prevent its exploitation under current conditions. (References: Stuart, 1963; Yeo, 1981).

Clear Hills deposits

The Clear Hills deposits are oolitic iron-formations occurring in the marine, Upper Cretaceous Kaskapau Formation at Clear Hills, Alberta. The largest deposit is estimated at more than 1 billion tons with an average grade of 31% Fe; another deposit of slightly higher grade exceeds 100 million tons. Other deposits are smaller and of lower grade. Possible processes for creating a marketable concentrate have been investigated. The main deposit, near Swift Creek, is 3 to 30 feet thick, underlies an area about 13 by 3 miles, and is covered by up to 200 feet of overburden (Kidd, 1959). The ferruginous sandstone contains ooids formed of concentric shells of goethite and quartz in a matrix of quartz, goethite, siderite, and chamosite, with a little apatite and sericite. (References: Kidd, 1959; Mellon, 1962).

REFERENCES

Aitken, J.D.
1981: Stratigraphy and sedimentology of the Upper Proterozoic Little Dal Group, Mackenzie Mountains, Northwest Territories; in Proterozoic Basins of Canada, F.H.A. Campbell (ed.); Geological Survey of Canada, Paper 81-10, p. 47-72.

Aitken, J.D. and McIlreath, I.A.
1984: The Cathedral Reef Escarpment, a Cambrian great wall with humble origins; Geos, v. 13, no. 1, p. 17-19.

Eisbacher, G.H.
1977: Tectono-stratigraphic framework of the Redstone Copper Belt, District of Mackenzie; in Report of Activities, Part A, Geological Survey of Canada, Paper 77-1A, p. 229-234.

Gabrielse, H. and Yorath, C.J. (ed.).
1991: Geology of the Cordilleran Orogen in Canada; Geological Survey of Canada, Geology of Canada no. 4. (Also Geological Society of America, The Geology of North America, v. G-2.).

Giusti, L.
1986: The morphology, mineralogy and behaviour of "fine-grained" gold from placer deposits of Alberta: sampling and implications for mineral exploration; Canadian Journal of Earth Sciences, v. 23, p. 1662-1672.

Halferdahl, L.B.
1965: The occurrence of gold in Alberta Rivers; Alberta Research Council, Economic Mineral Files, Open File Report 65-11, unpaged.

Hewton, R.S.
1982: Gayna River: a Proterozoic Mississippi Valley-type zinc-lead deposit; in Precambrian Sulphide Deposits, R.W. Hutchinson, C.D. Spence, and J.M. Franklin, (ed.), Geological Association of Canada, Special Paper 25, p. 667-700.

Jefferson, C.W.
1978: Stratigraphy and sedimentology, Upper Proterozoic Redstone Copper Belt, Mackenzie Mountains, Northwest Territories, a Preliminary Report; in Mineral Industry Report 1975, Northwest Territories, Indian and Northern Affairs, E.G.S. 1978-5, p. 157-175.

Jefferson, C.W. and Ruelle, J.C.L.
1986: The Late Proterozoic Redstone Copper Belt, Mackenzie Mountains, Northwest Territories; in Mineral Deposits of Northern Cordillera, J.A. Morin (ed.); Canadian Institute of Mining and Metallurgy, Special Volume 37, p. 154-168.

Kidd, D.J.
1959: Iron occurrences in the Peace River region, Alberta; Research Council of Alberta, Preliminary Report 59-3, 38 p.

Krebs, W. and Macqueen, R.W.
1984: Sequence of diagenetic and mineralization events, Pine Point lead-zinc property, Northwest Territories, Canada; Bulletin of Canadian Petroleum Geology, v. 32, no. 4, p. 434-464.

Macqueen, R.W. and Powell, T.G.
1983: Organic geochemistry of the Pine Point lead-zinc ore field and region, Northwest Territories, Canada; Economic Geology, v. 78, no. 1, p. 1-25.

Macqueen, R.W. and Thompson, I.A.
1978: Carbonate-hosted lead-zinc occurrences in northeastern British Columbia with emphasis on the Robb Lake deposit; Canadian Journal of Earth Sciences, v. 15, p. 1737-1762.

Mellon, G.B.
1962: Petrology of Upper Cretaceous oolitic iron-rich rocks from northern Alberta; Economic Geology, v. 57, p. 921-940.

Ney, C.S.
1954: Monarch and Kicking Horse Mines, Field, British Columbia; in Fourth Annual Field Conference Guidebook; Alberta Society of Petroleum Geologists, Calgary, p. 119-136.

Powell, T.G. and Macqueen, R.W.
1984: Precipitation of sulfide ores and organic matter: Sulfate reactions at Pine Point, Canada; Science, v. 224, p. 63-66.

Preto, V.A.
1971: Lode copper deposits of the Racing River - Gataga River area; in Geology, Exploration and Mining in British Columbia; British Columbia Department of Mines and Petroleum Resources, p. 75-107.

Rhodes, D., Lantos, E.A., Lantos, J.A., Webb, R.J., and Owens, D.C.
1984: Pine Point orebodies and their relationship to the stratigraphy, structure, dolomitization and karstification of the Middle Devonian barrier complex; Economic Geology, v. 79, p. 991-1055.

Ruelle, J.C.L.
1982: Depositional environments and genesis of stratiform copper deposits of the Redstone Copper Belt, Mackenzie Mountains, Northwest Territories; in Precambrian Sulphide Deposits, R.W. Hutchinson, C.D. Spence, and J.M. Franklin (ed.); Geological Association of Canada, Special Paper 25, p. 701-738.

Stuart, R.A.
1963: Geology of the Snake River iron deposit; Canada Department of Indian Affairs and Northern Development, Yellowknife, Open File, 18 p.

Yeo, G.M.
1981: The Late Proterozoic Rapitan glaciation in the northern Cordillera; in Proterozoic Basins of Canada, F.H.A. Campbell (ed.); Geological Survey of Canada, Paper 81-10, p. 25-46.

Authors' Addresses

J.D. Aitken
Institute of Sedimentary and Petroleum Geology
Geological Survey of Canada
3303 - 33 Street N.W.
Calgary, Alberta
T2L 2A7

D.F. Stott
Institute of Sedimentary and Petroleum Geology
Geological Survey of Canada
3303 - 33 Street N.W.
Calgary, Alberta
T2L 2A7

R.W. Macqueen
Institute of Sedimentary and Petroleum Geology
Geological Survey of Canada
3303 - 33 Street N.W.
Calgary, Alberta
T2L 2A7

Printed in Canada

Subchapter 6F

GROUNDWATER IN THE INTERIOR PLAINS REGION

D.H. Lennox

CONTENTS

HYDROGEOLOGICAL ENVIRONMENT

The Canadian Interior Plains lie between the Canadian Cordillera on the west and the Canadian Shield on the east and extend as far north as the Arctic Ocean. The Interior Plains Hydrogeological Region is defined as that part of the Canadian Interior Plains that lies to the south of the southern limit of discontinuous permafrost.

Physiography

The Interior Plains Hydrogeological Region and adjacent areas to the north were divided by Bostock (1970) into seven physiographic subregions (Fig. 6F.1). The three largest of these, Manitoba, Saskatchewan, and Alberta plains, are separated from one another by two pronounced west-to-east downward steps in the terrain, the Missouri Coteau and Manitoba Escarpment. To the north of Alberta Plain lie three somewhat smaller subregions: Alberta Plateau, Peace River Lowland, and Fort Nelson Lowland. The smallest of the seven subregions, Cypress Hills, rises 700 m above the surrounding Alberta Plain just north of the International Boundary.

Surface elevations in the region range from 1500 m next to the Rocky Mountain Foothills belt to 150 m at the eastern boundary. The Manitoba Plain is very gently undulating to flat, with an average elevation of about 250 m above sea level (Bostock, 1970). Average elevation increases successively in the Saskatchewan and Alberta plains, as does the amount of local relief. This trend continues into the Alberta Plateau to the northwest. In the Alberta Plain the average elevation is 750 m and river valleys are entrenched to 100 m or more beneath the surrounding terrain. Elevations in some upland areas, such as the Neutral Hills and Porcupine Hills, exceed 1000 m (Fig. 6F.2).

Elevations in the complex of subregions comprising the Alberta Plateau reach as much as 1300 m in the southwest. The topographic relief between major upland features such as the Swan Hills and the adjacent lowlands exceeds more than 600 m. Rivers in this part of the hydrogeological

Lennox, D.H.
1993: Groundwater in the Interior Plains region; Subchapter 6F in Sedimentary Cover of the Craton in Canada, D.F. Stott and J.D. Aitken (ed.); Geological Survey of Canada, Geology of Canada, no. 5, p. 616-641 (also Geological Society of America, The Geology of North America, v. D-1).

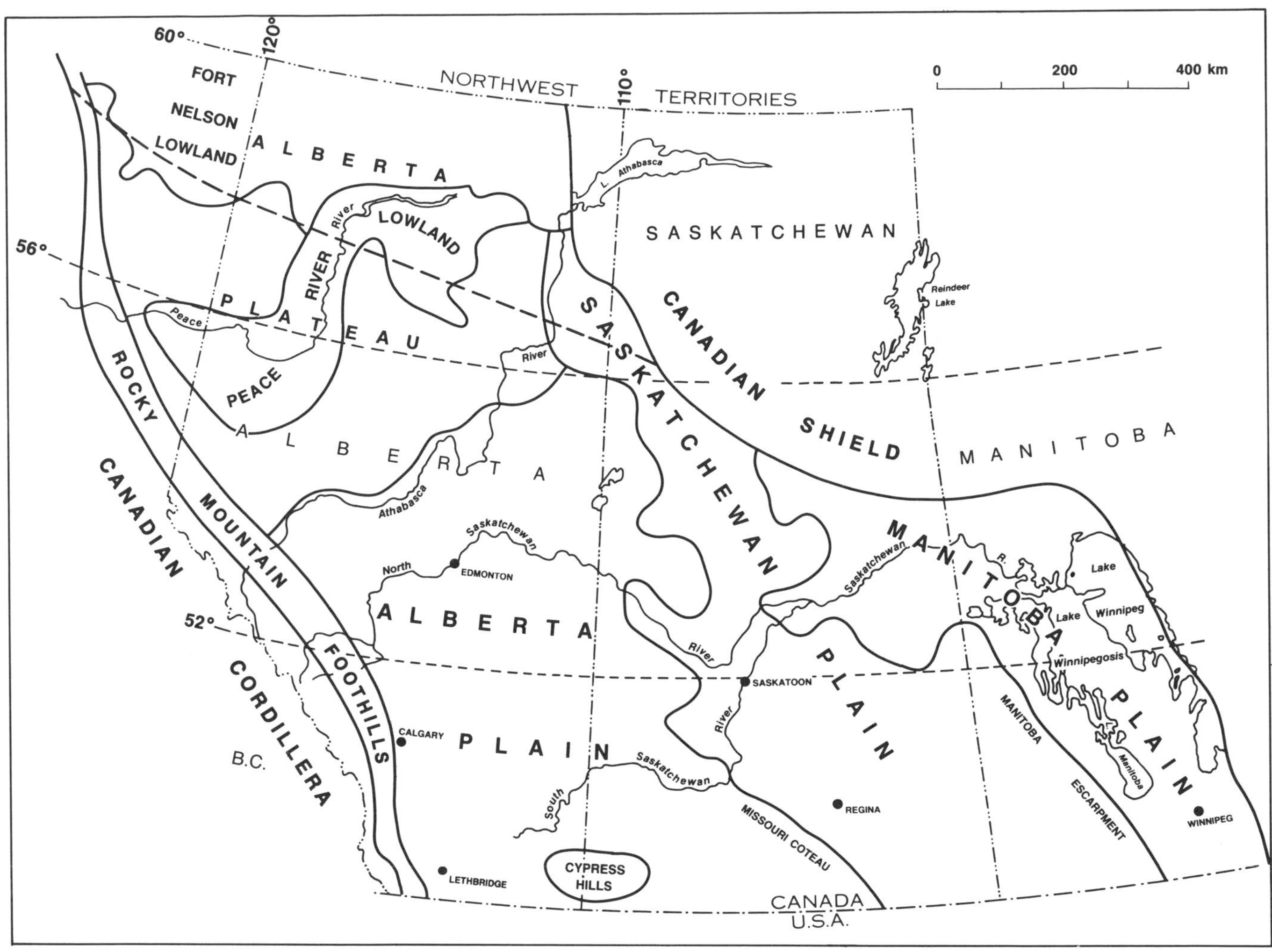

Figure 6F.1. Physiographic subdivisions of the Interior Plains Hydrogeologic Region (after Bostock, 1970).

region are deeply entrenched, to depths of 150 m below the surrounding terrain and as much as 300 m in the case of Peace River.

Climate

The climate of the Interior Plains Hydrogeological Region is subhumid continental, except for a southeastern portion of the Alberta Plain extending as far west as Lethbridge and north to about 52°N and for some adjoining sections of the Saskatchewan Plain (Fig. 6F.1). The climate in this latter area is arid to semi-arid, and average annual precipitation generally ranges from 300 to 400 mm, although in a few locations it is less (Fisheries and Environment Canada, 1978). For most of the region outside this central dry area, annual precipitation is between 400 and 500 mm, exceeding this range only in the extreme southeast and in a narrow strip along the Rocky Mountain Foothills. Throughout almost all of the region, most of the precipitation falls as rain, and local topographic relief has an important effect on its distribution. The distribution of average annual snowfall is similar to that of average annual precipitation, but average maximum snowfall accumulation follows a different pattern, increasing along a general southeast-to-northwest trend from less than 30 cm in southern Manitoba and Saskatchewan to about 60 cm in the Alberta Plateau and adjoining areas.

Acknowledgments

Studies by various Canadian scientists would not have been possible without the mass of information (on lithology, permeability, hydraulic head, subsurface temperature, and the chemical composition of formation waters) gathered by or for the petroleum industry during exploration for hydrocarbons in the Western Canada Basin. Much of the information is readily available from various provincial and federal agencies and provides a data base unequalled in any other country.

This subchapter has been reviewed by R.O. van Everdingen and B. Hitchon, whose comprehensive comments have been beneficial in making improvements to parts of the text.

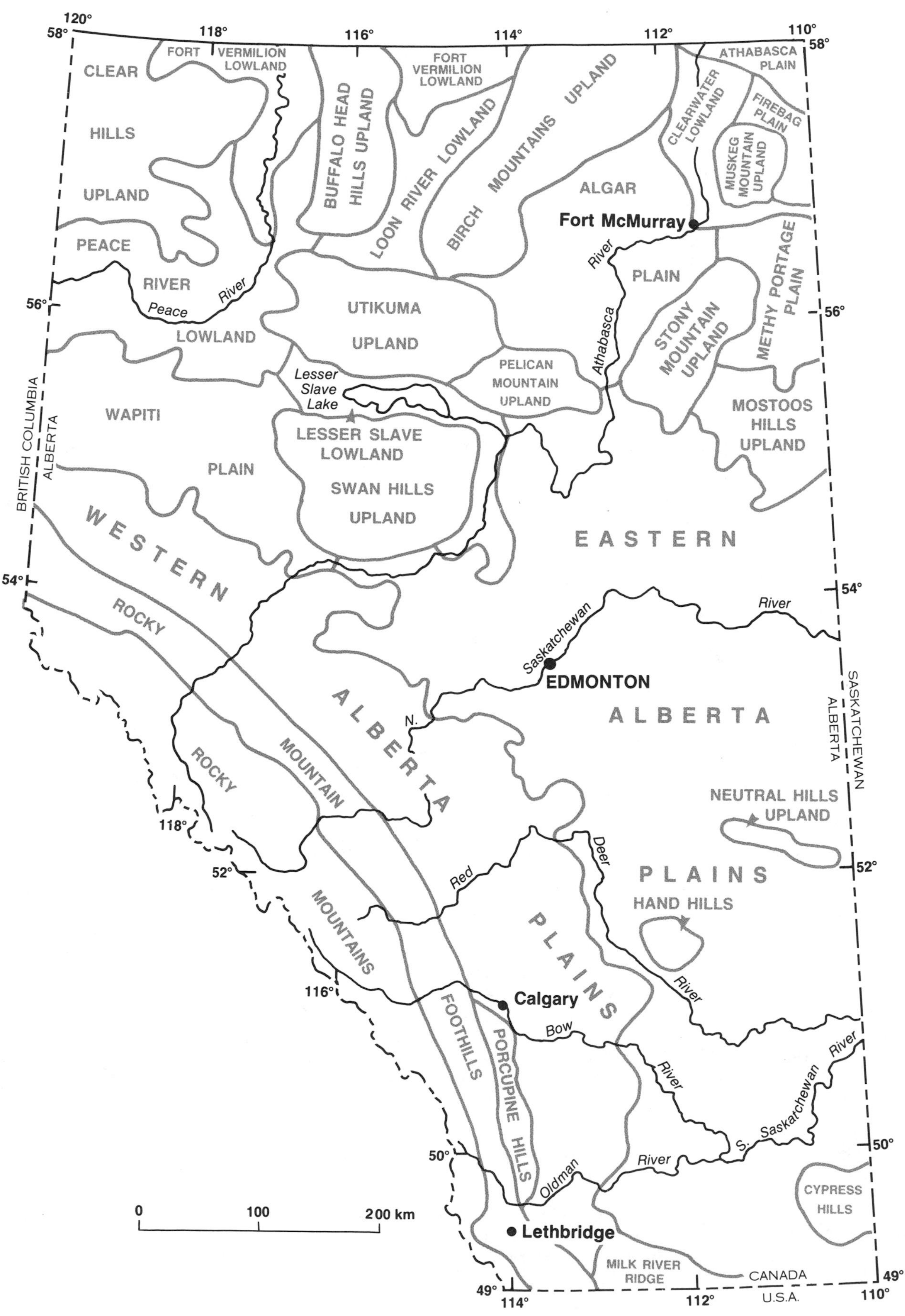

CLEAR HILLS UPLAND
FORT VERMILION LOWLAND
BUFFALO HEAD HILLS UPLAND
FORT VERMILION LOWLAND
LOON RIVER LOWLAND
BIRCH MOUNTAINS UPLAND
CLEARWATER LOWLAND
ATHABASCA PLAIN
FIREBAG PLAIN
MUSKEG MOUNTAIN UPLAND
ALGAR PLAIN
Fort McMurray
PEACE RIVER LOWLAND
Peace River
UTIKUMA UPLAND
PELICAN MOUNTAIN UPLAND
STONY MOUNTAIN UPLAND
METHY PORTAGE PLAIN
Athabasca River
Lesser Slave Lake
LESSER SLAVE LOWLAND
SWAN HILLS UPLAND
MOSTOOS HILLS UPLAND
WAPITI PLAIN
WESTERN ALBERTA PLAINS
EASTERN ALBERTA PLAINS
BRITISH COLUMBIA
ALBERTA
SASKATCHEWAN
ALBERTA
ROCKY MOUNTAIN FOOTHILLS
ROCKY MOUNTAINS
N. Saskatchewan River
EDMONTON
NEUTRAL HILLS UPLAND
HAND HILLS
Red Deer River
Calgary
Bow River
PORCUPINE HILLS
S. Saskatchewan River
Oldman River
CYPRESS HILLS
Lethbridge
MILK RIVER RIDGE
CANADA
U.S.A.
0
100
200 km
120°
118°
116°
114°
112°
110°
58°
56°
54°
52°
50°
49°

Table 6F.1. Estimated volumes of sandstone, shale, carbonate, and evaporite rocks in hydrostratigraphic units between 49° and 60°N

	Sandstone		Shale		Carbonate		Evaporite		Total	
Hydrostratigraphic Unit	10^5km^3	%	10^5km^3	%	10^5km^3	%	10^5km^3	%	10^5km^3	%
Upper Clastic	1.50	14.9	8.47	84.3	0.08	0.8	0.00	0.0	10.05	100.0
Middle Carbonate-Evaporite	0.24	2.3	3.26	31.3	5.66	54.2	1.27	12.2	10.43	100.0
Lower Clastic	0.74	37.9	1.04	53.4	0.17	8.7	0.00	0.0	1.95	100.0
Total	2.48	11.1	12.77	56.9	5.91	26.3	1.27	5.7	22.43	100.0

*Adapted from Tables I and II, Hitchon (1968)

HYDROSTRATIGRAPHIC UNITS

Bedrock in the Interior Plains Hydrogeological Region is almost everywhere concealed under a cover of Quaternary deposits, primarily till and glaciolacustrine and glaciofluvial sediments of Pleistocene age (Meyboom, 1967). This surficial cover forms the uppermost of five distinct hydrostratigraphic units. The second, third, and fourth of these units are broad subdivisions of the sedimentary sequence underlying the Pleistocene deposits. They are primarily gross units defined by lithology and are referred to here as the upper clastic, middle carbonate-evaporite, and lower clastic units, following the terminology of Kent and Simpson (1973).

The two clastic units are also readily distinguished from the intervening carbonate-evaporite unit on the basis of their geothermal properties (Jessop and Vigrass, 1984; Majorowicz et al., 1986). Their average thermal conductivities are appreciably less and their average geothermal gradients appreciably steeper than those in the carbonate-evaporite unit. Hydrogeologically, the picture is not as clear-cut, although Jessop and Vigrass (1984) suggested that the geothermal gradient change at the top of the carbonate-evaporite sequence might be partially attributed to a hydrodynamic effect.

The sedimentary rocks, which together make up the second, third, and fourth hydrostratigraphic units, range in age from Cambrian to Tertiary and form a wedge that increases in thickness southwesterly from the Canadian Shield (Fig. 6F.3). Strata within the sequence generally dip gently southwestward to the disturbed belt, as does the surface of the underlying Precambrian basement rocks, which form the fifth hydrostratigraphic unit. The sedimentary deposits reach a maximum thickness of 6000 m at the western edge of the region, where they pass into the strongly folded and faulted structures of the Canadian Cordillera.

Figure 6F.2. Physiographic features of Alberta (after Canadian Society of Petroleum Geologists, 1975).

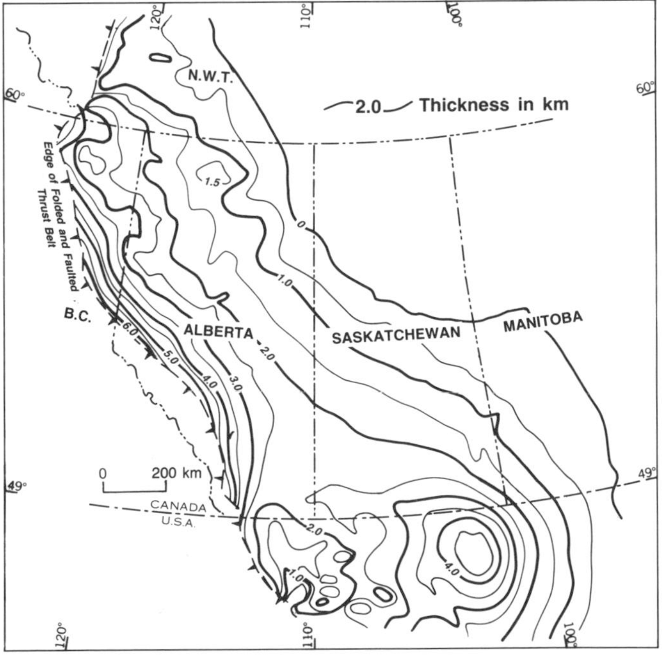

Figure 6F.3. Total preserved thickness of Phanerozoic rocks in the Interior Plains Hydrogeologic Region (after Porter et al., 1982).

The upper clastic hydrostratigraphic unit is a thick blanket of Tertiary, Cretaceous, and Jurassic rocks. In the area between latitudes 49° and 60°N, these rocks occupy an estimated volume of 10.05 x 10^5 km^3 (Table 6F.1), or about 45% of the total volume of sedimentary rocks in this part of the Western Canada Basin (Hitchon, 1968).

The upper clastic and middle carbonate-evaporite hydrostratigraphic units are separated by an unconformity that developed during Mesozoic uplift and erosion. The middle unit is a Paleozoic succession in which carbonates and evaporites predominate over sandstones and shales (Table 6F.1). Rocks in this unit range in age from Late Ordovician to Late Carboniferous. The estimated volume of the middle carbonate-evaporite unit between 49° and 60°N is 10.43 x 10^5 km^3 (Table 6F.1) or about 46.5% of the total.

Underlying the middle carbonate-evaporite unit through much of the region is the lower clastic hydrostratigraphic unit, consisting predominantly of Middle Cambrian to Middle Ordovician sandstones and shales but also including some carbonate rocks (Table 6F.1). The volume of this unit between 49° and 60°N is 1.95×10^5 km^3 (Table 6F.1) or about 8.5% of the total sedimentary cover in this part of the Western Canada Basin.

The lowest hydrostratigraphic unit is the Precambrian crystalline basement. As already noted, the basement surface generally dips gently southwestward. However, two major basement features in southern Alberta and southwestern Saskatchewan – Sweetgrass Arch and Williston Basin (Fig. 6F.4) – radically modify this uniform basement structural trend. Relief between the crest of Sweetgrass Arch and the deeper portions of the two adjoining basins is 1000 m. A third basement structural feature, North Battleford Arch, although more subdued, is of possible hydrogeological significance on a regional scale. The basement is considered to be impermeable, so that the upper surface of this unit is regarded as the lower boundary to the regional groundwater flow regime. The following discussion will concern only those hydrostratigraphic units in which significant flow takes place, that is, the Quaternary cover, the upper clastic sequence, the middle carbonate-evaporite sequence, and the lower clastic sequence. The Precambrian crystalline basement unit will be considered only in terms of the influence it exerts as a boundary on groundwater flow in the other hydrostratigraphic units.

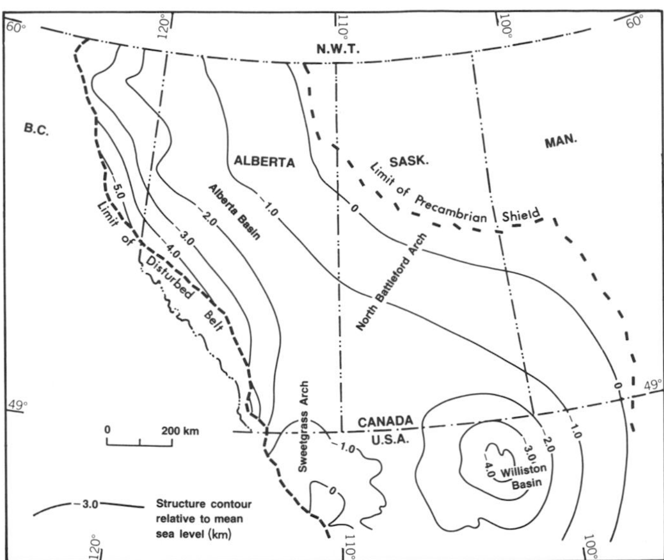

Figure 6F.4. Structure contours on Precambrian basement (after Porter et al., 1982).

Quaternary hydrostratigraphic unit

The surficial Pleistocene deposits are absent only in the Cypress Hills of southeastern Alberta and southwestern Saskatchewan (Westgate, 1968), Wood Mountain of southern Saskatchewan, and Del Bonita Upland of southwestern Alberta. Elsewhere, according to Meyboom (1967), about 60% of the region is covered by till, about 40% by glaciolacustrine sediments, and less than 1% by outwash deposits (Fig. 6F.5). The average thickness of surficial deposits tends to increase with distance from the mountain belt. In southern Alberta it is probably 30 m or less but becomes appreciably greater in Saskatchewan and farther north in Alberta. Maximum thicknesses are commonly observed along the courses of buried preglacial valleys, approaching 250 m in Alberta and exceeding 300 m at some locations in Saskatchewan.

The complex glacial and interglacial history of the Interior Plains Hydrogeological Region has resulted in an intricate and not readily predictable distribution of Pleistocene tills, clays, sands, and gravels. At the land surface, the probable presence of permeable, water-sorted sands and gravels is commonly made evident by the characteristic geomorphic forms of eskers, kames, outwash plains, beaches, and deltas. More deeply buried sands and gravels do not generally have any surface expression. For example, the courses of buried paleodrainage systems developed in the bedrock have only become relatively well known through systematic groundwater and geological investigations. However, information is often insufficient to predict whether a buried valley contains sand and gravel along a given reach or whether it is instead completely filled with glacial till, lake clays, or silts. The presence of isolated sand and gravel deposits interbedded with tills or lake sediments is equally unpredictable.

In Saskatchewan, the Quaternary stratigraphic succession has been subdivided into three groups. The oldest of these is the Empress Group of southern Saskatchewan and adjoining areas of Alberta. The Empress directly overlies the Cretaceous or Tertiary bedrock and comprises sand, gravel, silt, and clay of fluvial, lacustrine, and colluvial origin (Whitaker and Christiansen, 1972). This unit is thick, widespread, and permeable. The Empress Group is overlain by the Sutherland Group, which is in turn overlain by the Saskatoon Group (Christiansen, 1968, 1971). Both the Sutherland and Saskatoon groups, which are areally more extensive than the Empress, can be traced for hundreds of kilometres in southern Saskatchewan (Christiansen, 1971). They consist primarily of till but also contain layers of stratified drift made up of sand, gravel, silt, and clay. Contemporaneous with the Empress Group, and regarded by Meneley (1972) as part of that group, are blanket sand deposits. These are laterally extensive, range up to 30 m in thickness, and tend to be located along the flanks of major buried valleys.

Texture and mineralogy

Tills are both texturally and mineralogically variable. The variations are commonly related to the character of nearby underlying bedrock. In some areas they can also be related to source area modification by glacial erosion during a sequence of glacial advances and retreats. For instance, the tills of southeastern Alberta are less clayey than those of eastern Alberta, which are in turn less clayey than those of north-central and northwestern Alberta (Gravenor and Bayrock, 1961). This variation is related to a similar trend in clay content for the local bedrock. A similar correlation exists between till and bedrock textures for a more limited

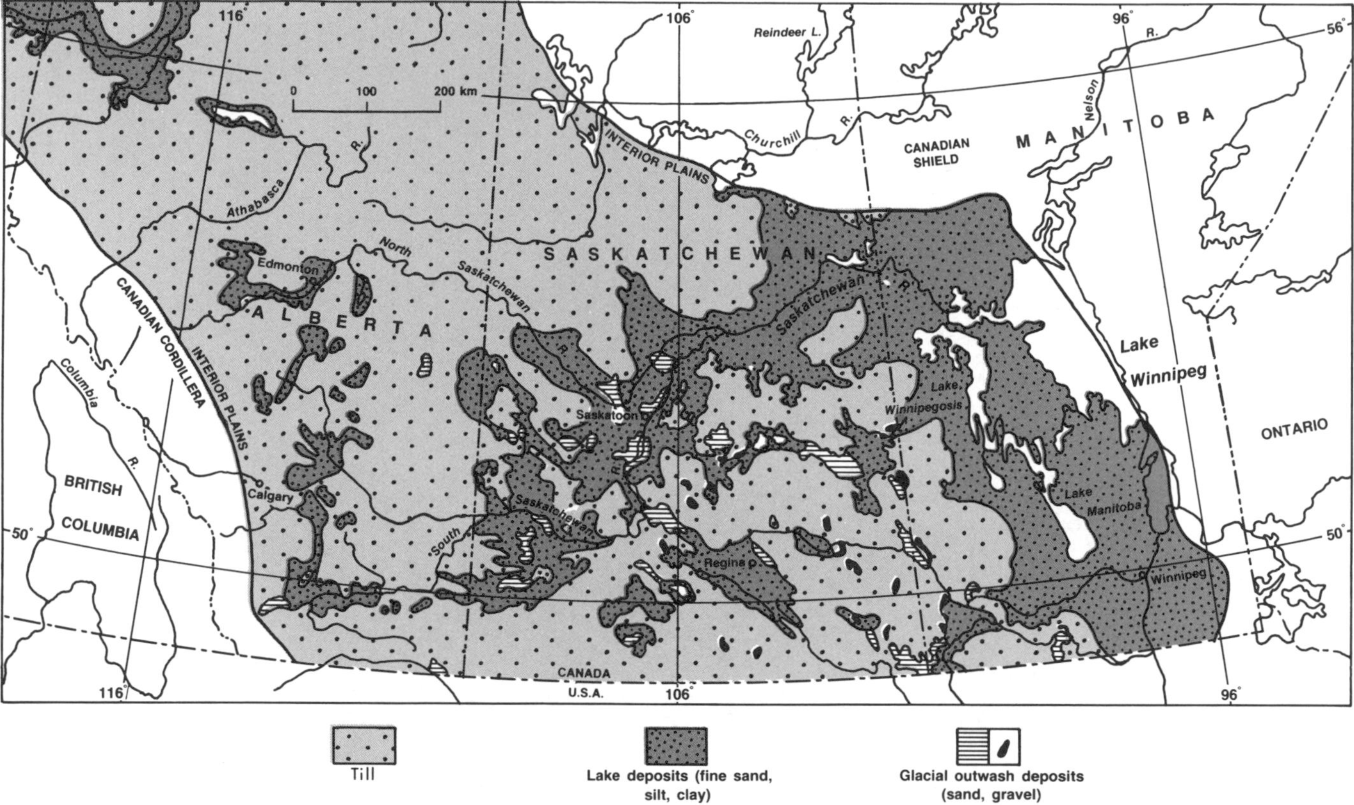

Figure 6F.5. Gross distribution of surficial deposits in the Interior Plains Hydrogeologic Region (after Meyboom, 1967).

area in the extreme southeast corner of Alberta (Westgate, 1968). The tills of the Saskatoon Group in southern Saskatchewan contain less clay and more carbonate than those of the underlying Sutherland Group, and this difference is related to the effects of glacial erosion on nearby source areas (Christiansen, 1971). Convincing mineralogical evidence in support of the concept of local bedrock sources for tills is provided by Christiansen (1971) and St. Arnaud (1976). Carbonate content was found to be high in those areas along the eastern and western margins of the hydrogeological region where Paleozoic carbonates come to the surface or subcrop beneath the surficial deposits (Fig. 6F.6; St. Arnaud, 1976). Lower carbonate values, in contrast, are associated with those areas where the Paleozoic rocks are deeply buried beneath the Tertiary and Mesozoic clastic rocks, which have a low carbonate content.

Glaciolacustrine deposits are principally calcareous, montmorillonitic, unoxidized clay and are found in the extensive basins once occupied by proglacial lakes, such as Lake Edmonton, Lake Regina, Lake Saskatchewan, and Lake Agassiz in southern Manitoba (Christiansen, 1979; Meyboom, 1967). These deposits are up to about 10 m thick in the central, deeper parts of the basins. Outwash deposits, consisting primarily of sand and fine gravel, are found along the shorelines of the former glacial lakes and at the mouths of meltwater channels.

Hydraulic conductivity of till

Hydraulic conductivity is a measure of the ease with which water can be induced to flow through a permeable medium. It is a function both of the geometric properties of the medium (i.e., of its intrinsic permeability or, for simplicity in what follows, its permeability) and of the density and viscosity of water at the temperature prevailing in the medium. Hydraulic conductivity increases as permeability and density increase and as viscosity decreases. The viscosity effect predominates, causing hydraulic conductivity for water flowing through a given medium to increase significantly as temperature increases.

The hydraulic conductivity of glacial tills in the Interior Plains Hydrogeological Region is an important factor in the rates of recharge through these materials to shallow and deep flow systems. Near the surface, where till structure is commonly modified by biological activities such as burrowing and root development, hydraulic conductivities can range up to 0.2 to 0.3 m/d (Meyboom, 1967). At greater depths, where these activities no longer exert an influence, hydraulic conductivity values characteristically decrease by a factor of 10 or more.

In general, for unfractured tills at depth, flows are intergranular and hydraulic conductivities are very low, of the order of 10^{-6} m/d. However, fracturing is believed to be common in tills of the Interior Plains Region, although it is not always evident visually (Keller et al., 1986). High conductivity values, such as those reported by Schwartz

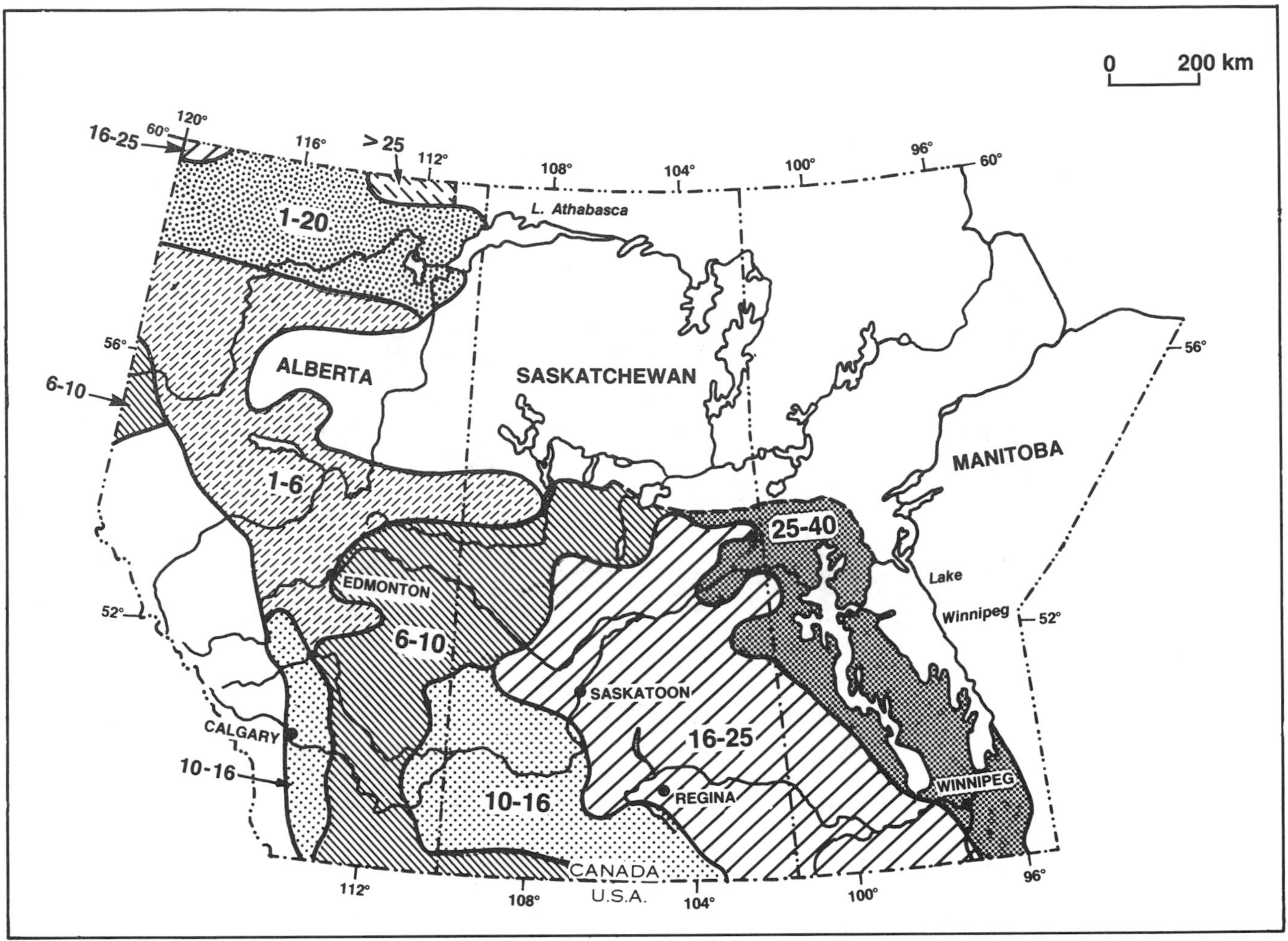

Figure 6F.6. Carbonate contents of glacial tills in the Interior Plains Hydrogeologic Region expressed as percentages calcium carbonate equivalent dry weight (after St. Arnaud, 1976).

(1975) for tills in southeastern Alberta, suggest that flow through such tills is actually strongly influenced by fractures.

Median bulk hydraulic conductivities for glacial till range as high as 2.7×10^{-2} m/d (Schwartz, 1975). A mean conductivity value of 1.7×10^{-2} m/d was determined for a southern Alberta till with large-scale fracturing (Hendry, 1982). For a till with small-scale fracturing, a mean value of 4.3×10^{-4} m/d was obtained. This is the same order of magnitude as the mean values for bulk conductivity of till determined by Grisak and Cherry (1975) and Keller et al. (1986). Other results, for fractured tills at southern prairie sites in all three provinces, give conductivity values in the 7×10^{-6} to 10^{-3} m/d range (Grisak et al., 1976). In contrast, determinations of hydraulic conductivity for tills in which flow is believed to be strictly intergranular give results that are approximately two orders of magnitude lower. Grisak and Cherry (1975) cited values ranging from 2.3×10^{-6} to 5.2×10^{-6} m/d; Hendry (1982) cited a range from 2.9×10^{-5} to 8.5×10^{-5} m/d; and Keller et al. (1986) suggested a best estimate for bulk hydraulic conductivity of unweathered till of 4×10^{-4} m/d. Mawson (1964) determined an average hydraulic conductivity of 0.8×10^{-2} m/d for a shallow Missouri Coteau till in central Saskatchewan. This value suggests that fractures and, perhaps, biological channels contribute significantly to flow within the till in that area.

Buried valleys

Prior to the onset of continental glaciation during the Pleistocene Epoch, drainage systems in the Interior Plains Hydrogeological Region were probably mature and well integrated (Farvolden, 1963; Fig. 6F.7). Subsequently, the bedrock surface underwent further fluvial dissection during interglacial periods, and additional valleys were formed which became intermingled with the preglacial set (Meyboom, 1967). At the end of the Pleistocene Epoch, the alluvial deposits at the base of the preglacial and interglacial valleys were overlain with glacial drift, which generally tended to fill in the valleys and to conceal the earlier drainage networks. Nevertheless, a number of major meltwater channels that were cut into bedrock during the final stages of deglaciation still remain open and from place to place are occupied by segments of the present-day drainage (Christiansen, 1963).

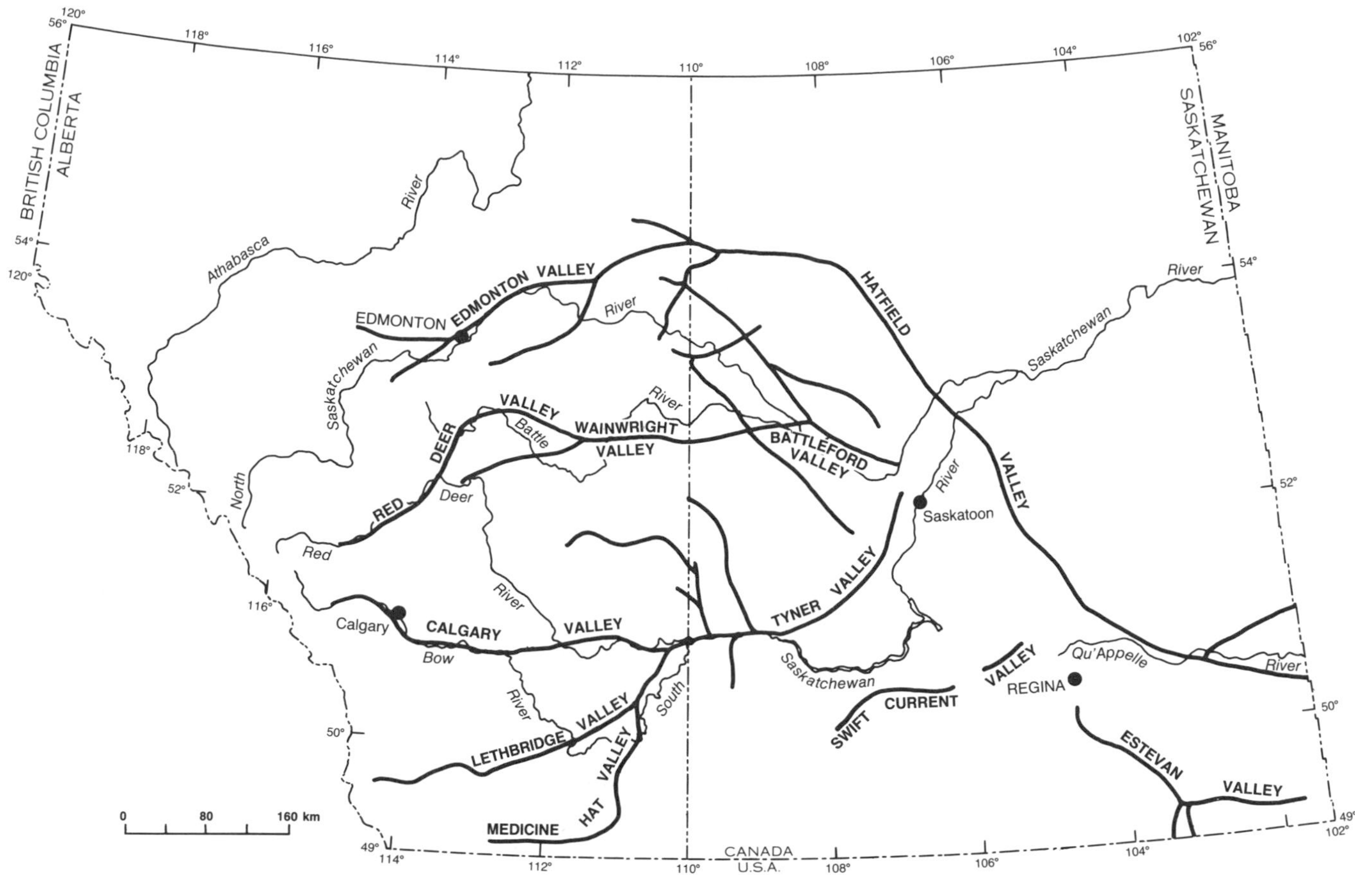

Figure 6F.7. Buried preglacial valleys of the Interior Plains Hydrogeologic Region (after Maathuis, 1986; Nielsen and Lorberg, 1971).

Sand and gravel are found in those preglacial valleys having their headwaters in the mountain areas or in areas of the plains where coarse materials were deposited during intervals of aggradation dating back to Oligocene time (Christiansen, 1963; Farvolden 1963). Sand and gravel were carried from these upland areas and deposited along the stream beds until gradients became too gentle for their continued transport. In Alberta, the regional erosional base level is now as low as, or lower than, it has been at any time since the Oligocene Epoch (Farvolden, 1963). Consequently, buried valleys in that province may intersect present-day streams, pass beneath them, or outcrop or subcrop above stream level along the present-day valley walls. In Saskatchewan, the preglacial erosional base level is from 30 to 90 m below the present level, and preglacial valleys are always deeper than present-day streams (Christiansen, 1963).

Upper clastic hydrostratigraphic unit

The upper clastic hydrostratigraphic unit is made up almost exclusively of sandstones, siltstones, and shales ranging from Tertiary to Jurassic in age (Table 6F.1). Jurassic rocks are found only in the southern and western parts of the hydrogeological region. In southern Saskatchewan and Manitoba, the lowermost Jurassic formations are primarily sandy, silty, or shaly, but they are intermixed with limestone, dolomite, and anhydrite. Elsewhere in the region, most of the formations comprising this hydrostratigraphic unit contain neither carbonate nor evaporite deposits. Exceptions to this general rule include limestone and calcareous shale and siltstone beds in the lower Mannville Group of southern and central Alberta and the calcareous First and Second White Speckled Shales of the Colorado Group of southern Alberta and Saskatchewan.

The upper clastic unit forms the bedrock surface throughout most of the region. The unit has been eroded, however, in a zone adjacent to the Canadian Shield as much as 200 km in width. There, Paleozoic rocks outcrop or subcrop beneath the Pleistocene deposits.

The upper clastic unit varies considerably in thickness, in a pattern similar to that shown in Figure 6F.3 for the total thickness of all Phanerozoic rocks. In general, thickness increases gradually toward the southwest from the erosional feather-edge in the northeast (Fig. 6F.8). Maximum thicknesses of 4000 m to slightly more than 5000 m are found close to the Foothills Belt. The approach to this belt is characterized by an appreciable increase in the rate of thickening. Outside the Foothills Belt, and away from the northeastern margin, the thickness of the clastic unit is 1000 to 2500 m.

Available information on hydraulic conductivities of strata of the upper clastic hydrostratigraphic unit is summarized in Table 6F.2. The data are biased towards higher conductivity values, which are more easily measured. The data in the table were obtained using a variety of field and laboratory techniques and are reported in some cases as ranges, in others as precise single values. Ranges, if given, commonly amount to an order of magnitude or more. Variations such as these are to be expected because of heterogeneity of the strata (Bachu et al., 1987). The dependence of measured conductivity values on measurement technique is also a factor, mean values from laboratory core analysis can be an order of magnitude greater than values based on drill-stem testing (Bachu et al., 1987).

Most of the values in Table 6F.2 are not given any directional significance, but a few are identified as vertical or horizontal conductivities. Because of these variations in sources, in measurement technology, in precision, and even in the very nature of the measured or estimated parameters, it is probably not particularly meaningful to compare individual values in the table, particularly those from different authors. Most of the data of the upper clastic hydrostratigraphic unit in Table 6F.2 are for sandstones, but a few conductivities are given for shales in the Upper Cretaceous Bearpaw and Oldman formations. As might be expected, the shale conductivities tend to be generally less than those for the coarse-grained strata of the upper clastic hydrostratigraphic unit. This trend is evident if the conductivity-map estimates taken from Schwartz et al. (1981) are omitted from the comparison.

Middle carbonate-evaporite hydrostratigraphic unit

The middle carbonate-evaporite hydrostratigraphic unit differs in a number of ways from the overlying and underlying clastic units. Carbonate and evaporites make up about two thirds of this unit (Table 6F.1). Dolomite formations predominate and limestone strata are common. Shale strata are also common but less so than in the two clastic units. Sandstones are relatively rare. Shale, sandstone, and siltstone units are in part calcareous.

As already noted, the middle hydrostratigraphic unit outcrops or immediately underlies the Pleistocene deposits in a narrow strip adjacent to the Canadian Shield. It is more uniform in thickness than the overlying clastic unit. The thickness is about 1000 m through Alberta and into western Saskatchewan (Fig. 6F.8). To the east, the middle carbonate-evaporite unit thins and pinches out.

Available information on hydraulic conductivities in the middle carbonate-evaporite unit is summarized in Table 6F.2. The information is primarily for carbonates. Conductivities for these strata range over similar sets of values to those that characterize sandstones of the upper hydrostratigraphic unit. In general, carbonates unmodified by fracturing, solution, or dolomitization can vary from dense, effectively pore-free rocks of very low permeability to extremely permeable rocks rich in voids (Matthes, 1982). High permeability can be developed in carbonate rocks through fracturing and dissolution of the carbonate minerals (Davis, 1969). Dolomitization can increase porosity and permeability because of the accompanying reduction in rock volume which, however, may be compensated for by diagenetic processes such as mechanical compaction, cementation, solution, and crystallization, and the formation of authigenic minerals (Matthes, 1982).

All this suggests that the middle carbonate-evaporite unit could contain some relatively permeable strata, in general appreciably more permeable than those strata that are characteristic of the overlying clastic unit, even though this is not evident from an examination of Table 6F.2. Hitchon (1969b), in his analysis of the effects of lithology on fluid flow in the Western Canada Basin, drew attention to the significance of the "relatively highly permeable Upper Devonian and Carboniferous carbonate rocks". In the same paper, however, he noted that considerably lower permeabilities are associated with the occurrence of anhydrite or halite and that thick evaporite beds in this unit may function as aquicludes, that is, may be virtually impermeable. Halites of the Lotsberg, Cold Lake, and Prairie formations are effective aquicludes (Bachu, 1985). Thus, hydraulic conductivities of the middle carbonate-evaporite hydrostratigraphic unit vary over a much wider range than is the case for either of the two clastic units.

Lower clastic hydrostratigraphic unit

The lower clastic hydrostratigraphic unit includes shales, siltstones, and sandstones of Cambrian to Middle Ordovician age. Overlying Upper Ordovician formations, beginning with the Red River Formation, are assigned to the carbonate-evaporite hydrostratigraphic unit. Limestones become prominent in Cambrian strata towards the western edge of the Interior Plains Hydrogeological Region.

Cambrian strata are missing from most of Manitoba except the southwestern part, from much of eastern Saskatchewan, and from most of northern Alberta (van Hees and North, 1966). They range in thickness from zero

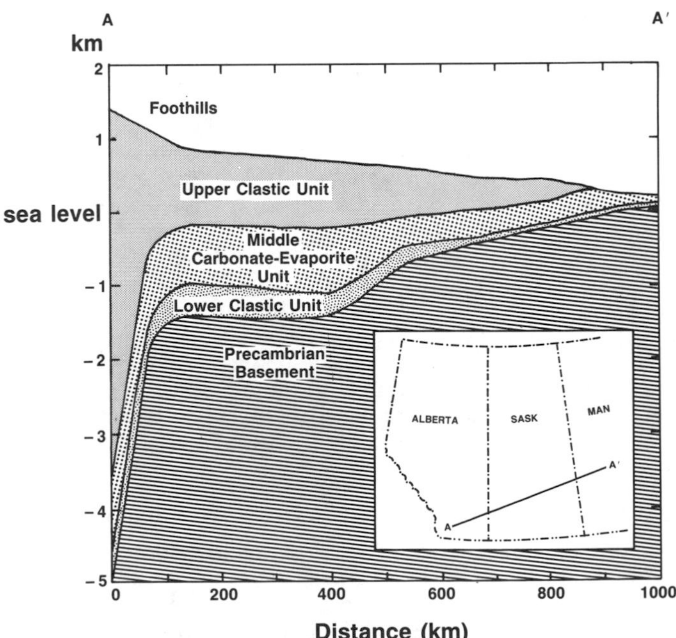

Figure 6F.8. Cross-section, southwestern Alberta to southern Manitoba, showing thicknesses of hydrostratigraphic units.

Table 6F.2. Hydraulic conductivities, sedimentary rocks of the Interior Plains Region

Stratigraphic Unit	Group or Formation	Lithology	Area	Hydraulic Conductivity K (m/d)*	Source	Remarks
Tertiary/Cretaceous	Ravenscrag, Frenchman, Eastend	Sandstone	S.Sask.	0.1 to 10	Meneley, 1983	
Upper Cretaceous	Bearpaw	Shale	Central Alta.	0.0017 (v)	Kunkle, 1962	45 m section
		Shale	Bethune, Sask.	0.04 0.00044	Meyboom, 1967	30 m below surface 90 m below surface
		Shale	Gardiner Dam Site	2.5×10^{-7} to 2.5×10^{-3}	Peterson, 1954	
		Sandstone	Riverhurst, Sask.	0.22	Meyboom, 1967	
	Oldman	Shale	SE Alta.	2.1×10^{-5} to 1.5×10^{-2}	Schwartz, 1975	Uppermost bedrock unit
	U. Belly River	Sandstone	Provost, Alta.	6.	Wallick, 1981	
	L. Belly River	Sandstone	Provost, Alta.	0.2	Wallick, 1981	
	Judith River	Sandstone	SW Sask.	0.2 to 2.0 (h)	Kewen and Schneider, 1979	
	Milk River	Sandstone, silstone, shale	SE Atla.	10^{-4} to 10	Schwartz et al., 1981	Range of contour values on conductivity map
	Medicine Hat	Sandstone	SE Alta.	10^{-6} to 10^{-4}	Schwartz et al., 1981	Range of contour values on conductivity map
Upper Cretaceous	Second White Speckled Shale	Sandstone	SE Alta.	10^{-4} to 10^{-2}	Schwartz et al., 1981	Range of contour values on conductivity map
Lower Cretaceous	Viking	Sandstone	Cold Lake, Alberta	(0.8 to 1.6) $\times 10^{-3}$	Bachu et al., 1987	Range cited is from harmonic to arithmetic mean for 187 K-values
		Sandstone	Swan Hills, Alberta	(1.3 to 3.9) $\times 10^{-3}$	Bachu et al., 1987	Range cited is from harmonic to arithmetic mean for 26 K-values
	Bow Island	Sandstone, shale	SE Alta.	10^{-3} to 10^{-1}	Schwartz et al., 1981	Range of contour values on conductivity map
	U. Manville	Sandstone	Cold Lake, Alberta	1.5×10^{-3} to 2.9×10^{-2}	Bachu et al., 1987	Range cited is from harmonic to arithmetic mean for 332 K-values
		Sandstone	Swan Hills, Alberta	1.5×10^{-3} to 5.4×10^{-2}	Bachu et al., 1987	Range cited is from harmonic to arithmetic mean for 65 K-values

Table 6F.2. Cont.

Stratigraphic Unit	Group or Formation	Lithology	Area	Hydraulic Conductivity K (m/d)*	Source	Remarks
	Peace River Spirit River	Sandstone, shale	N Alta.	10^{-2} 3×10^{-1} (v') 1.0 (h')	Toth, 1978	Notikewin is uppermost member of Spirit River Formation
	Mannville	Sandstone (?) Sandstone	Regina, Sask. NE Sask.	0.16 17	Lissey, 1962 Meneley et al., 1979	
	Wabiskaw, Bullhead	Shale, sandstone	N Alta.	10^{-2} (d) 6.6×10^{-2} (v') 2.1×10^{-1} (h')	Toth, 1978	
	L. Mannville	Sandstone	Cold Lake, Alberta	3.8×10^{-3} to 2.9×10^{-2}	Bachu et al., 1987	Range cited is from harmonic to arithmetic mean for 117 K-values
		Sandstone	Swan Hills, Alberta	1.4×10^{-3} to 8.6×10^{-2}	Bachu et al., 1987	Range cited is from harmonic to arithmetic mean for 141 K-values
	Glauconitic Sandstone	Shaly sandstone	SE Alta.	10^{-5} to 10^{-1}	Schwartz et al., 1981	Range of contour values on conductivity map
	Sunburst, Cutbank	Shaly sandstone	SE Alta.	10^{-3} to 10^{-1}	Schwartz et al., 1981	Range of contour values on conductivity map
Jurassic	Sawtooth	Sandstone, shale, limestone	SE Alta.	10^{-3} to 10^{-1}	Schwartz et al., 1981	Range of contour values on conductivity map
Lower Carboniferous	Livingstone	Limestone, dolomite	SE Alta.	10^{-3} to 10	Schwartz et al., 1981	Range of contour values on conductivity map
	Mission Canyon	Limestone	Weyburn, Sask.	2×10^{-3} 2×10^{-2}	Smith, 1980	Conversion of original author's intrinsic permeability estimates
		Limestone	SE Sask.	2×10^{-3} 2	Fuzesy, 1966	Conversion of original author's intrinsic permeability estimates
Permian/Lower Carboniferous	Belloy, Debolt, Upper Banff	Limestone, siltstone, shale	N Alta.	3×10^{-2} (d) 6.5×10^{-3} (v') 3.5×10^{-2} (h')	Toth, 1978	
Upper Devonian	Upper Wabamun, Calmar, Nisku	Limestone, dolomite, siltstone, shale	N Alta.	3×10^{-3} 1.0×10^{-1} (v') 7×10^{-1} (h')	Toth, 1978	
	Winterburn	Dolomite, siltstone	Cold Lake, Alberta	0.85	Bachu, 1985	Average, 10 K-values
	Birdbear	Limestone, dolomite anhydrite	SE Sask.	0.004 to 0.1	Nicholson, 1970	Conversion of original author's intrinsic permeability estimates

Table 6F.2. Cont.

Stratigraphic Unit	Group or Formation	Lithology	Area	Hydraulic Conductivity K (m/d)*	Source	Remarks
	Grosmont	Dolomite	Cold Lake, Alberta	0.04	Bachu, 1985	Average, 20 K-values
	Beaverhill Lake	Limestone, shale, dolomite	Cold Lake, Alberta	8.5×10^{-3}	Bachu, 1985	Average, 30 K-values
		Limestone, shale, dolomite	Swan Hills, Alberta	1.7×10^{-5} to 3.8×10^{-2}	Bachu et al., 1987	Range cited is from harmonic to arithmetic mean for 22 K-values
Upper/Middle Devonian	Slave Point, Watt Mountain	Limestone, shale, anhydrite, siltstone	N Alta.	7×10^{-4} (d) 3.6×10^{-4} (v') 1.3×10^{-3} (h')	Toth, 1978	
Middle Devonian	Gilwood	Sandstone, shale	Swan Hills, Alberta	8.4×10^{-4} to 8.4×10^{-2}	Bachu et al., 1987	Range cited is from harmonic to arithmetic mean for 48 K-values, corrected for in-situ aquifer conditions
	Keg River, Granite Wash	Dolomite, limestone, shale, siltstone, sandstone	N Alta.	2.1×10^{-2} (d)	Toth, 1978	
	Keg River	Dolomite, limestone	N Alta.	10^{-3} (v') 3×10^{-3} (h')	Toth, 1978	
	Winnipegosis	Dolomite	Cold Lake, Alberta	8.5×10^{-3}	Bachu, 1985	Average, 5 K-values
	Granite Wash	Shale, siltstone, sandstone	N. Alta.	3×10^{-2} (v') 1.2×10^{-1} (h')	Toth, 1978	
Silurian	Interlake	Limestone, dolomite	S Sask.	0.003 to 3.0	Jamieson, 1979	Conversion of original author's intrinsic permeability estimates
Ordovician	Winnipeg	Sandstone, shale	Regina	0.14	Vigrass and Jessop, 1984	
		Sandstone	E. Man.	0.28 to 100	Betcher, 1986	
Cambrian	Deadwood	Sandstone, conglomerate, minor shale	Regina	0.18 to 0.45	Vigrass and Jessop, 1984	
	Basal Cambrian	Coarse clastics	Cold Lake, Alberta	0.04	Bachu, 1985	Average, 5 K-values

* **Abbreviations:**

d - average conductivity from drill stem tests

h - horizontal conductivity

h' - average maximum horizontal conductivity from core measurements

v - vertical conductivity

v' - average vertical conductivity from core measurements

at the eastern erosional edge to more than 600 m in an area north of Calgary. Ordovician beds are found primarily in southern Manitoba and Saskatchewan but extend as well into eastern Alberta (Porter et al., 1966). The Middle Ordovician Winnipeg Formation of this region is the uppermost formation assigned to the lower clastic unit. It consists predominantly of arenaceous to nonarenaceous, siliceous to slightly calcareous shales and poorly consolidated to unconsolidated siliceous very fine- to coarse-grained sandstone (Baillie, 1952; Betcher, 1986). Its thickness in Manitoba ranges from zero in the north to more than 60 m in the extreme southwest (Betcher, 1986). Two widespread stratigraphic units characterize the Winnipeg Formation in Manitoba. The first is a basal sandstone, which is probably continuous throughout the area of deposition; the second is a shale layer at the top of the formation, which is a persistent feature only in southern Manitoba. Other and probably less extensive sandstone strata are found above the basal sandstone. In southern Manitoba, the shale layer is a very effective aquitard, separating groundwater flow systems in the Winnipeg sandstones from those in the overlying carbonates of the Red River Formation.

In Saskatchewan, the Winnipeg Formation and the Cambrian Deadwood Formation form the lower clastic unit, which ranges in thickness from 60 m in northeastern Saskatchewan to 600 m along the Alberta border (Paterson, 1975) (Fig. 6F.8).

Published information on hydraulic conductivities for this unit is limited to measurements made by Vigrass and Jessop (1984) in a test hole at Regina and to calculations by Betcher (1986) based on analyses of aquifer test data and of grain-size distributions in samples. The grain-size conductivity values are about two orders of magnitude larger than the aquifer-test values, probably because of the removal of fine intergranular clays and other cementing materials during sampling and preparation of the samples for analysis. Only Betcher's aquifer-test results are given in Table 6F.2. The results presented in the table are indicative of high permeabilities for the lower clastic hydrostratigraphic unit but, like other data in the table, these data are biased towards relatively coarse materials.

GROUNDWATER FLOW SYSTEMS

Flow in any hydrogeological system depends on the amount of recharge to the system from precipitation or other sources. It depends as well on the orientation and structure of the physical boundaries that limit the flow region and on the permeability variations within it. The most important boundary is the upper one, the water table, which is effectively the ground surface, through which all significant recharge to or discharge from the system is assumed to occur. Theoretical investigations of boundary effects on steady-state flow in hydrogeological systems have consequently been concerned principally with the influence of variations in surface topography. Other, subsurface boundaries have generally been assumed to be uniform vertical or horizontal planes. Intuitively, it is to be expected that major deviations from these idealized geometries will reveal themselves as significant perturbations of the flow regime.

The first analytical approach to the evaluation of topographic effects on regional groundwater flow was provided by Toth (1962, 1963). His solution for a homogeneous isotropic flow region bounded at the top by a sloping sinusoidal topography proved that topography alone can give rise to a complex set of local, intermediate, and regional groundwater flow systems. Subsequent numerical simulation studies by Freeze and Witherspoon (1966, 1967, 1968) demonstrated the degree to which flow regimes can be further complicated by the effects of geological heterogeneity.

These investigations built upon the theoretical foundation by Hubbert (1940), which proved groundwater motion to be a particular example of motion in a conservative force field. Such a field is characterized by a potential function, which varies from point to point. The work done in moving a particle from one position to another in the field is independent of the path taken and depends only on the difference in potential between the initial and final positions. In the case of groundwater motion, the function is known as the fluid potential and is equal to the mechanical energy of the fluid per unit mass. The fluid potential cannot be observed directly but is simply derived as the product of an observable quantity, the hydraulic head, and the acceleration due to gravity. Hydraulic head for a well tapping an aquifer at a given point is the height to which water rises in the well above some arbitrary datum. Hydraulic head is the sum of two components: elevation head and pressure head (Fig. 6F.9). Elevation head is the height of the observation point above datum. Pressure head is the height above the observation point to which water rises in the well because of hydraulic pressure in the aquifer. In hydrogeological practice, fluid potential is most commonly reported as hydraulic head, whereas the petroleum industry finds it more convenient to deal with formation pressures. These differences in approach must be reconciled when analyzing data from both sources.

The Western Canada Basin may be regarded conceptually as a vast hydrogeological flow region, geologically complex but generally permeable and hydraulically interconnected throughout (Hitchon, 1969b). The region is bounded below by the Precambrian basement, which may be considered, for all practical purposes, to be impermeable (Hitchon, 1969a). The eastern boundary of the flow region is the basement outcrops along the western edge of the Canadian Shield. Elevations along this boundary decrease from a maximum of more than 400 m near Peter Pond Lake in eastern Saskatchewan to slightly more than 200 m on Lake Winnipeg and to about 150 m near Great Slave Lake. The western boundary of the flow region cannot be indicated as precisely as the eastern boundary, but it lies somewhere within the Foothills belt of the Canadian Cordillera (Hitchon, 1969a, b). To the west of this belt, the rugged mountain topography gives rise to a number of discrete and relatively localized groundwater flow systems, which do not connect with or contribute to flow systems within the sedimentary basin proper (Toth, in Hitchon, 1969a).

Flow in the unsaturated zone

The upper boundary of the conceptual flow systems considered in the model studies by Toth (1962, 1963) and by Freeze and Witherspoon (1966, 1967, 1968) was assumed to be the water table. This boundary was assumed also to be a subdued replica of the ground surface, and groundwater recharge was considered to be a spatially and temporally uniform steady-state process. In the real world, however, recharge is due to an irregularly spaced sequence of discrete recharge events of varying magnitude. For practical purposes, this difference between concept and reality is probably of little or no consequence in most cases. However, it may acquire some significance in arid and semiarid areas such as the south-central portion of the Interior Plains Hydrogeological Region.

Recharge to the water table depends, in the first instance, on the intensity and the temporal and spatial distribution of rainfall and snowmelt. For these events to be translated undiminished into groundwater recharge, the infiltrating water would have to proceed downward without hindrance toward the water table. Downward flow in the unsaturated zone is, however, normally reduced to some degree by evaporation and by transpiration by plants. In humid areas, the infiltration process predominates and groundwater recharge usually results when there is rainfall or snowmelt at the surface. In drier areas only the most exceptional and intense events may lead to groundwater recharge.

Both physical and biological processes in the unsaturated zone may limit or even completely prevent flow to the water table. There have been no Canadian studies to gauge the significance of these processes in the Interior Plains Hydrogeological Region. However, the results from an investigation of water movement in till of east-central South Dakota (Cravens and Ruedisili, 1987) are probably transferable to sites in the drier parts of the Interior Plains Region. At the South Dakota site, where average annual precipitation is about 450 mm, it was concluded that discharge of water infiltrating the till is primarily through evapotranspiration, with no significant recharge to underlying aquifers or to the regional flow system. Because the conditions described by Cravens and Ruedisili (see also Houghton et al., 1987) can affect fairly large areas, they could give rise to significant modifications of conceptual groundwater flow models.

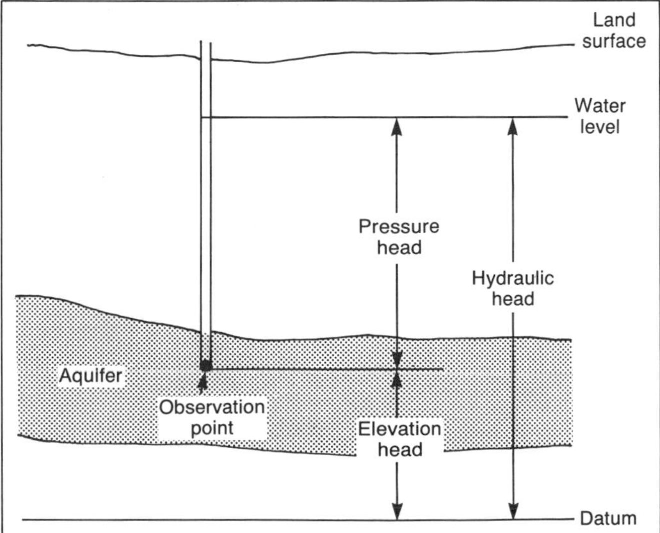

Figure 6F.9. Relation between hydraulic, elevation and pressure heads.

Shallow and intermediate flow systems

The existence in the Interior Plains Hydrogeologic Region of both shallow local groundwater flow systems and deeper, more extensive intermediate flow systems has been demonstrated in field studies by observers such as Meyboom (1963a) and Freeze (1969). These studies substantiated the predictions of the mathematical model of Toth (1963). The Assiniboia area cross-section of Freeze (1969; Fig.6F.10), for example, shows recharge through the glacial sediments, a local flow system near the Mulvaney slough, and an underlying intermediate flow system. Flow in the intermediate system is concentrated in a sand member of the Eastend Formation. A major permeability contrast (about five orders of magnitude) between the sand member and the underlying Bearpaw shale strongly suggests that the lower boundary of the intermediate flow system is the top of the Bearpaw Formation. The characteristics of shallow groundwater flow systems in southwestern Manitoba were investigated by Eilers (1973). He found evidence that an elevation difference of as little as 1 m over a lateral distance of about 60 m was sufficient to generate a small localized independent flow system.

Buried-valley aquifers can play a significant role in the shallow to intermediate flow regime (Lennox et al., 1988). Recharge to these aquifers is both from local precipitation that has infiltrated overlying aquitards and by lateral inflow from adjoining bedrock or surficial aquifers. Discharge is commonly into major streams of the present-day drainage network, which exert a strong controlling influence on the buried-valley flow systems. Where there are several distinct discharge areas along the length of a buried valley, as in the case of the Hatfield Valley aquifer (Fig. 6F.7), the valley is characterized by a number of discrete flow systems separated by hydraulic head divides. Where basal or intertill sand and gravel are sufficiently permeable and continuous and there is a good hydraulic connection with adjacent bedrock and surficial deposits, the buried valley also acts as a local drain (Le Breton, 1963; Meneley, 1972).

Deep flow and bedrock flow systems

The present level of understanding of deep flow systems could never have been achieved without access to the great mass of information on bedrock strata and formation waters, and their chemical and physical properties, that has been gathered in the course of oil and gas exploration in the Western Canada Basin. Investigations by Bachu, Hitchon, Jones, Majorowicz, Toth and van Everdingen, and their various co-workers referred to in this and later sections, have all relied heavily on this source of information.

Van Everdingen (1968), for example, used oil and gas exploration data in his investigation of the geochemistry of deep formation waters of the Western Canada Basin, which provided evidence for the existence of major basin-wide groundwater flow systems and for the influence of both topography and geology on those systems. The Rocky Mountains, the Cypress Hills, and other elevated areas were identified as the main recharge areas, although van Everdingen speculated that faults and high-angle thrusts along the eastern edge of the mountain belt may isolate groundwater flow systems in the belt from those in the sedimentary basin. This hypothesis fits with the suggestion of Hitchon (1969a) that the western boundary of the sedimentary basin flow system is somewhere in the Foothills Belt.

The groundwater flow regime is more precisely defined near the eastern boundary to the hydrogeologic region. A representative cross-section of the southeast portion of the region (Fig. 6F.11) shows both intermediate and regional flow systems. Toward the southwest of this area, these systems are isolated from each other by impermeable salt beds of the Elk Point Group. These beds, in particular those of the Prairie Formation, are found throughout much of southern Saskatchewan, eastern and northern Alberta, and in part of southern Manitoba (Grayston et al., 1966). Wherever they are found, they function as aquicludes to separate the upper and lower parts of the flow regime.

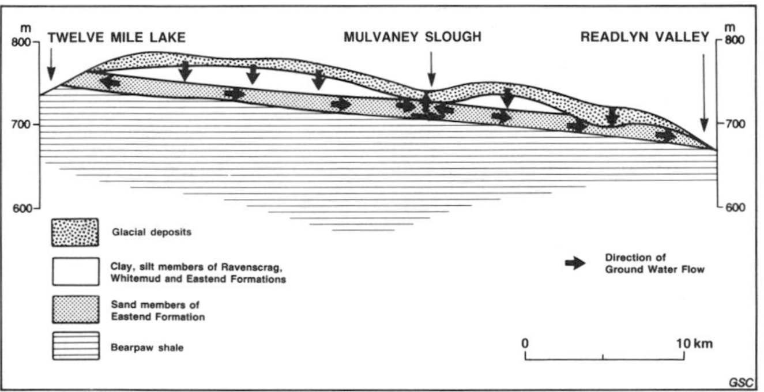

Figure 6F.10. Groundwater flow systems near Assiniboia, Saskatchewan (after Freeze, 1969).

In the updip direction, where the halite beds have been removed by salt solution (Bannantyne, 1960; de Mille et al., 1964; van Everdingen, 1968), intermediate and regional flow systems are no longer physically separated and their waters can mix. Relatively fresh waters of meteoric origin coming from the southwest dissolve salt as they move along the upper surface of the halite beds near the erosional edge. Some of these waters subsequently mix with saline Lower Paleozoic formation waters moving updip from deep in the basin and with other waters of meteoric origin moving downward from recharge areas near the Manitoba Escarpment. As a result, waters in the Lower Paleozoic rocks farther east near the western edge of the Canadian Shield can be relatively fresh (van Everdingen, 1968) and do not show the effects of salt solution (Hitchon et al., 1969). Undoubtedly, too, in this area, it is likely that flow systems become more limited in areal extent and controlled by local topography, as might be expected where the flow region pinches out over a relatively shallow impermeable boundary. This would also give rise to relatively fresh groundwaters.

Some saline waters resulting from halite solution obviously experience little or no mixing with waters from other sources. Salt springs emerging from Devonian outcrops directly overlying the Precambrian basement in northeastern Alberta have been shown to originate from halite solution by infiltrating waters of meteoric origin (Hitchon et al., 1969). To the southeast, in the area of the cross-section shown in Figure 6F.11, the Devonian rocks around Lake Winnipegosis are also noted for their saline springs (Cole, 1915; Cameron, 1949). Bannantyne (1960) first suggested that concentrated brines in Paleozoic formations of southwest Manitoba are due to dissolution of Prairie Formation salt beds. The same explanation was given by Van Everdingen (1968) for the saline springs in the Lake Winnipegosis area. He also suggested that high-permeability channels in the carbonates of the Middle Devonian Winnipegosis Formation likely favour the

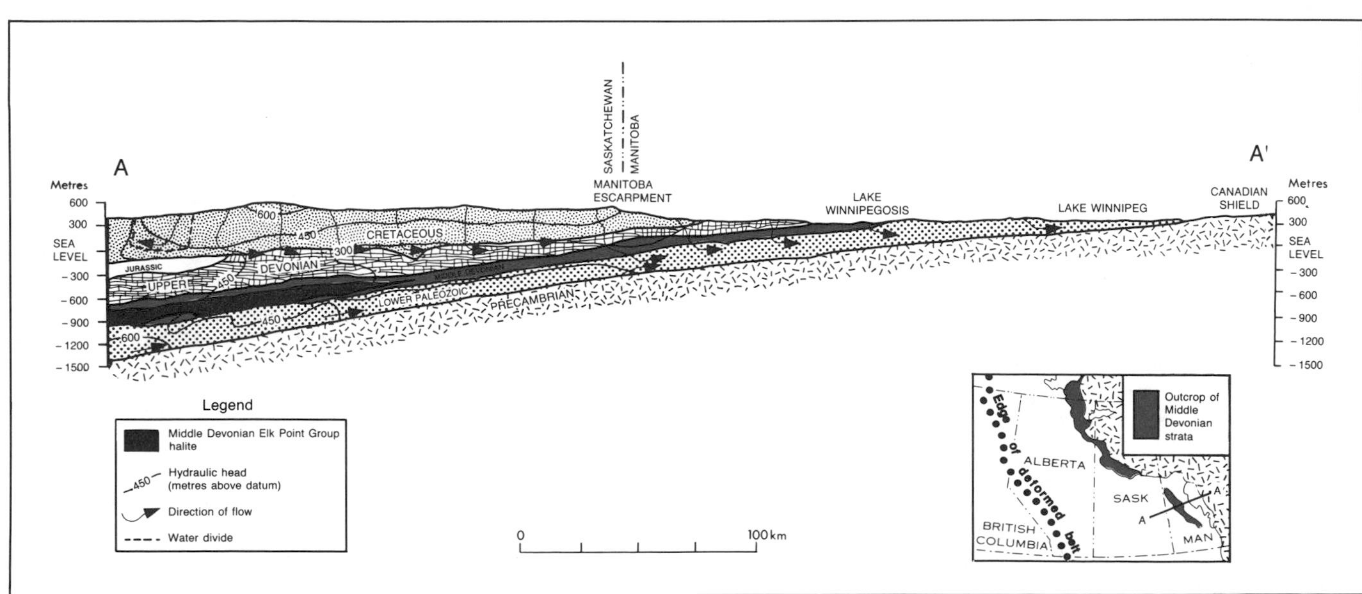

Figure 6F.11. Groundwater flow systems, southeastern portion of Interior Plains Hydrogeologic Region (after Hitchon et al., 1969).

movement of salt-solution brines to spring sites. They might also tend to reduce the effects of mixing with groundwaters from other sources.

Information on flow systems along the eastern margin of the regional flow regime is most abundant for southern Manitoba. Topographic features in southwestern Manitoba clearly exert a significant modifying effect on the basin-wide flow system. Both van Everdingen (1968) and Hitchon et al. (1969) have attributed the relatively low concentrations of total dissolved solids in formation waters from Paleozoic rocks in that area to dilution of deep formation waters by water entering along the Manitoba Escarpment.

In his review of the hydrogeology and hydrochemistry of the Lower Cretaceous Swan River Formation, Rutulis (1984) noted that deep regional flow predominates in this formation in the area west of the Manitoba Escarpment, whereas freshwater recharge contributes significantly to flow in the more northerly parts of the formation in those areas along and to the east of the escarpment. Freshwater recharge to this formation is also inferred for the Turtle Mountain area straddling the Manitoba-North Dakota boundary.

Betcher (1986) reached similar conclusions for the Middle Ordovician Winnipeg Formation in Manitoba. Waters from this formation in southwestern Manitoba have total dissolved solids approaching 300 000 mg/L. Total dissolved-solids concentrations decrease to the east and northeast, approximately in the direction of regional groundwater flow, to values of about 2000 mg/L. A freshwater zone in the formation in the extreme southeast of Manitoba has flow directed towards the northwest and total dissolved solids of less than 600 mg/L. These results, supplemented by geochemical analyses of a number of selected formation water samples, led Betcher to conclude that the brines encountered in the deeper southwestern portions of the Winnipeg Formation in southwestern Manitoba are progressively freshened by the influx of locally recharged "fresh" water. Isotopic (^{18}O) data tended to support this concept.

Render's (1970) detailed study of the hydrogeology of the metropolitan Winnipeg area demonstrated the existence of important groundwater flow systems strongly influenced by recharge at local topographic highs to the north and northwest of Winnipeg and by drainage into the valley of the Red River, which runs through the city from south to north. Render's study focused on the "Winnipeg Carbonate Aquifer", a 15- to 30-m section of karstic limestones and dolomites at the top of the Upper Ordovician Red River Formation. Solution acting over the past several thousand years has substantially increased the permeability of this section by enlarging the openings in a network of fractures, joints, and bedding planes. Render found flow in the carbonate aquifer to be dominated by a pronounced low in the piezometric surface, centred under central Winnipeg and due to extensive exploitation of the aquifer over a period of many decades. As a result, flow for much of Render's study area was directed towards this central low, and several well-defined flow systems could be clearly identified. Fritz et al. (1974) interpreted isotopic (^{18}O) and sulphate data for Winnipeg Carbonate Aquifer water samples to derive a much more detailed picture of flow towards the central low. They were able to demonstrate that recharge for flow systems in the area was in the uplands of the Red River basin. Unique isotopic and geochemical characteristics of individual flow systems were tentatively attributed to the effects of land clearing and agricultural development in recharge areas and to recharge of certain waters as long as 10 000 years ago, when the local climate had yet to recover from the effects of the last glaciation.

The analysis of fluid potential data for the Western Canada Basin (Hitchon, 1969a, 1969b) and a later study (Hitchon, 1984) of the relation between hydrodynamics and geothermal gradients also demonstrated the influence of major variations in topography and geology on hydraulic heads. A major finding was that, despite significant variations in basin geology, the fluid potential in any part of the basin corresponds closely to and is largely determined by the fluid potential at the topographic surface in that part of the basin. On a basin-wide scale, the analysis suggested that there is flow from recharge areas in the Foothills Belt (elevations ranging up to about 1800 m) towards low-lying discharge areas at elevations as low as 300 m adjacent to the Canadian Shield in northeastern Alberta and in Manitoba. These studies provide further evidence that upland areas not part of the Foothills Belt are significant sources of recharge to the hydrogeological flow system in the basin. Their importance as recharge sources is related to their physical size. The most important noted by Hitchon (1984) include the Caribou Mountains just south of the 60th parallel and the Swan Hills, both in Alberta, and the Cypress Hills in southeastern Alberta and southwestern Saskatchewan. Smaller upland features, such as the Whitemud Hills in the southeast portion of the Clear Hills Upland (Fig. 6F.2), also exert an effect to depths of well over 1000 m. Major rivers, such as the South Saskatchewan, act as local drains and their drawdown effects can be observed, in some areas, to depths of more than 1500 m. The investigation by Toth (1978) of gravity-induced cross-formational flow of formation waters in the 53 000 km^2 Red Earth region of northern Alberta provided further substantiating evidence. Topographic effects of such features as the Buffalo Head Hills, Birch Mountains Upland, and Utikuma Upland (Fig. 6F.2) persist to depths of about 1200 m or more. At these depths in this area flow has penetrated into the upper portion of the middle carbonate-evaporite hydrostratigraphic unit.

A detailed study of groundwater and terrestrial heat flow in the Cold Lake area of east-central Alberta was carried out by Bachu (1985). This 23 700 km^2 area adjacent to and north of the North Saskatchewan River lies mainly within the eastern Alberta Plains but includes a small part of the Mostoos Hills Upland (Fig. 6F.2). Information for the Cambrian rocks of the lower clastic hydrostratigraphic unit in this area is sparse, but the unit can be considered as an aquifer in which flow is regional and directed from south to north. The middle carbonate-evaporite unit, comprising the Watt Mountain, Beaverhill Lake, and Cooking Lake formations, constitutes another important regional aquifer in which groundwater flow is from the southwest to the northeast.

Although the Cold Lake area is relatively flat and low-lying, Bachu's (1985) analysis indicated that groundwater flow throughout most of the Cretaceous aquifers and aquitards of the upper clastic hydrostratigraphic unit is vertically downward, because the area as a whole is a groundwater recharge area. In the lowermost portion of this unit, the Wabiskaw and McMurray sandstones of the Mannville Group, groundwater flow is deflected laterally in accordance with the local hydrogeology of the underlying middle carbonate-evaporite hydrostratigraphic unit. Where the Mannville Group is underlain by shales of the Ireton Formation, flow at the base of the Mannville Group is to the west. In the eastern part of the Cold Lake area, where the Ireton Formation is absent, flow in the Mannville is eastward together with the flow in the immediately underlying Cooking Lake-Beaverhill Lake-Watt Mountain aquifer. To the northwest of the area, the Ireton Formation is overlain by carbonates and evaporites of the Grosmont Formation. This formation acts as a groundwater "drain", deflecting flow in the overlying Mannville Group toward the northwest.

Evidence was found by van Everdingen (1968), Hitchon (1969a, 1969b, 1984), and Toth (1978) for the modifying influence of major permeability variations on the regional groundwater flow system in the Western Canada Basin. Van Everdingen concluded that irregularities in the system are caused by major inhomogeneities in the geological framework. Hitchon's cross-sections and maps (1984) revealed two important and probably related fluid-potential features that cannot be explained by the effects of surface topography alone. The first of these is a ridge of high fluid potential extending from the Cypress Hills area to the Canadian Shield topographic high near Peter Pond Lake. To the west of, and adjoining, this feature is a low-potential trough, which extends northward from about 50°N latitude, perhaps as far as the Fort McMurray area (Hitchon, 1969a). Hitchon (1984) concluded that the Interior Plains Hydrogeological Region is characterized by two broad, regional flow systems separated from one another by the high-potential ridge (Fig. 6F.12).

The high-potential ridge may reflect the presence of the underlying basement high of the Sweetgrass Arch in southeastern Alberta. Its continuation to the northeast is likely similarly influenced by the topographic high at the boundary with the Canadian Shield and the rise in the basement toward this high. The adjoining low-potential trough may also be, in part, a consequence of the basement topography. Hitchon (1969b), however, suggested a geological origin for this feature related to preferential flow through the relatively high permeability Upper Devonian and Carboniferous rocks. It seems likely that some combination of basement topography and variations in permeability together provide the explanation for the Alberta low-potential trough.

Attention also was called by van Everdingen (1968) to the probable role of permeable reef and carbonate rocks as high-permeability paths that could have a significant effect on groundwater circulation in the Western Canada Basin. On a finer scale, Hitchon (1969b) noted evidence for a number of distinct drainage channels within the Upper Devonian and Carboniferous carbonate rocks, identifiable in graphs of the variation of formation pressure with depth. He also noted the influence of rapid lithological changes, such as a lateral change from dolomite to anhydrite and a fourfold lateral increase in shale percentage, on hydraulic head. Both of these changes were reflected by an appreciable localized steepening of the hydraulic gradient.

Summary

In summary, the studies done to date indicate that surface topography exerts a major controlling influence on groundwater flow in the Western Canada Basin, both at shallow depths and on a basin-wide scale. Major rivers act as local drains, notably at the discharge end of buried interglacial and preglacial valley systems, and their effects can be observed in some cases to depths of more than 1500 m. The basin-wide flow regime is, however, significantly modified in response to major structural features on the top of the impermeable Precambrian basement and as a result of permeability variations within the sedimentary rocks of the basin. The permeability contrasts having the greatest impact on the basin-wide flow regime are those within the middle carbonate-evaporite hydrostratigraphic unit. They include high-permeability carbonate zones, which serve as conduits for groundwater flow, and thick, effectively impermeable salt beds, which act as barriers to prevent flow between overlying and underlying strata. The eastern and western limits of the basin-wide flow regime are to some degree undefined. To the west there is a lack of precise knowledge of the effects of folding and faulting in the Foothills Belt on hydraulic continuity; to the east the thinning and pinching out of the flow region above the impermeable Precambrian basement leads to the intermixing of regional and local flow systems and leaves the boundaries between them uncertain.

HYDROGEOCHEMISTRY

The chemical composition of groundwater depends on several factors, including: (1) the chemical composition of the rainwater or snowmelt that provides recharge to hydrogeological flow systems; (2) the amount of rainwater or snowmelt that actually infiltrates; (3) the balance between infiltration and evaporation fluxes; (4) the chemical and biological reactions that take place as

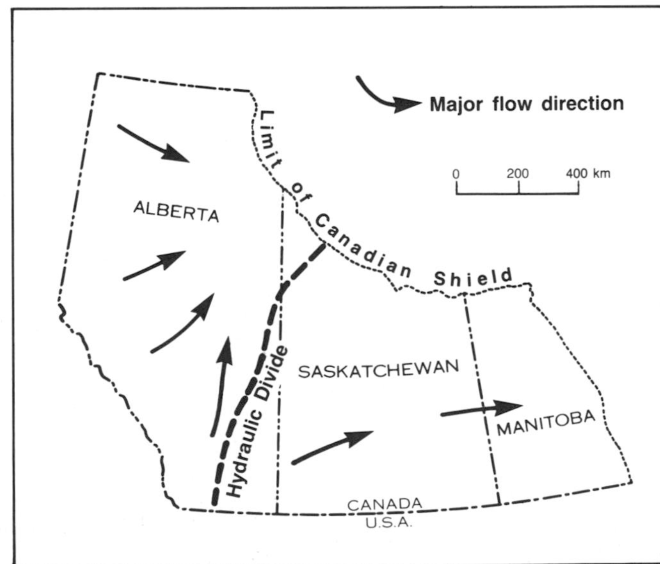

Figure 6F.12. Major regional flow systems in the Interior Plains Hydrogeologic Region (after Hitchon, 1984).

recharge water infiltrates the soil zone and percolates down to the water table; and (5) the path taken once the water has traversed the unsaturated zone and becomes part of the hydrogeological flow system beneath the water table. Changes in the chemical composition of the water along this path depend on the nature of the geological materials with which the water comes into contact, the sequence in which these materials are encountered, the time spent in contact with them, and the temperature, pressure, and other conditions during the period of contact. Freeze and Cherry (1979) reviewed these factors in considerable detail. The following summary is based to a large extent on their review, but reference is also made to the excellent detailed studies by Moran et al. (1978), Cravens and Ruedisili (1987), and Houghton et al. (1987) on the chemistry of waters in the unsaturated zone and in shallow groundwater systems.

Precipitation

Rain and snowmelt are characteristically very dilute solutions, generally containing at most a few tens of milligrams of dissolved solids per litre. Under natural conditions they are oxidizing and slightly to moderately acidic, with pH in the 5 to 6 range. Airborne emissions in industrialized and urbanized areas add sulphur and nitrogen compounds to the air, which react with atmospheric water, forming sulphate, nitrate, and hydrogen ions, thereby increasing the acidity of precipitation. Where such airborne contamination exerts an influence, the pH of precipitation can be reduced to less than 4 and the effect may extend for many hundreds of kilometres from the source. Acidic precipitation has not been identified as a major problem in the Interior Plains Hydrogeological Region, but industrial sources in the region, such as natural-gas processing plants and thermal generating stations, could affect the acidity of precipitation and might, ultimately, influence the quality of groundwaters in the region.

Regional variation in sulphate content for central Alberta precipitation and its dependence on sulphur dioxide emission from gas-processing plants in the area were investigated by Summers and Hitchon (1973). They found summer convective storms to be particularly efficient in scavenging sulphur dioxide from the atmosphere, causing one third to one half of the total emitted locally to be deposited as sulphate within 40 km of its source. Snow, in contrast, was found to be a very inefficient SO_2 scavenger. Highly localized deposition such as this could eventually give rise to correspondingly localized variations in the composition of shallow groundwaters.

Water in the unsaturated zone

Recharge water infiltrating and moving downward through the unsaturated zone is modified significantly during its passage towards the water table. Decaying organic matter reacts with the dissolved oxygen in the water to produce CO_2 and water. More CO_2 is added to the system by respiration of plant roots. Carbon dioxide and water combine to form carbonic acid (H_2CO_3). The acidity of the water increases appreciably as a result. Freeze and Cherry (1979) calculated a typical pH value of 4.3 to 4.5 for water in the soil zone, a value well below the characteristic values for uncontaminated precipitation.

The dissolved carbonic acid in the infiltrating water reacts with minerals in the unsaturated zone. Particularly important in the Interior Plans region are the reactions with calcite ($CaCO_3$) and dolomite ($CaMg(CO_3)_2$):

$$CaCO_3 + H_2CO_3 \rightarrow 2HCO_3^- + Ca^{2+}$$
$$CaMg(CO_3)_2 + 2H_2CO_3 \rightarrow 4HCO_3^- + Ca^{2+} + Mg^{2+}$$

Glacial tills from which many regional soils are derived contain carbonates in amounts varying from about 1 to about 40% expressed as percentage $CaCO_3$ (Fig. 6F.6; St. Arnaud, 1976). The main dissolution of calcite and dolomite probably takes place in the unsaturated zone; groundwater samples taken at or just below the water table are found to be saturated to supersaturated with respect to calcite and dolomite (Grisak et al., 1976). These groundwaters characteristically have pH values in the 7-8 range and HCO_3^- values in the 300-700 mg/L range.

Pyrite (FeS_2) oxidation possibly may be significant for the chemical composition of water in the unsaturated zone (Moran et al., 1978). R.O. van Everdingen (pers. comm., 1988) suggested that where pyrite is present and atmospheric oxygen available the following probable sequence of reactions, beginning with pyrite oxidation, is responsible for a dramatic increase in the acidity of the water, making possible further dissolution of calcite and dolomite and the production of additional bicarbonate.

$$2FeS_2 + 7O_2 + 2H_2O \rightarrow 2Fe^{2+} + 4SO_4^{2-} + 4H^+$$
$$2Fe^{2+} + 2H^+ + \tfrac{1}{2}O_2 \rightarrow 2Fe^{3+} + H_2O$$
$$2Fe^{3+} + 6H_2O \rightarrow 2Fe(OH)_3 + 6H^+$$

This chain of reactions is cited by Moran et al. (1978) as one of the strongest acid-producing sequences known to occur in natural geological systems. They identified pyrite oxidation as the cause of HCO_3^- levels in the 500-1500 mg/L range in some shallow groundwaters of west-central North Dakota. This area is part of the Western Glaciated Plains region of North America (Lennox et al., 1988), which includes the Interior Plains region and therefore shares many hydrogeological characteristics with it. Despite the similarity, Cherry (1972), Grisak et al. (1976), and Davison (1976) were able to explain the observed chemical properties of a selection of glacial-drift groundwaters in Manitoba and Saskatchewan without having to call on pyrite oxidation above or below the water table as a necessary step in their geochemical evolution. However, both Hackbarth (1978) and Wallick (1981) have suggested that this process has a role to play in some parts of Alberta. Hackbarth (1978) attributed anomalously high sulphate values for some groundwaters of the Grande Prairie-Beaverlodge area of Alberta to the oxidation of pyrite-rich till derived from bedrock of the Smoky Group. Sulphate concentrations in the most prominent anomalous area range up to and over 1000 mg/L. Bicarbonate concentrations in this area are also anomalously high, exceeding 1200 mg/L. Wallick (1981) identified finely disseminated pyrite in the soil and unsaturated zone as a source of sulphate in certain groundwaters of east-central Alberta. The pyrite in this area is also bedrock-derived, in this case from coal seams and clay shale of the Belly River Group. Observed sulphate and bicarbonate concentrations are relatively low, however, being less than 500 and less than 700 mg/L, respectively.

Climatic conditions also affect the composition of water in the unsaturated zone (Bouwer, 1978; Fig. 6F.13). In humid areas with abundant rainfall, minerals released from the soil through physical, chemical, and biological processes are regularly and frequently mobilized and carried down into the saturated groundwater zone by infiltrating precipitation. In more arid areas, where evaporation and transpiration predominate, salts accumulate in the soil water, and calcite and gypsum will precipitate in or near the root zone (Moran et al., 1978; Cravens and Ruedisili, 1987; Houghton et al., 1987). Eventually a recharge event of sufficient magnitude will leach them out of the soil and wash them down to the water table. The leachate, which then becomes part of the groundwater regime, is considerably enriched in dissolved constituents. Meyboom (1967) noted that regional variations in total dissolved solids for shallow till of the Interior Plains Hydrogeological Region can be correlated with regional variations in aridity. The most mineralized shallow groundwaters are found in till of the central, drier areas of southern Saskatchewan and Alberta.

The chemical characteristics of water in the unsaturated zone are also affected by exchange of cations in solution with cations adsorbed on soil or rock particles. Ions are attracted to these small particles because they have surfaces with small unbalanced electrical charges. Divalent cations, such as Ca^{2+}, are more strongly attracted than are monovalent cations, such as Na^{+}. Hence, when Ca^{2+} and Mg^{2+} ions are produced as a result of carbonate mineral dissolution, they tend to become adsorbed onto the soil or rock particles in exchange for Na^{+} ions, which go into solution. The removal of calcium and magnesium from solution by cation exchange reduces the saturation of the soil water with respect to calcite and dolomite and facilitates more carbonate mineral dissolution.

Groundwater

Soil water reaching the water table becomes part of the groundwater flow regime and its chemical properties are further modified as it moves through the saturated zone beneath the water table. Continuing contact with, and movement through, a variety of geological materials leads to an increase in total dissolved solids. The concentrations of the various major ions are determined by the minerals encountered, the order in which they are encountered, and their availability and relative solubility. Concentrations generally increase with distance along the flow path, but relative concentrations may increase or decrease. Thus one species may predominate along one segment of the flow path, only to become of lesser or even minor importance at some later stage. Chebotarev (1955) concluded that if flow paths are sufficiently long and deep, groundwater originating as precipitation tends to evolve towards a chemical composition similar to that of seawater. He determined that the anion sequence that commonly characterizes the evolution of the carbonate-rich waters filtering downward from the soil zone could be summarized as follows:

$$HCO_3^- \rightarrow HCO_3^- + SO_4^{2-} \rightarrow SO_4^{2-} + HCO_3^- \rightarrow$$

$$SO_4^{2-} + Cl^- \rightarrow Cl^- + SO_4^{2-} \rightarrow Cl^-$$

End members of the sequence can be highly saline brines, much more saline than seawater, if flow penetrates to great depth and there is contact with highly soluble chloride minerals, such as halite and sylvite (Freeze and Cherry, 1979).

The Chebotarev sequence is a valuable conceptual tool in understanding the chemical evolution of groundwaters originating as precipitation. Other origins are possible, however. In particular, deeper waters in a sedimentary basin may have originally been brackish or marine, trapped in the interstices of the sediments at the time of deposition. The observed chemical properties of formation waters in the Western Canada Basin can be explained on the basis of just such an origin, taking into account the subsequent modifying effects of a variety of physical, chemical, and bacteriological processes (Hitchon, 1966). In general, the concentration of dissolved solids in formation waters increases gradually with age and depth. The most important influence in chemical composition is the environment at the time of deposition. Modifying processes include osmosis, chemical reactions with the surrounding rocks, and dilution with fresher groundwaters. Highly saline springs at the discharge end of deeper flow systems with dissolved solids about 90% NaCl indicate that halite solution is one of the important chemical reactions. Such springs discharge from Devonian rocks in the Lake Winnipegosis and Lake Manitoba areas, about 200 km west of the boundary of the Canadian Shield (Cole, 1915; Cameron, 1949). Total dissolved solids for the spring waters were determined to range from about 26 000 to about 63 000 mg/L. Further evidence for discharge from these Devonian strata is provided by lake-water analyses (Thomas, 1959), which gave chloride concentrations up to about 600 mg/L.

Quaternary hydrostratigraphic unit

North American glacial-drift waters that have not been modified by passage through bedrock are classified into three main types (Freeze and Cherry, 1979). Type I waters are found in glacial deposits of the Canadian Shield and

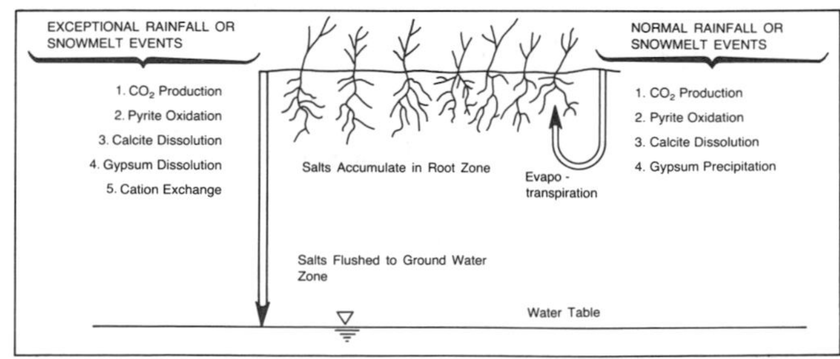

Figure 6F.13. Schematic representation of physical and chemical processes affecting the chemical properties of water in the unsaturated zone (after Moran et al., 1978).

other areas of igneous and metamorphic rocks. Type II and Type III both occur in the Interior Plains Hydrogeological Region, but Type III predominates.

Type II glacial-drift waters are hard to very hard and slightly alkaline. Total dissolved solids concentrations are generally less than 1000 mg/L, with Ca^{2+}, Mg^{2+}, and HCO_3^- being the dominant ions. These are the characteristics, as has already been noted, of waters reaching the water table from the unsaturated zone.

The Type II glacial-drift waters, found at or just below the water table in groundwater recharge areas, evolve into Type III waters as they migrate through the drift. Like Type II waters, Type III waters are slightly alkaline, but total dissolved solids are commonly in the 1000 to 10 000 mg/L range. These waters have major concentrations of Na^+, Ca^{2+}, Mg^{2+}, HCO_3^-, and SO_4^{2-}, with SO_4^{2-} as the dominant anion. Evolution to Type III is a consequence of dissolution of calcite, dolomite, gypsum, and anhydrite, and cation exchange. Dissolution of silicate minerals, despite their relative abundance in the drift, plays a comparatively minor role because they are much less soluble than the carbonates and sulphates.

Good representative water samples are difficult to obtain directly from glacial till, because of its low hydraulic conductivity. However, many samples have been collected from intertill aquifers or from other relatively permeable units in the Quaternary hydrostratigraphic unit. The chemical properties of waters in these aquifers must also reflect flow history because they are genetically related either to Type II or Type III glacial-drift waters (in the case of flows confined to the Quaternary unit) or to bedrock waters (in the case of flows returning to the surface from greater depths).

The freshest waters in the Quaternary hydrostratigraphic unit are found in surface and near-surface glaciofluvial sands and gravels, in sand dunes, and in the alluvial deposits of present-day rivers. In Saskatchewan, such aquifers typically contain quartz with minor amounts of carbonates and igneous and metamorphic rock fragments (Maathuis, 1986). Most of the aquifer minerals are relatively insoluble, and waters in these aquifers are generally of the calcium bicarbonate type with less than 1000 mg/L total dissolved solids. Total dissolved solids values of 200 to 1200 mg/L were found in alluvial terrace sands and gravels in the Peace River district of Alberta (Jones, 1966). Groundwaters from shallow, buried, preglacial sands and gravels with direct hydraulic connections to present-day rivers are also low in total dissolved solids. Total dissolved solids values for buried-valley aquifers at Edmonton and Medicine Hat, Alberta, are less than 500 mg/L (Stein, 1976; Meyboom 1963b). Water in glaciofluvial sands and gravels of the Assiniboine delta aquifer east of Brandon, Manitoba, seldom contains more than 700 mg/L total dissolved solids.

Information on total dissolved solids and ionic species for Alberta glacial-drift groundwaters clearly supports the Type II - Type III classification of Freeze and Cherry (1979) (see, for example, Barnes, 1978; Ozoray, 1974; Stein, 1976; Vogwill, 1978). A few total dissolved solids values (Le Breton, 1963; Borneuf, 1976) have been found, however, that significantly exceed the 10 000 mg/L upper limit assigned to Type III waters by Freeze and Cherry.

Trends indicative of geochemical evolution are generally difficult to discern in glacial-drift groundwaters. However, Borneuf (1976) noted that total dissolved solids increase downslope from the Cypress Hills, an observation in agreement with the conclusion drawn from flow-system studies by van Everdingen (1968) and Hitchon (1969a) that the Cypress Hills and other major topographic highs are important recharge areas. Analytical results for buried-valley aquifers in Saskatchewan (Maathuis and Schreiner, 1982a, b; Schreiner and Maathuis, 1982) are also, in some cases, suggestive of a similar geochemical evolution. This is particularly true for the Hatfield Valley aquifer (Fig. 6F.7) for which reaches of relatively fresh water alternate with reaches of more mineralized water throughout most of its 550-km course across the province. Buried-valley waters in Saskatchewan are mainly of Type III or transitional between Types II and III. Relatively fresh calcium bicarbonate waters are found in the Hatfield Valley near the Alberta border. Taken as a whole, the hydrogeochemical evidence supports the views that recharge to buried-valley aquifers is commonly derived from local precipitation and that discharge is commonly into major streams of the present-day drainage network.

Bedrock hydrostratigraphic units

Geochemical processes already considered for waters in the unsaturated zone and in the Quaternary hydrostratigraphic unit, such as dissolution of carbonate and sulphate minerals and cation exchange, also affect the chemical characteristics of groundwaters from shallow bedrock. Dissolution of halite, and alteration of feldspars, micas, and clay minerals may also be important. Cation exchange and CO_2 generation by biochemical reduction of sulphate are particularly significant processes in sedimentary rocks of the Interior Plains Hydrogeological Region (Freeze and Cherry, 1979; Wallick, 1981).

Cation exchange, as already indicated, removes calcium and magnesium ions from solution, making the groundwater undersaturated with respect to the carbonate minerals and allowing dissolution of these minerals to continue. If there is an abundant supply of hydrogen ions to facilitate carbonate dissolution, the linked processes of carbonate dissolution and cation exchange can yield groundwater with very low calcium and magnesium concentrations but with HCO_3^- levels of more than 1000 mg/L. Hydrogen ions necessary for the process are commonly available as a result of pyrite oxidation in the unsaturated zone.

At depths beyond a few hundred metres, temperatures and pressures increase significantly and these changes affect chemical reaction equilibrium points, dissociation products and constants, and mineral solubilities and solution rates. The solubilities of both $CaCO_3$ and $CaSO_4$ increase appreciably with dissolved NaCl concentration and pressure, and decrease with increasing temperature (van Everdingen, 1968). The chemical composition of water entrapped in the rock at the time of deposition also plays a role. About 85% of the sedimentary rocks of the Western Canada Basin were deposited in a marine environment. All these factors must be given consideration in any detailed analysis of the properties of deep bedrock waters.

Published tabulations and other compilations of geochemical information for bedrock waters of the Western Canada Basin were cited by Hitchon (1966). Major-ion concentrations and total dissolved solids were given by Meyboom (1967) for 65 water samples from formations ranging in age from Cambrian to Late Cretaceous. These results show total dissolved solids reaching a maximum of over 300 000 mg/L for waters from Devonian and older strata in Alberta and Saskatchewan. These deep waters (and many of the shallower ones as well) are predominantly of the sodium chloride type. Porter and Fuller (1959) constructed salinity maps for Cambrian to Silurian strata of the northern Williston Basin and adjacent areas. Hitchon (1966), on the basis of an examination and interpretation of analytical data for more than 10 000 formation-water samples, modified the maps of Porter and Fuller and produced additional chloride-content maps for stratigraphic units up to the Upper Cretaceous Belly River Formation. Chloride values on the maps exceed 190 000 mg/L for many Cambrian, Ordovician, Silurian, and Middle Devonian units. They are greater than 225 000 mg/L for parts of the Upper Devonian Beaverhill Lake Formation. Hitchon found the waters ranged from practically fresh to a total dissolved solids content of 675 000 mg/L for a sample from the Upper Devonian Nisku Formation. Sodium and chloride were the predominant dissolved ions, with calcium being next in importance.

Hitchon's maps indicate a gradual increase in total dissolved solids with depth, geologic age, and approach to the disturbed belt. They provide information as well on flow within and between stratigraphic units. Hitchon concluded that the most important factor governing observed regional variations in formation-water chemistry has been the environment of deposition. Post-depositional effects, such as the removal or addition of dissolved solids and mixing, were judged to have a somewhat lesser effect.

A later sequence of studies (Hitchon and Friedman, 1969; Billings et al., 1969; Hitchon et al., 1971) involved an in-depth analysis of the isotopic and chemical properties of a number of formation-water samples. These studies confirmed Hitchon's earlier conclusion (1966). In particular, factor analyses of the relations among 20 major and minor chemical components (Hitchon et al., 1971) clearly demonstrated that deeper formation waters were once seawaters, and that they have been subsequently modified by two major and competing processes: concentration due to membrane filtration, and dilution by fresh-water recharge. These formation waters are, on the average, about 1.3 times as saline as present-day seawater. Their increased salinity was shown to be primarily due to an increase in NaCl, and was associated with dissolution of salt from evaporite beds.

Other investigations of deep bedrock systems considered a number of chemical and physical processes that might influence the chemical evolution of deep western Canadian groundwaters, including, in particular, salt filtering by reverse osmosis through shales acting as semipermeable membranes (van Everdingen, 1968; Bredehoeft et al., 1963; Graf, 1982; Kharaka and Berry, 1973; Kharaka and Smalley, 1976). Van Everdingen concluded from his study that the chemical composition of deep formation waters in the Western Canada Basin is a reflection of the major regional groundwater flow systems, and that apparently anomalous pressures and salinities can probably in many cases be explained by the membrane effect. He also concluded that deep groundwater flow systems affect hydrocarbon accumulation in the basin.

Toth (1980) concluded that cross-formational groundwater flow is the principal agent in the transport and accumulation of hydrocarbons. Hydrocarbons migrate in the general direction of groundwater flow until they encounter and are held back by stratigraphic or structural traps. Formation and growth of oil and gas pools in these traps is favoured by low hydraulic gradients and flow velocities. Toth noted that particularly favourable hydrodynamic conditions exist in the regions of low-velocity flow that typically underlie areas of upward regional flow. Petroleum exploration data confirmed his theoretical expectations for these areas. Further confirmation has been provided as a result of studies of the interrelations between groundwater and terrestrial heat flow regimes (Bachu, 1985; Bachu et al., 1987) and their influence on hydrocarbon occurrences (Hitchon, 1984; Jones et al., 1986; Majorowicz et al., 1985, 1986).

According to Hitchon (1984), geologists have been aware for over a century that hydrocarbons occur more frequently where geothermal gradients locally exceed average regional values. Earlier evidence for this relation was discussed by Klemme (1975) and Roberts (1981). The explanation of the phenomenon lies in the distortion of the terrestrial heat flow regime caused by the convective transport of heat by moving groundwater. In recharge areas, this convective transport is downward, in opposition to the upward conductive flow of geothermal heat from the crust and upper mantle. In discharge areas, both convective and conductive heat flows are directed upward. As a result, isotherms for a uniform medium are displaced downward in recharge areas and upward in discharge areas (Domenico and Palciauskas, 1973). Thus, geothermal gradients are steeper in discharge areas where, as already noted, hydrodynamic conditions favour hydrocarbon accumulation.

Investigations for Alberta (Hitchon, 1984; Jones et al., 1986; Majorowicz et al., 1985) and for Saskatchewan (Majorowicz et al., 1986) demonstrated the general validity of theoretical expectations for the interrelations between fluid flow, terrestrial heat flow, and hydrocarbon occurrence. Jones et al. (1986) noted, however, that the geothermal gradient for one class of Cretaceous oil reservoirs is about 5% less than the gradient external to them, a result they explained by proposing that the source of these pools has been the surrounding shales, so that the oil in this case has undergone only a limited migration. Similarly, Majorowicz et al. (1986) observed that there is clear evidence for a significant contrast in thermal conductivity between the middle carbonate-evaporite hydrostratigraphic unit and the overlying and underlying clastic units. In particular, temperature measurements in a borehole at Regina gave a gradient of 36°C/km for the Paleozoic carbonate-evaporite sequence and 17°C/km for the overlying clastics (Jessop and Vigrass, 1984). The thermal-conductivity contrast, combined with the already-noted hydraulic-conductivity contrast for the three major hydrostratigraphic units, suggests that interpretation of the relations between fluid flow, temperature distribution, and hydrocarbon occurrence may require more than application of simple theory.

Table 6F.3. Consumptive use of groundwater, Prairie Provinces

Sector	Manitoba (10^3 m^3)	Saskatchewan (10^3 m^3)	Alberta (10^3 m^3)	Total (10^3 m^3)	Total Percentage[4]
Agriculture					
Livestock[2]	28 000	45 000	81 000	154 000	90
Irrigation	13 500	1 600	1 900	17 000	0.8
Subtotal	41 500	46 600	82 900	171 000	7.4
Rural[2]	14 000	22 000	35 000	71 000	90
Municipal	5 000	28 000	17 000	50 000	8
Industrial[3]	11 000	16 000	23 000	50 000	2.1
Total	71 500	112 600	157 900	342 000	9.8

1 Adapted from Hess (1986).

2 Estimated that only about 10% of the water consumed by livestock and for rural domestic purposes is drawn from surface-water sources.

3 Excludes water used for cooling purposes by thermal electrical power plants.

4 The total groundwater consumption in column 5 as a percentage of total consumption of water for a particular sector taken from all sources (groundwater and surface water).

A computer-based "dynamic basin analysis" methodology was described by Bachu et al. (1987) for the analysis of terrestrial heat and fluid-flow regimes in a sedimentary basin. The methodology involves the synthesis of available hydrogeological, geothermal, and geochemical information and its incorporation with information on basin geology and stratigraphy to develop a description of the interdependent fluid and heat regimes. The description so developed respects the physical realities that makes these two fields mutually dependent, as well as the other relations that must hold with regard to geochemistry and the geological framework. The methodology is particularly suitable for basins, such as the Alberta Basin, where large amounts of data are readily available.

The dynamic basin analysis approach was used by Bachu (1985) in his analysis of fluid flow and temperature distribution for the Cold Lake area of Alberta. The hydrogeological conclusions from this study have already been cited. The terrestrial heat-flow conclusion of interest is that convective heat transfer is negligible throughout the study area; geothermal heat is transferred from the basement to the land surface by thermal conduction through the stratigraphic column.

ECONOMIC SIGNIFICANCE OF GROUNDWATER

Groundwater in the Interior Plains Hydrogeological Region has been exploited almost exclusively as a source of water for consumptive use. Hess (1986) utilized estimates of groundwater use by various sectors of the economy to determine total Canadian groundwater use in 1981. Table 6F.3, based on his results, summarizes the findings for the three Prairie Provinces.

Estimated total groundwater use in Manitoba, Saskatchewan, and Alberta for 1981 was about $3.4 \times 10^8 m^3$, or about 10% of the total water use, exclusive of that used for cooling at power plants. About half of the groundwater total was used to satisfy agricultural needs, primarily livestock watering. Livestock can tolerate total dissolved solids up to about 7000 mg/L, depending on the species (U.S. Environmental Protection Agency, 1973), but the recommended maximum for farm animals, as given in a joint report by the U.S. National Academies of Sciences and Engineering, is only 3000 mg/L (Bouwer, 1978). The same source also gives recommended concentration limits for a number of elements, including one of 100 mg/L for total nitrogen, either as nitrates or nitrites. Over much of the Interior Plains Hydrogeological Region, groundwater of satisfactory quality is available in amounts sufficient to satisfy stock-watering needs.

Quality standards for irrigation water are more difficult to establish than for stock watering. Quality requirements in this case depend on the crop to be grown, the type and drainability of the soil, and the climate (Bouwer, 1978). Excess dissolved solids reduce crop yields. Excess sodium facilitates the exchange of sodium ions with calcium and magnesium ions attached to soil clays, thereby altering soil structure and reducing soil permeability. Guidelines for the interpretation of water quality for irrigation based on these and other factors are given by Ayers (1975) and reproduced by Bouwer (1978). Ayers suggested that total dissolved solids concentrations up to 480 mg/L should cause no problems for irrigation. Crop yields decrease as concentrations increase beyond this point and problems become severe when concentrations exceed 1920 mg/L. Ayers also listed similar critical values for the sodium absorption ratio (SAR), which is a measure of the relative abundance of sodium compared to calcium and magnesium.

Excess dissolved solids and high SAR values commonly make Alberta and Saskatchewan groundwaters unsuitable for irrigation use. In addition, aquifers in these provinces cannot generally yield water in the relatively large quantities usually required for irrigation purposes. This combination of factors limits groundwater use for irrigation to a small percentage of agricultural use in both Saskatchewan and Alberta. In Manitoba, in contrast, about one third of agricultural groundwater use is for irrigation. The difference reflects the recent development for irrigation purposes of the Assiniboine delta aquifer, described by Rutulis (1972). This aquifer, consisting of ancient glaciofluvial fine sands to coarse gravels, extends over about 2000 km^2. Its saturated thickness averages 15 m and may range up to 30 m. Total dissolved solids are generally in the 200-500 mg/L range and rarely exceed 700 mg/L. This combination of excellent water quality and high yield potential is seldom realized for other prairie aquifers, particularly in areas where soils and soil drainage characteristics also favour irrigated agriculture (L. Gray, pers. comm., 1987). The development of the aquifer for irrigation purposes, which began in the early 1980s, has now become so popular that consideration must be given to possible adverse effects such as nitrate contamination from overfertilization, declining water tables, and reduction of groundwater discharge into the Assiniboine River valley. These possible impacts are being studied to provide guidance on the permissible level of groundwater development (L. Gray, pers. comm., 1987).

A little over 20% of estimated groundwater use in the Prairie Provinces during 1981 satisfied rural domestic needs (Table 6F.3). Although the total amount used is small, groundwater is a vital resource in many interstream areas devoid of reliable surface-water alternatives. However, prairie groundwaters, with the exception of those taken from some Quaternary and shallow bedrock aquifers, are generally mineralized to the extent that they fail to meet established standards or guidelines for drinking-water quality. The 500 mg/L Canadian guideline for maximum concentration of total dissolved solids is rarely satisfied by prairie groundwaters, and the 250 and 150 mg/L limits for chloride and sulphate, respectively, are commonly exceeded. These particular limits are classed as aesthetic objectives (Health and Welfare Canada, 1987). More mineralized waters can be consumed without risk to human health and can, of course, be used for cooking, washing, sanitation, and other domestic needs. Dugouts can be an option but are subject to contamination by surface drainage. Consideration of the limited availability of slightly mineralized groundwaters has led provincial health authorities (see, e.g. LeBreton, 1963; Jones, 1966) to establish somewhat higher maximum concentration limits for total dissolved solids, chloride, sulphate, and some other dissolved constituents.

Municipal and industrial use of groundwater in the Interior Plains Hydrogeological Region during 1981 each amounted to about 15% of the total consumption (Table 6F.3). Larger municipalities are generally located on major rivers or streams and can obtain the large supplies they require from these sources. Many smaller municipalities are located in interstream areas and rely on aquifers for their water supplies. Regina is the largest prairie city that draws any part of its water supply from a groundwater source. In 1986, slightly more than 20% of the city's population of 163 000 was supplied from an aquifer developed in a highly complex intertill system (Environment Canada, 1987). The system includes sands and gravels of the Floral Formation within the Saskatoon Group, and sands and gravels of the Condie Moraine (Maathuis, 1986). Water analyses for the aquifer give average total dissolved solids, sulphate and chloride concentrations of about 1000 mg/L, about 300 mg/L, and less than 10 mg/L, respectively (Meneley, 1972).

A few other prairie municipalities with populations greater than 10 000 also rely in whole or in part on groundwater for their water supplies. Yorkton, Saskatchewan, with a population of about 15 000, takes all of its supply from a complex sequence of permeable glacial sands and gravels of the Sutherland Group (Meneley, 1972) at depths ranging from about 15 to about 60 m (U. Roeper, pers. comm., 1987). Total dissolved solids range from 700 to 2500 mg/L, and sulphate concentrations are considerably in excess of suggested limits in some cases (U. Roeper, pers. comm., 1987). North Battleford, Saskatchewan, with a population of about 14 000, obtains part of its water supply from alluvial deposits of the North Saskatchewan River. Well depths range from 8 to 45 m. The water in this case is primarily river water that has been filtered through the aquifer material. Total dissolved solids range from 1000 to 2500 mg/L. Until 1985, Lloydminster, on the Alberta-Saskatchewan border, obtained its water supply from glacial sands at a depth of about 35 m. Total dissolved solids in this supply were about 1300 mg/L. It was replaced by a surface-water supply because of drawdown effects on the formation and a nearby resort lake (U. Roeper, pers. comm., 1987). Selkirk, Manitoba, and Wetaskiwin, Alberta, also rely in part on groundwater for their municipal supplies, as do the majority of smaller towns and villages (Environment Canada, 1987).

Groundwater use in the prairie provinces for industrial purposes in 1981 was only about 9% of total industrial water use, exclusive of the large amounts of water used by power plants for cooling purposes (Hess, 1986). Detailed information on industrial groundwater use is provided in the Prairie Provinces Water Board study (1982) of water demands for the Saskatchewan-Nelson basin. Injection into oil and gas reservoirs for secondary recovery of hydrocarbons is a major industrial use of groundwater in both Alberta and Saskatchewan, amounting to nearly $15 \times 10^6 m^3$ in 1978, or about one quarter and one half, respectively, of total industrial groundwater use in these provinces. In Saskatchewan most of the injection water used is brackish to saline and derived primarily from the Mannville Group (Table 6F.3).

Other important industrial uses of groundwater in Saskatchewan are for process water in sodium sulphate recovery, and for dissolving and removing contaminants from potash ore. Potash mines near Esterhazy and Rocanville used 1.75 x $10^6 m^3$ of groundwater during 1978. Groundwater use in 1978 for sodium sulphate recovery was about half of this amount. Another industrial user in Saskatchewan, a refinery in Regina, consumes more than 1 x $10^6 m^3$ of groundwater annually. For Manitoba, only one major industrial groundwater user, the chemical manufacturing sector in Brandon, was identified in the Prairie Provinces Water Board study (1982). It used 3.6 x $10^6 m^3$ of groundwater in 1978.

Groundwater can also be exploited for the thermal energy it contains. Jessop (1984) stated ". . . there is a very large geothermal resource below the Prairies, the temperature of which makes it useful for direct-heating and other low-temperature applications, but not usually for generation of electrical power." He noted, however, the limitations imposed by the presence of dissolved solids and gases, which could cause corrosion and scaling problems. Lam and Jones (1984) considered waters warmer than 50°C as potentially useful for low-grade geothermal energy extraction and, on that basis, suggested that porous carbonates of the Lower Carboniferous Rundle Group, the Upper Devonian Leduc Formation, and the Middle Devonian Beaverhill Lake Group in the Hinton-Edson area of west-central Alberta could be exploited for geothermal energy. In Saskatchewan, the city of Moose Jaw once utilized artesian water from a Jurassic aquifer to supply heated water to an indoor swimming pool, and it is believed that there is potential there to use deeper formation waters for space heating (Jessop, 1985). Observations made in a 2215 m deep well drilled in Regina (Vigrass and Jessop, 1984) have demonstrated that ". . . there is, without any doubt, an exploitable geothermal resource in the Winnipeg and Deadwood sandstones at the base of the sedimentary sequence".

Heat pumps could be used to extract thermal energy from cooler, shallower, potable prairie groundwaters (MLM Groundwater Engineering, 1979). Water as cool as 5°C might be exploitable and is available for the southern three quarters of Alberta, the southern half of Saskatchewan, and the southern quarter of Manitoba. This application would be most viable economically in rural areas, particularly if the same well were used both for heat extraction and to provide a domestic water supply.

REFERENCES

Ayers, R.S.
1975: Quality of water for irrigation; Proceedings of Irrigation and Drainage Division, Specialty Conference, American Society of Civil Engineering, August 13-15, Logan, Utah, p. 24-56.

Bachu, S.
1985: Influence of lithology and fluid flow on the temperature distribution in a sedimentary basin: a case study from the Cold Lake area, Alberta, Canada; Tectonophysics, v. 120, p. 257-284.

Bachu, S., Sauveplane, C.M., Lytviak, A.T., and Hitchon, B.
1987: Analysis of fluid and heat regimes in sedimentary basins; techniques for use with large data bases; American Association of Petroleum Geologists Bulletin, v. 71, p. 822-843.

Baillie, A.D.
1952: Ordovician geology of Lake Winnipeg and adjacent areas, Manitoba; Manitoba Department of Mines and Natural Resources, Mines Branch, Publication 51-6, 64 p.

Bannantyne, B.B.
1960: Potash deposits, rock salt and brines in Manitoba; Manitoba Department of Mines and Natural Resources, Mines Branch, Publication 59-1, 30 p.

Barnes, R.G.
1978: Hydrogeology of the Brazeau-Canoe River area, Alberta; Alberta Research Council, Report 77-5, 32 p.

Betcher, R.N.
1986: Regional hydrogeology of the Winnipeg Formation in Manitoba; in Proceedings Third Canadian Hydrogeological Conference, G. van der Kamp and M. Madunicky (ed.), Saskatoon, Saskatchewan, p. 160-174.

Billings, G.K., Hitchon, B., and Shaw, D.R.
1969: Geochemistry and origin of formation waters in the Western Canada Sedimentary Basin, 2. Alkali metals; Chemical Geology, v. 4, p. 211-222.

Borneuf, D.M.
1976: Hydrogeology of the Foremost area, Alberta; Alberta Research Council, Report 74-4, 26 p.

Bostock, H.S.
1970: Physiographic subdivisions of Canada; in Geology and Economic Minerals of Canada; Geological Survey of Canada, Economic Geology Report 1, p. 10-30.

Bouwer, H.
1978: Groundwater Hydrology; McGraw-Hill Book Co., New York, N.Y., 480 p.

Bredehoeft, J.D., Blyth, C.R., White, W.A., and Maxey, G.B.
1963: Possible mechanisms for concentration of brines in subsurface formations; American Association of Petroleum Geologists Bulletin, v. 47, p. 257-269.

Cameron, E.L.
1949: Salt, potash and phosphate in Manitoba; Manitoba Department of Mines and Natural Resources, Mines Branch, Bulletin 48-9, 13 p.

Canadian Society of Petroleum Geologists
1975: Geological Highway Map of Alberta.

Chebotarev, I.I.
1955: Metamorphism of natural waters in the crust of weathering; Geochimica Cosmochimica Acta, v. 8, p. 22-48, 137-170, 198-212.

Cherry, J.A.
1972: Geochemical processes in shallow groundwater flow systems in five areas in southern Manitoba, Canada; Proceedings 24th International Geological Congress, Montreal, Canada, Sec. 11, p. 208-221.

Christiansen, E.A.
1963: Hydrogeology of surficial and bedrock valley aquifers in southern Saskatchewan; in Proceedings Hydrological Symposium No. 3; Groundwater: National Research Council of Canada, p. 49-66.
1968: Pleistocene stratigraphy of the Saskatoon area, Saskatchewan, Canada; Canadian Journal of Earth Sciences, v. 5, p. 1167-1173.
1971: Tills in southern Saskatchewan, Canada; in Tills: A Symposium, R.P. Goldthwait (ed.); Ohio State University Press, Columbus, Ohio, p. 167-183.
1979: The Wisconsinan deglaciation of southern Saskatchewan and adjacent areas; Canadian Journal of Earth Sciences, v. 16, p. 913-938.

Cole, L.H.
1915: The salt deposits of Canada and the salt industry; Canadian Department of Mines, Mines Branch, Report 325, 152 p.

Cravens, S.J. and Ruedisili, L.C.
1987: Water movement in till of east-central South Dakota; Ground Water, v. 25, p. 555-561.

Davis, S.N.
1969: Porosity and permeability of natural materials; in Flow Through Porous Media, R.J.M. DeWeist (ed.); Academic Press, New York, N.Y., p. 53-59.

Davison, C.C.
1976: A hydrogeochemcial investigation of a brine disposal lagoon-aquifer system, Esterhazy, Saskatchewan; M.Sc. thesis, University of Waterloo, Waterloo, Ontario.

de Mille, G., Shouldice, J.R., and Nelson, H.W.
1964: Collapse structures related to evaporites of the Prairie Formation, Saskatchewan; Geological Society of America Bulletin, v. 75, p. 307-316.

Domenico, P.A. and Palciauskas, V.V.
1973: Theoretical analysis of forced convective heat transfer in regional ground-water flow; Geological Society of America Bulletin, v. 84, p. 3803-3814.

Eilers, R.G.
1973: Relations between hydrology and soil characteristics near Deloraine, Manitoba; M.Sc thesis, University of Manitoba, Winnipeg, Manitoba.

Environment Canada
1987: National inventory of municipal waterworks and wastewater systems in Canada 1986; Department of Supply and Services, Canada, Catalogue No. E94-81/1987, 419 p.
Farvolden, R.N.
1963: Bedrock channels of southern Alberta; in Early Contributions to the Groundwater Hydrology of Alberta; Research Council of Alberta, Bulletin 12, p. 63-75.
Fisheries and Environment Canada
1978: Hydrological Atlas of Canada. (Text and 34 maps, scales 1:20 000 and 1:10 000 000).
Freeze, R.A.
1969: Regional groundwater flow – Old Wives Lake drainage basin, Saskatchewan; Canada, Department of Environment, Inland Waters Branch, Science Series 5, 245 p.
Freeze, R.A. and Cherry, J.A.
1979: Groundwater; Prentice-Hall, Inc., Englewood Cliffs, N.J., 604 p.
Freeze, R.A. and Witherspoon, P.A.
1966: Theoretical analysis of regional groundwater flow: 1. Analytical and numerical solutions to the mathematical model; Water Resources Research, v. 2, p. 641-656.
1967: Theoretical analysis of regional groundwater flow: 2. Effect of water-table configuration and subsurface permeability variation; Water Resources Research, v. 3, p. 623-634.
1968: Theoretical analysis of regional groundwater flow: 3. Quantitative interpretations; Water Resources Research, v. 4, p. 581-590.
Fritz, P., Drimmie, R.J., and Render, F.W.
1974: Stable isotope contents of a major prairie aquifer in central Manitoba, Canada; in Isotope Techniques in Groundwater Hydrology; International Atomic Energy Agency STI/PUB/373, v. 1, p. 379-398.
Fuzesy, L.M.
1966: Geology of the Frobisher-Alida beds, southeastern Saskatchewan; Saskatchewan Department of Mineral Resources, Report 104, 59 p.
Graf, D.L.
1982: Chemical osmosis, reverse chemical osmosis, and the origin of subsurface brines; Geochimica et Cosmochimica Acta, v. 46, p. 1431-1438.
Gravenor, C.P. and Bayrock, L.A.
1961: Glacial deposits of Alberta; in Soils in Canada; Royal Society of Canada, Special Publication 3, p. 33-50.
Grayston, L.D., Sherwin, D.F., and Allan, J.F.
1966: Middle Devonian; in Geological History of Western Canada (second edition), R.G. McCrossan and R.P. Glaister (ed.); Alberta Society of Petroleum Geologists, Calgary, p. 49-59.
Grisak, G.E. and Cherry, J.A.
1975: Hydrologic characteristics and response of fractured till and clay confining a shallow aquifer; Canadian Geotechnical Journal, v. 12, p. 23-43.
Grisak, G.E., Cherry, J.A., Vonhof, J.A., and Blumele, J.P.
1976: Hydrogeologic and hydrochemical properties of fractured till in the Interior Plains Region; in Glacial Till. An Inter-Disciplinary Study; Royal Society Canada, Special Publication 12, p. 304-355.
Hackbarth, D.A.
1978: Groundwater conditions in the Grande Prairie-Beaverlodge area, Alberta; Alberta Research Council, Report 77-4, 58 p.
Health and Welfare Canada
1987: Guidelines for Canadian drinking water quality; Supply and Services Canada, Catalogue No. H48-10/1987E, 20 p.
Hendry, J.J.
1982: Hydraulic conductivity of a glacial till in Alberta; Ground Water, v. 20, p. 162-169.
Hess, P.J.
1986: Ground-water use in Canada, 1981; Environment Canada, National Hydrological Research Institute, Paper 28, 43 p.
Hitchon, B.
1966: Formation fluids; in Geological History of Western Canada (second edition), R.G. McCrossan and R.P. Glaister (ed.); Alberta Society of Petroleum Geologists, Calgary, p. 201-217.
1968: Rock volume and pore volume data for plains region of Western Canada Sedimentary Basin between latitudes 49° and 60°N; American Association of Petroleum Geologists Bulletin, v. 52, p. 2318-2323.
1969a: Fluid flow in the Western Canada Sedimentary Basin. 1. Effect of topography; Water Resources Research, v. 5, p. 186-195.
1969b: Fluid flow in the Western Canada Sedimentary Basin. 2. Effect of geology; Water Resources Research, v. 5, p. 460-469.
1984: Geothermal gradients, hydrodynamics, and hydrocarbon occurrences, Alberta, Canada; American Association of Petroleum Geologists Bulletin, v. 68, p. 713-743.
Hitchon, B., Billings, G.K., and Klovan, J.E.
1971: Geochemistry and origin of formation waters in the Western Canada Sedimentary Basin – III. Factors controlling chemical composition; Geochimica et Cosmochimica Acta, v. 35, p. 567-598.
Hitchon, B. and Friedman, I.
1969: Geochemistry and origin of formation waters in the Western Canada Sedimentary Basin – I. Stable isotopes of hydrogen and oxygen; Geochimica et Cosmochimica Acta, v. 33, p. 1321-1349.
Hitchon, B., Levinson, A.A., and Reeder, S.W.
1969: Regional variations of river water composition resulting from halite solution, Mackenzie River drainage basin, Canada; Water Resources Research, v. 5, p. 1395-1403.
Houghton, R.L., Thorstenson, D.C., Fisher, D.W., and Groenewold, G.H.
1987: Hydrogeochemistry of the upper part of the Fort Union Group in the Gascoyne lignite strip-mining area, North Dakota; United States Geological Survey, Professional Paper 1340, 104 p.
Hubbert, M.K.
1940: The theory of groundwater motion; Journal of Geology, v. 48, p. 785-944.
Jamieson, E.R.
1979: Well data and lithologic description of the Interlake Group (Silurian) in southern Saskatchewan; Saskatchewan Department of Mineral Resources, Report 139, 67 p.
Jessop, A.M.
1984: Geothermal energy in Canada - A review of the resources; in Energy Developments: New Forms, Renewables, Conservation, F.A. Curtis (ed.); Pergamon Press, Inc., Toronto, p. 263-266.
1985: Coordination of geothermal research; Canada, Department of Energy, Mines and Resources, Earth Physics Branch, Internal Report 85-5, 21 p.
Jessop, A.M. and Vigrass, L.W.
1984: The Regina geothermal experiment - thermal aspects; in Energy Developments: New Forms, Renewables, Conservation, F.A. Curtis (ed.); Pergamon Press, Inc., Toronto, p. 315-318.
Jones, F.W., Majorowicz, J.A., Linville, A., and Osadetz, K.G.
1986: The relationship of hydrocarbon occurrences to geothermal gradients and time-temperature indices in Mesozoic formations of southern Alberta; Bulletin of Canada Petroleum Geology, v. 34, p. 226-239.
Jones, J.F.
1966: Geology and groundwater resources of the Peace River district, northwestern Alberta; Research Council of Alberta, Bulletin 16, 143 p.
Keller, C.K., van der Kamp, G., and Cherry, J.A.
1986: Fracture permeability and groundwater flow in clayey till near Saskatoon, Saskatchewan; Canadian Geotechnical Journal, v. 23, p. 229-240.
Kent, D.M. and Simpson, F.
1973: Outline of the geology of southern Saskatchewan; in An Excursion Guide to the Geology of Saskatchewan; Saskatchewan Geological Society, Special Publication 1, p. 103-125.
Kewen, T.J. and Schneider, A.T.
1979: Hydrogeologic evaluation of the Judith River Formation in west central Saskatchewan; Saskatchewan Research Council, Geology Division, Report G79-2, 78 p.
Kharaka, Y.K. and Berry, F.A.F.
1973: Simultaneous flow of water and solutes through geological membranes – I. Experimental investigation; Geochimica et Cosmochimica Acta, v. 37, p. 2577-2603.
Kharaka, Y.K. and Smalley, W.C.
1976: Flow of water and solutes through compacted clays; American Association of Petroleum Geologists Bulletin, v. 60, p. 973-980.
Klemme, H.D.
1975: Geothermal gradients, heat flow and hydrocarbon recovery; in Petroleum and Global Tectonics, A.G. Fischer and S. Judson (ed.); Princeton University Press, Princeton, N.J., p. 251-304.
Kunkle, G.R.
1962: Reconnaissance groundwater survey of the Oyen map-area, Alberta; Research Council of Alberta, Preliminary Report 62-3, 23 p.
Lam, H.L. and Jones, F.W.
1984: An assessment of the low-grade geothermal potential in a foothills area of west-central Alberta; in Energy Developments: New Forms, Renewables, Conservation, F.A. Curtis (ed.); Pergamon Press, Inc., Toronto, p. 273-278.
Le Breton, E.G.
1963: Groundwater geology and hydrology of east-central Alberta; Research Council of Alberta, Bulletin 13, 64 p.

Lennox, D.H., Maathuis, H., and Pederson, D.
1988: Region 13, Western Glaciated Plains; in Ground Water Hydrogeology, W. Back, J.S. Rosenshein, and P.R. Seaber (ed.); Geological Society of America, The Geology of North America, v. O-2.

Lissey, A.
1962: Groundwater resources of the Regina area, Saskatchewan; Regina Engineering Department, Hydrology Division, Report 1, 89 p.

Maathuis, H. and Schreiner, B.T.
1982a: Hatfield Valley aquifer system in the Wynyard region, Saskatchewan; Saskatchewan Research Council, Geology Division, Publication no. G-743-3-B-82, 61 p.
1982b: Hatfield Valley aquifer system in the Waterhen River area (73K), Saskatchewan; Saskatchewan Research Council, Geology Division, Publication no. G-744-7-C-82, 33 p.

Majorowicz, J.A., Jones, F.W., and Jessop, A.M.
1986: Geothermics of the Williston Basin in Canada in relation to hydrodynamics and hydrocarbon occurrences; Geophysics, v. 51, p. 767-779.

Majorowicz, J.A., Rahman, M., Jones, F.W., and McMillan, N.J.
1985: The paleogeothermal and present thermal regimes of the Alberta Basin and their significance for petroleum occurrences; Bulletin of Canadian Petroleum Geology, v. 33, p. 12-21.

Matthes, G.
1982: The Properties of Groundwater; John Wiley and Sons, New York, N.Y., 285 p.

Mawson, C.A.
1964: Survey of hydrological applications of tracers in Canada; Atomic Energy of Canada Ltd., AECL-2005, 14 p.

Meneley, W.A.
1972: Groundwater – Saskatchewan; in Water Supply for the Saskatchewan-Nelson Basin; Saskatchewan-Nelson Basin Board Report, Appendix 7, Section F, p. 673-723.

Meneley, W.A., Maathuis, H., Jaworski, E.J., and Allen, V.F.
1979: SRC observation wells in Saskatchewan, Canada: Introduction, design and discussion of accumulated data: Accumulated data for observation wells; Saskatchewan Research Council, Geology Division, Report 19, 3 volumes.

Meyboom, P.
1963a: Patterns of groundwater flow in the prairie profile; in Proceedings of Hydrology Symposium No. 3 – Groundwater; National Research Council of Canada, p. 5-20.
1963b: Induced infiltration, Medicine Hat, Alberta; in Early Contributions to the Groundwater Hydrology of Alberta; Research Council of Alberta, Bulletin 12, p. 88-97.
1967: Interior Plains Hydrogeological Region; in Groundwater in Canada; Geological Survey of Canada, Economic Geology Report 24, p. 131-158.

MLM Ground-Water Engineering
1979: Applicability of ground-water source heat pumps in western Canada; unpub. rep., National Research Council, Division of Building Research, DSS Contract No. ISX78-00103.

Moran, S.R., Groenewold, G.H., and Cherry, J.A.
1978: Geologic, hydrologic, and geochemical concepts and techniques in overburden characterization for mined-land reclamation; North Dakota Geological Survey, Report of Investigations 63, 152 p.

Nicholson, R.A.H.
1970: The petrology and economic geology of the Upper Devonian Birdbear Formation in southeastern Saskatchewan; Saskatchewan Department of Mineral Resources, Report 125, 94 p.

Nielsen, G.L. and Lorberg, E.
1971: Groundwater inventory of southern Alberta (south of 54 degrees); Alberta Environment, Earth Science and Licensing Division Report, 48 p.

Ozoray, G.F.
1974: Hydrogeology of the Waterways-Winefred Lake area, Alberta; Research Council of Alberta, Report 74-2, 18 p.

Paterson, D.F.
1975: Computer plotted isopach and structure maps of the Lower Paleozoic formations in Saskatchewan; Saskatchewan Department of Mineral Resources, Report 165, 15 p.

Peterson, R.
1954: Studies of the Bearpaw Shale at a dam site in Saskatchewan; Proceedings of American Society of Civil Engineers, Soil Mechanics and Foundation Division, v. 80, separate 476, 28 p.

Porter, J.W. and Fuller, J.G.C.M.
1959: Lower Paleozoic rocks of northern Williston Basin and adjacent areas; American Association of Petroleum Geologists Bulletin, v. 43, p. 124-189.

Porter, J.W., Fuller, J.G.C.M., and Norford, B.S.
1966: Ordovician-Silurian; in Geological History of Western Canada (second edition), R.G. McCrossan and R.P. Glaister (ed.); Alberta Society of Petroleum Geologists, Calgary, p. 34-48.

Porter, J.W., Price, R.A., and McCrossan, R.G.
1982: The Western Canada Sedimentary Basin; Philosophical Transactions, Royal Society of London A, v. 305, p. 169-192.

Prairie Provinces Water Board
1982: Municipal and industrial water uses; Appendix 2, Water Demand Study; Historical and Current Uses in the Saskatchewan-Nelson Basin; Regina, Saskatchewan, 147 p.

Render, F.W.
1970: Geohydrology of the metropolitan Winnipeg area as related to groundwater supply and construction; Canadian Geotechnical Journal, v. 7, p. 243-274.

Roberts, W.H., III
1981: Some uses of temperature data in petroleum exploration; in Proceedings of the Symposium on Unconventional Methods in Exploration for Petroleum and Natural Gas, B.M. Gottlieb (ed.), No. 2, p. 8-49.

Rutulis, M.
1972: Groundwater – Manitoba; in Water Supply for the Saskatchewan-Nelson Basin; Saskatchewan-Nelson Basin Board Report, Appendix 7, Section F., p. 725-742.

Rutulis, M. (cont.)
1984: Dakota aquifer system in the Province of Manitoba; in Geohydrology of the Dakota Aquifer; Proceedings First C.V. Theis Conference on Hydrology; National Water Well Association, p. 14-21.

Schreiner, B.T. and Maathuis, H.
1982: Hatfield Valley aquifer system in the Melville region, Saskatchewan; Saskatchewan Research Council, Geology Division, Publication no. G-743-3-B-82, 77 p.

St. Arnaud, R.J.
1976: Pedological aspects of glacial till; in Glacial Till; an Inter-Disciplinary Study; Royal Society of Canada, Publication 12, p. 133-155.

Schwartz, F.W.
1975: Hydrogeologic investigation of a radioactive waste management site in southern Alberta; Canadian Geotechnical Journal, v. 12, p. 349-361.

Schwartz, F.W., Muehlenbachs, K., and Chorley, D.W.
1981: Flow-system controls of the chemical evolution of groundwater; Journal of Hydrology, v. 54, p. 225-243.

Smith, S.R.
1980: Petroleum geology of the Mississippian Midale Beds, Benson oil field, southeastern Saskatchewan; Saskatchewan Department of Mineral Resources, Report 215, 98 p.

Stein, R.
1976: Hydrogeology of the Edmonton area (northeast segment), Alberta; Alberta Research Council, Report 76-1, 26 p.

Summers, P.W. and Hitchon, B.
1973: Source and budget of sulfate in precipitation from central Alberta; Journal of the Air Pollution Control Association, v. 23, p. 194-199.

Thomas, J.F.J.
1959: Industrial water resources of Canada, Water Survey Report No. 10, Nelson River drainage basin; Canadian Department of Mines and Technical Surveys, Mines Branch, Report 861, 147 p.

Toth, J.
1962: A theory of groundwater motion in small drainage basins in central Alberta; Journal of Geophysical Research, v. 67, p. 4375-4387.
1963: A theoretical analysis of groundwater flow in small drainage basins; Journal of Geophysical Research, v. 68, p. 4795-4812.
1978: Gravity-induced cross-formational flow of formation fluids, Red Earth region, Alberta, Canada: analysis, patterns and evolution; Water Resources Research, v. 14, p. 805-843.
1980: Cross-formational gravity flow of groundwater, a mechanism of the transport and accumulation of petroleum (generalized hydraulic theory of petroleum migration); in Problems of Petroleum Migration, W.M. Roberts III and C.J. Cordell (ed.); American Association of Petroleum Geologists, Studies in Geology 10, p. 121-167.

U.S. Environmental Protection Agency
1973: Water quality criteria 1972; U.S. Government Printing Office, Washington, D.C., EPA R3 73003, 606 p.

van Everdingen, R.O.
1968: Studies of formation waters in Western Canada: Geochemistry and hydrodynamics; Canadian Journal of Earth Sciences, v. 5, p. 523-543.

van Hees, H. and North, F.K.
1966: Cambrian; in Geological History of Western Canada (second edition), R.G. McCrossan and R.P. Glaister (ed.); Alberta Society of Petroleum Geologists, Calgary, p. 20-33.
Vigrass, L.W. and Jessop, A.M.
1984: Regina geothermal experiment-geological and hydrological aspects; in Energy Developments: New Forms, Renewables, Conservation, F.A. Curtis (ed.); Pergamon Press, Inc., Toronto, p. 303-308.
Vogwill, R.I.J.
1978: Hydrogeology of the Lesser Slave Lake area, Alberta; Alberta Research Council, Report 77-1, 30 p.
Wallick, E.I.
1981: Chemical evolution of groundwater in a drainage basin of Holocene age, east-central Alberta, Canada; Journal of Hydrology, v. 54, p. 245-283.
Westgate, J.A.
1968: Surficial geology of the Foremost-Cypress Hills area, Alberta; Research Council of Alberta, Bulletin 22, 122 p.
Whitaker, S.H. and Christiansen, E.A.
1972: The Empress Group in southern Saskatchewan; Canadian Journal of Earth Sciences, v. 9, p. 353-360.

Author's Address

D.H. Lennox
18 Glendenning St.
Nepean, Ontario
K2H 7Y9

Printed in Canada

Chapter 7

HUDSON PLATFORM – INTRODUCTION

CHAPTER 7

HUDSON PLATFORM - INTRODUCTION

A.W. Norris

GENERAL STATEMENT

Hudson Platform, in the central part of the Canadian Shield, covers an area of 971 250 km^2. Its margins have been greatly eroded, so that most of its sedimentary cover now resides in two linked cratonic basins, Hudson Bay Basin and Moose River Basin. Most of this area is covered by water of Hudson and James bays (608 650 km^2). Onshore in the south, parts of the platform form Hudson Bay Lowland (323 750 km^2) and in the north, Southampton, Coats, and Mansel islands form Southampton Plain (38 850 km^2).

A northeast-trending positive area, Cape Henrietta Maria Arch (Sanford et al., 1968; the Patricia Arch of Nelson and Johnson, 1966) divides Hudson Platform into two Phanerozoic sedimentary basins, Moose River Basin in the southeast and Hudson Bay Basin in the northwest (Fig. 7.1, 7.2). These basins are largely defined by post-Devonian movements; their largely common histories involve periods in which they were depocentres alternating with periods in which they were parts of a vast, undifferentiated platform. Thus, the two basins are in part depocentres, in part syndepositional and post-depositional structural basins, and in part basins of preservation. The most reliable thickness measurements are those based on drilling. In Moose River Basin, which developed within the platform somewhat later than Hudson Bay Basin, Ordovician, Silurian, Devonian, Middle Jurassic, and Lower Cretaceous rocks have a total thickness of about 760 m in the central part. In Hudson Bay Basin, Ordovician, Silurian, and Devonian rocks are at least 1575 m thick in the water-covered central part.

PHYSIOGRAPHY AND ACCESSIBILITY

The Hudson Bay Lowland (Lowlands of some authors; see Coombs, 1954) is a low, flat coastal plain, 160 to 420 km wide, bordering Hudson and James bays, that extends for a distance of 1350 km from Nottawa River in Quebec to North Knife River in Manitoba (Coombs, 1954). This physiographic division more or less corresponds to the area underlain by Phanerozoic sedimentary rocks, in contrast to the Canadian Shield, which is underlain by Precambrian crystalline rocks. Following the last glacial period, Hudson Bay and the bordering lowlands, which had been depressed by the load of ice, became temporarily inundated by the Tyrrell Sea. As the land rose by isostatic rebound, the Tyrrell Sea formed a succession of beaches, parallel to the present shorelines of Hudson and James bays, some of which are now far inland. Large parts of Hudson Bay Lowland are poorly drained and characterized by the presence of innumerable shallow lakes surrounded by muskeg. However, numerous major rivers that have their headwaters in the Canadian Shield flow across the lowlands to Hudson and James bays.

Three large islands, Southampton, Coats, and Mansel, are located in northern Hudson Bay. Large parts of these islands are underlain by Paleozoic rocks, and the region is referred to as Southampton Plain (Sanford and Grant, 1976). The Paleozoic terrain of these islands differs markedly from that of the Hudson Bay Lowland in that it is well drained and barren, resembling the Arctic Islands. In general, these islands are of low relief. Elevations of Southampton Plain rarely exceed 150 m on Southampton Island, and are generally less than 100 m on Coats and Mansel islands. Glacial deposits are relatively thin and consist largely of Paleozoic detritus, which has been reworked into a succession of raised beaches, bars, and terraces by the retreating Tyrrell Sea.

The seafloor beneath Hudson and James bays is irregular (Fig. 7.3). The bedrock and Quaternary surficial deposits have been shaped by glacial and current scouring (Pelletier, 1969). In places, the bathymetric configuration closely parallels structure and variations in the lithology of the bedrock surface. Linear cuestas and depressions are formed by the erosion of hard and soft rock units. Circular depressions, particularly in east-central Hudson Bay, are assumed to be of karst origin. The generalized geological map of Hudson Platform (Fig. 7.4) shows the onshore and offshore distribution of rock units based on bathymetric and all other available data.

HISTORY OF GEOLOGICAL INVESTIGATIONS

The earliest geological explorations of the Hudson Bay Lowland were mainly by personnel of the Geological Survey of Canada and the Ontario Department of Mines. Between 1870 and 1915, most of the bedrock exposures along the major rivers and tributaries and coastlines of Hudson and James bays had been described and delineated by a

Norris, A.W.
1993: Hudson Platform - Introduction; Chapter 7 in Sedimentary Cover of the Craton in Canada, D.F. Stott and J.D. Aitken (ed.); Geological Survey of Canada, Geology of Canada, no. 5, p. 643-651 (also Geological Society of America, The Geology of North America, v. D-1).

distinguished group of pioneers, including R. Bell, A.P. Low, W. McInnes, W.J. Wilson, J.B. Tyrrell, and F.J. Alcock (Nelson and Johnson, 1966; Sanford and Norris, 1975). These early reconnaissance surveys recognized the presence of Ordovician, Silurian, and Devonian rocks and established their approximate distribution in the Hudson Bay Lowland. However, the first stratigraphic synthesis of the area was by Savage and Van Tuyl (1919), who proposed a formational nomenclature for the Paleozoic succession and laid the foundation for all later work.

Dyer (1928) summarized all available information on the stratigraphy, paleontology, structural geology, and mineral resources of Moose River Basin. A report on the stratigraphy, structure, and hydrocarbon potential of the northern Ontario part of Moose River Basin was prepared by Martison (1953) based on a reconnaissance survey of the area in 1946.

Phanerozoic rocks of the Moose River Basin and bordering Precambrian rocks of the Quebec Embayment were mapped by Remick et al. (1963). The only other

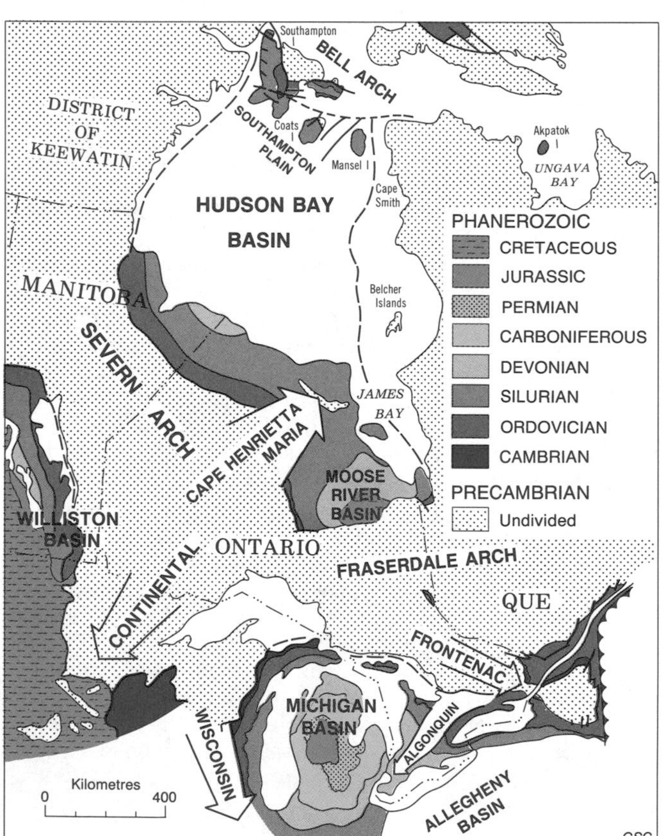

Figure 7.1. Sedimentary basins and cratonic arches of east-central Canada and adjacent area of United States (from Sanford and Norris, 1973).

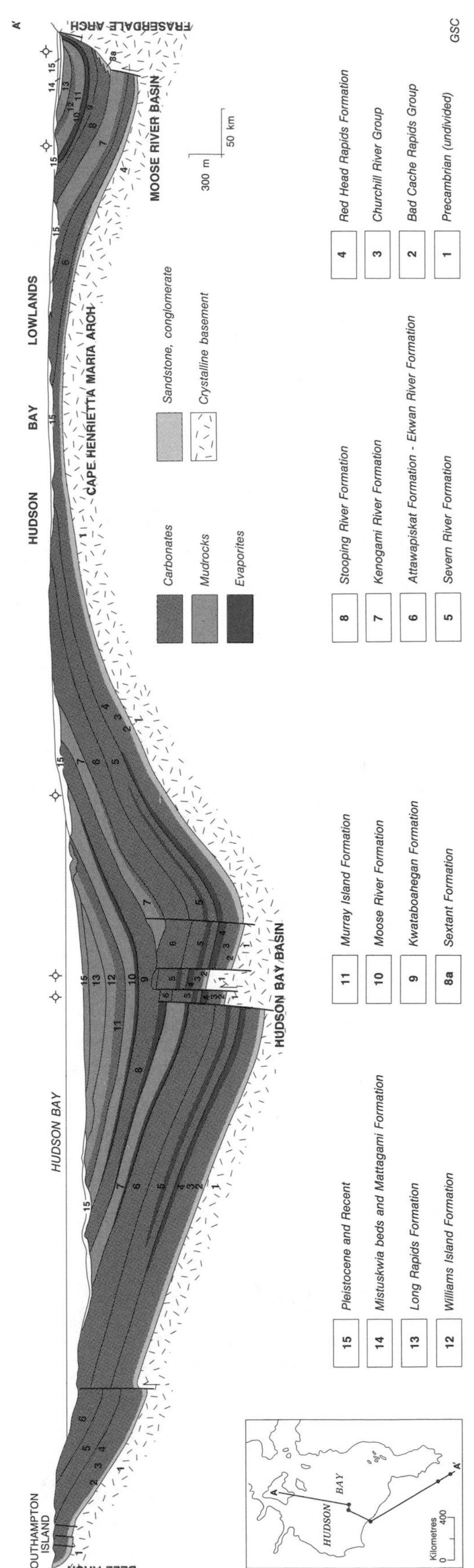

Figure 7.2. Schematic north-south cross-section of Hudson Platform showing gross lithologies and main structures (from B.V. Sanford, unpub. rep., 1980).

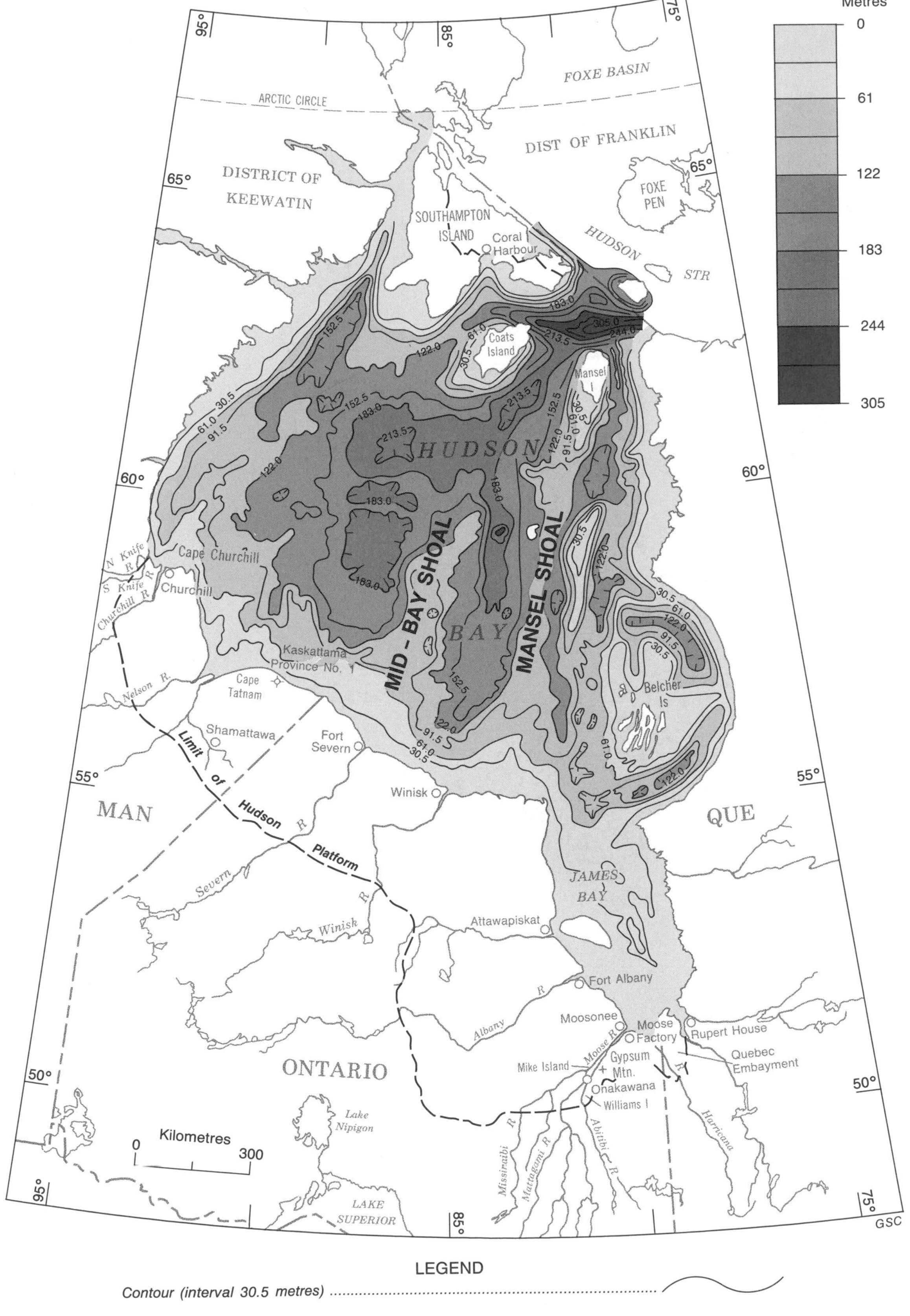

LEGEND

Contour (interval 30.5 metres) ....

Figure 7.3. Bathymetric chart of Hudson Bay (after Sanford and Norris, 1973).

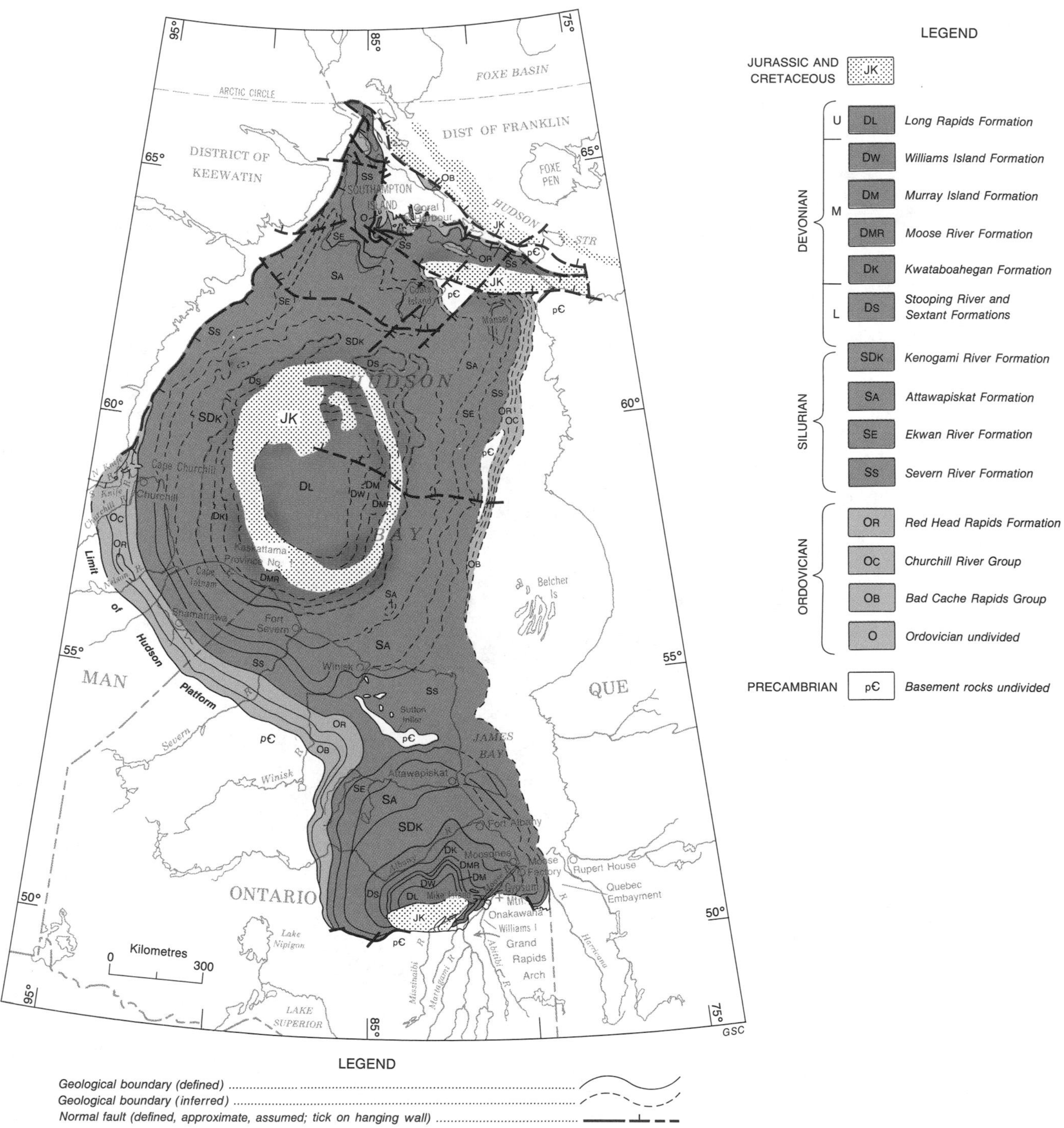

Figure 7.4. Geological map of Hudson Platform (after Sanford et al., 1979; areas of known and inferred Jurassic and Cretaceous rocks in Hudson Bay Basin added by B.V. Sanford and A.C. Grant, 1989).

reports covering this area are by Flower (1968) on Silurian cephalopods, and by Larsson (1984) on Silurian stratigraphy and paleontology.

Local aspects of Paleozoic stratigraphy and paleontology of Moose River Basin have been described in numerous reports published between 1904 and 1953, and summarized by Nelson and Johnson (1966), Sanford and Norris (1975), and others. Reports dealing with Mesozoic stratigraphy and coal deposits of the basin published between 1920 and 1953 have been summarized by Price (1978) and by Telford and Verma (1982).

Little comprehensive geological work had been done before 1950 on Phanerozoic rocks of the northern Hudson Bay Lowland. The more important early reports include those by Savage and Van Tuyl (1919), which described the geology of Nelson and Hayes rivers; by Miller and Youngquist (1947), which described Ordovician cephalopods collected near Churchill and along North Knife River; by Williams (1948), which outlined the geology of the Churchill area; and by Ethington and Furnish (1959), which described Ordovician conodonts from the Knife River delta area.

The first detailed reports on Ordovician stratigraphy and paleontology of the northern Hudson Bay Lowland were by Nelson (1963, 1964), based on field work in 1950 and 1951 along the Nelson, Churchill, South and North Knife rivers, and along the coast between Churchill and Cape Churchill.

Nelson and Johnson (1966) published a summary of the geology of Hudson Platform based on previous data and new observations. They demonstrated the presence of Devonian rocks in the Cape Tatnam area of the northern lowlands and in the offshore Mid-Bay Shoal area.

During Operation Kapuskasing (George et al., 1967a, b; Bennett et al., 1966, 1967) Ontario Department of Mines geologists mapped a southern part of the Hudson Bay Lowland, and their reports incorporated important new data from drilling by the Hydro-Electric Power Commission of Ontario.

In 1967 members of the Geological Survey of Canada carrying out Operation Winisk mapped the Hudson Bay Lowland (Norris and Sanford, 1968; Sanford et al., 1968). A symposium volume on Hudson Bay (Hood et al., 1969) includes reports on Precambrian rocks by Bostock (1969), Paleozoic and Mesozoic geology by Norris and Sanford (1969), Pleistocene geology by Craig (1969) and McDonald (1969), rivers of the Hudson Bay Lowland by Cumming (1969), and 17 other papers.

The most comprehensive account of background information on Hudson Bay is that edited by Beals (1968b) entitled, "Science, History and Hudson Bay".

Accounts of Paleozoic geology of the northern Manitoba part of Hudson Bay Lowland, which incorporate new data from the subsurface, include those by Cumming (1971) on the Ordovician, Norford (1971) on the Silurian, and Norris and Uyeno (1971) on the Devonian. A generalized geological map of Hudson Platform, constructed by Sanford (in Sanford and Norris, 1973), shows the offshore distribution of Ordovician, Silurian, and Devonian rock units based on observations made in 1971. Cumming (1975) provided faunal lists, descriptions, and locations of all Ordovician outcrops in the Hudson Bay Lowland that were examined by members of Operation Winisk. A detailed report on the Devonian geology of Hudson Platform by Sanford and Norris (1975) incorporates subsurface data from both basins acquired before 1972. The Precambrian and Paleozoic geology of Southampton, Coats, and Mansel islands was described in detail by Heywood and Sanford (1976).

Geophysical studies have contributed significantly toward establishing the distribution, thickness of sediments, and basement structures of both basins; these studies include sea magnetometer (Hood, 1964, 1966) and seismic refraction surveys (Hobson, 1964a, b; 1967). Reflection seismic data were obtained recently from the central part of the Hudson Bay Basin by the Aquitaine Arco Group, Shell Canada Resources (Dimian et al., 1983) and Mobil Oil Canada Limited. New information is provided by Grant and Sanford (in Chapter 8, this volume).

Information on the Mesozoic geology and mineral potential of Moose River Basin was based on geological and geophysical surveys and drilling programs between 1975 and 1978 (Telford and Verma, 1982). The presence of Jurassic sediments in the basin was established for the first time by G. Norris (1977; in Telford and Verma, 1982) on the basis of palynology. More recent information on Cretaceous rocks in Hudson Bay Basin has been obtained during investigations by the Geological Survey of Canada (see Sanford and Grant in Chapter 8, this volume).

Biostratigraphical reports on various fossil groups in Hudson Platform include those by Barnes and Munro (1973) on Ordovician conodonts; Le Fèvre et al. (1976) on Ordovician and Silurian conodonts; Norford (1970) on Ordovician and Silurian shelly fossils; Norford (1981) and Gass and Mikulic (1982) on Silurian trilobites; Boucot and Johnson (1979) on Silurian brachiopods; Hueber (1983) on a new plant species from the Lower Devonian Sextant Formation; McGregor (in McGregor et al., 1970) and McGregor and Camfield (1976) on Upper Silurian? to Middle Devonian spores; Playford (1977) on Lower to Middle Devonian acritarchs; G. Norris (1977) on Middle Jurassic palynofloral assemblages; Hopkins and Sweet (1976), and G. Norris (in Telford and Verma, 1982) on Lower Cretaceous (Albian) spore-pollen assemblages.

NATURE OF THE DATA BASE

The main Phanerozoic rock outcrops of the Hudson Bay Lowland are confined largely to the banks of rivers and tributary streams, and along some stretches of the coastline and adjacent tidal flats. These outcrops, as well as the locations of the main drillholes, are indicated on Map 17-1967 of Sanford et al. (1968), and in Figures 3 and 5 of Sanford and Norris (1975). The general paucity of Phanerozoic outcrops is due largely to a relatively thick blanket of Recent and Pleistocene deposits.

In contrast to the tree-and-muskeg-covered Hudson Bay Lowland, the Paleozoic terrain of Southampton, Coats, and Mansel islands in northern Hudson Bay is relatively well drained, and Recent and Pleistocene deposits are relatively thin. As a result, bedrock exposures are numerous on these islands and are present along the shorelines, rivers, and interior uplands (Map 1404A, Heywood and Sanford, 1976).

Since 1930 the Phanerozoic rocks of the more intensely studied Moose River Basin have been penetrated by about 352 boreholes. Of these, about 35 were targeted on Paleozoic rocks (Sanford and Norris, 1975), and the remainder were concerned with delineating the Mesozoic lignite deposits in the Onakawana area (Price, 1978) and three other small lignite deposits (Verma, in Telford and Verma, 1982).

Considerably less drilling has been done in the Hudson Bay Basin. Up to 1972, 11 boreholes penetrating Paleozoic rocks had been completed on the mainland at the southern end of the basin (Sanford and Norris, 1975). Between 1969 and 1974 three offshore wells, two of which penetrated to basement, were drilled in the south-central part (Dimian et al., 1983).

ACKNOWLEDGMENTS

Included in Chapter 8 are new data and interpretations from an unpublished report by B.V. Sanford entitled, "Evaporite Deposits of the Hudson Bay Region". Drafts of Chapters 8 and 9 for this volume were critically read by R.D. Johnson, B.S. Norford, B.V. Sanford, and T.T. Uyeno. Most of their comments have been incorporated in the text and are gratefully acknowledged.

The text for Chapter 7 and that portion of Chapter 8 dealing with Paleozoic and Mesozoic biostratigraphy and lithostratigraphy were prepared by A.W. Norris. The section on Crustal Geophysics was contributed by A.C. Grant and B.V. Sanford, and they have also provided a summary of new information on the geology of Hudson Platform, obtained prior to the end of 1988. The summary of Quaternary stratigraphy and history is by W.R. Cowan and is based in part on material included in the companion volume on the Quaternary Geology of Canada and Greenland (Fulton, 1989).

REFERENCES

Barnes, C.R. and Munro, I.
1973: Middle and Upper Ordovician conodont faunas from Manitoba, Hudson Bay, and Canadian Shield outliers (abstract); Geological Survey of America, Abstracts with Programs, v. 5, no. 4, p. 297.

Beals, C.S. (ed.)
1968: Science, History and Hudson Bay; Department of Energy, Mines and Resources, Ottawa, Canada, 2 volumes, 1057 p.

Bennett, G., Brown, D.D., and George, P.T.
1966: Operation Kapuskasing, District of Cochrane; Ontario Department of Mines, Summary of Field Work, 1966, by the Geological Branch, E.G. Pye (ed.); Preliminary Report 1966-1, p. 42-47.

Bennett, G., Brown, D.D., George, P.T., and Leahy, E.J.
1967: Operation Kapuskasing; Ontario Department of Mines, Miscellaneous Paper 10, 73 p.

Bostock, H.H.
1969: Precambrian sedimentary rocks of the Hudson Bay Lowlands; in Earth Science Symposium on Hudson Bay, Ottawa, February 1968, P.J. Hood et al. (ed.); Geological Survey of Canada, Paper 68-53, p. 206-214.

Boucot, A.J. and Johnson, J.G.
1979: Pentamerinae (Silurian Brachiopods); Palaeontolographica, Abteilung A, v. 163, Lieferung 4-6, p. 87-129.

Coombs, D.B.
1954: The physiographic subdivisions of the Hudson Bay Lowlands south of 60 degrees North; Department of Mines and Technical Surveys, Canada, Geographical Bulletin No. 6, p. 1-16.

Craig, B.G.
1969: Late glacial and post-glacial history of the Hudson Bay region; in Earth Science Symposium on Hudson Bay, Ottawa, February 1968, P.J. Hood et al. (ed.); Geological Survey of Canada, Paper 68-53, p. 63-77.

Cumming, L.M.
1969: Rivers of the Hudson Bay Lowlands; in Earth Science Symposium on Hudson Bay, Ottawa, February 1968, P.J. Hood et al. (ed.); Geological Survey of Canada, Paper 68-53, p. 144-168.
1971: Ordovician strata of the Hudson Bay Lowlands in northern Manitoba; in Geoscience Studies in Manitoba, A.C. Turnock (ed.); Geological Association of Canada, Special Paper No. 9, p. 189-197.
1975: Ordovician strata of the Hudson Bay Lowlands; Geological Survey of Canada, Paper 74-28, 91 p.

Dimian, M.V., Gray, R., Stout, J., and Wood, B.
1983: Hudson Bay Basin; in Seismic Expression of Structural Styles; a Picture and Work Atlas, A.W. Bally (ed.); American Association of Petroleum Geologists, Studies in Geology Series, no. 15, v. 2, p. 2.2.4-1 - 2.2.4-4.

Dyer, W.S.
1928: Geology and economic deposits of the Moose River Basin; Ontario Department of Mines, Annual Report, v. 37, Pt. 6, p. 1-69.

Ethington, R.L. and Furnish, W.M.
1959: Ordovician conodonts from northern Manitoba; Journal of Paleontology, v. 33, no. 4, p. 540-546.

Flower, R.H.
1968: Silurian cephalopods from James Bay Lowland, with a revision of the Family Narthecoceratidae; Geological Survey of Canada, Bulletin 164, 88 p.

Fulton, R.J. (ed.)
1989: Quaternary Geology of Canada and Greenland; Geological Survey of Canada, Geology of Canada, no. 1. (Also Geological Society of America, The Geology of North America, v. K-1).

Gass, K.C. and Mikulic, D.G.
1982: Observations on the Attawapiskat Formation (Silurian) trilobites of Ontario, with description of a new encrinurine; Canadian Journal of Earth Sciences, v. 19, no. 3, p. 589-596.

George, P.T., Brown, D.D., and Bennett, G.
1967a: Operation Kapuskasing; Ontario Department of Mines preprint of paper presented at 35th Annual Meeting, Prospectors and Developers Association, Toronto, March 8, 1967.

George, P.T., Brown, D.D., Bennett, G., and Leahy, E.J.
1967b: Operation Kapuskasing; Canadian Mining Journal, v. 88, no. 4, p. 129-136.

Heywood, W.W. and Sanford, B.V.
1976: Geology of Southampton, Coats, and Mansel Islands, District of Keewatin, Northwest Territories; Geological Survey of Canada, Memoir 382, 35 p.

Hobson, G.D.
1964a: Ontario-Hudson Bay Lowlands, thickness of sedimentary section (Paleozoic to Cretaceous) from reconnaissance seismic refraction survey, March and April, 1964; Ontario Department of Mines Preliminary Map P243, 1 inch = 31 miles.
1964b: Nine reversed refraction seismic profiles, Hudson Bay Lowlands, Manitoba; Geological Survey of Canada, Paper 64-2, p. 33-40.
1967: Reconnaissance seismic refraction survey of Hudson Bay, Canada; reprinted from Proceedings of the 7th World Petroleum Congress, p. 813-826.

Hood, P.J.
1964: Sea magnetometer reconnaissance of Hudson Bay; Geophysics, v. 29, p. 916-921.
1966: Geophysical reconnaissance of Hudson Bay, Part I: Sea magnetometer survey, Part II: Sub-bottom depth recorder survey; Geological Survey of Canada, Paper 65-32, 42 p.

Hood, P.J., Hobson, G.D., Norris, A.W., and Pelletier, B.R. (ed.)
1969: Earth Science Symposium on Hudson Bay, Ottawa, February, 1968; Geological Survey of Canada, Paper 68-53, 386 p.

Hopkins, W.S. and Sweet, A.R.
1976: Miospores and megaspores from the Lower Cretaceous Mattagami Formation of Ontario; in Contributions to Canadian Paleontology; Geological Survey of Canada, Bulletin 256, p. 55-71.

Hueber, F.M.
1983: A new species of *Baragwanathia* from the Sextant Formation (Emsian), northern Ontario, Canada; Botanical Journal of the Linnean Society, v. 86, p. 57-79.

Larsson, S.Y.
1984: Silurian paleontology and stratigraphy of the Hudson Bay Lowlands in western Quebec; M.Sc. thesis, McGill University, Montreal, Quebec, 188 p.

Le Fèvre, J., Barnes, C.R., and Tixier, M.
1976: Paleoecology of Late Ordovician and Early Silurian conodontophorids, Hudson Bay Basin; in Conodont Paleoecology, C.R. Barnes (ed.); Geological Association of Canada, Special Paper No. 15, p. 69-89.

Martison, N.W.
1953: Petroleum possibilities of the James Bay Lowland area, Ontario; Department of Mines, v. 61, Pt. 6, 1952, p. 1-58.

McDonald, B.C.
1969: Glacial and interglacial stratigraphy, Hudson Bay Lowland, p. 78-99; in Earth Science Symposium on Hudson Bay, Ottawa, February, 1968, P.J. Hood et al. (ed.); Geological Survey of Canada, Paper 68-53, 386 p.

McGregor, D.C. and Camfield, M.
1976: Upper Silurian? to Middle Devonian spores of the Moose River Basin; Geological Survey of Canada, Bulletin 263, 63 p.

McGregor, D.C., Sanford, B.V., and Norris, A.W.
1970: Palynology and correlation of Devonian formations in the Moose River Basin, northern Ontario; Geological Association of Canada, Proceedings, v. 22, p. 45-54.

Miller, A.K. and Youngquist, W.
1947: Ordovician cephalopods from the west-central shore of Hudson Bay; Journal of Paleontology, v. 21, no. 5, p. 409-410.

Nelson, S.J.
1963: Ordovician paleontology of northern Hudson Bay Lowland; Geological Society of America, Memoir 90, 152 p.
1964: Ordovician stratigraphy of northern Hudson Bay Lowland; Geological Survey of Canada, Bulletin 108, 36 p.

Nelson, S.J. and Johnson, R.D.
1966: Geology of Hudson Bay Basin; Bulletin of Canadian Petroleum Geology, v. 14, no. 4, p. 520-578.

Norford, B.S.
1970: Ordovician and Silurian biostratigraphy of the Sogepet-Aquitaine Kaskattama Province No. 1 well, northern Manitoba; Geological Survey of Canada, Paper 69-8, 36 p.
1971: Silurian stratigraphy of northern Manitoba; in Geoscience Studies in Manitoba, A.C. Turnock (ed.); Geological Association of Canada, Special Paper No. 9, p. 199-207.
1981: The trilobite fauna of the Silurian Attawapiskat Formation, northern Ontario and northern Manitoba; Geological Survey of Canada, Bulletin 327, 37 p.

Norris, A.W. and Sanford, B.V.
1968: Operation Winisk - An air-supported geological reconnaissance survey of the Hudson Bay Lowlands; Ontario Petroleum Institute Inc., Seventh Annual Conference, Windsor, Ontario, October 30 to November 1, 1968, p. 1-33.
1969: Paleozoic and Mesozoic geology of the Hudson Bay Lowlands; in Earth Science Symposium on Hudson Bay, P.J. Hood et al. (ed.); Geological Survey of Canada, Paper 68-53, p. 169-205.

Norris, A.W. and Uyeno, T.T.
1971: Stratigraphy and conodont faunas of Devonian outcrop belts, Manitoba; in Geoscience Studies in Manitoba, A.C. Turnock (ed.); Geological Association of Canada, Special Paper No. 9, p. 209-223.

Norris, G.
1977: Palynofloral evidence for terrestrial Middle Jurassic in the Moose River Basin, Ontario; Canadian Journal of Earth Sciences, v. 14, no. 2, p. 153-158.

Pelletier, B.R.
1969: Submarine physiography, bottom sediments and models of sediment transport in Hudson Bay; in Earth Science Symposium on Hudson Bay, Ottawa, February 1968, P.J. Hood et al. (ed.); Geological Survey of Canada, Paper 68-53, p. 100-135.

Playford, G.
1977: Lower and Middle Devonian acritarchs of the Moose River Basin, Ontario; Geological Survey of Canada, Bulletin 279, 87 p.

Price, L.L.
1978: Mesozoic deposits of the Hudson Bay Lowlands and coal deposits of the Onakawana area, Ontario; Geological Survey of Canada, Paper 75-13, 39 p.

Remick, J.H., Gillain, R.R., and Durden, C.J.
1963: Geology of Rupert Bay, Missisicabi River area; Quebec Department of Natural Resources, Preliminary Report 498, 20 p.

Sanford, B.V. and Grant, G.M.
1976: Physiography (of) eastern Canada and adjacent areas; Geological Survey of Canada, Map 1399A (4 sheets), scale 1:2 000 000.

Sanford, B.V. and Norris, A.W.
1973: The Hudson Platform; in The Future Petroleum Provinces of Canada - Their Geology and Potential, R.G. McCrossan (ed.); Canadian Society of Petroleum Geologists, Memoir 1, p. 387-409.
1975: Devonian stratigraphy of the Hudson Platform - Part I: Stratigraphy and economic geology, Part II; Outcrop and subsurface sections; Geological Survey of Canada, Memoir 379, 372 p.

Sanford, B.V., Norris, A.W., and Bostock, H.H.
1968: Geology of the Hudson Bay Lowlands (Operation Winisk); Geological Survey of Canada, Paper 67-60, p. 1-45, and Geological Map 17-1967.

Savage T.E. and Van Tuyl, F.M.
1919: Geology and stratigraphy of the area of Paleozoic rocks in the vicinity of Hudson and James bays; Geological Society of America, Bulletin, v. 30, p. 339-378.

Telford, P.G. and Verma, H.M. (ed.)
1982: Mesozoic geology and mineral potential of the Moose River Basin; Ontario Geological Survey, Study 21, 193 p.

Williams, M.Y.
1948: The geological history of Churchill, Manitoba; Western Miner, v. 21, no. 6, p. 39-42.

Author's Address

A.W. Norris
Institute of Sedimentary and Petroleum Geology
Geological Survey of Canada
3303 - 33rd Street N.W.
Calgary, Alberta
T2L 2A7

Printed in Canada

Chapter 8

HUDSON PLATFORM - GEOLOGY

CHAPTER 8

HUDSON PLATFORM - GEOLOGY

A.W. Norris

With contributions by A.C. Grant, B.V. Sanford, and W.R. Cowan

TECTONIC FRAMEWORK

A.W. Norris

The Phanerozoic rocks of Moose River Basin are separated from those of Hudson Bay Basin by a northeast-trending Precambrian basement high, the Cape Henrietta Maria Arch (Fig. 8.1; see also Fig. 7.1). Because of the close similarity of facies and faunas, both basins were almost certainly connected with the Williston, Michigan, and Appalachian basins to the south during various intervals of Paleozoic time. The character of Ordovician and Silurian outliers on the Canadian Shield (Caley and Liberty, 1957), at Clearwater Lake (Kranck and Sinclair, 1963), Waswanipi (Clark and Blake, 1952), Lake Timiskaming (Hume, 1925), Mattawa, Lake Nippissing, and others, tends to support this hypothesis.

The northern boundary of Hudson Bay Basin is Bell Arch, trending northwestward through Southampton Island and bordering Foxe Basin of the southeastern Arctic Platform (see Fig. 7.1). The Phanerozoic rocks of Hudson Platform are separated from Williston and Elk Point basins to the southwest by Precambrian rocks of Severn Arch. To the south, Fraserdale Arch separates Moose River Basin from St. Lawrence Platform and Michigan Basin.

A basin concept for the offshore regions of Hudson Bay, inferred from sea-magnetometer profiles, was firmly established by Hood (1964). On the basis of refraction seismic data, Hobson (1964a, b) confirmed a basin of major proportions beneath Hudson Bay. Seismic profiles suggest that about 2000 m of Phanerozoioc strata are present in Hudson Bay Basin, and 600 to 800 m in the deeper parts of Moose River Basin.

Two structural provinces of the Canadian Shield, Superior and Churchill, border and extend beneath Hudson Platform. Their boundary extends from the vicinity of Nelson River, Manitoba, eastward and southeastward beneath the Paleozoic cover of Hudson Bay Basin to reappear in the Sutton Inlier on Cape Henrietta Maria Arch; from there it extends beneath Paleozoic rocks of Moose River Basin, and across James Bay to the east side of Hudson Bay (Wanless et al., 1967). Within Superior Province, plutonic rocks of Archean age form most of the Precambrian basement. Rocks of Proterozoic age dominate Churchill Province; some are plutonic crystalline rocks and others, as in the Sutton Inlier, are slightly altered and little deformed Aphebian sedimentary rocks (see the companion volume, "Precambrian Geology of the Craton in Canada and Greenland" (Geology of Canada, no. 7, Hoffman et al., in prep.).

Configuration of the Precambrian basement underlying Hudson Platform is shown in a generalized way by Figure 8.1. This figure also portrays the approximate variation in thickness of the Phanerozoic sediments in various parts of both basins. A reflection seismic profile across the central part of the Hudson Bay (Fig. 8.2) shows that the Precambrian basement as well as some of the overlying Phanerozoic rocks has been affected by block faulting.

A number of fracture systems can be seen in outcrop, and others are inferred from bathymetric charts and seismic data. A pronounced east-west lineament extending across east-central Hudson Bay abruptly terminates the Mid-Bay Shoal in the north and is interpreted as a normal fault dipping steeply to the north. On the basis of fault systems and unconformities mapped on the surface and detected in subsurface by drilling in both basins, three periods of epeirogenic uplift are suggested: (1) Late Ordovician (late Gamachian) to Early Silurian (early Llandovery) inclusive, (2) Early Devonian (Siegenian to Emsian) inclusive, and (3) Early Cretaceous.

CRUSTAL GEOPHYSICS

A.C. Grant and B.V. Sanford

The coast of Hudson Bay east of the Belcher Islands forms a remarkable arc, which Beals (1968a) and Halliday (1968) suggested may be the impact scar of a large meteorite. Wilson (1968) proposed an alternative interpretation, namely, that the east coast of Hudson Bay may be a series of secondary sedimentary basins formed along a Precambrian suture. Gibb and Walcott (1971) extended this interpretation to the length of the boundary between Superior and Churchill provinces, on the explicit assumption that plate tectonic mechanisms were active in Early Proterozoic time. The most recent synthesis of the

Norris, A.W.
1993: Hudson Platform - Geology; Chapter 8 in Sedimentary Cover of the Craton in Canada, D.F. Stott and J.D. Aitken (ed.); Geological Survey of Canada, Geology of Canada, no. 5, p. 653-700 (also Geological Society of America, The Geology of North America, v. D-1).

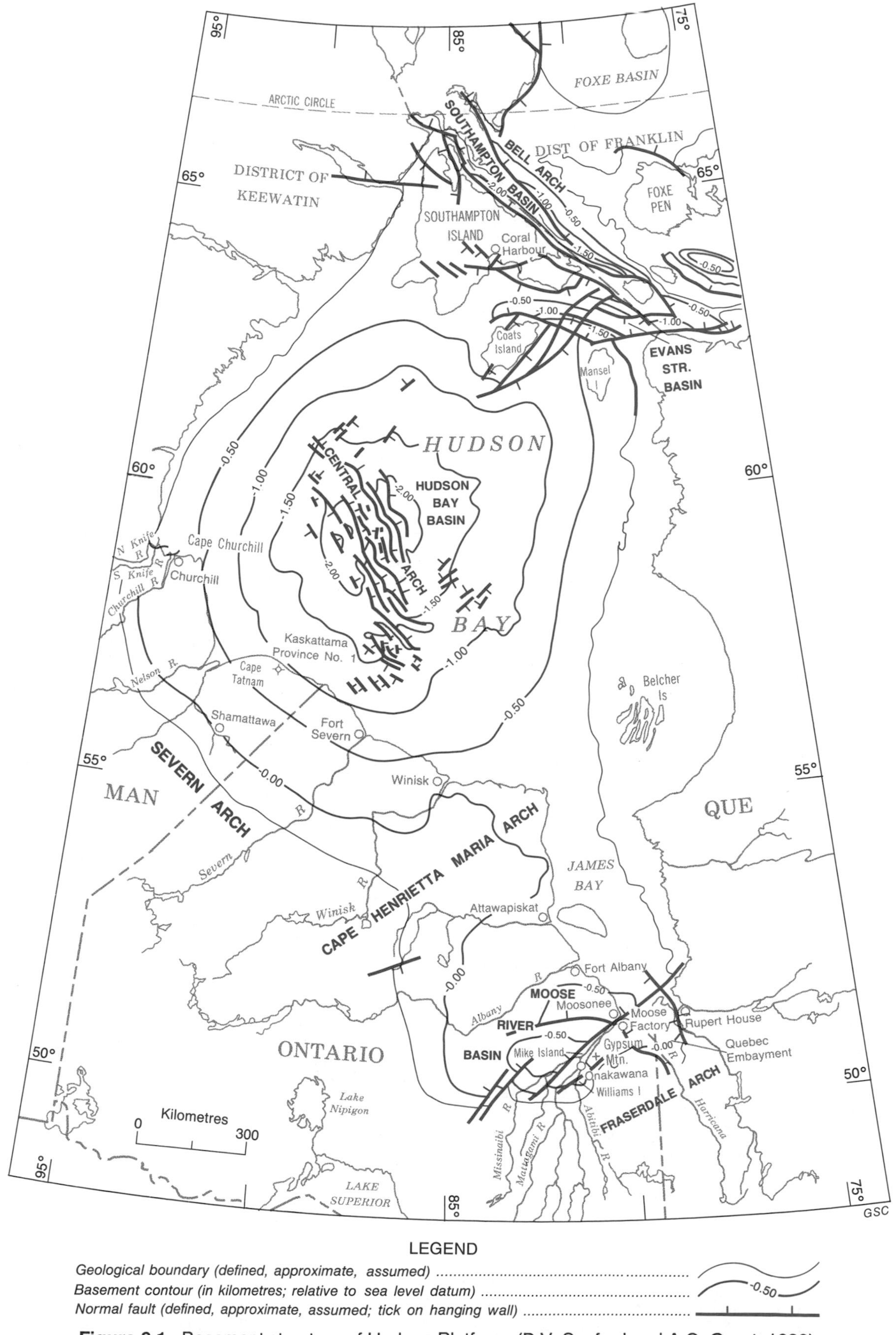

Figure 8.1. Basement structure of Hudson Platform (B.V. Sanford and A.C. Grant, 1989).

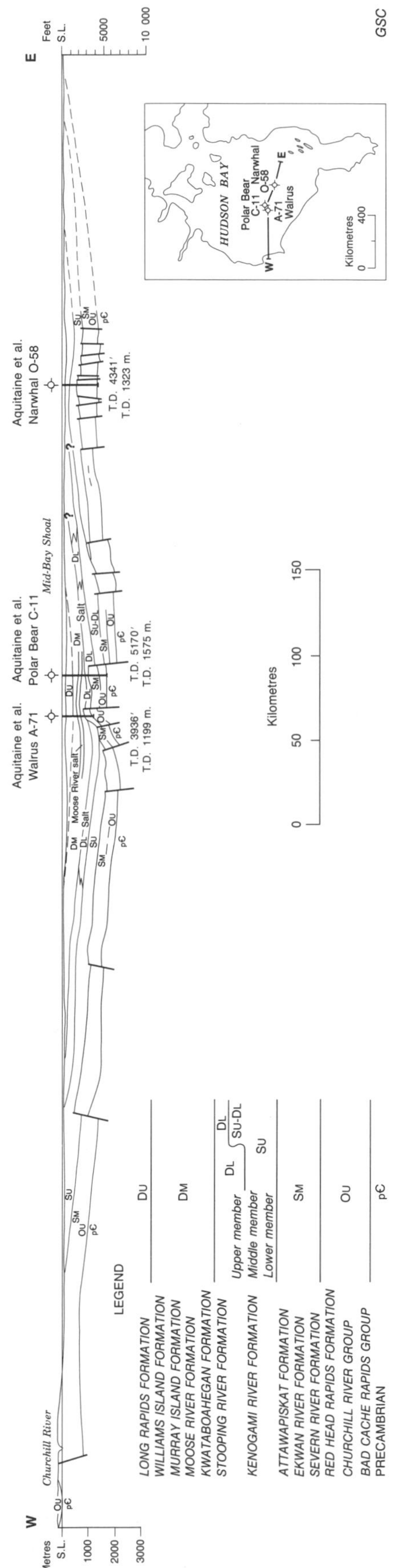

Figure 8.2. East-west structure section based on a reflection seismic profile across Hudson Bay (after Wood in Dimian et al., 1983).

assembly of the Canadian Shield (Hoffman, 1988) follows a plate-tectonic scenario, but suture-straddling basins play no important role therein.

Large-scale features of the Canadian Shield, especially the boundary between Superior and Churchill structural provinces, which lies within the Trans-Hudson Orogen (Fig. 8.3; Hoffman, 1988), can be traced in part beneath Hudson Platform with the use of maps of geophysical fields. Gravity and magnetic anomalies correlated with lithological and tectonic belts of the exposed shield can be traced, especially from the country west of Hudson Bay eastward beneath the sedimentary platform. Because of obvious "oroclinal" bends and discontinuities, the tracing of corresponding elements westward from northern Quebec

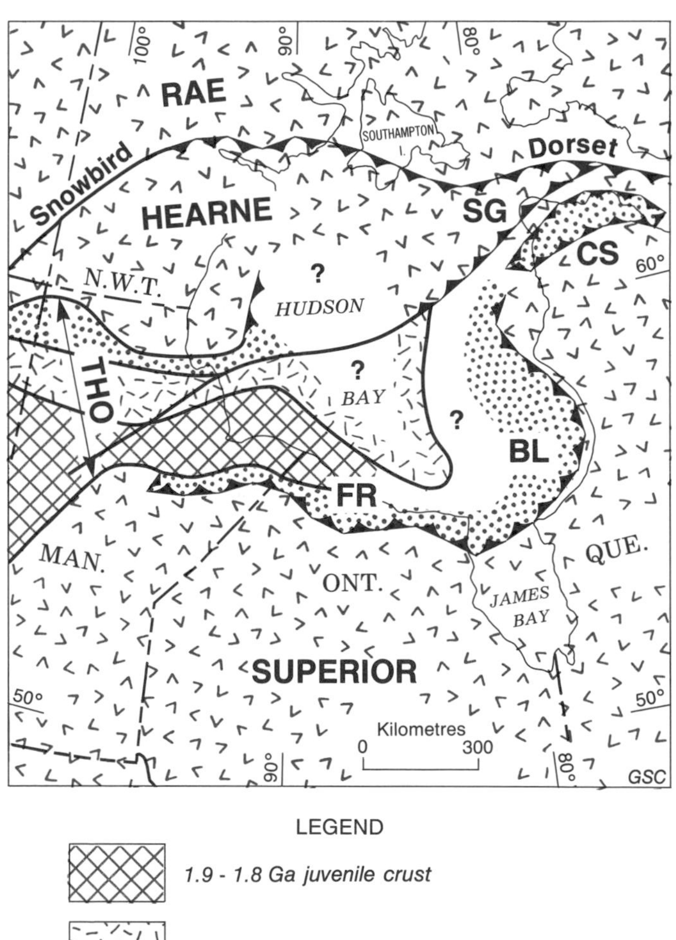

Figure 8.3. Basement structure beneath Hudson Plaform. Upper case Archaean provinces; - Lower case Proterozoic orogens. BL - Belcher Fold Belt; CS - Cape Smith Fold Belt; FR - Fox River Belt; SG - Suglik Terrane; THO - Trans-Hudson Orogen (after Hoffman 1988).

is highly conjectural (Geological Survey of Canada, 1980, 1987; Hoffman, 1988). Despite gaps in the observational network, it is clear that the principal trends of magnetic and gravity anomalies in Hudson Bay are similar, and both display a marked "kink" in the Trans-Hudson Orogen beneath the southeastern part of the bay (Fig. 8.3). Recent papers that provide syntheses of the geology of Hudson Platform from a geophysical standpoint include those of Baragar and Scoates (1981), Gibb et al. (1983), Green et al. (1985), Hoffman (1988), and Thomas and Gibb (1985).

Inferences from magnetic surveys

Regionl magnetic profiles, as interpreted by Bower (1960) and Hood (1964), indicated a substantial thickening of the Paleozoic section beneath Hudson Bay and stimulated the interest of the petroleum exploration industry. Compilations of data from airborne and shipborne magnetic surveys in the Hudson Bay region by government and industry have been published by Morley et al. (1968) and McGrath et al. (1977). An extensive shipborne survey in the summers of 1975 to 1978 provided the first overview of the principal magnetic anomalies (Coles and Haines, 1982). Gaps remain in the Magnetic Anomaly Map of Canada (Geological Survey of Canada, 1987), but the most prominent and most important features, the anomalies outlining the interior zone of the Trans-Hudson Orogen, can be traced from the west, beneath southern Hudson Bay to the aforementioned "kink". At the "kink", the anomalies turn north for about 300 km before encountering a region of confusion, beyond which their trace is conjectural (Fig. 8.3).

Inferences from gravity surveys

Regional gravity coverage is complete over Hudson Bay, James Bay, and adjacent regions (Geological Survey of Canada, 1980). Thomas and Gibb (1977) and Mukhopadhyay and Gibb (1981) modelled gravity profiles of the circum-Superior suture from the Cape Smit Fold Belt and offshore in eastern Hudson Bay. Their interpretation suggested that the crust of Churchill Province is thicker and denser than of Superior Province. Hoffman (1985) used gravity data to support an interpretation that the Cape Smith Fold Belt is a klippe, rather than a suture. Green et al. (1985) have discussed the geological significance of the gravity field south of Hudson Platform.

Structural trends beneath Hudson Platform as expressed by gravity probably are shown most clearly by maps of the horizontal gradient of Bouger gravity (Sharpton et al., 1987). Thomas et al. (1987, 1988) have defined "gravity domains" of North America on this basis and shown (1988) that the Trans-Hudson Orogen (Hoffman, 1988) underlies much of Hudson Platform.

Inferences from seismic refraction surveys

The earliest seismic-refraction work in the Hudson Bay region was carried out in the Hudson Bay Lowland in 1963 and 1964 (Hobson, 1964a). Further seismic studies, including a large-scale crustal refraction experiment, were conducted in Hudson Bay in 1965. These revealed the general configuration and thickness of the Paleozoic basin-filling sediments beneath Hudson Bay (Hobson, 1967).

Ruffman and Keen (1967) published a time-term analysis of the crustal seismic data from Hudson Bay that indicated a single-layered crust with velocity of 6.33 km/s and an upper-mantle velocity of 8.27 km/s. They reported that the depth to mantle ranged from 26 to 42.7 km, and attributed crustal thickening from 30 km to 42 km, in eastern Hudson Bay to the boundary between Churchill and Superior provinces. Overton (1969) reviewed several analyses of Hudson Bay refraction data (Hobson et al., 1967; Hunter and Mereu, 1967; Ruffman and Keen, 1967), and concluded that the data should be interpreted more generally as showing an upper mantle velocity of 8.1 to 8.3 km/s and average crustal thickness of about 35 km. Barr (1969) concluded that the mantle velocity in northern Hudson Bay might be higher (8.4 to 8.5 km/s) than the aforementioned values. Hajnal (1969) suggested that the Hudson Bay data might be best interpreted in terms of a two-layer model. These differences in interpretation relate in part to the quality of seismic refraction data, and in part to sparse coverage in a region of complex crystalline basement.

Inferences from seismic reflection surveys

Various oil companies began seismic exploration in Hudson Bay in the early 1960s, and experimented with both reflection and refraction techniques. Seismic surveys in Hudson Bay are difficult because of generally shallow water and "hard bottom" of Paleozoic carbonate rocks. It can be difficult, in some areas, to map the basement surface, because the seismic velocities of the carbonates can be similar to those of the basement rocks.

Reflection surveys have confirmed and detailed the relatively shallow, saucer shape of the Hudson Bay Paleozoic basin, which is interrupted by a linear, NNW-trending, central horst block with relief of about 1 km (Fig. 7.4, 8.1). Seismic reflection data acquired by the petroleum industry from much of the eastern side of Hudson Bay show basement reflectors, presumably Proterozoic volcanic and sedimentary layers of the interior zone of the Trans-Hudson Orogen. Apparent dips in these rocks generally are westward and can be interpreted in terms of eastward-directed folding and thrusting, as discussed by Roksandic (1987).

Conclusion

The nature of the Precambrian crust beneath the northern Hudson Platform is largely unknown, because of the lack of magnetic, gravity, and seismic refraction surveys there. Magnetic and gravity data indicate that the southern part of Hudson Platform overlies a complex embayment in the margin of the Superior Province of the Canadian Shield, and a corresponding "kink" (orocline?) in the Trans-Hudson Orogen (Fig. 8.3).

Sanford (1987) has pointed to instances of synchrony of subsidence among intracratonic basins and of uplift of their delimiting arches, and between these and convergent activity at the margins of the North American Plate. Nevertheless, as remains the case with most intracratonic basins, neither geology nor geophysics has yet provided an insight into the fundamental cause(s) of subsidence of Hudson Platform and Hudson and Moose River basins.

TECTONOSTRATIGRAPHIC ASSEMBLAGES

A.W. Norris

The oldest Phanerozoic marine sediments of Hudson Platform are of Middle (Caradoc) and Late (Ashgill) Ordovician age. In terms of sequence and sub-sequence terminology of Sloss (1963) and Vail et al. (1977), these Ordovician sediments are within the Tippecanoe I sub-sequence (Fig. 8.4). Older sediments, equivalent to the Sauk sequence, are missing. The succeeding unconformity-bounded sequence of sediments ranges in age from Middle Llandovery of Early Silurian time to about mid Siegenian of the Early Devonian Epoch. This succession is within the Tippecanoe II sub-sequence. The youngest Paleozoic succession of sediments in Hudson Platform ranges in age from about mid-Siegenian of the Early Devonian to within the Famennian of the Late Devonian Epoch. It belongs to the Kaskaskia I sub-sequence, and is the most nearly complete sub-sequence in Hudson Platform.

Small areas of thin continental sediments of Middle Jurassic and Early Cretaceous ages are present only in Moose River Basin (but see "Recent Advances in the Geology of Hudson Platform", below). These belong to the Zuni I and Zuni II sub-sequences, respectively.

BIOSTRATIGRAPHY

A.W. Norris

Ordovician

State-of-the-art

Correlation of Ordovician strata in Hudson Platform and adjacent areas is based mainly on shelly benthonic fossils and conodonts. Graptolites occur in only two rock units. Although zonal schemes based on conodonts, graptolites, and shelly fossils (Fig. 8.5) have proven to be the most useful for correlation, other fossil groups including ostracodes and cephalopods are receiving increasing attention.

Systemic and series boundaries

The historical three-fold terminology for the Ordovician System of North America has been Lower, Middle, and Upper, corresponding to the Canadian, Champlainian, and Cincinnatian Series (Barnes et al., 1981). The series were defined primarily in carbonate facies with shelly fossils and conodonts, and precise correlation with the standard British series, which are mainly graptolitic, is difficult. The British series for the Ordovician, which are commonly used internationally, are in ascending sequence: Tremadoc, Arenig, Llanvirn, Llandeilo, Caradoc, and Ashgill. The base of the Ordovician System recognized by Barnes et al. (1981) for Canada is the base of the trilobite *Missisquoia* Zone, recognized in the Rocky Mountains and western Newfoundland. The base of the Middle Ordovician Series is the base of the brachiopod *Orthidiella* Zone in the southern Rocky Mountains and coincides with the base of the Whiterockian Stage. The base of this brachiopod zone falls within the graptolite *Isograptus victoriae* Zone of western Canada. The base of the Upper Ordovician Series, as selected by Barnes et al. (1981), is the base of the Edenian Stage, which corresponds approximately to the base of the graptolite *Climacograptus spiniferus* Zone of the St. Lawrence Lowlands.

Faunal realms

Ordovician biofacies realms on the North American Plate have a concentric spatial distribution somewhat analogous to the Middle and Late Cambrian biofacies realms described by Palmer (1973) and others. The inner, mid-continent faunas inhabited a wide carbonate platform in the continental interior of North America, whereas the peripheral faunas (North Atlantic) occupied the outer, oceanward margin of the platform. Faunas of the Mid-continent Province were largely restricted to regions characterized by raised salinity and temperatures, whereas the cosmopolitan North Atlantic faunas represent a normal marine province (Barnes and Fåhraeus, 1975).

Biostratigraphic framework

Conodonts in the Ordovician succession of Hudson Platform belong to the Mid-continent Province for which no formal zonation exists (Fig. 8.5), although the faunas of the succession have been described by numerous workers. Faunas 12 and 13 of the succession are represented in the Churchill River Group and the Red Head Rapids Formation, respectively (Barnes, 1974; Barnes et al., 1981).

Two shelly faunas, the Red River and *Bighornia-Thaerodonta* (Fig. 8.5), are represented in the Bad Cache and Churchill River groups, respectively (Barnes et al., 1981); both faunas are widely distributed in the central cratonic regions of Canada.

The graptolites, trilobites, and conodonts present in the locally occurring Boas River Shale on Southampton Island have a wide distribution in equivalent shale units in eastern Canada.

Silurian

State-of-the-art

Worldwide correlation of Silurian marine beds is presently based almost entirely on brachiopods, graptolites, and conodonts (Boucot, 1979). These are supplemented locally by chitinozoans, tetracorals, tabulate corals, stromatoporoids, and ostracodes. Other potentially useful fossil groups in Hudson Platform include trilobites, and in the upper part of the succession, spores.

Systemic and series boundaries

The standard series for the Silurian System recommended by Berry and Boucot (1970) for use in North America are the Llandovery, Wenlock, Ludlow, and Pridoli (Fig. 8.6). In current usage the Lower Silurian includes the Llandovery and Wenlock Series, and the Upper Silurian includes the Ludlow and Pridoli Series.

The base of the type Silurian in Britain is defined as the base of the "Lower Llandovery beds" (Berry and Boucot, 1970). This level in Canada corresponds approximately to the base of the graptolite *Glyptograptus persculptus* Zone, or the base of the conodont *Oulodus? nathani* Zone on Anticosti Island, Quebec. Recently, Holland et al. (1984) have recommended that the base of the Silurian System be selected at a level that coincides with the base of the graptolite *Akidograptus acuminatus* Zone, a level above that of the *Glyptograptus persculptus* Zone.

Both the Llandovery and Wenlock have been subdivided on the basis of brachiopods into Lower, Middle, and Upper units (Berry and Boucot, 1970).

Recognition of the Ludlow Series in North America by Berry and Boucot (1970) is based largely on graptolites, but it can be identified also by brachiopods and conodonts.

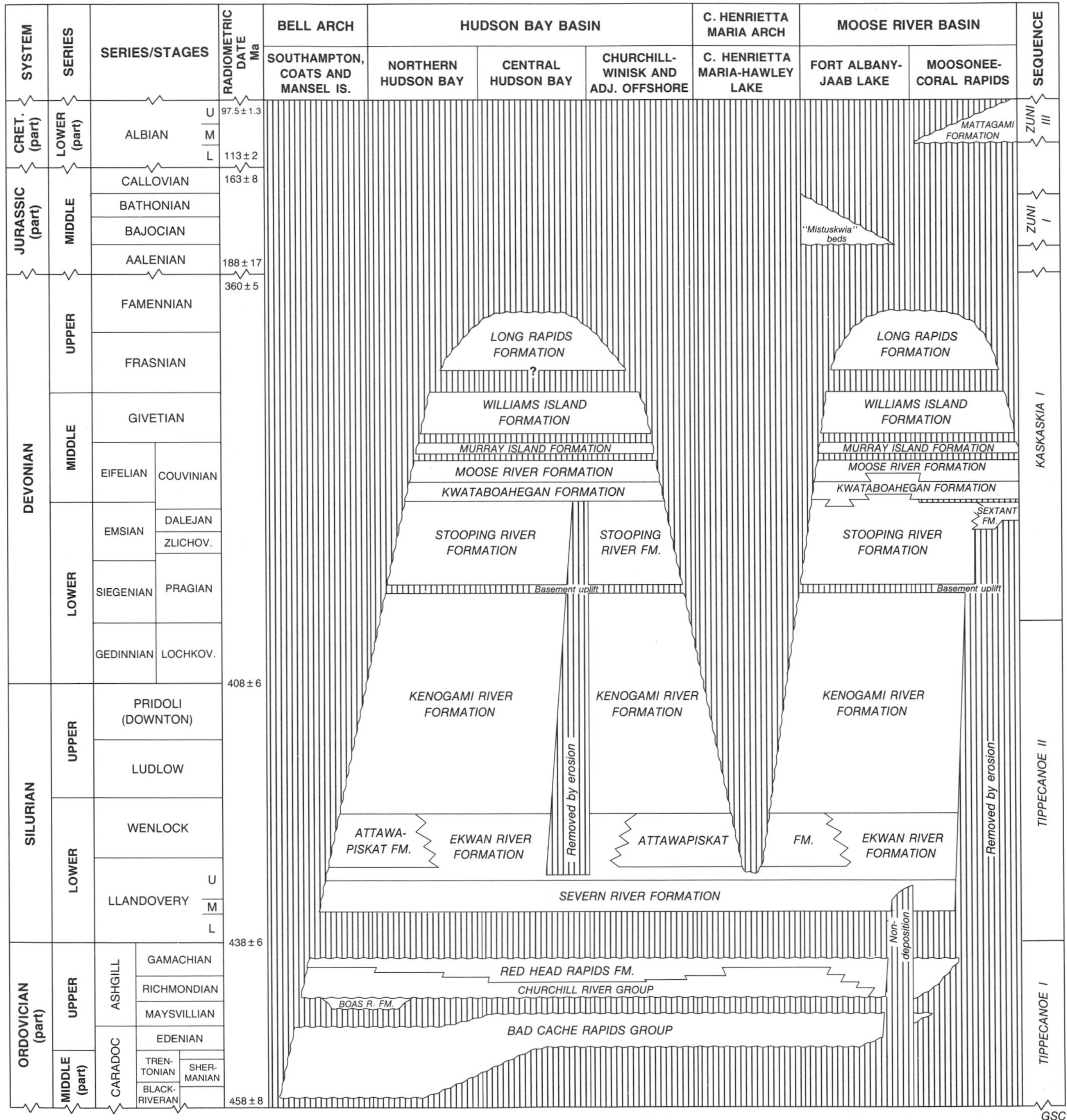

Figure 8.4. Schematic outline of Paleozoic and Mesozoic formations of Hudson Platform.

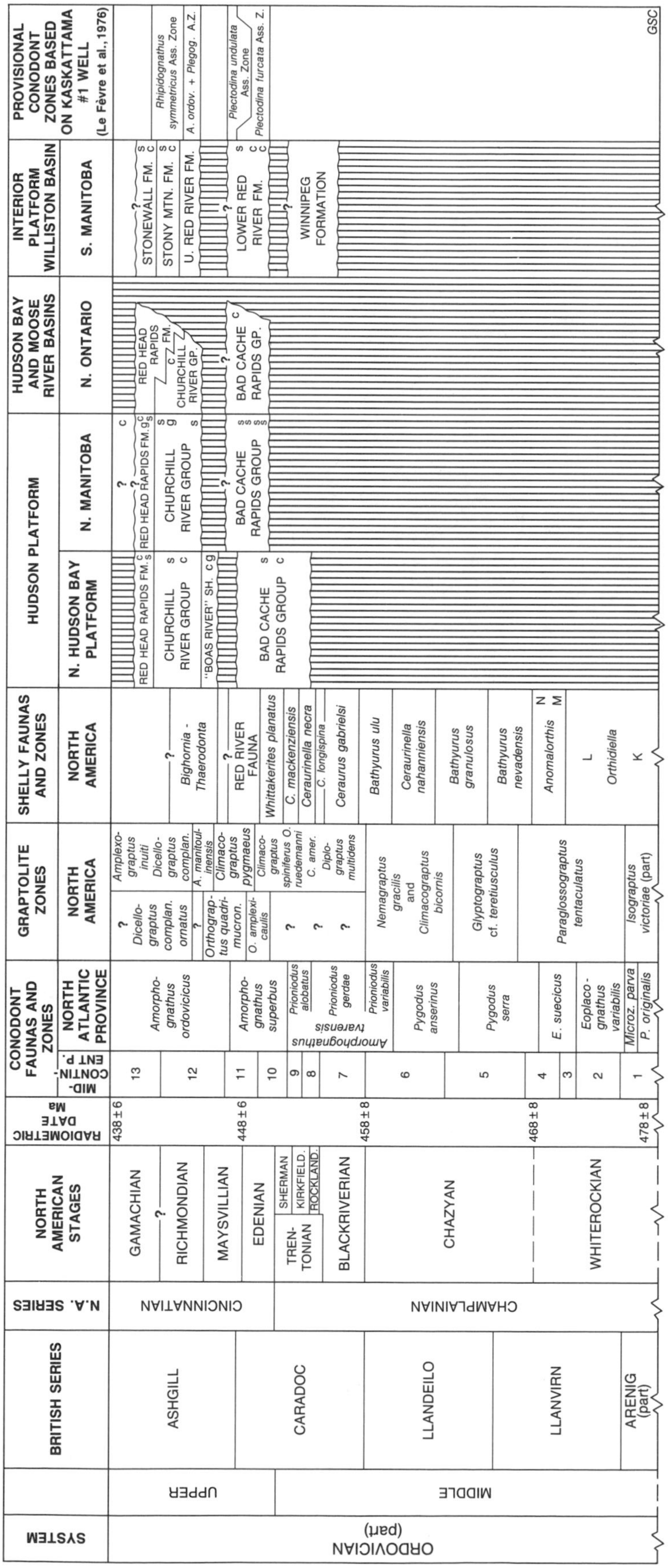

SYSTEM: ORDOVICIAN (part)

(Subdivision): UPPER; MIDDLE

BRITISH SERIES: ASHGILL; CARADOC; LLANDEILO; LLANVIRN; ARENIG (part)

N.A. SERIES: CINCINNATIAN; CHAMPLAINIAN

NORTH AMERICAN STAGES: GAMACHIAN; ? ; RICHMONDIAN; MAYSVILLIAN; EDENIAN; TRENTONIAN (SHERMAN, KIRKFIELD, ROCKLAND); BLACKRIVERIAN; CHAZYAN; WHITEROCKIAN

RADIOMETRIC DATE Ma: 438 ± 6; 448 ± 6; 458 ± 8; 468 ± 8; 478 ± 8

CONODONT FAUNAS AND ZONES

- MID-CONTINENT P.: 13; 12; 11; 10; 9; 8; 7; 6; 5; 4; 3; 2; 1
- NORTH ATLANTIC PROVINCE: *Amorphognathus ordovicicus*; *Amorphognathus superbus*; *Amorphognathus tvarensis* (*Prioniodus alobatus*; *Prioniodus gerdae*; *Prioniodus variabilis*); *Pygodus anserinus*; *Pygodus serra*; *E. suecicus*; *Eoplacognathus variabilis*; *Microz. parva*; *P. originalis*

GRAPTOLITE ZONES – NORTH AMERICA: ? *Dicellograptus complan. ornatus*; *Amplexograptus inuiti*; *Dicellograptus complan.*; ? ; *A. manitoulinensis*; *Orthograptus quadrimucron.*; *Climacograptus pygmaeus*; *O. amplexicaulis*; *Climacograptus spiniferus*; *O. ruedemanni*; *C. amer.*; *Diplograptus multidens*; ? ; ? ; ? ; *Nemagraptus gracilis* and *Climacograptus bicornis*; *Glyptograptus* cf. *teretiusculus*; *Paraglossograptus tentaculatus*; *Isograptus victoriae* (part)

SHELLY FAUNAS AND ZONES – NORTH AMERICA: ? ; *Bighornia - Thaerodonta*; ? ; RED RIVER FAUNA; *Whittakerites planatus*; *C. mackenziensis*; *Ceraurinella necra*; *C. longispina*; *Ceraurus gabrielsi*; *Bathyurus ulu*; *Ceraurinella nahanniensis*; *Bathyurus granulosus*; *Bathyurus nevadensis*; *Anomalorthis* (N, M); L; *Orthidiella*; K

HUDSON PLATFORM

- N. HUDSON BAY PLATFORM: RED HEAD RAPIDS FM. (c, s); CHURCHILL RIVER GROUP (s, c); "BOAS RIVER" SH. (c g); BAD CACHE RAPIDS GROUP (s, c)
- N. MANITOBA: ? (c); ? ; RED HEAD RAPIDS FM. (g, c, s); CHURCHILL RIVER GROUP (s, g, s); ? ; BAD CACHE RAPIDS GROUP (s, s, s)

HUDSON BAY AND MOOSE RIVER BASINS – N. ONTARIO: RED HEAD RAPIDS FM. (c); CHURCHILL RIVER GP.; ? ; BAD CACHE RAPIDS GP. (c)

INTERIOR PLATFORM WILLISTON BASIN – S. MANITOBA: ? ; STONEWALL FM. (s, c); STONY MTN. FM. (s, c); U. RED RIVER FM.; ? ; LOWER RED RIVER FM. (s, c, c); ? ; WINNIPEG FORMATION

PROVISIONAL CONODONT ZONES BASED ON KASKATTAMA #1 WELL (Le Fèvre et al., 1976): *Rhipidognathus symmetricus* Ass. Zone; *A. ordov.* + *Plegog.* A.Z.; *Plectodina undulata* Ass. Zone; *Plectodina furcata* Ass. Z.

GSC

Figure 8.5. Faunal zonation schemes for the Middle and Upper Ordovician. c - conodonts; s - shelly fossils; g - graptolites (modified from Barnes et al., 1981).

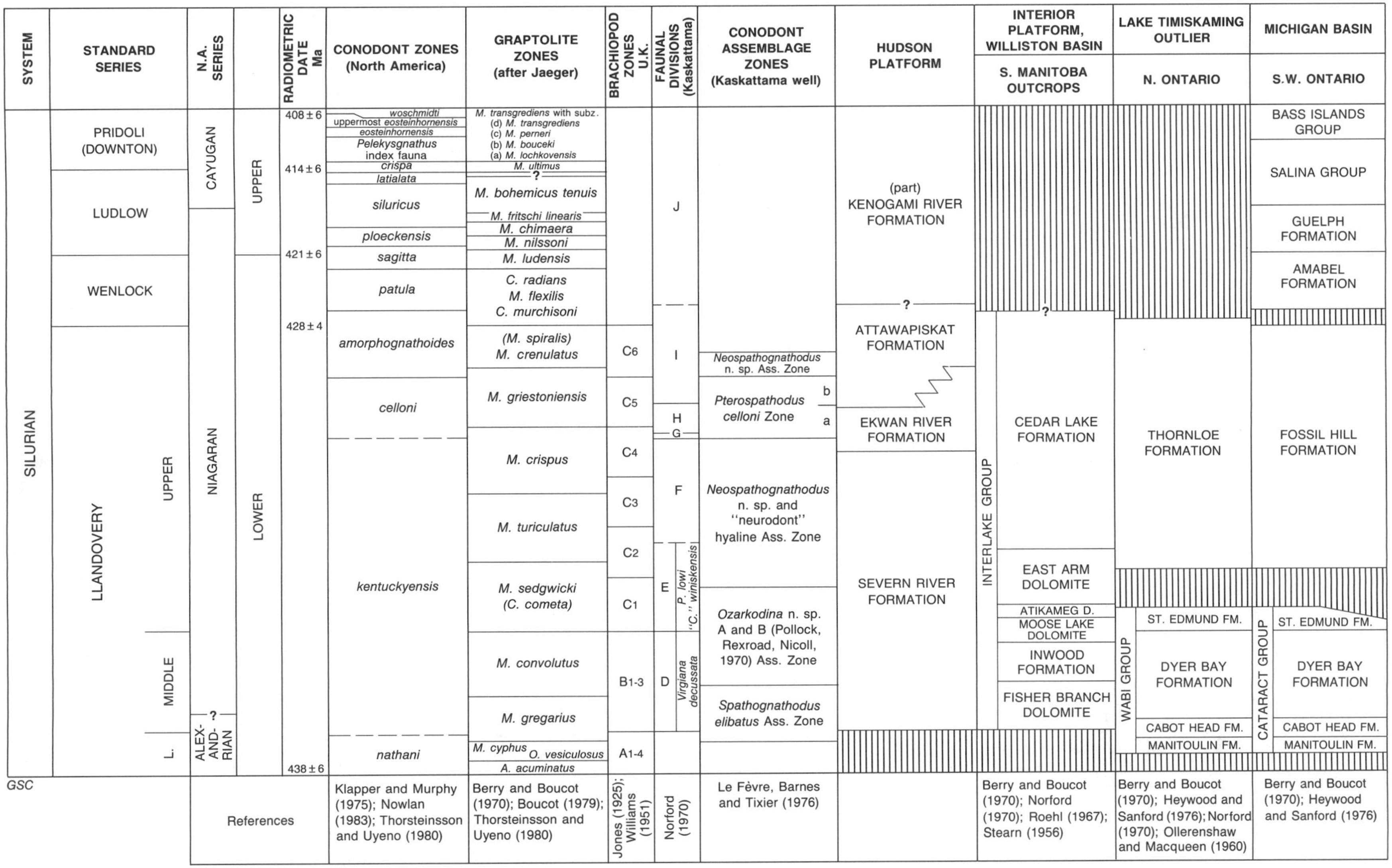

SYSTEM	STANDARD SERIES		N.A. SERIES		RADIOMETRIC DATE Ma	CONODONT ZONES (North America)	GRAPTOLITE ZONES (after Jaeger)	BRACHIOPOD ZONES U.K.	FAUNAL DIVISIONS (Kaskattama)	CONODONT ASSEMBLAGE ZONES (Kaskattama well)	HUDSON PLATFORM	INTERIOR PLATFORM, WILLISTON BASIN: S. MANITOBA OUTCROPS	LAKE TIMISKAMING OUTLIER: N. ONTARIO	MICHIGAN BASIN: S.W. ONTARIO
SILURIAN	PRIDOLI (DOWNTON)		CAYUGAN	UPPER	408 ± 6	woschmidti; uppermost eosteinhornensis; eosteinhornensis; Pelekysgnathus index fauna	M. transgrediens with subz. (d) M. transgrediens (c) M. perneri (b) M. bouceki (a) M. lochkovensis		J		(part) KENOGAMI RIVER FORMATION	(hiatus)	(hiatus)	BASS ISLANDS GROUP
					414 ± 6	crispa	M. ultimus							
							?							
	LUDLOW					latialata; siluricus	M. bohemicus tenuis							SALINA GROUP
						ploeckensis	M. fritschi linearis; M. chimaera; M. nilssoni							
	WENLOCK		NIAGARAN	LOWER	421 ± 6	sagitta	M. ludensis							GUELPH FORMATION
						patula	C. radians; M. flexilis; C. murchisoni				?	?		AMABEL FORMATION
	LLANDOVERY	UPPER			428 ± 4	amorphognathoides	(M. spiralis) M. crenulatus	C6	I	Neospathognathodus n. sp. Ass. Zone	ATTAWAPISKAT FORMATION	INTERLAKE GROUP: CEDAR LAKE FORMATION	THORNLOE FORMATION	FOSSIL HILL FORMATION
						celloni	M. griestoniensis	C5	H	Pterospathodus celloni Zone (b, a)	EKWAN RIVER FORMATION			
									G					
						kentuckyensis	M. crispus	C4	F	Neospathognathodus n. sp. and "neurodont" hyaline Ass. Zone	SEVERN RIVER FORMATION			
							M. turiculatus	C3						
								C2				EAST ARM DOLOMITE	(hiatus)	(hiatus)
							M. sedgwicki (C. cometa)	C1	E: P. lowi; "C." winiskensis	Ozarkodina n. sp. A and B (Pollock, Rexroad, Nicoll, 1970) Ass. Zone		ATIKAMEG D.; MOOSE LAKE DOLOMITE	WABI GROUP: ST. EDMUND FM.	CATARACT GROUP: ST. EDMUND FM.
		MIDDLE					M. convolutus	B1-3	D: Virgiana decussata			INWOOD FORMATION	DYER BAY FORMATION	DYER BAY FORMATION
			?				M. gregarius			Spathognathodus elibatus Ass. Zone		FISHER BRANCH DOLOMITE	CABOT HEAD FM.	CABOT HEAD FM.
		L.	ALEX-AND-RIAN		438 ± 6	nathani	M. cyphus; O. vesiculosus; A. acuminatus	A1-4			(hiatus)	(hiatus)	MANITOULIN FM.	MANITOULIN FM.
			References			Klapper and Murphy (1975); Nowlan (1983); Thorsteinsson and Uyeno (1980)	Berry and Boucot (1970); Boucot (1979); Thorsteinsson and Uyeno (1980)	Jones (1925); Williams (1951)	Norford (1970)	Le Fèvre, Barnes and Tixier (1976)		Berry and Boucot (1970); Norford (1970); Roehl (1967); Stearn (1956)	Berry and Boucot (1970); Heywood and Sanford (1976); Norford (1970); Ollerenshaw and Macqueen (1960)	Berry and Boucot (1970); Heywood and Sanford (1976)

GSC

Figure 8.6. Silurian faunal zonation schemes and correlation of the Silurian sequence of Hudson Platform with other areas.

The base of the Pridoli is marked by the graptolite *Monograptus ultimus* Zone, which occurs in many parts of the world including Canada. In nonmarine facies containing fish remains and spores, the approximately equivalent Downton Series is used by some workers in place of the Pridoli.

Faunal realms

According to some authors, notably Holland (1971), the Silurian Period was characterized by relatively cosmopolitan marine faunas. In more recent analyses, Oliver (1977), Boucot (1979), and others have recognized moderate provincialism. The marine shelf and platform biota of the Silurian is characterized by a Northern Silurian Realm of moderate diversity and a southern Malvinokaffric Realm of low diversity, which record "warm" and "cold" climates, respectively. North America lies within the Northern Silurian Realm.

The initial Silurian faunas of the Lower Llandovery are relicts of Late Ordovician (Ashgill), Old World Realm faunas. New taxa characteristic of southeastern Kazakhstan spread to other areas near the beginning of Late Llandovery time (Boucot, 1975). A marked enrichment of the Silurian marine fauna during the later Wenlock Epoch was coincident with the rapid spread and diversification of reef biotas. At the end of the Ludlow Epoch, a marked reduction in taxonomic diversity coincided with the diminution of the reef environment (Boucot, 1979).

Biostratigraphic framework

Shelly bethonic fossils, conodonts, and spores, the latter in the uppermost part of the sequence, have been the most useful taxa in dating and correlating the Silurian rocks of Hudson Platform. Informal faunal divisions (Fig. 8.6) based on shelly fossils in the Silurian sequence, penetrated by the Sogepet Aquitaine Kaskattama No. 1 well (59°04′18.5"N, 90°10′29.4"W) at the southern end of Hudson Bay Basin, have been proposed by Norford (1970). Faunal division D corresponds to the interval containing the widespread brachiopod *Virgiana decussata*, which is aligned by Berry and Boucot (1970) with the Middle (B_{1-3}) Landovery, and marks the hiatus at the base of the Silurian sequence. The succeeding faunal division E is characterized by the equally widespread brachiopods *Plectatrypa lowi* and "*Camarotoechia*" *winiskensis*, which is aligned with the lower Upper (C_{1-2}) Llandovery. Norford's (1970) faunal divisions F, G, H, and I, based primarily on diagnostic brachiopods, an ostracode, and diagnostic corals, are aligned with the Upper Landovery to lower Wenlock. His faunal division J corresponds to the thick, largely unfossiliferous Kenogami River Formation, which spans the upper Wenlock, Ludlow, and Pridoli of the Silurian, and the Gedinnian and part of the Siegenian of the Devonian.

Four provisional conodont assemblage zones and one formal conodont zone have been recognized by Le Fèvre et al. (1976) in the Silurian sequence penetrated by the Kaskattama No. 1 and nearby wells. The one assemblage tied in with the standard conodont zonal scheme is the *Pterospathodus celloni* Zone, which is aligned approximately with the upper part of the Upper Llandovery (McCracken and Barnes, 1981).

Devonian

State-of-the-art

Fossil groups most useful for dating and correlating Devonian rocks in Hudson Platform include corals, brachiopods, and conodonts (Fig. 8.7). Other groups of lesser importance include acritarchs from the Lower and Middle Devonian of Moose River Basin (Playford, 1977); stromatoporoids from the Kwataboahegan Formation (Fritz and Waines, 1956); cephalopods from the Moose River Basin (Foerste, 1929); and ammonoids from the Upper Devonian Long Rapids Formation (Foerste, 1928; Miller, 1938). Other fossil groups known to be present and potentially useful include the tentaculitids and ostracodes.

Systemic and series boundaries

Terminology used for Devonian series and stages in North America has tended to follow that used in the classic Devonian areas of western Europe. An exception is the series and stage terminology introduced by Cooper et al. (1942) for the Devonian succession in New York State and adjacent areas, where the invertebrate fossils for the Lower and Middle Devonian are highly endemic and distinct from the Old World faunas found elsewhere. Two sets of stages used for the Lower Devonian are the Gedinnian, Siegenian, and Emsian, defined in nearshore clastic (Rhenish) facies of Belgium and West Germany; and the Lochkovian, Pragian, Zlichovian, and Dalejan, defined in more carbonate-rich (Hercynian) facies of Czechoslovakia. Both the Gedinnian and Siegenian contain sparse marine faunas and are difficult to correlate, and for this reason the Lochkovian and Pragian have been adopted as stage standards for marine faunas. The Middle Devonian Series includes the Couvinian and Givetian stages, as used in France and Belgium, and the Eifelian and Givetian stages, as used in Germany, with the latter terminology currently preferred. The Upper Devonian Series encompasses the Frasnian and Famennian stages.

The Silurian-Devonian boundary is defined (McLaren, 1972) as lying directly below the first occurrence of *Monograptus uniformis*; and this level is close to the base of the conodont *Icriodus woschmidti* Zone. The Lower-Middle Devonian boundary is selected at the base of the conodont *partitus* Zone; and the Middle-Upper Devonian boundary at the base of the conodont Lower *asymmetricus* Zone. The latter level is slightly higher than that traditionally used in North America.

The top of the Devonian succession (base of the Carboniferous) is marked by the base of the conodont *Siphonodella sulcata* Zone (Paproth, 1980). This level is very slightly below the ammonoid *Wocklumeria-Gattendorfia* "Stufen" boundary previously recommended.

Faunal realms

Three major faunal provinces or realms have been recognized in the Devonian, based mainly on the world distribution of brachiopods (Johnson and Boucot, 1973;

Boucot, 1975). Other shelly fossils, such as corals (Oliver, 1977), and trilobites (Ormiston, 1972), have a somewhat similar pattern of distribution. The three major faunal realms are referred to as the Old World, Eastern Americas, and Malvinokaffric. In North America, the faunas characterizing the Old World Realm occur in the western part of the continent and in the Canadian Arctic; whereas the faunas of the Eastern Americas Realm are found in a belt along the eastern and southern parts of the continent. The principal cause of provinciality between eastern and western North America during Early and part of Middle Devonian time was the presence of a northeast-trending land barrier referred to as the Transcontinental Arch or Continental Backbone. This land barrier was breached during the latest Givetian and early Frasnian ages so that the Eastern Americas Realm finally disappeared as a

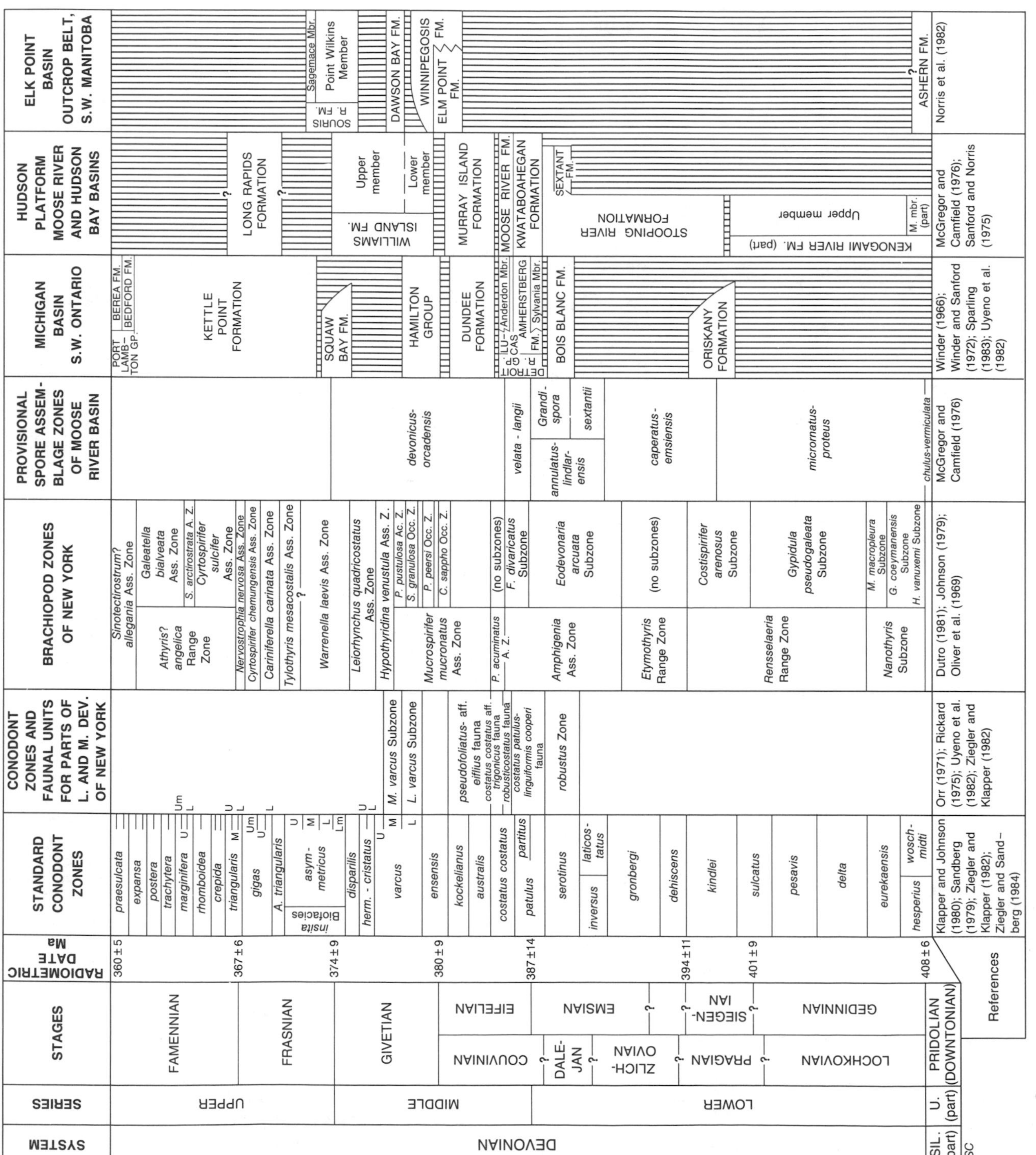

Figure 8.7. Conodont, brachiopod and spore zonation schemes for the Devonian, and correlation of the Devonian succession of Hudson Platform with other areas.

distinct entity and was replaced by a cosmopolitan Old World Realm fauna. The Malvinokaffric Realm, which developed in Emsian time, comprises certain parts of the southern hemisphere, specifically in southern South America, South Africa, and Antarctica.

Early Devonian marine faunas are absent from Hudson Platform where the interval (Gedinnian and Siegenian) is represented by terrigenous clastics and restricted marine carbonates and evaporites. Hudson Platform during that time appears to have been part of the 'Old Red Sandstone Continent' as depicted by House (1968).

The brachiopods, corals, and conodonts of late Emsian, Eifelian, and Givetian ages indicate that Hudson Platform was a part of the Eastern Americas Realm during those times (Johnson and Boucot, 1973; Oliver, 1977; Oliver and Pedder, 1979). The brachiopods and corals show maximum provinciality during the Emsian age. However, a few brachiopods of Old World Realm affinities indicate that there were minor marine connections between Moose River Basin and Elk Point Basin of the Interior Platform in late Eifelian, and between Hudson Bay and Elk Point basins in mid-Givetian time. Widespread goniatite occurrences and cosmopolitan brachiopods and corals reflect a pronounced marine transgression during Frasnian time.

Biostratigraphic framework

Biostratigraphical zonal schemes that are important in dating and correlating the Devonian rock units of Hudson Platform are indicated in Figure 8.7. These include the standard worldwide conodont zones that have been developed and refined by numerous workers over the past 20 years; the local conodont zones and faunal units for parts of the Lower and Middle Devonian of New York and adjacent areas; the brachiopod zones for the New York area; and the provisional spore assemblage zones proposed for Moose River Basin. Correlation between the zonal schemes is only approximate.

Of special significance is the presence in Hudson Bay Platform of Lower and Middle Devonian shelly benthonic fossils that belong to the Eastern Americas Realm and are distinct from Old World Realm faunas. Correlation between these two faunal realms has gradually become established mainly through the use of conodonts.

The provisional spore assemblage zones proposed by McGregor and Camfield (1976) for the Devonian succession in Moose River Basin are important because the basin is one of the few areas in the world where spores derived from adjacent land areas can be tied in with a marine succession. Parts of the evaporitic and continental Kenogami River Formation were dated for the first time by means of spores.

Mesozoic

Biostratigraphic framework

The continental Lower Cretaceous Mattagami Formation in Moose River Basin was first recognized on the basis of plants (Keele, 1920a; Bell, 1928). Spore-pollen analyses of drill samples later refined the dating of the Mattagami Formation and indicated a new continental rock unit, the Mistuskwia Beds, of Middle Jurassic age (Norris et al., 1976; Norris, 1977).

LITHOSTRATIGRAPHY

A.W. Norris

Ordovician

The oldest known Paleozoic rocks of Hudson Platform are of late Middle Ordovician (Caradoc) and Late Ordovician (Ashgill) age, which have been subdivided by Nelson (1963, 1964) into the Bad Cache Rapids and Churchill River groups (Fig. 8.8) and the Red Head Rapids Formation, in ascending sequence. They form narrow outcrop belts along the southwest margin of Hudson Bay Lowland, in Quebec Embayment of Moose River Basin, and on Southampton and Coats islands at the north end of Hudson Bay Basin (see Fig. 7.3). Maximum thickness of Ordovician strata in the Manitoba part of Hudson Bay Basin is about 180 m (Cumming, 1971), in Moose River Basin about 83 m (Sanford et al., 1968), and on Southampton Island about 160 m (Heywood and Sanford, 1976; Fig. 8.9, 8.11). The Ordovician succession thickens slightly towards the central offshore part of Hudson Bay Basin, and thins markedly over the Cape Henrietta Maria Arch.

Bad Cache Rapids Group

The Bad Cache Rapids Group (Nelson, 1963, 1964) includes Ordovician strata of the northern Hudson Bay Lowland that rest nonconformably on the peneplaned surface of the Precambrian basement (Fig. 8.10) and are succeeded disconformably by limestone of the Churchill River Group. The Bad Cache Rapids was divided into the Portage Chute and Surprise Creek formations, in ascending sequence. Although these are useful formational terms in the outcrop belt of the type area, they are difficult to apply in the subsurface and are seldom, if ever, used regionally. The Bad Cache Rapids Group occurs throughout Hudson Bay Basin, but thins to a feather edge over Cape Henrietta Maria Arch, and is not represented in Moose River Basin.

Several drill holes penetrating the Bad Cache Rapids Group in the Nelson River area of the outcrop belt show thicknesses from 16 to 30 m. In the Kaskattama No. 1 well, located on the coast near the mouth of Kaskattama River, the thickness is 48 m. On Southampton Island, a thickness of 46 m is estimated. Two wells in the offshore central part of the basin penetrated 71 and 90.5 m of Bad Cache Rapids strata (Aquitaine Company of Canada Ltd., 1974a, b).

The lower, Portage Chute Formation, consists of a basal calcareous quartz sandstone, overlain by microcrystalline dolomite and bioclastic limestone commonly exhibiting nodular bedding (Cumming, 1971). The overlying Surprise Creek Formation consists of finely crystalline, cherty, dolomitic limestone. In the subsurface some anhydrite is also evident (Le Fèvre et al., 1976).

On Southampton Island, the Bad Cache Rapids Group is a relatively uniform sequence of micritic limestone with extensive light yellowish orange mottling associated with incipient dolomitization (Heywood and Sanford, 1976). The basal metre is generally sandy and conglomeratic. Anastomosing structures, presumably of algal origin, and dark grey shale in laminae and thin interbeds are common in the lower part of the group.

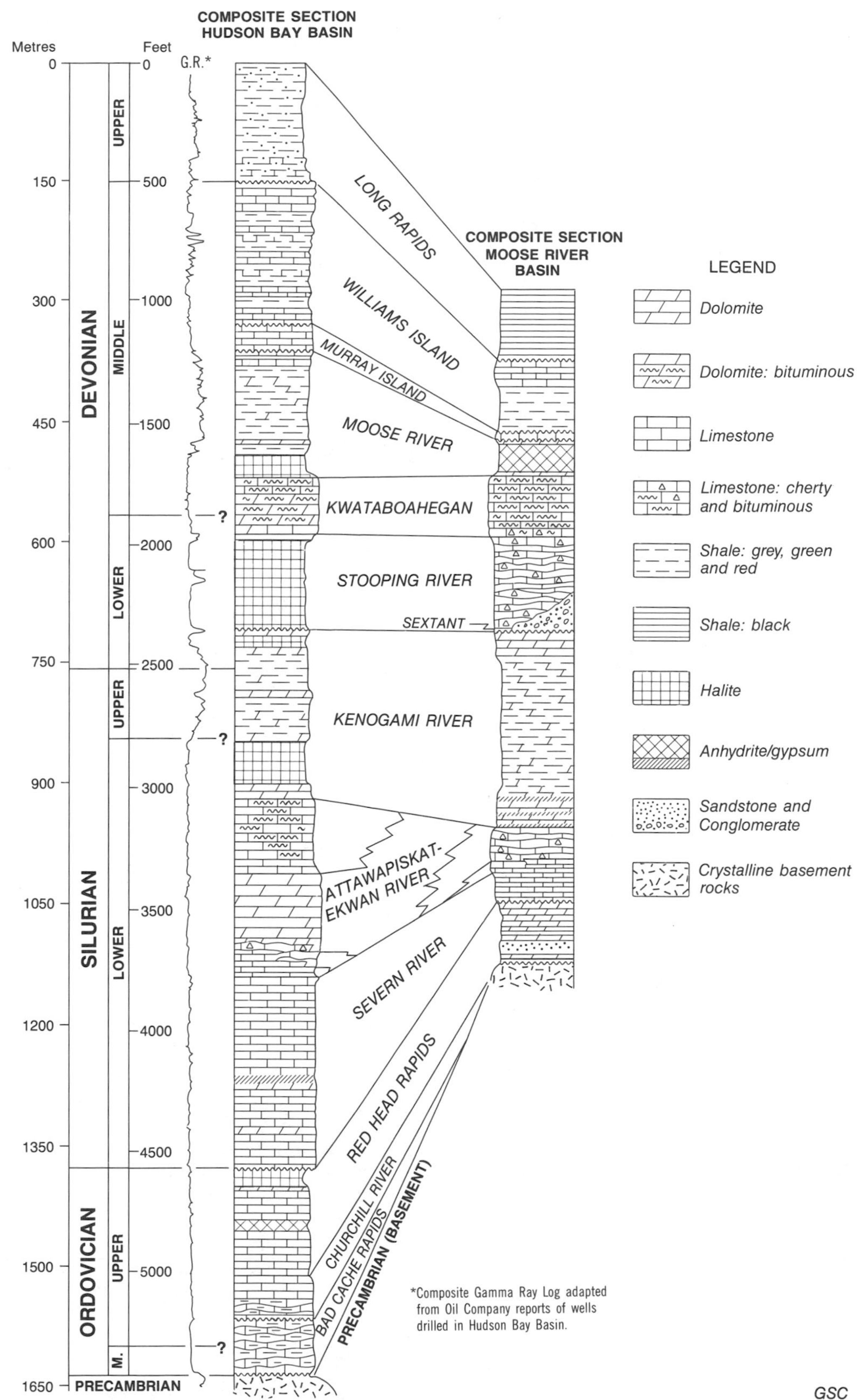

Figure 8.8. Composite Paleozoic stratigraphic successions for Moose River and Hudson Bay basins (from B.V Sandford, unpub. rep., 1980).

The sandy beds at the base of the Portage Chute Formation mark the initial marine transgression over the Canadian Shield. Succeeding beds with abundant shelly fossils reflect a normal marine subtidal environment. Decrease in shelly fossils, presence of leperditiid ostracodes, and gradual upward increase in evaporites and dolomite in the Surprise Creek Formation reflect an intertidal environment of deposition (Le Fèvre et al., 1976).

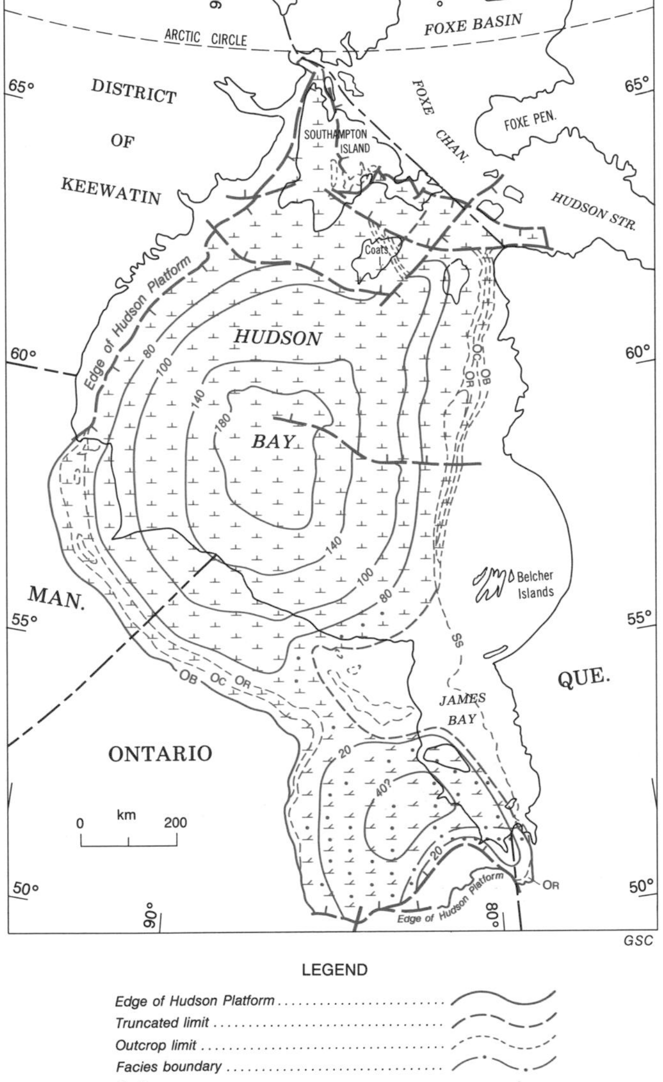

Figure 8.9. Facies and isopachs of the Bad Cache Rapids and Churchill River groups (upper Middle and Upper Ordovician) of Hudson Platform.

Corals, brachiopods, gastropods, and cephalopods are sporadically abundant in the Portage Chute and Surprise Creek formations in the type area (Nelson, 1963, 1964). A similar but less abundant assemblage of fossils has been reported (Heywood and Sanford, 1976) from the Bad Cache Rapids Group on Southampton Island. The fossils of the Portage Chute and Surprise Creek formations suggest correlation with the Dog Head and Cat Head members, respectively, of the Red River Formation of southern Manitoba (Nelson and Johnson, 1966). Sinclair (in Heywood and Sanford, 1976) suggested that the fossils of the Bad Cache Rapids Group on Southampton Island correlate with the Farr Formation of Lake Timiskaming and the upper Cobourg beds of the Trenton Group in south-central and eastern Ontario.

Conodonts from the Bad Cache Rapids Group on Southampton and Coats islands indicate a late Middle Ordovician (post-Chazyan) or early Late Ordovician age (Barnes, 1974). Conodonts from the Portage Chute Formation of northern Manitoba are placed in a provisional *Plectodina furcata* Assemblage Zone dated as Edenian of the Late Ordovician. Conodonts from the succeeding Surprise Creek Formation are placed in the *Plectodina undulata* Assemblage Zone, dated as possibly Maysvillian of the Late Ordovician (Le Fèvre et al., 1976).

Boas River shale

The informal name Boas River shale was introduced by Sanford (in Heywood and Sanford, 1976) for a distinctive petroliferous shale overlying the Bad Cache Rapids Group on Southampton Island. The "oil shale interval" of Nelson and Johnson (1966, 1976) on Southampton Island is younger than the Boas River shale and is considered by Sanford (in Heywood and Sanford, 1976) to be a part of the Red Head Rapids Formation. The Boas River shale may

Figure 8.10. Nonconformable contact between Aphebian granitic rocks and basal beds of the Ordovician Portage Chute Formation of the Bad Cache Rapids Group, on Churchill River (photo - B.V. Sanford, GSC 200481B).

have had a wider distribution, but either may be covered by drift or have been removed by erosion during a mid-Late Ordovician (Maysvillian) hiatus. It consists of black calcareous and petroliferous shale. Very thin, uniform bedding, dark colour, and pelagic fossils suggest that it was deposited in a relatively deep water, marine environment.

A graptolite species from the Boas River shale has been dated (Riva in Heywood and Sanford, 1976) as equivalent to the *Amplexograptus manitoulensis* Zone of eastern North America (Fig. 8.5). A trilobite, *Pseudogygites* sp. cf. *P. latimarginata* (Hall), from the same beds, is a typical element of the upper Cobourg and Collingwood faunas of southern Ontario, which are of late Middle to Late Ordovician age (Sinclair in Heywood and Sanford, 1976). Barnes (1974) has indicated that the conodonts from the Boas River shale are more closely related to the overlying Churchill River Group than to the underlying Bad Cache Rapids Group.

A brown, bituminous and argillaceous limestone containing a *Receptaculites-Maclurites-Hormotoma* faunal assemblage on Akpatok Island has been correlated by Workum et al. (1976) with the Boas River shale of Southampton Island.

Churchill River Group

The Churchill River Group (Nelson, 1963, 1964) disconformably overlies the Bad Cache Rapids Group and is, in turn, conformably succeeded by the Red Head Rapids Formation. In the type area in northern Hudson Bay Lowland, it has been divided, in ascending order, into the Caution Creek and Chasm Creek formations.

Rocks of the group occur in a narrow outcrop belt in western Hudson Bay Lowland but are concealed by faults along the southern margin of Moose River Basin. They appear again directly south of James Bay in the Quebec Embayment, and on Southampton and Coats islands. On the southern flank of Cape Henrietta Maria Arch they overlap the Bad Cache Rapids Group to rest directly on Precambrian rocks in Moose River Basin.

Thicknesses for the Caution Creek and Chasm Creek formations, based on outcrop data in northern Hudson Bay Lowland, are 13 and 55 m, respectively (Nelson and Johnson, 1966). In the nearby Kaskattama No. 1 well, the Churchill River Group is 90 m thick (Norford, 1970). Nearly complete outcrop sections of the group on Southampton Island show a thickness of about 53 m (Heywood and Sanford, 1976). Maximum recorded thickness of the group in the offshore central part of the Hudson Bay Basin is about 96 m (Aquitaine Company of Canada Ltd., 1974a).

The lower Caution Creek Formation, in the type area, consists of microcrystalline, dolomitic limestone, with skeletal fragments common in some beds (Nelson and Johnson, 1966). These sediments containing abundant shelly benthonic fossils are interpreted by Le Fèvre et al. (1976) as indicating an open marine, subtidal environment of deposition. The conformably overlying Chasm Creek Formation consists mainly of resistant, microcrystalline, iron-rich dolomite and slightly dolomitic limestone, commonly with indeterminate trace fossil markings, and in part containing fragments of skeletal material (Nelson and Johnson, 1966). With a marked reduction in shelly benthonic fossils and traces of anhydrite in the subsurface, it records a slightly to moderately restricted intertidal environment of deposition (Le Fèvre et al., 1976). On Southampton Island, the group is composed of argillaceous, micritic limestone. Anastomosing structures of presumed algal origin appear in a thin interval at the top (Heywood and Sanford, 1976).

The Caution Creek Formation, on the basis of shelly fossils, is correlated with the Gunn and lower Penitentiary members, and the Chasm Creek Formation is correlated with the upper Penitentiary and Gunton members of the Stony Mountain Formation of southern Manitoba (Nelson and Johnson, 1966). Shelly fossils in the Churchill River

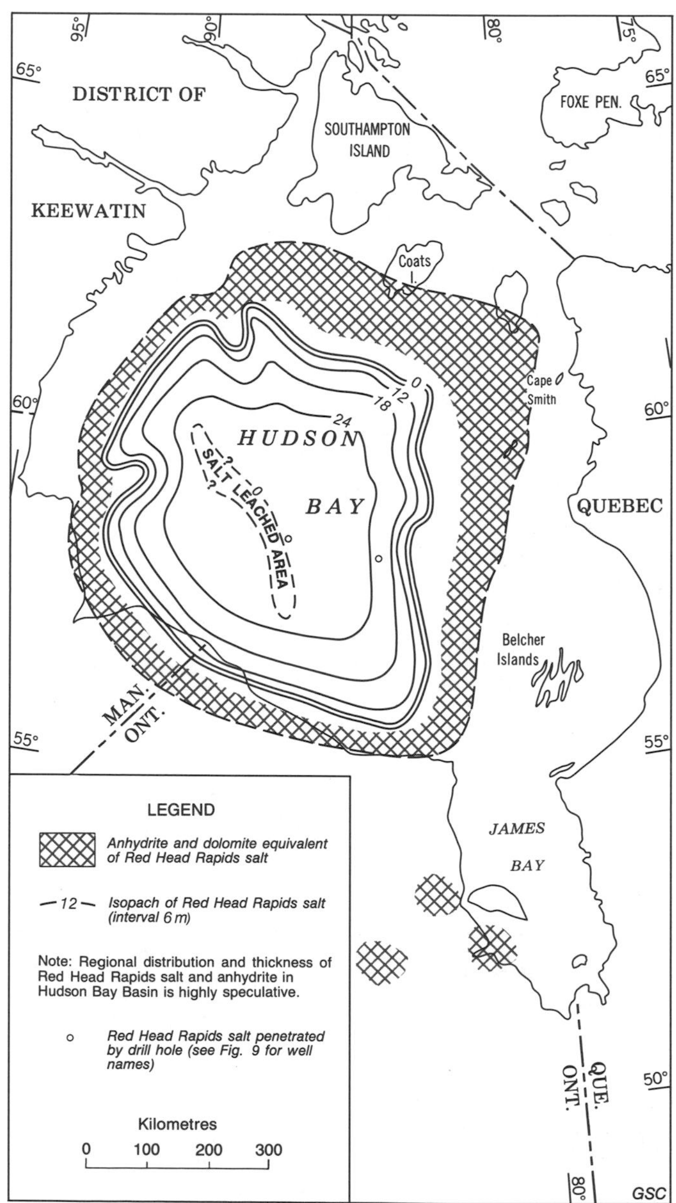

Figure 8.11. Isopachs of evaporites of Red Head Rapids Formation (Upper Ordovician, Gamachian) of Hudson Platform (from B.V. Sanford, unpub. rep., 1980).

Group on Southampton Island consist mainly of corals and cephalopods. The presence of *Favosites* and *Paleofavosites*, and the absence of such genera as *Receptaculites, Hormotoma*, and *Rafinesquina*, distinguish this fauna from that of the underlying Bad Cache Rapids Group.

Conodonts from the Caution Creek Formation of northern Manitoba have been placed (Le Fèvre et al., 1976) in the *Amorphognathus ordovicicus* and *Plegognathus* Assemblage Zone, dated as early Richmondian. Conodonts from the overlying Chasm Creek Formation are placed in the lower part of the *Rhipidognathus symmetricus* Assemblage Zone, dated as mid-Richmondian. Conodonts from the Churchill River Group on Southampton Island belong to Fauna 12 of Sweet et al. (1971) and indicate a late Maysvillian to Richmondian age (Barnes, 1974; Heywood and Sanford, 1976).

Red Head Rapids Formation

The Red Head Rapids Formation (Nelson, 1963, 1964) comprises dolomite and calcareous dolomite that overlie the Churchill River Group and is succeeded unconformably by limestone of the Silurian Severn River Formation in the northern Hudson Bay Lowland. The unit is now recognized in Hudson Bay and Moose River basins (Fig. 8.11). In Moose River Basin, it overlaps the Churchill River Group to rest nonconformably on Precambrian basement rocks.

In the subsurface of northern Hudson Bay Lowland, where it was referred to by Norford (1970) as "Stonewall equivalent", the thickness of the Red Head Rapids Formation is 32 m (Fig. 8.10). From outcrop sections on Southampton and Coats islands the estimated thickness is about 61 m (Heywood and Sanford, 1976). A noticeable thickening of the formation occurs in the offshore central part of the basin, where a maximum of 98 m was recorded (Aquitaine Company of Canada Ltd., 1974a).

In the type area of the Churchill River, the Red Head Rapids Formation consists of microcrystalline and slightly calcareous dolomite (Nelson, 1963). In the Kaskattama No. 1 and other nearby wells, inclusions of anhydrite and some shale are also evident (Cumming, 1971). On Southampton Island three distinct lithological units are recognized, which are referred to informally as laminated, biostromal, and biohermal beds, in ascending sequence (Heywood and Sanford, 1976). The lower unit is composed of microcrystalline to microgranular, laminated limestone and dolomite, in which stromatolites and flat pebble conglomerate beds occur sporadically. The succeeding unit is a massive, stromatolitic, biostromal carbonate (Fig. 8.12), consisting of vuggy, microcrystalline limestone, in places variably dolomitized. The uppermost unit is composed of a biohermal facies of algal limestone and an inter-reef facies of thin bedded, microcrystalline limestone. The biohermal structures are up to 23 m high and up to 1.5 km wide. Vuggy porosity is present in some of the reefs, and some of the vugs contain dead oil.

Evaporites up to 20 m thick in the Red Head Rapids, including anhydrite near the middle and salt at the top of the formation, have been penetrated by two of the offshore wells in Hudson Bay Basin. Very thin intervals of anhydrite have been encountered in several wells on the mainland in the southern peripheral part of the basin, and in two wells in Moose River Basin (B.V. Sanford, pers. comm., 1980).

The general lack of marine faunas and burrowing organisms suggest that most of the carbonates originated in a regressive, restricted, highly saline environment of deposition. Biohermal algal carbonates on parts of the flanks of the basins, and evaporites in the centres, suggest bioherm-silled central basins, with sea-water inflow and evaporation.

The exceedingly scarce shelly fossils of the formation indicate a correlation with the Stonewall Formation of southern Manitoba (Nelson and Johnson, 1966; Norford, 1970). The few conodonts recovered from various localities are closely similar to those from the underlying Churchill River Group and were originally assigned to Fauna 12 of Sweet et al. (1971), dated as probably Richmondian in age (Barnes, 1974). More recently, on the basis of the work of McCracken and Barnes (1981) on Anticosti Island, conodonts from the Red Head Rapids Formation have been aligned by Barnes et al. (1981) with Fauna 13 of Gamachian age. Conodonts from the lower part of the formation are assigned by Le Fèvre et al. (1976) to the upper part of their *Rhipidognathus symmetricus* Assemblage Zone.

Silurian

Silurian rocks of Hudson Platform include, in ascending sequence, the Severn River, Ekwan River, and Attawapiskat formations, and the lower and middle members of the Kenogami River Formation. Part of the Gamachian Stage of the Ordovician System and all of the lower Llandovery Series of the Silurian System are missing; Silurian and Ordovician sequences are separated by a hiatus of some magnitude. Silurian rocks are widely distributed in the outcrop belts and subsurface of both

Figure 8.12. Domal stromatolites in the Upper Ordovician Red Head Rapids Formation, south coast of Bell Peninsula, Southampton Island, northern Hudson Bay Basin (photo - B.V. Sanford, GSC 161935).

basins. The aggregate thickness of the Silurian rocks is about 303 m in Moose River Basin; about 537 m in the subsurface of northern Hudson Bay Lowland; about 617 m in the offshore central part of Hudson Bay Basin; and between 298 and 305 m (estimated) for the incomplete succession on Southampton Island.

Severn River Formation

The Severn River Formation (Savage and Van Tuyl, 1919) comprises fine-grained, thin-bedded limestone and some dolomite, typically exposed along Severn River near the junction with Fawn River. Some authors include within the Severn River Formation beds named the Port Nelson

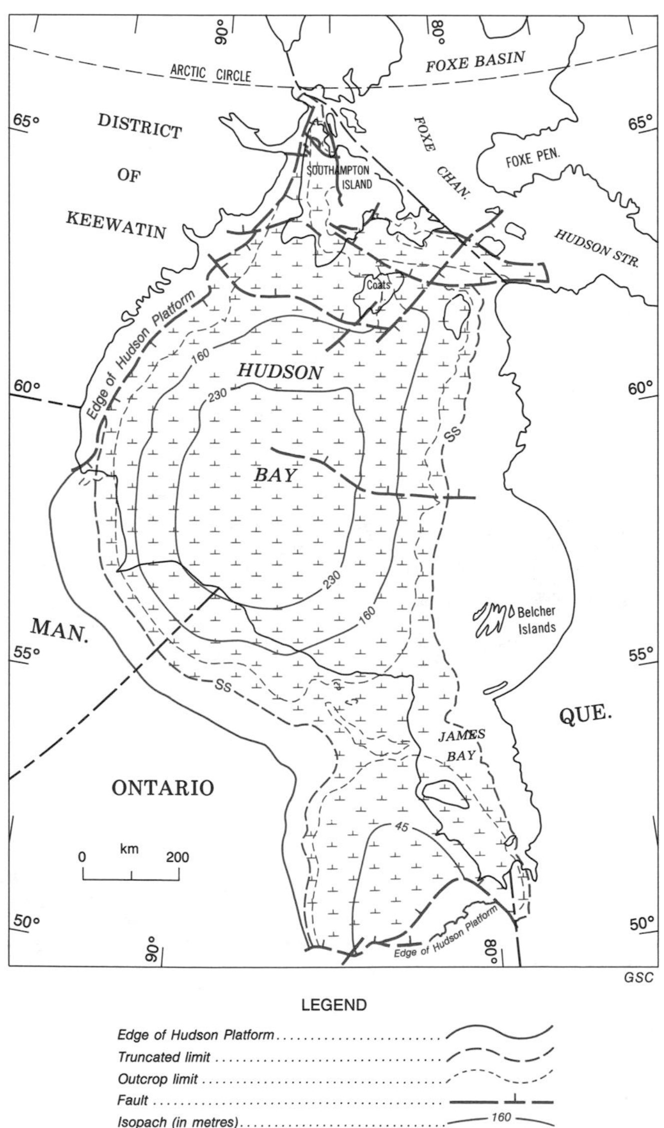

Figure 8.13. Facies and isopachs of the Severn River Formation (Lower Silurian, Middle and Upper Llandoverian) of Hudson Platform.

Formation by Savage and Van Tuyl (1919). As currently defined, the Severn River applies to basal Silurian beds that unconformably overlie strata of the Ordovician Red Head Rapids Formation and are, in turn, conformably succeeded by strata of the Ekwan River Formation. Locally, as in the Churchill and Hawley Lake areas of Hudson Bay Basin, and in the subsurface of Moose River Basin, Severn River strata overlap Ordovician rocks to rest nonconformably on Precambrian rocks.

Thickness of the Severn River Formation in the subsurface of the southern Hudson Bay Basin is about 230 m (Fig. 8.13), in the outcrop belt of Southampton Island about 152 m (estimated), in the subsurface of the offshore central Hudson Bay Basin about 230 m, and in the subsurface of northeastern Moose River Basin about 45 m.

The Severn River Formation in the outcrop belt of northern Hudson Bay Lowland consists of a heterogeneous assemblage of limestone, dolomitic limestone, and dolomite (Norford, 1971). Some limestone beds are burrowed and mottled, whereas others contain layered and mounded stromatolites and flat-pebble conglomerates. Minor inclusions of anhydrite occur in the subsurface of northern Hudson Bay Lowland, and anhydrite forms a thin interval near the middle of the formation in the offshore region of Hudson Bay. Where the formation rests on Precambrian rocks, the basal beds are sandy and conglomeratic.

On Southampton Island, the formation consists primarily of microcrystalline limestone and dolomite (Heywood and Sanford, 1976). Calcareous structures of presumed cryptalgal origin, producing irregular and wavy bedding surfaces, are present in parts of the succession. At some localities, large parts of the Severn River are dolomitized.

A lower part of the Severn River Formation contains locally abundant shelly fossils alternating with leperditiid and beyrichiid ostracodes, suggesting open marine conditions in a shallow subtidal environment, which periodically became slightly restricted (Le Fèvre et al., 1976). A middle part with rare benthonic fossils, leperditiid ostracodes, and minor evaporites, suggests more pronounced restricted marine conditions and deposition in an intertidal environment. A major upper part of the formation with abundant benthonic fossils, minor leperditiid ostracodes, reflects normal salinity and deposition in a shallow subtidal environment.

In Kaskattama No. 1 well, faunal division D of Norford (1970) occurs in the lower third of the formation, which is characterized by the brachiopod *Virgiana decussata* (Whiteaves) (Fig. 8.6). This faunal division is dated as Middle Llandovery and correlated with the Dyer Bay Member of the Cabot Heat Formation of southern Ontario, with the Lime Island Dolomite of northern Michigan, and with the Fisher Branch Dolomite of southern Manitoba. Faunal division E, in the middle third of the formation, contains elements of the "*Camarotoechia*" cf. "*C.*" *winiskensis* Whiteaves and *Plectatrypa lowi* (Whiteaves) Zone of Berry and Boucot (1970), and is dated as early Late Llandovery and correlated with the Wabi Formation of Lake Timiskaming, with the Mindemoya Formation of southern Ontario, with part of the Hendricks Dolomite of northern Michigan, and with the Atikameg Dolomite of southern Manitoba. Faunal division F, in the upper third of the formation, contains *Pteroleperditia* and ?*Glassia* cf.

?*G. variabilis* Whiteaves amongst other forms, but does not contain *Pentamerus*. It is dated as early Late Llandovery and correlated with the upper part of the Hendricks Formation of northern Michigan, with the lower part of the Fossil Hill Formation of southern Ontario, with the Thornloe Formation of Lake Timiskaming, and possibly with the East Arm Dolomite of southern Manitoba.

Some of the more diagnostic shelly fossils from the Severn River Formation on Southampton Island (Bolton in Heywood and Sanford, 1976) include the brachiopods *Virgiana* sp., *Stegerhynchus winiskensis* (Whiteaves), and *Costistricklandia* sp., and the ostracode *Dihogmochilina latimarginata* (Jones), dated as Middle to Late Llandovery.

Three provisional conodont assemblage zones recognized by Le Fèvre et al. (1976) in the formation from the Kaskattama No. 1 and nearby wells are as follows in ascending sequence: *Spathognathodus elibatus*, dated as Middle Llandovery; *Ozarkodina* n. sp. A and B of Pollock et al. (1970), dated as late Middle and early Late Llandovery; and *Neospathognathodus* n. sp. and "Neurodont" hyaline forms, dated as mid-Late Llandovery in age.

Ekwan River Formation

The Ekwan River Formation (Savage and Van Tuyl, 1919) consists of strata that conformably succeed the Severn River Formation and are, in turn, conformably overlain by reef-bearing carbonate of the Attawapiskat Formation. Where the Attawapiskat Formation is not developed, the Ekwan River is overlain by the lower member of the Kenogami River Formation. Parts of the Ekwan River Formation are well exposed along some of the rivers cutting across the southern Hudson Bay Lowland, but fewer exposures occur in the northern lowland. The formation is widely exposed on Southampton, Coats, and Mansel islands (Fig. 8.14). Representative thicknesses of the formation are as follows: 40 m in the Kaskattama No. 1 well, up to 235 m in the offshore central part of Hudson Bay Basin, and up to 90 m (estimated) on the islands to the north (Fig. 8.15).

In the Hudson Bay Lowland the Ekwan River consists of well-bedded, skeletal and pelletoidal limestone and finely crystalline dolomite, which locally swell into

Figure 8.14. Dolomite beds of the Ekwan River Formation near Cape Southampton, southern Coats Island, northern Hudson Bay Basin. Tilting of beds is related to normal faulting (photo - B.V. Sanford, GSC 149652).

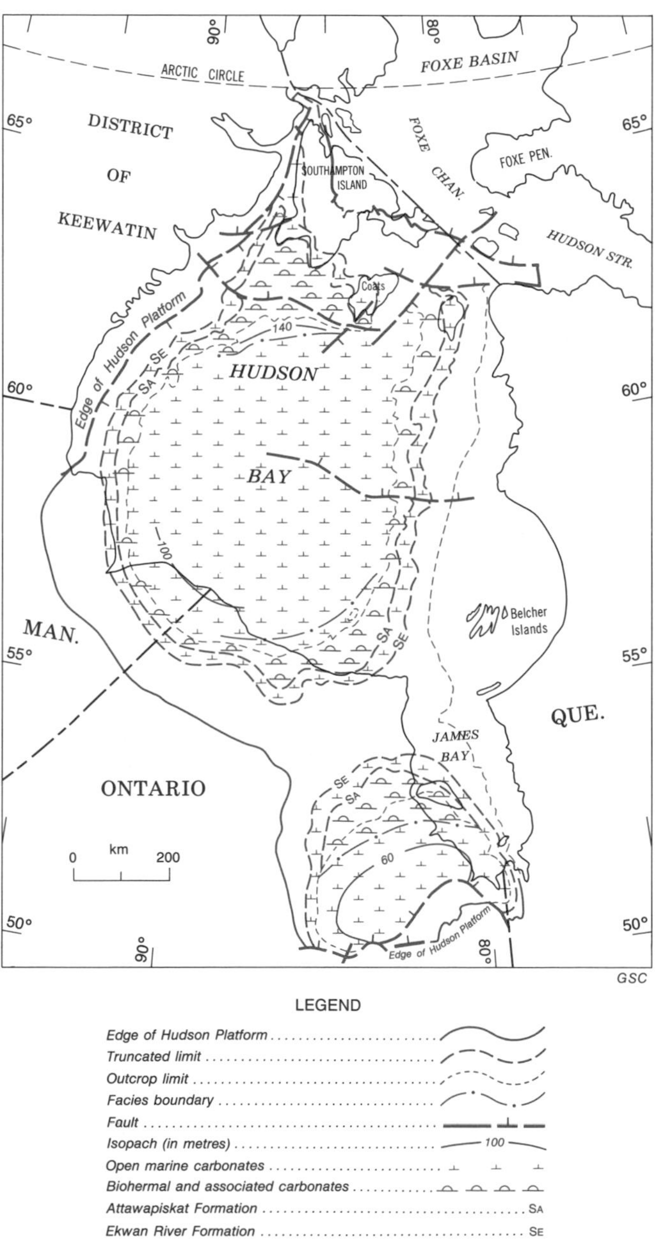

Figure 8.15. Facies and isopachs of the Ekwan River and Attawapiskat formations (Lower Silurian, Upper Llandoverian and Wenlock) of Hudson Platform.

irregular massive biostromal lenses. Grainstone carbonate and skeletal fragments form a high percentage of the sequence in some places. Nodular chert is also common.

On Southampton, Coats, and Mansel islands, the formation is divisible into three rock units (Heywood and Sanford, 1976). The lower unit is composed of limestone that locally contains columnar stromatolitic zones up to 9 m thick. Laterally equivalent strata consist of laminated limestone with sporadic flat-pebble conglomerate. The middle unit consists of resistant, planar- to lenticular-bedded limestone with stromatolite and stromatoporoid-coral biostromes. Chert nodules are locally abundant. The upper unit is generally a resistant, thickly bedded sequence of fine- to medium-grained crinoidal limestone and dolomite.

The abundant shelly fossils and development of local biostromes indicate normal open marine conditions and deposition in a shallow subtidal environment.

Ekwan River fossils include stromatoporoids, corals, brachiopods, and cephalopods. Faunal division G of Norford (1970) occupies a thin interval in the lower part of the formation in the Kaskattama No. 1 well (Fig. 8.6). It is characterized by the ostracode *Dihogmochilina latimarginata* (Jones) and the brachiopod *Pentamerus* sp., dated as Late Llandovery and correlated with the upper part of the East Arm Dolomite and the Cedar Lake Dolomite of southern Manitoba, and with the upper part of the Hendricks Dolomite of northern Michigan. Faunal division H occurs in the upper three-quarters of the formation and includes a coral fauna and the brachiopod *Pentamerus* sp., indicating a Late Llandovery age. It is correlated with the upper part of the Fossil Hill and Thornloe formations of southern Ontario and Lake Timiskaming, and the Schoolcraft Dolomite of northern Michigan.

Figure 8.16. Reefal carbonate of the Attawapiskat Formation (Upper Llandoverian – lower Wenlock) at the southeast end of a small island, Attawapiskat River, at 52°53.7′N; 83°31′W, northern Moose River Basin, northern Ontario. Note steep dip of flanking beds. Jacob's staff graduated in feet and tenths (photo - A.W. Norris, GSC 5-6-67).

Conodonts of the *Pterospathodus celloni* Zone of Late but not latest Llandovery age were obtained from outcrops of Ekwan River Formation along the Severn and Attawapiskat rivers (Uyeno in Norford, 1981). Conodonts from the formation in the Kaskattama No. 1 well were assigned to the lower part of the *celloni* Zone (Le Fèvre et al., 1976).

Attawapiskat Formation

Conspicuous exposures of reefal carbonates occur along the lower stretches of the Attawapiskat River (Fig. 8.16). These reefal carbonates, assigned to the Attawapiskat Formation (Savage and Van Tuyl, 1919), appear to incompletely surround and cover the flanks of Moose River and Hudson Bay basins. In outcrops they are most fully developed on the northwestern and southeastern flanks of Cape Henrietta Maria Arch and have been mapped also on Southampton, Coats, and Mansel islands on the southern flank of Bell Arch. The Attawapiskat appears to be in part laterally equivalent to the upper part of the Ekwan River Formation.

The thickness of the Attawapiskat Formation in Kaskattama No. 1 well is about 62 m (Norford, 1970; Le Fèvre et al., 1976). The thickness in the outcrop belts of Southampton, Coats, and Mansel islands is estimated at about 53 m (Heywood and Sanford, 1976).

Two predominant lithofacies, reef and inter-reef, are present. Most conspicuous in the outcrop belts are the swarms of bioherms that are scores of metres wide and up to 10 m high, consisting of variably textured limestone, commonly microcrystalline, and commonly with fragmental texture throughout. Organic remains within the bioherms appear to consist mainly of calcareous algae, bulbous stromatoporoids, and favositid and halysitid corals. Coarse vugs are locally present within the bioherms. The inter-reef facies is more uniformly bedded and appears to overlie and flank the reefs. This facies consists generally of lime mudstone and dolomite with numerous coarse, granular textured detrital beds, commonly with excellent intergranular and pinpoint porosity. Concentric flanking beds of detrital carbonate show inclined bedding with dips as high as 30° (Fig. 8.16).

On Southampton, Coats, and Mansel islands, the bioherms are highly variable in size, but they are generally small with diameters and heights of less than 1.5 m. They are composed of a framework of calcareous algae, stromatoporoids, and corals. The inter-reef beds form the predominant lithofacies in this area, and consist of dolomite and dolomitic limestone that are commonly cherty.

The thickness of the Attawapiskat Formation in Kaskattama No. 1 well is about 62 m (Norford, 1970; Le Fèvre et al., 1976). The thickness in the outcrop belts of Southampton, Coats, and Mansel islands is estimated at about 53 m (Heywood and Sanford, 1976).

Two predominant lithofacies, reef and inter-reef, are present. Most conspicuous in the outcrop belts are the swarms of bioherms that are scores of metres wide and up to 10 m high, consisting of variably textured limestone, commonly microcrystalline, and commonly with fragmental texture throughout. Organic remains within the bioherms appear to consist mainly of calcareous algae, bulbous stromatoporoids, and favositid and halysitid corals. Coarse vugs are locally present within the bioherms. The inter-reef facies is more uniformly bedded and appears to overlie and flank the reefs. This facies consists generally of lime mudstone and dolomite with numerous coarse, granular textured detrital beds, commonly with excellent intergranular and pinpoint porosity. Concentric flanking beds of detrital carbonate show inclined bedding with dips as high as 30° (Fig. 8.16).

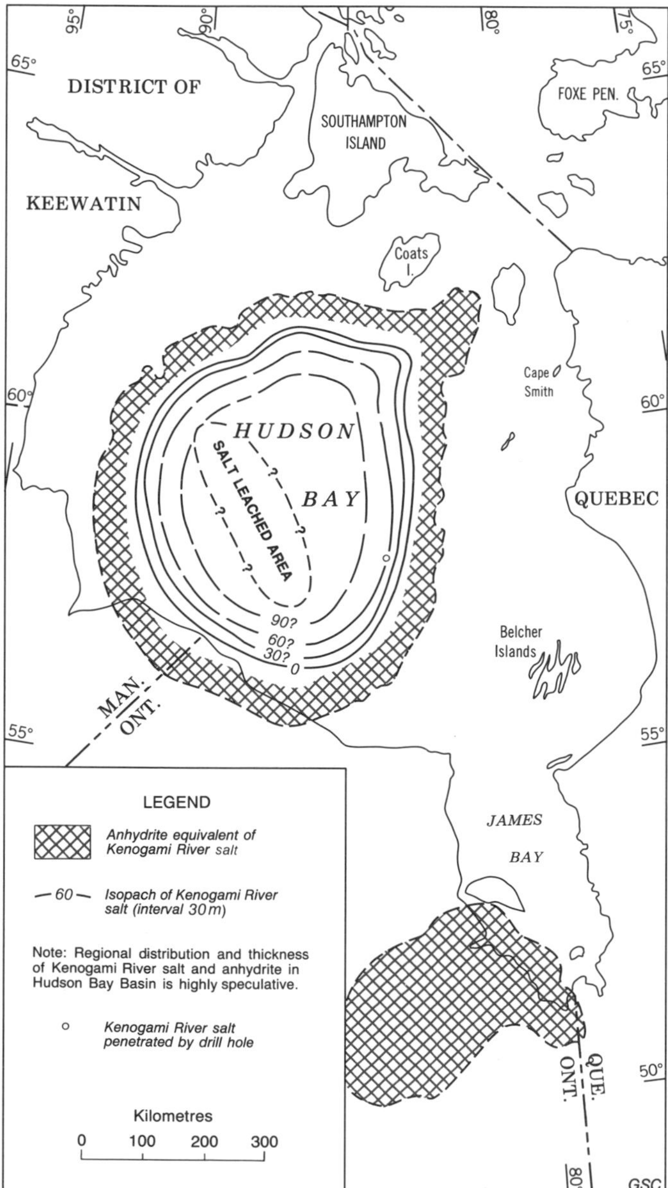

Figure 8.17. Isopachs of evaporites of the Kenogami River Formation (Upper Silurian and lower Lower Devonian) of Hudson Platform (from B.V. Sanford, unpub. rep., 1980).

On Southampton, Coats, and Mansel islands, the bioherms are highly variable in size, but they are generally small with diameters and heights of less than 1.5 m. They are composed of a framework of calcareous algae, stromatoporoids, and corals. The inter-reef beds form the predominant lithofacies in this area, and consist of dolomite and dolomitic limestone that are commonly cherty.

The development of reef-like facies rich in stromatoporoids and corals and their debris indicate (Le Fèvre et al., 1976) an open marine, subtidal environment on the higher peripheral parts of both basins.

Shelly fossils from the Attawapiskat Formation (Savage and Van Tuyl, 1919; Nelson and Johnson, 1966) include stromatoporoids, corals, brachiopods, and trilobites. Fossils from the formation in Kaskattama No. 1 well, assigned by Norford (1970, p. 6) to faunal division I, include *Palaeocyclus, Solenohalysites*, and *Pentameroides* of latest Llandovery or early Wenlock age, and suggest correlation with the Cordell Dolomite of Michigan. Trilobites from the formation have been described by Norford (1971) and Gass and Mikulic (1982), and a new pentamerid brachiopod species has been described by Boucot and Johnson (1979).

The few conodonts from the lower half of the Attawapiskat Formation are assigned to the upper part of the *Pterospathodus celloni* Zone of late Llandovery age (Le Fèvre et al., 1976). Higher conodonts are assigned to a *Neospathognathodus* n. sp. Assemblage Zone.

Silurian and Devonian

Kenogami River Formation

The Kenogami River Formation (Dyer, 1930) comprises a barren sequence of shale and dolomite exposed along the Kenogami, Pagwachuan, and Albany rivers in the Moose River Basin (Fig. 8.17). This formation overlies the Attawapiskat Formation, and the Ekwan River Formation in places where the Attawapiskat is not developed, and is succeeded by the Lower Devonian Stooping River Formation (Sanford et al., 1968). Locally, as in the southern Moose River Basin, the Kenogami River Formation is overlain by clastic continental beds of the Lower Devonian Sextant Formation. In other areas in both basins, Kenogami River strata have been uplifted by block faulting and removed by erosion (see Fig. 7.2, 8.2).

Martison (1953) outlined five rock units within the Kenogami River Formation. In a revision by Sanford et al. (1968), three major lithological units were recognized, informally designated as lower, middle, and upper members.

Tillement et al. (1976) introduced the name Hudson Formation to apply to beds here included in the Kenogami River Formation in Aquitaine Narwhal South O-48 well (58°07′56″28N, 84°08′16″78′W) in east-central Hudson Bay.

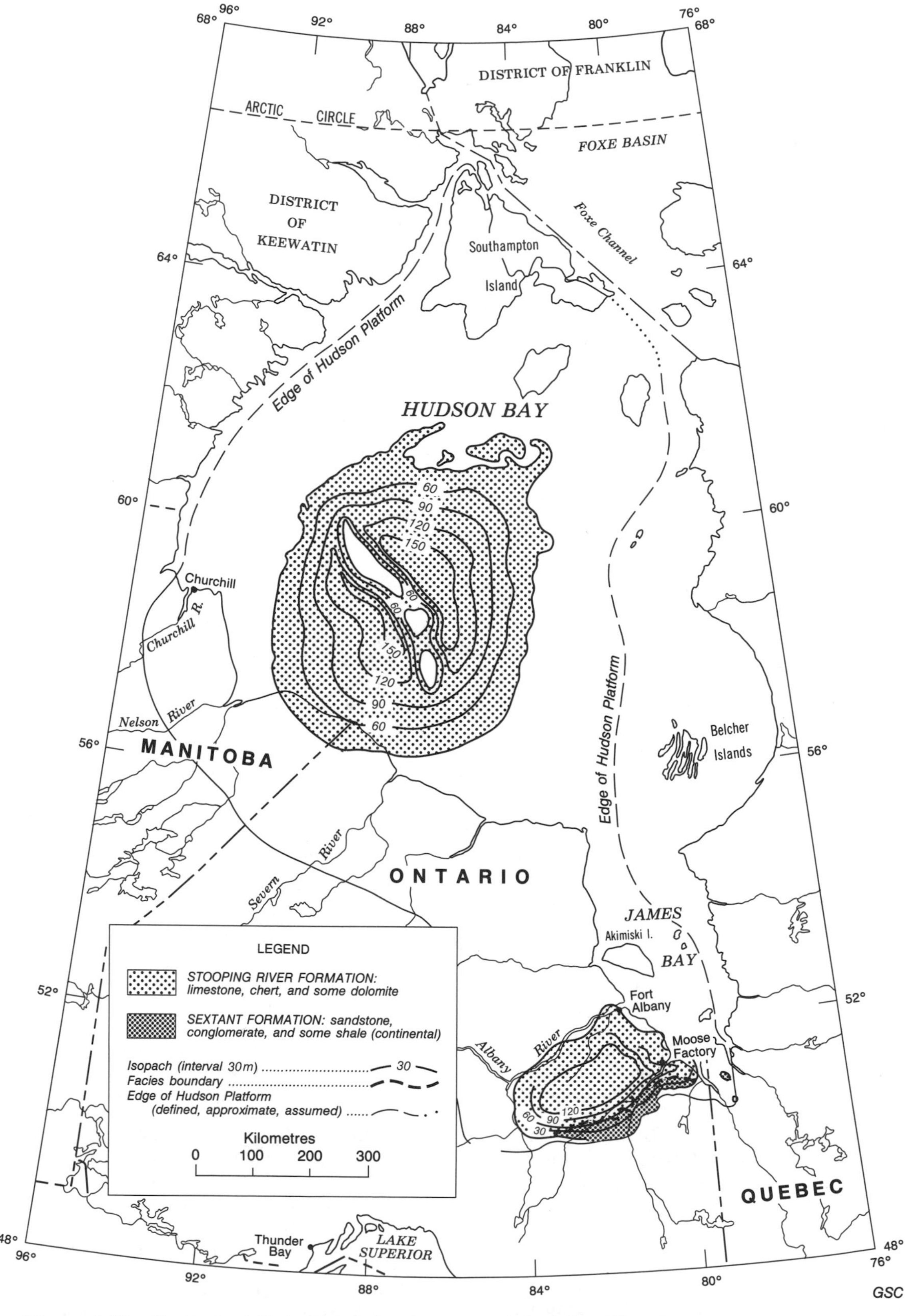

Figure 8.18. Facies and isopachs of the Sextant and Stooping River formations (upper Lower Devonian) of Hudson Platform (from Sanford and Norris, 1975).

They recorded palynomorphs of Pennsylvanian (Westphalian) age from drilling residue from the Kenogami River interval. Examination of the samples by palynologists of the Geological Survey of Canada showed a mixture of spores and pollen of Pennsylvanian (Westphalian), Mesozoic, and Cenozoic ages, indicating contamination of the samples (B.V. Sanford, pers. comm., 1980). The lithology and the section based on the seismic profile across Hudson Bay (Fig. 8.2) strongly support the contention that the interval in question is a part of the Kenogami River Formation.

In Moose River Basin, the thickness of the lower member ranges from 23 to 53 m, the middle member from 145 to 168 m, and the upper member from 11 to 33 m (Sanford et al., 1968). In Kaskattama No. 1 well at the southern end of Hudson Bay Basin, the three members are 36, 158, and 8 m thick, respectively.

The lower member consists of uniform microcrystalline dolomite. The middle member succeeds the lower member gradationally and consists of gypsiferous, and in part mottled mudstone, siltstone, sandstone, minor argillaceous dolomite, and coarsely vuggy limestone. The upper member consists of fine to microcrystalline dolomite. In Moose River Basin, the lower member contains thin interbeds of anhydrite, and the upper member contains a brecciated zone near the top, which probably resulted from leaching of evaporites. In Hudson Bay Basin, salt units occur in the lower member and near the top of the middle member.

The lower member was deposited under supratidal conditions, and clastic sediments and red colouration of the middle member suggest subaerial exposure. The upper member reflects a return to shallow water tidal-flats conditions. In Hudson Bay Basin a somewhat more complete restriction behind a carbonate barrier provided conditions for the precipitation of halite, evaporitic carbonates, and associated redbeds during a time of widespread aridity.

The only shelly fossils reported from the Kenogami River Formation are from the upper member in Moose River Basin (Wilson, 1953), and these are poorly preserved and not diagnostic. On the basis of stratigraphic position and lithological similarity to the Salina and Bass Islands formations of the Great Lakes region, the formation was assumed to be of Late Silurian age (Sanford et al., 1968). Spores described by McGregor and Camfield (1976) from two wells in Moose River Basin indicate a Late Silurian (?Downton) or Early Devonian age for the upper part of the middle member, and an Early Devonian (Gedinnian and Siegenian) age for the upper member. The lower member remains undated but is limited by the underlying Attawapiskat Formation of latest Llandovery or early Wenlock age (Norford, 1970). The Silurian-Devonian boundary has not been determined but is probably within the middle member of the formation.

Devonian

The Devonian succession of Hudson Platform, preserved mainly in the centres of the two basins, comprises, in ascending sequence, the upper member of the Kenogami River Formation and Sextant, Stooping River, Kwataboahegan, Moose River, Murray Island, Williams Island, and Long Rapids formations. In Moose River Basin, Devonian rocks are partly covered by Mesozoic rocks, and in Hudson Bay Basin Devonian strata are largely covered by water, with only a minor part present on the mainland in the Cape Tatnam area. The composite thickness in Moose River Basin is about 396 m, and in Hudson Bay Basin about 572 m.

Sextant Formation

The Sextant Formation contains the continental clastic beds that overlie Precambrian crystalline rocks along the southern margin of Moose River Basin (Sanford and Norris, 1975). These beds are overlain by, and change northward into carbonate beds of the Stooping River Formation. In general, the thicker sequences of the Sextant Formation, about 45 m, occur along and near the southern margin of Moose River Basin (Fig. 8.18). From the margin of the Canadian Shield the formation thins northward to a feather edge as the continental Sextant beds are overlapped and abruptly replaced by marine carbonate beds of the Stooping River Formation.

The Sextant Formation consists of nonmarine sandstone, shale, clay, siltstone, and conglomeratic sandstone; friable arkosic sandstone is most abundant. The sandstones occur as thin to medium lensing beds that are commonly crossbedded. The Sextant Formation presumably was deposited by a series of streams debouching from the nearby Precambrian highlands uplifted by faulting.

Plant remains are common in the formation, particularly in some shale lenses. These have been noted by numerous workers, including Wilson (1903), Bell (in Dyer, 1928; in Martison, 1953), Lemon (1953, 1954), and Hueber (1983). On the basis of plants, the Sextant was dated as Early Devonian.

Spores from the Sextant Formation indicate a mid to late Emsian age, and correlate with the lower part of the Battery Point Formation of eastern Gaspésie (McGregor et al., 1970; McGregor in Sanford and Norris, 1975; McGregor and Camfield, 1976).

Stooping River Formation

The Stooping River Formation (Sanford et al., 1968) includes Lower Devonian limestone and dolomite that normally succeed the Kenogami River Formation. However, where older Paleozoic rocks have been eroded, the Stooping River or its nonmarine equivalent (Sextant Formation) rests on older Silurian, Ordovician, or Precambrian rocks. The Stooping River in both basins is overlain by carbonate of the Kwataboahegan Formation. Thickness of the formation in the subsurface of Moose River Basin ranges from a maximum of 143 m to a minimum of 12 m. Thickness in the subsurface of southern Hudson Bay Basin ranges from 78 to 86 m (Fig. 8.19).

The Stooping River Formation in Moose River Basin consists of nodular, thin bedded, finely crystalline and locally fragmental, cherty limestone and dolomite. Where these beds overlap the clastic sediments of the Sextant Formation in the southern part of the basin, the basal beds are sandy and dolomitic. In wells drilled onshore in the southern part of Hudson Bay Basin, the formation consists of sparingly fossiliferous, finely crystalline to aphanitic limestone and dolomite. These carbonate beds are free of chert, but contain minor lenses of anhydrite. Traced offshore into the central part of the basin, the evaporitic carbonate changes to a unit consisting mainly of halite.

The abundant shelly fossils and carbonate rocks of the Stooping River Formation in Moose River Basin indicate an open marine, subtidal environment of deposition. A marked reduction of fossils and presence of minor evaporites suggest slightly restricted marine conditions for the southern end of Hudson Bay Basin, which change to markedly restricted conditions in the central part of the basin.

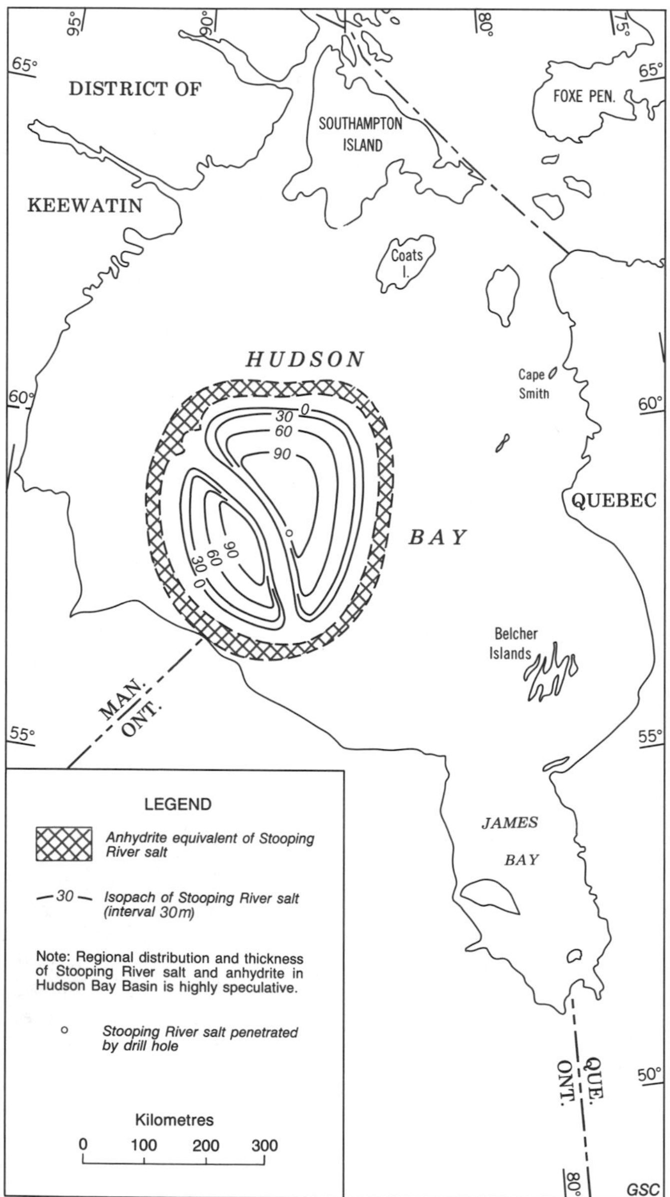

Figure 8.19. Isopachs of evaporites of Stooping River Formation (upper Lower Devonian) of Hudson Platform (from B.V. Sanford, unpub. rep., 1980).

Fossils in the Stooping River of Moose River Basin are abundant and diverse (Sanford and Norris, 1975). The brachiopods include many elements that occur typically in the Schoharie and Bois Blanc and equivalent formations of eastern North America, which are dated as Emsian (Boucot and Johnson in Sanford and Norris, 1975). Conodonts from the upper two-thirds of the formation range in age from mid-Emsian to possibly early Eifelian (Uyeno in Sanford and Norris, 1975). Spores from the lower two-thirds of the formation are assigned by McGregor and Camfield (1976) to the *caperatus-emsiensis* Assemblage Zone, dated as possibly Siegenian at the base to mid-Emsian at the top. Spores from the uppermost third of the formation are assigned to the lower part of the *annulatus-lindlarensis* Assemblage Zone, dated as late Emsian.

Kwataboahegan Formation

Resistant, thick-bedded coral limestone of the Kwataboahegan Formation (Sanford et al., 1968) is well exposed along the lower reaches of the Kwataboahegan River. The contact with the underlying Stooping River Formation is generally conformable, but in southeastern Moose River Basin, where underlying rock units have been affected by block faulting, the contact is disconformable. In both basins, the Kwataboahegan is conformably overlain by evaporite and carbonate beds of the Moose River Formation.

In the subsurface of Moose River Basin, thickness of the Kwataboahegan ranges from a minimum of 24 m to a maximum of 77 m (Fig. 8.20). In the subsurface of the southern mainland area of Hudson Bay Basin, thicknesses range from 39 to 43 m (Sanford and Norris, 1975). Offshore, in the central part of the latter basin the thickness is about 70 m.

The Kwataboahegan Formation in Moose River Basin consists mainly of resistant, thick bedded to massive, medium-grained, biostromal limestone with abundant stromatoporoids and corals (Fig. 8.21). Some of the beds are dolomitic, and these are recessive weathering. Angular clasts of quartz and feldspar are present in basal beds in the southern part of the basin. The biostromal carbonate buildups appear to be related to Precambrian knobs projecting through Paleozoic sediments. Away from the Precambrian knobs the formation is thinner bedded, highly bituminous, and generally less fossiliferous.

In the subsurface of the southern mainland area of Hudson Bay Basin, the formation consists of fine-grained and partly saccharoidal limestone containing abundant skeletal fragments. Also present are interbeds of argillaceous limestone, and limestone and calcareous dolomite with closely spaced dark brown bituminous laminae.

The biostromal carbonates with abundant fossils indicate deposition in a normal marine, shallow subtidal environment on topographically high areas, and the bituminous carbonates suggest deposition in a slightly deeper subtidal environment in other parts of the basins.

The Kwataboahegan Formation is by far the most abundantly fossiliferous unit of the Devonian succession in Hudson Platform. The most abundant shelly fossils are corals (Cranswick and Fritz, 1958), stromatoporoids (Fritz and Waines, 1956), and brachiopods (Sanford and Norris, 1975).

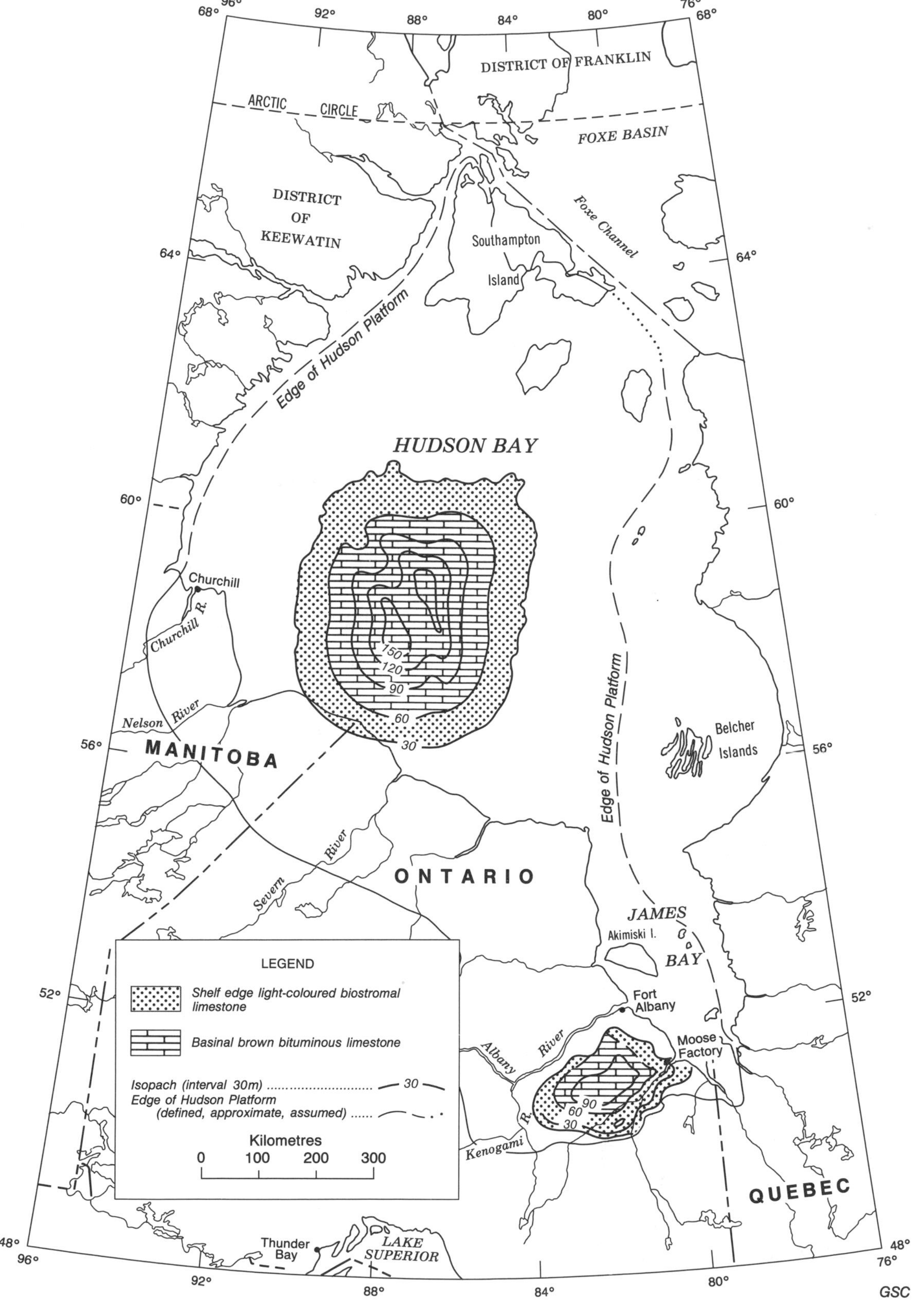

Figure 8.20. Facies and isopachs of the Kwataboahegan Formation (lower Middle Devonian) of Hudson Platform (from Sanford and Norris, 1975).

According to Oliver (in Sanford and Norris, 1975), the corals are a mixed assemblage, with most of the forms indicating a Schoharie-Bois Blanc (Emsian) age, and others indicating a younger (late Emsian to Eifelian) Onondaga age. The presence of *Amphigenia* among the brachiopods suggests an alignment with the upper part of Dutro's (1981) *Amphigenia* Assemblage Zone of New York State. The conodonts from the lower third of the formation are similar to those from the Edgecliff Member of the Onondaga Limestone of New York dated as late Emsian (Uyeno et al., 1982). Spores from the Kwataboahegan are assigned (McGregor and Camfield, 1976) to the upper part of the *annulatus-lindlarensis* and lower part of the *velata-langii* Assemblage Zones, dated as late Emsian and early Eifelian, respectively.

Moose River Formation

The Moose River Formation (Dyer, 1928) comprises unfossiliferous, generally brecciated limestone and associated gypsum beds in Moose River Basin. It overlies the Kwataboahegan Formation, and is, in turn, succeeded by limestone of the Murray Island Formation (Sanford et al., 1968). The Moose River is present also in Hudson Bay Basin, where it has been penetrated by a number of drillholes.

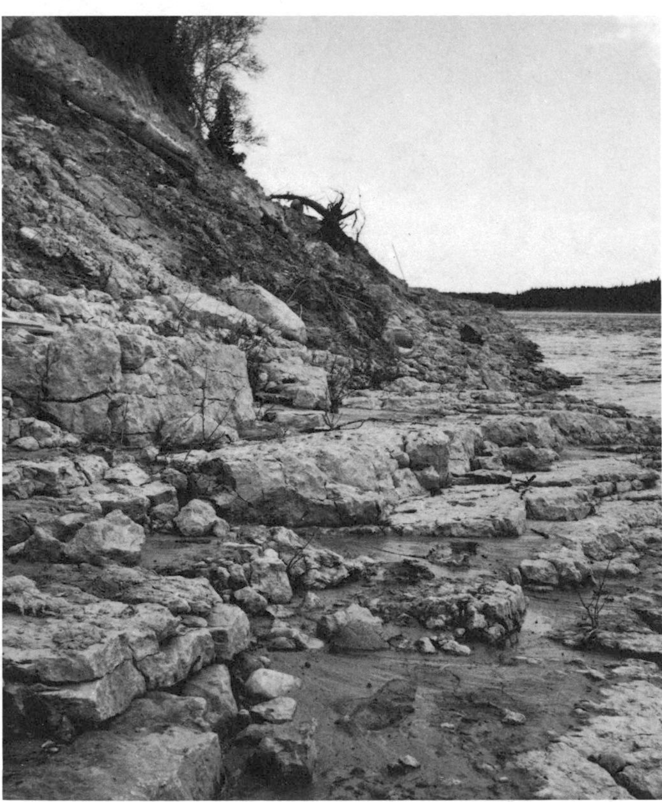

Figure 8.21. North-dipping limestone beds of Devonian Kwataboahegan Formation on west bank of Mattagami River at downstream end of Grand Rapids, southern Moose River Basin, northern Ontario (photo - A.W. Norris, GSC 1-8-67).

The maximum recorded thickness of the formation is 89 m in the central part of Moose River Basin, where the sequence is unbrecciated and where there is little or no removal of evaporites by solution (Fig. 8.22). In central Hudson Bay Basin, the maximum recorded thickness is about 160 m.

In the central part of Moose River Basin, the formation consists mainly of gypsum, 45 to 61 m thick (Fig. 8.23), occurring between thin carbonate units of limestone and/or dolomite at the base and top of the formation. Thin shale beds are found within the sequence. Carbonate breccias of collapse origin are common in outcrops in the southern parts where the evaporites have been dissolved. The gypsum pinches out in the northern part, where the formation changes entirely to a thin sequence of limestone and dolomite with minor shale beds.

In the subsurface of the southern mainland part of Hudson Bay Basin, the Moose River Formation consists of variably dolomitic limestone, argillaceous limestone, dolomite, and argillaceous dolomite. Some of the carbonate beds contain inclusions of anhydrite and gypsum, some bituminous laminae, and minor chert. A few thin beds of brick-red shale are also present. In central Hudson Bay Basin, the formation changes to a basal unit composed mainly of halite, succeeded by a thick sequence of interbedded carbonate and red shale.

The evaporitic deposits of the Moose River Formation were produced by restricted marine conditions coupled with an arid climate that also affected other basins of deposition across North America at this time, for example, the Michigan and Elk Point basins.

Fossils are exceedingly scarce in the formation because of the restricted marine environment. Lithologically and stratigraphically, the Moose River is comparable to the evaporitic and dolomitic Lucas Formation of southwestern Ontario. On the basis of stratigraphic position and the few spores recorded by McGregor and Camfield (1976), the Moose River Formation is dated approximately as mid-Eifelian of the Middle Devonian.

Murray Island Formation

The Murray Island Formation (Sanford et al., 1968), a relatively thin unit of fossiliferous limestone, disconformably overlies the Moose River Formation and is disconformably overlain by shale and limestone of the Williams Island Formation. Outcrops of the formation are limited to the southeastern part of Moose River Basin.

The formation in the subsurface of Moose River Basin is 6 to 20 m thick (Fig. 8.24). In central Hudson Bay Basin, its maximum thickness is 53 m.

In the type area, the Murray Island Formation consists of a cliff-forming succession of banded, bituminous, highly calcareous dolomite; fine- to very coarse-grained fragmental limestone; and thin- to medium-bedded argillaceous limestone. The carbonate beds are generally jointed and fractured and in places slightly brecciated by solution and subsidence of the underlying evaporitic Moose River beds. The basal beds of the formation commonly contain rounded carbonate pebbles derived from the underlying Moose River Formation, suggesting that they rest on an erosional surface. In the subsurface of central

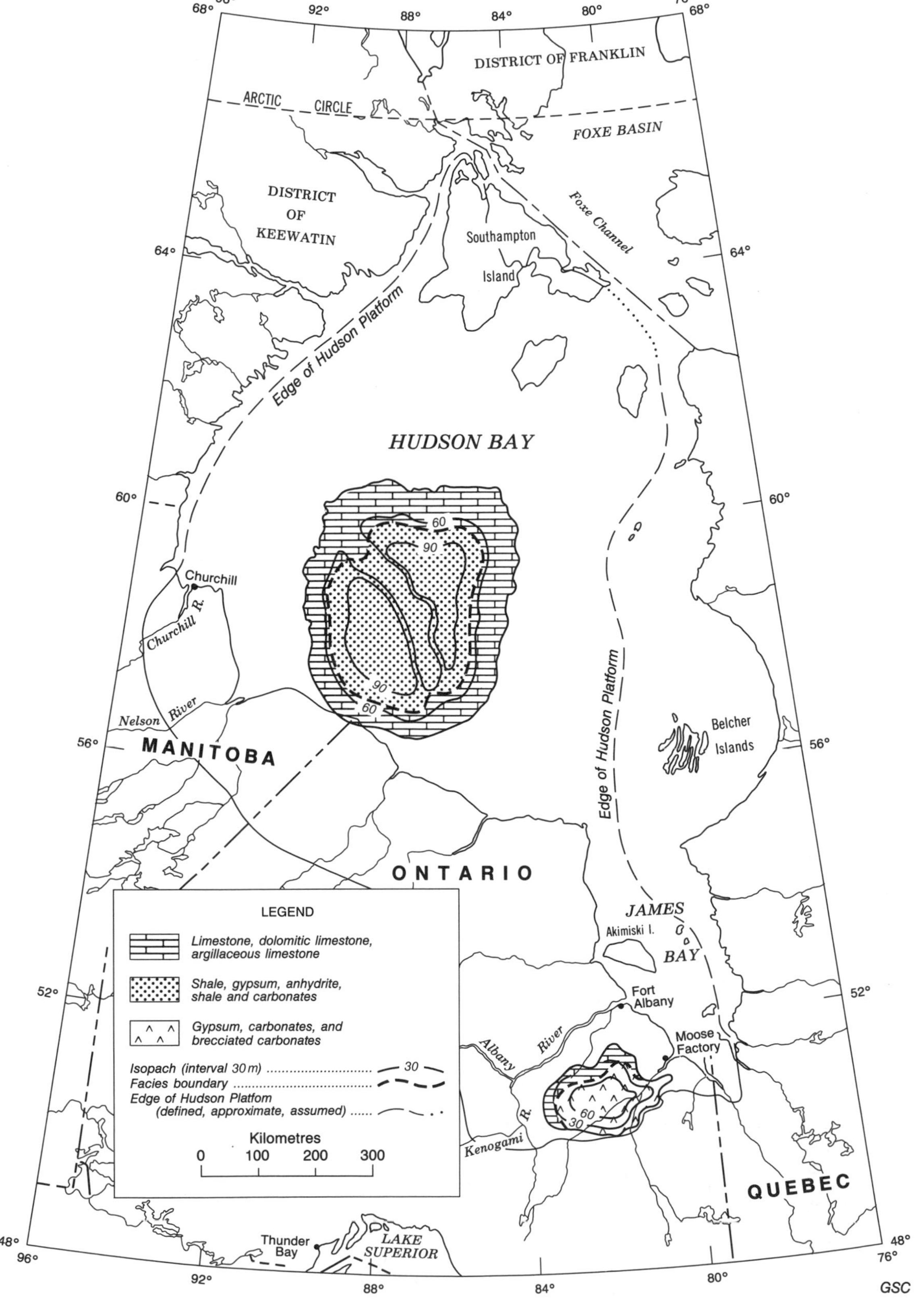

Figure 8.22. Facies and isopachs of the Moose River Formation (lower Middle Devonian) of Hudson Platform (from Sanford and Norris, 1975).

Hudson Bay Basin the Murray Island Formation consists of fine- to medium-grained, fragmental limestone, associated with argillaceous limestone.

The Murray Island sediments and shelly fossils reflect a return to normal marine salinities after the preceding evaporitic phase.

Shelly fossils are not abundant but include a diverse assemblage of brachiopods, including elements from the Appalachian, Mid-continent, and Cordilleran Faunal Provinces. The highest occurrence of *Amphigenia* cf. *A. elongata* (Vanuxem) in Hudson Bay Lowland is within the Murray Island Formation. This is a typical Appalachian element, which in New York ranges up into the Moorehouse Member of the Onondaga Limestone below the Tioga Bentonite. Mid-continent elements include *Spinatrypa costata* (Bassett) and *Spinatrypa ehlersi* (Bassett), which occur typically in the Dundee Limestone of southeastern Michigan (Bassett, 1935). A Cordilleran element, *Desquamatia arctica* (Warren), occurs also in the lower Rogers City Formation of Michigan, Elm Point Formation of southern Manitoba, lower Methy Formation of northwestern Saskatchewan and northeastern Alberta, as well as many equivalent rock units in northwestern Canada.

Conodonts in the Murray Island Formation indicate a late Eifelian age (Uyeno in Sanford and Norris, 1975).

Spores in formations immediately underlying and overlying the Murray Island are placed (McGregor and Camfield, 1976) in the *devonicus-orcadensis* Assemblage Zone.

Williams Island Formation

Williams Island (Kindle, 1924) carbonate and shale are exposed on Williams Island and on adjacent banks of Abitibi River. Throughout Moose River Basin, the Williams Island Formation consists of a lower, recessive shale member and an upper, more resistant carbonate member. In the subsurface of Moose River Basin, the lower member is 36 to 47 m thick, and the upper member is 33 to 45 m thick (Fig. 8.25). In the offshore central part of the Hudson Bay Basin, the formation is about 177 m.

The lower member, exposed on Mike Island, in Moose River, consists of soft grey shale, irregularly bedded, soft sandstone, gypsiferous sandy shale, gypsiferous siltstone and sandstone, soft limestone, and some brecciated limestone. At several localities in Moose River Basin, the exposed basal beds of the formation consist of abundantly fossiliferous, calcareous shale. Terrigenous detritus in association with marine beds in the lower member reflect uplift in the highland areas. Evaporitic beds probably reflect supratidal conditions.

The upper member, exposed on and near Williams Island in Abitibi River, consists of thin- to medium-bedded argillaceous limestone and calcareous shale; medium- to coarse-grained saccharoidal and oolitic limestone; platy, argillaceous limestone; and partly brecciated, vuggy, oolitic limestone. These rocks are an alternation of subtidal open marine deposits and marine shoreline deposits, with brecciation resulting from solution of evaporites of possible lagoonal origin.

Figure 8.23. Cliffs of massive to thick-bedded gypsum of lower Middle Devonian Moose River Formation exposed along Cheepash River, southern Moose River Basin, northern Ontario (photo - B.V. Sanford, GSC 200841).

In offshore central Hudson Bay Basin, the Williams Island Formation consists of dusky red plastic clay, micritic limestone, and argillaceous limestone, all apparently of deep water origin.

Fossils from the lower shaly member in Moose River Basin consist mainly of brachiopods and horn corals that are closely similar to forms occurring in the Hamilton Group of southwestern Ontario and New York dated as late Middle Devonian (Givetian). Fossils in a red calcareous clay dredged from the southern end of the Mid-Bay Shoal in Hudson Bay (57°40'N, 85°17'W) are closely similar to species in the Potter Farm and Norway Point formations of the Traverse Group of northern Michigan (Sanford and Norris, 1975).

A fauna consisting largely of corals (Fritz et al., 1957), from the upper member of the type section of the formation contains many elements in common with the Traverse Group of Michigan, dated as late Middle Devonian (Givetian).

A diagnostic brachiopod, closely similar to but distinct from *Leiorhynchus quadracostatus* (Vanuxem), occurs in argillaceous limestone beds at or near the top of the formation near Williams Island. *L. quadracostatus* occurs typically in the upper Geneseo Shale and lower overlying Sherburne Sandstone of New York (Sartenaer in Sanford and Norris, 1975). These units are aligned with the conodont Lowermost *asymmetricus* Zone (Klapper, 1981; Klapper in Oliver and Klapper, 1981) of late Givetian, late Middle Devonian age.

Spores recovered from the lower part of the formation in the subsurface of Moose River Basin are assigned to the *devonicus-orcadensis* Assemblage Zone, dated as early to middle Givetian (McGregor and Camfield, 1976).

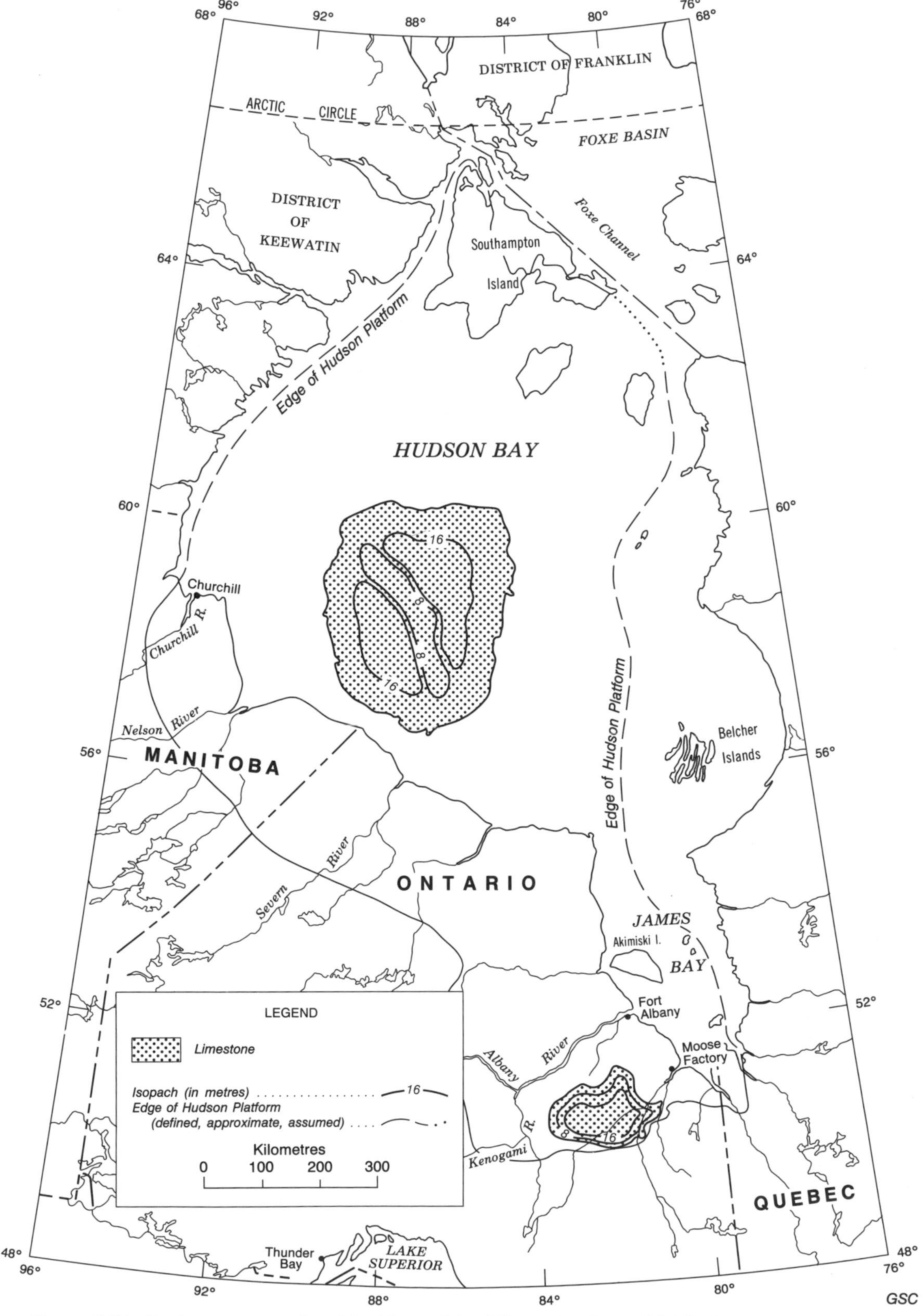

Figure 8.24. Facies and isopachs of the Murray Island Formation (lower Middle Devonian) of Hudson Platform (from Sanford and Norris, 1975).

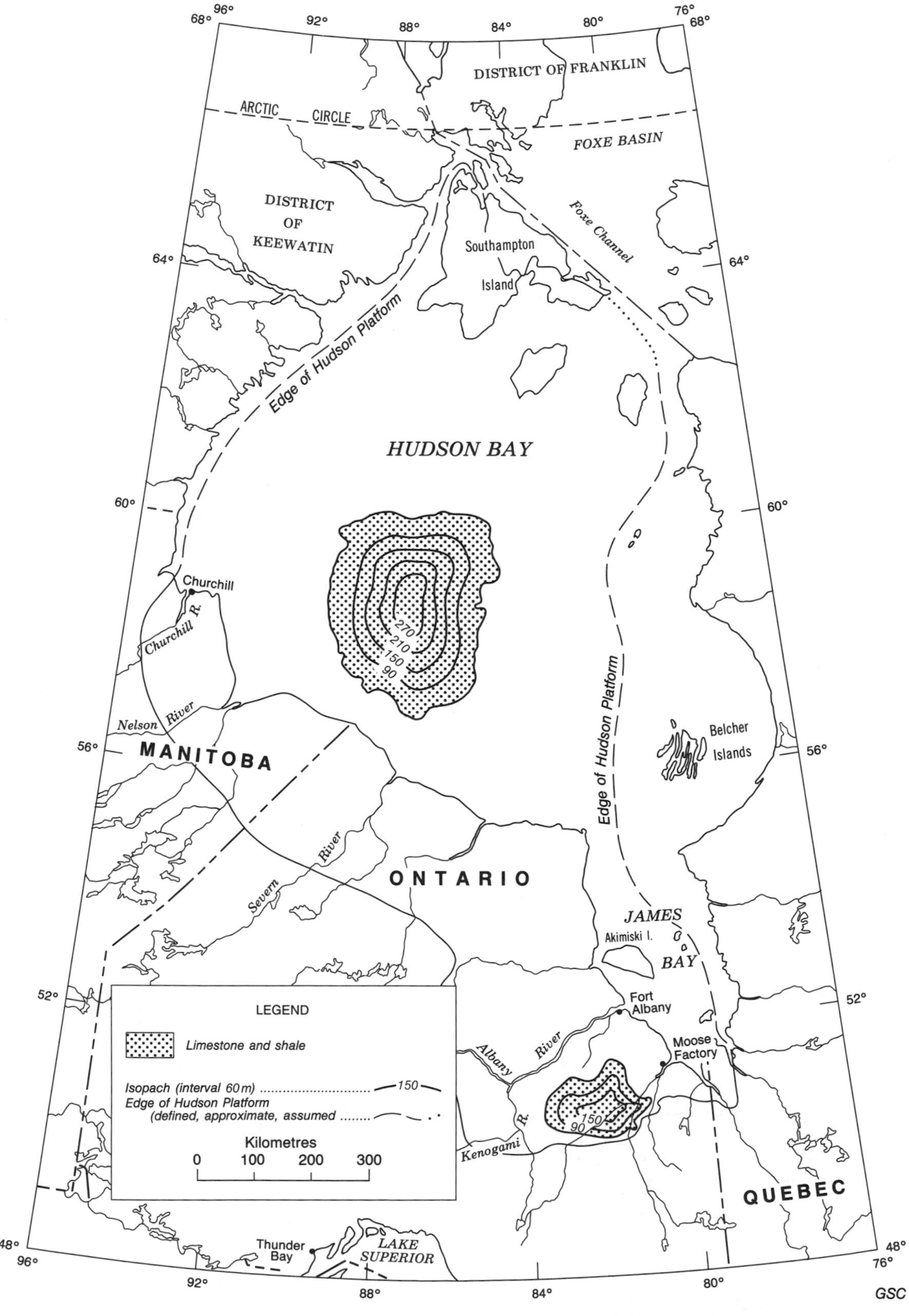

Figure 8.25. Facies and isopachs of Williams Island Formation (upper Middle Devonian) of Hudson Platform (from Sanford and Norris, 1975).

Long Rapids Formation

The Long Rapids shale (Savage and Van Tuyl, 1919) is exposed along Abitibi River. It disconformably overlies the upper carbonate member of the Williams Island Formation and is unconformably overlain by continental beds of Mesozoic age in Moose River Basin.

Recorded thicknesses of the formation in the subsurface of Moose River Basin are generally less than 30 m, but they are up to 87 m in the Onakawana area where less erosion has occurred at the top of the sequence (Fig. 8.26). Thickness of the Long Rapids in central Hudson Bay is about 150 m (Sanford and Norris, 1975).

The Long Rapids Formation in Moose River Basin consists of dark shales of varying composition, some mudstone, minor scattered thin beds of limestone and dolomite, and some clay ironstone concretions. These dark shales and sparse pelagic fossils indicate starved basin, anoxic conditions and are similar to coeval shales in the Michigan and Allegheny basins, suggesting interconnection of the three basins.

Beds tentatively assigned to the Long Rapids Formation in central Hudson Bay Basin are significantly different from those in Moose River Basin. They consist of red, pink, and salt-and-pepper grey mudstone, shale, claystone, siltstone, and sandstone, reflecting restriction of the basin and intermittent supratidal conditions.

With the exception of the large sporomorph, *Tasmanites huronensis* (Dawson), fossils in the Long Rapids of the type area are exceedingly scarce. The few fossils recorded include tentaculitids, brachiopods, ammonoids, and fragments of fish plates and plant tissue. The beds in Hudson Bay Basin appear to be barren.

Manticoceras cf. *M. sinuosum* (Hall) collected from near the base of the formation on Abitibi River, is a species that occurs in the Cashaqua Shale of the Sonyea Group of New York. Most of the Cashaqua Shale, excluding the basal beds, is assigned by Klapper (1981) to the conodont *Ancyrognathus triangularis* Zone of mid Frasnian age.

Conodonts (Telford, 1985) indicate that the Long Rapids Formation in Moose River Basin ranges from earliest Frasnian Lower *asymmetricus* Zone at the base to middle Famennian Lower *rhomboidea* Zone at the top. The Long Rapids Formation is comparable in lithology to the Kettle Point Formation of southwestern Ontario, but is more restricted in age.

Mesozoic

Rocks of Mesozoic age in Hudson Platform have been recognized only in the Moose River Basin (but see below, under "Recent Advances in the Geology of Hudson Platform"). They include continental beds referred to informally as the Mistuskwia Beds (Telford and Verma, 1982); continental beds of Early Cretaceous age named the Mattagami Formation (Dyer, 1928; see also Price, 1978); and lamprophyric and kimberlitic dykes and sills of Late Jurassic-Early Cretaceous age, which intruded the Devonian succession near the southern margin of the basin (Sanford and Norris, 1975).

Mistuskwia Beds

Mistuskwia nonlithified sand and clay unconformably overlie Devonian rocks, and are succeeded by continental beds of the Lower Cretaceous Mattagami Formation or Quaternary deposits in Moose River Basin. The Mistuskwia Beds have been recognized in only two wells and appear to be confined to the south-central part of the basin. Their maximum known thickness is 19.4 m.

The Mistuskwia Beds consist of varicoloured calcareous clays and thin interbeds of fine- to medium-grained, calcareous quartz sands. The sands are characterized by high rounding and sphericity, include limestone (Paleozoic) and igneous (Precambrian) granules, and have a very restricted heavy-mineral assemblage.

The deposit is interpreted as deltaic and likely built out from the northwest from a dominantly sedimentary source with minor input from metamorphosed igneous rocks (Hamblin in Telford and Verma, 1982).

Palynofloral assemblages from the Mistuskwia Beds (G. Norris, 1977; Telford and Verma, 1982) were dated as Bajocian-Bathonian of the Middle Jurassic and were correlated with the upper Gravelbourg Formation of Alberta and Saskatchewan, and the Shaunavon and Sawtooth formations of the southern Canadian Plains.

Mattagami Formation

The Mattagami Formation (Keele, 1920a; Dyer, 1928) includes fireclays with accompanying lignite and quartzose sand that overlie the Upper Devonian on Mattagami River near Onakawana. The Mattagami rests unconformably on the Upper Devonian Long Rapids Formation, but elsewhere it rests locally on Paleozoic rocks ranging from Ordovician to Middle Devonian in age. It nonconformably overlies Precambrian rocks on the flanks of inliers, and also along the faulted, southern margin of the basin. Although the lateral extent of the formation is not delineated precisely because of a thick cover of Quaternary deposits, it appears to extend continuously around the southeastern margin of Moose River Basin.

Thickness of the formation varies greatly. The maximum known thickness occurs in the south-central part of the basin where 121.4 m was penetrated (Telford in Telford and Verma, 1982). In the intensively drilled Onakawana lignite field, the maximum thickness is 51 m. The Mattagami Formation consists of an extremely variable sequence of nonlithified varicoloured clays and silts, quartz sands, and seams of lignite.

Two sediment associations are present in the formation, one typified by dark grey or black clay and silt commonly associated with abundant detrital or *in situ* carbonaceous material, and the other characterized by thick sections of interbedded sand, oxidized reddish to light brown clays, and white kaolinitic clays.

The Mattagami Formation represents a high-constructive segment of a major river system, which drained an extensive tract of the southern Canadian Shield, eventually debouching into the Albian epeiric seas of western Canada (Try et al., 1984).

Plant remains from the Mattagami were described by Bell (1928), who suggested a Cretaceous or possibly Late Jurassic age for the formation. According to G. Norris

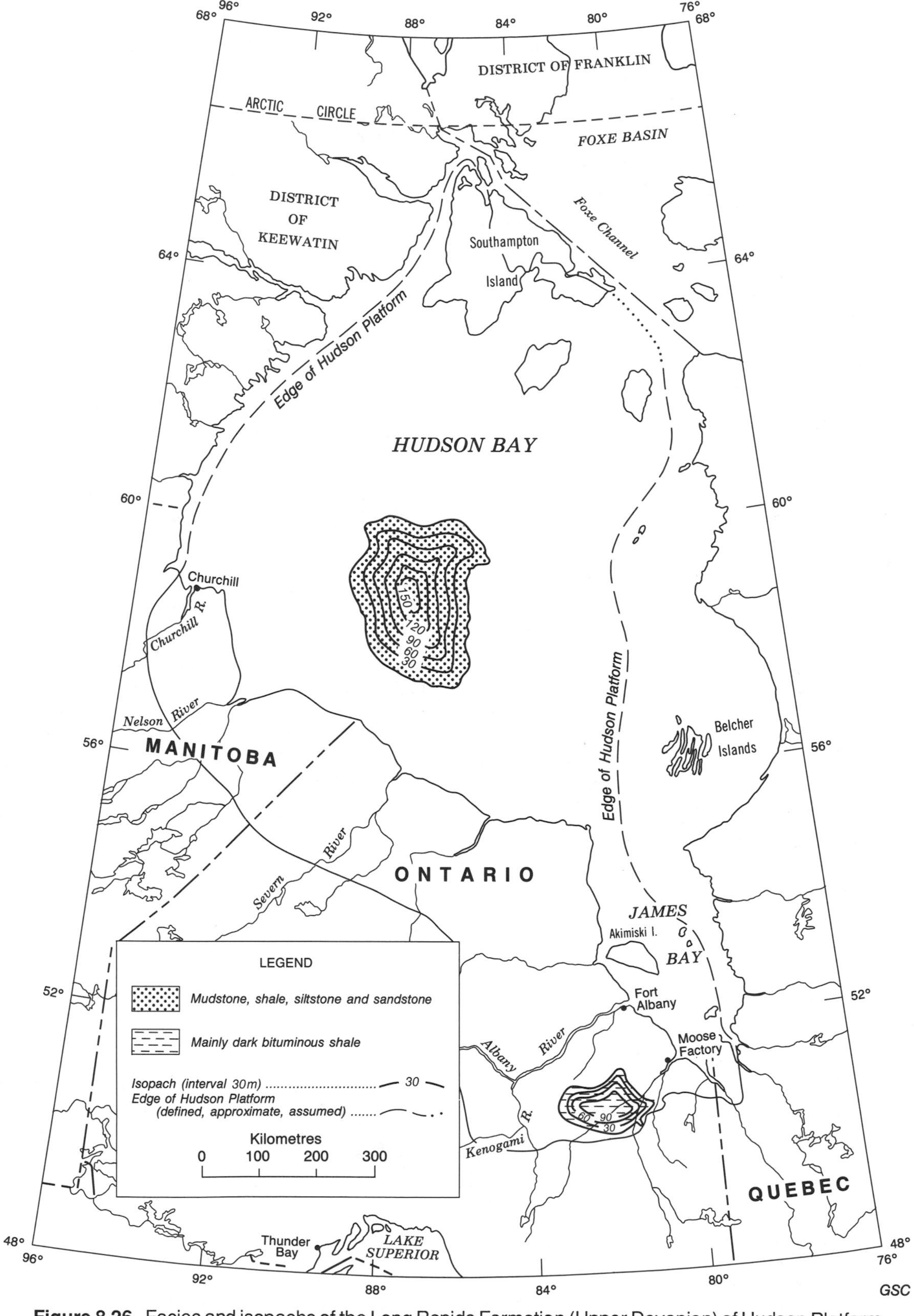

Figure 8.26. Facies and isopachs of the Long Rapids Formation (Upper Devonian) of Hudson Platform (from Sanford and Norris, 1975).

(in Telford and Verma, 1982) the formation, on the basis of contained spores, ranges in age from Middle to Late Albian (Early Cretaceous) and is correlated specifically with the Loon River, Peace River, and lower Shaftesbury formations of northeast Alberta, and the Swan River and Ashville (in part) groups of Manitoba and Saskatchewan.

Intrusive dykes and sills

Dykes and sills of lamprophyric and kimberlitic composition have intruded Devonian sedimentary rocks at the southern margin of Moose River Basin. At Sextant Rapids, the main intrusive body is a lamprophyre sill about 18 m thick occurring within the Sextant Formation. At the head of Coral Rapids, 3.2 km downstream, a lamprophyre sill, about 18 m thick, occurs between the Sextant and Stooping River formations. A composite dyke at the foot of Coral Rapids consists of lamprophyre and kimberlitic rock. Drilling in the area has demonstrated that the intrusions cut the Sextant, Stooping River, and Kwataboahegan formations, a relationship that indicates that the time of intrusion was post-early Middle Devonian (post-Eifelian). A potassium-argon age determination on lamprophyre collected from the foot of Coral Rapids indicated 128 ± 18 Ma (Sanford and Norris, 1975). This date suggests that the intrusion occurred during Late Jurassic to Early Cretaceous time, about the same age as the Monteregian Intrusions in Quebec (Sanford and Norris, 1975; Cook and Bally, 1975).

BASIN HISTORY

A.W. Norris

Ordovician

Paleogeography and tectonic history

Although the first widespread marine transgression of the southern Canadian Shield occurred during the Middle Ordovician Epoch (Black Riveran-Trentonian), it was possibly not until the Late Ordovician Epoch (Edenian) that the seas finally transgressed the central part of the Shield to initiate the Hudson Platform sequence. The Precambrian basement over which this marine transgression took place was largely a peneplain with monadnocks of 200 to 300 m relief.

In the initial sequence of Hudson Platform, the Portage Chute Formation of the Bad Cache Rapids Group is of subtidal origin, whereas the overlying Surprise Creek Formation reflects a rapid transition from subtidal to supratidal conditions (Le Fèvre et al., 1976).

In the succeeding Maysvillian Stage, a black shale facies, a possible product of the Taconian Orogeny to the southeast, blanketed the carbonate over a large segment of the southern, central, and eastern parts of the Canadian Shield, where it is now preserved as erosional remnants at Lake Timiskaming (Dawson Point Formation), on Southampton Island (Boas River Shale), and on Akpatok Island (unnamed beds) in Ungava Bay. The absence of this unit on a regional scale in Hudson Platform is presumably due to its softness and to extensive erosion that prevailed during part of Maysvillian time following withdrawal of the Edenian sea.

Succeeding rocks of Richmondian-Gamachian age in Hudson Platform are lithologically and faunally similar to equivalent rocks of the eastern St. Lawrence, Arctic, and Interior platforms with which they are presumed to have been originally connected. In Hudson Platform two major rock units are recognized that are partial facies equivalents of one another. The lower, Churchill River Group (Fig. 8.9), consists of open marine limestone. The upper, Red Head Rapids Formation (Fig. 8.11), consists of restricted marine dolomite and limestone that contain abundant stromatolitic bioherms on Southampton Island and evaporite (gypsum/anhydrite and salt) deposits in the central Hudson Bay Basin. The concentric distribution of these lithofacies is the first suggestion of differentiation of Hudson Platform: Hudson Bay Basin had begun to subside differentially, while Moose River Basin had yet to do so. The Fraserdale Arch was apparently emergent during this period, as indicated by the presence of quartz and feldspathic sandstone, conglomerate, and minor red shale, which were deposited in the basal part of the Red Head Rapids and sporadically at higher levels in the formation along the southern margin of Moose River Basin. At or near the close of the Ordovician Period, the sea again withdrew from Hudson Platform, initiating a period of subaerial erosion that continued well into the succeeding Silurian Llandovery Series.

This cycle of open marine carbonate followed by an evaporitic phase and emergence is similar to the earlier Ordovician cycle. Both cycles are characterized by relatively rapid marine transgressions followed by slower regressions, and both regressions are thought to be related to climaxes of continental glaciation in the southern hemisphere (Le Fèvre et al., 1976).

Paleolatitudes and paleoclimatology

Paleomagnetic data (Jaanusson, 1979) indicate that the Ordovician equator extended diagonally across North America in a north-northeast direction from approximately the north end of the Gulf of California to the northwest end of Baffin Island. Thus, the Ordovician succession of Hudson Platform would have been deposited at a low latitude, in warm to tropical seas.

Silurian

Paleogeography and tectonic history

A period of subaerial erosion followed withdrawal of the sea from Hudson Platform at or near the end of the Ordovician Period. The sea did not return to the area until Middle Llandovery (Early Silurian) time. In Hudson Platform, the Middle and Upper Llandovery and Wenlock series of the Lower Silurian are represented by thick carbonate units of the Severn River, Ekwan River, and Attawapiskat formations (Fig. 8.13, 8.15), which at one time were continuous with similar strata in the Arctic, Interior, and St. Lawrence platforms to the north, west, and southeast, respectively. In contrast to the uniform Severn River limestone and dolomite, the succeeding Ekwan River and Attawapiskat formations contain a wide variety of limestone and dolomite facies, which were deposited in environments ranging from tidal flat to reef platform. The conspicuous development of the latter facies around the peripheries of the basins, particularly on the flanks of Cape

Henrietta Maria and Bell arches, indicates differential subsidence of the basins and relative uplift of the two arches. At this time, Silurian reef development in North America was relatively widespread within low latitudes. At or near the close of the Early Silurian (Wenlock) Epoch, Attawapiskat reef development was terminated by a pronounced and widespread drop in sea level, which resulted in deposition of evaporites and carbonates in association with terrigenous redbeds of the Kenogami River Formation in both basins (Fig. 8.17). The restricted marine conditions persisted throughout the remaining part of the Silurian Period and continued to Early Devonian (Siegenian) time.

Paleolatitudes and paleoclimatology

The paleoequator for the Silurian is shown by most authors (e.g., Irving, 1964; Holland, 1971; Oliver, 1977) as extending obliquely across North America from southwest to northeast to intersect Hudson Bay. The reefs of the Attawapiskat Formation and evaporites of the Kenogami River Formation are typical low-latitude sediments consistent with the paleomagnetic data.

Devonian

Paleogeography and tectonic history

Within the Siegenian Stage of the Lower Devonian, Hudson Platform underwent epeirogeny, resulting in uplift of Fraserdale, Severn, and Bell arches, and block faulting and erosion within both basins (see Fig. 7.2, 8.2, 8.4). Salt solution was most pronounced in those parts of the basins affected by uplift, faults, and erosion.

Following Early Devonian epeirogeny, the uplifted Bell and Severn arches created land barriers separating Hudson Platform from the Arctic and Interior platforms lying to the north and southwest, respectively. Marine connection throughout the remainder of Early and Middle Devonian time was thus with the St. Lawrence Platform to the south and southeast.

In response to block faulting and uplift along the northern margin of Fraserdale Arch in late Early Devonian time, clastic sediments (Sextant) were deposited in a continental environment along the adjacent southern part of Moose River Basin (Fig. 8.18). The nonmarine conglomerates and finer clastics abruptly change northward to carbonates (Stooping River), the latter being deposited on a highly deformed and weathered surface. At the same time, evaporites (gypsum/anhydrite and salt) were deposited in the nearly enclosed Hudson Bay Basin (Fig. 8.19). The succeeding open marine carbonate (Kwataboahegan Formation) marks a widespread marine transgression in early Eifelian time. The lithological and faunal similarities of the Kwataboahegan Formation with the Onondaga Limestone of Ontario and New York indicate a marine connection with these areas to the southeast. The development of thick biostromal carbonate on the marginal parts of both Hudson Bay and Moose River basins (Fig. 8.20) indicates further differential subsidence, and set the stage for deposition of the succeeding evaporites of the Moose River Formation (Fig. 8.22). Thick shale deposits in Hudson Bay Basin and minor shale in Moose River Basin associated with evaporites and carbonates suggest a rejuvenation of land areas at this time (Bell or Severn Arch, or possibly both). Following deposition of the regressive Moose River sequence, the sea retreated from Hudson Platform for a brief interval. In late Eifelian time it returned and deposited limestone (Murray Island Formation) (Fig. 8.24). The fossils in this unit indicate marine connections with Michigan Basin, the mid-continent area, and Elk Point Basin of the Interior Platform.

The Givetian Stage of the upper Middle Devonian in Hudson Platform is represented by the Williams Island shale, argillaceous limestone, limestone, and some scattered evaporites (Fig. 8.25). The shelly fossils in these sediments, of normal and restricted marine origin, indicate a strong marine connection with Michigan Basin to the south and a weak connection with Elk Point Basin to the southwest. A hiatus of short duration with possible subaerial exposure separates the top of the Williams Island Formation from succeeding beds in Moose River Basin.

The Frasnian and Famennian black shales in Moose River Basin are lithologically so similar to Upper Devonian black shale of the Michigan and Allegheny basins, and other shale sequences of the eastern and central United States, that they are thought to have been deposited under similar conditions possibly related to the Acadian Orogeny. The black shale is interpreted as being deposited in an enclosed sea analogous to the modern Black Sea where black organic mud is accumulating in the deeper parts. Fine detritus from the distal part of the Catskill Delta and partly decomposed organic matter from pelagic organisms are important components of the sequence (Fig. 8.26). In contrast, the sequence in Hudson Bay Basin of red and grey shale, siltstone, and sandstone, associated with minor gypsum, suggest restricted marine, supratidal conditions, and derivation of terrigenous detritus from a more local source, either the Bell Arch or Severn Arch, or both.

In late Famennian time, Hudson Platform was elevated above sea level and remained so, as far as any record exists, throughout the remainder of the Paleozoic, Mesozoic, and Tertiary, until the time of the Wisconsinan and Holocene ice sheet, subsequent to which it was submerged beneath the Tyrrell Sea. At present a large part of the platform remains covered by water of an epeiric sea (Hudson and James bays).

Paleolatitudes and paleoclimatology

Global reconstructions for the Devonian Period by House (1979), Oliver (1977), and others, based on paleomagnetic and other data, show North America, western Europe, and northwest Africa in juxtaposition. The paleoequator for the Devonian Period is shown as extending diagonally across North America from the north end of California in the west to the central part of Hudson Bay in the east. On the other hand, a Devonian world reconstruction by Heckel and Witzke (1979), based on paleoclimatic criteria, shows the paleoequator about 25° north of the paleomagnetic equator. These authors suggest that the rotational poles did not coincide with the paleomagnetic poles for much of Devonian time. The reconstruction by Heckel and Witzke (1979) is, perhaps, slightly favoured over some of the others.

Mesozoic

Paleogeography and tectonic history

Rejuvenation of block faulting and deposition of continental deposits along the southern margin of Moose River Basin were probably coeval with the intrusion of lamprophyric and kimberlitic rocks dated as Late Jurassic-Early Cretaceous. Radiometric data (Cook and Bally, 1975) for Jurassic and Early Cretaceous times show widespread igneous activity in eastern Canada and the northeastern United States, associated with widespread regional faulting. The apparently synchronous igneous activity in the Moose River Basin is, perhaps, related to that in eastern Canada, to a prevailing regime of extension and the opening of the Atlantic Ocean. Although the southern intrusions may be related to a hot spot track, no clear, consistent sequence has been determined for those of Hudson Platform.

Three Mesozoic depositional units of continental origin are recognized in the Moose River Basin. Tectonic factors affecting development of the Mesozoic units were the differential subsidence of the Precambrian and Paleozoic basement and the reactivated fault-controlled Precambrian escarpment along the southern margin of the Moose River Basin (Sanford and Norris, 1975). This escarpment limited the southern extent of Mesozoic deposition and provided an upland source of terrigenous detritus. A broad arch, probably related to the northwest-trending Grand Rapids Arch, evident on the Mattagami River at and near Grand Rapids (see Sanford and Norris, 1975), appears to subdivide the Mesozoic deposits into two subsidiary basins.

The Middle Jurassic Mistuskwia Beds appear to be restricted to the central part of the basin on the north side of the Grand Rapids Arch (Hamblin in Telford and Verma, 1982). Sphericity and rounding of sand grains derived from Paleozoic limestone and Precambrian rocks, and a restricted and sparse heavy-mineral assemblage, indicate a dominantly sedimentary source with minor input from metamorphosed igneous rocks. The beds are of deltaic origin, with provenance from the northwest.

The early phase of deposition of the deltaic Lower Cretaceous Mattagami Formation was restricted to the southern and southeastern marginal areas of both subbasins of the Moose River Basin. The terrigenous detritus was derived from the Precambrian upland to the south.

The second phase of deposition of the Mattagami Formation was a more widespread event with sediments extending over the former barrier of the Grand Rapids Arch to partly overlap the older Mistuskwia Beds in the northern sub-basin. Several deltaic systems, each with distinct provenance, drained rejuvenated Precambrian uplands to the south (Hamblin in Telford and Verma, 1982).

Paleolatitudes and paleoclimatology

Paleolatitudes of North America for the Cretaceous Period, based on paleomagnetic observations, indicate that the magnetic pole was located just west of Alaska (Couillard and Irving, 1975). This pole position suggests that paleolatitudes in northern and western parts of the continent were about 10° higher, and those in the eastern parts were about 10° lower in the Cretaceous than at present. Hudson Platform during this time would have been between the paleolatitudes of about 44°N and 56°N, in a temperate climate.

Palynofloras from the Middle Jurassic Mistuskwia Beds, dominated by coniferous pollen, are comparable with Middle Jurassic palynofloras of the western Canadian plains (Norris in Telford and Verma, 1982). Palynofloras from the Lower Cretaceous Middle and Upper Albian Mattagami Formation are much more diverse than those from the Mistuskwia Beds, with many pteridophyte species represented. The hiatus between the Middle Jurassic Mistuskwia Beds and the Albian Mattagami Formation in the Moose River Basin corresponds to that in Saskatchewan and Manitoba between the Jurassic and the overlying Albian Swan River Group.

RECENT ADVANCES IN THE GEOLOGY OF HUDSON PLATFORM

B.V. Sanford and A.C. Grant

Introduction

Since the preparation of the preceding text by A.W. Norris, new information concerning the stratigraphy and structure of Hudson Platform has become available. The new data came from marine and onshore surveys by the Geological Survey of Canada, new exploration drilling, and seismic-reflection profiles recently released by the petroleum industry. They have provided an improved appreciation of the style and timing of basement faulting and improved definition of the distribution of stratigraphic units, particularly those of Mesozoic strata previously unknown in most of the region.

New seismic-reflection profiles reveal Cretaceous movements along some of the major arches (Fig. 7.4, 8.1). Block faulting was accompanied by marine submergence and deposition of thick siliciclastic strata in the grabens. Continental sedimentation prevailed in Moose River Basin, as described earlier, but sediments deposited in Hudson Bay, Southampton and Evans Strait basins, although mostly undrilled, are believed to be of marine origin. They probably are related to equivalent Cretaceous strata in the Ungava Basin and on the Labrador Shelf and Baffin Bay to the east.

Hudson Bay Basin

Bathymetric and single and multichannel seismic data recently acquired, and the findings of two deep stratigraphic test holes offshore, have added much to the understanding of the Hudson Bay Basin (Dingwall, 1986; Roksandic, 1987; Sanford, 1987; Grant and Sanford, 1988). These data, tied to earlier stratigraphic tests, have permitted more precise mapping of the fractures associated with the Central Hudson Bay Arch, and a more accurate delineation of the structure of the basin generally.

The new seismic profiles and the two basement tests, Netsiq and Beluga, completed in 1985, indicate an important episode of block faulting near the close of the Early Silurian (Llandovery to Wenlock) Epoch and the termination of Ekwan River/Attawapiskat deposition. The response was acceleration of deposition in two sub-basins, one southeast, the other northwest of the axis of the central arch (Fig. 7.4, 8.1). After their filling with Upper Silurian

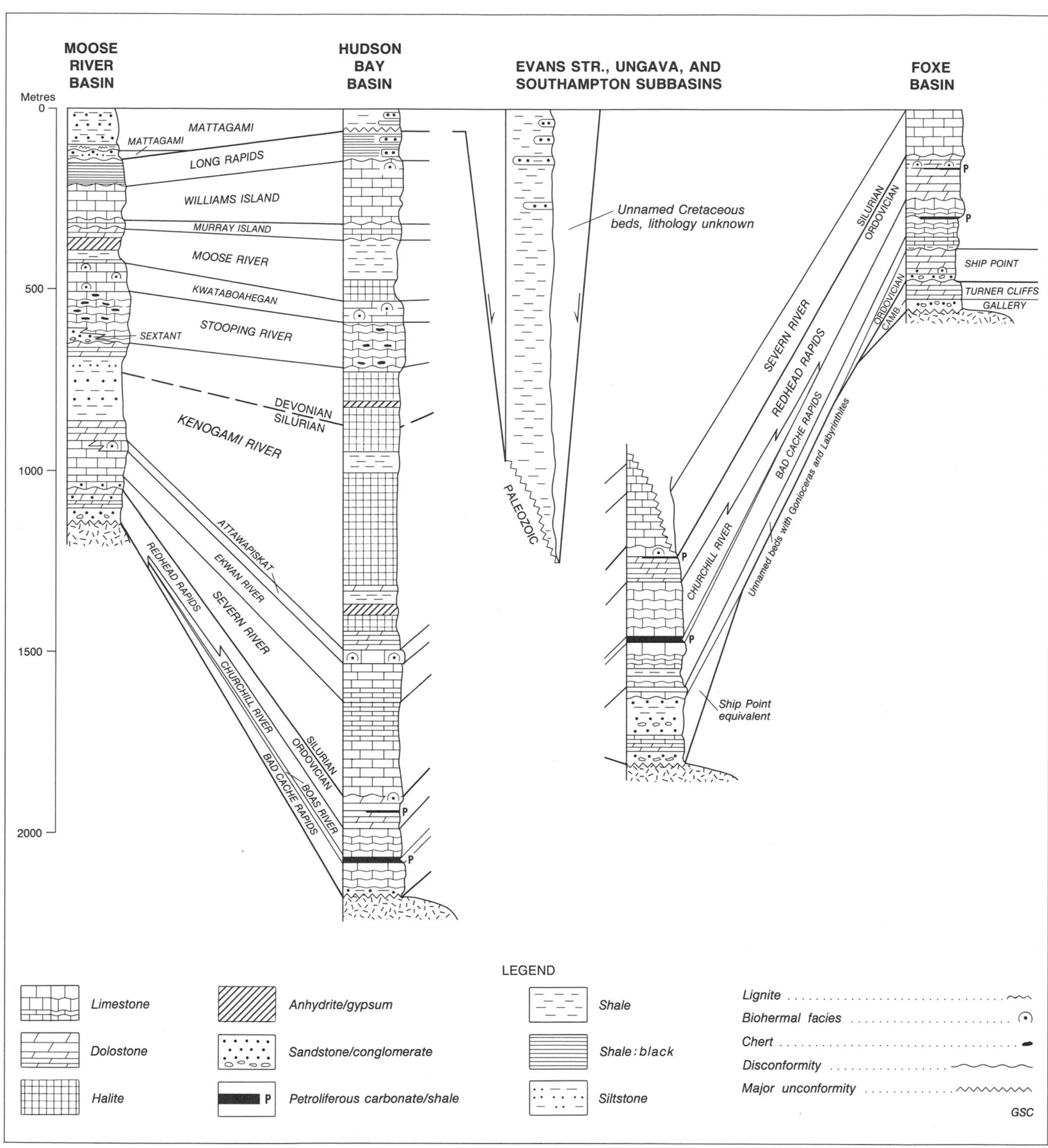

Figure 8.27. Stratigraphic/structural cross-section, Moose River Basin to Foxe Basin (B.V. Sanford and A.C. Grant, 1989).

to Lower Devonian sediments, the basins ceased to develop separately but continued as a single basin through the rest of the Devonian Period.

One of the stratigraphic tests, I.G.C. Sogepet et al. Netsiq N-01, was located on the highest part of the Central Hudson Bay Arch. There, upper Middle Devonian Williams Island Formation rests unconformably on Lower Silurian Ekwan River strata. The absence of some 1200 m of Upper Silurian to lower Middle Devonian strata, present elsewhere in the basin, demonstrates the structural relief of the arch.

Important to the understanding of the timing and depositional history of the sub-basins bordering Central Hudson Bay Arch is the log of the Trillium et al. Beluga O-23 test. There, in contrast to the successions drilled on the arch, more than 800 m of halite and minor shale and carbonates of the Upper Silurian to Lower Devonian Kenogami River Formation were encountered. The accelerated subsidence of the sub-basins corresponds to synchronous developments in the Michigan and Allegheny basins (B.V. Sanford, Chapter 11, this volume), and basins flanking the Boothia Arch (Okulitch et al., 1986).

The Boas River oil shale of Ordovician age (Sanford in Heywood and Sanford, 1976; Dewing et al., 1987; Macauley, 1986), long thought to be restricted to Southampton Island, has now been identified in many drillholes in the Hudson Bay Basin. Only a thin veneer is recognized in the wells drilled offshore on the higher structures, and the unit may be best developed in the deeper parts of the basin. Its widespread distribution is now known, however, even as far south as Winisk River in the Hudson Bay Lowland. There, drillholes completed by the International Nickel Company of Canada encountered 10 m and more of dark brown and black, petroliferous carbonate. The very thin "oil shale" interbeds that occur locally in the Red Head Rapids Formation of Southampton Island (Nelson and Johnson, 1976; Dewing et al., 1987) have not been recognized elsewhere in the Hudson Bay Basin.

Marine surveys of Hudson Bay conducted by the Geological Survey of Canada in 1986 and 1987 acquired single-channel, seismic-reflection data of high quality over wide areas (Grant and Sanford, 1988). These, combined with shallow drilling, and detailed bathymetry from the Canadian Hydrographic Service, have permitted more accurate geological mapping. The seismic data revealed strata up to 150 m thick, suspected to be of Cretaceous age, unconformably overlying the Paleozoic strata. Analysis of borehole cuttings from the suspect strata in the north-central part of the basin confirmed a Cretaceous age (D.C. McGregor, unpub. rep., 1987), later more precisely determined as from Aptian to Cenomanian (R.A. Fensome, pers. comm., 1988).

The presence of Cretaceous strata (as erosional remnants?) in Hudson Bay may be of direct significance to exploration for hydrocarbons. It may record previously unrecognized sedimentary loading and consequent maturation of hydrocarbons in underlying Devonian rocks. Thus, Devonian strata, previously largely disregarded as potentially oil and gas bearing, may be prospective.

A further recent discovery, based mainly on shallow seismic-reflection profiles, is evidence of widespread dissolution of Ordovician, Silurian, and Devonian salt, resulting in the formation of collapse structures. Some of these closely resemble productive Devonian structures of the Michigan Basin. Important also from the perspective of oil and gas potential is the appearance of anomalous structures, tentatively interpreted as bioherms, in the Williams Island Formation.

Southampton and Evans Strait Basins

Block faulting associated with uplift of the Bell Arch has given rise to graben and half-graben basins (Fig. 7.4, 8.1). Movements of the arch system took place in late Early Silurian time, but the greatest displacements apparently occurred in Cretaceous and/or post-Cretaceous time. Recent investigations in Hudson Strait (MacLean et al., 1986), and in the Evans Strait and southeastern Southampton basins (Grant and Sanford, 1988) have established the presence of thick wedges of sand and shale of probable Cretaceous age (Fig 8.27). No biostratigraphical data are yet available; the distribution and Cretaceous designation of these clastics is based on their seismic signature, seafloor configuration, and profoundly unconformable relationship to underlying Paleozoic and Precambrian rocks. The newly recognized strata of probable Cretaceous age are more than 1500 m thick in the southern Southampton Basin, and 1000 m thick in the deeper parts of the Evans Strait Basin.

QUATERNARY

W.R. Cowan

The Hudson Platform area is well known for its preserved Quaternary sediments, which are complexly interrelated with the sediments of the surrounding Canadian Shield regions. The stratigraphy of the region is important to understanding the Quaternary chronology of central North America, with which many North American geologists are familiar. Readers are referred to more complete summaries of this and the surrounding Shield regions by Andrews (1989), Dredge and Cowan (1989), Dyke and Dredge (1989), and Vincent (1989); all are found within the companion volume, "Quaternary Geology of Canada and Greenland" (R.J. Fulton (ed.), 1989).

Review literature or major works on the region include those of Terasmae and Hughes (1960), Lee (1968), McDonald (1969, 1971), Prest (1970), Skinner (1973), Shilts (1982, 1984), Andrews et al. (1983), and Dredge and Nielsen (1985).

Within Hudson Platform, only the latest two of the main North American glaciations, Illinoian and Wisconsinan, are well documented (Fig. 8.28), although pre-Illinoian deposits are recognized by some authors. Sediments of the two glaciations, together with those of the interglacial Sangamonian interval, have been studied in two main areas of the Hudson Bay Lowland: Moose River Basin and Hudson Bay Basin. Moose River Basin in the south borders James Bay. Hudson Bay Basin (Manitoba Lowland) borders Hudson Bay north of the Cape Henrietta Maria Arch, which is expressed at the surface by the Sutton Ridge.

Stratigraphy of the Hudson Platform region records numerous glacial events, represented by tills and associated sediments, and several nonglacial events of debatable rank and significance. Key marker horizons include late-glacial marine clays deposited in the Tyrrell Sea and terrestial and marine sediments of the Missinaibi Formation dated at more than 72.5 ka BP (Stuiver et al., 1978).

These marine sediments indicate the onset of interglacial episodes, such as the present interglaciation, when eustatic rises in sea level exceed the rate of isostatic rebound of the glacially depressed region. In addition, marine mollusc fragments, commonly reworked, have been found at various intervals below and above the Missinaibi Formation; the meaning of many of these remains conjectural.

Pre-Missinaibi sediments

Though interglacial or interstadial deposits have long been recognized in Hudson Bay Lowland (see summary by Prest, 1970), it was not until the 1950s that studies of pre-Missinaibi sediments were initiated by Terasmae and Hughes. Subsequent studies have revealed a complex array of pre-Missinaibi tills, related sediments, and possible nonglacial sediments, as briefly described below.

Moose River Basin

Terasmae and Hughes (1960) identified two pre-Missinaibi drift packages in the Moose River Basin. The lower drift is a till unit exposed at only one site. The middle drift, exposed directly beneath the Missinaibi beds (now Missinaibi Formation), consists of a lower clay unit overlain by a sequence of gravels and interbedded till and silt.

Skinner (1973) identified three sub-Missinaibi tills in the Moose River Basin (Table 8.1). His conclusion, based on till fabrics, till lithology, current bedding directions for intertill sediments, and the lack of intertill organic remains, was that the tills represent a series of local advances into and retreats from a proglacial lake by a southwesterly flowing (or northeasterly retreating) ice sheet. More recently, Shilts (1984) has located a still "older?" till beneath Skinner's (1973) lowest till (Till 1, Table 8.1). Furthermore, he suggested that these four tills may represent glacial events of a more significant scale than recognized by Skinner. At present the writer believes that there is insufficient evidence provided by these till deposits to conclude that more than two pre-Missinaibi glaciations occurred.

Hudson Bay Basin

Within Hudson Bay Basin, stratigraphic work has been carried out principally by McDonald (1969), Netterville (1974), Nielsen and Dredge (1982), and Dredge and Nielsen (1985). From this work, several pre-Missinaibi tills have been recognized, principally in the Manitoba Lowland, where recent detailed work has been carried out by L.A. Dredge and E. Nielsen (Fig. 8.28a, Table 8.2). There they have identified as many as four till units beneath the interglacial marker beds (Missinaibi equivalents), but the significance and rank of the lowermost two units have not been determined. The older of the two better known tills is referred to as the Sundance Till. Striae, fabric, and lithological data demonstrate that it was deposited by Keewatin ice flowing from north to south. It has been correlated with Skinner's (1973) Till II by Dredge (Dredge and Cowan, 1989) and is considered to be pre-Illinoian in age. An oxidized and bleached zone at the top of this till is believed to be a paleosol; its pollen content suggests tundra conditions during the soil-forming period. Overlying this

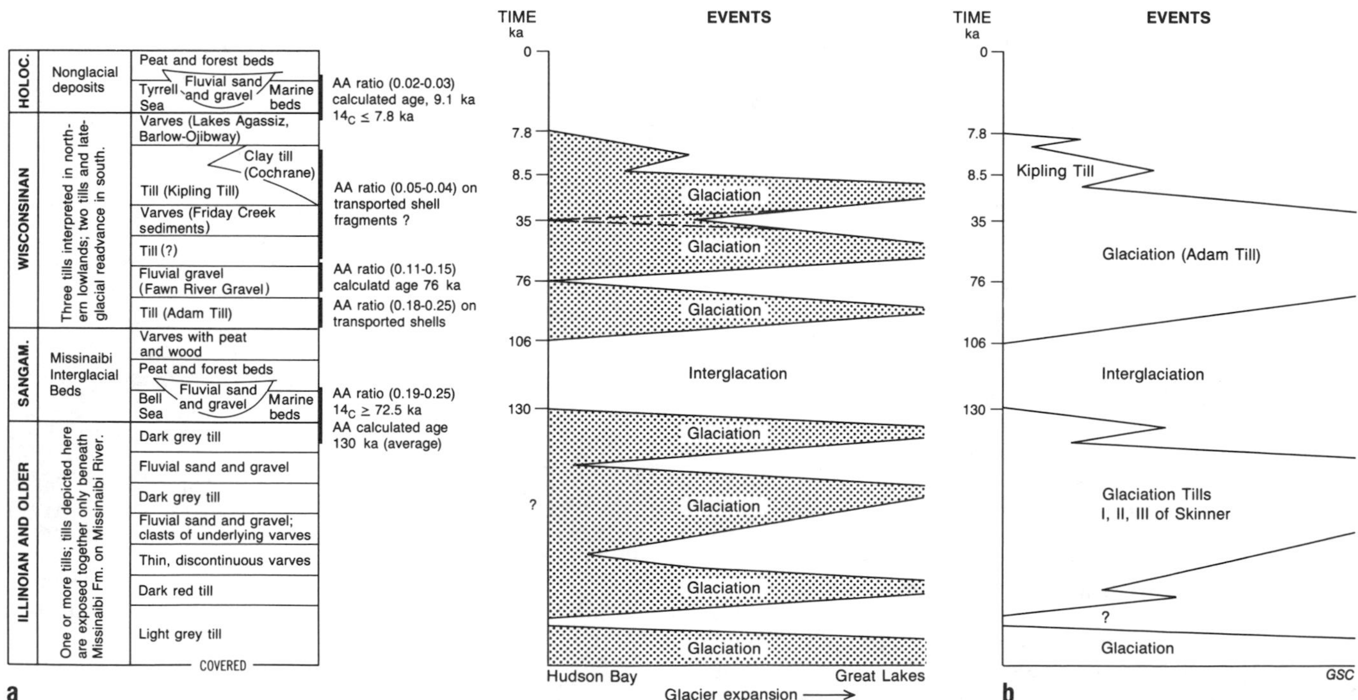

Figure 8.28. (a) Possible stratigraphic interpretation of glacial sediments presently exposed in Hudson Bay Lowland. Ages (other than for Tyrrell Sea) are approximate and speculative, based on recent amino acid (AA) data (total ratios) (from Shilts, 1984). This interpretation maximizes the number of events, (b) a more conservative interpretation of the data presented in Figure 8.28a.

Table 8.1. Quaternary lithostratigraphic units and inferred events, Moose River Basin (Modified from Skinner, 1973)

	Rock-Stratigraphic Unit	Inferred Event	Age ^{14}C Years BP
WISCONSINAN	Terrestrial unit	weathering; peat and forest growth eolian activity	
	Marine unit	Stream incision and deposition marine recession	
	Glaciolacustrine unit	marine incursion glacial retreat	8000
	KIPLING TILL	glacial advance	
	Friday Creek sediments	retreat	
	ADAM TILL	glacial advance	
SANGAMONIAN MISSINAIBI FORMATION	Lacustrine member	lacustrine transgression	
	Forest-peat member (buried soil)	weathering; peat and forest growth	
	Fluvial member		
	Marine member	stream incision and deposition marine recession marine incursion glacial retreat	>72 500
ILLINONIAN and OLDER	Till III	advance (glaciation)	
	Intertill sediments II-III	retreat (glaciation)	
	Till II	advance (glaciation)	
	Intertill sediments I-II	retreat (glaciation)	
	Till I	advance (glaciation)	
	OLDER TILL	advance	

weathered horizon is a calcareous till (Amery Till), containing shell chips, which was deposited by ice flowing from or across Hudson Bay from the east and northeast. This is correlated by Dredge with Skinner's (1973) Till III and represents a pre-Sangamon glaciation, probably Illinoian.

Correlations from the Manitoba Lowland to the Moose River Basin are tenuous at best. However, it is clear that several scientists now recognize pre-Illinoian glaciations within the region.

Missinaibi Formation

Nonglacial sediments within the stratigraphic sequence of Hudson Bay Lowland include fluvial sands and gravels, peat and wood, lacustrine silts and clays, and marine deposits (Fig. 8.28a). Of these, the peat and wood have been well studied for pollen composition and radiocarbon age. In addition, the recognition of marine beds is critical, in that their presence is indicative of high relative sea levels and nonglacial events.

Modern study of the nonglacial deposits dates to the 1950s when Terasmae and Hughes (Terasmae, 1958; Terasmae and Hughes, 1960) reported on nonglacial sediments along the Missinaibi and Opasatika rivers. They concluded that the nonglacial Missinaibi beds (now Missinaibi Formation, Skinner, 1973) probably represented an interstadial episode correlative with the St. Pierre Interstade of Quebec. Subsequently, McDonald (1969) and Skinner (1973) carried out regional and detailed studies respectively and concluded from various criteria that the Missinaibi beds were probably deposited during the Sangamon Interglacial Stage. Principal criteria were 1) evidence of relatively high sea level (Bell Sea) at the base of the sequence; 2) palynological and botanical evidence that the climate was similar to or even warmer than the present (Terasmae, 1958; Skinner, 1973; Lichti-Federovich, 1974); 3) similarity of the Interglacial sedimentation sequences to those of postglacial sediments; and 4) current bedding in gravels indicating flow directions toward Hudson Bay. Though there has been debate as to the rank of the Missinaibi Formation, it is the opinion of the present author and many others (e.g., Shilts, 1984; Andrews et al., 1983; Fulton, 1983; Prest, 1970) that the Missinaibi Formation represents the last interglacial stage, the Sangamonian. Radiocarbon dates on this material are nonfinite, the oldest exceeding 72.5 ka (Stuiver et al., 1978).

In the Manitoba Lowland, sediments of similar stratigraphic position and geologic composition, such as the Gods River sediments (Netterville, 1974) and sediments along the Nelson River (Dredge and Nielsen, 1985), are correlated with the Missinaibi Formation.

Post Missinaibi deposits

Post-Missinaibi deposits include tills and intertill sediments of various origins. The latter have become the subject of considerable debate with respect to the magnitude of the "nonglacial" events they represent.

Moose River Basin

The Adam Till (Fig. 8.28a, Table 8.1) of Moose River Basin was described in detail by Skinner (1973) and interpreted as representing a southwestward-flowing glaciation. Skinner was unable to determine how much of the Wisconsinan Stage is spanned by Adam Till because the

Table 8.2. Quaternary stratigraphic record in the Hudson Bay Basin area of Manitoba (compiled by L.A. Dredge; data from Dredge and Cowan, 1989; Dredge and Neilson, 1985; and R.W. Klassen).

Unit	Subunit	Inferred Event
Terrestrial sediments	(a) river deposits and (b) peat	
	(d) estuarine facies (c) shallow marine facies (b) deep-water marine facies (a) glaciomarine facies	Marine incursion and regression
Lake Agassiz sediments	(c) surge till (b) distal varves (a) sandy glaciolacustrine facies (receding ice)	Ice recession accompanied by Lake Agassiz
Wigwam Creek Formation (Tills and sand beds)	(e) unoxidized brown till (Keewatin and Hudson ice) (d) glaciofluvial relict (c) oxidized brown till (Hudson ice) (b) glaciofluvial relict (a) oxidized grey-brown till (Hudson Ice)	Wisconsin glacial deposits (shifting centres of outflow)
Nelson River, Gods River sediments	(g) local glacial diamicton (f) laminated lacustrine - facies (e) peat and wood, sand (d) freshwater marl (c) blocky marine (?) facies (b) varved glaciolacustrine facies (a) glaciofluvial facies (oxidized gravel)	Sangamon Interglacial
Amery Till	grey till (Hudson ice)	pre-Sangamon glaciation
(bleached zone)	paleosol	interglaciation (ice-free Hudson Bay)
Sundance Till	grey till, deep red oxidized (Keewatin ice)	glaciation
Mountain Rapids diamictons	(b) sand and gravel facies (a) dark grey till	non-glacial interval glacial interval

age and rank of the overlying sediments (Friday Creek) are unknown. The Friday Creek sediments consist of glaciolacustrine materials containing little or no organic matter suitable for radiocarbon dating. A sample of Friday Creek silt and clay from the junction of Adam Creek and Mattagami River contains pollen very similar to that of Erie Interstadial and Mackinaw Interstadial silts of southwestern Ontario. Typically these pollen are representative of deposition in deep, sterile proglacial lakes, the pollen being largely derived from glacial meltwater (J.H. McAndrews, unpublished report, 1979). There is little evidence to suggest that the Friday Creek sediments represent events very different from the late glacial Barlow-Ojibway glaciolacustrine sediments; they may be in part correlative. If this is the case, the Adam Till spans most of Wisconsinan time.

The Kipling Till, which overlies Friday Creek sediments, represents a glacial event of unknown magnitude; some writers (e.g., Shilts, 1982) believe it may represent a major Late Wisconsinan event, whereas others are not convinced that it represents more than a fluctuation during deglaciation. The stratigraphic relationship between the Kipling and Cochrane tills (Hughes, 1965) is unknown; they are lithologically very similar and may well be correlative (R.G. Skinner, pers. comm., 1985, believes this may be true). Certainly the Friday Creek-Kipling Till sequence of the type area is comparable to materials representing short-lived glacial events within the Great Lakes region.

Hudson Bay Basin

McDonald (1969) recognized two post-Missinaibi till units throughout Hudson Bay Lowland; these are commonly separated by intervening glacial lacustrine or fluvial sediments, although in much of the central part of the Lowland the two tills are in contact with each other. The upper till is distinct from the lower in that it is generally finer grained and noticeably different in colour. McDonald (1971) further reported that the lower till overlying the Missinaibi Formation was deposited by ice flowing from the northeast, whereas the upper till was deposited from the northwest. Only at one site did he recognize a possible third upper till that could relate to the Cochrane events. McDonald also recognized numerous outcrops of gravel containing water-worn shell fragments, with current bedding indicating probable interglacial conditions and stream flow toward Hudson Bay.

More recent work by Klassen (1985), Nielsen and Dredge (1982), and Dredge and Nielsen (1985) has identified a package of till and glaciofluvial sediments in Manitoba Lowland referred to as the Wigwam Creek Formation (Table 8.2). This formation consists of three or more tills separated by glaciofluvial sands and gravels or in direct contact. Lithologic composition and ice-flow indicators suggest that the lower two tills in this package resulted from ice flowing out of or across Hudson Bay, whereas the uppermost till reflects sources both in Keewatin and in the Hudson Bay region. The interbedded sediments, interpreted as meltout sediments, indicate that the Wigwam Creek Formation represents continuous Wisconsinan ice cover, accompanied by several shifts of ice flow, which account for the variation in till composition. The Wigwam Creek Formation has been correlated by Dredge (Dredge and Cowan, 1989) with the Adam and Kipling till sequences of Moose River Basin.

Late Glacial Events

During Late Wisconsinan and Early Holocene retreat into Hudson Bay Lowland, the ice front was bounded by Glacial Lake Agassiz on the west and Lake Barlow-Ojibway along a broad southern margin. Glaciolacustrine sedimentation is represented throughout much of the area by varved silt, sand, and clay. Buoyancy of the ice fronts in these lakes resulted in surging or enhanced glacial flow, with consequent deposition of fine-grained tills derived from the glaciolacustrine sediments. Numerous surges of varying magnitude probably occurred and deposition was not continuous. These tills are generally correlative with the Cochrane Till to the south (Hughes, 1965), which is estimated at about 8.2 ka BP, and possibly with all or part of the Kipling Till in Moose River Basin.

Recession and thinning of the ice was accompanied by rising sea levels, which eventually resulted in the sea entering Hudson Bay along its eastern margin; this was probably accompanied by calving bays in Hudson Strait and elsewhere, with a rapid drawdown of residual ice in eastern Hudson Bay. The sea also probably entered western Hudson Bay along a zone of weakness between residual ice centred in Keewatin and that in Hudson Bay. Eventually, Lakes Agassiz and Barlow-Ojibway drained into the marine environment. The catastrophic drainage of Lake Barlow-Ojibway is marked by "drainage horizons" at the base of marine sediments. These consist of clay-pebble gravel and pebbly sands (Hardy, 1982; Skinner, 1973) created by density underflows. The incursion of the Tyrrell Sea, now estimated to have occurred above 8.0 ka BP, deposited a blanket of marine sands and clays over much of Hudson Bay Lowland. Marine limits following isostatic adjustment indicate that the lowest levels occur in the Nelson River area (122 m) and the highest on Sutton Ridge at 180 m a.s.l., and near Seal River, also at 180 m a.s.l. The two high marine limit areas may represent loading by residual ice in Hudson Bay and Keewatin respectively. Rapid emergence through isostatic rebound resulted in uncovering of the present land surface. Present-day rate of emergence is about 39 cm per century at Churchill, Manitoba (Barnett, 1970) and 90-120 cm per century at Fort Albany in the James Bay area (Hunter, 1970).

Summary

Until recently, lithostratigraphic studies on tills and other sediments were inadequate, rendering correlation of glacial sediments across Hudson Platform tenuous. Counting up or down from a known marker, in this case the Missinaibi marker beds, was common procedure as it is in many geological studies; variations in sediment packages over very short distances hampered correlations. Attempts have been made to relate intertill sediments to nonglacial events through facies interpretation, weathering or oxidation of gravels, and the search for marine events that indicate high relative sea levels and minimum ice cover in the Hudson Bay region.

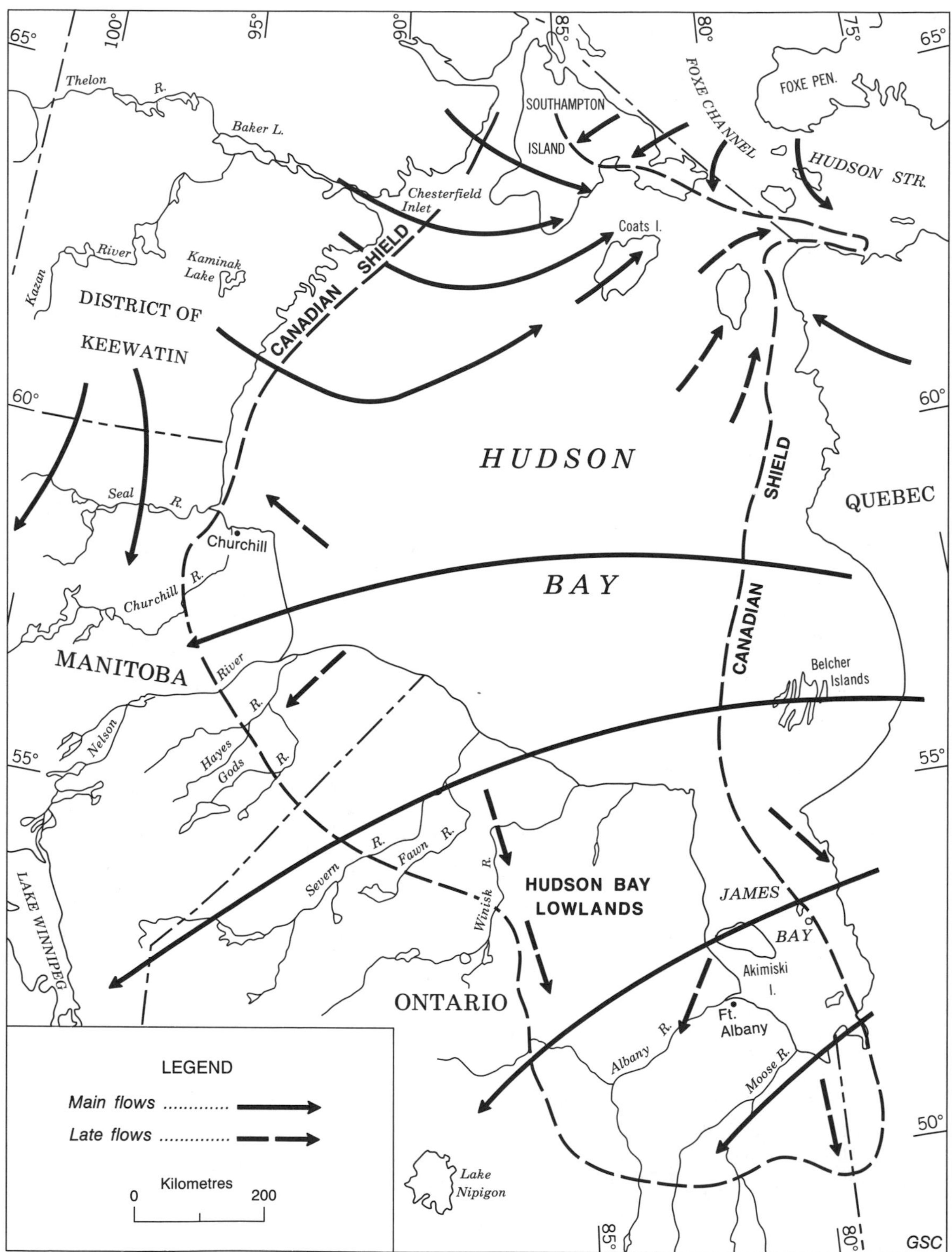

Figure 8.29. Ice flow directions in Hudson Platform area (after Shilts, 1982; Dyke et al., 1982; Prest et al., 1972).

The presence of reworked marine shells in many till units and interglacial sands and gravels, and *in situ* marine shells in the Sangamonian (Bell Sea) and Postglacial (Tyrrell Sea) marine sediments led Andrews et al. (1983), Shilts et al. (1981), and Shilts (1982, 1984) to use cluster analysis of amino acid ratios of both erratic and *in situ* mollusc fragments to determine a possible relative chronology, the Bell and Tyrrell Sea material providing end-member control. Results of this work, using values of 130 ka for the Bell Sea sediments and 8 ka for Tyrrell Sea sediments, led to the prediction that Hudson Bay was free of ice along its southern shore about 35, 75, and 105 ka BP (Fig. 8.28a). Of these, the 105 ka cluster is believed to represent erosion of Hudson Bay sediments at the end of the last interglacial stage (Sangamon; Shilts, 1982).

The 75 ka period was determined from shells in the Fawn River Gravels (flow direction toward Hudson Bay) and the 35 ka material from upper tills in the lowlands. To date, these results serve to put forward an hypothesis for testing; the challenge is to find sites containing *in situ* shells to verify the hypothesis.

On the basis of the foregoing data, Shilts (1984) put forward an interpretation of the glacial history of the lowlands, which included four major pre-Missinaibi glaciations and two major nonglacial episodes within the post-Missinaibi sediments, approximately equivalent to the St. Pierre and Port Talbot Interstades in the St. Lawrence – Great Lakes region. This interpretation, presented here as Figure 8.28a, may be considered as maximizing the numbers of major glacial and nonglacial events displayed by the known sediments. Conversely, Dredge and Nielsen (1985) interpreted two major pre-Missinaibi glaciations and postulated continuous Wisconsinan glaciation in the Manitoba Lowland, though recognizing shifting centres of ice dispersal. Figure 8.28b summarizes the consequences of minimizing the number of events.

The following areas need to be investigated to clarify some of the questions in doubt.

1) The writer favours Skinner's (1973) conclusion that pre-Missinaibi tills I, II, and III probably constitute one sediment package. More work is required on these tills and intervening sediments.
2) More lithological data must be obtained for tills in the region to facilitate correlation.
3) The stratigraphy and interpretation of the Fawn River Gravels need to be studied in detail.
4) The origin and rank of the Friday Creek sediments need to be studied in greater detail, given the major significance accorded them by several writers. Limited experience by the writer leads to the belief that these represent proglacial lacustrine sediments of relatively minor chronological significance, i.e., they may be part of the Barlow-Ojibway sediments (broad sense).
5) The relationship of Kipling Till to Cochrane events requires thorough examination. Are the Kipling and Cochrane tills part of one sediment package? Is the retreat and readvance to deposit Kipling Till a major Middle – Late Wisconsinan event or a late recessional phase during Late Wisconsinan retreat?

Ice flow directions

The significance of the Hudson Bay area as a centre of glacial dispersion has been (and is) a topic of much commentary. Recent discussion on this topic includes work of Ives et al. (1975), Andrews and Barry (1978), Denton and Hughes (1981), Andrews (1982), Dyke et al. (1982), Shilts (1982), Prest (1983, 1984), and Dredge and Cowan (1989). Parts of the Hudson Platform area have been affected directly or indirectly by ice flowing from central Quebec and Labrador, Keewatin, Foxe Basin, and from thick ice over south-central Hudson Bay (Fig. 8.29).

Labrador ice

Laurentide ice centred in central Quebec and Labrador flowed westerly and southwesterly across the southern part of Hudson Platform during the Wisconsin glaciation and probably during preceding events. Evidence is provided by 1) striations and streamlined drift forms (Prest et al., 1968); 2) a dispersion of dark erratics from the Omarolluk Formation in the Belcher Islands (eastern Hudson Bay) and perhaps Sutton Ridge (Fig. 8.29), westward across the lowlands to at least Lethbridge, Alberta and southward to at least Leamington, Ontario (Prest, 1963; E.V. Sado, pers. comm., 1985; E. Nielsen, pers. comm., 1985; Shilts et al., 1979); and 3) a large fan of Paleozoic limestone and dolostone erratics, which were dispersed southward and westward from the Hudson Bay Lowland.

Keewatin ice

Tyrrell (1898) proposed that a centre of glaciation existed in central Keewatin, Northwest Territories. Lee (1959; Lee et al., 1957) subsequently determined that Keewatin was an ice divide towards which late glacial ice retreated. More recently, Shilts (1980, 1982) and Shilts et al. (1979) have demonstrated that ice centred in Keewatin dispersed red erratics of the Dubawnt Group easterly and southeasterly into northwestern Hudson Bay, then northeasterly across Coats Island in north-central Hudson Bay; although these erratics are not reported on Southampton Island, Keewatin ice also affected the southern part of this island, though on a different flow trajectory.

Foxe Basin ice

Wisconsinan glacier ice centred in Foxe Basin (Ives and Andrews, 1963; Prest, 1984; Dyke et al., 1982) flowed radially outward, becoming confluent with Keewatin and Labrador ice. Southward flow from this ice mass affected the northern part of the Hudson Platform area, particularly Southampton Island.

Hudson Bay ice

Tyrrell (1913, 1914) proposed a centre of glacial dispersion from an area southwest of Hudson Bay — the Patrician centre of glaciation. Flint (1943) proposed a major single domed ice sheet over Hudson Bay, which became a standard model for many years.

Evidence of Tyrrell's Patrician centre of dispersal is generally lacking; however, numerous writers have proposed or accepted a major ice dome to the northeast of this centre, notably in the southwestern region of Hudson Bay. This concept is based on glacio-isostatic considerations (e.g., Andrews and Peltier, 1976), ice sheet modelling (e.g., Paterson, 1972; Sugden, 1977; Denton and Hughes, 1981), and multiple criteria (e.g., Dyke et al., 1982). Several of the authors cited believe that at some time during the maximum phase of the last glaciation, at least, there was a major concentration of ice in the southern Hudson Bay region. Ice-flow data indicate that ice at one time flowed northeasterly across the Ottawa Islands (northeastern Hudson Bay) from central Hudson Bay (Andrews and Falconer, 1969), whereas late glacial events such as the Cochrane surges were in many instances related to residual ice in the Hudson and James Bay basins. The writer agrees with Prest (1983, 1984) that the confluence of Keewatin and Labrador ice over Hudson Bay resulted in thick ice in that region; this confluence was aided by a blockage of the Hudson Strait outlet by Foxe-Baffin Island ice. During recession of the Keewatin and Labrador ice, a large residual mass of ice remained over much of Hudson Bay, acting independently for perhaps a few thousand years. Contact areas between Keewatin ice and Labrador ice are marked by the Burntwood-North Knife and Harricana moraines respectively. In addition to Cochrane Till, this ice mass may also be responsible for Kipling Till (Skinner, 1973) in the Moose River Basin as well as the upper brown till reported in Hudson Bay Basin by Dredge (Dredge and Cowan, 1989).

REFERENCES

Andrews, J.T.

1982: On the reconstruction of Pleistocene Ice sheets: a review; Quaternary Science Reviews, v. 1, p. 1-30.

1989: Quaternary geology of the northeastern Canadian Shield region; in Quaternary Geology of Canada and Greenland, R.J. Fulton (ed.); Geological Survey of Canada, Geology of Canada, no. 1. (Also Geological Society of America, The Geology of North America, vol. K-1).

Andrews, J.T. and Barry, R.G.

1978: Glacial inception and disintegration during the last glaciation; Annual Review of Earth Planetary Science, v. 6, p. 205-228.

Andrews, J.T. and Falconer, G.

1969: Late glacial and post-glacial history and emergence of the Ottawa Islands, Hudson Bay, Northwest Territories: evidence of the deglaciation of Hudson Bay; Canadian Journal of Earth Sciences, v. 6, p. 1263-1276.

Andrews, J.T. and Peltier, W.R.

1976: Collapse of the Hudson Bay ice center and glacio-isostatic rebound; Geology, v. 4, p. 73-75.

Andrews, J.T., Shilts, W.W., and Miller, G.H.

1983: Multiple deglaciations of the Hudson Bay Lowlands, Canada, since deposition of the Missinaibi (last-interglacial?) Formation; Quaternary Research, v. 19, p. 18-37.

Aquitaine company of Canada Ltd.

1974a: Well history report (on) Aquitaine et al. Narwhal South O-58; available at Geological Survey of Canada, Calgary (Release date September 1, 1976).

Auitaine company of Canada Ltd. (cont.)

1974b: Well history report (on) Aquitaine et al. Polar Bear C-11; available at Geological Survey of Canada, Calgary (Release date October 20, 1976).

Baragar, W.R.A. and Scoates, R.F.J.

1981: The Circum-Superior Belt: a Proterozoic plate margin?; in Precambrian Plate Tectonics, A. Kroner (ed.); Developments in Precambrian Geology, v. 4, p. 297-330.

Barnes, C.R.

1974: Ordovician conodont biostratigraphy of the Canadian Arctic; in Proceedings, Symposium on the Geology of the Canadian Arctic, J.D. Aitken and D.J. Glass (ed.); Geological Association of Canada and Canadian Society of Petroleum Geologists, May, 1973, p. 221-240.

Barnes, C.R. and Fåhraeus, L.E.

1975: Provinces, communities, and proposed nektobenthic habit of conodontophorids; Lethaia, v. 8, p. 133-149.

Barnes, C.R., Norford, B.S., and Skevington, D.

1981: The Ordovician System in Canada: Correlation chart and explanatory notes; International Union of Geological Sciences, Publication No. 8, 26 p.

Barnett, D.M.

1970: An amendment and extension of tide gauge data analysis for Churchill, Manitoba; Canadian Journal of Earth Sciences, v. 7, p. 626-627.

Barr, K.G.

1969: Evidence for variations in upper mantle velocity in the Hudson Bay area; in Earth Science Symposium on Hudson Bay, Ottawa, February 1968, P.J. Hood et al. (ed.); Geological Survey of Canada, Paper 68-53, p. 365-376.

Bassett, C.F.

1935: Stratigraphy and paleontology of the Dundee Limestone of southeastern Michigan; Bulletin of the Geological Society of America, v. 46, Pt. 1, p. 425-462.

Beals, C.S.

1968: On a possibility of a catastrophic origin for the great arc of eastern Hudson Bay; in Science, History and Hudson Bay, C.S. Beals (ed.); Department of Energy, Mines and Resources, Ottawa, v. 2, p. 988-995.

Bell, W.A.

1928: Mesozoic plants from the Mattagami Series, Ontario; National Museum of Canada, Bulletin 49, Contributions to Canadian Palaeontology, Geology Series No. 48, p. 27-30, 59-67.

Berry, W.B.N. and Boucot, A.J.

1970: Correlation of the North American Silurian rocks; Geological Society of America, Special Paper 102, 289 p.

Bostock, H.H.

1969: Precambrian sedimentary rocks of the Hudson Bay Lowlands; in Earth Science Symposium on Hudson Bay, Ottawa, February 1968, P.J. Hood et al. (ed.); Geological Survey of Canada, Paper 68-53, p. 206-214.

Boucot, A.J.

1975: Evolution and extinction rate controls; developments in palaeontology and stratigraphy, v. 1; Elsevier Scientific Publishing Company, 426 p.

1979 Silurian; in Treatise on Invertebrate Paleontology, Part A – Introduction, Fossilization (Taphonomy), Biogeography and Biostratigraphy, R.A. Robison and C. Teichert (ed.); The Geological Society of America, Inc., and the University of Kansas, Boulder, Colorado and Lawrence, Kansas, p. A167-A182.

Boucot, A.J. and Johnson, J.G.

1979: Pentamerinae (Silurian Brachiopods); Palaeontolographica, Abteilung A, v. 163, Lieferung 4-6, p. 87-129.

Bower, M.E.

1960: Aeromagnetic surveys across Hudson Bay from Churchill to Coral Harbour and Churchill to Great Whale River; Geological Survey of Canada, Paper 59-13, 32 p.

Caley, J.F. and Liberty, B.A.

1957: The St. Lawrence and Hudson Bay Lowlands, and Palaeozoic outliers; in Geology and Economic Minerals of Canada (4th Edition); Geological Survey of Canada, p. 207-246.

Clark, T.H. and Blake, D.A.W.

1952: Ordovician fossils from Waswanipi Lake, Quebec; Canadian Field-Naturalist, v. 66, no. 5, p. 119-121.

Coles, R.L. and Haines, G.V.

1982: Regional patterns in magnetic anomalies over Hudson Bay; Canadian Journal of Earth Sciences, v. 19, p. 1116-1121.

Cook, T.D. and Bally, A.W. (ed.)

1975: Stratigraphic Atlas of North and Central America; prepared by the Exploration Department of Shell Oil Company, Houston, Texas; Princeton University Press, Princeton, New Jersey, 272 p.

Cooper, G.A., Butts, C., Caster, K.E., Chadwick, G.M., Goldring, W., Kindle, E.M., Kirk, E., Merriam, C.W., Swartz, F.M., Warren, P.S., Warthin, A.S., and Willard, B.

1942: Correlation of Devonian sedimentary formations of North America; Bulletin of the Geological Society of America, v. 53, p. 1729-1794.

Couillard, R. and Irving, E.

1975: Palaeolatitude and reversals: Evidence from the Cretaceous Period; in The Cretaceous System in the Western Interior of North America, W.G.E. Caldwell (ed.); Geological Association of Canada, Special Paper No. 13, p. 21-29.

Cranswick, J.S. and Fritz, M.A.

1958: Coral fauna of the Upper Abitibi Limestone; Geological Association of Canada, v. 10, p. 31-81.

Cumming, L.M.

1971: Ordovician strata of the Hudson Bay Lowlands in northern Manitoba; in Geoscience Studies in Manitoba, A.C. Turnock (ed.); Geological Association of Canada, Special Paper No. 9, p. 189-197.

Denton, G.H. and Hughes, T.J. (ed.)

1981: The Last Great Ice Sheets; John Wiley and Sons, New York, 484 p.

Dewing, K., Copper, P., and Hamilton, S.M.
1987: Preliminary report on the stratigraphic position of oil shales in the lower Paleozoic of Southampton Island, Northwest Territories; in Current Research, Part A, Geological Survey of Canada, Paper 87-1A, p. 883-888.

Dimian, M.V., Gray, R., Stout, J., and Wood, B.
1983: Hudson Bay Basin; in Seismic Expression of Structural Styles; a Picture and Work Atlas, A.W. Bally (ed.); American Association of Petroleum Geologists, Studies in Geology Series, no. 15, v. 2, p. 2.2.4-1 - 2.2.4-4.

Dingwall, R.G.
1986: The exploration of Hudson Bay; in Proceedings of The Ontario Petroleum Institute Annual Meeting, London, Ontario, p. 1-41.

Dredge, L.A. and Cowan, W.R.
1989: Quaternary geology of the southwestern Canadian Shield region; in Quaternary Geology of Canada and Greenland, R.J. Fulton (ed.); Geological Survey of Canada, Geology of Canada, no. 1 (Also Geological Society of America, The Geology of North America, vol. K-1).

Dredge, L.A. and Nielsen, E.
1985: Glacial and interglacial deposits in the Hudson Bay Lowlands: a summary of sites in Manitoba; in Current Research, Part A, Geological Survey of Canada, Paper 85-1A, p. 247-257.

Dutro, J.T., Jr.
1981: Devonian brachiopod biostratigraphy; in Devonian Biostratigraphy of New York, Part 1, W.A. Oliver, Jr. and G. Klapper (ed.); International Union of Geological Sciences, Subcommission on Devonian Stratigraphy, p. 67-82.

Dyer, W.S.
1928: Geology and economic deposits of the Moose River Basin; Ontario Department of Mines, Annual Report, v. 37, Pt. 6, p. 1-69.
1930: The Onakawana lignite deposit, Moose River Basin: Ontario Department of Mines, v. 39, Pt. 6, p. 1-14; also published as The Lignite Deposits of Onakawana, Moose River Basin, northern Ontario; Canadian Institute of Mining and Metallurgy, Bulletin, v. 33, p. 450-472.

Dyke, A.S. and Dredge, L.A.
1989: Quaternary geology of the northwestern Canadian Shield region; in Quaternary Geology of Canada and Greenland, R.J. Fulton (ed.); Geological Survey of Canada, Geology of Canada, no. 1. (Also Geological Society of America, The Geology of North America, vol. K-1).

Dyke, A.S., Dredge, L.A., and Vincent, J.-S.
1982: Configuration and dynamics of the Laurentide ice sheet during the Late Wisconsin maximum; Geographie physique et Quaternaire, v. 26, no. 1-2, p. 5-14.

Flint, R.F.
1943: Growth of the North American ice sheet during the Wisconsin age; Bulletin of the Geological Society of America, v. 54, p. 352-362.

Foerste, A.F.
1928: American Arctic and related cephalopods; Science Laboratory, Journal, Denison University, Granville, Ohio, v. 23, p. 1-110.
1929: Devonian cephalopods from the Moose River Basin; Ontario Department of Mines, v. 37, Pt. 6, p. 70-79.:

Fritz, M.A., Lemon, R.R.H., and Norris, A.W.
1957: Stratigraphy and palaeontology of the Williams Island Formation; Geological Association of Canada, Proceedings, v. 9, p. 21-40.

Fritz, M.A. and Waines, R.H.
1956: Stromatoporoids from the Upper Abitibi Limestone; Geological Association of Canada, Proceedings, v. 8, Pt. 1, p. 87-126.

Fulton, R.J.
1983: Correlation of Quaternary events in Canada; p. 70-89 in Quaternary Glaciations in the Northern Hemisphere, Report No. 9 of International Geological Correlation Project 73/1/24, Prague, 248 p.

Fulton, R.J. (ed.)
1989: Quaternary Geology of Canada and Greenland; Geological Survey of Canada, Geology of Canada, no. 1. (Also Geological Society of America, The Geology of North America, vol. K-1).

Gass, K.C. and Mikulic, D.G.
1982: Observations on the Attawapiskat Formation (Silurian) trilobites of Ontario, with description of a new encrinurine; Canadian Journal of Earth Sciences, v. 19, no. 3, p. 589-596.

Geological Survey of Canada
1980: Gravity Map of Canada; Geological Survey of Canada, Gravity Map Series No. 80-1, scale 1:5 000 000.
1987: Magnetic Anomaly Map of Canada, fifth edition; Geological Survey of Canada, Map 1255A, scale 1:5 000 000.

Gibb, R.A. and Walcott, R.I.
1971: A Precambrian suture in the Canadian Shield; Earth and Planetary Science Letters, v. 10, p. 417-422.

Gibb, R.A., Thomas, M.D., LaPointe, P.L., and Mukhopadyay, M.
1983: Geophysics of proposed Proterozoic sutures in Canada; Precambrian Research, v. 19, p. 349-384.

Grant, A.C. and Sanford, B.V.
1988: Bedrock geological mapping and basin studies in the Hudson Bay region; in Current Research, Part B, Geological Survey of Canada, Paper 88-1B, p. 287-296.

Green, A.G., Hajnal, Z., and Weber, W.
1985: An evolutionary model of the western Churchill Province and western margin of the Superior Province in Canada and the North-central United States; Tectonophysics, v. 116, p. 281-322.

Hajnal, Z.
1969: A two-layer model for the earth's crust under Hudson Bay; in Earth Science Symposium on Hudson Bay, Ottawa, February 1968, P.J. Hood et al. (ed.); Geological Survey of Canada, Paper 68-53, p. 326-336.

Halliday, I.
1968: Supporting astronomical evidence from three members of the solar system; in Science, History and Hudson Bay, C.S. Beals (ed.); Department of Energy, Mines and Resources, Ottawa, v. 2, p. 999-1015.

Hardy, L.
1982: Le Wisconinien Superieur a l'est de la Baie James (Quebec); Le Naturaliste Canadienne, v. 109, p. 333-351.

Heckel, P.H. and Witzke, B.J.
1979: Devonian world palaeogeography determined from distribution of carbonates and related lithic palaeoclimatic indicators; in The Devonian System, M.R. House, C.T. Scrutton, and M.G. Bassett (ed.); Special Papers in Palaeontology, no. 23, p. 99-123.

Heywood, W.W. and Sanford, B.V.
1976: Geology of Southampton, Coats, and Mansel Islands, District of Keewatin, Northwest Territories; Geological Survey of Canada, Memoir 382, 35 p.

Hobson, G.D.
1964a: Ontario-Hudson Bay Lowlands, thickness of sedimentary section (Paleozoic to Cretaceous) from reconnaissance seismic refraction survey, March and April, 1964; Ontario Department of Mines Preliminary Map P243, 1 inch = 31 miles.
1964b: Nine reversed refraction seismic profiles, Hudson Bay Lowlands, Manitoba; Geological Survey of Canada, Paper 64-2, p. 33-40.

Hobson, G.D.
1967: Reconnaissance seismic refraction survey of Hudson Bay, Canada; reprinted from Proceedings of the 7th World Petroleum Congress, p. 813-826.

Hobson, G.D., Overton, A., Clay, D.N., and Thatcher, W.
1967: Crustal structure under Hudson Bay; Canadian Journal of Earth Sciences, v. 4, p. 929-947.

Hoffman, P.F.
1985: Is the Cape Smith Belt (northern Quebec) a klippe? Canadian Journal of Earth Sciences, v. 22, p. 1361-1369.
1988: United plates of America, the birth of a craton; Early Proterozoic assembly and growth of Laurentia; Annual Reviews of Earth and Planetary Sciences, v. 16, p. 543-603.

Holland, C.H.
1971: Silurian faunal provinces?; in Faunal Provinces in Space and time, F.A. Middlemiss et al. (ed.); Seel House Press, Liverpool, p. 61-76.

Holland, C.H., Ross, R.L., and Cocks, L.R.M.
1984: Ordovician-Silurian boundary; Lethaia, v. 17, p. 184.

Hood, P.J.
1964: Sea magnetometer reconnaissance of Hudson Bay; Geophysics, v. 29, p. 916-921.

House, H.R.
1968: Continental drift and the Devonian System: An inaugural lecture delivered in the University of Hull, December 1967; University of Hull, 24 p.
1979: Devonian in the eastern hemisphere; in Treatise on Invertebrate Paleontology, Part A, Introduction, Fossilization (Taphonomy), Biogeography and Biostratigraphy, R.A. Robison and C. Teichert (ed.); The Geological Society of America, Inc. and University of Kansas, Boulder, Colorado and Lawrence, Kansas, p. A183-A217.

Hueber, F.M.
1983: A new species of *Baragwanathia* from the Sextant Formation (Emsian), northern Ontario, Canada; Botanical Journal of the Linnean Society, v. 86, p. 57-79.

Hughes, O.L.
1965: Surficial geology of part of the Cochrane District, Ontario, Canada; p. 535-565 in International Studies of the Quaternary, H.E. Wright, Jr., and D.G. Frey (ed.); 7th INQUA Congress, Boulder, Colorado; Geological Society of America, Special Paper 84, p. 535-565.

Hume, G.S.
1925: The Paleozoic outlier of Lake Timiskaming, Ontario and Quebec; Geological Survey of Canada, Memoir 145, 129 p.

Hunter, G.T.
1970: Postglacial uplift at Fort Albany, James Bay; Canadian Journal of Earth Sciences, v. 7, p. 547-548.

Hunter, J.A. and Mereu, R.F.
1967: The crust of the earth under Hudson Bay; Canadian Journal of Earth Sciences, v. 4, p. 929-947.

Irving, E.
1964: Paleomagnetism; John Wiley and Sons, New York, 399 p.

Ives, J.D. and Andrews, J.T.
1963: Studies in the physical geography of north central Baffin Island; Geographical Bulletin, v. 19, p. 5-48.

Ives, J.D., Andrews, J.T., and Barry, R.G.
1975: Growth and decay of the Laurentide ice sheet and comparison with Fenno-Scandia; Naturwissenschaften, v. 62, p. 118-125.

Jaanusson, V.
1979: Ordovician; in Treatise on Invertebrate Paleontology, Part A, Introduction, Fossilization (Taphonomy), R.A. Robison and C. Teichert (ed.); Biogeography and Biostratigraphy, p. A136-A166.

Johnson, J.G. and Boucot, A.J.
1973: Devonian brachiopods; in Atlas of Palaeobiogeography, A. Hallam (ed.); Elsevier Scientific Publishing Company, Amsterdam, London, New York, p. 89-96.

Keele, J.
1920: Mesozoic clays in northern Ontario; Geological Survey of Canada, Summary Report, 1919, Pt. G, p. 13-19.

Kindle, E.M.
1924: Geology of a portion of the northern part of the Moose River Basin, Ontario; Geological Survey of Canada, Summary Report, 1923, Pt. CI, p. 21-41.

Klapper, G.
1981: Review of New York Devonian conodont biostratigraphy; in Devonian Biostratigraphy of New York, Part I, Text, W.A. Oliver, Jr. and G. Klapper (ed.); International Union of Geological Sciences, Subcommission on Devonian Stratigraphy, Washington, D.C., p. 57-66.

Klassen, R.W.
1985: Surficial geology of north-central Manitoba; Geological Survey of Canada, Memoir 419, 57 p.

Kranck, S.H. and Sinclair, G.W.
1963: Clearwater Lake, New Quebec; Geological Survey of Canada, Bulletin 100, 25 p.

Lee, H.A.
1959: Surficial geology of southern District of Keewatin and the Keewatin ice divide, Northwest Territories; Geological Survey of Canada, Bulletin 51, 42 p.
1968 Quaternary geology; in Science, History and Hudson Bay, C.S. Beals (ed.); Department of Energy, Mines and Resources, Ottawa, p. 503-543.

Lee, H.A., Craig, B.G., and Fyles, J.G.
1957: Keewatin ice divide (abstract); Bulletin of the Geological Society of America, v. 68, p. 1760-1761.

Le Fèvre, J., Barnes, C.R., and Tixier, M.
1976: Paleoecology of Late Ordovician and Early Silurian conodontophorids, Hudson Bay Basin; in Conodont Paleoecology, C.R. Barnes (ed.); Geological Association of Canada, Special Paper No. 15, p. 69-89.

Lemon, R.R.H.
1953: The Sextant Formation and its flora; M.A. thesis, University of Toronto, Toronto, Ontario.
1954: The Sextant Formation and its flora (abstract); Canadian Mining Journal, v. 75, no. 6, p. 102.

Lichti-Federovich, S.
1974: *Najas Guadalupensis* (Spreng) Morong. in the Missinaibi Formation, northern Ontario; in Report of Activities, Part A, Geological Survey of Canada, Paper 74-1A, p. 201.

Macauley, G.
1986: Geochemistry of the Ordovician Boas oil shale, Southampton Island, Northwest Territories; Geological Survey of Canada, Open File 1285, 15 p.

MacLean, B., Williams, G.L., Sanford, B.V., Klassen, R.A., Blakeney, C., and Jennings, A.
1986: A reconnaissance study of the bedrock and surficial geology of Hudson Strait, Northwest Territories; in Current Research, Part B, Geological Survey of Canada, Paper 86-1B, p. 617-635.

Martison, N.W.
1953: Petroleum possibilities of the James Bay Lowland area, Ontario; Department of Mines, v. 61, Pt. 6, 1952, p. 1-58.

McCracken, A.D. and Barnes, C.R.
1981: Conodont biostratigraphy and paleoecology of the Ellis Bay Formation, Anticosti Island, Quebec, with special reference to Late Ordovician-Early Silurian chronostratigraphy and the systemic boundary; Geological Survey of Canada, Bulletin 329, Pt. 2, p. 51-134.

McDonald, B.C.
1969: Glacial and interglacial stratigraphy, Hudson Bay Lowland, p. 78-99; in Earth Science Symposium on Hudson Bay, Ottawa, February, 1968, P.J. Hood et al. (ed.); Geological Survey of Canada, Paper 68-53, 386 p.
1971: Late Quaternary stratigraphy and deglaciation in eastern Canada; in The Late Cenozoic Glacial Ages, K.K. Turekian (ed.); Yale University Press, New Haven, p. 331-353.

McGrath, P.M., Hood, P.J., and Darnley, A.G.
1977: Magnetic Anomaly Map of Canada; Geological Survey of Canada, Map 1255A (3rd. ed.), 1:5 000 000.

McGregor, D.C. and Camfield, M.
1976: Upper Silurian? to Middle Devonian spores of the Moose River Basin; Geological Survey of Canada, Bulletin 263, 63 p.

McGregor, D.C., Sanford, B.V., and Norris, A.W.
1970: Palynology and correlation of Devonian formations in the Moose River Basin, northern Ontario; Geological Association of Canada, Proceedings, v. 22, p. 45-54.

McLaren, D.J.
1972: Report from the Committee on the Silurian-Devonian Boundary and stratigraphy to the president of the Commission on Stratigraphy; Geological Newsletter, v. 1972, p. 268-288.

Miller, A.K.
1938: Devonian ammonoids of America; Geological Society of America, Special Paper 14, 262 p.

Morley, L.W., McLaren, A.S., and Hood, P.J.
1968: Low-level aeromagnetic and ship magnetometer observations; in Science, History and Hudson Bay, C.S. Beals (ed.); Department of Energy, Mines and Resources, Ottawa, v. 2, p. 687-702.

Mukhopadhyay, M. and Gibb, R.A.
1981: Gravity anomalies and deep structure of eastern Hudson Bay; Tectonophysics, v. 72, p. 43-60.

Nelson, S.J.
1963: Ordovician paleontology of northern Hudson Bay Lowland; Geological Society of America, Memoir 90, 152 p.
1964: Ordovician stratigraphy of northern Hudson Bay Lowland; Geological Survey of Canada, Bulletin 108, 36 p.

Nelson, S.J. and Johnson, R.D.
1966: Geology of Hudson Bay Basin; Bulletin of Canadian Petroleum Geology, v. 14, p. 520-578.
1976: Oil shales on Southampton Island, northern Hudson Bay; Bulletin of Canadian Petroleum Geology, v. 24, p. 70-91.

Netterville, J.J.
1974: Quaternary stratigraphy of the lower Gods River region, Hudson Bay Lowlands, Manitoba; M.Sc. thesis, University of Calgary, Alberta, 79 p.

Nielsen, E. and Dredge, L.
1982: Quaternary stratigraphy and geomorphology of a part of the lower Nelson River; Geological Association of Canada Field Trip Guidebook, no. 5, Winnipeg, Manitoba, 56 p.

Norford, B.S.
1970: Ordovician and Silurian biostratigraphy of the Sogepet-Aquitaine Kaskattama Province No. 1 well, northern Manitoba; Geological Survey of Canada, Paper 69-8, 36 p.
1971: Silurian stratigraphy of northern Manitoba; in Geoscience Studies in Manitoba, A.C. Turnock (ed.); Geological Association of Canada, Special Paper No. 9, p. 199-207.
1981: The trilobite fauna of the Silurian Attawapiskat Formation, northern Ontario and northern Manitoba; Geological Survey of Canada, Bulletin 327, 37 p.

Norris, G.
1977: Palynofloral evidence for terrestrial Middle Jurassic in the Moose River Basin, Ontario; Canadian Journal of Earth Sciences, v. 14, no. 2, p. 153-158.

Norris, G., Telford, P.G., and Vos, M.A.
1976: An Albian microflora from the Mattagami Formation, James Bay Lowlands, Ontario; Canadian Journal of Earth Sciences, v. 13, p. 400-403.

Okulitch, A.V., Packard, J.J., and Zolnai, A.I.
1986: Evolution of the Boothia Uplift, Arctic Canada; Canadian Journal of Earth Sciences, v. 23, p. 350-358.

Oliver, W.A., Jr.
1977: Biogeography of Late Silurian and Devonian rugose corals; Palaeogeography, Palaeoclimatology, Palaeoecology, v. 22, p. 85-135.

Oliver, W.A., Jr. and Klapper, G. (ed.)
1981: Devonian Biostratigraphy of New York, Part 2 – Stop Descriptions; International Union of Geological Sciences, Subcommission on Devonian Stratigraphy, Washington, D.C., 69 p.

Oliver, W.A., Jr. and Pedder, A.E.H.
1979: Biogeography of Late Silurian and Devonian rugose corals in North America; in Historical Biogeography, Plate Tectonics, and the Changing Environment, J. Gray and A.J. Boucot (ed.); Oregon State University Press, p. 131-145.

Ormiston, A.R.
1972: Lower and Middle Devonian trilobite zoogeography in northern North America; 24th International Geological Congress, Montreal, Canada, 1972, sec. 7, Paleontology, p. 594-604.

Overton, A.
1969: An alternative interpretation of the 1985 Hudson Bay crustal seismic data; in Earth Science Symposium on Hudson Bay, Ottawa, February 1968, P.J. Hood et al. (ed.); Geological Survey of Canada, Paper 68-53, p. 292-306.

Palmer, A.R.
1973: Cambrian trilobites; in Atlas of Palaeobiogeography, A. Hallam (ed.); Elsevier Scientific Publishing Company, Amsterdam, London, New York, p. 3-11.

Paproth, E.
1980: Devonian-Carboniferous boundary resolution; Episodes, v. 1980, no. 3, p. 27.

Paterson, W.J.B.
1972: Laurentide ice sheet: estimated volumes during Late Wisconsin; Reviews of Geophysics and Space Physics, v. 10, p. 885-917.

Playford, G.
1977: Lower and Middle Devonian acritarchs of the Moose River Basin, Ontario; Geological Survey of Canada, Bulletin 279, 87 p.

Pollock, C.A., Rexroad, C.B., and Nicoll, R.S.
1970: Lower Silurian conodonts from northern Michigan and Ontario; Journal of Paleontology, v. 44, p. 743-764.

Prest, V.K.
1963: Surficial geology, Red Lake-Lansdowne House area, northwestern Ontario (parts of 42, 43, 52 and 53); Geological Survey of Canada, Paper 63-6, 23 p.
1970: Quaternary geology of Canada; in Geology and Economic Minerals of Canada, R.J.W. Douglas (ed.); Geological Survey of Canada, Economic Geology Report No. 1, 5th edition, p. 676-764.
1983: The Wisconsinan glacier complex; p. 90-102 in Quaternary glaciations in the northern hemisphere, Report no. 9 of International Geological Correlation Project 73/1/24, Prague, 284 p.
1984: The Late Wisconsinan glacier complex; p. 21-36 in Quaternary Stratigraphy of Canada – A Canadian Contribution to IGCP Project 24, R.J. Fulton (ed.); Geological Survey of Canada, Paper 84-10, 210 p.

Prest, V.K., Grant, D.R., and Rampton, V.N.
1968: Glacial map of Canada; Geological Survey of Canada, Map 1253A, scale 1:5 000 000.

Price, L.L.
1978: Mesozoic deposits of the Hudson Bay Lowlands and coal deposits of the Onakawana area, Ontario; Geological Survey of Canada, Paper 75-13, 39 p.

Roksandic, M.M.
1987: The tectonics and evolution of the Hudson Bay Region; in Sedimentary Basins and Basin-forming Mechanisms, C. Beaumont and A.J. Tankard (ed.); Canadian Society of Petroleum Geologists, Memoir 12, p. 507-518.

Ruffman, A. and Keen, M.J.
1967: A time-term analysis of the first arrival data from the seismic experiment in Hudson Bay, 1965; Canadian Journal of Earth Sciences, v. 4, p. 901-928.

Sanford, B.V.
1987: Paleozoic geology of the Hudson Platform; in Sedimentary Basins and Basin Forming Mechanisms, C. Beaumont and A.J. Tankard (ed.); Canadian Society of Petroleum Geologists, Memoir 12, p. 483-505.

Sanford, B.V. and Norris, A.W.
1973: The Hudson Platform; in The Future Petroleum Provinces of Canada – Their Geology and Potential, R.G. McCrossan (ed.); Canadian Society of Petroleum Geologists, Memoir 1, p. 387-409.
1975: Devonian stratigraphy of the Hudson Platform – Part I: Stratigraphy and economic geology; Part II: Outcrop and subsurface sections; Geological Survey of Canada, Memoir 379, 372 p.

Sanford, B.V., Norris, A.W., and Bostock, H.H.
1968: Geology of the Hudson Bay Lowlands (Operation Winisk); Geological Survey of Canada, Paper 67-60, p. 1-45, and Geological Map 17-1967.

Savage, T.E. and Van Tuyl, F.M.
1919: Geology and stratigraphy of the area of Paleozoic rocks in the vicinity of Hudson and James bays; Geological Society of America, Bulletin, v. 30, p. 339-378.

Sharpton, V.L., Grieve, R.A.F., Thomas, M.D., and Halpenny, J.F.
1987: Horizontal gravity gradient: an aid to the definition of crustal structure in North America; Geophysical Research Letters, v. 14, p. 808-811.

Shilts, W.W.
1980: Flow patterns in the central North American ice sheet; Nature, v. 286, p. 213-218.
1982: Quaternary evolution of the Hudson/James Bay Region; Le Naturaliste Canadien, v. 109, p. 309-332.
1984: Quaternary events – Hudson Bay Lowland and southern District of Keewatin, p. 117-126; in Quaternary Stratigraphy of Canada – A Canadian Contribution to IGCP Project 24, R.J. Fulton (ed.); Geological Survey of Canada, Paper 84-10, 210 p.

Shilts, W.W., Cunningham, C.M., and Kaszycki, C.A.
1979: Keewatin Ice Sheet – re-evaluation of the traditional concept of the Laurentide Ice Sheet; Geology, v. 7, p. 537-541.

Shilts, W.W., Miller, G.H., and Andrews, J.T.
1981: Glacial flow indicators and Wisconsin glacial chronology, Hudson Bay/James Bay Lowlands: evidence against a Hudson Bay ice divide (abstract); Geological Society of America, Abstracts with Programs, v. 13, no. 7, p. 553.

Skinner, R.G.
1973: Quaternary stratigraphy of the Moose River Basin, Ontario; Geological Survey of Canada, Bulletin 225, 77 p.

Sloss, L.L.
1963: Sequences in the cratonic interior of North America; Geological Society of America, Bulletin, v. 74, p. 93-113, 6 figs.

Stuiver, M., Huesser, C.J., and Yang, In Che.
1978: North American glacial history extended to 75,000 years ago; Science, v. 200, p. 16-21.

Sugden, D.E.
1977: Reconstruction of the morphology, dynamics, and thermal characteristics of the Laurentide ice sheet at its maximum; Arctic and Alpine Research, v. 9, p. 21-47.

Sweet, W.C., Ethington, R.L., and Barnes, C.R.
1971: North American Middle and Upper Ordovician conodont faunas; in Symposium on conodont biostratigraphy, W.C. Sweet and S.M. Bergström (ed.); Geological Society of America, Memoir 127, p. 163-193.

Telford, P.G.
1985: Biostratigraphy of the Long Rapids Formation (Upper Devonian), Moose River Basin, Ontario; in Canadian Paleontology and Biostratigraphy Seminar, J.F. Riva (ed.); Programme with Abstracts, Quebec City, September 27-29, 1985 (pages not numbered).

Telford, P.G. and Verma, H.M. (ed.)
1982: Mesozoic geology and mineral potential of the Moose River Basin; Ontario Geological Survey, Study 21, 193 p.

Terasmae, J.
1958: Non-glacial deposits along Missinaibi River, Ontario; Part III, p. 29-34; in Contributions to Canadian Palynology; Geological Survey of Canada, Bulletin 46, 35 p.

Terasmae, J. and Hughes, O.L.
1960: A palynological and geological study of Pleistocene deposits in the James Bay Lowlands, Ontario (42N1/2); Geological Survey of Canada, Bulletin 62, 15 p.

Thomas, M.D. and Gibb, R.A.
1977: Gravity anomalies and deep structure of the Cape Smith Fold Belt, Northern Ungava, Quebec; Geology, v. 5, p. 169-172.
1985: Proterozoic plate subduction and collision processes for reactivation of Archean crust; in Evolution of Archean Supracrustal Sequences, L.D. Ayres, P.C. Thurston, K.D. Card, and W. Weber (ed.); Geological Association of Canada, Special Paper 28, p. 263-279.

Thomas, M.D., Grieve, R.A.F., and Sharpton, V.L.
1988: Gravity domains and assembly of the North American continent by collisional tectonics; Nature, v. 331, p. 333-334.

Thomas, M.D., Sharpton, V.L., and Grieve, R.A.F.
1987: Gravity patterns and Precambrian structure in the North American central plains; Geology, v. 15, p. 489-492.

Tillement, B.A., Peniguel, G., and Fuillemin, J.P.
1976: Marine Pennsylvanian rocks in Hudson Bay; Bulletin of Canadian Petroleum Geology, v. 24, p. 418-439.

Try, C.F., Long, D.G.F., and Winder, C.G.
1984: Sedimentology of the Lower Cretaceous Mattagami Formation, Moose River Basin, James Bay Lowlands, Ontario, Canada; in The Mesozoic of Middle North America, D.F. Stott and D.J. Glass (ed.); Canadian Society of Petroleum Geologists, Memoir 9, p. 345-359.

Tyrrell, J.B.
1898: The glaciation of north-central Canada; Journal of Geology, v. 6, p. 147-160.
1913: Hudson Bay Exploring Expedition, 1912; Twenty-second Annual Report of the Ontario Bureau of Mines, v. 22, pt. 1, p. 161-209.
1914: The Patrician glacier south of Hudson Bay; Twelfth International Geological Congress (1913), Comptes Rendus, Ottawa, Canada, p. 523-534.

Uyeno, T.T., Telford, P.G., and Sanford, B.V.
1982: Devonian conodonts and stratigraphy of southwestern Ontario; Geological Survey of Canada, Bulletin 332, 55 p.
Vail, P.R., Mitchum, R.M., and Thompson, S., III
1977: Seismic stratigraphy and global changes of sea level, Part 4: Global cycles of relative changes of sea level; in Seismic Stratigraphy – Applications to Hydrocarbon Exploration, C.E. Payton (ed.); American Association of Petroleum Geologists, Memoir 26, p. 83-98.
Vincent, J.S.
1989: Quaternary geology of the southeastern Canadian Shield; in Quaternary Geology of Canada and Greenland, R.J. Fulton (ed.); Geological Survey of Canada, Geology of Canada, no. 1. (Also Geological Society of America, The Geology of North America, v. K-1).
Wanless, R.K., Stevens, R.D., Lachance, G.R., and Edmonds, C.M.
1967: Age determinations and geological studies, K-Ar Isotopic Ages, Report 7; Geological Survey of Canada, Paper 66-17, 120 p.
Wilson, A.E.
1953: A report on fossil collections from James Bay Lowland; Ontario Department of Mines, v. 61, Pt. 6, p. 59-81.
Wilson, J.T.
1968: Theories on the origin of Hudson Bay, Part III: Comparisons of the Hudson Bay arc with some other features; in Science, History and Hudson Bay, C.S. Beals (ed.); Department of Energy, Mines and Resources, Ottawa, v. 2, p. 1015-1033.
Wilson, W.J.
1903: Reconnaissance surveys of four rivers south-west of James Bay; Geological Survey of Canada, Summary Report, 1902, n.s., v. 15 (Annual Report 1902-03), Pt. A, p. 222-243.
Workum, R.H., Bolton, T.E., and Barnes, C.R.:
1976: Ordovician geology of Akpatok Island, Ungava Bay, District of Franklin; Canadian Journal of Earth Sciences, v. 13, p. 157-178.

ADDENDUM

The most notable advances and revisions of the geology of the Hudson Platform that have been made since the above report was prepared are as follows:

Updated bathymetric charts of Hudson Bay compiled from many sources are illustrated by Sanford (1987, fig. 4, p. 485) and Grant and Sanford (1988, fig. 3, p. 290). The distribution of recently discovered Cretaceous rocks in the Hudson Bay Basin is depicted by Sanford and Grant (1990, fig. 4, p.21). Conodont zones for the upper Givetian of the late Middle Devonian and for the Frasian of early Late Devonian have been revised by Klapper and Johnson (1990) and Klapper (1988), respectively. Brachiopods of the lower and brachiopods and conodonts of the uppermost shale members of the Williams Island Formation (late Middle Devonian) and lower Long Rapids Formation (early Late Devonian) have been described and illustrated by Norris (manuscript in preparation) and Norris et al. (1992), respectively.

Grant, A.C. and Sanford, B.V.
1988: Bedrock geological mapping and basin studies in the Hudson Bay region; in Current Research, Part B; Geological Survey of Canada, Paper 88-1B, p. 287-296.
Klapper, G.
1988: The Montagne Noire (Upper Devonian) conodont succession; in Devonian of the World, N.J. McMillan, A.F. Embry and D.J. Glass (ed.); Canadian Society of Petroleum Geologists, Memoir 14, v. 3, p. 449-468.
Klapper, G. and Johnson, J.G.
1990: Revisions of Middle Devonian conodont zones; in Lower and Middle Devonian brachiopod-dominated communities of Nevada, and their position in a biofacies-province-realm model; Journal of Paleontology, v. 64, no. 6, p. 902-941.
Norris, A.W., Uyeno, T.T., Sartenaer, P., and Telford, P.G.
1992: Brachiopod and conodont faunas from the uppermost Williams Island Formation and lower Long Rapids Formation (Middle and Upper Devonian), Moose River Basin, northern Ontario; Geological Survey of Canada, Bulletin 434, 133 p.
Sanford, B.V.
1987: Paleozoic geology of the Hudson Platform; in Sedimentary Basins and Basin-Forming Mechanisms, C.Beaumont and A.J. Tankard (ed.); Canadian Society of Petroleum Geologists, Memoir 12, p. 483-505.
Sanford, B.V. and Grant, A.C.
1990: New findings relating to the stratigraphy and structure of the Hudson Platform; in Current Research, Part D; Geological Survey of Canada, Paper 90-1D, p. 17-30.

Authors' Addresses

A.W. Norris
Institute of Sedimentary and Petroleum Geology
Geological Survey of Canada
3303 - 33rd Street N.W.
Calgary, Alberta
T2L 2A7

B.V. Sanford
17 Meadowglade Gardens
Nepean, Ontario
K2G 5J4

A.C. Grant
Atlantic Geoscience Centre
Geological Survey of Canada
Bedford Institute of Oceanography
P.O. Box 1006
Dartmouth, Nova Scotia
B2Y 4A2

W.L. Cowan
Ontario Geological Survey
Ministry of Northern Development and Mines
77 Grenville Street, 7th Floor
Toronto, Ontario
M5S 1W4

Printed in Canada

Chapter 9

HUDSON PLATFORM – ECONOMIC GEOLOGY

Chapter 9

HUDSON PLATFORM - ECONOMIC GEOLOGY

B.V. Sanford, A.W. Norris, and A.R. Cameron

Hudson Platform underlies an area of 971 250 km^2, more than half of which is covered by the waters of Hudson Bay, James Bay, Hudson Strait, and Foxe Basin. The region has potential for a variety of economic commodities, including petroleum and natural gas, oil shale, lignite, lead and zinc, and industrial minerals such as salt, gypsum, and limestone.

The principal emphasis over the years has been placed on exploration for oil and gas, first in the onshore areas of Moose River Basin, and in more recent years in the onshore and offshore regions of Hudson Bay Basin. Lignite deposits that have long been known to occur along the extreme southern margins of Moose River Basin have also been investigated from time to time for about a century, but have not been exploited commercially. Promising industrial mineral deposits including salt, gypsum, clay, sand, and limestone are known in widely separated parts of the platform, but have remained undeveloped owing to their great distance from densely populated regions of the country.

PETROLEUM AND NATURAL GAS

B.V. Sanford and A.W. Norris

Hudson Bay and Moose River basins of Hudson Platform, underlain by lower and middle Paleozoic carbonate strata, have undergone only limited exploration. The most comprehensive assessments of the oil and gas possibilities of the platform are those of Sanford and Norris (1973) and Johnson et al. (1986). Several authors have commented on the close similarity of the Devonian succession of Hudson Platform to that of the Michigan Basin, pointed out the long record of Devonian production from the latter area, and suggested that the potential for similar production from Hudson Platfor.

The Moose River Basin at the southern end of Hudson Platform was tested unsuccessfully by several early wells drilled by the Ontario Department of Mines and by two exploratory wells in 1971. The stratigraphic section is similar to that of southern Ontario and has many similarities as to potential sources, reservoirs, and trapping mechanisms. The sedimentary section is only about 760 m thick.

Sanford, B.V., Norris, A.W., and Cameron, A.R.
1993: Hudson Platform - Economic geology; Chapter 9 in Sedimentary Cover of the Craton in Canada, D.F. Stott and J.D. Aitken (ed.); Geological Survey of Canada, Geology of Canada, no. 5, p. 701-707 (also Geological Society of America, The Geology of North America, v. D-1).

Hudson Bay Basin, viewed as more promising, is covered by water except for the southwest flank. The main basin is roughly circular, and the area containing more than a kilometre of sediments has a diameter of 500 km, similar to those of Williston, Michigan, and Illinois basins. The maximum thickness of sediments is uncertain because of lack of seismic definition, but is probably in the range of 2500 to 3000 m (Wade et al., 1977; Hood, 1964). The stratigraphic section consists of Ordovician, Silurian, and Devonian rocks, with some recently (1988) identified Cretaceous rocks beneath Hudson Bay.

Studies by Tillement (1975) and others on the quantity of organic material, its quality, and its state of maturation for the Ordovician, Silurian, and Devonian strata of Hudson Platform concluded that these rocks have a relatively low degree of maturation. Specific data on Silurian conodonts from the Quebec Embayment indicate a colour alteration index of 1 (T.T. Uyeno, pers. comm., 1984), which is within the lower part of the range favourable for hydrocarbon generation (Epstein et al., 1977).

Northeast- and northwest-trending conjugate fault sets occur within Hudson Bay Basin (Dimian et al., 1983). A complex, north-trending, regional horst, the Mid-Bay "High" or Shoal, is 400 km long and 60 km wide (see Fig. 7.3, 8.1; Wade et al., 1977). Vertical displacements during Late Silurian (?) and Early Devonian times were of the order of 300 m. The horst complex essentially bisected the Late Silurian basin, and the eastern and western sub-basins continued to receive additional sediments during the regressive phase of the Late Silurian and Early Devonian epochs. Subsequent Middle and Upper Devonian strata onlapped and buried the horst. Fault traps and related drapes and folds may have developed around the Mid-Bay "High", and may also occur along regional faults.

Some of the potential traps have been tested by exploratory drilling. Prior to 1969 only five tests had been drilled onshore in Hudson Bay Basin. Drilling between 1969 and 1974 was in the southern half of the basin, with three tests offshore – two on the Mid-Bay "High" Basin, and one on the southeastern flank. Two tests of the northern half of the basin were completed in 1985, one on the "High" and the second in the western sub-basin. None of the wells found commercial quantities of oil or gas.

Reflection seismic data in Hudson Bay Basin (Dimian et al., 1983; B.V. Sanford, unpubl.) indicate the widespread presence of block-faulted and salt-dissolution structures and biohermal facies, some of which could conceivably be favourable for the entrapment of oil and gas. Traps of these types hold oil and gas in the Michigan and Allegheny basins (Sanford et al., 1985) and in the Williston and Elk Point basins of the Interior Platform (Gwynn, 1964; Little et al., 1970).

From a lithological viewpoint, rough comparisons of Paleozoic sequences and their hydrocarbon potential can also be made between the Hudson, Interior, and St. Lawrence platforms.

Cambrian to Lower Ordovician (Canadian)

Although Cambrian to Lower Ordovician orthoquartzitic sandstone and dolostone are present in Foxe Basin directly northeast of the Hudson Bay Basin, equivalent strata are absent in the Hudson Bay and Moose River basins.

Middle Ordovician (Whiterockian to Trentonian)

Middle Ordovician strata of Hudson Platform have lithological counterparts in the St. Lawrence Platform. The upper Middle Ordovician (Trentonian) shales and interbedded shaly limestones of the lower Bad Cache Rapids Group lack reservoir characteristics in this region, but have important source-rock potential.

Upper Ordovician to Lower Silurian (Edenian to Wenlock)

Upper Ordovician to Lower Silurian carbonates of the upper Bad Cache Rapids, Churchill River, Red Head Rapids, Severn River, Ekwan River, and Attawapiskat formations are lithologically similar to their counterparts in Williston Basin (Sanford and Norris, 1973). Hydrocarbons in the latter region (Gwynn, 1964) have been produced from reservoirs associated with block faulting along the Nesson and Cedar Creek anticlines of North Dakota. Significant oil and gas production has also been obtained in the Great Lakes region of Canada and United States from the partly equivalent upper Middle (Trentonian) to lower Upper Ordovician (Edenian) carbonates that have been repeatedly fractured and dolomitized. Considerable potential lies in the block-faulted structures in Hudson Bay Basin, including those along the central Hudson Bay "high" and particularly on the downthrown side of the tilted blocks where dolomitization is most likely to have occurred.

The Attawapiskat reef development in Hudson Platform has its counterpart in the upper Llandovery and (?)Wenlock bioherms bordering Michigan Basin. The latter are largely devoid of hydrocarbons and lack source rock. Similar conditions appear to have prevailed for the Attawapiskat facies of Hudson Bay Basin. The oil- and gas-bearing reefs of Late Silurian (early Ludlow) age that occur in great profusion around the margins of Michigan Basin (Sanford et al., 1985) apparently have no counterpart in Hudson Platform.

Upper Ordovician Boas River oil shales on Southampton Island may be an important petroleum resource and may have source-rock potential in the deep parts of Hudson Bay Basin. On Southampton Island, the shales, up to 2.1 m thick, outcrop at a single locality along the upper reaches of the Boas River (Sanford, in Heywood and Sanford, 1976). There, the unit overlies the Bad Cache Rapids Group with gradational contact, and is in turn overlain by the Churchill River Group. Its distribution elsewhere is generally unknown, but traces of the shale in deep wells in the central part of Hudson Bay indicate it may be widespread.

A second, albeit thin, Upper Ordovician oil shale occurs near the top of the Red Head Rapids Formation on Southampton Island. The shales are interbedded with thin, platy carbonates, and form a relatively minor constituent of the formation.

Upper Silurian to Lower Devonian (Ludlow to Gedinnian)

The Upper Silurian and Lower Devonian terrigenous redbed and evaporitic deposits of the Kenogami River Formation have certain lithological characteristics in common with the Salina Formation of the Michigan Basin. However, the Kenogami River appears to lack the reservoir characteristics and hydrocarbon source-rock potential of the Salina, and may have limited potential for oil and gas accumulation in Hudson Platform.

Lower to Upper Devonian (Siegenian to Famennian)

Lower and Middle Devonian rocks of the Moose River and Hudson Bay basins have lithological counterparts in the Michigan Basin. In the latter area, equivalents of the Moose River, Murray Island, and Williams Island formations have produced substantial volumes of oil for more than a century. Biostromal carbonate banks with reservoir potential occur in the Kwataboahegan Formation, particularly in the southeastern part of Moose River Basin. Parts of the Moose River and Murray Island formations are fractured, brecciated, and porous due to solution of evaporites.

Salt-leached structures, similar to traps in Michigan Basin, are present over a wide region of Hudson Bay Basin, as determined by bathymetry and shallow high-resolution seismic data (B.V. Sanford, unpub. rep., 1980). Most of the Devonian oil discovered to date in the Michigan Basin is from carbonate rocks in structural traps that were formed as a result of selective dissolution of underlying Silurian and/or Devonian salts along rejuvenated fractures. Similarly, some of the structures in Hudson Bay Basin may offer good possibilities for hydrocarbon entrapment in the Devonian rocks.

Resources

No commercial quantities of oil or gas have been discovered by the limited drilling undertaken within Hudson Platform.

OIL SHALE

A.W. Norris

Oil shales (Macauley, 1984) occur within three stratigraphic units in Hudson Platform: the Upper Ordovician Boas River Shale on Southampton Island (Heywood and Sanford, 1976); the uppermost Ordovician 'oil shale interval' on Southampton Island (Nelson and Johnson, 1966, 1976); and the Upper Devonian Long Rapids Formation in Moose River Basin. Analyses of 95 samples of the 'oil shale interval' by Nelson and Johnson (1976) indicate an average yield of 50 litres of oil per tonne. The few analyses of oil shale from the Long Rapids Formation were summarized by Macauley (1984); these are insufficient to properly evaluate the potential net oil content.

COAL

A.R. Cameron

Moose River Basin, located at the south end of Hudson Bay Platform, contains the only known coal deposits in Ontario. An oval-shaped area (approximately 80 x 172 km) at the south end of the basin is underlain by known or possible occurrences of Jurassic and lignite-bearing Cretaceous sediments (Fig. 9.1). The region is topographically low, with extensive swampy areas, and ground access is difficult.

The coal occurs in the Lower Cretaceous Mattagami Formation, which ranges in thickness from 0 to 166 m, thinning northward (Winder et al., 1982; Try et al., 1984). In part of the area, the Mattagami is underlain disconformably by Jurassic Mistuskwia Beds and overlain by glacial deposits and marine clays of the post-glacial Tyrrell Sea (Telford, 1982). Mattagami sediments are largely unconsolidated clays, with lesser amounts of silica sand deposited in a multiple-channel fluvial environment with peat formation in channels and on interfluve areas.

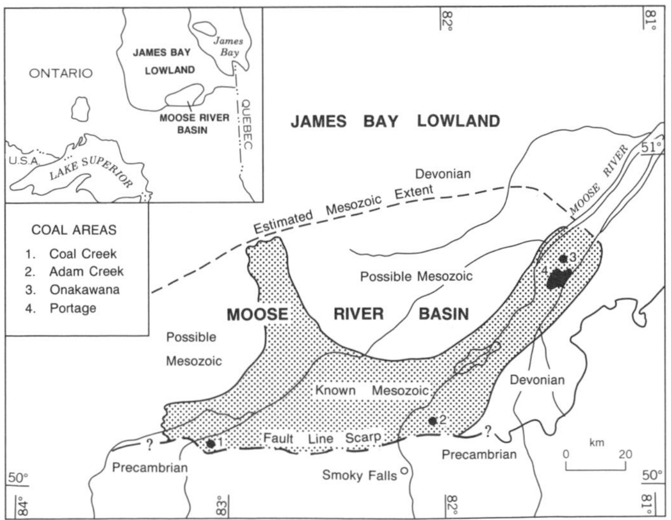

Figure 9.1. Area of Mesozoic beds and coal occurrences in Moose River Basin (modified from Telford, 1982).

Table 9.1. Composition data on Onakawana lignites (from Vos, 1982)

Average analysis lignites:		
		Per cent
Moisture		46.00
Ash		8.83
Volatiles		21.92
Fixed carbon		23.25
Calorific value		12.20 MJ/kg
These averages correspond to the following ultimate analysis:		
		Dry, ash free
Carbon	32.23	70.63
Hydrogen	2.20	4.82
Sulphur	0.51	1.12
Nitrogen	0.23	0.50
Ash	8.37	-
Oxygen	10.37	22.73
Chlorine	0.09	0.20
Moisture	46.00	-

Lignitic beds in the Mattagami Formation have been reported at four locations (Fig. 9.1; Price, 1978). Of these, the most important is near Onakawana, though recent drilling 40 to 60 km northwest of Onakawana indicated at least one lignite bed 5-6 m thick (Telford and Sawacki, 1983). At Onakawana, Price (1978) reported two lignite beds with average thicknesses of 4.2 m (lower seam) and 5.4 m (upper seam). According to Trusler et al. (1974), the lower seam is continuous over a 27 km^2 area. The upper seam appears to be less persistent. Spotty occurrences of a third seam higher in the section were also reported.

Much of the structure within the Mattagami Formation appears to be the result of slumping and related deformation of unconsolidated beds, although ice thrusting has taken place also (Baker, 1911). The southern limit of Mesozoic beds and Moose River Basin is marked by a prominent fault-line scarp of Precambrian rocks (Fig. 9.1). This structure appears to have been active after deposition of the Mesozoic beds. Within Moose River Basin, a northwest-trending basement ridge, Grand Rapids Arch, likely influenced the deposition of the Mesozoic beds.

Comparison of values of Mattagami lignites (Table 9.1) with those of Saskatchewan lignites shows that the Moose River lignites are lower in rank and would classify as lignite B in the ASTM system. The lignite consists of flattened, carbonized wood, partly as recognizable tree trunks and branches, with brown earthy layers containing spores (?) and resinous lenses.

No lignite has been produced commercially from Moose River Basin, although since 1900 there have been several periods of relatively intense exploration (1926-32, 1943-45, 1966-67, 1975-present). Between 1945 and 1946 at Onakawana, near Abitibi River, some 3000 tonnes of lignite were mined and a small test boiler unit installed. Shortly after 1946, this development was discontinued.

Bielenstein et al. (1972) estimated the lignite resources at 218 million tonnes, all in the measured category. Vos (1982) reported recoverable resources in the Onakawana area as 189 million tonnes, based on an average stripping ratio of 6.56.

INDUSTRIAL MINERALS

A.W. Norris

Limestone

Various types of limestone are readily accessible along the railway lines crossing Hudson Bay Lowland in northern Ontario and northern Manitoba. The more promising and accessible high-calcium limestone deposits occur in the Devonian strata of Moose River Basin and include the outcrops of the Kwataboahegan Formation near Coral Rapids on Abitibi River and at Grand Rapids on Mattagami River; the Murray Island Formation near the upper end of Long Rapids on Abitibi River, and on Moose River near Moose River Crossing; and the upper member of the Williams Island Formation near Williams Island.

Analyses of some of the Devonian limestones of Moose River Basin were provided by Malcolm (1926), Dyer (1929), and Goudge (1938).

Gypsum

The gypsum deposits of Moose River Basin have attracted attention because of their high purity, conspicuous outcrops, and location near transportation routes. They are part of the Middle Devonian Moose River Formation, and were described most recently by Sanford and Norris (1975). The main outcrop areas are on the Cheepash and Moose rivers, a slightly elevated area known as Gypsum Mountain, and on Wakwayokastic River (see Sanford and Norris, 1975). These outcrops occur in a northwest-trending belt as much as 17 km wide and about 72 km long. Calculations by Vos (1982) suggest that 160 million tonnes of gypsum are available for commercial exploitation in the areas underlying the outcrops. If the areas between outcrops are included, a total of 5.5 billion tonnes of gypsum may be present.

Salt

Salt deposits are widely distributed in Hudson Bay Basin, occurring within the Ordovician Red Head Rapids Formation, the Silurian-?Devonian Kenogami River Formation, and the Devonian Stooping River Formation. These occurrences are in the subsurface and located mainly offshore beneath Hudson Bay, and are thus beyond the capability of conventional exploitation by brining or mining methods of recovery. The most accessible of these occurrences is the 12-m interval of Red Head Rapids salt that lies in a relatively narrow onshore coastal belt that straddles the Manitoba-Ontario border. The salt lies from 669 to 790 m below the surface, is relatively pure, and may be of sufficient thickness for commercial use (B.V. Sanford, unpub. rep., 1980).

Silica sand

Deposits of quartz sand in the Lower Cretaceous Mattagami Formation of Moose River Basin were described by Price (1978), Vos (1982), and others. The sands are remarkably pure, non-indurated, and quartzose, with grains that are angular to subangular (Price, 1978). Analyses of the sands (Vos, 1982) indicates that, with minor beneficiation to reduce iron and clay content, the sand is suitable for glass production.

Clays

Clays within the Lower Cretaceous Mattagami Formation of Moose River Basin have much the same distribution as silica sand. Price (1978) reported important outcrop localities along the Missinaibi, Pivabiska, Mattagami, and Abitibi rivers, and Vos (1982) described their distribution in the subsurface, as well as summarizing all the previous work on refractory clays in Moose River Basin. The clays have a wide range of composition and include kaolinitic clay and illite. The large volume and high quality of the clays make them attractive prospects for future commercial development.

Siderite and limonite

Siderite and limonite occur as a replacement deposit in the Devonian Stooping River and Kwataboahegan formations at Grand Rapids on Mattagami River (Sanford and Norris, 1975). The mineralization occurs in solution cavities, and forms three small bodies, one near the lower, and two near the upper end of the rapids. The iron minerals consists mainly of siderite and a lesser amount of limonite (Cross, 1920). The siderite is of good quality, but the tonnage is insufficient for commercial development at the present time.

REFERENCES

Baker, M.B.
1911: Iron and lignite in the Mattagami basin; Ontario Bureau of Mines, Annual Report, v. 20, Part 1, p. 214-246.

Bielenstein, H.U., Chrismas, L.P., Latour, B.A., and Tibbetts, T.E.
1979: Coal resources and reserves of Canada; Department of Energy, Mines and Resources, Canada, Report ER 79-9, 37 p.

Cross, J.G.
1920: Pre-Cambrian rocks and iron ore deposits in the Abitibi-Mattagami area; Ontario Department of Mines, v. 29, Pt. 2, p. 1-18.

Dimian, M.V., Gray, R., Stout, J., and Wood, B.
1983: Hudson Bay Basin; in Seismic Expression of Structural Styles, a Picture and Work Atlas, A.W. Bally (ed.); American Association of Petroleum Geologists, Studies in Geology Series, No. 5, v. 2, p. 2.2.2-1 to 2.2.4-4.

Dyer, W.S.
1929: Limestone of the Moose River and Albany River basins; Ontario Department of Mines, v. 38, Pt. 4, p. 31-33.

Epstein, A.G., Epstein, J.B., and Harris, L.D.
1977: Conodont colour alteration – An index to organic metamorphism; United States Geological Survey, Professional Paper 995, 27 p.

Goudge, M.F.
1938: Limestones of Canada, Part IV, Ontario; Department of Mines and Resources, Bureau of Mines, No. 781, 362 p.

Gwynn, T.A.
1964: The Cedar Creek Anticline, 43 years of history and development, 1921 to 1964; Third International Williston Basin Symposium, Regina, Saskatchewan; Billings Geological Society, Billings, Montana, p. 192-199.

Heywood, W.W. and Sanford, B.V.
1976: Geology of Southampton, Coats, and Mansel Islands, District of Keewatin, Northwest Territories; Geological Survey of Canada, Memoir 382, 35 p.

Hood, P.J.
1964: Sea magnetometer reconnaissance of Hudson Bay; Geophysics, v. 29, p. 916-921.

Johnson, R.D., Joubin, F.R., Nelson, S.J., and Olsen, E.
1986: Mineral Resources; Chapter 19 in Canadian Inland Seas, I.P. Martini, (ed.); Elsevier Oceanography Series 44, p. 387-402.

Little, H.W., Belyea, H.R., Stott, D.F., Latour, B.A., and Douglas, R.J.W.
1970: Economic minerals of western Canada; in Geology and Economic Minerals of Canada, R.J.W. Douglas (ed.); Geological Survey of Canada, Economic Geology Report No. 1, p. 490-546.

Macauley, G.
1984: Geology of the oil shale deposits of Canada; Geological Survey of Canada, Paper 81-25, 63 p.

Malcolm, W.
1926: Limestone on Abitibi and Mattagami rivers, Ontario; Geological Survey of Canada, Summary Report, 1924, Pt. C, p. 26-98.

Nelson, S.J. and Johnson, R.D.
1966: Geology of Hudson Bay Basin; Bulletin of Canadian Petroleum Geology, v. 14, no. 4, p. 520-578.
1976: Oil shales on Southampton Island, northern Hudson Bay; Bulletin of Canadian Petroleum Geology, v. 24, no. 1, p. 70-91.

Price, L.L.
1978: Mesozoic deposits of the Hudson Bay Lowlands and coal deposits of the Onakawana area, Ontario; Geological Survey of Canada, Paper 75-13, 39 p.

Sanford, B.V. and Norris, A.W.
1973: The Hudson Platform; in The Future Petroleum Provinces of Canada – Their Geology and Potential, R.G. McCrossan (ed.); Canadian Society of Petroleum Geologists, Memoir 1, p. 387-409.
1975: Devonian stratigraphy of the Hudson Platform – Part I: Stratigraphy and economic geology; Part II: Outcrop and subsurface sections; Geological Survey of Canada, Memoir 379, 372 p.

Sanford, B.V., Thompson, F.J., and McFall, G.H.
1985: Plate tectonics – a possible controlling mechanism in the development of hydrocarbon traps in southwestern Ontario; Bulletin of Canadian Petroleum Geology, v. 33, no. 1, p. 52-71.

Telford, P.G.
1982: Mesozoic stratigraphy of the Moose River Basin; in Mesozoic Geology and Mineral Potential of the Moose River Basin, P.G. Telford and H.M. Verma (ed); Ontario Geological Survey, Study 21, p. 21-50.

Telford, P.G. and Sawacki, D.G.
1983: No. 529 Lignite assessment project, Moose River Basin, James Bay Lowland; Ontario Geological Survey, Miscellaneous Paper 116, p. 126-127.

Tillement, B.A.
1975: Hydrocarbon potential of Hudson Bay Basin–1975 evaluation; Report prepared by Aquitaine Company of Canada Ltd., on behalf of the Aquitaine-Acro Group. Available from Information Services, Canterra Energy Ltd., 202, 505-4th Anvenue S.W., Calgary, 35 p.

Try, C.F., Long, D.G.F., and Winder, C.F.
1984: Sedimentology of the Lower Cretaceous Mattagami Formation, Moose River Basin, James Bay Lowlands, Ontario, Canada; in The Mesozoic of Middle North America, D.F. Stott and D.J. Glass (ed.); Canadian Society of Petroleum Geologists, Memoir 9, p. 345-359.

Trusler, J.R., and others.
1974: Onakawana lignite area, District of Cochrane; Ontario Division of Mines, Open File Report 5111.

Vos, M.A.
1982: Lignite and industrial mineral resources of the Moose River Basin; in Mesozoic Geology and Mineral Potential of the Moose River Basin, P.G. Telford and H.M. Verma (ed.); Ontario Geological Survey, Study 21, p. 135-190.

Wade, J.A., Grant, A.C., Sanford, B.V., and Barss, M.S.
1977: Basement structure (of) eastern Canada and adjacent areas; Geological Survey of Canada, Map 1400A (4 sheets), scale 1:2 000 000.

Winder, C.G., Telford, P.G., Verma, H., Fyfe, W.S., and Long, D.G.F.
1982: Flyluvial model for Lower Cretaceous lignite, northern Ontario; American Association of Petroleum Geologists, v. 66, no. 5, p. 643.

Authors' Addresses

B.V. Sanford (Retired)
17 Meadowglade Gardens
Nepean, Ontario
K2G 5J4

A.W. Norris
Institute of Sedimentary and Petroleum Geology
3303 – 33rd Street N.W.
Calgary, Alberta
T2L 2A7

A.R. Cameron (Retired)
Institute of Sedimentary and Petroleum Geology
3303 – 33rd Street N.W.
Calgary, Alberta
T2L 2A7

Printed in Canada

CHAPTER 10

ST. LAWRENCE PLATFORM – INTRODUCTION

Chapter 10

ST. LAWRENCE PLATFORM - INTRODUCTION

B.V. Sanford

GENERAL STATEMENT

The St. Lawrence Platform borders the Canadian Shield in southeastern Canada and adjacent areas of the United States, and is in turn bounded on the south by the Appalachian Orogen (Fig. 10.1). The Upper Precambrian to Carboniferous (and locally Jurassic) rocks within the platform consist of carbonates, shales, evaporites, and sandstones that dip at low angles southward, either into Michigan Basin, or into Allegheny, Quebec, and Anticosti foreland basins that parallel the Appalachian structural front (Fig. 10.2, see also Sanford et al., 1979). The strata are for the most part undeformed but are locally folded and faulted. Deformation is most evident near the southeastern platform margin where the beds were complexly faulted and folded by overriding thrust slices during various episodes of Paleozoic orogeny.

This account provides a summary of the depositional and tectonic history of the region. Brief reference is made to events in the Appalachian Orogen, Hudson Platform, Interior Platform, and adjoining regions of the United States. For complete treatment of these regions, the reader is referred to the appropriate chapters of this volume, and to the companion volumes of the Decade of North American Geology series (Williams and Neale, in prep.; Sloss, 1988).

Acknowledgments

Grateful acknowledgment is extended to M.J. Copeland, T.E. Bolton, and T.T. Uyeno for their contribution "Diagnostic Shelly Fauna and Conodont Biostratigraphy" in Chapter 11. Early drafts of the manuscripts benefitted substantially from critical reading by F. Frey and T.T. Uyeno.

REGIONAL ELEMENTS

The Paleozoic-Precambrian contact extending from northern Michigan to the Strait of Belle Isle is a highly irregular feature reflecting the northwest- and northeast-trending basement arches that were rejuvenated from time to time during the Phanerozoic Eon. The periodic arch movements appear to have coincided in many instances with events at the Taconic/Appalachian continental margin, and may have been causally related to them.

St. Lawrence Platform is limited on the southeast by the Appalachian Orogen, which extends from the southeastern United States to Newfoundland. A major sag occurs beneath the northern Gulf of St. Lawrence where Appalachian deformed rocks are buried beneath Carboniferous rocks of Magdalen Basin.

The original depositional limits of St. Lawrence Platform during the development of the Appalachian Miogeocline and subsequent foreland basins are imprecisely known. The early Paleozoic continent-ocean transition may coincide with a major gravity gradient, negative on the continental side, positive on the oceanic (see Fig. 11.14; Haworth et al., 1980).

Paleozoic outliers at widely separated localities of the Canadian Shield indicate that St. Lawrence Platform is now but an erosional remnant of a much broader cratonic cover that once connected with the Hudson and Interior platforms to the north and west respectively. Broad uplift of the Canadian Shield at various times during the Phanerozoic Eon has resulted in widespread erosion of lower and middle Paleozoic strata, reducing their areal distribution to three principal divisions:

i. Western St. Lawrence Platform,
ii. Central St. Lawrence Platform, and
iii. Eastern St. Lawrence Platform.

The sedimentary record of Late Proterozoic to Carboniferous times preserved in St. Lawrence Platform consists of nine depositional cycles. The distribution of these nine stratigraphic packages and locally preserved Jurassic rocks is shown in Figure 10.1

Western St. Lawrence Platform

Western St. Lawrence Platform is separated from the central division by the southeast-trending Frontenac Arch, a positive basement feature connecting the Canadian Shield with the Adirondack Inlier of New York. Contained within the western division are two sedimentary basins, the foreland Allegheny Basin on the south, and the intracratonic Michigan Basin to the northwest (Fig. 10.2). The two basins are separated by the structurally prominent, northeast-trending Findlay and Algonquin arches. Bordering and overlying the arches in southwestern

Sanford, B.V.
1993: St. Lawrence Platform - Introduction; Chapter 10 in Sedimentary Cover of the Craton in Canada, D.F. Stott and J.D. Aitken (ed.); Geological Survey of Canada, Geology of Canada, no. 5, p. 709-722 (also Geological Society of America, The Geology of North America, v. D-1).

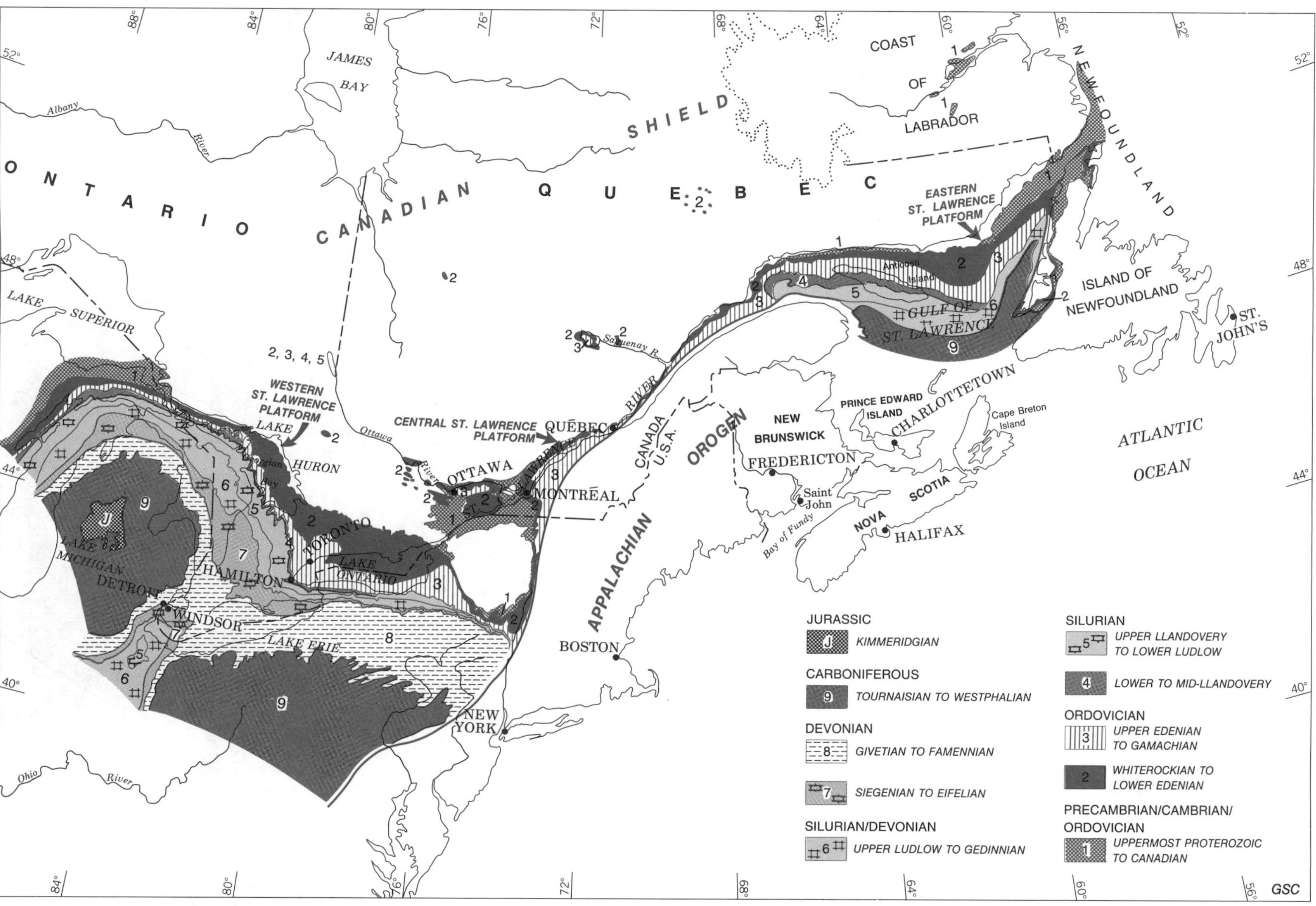

Figure 10.1. Geology of St. Lawrence Platform, showing depositional cycles.

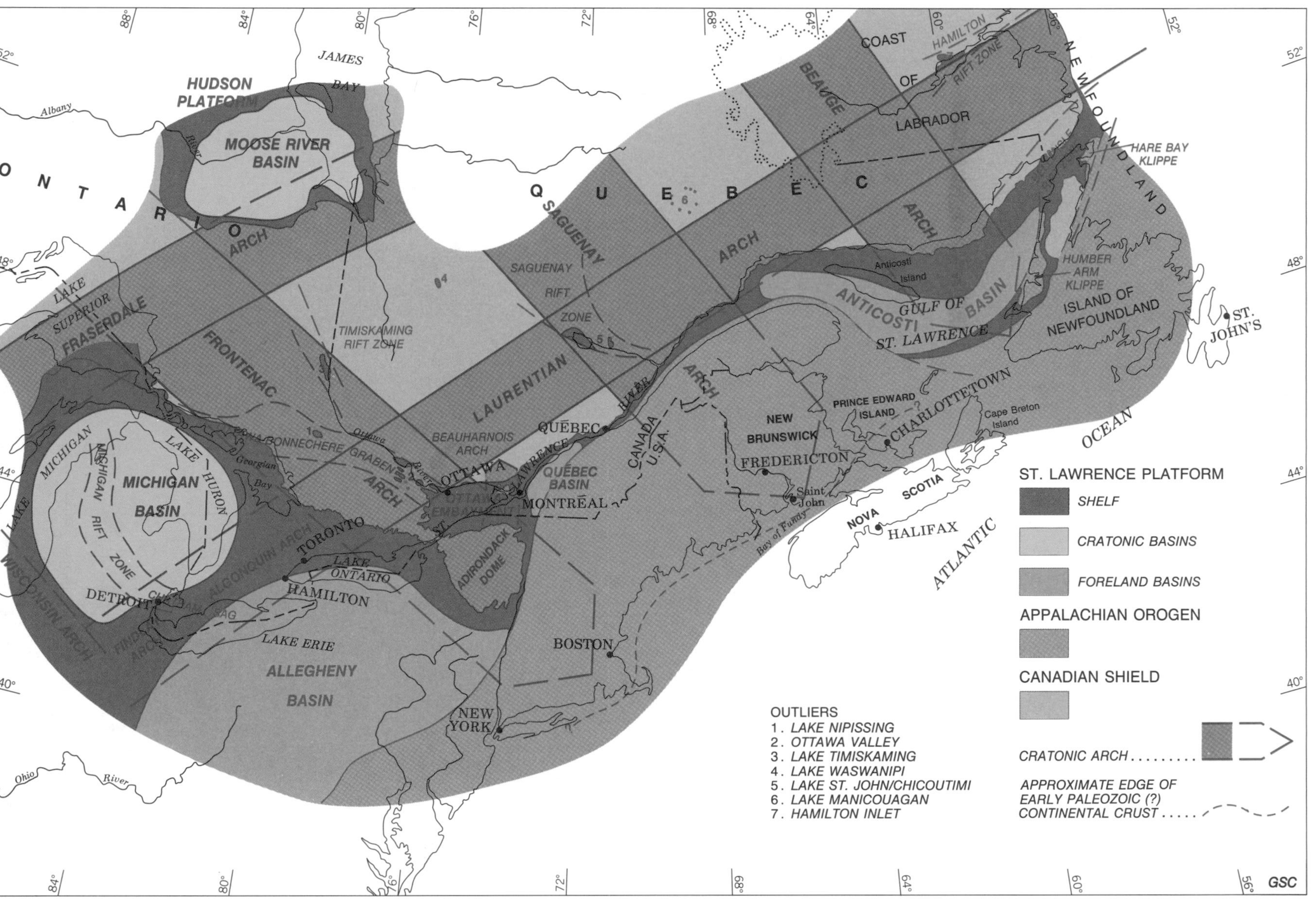

Figure 10.2. Principal tectonic elements of St. Lawrence Platform.

Ontario, Cambrian to Devonian rocks are generally less than 1500 m thick, but these, plus younger strata including Carboniferous, increase in thickness to more than 4500 m at the centre of the epicratonic Michigan Basin, and to 7000 m in the foreland Allegheny Basin near the Appalachian structural front.

In the Canadian sector of Western St. Lawrence Platform (southwestern Ontario), a comprehensive geoscience data base has been assembled as a result of longterm studies of natural bedrock exposures and rock quarries, borehole drilling, and geophysical surveys. Natural bedrock outcrops occur along many of the stream systems and the shores of Great Lakes Huron, Erie, and Ontario, and rock quarries in large numbers are scattered throughout the countryside. The best natural exposures are along the Niagara Escarpment (Fig. 10.3) and certain segments of the partly buried escarpments that trend northwesterly across southwestern Ontario and beneath the adjacent Great Lakes. Such prominent features in the latter areas are readily recognizable on bathymetric charts and marine seismic profiles, and permit the extrapolation and regional mapping of rock units offshore. Surface bedrock studies coupled with data from several thousand boreholes drilled during exploration for oil and gas are the principal sources of information for southwestern Ontario and adjacent areas of the United States.

Magnetic, gravity, and seismic surveys have contributed to knowledge of the subsurface geology of southwestern Ontario. Such information is important in the interpretation of the structural geology, particularly in fracture mapping, delineation of collapse structures formed by salt leaching, and exploration for oil- and gas-bearing pinnacle reefs and related structures.

Central St. Lawrence Platform

Central St. Lawrence Platform is bounded on the west by Frontenac Arch, on the east by impingement of the Appalachian front against Saguenay Arch, and on the north by the northeast-trending Laurentian Arch. The southeastern boundary of the platform is Logan's Line, the northwestern edge of allochthonous Ordovician rocks of the Appalachian Orogen.

Figure 10.3. Resistant Silurian carbonates form the lip of Niagara Falls. The deep gorge below is carved in shale, sandstone and carbonates of Early Silurian and Late Ordovician age (photo - R.D. Hutt, GSC 200934).

The stratigraphic sequences underlying the region are contained in the foreland Quebec Basin and its extension, Ottawa Embayment. Post-Ordovician uplift of Beauharnois Arch, possibly a northeast-trending arm of the Adirondack Dome, elevated the eastern margin of Ottawa Embayment to form what is now a closed structural depression (Fig. 10.2, 11.4).

Upper Proterozoic(?), Cambrian, and Ordovician rocks within Ottawa Embayment have a maximum thickness of about 1000 m. Quebec Basin, now in the form of a northeast-trending syncline, contains a similar, but much thicker sequence, perhaps in the order of 3000 m or more in the south.

Lower Devonian (Gedinnian and Siegenian) and Middle Devonian (Givetian) limestone and sandstone blocks occur in a diatreme breccia at St. Helen's Island in Montreal. Although the youngest mappable rock units in Quebec Basin are of Late Ordovician age, the brecciated remnants clearly indicate that Lower and Middle Devonian strata, and possibly even younger Devonian rocks, were also deposited there.

Geological knowledge of Ottawa Embayment and Quebec Basin is based on the numerous bedrock exposures that occur along the river systems, and the operating and abandoned quarries scattered throughout the region. Widespread normal faulting, particularly along the margins of Quebec Basin and in Ottawa Embayment, has rotated large blocks of Paleozoic strata; the uptilted edges of these blocks form low escarpments above the lowland terrain. These escarpments provide much additional control for the lithological, biostratigraphical, and structural framework of the region.

Because of the low potential for oil and gas in Ottawa Embayment, relatively few wells have been drilled. More encouraging showings of oil and gas have been encountered in Quebec Basin, where exploration drilling has been somewhat more systematic. Even so, reliable stratigraphic information is available for relatively few boreholes completed to 150 m or more (see Quebec Department of Natural Resources, 1974). However, some boreholes do penetrate the complete Paleozoic succession in the deeper parts of the basin and provide invaluable information.

Regional gravity and aeromagnetic coverage of the central platform facilitate interpretation of the composition and structure of the underlying basement rocks. Magnetic maps have practical application also in the recognition of fractures and fracture patterns in the Paleozoic strata. Reflection seismic data are particularly helpful in the mapping of subsurface structure, but only a limited amount of this information is publicly available.

Eastern St. Lawrence Platform

Eastern St. Lawrence Platform is separated from the central division by Saguenay Arch and is bounded on the north by Laurentian Arch. The platform is open to the east (Labrador Shelf) but narrows at the Strait of Belle Isle and in adjacent Newfoundland, where it has been deformed and overridden by Appalachian allochthons. The southwestern boundary can be roughly mapped offshore of Gaspé coast on hydrographic charts. To the east beneath the Gulf of St. Lawrence, deformed rocks are present beneath a Carboniferous cover. In this region, the southeastern margin can only be inferred from regional gravity and magnetic anomaly maps.

Much of Eastern St. Lawrence Platform is occupied by the foreland Anticosti Basin, the latter extending from the northwest extremity of the Gulf of St. Lawrence to western Newfoundland (Fig. 10.2). The basin-fill ranges in age from Late Proterozoic(?) and Early Cambrian to Late Silurian.

The southeast-trending Beaugé Arch divides Anticosti Basin into two major segments. The deeper, western part contains Cambrian to Carboniferous strata probably in excess of 6000 m, whereas strata of the same age-span in the eastern segment off southwestern Newfoundland are up to 4000 m thick.

Paleozoic rocks in Eastern St. Lawrence Platform have been studied extensively at four principal localities:

i. Southeastern Labrador bordering the Strait of Belle Isle,
ii. West coast of Newfoundland
iii. Anticosti Island, and
iv. Mingan Islands and the adjacent coast of mainland Quebec.

A fairly complete Paleozoic succession ranging in age from Early Cambrian to Late Silurian has been pieced together from the excellent coastal exposures in southeastern Labrador and western Newfoundland, and in uplifted blocks bordering the Great Northern Highlands of Newfoundland. Lower Ordovician to Lower Silurian rocks have been studied in detail on Mingan Islands and Anticosti Island. Wells drilled during oil and gas exploration on Anticosti Island are another major source of geological data. Of the eight holes completed to date, all but one completely penetrated the Paleozoic succession.

In the offshore region between Anticosti Island, western Newfoundland, and the Strait of Belle Isle, marine surveys by government and private agencies have generated a comprehensive gridwork of reflection seismic, magnetic, gravity, and shallow borehole data. This information, in conjunction with bathymetric charts, has greatly facilitated the extrapolation of the onshore geology into the broad offshore regions of St. Lawrence Platform.

Paleozoic outliers

Outliers of Paleozoic rocks rest upon rocks of the Canadian Shield north of St. Lawrence Platform (Fig. 10.1). Most are small and exhibit only a few feet of flat-lying to gently tilted strata. Several, however, including the Lake Timiskaming, Lac St. Jean, and Ottawa Valley outliers, are structurally preserved in down-faulted blocks, and consequently contain relatively thick successions. The Lake Timiskaming outlier, perhaps the best known, measures 40 by 15 km and contains more than 250 m of strata ranging from Middle Ordovician to Lower Silurian.

Most of the outliers contain good bedrock exposures, thus contributing to lower Paleozoic paleogeographical reconstructions and regional correlation of rock units between St. Lawrence and Hudson platforms. Core from a deep borehole drilled through the complete Paleozoic succession at Lake Timiskaming has provided much new information, especially about certain rock units that are nowhere exposed at surface in the outlier.

HISTORY OF GEOLOGICAL EXPLORATION

Preliminary interpretation of the geology of St. Lawrence Platform began in the early 19th century. In 1829, Major General Baddeley of the Royal Engineers was the first to describe the Paleozoic limestones of Lac St. Jean and Murray Bay. At about this time also, Dr. J.J. Bigsby, Secretary to the Boundary Commission, carried out investigations between Quebec City and Lake Superior. His "Sketch of the geology of the island of Montreal" published in 1825, for example, is an important and accurate contribution.

With the founding of the Geological Survey of Canada in 1842, a program for the systematic mapping of Canada was begun. Sir William Logan, the Survey's first director, was assisted in his initial investigations by Alexander Murray, T. Sterry Hunt, James Richardson, and Robert Bell, all of whom made significant contributions to the knowledge of St. Lawrence Platform. In 1843, Alexander Murray investigated the region between Lake Erie and Georgian Bay and four years later began studies of Manitoulin Island. The New York Geological Survey, having been founded prior to the Geological Survey of Canada, already had established a classification for Paleozoic rocks. Alexander Murray successfully applied the New York classification to Ontario, setting up ten major stratigraphic divisions, most of which still apply.

Another of Logan's colleagues, the chemist and mineralogist T. Sterry Hunt, is remembered for his scientific contribution to petroleum geology. From his knowledge of oil seepages in the Gaspésie, he established the "Anticlinal Theory of Accumulation" in 1861. This basic concept was soon widely accepted and successfully applied to early oil exploration in Ontario and adjacent regions of the United States.

James Richardson, another of Logan's staff, visited Anticosti and Mingan Islands in 1857. He later investigated the Ottawa River from Pembroke to Grenville, and examined the north shore of the St. Lawrence River between Montreal and St. Maurice. Richardson also examined Paleozoic exposures in the Eastern Townships of Quebec, Gaspésie, and along the west coast of Newfoundland. He thus helped to establish the first regional correlation of rock units over a wide region of St. Lawrence Platform and elsewhere in eastern Canada.

Logan himself examined much of southwestern Ontario, the Ottawa-Quebec Lowland, and Mingan Islands. In 1863 he published "Geology of Canada", a monumental and still valuable work. Credit for Logan's contribution to the dating and regional correlation of Phanerozoic rocks must be shared with Elkanah Billings, Survey paleontologist, who had the important and laborious task of identifying and describing the huge fossil collections.

During the latter part of the 19th century, geological investigations of a systematic nature in St. Lawrence Platform were curtailed, although significant contributions were made by Robert Bell on Manitoulin Island in 1865 (Bell, 1869), R.W. Ells in the Eastern Townships of Quebec (Ells, 1896) and in various parts of Ontario during the years 1886 to 1906, and A.P. Low (1892) in parts of Portneuf and Montmorency counties, Quebec, in 1890 and 1891. In 1876, Professor E.J. Chapman published "Outline of the Geology of Canada", which included important observations on the geology of St. Lawrence Platform.

Since the beginning of this century extensive geological investigations have been conducted in St. Lawrence Platform. In central Ontario east of the Niagara Escarpment, W.A. Johnson (1912, 1914), W.A. Parks (1925, 1928), and V.J. Okulitch (1938) provided detailed lithological and biostratigraphical classification of the Middle and Upper Ordovician sequences respectively, which have for the most part remained valid to the present day. In southwestern Ontario (Niagara Escarpment and westward), M.Y. Williams (1919) and C.R. Stauffer (1915) investigated the Silurian and Devonian systems respectively, and published the first detailed geological maps and reports.

From 1936 to 1947, systematic mapping of the Paleozoic rocks of southwestern Ontario was carried out by J.F. Caley. The results of these surface and subsurface investigations (Caley, 1940, 1941, 1943, 1945) were a valuable addition to the geology of Ontario and a useful guide to economic minerals and fuels.

The studies initiated by Caley were continued by B.V. Sanford from 1949 to 1985, incorporating new information from oil and gas exploration activities. This work has resulted in the publication of revised geological and oil and gas field maps of southwestern Ontario, and reports describing the subsurface geology and economic potential of the Cambrian, Ordovician, Silurian, and Devonian systems (Sanford, 1961, 1965, 1968, 1969, 1977; Sanford and Quillian, 1959; and Sanford et al., 1985). Comprehensive stratigraphic investigations have also been carried out in Ontario on the Ordovician System by B.A. Liberty (1969), on the Silurian by T.E. Bolton (1957) and on the Devonian by T.T. Uyeno (1982).

In 1975, P.G. Telford and colleagues of the Ontario Geological Survey began a program of systematic remapping of the Paleozoic rocks of Ontario, and that program is continuing.

In the central St. Lawrence Platform, contributions to Ordovician stratigraphy and biostratigraphy were made by A.E. Wilson and T.H. Clark. Wilson made an exhaustive lifetime study of the Ottawa region, culminating (1964) in publication of "Geology of the Ottawa-St. Lawrence Lowland, Ontario and Quebec". In 1938, T.H. Clark began a thirty-year study of the Quebec Lowland for the Quebec Department of Mines. His investigations resulted in the present comprehensive classification of the Cambrian(?) and Ordovician systems, and modern detailed maps and reports (1965, 1972). In addition to Clark's studies, significant lithological and biostratigraphical contributions were made by H.R. Belyea (1952), H.J. Hofman (1963, 1972, 1979), F.F. Osborne (1956), J. Riva (1969, 1972), Y. Globensky and J.C. Jauffred (1971), A.E. Foerste (1924), and others.

In the eastern platform, the first comprehensive investigations on Anticosti and Mingan Islands during this century were begun in 1909 by W.H. Twenhofel. His studies resulted in the publication (1928) of "Geology of Anticosti Island", followed (1938) by "Geology and Paleontology of the Mingan Islands, Quebec". No further detailed field investigations on Anticosti Island were carried out until 1957, when Bolton (1961) began a revision of the stratigraphy and paleontology of the Ordovician and Silurian formations. More recent investigations of Anticosti Island include those of Petryk (1976, 1981a, b) on the stratigraphic and sedimentology framework and oil and gas potential. Additional information on the biostratigraphic framework of the Ordovician and Silurian formations has been provided by Barnes et al. (1978, 1981a, b), Nowlan (1980, 1981), Nowlan and Barnes (1981), Bolton (1966, 1972, 1981a, b), Bolton and Copeland (1972), Bolton and Nowlan (1979), Uyeno and Barnes (1983), McCracken (1981), McCracken and Barnes (1981), Copeland (1970, 1973, 1974, 1976, 1981), and Copeland and Bolton (1975).

Paleozoic and older rocks of western Newfoundland and Labrador have been examined at various times since the first mapping of the region, begun by James Richardson in 1860. No further comprehensive mapping was undertaken until 1910, when C. Schuchert and W.H. Twenhofel carried out a two-month survey of western Newfoundland. The work was continued in 1918 and 1920 by Schuchert and C.O. Dunbar, and published in 1934.

Since that time, a host of individual projects have been carried out along the west coast of Newfoundland, southeast Labrador and on Belle Isle. Some of the more recent and comprehensive of these investigations are those of Bergström et al. (1974), Boyce (1978, 1979, 1983, 1985), Collins and Smith (1975), Copeland and Bolton (1977), Cumming (1983), Fåhraeus (1973, 1974), Fåhraeus and Nowlan (1978), Flower (1978), Fortey (1979), James and Stevens (1979, 1982), Klappa et al. (1980), Kluyver (1975), Knight (1977, 1983), Lane (1984), Nelson (1955), Pratt and James (1982), Riley (1962), Rodgers (1965, 1968, 1971), Stouge (1982, 1984), Whittington and Kindle (1963, 1966, 1969), Williams (1984), and Williams and Smyth (1983).

Offshore mapping of Paleozoic rocks in the northeastern Gulf of St. Lawrence, based on shallow high-resolution seismic reflection surveys, was carried out by Shearer (1973). More recent mapping, using similar seismic techniques augmented by borehole drilling was conducted by Haworth and Sanford (1976).

Geological investigations of Paleozoic outliers (Lac St. Jean and Chicoutimi) were first published by Frederick Baddeley (1829). These outliers have since been investigated by a number of workers including James Richardson (1857), Dresser (1916), Denis (1933), Sinclair (1953), and Harland et al. (1985). The Lake Timiskaming outlier was first investigated by Robert Bell (1894) and later mapped by Barlow (1897). In 1916 G.S. Hume began a comprehensive study of the region culminating in publication of "The Paleozoic outlier of Lake Timiskaming, Ontario and Quebec" (1925). Some recent publications pertaining to the stratigraphy and paleontology of the outlier are by Bolton and Copeland (1972), Ollerenshaw and MacQueen (1960), Copeland (1965), and Sinclair (1965). The most recent geological mapping of the Lake Timiskaming outlier is by Russell (1984). Perhaps the most significant single contribution dealing with the Ottawa Valley outliers is by Kay (1942). Contributions pertaining to the age and correlation of the Waswanipi and

Manicouagan outliers are included in Quebec Department of Mines Geological Report 59 (Blake, 1953) and Geological Survey of Canada Bulletin 134 (Bolton, 1965).

PHYSIOGRAPHIC ELEMENTS

The main physiographic elements of St. Lawrence Platform are: Western St. Lawrence Lowland and adjacent Allegheny Plateau; Central St. Lawrence Lowland; and Eastern St. Lawrence Lowland and adjacent Blow Me Down Highlands, St. John Highlands, and Great Northern Highlands (Fig. 10.4).

The more widespread and consistent of these elements are the Western, Central, and Eastern St. Lawrence Lowlands, which extend from Michigan and the Great Lakes region northeastward along the St. Lawrence River and northern Gulf of St. Lawrence to western Newfoundland, a distance of some 2400 km. All St. Lawrence Lowland segments have many physiographic characteristics in common, chief among them are large expanses of low, uniformly flat terrain that reflect the relatively undisturbed character of the underlying sedimentary strata. Local and regional differences in topographical relief are due to one or more of the following: variations in weathering characteristics of the underlying Paleozoic formations; faulting and other structural phenomena; glacial scouring; stream erosion; variations in composition and thickness of glacial and interglacial deposits; and to some small degree, modification of terrain in the western and central lowlands by man, through prolonged agricultural use and industrial development, during the past three hundred years.

In sharp contrast to the St. Lawrence Lowlands are plateaux and highland terrains bordering the St. Lawrence Platform on the northwest and west (Canadian Shield) and the southeast and east (frontal highland and plateaux of the Appalachian Orogen).

Although Quaternary deposits, modified by fluvial and marine deposition and by erosion, have strongly affected the shape of the present landscape, the larger scale landforms are controlled by bedrock formations. A description of the Quaternary deposits of the St. Lawrence Lowlands and adjacent physiographic elements is provided in Chapter 11. The treatment of physiographic elements in this chapter is restricted to a brief summary of the role of Paleozoic formations in shaping the more regional aspects of the topography, including the offshore regions of the Great Lakes and the northern Gulf of St. Lawrence.

Western St. Lawrence Lowland

The northeastern boundary of Western St. Lawrence Lowland essentially parallels the erosional limit of relatively flat-lying Paleozoic strata where they lap onto the southern margins of the Precambrian Penokean Hills and Laurentian Highlands.

The configuration of Western St. Lawrence Lowland is controlled in large part by the Algonquin Arch, which extends from the Lake Simcoe region southwestward to the vicinity of Lake St. Clair. Surface elevations over the higher parts of the arch near Georgian Bay reach 518 m, decreasing to the northwest and southeast to 177 m and 174 m, the levels of Lake Huron and Lake Erie, respectively.

The surface expression of Findlay Arch, which is southwest of the Algonquin Arch, is much less pronounced in Ontario than in Ohio and Michigan, but is reflected by shallow water in Lake Erie and slightly elevated topography onshore.

The effects of preglacial erosion of the Paleozoic formations are readily apparent in Ontario and adjacent Michigan. Spectacular features are the north- and northeast-facing escarpments, most of which formed in resistant carbonate rocks underlain by recessive-weathering shales. The five principal escarpments and one lesser one are as follows (Fig. 4.10):

i. Black River Escarpment, marking the Paleozoic-Precambrian boundary from the head of St. Lawrence River to Georgian Bay and continuing beneath the waters of Lake Huron;

ii. Georgian Bay Escarpment, extending intermittently from southern Lake Ontario to Manitoulin Island;

iii. Niagara Escarpment, beginning at Rochester, New York and extending westward to Hamilton, thence northwestward to Bruce Peninsula and Manitoulin Island;

iv. Bois Blanc Escarpment (also known as the Onondaga Escarpment), marking the Silurian- Devonian boundary from Niagara River to Lake Huron, thence beneath the lake where it is known as Six Fathom Scarp;

v. Hamilton Escarpment, beginning beneath eastern Lake Erie and extending to Lake Huron, thence beneath the lake where it becomes Ipperwash Scarp; and

vi. An escarpment carved into Hamilton Group shales and carbonates, which occurs intermittently across the crest of the Findlay Arch.

The conspicuous Black River and Niagara escarpments extend across the Paleozoic terrain of Ontario. The Niagara Escarpment, in particular, forms a spectacular geological feature up to 77 m high, dividing the terrain into two: the southwestern and south-central lowlands. All other escarpments are partly buried beneath Quaternary deposits, but are prominently exposed at some onshore localities. Several escarpments also form major physiographic elements beneath the waters of Lakes Ontario and Huron.

The bedrock surface in southwestern and south-central Ontario has been significantly modified locally by glacial scouring. In the Great Lakes are deep basins presumably formed as a result of ice scouring. Some of the larger scoured features illustrated in Figure 10.4 are: Long Point Basin in Lake Erie; Ontario Basin in Lake Ontario; and Georgian and Manitoulin basins in Lake Huron.

Allegheny Plateau

Allegheny Plateau, adjoining Western St. Lawrence Lowland on the south, rises abruptly from 200 m along the south shore of Lake Erie to elevations of over 700 m, and is extensively dissected by major rivers. The rocks of the plateau are deformed and are part of the Appalachian Orogen.

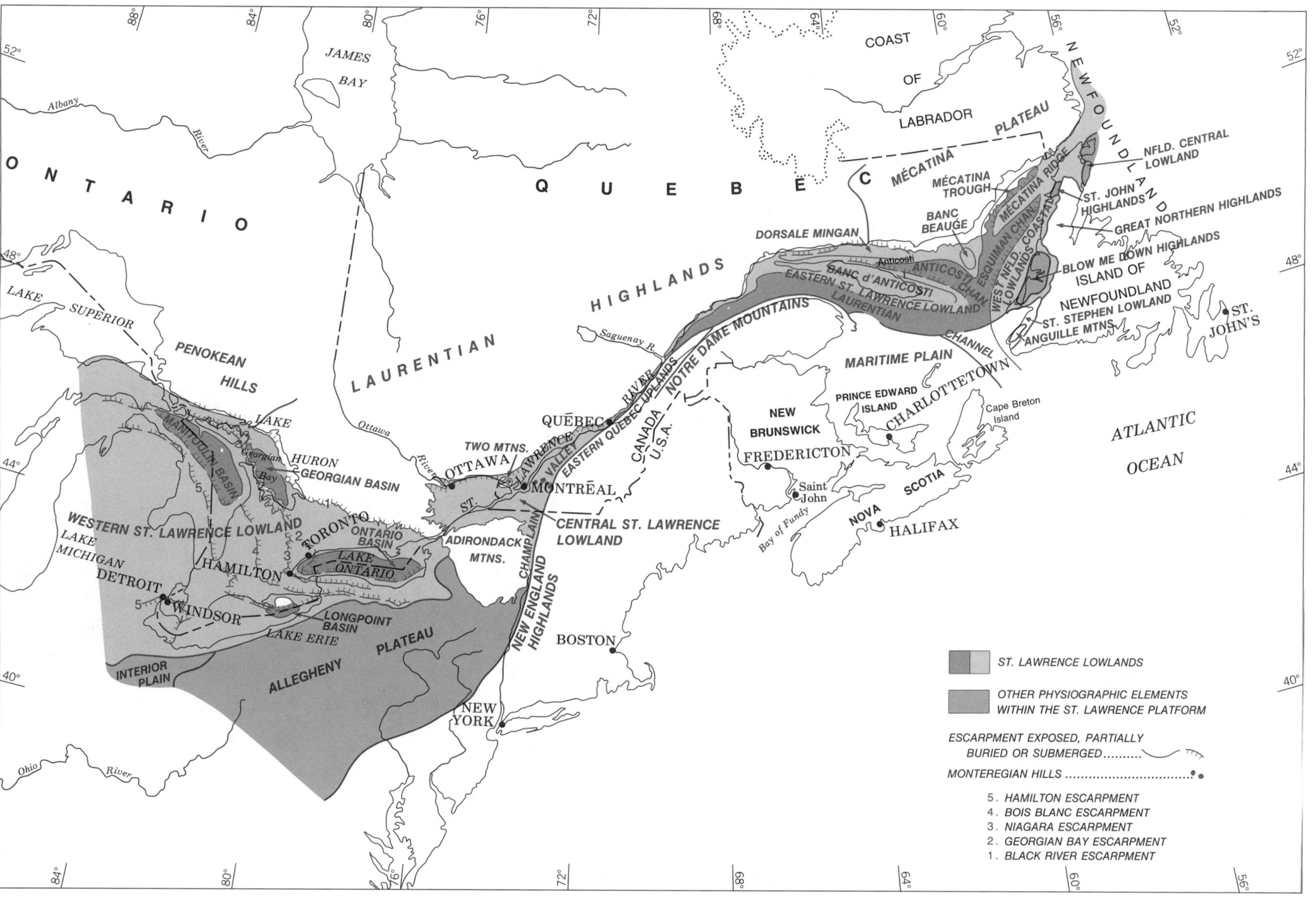

Figure 10.4. Physiographic elements of St. Lawrence Platform.

Central St. Lawrence Lowland

Central St. Lawrence Lowland is bounded on the north and west by Precambrian rocks of the Laurentian Highlands; on the southwest by the Adirondack Mountains; and on the southeast by the New England Highlands and Eastern Quebec Uplands of the Appalachian Orogen (Fig. 10.4).

The Ottawa-St. Lawrence Lowland, coinciding with the Ottawa Embayment tectonic sub-province, varies in elevation from 45 to 90 m and is almost completely surrounded by Precambrian terrain. Quebec Lowland and Champlain Valley to the east, largely coincident with Quebec Basin, extend from the vicinity of Two Mountains (Precambrian rocks of Beauharnois Arch) and Adirondack Mountains eastward to Quebec City. With the exception of six Monteregian Hills (denuded intrusions), some of which exceed 300 m in elevation, the lowland area of Quebec lies at less than 30 m along the St. Lawrence, rising to 90 m or more north and south of the river. In this region, as well as the Ottawa-St. Lawrence Lowland, Paleozoic rocks are commonly in fault contact with Precambrian rocks of the Canadian Shield to the north. In such areas, Precambrian rocks form escarpments above the lowland areas.

Widespread normal faulting that has occurred throughout Central St. Lawrence Lowland has rotated large blocks of Paleozoic strata so that their uptilted edges form low escarpments and other interruptions of the otherwise flat terrain. Many of the bedrock irregularities, reflecting tectonic structures, igneous intrusions, pre-glacial erosion, and glacial scouring have been masked, in part at least, by Quaternary glacial and marine (Champlain Sea) deposits. The Quaternary sediments in turn have undergone widespread erosional modification by Recent and contemporary stream erosion.

Eastern St. Lawrence Lowland

For more than 400 km along St. Lawrence River and its estuary below Quebec City, autochthonous Paleozoic strata are confined to a narrow strip 10 to 30 km wide, largely concealed beneath the river. Where the river widens into the northern Gulf of St. Lawrence, Paleozoic platformal strata also widen to form Eastern St. Lawrence Lowland. This terrain continues eastward beneath the gulf, rising to form Anticosti Island, Mingan Islands, parts of western Newfoundland, and a narrow belt along the southeast coast of Labrador bordering the Strait of Belle Isle (Fig. 10.4).

Eastern St. Lawrence Lowland is bordered on the northwest by Precambrian terrain of the Laurentian Highlands and Mecatina Plateau, on the southwest by the Notre Dame Mountains of northern Gaspésie, on the south by the Maritime Plain and St. Stephen Lowland, and on the east by the Blow Me Down Highlands, St. John Highlands, and Great Northern Highlands of Newfoundland.

The Paleozoic platformal succession underlying the northern Gulf of St. Lawrence and adjacent onshore regions is composed for the most part of alternating units of carbonate, shale, and sandstone. The carbonate and sandstone units, being the more resistant, form the offshore ridges and banks, and the shale units the major elongated depressions on the seafloor, herein called channels and troughs (Fig. 10.4).

In the western part of the lowland are two prominent positive bathymetric features, Dorsale Mingan and Banc d'Anticosti. Dorsale Mingan, some 350 km long, consists of resistant Lower and Middle Ordovician carbonate rocks. These rise above the surface of the Gulf at the mid-point of the ridge, to form the Mingan Islands and adjacent north shore of Quebec. The Mingan Islands are some twenty-two in number. The strata underlying the islands dip gently south, so that most have steep cliffs on their landward side.

South of Dorsale Mingan, the southeast-trending Banc d'Anticosti extends for some 425 km across the northern Gulf of St. Lawrence. The Ordovician and Silurian carbonates that support the bank rise above the Gulf to form Anticosti Island. The carbonates forming the bank are underlain by a thick Ordovician shale succession whose erosion accounts for the northeast-facing, submarine escarpment immediately to the north. The conspicuously terraced terrain of Anticosti Island reaches elevations above 300 m.

The conspicuous, southeast-elongated depressions on the seafloor north and south of Banc d'Anticosti are named Anticosti Channel and Laurentian Channel respectively. Both are undoubtedly pre-glacial stream valleys, modified by glacial scouring.

The West Newfoundland Coastal Lowland and the coast of Labrador bordering the Strait of Belle Isle are the only other land areas in the Eastern St. Lawrence Lowland, They are both dissected by faults, particularly the West Newfoundland Coastal Lowland, where it is bounded on the east by Blow Me Down Highlands, St. John Highlands, and Great Northern Highlands.

In the adjacent offshore region of the northeastern Gulf of St. Lawrence, resistant Cambrian and Ordovician carbonates and sandstones form a southwest-trending prominence, Mecatina Ridge. Elongated depressions to the north and south are floored by less resistant sandstones and shales and are known as Mecatina Trough and Esquiman Channel, respectively. These, like the seafloor depressions in the western gulf, are assumed to be pre-glacial valleys modified by glacial scouring.

Banc Beaugé in the north-central Gulf of St. Lawrence is the subsea expression of the southeast-trending Beaugé Arch. Its configuration is partly attributable to resistant Middle Ordovician carbonates, and partly to the structural drape of these over a basement structural high.

Blow Me Down Highlands, St. John Highlands, and Great Northern Highlands

West Newfoundland Coastal Lowland, a part of Eastern St. Lawrence Lowland, is bounded abruptly along its southeast margin by the Blow Me Down Highlands, St. John Highlands, and Great Northern Highlands. These highland areas were of platformal character until incorporated into the orogen by emplacement of the Humber Arm allochthon in early Middle Ordovician (Chazyan) time.

REFERENCES

Baddeley, F.
1829: On the geognosy of a part of the Saguenay country; Quebec Literary and Historical Society Translations, v. 1, p. 79-166.

Barlow, A.E.
1897: Report on the geology and natural resources of the area included in the Nipissing and Timiskaming map sheets, emprising portions of the district of Nipissing, Ontario, and of the county of Pontiac, Quebec; Geological Survey of Canada, Annual Report, v. X, pt. 1, 302 p.

Barnes, C.R., Norford, B.S., and Skevington, D.
1981a: The Ordovician System in Canada; International Union of Geological Sciences, Publication No. 8, 26 p.

Barnes, C.R., Petryk, A.A., and Bolton, T.E.
1981b: Anticosti Island, Quebec; in Volume I: Stratigraphy and Paleontology, P.J. Lespérance (ed.), IUGS Subcommission on Silurian Stratigraphy, Ordovician-Silurian Boundary Working Group; Field Meeting, Anticosti-Gaspé, Quebec, 1981, Guidebook, p. 1-24.

Barnes, C.R., Telford, P.G., and Tarrant, G.A.
1978: Ordovician and Silurian conodont biostratigraphy, Manitoulin Island and Bruce Peninsula, Ontario; Michigan Basin Geological Society, Special Paper No. 3, p. 63-71.

Belle, R.N.
1869: Report on the Manitoulin Islands; Geological Survey of Canada, Report of Progress 1866-1869.
1894: Pre-Paleozoic decay of crystalline rocks north of Lake Huron; Geological Society of America Bulletin, v. 5, p. 357-366.

Belyea, H.R.
1952: Deep wells and subsurface stratigraphy of part of the St. Lawrence Lowlands, Quebec; Geological Survey of Canada, Bulletin 22, 113 p.

Bergström, S.M., Riva, J., and Kay, M.
1974: Significance of conodonts, graptolites and shelly faunas from the Ordovician of western and north-central Newfoundland; Canadian Journal of Earth Sciences, v. 11, p. 1625-1660.

Bigsby, J.J.
1825: A sketch of the geology of the Island of Montreal; Lyceum Natural History of New York, Annals, v. 1, p. 198-219.

Blake, D.A.W.
1953: Waswanipi Lake area; Quebec Department of Mines, Geological Report 59, 23 p.

Bolton, T.E.
1957: Silurian stratigraphy and paleontology of the Niagara Escarpment in Ontario; Geological Survey of Canada, Memoir 289, 145 p.
1961: Ordovician and Silurian formations of Anticosti Island, Quebec; Geological Survey of Canada, Paper 61-26, 18 p.
1965: Ordovician and Silurian Tabulata corals *Labrinthites, Arcturia, Troedssonites, Muetisolenia* and *Boreaster*; in Contributions to Canadian Paleontology; Geological Survey of Canada, Bulletin 134, pt. II, p. 15-33.
1966: Illustrations of Canadian fossils: Silurian faunas of Ontario; Geological Survey of Canada, Paper 66-5, 7 p.
1972: Geological map and notes on the Ordovician and Silurian litho- and biostratigraphy, Anticosti Island, Quebec; Geological Survey of Canada, Paper 71-19, 44 p.
1981a: Late Ordovician and Early Silurian Anthozoa of Anticosti Island, Quebec; in Volume II: Stratigraphy and Paleontology, P.J. Lespérance (ed.), IUGS Subcommission on Silurian Stratigraphy, Ordovician-Silurian Boundary Working Group; Field Meeting, Anticosti-Gaspé, Quebec, 1981, p. 107-109.
1981b: Ordovician and Silurian Stratigraphy, Ordovician-Silurian Boundary Working Group; Field Meeting, Anticosti-Gaspé, Quebec, 1981, p. 41-46.

Bolton, T.E. and Copeland, M.J.
1972: Paleozoic formations and Silurian biostratigraphy, Lake Timiskaming region, Ontario and Quebec; Geological Survey of Canada, Paper 72-15, 48 p.

Bolton, T.E. and Nowlan, G.S.
1979: A Late Ordovician fossil assemblage from an outlier north of Aberdeen Lake, District of Keewatin; Geological Survey of Canada, Bulletin 321, p. 1-26.

Boyce, D.H.
1978: Recent developments in western Newfoundland Cambro-Ordovician trilobite biostratigraphy; Newfoundland Department of Mines and Energy, Report 78-1, p. 80-84.
1979: Further developments in western Newfoundland Cambro-Ordovician biostratigraphy; Newfoundland Department of Mines and Energy, Report 79-1, p. 7-10.
1983: Preliminary Ordovician trilobite biostratigraphy of the Eddies Cove west - Port au Choix area, western Newfoundland; in Current Research, Department of Mines and Energy, Government of Newfoundland and Labrador, Report 85-1, p. 60-70.

Caley, J.F.
1940: Paleozoic geology of the Toronto-Hamilton area, Ontario; Geological Survey of Canada, Memoir 224, 284 p.
1941: Paleozoic geology of the Brantford area, Ontario; Geological Survey of Canada, Memoir 226, 176 p.
1943: Paleozoic geology of the London area, Ontario; Geological Survey of Canada, Memoir 237, 171 p.
1945: Paleozoic geology of the Windsor-Sarnia area, Ontario; Geological Survey of Canada, Memoir 240, 227 p.

Chapman, E.J.
1876: An outline of the geology of Canada: based on a subdivision of the provinces into natural areas; Copp, Clark and Company, Toronto, 104 p.

Clark, T.H.
1965: Oil and gas potential of Quebec and the Maritime Provinces; Ontario Petroleum Institute, 4th Annual Conference, Technical Paper 11, 5 p.
1972: Stratigraphy and structure of the St. Lawrence Lowland of Quebec; XXIV International Geological Congress, Montreal, Quebec, Excurison C-52, 59 p.

Collins, J.A. and Smith, L.
1975: Zinc deposits related to diagenesis and intrakarstic sedimentation in the Lower Ordovician St. George Formation, western Newfoundland; Bulletin of Canadian Petroleum Geology, v. 23, p. 393-427.

Copeland, M.J.
1965: Ordovician Ostracoda from Lake Timiskaming, Ontario; Geological Survey of Canada, Bulletin 127, 5 p.
1970: Ostracoda from the Vauréal Formation (Upper Ordovician) of Anticosti Island, Quebec; in Contributions to Canadian Paleontology; Geological Survey of Canada, Bulletin 187, p. 15-29.
1973: Ostracoda from the Ellis Bay Formation (Ordovician), Anticosti Island, Quebec; Geological Survey of Canada, Paper 72-43, 28 p.
1974: Silurian Ostracoda from Anticosti Island, Quebec; Geological Survey of Canada, Bulletin 241, 73 p.
1976: Leperditicopid ostracodes as Silurian biostratigraphic indices; in Report of Activities, Part B, Geological Survey of Canada, Paper 76-1B, p. 83-88.
1981: Latest Ordovician and Silurian ostracodes from Anticosti Island, Quebec; in Volume II: Stratigraphy and Paleontology, Lespérance, P.J. (ed.); IUGS Subcommission on Silurian Stratigraphy, Ordovician-Silurian Boundary Working Group; Field Meeting, Anticosti-Gaspé, Quebec, 1981, p. 185-195.

Copeland, M.J. and Bolton, T.E.
1975: Geology of the central part of Anticosti Island, Quebec; in Report of Activities, Part A, Geological Survey of Canada, Paper 75-1A, p. 519-523.
1977: Additional paleontological observations bearing on the age of the Lourdes Formation (Ordovician), Port au Port Peninsula, western Newfoundland; in Report of Activities, Part B, Geological Survey of Canada, Paper 77-1B, p. 1-13.

Cumming, L.M.
1983: Lower Paleozoic autochthonous strata of the Strait of Belle Isle area; in Geology of the Strait of Belle Isle area, northwestern insular Newfoundland, southern Labrador and adjacent Quebec; Geological Survey of Canada, Memoir 400, part 2, p. 75-108.

Denis, B.T.
1933: The Simard map area; Quebec Bureau of Mines Report for 1932, part D, p. 53-81.

Dresser, J.A.
1916: Part of the district of Lake St. John; Geological Survey of Canada, Memoir 92, 88 p.

Ells, R.W.
1896: Report of a portion of the Province of Quebec comprised in the southwest sheet of the "Eastern Townships" map (Montreal Sheet); Geological Survey of Canada, Annual Report 7, part J, p. 1-92.

Eyles, C.H. and Eyles, N.
1983: Sedimentation in a large lake; a reinterpretation of the late Pleistocene stratigraphy at Scarborough Bluffs, Ontario, Canada; Geology, v. 11, p. 146-152.

Fåhraeus, L.E.
1973: Depositional environments and conodont-based correlation of the Long Point Formation (Middle Ordovician), western Newfoundland; Canadian Journal of Earth Sciences, v. 10, p. 1822-1833.
1974: Lower Paleozoic stratigraphy of the Port au Port area, west Newfoundland; in Geological Association of Canada/Mineralogical Association of Canada, R.K. Stevens (ed.); Field Trip Manual, 13 p.

Fåhraeus, L.E. and Nowlan, G.S.
1978: Franconian (Late Cambrian) to early Champlainian (Middle Ordovician) conodonts from the Cow Head Group, Western Newfoundland; Journal of Paleontology, v. 52, no. 2, p. 444-471.

Flower, R.H.
1978: St. George and Table Head cephalopod zonation in western Newfoundland; in Current Research, Part A, Geological Survey of Canada, Paper 78-1A, p. 217-224.

Foerste, A.F.
1924: Upper Ordovician faunas of Ontario and Quebec; Geological Survey of Canada, Memoir 138, 255 p.

Fortey, R.A.
1979: Early Ordovician trilobites from the Catoche Formation (St. George Group), western Newfoundland; in Contributions to Canadian Paleontology; Geological Survey of Canada, Bulletin 321, p. 61-114.

Globensky, Y. and Jauffred, J.C.
1971: Stratigraphic distribution of conodonts in the Middle Ordovician Neuville section of Quebec; Geological Association of Canada Proceedings, v. 23, p. 43-68.

Harland, T.L., Pickerill, R.K., and Fillion, D.
1985: Ordovician intracratonic sediments from the Lac-St. Jean and Chicoutimi areas, Quebec, eastern Canada; Canadian Journal of Earth Sciences, v. 22, p. 240-255.

Haworth, R.T. and Sanford, B.V.
1976: Paleozoic geology of northeast Gulf of St. Lawrence; in Report of Activities, Part A; Geological Survey of Canada, Paper 76-1A, p. 1-6.

Haworth, R.T., Daniels, D.L., Williams, H., and Zietz, I.
1980: Bouguer gravity anomaly map of the Appalachian Orogen; Memorial University of Newfoundland, St. Johns, Newfoundland, Map No. 3a, scale 1: 2 000 000.

Hofmann, H.J.
1963: Ordovician Chazy Group in southern Quebec; Bulletin of American Association of Petroleum Geologists, v. 47, p. 270-301.
1972: Stratigraphy of the Montreal area; XXIV International Geological Congress, Montreal, Quebec, Excursion B-03, p. 27-59.
1979: Chazy (Middle Ordovician) trace fossils in the Ottawa-St. Lawrence Lowlands; in Contributions to Canadian Paleontology; Geological Survey of Canada, Bulletin 321, p. 27-59.

Hume, G.S.
1925: The Paleozoic outlier of Lake Timiskaming, Ontario and Quebec; Geological Survey of Canada, Memoir 145, 129 p.

Hunt, T.S.
1861: Notes on the history of petroleum or rock oil; Canadian Naturalist, v. 6, p. 241-255.

James, N.P. and Stevens, R.K.
1979: Correlation and timing of platform-margin megabreccia deposition, Cow Head and related groups, western Newfoundland (abst.); in American Association of Petroleum Geologists, Bulletin, v. 63, p. 474.
1982: Anatomy and evolution of a lower Paleozoic continental margin, western Newfoundland; International Association of Sedimentologists, 11th International Congress on Sedimentology, McMaster University, Excursion 2B, 75 p.

Johnson, W.A.
1912: Geology of the Lake Simcoe area, Ontario; Brechin and Kirkfield sheets; Geological Survey of Canada, Summary Report 1911, p. 253-261.
1914: Geology of the Lake Simcoe area, Ontario; Beaverton, Sutton and Barrie sheets; Geological Survey of Canada, Summary Report 1912, p. 294-300.

Kay, G.M.
1942: Ottawa-Bonnechère Graben – Lake Ontario homocline; Geological Society of America, Bulletin, v. 53, p. 585-646.

Klappa, C.F., Opalinski, P.R., and James, N.P.
1980: Middle Ordovician Table Head Group of western Newfoundland: a revised stratigraphy; Canadian Journal of Earth Sciences, v. 17, p. 1007-1019.

Kluyver, H.M.
1975: Stratigraphy of the Ordovician St. George Group in the Port-au Choix Area, western Newfoundland; Canadian Journal of Earth Sciences, v. 12, p. 589-594.

Knight, I.
1977: Cambro-Ordovician platformal rocks of the northern peninsula, Newfoundland; Department of Mines and Energy, Government of Newfoundland and Labrador, Report 77-6, 27 p.
1983: Geology of Cambro-Ordovician rocks in parts of the Castors River, St. John Island and Port Saunders map sheets; in Current Research, Department of Mines and Energy, Government of Newfoundland and Labrador, Report 83-1, p. 1-10.

Lane, T.E.
1984: Preliminary classification of carbonate breccias, Newfoundland zinc mines, Daniel's Harbour, Newfoundland; in Current Research, Part A, Geological Survey of Canada, Paper 84-1A, p. 505-512.

Liberty, B.A.
1969: Paleozoic geology of the Lake Simcoe area, Ontario; Geological Survey of Canada, Memoir 355, 201 p.

Logan, W.E.
1863: Report on the geology of Canada; Geological Survey of Canada, Report of Progress from its commencement to 1863, 983 p.

Low, A.P.
1892: Report on the geology and economic minerals of the southern portion of Portneuf, Quebec and Montmorency counties, Province of Quebec; Geological Survey of Canada, Summary Report, v. 5, part L, 82 p.

McCracken, A.D.
1981: Conodont biostratigraphy across the Ordovician-Silurian boundary, Ellis Bay Formation, Anticosti Island, Quebec; in Volume II: Stratigraphy and Paleontology, P.J. Lespérance (ed.); IUGS Subcommission on Silurian Stratigraphy, Ordovician-Silurian Boundary Working Group, Field Meeting, Anticosti-Gaspé, Quebec, 1981, p. 61-69.

McCracken, A.D. and Barnes, C.R.
1981: Conodont biostratigraphy and paleontology of the Ellis Bay Formation Anticosti Island, Quebec, with special reference to Late Ordovician-Early Silurian chronostratigraphy and the systemic boundary; Geological Survey of Canada, Bulletin 329, Part 2, p. 51-134.

Murray, A.
1843: On the geology of the district between Georgian Bay and the lower extremity of Lake Erie; Geological Survey of Canada; Report of Progress, p. 51-159.

Nelson, S.J.
1955: Geology of Portland Creek – Port Saunders area, west coast; Department of Mines and Resources, Newfoundland Geological Survey Report No. 7, 58 p.

Nowlan, G.S.
1980: Early and Middle Ordovician conodonts from the Mingan Islands, Quebec (Abstract); Geological Society of America, North-Central Section, Abstracts with Programs, p. 253.
1981: Stratigraphy and conodont faunas of the Lower and Middle Ordovician Romaine and Mingan formations, Mingan Islands, Quebec (Abstract); Maritime Sediments and Atlantic Geology, v. 17, p. 67.

Nowlan, G.S. and Barnes, C.R.
1981: Late Ordovician conodonts from the Vauréal Formation, Anticosti Island; Geological Survey of Canada; Bulletin 329, Part 1, p. 1-49.

Okulitch, V.J.
1938: The Ordovician section at Coboconk, Ontario; Transactions of Royal Canadian Institute, v. 22, part 2, p. 319-339.

Ollerenshaw, N.C. and Macqueen, R.W.
1960: Ordovician and Silurian of the Lake Timiskaming area; Geological Association of Canada, Proceedings, v. 12, p. 105-115.

Osborne, F.F.
1956: Geology near Quebec City; Naturalists Canada, v. 83, p. 157-223.

Parks, W.A.
1925: The stratigraphy and correlation of the Dundas Formation; Ontario Department of Mines, Annual Report, 1923, v. 32, part VII, p. 99-116.
1928: Faunas and stratigraphy of the Ordovician black shales and related rocks in southern Ontario; Royal Society of Canada, Transcript, 3rd Series, v. 22, Section 4, p. 39-90.

Petryk, A.A.
1976: Geology and oil and gas exploration of Anticosti Island, Gulf of St. Lawrence Quebec; preliminary reconnaissance; Quebec Department of Natural Resources, Preliminary Report no. 1, 25 p.
1981a: Stratigraphy, sedimentology and paleogeography of the Upper Ordovician-Lower Silurian of Anticosti Island, Quebec; in Volume II: Stratigraphy and Paleontology, P.J. Lespérance (ed.), IUGS Subcommission of Silurian stratigraphy, Ordovician-Silurian Boundary Working Group, Field Meeting, Anticosti-Gaspé, Quebec, 1981, Guidebook, p. 11-39.
1981b: Lithostratigraphy, paleogeography and hydrocarbon potential of Anticosti Island; Ministère de l'Énergie et des Ressources du Québec, Publication DPV-817, 133 p.

Pratt, B.R. and James, N.P.
1982: Cryptalgal-metazoan bioherms of Early Ordovician age in the St. George Group, Western Newfoundland; Sedimentology, v. 29, p. 543-569.

Quebec Department of Natural Resources
1974: Data on wells drilled for petroleum and natural gas in the St. Lawrence Lowlands area; Part II, wells more than 500 feet in depth.

Richardson, J.
1857: Report for the year 1856; Geological Survey of Canada; Report of Progress 1853-1856, p. 191-245.

Riley, G.C.
1962: Stephenville map area, Newfoundland; Geological Survey of Canada, Memoir 323, 72 p.

Riva, J.
1969: Middle and Upper Ordovician graptolite faunas of the St. Lawrence Lowlands of Quebec and of Anticosti Island; in North Atlantic – Geology and Continental Drift; American Association of Petroleum Geologists, Memoir 12, p. 513-556.
1972: Geology of the environs of Quebec City; XXIV International Geological Congress, Montreal, Quebec, Excursion B-19, 53 p.
Rodgers, J.
1965: Long Point and Clam Bank formations, western Newfoundland; Geological Association of Canada, Proceedings, v. 16, p. 83-94.
1968: The eastern edge of the North American continent during the Cambrian and Early Ordovician, in Studies of Appalachian geology, northern and maritime; E-an Zen, W.S. Shite, J.B. Hadley, and J.B. Thompson Jr. (ed.); Interscience New York, p. 141-149.
1971: The Taconic orogeny; Geological Society of America, Bulletin, v. 82, p. 1141-1178.
Russell, D.J.
1984: Paleozoic geology of the Lake Timiskaming area, Timiskaming District; Ontario Geological Survey, Map P.2700, Geological Series – Preliminary Map, scale 1:50 000.
Sanford, B.V.
1961: Subsurface stratigraphy of Ordovician rocks in southwestern Ontario; Geological Survey of Canada, Paper 60-26, 54 p.
1965: Salina salt beds of southwestern Ontario; Geological Survey of Canada, Paper 65-9, 7 p.
1968: Devonian of Ontario and Michigan; in International Symposium on the Devonian System; Alberta Society of Petroleum Geologists, v. 1, p. 973-999. (1967 print).
1969: Silurian of southwestern Ontario; 8th Annual Conference of the Ontario Petroleum Institute, Proceedings, 44 p.
1977: Three-dimensional geometry of salt deposits, southwestern Ontario, maps and structure sections; Geological Survey of Canada, Open File 401, 6 maps.
Sanford, B.V. and Quillian, R.G.
1959: Subsurface stratigraphy of Upper Cambrian rocks in southwestern Ontario; Geological Survey of Canada, Paper 58-12, 16 p.
Sanford, B.V., Thompson, F.J., and McFall, G.H.
1985: Plate tectonics – a possible controlling mechanism in the development of hydrocarbon traps in southwestern Ontario; Bulletin of Canadian Petroleum Geology, v. 33, no. 1, p. 52-71.
Schuchert, C. and Dunbar, C.O.
1934: Stratigraphy of western Newfoundland; Geological Society of America, Memoir 1, 123 p.
Shearer, J.M.
1973: Bedrock and surficial geology of the northern Gulf of St. Lawrence as interpreted from continuous seismic reflection profiles; in Earth Sciences Symposium on offshore Eastern Canada, P.J. Hood (ed.); Geological Survey of Canada, Paper 71-23, p. 285-303.
Sinclair, G.W.
1953: Middle Ordovician beds in the Saguenay Valley, Quebec; American Journal of Science, v. 251, p. 841-854.
1965: Succession of Ordovician rocks at Lake Timiskaming; Geological Survey of Canada, Paper 65-34, 6 p.
Sloss, L.L.
1963: Sequences in the cratonic interior of North America; Geological Society of America, Bulletin, v. 74, p. 93-114.
Sloss, L.L. (ed.)
1988: Sedimentary Cover – North American Craton: United States; Geological Society of America, The Geology of North America, v. D-2, 506 p.
Stauffer, C.R.
1915: Devonian of southwestern Ontario; Geological Survey of Canada, Memoir 63, 391 p.
Stouge, S.S.
1982: Preliminary conodont biostratigraphy and correlation of Lower to Middle Ordovician carbonates of the St. George Group, Great Northern Peninsula, Newfoundland; Department of Mines and Energy, Government of Newfoundland and Labrador, Report 82-3, 59 p.
1984: Conodonts of the Middle Ordovician Table Head Formation, Western Newfoundland; Fossils and Strata, no. 16, 145 p.
Twenhofel, W.H.
1928: Geology of Anticosti Island; Geological Survey of Canada, Memoir 154, 481 p.
1938: Geology and paleontology of the Mingan Islands, Quebec; Geological Society of America, Special Paper No. 11, 132 p.
Uyeno, T.T. and Barnes, C.R.
1983: Conodonts of the Jupiter and Chicotte formations (Lower Silurian), Anticosti Island, Quebec; Geological Survey of Canada, Bulletin 355, 49 p.
Uyeno, T.T., Telford, P.G., and Sanford, B.V.
1982: Devonian conodonts and stratigraphy of southwestern Ontario; Geological Survey of Canada, Bulletin 332, 45 p.
Whittington, H.B. and Kindle, C.H.
1963: Middle Ordovician Table Head Formation, western Newfoundland; Geological Society of America, Bulletin, v. 74, p. 745-758.
1966: Middle Cambrian strata of the Strait of Belle Isle, Newfoundland, Canada (Abstract); Geological Society of America, Northeastern Section, Program 1966 Annual Meeting, p. 46.
1969: Cambrian and Ordovician stratigraphy of western Newfoundland; in North Atlantic Geology and Continental Drift, M. Kay (ed.); American Association of Petroleum Geologists, Memoir 12, p. 655-664.
Williams, H.
1984: Miogeosynclines and suspect terranes of the Caledonian-Appalachian Orogen: tectonic patterns in the North Atlantic region; Canadian Journal of Earth Sciences, v. 21, p. 887-901.
Williams, H. and Neale, E.R.W. (ed.)
in prep: Geology of the Appalachian-Caledonian orogen in Canada and Greenland; Geological Survey of Canada, Geology of Canada no. 5. (Also Geological Society of America, The Geology of North America, v. F-1.)
Williams, H. and Smyth, W.R.
1983: Geology of the Hare Bay Allochthon; in Geology of the Strait of Belle Isle area, northwestern insular Newfoundland, southern Labrador, and adjacent Quebec; Geological Survey of Canada, Memoir 400, pt. 3, p. 109-132.
Williams, M.Y.
1919: The Silurian geology and faunas of Ontario Peninsula and Manitoulin and adjacent islands; Geological Survey of Canada, Memoir 111, 195 p.
Wilson, A.E.
1964: Geology of the Ottawa-St. Lawrence Lowland, Ontario and Quebec; Geological Survey of Canada, Memoir 241, 66 p.

Authors' Addresses

B.V. Sanford
17 Meadowglade Gardens
Nepean, Ontario
K2G 5J4

Printed in Canada

Chapter 11

ST. LAWRENCE PLATFORM — GEOLOGY

CHAPTER 11

ST. LAWRENCE PLATFORM - GEOLOGY

B.V. Sanford

with contributions by W.R. Cowan and K.L. Currie

EVOLUTION OF TECTONIC FRAMEWORK

Precambrian rocks of the Canadian Shield extend into the subsurface to form the basement beneath Paleozoic rocks of St. Lawrence Platform and St. Lawrence Trough (Fig. 11.1), where they were in part involved in Taconian, Acadian, and Alleghenian orogenic deformation. The southeastern boundary of the Precambrian basement beneath the Appalachian Orogen is unknown, but is assumed to coincide roughly with the southeastern limit of anomalous concentrations of seismic events (Sanford et al., 1985; Fig. 10.2).

Evolution of St. Lawrence Platform and its periodic connections with the Hudson and Interior platforms were linked with Paleozoic plate-tectonic events at the eastern margins of the North American continent (which at that time included Greenland).

Two major tectonic cycles are recorded: **Tectonic Cycle I**, extending from late Precambrian to late Paleozoic time, and **Tectonic Cycle II**, extending from early Mesozoic to the present (Fig. 11.2).

Tectonic Cycle I began with rifting in Late Proterozoic time and eventual separation of the continental masses (opening of the Iapetus Ocean) in Early Cambrian and continuing through the Cambrian and Early Ordovician (Williams and Hiscott, 1987). In late Early and Middle Ordovician time, the Iapetus Ocean began to close; final closing and continental collision occurred in the middle to late Paleozoic. Tectonic Cycle II began with Triassic rifting, followed by diachronous continental separation in mid-Jurassic to mid-Cretaceous time, and subsequent passive-margin subsidence continuing to the present.

Rifting and separation of the continents at the onset of each of the two Phanerozoic tectonic cycles was attended by uplift and faulting, continental redbed deposition, and dyke and pluton emplacement, over a wide region of the platform and adjacent craton. However, the more intense epeirogeny on the platform and exposed craton, in the form of arch rejuvenation, basin inception and development, and fault-block rotation, was apparently associated with the active margin phase of Tectonic Cycle I, during Taconian, Acadian, and Alleghenian orogenies, notably the first two. Normal faults that had been active since the late Proterozoic display greatest displacement in Middle and Late Ordovician (late Trentonian and early Edenian; St. Julien and Hubert, 1975), and Late Silurian to Early Devonian (Ludlow to Gedinnian) time.

Initially (in the Cambrian and Early Ordovician) craton-derived clastics were deposited in association with platform carbonates on a broad continental shelf open to the southeast. With the onset of convergence, first documented in early Middle Ordovician (Chazyan) time, destruction of the outer shelf and its adjoining continental slope, and building of the Appalachian Orogen began. Remnants of the slope deposits are preserved only in tectonic slices transported cratonward during the Taconian Orogeny.

Uplift and faulting followed by subaerial erosion took place along much of St. Lawrence Platform near the close of the Early Ordovician. The most pronounced movements during this period are recorded in the central and western platform, in uplift of the Saguenay, Frontenac, Wisconsin, and Algonquin arches, and inception of Michigan Basin in Early to early Middle Ordovician (Canadian to early Whiterockian) time (see Fisher and Barrett, 1985).

The first widespread transport of orogen-derived clastics toward the craton was in the early Middle Ordovician (Whiterockian/Chazyan). Flysch sediments containing ophiolitic detritus derived from an emerging Taconian welt along the southeastern margin of the continent were deposited in the rapidly subsiding foreland basin, St. Lawrence Trough, to the northwest. Remnants of this sedimentary wedge are preserved at the northeastern end of St. Lawrence Platform in western Newfoundland, and elsewhere along the margins of the platform to the southwest. During the same interval (Chazyan), the Taconian structural front migrated northwestward onto the shelf. Remnants of the slices that underwent early movement are the Cow Head carbonate breccias and the flysch and ophiolite of the Humber Arm

Sanford, B.V.
1993: St. Lawrence Platform - Geology; Chapter 11 in Sedimentary Cover of the Craton in Canada, D.F. Stott and J.D. Aitken (ed.); Geological Survey of Canada, Geology of Canada, no. 5, p. 723-786 (also Geological Society of America, The Geology of North America, D-1).

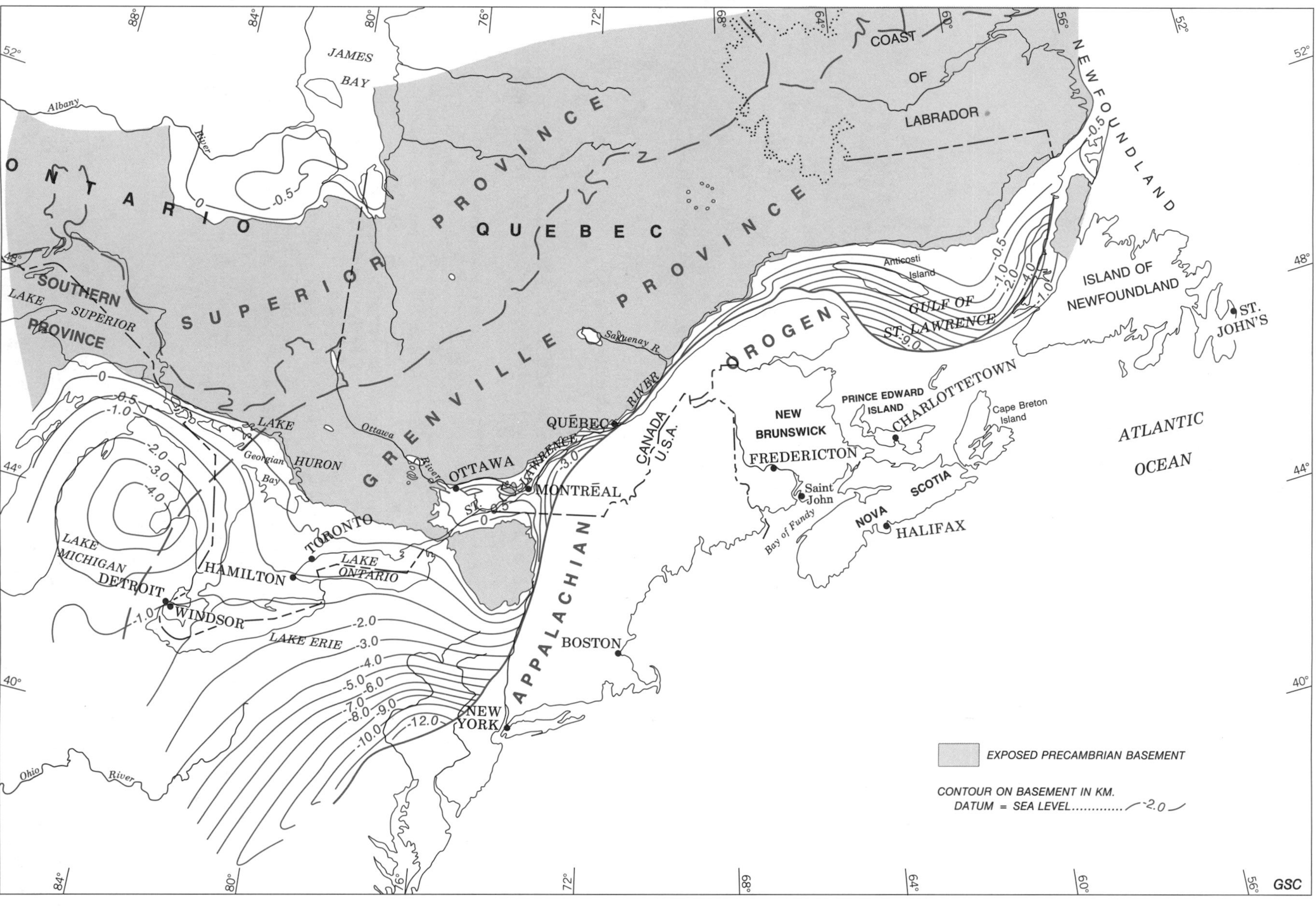

Figure 11.1. Configuration of Precambrian basement rocks beneath Phanerozoic cover, St. Lawrence Platform.

and Hare Bay allochthons (klippen) in western Newfoundland. This tectonic loading initiated subsidence of the foreland Anticosti Basin, which persisted until Late Silurian time.

Continued northwestward migration of the Taconian structural front in late Middle and Late Ordovician time resulted in progressive narrowing of St. Lawrence Trough and transport of orogen-derived clastics farther and farther onto the platform. In the late Middle Ordovician (late Trentonian and early Edenian), orogen-derived clastics intertongued with carbonates and minor craton-derived shales along the platform edges, eventually spilling over the platform in late Edenian and Maysvillian time to extend far onto the southern margins of the craton.

In Late Ordovician (Richmondian) time the diachronous Taconian structural front reached its present position southeast of St. Lawrence River. This late Taconian deformation is not recorded in the extreme eastern segment of the platform, where clastic deposition abruptly gave way to carbonate. Along the southern margins of the central and western platform, however, newly elevated tectonic lands provided fluvio-deltaic clastic redbeds, which were transported onto and across the adjacent platform.

Renewed uplift of the cratonic arches, Beaugé, Saguenay, Frontenac, Fraserdale and Findlay- Algonquin-Laurentian, coincided broadly with the Taconian orogeny. Initial differential subsidence of Michigan Basin in Middle Ordovician (early Blackriveran-early Caradocian) time also coincided with Taconian (but not earliest Taconian) movements.

As the Taconian front encountered the southeast-trending Frontenac and Saguenay arches during the Richmondian Age, the previously narrowed St. Lawrence Trough became segmented into the three foreland basins shown in Figures 10.2 and 11.1, the Anticosti, Quebec, and Allegheny basins.

Marine seaways were confined largely to the southern margins of the craton in earliest Silurian (early Llandovery), but expanded onto the craton in Mid-Silurian (mid- to late Llandovery and Wenlock) times. The reasons for the extensive flooding that led to interconnection of the seaways between the St. Lawrence, Hudson, Interior, and Arctic platforms from early Late Ordovician to Early Silurian times inclusive, are poorly understood. Nevertheless, widespread transgression, in part at least eustatic, occurred in most of North America.

The end of maximum flooding of the craton in Late Silurian (Ludlow) time was accompanied by uplift and widespread block faulting that prevailed throughout the Late Silurian to Early Devonian. This epeirogenic event is recorded in most of North America, and may correspond to an early phase of continental collision (Acadian Orogeny). It established the framework in which, following renewed marine transgression, deposition took place during the remainder of the Paleozoic Era. Continued subsidence and

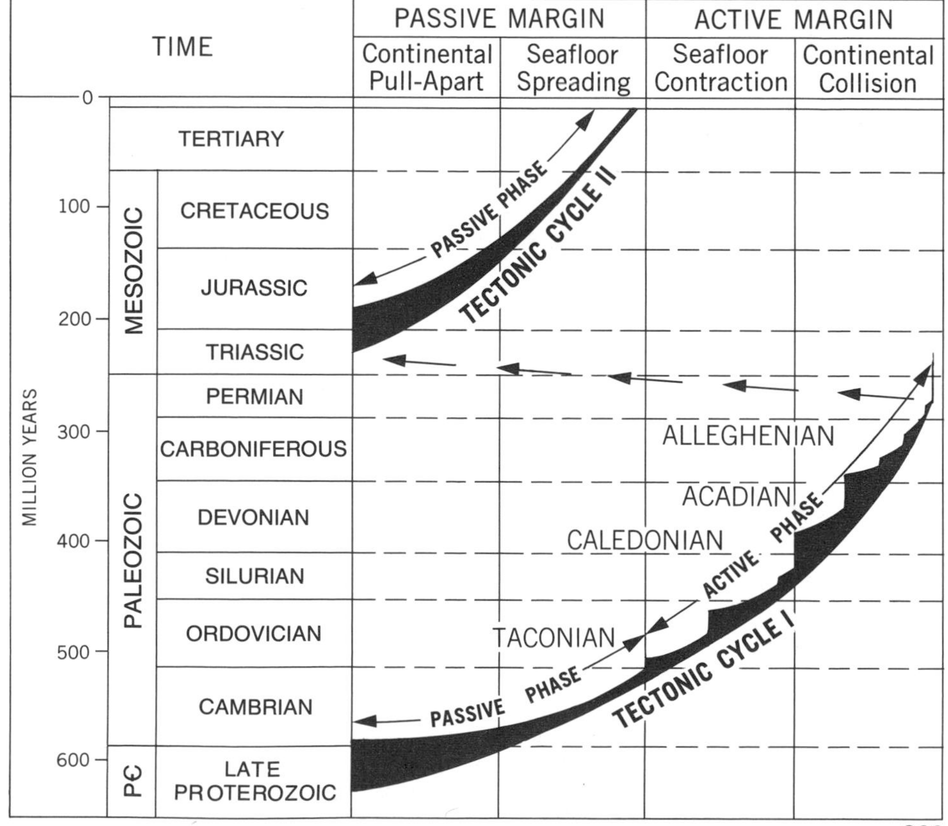

Figure 11.2. Phanerozoic tectonic cycles (broader segments of curve indicate more intense levels of tectonism).

coeval positive behaviour of the arches characterized the St. Lawrence and Hudson platforms. Semi-continental redbed and marine deposits accumulated in the foreland basins along the length of St. Lawrence Platform, giving place to marine carbonates and evaporites farther onto the platform. Michigan Basin underwent accelerated subsidence, and in Hudson Platform, uplift of Cape Henrietta Maria Arch and bordering arches was accompanied by subsidence of Hudson Bay Basin on the north and Moose River Basin on the south. The seaways had now become greatly reduced in extent and this, in combination with the increased expression of the arches, provided the necessary marine restriction for deposition of evaporites, conditions that continued to prevail intermittently during the succeeding Devonian period.

In Middle and Late Devonian time, Anticosti Basin and the adjacent craton were emergent; sedimentation was confined to the central(?) and western platform. The Acadian Orogeny is registered in the latter regions by a clastic wedge giving place to marine shales and carbonates farther onto the platform.

Paleozoic rocks have been stripped from the broad expanse of the Canadian Shield, except for widely separated outliers. Their erosion must be a result of uplift that undoubtedly began at the close of the Early Silurian, and continued intermittently through the remainder of the Paleozoic Era.

Following stabilization of the Appalachian Orogen in Carboniferous-Permian time, a new tectonic cycle was born with Triassic rifting and Late Jurassic to mid-Cretaceous continental separation. The Atlantic margin remains passive to the present (Tectonic Cycle II, Fig. 11.2). The preserved, undeformed St. Lawrence Platform is relatively remote from the Atlantic margin, and accordingly shows only subdued effects of the initiation of Cycle II. Nevertheless, there is clear evidence of widespread, early

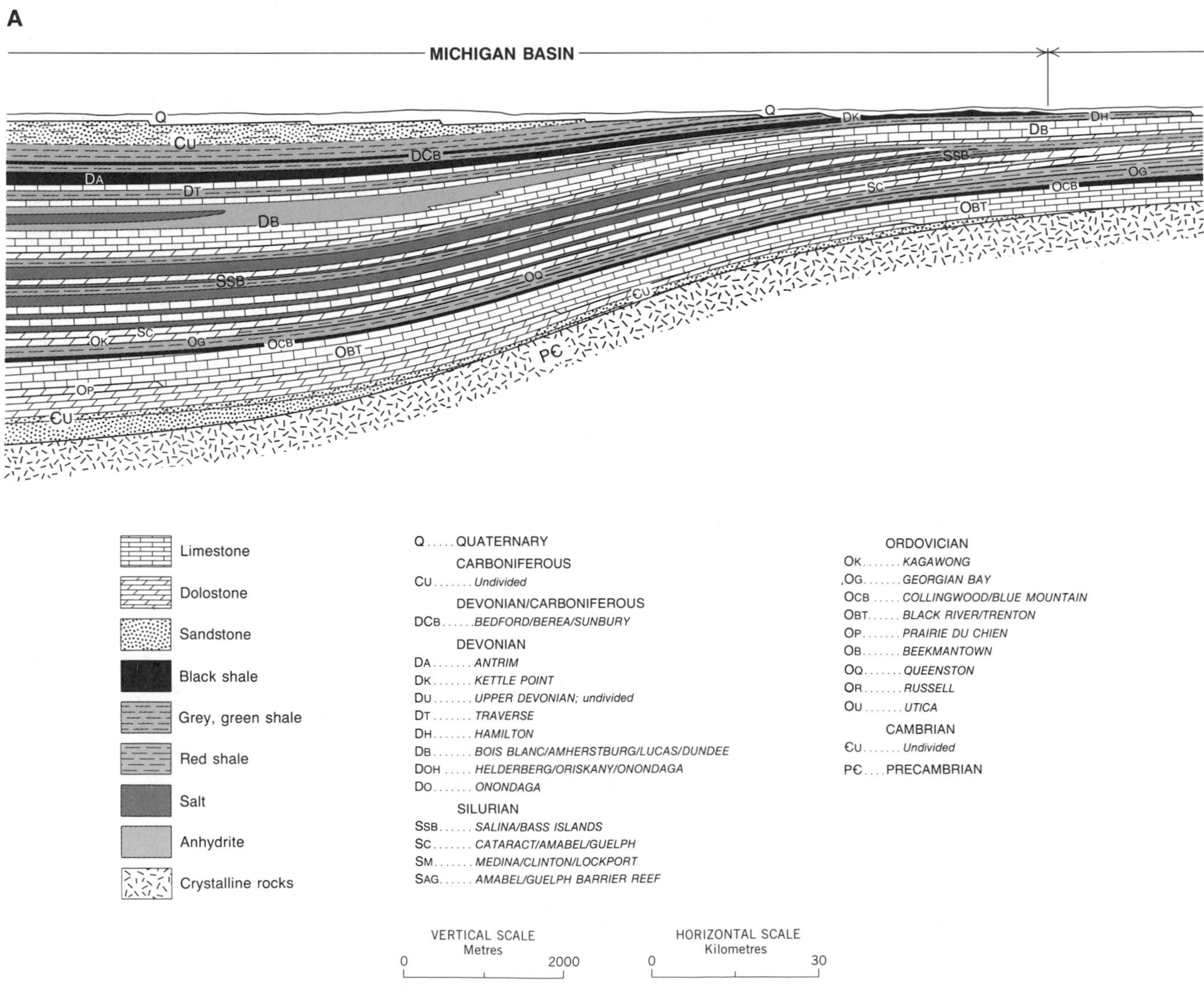

Figure 11.3. Section A-A′ Michigan Basin to Allegheny Basin.

Mesozoic basement rejuvenation in the form of block faulting and dyke and pluton emplacement. The presence of Jurassic terrigenous clastics in Michigan Basin, and Jurassic-Cretaceous clastics of continental origin in southern Hudson Platform, point to reactivation of the Frontenac and Fraserdale arches at that time.

The basement arches, most of which have risen intermittently throughout the Phanerozoic, are still active, as evidenced by the concentrations of earthquake epicentres along the axes or the margins of several of them (Sanford et al., 1985), especially the Frontenac, Saguenay, Algonquin, and Laurentian arches. This observation has important economic and environmental ramifications.

The highly generalized stratigraphic and structural framework of the several segments of the platform is illustrated in Figures 11.3 to 11.6. Uplift of the cratonic arches and basement block movements have resulted in extensive faulting of the Paleozoic rocks. Only the more significant faults are shown in the cross-sections.

GEOPHYSICAL CHARACTERISTICS

The boundaries of the structural provinces of St. Lawrence Platform and adjoining regions are traceable on regional magnetic and gravity maps (Haworth et al., 1980; Earth Physics Branch, 1980; Dods et al., 1984).

The northwestern margin of the Appalachian Orogen is delineated by an arc of northeast-trending, positive gravity anomalies that border Allegheny Basin and Adirondack Mountains in the United States and continue into Canada along the southern margin of Quebec Basin. From Quebec City the anomalies continue east along northern Gaspésie, thence beneath the northern Gulf of St. Lawrence, where they arc southeastward toward the Port au Port Peninsula of western Newfoundland.

Magnetic and gravity maps provide only limited information pertaining to the geology of St. Lawrence Platform. Magnetic and gravity anomalies appear for the most part to reflect the composition and structure of the

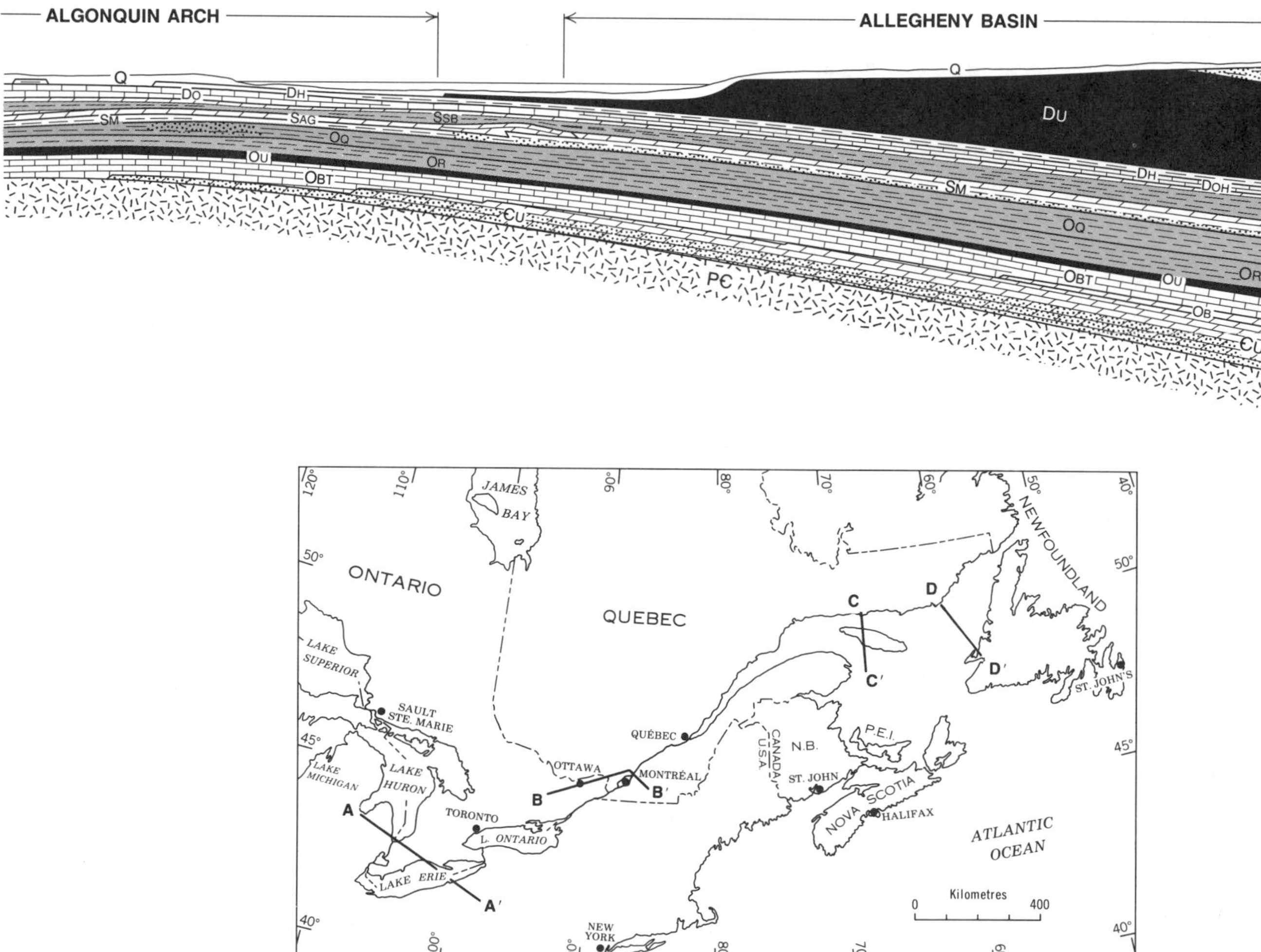

Figure 11.3. Cont.

underlying Precambrian basement. However, many of the northeast and northwest linear magnetic anomalies presumably indicate reduced magnetic susceptibility due to alteration along basement faults. In such cases, fracturing of the Paleozoic rocks should also be suspected, as many of the known faults in Paleozoic rocks of the St. Lawrence Platform are rejuvenated Precambrian faults.

Detailed gravity surveys were used successfully for many years in southwestern Ontario in exploration for pinnacle reefs and salt-collapse structures in Paleozoic rocks. Gravity interpretation has gradually given way to the use of high-resolution seismic surveys as record quality has improved with greatly advanced survey and data processing techniques. The seismic profiles reveal faults, salt-collapse structures, pinnacle reefs, and other structural and stratigraphic phenomena that may have affected accumulation of oil and gas.

DIAGNOSTIC SHELLY FAUNAS AND CONODONT BIOSTRATIGRAPHY

M.J. Copeland, T.E. Bolton and T.T. Uyeno

Shelly faunas and microfossils are used extensively in the dating and correlation of Paleozoic rocks of St. Lawrence Platform (Fig. 11.7, 11.8). Shelly faunas have been studied in detail and have provided a biostratigraphic zonation for the region. Paleozoic shelly-fauna biostratigraphy of the platform is based primarily on brachiopods and trilobites, but locally and in parts of the succession, other shelly elements such as corals, cephalopods, and ostracodes have proved useful. Microfossils, particularly conodonts, although much less studied, have become increasingly important recently for purposes of dating and correlation.

Faunas diagnostic of the Early, Middle, and Late Cambrian are confined to eastern Anticosti Basin (western Newfoundland and southeastern Labrador). Shelly faunas are common in Ordovician rocks in all parts of the platform, and in the Silurian in the eastern and western divisions. Devonian shelly faunas are known only in the western platform, with the exception of Lower and Middle Devonian blocks preserved in the St. Helen's Island diatreme breccia at Montreal.

Conodont-based biostratigraphic zonation has been carried out in the Ordovician, Silurian, and Devonian systems in all three divisions of the St. Lawrence platform, although the level of sampling and accuracy of the zonation is variable.

Cambrian

Strata assumed to be Cambrian occur throughout the platform, but only the rocks exposed in western Newfoundland and southeastern Labrador contain fossils of biostratigraphic significance. Assemblages of late Early, late Middle, and Late Cambrian age have been identified (Fig. 11.7). These include archaeocyathids (Debrenne and James, 1981) near the base and trilobite *Bonnia-Olenellus* to *Cedaria-Crepicephalus* zones of parts of the Lower, Middle, and Upper Cambrian (Boyce, 1978, 1979, 1983; Stouge and Boyce, 1983). Elsewhere on the platform diagnostic Cambrian fossils are absent in the basal pre-Middle Ordovician sandstones and carbonates. The Late Cambrian designation of the rocks in the central and western platform is thus speculative and based largely on stratigraphic position and lithological correlation with known Upper Cambrian rocks in adjacent regions of the United States.

B

OTTAWA

Q ..... QUATERNARY
ORDOVICIAN
OQ .... QUEENSTON
OR .... RUSSELL
OC .... CARLSBAD
OEB .... EASTVIEW/BILLINGS
OBT ...... BLACK RIVER/TRENTON
ORC ...... ROCKCLIFFE
OO .... OXFORD
OM .... MARCH
OBR ...... BÉCANCOUR RIVER
OPR ...... PONTGRAVÉ RIVER
ONR ...... NICOLET RIVER
OL .... LACHINE
OU .... UTICA
OLa ...... LAVAL
OB .... BEAUHARNOIS
CAMBRIAN/ORDOVICIAN
ЄOT ...... THERESA
ЄOCN .... COVEY HILL/NEPEAN
ЄOCC .... COVEY HILL/CHATEAUGUAY
ЄOU ..... Allochthonous rocks; undivided
PЄ .... PRECAMBRIAN

VERTICAL SCALE
Metres
0 2000

HORIZONTAL SCALE
Kilometres
0 30

Figure 11.4. Section B-B′ Ottawa Embayment to Quebec Basin. See Figure 11.3 for location of section.

Ordovician

The Ordovician System in St. Lawrence Platform is complex in terms of diversity and abundance of faunas. Much of the early biostratigraphic work began in southern Ontario and adjacent New York, and those regions contain what is considered to be the type sequence for the Middle Ordovician of North America. A virtually complete succession of Ordovician rocks occurs in Anticosti Basin, whereas the lower Middle Ordovician (Whiterockian) is missing from the central region (Quebec Basin and Ottawa Embayment), and the Lower Ordovician (Canadian) as well as the lower Middle Ordovician (Whiterockian-Chazyan) are missing in the western platform.

The chief shelly faunal elements of biostratigraphic significance are trilobites (Lespérance and Bertrand, 1976; Ludvigsen, 1978a, b; Dean, 1979), brachiopods (Cooper, 1956), cephalopods (Flower, 1978), corals (Copper and Morrison, 1978; Bolton, 1981a; Elias, 1982), and ostracodes (Copeland, 1965, 1970, 1973; Copeland and Bolton, 1977).

Some shelly faunal elements are common throughout the platform. The Lower Ordovician (Canadian) is marked near the base by the occurrence of the cephalopod *Diphragmoceras*, and at the top by the trilobites

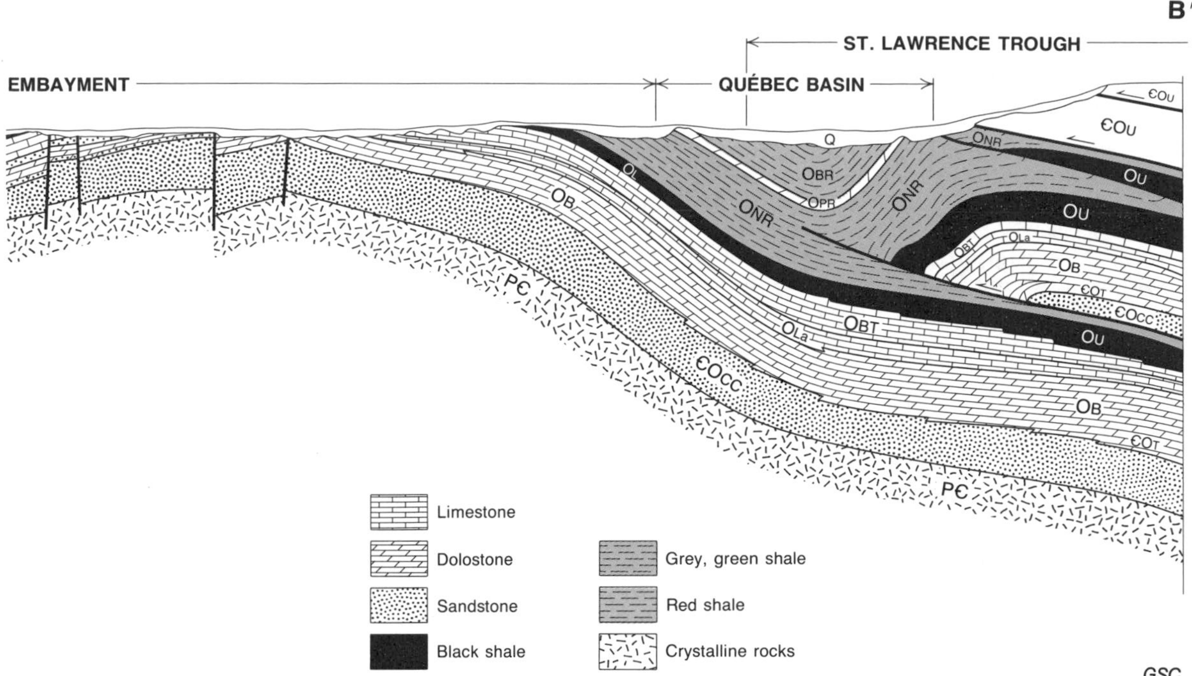

Figure 11.4. Cont.

Bolbocephalus and *Goniotelina*. In the lower part of the Middle Ordovician (Chazyan), the brachiopod *Rostricellula*, the coral *Eofletcheria*, and the cystoid *Bolboporites* are common faunal elements. The Blackriveran cephalopod *Gonioceras* and the Trentonian trilobite *Cryptolithus* are distinctive components of the later Middle Ordovician. The trilobites *Pseudogygites* and *Triarthrus* (with graptolites) are present in the Edenian Stage of the Upper Ordovician. The upper part of the Upper Ordovician (Maysvillian and Richmondian) is marked throughout the platform by the presence of the solitary coral *Grewingkia* and the stromatoporoid *Beatricea*. The uppermost Ordovician (Gamachian) is represented only on Anticosti Island; diagnostic fossils include the ostracode *Jonesites* and the cephalopods *Billingsites* and *Schuchertoceras*, associated with the conodont *Gamachignathus*.

Conodonts are generally abundant in Ordovician strata throughout the platform (Bergström et al., 1974; Uyeno, 1974; Barnes et al., 1978; Nowlan and Barnes, 1981; McCracken and Barnes, 1981; Stouge, 1982, 1984). Graptolites are prominent only in the Lower to lower Middle Ordovician St. George and Table Head groups of western Newfoundland (Whittington and Kindle, 1963; Boyce, 1985), and in the Edenian in all three divisions of the platform (Riva, 1969).

The Lower Ordovician (Tremadoc and Arenig) in western Newfoundland has been divided into informal conodont faunal units (Stouge, 1982), which correspond approximately with the Mid-continental Realm faunal zones C to E and 1 to 3 (Fig. 11.7). Strata of the St. George Group may range through zones C to E and 1, with elements of faunas E and 1 to 3 missing from the sequence. In the Mingan Islands, faunas D and E and elements possibly as high as early Whiterockian were reported by Nowlan (1980, 1981) from the Romaine Formation.

In the central platform (Ottawa Embayment), the Lower Ordovician (Tremadoc and Arenig) is represented by conodont faunas C and D (Greggs and Bond, 1971; Bond and Greggs, 1973, 1976).

A Middle Ordovician conodont fauna (Fauna 4 and possibly older) occurs in the Whiterockian Table Head Group in eastern Anticosti Basin (southwestern Newfoundland), succeeded by elements representative of the Chazyan Stage. In the central platform, the Whiterockian is not represented, and rocks containing late Chazyan fauna rest directly on Lower Ordovician strata.

Blackriveran and Trentonian rocks have wide distribution in the platform. In the eastern division, the beds are known only in the southwestern part of Newfoundland, where they contain conodonts representative of *Prioniodus gerdae* Subzone of the *Amorphognathus tvaerensis* Zone of the North Atlantic faunal realm (Fåhraeus, 1973). In the central and western divisions, Faunas 7 to 10 of the Midcontinental Realm are represented (Schopf, 1966; Uyeno, 1974; Winder, 1966a; and Barnes et al., 1978).

Upper Ordovician rocks (Edenian to Richmondian) containing Faunas 10, 11, and 12 are also widely distributed. On Anticosti Island, the Vauréal Formation containing Fauna 12 is overlain by the Ellis Bay Formation containing Fauna 13 of Gamachian age (Nowlan and Barnes, 1981). Overlying Fauna 13 is the *nathani* Zone of Llandovery age. This transition marks the Ordovician-Silurian boundary (McCracken and Barnes, 1981).

Silurian

Silurian shelly fossils occur in Anticosti Basin (Copeland, 1974, 1981; Bolton, 1981b; Cooper, 1981; Lespérance, 1985), and in the Western Platform (Bolton, 1966; Chiang, 1971; Copeland, 1976; Cooper, 1978, 1982; Bolton and Copeland, 1972). Lower Llandovery strata contain the brachiopod *Zygospiraella*, which is succeeded in the mid-Llandovery by a succession of brachiopods, *Virgiana, Stricklandia, Eocoelia, Pentamerus, Costistricklandia,* and *Pentameroides*. Accompanying these fossils are the ostracodes *Zygocosta* and *Zygobursa* (with *Virgiana*) and *Zygobolba* (with *Stricklandia* to *Costistricklandia*). The button coral, *Palaeocyclus*, is prominent near the top of the Llandovery beds.

Wenlock faunas of southern Ontario are represented in ascending order by the crinoid *Stephanocrinus*, the brachiopod *Plicostricklandia*, and the ostracode *Drepenallina*. Late Silurian (early Ludlow) faunas of southern Ontario contain the diagnostic coral *Fletcheria*, pelecypod *Megalomus*, and brachiopods *Trimerella* and *Conchidium*; and the Pridoli Clam Bank Formation of southwestern Newfoundland contains the crinoid hold-fast *Camarocrinus*.

Silurian strata of Anticosti Island contain one of the most complete Llandovery conodont successions in the world. The Ordovician-Silurian boundary is determined approximately by the contact of conodont Fauna 13 (Gamachian) with the *nathani* Zone (lower Llandovery),

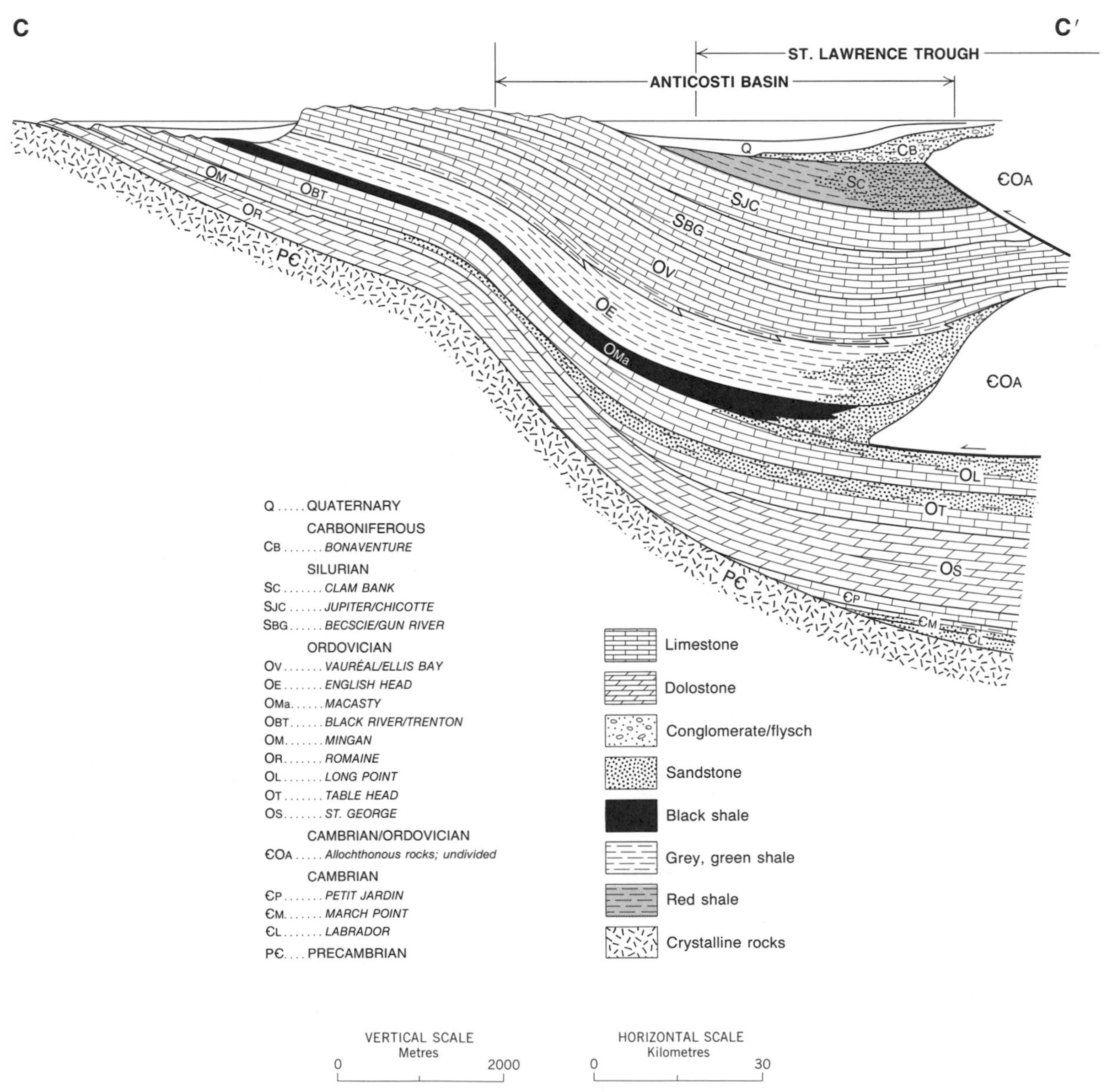

Figure 11.5. Section C-C′ West Anticosti Basin. See Figure 11.3 for location of section.

and occurs within the Ellis Bay Formation (McCracken and Barnes, 1981). The *kentuckyensis* Zone ranges from within the Becscie Formation to high in the Jupiter Formation.

Higher rock units on Anticosti Island can be more appropriately referred to in terms of British conodont nomenclature; thus most of the Jupiter Formation is referred to the *staurognathoides* Zone (2 to 4 Llandovery), with the succeeding *inconstans (=celloni)* Zone in the upper Jupiter and the Chicotte Formation. The higher beds of the Chicotte contain the *amorphognathoides* Zone (upper Llandovery-lowermost Wenlock).

In the western platform and Lake Timiskaming outlier, only the Llandovery and Wenlock Series have been zoned (Rexroad and Rickard, 1965; Barnes et al., 1978; T.T. Uyeno in Bolton and Copeland, 1972; Uyeno and Barnes, 1983). The zones *kentuckyensis* and *discreta-deflecta* through *celloni, amorphognathoides, patula, sagitta* to *crassa* have been identified.

Devonian

With the exception of Lower and Middle Devonian exotic blocks found in the St. Helen's Island diatreme breccia, Devonian strata are confined to the Western St. Lawrence Platform, where Siegenian, Emsian, Eifelian, Givetian, Frasnian, and Famennian rocks are represented.

Brachiopods (Fagerstrom, 1971; Uyeno et al., 1982), corals (Oliver, 1976), and stromatoporoids (Fagerstrom, 1982) are the most useful shelly fauna. The mid-Early Devonian (Siegenian) brachiopod *Costispirifer arenosus* is present in Oriskany strata in southern Ontario (Winder and Sanford, 1972) and in the St. Helen's Island Breccia (Hofmann, 1972). A succession of brachiopods of late Early and Middle Devonian (Emsian to Givetian) age, *Amphigenia, Prosserella, Paraspirifer, Leiorhynchus* and *Mucrospirifer*, occur in southwestern Ontario in Bois Blanc, Amherstburg, Lucas, Dundee, and Hamilton strata. These are succeeded by Upper Devonian (Frasnian and Famennian) Kettle Point strata, which have no diagnostic shelly fauna but contain the sporomorph *Tasmanites huronensis* and the microfossil *Foerstia (Protosalvinia)* (Russell, 1985) and conodonts.

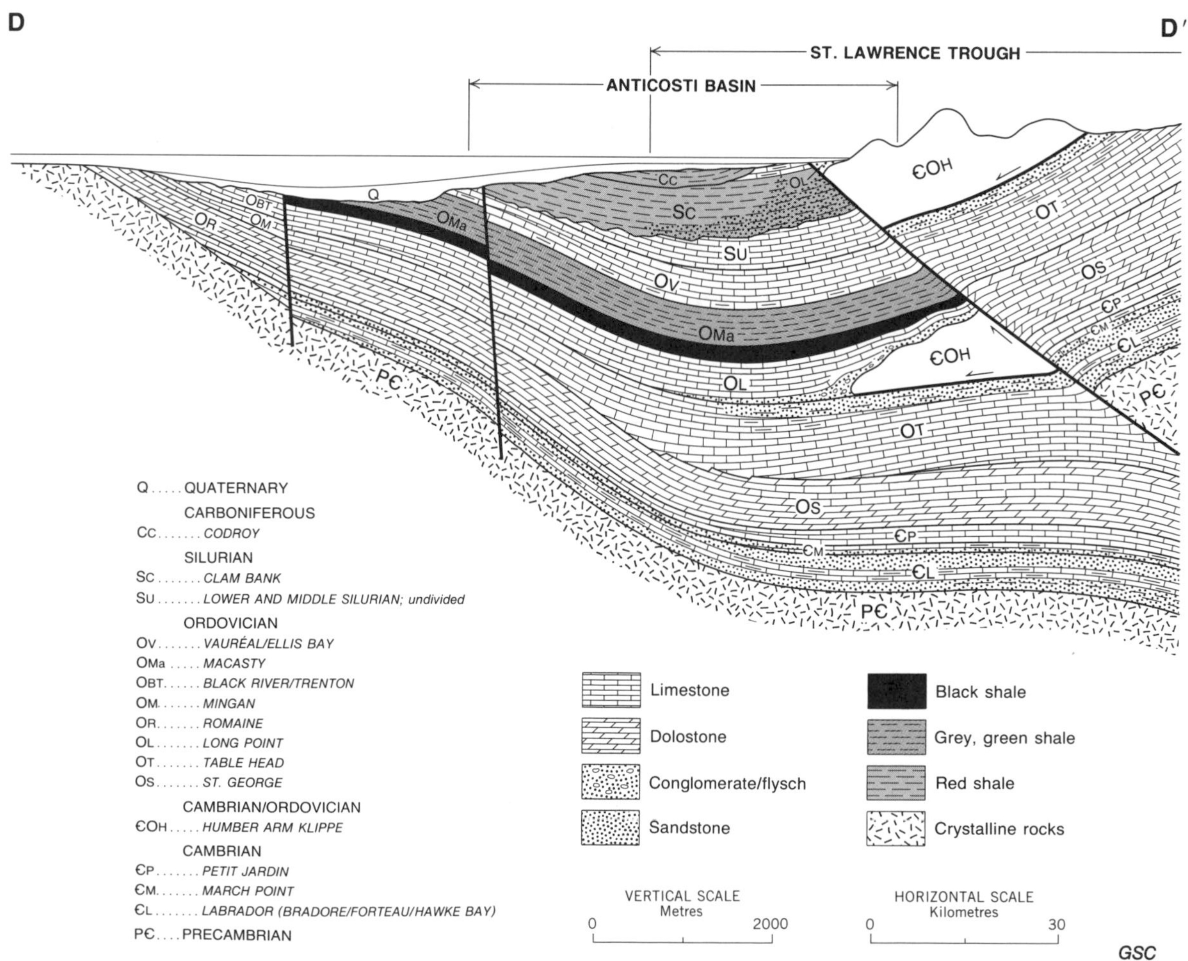

Figure 11.6. Section D-D′ East Anticosti Basin. See Figure 11.3 for location of section.

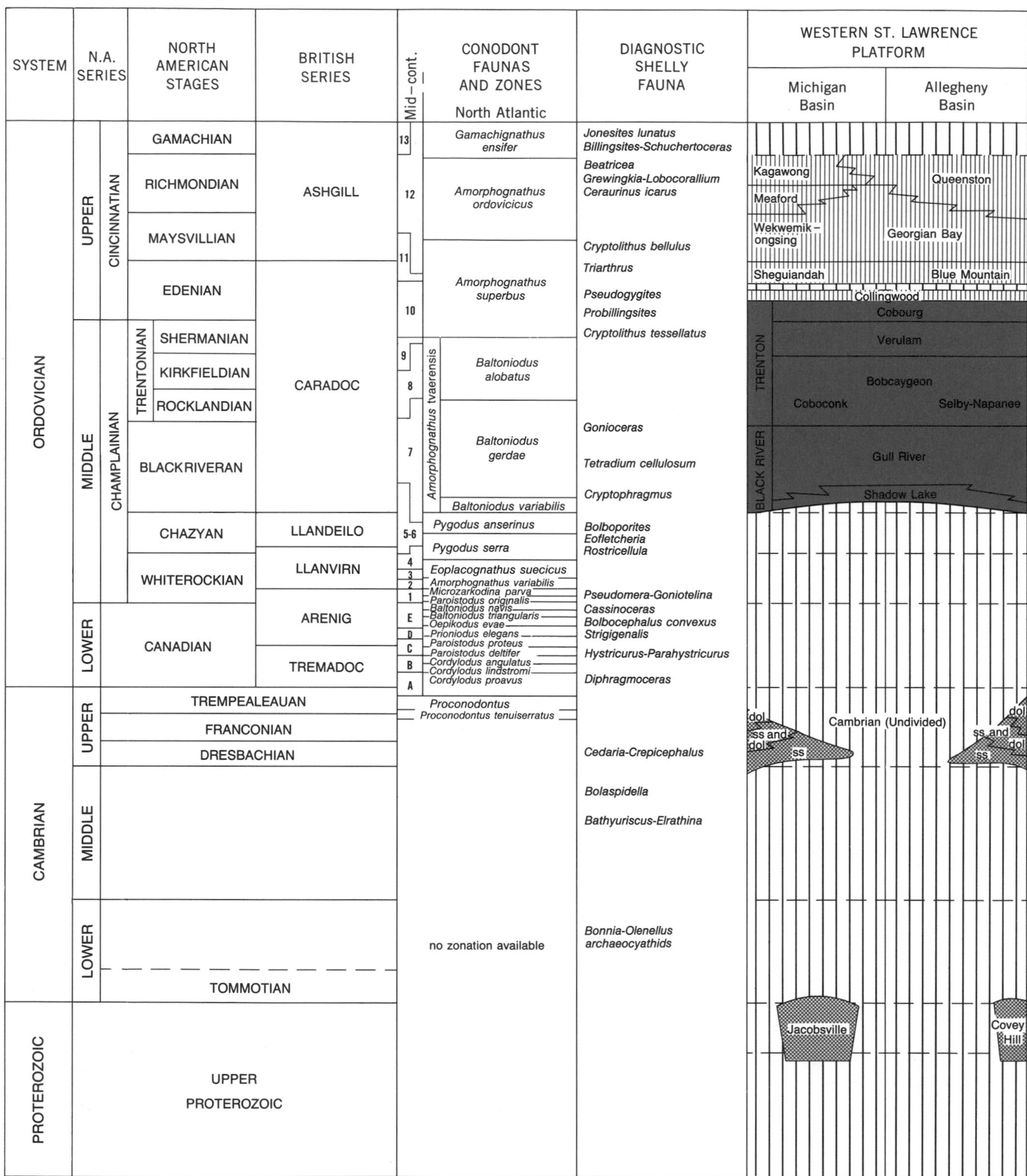

Figure 11.7. Cambrian and Ordovician biostratigraphy and regional correlation.

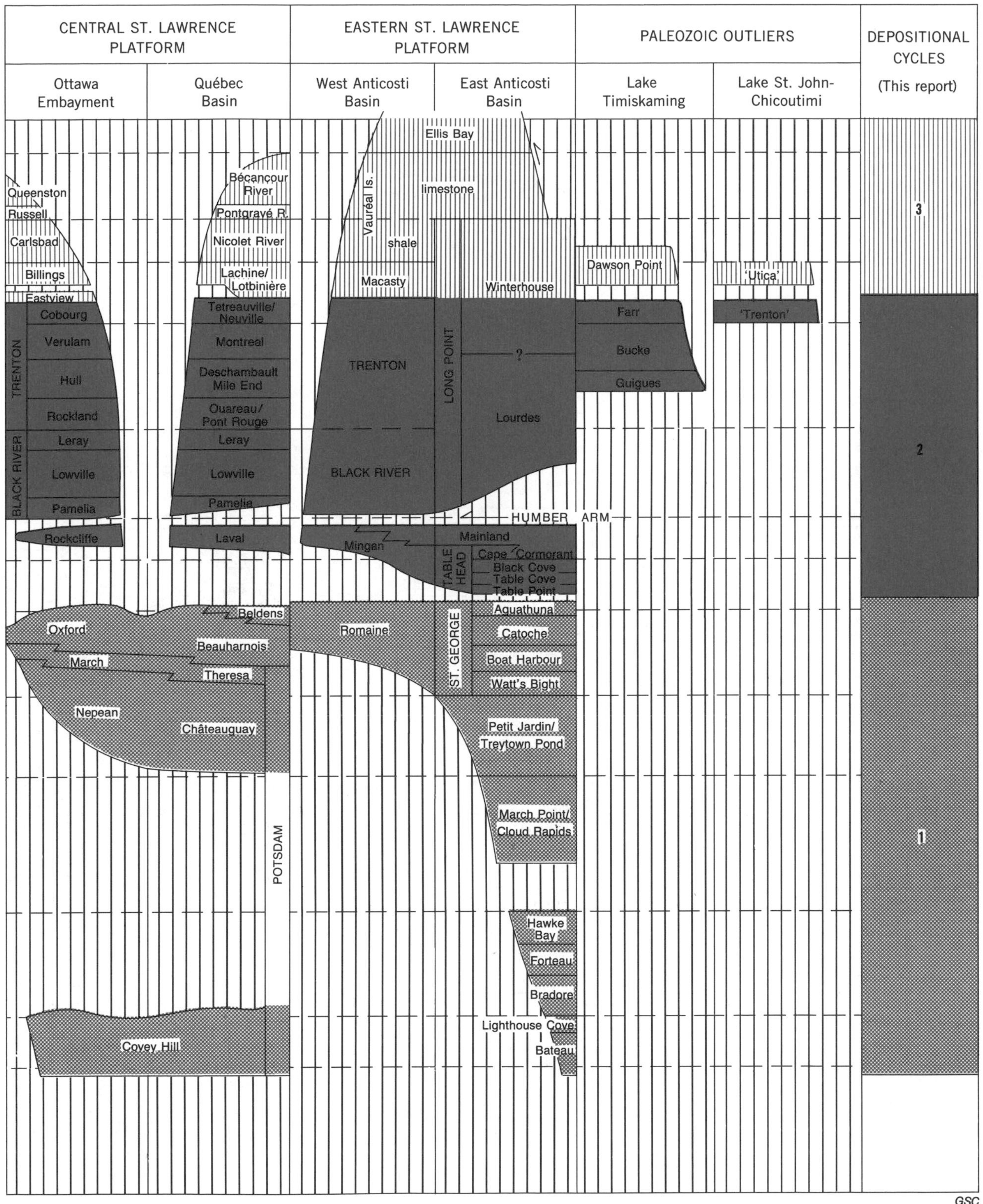

Figure 11.7. Cont.

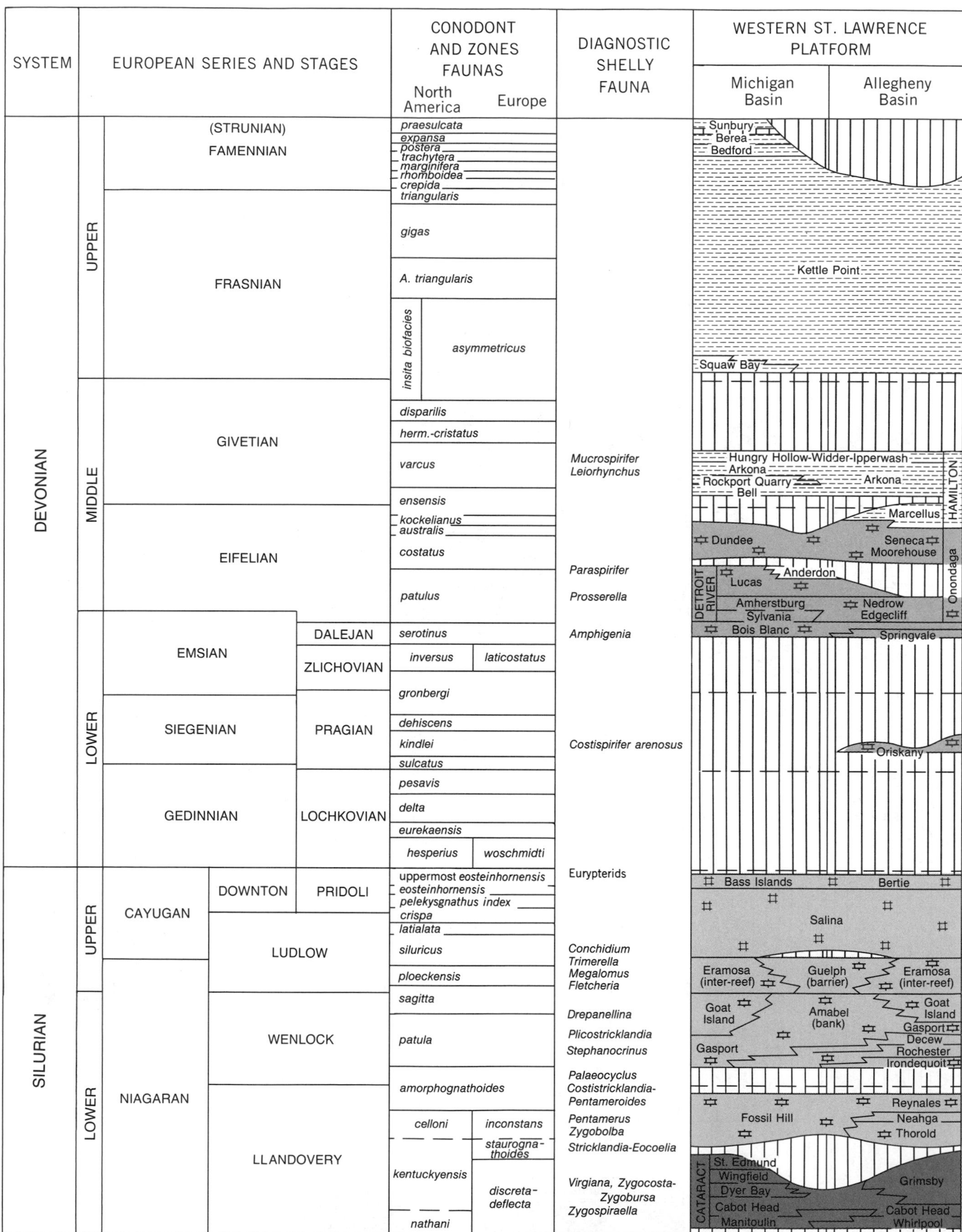

Figure 11.8. Silurian and Devonian biostratigraphy and regional correlation.

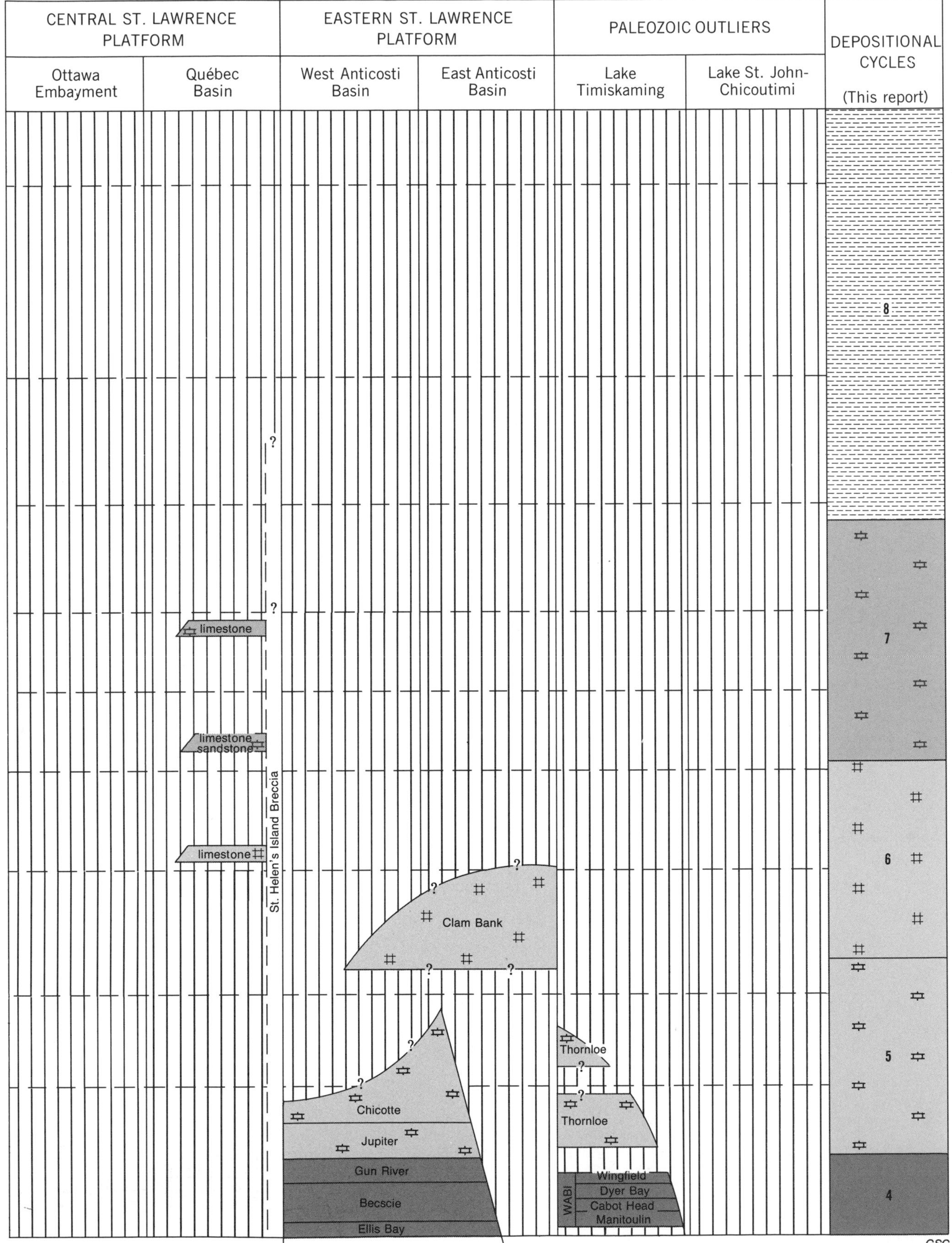

Figure 11.8. Cont.

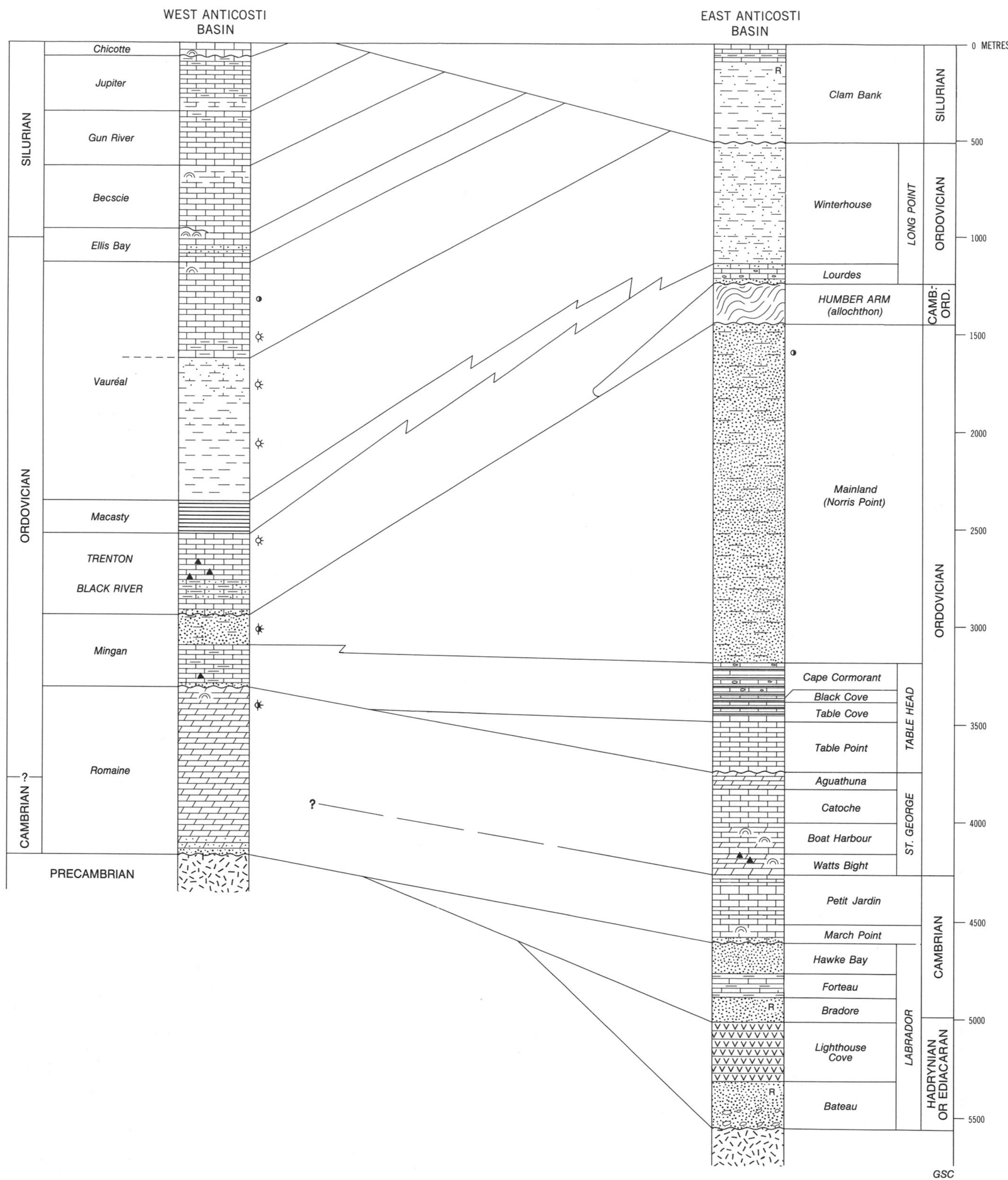

Figure 11.9. Composite stratigraphic succession - Eastern St. Lawrence Platform. For legend see Figure 11.10.

The youngest Paleozoic formations in southwestern Ontario, the Bedford, Berea, and Sunbury, have been assigned traditionally to the Tournaisian and thus to the Lower Carboniferous. The Tournaisian designation still stands, but the lower part of that succession ("Strunian") is now included in the Upper Devonian. The presence of the spore *Hymenozonotriletes lepidophytus*, a "Strunian" form, in these three formations indicates that they are uppermost Devonian (McGregor, 1970; Sanford, 1968).

The Devonian conodont succession of southwestern Ontario is summarized in Uyeno et al. (1982), and Sparling (1983). The Bois Blanc is assigned to the *serotinus* Zone (late Emsian), and the succeeding Detroit River Group and Dundee Formation range from the *patulus* Zone through the *costatus* and *kockelianus* zones (Eifelian). The upper strata of the Hamilton Group (Arkona and Ipperwash formations) are assigned to the lower *varcus* Subzone (lower Givetian), and the lower beds (Bell and Rockport Quarry formations) may belong in the *ensensis* Zone (straddling the Eifelian-Givetian boundary). Conodonts of the Kettle Point Formation range from the middle *asymmetricus* Zone (lower Frasnian) to at least as high as the *marginifera* Zone (Winder, 1966b).

DEPOSITIONAL CYCLES AND LITHOSTRATIGRAPHIC FRAMEWORK

The preserved geological record of the St. Lawrence Platform evolved through a long and complex series of tectonic and depositional events that began in late Proterozoic time and continued through much of the Paleozoic Era. Jurassic strata are preserved only locally. The pre-Mesozoic deposits are divided into nine tectonostratigraphic cycles as follows:

Cycle 9 — upper Tournaisian to Stephanian

Cycle 8 — Givetian to lower Tournaisian ("Strunian")

Cycle 7 — Siegenian to Eifelian

Cycle 6 — upper Ludlow to Gedinnian

Cycle 5 — upper Llandovery to lower Ludlow

Cycle 4 — lower to mid-Llandovery

Cycle 3 — upper Edenian to Gamachian

Cycle 2 — Whiterockian to lower Edenian

Cycle 1 — uppermost Proterozoic to Canadian

Most of the following discussion deals with Cycles 1 to 8, whose biostratigraphy and regional correlations are summarized in Figures 11.7 to 11.11. For each of these cycles, a map summarizing distribution, depositional facies, and thickness is presented (Fig. 11.12, 14, 19, 22, 24, 29, 31, 32). Cycle 9 and Jurassic (Kimmeridgian) strata are confined to small areas at the eastern and western extremities of the platform (Fig. 10.1) and receive only brief mention here.

Most, but not all, of the tectonostratigraphic cycles are bounded by disconformities. They represent the most logical units through which the succession of depositional and tectonic events can be demonstrated. These events are clearly linked to the origin and changing character of the Appalachian/Atlantic margin of Laurentia, and to orogeny resulting from interaction of that continental plate with neighbouring plates.

Cycle 1 – Uppermost Proterozoic to Canadian

Sedimentation on St. Lawrence Platform during late Proterozoic to Canadian time (Sauk Sequence of Sloss, 1963) took place under passive-margin (divergent) conditions. Precambrian and Lower Cambrian redbed clastics and volcanics were deposited in rift-basins from northern Michigan to western Newfoundland and southeastern Labrador prior to continental separation. Dyke swarms record the tensional regime (Sanford et al., 1985).

The opening of the Iapetus Ocean at the onset of seafloor spreading in Early Cambrian time (Williams and Hiscott, 1987) produced a broad continental shelf along the eastern margin of North America. This shelf persisted through Early Ordovician time, then was destroyed when the margin became convergent. Deposits of the Early and Middle Cambrian continental shelf are exposed only in the eastern platform and a narrow belt to the southwest (Fig. 11.12), the latter now largely obscured beneath allochthonous tectonic slices. In Late Cambrian and Early Ordovician time the shelf broadened, so that shelf strata of these ages now outcrop widely in the central and western parts of the platform also.

The Cambrian and Lower Ordovician sediments of the continental shelf were quartz-carbonate facies that gave way to shale and siltstone and carbonate breccias along the continental slope and rise. From their present erosional edges around the inner margins of the platform, rocks of this cycle thicken southeastward to more than 2000 m toward the reconstructed edge of the continental shelf, and to 1000 m or more in the Michigan Embayment.

During this time, uplift of the cratonic arches is recorded, involving Fraserdale, Algonquin, and Frontenac arches in Ontario and adjacent areas of the United States, Saguenay and Beaugé arches in Quebec, and probably parts also of the northeast-trending Laurentian Arch (Fig. 10.2). Movement of the Fraserdale, Algonquin, and Frontenac arches defined the Ottawa Embayment and Michigan Basin. The Algonquin and Frontenac arches also are known to have undergone subaerial erosion following Late Cambrian and Early Ordovician deposition prior to readvance of the sea and onset of Cycle 2 deposition in Middle Ordovician time.

Much of the continental shelf is inferred to be preserved but obscured beneath the Taconic allochthons and the foreland thrust and fold belt of the Appalachians, whereas the outer shelf, slope, and rise were tectonically fragmented and largely obliterated. Thus, the only record now available of the latter are remnants (i.e., Cow Head Breccia and Quebec Group breccias) preserved in some of the early thrust slices that now rest on the southeastern margin of St. Lawrence Platform.

The redbed clastics and volcanics that form the initial deposits of Cycle 1 are referred to the Bateau, Lighthouse Cove, and Bradore formations in western Newfoundland, Covey Hill Formation in Quebec Basin and Ottawa Embayment, and Jacobsville Formation in Michigan Basin and adjacent Lake Superior region (Fig. 11.9, 11.10, 11.11). The succeeding marine sequences of

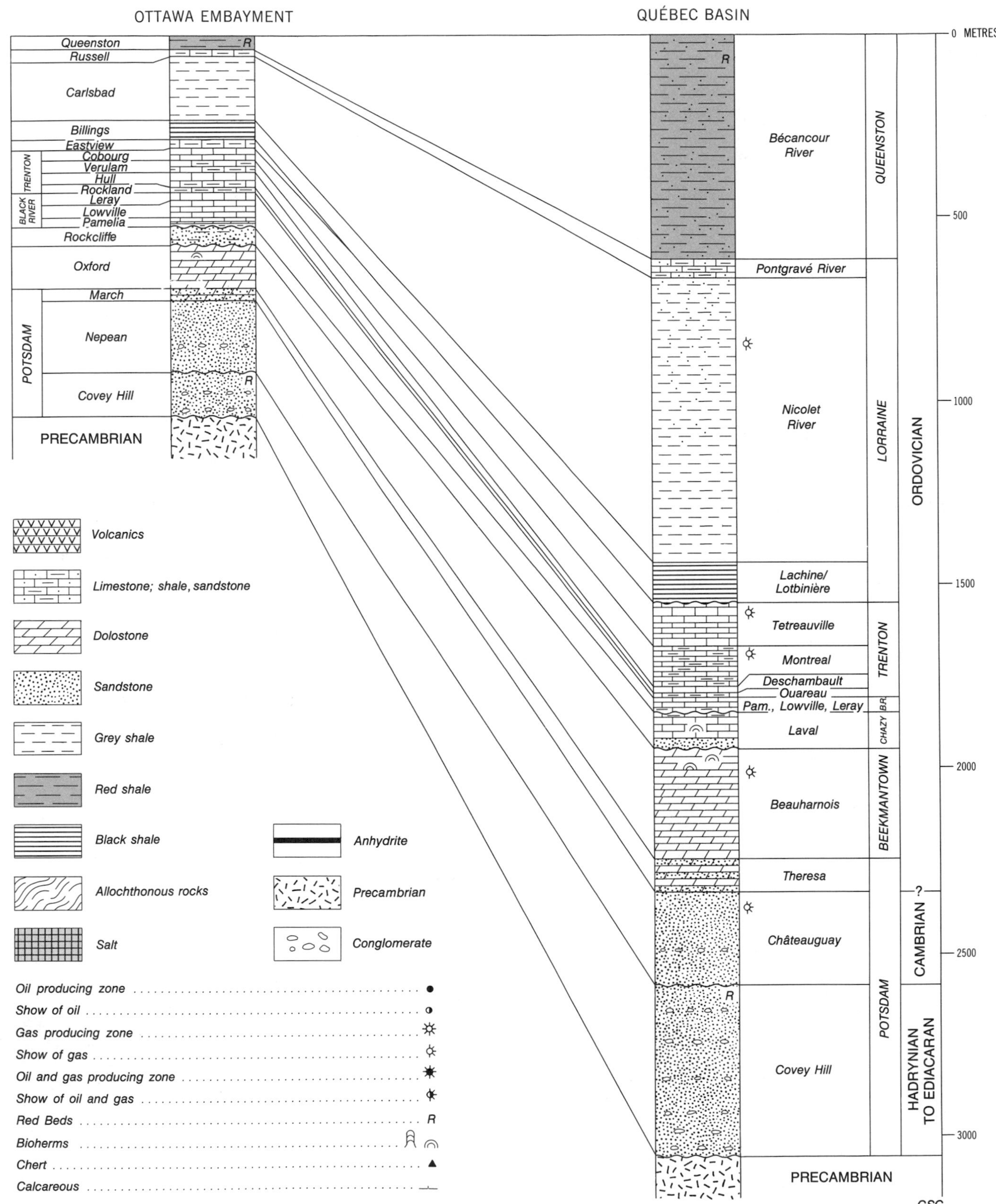

Figure 11.10. Composite stratigraphic succession - Central St. Lawrence Platform.

Cycle 1 are the Forteau, Hawke Bay, March Point, and Petit Jardin formations and St. George Group-Romaine Formation in Anticosti Basin, Châteauguay-Nepean, Theresa-March, and Beauharnois-Oxford formations in Quebec Basin and Ottawa Embayment respectively, and Upper Cambrian undivided in southwestern Ontario bordering the Michigan and Allegheny basins (Fig. 11.11).

Eastern St. Lawrence Platform

Initial deposits of depositional Cycle 1 are purplish conglomerate and quartzite of the Bateau Formation, up to 244 m thick (Fig. 11.9). They are confined to the extreme northern part of the Great Northern Peninsula of Newfoundland and adjacent regions of Belle Island and the southeast coast of Labrador, where they form discontinuous patches on an irregular, faulted basement surface.

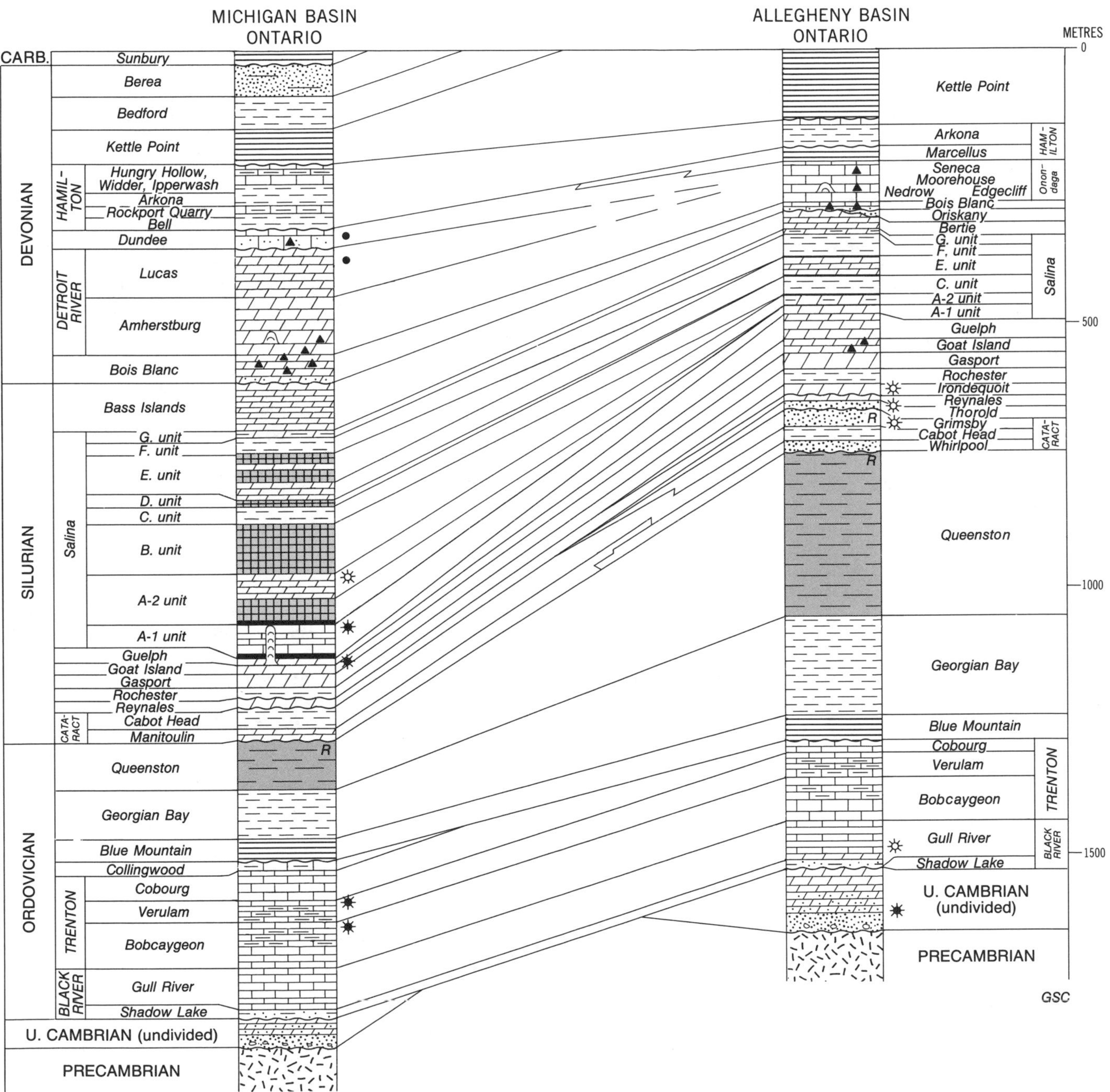

Figure 11.11. Composite stratigraphic succession - Western St. Lawrence Platform. For legend see Figure 11.10.

Black to dark green and purple to reddish-brown basalts of the Lighthouse Cove Formation, up to 310 m thick, overlie the Bateau. Their distribution coincides with that of the Bateau Formation, except that the basalts locally, on horst blocks, overlap the basal siliciclastics to lie directly on Precambrian basement.

The Bradore Formation, completing the basal nonmarine(?) sequence of Cycle 1, comprises red, pink, and grey, massive weathering, orthoquartzitic sandstones, arkoses, and conglomerates, up to 120 m thick. These overlie Lighthouse Cove volcanics in the Northern Peninsula of Newfoundland but overlap the latter to the southwest and northeast to lie directly on basement.

The radiometric age ($^{40}Ar/^{39}Ar$ method) of dykes that may have been feeders to the Lighthouse Cove volcanics is 605±10 Ma (Stukas and Reynolds, 1974). Carbonates conformably overlying the Bradore Formation contain a late Early Cambrian marine fauna. The Bateau, Lighthouse Cove, and possibly the lower part of the Bradore formations can thus be regarded as latest Proterozoic to early Early Cambrian.

The oldest marine Paleozoic strata in the eastern platform are the Forteau carbonates and shales. Remnants of the formation are present in southeastern Labrador bordering the Strait of Belle Isle, and a more continuous subsea outcrop belt extending to the southwest beneath the northeastern Gulf of St. Lawrence (Haworth and Sanford, 1976). To the southwest, the subsea Forteau outcrop belt is overlapped by progressively younger strata, reaching the St. George Group and its equivalent the Romaine Formation (in western Anticosti Basin) on approach to Beaugé Arch. From the Quebec-Labrador coast, the Forteau Formation dips to the southeast beneath Hawke Bay sandstones and younger Cambrian, Ordovician, and Silurian strata, but resurfaces in a thrust belt along the western and northern margins of the uplifted Great Northern Highlands of western Newfoundland.

The Forteau Formation consists of grey and bluish-grey, shaly, even- to nodular-bedded limestones, containing thick interbeds of dark grey and blue-grey shale. It thickens from about 119 m in the Strait of Belle Isle region to 224 m near Canada Bay. Much of the thickening and increased fine-clastic content presumably is due to increased water depth and more rapid subsidence toward the margin of the Cambrian continental shelf.

A common characteristic of the Forteau Formation is the presence of oncolites (button algae), which invariably occur in the upper beds, but may also occur throughout the formation. The Forteau is richly fossiliferous; coral-like archeocyathid reef mounds are characteristic in southeastern Labrador, and occur locally in western Newfoundland. An *Olenellus* trilobite fauna confirms a late Early Cambrian age for the formation.

Sandstones of the Hawke Bay Formation complete the Lower Cambrian succession in the eastern half of Anticosti Basin. The rocks parallel the distribution of the Forteau, outcropping on the seafloor beneath the northeastern Gulf of St. Lawrence and Strait of Belle Isle and along the northern and western margins of the Great Northern Highlands of western Newfoundland. In the latter area they extend as far south as Port au Port Peninsula (James and Stevens, 1982).

The Hawke Bay Formation is composed of pink and white, glauconitic, fine- to medium-grained orthoquartzitic sandstone up to 153 m thick. It also contains varying amounts of interbedded dark grey to black and locally green or red sandy shale and minor quartz pebble conglomerate. A variety of "shallow-water" sedimentary structures such as crossbedding, ripple-marks, and desiccation cracks are present. The well abraded detritus of the Hawke Bay was presumably derived from the Canadian Shield to the northwest and deposited in a nearshore peritidal environment. Fragments of *Olenellus* and other organisms confirm a late Early Cambrian age and point to a marine origin.

Hawke Bay deposition was followed by a period of subaerial erosion that continued well into early Middle Cambrian time. Because this break is widely recorded in North America and elsewhere (Palmer and James, 1980; Aitken, 1981), it probably records a eustatic drop in sea level.

Carbonates, shales, and sandstones of the March Point Formation (Knight, 1977, 1983; James, 1981; Lochman, 1938) succeed the Hawke Bay sandstones with disconformable contact. They form the bedrock surface beneath the southern Strait of Belle Isle and along the adjacent coast of western Newfoundland, again rising to surface around the margins of the Great Northern Highlands and in Port au Port Peninsula. The formation, 270 m or more thick, consists largely of nodular to uniformly bedded, grey to black dolostones and dolomitic limestones, all of shallow marine origin. The lower strata, consisting of sandy dolostones and interbedded glauconitic sandstones and shales, were presumably derived from the underlying Hawke Bay Formation. Shale partings and interbeds are common also in the upper part of the formation, as are complex, stromatolitic and thrombolitic biohermal mounds. Intraformational conglomerates occur throughout the March Point Formation. The strata are sparsely fossiliferous, but locally contain trilobite faunas that confirm a late Middle Cambrian age (Whittington and Kindle, 1966, 1969).

Carbonates of the Upper Cambrian Petit Jardin Formation succeed the March Point and have wide distribution along the northern and western margins of the Great Northern Highlands and in Port au Port Peninsula. The rocks are relatively unfossiliferous, light grey, yellow-weathering dolostones containing grey, red, and black dolomitic shale and minor blue-grey limestone (Knight, 1977, 1983; James and Stevens, 1982). Stromatolites occur sporadically throughout the formation, commonly associated with crossbedded oolitic dolostone or limestone. The formation is complexly faulted along the Strait of Belle Isle, and its apparent thickness of 270 m in that region is estimated.

Cycle 1 in the Eastern St. Lawrence Platform terminates with limestones and dolostones of the St. George Group and equivalents in the Romaine Formation. The St. George Group, with wide outcrop distribution beneath the northeast Gulf of St. Lawrence and bordering the Great Northern Highlands and in Port au Port Peninsula, is divisible into the following four formations (Knight, 1983), in ascending order:

i. Watts Bight Formation, 90 to 100 m: massive-bedded, dark grey to black cherty dolostones. Stromatolite mounds, gastropods, and cephalopods are common throughout;
ii. Boat Harbour Formation, 135 to 155 m: a lower unit of light grey dolostone with extensive collapse breccias, interbedded with blue-grey and white limestone; and an upper unit of grey limestone with interbedded light grey dolostone or dolomitic limestone;
iii. Catoche Formation, 105 to 165 m: grey, rubbly weathering limestone in the lower part, grading upward to light and dark grey bituminous dolostone. Secondary white sparry dolomite is widespread, forming pseudobreccia textures;
iv. Aguathuna Formation, 70 to 90 m: massive to laminated, yellow-weathering dolostone and interbedded limestone.

Northwest of Beaugé Arch, the St. George Group merges with the Romaine Formation. In western Anticosti Basin, the Romaine overlaps older strata to rest on basement.

The Romaine Formation outcrops in the Mingan Islands and the immediately adjacent area (north shore) of the Quebec mainland. The beds dip southwest beneath Jacques Cartier Channel and reach a thickness of 365 m beneath the west-central part of Anticosti Island, and to 855 m beneath the south shore. Some of the lower beds, although mostly Lower Ordovician, could conceivably include Middle and Upper Cambrian beds equivalent to the March Point and Petit Jardin formations of western Newfoundland.

The Romaine consists of light grey to dark brownish-grey dolostones with minor interbeds of dark brown limestone. Thin sandstone usually forms the base. The yellow-weathering, massive dolostones contain numerous stromatolites, which locally swell into small biohermal mounds. They have many lithological and biostratigraphical features in common with the St. George Group (particularly Catoche Formation) of western Newfoundland, and the Beauharnois and Oxford formations of the Quebec Basin and Ottawa Embayment, and undoubtedly formed a part of the same depositional sequence. The St. George Group and Romaine Formation are largely Lower Ordovician (Canadian), but apparently bridge the Lower-Middle Ordovician boundary to extend upward into the Whiterockian Stage (Nowlan, 1981; Stouge, 1982).

Thinning of Cycle 1 over Beaugé Arch (Fig. 11.12) indicates that the arch was active during and immediately following deposition. Elsewhere on St. Lawrence Platform, the evidence suggests that most of the epeirogenic movement affecting Cycle 1 deposition took place at or near the close of the Early Ordovician, and resulted in widespread emergence of the platform and subaerial erosion.

Central St. Lawrence Platform

The basal strata of Cycle 1 (Fig. 11.10) are redbeds known as the Covey Hill Formation. The formation has many lithological and structural characteristics in common with the Bateau and Bradore formations of the Strait of Belle Isle region, and the Jacobsville Formation of Michigan Basin and adjacent Lake Superior region. The beds occur sporadically on and marginal to Frontenac Arch, but are continuous in subsurface to the east, beneath much of Ottawa Embayment and Quebec Basin. The rocks are red, pink, and white, crossbedded, orthoquartzitic sandstones, arkosic and conglomeratic at the base. Quartz-pebble conglomerate lenses occur in abundance in the lower beds and sporadically higher in the formation. The formation, highly variable in thickness, probably reaches its maximum development of more than 517 m near the type locality in southern Quebec bordering the Adirondack Mountains (Clark, 1972).

Covey Hill detritus was derived mainly from the newly rejuvenated Frontenac and Laurentian arches bordering the central platform on the west and northwest respectively. Thick, locally derived conglomeratic wedges here and there along the axis and margins of Frontenac Arch indicate reactivation of some of the Precambrian faults, which mainly trend northeasterly.

The Covey Hill, like the Bateau-Lighthouse Cove-Bradore sequence, is continental, and distinct from the marine sandstones and carbonates of the overlying Upper Cambrian and Lower Ordovician sequences. It is structurally different also, having been deposited on a horst-and-graben basement surface created by rifting prior to and during continental separation. The formation is thus assumed to be uppermost Proterozoic to lowest Cambrian, as are similar rocks in the extreme eastern part of the St. Lawrence Platform.

In the central platform, Lower and Middle Cambrian marine strata were not deposited. The marine facies of Cycle 1 are represented by Upper Cambrian and Lower Ordovician strata of the Chateauguay-Nepean, Theresa-March, and Beauharnois-Oxford formations, in ascending order.

Sandstones of the Chateauguay Formation (Quebec Basin) and equivalent Nepean Formation (Ottawa Embayment) overlie Covey Hill redbeds with profound unconformity, and locally overlap these to rest on basement rocks. Both formations are typically white and grey-buff to rusty yellowish orange, orthoquartzitic sandstone of exceptional maturity (Fig. 11.13). Quartz pebble conglomerate, arkose, and shaly interbeds are common at the base, and occur less commonly higher in the sequence. From its erosional edge, the Nepean thickens eastward from 180 m in Ottawa Embayment to more than 245 m in western Quebec Basin where it is known as the Chateauguay Formation. Primary sedimentary structures in the Nepean and Chateauguay formations point to shallow marine and fluvial depositional environments. The source of detritus was rejuvenated highlands of the Frontenac and Laurentian arches. The Nepean and Chateauguay are sparsely fossiliferous but are lithologically similar and presumably equivalent at least in part to the Potsdam Formation of known Late Cambrian age in northern New York.

Theresa and March formations gradationally succeed the Chateauguay and Nepean respectively, and in some localities around the margins of Ottawa Embayment and Quebec Basin overlap them to rest on basement. The rocks are light grey to white orthoquartzitic sandstone

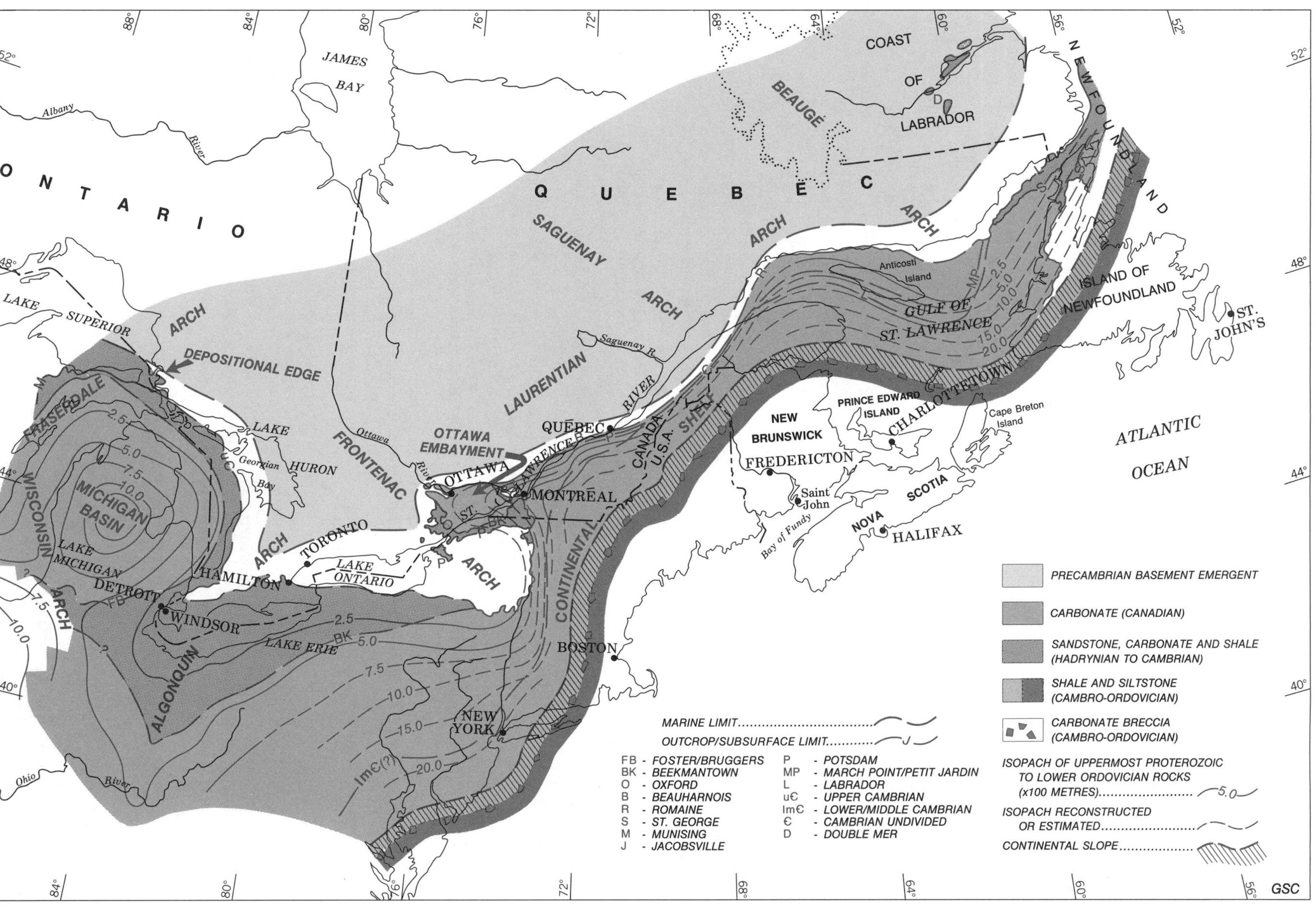

Figure 11.12. Depositional Cycle 1 - Upper Proterozoic/Ediacaran to Canadian.

Figure 11.13. Upper Cambrian to Lower Ordovician Nepean Formation, Highway 15 near Elgin, Ontario. Flat-lying orthoquartzitic sandstone and conglomerate lie unconformably on steeply dipping Precambrian gneiss (photo - B.V. Sanford, GSC 9-2-68).

interbedded with light to medium grey dolostone. Dolostone is generally subordinate in the lower part, increasing upward. With the upward disappearance of sandstone, the Theresa and March give way to Beauharnois and Oxford carbonates. The March Formation is 10 m thick or less around the margins of Ottawa Embayment, increasing to about 60 m in the centre. Equivalent Theresa strata to the east thicken from 75 m in the Montreal area to 150 m or more in the southern Quebec Basin.

Cycle 1 in the central platform ends with the Beauharnois and Oxford formations. The rocks along the margins of the platform are light to dark greyish-brown, stromatolitic dolostones, giving place basinward to dark brownish-grey to black micritic dolostones with interbeds of limestone and shale. They thicken eastward, from 125 m in the Ottawa Embayment to 245 m in the Montreal area and 460 m in the southeastern Quebec Basin.

The Theresa/March and Beauharnois/Oxford formations are characteristically dolomitic and sparsely fossiliferous, except for the widespread occurrence of algal stromatolites. In combination with intraformational breccias, gypsum lenses, halite casts, and mudcracks, these characteristics point to a variety of depositional environments varying through restricted marine to lagoonal, intertidal, and supratidal (Hofmann, 1972).

The Theresa/March and Beauharnois/Oxford formations where exposed along the margins of Quebec Basin and Ottawa Embayment are of Early Ordovician age. The March Formation along the eastern margin of Frontenac Arch is largely of Tremadoc age, on the basis of its conodont fauna (Bond and Greggs, 1973), but may range as young as Arenig (T.T. Uyeno, pers. comm.). The succeeding Oxford and Beauharnois formations are also of Arenig age. In the deeper parts of the Quebec Basin, however, the Theresa and lower beds of the Beauharnois may be as old as Late Cambrian.

Western St. Lawrence Platform

The basal redbeds of Cycle 1 in the western platform are the sandstones and conglomerates of the Jacobsville Formation. In southwestern Ontario, they occur as small subsurface outliers up to 6 m thick between Sault St. Marie and the western extremity of Manitoulin Island. The formation is continuous from the Sault St. Marie region to the Keweenawan Peninsula of northwestern Michigan. It consists of red, reddish-brown, pink or white, crossbedded orthoquartzitic sandstone containing intervals of red siltstone and shale. Lenses of conglomerate occur throughout the Jacobsville, but are more abundant near the basal contact with Aphebian or older, crystalline basement. The formation is more than 300 m thick only in a narrow outcrop belt along the northern peninsula of Michigan, where it dips north beneath Lake Superior. Jacobsville detritus was derived from a "Northern Michigan Highland" (Hamblin, 1958), which was presumably part of Fraserdale Arch connected with Wisconsin Dome beneath what is now the northern peninsula of Michigan.

The age of the Jacobsville is uncertain. It has been variously classified as Proterozoic, Early, Middle, and Late Cambrian. Inasmuch as the formation is structurally discordant with overlying rocks of Late Cambrian age, it is assumed to be significantly older. Similar rocks in the eastern platform that have been dated as Late Proterozoic to early Early Cambrian record rifting, prior to continental separation, that affected the entire St. Lawrence Platform. The Jacobsville is probably a product of the same rifting event.

Succeeding marine deposits in southwestern Ontario are the undivided Upper Cambrian rocks confined to subsurface around the margins of the Algonquin Arch (Fig. 11.12), where their bevelled edges are overlapped disconformably by Middle Ordovician (Black River) limestones. These rocks are erosionally reduced from their former much greater extent by rejuvenation of Algonquin Arch after deposition of Cycle 1. To the south and west, the section is progressively thicker and includes younger Cambrian and Lower Ordovician strata toward the centres of Allegheny and Michigan basins.

In southwestern Ontario, the Upper Cambrian is represented by three principal facies: a lower facies of white to buff orthoquartzitic sandstone; a middle facies of glauconitic, orthoquartzitic sandstone alternating with dolostone and grey to tan oolitic dolostone; and an upper facies of tan to light brown dolostone. From their erosional edges along the margins of Algonquin Arch, the strata thicken from 150 m beneath central Lake Erie to more than 2000 m in Allegheny Basin. On the western flank of the arch, approximately 75 m of strata are preserved; these thicken to more than 1300 m in Michigan Basin.

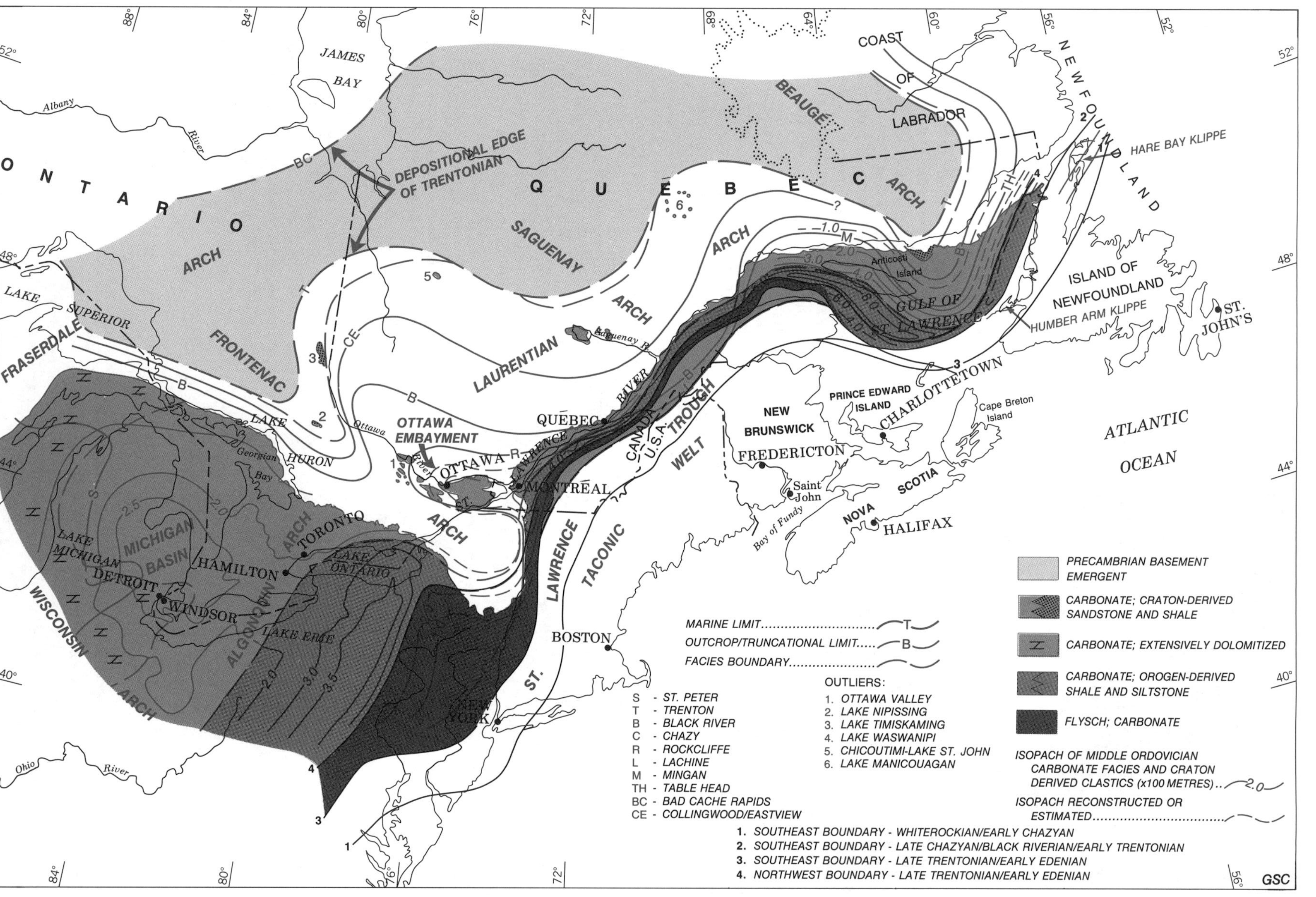

Figure 11.14. Depositional Cycle 2 - Whiterockian to early Edenian.

Figure 11.15. Middle Ordovician Lourdes Formation of the Long Point Group on Long Point Peninsula, western Newfoundland, view southwestward. The southeast-facing escarpment of resistant fossiliferous limestone forms the peninsula (photo - B.V. Sanford, GSC 2-11-68).

Cycle 2 – Whiterockian to lower Edenian

In early Middle Ordovician time, the broad continental shelf at the divergent, Iapetus margin of North America, upon which Cycle 1 had accumulated, was destroyed by the onset of convergent tectonism. The load of thickened and stacked tectonic sheets at the northwestward-moving front of the Taconian Orogen depressed the earlier shelf to form a foreland trough (Fig. 11.14). Southeastward transport of craton-derived detritus onto and across a shallow shelf gave way to cratonward transport of detritus from new tectonic highlands formed as Iapetus began to close. Depositional Cycle 2 comprises the earliest deposits of the newly formed trough and the foreland platform separating it from the Canadian Shield. Anticosti Basin was established at this time, its eastern limit being the structural salient of the Humber Arm and Hare Bay allochthons.

Cycle 2 commences at an unconformity widely expressed in North America and marking the base of the Tippecanoe Sequence of Sloss (1963). The ensuing transgression, first recorded only in the Eastern and Central St. Lawrence Platform, reached the western platform in late Middle Ordovician time, and progressed to submerge most of the Canadian Shield during the Trentonian and early Edenian (Fig. 11.14). This eustatic rise of sea level coincided with Taconian orogeny.

Rocks of Cycle 2 thicken from their erosional edge along the northern margins of the platform toward the marginal trough. Carbonates with minor sandstone and shale, the principal facies on the platform, reach their maximum thickness in Michigan Basin and along the extreme southern margin of the platform, before giving way abruptly to orogen-derived flysch (shale, siltstone) in the trough. The thickness of the latter facies, now deformed within the orogen, is unknown. Depositional thinning and local non-deposition of Cycle 2 in various parts of the platform (Fig. 11.14) record activity of the Frontenac, Saguenay, and Beaugé arches.

Cycle 2 is divisible into two depositional sub-sequences (Fig. 11.7, 11.9, 11.10, 11.11): (1) a lower sub-sequence, missing in the western platform, that comprises the Table Head Group, Mainland (Norris Point), Goose Tickle, and Mingan formations in the eastern platform, and Laval and Rockcliffe formations in the central platform; and (2) an upper sequence referred to the Black River and Trenton groups throughout the platform, except in the eastern Anticosti Basin (western Newfoundland), where equivalent strata form the lower Long Point Group (Lourdes and basal Winterhouse formations).

Upper Ordovician rocks have been dated on the basis of graptolites on the one hand, and conodonts on the other. Depending on which zonal scheme is followed, a discrepancy of as much as a stage may occur at the Edenian-Maysvillian level. The problem is unresolved and explains the differences in age assignments between the text and the Correlation Chart (in pocket).

Eastern St. Lawrence Platform

The initial deposits of Cycle 2 are the limestones, shales, and carbonate breccias of the Table Head Group. These form the bedrock surface beneath the northeast Gulf of St. Lawrence and around the margins of the Humber Arm and Hare Bay allochthons (see Fig. 10.1). Their reconstructed depositional edges in the southern part of the gulf describe an arc around the prominent nose of the Beaugé Arch in subsurface, to merge with the Mingan Formation in western Anticosti Basin (Fig. 11.14).

The Table Head Group is divisible into four principal stratigraphic units as follows, in ascending order (Klappa et al., 1980; Cumming, 1983):

i. Table Point Formation, 256 m: dark brownish-grey, finely crystalline limestone with minor interbeds of dolostone. Thin interbeds and partings of dark grey to black shale are common throughout, increasing in number toward the top. The contact with the underlying St. George Group is disconformable in most localities in western Newfoundland;

ii. Table Cove Formation, 94 m: crinoidal and shaly bioclastic limestones at the base, grading upward to turbiditic limestones and interbedded shales. Lenses of limestone breccia occur in the upper part of the formation, and some of the beds have undergone large-scale slump folding;

iii. Black Cove Formation, 9 m: black graptolitic shales and mudstones, representing the transition from dominant limestone facies of the Table Point/Table Cove Formation to the megabreccias and flysch deposits of the succeeding Cape Cormorant Formation;

iv. Cape Cormorant Formation, 180 m: dark grey to black calcareous shales and green siltstones with interbedded turbiditic limestones and limestone breccias.

Gradationally succeeding the Table Head Group in western Newfoundland is a unit of grey micaceous sandstones and shales containing interbeds of grey limestone, up to 1700 m thick (James and Stevens, 1982). This unit forms the top of the autochthonous sequence on the Port au Port Peninsula and is informally referred to as the Mainland sandstone. Strata approximately equivalent to the upper Table Head Group and Mainland sandstone in the Canada Bay-Hare Bay area of western Newfoundland are the black shales, siltstones, greywackes, and chaotic conglomerates of the Goose Tickle Formation (Williams and Smyth, 1983).

West of Beaugé Arch, initial deposits of Cycle 2 are represented by limestones of the Mingan Formation. These fringe the Quebec shoreline, form Mingan Islands, and dip south beneath Anticosti Island and the northwestern Gulf of St. Lawrence. The basal strata, resting disconformably on the Romaine Formation, consist of carbonate conglomerate, sandstone, and shale. These are succeeded by brown, bioclastic and calcarenitic limestones grading upward in turn to brown micritic limestones. In boreholes on Anticosti Island, the upper beds of the Mingan consist of grey micaceous sandstones containing interbeds of limestone and shale. The formation is 47 m thick at the Mingan Islands, 136 m in west-central Anticosti Island, and 366 m beneath the south shore of the island. Its absence beneath the eastern end of Anticosti Island is presumably due to erosion following reactivation of the Beaugé Arch at the close of Chazyan time.

The sandstone beds of the upper Mingan Formation beneath Anticosti Island, up to 155 m thick, form a wedge that thickens rapidly to the south. They are lithologically similar to the Mainland sandstone of Port au Port Peninsula, and are assumed to be equivalent, in part at least, and derived from the orogen. Both the Mingan and Mainland have been dated as Chazyan (Barnes et al., 1981b). The Table Head Group, on the other hand, is Whiterockian (Klappa et al., 1980; Barnes et al., 1981a) and is thus older in part than the type Mingan. It is conceivable, however, that some or all of the Mingan limestones in the subsurface beneath the south shore of Anticosti Island may also be Whiterockian, and equivalent in part to the limestones and shales of the Table Head Group.

The cratonward-migrating, Taconian structural front reached its present position in western Newfoundland at or near the close of Chazyan time, with the emplacement of the Humber Arm and Hare Bay allochthons over the eastern flank of the St. Lawrence Trough. Subsequently, thick sequences of Blackriveran, Trentonian, and lower Edenian clastic and carbonate rocks continued to be deposited in the trough and the foreland platform, burying the leading edge of the allochthons (Fig. 11.6). In western Newfoundland, these are the limestones, shales, and sandstones of the Long Point Group (Rodgers and Neale, 1963; Bergström et al., 1974; Fåhraeus, 1973; Copeland and Bolton, 1977). They rest with profound unconformity on allochthonous rocks of the Humber Arm Supergroup on Port au Port Peninsula, and with contact of unknown character on the Mainland sandstone off the edges of the allochthon beneath the northeast Gulf of St. Lawrence. The contact between the Mainland sandstone and Long Point Group thus dates the emplacement of the Humber Arm and Hare Bay allochthons.

Where exposed on Port au Port Peninsula, the Long Point Group consists of two units, the Lourdes and overlying Winterhouse formations. The Lourdes Formation (Fig. 11.15), 85 m thick, consists of three limestone subunits (members I, II, and III), each with characteristic sandstone and shale layers at the base. Small coral (*Labrinthites*) bioherms are a characteristic feature of the uppermost beds of member II. Alternating shales, siltstones, and sandstones of the Winterhouse, up to 300 m thick, conformably succeed the Lourdes.

The Lourdes Formation has been referred to the Porterfieldian-Bolarian (Chazyan-Blackriveran) on the basis of shelly and conodont faunas (Bergström et al., 1974). On the other hand, an ostracode assemblage from the shaly beds at the base of member III of the Lourdes points to a Wilderness-Barneveld (Trentonian) age for that part (Copeland and Bolton, 1977). The lower beds of the overlying Winterhouse Formation may also be Middle Ordovician (Kirkfieldian and Shermanian) on the basis of shelly (Bergström et al., 1974) and conodont faunas (Fåhraeus, 1973). The upper Winterhouse is considered Cincinnatian (Edenian to Maysvillian) on graptolite evidence (J. Riva, in Bergström et al., 1974).

Limestones of the Black River and Trenton groups, undivided on the west side of Anticosti Basin, are coeval with the Lourdes and lower Winterhouse formations. These overlie the Mingan Formation, probably disconformably, and consist of tan to medium and dark brown limestone, with minor sandstone, siltstone, and shale at base, and grey to greenish-grey shale partings and interbeds throughout. Black River and Trenton strata dip southwest from the subsea outcrop belt south of the Mingan

Figure 11.16. Middle Ordovician Laval Formation (Ste. Therese Member), Legace quarry, Laval, Quebec. The lower part is interbedded sandstone and shale, and the upper part sandy calcarenite and calcarenite (photo - H.J. Hofmann).

Islands. They are about 300 m thick in the subsurface beneath west-central Anticosti Island and more than 400 m along the south shore. Farther south and southeast, the carbonates of the Trenton are continuous with the Lourdes Formation and the shales, siltstones, and sandstones of the lower Winterhouse Formation of the Long Point Group. From western and central Anticosti Island, Black River and Trenton carbonates thin to 200 m or less toward Beaugé Arch, where they overlap the Mingan to rest disconformably on the Lower Ordovician Romaine Formation, and ultimately on basement over the crest of the arch (Fig. 11.14).

Deep marine conditions continued in the greatly narrowed St. Lawrence Trough following emplacement of the Humber Arm and Hare Bay allochthons. On the other hand, the carbonates and sands of the Long Point Group reflect shallower, higher energy conditions. The Taconian structural front had no relief during Blackriveran, Trentonian, and early Edenian times.

Central St. Lawrence Platform

Deposition of Cycle 2 in the central platform was confined to the Chazyan, Blackriveran, Trentonian, and early Edenian ages. The initial deposits are the Chazy Group, more commonly referred to the Rockcliffe Formation in the Ottawa Embayment and the Laval Formation in the Quebec Basin (Fig. 11.10).

The Rockcliffe Formation, now separated from the Laval by the Beauharnois Arch (Fig. 10.2), is exposed along the north, east, and south margins of the Ottawa Embayment and is overlapped northwest of Ottawa by carbonates of the Black River and Trenton groups. In the western embayment, the rocks are dark grey and olive green, and locally red shales, siltstones, and greenish-grey sandstones, up to 50 m thick (Wilson, 1964). In the eastern embayment, the upper sandstone beds of the Rockcliffe intertongue with calcarenites once continuous with the St. Martin Member of the Laval Formation in the Quebec Basin to the east.

Figure 11.17. Middle Ordovician Gull River Formation, roadcut on Highway 401 near Kingston, Ontario. The limestones lie unconformably on Precambrian (Grenvillian) granites and dip gently southwest (photo - B.V. Sanford, GSC 200841P).

The Laval Formation outcrops along the west and north margins of Quebec Basin, but is overlapped to the east by limestones of the Black River Group and ultimately the Trenton Group, near Saguenay Arch. The formation, up to 95 m thick, consists of a basal, Ste. Thérèse, member of sandstone and shale (Fig. 11.16), gradationally succeeded by shaly limestones with interbeds of calcarenite, the St. Martin Member, and lime mudstones of the Beaconsfield Member (Clark, 1972; Hofmann, 1972). Within the St. Martin facies are small biohermal mounds with widespread coral and bryozoan framework.

The Rockcliffe Formation and Ste. Thérèse Member of the Laval Formation form a terrigenous wedge thickening to the west and northwest. This wedge apparently records rejuvenation of the Frontenac and Laurentian arches in Chazyan time.

The Rockcliffe and Ste. Thérèse sandstones contain a variety of primary sedimentary features, such as ripple-marks, desiccation cracks, and cut-and-fill structures, presumably recording fluvial and marine-deltaic deposition. The equivalent carbonates in the eastern Ottawa Embayment and Quebec Basin record marine, high-energy to quiescent environments representative of shelf edge and open shelf respectively.

Carbonates of the Black River and Trenton groups succeed the Rockcliffe and Laval formations in the central platform and represent a second phase of Cycle 2 deposition. The rocks, now confined to small remnants in Ottawa Embayment and around the margins of Quebec Basin, once broadly overlapped the Chazy and older strata of Cycle 1 to extend far across the craton to the northwest.

Combined Black River and Trenton strata are 213 m thick in the central part of Ottawa Embayment and along the northern margin of Quebec Basin, and thicken to more than 300 m in the central basin before their upper part (upper Trenton) gives way abruptly to black shales of the Utica Group. They thin eastward toward Saguenay Arch, where the Black River and lower Trenton were not deposited. On the crest of the arch, from Quebec City to the entrance of the Gulf of St. Lawrence, a relatively thin Trenton Group rests on basement. Trenton rocks underlie a narrow belt beneath St. Lawrence River and locally outcrop on the north shore of the river. As elsewhere in Quebec Basin, they dip southward beneath Appalachian allochthons.

The Black River Group (Clark, 1972; Wilson, 1964) consists of three formations, in ascending order (thicknesses in parenthesis refer to Ottawa Embayment and Quebec Basin respectively):

i. Pamelia Formation (23 m, 3 m): buff-weathering dolostone with interbeds of shale, limestone, and minor sandstone;

ii. Lowville Formation (52 m, 5 m): largely thin bedded, tan to dark brown micritic limestone;

iii. Leray Formation (8 m, 9 m): massive and thick-bedded, dark brownish-grey fossiliferous limestone.

The cause of thinning of the Black River Group, from 83 m in Ottawa Embayment to 17+ m in Quebec Basin, and parallel thinning of the lower units of the succeeding Trenton, is not well understood.

The formations of the Trenton Group, unlike those of the Black River, have long been identified by different stratigraphic nomenclature in Ottawa Embayment and Quebec Basin. The rock units are reasonably similar and readily correlated, however, and are combined here for brevity, as follows:

i. Rockland (13 m) and Ouareau (5 m) formations: medium- to thick-bedded, dark brownish-grey, argillaceous, fine-grained limestones, with dark grey shale partings and interbeds throughout.

   The Rockland and Ouareau are characterized by important facies changes as they are traced across Ottawa Embayment and Quebec Basin respectively. Rockland beds, as described above in the central and eastern part of the Ottawa Embayment, change abruptly to coarse-grained, crossbedded calcarenites, obviously the product of a high-energy, marine (littoral?) environment, on approach to the margin of Frontenac Arch. The Ouareau, applicable mainly to the western part of Quebec Basin, also changes to quite different and distinctive rock units, variously referred to as Fontaine, St. Alban, and Pont Rouge formations, northeastward toward Saguenay Arch;

ii. Hull (37 m) and Mile End/Deschambault (8 m) formations: dark grey, argillaceous limestone, with dark grey to black shale interbeds (lower Hull-Mile End), succeeded by light grey, massive-bedded calcarenite and bioclastic limestone (upper Hull-Deschambault);

iii. Verulam (30 m) and Montreal (114 m) formations: thin-bedded, nodular-bedded, dark grey, shaly limestone, with numerous interbeds of dark grey to black shale;

iv. Cobourg (55 m) and Tetreauville (122 m) formations: dark brown, micro-crystalline limestone. The Tetreauville, as it is known in western Quebec Basin, gives place northeastward, toward Quebec City, to rhythmically bedded lithographic limestones with interbeds of dark grey shale, locally referred to as the Neuville Formation.

In contrast to the eastward thinning of Rockland/Ouareau, and Hull/Mile End-Deschambault (seen also in the underlying Black River Group), upper Trenton strata increase in thickness southeastward across Quebec Basin and acquire greatly increased shale and siltstone contents. These changes record accelerated subsidence of the St. Lawrence Trough, presumably caused by intensified orogeny and, possibly, advance of the Taconian front, in late Trentonian time (boundary 3 in Fig. 11.14).

Western St. Lawrence Platform

In early Middle Ordovician, Chazyan time, the western platform was emergent. The carbonates of Blackriveran, Trentonian, and early Edenian ages were deposited directly on Upper Cambrian strata, overlapping the bevelled edges of the latter to rest directly on Precambrian basement along the axis of Algonquin Arch.

Figure 11.18. Ordovician limestones and minor shales, approximately equivalent to the Cobourg, Tetreauville, and Neuville formations of the St. Lawrence Lowland, exposed at Chute aux Galets, Shipshaw River, Quebec (Chicoutimi outlier) (photo - B.V. Sanford, GSC 200841R).

In southwestern Ontario, Black River and Trenton groups are similar to their counterparts in Ottawa Embayment and Quebec Basin. They outcrop across southwestern Ontario in a broad belt extending from eastern Lake Ontario to Georgian Bay, thence northwestward beneath Lake Huron to Manitoulin Island and the northern peninsula of Michigan (Sanford and Baer, 1981). The initial deposits of this sequence, the Black River Group, form two formations (Sanford, 1961; Liberty, 1969), in ascending order:

i. Shadow Lake Formation, 0 to 15 m: grey-green and locally red, dolomitic shale, siltstone, and sandstone, with minor interbeds of tan and brown limestone or dolostone;

ii. Gull River Formation, 23 to 122 m: thin-bedded, brown to grey-tan and cream coloured, finely crystalline to lithographic limestones; interbeds of light tan dolostone and minor shale are common in the lower part (Fig. 11.17).

The Black River Group thins from a maximum of 135 m in Ontario, bordering Michigan and Allegheny basins, to less than 30 m over the structurally higher part of Algonquin Arch. Toward the southern margin of Allegheny Basin in New York and Pennsylvania, the group thins, and eventually disappears near the Taconian front, posing an unresolved tectonic problem. Black River strata, lithologically consistent over broad areas of the western platform, are fine in texture, sparsely fossiliferous and commonly stromatolitic, and may have been deposited in a hypersaline, lagoonal environment.

Conformably succeeding the Black River are the shaly limestones and shales of the Trenton Group as follows:

i. Bobcaygeon Formation, 57 to 91 m, with lower, Coboconk-Selby/Napanee members and an upper, Kirkfield Member. The Coboconk Member consists of tan and light brown calcisiltite and calcarenite on the Algonquin Arch and adjacent areas bordering Michigan Basin. These grade to grey and grey-tan, thin-bedded, argillaceous limestone and dark grey shale of the

Selby-Napanee members off the arch to the southeast. The overlying Kirkfield Member consists of dark grey, thin-bedded limestone and interbedded shale, the shaly units decreasing in number upward, giving place to greyish brown, finely crystalline and bioclastic limestones at the top;

ii. Verulam Formation, 27 to 37 m: grey to dark greyish brown, finely crystalline to fragmental limestone with shale partings, and interbedded dark grey shale. The upper few metres of the formation is white and grey mottled, fragmental or bioclastic limestone;

iii. Cobourg Formation (Lindsay Formation of Liberty, 1969), 12 to 61 m: brown to dark brown and greyish brown, evenly thin-bedded microcrystalline limestone with a few shale partings and minor interbeds of bioclastic limestone at or near the base.

Trenton strata thin eastward from about 150 m in Michigan Basin to 120 m over Algonquin Arch, and thicken to 160 m in Allegheny Basin. Southeastward across Allegheny Basin, the Cobourg, Verulam, and Bobcaygeon carbonates gradually give place (in that order) to black shales of the Utica Group. The Utica, in turn, ultimately gives place to flysch and black shale facies of the Canajoharie Formation toward the Taconican front.

Trenton strata, in contrast to Black River, contain a profuse shelly fauna that points to widespread open marine conditions.

Paleozoic outliers

In contrast to the Chazyan seas, which did not extend far outside St. Lawrence Trough, Blackriveran, Trentonian, and early Edenian seas transgressed widely over the southern Canadian Shield, as evidenced by the widely separated Paleozoic outliers illustrated in Figure 11.14.

Blackriveran deposition was confined largely to the outer margins of the craton (Ottawa Valley outliers), whereas overlapping Trentonian and early Edenian deposits extended far to the north and are preserved in outliers at Chicoutimi and Lakes Timiskaming, Nipissing, Waswanipi, Manicouagan, and Lac St. Jean.

The Lake Timiskaming outlier is the largest of the group and contains the following succession representing Cycle 2 (Russell, 1984):

i. Guigues Formation, 30 m: red and grey-green orthoquartzitic and arkosic sandstones;

ii. Bucke Formation, 24 m: grey and grey-green, sandy shales, containing medium-grey, argillaceous, microcrystalline limestone beds, the latter increasing in numbers upward to grade into the overlying;

iii. Farr Formation, 30 m: medium to dark brown, nodular, microcrystalline limestone. The beds contain dolomitized burrows that produce a characteristic yellowish orange mottled effect. The Farr limestones, of early Edenian age, are lithologically and faunally similar to the Bad Cache Rapids Group of the Hudson Platform and the Red River Formation of the Interior Platform, with which they were undoubtedly once continuous. The Farr and the underlying Guigues and Bucke formations are in turn roughly equivalent to the upper Bobcaygeon-Hull, Verulam, and Cobourg formations of the Trenton Group in the south-central and Ottawa Valley regions of Ontario.

The Paleozoic rocks at Chicoutimi and Lac St. Jean, as at Lake Timiskaming, have long been a subject of interest and curiosity. Sinclair (1953) subdivided the Cycle 2 limestones, 36 m thick, into four informal rock units, which he tentatively correlated with formations comprising the Black River and Trenton groups in the Central and Western St. Lawrence Platform. Sinclair's units are readily identifiable in the field, but may not embrace the long interval of time (Blackriveran to early Edenian) that he proposed.

More recent investigations of Chicoutimi and Lac St. Jean outliers by Harland et al. (1985), although largely of a sedimentological nature, suggest that only the upper part of the Trenton Group (Cobourg equivalent) is represented (Fig. 11.18). The more recent dating by C.R. Barnes, based on conodonts, restricts the sequence to early Edenian and fits well with the paleogeological reconstruction outlined in Figure 11.14 (as previously noted, discrepancies exist at the Edenian-Maysvillian level between text and correlation charts).

Cycle 3 – upper Edenian to Gamachian

Central and western St. Lawrence Trough were narrowed further in late Edenian to Richmondian time. Initially, about mid-Edenian, the northern part of the platform experienced a brief regression. With the ensuing transgression, flysch deposition, formerly restricted to the trough, expanded across the upper Trentonian carbonates of the southern margin of the platform (Fig. 11.19). Farther northwest, equivalent (distal) black and grey shales covered broad expanses of the southern Canadian Shield, and marine connections with the Hudson and Interior platforms were re-established. The initial, onlapping deposits were black shales and flysch (late Edenian), succeeded by open marine shales, siltstones, and sandstones (Maysvillian); both facies are remarkably persistent along the length of the trough and adjacent platform. In the early Richmondian, following an apparent pause, thrusting and further narrowing of the St. Lawrence Trough took place. Where the allochthons overrode the southeastern extremities of Frontenac and Saguenay arches during the waning phase of the Taconian orogeny, the trough became segmented into separate Anticosti, Quebec, and Allegheny basins.

Following emplacement of the youngest allochthons and final definition of the Taconian front, redbed deposition ensued almost immediately in the Quebec and Allegheny basins, but carbonate sedimentation was re-established over the craton farther northwest (Fig. 11.19).

The eastern platform (Anticosti Basin) escaped significant deformation during late Edenian to Richmondian time. Maysvillian clastics abruptly give way, not to thick redbeds as elsewhere, but to limestone, the latter facies continuing uninterrupted through Richmondian and Gamachian, and Silurian Llandovery and possibly Wenlock times. It is not clear whether the absence of deformation and orogenic clastics is due to the north-to-south diachroneity of the Taconian orogeny, to the

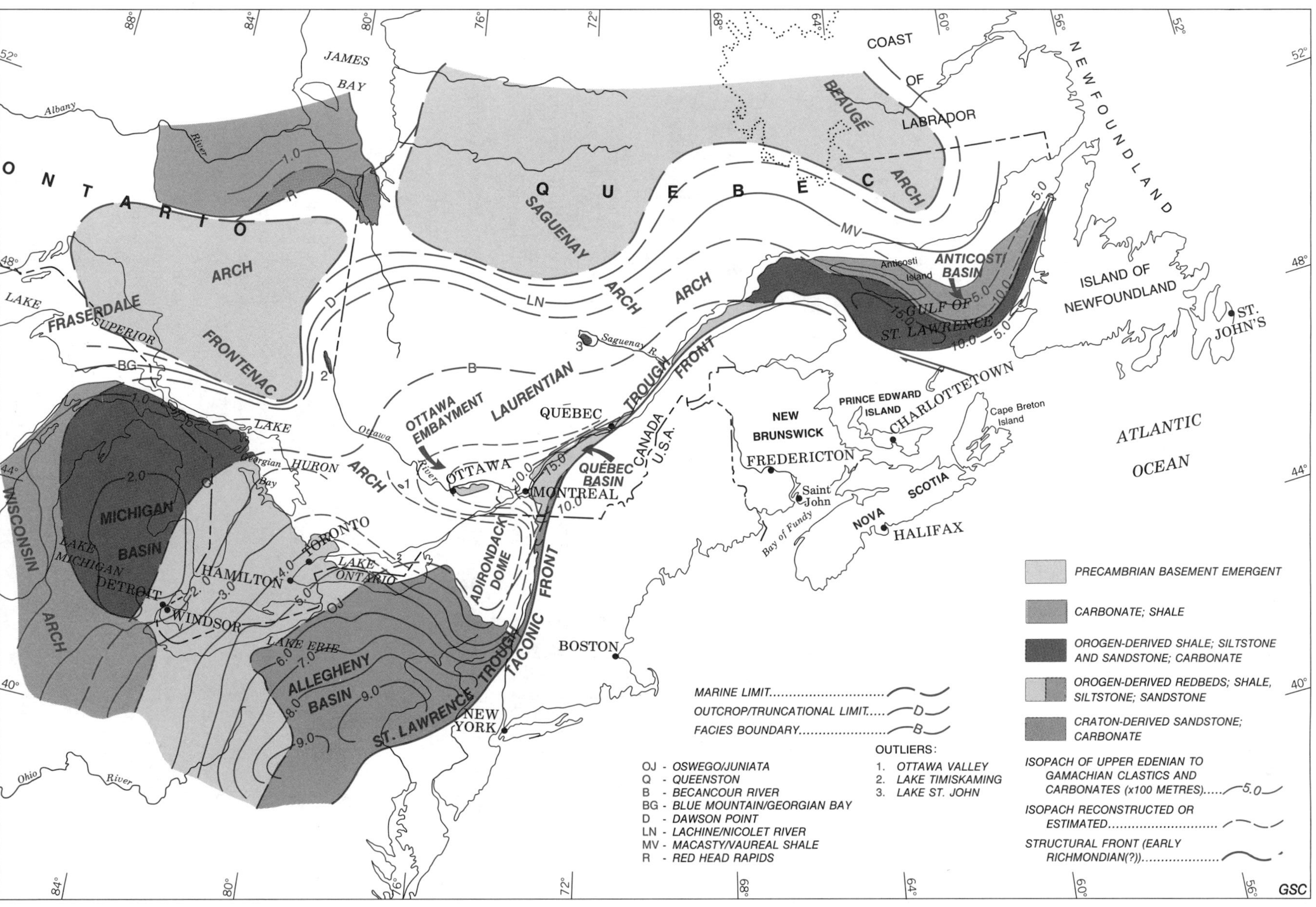

Figure 11.19. Depositional Cycle 3 - late Edenian to Gamachian,.

development of a transcurrent southern margin of Anticosti Basin, or to the position of the basin adjacent to an embayment in the continental margin.

From their erosional edges along the northern margins of the platform, late Edenian to Richmondian deposits (and Gamachian in Anticosti Basin) thicken generally southeastward into the foredeeps bordering the Appalachian Orogen. The deposits reach thicknesses of 1500 m in the Anticosti and Quebec basins and 900 m in the Allegheny Basin. Reconstructed isopachs in Figure 11.19 indicate that the Frontenac, Saguenay, and Beaugé arches were active during the Late Ordovician Epoch. A closed, 200 m isopach defining the centre of Michigan Basin indicates it underwent differential subsidence during the same period.

Although the lithostratigraphic succession of Cycle 3 is common to the entire platform, each division and subdivision of the platform has its own nomenclature. In western Anticosti Basin the cycle comprises, from base to top, Macasty, Vauréal, and Ellis Bay formations, whereas in the eastern part it is the upper beds of the Winterhouse Formation of the Long Point Group. In Central St. Lawrence Platform, the cycle comprises Lachine (Lotbinière; Utica), Nicolet River, Pontgravé River, and Bécancour River formations; and in Ottawa Embayment, Eastview, Billings, Carlsbad, Russell, and Queenston formations. Equivalent strata in the western platform (southwestern Ontario) are referred to as the Collingwood, Blue Mountain, Georgian Bay, and Queenston formations. From Bruce Peninsula west to Manitoulin Island and northern Michigan Basin generally, facies changes lead to different terminology, in stratigraphic order: Collingwood, Sheguiandah, Wekwemikongsing, Meaford, and Kagawong formations.

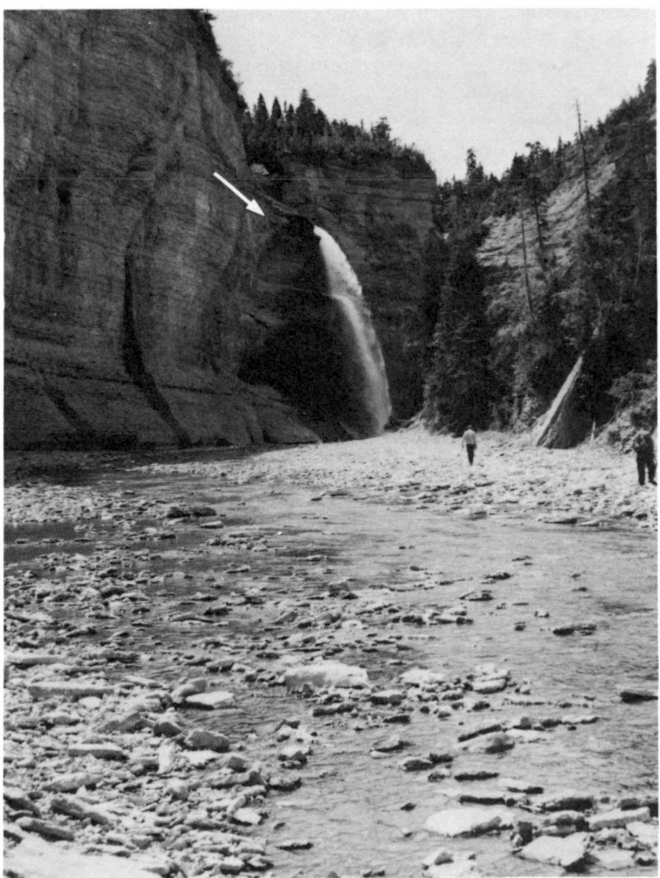

Figure 11.20. Upper Ordovician Vauréal Formation and Ellis Bay Formation, Vauréal River, Anticosti Island, Quebec. Arrow at base of Ellis Bay (photo - T.E. Bolton, GSC 158944).

Eastern St. Lawrence Platform

Trenton limestones in Anticosti Basin are succeeded by dark grey and black bituminous shales with minor limestone interbeds, the Macasty Formation. In the western part of the basin, the Macasty forms the bedrock surface beneath Jacques Cartier Channel off the northeast shore of Anticosti Island and dips beneath the island and adjacent regions of the western Gulf of St. Lawrence. Near the eastern end of the island, the subsea outcrop veers sharply southeast on approach to Beaugé Arch, where the Macasty is overlapped by the lower shale beds of the Vauréal Formation. Macasty shales are also present on the seafloor east of Beaugé Arch, where they dip southeast beneath Vauréal shales. The Macasty black shale facies does not occur in the shale-siltstone-sandstone sequence of the Winterhouse Formation in any of the exposures on Long Point, western Newfoundland. As the upper Winterhouse is thought to range in age from late Trentonian to Maysvillian (Bergström et al., 1974), beds equivalent to the Macasty are assumed to be present within that sequence.

Macasty strata are 100 m thick beneath the western part of Anticosti Island and about 175 m along the southwest shore. They thin to 42 m in the central part and zero in the eastern parts of the island, on the flank of Beaugé Arch. Rapid thinning and apparent non-deposition across the nose of the arch are presumably due to uplift of the arch in mid-Edenian time, a tectonic event that affected the entire St. Lawrence Platform.

Medium to dark grey micaceous shales and interbedded limestones and sandstones of the lower Vauréal Formation succeed the black shales of the Macasty Formation (Fig. 11.20). The beds are nowhere exposed onshore, but outcrop beneath Jacques Cartier Channel immediately north and southeast of Anticosti Island, thence forming an arc around the nose of Beaugé Arch to trend northeastward beneath the northeast Gulf of St. Lawrence. They dip southwest and southeast beneath carbonate rocks of the upper Vauréal Formation. Lower Vauréal strata are lithologically similar to the Winterhouse Formation of southwestern Newfoundland and are undoubtedly continuous with that formation beneath the intervening Gulf of St. Lawrence. The sandstone interbeds within the predominantly shale sequence appear to increase in number southward across Anticosti Island, and like sandstones in the Winterhouse Formation were presumably derived from a southeastern, orogenic source.

The Vauréal Formation is thickest beneath the western half of Anticosti Island, approximately 465 m along the north shore, and 720 m along the southwest shore. The beds thin toward Beaugé Arch, from 278 m in the east-central island to 253 m beneath the extreme southern end. In sharp contrast to the thick, orogen-derived redbed clastics that terminated Late Ordovician deposition in the central and western platform, carbonates and siliciclastics (possibly craton derived) were deposited in Anticosti Basin during Richmondian and Gamachian time. These are the upper Vauréal and Ellis Bay formations, which outcrop across the northern part of Anticosti Island and continue offshore (Bolton, 1972; Petryk, 1981). Petryk (1981) presented evidence of a cratonic source for the Vauréal sandstones, but equivalent strata in the Quebec and Allegheny basins are derived from the orogen.

The Vauréal carbonates that gradationally succeed the shale facies are up to 482 m thick and consist of grey and bluish grey, finely crystalline to semi-lithographic limestone. Alternating grey and greenish grey shale units are abundant in the lower beds, decreasing in number upward. Small patch reefs occur near the top of the formation at widely separated localities. Intraformational limestone conglomerates and detrital fragments of shale are common, as are minor detrital quartz silt and sand, the latter occurring as interbeds in the upper part of the formation.

The youngest Ordovician rocks in Anticosti Basin, argillaceous limestones with interbedded shales and sandstones of the Ellis Bay Formation (Fig. 11.21), are Gamachian in age. Biohermal mounds with a framework of corals, stromatoporoids, and bryozoans, resting on a thin tabular carbonate platform, occur at the top of the formation. Unlike the preceding rock units, the Ellis Bay thickens eastward, from 75 m in western Anticosti Island to about 100 m at the eastern end. The eastward thickening is presumably attributable in part to intertonguing sandstones that form a greater proportion of the formation over the crest of Beaugé Arch. In contrast to the orogen-derived clastics of the Macasty and lower Vauréal formations, the sandstones that occur as interbeds and units within the upper Vauréal and Ellis Bay formations were probably derived from the craton, specifically Beaugé Arch, rejuvenated near the close of the Ordovician Period.

Figure 11.21. Upper Ordovician to Lower Silurian Ellis Bay Formation, west bank of Salmon River below 10-Mile Lodge, Anticosti Island, Quebec. Massive carbonates in centre of photo are biohermal (photo - T.E. Bolton, GSC 127610).

Cycle 3 deposition in the Eastern St. Lawrence Platform was characterized by a variety of depositional environments, which in general reflected the changing tectonics of the Appalachian Orogen and its foreland trough. Orogenic activity, renewed in the late Trentonian and intensified in the mid-Edenian, led to accelerated subsidence and deep semi-restricted to open marine depositional conditions throughout much of the marginal trough and foreland platform from late Edenian to Maysvillian time. Bypassing of the eastern platform in early Richmondian time by northwest-advancing allochthons permitted stabilization of Anticosti Basin. This resulted in the return of low- to medium-energy open marine conditions conducive to deposition of carbonates, growth of bioherms, and spread of craton-derived sands.

Central St. Lawrence Platform

The initial deposits of Cycle 3 in the Ottawa Embayment and Quebec Basin are the Eastview-Billings and Lachine formations respectively (Wilson, 1964; Clark, 1972). Only remnants of Eastview-Billings are preserved in down-faulted blocks, whereas the Lachine and its lateral equivalent the Lotbinière Formation border Quebec Basin in the north and west, and dip southeastward beneath younger Ordovician strata.

The Eastview Formation, up to 8 m thick, consists of alternating dark brown, bituminous limestones and black shales, and represents the transition from the limestones of the Cobourg (Trenton Group) below, to the black and dark grey shales, up to 52 m thick, of the Billings Formation above. Rocks representative of the Eastview interval are apparently absent along the northern margin of Quebec Basin, where dark brown and black shales 90 to 100 m thick of the Lachine Formation rest with sharp and disconformable(?) contact on Tetreauville limestones of the Trenton Group. In the northeastern part of Quebec Basin, on the other hand, the Lotbinière Formation, coeval dark grey to black shales with abundant dolomitic limestone interbeds, may in places represent continuous sedimentation between Cycles 2 and 3. South of St. Lawrence River, upper Trentonian (Shermanian) to upper Edenian black shales of the Utica Group, up to 350 m thick, intertongue with upper Trenton limestones and succeed the latter.

The Carlsbad Formation, 185 m of dark grey and greenish grey shales with minor limestone interbeds near the top, succeeds the Billings Formation in Ottawa Embayment. These beds, like all other Upper Ordovician units of Cycle 3, are confined to small pockets preserved along the downthrown sides of tilted fault-blocks. Equivalent rocks in central Quebec Basin are the dark grey shales, siltstones, and fine-grained sandstones, 762 m thick, of the Nicolet River Formation. The upper parts of both the Nicolet River and Carlsbad contain silty limestone interbeds that increase in number upward to give place ultimately to dolomitic limestones and interbedded siltstones of the Russell Formation. The Russell is 12 m thick in Ottawa Embayment, and the equivalent Pontgravé River Formation is 48 m thick in Quebec Basin.

The reasons for the increase in carbonate deposition in an otherwise continuous clastic sequence beginning in early Richmondian time are not clearly understood. If diminished orogenic tectonic activity was the cause, the pause was brief, because redbed clastics again invaded the central platform immediately following deposition of the Russell and Pontgravé River carbonates. Alternatively, a brief sea-level rise may have ponded the orogenic clastics near their source.

The westward transport of Taconian allochthons in early Richmondian time, resulting in segmentation of the St. Lawrence Trough between cratonic arches, presumably also led to inception of the Queenston delta, and ultimately the final regression of Ordovician seas from the central and western platform.

In the Ottawa Embayment, a small outlier of the Queenston Formation, consisting of red and grey-green mottled shales 13 m thick, is preserved in a downfaulted block. A much more complete sequence of the same age is preserved in the central part of Quebec Basin. It consists of 610 m of red siltstone, sandstone, and shale of the Bécancour River Formation.

In the extreme southern part of Quebec Basin, Utica and Nicolet River strata dip beneath allochthonous rocks of the Appalachian Orogen. In this region, the complexly folded and faulted beds are undivided, and collectively referred to as the St. Germain Complex.

The semi-restricted to open marine conditions that had prevailed from late Edenian to Maysvillian time throughout the trough and platform ended in the early Richmondian, with the renewed advance of the Taconian front and isolation of the Quebec and Allegheny foreland basins. Initiation of the Queenston delta was marked by the deposition of fluvial redbed and fluvial-to-marine deltaic strata that intertongued with marine clastics, initially along the structural front, then progressively onto the craton (Fig. 11.19).

Western St. Lawrence Platform

In southwestern Ontario, the initial beds of Cycle 3 are composed of alternating brown bituminous limestones and black shales, 5 to 7 m thick, of the Collingwood Formation (Russell and Telford, 1983). Along their outcrop from Lake Ontario to Georgian Bay, and in the subsurface to the west, the beds are discontinuous. They were undoubtedly deposited throughout the entire region, but were eroded along the uptilted margins of fault-bounded blocks during a brief hiatus in about mid-Edenian time. Thus, the succeeding, dark grey, non-calcareous shales of the Blue Mountain Formation (Shequiandah Formation on Manitoulin Island) rest disconformably on the Collingwood or, in the absence of the latter, on the Cobourg limestones of the Trenton Group.

Blue Mountain strata in turn grade upward to medium grey shale with siltstone and silty dolostone interbeds of the Georgian Bay Formation, 100 to 140 m thick. The dolostone interbeds increase in number toward the northeastern Michigan Basin in Ontario (Manitoulin Island and adjacent islands), where they form a distinctive unit 20 m thick, the Meaford Formation. In this region, the soft grey shales, 45 m thick, between the Shequiandah below and the Meaford above have been referred traditionally to the Wekwemikongsing Formation.

The youngest Ordovician strata in southwestern Ontario (Richmondian) are the red, maroon, and grey-green mottled shales and siltstones, 23 to 335 m thick, of the Queenston Formation. Silty carbonate interbeds occur at the base, and sporadically throughout the formation, but are most prominent near the pinchout, where the red shales give place to grey-brown and tan coloured dolostones, 27 m thick, of the Kagawong Formation (Fig. 11.3).

Depositional environments during Cycle 3 deposition in Western St. Lawrence Platform were similar in many respects to those in Ottawa Embayment and Quebec Basin. One important difference, however, was that Allegheny Basin and the adjacent region were more stable and slowly subsiding. Instead of being trapped near its source, orogenic detritus was spread widely over the foreland basin and platform. In addition, the sparse, normal marine faunas of the Meaford and Kagawong formations, and the presence of stromatolites, suggest Richmondian marine restriction in Michigan Basin.

Paleozoic outliers

The lower, Eastview-Collingwood strata of Cycle 3 are known at only one locality beyond the limits of St. Lawrence Platform, namely the Clear Lake outlier (Ottawa valley outliers). There, 6 m of black shale with bituminous limestone interbeds lies conformably on Cobourg limestones of the Trenton Group.

At Lake Timiskaming outlier, the Collingwood-Eastview interval is missing, and the Dawson Point Formation, 30 m of dark grey shales with a thin basal limestone conglomerate, rests with sharp contact on the Farr Formation (Cobourg equivalent). The Dawson Point shales contain *Triarthrus* cf. *T. rougensis* Parks and *Leptobolus insignis* Hall, as reported by Sinclair (1965), and are thus late Edenian in age and equivalent to the Blue Mountain and Billings formations of the central and western St. Lawrence Platform respectively. Some of the higher beds of the Dawson Point may be even younger, that is, Maysvillian, as suggested in Figure 11.7.

Collingwood-Eastview equivalents are apparently missing also at Lac St. Jean and Chicoutimi outliers. There, dark grey to black shales 8 m or more thick, informally known as "Gloucester" beds (Sinclair, 1953), rest with uneven contact on carbonates (Tetreauville-Cobourg equivalent) of the upper Trenton Group. These beds are lithologically identical to the Lachine and Billings formations of Quebec Basin and Ottawa Embayment respectively, contain a similar fauna, and are assumed to be equivalent.

Cycle 4 – lower to mid-Llandovery

Except on the eastern platform, where sedimentation was continuous from Ordovician to Silurian times, the sea regressed at the close of the Richmondian Age, initiating a period of erosion that continued through Gamachian time. In the early Silurian (Llandovery), the sea again transgressed the platform. Initial submergence was confined to St. Lawrence Platform and the adjacent Canadian Shield (see depositional limit of Manitoulin-Cabot Head formations in Fig. 11.22), but eventually the sea progressed across the shield to the Hudson, Arctic, and Interior platforms in mid-Llandovery time.

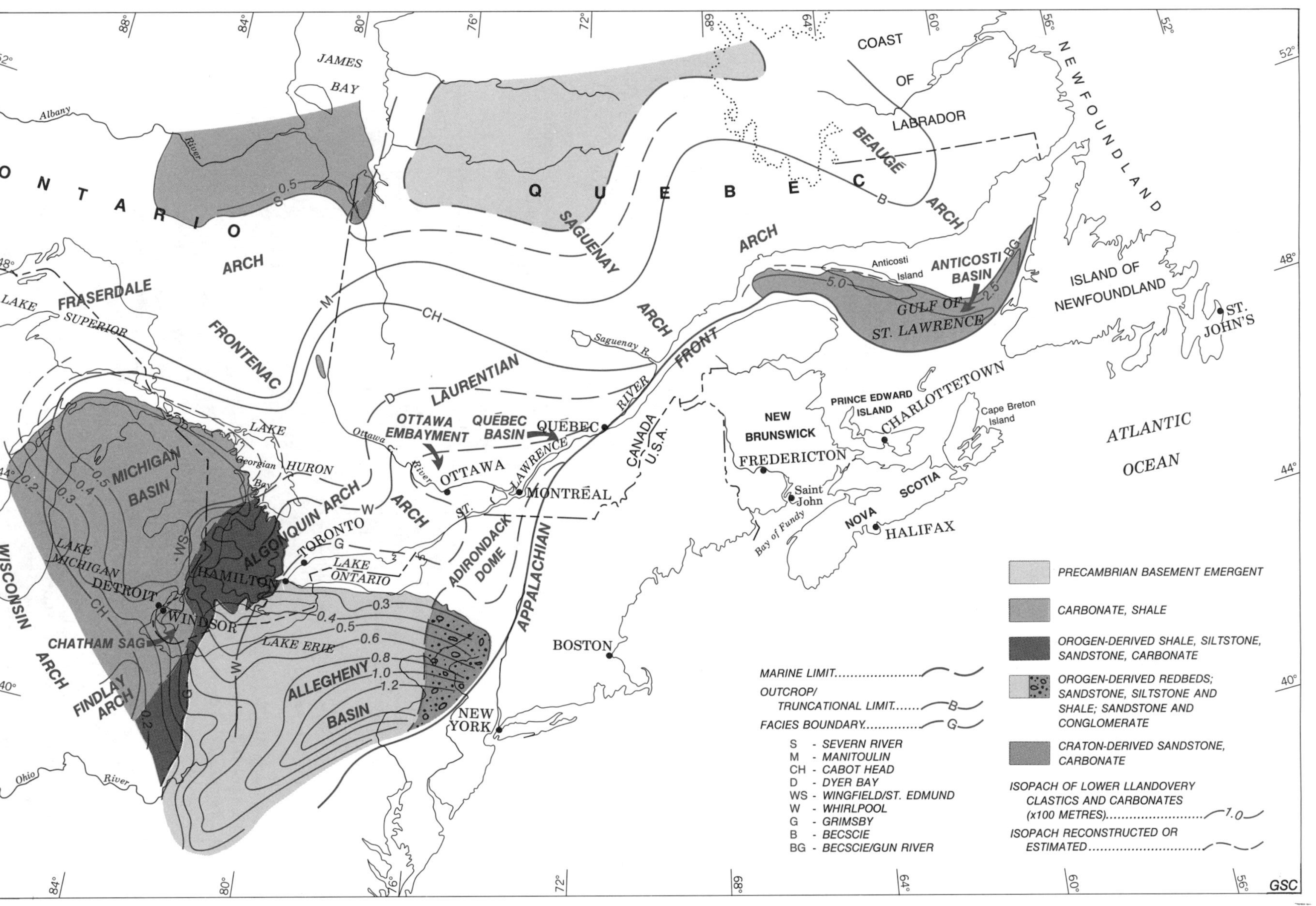

Figure 11.22. Depositional Cycle 4 - early to mid-Llandovery.

Rocks of Cycle 4 are preserved in the eastern and western platform, and at Lake Timiskaming outlier. An equivalent sedimentary sequence was undoubtedly deposited also in the central platform (Fig. 11.22) but eroded in post-Devonian times.

The lower- to mid-Llandovery strata of Anticosti Basin, mainly limestones, record neither contemporaneous orogeny nor nearby land. On the other hand, the relatively large volumes of nonmarine and marine, fine to coarse clastics deposited in the Allegheny and Quebec basins and the adjacent craton at the same time record either tectonism or persisting orogenic highlands to the southeast.

From their erosional edges along the northern margins of the platform, lower to mid-Llandovery beds thicken to 500 and 120 m in Anticosti and Allegheny basins respectively. A closed, 50 m isopach identifies the centre of Michigan Basin.

Rocks of Cycle 4 in Anticosti Basin consist of two lithological units, the Becscie and Gun River formations. Equivalent rocks in southwestern Ontario and Lake Timiskaming outlier are known as Cataract and Wabi groups respectively.

Eastern St. Lawrence Platform

The Becscie and Gun River formations outcrop along central Anticosti Island (Fig. 10.1) and dip southwest beneath younger Silurian rocks on the south shore and in the adjacent Gulf of St. Lawrence. The distribution of these rocks elsewhere in the Anticosti Basin is presently unknown.

The initial beds of Cycle 4, the Becscie Formation, are 131 to 173 m thick and consist of whitish grey to blue-grey, finely crystalline limestone, grading upward to green shale and nodular limestone. Overlying the Becscie are 146 m of ash-grey to yellowish white limestone, and alternating limestone and shale of the Gun River Formation. Common to both formations are intraformational conglomerates, current structures, and bioherms, most of the latter occurring near the top of each formation.

Western St. Lawrence Platform

The lower to mid-Llandovery rocks of Allegheny Basin in Ontario are a complex assortment of siliciclastics that border the Algonquin Arch on the south and intertongue with carbonate units in Michigan Basin. The dominant

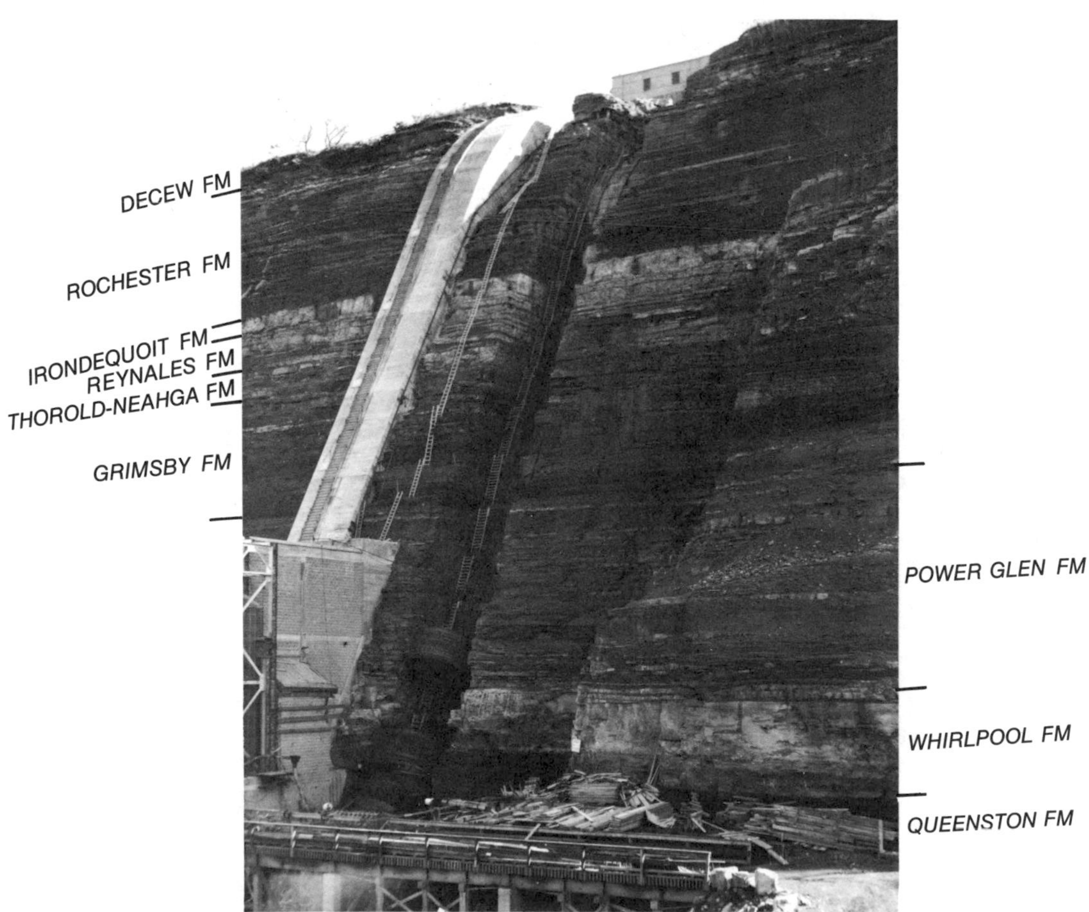

Figure 11.23. Silurian section, Decew Falls, Ontario. The strata, representative of Allegheny Basin, are largely shales and sandstones derived from the Appalachian Orogen to the southeast (photo - T.E. Bolton, GSC)

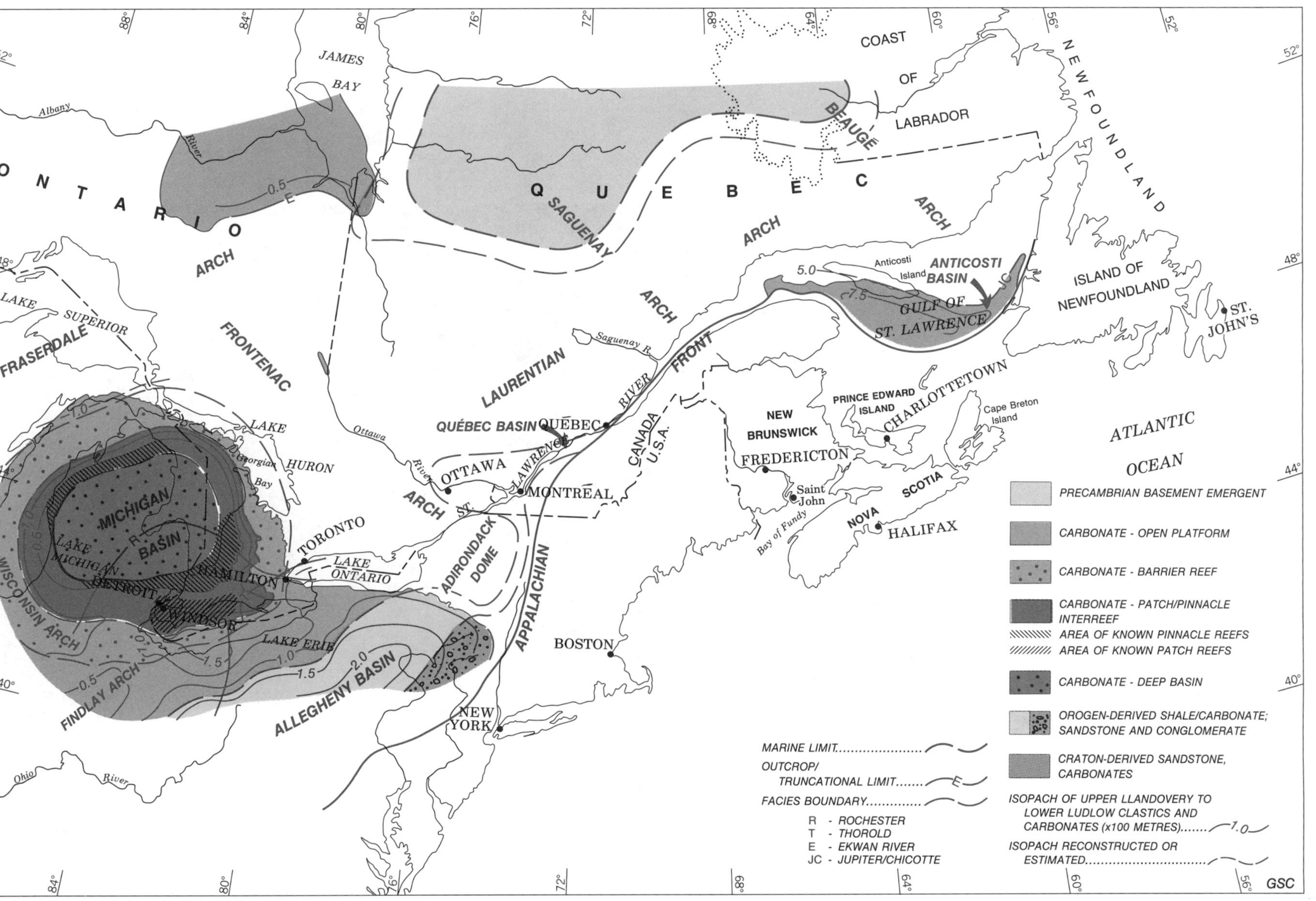

Figure 11.24. Depositional Cycle 5 - late Llandovery to early Ludlow.

facies on the south are sandstones derived in part from underlying Upper Ordovician strata and presumably also from rejuvenated highlands along the Appalachian Front. There, the Cataract Group is divided into three formations: a basal unit, 0 to 7 m thick, of white and greenish grey, orthoquartzitic sandstone, the Whirlpool Formation; an intermediate unit, 15 m thick, of medium grey to greenish grey and locally red shale, siltstone, and sandstone, the Cabot Head Formation; and an upper unit, 0 to 20 m thick, of red and greyish green sandstone with interbeds of red shale and siltstone, the Grimsby Formation.

As the Cataract Group is traced across Algonquin Arch and into Michigan Basin, carbonates become the dominant rocks (Fig. 11.23). Intertonguing with the Whirlpool and lower Cabot Head shales are brown and blue-grey shaly dolostones, 0 to 10 m thick, of the Manitoulin Formation. Intertonguing with the upper Cabot Head shale farther into the basin (i.e., Bruce Peninsula and Manitoulin Island) are the blue-grey argillaceous dolostones (12 m) of the Dyer Bay Formation, grey-tan micritic dolostones and interbedded green and red shales of the Wingfield (12 m), and medium to dark brown, semi-lithographic dolostones (27 m) of the St. Edmund.

Lake Timiskaming outlier

The Wabi Group at Lake Timiskaming, 46 m thick, is divisible into rock units remarkably similar to the Manitoulin, Cabot Head, Dyer Bay, and Wingfield formations bordering Michigan Basin in Ontario, and that nomenclature is used (B.V. Sanford, in Poole et al., 1970; Russell, 1984). The basal beds of the Wabi Group, 16 m thick, are grey limestone with interbedded dolostone of the Manitoulin Formation, lying disconformably on the Upper Ordovician Dawson Point shales. Red and green shales 10 m thick, with nodular gypsum, gradationally succeed the Manitoulin, and represent a feather-edge of the Cabot Head Formation. Completing the Wabi Group are the grey and blue-grey limestones of the Dyer Bay Formation, 7 m thick, and the grey and grey-tan, semi-lithographic dolostones with green and red nodular shale interbeds of the Wingfield Formation, 12 m thick.

Figure 11.25. Lower Silurian Jupiter Formation, south coast of Anticosti Island, Quebec; view eastward from Cape Jupiter to Cape Ottawa. The limestones dip gently south beneath the Gulf of St. Lawrence (photo - T.E. Bolton, GSC 201574D).

On the central(?) and western platform, regression is recorded between middle Llandovery and upper Llandovery strata. The eastern platform (Anticosti Basin) on the other hand remained submerged.

Depositional environments of Cycle 4 exhibit many similarities to those of the preceding cycle, inasmuch as both cycles began with transgression and open marine conditions, and ended with deposition of regressive redbeds in northwestward migrating deltas. Unique to the lower to mid-Llandovery, however, particularly in the western platform, are the thin units with ripple-marks, cross-bedding, and cut-and-fill (shoestring) sands suggestive of shallow subtidal to intertidal deposition. The upper deltaic redbeds of the sequence (Grimsby Formation) were probably deposited in a variety of environments from fluvial to marine.

Cycle 5 – upper Llandovery to lower Ludlow

Renewed transgression over St. Lawrence Platform in Llandovery to early Ludlow time resulted in the submergence of the platform and wide regions of the Canadian Shield, the seas connecting over the Hudson, Arctic, and Interior platforms as in the previous cycle (Fig. 11.24). Rocks of Cycle 5 are preserved in the eastern and western platform and in Lake Timiskaming outlier. They were presumably deposited also in the central division, and eroded in post-Devonian time.

Carbonate rocks with minor bioherms dominate the Anticosti Basin succession. As in the preceding cycle, the general lack of clastic deposits in Cycle 5 suggests tectonic stability along the southern margins there. Minor movements along the margins of the central and western platform, or persistent orogenic highlands there, are recorded by coarse to fine clastics in Allegheny Basin. The marine sandstones and shales that dominate the southern and central Allegheny Basin change abruptly to carbonates northward. The carbonates fringing Michigan Basin are complex bank and barrier reef deposits; these give place to patch and pinnacle reef belts toward the basin centre.

From their erosional edge along the northern margins of the platform, the rocks of Cycle 5 thicken to the south, reaching 750 m in Anticosti Basin and 200 m in Allegheny Basin. Carbonate deposits bordering Michigan Basin vary in thickness from 100 to 150 m, but thin to 50 m or less in the centre of the basin.

In Anticosti Basin, late Llandovery and possibly early Wenlock times are represented by two lithological units, the Jupiter and Chicotte formations. In southwestern Ontario a more complete and varied succession is represented by the Thorold, Neahga, Reynales, Fossil Hill, Irondequoit, Rochester, Decew, Gasport, Goat Island, Amabel, Eramosa, and Guelph formations, in stratigraphic order (Fig. 11.8).

Strata approximately equivalent to the Fossil Hill and lower Amabel, locally called Thornloe Formation, complete the Paleozoic succession at Lake Timiskaming.

Eastern St. Lawrence Platform

The Jupiter and Chicotte formations outcrop in southern Anticosti Island, and continue offshore beneath the Gulf of St. Lawrence. Southeast of Anticosti Island, the Chicotte dips beneath redbeds of the Clam Bank Formation and is presumably continuous in subsurface beneath the extreme southern and eastern parts of the basin.

The Jupiter Formation, 198 m thick, consists of ash-grey, finely crystalline limestone with grey shale intervals, grading upward to light green or grey calcareous shale and argillaceous limestone (Fig. 11.25). The Chicotte Formation, resting with abrupt contact on the Jupiter, is distinct from all underlying Silurian or Ordovician formations. It is largely light grey and white, bioclastic limestone, composed mainly of fragments of crinoid columns and corals, which locally form into small mounds. The limestones, up to 75 m thick, closely resemble the Wiarton Member of the Amabel Formation, as developed along the eastern and northern margins of Michigan Basin in Ontario.

The Jupiter Formation, like all underlying Silurian and Ordovician units, is a product of subtidal deposition on a moderately subsiding, open-marine platform. A marked change in physical and biological conditions, recorded also in the eastern platform, Lake Timiskaming outlier, and southern Hudson Platform, took place at the onset of deposition of the Chicotte Formation, which represents a complex of crinoid banks unlike any formation seen earlier.

Western St. Lawrence Platform

The late Llandovery to early Ludlow was a time of major reef development throughout the Great Lakes region of Canada and United States. Carbonate bank and barrier reef facies were forming around the stable shallow margins of the Michigan Basin. Basinward, patch and pinnacle reefs flourished during early to early late Ludlow time.

Initial bank and barrier reef growth began in late Llandovery time along the northern and eastern margins of Michigan Basin in Ontario, with deposition of tan and light brown, medium crystalline dolostones, 12 to 24 m thick, of the Fossil Hill Formation. Bioherms, although abundant, were small. Equivalent inter-reef facies in the Michigan and Allegheny basins are the grey-brown micritic dolostones (3 to 9 m) of the Reynales Formation. The basal transgressive facies of the Reynales, which borders the Allegheny Basin, are light grey and greenish grey orthoquartzitic sandstone (7 m) and overlying dark grey fissile shale of the Thorold and Neahga formations. Deposition of this sequence was followed by regression and brief subaerial erosion.

Marine readvance during early Wenlock time led to deposition of a facies mosaic in the Great Lakes region. The initial strata are the thick crinoid bank deposits of the Colpoy Bay and Wiarton members of the Amabel Formation (Fig. 11.26). These beds, consisting of white to bluish grey and beige crinoidal dolostones up to 46 m thick, formed along the northern, eastern, and southern margins of Michigan Basin in Ontario, giving place to grey, fine-grained, cherty dolostones (18 m) of the Lions Head toward the basin centre. The crinoid bank bordering Allegheny Basin in Ontario intertongues with the grey shales (0 to 24 m) of the Rochester Formation, grey shaly dolostones (0 to 9 m) of the Decew Formation, and grey cherty dolostones (0 to 18 m) of the Goat Island Formation. The Amabel crinoid bank migrated periodically across the northern Allegheny Basin, to emplace wedges of clean crinoidal dolostones of the Irondequoit (3 m) and Gasport (10 m) formations into an otherwise basinal sequence. Bioherms, usually small, occur within the Wiarton and Gasport formations.

The most important of the reef-bearing sequences, and the youngest, is the Guelph Formation (Fig. 11.27), which contains the oil- and gas-bearing reefs in southwestern Ontario and Michigan. In contrast to the broadly distributed crinoid banks of the Amabel Formation, the

Figure 11.26. White-weathering, resistant dolostones of the Silurian Amabel Formation form cliffs on Niagara Escarpment, Bruce Peninsula, Ontario (photo - B.V. Sanford, GSC 2-17-70).

Figure 11.27. Lower Silurian Guelph Formation, Elora, Ontario. Fossiliferous dolostones (concentric casts and cavities are remains of *Megalomus*) are part of the barrier reef complex that encircled the Michigan Basin in Silurian time (photo - B.V. Sanford, GSC 200874C).

barrier deposits of the Guelph formed a band 20 to 60 km wide around what had become a rapidly subsiding Michigan Basin. Cream to white dolostone up to 100 m thick that forms the barrier changes abruptly to dark brown and locally black, bituminous dolostones of the Eramosa inter-reef facies, thinning from 50 m against the barrier reef to 6 m or less in the basin centre.

Immediately basinward of the barrier is the belt of large patch reefs, with vertical relief of up to 49 m on bases of several thousand hectares. Farther into the basin (i.e., in Ontario bordering Lake Huron and southeastern Michigan, and along the northern part of the southern peninsula of Michigan) are numerous pinnacle reefs, with vertical relief of 90 to 150 m on bases of 125 to 200 ha. Bioherms occur sporadically within the Guelph barrier reef complex along the outer margins of Michigan Basin, but nowhere is the relief of individual mounds known to exceed a few metres.

Shallowing of the seas towards the close of Llandovery time affected all of the St. Lawrence Platform. This records, at least in part, rejuvenation of the cratonic arches; a eustatic component may be involved as well. During part of this time also (Wenlock to early Ludlow), Michigan Basin underwent accelerated differential subsidence, leading to extensive reef development.

Lake Timiskaming outlier

The youngest Paleozoic rocks of the Lake Timiskaming outlier are reefal carbonates, 60 m thick, of the upper Llandovery to lower Wenlock Thornloe Formation (Fig. 11.28). They rest with abrupt, disconformable contact on the Wabi Group and consist of two units, a lower unit similar to the Fossil Hill Formation of southwestern Ontario, and an upper unit that more closely resembles the Amabel Formation of the same region. The lower beds are alternating grey-buff to light brown limestones and dolostones 39 m thick, with a thin, basal, limestone conglomerate. These are succeeded, with possible disconformable contact, by somewhat lighter coloured, more massive and reefal, light grey and grey-tan dolostones 21 m thick, which are undoubtedly the northern equivalent of the Amabel.

The late Llandovery to Wenlock successions of the eastern and western platforms, the Lake Timiskaming outlier, and southern Hudson Platform point to not only widely uniform but also unusual marine conditions. The relatively coarse-grained, crinoidal-bank limestones (Chicotte) of the eastern platform might be attributable to a higher energy environment, and that in turn to shallowing. Such an explanation fails in the case of the similar limestones (Amabel) of Michigan Basin, which was undergoing accelerated subsidence, or those of the Lake Timiskaming outlier (upper Thornloe), deposited during a period of maximum inundation of the Canadian Shield. Perhaps, taking a clue from the maximum development of coralline reefs in Michigan Basin (Guelph), and Hudson Platform (Attawapiskat) immediately following, the explanation may lie in a climatic/biological event, an extraordinary flourishing of crinoids, followed by a similar event for reef-building corals.

During Llandovery to Wenlock time, maximum expansion of the seaways, an undoubted eustatic event, took place. The Canadian Shield was widely inundated, and on the evidence of similarity of rock-types as well as faunas, St. Lawrence, Hudson, and Interior platforms were interconnected. Uniform open-platform marine environments prevailed over wide regions of the craton until early Ludlow time, when much of the southern craton suddenly became emergent.

Cycle 6 – upper Ludlow to Gedinnian

Widespread regression beginning in early Ludlow time set the stage for deposition of Cycle 6 (Fig. 11.29). The tectonics of the time are puzzling. In view of the existence in the Appalachians of latest Silurian granites and some evidence for latest Silurian deformation (see Osberg, 1983), the rapid subsidence of the Michigan and Allegheny basins might be thought to be of foredeep type, and indeed, red clastics of the Bloomsburg delta occur at the southern margin of the latter, but the rocks of the western and central St. Lawrence Platform are evaporites and carbonates. Only in the eastern platform were deltaic redbeds deposited, evidence of nearby orogenic(?) highlands. The cratonic arches were again active.

Rocks of Cycle 6 are preserved only in the eastern and western divisions of the platform. Their former presence in the central division is recorded only by blocks of Gedinnian (Helderbergian) age that are locally preserved in a diatreme breccia.

Cycle 6 thickens southward from its erosional edge along the eastern and western platform to more than 600 m in Anticosti Basin and 700 m in Allegheny Basin. Maximum thickness is approximately 1300 m, in the centre of Michigan Basin.

Rocks of Cycle 6 in Anticosti Basin are a single lithological unit, the Clam Bank Formation. In Quebec Basin, the breccia blocks of Gedinnian age are unnamed, other than having been associated on fossil evidence with the Helderberg limestone of Allegheny Basin (Hofmann, 1972).

Figure 11.28. Lower Silurian Thornloe Formation, Dawson Point, Ontario (Lake Timiskaming outlier). The thin- to thick-bedded limestones and dolostones are lithologically and faunally similar to the Fossil Hill Formation of southern Ontario. The two units presumably were continuous across the southern Canadian Shield (photo - B.V. Sanford, GSC 200841X).

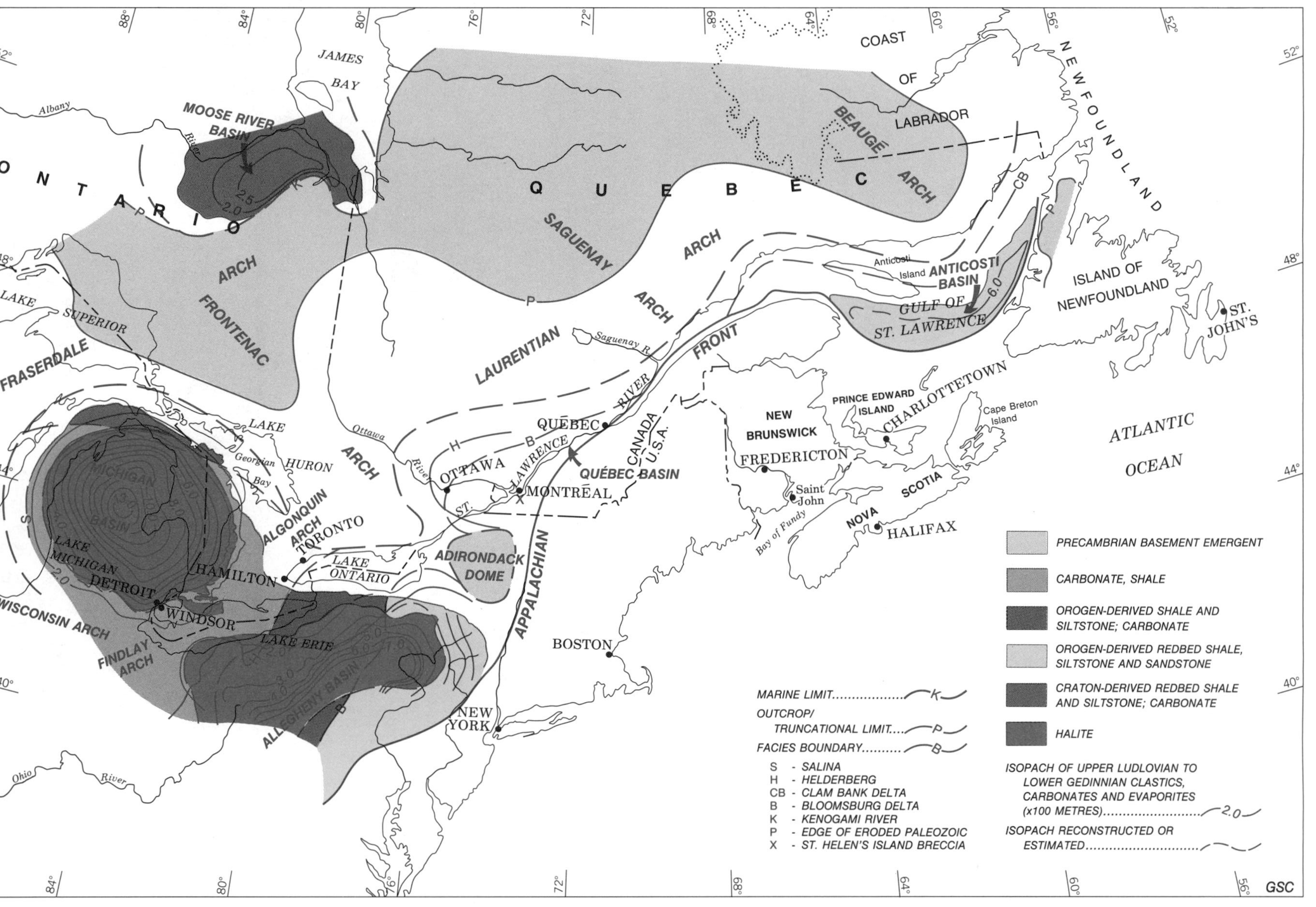

Figure 11.29. Depositional Cycle 6 - late Ludlow to Gedinnian.

In southwestern Ontario, the sequence is divided into two formations, a lower, Salina, and an upper, Bass Islands (Bertie).

Eastern St. Lawrence Platform

The Clam Bank Formation is exposed at a single locality in Anticosti Basin, Long Point Peninsula in southwestern Newfoundland. Its broader distribution in the immediate offshore regions of the Gulf of St. Lawrence has been confirmed by Shearer (1973) and Haworth and Sanford (1976). Where exposed in western Newfoundland, the beds are alternating red sandstones, siltstones, and shales 500 m thick, containing light grey fossiliferous limestone interbeds in the upper part (Fig. 11.30). The carbonates contain a rich invertebrate fauna indicating a Late Silurian (Pridoli) age (Berry and Boucot, 1970), a dating confirmed by study of the conodonts (Fåhraeus, 1974).

On Long Point, deformed beds of the Clam Bank overlie the Middle to Late Ordovician Long Point Group with possible fault contact (Rodgers, 1965). In the offshore southeast of Anticosti Island, Clam Bank beds overlie Chicotte carbonates and in this context must be considered Wenlock to Ludlow in age. They probably represent a remnant of the Bloomsburg delta, erosionally separated from the main part exposed along the southern margins of Allegheny Basin.

Central St. Lawrence Platform

Rocks of Cycle 6 are known at only one locality in Quebec Basin (St. Helen's Island, Montreal), where they form a part of St. Helen's Island Breccia. Much of St. Helen's Island is formed of a diatreme breccia consisting of blocks of all sizes up to 6 m across, imbedded in a matrix of smaller fragments of the same rock types (Hofmann, 1972; Clark, 1972). The blocks consist of Precambrian crystalline rocks, Potsdam sandstone, and several of the Ordovician limestones, as well as Lower and Middle Devonian limestones and sandstones. Fossils contained in the Lower Devonian blocks include Gedinnian and Siegenian forms that are common to the Helderberg and Oriskany formations of New York state. The Helderberg limestone blocks are the only remaining evidence that marine waters transgressed the Central St. Lawrence Platform during deposition of Cycle 6.

Western St. Lawrence Platform

Marine restriction caused by (1) a major landmass, the Appalachian Orogen, to the south, (2) a barrier reef complex paralleling the present Great Lakes system, and (3) the Frontenac Arch led to the deposition of upper Ludlow and Pridoli evaporites and associated sediments in the Allegheny and Michigan basins.

In southwestern Ontario, Cycle 6 is represented by the Salina Formation, up to 427 m thick, and the Bass Islands and Bertie formations, 100 and 10 m thick respectively. The Salina is further divided into units A-1, A-2, B, C, D, E, F, and G from the base upward. Units A-1, A-2, E, and G are mainly carbonate, and units C and F are shale. Units A-1, A-2, B, D, and E contain thick halite and anhydrite intervals along the eastern rim of Michigan Basin (see Fig. 11.11). Near the foot of Lake Huron, Salina halite totals 223 m, the maximum in Ontario. The B unit, the most widespread and thickest of the salt units, originally extended across the Guelph barrier reef in west-central Lake Erie to connect with the Salina B salt in Allegheny Basin. Younger Salina salt units (i.e., D and E salts) are also present in Allegheny Basin, but were probably not continuous across southwestern Ontario with their counterparts in Michigan Basin. Stratigraphic continuity of thin anhydrite layers within halite units along depositional strike in the eastern Michigan Basin, revealed by geophysical logs, strongly suggests deposition in deep water. The rapid facies change to thin anhydrite and carbonate units is a response to slower subsidence and shallower environments toward the basin rim.

The Salina is overlain by cream and tan-coloured, finely crystalline dolostones of the Bass Islands (Michigan Basin) and Bertie (Allegheny Basin) formations, locally deformed by dissolution of the underlying salt. Removal of salt units along the eastern margins of Michigan Basin began on a very large scale in late Pridoli time. Dissolution began above such features as Guelph patch and pinnacle reefs, and along some of the major fracture systems (faults) that were undergoing rejuvenation at that time. The numerous depressions that were thus forming on the seafloor became the receptacles for anomalous thicknesses of Bass Islands (Bertie) carbonates.

At the close of Pridoli time, the sea withdrew from St. Lawrence Platform, initiating an erosional interval that lasted until mid-Siegenian time in the central and western platform, and Viséan time in the eastern platform.

Figure 11.30. Upper Silurian (Pridoli) Clam Bank Formation, north shore of Long Point Peninsula, western Newfoundland; dominant redbed shales, siltstones, and sandstones with minor fossiliferous, light-coloured limestones (photo - B.V. Sanford, GSC 2-11-68).

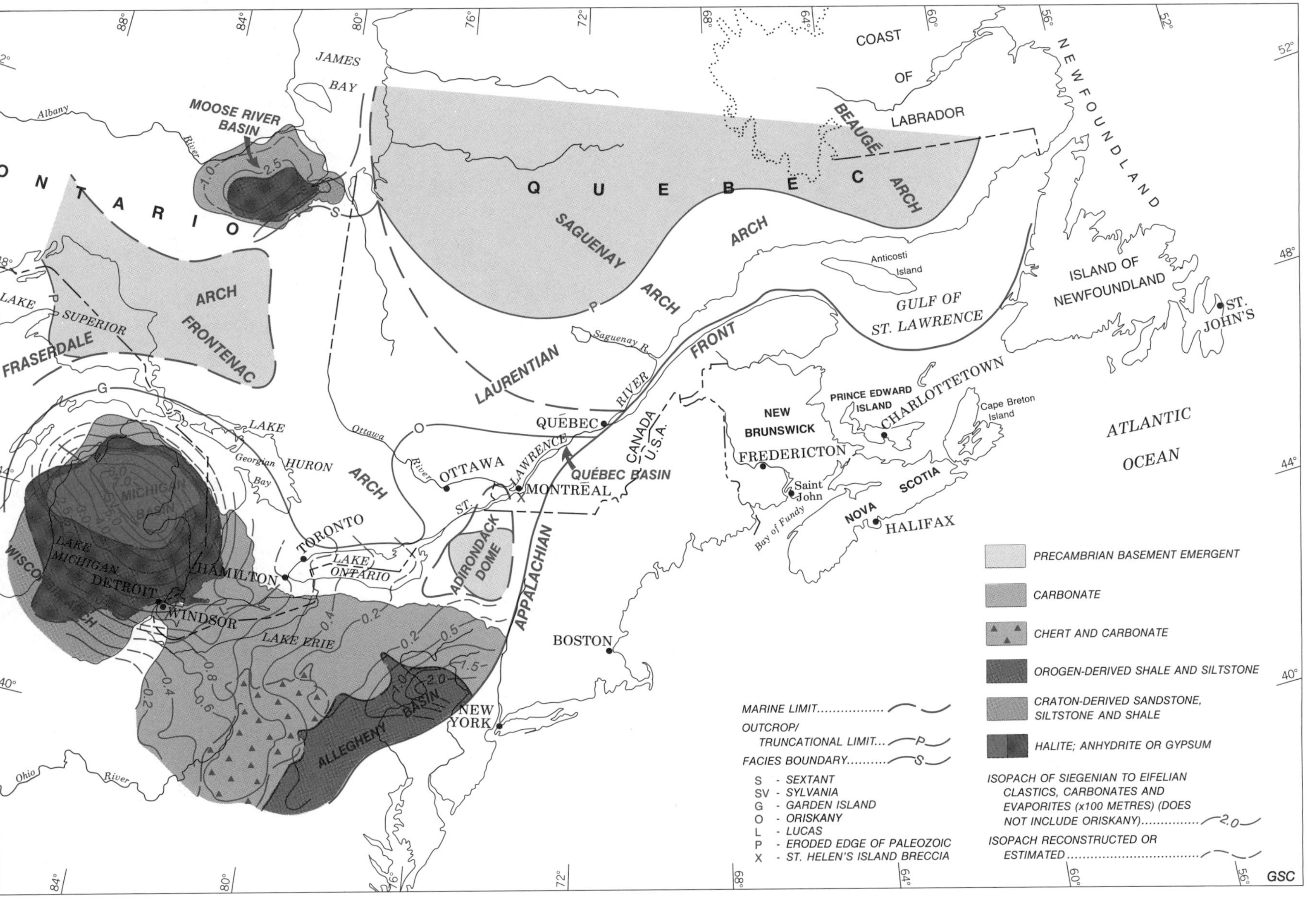

Figure 11.31. Depositional Cycle 7 - Siegenian to Eifelian.

Cycle 7 – Siegenian to Eifelian

Cycle 7 is the product of three marine transgressions of the Central and Western St. Lawrence Platform and adjacent craton. The first, in mid-Siegenian time, was confined to the platform, whereas the second and third transgressions, in late Emsian to mid-Eifelian and late Eifelian times respectively, continued northward across the southern Canadian Shield to unite the waters covering St. Lawrence and Hudson platforms (Fig. 11.31).

Rocks of Cycle 7 are unknown in Anticosti Basin, and in Quebec Basin are known only from the St. Helen's Island Breccia. In the western platform, however, they are widely preserved at the surface, and down-dip in the subsurface of Michigan and Allegheny basins (Fig. 10.1).

Sandstones, shales, and carbonates make up the sequence in Allegheny Basin, while carbonates, evaporites, and minor sandstones dominate in Michigan Basin. Epeirogenic movements of the craton during deposition of Cycle 6 (particularly during Pridoli time) apparently continued into Gedinnian time, resulting in partial denudation of the Wisconsin, Fraserdale, and Frontenac arches and the emergence of source areas for the thin siliciclastic units that were deposited here and there in the western platform. These movements also caused marine restriction and the ensuing deposition of evaporites in Michigan Basin in mid-Eifelian time.

Rocks of Cycle 7 reach a thickness of 700 m in Michigan Basin and 200 m along the extreme southern margin of Allegheny Basin. A northeast-trending "thin" or arch, along which the sequence thins to 20 m or less (Fig. 11.31) is not seen in any other sequence. If due to renewed uplift of Adirondack Dome, it represents both an expansion of the area of the dome and extension along a trend not seen at other times.

Blocks in the St. Helen's Island Breccia are unnamed, except for their association with the Oriskany Formation of New York State on the basis of fossil evidence (Hofmann, 1972). The assumed, depositional distribution of these strata on the central platform is sketched in Figure 11.31. The rock units comprising the cycle in southwestern Ontario bordering Michigan and Allegheny basins are, in upward order, the Oriskany, Bois Blanc, Amherstburg, Lucas, and Dundee formations and the partly equivalent Onondaga Formation.

Western St. Lawrence Platform

The oldest Devonian rocks in southwestern Ontario are of mid-Siegenian age. They are preserved mainly as subsurface erosional remnants, none more than 6 m thick, in the Niagara Peninsula and adjacent offshore regions of Lake Erie. They are light grey, orange-stained, glauconitic, medium- to coarse-grained orthoquartzic sandstones with a thin basal conglomerate and are outliers of the Oriskany Formation of Allegheny Basin (Sanford, 1968). Where exposed at a single locality in Haldimand County, Ontario, the beds overlie the Upper Silurian Bertie Formation with abrupt, disconformable contact. Sandy dolostones (Garden Island Formation) of the same age are exposed at a single locality off the northwest shore of the southern peninsula of Michigan. These occurrences suggest the broad extent of the Oriskany sea and of Oriskany sandstones prior to late Siegenian to late Emsian erosion.

Prolonged, post-Oriskany exposure led to deformation and brecciation caused by extensive leaching and removal of underlying Salina salt units along the outer margins of Michigan Basin. The anomalous masses of carbonate that had formed in seafloor depressions during the deposition of the Bass Islands and Bertie formations in Pridoli time remained to form a higher irregular surface topography, with mounds rising from 60 to 120 m above the lowland.

The late Emsian to mid-Eifelian sea transgressed the highly irregular surface northward across the Shield to Hudson Platform. The initial deposits are the blue-grey cherty limestones and dolostones of the Bois Blanc Formation. At places along the northern margin of the Allegheny Basin in Ontario, orthoquartzitic sandstone (Springvale Member), derived from the underlying Oriskany Formation, was deposited as a basal unit of the Bois Blanc and as layers interbedded with the limestones above. The highly irregular and unpredictable thickness of the Bois Blanc reflects the character of the underlying surface. The Bois Blanc Formation was not deposited on the higher mounds and ridges of Bass Islands (Bertie) carbonates. Elsewhere, it may be 70 to 100 m thick, in post-Bass Islands solution depressions. Where the underlying surface is more regular, the Bois Blanc varies from 36 m in the Ontario part of Michigan Basin to 5 m or less at the border of Allegheny Basin.

Brown, sucrosic, bituminous and cherty dolostones and limestones of the Amherstburg Formation, up to 100 m thick, gradationally succeed the Bois Blanc Formation. Where the Amherstburg overlaps the Bois Blanc along the margins of Michigan Basin (high on Wisconsin and Findlay arches), its basal beds give place to white, orthoquartzitic sandstones and light grey, sandy, crinoidal dolostones of the Sylvania Member. Eastward toward Allegheny Basin, the Amherstburg dolostones change to crinoidal and coral limestones and dark greyish brown, argillaceous, cherty limestones of the Edgecliff and Nedrow members of the Onondaga Formation. Bioherms are common in the Amherstburg and lower Onondaga (Edgecliff) formations in certain regions of Ontario. Abundant patch reefs, some more than 27 m thick, with a stromatoporoid framework (Formosa reef facies) occur in the lower Amherstburg in an area of 165 km^2 bordering Lake Huron. Similar bioherms containing a rich open marine biota occur in the Edgecliff Member of the Onondaga Formation at scattered localities of the Niagara Peninsula and beneath the eastern part of Lake Erie. Most of these latter are restricted in area and height and reflect a relatively slowly subsiding and undifferentiated seafloor.

In the central parts of Michigan Basin, halite, anhydrite, limestone, and dolostone of the Lucas Formation conformably succeed the Amherstburg Formation. The salt and anhydrite units that dominate the Michigan succession wedge out to the east in Ontario and are replaced by tan and light brown, microsucrosic dolostones with minor interbeds of anhydrite and/or gypsum. These latter thin markedly toward the basin margin and give way in turn to high-calcium, sublithographic limestones and sandy limestones of the Anderdon Member. Whether or not the Anderdon Member of the Lucas Formation is represented by lower beds of the Moorehouse Member of the Onondaga Formation in Niagara Peninsula and beneath eastern Lake Erie, or is absent in the latter regions, has not been firmly

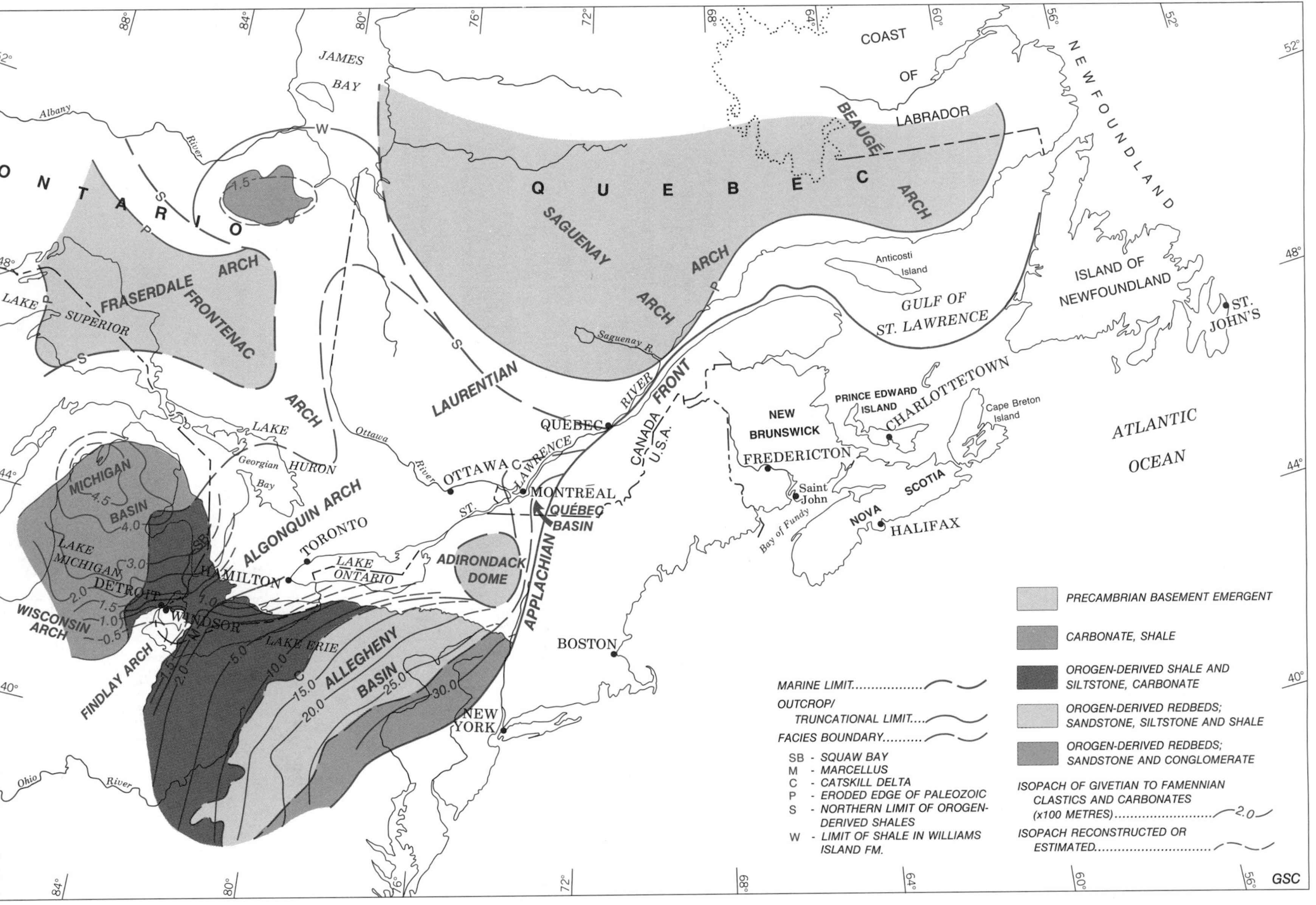

Figure 11.32. Depositional Cycle 8 - Givetian to Famennian.

established. On the other hand, a regional disconformity at the top of the Lucas in Ontario has been traced eastward as far as Norfolk and Haldimand counties in the western part of the Niagara Peninsula. A corresponding break is probably present within the Onondaga of the northern Allegheny Basin, possibly between the Nedrow and Moorehouse members, and Lucas equivalents are probably absent there.

The third and final transgression of Cycle 7 was in late Eifelian time. As in the preceding, Bois Blanc-to-Lucas invasion, transgression across the southern part of the Canadian Shield effected marine interconnection with the Hudson Platform to the north, and possibly also with the Interior Platform to the west. Rocks of this interval bordering Michigan Basin in Ontario are the Dundee Formation (30 m), fine- to medium-grained, crinoidal and calcarenitic limestones below, grading upward to medium and dark brown, argillaceous limestones. Minor sandstone and carbonate conglomerate characterize the basal beds of the formation locally. These beds disconformably overlie the Lucas Formation everywhere, and overlap the truncated edges of the latter along the limits of Michigan Basin to lie on the Nedrow Member of the Onondaga Formation. Where the Dundee beds overlap the Lucas they become darker brown, more argillaceous, thin bedded, and cherty, and as such are more readily associated with the Moorehouse and Seneca members (45 m) of the Onondaga.

The upper Emsian to Eifelian succession of Cycle 7 is a highly variable sequence of carbonates and evaporites that reflects contrasting depositional environments across Algonquin and Findlay arches between Michigan and Allegheny basins. The carbonates in the central part of Michigan Basin were mainly dolostones, products of restriction and hypersaline conditions persisting from the basin configuration characteristics of the preceding cycle. In contrast, open marine conditions, a flourishing biota and bioherm growth prevailed along the hingelines of Algonquin and Findlay arches and the adjacent northern margin of Allegheny Basin.

Cycle 8 – Givetian to lower Tournaisian ("Strunian")

In the Givetian to early Tournaisian cycle, two major marine transgressions inundated Central and Western St. Lawrence Platform, both effecting marine connections with Hudson Platform. Rocks of Cycle 8 in Quebec Basin are confined to the St. Helen's Island Breccia. In the western platform, Cycle 8 is widely preserved around the margins of Allegheny and Michigan basins and includes the youngest rocks in southwestern Ontario (Fig. 11.32).

The thick marine clastics of the Catskill Delta deposited in Allegheny Basin and possibly Quebec Basin in Givetian, Frasnian, Famennian, and early Tournaisian times reflect the Acadian orogeny along the central and southwestern Appalachian front. In southern Allegheny Basin, the rocks are continental redbeds. These intertongue with marine sandstones, siltstones, shales and minor carbonates in northwestern Allegheny Basin, and ultimately give way to shales and carbonates in Michigan Basin (Fig. 11.32).

The thickness of upper Middle and Upper Devonian strata in the central platform is unknown. In the southern part of Allegheny Basin, it exceeds 3000 m. The strata thin northwestward to 100 m on Algonquin and Findlay arches in Ontario, thickening again to 450 m in Michigan Basin.

Blocks representative of Cycle 8 in the St. Helen's Island Breccia are unnamed, except for their association on fossil evidence with the Hamilton Group of New York State (Boucot and Johnson, 1967; Clark, 1972). Cycle 8 in the Ontario part of the western platform comprises, in upward succession, the Hamilton Group and Kettle Point, Bedford, Berea, and Sunbury formations. Bedford, Berea, and Sunbury formations have been traditionally dated as Tournaisian and thus Early Carboniferous. The lower Tournaisian designation still applies, but the three formations are now assigned to the Upper Devonian, on the basis of the presence of *Hymenozonotriletes lepidophytus*, a species common to the "Strunian" (latest Devonian) of the Belgian succession (McGregor, 1970; Sanford, 1968).

Central St. Lawrence Platform

Sandstone blocks containing a Middle Devonian (Givetian) Hamilton fauna form a part of the St. Helen's Breccia at Montreal. The original depositional extent of these otherwise unknown rocks is assumed to be wide, as speculatively sketched in Figure 11.32.

Western St. Lawrence Platform

Shales and limestones of the Hamilton Group form the lower part of Cycle 8. Along the northwest flank of the Allegheny Basin in Ontario these are divided into three units: at the base, the Marcellus Formation (15 m), dark grey and black bituminous shales, containing interbeds of dark brown argillaceous limestone in the lower part where it gradationally succeeds the Seneca (Dundee equivalent) limestones of the Onondaga Formation; a succeeding unit of medium grey shales (30 m), roughly equivalent to the combined Bell and Arkona shales bordering the Michigan Basin to the west; and an upper, dark greyish brown argillaceous limestone unit (2 to 3 m), possibly equivalent to the Ipperwash limestones of the Michigan Basin succession.

The Hamilton Group in Ontario thickens northwestward toward Michigan Basin, where it consists of the following formations in upward succession: Bell shales (18 m); Rockport Quarry limestones (6 m); Arkona shales (37 m); Hungry Hollow limestone and shale (1 m); Widder shales with limestone interbeds (14 m); and Ipperwash limestones (15 m).

The Hamilton Group is succeeded by the Kettle Point reddish brown and black, bituminous shales with interspersed grey-green shale and siltstone interbeds. At the type section on Lake Huron, the lower shales of the Kettle Point Formation contain numerous concentric limestone concretions commonly referred to as "kettles", some reaching more than a metre in diameter (Fig. 11.33). In Ontario, the Kettle Point reaches its maximum development of 120 m beneath Lake Erie. Onshore the formation is highly variable in thickness (30-60 m), due to post-Devonian dissolution of the underlying salt and consequent irregular subsidence, followed by erosional bevelling of the Kettle Point at subsequent unconformities. The Squaw Bay Formation, 1 to 2 m of brown bituminous limestone, intertongues with lower Kettle Point shales at the margin of Michigan Basin. In Ontario, it is limited to an area bordering the St. Clair River, but in adjacent areas of Michigan, where it represents a basinal facies of the Antrim (Kettle Point), it is widely preserved.

In eastern Michigan Basin, the Kettle Point is locally succeeded by shales and sandstones of the Bedford, Berea, and Sunbury formations. Erosion across the intervening Algonquin and Findlay arches has severed the connections between these and their type localities in Ohio. The Bedford (30 m) consists of soft grey fissile shale that is silty and sandy in its upper part. The overlying Berea (up to 60 m) consists of friable, shaly and dolomitic sandstones, which locally cut through the underlying Bedford shales as shoestring (channel) sands to rest directly on the Kettle Point Formation. The upper unit is the Sunbury Formation (25 m), consisting of black fissile shales, the youngest Devonian strata in Michigan Basin. Its distribution in Ontario, like those of the Bedford and Berea, is confined to a few square kilometres bordering the St. Clair River.

Cycle 8, whose distribution was greatly reduced by erosion following post-Devonian uplift of the cratonic arches, records a waning, Acadian landmass to the southeast. Continental deposition prevailed in southern, marine-deltaic deposition in central Allegheny Basin (and possibly also in Quebec Basin). Initially, the northern part of Allegheny Basin and Michigan Basin were open marine platforms. The causes for the subsequent change to deposition of thick, anaerobic muds (black shales) are obscure; it is worth noting that this change occurred over much of the North American Craton near the Devonian-Carboniferous boundary.

Cycle 9 – Upper Tournaisian to Stephanian

Rocks of Carboniferous age are widely distributed in the eastern and western St. Lawrence Platform and formerly may have covered the central platform also.

In the eastern platform (Anticosti Basin), Carboniferous strata are products of a depositional regime unrelated to the regimes of Cycles 1 through 8. They belong instead to the Magdalen successor basin centred beneath the Gulf of St. Lawrence. From the central gulf, the deposits of the Magdalen Basin extend northward across the deformed, lower and middle Paleozoic orogen and its structural front, to lie upon the southern margin of the Anticosti Basin.

Carboniferous rocks do not occur in southwestern Ontario, but are present in the immediately adjacent regions of Michigan and Allegheny basins. They were undoubtedly once continuous across Algonquin and Findlay arches in Ontario and adjacent western Ohio and southern Michigan, but the connection has been eroded.

Eastern St. Lawrence Platform

In the offshore Anticosti Basin, the Lower Carboniferous Codroy Formation disconformably succeeds the Upper Silurian Clam Bank Formation. On Port au Port Peninsula, however, it lies unconformably on the Cambrian (?) to Ordovician platformal succession and on rocks of the Humber Arm allochthon that are locally highly deformed. The beds onshore are red siltstone, sandstone, and conglomerate, limestone and gypsum, ranging in age from late Tournaisian to Viséan. The composition and age of the Codroy offshore are uncertain.

Western St. Lawrence Platform

Upper Carboniferous rocks underlie the central Michigan and Allegheny basins. The strata, largely nonmarine, are shale, siltstone, and sandstone, with minor limestone, and thick accumulations of coal, particularly in Allegheny Basin. The Michigan Basin sequence is thin and largely Westphalian, while that in Allegheny Basin ranges from late Namurian to Stephanian in age.

The thick clastic wedges of Allegheny and Michigan basins were largely derived from the Appalachian Orogen, whose relief was renewed during the Alleghanian Orogeny. Some of the Michigan Basin sediments may, however, be derived in part from the Canadian Shield.

THE MONTEREGIAN INTRUSIONS

K.L. Currie

The Monteregian Hills Petrographic Province (Adams, 1903) comprises seven major alkaline intrusions and numerous minor bodies of Early Cretaceous age, which form an east-west belt 120 km long by 25 km wide (Fig. 11.34), extending from Oka on the west to Bromont on the east. The three eastern major bodies, Yamaska, Shefford, and Brome, intrude the Appalachian Orogen. An eighth body, Mont Megantic, is sometimes included with the Monteregian Hills, but may belong to the less alkaline White Mountains Magma Series (Bedard et al., 1987). The major intrusions form prominent steep wooded hills rising hundreds of metres above the surrounding flat farm lands. Minor intrusions (sills and dykes) outcrop mainly in rock cuttings or in rapids in streams.

Figure 11.33. Upper Devonian Kettle Point Formation, Kettle Point on Lake Huron, Ontario. The locality is named for the profusion of spheroidal limestone concretions, called kettles", in the black shales (photo - B.V. Sanford, GSC 4-11-70).

Monteregian alkaline rocks belong mainly to one of three suites, namely (1) heterogeneous gabbro-to-pyroxenite suites lacking feldspathoids, (2) feldspathoid-bearing dioritic to syenitic suites, and (3) silica-saturated suites thought to be due to hybridization of alkaline magmas with siliceous wall rocks. Petrographic and chemical data are given by Eby (1984b, 1985, 1987). In general the rocks become increasingly undersaturated with silica from east to west. West of Montreal the rocks are so undersaturated that they lack feldspar. Major and minor intrusions consist of carbonatite, jacupirangite, ijolite, and melilite-bearing rocks (okaite, alnoite) in varying degrees of brecciation. The central plutons (Mont Royal, Mont Saint Bruno, Mont Saint Hilaire, Yamaska) all contain a large component of melagabbro and pyroxenite with titanaugite and kaersutite as essential components and varying amounts of olivine and plagioclase (An_{85-44}). Mount Johnson consists entirely of a younger component of nepheline-bearing diorite and monzonite, and Mont Royal, Saint Hilaire, and Yamaska contain similar rocks cutting the older mafic sequence. The dyke suite associated with all these plutons is markedly bimodal, with salic members exhibiting a trend to peralkalinity, as evidenced by sodalite and hauyne-bearing varieties, although none of the dyke suites exhibit the spectacular peralkaline phonolites and syenites found in the Mont Saint Hilaire intrusion (Currie et al., 1986). All of the central plutons contain minor amounts of hypersthene-bearing phases, thought to be due to reaction of the magma with siliceous wall rocks. The eastern plutons, Shefford, Brome, and Megantic tend to be larger and less undersaturated than those to the west, and consist of arcuate diorite to gabbro bodies and essentially saturated leucocratic syenite bodies. At Shefford and Brome nepheline-bearing diorites and monzonites intrude older syenite. At Megantic a large central plug of biotite granites resembles granite in the White Mountains. The bimodal dyke suite comprises alkali olivine basalts and saturated comptonites on the mafic side and quartz-bearing felsic dykes.

Emplacement ages of the Monteregian igneous rocks as measured by Rb-Sr isochron, K-Ar mineral ages, Ar-Ar ages, and fission-track dating fall in the range 115-140 Ma. Eby (1984a) noted that the older mafic to ultramafic units all gave ages of 130 to 140 Ma, while the younger, strongly

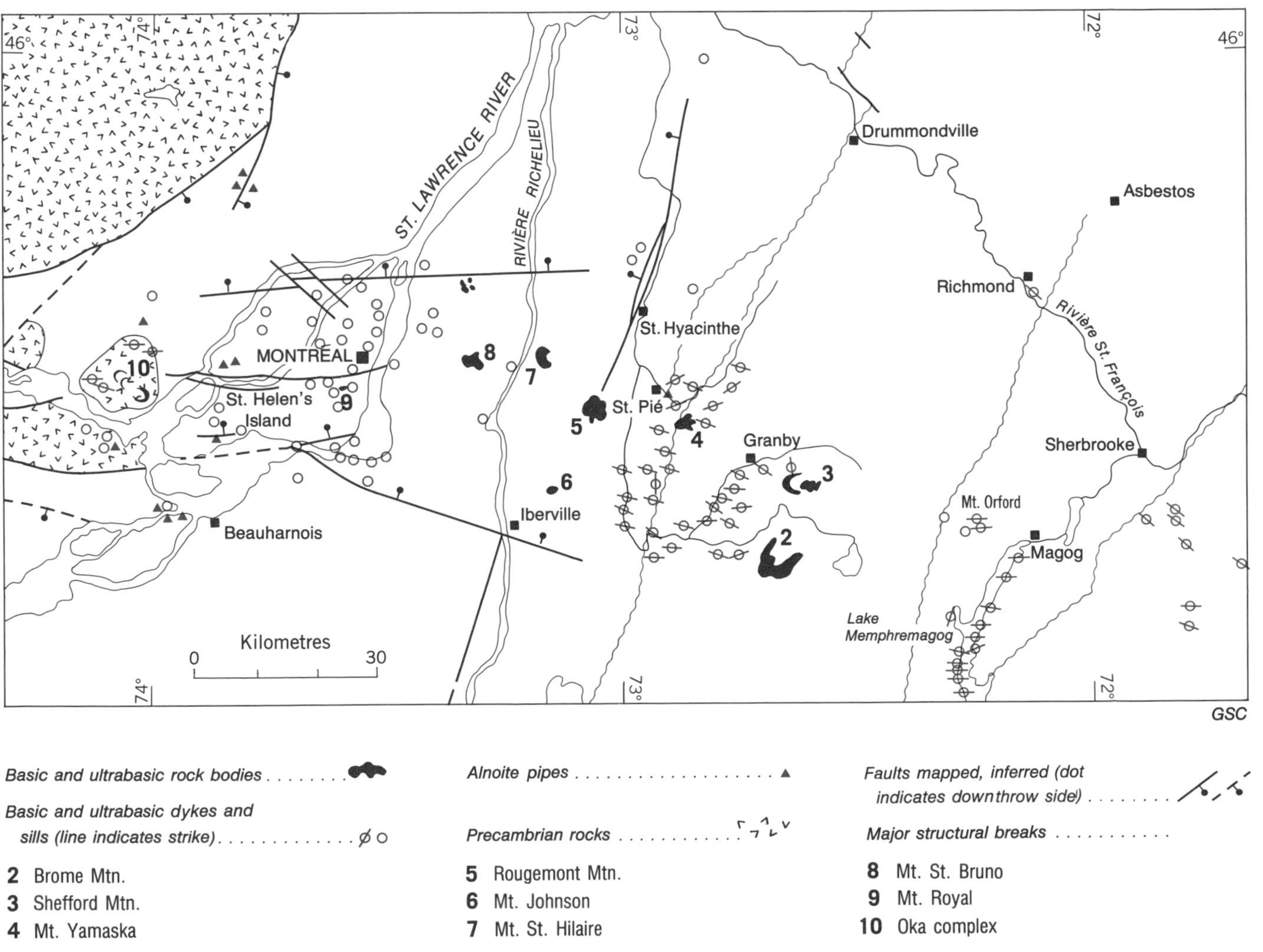

Figure 11.34. Sketch-map showing the distribution of the Monteregian intrusions (after Gold, 1967).

undersaturated units, gave ages of 118-122 Ma. The phases thought to have reacted with wall rocks may give either type of age. The dyke suite shows the same pattern as the major intrusions, with strongly undersaturated compositions giving younger ages. A small number of monchiquites and alnoites give ages between 100-115 Ma. The age pattern implies that emplacement of the composite plutons must have been a two stage process extending over 10-15 million years. However, Gilbert and Foland (1986) claimed a short intrusion history for Mont Saint Hilaire, an interpretation disputed by Currie et al. (1986). Whatever the mechanism, geophysical studies suggest the major intrusions are essentially vertical cylinders, or sets of cylinders. Their emplacement had little effect on the country rocks more than 20-100 m from the contact. In this inner zone inward dips of bedding have been ascribed to partial melting of the wall-rocks (Pouliot, 1969), and metasomatism is seen at Oka and Mont Saint Hilaire.

Petrogenetic modelling indicates that the differing rock series in the Monteregians must be derived from different protoliths. Most of the mafic rocks are partly of cumulate origin so that liquid lines of descent cannot be constructed. However, Eby (1984b, 1985) has used single melt compositions together with the compositions of cumulate minerals to reconstruct a magma composition in equilibrium with mantle olivine. Calculations suggest the mafic, feldspathoid-free series crystallized from a magma similar to alkaline picrite, derived from a garnet lherzolite protolith. Such compositions occur in the dyke suite. Pb and Sr isotopic data (Eby, 1987) indicate this protolith had undergone a Precambrian melting event, and high incompatible element contents suggest subsequent mantle metasomatism. The alnoitic and carbonatitic rocks of the Oka district could be derived from a similar parent by small degrees of partial melting if the parent was slightly carbonated. Compositions of the strongly undersaturated rocks suggest derivation from metasomatised, depleted spinel lherzolite. The data as a whole suggest a progressive eastward increase in degree of melting and decrease in depth of melting.

The tectonic setting of the Monteregian Hills has been debated for many years. Some of the suggested possibilities are leaky transform faults (Foland and Faul, 1977), mantle plumes (Crough, 1981), and southeast-propagating tensional events (Bedard, 1985). No single model has overwhelming support at present, and elements of all three may be involved. The location of the plutons seems strongly influenced by older structures (Kumarapeli, 1978), but the trend of the province cuts across the Saint Lawrence Trough. The complex nature of the protolith implies that the Monteregian province was derived from subcontinental mantle. A possible scenario involves impingement of a deep mantle diapir on the base of continental lithosphere, bringing heat and volatiles. At higher levels, stretching and reactivation of old zones of weakness provide channels for fluids. The pattern of lithologies suggests greater melting at shallower depths from west to east, consistent with westward motion of the continental plate and subjacent lithosphere over underlying heat source. Such a pattern is consistent with opening of the Atlantic Ocean, known to have been in progress at this time. On this model the Monteregian hills are a minor side effect of ocean opening, possibly associated with transform faulting or local tensional effects.

SUMMARY OF BASIN HISTORY

B.V. Sanford

The undeformed St. Lawrence Platform is the inboard remnant of a much broader platform that originated as a passive continental margin through continental separation in Early Cambrian time, as the Iapetus Ocean began to open. The initial deposits were Upper Proterozoic nonmarine redbed clastics, and volcanics locally, laid down in pre-separation rift-depressions here and there from western Newfoundland to northern Michigan (Kumarapeli, 1985).

From Early Cambrian through Early Ordovician times, the passive margin was a broad continental shelf (Fig. 11.12), whose sediments were largely carbonates of shallow marine origin with intercalations of orthoquartzitic sandstone derived from the Canadian Shield to the northwest. Lower and Middle Cambrian deposits are confined to outboard areas (eastern platform), but Upper Cambrian and Lower Ordovician strata cover much of the central and western platforms as well.

During Late Cambrian and Early Ordovician times, two principal negative tectonic elements were in evidence, (i) an ancestral Michigan Basin (embayment) at the western end of the St. Lawrence Platform, and (ii) the Ottawa Embayment in the central part, a gently subsiding element, initiated as an aulacogen in latest Proterozoic time (Kumarapeli, 1985), which would later (Late Ordovician to Late Devonian?) become an appendage of the Quebec Basin.

In early Middle Ordovician (Whiterockian to Chazyan) time, the passive (divergent) margin became active, as the Iapetus seafloor began to converge against North America; this was the beginning of the Taconian orogeny. The rise of the orogen provided both a new source of detritus in the southeast, and a load that depressed the lithosphere to form a linear, northeast-trending, foreland basin, St. Lawrence Trough.

During the initial development of the trough, carbonates and fine-to-coarse clastics derived from the Canadian Shield were the dominant facies bordering the craton. In the central and eastern trough, these passed basinward to flysch derived from a rising landmass to the southeast and east, and black shales. Near the close of Chazyan time, allochthonous slices of slope, rise, and abyssal-plain sediments and oceanic crust were emplaced into the eastern trough. Remnants of the allochthons, namely the Humber Arm and Hare Bay klippen of western Newfoundland, formed the eastern structural margin of Anticosti Basin from late Chazyan time onward.

Continuation of convergent orogeny in late Middle Ordovician (Blackriveran and Trentonian) time caused further northwestward migration of the Taconian front south of the Gulf of St. Lawrence, attendant migration of the axis of the St. Lawrence Trough, and narrowing of its central and western divisions. Shale and flysch, derived from the advancing orogen, initially intertongued with Trentonian and lower Edenian carbonates, but later, in late Edenian and Maysvillian times, spilled over the carbonates to extend widely across the southern craton. By the Edenian, eustatic sea-level rise had effected marine conditions between the St. Lawrence, Hudson, and Interior platforms.

Vertical movements of structural elements on the craton coincided with the Taconian orogeny; the cratonal arches rose and Hudson Platform began to subside and receive marine sediments (Sanford and Norris, 1973). Faults associated with the arches were most active in late Trentonian and early Edenian time.

Orogenic activity was intense during the Late Ordovician (late Edenian to Richmondian). Large volumes of clastic sediments (euxinic and open-marine muds) were transported into the St. Lawrence Trough, across the trough and across vast areas of the Canadian Shield (Fig. 11.19). In Richmondian time, the Taconian structural front reached its final and present position in Canada, marking the southeastern limit of the St. Lawrence Platform. Impingement of the front against the southeast-trending Frontenac, Saguenay, and Beaugé arches segmented the foreland basin, forming the independent Allegheny, Quebec, and Anticosti basins.

From the inception of convergent tectonism until the end of Chazyan time, the eastern platform responded, as did the central and western parts, to an approaching tectonic front. The Richmondian and Gamachian strata of Anticosti Basin are limestones, however, bearing no witness to nearby tectonic lands, whereas those of the central and western platform are thick, orogen-derived redbeds. The lack of tectonism in the eastern sector may be attributable to cessation of convergence there, to by-passing of convergence along a northwest-trending transform fault, to the position of Anticosti Basin adjacent to an embayment in the continental margin, or to some combination of these.

The Michigan Basin continued to form a gently subsiding depocentre beneath the southern peninsula of Michigan throughout the Late Ordovician (late Edenian to Richmondian). Ottawa Embayment, largely an appendage of Quebec Basin, also persisted as a negative tectonic element during that time.

Early and early Middle Silurian (early Llandovery) were presumably times of relatively subdued orogenic activity along the Appalachian front, although siliciclastic redbed and marine shale deposition continued throughout Allegheny Basin, and presumably also Quebec Basin (Fig. 11.22). Much of the coarse detritus is of second-cycle origin, and apparently derived from the older sediments exposed along the structural front. Lower Llandovery coarse and fine clastics intertongued with carbonates along the eastern rim of Michigan Basin and elsewhere along the southern margin of the Canadian Shield. In Anticosti Basin, lower Llandovery carbonates succeeded Ordovician carbonates with no apparent break in deposition.

Late Llandovery to early Ludlow was a time of prolific reef growth in the Great Lakes region of Canada and the United States, which reached a climax in late Wenlock and early Ludlow times with the establishment, in the Michigan Basin, of a barrier reef complex and, basinward of the barrier, many patch and pinnacle reefs (Fig. 11.24). These developments coincided with rejuvenation of the cratonic arches, and accelerated subsidence of the Michigan Basin.

Depositional effects of the Acadian orogeny probably were felt first in the eastern platform (Anticosti Basin), with deposition of redbed deposits beginning as early as Wenlock time, and continuing through Ludlow and into Pridoli time. In late Ludlow and Pridoli times, these effects were felt in the central and western platform. These redbeds, parts of the Bloomsburg delta, were deposited along the southern margin of Allegheny Basin and possibly also Quebec Basin (Fig. 11.29). They pass northwestward to marine shales, carbonates, and thick evaporites in the central and northern Allegheny Basin and Michigan Basin.

In contrast to the widespread marine transgressions of late Middle Ordovician to Middle Silurian times, the late Silurian (late Ludlow to Pridoli) was a time of regression resulting in restriction and the deposition of evaporites. Restriction was intensified by rejuvenation of arches and downwarping of basins, effects recorded as far distant as the central Canadian Shield, where reactivation of the Cape Henrietta Maria Arch, for example, segmented Hudson Platform into two depocentres.

The youngest Paleozoic series was apparently deposited in the Eastern St. Lawrence Platform in Late Silurian (Wenlock? to Pridoli) time, except for Carboniferous deposits that form a part of the Magdalen Basin to the south and overlap the southern margin of Anticosti Basin. Thus, during Devonian deposition in the central and western platform, Anticosti Basin, again aberrant with respect to Appalachian tectonics, was apparently emergent.

In contrast to the limited Late Silurian seaways, Devonian (Emsian to Famennian) seas extended across much of the southern Canadian Shield, connecting St. Lawrence and Hudson platforms (Sanford and Norris, 1975). In late Emsian to Eifelian time, carbonates and evaporites were the dominant facies. This was particularly so in Michigan Basin, where thick evaporites (halite and anhydrite) were deposited in association with carbonates in a regime of restriction and rapid subsidence (Fig. 11.31). Thin, widespread, orthoquartzitic sandstones, derived partly from the orogen and partly from rejuvenated Precambrian basement arches, punctuate the evaporite-carbonate succession.

The onset of the main phase of the Acadian orogeny in late Middle to Late Devonian (Givetian to Famennian) time led to deposition of thick molasse and deltaic redbeds in the central and western platforms near the deformation front and euxinic marine shales and open-marine carbonates to the northwest (Fig. 11.32). In Michigan Basin and southern Hudson Platform (Moose River Basin), carbonates are the dominant Givetian lithofacies, but orogen-derived shales are prominent in the Frasnian and Famennian stages.

Carboniferous sediments, later largely eroded, may have covered a substantial part of the St. Lawrence Platform. They are widely preserved in the central and southern Gulf of St. Lawrence (Magdalen Basin), and overlap the Taconian structural front and the southern margin of Anticosti Basin beneath the northern Gulf of St. Lawrence. A large remnant of Carboniferous rocks in Michigan Basin was clearly once continuous across the Algonquin-Findlay arch trend in Ontario with its counterpart in Allegheny Basin.

The repeatedly observed synchroneity of epeirogenic movements and tectonism at the Appalachian plate margin points to a causative relationship. Most of the epeirogenic movements of the craton were in the form of basement arch rejuvenation and complementary basin subsidence. Uplift of the arches was for the most part accomplished through movement of fault-bounded megablocks, the principal times of uplift and block-rotation being:

i. Near the close of the Early Ordovician Epoch;
ii. At the end of Collingwood deposition in Late Ordovician (mid-Edenian) time;
iii. At the close of Early Silurian (Wenlock) time;
iv. Near the close of Late Silurian (Pridoli) time; and
v. At various times between late Middle and Late Devonian, and in post-Devonian times.

The most marked and widespread movements were those near the close of Wenlock and Pridoli times; these were presumably caused in some way by collisional tectonism.

The Paleozoic formations of the St. Lawrence Platform have been deeply and extensively bevelled by erosion. The present distribution and highly generalized stratigraphic and structural framework of the various divisions of the platform are illustrated in Figures 11.3 to 11.6. Only a few of the more significant faults associated with vertical movements of the arches and basins can be shown in these figures.

QUATERNARY PERIOD

W.R. Cowan

General statement

The Quaternary Period witnessed deposition of the glacial and intervening nonglacial deposits that mantle the bedrock. It commonly is divided into Pleistocene, the main glacial episode, and Recent, the interval following dissipation of the major mainland ice sheet. Only the Pleistocene is considered here. Traditionally, in central North America, four major glaciations were recognized,

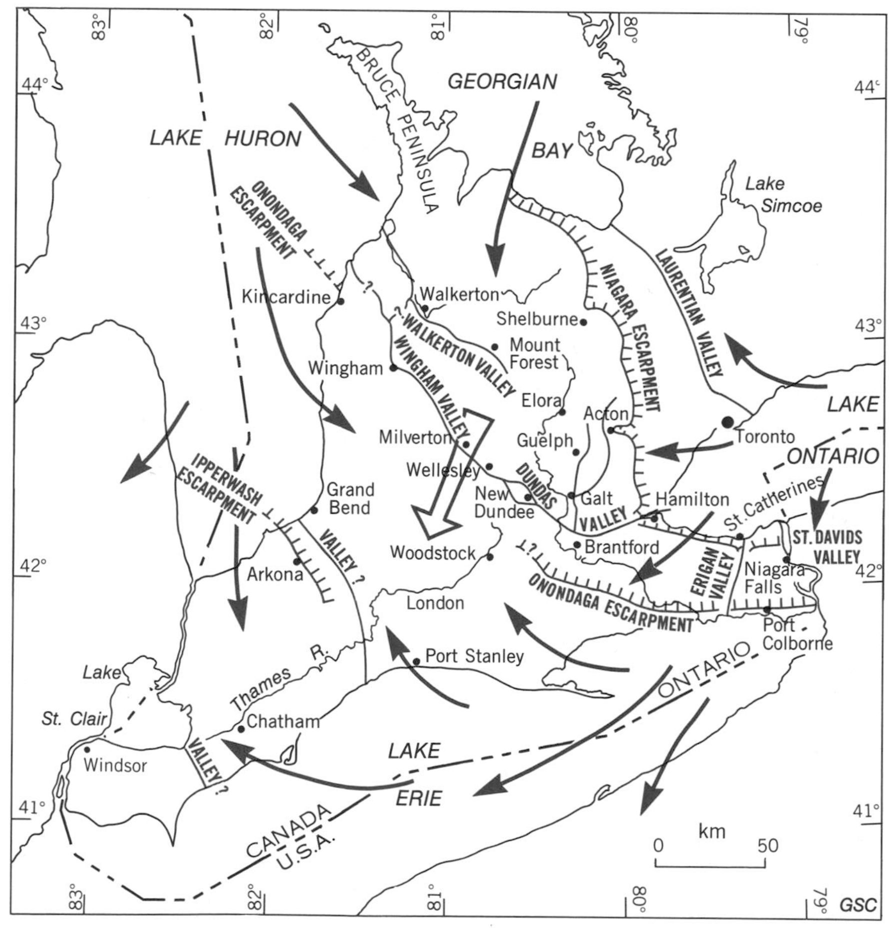

Figure 11.35. Major features of bedrock topography in southern Ontario (after Karrow, 1973). Small arrows indicate early and late ice-flow patterns influenced by the Great Lakes basins. Large arrow indicates main direction of flow during full glaciation.

from oldest to youngest, Nebraskan, Kansan, Illinoian, and Wisconsinan. These were separated by interglacial intervals – Aftonian, Yarmouthian, and Sangamonian – when climates were warm and the continent was largely ice-free. This nomenclature is now thought to be oversimplified, and some of these terms should be abandoned. In the St. Lawrence Lowland, only deposits of Wisconsinan age are well developed, although older sediments, assigned to the Illinoian and Sangamonian, are present locally.

For discussion of Quaternary deposits, the St. Lawrence Lowland is divided into two regions, western and central, separated by Frontenac Arch; each is characterized by thick glacial deposits. Morainal belts in the western region provide evidence of the advance and retreat of several ice lobes, accompanied by the development of the precursors of the Great Lakes. The central St. Lawrence Lowland is noted for late glacial marine sediments, which do not extend into the western area.

Western St. Lawrence Lowland

The Quaternary geological history of the Western St. Lawrence Lowland area in Ontario is complex and by no means fully understood. This brief synopsis presents the current knowledge and thought; for a more thorough review the reader is referred to the companion volume in this series (Fulton, 1989) and to Karrow (1984). Literature referred to herein is primarily of a synoptic nature and will provide readers with an overall understanding of trends of research and thought over the past 25 years.

Additional key review literature includes Chapman and Putnam (1984), Dreimanis (1977a), Dreimanis et al. (1966), Dreimanis and Karrow (1972), Dreimanis and Goldthwait (1973), Fullerton (1980), Karrow (1969, 1974, 1989), Karrow and Calkin (1985), and Prest (1970).

Bedrock surface

The preglacial surface of the southern Ontario area was probably flat with a few significant perturbations such as bedrock escarpments and buried river valleys. The best known of these are the Niagara Escarpment, the Onondaga Escarpment (also known as Bois Blanc Escarpment), and the buried Laurentian Valley (Fig. 11.35) with its associated delta at Toronto. The locations of prominent bedrock topographic features as outlined by Karrow (1973) are shown by Figure 11.35; in addition, this figure shows the major directions of glacier flow during Quaternary glaciations. Many of these bedrock surface features are known only from the compilation of well-record data, as the features are completely filled or subdued with glacial drift.

The Niagara Escarpment is formed from resistant dolostones of the Silurian Lockport Formation, which overlies less resistant Silurian and Ordovician shales, sandstones, and siltstones. The Onondaga Escarpment is formed by resistant Silurian dolostone (Bass Islands Formation) and limestones of Devonian age (Onondaga and Bois Blanc formations); these overlie less resistant shales and dolostones of the Salina Formation. This escarpment is exposed at the surface for only a short distance near Port Colborne, Ontario. A small, less well-known escarpment near Arkona, Ontario, Ipperwash Escarpment, appears to be the expression of a resistant limestone member within the Devonian Hamilton shales. These escarpments were undoubtedly present in preglacial times and probably exerted a considerable influence on patterns of glacial deposition. The degree to which the escarpments were affected by glacial erosion is poorly understood and the subject of debate. However, it seems that, although considerable modifications may have occurred, the escarpments are essentially preglacial landforms resulting from mass wasting and fluvial processes.

The buried river valleys were formed during preglacial, interglacial, and interstadial erosional episodes; in many instances they were re-occupied during successive events. The Laurentian and Erigan valleys (Fig. 11.35) are generally believed to be preglacial features that carried the waters of Lakes Huron and Erie to the Lake Ontario-St. Lawrence River system. The age of the remaining valleys is unknown.

Finally, the origin of the Great Lakes basins themselves has been a question for some time. The long axis of Lake Huron tends to follow the structure of underlying bedrock; Lake Erie is mainly underlain by Devonian shales; and Lake Ontario is underlain by Ordovician shales and shaly limestones. In general, it is believed that the lake basins originated as part of a large river system or systems. Subsequent glacial erosion enlarged them.

WISCONSINAN	LATE (?)	North Bay Interstade	
		Greatlakean Stade	
		Two Creeks Interstade	11.85 ka
		Port Huron Stade	
		Mackinaw Interstade	13.3 ka
		Port Bruce Stade	
		Erie Interstade	
		Nissouri Stade	
	MIDDLE	Plum Point Interstade	28 ka
		Cherrytree Stade	
		Port Talbot Interstade II	45 ka
		unnamed stadial (Dunwich Till)	
		Port Talbot I	
	EARLY	Guildwood Stade	
		St. Pierre Interstade	75 ka
		Nicolet Stade	
SANGAMONIAN			
ILLINOIAN			

GSC

Figure 11.36. Time classification of Quaternary Period for Great Lakes-St. Lawrence region. Ages from significant dated sites are indicated. Adapted from Dreimanis and Karrow (1972) by Karrow (1984).

Quaternary geology

To date only two major glacial episodes have been recognized within the region – the Illinoian and the Wisconsinan. Of these, the Illinoian has been proven only in the Toronto area where sediments representing the Sangamonian Interglacial Stage overlie glacial sediment. All pre-Illinoian glacial and nonglacial sediments have either been eroded, have not yet been found, or have been found but not recognized because of a lack of biotic remains or materials suitable for dating with radiocarbon or other methods.

The stratigraphic system currently in use in the region generally follows that of Dreimanis and Karrow (1972). They divided the Wisconsinan Stage into three substages (Early, Middle, and Late), which were in turn subdivided into stades and interstades representing short periods of general glacial advance or retreat within a region or ice lobe (Fig. 11.36). In general the Illinoian Stage ended at about 130 ka, the Sangamonian at 75 ka, and the Wisconsinan at about 10 ka; significant dated nonglacial events are shown in Figure 11.36.

The stades and interstades within this system are based on the presence of glacial sediment (tills) representing advances, and sediments, such as lacustrine sediments (commonly of glacial origin), which indicate glacial withdrawal (sometimes over very short distances).

The tills within the system are treated as lithostratigraphic units based on their areal continuity, textural composition, clast composition, and the inferred source of erratic clasts. As shown in Figure 11.35, early phases of glaciation resulted in the channelling of ice into the Great Lakes basins, with subsequent expansion out of the basins; a reverse sequence occurred during glacier retreats. At times of maximum ice thickness the overall ice flow within the region was from northeast to southwest. This series of events gave rise to till sheets related areally and lithologically to the various lake basins and direction of

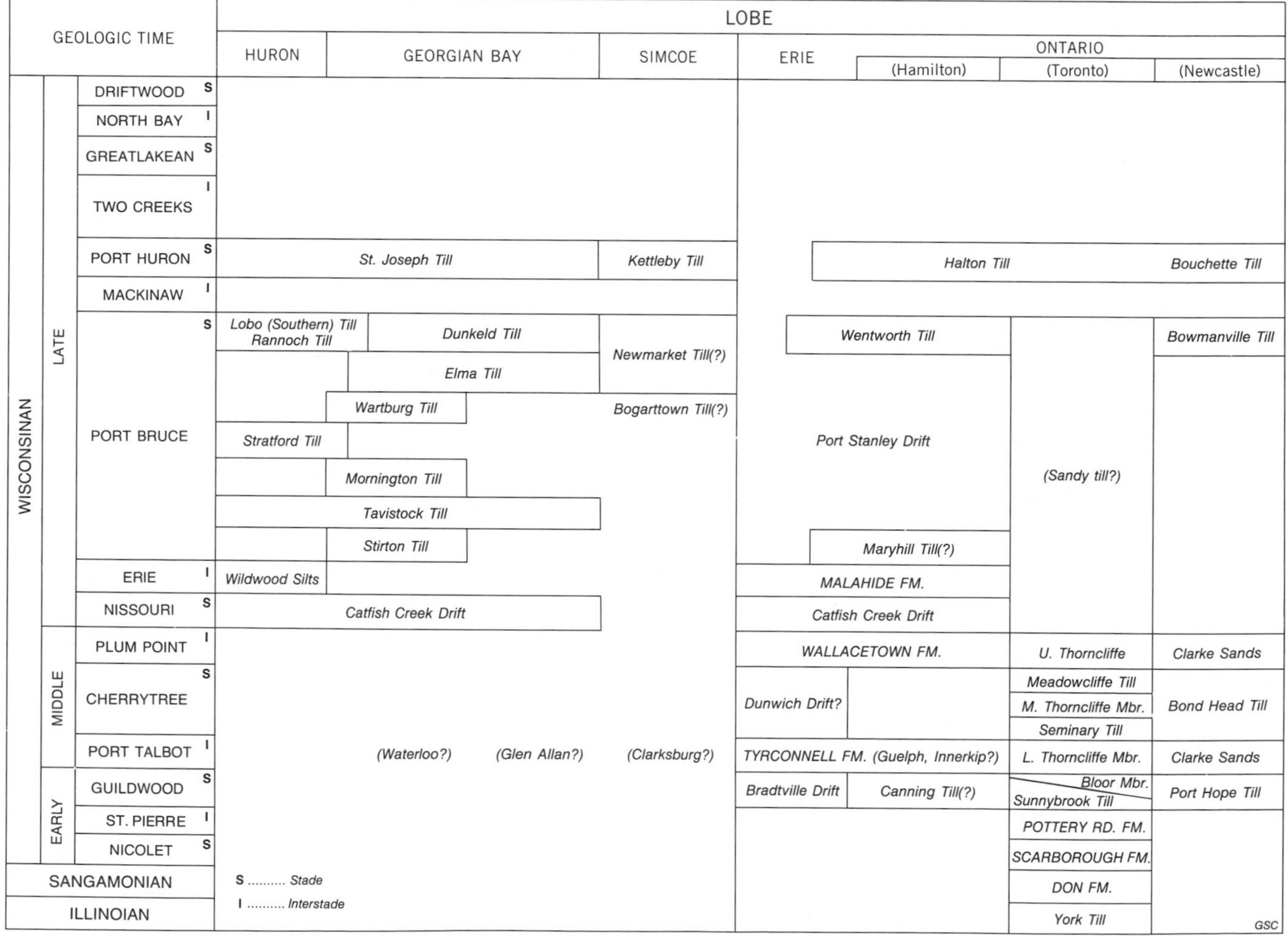

Figure 11.37. Correlation chart for glacial and non-glacial events in southern Ontario. (From Karrow, 1989.)

ice flow. Through detailed lithostratigraphic mapping, primarily by the Ontario Geological Survey, a detailed stratigraphy has been developed for each lobe and a cross-correlation, based on overlap of sediments from the various basins, proposed (Fig. 11.37; Karrow, 1989). It is noteworthy that for the Late Wisconsinan Substage, the Catfish Creek Drift is common to several lobes or basins; it represents the southwesterly flow during maximum ice cover.

Illinoian Stage

The oldest recognized glacial material in southern Ontario consists of the York Till deposited by ice flowing out of the Lake Ontario Basin in the Toronto area. It is recognized as being pre-Sangamonian on the basis of overlying organic sediments of Sangamonian age. North of Toronto, deep wells drilled into the buried Laurentian River channel have encountered tills that could eventually prove to be Illinoian or older in age (Karrow, 1984).

Sangamonian Stage

Interglacial beds have been recognized in the Toronto region for more than a half century. The most famous of these comprise the Don Formation, an assemblage of fluvial and estuarine sediments that occurs throughout much of the City of Toronto and extends northward into the buried Laurentian channel (Fig. 11.35). Fossil material indicating a climate perhaps 2°C warmer than present (Terasmae, 1960) leads to the interglacial assignment of these beds. Numerous biological studies of these materials are referenced in Karrow (1984), but there is currently no substantial overview of this formation and its contained fossil material.

Wisconsinan Stage

The three-fold division of the Wisconsinan Stage in Ontario followed the discovery by Dreimanis (1958) of the Port Talbot Interstadial beds in the bluffs along the north shore of Lake Erie. Subsequently, sub-till organic materials have been located at twelve more sites within the region. Approximate age ranges for these divisions are as follows: Early Wisconsinan – 75 to 65 ka; Middle Wisconsinan – 65 to 23 ka; and Late Wisconsinan – 23 to 10 ka.

Early Wisconsinan

The Early Wisconsinan Substage is subdivided into two glacial stades with an intervening interstade (Fig. 11.36, 11.37). In southern Ontario, the first of these, the Nicolet Stade, is marked by a cooling trend and deposits of the Scarborough deltaic sands and clays in a high-level lake in the Lake Ontario Basin.

Glacier recession in Lake Ontario resulted in lowering of Lake Scarborough, entrenchment of subaerial streams, and deposition of the Pottery Road fluvial sediments, which are correlated with the St. Pierre Interstadial sediments of the St. Lawrence Lowland. Fossil remains indicate a cool climate (Karrow, 1984). Subsequently, a major glacier readvance occurred with the deposition of the Sunnybrook Till at Toronto, the Bradtville Till at Lake Erie, and a series of tills in the United States including the Titusville Till (Fullerton, 1986).

Middle Wisconsinan

Retreat of the Early Wisconsinan glaciers was followed by a long period (65-23 ka) of cool climate with but little glacier activity in the Great Lakes region – the Middle Wisconsinan Substage. This substage consisted of an early, cool, nonglacial phase, which has been dated by numerous radiocarbon analyses as generally exceeding 40 ka, a middle phase during which some glacier expansion into the Great Lakes region occurred, and a late phase of very cool nonglacial conditions.

The early phase, the Port Talbot Interstade, has yielded radiocarbon dates on organic materials from throughout the central area of southwestern Ontario, which generally imply little or no ice cover in the region between 40 and 65 ka. Evidence consists of peat beds, weathering profiles, low water levels in Lake Erie and cool climate pollen spectra (Berti, 1975; Dreimanis et al., 1966; Karrow, 1984).

In the middle phase, between 40 and 35 ka, glacier expansion into Lake Ontario resulted in deposition of glaciogene sediments within the Toronto area. These are referred to as the Seminary and Meadowcliffe tills (Karrow, 1967) and appear to be of very limited extent; indeed, Eyles and Eyles (1983) suggested that these diamictons may not represent deposits associated with grounded ice but perhaps are glaciolacustrine sediments deposited below floating ice. The readvance is termed the Cherrytree Stade. At Lake Erie, Dreimanis (de Vries and Dreimanis, 1960) has described Dunwich Till outcrops of northern (Georgian Bay lobe) affinity, which appear to fit within this time span. However, evidence is lacking elsewhere to verify that this Dunwich Till does indeed represent this glacial event; if it does then it would appear to be of very limited extent.

During the late phase, from about 23-35 ka, very cool nonglacial conditions again prevailed (Berti, 1975); this episode is termed the Plum Point Interstade. Sediments containing organic remains representative of this interstade occur at Plum Point on Lake Erie, at Toronto, and at Niagara Falls. Karrow (1984) has concluded that ice cover during the Plum Point Interstade exceeded that of the Port Talbot Interstade; his conclusion is based on the relative frequencies of radiocarbon-dated sediment representing each episode.

Late Wisconsinan

Following the Plum Point Interstade, major glacier expansion in the Great Lakes region resulted in complete coverage of the area by ice; this onset was initiated at 23 ka and culminated in Ohio about 18 ka. Detailed mapping of surficial sediments indicates numerous fluctuations of the various ice lobes, especially during retreat from the maximum cover (Fig. 11.37). If the number of events that occurred between 23 and 12 ka is representative of Pleistocene glaciation in general, then knowledge of older glacial events in southern Ontario is very cursory indeed.

The **Nissouri Stade** marks the onset of Late Wisconsinan glacial events. It is represented throughout southwestern Ontario by a stony, very stiff, dense carbonate-rich, sandy silt till known as Catfish Creek Drift (de Vries and Dreimanis, 1960). Though it generally represents regional flow toward the southwest, early and late flow out of the lake basins is assumed but poorly documented except perhaps at Lake Erie.

By about 16 ka, retreat from the Nissouri stadial maximum ice-frontal positions was well underway and the ice was thinning rapidly. The melting gave rise to numerous glacial lakes through much of central southwestern Ontario and a large glacial lake within the Erie Basin where these lacustrine sediments are best known (Dreimanis, 1958). This episode of glacier recession is referred to as the **Erie Interstade** (Fig. 11.36).

Rejuvenation of the glaciers (**Port Bruce Stade**) resulted in readvances across much of the area in southern Ontario that was mantled by lacustrine sediments; these sediments were incorporated and redeposited as stone-poor, clay-rich tills. Where lacustrine sediments were coarse grained or absent, sandy silt tills were the result. This event or series of events resulted in the Erie-Ontario lobes advancing to the Orangeville Moraine in the north, the Ingersoll Moraine in the south, and perhaps to the Leamington and Windsor areas in the extreme southwest. The Georgian Bay-Huron lobe advanced southeasterly to the Orangeville Moraine in the north and to east of Woodstock in the south. The principal sediments related to the Ontario-Erie Lobe are the Port Stanley Drift (Fig. 11.37) (associated with the Ingersoll, Westminster, St. Thomas and Norwich moraines) and the sandy Wentworth Till associated with the Paris and Galt moraines; the Paris and Galt moraines represent fluctuation near the end of the Port Bruce Stade.

The Georgian Bay-Huron lobe produced the Tavistock Till, which has been mapped widely; associated lesser fluctuations created the Stirton and Mornington tills (Fig. 11.37). Ice retreat after their deposition resulted in a break between the Huron and Georgian Bay ice lobes. Several late Port Bruce Stadial fluctuations created additional tills and moraines on a lobal basis, including the Rannoch Till, which is associated with several end moraines (Mitchell, Dublin, Lucan, and Seaforth).

The smaller Simcoe Lobe became independent at this time, readvancing to deposit the Newmarket Till along the northern part of the Niagara Escarpment.

Much of this complex history occurred between 15 and 13.5 ka and records unstable ice masses advancing and retreating rapidly, often into proglacial lakes, which enhanced both flow and retreat rates. Some of the early Glacial Great Lakes came into being at this time (Fig. 11.38), principally phases of Lake Maumee, which are associated with retreat from ice marginal positions recorded by Port Stanley Till, and Lake Arkona phases, which may possibly be associated with Wentworth Till and the Paris Moraine; Maumee lakes drained either southerly or westerly through the United States, while Arkona lakes drained westerly through the United States.

The **Mackinaw Interstade** followed the Port Bruce Stade. Though most of southwestern Ontario was ice-free at this time, no organic beds have been observed beneath younger glacial sediments in Ontario. In Michigan, organic materials of this age have been dated at 13.3 ka. Wood in younger till has been dated at 13.1 ka on the Lake Huron shore (GSC-2213; Lowdon and Blake, 1976); this agrees with the estimated age of 13 ka for the succeeding Port Huron Stade. The low-level Lake Ypsilanti formed in the Erie basin at this time and drained easterly.

The western Lake Ontario basin was ice-free at this time (Karrow, 1969). Glaciolacustrine sedimentation occurred within the Huron Basin and locally around ice margins.

The Mackinaw Interstadial was followed by significant readvances of the Ontario, Huron, Georgian Bay, and Simcoe ice lobes – the well known **Port Huron Stade**. This advance resulted in deposition of the fine-grained St. Joseph Till in the Huron basin where the ice advanced to the Wyoming Moraine (Port Huron Moraine System); an equivalent till of the same name was deposited by the Georgian Bay Lobe, which advanced to the Banks Moraine. On the Simcoe and Ontario lobe fronts, the ice advanced to the south and west, depositing the clayey Kettleby and Halton tills respectively. These two lobes contributed much fine-grained glaciolacustrine sediment to the massive Oak Ridges Moraine.

Blockage of easterly drainage resulted in high-level Lake Whittlesey (Fig. 11.38), which has a well developed shoreline and suite of deltaic sediments in southern Ontario. The lake drained westerly through Michigan.

Retreat of the Port Huron ice lobes to their respective basins essentially completed the glacial sculpturing of the region, initiating the **Two Creeks Interstade**, which has been dated at about 11.9 ka at its type area in Wisconsin. However, no organic material of this age has been found beneath glacial sediment in southern Ontario. There, vegetation rapidly colonized the upland areas, while in the Huron-Erie basins final retreat was accompanied by falling lake levels through the Warren, Grassmere, and Lundy stages (Fig. 11.38). The several levels of Lake Warren (I, II, and III) drained westward through Michigan, whereas Lundy and Grassmere may have drained eastward. Further retreat into the Ontario Basin opened an isostatically depressed outlet near Buffalo, causing Lake Erie to drain to very low levels (Fig. 11.38).

Continuing retreat in the Ontario Basin led to the formation of Lake Iroquois at about 12 ka (Karrow, 1969). This lake initially had its outlet at Rome, New York, later draining into the St. Lawrence basin as it became ice-free. Early Lake Ontario was much lower than present, owing to the isostatic depression of the St. Lawrence Valley. Subsequently, it rose to present levels; ongoing uplift at the east end of the lake currently is causing a minor increase in water level at the west end.

Gradual uplift of the eastern outlet of Lake Erie created rising water levels; the present day level of Lake Erie was attained only within the last 2 ka (Barnett, 1985).

The **Great Lakean Stade** (Fig. 11.37) represents a glacial readvance from the Superior basin into Michigan and Wisconsin, but no positive evidence of this advance is found in the central Ontario region. However, it is assumed that a rapidly downwasting, retreating ice mass must have had its margin somewhere in this region about that time (Fig. 11.39). Uplift of the Huron basin outlet near Kirkfield, Ontario caused water levels to rise to the Main Algonquin level, which developed very strong shoreline features in the southern part of its basin. Northward retreat of the ice allowed the opening of several channels across the Algonquin Upland, causing a series of lower Post Algonquin levels to form. Opening of the depressed North Bay Outlet allowed the Lake Huron Basin to drain to a very low Lake Stanley level dated at between 10.5 and 10.0 ka; a contemporaneous low-level lake in Georgian Bay is called Lake Hough. These latter events are referred to as the **North Bay Interstade**. Uplift of the North Bay Outlet caused rising lake levels in the Huron basin, giving rise to the Nipissing Great Lakes between 6 and 4 ka. Eventually downcutting of the outlet at Port Huron caused lowering to the Lake Algoma level, and subsequently, the current Lake Huron level was attained.

Central St. Lawrence Lowland

The Central St. Lawrence Lowland region (Sanford and Grant, 1976) is bounded by the Precambrian Frontenac Axis on the west, the Canadian Shield to the north and the Appalachian Mountains on the south; to the east the region is limited by the converging Appalachian and Shield regions just west of the Saguenay River.

The major Quaternary elements in the lowland were deciphered by N.R. Gadd and P.F. Karrow during the 1950s; this work was summarized by Gadd in 1971. Subsequent syntheses include papers by Gadd (1976), Gadd et al. (1972), Dreimanis and Karrow (1972), Dreimanis (1977b), McDonald (1971), Occhietti (1977, 1980), and Prest (1970); current summary works include

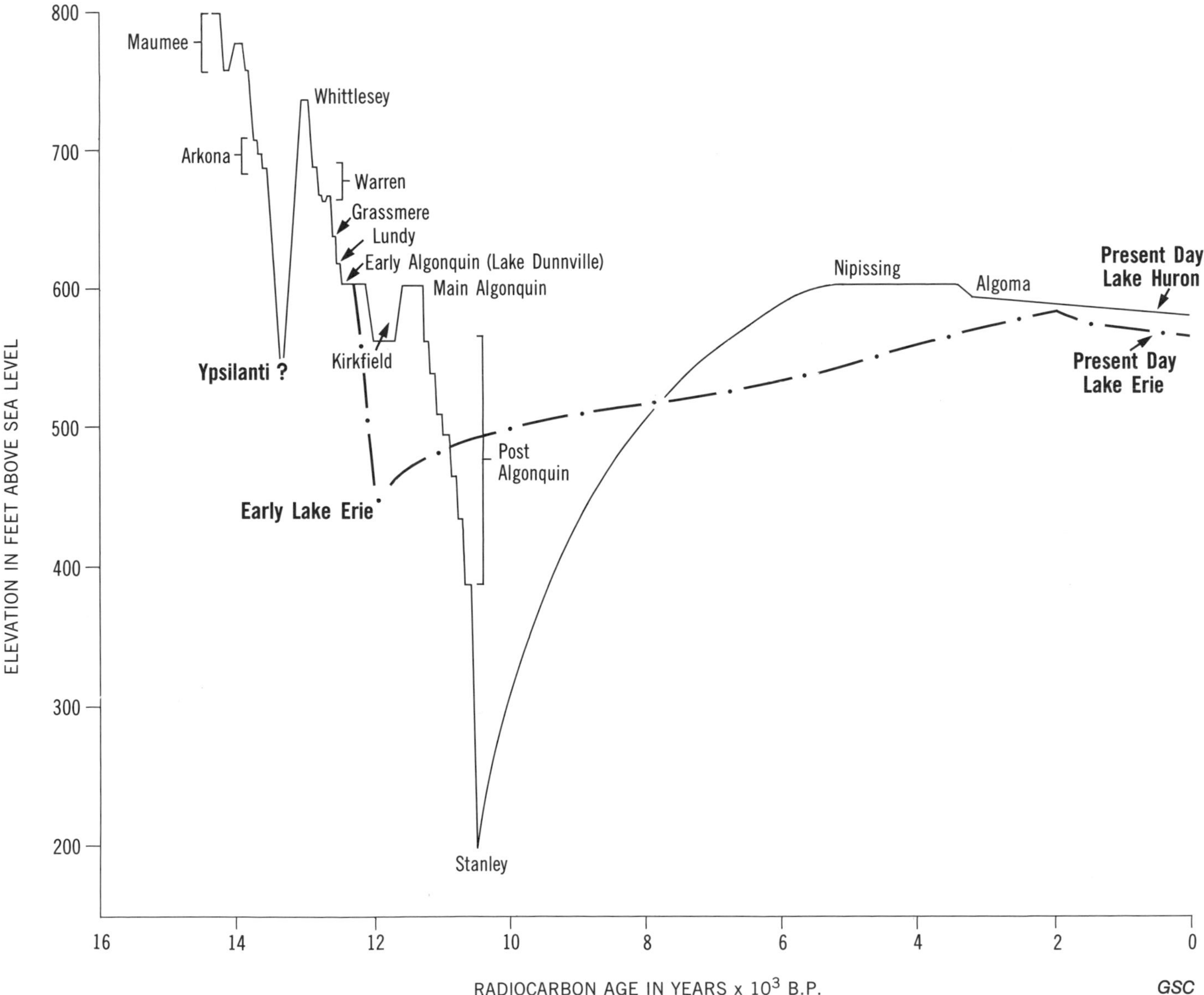

Figure 11.38. Sequence of former lakes of the Huron and Erie basins (Chapman and Putnam, 1984 after Lewis, 1969; Fullerton, 1980; Barnett, 1985).

Karrow (1984), LaSalle (1984, Fulton et al. (1984), Occhietti (1989), and Vincent (1989). For related information on the Appalachian region to the south, papers by McDonald (1969), Shilts (1981), and McDonald and Shilts (1971) provide considerable information.

Bedrock surface

Much of the physiography of the region is structurally controlled, and the major landscape elements were in place in pre-Quaternary time. Paleozoic rocks underlying the lowland are largely covered by glacial and postglacial sediments. Gadd (1971) believed that Pleistocene glaciation of the area did not significantly alter the bedrock surface, removing and redistributing little bedrock material. This interpretation is supported by a growing list of saprolite occurrences, presumably of Tertiary age, in Quebec (LaSalle, 1984). Northeasterly drainage along the St. Lawrence Valley has been in place for at least that part of the Quaternary for which there is a sedimentary record.

Quaternary geology

Pre-Wisconsinan

Though no biotic data are available to verify the presence of pre-Wisconsinan glacial or interglacial sediments in the St. Lawrence Lowlands, fragmentary data suggest that such sediments may someday be better known through drilling or the location of new outcrops. The saprolites mentioned above may represent pre-Quaternary weathering or Quaternary interglacial weathering. As well, weathered gravels beneath Lower Wisconsinan tills in the Appalachian area of Quebec indicate an ice-free lowland following an earlier glacial episode (McDonald, 1971). Occhietti (1989) suggested that as many as three glaciations prior to the deposition of the St. Pierre sediments may be documented by scattered data. Though this is plausible, the data to date are ambiguous.

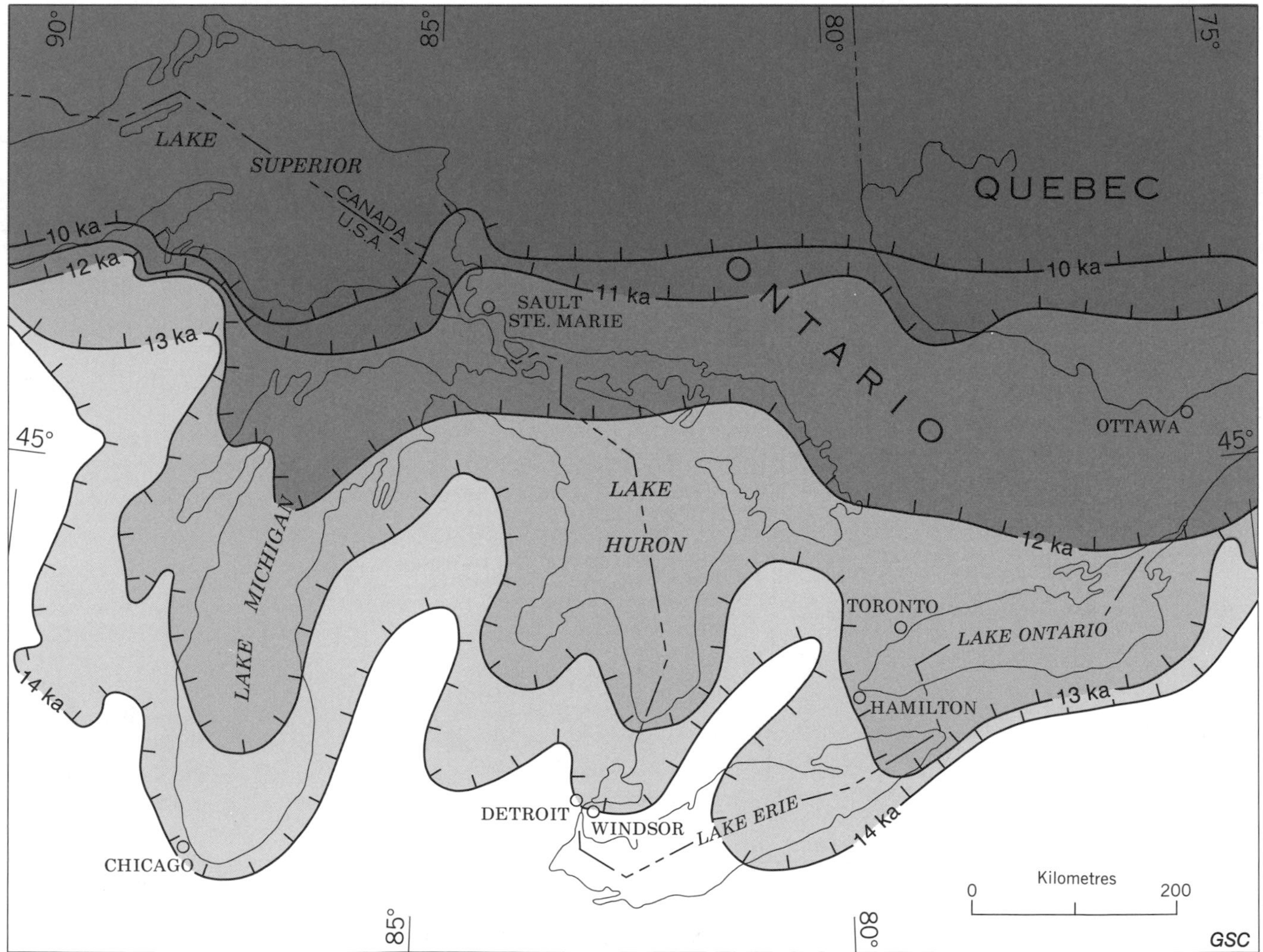

Figure 11.39. Ice-margin positions in the Michigan Basin area of Ontario from 14 ka to 10 ka. (Adapted from P.F. Karrow, 1989.)

Wisconsinan Stage

Stratigraphic correlation within the region is based on the sequence developed by Gadd (1971). In addition, the sequence developed by McDonald and Shilts (1971) for southeastern Quebec is critical as it relates to events in the Appalachian and Appalachian front region. Information from other areas is correlated with these sections and is described by LaSalle (1984) and Occhietti (1989). The lithological sequences for the central St. Lawrence Lowland and southeastern Quebec are illustrated in Figure 11.40.

Early Wisconsinan events within the lowland include those related to the Nicolet Stade, the St. Pierre Interstade and the lower part of the Gentilly Stade (Guildwood Stade, Fig. 11.40; Dreimanis and Karrow, 1972). The Nicolet Stade and St. Pierre Interstade were previously referred to as the Bécancour Stade and St. Pierre Interval respectively (Gadd, 1960, 1971). The Gentilly Stade (Gadd, 1960, 1971) relates to the period spanning the Guildwood Stade through Port Huron Stade of Dreimanis and Karrow (1972).

The **Nicolet Stade** is represented by the Bécancour Till, which was deposited by ice moving in a generally southerly direction. Blockage of St. Lawrence drainage resulted in deposition of proglacial varved sediments. Retreat prior to the beginning of the St. Pierre Interstade was accompanied by a later phase of glaciolacustrine deposition represented by the Pierreville varves. Although these varves are generally accepted as being related to the Bécancour Till, Occhietti (1989) discussed the possibility that they may relate to older events. In southeastern

	SOUTHERN QUEBEC	CENTRAL ST. LAWRENCE LOWLAND Gadd, 1971; Occhietti, 1980	SOUTHEASTERN QUEBEC McDonald and Shilts, 1971	TIME	GREAT LAKES-ST. LAWRENCE Dreimanis and Karrow, 1972	
WISCONSINAN	CHAMPLAIN SEA EPISODE	LAMPSILIS LAKE SEDIMENTS CHAMPLAIN SEA CLAY AND SAND	CHAMPLAIN SEA CLAY AND SAND	12 Ka	TWO CREEKS INTERSTADE	LATE WISCONSINAN
	GENTILLY STADE	VARVED SEDIMENTS	LENNOXVILLE TILL		PORT HURON STADE	
		GENTILLY TILL		13 Ka	MACKINAW INTERSTADE	
					PORT BRUCE STADE	
					ERIE INTERSTADE	
					NISSOURI STADE	
				25 Ka	PLUM POINT INTERSTADE	MIDDLE WISCONSINAN
			GAYHURST FORMATION		CHERRYTREE STADE	
				42-54 Ka	PORT TALBOT INTERSTADE	
		DESCHAILLONS FORMATION, varved clay and silt	CHAUDIÈRE TILL		GUILDWOOD STADE	EARLY WISCONSINAN
	ST. PIERRE STADE	ST. PIERRE SEDIMENTS	*MASSAWIPPI FORMATION*	80 Ka	ST. PIERRE INTERSTADE	
	NICOLET STADE	PIERREVILLE VARVES BECANCOUR TILL VARVES	JOHNVILLE TILL		NICOLET STADE	
PRE-WISCONSINAN			PRE-WISCONSINAN OR TERTIARY cemented river gravels subtill saprolite	125 Ka	SANGAMONIAN	

GSC

Figure 11.40. Upper Tertiary (?) and Quaternary stratigraphy of Central St. Lawrence Lowland and southeastern Quebec. (From LaSalle, 1984.)

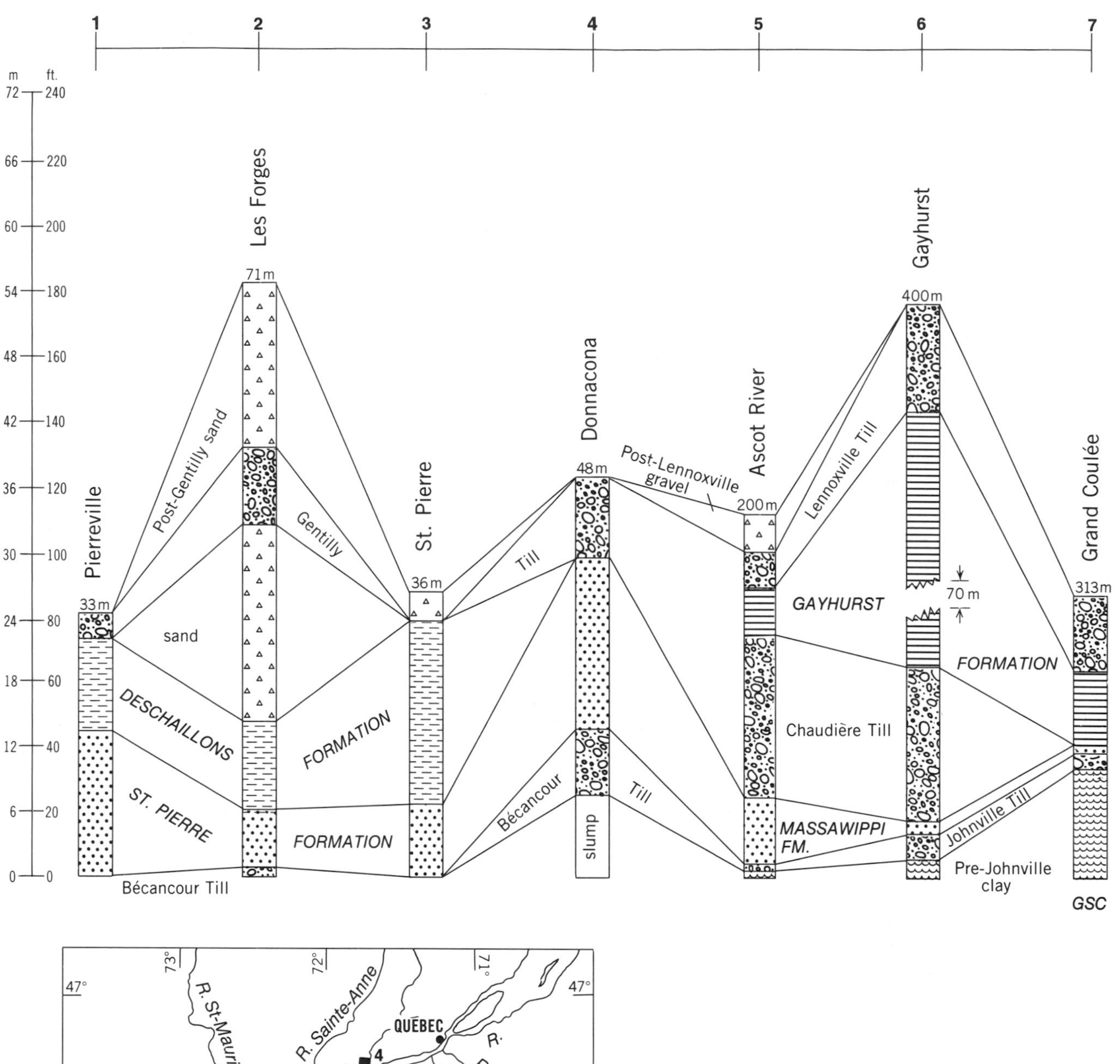

Figure 11.41. Principal sections of Quaternary deposits in southern Quebec (Gadd, 1971; McDonald and Shilts, 1971; Occhietti, 1980). Numbers at top of section are elevations above sea level. (From Karrow, 1984.)

Quebec the Nicolet Stade is represented by the Johnville Till and associated sediments (McDonald, 1971), whereas at Toronto it is represented by the Scarborough Formation, which consists of sediments deposited within an ice-dammed lake in the Lake Ontario Basin.

The **St. Pierre Interstade** is a nonglacial interval during which Early Wisconsinan ice retreated north of the St. Lawrence Valley, perhaps well into north-central Quebec (McDonald, 1971). The St. Pierre sediments (Gadd, 1960) deposited in the St. Lawrence Lowland consist of fluvial sediments, with current directions similar to the present, and sphagnum and carex peat. The latter appear to indicate a climatic condition suited to boreal forest development, i.e., colder than present. Wood from this interval has been dated at 74.7+2.7/-2.0 ka BP (Stuiver et al., 1978). This interval is represented in southeastern Quebec by the Masawippi Formation, which consists of peat beds and plant-bearing lake sediments (McDonald, 1971) dated at >40 ka (Shilts, 1981). Relative sea level was not as high as during the Late Wisconsinan Champlain Sea episode, as apparently no marine sediments are associated with the St. Pierre sediments.

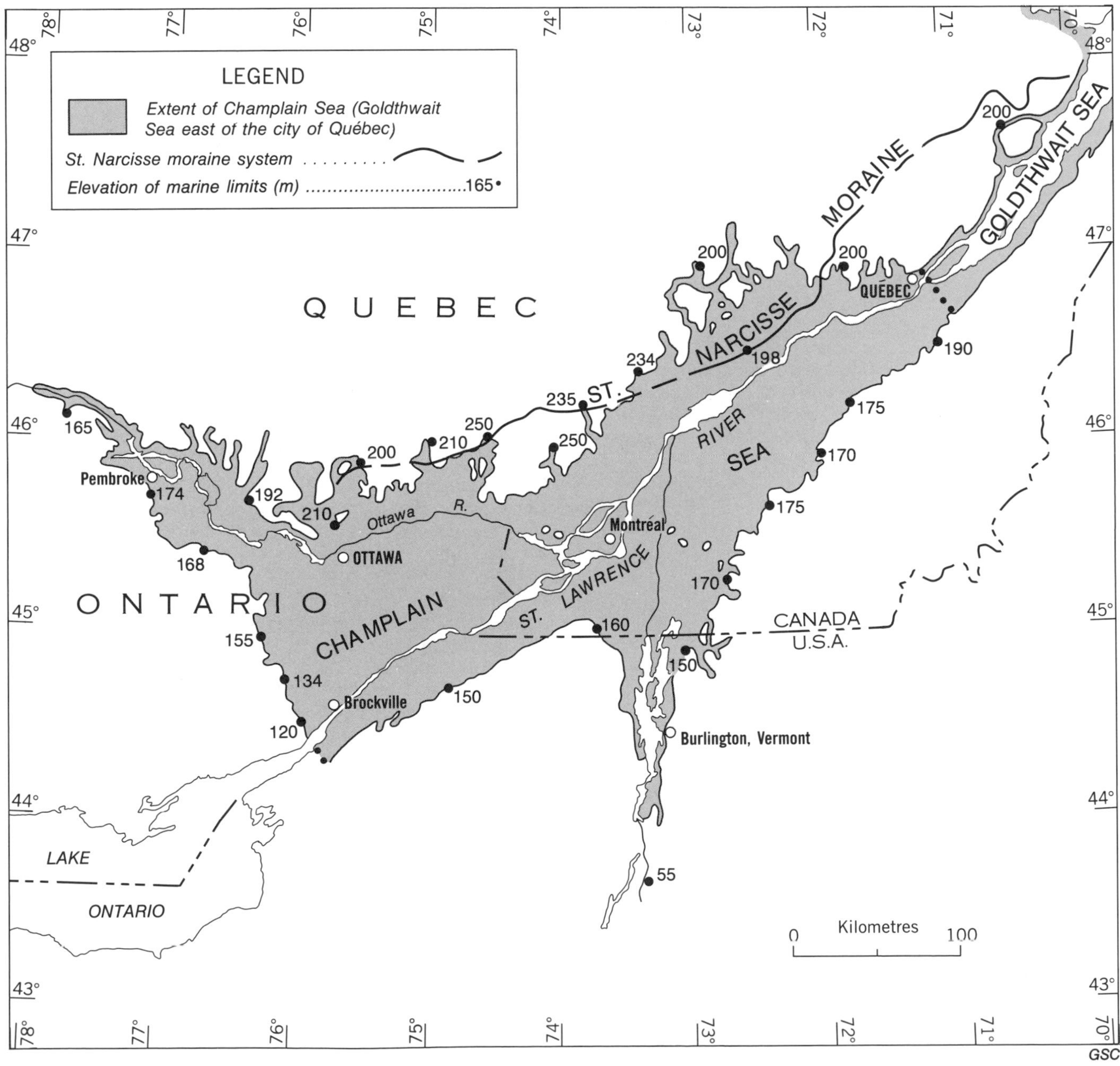

Figure 11.42. Extent of Champlain Sea; elevation of marine limits, and location of St. Narcisse Moraine System (from Occhietti, 1989.)

The St. Pierre Interstade terminated when the St. Lawrence River system was blocked by Labrador-based ice flowing from northeast to southwest, creating a large glacial lake in the St. Lawrence Basin. This event marked the onset of the **Gentilly Stade** and was accompanied by deposition of varved sediments of the Deschaillons Formation. Continued advance resulted in deposition of the Gentilly Till in the St. Lawrence Basin. This ice is believed to have covered much of the St. Lawrence Basin from its initial advance during Early Wisconsinan time until deglaciation at the end of Late Wisconsinan time (about 12 ka), because dated intervening sediments of Middle Wisconsinan age are generally unknown.

In southeastern Quebec, the early part of the Gentilly Stade is marked by the Chaudiere Till, which initially was deposited by ice flowing from a northeasterly direction; however, lithological and till-fabric data indicate that the centre of dispersion shifted westward during deposition of this till (McDonald, 1971). In this region, a limited glacier retreat occurred during Middle Wisconsinan time and glacial Lake Gayhurst formed between the ice front and the Appalachians. Approximately 4000 varves were laid down during this episode, which has been dated at >20 ka (McDonald and Shilts, 1971). St. Lawrence drainage remained blocked by ice. Lake sediments occupying a similar stratigraphic position have been identified in the Montreal region (Prest and Hode-Keyser, 1977; LaSalle, 1984), where underlying till is known as Malone Till. The Gayhurst sediments appear to be comparable in age and stratigraphic position with sediments of Port Talbot Interstadial age (*sensu lato*) of the western St. Lawrence Lowland region (McDonald and Shilts, 1971). Readvance from this position resulted in deposition of the Lennoxville Till in southeastern Quebec (Fig. 11.40, 11.41). This till appears to represent the entire Late Wisconsinan in southeastern Quebec (McDonald, 1971), though this interpretation is somewhat controversial (LaSalle, 1984); it is equivalent to the upper part of the Gentilly Till in most of the St. Lawrence River area and perhaps the Fort Covington Till in the Montreal/Cornwall region (Prest and Hode-Keyser, 1977).

Deglaciation

During the Late Wisconsinan maximum, the entire region was covered by thick ice. Retreat of the ice margin and glacier thinning was accompanied by several complex events. Retreat from Maine may have begun as early as 15 ka (Gadd et al., 1972) and perhaps from the Quebec Appalachians about 13.5 ka. A calving bay in the Gulf of St. Lawrence is believed to have caused a considerable drawdown of ice in the St. Lawrence Valley, leaving an active mass of residual ice in the Thetford Mines area; the latter is indicated by radial outward flow features.

Early retreat of the Laurentide ice from the Appalachian area was accompanied by deposition of the Highland Front moraine system before 12.5 ka (Gadd et al., 1972), and perhaps before 13 ka (Shilts, 1981). Several of the Monteregian Hills became nunataks, and glacial lakes formed along the retreating ice margin, the largest being Lake Vermont in the Lake Champlain Valley. These lakes are represented by varved lacustrine sediments overlying Gentilly Till. Farther northward, the Drummondville Moraine was deposited prior to the invasion of the Champlain Sea, i.e., prior to 12 ka. About this time the narrows at Quebec City became ice free and early phases of the Champlain Sea invaded the southern margin of the St. Lawrence Lowland between the retreating ice front and the Appalachian region. Ice retreat and disintegration was rapid along the Champlain Sea front and the sea soon expanded to occupy the Ottawa-St. Lawrence Lowland as far north as Pembroke (Fig. 11.42). Radiocarbon dates on molluscs have created some controversy over the routes, means, and timing of marine invasions, the relationship of Champlain Sea to Lake Iroquois in the Lake Ontario Basin, and several other topics, which are discussed in detail by Occhietti (1989). It is known, however, that the invasion occurred before 12 ka and that by 9.5 or 10.0 ka isostatic rebound had caused much of the St. Lawrence Lowland to be drained of marine waters, which left thick fossiliferous deep-water clay and near-shore sand as their legacy. Toward the end of the drainage phase, fresh water predominated as marine waters were shut off by uplift of the sill at Quebec City; Elson (1969) termed this the Lampsilis Lake phase and postulated that large volumes of fresh water derived from Lake Agassiz to the west were entering the system.

Perhaps the last major significant glacial event to directly affect much of the St. Lawrence Valley occurred at about 11.0 ka, when Laurentide ice advanced into the Champlain Sea along parts of the 300 km+ St. Narcisse morainic system (LaSalle and Elson, 1975; LaSalle, 1984). Though the geological and climatic significance of this event has been debated (LaSalle, 1984), the moraine provides an important chronological marker by its continuity over such a long distance (Fig. 11.42).

Summary of Quaternary history

The Quaternary history of the western St. Lawrence Lowland is very complex, and knowledge of pre-Late Wisconsinan events is very general. A great deal is known about Late Wisconsinan events, though greater radiometric chronological control is desired. A generalized summary of late glacial retreat is shown in Figure 11.39.

Only a cursory and conventional overview of the Quaternary glacial history of the central St. Lawrence Lowland has been provided. Recent correlations with deep-sea stratigraphy and the major chronological modifications implied have not been discussed; readers are referred to Fulton et al. (1984) for discussion of this and other topics. Other areas of debate not discussed in detail are presented by Karrow (1984), LaSalle (1984), and Occhietti (1989).

REFERENCES

Adams, F.D.

1903: The Monteregian Hills: a Canadian petrographical province; Journal of Geology, v. 11, p. 239-282.

Aitken, J.D.

1981: Cambrian System in the southern Canadian Rocky Mountains, Alberta and British Columbia; in Second International Symposium on the Cambrian System, M.E. Taylor (ed.); IUGS, United States Geological Survey, Subcommission on Cambrian Stratigraphy and Geological Survey of Canada; Guidebook for Field Trip 2, August 14-20, 1981, 61 p.

Barnes, C.R., Norford, B.S., and Skevington, D.

1981a: The Ordovician System in Canada; International Union of Geological Sciences, Publication No. 8, 26 p.

Barnes, C.R., Petryk, A.A., and Bolton, T.E.
1981b: Anticosti Island, Quebec; in Volume I: Stratigraphy and Paleontology, P.J. Lespérance (ed.); IUGS Subcommission on Silurian Stratigraphy, Ordovician-Silurian Boundary Working Group, Field Meeting, Anticosti-Gaspé, Quebec, 1981, Guidebook, p. 1-24.

Barnes, C.R., Telford, P.G., and Tarrant, G.A.
1978: Ordovician and Silurian conodont biostratigraphy, Manitoulin Island and Bruce Peninsula, Ontario; Michigan Basin Geological Society, Special Paper No. 3, p. 63-71.

Barnett, P.J.
1985: Glacial retreat and lake levels, north-central Lake Erie Basin, Ontario; in Quaternary History of the Great Lakes, P.F. Karrow (ed.); Geological Association of Canada, Special Paper No. 30, p. 185-194.

Bedard, J.H.
1985: The opening of the Atlantic, the Mesozoic New England igneous province, and mechanisms of continental breakup; Tectonophysics, v. 113, p. 209-232.

Bedard, J.H., Ludden, J.N., and Francis, D.M.
1987: The Megantic intrusive complex, Quebec: a study of the derivation of silica-saturated anorogenic magma of alkaline affinity; Journal of Petrology, v. 28, p. 355-388.

Bergström, S.M., Riva, J., and Kay, M.
1974: Significance of conodonts, graptolites and shelly faunas from the Ordovician of western and north-central Newfoundland; Canadian Journal of Earth Sciences, v. 11, p. 1625-1660.

Berry, W.B.N. and Boucot, A.J.
1970: Correlation of the North American Silurian Rocks; Geological Society of America, Special Paper 102, 289 p.

Berti, A.A.
1975: Paleobotany of Wisconsin interstadials, eastern Great Lakes Region, North America; Quaternary Research, v. 5, p. 591-612.

Bolton, T.E.
1966: Illustrations of Canadian fossils, Silurian faunas of Ontario; Geological Survey of Canada, Paper 66-5, 7 p.
1972: Geological map and notes on the Ordovician and Silurian litho- and biostratigraphy, Anticosti Island, Quebec; Geological Survey of Canada, Paper 71-19, 44 p.
1981a: Late Ordovician and Early Silurian Anthozoa of Anticosti Island, Quebec; in Volume II: Stratigraphy and Paleontology, P.J. Lespérance (ed.); IUGS Subcommission on Silurian Stratigraphy, Ordovician-Silurian Boundary Working Group, Field Meeting, Anticosti-Gaspé, Quebec, 1981, p. 107-109.
1981b: Ordovician and Silurian biostratigraphy, Anticosti Island, Quebec; in Volume II: Stratigraphy and Paleontology, P.J. Lespérance (ed.); IUGS Subcommission on Silurian Stratigraphy, Ordovician-Silurian Boundary Working Group, Field Meeting, Anticosti-Gaspé, Quebec, 1981, p. 41-46.

Bolton, T.E. and Copeland, M.J.
1972: Paleozoic formations and Silurian biostratigraphy, Lake Timiskaming region, Ontario and Quebec; Geological Survey of Canada, Paper 72-15, 48 p.

Bond, I.J. and Greggs, R.G.
1973: Revision of the March Formation (Tremadocian) in southeastern Ontario; Canadian Journal of Earth Sciences, v. 10, p. 1140-1155.
1976: Revision of the Oxford Formation (Arenig) of southeastern Ontario; Canadian Journal of Earth Sciences, v. 13, p. 19-26.

Boucot, A.J. and Johnson, J.G.
1967: Paleogeography and correlation of Appalachian Province Lower Devonian sedimentary rocks; Tulsa Geological Society Digest, v. 35, p. 35-87.

Boyce, D.H.
1978: Recent developments in western Newfoundland Cambro-Ordovician trilobite biostratigraphy; in Report of Activities for 1977, R.V. Gibbons (ed.); Newfoundland Department of Mines and Energy, Report 78-1, p. 80-84.
1979: Further developments in western Newfoundland Cambro-Ordovician biostratigraphy; in Report of Activities for 1977, R.V. Gibbons (ed.); Newfoundland Department of Mines and Energy, Report 79-1, p. 7-10.
1983: Preliminary Ordovician trilobite biostratigraphy of the Eddies Cove west – Port au Choix area, western Newfoundland; in Current Research, Department of Mines and Energy, Government of Newfoundland and Labrador, Report 83-1, p. 11-15.
1985: Cambrian-Ordovician biostratigraphic investigations, Great Northern Peninsula, western Newfoundland; in Current Research, Department of Mines and Energy, Government of Newfoundland and Labrador, Report 85-1, p. 60-70.

Chapman, L.J. and Putman, D.F.
1984: The physiography of southern Ontario; Ontario Geological Survey, Special Volume 2, 270 p. (see also earlier 1951 and 1966 editions).

Chiang, K.K.
1971: Silurian pentameracean brachiopods of the Fossil Hill Formation, Ontario; Journal of Paleontology, v. 45, p. 849-861.

Clark, T.H.
1972: Stratigraphy and structure of the St. Lawrence Lowland of Quebec; XXIV International Geological Congress, Montreal, Quebec, Excursion C-52, 59 p.

Cooper, G.A.
1956: Chazyan and related brachiopods; Smithsonian Miscellaneous Collections, v. 127, parts I and II, 1017 p.

Cooper, P.
1978: Paleoenvironments and paleocommunities in the Ordovician-Silurian sequence of Manitoulin Island; Michigan Basin Geological Society, Special Paper No. 3, p. 47-61.
1981: Atrypoid brachiopods and their distribution in the Ordovician-Silurian sequence of Anticosti Island; in Volume II: Stratigraphy and Paleontology, P.J. Lespérance (ed.); IUGS Subcommission on Silurian Stratigraphy, Ordovician-Silurian Boundary Working Group; Field Meeting, Anticosti-Gaspé, Quebec, 1981, p. 137-141.
1982: Early Silurian atrypoids from Manitoulin Island and Bruce Peninsula, Ontario; Journal of Paleontology, v. 56, p. 680-702.

Copeland, M.J.
1965: Ordovician Ostracoda from Lake Timiskaming, Ontario; Geological Survey of Canada, Bulletin 127, 5 p.
1970: Ostracoda from the Vauréal Formation (Upper Ordovician) of Anticosti Island, Quebec; in Contributions to Canadian Paleontology; Geological Survey of Canada, Bulletin 187, p. 15-29.
1973: Ostracoda from the Ellis Bay Formation (Ordovician), Anticosti Island, Quebec; Geological Survey of Canada, Paper 72-43, 28 p.
1974: Silurian Ostracoda from Anticosti Island, Quebec; Geological Survey of Canada, Bulletin 241, 73 p.
1976: Leperditicopid ostracoda as Silurian biostratigraphic indices; in Report of Activities, Part B, Geological Survey of Canada, Paper 76-1B, p. 83-88.
1981: Latest Ordovician and Silurian ostracoda from Anticosti Island, Quebec; in Volume II: Stratigraphy and Paleontology, Lespérance, P.J. (ed.); IUGS Subcommission on Silurian Stratigraphy, Ordovician-Silurian Boundary Working Group, Field Meeting, Anticosti-Gaspé, Quebec, 1981, p. 185-195.

Copeland, M.J. and Bolton, T.E.
1977: Additional paleontological observations bearing on the age of the Lourdes Formation (Ordovician), Port au Port Peninsula, western Newfoundland; in Report of Activities, Part B, Geological Survey of Canada, Paper 77-1B, p. 1-13.

Copper, P. and Morrison, R.
1978: Morphology and paleoecology of Ordovician tetradiid corals from the Manitoulin District, Ontario; Canadian Journal of Earth Sciences, v. 15, p. 2006-2020.

Crough, S.T.
1981: Mesozoic hotspot epeirogeny in eastern North America; Geology, v. 9, p. 2-6.

Cumming, L.M.
1983: Lower Paleozoic autochthonous strata of the Strait of Belle Isle area; in Geology of the Strait of Belle Isle area, northwestern insular Newfoundland, southern Labrador and adjacent Quebec; Geological Survey of Canada, Memoir 400, part 2, p. 75-108.

Currie, K.L., Eby, G.N., and Gittins, J.
1986: The petrology of the Mont Saint Hilaire complex, southern Quebec; an alkaline gabbro-peralkaline syenite association; Lithos, v. 19, p. 65-81.

de Vries, H. and Dreimanis, A.
1960: Finite radiocarbon dates of the Port Talbot Interstadial deposits in southern Ontario; Science, Volume 131, p. 1738-1739.

Dean, W.T.
1979: Trilobites from the Long Point Group (Ordovician), Port au Port Peninsula, southwestern Newfoundland; Geological Survey of Canada, Bulletin 29, 23 p.

Debrenne, F. and James, N.P.
1981: Reef associated archaeocyathans from the Lower Cambrian of Labrador and Newfoundland; Paleontology, v. 24, part 2, p. 343-378.

Dods, S.D., Hood, P.J., Teskey, D.J., and McGrath, P.H.
1984: Magnetic anomaly map of Canada; Geological Survey of Canada, Map 1255A, 4th Edition, scale 1:5 000 000.

Dreimanis, A.
1958: Wisconsin stratigraphy at Port Talbot on the north shore of Lake Erie, Ontario; The Ohio Journal of Science, v. 58, p. 55-84.
1977a: Late Wisconsin glacial retreat in the Great Lakes region, North America; Annals of the New York Academy of Sciences, v. 288, p. 70-89.
1977b: Correlation of Wisconsin glacial events between the eastern Great Lakes and the St. Lawrence Lowlands; Geographie Physique et Quaternaire, v. 21, p. 37-51.

Dreimanis, A. and Goldthwait, R.P.
1973: Wisconsin glaciation in the Huron, Erie, and Ontario lobes; Geological Society of America, Memoir 36, p. 71-106.

Dreimanis, A. and Karrow, P.F.
1972: Glacial history of the Great Lakes-St. Lawrence region, the classification of the Wisconsin(an) Stage and its correlatives; 24th International Geological Congress (Montreal), Section 12, p. 5-15.
Dreimanis, A., Terasmae, J., and McKenzie, G.D.
1966: The Port Talbot Interstade of the Wisconsin glaciation; Canadian Journal of Earth Sciences, v. 3, p. 305-325.
Earth Physics Branch
1980: Gravity map of Canada; Earth Physics Branch, Department of Energy, Mines and Resources, Gravity Map Series No. 80-1, scale 1:5 000 000.
Eby, G.N.
1984a: Geochronology of the Monteregian Hills alkaline igneous province, Quebec; Geology, v. 12, p. 468-470.
1984b: Monteregian Hills I; Petrography, major and trace element geochemistry and strontium isotope chemistry of the western intrusions, Mounts Royal, Saint Bruno and Johnson; Journal of Petrology, v. 25, p. 421-452.
1985: Monteregian Hills II; Petrography, major and trace element geochemistry and strontium isotope chemistry of the eastern intrusions, Mounts Shefford, Brome and Megantic; Journal of Petrology, v. 26, p. 418-448.
1987: The Monteregian Hills and White Mountain alkaline igneous provinces, eastern North America; in Alkaline Igneous Rocks, J.G. Fitton and B.J.G. Upton (ed.); Geological Society of London Special Publication, v. 30, p. 433-447.
Elias, R.J.
1982: Latest Ordovician solitary rugose corals of eastern North America; Bulletin of American Paleontology, v. 81, no. 314, 93 p.
Elson, J.A.
1969: Late Quaternary marine submergence of Quebec; Revue de Geographie de Montreal, v. 23, p. 247-250.
Eyles, C.H. and Eyles, N.
1983: Sedimentation in a large lake; a reinterpretation of the late Pleistocene stratigraphy at Scarborough Bluffs, Ontario, Canada; Geology, v. 11, p. 146-152.
Fagerstrom, J.A.
1971: Brachiopods of the Detroit River Group (Devonian) from southwestern Ontario and adjacent areas of Michigan and Ohio; Geological Survey of Canada, Bulletin 204, 112 p.
1982: Stromatoporoids of the Detroit River Group and adjacent rocks (Devonian) in the vicinity of the Michigan Basin; Geological Survey of Canada, Bulletin 339, 64 p.
Fåhraeus, L.E.
1973: Depositional environments and conodont-based correlation of the Long Point Formation (Middle Ordovician), western Newfoundland; Canadian Journal of Earth Sciences, v. 10, p. 1822-1833.
1974: Lower Paleozoic stratigraphy of the Port au Port area, west Newfoundland; in Geological Association of Canada/Mineralogical Association of Canada, R.K. Stevens (ed.); Field Trip Manual, 13 p.
Fisher, J.H. and Barrett, M.W.
1985: Exploration in Ordovician of central Michigan; American Association of Petroleum Geologists, Bulletin, v. 69, p. 2065-2076.
Flower, R.H.
1978: St. George and Table Head cephalopod zonation in western Newfoundland; in Current Research, Part A, Geological Survey of Canada, Paper 78-1A, p. 217-224.
Foland, K.A. and Faull, F.
1977: Ages of the White Mountain intrusives, New Hampshire, Vermont and Maine; American Journal of Science, v. 277, p. 888-904.
Fullerton, D.S.
1980: Preliminary correlation of Post-Erie Interstadial events (16 000-10 000 radiocarbon years before present), central and eastern Great Lakes region, and Hudson, Champlain and St. Lawrence lowlands, United States and Canada; United States Geological Survey, Professional Paper 1089, 52 p.
1986: Stratigraphy and correlation of glacial deposits from Indiana to New York and New Jersey; in Quaternary glaciations in the Northern Hemisphere, V. Sibrava, D.Q. Bowen, and G.M. Richmond (ed.); Quaternary Science Reviews, v. 5, p. 197-200 (International Geological Correlation Project No. 24).
Fulton, R.J. (ed.)
1989: Quaternary Geology of Canada and Greenland; Geological Survey of Canada, Geology of Canada, No. 1 (Also Geological Society of America, The Geology of North America, v. K-1).
Fulton, R.J., Karrow, P.F., LaSalle, P., and Grant, D.R.
1984: Summary of Quaternary stratigraphy and history, Eastern Canada; in Quaternary Stratigraphy of Canada – A Canadian Contribution to I.G.C.P. Project 24, R.J. Fulton (ed.); Geological Survey of Canada, Paper 84-10, p. 193-210.
Gadd, N.R.
1960: Surficial geology of the Bécancour map-area, Quebec; Geological Survey of Canada, Paper 59-8, 34 p.
Gadd, N.R. (cont.)
1971: Pleistocene geology of the central St. Lawrence Lowland; Geological Survey of Canada, Memoir 359, 153 p.
1976: Quaternary stratigraphy in southern Quebec; in Quaternary Stratigraphy of North America, W.C. Mahoney (ed.); Dowden, Hutchinson and Ross, Stroudsburg, Pennsylvania, U.S.A., p. 37-50.
Gadd, N.R., McDonald, B.C., and Shilts, W.W.
1972: Deglaciation of southern Quebec; Geological Survey of Canada, Paper 71-47, 19 p.
Gilbert, L.A. and Foland, K.A.
1986: The Mount Saint Hilaire plutonic complex; occurrence of excess ^{40}Ar and short intrusion history; Canadian Journal of Earth Sciences, v. 23, p. 943-958.
Greggs, R.G. and Bond, I.J.
1971: Conodonts from the March and Oxford formations in the Brockville area, Ontario; Canadian Journal of Earth Sciences, v. 8, p. 1455-1471.
Hamblin, W.K.
1958: The Cambrian sandstones of northern Michigan; Geological Survey Division, Michigan Department of Conservation, Publication 51, 146 p.
Harland, T.L., Pickerill, R.K., and Fillion, D.
1985: Ordovician intracratonic sediments from the Lac-St. Jean and Chicoutimi areas, Quebec, eastern Canada; Canadian Journal of Earth Sciences, v. 22, p. 240-255.
Haworth, R.T., Daniels, D.L., Williams, H., and Zietz, I.
1980: Bouguer gravity anomaly map of the Appalachian Orogen; Memorial University of Newfoundland, St. Johns, Newfoundland, Map No. 3a, scale 1:2 000 000.
Haworth, R.T. and Sanford, B.V.
1976: Paleozoic geology of northeast Gulf of St. Lawrence; in Report of Activities, Part A, Geological Survey of Canada, Paper 76-1A, p. 1-6.
Hofmann, H.J.
1972: Stratigraphy of the Montreal area; XXIV International Geological Congress, Montreal, Quebec, Excursion B-03, p. 27-59.
James, N.P.
1981: Megablocks of calcified algae in the Cow Head Breccia, western Newfoundland; vestiges of Cambro-Ordovician platform margin; Geological Society of America, Bulletin, v. 92, p. 797-811.
James, N.P. and Stevens, R.K.
1982: Anatomy and evolution of a lower Paleozoic continental margin, western Newfoundland; International Association of Sedimentologists, 11th International Congress on Sedimentology, McMaster University, Excursion 2B, 75 p.
Karrow, P.F.
1967: Pleistocene geology of the Scarborough area; Ontario Department of Mines, Geological Report 46, 108 p.
1969: Stratigraphic studies in the Toronto Pleistocene; Geological Association of Canada Proceedings, v. 20, p. 4-16.
1973: Bedrock topography in southwestern Ontario: a progress report; Geological Association of Canada Proceedings, v. 25, p. 67-77.
1974: Till stratigraphy in parts of southwestern Ontario; Geological Society of America, Bulletin, v. 85, p. 761-768.
1984: Quaternary stratigraphy and history, Great Lakes-St. Lawrence region; in Quaternary Stratigraphy of Canada – a Canadian Contribution to I.G.C.P. Project 24, R.J. Fulton (ed.); Geological Survey of Canada, Paper 84-10, p. 127-153.
1989: Quaternary Geology of the Great Lakes subregion; in Quaternary Geology of Canada and Greenland, R.J. Fulton (ed.); Geological Survey of Canada, Geology of Canada no. 1. (Also Geological Society of America, The Geology of North America, v. K-1.)
Karrow, P.F. and Calkin, P.E. (ed.)
1985: Quaternary evolution of the Great Lakes; Geological Association of Canada, Special Paper 30, 230 p.
Klappa, C.F., Opalinski, P.R., and James, N.P.
1980: Middle Ordovician Table Head Group of western Newfoundland: a revised stratigraphy; Canadian Journal of Earth Sciences, v. 17, p. 1007-1019.
Knight, I.
1977: Cambro-Ordovician platformal rocks of the northern peninsula, Newfoundland; Department of Mines and Energy, Government of Newfoundland and Labrador, Report 77-6, 29 p.
1983: Geology of Cambro-Ordovician rocks in parts of the Castors River, St. John Island and Port Saunders map sheets; in Current Research, Department of Mines and Energy, Government of Newfoundland and Labrador, Report 83-1, p. 1-10.
Kumarapeli, P.S.
1978: The Saint Lawrence paleorift system; in Tectonics and geophysics of continental rifts, I.B. Ramberg and E.R. Neumann (ed.); Reidel, Dordrecht, p. 367-384.
1985: Vestiges of Iapetan rifting in the craton west of the Northern Appalachians; Geoscience Canada, v. 12, p. 54-59.

LaSalle, P.
1984: Quaternary stratigraphy of Quebec; a review; in Quaternary Stratigraphy of Canada – a Canadian Contribution to IGCP Project 24, R.J. Fulton (ed.); Geological Survey of Canada, Paper 84-10, p. 155-171.

LaSalle, P. and Elson, J.A.
1975: Emplacement of the St. Narcisse Moraine as a climatic event in eastern Canada; Quaternary Research, v. 5, p. 621-625.

Lespérance, P.J.
1985: Faunal distributions across the Ordovician-Silurian boundary, Anticosti Island and Percé, Quebec, Canada; Canadian Journal of Earth Sciences, v. 22, p. 838-849.

Lespérance, P.J. and Bertrand, R.
1976: Population systematics of the Middle and Upper Ordovician trilobite *Cryptolithus* from the St. Lawrence Lowlands and areas of Quebec; Journal of Paleontology, v. 50, p. 598-613.

Liberty, B.A.
1969: Paleozoic geology of the Lake Simcoe area, Ontario; Geological Survey of Canada, Memoir 355, 201 p.

Lochman, C.
1938: Middle and Upper Cambrian faunas from western Newfoundland; Journal of Paleontology, v. 12, p. 461-477.

Lowdon, J.A. and Blake, W., Jr.
1976: Geological Survey of Canada radiocarbon dates XVI; Geological Survey of Canada, Paper 76-7, 21 p.

Ludvigsen, R.
1978a: Lower Ordovician trilobites of the Oxford Formation, eastern Ontario; Canadian Journal of Earth Sciences, v. 16, p. 859-865.
1978b: Towards an Ordovician trilobite biostratigraphy of southern Ontario; Michigan Basin Geological Society, Special Paper No. 3, p. 73-84.

McCracken, A.D. and Barnes, C.R.
1981: Conodont biostratigraphy and paleontology of the Ellis Bay Formation, Anticosti Island, Quebec, with special reference to Late Ordovician-Early Silurian chronostratigraphy and the systemic boundary; Geological Survey of Canada, Bulletin 329, Part 2, p. 51-134.

McDonald, B.C.
1969: Surficial geology of La Patrie-Sherbrooke area, Quebec, including Eaton River Watershed; Geological Survey of Canada, Paper 67-52, 21 p.
1971: Late Quaternary stratigraphy and deglaciation in eastern Canada; in The Late Cenozoic Glacial Ages, K.K. Turekian et al. (ed.); Yale University Press, p. 331-353.

McDonald, B.C. and Shilts, W.W.
1971: Quaternary stratigraphy and events in southeastern Quebec; Geological Society of America, Bulletin, v. 31, p. 636-698.

McGregor, D.C.
1970: *Hymenozonotriletes lepidophytus* Kedo and associated spores from the Devonian of Canada; Colloque sur la stratigraphie du Carbonifère, Université de Liège, v. 55, p. 315-326.

Nowlan, G.S.
1980: Early and Middle Ordovician conodonts from the Mingan Islands, Quebec (Abstract); Geological Society of America, North-Central Section, Abstracts with Programs, p. 253.
1981: Stratigraphy and conodont faunas of the Lower and Middle Ordovician Romaine and Mingan formations, Mingan Islands, Quebec; Maritime Sediments and Atlantic Geology, v. 17, p. 67.

Nowlan, G.S. and Barnes, C.R.
1981: Late Ordovician conodonts from the Vauréal Formation, Anticosti Island, Quebec; Geological Survey of Canada, Bulletin 329, Part 1, p. 1-49.

Occhietti, S.
1977: Stratigraphie du Wisconsinien de la region de Trois Rivières-Shawinigan, Québec; Geographie Physique et Quaternaire, v. 31, p. 307-322.
1980: Le Quaternaire de la region de Trois Rivières-Shawinigan, Québec; Contribution à la paleogéographie de la vallée moyenne du Saint Laurent et correlations stratigraphiques, Université du Québec à Trois Rivières, Paleo-Québec, v. 10, 227 p.
1989: Quaternary geology of the St. Lawrence Valley and adjacent Appalachian Subregion; in Quaternary Geology of Canada and Greenland, R.J. Fulton (ed.); Geological Survey of Canada, Geology of Canada No. 1. (Also Geological Society of America, The Geology of North America, v. K-1.)

Oliver, W.A.
1976: Noncystimorph colonial rugose corals of the Onesquethaw and Lower Cazenova Stages (Lower and Middle Devonian) in New York and adjacent areas; United States Geological Survey, Professional Paper 869, 152 p.

Palmer, A.R. and James, N.P.
1980: The Hawke Bay event: a circum-Iapetus regression near the Lower-Middle Cambrian boundary; in Proceedings of "The Caledonides in the U.S.A.", D.R. Wones (ed.); Virginia Polytechnical Institute, Department of Geological Sciences, Memoir no. 2, p. 15-18 (International Geological Correlation Project No. 27).

Petryk, A.A.
1981: Stratigraphy, sedimentology and paleogeography of the Upper Ordovician-Lower Silurian of Anticosti Island, Quebec; in Volume II: Stratigraphy and Paleontology, P.J. Lespérance (ed.), IUGS Subcommission on Silurian Stratigraphy, Ordovician-Silurian Boundary Working Group, Field Meeting, Anticosti-Gaspé, Quebec, 1981, p. 11-39.

Poole, W.H., Sanford, B.V., Williams, H., and Kelley, D.G.
1970: Geology of southeastern Canada; in Geology and Economic Minerals of Canada, R.J.W. Douglas (ed.); Geological Survey of Canada, Economic Geology Report No. 1, p. 229-304.

Pouliot, G.
1969: Guidebook for the Geology of Monteregian Hills; Mineralogical Association of Canada, Montreal, 169 p.

Prest, V.K.
1970: Quaternary geology of Canada; in Geology and Economic Minerals of Canada, R.J.W. Douglas (ed.); Geological Survey of Canada, Economic Geology Report No. 1, 5th Edition, p. 676-764.

Prest, V.K. and Hode-Keyser, J.
1977: Geology and engineering characteristics of surficial deposits, Montreal Island and vicinity, Quebec; Geological Survey of Canada, Paper 75-27, 29 p.

Rexroad, C.B. and Rickard, L.V.
1965: Zonal conodonts from the Silurian strata of the Niagara gorge; Journal of Paleontology, v. 39, p. 1217-1220.

Riva, J.
1969: Middle and Upper Ordovician graptolite faunas of the St. Lawrence Lowlands of Quebec and of Anticosti Island; in North Atlantic – Geology and Continental Drift; American Association of Petroleum Geologists, Memoir 12, p. 513-556.

Rodgers, J.
1965: Long Point and Clam Bank formations, western Newfoundland; Geological Association of Canada, Proceedings, v. 16, p. 83-94.

Rodgers, J. and Neale, E.R.W.
1963: Possible "Taconic" klippen in western Newfoundland; American Journal of Science, v. 261, p. 713-730.

Russell, D.J.
1984: Paleozoic geology of the Lake Timiskaming area, Timiskaming District; Ontario Geological Survey, Geological Series – Preliminary Map P.2700, scale 1:50 000.
1985: Depositional analysis of a black shale by using gamma-ray stratigraphy; the Upper Devonian Kettle Point Formation of Ontario; Bulletin of Canadian Petroleum Geology, v. 33, no. 2, p. 236-253.

Russell, D.J. and Telford, P.G.
1983: Revisions to the stratigraphy of the Upper Ordovician Collingwood beds of Ontario – a potential oil shale; Canadian Journal of Earth Sciences, v. 20, p. 1780-1790.

Sanford, B.V.
1961: Subsurface stratigraphy of Ordovician rocks in southwestern Ontario; Geological Survey of Canada, Paper 60-26, 54 p.
1968: Devonian of Ontario and Michigan; in International Symposium on the Devonian System; Alberta Society of Petroleum Geologists, v. 1, p. 973-999, (1967 imprint).

Sanford, B.V. and Baer, A.J.
1981: Southern Ontario, Ontario; Geological Survey of Canada, Map 1335A (Sheet 30S Geological Atlas), 1:1 000 000 scale.

Sanford, B.V. and Grant, G.M. (compilers)
1976: Physiography, eastern Canada and adjacent areas; Geological Survey of Canada, Map 1399A (4 sheets), 1:2 000 000 scale.

Sanford, B.V. and Norris, A.W.
1973: The Hudson Platform; in Future Petroleum Provinces of Canada, R.G. McCrossan (ed.); Canadian Society of Petroleum Geologists, Memoir 1, p. 387-409.
1975: Devonian stratigraphy of the Hudson Platform; Geological Survey of Canada, Memoir 379, 124 p.

Sanford, B.V., Thompson, F.J., and McFall, G.H.
1985: Plate tectonics – a possible controlling mechanism in the development of hydrocarbon traps in southwestern Ontario; Bulletin of Canadian Petroleum Geology, v. 33, no. 1, p. 52-71.

Schopf, T.J.M.
1966: Conodonts of the Trenton Group (Ordovician) in New York, southern Ontario and Quebec; New York State Museum and Science Service, Bulletin 405, 105 p.

Shearer, J.M.
1973: Bedrock and surficial geology of the northern Gulf of St. Lawrence as interpreted from continuous seismic reflection profiles; in Earth Sciences Symposium on offshore Eastern Canada, P.J. Hood (ed.); Geological Survey of Canada, Paper 71-23, p. 285-303.

Shilts, W.W.
1981: Surficial geology of the Lac Megantic area, Quebec; Geological Survey of Canada, Memoir 397, 102 p.

Sinclair, G.W.
1953: Middle Ordovician beds in the Saguenay Valley, Quebec; American Journal of Science, v. 251, p. 841-854.
1965: Succession of Ordovician rocks at Lake Timiskaming; Geological Survey of Canada, Paper 65-34, 6 p.

Sloss, L.L.
1963: Sequences in the cratonic interior of North America; Geological Society of America, Bulletin, v. 74, p. 93-114.

Sparling, D.R.
1983: Conodont biostratigraphy and biofacies of Lower/Middle Devonian limestone of north-central Ohio; Journal of Paleontology, v. 57, p. 825-864.

St. Julien, P. and Hubert, C.
1975: Evolution of the Taconian Orogen in the Quebec Appalachians; American Journal of Science, v. 275-A, p. 337-362.

Stouge, S.S.
1982: Preliminary conodont biostratigraphy and correlation of Lower to Middle Ordovician carbonates of the St. George Group, Great Northern Peninsula, Newfoundland; Department of Mines and Energy, Government of Newfoundland and Labrador, Report 82-3, 59 p.
1984: Conodonts of the Middle Ordovician Table Head Formation Western Newfoundland; Fossils and Strata, no. 16, 145 p.

Stouge, S.S. and Boyce, W.D.
1983: Fossils of northwestern Newfoundland and southeastern Labrador: conodonts and trilobites; Department of Mines and Energy, Government of Newfoundland and Labrador, Report 83-3, 23 p.

Stuiver, M., Heusser, C.J., and Yange, In Che.
1978: North American glacial history extended to 75 000 years ago; Science, v. 200, p. 16-21.

Stukas, V. and Reynolds, P.H.
1974: $^{40}Ar/^{39}Ar$ dating of the Long Range dikes, Newfoundland; Earth and Planetary Science Letters, v. 22, p. 256-266.

Terasmae, J.
1960: Contributions to Canadian palynology no. 2; Geological Survey of Canada, Bulletin 56, 41 p.

Uyeno, T.T.
1974: Conodonts of the Hull Formation; Ottawa Group (Middle Ordovician) of the Ottawa-Hull area, Ontario and Quebec; Geological Survey of Canada, Bulletin 332, 25 p.

Uyeno, T.T. and Barnes, C.R.
1983: Conodonts of the Jupiter and Chicotte formations (Lower Silurian), Anticosti Island, Quebec; Geological Survey of Canada, Bulletin 355, 49 p.

Uyeno, T.T., Telford, P.G., and Sanford, B.V.
1972: Devonian conodonts and stratigraphy of southwestern Ontario; Geological Survey of Canada, Bulletin 332, 45 p.

Vincent, J.S.
1989: Quaternary geology of the southwestern Canadian Shield; in Quaternary Geology of Canada and Greenland, R.J. Fulton (ed.); Geological Survey of Canada, Geology of Canada no. 1. (Also Geological Society of America, The Geology of North America, v. K-1.)

Whittington, H.B. and Kindle, C.H.
1963: Middle Ordovician Table Head Formation, western Newfoundland; Geological Society of America, Bulletin, v. 74, p. 745-758.
1966: Middle Cambrian strata of the Strait of Belle Isle, Newfoundland, Canada (Abstract); Geological Society of America, Northeastern Section, Program 1966 Annual Meeting, p. 46.
1969: Cambrian and Ordovician stratigraphy of western Newfoundland; in North Atlantic Geology and Continental Drift, M. Kay (ed.); American Association of Petroleum Geologists, Memoir 12, p. 655-664.

Williams, H. and Hiscott, R.N.
1987: Definition of the Iapetus rift-drift transition in western Newfoundland; Geology, v. 15, p. 1044-1047.

Williams, H. and Smyth, W.R.
1983: Geology of the Hare Bay Allochthon; in Geology of the Strait of Belle Isle area, northwestern insular Newfoundland, southern Labrador, and adjacent Quebec; Geological Survey of Canada, Memoir 400, Part 3, p. 109-132.

Wilson, A.E.
1964: Geology of the Ottawa-St. Lawrence Lowland, Ontario and Quebec; Geological Survey of Canada, Memoir 421, 66 p.

Winder, C.G.
1966a: Conodonts from the upper Cobourg Formation (late Middle Ordovician) at Colborne, Ontario; Journal of Paleontology, v. 40, p. 46-63.
1966b: Conodont zones and stratigraphic variability in Upper Devonian rocks, Ontario; Journal of Paleontology, v. 40, p. 1275-1293.

Winder, C.G. and Sanford, B.V.
1972: Stratigraphy and paleontology of the Paleozoic rocks of southern Ontario; XXIV International Geological Congress, Montréal, Quebec, Excursion A45-C45, 74 p.

ADDENDUM

Since completion of this manuscript, many new important findings have come to light concerning the origin and development of the St. Lawrence Platform. Plate tectonic reconstructions and distribution of depositional cycles in some parts of the region are thus outdated. This is particularly true of the eastern segment of the St. Lawrence Platform where marine surveys were carried out by the Geological Survey of Canada in 1989, 1990 and 1991, resulting in new concepts pertaining to the origin and development of the Anticosti Basin (Sanford and Grant, 1990a,b)

Sanford, B.V. and Grant, A.C.
1990a: Bedrock geological mapping and basin studies in the Gulf of St. Lawrence; in Current Research, Part B; Geological Survey of Canada, Paper 90-1B, p. 33-42.

Sanford, B.V. and Grant, A.C. (cont.)
1990b: Geology and petroleum potential of the Gulf of St. Lawrence region; Program with Abstracts for Eastern Section of AAPG Annual Meeting, London, Ontario, Sept. 10-12, 1990.

Authors' Addresses

B.V. Sanford
17 Meadowglade Gardens
Nepean, Ontario
K2G 5J4

K.L. Currie
Continental Geoscience Division
Geological Survey of Canada
601 Booth Strert
Ottawa, Ontario
K1A 0E8

W.R. Cowan
Ontario Geological Survey
Ministry of Northern Development and Mines
77 Grenville Street, 7th Floor
Toronto, Ontario
M5S 1W4

Printed in Canada

Chapter 12

ST. LAWRENCE PLATFORM – ECONOMIC GEOLOGY

Chapter 12

ST. LAWRENCE PLATFORM - ECONOMIC GEOLOGY

B.V. Sanford

The principal economic deposits of the Paleozoic formations of St. Lawrence Platform are non-metallic, the most important being limestone and dolostone, salt, oil and gas, shale, gypsum, and sandstone in that order. Carbonate rocks also have some potential as hosts to metallic deposits, such as lead, zinc, and cadmium.

PETROLEUM

Although good showings of petroleum have been encountered and small quantities of oil and gas produced in Quebec and Anticosti basins, production of commercial significance has thus far been confined to the western end of the platform in southwestern Ontario, where the Algonquin-Findlay Arch separates the Michigan and Allegheny basins.

Southwestern Ontario - Michigan and Allegheny Basins

The oil industry in southwestern Ontario began with the utilization of oil from seepages in glacial deposits along Black Creek in Lambton County. The first wells drilled into bedrock were at Oil Springs, in 1858. They were highly successful, and many flowing wells were soon completed. After the discovery of Petrolia in 1861 and Bothwell in 1862, the output of oil steadily increased until a peak of 132.8 x 10^3 m^3 was recorded in 1895. The annual yield gradually declined to 18.0 x 10^3 m^3 in 1945, but more recent discoveries, particularly in the Trenton and Black River groups, and the application of secondary recovery techniques have increased the production to 190.6 x 10^3 m^3 in 1988.

In 1889, Ontario's natural gas industry was born with discovery of the Kingsville field in Essex County. Since then, numerous accumulations, extending from southern Lake Huron to the Niagara River, have been discovered. Several onshore fields have been extended beneath Lake Erie, and many new discoveries have been made beneath the lake itself. In 1988, 501.7 x 10^6 m^3 of gas were produced in southwestern Ontario.

The oil and gas pools discovered to date in southwestern Ontario number some two hundred and twenty and include some of the oldest in North America. Although these now produce only a small fraction of the energy requirements of eastern Canada, they are important because of their proximity to markets. This is particularly true of the gas fields that produced from pinnacle reefs in the Sarnia area. The use of some of these depleted fields as storage reservoirs has greatly increased the practicability of transporting large quantities of natural gas from western Canada to the densely populated regions of southwestern Ontario and western Quebec.

The oil and gas occur in Cambrian, Ordovician, Silurian, and Devonian strata in a variety of stratigraphic and structural traps.

Cambrian

The Cambrian hydrocarbon pools are in tilted fault-blocks (Fig. 12.1). The oil and gas are trapped in orthoquartzitic sandstones and dolostones, in blocks that are uptilted so that the reservoir rocks lie against impermeable Middle Ordovician, Black River limestones. Thus far, fourteen oil and gas fields have been found. As of 1988, these had yielded 731.0 x 10^3 m^3 of oil, and 171.0 x 10^6 m^3 of gas.

Ordovician

In contrast to the fault traps of Upper Cambrian pools, the traps in the Middle Ordovician Black River and Trenton groups are permeability pinchouts in carbonate rocks (Fig. 12.2). Probably the two most important prerequisites for these traps are: (i) intersecting fracture systems, with one or more of the systems having undergone repeated rejuvenation, and (ii) substantial, pre-accumulation fluid migration to dolomitize the limestones along and adjacent to the fractures, thus providing suitable porosity and permeability.

Up until 1983, exploration for Trenton and Black River fields met with only moderate success. Two new pools were discovered in 1983, leading to a series of new discoveries. As of 1988, twenty-seven pools had been discovered, and these have yielded 484.7 x 10^3 m^3 of oil and 434.5 x 10^6 m^3 of gas.

Sanford, B.V.
1993: St. Lawrence Platform - Economic geology; Chapter 12 in Sedimentary Cover of the Craton in Canada, D.F. Stott and J.D. Aitken (ed.); Geological Survey of Canada, Geology of Canada, no. 5, p. 787-798 (also Geological Society of America, The Geology of North America, v. D-1).

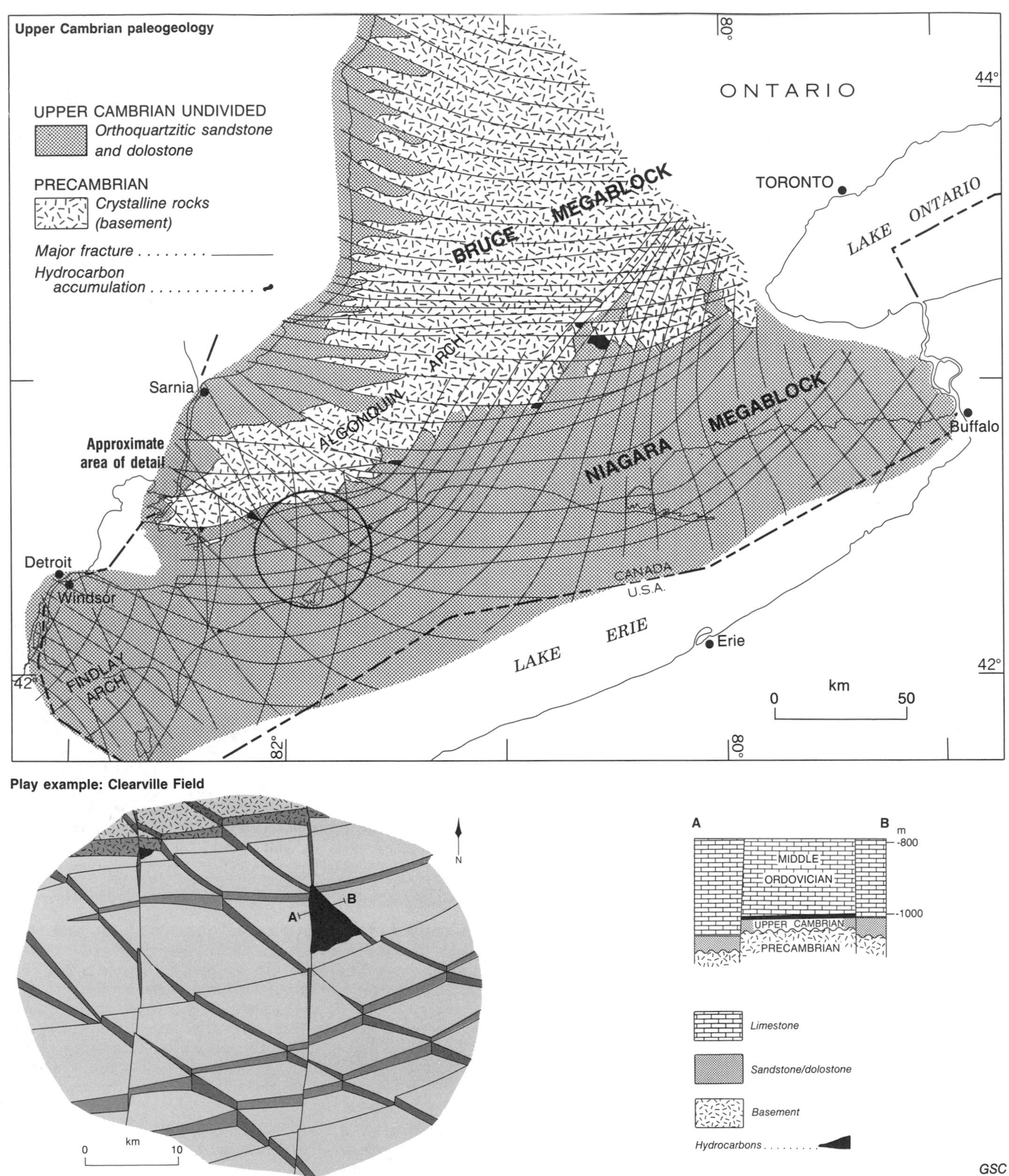

Figure 12.1. Upper Cambrian block rotation play.

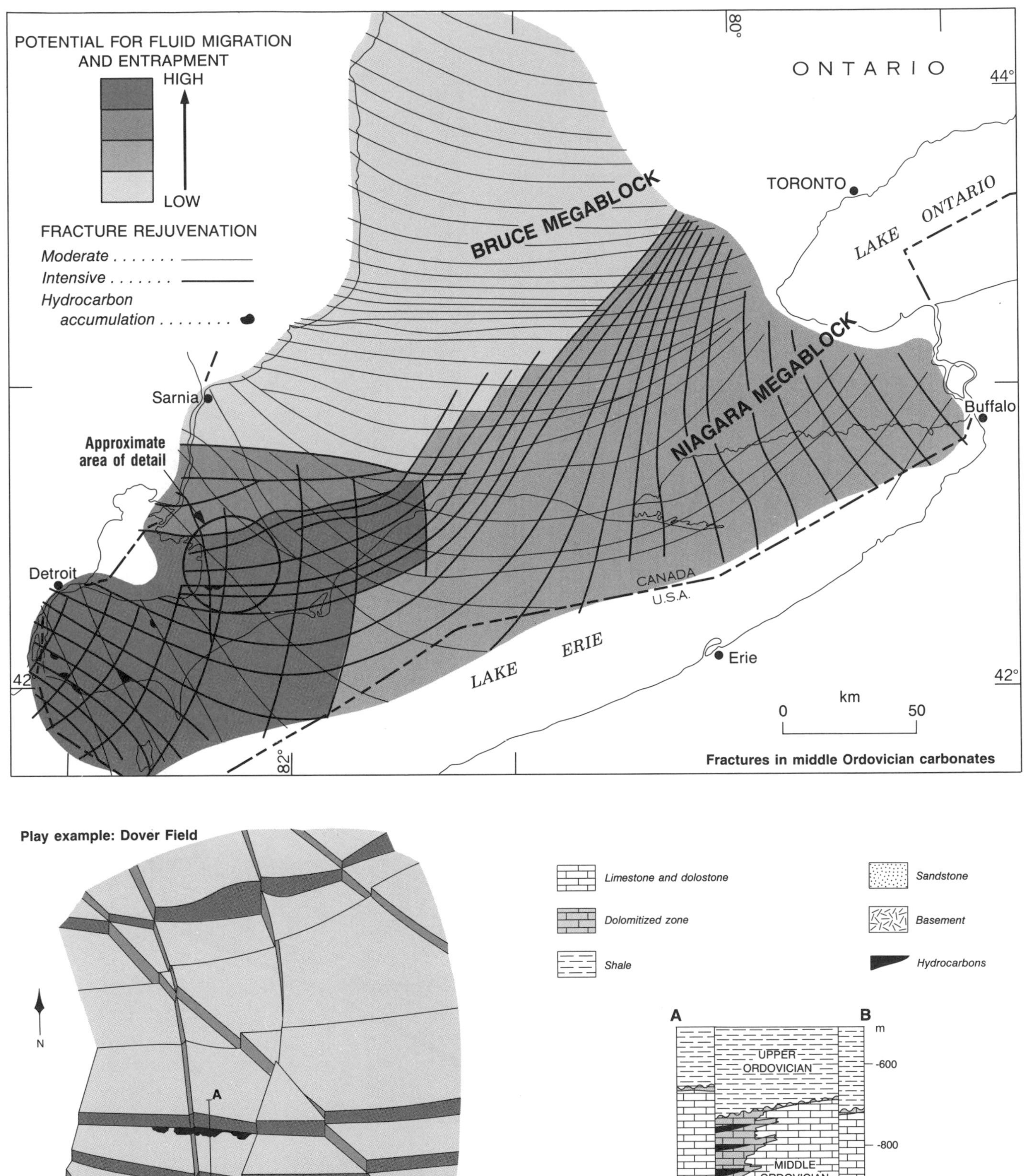

Figure 12.2. Middle Ordovician permeability pinchout play.

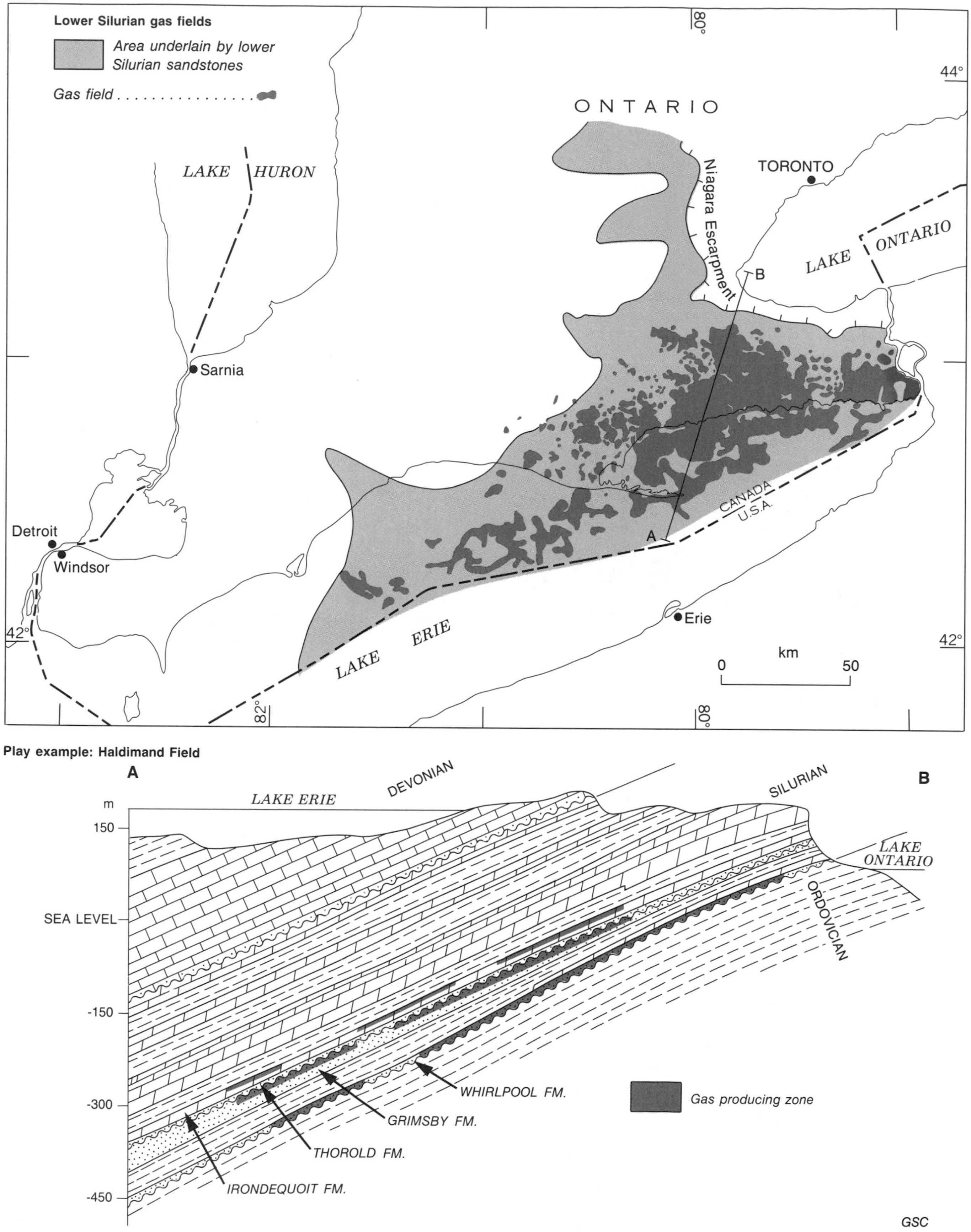

Play example: Haldimand Field

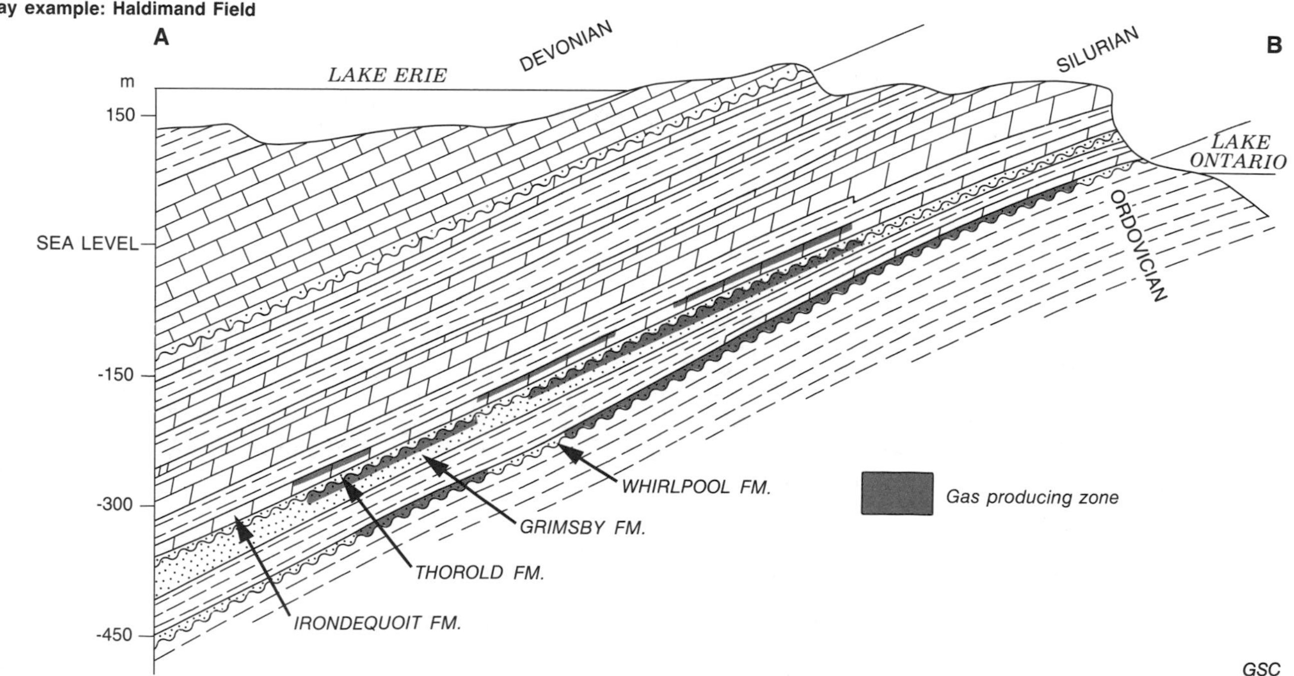

Figure 12.3. Lower Silurian permeability pinchout play.

Lower Silurian

The Silurian Whirlpool, Grimsby, and Thorold sandstones and Irondequoit dolostones are widely present in Niagara Peninsula and beneath central and eastern Lake Erie, and account for a substantial amount of the gas produced in southwestern Ontario. The trapping mechanism is not fully understood, but is assumed to be permeability pinchout, due in part to textural and compositional variations in the sandstones, and to diagenetic changes involving quartz infilling of early porosity (Fig. 12.3). Faulting may also account for some of the traps.

As of 1988, twenty-one pools had been discovered, and these had yielded a total of 10.2×10^9 m^3 of gas. In addition, a small amount of oil (6.6×10^3 m^3) was recovered in the earlier part of the century from the Whirlpool sandstone in the Haldimand field.

Upper Silurian

The most productive exploration plays in southwestern Ontario are the Upper Silurian, Guelph patch and pinnacle reefs and overlying Salina A-1 and A-2 carbonate traps. A regional facies reconstruction (see Fig. 11.24) of the Guelph Formation implies a relatively stable basin margin on which developed a barrier reef enclosing the basin centre on the north, east, and southeast, flanked on the basin side by a belt of patch and pinnacle reefs. The controls of bioherm growth during Guelph time have been a long-lived subject of speculation and debate. Recent studies by Sanford et al. (1985) show that many of the Guelph reefs and superposed Salina A-1 and A-2 structures occur mainly along the crests of tilted blocks (Fig. 12.4). Similarly, pinnacle reef growth farther into Michigan Basin appears also to have favoured fault-line scarps.

As of 1988, some one hundred and thirty-two Guelph and Salina pools had been discovered, and these had yielded a total of 1.9×10^6 m^3 of oil, and 17.2×10^6 m^3 of gas.

Devonian

Hydrocarbon traps in Devonian rocks in southwestern Ontario, like those in the Cambrian and Ordovician rocks, are structural. They differ, however, in that the oil occurs in structures in which the reservoir carbonates of the Dundee and Detroit River groups have been deformed by selective dissolution of the underlying Salina salts (Fig. 12.5). Generally, salt leaching occurred along northwest and east-west trending faults, and resulted in local reversal of regional dip to form domal structures over the thicker salt preserved between faults.

Numerous salt-dissolution structures occur in the Upper Silurian and Devonian rocks of southwestern Ontario. Hydrocarbons have been found in Devonian strata in a least twenty-three of the structures, which as of 1988 had yielded 6.6×10^6 m^3 of oil.

Ottawa Embayment and Quebec Basin

Approximately two hundred and fifty wells to depths greater than 150 m have been drilled in the Ottawa Embayment and Quebec Basin. None was successful in finding commercial quantities of oil or natural gas. The relatively few deep stratigraphic tests completed in the central part of the Ottawa Embayment encountered small showings of gas in various parts of the stratigraphic column, mainly the Trenton Group and overlying Billings and Carlsbad formations. Showings of gas were also locally encountered in glacial deposits, in areas underlain by bituminous shales of the Eastview and Billings formations.

In the Quebec Basin, oil and gas shows have been reported in the Potsdam, Beauharnois, Black River, Trenton, Utica, and Nicolet River formations. In addition, many of the shallow boreholes drilled throughout the basin for water encountered substantial gas flows in the Quaternary deposits overlying bedrock. The gas commonly occurs in porous sand layers at or near the bedrock (Utica or Lorraine) contact. One such accumulation of marginal commercial significance is the Pointe du Lac gas field in St. Maurice County.

Much of the deep drilling to date in the Quebec Basin has been concentrated along St. Lawrence River between Montreal and Quebec City. Recent drilling has been carried out in the overthrust belt at the southern margin of the basin. Shows of oil and gas have been encountered in several places in Cambrian and Ordovician strata, with the most encouraging flows obtained from the Beauharnois Formation at a locality near St. Flavien in Lotbinière County.

Anticosti Basin

Natural oil seepages near Parson's Pond and on St. Paul's Inlet and Port au Port Peninsula, all on the west coast of Newfoundland, were known as early as 1812. About forty wells, ranging in depth from 90 to 1220 m, have been drilled near these seepages. Some reported production of 0.16 to 1 m^3 daily. Total production was 400 m^3. The oil at Port au Port was obtained from porous limestones interbedded with tightly folded shales of the Humber Arm Supergroup. The oil-bearing strata near Parson's Pond and St. Paul's Inlet are dolostones of the St. George Group.

Since 1963, eight test holes, from 988 to 3848 m deep, have been drilled on Anticosti Island. Although most of the wells are in the northwestern part of the island, others were drilled on the south coast and at the extreme eastern end of the island. With one exception, the wells completely penetrated the Paleozoic succession and reached crystalline rocks of the Precambrian basement. Shows of oil and/or gas were encountered in the Romaine, Mingan, and Vauréal formations in some of the holes, but were insufficient to justify completion as producers.

INDUSTRIAL MINERALS

Limestone and dolostone

The limestones and dolostones of Cycles 1, 2, 5, and 7 (see Fig. 10.1) are a widely accessible source of raw material for the production of a variety of commodities for the building and heavy construction and chemical industries. The chief commodities are crushed stone for road construction and concrete aggregate, railway ballast, fluxstone, building and ornamental stone, terazzo, stucco dash, riprap (armourstone), and pulverized stone used as asphalt sealer. During 1988, approximately 95 million tonnes of stone valued at about $507.5 million were quarried in various

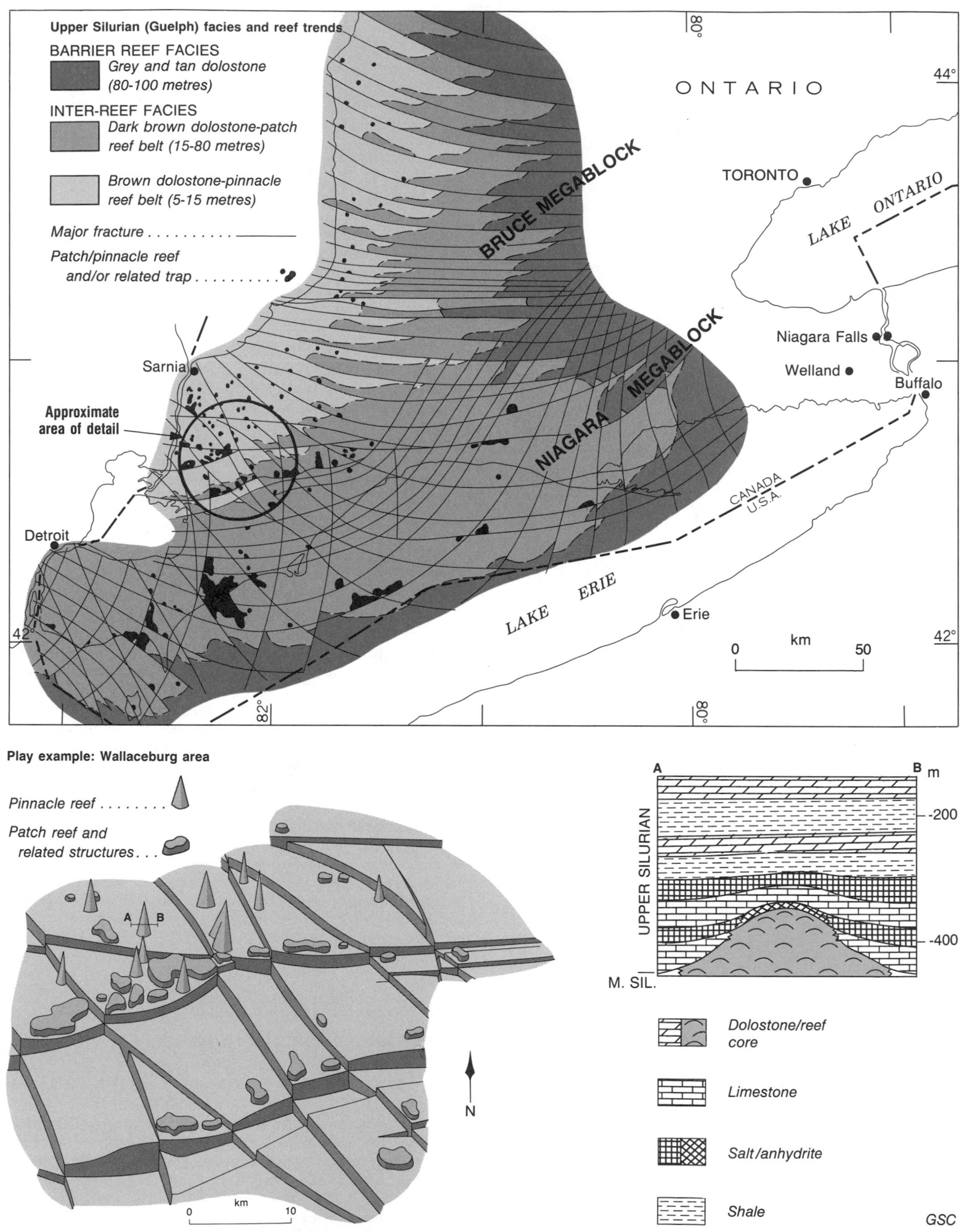

Upper Silurian (Guelph) facies and reef trends
BARRIER REEF FACIES
Grey and tan dolostone (80-100 metres)
INTER-REEF FACIES
Dark brown dolostone-patch reef belt (15-80 metres)
Brown dolostone-pinnacle reef belt (5-15 metres)
Major fracture
Patch/pinnacle reef and/or related trap
ONTARIO
TORONTO
LAKE ONTARIO
BRUCE MEGABLOCK
NIAGARA MEGABLOCK
Niagara Falls
Welland
Buffalo
Sarnia
Approximate area of detail
Detroit
CANADA
U.S.A.
LAKE ERIE
Erie
km
0
50
80°
44°
42°
82°
Play example: Wallaceburg area
Pinnacle reef
Patch reef and related structures
A
B
N
0
km
10
m
-200
-400
UPPER SILURIAN
M. SIL.
Dolostone/reef core
Limestone
Salt/anhydrite
Shale
GSC

Figure 12.4. Upper Silurian patch and pinnacle reef play.

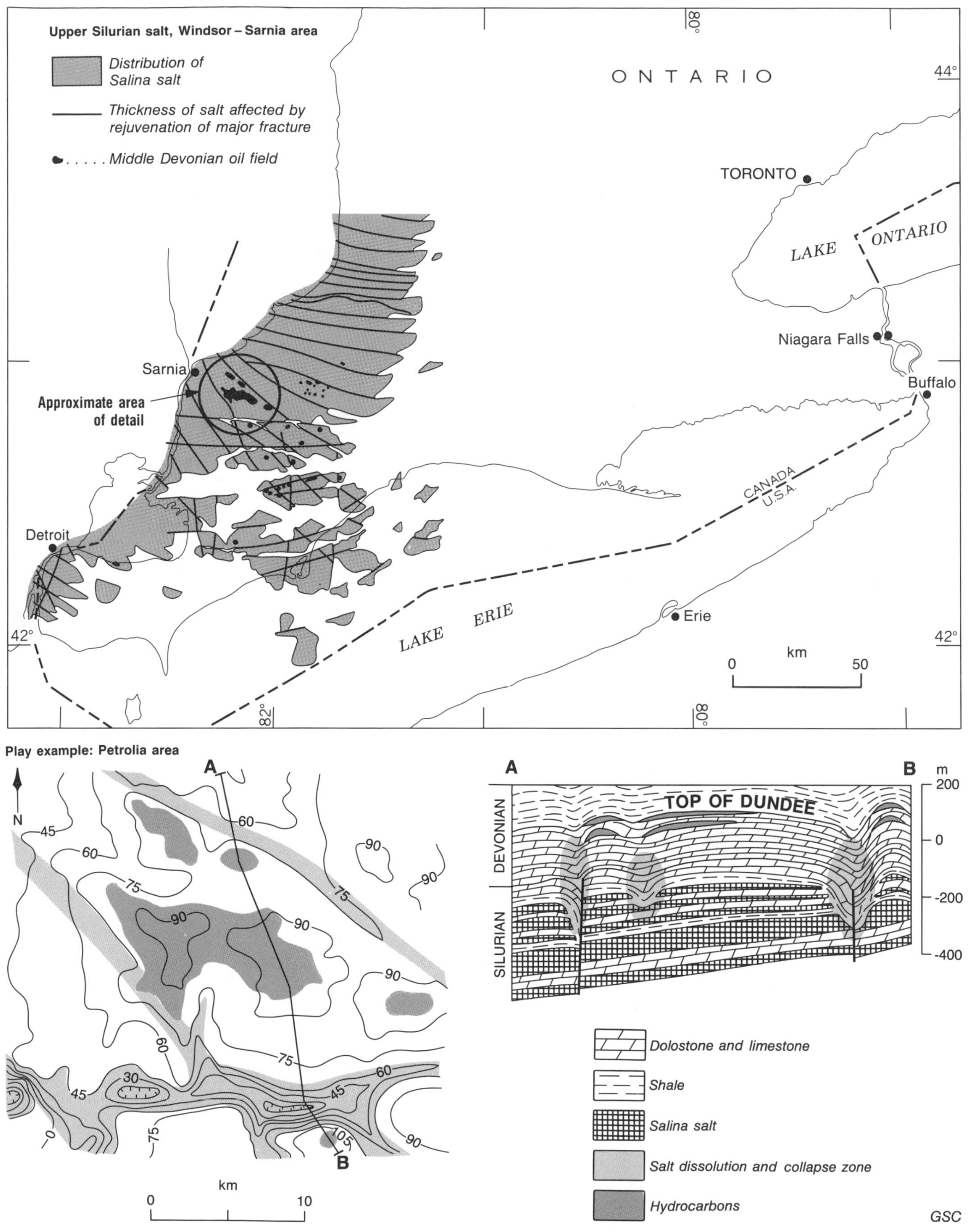

Figure 12.5. Middle Devonian salt dissolution and collapse play.

parts of the St. Lawrence Platform for the foregoing purposes. In addition, about 15 million tonnes of limestone were processed into cement and chemical lime valued at $999.7 million. Both of these commodities are consumed in quantity by the building trades, the latter also being used in the iron and steel industry, uranium processing, pulp and paper manufacture, glass making, gold milling, sugar refining, smelting, and in agriculture for conditioning soils.

More than half of the production of cement, lime, and stone from the platform is from limestones and dolostones quarried in Ontario. Production is obtained from most of the Ordovician, Silurian, and Devonian carbonate units, including the Oxford Formation, Black River and Trenton groups, Reynales-Fossil Hill, Irondequoit, Gasport, Goat Island, Amabel, Eramosa, Guelph, Bertie, Bois Blanc, Amherstburg, Lucas, and Dundee formations.

Most of the remaining production of limestone and dolostone takes place in Quebec, from the Beauharnois and Laval formations and Black River and Trenton groups of Early and Middle Ordovician age.

In western Newfoundland, limestones are quarried for the manufacture of cement. Production is from the Corner Brook Formation (Table Head?) of probable Middle Ordovician age. The St. George and Table Head groups offer good potential as raw material for the stone, lime, and cement industries, and both units have been extensively used for this purpose at one time or another in the Port au Port region of southwestern Newfoundland.

Salt (Halite)

Salt (halite) deposits are confined to the western platform, where they form thick units within the Salina Formation of Late Silurian age (Fig. 12.6). Salt was formerly produced (by the artificial brine and evaporation method) at a number of places in Ontario, but is now concentrated at four localities: Goderich, Sarnia, Windsor, and Amherstburg. Brine is taken from the B, D, and E units of the Salina Formation.

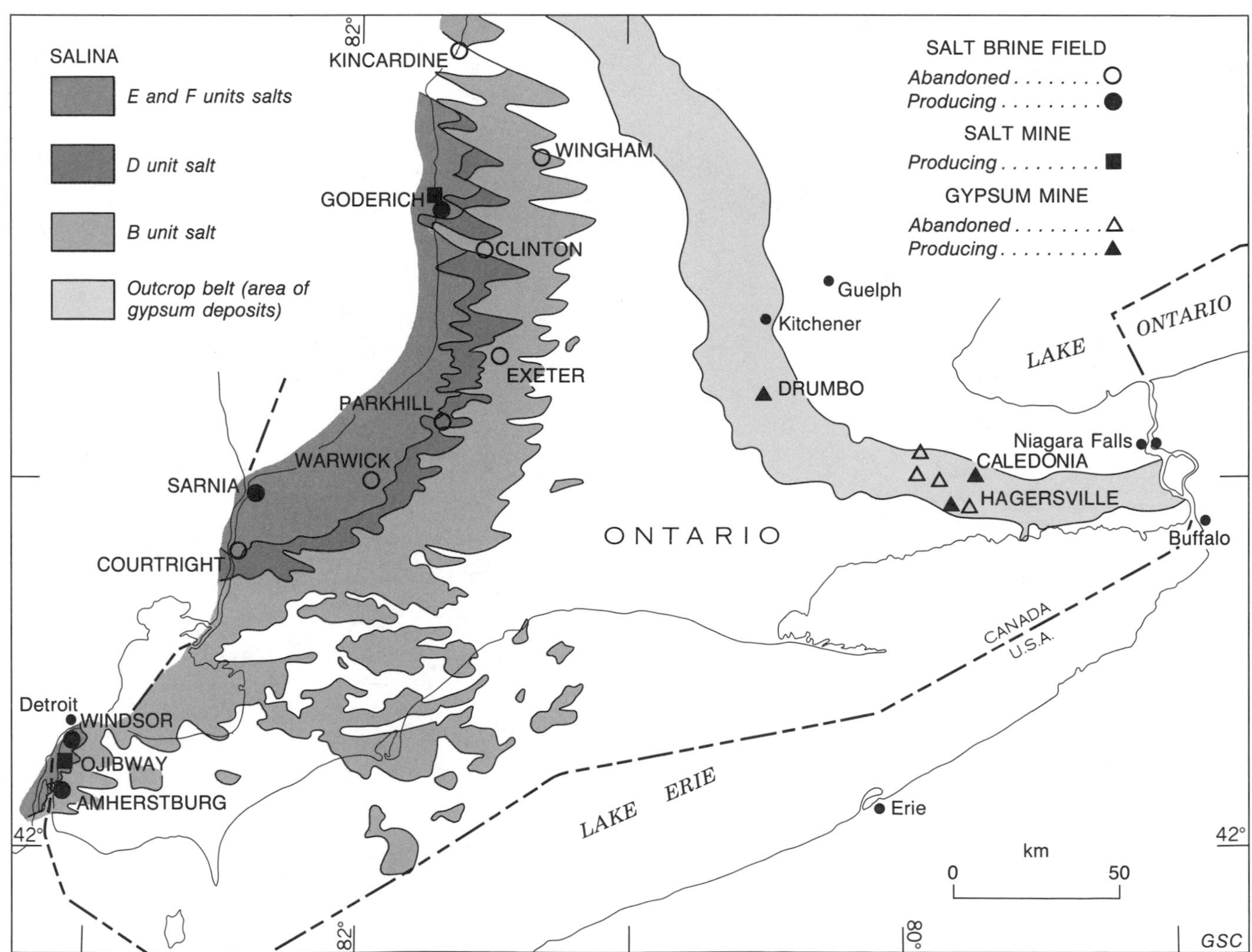

Figure 12.6. Salt and gypsum deposits in southwestern Ontario.

Figure 12.7. Salt mine of Canadian Rock Salt Company Limited at Ojibway, Ontario, in the Silurian Salina Formation (photo - Canadian Rock Salt Company Limited.)

In recent years, two salt mines have come into production in Ontario, at Ojibway and Goderich. The mine at Ojibway near Windsor began operations in 1955 (Fig. 12.7). Production is from a 6.7-m salt interval of the Salina E unit, at a depth of 289 m. The mine at Goderich began operations in 1959, with production from a 24-m salt unit at the base of the A-2 unit of the Salina at a depth of 510 m. These two operations have greatly increased the annual salt production in Ontario, and meet the demand for a coarse product for road de-icing, a purpose for which the evaporated salt was not well suited. In 1988, combined production from mining and brining operations in Ontario totalled 6.9 x 10^6 tonnes, valued at $158.4 million.

Large quantities of salt are used in Ontario during the winter months for de-icing streets and highways. Other large consumers of salt are the food processing industries and the chemical industries, which manufacture soda ash, calcium chloride, and chlorine.

In addition to its commercial use, salt has become increasingly important as a hostrock for the storage of liquid natural gases and refined petroleum products. According to Guillet (1984), sixty-three active storage caverns were in use in 1980 for this purpose at Sarnia and three at Windsor, with three more under development, the total storage capacity being 2.2 x 10^6 m^3.

Shale

Shales of Cycles 3, 4, and 8 (see Fig. 10.1) are widespread at the surface in the St. Lawrence Platform and are raw material for the manufacture of a variety of products. Brick and drain and industrial tile are the chief products in southwestern, central, and eastern Ontario and western Quebec. In Ontario, the Upper Ordovician Georgian Bay and Queenston formations east of the Niagara Escarpment and the Queenston Formation near Ottawa are used extensively for manufacturing brick and tile, as are the Middle Devonian Hamilton shales in the Thedford and Arkona areas in the extreme southwest. The Georgian Bay and Queenston shales are also used in the manufacture of pottery and cement. The total value of products manufactured from shales in Ontario was estimated at $122 million for 1988.

In Quebec, shales of the Upper Ordovician Nicolet River Formation provide a good source of raw material for the manufacture of brick and drain and chimney tile. In 1988, these commodities had a value of approximately $24.7 million.

In Western Newfoundland, shale is quarried near Corner Brook, where it is used in the manufacture of cement. In 1988 shale production was valued at $1.2 million.

Gypsum

Commercial gypsum deposits are confined to the western St. Lawrence Platform, where they occur in Cycle 6 (see Fig. 10.1, Fig. 12.6). Occurrences in southwestern Ontario have been known for more than 150 years, the first mine being opened about 1822 at a site near the present town of Paris, Ontario. Since then, about fifteen deposits have been worked, the most important being in Haldimand, Brant, and Oxford counties. The deposits, belonging to the Salina Formation, are lenticular, up to 3.3 m thick, and occur at sporadic horizons.

In 1988, three mines were operating, at Caledonia and Hagersville in Haldimand County, and at Drumbo in Oxford County. Their combined output during the year totalled 1.5 million tonnes, valued at $19.7 million.

During the early development of the industry in Ontario, gypsum was used solely as fertilizer and the markets were local. Gypsum now has multiple uses, principally in the construction industry, as in the manufacture of wallboard, plaster of paris and gypsum lath, as a setting control in portland cement, and as a filler in paint and paper.

Gravel, sand, and clay

Unconsolidated Quaternary deposits of the St. Lawrence Lowland, in addition to providing a young, unleached substrate from which fertile soils have developed, are of industrial importance. Although less abundant on the lowland than on the Canadian Shield, glacial gravels are locally important as road metal and concrete aggregate. Proglacial and glaciolacustrine sands, suitable for the latter purpose, are fairly widespread. Glaciolacustrine clays have been widely used for brick-making, especially in Quebec. Because provincial summaries do not separate lowland and shield, production figures for the lowland are unavailable.

Peat

Cool, moist climate and abundant closed depressions on the post-glacial surface have provided favourable conditions for peat accumulation. Although the peat deposits of the lowland are less generously distributed than those of the shield, their proximity to points of consumption has made them valuable, mainly as a conditioner of soils. Production figures are available only on a provincial basis and not on the basis of geological region.

Sandstone

The orthoquartzitic sandstone deposits of Cycles 1, 4, and 7 (Fig. 10.1) in Ontario and Quebec have long been utilized for building and ornamental stones. In southwestern Ontario, the Silurian Whirlpool and Devonian Oriskany formations have been used extensively for this purpose. The Upper Cambrian to Lower Ordovician Nepean sandstone of the Ottawa Valley has long been famous for its pleasing qualities as a building and ornamental stone, having been used in the construction of the Houses of Parliament and other government buildings in Ottawa.

Because of its exceptional purity where exposed along the Beauharnois uplift west of Montreal, the Potsdam sandstone is quarried for use in glass-making and the manufacture of silicon carbide, autoclaved concrete, asbestos pipe, and ceramic products.

METALS

Lead and zinc

The carbonate strata of the St. Lawrence Platform (mainly the dolostones of Cycles 1 and 4, see Fig. 10.1 and 11.10 to 12) have potential as hosts to lead and zinc deposits, but thus far production (of zinc) is confined to western Newfoundland. The Silurian Amabel and Guelph formations between Hamilton and Bruce Peninsula contain scattered showings of galena and sphalerite. Some mineralization also occurs at the surface and in the subsurface in the barrier reef facies of the Guelph Formation in the Bruce Peninsula.

In western Newfoundland, zinc deposits of commercial grade (8.55%) and cadmium occur near Daniels Harbour. Production of about 1600 tonnes of ore a day is from dolostones of the St. George Group.

ECONOMIC ASPECTS OF MONTEREGIAN INTRUSIONS

K.L. Currie

No significant metallic deposits are known to be associated with Monteregian intrusions of the St. Lawrence Lowland. The St. Lawrence niobium deposit occurs in the Oka carbonatite complex, which lies within a Precambrian crystalline inlier. However, minor intrusions of carbonatitic affinity on the northwest of Montreal Island contain pyrochlore (St. Scholastique, St. Andre, St. Michel; Dawson, 1974). Because they show exposed rock faces, most of the major plutons have been quarried at various times for aggregate and road metal. The quarries at Mont St. Hilaire (De Mix and Poudrette quarries) are world famous for the variety (140 species) and beauty of mineral specimens found in large vugs. Mount Johnson (Mont St. Gregoire, Monoir) has been extensively quarried for dimension stone.

REFERENCES

Dawson, K.R.
1974: Niobium (Columbium) and tantalum in Canada; Geological Survey of Canada, Economic Geology Report 29, 76 p.

Guillet, G.R.
1984: Salt in Ontario; in The Geology of Industrial Minerals of Canada, G.R. Guillet and W. Martin (ed.); The Canadian Institute of Mining and Metallurgy, Special Volume 29, p. 143-147.

Sanford, B.V., Thompson, F.J., and McFall, G.H.
1985: Plate tectonics – a possible controlling mechanism in the development of hydrocarbon traps in southwestern Ontario; Bulletin of Canada Petroleum Geology, v. 33, p. 52-71.

Authors' Addresses

B.V. Sanford
17 Meadowlade Gardens
Nepean, Ontario
K2G 5J4

K.L. Currie
Continental Geoscience Division
Geological Survey of Canada
601 Booth Street
Ottawa, Ontario
K1A 0E8

Printed in Canada

CHAPTER 13

EVOLUTIONARY MODELS AND TECTONIC COMPARISONS

Chapter 13

EVOLUTIONARY MODELS AND TECTONIC COMPARISONS

J.D. Aitken

INTRODUCTION

This chapter attempts comparisons among the various sectors of the Phanerozoic platforms and basins of North America (Fig. 13.1) and among the various sectors of the frontal parts of the outboard orogens, whose behaviour has profoundly influenced the sedimentary record of the former. The syntactical difficulties of making comparisons among more than two entities are formidable; accordingly, a graphical summary is also provided (Fig. 13.2). Each column of the graphical summary is a generalization of an entire transect of a platform, basin, or frontal orogen; it thus contains numerous half-truths and serves mainly to highlight coincident and non-coincident events. The graphical summary should not be treated as a source of data.

PLATFORMS AND BASINS OF THE PASSIVE PHASES

The vast Paleozoic depositional platforms of Canada have no exact modern counterparts; furthermore, few areas of the modern earth display even a few of the essential characteristics of those ancient, tectonically stable regions.

The closest analogue is the modern Hudson Bay itself. Vast, landlocked, and shallow by oceanic standards, it is a fair model for the Paleozoic Hudson Platform. Like the ancient scene of sedimentation, the modern basin receives only modest inputs of detritus from the surrounding Canadian Shield, but for different reasons. Sedimentation on the ancient platform took place mainly during long periods of very high relative sea level, so that the surrounding arches were much reduced in area, and repeatedly submerged. Today, the surrounding arches stand high, and although several large rivers flow to the bay, they drain regions from which all mature soils were swept by Quaternary glaciers, and in which a cold climate allows minimal chemical weathering and hence yields mainly coarse detritus. Furthermore, thousands of glacial depressions are occupied by lakes and bogs that form efficient sediment traps. Unlike the warm, carbonate-depositing Paleozoic seas of Hudson Platform, the modern sea is cold, and significant carbonate sedimentation, to take the place of the sparse detrital input, is forbidden. If the modern Hudson and James bays lay today at latitudes below 35°, they would undoubtedly be receiving carbonate sediments, probably at a rapid rate.

Other regions with sedimentological similarities to the Paleozoic platforms are the Yucatan Platform (at present, largely subaerial) and the Bonaparte Gulf off northern Australia. Both are tropical or subtropical areas of carbonate sedimentation, isolated by distance and deep water from highlands, but their scale does not compare with that of the Paleozoic examples. In their marginal positions they more closely resemble the early St. Lawrence and Interior platforms than the landlocked Hudson Platform. At various times during the Paleozoic, the Canadian platforms contained vast, featureless areas that were subaerial except for rare and brief marine floodings, and were either receiving sediment at minimal rates or undergoing slow net erosion. The Rann of Kutch, at the margin of the Indian continent, is a modern example of such conditions.

Hudson Platform, although sharing some characteristics with the Williston and Michigan basins, is a unique tectonic feature in North America. Situated near the centre of the North American craton, and broader and of flatter profile than the Williston and Michigan basins, it alternated, during its Late Ordovician to Late Devonian history, between being a nearly land-locked basin and being a slightly subsiding part of an epeiric sea that covered all of North America except the Appalachians. Remote from the continent-rimming orogens, it received virtually none of their detritus, and was deformed only by localized, steep faults.

The histories of the Interior, Arctic, and St. Lawrence platforms and their associated miogeoclines, although differing in many particulars, share some common themes. One important difference, however, is revealed by the Cambrian sediments of the three regions. From the time that the horsts resulting from Late Proterozoic rifting subsided (coincident with the onset of continental separation?), there is no sign of source lands offshore the Cordilleran margin (except the ?Lower Ordovician Broadview Formation of Kootenay Terrane), and none offshore the Innuitian margin until the Silurian Period. In contrast, the Cambrian miogeoclinal limestones and shales of the Appalachian outer platform are replaced eastward by greywackes indicative of an off-platform source of detritus. In a general way, however, all faced open oceans from the time of their inception as platforms inboard of passive margins near the beginning of the Cambrian, until

Aitken, J.D.
1993: Evolutionary models and tectonic comparisons; Chapter 13 in Sedimentary Cover of the Craton in Canada, D.F. Stott and J.D. Aitken (ed.); Geological Survey of Canada, Geology of Canada, no. 5, p. 799-808 (also Geological Society of America, The Geology of North America, v. D-1).

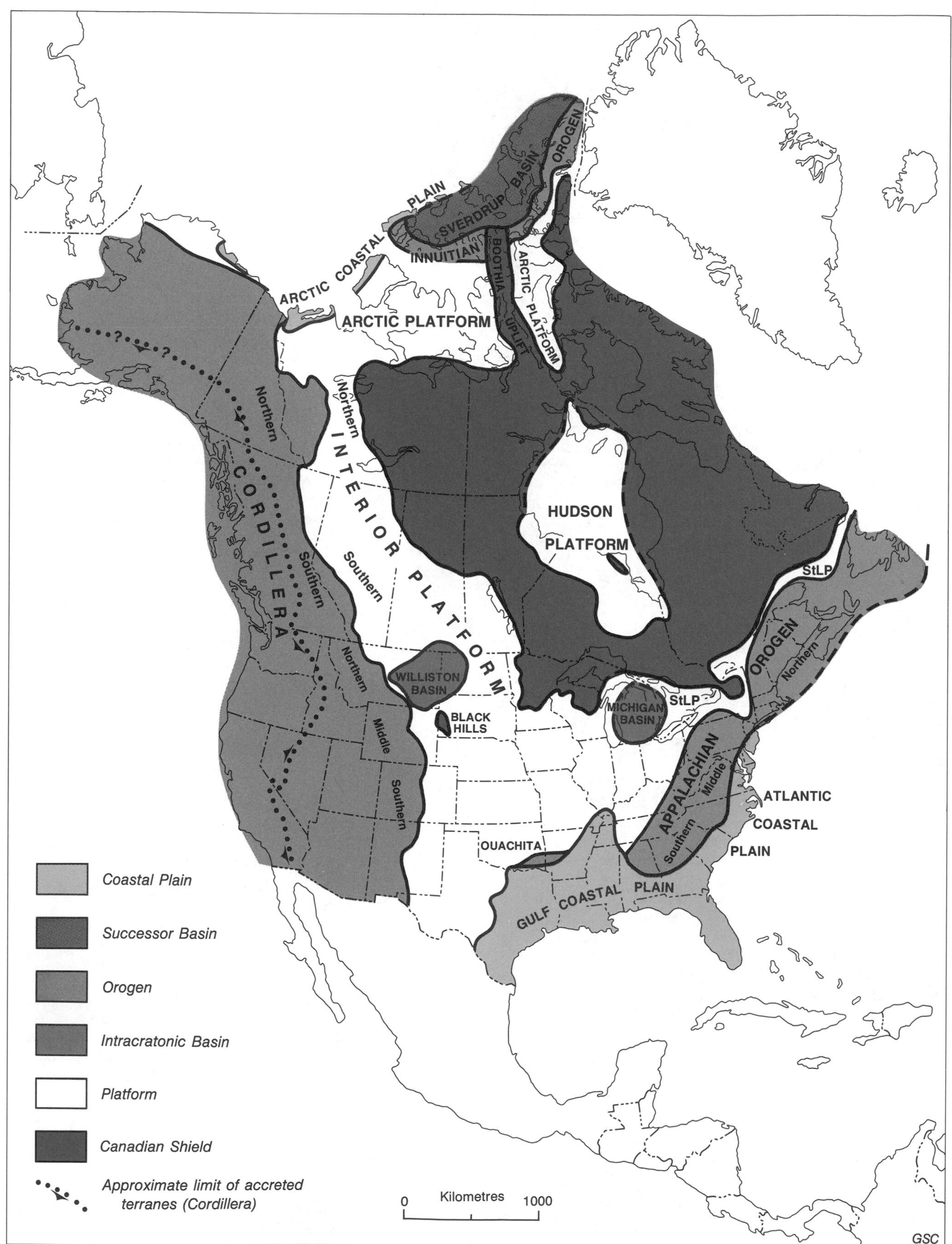

Figure 13.1. Major tectonic elements of St. Lawrence Platform (StLP).

the first tectonic disruptions of those margins (beginning of Middle Ordovician time for the Appalachian margin, no earlier than mid-Silurian for the Cordilleran, Late Silurian for the Innuitian). There seem to be no modern examples of similar continental margins at which broad peneplains (whether sediment-mantled or not) pass with imperceptible slope beneath the ocean.

Another important difference is that of basement age and character (Beaumont, 1981). The ancient Appalachian continental margin was formed on relatively young (thin, warm), Grenvillian lithosphere (about 400 million years old), and was parallel to the gross structure of that basement. The ancient Cordilleran margin was formed on relatively old (thick, cool), "Hudsonian" lithosphere (about 1200 million years old), with a structural grain mostly non-parallel to the margin. The ancient Innuitian margin had basement rocks similar in age to those of the Cordilleran margin; basement "grain" was in part sub-parallel to the margin and in part near-perpendicular. These differences may account for the relatively "active" basement of the St. Lawrence Platform [see, for example, Sanford (Chapter 11 in this volume) and Sanford et al. (1985)], as compared with that of the Interior and Arctic platforms [Chapter 4 in this volume and Trettin (1991)]. The "active" basement is seen, for example, in the Paleozoic outliers (effectively graben) of the southeastern part of the Canadian Shield and the normal faults that so commonly mark the northwestern boundary of today's St. Lawrence Platform. In contrast, the faulted epicratonic arches of the Interior and Arctic platforms are widely separated anomalies.

Basement age may account for another important difference among the pre-orogenic Appalachian, Cordilleran, and Innuitian miogeoclines and platforms. Basement of the first-named has no unmetamorphosed sedimentary cover pre-dating the Late Proterozoic rifting and sedimentation common to all. The Cordilleran margin has one such cover succession in the south (Purcell, 1500-1300 Ma), and at least two in the north (Wernecke and Hornby Bay - Dismal Lakes, 1700-1200 Ma; "Mackenzie Mountains supergroup", 880-770 Ma). Similar Proterozoic successions lie along the Innuitian margin where, as in the Cordillera, they occur partly in extensive blankets and partly in rift-bounded basins. This suggests that the younger, warmer, Grenvillian basement of the Appalachians stood high until very late Proterozoic time. New evidence yet to be assimilated is the presence of a thick, faulted and broadly folded succession of layered rocks (Wernecke - Hornby Bay - Dismal Lakes and, very likely, Coppermine Lava equivalents) unconformably beneath "Mackenzie Mountains supergroup" strata in the Mackenzie Platform (Cook, 1988a, b). This ancient deformation, whether of orogenic character or not, clearly seen on seismic-reflection profiles, is startling in a region hitherto considered to have been platformal since the cessation of terminal Early Proterozoic orogeny.

Basement uplifts or epicratonic arches (excluding orogenic forebulges) can generally be categorized as either sub-parallel to continental margins or quasi-radial with respect to the Canadian Shield. They are generally dated with reference to onlapping formations; there is thus more uncertainty in the date(s) of their uplift, once or several times, than is generally acknowledged. The quasi-radial arches of the St. Lawrence Platform (Frontenac, Saguenay, Beaugé), oriented across the basement grain, are narrow and closely spaced as compared with the single, early, quasi-radial arch of the Interior Platform (Peace-Athabasca), aligned near the basement grain. Both regions also had margin-parallel uplifts during the early passive phase; Vermont-Quebec and Stoke Mountain geanticlines in the east (Cady, 1970), the Mackenzie Arch in the west. The anorogenic arches of the St. Lawrence Platform are largely recognizable from the onset of passive-margin sedimentation and continued in existence through the subsequent active-margin phase, although new arches (broad and variously labelled) became evident after the onset of the Taconian orogeny and the more or less coincident inception of subsidence in the Hudson Bay, Michigan and Williston basins. Sanford (Chapter 11 in this volume) assumes a cause-and-effect relationship, but the case seems far from settled. In the west, new arches appeared in mid-Ordovician (Sweetgrass, radial), Devonian (Peace River, radial; Keele, oblique; West Alberta, parallel), and Triassic (Blackwater, radial) times. These later arches post-date dimly perceived Devonian or earlier, Carboniferous, and Triassic tectonic and igneous activity at the western margin, and thus may not belong to a strictly passive regime. Their subsequent histories are varied: briefly positive then neutral; intermittently positive to the present; positive then negative (Fig. 2D.4, 5.2). Like typical arches of the St. Lawrence Platform, some of these western arches underwent uplift during both passive (divergent) or pseudo-passive, and active (convergent) margin phases.

The rise and fall of cratonic arches and basins has been claimed by some to be a response to the forces involved in contractional orogeny at the continental margins. The basis for the claim is approximate or partial synchrony between the activity of the cratonic features and contractional activity at the margins. The hypothesis may be valid, but if so, it fails to explain the varied orientations of the cratonic features relative to the margins, and their periods of activity not coinciding with marginal orogenies. Alternative hypotheses should be tested.

An alternative is suggested by the extraordinary development of basins and arches, in the interior United States, during deposition of the Absaroka Sequence (latest Early Carboniferous to Early Permian). "Absaroka stratigraphy and structure .... indicate vertical movements on scales of kilometres, commonly involving fracturing and faulting at positions as central as Illinois and Kansas, and oscillations with respect to baselevel with periods measured in 10^5 years. A number of the older cratonic arches and basins renewed their differential rates of subsidence, and these were supplemented and, in places, grossly modified by greatly amplified vertical displacements and "new" additions to the tectonic geography of the craton" (Sloss, 1982). Budnik (1986) made a case for the rise, in the Late Carboniferous, of the "ancestral Rocky Mountains" of the west-central United States as a response to sinistral transcurrent movement on the Wichita Megashear. That rise is one of the Absaroka events summarized by Sloss (above), and suggests the hypothesis of intraplate transcurrent faulting, whatever its relationship to activity at plate margins, as the cause of all the extraordinary vertical movements of that period. Such

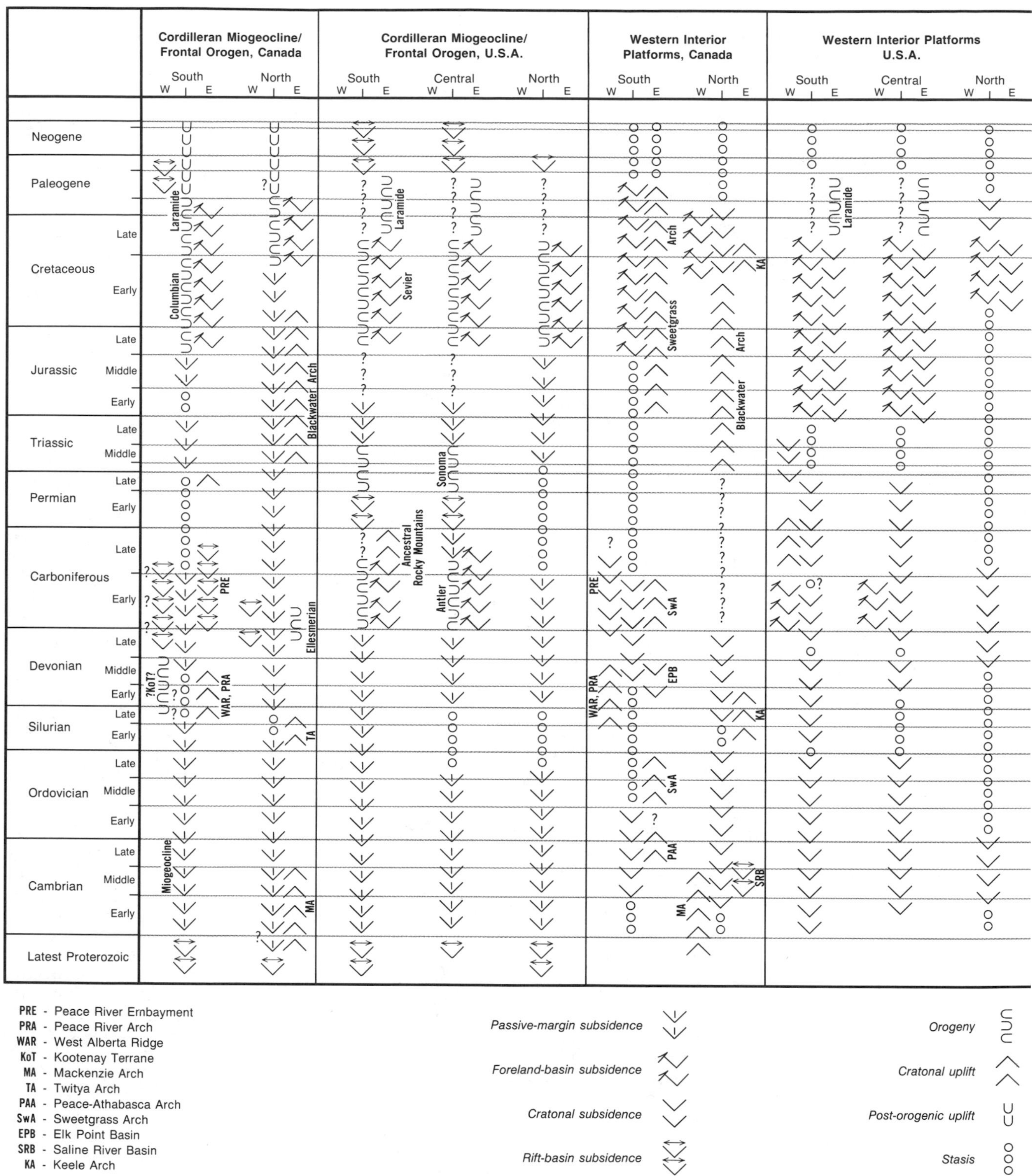

Figure 13.2. Graphical comparison of the generalized histories of continental margins, basins, and platforms, arches and orogens of North America.

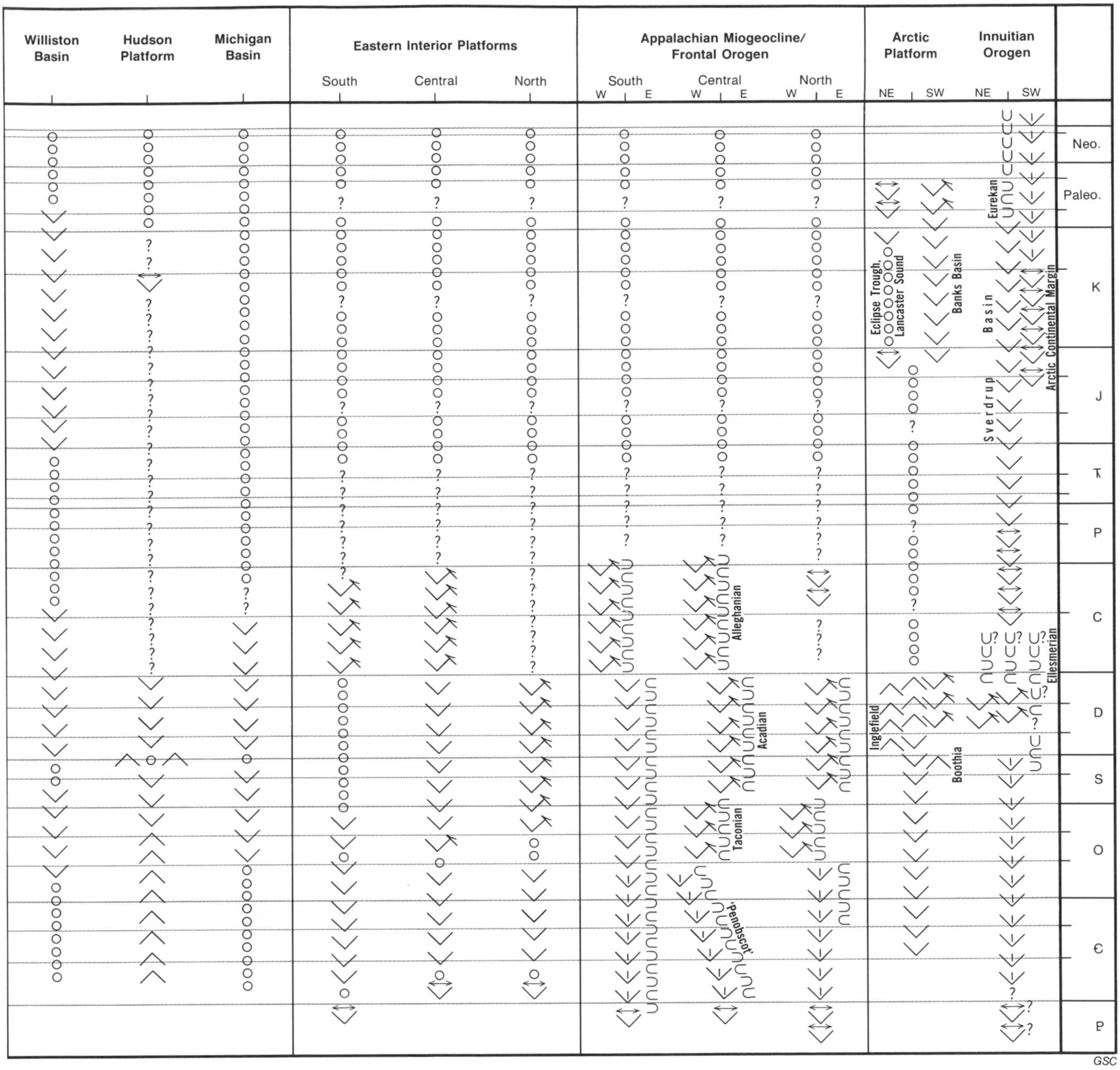

Williston Basin
Hudson Platform
Michigan Basin
Eastern Interior Platforms
South Central North
Appalachian Miogeocline/ Frontal Orogen
South W E Central W E North W E
Arctic Platform NE SW
Innuitian Orogen NE SW
Neo.
Paleo.
K
J
T̄R
P
C
D
S
O
Є
P
Eclipse Trough, Lancaster Sound
Banks Basin
Sverdrup Basin
Eurekan
Arctic Continental Margin
Alleghanian
Ellesmerian
Acadian
Inglefield
Boothia
Taconian
'Penobscot'
GSC

Figure 13.2. Cont.

a hypothesis would explain the diversity of trends and the different up/down histories of cratonic features of Absaroka times, including subsidence of the formerly elevated Peace River Embayment, in which steep basement faults played an important role. The hypothesis of intraplate transcurrent faulting may have more general application in Canada, in explaining the vertical movements of cratonal arches and basins, especially the widely separated and variously oriented cratonic structures, mainly arches, of Devonian times.

The Hudson Bay and Moose River basins of the Hudson Platform, the Williston Basin of the Interior Platform, and the Michigan Basin of the St. Lawrence Platform have never undergone orogenic deformation. Despite the near-coincidence in their dates of inception (as **differentially** subsiding elements) with the onset of the Taconic orogeny in the east at about the beginning of the Middle Ordovician, they appear to have no essential relationship to any coexisting orogenic belt, and indeed, received at most the finest and farthest travelled detritus of the orogenic clastic wedges. In contrast, the St. Lawrence, Arctic, and Interior platforms suffered orogenic destruction of their outboard reaches and transformation, in part, to foreland basins caught between the Canadian Shield and young mountain ranges, with consequent revolutionary changes in subsidence mechanisms and the source and character of their sediments.

PLATFORMS AND BASINS OF THE ACTIVE PHASES

Price and Hatcher (1983) have noted numerous similarities between the tectonic histories of the southern-central Appalachian Mountains and the southern Canadian Cordillera. Some differences (including several cited by Price and Hatcher) have been important in determining the character and distribution of both pre-orogenic and orogenic sediments on the former platforms inboard. The present discussion is expanded to include consideration of the northern Appalachians, the northern Canadian Cordillera, the Innuitian Orogen, and the Cordillera in the United States.

Until recently, one of the principal apparent differences between the Appalachian and Cordilleran orogens has been the duration of the convergent phase in each. The Appalachians record about 200 million years of convergence. Osberg (1983) emphasized the episodicity of this tectonism, but his data, combined for the orogen as a whole, can as readily be viewed as recording almost continuous convergence with peaks of intensity (it remains difficult for a non-Appalachian geologist to sort out, from the literature, which deformations were North American and which took place in allochthonous terranes). The Cordilleran record has appeared to be about 100 million years of relatively continuous convergence. The short western history arises in part from the seemingly continuous record of passive-margin subsidence in the west, from Cambrian through Middle Jurassic times, but new evidence of pre-Jurassic deformation and plutonism to the west is blurring this apparent simplicity. If the first folding of Cordilleran, miogeoclinal rocks, recorded in Kootenay Terrane and dated only as Paleozoic, pre-Late Devonian, be viewed as the onset of "Cordilleran" deformation, then, taking note of evidence for Early Carboniferous, Permian, and Triassic convergence (as discussed in Chapter 5), Cordilleran orogenic history would be as long as Appalachian. A similar approach was taken by Burchfiel and Davis (1975). They pointed out that the variously named orogenies recorded in the Cordillera of the western U.S.A., Antler (Late Devonian or earlier, and early Early Carboniferous), Sonoma (Permian and Triassic), Sevier (Middle Jurassic or earlier to Late Cretaceous), and Laramide (latest Cretaceous to Eocene) could be viewed as a continuous record of convergence of changing style. In the Canadian Cordillera, in addition to evidence for Antler (Ellesmerian) deformation and igneous activity, evidence for Sonoman deformation and vulcanism is increasingly coming to light. In Canada, however, the weight of evidence indicates that Sonoman deformation took place in offshore terranes prior to suturing to North America (Gabrielse and Yorath, 1991). A similar approach may be fruitfully applied to the question of the Columbian and Laramide orogenies in Canada, as follows. The Cordillera-long extent of mid-Cretaceous (late Columbian) plutons is in marked contrast with the feeble size and number of Late Cretaceous/Tertiary plutons and their absence from the eastern Cordillera (Woodsworth et al., 1991). Clearly, Columbian and Laramide orogenies, however poorly understood, were different processes. The puzzling and unexplained fact, given evidence for pre-Jurassic tectonic activity, is that the only established, pre-Late Jurassic, westerly derived clastic sediments of autochthonous Canadian Cordilleran terranes are Devono-Carboniferous. With one exception, these occur far outboard, were mainly deposited in rift basins, and do not extend as clastic wedges into the Interior Platform. The favoured interpretation for the tectonic environment of this sedimentation is extensional-transcurrent (Gordey et al., 1987), though other interpretations are possible. Only in the northern part of Yukon Territory and adjacent Alaska is a coarse and far-reaching clastic wedge (Ellesmerian) known that is clearly related to "outboard" contractional tectonism. The older, Upper Devonian part of this wedge may indeed be related to the non-contractional tectonism just referred to, but the younger, Lower Carboniferous part, and possibly the entire wedge, resulted from the contractional Ellesmerian Orogeny experienced by most of the Arctic Archipelago but not clearly established in the Canadian Cordillera outside northern Yukon Territory.

Episodes of tectonism in the Innuitian Orogen appear to be more clearly separated in time. A Late Silurian-Early Devonian episode, recorded by a clastic wedge derived from the north and coinciding with initiation of the Boothia Uplift, is attributed to sinistral transcurrent movements that brought the terrane Pearya (northernmost Ellesmere Island) to its present relationship with the miogeocline (Trettin, in press). The subsequent, Late Devonian-Early Carboniferous contractional orogeny, the type Ellesmerian, was experienced throughout the Innuitian Belt and shed a clastic wedge that appears to have reached as far as the Mackenzie Platform and northern Yukon Territory. Much of the Innuitian Orogen and some of the Arctic Platform then subsided as the rift-initiated Sverdrup Basin (in mid-Early Carboniferous to Late Cretaceous times). Contractional orogeny was not experienced again until the Late Cretaceous-Early Oligocene Eurekan Orogeny, an event whose effects diminish southwestward along the orogen. Interestingly, extensional tectonism was experienced concurrently in the southeastern part of the orogen (Trettin, 1991).

Because of the lack of pre-Jurassic clastic wedges in the Interior Platform, it has been customary to treat the Cambrian through Middle Jurassic sediments of the platform as a single passive-margin ("miogeoclinal") succession. This familiar approach may yet be justified; if the Columbian-Laramide deformation in the Cordillera is a consequence of the docking of an exotic superterrane against North America, as generally considered, then it is clear from their limited extent that, excepting Ellesmerian orogeny in the Arctic regions, the pre-Jurassic deformations, Antler and Sonoma, had some different cause. Transcurrent faulting along the margin has been suggested. At other times since the Silurian, the margin appears to have been extensional (passive) into a back-arc basin, the arc itself being then well offshore, but since accreted to North America.

The spatial association between the deformed belts and the "miogeoclinal" belts is strong, but not total. Burchfiel and Davis (1975) pointed out that in southeastern California and southwestern Nevada, Jura-Cretaceous contractional deformation affected regions that were thoroughly cratonal, as did Cretaceous-Tertiary ("Classical Laramide") deformation in Colorado and Wyoming. In Canada, most of the cover of the Porcupine Platform was deformed in mid-Cretaceous to Paleocene times. The folds and thrusts of the Franklin Mountains represent an extraordinary salient of the deformation front toward the craton; this was facilitated by the presence of Cambrian evaporites as a detachment horizon. The Appalachian Orogen appears to lack these excursions of orogenic deformation beyond the miogeocline.

The foreland-basin succession of the Interior Platform differs from that of the St. Lawrence Platform in several important ways related to climate and tectonic style in the orogens. The first difference is in the proportion of carbonate rocks, high in the Appalachian case and almost negligible in the Cordilleran, especially in Canada (the Mesozoic carbonate rocks of the central United States are not regarded here as of the foreland basin). This difference relates in large part to climate, reflecting paleolatitude, and the globally warm conditions of most of the Paleozoic as compared with Jurassic to Eocene times. It also relates to tectonic style. Cordilleran deformation is generally interpreted to have progressed toward the foreland, with few out-of-sequence structures, so that the easternmost contractional deformation and eroding highlands were the youngest. In contrast, early Taconic thrust-sheets of the northern Appalachians reached positions near the western limit of the subsequent Acadian deformation and well west of the limit of the final, Alleghenian, deformation, and were soon buried, thus ceasing to be sources of detritus. In this latter instance, the distal thrusts did not create highlands of any significant height or duration, and the source of clastics actually retreated from the foreland. Furthermore, there are numerous instances in the frontal Appalachians of synclinoria of folded miogeoclinal strata within which orogenic sediments were subsequently deposited and refolded (Rodgers, 1970). These synclinoria appear to have been sediment traps. Although foredeep deposits were indeed caught up by the progressive eastward advance of the deformation-front in the Cordillera (Price and Mountjoy, 1970), examples of Columbian-Laramide orogenic sediments that were deposited on folded miogeoclinal sediments and subsequently were themselves folded are unknown. Another aspect of tectonic style affecting the dispersal of orogenic detritus relates to the expected flexural bulges on the foreland sides of the foredeeps. In the Appalachian case, the successive forebulges appear to be pronounced, and represented by the several phases of uplift of the Findlay- Algonquin-Laurentian arch system. This hypothesis was quantitatively evaluated from a geophysical point of view by Quinlan and Beaumont (1984), with positive results. On the other hand, although there is a good correlation between episodes of Appalachian convergent tectonism and episodes of uplift on the arches named, the forebulge interpretation plays no role in Sanford's (this volume) scenario for the evolution of the St. Lawrence Lowland. In the Cordilleran case, the forebulge is elusive (Stott, Chapter 4I in this volume); the example of Sweetgrass Arch as a forebulge (Beaumont, 1981) is flawed, because the structure existed prior to Mesozoic convergent orogeny. The differences among orogens in expression of the forebulge may again lie in the character of the lithosphere (Beaumont, 1981), relatively young, thin, and warm beneath the eastern platform, and old, thick, and cool beneath the western.

In none of the continent-bounding orogens were pulses of orogeny felt equally along the mountain system, and in each, one or other of the orogenic pulses appears to have spared some areas. Thus, the clastic wedges are not orogen-long. Examples are the absence of an Acadian clastic wedge in the southern Appalachians and the absence or feeble expression of Alleghenian orogeny in the maritime provinces and Newfoundland (Fig. 13.2). Antler (Devono-Carboniferous) and Sonoma (Permo-Triassic) events, which produced some large-scale contractional structures and clastic wedges in the western U.S.A., are only feebly expressed in British Columbia. Contractional deformation and clastic wedges of Antler age reappear in northern Yukon Territory and the Arctic Islands as manifestations of Ellesmerian orogeny. The intensity of Jurassic to Paleocene convergence in the Canadian Cordillera was similarly non-uniform along the length of the orogen. Early Columbian (pre-Late Aptian) convergence is recorded by thick clastic wedges south of latitude 60°, but the earliest wedge around and north of the Mackenzie Arc is Late Aptian-Albian. In the western U.S.A., the "province of Laramide Block uplifts", sometimes referred to as the Wyoming Province and now generally accepted to be of contractional origin, is a latitudinally limited excursion of the Cordilleran deformation front into a thoroughly cratonal region. This "type Laramide" deformation was entirely younger than the Late Jurassic-Cretaceous deformation and extended into the Eocene. There is no comparable tectonic province in Canada. In the Arctic, Eurekan (Late Cretaceous-Tertiary) contractional structures are most strongly developed in the northeast, and fade to insignificance southwestward.

Some of the examples given above introduce the question of whether the already-deformed hinterlands took part in later deformations. The Sevier Belt did not undergo significant Laramide deformation (Burchfiel and Davis, 1975). Similarly, mid-Cretaceous plutons in the eastern Cordillera of Canada are post-tectonic and undeformed (Woodsworth et al., 1991), although the final deformation in the frontal ranges extended into the Paleocene. This apparent passivity of the earlier fold belts may have sedimentological consequences; if young tectonic relief was being created only in the exterior ranges of the particular epoch, the young, frontal ranges commonly or generally may have contained the continental drainage divide. In such case, only the frontal ranges, and not the entire orogen, would have been the source lands for the foreland basin. The petrology of Cretaceous foreland-basin sediments of the Canadian Cordillera supports the notion. This idea might be applicable to the southern Appalachians, where Acadian deformation, but no Acadian clastic wedge, is recorded.

Examples of extensional tectonism that coincided with contractional effects are widespread. A chain of Early Carboniferous rift-basins in the Maritime Provinces and Newfoundland ("Fundy Epieugeosyncline" of Poole et al., 1970) apparently developed during the span of the Alleghanian Orogeny. Rodgers (1970) suggested that the extension occurred in a transcurrent regime. Similarly, small Devono-Carboniferous extensional basins and companion high blocks developed in Selwyn Basin, either concurrently with, or immediately preceding Ellesmerian (Brookian) folding in northern Yukon Territory. Again, a transcurrent regime has been suggested (Gordey et al., 1987). In the Arctic, Eurekan outboard contraction was accompanied by inboard extension.

Sverdrup Basin is unique in North America. As successor basins, the Carboniferous basins of the Canadian Appalachians bear some similarities, including thick, early accumulations of evaporites. They appear, however, to be transtensional in origin and not to have developed much beyond the early rift phase; they do not approach the extent or depth of subsidence of the Sverdrup Basin. The Gulf Coast Basin also bears some similarities, as pointed out by Douglas et al. (1963) and Thorsteinsson (1974). It is not clear, however, to what extent initial rifting of the Gulf Coast Basin took place within the Appalachian-Ouachita orogen, nor to what extent the initial basin was a continental-margin rather than a successor basin. The nearest well-studied analogue of Sverdrup Basin may be the Zechstein-North Sea Basin of western Europe.

Few modern zones of crustal convergence are good "models" for the North American orogens, partly because most are oceanic or at continent-ocean boundaries. A marine and a subaerial foreland basin can, however, be studied today.

In the vicinity of the Persian Gulf, the Arabian Shield (African Plate) is being obliquely subducted beneath Asia. The Zagros Mountains forming the northeastern coast of the gulf are a foreland fold and thrust belt closely analogous to the eastern Cordillera or the Valley and Ridge Province. They extend southwestward hundreds of kilometres beyond the intercontinental suture, involving the detached cover of the platform. The foredeep is the Persian Gulf, a site of both carbonate and clastic sedimentation.

The Cordillera Oriental and Llanos (plains) of Columbia demonstrate a linked orogen and subaerial foreland basin. The mountain front is seismically active, and young river terraces are folded and faulted. From a spectacular mountain front, continuously linked and highly active alluvial fans extend across the foreland basin more than 450 km; their streams are tributaries to the longitudinal Rio Guaviare-Rio Orinoco system, which is pushed against the Guyana Shield by this massive, one-sided input of detritus. Inter-fan lakes and swamps are small potential sites of coal deposition. Tongues of marine shale in the Llanos subsurface demonstrate that the foreland was inundated as recently as Oligocene time.

REFERENCES

Beaumont, C.
1981: Foreland basins; Geophysical Journal of the Royal Astronomical Society, v. 65, p. 291-329.

Budnick, R.T.
1986: Left-lateral intraplate deformation along the Ancestral Rocky Mountains: implications for late Paleozoic plate motions; Tectonophysics, v. 132, p. 195-214.

Burchfiel, B.C. and Davis, G.A.
1975: Nature and controls of Cordilleran orogenesis, western United States: extensions of an earlier synthesis; American Journal of Science, v. 275-A, p. 363-396.

Cady, W.M.
1970: The lateral transition from the miogeosynclinal to the eugeosynclinal zone in northwestern New England and adjacent Quebec; in Studies of Appalachian Geology, Northern and Maritime; Interscience, New York, p. 151-161.

Cook, F.A.
1988a: Proterozoic thin-skinned thrust and fold belt beneath the Interior Platform in northwest Canada; Geological Society of America Bulletin, v. 100, p. 877-890.
1988b: Middle Proterozoic compressional orogen in northwestern Canada; Journal of Geophysical Research, v. 93, p. 8985-9005.

Douglas, R.J.W., Norris, D.K., Thorsteinsson, R., and Tozer, E.T.
1963: Geology and petroleum potentialities of northern Canada; Sixth World Petroleum Congress, Proceedings, Section 1, p. 1-52.

Gabrielse, H. and Yorath, C.J. (ed.)
1991: Geology of the Cordilleran Orogen in Canada; Geological Survey of Canada, Geology of Canada, no. 4. (Also Geological Society of America, The Geology of North America, v. G-2.)

Gordey, S.P., Abbot, J.G., Tempelman-Kluit, D.J., and Gabrielse, H.
1987: "Antler" clastics in the Canadian Cordillera; Geology, v. 15, p. 103-107.

Osberg, P.H.
1983: Timing of orogenic events in the U.S. Appalachians; in Regional Trends in the Geology of the Appalachian- Caledonian-Hercynian-Mauritanide Orogen, P.E. Schenk (ed.); D. Reidel, Boston, p. 315-337.

Poole, W.H., Sanford, B.V., Williams, H., and Kelley, D.G.
1970: Geology of southeastern Canada; in Geology and Economic Minerals of Canada, R.J.W. Douglas (ed.); Geological Survey of Canada, Economic Geology Report No. 1, p. 229-306.

Price, R.A., and Hatcher, R.D., Jr.
1983: Tectonic similarities in the evolution of the Alabama-Pennsylvania Appalachians and the Alberta-British Columbia Canadian Cordillera; in Contributions to the Tectonics and Geophysics of Mountain Chains, R.D. Hatcher, Jr., H. Williams, and I. Zietz (ed.); Geological Society of America, Memoir 158, p. 149-160.

Price, R.A., and Mountjoy, E.W.
1970: Geologic structure of the Canadian Rocky Mountains between Bow and Athabasca Rivers - a progress report; in Structure of the Southern Canadian Cordillera, J.O. Wheeler (ed.); Geological Association of Canada, Special Paper no. 6, p. 7-25.

Quinlan, G.M., and Beaumont, C.
1984: Appalachian thrusting, lithospheric flexure, and the Paleozoic stratigraphy of the eastern interior of North America; Canadian Journal of Earth Sciences, v. 21, p. 973-996.

Rodgers, J.
1970: The tectonics of the Appalachians; Wiley-Interscience, New York, 271 p.

Sanford, B.V., Thompson, F.J., and McFall, G.H.
1985: Plate tectonics - a possible controlling mechanism in the development of hydrocarbon traps in southwestern Ontario; Bulletin of Canadian Petroleum Geology, v. 33, p. 52-71.

Sloss, L.L.
1982: The Midcontinent Province: United States; in Perspectives in Regional Geological Synthesis, A.R. Palmer (ed.); Geological Society of America, DNAG Special Publication I, p. 27-40.

Thorsteinsson, R.
1974: Carboniferous and Permian stratigraphy of Axel Heiberg Island and western Ellesmere Island, Canadian Arctic Archipelago; Geological Survey of Canada, Bulletin 224, 115 p.

Trettin, H.P. (ed.)
1991: Geology of the Innuitian Orogen and Arctic Platform of Canada and Greenland; Geological Survey of Canada, Geology of Canada, no. 3. (also Geological Society of America, The Geology of North America, v. E.)

Woodsworth, G.J., Anderson, R.G., and Armstrong, R.L.
1991: Plutonic regimes; Chapter 15 in Geology of the Cordilleran Orogen in Canada, H. Gabrielse and C.J. Yorath (ed.); Geological Survey of Canada, Geology of Canada no. 4. (Also Geological Society of America, The Geology of North America, v. G-2.)

Author's Address

J.D. Aitken
2676 Jemima Road
Denman Island, British Columbia
V0R 1T0

Printed in Canada

INDEX

The letters f and t refer to figures or tables respectively

Survey Peak Formation, 102f, 109f, 118f, 120, 121f, 131, 135f, 136f, 141f